P. R. Kumar

John Baillieul

Kirsten Morris

Roberto Tempo

Christos G. Cassandras

PJ Antsaklis

Hassan Khalil

Andrew R. Teel

Bozenna Pasik-Duncan

Control System Advanced Methods

The Electrical Engineering Handbook Series

Series Editor
Richard C. Dorf
University of California, Davis

Titles Included in the Series

The Avionics Handbook, Second Edition, Cary R. Spitzer
The Biomedical Engineering Handbook, Third Edition, Joseph D. Bronzino
The Circuits and Filters Handbook, Third Edition, Wai-Kai Chen
The Communications Handbook, Second Edition, Jerry Gibson
The Computer Engineering Handbook, Vojin G. Oklobdzija
The Control Handbook, Second Edition, William S. Levine
CRC Handbook of Engineering Tables, Richard C. Dorf
Digital Avionics Handbook, Second Edition, Cary R. Spitzer
The Digital Signal Processing Handbook, Vijay K. Madisetti and Douglas Williams
The Electric Power Engineering Handbook, Second Edition, Leonard L. Grigsby
The Electrical Engineering Handbook, Third Edition, Richard C. Dorf
The Electronics Handbook, Second Edition, Jerry C. Whitaker
The Engineering Handbook, Third Edition, Richard C. Dorf
The Handbook of Ad Hoc Wireless Networks, Mohammad Ilyas
The Handbook of Formulas and Tables for Signal Processing, Alexander D. Poularikas
Handbook of Nanoscience, Engineering, and Technology, Second Edition, William A. Goddard, III, Donald W. Brenner, Sergey E. Lyshevski, and Gerald J. Iafrate
The Handbook of Optical Communication Networks, Mohammad Ilyas and Hussein T. Mouftah
The Industrial Electronics Handbook, J. David Irwin
The Measurement, Instrumentation, and Sensors Handbook, John G. Webster
The Mechanical Systems Design Handbook, Osita D.I. Nwokah and Yidirim Hurmuzlu
The Mechatronics Handbook, Second Edition, Robert H. Bishop
The Mobile Communications Handbook, Second Edition, Jerry D. Gibson
The Ocean Engineering Handbook, Ferial El-Hawary
The RF and Microwave Handbook, Second Edition, Mike Golio
The Technology Management Handbook, Richard C. Dorf
Transforms and Applications Handbook, Third Edition, Alexander D. Poularikas
The VLSI Handbook, Second Edition, Wai-Kai Chen

The Control Handbook

Second Edition

Edited by

William S. Levine

University of Maryland

College Park, MD, USA

Control System Fundamentals

Control System Applications

Control System Advanced Methods

Control System Advanced Methods

Edited by

William S. Levine

University of Maryland

College Park, MD, USA

CRC Press
Taylor & Francis Group
Boca Raton London New York

CRC Press is an imprint of the
Taylor & Francis Group, an **informa** business

CRC Press
Taylor & Francis Group
6000 Broken Sound Parkway NW, Suite 300
Boca Raton, FL 33487-2742

Printed in the United States of America on acid-free paper
10 9 8 7 6 5 4 3 2 1

International Standard Book Number: 978-1-4200-7364-5 (Hardback)

Library of Congress Cataloging-in-Publication Data

Control system advanced methods / edited by William S. Levine. -- 2nd ed.
p. cm. -- (The electrical engineering handbook series)
Includes bibliographical references and index.
ISBN 978-1-4200-7364-5
1. Automatic control. I. Levine, W. S. II. Title. III. Series.

TJ212.C685 2011
629.8--dc22 2010026367

Visit the Taylor & Francis Web site at
http://www.taylorandfrancis.com

and the CRC Press Web site at
http://www.crcpress.com

Contents

SECTION II Kalman Filter and Observers

SECTION III Design Methods for MIMO LTI Systems

SECTION IV Analysis and Design of Hybrid Systems

SECTION V Adaptive Control

SECTION VI Analysis and Design of Nonlinear Systems

SECTION VII Stability

SECTION VIII Design

SECTION IX System Identification

SECTION X Stochastic Control

SECTION XI Control of Distributed Parameter Systems

SECTION XII Networks and Networked Controls

Preface to the Second Edition

As you may know, the first edition of *The Control Handbook* was very well received. Many copies were sold and a gratifying number of people took the time to tell me that they found it useful. To the publisher, these are all reasons to do a second edition. To the editor of the first edition, these same facts are a modest disincentive. The risk that a second edition will not be as good as the first one is real and worrisome. I have tried very hard to insure that the second edition is at least as good as the first one was. I hope you agree that I have succeeded.

I have made two major changes in the second edition. The first is that all the *Applications* chapters are new. It is simply a fact of life in engineering that once a problem is solved, people are no longer as interested in it as they were when it was unsolved. I have tried to find especially inspiring and exciting applications for this second edition.

Secondly, it has become clear to me that organizing the *Applications* book by academic discipline is no longer sensible. Most control applications are interdisciplinary. For example, an automotive control system that involves sensors to convert mechanical signals into electrical ones, actuators that convert electrical signals into mechanical ones, several computers and a communication network to link sensors and actuators to the computers does not belong solely to any specific academic area. You will notice that the applications are now organized broadly by application areas, such as automotive and aerospace.

One aspect of this new organization has created a minor and, I think, amusing problem. Several wonderful applications did not fit into my new taxonomy. I originally grouped them under the title Miscellaneous. Several authors objected to the slightly pejorative nature of the term "miscellaneous." I agreed with them and, after some thinking, consulting with literate friends and with some of the library resources, I have renamed that section "Special Applications." Regardless of the name, they are all interesting and important and I hope you will read those articles as well as the ones that did fit my organizational scheme.

There has also been considerable progress in the areas covered in the *Advanced Methods* book. This is reflected in the roughly two dozen articles in this second edition that are completely new. Some of these are in two new sections, "Analysis and Design of Hybrid Systems" and "Networks and Networked Controls."

There have even been a few changes in the *Fundamentals*. Primarily, there is greater emphasis on sampling and discretization. This is because most control systems are now implemented digitally.

I have enjoyed editing this second edition and learned a great deal while I was doing it. I hope that you will enjoy reading it and learn a great deal from doing so.

William S. Levine

Acknowledgments

The people who were most crucial to the second edition were the authors of the articles. It took a great deal of work to write each of these articles and I doubt that I will ever be able to repay the authors for their efforts. I do thank them very much.

The members of the advisory/editorial board for the second edition were a very great help in choosing topics and finding authors. I thank them all. Two of them were especially helpful. Davor Hrovat took responsibility for the automotive applications and Richard Braatz was crucial in selecting the applications to industrial process control.

It is a great pleasure to be able to provide some recognition and to thank the people who helped bring this second edition of *The Control Handbook* into being. Nora Konopka, publisher of engineering and environmental sciences for Taylor & Francis/CRC Press, began encouraging me to create a second edition quite some time ago. Although it was not easy, she finally convinced me. Jessica Vakili and Kari Budyk, the project coordinators, were an enormous help in keeping track of potential authors as well as those who had committed to write an article. Syed Mohamad Shajahan, senior project executive at Techset, very capably handled all phases of production, while Richard Tressider, project editor for Taylor & Francis/CRC Press, provided direction, oversight, and quality control. Without all of them and their assistants, the second edition would probably never have appeared and, if it had, it would have been far inferior to what it is.

Most importantly, I thank my wife Shirley Johannesen Levine for everything she has done for me over the many years we have been married. It would not be possible to enumerate all the ways in which she has contributed to each and everything I have done, not just editing this second edition.

William S. Levine

Editorial Board

Editor

William S. Levine received B.S., M.S., and Ph.D. degrees from the Massachusetts Institute of Technology. He then joined the faculty of the University of Maryland, College Park where he is currently a research professor in the Department of Electrical and Computer Engineering. Throughout his career he has specialized in the design and analysis of control systems and related problems in estimation, filtering, and system modeling. Motivated by the desire to understand a collection of interesting controller designs, he has done a great deal of research on mammalian control of movement in collaboration with several neurophysiologists.

He is co-author of *Using MATLAB to Analyze and Design Control Systems*, March 1992. Second Edition, March 1995. He is the coeditor of *The Handbook of Networked and Embedded Control Systems*, published by Birkhauser in 2005. He is the editor of a series on control engineering for Birkhauser. He has been president of the IEEE Control Systems Society and the American Control Council. He is presently the chairman of the SIAM special interest group in control theory and its applications.

He is a fellow of the IEEE, a distinguished member of the IEEE Control Systems Society, and a recipient of the IEEE Third Millennium Medal. He and his collaborators received the Schroers Award for outstanding rotorcraft research in 1998. He and another group of collaborators received the award for outstanding paper in the *IEEE Transactions on Automatic Control*, entitled "Discrete-Time Point Processes in Urban Traffic Queue Estimation."

Contributors

Eyad H. Abed
Department of Electrical Engineering
and the Institute for Systems Research
University of Maryland
College Park, Maryland

Andrew G. Alleyne
Department of Mechanical Science
and Engineering
University of Illinois at Urbana–Champaign
Urbana, Illinois

Albert N. Andry, Jr.
Teledyne Reynolds Group
Los Angeles, California

Panos J. Antsaklis
Department of Electrical Engineering
University of Notre Dame
Notre Dame, Indiana

Murat Arcak
Department of Electrical Engineering
and Computer Sciences
University of California, Berkeley
Berkeley, California

Karl J. Åström
Department of Automatic Control
Lund Institute of Technology
Lund, Sweden

Michael Athans
Department of Electrical Engineering
and Computer Science
Massachusetts Institute of
Technology
Cambridge, Massachusetts

J. Baillieul
Department of Electrical & Computer
Engineering
Boston University
Boston, Massachusetts

Gary J. Balas
Department of Aerospace Engineering
and Mechanics
University of Minnesota
Minneapolis, Minnesota

Simone Baldi
Department of Systems and Computer Science
University of Florence
Florence, Italy

Kira L. Barton
Department of Mechanical Science
and Engineering
University of Illinois at Urbana–Champaign
Urbana, Illinois

Sorin Bengea
Control Systems Group
United Technologies Research Center
East Hartford, Connecticut

Franco Blanchini
Department of Mathematics and
Computer Science
University of Udine
Udine, Italy

Anthony M. Bloch
Department of Mathematics
University of Michigan
Ann Arbor, Michigan

Julio H. Braslavsky
School of Electrical Engineering and Computer Science
The University of Newcastle
Callaghan, New South Wales, Australia

Douglas A. Bristow
Department of Mechanical and Aerospace Engineering
Missouri University of Science and Technology
Rolla, Missouri

Christopher I. Byrnes
Department of Systems Sciences and Mathematics
Washington University
St. Louis, Missouri

Alan Chao
Laboratory for Information and Decision Systems
Massachusetts Institute of Technology
Cambridge, Massachusetts

David W. Clarke
Department of Engineering Science
University of Oxford
Oxford, United Kingdom

Fabrizio Dabbene
Institute for Electronics, Engineering, Information and Telecommunications–National Research Council
Polytechnic University of Turin
Turin, Italy

Munther A. Dahleh
Laboratory for Information and Decision Systems
Massachusetts Institute of Technology
Cambridge, Massachusetts

Edward J. Davison
Department of Electrical and Computer Engineering
University of Toronto
Toronto, Ontario, Canada

Raymond A. DeCarlo
Department of Electrical and Computer Engineering
Purdue University
West Lafayette, Indiana

David F. Delchamps
School of Electrical and Computer Engineering
Cornell University
Ithaca, New York

Maria Domenica Di Benedetto
Department of Electrical Engineering
University of L'Aquila
L'Aquila, Italy

Joel Douglas
Department of Electrical Engineering and Computer Science
Massachusetts Institute of Technology
Cambridge, Massachusetts

T.E. Duncan
Department of Mathematics
University of Kansas
Lawrence, Kansas

Sergey V. Drakunov
Department of Physical Sciences
Embry-Riddle Aeronautical University
Daytona Beach, Florida

Jay A. Farrell
Department of Electrical Engineering
University of California, Riverside
Riverside, California

Bruce Francis
Department of Electrical and Computer Engineering
University of Toronto
Toronto, Ontario, Canada

Randy A. Freeman
University of California, Santa Barbara
Santa Barbara, California

James S. Freudenberg
Department of Electrical Engineering and Computer Science
University of Michigan
Ann Arbor, Michigan

Bernard Friedland
Department of Electrical and Computer Engineering
New Jersey Institute of Technology
Newark, New Jersey

T.T. Georgiou
Department of Electrical Engineering
University of Minnesota
Minneapolis, Minnesota

Simon Grocott
Space Engineering Research Center
Massachusetts Institute of Technology
Cambridge, Massachusetts

John A. Gubner
College of Engineering
University of Wisconsin–Madison
Madison, Wisconsin

Vijay Gupta
Department of Electrical Engineering
University of Notre Dame
Notre Dame, Indiana

Tore Hägglund
Department of Automatic Control
Lund Institute of Technology
Lund, Sweden

Fumio Hamano
Department of Electrical Engineering
California State University
Long Beach, California

Constantine H. Houpis
Department of Electrical and Computer Engineering
Air Force Institute of Technology
Wright-Patterson Air Force Base, Ohio

Petros Ioannou
Department of Electrical Engineering
University of Southern California
Los Angeles, California

Alberto Isidori
Department of Computer Science and Systemics
Sapienza University of Rome
Rome, Italy

Ulf T. Jönsson
Department of Mathematics
Royal Institute of Technology
Stockholm, Sweden

V. Jurdjevic
Department of Mathematics
University of Toronto
Toronto, Ontario, Canada

Edward W. Kamen
School of Electrical and Computer Engineering
Georgia Institute of Technology
Atlanta, Georgia

Chung-Yao Kao
Department of Electrical Engineering
National Sun Yat-Sen University
Kaohsiung, Taiwan

Hassan K. Khalil
Department of Electrical & Computer Engineering
Michigan State University
East Lansing, Michigan

Petar V. Kokotović
Department of Electrical and Computer Engineering
University of California, Santa Barbara
Santa Barbara, California

Miroslav Krstić
Department of Mechanical and Aerospace Engineering
University of California, San Diego
San Diego, California

Vladimír Kučera
Czech Technical University
Prague, Czech Republic

and

Institute of Information Theory and Automation
Academy of Sciences
Prague, Czech Republic

P.R. Kumar
Department of Electrical and Computer Engineering and Coordinated Science Laboratory
University of Illinois at Urbana–Champaign
Urbana, Illinois

Alex A. Kurzhanskiy
Department of Electrical Engineering and Computer Sciences
University of California, Berkeley
Berkeley, California

Françoise Lamnabhi-Lagarrigue
Laboratory of Signals and Systems
National Center for Scientific Research
Supélec
Gif–sur–Yvette, France

J.E. Lagnese
Department of Mathematics
Georgetown University
Washington, D.C.

A.J. Laub
Electrical Engineering Department
University of California, Los Angeles
Los Angeles, California

Jay H. Lee
Korean Advanced Institute of Science and Technology
Daejeon, South Korea

B. Lehman
Department of Electrical & Computer Engineering
Northeastern University
Boston, Massachusetts

G. Leugering
Faculty for Mathematics, Physics, and Computer Science
University of Bayreuth
Bayreuth, Germany

William S. Levine
Department of Electrical & Computer Engineering
University of Maryland
College Park, Maryland

Frank L. Lewis
Automation and Robotics Research Institute
The University of Texas at Arlington
Arlington, Texas

Hai Lin
Department of Electrical and Computer Engineering
National University of Singapore
Singapore

Shu-Jun Liu
Department of Mathematics
Southeast University
Nanjing, China

Lennart Ljung
Department of Electrical Engineering
Linköping University
Linköping, Sweden

Douglas P. Looze
Department of Electrical and Computer Engineering
University of Massachusetts Amherst
Amherst, Massachusetts

Kenneth A. Loparo
Department of Electrical Engineering
and Computer Science
Case Western Reserve University
Cleveland, Ohio

Leonard Lublin
Space Engineering Research Center
Massachusetts Institute of Technology
Cambridge, Massachusetts

Kevin M. Lynch
Mechanical Engineering Department
Northwestern University
Evanston, Illinois

Lorenzo Marconi
Department of Electronics, Computer
Sciences and Systems
University of Bologna
Bologna, Italy

Nuno C. Martins
Department of Electrical & Computer
Engineering
University of Maryland
College Park, Maryland

Alexandre Megretski
Laboratory for Information and
Decision Systems
Massachusetts Institute of Technology
Cambridge, Massachusetts

Richard H. Middleton
The Hamilton Institute
National University of Ireland,
Maynooth
Maynooth, Ireland

Manfred Morari
Swiss Federal Institute of Technology
Zurich, Switzerland

Kirsten Morris
Department of Applied Mathematics
University of Waterloo
Waterloo, Ontario, Canada

Andy Packard
Department of Mechanical Engineering
University of California, Berkeley
Berkeley, California

B. Pasik-Duncan
Department of Mathematics
University of Kansas
Lawrence, Kansas

Kevin M. Passino
Department of Electrical Engineering
The Ohio State University
Columbus, Ohio

Stephen D. Patek
Laboratory for Information and
Decision Systems
Massachusetts Institute of Technology
Cambridge, Massachusetts

R.V. Patel
Department of Electrical and
Computer Engineering
University of Western Ontario
London, Ontario, Canada

Marios M. Polycarpou
Department of Electrical and
Computer Engineering
University of Cyprus
Nicosia, Cyprus

L. Praly
Systems and Control Centre
Mines Paris Institute of Technology
Paris, France

Jörg Raisch
Department of Electrical Engineering and
Computer Science
Technical University of Berlin
Berlin, Germany

Anders Rantzer
Department of Automatic Control
Lund University
Lund, Sweden

Mahmut Reyhanoglu
Physical Sciences Department
Embry-Riddle Aeronautical University
Daytona Beach, Florida

Michael C. Rotkowitz
Department of Electrical and Electronic Engineering
The University of Melbourne
Melbourne, Australia

David L. Russell
Department of Mathematics
Virginia Polytechnic Institute and State University
Blacksburg, Virginia

Carsten Scherer
Department of Mathematics
University of Stuttgart
Stuttgart, Germany

Jacquelien M.A. Scherpen
Faculty of Mathematics and Natural Sciences
University of Groningen
Groningen, the Netherlands

Thomas I. Seidman
Department of Mathematics and Statistics
University of Maryland, Baltimore County
Baltimore, Maryland

M.E. Sezer
Department of Electrical and Electronics Engineering
Bilkent University
Ankara, Turkey

Eliezer Y. Shapiro (deceased)
HR Textron
Valencia, California

Adam Shwartz
Department of Electrical Engineering
Technion–Israel Institute of Technology
Haifa, Israel

D.D. Šiljak
Department of Electrical Engineering
Santa Clara University
Santa Clara, California

Andrey Smyshlyaev
Department of Mechanical and Aerospace Engineering
University of California, San Diego
La Jolla, California

Kenneth M. Sobel
Department of Electrical Engineering
The City College of New York
New York, New York

Torsten Söderström
Department of Information Technology
Uppsala University
Uppsala, Sweden

Eduardo D. Sontag
Department of Mathematics
Rutgers University
New Brunswick, New Jersey

A.R. Teel
Department of Electrical Engineering
University of Minnesota
Minneapolis, Minnesota

Roberto Tempo
Institute for Electronics, Engineering, Information and Telecommunications–National Research Council
Polytechnic University of Turin
Turin, Italy

Alberto Tesi
Department of Systems and Computer Science
University of Florence
Florence, Italy

Kasemsak Uthaichana
Department of Electrical Engineering
Chiang Mai University
Chiang Mai, Thailand

P.M. Van Dooren
Department of Mathematical Engineering
Catholic University of Leuven
Leuven, Belgium

Pravin Varaiya
Department of Electrical Engineering
and Computer Sciences
University of California, Berkeley
Berkeley, California

Rafael Vazquez
Department of Aerospace Engineering
University of Seville
Seville, Spain

Draguna Vrabie
Automation and Robotics Research Institute
The University of Texas at Arlington
Arlington, Texas

Hua O. Wang
United Technologies Research Center
East Hartford, Connecticut

Siep Weiland
Department of Electrical Engineering
Eindhoven University of Technology
Eindhoven, the Netherlands

Trevor Williams
Department of Aerospace Engineering
and Engineering Mechanics
University of Cincinnati
Cincinnati, Ohio

Stephen Yurkovich
Department of Electrical Engineering
The Ohio State University
Columbus, Ohio

S.H. Żak
School of Electrical and
Computer Engineering
Purdue University
West Lafayette, Indiana

Milos Žefran
Department of Electrical Engineering
University of Illinois at Chicago
Chicago, Illinois

Dmitry Zenkov
Department of Mathematics
North Carolina State University
Raleigh, North Carolina

I

Analysis Methods for MIMO Linear Systems

1

Numerical and Computational Issues in Linear Control and System Theory*

A.J. Laub
University of California, Los Angeles

R.V. Patel
University of Western Ontario

P.M. Van Dooren
Catholic University of Leuven

1.1 Introduction

This chapter provides a survey of various aspects of the numerical solution of selected problems in linear systems, control, and estimation theory. Space limitations preclude an exhaustive survey and extensive list of references. The interested reader is referred to [1,4,10,14] for sources of additional detailed information.

* This material is based on a paper written by the same authors and published in Patel, R.V., Laub, A.J., and Van Dooren, P.M., Eds., *Numerical Linear Algebra Techniques for Systems and Control*, Selected Reprint Series, IEEE Press, New York, pp. 1–29 1994, copyright 1994 IEEE.

Many of the problems considered in this chapter arise in the study of the "standard" linear model

$$\dot{x}(t) = Ax(t) + Bu(t), \tag{1.1}$$

$$y(t) = Cx(t) + Du(t). \tag{1.2}$$

Here, $x(t)$ is an n-vector of states, $u(t)$ is an m-vector of controls or inputs, and $y(t)$ is a p-vector of outputs. The standard discrete-time analog of Equations 1.1 and 1.2 takes the form

$$x_{k+1} = Ax_k + Bu_k, \tag{1.3}$$

$$y_k = Cx_k + Du_k. \tag{1.4}$$

Of course, considerably more elaborate models are also studied, including time-varying, stochastic, and nonlinear versions of the above, but these are not discussed in this chapter. In fact, the above linear models are usually derived from linearizations of nonlinear models regarding selected nominal points.

The matrices considered here are, for the most part, assumed to have real coefficients and to be small (of order a few hundred or less) and dense, with no particular exploitable structure. Calculations for most problems in classical single-input, single-output control fall into this category. Large sparse matrices or matrices with special exploitable structures may significantly involve different concerns and methodologies than those discussed here.

The systems, control, and estimation literature is replete with *ad hoc* algorithms to solve the computational problems that arise in the various methodologies. Many of these algorithms work quite well on some problems (e.g., "small-order" matrices) but encounter numerical difficulties, often severe, when "pushed" (e.g., on larger order matrices). The reason for this is that little or no attention has been paid to the way algorithms perform in "finite arithmetic," that is, on a finite word length digital computer.

A simple example by Moler and Van Loan [14, p. 649]* illustrates a typical pitfall. Suppose it is desired to compute the matrix e^A in single precision arithmetic on a computer which gives six decimal places of precision in the fractional part of floating-point numbers. Consider the case

$$A = \begin{bmatrix} -49 & 24 \\ -64 & 31 \end{bmatrix}$$

and suppose the computation is attempted with the Taylor series formula

$$e^A = \sum_{k=0}^{+\infty} \frac{1}{k!} A^k. \tag{1.5}$$

This is easily coded and it is determined that the first 60 terms in the series suffice for the computation, in the sense that the terms for $k \geq 60$ of the order 10^{-7} no longer add anything significant to the sum. The resulting answer is

$$\begin{bmatrix} -22.2588 & -1.43277 \\ -61.4993 & -3.47428 \end{bmatrix}.$$

Surprisingly, the true answer is (correctly rounded)

$$\begin{bmatrix} -0.735759 & 0.551819 \\ -1.47152 & 1.10364 \end{bmatrix}.$$

What happened here was that the intermediate terms in the series became very large before the factorial began to dominate. The 17th and 18th terms, for example, are of the order of 10^7 but of opposite signs so

* The page number indicates the location of the appropriate reprint in [14].

that the less significant parts of these numbers, while significant for the final answer, are "lost" because of the finiteness of the arithmetic.

For this particular example, various fixes and remedies are available. However, in more realistic examples, one seldom has the luxury of having the "true answer" available so that it is not always easy to simply inspect or test a computed solution and determine that it is erroneous. Mathematical analysis (truncation of the series, in the example above) alone is simply not sufficient when a problem is analyzed or solved in finite arithmetic (truncation of the arithmetic). Clearly, a great deal of care must be taken.

The finiteness inherent in representing real or complex numbers as floating-point numbers on a digital computer manifests itself in two important ways: floating-point numbers have only finite precision and finite range. The degree of attention paid to these two considerations distinguishes many reliable algorithms from more unreliable counterparts.

The development in systems, control, and estimation theory of stable, efficient, and reliable algorithms that respect the constraints of finite arithmetic began in the 1970s and still continues. Much of the research in numerical analysis has been directly applicable, but there are many computational issues in control (e.g., the presence of hard or structural zeros) where numerical analysis does not provide a ready answer or guide. A symbiotic relationship has developed, especially between numerical linear algebra and linear system and control theory, which is sure to provide a continuing source of challenging research areas.

The abundance of numerically fragile algorithms is partly explained by the following observation:

> If an algorithm is amenable to "easy" manual calculation, it is probably a poor method if implemented in the finite floating-point arithmetic of a digital computer.

For example, when confronted with finding the eigenvalues of a 2×2 matrix, most people would find the characteristic polynomial and solve the resulting quadratic equation. But when extrapolated as a general method for computing eigenvalues and implemented on a digital computer, this is a very poor procedure for reasons such as roundoff and overflow/underflow. The preferred method now would generally be the double Francis QR algorithm (see [17] for details) but few would attempt that manually, even for very small-order problems.

Many algorithms, now considered fairly reliable in the context of finite arithmetic, are not amenable to manual calculations (e.g., various classes of orthogonal similarities). This is a kind of converse to the observation quoted above. Especially in linear system and control theory, we have been too easily tempted by the ready availability of closed-form solutions and numerically naive methods to implement those solutions. For example, in solving the initial value problem

$$\dot{x}(t) = Ax(t); \quad x(0) = x_0, \tag{1.6}$$

it is not at all clear that one should explicitly compute the intermediate quantity e^{tA}. Rather, it is the vector $e^{tA}x_0$ that is desired, a quantity that may be computed by treating Equation 1.6 as a system of (possibly stiff) differential equations and using an implicit method for numerically integrating the differential equation. But such techniques are definitely not attractive for manual computation.

The awareness of such numerical issues in the mathematics and engineering community has increased significantly in the last few decades. In fact, some of the background material well known to numerical analysts has already filtered down to undergraduate and graduate curricula in these disciplines. This awareness and education has affected system and control theory, especially linear system theory. A number of numerical analysts were attracted by the wealth of interesting numerical linear algebra problems in linear system theory. At the same time, several researchers in linear system theory turned to various methods and concepts from numerical linear algebra and attempted to modify them in developing reliable algorithms and software for specific problems in linear system theory. This cross-fertilization has been greatly enhanced by the widespread use of software packages and by developments over the last couple of decades in numerical linear algebra. This process has already begun to have a significant impact on the future directions and development of system and control theory, and on applications, as evident from the growth of computer-aided control system design (CACSD) as an intrinsic tool. Algorithms implemented as mathematical software are a critical "inner" component of a CACSD system.

In the remainder of this chapter, we survey some results and trends in this interdisciplinary research area. We emphasize numerical aspects of the problems/algorithms, which is why we also spend time discussing appropriate numerical tools and techniques. We discuss a number of control and filtering problems that are of widespread interest in control.

Before proceeding further, we list here some notations to be used:

$\mathbb{F}^{n\times m}$	the set of all $n \times m$ matrices with coefficients in the field $\mathbb{F}$ ($\mathbb{F}$ is generally $\mathbb{R}$ or $\mathbb{C}$)
A^T	the transpose of $A \in \mathbb{R}^{n\times m}$
A^H	the complex-conjugate transpose of $A \in \mathbb{C}^{n\times m}$
A^+	the Moore–Penrose pseudoinverse of A
$\|A\|$	the spectral norm of A (i.e., the matrix norm subordinate to the Euclidean vector norm: $\|A\| = \max_{\|x\|_2=1} \|Ax\|_2$)
diag $(a_1, \ldots, a_n)$	the diagonal matrix $\begin{bmatrix} a_1 & & 0 \\ & \ddots & \\ 0 & & a_n \end{bmatrix}$
$\Lambda(A)$	the set of eigenvalues $\lambda_1, \ldots, \lambda_n$ (not necessarily distinct) of $A \in \mathbb{F}^{n\times n}$
$\lambda_i(A)$	the ith eigenvalue of A
$\Sigma(A)$	the set of singular values $\sigma_1, \ldots, \sigma_m$ (not necessarily distinct) of $A \in \mathbb{F}^{n\times m}$
$\sigma_i(A)$	the ith singular value of A

Finally, let us define a particular number to which we make frequent reference in the following. The *machine epsilon* or *relative machine precision* is defined, roughly speaking, as the smallest positive number ϵ that, when added to 1 on our computing machine, gives a number greater than 1. In other words, any machine representable number δ less than ϵ gets " rounded off" when (floating-point) added to 1 to give exactly 1 again as the rounded sum. The number ϵ, of course, varies depending on the kind of computer being used and the precision of the computations (single precision, double precision, etc.). But the fact that such a positive number ϵ exists is entirely a consequence of finite word length.

1.2 Numerical Background

In this section, we give a very brief discussion of two concepts fundamentally important in numerical analysis: *numerical stability* and *conditioning*. Although this material is standard in textbooks such as [8], it is presented here for completeness and because the two concepts are frequently confused in the systems, control, and estimation literature.

Suppose we have a mathematically defined problem represented by f which acts on data d belonging to some set of data $\mathcal{D}$, to produce a solution $f(d)$ in a solution set $\mathcal{S}$. These notions are kept deliberately vague for expository purposes. Given $d \in \mathcal{D}$, we desire to compute $f(d)$. Suppose d^* is some approximation to d. If $f(d^*)$ is "near" $f(d)$, the problem is said to be well conditioned. If $f(d^*)$ may potentially differ greatly from $f(d)$ even when d^* is near d, the problem is said to be ill-conditioned. The concept of "near" can be made precise by introducing norms in the appropriate spaces. We can then define the condition of the problem f with respect to these norms as

$$\kappa[f(d)] = \lim_{\delta\to 0} \sup_{d_2(d,d^*)=\delta} \left[\frac{d_1(f(d),f(d^*))}{\delta}\right], \tag{1.7}$$

where $d_i\,(\cdot\,,\cdot)$ are distance functions in the appropriate spaces. When $\kappa[f(d)]$ is infinite, the problem of determining $f(d)$ from d is *ill-posed* (as opposed to *well-posed*). When $\kappa[f(d)]$ is finite and *relatively large* (or *relatively small*), the problem is said to be *ill-conditioned* (or *well-conditioned*).

A simple example of an ill-conditioned problem is the following. Consider the $n \times n$ matrix

$$A = \begin{bmatrix} 0 & 1 & 0 & \cdot & \cdot & \cdot & 0 \\ \cdot & \cdot & \cdot & \cdot & \cdot & \cdot & \cdot \\ \cdot & & \cdot & \cdot & \cdot & \cdot & \cdot \\ \cdot & & & \cdot & \cdot & \cdot & 0 \\ \cdot & & & & \cdot & \cdot & 1 \\ 0 & \cdot & \cdot & \cdot & \cdot & \cdot & 0 \end{bmatrix},$$

with n eigenvalues at 0. Now, consider a small perturbation of the data (the n^2 elements of A) consisting of adding the number 2^{-n} to the first element in the last (nth) row of A. This perturbed matrix then has n distinct eigenvalues $\lambda_1, \ldots, \lambda_n$ with $\lambda_k = 1/2\exp(2k\pi j/n)$, where $j := \sqrt{-1}$. Thus, we see that this small perturbation in the data has been magnified by a factor on the order of 2^n resulting in a rather large perturbation in solving the problem of computing the eigenvalues of A. Further details and related examples can be found in [9,17].

Thus far, we have not mentioned how the problem f above (computing the eigenvalues of A in the example) was to be solved. Conditioning is a function solely of the problem itself. To solve a problem numerically, we must implement some numerical procedures or algorithms which we denote by f^*. Thus, given d, $f^*(d)$ is the result of applying the algorithm to d (for simplicity, we assume d is "representable"; a more general definition can be given when some approximation d^{**} to d must be used). The algorithm f^* is said to be numerically (backward) stable if, for all $d \in \mathcal{D}$, there exists $d^* \in \mathcal{D}$ near d so that $f^*(d)$ is near $f(d^*)$, ($f(d^*)$ = the exact solution of a nearby problem). If the problem is well-conditioned, then $f(d^*)$ is near $f(d)$ so that $f^*(d)$ is near $f(d)$ if f^* is numerically stable. In other words, f^* does not introduce any more sensitivity to perturbation than is inherent in the problem. Example 1.1 further illuminates this definition of stability which, on a first reading, can seem somewhat confusing.

Of course, one cannot expect a stable algorithm to solve an ill-conditioned problem any more accurately than the data warrant, but an unstable algorithm can produce poor solutions even to well-conditioned problems. Example 1.2, illustrates this phenomenon. There are thus two separate factors to consider in determining the accuracy of a computed solution $f^*(d)$. First, if the algorithm is stable, $f^*(d)$ is near $f(d^*)$, for some d^*, and second, if the problem is well conditioned, then, as above, $f(d^*)$ is near $f(d)$. Thus, $f^*(d)$ is near $f(d)$ and we have an "accurate" solution.

Rounding errors can cause unstable algorithms to give disastrous results. However, it would be virtually impossible to account for every rounding error made at every arithmetic operation in a complex series of calculations. This would constitute a *forward* error analysis. The concept of *backward* error analysis based on the definition of numerical stability given above provides a more practical alternative. To illustrate this, let us consider the singular value decomposition (SVD) of an arbitrary $m \times n$ matrix A with coefficients in $\mathbb{R}$ or $\mathbb{C}$ [8] (see also Section 1.3.3),

$$A = U\Sigma V^H\,. \tag{1.8}$$

Here U and V are $m \times m$ and $n \times n$ unitary matrices, respectively, and Σ is an $m \times n$ matrix of the form

$$\Sigma = \left[\begin{array}{c|c} \Sigma_r & 0 \\ \hline 0 & 0 \end{array}\right]; \quad \Sigma_r = \mathrm{diag}\{\sigma_1, \ldots, \sigma_r\} \tag{1.9}$$

with the *singular values* σ_i positive and satisfying $\sigma_1 \geq \sigma_2 \cdots \geq \sigma_r > 0$. The computation of this decomposition is, of course, subject to rounding errors. Denoting computed quantities by an overbar, for some *error matrix* E_A,

$$\overline{A} = A + E_A = \overline{U\Sigma V}^H. \tag{1.10}$$

The computed decomposition thus corresponds exactly to a *perturbed* matrix $\overline{A}$. When using the SVD algorithm available in the literature [8], this perturbation can be bounded by

$$\| E_A \| \leq \pi\epsilon \| A \|, \tag{1.11}$$

where ϵ is the machine precision and π is some quantity depending on the dimensions m and n, but reasonably close to 1 (see also [14, p. 74]). Thus, the *backward error* E_A induced by this algorithm has roughly the same norm as the *input error* E_i resulting, for example, when reading the data A into the computer. Then, according to the definition of numerical stability given above, when a bound such as that in Equation 1.11 exists for the error induced by a numerical algorithm, the algorithm is said to be *backward stable* [17]. Note that backward stability does not guarantee any bounds on the errors in the result $\overline{U}, \overline{\Sigma}$, and $\overline{V}$. In fact, this depends on how perturbations in the data (namely, $E_A = \overline{A} - A$) affect the resulting decomposition (namely, $E_U = \overline{U} - U, E_\Sigma = \overline{\Sigma} - \Sigma$, and $E_V = \overline{V} - V$). This is commonly measured by the condition $\kappa[f(A)]$.

Backward stability is a property of an algorithm, and the condition is associated with a problem and the specific data for that problem. The errors in the result depend on the stability of the algorithm used and the condition of the problem solved. A *good* algorithm should, therefore, be backward stable because the size of the errors in the result is then mainly due to the condition of the problem, not to the algorithm. An unstable algorithm, on the other hand, may yield a large error even when the problem is well conditioned.

Bounds of the type Equation 1.11 are obtained by an error analysis of the algorithm used, and the condition of the problem is obtained by a sensitivity analysis; for example, see [9,17].

We close this section with two simple examples to illustrate some of the concepts introduced.

Example 1.1:

Let x and y be two floating-point computer numbers and let $fl(x * y)$ denote the result of multiplying them in floating-point computer arithmetic. In general, the product $x * y$ requires more precision to be represented exactly than was used to represent x or y. But for most computers

$$fl(x * y) = x * y(1 + \delta), \tag{1.12}$$

where $|\delta| < \epsilon$ (= relative machine precision). In other words, $fl(x * y)$ is $x * y$ correct to within a unit in the last place. Another way to write Equation 1.12 is as follows:

$$fl(x * y) = x(1 + \delta)^{1/2} * y(1 + \delta)^{1/2}, \tag{1.13}$$

where $|\delta| < \epsilon$. This can be interpreted as follows: the computed result $fl(x * y)$ is the exact product of the two slightly perturbed numbers $x(1 + \delta)^{1/2}$ and $y(1 + \delta)^{1/2}$. The slightly perturbed data (not unique) may not even be representable as floating-point numbers. The representation of Equation 1.13 is simply a way of accounting for the roundoff incurred in the algorithm by an initial (small) perturbation in the data.

Example 1.2:

Gaussian elimination with no pivoting for solving the linear system of equations

$$Ax = b \tag{1.14}$$

is known to be numerically unstable; see for example [8] and Section 1.3. The following data illustrate this phenomenon. Let

$$A = \begin{bmatrix} 0.0001 & 1.000 \\ 1.000 & -1.000 \end{bmatrix}, \quad b = \begin{bmatrix} 1.000 \\ 0.000 \end{bmatrix}.$$

All computations are carried out in four-significant-figure decimal arithmetic. The "true answer" $x = A^{-1}b$ is

$$\begin{bmatrix} 0.9999 \\ 0.9999 \end{bmatrix}.$$

Using row 1 as the "pivot row" (i.e., subtracting 10,000 $\times$ row 1 from row 2) we arrive at the equivalent triangular system

$$\begin{bmatrix} 0.0001 & 1.000 \\ 0 & -1.000 \times 10^4 \end{bmatrix} \begin{bmatrix} x_1 \\ x_2 \end{bmatrix} = \begin{bmatrix} 1.000 \\ -1.000 \times 10^4 \end{bmatrix}.$$

The coefficient multiplying x_2 in the second equation should be $-10,001$, but because of roundoff, becomes $-10,000$. Thus, we compute $x_2 = 1.000$ (a good approximation), but back substitution in the equation

$$0.0001x_1 = 1.000 - fl(1.000 * 1.000)$$

yields $x_1 = 0.000$. This extremely bad approximation to x_1 is the result of numerical instability. The problem, it can be shown, is quite well conditioned.

1.3 Fundamental Problems in Numerical Linear Algebra

In this section, we give a brief overview of some of the fundamental problems in numerical linear algebra that serve as building blocks or "tools" for the solution of problems in systems, control, and estimation.

1.3.1 Linear Algebraic Equations and Linear Least-Squares Problems

Probably the most fundamental problem in numerical computing is the calculation of a vector x which satisfies the linear system

$$Ax = b, \tag{1.15}$$

where $A \in \mathbb{R}^{n\times n}$(or $\mathbb{C}^{n\times n}$) and has rank n. A great deal is now known about solving Equation 1.15 in finite arithmetic both for the general case and for a large number of special situations, for example, see [8,9].

The most commonly used algorithm for solving Equation 1.15 with general A and small n (say $n \leq 1000$) is Gaussian elimination with some sort of pivoting strategy, usually "partial pivoting." This amounts to factoring some permutation of the rows of A into the product of a unit lower triangular matrix L and an upper triangular matrix U. The algorithm is effectively stable, that is, it can be proved that the computed solution is near the exact solution of the system

$$(A + E)x = b \tag{1.16}$$

with $|e_{ij}| \leq \phi(n)\,\gamma\,\beta\,\epsilon$, where $\phi(n)$ is a modest function of n depending on details of the arithmetic used, γ is a "growth factor" (which is a function of the pivoting strategy and is usually—but not always—small), β behaves essentially like $\|A\|$, and ϵ is the machine precision. In other words, except for moderately pathological situations, E is "small"—on the order of $\epsilon\,\|A\|$.

The following question then arises. If, because of rounding errors, we are effectively solving Equation 1.16 rather than Equation 1.15, what is the relationship between $(A+E)^{-1}b$ and $A^{-1}b$? To answer this question, we need some elementary perturbation theory and this is where the notion of condition number arises. A condition number for Equation 1.15 is given by

$$\kappa(A) := \|A\| \, \|A^{-1}\|. \tag{1.17}$$

Simple perturbation results can show that perturbation in A and/or b can be magnified by as much as $\kappa(A)$ in the computed solution. Estimating $\kappa(A)$ (since, of course, A^{-1} is unknown) is thus a crucial aspect of assessing solutions of Equation 1.15 and the particular estimating procedure used is usually the principal difference between competing linear equation software packages. One of the more sophisticated and reliable condition estimators presently available is implemented in LINPACK [5] and its successor LAPACK [2]. LINPACK and LAPACK also feature many codes for solving Equation 1.14 in case A has certain special structures (e.g., banded, symmetric, or positive definite).

Another important class of linear algebra problems, and one for which codes are available in LINPACK and LAPACK, is the linear least-squares problem

$$\min \|Ax - b\|_2, \tag{1.18}$$

where $A \in \mathbb{R}^{m\times n}$ and has rank k, with (in the simplest case) $k = n \le m$, for example, see [8]. The solution of Equation 1.18 can be written formally as $x = A^+b$. The method of choice is generally based on the QR factorization of A (for simplicity, let $\text{rank}(A) = n$)

$$A = QR, \tag{1.19}$$

where $R \in \mathbb{R}^{n\times n}$ is upper triangular and $Q \in \mathbb{R}^{m\times n}$ has orthonormal columns, that is, $Q^TQ = I$. With special care and analysis, the case $k < n$ can also be handled similarly. The factorization is effected through a sequence of Householder transformations H_i applied to A. Each H_i is symmetric and orthogonal and of the form $I - 2uu^T/u^Tu$, where $u \in \mathbb{R}^m$ is specially chosen so that zeros are introduced at appropriate places in A when it is premultiplied by H_i. After n such transformations,

$$H_nH_{n-1}\dots H_1A = \begin{bmatrix} R \\ 0 \end{bmatrix},$$

from which the factorization Equation 1.19 follows. Defining c and d by

$$\begin{bmatrix} c \\ d \end{bmatrix} := H_nH_{n-1}\dots H_1b,$$

where $c \in \mathbb{R}^n$, it is easily shown that the least-squares solution x of Equation 1.18 is given by the solution of the linear system of equations

$$Rx = c\,. \tag{1.20}$$

The above algorithm is numerically stable and, again, a well-developed perturbation theory exists from which condition numbers can be obtained, this time in terms of

$$\kappa(A) := \|A\| \, \|A^+\|.$$

Least-squares perturbation theory is fairly straightforward when $\text{rank}(A) = n$, but is considerably more complicated when A is rank deficient. The reason for this is that, although the inverse is a continuous function of the data (i.e., the inverse is a continuous function in a neighborhood of a nonsingular matrix), the pseudoinverse is discontinuous. For example, consider

$$A = \begin{bmatrix} 1 & 0 \\ 0 & 0 \end{bmatrix} = A^+$$

and perturbations

$$E_1 = \begin{bmatrix} 0 & 0 \\ \delta & 0 \end{bmatrix} \quad \text{and} \quad E_2 = \begin{bmatrix} 0 & 0 \\ 0 & \delta \end{bmatrix}$$

with δ being small. Then

$$(A+E_1)^+ = \begin{bmatrix} \dfrac{1}{1+\delta^2} & \dfrac{\delta}{1+\delta^2} \\ 0 & 0 \end{bmatrix},$$

which is close to A^+ but

$$(A+E_2)^+ = \begin{bmatrix} 1 & 0 \\ 0 & \dfrac{1}{\delta} \end{bmatrix},$$

which gets arbitrarily far from A^+ as δ is decreased toward 0.

In lieu of Householder transformations, Givens transformations (elementary rotations or reflections) may also be used to solve the linear least-squares problem [8]. Givens transformations have received considerable attention for solving linear least-squares problems and systems of linear equations in a parallel computing environment. The capability of introducing zero elements selectively and the need for only local interprocessor communication make the technique ideal for "parallelization."

1.3.2 Eigenvalue and Generalized Eigenvalue Problems

In the algebraic eigenvalue/eigenvector problem for $A \in \mathbb{R}^{n\times n}$, one seeks nonzero solutions $x \in \mathbb{C}^n$ and $\lambda \in \mathbb{C}$, which satisfy

$$Ax = \lambda x. \tag{1.21}$$

The classic reference on the numerical aspects of this problem is Wilkinson [17]. A briefer textbook introduction is given in [8].

Quality mathematical software for eigenvalues and eigenvectors is available; the EISPACK [7,15] collection of subroutines represents a pivotal point in the history of mathematical software. The successor to EISPACK (and LINPACK) is LAPACK [2], in which the algorithms and software have been restructured to provide high efficiency on vector processors, high-performance workstations, and shared memory multiprocessors.

The most common algorithm now used to solve Equation 1.21 for general A is the QR algorithm of Francis [17]. A shifting procedure enhances convergence and the usual implementation is called the double-Francis-QR algorithm. Before the QR process is applied, A is initially reduced to upper Hessenberg form A_H ($a_{ij} = 0$ if $i - j \geq 2$). This is accomplished by a finite sequence of similarities of the Householder form discussed above. The QR process then yields a sequence of matrices orthogonally similar to A and converging (in some sense) to a so-called quasi-upper triangular matrix S also called the real Schur form (RSF) of A. The matrix S is block upper triangular with 1×1 diagonal blocks corresponding to real eigenvalues of A and 2×2 diagonal blocks corresponding to complex-conjugate pairs of eigenvalues. The quasi-upper triangular form permits all arithmetic to be real rather than complex as would be necessary for convergence to an upper triangular matrix. The orthogonal transformations from both the Hessenberg reduction and the QR process may be accumulated in a single orthogonal transformation U so that

$$U^T A U = R \tag{1.22}$$

compactly represents the entire algorithm. An analogous process can be applied in the case of symmetric A, and considerable simplifications and specializations result.

Closely related to the QR algorithm is the QZ algorithm for the generalized eigenvalue problem

$$Ax = \lambda Mx, \tag{1.23}$$

where $A, M \in \mathbb{R}^{n \times n}$. Again, a Hessenberg-like reduction, followed by an iterative process, is implemented with orthogonal transformations to reduce Equation 1.23 to the form

$$QAZy = \lambda QMZy, \tag{1.24}$$

where QAZ is quasi-upper triangular and QMZ is upper triangular. For a review and references to results on stability, conditioning, and software related to Equation 1.23 and the QZ algorithm, see [8]. The generalized eigenvalue problem is both theoretically and numerically more difficult to handle than the ordinary eigenvalue problem, but it finds numerous applications in control and system theory [14, p. 109].

1.3.3 The Singular Value Decomposition and Some Applications

One of the basic and most important tools of modern numerical analysis, especially numerical linear algebra, is the SVD. Here we make a few comments about its properties and computation as well as its significance in various numerical problems.

Singular values and the SVD have a long history, especially in statistics and numerical linear algebra. These ideas have found applications in the control and signal processing literature, although their use there has been overstated somewhat in certain applications. For a survey of the SVD, its history, numerical details, and some applications in systems and control theory, see [14, p. 74].

The fundamental result was stated in Section 1.2 (for the complex case). The result for the real case is similar and is stated below.

Theorem 1.1:

Let $A \in \mathbb{R}^{m \times n}$ with rank(A) $= r$. Then there exist orthogonal matrices $U \in \mathbb{R}^{m \times m}$ and $V \in \mathbb{R}^{n \times n}$ so that

$$A = U\Sigma V^T, \tag{1.25}$$

where

$$\Sigma = \begin{bmatrix} \Sigma_r & 0 \\ 0 & 0 \end{bmatrix}$$

and $\Sigma_r = diag\,\{\sigma_1, \ldots, \sigma_r\}$ with $\sigma_1 \geq \cdots \geq \sigma_r > 0$.

The proof of Theorem 1.1 is straightforward and can be found, for example, in [8]. Geometrically, the theorem says that bases can be found (separately) in the domain and codomain spaces of a linear map with respect to which the matrix representation of the linear map is diagonal. The numbers $\sigma_1, \ldots, \sigma_r$, together with $\sigma_{r+1} = 0, \ldots, \sigma_n = 0$, are called the singular values of A, and they are the positive square roots of the eigenvalues of A^TA. The columns $\{u_k, k = 1, \ldots, m\}$ of U are called the left singular vectors of A (the orthonormal eigenvectors of AA^T), while the columns $\{v_k, k = 1, \ldots, n\}$ of V are called the right singular vectors of A (the orthonormal eigenvectors of A^TA). The matrix A can then be written (as a dyadic expansion) also in terms of the singular vectors as follows:

$$A = \sum_{k=1}^{r} \sigma_k u_k v_k^T.$$

The matrix A^T has m singular values, the positive square roots of the eigenvalues of AA^T. The r [= rank(A)] nonzero singular values of A and A^T are, of course, the same. The choice of A^TA rather than

AA^T in the definition of singular values is arbitrary. Only the nonzero singular values are usually of any real interest and their number, given the SVD, is the rank of the matrix. Naturally, the question of how to distinguish nonzero from zero singular values in the presence of rounding error is a nontrivial task.

It is not generally advisable to compute the singular values of A by first finding the eigenvalues of A^TA, tempting as that is. Consider the following example, where μ is a real number with $|\mu| < \sqrt{\epsilon}$ (so that $fl(1+\mu^2) = 1$, where $fl(\cdot)$ denotes floating-point computation). Let

$$A = \begin{bmatrix} 1 & 1 \\ \mu & 0 \\ 0 & \mu \end{bmatrix}.$$

Then

$$fl(A^TA) = \begin{bmatrix} 1 & 1 \\ 1 & 1 \end{bmatrix}.$$

So we compute $\hat{\sigma}_1 = \sqrt{2}$, $\hat{\sigma}_2 = 0$ leading to the (erroneous) conclusion that the rank of A is 1. Of course, if we could compute in infinite precision, we would find

$$A^TA = \begin{bmatrix} 1+\mu^2 & 1 \\ 1 & 1+\mu^2 \end{bmatrix}$$

with $\sigma_1 = \sqrt{2+\mu^2}, \sigma_2 = |\mu|$ and thus $\text{rank}(A) = 2$. The point is that by working with A^TA we have unnecessarily introduced μ^2 into the computations. The above example illustrates a potential pitfall in attempting to form and solve the normal equations in a linear least-squares problem and is at the heart of what makes square root filtering so attractive numerically. Very simplistically, square root filtering involves working directly on an "A-matrix," for example, updating it, as opposed to updating an "A^TA-matrix."

Square root filtering is usually implemented with the QR factorization (or some closely related algorithm) as described previously rather than SVD. Moreover, critical information may be lost irrecoverably by simply forming A^TA.

Returning now to the SVD, two features of this matrix factorization make it attractive in finite arithmetic: first, it can be computed in a numerically stable way, and second, singular values are well conditioned. Specifically, there is an efficient and numerically stable algorithm by Golub and Reinsch [8] which works directly on A to give the SVD. This algorithm has two phases. In the first phase, it computes orthogonal matrices U_1 and V_1 so that $B = {U_1}^TAV_1$ is in bidiagonal form, that is, only the elements on its diagonal and first superdiagonal are nonzero. In the second phase, the algorithm uses an iterative procedure to compute orthogonal matrices U_2 and V_2 so that ${U_2}^TBV_2$ is diagonal and nonnegative. The SVD defined in Equation 1.25 is then $\Sigma = U^TBV$, where $U = U_1U_2$ and $V = V_1V_2$. The computed U and V are orthogonal approximately to the working precision, and the computed singular values are the exact σ_i's for $A+E$, where $\|E\|/\|A\|$ is a modest multiple of ϵ. Fairly sophisticated implementations of this algorithm can be found in [5,7]. The well-conditioned nature of the singular values follows from the fact that if A is perturbed to $A+E$, then it can be proved that

$$\|\sigma_i(A+E) - \sigma_i(A)\| \leq \|E\|.$$

Thus, the singular values are computed with small absolute error although the relative error of sufficiently small singular values is not guaranteed to be small.

It is now acknowledged that the singular value decomposition is the most generally reliable method of determining rank numerically (see [14, p. 589] for a more elaborate discussion). However, it is considerably more expensive to compute than, for example, the QR factorization which, with column pivoting [5], can usually give equivalent information with less computation. Thus, while the SVD is a useful theoretical tool, its use for actual computations should be weighed carefully against other approaches.

The problem of numerical determination of rank is now well understood. The essential idea is to try to determine a "gap" between "zero" and the "smallest nonzero singular value" of a matrix A. Since the computed values are exact for a matrix near A, it makes sense to consider the ranks of all matrices in some δ-ball (with respect to the spectral norm $\|\cdot\|$, say) around A. The choice of δ may also be based on measurement errors incurred in estimating the coefficients of A, or the coefficients may be uncertain because of rounding errors incurred in a previous computation. However, even with SVD, numerical determination of rank in finite arithmetic is a difficult problem.

That other methods of rank determination are potentially unreliable is demonstrated by the following example. Consider the Ostrowski matrix $A \in \mathbb{R}^{n\times n}$ whose diagonal elements are all -1, whose upper triangle elements are all $+1$, and whose lower triangle elements are all 0. This matrix is clearly of rank n, that is, is invertible. It has a good "solid" upper triangular shape. All of its eigenvalues (-1) are well away from zero. Its determinant $(-1)^n$ is definitely not close to zero. But this matrix is, in fact, very nearly singular and becomes more nearly so as n increases. Note, for example, that

$$\begin{bmatrix} -1 & +1 & \cdots & & \cdots & +1 \\ 0 & \ddots & \ddots & & & \vdots \\ \vdots & \ddots & \ddots & \ddots & & \\ & & & \ddots & \ddots & \vdots \\ \vdots & & & \ddots & \ddots & +1 \\ 0 & \cdots & & \cdots & 0 & -1 \end{bmatrix} \begin{bmatrix} 1 \\ 2^{-1} \\ \vdots \\ 2^{-n+1} \end{bmatrix} = \begin{bmatrix} -2^{-n+1} \\ -2^{-n+1} \\ \vdots \\ -2^{-n+1} \end{bmatrix} \to \begin{bmatrix} 0 \\ 0 \\ \vdots \\ 0 \end{bmatrix} \quad (n \to +\infty).$$

Moreover, adding 2^{-n+1} to every element in the first column of A gives an exactly singular matrix. Arriving at such a matrix by, say, Gaussian elimination would give no hint as to the near singularity. However, it is easy to check that $\sigma_n(A)$ behaves as 2^{-n+1}. A corollary for control theory is that eigenvalues do not necessarily give a reliable measure of "stability margin." It is useful to note that in this example of an invertible matrix, the crucial quantity, $\sigma_n(A)$, which measures nearness to singularity, is simply $1/\|A^{-1}\|$, and the result is familiar from standard operator theory. There is nothing intrinsic about singular values in this example and, in fact, $\|A^{-1}\|$ might be more cheaply computed or estimated in other matrix norms.

Because rank determination, in the presence of rounding error, is a nontrivial problem, the same difficulties naturally arise in any problem equivalent to, or involving, rank determination, such as determining the independence of vectors, finding the dimensions of certain subspaces, etc. Such problems arise as basic calculations throughout systems, control, and estimation theory. Selected applications are discussed in more detail in [14, p. 74] and in [1,4,10].

Finally, let us close this section with a brief example illustrating a totally inappropriate use of SVD. The rank condition

$$\operatorname{rank}\,[B, AB, \ldots, A^{n-1}B] = n \tag{1.26}$$

for the controllability of Equation 1.1 is too well known. Suppose

$$A = \begin{bmatrix} 1 & \mu \\ 0 & 1 \end{bmatrix}, \quad B = \begin{bmatrix} 1 \\ \mu \end{bmatrix}$$

with $|\mu| < \sqrt{\epsilon}$. Then

$$fl[B, AB] = \begin{bmatrix} 1 & 1 \\ \mu & \mu \end{bmatrix},$$

and now even applying SVD, the erroneous conclusion of uncontrollability is reached. Again the problem is in just forming AB; not even SVD can come to the rescue after that numerical *faux pas*.

1.4 Applications to Systems and Control

A reasonable approach to developing numerically reliable algorithms for computational problems in linear system theory would be to reformulate the problems as concatenations of subproblems for which numerically stable algorithms are available. Unfortunately, one cannot ensure that the stability of algorithms for the subproblems results in the stability of the overall algorithm. This requires separate analysis that may rely on the sensitivity or condition of the subproblems. In the next section, we show that delicate (i.e., badly conditioned) subproblems should be avoided whenever possible; a few examples are given where a possibly badly conditioned step is circumvented by carefully modifying or completing existing algorithms; see, for example, [14, p. 109].

A second difficulty is the ill-posedness of some of the problems occurring in linear system theory. Two approaches can be adopted. One can develop an acceptable perturbation theory for such problems, using a concept such as *restricted condition*, which is, the condition under perturbations for which a certain property holds, for example, fixed rank [14, p. 109]. One then looks for restricting assumptions that make the problem well posed. Another approach is to delay any such *restricting choices* to the end and leave it to the user to decide which choice to make by looking at the results. The algorithm then provides quantitative measures that help the user make this choice; see, for example, [14, p. 171, 529]. By this approach, one may avoid artificial restrictions of the first approach that sometimes do not respect the practical significance of the problem.

A third possible *pitfall* is that many users almost always prefer fast algorithms to slower ones. However, slower algorithms are often more reliable.

In the subsections that follow, we survey a representative selection of numerical linear algebra problems arising in linear systems, control, and estimation theory, which have been examined with some of the techniques described in the preceding sections. Many of these topics are discussed briefly in survey papers such as [11] and [16] and in considerably more detail in the papers included or referenced in [14] and in [1,4,10]. Some of the scalar algorithms discussed here do not extend trivially to the matrix case. When they do, we mention only the matrix case. Moreover, we discuss only the numerical aspects here; for the system-theoretical background, we refer the reader to the control and systems literature.

1.4.1 Some Typical Techniques

Most of the reliable techniques in numerical linear algebra are based on the use of orthogonal transformations. Typical examples of this are the QR decomposition for least-squares problems, the Schur decomposition for eigenvalue and generalized eigenvalue problems, and the SVD for rank determinations and generalized inverses. Orthogonal transformations also appear in most of the reliable linear algebra techniques for control theory. This is partly due to the direct application of existing linear algebra decompositions to problems in control. Obvious examples of this are the Schur approach for solving algebraic Riccati equations, both continuous- and discrete-time [14, p. 529, 562, 573], for solving Lyapunov equations [14, p. 430] and for performing pole placement [14, p. 415]. New orthogonal decompositions have also been introduced that rely heavily on the same principles but were specifically developed for problems encountered in control. Orthogonal state-space transformations on a system $\{A, B, C\}$ result in a new state-space representation $\{U^H AU, U^H B, CU\}$, where U performs some kind of decomposition on the matrices A, B, and C. These special forms, termed "condensed forms," include

- The state Schur form [14, p. 415]
- The state Hessenberg form [14, p. 287]
- The observer Hessenberg form [14, p. 289, 392]
- The controller Hessenberg form [14, p. 128, 357].

Staircase forms or block Hessenberg forms are other variants of these condensed forms that have proven useful in dealing with MIMO systems [14, p. 109, 186, 195].

There are two main reasons for using these orthogonal state-space transformations:

- The numerical sensitivity of the control problem being solved is not affected by these transformations because sensitivity is measured by norms or angles of certain spaces and these are unaltered by orthogonal transformations.
- Orthogonal transformations have minimum condition number, essential in proving bounded error propagation and establishing numerical stability of the algorithm that uses such transformations.

More details on this are given in [14, p. 128] and in subsequent sections where some of these condensed forms are used for particular applications.

1.4.2 Transfer Functions, Poles, and Zeros

In this section, we discuss important structural properties of linear systems and the numerical techniques available for determining them. The transfer function $R(\lambda)$ of a linear system is given by a polynomial representation $V(\lambda)T^{-1}(\lambda)U(\lambda)+W(\lambda)$ or by a state-space model $C(\lambda I-A)^{-1}B+D$. The results in this subsection hold for both the discrete-time case (where λ stands for the shift operator z) and the continuous-time case (where λ stands for the differentiation operator D).

1.4.2.1 The Polynomial Approach

One is interested in a number of structural properties of the transfer function $R(\lambda)$, such as poles, transmission zeros, decoupling zeros, etc. In the scalar case, where $\{T(\lambda), U(\lambda), V(\lambda), W(\lambda)\}$ are scalar polynomials, all of this can be found with a greatest common divisor (GCD) extraction routine and a rootfinder, for which reliable methods exist. In the matrix case, the problem becomes much more complex and the basic method for GCD extraction, the Euclidean algorithm, becomes unstable (see [14, p. 109]). Moreover, other structural elements (null spaces, etc.) come into the picture, making the polynomial approach less attractive than the state-space approach [14, p. 109, and references therein].

1.4.2.2 The State-Space Approach

The structural properties of interest are poles and zeros of $R(\lambda)$, decoupling zeros, controllable and unobservable subspaces, supremal (A, B)-invariant and controllability subspaces, factorizability of $R(\lambda)$, left and right null spaces of $R(\lambda)$, etc. These concepts are fundamental in several design problems and have received considerable attention over the last few decades; see, for example, [14, p. 74, 109, 174, 186, 529]. In [14, p. 109], it is shown that all the concepts mentioned above can be considered generalized eigenstructure problems and that they can be computed via the Kronecker canonical form of the pencils

$$\begin{array}{cc} [\lambda I - A] & [\lambda I - A \mid B] \\ \left[\begin{array}{c} \lambda I - A \\ \hline -C \end{array}\right] & \left[\begin{array}{c|c} \lambda I - A & B \\ \hline -C & D \end{array}\right] \end{array} \tag{1.27}$$

or from other pencils derived from these. Backward stable software is also available for computing the Kronecker structure of an arbitrary pencil. A remaining problem here is that determining several of the structural properties listed above may be ill-posed in some cases in which one has to develop the notion of restricted condition (see [14, p. 109]). A completely different approach is to reformulate the problem as an approximation or optimization problem for which *quantitative measures* are derived, leaving the final choice to the user. Results in this vein are obtained for controllability, observability [14, p. 171, 186, 195] (almost) (A, B)-invariant, and controllability subspaces.

1.4.3 Controllability and Other "Abilities"

The various "abilities" such as controllability, observability, reachability, reconstructibility, stabilizability, and detectability are basic to the study of linear control and system theory. These concepts can also be viewed in terms of decoupling zeros, controllable and unobservable subspaces, controllability subspaces, etc. mentioned in the previous section. Our remarks here are confined, but not limited, to the notion of controllability.

A large number of algebraic and dynamic characterizations of controllability have been given; see [11] for a sample. But each one of these has difficulties when implemented in finite arithmetic. For a survey of this topic and numerous examples, see [14, p. 186]. Part of the difficulty in dealing with controllability numerically lies in the intimate relationship with the invariant subspace problem [14, p. 589]. The controllable subspace associated with Equation 1.1 is the smallest A-invariant subspace (subspace spanned by eigenvectors or principal vectors) containing the range of B. Since the A-invariant subspaces can be extremely sensitive to perturbation, it follows that, so too, is the controllable subspace. Similar remarks apply to the computation of the so-called controllability indices. The example discussed in the third paragraph of Section 1.2 dramatically illustrates these remarks. The matrix A has but one eigenvector (associated with 0), whereas the slightly perturbed A has n eigenvectors associated with the n distinct eigenvalues.

Attempts have been made to provide numerically stable algorithms for the pole placement problem discussed in a later section. It suffices to mention here that the problem of pole placement by state feedback is closely related to controllability. Work on developing numerically stable algorithms for pole placement is based on the reduction of A to a Hessenberg form; see, for example, [14, p. 357, 371, 380]. In the single-input case, a good approach is the controller Hessenberg form mentioned above where the state matrix A is upper Hessenberg and the input vector B is a multiple of $(1, 0, \ldots, 0)^T$. The pair (A, B) is then controllable if, and only if, all $(n-1)$ subdiagonal elements of A are nonzero. If a subdiagonal element is 0, the system is uncontrollable, and a basis for the uncontrollable subspace is easily constructed. The transfer function gain or first nonzero Markov parameter is also easily constructed from this "canonical form." In fact, the numerically more robust system Hessenberg form, playing an ever-increasing role in system theory, is replacing the numerically more fragile special case of the companion or rational canonical or Luenberger canonical form.

A more important aspect of controllability is a topological notion such as "near uncontrollability." But there are numerical difficulties here also, and we refer to Parts 3 and 4 of [14] for further details. Related to this is an interesting system-theoretic concept called "balancing" discussed in Moore's paper [14, p. 171]. The computation of "balancing transformations" is discussed in [14, p. 642].

There are at least two distinct notions of near uncontrollability [11] in the parametric sense and in the energy sense. In the parametric sense, a controllable pair (A, B) is said to be near uncontrollable if the parameters of (A, B) need be perturbed by only a relatively small amount for (A, B) to become uncontrollable. In the energy sense, a controllable pair is near-uncontrollable if large amounts of control energy ($\int u^T u$) are required for a state transfer. The pair

$$A = \begin{bmatrix} 0 & 1 & & \cdots & 0 \\ \vdots & \ddots & \ddots & & \vdots \\ & & & & \\ \vdots & & & \ddots & 1 \\ 0 & \cdots & & \cdots & 0 \end{bmatrix}, \quad B = \begin{bmatrix} 0 \\ \vdots \\ 0 \\ 1 \end{bmatrix}$$

is very near-uncontrollable in the energy sense but not as badly as in the parametric sense. Of course, both measures are coordinate dependent and "balancing" is one attempt to remove this coordinate bias. The pair (A, B) above is in "controllable canonical form." It is now known that matrices in this form (specifically, the A matrix in rational canonical form) almost always exhibit poor numerical behavior and are "close to" uncontrollable (unstable, etc.) as the size n increases. For details, see [14, p. 59].

1.4.4 Computation of Objects Arising in the Geometric Theory of Linear Multivariable Control

A great many numerical problems arise in the geometric approach to control of systems modeled as Equations 1.1 and 1.2. Some of these are discussed in the paper by Klema and Laub [14, p. 74]. The power of the geometric approach derives in large part from the fact that it is independent of specific coordinate systems or matrix representations. Numerical issues are a separate concern.

A very thorough numerical treatment of numerical problems in linear system theory has been given by Van Dooren [14, p. 109]. This work has applications for most calculations with linear state-space models. For example, one by-product is an extremely reliable algorithm (similar to an orthogonal version of Silverman's structure algorithm) for the computation of multivariable system zeros [14, p. 271]. This method involves a generalized eigenvalue problem (the Rosenbrock pencil), but the "infinite zeros" are first removed by deflating the given matrix pencil.

1.4.5 Frequency Response Calculations

Many properties of a linear system such as Equations 1.1 and 1.2 are known in terms of its frequency response matrix

$$G(j\omega) := C(j\omega I - A)^{-1}B + D; \quad (\omega \geq 0) \tag{1.28}$$

(or $G(e^{j\theta})$; $\theta \in [0, 2\pi]$ for Equations 1.3 and 1.4). In fact, various norms of the return difference matrix $I + G(j\omega)$ and related quantities have been investigated in control and system theory to providing robust linear systems with respect to stability, noise response, disturbance attenuation, sensitivity, etc.

Thus it is important to compute $G(j\omega)$ efficiently, given A, B, and C for a (possibly) large number of values of ω (for convenience we take D to be 0, because if it is nonzero it is trivial to add to G). An efficient and generally applicable algorithm for this problem is presented in [14, p. 287]. Rather than solving the linear equation $(j\omega I - A)X = B$ with dense unstructured A, which would require $O(n^3)$ operations for each successive value of ω, the new method initially reduces A to upper Hessenberg form H. The orthogonal state-space coordinate transformations used to obtain the Hessenberg form of A are incorporated into B and C giving $\tilde{B}$ and $\tilde{C}$. As ω varies, the coefficient matrix in the linear equation $(j\omega I - H)X = \tilde{B}$ remains in upper Hessenberg form. The advantage is that X can now be found in $O(n^2)$ operations rather than $O(n^3)$ as before, a substantial saving. Moreover, the method is numerically very stable (via either LU or QR factorization) and has the advantage of being independent of the eigenstructure (possibly ill-conditioned) of A. Another efficient and reliable algorithm for frequency response computation [14, p. 289] uses the observer Hessenberg form mentioned in Section 1.4.1 together with a determinant identity and a property of the LU decomposition of a Hessenberg matrix.

The methods above can also be extended to state-space models in implicit form, that is, where Equation 1.1 is replaced by

$$E\dot{x} = Ax + Bu. \tag{1.29}$$

Then Equation 1.28 is replaced with

$$G(j\omega) = C(j\omega E - A)^{-1}B + D, \tag{1.30}$$

and the initial triangular/Hessenberg reduction [8] can be employed again to reduce the problem to updating the diagonal of a Hessenberg matrix and consequently an $O(n^2)$ problem.

An improvement for the frequency response evaluation problem is using matrix interpolation methods to achieve even greater computational efficiency.

1.4.6 Numerical Solution of Linear Ordinary Differential Equations and Matrix Exponentials

The "simulation" or numerical solution of linear systems of ordinary differential equations (ODEs) of the form

$$\dot{x}(t) = Ax(t) + f(t), \quad x(0) = x_0, \tag{1.31}$$

is a standard problem that arises in finding the time response of a system in state-space form. However, there is still debate as to the most effective numerical algorithm, particularly when A is defective (i.e., when A is $n \times n$ and has fewer than n linearly independent eigenvectors) or nearly defective. The most common approach involves computing the matrix exponential e^{tA}, because the solution of Equation 1.31 can be written simply as

$$x(t) = e^{tA}x_0 + \int_0^t e^{(t-s)A}f(s)\,ds.$$

A delightful survey of computational techniques for matrix exponentials is given in [14, p. 649]. Nineteen "dubious" ways are explored (there are many more ways not discussed) but no clearly superior algorithm is singled out. Methods based on Padé approximation or reduction of A to RSF are generally attractive while methods based on Taylor series or the characteristic polynomial of A are generally found unattractive. An interesting open problem is the design of a special algorithm for the matrix exponential when the matrix is known *a priori* to be stable ($\Lambda(A)$ in the left half of the complex plane).

The reason for the adjective "dubious" in the title of [14, p. 649] is that in many (maybe even most) circumstances, it is better to treat Equation 1.31 as a system of differential equations, typically stiff, and to apply various ODE techniques, specially tailored to the linear case. ODE techniques are preferred when A is large and sparse for, in general, e^{tA} is unmanageably large and dense.

1.4.7 Lyapunov, Sylvester, and Riccati Equations

Certain matrix equations arise naturally in linear control and system theory. Among those frequently encountered in the analysis and design of continuous-time systems are the Lyapunov equation

$$AX + XA^T + Q = 0, \tag{1.32}$$

and the Sylvester equation

$$AX + XF + Q = 0. \tag{1.33}$$

The appropriate discrete-time analogs are

$$AXA^T - X + Q = 0 \tag{1.34}$$

and

$$AXF - X + Q = 0. \tag{1.35}$$

Various hypotheses are posed for the coefficient matrices A, F, and Q to ensure certain properties of the solution X.

The literature in control and system theory on these equations is voluminous, but most of it is ad hoc, at best, from a numerical point of view, with little attention to questions of numerical stability, conditioning, machine implementation, and the like.

For the Lyapunov equation, the best overall algorithm in terms of efficiency, accuracy, reliability, availability, and ease of use is that of Bartels and Stewart [14, p. 430]. The basic idea is to reduce A to quasi-upper triangular form (or RSF) and to perform a back substitution for the elements of X.

An attractive algorithm for solving Lyapunov equations has been proposed by Hammarling [14, p. 500]. This algorithm is a variant of the Bartels–Stewart algorithm but instead solves directly for the Cholesky factor Y of X: $Y^TY = X$ and Y is upper triangular. Clearly, given Y, X is easily recovered if necessary. But in many applications, for example, [14, p. 642] only the Cholesky factor is required.

For the Lyapunov equation, when A is stable, the solutions of the equations above are also equal to the reachability and observability Grammians $P_r(T)$ and $P_o(T)$, respectively, for $T = +\infty$ for the system $\{A, B, C\}$:

$$\begin{aligned} P_r(T) &= \int_0^T e^{tA}BB^T e^{tA^T}\,dt; \quad P_o(T) = \int_0^T e^{tA^T}C^TCe^{tA}\,dt \\ P_r(T) &= \sum_{k=0}^{T} A^kBB^T(A^T)^k; \quad P_o(T) = \sum_{k=0}^{T}(A^T)^kC^TCA^k. \end{aligned} \tag{1.36}$$

These can be used along with some additional transformations (see [14, p. 171, 642]) to compute the so-called *balanced* realizations $\{\tilde{A}, \tilde{B}, \tilde{C}\}$. For these realizations, both P_o and P_r are equal and diagonal. These realizations have some nice sensitivity properties with respect to poles, zeros, truncation errors in digital filter implementations, etc. [14, p. 171]. They are, therefore, recommended whenever the choice of a realization is left to the user. When A is not stable, one can still use the *finite range* Grammians Equation 1.36, for $T < +\infty$, for balancing [14, p. 171]. A reliable method for computing integrals and sums of the type Equation 1.36 can be found in [14, p. 681]. It is also shown in [14, p. 171] that the reachable subspace and the unobservable subspace are the image and the kernel of $P_r(T)$ and $P_o(T)$, respectively. From these relationships, sensitivity properties of the spaces under perturbations of $P_r(T)$ and $P_o(T)$ can be derived.

For the Sylvester equation, the Bartels–Stewart algorithm reduces both A and F to RSF and then a back substitution is done. It has been demonstrated in [14, p. 495] that some improvement in this procedure is possible by only reducing the larger of A and F to upper Hessenberg form. The stability of this method has been analyzed in [14, p. 495]. Although only *weak stability* is obtained, this is satisfactory in most cases.

Algorithms are also available in the numerical linear algebra literature for the more general Sylvester equation

$$A_1XF_1{}^T + A_2XF_2{}^T + Q = 0$$

and its symmetric Lyapunov counterpart

$$AXF^T + FXA^T + Q = 0.$$

Questions remain about estimating the condition of Lyapunov and Sylvester equations efficiently and reliably in terms of the coefficient matrices. A deeper analysis of the Lyapunov and Sylvester equations is probably a prerequisite to at least a better understanding of condition of the Riccati equation for which, again, there is considerable theoretical literature but not as much known from a purely numerical point of view. The symmetric $n \times n$ algebraic Riccati equation takes the form

$$Q + AX + XA^T - XGX = 0 \tag{1.37}$$

for continuous-time systems and

$$A^TXA - X - A^TXG_1(G_2 + G_1^TXG_1)^{-1}G_1^TXA + Q = 0 \tag{1.38}$$

for discrete-time systems. These equations appear in several design/analysis problems, such as optimal control, optimal filtering, spectral factorization, for example, see the papers in Part 7 of [14] and references therein. Again, appropriate assumptions are made on the coefficient matrices to guarantee the existence and/or uniqueness of certain kinds of solutions X. Nonsymmetric Riccati equations of the form

$$Q + A_1X + XA_2 - XGX = 0 \tag{1.39}$$

for the continuous-time case (along with an analog for the discrete-time case) are also studied and can be solved numerically by the techniques discussed below.

Several algorithms have been proposed based on different approaches. One of the more reliable general-purpose methods for solving Riccati equations is the Schur method [14, p. 529]. For the case of Equation 1.37, for example, this method is based on reducing the associated $2n \times 2n$ Hamiltonian matrix

$$\begin{bmatrix} A & -G \\ -Q & -A^T \end{bmatrix} \tag{1.40}$$

to RSF. If the RSF is ordered so that its stable eigenvalues (there are exactly n of them under certain standard assumptions) are in the upper left corner, the corresponding first n vectors of the orthogonal matrix, which effects the reduction, forms a basis for the stable eigenspace from which the nonnegative definite solution X is then easily found.

Extensions to the basic Schur method have been made [14, p. 562, 573], which were prompted by the following situations:

- G in Equation 1.37 is of the form $BR^{-1}B^T$, where R may be nearly singular, or G_2 in Equation 1.38 may be exactly or nearly singular.
- A in Equation 1.38 is singular (A^{-1} is required in the classical approach involving a symplectic matrix that plays a role analogous to Equation 1.40).

This resulted in the generalized eigenvalue approach requiring the computation of a basis for the deflating subspace corresponding to the stable generalized eigenvalues. For the solution of Equation 1.37, the generalized eigenvalue problem is given by

$$\lambda \begin{bmatrix} I & 0 & 0 \\ 0 & I & 0 \\ 0 & 0 & 0 \end{bmatrix} - \begin{bmatrix} A & 0 & B \\ -Q & -A^T & 0 \\ 0 & B^T & R \end{bmatrix}; \tag{1.41}$$

for Equation 1.38, the corresponding problem is

$$\lambda \begin{bmatrix} I & 0 & 0 \\ 0 & A^T & 0 \\ 0 & {G_1}^T & 0 \end{bmatrix} - \begin{bmatrix} A & 0 & -G_1 \\ -Q & I & 0 \\ 0 & 0 & G_2 \end{bmatrix}. \tag{1.42}$$

The extensions above can be generalized even further, as the following problem illustrates. Consider the optimal control problem

$$\min \frac{1}{2} \int_0^{+\infty} [x^T Q x + 2x^T S u + u^T R u]\, dt \tag{1.43}$$

subject to

$$E\dot{x} = Ax + Bu. \tag{1.44}$$

The Riccati equation associated with Equations 1.43 and 1.44 then takes the form

$$E^T X B R^{-1} B^T X E - (A - BR^{-1}S^T)^T X E - E^T X (A - BR^{-1}S^T) - Q + SR^{-1}S^T = 0 \tag{1.45}$$

or

$$(E^T X B + S) R^{-1} (B^T X E + S^T) - A^T X E - E^T X A - Q = 0. \tag{1.46}$$

This so-called "generalized" Riccati equation can be solved by considering the associated matrix pencil

$$\begin{bmatrix} A & 0 & B \\ -Q & -A^T & -S \\ S^T & B^T & R \end{bmatrix} - \lambda \begin{bmatrix} E & 0 & 0 \\ 0 & E^T & 0 \\ 0 & 0 & 0 \end{bmatrix}. \tag{1.47}$$

Note that S in Equation 1.43 and E in Equation 1.44 are handled directly and no inverses appear. The presence of a nonsingular E in state-space models of the form Equation 1.44 adds no particular difficulty to the solution process and is numerically the preferred form if E is, for example, near singular or even sparse. Similar remarks apply to the frequency response problem in Equations 1.29 and 1.30 and, indeed, throughout all of linear control and system theory. The stability and conditioning of these approaches are discussed in [14, p. 529, 573]. Other methods, including Newton's method and iterative refinement, have been analyzed in, for example, [14, p. 517]. Numerical algorithms for handling Equations 1.41, 1.42, and 1.47 and a large variety of related problems are described in [14, p. 421, 573]. A thorough survey of the Schur method, generalized eigenvalue/eigenvector extensions, and the underlying algebraic structure in terms of "Hamiltonian pencils" and "symplectic pencils" is included in [3,12].

Schur techniques can also be applied to Riccati differential and difference equations and to nonsymmetric Riccati equations that arise, for example, in invariant imbedding methods for solving linear two-point boundary value problems.

As with the linear Lyapunov and Sylvester equations, satisfactory results have been obtained concerning condition of Riccati equations, a topic of great interest independent of the solution method used, be it a Schur-type method or one of numerous alternatives. We refer to [1] for a further discussion on this.

A very interesting class of invariant-subspace-based algorithms for solving the Riccati equation and related problems uses the so-called matrix sign function. These methods, which are particularly attractive for very large-order problems, are described in detail in [14, p. 486] and the references therein. These algorithms are based on Newton's method applied to a certain matrix equation. A new family of iterative algorithms of arbitrary order convergence has been developed in [14, p. 624]. This family of algorithms can be parallelized easily and yields a viable method of solution for very high-order Riccati equations.

1.4.8 Pole Assignment and Observer Design

Designing state or output feedback for a linear system, so that the resulting closed-loop system has a desired set of poles, can be considered an inverse eigenvalue problem. The state feedback pole assignment problem is as follows: Given a pair (A, B), one looks for a matrix F so that the eigenvalues of the matrix

$$A_F = A + BF$$

lie at specified locations or in specified regions. Many approaches have been developed for solving this problem. However, the emphasis is on numerically reliable methods and consideration of the numerical sensitivity of the problem, for example, see the papers in Part 6 of [14]. Special cases of the pole assignment problem arise in observer design [14, p. 407], and in deadbeat control for discrete-time systems (where $A + BF$ is required to be nilpotent) [14, p. 392]. The numerically reliable methods for pole assignment are based on reducing A to either an RSF, [14, p. 415], or to a Hessenberg or block Hessenberg (staircase) form [14, p. 357, 380]. The latter may be regarded a numerically robust alternative to the controllable or Luenberger canonical form whose computation is known to be numerically unreliable [14, p. 59]. For multi-input systems, the additional freedom available in the state-feedback matrix can be used for eigenvector assignment and sensitivity minimization for the closed-loop poles [14, p. 333]. There the resulting matrix A_F is not computed directly, but instead the matrices Λ and X of the decomposition

$$A_F = X \Lambda X^{-1}$$

are computed via an iterative technique. The iteration aims to minimize the sensitivity of the placed eigenvalues λ_i or to maximize the orthogonality of the eigenvectors x_i.

Pole assignment by output feedback is more difficult, theoretically as well as computationally. Consequently, there are a few numerically reliable algorithms available [14, p. 371]. Other works on pole assignment have been concerned with generalized state-space or descriptor systems.

The problem of observer design for a given state-space system $\{A, B, C\}$ is finding matrices T, A_K, and K so that

$$TA_K - AT = KC \tag{1.48}$$

whereby the spectrum of A_K is specified. Because this is an underdetermined (and nonlinear) problem in the unknown parameters of T, A_K, and K, one typically sets $T = I$ and Equation 1.48 then becomes

$$A_K = A + KC,$$

which is a transposed pole placement problem. In this case, the above techniques of pole placement automatically apply here. In reduced order design, T is nonsquare and thus cannot be equated to the identity matrix. One can still solve Equation 1.48 via a recurrence relationship when assuming A_K in Schur form [14, p. 407].

1.4.9 Robust Control

In the last decade, there has been significant growth in the theory and techniques of robust control; see, for example, [6] and the references therein. However, the area of robust control is still evolving and its numerical aspects have just begun to be addressed [13]. Consequently, it is premature to survey reliable numerical algorithms in the area. To suggest the flavor of the numerical and computational issues involved, in this section we consider a development in robust control that has attracted a great deal of attention, the so-called H_∞ approach. H_∞ and the related structured singular value approach have provided a powerful framework for synthesizing *robust* controllers for linear systems. The controllers are robust, because they achieve desired system performance despite a significant amount of uncertainty in the system.

In this section, we denote by $\mathbb{R}(s)^{n\times m}$ the set of proper real rational matrices of dimension $n \times m$. The H_∞ norm of a stable matrix $G(s) \in \mathbb{R}(s)^{n\times m}$ is defined as

$$\|G(s)\|_\infty := \sup_{\omega\in\mathbb{R}} \sigma_{\max}[G(j\omega)], \tag{1.49}$$

where $\sigma_{\max}[\cdot]$ denotes the largest singular value of a (complex) matrix. Several iterative methods are available for computing this norm. In one approach, a relationship is established between the singular values of $G(jw)$ and the imaginary eigenvalues of a Hamiltonian matrix obtained from a state-space realization of $G(s)$. This result is then used to develop an efficient bisection algorithm for computing the H_∞ norm of $G(s)$.

To describe the basic H_∞ approach, consider a linear, time-invariant system described by the state-space equations

$$\begin{aligned}\dot{x}(t) &= Ax(t) + B_1w(t) + B_2u(t),\\ z(t) &= C_1x(t) + D_{11}w(t) + D_{12}u(t),\\ y(t) &= C_2x(t) + D_{21}w(t) + D_{22}u(t),\end{aligned} \tag{1.50}$$

where $x(t) \in \mathbb{R}^n$ denotes the state vector; $w(t) \in \mathbb{R}^{m_1}$ is the vector of disturbance inputs; $u(t) \in \mathbb{R}^{m_2}$ is the vector of control inputs; $z(t) \in \mathbb{R}^{p_1}$ is the vector of error signals; and $y(t) \in \mathbb{R}^{p_2}$ is the vector of measured variables. The transfer function relating the inputs $\begin{bmatrix} w \\ u \end{bmatrix}$ to the outputs $\begin{bmatrix} z \\ y \end{bmatrix}$ is

$$G(s) := \begin{bmatrix} G_{11}(s) & G_{12}(s) \\ G_{21}(s) & G_{22}(s) \end{bmatrix} \tag{1.51}$$

$$= \begin{bmatrix} D_{11} & D_{12} \\ D_{21} & D_{22} \end{bmatrix} + \begin{bmatrix} C_1 \\ C_2 \end{bmatrix} (sI - A)^{-1} \begin{bmatrix} B_1 & B_2 \end{bmatrix}. \tag{1.52}$$

Implementing a feedback controller defined by

$$u = K(s)y \tag{1.53}$$

where $K(s) \in \mathbb{R}(s)^{m_2 \times p_2}$, we obtain the closed-loop transfer matrix $T_{zw}(s) \in \mathbb{R}(s)^{p_1 \times m_1}$ from the disturbance w to the regulated output z

$$T_{zw} := G_{11} + G_{12}K(I - G_{22}K)^{-1}G_{21}. \tag{1.54}$$

Next, define the set $\mathcal{K}$ of all internally stabilizing feedback controllers for the system in Equation 1.50, that is,

$$\mathcal{K} := \{K(s) \in \mathbb{R}(s)^{m_2 \times p_2} : T_{zw}(s) \text{ is internally stable}\}.$$

Now let $K(s) \in \mathcal{K}$, and define

$$\gamma := \|T_{zw}(s)\|_\infty. \tag{1.55}$$

Then the H_∞ control problem is to find a controller $K(s) \in \mathcal{K}$ that minimizes γ. The optimal value of γ is defined as

$$\gamma_{\text{opt}} := \inf_{K \in \mathcal{K}} \|T_{zw}(s)\|_\infty. \tag{1.56}$$

The original formulation of this problem was in an input–output setting, and the early methods for computing γ_{opt} used either an iterative search involving spectral factorization and solving the resulting Nehari problem or computed the spectral norm of the associated Hankel plus Toeplitz operator. In a state-space formulation for computing γ_{opt}, promising from the viewpoint of numerical computation, the problem is formulated in terms of two algebraic Riccati equations that depend on a gain parameter γ. Then, under certain assumptions [13], it can be shown that for a controller $K(s) \in \mathcal{K}$ to exist so that $\|T_{zw}\|_\infty < \gamma$, three conditions have to be satisfied, namely, stabilizing solutions exist for the two Riccati equations, and the spectral radius of the product of the solutions is bounded by γ^2. If these conditions are satisfied for a particular value of γ, the corresponding controller $K(s)$ can be obtained from the solutions of the Riccati equations. The optimal gain, γ_{opt}, is the infimum over all suboptimal values of γ such that the three conditions are satisfied.

The approach above immediately suggests a bisection-type algorithm for computing γ_{opt}. However, such an algorithm can be very slow in the neighborhood of the optimal value. To obtain speedup near the solution, a gradient approach is proposed in [13]. The behavior of the Riccati solution as a function of γ is used to derive an algorithm that couples a gradient method with bisection. It has been pointed out in [13] that the Riccati equation can become ill-conditioned as the optimal value of γ is approached. It is therefore recommended in [13] that, instead of computing the Riccati solutions explicitly, invariant subspaces of the associated Hamiltonian matrices should be used.

1.5 Mathematical Software

1.5.1 General Remarks

The previous sections have highlighted some topics from numerical linear algebra and their application to numerical problems arising in systems, control, and estimation theory. These problems represent only a very small subset of numerical problems of interest in these fields but, even for problems apparently "simple" from a mathematical viewpoint, the myriad of details that constitute a sophisticated implementation become so overwhelming that the only effective means of communicating an algorithm is through mathematical software. Mathematical or numerical software is an implementation on a computer of an algorithm for solving a mathematical problem. Ideally, such software would be reliable, portable, and unaffected by the machine or system environment.

The prototypical work on reliable, portable mathematical software for the standard eigenproblem began in 1968. EISPACK, Editions I and II [7,15], were an outgrowth of that work. Subsequent efforts of interest to control engineers include LINPACK [5] for linear equations and linear least-squares problems, FUNPACK (Argonne) for certain function evaluations, MINPACK (Argonne) for certain optimization problems, and various ODE and PDE codes. High-quality algorithms are published regularly in the

ACM Transactions on Mathematical Software. LAPACK, the successor to LINPACK and EISPACK, is designed to run efficiently on a wide range of machines, including vector processors, shared-memory multiprocessors, and high-performance workstations.

Technology to aid in developing mathematical software in Fortran has been assembled as a package called TOOLPACK. Mechanized code development offers other advantages with respect to modifications, updates, versions, and maintenance.

Inevitably, numerical algorithms are strengthened when their mathematical software is portable, because they can be used widely. Furthermore, such a software is markedly faster, by factors of 10 to 50, than earlier and less reliable codes.

Many other features besides portability, reliability, and efficiency characterize "good" mathematical software, for example,

- High standards of documentation and style so as to be easily understood and used
- Ease of use; ability of the user to interact with the algorithm
- Consistency/compatibility/modularity in the context of a larger package or more complex problem
- Error control, exception handling
- Robustness in unusual situations
- Graceful performance degradation as problem domain boundaries are approached
- Appropriate program size (a function of intended use, e.g., low accuracy, real-time applications)
- Availability and maintenance
- "Tricks" such as underflow-/overflow-proofing, if necessary, and the implementation of columnwise or rowwise linear algebra

Clearly, the list can go on.

What becomes apparent from these considerations is that evaluating mathematical software is a challenging task. The quality of software is largely a function of its operational specifications. It must also reflect the numerical aspects of the algorithm being implemented. The language used and the compiler (e.g., optimizing or not) for that language have an enormous impact on quality, perceived and real, as does the underlying hardware and arithmetic. Different implementations of the same algorithm can have markedly different properties and behavior.

One of the most important and useful developments in mathematical software for most control engineers has been very high-level systems such as MATLAB® and Scilab. These systems spare the engineer the drudgery of working at a detailed level with languages such as Fortran and C, and they provide a large number of powerful computational "tools" (frequently through the availability of formal "toolboxes"). For many problems, the engineer must still have some knowledge of the algorithmic details embodied in such a system.

1.5.2 Mathematical Software in Control

Many aspects of systems, control, and estimation theory are at the stage from which one can start the research and design necessary to produce reliable, portable mathematical software. Certainly, many of the underlying linear algebra tools (e.g., in EISPACK, LINPACK, and LAPACK) are considered sufficiently reliable to be used as black, or at least gray, boxes by control engineers. An important development in this area is the SLICOT library, which is described in [1, p. 60]. Much of that theory and methodology can and has been carried over to control problems, but this applies only to a few basic control problems. Typical examples are Riccati equations, Lyapunov equations, and certain basic state-space transformations and operations. Much of the work in control, particularly design and synthesis, is simply not amenable to nice, "clean" algorithms. The ultimate software must be capable of enabling a dialogue between the computer and the control engineer, but with the latter probably still making the final engineering decisions.

1.6 Concluding Remarks

Several numerical issues and techniques from numerical linear algebra together with a number of important applications of these ideas have been outlined. A key question in these and other problems in systems, control, and estimation theory is what can be computed reliably and used in the presence of parameter uncertainty or structural constraints (e.g., certain "hard zeros") in the original model, and rounding errors in the calculations. However, because the ultimate goal is to solve real problems, reliable tools (mathematical software) and experience must be available to effect real solutions or strategies. The interdisciplinary effort during the last few decades has significantly improved our understanding of the issues involved in reaching this goal and has resulted in some high-quality control software based on numerically reliable and well-tested algorithms. This provides clear evidence of the fruitful symbiosis between numerical analysis and numerical problems from control. We expect this symbiotic relationship to flourish, as control engineering realizes the full potential of the computing power becoming more widely available in multiprocessing systems and high-performance workstations. However, as in other applications areas, software continues to act as a constraint and a vehicle for progress. Unfortunately, high-quality software is very expensive.

In this chapter, we have focused only on dense numerical linear algebra problems in systems and control. Several related topics that have not been covered here are, for example, parallel algorithms, algorithms for sparse or structured matrices, optimization algorithms, ODE algorithms, algorithms for differential-algebraic systems, and approximation algorithms. These areas are well established in their own right, but for control applications a lot of groundwork remains undone. The main reason we have confined ourselves to dense numerical linear algebra problems in systems and control is that, in our opinion, this area has reached a mature level where definitive statements and recommendations can be made about various algorithms and other developments.

References

1. Varga, A. et al. The amazing power of numerical awareness in Control, *IEEE Control Systems Magazine* pp. 14–76, 24, February 2004.
2. Anderson, E., Bai, Z., Bischof, C., Blackford, S., Demmel, J., Dongarra, J., DuCroz, J., et al. *LAPACK Users' Guide* (3rd edition), SIAM, Philadelphia, PA, 1999.
3. Bunse-Gerstner, A., Byers, R., and Mehrmann, V., Numerical methods for algebraic Riccati equations, in *Proceedings of Workshop on the Riccati Equation in Control, Systems, and Signals* (Como, Italy) Bittanti, S., Ed., Pitagora, Bologna, Italy, pp. 107–116, 1989.
4. Datta, B.N., *Linear Algebra and Applications* (2nd edition), SIAM, Philadelphia, PA, 2010.
5. Dongarra, J.J., Bunch, J.R., Moler, C.B., and Stewart, G.W., *LINPACK Users' Guide*, SIAM, Philadelphia, PA, 1979.
6. Dorato, P. and Yedavalli, R.K., Eds., *Recent Advances in Robust Control*, Selected Reprint Series, IEEE Press, New York, 1990.
7. Garbow, B.S., Boyle, J.M., Dongarra, J.J., and Moler, C.B., *Matrix Eigensystem Routines—EISPACK Guide Extension*, in *Lecture Notes in Computer Science*, Springer (vol. 51) New York, 1977.
8. Golub, G.H. and Van Loan, C.F., *Matrix Computations* (3rd edition), Johns Hopkins University Press, Baltimore, MD, 1996.
9. Higham, N.J., *Accuracy and Stability of Numerical Algorithms* (2nd edition), SIAM, Philadelphia, PA, 2002.
10. Higham, N.J., *Functions of Matrices. Theory and Computation*, SIAM, Philadelphia, PA, 2008.
11. Laub, A.J., Survey of computational methods in control theory, in *Electric Power Problems: The Mathematical Challenge*, Erisman, A.M., Neves, K.W., and Dwarakanath, M.H., Eds., SIAM, Philadelphia, PA, pp. 231–260, 1980.
12. Laub, A.J., Invariant subspace methods for the numerical solution of Riccati equations, in *The Riccati Equation*, Bittanti, S., Laub, A.J., and Willems, J.C., Eds., Springer, Berlin, pp. 163–196, 1991.
13. Pandey, P. and Laub, A.J., Numerical issues in robust control design techniques, in *Control and Dynamic Systems—Advances in Theory and Applications: Digital and Numeric Techniques and Their Applications in Control Systems*, vol. 55, Leondes, C.T., Ed., Academic, San Diego, CA, pp. 25–50, 1993.

14. Patel, R.V., Laub, A.J., and Van Dooren, P.M., Eds., *Numerical Linear Algebra Techniques for Systems and Control*, Selected Reprint Series, IEEE Press, New York, 1994.
15. Smith, B.T, Boyle, J.M., Dongarra, J.J., Garbow, B.S., Ikebe, Y., Klema, V.C., and Moler, C.B., *Matrix Eigensystem Routines—EISPACK Guide*, in Lecture Notes in Computer Science, vol. 6, Springer, New York, 1976.
16. Van Dooren, P., Numerical aspects of system and control algorithms, *Journal A*, 30, pp. 25–32, 1989.
17. Wilkinson, J.H., *The Algebraic Eigenvalue Problem*, Oxford University Press, Oxford, England, 1965.

2

Multivariable Poles, Zeros, and Pole-Zero Cancellations

Joel Douglas
Massachusetts Institute of Technology

Michael Athans
Massachusetts Institute of Technology

2.1 Introduction

In this chapter we will introduce the basic building blocks necessary to understand linear time-invariant, multivariable systems. We will examine solutions of linear systems in both the time domain and frequency domain. An important issue is our ability to change the system's response by applying different inputs. We will thus introduce the concept of controllability. Similarly, we will introduce the concept of observability to quantify how well we can determine what is happening internally in our model when we can observe only the outputs of the system. An important issue in controllability and observability is the role of zeros. We will define them for multivariable systems and show their role in these concepts. Throughout, we will introduce the linear algebra tools necessary for multivariable systems.

2.2 Unforced Linear Systems

2.2.1 Eigenvectors and Eigenvalues

Given a matrix $A \in \mathcal{R}^{n\times n}$, the eigenstructure of A is defined by n complex numbers λ_i. When the λ_i are all different, each λ_i has corresponding vectors $v_i \in \mathcal{C}^n$ and $w_i \in \mathcal{C}^n$ so that

$$Av_i = \lambda_i v_i; \quad w_i^H A = \lambda_i w_i^H; \quad i = 1 \dots n \tag{2.1}$$

where w^H is the complex conjugate transpose of w. The complex numbers λ_i are called the eigenvalues of A. The vectors v_i are called the right eigenvectors, and the vectors w_i are called the left eigenvectors. Notice that any multiple of an eigenvector is also an eigenvector.

The left and right eigenvectors are mutually orthogonal, that is, they satisfy the property

$$w_i^H v_j = \delta_{ij} \triangleq \begin{cases} 1 & \text{if } i = j \\ 0 & \text{if } i \neq j \end{cases} \tag{2.2}$$

We assume throughout that the eigenvalues are distinct; that is, they are all different. The case where eigenvalues repeat is much more complicated, both theoretically and computationally. The case of repeated eigenvalues is covered in Kailath [1].

One other formula useful for describing linear systems is the dyadic formula. This formula shows how the matrix A can be formed from its eigenvalues and eigenvectors. It is given by

$$A = \sum_{i=1}^{n} \lambda_i v_i w_i^H \tag{2.3}$$

2.2.2 The Matrix Exponential

The matrix exponential, e^A, is defined by

$$e^A = I + A + \frac{1}{2}A^2 + \frac{1}{6}A^3 + \cdots \tag{2.4}$$

$$= \sum_{k=0}^{\infty} \frac{1}{k!}A^k \tag{2.5}$$

The matrix exponential solves the following matrix differential equation

$$\dot{x}(t) = Ax(t), \quad x(0) = \xi \tag{2.6}$$

The solution is

$$x(t) = e^{At}\xi \tag{2.7}$$

The matrix exponential can be calculated from the eigenstructure of the matrix A. If A has the eigenstructure as in Equation 2.1, then

$$e^A = \sum_{i=1}^{n} e^{\lambda_i} v_i w_i^H \quad \text{and} \quad e^{At} = \sum_{i=1}^{n} e^{\lambda_i t} v_i w_i^H \tag{2.8}$$

This can be seen by writing the matrix exponential using the infinite sum, substituting in the dyadic formula (Equation 2.3), and using Equation 2.2.

Taking the Laplace transform of Equation 2.6,

$$sx(s) - \xi = Ax(s) \tag{2.9}$$

where we have used the initial condition $x(0) = \xi$. Thus, the solution is

$$x(s) = (sI - A)^{-1}\xi \tag{2.10}$$

Therefore, $(sI - A)^{-1}$ is the Laplace transform of e^{At}.

2.2.3 Definition of Modes

The solution to the unforced system Equation 2.6 can be written in terms of the eigenstructure of A as

$$x(t) = e^{At}\xi = \sum_{i=1}^{n} e^{\lambda_i t} v_i (w_i^H \xi) \tag{2.11}$$

The ith mode is $e^{\lambda_i t} v_i$, defined by the direction of the right eigenvector v_i, and the exponential associated with the eigenvalue λ_i. $w_i^H \xi$ is a scalar, specifying the degree to which the initial condition ξ excites the ith mode.

Taking Laplace transforms,

$$x(s) = \sum_{i=1}^{n} \frac{1}{s - \lambda_i} v_i w_i^H \xi \tag{2.12}$$

From this equation,

$$x(s) = (sI - A)^{-1}\xi = \sum_{i=1}^{n} \frac{1}{s - \lambda_i} v_i w_i^H \xi \tag{2.13}$$

When the initial condition is equal to one of the right eigenvectors, only the mode associated with that eigenvector is excited. To see this, let $\xi = v_j$. Then, using Equation 2.2,

$$x(t) = \sum_{i=1}^{n} e^{\lambda_i t} v_i (w_i^H v_j) = e^{\lambda_j t} v_j \tag{2.14}$$

A mode is called stable if the dynamics of the modal response tend to zero asymptotically. This is, therefore, equivalent to

$$\mathrm{Re}(\lambda_i) < 0 \tag{2.15}$$

A system is said to be stable if all of its modes are stable. Thus, for any initial condition, $x(t) \to 0$ if the system is stable.

2.2.4 Multivariable Poles

Consider the system

$$\dot{x}(t) = Ax(t) + Bu(t) \tag{2.16}$$

$$y(t) = Cx(t) + Du(t) \tag{2.17}$$

The multivariable poles are defined as the eigenvalues of A. Thus, the system is stable if all of its poles are strictly in the left half of the s-plane. The poles are therefore the roots of the equation

$$\det(sI - A) = 0 \tag{2.18}$$

2.3 Forced Linear Time-Invariant Systems

2.3.1 Solution to Forced Systems

We consider the dynamical system

$$\dot{x}(t) = Ax(t) + Bu(t), \quad x(0) = \xi \tag{2.19}$$

$$y(t) = Cx(t) + Du(t) \tag{2.20}$$

The solution $x(t)$ to this equation is

$$x(t) = e^{At}\xi + \int_0^t e^{A(t-\tau)}Bu(\tau)\,d\tau \tag{2.21}$$

This can be seen by differentiating the solution, and substituting in Equation 2.19. The output $y(t)$ is thus

$$y(t) = Cx(t) + Du(t) = Ce^{At}\xi + \int_0^t Ce^{A(t-\tau)}Bu(\tau)\,d\tau + Du(t) \tag{2.22}$$

Using the eigenstructure of A, we can write

$$y(t) = \sum_{i=1}^{n} e^{\lambda_i t}(Cv_i)(w_i^H\xi) \tag{2.23}$$

$$+\sum_{i=1}^{n} e^{\lambda_i t}(Cv_i)(w_i^H B)\int_0^t e^{-\lambda_i\tau}u(\tau)\,d\tau + Du(t) \tag{2.24}$$

Applying the Laplace Transform to the system Equations 2.19 and 2.20,

$$y(s) = C(sI - A)^{-1}Bu(s) + Du(s) \tag{2.25}$$

Using the eigenstructure of A, we can substitute for $(sI - A)^{-1}$ to get the Laplace Transform equation

$$y(s) = \sum_{i=1}^{n} \frac{Cv_i w_i^H B}{s - \lambda_i}u(s) + Du(s) \tag{2.26}$$

The matrix $\frac{Cv_i w_i^H B}{(s-\lambda_i)}$ is called the residue matrix at the pole $s = \lambda_i$.

We can see that $w_i^H B$ is an indication of how much the ith mode is exited by the inputs, and Cv_i indicates how much the ith mode is observed in the outputs. This is the basis for the concepts of controllability and observability, respectively.

2.3.2 Controllability and Observability

Controllability is concerned with how much an input affects the states of a system. The definition is general enough to handle nonlinear systems and time-varying systems.

Definition 2.1:

The nonlinear time-invariant (LTI) system

$$\dot{x}(t) = f[x(t), u(t)], \quad x(0) = \xi, \tag{2.27}$$

is called completely controllable if, for any initial state ξ and any final state θ, we can find a piecewise, continuous bounded function $u(t)$ for $0 \le t \le T$, $T < \infty$, so that $x(T) = \theta$. A system which is not completely controllable is called uncontrollable.

Observability is defined similarly.

Definition 2.2:

The nonlinear time-invariant system

$$\dot{x}(t) = f[x(t), u(t)], \quad x(0) = \xi \tag{2.28}$$

$$y(t) = g[x(t), u(t)] \tag{2.29}$$

is observable if one can calculate the initial state ξ *based upon measurements of the input* $u(t)$ *and output* $y(t)$ *for* $0 \leq t \leq T$, $T < \infty$. *A system which is not observable is called unobservable.*

For LTI systems, there are simple tests to determine if a system is controllable and observable. The ith mode is uncontrollable if, and only if,

$$w_i^H B = 0 \tag{2.30}$$

Thus a mode is called uncontrollable if none of the inputs can excite the mode. The system is uncontrollable if it is uncontrollable from any mode. Thus, a system is controllable if

$$w_i^H B \neq 0 \quad i = 1, \ldots, n \tag{2.31}$$

Similarly, the ith mode is unobservable if, and only if,

$$Cv_i = 0 \tag{2.32}$$

Thus a mode is called unobservable if we can not see its effects in any of the outputs. The system is unobservable if it is unobservable from any mode. Thus, a system is unobservable if

$$Cv_i \neq 0, \quad i = 1, \ldots, n \tag{2.33}$$

2.3.3 Other Tests for Controllability and Observability

There is a simple algebraic test for controllability and observability. Let us define the controllability matrix M_c as

$$M_c = \begin{bmatrix} B & AB & A^2B & \cdots & A^{n-1}B \end{bmatrix} \tag{2.34}$$

The system is controllable if, and only if, $rank(M_c) = n$. Note that if the rank of M_c is less than n, we do not know anything about the controllability of individual modes. We only know that at least one mode is uncontrollable.

There is a similar test to determine the observability of the system. We form the observability matrix M_o as

$$M_o = \begin{bmatrix} C \\ CA \\ CA^2 \\ \vdots \\ CA^{n-1} \end{bmatrix}. \tag{2.35}$$

The system is observable if, and only if, $rank(M_o) = n$. Again, this test provides no insight into the observability of the modes of the system. It only determines whether or not the whole system is observable.

We now have tests for controllability and observability. We will now relate the loss of controllability and observability to pole-zero cancellations. First, we need to define the concept of a zero for a multi-input multioutput system.

2.4 Multivariable Transmission Zeros

There are several ways in which zeros can be defined for multivariable systems. The one we will examine is based on a generalized eigenvalue problem and has an interesting physical interpretation. Another approach to defining and calculating the zeros of MIMO systems can be found in Kailath [1].

2.4.1 Definition of MIMO Transmission Zeros

To define the multi-input, multi-output (MIMO) transmission zeros, we will first assume that we have a system with the same number of inputs and outputs. This is referred to as a square system. We will later extend the definition to nonsquare systems. For square systems, we can represent the system in the time domain as

$$\dot{x}(t) = Ax(t) + Bu(t) \tag{2.36}$$

$$y(t) = Cx(t) + Du(t) \tag{2.37}$$

where $x(t) \in \mathcal{R}^n$, $u(t) \in \mathcal{R}^m$, and $y(t) \in \mathcal{R}^m$. We can also write the transfer function matrix as

$$G(s) = C(sI - A)^{-1}B + D \tag{2.38}$$

where $G(s) \in \mathcal{C}^{m \times m}$. Given this system, we have the following definition:

Definition 2.3:

The plant has a zero at the (complex) value z_k if vectors $\xi_k \in \mathcal{C}^n$ and $u_k \in \mathcal{C}^m$ exist which are not both zero, so that the solution to the equations

$$\dot{x}(t) = Ax(t) + Bu_k e^{z_k t}, \quad x(0) = \xi_k \tag{2.39}$$

$$y(t) = Cx(t) + Du(t) \tag{2.40}$$

has the property that

$$y(t) \equiv 0 \quad \forall t > 0 \tag{2.41}$$

This property of transmission zeros is sometimes called transmission blocking. When zeros repeat, this definition still holds but a more complicated transmission blocking property also holds [2].

As an example, let us show that this definition is consistent with the standard definition of zeros for single-input, single-output systems.

Example 2.1:

Let us consider the following plant:

$$g(s) = \frac{s+1}{s(s-1)} \tag{2.42}$$

Then a state-space representation of this is

$$\dot{x}_1(t) = x_2(t) \tag{2.43}$$

$$\dot{x}_2(t) = x_2(t) + u(t) \tag{2.44}$$

$$y(t) = x_1(t) + x_2(t) \tag{2.45}$$

Let us define $u(t) = 2e^{-t}$, so that $u_k = 2$ and $z_k = -1$. Let us also define the vector ξ_k as

$$\xi_k = \begin{bmatrix} 1 \\ -1 \end{bmatrix} \tag{2.46}$$

Then

$$\dot{x_2}(t) = x_2(t) + 2e^{-t}, \quad x_2(0) = -1 \tag{2.47}$$

$$\Rightarrow x_2(t) = -e^t + e^t \int_0^t e^{-\tau} 2e^{-\tau} \partial\tau = -e^{-t} \tag{2.48}$$

$$\dot{x}_1(t) = x_2(t), \quad x_1(0) = 1 \tag{2.49}$$

$$\Rightarrow x_1(t) = 1 + e^{-t} - 1 = e^{-t} \tag{2.50}$$

Thus,

$$y(t) = x_1(t) + x_2(t) = 0 \tag{2.51}$$

So we have confirmed that $z = -1$ is a transmission zero of the system.

From the definition, we can see how transmission zeros got their name. If an input is applied at the frequency of the transmission zero in the correct direction (u_k), and the initial condition is in the correct direction ξ_k, then nothing is transmitted through the system.

2.4.2 Calculation of Transmission Zeros

To calculate the transmission zero, we can rewrite the definition of the transmission zero in matrix form as

$$\begin{bmatrix} z_k I - A & -B \\ -C & -D \end{bmatrix} \begin{bmatrix} \xi_k \\ u_k \end{bmatrix} = \begin{bmatrix} 0 \\ 0 \end{bmatrix} \tag{2.52}$$

This is in the form of a generalized eigenvalue problem, typically written as

$$G v_i = z_i M v_i \tag{2.53}$$

$$w_i^H G = z_i w_i^H M \tag{2.54}$$

where z_i is a generalized eigenvalue, with right and left generalized eigenvectors v_i and w_i, respectively. The generalized eigenvalues are the roots of the equation

$$\det(zM - G) = 0 \tag{2.55}$$

Note that if M is invertible, there are n generalized eigenvalues. Otherwise, there are less than n.

Equation 2.52 is a generalized eigenvalue problem with

$$G = \begin{bmatrix} A & B \\ C & D \end{bmatrix} \quad M = \begin{bmatrix} I & 0 \\ 0 & 0 \end{bmatrix} \tag{2.56}$$

Let us look at the implication of the generalized eigenvalue problem. Let z_k be a transmission zero. Then it must be true that

$$0 = \det \begin{bmatrix} z_k I - A & -B \\ -C & -D \end{bmatrix} \tag{2.57}$$

If there are no common poles and zeros, then $z_k I - A$ is invertible, and we can write

$$0 = \det(z_k I - A) \det[C(z_k I - A)^{-1} B + D] \tag{2.58}$$

$$= \det(z_k I - A) \det[G(z_k)] \tag{2.59}$$

Since we assume there are no common poles and zeros in the system, then $\det(z_k I - A) \neq 0$, and so it must be true that

$$\det(G(z_k)) = 0 \tag{2.60}$$

Thus, in the case that there are no common poles and zeros, the MIMO transmission zeros are the roots of Equation 2.60. To check for transmission zeros at the same frequencies as the poles, we must use the generalized eigenvalue problem.

Let us now give a multivariable example.

Example 2.2:

Consider the system given by Equations 2.36 and 2.37, with the matrices given by

$$A = \begin{bmatrix} -2 & 0 & 0 \\ 0 & -3 & 0 \\ 0 & 0 & -4 \end{bmatrix} \quad B = \begin{bmatrix} -1 & 0 \\ -1 & 0 \\ 0 & 1 \end{bmatrix} \tag{2.61}$$

$$C = \begin{bmatrix} 1 & 0 & 0 \\ 0 & 1 & 1 \end{bmatrix} \quad D = \begin{bmatrix} 1 & 0 \\ 1 & 0 \end{bmatrix} \tag{2.62}$$

Solving the generalized eigenvalue problem Equation 2.52, we find that there are two zeros, given by

$$z_1 = -1 \quad z_2 = -3 \tag{2.63}$$

$$\xi_1 = \begin{bmatrix} 2 \\ 1 \\ 1 \end{bmatrix} \quad \xi_2 = \begin{bmatrix} 0 \\ 1 \\ -1 \end{bmatrix} \tag{2.64}$$

$$u_1 = \begin{bmatrix} -2 \\ 3 \end{bmatrix} \quad u_2 = \begin{bmatrix} 0 \\ -1 \end{bmatrix} \tag{2.65}$$

Let us also check the transfer function calculation. The transfer function matrix for this system is given by

$$G(s) = \begin{bmatrix} \dfrac{s+1}{s+2} & 0 \\ \dfrac{s+2}{s+3} & \dfrac{1}{s+4} \end{bmatrix} \tag{2.66}$$

To find transmission zeros at locations other than the poles of the system, we take the determinant and set it equal to zero to find

$$0 = \det(G(z)) = \frac{s+1}{(s+2)(s+4)} \tag{2.67}$$

Thus we find a transmission zero at $z = -1$. Notice that we have correctly found the transmission zero which is not at the pole frequency, but in this case we did not find the transmission zero at the same frequency as the pole at $s = -3$. It is also important to realize that, although one of the SISO transfer functions has a zero at $s = -2$, this is not a transmission zero of the MIMO system.

Also notice that this frequency domain method does not give us the directions of the zeros. We need to use the generalized eigenvalue problem to determine the directions.

2.4.3 Transmission Zeros for Nonsquare Systems

When we have a nonsquare system, we differentiate between right zeros and left zeros. In the following, we will assume we have a system with n states, m inputs, and p outputs.

Definition 2.4:

z_k is a right zero of the system,

$$\dot{x}(t) = Ax(t) + Bu(t) \tag{2.68}$$

$$y(t) = Cx(t) + Du(t) \tag{2.69}$$

if vectors $\xi_k \in C^n$ and $u_k \in C^m$ exist, both not zero, so that

$$\begin{bmatrix} z_k I - A & -B \\ -C & -D \end{bmatrix} \begin{bmatrix} \xi_k \\ u_k \end{bmatrix} = \begin{bmatrix} 0 \\ 0 \end{bmatrix} \tag{2.70}$$

Definition 2.5:

z_k is a left zero of the system,

$$\dot{x}(t) = Ax(t) + Bu(t) \tag{2.71}$$

$$y(t) = Cx(t) + Du(t) \tag{2.72}$$

if it is a right zero of the system,

$$\dot{x}(t) = A^T x(t) + C^T u(t) \tag{2.73}$$

$$y(t) = B^T x(t) + D^T u(t) \tag{2.74}$$

In other words, z_k is a left zero of our system if vectors $\eta_k \in C^n$ and $\gamma_k \in C^p$ exist, both not zero, so that

$$\begin{bmatrix} \eta_k & \gamma_k \end{bmatrix} \begin{bmatrix} z_k I - A & -B \\ -C & -D \end{bmatrix} = \begin{bmatrix} 0 & 0 \end{bmatrix} \tag{2.75}$$

For square systems, the left and right zeros coincide; any frequency which is a right zero is also a left zero.

2.5 Multivariable Pole-Zero Cancellations

Now that we have defined the zeros of a multivariable system, we are in a position to describe pole-zero cancellations and what they imply in terms of controllability and observability.

First, we give a SISO example, which shows that a pole-zero cancellation implies loss of controllability or observability.

Example 2.3:

Consider the system given by the equations

$$\begin{bmatrix} \dot{x}_1(t) \\ \dot{x}_2(t) \end{bmatrix} = \begin{bmatrix} 0 & 1 \\ 1 & 0 \end{bmatrix} \begin{bmatrix} x_1(t) \\ x_2(t) \end{bmatrix} + \begin{bmatrix} 0 \\ 1 \end{bmatrix} u(t) \tag{2.76}$$

$$y(t) = \begin{bmatrix} 1 & -1 \end{bmatrix} \begin{bmatrix} x_1(t) \\ x_2(t) \end{bmatrix} \tag{2.77}$$

The transfer function for this system is given by

$$g(s) = \frac{-s+1}{s^2-1} = -\frac{s-1}{(s+1)(s-1)} \tag{2.78}$$

Thus, there is a pole-zero cancellation at $s = 1$. To check for loss of controllability and observability, we will perform an eigenvalue decomposition of the system. The eigenvalues and associated left and right eigenvalues are given by

$$\lambda_1 = -1 \quad v_1 = \begin{bmatrix} 1 \\ -1 \end{bmatrix} \quad w_1 = \begin{bmatrix} 0.5 & -0.5 \end{bmatrix} \tag{2.79}$$

$$\lambda_2 = 1 \quad v_2 = \begin{bmatrix} 1 \\ 1 \end{bmatrix} \quad w_2 = \begin{bmatrix} 0.5 & 0.5 \end{bmatrix} \tag{2.80}$$

It is easy to verify that the dyadic decomposition Equation 2.3 holds. We now can use the modal tests for controllability and observability. We can see that

$$Cv_1 = 2 \quad w_1^T B = -0.5 \tag{2.81}$$

This tells us that the first mode is both controllable and observable, as expected. For the second mode,

$$Cv_2 = 0 \quad w_2^T B = 0.5 \tag{2.82}$$

We see that the second mode is controllable, but unobservable. The conclusion is that this particular state-space realization has an unobservable mode, but is controllable.

Now let us examine what happens when we lose observability in more detail. We will assume we have the system described by Equations 2.36 through 2.38. Let us assume a pole with frequency λ_k and direction v_k, that is, A has an eigenvalue λ_k with associated eigenvalue v_k. Thus

$$(\lambda_k I - A)v_k = 0 \tag{2.83}$$

Let us also assume a (right) zero at frequency z_k with direction given by $\begin{bmatrix} \xi_k^T & u_k^T \end{bmatrix}^T$, that is,

$$(z_k I - A)\xi_k - Bu_k = 0 \tag{2.84}$$

$$C\xi_k + Du_k = 0 \tag{2.85}$$

If it is true that $\lambda_k = z_k$ and $\xi_k = \beta v_k$ with β any scalar, then

$$Bu_k = 0 \tag{2.86}$$

If we assume that the columns of B are linearly independent (which says we don't have redundant controls), then it must be true that $u_k = 0$. However, with $u_k = 0$,

$$Cv_k = 0 \Rightarrow k\text{th mode is unobservable} \tag{2.87}$$

On the other hand, let us assume that the kth mode is unobservable. Then we know that

$$Cv_k = 0 \tag{2.88}$$

where v_k is the eigenvector associated with the kth mode. We will show that there must be a pole-zero cancellation. Let us choose $u_k = 0$. If λ_k is the eigenvalue associated with the kth mode, then we know

$$(\lambda_k I - A)v_k = 0 \Rightarrow (\lambda_k I - A)v_k - Bu_k = 0 \tag{2.89}$$

$$Cv_k + Du_k = 0 \tag{2.90}$$

Thus, λ_k is also a zero with direction

$$\begin{bmatrix} v_k \\ 0 \end{bmatrix} \tag{2.91}$$

We have now proven the following theorem:

Theorem 2.1:

The kth mode, with eigenvalue λ_k and eigenvector v_k, is unobservable if, and only if, λ_k is a right transmission zero with direction given by Equation 2.91.

Following the exact same steps, we show that loss of controllability implies a pole-zero cancellation. Using the same steps as before, we can prove the following theorem.

Theorem 2.2:

Let the kth mode have eigenvalue λ_k and left eigenvector w_k. Then the kth mode is uncontrollable if, and only if, λ_k is a left transmission zero with direction given by

$$\begin{bmatrix} w_k^T & 0 \end{bmatrix} \tag{2.92}$$

We conclude with an example.

Example 2.4:

Consider the system Equations 2.19 and 2.20, with matrices

$$A = \begin{bmatrix} -2.5 & -0.5 & 0.5 \\ 0.5 & -1.5 & 0.5 \\ 1 & 1 & -2 \end{bmatrix} \quad B = \begin{bmatrix} 1 \\ 1 \\ -2 \end{bmatrix} \tag{2.93}$$

$$C = \begin{bmatrix} 0 & 0 & 1 \\ 0 & 1 & 0 \end{bmatrix} \quad D = \begin{bmatrix} 0 \\ 0 \end{bmatrix} \tag{2.94}$$

A has the following eigenstructure:

$$\lambda_1 = -1 \quad v_1 = \begin{bmatrix} 0 \\ 1 \\ 1 \end{bmatrix} \quad w_1^T = \begin{bmatrix} \frac{1}{2} & \frac{1}{2} & \frac{1}{2} \end{bmatrix} \tag{2.95}$$

$$\lambda_2 = -2 \quad v_2 = \begin{bmatrix} 1 \\ -1 \\ 0 \end{bmatrix} \quad w_2^T = \begin{bmatrix} \frac{1}{2} & -\frac{1}{2} & \frac{1}{2} \end{bmatrix} \tag{2.96}$$

$$\lambda_3 = -3 \quad v_1 = \begin{bmatrix} 1 \\ 0 \\ -1 \end{bmatrix} \quad w_1^T = \begin{bmatrix} \frac{1}{2} & \frac{1}{2} & -\frac{1}{2} \end{bmatrix} \tag{2.97}$$

Let us check for controllability and observability.

$$Cv_1 = \begin{bmatrix} 1 \\ 1 \end{bmatrix} \quad w_1^T B = 0 \tag{2.98}$$

$$Cv_2 = \begin{bmatrix} 0 \\ -1 \end{bmatrix} \quad w_2^T B = -1 \tag{2.99}$$

$$\text{and} \quad Cv_3 = \begin{bmatrix} -1 \\ 0 \end{bmatrix} \quad w_3^T B = 2. \tag{2.100}$$

The conclusion is, therefore, that all modes are observable and that the first mode is uncontrollable, but the second and third modes are controllable. It is easy to verify that $z = -1$ is a left zero of this system, with direction

$$[\eta^T \quad \gamma^T] = \begin{bmatrix} \frac{1}{2} & \frac{1}{2} & \frac{1}{2} & 0 & 0 \end{bmatrix}. \tag{2.101}$$

Thus there is a pole-zero cancellation at $z = -1$, which makes the system uncontrollable.

References

1. Kailath, T., *Linear Systems*, Prentice Hall, Englewood Cliffs, NJ, 1980.
2. MacFarlane, A.G.J. and Karcanias, N., Poles and Zeros of Linear Multivariable Systems: A Survey of the Algebraic, Geometric, and Complex Variable Theory, *Int. J. Control*, 24, 33–74, 1976.
3. Rosenbrock, H.H., *State-Space and Multivariable Theory*, John Wiley & Sons, New York, 1973.
4. Sain, M.K. and Schrader, C.B., The Role of Zeros in the Performance of Multiinput, Multioutput Feedback Systems, *IEEE Trans. Education*, 33(3), 1990.
5. Schrader, C.B. and Sain, M.K., Research on System Zeros: A Survey, *Int. J. Control*, 50(4), 1989.

3

Fundamentals of Linear Time-Varying Systems

Edward W. Kamen
Georgia Institute of Technology

3.1 Introduction

In this chapter, various fundamental elements of the theory of linear time-varying systems are studied in both the continuous-time and discrete-time cases. The chapter is a revised and updated version of Chapter 25 that appeared in the first edition of *The Control Handbook*. The revision includes material on the existence of coordinate transformations that transform time-varying system matrices into diagonal forms, and the connection to the concept of "dynamic" eigenvalues and eigenvectors. Also included in the revision is the use of a simplified expression derived in [11] for the feedback gain in the design of a state feedback controller in the single-input case based on the control canonical form. Generalizations to the multi-input multi-output case can be carried out using the results in [12]. The theory begins in Section 3.2 after the following comments.

The distinguishing characteristic of a time-varying system is that the values of the output response depend on when the input is applied to the system. Time variation is often a result of system parameters changing as a function of time, such as aerodynamic coefficients in high-speed aircraft, circuit parameters in electronic circuits, mechanical parameters in machinery, and diffusion coefficients in chemical processes. Time variation may also be a result of linearizing a nonlinear system about a family of operating points and/or about a time-varying operating point.

The values of the time-varying parameters in a system are often not known *a priori*, that is, before the system is put into operation, but can be measured or estimated during system operation. Systems whose parameters are not known *a priori* are often referred to as parameter varying systems, and the control of such systems is referred to as gain scheduling. Parameter varying systems and gain scheduling are considered in another chapter of this handbook, and are not pursued here. In some applications, time variations in the coefficients of the system model are known *a priori*. For example, this may be the case if the system model is a linearization of a nonlinear system about a known time-varying nominal trajectory. In such cases, the theory developed in this chapter can be directly applied.

In this chapter, the study of linear time-varying systems is carried out in terms of input/output equations and the state model. The focus is on the relationships between input/output and state models, the construction of canonical forms, the study of the system properties of stability, controllability, and observability, and the design of controllers. Both the continuous-time and the discrete-time cases are considered. In the last part of the chapter, an example is given on the application to state observers and state feedback control, and examples are given on checking for stability. Additional references on the theory of linear time-varying systems are given in Further Reading.

3.2 Analysis of Continuous-Time Causal Linear Time-Varying Systems

Consider the single-input single-output continuous-time system given the input/output relationship

$$y(t) = \int_{-\infty}^{t} h(t,\tau)u(\tau)\,d\tau \tag{3.1}$$

where t is the continuous-time variable, $y(t)$ is the output response resulting from input $u(t)$, and $h(t,\tau)$ is a real-valued continuous function of t and τ. It is assumed that there are conditions on $h(t,\tau)$ and/or $u(t)$, which insure that the integral in Equation 3.1 exists. The system given by Equation 3.1 is causal since the output $y(t)$ at time t depends only on the input $u(\tau)$ for $\tau \le t$. The system is also linear since integration is a linear operation. Linearity means that if $y_1(t)$ is the output response resulting from input $u_1(t)$, and $y_2(t)$ is the output response resulting from input $u_2(t)$, then for any real numbers a and b, the output response resulting from input $au_1(t) + bu_2(t)$ is equal to $ay_1(t) + by_2(t)$.

Let $\delta(t)$ denote the unit impulse defined by $\delta(t) = 0$, $t \neq 0$ and $\int_{-\varepsilon}^{\varepsilon} \delta(\lambda)\,d\lambda = 1$ for any real number $\varepsilon > 0$. For any real number t_1, the time-shifted impulse $\delta(t - t_1)$ is the unit impulse located at time $t = t_1$. Then from Equation 3.1 and the sifting property of the impulse, the output response $y(t)$ resulting from input $u(t) = \delta(t - t_1)$ is given by

$$y(t) = \int_{-\infty}^{t} h(t,\tau)u(\tau)\,d\tau = \int_{-\infty}^{t} h(t,\tau)\delta(\tau - t_1)\,d\tau = h(t,t_1)$$

Hence, the function $h(t,\tau)$ in Equation 3.1 is the *impulse response function* of the system, that is, $h(t,\tau)$ is the output response resulting from the impulse $\delta(t-\tau)$ applied to the system at time τ.

The linear system given by Equation 3.1 is *time invariant* (or *constant*) if and only if

$$h(t+\gamma, \tau+\gamma) = h(t,\tau), \quad \text{for all real numbers } t, \tau, \gamma \tag{3.2}$$

Time invariance means that if $y(t)$ is the response to $u(t)$, then for any real number t_1, the time-shifted output $y(t-t_1)$ is the response to the time-shifted input $u(t-t_1)$. Setting $\gamma = -\tau$ in Equation 3.2 gives

$$h(t-\tau, 0) = h(t,\tau), \quad \text{for all real numbers } t, \tau \tag{3.3}$$

Hence, the system defined by Equation 3.1 is time invariant if and only if the impulse response function $h(t,\tau)$ is a function only of the difference $t-\tau$. In the time-invariant case, Equation 3.1 reduces to the convolution relationship

$$y(t) = h(t) * u(t) = \int_{-\infty}^{t} h(t-\tau)u(\tau)\, d\tau \tag{3.4}$$

where $h(t) = h(t, 0)$ is the impulse response (i.e., the output response resulting from the impulse $\delta(t)$ applied to the system at time 0).

The linear system defined by Equation 3.1 is *finite-dimensional* or *lumped* if the input $u(t)$ and the output $y(t)$ are related by the nth-order input/output differential equation

$$y^{(n)}(t) + \sum_{i=0}^{n-1} a_i(t) y^{(i)}(t) = \sum_{i=0}^{m} b_i(t) u^{(i)}(t) \tag{3.5}$$

where $y^{(i)}(t)$ is the ith derivative of $y(t)$, $u^{(i)}(t)$ is the ith derivative of $u(t)$, and $a_i(t)$ and $b_i(t)$ are real-valued functions of t. In Equation 3.5, it is assumed that $m \le n$. The linear system given by Equation 3.5 is time invariant if and only if all coefficients in Equation 3.5 are constants, that is, $a_i(t) = a_i$ and $b_i(t) = b_i$ for all i, where a_i and b_i are real constants.

3.2.1 State Model Realizations

A state model for the system given by Equation 3.5 can be constructed as follows. First, suppose that $m = 0$ so that Equation 3.5 becomes

$$y^{(n)}(t) + \sum_{i=0}^{n-1} a_i(t) y^{(i)}(t) = b_0(t) u(t). \tag{3.6}$$

Then defining the state variables

$$x_i(t) = y^{(i-1)}(t), \quad i = 1, 2, \ldots, n, \tag{3.7}$$

the system defined by Equation 3.6 has the state model

$$\dot{x}(t) = A(t)x(t) + B(t)u(t) \tag{3.8}$$

$$y(t) = Cx(t) \tag{3.9}$$

where the coefficient matrices $A(t)$, $B(t)$, and C are given by

$$A(t) = \begin{bmatrix} 0 & 1 & 0 & \cdots & 0 & 0 \\ 0 & 0 & 1 & \cdots & 0 & 0 \\ 0 & 0 & 0 & \cdots & 0 & 0 \\ \vdots & & & \cdots & & \vdots \\ 0 & 0 & 0 & \cdots & 0 & 1 \\ -a_0(t) & -a_1(t) & -a_2(t) & \cdots & -a_{n-2}(t) & -a_{n-1}(t) \end{bmatrix} \tag{3.10}$$

$$B(t) = \begin{bmatrix} 0 \\ 0 \\ 0 \\ \vdots \\ 0 \\ b_0(t) \end{bmatrix} \tag{3.11}$$

$$C = \begin{bmatrix} 1 & 0 & 0 & \cdots & 0 & 0 \end{bmatrix} \tag{3.12}$$

and $x(t)$ is the n-dimensional state vector given by

$$x(t) = \begin{bmatrix} x_1(t) \\ x_2(t) \\ \vdots \\ x_{n-1}(t) \\ x_n(t) \end{bmatrix}$$

When $m \geq 1$ in Equation 3.5, the definition in Equation 3.7 of the state variables will not yield a state model of the form given in Equations 3.8 and 3.9. If $m < n$, a state model can be generated by first rewriting Equation 3.5 in the form

$$D^n y(t) + \sum_{i=0}^{n-1} D^i \left[\alpha_i(t) y(t)\right] = \sum_{i=0}^{m} D^i \left[\beta_i(t) u(t)\right] \tag{3.13}$$

where D is the derivative operator and $\alpha_i(t)$ and $\beta_i(t)$ are real-valued functions of t. The form of the input/output differential equation given by Equation 3.13 exists if $a_i(t)$ and $b_i(t)$ are differentiable a suitable number of times. If $a_i(t)$ are constants so that $a_i(t) = a_i$ for all t, then $\alpha_i(t)$ are constants and $\alpha_i(t) = a_i$, $i = 0, 1, \ldots, n-1$. If $b_i(t)$ are constants so that $b_i(t) = b_i$ for all t, then $\beta_i(t)$ are constants and $\beta_i(t) = b_i$, $i = 0, 1, \ldots, m$.

When $a_i(t)$ and $b_i(t)$ are not constants, $\alpha_i(t)$ is a linear combination of $a_i(t)$ and the derivatives of $a_j(t)$ for $j = n-i-1, n-i-2, \ldots 1$, and $\beta_i(t)$ is a linear combination of $b_i(t)$ and the derivatives of $b_j(t)$ for $j = m-i-1, m-i-2, \ldots 1$. For example, when $n = 2$, $\alpha_0(t)$ and $\alpha_1(t)$ are given by

$$\alpha_0(t) = a_0(t) - \dot{a}_1(t) \tag{3.14}$$

$$\alpha_1(t) = a_1(t) \tag{3.15}$$

When $n = 3$, $\alpha_i(t)$ are given by

$$\alpha_0(t) = a_0(t) - \dot{a}_1(t) + \ddot{a}_2(t) \tag{3.16}$$

$$\alpha_1(t) = a_1(t) - 2\dot{a}_2(t) \tag{3.17}$$

$$\alpha_2(t) = a_2(t) \tag{3.18}$$

In the general case (n arbitrary), there is a one-to-one and onto correspondence between the coefficients $a_i(t)$ in the left-hand side of Equation 3.5 and the coefficients $\alpha_i(t)$ in the left-hand side of Equation 3.13. Similarly, there is a one-to-one and onto correspondence between the coefficients in the right-hand side of Equation 3.5 and the right-hand side of Equation 3.13.

Now given the system defined by Equation 3.5 written in the form of Equation 3.13, define the state variables

$$\begin{aligned} x_n(t) &= y(t) \\ x_{n-1}(t) &= Dx_n(t) + \alpha_{n-1}(t)x_n(t) - \beta_{n-1}(t)u(t) \\ x_{n-2}(t) &= Dx_{n-1}(t) + \alpha_{n-2}(t)x_n(t) - \beta_{n-2}(t)u(t) \\ &\vdots \\ x_1(t) &= Dx_2(t) + \alpha_1(t)x_n(t) - \beta_1(t)u(t) \end{aligned} \tag{3.19}$$

where $\beta_i(t) = 0$ for $i > m$. Then with the state variables defined by Equations 3.19, the system given by Equation 3.5 has the state model

$$\dot{x}(t) = A(t)x(t) + B(t)u(t) \tag{3.20}$$

$$y(t) = Cx(t) \tag{3.21}$$

where the coefficient matrices $A(t)$, $B(t)$, and C are given by

$$A(t) = \begin{bmatrix} 0 & 0 & 0 & \cdots & 0 & -\alpha_0(t) \\ 1 & 0 & 0 & \cdots & 0 & -\alpha_1(t) \\ 0 & 1 & 0 & \cdots & 0 & -\alpha_2(t) \\ \vdots & & & \cdots & & \vdots \\ 0 & 0 & 0 & \cdots & 0 & -\alpha_{n-2}(t) \\ 0 & 0 & 0 & \cdots & 1 & -\alpha_{n-1}(t) \end{bmatrix} \tag{3.22}$$

$$B(t) = \begin{bmatrix} \beta_0(t) \\ \beta_1(t) \\ \beta_2(t) \\ \vdots \\ \beta_{n-2}(t) \\ \beta_{n-1}(t) \end{bmatrix} \tag{3.23}$$

and

$$C = \begin{bmatrix} 0 & 0 & \cdots & 0 & 1 \end{bmatrix} \tag{3.24}$$

The state model with $A(t)$ and C specified by Equations 3.22 and 3.24 is said to be in *observer canonical form*, which is observable as discussed below. Note that for this particular state realization, the row vector C is constant (independent of t).

In addition to the state model defined by Equations 3.22 through 3.24, there are other possible state models for the system given by Equation 3.5. For example, another state model can be constructed in the case when the left-hand side of Equation 3.5 can be expressed in a Floquet factorization form (see [1]), so that Equation 3.5 becomes

$$(D - p_1(t))(D - p_2(t))\ldots(D - p_n(t))y(t) = \sum_{i=0}^{m} b_i(t)u^{(i)}(t) \tag{3.25}$$

where again D is the derivative operator and the $p_i(t)$ are real-valued or complex-valued functions of the time variable t. For example, consider the $n = 2$ case for which

$$\begin{aligned} (D - p_1(t))(D - p_2(t))y(t) &= (D - p_1(t))[\dot{y}(t) - p_2(t)y(t)] \\ (D - p_1(t))(D - p_2(t))y(t) &= \ddot{y}(t) - [p_1(t) + p_2(t)]\dot{y}(t) + [p_1(t)p_2(t) - \dot{p}_2(t)]y(t) \end{aligned} \tag{3.26}$$

With $n = 2$ and $m = 1$, the state variables may be defined by

$$x_1(t) = \dot{y}(t) - p_2(t)y(t) - b_1(t)u(t) \tag{3.27}$$

$$x_2(t) = y(t) \tag{3.28}$$

which results in the state model given by Equations 3.20 and 3.21 with

$$A(t) = \begin{bmatrix} p_1(t) & 0 \\ 1 & p_2(t) \end{bmatrix} \tag{3.29}$$

$$B(t) = \begin{bmatrix} b_0(t) - \dot{b}_1(t) + p_1(t)b_1(t) \\ b_1(t) \end{bmatrix} \tag{3.30}$$

and

$$C = \begin{bmatrix} 0 & 1 \end{bmatrix} \tag{3.31}$$

In the general case given by Equation 3.25, there is a state model in the form of Equations 3.20 and 3.21 with $A(t)$ and C given by

$$A(t) = \begin{bmatrix} p_1(t) & 0 & 0 & 0 & \cdots & 0 & 0 \\ 1 & p_2(t) & 0 & 0 & \cdots & 0 & 0 \\ 0 & 1 & p_3(t) & 0 & \cdots & 0 & 0 \\ \vdots & & & & & \vdots & \vdots \\ 0 & 0 & 0 & 0 & \cdots & p_{n-1}(t) & 0 \\ 0 & 0 & 0 & 0 & \cdots & 1 & p_n(t) \end{bmatrix} \tag{3.32}$$

$$C = \begin{bmatrix} 0 & 0 & 0 & \cdots & 1 \end{bmatrix} \tag{3.33}$$

A very useful feature of the state model with $A(t)$ given by Equation 3.32 is the lower triangular form of $A(t)$. In particular, as discussed below, stability conditions can be specified in terms of $p_i(t)$, which can be viewed as "time-varying poles" of the system. However, in the time-varying case the computation of $p_i(t)$ based on the Floquet factorization given in the left-hand side of Equation 3.25 requires that nonlinear Riccati-type differential equations must be solved (see, e.g., [2,13]). The focus of [2] is on the computation of poles and zeros and the application to stability and transmission blocking, with the emphasis on time-varying difference equations. In [4], the authors generate a Floquet factorization given by the left-hand side of Equation 3.25 by utilizing successive Riccati transformations in a state-space setting.

A significant complicating factor in the theory of factoring polynomial operators with time-varying coefficients is that in general there are an infinite number of different "pole sets"$\{p_1(t), p_2(t), \ldots, p_n(t)\}$. This raises the question as to whether one pole set may be "better" in some sense than another pole set. In the case of linear difference equations with time-varying coefficients, it is shown in [2] that poles can be computed using a nonlinear recursion and that uniqueness of pole sets can be achieved by specifying initial values for the poles. For recent work on the factorization of time-varying differential equations by using ground field extensions, see [6]. As discussed below, in [7] the definition and computation of time-varying poles is pursued by using Lyapunov coordinate transformations in a state-space setting.

3.2.2 The State Model

For an m-input p-output linear n-dimensional time-varying continuous-time system, the general form of the state model is

$$\dot{x}(t) = A(t)x(t) + B(t)u(t) \tag{3.34}$$

$$y(t) = C(t)x(t) + D(t)u(t) \tag{3.35}$$

where Equation 3.34 is the *state equation* and Equation 3.35 is the *output equation*. In Equations 3.34 and 3.35, $A(t)$ is the $n \times n$ *system matrix*, $B(t)$ is the $n \times m$ *input matrix*, $C(t)$ is the $p \times n$ *output matrix*, $D(t)$ is the $p \times m$ *direct feed matrix*, $u(t)$ is the m-dimensional input vector, $x(t)$ is the n-dimensional state vector, and $y(t)$ is the p-dimensional output vector. The term $D(t)u(t)$ in Equation 3.35 is of little significance in the theory, and thus $D(t)u(t)$ is usually omitted from Equation 3.35, which will be done here.

To solve Equation 3.34, first consider the homogeneous equation

$$\dot{x}(t) = A(t)x(t), \quad t > t_0 \tag{3.36}$$

with the initial condition $x(t_0)$ at initial time t_0. For any $A(t)$ whose entries are continuous functions of t, it is well known (see, e.g., [8]) that for any initial condition $x(t_0)$, there is a unique continuously differentiable solution of Equation 3.36 given by

$$x(t) = \Phi(t, t_0)x(t_0), \quad t > t_0 \tag{3.37}$$

where $\Phi(t, t_0)$ is an $n \times n$ matrix function of t and t_0, called the *state-transition matrix*. The state-transition matrix has the following fundamental properties:

$$\Phi(t, t) = I = n \times n \text{ identity matrix}, \quad \text{for all } t \tag{3.38}$$

$$\Phi(t, \tau) = \Phi(t, t_1)\Phi(t_1, \tau), \quad \text{for all } t_1, t, \tau \tag{3.39}$$

$$\Phi^{-1}(t, \tau) = \Phi(\tau, t), \quad \text{for all } t, \tau \tag{3.40}$$

$$\frac{\partial}{\partial t}\Phi(t, \tau) = A(t)\Phi(t, \tau), \quad \text{for all } t, \tau \tag{3.41}$$

$$\frac{\partial}{\partial \tau}\Phi(t, \tau) = -\Phi(t, \tau)A(\tau), \quad \text{for all } t, \tau \tag{3.42}$$

$$\det\Phi(t, \tau) = \exp\left[\int_\tau^t \operatorname{tr}[A(\sigma)]\, d\sigma\right] \tag{3.43}$$

In Equation 3.43, "det" denotes the determinant and "tr" denotes the trace. Equation 3.39 is called the *composition property*. It follows from this property that $\Phi(t, \tau)$ can be written in the factored form

$$\Phi(t, \tau) = \Phi(t, 0)\Phi(0, \tau), \quad \text{for all } t, \tau \tag{3.44}$$

Another important property of the state-transition matrix $\Phi(\tau, t)$ is that it is a continuously differentiable matrix function of t and τ.

It follows from Equation 3.42 that the *adjoint equation*

$$\dot{\gamma}(t) = -A^T(t)\gamma(t) \tag{3.45}$$

has state-transition matrix equal to $\Phi^T(\tau, t)$, where again $\Phi(t, \tau)$ is the state-transition matrix for Equation 3.36 and superscript "T" denotes the transpose operation.

If the system matrix $A(t)$ is constant over the interval $[t_1, t_2]$, that is, $A(t) = A$, for all $t \in [t_1, t_2]$, then the state-transition matrix is equal to the matrix exponential over $[t_1, t_2]$:

$$\Phi(t, \tau) = e^{A(t-\tau)} \quad \text{for all } t, \tau \in [t_1, t_2] \tag{3.46}$$

If $A(t)$ is time varying and $A(t)$ commutes with its integral over the interval $[t_1, t_2]$, that is,

$$A(t)\left[\int_\tau^t A(\sigma)\, d\sigma\right] = \left[\int_\tau^t A(\sigma)\, d\sigma\right]A(t), \quad \text{for all } t, \tau \in [t_1, t_2] \tag{3.47}$$

then $\Phi(t, \tau)$ is given by

$$\Phi(t, \tau) = \exp\left[\int_\tau^t A(\tau)\, d\tau\right], \quad \text{for all } t, \tau \in [t_1, t_2] \tag{3.48}$$

Note that the commutativity condition in Equation 3.47 is always satisfied in the time-invariant case. It is also always satisfied in the one-dimensional ($n = 1$) time-varying case since scalars commute. Thus, $\Phi(t, \tau)$ is given by the exponential form in Equation 3.48 when $n = 1$. Unfortunately, the exponential form for $\Phi(t, \tau)$ does not hold for an arbitrary time-varying matrix $A(t)$ when $n > 1$, and, in general, there is no known closed-form expression for $\Phi(t, \tau)$ when $n > 1$. However, approximations to $\Phi(t, \tau)$ can be readily computed from $A(t)$ by numerical techniques, such as the method of successive approximations (see, e.g., [8]). Approximations to $\Phi(t, \tau)$ can also be determined by discretizing the time variable t as shown next.

Let k be an integer-valued variable, let T be a positive number, and let $a_{ij}(t)$ denote the i,j entry of the matrix $A(t)$. Suppose that by choosing T to be sufficiently small, the absolute values $|a_{ij}(t) - a_{ij}(kT)|$ can be made as small as desired over the interval $t \in [kT, kT + T]$ and for all integer values of k. Then for a suitably small $T, A(t)$ is approximately equal to $A(kT)$ for $t \in [kT, kT + T]$, and from Equation 3.46 the state-transition matrix $\Phi(t, kT)$ is given approximately by

$$\Phi(t, kT) = e^{A(kT)(t-kT)} \quad \text{for all } t \in [kT,\ kT + T] \quad \text{and} \quad \text{all } k \tag{3.49}$$

Setting $t = kT + T$ in Equation 3.49 yields

$$\Phi(kT + T, kT) = e^{A(kT)T} \quad \text{for all } k \tag{3.50}$$

The state-transition matrix $\Phi(kT + T, kT)$ given by Equation 3.50 can be computed by using a parameterized Laplace transform. To show this, first define the $n \times n$ matrix

$$q(k, t) = e^{A(kT)t} \tag{3.51}$$

Then $q(k, t)$ is equal to the inverse Laplace transform of

$$Q(k, s) = [sI - A(kT)]^{-1} \tag{3.52}$$

Note that the transform $Q(k, s)$ is parameterized by the integer-valued variable k. A closed-form expression for $q(k, t)$ as a function of t can be obtained by expanding the entries of $Q(k, s)$ into partial fraction expansions and then using standard Laplace transform pairs to determine the inverse transform of the terms in the partial faction expansions. Then from Equation 3.51, setting $t = T$ in $q(k, t)$ yields $q(k, T) = \Phi(kT + T, kT)$.

Again consider the general case with the state equation given by Equation 3.34. Given the state-transition matrix $\Phi(t, \tau)$, for any given initial state $x(t_0)$ and input $u(t)$ applied for $t \geq t_0$, the complete solution to Equation 3.34 is

$$x(t) = \Phi(t, t_0)x(t_0) + \int_{t_0}^{t} \Phi(t, \tau)B(\tau)u(\tau)\, d\tau, \quad t > t_0 \tag{3.53}$$

Then, when $y(t) = C(t)x(t)$, the output response $y(t)$ is given by

$$y(t) = C(t)\Phi(t, t_0)x(t_0) + \int_{t_0}^{t} C(t)\Phi(t, \tau)B(\tau)u(\tau)\, d\tau, \quad t > t_0 \tag{3.54}$$

If the initial time t_0 is taken to be $-\infty$ and the initial state is zero, Equation 3.54 becomes

$$y(t) = \int_{-\infty}^{t} C(t)\Phi(t, \tau)B(\tau)u(\tau)\, d\tau \tag{3.55}$$

Comparing Equation 3.55 with the m-input, p-output version of Equation 3.1 reveals that

$$H(t, \tau) = \begin{cases} C(t)\Phi(t, \tau)B(\tau) & \text{for } t \geq \tau \\ 0 & \text{for } t < \tau \end{cases} \tag{3.56}$$

where $H(t, \tau)$ is the $p \times m$ impulse response function matrix. Inserting Equation 3.44 into Equation 3.56 reveals that $H(t, \tau)$ can be expressed in the factored form,

$$H(t, \tau) = H_1(t)H_2(\tau), \quad t \geq \tau \tag{3.57}$$

where

$$H_1(t) = C(t)\Phi(t, 0) \quad \text{and} \quad H_2(\tau) = \Phi(0, \tau)B(\tau) \tag{3.58}$$

It turns out [8] that a linear time-varying system with impulse response matrix $H(t, \tau)$ has a state realization given by Equations 3.34 and 3.35 with $D(t) = 0$ if and only if $H(t, \tau)$ can be expressed in the factored form given in Equation 3.57.

3.2.3 Change of State Variables

Suppose that the system under study has the n-dimensional state model

$$\dot{x}(t) = A(t)x(t) + B(t)u(t) \tag{3.59}$$

$$y(t) = C(t)x(t) \tag{3.60}$$

In the following development, the system given by Equations 3.59 and 3.60 will be denoted by the triple $[A(t), B(t), C(t)]$.

Given any $n \times n$ invertible continuously differentiable matrix $P(t)$, which is real or complex valued, another state model can be generated by defining the new state vector $z(t) = P^{-1}(t)x(t)$, where $P^{-1}(t)$ is the inverse of $P(t)$. The matrix $P^{-1}(t)$ (or the matrix $P(t)$) is called the transformation matrix and the process of going from $x(t)$ to $z(t)$ is referred to as a coordinate transformation. The state variables for the new state model (i.e., the $z_i(t)$) are linear combinations with time-varying coefficients of the state variables of the given state model. To determine the state equations for the new state model, first note that $P(t)z(t) = x(t)$. Then taking the derivative of both sides of $P(t)z(t) = x(t)$ gives

$$P(t)\dot{z}(t) + \dot{P}(t)z(t) = \dot{x}(t) \tag{3.61}$$

Inserting the expression for $\dot{x}(t)$ given by Equation 3.59 yields

$$P(t)\dot{z}(t) + \dot{P}(t)z(t) = A(t)x(t) + B(t)u(t) \tag{3.62}$$

and replacing $x(t)$ by $P(t)z(t)$ in Equation 3.62 and rearranging terms results in

$$P(t)\dot{z}(t) = [A(t)P(t) - \dot{P}(t)]z(t) + B(t)u(t) \tag{3.63}$$

Finally, multiplying both sides of Equation 3.63 on the left by $P^{-1}(t)$ gives

$$\dot{z}(t) = [P^{-1}(t)A(t)P(t) - P^{-1}(t)\dot{P}(t)]z(t) + P^{-1}(t)B(t)u(t) \tag{3.64}$$

and replacing $x(t)$ by $P(t)z(t)$ in Equation 3.60 yields

$$y(t) = C(t)P(t)z(t) \tag{3.65}$$

The state equations for the new state model are given by Equations 3.64 and 3.65. This new model will be denoted by the triple $[\bar{A}(t), \bar{B}(t), \bar{C}(t)]$ where

$$\bar{A}(t) = P^{-1}(t)A(t)P(t) - P^{-1}(t)\dot{P}(t) \tag{3.66}$$

$$\bar{B}(t) = P^{-1}(t)B(t) \tag{3.67}$$

$$\bar{C}(t) = C(t)P(t) \tag{3.68}$$

Multiplying both sides of Equation 3.66 on the left by $P(t)$ and rearranging the terms yield the result that the transformation matrix $P(t)$ satisfies the differential equation

$$\dot{P}(t) = A(t)P(t) - P(t)\bar{A}(t) \tag{3.69}$$

The state models $[A(t), B(t), C(t)]$ and $[\overline{A}(t), \overline{B}(t), \overline{C}(t)]$ with $\bar{A}(t), \bar{B}(t), \bar{C}(t)$ given by Equations 3.66 through 3.68 are said to be *algebraically equivalent*. The state-transition matrix $\bar{\Phi}(t, \tau)$ for $[\bar{A}(t), \bar{B}(t), \bar{C}(t)]$ is given by

$$\bar{\Phi}(t, \tau) = P^{-1}(t)\Phi(t, \tau)P(\tau) \tag{3.70}$$

where $\Phi(t, \tau)$ is the state-transition matrix for $[A(t), B(t), C(t)]$.

Given an n-dimensional state model $[A(t), B(t), C(t)]$ and any $n \times n$ continuous matrix function $\Gamma(t)$, there is a transformation matrix $P(t)$ that transforms $A(t)$ into $\Gamma(t)$, that is, $\bar{A}(t) = \Gamma(t)$. To show this, define $P(t)$ by

$$P(t) = \Phi(t, 0)\bar{\Phi}(0, t) \tag{3.71}$$

where $\bar{\Phi}(t, \tau)$ is the state-transition matrix for $\dot{z}(t) = \Gamma(t)z(t)$. Note that $P(t)$ is continuously differentiable since $\Phi(t, 0)$ and $\bar{\Phi}(0, t)$ are continuously differentiable, and $P(t)$ is invertible since state-transition matrices are invertible. Then taking the derivative of both sides of Equation 3.71 gives

$$\dot{P}(t) = \Phi(t, 0)\dot{\bar{\Phi}}(0, t) + \dot{\Phi}(t, 0)\bar{\Phi}(0, t) \tag{3.72}$$

Using Equations 3.41 and 3.42 in Equation 3.72 yields

$$\dot{P}(t) = -\Phi(t, 0)\bar{\Phi}(0, t)\Gamma(t) + A(t)\Phi(t, 0)\bar{\Phi}(0, t) \tag{3.73}$$

Finally, inserting Equation 3.71 into Equation 3.73 results in Equation 3.69 with $\bar{A}(t) = \Gamma(t)$, and thus, $P(t)$ transforms $A(t)$ into $\Gamma(t)$.

This result shows that via a change of state, $A(t)$ can be transformed to any desired continuous matrix $\Gamma(t)$. The fact that any continuous system matrix $A(t)$ can be transformed to any other continuous matrix raises some interesting issues. For example, suppose that the transformation $z(t) = P^{-1}(t)x(t)$ is defined so that the new system matrix $\bar{A}(t)$ is equal to a diagonal matrix $\Lambda(t)$ with real or complex-valued functions $\lambda_1(t), \lambda_2(t), \ldots, \lambda_n(t)$ on the diagonal. Then from Equation 3.66,

$$\Lambda(t) = P^{-1}(t)A(t)P(t) - P^{-1}(t)\dot{P}(t) \tag{3.74}$$

and multiplying both sides of Equation 3.74 on the left by $P(t)$ gives

$$P(t)\Lambda(t) = A(t)P(t) - \dot{P}(t) \tag{3.75}$$

Now for $i = 1, 2, \ldots, n$, let $\gamma_i(t)$ denote the ith column of the transformation matrix $P(t)$. Then from Equation 3.75, it follows that

$$\lambda_i(t)\gamma_i(t) = A(t)\gamma_i(t) - \dot{\gamma}_i(t), \quad i = 1, 2, \ldots, n \tag{3.76}$$

Rearranging terms in Equation 3.76 gives

$$[A(t) - \lambda_i(t)I]\gamma_i(t) = \dot{\gamma}_i(t), \quad i = 1, 2, \ldots, n \tag{3.77}$$

where I is the $n \times n$ identity matrix.

Equation 3.77 has appeared in the literature on linear time-varying systems, and has sometimes been used as the justification for referring to the $\lambda_i(t)$ as "dynamic" eigenvalues or poles, and the corresponding $\gamma_i(t)$ as eigenvectors, of the time-varying system having system matrix $A(t)$. But since Equation 3.77 holds for any continuous scalar-valued functions $\lambda_1(t), \lambda_2(t), \ldots, \lambda_n(t)$, or for any constants $\lambda_1, \lambda_2, \ldots, \lambda_n$, the existence of $\lambda_i(t)$ and $\gamma_i(t)$ satisfying Equation 3.77 is not useful unless further conditions are placed on the transformation matrix $P(t)$. For example, one can consider stronger notions of equivalence such as *topological equivalence*. This means that, in addition to being algebraically equivalent, the transformation matrix $P(t)$ has the properties

$$|\det P(t)| \geq c_0 \quad \text{for all } t \tag{3.78}$$

$$|p_{ij}(t)| \leq c_1 \quad \text{for all } t \text{ and } i, j = 1, 2, \ldots, n \tag{3.79}$$

where $p_{ij}(t)$ is the i, j entry of $P(t)$, and c_0 and c_1 are finite positive constants. The conditions given in Equations 3.78 and 3.79 are equivalent to requiring that $P(t)$ and its inverse $P^{-1}(t)$ be bounded matrix functions of t. A transformation $z(t) = P^{-1}(t)x(t)$ with $P(t)$ satisfying Equations 3.78 and 3.79 is called a *Lyapunov transformation*. As noted in Section 3.2.4, stability is preserved under a Lyapunov transformation.

If the transformation $z(t) = P^{-1}(t)x(t)$ is a Lyapunov transformation and puts the system matrix $A(t)$ into the diagonal form $\Lambda(t) = \text{diag}(\lambda_1(t), \lambda_2(t), \ldots, \lambda_n(t))$, then the $\lambda_i(t)$ are of interest since they determine the stability of the system. However, in the general linear time-varying case, the existence of a Lyapunov transformation that puts $A(t)$ into a diagonal form is a strong condition. A much weaker condition is the existence of a Lyapunov transformation that puts $A(t)$ into an upper or lower triangular form (see, e.g., [7]). This is briefly discussed in Section 3.2.4.

3.2.4 Stability

Given a system with n-dimensional state model $[A(t), B(t), C(t)]$, again consider the homogeneous equation

$$\dot{x}(t) = A(t)x(t), \quad t > t_0 \tag{3.80}$$

with solution

$$x(t) = \Phi(t, t_0)x(t_0), \quad t > t_0 \tag{3.81}$$

The system is said to be *asymptotically stable* if for some initial time t_0, the solution $x(t)$ satisfies the condition $\|x(t)\| \to 0$ as $t \to \infty$ for any initial state $x(t_0)$ at initial time t_0. Here $\|x(t)\|$ denotes the *Euclidean norm* of the state $x(t)$ given by

$$\|x(t)\| = \sqrt{x^T(t)x(t)} = \sqrt{x_1^2(t) + x_2^2(t) + \cdots + x_n^2(t)} \tag{3.82}$$

where $x(t) = [x_1(t)\ x_2(t) \ldots x_n(t)]^T$. A system is asymptotically stable if and only if

$$\|\Phi(t, t_0)\| \to 0 \quad \text{as}\ t \to \infty \tag{3.83}$$

where $\|\Phi(t, t_0)\|$ is the matrix norm equal to the square root of the largest eigenvalue of $\Phi^T(t, t_0)\Phi(t, t_0)$, where t is viewed as a parameter.

A stronger notion of stability is *exponential stability*, which requires that for some initial time t_0, there exist finite positive constants c and λ such that for any $x(t_0)$, the solution $x(t)$ to Equation 3.80 satisfies

$$\|x(t)\| \leq ce^{-\lambda(t-t_0)} \|x(t_0)\|, \quad t \geq t_0 \tag{3.84}$$

If the condition in Equation 3.84 holds for all t_0 with the constants c and λ fixed, the system is said to be *uniformly exponential stable*. Uniform exponential stability is equivalent to requiring that there exists finite positive constants γ and λ such that

$$\|\Phi(t, \tau)\| \leq \gamma e^{-\lambda(t-\tau)} \quad \text{for all}\ t, \tau\ \text{such that}\ t \geq \tau \tag{3.85}$$

Uniform exponential stability is also equivalent to requiring that, given any positive constant δ, there exists a positive constant T such that for any t_0 and $x(t_0)$, the solution $x(t)$ to Equation 3.80 satisfies

$$\|x(t)\| \leq \delta \|x(t_0)\| \quad \text{for}\ t \geq t_0 + T \tag{3.86}$$

It follows from Equation 3.85 that uniform exponential stability is preserved under a Lyapunov transformation $z(t) = P^{-1}(t)x(t)$. To see this, let $\bar{\Phi}(t, \tau)$ denote the state-transition matrix for $\dot{z}(t) = \bar{A}(t)z(t)$,

where

$$\bar{A}(t) = P^{-1}(t)A(t)P(t) - P^{-1}(t)\dot{P}(t) \tag{3.87}$$

Then using Equation 3.70 gives

$$\left\|\bar{\Phi}(t,\tau)\right\| \le \left\|P^{-1}(t)\right\| \, \|\Phi(t,\tau)\| \, \|P(\tau)\| \tag{3.88}$$

But Equations 3.78 and 3.79 imply that

$$\|P(\tau)\| \le M_1 \quad \text{and} \quad \left\|P^{-1}(t)\right\| \le M_2 \tag{3.89}$$

for some finite constants M_1 and M_2. Then inserting Equations 3.85 and 3.89 into Equation 3.88 yields

$$\left\|\bar{\Phi}(t,\tau)\right\| \le \gamma M_1 M_2 e^{-\lambda(t-\tau)} \quad \text{for all } t, \tau \text{ such that } t \ge \tau \tag{3.90}$$

which verifies that $\dot{z}(t) = \bar{A}(t)z(t)$ is uniformly exponentially stable.

If $P(t)$ and $P^{-1}(t)$ are bounded only for $t \ge t_1$ for some finite t_1, the coordinate transformation $z(t) = P^{-1}(t)x(t)$ preserves exponential stability of the given system. This can be proved using constructions similar to those given above. The details are omitted.

When $n = 1$ so that $A(t) = a(t)$ is a scalar-valued function of t, as seen from Equation 3.48,

$$\Phi(t, t_0) = \exp\left[\int_{t_0}^{t} a(\tau)\, d\tau\right]. \tag{3.91}$$

It follows that the system is asymptotically stable if and only if

$$\int_{t_0}^{t} a(\sigma)\, d\sigma \to -\infty \text{ as } t \to \infty \tag{3.92}$$

and the system is uniformly exponentially stable if a positive constant λ exists so that

$$\int_{\tau}^{t} a(\sigma)\, d\sigma \le -\lambda(t-\tau) \quad \text{for all } t, \tau \text{ such that } t \ge \tau \tag{3.93}$$

When $n > 1$, if $A(t)$ commutes with its integral for $t > t_1$ for some $t_1 \ge t_0$ (see Equation 3.47), then

$$\Phi(t, t_1) = \exp\left[\int_{t_1}^{t} A(\sigma)\, d\sigma\right], \quad t > t_1 \tag{3.94}$$

and a sufficient condition for exponential stability (not uniform in general) is that the matrix function

$$\frac{1}{t}\int_{t_2}^{t} A(\sigma)\, d\sigma \tag{3.95}$$

be a bounded function of t and its *pointwise eigenvalues* have real parts $\le -\beta$ for $t > t_2$ for some $t_2 \ge t_1$, where β is a positive constant. The pointwise eigenvalues of a time-varying matrix $M(t)$ are the eigenvalues of the constant matrix $M(\tau)$ for each fixed value of τ viewed as a time-independent parameter. If $A(t)$ is upper or lower triangular with $p_1(t), p_2(t), \ldots, p_n(t)$ on the diagonal, then a sufficient condition for uniform exponential stability is that the off-diagonal entries of $A(t)$ be bounded and the scalar systems

$$\dot{x}_i(t) = p_i(t)x_i(t) \tag{3.96}$$

be uniformly exponentially stable for $i = 1, 2, \ldots, n$. Note that the system matrix $A(t)$ given by Equation 3.32 is lower triangular, and thus, in this case, the system with this particular system matrix is uniformly exponentially stable if the poles $p_i(t)$ in the Floquet factorization given in the left-hand side of

Equation 3.25 are stable in the sense that the scalar systems in Equation 3.96 are uniformly exponentially stable.

If there exists a Lyapunov transformation $z(t) = P^{-1}(t)x(t)$ that puts the system matrix $A(t)$ into an upper or lower triangular form with $p_1(t), p_2(t), \ldots, p_n(t)$ on the diagonal, then sufficient conditions for uniform exponential stability are that the off-diagonal entries of the triangular form be bounded and the scalar systems defined by Equation 3.96 be uniformly exponentially stable. The construction of a Lyapunov transformation that puts $A(t)$ into an upper triangular form is given in [7]. In that paper, the authors define the set $\{p_1(t), p_2(t), \ldots, p_n(t)\}$ of elements on the diagonal of the upper triangular form to be a pole set of the linear time-varying system with system matrix $A(t)$.

Another condition for uniform exponential stability is that a symmetric positive-definite matrix $Q(t)$ exists with $c_1 I \leq Q(t) \leq c_2 I$ for some positive constants c_1, c_2, such that

$$A^T(t)Q(t) + Q(t)A(t) + \dot{Q}(t) \leq -c_3 I \tag{3.97}$$

for some positive constant c_3. Here $F \leq G$ means that F and G are symmetric matrices and $G - F$ is positive semidefinite (all pointwise eigenvalues are ≥ 0). This stability test is referred to as the Lyapunov criterion. For more details, see [8].

Finally, it is noted that if the entries of $B(t)$ and $C(t)$ are bounded, then uniform exponential stability implies that the system is *bounded-input, bounded-output (BIBO) stable*, that is, a bounded input $u(t)$ always results in a bounded output response $y(t)$.

3.2.5 Controllability and Observability

Given a system with the n-dimensional state model $[A(t), B(t), C(t)]$, it is now assumed that the entries of $A(t)$, $B(t)$, and $C(t)$ are at least continuous functions of t. The system is said to be *controllable* on the interval $[t_0, t_1]$, where $t_1 > t_0$, if for any states x_0 and x_1, a continuous input $u(t)$ exists that drives the system to the state $x(t_1) = x_1$ at time $t = t_1$ starting from the state $x(t_0) = x_0$ at time $t = t_0$.

Define the *controllability Gramian* which is the $n \times n$ matrix given by

$$W(t_0, t_1) = \int_{t_0}^{t_1} \Phi(t_0, t)B(t)B^T(t)\Phi^T(t_0, t)\, dt \tag{3.98}$$

The controllability Gramian $W(t_0, t_1)$ is symmetric positive semidefinite and is the solution to the matrix differential equation

$$\begin{aligned} \frac{d}{dt}W(t, t_1) &= A(t)W(t, t_1) + W(t, t_1)A^T(t) - B(t)B^T(t), \\ W(t_1, t_1) &= 0 \end{aligned} \tag{3.99}$$

Then the system is controllable on $[t_0, t_1]$ if and only if $W(t_0, t_1)$ is invertible, in which case a continuous input $u(t)$ that drives the system from $x(t_0) = x_0$ to $x(t_1) = x_1$ is

$$u(t) = -B^T(t)\Phi^T(t_0, t)W^{-1}(t_0, t_1)\left[x_0 - \Phi(t_0, t_1)x_1\right], \quad t_0 \leq t \leq t_1 \tag{3.100}$$

There is a sufficient condition for controllability that does not require that the controllability Gramian be computed: Given a positive integer q, suppose that the entries of $B(t)$ are $q - 1$ times continuously differentiable and the entries of $A(t)$ are $q - 2$ times continuously differentiable, and define the $n \times m$ matrices

$$K_0(t) = B(t) \tag{3.101}$$

$$K_i(t) = -A(t)K_{i-1}(t) + \dot{K}_{i-1}(t), \quad i = 1, 2, \ldots, q - 1 \tag{3.102}$$

Finally, let $K(t)$ denotes the $n \times mq$ matrix whose ith block column is equal to $K_{i-1}(t)$, that is

$$K(t) = [K_0(t)\ K_1(t) \ldots K_{q-1}(t)] \tag{3.103}$$

Then a sufficient condition for the system $[A(t), B(t), C(t)]$ to be controllable on the interval $[t_0, t_1]$ is that the $n \times mq$ matrix $K(t)$ defined by Equation 3.103 has rank n for at least one value of $t \in [t_0, t_1]$. This condition was first derived in [9].

The rank condition on the matrix $K(t)$ is preserved under a change of state variables. To show this, suppose the coordinate transformation $z(t) = P^{-1}(t)x(t)$ results in the state model $[\overline{A}(t), \overline{B}(t), \overline{C}(t)]$. For this model, define

$$\bar{K}(t) = [\bar{K}_0(t)\ \bar{K}_1(t)\ \ldots\ \bar{K}_{q-1}(t)] \tag{3.104}$$

where the $\bar{K}_i(t)$ are given by Equations 3.101 and 3.102 with $A(t)$ and $B(t)$ replaced by $\bar{A}(t)$ and $\bar{B}(t)$, respectively. It is assumed that $\bar{A}(t)$ and $\bar{B}(t)$ satisfy the same differentiability requirements as given above for $A(t)$ and $B(t)$ so that $\bar{K}(t)$ is well defined. Then it follows that

$$P^{-1}(t)K(t) = \overline{K}(t) \tag{3.105}$$

Now since $P^{-1}(t)$ is invertible for all $t, P^{-1}(t)$ has rank n for all $t \in [t_0, t_1]$, and thus $\bar{K}(t)$ has rank n for some $t_c \in [t_0, t_1]$ if and only if $K(t_c)$ has rank n. Therefore, the system defined by $[A(t), B(t), C(t)]$ is controllable on the interval $[t_0, t_1]$ if and only if the transformed system $[\overline{A}(t), \overline{B}(t), \overline{C}(t)]$ is controllable on $[t_0, t_1]$.

If the matrix $K(t)$ has rank n for all t, the $n \times n$ matrix $K(t)K^T(t)$ is invertible for all t, and in this case Equation 3.105 can be solved for $P^{-1}(t)$. This gives

$$P^{-1}(t) = \overline{K}(t)K^T(t)\left[K(t)K^T(t)\right]^{-1} \tag{3.106}$$

Note that Equation 3.106 can be used to compute the transformation matrix $P^{-1}(t)$ directly from the coefficient matrices of the state models $[A(t), B(t), C(t)]$ and $[\overline{A}(t), \overline{B}(t), \overline{C}(t)]$.

Now suppose that the system input $u(t)$ is zero, so that the state model is given by

$$\dot{x}(t) = A(t)x(t) \tag{3.107}$$

$$y(t) = C(t)x(t) \tag{3.108}$$

Inserting Equation 3.108 into the solution of Equation 3.107 results in the output response $y(t)$ resulting from initial state $x(t_0)$:

$$y(t) = C(t)\Phi(t, t_0)x(t_0), \quad t > t_0 \tag{3.109}$$

The system is said to be observable on the interval $[t_0, t_1]$ if any initial state $x(t_0)$ can be determined from the output response $y(t)$ given by Equation 3.109 for $t \in [t_0, t_1]$. Define the *observability Gramian* which is the $n \times n$ matrix given by

$$M(t_0, t_1) = \int_{t_0}^{t_1} \Phi^T(t, t_0)C^T(t)C(t)\Phi(t, t_0)\, dt \tag{3.110}$$

The observability Gramian $M(t_0, t_1)$ is symmetric positive semidefinite and is the solution to the matrix differential equation

$$\begin{aligned} \frac{d}{dt}M(t, t_1) &= -A^T(t)M(t, t_1) - M(t, t_1)A(t) - C^T(t)C(t), \\ M(t_1, t_1) &= 0 \end{aligned} \tag{3.111}$$

Then the system is *observable* on $[t_0, t_1]$ if and only if $M(t_0, t_1)$ is invertible, in which case the initial state $x(t_0)$ is given by

$$x(t_0) = M^{-1}(t_0, t_1) \int_{t_0}^{t_1} \Phi^T(t, t_0) C^T(t) y(t)\, dt \tag{3.112}$$

Again given an m-input p-output n-dimensional system with state model $[A(t), B(t), C(t)]$, the *adjoint system* is the p-input m-output n-dimensional system with state model $[-A^T(t), C^T(t), B^T(t)]$. The adjoint system is given by the state equations

$$\dot{\gamma}(t) = -A^T(t)\gamma(t) + C^T(t)\nu(t) \tag{3.113}$$

$$\eta(t) = B^T(t)\gamma(t) \tag{3.114}$$

where $\gamma(t)$, $\nu(t)$, and $\eta(t)$ are the state, input, and output, respectively, of the adjoint system. As noted above in Section 3.2.2, the state-transition matrix of the adjoint system $[-A^T(t), C^T(t), B^T(t)]$ is equal to $\Phi^T(\tau, t)$, where $\Phi(t, \tau)$ is the state-transition matrix for the given system $[A(t), B(t), C(t)]$. Using this fact and the definition of the input and output coefficient matrices of the adjoint system given by Equations 3.113 and 3.114, it is clear that the controllability Gramiam of the adjoint system is identical to the observability Gramiam of the given system. Thus, the given system is observable on $[t_0, t_1]$ if and only if the adjoint system is controllable on $[t_0, t_1]$. In addition, the observability Gramiam of the adjoint system is identical to the controllability Gramiam of the given system. Hence, the given system is controllable on $[t_0, t_1]$ if and only if the adjoint system is observable on $[t_0, t_1]$.

A sufficient condition for observability is given next in terms of the system matrix $A(t)$ and the output matrix $C(t)$: Consider the system with state model $[A(t), B(t), C(t)]$, and define the $p \times n$ matrices

$$L_0(t) = C(t) \tag{3.115}$$

$$L_i(t) = L_{i-1}(t)A(t) + \dot{L}_{i-1}(t), \quad i = 1, 2, \ldots, q-1 \tag{3.116}$$

where q is a positive integer. Consider the $pq \times n$ matrix $L(t)$ whose ith block row is equal to $L_{i-1}(t)$, that is,

$$L(t) = \begin{bmatrix} L_0(t) \\ L_1(t) \\ \vdots \\ L_{q-1}(t) \end{bmatrix} \tag{3.117}$$

It is assumed that the entries of $C(t)$ are $q-1$ times differentiable and the entries of $A(t)$ are $q-2$ times differentiable, so that $L(t)$ is well defined. Then as first shown in [9], the system is observable on $[t_0, t_1]$ if the $pq \times n$ matrix $L(t)$ defined by Equations 3.115 through 3.117 has rank n for at least one value of $t \in [t_0, t_1]$.

Now consider the $n \times pq$ matrix $U(t) = [U_0(t)\, U_1(t) \ldots U_{q-1}(t)]$ generated from the adjoint system $[-A^T(t), C^T(t), B^T(t)]$, where

$$U_0(t) = C^T(t) \tag{3.118}$$

$$U_i(t) = A^T(t)U_{i-1}(t) + \dot{U}_{i-1}(t), \quad i = 1, 2, \ldots, q-1 \tag{3.119}$$

Then from Equations 3.115 through 3.119, it is seen that the transpose $U^T(t)$ of the $n \times pq$ matrix $U(t)$ is equal to the $pq \times n$ matrix $L(t)$ of the given system with state model $[A(t), B(t), C(t)]$. Since the transpose operation does not affect the rank of a matrix, the sufficient condition rank $L(t) = n$ for observability of the system $[A(t), B(t), C(t)]$ implies that the adjoint system $[-A^T(t), C^T(t), B^T(t)]$ is controllable. In addition, the rank condition for observability of the adjoint system implies that the given system is controllable. It also follows from the above constructions that the rank condition for observabiliy of the system $[A(t), B(t), C(t)]$ is preserved under a coordinate transformation.

3.2.6 Control Canonical Form and Controller Design

Now suppose that the system with state model $[A(t), B(t), C(t)]$ has a single input ($m = 1$) so that the input matrix $B(t)$ is an n-element column vector. Assuming that $B(t)$ and $A(t)$ can be differentiated an appropriate number of times, let $R(t)$ denote the $n \times n$ matrix whose columns $r_i(t)$ are defined by

$$r_1(t) = B(t) \tag{3.120}$$

$$r_{i+1}(t) = A(t)r_i(t) - \dot{r}_i(t), \quad i = 1, 2, \ldots, n-1 \tag{3.121}$$

Note that $R(t)$ is a minor variation of the $n \times n$ matrix $K(t)$ defined by Equations 3.101 through 3.103 with $q = n$. In fact, since $R(t)$ is equal to $K(t)$ with a sign change in the columns, $R(t)$ has rank n for all t if and only if $K(t)$ has rank n for all t. Thus, $R(t)$ is invertible for all t if and only if the matrix $K(t)$ has rank n for all t.

In some textbooks, such as [1], the matrix $R(t)$ is called the *controllability matrix* of the system with state model $[A(t), B(t), C(t)]$.

Assuming that $R(t)$ is invertible for all t, define the n-element column vector

$$\eta(t) = -R^{-1}(t)r_{n+1}(t) \tag{3.122}$$

where $r_{n+1}(t)$ is defined by Equation 3.121 with $i = n$. The vector $\eta(t)$ is invariant under any change of state $z(t) = P^{-1}(t)x(t)$. In other words, if Equations 3.120 through 3.122 are evaluated for the new state model $[\overline{A}(t), \overline{B}(t), \overline{C}(t)]$ resulting from the transformation $z(t) = P^{-1}(t)x(t)$, Equation 3.122 will yield the same result for $\eta(t)$. In addition, if $A(t) = A$ and $B(t) = B$ where A and B are constant matrices, the vector η is constant and is given by

$$\eta = [a_0 \; a_1 \; \ldots \; a_{n-1}]^T \tag{3.123}$$

where a_i are the coefficients of the characteristic polynomial of A, that is,

$$\det(sI - A) = s^n + \sum_{i=0}^{n-1} a_i s^i \tag{3.124}$$

Given the analogy with the time-invariant case, the vector $\eta(t)$ given by Equation 3.122 can be viewed as a time-varying version of the *characteristic vector* of the system.

If the $n \times n$ matrix $R(t)$ with columns defined by Equations 3.120 and 3.121 is invertible for all t, there is a coordinate transformation $z(t) = P^{-1}(t)x(t)$, which converts $[A(t), B(t), C(t)]$ into the *control canonical form* $[\bar{A}(t), \bar{B}, \bar{C}(t)]$ with $\bar{A}(t)$ and $\bar{B}$ given by

$$\bar{A}(t) = \begin{bmatrix} 0 & 1 & 0 & \cdots & 0 & 0 \\ 0 & 0 & 1 & \cdots & 0 & 0 \\ 0 & 0 & 0 & \cdots & 0 & 0 \\ \vdots & & & \cdots & & \vdots \\ 0 & 0 & 0 & \cdots & 0 & 1 \\ -\psi_0(t) & -\psi_1(t) & -\psi_2(t) & \cdots & -\psi_{n-2}(t) & -\psi_{n-1}(t) \end{bmatrix} \tag{3.125}$$

$$\overline{B} = \begin{bmatrix} 0 \\ 0 \\ 0 \\ \vdots \\ 0 \\ 1 \end{bmatrix} \tag{3.126}$$

This form was first derived in [10].

A recursive procedure for computing the rows $c_i(t)$ of the matrix $P^{-1}(t)$ in the transformation $z(t) = P^{-1}(t)x(t)$ to the control canonical form was derived in [11], and is given by

$$c_1(t) = (e_n)^T R^{-1}(t) \tag{3.127}$$

$$c_{i+1}(t) = c_i(t)A(t) + \dot{c}_i(t), \quad i = 1, 2, \ldots, n-1 \tag{3.128}$$

where

$$(e_n)^T = \begin{bmatrix} 0 & 0 & 0 & \cdots & 0 & 1 \end{bmatrix} \tag{3.129}$$

As shown in [11], the entries $\psi_i(t)$ in the bottom row of $\bar{A}(t)$ are given by

$$\psi(t) = [\psi_0(t)\ \psi_1(t)\ \ldots\ \psi_{n-1}(t)] = -(c_n(t)A(t) + \dot{c}_n(t))P(t) \tag{3.130}$$

Now consider the system in the control canonical form given by the state equation

$$\dot{z}(t) = \bar{A}(t)z(t) + \bar{B}u(t) \tag{3.131}$$

where $\bar{A}(t)$ and $\bar{B}$ are defined by Equations 3.125 and 3.126. Then with the *state feedback control* $u(t) = -\bar{g}(t)z(t)$, where $\bar{g}(t)$ is an n-element row vector, from Equation 3.131 the state equation for the resulting closed-loop system is

$$\dot{z}(t) = [\bar{A}(t) - \bar{B}\bar{g}(t)]z(t) \tag{3.132}$$

Let $d = [d_0\ d_1\ \ldots\ d_{n-1}]$ be an n-element row vector with any desired constants $d_0, d_1, \ldots, d_{n-1}$. Then from the form of $\bar{A}(t)$ and $\bar{B}$, it is clear that if $\bar{g}(t)$ is taken to be $\bar{g}(t) = d - \psi(t)$, then the resulting closed-loop system matrix $\bar{A}(t) - \bar{B}\bar{g}(t)$ is equal to the matrix in the right-hand side of Equation 3.125 with the elements $-d_0, -d_1, \ldots, -d_{n-1}$ in the bottom row. Thus, the matrix $\bar{A}(t) - \bar{B}\bar{g}(t)$ is constant and its characteristic polynomial is equal to $s^n + d_{n-1}s^{n-1} + \cdots d_1 s + d_0$. The coefficients of the characteristic polynomial can be assigned to have any desired values $d_0, d_1, \ldots, d_{n-1}$.

The state feedback $u(t) = -\bar{g}(t)z(t)$ in the control canonical form can be expressed in terms of the state $x(t)$ of the given system $[A(t), B(t), C(t)]$ by using the coordinate transformation $z(t) = P^{-1}(t)x(t)$. This results in

$$u(t) = -\bar{g}(t)z(t) = -\bar{g}(t)P^{-1}(t)x(t) = -[d - \psi(t)]P^{-1}(t)x(t) \tag{3.133}$$

Inserting the feedback $u(t)$ given by Equation 3.133 into the state equation $\dot{x}(t) = A(t)x(t) + B(t)u(t)$ for the system $[A(t), B(t), C(t)]$ yields the following equation for the resulting closed-loop system:

$$\dot{x}(t) = (A(t) - B(t)g(t))x(t) \tag{3.134}$$

where the feedback gain vector $g(t)$ is given by

$$g(t) = \bar{g}(t)P^{-1}(t) = [d - \psi(t)]P^{-1}(t) \tag{3.135}$$

By the above constructions, the coordinate transformation $z(t) = P^{-1}(t)x(t)$ transforms the system matrix $A(t) - B(t)g(t)$ of the closed-loop system defined by Equation 3.134 into the system matrix $\bar{A}(t) - \bar{B}\bar{g}(t)$ of the closed-loop system given by Equation 3.132.

As shown in [11], the gain vector $g(t)$ defined by Equation 3.135 can be computed without having to determine $\bar{A}(t)$ by using the following formula:

$$g(t) = -\left[c_{n+1}(t) + \sum_{i=0}^{n-1} d_i c_{i+1}\right] \tag{3.136}$$

where the $c_i(t)$ are given by Equations 3.127 and 3.128 with $i = 1, 2, \ldots, n$. If the zeros of the characteristic polynomial of $\bar{A}(t) - \bar{B}\bar{g}(t)$ are chosen so that they are located in the open left half of the complex plane and if the coordinate transformation $z(t) = P^{-1}(t)x(t)$ is a Lyapunov transformation, then the closed-loop system given by Equation 3.134 is uniformly exponentially stable. If $P(t)$ and $P^{-1}(t)$ are bounded only for $t \geq t_1$ for some finite t_1, then the closed-loop system is exponentially stable. In Section 3.4, an example is given which illustrates the computation of the feedback gain $g(t)$.

3.3 Discrete-Time Linear Time-Varying Systems

A discrete-time causal linear time-varying system with single-input $u(k)$ and single-output $y(k)$ can be modeled by the input/output relationship

$$y(k) = \sum_{j=-\infty}^{k} h(k, j)u(j) \tag{3.137}$$

where k is an integer-valued variable (the discrete-time index) and $h(k,j)$ is the output response resulting from the unit pulse $\delta(k-j)$ (where $\delta(k-j)=1$ for $k=j$ and $=0$ for $k\neq j$) applied at time j. It is assumed that $u(k)$ and/or $h(k,j)$ is constrained so that the summation in Equation 3.137 is well defined. The system defined by Equation 3.137 is time invariant if and only if $h(k,j)$ is a function of only the difference $k-j$, in which case Equation 3.137 reduces to the convolution relationship

$$y(k) = h(k) * u(k) = \sum_{j=-\infty}^{k} h(k-j)u(j) \tag{3.138}$$

where $h(k-j) = h(k-j, 0)$.

The system defined by Equation 3.138 is finite dimensional if the input $u(k)$ and the output $y(k)$ are related by the nth-order difference equation

$$y(k+n) + \sum_{i=0}^{n-1} a_i(k)y(k+i) = \sum_{i=0}^{m} b_i(k)u(k+i) \tag{3.139}$$

where $m \leq n$ and the $a_i(k)$ and the $b_i(k)$ are real-valued functions of the discrete-time variable k. The system given by Equation 3.139 is time invariant if and only if all coefficients in Equation 3.139 are constants, that is, $a_i(k) = a_i$ and $b_i(k) = b_i$ for all i, where a_i and b_i are constants.

When $m < n$ the system defined by Equation 3.139 has the n-dimensional state model

$$x(k+1) = A(k)x(k) + B(k)u(k) \tag{3.140}$$

$$y(k) = Cx(k) \tag{3.141}$$

where

$$A(k) = \begin{bmatrix} 0 & 0 & 0 & \cdots & 0 & -a_0(k) \\ 1 & 0 & 0 & \cdots & 0 & -a_1(k-1) \\ 0 & 1 & 0 & \cdots & 0 & -a_2(k-2) \\ \vdots & & & \cdots & & \vdots \\ 0 & 0 & 0 & \cdots & 0 & -a_{n-2}(k-n+2) \\ 0 & 0 & 0 & \cdots & 1 & -a_{n-1}(k-n+1) \end{bmatrix} \tag{3.142}$$

$$B(k) = \begin{bmatrix} b_0(k) \\ b_1(k-1) \\ b_2(k-2) \\ \vdots \\ b_{n-2}(k-n+2) \\ b_{n-1}(k-n+1) \end{bmatrix} \tag{3.143}$$

and

$$C = \begin{bmatrix} 0 & 0 & \cdots & 0 & 1 \end{bmatrix}$$

where $b_i(k) = 0$ for $i > m$. This particular state model is referred to as the observer canonical form. As in the continuous-time case, there are other possible state realizations of Equation 3.139, but these will not be considered here. It is interesting to note that the entries of $A(k)$ and $B(k)$ in the observer canonical form are simply time shifts of the coefficients of the input/output difference Equation 3.139, whereas as shown above, in the continuous-time case this relationship is rather complicated.

3.3.1 State Model in the General Case

For an m-input p-output linear n-dimensional time-varying discrete-time system, the general form of the state model is

$$x(k+1) = A(k)x(k) + B(k)u(k) \tag{3.144}$$

$$y(k) = C(k)x(k) + D(k)u(k) \tag{3.145}$$

where the system matrix $A(k)$ is $n \times n$, the input matrix $B(k)$ is $n \times m$, the output matrix $C(k)$ is $p \times n$, and the direct feed matrix $D(k)$ is $p \times m$. The state model given by Equations 3.144 and 3.145 may arise as a result of sampling a continuous-time system given by the state model

$$\dot{x}(t) = A(t)x(t) + B(t)u(t) \tag{3.146}$$

$$y(t) = C(t)x(t) + D(t)u(t) \tag{3.147}$$

If the sampling interval is equal to T, then setting $t = kT$ in Equation 3.147 yields an output equation of the form in Equation 3.145, where $C(k) = C(t)|_{t=kT}$ and $D(k) = D(t)|_{t=kT}$. To "discretize" Equation 3.146, first recall (see Equation 3.53) that the solution to Equation 3.146 is

$$x(t) = \Phi(t, t_0)x(t_0) + \int_{t_0}^{t} \phi(t, \tau)B(\tau)u(\tau)\, d\tau, \quad t > t_0 \tag{3.148}$$

Then setting $t = kT + T$ and $t_0 = kT$ in Equation 3.148 yields

$$x(kT+T) = \Phi(kT+T, kT)x(kT) + \int_{kT}^{kT+T} \Phi(kT+T, \tau)B(\tau)u(\tau)\, d\tau \tag{3.149}$$

The second term on the right-hand side of Equation 3.149 can be approximated by

$$\left[\int_{kT}^{kT+T} \Phi(kT+T, \tau)B(\tau)\, d\tau\right] u(kT)$$

and thus Equation 3.149 is in the form of Equation 3.144 with

$$A(k) = \Phi(kT+T, kT) \tag{3.150}$$

$$B(k) = \int_{kT}^{kT+T} \Phi(kT+T, \tau)B(\tau)\, d\tau \tag{3.151}$$

Note that the matrix $A(k)$ given by Equation 3.150 is always invertible since $\Phi(kT+T, kT)$ is always invertible (see the property given by Equation 3.40). As discussed below, this implies that discretized or sampled data systems are "reversible."

From Equations 3.150 and 3.151 it is seen that the computation of $A(k)$ and $B(k)$ requires knowledge of the state-transition matrix $\Phi(t, \tau)$ for $t = kT + T$ and $\tau \in [kT, kT+T)$. As discussed in Section 3.3, if $A(t)$ in Equation 3.146 is a continuous function of t and the variation of $A(t)$ over the intervals

$[kT, kT+T]$ is sufficiently small for all k, then $\Phi(kT+T,\tau)$ can be approximated by

$$\Phi(kT+T,\tau) = e^{A(kT)(kT+T-\tau)} \quad \text{for } \tau \in [kT, kT+T) \tag{3.152}$$

and hence, $A(k)$ and $B(k)$ can be determined using

$$A(k) = e^{A(kT)T} \tag{3.153}$$

$$B(k) = \int_{kT}^{kT+T} e^{A(kT)(kT+T-\tau)} B(\tau)\,\mathrm{d}\tau \tag{3.154}$$

Given the discrete-time system defined by Equations 3.144 and 3.145, the solution to Equation 3.144 is

$$x(k) = \Phi(k,k_0)x(k_0) + \sum_{j=k_0}^{k-1} \Phi(k,j+1)B(j)u(j), \quad k > k_0 \tag{3.155}$$

where the $n \times n$ state-transition matrix $\Phi(k,j)$ is given by

$$\Phi(k,k_0) = \begin{cases} \text{not defined for} & k < k_0 \\ I, & k = k_0 \\ A(k-1)A(k-2)\ldots A(k_0), & k > k_0 \end{cases} \tag{3.156}$$

It follows directly from Equation 3.156 that $\Phi(k,k_0)$ is invertible for $k > k_0$ only if $A(k)$ is invertible for $k \geq k_0$. Thus, in general, the initial state $x(k_0)$ cannot be determined from the relationship $x(k) = \Phi(k,k_0)x(k_0)$. In other words, a discrete-time system is not necessarily *reversible,* although any continuous-time system given by Equations 3.146 and 3.147 is reversible since $\Phi(t,t_0)$ is always invertible. However, as noted above, any sampled data system is reversible.

The state-transition matrix $\Phi(k,k_0)$ satisfies the composition property:

$$\Phi(k,k_0) = \Phi(k,k_1)\Phi(k_1,k_0), \quad \text{where } k_0 \leq k_1 \leq k \tag{3.157}$$

and in addition,

$$\Phi(k+1,k_0) = A(k)\Phi(k,k_0), \quad k \geq k_0 \tag{3.158}$$

If $A(k)$ is invertible for all k, $\Phi(k,k_0)$ can be written in the factored form

$$\Phi(k,k_0) = \Phi_1(k)\Phi_2(k_0), \quad k \geq k_0 \tag{3.159}$$

where

$$\Phi_1(k) = \begin{cases} A(k-1)A(k-2)\ldots A(0), & k \geq 1 \\ I, & k = 0 \\ A^{-1}(k-2)A^{-1}(k-3)\ldots A^{-1}(-1), & k < 0 \end{cases} \tag{3.160}$$

$$\Phi_2(k_0) = \begin{cases} A^{-1}(0)A^{-1}(1)\ldots A^{-1}(k_0-1), & k_0 > 0 \\ I, & k_0 = 0 \\ A(-1)A(-2)\ldots A(k_0), & k_0 < 0 \end{cases} \tag{3.161}$$

When the direct feed matrix $D(k)$ in Equation 3.145 is zero, so that $y(k) = C(k)x(k)$, the output response $y(k)$ is given by

$$y(k) = C(k)\Phi(k,k_0)x(k_0) + \sum_{j=k_0}^{k-1} C(k)\Phi(k,j+1)B(j)u(j), \quad k > k_0 \tag{3.162}$$

If the initial time k_0 is set equal to $-\infty$ and the initial state is zero, Equation 3.162 becomes

$$y(k) = \sum_{j=-\infty}^{k-1} C(k)\Phi(k,j+1)B(j)u(j) \tag{3.163}$$

Comparing Equation 3.163 with the m-input p-output version of the input/output Equation 3.137 reveals that

$$H(k,j) = \begin{cases} C(k)\Phi(k,j+1)B(j), & k > j \\ 0, & k \leq j \end{cases} \tag{3.164}$$

where $H(k,j)$ is the $p \times m$ unit-pulse response function matrix. Note that if $A(k)$ is invertible so that $\Phi(k,k_0)$ has the factorization given in Equation 3.159, then $H(k,j)$ can be expressed in the factored form as

$$H(k,j) = [C(k)\Phi_1(k)][\Phi_2(j+1)B(j)] \quad \text{for } k > j. \tag{3.165}$$

As in the continuous-time case, this factorization is a fundamental property of unit-pulse response matrices $H(k,j)$ that are realizable by a state model (with invertible $A(k)$).

3.3.2 Stability

Given an n-dimensional discrete-time system defined by Equations 3.144 and 3.145, consider the homogeneous equation

$$x(k+1) = A(k)x(k), \quad k \geq k_0. \tag{3.166}$$

The solution is

$$x(k) = \Phi(k,k_0)x(k_0), \quad k > k_0 \tag{3.167}$$

where $\Phi(k,k_0)$ is the state-transition matrix defined by Equation 3.156.

The system is said to be asymptotically stable if for some initial time k_0, the solution $x(k)$ satisfies the condition $\|x(k)\| \to 0$ as $k \to 0$ for any initial state $x(k_0)$ at time k_0. This is equivalent to requiring that

$$\|\Phi(k,k_0)\| \to 0, \quad \text{as } k \to \infty \tag{3.168}$$

The system is exponentially stable if for some initial time k_0, there exist finite positive constants c and ρ with $\rho < 1$, such that for any $x(k_0)$ the solution $x(k)$ satisfies

$$\|x(k)\| \leq c\rho^{k-k_0} \|x(k_0)\|, \quad k > k_0 \tag{3.169}$$

If Equation 3.169 holds for all k_0 with the constants c and ρ fixed, the system is said to be uniformly exponentially stable. This is equivalent to requiring that there exist a finite positive constant γ and a nonnegative constant ρ with $\rho < 1$ such that

$$\left\|\Phi(k,j)\right\| \leq \gamma\rho^{k-j}, \quad \text{for all } k,j \text{ such that } k \geq j \tag{3.170}$$

Uniform exponential stability is also equivalent to requiring that given any positive constant δ, there exists a positive integer q such that for any k_0 and $x(k_0)$, the solution to Equation 3.166 satisfies

$$\|x(k)\| \leq \delta \|x(k_0)\|, \quad k \geq k_0 + q \tag{3.171}$$

Uniform exponential stability is also equivalent to the existence of a finite positive constant β such that

$$\sum_{i=j+1}^{k} \|\Phi(k,i)\| \leq \beta \quad \text{for all } k,j \text{ such that } k \geq j+1 \tag{3.172}$$

Another necessary and sufficient condition for uniform exponential stability is that a symmetric positive-definite matrix $Q(k)$ exists with $c_1 I \leq Q(k) \leq c_2 I$ for some positive constants c_1 and c_2 so that

$$A^T(k)Q(k+1)A(k) - Q(k) \leq -c_3 I \tag{3.173}$$

for some positive constant c_3.

3.3.3 Controllability and Observability

The discrete-time system defined by Equations 3.144 and 3.145 with $D(k) = 0$ will be denoted by the triple $[A(k), B(k), C(k)]$. The system is said to be controllable on the interval $[k_0, k_1]$ with $k_1 > k_0$ if, for any states x_0 and x_1, an input $u(k)$ exists that drives the system to the state $x(k_1) = x_1$ at time $k = k_1$ starting from the state $x(k_0) = x_0$ at time $k = k_0$. To determine a necessary and sufficient condition for controllability, first solve Equation 3.144 to find the state $x(k_1)$ at time $k = k_1$ resulting from state $x(k_0)$ at time $k = k_0$ and the input sequence $u(k_0), u(k_0 + 1), \ldots, u(k_1 - 1)$. The solution is

$$\begin{aligned} x(k_1) = \Phi(k_1, k_0)x(k_0) + \Phi(k_1, k_0 + 1)B(k_0)u(k_0) + \Phi(k_1, k_0 + 2)B(k_0 + 1)u(k_0 + 1) \\ + \cdots \Phi(k_1, k_1 - 1)B(k_1 - 2)u(k_1 - 2) + B(k_1 - 1)u(k_1 - 1) \end{aligned} \tag{3.174}$$

Let $R(k_1, k_0)$ be the *controllability (or reachability) matrix* with n rows and $(k_1 - k_0)m$ columns defined by

$$R(k_0, k_1) = [B(k_1 - 1)\ \Phi(k_1, k_1 - 1)B(k_1 - 2) \ldots \Phi(k_1, k_0 + 2)B(k_0 + 1)\ \Phi(k_1, k_0 + 1)B(k_0)] \tag{3.175}$$

Then Equation 3.174 can be written the form

$$x(k_1) = \Phi(k_1, k_0)x(k_0) + R(k_0, k_1)U(k_0, k_1) \tag{3.176}$$

where $U(k_0, k_1)$ is the $(k_1 - k_0)m$-element column vector of inputs given by

$$U(k_0, k_1) = \left[u^T(k_1 - 1)\ u^T(k_1 - 2)\ \cdots\ u^T(k_0 + 1)\ u^T(k_0)\right]^T \tag{3.177}$$

Now for any states $x(k_0) = x_0$ and $x(k_1) = x_1$, from Equation 3.176, there is a sequence of inputs given by $U(k_0, k_1)$ that drives the system from x_0 to x_1 if and only if the matrix $R(k_0, k_1)$ has rank n. If this is the case, Equation 3.176 can be solved for $U(k_0, k_1)$, giving

$$U(k_0, k_1) = R^T(k_0, k_1)\left[R(k_0, k_1)R^T(k_0, k_1)\right]^{-1} [x_1 - \Phi(k_1, k_0)x_0] \tag{3.178}$$

Hence, rank $R(k_0, k_1) = n$ is a necessary and sufficient condition for controllability over the interval $[k_0, k_1]$.

Given a fixed positive integer N, set $k_0 = k - N + 1$ and $k_1 = k + 1$ in $R(k_0, k_1)$, which results in the matrix $R(k - N + 1, k + 1)$. Note that $R(k - N + 1, k + 1)$ is a function of only the integer variable k and that the size of the matrix $R(k - N + 1, k + 1)$ is equal to $n \times Nm$ since $k_1 - k_0 = N$. The $n \times Nm$ matrix $R(k - N + 1, k + 1)$ will be denoted by $R(k)$. By definition of the state-transition matrix $\Phi(k, k_0)$, $R(k)$ can be written in the form

$$R(k) = [R_0(k)\ R_1(k)\ \ldots\ R_{N-1}(k)] \tag{3.179}$$

where the block columns $R_i(k)$ of $R(k)$ are given by

$$R_0(k) = B(k) \tag{3.180}$$

$$R_i(k) = A(k)R_{i-1}(k - 1), \quad i = 1, 2, \ldots, N - 1 \tag{3.181}$$

The system is said to be *uniformly N-step controllable* if rank $R(k) = n$ for all k. Uniformly N-step controllable means that the system is controllable on the interval $[k - N + 1,\ k + 1]$ for all k.

Now suppose that the system input $u(k)$ is zero, so that the state model is given by

$$x(k + 1) = A(k)x(k) \tag{3.182}$$

$$y(k) = C(k)x(k) \tag{3.183}$$

From Equations 3.182 and 3.183, the output response $y(k)$ resulting from initial state $x(k_0)$ is given by

$$y(k) = C(k)\Phi(k, k_0)x(k_0), \quad k \geq k_0 \tag{3.184}$$

Then the system is said to be observable on the interval $[k_0, k_1]$ if any initial state $x(k_0) = x_0$ can be determined from the output response $y(k)$ given by Equation 3.184 for $k = k_0, k_0 + 1, \ldots, k_1 - 1$. Using

Equation 3.184 for $k = k_0$ to $k = k_1 - 1$ yields

$$\begin{bmatrix} y(k_0) \\ y(k_0+1) \\ \vdots \\ y(k_1-2) \\ y(k_1-1) \end{bmatrix} = \begin{bmatrix} C(k_0)x_0 \\ C(k_0+1)\Phi(k_0+1,k_0)x_0 \\ \vdots \\ C(k_1-2)\Phi(k_1-2,k_0)x_0 \\ C(k_1-1)\Phi(k_1-1,k_0)x_0 \end{bmatrix} \tag{3.185}$$

The right-hand side of Equation 3.185 can be written in the form $O(k_0,k_1)x_0$ where $O(k_0,k_1)$ is the $(k_1-k_0)p \times n$ *observability matrix* defined by

$$O(k_0,k_1) = \begin{bmatrix} C(k_0) \\ C(k_0+1)\Phi(k_0+1,k_0) \\ \vdots \\ C(k_1-2)\Phi(k_1-2,k_0) \\ C(k_1-1)\Phi(k_1-1,k_0) \end{bmatrix} \tag{3.186}$$

Equation 3.185 can be solved for any initial state x_0 if and only if rank $O(k_0,k_1) = n$, which is a necessary and sufficient condition for observability on $[k_0,k_1]$. If the rank condition holds, the solution of Equation 3.185 for x_0 is

$$x_0 = \left[O^T(k_0,k_1)O(k_0,k_1)\right]^{-1} O^T(k_0,k_1)Y(k_0,k_1) \tag{3.187}$$

where $Y(k_0,k_1)$ is the $(k_1-k_0)p$-element column vector of outputs given by

$$Y(k_0,k_1) = \left[y^T(k_0)\; y^T(k_0+1) \cdots y^T(k_1-2)\; y^T(k_1-1)\right]^T \tag{3.188}$$

Given a positive integer N, setting $k_0 = k$ and $k_1 = k+N$ in $O(k_0,\ k_1)$ yields the $Np \times n$ matrix $O(k,k+N)$, which will be denoted by $O(k)$. By definition of the state-transition matrix $\Phi(k,k_0)$, $O(k)$ can be written in the form

$$O(k) = \begin{bmatrix} O_0(k) \\ O_1(k) \\ \vdots \\ O_{N-1}(k) \end{bmatrix} \tag{3.189}$$

where the block rows $O_i(k)$ of $O(k)$ are given by

$$O_0(k) = C(k) \tag{3.190}$$

$$O_i(k) = O_{i-1}(k+1)A(k), \quad i = 1,2,\ldots,N-1 \tag{3.191}$$

The system is said to be *uniformly N-step observable* if rank $O(k) = n$ for all k. Uniformly N-step observable means that the system is observable on the interval $[k,k+N]$ for all k.

3.3.4 Change of State Variables and Canonical Forms

Again, consider the discrete-time system with state model $[A(k),B(k),C(k)]$. For any $n \times n$ invertible matrix $P(k)$, another state model can be generated by defining the new state vector $z(k) = P^{-1}(k)x(k)$.

The new state model is given by

$$z(k+1) = \bar{A}(k)z(k) + \bar{B}(k)u(k) \tag{3.192}$$

$$y(k) = \bar{C}(k)z(k) \tag{3.193}$$

where

$$\bar{A}(k) = P^{-1}(k+1)A(k)P(k) \tag{3.194}$$

$$\bar{B}(k) = P^{-1}(k+1)B(k) \tag{3.195}$$

$$\bar{C}(k) = C(k)P(k) \tag{3.196}$$

The state-transition matrix $\bar{\Phi}(k, k_0)$ for the new state model is given by

$$\bar{\Phi}(k, k_0) = P^{-1}(k)\Phi(k, k_0)P(k_0) \tag{3.197}$$

where $\Phi(k, k_0)$ is the state-transition matrix for $[A(k), B(k), C(k)]$. The new state model, which will be denoted by $[\bar{A}(k), \bar{B}(k), \bar{C}(k)]$, and the given state model $[A(k), B(k), C(k)]$ are said to be algebraically equivalent.

Given an n-dimensional state model $[A(k), B(k), C(k)]$ and any $n \times n$ invertible matrix function $\Gamma(k)$, if $A(k)$ is invertible, there is an invertible coordinate transformation matrix $P(k)$ for $k \geq k_0$, which transforms $A(k)$ into $\Gamma(k)$ for $k \geq k_0$, that is, $\bar{A}(k) = \Gamma(k)$, $k \geq k_0$. To show this, define $P(k)$ by the matrix difference equation

$$P(k+1) = A(k)P(k)\Gamma^{-1}(k), \quad k \geq k_0 \tag{3.198}$$

with initial condition $P(k_0) = I = n \times n$ identity matrix. Then multiplying both sides of Equation 3.198 on the right by $\Gamma(k)$ and multiplying the resulting equation on the left by $P^{-1}(k+1)$ yields Equation 3.194 with $\bar{A}(k) = \Gamma(k)$ for $k \geq k_0$. This result shows that an invertible matrix $A(k)$ can be put into a diagonal form via a coordinate transformation with any desired nonzero functions or constants on the diagonal. Thus, as in the continuous-time case, there is no useful generalization of the notion of eigenvalues and eigenvectors in the time-varying case, unless additional conditions are placed on the coordinate transformation. For example, one can require that the transformation be a Lyapunov transformation, which means that both $P(k)$ and its inverse $P^{-1}(k)$ are bounded matrix functions of the integer variable k. It follows from Equation 3.197 that uniform exponential stability is preserved under a Lyapunov transformation.

Suppose that the systems $[A(k), B(k), C(k)]$ and $[\bar{A}(k), \bar{B}(k), \bar{C}(k)]$ are algebraically equivalent and let $R(k)$ denote the $n \times Nm$ controllability matrix for the system $[A(k), B(k), C(k)]$, where $R(k)$ is defined by Equations 3.179 through 3.181. Similarly, for the system given by $[\bar{A}(k), \bar{B}(k), \bar{C}(k)]$ define

$$\bar{R}(k) = [\bar{R}_0(k)\, \bar{R}_1(k)\, \ldots\, \bar{R}_{n-1}(k)] \tag{3.199}$$

where the $\bar{R}_i(k)$ are given by Equations 3.180 and 3.181 with $A(k)$ and $B(k)$ replaced by $\bar{A}(k)$ and $\bar{B}(k)$, respectively. Then the coordinate transformation $P^{-1}(k)$ is given by

$$P^{-1}(k+1)R(k) = \bar{R}(k) \tag{3.200}$$

If the system $[A(k), B(k), C(k)]$ is uniformly N-step controllable, $R(k)$ has rank n for all k, and the $n \times n$ matrix $R(k)R^T(k)$ is invertible. Thus, Equation 3.200 can be solved for $P^{-1}(k+1)$, which gives

$$P^{-1}(k+1) = \bar{R}(k)R^T(k)[R(k)R^T(k)]^{-1} \tag{3.201}$$

It follows from Equation 3.200 that uniform N-step controllability is preserved under a change of state variables.

Now suppose that the n-dimensional system with state model $[A(k), B(k), C(k)]$ is uniformly N-step controllable with $N = n$ and that the system has a single input ($m = 1$) so that $B(k)$ is an n-element column vector and $R(k)$ is a $n \times n$ invertible matrix. Define

$$R_n(k) = A(k)R_{n-1}(k-1) \tag{3.202}$$

$$\eta(k) = -R^{-1}(k)R_n(k) \tag{3.203}$$

where $R_{n-1}(k)$ is the column vector defined by Equation 3.181 with $i = n-1$. The n-element column vector $\eta(k)$ defined by Equation 3.203 is invariant under any change of state $z(k) = P^{-1}(k)x(k)$, and in the time-invariant case, η is constant and is given by

$$\eta = \begin{bmatrix} a_0 & a_1 & \cdots & a_{n-1} \end{bmatrix}^T \tag{3.204}$$

where the a_i are the coefficients of the characteristic polynomial of A.

Given $\eta(k)$ defined by Equation 3.203, write $\eta(k)$ in the form

$$\eta(k) = \begin{bmatrix} \eta_0(k) & \eta_1(k) & \cdots & \eta_{n-1}(k) \end{bmatrix} \tag{3.205}$$

Then as proved in [3], there is a transformation $P(k)$ which converts $[A(k), B(k), C(k)]$ into the control canonical form $[\bar{A}(k), \bar{B}, \bar{C}(k)]$, with $\bar{A}(k)$ and $\bar{B}$ given by

$$\bar{A}(k) = \begin{bmatrix} 0 & 1 & 0 & \cdots & 0 & 0 \\ 0 & 0 & 1 & \cdots & 0 & 0 \\ 0 & 0 & 0 & \cdots & 0 & 0 \\ \vdots & & & \cdots & & \vdots \\ 0 & 0 & 0 & \cdots & 0 & 1 \\ -\eta_0(k) & -\eta_1(k+1) & -\eta_2(k+2) & \cdots & -\eta_{n-2}(k+n-2) & -\eta_{n-1}(k+n-1) \end{bmatrix} \tag{3.206}$$

$$\bar{B} = \begin{bmatrix} 0 \\ 0 \\ \vdots \\ 0 \\ 1 \end{bmatrix} \tag{3.207}$$

The transformation matrix $P^{-1}(k)$ that yields the control canonical form can be determined using Equation 3.201. As in the continuous-time case, the control canonical form can be used to design a state feedback control which results in a uniformly exponentially stable closed-loop system if the coordinate transformation defined by $P^{-1}(k)$ is a Lyapunov transformation. If $P(k)$ and $P^{-1}(k)$ are bounded only for $k \geq k_1$ for some finite k_1, the resulting closed-loop system is exponentially stable. The details are a straightforward modification of the continuous-time case, and thus are not pursued here.

3.4 Applications and Examples

In Section 3.4.1, an example is given on the construction of canonical forms and the design of observers and controllers.

3.4.1 Observer and Controller Design

Consider the single-input single-output linear time-varying continuous-time system given by the input/output differential equation

$$\ddot{y}(t) + e^{-t}\dot{y}(t) + y(t) = \dot{u}(t) \tag{3.208}$$

To determine the state model, which is in observer canonical form, write Equation 3.208 in the form

$$\ddot{y}(t) + D[\alpha_1(t)y(t)] + \alpha_0(t)y(t) = \dot{u}(t) \tag{3.209}$$

In this case,

$$\alpha_1(t) = e^{-t} \quad \text{and} \quad \alpha_0(t) = 1 + e^{-t} \tag{3.210}$$

Then from Equations 3.22 through 3.24, the observer canonical form of the state model is

$$\begin{bmatrix} \dot{x}_1(t) \\ \dot{x}_2(t) \end{bmatrix} = \begin{bmatrix} 0 & -1-e^{-t} \\ 1 & -e^{-t} \end{bmatrix} \begin{bmatrix} x_1(t) \\ x_2(t) \end{bmatrix} + \begin{bmatrix} 0 \\ 1 \end{bmatrix} u(t) \tag{3.211}$$

$$y(t) = \begin{bmatrix} 0 & 1 \end{bmatrix} \begin{bmatrix} x_1(t) \\ x_2(t) \end{bmatrix} \tag{3.212}$$

The state variables $x_1(t)$ and $x_2(t)$ in this state model are given by

$$x_1(t) = \dot{y}(t) + e^{-t}y(t) - u(t) \tag{3.213}$$

$$x_2(t) = y(t) \tag{3.214}$$

If the output $y(t)$ of the system can be differentiated, then $x_1(t)$ and $x_2(t)$ can be directly determined from the input $u(t)$ and the output $y(t)$ by using Equations 3.213 and 3.214. In practice, however, differentiation of signals should be avoided, and thus directly determining $x_1(t)$ using Equation 3.213 is usually not viable. As discussed next, by using a state observer, the state $x(t)$ can be estimated without having to differentiate signals. Actually, in this particular example it is necessary to estimate only $x_1(t)$ since the output $y(t) = x_2(t)$ is known, and thus a reduced-order observer could be used, but this is not considered here.

An observer for the state $x(t)$ is given by

$$\frac{d}{dt}\hat{x}(t) = A(t)\hat{x}(t) + H(t)[y(t) - C(t)\hat{x}(t)] + B(t)u(t) \tag{3.215}$$

where $H(t)$ is the n-element observer gain vector and $\hat{x}(t)$ is the estimate of $x(t)$. With the estimation error $e(t)$ defined by $e(t) = x(t) - \hat{x}(t)$, the error is given by the differential equation

$$\dot{e}(t) = [A(t) - H(t)C(t)]\, e(t), \quad t > t_0 \tag{3.216}$$

with initial error $e(t_0)$ at initial time t_0. The objective is to choose the gain vector $H(t)$ so that, for any initial error $e(t_0)$, $\|e(t)\| \to 0$ as $t \to \infty$, with some desired rate of convergence.

For the system given by Equations 3.211 and 3.212, the error Equation 3.216 is

$$\dot{e}(t) = \begin{bmatrix} 0 & -1-e^{-t}-h_1(t) \\ 1 & -e^{-t}-h_2(t) \end{bmatrix} e(t) \tag{3.217}$$

where $H(t) = [h_1(t) \quad h_2(t)]^T$. From Equation 3.217, it is obvious that by setting

$$h_1(t) = m_0 - 1 - e^{-t} \tag{3.218}$$

$$h_2(t) = m_1 - e^{-t} \tag{3.219}$$

where m_0 and m_1 are constants, the coefficient matrix on the right-hand side of Equation 3.217 is constant, and its eigenvalues can be assigned by choosing m_0 and m_1. Hence, any desired rate of convergence to zero can be achieved for the error $e(t)$.

The estimate $\hat{x}(t)$ of $x(t)$ can then be used to realize a feedback control law of the form

$$u(t) = -g(t)\hat{x}(t) \tag{3.220}$$

where $g(t)$ is the feedback gain vector. The first step in pursuing this is to consider the extent to which the system can be controlled by state feedback of the form given in Equation 3.220 with $\hat{x}(t)$ replaced by $x(t)$; in other words, the true system state $x(t)$ is assumed to be available. In particular, we can ask whether or not there is a gain vector $g(t)$ so that with $u(t) = -g(t)x(t)$, the state of the resulting closed-loop system decays to zero exponentially with some desired rate of convergence. This can be answered by attempting to transform the state model given by Equations 3.211 and 3.212 to control canonical form. Following the procedure given in Section 3.2, the steps are as follows.

Let $R(t)$ denote the 2×2 matrix whose columns $r_i(t)$ are defined by Equations 3.120 and 3.121 with $n = 2$. This yields

$$r_1(t) = B(t) = \begin{bmatrix} 0 \\ 1 \end{bmatrix} \tag{3.221}$$

$$r_2(t) = A(t)r_1(t) + \dot{r}_1(t) = \begin{bmatrix} -1 - e^{-t} \\ -e^{-t} \end{bmatrix} \tag{3.222}$$

Then

$$R(t) = [r_1(t) \quad r_2(t)] = \begin{bmatrix} 0 & -1 - e^{-t} \\ 1 & -e^{-t} \end{bmatrix} \tag{3.223}$$

and

$$\det[R(t)] = 1 + e^{-t} \quad \text{for all } t \tag{3.224}$$

where "det" denotes the determinant. Since $\det[R(t)] \neq 0$ for all t, $R(t)$ has rank 2 for all t, and thus the control canonical form exists for the system given by Equations 3.211 and 3.212.

The rows $c_i(t)$ of the matrix $P^{-1}(t)$ in the transformation $z(t) = P^{-1}(t)x(t)$ to the control canonical form are computed using Equations 3.127 and 3.128. This yields

$$c_1(t) = [-(1+e^{-t})^{-1} \quad 0] \quad \text{and} \quad c_2(t) = [-e^{-t}(1+e^{-t})^{-2} \quad 1] \tag{3.225}$$

and thus

$$P^{-1}(t) = \begin{bmatrix} -(1+e^{-t})^{-1} & 0 \\ -e^{-t}(1+e^{-t})^{-2} & 1 \end{bmatrix} \quad \text{and} \quad P(t) = \begin{bmatrix} -(1+e^{-t}) & 0 \\ -e^{-t}(1+e^{-t})^{-1} & 1 \end{bmatrix} \tag{3.226}$$

Now choose the coefficients of the characteristic polynomial of the closed-loop system matrix in the control canonical form to be d_0 and d_1. Then from Equation 3.136, the feedback gain vector $g(t)$ back in the original state coordinates is given by

$$g(t) = -\left(c_3(t) + d_0c_1(t) + d_1c_2(t)\right) \tag{3.227}$$

where

$$c_3(t) = c_2(t)A(t) + \dot{c}_2(t) \tag{3.228}$$

$$c_3(t) = [1 - 2e^{-2t}(1+e^{-t})^{-3} + e^{-t}(1+e^{-t})^{-2} \quad e^{-t}(1+e^{-t})^{-1} - e^{-t}] \tag{3.229}$$

Inserting the expressions for $c_1(t)$, $c_2(t)$, and $c_3(t)$ given by Equations 3.225 and 3.229 into Equation 3.227 results in the feedback gain vector $g(t)$.

Since $P(t)$ and $P^{-1}(t)$ given by Equation 3.226 are bounded for $t \geq t_1$ for any finite t_1, by choosing appropriate values for d_0 and d_1 it follows that via the state feedback control given by $u(t) = -g(t)x(t)$, the resulting closed-loop system is exponentially stable with any desired rate λ of convergence to zero. It also follows that with the state feedback control $u(t) = -g(t)\hat{x}(t)$, where $\hat{x}(t)$ is the estimated state, the resulting closed-loop system is also exponentially stable with any desired rate of convergence to zero.

3.4.2 Exponential Systems

A system with n-dimensional state model $[A(t), B(t), C(t)]$ is said to be an *exponential system* if its state-transition matrix $\Phi(t, \tau)$ can be written in the matrix exponential form

$$\Phi(t,\tau) = e^{\Gamma(t,\tau)} \tag{3.230}$$

where $\Gamma(t, \tau)$ is a $n \times n$ matrix function of t and τ. The form given in Equation 3.230 is valid (at least locally, that is, when t is close to τ) for a large class of time-varying systems. In fact, as noted above, for $t \in [\tau, \tau + T]$, $\Phi(t, \tau)$ can be approximated by the matrix exponential $e^{A(kT)(t-\tau)}$ if $A(t)$ is approximately equal to $A(\tau)$ for $t \in [\tau, \tau + T]$. For a mathematical development of the matrix exponential form, see [5].

As noted previously, the exponential form in Equation 3.230 is valid for any system where $A(t)$ commutes with its integral (see Equation 3.47), in which case

$$\Gamma(t,\tau) = \int_\tau^t A(\sigma)\,d\sigma \tag{3.231}$$

The class of systems for which $A(t)$ commutes with its integral is actually fairly large; in particular, this is the case for any $A(t)$ given by

$$A(t) = \sum_{i=1}^{r} f_i(t) A_i \tag{3.232}$$

where $f_i(t)$ are arbitrary real-valued functions of t and the A_i are arbitrary constant $n \times n$ matrices that satisfy the commutativity conditions

$$A_i A_j = A_j A_i, \quad \text{for all integers } 1 \le i, j \le r \tag{3.233}$$

For example, suppose that

$$A(t) = \begin{bmatrix} f_1(t) & c_1 f_2(t) \\ c_2 f_2(t) & f_1(t) \end{bmatrix} \tag{3.234}$$

where $f_1(t)$ and $f_2(t)$ are arbitrary real-valued functions of t and c_1 and c_2 are arbitrary constants. Then

$$A(t) = f_1(t) A_1 + f_2(t) A_2 \tag{3.235}$$

where $A_1 = I$ and

$$A_2 = \begin{bmatrix} 0 & c_1 \\ c_2 & 0 \end{bmatrix} \tag{3.236}$$

Obviously, A_1 and A_2 commute, and thus $\Phi(t, \tau)$ is given by Equations 3.230 and 3.231. In this case, $\Phi(t, \tau)$ can be written in the form

$$\Phi(t,\tau) = \exp\left[\left(\int_\tau^t f_1(\sigma)\,d\sigma\right) I\right] \exp\left[\left(\int_\tau^t f_2(\sigma)\,d\sigma\right) A_2\right] \tag{3.237}$$

Given an n-dimensional system with exponential state-transition matrix $\Phi(t, \tau) = e^{\Gamma(t,\tau)}$, $\Phi(t, \tau)$ can be expressed in terms of scalar functions using the Laplace transform as in the time-invariant case.

In particular, let

$$\Phi(t,\beta,\tau) = \text{inverse transform of } [sI - (1/\beta)\Gamma(\beta,\tau)]^{-1} \tag{3.238}$$

where $\Gamma(\beta,\tau) = \Gamma(t,\tau)|_{t=\beta}$ and β is viewed as a parameter. Then

$$\Phi(t,\tau) = \Phi(t,\beta,\tau)|_{\beta=t} \tag{3.239}$$

For example, suppose that

$$A(t) = \begin{bmatrix} f_1(t) & f_2(t) \\ -f_2(t) & f_1(t) \end{bmatrix} \tag{3.240}$$

where $f_1(t)$ and $f_2(t)$ are arbitrary functions of t with the constraint that $f_2(t) \geq 0$ for all t. Then

$$\Gamma(t,\tau) = \int_\tau^t A(\sigma)\,d\sigma \tag{3.241}$$

and $\Phi(t,\beta,\tau)$ is equal to the inverse transform of

$$\Phi(s,\beta,\tau) = \begin{bmatrix} s-\gamma_1(\beta,\tau) & -\gamma_2(\beta,\tau) \\ \gamma_2(\beta,\tau) & s-\gamma_1(\beta,\tau) \end{bmatrix}^{-1} \tag{3.242}$$

where

$$\gamma_1(\beta,\tau) = (1/\beta)\int_\tau^\beta f_1(\sigma)\,d\sigma \tag{3.243}$$

$$\gamma_2(\beta,\tau) = (1/\beta)\int_\tau^\beta f_2(\sigma)\,d\sigma \tag{3.244}$$

Computing the inverse Laplace transform of $\Phi(s,\beta,\tau)$ and using Equation 3.239 give (for $t > 0$)

$$\Phi(t,\tau) = \begin{bmatrix} e^{\gamma_1(t,\tau)t}\cos[\gamma_2(t,\tau)t] & e^{\gamma 1(t,\tau)t}\sin[\gamma_2(t,\tau)t] \\ -e^{\gamma_1(t,\tau)t}\sin[\gamma_2(t,\tau)t] & e^{\gamma_1(t,\tau)t}\cos[\gamma_2(t,\tau)t] \end{bmatrix} \tag{3.245}$$

3.4.3 Stability

Again consider an n-dimensional exponential system $[A(t), B(t), C(t)]$ with state-transition matrix $\Phi(t,\tau) = e^{\Gamma(t,\tau)}$. A sufficient condition for exponential stability of the differential equation $\dot{x}(t) = A(t)x(t)$is that the $n \times n$ matrix $(1/t)\Gamma(t,\tau)$ be bounded as a function of t and its pointwise eigenvalues have real parts $\leq -v$ for some $v > 0$ and all $t > \tau$ for some finite τ. For example, suppose that $A(t)$ is given by Equation 3.234 so that

$$(1/t)\Gamma(t,\tau) = \begin{bmatrix} \gamma_1(t,\tau) & c_1\gamma_2(t,\tau) \\ c_2\gamma_2(t,\tau) & \gamma_1(t,\tau) \end{bmatrix} \tag{3.246}$$

where $\gamma_1(t,\tau)$ and $\gamma_2(t,\tau)$ are given by Equations 3.243 and 3.244 with $\beta = t$. Then

$$\det[sI - (1/t)\Gamma(t,\tau)] = s^2 - 2\gamma_1(t,\tau)s + \gamma_1^2(t,\tau) - c_1c_2\gamma_2^2(t,\tau) \tag{3.247}$$

and the pointwise eigenvalues of $(1/t)\Gamma(t,\tau)$ have real parts $\leq -v$ for all $t > \tau$ for some $v > 0$ and τ if

$$\gamma_1(t,\tau) \leq \nu_1, \quad \text{for all } t > \tau \quad \text{and} \quad \text{some } \nu_1 < 0, \tag{3.248}$$

$$\gamma_1^2(t,\tau) - c_1c_2\gamma_2^2(t,\tau) \geq \nu_2 \quad \text{for all } t > \tau \quad \text{and} \quad \text{some } \nu_2 > 0 \tag{3.249}$$

Therefore, if Equations 3.248 and 3.249 are satisfied and $\gamma_1(t,\tau)$ and $\gamma_2(t,\tau)$ are bounded functions of t, the solutions to $\dot{x}(t) = A(t)x(t)$ with $A(t)$ given by Equation 3.234 decay to zero exponentially.

It is well known that in general there is no pointwise eigenvalue condition on the system matrix $A(t)$ that insures exponential stability, or even asymptotic stability. For an example (taken from [8]), suppose that

$$A(t) = \begin{bmatrix} -1+\alpha(\cos^2 t) & 1-\alpha(\sin t)(\cos t) \\ -1-\alpha(\sin t)(\cos t) & -1+\alpha(\sin^2 t) \end{bmatrix} \tag{3.250}$$

where α is a real parameter. The pointwise eigenvalues of $A(t)$ are equal to

$$\frac{\alpha - 2 \pm \sqrt{\alpha^2 - 4}}{2} \tag{3.251}$$

which are strictly negative if $0 < \alpha < 2$. But

$$\Phi(t, 0) = \begin{bmatrix} e^{(\alpha-1)t}(\cos t) & e^{-t}(\sin t) \\ -e^{(\alpha-1)t}(\sin t) & e^{-t}(\cos t) \end{bmatrix} \tag{3.252}$$

and thus, the system is obviously not asymptotically stable if $\alpha > 1$.

3.4.4 The Lyapunov Criterion

By using the Lyapunov criterion (see Equation 3.97), it is possible to derive sufficient conditions for uniform exponential stability without computing the state-transition matrix. For example, suppose that

$$A(t) = \begin{bmatrix} 0 & 1 \\ -1 & -a(t) \end{bmatrix} \tag{3.253}$$

where $a(t)$ is a real-valued function of t with $a(t) \geq c$ for all $t > t_1$, for some t_1 and some constant $c > 0$. Now in Equation 3.97, choose

$$Q(t) = \begin{bmatrix} a(t) + \dfrac{2}{a(t)} & 1 \\ 1 & \dfrac{2}{a(t)} \end{bmatrix} \tag{3.254}$$

Then, $c_1 I \leq Q(t) \leq c_2$, for all $t > t_1$ for some constants $c_1 > 0$ and $c_2 > 0$. Now

$$Q(t)A(t) + A^T(t)Q(t) + \dot{Q}(t) = \begin{bmatrix} -2 + \dot{a}(t) - \dfrac{\dot{a}(t)}{a^2(t)} & 0 \\ 0 & -1 - \dfrac{\dot{a}(t)}{a^2(t)} \end{bmatrix} \tag{3.255}$$

Hence, if

$$-2 + \dot{a}(t) - \frac{\dot{a}(t)}{a^2(t)} \leq -c_3 \quad \text{for } t > t_1 \quad \text{for some } c_3 > 0 \tag{3.256}$$

and

$$-1 - \frac{\dot{a}(t)}{a^2(t)} \leq -c_4 \quad \text{for } t > t_1 \quad \text{for some } c_4 > 0 \tag{3.257}$$

the system is uniformly exponentially stable. For instance, if $a(t) = b - \cos t$, then Equations 3.256 and 3.257 are satisfied if $b > 2$, in which case the system is uniformly exponentially stable.

Now suppose that

$$A(t) = \begin{bmatrix} 0 & 1 \\ -a_1(t) & -a_2(t) \end{bmatrix} \tag{3.258}$$

As suggested in [8, p. 109], sufficient conditions for uniform exponential stability can be derived by taking

$$Q(t) = \begin{bmatrix} a_1(t) + a_2(t) + \dfrac{a_1(t)}{a_2(t)} & 1 \\ 1 & 1 + \dfrac{1}{a_2(t)} \end{bmatrix} \tag{3.259}$$

3.5 Defining Terms

State model: For linear time-varying systems, this is a mathematical representation of the system in terms of state equations of the form $\dot{x}(t) = A(t)x(t) + B(t)u(t), y(t) = C(t)x(t) + D(t)u(t)$.

State-transition matrix: The matrix $\Phi(t, t_0)$ where $\Phi(t, t_0)x(t_0)$ is the state at time t starting with state $x(t_0)$ at time t_0 and with no input applied for $t \geq t_0$.

Exponential system: A system whose state-transition matrix $\Phi(t, \tau)$ can be written in the exponential form $e^{\Gamma(t,\tau)}$ for some $n \times n$ matrix function $\Gamma(t, \tau)$.

Reversible system: A system whose state-transition matrix is invertible.

Sampled data system: A discrete-time system generated by sampling the inputs and outputs of a continuous-time system.

Change of state: A transformation $z(t) = P^{-1}(t)x(t)$ from the state vector $x(t)$ to the new state vector $z(t)$.

Algebraic equivalence: Refers to two state models of the same system related by a change of state.

Lyapunov transformation: A change of state $z(t) = P^{-1}(t)x(t)$ where $P(t)$ and its inverse $P^{-1}(t)$ are both bounded functions of t.

Canonical form: A state model $[A(t), B(t), C(t)]$ with one or more of the coefficient matrices $A(t), B(t), C(t)$ in a special form.

Control canonical form: In the single-input case, a canonical form for $A(t)$ and $B(t)$ that facilitates the study of state feedback control.

Observer canonical form: In the single-output case, a canonical form for $A(t)$ and $C(t)$ that facilitates the design of a state observer.

Characteristic vector: A time-varying generalization corresponding to the vector of coefficients of the characteristic polynomial in the time-invariant case.

Asymptotic stability: Convergence of the solutions of $\dot{x}(t) = A(t)x(t)$ to zero for any initial state $x(t_0)$.

Exponential stability: Convergence of the solutions of $\dot{x}(t) = A(t)x(t)$ to zero at an exponential rate.

Uniform exponential stability: Convergence of the solutions of $\dot{x}(t) = A(t)x(t)$ to zero at an exponential rate uniformly with respect to the initial time.

Pointwise eigenvalues: The eigenvalues of a $n \times n$ time-varying matrix $M(t)$ with t replaced by τ, where τ is viewed as a time-independent parameter.

Controllability: The existence of inputs that drive a system from any initial state to any desired state.

Observability: The ability to compute the initial state $x(t_0)$ from knowledge of the output response $y(t)$ for $t \geq t_0$.

State feedback control: A control signal of the form $u(t) = -F(t)x(t)$ where $F(t)$ is the feedback gain matrix and $x(t)$ is the system state.

Observer: A system which provides an estimate $\hat{x}(t)$ of the state $x(t)$ of a system.

Acknowledgment

The author wishes to thank Professor Wilson J. Rugh of the Johns Hopkins University for his comments regarding a technical issue involving the concept of uniform exponential stability in the time-varying case.

References

1. Chen, C. T., *Linear System Theory and Design*, Holt, Rinehart, and Winston, New York, 1984.
2. Kamen, E.W., The poles and zeros of a linear time-varying system, *Linear Algebra Appl.*, 98, 263–289, 1988.

3. Kamen, E.W. and Hafez, K.M., Algebraic theory of linear time-varying systems, *SIAM J. Control Optim.*, 17, 500–510, 1979.
4. Kloet, P. van der and Neerhoff, F.L., The Cauchy–Floquet factorization by successive Riccati transformations, *Proc. IEEE Int. Sym. Circuits Systems*, Phoenix, AZ, 257–260, 2002.
5. Magnus, W., On the exponential solution of differential equations for a linear operator, *Commun. Pure Appl. Math.*, VII, 649–673, 1954.
6. Marinescu, B. and Bourlès, H., An intrinsic algebraic setting for poles and zeros of linear time-varying systems, *Systems Control Lett.*, 58, 248–253, 2009.
7. O'Brien, R.T. and Iglesias, P. A., On the poles and zeros of linear, time-varying systems, *IEEE Trans. Circuits Systems-I: Fundam. Theory Appl.*, 48, 565–577, 2001.
8. Rugh, W.J., *Linear System Theory*, 2nd edn, Prentice-Hall, Englewood Cliffs, NJ, 1996.
9. Silverman, L.M. and Meadows, H.E., Controllability and observability in time-variable linear systems, *SIAM J. Control Optim.*, 5, 64–73, 1967.
10. Silverman, L.M., Transformation of time-variable systems to canonical (Phase-variable) form, *IEEE Trans. Automat. Control*, AC-11, 300, 1966.
11. Valasek, M. and Olgac, N., An efficient pole placement technique for linear time-variant SISO systems, *IEE Proc. Control Theory Appl. D*, 451–458, 1995.
12. Valasek, M. and Olgac, N., Pole placement for linear time-varying non-lexicographically fixed MIMO systems, *Automatica*, 35, 101–108, 1999.
13. Zhu, J.J., Well-defined series and parallel d-spectra for linear time-varying systems, *Proc. Am. Control Conf.*, Baltimore, MD, 734–738,1994.

Further Reading

There are a large number of research papers and textbooks that contain results on the theory of linear time-varying systems. Only a small portion of the existing work is mentioned here, with the emphasis on textbooks: In addition to [1] and [8] listed in the references above, textbooks that contain material on the fundamentals of linear time-varying systems include the following:

14. Zadeh, L.A. and Desoer, C.A., *Linear System Theory*, McGraw-Hill, New York, 1963.
15. Brockett, R.W., *Finite Dimensional Linear Systems*, Wiley & Sons, New York, 1970.
16. D'Angelo, H., *Linear Time-Varying Systems*, Allyn and Bacon, Boston, MA, 1970.
17. Kailath, T., *Linear Systems*, Prentice-Hall, Englewood Cliffs, NJ, 1980.
18. Sontag, E.D., *Mathematical Control Theory*, Springer-Verlag, New York, 1990.
19. Antsaklis, P.J. and Michel, A.N., *Linear Systems*, Birkhauser, Boston, MA, 2006.

For textbooks on H-infinity and $H2$ control of time-varying systems, see

20. Peters, M.A. and Iglesias, P.A., *Minimum Entropy Control for Time-Varying Systems*, Birkhauser, Boston, MA, 1997.
21. Ichikawa, A. and Katayama, H., *Linear Time-Varying and Sampled-Data Systems, Lecture Notes in Control and Information Science*, 265, Springer-Verlag, London, 2001.

An approach to linear time-varying systems given in terms of matrix algebra and analytic function theory is developed in the following textbook:

22. Dewilde, P. and Veen, A. van der, *Time-Varying Systems and Computations*, Kluwer, Boston, MA, 1998.

For textbooks on the adaptive control of time-varying systems and observers for time-varying systems, see

23. Tsakalis, K.S. and Ioannou, P.A., *Linear Time-Varying Plants: Control and Adaptation*, Prentice-Hall, Englewood Cliffs, NJ, 1993.
24. O'Reilly, J., *Observers for Linear* Systems, Academic Press, New York, 1983.

Many textbooks exist on the stability of time-varying differential equations and systems. Examples are

25. Bellman, R., *Stability of Differential Equations*, McGraw-Hill, New York, 1953.
26. Hahn, W., *Stability of Motion*, Springer-Verlag, New York, 1967.
27. Harris, C.J. and Miles, J.F., *Stability of Linear* Systems, Academic Press, New York, 1980.
28. Lukes, D.L., *Differential Equations: Classical to Controlled*, Academic Press, New York, 1982.
29. Miller, R.K. and Michel, A.N., *Ordinary Differential Equations*, Academic Press, New York, 1982.

30. Michel, A.N., Liu, D., and Hou, L., *Stability of Dynamical Systems: Continuous, Discontinuous, and Discrete Systems*, Birkhauser Verlag, Boston, MA, 2008.

For an in-depth treatment of the stability of second-order linear time-varying differential equations, see

31. Duc, L.H., Ilchmann, A., Siegmund, S., and Taraba, P., On stability of linear time-varying second-order differential equations, *Quart. Appl. Math.* 64, 137–151, 2006.

There are a number of papers on the study of time-varying systems given in terms of rings of differential or difference polynomials. In addition to [2], [3], and [6] in the references above, examples of this work are as follows:

32. Kamen, E.W., Khargonekar, P.P., and Poolla, K.R., A transfer function approach to linear time-varying discrete-time systems, *SIAM J. Control Optim.*, 23, 550–565, 1985.
33. Poolla, K. R. and Khargonekar, P.P., Stabilizability and stable proper factorizations for linear time-varying systems, *SIAM J. Control Optim.*, 25, 723–736, 1987.
34. Fliess, M., Some basic structural properties of generalized linear systems, *Systems Control Lett.*, 15, 391–396, 1990.

4

Balanced Realizations, Model Order Reduction, and the Hankel Operator

Jacquelien M.A. Scherpen
University of Groningen

4.1 Introduction

In many engineering applications, processes are described by increasingly complex models that are difficult to analyze and difficult to control. Reduction of the order of the model may overcome some of these difficulties, but it is quite possible that model reduction incurs a significant loss of accuracy. Therefore, the system has to be analyzed in a manner that is useful for the application purpose. Simplification of the model based on this analysis usually results in a model of lower complexity which is easier to handle, and in a corresponding simplification of synthesis procedures for control and filtering problems. Furthermore, the simplification decreases the computational effort. Every application has its own demands, and different model reduction methods have different properties.

In [1] two types of approximation methods for large-scale linear systems are discussed, namely singular-value decomposition (SVD) and Krylov methods. The Krylov methods are based on matching the moments of the impulse response of the linear system up to a certain level. This can be interpreted in terms of the series expansions of the transfer function; see, for example, [1] and the references therein. The key to the success of these methods is that they can be implemented iteratively and are able to handle much higher-order systems than the SVD methods. Nevertheless, they are not always resulting in models that are useful for control, for example, stability may not be preserved, and no *a priori* error bounds for the reduced order system can be given. Furthermore, an extension of these methods to nonlinear systems is still largely unexplored. Only recently, the first steps toward such extension is presented by [2]. The

SVD methods based on balanced realizations, on the other hand, offer a clear structure for analysis of the system based on controllability and observability properties, and model reduction by balanced truncation does preserve stability and other properties and does have an *a priori* error bound for the reduced order system. Furthermore, the extension of balancing to nonlinear systems has been studied in the past two decades as well.

Kalman's minimal realization theory (e.g., [27]) offers a clear picture of the structure of linear systems. However, the accompanying algorithms are not very satisfactory, since they are only textbook algorithms, which are numerically deficient. Moore [32] showed that there are very useful tools that may be used to cope with this problem. He used the principal component analysis which was introduced in statistics in the 1930s to analyze a linear system, and specifically to apply it to model reduction. The most important contribution of [32] is the introduction of balancing for stable minimal linear systems. The balancing method offers a tool to measure the contribution of the different state components to the past input and future output energy of the system, which are measures of controllability and observability. The algorithmic methods corresponding to the balancing theory nowadays are standard toolboxes in simulation packages like MATLAB®.

In the theory of continuous-time linear systems, the system Hankel operator plays an important role in a number of realization problems. For example, when viewed as a mapping from past inputs to future outputs, it plays a direct role in the abstract definition of *state*. It also plays a central role in minimality theory, in model reduction problems, and related to these, in linear identification methods. Specifically, the Hankel operator supplies a set of similarity invariants, the so-called Hankel singular values, which can be used to quantify the importance of each state in the corresponding input–output system. The Hankel operator can also be factored into the composition of an observability and controllability operator, from which Gramian matrices can be defined and the notion of a balanced realization follows. The Hankel singular values are most easily computed in a state-space setting using the product of the Gramian matrices, though intrinsically they depend only on the given input–output mapping. For linear systems, the Hankel operator offers an immediate relation with the frequency domain setting of balancing for linear systems, for example, [63].

Furthermore, these methods have proved to be very useful for application purposes. To mention a few older applications, we refer to [16,62]. Reference [62] successfully applies methods based on balanced realizations on the controller design of the Philips CD player. In [16], several balanced controller designs and algorithms for sensor and actuator placement based on balanced realizations are given and demonstrated for the NASA Deep Space Network Antenna.

For nonlinear systems, the first step toward extension of the linear balancing methods has been set in [45], where a balancing formalism is developed for stable nonlinear continuous-time state-space systems based on the idea that state components that correspond to low control costs and high output energy generation (in the sense of L_2 energy in a signal) are important for the description of the dynamics of the input–output behavior of the system, while state components with high control costs and low output energy generation can be left out of the description. Since then, many results on state-space balancing, modifications based on sliding time windows, and modifications based on proper orthogonal decomposition (POD), and computational issues for model reduction and related minimality considerations for nonlinear systems have appeared in the literature; for example, [18,20,28,37,38,58,59,64]. The relations of the original nonlinear balancing method of [45] with minimality are later explored in [47], and a more constructive approach that includes a strong relation with the nonlinear input–output operators, the nonlinear Hankel operator, and the Hankel norm of the system is presented in [10,12,48]. Model reduction based on these studies is recently treated in [13,14].

Here we first review the Hankel operator, balancing and balanced truncation for stable linear systems. Then unstable systems and closed-loop balancing are reviewed for linear systems. Finally, the extension of the linear theory to nonlinear systems is treated, based on balancing procedures in a neighborhood (possibly large, almost global) of an equilibrium point.

4.2 Linear Systems

In this section, we briefly review the well-known linear system definitions of the system Hankel matrix; the Hankel operator; the controllability and observability operators, Gramians and functions; and the balanced realization and the corresponding model reduction procedure.

4.2.1 The Hankel, Controllability, and Observability Operator

Consider a continuous-time, causal linear input–output system $\Sigma : u \rightarrow y$, with m-dimensional input space $u \in U$ and p-dimensional output space $y \in Y$, and with impulse response $H(t)$. Let

$$H(t) = \sum_{k=0}^{\infty} H_{k+1} \frac{t^k}{k!}, \quad t \geq 0 \tag{4.1}$$

denote its Taylor series expansion about $t = 0$, where $H_k \in \mathbb{R}^{p\times m}$ for each k. The system Hankel matrix is defined as $\hat{\mathcal{H}} = [\hat{\mathcal{H}}_{i,j}]$, where $\hat{\mathcal{H}}_{i,j} = H_{i+j-1}$ for $i, j \geq 1$. If Σ is also (bounded input bounded output) BIBO stable, then the system Hankel operator is the well-defined mapping

$$\begin{aligned} \mathcal{H} &: L_2^m[0, +\infty) \rightarrow L_2^p[0, +\infty), \\ &: \hat{u} \rightarrow \hat{y}(t) = \int_0^{\infty} H(t + \tau)\hat{u}(\tau)\, d\tau. \end{aligned} \tag{4.2}$$

mapping the past inputs to the future outputs. If we define the *time flipping* operator as

$$\begin{aligned} \tilde{\mathcal{F}} &: L_2^m[0, +\infty) \rightarrow L_2^m(-\infty, 0] \\ &: \hat{u} \rightarrow u(t) = \begin{cases} \hat{u}(-t) : t < 0 \\ 0 : t \geq 0, \end{cases} \end{aligned}$$

then clearly $\mathcal{H}(\hat{u}) = (\Sigma \circ \tilde{\mathcal{F}})(\hat{u})$; see the illustration in Figure 4.1. The lower side of the figure depicts the input–output behavior of the original operator Σ. The upper side depicts the input–output behavior of the Hankel operator of Σ, where the signal in the upper left side is the time-flipped signal of the lower left side signal. The flipping operator $\mathcal{F}$ is defined by $\mathcal{F}(u(t)) := u(-t)$. The upper right side signal is the truncated signal (to the space $L_2[0, \infty)$) of the lower left side signal. The corresponding truncation operator is given by

$$\mathcal{T}(y(t)) := \begin{cases} 0 & (t < 0) \\ y(t) & (t \geq 0) \end{cases},$$

and $\tilde{\mathcal{F}} = \mathcal{T} \circ \mathcal{F}$. The definition of a Hankel operator implies that it describes the mapping from the input to the output generated by the state at $t = 0$. Hence, we can analyze the relationship between the state and the input–output behavior of the original operator Σ by investigating its Hankel operator.

When $\mathcal{H}$ is known to be a compact operator, then its (Hilbert) adjoint operator, $\mathcal{H}^*$, is also compact, and the composition $\mathcal{H}^*\mathcal{H}$, is a self-adjoint compact operator with a well-defined spectral decomposition:

$$\mathcal{H}^*\mathcal{H} = \sum_{i=1}^{\infty} \sigma_i^2 \, \langle \cdot, \psi_i \rangle_{L_2} \, \psi_i, \quad \sigma_i \geq 0, \tag{4.3}$$

$$\langle \psi_i, \psi_j \rangle_{L_2} = \delta_{ij}, \quad \langle \psi_i, (\mathcal{H}^*\mathcal{H})(\psi_i) \rangle_{L_2} = \sigma_i^2, \tag{4.4}$$

where σ_i^2 is an eigenvalue of $\mathcal{H}^*\mathcal{H}$ with the corresponding eigenvector ψ_i, ordered as $\sigma_1 \geq \cdots \geq \sigma_n > \sigma_{n+1} = \sigma_{n+2} = 0$, and called the *Hankel singular values* for the input–output system Σ.

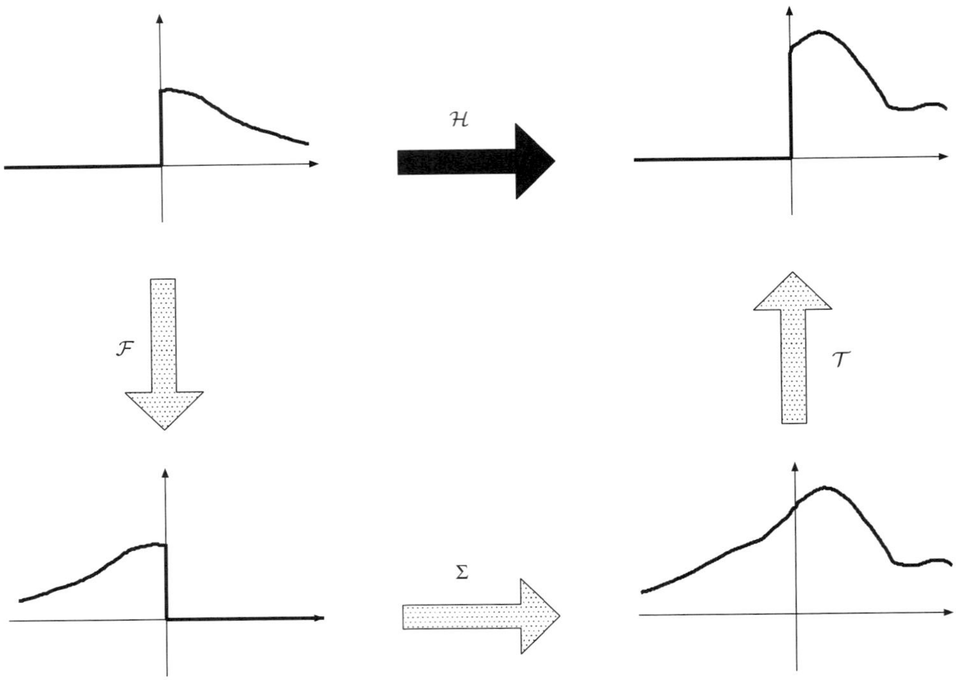

FIGURE 4.1 Hankel operator $\mathcal{H}$ of Σ.

Let (A, B, C) be a state-space realization of Σ with dimension n, that is, consider a linear system:

$$\begin{aligned} \dot{x} &= Ax + Bu, \\ y &= Cx, \end{aligned} \tag{4.5}$$

where $u \in \mathbb{R}^m$, $x \in \mathbb{R}^n$, and $y \in \mathbb{R}^p$. We assume that Equation 4.5 is *stable and minimal*, that is, controllable and observable. Any such realization induces a factorization of the system Hankel matrix into the form $\hat{\mathcal{H}} = \hat{\mathcal{O}}\hat{\mathcal{C}}$, where $\hat{\mathcal{O}}$ and $\hat{\mathcal{C}}$ are the (extended) observability and controllability matrices. If the realization is asymptotically stable, then the Hankel operator can be written as the composition of uniquely determined observability and controllability operators; that is, $\mathcal{H} = \mathcal{O}\mathcal{C}$, where the controllability and observability operators are defined as

$$\begin{aligned} \mathcal{C} &: L_2^m[0, +\infty) \to \mathbb{R}^n : \hat{u} \to \int_0^\infty e^{At} B\hat{u}(t)\, dt, \\ \mathcal{O} &: \mathbb{R}^n \to L_2^p[0, +\infty) : x \to \hat{y}(t) = Ce^{At}x. \end{aligned}$$

Since $\mathcal{C}$ and $\mathcal{O}$ have a finite dimensional range and domain, respectively, they are compact operators; and the composition $\mathcal{O}\mathcal{C}$ is also a compact operator. From the definition of the (Hilbert) adjoint operator, it is easily shown that $\mathcal{C}$ and $\mathcal{O}$ have corresponding adjoints

$$\begin{aligned} \mathcal{C}^* &: \mathbb{R}^n \to L_2^m[0, +\infty) : x \to B^T e^{A^T t} x, \\ \mathcal{O}^* &: L_2^p[0, +\infty) \to \mathbb{R}^n : y \to \int_0^\infty e^{A^T t} C^T y(t)\, dt. \end{aligned}$$

4.2.2 Balanced State-Space Realizations

The above input–output setting can be related with the well-known Gramians that are related to the state-space realization. In order to do so, we consider the energy functions given in the following definition.

Definition 4.1:

The controllability and observability functions of a smooth state-space system are defined as

$$L_c(x_0) = \min_{\substack{u \in L_2(-\infty,0) \\ x(-\infty)=0, x(0)=x_0}} \frac{1}{2} \int_{-\infty}^{0} \| u(t) \|^2 \, dt \tag{4.6}$$

and

$$L_o(x_0) = \frac{1}{2} \int_{0}^{\infty} \| y(t) \|^2 \, dt, \quad x(0) = x_0, \quad u(t) \equiv 0, \quad 0 \le t < \infty, \tag{4.7}$$

respectively.

The value of the controllability function at x_0 is the minimum amount of input energy required to reach the state x_0 from the zero state, and the value of the observability function at x_0 is the amount of output energy generated by the state x_0. The following results are well known:

Theorem 4.1:

Consider the system (Equation 4.5). Then $L_c(x_0) = \frac{1}{2} x_0^T P^{-1} x_0$ and $L_o(x_0) = \frac{1}{2} x_0^T Q x_0$, where $P = \int_0^\infty e^{At} BB^T e^{A^T t} dt$ is the controllability Gramian and $Q = \int_0^\infty e^{A^T t} C^T C e^{At} dt$ is the observability Gramian. Furthermore, P and Q are symmetric and positive definite, and are unique solutions of the Lyapunov equations

$$AP + PA^T = -BB^T \tag{4.8}$$

and

$$A^T Q + QA = -C^T C, \tag{4.9}$$

respectively.

From the form of the Gramians in this theorem, it follows immediately that for any $x_1, x_2 \in \mathbb{R}^n$,

$$\begin{aligned} < x_1, \mathcal{C}\mathcal{C}^* x_2 > &= x_1^T \int_0^\infty e^{At} BB^T e^{A^T t} dt \, x_2, \\ &= x_1^T P x_2 \end{aligned} \tag{4.10}$$

$$\begin{aligned} < x_1, \mathcal{O}^*\mathcal{O} x_2 > &= x_1^T \int_0^\infty e^{A^T t} C^T C e^{At} dt \, x_2, \\ &= x_1^T Q x_2 \end{aligned} \tag{4.11}$$

and the relation with the energy functions is given as

$$L_c(x) = \frac{1}{2} x^T P^{-1} x = \frac{1}{2} \langle x, (\mathcal{C}\mathcal{C}^*)^{-1} x \rangle, \tag{4.12}$$

$$L_o(x) = \frac{1}{2} x^T Q x = \frac{1}{2} \langle x, (\mathcal{O}^*\mathcal{O}) x \rangle. \tag{4.13}$$

The following (balancing) theorem is originally due to [32].

Theorem 4.2: [32]

The eigenvalues of QP are similarity invariants, that is, they do not depend on the choice of the sate-space coordinates. There exists a state-space representation where

$$\Sigma := Q = P = \begin{pmatrix} \sigma_1 & & 0 \\ & \ddots & \\ 0 & & \sigma_n \end{pmatrix}, \tag{4.14}$$

with $\sigma_1 \geq \sigma_2 \geq \cdots \geq \sigma_n > 0$ *the square roots of the eigenvalues of QP. Such representations are called* balanced, *and the system is in* balanced form. *Furthermore, the* σ_i*'s,* $i = 1, \ldots, n$, *equal the Hankel singular values, that is, the singular values of the Hankel operator (Equation 4.2).*

Two other representations that may be obtained from Equation 4.14 by coordinate transformations $x = \Sigma^{-\frac{1}{2}}\bar{x}$ and $x = \Sigma^{\frac{1}{2}}\bar{x}$, respectively, follow easily from the above theorem.

Definition 4.2: [32]

A state-space representation is an input-normal/output-diagonal *representation if* $P = I$ *and* $Q = \Sigma^2$, *where* Σ *is given by Equation 4.14. Furthermore, it is an output-normal/input-diagonal representation if* $P = \Sigma^2$ *and* $Q = I$.

The largest Hankel singular value is equal to the Hankel norm of the system, that is,

$$\| G \|_H^2 = \max_{\substack{x \in \mathbb{R}^n \\ x \neq 0}} \frac{L_o(x)}{L_c(x)} = \max_{\substack{x \in \mathbb{R}^n \\ x \neq 0}} \frac{x^T Q x}{x^T P^{-1} x} = \max_{\substack{x \in \mathbb{R}^n \\ x \neq 0}} \frac{\bar{x}^T \Sigma^2 \bar{x}}{\bar{x}^T \bar{x}} = \sigma_1^2, \tag{4.15}$$

where $G = C(sI - A)^{-1}B$ is the transfer matrix of the system. This gives a characterization of the largest Hankel singular value. The other Hankel singular values may be characterized inductively in a similar way; we refer to [6,15].

So far, we have assumed the state-space representation to be minimal. However, if we consider nonminimal state-space realizations, that is, the system is not controllable and/or observable, then we obtain σ_i's that are zero, corresponding to the noncontrollable or nonobservable part of the system, and thus, to the nonminimal part of the system. Related to this observation, we have that the minimal realization of a linear input–output system has a dimension n that is equal to the Hankel rank, or in other words, it equals the rank of the Hankel matrix. A well-known result related to the latter is the following theorem for example, [63].

Theorem 4.3:

If (A, B, C) *is asymptotically stable, then the realization is minimal if and only if* $P > 0$ *and* $Q > 0$.

4.2.3 Model Reduction

Once the state-space system is in balanced form, an order reduction procedure based on this form may be applied. Thus, in order to proceed, we assume that the system (Equation 4.5) is in balanced form. Then the controllability and observability function are $\bar{L}_c(\bar{x}_0) = \frac{1}{2}\bar{x}_0^T \Sigma^{-1}\bar{x}_0$ and $\bar{L}_c(\bar{x}_0) = \frac{1}{2}\bar{x}_0^T \Sigma\bar{x}_0$, respectively. For small σ_i, the amount of control energy required to reach the state $\bar{x} = (0, \dots, 0, x_i, 0, \dots, 0)$ is large while the output energy generated by this state $\bar{x}$ is small. Hence, if $\sigma_k \gg \sigma_{k+1}$, the state components x_{k+1} to x_n are far less important from this energy point of view and may be removed to reduce the number of state components of the model. We partition the system (Equation 4.5) in a corresponding way as follows:

$$A = \begin{pmatrix} A_{11} & A_{12} \\ A_{21} & A_{22} \end{pmatrix}, \quad B = \begin{pmatrix} B_1 \\ B_2 \end{pmatrix}, \quad C = \begin{pmatrix} C_1 & C_2 \end{pmatrix}, \tag{4.16}$$

$$x^1 = (x_1, \dots, x_k)^T, \quad x^2 = (x_{k+1}, \dots, x_n)^T, \quad \Sigma = \begin{pmatrix} \Sigma_1 & 0 \\ 0 & \Sigma_2 \end{pmatrix},$$

where $\Sigma_1 = \text{diag}(\sigma_1, \dots, \sigma_k)$ and $\Sigma_2 = \text{diag}(\sigma_{k+1}, \dots, \sigma_n)$.

Theorem 4.4:

Both subsystems (A_{ii}, B_i, C_i), $i = 1, 2$, are again in balanced form, and their controllability and observability Gramians are equal to Σ_i, $i = 1, 2$.

The following result has been proved by [43].

Theorem 4.5:

Assume that $\sigma_k > \sigma_{k+1}$. Then both subsystems (A_{ii}, B_i, C_i), $i = 1, 2$, are asymptotically stable.

The subsystem (A_{11}, B_1, C_1) may be used as an approximation of the full order system (Equation 4.5). The optimality of this approximation in the Hankel and $\mathcal{H}_\infty$-norm has been studied by [17], and an upper bound for the error is given. The $\mathcal{H}_\infty$-norm of $G(s) = C(sI - A)^{-1}B$ is defined as

$$\|G\|_\infty = \sup_{\omega \in \mathbb{R}} \lambda_{\max}^{\frac{1}{2}}(G(-j\omega)^T G(j\omega)),$$

where $\lambda_{\max}^{\frac{1}{2}}(G(-j\omega)^T G(j\omega))$ is the square root of the maximum eigenvalue of $G(-j\omega)^T G(j\omega)$. Denote the transfer matrix of the reduced order system (A_{11}, B_1, C_1) by $\tilde{G}(s) = C_1(sI - A_{11})^{-1}B_1$. The following error bound was originally proved in [17]. A number of proofs can be found in [63].

Theorem 4.6: [17]

$$\|G - \tilde{G}\|_H \le \|G - \tilde{G}\|_\infty \le 2(\sigma_{k+1} + \dots + \sigma_n).$$

Hence, if we remove the state components $x_{k+1}, \dots, x_n$ that correspond to small Hankel singular values $\sigma_{k+1}, \dots, \sigma_n$ (small compared to the rest of the singular values, that is, $\sigma_k \gg \sigma_{k+1}$), then the error is small, and the reduced order system (A_{11}, B_1, C_1) constitutes a good approximation in terms of the Hankel norm to the full order system.

The model reduction method that we gave above consists of simply truncating the model. It is also possible to reduce the model in a different way see, for example, [8,21]. Instead of setting

$x^2 = (x_{k+1}, \ldots, x_n) = 0$ we approximate the system by setting $\dot{x}^2 = 0$ (thus interpreting x^2 as a very fast stable state, which may be approximated by a constant function of x^1 and u). The resulting algebraic equation can be solved for x^2 as (note that A_{22}^{-1} exists by Theorem 4.5)

$$x^2 = -A_{22}^{-1}\left(A_{21}x^1 + B_2 u\right).$$

Substitution in Equation 4.5 leads to a reduced order model $(\hat{A}, \hat{B}, \hat{C})$ defined as

$$\begin{aligned}\hat{A} &:= A_{11} - A_{12}A_{22}^{-1}A_{21},\\ \hat{B} &:= B_1 - A_{12}A_{22}^{-1}B_2,\\ \hat{C} &:= C_1 - C_2A_{22}^{-1}A_{21},\end{aligned}$$

The system $(\hat{A}, \hat{B}, \hat{C})$ also gives an approximation to the full order system (Equation 4.5). Theorems 4.4 through 4.6 also hold if we replace the system (A_{11}, B_1, C_1) by $(\hat{A}, \hat{B}, \hat{C})$.

4.2.4 Unstable Systems, Closed-Loop Balancing

4.2.4.1 Linear Quadratic Gaussian Balancing

A major drawback of the original balancing method as described in Section 4.2.2 is that it only applies to stable systems. Furthermore, the method emphasizes the (open-loop) input–output characteristics of the system, while it is *a priori* not clear if it yields good approximations in closed-loop configurations. In this section, we treat (linear quadratic Gaussian) LQG balancing for linear systems, which was introduced by [25,26] (see also [57]). In [41], this concept is further developed. LQG balancing was introduced with the aim of finding a model reduction method for a system (not necessarily stable) together with its corresponding LQG compensator. LQG balancing has been treated from another point of view in [61]. First, we give a review of the formulation of [26,41].

LQG compensation is formulated for a minimal state-space system

$$\begin{aligned}\dot{x} &= Ax + Bu + Bd,\\ y &= Cx + v,\end{aligned} \tag{4.17}$$

where $u \in \mathbb{R}^m$, $x \in \mathbb{R}^n$, $y \in \mathbb{R}^p$, and d and v are independent Gaussian white-noise processes with covariance functions $I\delta(t - \tau)$. The criterion

$$J(x_0, u(\cdot)) = E \lim_{T\to\infty} \frac{1}{T}\int_0^T [x^T(t)C^TCx(t) + u^T(t)u(t)]\,dt \tag{4.18}$$

is required to be minimized. The resulting optimal compensator is given by

$$\begin{aligned}\dot{z} &= Az + Bu + SC^T(y - Cz),\\ u &= -B^TPz.\end{aligned} \tag{4.19}$$

Here S is the stabilizing solution (i.e., $\sigma(A - SC^TC) \subset \mathbb{C}^-$) to the filter algebraic Riccati equation (FARE)

$$AS + SA^T + BB^T - SC^TCS = 0, \tag{4.20}$$

and P is the stabilizing solution (i.e., $\sigma(A - BB^TP) \subset \mathbb{C}^-$) to the control algebraic Riccati equation (CARE)

$$A^TP + PA + C^TC - PBB^TP = 0. \tag{4.21}$$

Theorem 4.7: [26,41]

The eigenvalues of PS are similarity invariants and there exists a state-space representation where

$$M := P = S = \begin{pmatrix} \mu_1 & & 0 \\ & \ddots & \\ 0 & & \mu_n \end{pmatrix}, \tag{4.22}$$

with $\mu_1 \geq \mu_2 \geq \cdots \geq \mu_n > 0$. *This is called an* LQG balanced representation *or* LQG balanced form.

Jonckheere and Silverman [26] and Opdenacker and Jonckheere [41] argue that if $\mu_k \gg \mu_{k+1}$ then the state components x_1 up to x_k are more difficult both to control and to filter than x_{k+1} up to x_n and a synthesis based only on the state components x_1 up to x_k probably retains the essential properties of the system in a closed-loop configuration. Corresponding to the partitioning of the state in the first k components and the last $n-k$ components, the partitioning of the matrices is done as in (4.16), and the reduced order system is

$$\begin{aligned} \dot{x} &= A_{11}x + B_1u + B_1d, \\ y &= C_1x + v. \end{aligned} \tag{4.23}$$

Theorem 4.8: [26,41]

Assume $\mu_k > \mu_{k+1}$. *Then* (A_{11}, B_1, C_1) *is minimal, the reduced order system (Equation 4.23) is again LQG balanced and the optimal compensator for system (Equation 4.23) is the reduced order optimal compensator of the full order system (Equation 4.17).*

As explained in Section 4.2.2, the original idea of balancing stable linear systems, as introduced by [32], considers the Hankel singular values σ_i, $i = 1, \ldots, n$, which are a measure for the importance of a state component in a balanced representation. This balancing technique is based on the input energy which is necessary to reach this state component and the output energy which is generated by this state component. A similar kind of reasoning, using a different pair of energy functions, may be used to achieve the similarity invariants μ_i, $i = 1, \ldots, n$, as above; see [61]. To follow this reasoning, we consider the minimal system (Equation 4.17) *without noise*, that is,

$$\begin{aligned} \dot{x} &= Ax + Bu, \\ y &= Cx, \end{aligned} \tag{4.24}$$

where $u \in \mathbb{R}^m$, $x \in \mathbb{R}^n$, and $y \in \mathbb{R}^p$. We define the energy functions

$$K^-(x_0) := \min_{\substack{u \in L_2(-\infty,0) \\ x(-\infty)=0, x(0)=x_0}} \frac{1}{2}\int_{-\infty}^{0} (\| y(t) \|^2 + \| u(t) \|^2)\, dt,$$

$$K^+(x_0) := \min_{\substack{u \in L_2(0,\infty) \\ x(\infty)=0, x(0)=x_0}} \frac{1}{2}\int_{0}^{\infty} (\| y(t) \|^2 + \| u(t) \|^2)\, dt.$$

$K^-(x_0)$ is called the *past energy* and $K^+(x_0)$ the *future energy* of the system in the state x_0.

Theorem 4.9: [61]

$K^-(x_0) = \frac{1}{2}x_0^T S^{-1} x_0$ and $K^+(x_0) = \frac{1}{2}x_0^T P x_0$, where S and P are the stabilizing solutions of the FARE and the CARE, Equations 4.20 and 4.21, respectively.

For the LQG balanced representation of Theorem 4.7, the past and future energy function are $K^-(x_0) = \frac{1}{2}x_0^T M^{-1} x_0$ and $K^+(x_0) = \frac{1}{2}x_0^T M x_0$, respectively, where M is diagonal. The importance of the state $\tilde{x} = (0, \ldots, 0, x_i, 0, \ldots, 0)$ in terms of past and future energy may be measured by the similarity invariant μ_i. For large μ_i the influence of the state $\tilde{x}$ on the future energy is large while the influence on the past energy is small. Hence, if $\mu_k \gg \mu_{k+1}$, the state components x_{k+1} to x_n are "not important" from this energy point of view and may be removed to reduce the number of state components of the model.

4.2.4.2 Balancing of the Normalized Coprime Representation

In [29,39], balancing of the normalized coprime representation of a linear system is treated. Balancing of the normalized coprime representation was introduced with the aim of finding a model reduction method for unstable linear systems. In [29], balancing of the normalized *right* coprime factorization is treated, while in [39], balancing of the normalized *left* coprime factorization is treated. Here we provide a brief review on this subject.

We consider the system (Equation 4.24) and its transfer function $G(s) = C(sI - A)^{-1}B$. Furthermore, we consider the stabilizing solution P to the CARE (Equation 4.21) and the stabilizing solution S to the FARE (Equation 4.20), leading to the stable matrices $\hat{A} := A - BB^T P$ and $\tilde{A} := A - SC^T C$. First we treat normalized right coprime factorizations and then normalized left coprime factorizations.

We may write any transfer matrix $G(s) = C(sI - A)^{-1}B$ as a right fraction $G(s) = N(s)D(s)^{-1}$ of stable transfer matrices $N(s)$ and $D(s)$. If we choose (e.g., [36]),

$$N(s) := C(sI - \hat{A})^{-1}B,$$
$$D(s) := I - B^T P(sI - \hat{A})^{-1}B,$$

then the *factorization* is *right coprime*, that is, $N(s)$ and $D(s)$ have no common zeros at the same place in the closed right half plane. A state-space realization of the transfer matrix (the so-called graph operator)

$$\begin{pmatrix} N(s) \\ D(s) \end{pmatrix}$$

is

$$\begin{aligned} \dot{x} &= (A - BB^T P)x + Bw, \\ \begin{pmatrix} y \\ u \end{pmatrix} &= \begin{pmatrix} C \\ -B^T P \end{pmatrix} x + \begin{pmatrix} 0 \\ I \end{pmatrix} w, \end{aligned} \tag{4.25}$$

with w a (fictitious) input variable. Furthermore, we are able to find stable transfer matrices $U(s)$ and $V(s)$, such that the Bezout identity

$$U(s)N(s) + V(s)D(s) = I. \tag{4.26}$$

is fulfilled. Indeed, take $U(s) = B^T P(sI - \tilde{A})^{-1}SC^T$ and $V(s) = I + B^T P(sI - \tilde{A})^{-1}B$ (see, e.g., [36,60]). The fact that we are able to find a stable left inverse of the graph operator, that is, we can find the solutions $U(s)$ and $V(s)$ to the Bezout identity (Equation 4.26), is equivalent to the factorization being right coprime.

Furthermore, the graph operator is inner, that is,

$$\| \begin{pmatrix} N \\ D \end{pmatrix} w \|_2 = \|w\|_2$$

or

$$N(-s)^T N(s) + D(-s)^T D(s) = I.$$

Therefore, the factorization is called *normalized*. It is easily checked that the observability Gramian of the system (Equation 4.25) is P. Denote its controllability Gramian by R.

In a similar way, we may write the transfer matrix $G(s)$ as a left fraction $G(s) = \tilde{D}(s)^{-1}\tilde{N}(s)$ of stable transfer matrices $\tilde{D}(s)$ and $\tilde{N}(s)$. If we choose (e.g., [36])

$$\begin{aligned} \tilde{N}(s) &:= C(sI - \tilde{A})^{-1}B, \\ \tilde{D}(s) &= C(sI - \tilde{A})^{-1}SC^T - I, \end{aligned}$$

then this is a *left factorization*. Obviously, $\hat{y}(s) = G(s)\hat{u}(s)$ is equivalent with $0 = \tilde{N}(s)\hat{u}(s) - \tilde{D}(s)\hat{y}(s)$. Moreover, a state-space realization of the transfer matrix

$$\begin{pmatrix} \tilde{N}(s) & \tilde{D}(s) \end{pmatrix}$$

is

$$\begin{aligned} \dot{x} &= (A - SC^T C)x + \begin{pmatrix} B & SC^T \end{pmatrix} \tilde{w}, \\ z &= Cx + \begin{pmatrix} 0 & -I \end{pmatrix} \tilde{w}. \end{aligned} \tag{4.27}$$

If we take

$$\tilde{w} = \begin{pmatrix} u \\ y \end{pmatrix}$$

as the input variable, then the dynamics resulting from setting $z = 0$ in Equation 4.27 is a state-space representation of $G(s)$. We are able to find stable transfer matrices such that the Bezout Identity is fulfilled, that is, there exist stable transfer matrices $\tilde{U}(s)$ and $\tilde{V}(s)$, such that

$$\tilde{N}(s)\tilde{U}(s) + \tilde{D}(s)\tilde{V}(s) = I. \tag{4.28}$$

Indeed, we may take $\tilde{U}(s) = B^T P(sI - \hat{A})^{-1}SC^T$ and $\tilde{V}(s) = I + C(sI - \hat{A})^{-1}SC^T$ (see, e.g., Vid,Ne). This proves that the factorization is *left coprime*. Furthermore $\begin{pmatrix} \tilde{N}(s) & \tilde{D}(s) \end{pmatrix}$ is co-inner, that is,

$$\tilde{N}(s)\tilde{N}(-s)^T + \tilde{D}(s)\tilde{D}(-s)^T = I,$$

which means that the factorization is *normalized*. Hence $\begin{pmatrix} \tilde{N}(s) & \tilde{D}(s) \end{pmatrix}$ represents the normalized left coprime factorization of system (Equation 4.24). The system (Equation 4.27) has as controllability Gramian the positive-definite matrix S and we denote its observability Gramian by the matrix Q. Note that the right factorization

$$\begin{pmatrix} N(s) \\ D(s) \end{pmatrix}$$

can be seen as an *image* representation of $G(s)$, while the left factorization

$$\begin{pmatrix} \tilde{N}(s) & \tilde{D}(s) \end{pmatrix}$$

can be regarded as a *kernel* representation of $G(s)$.

The following result follows rather straightforwardly.

Theorem 4.10:

The Hankel singular values of the right and left factorization (Equations 4.25 and 4.27), respectively, are the same.

Proof. It follows from the Lyapunov Equations 4.8 and 4.9 for the systems (Equations 4.25 and 4.27), that $R = (I + SP)^{-1}S$ and $Q = (I + PS)^{-1}P$. Now, it is easily obtained that PR and SQ have the same eigenvalues.

The Hankel singular values of Equation 4.25 (and, hence, of Equation 4.27) are called the *graph Hankel singular values* of the original system (Equation 4.24). These singular values have the following property:

Theorem 4.11: [29,39]

The graph Hankel singular values of system (Equation 4.24) are strictly less than one.

Denote the graph Hankel singular values by τ_i, $i = 1, \ldots, n$, and assume $\tau_1 \geq \cdots \geq \tau_n$. The relation between τ_i, $i = 1, \ldots, n$, and the similarity invariants μ_i, $i = 1, \ldots, n$, of Theorem 4.7 is given by the following theorem:

Theorem 4.12: [39,61]

$$\mu_i = \tau_i(1 - \tau_i^2)^{-\frac{1}{2}} \quad \textit{for } i = 1, \ldots, n.$$

This implies that the reduced model that is obtained by model reduction based on balancing the (left or right) normalized coprime factorization is the same as the reduced model that is obtained by model reduction based on LQG balancing. Consider the normalized right coprime representation (Equation 4.25) and assume that it is in balanced form with

$$\Lambda := P = R = \begin{pmatrix} \tau_1 & & 0 \\ & \ddots & \\ 0 & & \tau_n \end{pmatrix}.$$

Furthermore, assume that $\tau_k > \tau_{k+1}$ and define correspondingly $\Lambda =: \text{diag}\{\Lambda_1, \Lambda_2\}$. It follows, [29], that reducing the order of Equation 4.25 by truncating the system to the first k state components (the partitioning is done corresponding to Equation 4.16) again gives a normalized right coprime representation.

Theorem 4.13: [29]

The reduced order system of Equation 4.25 is of the form

$$\left(\left(A_{11} - B_1 B_1^T D_1\right), B_1, \begin{pmatrix} C_1 \\ -B_1^T D_1 \end{pmatrix}, \begin{pmatrix} 0 \\ I \end{pmatrix}\right),$$

with controllability and observability Gramian Λ_1. This system is the normalized right coprime representation of the system (A_{11}, B_1, C_1), which is minimal.

4.2.4.3 Extensions to Other Types of Balancing

The methods described above can be put in a more general setting, and extended to the H_∞ case. $H\infty$ balancing for linear systems is introduced in [33–35]. For details, we refer to the latter references. In the H_∞ case, balancing is performed on Q_γ^- and Q_γ^+, [46], which are defined as

$$Q_\gamma^-(x_0) = \min_{u\in L_2(-\infty,0)} \frac{1}{2}\int_{-\infty}^{0} (1-\gamma^{-2})||y(t)||^2 + ||u(t)||^2\, dt, \quad x(-\infty)=0,\ x(0)=x_0, \forall\gamma$$

and

$$Q_\gamma^+(x_0) = \min_{u\in L_2(0,\infty)} \frac{1}{2}\int_0^\infty ||y(t)||^2 + \frac{1}{1-\gamma^{-2}}||u(t)||^2\, dt, \quad x(\infty)=0,\ x(0)=x_0, \quad \text{for } \gamma>1,$$

while if $\gamma < 1$, then

$$Q_\gamma^+(x_0) = \max_{u\in L_2(0,\infty)} \frac{1}{2}\int_0^\infty ||y(t)||^2 + \frac{1}{1-\gamma^{-2}}||u(t)||^2\, dt, \quad x(\infty)=0,\ x(0)=x_0.$$

There is an immediate relation with the solutions to the H_∞ Filter and Control Algebraic Riccati equations, see, for example, [46].

Positive real and bounded real balancing can be done by considering dissipativity with respect to a quadratic supply rate that depends on the input and the output of the system:

$$s(u,y) = \frac{1}{2}[u^T\ y^T]J\begin{bmatrix} u \\ y \end{bmatrix}, \tag{4.29}$$

with $u \in \mathbb{R}^m$, $y \in \mathbb{R}^p$, and $J \in \mathbb{R}^{(m+p)\times(m+p)}$, such that $J = J^T$; see, for example, [23,52]. The bounded real balancing case , that is, $J = \begin{bmatrix} I & 0 \\ 0 & -I \end{bmatrix}$ is treated in [1,22,39,42]. The positive real balancing case, that is balancing of strictly passive, asymptotically stable, minimal systems, that is $J = \begin{bmatrix} 0 & I \\ I & 0 \end{bmatrix}$, is treated in [1,7,19]. For all these methods it holds that truncation preserves the original balanced structure, and thus truncation based on bounded real or positive real balancing preserves bounded and positive realness, respectively. Furthermore, error bounds for model reduction based on several of the above mentioned methods are available; see, for example, [52].

Additionally, model reduction methods based on balancing methods that preserve some internal structure of the system currently receive a lot of interest due to applications that require this, such as electrical circuit simulators. For some results on Hamiltonian structure preservation, and second-order system structure preservation; see, for example, [30,54,55].

4.3 Balancing for Nonlinear Systems

Balanced realizations and the related model order reduction technique rely on singular-value analysis. The analysis is important since it extracts the gain structure of the operator, that is, it characterizes the largest input–output ratio and the corresponding input [51]. Since linear singular values are defined as eigenvalues of the composition of the given operator and its adjoint, it is natural to introduce a nonlinear version of adjoint operators to obtain a nonlinear counterpart of a singular value. There has been done quite some research on the nonlinear extension of adjoint operators, for example, [3,10,48], and the references therein. Here we do not explicitly use these definitions of nonlinear adjoint operators. We rely on a characterization of singular values for nonlinear operators based on the gain structure as studied in [9]. The balanced realization based on this analysis yields a realization that is based on the singular values of the corresponding Hankel operator, and results in a method which can be viewed as a complete extension of the linear methods, both from an input–output and a state-space point of view [12].

The related model order reduction technique, nonlinear balanced truncation, preserves several important properties of the original system and the corresponding input–output operator, such as stability, controllability, observability, and the gain structure [11,14].

This section gives a very brief overview of the series of research on balanced realization and the related model order reduction method based on nonlinear singular-value analysis. We refer to [13] for more details.

4.3.1 Basics of Nonlinear Balanced Realizations

This section gives a nonlinear extension of balanced realization introduced in the Section 4.2. Let us consider the following asymptotically stable input-affine nonlinear system

$$\Sigma : \begin{cases} \dot{x} = f(x) + g(x)u \quad x(0) = x^0, \\ y = h(x), \end{cases} \tag{4.30}$$

where $x(t) \in \mathbb{R}^n$, $u(t) \in \mathbb{R}^m$, and $y(t) \in \mathbb{R}^p$. The controllability operator $\mathcal{C} : U \to X$ with $X = \mathbb{R}^n$ and $U = L_2^m[0,\infty)$, and the observability operator $\mathcal{O} : X \to Y$ with $Y = L_2^p[0,\infty)$ for this system are defined by

$$\mathcal{C} : u \mapsto x^0 : \begin{cases} \dot{x} = -f(x) - g(x)u \quad x(\infty) = 0 \\ x^0 = x(0) \end{cases}$$

$$\mathcal{O} : x^0 \mapsto y : \begin{cases} \dot{x} = f(x) \quad x(0) = x^0 \\ y = h(x) \end{cases}$$

This definition implies that the observability operator $\mathcal{O}$ is a map from the initial condition $x(0) = x^0$ to the output L_2 signal when no input is applied. To interpret the meaning of $\mathcal{C}$, let us consider a time-reversal behavior of the $\mathcal{C}$ operator as

$$\mathcal{C} : u \mapsto x^0 : \begin{cases} \dot{x} = f(x) + g(x)u(-t) \quad x(-\infty) = 0 \\ x^0 = x(0) \end{cases}$$

Then the controllability operator $\mathcal{C}$ can be regarded as a mapping from the input L_2 signal to the terminal state $x(0) = x^0$ when the initial state is $x(-\infty) = 0$. Therefore, as in the linear case, $\mathcal{C}$ and $\mathcal{O}$ represent the input-to-state behavior and the state-to-output behavior, respectively, and the Hankel operator for the nonlinear system Σ in Equation 4.30 is given by the composition of $\mathcal{C}$ and $\mathcal{O}$

$$\mathcal{H} := \mathcal{O} \circ \mathcal{C}. \tag{4.31}$$

To relate the Hankel, controllability and observability operator with the observability and controllability functions defined in Equations 4.6 and 4.7, we first introduce a norm-minimizing inverse $\mathcal{C}^\dagger : X \to U$ of $\mathcal{C}$.

$$\mathcal{C}^\dagger : x^0 \mapsto u := \arg \min_{\mathcal{C}(u)=x^0} \|u\|$$

The operators $\mathcal{C}^\dagger$ and $\mathcal{O}$ yield the definitions of the controllability function $L_c(x)$ and the observability function $L_o(x)$ that are generalizations of the controllability and observability Gramians, respectively that is,

$$L_c(x^0) := \frac{1}{2} \parallel C^\dagger(x^0) \parallel^2 = \min_{\substack{u\in L_2(-\infty,0] \\ x(-\infty)=0, x(0)=x_0}} \frac{1}{2}\int_{-\infty}^{0} \parallel u(t) \parallel^2 \, dt, \tag{4.32}$$

$$L_o(x^0) := \frac{1}{2} \parallel \mathcal{O}(x^0) \parallel^2 = \frac{1}{2}\int_{0}^{\infty} \parallel y(t) \parallel^2 \, dt, \quad x(0) = x^0, \quad u(t) \equiv 0, \quad 0 \le t < \infty. \tag{4.33}$$

For linear systems, the relation with the Gramians is given in Theorem 4.1. Here the inverse of P appears in the equation for L_c because $C^\dagger$ appears in the definition (Equation 4.32), whereas C can be used in the linear case. In order to obtain functions $L_c(x)$ and $L_o(x)$, we need to solve a Hamilton–Jacobi equation and a Lyapunov equation.

Theorem 4.14: [45]

Consider the system of Equation 4.30. Suppose that 0 *is an asymptotically stable equilibrium point and that a smooth observability function* $L_o(x)$ *exists. Then* $L_o(x)$ *is the unique smooth solution of*

$$\frac{\partial L_o(x)}{\partial x} f(x) + \frac{1}{2}\, h(x)^{\mathrm{T}} h(x) = 0,$$

with $L_o(0) = 0$. *Furthermore, assume that a smooth controllability function* $L_c(x)$ *exists. Then* $L_c(x)$ *is the unique smooth solution of*

$$\frac{\partial L_c(x)}{\partial x} f(x) + \frac{1}{2}\frac{\partial L_c(x)}{\partial x} g(x) g(x)^{\mathrm{T}} \frac{\partial L_c(x)}{\partial x}^{\mathrm{T}} = 0,$$

with $L_c(0) = 0$ *such that* 0 *is an asymptotically stable equilibrium point of* $\dot{x} = -f(x) - g(x)g(x)^{\mathrm{T}} (\partial L_c(x)/\partial x)^{\mathrm{T}}$.

Similar to the linear case, the positive definiteness of the controllability and observability functions implies strong reachability and zero-state observability of the system Σ in Equation 4.30, respectively. Combining these two properties, we can obtain the following result on the minimality of the system.

Theorem 4.15: [47]

Consider the system of Equation 4.30, and assume it is analytic. Suppose that

$$0 < L_c(x) < \infty,$$
$$0 < L_o(x) < \infty,$$

hold for all $x \neq 0$. *Then the system is a minimal realization as defined in and under the conditions from, [24].*

$L_c(x)$ and $L_o(x)$ can be used to "measure the minimality" of a nonlinear dynamical system. Furthermore, a basis for nonlinear balanced realizations is obtained as a nonlinear generalization of Definition 4.2 in the linear case. For that, a factorization of $L_o(x)$ into a semiquadratic form needs to be done, that is, in a convex neighborhood of the equilibrium point 0, we can write

$$L_o(x) = \frac{1}{2} x^{\mathrm{T}} M(x) x, \quad \text{with } M(0) = \frac{\partial^2 L_o}{\partial x^2}(0).$$

Now, an input-normal/output-diagonal form can be obtained.

Theorem 4.16: [45]

Consider the system of Equation 4.30 on a neighborhood W of 0. Suppose that 0 is an asymptotically stable equilibrium point, that it is zero-state observable, that smooth controllability and observability functions $L_c(x)$ and $L_o(x)$ exist on W, and that $(\partial^2 L_c/\partial x^2)(0) > 0$ and $(\partial^2 L_o/\partial x^2)(0) > 0$ hold. Furthermore, assume that the number of distinct eigenvalues of $M(x)$ is constant on W. Then there exists coordinates such that the controllability and observability functions $L_c(x)$ and $L_o(x)$ satisfy

$$L_c(x) = \frac{1}{2}\sum_{i=1}^{n} x_i^2, \tag{4.34}$$

$$L_o(x) = \frac{1}{2}\sum_{i=1}^{n} x_i^2 \tau_i(x), \tag{4.35}$$

where $\tau_1(x) \geq \tau_2(x) \geq \cdots \geq \tau_n(x)$.

A state-space realization satisfying the conditions (Equations 4.34 and 4.35) is called an *input-normal form*, and the functions $\tau_i(x)$, $i = 1, 2, \ldots, n$ are called singular-value functions. We refer to [45] for the construction of the coordinate transformation that brings the system in the form of Theorem 4.16. If a singular value function $\tau_i(x)$ is larger than $\tau_j(x)$, then the coordinate axis x_i plays a more important role than the coordinate axis x_j does. Thus this realization is similar to the linear input-normal/output-diagonal realization of Definition 4.2, and it directly yields a tool for model order reduction of a nonlinear systems. However, a drawback of the above realization is that the the singular-value functions $\tau_i(x)$'s and consequently, the corresponding realization are not unique, for example, [18]. For example, if the observability function is given by

$$L_o(x) = \frac{1}{2}(x_1^2\tau_1(x) + x_2^2\tau_2(x)) = \frac{1}{2}(2x_1^2 + x_2^2 + x_1^2x_2^2),$$

with the state-space $x = (x_1, x_2)$, then the corresponding singular-value functions are

$$\tau_1(x) = 2 + kx_2^2,$$
$$\tau_2(x) = 1 + (1-k)x_1^2,$$

with an arbitrary scalar constant k. This example reveals that the singular value function are not uniquely determined by this characterization. To overcome these problems, balanced realizations based on nonlinear singular value analysis is presented in the following section.

4.3.2 Balanced Realizations Based on Singular-Value Analysis of Hankel Operators

In this section, application of singular-value analysis to nonlinear Hankel operators determines a balanced realization with a direct input–output interpretation, whereas the balanced realization of Theorem 4.16 is completely determined based on state-space considerations only. To this end, we consider the Hankel operator $\mathcal{H} : U \to Y$ as defined in Equation 4.31 with $U = L_2^m[0, \infty)$ and $Y = L_2^p[0, \infty)$. Then a singular-value analysis based on the differential form, [12], is given by

$$(\mathrm{d}\mathcal{H}(v))^* \, \mathcal{H}(v) = \lambda \, v, \quad \lambda \in \mathbb{R}, \quad v \in U, \tag{4.36}$$

with λ and v the eigenvalues and corresponding eigenvectors, respectively. Since we consider a singular-value analysis problem on L_2 spaces, we need to find state trajectories of certain Hamiltonian dynamics;

see, for example, [12]. In the linear case, we only need to solve an eigenvalue problem on a finite dimensional space $X = \mathbb{R}^n$ to obtain the singular values and singular vectors of the Hankel operator $\mathcal{H}$. Here we provide the nonlinear counterpart as follows.

Theorem 4.17: [12]

Consider the Hankel operator defined by Equation 4.31. Suppose that the operators $\mathcal{C}^\dagger$ and $\mathcal{O}$ exist and are smooth. Moreover, suppose that $\lambda \in \mathbb{R}$ and $\xi \in X$ satisfy the following equation:

$$\frac{\partial L_o(\xi)}{\partial \xi} = \lambda \, \frac{\partial L_c(\xi)}{\partial \xi}, \quad \lambda \in \mathbb{R}, \quad \xi \in X. \tag{4.37}$$

Then λ is an eigenvalue of $(d\mathcal{H}(u)^*\mathcal{H}(u)$, and

$$v := \mathcal{C}^\dagger(\xi). \tag{4.38}$$

That is, v defined above is a singular vector of $\mathcal{H}$.

Although the singular-value analysis problem, [13], is a nonlinear problem on an infinite dimensional signal space $U = L_2^m[0, \infty)$, the problem to be solved in the above theorem is a nonlinear algebraic equation on a finite dimensional space $X = \mathbb{R}^n$ which is also related to a nonlinear eigenvalue problem on X; see [9].

In the linear case, Equation 4.37 reduces to

$$\xi^{\mathrm{T}} Q = \lambda \, \xi^{\mathrm{T}} P^{-1}$$

where P and Q are the controllability and observability Gramians, and λ and ξ are an eigenvalue and an eigenvector of PQ. Furthermore, Equation 4.38 characterizes the relationship between a singular vector v of $\mathcal{H}$ and an eigenvector ξ of PQ. Also, in the linear case, there always exist n independent pairs of eigenvalues and eigenvectors of PQ. What happens in the nonlinear case? The answer is provided in the following theorem.

Theorem 4.18: [12]

Consider the system Σ in Equation 4.30 and the Hankel operator $\mathcal{H}$ in Equation 4.31 with $X = \mathbb{R}^n$. Suppose that the Jacobian linearization of the system has n distinct Hankel singular values. Then Equation 4.37 has n independent solution curves $\xi = \xi_i(s)$, $s \in \mathbb{R}$, $i = 1, 2, \ldots, n$ intersecting to each other at the origin and satisfying the condition

$$\|\xi_i(s)\| = |s|.$$

For linear systems, the solutions of Equation 4.37 are lines (orthogonally) intersecting at the origin. The above theorem shows that instead of these lines, in the nonlinear case n independent curves $x = \xi_i(s)$, $i = 1, 2, \ldots, n$ exist. For instance, if the dimension of the state is $n = 2$, the solution of Equation 4.37 is illustrated in Figure 4.2.

We can relate the solutions $\xi_i(s)$ to the singular values of the Hankel operator $\mathcal{H}$. Let $v_i(s)$ and $\sigma_i(s)$ denote the singular vector and the singular-value parameterized by s corresponding to $\xi_i(s)$. Then we have

$$v_i(s) := \mathcal{C}^\dagger(\xi_i(s)),$$

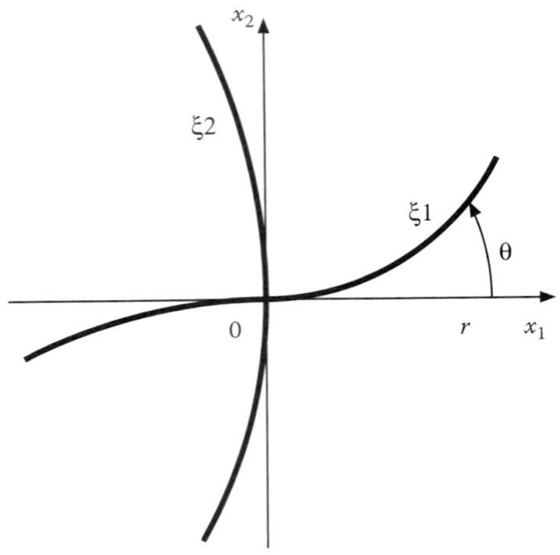

FIGURE 4.2 Configuration of $\xi_1(s)$ and $\xi_2(s)$ in the case $n = 2$.

$$\sigma_i(s) := \frac{\|\mathcal{H}(v_i(s))\|_{L_2}}{\|v_i(s)\|_{L_2}} = \frac{\|\mathcal{O}(\xi_i(s))\|_{L_2}}{\|\mathcal{C}^{\dagger}(\xi_i(s))\|_{L_2}} = \sqrt{\frac{L_o(\xi_i(s))}{L_c(\xi_i(s))}}.$$

This expression yields an explicit expression of the singular values $\sigma_i(s)$'s of the Hankel operator $\mathcal{H}$. These functions $\sigma_i(s)$'s are called *Hankel singular values*. Without loss of generality we assume that the following equation holds for $i = 1, 2, \ldots, n$ in a neighborhood of the origin

$$\min\{\sigma_i(s), \sigma_i(-s)\} > \max\{\sigma_{i+1}(s), \sigma_{i+1}(-s)\}. \tag{4.39}$$

As in the linear case, the solution curves $\xi_i(s)$ play the role of the coordinate axes of a balanced realization. By applying an isometric coordinate transformation which maps the solution curves $\xi_i(s)$'s into the coordinate axes, we obtain a realization whose (new) coordinate axes x_i are the solution of Equation 4.37, that is,

$$\left.\frac{\partial L_o(x)}{\partial x}\right|_{x=(0,\ldots,0,x_i,0,\ldots,0)} = \lambda \left.\frac{\partial L_c(x)}{\partial x}\right|_{x=(0,\ldots,0,x_i,0,\ldots,0)}, \tag{4.40}$$

$$\sigma_i(x_i) = \sqrt{\frac{L_o(0,\ldots,0,x_i,0,\ldots,0)}{L_c(0,\ldots,0,x_i,0,\ldots,0)}}. \tag{4.41}$$

Equation 4.41 implies that the new coordinate axes x_i, $i = 1, \ldots, n$ are the solutions of Equation 4.37 for Hankel singular-value analysis. Therefore, the Hankel norm can be obtained by

$$\begin{aligned}\|\Sigma\|_H &= \sup_{u\neq 0} \frac{\|\mathcal{H}(u)\|_{L_2}}{\|u\|_{L_2}} = \sup_{s\in\mathbb{R}} \max_i \sigma_i(s) \\ &= \sup_{x_1\in\mathbb{R}} \sqrt{\frac{L_o(x_1,0,\ldots,0)}{L_c(x_1,0,\ldots,0)}},\end{aligned}$$

provided the ordering condition (Equation 4.39) holds for all $s \in \mathbb{R}$. Furthermore, apply this coordinate transformation recursively to all lower dimensional subspaces such as $(x_1, x_2, \ldots, x_k, 0, \ldots, 0)$, then we obtain a state-space realization satisfying Equation 4.41 and

$$x_i = 0 \iff \frac{\partial L_o(x)}{\partial x_i} = 0 \iff \frac{\partial L_c(x)}{\partial x_i} = 0. \tag{4.42}$$

This property is crucial for balanced realization and model order reduction. Using tools from differential topology, for example, [31], we can prove that the obtained realization is diffeomorphic to the following *precise* input-normal/output-diagonal realization.

Theorem 4.19: [11,14]

Consider the system Σ in Equation 4.30. Suppose that the assumptions in Theorem 4.18 hold. Then there exists coordinates in a neighborhood of the origin such that the system is in input-normal/output-diagonal form satisfying

$$L_c(x) = \frac{1}{2}\sum_{i=1}^{n} x_i^2,$$

$$L_o(x) = \frac{1}{2}\sum_{i=1}^{n} x_i^2\, \sigma_i(x_i)^2.$$

This realization is more precise than the realization in Theorem 4.16 in the following sense: (a) The solutions of Equation 4.37 coincide with the coordinate axes, that is, Equation 4.40 holds. (b) The ratio of the observability function L_o to the controllability function L_c equals the singular values $\sigma_i(x_i)$'s on the coordinate axes, that is, Equation 4.41 holds. (c) An exact balanced realization can be obtained by a coordinate transformation

$$z_i = \phi_i(x_i) := x_i \sqrt{\sigma_i(x_i)}, \tag{4.43}$$

which is well-defined in a neighborhood of the origin.

Corollary 4.1: [11,14]

The coordinate change (Equation 4.43) transforms the input-normal realization in Theorem 4.19 into the following balanced form:

$$L_c(z) = \frac{1}{2}\sum_{i=1}^{n} \frac{z_i^2}{\sigma_i(z_i)},$$

$$L_o(z) = \frac{1}{2}\sum_{i=1}^{n} z_i^2 \sigma_i(z_i).$$

Since we only use the coordinate transformation (Equation 4.43) preserving the coordinate axes, the realization obtained here also satisfies the properties (a) and (b) explained above. The controllability and observability functions can be written as

$$L_c(z) = \frac{1}{2} z^{\mathrm{T}} \underbrace{\mathrm{diag}(\sigma_1(z_1), \ldots, \sigma_n(z_n))^{-1}}_{P(z)} z,$$

$$L_o(z) = \frac{1}{2}\, z^{\mathrm{T}} \underbrace{\mathrm{diag}(\sigma_1(z_1), \ldots, \sigma_n(z_n))}_{Q(z)}\, z.$$

Here $P(z)$ and $Q(z)$ can be regarded as nonlinear counterparts of the balanced controllability and observability Gramians, since

$$P(z) = Q(z) = \text{diag}(\sigma_1(z_1), \sigma_2(z_2), \ldots, \sigma_n(z_n)).$$

The axes of this realization are uniquely determined. We call this state-space realization a *balanced realization* of the nonlinear system Σ in Equation 4.30. As in the linear case, both the relationship between the input-to-state and state-to-output behavior and that among the coordinate axes are balanced.

4.3.3 Model Order Reduction

An important application of balanced realizations is that it is a tool for model order reduction called *balanced truncation*. Here, a model order reduction method preserving the Hankel norm of the original system is proposed. Suppose that the system of Equation 4.30 is balanced in the sense that it satisfies Equations 4.41 and 4.42. Note that the realizations in Theorem 4.19 and Corollary 4.1 satisfy these conditions. Suppose, that

$$\min\{\sigma_k(s), \sigma_k(-s)\} \gg \max\{\sigma_{k+1}(s), \sigma_{k+1}(-s)\}$$

holds with a certain k $(1 \leq k < n)$. Divide the state into two vectors $x = (x^a, x^b)$

$$x^a := (x_1, \ldots, x_k) \in \mathbb{R}^k,$$
$$x^b := (x_{k+1}, \ldots, x_n) \in \mathbb{R}^{n-k},$$

and the vector field into two vector fields accordingly

$$f(x) = \begin{pmatrix} f^a(x) \\ f^b(x) \end{pmatrix},$$
$$g(x) = \begin{pmatrix} g^a(x) \\ g^b(x) \end{pmatrix},$$

and truncate the state by substituting $x^b = 0$. Then we obtain a k-dimensional state-space model Σ^a with the state x^a (with a $(n-k)$-dimensional residual model Σ^b with the state x^b).

$$\Sigma^a : \begin{cases} \dot{x}^a = f^a(x^a, 0) + g^a(x^a, 0)u^a \\ y^a = h(x^a, 0) \end{cases} \tag{4.44}$$

$$\Sigma^b : \begin{cases} \dot{x}^b = f^b(0, x^b) + g^b(0, x^b)u^b \\ y^b = h(0, x^b) \end{cases} \tag{4.45}$$

This procedure is called *balanced truncation*. The obtained reduced order models have preserved the following properties.

Theorem 4.20: [13,14]

Suppose that the system Σ satisfies Equations 4.41 and 4.42 and apply the balanced truncation procedure explained above. Then the controllability and observability functions of the reduced order models Σ^a and

Σ^b denoted by L_c^a, L_c^b, L_o^a, and L_o^b, respectively, satisfy the following equations:

$$L_c^a(x^a) = L_c(x^a, 0), \quad L_o^a(x^a) = L_o(x^a, 0),$$
$$L_c^b(x^b) = L_c(0, x^b), \quad L_o^b(x^b) = L_o(0, x^b),$$

which implies

$$\sigma_i^a(x_i^a) = \sigma_i(x_i^a), \quad i = 1, 2, \ldots, k,$$
$$\sigma_i^b(x_i^b) = \sigma_{i+k}(x_i^b), \quad i = 1, 2, \ldots, n-k,$$

with the singular values σ^a's of the system Σ^a and the singular values σ^b of the system Σ^b. In particular, if σ_1 is defined globally, then

$$\|\Sigma^a\|_H = \|\Sigma\|_H.$$

Theorem 4.20 states that the important characteristics of the original system such as represented by the controllability and observability functions and Hankel singular values are preserved. Moreover, by Theorem 4.15, this implies that the controllability, observability, minimality, and the gain property is preserved under the model reduction. These preservation properties hold for truncation of any realization satisfying the conditions (Equations 4.41 and 4.42), such as the realizations in Theorem 4.19 and Corollary 4.1 [13,14]. Furthermore, concerning the stability, (global) Lyapunov stability and local asymptotic stability are preserved with this procedure as well. Note that this theorem is a natural nonlinear counterpart of the linear theory. However, a nonlinear counterpart of the error bound of the reduced order model has not been found yet.

4.3.4 Other Types of Balancing for Nonlinear Systems

As for linear systems, there exist extensions of LQG, and coprime balancing [49], H_∞ or L_2-gain balancing [46], and positive/bounded real and dissipativity-based balancing [23]. In fact, in [23] a direct relation is obtained with Hankel operator analysis for augmented systems.

The presented work treats balanced realizations for nonlinear systems based on balancing in a (possibly large) region around an equilibrium point, where a relation with the Hankel operator, observability and controllability operators and functions, and minimality of the nonlinear system is obtained. A drawback of these methods is the computational effort that is needed to compute the balanced realization. As mentioned in the introduction, other extensions of the linear notion of balancing can be found in for example, [20,28,58,59].

4.4 Concluding Remarks

In this Chapter, balanced realizations for linear and nonlinear systems and model reduction based on these realizations are treated. There exists a vast amount of literature on this topic, and the reference list in this paper is certainly not complete. For example, some of the basics for linear balanced realizations can be found in [40], and we did not pay any attention to the balancing methods treating uncertain, and time- and parameter-varying linear systems; for example, [4,44], behavioral balancing, for example, [53], or the numerical side of balancing, for example, [5].

Recently, a lot of interest is taken in structure preserving order techniques for both linear and nonlinear systems, where the additional structure to be preserved is a physical structure, such as port-Hamiltonian structure, [56], and other physical structures, as mentioned in our linear systems section. For example, for circuit simulators with continuously growing orders of the models, a need for interpreting a reduced order model as a circuit is important, such as, [50]. Due to the explicit interconnection (input/output like) structure of the circuits, order reduction methods based on balancing are attractive to apply to these circuits. However, structure preservation and circuit interpretation of the corresponding reduced order models is not possible yet, and is one of the motivators for further research to both linear and nonlinear structure preserving order reduction methods.

References

1. A.C. Antoulas, *Approximation of Large-Scale Dynamical Systems*, SIAM, Philadelphia, 2005.
2. A. Astolfi, Model reduction by moment matching, *Proc. 7th IFAC NOLCOS*, Pretoria, South Africa, 95–102, August 2007.
3. J. Batt, Nonlinear compact mappings and their adjoints, *Math. Ann.*, 189, 5–25, 1970.
4. C. L. Beck, J. Doyle, and K. Glover, Model reduction of multi-dimensional and uncertain systems, *IEEE Trans. Automat. Control*, AC-41, 10, 1466–1477, 1996.
5. P. Benner, V. Mehrmann, and D.C. Sorensen, Eds., *Dimension Reduction of i Large-Scale Systems*, Springer-Verlag, Berlin, 2005.
6. R. Courant and D. Hilbert, *Methods of Mathematical Physics*, Vol. 1, Interscience Publishers, New York, 1953.
7. U. B. Desai and D. Pal, A transformation approach to stochastic model reduction, *IEEE Trans. Automat. Control*, 29, 1097–1100, 1984.
8. K.V. Fernando and H. Nicholson, Singular pertubational model reduction of balanced systems, *IEEE Trans. Automat. Control*, AC-27, 466–468, 1982.
9. K. Fujimoto, What are singular values of nonlinear operators?, *Proc. 43rd IEEE Conf. on Decision and Control*, The Bahamas, 1623–1628, 2004.
10. K. Fujimoto, J. M. A. Scherpen, and W. S. Gray, Hamiltonian realizations of nonlinear adjoint operators, *Automatica*, 38, 10, 1769–1775, 2002.
11. K. Fujimoto and J.M.A. Scherpen, Nonlinear balanced realization based on singular value analysis of Hankel operators, *Proc. 42nd IEEE Conf. on Decision and Control*, Maui, HI, 6072–6077, 2003.
12. K. Fujimoto and J.M.A. Scherpen, Nonlinear input-normal realizations based on the differential eigen-structure of Hankel operators, *IEEE Trans. Automat. Control*, AC-50, 1, 2–18, 2005.
13. K. Fujimoto and J.M.A. Scherpen, Singular value analysis and balanced realizations for nonlinear systems, in: *Model Order Reduction: Theory, Research Aspects and Applications*, Eds. W. Schilders, H. van der Vorst, and J. Rommes, Springer-Verlag, Berlin, 251–272, 2008.
14. K. Fujimoto and J.M.A. Scherpen, Model reduction for nonlinear systems based on the balanced realization, *SIAM J. Control Optim.*, 48 (7), 4591–4623, 2010.
15. F.R. Gantmacher, *The Theory of Matrices*, Chelsea, New York, 1960.
16. W. Gawronski, *Balanced Control of Flexible Structures, Lecture Notes in Contr. Inf. Sc.* 211, Springer-Verlag, Berlin, 1996.
17. K. Glover, All optimal Hankel-norm approximations of linear multivariable systems and their L^{∞}-error bounds, *Int. J. Control*, 39, 1115–1193, 1984.
18. W.S. Gray and J.M.A. Scherpen, State dependent matrices in quadratic forms, *Systems Control Lett.*, 44, 3, 219–232, 2001.
19. S. Gugercin and A. C. Antoulas, A survey of model reduction by balanced truncation and some new results, *Int. J. Control*, 77 (8), 748–766, 2004.
20. J. Hahn and T.F. Edgar, An improved method for nonlinear model reduction using balancing of empirical Gramians, *Comp. Chem. Eng.*, 26, 10, 1379–1397, 2002.
21. P. Heuberger, A family of reduced order models, based on open-loop balancing, *Ident. Model. Contr.*, Delft University Press, 1, 1–10, 1990.
22. J. W. Hoffmann, Normalized coprime factorizations in continuous and discrete time—a joint state-space approach, *IMA J. Math. Contr. Info.* 13, 359–384, 1996.
23. T.C. Ionescu, K. Fujimoto and J.M.A. Scherpen, Dissipativity preserving balancing for nonlinear systems—a Hankel operator approach, *Systems and Control Letters*, 59, 180–194, 2010.
24. A. Isidori, *Nonlinear Control Systems* (3rd edition), Springer-Verlag, Berlin, 1995.
25. E.A. Jonckheere and L.M. Silverman, Singular value analysis of deformable systems, *J. Circ., Syst. Sign. Proc.*, 1, 447–470, 1982.
26. E.A. Jonckheere and L.M. Silverman, A new set of invariants for linear systems—Applications to reduced order compensator design, *IEEE Trans. Automat. Control*, AC-28, 953–964, 1983.
27. T. Kailath, *Linear Systems*, Prentice-Hall, Englewood Cliffs, NJ, 1980.

28. S. Lall, J.E. Marsden, and S. Glavaski, A subspace approach to balanced truncation for model reduction of nonlinear control systems, *Int. J. Robust Nonlinear Control*, 12, 6, 519–535, 2002.
29. D.G. Meyer, Fractional balanced reduction—model reduction via a fractional representation, *IEEE Trans. Automat. Control*, AC-26, 1341–1345, 1990.
30. D.G. Meyer and S. Srinivasan, Balancing and model reduction for second order form linear systems, *IEEE Trans. Automat. Control*, AC-41, 11, 1632–1644, 1996.
31. J. W. Milnor, *Topology from the Differential Viewpoint*, Princeton University Press, Princeton, NJ, 1965.
32. B.C. Moore, Principal component analysis in linear systems: Controllability, observability and model reduction, *IEEE Trans. Automat. Control*, AC-26, 17–32, 1981.
33. D. Mustafa, $\mathcal{H}_\infty$-characteristic values, *Proc. 28th IEEE Conf. on Decision and Control*, Tampa, FL, 1483–1487, 1989.
34. D. Mustafa and K. Glover, *Minimum Entropy $\mathcal{H}_\infty$ Control, Lect. Notes Contr. Inf. Sci.* No. 146, Springer-Verlag, Berlin, 1990.
35. D. Mustafa and K. Glover, Controller reduction by $\mathcal{H}_\infty$-balanced truncation, *IEEE Trans. Automat. Control*, AC-36, 668–682, 1991.
36. C.N. Nett, C.A. Jacobson, and M.J. Balas, A connection between state-space and doubly coprime fractional representations, *IEEE Trans. Autom. Control*, AC-29, 831–832, 1984.
37. A.J. Newman and P.S. Krishnaprasad, Computation for nonlinear balancing, *Proc. 37th IEEE Conf. on Decision and Control*, Tampa, FL, 4103–4104, 1998.
38. A. Newman and P. S. Krishnaprasad, Computing balanced realizations for nonlinear systems, *Proc. Symp. Mathematical Theory of Networks and Systems*, 2000.
39. R. Ober and D. McFarlane, Balanced canonical forms for minimal systems: A normalized coprime factor approach, *Linear Algebra and its Applications*, 122–124, 23–64, 1989.
40. G. Obinata and B.D.O. Anderson, *Model Reduction for Control System Design*, Springer-Verlag, London, 2001.
41. P.C. Opdenacker and E.A. Jonckheere, LQG balancing and reduced LQG compensation of symmetric passive systems, *Int. J. Control*, 41, 73–109, 1985.
42. P. C. Opdenacker and E. A. Jonckheere, A contraction mapping preserving balanced reduction scheme and its infinity norm error bounds, *IEEE Trans. Circuits Systems* 35 (2), 184–189, 1988.
43. L. Pernebo and L.M. Silverman, Model reduction via balanced state space representations, *IEEE Trans. Automat. Control*, AC-27, 382–387, 1982.
44. H. Sandberg and A. Rantzer, Balanced truncation of linear time-varying systems, *IEEE Trans. Automat. Control*, 49, 2, 217—229, 2004.
45. J.M.A. Scherpen, Balancing for nonlinear systems, *Systems Control Lett.*, 21, 143–153, 1993.
46. J.M.A. Scherpen, $\mathcal{H}_\infty$ balancing for nonlinear systems, *Int. J. Robust Nonlinear Control*, 6, 645–668, 1996.
47. J.M.A. Scherpen and W.S. Gray, Minimality and local state decompositions of a nonlinear state space realization using energy functions, *IEEE Trans. Automat. Control*, AC-45, 11, 2079–2086, 2000.
48. J.M.A. Scherpen and W.S. Gray, Nonlinear Hilbert adjoints: Properties and applications to Hankel singular value analysis, *Nonlinear Anal., Theory, Methods Appl.*, 51, 5, 883–901, 2002.
49. J.M.A. Scherpen and A.J. van der Schaft, Normalized coprime factorizations and balancing for unstable nonlinear systems, *Int. J. Control*, 60, 6, 1193–1222, 1994.
50. W.H.A. Schilder, H.A. van der Vorst, and J. Rommes, Eds., *Model Order Reduction: Theory, Research Aspects and Applications*, Mathematics in Industry, Vol. 13, Springer, Berlin, 2008.
51. G.W. Stewart, On the early history of the singular value decomposition, *SIAM Rev.*, 35, 4, 551–566, 1993.
52. H.L. Trentelman, Bounded real and positivey real balanced truncation using Σ normalized coprime factor, *System Control Lett.*, 58, 12, 871–879, 2009.
53. H.L. Trentelman and P. Rapisarda, A behavioral approach to passivity and bounded realness preserving balanced truncation with error bounds, *Proc. 47th IEEE Conf. on Decision and Control*, Shanghai, China, 2009.
54. A.J. van der Schaft, Balancing of lossless and passive systems, *IEEE Trans. Automat. Control*, AC-53, 2153–2157, 2008.
55. A.J. van der Schaft and J.E. Oeloff, Model reduction of linear conservative mechanical systems, *IEEE Trans. Automat. Control*, AC-35, 729–733, 1990.
56. A.J. van der Schaft and R. Polyuga, Structure-preserving model reduction of complex physical systems, *Proc. 48th IEEE Conf. on Decision and Control*, Shanghai, China, December 16–18, 2009.
57. E. Verriest, Low sensitivity design and optimal order reduction for the LQG-problem, *Proc. 24th Midwest Symp. Circ. Syst. Albuquerque*, NM, 1981, 365–369.
58. E.I. Verriest and W.S. Gray, Flow balancing nonlinear systems, *Proc. 2000 Int. Symp. Math. Th. Netw. Syst. (MTNS)*, 2000.

59. E.I. Verriest and W.S. Gray, Balanced nonlinear realizations, *Proc. 43rd IEEE Conf. on Decision and Control*, The Bahamas, 1164–1169, 2004.
60. M. Vidyasagar, *Control System Synthesis: A Factorization Approach*, MIT Press, Cambridge, 1985.
61. S. Weiland, *Theory of Approximation and Disturbance Attenuation for Linear Systems*, Doctoral dissertation, University of Groningen, 1991.
62. P. Wortelboer, *Frequency-Weighted Balanced Reduction of Closed Loop Mechanical Servo Systems: Theory and Tools*, doctoral dissertation, Delft University of Technology, 1994.
63. K. Zhou, J. C. Doyle, and K. Glover, *Robust and Optimal Control*, Prentice-Hall, Inc., Upper Saddle River, NJ, 1996.
64. A. J. Krener, Reduced order modeling of nonlinear control systems, in *Analysis and Design of Nonlinear Control Systems*, A. Astolfi and L. Marconi, eds., Springer-Verlag, 41–62, 2008.

5

Geometric Theory of Linear Systems*

Fumio Hamano
California State University

5.1 Introduction

In the late 1960s, Basile and Marro (1969) (and later Wonham and Morse, 1970) discovered that the behavior of time-invariant linear control systems can be seen as a manifestation of the subspaces similar to the invariant subspaces characterized by the system matrices. As a result, the system behavior can be predicted and the solvability of many control problems can be tested by examining the properties of such subspaces. In many instances, one can understand essential issues intuitively in geometric terms. Moreover, thanks to good algorithms and software available in the literature (see Basile and Marro, 1992), the above subspaces can be generated and the properties can be readily examined by using personal computers. Thus, a large class of problems involving feedback control laws and observability of linear systems can be solved effectively by this geometric method, for example, problems of disturbance localization, decoupling, unknown input observability and system inversion, observer design, regulation

* This chapter is dedicated to Professors G. Basile and G. Marro for their pioneering contributions in the development of geometric methods for linear systems.

and tracking, model following, robust control, and so on. Comprehensive treatments of the basic theory and many applications, including the ones mentioned above, can be found in the excellent books by Wonham (1985), Basile and Marro (1992), and Trentelman et al. (2002), with additional later results found in the newer books, for example, controlled and conditioned invariant subspaces and duality are treated in an organized fashion in the 2nd book and, in addition to the duality, an extension of the theory to include distributions as inputs and the relation between disturbance localization and H_2 optimal control problem are also given in the 3rd book. The method is also useful in the analysis and design of decentralized control systems (Hamano and Furuta, 1975), perfect and near-perfect signal decoupling (and its applications to other control problems) for nonminimum-phase systems (Marro and Zattoni, 2006), and failure/fault detection (Massoumnia et al., 1989). This chapter serves as an introduction to the subject. Regrettably, citations in this chapter are rather limited due to editorial restrictions. Extensive references can be found in the above-mentioned books as well as in the chapter by Marro (2008). To prepare the present chapter, the book by Basile and Marro (1992) has been used as the primary reference, and the majority of the proofs omitted in this chapter can be found in this reference.

Section 5.2 gives a review of elementary notions including invariant subspaces, reachability, controllability, observability, and detectability. It also provides convenient formulae for subspace calculations. Sections 5.3 through 5.7 describe the basic ingredients of the geometric theory (or approach). More specifically, Section 5.3 introduces the fundamental notions of (A, im B)-controlled and (A, ker C)-conditioned invariants (which are subspaces of the state space), and Section 5.4 provides some algebraic properties of these invariants. In Section 5.5, "largest" controlled and "smallest" conditioned invariants are presented with respect to certain subspaces and Section 5.6 discusses well-structured special classes of controlled and conditioned invariants. Section 5.7 analyzes the above invariants in relation to stabilization. Sections 5.8 through 5.10 describe applications to demonstrate the use of the basic tools developed in the previous sections. For this goal, the disturbance localization problem is chosen and it is discussed in three different situations with varying degrees of sophistication. The disturbance localization problems are chosen since the methods used to solve the problems can be used or extended to solve other more involved problems. It also has historical significance as one of the first problems for which the geometric method was used. Section 5.11 provides the conclusion and brief statements about practical applications and about extensions of geometric notions and approach to more general or different systems.

Notation: Capital letters A, B, and so on denote the matrices (or linear maps) with I and I_n reserved, respectively, for an identity matrix (of appropriate dimension) and an $n \times n$ identity matrix. The transpose of a matrix A is denoted by A'. Capital script letters such as $\mathcal{V}, \mathcal{W}$ represent vector spaces or subspaces. Small letters x, y, and so on are column vectors. Scalars are also denoted by small letters. The letter 0 is used for a zero matrix, vector, or scalar depending on the context. Notation ":=" means "(the left-hand side, i.e., ':' side) is defined by (the right-hand side, i.e., '=' side)." Similarly for "=:" where the roles of the left- and right-hand sides are reversed. The image (or range) and the kernel (or null space) of M are denoted by im M and ker M, respectively. The expression $\mathcal{V} + \mathcal{W}$ represents the sum of two subspaces $\mathcal{V}$ and $\mathcal{W}$, that is, $\mathcal{V} + \mathcal{W} := \{v + w : v \in \mathcal{V} \text{ and } w \in \mathcal{W}\}$. If $\mathcal{V}$ is a subspace of $\mathcal{W}$, we write $\mathcal{V} \subset \mathcal{W}$. If $\mathcal{V} \subset \mathcal{X}$, we use $A^{-1}\mathcal{V} := \{x \in \mathcal{X} : Ax \in \mathcal{V}\}$, that is, the set of all $x \in \mathcal{X}$ satisfying $Ax \in \mathcal{V}$. Similarly, $A^{-k}\mathcal{V} := \left\{x \in \mathcal{X} : A^k x \in \mathcal{V}\right\}, k = 1, 2, \ldots$.

5.2 Review of Elementary Notions

In this section we will review invariant subspaces and some of their basic roles in the context of the linear systems.

Definition 5.1:

A subspace $\mathcal{V}$ of $\mathcal{X} := R^n$ is said to be A-invariant if

$$A\mathcal{V} \subset \mathcal{V}, \tag{5.1}$$

that is, $x \in \mathcal{V} \Rightarrow Ax \in \mathcal{V}$.

An A-invariant subspace plays the following obvious but important role for an autonomous linear system. Consider the autonomous linear system described by

$$\dot{x}(t) = Ax(t), \quad x(0) = x_o, \tag{5.2}$$

where the column vectors $x(t) \in \mathcal{X} := R^n$ and $x_o \in \mathcal{X}$ are, respectively, the state of the system at time $t \geq 0$ and the initial state, and A is an $n \times n$ real matrix. Now, suppose a subspace $\mathcal{V}$ is A-invariant. Clearly, if $x(t) \in \mathcal{V}$, then the rate of change $\dot{x}(t) \in \mathcal{V}$, which implies that the state remains in $\mathcal{V}$. More strongly, we have

---

Lemma 5.1:

Let $\mathcal{V} \subset \mathcal{X}$. For the autonomous linear system described by Equation 5.2, $x_o \in \mathcal{V}$ implies $x(t) \in \mathcal{V}$ for all $t \geq 0$ if and only if $\mathcal{V}$ is A-invariant.

Let $\mathcal{V}$ be A-invariant, and introduce a new basis $\{e_1, \ldots, e_n\}$, such that

$$span\{e_1, \ldots, e_\nu\} = \mathcal{V}, \quad \nu \leq n. \tag{5.3}$$

Define a coordinate transformation by

$$x = [e_1 \ldots e_n] \begin{bmatrix} \tilde{x}_1 \\ \tilde{x}_2 \end{bmatrix}, \quad x \in R^n, \quad \tilde{x}_1 \in R^\nu, \quad \tilde{x}_2 \in R^{n-\nu}. \tag{5.4}$$

Then, it is easy to see that, with respect to the new basis, the state Equation 5.2 can be rewritten as

$$\begin{bmatrix} \dot{\tilde{x}}_1(t) \\ \dot{\tilde{x}}_2(t) \end{bmatrix} = \begin{bmatrix} \tilde{A}_{11} & \tilde{A}_{12} \\ 0 & \tilde{A}_{22} \end{bmatrix} \begin{bmatrix} \tilde{x}_1(t) \\ \tilde{x}_2(t) \end{bmatrix}, \tag{5.5}$$

$$\tilde{x}_1(0) = \tilde{x}_{10}, \quad \tilde{x}_2(0) = \tilde{x}_{20}. \tag{5.6}$$

Clearly, if $\tilde{x}_{20} = 0$, then $\tilde{x}_2(t) = 0$ for all $t \geq 0$, that is, $x_o \in \mathcal{V}$ implies $x(t) \in \mathcal{V}$ for all $t \geq 0$ (which has been stated in Lemma 5.1).

Let $\mathcal{V}$ be an A-invariant subspace in $\mathcal{X}$. The *restriction* $A|\,\mathcal{V}$ of a linear map $A : \mathcal{X} \to \mathcal{X}$ (or an $n \times n$ real matrix A) to a subspace $\mathcal{V}$ is a linear map from $\mathcal{V}$ to $\mathcal{V}$, mapping $v \mapsto Av$ for all $v \in \mathcal{V}$. For $x \in \mathcal{X}$, we write $x + \mathcal{V} := \{x + v : v \in \mathcal{V}\}$ called the *coset* of x modulo $\mathcal{V}$, which represents a hyperplane passing through a point x. The set of cosets modulo $\mathcal{V}$ is a vector space called the *factor space* (or *quotient space*) and it is denoted by $\mathcal{X}/\mathcal{V}$. An *induced map* $A|\mathcal{X}/\mathcal{V}$ is a linear map defined by $x + \mathcal{V} \mapsto Ax + \mathcal{V}$, $x \in \mathcal{X}$.

An A-invariant subspace $\mathcal{V}$ is said to be *internally stable* if $\tilde{A}_{11}$ in Equation 5.5 is stable, that is, all the eigenvalues have negative real parts, or equivalently, if $A|\,\mathcal{V}$ is stable. Therefore, $x(t)$ converges to the zero state as $t \to \infty$ whenever $x_o \in \mathcal{V}$ if and only if $\mathcal{V}$ is internally stable. Also, an A-invariant subspace $\mathcal{V}$ is said to be *externally stable* if $\tilde{A}_{22}$ is stable, that is, if $A|\,\mathcal{X}/\mathcal{V}$ is stable. Clearly, $\tilde{x}_2(t)$ converges to zero as

$t \to \infty$, that is, $x(t)$ converges to $\mathcal{V}$ as $t \to \infty$ if and only if $\mathcal{V}$ is externally stable. Note that the eigenvalues of $\tilde{A}_{11}$and $\tilde{A}_{22}$ do not depend on a particular choice of coordinates (as long as Equation 5.3 is satisfied).

Let us now consider a continuous time, time-invariant linear control system $\Sigma := [A,B,C]$ described by

$$\dot{x}(t) = Ax(t) + Bu(t), \quad x(0) = x_o, \tag{5.7}$$

$$y(t) = Cx(t), \tag{5.8}$$

where the column vectors $x(t) \in \mathcal{X} := R^n, u(t) \in R^m$, and $t(t) \in R^p$ are, respectively, the state, input, and output of the system at time $t \geq 0, x_o \in R^n$ is the initial state, and A, B, and C are real matrices with consistent dimensions. We assume that $u(t)$ is piecewise continuous. We are also interested in the closed-loop system, namely, we apply a linear state-feedback law of the form

$$u(t) = Fx(t), \tag{5.9}$$

then Equation 5.7 becomes

$$\dot{x}(t) = (A + BF)x(t), \quad x(0) = x_o, \tag{5.10}$$

where F is an $m \times n$ real matrix.

Some invariant subspaces are associated with reachability (controllability) and observability. A state $\tilde{x}$ is said to be *reachable* (or *controllable*) if there is a control which drives the zero state to $\tilde{x}$ (or, respectively, $\tilde{x}$ to the zero state) in finite time, that is, if there is $u(t), 0 \leq t \leq t_f$ such that $x(0) = 0$ (or, respectively, $\tilde{x}$) and $x(t_f) = \tilde{x}$ (or, respectively 0) for some $0 < t_f < \infty$ (see Kalman et al., 1969, Chapter 2). The set of reachable (or controllable) states forms a subspace and it is called the *reachable* (or, respectively, *controllable*) *subspace*, which will be denoted as $\mathcal{V}_{reach}$ (or, respectively, $\mathcal{V}_{contr}$). For an $n \times n$ matrix M and a subspace $\mathcal{I} \subset \mathcal{X}$, define

$$\mathcal{R}(M, \mathcal{I}) := \mathcal{I} + M\mathcal{I} + \cdots + M^{n-1}\mathcal{I}.$$

The reachable and controllable subspaces are characterized by the following.

Theorem 5.1:

For the continuous-time system $\Sigma := [A, B, C]$,

$$\begin{aligned}\mathcal{V}_{reach} &= \mathcal{R}(A, \operatorname{im} B) = \operatorname{im}\left[B\; AB\; \ldots\; A^{n-1}B\right], \\ &= \mathcal{R}(A + BF, \operatorname{im} B) = \operatorname{im}\left[B\; (A+BF)B\; \ldots\; (A+BF)^{n-1}B\right]\end{aligned} \tag{5.11}$$

for any $m \times n$ real matrix F. Furthermore,

$$\mathcal{V}_{reach} = \mathcal{V}_{contr}. \tag{5.12}$$

Remark

The subspace $\mathcal{V}_{reach} = \mathcal{R}(A, \operatorname{im} B)$ is the minimum A-invariant subspace containing $\operatorname{im} B$. It is also $(A + BF)$-invariant for any $m \times n$ real matrix F. (For a discrete-time system, Equation 5.12 does not hold; instead we have $\mathcal{V}_{reach} \subset \mathcal{V}_{contr}$.)

The pair (A, B) or system $\Sigma := [A, B, C]$ is said to be *reachable* (or *controllable*) if $\mathcal{V}_{reach} = \mathcal{X}$ (or, respectively, $\mathcal{V}_{contr} = \mathcal{X}$). The set $\Lambda := \{\lambda_1, \ldots, \lambda_n\}$ of complex numbers is called a *symmetric* set if, whenever λ_i is not a real number, $\lambda_j = \lambda_i^*$ for some $j = 1, \ldots, n$ where λ_i^* is the complex conjugate of λ_i. Denote by $\sigma(A + BF)$ the spectrum (or the eigenvalues) of $A + BF$. Then, we have

Theorem 5.2:

For any symmetric set $\Lambda := \{\lambda_1, \ldots, \lambda_n\}$ *of complex numbers* $\lambda_1, \ldots, \lambda_n$ *there is an* $m \times n$ *real matrix* F*, such that* $\sigma(A + BF) = \Lambda$ *if and only if the pair* (A, B) *is reachable (or controllable).*

Proof. See Wonham (1985); Basile and Marro (1992); and Heymann (1968).

Remark

Let $\dim \mathcal{V}_{reach} = r \leq n$. For any symmetric set $\Lambda_r := \{\lambda_1, \ldots, \lambda_r\}$ of complex numbers $\lambda_1, \ldots, \lambda_r$, there is an $m \times n$ real matrix F, such that $\sigma(A + BF|\,\mathcal{V}_{reach}) = \Lambda_r$. In fact, since $\mathcal{V}_{reach}$ is A-invariant, using the same state coordinate transformation given by Equations 5.3 and 5.4 with $\mathcal{V} = \mathcal{V}_{reach}$ and $\nu = r$, we obtain, from Equation 5.7,

$$\begin{bmatrix} \dot{\tilde{x}}_1(t) \\ \dot{\tilde{x}}_2(t) \end{bmatrix} = \begin{bmatrix} \tilde{A}_{11} & \tilde{A}_{12} \\ 0 & \tilde{A}_{22} \end{bmatrix} \begin{bmatrix} \tilde{x}_1(t) \\ \tilde{x}_2(t) \end{bmatrix} + \begin{bmatrix} \tilde{B}_1 \\ 0 \end{bmatrix} u(t).$$

It is easy to show $\dim \operatorname{im} \begin{bmatrix} \tilde{B}_1 \; \tilde{A}_{11}\tilde{B}_1 \; \ldots \; \tilde{A}_{11}^{r-1}\tilde{B}_1 \end{bmatrix} = r$, that is, the pair $(\tilde{A}_{11}, \tilde{B}_1)$ of the r-dimensional subsystem is reachable. The eigenvalue assignment for $\mathcal{V}_{reach}$ follows by applying Theorem 5.2.

The pair (A, B) is said to be *stabilizable* if there is a real matrix F, such that the eigenvalues of $A + BF$ have negative real parts. We have (see Basile and Marro, 1992).

Corollary 5.1:

Pair (A, B) *is stabilizable if and only if* $\mathcal{V}_{reach}$ *is externally stable.*

The state $\bar{x}$ of the system Σ is said to be *unobservable* if it produces zero output when the input is not applied, that is, if $x(0) = \bar{x}$ and $u(t) = 0$ for all $t \geq 0$ implies $y(t) = 0$ for all $t \geq 0$. The set of unobservable states forms a subspace which is called the *unobservable subspace*. This will be denoted by $\mathcal{S}_{unobs}$.

Theorem 5.3:

$$\begin{aligned} \mathcal{S}_{unobs} &= \ker \begin{bmatrix} C \\ CA \\ \vdots \\ CA^{n-1} \end{bmatrix} = \ker \begin{bmatrix} C \\ C(A+GC) \\ \vdots \\ C(A+GC)^{n-1} \end{bmatrix} \\ &= \ker C \cap A^{-1} \ker C \cap \cdots \cap A^{-(n-1)} \ker C \\ &= \ker C \cap (A+GC)^{-1} \ker C \cap \cdots \cap (A+GC)^{-(n-1)} \ker C \end{aligned} \tag{5.13}$$

for any $n \times p$ *real matrix* G.

Remark

The subspace $\mathcal{S}_{unobs}$ is A-invariant. It is also $(A + GC)$-invariant for any $n \times p$ real matrix G.

The pair (A, C) or system Σ is said to be *observable* if $\mathcal{S}_{unobs} = \{0\}$, and the pair (A, C) is said to be *detectable* if $A + GC$ is stable for some real matrix G. For observability and detectability we have the following facts.

Theorem 5.4:

For any symmetric set $\Lambda := \{\lambda_1, \ldots, \lambda_n\}$ of complex numbers $\lambda_1, \ldots, \lambda_n$, there is an $n \times p$ real matrix G such that $\sigma(A + GC) = \Lambda$ if and only if the pair (A, C) is observable.

Corollary 5.2:

The pair (A, C) is detectable if and only if $\mathcal{S}_{unobs}$ is internally stable.

The following formulae are useful for subspace calculations.

Lemma 5.2:

Let $\mathcal{V}$, $\mathcal{V}_1$, $\mathcal{V}_2$, $\mathcal{V}_3 \subset \mathcal{X}$. Then, letting $\mathcal{V}^\perp := \{x \in \mathcal{X} : x'v = 0 \text{ for all } v \in \mathcal{V}\}$,

$$
\begin{aligned}
(\mathcal{V}^\perp)^\perp &= \mathcal{V}, \\
(\mathcal{V}_1 + \mathcal{V}_2)^\perp &= \mathcal{V}_1^\perp \cap \mathcal{V}_2^\perp, \\
(\mathcal{V}_1 \cap \mathcal{V}_2)^\perp &= \mathcal{V}_1^\perp + \mathcal{V}_2^\perp, \\
A(\mathcal{V}_1 + \mathcal{V}_2) &= A\mathcal{V}_1 + A\mathcal{V}_2, \\
A(\mathcal{V}_1 \cap \mathcal{V}_2) &\subset A\mathcal{V}_1 \cap A\mathcal{V}_2,
\end{aligned}
$$

$(A_1 + A_2)\mathcal{V} = A_1\mathcal{V} + A_2\mathcal{V}$, where A_1 and A_2 are $n \times n$ matrices,

$$
\begin{aligned}
(A\mathcal{V})^\perp &= A'^{-1}\mathcal{V}^\perp, \\
\mathcal{V}_1 + (\mathcal{V}_2 \cap \mathcal{V}_3) &\subset (\mathcal{V}_1 + \mathcal{V}_2) \cap (\mathcal{V}_1 + \mathcal{V}_3), \\
\mathcal{V}_1 \cap (\mathcal{V}_2 + \mathcal{V}_3) &\supset (\mathcal{V}_1 \cap \mathcal{V}_2) + (\mathcal{V}_1 \cap \mathcal{V}_3),
\end{aligned}
$$

$\mathcal{V}_1 \cap (\mathcal{V}_2 + \mathcal{V}_3) = \mathcal{V}_2 + (\mathcal{V}_1 \cap \mathcal{V}_3)$, provided $\mathcal{V}_1 \supset \mathcal{V}_2$.

5.3 $(A, \operatorname{im} B)$-Controlled and $(A, \ker C)$-Conditioned Invariant Subspaces and Duality

In this section, we introduce important subspaces associated with system $\Sigma := [A, B, C]$ described by Equation 5.7 (or Equation 5.10) and Equation 5.8 with a state feedback law (Equation 5.9). According to Lemma 5.1, for an autonomous linear system described by Equation 5.2, an A-invariant subspace is a subspace having the property that any state trajectory starting in the subspace remains in it. However, for a linear system $\Sigma := [A, B, C]$ with input, A-invariance is not necessary in order for a subspace to have the above trajectory confinement property. In fact, let $\mathcal{V} \subset \mathcal{X}$. A state trajectory can be confined in $\mathcal{V}$ if and only if $\dot{x}(t) \in \mathcal{V}$, and to produce $\dot{x}(t) \in \mathcal{V}$ whenever $x(t) \in \mathcal{V}$, we need $\dot{x}(t) = Ax(t) + Bu(t) = v(t) \in \mathcal{V}$, that is, $Ax(t) = v(t) - Bu(t)$ for some $u(t)$ and $v(t)$, which implies $A\mathcal{V} \subset \mathcal{V} + \operatorname{im} B$. The converse also holds. Summarizing, we have

---

Lemma 5.3:

Consider the system described by Equation 5.7. For each initial state $x_o \in \mathcal{V}$, there is an (admissible) input $u(t), t \geq 0$ such that the corresponding $x(t) \in \mathcal{V}$ for all $t \geq 0$ if and only if

$$A\mathcal{V} \subset \mathcal{V} + \text{im}\, B. \tag{5.14}$$

Subspaces satisfying Equation 5.14 play a fundamental role in the geometric approach, and we introduce

---

Definition 5.2:

A subspace $\mathcal{V}$ is said to be an $(A, \text{im}\, B)$-controlled invariant (subspace) (Basile and Marro, 1969, 1992) or an (A, B)-invariant subspace (Wonham and Morse, 1970; Wonham, 1985) if it is A-invariant modulo imB, that is, if Equation 5.14 holds.

An important property of the above subspace is described by

Theorem 5.5:

Let $\mathcal{V} \subset \mathcal{X}$. Then, there exists an $m \times n$ real matrix F, such that

$$(A + BF)\mathcal{V} \subset \mathcal{V}, \tag{5.15}$$

if and only if $\mathcal{V}$ is an (A, im B)-controlled invariant.

Remark

If the state feedback control law (Equation 5.9) is applied to the system $\Sigma := [A, B, C]$, the corresponding state equation is described by Equation 5.10. Therefore, recalling Lemma 5.1, if $\mathcal{V}$ is an $(A, \text{im}\, B)$-controlled invariant, then there is an F for Equation 5.10 such that $x(t) \in \mathcal{V}$ for all $t \geq 0$, provided $x_o \in \mathcal{V}$.

Another class of important subspaces is now introduced (Basile and Marro 1969, 1992).

Definition 5.3:

A subspace $\mathcal{S}$ of $\mathcal{X}$ is said to be an $(A, \ker C)$-conditioned invariant (subspace), if

$$A\left(\mathcal{S} \cap \ker C\right) \subset \mathcal{S}. \tag{5.16}$$

There is a duality between controlled invariants and conditioned invariants in the following sense. By taking the orthogonal complements of the quantities on both sides of Equation 5.16, we see that Equation 5.16 is equivalent to $\{A\left(\mathcal{S} \cap \ker C\right)\}^{\perp} \supset \mathcal{S}^{\perp}$, which, in turn, is equivalent to $A'^{-1}\left(\mathcal{S}^{\perp} + \text{im}\, C'\right) \supset \mathcal{S}^{\perp}$, that is, $A'\mathcal{S}^{\perp} \subset \mathcal{S}^{\perp} + \text{im}\, C'$. Similarly, Equation 5.14 holds if and only if $A'(\mathcal{V}^{\perp} \cap \ker B') \subset \mathcal{V}^{\perp}$. Thus, we have

Lemma 5.4:

A subspace $\mathcal{S}$ is an $(A, \ker C)$-conditioned invariant if and only if $\mathcal{S}^{\perp}$ is an $(A', \text{im } C')$-controlled invariant. Also, a subspace $\mathcal{V}$ is an $(A, \text{im } B)$-controlled invariant if and only if $\mathcal{V}^{\perp}$ is an $(A', \ker B')$-conditioned invariant.

Due to the above lemma, the previous theorem can be translated easily into the following property (Basile and Marro, 1969, 1992).

Theorem 5.6:

Let $\mathcal{S} \subset \mathcal{X}$. Then, there exists an $n \times p$ real matrix G satisfying

$$(A + GC)\mathcal{S} \subset \mathcal{S}, \tag{5.17}$$

if and only if $\mathcal{S}$ is an $(A, \ker C)$-conditioned invariant.

The conditioned invariants are naturally associated with dynamic observers (not necessarily asymptotic). The following Lemma 5.5 summarizes the meaning of the invariants in this context.

Consider the system $\Sigma := [A, B, C]$ described by Equations 5.7 and 5.8. To focus on the state observation aspect of the system, we assume that $u(t) = 0$ for all $t \geq 0$. (It is straightforward to extend the result to include the nonzero measurable input.) We define the full-order and reduced order observers Σ_{obs} and $\Sigma_{obs,red}$, respectively, as follows:

$$\Sigma_{obs} : \dot{\hat{x}}(t) = \left(A + GC\right)\hat{x}(t) - Gy(t) = \left(A + GC\right)\hat{x}(t) - GCx(t),$$
$$\Sigma_{obs,red} : \dot{z}(t) = Nz(t) - My(t) = Nz(t) - MCx(t).$$

Here, $\hat{x}(t)$ and $z(t)$ are column vectors with dimensions n and n_z (to be determined), respectively, and N and M are, respectively, $n_z \times n_z$ and $n_z \times p$ real matrices. Then, we have

Lemma 5.5:

Let $\mathcal{S} \subset \mathcal{X} := R^n$. The following statements are equivalent:

i. *$\mathcal{S}$ is $(A, \ker C)$-conditioned invariant, that is, Equation 5.16 holds.*
ii. *There exists a real matrix G with an appropriate dimension and an observer Σ_{obs} such that*

$$\hat{x}(0) = x(0)(\text{mod } \mathcal{S}), \quad \text{that is } \hat{x}(0) - x(0) \in \mathcal{S}$$

implies

$$\hat{x}(t) = x(t)(\text{mod } \mathcal{S}), \quad \text{that is } \hat{x}(t) - x(t) \in \mathcal{S}, \quad \text{for all } t > 0.$$

iii. *There exist real matrices M, N, and H, satisfying $\ker H = \mathcal{S}$, with suitable dimensions and an observer $\Sigma_{obs,red}$ such that*

$$z(0) = Hx(0) \Rightarrow z(t) = Hx(t), \quad \text{for all } t > 0.$$

Remark

The statement (iii) in the above lemma is used by Willems (1981) as the definition of the conditioned invariant. Also, see Trentelman et al. (2001, Section 5.1).

Proof of Lemma 5.5. (i) $\Rightarrow$ (ii) : Suppose Equation 5.16 holds. In view of Theorem 5.6 choose G, so that Equation 5.17 holds. Define the observer error by $e_{rr}(t) := \hat{x}(t) - x(t)$. The equations for Σ and Σ_{obs} lead to $\dot{e}_{rr}(t) = (A + GC)e_{rr}(t)$. Since $\mathcal{S}$ is $(A + GC)$-invariant, $e_{rr}(0) \in \mathcal{S} \Rightarrow e_{rr}(t) \in \mathcal{S}$ for all $t > 0$. (ii) $\Rightarrow$ (iii): Basically, a modulo $\mathcal{S}$ version of Σ_{obs} in (ii) is a reduced order observer $\Sigma_{obs,red}$. To be explicit, choose a new basis $\{e_1, \ldots, e_{n_\mathcal{S}}, \ldots, e_n\}$ for $\mathcal{X} := R^n$ so that $\{e_1, \ldots, e_{n_\mathcal{S}}\}$ is a basis of subspace $\mathcal{S} \subset \mathcal{X} := R^n$, where $n_\mathcal{S} := \dim \mathcal{S}$. Define $T := [e_1 \ldots e_{n_\mathcal{S}} \ldots e_n]$ and apply the state coordinate transformation $x(t) = T\tilde{x}(t)$ to Σ (with zero inputs) to obtain $\tilde{\Sigma} : \dot{\tilde{x}}(t) = T^{-1}AT\tilde{x}(t), y(t) = CT\tilde{x}(t)$, where $\tilde{x}(t) = [\tilde{x}_1'(t) \quad \tilde{x}_2'(t)]', \tilde{x}_1(t) \in R^{n_\mathcal{S}}$, and $\tilde{x}_2(t) \in R^{n-n_\mathcal{S}}$. We apply the same coordinate transformation to Σ_{obs}, that is, $\hat{x}(t) = T\tilde{\hat{x}}(t)$, to obtain $\tilde{\Sigma}_{obs} : \dot{\tilde{\hat{x}}}(t) = T^{-1}(A + GC)T\tilde{\hat{x}}(t) - T^{-1}Gy(t)$, which has the following structure:

$$\tilde{\Sigma}_{obs} : \begin{bmatrix} \dot{\tilde{\hat{x}}}_1(t) \\ \dot{\tilde{\hat{x}}}_2(t) \end{bmatrix} = \begin{bmatrix} \tilde{A}_{G11} & \tilde{A}_{G12} \\ 0 & \tilde{A}_{G22} \end{bmatrix} \begin{bmatrix} \tilde{\hat{x}}_1(t) \\ \tilde{\hat{x}}_2(t) \end{bmatrix} - \begin{bmatrix} \tilde{G}_1 \\ \tilde{G}_2 \end{bmatrix} y(t),$$

where $\tilde{\hat{x}}_1(t) \in R^{n_\mathcal{S}}, \tilde{\hat{x}}_2(t) \in R^{n-n_\mathcal{S}}, \tilde{A}_{G22} \in R^{(n-n_\mathcal{S})\times(n-n_\mathcal{S})}, \tilde{G}_2 \in R^{(n-n_\mathcal{S})\times p}$, and $\tilde{A}_{G11}$, $\tilde{A}_{G12}$, and $\tilde{G}_1$ with consistent dimensions. Note that the 0 block in matrix $T^{-1}(A + GC)T$ is due to $(A + GC)-$ invariance of $\mathcal{S}$. The corresponding observer error $[\tilde{e}_{rr1}'(t) \quad \tilde{e}_{rr2}'(t)]' := [\tilde{\hat{x}}_1'(t) \quad \tilde{\hat{x}}_2'(t)]' - [\tilde{x}_1'(t) \quad \tilde{x}_2'(t)]'$ satisfies

$$\begin{bmatrix} \dot{\tilde{e}}_{rr1}(t) \\ \dot{\tilde{e}}_{rr2}(t) \end{bmatrix} = \begin{bmatrix} \tilde{A}_{G11} & \tilde{A}_{G12} \\ 0 & \tilde{A}_{G22} \end{bmatrix} \begin{bmatrix} \tilde{e}_{rr1}(t) \\ \tilde{e}_{rr2}(t) \end{bmatrix}.$$

The reduced order observer $\Sigma_{obs,red}$ can be chosen as follows:

$$\Sigma_{obs,red} : \dot{\tilde{\hat{x}}}_2(t) = \tilde{A}_{G22}\tilde{\hat{x}}_2(t) - \tilde{G}_2 y(t).$$

Define H by $H := [0 \quad I_{n-n_\mathcal{S}}]\, T^{-1}$. Then,

$$\tilde{\hat{x}}_2(0) - Hx(0) = \tilde{\hat{x}}_2(0) - \tilde{x}_2(0) = \tilde{e}_{rr2}(0) = 0$$

implies

$$\tilde{\hat{x}}_2(t) - Hx(t) = \tilde{\hat{x}}_2(t) - \tilde{x}_2(t) = \tilde{e}_{rr2}(t) = e^{\tilde{A}_{G22}t}\, \tilde{e}_{rr2}(0) = 0.$$

(iii) $\Rightarrow$ (i) : Let $x(0) = x_o \in \mathcal{S} \cap \ker C$ and choose $z(0) = Hx_o = 0$. Then, we have $\dot{z}(0) = Nz(0) - My(0) = Nz(0) - MCx(0) = 0$. Since $z(t) = Hx(t)$ this implies $H\dot{x}(0) = HAx(0) = HAx_o = 0$, that is, $Ax_o \in \ker H = \mathcal{S}$ (for every $x_o \in \mathcal{S} \cap \ker C$). QED

A subspace may be both controlled and conditioned invariant. In such a case, we have the following (Basile and Marro, 1992; Hamano and Furuta, 1975):

Lemma 5.6:

There exists an $m \times p$ real matrix K, such that

$$(A + BKC)\mathcal{V} \subset \mathcal{V}, \tag{5.18}$$

if and only if $\mathcal{V}$ is both an $(A, \operatorname{im} B)$-controlled invariant and an $(A, \ker\ C)$-conditioned invariant.

Proof. [*Only if part*]. The controlled invariance and conditioned invariance follow trivially by Theorems 5.5 and 5.6, respectively.

[If part]. For $C = 0$ or $\ker C = 0$ or $\mathcal{V} = 0$, the statement of the lemma trivially holds. So we prove the lemma assuming such trivialities do not occur. Since by assumption $\mathcal{V}$ is an $(A, \operatorname{im} B)$-controlled invariant, there is an $m \times n$ real matrix $\bar{F}$ satisfying $(A+B\bar{F})\mathcal{V} \subset \mathcal{V}$. Assuming $\mathcal{V} \cap \ker C \neq \{0\}$, let $\{v_1 \dots v_\mu\}$ be a basis of $\mathcal{V} \cap \ker C$. Complete the basis $\{v_1 \dots v_\mu \dots v_\nu \dots v_n\}$ of $\mathcal{X}$ in such a way that $\{v_1 \dots v_\mu \dots v_\nu\}$ is a basis of $\mathcal{V}$ where $1 \leq \mu \leq \nu \leq n$. Define an $m \times n$ real matrix $\tilde{F}$ by

$$\tilde{F}v_i = \begin{cases} \bar{F}v_i, & i = \mu+1, \dots, \nu \\ 0, & i = 1, \dots, \mu \\ \text{arbitrary}, & \text{otherwise} \end{cases} \tag{5.19}$$

Choose K so that $\tilde{F} = KC$, that is, $\tilde{F}\left[v_{\mu+1} \dots v_\nu\right] = K\left[Cv_{\mu+1} \dots Cv_\nu\right]$. Note that due to the particular choice of basis, the columns of $\left[Cv_{\mu+1} \dots Cv_\nu\right]$ are linearly independent, and so the above K certainly exists. To show that Equation 5.18 holds, let $v \in \mathcal{V}$. Then, $v = \sum_{i=1}^{\mu} \alpha_i v_i + \sum_{i=\mu+1}^{\nu} \alpha_i v_i$ for some numbers α_i's. Therefore, by the above choice of K,

$$(A+BKC)v = A\sum_{i=1}^{\mu} \alpha_i v_i + (A+B\bar{F})\sum_{i=\mu+1}^{\nu} \alpha_i v_i.$$

But, the first sum belongs to $\mathcal{V}$ by Equation 5.16 (where $\mathcal{S} = \mathcal{V}$) and the second sum is an element of $\mathcal{V}$ by the choice of $\bar{F}$. Thus, $(A+BKC)v \in \mathcal{V}$.

If $\mathcal{V} \cap \ker C = \{0\}$, the basis for $\mathcal{V} \cap \ker C$ is empty and Equation 5.19 can be modified by dropping the second row and setting $\mu = 0$. Equation 5.18 follows similarly to the above. QED

5.4 Algebraic Properties of Controlled and Conditioned Invariants

An $(A, \operatorname{im} B)$-controlled invariant has the following properties in addition to the ones discussed in the previous section. The proofs are omitted. They can be found in Basile and Marro (1992, Chapter 4).

Lemma 5.7:

If $\mathcal{V}_1$ and $\mathcal{V}_2$ are $(A, \operatorname{im} B)$-controlled invariants, so is $\mathcal{V}_1 + \mathcal{V}_2$.

Remark

But, the intersection of two $(A, \operatorname{im} B)$-controlled invariants is not in general an $(A, \operatorname{im} B)$-controlled invariant.

Lemma 5.8:

Let $\mathcal{V}_1$ and $\mathcal{V}_2$ be $(A, \operatorname{im} B)$-controlled invariants. Then, there is an $m \times n$ real matrix F satisfying

$$(A+BF)\mathcal{V}_i \subset \mathcal{V}_i, \quad i = 1, 2, \tag{5.20}$$

if and only if $\mathcal{V}_1 \cap \mathcal{V}_2$ is an $(A, \operatorname{im} B)$-controlled invariant.

By duality, we have

Lemma 5.9:

If $\mathcal{S}_1$ and $\mathcal{S}_2$ are $(A, \ker C)$-conditioned invariants, so is $\mathcal{S}_1 \cap \mathcal{S}_2$.

Remark

$\mathcal{S}_1 + \mathcal{S}_2$ is not necessarily an (*A*, ker *C*)-conditioned invariant.

Lemma 5.10:

Let $\mathcal{S}_1$ and $\mathcal{S}_2$ be $(A, \ker C)$-conditioned invariants. Then, there is an $n \times p$ real matrix G satisfying

$$(A + GC)\mathcal{S}_i \subset \mathcal{S}_i, \quad i = 1, 2, \tag{5.21}$$

if and only if $\mathcal{S}_1 + \mathcal{S}_2$ is an $(A, \ker C)$-conditioned invariant.

5.5 Maximum-Controlled and Minimum-Conditioned Invariants

Let $\mathcal{K} \subset \mathcal{X}$, and consider the set of (*A*, im *B*)-controlled invariants contained in $\mathcal{K}$ (by subspace inclusion). Lemma 5.7 states that the set of (*A*, im *B*)-controlled invariants is closed under subspace addition. As a result, the set of (*A*, im *B*)-controlled invariants contained in $\mathcal{K}$ has a largest element or supremum. This element is a unique subspace that contains (by subspace inclusion) any other (*A*, im *B*)-controlled invariants contained in $\mathcal{K}$, and is called the *maximum* (or *supremum*) (*A*, im *B*)-*controlled invariant* contained in $\mathcal{K}$. This is denoted in the sequel as $\mathcal{V}_{\max}(A,\ \operatorname{im} B, \mathcal{K})$. Similarly, let $\mathcal{I} \subset \mathcal{X}$. Then, owing to Lemma 5.9, it can be shown that the set of (*A*, ker *C*)-conditioned invariants containing $\mathcal{I}$ has a smallest element or infimum. This is a unique (*A*, ker *C*)-conditioned invariant contained in all other (*A*, ker *C*)-conditioned invariants containing $\mathcal{I}$, and is called the *minimum* (or *infimum*) (*A*, ker *C*)-*conditioned invariant* containing $\mathcal{I}$. This subspace will be denoted as $\mathcal{S}_{\min}(A, \ker C, \mathcal{I})$. The subspaces $\mathcal{V}_{\max}(A,\ \operatorname{im} B, \mathcal{K})$ and $\mathcal{S}_{\min}(A, \ker C, \mathcal{I})$ are important because they can be computed in a finite number of steps (in at most *n* iterations) and because testing the solvability of control problems typically reduces to checking the conditions involving these subspaces. The geometric algorithms to compute $\mathcal{V}_{\max}(A,\ \operatorname{im} B, \mathcal{K})$ and $\mathcal{S}_{\min}(A, \ker C, \mathcal{I})$ are given below.

Algorithm to compute $\mathcal{V}_{\max}(A,\ \operatorname{im} B, \mathcal{K})$:

$$\mathcal{V}_{\max}(A,\ \operatorname{im} B, \mathcal{K}) = \mathcal{V}_{\dim \mathcal{K}}, \tag{5.22a}$$

where

$$\mathcal{V}_o = \mathcal{K}, \tag{5.22b}$$

$$\mathcal{V}_i = \mathcal{K} \cap A^{-1}(\mathcal{V}_{i-1} + \operatorname{im} B), \quad i = 1, \ldots, \dim \mathcal{K}. \tag{5.22c}$$

Proof of Equation 5.22a. See Basile and Marro (1992) or Wonham (1985).

Remark

The algorithm defined by Equations 5.22 has the following properties:

i. $\mathcal{V}_1 \supset \mathcal{V}_2 \supset \cdots \supset \mathcal{V}_{\dim \mathcal{K}}$.
ii. If $\mathcal{V}_\ell = \mathcal{V}_{\ell+1}$, then $\mathcal{V}_\ell = \mathcal{V}_{\ell+1} = \cdots = \mathcal{V}_{\dim \mathcal{K}}$.

Remark

For an algorithm to compute *F* such that $(A + BF)\mathcal{V}_{\max}(A,\ \operatorname{im} B, \mathcal{K}) \subset \mathcal{V}_{\max}(A,\ \operatorname{im} B, \mathcal{K})$, see Basile and Marro (1992).

The following is the dual of the above algorithm (see Basile and Marro, 1992).

Algorithm to calculate $\mathcal{S}_{\min}(A, \ker C, \mathcal{I})$:

$$\mathcal{S}_{\min}(A, \ker C, \mathcal{I}) = \mathcal{S}_{n-\dim \mathcal{I}}, \tag{5.23a}$$

where

$$\mathcal{S}_o = \mathcal{I}, \tag{5.23b}$$

$$\mathcal{S}_i = \mathcal{I} + A(\mathcal{S}_{i-1} \cap \ker C), \quad i = 1, \ldots, n - \dim \mathcal{I}. \tag{5.23c}$$

Remark

The algorithm generates a monotonically nondecreasing sequence:

(i) $\mathcal{S}_1 \subset \mathcal{S}_2 \subset \cdots \subset \mathcal{S}_{n-\dim \mathcal{I}}$.
(ii) If $\mathcal{S}_\ell = \mathcal{S}_{\ell+1}$, then $\mathcal{S}_\ell = \mathcal{S}_{\ell+1} = \cdots = \mathcal{S}_{n-\dim \mathcal{I}}$.

5.6 Self-Bounded-Controlled and Self-Hidden-Conditioned Invariants and Constrained Reachability and Observability

Let $\mathcal{V}_o$ be an $(A, \text{im } B)$-controlled invariant contained in $\mathcal{K}$, and consider all possible state trajectories (with different controls) starting at x_o in $\mathcal{V}_o$ and confined in $\mathcal{K}$. We know that there is at least one control for which the state trajectory remains in $\mathcal{V}_o$, but that there may be another control for which the state trajectory goes out of $\mathcal{V}_o$ while remaining in $\mathcal{K}$. However, some $(A, \text{im } B)$-controlled invariant contained in $\mathcal{K}$, say $\mathcal{V}$, has a stronger property that, for any initial state in $\mathcal{V}$, there is no control which drives the state (initially in $\mathcal{V}$) out of $\mathcal{V}$ while maintaining the state trajectory in $\mathcal{K}$, that is, the state trajectory must go out of $\mathcal{K}$ if it ever goes out of $\mathcal{V} \subset \mathcal{K}$. The subspace $\mathcal{V}_{\max}(A, \text{ im } B, \mathcal{K})$ is one of those since it contains all states that can be controlled to remain in $\mathcal{K}$. Such an $(A, \text{im } B)$-controlled invariant $\mathcal{V}$ contained in $\mathcal{K}$ is in general characterized by

$$\mathcal{V}_{\max}(A, \text{ im } B, \mathcal{K}) \cap \text{ im } B \subset \mathcal{V}, \tag{5.24}$$

and we have (Basile and Marro, 1982, 1992)

Definition 5.4:

An $(A, \text{im } B)$-controlled invariant $\mathcal{V}$ contained in $\mathcal{K}$ is said to be self-bounded with respect to $\mathcal{K}$ if Equation 5.24 holds.

Remark

The left-hand side of inclusion 5.24 represents the set of all possible influences of control on the state velocity at each instant of time that do not pull the state out of $\mathcal{K}$.

Self-bounded $(A, \text{im } B)$-controlled invariants have the following properties.

For each $\mathcal{K}$ we can choose a single state-feedback control law, which works for all the self-bounded $(A, \text{im } B)$-controlled invariants with respect to $\mathcal{K}$. More precisely,

Lemma 5.11:

Let F be an $m \times n$ real matrix satisfying

$$(A + BF)\mathcal{V}_{\max}(A, \text{ im } B, \mathcal{K}) \subset \mathcal{V}_{\max}(A, \text{ im } B, \mathcal{K}). \tag{5.25}$$

Then, every self-bounded $(A, \text{im}\, B)$*-controlled invariant* $\mathcal{V}$ *with respect to* $\mathcal{K}$ *satisfies*

$$(A + BF)\mathcal{V} \subset \mathcal{V}. \tag{5.26}$$

Proof. See Basile and Marro (1992).

It can be shown that the set of self-bounded $(A, \text{im}\, B)$-controlled invariants in $\mathcal{K}$ is closed under subspace intersection, that is, if $\mathcal{V}_1$ and $\mathcal{V}_2$ are self-bounded $(A, \text{im}\, B)$-controlled invariants with respect to $\mathcal{K}$, so is $\mathcal{V}_1 \cap \mathcal{V}_2$. Therefore, the above set has a minimum element which is called *the minimum self-bounded* $(A, \text{im}\, B)$*-controlled invariant with respect to* $\mathcal{K}$ (Basile and Marro, 1982, 1992), denoted in this chapter by $\mathcal{V}_{sb,\min}(A, \text{im}\, B, \mathcal{K})$. The subspace $\mathcal{V}_{sb,\min}(A, \text{im}\, B, \mathcal{K})$ is related to $\mathcal{V}_{\max}(A, \text{im}\, B, \mathcal{K})$ and $\mathcal{S}_{\min}(A, \mathcal{K}, \text{im}\, B)$ as follows.

Theorem 5.7:

$$\mathcal{V}_{sb,\min}(A, \text{im}\, B, \mathcal{K}) = \mathcal{V}_{\max}(A, \text{im}\, B, \mathcal{K}) \cap \mathcal{S}_{\min}(A, \mathcal{K}, \text{im}\, B). \tag{5.27}$$

Proof. See Basile and Marro (1992, Chapter 4) and Morse (1973).

The minimum self-bounded $(A, \text{im}\, B)$-controlled invariant is closely related to constrained reachability. The set of all states that can be reached from the zero state through state trajectories constrained in $\mathcal{K}$ in finite time is called the *reachable set* (or *reachable subspace*, since the set forms a subspace) *in* $\mathcal{K}$ (Basile and Marro, 1992). It is also called the *supremal* (A, B)*-controllability subspace contained in* $\mathcal{K}$ (Wonham, 1985). This set will be denoted by $\mathcal{V}_{reach}(\mathcal{K})$. Clearly, it has the following properties.

Lemma 5.12:

$$\mathcal{V}_{reach}(\mathcal{K}) \subset \mathcal{V}_{\max}(A, \text{im}\, B, \mathcal{K}) \subset \mathcal{K}, \tag{5.28}$$

$$\mathcal{V}_{reach}(\mathcal{V}_{\max}(A, \text{im}\, B, \mathcal{K})) = \mathcal{V}_{reach}(\mathcal{K}). \tag{5.29}$$

To examine $\mathcal{V}_{reach}(\mathcal{K})$ further, let F be a matrix satisfying Equation 5.25. It is easy to observe that the set of $\dot{x}(t)$ (and so the trajectories) in $\mathcal{V}_{\max}(A, \text{im}\, B, \mathcal{K})$ (or in $\mathcal{K}$) that can be generated by the state Equation 5.7, when $x(t) \in \mathcal{V}_{\max}(A, \text{im}\, B, \mathcal{K})$, is identical to those generated by $\dot{x}(t) = (A + BF)x(t) + BL\tilde{u}(t)$ for a matrix L satisfying $\text{im}\, BL = \text{im}\, B \cap \mathcal{V}_{\max}(A, \text{im}\, B, \mathcal{K})$ and the admissible inputs $\tilde{u}(t)$. Then, by Equation 5.11, $\mathcal{V}_{reach}(\mathcal{K}) = \mathcal{R}(A + BF, \text{im}\, B \cap \mathcal{V}_{\max}(\mathcal{K}))$, where $\mathcal{V}_{\max}(\mathcal{K}) := \mathcal{V}_{\max}(A, \text{im}\, B, \mathcal{K})$. Clearly, $\mathcal{V}_{reach}(\mathcal{K}) \supset \text{im}\, B \cap \mathcal{V}_{\max}(\mathcal{K})$ and the subspace $\mathcal{V}_{reach}(\mathcal{K})$ is the minimum $(A + BF)$-invariant satisfying this inclusion (see the remark immediately after Theorem 5.1), that is, it is the minimum $(A, \text{im}\, B)$-controlled invariant self-bounded with respect to $\mathcal{K}$. Thus, we have

---

Theorem 5.8:

Let F *be a real matrix satisfying* $(A + BF)\mathcal{V}_{\max}(\mathcal{K}) \subset \mathcal{V}_{\max}(\mathcal{K})$, *where* $\mathcal{V}_{\max}(\mathcal{K}) := \mathcal{V}_{\max}(A, \text{im}\, B, \mathcal{K})$. *Then,*

$$\mathcal{V}_{reach}(\mathcal{K}) = \mathcal{R}(A + BF, \text{im}\, B \cap \mathcal{V}_{\max}(\mathcal{K})) = \mathcal{V}_{sb,\min}(A, \text{im}\, B, \mathcal{K}). \tag{5.30}$$

Dual to the above results we have

Definition 5.5:

An $(A, \ker\ C)$-conditioned invariant $\mathcal{S}$ containing $\mathcal{I}$ is said to be self-hidden with respect to $\mathcal{I}$ if

$$\mathcal{S} \subset \mathcal{S}_{\min}(A,\ \ker C, \mathcal{I}) + \ker C. \tag{5.31}$$

Lemma 5.13:

Let G be an $n \times p$ real matrix satisfying

$$(A + GC)\mathcal{S}_{\min}(A,\ \ker C, \mathcal{I}) \subset \mathcal{S}_{\min}(A,\ \ker C, \mathcal{I}). \tag{5.32}$$

Then, every $(A, \ker\ C)$-conditioned invariant $\mathcal{S}$ (containing $\mathcal{I}$) self-hidden with respect to $\mathcal{I}$ satisfies

$$(A + GC)\mathcal{S} \subset \mathcal{S}. \tag{5.33}$$

If $(A, \ker C)$-conditioned invariants $\mathcal{S}_1$ and $\mathcal{S}_2$ containing $\mathcal{I}$ are self-hidden with respect to $\mathcal{I}$, so is $\mathcal{S}_1 + \mathcal{S}_2$. Therefore, the above set has a maximum element which is called *the maximum self-hidden* $(A, \ker C)$-*conditioned invariant with respect to* $\mathcal{I}$ denoted by $\mathcal{S}_{sh,\max}(A, \ker C, \mathcal{I})$. The subspace $\mathcal{S}_{sh,\max}(A,\ \ker C, \mathcal{I})$ is related to $\mathcal{V}_{\max}(A, \mathcal{I},\ \ker C)$ and $\mathcal{S}_{\min}(A, \ker C, \mathcal{I})$ as follows.

Theorem 5.9:

$$\mathcal{S}_{sh,\max}(A, \ker C, \mathcal{I}) = \mathcal{S}_{\min}(A, \ker C, \mathcal{I}) + \mathcal{V}_{\max}(A, \mathcal{I}, \ker C). \tag{5.34}$$

Furthermore, we have

Theorem 5.10:

$$\begin{aligned} \mathcal{S}_{sh,\max}(A, \ker C, \mathcal{I}) = (\ker C + \mathcal{S}_{\min}(\mathcal{I})) \cap (A + GC)^{-1}(\ker C + \mathcal{S}_{\min}(\mathcal{I})) \\ \cap \cdots \cap (A + GC)^{-(n-1)}(\ker C + \mathcal{S}_{\min}(\mathcal{I})), \end{aligned} \tag{5.35}$$

where $\mathcal{S}_{\min}(\mathcal{I}) := \mathcal{S}_{\min}(A, \ker C, \mathcal{I})$ and the matrix G satisfies Equation 5.32, that is, $(A + GC)\mathcal{S}_{\min}(\mathcal{I}) \subset \mathcal{S}_{\min}(\mathcal{I})$.

Remark

The right-hand side of Equation 5.35 represents the largest $A + GC$ invariant subspace contained in $\ker C + \mathcal{S}_{\min}(\mathcal{I})$.

Remark

Equation 5.35 indicates that $\mathcal{S}_{sh,\max}(A, \ker C, \mathcal{I})$ represents the set of unobservable states through a dynamic observer when unknown disturbances are present if $\mathcal{I}$ represents the disturbance channel. See Basile and Marro (1992, Section 4.1.3) for details on the topic.

5.7 Internal, External Stabilizability

We now introduce the notions of stability associated with controlled and conditioned invariants.

Definition 5.6:

An (A, im B)-controlled invariant $\mathcal{V}$ is said to be internally stabilizable if, for any initial state $x_o \in \mathcal{V}$, there is a control $u(t)$ such that $x(t) \in \mathcal{V}$ for all $t \geq 0$ and $x(t)$ converges to the zero state as $t \to \infty$, or, alternatively (and, in fact, equivalently), if there exists an $m \times n$ real matrix F satisfying $(A + BF)\mathcal{V} \subset \mathcal{V}$ and $(A + BF)|\, \mathcal{V}$ is stable.

Definition 5.7:

An (A, im B)-controlled invariant $\mathcal{V}$ is said to be externally stabilizable if, for any initial state $x_o \in \mathcal{X}$, there is a control $u(t)$ such that $x(t)$ converges to $\mathcal{V}$ as $t \to \infty$, or, alternatively (and, equivalently), if there exists an $m \times n$ real matrix F satisfying $(A + BF)\mathcal{V} \subset \mathcal{V}$ and $(A + BF)|\, \mathcal{X}/\mathcal{V}$ is stable.

Internal and external stabilizabilities can be easily examined by applying appropriate coordinate transformations (consisting of new sets of basis vectors) for the state and input spaces. For this, let $\mathcal{V}$ be an (A, im B)-controlled invariant. Also, let $n_r := \dim \mathcal{V}_{reach}(\mathcal{V})$, $n_{\mathcal{V}} := \dim \mathcal{V}$, and $n_{\mathcal{S}} := \dim \mathcal{S}_{\min}(A, \mathcal{V}, \text{im}\, B)$. Define $n \times n$ and $m \times m$ nonsingular matrices $T = [T_1\ T_2\ T_3\ T_4]$ and $U = [U_1\ U_2]$, respectively, as follows: Choose U_1, T_1, and T_2, so that $\text{im}\, BU_1 = \mathcal{V} \cap \text{im}\, B$, $\text{im}\, T_1 = \mathcal{V}_{reach}(\mathcal{V})$, and $\text{im}\begin{bmatrix} T_1 & T_2 \end{bmatrix} = \mathcal{V}$. Also, choose U_2, so that $\text{im}\, B\,[U_1\ U_2] = \text{im}\, B$. Noting that $\mathcal{V}_{reach}(\mathcal{V}) = \mathcal{V} \cap \mathcal{S}_{\min}(A, \mathcal{V}, \text{im}\, B)$ (see Equations 5.27 and 5.30; let $\mathcal{K} = \mathcal{V}$), select T_3 satisfying $\text{im}\, [T_1\ T_3] = \mathcal{S}_{\min}(A, \mathcal{V}, \text{im}\, B)$ with $\text{im}\, BU_2 \subset \text{im}\, T_3$. Then, we have

$$\tilde{A} := T^{-1}AT = \begin{bmatrix} \tilde{A}_{11} & \tilde{A}_{12} & \tilde{A}_{13} & \tilde{A}_{14} \\ 0 & \tilde{A}_{22} & \tilde{A}_{23} & \tilde{A}_{24} \\ \tilde{A}_{31} & \tilde{A}_{32} & \tilde{A}_{33} & \tilde{A}_{34} \\ 0 & 0 & \tilde{A}_{43} & \tilde{A}_{44} \end{bmatrix}, \tag{5.36}$$

$$\tilde{B} := T^{-1}BU = \begin{bmatrix} \tilde{B}_{11} & 0 \\ 0 & 0 \\ 0 & \tilde{B}_{32} \\ 0 & 0 \end{bmatrix}. \tag{5.37}$$

Here, $\tilde{A}$ is divided into blocks in accordance with the ranks of $T_1,\ T_2,\ T_3,\ T_4$, that is, $\tilde{A}_{11},\ \tilde{A}_{12}, \ldots, \tilde{A}_{44}$ respectively have dimensions $n_1 \times n_1,\ n_1 \times n_2, \ldots, n_4 \times n_4$, where $n_1 := rank T_1 = n_r, n_2 := rank T_2 = n_{\mathcal{V}} - n_r, \ldots$, and $\tilde{B}$ is divided similarly, for example, $\tilde{B}_{11}$ and $\tilde{B}_{32}$ have dimensions $n_1 \times m_1$ and $n_3 \times m_2$, where m_1 and m_2 are, respectively, the numbers of columns of U_1 and U_2. Note that the zero matrix in the second block row of $\tilde{A}$ is due to the following facts: $\mathcal{V}_{reach}(\mathcal{V})$ is $(A + \text{BF})$-invariant for every F satisfying $(A + BF)\mathcal{V} \subset \mathcal{V}$, but the second block row of $\tilde{B}$ is zero (which makes F ineffective in the block). The fourth block row of $\tilde{A}$ has zero blocks since $\mathcal{V}$ is $(A + BF)$-invariant for any F defined above, but the fourth block row of $\tilde{B}$ is zero. Now, using F satisfying $(A + BF)\mathcal{V} \subset \mathcal{V}$, consider $A + BF$ with respect to the new bases. Noting that $\tilde{A}_{31} + \tilde{B}_{32}\tilde{F}_{21} = 0$ and $\tilde{A}_{32} + \tilde{B}_{32}\tilde{F}_{22} = 0$ by construction, we obtain

$$\tilde{A} + \tilde{B}\tilde{F} = T^{-1}(A + BF)T = \begin{bmatrix} \tilde{A}_{11} + \tilde{B}_{11}\tilde{F}_{11} & \tilde{A}_{12} + \tilde{B}_{11}\tilde{F}_{12} & \tilde{A}_{13} + \tilde{B}_{11}\tilde{F}_{13} & \tilde{A}_{14} + \tilde{B}_{11}\tilde{F}_{14} \\ 0 & \tilde{A}_{22} & \tilde{A}_{23} & \tilde{A}_{24} \\ 0 & 0 & \tilde{A}_{33} + \tilde{B}_{32}\tilde{F}_{23} & \tilde{A}_{34} + \tilde{B}_{32}\tilde{F}_{24} \\ 0 & 0 & \tilde{A}_{43} & \tilde{A}_{44} \end{bmatrix} \tag{5.38}$$

where

$$\tilde{F} = \begin{bmatrix} \tilde{F}_{11} & \tilde{F}_{12} & \tilde{F}_{13} & \tilde{F}_{14} \\ \tilde{F}_{21} & \tilde{F}_{22} & \tilde{F}_{23} & \tilde{F}_{24} \end{bmatrix}, \quad \tilde{A} := T^{-1}AT, \ \ \tilde{B} := T^{-1}BU, \quad \text{and} \quad \tilde{F} := U^{-1}FT.$$

Note that $\tilde{A}_{22}$ cannot be altered by any linear state feedback satisfying $(A+BF)\mathcal{V} \subset \mathcal{V}$, that is, the (2,2)-block of $\tilde{A}+\tilde{B}\tilde{F}$ remains $\tilde{A}_{22}$ for any real $\tilde{F}$ satisfying

$$(\tilde{A}+\tilde{B}\tilde{F})\text{im}\begin{bmatrix} I_{n_1} & 0 \\ 0 & I_{n_2} \\ 0 & 0 \\ 0 & 0 \end{bmatrix} \subset \text{im}\begin{bmatrix} I_{n_1} & 0 \\ 0 & I_{n_2} \\ 0 & 0 \\ 0 & 0 \end{bmatrix}, \tag{5.39}$$

where 0's are zero matrices with suitable dimensions. Thus,

Lemma 5.14:

$$\sigma((A+BF)|\mathcal{V}/\mathcal{V}_{reach}(\mathcal{V})) = \sigma(\tilde{A}_{22}) \tag{5.40}$$

(i.e., fixed) for all F satisfying $(A+BF)\mathcal{V} \subset \mathcal{V}$. Here, $(A+BF)|\,\mathcal{V}/\mathcal{V}_{reach}(\mathcal{V})$ is the induced map of A+BF restricted to $\mathcal{V}/\mathcal{V}_{reach}(\mathcal{V})$.

Note also that by construction the pair (A_{11}, B_{11}) is reachable. Hence, by Theorem 5.2, we have

---

Lemma 5.15:

$\sigma((A+BF)|\,\mathcal{V}_{reach}(\mathcal{V})) = \sigma(\tilde{A}_{11}+\tilde{B}_{11}\tilde{F}_{11})$ can be freely assigned by an appropriate choice of F satisfying $(A+BF)\mathcal{V} \subset \mathcal{V}$ (or $\tilde{F}$ satisfying Equation 5.39).

The eigenvalues of $(A+BF)|\,\mathcal{V}/\mathcal{V}_{reach}(\mathcal{V})$ are called the *internal unassignable eigenvalues of* $\mathcal{V}$. The internal unassignable eigenvalues of $\mathcal{V}_{\max}(A, \text{im}\, B, \ker C)$ are called *invariant zero's* of the system Σ:=$[A, B, C]$ (or triple (A, B, C)), which are equal to the transmission zero's of $C(sI{-}A)^{-1}B$ if (A, B) is reachable and (A, C) is observable. Table 5.1 shows how freely the eigenvalues can be assigned for $A+BF$ by choosing F satisfying $(A+BF)\mathcal{V} \subset \mathcal{V}$ given an (A, im B)-controlled invariant. Theorems 5.11 and 5.12 given below easily follow from the above lemmas and Table 5.1.

Theorem 5.11:

An (A, im B)-controlled invariant $\mathcal{V}$ is internally stabilizable if and only if all its internal unassignable eigenvalues have negative real parts.

TABLE 5.1 Spectral Assignability of $(A+BF)\mid\mathcal{W}$, Given an (A, im B)-Controlled Invariant $\mathcal{V}$

$\mathcal{W}$	$\mathcal{X}/(\mathcal{V}+\mathcal{V}_{reach})$	$(\mathcal{V}+\mathcal{V}_{reach})/\mathcal{V}$	$\mathcal{V}/\mathcal{V}_{reach}(\mathcal{V})$	$\mathcal{V}_{reach}(\mathcal{V})$
Assignability	Fixed	Free	Fixed	Free

Note: The table indicates that $\sigma((A+BF)\mid\mathcal{X}/(\mathcal{V}+\mathcal{V}_{reach}))$ is fixed for all F satisfying $(A+BF)\mathcal{V} \subset \mathcal{V}$, $\sigma((A+BF)\mid(\mathcal{V}+\mathcal{V}_{reach})/\mathcal{V})$ is freely assignable (up to a symmetric set) by choosing an appropriate F satisfying $(A+BF)\mathcal{V} \subset \mathcal{V}$, and so on.

Theorem 5.12:

An (A, im B)-controlled invariant $\mathcal{V}$ is externally stabilizable if and only if $\mathcal{V} + \mathcal{V}_{reach}$ is externally stable.

Remark

$\mathcal{V} + \mathcal{V}_{reach}$ is A-invariant since $A(\mathcal{V} + \mathcal{V}_{reach}) = A\mathcal{V} + A\mathcal{V}_{reach} = \mathcal{V} + \text{im } B + \mathcal{V}_{reach} = \mathcal{V} + \mathcal{V}_{reach}$.

Lemma 5.16:

If the pair (A, B) is stabilizable, then all (A, im B)-controlled invariants are externally stabilizable.

Remark

The matrix F can be defined independently on $\mathcal{V}$ and on $\mathcal{X}/\mathcal{V}$.

Dual to the above internal and external stabilizabilities for a controlled invariant are, respectively, external and internal stabilizabilities for a conditioned invariant which will be defined below.

Definition 5.8:

An $(A, \ker C)$-conditioned invariant $\mathcal{S}$ is said to be externally stabilizable if there exists an $n \times p$ real matrix G satisfying $(A + GC)\mathcal{S} \subset \mathcal{S}$ and $(A + GC)|\ \mathcal{X}/\mathcal{S}$ is stable.

Definition 5.9:

An $(A, \ker C)$-conditioned invariant $\mathcal{S}$ is said to be internally stabilizable if there exists an $n \times p$ real matrix G satisfying $(A + GC)\mathcal{S} \subset \mathcal{S}$ and $(A + GC)|\ \mathcal{S}$ is stable.

How freely the eigenvalues can be chosen for $A + GC$ by means of G is given in Table 5.2, from which necessary and sufficient conditions of the stabilizabilities follow easily. (See Basile and Marro (1992) for further discussions on the topic of the internal and external stabilizabilities.)

5.8 Disturbance Localization (or Decoupling) Problem

One of the first problems to which the geometric notions were applied is the problem of disturbance localization (or disturbance decoupling). As we see below, the solution to the problem is remarkably simple in geometric terms. The solution and analysis of this problem can also be used to solve other

TABLE 5.2 Spectral Assignability of $(A + GC) \mid \mathcal{W}$, Given an $(A, \ker C)$-Conditioned Invariant $\mathcal{S}$

$\mathcal{W}$	$\mathcal{X}/\mathcal{S}_{sh,\max}(A, \ker C, \mathcal{S})$	$\mathcal{S}_{sh,\max}(A, \ker C, \mathcal{S})/\mathcal{S}$	$\mathcal{S}/\mathcal{S} \cap \mathcal{S}_{unobs}$	$\mathcal{S} \cap \mathcal{S}_{unobs}$
Assignability	Free	Fixed	Free	Fixed

Note: The table indicates that $\sigma((A+BF)|\mathcal{X}/\mathcal{S}_{sh,\max}(A, \ker C, \mathcal{S}))$ is freely assignable (up to a symmetric set) by choosing an appropriate G satisfying $(A + GC)\mathcal{S} \subset \mathcal{S}$, $\sigma((A + BF)|\mathcal{S}_{sh,\max}(A, \ker C, \mathcal{S})/\mathcal{S})$ is fixed for all G satisfying $(A + GC)\mathcal{S} \subset \mathcal{S}$, and so on.

more involved problems (e.g., model following, decoupling, and disturbance decoupling in decentralized systems).

We will be concerned with a time-invariant linear control system Σ_d:=[A, B, D, C] described by

$$\dot{x}(t) = Ax(t) + Bu(t) + Dw(t), \quad x(0) = x_o, \tag{5.41}$$

$$y(t) = Cx(t), \tag{5.42}$$

where the column vectors $x(t) \in \mathcal{X} := R^n, u(t) \in R^m, y(t) \in R^p$, and $w(t) \in R^{m_d}$ are, respectively, the state, input, output, and unknown disturbance of the system at time $t \geq 0$, the vector $x_o \in R^n$ is the initial state and the real matrices A, B, C, and D have consistent dimensions. We assume that $u(t)$ and $w(t)$ are piecewise continuous. Here, $w(t)$ is called a "disturbance" and it can neither be measured nor controlled. The above notation will be standard in the sequel. If we apply the linear state feedback control law

$$u(t) = Fx(t) \tag{5.43}$$

to the above system, we obtain the state equation

$$\dot{x}(t) = (A + BF)x(t) + Dw(t), \quad x(0) = x_o, \tag{5.44}$$

where F is an $m \times n$ real matrix. Our problem is to choose the control law (Equation 5.43) so that the disturbance doses not affect the output in the resulting closed-loop system given by Equations 5.44 and 5.42, that is, so that $y(t) = 0$ for all $t \geq 0$ for any $w(t)$, $t \geq 0$, provided $x(0) = x_o = 0$. By virtue of Theorem 5.1, note that at any time $t \geq 0$ all the possible states due to all the admissible $w(\tau)$, $0 \leq \tau < t$ are characterized by

$$\mathcal{R}(A + BF,\ \text{im}\, D) := \text{im}\, D + (A + BF)\ \text{im}\, D + \cdots + (A + BF)^{n-1}\ \text{im}\, D. \tag{5.45}$$

Therefore, in algebraic terms, the above problem can be restated as

---

Problem 5.1: Disturbance Localization (or Disturbance Decoupling) Problem

Given $n \times n, n \times m, n \times m_d$, and $p \times n$ real matrices A, B, D, and C, find an $m \times n$ real matrix F satisfying

$$\mathcal{R}(A + BF,\ \text{im}\, D) \subset \ker C. \tag{5.46}$$

It is not always possible to find F satisfying Inclusion 5.46. The following theorem gives the necessary and sufficient condition for the existence of such an F in terms of given matrices. (See Basile and Marro (1992, Section 4.2) and Wonham (1985, Section 4.3).)

Theorem 5.13:

There exists an $m \times n$ real matrix F satisfying Inclusion 5.46 if and only if

$$\text{im}\, D \subset \mathcal{V}_{\max}(A,\ \text{im}\, B, \ker C) \tag{5.47}$$

Proof. [*Only if part.*] Trivially, $\text{im}\, D \subset \mathcal{R}(A + BF, \text{im}\, D)$. Since $\mathcal{R}(A + BF, \text{im}\, D)$ is an $(A + BF)$-invariant, it is an (A, im B)-controlled invariant in view of Theorem 5.5, and, by assumption, is contained in ker C. Therefore, $\mathcal{R}(A + BF, \text{im}\, D) \subset \mathcal{V}_{\max}(A, \text{im}\, B, \ker C)$. Thus, Equation 5.47 holds.

[If part] Let F be such that

$$(A+BF)\mathcal{V}_{\max}(A, \operatorname{im} B, \ker C) \subset \mathcal{V}_{\max}(A, \operatorname{im} B, \ker C). \tag{5.48}$$

Inclusion 5.47 implies, consecutively, $(A+BF)\operatorname{im} D \subset \mathcal{V}_{\max}(A, \operatorname{im} B, \ker C), \ldots, (A+BF)^{n-1}\operatorname{im} D \subset \mathcal{V}_{\max}(A, \operatorname{im} B, \ker C)$. Therefore, $\mathcal{R}(A+BF, \operatorname{im} D) \subset \mathcal{V}_{\max}(A, \operatorname{im} B, \ker C) \subset \ker C$. QED

Remark

Let $n^* := \dim \mathcal{V}_{\max}(A, \operatorname{im} B, \ker C)$, and let T be a real nonsingular matrix for which the first n^* columns form a basis of $\mathcal{V}_{\max}(A, \operatorname{im} B, \ker C)$. Then, Inclusion 5.47 means that, with an $m \times n$ real matrix F satisfying Inclusion 5.48, the coordinate transformation $x(t) = T\tilde{x}(t)$ transforms Equations 5.44 and 5.42 to

$$\begin{bmatrix} \dot{\tilde{x}}_1(t) \\ \dot{\tilde{x}}_2(t) \end{bmatrix} = \begin{bmatrix} \tilde{A}_{11}+\tilde{B}_1\tilde{F}_1 & \tilde{A}_{12}+\tilde{B}_1\tilde{F}_2 \\ 0 & \tilde{A}_{22}+\tilde{B}_2\tilde{F}_2 \end{bmatrix} \begin{bmatrix} \tilde{x}_1(t) \\ \tilde{x}_2(t) \end{bmatrix} + \begin{bmatrix} \tilde{D}_1 \\ 0 \end{bmatrix} w(t) \tag{5.49}$$

and

$$y(t) = \begin{bmatrix} 0 & \tilde{C}_2 \end{bmatrix} \begin{bmatrix} \tilde{x}_1(t) \\ \tilde{x}_2(t) \end{bmatrix}, \tag{5.50}$$

where $\begin{bmatrix} \tilde{F}_1 & \tilde{F}_2 \end{bmatrix} = FT$. Clearly, $w(t)$ does not affect $\tilde{x}_2(t)$; hence, no effect on $y(t)$.

5.9 Disturbance Localization with Stability

In the previous section, the disturbance localization problem without additional constraints was solved. In this section, the problem is examined with an important constraint of stability. For the system Σ_d:=[A,B,D,C] described by Equations 5.44 and 5.42, or for the matrices A, B, D, and C, we have the following.

Problem 5.2: Disturbance Localization Problem with Stability

Find (if possible) an $m \times n$ real matrix F, such that (1) Inclusion 5.46 holds and (2) $A+BF$ is stable, that is, the eigenvalues of $A+BF$ have negative real parts.

Trivially, for condition (2) to be true, it is necessary that the pair (A, B) is stabilizable. We have (see Basile and Marro, 1992, Section 4.2)

Theorem 5.14:

Let A, B, D, and C be as before, and assume that the pair (A, B) is stabilizable. Then, the disturbance localization problem with stability has a solution if and only if (1) Inclusion 5.47 holds and (2) $\mathcal{V}_{sb,\min}(A, \operatorname{im} B + \operatorname{im} D, \ker C)$ is an internally stabilizable $(A, \operatorname{im} B)$-controlled invariant.

Remark

An alternative condition can be found in Wonham (1985, Theorem 5.8).

We will first prove that the above two conditions are sufficient. This part of the proof leads to a constructive procedure to find a matrix F which solves the problem.

Sufficiency proof of Theorem 5.14. The condition (2) means that there is a real matrix F, such that $(A+BF)$ $\mathcal{V}_{sb,\min}(A, \operatorname{im} B + \operatorname{im} D, \ker C) \subset \mathcal{V}_{sb,\min}(A, \operatorname{im} B + \operatorname{im} D, \ker C)$ and $\mathcal{V}_{sb,\min}(A, \operatorname{im} B + \operatorname{im} D, \ker C)$ is internally stable. Since the pair (A, B) is stabilizable, $\mathcal{V}_{sb,\min}(A, \operatorname{im} B + \operatorname{im} D, \ker C)$ is externally stabilizable by virtue of Lemma 5.16. Since the internal stability of $\mathcal{V}_{sb,\min}(A, \operatorname{im} B + \operatorname{im} D, \ker C)$ (with respect to $A+BF$) is determined by $F|\,\mathcal{V}_{sb,\min}(A, \operatorname{im} B + \operatorname{im} D, \ker C)$ and since the external stability depends only on $F|\,\mathcal{X}/\mathcal{V}_{sb,\min}(A, \operatorname{im} B + \operatorname{im} D, \ker C)$, the matrix F can be chosen so that the controlled invariant is both internally and externally stable, hence, $A + BF$ is stable. It remains to show that Equation 5.46 holds. Since $\mathcal{V}_{sb,\min}(A, \operatorname{im} B + \operatorname{im} D, \ker C)$ is self-bounded with respect to $\ker C$, and since Equation 5.47 holds, we have

$$\begin{aligned}\mathcal{V}_{sb,\min}(A, \operatorname{im} B + \operatorname{im} D, \ker C) &\supset (\operatorname{im} B + \operatorname{im} D) \cap \mathcal{V}_{\max}(A, \operatorname{im} B + \operatorname{im} D, \ker C)\\ &\supset (\operatorname{im} B + \operatorname{im} D) \cap \mathcal{V}_{\max}(A, \operatorname{im} B, \ker C)\\ &= (\operatorname{im} B \cap \mathcal{V}_{\max}(A, \operatorname{im} B, \ker C)) + \operatorname{im} D \supset \operatorname{im} D.\end{aligned}$$

This inclusion and the $(A+BF)$-invariance of $\mathcal{V}_{sb,\min}(A, \operatorname{im} B + \operatorname{im} D, \ker C)$ imply $\mathcal{R}(A+BF, \operatorname{im} D) \subset \mathcal{V}_{sb,\min}(A, \operatorname{im} B + \operatorname{im} D, \ker C) \subset \ker C$. QED

Remark

In the remark after Theorem 5.13, redefine $n^* := \dim \mathcal{V}_{sb,\min}(A, \operatorname{im} B + \operatorname{im} D, \ker C)$. Then, $\sigma(A + BF|\mathcal{V}_{sb,\min}(A, \operatorname{im} B + \operatorname{im} D, \ker C)) = \sigma\left(\tilde{A}_{11} + \tilde{B}_1\tilde{F}_1\right)$ and $\sigma(A + BF|\mathcal{X}/\mathcal{V}_{sb,\min}(A, \operatorname{im} B + \operatorname{im} D, \ker C)) = \sigma\left(\tilde{A}_{22} + \tilde{B}_2\tilde{F}_2\right)$. Therefore, we choose $\tilde{F}_1$ and $\tilde{F}_2$ so that $\tilde{A}_{11} + \tilde{B}_1\tilde{F}_1$ and $\tilde{A}_{22} + \tilde{B}_2\tilde{F}_2$ are stable and $\tilde{A}_{21} + \tilde{B}_2\tilde{F}_1 = 0$. Then, the desired F is given by $F = \begin{bmatrix}\tilde{F}_1 & \tilde{F}_2\end{bmatrix} T^{-1}$.

To prove the necessity we use the following

Lemma 5.17:

If there exists an internally stabilizable $(A, \operatorname{im} B)$-controlled invariant $\mathcal{V}$ satisfying $\mathcal{V} \subset \ker C$ and $\operatorname{im} D \subset \mathcal{V}$, then $\mathcal{V}_{sb,\min}(A, \operatorname{im} B + \operatorname{im} D, \ker C)$ is an internally stabilizable $(A, \operatorname{im} B)$-controlled invariant, that is, it is $(A+BF)$-invariant and $A + BF|\,\mathcal{V}_{sb,\min}(A, \operatorname{im} B + \operatorname{im} D, \ker C)$ is stable for some real matrix F.

Proof. The lemma follows from Schumacher (1983, Propositions 3.1 and 4.1). It is also an immediate consequence of Lemma 4.2.1 in Basile and Marro (1992).

Necessity proof of Theorem 5.14. Suppose that the problem has a solution, that is, there is a real matrix F such that $A + BF$ is stable and $\mathcal{V} := \mathcal{R}(A+BF, \operatorname{im} D) \subset \ker C$. Clearly, $\operatorname{im} D \subset \mathcal{V}$. Also, $\mathcal{V}$ is an $(A+BF)$-invariant contained in $\ker C$. Hence, $\mathcal{V} \subset \mathcal{V}_{\max}(A, \operatorname{im} B, \ker C)$. Therefore, condition (i) holds. To show condition (ii), note that $\mathcal{V}$ is internally (and externally) stable with respect to $A + BF$, $\mathcal{V} \subset \ker C$, and $\operatorname{im} D \subset \mathcal{V}$. Condition (ii) follows by Lemma 5.17. QED

5.10 Disturbance Localization by Dynamic Compensator

In the previous section, we used the (static) state-feedback control law $u(t) = Fx(t)$ to achieve disturbance rejection and stability. In this section, we use measurement output feedback with a dynamic compensator Σ_c placed in the feedback loop to achieve the same objectives.

More specifically, we are concerned with the system $\Sigma_{dm} := [A, B, D, C, C_{meas}]$ described by

$$\dot{x}(t) = Ax(t) + Bu(t) + Dw(t), \quad x(0) = x_o \tag{5.41}$$

$$y(t) = Cx(t), \tag{5.42}$$

$$y_{meas}(t) = C_{meas}x(t), \tag{5.51}$$

where $y_{meas}(t) \in \mathcal{Y}_{meas} := R^{p_m}$ stands for a measurement output and C_{meas} is a $p_m \times n$ real matrix. Define the dynamic compensator $\Sigma_c := [A_c, B_c, C_c, K_c]$ by

$$\dot{x}_c(t) = A_c x_c(t) + B_c y_{meas}(t), \quad x_c(0) = x_{co}, \tag{5.52}$$

$$u(t) = C_c x_c(t) + K_c y_{meas}(t), \tag{5.53}$$

where $x_c(t),\ x_{co} \in \mathcal{X}_c := R^{n_c}$, and A_c, B_c, C_c, and K_c are, respectively, $n_c \times n_c, n_c \times p_m, m \times n_c, m \times p_m$ real matrices. We wish to determine A_c, B_c, C_c, and K_c such that (1) $y(t) = 0$ for all $t \geq 0$ for any (admissible) $w(t), t \geq 0$ provided $x_o = 0$ and $x_{co} = 0$, and (2) $x(t)$ and $x_c(t)$ converges to zero as $t \to +\infty$ for all $x_o \in \mathcal{X}$ and $x_{co} \in \mathcal{X}_c$. For this, it is convenient to introduce the *extended state* defined by

$$\hat{x} := \begin{bmatrix} x \\ x_c \end{bmatrix} \in \hat{\mathcal{X}} := \mathcal{X} \times \mathcal{X}_c = R^{n+n_c},$$

where $x \in \mathcal{X}$ and $x_c \in \mathcal{X}_c$. Then, the overall system (including Σ_{dm} and Σ_c) can be rewritten as

$$\dot{\hat{x}}(t) = \hat{A}\hat{x}(t) + \hat{D}w(t), \quad \hat{x}(0) = \hat{x}_o, \tag{5.54}$$

$$y(t) = \hat{C}\hat{x}(t), \tag{5.55}$$

where $\hat{x}_o := \begin{bmatrix} x_o' & x_{co}' \end{bmatrix}'$ and the matrices are defined by

$$\hat{A} := \begin{bmatrix} A + BK_cC_{meas} & BC_c \\ B_cC_{meas} & A_c \end{bmatrix}, \quad \hat{D} := \begin{bmatrix} D \\ 0 \end{bmatrix}, \quad \hat{C} := \begin{bmatrix} C & 0 \end{bmatrix}. \tag{5.56}$$

Remark

Define

$$\hat{A}_o := \begin{bmatrix} A & 0 \\ 0 & 0 \end{bmatrix}, \quad \hat{B}_o := \begin{bmatrix} B & 0 \\ 0 & I_{n_c} \end{bmatrix}, \quad \hat{C}_{meas} := \begin{bmatrix} C_{meas} & 0 \\ 0 & I_{n_c} \end{bmatrix},$$

$$\hat{K}_c := \begin{bmatrix} K_c & C_c \\ B_c & A_c \end{bmatrix}. \tag{5.57}$$

Then, it is easy to verify that

$$\hat{A} = \hat{A}_o + \hat{B}_o\hat{K}_c\hat{C}_{meas}. \tag{5.58}$$

(See Basile and Marro (1992, Section 5.1) and Willems and Commault (1981).)

Remark

An important special case of Σ_c is a state estimate feedback through an observer. More specifically, let L_1 and L_2satisfy

$$L_1C_{meas} + L_2 = I_n, \tag{5.59}$$

and consider an asymptotic observer (or state estimator) described by

$$\dot{x}_c(t) = Ax_c(t) + Bu(t) + G\{C_{meas}x_c(t) - y_{meas}(t)\}, x_c(0) = x_{co}, \tag{5.60}$$

$$x_{est}(t) = L_2x_c(t) + L_1y_{meas}(t), \tag{5.61}$$

where $x_{est}(t) \in R^n$ is an "estimate" of $x(t)$ (under proper conditions) and G, L_1, and L_2 are real matrices with appropriate dimensions. If we use a state estimate feedback law

$$u(t) = Fx_{est}(t) \tag{5.62}$$

with an $m \times n$ real matrix F, then the overall system reduces to Equations 5.54 and 5.55, where $\hat{A}$, $\hat{D}$, and $\hat{C}$ are given, instead of Equation 5.56, by

$$\hat{A} := \begin{bmatrix} A + BFL_1C_{meas} & BFL_2 \\ (BFL_1 - G)C_{meas} & A + BFL_2 + GC_{meas} \end{bmatrix}, \quad \hat{D} := \begin{bmatrix} D \\ 0 \end{bmatrix}, \quad \hat{C} := \begin{bmatrix} C & 0 \end{bmatrix}. \tag{5.63}$$

It should also be noted that, if we apply the coordinate transformation

$$\hat{x} = \begin{bmatrix} I_n & 0 \\ I_n & -I_n \end{bmatrix} \bar{x},$$

Equations 5.54 and 5.63 can be rewritten as

$$\dot{\bar{x}}(t) = \bar{A}\bar{x}(t) + \bar{D}w(t), \quad \bar{x}(0) = \bar{x}_o, \tag{5.64}$$

where

$$\bar{A} := \begin{bmatrix} A + BF & -BFL_2 \\ 0 & A + GC_{meas} \end{bmatrix}, \quad \bar{D} := \begin{bmatrix} D \\ D \end{bmatrix}, \tag{5.65}$$

from which it is apparent that F and G can be selected independently (based on Theorem 5.2 and its dual) to make $\bar{A}$ stable, that is, the separation property holds. (See Basile and Marro (1992, Sections 3.4.2, 5.1.2) for further discussions.)

With the above notation the problem can be restated in geometric terms as:

Problem 5.3: Disturbance Localization Problem with Stability by Dynamic Compensator

Find (if possible) a number $n_c(= \dim \mathcal{X}_c)$ and real matrices A_c, B_c, C_c, and K_c of dimensions $n_c \times n_c, n_c \times p_m, m \times n_c$, and $m \times p_m$, respectively, such that

$$(i) \quad \mathcal{R}(\hat{A}, \text{im}\,\hat{D}) \subset \ker \hat{C}, \tag{5.66}$$

and

$$(ii) \quad \hat{A} \text{ is stable.} \tag{5.67}$$

Remark

Noting that $\mathcal{R}(\hat{A}, \text{im}\,\hat{D})$ is the minimum $\hat{A}$-invariant containing im $\hat{D}$, it is easy to see that condition (i) is equivalent to the following condition:

$(i)'$ there is an $\hat{A}$-invariant $\hat{\mathcal{V}}$ satisfying

$$im\,\hat{D} \subset \hat{\mathcal{V}} \subset \ker \hat{C}. \tag{5.68}$$

The conditions under which the above problem is solvable are given by the following theorems.

Theorem 5.15:

Let (A, B) be stabilizable, and also let (A, C_{meas}) be detectable. Then, the disturbance localization problem with stability by dynamic compensator has a solution if and only if there exists an internally stabilizable

$(A, \text{im } B)$-controlled invariant $\mathcal{V}$ and an externally stabilizable $(A, \ker C_{meas})$-conditioned invariant $\mathcal{S}$ satisfying the condition

$$\text{im } D \subset \mathcal{S} \subset \mathcal{V} \subset \ker C. \tag{5.69}$$

Proof. See Basile and Marro (1992, Section 5.2) and Willems and Commault (1981). QED

The above condition is an existence condition and it is not convenient for testing. The following theorem (see Theorem 5.2–2 in Basile and Marro, 1992) gives constructive conditions and they can be readily tested.

Theorem 5.16:

Assume that (A, B) is stabilizable and (A, C_{meas}) is detectable. The disturbance localization problem with stability by dynamic compensator has a solution if and only if the following conditions hold:

i.

$$\mathcal{S}_{\min}(A, \ker C_{meas}, \text{im } D) \subset \mathcal{V}_{\max}(A, \text{im } B, \ker C), \tag{5.70}$$

ii. $\mathcal{S}_{\min}(A, \ker C_{meas}, \text{im } D) + \mathcal{V}_{\max}(A, \text{im } D, \ker C \cap \ker C_{meas})$ *is an externally stabilizable $(A, \ker C_{meas})$-conditioned invariant*
iii. $\mathcal{V}_{sb,\min}(A, \text{im } B + \text{im } D, \ker C) + \mathcal{V}_{\max}(A, \text{im } D, \ker C \cap \ker C_{meas})$ *is an internally stabilizable $(A, \text{im } B)$-controlled invariant.*

Remark

The subspaces $\mathcal{S}_{\min}(A, \ker C_{meas}, \text{im } D) + \mathcal{V}_{\max}(A, \text{im } D, \ker C \cap \ker C_{meas})$ and $\mathcal{V}_{sb,\min}(A, \text{im } B + \text{im } D, \ker C) + \mathcal{V}_{\max}(A, \text{im } D, \ker C \cap \ker C_{meas})$ defined above are, respectively, an $(A, \ker C_{meas})$-conditioned invariant and an $(A, \text{im } B)$-controlled invariant. (See Lemma 5.18 at the end of this section for the proof the above remark.)

Proof of Necessity of Theorem 5.16. See Basile and Marro (1992, Section 5.2) and Basile et al. (1986). QED

Proof of Sufficiency for Theorem 5.16. Let

$$\mathcal{S} := \mathcal{S}_{\min}(A, \ker C_{meas}, \text{im } D) + \mathcal{V}_{\max}(A, \text{im } D, \ker C \cap \ker C_{meas}), \tag{5.71}$$

$$\mathcal{V} := \mathcal{V}_{sb,\min}(A, \text{im } B + \text{im } D, \ker C) + \mathcal{V}_{\max}(A, \text{im } D, \ker C \cap \ker C_{meas}). \tag{5.72}$$

We will show that the above $\mathcal{S}$ and $\mathcal{V}$ satisfy the conditions of Theorem 5.15. By assumptions (ii) and (iii), $\mathcal{S}$ is externally stabilizable and $\mathcal{V}$ is internally stabilizable. By Inclusion 5.70 and Equation 5.71, trivially $\text{im } D \subset \mathcal{S}$. By Equation 5.72, $\mathcal{V} \subset \ker C$ trivially also. To see the validity of the middle inclusion of Inclusion 5.69, that is, $\mathcal{S} \subset \mathcal{V}$, it suffices to show $\mathcal{S}_{\min}(A, \ker C_{meas}, \text{im } D) \subset \mathcal{V}_{sb,\min}(A, \text{im } B + \text{im } D, \ker C)$. For this, first observe that, by Inclusion 5.70, we have $\text{im } D \subset \mathcal{V}_{\max}(A, \text{im } B, \ker C)$, which implies $\mathcal{V}_{\max}(A, \text{im } B + \text{im } D, \ker C) = \mathcal{V}_{\max}(A, \text{im } B, \ker C)$ since trivially $\mathcal{V}_{\max}(A, \text{im } B + \text{im } D, \ker C) \supset \mathcal{V}_{\max}(A, \text{im } B, \ker C)$ and $A\mathcal{V}_{\max}(A, \text{im } B + \text{im } D, \ker C) \subset \mathcal{V}_{\max}(A, \text{im } B + \text{im } D, \ker C) + \text{im } B + \text{im } D = \mathcal{V}_{\max}(A, \text{im } B + \text{im } D, \ker C) + \text{im } B$. In addition, Inclusion 5.70 implies $\mathcal{S}_{\min}(A, \ker C_{meas}, \text{im } D) \subset \ker C$, which leads to $\mathcal{S}_{\min}(A, \ker C_{meas}, \text{im } D) = \mathcal{S}_{\min}(A, \ker C_{meas} \cap \ker C, \text{im } D) \subset \mathcal{S}_{\min}(A, \ker C, \text{im } D) \subset \mathcal{S}_{\min}(A, \ker C, \text{im } B + \text{im } D)$.

By using Inclusion 5.70, the above facts, and Theorem 5.7, we have

$$\begin{aligned}\mathcal{S}_{\min}(A, \ker C_{meas}, \operatorname{im} D) &\subset \mathcal{V}_{\max}(A, \operatorname{im} B, \ker C) \cap \mathcal{S}_{\min}(A, \ker C, \operatorname{im} B + \operatorname{im} D)\\ &= \mathcal{V}_{\max}(A, \operatorname{im} B + \operatorname{im} D, \ker C) \cap \mathcal{S}_{\min}(A, \ker C, \operatorname{im} B + \operatorname{im} D)\\ &= \mathcal{V}_{sb,\min}(A, \operatorname{im} B + \operatorname{im} D, \ker C).\end{aligned} \tag{5.73}$$

Thus, Inclusion 5.69 holds. QED

Procedure for Finding A_c, B_c, C_c, and K_c

Define $\mathcal{S}$ and $\mathcal{V}$ by Equations 5.71 and 5.72. Let $\mathcal{L}$ be a subspace of R^n satisfying

$$\mathcal{L} \oplus (\mathcal{S} \cap \ker C_{meas}) = \mathcal{S}. \tag{5.74}$$

Clearly,

$$\mathcal{L} \cap \ker C_{meas} = \{0\}. \tag{5.75}$$

Now let $\mathcal{L}_{comp}$ be a subspace satisfying $\mathcal{L} \oplus \mathcal{L}_{comp} = R^n$ and $\mathcal{L}_{comp} \supset \ker C_{meas}$, and define the projection L_2 on $\mathcal{L}_{comp}$ along $\mathcal{L}$. Select L_1 such that

$$L_1 C_{meas} = I_n - L_2. \tag{5.76}$$

Such an L_1 exists since $\ker(I_n - L_2) = \mathcal{L}_{comp} \supset \ker C_{meas}$.

Choose real matrices F and G so that

$$(A + BF)\mathcal{V} \subset \mathcal{V}, \tag{5.77}$$

$$(A + GC_{meas})\mathcal{S} \subset \mathcal{S}, \tag{5.78}$$

and $A + BF$ and $A + GC_{meas}$ are stable. Then, use Equations 5.60 through 5.62 as a dynamic compensator, or equivalently, set $A_c := A + BFL_2 + GC_{meas}$, $B_c := BFL_1 - G$, $C_c := FL_2$ and $K_c := FL_1$ in Equations 5.52 and 5.53. From Equation 5.65, the overall system is clearly stable (and so is $\hat{A}$). It can also be shown (see Basile and Marro, 1992; Lemmas 5.1.1 and 5.1.2, and Section 5.2.1) that

$$\hat{\mathcal{V}} := \left\{ \begin{bmatrix} v \\ v - s \end{bmatrix} : v \in \mathcal{V},\ s \in \mathcal{S} \right\}$$

is $\hat{A}$-invariant and satisfies Equation 5.68.

Lemma 5.18:

The subspaces $\mathcal{S}$ defined by Equations 5.71 is an $(A, \ker C_{meas})$-conditioned invariant; and, if Inclusion 5.70 holds, then $\mathcal{V}$ defined by Equation 5.72 is an $(A, \operatorname{im} B)$-controlled invariant.

Proof. Since $\mathcal{V}_{\max}(A, \operatorname{im} D, \ker C \cap \ker C_{meas}) \subset \ker C_{meas}$, by the last equality of Lemma 5.2 we have $\mathcal{S} \cap \ker C_{meas} = \mathcal{S}_{\min}(A, \ker C_{meas}, \operatorname{im} D) \cap \ker C_{meas} + \mathcal{V}_{\max}(A, \operatorname{im} D, \ker C \cap \ker C_{meas})$. Then, it is easy to see $A(\mathcal{S} \cap \ker C_{meas}) \subset \mathcal{S} + \operatorname{im} D$. But, $\operatorname{im} D \subset \mathcal{S}_{\min}(A, \ker C_{meas}, \operatorname{im} D) \subset \mathcal{S}$. Thus, $A(\mathcal{S} \cap \ker C_{meas}) \subset \mathcal{S}$. To prove that $\mathcal{V}$ is an $(A, \operatorname{im} B)$-controlled invariant, first it is easy to show $A\mathcal{V} \subset \mathcal{V} + \operatorname{im} B + \operatorname{im} D$. But, by Inclusion 5.73, $\operatorname{im} D \subset \mathcal{V}_{sb,\min}(A, \operatorname{im} B + \operatorname{im} D, \ker C) \subset \mathcal{V}$. Thus, $A\mathcal{V} \subset \mathcal{V} + \operatorname{im} B$. QED

5.11 Conclusion

An introduction to the geometric method (often called geometric approach) has been provided. The basic tools have been described and applied to the disturbance localization (also called disturbance decouping) problems to demonstrate the use of the tools. Numerous research articles were published on the subject for the last 40 years. Some are included in the list of references in this chapter and a great many others can be found in the books and articles already referenced in this chapter. Reports for practical applications can also be found in the literature, for example, Takamatsu et al. (1979); Massoumnia et al. (1989); Prattichizzott et al. (1997); Barbagli et al. (1998); and Marro and Zattoni (2004). This chapter has dealt with continuous-time systems. Many articles in the literature treat discrete-time systems, for example, Akashi and Imai (1979), Hamano and Basile (1983), and Marro and Zattoni (2006).

The notion of controlled invariant subspaces has been extended to "almost" controlled invariant subspaces, to which the state trajectory can stay arbitrary close by a suitable choice of inputs, possibly with high-gain feedback (Willems, 1981). The theory involves use of distributions for inputs. The dual notion of "almost" conditionally invariant subspaces has also been introduced (Willems, 1982). The geometric approach has also been extended to more general or different classes of linear systems, for example, 2D systems (Conte and Perdon, 1988), systems over a ring including delay differential systems (Conte and Perdon, 1998), and periodic systems (Longhi and Monteriù, 2007). The geometric notions for linear systems have been extended to nonlinear systems (controlled invariant distributions, in particular) using differential geometric tools and problems such as disturbance localization and output decoupling have been studied in the literature (see Nijmeijer and van der Schaft (1990), Isidori (1995), and Mattone and De Luca (2006), for instance). A 40-year historical perspective of the development of geometric methods for linear and nonlinear systems with an insight into the future can be found in Marro (2008).

References

Akashi, H. and Imai, H. 1979. Disturbance localization and output deadbeat control through an observer in discrete-time linear multivariable systems. *IEEE Trans. Automat. Control,* AC-24: 621–627.

Barbagli, F., Marro, G., Mercorelli, P., and Prattichizzo, D. 1998. Some results on output algebraic feedback with applications to mechanical systems. *Proc. 37 th IEEE Conf. Decision Control (CDC 1998),* Tampa, Florida, USA, 3545–3550.

Basile, G. and Marro, G. 1969. Controlled and conditioned invariant subspaces in linear system theory. *J. Optim. Theory Appl.,* 3(5): 305–315.

Basile, G. and Marro, G. 1982. Self-bounded controlled invariant subspaces. *J. Optim. Theory Appl.,* 38(1):71–81.

Basile, G. and Marro, G. 1992. *Controlled and Conditioned Invariants in Linear System Theory,* Prentice-Hall, Englewood Cliffs, NJ.

Basile, G., Marro, G., and Piazzi, A. 1986. Stability without eigenspaces in the geometric approach: some new results. In *Frequency Domain and State Space Methods for Linear Systems,* C. A. Byrnes and A. Lindquist (Eds). North-Holland (Elsevier), Amsterdam, pp. 441–450.

Conte, G. and Perdon, A. M. 1988. A geometric approach to the theory of 2-D systems. *IEEE Trans. Automat. Control,* 33: 946–950.

Conte, G. and Perdon, A. M. 1998. The block decoupling problem for systems over a ring. *IEEE Trans. Automat. Control,* AC-43: 1600–1604.

Hamano, F. and Basile, G. 1983. Unknown-input present-state observability of discrete-time linear systems. *J. Optim. Theory Appl.,* 40(2): 293–307.

Hamano, F. and Furuta, K. 1975. Localization of disturbances and output decomposition of linear multivariable systems. *Int. J. Control,* 22(4): 551–562.

Heymann, M. 1968. Comments on Pole assignment in multi-input controllable linear systems. *IEEE Trans. Automat. Control,* AC-13: 748–749.

Isidori, A. 1995. *Nonlinear Control Systems* (3rd edition), Springer-Verlag, New York.

Kalman, R. E., Falb, P. L., and Arbib, M. A. 1969. *Topics in Mathematical System Theory,* McGraw-Hill, New York.

Longhi, S. and Monteriù, A. 2007. A geometric approach to fault detection of periodic systems. *Proc. 46th IEEE Conf. Decision Control,* New Orleans, USA.

Marro, G. 2008. The geometric approach to control: a light presentation of theory and applications. In *Control Science Evolution*, S. Bittanti (Ed.), CNR Publications, Rome, pp. 157–204.

Marro, G. and Zattoni, E. 2004. Detection of incipient failures by using geometric approach techniques: An application to railway switching points. *Proceedings of the 2nd IFAC Symposium on System, Structure and Control,* Oaxaca, Mexico. Marro, G. and Zattoni, E. 2006. Signal decoupling with preview in the geometric context: Exact solution for nonminimum-phase systems. *J. Optim. Theory Appl.,* 129(1): 165–183.

Massoumnia, M. A., Verghese, G. C., and Willski, A. S. 1989. Failure detection and identification. *IEEE Trans. Automat. Control,* 34: 316–321.

Mattone, R. and De Luca, A. 2006. Nonlinear fault detection and isolation in a three-tank heating system. *IEEE Trans. Control Systems Tech.,*14: 1158–1166.

Morse, A. S. 1973. Structural invariants of linear multivariable systems. *SIAM J. Control Optim.*, 11(3): 446–465.

Nijmeijer, H. and van der Schaft, A. J. 1990. *Nonlinear Dynamical Control Systems,* Springer-Verlag, New York.

Prattichizzott, D., Mercorellit, P., Bicchit, A., and Vicinol, A. 1997. On the geometric control of internal forces in power grasps. *Proceedings of the 36th Conf Decision Control,* San Diego, USA, pp. 1942–1947.

Schumacher, J.M. 1983. On a conjecture of Basile and Marro. *J. Optim. Theory Appl.,* 41(2): 371–376.

Takamatsu, T., Hashimoto, I., and Nakai, Y. 1979. A geometric approach to multivariable control system design of a distillation column. *Automatica,* 15: 387–402.

Trentelman, H. L., Stoorvogel, A. A., and Hautus, M. 2002. *Control Theory for Linear Systems.* Springer-Verlag, London.

Willems, J. C. 1981. Almost invariant subspaces: An approach to high gain feedback design—Part I: Almost controlled invariant subspaces. *IEEE Trans. Automat. Control* AC-26: 235–252.

Willems, J. C. 1982. Almost invariant subspaces: An approach to high gain feedback design—Part II: Almost conditionally invariant subspaces. *IEEE Trans. Automat. Control* AC-27: 1071–1085.

Willems, J. C. and Commault, C. 1981. Disturbance decoupling by measurement feedback with stability or pole placement. *SIAM J. Control Optim.*, 19(4): 490–504.

Wonham, W. M. 1985. *Linear Multivariable Control, A Geometric Approach* (3rd edition), Springer-Verlag, New York.

Wonham, W. M. and Morse, A. S. 1970. Decoupling and pole assignment in linear multivariable systems: A geometric approach. *SIAM J. Control Optim.*, 8(1): 1–18.

6

Polynomial and Matrix Fraction Descriptions

David F. Delchamps
Cornell University

6.1 Introduction

For control system design, it is useful to characterize multi-input, multi-output, time-invariant linear systems in terms of their transfer function matrices. The transfer function matrix of a real m-input, p-output continuous-time system is a $(p \times m)$ matrix-valued function $G(s)$, where s is the Laplace transform variable; the corresponding object in discrete time is a $(p \times m)$ matrix-valued function $G(z)$, where z is the z-transform variable. Things are particularly interesting when $G(s)$ or $G(z)$ is a *proper rational matrix* function of s or z, that is, when every entry in $G(s)$ or $G(z)$ is a ratio of two real polynomials in s or z whose denominator's degree is at least as large as its numerator's degree. In this case, the system has state space realizations of the form

$$\begin{aligned} \dot{x}(t) &= Ax(t) + Bu(t) \\ y(t) &= Cx(t) + Du(t) \end{aligned}$$

or

$$\begin{aligned} x(k+1) &= Ax(k) + Bu(k) \\ y(k) &= Cx(k) + Du(k), \end{aligned}$$

where the state vector x takes values in $\mathbf{R}^n$. Any such realization defines a decomposition $G(s) = C(sI_n - A)^{-1}B + D$ or $G(z) = C(zI_n - A)^{-1}B + D$ for the system's transfer function matrix. A realization is minimal when the state vector dimension n is as small as it can be; the *MacMillan degree* $\delta_M(G(s))$ or $\delta_M(G(z))$ is the value of n in a minimal realization. A system's MacMillan degree is a natural candidate for the "order of the system."

It is not easy to construct minimal realizations for a multi-input, multi-output system with an arbitrary proper rational transfer function matrix; even determining such a system's MacMillan degree requires some effort. Circumstances are simpler for single-input, single-output (SISO) systems. Consider, for example, a real SISO continuous-time system with proper rational transfer function $g(s)$. We can express

$g(s)$ as a ratio

$$g(s) = \frac{p(s)}{q(s)}, \tag{6.1}$$

where $q(s)$ is a polynomial of degree n, $p(s)$ is a polynomial of degree at most n, and $p(s)$ and $q(s)$ are coprime, which is to say that their only common factors are nonzero real numbers. The coprimeness of $p(s)$ and $q(s)$ endows the fractional representation Equation 6.1 with a kind of irreducibility; furthermore, Equation 6.1 is "minimal" in the sense that any other factorization $g(s) = \hat{p}(s)/\hat{q}(s)$ features a denominator polynomial $\hat{q}(s)$ whose degree is at least n.

The MacMillan degree $\delta_M(g(s))$ is precisely n, the degree of $q(s)$ in Equation 6.1. To see that $n \leq \delta_M(g(s))$, suppose (A, B, C, D) are the matrices in a minimal realization for $g(s)$. Set $\hat{n} = \delta_M(g(s))$, so A is $(\hat{n} \times \hat{n})$. The matrix $(sI_{\hat{n}} - A)^{-1}$ has rational entries whose denominator polynomials have degrees at most $\hat{n}$. Multiplying on the left by $\hat{C}$ and on the right by $\hat{B}$ and finally adding $\hat{D}$ results in a rational function whose denominator degree in lowest terms is at most $\hat{n}$. This rational function, however, is $g(s)$, whose lowest-terms denominator degree is n, whence it follows that $n \leq \hat{n}$.

To finish showing that $\delta_M(g(s)) = n$, it suffices to construct a realization whose A-matrix is $(n \times n)$. There are many ways to do this; here is one. Begin by setting $d = \lim_{|s| \to \infty} g(s)$; d is well-defined because $g(s)$ is proper. Then

$$g(s) = \frac{\tilde{p}(s)}{q(s)} + d,$$

where the degree of $\tilde{p}(s)$ is at most $n - 1$. Cancel the coefficient of s^n from $q(s)$ and $\tilde{p}(s)$ so that

$$q(s) = s^n + q_1 s^{n-1} + \cdots + q_{n-1}s + q_n$$

and

$$\tilde{p}(s) = \gamma_1 s^{n-1} + \gamma_2 s^{n-2} + \cdots + \gamma_{n-1}s + \gamma_n.$$

It is not hard to verify that $g(s) = C(sI_n - A)^{-1}B + D$ when

$$A = \begin{bmatrix} 0 & 1 & 0 & . & . & . & 0 \\ 0 & 0 & 1 & 0 & . & . & 0 \\ 0 & 0 & 0 & 1 & . & . & 0 \\ . & . & . & . & . & . & . \\ . & . & . & . & . & 1 & 0 \\ 0 & 0 & . & . & . & 0 & 1 \\ -q_n & -q_{n-1} & . & . & . & -q_2 & -q_1 \end{bmatrix}, \quad B = \begin{bmatrix} 0 \\ 0 \\ . \\ . \\ . \\ 0 \\ 1 \end{bmatrix}, \tag{6.2}$$

and

$$C = \begin{bmatrix} \gamma_n & \gamma_{n-1} & . & . & . & \gamma_1 \end{bmatrix}, \quad D = d. \tag{6.3}$$

To summarize the foregoing discussion, if $g(s)$ is the transfer function of a real time-invariant continuous-time SISO linear system, and $g(s)$ is proper rational, then $g(s)$ possesses a fractional representation Equation 6.1 that

- is irreducible because $p(s)$ and $q(s)$ are coprime, or, equivalently.
- is minimal because the degree of $q(s)$ is as small as it can be in such a representation.

Moreover,

- the degree of $q(s)$ in any minimal and irreducible representation Equation 6.1 is equal to the MacMillan degree of $g(s)$.

Replacing s with z results in identical assertions about discrete-time SISO systems.

Our goal in what follows will be to generalize these results to cover continuous-time, multi-input, multi-output (MIMO) systems with $(p \times m)$ proper rational transfer function matrices $G(s)$. The development is purely algebraic and applies equally well to discrete-time MIMO systems with proper rational transfer function matrices $G(z)$; simply replace s with z throughout. In Section 6.2, we present MIMO versions of the fractional representation Equation 6.1, complete with appropriate matrix analogues of irreducibility and minimality. In Section 6.3, we relate the results of Section 6.2 to the MacMillan degree of $G(s)$. We obtain as a bonus a minimal realization of $G(s)$ that turns out to be the MIMO analogue of the realization in Equations 6.2 and 6.3. In Section 6.4 we prove a well-known formula for the MacMillan degree of $G(s)$ and discuss briefly some connections between the material in Section 6.2 and the theory of ARMA models for MIMO systems, with a nod toward some modern results about stable coprime factorization and robust control. The algebra we will be employing throughout, while tedious at times, is never terribly abstract, and much of it is interesting in its own right.

6.2 Polynomial Matrix Fraction Descriptions

We begin by establishing some basic terminology. A *real polynomial matrix* is a matrix each of whose entries is a polynomial in s with real coefficients. We often omit the adjective "real" but assume, unless stated otherwise, that all polynomial matrices are real. One can perform elementary operations on polynomial matrices such as addition and multiplication using the rules of polynomial algebra along with the standard formulas that apply to matrices of numbers. Likewise, one can compute the determinant $\det F(s)$ of a square polynomial matrix $F(s)$ by any of the usual formulas or procedures. All the familiar properties of the determinant hold for polynomial matrices, for example, the product rule

$$\det[F_1(s)F_2(s)] = [\det F_1(s)][\det F_2(s)], \tag{6.4}$$

which applies when $F_1(s)$ and $F_2(s)$ are the same size.

There are two shades of invertibility for square polynomial matrices.

Definition 6.1:

A square polynomial matrix $F(s)$ is said to be nonsingular if, and only if, $\det F(s)$ *is a nonzero polynomial and unimodular if, and only if,* $\det F(s)$ *is a nonzero real number.*

Thus if $F(s)$ is *nonsingular*, we are free to compute $F^{-1}(s)$ using the adjugate-determinant formula for the inverse [12], namely,

$$F^{-1}(s) = \frac{1}{\det F(s)} \operatorname{adj} F(s).$$

The (i, j) entry of $\operatorname{adj} F(s)$ is $(-1)^{i+j}$ times the determinant of the matrix obtained from $F(s)$ by eliminating its jth row and ith column. $F^{-1}(s)$ is in general a rational matrix function of s. If $F(s)$ is not just nonsingular but also *unimodular*, then $F^{-1}(s)$ is actually a polynomial matrix; thus unimodular polynomial matrices are "polynomially invertible."

From Definition 6.1 along with the product rule Equation 6.4, the product of two nonsingular polynomial matrices is nonsingular and the product of two unimodular polynomial matrices is unimodular. Furthermore, we have the following "pointwise" characterization of unimodularity: a square polynomial matrix $F(s)$ is unimodular precisely when $F(s_o)$ is an invertible matrix of complex numbers for every complex number s_o. This statement follows from the fundamental theorem of algebra, which implies that if det $F(s)$ is a polynomial of nonzero degree, then det $F(s_o) = 0$ for at least one complex number s_o.

As an example, consider the two polynomial matrices

$$F_1(s) = \begin{bmatrix} s+2 & s+2 \\ s+1 & s+2 \end{bmatrix} \quad \text{and} \quad F_2(s) = \begin{bmatrix} s+2 & s+4 \\ s & s+2 \end{bmatrix}.$$

$F_2(s)$ is unimodular and $F_1(s)$ is merely nonsingular; in fact,

$$F_1^{-1}(s) = \begin{bmatrix} \dfrac{1}{s+1} & -1 \\ -\dfrac{s+1}{s+2} & 1 \end{bmatrix} \quad \text{and} \quad F_2^{-1}(s) = \frac{1}{4}\begin{bmatrix} s+2 & -s-2 \\ -s & s+2 \end{bmatrix}.$$

We are ready to define the matrix generalization(s) of the fractional representation Equation 6.1.

Definition 6.2:

Let $G(s)$ be a real $(p \times m)$ proper rational matrix function of s.

- *A right matrix fraction description, or right MFD, of $G(s)$ is a factorization of the form $G(s) = P(s)Q^{-1}(s)$, where $P(s)$ is a real $(p \times m)$ polynomial matrix and $Q(s)$ is a real $(m \times m)$ nonsingular polynomial matrix.*
- *A left matrix fraction description, or left MFD, of $G(s)$ is a factorization of the form $G(s) = Q_L^{-1}(s)P_L(s)$, where $P_L(s)$ is a real $(p \times m)$ polynomial matrix and $Q_L(s)$ is a real $(p \times p)$ nonsingular polynomial matrix.*

It is easy to construct left and right matrix fraction descriptions for a given $(p \times m)$ proper rational matrix $G(s)$. For instance, let $q(s)$ be the lowest common denominator of the entries in $G(s)$. Set $P(s) = q(s)G(s)$; then $P(s)$ is a polynomial matrix. Setting $Q(s) = q(s)I_m$ makes $P(s)Q^{-1}(s)$ a right MFD for $G(s)$; setting $Q_L(s) = q(s)I_p$ makes $Q_L^{-1}(s)P(s)$ a left MFD of $G(s)$.

Next we introduce matrix versions of the notions of irreducibility and minimality associated with the fractional representation Equation 6.1. First consider minimality. Associated with each right (or left) MFD of $G(s)$ is the "denominator matrix" $Q(s)$ (or $Q_L(s)$); the degree of the nonzero polynomial det $Q(s)$ (or det $Q_L(s)$) plays a role analogous to that of the degree of the polynomial $q(s)$ in Equation 6.1.

Definition 6.3:

Let $G(s)$ be a real $(p \times m)$ proper rational matrix.

- *A right MFD $P(s)Q^{-1}(s)$ of $G(s)$ is minimal if, and only if, the degree of* det $\hat{Q}(s)$ *in any other right MFD $\hat{P}(s)\hat{Q}^{-1}(s)$ of $G(s)$ is at least as large as the degree of* det $Q(s)$.
- *A left MFD $Q_L^{-1}(s)P_L(s)$ of $G(s)$ is minimal if, and only if, the degree of* det $\hat{Q}_L(s)$ *in any other left MFD $\hat{Q}_L^{-1}(s)\hat{P}_L(s)$ of $G(s)$ is at least as large as the degree of* det $Q_L(s)$.

To formulate the matrix version of irreducibility, we require the following analogues of the coprimeness condition on the polynomials $p(s)$ and $q(s)$ appearing in Equation 6.1.

Definition 6.4:

- *Two polynomial matrices $P(s)$ and $Q(s)$ possessing m columns are said to be right coprime if, and only if, the following condition holds: If $F(s)$ is an $(m \times m)$ polynomial matrix so that for some polynomial matrices $\hat{P}(s)$ and $\hat{Q}(s)$, $P(s) = \hat{P}(s)F(s)$ and $Q(s) = \hat{Q}(s)F(s)$, then $F(s)$ is unimodular.*
- *Two polynomial matrices $P_L(s)$ and $Q_L(s)$ possessing p rows are said to be left coprime if, and only if, the following condition holds: If $F(s)$ is a $(p \times p)$ polynomial matrix so that for some polynomial matrices $\hat{P}_L(s)$ and $\hat{Q}_L(s)$, $P(s) = F(s)\hat{P}_L(s)$ and $Q_L(s) = F(s)\hat{Q}_L(s)$, then $F(s)$ is unimodular.*

Saying that two polynomial matrices are right (or left) coprime is the same as saying that they have no right (or left) common square polynomial matrix factors other than unimodular matrices. In this way, unimodular matrices play a role similar to that of nonzero real numbers in the definition of coprimeness for scalar polynomials. Any two unimodular matrices of the same size are trivially right and left coprime. Similarly, any $(r \times r)$ unimodular matrix is right coprime with any polynomial matrix that has r columns and left coprime with any polynomial matrix that has r rows. Equipped with Definition 6.4, we can explain what it means for a right or left MFD to be irreducible.

Definition 6.5:

Let $G(s)$ be a real $(p \times m)$ proper rational matrix.

- *A right MFD $P(s)Q^{-1}(s)$ of $G(s)$ is irreducible if, and only if, $P(s)$ and $Q(s)$ are right coprime.*
- *A left MFD $Q_L^{-1}(s)P_L(s)$ of $G(s)$ is irreducible if, and only if, $P_L(s)$ and $Q_L(s)$ are left coprime.*

It is time to begin reconciling Definition 6.3, Definition 6.5, and the scalar intuition surrounding irreducibility and minimality of the fractional representation Equation 6.1. The proof of the following central result will occupy most of the remainder of this section; the finishing touches appear after Lemmas 6.1 and 6.2 below.

Theorem 6.1:

Let $G(s)$ be a real $(p \times m)$ proper rational matrix.

- *A right MFD $P(s)Q^{-1}(s)$ of $G(s)$ is minimal if, and only if, it is irreducible. Furthermore, if $P(s)Q^{-1}(s)$ and $\hat{P}(s)\hat{Q}^{-1}(s)$ are two minimal right MFDs of $G(s)$, then an $(m \times m)$ unimodular matrix $V(s)$ exists so that $\hat{P}(s) = P(s)V(s)$ and $\hat{Q}(s) = Q(s)V(s)$.*
- *A left MFD $Q_L^{-1}(s)P_L(s)$ of $G(s)$ is minimal if, and only if, it is irreducible. Furthermore, if $Q_L^{-1}(s)P_L(s)$ and $\hat{Q}_L^{-1}(s)\hat{P}_L(s)$ are two minimal left MFDs of $G(s)$, then a $(p \times p)$ unimodular matrix $V(s)$ exists so that $\hat{P}_L(s) = V(s)P(s)$ and $\hat{Q}_L(s) = V(s)Q(s)$.*

Some parts of Theorem 6.1 are easier to prove than others. In particular, it is straightforward to show that a minimal right or left MFD is irreducible. To see this, suppose $G(s) = P(s)Q^{-1}(s)$ is a right MFD that is not irreducible. Since $P(s)$ and $Q(s)$ are not right coprime, factorizations $P(s) = \hat{P}(s)F(s)$ and $Q(s) = \hat{Q}(s)F(s)$ exist with $F(s)$ not unimodular. $F(s)$ and $\hat{Q}(s)$ are evidently nonsingular because $Q(s)$ is.

Furthermore,

$$G(s) = P(s)Q^{-1}(s) = \hat{P}(s)F(s)F^{-1}(s)\hat{Q}^{-1}(s) = \hat{P}(s)\hat{Q}^{-1}(s),$$

which shows how to "reduce" the original MFD $P(s)Q^{-1}(s)$ by "canceling" the right common polynomial matrix factor $F(s)$ from $P(s)$ and $Q(s)$. Because $F(s)$ is not unimodular, det $F(s)$ is a polynomial of positive degree, and the degree of det $\hat{Q}(s)$ is therefore strictly less than the degree of det $Q(s)$. Consequently, the original right MFD $P(s)Q^{-1}(s)$ is not minimal. The argument demonstrating that minimality implies irreducibility for left MFDs is essentially identical.

Proving the converse parts of the assertions in Theorem 6.1 is substantially more difficult. The approach we adopt, based on the so-called Smith form for polynomial matrices, is by no means the only one. Rugh [10] follows a quite different route, as does Vidyasagar [13]. We have elected to center our discussion on the Smith form partly because the manipulations we will be performing are similar to the things one does to obtain LDU decompositions [12] for matrices of numbers via Gauss elimination.

We will be employing a famous result from algebra known as the *Euclidean algorithm*. It says that if $f(s)$ is a real polynomial and $g(s)$ is another real polynomial of lower degree, then unique real polynomials $\kappa(s)$ and $\rho(s)$ exist, where $\rho(s)$ has lower degree than $g(s)$, for which

$$f(s) = g(s)\kappa(s) + \rho(s).$$

It is customary to regard the zero polynomial as having degree $-\infty$ and nonzero constant polynomials as having degree zero. If $g(s)$ divides $f(s)$, then $\rho(s) = 0$; otherwise, one can interpret $\kappa(s)$ and $\rho(s)$ respectively as the quotient and remainder obtained from dividing $f(s)$ by $g(s)$.

The Smith form of a real $(k \times r)$ polynomial matrix $F(s)$ is a matrix obtained by performing elementary row and column operations on $F(s)$. The row and column operations are of three kinds:

- Interchanging the ith and jth rows (or the ith and jth columns).
- Replacing row i with the difference (row i) $-\kappa(s)\times$ (row j), where $\kappa(s)$ is a polynomial (or replacing column j with the difference (column j) $-\kappa(s)\times$ (column i)).
- Replacing row (or column) i with its multiple by a nonzero real number γ.

One can view each of these operations as the result of multiplying $F(s)$ on the left or the right by a unimodular matrix. To interchange the ith and jth rows of $F(s)$, multiply on the left by the permutation matrix Π^{ij}; Π^{ij} is a $(k \times k)$ identity matrix with the ith and jth rows interchanged. To replace row i with itself minus $\kappa(s)$ times row j, multiply on the left by the matrix $E^{ij}[-\kappa(s)]$, a $(k \times k)$ identity matrix except with $-\kappa(s)$ in the ij position. Because det $\Pi^{ij} = -1$ and det $E^{ij}[-\kappa(s)] = 1$, both matrices are unimodular. To multiply row i by γ, multiply $F(s)$ on the left by the $(k \times k)$ diagonal matrix with γ in the ith diagonal position and ones in all the others; this matrix is also unimodular, because its determinant is γ. The column operations listed above result from multiplying on the right by $(r \times r)$ unimodular matrices of the types just described.

The following result defines the Smith form of a polynomial matrix. It is a special case of a more general theorem [6, pp. 175–180].

Theorem 6.2: Smith Form

Let $F(s)$ be a real $(k \times r)$ polynomial matrix with $k \geq r$. A $(k \times k)$ unimodular matrix $U(s)$ and an $(r \times r)$

unimodular matrix $R(s)$ exist so that $U(s)F(s)R(s)$ takes the form

$$\Lambda(s) = \begin{bmatrix} d_1(s) & 0 & 0 & . & . & 0 \\ 0 & d_2(s) & 0 & . & . & 0 \\ 0 & 0 & d_3(s) & 0 & . & 0 \\ . & . & . & . & . & . \\ . & . & . & . & . & 0 \\ 0 & . & . & . & 0 & d_r(s) \\ 0 & . & . & . & . & 0 \\ . & . & . & . & . & . \\ . & . & . & . & . & . \\ 0 & . & . & . & . & 0 \end{bmatrix}. \tag{6.5}$$

Furthermore, each nonzero $d_j(s)$ is monic, and $d_j(s)$ divides $d_{j+1}(s)$ for all j, $1 \leq j \leq r-1$.

Proof 6.1. Let $\delta_{\min}$ be the smallest of the degrees of the nonzero polynomials in $F(s)$. Now invoke the following procedure:

Reduction Step

Pick an entry in $F(s)$ whose degree is $\delta_{\min}$ and use row and column exchanges to bring it to the $(1,1)$ position; denote the resulting matrix by $G(s)$. For each i, $2 \leq i \leq k$, use the Euclidean algorithm to find $\kappa_i(s)$ so that $\rho_i(s) = [G(s)]_{i1} - \kappa_i(s)[G(s)]_{11}$ has strictly lower degree than $[G(s)]_{11}$, which has degree $\delta_{\min}$. Observe that $\rho_i(s) = 0$ if $[G(s)]_{11}$ divides $[G(s)]_{i1}$. Multiply $G(s)$ on the left by the sequence of matrices $E^{i1}[-\kappa_i(s)]$; this has the effect of replacing each $[G(s)]_{i1}$ with $\rho_i(s)$. Multiplying on the right by a similar sequence of E^{1j}-matrices replaces each $[G(s)]_{1j}$ with a polynomial whose degree is lower than $\delta_{\min}$.

The net result is a new matrix whose $\delta_{\min}$ is lower than the $\delta_{\min}$ of $F(s)$. Repeating the reduction step on this new matrix results in a matrix whose $\delta_{\min}$ is still lower. Because we cannot continue reducing $\delta_{\min}$ forever, iterating the reduction step on the successor matrices of $F(s)$ leads eventually to a matrix of the form

$$F_1(s) = \begin{bmatrix} d_1(s) & 0 & 0 & . & . & 0 \\ 0 & \pi & \pi & . & . & \pi \\ 0 & \pi & \pi & . & . & \pi \\ . & . & . & . & . & . \\ . & . & . & . & . & . \\ 0 & \pi & . & . & . & \pi \end{bmatrix}, \tag{6.6}$$

where each of the πs is a polynomial.

Divisibility Check

If $d_1(s)$ does not divide all the πs in Equation 6.6, find a π that $d_1(s)$ does not divide and multiply $F_1(s)$ on the left by an E^{i1}-matrix so as to add the row containing the offending π to the first row of $F_1(s)$. Now repeat the reduction step on this new matrix. What results is a matrix of the form Equation 6.6 with a lower $\delta_{\min}$.

Repeat the divisibility check on this new matrix. The process terminates eventually in a matrix of the form Equation 6.6 whose $d_1(s)$ divides all the polynomials π. Note that this newest $F_1(s)$ can be written in the form $U_1(s)F(s)R_1(s)$ for some unimodular matrices $U_1(s)$ and $R_1(s)$. The next phase of the Smith form computation entails performing the reduction step and divisibility check on the $(k-1 \times r-1)$ matrix of

πs in $F_1(s)$. What results is a matrix

$$F_2(s) = \begin{bmatrix} d_1(s) & 0 & 0 & . & . & 0 \\ 0 & d_2(s) & 0 & . & . & 0 \\ 0 & 0 & \pi & . & . & \pi \\ 0 & 0 & \pi & . & . & \pi \\ . & . & . & . & . & . \\ . & . & . & . & . & . \\ 0 & 0 & \pi & . & . & \pi \end{bmatrix}$$

in which $d_2(s)$ divides all of the πs. Moreover, $d_1(s)$ divides $d_2(s)$ because $d_2(s)$ is a polynomial linear combination of the πs in $F_1(s)$. Furthermore, there are unimodular matrices $U_2(s)$ and $R_2(s)$ so that $F_2(s) = U_2(s)F_1(s)R_2(s)$, whereby $F_2(s) = U_2(s)U_1(s)F(s)R_1(s)R_2(s)$.

Continuing in this fashion leads successively to $F_3(s), \ldots,$ and finally $F_r(s)$, which has the same form as $\Lambda(s)$. To get $\Lambda(s)$, modify $F_r(s)$ by scaling all of the rows of $F_r(s)$ so that the nonzero $d_j(s)$-polynomials are monic. It is evident that there are unimodular matrices $U(s)$ and $R(s)$ so that $\Lambda(s) = U(s)F(s)R(s)$.

A few comments are in order. First, if $F(s)$ is a real $(k \times r)$ polynomial matrix with $k \leq r$, then applying Theorem 6.2 to $F^T(s)$ and transposing the result yields unimodular matrices $U(s)$ and $R(s)$ so that $U(s)F(s)R(s)$ takes the form

$$\Lambda(s) = \begin{bmatrix} d_1(s) & 0 & 0 & . & . & 0 & 0 & . & 0 \\ 0 & d_2(s) & 0 & . & . & . & . & . & 0 \\ 0 & 0 & d_3(s) & 0 & . & . & . & . & 0 \\ . & . & . & . & . & . & . & . & . \\ . & . & . & . & . & . & . & . & 0 \\ 0 & . & . & . & 0 & d_k(s) & 0 & . & 0 \end{bmatrix}; \tag{6.7}$$

this is the Smith form of an $F(s)$ with more columns than rows. Next, consider the polynomials $\{d_j(s)\}$ in Equation 6.5; these are called the *elementary divisors* of $F(s)$. It might not appear prima facie as if the proof of Theorem 6.2 specified them uniquely, but it does. In fact, we have the following explicit characterization.

Fact 6.1:

For each j, $1 \leq j \leq \min(k, r)$, the product $d_1(s) \ldots d_j(s)$ is the monic greatest common divisor of all of the $(j \times j)$ minors in $F(s)$.

Proof 6.2. The divisibility conditions on the $\{d_j(s)\}$ make it obvious that $d_1(s) \cdots d_j(s)$ is the monic greatest common divisor of the $(j \times j)$ minors in $\Lambda(s)$. Fact 6.1 follows immediately from the observation that the reduction procedure leading from $F(s)$ to $\Lambda(s)$ is such that for each i, the set of $(j \times j)$ minors of $F_i(s)$ has the same family of common divisors as the set of $(j \times j)$ minors of $F_{i+1}(s)$. (See also the proof of Corollary 6.1 below.) As a result, the monic greatest common divisor of the $(j \times j)$ minors does not change over the course of the procedure.

One final comment: the sequence of elementary divisors, in general, might start out with one or more 1s and terminate with one or more 0s. In fact, because every nonzero elementary divisor is a monic polynomial, each constant elementary divisor is either a 1 or a 0. If $F(s)$ is a square matrix, then $\det F(s)$ is a nonzero real multiple of the product of its elementary divisors; this follows from the unimodularity of $U(s)$ and $R(s)$ in Theorem 6.2. Hence if $F(s)$ is nonsingular, none of its elementary divisors is zero.

If $F(s)$ is unimodular, then all of its elementary divisors are equal to 1. In other words, the Smith form of a $(k \times k)$ unimodular matrix is simply the $(k \times k)$ identity matrix I_k.

The Smith form is the key to finishing the proof of Theorem 6.1. The following lemma, which offers useful alternative characterizations of coprimeness, is the first step in that direction.

Lemma 6.1:

Let $P(s)$ and $Q(s)$ be real polynomial matrices having respective sizes $(p \times m)$ and $(m \times m)$. The following three conditions are equivalent:

1. *$P(s)$ and $Q(s)$ are right coprime.*
2. *There exist real polynomial matrices $X(s)$ and $Y(s)$ with respective sizes $(m \times m)$ and $(m \times p)$ so that $X(s)Q(s) + Y(s)P(s) = I_m$.*
3. *The $(m + p \times m)$ matrix*

$$\begin{bmatrix} Q(s_o) \\ P(s_o) \end{bmatrix}$$

has full rank m for every complex number s_o.

Proof 6.3. We show that $(1) \Longrightarrow (2) \Longrightarrow (3) \Longrightarrow (1)$. Set

$$F(s) = \begin{bmatrix} Q(s) \\ P(s) \end{bmatrix}$$

and, using the notation of Theorem 6.2, let $\Lambda(s) = U(s)F(s)R(s)$ be the Smith form of $F(s)$. Denote by $\Delta(s)$ the $(m \times m)$ matrix comprising the first m rows of $\Lambda(s)$; the diagonal elements of $\Delta(s)$ are the elementary divisors of $F(s)$.

If (1) holds, we need $\Delta(s) = I_m$. To see this, partition $U^{-1}(s)$ as

$$U^{-1}(s) = W(s) = \begin{bmatrix} W_1(s) & W_2(s) \\ W_3(s) & W_4(s) \end{bmatrix};$$

then

$$F(s) = \begin{bmatrix} Q(s) \\ P(s) \end{bmatrix} = \begin{bmatrix} W_1(s) \\ W_3(s) \end{bmatrix} \Delta(s)R^{-1}(s),$$

so that $P(s)$ and $Q(s)$ have $\Delta(s)R^{-1}(s)$ is an $(m \times m)$ as a right common polynomial matrix factor. For (1) to hold, this last matrix must be unimodular, and that happens when $\Delta(s) = I_m$.

Hence, right coprimeness of $P(s)$ and $Q(s)$ implies that $\Delta(s) = I_m$. It follows immediately in this case that if we partition the first m rows of $U(s)$ as $\begin{bmatrix} U_1(s) & U_2(s) \end{bmatrix}$, then (2) holds with $X(s) = R(s)U_1(s)$ and $Y(s) = R(s)U_2(s)$. (3) is a straightforward consequence of (2) because, if $F(s_o)$ were rank-deficient for some s_o, then $X(s_o)Q(s_o) + Y(s_o)P(s_o)$ could not be I_m. Finally, (3) implies (1) because, if (3) holds and

$$F(s) = \begin{bmatrix} \hat{P}(s) \\ \hat{Q}(s) \end{bmatrix} \hat{R}(s)$$

is a factorization with $\hat{R}(s)$ $(m \times m)$ but not unimodular, then $\det \hat{R}(s_o) = 0$ for at least one complex number s_o, which contradicts (3).

Not surprisingly, Lemma 6.1 has the following left-handed analogue.

Lemma 6.2:

Let $P_L(s)$ and $Q_L(s)$ be real polynomial matrices with respective sizes $(p \times m)$ and $(p \times p)$. The following three conditions are equivalent:

1. *$P_L(s)$ and $Q_L(s)$ are left coprime.*
2. *Real polynomial matrices $X(s)$ and $Y(s)$ exist with respective sizes $(p \times p)$ and $(m \times p)$ so that $Q_L(s)X(s) + P_L(s)Y(s) = I_p$.*
3. *The $(p \times p + m)$ matrix*

$$\begin{bmatrix} Q_L(s_o) & P_L(s_o) \end{bmatrix}$$

has full rank p for every complex number s_o.

We are ready now to finish proving Theorem 6.1.

Proof of Theorem 6.1: We prove only the assertions about right MFDs. We have shown already that a minimal right MFD of $G(s)$ is irreducible. As for the converse, suppose $G(s) = P(s)Q^{-1}$ is an irreducible right MFD and that $G(s) = \hat{P}(s)\hat{Q}^{-1}(s)$ is a minimal (hence irreducible) right MFD. By Lemma 6.1, there exist $X(s)$, $Y(s)$, $\hat{X}(s)$, and $\hat{Y}(s)$ of appropriate sizes satisfying

$$X(s)Q(s) + Y(s)P(s) = I_m$$
$$\hat{X}(s)\hat{Q}(s) + \hat{Y}(s)\hat{P}(s) = I_m.$$

Invoke the fact that $P(s)Q^{-1}(s) = \hat{P}(s)\hat{Q}^{-1}(s)$; a straightforward manipulation yields

$$X(s)\hat{Q}(s) + Y(s)\hat{P}(s) = Q^{-1}(s)\hat{Q}(s),$$
$$\hat{X}(s)Q(s) + \hat{Y}(s)P(s) = \hat{Q}^{-1}(s)Q(s).$$

The matrices on the right-hand sides are inverses of each other and are polynomial matrices; hence they must be unimodular. This implies, in particular, that the degrees of the determinants of $Q(s)$ and $\hat{Q}(s)$ are the same, and $P(s)Q^{-1}(s)$ is, therefore, also a minimal right MFD.

Finally, setting $V(s) = Q^{-1}(s)\hat{Q}(s)$ reveals that $\hat{P}(s) = P(s)V(s)$ and $\hat{Q}(s) = Q(s)V(s)$, proving that the two minimal right MFDs are related as in the theorem statement.

Theorem 6.1 establishes the equivalence of minimality and irreducibility for MFDs. It is worth remarking that its proof furnishes a means for reducing a nonminimal MFD to a minimal one. The argument supporting Lemma 6.1 provides the key. Let $G(s) = P(s)Q^{-1}(s)$ be a nonminimal right MFD. Employing the same notation as in the proof of Lemma 6.1,

$$F(s) = \begin{bmatrix} Q(s) \\ P(s) \end{bmatrix} = \begin{bmatrix} W_1(s) \\ W_3(s) \end{bmatrix} \times \Delta(s)R^{-1}(s) \stackrel{\text{def}}{=} \hat{F}(s)\Delta(s)R^{-1}(s).$$

Setting $\hat{Q}(s) = W_1(s)$ and $\hat{P}(s) = W_2(s)$ makes $\hat{P}(s)\hat{Q}^{-1}(s)$ a right MFD of $G(s)$. By definition of $W(s)$ and $U(s)$,

$$U(s)\hat{F}(s) = \begin{bmatrix} I_m \\ 0 \end{bmatrix};$$

taken in conjunction with Theorem 6.2, this implies that $\hat{F}(s)$ has ones as its elementary divisors. As in the proof of Lemma 6.1, we conclude that $\hat{P}(s)\hat{Q}^{-1}(s)$ is an irreducible (hence minimal) right MFD.

Theorem 6.1 also makes an important assertion about the "denominator matrices" appearing in minimal MFDs. The theorem states that if $Q(s)$ and $\hat{Q}(s)$ are denominator matrices in two minimal right MFDs of a proper rational $G(s)$, then $Q(s)$ and $\hat{Q}(s)$ differ by a right unimodular matrix factor. It follows that the determinants of $Q(s)$ and $\hat{Q}(s)$ are identical up to a nonzero real multiple; in particular, det $Q(s)$ and det $\hat{Q}(s)$ have the same roots including multiplicities. The same is true about the determinants of the denominator matrices $Q_L(s)$ and $\hat{Q}_L(s)$ from two minimal left MFDs. It is clear that the poles of $G(s)$ must lie among the roots of det $Q(s)$ and det $Q_L(s)$. What is not so obvious is the fact, whose proof we postpone until the next section, that these last two polynomials are actually nonzero real multiples of each other.

6.3 Fractional Degree and MacMillan Degree

Conspicuously absent from Section 6.2 is any substantive discussion of relationships between right and left MFDs for a proper rational matrix $G(s)$. The most important connection enables us to close the circle of ideas encompassing the minimal MFDs of Section 6.2 and the irreducible fractional representation Equation 6.1 for a scalar rational function. A crucial feature of Equation 6.1 is that the degree of $q(s)$ is the same as the MacMillan degree of $g(s)$. Our principal goal in what follows is to prove a similar assertion about the MacMillan degree of $G(s)$ and the degrees of the determinants of the denominator matrices appearing in minimal right and left MFDs of $G(s)$.

Our first task is to demonstrate that, when $P(s)Q^{-1}(s)$ and $Q_L^{-1}(s)P_L(s)$ are right and left MFDs of a proper rational matrix $G(s)$, the degrees of the polynomials det $Q(s)$ and det $Q_L(s)$ are the same. To that end, we need the following technical lemma.

Lemma 6.3:

Suppose that the $(m+p \times m+p)$ polynomial matrix $W(s)$ is nonsingular and that the $(m \times m)$ submatrix $W_1(s)$ is also nonsingular, where

$$W(s) = \begin{bmatrix} W_1(s) & W_2(s) \\ W_3(s) & W_4(s) \end{bmatrix}.$$

Then,

1. $H(s) = W_4(s) - W_3(s)W_1^{-1}(s)W_2(s)$ *is also nonsingular*
2. $\det W(s) = \det W_1(s) \det H(s)$
3. $W^{-1}(s)$ *is given by*

$$\begin{bmatrix} W_1^{-1}(s) + W_1^{-1}(s)W_2(s)H^{-1}(s)W_3(s)W_1^{-1}(s) & -W_1^{-1}(s)W_2(s)H^{-1}(s) \\ -H^{-1}(s)W_3(s)W_1^{-1}(s) & H^{-1}(s) \end{bmatrix}$$

Proof 6.4. Statements (1) and (2) follow from the identity

$$\begin{bmatrix} I_m & 0 \\ -W_3(s)W_1^{-1}(s) & I_p \end{bmatrix} W(s) = \begin{bmatrix} W_1(s) & W_2(s) \\ 0 & H(s) \end{bmatrix},$$

because the matrix multiplying $W(s)$ has determinant 1. Multiplying the last equation on the left by

$$\begin{bmatrix} W_1^{-1}(s) & 0 \\ 0 & H^{-1}(s) \end{bmatrix} \begin{bmatrix} I_m & -W_2(s)H^{-1}(s) \\ 0 & I_p \end{bmatrix}$$

yields statement (3).

A direct consequence of Lemma 6.3 is the advertised relationship between determinants of the denominator matrices in right and left MFDs.

Theorem 6.3:

Let $P(s)Q^{-1}(s)$ and $Q_L^{-1}(s)P_L(s)$, respectively, be minimal right and left MFDs of a real $(p \times m)$ proper rational matrix $G(s)$. Then $\det Q_L(s)$ is a nonzero real multiple of $\det Q(s)$; in particular, the two polynomials have the same degree and the same roots.

Proof 6.5. We learned in the proof of Lemma 6.1 that unimodular matrices $U(s)$ and $R(s)$ exist so that

$$U(s)\begin{bmatrix} Q(s) \\ P(s) \end{bmatrix} R(s) = \begin{bmatrix} I_m \\ 0 \end{bmatrix}.$$

Set

$$M(s) = \begin{bmatrix} R(s) & 0 \\ 0 & I_p \end{bmatrix} U(s)$$

and define $W(s) = M^{-1}(s)$; observe that $M(s)$ and $W(s)$ are unimodular. Partition $M(s)$ and $W(s)$ conformably as follows:

$$M(s) = \begin{bmatrix} M_1(s) & M_2(s) \\ M_3(s) & M_4(s) \end{bmatrix}; \quad W(s) = \begin{bmatrix} W_1(s) & W_2(s) \\ W_3(s) & W_4(s) \end{bmatrix}.$$

It follows that $W_1(s) = Q(s)$ and $W_3(s) = P(s)$. In particular, $W_1(s)$ is nonsingular. By Lemma 6.3, $M_4(s)$ is nonsingular and equals $H^{-1}(s)$, where $H(s) = W_4(s) - W_3(s)W_1^{-1}(s)W_2(s)$. Because $M(s)W(s) = I_{m+p}$,

$$M_3(s)W_2(s) + M_4(s)W_4(s) = I_p;$$

hence $M_3(s)$ and $M_4(s)$ are left coprime by Lemma 6.1. Furthermore,

$$M_3(s)W_1(s) + M_4(s)W_3(s) = 0;$$

the fact that $G(s) = P(s)Q^{-1}(s) = W_3(s)W_1^{-1}(s)$ makes

$$G(s) = M_4^{-1}(s)[-M_3(s)]$$

a minimal left MFD of $G(s)$. By Theorem 6.1, $\det M_4(s)$ is a nonzero real multiple of the determinant of the matrix $Q_L(s)$ appearing in any minimal left MFD $Q_L^{-1}(s)P_L(s)$ of $G(s)$.

By item (2) in Lemma 6.3, $\det W(s) = \det W_1(s) \det H(s)$. Because $W_1(s) = Q(s)$ and $M_4(s) = H^{-1}(s)$,

$$\det Q(s) = \det W(s) \det M_4(s).$$

Unimodularity of $W(s)$ implies that $\det Q(s)$ and $\det M_4(s)$ differ by a nonzero real multiple.

Theorem 6.3 makes possible the following definition.

Definition 6.6:

The fractional degree $\delta_F[G(s)]$ of a real $(p \times m)$ proper rational matrix $G(s)$ is the degree of $\det Q(s)$ (or of $\det Q_L(s)$), where $Q(s)$ (or $Q_L(s)$) comes from a minimal right (or left) MFD $P(s)Q^{-1}(s)$ (or $Q_L^{-1}(s)P_L(s)$) of $G(s)$.

As promised, we will demonstrate below that the fractional degree of a proper rational matrix $G(s)$ is the same as its MacMillan degree $\delta_M[G(s)]$. Our approach will be first to show that $\delta_F \leq \delta_M$ and then to construct a state space realization for $G(s)$ with state vector dimension $\delta_F[G(s)]$. The existence of such a realization guarantees that $\delta_F \geq \delta_M$, from which it follows that the two degrees are equal.

Accordingly, let $G(s)$ be a given real $(p \times m)$ proper rational matrix. Let $D = \lim_{s\to\infty} G(s)$; set $Z(s) = G(s) - D$. If $P(s)Q^{-1}(s)$ is a minimal right MFD of $G(s)$, then $\tilde{P}(s)Q^{-1}(s)$ is a minimal right MFD for $Z(s)$, where $\tilde{P}(s) = P(s) - DQ(s)$. To see why $\tilde{P}(s)Q^{-1}(s)$ is minimal, note that the two matrices

$$F(s_o) = \begin{bmatrix} Q(s_o) \\ P(s_o) \end{bmatrix}, \quad \tilde{F}(s_o) = \begin{bmatrix} Q(s_o) \\ P(s_o) - DQ(s_o) \end{bmatrix}$$

have the same nullspace (and hence the same rank) for every complex number s_o. It follows from item 3 in Lemma 6.1 that $\tilde{P}(s)Q^{-1}(s)$ is irreducible and hence minimal. In addition, we can conclude that the fractional degrees of $G(s)$ and $Z(s)$ are the same.

Suppose that $n = \delta_M(G(s))$ and that (A, B, C, D) is a minimal realization of $G(s)$. The Popov-Belevitch-Hautus test for reachability [11] implies that the $(n \times n + m)$ matrix,

$$K(s_o) = \begin{bmatrix} (s_o I_n - A) & B \end{bmatrix}$$

has full rank n for every complex number s_o. By Lemma 6.1, $(sI_n - A)^{-1}B$ is an irreducible (hence minimal) left MFD of $K(s)$. Thus $K(s)$ has fractional degree n. Now,

$$Z(s) = C(sI_n - A)^{-1}B = CF(s)\,;$$

it follows that the fractional degree of $Z(s)$ is at most equal to n. To see this, suppose $K(s) = P(s)Q^{-1}(s)$ is a minimal right MFD of $K(s)$; then $Z(s) = [CP(s)]Q^{-1}(s)$ is a right MFD of $Z(s)$, and this MFD need not be minimal. Hence $\delta_F(Z) \leq \delta_F(K) = n$. The upshot is that the fractional degree of $G(s)$, which is the same as the fractional degree of $Z(s)$, is bounded from above by the MacMillan degree of $G(s)$. In other words,

$$\delta_F[G(s)] \leq \delta_M[G(s)].$$

Proving the reverse inequality requires a bit more effort. Suppose we have a minimal right MFD $G(s) = P(s)Q^{-1}(s)$ and corresponding right MFD $Z(s) = \tilde{P}(s)Q^{-1}(s)$. The fractional degree of $G(s)$ is the same as the degree of det $Q(s)$. To show how the entries in $Q(s)$ combine to determine the degree of det $Q(s)$, we need the following definition.

Definition 6.7:

Let $Q(s)$ be a real nonsingular $(m \times m)$ polynomial matrix. The jth column degree of $Q(s)$, $\delta_j[Q(s)]$, is the highest of the degrees of the polynomials in the jth column of $Q(s)$, $1 \leq j \leq m$. The high-order coefficient matrix of $Q(s)$, Q_H, is the real $(m \times m)$ matrix whose (i, j) entry is the coefficient of $s^{\delta_j[Q(s)]}$ in $[Q(s)]_{ij}$.

The nonsingularity of $Q(s)$ guarantees that all of the column degrees are nonnegative integers. The classic expansion for the determinant [12, page 157] reveals det $Q(s)$ as the sum of m-fold products each of which contains precisely one element from each column of $Q(s)$. It follows that the degree of det $Q(s)$ is bounded from above by the sum of all the column degrees of $Q(s)$. In fact, if $\delta_1, \ldots, \delta_m$ are the column degrees of $Q(s)$, then the coefficient of the $s^{(\delta_1+\cdots+\delta_m)}$ in the expansion for det $Q(s)$ is exactly det Q_H, the determinant of the high-order coefficient matrix. In other words, det $Q(s)$ has degree *equal* to the sum of the column degrees when det $Q_H \neq 0$, which is the same as saying that the high-order coefficient matrix is invertible.

Our method for constructing for $G(s)$ a realization whose state dimension is $\delta_F[G(s)]$ hinges crucially on having a minimal right MFD $P(s)Q^{-1}(s)$ of $G(s)$ whose $Q(s)$ has an invertible high-order coefficient matrix. It turns out that such an MFD always exists. The idea is to start with an arbitrary minimal right MFD $\hat{P}(s)\hat{Q}^{-1}(s)$ and "operate on it" with a unimodular $V(s)$ via

$$P(s) = \hat{P}(s)V(s), \quad Q(s) = \hat{Q}(s)V(s)$$

to get another minimal right MFD $P(s)Q^{-1}(s)$ with an invertible Q_H. We construct $V(s)$ by looking at $\hat{Q}(s)$ only. Specifically, we prove the following assertion.

Lemma 6.4:

If $\hat{Q}(s)$ is a real nonsingular $(m \times m)$ polynomial matrix, a unimodular matrix $V(s)$ exists so that

- *$Q(s) = \hat{Q}(s)V(s)$ has an invertible high-order coefficient matrix Q_H*
- *The column degrees $\{\delta_j(Q(s))\}$ are in decreasing order, i.e.,*

$$\delta_1(Q(s)) \geq \delta_2(Q(s)) \cdots \geq \delta_m(Q(s)). \tag{6.8}$$

Proof 6.6. If $\hat{Q}_H$ is already invertible, let $V(s)$ be a permutation matrix Π so that the columns of $\hat{Q}(s)\Pi$ are lined up in decreasing order of column degree. If $\hat{Q}_H$ is not invertible, after finding Π as above, choose a nonzero $w \in \mathbf{R}^m$ satisfying $\hat{Q}_H \Pi w = 0$. Assume without loss of generality that the first nonzero element in w is a 1 and occurs in the kth position. Let $E(s)$ be the $(m \times m)$ polynomial matrix all of whose columns except the kth are the same as those in the $(m \times m)$ identity matrix I_m; as for the the kth column, let $[E(s)]_{ik}$ be 0 when $i < k$ and $w_i s^{\delta_k - \delta_i}$ when $i > k$, where δ_j denotes the jth column degree of $\hat{Q}(s)\Pi$. $E(s)$ has determinant 1 and is therefore unimodular; furthermore, $\hat{Q}(s)\Pi E(s)$ has the same columns as $\hat{Q}(s)\Pi$ except for the kth column, and the choice of $E(s)$ guarantees that the kth column degree of $\hat{Q}(s)\Pi E(s)$ is lower than the kth column degree of $\hat{Q}(s)\Pi$.

The preceding paragraph describes a technique for taking a $\hat{Q}(s)$ with singular $\hat{Q}_H$ and finding a unimodular matrix $\Pi E(s)$ so that $\hat{Q}(s)\Pi E(s)$ has a set of column degrees whose sum is less than the sum of the column degrees of $\hat{Q}(s)$. If $Q(s) = \hat{Q}(s)\Pi E(s)$ still fails to have an invertible high-order coefficient matrix, we can repeat the column-permutation-and-reduction procedure on $\hat{Q}(s)\Pi E(s)$, and so on. This iteration must terminate after a finite number of steps because we cannot reduce the sum of the column degrees forever. When all is said and done, we will have a unimodular matrix $V(s)$ so that $Q(s) = \hat{Q}(s)V(s)$ has an invertible high-order coefficient matrix Q_H with columns arrayed in decreasing order of column degree, so that the column degrees of $Q(s)$ satisfy Equation 6.8.

Thus we can take an arbitrary minimal right MFD $G(s) = \hat{P}(s)\hat{Q}^{-1}(s)$ and form a new minimal right MFD $P(s)Q^{-1}(s)$ using $P(s) = \hat{P}(s)V(s)$ and $Q(s) = \hat{Q}(s)V(s)$; choosing $V(s)$ appropriately, using Lemma 6.4, makes Q_H invertible, ensuring in turn that

$$\delta_F[G(s)] = \text{degree}[\det Q(s)] = \delta_1[Q(s)] + \cdots + \delta_m[Q(s)].$$

Furthermore, we can assume that the column degrees of $Q(s)$ satisfy the ordering of Equation 6.8.

Our final step is to produce a realization (A, B, C, D) of $G(s)$ where A is $(\delta_F \times \delta_F)$. This will confirm that $\delta_F(G(s)) \geq \delta_M(G(s))$ and, hence, that the two degrees are equal. First define D and $Z(s) = G(s) - D$ as before, and let $Z(s) = \tilde{P}(s)Q^{-1}(s)$ be the corresponding minimal right MFD for $Z(s)$. Because

$$[\tilde{P}(s)]_{ij} = \sum_{k=1}^{m} [Z(s)]_{ik}[Q(s)]_{kj},$$

and because $Z(s)$ is strictly proper, the degree of $[\tilde{P}(s)]_{ij}$ is strictly less than δ_j for all i and j. In particular, if $\delta_j = 0$, then $[\tilde{P}(s)]_{ij} = 0$ for all i, $1 \leq i \leq p$ (as usual, the zero polynomial has degree $-\infty$).

Next define, for each $k > 0$, the polynomial k-vector $\mathbf{s}_k$ by

$$\mathbf{s}_k = \begin{bmatrix} 1 \\ s \\ s^2 \\ \cdot \\ \cdot \\ s^{k-1} \end{bmatrix}.$$

We now form a matrix $\mathbf{S}$ that has m columns and number of rows equal to δ_F, which is equal in turn to the sum of the column degrees $\{\delta_j\}$. The jth column of $\mathbf{S}$ has $\delta_1 + \cdots + \delta_{j-1}$ 0s at the top, the vector $\mathbf{s}_{\delta_j}$ in the next δ_j positions, and 0s in the remaining positions. For example, if $m = 3$ and $\delta_1 = 3$, $\delta_2 = 2$, and $\delta_3 = 0$, then

$$\mathbf{S} = \begin{bmatrix} 1 & 0 & 0 \\ s & 0 & 0 \\ s^2 & 0 & 0 \\ 0 & 1 & 0 \\ 0 & s & 0 \end{bmatrix}.$$

Our observation above concerning the degrees of the entries in $\tilde{P}(s)$ ensures that a real $(p \times \delta_F)$ matrix C exists so that $\tilde{P}(s) = C\mathbf{S}$. This C will be the C-matrix in our realization of $G(s)$. Since we want

$$Z(s) = \tilde{P}(s)Q^{-1}(s) = C\mathbf{S}Q^{-1}(s)B = C(sI_{\delta_F} - A)^{-1}B,$$

we will construct A and B so that

$$\mathbf{S}Q^{-1}(s) = (sI_{\delta_F} - A)^{-1}B,$$

or, equivalently,

$$s\mathbf{S} = A\mathbf{S} + BQ(s). \tag{6.9}$$

Recall first that Q_H, the high-order coefficient matrix of $Q(s)$, is invertible; by definition of Q_H and the column degrees $\{\delta_j\}$,

$$Q(s) = Q_H \begin{bmatrix} s^{\delta_1} & 0 & \cdot & \cdot & \cdot & 0 \\ 0 & s^{\delta_2} & 0 & \cdot & \cdot & \cdot \\ \cdot & 0 & \cdot & \cdot & \cdot & \cdot \\ \cdot & \cdot & \cdot & \cdot & 0 & \cdot \\ \cdot & \cdot & \cdot & 0 & s^{\delta_{m-1}} & 0 \\ \cdot & \cdot & \cdot & \cdot & 0 & s^{\delta_m} \end{bmatrix} + \tilde{Q}(s), \tag{6.10}$$

where $\tilde{Q}(s)$ satisfies the same constraints as $\tilde{P}(s)$ on the degrees of its entries. We may write $\tilde{Q}(s) = G\mathbf{S}$ for some real $(m \times \delta_F)$ matrix G. Denote by $\Sigma(s)$ the diagonal matrix on the right-hand side of Equation 6.10. Then A and B satisfy Equation 6.9 when

$$s\mathbf{S} = (A + BQ_H G)\mathbf{S} + BQ_H \Sigma(s). \tag{6.11}$$

Define B as follows: $B = \tilde{B}Q_H^{-1}$, where $\tilde{B}$ is the real $(\delta_F \times m)$ matrix whose jth column is zero when $\delta_j = 0$ and contains a single 1 at the $\delta_1 + \cdots + \delta_j$ position if $\delta_j \neq 0$. Observe that if $i = \delta_1 + \cdots + \delta_j$ for some j, then the ith row of $BQ_H G$ is the same as the jth row of G; for all other values of i, the ith row of $BQ_H G$ is zero.

Finally, define A. If $i = \delta_1 + \cdots + \delta_j$ for some j, then the ith row of A is the negative of the ith row of $BQ_H G$. For other values of i, the ith row of A contains a 1 in the $(i, i+1)$ position (just above the diagonal) and 0s elsewhere.

Verifying that A and B so defined satisfy the required relationship Equation 6.11 is straightforward. The important consequence of the construction is that $Z(s) = C(sI_{\delta_F} - A)^{-1}B$, so that (A, B, C, D) is a realization of $G(s)$ whose A-matrix has size $(\delta_F \times \delta_F)$. The following theorem summarizes the results of this section so far.

Theorem 6.4:

If $G(s)$ is a real $(p \times m)$ proper rational matrix, then the MacMillan degree of $G(s)$ is the same as the fractional degree of $G(s)$. The procedure outlined above yields a minimal realization (A, B, C, D) of $G(s)$.

The central role of Lemma 6.4 in the proof of Theorem 6.4 merits a closer look. In essence, Lemma 6.4 guarantees the existence of a minimal right MFD $G(s) = P(s)Q^{-1}(s)$ wherein the column degrees of $Q(s)$ sum to the degree of the determinant of $Q(s)$. The crucial enabling feature of $Q(s)$ is the invertibility of its high-order coefficient matrix Q_H. We will see presently that any minimal right MFD whose denominator matrix possesses this last property will have the same column degrees as $Q(s)$ up to a reordering. As a result, these special column degrees are a feature of certain right MFDs and of the transfer function matrix $G(s)$ itself.

Definition 6.8:

The ordered column indices $l_1[Q(s)], \ldots, l_m[Q(s)]$ of a real nonsingular $(m \times m)$ polynomial matrix $Q(s)$ are the column degrees of $Q(s)$ arranged in decreasing order.

Theorem 6.5:

Let $Q(s)$ and $\hat{Q}(s)$ be real nonsingular $(m \times m)$ polynomial matrices appearing in minimal right MFDs $P(s)Q^{-1}(s)$ and $\hat{P}(s)\hat{Q}^{-1}(s)$ of a real $(p \times m)$ proper rational matrix $G(s)$. If the high-order coefficient matrices Q_H and $\hat{Q}_H$ are both invertible, then the ordered column indices of $Q(s)$ and $\hat{Q}(s)$ are identical.

Proof 6.7. Assume first that $Q(s)$ and $\hat{Q}(s)$ satisfy the second item in Lemma 6.4, i.e., have respective column degrees $\{\delta_j\}$ and $\{\hat{\delta}_j\}$ that are decreasing in j. Write Equation 6.10 along with a similar equation for $\hat{Q}(s)$ as follows:

$$Q(s) = Q_H \Sigma(s) + \tilde{Q}(s)$$
$$\hat{Q}(s) = \hat{Q}_H \hat{\Sigma}(s) + \tilde{\hat{Q}}(s).$$

By construction,

$$\lim_{|s|\to\infty} \tilde{Q}(s)\Sigma^{-1}(s) \stackrel{\text{def}}{=} \lim_{|s|\to\infty} \Delta(s) = 0,$$
$$\lim_{|s|\to\infty} \tilde{\hat{Q}}(s)\hat{\Sigma}^{-1}(s) \stackrel{\text{def}}{=} \lim_{|s|\to\infty} \hat{\Delta}(s) = 0.$$

Because $Q(s)$ and $\hat{Q}(s)$ both come from minimal right MFDs of $G(s)$, by Theorem 6.1 an $(m \times m)$ unimodular matrix $V(s)$ exists so that $\hat{Q}(s) = Q(s)V(s)$. Manipulation yields

$$\Sigma(s)U(s)\hat{\Sigma}^{-1}(s) = [I_m + \Delta(s)]^{-1}Q_H^{-1}\hat{Q}_H[I_m + \hat{\Delta}(s)]. \tag{6.12}$$

The right-hand side of Equation 6.12 approaches a constant limit as $|s| \to \infty$; note, in particular, that $I_m + \Delta(s)$ is nonsingular for $|s|$ large enough. Meanwhile, the (i, j) entry of the matrix on the left-hand side of Equation 6.12 is simply $s^{\delta_i - \hat{\delta}_j}[U(s)]_{ij}$. Hence we need $[U(s)]_{ij} = 0$ whenever $\delta_i > \hat{\delta}_j$. One by one we will show that $\delta_j \leq \hat{\delta}_j$. If $\delta_1 > \hat{\delta}_1$, then, by the ordering on the $\hat{\delta}$s, $\delta_1 > \hat{\delta}_j$ for all j, and the entire first row of $U(s)$ must be zero, contradicting nonsingularity and, a fortiori, unimodularity of $U(s)$. Assume inductively that $\delta_j \leq \hat{\delta}_j$ for $j < k$ but that $\delta_k > \hat{\delta}_k$. In this case, the orderings on the δs and $\hat{\delta}$s imply that $\delta_i > \hat{\delta}_k \geq \hat{\delta}_j$ for all $i \leq k$ and all $j \geq k$; hence the entire upper right-hand $(k \times m - k)$ corner of $U(s)$ must be zero, which contradicts unimodularity of $U(s)$ once again.

Thus $\delta_j \leq \hat{\delta}_j$ for all j, $1 \leq j \leq m$. It follows that $\delta_j = \hat{\delta}_j$ for every j because the sum of the δs and the sum of the $\hat{\delta}$s must both equal the fractional degree of $G(s)$, which is the common degree of the determinants of $Q(s)$ and $\hat{Q}(s)$. Our initial assumption that the columns of $Q(s)$ and $\hat{Q}(s)$ were arrayed in decreasing order of column degree means, in terms of Definition 6.8, that $Q(s)$ and $\hat{Q}(s)$ have the same ordered column indices. Finally, it is easy to eliminate this initial assumption; simply precede the argument with right multiplications by permutation matrices Π and $\hat{\Pi}$ that reorder the matrices' columns appropriately. In any event, the ordered column indices of $Q(s)$ and $\hat{Q}(s)$ are identical.

The principal consequence of Theorem 6.5 is that any two $Q(s)$-matrices with invertible Q_Hs appearing in minimal right MFDs of $G(s)$ have the same set of ordered column indices. These special ordered column indices are sometimes called the *Kronecker controllability indices* of $G(s)$. They have other names, as well; Forney [4], for example, calls them *invariant dynamical indices.* They relate to controllability because the realization (A, B, C, D) we constructed *en route* to Theorem 6.4 is precisely the MIMO analogue to the SISO realization given in Equations 6.2 and 6.3. The realizations are called *controllable canonical forms* [2,7,10]. Interested readers can verify that applying the realization procedure following Lemma 6.4 to a scalar irreducible fractional representation Equation 6.1 leads exactly to Equations 6.2 and 6.3.

It is worth making one final observation. Our proof of Theorem 6.4 relied on constructing a minimal realization of a proper rational transfer matrix $G(s)$ starting from a minimal right MFD $P(s)Q^{-1}(s)$ of $G(s)$. We could have worked instead with a minimal left MFD $G(s) = Q_L^{-1}(s)P_L(s)$, in which case we would have considered the *row degrees* and high-order coefficient matrix of $Q_L(s)$. Perhaps the simplest way to view this is to realize $G^T(s)$ by following the route we have already laid out beginning with the right MFD $P_L^T(s)(Q_L^T(s))^{-1}$ of $G^T(s)$ and subsequently transposing the end result.

6.4 Smith–MacMillan Form, ARMA Models, and Stable Coprime Factorization

The aim of this section is to tie up some loose ends and to point the reader toward some important modern control theoretic developments that rest heavily on the theory of polynomial MFDs described in the foregoing sections. First we discuss the so-called Smith–MacMillan form for proper rational matrices; many of the results detailed in Sections 6.2 and 6.3 have alternative derivations based on the Smith–MacMillan form. Next, we describe briefly the connection between MFDs and ARMA models for MIMO linear systems. We close with a quick introduction to stable coprime factorization and mention briefly its generalizations and applications in robust control theory.

The Smith–MacMillan form of a real $(p \times m)$ proper rational matrix $G(s)$ is basically a rational version of the Smith form of a polynomial matrix $F(s)$. It was introduced originally by MacMillan [9] and later exploited by Kalman in an important paper [8] that demonstrated correspondences between several notions of rational matrix degree. Given $G(s)$, begin by letting $q(s)$ be the monic lowest common denominator of its entries. Set $F(s) = q(s)G(s)$; by Theorem 6.2, we can find unimodular matrices $U(s)$ and $R(s)$ of respective sizes $(p \times p)$ and $(m \times m)$ so that $\Lambda(s) = U(s)F(s)R(s)$ has the form of Equation 6.5 or 6.7 depending on whether $p \geq m$ or $p \leq m$, respectively.

Assuming temporarily that $p \geq m$, the matrix $U(s)G(s)R(s) = \frac{1}{q(s)}F(s) \overset{\text{def}}{=} \Lambda_{SM}(s)$ takes the form

$$\begin{bmatrix} \gamma_1(s)/\phi_1(s) & 0 & 0 & . & . & 0 \\ 0 & \gamma_2(s)/\phi_2(s) & 0 & . & . & 0 \\ 0 & 0 & \gamma_3(s)/\phi_3(s) & 0 & . & 0 \\ . & . & . & . & . & . \\ . & . & . & . & . & 0 \\ 0 & . & . & . & 0 & \gamma_m(s)/\phi_m(s) \\ 0 & . & . & . & . & 0 \\ . & . & . & . & . & . \\ . & . & . & . & . & . \\ 0 & . & . & . & . & 0 \end{bmatrix}, \tag{6.13}$$

where $\gamma_k(s)/\phi_k(s)$ is the fraction $d_k(s)/q(s)$ expressed in lowest terms. If $d_k(s) = 0$, set $\phi_k(s) = 1$. The divisibility conditions on $\{d_k(s)\}$ guaranteed by Theorem 6.2 ensure that $\gamma_k(s)$ divides $\gamma_{k+1}(s)$ and $\phi_{k+1}(s)$ divides $\phi_k(s)$ for all k, $1 \leq k < m$.

Furthermore, if we set

$$\begin{aligned} P(s) = U^{-1}(s) &\begin{bmatrix} \gamma_1(s) & 0 & 0 & . & . & 0 \\ 0 & \gamma_2(s) & 0 & . & . & 0 \\ 0 & 0 & \gamma_3(s) & 0 & . & 0 \\ . & . & . & . & . & . \\ . & . & . & . & . & 0 \\ 0 & . & . & . & 0 & \gamma_m(s) \\ 0 & . & . & . & . & 0 \\ . & . & . & . & . & . \\ . & . & . & . & . & . \\ 0 & . & . & . & . & 0 \end{bmatrix} \\ &\overset{\text{def}}{=} U^{-1}(s)\Gamma(s) \end{aligned}$$

and

$$\begin{aligned} Q(s) = R(s) &\begin{bmatrix} \phi_1(s) & 0 & 0 & . & . & 0 \\ 0 & \phi_2(s) & 0 & . & . & 0 \\ 0 & 0 & \phi_3(s) & 0 & . & 0 \\ . & . & . & . & . & . \\ . & . & . & . & . & 0 \\ 0 & . & . & . & 0 & \phi_m(s) \end{bmatrix} \\ &\overset{\text{def}}{=} R(s)\Phi(s), \end{aligned}$$

then $P(s)Q^{-1}(s)$ is a right MFD of $G(s)$.

$P(s)Q^{-1}(s)$ is minimal because of the divisibility conditions on the $\{\gamma_j\}$ and $\{\phi_j\}$. The easiest way to see this is by checking that the $(m + p \times m)$ matrix

$$\begin{bmatrix} Q(s_o) \\ P(s_o) \end{bmatrix} = \begin{bmatrix} R(s_o) & 0 \\ 0 & U^{-1}(s_o) \end{bmatrix} \begin{bmatrix} \Phi(s_o) \\ \Gamma(s_o) \end{bmatrix}$$

has full rank m for every complex number s_o. The idea is that if, for example, $\gamma_k(s_o) = 0$ for some smallest value of k, then $\gamma_j(s_o) = 0$ for all $j \geq k$; hence $\phi_j(s_o) \neq 0$ for $j \geq k$ which, coupled with $\gamma_j(s_o) \neq 0$ for $j < k$, means that m of the $\{\gamma_j(s_o)\}$ and $\{\phi_j(s_o)\}$ are nonzero. Because $P(s)Q^{-1}(s)$ is a minimal right MFD of $G(s)$, it follows from Theorem 6.4 that the MacMillan degree of $G(s)$ is the degree of det $Q(s)$, which is the sum of the degrees of the polynomials $\{\phi_j(s)\}$.

The same sort of analysis works when $p \leq m$; in that case, $\Lambda_{SM}(s) = U(s)G(s)R(s)$ looks like a rational version of the matrix in Equation 6.7. In either case, $\Lambda_{SM}(s)$ is called the Smith–MacMillan form of $G(s)$. To summarize,

Theorem 6.6: Smith–MacMillan Form

Let $G(s)$ be a real $(p \times m)$ proper rational matrix and let $m \wedge p$ be the minimum of m and p. Unimodular matrices $U(s)$ and $R(s)$ exist so that $\Lambda_{SM}(s) = U(s)G(s)R(s)$, where

- $[\Lambda_{SM}(s)]_{ij} = 0$, *except when* $i = j$ *and* $[\Lambda_{SM}(s)]_{jj}(s) = \gamma_j(s)/\phi_j(s)$, $1 \leq j \leq m \wedge p$, *where the* $\{\gamma_j(s)/\phi_j(s)\}$ *are ratios of coprime monic polynomials,*
- $\gamma_j(s)$ *divides* $\gamma_{j+1}(s)$ *and* $\phi_{j+1}(s)$ *divides* $\phi_j(s)$, $1 \leq j < m \wedge p$,
- *The MacMillan degree of $G(s)$ is the sum of the degrees of the $\{\phi_j(s)\}$.*

An interesting consequence of Theorem 6.6 is the following characterization of MacMillan degree [9].

Corollary 6.1:

The MacMillan degree of a real $(p \times m)$ proper rational matrix $G(s)$ is the degree of the lowest common denominator of all of the minor subdeterminants of $G(s)$.

Proof 6.8. Assume $p \geq m$; the argument is similar when $p \leq m$. A glance at $\Lambda_{SM}(s)$ in Equation 6.13 and the divisibility conditions on the $\{\phi_j(s)\}$ in Theorem 6.6 reveals that for each k, $1 \leq k \leq m$, the product $\phi_1(s) \cdots \phi_k(s)$ is the monic lowest common denominator of all of the $(j \times j)$ minors in $\Lambda_{SM}(s)$ of order $j \leq k$. Hence the product of all of the $\phi_j(s)$ is the monic lowest common denominator of all of the minors of $\Lambda_{SM}(s)$. Now, $G(s)$ and $\Lambda_{SM}(s)$ are related via pre- and postmultiplication by unimodular matrices; hence the $(j \times j)$ minor determinants of each matrix are polynomial linear combinations of the $(j \times j)$ minors of the other matrix. It follows that any common denominator for one set of minors is a common denominator for the other set, and the proof is complete.

Observe that we could have generated many of the results in Sections 6.2 and 6.4 by appealing to the Smith–MacMillan form. A disadvantage of this approach is that Theorem 6.6 produces only right MFDs for $(p \times m)$ rational matrices when $p > m$ and only left MFDs when $p < m$.

We consider next some relationships between MFDs and some well-known time-domain representations of input–output linear systems. Let $G(s)$ be a real $(p \times m)$ proper rational matrix that is the transfer function of a real m-input, p-output, time-invariant linear system with input $u : \mathbf{R} \rightarrow \mathbf{R}^m$ and output $y : \mathbf{R} \rightarrow \mathbf{R}^p$. Let $\mathcal{L}\{y\}(s)$ and $\mathcal{L}\{u\}(s)$ be the Laplace transforms of y and u, so $\mathcal{L}\{y\}(s) = G(s)\mathcal{L}\{u\}(s)$. If $G(s) = Q_L^{-1}(s)P_L(s)$ is a left MFD of $G(s)$, then

$$Q_L(s)\mathcal{L}\{y\}(s) = P_L(s)\mathcal{L}\{u\}(s);$$

in the time domain, this last equation corresponds with the vector differential equation

$$Q_L(D)y(t) = P_L(D)u(t), \tag{6.14}$$

where D is the differential operator $\frac{d}{dt}$. Equation 6.14 is an *ARMA (autoregressive moving-average) representation* for the system's input–output relationship.

Similarly, if $G(s) = P(s)Q^{-1}(s)$ is a right MFD of $G(s)$, we can define $w : \mathbf{R} \to \mathbf{R}^m$ by means of the autoregressive (AR) differential equation

$$Q(D)w(t) = u(t) \tag{6.15}$$

and use w as the input to a moving-average (MA) specification of y, namely,

$$y(t) = P(D)w(t). \tag{6.16}$$

Because

$$\mathcal{L}\{y\}(s) = P(s)\mathcal{L}\{w\}(s) \text{ and } \mathcal{L}\{w\}(s) = Q^{-1}(s)\mathcal{L}\{u\}(s),$$
$$\mathcal{L}\{y\}(s) = P(s)Q^{-1}(s)\mathcal{L}\{u\}(s) = G(s)\mathcal{L}\{u\}(s),$$

so Equations 6.15 and 6.16 together constitute another time-domain description of the input–output behavior of the system. Whereas Equation 6.14 gives an ARMA description for the system, Equations 6.15 and 6.16 split the input–output relation into autoregressive and moving-average parts. For a SISO system, any fractional representation of the form Equation 6.1 acts as a left and right "MFD," so that the two time-domain characterizations are identical.

We close by presenting a very brief introduction to some of the ideas underlying stable coprime factorization, which is the single most important off-shoot of the theory of MFDs for input–output systems. We call a rational matrix $H(s)$ *stable* if, and only if, the poles of the entries in $H(s)$ lie in the open left half-plane $\text{Re}\{s\} < 0$. As usual, let $G(s)$ be a real $(p \times m)$ proper rational matrix. It turns out to be possible to write $G(s)$ in the form $G(s) = H_1(s)H_2^{-1}(s)$, where

- $H_1(s)$ and $H_2(s)$ are stable proper rational matrices of respective sizes $(p \times m)$ and $(m \times m)$,
- $H_1(s)$ and $H_2(s)$ are *right coprime over the ring of stable rational functions,* that is, the only common right stable $(m \times m)$ proper rational matrix factors of $H_1(s)$ and $H_2(s)$ have inverses that are also stable and proper.

Any such representation $G(s) = H_1(s)H_2^{-1}(s)$ is called a *stable right coprime factorization* of $G(s)$. One can define stable left coprime factorizations similarly.

It is not difficult to show that stable coprime factorizations exist. One approach, patterned after a technique due originally to Vidyasagar [14], goes as follows. Given $G(s)$, choose $\alpha > 0$ so that $-\alpha$ is not a pole of any entry in $G(s)$. Let $\sigma = 1/(s+\alpha)$, so that $s = (1-\alpha\sigma)/\sigma$. Define $\tilde{G}(\sigma) = G((1-\alpha\sigma)/\sigma)$. It follows that $\tilde{G}(\sigma)$ is a proper rational matrix function of σ. To see why it is proper, observe that the condition $\sigma \to \infty$ is the same as $s \to -\alpha$, and $-\alpha$ is not a pole of $G(s)$.

Now invoke the theory of Section 6.2 and find a minimal right MFD $\tilde{G}(\sigma) = P(\sigma)Q^{-1}(\sigma)$ of $\tilde{G}(\sigma)$. Finally, set $H_1(s) = P[1/(s+\alpha)]$ and $H_2(s) = Q[1/(s+\alpha)]$. Then

$$G(s) = \tilde{G}\left(\frac{1}{s+\alpha}\right) = H_1(s)H_2^{-1}(s).$$

Because $P(\sigma)$ and $Q(\sigma)$ are polynomial in σ, $H_1(s)$ and $H_2(s)$ are proper rational matrix functions of s. Moreover, all of the poles of $H_1(s)$ and $H_1(s)$ are at $-\alpha$, which means that $H_1(s)$ and $H_2(s)$ are stable. Furthermore, any stable $(m \times m)$ proper rational right common factor $H(s)$ of $H_1(s)$ and $H_2(s)$ defines, via $V(s) = H((1-\alpha\sigma)/\sigma)$, a polynomial right common factor of $P(\sigma)$ and $Q(\sigma)$, which must have a polynomial inverse by minimality of $P(\sigma)Q^{-1}(\sigma)$. It follows that $H^{-1}(s) = V^{-1}[1/(s+\alpha)]$ is stable and proper, implying that $H_1(s)$ and $H_2(s)$ are right coprime over the ring of stable proper rational functions.

The principal application of stable coprime factorizations is to robust control system design. At the heart of such applications is the notion of the H^∞ *norm* of a stable proper rational matrix $H(s)$. Given such an $H(s)$, the H^∞ norm of $H(s)$ is the supremum over $\omega \in \mathbf{R}$ of the largest singular value of $H(i\omega)$. Given two possibly unstable $(p \times m)$ transfer function matrices $G_a(s)$ and $G_b(s)$, one can define the distance between $G_a(s)$ and $G_b(s)$ in terms of the H^∞ norm of the *stable* rational matrix

$$\begin{bmatrix} H_{a1}(s) \\ H_{a2}(s) \end{bmatrix} - \begin{bmatrix} H_{b1}(s) \\ H_{b2}(s) \end{bmatrix},$$

where $G_a(s) = H_{a1}(s)H_{a2}^{-1}(s)$ and $G_b(s) = H_{b1}(s)H_{b2}^{-1}(s)$ are stable coprime factorizations of $G_a(s)$ and $G_b(s)$.

Interested readers can consult [1,3,5,13], and the references therein for a through development of the ideas underlying robust control system design and their dependence on the theory of stable coprime factorization. A by-product of Vidyasagar's approach [13] is a framework for understanding MFDs and stable coprime factorizations in terms of more general themes from abstract algebra, notably ring theory. This framework reveals that many of our results possess natural generalizations that apply in contexts broader than those considered here.

6.5 Defining Terms

Proper rational matrix: A matrix whose entries are proper rational functions, i.e., ratios of polynomials each of whose numerator degrees is less than or equal to its denominator degree.

MacMillan degree: The dimension of the state in a minimal realization of a proper rational transfer function matrix.

Real polynomial matrix: A matrix whose entries are polynomials with real coefficients.

Nonsingular: A real square polynomial matrix is nonsingular if its determinant is a nonzero polynomial.

Unimodular: A real square polynomial matrix is unimodular if its determinant is a nonzero real number.

References

1. Boyd, S. P. and Barratt, C., *Linear Controller Design: Limits of Performance,* Prentice Hall, Englewood Cliffs, NJ, 1991.
2. Delchamps, D. F., *State Space and Input-Output Linear Systems,* Springer, New York, 1988.
3. Doyle, J. C., Francis, B.A., and Tannenbaum, A., *Feedback Control Theory,* MacMillan, New York, 1992.
4. Forney, G. D., Minimal bases of rational vector spaces with applications to multivariable linear systems, *SIAM J. Control,* 13, 493–520, 1975.
5. Francis, B., *A Course in H_∞ Control Theory,* Volume 88 In Lecture Notes in Control and Information Sciences, Springer, 1987.
6. Jacobson, N., *Basic Algebra I,* W. H. Freeman, San Francisco, 1974.
7. Kailath, T., *Linear Systems,* Prentice Hall, Englewood Cliffs, NJ, 1980.
8. Kalman, R. E., Irreducible realizations and the degree of a rational matrix, *J. SIAM,* 13, 520–544, 1965.
9. MacMillan, B., An introduction to formal realizability theory parts I and II, *Bell Syst. Tech. J.,* 31, 217–279, 591-600, 1952.
10. Rugh, W. J., *Linear System Theory,* Prentice Hall, Englewood Cliffs, NJ, 1993.
11. Sontag, E., *Mathematical Control Theory,* Springer, New York, 1990.
12. Strang, G., *Linear Algebra and Its Applications,* Academic, New York, 1976.
13. Vidyasagar, M., *Control System Synthesis: A Factorization Approach,* MIT, Cambridge, MA, 1985.
14. Vidyasagar, M., On the use of right-coprime factorizations in distributed feedback systems containing unstable subsystems, *IEEE Trans. Circuits Syst. CAS-25,* 916–921, 1978.

7

Robustness Analysis with Real Parametric Uncertainty

Roberto Tempo
Polytechnic University of Turin

Franco Blanchini
University of Udine

7.1 Motivations and Preliminaries

Over the last decades, stability and performance of control systems affected by bounded perturbations have been studied in depth. The attention of researchers and control engineers concentrated on robustness tools in the areas H_∞, Kharitonov (or real parametric uncertainty), L_1, Lyapunov, μ, and quantitative feedback control (QFT). For further discussions on these topics and on the exposition of the main technical results, the reader may consult different sections of this volume and the special issue on robust control of *Automatica* (1993).

One of the key features of this chapter is the concept of *robustness*. To explain, instead of a single (nominal) system, we study a family of systems and we say that a certain property (e.g., stability or performance) is robustly satisfied if it is satisfied for all members of the family. In particular, we focus on linear, time-invariant, single-input, single-output systems affected by real parametric uncertainty. Stability of interval polynomials (i.e., polynomials whose coefficients lie within given intervals) and the well-known *Theorem of Kharitonov* (Kharitonov, 1978) are at the core of this research area. This theorem states that an interval polynomial has all its roots in the open left half-plane if and only if four specially constructed polynomials have roots in the open left half-plane. Subsequently, the *Edge Theorem* (Bartlett et al., 1988) studied the problem of affine dependence between coefficients and uncertain parameters, and more general regions than the open left half-plane. This result provides an elegant solution proving that it suffices to check stability of the so-called one-dimensional exposed edges. We refer to the books (Ackermann, 1993; Barmish, 1994; Bhattacharyya et al., 1995; Djaferis, 1995; Kogan, 1995) for a discussion of the extensive literature on this subject.

To explain robustness analysis more precisely with real parametric uncertainty, we consider a family of polynomials $p(s,q)$ of degree n whose real coefficients $a_i(q)$ are continuous functions of an ℓ-dimensional vector of real uncertain parameters q, each bounded in the interval $[q_i^-, q_i^+]$. More formally, we define

$$p(s,q) \doteq a_0(q) + a_1(q)s + a_2(q)s^2 + \cdots + a_n(q)s^n,$$
$$q \doteq [q_1, q_2, \ldots, q_\ell],$$

and the set

$$Q \doteq \{q : q_i^- \leq q_i \leq q_i^+,\ i = 1, 2, \ldots, \ell\}.$$

We assume that $p(s,q)$ is of degree n for all $q \in Q$—that is, we assume that $a_n(q) \neq 0$ for all $q \in Q$. Whenever the relations between the polynomial coefficients $a_i(q)$ and the vector q are specified, we study the root location of $p(s,q)$ for all $q \in Q$. Within this framework, the basic property we need to guarantee is robust stability. In particular, we say that $p(s,q)$ is robustly stable if $p(s,q)$ has roots in the open left half-plane for all $q \in Q$.

The real parametric approach can be also formulated for control systems. In this case, we deal with a family of plants denoted by $P(s,q)$. More precisely, we concentrate on robust stability or performance of a proper plant

$$P(s,q) \doteq \frac{N_P(s,q)}{D_P(s,q)},$$

where $N_P(s,q)$ and $D_P(s,q)$ are the numerator and denominator polynomials whose real coefficients are continuous functions of q. We assume that $D_P(s,q)$ has invariant degree for all $q \in Q$. We also assume that there is no unstable pole–zero cancellation for all $q \in Q$; the reader may refer to Chockalingam and Dasgupta (1993) for further discussions.

Robustness analysis is clearly of interest when the plant requires compensation. In practice, if the compensator is designed on the basis of the nominal plant, then, robustness analysis can be performed by means of the closed-loop polynomial. That is, given a compensator transfer function

$$C(s) \doteq \frac{N_C(s)}{D_C(s)}$$

connected in a feedback loop with $P(s,q)$, we immediately write the closed-loop polynomial

$$p(s,q) = N_P(s,q)N_C(s) + D_P(s,q)D_C(s)$$

whose root location determines closed-loop stability.

To conclude this preliminary discussion, we remark that one of the main technical tools described here is the so-called value set (or template in the QFT jargon, see Horowitz, 1991; see Barmish, 1994 for a detailed exposition of its properties). In particular, we show that if the polynomial or plant coefficients are affine functions of q, the value set can be easily constructed with 2D graphics. Consequently, robustness tests in the frequency domain can be readily performed. Finally, in this chapter, since the main goal is to introduce the basic concepts and tools available for robustness analysis with real parametric uncertainty, we do not provide formal proofs but we make reference to the specific literature on the subject.

7.1.1 Motivating Example: DC Electric Motor with Uncertain Parameters

For the sake of illustrative purposes, an example of a DC electric motor is formulated and carried out throughout this chapter in various forms. Consider the system represented in Figure 7.1 of an armature-controlled DC electric motor with independent excitation. The voltage to angle transfer function $P(s) = \Theta(s)/V(s)$ is given by

$$P(s) = \frac{K}{LJs^3 + (RJ + BL)s^2 + (K^2 + RB)s},$$

where L is the armature inductance, R the armature resistance, K the motor electromotive force-speed constant, J the moment of inertia, and B the mechanical friction. Clearly, the values of some of these parameters may be uncertain. For example, the moment of inertia and the mechanical friction are functions of the load. Therefore, depending on the specific application, if the load is not fixed, the values of J and B are not precisely known. Similarly, the armature resistence R is a parameter that can be measured very accurately but is subject to temperature variations, and the motor constant K is a function of the field magnetic flow which may vary.

To summarize, it is reasonable to say that the motor parameters, or a subset of them, may be unknown but bounded within given intervals. More precisely, we can identify

$$q_1 = L; \quad q_2 = R; \quad q_3 = K; \quad q_4 = J; \quad q_5 = B$$

and specify a given interval $[q_i^-, q_i^+]$ for each q_i, $i = 1, 2, \ldots, 5$. Then, instead of $P(s)$, we write

$$P(s, q) = \frac{q_3}{q_1 q_4 s^3 + (q_2 q_4 + q_1 q_5)s^2 + (q_3^2 + q_2 q_5)s}.$$

7.2 Description of the Uncertainty Structures

As discussed in the preliminaries in Section 7.1, we consider a proper plant $P(s, q)$ whose coefficients are continuous functions of the uncertainty q which is confined to the set Q. Depending on the specific problem under consideration, the coefficients of $N_P(s, q)$ and $D_P(s, q)$ may be linear or nonlinear functions of q. To explain more precisely, we consider the example of Section 7.1.1. Assume that the armature

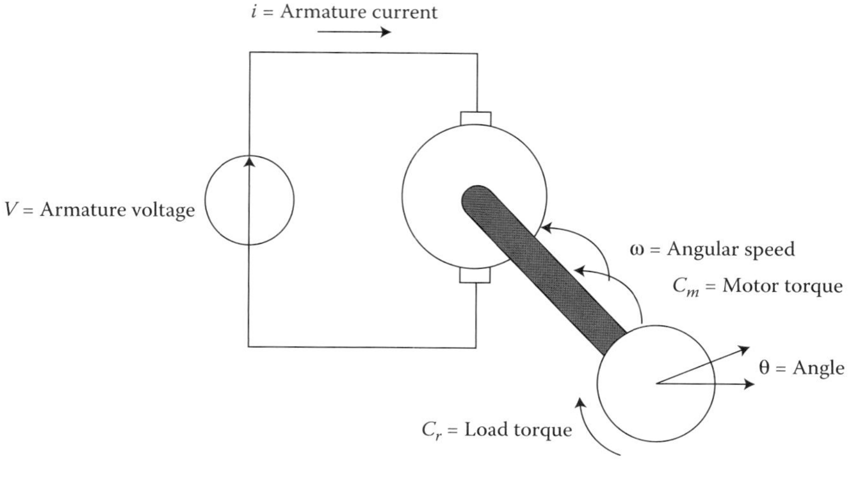

FIGURE 7.1 DC electric motor.

inductance L, the armature resistance R, and the constant K are fixed while the moment of inertia J and the mechanical friction B are unknown. Then, we take $q_1 = J$ and $q_2 = B$ as uncertain parameters; the resulting set Q is a two-dimensional rectangle. In this case, the plant coefficients are *affine** functions of q_1 and q_2

$$N_P(s,q) = K;$$
$$D_P(s,q) = Lq_1s^3 + (Rq_1 + Lq_2)s^2 + (K^2 + Rq_2)s$$

and we say that the plant has an affine uncertainty structure. This situation arises in practice whenever, for example, the load conditions are not known. Other cases, however, may be quite different from the point of view of the uncertainty description. For example, if L, K, and B are fixed and R and J are uncertain, we identify q_1 and q_2 with R and J, respectively. We observe that the denominator coefficient of s^2 contains the product of the two uncertain parameters q_1 and q_2

$$N_P(s,q) = K;$$
$$D_P(s,q) = Lq_2s^3 + (q_1q_2 + BL)s^2 + (K^2 + Bq_1)s.$$

In this case, the plant coefficients are no longer affine functions of the uncertainties but they are *multiaffine* functions† of q. This discussion can be further generalized. It is well known that the motor constant K is proportional to the magnetic flow. In an ideal machine with independent excitation, such a flow is constant, but in a real machine, the armature reaction phenomenon causes magnetic saturation with the consequence that the constant K drops when the armature current exceeds a certain value. Hence, we consider K as an uncertain parameter and we set $q_1 = K$. In turn, this implies that the plant coefficients are *polynomial* functions of the uncertainties. In addition, since q_1 enters in $N_P(s,q)$ and $D_P(s,q)$, we observe that there is coupling between numerator and denominator coefficients. In different situations when this coupling is not present, we say that the numerator and denominator uncertainties are *independent*. An important class of independent uncertainties, in which all the coefficients of the numerator and denominator change independently within given intervals, is the so-called interval plant; for example, see Barmish et al. (1992). In other words, an interval plant is the ratio of two independent interval polynomials; recall that an interval polynomial

$$p(s,q) = q_0 + q_1s + q_2s^2 + \cdots + q_ns^n$$

has independent coefficients bounded in given intervals $q_i^- \leq q_i \leq q_i^+$ for $i = 0, 1, 2, \ldots, n$.

The choice of the uncertain parameters for a control system is a modeling problem, but robustness analysis is of increasing difficulty for more general uncertainty structures. In the following sections, we show that this analysis can be easily performed if the structure is affine and we demonstrate that a "tight" approximate solution can be readily computed in the multiaffine case.

7.3 Uncertainty Structure Preservation with Feedback

In the previous sections, we described the classes of uncertainty structures entering into the open-loop plant. From the control system point of view, an important and closely related question arises: what are the conditions under which a certain uncertainty structure is preserved with feedback? To answer this question, we consider a plant $P(s,q)$ and a compensator $C(s)$ connected with the feedback structure shown in Figure 7.2.

* An affine function $f: Q \to \mathbf{R}$ is the sum of a linear function and a constant. For example, $f(q) = 3q_1 + 2q_2 - 4$ is affine.

† A function $f: Q \to \mathbf{R}$ is said to be multiaffine if the following condition holds: If all components $q_1, \ldots, q_\ell$ except for one are fixed, then f is affine. For example, $f(q) = 3q_1q_2q_3 - 6q_1q_3 + 4q_2q_3 + 2q_1 - 2q_2 + q_3 - 1$ is multiaffine.

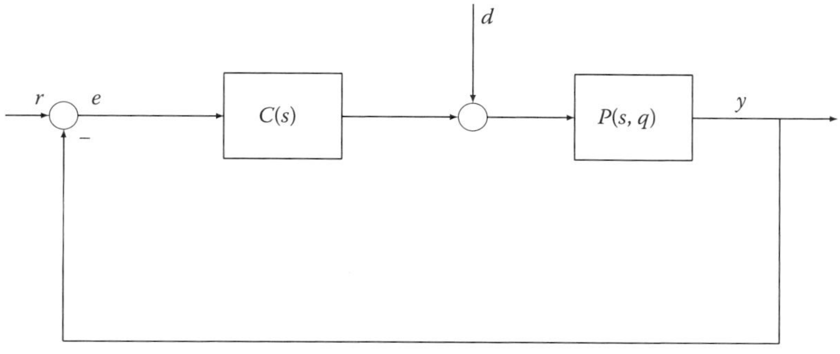

FIGURE 7.2 Closed-loop system.

Depending on the specific problem under consideration (e.g., disturbance attenuation or tracking), we study sensitivity, complementary sensitivity, and output-disturbance transfer functions (e.g., see Doyle et al., 1992)

$$S(s,q) \doteq \frac{1}{1+P(s,q)C(s)}; \quad T(s,q) \doteq \frac{P(s,q)C(s)}{1+P(s,q)C(s)}; \quad R(s,q) \doteq \frac{P(s,q)}{1+P(s,q)C(s)}.$$

For example, it is immediate to show that the sensitivity function $S(s,q)$ takes the form

$$S(s,q) = \frac{D_P(s,q)D_C(s)}{N_P(s,q)N_C(s) + D_P(s,q)D_C(s)}.$$

If the uncertainty q enters affinely into the plant numerator and denominator, q also enters affinely into $S(s,q)$. We conclude that the affine structure is preserved with feedback. The same fact also holds for $T(s,q)$ and $R(s,q)$. Next, we consider the interval plant structure; recall that an interval plant has independent coefficients bounded in given intervals. It is easy to see that, in general, this structure is not preserved with compensation. Moreover, if the plant is affected by uncertainty entering independently into numerator and denominator coefficients, this decoupling is destroyed for all transfer functions $S(s,q)$, $T(s,q)$, and $R(s,q)$. Finally, it is important to note that the multiaffine and polynomial uncertainty structures are preserved with feedback. Table 7.1 summarizes this discussion. In each entry of the first row of the table we specify the structure of the uncertain plant $P(s,q)$ and in the entry below the corresponding structure of $S(s,q)$, $T(s,q)$, and $R(s,q)$.

7.4 Overbounding with Affine Uncertainty: The Issue of Conservatism

As briefly mentioned at the end of Section 7.3, the affine structure is very convenient for performing robustness analysis. However, in several real applications, the plant does not have this form; for example, see Abate et al. (1994). In such cases, the nonlinear uncertainty structure can always be embedded into an affine structure by replacing the original family by a "larger" one. Even though this process has the advantage that it handles much more general robustness problems, it has the obvious drawback that it gives only an approximate but guaranteed solution. Clearly, the goodness of the approximation depends on the

TABLE 7.1 Uncertainty Structure with Feedback

$P(s,q)$	Independent	Interval	Affine	Multiaffine	Polynomial
$S(s,q), T(s,q), R(s,q)$	Dependent	Affine	Affine	Multiaffine	Polynomial

specific problem under consideration. To illustrate this simple overbounding methodology, we consider the DC-electric motor transfer function with two uncertain parameters and take $q_1 = R$ and $q_2 = J$. As previously discussed, with this specific choice, the plant has a multiaffine uncertainty structure. That is,

$$P(s,q) = \frac{K}{Lq_2s^3 + (q_1q_2 + BL)s^2 + (K^2 + Bq_1)s}.$$

To overbound $P(s,q)$ with an affine structure, we set $q_3 = q_1q_2$. Given bounds $[q_1^-, q_1^+]$ and $[q_2^-, q_2^+]$ for q_1 and q_2, the range of variation $[q_3^-, q_3^+]$ for q_3 can be easily computed:

$$\begin{aligned} q_3^- &= \min\{q_1^- q_2^-, q_1^- q_2^+, q_1^+ q_2^-, q_1^+ q_2^+\}; \\ q_3^+ &= \max\{q_1^- q_2^-, q_1^- q_2^+, q_1^+ q_2^-, q_1^+ q_2^+\}. \end{aligned}$$

Clearly, the new uncertain plant

$$P(s,\tilde{q}) = \frac{K}{Lq_2s^3 + (q_3 + BL)s^2 + (K^2 + Bq_1)s}$$

has three uncertain parameters $\tilde{q} = (q_1, q_2, q_3)$ entering affinely into $P(s,\tilde{q})$. This new parameter is not independent, because $q_3 = q_1q_2$ and not all values of $[q_3^-, q_3^+]$ are physically realizable. However, since we assume that the coefficients q_i are independent, this technique leads to an affine overbounding of $P(s,q)$ with $P(s,\tilde{q})$. We conclude that if a certain property is guaranteed for $P(s,\tilde{q})$, then, this same property is also guaranteed for $P(s,q)$. Unfortunately, the converse is not true. The control systems interpretation of this fact is immediate: Suppose that a certain compensator $C(s)$ does not stabilize $P(s,\tilde{q})$. It may turn out that this same compensator does stabilize the family $P(s,q)$. Figure 7.3 illustrates the overbounding procedure for $q_1^- = 1.2, q_1^+ = 1.7, q_2^- = 1.7, q_2^+ = 2.2, q_3^- = 2.04$ and $q_3^+ = 3.74$.

To generalize this discussion, we restate the overbounding problem as follows: Given a plant $P(s,q)$ having nonlinear uncertainty structure and a set Q, find a new uncertain plant $P(s,\tilde{q})$ with affine uncertainty structure and a new set $\tilde{Q}$. In general, there is no unique solution and there is no systematic procedure to construct an "optimal" overbounding. In practice, however, the control engineer may find via heuristic considerations a reasonably good method to perform it. The most natural way to obtain this bound may be to compute an interval overbounding for each coefficient of the numerator and denominator coefficients—that is, an interval plant overbounding. To illustrate, letting $a_i(q)$ and $b_i(q)$

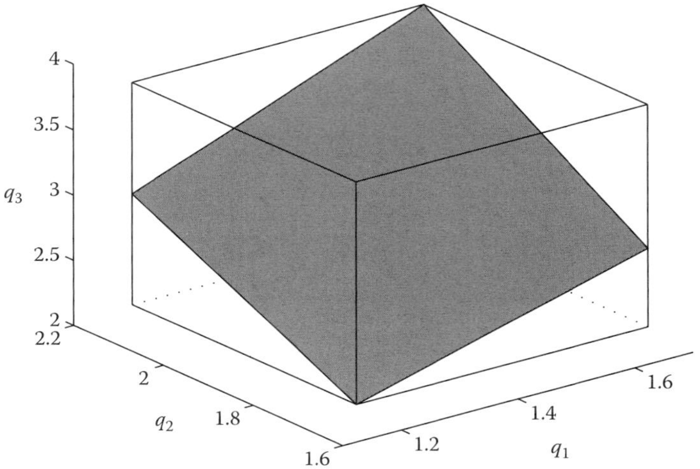

FIGURE 7.3 Overbounding procedure.

denote the numerator and denominator coefficients of $P(s, q)$, lower and upper bounds are given by

$$a_i^- = \min_{q \in Q} a_i(q); \quad a_i^+ = \max_{q \in Q} a_i(q)$$

and

$$b_i^- = \min_{q \in Q} b_i(q); \quad b_i^+ = \max_{q \in Q} b_i(q).$$

If $a_i(q)$ and $b_i(q)$ are affine or multiaffine functions and q lies in a rectangular set Q, these minimizations and maximizations can be easily performed. That is, if we denote by $q^1, q^2, \ldots, q^L \doteq q^{2^\ell}$ the vertices of Q, then

$$a_i^- = \min_{q \in Q} a_i(q) = \min_{k=1,2,\ldots,L} a_i(q^k); \quad a_i^+ = \max_{q \in Q} a_i(q) = \max_{k=1,2,\ldots,L} a_i(q^k)$$

and

$$b_i^- = \min_{q \in Q} b_i(q) = \min_{k=1,2,\ldots,L} b_i(q^k); \quad b_i^+ = \max_{q \in Q} b_i(q) = \max_{k=1,2,\ldots,L} b_i(q^k).$$

To conclude this section, we remark that for more general uncertainty structures than multiaffine, a tight interval plant overbounding may be difficult to construct.

7.5 Robustness Analysis for Affine Plants

In this section, we study robustness analysis of a plant $P(s, q)$ affected by affine uncertainty $q \in Q$. The approach taken here is an extension of the classical Nyquist criterion and requires the notion of *value set*.

For fixed frequency $s = j\omega$, we define the value set $P(j\omega, Q) \subset \mathbf{C}$ as

$$P(j\omega, Q) \doteq \{P(j\omega, q) : D_P(j\omega, q) \neq 0, q \in Q\}.$$

Roughly speaking, $P(j\omega, Q)$ is a set in the complex plane which graphically represents the uncertain plant. Without uncertainty, $P(j\omega)$ is a singleton and its plot for a range of frequencies is the Nyquist diagram. The nice feature is that this set is two-dimensional even if the number of uncertain parameters is large. Besides the issue of the value set construction (which is relegated to the next subsection for the specific case of affine plants), we now formally state a robustness criterion. This is an extension of the classical Nyquist criterion and holds for more general uncertainty structures than affine—continuity of the plant coefficients with respect to the uncertain parameters suffices. However, for more general classes of plants than affine, the construction of the value set is a hard problem.

Criterion 7.1: Robustness Criterion for Uncertain Plants

The plant $P(s, q)$ is robustly stable for all $q \in Q$ if and only if the Nyquist stability criterion is satisfied for some $q \in Q$ and $-1 + j0 \notin P(j\omega, Q)$ for all $\omega \in \mathbf{R}$.

This criterion can be proved using continuity arguments; see Fu (1990). To detect robustness, one should check if the Nyquist stability criterion holds for some $q \in Q$; without loss of generality, this check can be performed for the nominal plant. Second, it should be verified that the value set does not go through the point $-1 + j0$ for all $\omega \in \mathbf{R}$. In practice, however, one can discretize a bounded interval $\Omega \subset \mathbf{R}$ with a "sufficiently" large number of samples—continuity considerations guarantee that the intersampling is not a critical issue. Finally, by drawing the value set, the gain and phase margins can be graphically evaluated; similarly, the resonance peak of the closed-loop system can be computed using the well-known constant M-circles.

7.5.1 Value Set Construction for Affine Plants

In this section, we discuss the generation of the value set $P(j\omega, Q)$ in the case of affine plants. The reader familiar with Nyquist-type analysis and design is aware of the fact that a certain range of frequencies, generally close to the crossover frequencies, can be specified *a priori*. That is, a range $\Omega = [\omega^-, \omega^+]$ may be imposed by design specifications or estimated by performing a frequency analysis of the nominal system under consideration. In this section, we assume that $D_P(j\omega, q) \neq 0$ for all $q \in Q$ and $\omega \in \Omega$. We remark that if the frequency $\omega = 0$ lies in the interval Ω, this assumption is not satisfied for type 1 or 2 systems—however, these systems can be easily handled with contour indentation techniques as in the classical Nyquist analysis. We also observe that the assumption that $P(s, q)$ does not have poles in the interval $[\omega^-, \omega^+]$ simply implies that $P(j\omega, Q)$ is bounded.

To proceed with the value set construction, we first need a preliminary definition. The one-dimensional *exposed edge* e^{ik} is a convex combination of the adjacent vertices* q^i and q^k of Q

$$e^{ik} \doteq \lambda q^i + (1-\lambda)q^k$$

for $\lambda \in [0, 1]$. Denote by E the set of all $q \in e^{ik}$ for some i, k, and $\lambda \in [0, 1]$. This set is the collection of all one-dimensional exposed edges of Q.

Under our assumption $\partial P(j\omega, Q) \neq 0$, for $q \in Q$ and all ω, the set $P(j\omega, Q)$ is compact. Then, for fixed $\omega \in \Omega$, it can be shown (see Fu, 1990) that

$$\partial P(j\omega, Q) \subseteq P(j\omega, E) \doteq \{P(j\omega, q) : q \in E\}$$

where $\partial P(j\omega, Q)$ denotes the boundary of the value set and $P(j\omega, E)$ is the image in the complex plane of the exposed edges (remember that we assumed $\partial P(j\omega, Q) \neq 0$, for $q \in Q$ and all ω). This says that the construction of the value set only requires computations involving the one-dimensional exposed edges. The second important fact observed in Fu (1990) is that the image of the edge e^{ik} in the complex plane is an arc of a circle or a line segment. To explain this claim, in view of the affine dependence of both $N(s, q)$ and $D(s, q)$ versus q, we write the uncertain plant corresponding to the edge e^{ik} in the form

$$P(s, e^{ik}) = \frac{N_P(s, \lambda q^i + (1-\lambda)q^k)}{D_P(s, \lambda q^i + (1-\lambda)q^k)} = \frac{N_P(s, q^k) + \lambda N_P(s, (q^i - q^k))}{D_P(s, q^k) + \lambda N_P(s, (q^i - q^k))}$$

for $\lambda \in [0, 1]$. For fixed $s = j\omega$, it follows that the mapping from the edge e^{ik} to the complex plane is bilinear. Then, it is immediate to conclude that the image of each edge is an arc of a circle or a line segment; the center and the radius of the circle and the extreme points of the segment can be also computed. Even though the number of one-dimensional exposed edges of the set Q is $\ell 2^{\ell-1}$, the set E is one dimensional. Therefore, a fast computation of $P(j\omega, E)$ can be easily performed and the boundary of the value set $P(j\omega, Q)$ can be efficiently generated.

Finally, an important extension of this approach is robustness analysis of systems with time delay. That is, instead of $P(s, q)$, we consider

$$P_\tau(s, q) \doteq \frac{N(s, q)}{D(s, q)} e^{-\tau s}$$

where $\tau \geq 0$ is a delay. It is immediate to see that the value set of $P_\tau(s, q)$ at frequency $s = j\omega$ is given by the value set of the plant $N(s, q)/D(s, q)$ rotated with respect to the origin of the complex plane of an angle $\tau\omega$ in clockwise direction. Therefore, Criterion 7.1 still applies; see Barmish and Shi (1990) for further details.

* Two vertices are adjacent if they differ for only one component. For example, in Figure 7.3 the vertices $q^1 = (1.2, 2.2, 2.04)$ and $q^2 = (1.2, 2.2, 3.74)$ are adjacent.

7.5.2 The DC-Electric Motor Example Revisited

To illustrate the concepts discussed in this section, we revisit the DC-electric motor example. We take two uncertain parameters

$$q_1 = J; \quad q_2 = B,$$

where $q_1 \in [0.03, 0.15]$ and $q_2 \in [0.001, 0.03]$ with nominal values $J = 0.042\,\text{kg}\,\text{m}^2$ and $B = 0.01625\,\text{N m/rps}$. The remaining parameters take values $K = 0.9$ V/rps, $L = 0.025$ H, and $R = 5\Omega$. The voltage to angle uncertain plant is

$$P(s,q) = \frac{0.9}{0.025q_1s^3 + (5q_1 + 0.025q_2)s^2 + (0.81 + 5q_2)s}.$$

To proceed with robustness analysis, we first estimate the critical range of frequencies obtaining $\Omega = [10, 100]$. We note that the denominator of $P(s,q)$ is nonvanishing for all $q \in Q$ in this range. Then, we study robust stability of the plant connected in a feedback loop with a PID compensator

$$C(s) = K_P + \frac{K_I}{s} + K_D s.$$

For closed-loop stability, we recall that the Nyquist criterion requires that the Nyquist plot of the open-loop system does not go through the point $-1 + j0$ and that it does encircle this point (in counter-clockwise direction) a number of times equal to the number of unstable poles; for example, see Horowitz (1963). In this specific case, setting $K_P = 200$, $K_I = 5120$, and $K_D = 20$, we see that the closed-loop nominal system is stable with a phase margin $\phi \approx 63.7°$ and a crossover frequency $\omega_c \approx 78.8$ rad/s. As a result of the analysis carried out by sweeping the frequency, it turns out that the closed-loop system is not robustly stable, since at the frequency $\omega \approx 16$ rad/s the value set includes the point $-1 + j0$. Figure 7.4 shows the Nyquist plot of the nominal plant and the value set for 12 equispaced frequencies in the range (12,34).

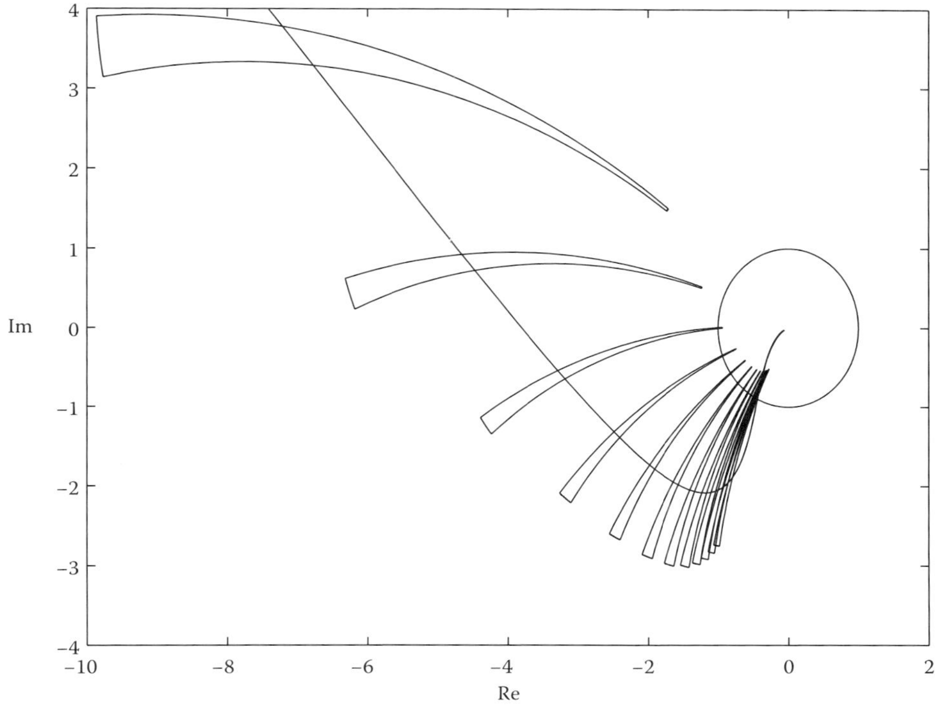

FIGURE 7.4 Nyquist plot and value set.

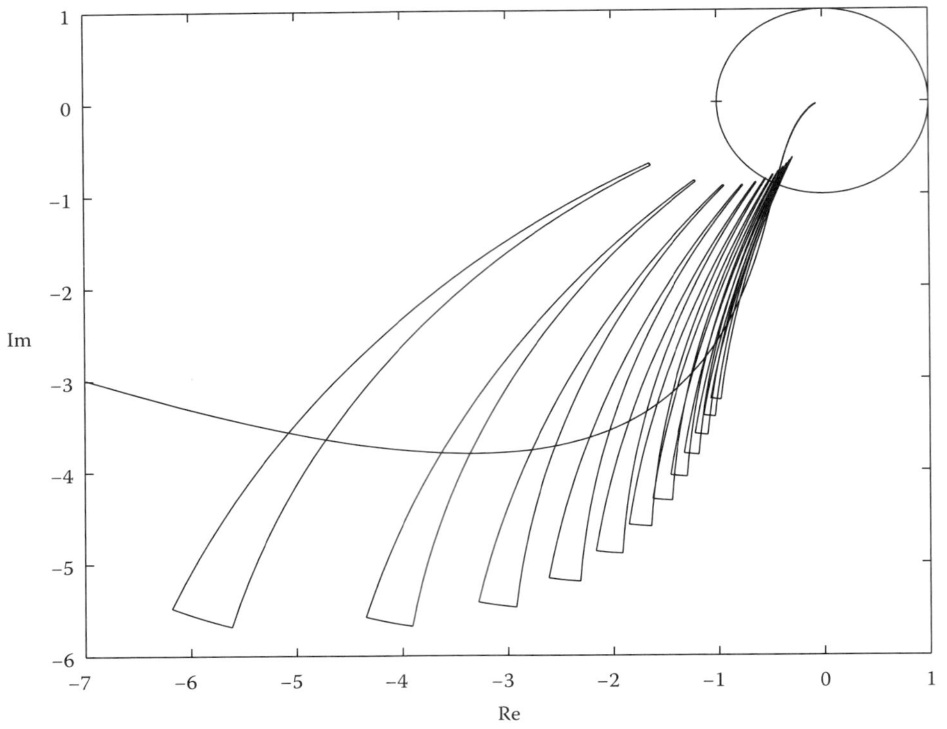

FIGURE 7.5 Nyquist plot and value set.

To robustly stabilize $P(s,q)$, we take $K_I = 2000$ and the same values of K_P and K_D as before. The reasons for choosing this value of K_I can be explained as follows: The compensator transfer function has phase zero at the frequency $\overline{\omega} = \sqrt{K_I/K_D}$. Thus, reducing K_I from 5120 to 2000 causes a drop of $\overline{\omega}$ from 16 to 10 rad/s. This implies that the phase lead effect begins at lower frequencies "pushing" the value set out of the critical point $-1+j0$. Since the nominal system is stable with a phase margin $\phi \approx 63.7°$ and a crossover frequency $\omega_c \approx 80.8$ rad/s, this new control system has nominal performance very close to the previous one. However, with this new PID compensator, the system becomes robustly stable. To see this, we generated the value sets for the same frequencies as before; see Figure 7.5.

From this figure, we observe that $P(j\omega,q)$ does not include the point $-1+j0$. We conclude that the closed-loop system is now robustly stable; the worst-case phase margin is $\phi \approx 57.9°$.

7.6 Robustness Analysis for Affine Polynomials

In this section, we study robustness analysis of the closed-loop polynomial

$$p(s,q) = N_P(s,q)N_C(s) + D_P(s,q)D_C(s)$$

when the coefficients of $N_P(s,q)$ and $D_P(s,q)$ are affine functions of q. The main goal is to provide an alternative criterion for polynomials instead of plants. With this approach, we do not need the nonvanishing condition about $D_P(s,q)$; furthermore, unstable pole–zero cancellations are not an issue. In this case, however, we lose the crucial insight given by the Nyquist plot. For fixed frequency $s = j\omega$, we define the value set $p(j\omega, Q) \subset \mathbf{C}$ as

$$p(j\omega, Q) \doteq \{p(j\omega, q) : q \in Q\}.$$

As in the plant case, $p(j\omega, Q)$ is a set in the complex plane which moves with frequency and which graphically represents the uncertain polynomial.

Criterion 7.2: Zero-Exclusion Condition for Uncertain Polynomials

The polynomial $p(s, q)$ is robustly stable for all $q \in Q$ if and only if $p(s, q)$ is stable for some $q \in Q$ and $0 \notin p(j\omega, Q)$ for all $\omega \in \mathbf{R}$.

The proof of this criterion requires elementary facts and, in particular, continuity of the roots of $p(s, q)$ versus its coefficients; see Frazer and Duncan (1929). Similar to the discussion in Section 7.5, we note that Criterion 7.2 is easily implementable—at least whenever the value set can be efficiently generated. That is, given an affine polynomial family, we take any element in this family and we check its stability. This step is straightforward using the Routh table or any root finding routine. Then, we sweep the frequency ω over a selected range of critical frequencies $\Omega = [\omega^-, \omega^+]$. This interval can be estimated, for example, using some *a priori* information on the specific problem or by means of one of the bounds given in Marden (1966). If there is no intersection of $p(j\omega, Q)$ with the origin of the complex plane for all $\omega \in \Omega$, then $p(s, q)$ is robustly stable.

Remark 7.1

A very similar zero exclusion condition can be stated for more general regions $\mathcal{D}$ than the open left half-plane. Meaningful examples of $\mathcal{D}$ regions are the open unit disk, a shifted left half-plane and a damping cone*. In this case, instead of sweeping the imaginary axis, we need to discretize the boundary of $\mathcal{D}$.

7.6.1 Value Set Construction for Affine Polynomials

In this section, we discuss the generation of the value set $p(j\omega, Q)$. Whenever the polynomial coefficients are affine functions of the uncertain parameters, the value set can be easily constructed. To this end, two key facts are very useful. First, for fixed frequency, we note that $p(j\omega, Q)$ is a two-dimensional convex polygon. Second, letting $q^1, q^2, \ldots, q^L$ denote the vertices of Q as in Section 7.4, we note that the vertices of the value set are a subset of the complex numbers $p(j\omega, q^1), p(j\omega, q^2), \ldots, p(j\omega, q^L)$. These two observations follow from the fact that, for fixed frequency, real and imaginary parts of $p(s, q)$ are both affine functions of q. Then, the value set is a two-dimensional affine mapping of the set Q and its vertices are generated by vertices of Q. Thus, for fixed ω, it follows that

$$p(j\omega, Q) = \text{conv}\,\{p(j\omega, q^1), p(j\omega, q^2), \ldots, p(j\omega, q^L)\}$$

where conv denotes the convex hull†. The conclusion is then immediate: For fixed frequency, one can generate the points $p(j\omega, q^1), p(j\omega, q^2), \ldots, p(j\omega, q^L)$ in the complex plane. The value set can be constructed by taking the convex hull of these points—this can be readily done with 2D graphics. From the computational point of view, we observe that the number of edges of the polygon is at most 2ℓ at each frequency. This follows from the observations that any edge of the value set is the image of an exposed edge of Q. In addition, parallel edges of Q are mapped into parallel edges in the complex plane and the edges of Q have only ℓ distinct directions. These facts can be used to efficiently compute $p(j\omega, Q)$. We now provide an example which illustrates the value set generation.

---

* A damping cone is a subset of the complex plane defined as $\{s : Re(s) \leq -\alpha|Im(s)|\}$ for $\alpha > 0$.

† The convex hull conv S of a set S is the smallest convex set containing S.

7.6.2 Example of Value Set Generation

Using the same data as in the example of Section 7.5.2 and a PID controller with gains $K_P = 200$, $K_I = 5120$, and $K_D = 20$, we study the closed-loop polynomial

$$p(s,q) = 0.025q_1s^4 + s^3(5q_1 + 0.025q_2) + s^2(5q_2 + 18.81) + 180s + 4608.$$

Robustness analysis is performed for 29 equispaced frequencies in the range (2,30). Figure 7.6 shows the polygonality of the value set and zero inclusion for $\omega \approx 16$ rad/s which demonstrates instability. This conclusion is in agreement with that previously obtained in Section 7.5.2.

7.6.3 Interval Polynomials: Kharitonov's Theorem and Value Set Geometry

In the special case of interval polynomials, robustness analysis can be greatly facilitated via Kharitonov's Theorem (Kharitonov, 1978). We now recall this result. Given an interval polynomial

$$p(s,q) = q_0 + q_1s + q_2s^2 + \cdots + q_ns^n$$

of order n (i.e., $q_n \neq 0$) and bounds $[q_i^-, q_i^+]$ for each coefficient q_i, define the following four polynomials:

$$\begin{aligned}
p_1(s) &\doteq q_0^+ + q_1^+s + q_2^-s^2 + q_3^-s^3 + q_4^+s^4 + q_5^+s^5 + q_6^-s^6 + q_7^-s^7 + \cdots;\\
p_2(s) &\doteq q_0^- + q_1^-s + q_2^+s^2 + q_3^+s^3 + q_4^-s^4 + q_5^-s^5 + q_6^+s^6 + q_7^+s^7 + \cdots;\\
p_3(s) &\doteq q_0^+ + q_1^-s + q_2^-s^2 + q_3^+s^3 + q_4^+s^4 + q_5^-s^5 + q_6^-s^6 + q_7^+s^7 + \cdots;\\
p_4(s) &\doteq q_0^- + q_1^+s + q_2^+s^2 + q_3^-s^3 + q_4^-s^4 + q_5^+s^5 + q_6^+s^6 + q_7^-s^7 + \cdots.
\end{aligned}$$

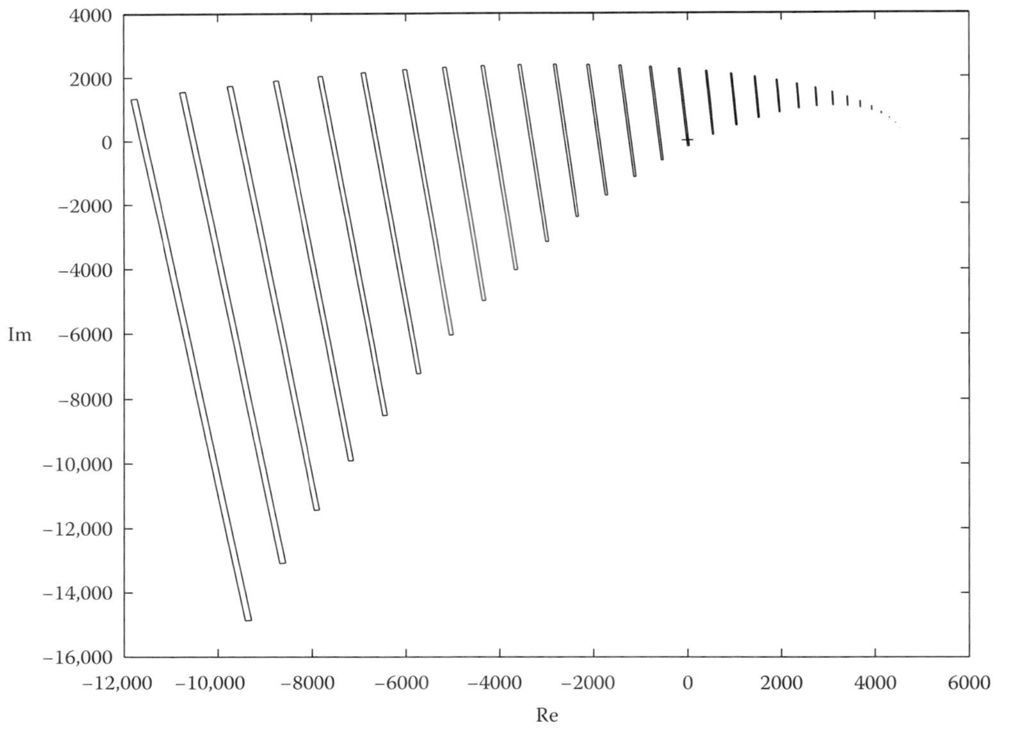

FIGURE 7.6 Value set plot.

Then, $p(s,q)$ is stable for all $q \in Q$ if and only if the four Kharitonov polynomials $p_1(s), p_2(s), p_3(s)$ and $p_4(s)$ are stable. To provide a geometrical interpretation of this result, we observe that the value set for fixed frequency $s = j\omega$ is a rectangle with level edges parallel to real and imaginary axis. The four vertices of this set are the complex numbers $p_1(j\omega), p_2(j\omega), p_3(j\omega)$, and $p_4(j\omega)$; see Dasgupta (1988). If the four Kharitonov polynomials are stable, due to the classical Mikhailov's criterion (Mikhailov, 1938), their phase is strictly increasing for ω increasing. In turn, this implies that the level rectangular value set moves in a counterclockwise direction around the origin of the complex plane. Next, we argue that the strictly increasing phase of the vertices and the parallelism of the four edges of the value set with real or imaginary axis guarantee that the origin does not lie on the boundary of the value set. By continuity, we conclude that the origin is outside the value set and the zero exclusion condition is satisfied; see Minnichelli et al. (1989).

7.6.4 From Robust Stability to Robust Performance

In this section, we point out the important fact that the polynomial approach discussed in this chapter can be also used for robust performance. To explain, we take an uncertain plant with affine uncertainty and we show how to compute the largest peak of the Bode plot magnitude for all $q \in Q$—that is, the worst-case H_∞ norm. Formally, for a stable strictly proper plant $P(s,q)$, we define

$$\max_{q \in Q} ||P(s,q)||_\infty \doteq \max_{q \in Q} \sup_{\omega} |P(j\omega, q)|.$$

Given a performance level $\gamma > 0$, then

$$\max_{q \in Q} ||P(s,q)||_\infty < \gamma$$

if and only if

$$\left| \frac{N_P(j\omega, q)}{D_P(j\omega, q)} \right| < \gamma$$

for all $\omega \geq 0$ and $q \in Q$. Since $P(j\omega, q) \to 0$ for $\omega \to \infty$, this is equivalent to check if the zero-exclusion condition

$$N_P(j\omega, q) - \gamma D_P(j\omega, q) e^{j\phi} \neq 0$$

is satisfied for all $\omega \in \mathbf{R}, q \in Q$ and $\phi \in [0, 2\pi]$. In turn, this implies that the polynomial with complex coefficients

$$p_\phi(s,q) = N_P(s,q) - \gamma D_P(s,q) e^{j\phi}$$

has roots in the open left half-plane for all $q \in Q$ and $\phi \in [0, 2\pi]$. Clearly, for fixed $\phi \in [0, 2\pi]$, Criterion 7.2 can be readily used; however, since $p_\phi(s,q)$ has complex coefficients, it should be necessarily checked for all $\omega \in \mathbf{R}$, including negative frequencies.

7.6.5 Algebraic Criteria for Robust Stability

If the uncertain polynomial under consideration is affine, the well-known Edge Theorem applies (Bartlett et al., 1988). This algebraic criterion is alternative to the frequency domain approach studied in this section. Roughly speaking, this result says that an affine polynomial family is robustly stable if and only if all the polynomials associated with the one-dimensional exposed edges of the set Q are stable. Even though this result is of algebraic nature, it can be explained by means of value set arguments. For affine polynomials and for fixed ω, we have already observed in Section 7.6.3 that the boundary of the value set $p(j\omega, Q)$ is the image of the one-dimensional exposed edges of Q. Thus, to guarantee the zero-exclusion condition, we need to guarantee that all edge polynomials are nonvanishing for all $\omega \in \mathbf{R}$—otherwise an instability occurs. We conclude that stability detection for affine polynomials requires the solution of a number of one-dimensional stability problems. Each of these problems can be stated as follows: Given polynomials $p_0(s)$ and $p_1(s)$ of order n and $m < n$, respectively, we need to study the stability of

$$p(s, \lambda) = p_0(s) + \lambda p_1(s)$$

for all $\lambda \in [0, 1]$. A problem of great interest is to ascertain when the robust stability of $p(s, \lambda)$ can be deduced from the stability of the extreme polynomials $p(s, 0)$ and $p(s, 1)$. This problem can be formulated in more general terms: To construct classes of uncertain polynomials for which the stability of the vertex polynomials (or a subset of them) implies stability of the family. Clearly, the edges associated with an interval polynomial is one such class. Another important example is given by the edges of the closed-loop polynomial of a control system consisting of a first-order compensator and an interval plant; see Barmish et al. (1992). Finally, see Rantzer (1992) for generalizations and for the concept of convex directions polynomials.

7.6.6 Further Extensions: The Spectral Set

In some cases, it is of interest to generate the entire root location of a family of polynomials. This leads to the concept of *spectral set*; see Barmish and Tempo (1991). Given a polynomial $p(s, q)$, we define the spectral set as

$$\sigma \doteq \{s \in \mathbf{C} : p(s, q) = 0 \text{ for some } q \in Q\}.$$

The construction of this set is quite easy for affine polynomials. Basically, the key idea can be explained as follows: For fixed $s \in \mathbf{C}$, checking if s is a member of the spectral set can be accomplished by means of the zero-exclusion condition; see also Remark 7.6.2. Next, it is easy to compute a bounded root confinement region $\bar{\sigma} \supseteq \sigma$; for example, see Marden (1966). Then, the construction of the spectral set σ amounts to a two-dimensional gridding of $\bar{\sigma}$ and, for each grid point, checking zero exclusion.

The spectral set concept can be further extended to control systems consisting of a plant $P(s, q)$ with a feedback gain K_P which needs tuning. In this case, we deal with the so-called robust root locus (Barmish and Tempo, 1990)—that is, the generation of all the roots of the closed-loop polynomial when K_P ranges in a given interval. To illustrate, we consider the same data as in Section 7.6.3 and a feedback gain K_P, thus obtaining the closed loop polynomial

$$p(s, q) = 0.025q_1s^3 + (5q_1 + 0.025q_2)s^2 + (0.81 + 5q_2)s + 0.9K_P,$$

where $q_1 \in [0.03, 0.15]$ and $q_2 \in [0.001, 0.03]$. In Figure 7.7, we show the spectral set for $K_P = 200$. Actually, only the portion associated with the dominant roots is visible since the spectral set, obviously, includes a real root in the interval $[-200.58, -200.12]$ which is out of the plot.

7.7 Multiaffine Uncertainty Structures

In this section, we discuss the generation of the value set for polynomials with more general uncertainty structures than affine. In particular, we study the case when the polynomial coefficients are multiaffine functions of the uncertain parameters. Besides the motivations provided in Section 7.3, we recall that this uncertainty structure is quite important for a number of reasons. For example, consider a linear state-space system of the form $\dot{x}(t) = A(q)x(t)$ where each entry of the matrix $A(q) \in \mathbf{R}^{m \times m}$ lies in a bounded interval, that is, it is an interval matrix. Then, the characteristic polynomial required for stability considerations has a multiaffine uncertainty structure.

In the case of multiaffine uncertainty, the value set is generally not convex and its construction cannot be easily performed. However, we can easily generate a "tight" convex approximation of $p(j\omega, Q)$—this approximation being its convex hull $\operatorname{conv} p(j\omega, Q)$. More precisely, the following fact, called the Mapping

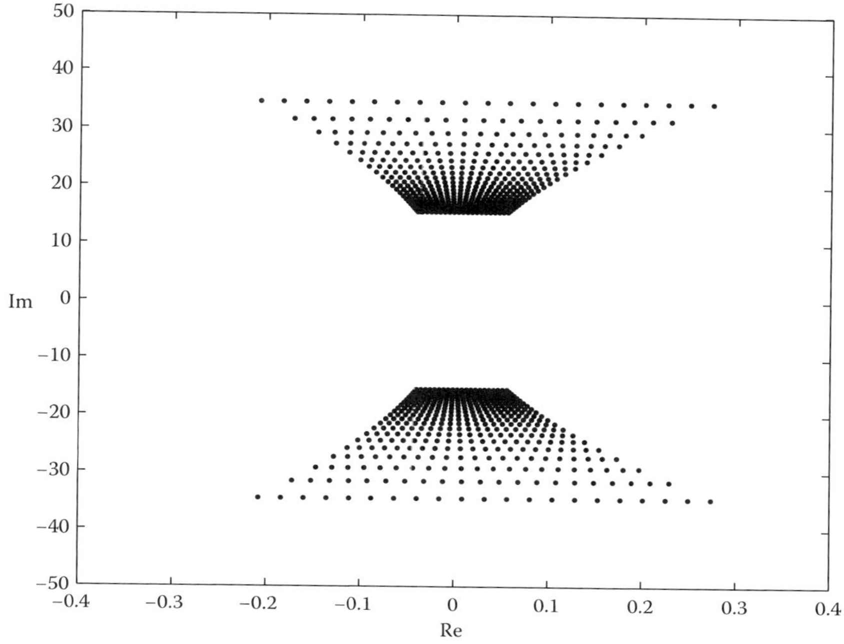

FIGURE 7.7 Spectral set.

Theorem, holds: The convex hull of the value set conv $p(j\omega, Q)$ is given by the convex hull of the vertex polynomials $p(j\omega, q^1), p(j\omega, q^2), \ldots, p(j\omega, q^L)$. In other words, the parts of the boundary of $p(j\omega, Q)$ which are not line segments are always contained inside this convex hull. Clearly, if conv $p(j\omega, Q)$ is used instead of $p(j\omega, Q)$ for robustness analysis through zero exclusion, only a sufficient condition is obtained. That is, if the origin of the complex plane lies inside the convex hull, we do not know if it is also inside the value set. We now formally state the Mapping Theorem; see Zadeh and Desoer (1963).

Theorem 7.1: Mapping Theorem

For fixed frequency $\omega \in \mathbf{R}$,

$$\mathit{conv}\ p(j\omega, Q) = \mathit{conv}\ \{p(j\omega, q^1), p(j\omega, q^2), \ldots, p(j\omega, q^L)\}.$$

With regard to applicability and usefulness of this result, comments very similar to those made in Section 7.7 about the construction of the value set for affine polynomials can be stated. Figure 7.8 illustrates the Mapping Theorem for the polynomial

$$p(s, q) = s^3 + (q_2 + 4q_3 + 2q_1q_2)s^2 + (4q_2q_3 + q_1q_2q_4)s + q_3 + 2q_1q_3 - q_1q_2(q_4 - 0.5)$$

with four uncertain parameters q_1, q_2, q_3, and q_4 each bounded in the interval [0, 1] and frequency $\omega = 1$ rad/s. The "true" value set shown in this figure is obtained via random generation of 10,000 samples uniformly distributed in the set Q.

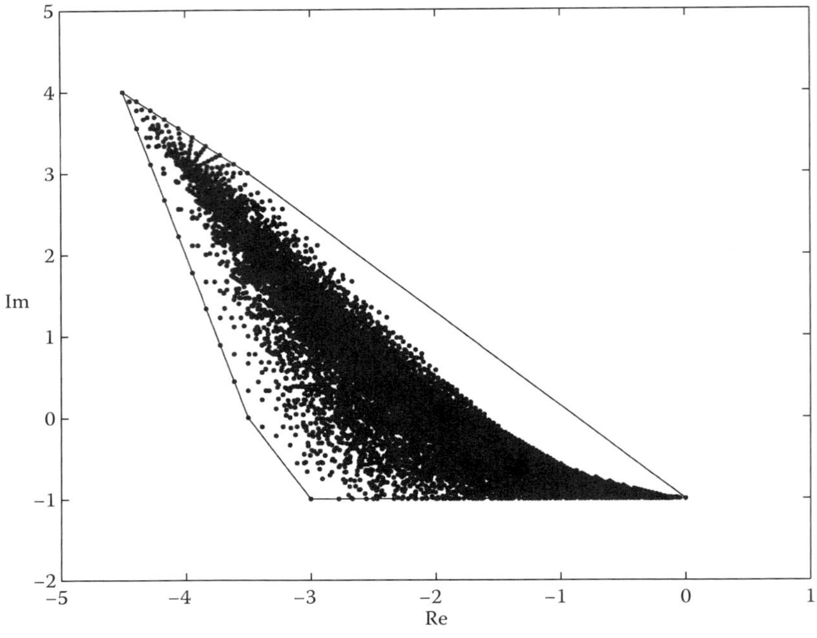

FIGURE 7.8 Value set and its convex hull.

7.8 General Uncertainty Structures and Controller Synthesis

For general uncertainty structures there is no analytical tool that enables us to construct the value set. For systems with a very limited number of uncertain parameters, a brute force approach, as in the example of Theorem 7.1, can be taken by simply gridding the set Q with a sufficient number of points. With this procedure, one can easily obtain a "cloud" in the complex plane which approximates the value set. Obviously, this method is not practical for a large number of uncertain parameters and provides no guarantee that robustness is satisfied outside the grid points.

Several attempts in the literature aimed at solving this general problem. Among the pioneering contributions along this line we recall the parameter space approach (Ackermann, 1980), techniques for the multivariable gain margin computation (de Gaston and Safonov, 1988) and (Sideris and Sanchez Pena, 1989) and the geometric programming approach (Vicino et al., 1990).

As far as the robust synthesis problem is concerned, it is worth mentioning that the methods described in this chapter are suitable for synthesis in a trial-and-error fashion. A practical use of the techniques can be summarized in the following two steps (as illustrated in Section 7.5.2):

- Synthesize a controller for the nominal plant with any proper technique.
- Perform robustness analysis taking into account the uncertainties and go back to the previous step if necessary.

In the special case of low complexity compensators, the two steps can be combined. This is the case of PID synthesis for which efficient robust design tools have been proposed (Ho et al., 2000).

To facilitate robustness analysis for general parametric uncertainty structures and for nonparametric uncertainty, different approaches have been proposed in the literature. Among the others, we recall the μ approach (Zhou et al., 1996; Sanchez Pena and Sznaier, 1998) and the probabilistic methods (Tempo et al., 2005). These approaches can be also used for designing a controller, but μ-synthesis is based on a procedure which is not guaranteed to converge to a global solution and probabilistic methods provide a controller which satisfies the required specification only with a given probability. However, both methods are useful tools for many robust control problems.

As a final remark, we emphasize that in this chapter we considered only uncertainties which are constant in time. It is known that parameter variations or, even worse, switching, may have a destabilizing effect. There are, indeed, examples of very simple systems that are (Hurwitz) stable for any fixed value of the parameters but can be destabilized by parameter variations. A class of systems which has been investigated is that of Linear Parameter-Varying (LPV) systems. For these systems, it has been established that the Lyapunov approach is crucial. For an overview of the problem of robustness of LPV systems and the link with switching systems, the reader is referred to Blanchini and Miani (2008).

References

Abate, M., Barmish, B.R., Murillo-Sanchez, C., and Tempo, R. 1994. Application of some new tools to robust stability analysis of spark ignition engines: A case study. *IEEE Transactions on Control Systems Technology*, CST-2:22–30.

Ackermann, J.E. 1980. Parameter space design of robust control systems. *IEEE Transactions on Automatic Control*, AC-25:1058–1072.

Ackermann, J. in cooperation with Bartlett, A., Kaesbauer, D., Sienel, W., and Steinhauser, R. 1993. *Robust Control, Systems with Uncertain Physical Parameters*, Springer-Verlag, London.

Ackermann, J., Curtain, R. F., Dorato, P., Francis, B. A., Kimura, H., and Kwakernaak, H. (Editors). 1993. Special Issue on robust control. *Automatica*, 29:1–252.

Barmish, B.R. 1994. *New Tools for Robustness of Linear Systems*, Macmillan, New York.

Barmish, B.R., Hollot, C.V., Kraus, F., and Tempo, R. 1992. Extreme point results for robust stabilization of interval plants with first-order compensators. *IEEE Transactions on Automatic Control*, AC-37:707–714.

Barmish, B.R. and Shi, Z. 1990. Robust stability of perturbed systems with time delays. *Automatica*, 25:371–381.

Barmish, B.R. and Tempo, R. 1990. The robust root locus. *Automatica*, 26:283–292.

Barmish, B.R. and Tempo, R. 1991. On the spectral set for a family of polynomials. *IEEE Transactions on Automatic Control*, AC-36:111–115.

Bartlett, A.C., Hollot, C.V., and Huang, L. 1988. Root locations of an entire polytope of polynomials: It suffices to check the edges. *Mathematics of Control, Signals and Systems*, 1:61–71.

Bhattacharyya, S.P., Chapellat, H., and Keel, L.H. 1995. *Robust Control: The Parametric Approach*, Prentice-Hall, Englewood Cliffs, NJ.

Blanchini, F. and Miani, S. 2008. *Set-Theoretic Methods in Control*, Birkhäuser, Boston.

Chockalingam, G. and Dasgupta, S. 1993. Minimality, stabilizability and strong stabilizability of uncertain plants. *IEEE Transactions on Automatic Control*. AC-38:1651–1661.

Dasgupta, S. 1988. Kharitonov's theorem revisited. *Systems and Control Letters*. 11:381–384.

de Gaston, R.R.E. and Safonov, M.G. 1988. Exact calculation of the multiloop stability margin. *IEEE Transactions on Automatic Control*, AC-33:156–171.

Djaferis, T.E. 1995. *Robust Control Design: A Polynomial Approach*, Kluwer Academic Publishers, Boston.

Doyle, J.C., Francis, B.A., and Tannenbaum, A.R. 1992. *Feedback Control Theory*, Macmillan, New York.

Frazer, R.A. and Duncan, W.J. 1929. On the criteria for the stability of small motion. *Proceedings of the Royal Society*, A, 124:642–654.

Fu, M. 1990. Computing the frequency response of linear systems with parametric perturbations. *Systems and Control Letters*, 15:45–52.

Ho, M.-T., Datta, A., and Bhattacharyya, S.P. 2000, *Structure and Synthesis of PID Controllers*, Springer-Verlag, London.

Horowitz, I.M. 1963. *Synthesis of Feedback Systems*, Academic Press, New York.

Horowitz, I. 1991. Survey of quantitative feedback theory (QFT). *International Journal of Control*, 53:255–291.

Kharitonov, V.L. 1978. Asymptotic stability of an equilibrium position of a family of systems of linear differential equations. *Differential'nye Uraveniya*, 14:1483–1485.

Kogan, J. 1995. *Robust Stability and Convexity*. Lecture Notes in Control and Information Sciences, Springer-Verlag, London.

Marden, M. 1966. *Geometry of Polynomials*, Mathematical Surveys, No. 3, American Mathematical Society, Providence, RI.

Mikhailov, A.W. 1938. Methods of harmonic analysis in control theory. *Avtomatika i Telemekhanika*, 3:27–81.

Minnichelli, R.J., Anagnost, J.J., and Desoer, C.A. 1989. An elementary proof of Kharitonov's theorem with extensions. *IEEE Transactions on Automatic Control*. AC-34:995–998.

Rantzer, A. 1992. Stability conditions for polytopes of polynomials, *IEEE Transactions on Automatic Control*, AC-37:79–89.

Sanchez Pena, R. and Sznaier, M. 1998. *Robust Systems, Theory and Applications*, John Wiley and Sons, New York.

Sideris, A. and Sanchez Pena, R.S. 1989. Fast calculation of the multivariable stability margin for real interrelated uncertain parameters. *IEEE Transactions on Automatic Control*, AC-34:1272–1276.

Tempo, R., Calafiore, G., and Dabbene, F. 2005. *Randomized Algorithms for Analysis and Control of Uncertain Systems*, Springer-Verlag, London.

Vicino, A., Tesi, A., and Milanese, M. 1990. Computation of nonconservative stability perturbation bounds for systems with nonlinearly correlated uncertainties. *IEEE Transactions on Automatic Control*, AC-35:835–841.

Zadeh, L.A. and Desoer, C.A., 1963. *Linear System Theory*, McGraw-Hill, New York.

Zhou, K., Doyle, J.C., and Glover, K. 1996. *Robust and Optimal Control*, Prentice-Hall, Upper Saddle River, NJ.

8

MIMO Frequency Response Analysis and the Singular Value Decomposition

Stephen D. Patek
Massachusetts Institute of Technology

Michael Athans
Massachusetts Institute of Technology

8.1 Modeling MIMO Linear Time-Invariant Systems in Terms of Transfer Function Matrices

Any multivariable linear time invariant (LTI) system is uniquely described by its impulse response matrix. The Laplace transform of this matrix function of time gives rise to the system's transfer function matrix (TFM). We assume that all systems in the sequel have TFM representations; frequency response analysis is always applied to TFM descriptions of systems. For the case of finite dimensional LTI systems, when we have a state-space description of the system, closed-form evaluation of the TFM is particularly easy. This is discussed in Example 8.1. More generally, *infinite*-dimensional LTI systems also have TFM representations, although closed-form evaluation of the TFM of these systems is less straightforward. Stability (in whatever sense) is *presumed*; without stability we cannot talk about the steady-state response of a system to sinusoidal inputs. For such systems, the TFM can be measured by means of sinusoidal inputs. We will not discuss the use of frequency-domain techniques in robustness analysis.

Example 8.1: Finite-Dimensional LTI Systems

Suppose that we have a finite-dimensional LTI system G, with the following state equations

$$\dot{x}(t) = Ax(t) + Bu(t),$$
$$y(t) = Cx(t) + Du(t).$$

Taking the Laplace transform of the state equation, we see that,

$$x(s) = (sI - A)^{-1}Bu(s)$$

where we have abused notation slightly with $x(s)$ and $u(s)$ representing the Laplace transforms of $x(t)$ and $u(t)$, respectively. Similarly,

$$y(s) = [C(sI - A)^{-1}B + D]u(s) \equiv G(s)u(s).$$

The matrix quantity $G(s)$ is the transfer function matrix associated with the system G.

As a matter of convention, we will refer to the input of the system generically as u. The more specific notation $u(t)$ and $u(s)$ will be used in reference to u represented as time-domain and frequency-domain (Laplace-transform) signals, respectively. A similar convention applies to the output, y. We will assume that u is a vector with m components and that the output y is a vector with p components. This makes the TFM a $p \times m$ matrix. To make this explicit,

$$G(s) = \begin{bmatrix} g_{11}(s) & \cdots & g_{1m}(s) \\ \vdots & & \vdots \\ g_{p1}(s) & \cdots & g_{pm}(s) \end{bmatrix} \tag{8.1}$$

$$u(s) = [u_1(s), \cdots, u_m(s)]^T \tag{8.2}$$

$$y(s) = [y_1(s), \cdots, y_p(s)]^T \tag{8.3}$$

Componentwise,

$$y_k(s) = \sum_{j=1}^{m} g_{kj}(s)u_j(s), \quad k = 1, \ldots, p. \tag{8.4}$$

As Laplace transforms, $u(s), y(s)$, and $G(s)$ are generally complex-valued quantities. In more formal mathematical notation, $u(s) \in C^m, y(s) \in C^p$, and $G(s) \in C^{p\times m}$.

8.2 Frequency Response for MIMO Plants

In discussing the frequency response of LTI systems, we focus our attention on systems which are *strictly* stable. This allows us to envision applying sinusoidal inputs to the system and measuring steady-state output signals which are appropriately scaled and phase-shifted sinusoids of the same frequency. Because we are dealing with MIMO systems, there are now additional factors which affect the nature of the frequency response, particularly the relative magnitude and phase of each of the components of the input vector u. These considerations will be discussed in detail below.

Suppose that we have in mind a complex exponential input, as below,

$$u(t) = \tilde{u}e^{j\omega t} \tag{8.5}$$

where $\tilde{u} = (\tilde{u}_1, \ldots, \tilde{u}_m)^T$ is a fixed (complex) vector in C^m. Note that by allowing $\tilde{u}$ to be complex, the individual components of $u(t)$ can have different phases relative to one another.

Example 8.2:

Suppose that $m = 2$ and that $\tilde{u} = ((1 + j1), (1 - j1))^T$, then,

$$u(t) = \begin{pmatrix} (1 + j1) \\ (1 - j1) \end{pmatrix} e^{j\omega t} = \begin{pmatrix} \sqrt{2}e^{j\pi/4} \\ \sqrt{2}e^{-j\pi/4} \end{pmatrix} e^{j\omega t} = \begin{pmatrix} \sqrt{2}e^{j(\omega t + \pi/4)} \\ \sqrt{2}e^{j(\omega t - \pi/4)} \end{pmatrix}$$

Thus, the two components of $u(t)$ are phase shifted by $\pi/2$ radians (or 90 degrees).

Suppose that this input is applied to our stable LTI system G (of compatible dimension). We know from elementary linear systems theory that each component of the output of G can be expressed in terms of G's *frequency response* $G(j\omega)$. (We obtain the frequency response from $G(s)$, literally, by setting $s = j\omega$.) Thus, at steady state,

$$y_k(t) = \sum_{j=1}^{m} g_{kj}(j\omega)\tilde{u}_j e^{j\omega t}; \quad k = 1, \ldots, p. \tag{8.6}$$

We may now express the vector output $y(t)$ at steady state as follows:

$$y(t) = \tilde{y}e^{j\omega t}, \quad \tilde{y} = (\tilde{y}_1, \ldots, \tilde{y}_p)^T \in C^p, \tag{8.7}$$

where

$$\tilde{y}_k = \sum_{j=1}^{m} g_{kj}(j\omega)\tilde{u}_j, \quad k = 1, \ldots, p. \tag{8.8}$$

Putting all of this together,

$$\tilde{y} = G(j\omega)\tilde{u}. \tag{8.9}$$

Just as in the SISO case, the MIMO frequency response $G(j\omega)$ provides a convenient means of computing the output of an LTI system driven by a complex exponential input. *Analysis* of the frequency response, however, is now complicated by the fact that $G(j\omega)$ is a matrix quantity. A simple way is needed to characterize the "size" of the frequency response as a function of ω. The effect of $G(j\omega)$ on complex exponential input signals depends on the direction of $\tilde{u} \in C^m$, including the relative phase of the components. In fact, a whole range of "sizes" of $G(j\omega)$ exists, depending on the directional nature of $\tilde{u}$. Our characterization of size should thus provide both upper and lower bounds on the magnitude gain of the frequency response matrix. The mathematical tool we need here is the Singular Value Decomposition (SVD) discussed briefly in the following section.

8.3 Mathematical Detour

We present here some mathematical definitions and basic results from linear algebra germane to our discussion of MIMO frequency response analysis. An excellent reference for this material can be found in [5].

8.3.1 Introduction to Complex Vectors and Complex Matrices

Given a complex-valued column vector $x \in C^n$, we may express x in terms of its real and imaginary components,

$$x = \begin{pmatrix} x_1 \\ \vdots \\ x_n \end{pmatrix} = \begin{pmatrix} a_1 + jb_1 \\ \vdots \\ a_n + jb_n \end{pmatrix} = a + jb$$

where a and b are both purely real-valued vectors in C^n. We define the row vector x^H as the complex-conjugate transpose of x, i.e.,

$$x^H = (x_1^*, \ldots, x_n^*) = a^T - jb^T.$$

The superscript asterix above denotes the complex-conjugate operation.

If we have two vectors x and y, both elements of C^n, then the *inner product of x and y* is given by the (complex) scalar $x^H y$. Two vectors x and y in C^n, which satisfy $x^H y = 0$, are said to be *orthogonal.*

The *Euclidean norm* of x, denoted by $\|x\|_2$, is given by the square root of the inner product of x with itself,

$$\|x\|_2 = \sqrt{x^H x} = \sqrt{a^T a + b^T b} = \sqrt{\sum_{i=1}^{n} (a_i{}^2 + b_i{}^2)}.$$

It is important to note that $\| \cdot \|_2$ is a real scalar function chosen arbitrarily to reflect the "size" of vectors in C^n. (It is true that, as a norm, $\| \cdot \|_2$ has to satisfy certain mathematical requirements, particularly positivity, scaling, and the triangle inequality. Aside from this, our definition of $\| \cdot \|_2$ is arbitrary.) Because all of the components of x are taken into account simultaneously, the value (and interpretation) of $\|x\|_2$ will depend on the *units* in which the components of x are expressed.

Example 8.3:

Suppose that we are dealing with a high-power (AC) electronic device and that the state of the device is determined by a vector $x \in C^2$ made up of phased voltages at two distinct points in the circuitry. Suppose first that both quantities are expressed in terms of kilovolts (kV). For example,

$$x = (1 + j2, 2 - j3)^T \quad kV$$

then,

$$\begin{aligned} \|x\|_2{}^2 &= [(1 - j2), (2 + j3)][(1 + j2), (2 - j3)]^T \\ &= (1 + 4) + (4 + 9) \\ &= 18 \end{aligned}$$

If, however, the first component is expressed in terms of Volts (V), then

$$\begin{aligned} \|x\|_2{}^2 &= [(1000 - j2000), (2 + j3)][(1000 + j2000), (2 - j3)]^T, \\ &= (10^6 + 4 \times 10^6) + (4 + 9) \\ &= 5000013, \end{aligned}$$

which is a far cry from what we had before! Note that this is not an entirely unreasonable example. In general, the components of x can consist of entirely different types of physical quantities, such as voltage, current, pressure, concentration, etc. The choice of units is arbitrary and will have an important impact on the "size" of x when measured in terms of the Euclidean norm.

A complex-valued matrix $M \in C^{p\times m}$ is a matrix whose individual components are complex valued. Since M is complex valued, we can express it as the sum of its real and imaginary matrix parts, i.e., $M = A + jB$, where A and B are both purely real-valued matrices in $C^{p\times m}$.

We define the *Hermitian of a complex matrix M* as the complex-conjugate transpose of M, that is, M^H is computed by taking the complex conjugate of the transpose of M. Mathematically,

$$M^H = A^T - jB^T \tag{8.10}$$

The following will play an important role in the next subsection.

8.3.1.1 Important Fact

Both M^HM and MM^H have eigenvalues that are purely real valued and nonnegative. Moreover, their nonzero eigenvalues are identical even though $M^HM \neq MM^H$.

Example 8.4:

Let

$$M = \begin{bmatrix} 1 & 1+j \\ -j & 2+j \end{bmatrix}$$

Then,

$$M^H = \begin{bmatrix} 1 & j \\ 1-j & 2-j \end{bmatrix}$$

$$M^HM = \begin{bmatrix} 2 & j3 \\ -j3 & 7 \end{bmatrix}$$

and

$$MM^H = \begin{bmatrix} 3 & 3+j2 \\ 3-j2 & 6 \end{bmatrix}$$

Although M^HM and MM^H are clearly not equal, a simple calculation easily reveals that both products have the same characteristic polynomial,

$$\det(\lambda I - M^HM) = \det(\lambda I - MM^H) = \lambda^2 - 9\lambda + 5$$

This implies that M^HM and MM^H share the same eigenvalues.

A complex-valued matrix M is called *Hermitian* if $M = M^H$. A nonsingular, complex-valued matrix is called *unitary* if $M^{-1} = M^H$. Stated another way, a complex-valued matrix M is unitary if its column-vectors are mutually orthonormal.

The *spectral norm of a matrix* $M \in C^{p\times m}$, denoted $\|M\|_2$, is defined by

$$\|M\|_2 = \max_{\|x\|_2=1} \|Mx\|_2 \tag{8.11}$$

The best way to interpret this definition is to imagine the vector x rotating on the surface of the unit hypersphere in C^m, generating the vector Mx in C^p. The size of Mx, i.e., its Euclidean norm, will depend on the direction of x. For some direction of x, the vector Mx will attain its maximum value. This value is equal to the spectral norm of M.

8.3.2 The Singular Value Decomposition

In this subsection we give a quick introduction to the singular value decomposition. This will be an essential tool in analyzing MIMO frequency response. For more details, the reader is referred to [5].

8.3.2.1 The Singular Values of a Matrix

Suppose that M is a $p \times m$ matrix, real or complex. Assume that the rank of M is k. We associate with M a total of k positive constants, denoted $\sigma_i(M)$, or simply $\sigma_i, i = 1, \ldots, k$. These are the *singular values of* M, computed as the positive square roots of the nonzero eigenvalues of either $M^H M$ or MM^H, that is,

$$\sigma_i(M) = \sqrt{\lambda_i(M^H M)} = \sqrt{\lambda_i(MM^H)} \;> 0, \quad i = 1, \ldots, k \tag{8.12}$$

where $\lambda_i(\cdot)$ is a shorthand notation for "the ith nonzero eigenvalue of". Note that the matrices $M^H M$ and MM^H may have one or more zero valued eigenvalues in addition to the ones used to compute the singular values of M. It is common to index and rank the singular values as follows:

$$\sigma_1(M) \geq \sigma_2(M) \geq \cdots \geq \sigma_k(M) > 0 \tag{8.13}$$

The largest singular value of M, denoted $\sigma_{max}(M)$, is thus equal to $\sigma_i(M)$. Similarly, $\sigma_{min}(M) = \sigma_k(M)$.

While it is tricky, in general, to compute the eigenvalues of a matrix numerically, reliable and efficient techniques for computing singular values are available in commercial software packages, such as MATLAB®.

8.3.2.2 The Singular Value Decomposition

The SVD is analogous to matrix diagonalization. It allows one to write the matrix M in terms of its singular values and involves the definition of special directions in both the range and domain spaces of M.

To begin the definition of the SVD, we use the k nonzero singular values σ_i of M, computed above. First form a square matrix with the k singular values along the main diagonal. Next, add rows and columns of zeros until the resulting matrix Σ is $p \times m$. Thus,

$$\Sigma = \left[\begin{array}{cccc|c} \sigma_1 & 0 & \cdots & 0 & \\ 0 & \sigma_2 & \cdots & 0 & 0_{k\times(m-k)} \\ \cdots & \cdots & \cdots & \cdots & \\ 0 & 0 & \cdots & \sigma_k & \\ \hline & 0_{(p-k)\times k} & & & 0_{(p-k)\times(m-k)} \end{array}\right] \tag{8.14}$$

By convention, assume that

$$\sigma_1 \geq \sigma_2 \geq \cdots \geq \sigma_k$$

Theorem 8.1: Singular Value Decomposition

A $p \times p$ unitary matrix U (with $U = U^H$) and an $m \times m$ unitary matrix V (with $V = V^H$) exist so that

$$M = U\Sigma V^H \qquad \Sigma = U^H M V \tag{8.15}$$

The p-dimensional column vectors $u_i, i = 1, \ldots, p$, of the unitary matrix U are called the *left singular vectors* of M. The m-dimensional column vectors $v_i, i = 1, \ldots, m$ of V are called the *right singular vectors* of M. Thus, we can visualize,

$$U = [u_1\; u_2\; \ldots\; u_p], \quad V = [v_1\; v_2\; \ldots\; v_m]$$

Since U and V are each unitary,

$$u_i^H u_k = \delta_{ik}, \quad i, k = 1, \ldots, p$$
$$v_i^H v_k = \delta_{ik}, \quad i, k = 1, \ldots, m$$

where δ_{ik} is the Kronecker delta. Because the left and right singular vectors are linearly independent, they can serve as basis vectors for C^p and C^m, respectively. Moreover, the left and right singular vectors associated with the (nonzero) singular values of M span the range and left-null spaces of the matrix M, respectively. Finally, a simple calculation shows that the left singular vectors of M are the normalized right eigenvectors of the $p \times p$ matrix MM^H. Similarly, the right singular vectors of M are the normalized left eigenvectors of the $p \times p$ matrix $M^H M$.

8.3.2.3 Some Properties of Singular Values

We list here some important properties of singular values. We leave the proofs to the reader. Some of the properties require that the matrix be square and nonsingular.

1. $\sigma_{max}(M) = \max_{\|x\|_2=1} \|Mx\|_2 = \|M\|_2 = \frac{1}{\sigma_{min}(M^{-1})}$.
2. $\sigma_{min}(M) = \min_{\|x\|_2=1} \|Mx\|_2 = \frac{1}{\|M^{-1}\|_2} = \frac{1}{\sigma_{max}(M^{-1})}$.
3. $\sigma_i(M) - 1 \leq \sigma_i(I + M) \leq \sigma_i(M) + 1, \quad i = 1, \dots, k$.
4. $\sigma_i(\alpha M) = |\alpha| \sigma_i(M)$ for all $\alpha \in C, i = 1, \dots, k$.
5. $\sigma_{max}(M_1 + M_2) \leq \sigma_{max}(M_1) + \sigma_{max}(M_2)$.
6. $\sigma_{max}(M_1 M_2) \leq \sigma_{max}(M_1) \cdot \sigma_{max}(M_2)$.

Property 1 indicates that maximum singular value $\sigma_{max}(M)$ is identical to the spectral norm of M. Thus, Properties 5 and 6 are restatements of the triangle inequality and submultiplicative property, respectively.

8.3.2.4 The SVD and Finite Dimensional Linear Transformations

We shall now present some geometric interpretations of the SVD result. Consider the linear transformation

$$y = Mu, \quad u \in C^m, \quad y \in C^p. \tag{8.16}$$

Let M have the singular value decomposition discussed above, that is, $M = U \Sigma V^H$. It may help the reader to think of u as the input to a static system M with output y. From the SVD of M,

$$y = Mu = U \Sigma V^H u.$$

Suppose we choose u to be one of the right singular vectors, say v_i, of M. Let y_i denote the resulting "output" vector. Then,

$$y_i = Mv_i = U \Sigma V^H v_i.$$

Because the right singular vectors of M are orthonormal,

$$V^H v_i = \left(0, \dots, 0, 1, 0, \dots, 0\right)^T,$$

where the ith component only takes on the value of 1. In view of the special structure of the matrix of singular values Σ,

$$\Sigma V^H v_i = \left(0, \dots, 0, \sigma_i, 0, \dots, 0\right)^T$$

where, again, only the ith component is potentially nonzero. Thus, finally,

$$y_i = U \Sigma V^H v_i = \sigma_i u_i. \tag{8.17}$$

Equation 8.17 interprets the unique relationship between singular values and singular vectors. In the context of M as a "static" system, when the input u is equal to a right singular vector v_i, the output direction is fixed by the corresponding left singular vector u_i. Keeping in mind that both u_i and v_i have unit magnitudes (in the Euclidean sense), the amplification (or attenuation) of the input is measured by the associated singular value σ_i. If we choose $u = v_i$, where $i > k$, then the corresponding output vector is zero because the matrix is not full rank and there are no more (nonzero) singular values.

Because Equation 8.17 holds for $i = 1, \dots, k$, it is true in particular for the maximum and minimum singular values and associated singular vectors. By abuse of notation, we shall refer to these left and right

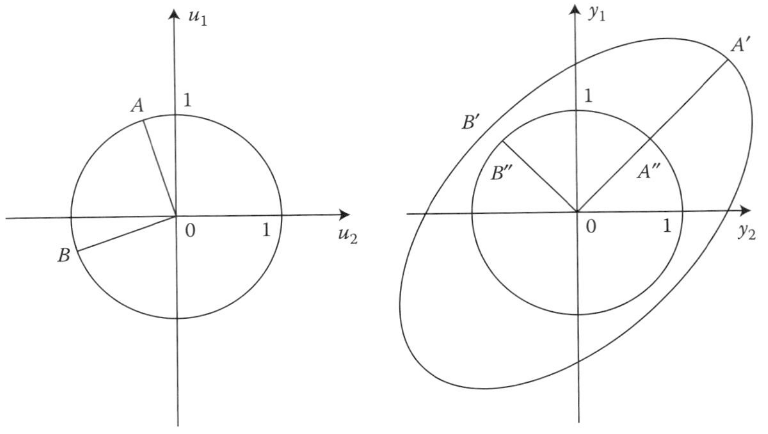

FIGURE 8.1 Visualization of SVD quantities.

singular vectors as *maximum and minimum singular vectors*, respectively, and use the subscripts "max" and "min" to distinguish them. Within the context of "static" systems, inputs along the *maximum* right singular vector generate the *largest* output along the direction of the *maximum* left singular vector. Similar comments apply to the case where inputs are in the direction of the minimum left singular vector.

Example 8.5: The Case Where M Is Real and 2 × 2

Let us suppose that M is a real-valued, nonsingular matrix mapping $u \in R^2$ to $y = Mu \in R^2$. Let us suppose further that u rotates on the circumference of the unit circle. The image of u under the transformation of M will then trace an ellipse in the (output) plane, as illustrated in Figure 8.1.

Because M is a real, nonsingular, 2 × 2 matrix, the SVD analysis will give

$$\Sigma = \begin{bmatrix} \sigma_{max} & 0 \\ 0 & \sigma_{min} \end{bmatrix} \quad U = [u_{max}\ u_{min}] \quad V = [v_{max}\ v_{min}]$$

where Σ, U, and V are all real valued.

Suppose that when u equals the vector OA, the output y is the vector OA'. Suppose further that, $y = y_{max} \equiv \sigma_{max} u_{max}$. Thus, the maximum right singular vector v_{max} equals the (unit length) vector OA, and the maximum left singular vector u_{max} equals the (unit length) vector OA". Moreover, the maximum singular value, σ_{max} equals the length of the vector OA'.

Similarly, suppose that when u equals the vector OB, the output y is the vector OB'. Suppose further that, $y = y_{min} \equiv \sigma_{min} u_{min}$. Thus, the minimum right singular vector v_{min} equals the (unit length) vector OB, and the minimum left singular vector u_{min} equals the (unit length) vector OB". Moreover, the minimum singular value, σ_{min} equals the length of the vector OB'.

Notice in Figure 8.1 that the left singular vectors are normal to each other, as are the right singular vectors.

As the minimum singular value decreases, so does the semiminor axis of the ellipse. As this happens, the ellipse becomes more and more elongated. In the limit, as $\sigma_{min} \to 0$, the ellipse degenerates into a straight line segment, and the matrix M becomes singular. In this limiting case, there are directions in the output space that we cannot achieve.

If the matrix M were a 3 × 3 real nonsingular matrix, then we could draw a similar diagram, illustrating the unit sphere mapping into a three-dimensional ellipsoid. Unfortunately, diagrams for higher dimensional matrices are impossible. Similarly, diagrams for *complex* matrices (even 2 × 2 matrices) are impossible, because we need a plane to represent each complex number.

Using these geometric interpretations of SVD quantities, it is possible to be precise about the meaning of the "size" of a real or complex matrix. From an intuitive point of view, if we consider the "system" $y = Mu$, and if we restrict the input vector u to have unit length, then

1. The matrix M is "large" if $\|y\|_2 \gg 1$, independent of the direction of the unit input vector u.
2. The matrix M is "small" if $\|y\|_2 \ll 1$, independent of the direction of the unit input vector u.

If we accept these definitions, then we can quantify size as follows:

1. The matrix M is "large" if its minimum singular value is large, i.e., $\sigma_{min}(M) \gg 1$.
2. The matrix M is "small" if its maximum singular value is small, i.e., $\sigma_{max}(M) \ll 1$.

8.3.2.5 More Analytical Insights

Once we have computed an SVD for a matrix M, in $y = Mu$, then we can compute many other important quantities. In particular, suppose that M is $m \times m$ and nonsingular. It follows that M has m nonzero singular values. We saw earlier (in Equation 8.17) that

$$y_i = \sigma_i u_i$$

when $u = v_i, i = 1, \ldots, m$. Because the left singular vectors are orthonormal, they form a basis for the m-dimensional input space, so that we can write any (input) vector in C^m as a linear combination of the v_is. For example, let u be given as follows:

$$u = \gamma_1 v_1 + \gamma_2 v_2 + \cdots + \gamma_m v_m$$

where the γ_i are real or complex scalars. From the linearity of the transformation M,

$$y = Mu = \gamma_1 \sigma_1 u_1 + \gamma_2 \sigma_2 u_2 + \cdots + \gamma_m \sigma_m v_m.$$

Using the SVD, we can also gain insight on the inverse transformation $u = M^{-1}y$. From the SVD theorem, we know that $M = U\Sigma V^H$. Using the fact that U and V are unitary, $M^{-1} = V\Sigma^{-1}U^H$. Notice that

$$\Sigma^{-1} = \text{diag}\left\{\frac{1}{\sigma_1}, \frac{1}{\sigma_2}, \ldots, \frac{1}{\sigma_m}\right\}$$

Thus, if

$$y = \delta_1 u_1 + \delta_2 u_2 + \cdots + \delta_m u_m$$

then

$$u = M^{-1}y = \delta_1 \frac{1}{\sigma_1} v_1 + \delta_2 \frac{1}{\sigma_2} v_2 + \cdots + \delta_m \frac{1}{\sigma_m} v_m$$

This implies that the information in the SVD of M can be used to solve systems of linear equations without computing the inverse of M.

8.4 The SVD and MIMO Frequency Response Analysis

We now return to our discussion of MIMO frequency response with the full power of the SVD at our disposal. Once again, we shall focus our attention on the transfer function matrix (TFM) description of the strictly stable LTI system $G(s)$. As before, we will assume that G has m inputs and p outputs, making $G(s)$ a $p \times m$ matrix. In general we shall assume that $p \geq m$, so that, unless the rank k of $G(s)$ becomes less than m, the response of the system to a non-zero input is always non-zero.

Recall that if the input vector signal $u(t)$ is a complex exponential of the form $u(t) = \tilde{u}e^{j\omega t}$, with $\tilde{u}$ fixed in C^m, then at steady state, the output vector $y(t)$ will also be a complex exponential function,

$y(t) = \tilde{y}e^{j\omega t}$, for some $\tilde{y} \in C^p$. Recall, also, that the complex vectors $\tilde{u}$ and $\tilde{y}$ are related by $G(s)$ evaluated at $s = j\omega$, that is,

$$\tilde{y} = G(j\omega)\tilde{u}.$$

It is important to note that $G(j\omega)$ is a complex matrix that changes with frequency ω. For any given fixed frequency, we can calculate the SVD of $G(j\omega)$:

$$G(j\omega) = U(j\omega)\Sigma(\omega)V^H(j\omega)$$

Note that, in general, all of the factors in the SVD of $G(j\omega)$ are explicitly dependent on omega:

1. The matrix $\Sigma(\omega)$ is a $p \times m$ matrix whose main diagonal is composed of the singular values of $G(j\omega)$,
$$\sigma_{max}(\omega) = \sigma_1(\omega), \sigma_2(\omega), \ldots, \sigma_{k_\omega}(\omega) = \sigma_{min}(\omega)$$
where k_ω is the rank of $G(j\omega)$.
2. The matrix $U(j\omega)$ is an $m \times m$ complex-valued matrix whose column vectors $\{u_j(j\omega)\}$ are the left singular vectors of $G(j\omega)$.
3. The matrix $V(j\omega)$ is a $p \times p$ complex-valued matrix whose column vectors $\{v_j(j\omega)\}$ are the right singular vectors of $G(j\omega)$.

8.4.1 Singular Value Plots (SV Plots)

Once we calculate the maximum and minimum singular values of $G(j\omega)$ for a range of frequencies ω, we can plot them together on a Bode plot (decibels versus rad/sec in log-log scale). Figure 8.2 shows a hypothetical SV plot.

With the proper interpretation, the SV plot can provide valuable information about the properties of the MIMO system G. In particular, it quantifies the "gain-band" of the plant at each frequency, and shows how this changes with frequency. It is a natural generalization of the information contained in the classical Bode magnitude plot for SISO plants. One main difference here is that, in the multivariable case, this "gain-band" is described by two curves, not one.

It is crucial to interpret the information contained in the SV plot correctly. At each frequency ω we assume that the input is a *unit* complex exponential, $u(t) = \tilde{u}e^{j\omega t}$. Then, assuming that we have reached steady state, we know that the output is also a complex exponential with the same frequency, $y(t) = \tilde{y}e^{j\omega t}$,

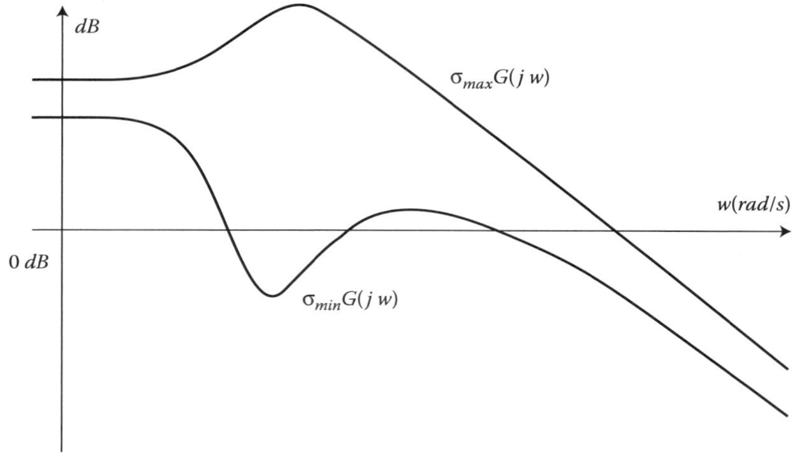

FIGURE 8.2 A hypothetical SV plot.

where $\tilde{y} = G(j\omega)\tilde{u}$. The magnitude $\|\tilde{y}\|_2$ of the output complex exponential thus depends on the direction of the input as well as on the frequency ω. Now, by looking at an SV plot, we can say that, at a given frequency:

1. The largest output size is $\|\tilde{y}\|_{2,max} = \sigma_{max}G(j\omega)$, for $\|\tilde{u}\|_2 = 1$.
2. The smallest output size is $\|\tilde{y}\|_{2,min} = \sigma_{min}G(j\omega)$, for $\|\tilde{u}\|_2 = 1$.

This allows us to discuss qualitatively the size of the plant gain as a function of frequency:

1. The plant has large gain at ω if $\sigma_{min}G(j\omega) \gg 1$.
2. The plant has small gain at ω if $\sigma_{max}G(j\omega) \ll 1$.

8.4.2 Computing Directional Information

In addition to computing system gain as a function of frequency, we can also use the SVD to compute "directional information" about the system. In particular, we can compute the direction of maximum and minimum amplification of the unit, real-valued sinusoidal input. In the following, we present a step-by-step methodology for maximum amplification direction analysis. Minimum amplification direction analysis is completely analogous, and will therefore not be presented explicitly.

8.4.2.1 Maximum Amplification Direction Analysis

1. Select a specific frequency ω.
2. Compute the SVD of $G(j\omega)$, i.e., find $\Sigma(\omega), U(j\omega)$, and $V(j\omega)$ such that $G(j\omega) = U(j\omega)\Sigma(\omega)V^H(j\omega)$ where U and V are unitary and Σ is the matrix of singular values.
3. In particular, find the maximum singular value $\sigma_{max}(\omega)$ of $G(j\omega)$.
4. Find the maximum right singular vector $v_{max}(\omega)$. This is the first column of the matrix $V(j\omega)$ found in the SVD. Note that $v_{max}(\omega)$ is a complex vector with m elements. Write the elements of $v_{max}(\omega)$ in polar form, i.e.,

   $$[v_{max}(\omega)]_i = |a_i|e^{j\psi_i}, \quad i = 1, 2, \ldots, m.$$

   Notice that a_i and ψ_i are really functions of ω; we suppress this frequency dependence for clarity.
5. Find the maximum left singular vector $u_{max}(\omega)$. This is the first column of the matrix $U(j\omega)$ found in the SVD. Note that $u_{max}(\omega)$ is a complex vector with p elements. Write the elements of $u_{max}(\omega)$ in polar form, i.e.,

   $$[u_{max}(\omega)]_i = |b_i|e^{j\phi_i}, \quad i = 1, 2, \ldots, p.$$

   Notice that b_i and ϕ_i are functions of ω; we suppress this frequency dependence for clarity.
6. We are now in a position to construct the real sinusoidal input signals that correspond to the direction of maximum amplification and to predict the output sinusoids that are expected at steady state. The input vector $u(t)$ is defined componentwise by

   $$u_i(t) = |a_i| \sin(\omega t + \psi_i), \quad i = 1, 2, \ldots, m$$

   where the parameters a_i and ψ_i are those determined above. Note that the amplitude and phase of each component sinusoid is distinct. We can utilize the implications of the SVD to predict the steady-state output sinusoids as

   $$y_i(t) = \sigma_{max}(\omega)|b_i| \sin(\omega t + \phi_i), \quad i = 1, 2, \ldots, p.$$

   Notice that all parameters needed to specify the output sinusoids are already available from the SVD.

When we talk about the "directions of maximum amplification," we mean input *sinusoids* of the form described above with very precise magnitude and phase relations to one another. The resulting output sinusoids also have very precise magnitude and phase relations, all as given in the SVD of $G(j\omega)$. Once again, a completely analogous approach can be taken to compute the minimum amplification direction associated with $G(j\omega)$.

It is important to remember that the columns of $U(j\omega)$ and $V(j\omega)$ are orthonormal. This means we can express *any* sinusoidal input vector as a linear combination of the right singular vectors of $G(j\omega)$ at a particular value of ω. The corresponding output sinusoidal vector will be a linear combination of the left singular vectors, after being scaled by the appropriate singular values.

Finally, because we measure system "size" in terms of the ratio of output Euclidean norm to input Euclidean norm, the "size" of the system is heavily dependent on the units of the input and output variables.

8.5 Frequency Response Analysis of MIMO Feedback Systems

In this section, we look at frequency domain-analysis for various control system configurations. We will pay particular attention to the classical unity-feedback configuration, where the variables to be controlled are used as feedback. Next we will look at a broader class of control system configurations relevant for some of the more modern controller design methodologies such as H_∞ and l_1 synthesis, as well as in robustness analysis and synthesis. MIMO frequency-domain analysis as discussed above will be pivotal throughout.

8.5.1 Classical Unity-Feedback Systems

Consider the unity-feedback system in the block diagram of Figure 8.3. Recall that the *loop* transfer function matrix is defined as

$$T(s) = G(s)K(s)$$

The *sensitivity* $S(s)$ and *complementary-sensitivity* $C(s)$ transfer function matrices are, respectively, defined as

$$S(s) = [I + T(s)]^{-1}$$
$$C(s) = [I + T(s)]^{-1}G(s)K(s) = S(s)T(s)$$

With these definitions,

$$e(s) = S(s)[r(s) - d(s)] + C(s)n(s) \tag{8.18}$$

The objective in control system design is to keep the error signal e "small". This means the *transfer* from the various disturbances to e must be small. Because it is always true that $S(s) + C(s) = I$, there is a

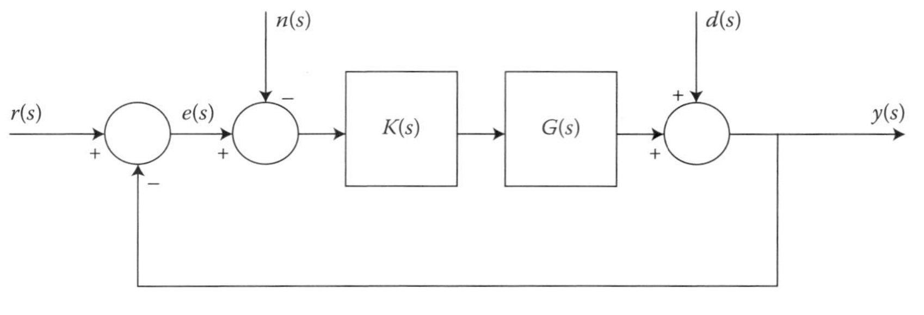

FIGURE 8.3 The unity feedback control system configuration.

trade-off involved. From Equation 8.18, we would like both $S(s)$ and $C(s)$ to be "small" for all s; but this is impossible because $S(s) + C(s) = I$. SVD analysis of the MIMO frequency response can be important in quantifying these issues.

8.5.1.1 Command Following

Suppose that the reference (command) signal $r(t)$ is sinusoidal, $r(t) = \tilde{r}e^{j\omega t}$. Then, as long as $d(s) = 0$ and $n(s) = 0$,

$$e(t) = \tilde{e}e^{j\omega t}$$

where $\tilde{e} = S(j\omega)\tilde{r}$. Thus,

$$\|\tilde{e}\|_2 \leq \sigma_{max}[S(j\omega)] \cdot \|\tilde{r}\|_2$$

Now suppose that $r(t)$ is the superposition of more than one sinusoid. Let Ω_r be the range of frequencies at which the input $r(t)$ has its energy. Then, in order to have good command following, we want

$$\sigma_{max}[S(j\omega)] \ll 1 \quad \forall \omega \in \Omega_r \tag{8.19}$$

Our objective now is to express this prescription for command following in terms of the loop-transfer function $T(s) = G(s)K(s)$. From our earlier discussion

$$\begin{aligned}\sigma_{max}[S(j\omega)] &= \sigma_{max}\{[I + T(j\omega)]^{-1}\}\\ &= \frac{1}{\sigma_{min}[I + T(j\omega)]}\end{aligned}$$

This implies that, for good command following, we must have $\sigma_{min}[I + T(j\omega)] \gg 1$ for all $\omega \in \Omega_r$. However, as we saw earlier,

$$\sigma_{min}[I + T(j\omega)] \geq \sigma_{min}[T(j\omega)] - 1$$

so it is sufficient that, for all $\omega \in \Omega_r, \sigma_{min}[T(j\omega)] \gg 1$ for good command following.

8.5.1.2 Disturbance Rejection

Suppose that the disturbance signal $d(t)$ is sinusoidal, $d(t) = \tilde{d}e^{j\omega t}$. Then, as long as $r(s) = 0$ and $n(s) = 0$,

$$e(t) = \tilde{e}e^{j\omega t}$$

where $\tilde{e} = -S(j\omega)\tilde{d}$. Thus,

$$\|\tilde{e}\|_2 \leq \sigma_{max}[S(j\omega)] \cdot \|\tilde{d}\|_2$$

Now suppose that $d(t)$ is the superposition of more than one sinusoid. Let Ω_d be the range of frequencies at which the input $d(t)$ has its energy. Then, just as with command following, for good disturbance rejection, we want

$$\sigma_{max}[S(j\omega)] \ll 1 \quad \forall \omega \in \Omega_d \tag{8.20}$$

Using the same argument given earlier, this prescription for disturbance rejection makes it sufficient that, for all $\omega \in \Omega_d, \sigma_{min}[T(j\omega)] \gg 1$.

8.5.1.3 Relationships to C(s)

Here we wish to determine in a precise quantitative way the consequences of obtaining command following and disturbance rejection. As we shall see, a price is paid in constraints on the complementary-sensitivity function $C(s)$.

Theorem 8.2:

Let $\Omega_p = \Omega_r \bigcup \Omega_d$. (Here "p" refers to "performance".) Consider δ so that $0 < \delta \ll 1$. If

$$\sigma_{max}[S(j\omega)] \leq \delta \ll 1$$

for all $\omega \in \Omega_p$, then

$$1 \ll \frac{1-\delta}{\delta} \leq \sigma_{min}[T(j\omega)]$$

and

$$1 - \delta \leq \sigma_{min}[C(j\omega)] \leq \sigma_{max}[C(j\omega)] \leq 1 + \delta$$

for all $\omega \in \Omega_p$.

Thus, in obtaining a performance level of δ, it is necessary that all of the singular values of $C(j\omega)$ are within δ of 1. In fact, because $S(s) + C(s) = I$, we must have $C(j\omega) \approx I$. (We shall discuss below why this can be a problem.)

Proof 8.1. We start by using the definition of $S(s)$,

$$\begin{aligned}\sigma_{max}[S(j\omega)] &= \sigma_{max}[(I + T(j\omega))^{-1}] \\ &= \frac{1}{\sigma_{min}[I + T(j\omega)]} \\ &\geq \frac{1}{1 + \sigma_{min}[T(j\omega)]}\end{aligned}$$

Using the hypothesis that $\sigma_{max}[S(j\omega)] \leq \delta$,

$$\frac{1}{1 + \sigma_{min}[T(j\omega)]} \leq \delta$$

which by solving for $\sigma_{min}[T(j\omega)]$ yields the first inequality. By cross-multiplying, we obtain the following useful expression:

$$\frac{1}{\sigma_{min}[T(j\omega)]} \leq \frac{\delta}{1-\delta} \ll 1 \tag{8.21}$$

Now consider the complementary-sensitivity function $C(s)$. $C(s) = [I + T(s)]^{-1}T(s)$. By taking the inverse of both sides, $C^{-1}(s) = T^{-1}(s)[I + T(s)] = I + T^{-1}(s)$. Thus

$$\frac{1}{\sigma_{min}[C(j\omega)]} = \sigma_{max}[C^{-1}(j\omega)] = \sigma_{max}[I + T^{-1}(j\omega)]$$

which implies

$$\begin{aligned}\frac{1}{\sigma_{min}[C(j\omega)]} &\leq 1 + \sigma_{max}[T^{-1}(j\omega)] \\ &= 1 + \frac{1}{\sigma_{min}[T(j\omega)]} \\ &\leq 1 + \frac{\delta}{1-\delta} = \frac{1}{1-\delta}\end{aligned}$$

(Notice that the second inequality follows from Equation 8.21.) Now,

$$1-\delta \le \sigma_{min}[C(j\omega)] \le \sigma_{max}[C(j\omega)]$$
$$= \sigma_{max}[I - S(j\omega)] \le 1 + \sigma_{max}[S(j\omega)] \le 1+\delta$$

which is the second desired inequality.

8.5.1.4 Measurement Noise Insensitivity: A Conflict!

Suppose that the measurement noise $n(t)$ is sinusoidal, $n(t) = \tilde{n}e^{j\omega t}$. Then, as long as $r(s) = 0$ and $d(s) = 0$,

$$e(t) = \tilde{e}e^{j\omega t}$$

where $\tilde{e} = C(j\omega)\tilde{d}$. Thus,

$$\|\tilde{e}\|_2 \le \sigma_{max}[C(j\omega)] \cdot \|\tilde{n}\|_2$$

Now suppose that $n(t)$ is the superposition of more than one sinusoid. Let Ω_n be the range of frequencies at which the input $n(t)$ has its energy. Then, in order to be insensitive to measurement noise, we want

$$\sigma_{max}[C(j\omega)] \ll 1 \quad \forall \omega \in \Omega_n \tag{8.22}$$

Theorem 8.3:

Let γ be such that $0 < \gamma \ll 1$. If

$$\sigma_{max}[C(j\omega)] \le \gamma$$

for all $\omega \in \Omega_n$, then

$$\sigma_{min}[T(j\omega)] \le \sigma_{max}[T(j\omega)] \le \frac{\gamma}{1-\gamma} \approx \gamma \ll 1$$

and

$$1 \approx 1-\gamma \le \sigma_{min}[S(j\omega)] \le \sigma_{max}[S(j\omega)]$$

for all $\omega \in \Omega_n$.

Thus, if the complementary-sensitivity function $C(j\omega)$ has low gain on Ω_n, then so does the loop-transfer function $T(j\omega)$. This in turn implies that the sensitivity transfer function $S(j\omega)$ has nearly unity gain on Ω_n. In other words, wherever (in frequency) we are insensitive to measurement noise we are necessarily prone to poor command following *and* disturbance rejection. This is primarily a consequence of the fact that $C(s) + S(s) = I$.

Proof 8.2. To prove the first relationship we use the fact that $C(s) = [I + T(s)]^{-1}T(s) = [T^{-1}(s) + I]^{-1}$ (proved using a few algebraic manipulations). This gives

$$\sigma_{max}[C(j\omega)] = \frac{1}{\sigma_{min}[T^{-1}(j\omega) + I]}$$
$$\ge \frac{1}{\sigma_{min}[T^{-1}(j\omega)] + 1}$$
$$= \frac{\sigma_{max}[T(j\omega)]}{1 + \sigma_{max}[T(j\omega)]}$$

Thus,

$$\sigma_{max}[T(j\omega)] \le \sigma_{max}[C(j\omega)] + \sigma_{max}[C(j\omega)]\sigma_{max}[T(j\omega)]$$
$$\le \gamma + \gamma\sigma_{max}[T(j\omega)]$$

which yields the desired inequality.

To prove the second relationship, observe that

$$\begin{aligned}\sigma_{max}[S(j\omega)] &\geq \sigma_{min}[S(j\omega)] \\ &= \sigma_{min}[(I+T(j\omega))^{-1}] \\ &= \frac{1}{\sigma_{max}[I+T(j\omega)]} \\ &\geq \frac{1}{\sigma_{max}[T(j\omega)]+1} \\ &\geq 1-\gamma\end{aligned}$$

where the last inequality comes from the first relationship proved above.

8.5.1.5 Design Implications

Achievable control design specifications must have a wide separation (in frequency) between the sets $\Omega_p = \Omega_r \bigcup \Omega_d$ and Ω_n. We cannot obtain good command following and disturbance rejection when we have sensors that are noisy on Ω_p. Figure 8.4 illustrates a problem that is well-posed in terms of these constraints.

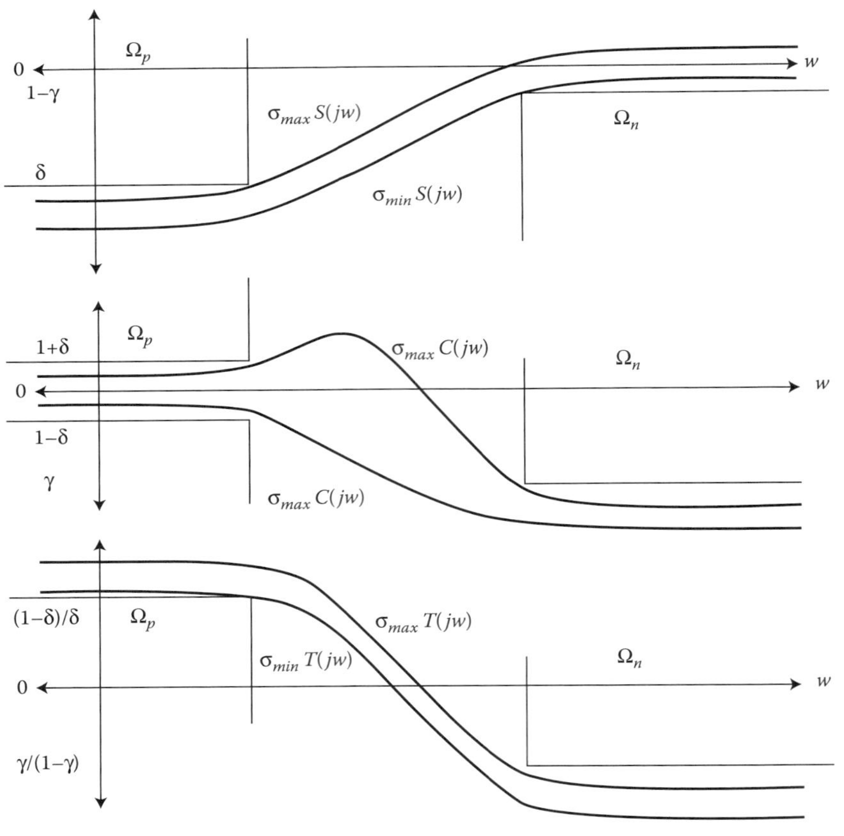

FIGURE 8.4 A well-posed problem: the singular value traces fall within the regions defined by Ω_p and δ (for performance) and Ω_n and γ (for noise insensitivity).

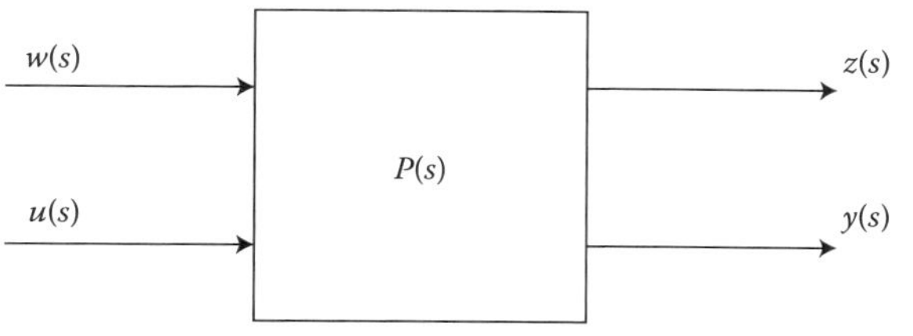

FIGURE 8.5 A generalized plant.

8.5.2 A More General Setting

Recent control design methodologies are convenient for more generalized problem descriptions. Consider the LTI system P shown in Figure 8.5, where u denotes the control vector input to the plant, y represents the measurement vector, w is a generalized disturbance vector, and z is a generalized performance vector. The assignment of physical variables to w and z here is arbitrary and is left to the control system analyst. One illustration is given in Example 8.6. We see that the transfer function matrix for this system can be partitioned as follows:

$$P(s) = \begin{bmatrix} P_{11}(s) & P_{12}(s) \\ P_{21}(s) & P_{22}(s) \end{bmatrix}$$

From this partition,

$$\begin{aligned} z(s) &= P_{11}(s)w(s) + P_{12}(s)u(s) \\ y(s) &= P_{21}(s)w(s) + P_{22}(s)u(s) \end{aligned}$$

Let K be a feedback controller for the system so that $u(s) = K(s)y(s)$. This feedback interconnection is shown in Figure 8.6. It is simple to verify that

$$\begin{aligned} y(s) &= [I - P_{22}(s)K(s)]^{-1}P_{21}(s)w(s), \\ z(s) &= \{P_{11}(s) + P_{12}(s)K(s)[I - P_{22}(s)K(s)]^{-1}P_{21}(s)\}w(s), \\ &\equiv F(P,K)(s)w(s). \end{aligned}$$

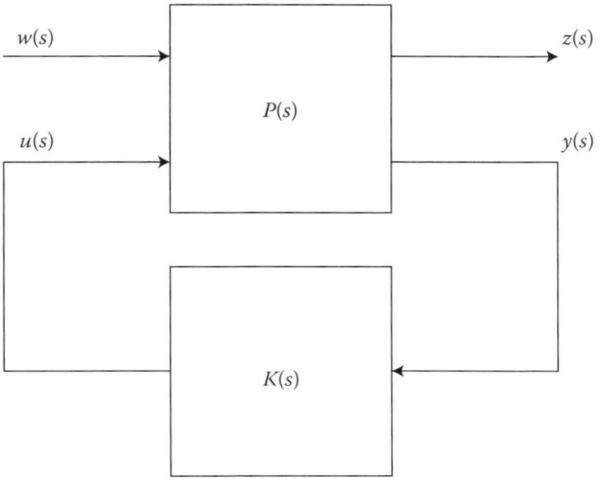

FIGURE 8.6 A feedback interconnection.

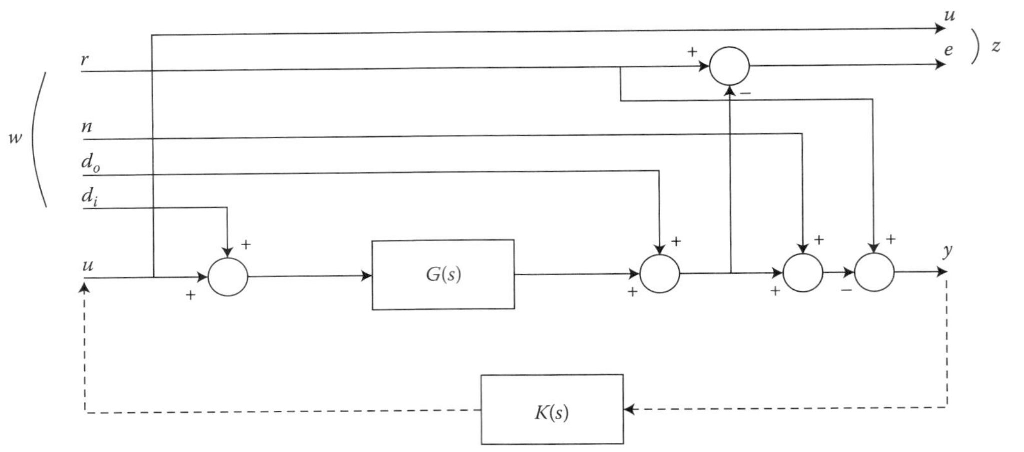

FIGURE 8.7 Unity feedback example.

The closed-loop transfer function from $w(s)$ to $z(s)$, denoted $F(P, K)(s)$, is called the (lower) linear fractional transformation (of the plant P and K). This type of mathematical object is often very useful in describing feedback interconnections.

Example 8.6: Relationships

In this example we show how the classical unity-feedback setup can be mapped into the general formulation of this section. Consider the block diagram shown in Figure 8.7. With four generalized disturbance inputs, a command input r, a sensor noise input n, a system-output disturbance d_o, and a system-input disturbance d_i. (These variables are lumped into the generalized disturbance vector w.) There are two generalized performance variables, control (effort) u and tracking error e. (These are lumped into the generalized performance vector z.) The variables u and y are the control inputs and sensor outputs of the generalized plant P.

8.5.2.1 Frequency Weights

In control design or analysis it is often necessary to incorporate extra information about the plant and its environment into the generalized description P. For example, we may know that system-input disturbances are always low-frequency in nature, and/or we may only care about noise rejection at certain high frequencies. This kind of information can be incorporated as "frequency weights" augmenting the generalized inputs and outputs of the plant P. Such weighting functions $W_d(s)$ and $W_p(s)$ are included in the block diagram of Figure 8.8. Examples are given below.

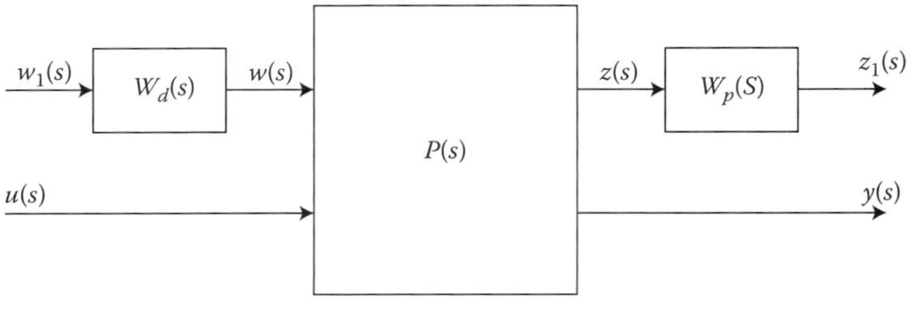

FIGURE 8.8 Weighting functions and the generalized plant.

With frequency-dependent weightings we can define a new generalized plant P_w. Referring to Figure 8.8 and using the block partition of P discussed earlier,

$$P_w(s) = \begin{bmatrix} W_p(s)P_{11}(s)W_d(s) & W_p(s)P_{12}(s) \\ P_{21}(s)W_d(s) & P_{22}(s) \end{bmatrix}.$$

With the weighting functions augmenting the plant description, P_w is ready (at least conceptually) for control design or analysis. If state-space techniques are being employed, the *dynamics* (i.e., the state variables) for W_d and W_p must be included in the state-space representation of P_w. This can be accomplished by state augmentation.

Typically, weights W_d on the generalized disturbance vector emphasize the frequency regions of most significant disturbance strength. Here $w(s) = W_d(s)w_1(s)$. We choose W_d so that a complete description of the generalized disturbance is obtained with $\|w_1(j\omega)\| = 1$ for all ω. Thus, we may think of W_d as an active filter which emphasizes (or de-emphasizes) certain variables in specific frequency domains consistent with our understanding of the system's environment. Reference inputs and system disturbances are most often low frequency, so these parts of W_d are usually low-pass filters. On the other hand, certain noise inputs are notched (like 60 Hz electrical hum) or high frequency. These parts of $W_d(s)$ should be band-pass or high-pass filters, respectively.

Weights W_p on the generalized performance vector emphasize the *importance* of good performance in different frequency regions. For example, we may be very interested in good tracking at low frequencies (but not at high frequencies), so that the weighting on this part of $W_p(s)$ should be a low-pass filter. On the other hand, we may not want to use control energy at high frequencies, so we would choose this part of $W_p(s)$ as high-pass. The various components of $W_p(s)$ must be consistent with one another in gain. The relative weights on variables in z should make sense as a whole.

It is important to note that we must still operate within constraints to mutually achieve various types of performance, as was the case with the classical unity-feedback formulation. Because we are no longer constrained to that rigid type of feedback interconnection, general results are difficult to state. The basic idea, however, remains that there is an essential conflict between noise-rejection and command-following/disturbance-rejection. In general there must be a separation in frequency between the regions in which the respective types of performance are important.

8.5.2.2 Weighted Sensitivity

We consider here the case of only one generalized input variable, d, a system-output disturbance. We are also interested in only one performance variable y, the disturbed output of the plant. Specifically, we have in mind a weighting function W_p which reflects our specifications for y. The feedback configuration is shown in Figure 8.9.

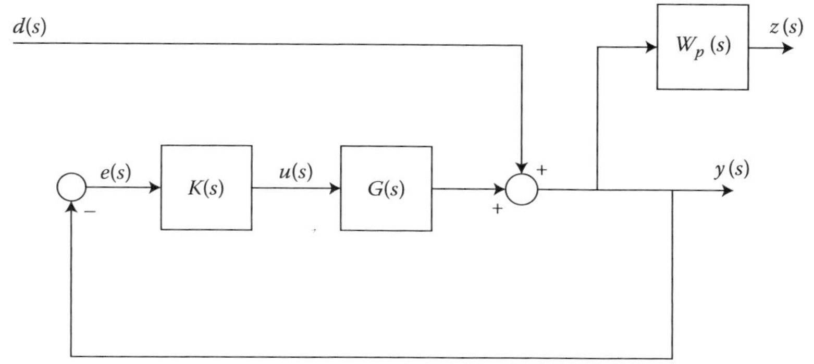

FIGURE 8.9 Feedback interconnection for weighted sensitivity.

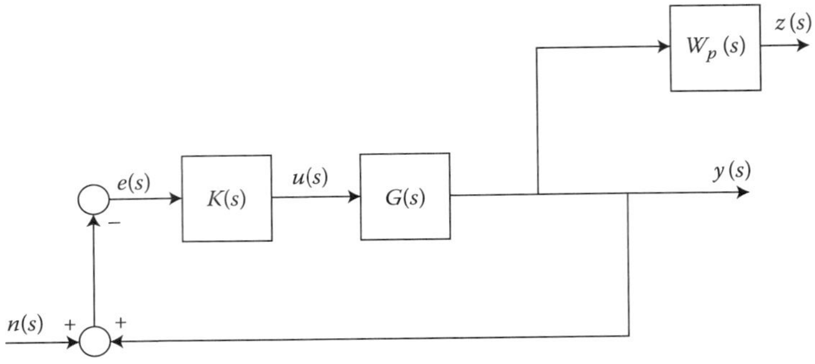

FIGURE 8.10 Feedback interconnection for weighted complementary sensitivity.

It is not hard to see that $y(s) = S(s)d(s)$. Thus, $z(s) = W_p(s)S(s)d(s)$. We refer to $W_p(s)S(s)$ as the "weighted" sensitivity transfer function for the feedback system. If $\sigma_{max}[W_p(j\omega)\ S(j\omega)] < 1$ for all ω, then,

$$\begin{aligned}\sigma_{max}[S(j\omega)] &= \sigma_{max}[W_p^{-1}(j\omega)W_p(j\omega)S(j\omega)]\\ &\leq \sigma_{max}[W_p^{-1}(j\omega)]\sigma_{max}[W_p(j\omega)S(j\omega)]\\ &< \sigma_{max}[W_p^{-1}(j\omega)]\end{aligned}$$

To interpret this, when $\sigma_{max}[W_p(j\omega)S(j\omega)] < 1$ for all ω, then the largest singular value of the sensitivity transfer function is strictly less than the largest singular value of the inverse of the weighting function.

8.5.2.3 Weighted Complementary Sensitivity

We consider here the case of only one generalized input variable, n, a sensor noise input. We are also only interested in one performance variable y, the output of the plant. Once again, we have in mind a weighting function W_p which reflects our specifications for y. The feedback configuration is shown in Figure 8.10.

It is not hard to see that $y(s) = C(s)d(s)$. Thus, $z(s) = W_p(s)C(s)d(s)$. We refer to $W_p(s)C(s)$ as the "weighted" complementary-sensitivity transfer function for the feedback system. Notice that if $\sigma_{max}[W_p(j\omega)C(j\omega)] < 1$ for all ω then,

$$\begin{aligned}\sigma_{max}[C(j\omega)] &= \sigma_{max}[W_p^{-1}(j\omega)W_p(j\omega)C(j\omega)]\\ &\leq \sigma_{max}[W_p^{-1}(j\omega)]\sigma_{max}[W_p(j\omega)C(j\omega)]\\ &< \sigma_{max}[W_p^{-1}(j\omega)]\end{aligned}$$

To interpret this, when $\sigma_{max}[W_p(j\omega)C(j\omega)] < 1$ for all ω, then the largest singular value of the complementary-sensitivity transfer function is strictly less than the largest singular value of the inverse of the weighting function.

References

1. Athans, M., Lecture Notes for Multivariable Control Systems I and II, Massachusetts Institute of Technology, 1994. (This reference may not be generally available.)
2. Freudenberg, J.S. and Looze, D.P., *Frequency Domain Properties of Scalar and Multivariable Feedback Systems,* Springer, Berlin, 1987.
3. Kailath, T., *Linear Systems,* Prentice Hall, Englewood Cliffs, NJ, 1980.
4. Maciejowski, J.M., *Multivariable Feedback Design,* Addison-Wesley, Wokingham, 1989.
5. Strang, G., *Linear Algebra and Its Applications,* Harcourt Brace Jovanovich, San Diego, 1988.

9

Stability Robustness to Unstructured Uncertainty for Linear Time Invariant Systems

Alan Chao
Massachusetts Institute of Technology

Michael Athans
Massachusetts Institute of Technology

9.1 Introduction

In designing feedback control systems, the stability of the resulting closed-loop system is a primary objective. Given a finite dimensional, linear time-invariant (FDLTI) model of the plant, $G(s)$, the stability of the nominal closed-loop system based on this model, Figure 9.1, can be guaranteed through proper design: in the Nyquist plane for single-input, single-output (SISO) systems or by using well-known design methodologies such as LQG and $\mathcal{H}_\infty$ for multi-input, multi-output (MIMO) systems. In any case, since the mathematical model is FDLTI, this nominal stability can be analyzed by explicitly calculating the closed-loop poles of the system. It is clear, however, that nominal stability is never enough since the model is never a true representation of the actual plant. That is, there are always modeling errors or uncertainty. As a result, the control engineer must ultimately ensure the stability of the actual closed-loop system, Figure 9.2. In other words, the designed controller, $K(s)$, must be robust to the model uncertainty. In this article, we address this topic of stability robustness. We present a methodology to analyze the stability of the actual closed-loop system under nominal stability to a given model and a certain representation of the uncertainty.

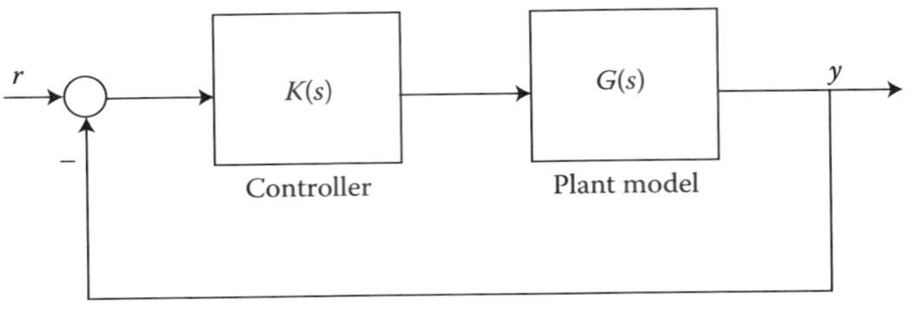

FIGURE 9.1 Block diagram of nominal feedback loop.

The outline of the chapter is as follows. We first establish a representation of the uncertainty on which we will base our analysis. We then proceed to derive conditions that guarantee the stability of the actual closed-loop system. First, we concentrate on SISO systems where we use the familiar Nyquist stability criterion to derive the stability robustness condition. We then derive this same condition using the small gain theorem. The purpose of this is to provide a simple extension of our analysis to MIMO systems, which we present next. We then interpret the stability robustness conditions and examine their impact on attainable closed-loop performance, such as disturbance rejection and command following. Finally, we present a discussion on other possible representations of uncertainty and their respective stability robustness conditions. Examples are presented throughout the discussion.

9.2 Representation of Model Uncertainty

9.2.1 Sources of Uncertainty

Before we can analyze the stability of a closed-loop system under uncertainty, we must first understand the causes of uncertainty in the model so as to find a proper mathematical representation for it. Throughout the discussion we will assume that the actual plant, $G_a(s)$, is linear time-invariant (LTI) and that we have a nominal LTI model, $G(s)$. Although this assumption may seem unreasonable since actual physical systems are invariably nonlinear and since a cause of uncertainty is that we model them as LTI systems, we need this assumption to obtain simple, practical results. In practice, these results work remarkably well for a large class of engineering problems, because many systems are designed to be as close to linear time-invariant as possible.

The sources of modeling errors are both intentional and unintentional. Unintentional model errors arise from the underlying complexity of the physical process, the possible lack of laws for dynamic cause and effect relations, and the limited opportunity for physical experimentation. Simply put, many physical processes are so complex that approximations are inevitable in deriving a mathematical model. On the other hand, many modeling errors are intentionally induced. In the interest of reducing the complexity and cost of the control design process, the engineer will often neglect "fast" dynamics in an effort to reduce the order or the dimension of the state-space representation of the model. For example, one may

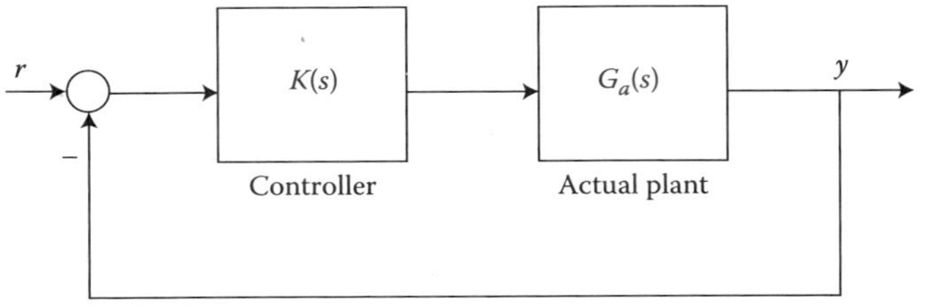

FIGURE 9.2 Block diagram of actual feedback loop.

neglect "fast" actuator and sensor dynamics, "fast" bending and/or torsional modes, and "small" time delays. In addition, the engineer will often use nominal values for the parameters of his model, such as time constants and damping ratios, even though he knows that the actual values will be different because of environmental and other external effects.

9.2.2 Types of Uncertainty

The resulting modeling errors can be separated into two types. The first type is known as parametric uncertainty. Parametric uncertainty refers to modeling errors, under the assumption that the actual plant is of the same order as the model, where the numerical values of the coefficients to the differential equation, which are related to the physical parameters of the system, between the actual plant and the model are different. The second type of uncertainty is known as unstructured uncertainty. In this case, the modeling errors refer to the difference in the dynamics between the finite dimensional model and the unknown and possibly infinite dimensional actual process.

In this chapter, we limit ourselves to addressing stability robustness with respect to unstructured uncertainty. We do this for two main reasons. First, we can capture the parametric errors in terms of the more general definition of unstructured uncertainty. Second and more importantly, unstructured uncertainty allows us to capture the effects of unmodeled dynamics. From our discussion above, we admit that we often purposely neglect "fast" dynamics in an effort to simplify the model. Furthermore, it can be argued that all physical processes are inherently distributed systems and that the modeling process acts to lump the dynamics of the physical process into a system that can be defined in a finite dimensional state-space. Therefore, unmodeled dynamics are always present and need to be accounted for in terms of stability robustness.

9.2.3 Multiplicative Representation of Unstructured Uncertainty

What we need now is a mathematical representation of unstructured uncertainty. The difficulty is that since the actual plant is never exactly known, we cannot hope to model the uncertainty to obtain this representation, for otherwise, it would not be uncertain. On the other hand, in practice, the engineer is never totally ignorant of the nature and magnitude of the modeling error. For example, from our arguments above, it is clear that unstructured uncertainty cannot be captured by a state-space representation, since the order of the actual plant is unknown, and thus we are forced to find a representation in terms of input–output relationships. In addition, if we choose to neglect "fast" dynamics, then we would expect that the magnitude of the uncertainty will be large at high frequencies in the frequency domain. In any case, the key here is to define a representation that employs the minimal information regarding the modeling errors and that is, in turn, sufficient to address stability robustness.

9.2.3.1 Set Membership Representation for Uncertainty

In this article we adopt a set membership representation of unstructured uncertainty. The idea is to define a bounded set of transfer function matrices, $\mathcal{G}$, which contains $G_a(s)$. Therefore, if $\mathcal{G}$ is properly defined such that we can show stability for all elements of $\mathcal{G}$, then we would have shown stability robustness. Towards this end, we define $\mathcal{G}$ as

$$\mathcal{G} = \{\tilde{G}(s) \mid \tilde{G}(s) = (I + w(s)\Delta(s))G(s), \|\Delta(j\omega)\|_{\mathcal{H}_\infty} \leq 1\} \tag{9.1}$$

where

1. $w(s)$ is a fixed, proper, and strictly stable scalar transfer function
2. $\Delta(s)$ is a strictly stable transfer function matrix (TFM)
3. No unstable or imaginary axis poles of $G(s)$ are cancelled in forming $\tilde{G}(s)$

This is known as the multiplicative representation for unstructured uncertainty. Since it is clear that the nominal model $G(s)$ is contained in $\mathcal{G}$, we view $\mathcal{G}$ as a set of TFMs perturbed from $G(s)$ that covers the actual plant. Our requirements that $w(s)\Delta(s)$ is strictly stable and that there are no unstable or imaginary axis pole-zero cancellations mean that the unstable and imaginary axis poles of our model and any $\tilde{G}(s) \in \mathcal{G}$ coincide. This assumes that the modeling effort is at least adequate enough to capture the unstable dynamics of $G_a(s)$.

In Equation 9.1, the term $w(s)\Delta(s)$ is known as the multiplicative error. We note that since the $\mathcal{H}_\infty$ norm of $\Delta(j\omega)$ varies between 0 and 1 and since the phase and direction of $\Delta(j\omega)$ are allowed to vary arbitrarily, the multiplicative error for any $\tilde{G}(s) \in \mathcal{G}$ is contained in a bounded hypersphere of radius $|w(j\omega)|$ at each frequency. Therefore, our representation of unstructured uncertainty is one in which we admit total lack of knowledge of the phase and direction of the actual plant with respect to the model, but that we have a bound on the magnitude of this multiplicative error. This magnitude-bound information is frequency dependent and is reflected in the fixed transfer function $w(s)$, which we will refer to as the weight. We note that we could have used a TFM $W(s)$ instead of a scalar $w(s)$ in our representation. However, in that case, the multiplicative error will reflect directional information of $W(s)$. Since we have presumed total lack of knowledge regarding directional information of the actual plant relative to the model, it is common in practice to choose $W(s)$ to be scalar.

9.2.3.2 SISO Interpretations for $\mathcal{G}$

To get a better feel for $\mathcal{G}$ and our representation for unstructured uncertainty, we first specialize to the SISO case in which our definition for $\mathcal{G}$ gives

$$\tilde{g}(s) = (1 + w(s)\Delta(s))g(s) \tag{9.2}$$

where

$$\|\Delta(j\omega)\|_{\mathcal{H}_\infty} = \sup_{\omega} |\Delta(j\omega)| \leq 1 \tag{9.3}$$

Since the phase of $\Delta(j\omega)$ is allowed to vary arbitrarily and its magnitude varies from 0 to 1 at all frequencies, the set $\mathcal{G}$ is the set of transfer functions whose magnitude bode plot lies in an envelope surrounding the magnitude plot of $g(s)$, as shown in Figure 9.3. Therefore, the size of the unstructured uncertainty

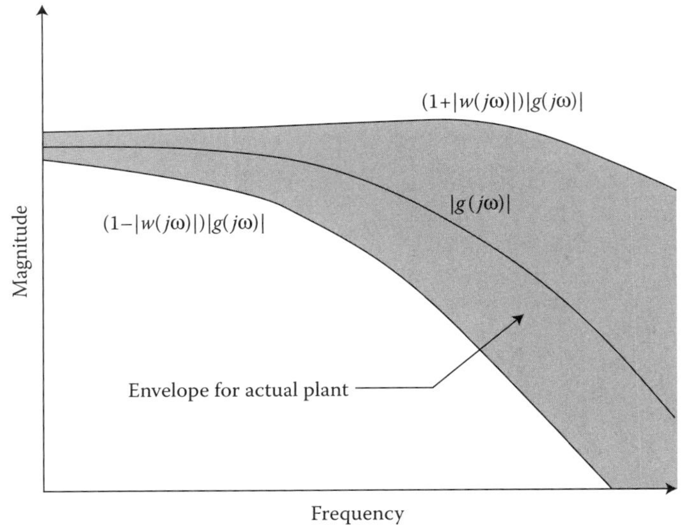

FIGURE 9.3 Bode plot interpretation of multiplicative uncertainty.

is represented by the size of this envelope. From the figure, the upper edge of the envelope corresponds to the plot of $(1+|w(j\omega)|)|g(j\omega)|$ while the lower edge corresponds to the plot of $(1-|w(j\omega)|)|g(j\omega)|$. Therefore, $|w(j\omega)|$ is seen as a frequency-dependent magnitude bound on the uncertainty. As mentioned beforehand, the size of the unstructured uncertainty typically increases with increasing frequency. Therefore, we would typically expect the size of the envelope containing $\mathcal{G}$ to increase with increasing frequency and also $|w(j\omega)|$ to increase with increasing frequency, as shown in Figure 9.3. Furthermore, we want to stress that since the phase of $\Delta(j\omega)$ is allowed to vary arbitrarily, the phase difference between any $\tilde{g}(j\omega) \in \mathcal{G}$ and $g(j\omega)$ can be arbitrarily large at any frequency.

For another interpretation, we can look at the multiplicative error $w(s)\Delta(s)$ in the SISO case. Solving for $w(s)\Delta(s)$ in Equation 9.2 gives

$$\frac{\tilde{g}(s)-g(s)}{g(s)} = w(s)\Delta(s) \tag{9.4}$$

which shows that $w(s)\Delta(s)$ is the normalized error in the transfer function of the perturbed system with respect to the nominal model. Using Equation 9.3 and noting that everything is scalar, we take magnitudes on both sides of (Equation 9.4) for $s=j\omega$ to get

$$|\tilde{g}(j\omega)-g(j\omega)| \leq |w(j\omega)||\Delta(j\omega)||g(j\omega)| \leq |w(j\omega)||g(j\omega)| \quad \forall\, \omega \tag{9.5}$$

As shown in Figure 9.4, for each ω, this inequality describes a closed disk in the complex plane of radius $|w(j\omega)||g(j\omega)|$ centered at $g(j\omega)$ which contains $\tilde{g}(j\omega)$. Since Equation 9.5 is valid for any $\tilde{g}(s) \in \mathcal{G}$, the set $\mathcal{G}$ is contained in that closed disk for each ω. In this interpretation, the unstructured uncertainty is represented by the closed disk, and therefore, we see that the direction and phase of the uncertainty is left arbitrary. However, we note that the radius of the closed disk does not necessarily increase with increasing frequency because it depends also on $|g(j\omega)|$, which typically decreases with increasing frequency at high frequencies due to roll-off.

9.2.3.3 Choosing $w(s)$

From our discussion, it is clear that our representation of the uncertainty only requires a nominal model, $G(s)$, and a scalar weight, $w(s)$, which reflects our knowledge on the magnitude bound of the uncertainty. The next logical question is how to choose $w(s)$ in the modeling process. From the definition

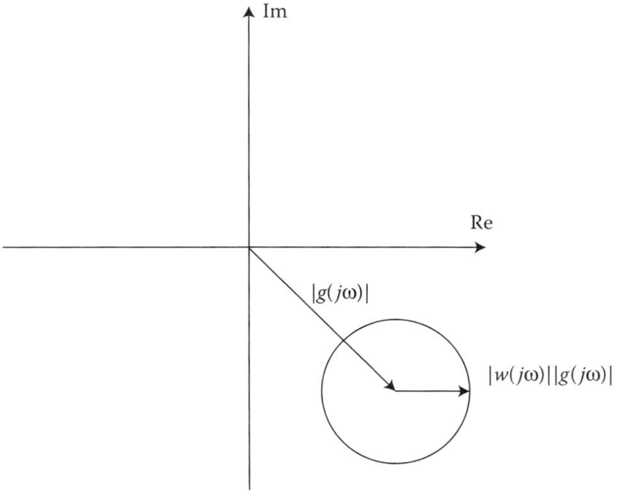

FIGURE 9.4 Interpretation of multiplicative uncertainty in the complex plane.

in Equation 9.1, we know that we must choose $w(s)$ so that the actual plant is contained in $\mathcal{G}$. In the course of modeling, whether through experimentation, such as frequency response, and/or model reduction, we will arrive at a set of transfer function matrices. It is assumed that our modeling effort is thorough enough to adequately capture the actual process so that this set will cover the TFM of the actual plant. From this set we will choose a nominal model, $G(s)$. We assume that $G(s)$ is square, same number of inputs as outputs, and nonsingular along the $j\omega$-axis in the s-plane. With this assumption, we can calculate, at each frequency, the multiplicative error for each TFM $G_i(s)$ in our set using

$$w(j\omega)\Delta(j\omega) = G_i(j\omega)G^{-1}(j\omega) - I \tag{9.6}$$

Taking maximum singular values on both sides of the above equation gives

$$\sigma_{\max}[w(j\omega)\Delta(j\omega)] = \sigma_{\max}[G_i(j\omega)G^{-1}(j\omega) - I] \tag{9.7}$$

Since $\|\Delta(j\omega)\|_{\mathcal{H}_\infty} \leq 1$ and $w(j\omega)$ is a scalar, it is clear that we must choose

$$|w(j\omega)| \geq \sigma_{\max}[G_i(j\omega)G^{-1}(j\omega) - I] \quad \forall\, \omega \geq 0 \tag{9.8}$$

to include the TFM $G_i(s)$ in $\mathcal{G}$. Therefore, we must choose a stable, proper $w(s)$ such that

$$|w(j\omega)| \geq \max_i \sigma_{\max}[G_i(j\omega)G^{-1}(j\omega) - I] \quad \forall\, \omega \geq 0 \tag{9.9}$$

to ensure that we include all $G_i(s)$ and, thus, the actual plant in $\mathcal{G}$. This process is illustrated in the following example.

Example 9.1: Integrator with Time Delay

Consider the set of SISO plants

$$g_\tau(s) = \frac{1}{s}\exp^{-s\tau}, \quad 0 \leq \tau \leq 0.2 \tag{9.10}$$

which is the result of our modeling process on the actual plant. Since we cannot fully incorporate the delay in a state-space model, we choose to ignore it. As a result, our model of the plant is

$$g(s) = \frac{1}{s} \tag{9.11}$$

To choose the appropriate $w(s)$ to cover all $g_\tau(s)$ in $\mathcal{G}$, we have to satisfy Equation 9.9, which in the SISO case is

$$|w(j\omega)| \geq \max_\tau \left|\frac{g_\tau(j\omega)}{g(j\omega)} - 1\right| \quad \forall\, \omega \geq 0 \tag{9.12}$$

Therefore, we need to choose $w(s)$ such that

$$|w(j\omega)| \geq \max_\tau \left|e^{-j\omega\tau} - 1\right| \quad \forall\, \omega \geq 0 \tag{9.13}$$

Using $e^{-j\omega\tau} = \cos(\omega\tau) - j\sin(\omega\tau)$ and a few trigonometry identities, the above inequality can be simplified as

$$|w(j\omega)| \geq \max_\tau \left|2\sin\left(\frac{\omega\tau}{2}\right)\right| \quad \forall\, \omega \geq 0 \tag{9.14}$$

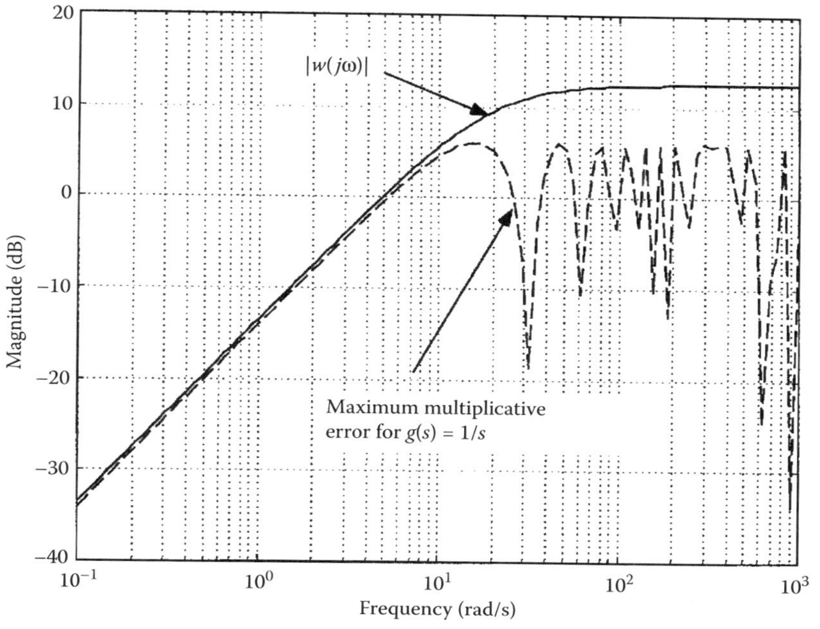

FIGURE 9.5 $w(s)$ for time delay uncertainty.

A simple $w(s)$ that will satisfy Equation 9.14 is

$$w(s) = \frac{0.21s}{0.05s + 1} \tag{9.15}$$

This is shown in Figure 9.5, where $|w(j\omega)|$ and $2\left|\sin(\frac{\omega\tau}{2})\right|$ are plotted together on a magnitude bode plot for $\tau = 0.2$, which is the worst-case value. We note that $w(s)$ is proper and strictly stable as required.

Now, let us suppose that we seek a better model by approximating the delay with a first-order Pade approximation

$$e^{-s\tau} \approx \frac{1 - \frac{s\tau}{2}}{1 + \frac{s\tau}{2}} \tag{9.16}$$

Our model becomes

$$g(s) = \frac{1}{s}\left(\frac{1 - \frac{0.1s}{2}}{1 + \frac{0.1s}{2}}\right) \tag{9.17}$$

where we approximate the delay at its midpoint value of 0.1 s. To choose $w(s)$ in this case, we again need to satisfy Equation 9.12 which becomes

$$|w(j\omega)| \geq \max_{\tau}\left|-\frac{e^{-j\omega\tau}(j\omega + 20)}{(j\omega - 20)} - 1\right| \quad \forall\, \omega \geq 0 \tag{9.18}$$

A simple $w(s)$ that will satisfy Equation 9.18 is

$$w(s) = \frac{0.11s}{0.025s + 1} \tag{9.19}$$

This is shown in Figure 9.6. We again note that $w(s)$ is proper and strictly stable as required.

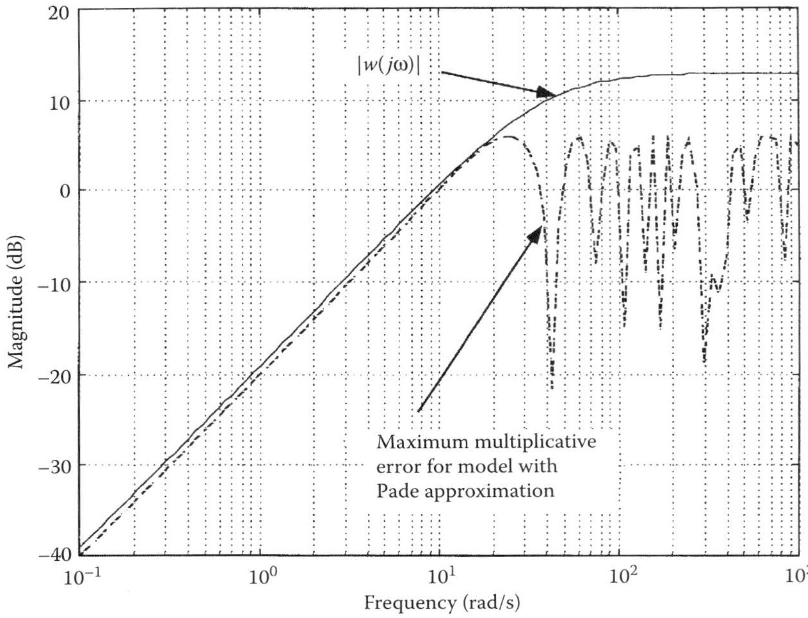

FIGURE 9.6 $w(s)$ for time delay uncertainty with Pade approximation.

Comparing the magnitudes of $w(s)$ for the two models in Figure 9.7, we note that, for the model with the Pade approximation, $|w(j\omega)|$ is less than that for the original model for $\omega \leq 100$ rad/s. Since $|w(j\omega)|$ is the magnitude bound on the uncertainty, Figure 9.7 shows that the uncertainty for the model with the Pade approximation is smaller than that for the original model in this frequency range. Physically, this is

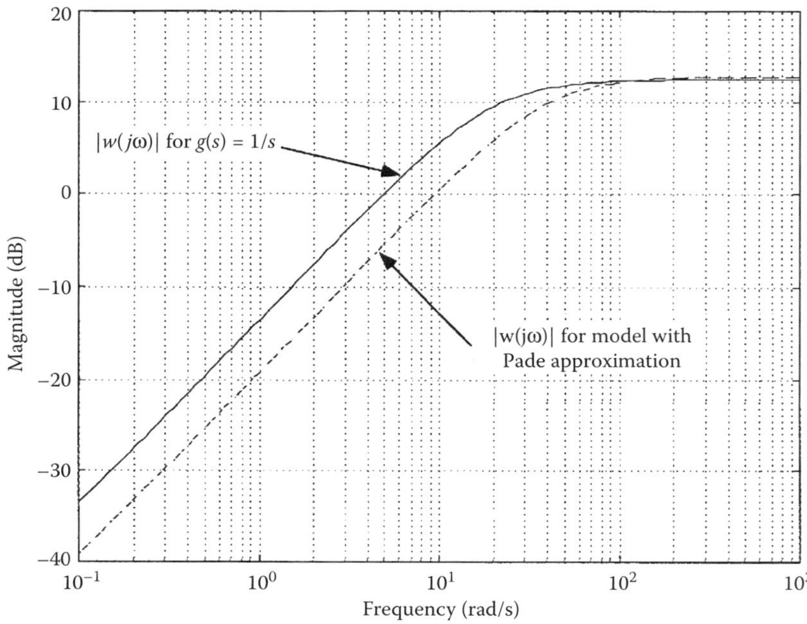

FIGURE 9.7 Comparing $w(s)$ for the two models.

because the model with the Pade approximation is a better model of the actual plant for $\omega \leq 100$ rad/s. As a result, its uncertainty is smaller.

We note that our choice of representing the uncertainty as being bounded in magnitude but arbitrary in phase and direction is, in general, an overbound on the set of TFMs that we obtained from the modeling process. That is, the set of TFMs from the modeling process may only be a small subset of $\mathcal{G}$. The benefit of using this uncertainty representation is that it allows for simple analysis of the stability robustness problem using minimal information regarding the modeling errors. However, the cost of such a representation is that the stability robustness results obtained may be conservative. This is because these results are obtained by showing stability with respect to the larger set $\mathcal{G}$ instead of the smaller set from our modeling process. Another way of looking at it is that the resulting set of stabilizing controllers for the larger set $\mathcal{G}$ will be smaller. As a result, the achievable performance may be worse. This conservatism will be discussed throughout as we develop our analysis for stability robustness, and the impact of stability robustness and this conservatism on closed-loop performance will be discussed in Section 9.4.

9.2.3.4 Reflecting Modeling Errors to the Input of the Plant

Finally, we note that in our representation of multiplicative unstructured uncertainty, we choose to lump the uncertainty at the plant output. In other words, we assume that the actual plant is of the form $(I + w(s)\Delta(s))G(s)$ where the uncertainty, $(I + w(s)\Delta(s))$ multiplying the model is at the plant output. To be more precise notationally, we should rewrite our definition for $\mathcal{G}$ as

$$\mathcal{G} = \{\tilde{G}(s) \mid \tilde{G}(s) = (I + w_o(s)\Delta(s))G(s), \|\Delta(j\omega)\|_{\mathcal{H}_\infty} \leq 1\} \tag{9.20}$$

where we have added the subscript "o" to $w(s)$ to distinguish it as the weight corresponding to uncertainty at the plant output. Alternatively, we can instead choose to lump the uncertainty at the plant input. In this case, we assume that the actual plant is of the form $G(s)(I + w_i(s)\Delta(s))$ where the uncertainty multiplying the model is at the plant input. As a result, the definition for $\mathcal{G}$ in this case is

$$\mathcal{G} = \{\tilde{G}(s) \mid \tilde{G}(s) = G(s)(I + w_i(s)\Delta(s)), \|\Delta(j\omega)\|_{\mathcal{H}_\infty} \leq 1\} \tag{9.21}$$

where $w_i(s)$ is the weight corresponding to the uncertainty at the plant input. Comparing the two representations reveals that, in general, the two sets defined by Equations 9.20 and 9.21 are not the same because matrix multiplication does not commute. Since $\Delta(s)$ is constrained similarly and we must choose the weight so that $\mathcal{G}$ covers the actual plant in both cases, we note that, in general, $w_o(s)$ is not the same as $w_i(s)$.

Of course, we do not physically lump the modeling errors to the plant input or output. Instead, in the course of modeling, we lump the model so that the modeling errors are reflected either to the input of the model or to the output. To be sure, modeling errors associated with the plant actuators and sensors, such as neglected actuator and sensor dynamics, are reflected more naturally to the model's input and output, respectively. However, modeling errors associated with internal plant dynamics, such as neglected flexible modes, are not naturally reflected to either the model's input or output. In this case, we have a choice as to where we wish to reflect these errors. As a final note, for the SISO case, the two representations are equivalent since scalar multiplication does commute. As a result, we can choose $w_o(s) = w_i(s) = w(s)$ so that it does not make a difference where we reflect our modeling errors.

9.3 Conditions for Stability Robustness

Having established a multiplicative representation for unstructured uncertainty, we proceed to use it to analyze the stability robustness problem. As mentioned beforehand, since the actual modeling error and, thus, the actual plant is never known, we cannot hope to simply evaluate the stability of the actual closed-loop system by using the Nyquist stability criterion or by calculating closed-loop poles. Instead, we must rely on our representation of the uncertainty to arrive at conditions or tests in the frequency domain that guarantee stability robustness. Throughout the discussion, we assume that we have a controller, $K(s)$, that gives us nominal stability for the feedback loop in Figure 9.1 and that we use this controller in the actual feedback loop, Figure 9.2. In addition, since we are interested in internal stability of the feedback loop, we assume throughout that for SISO systems, the transfer function $k(s)g(s)$ is stabilizable and detectable; that is, there are no unstable pole-zero cancellations in forming $k(s)g(s)$. For MIMO systems, the corresponding conditions are that both $K(s)G(s)$ and $G(s)K(s)$ are stabilizable and detectable.

9.3.1 Stability Robustness for SISO Systems

In this section we analyze the stability robustness of SISO feedback systems to multiplicative unstructured uncertainty. We first derive a sufficient condition for stability robustness using the familiar Nyquist stability criterion. We then derive the same condition using another method: the small gain theorem. The goal here is not only to show that one can arrive at the same answer but also to present the small gain approach to analyzing stability robustness, which can be easily extended to MIMO systems. Finally, we compare our notion of stability robustness to more traditional notions of robustness such as gain and phase margins.

9.3.1.1 Stability Robustness Using the Nyquist Stability Criterion

We begin the analysis of stability robustness to unstructured uncertainty for SISO systems with the Nyquist stability criterion. We recall from classical control theory that the Nyquist stability criterion is a graphical representation of the relationship between the number of unstable poles of an open-loop transfer function, $l(s)$, and the unstable zeros of the return difference transfer function, $1 + l(s)$. Since the zeros of $1 + l(s)$ correspond to the closed-loop poles of $l(s)$ under negative unity feedback, the Nyquist stability criterion is used to relate the number of unstable poles of $l(s)$ to the stability of the resulting closed-loop system. Specifically, the Nyquist stability criterion states that the corresponding closed-loop system is stable if and only if the number of positive clockwise encirclements (or negative counterclockwise encirclements) of the point $(-1, 0)$ in the complex plane by the Nyquist plot of $l(s)$ is equal to $-P$, where P is the number of unstable poles of $l(s)$. Here, the Nyquist plot is simply the plot in the complex plane of $l(s)$ evaluated along the closed Nyquist contour D_r, which is defined in the usual way with counterclockwise indentations around the imaginary axis poles of $l(s)$ so that they are excluded from the interior of D_r. Notationally, we express the Nyquist stability criterion as

$$\aleph(-1, l(s), D_r) = -P \tag{9.22}$$

In our analysis we are interested in the stability of the two feedback loops given in Figures 9.1 and 9.2. For the nominal feedback loop, we define the nominal loop transfer function as

$$l(s) = g(s)k(s) \tag{9.23}$$

Since we assume that the nominal closed loop is stable, we have that

$$\aleph(-1, g(s)k(s), D_r) = -P \tag{9.24}$$

where P is the number of unstable poles of $g(s)k(s)$. For the actual feedback loop, we define the actual loop transfer function as

$$l_a(s) = g_a(s)k(s) = (1 + w(s)\Delta(s))g(s)k(s) \tag{9.25}$$

where the second equality holds for some $\|\Delta(j\omega)\|_{\mathcal{H}_\infty} \leq 1$ since we assumed that $g_a(s) \in \mathcal{G}$ in our representation of the unstructured uncertainty. In addition, since we assumed that the unstable poles of $g(s)$ and $g_a(s)$ coincide and since the same $k(s)$ appears in both loop transfer functions, the number of unstable poles of $g_a(s)k(s)$ and thus $l_a(s)$ is also P. Similarly, since we assumed that the imaginary axis poles of $g(s)$ and $g_a(s)$ coincide, the same Nyquist contour D_r can be used to evaluate the Nyquist plot of $l_a(s)$. Therefore, by the Nyquist stability criterion, the actual closed-loop system is stable if and only if

$$\begin{aligned}\aleph(-1, g_a(s)k(s), D_r) &= \aleph(-1, (1 + w(s)\Delta(s))g(s)k(s), D_r) \\ &= -P\end{aligned} \tag{9.26}$$

Since we do not know $g_a(s)$, we can never hope to use Equation 9.26 to evaluate or show the stability of the actual system. However, we note that Equation 9.26 implies that the actual closed-loop system is stable if and only if the number of counterclockwise encirclements of the critical point, $(-1, 0)$, is the same for the Nyquist plot of $l_a(s)$ as that for $l(s)$. Therefore, if we can show that for the actual loop transfer function, we do not change the number of counterclockwise encirclements from that of the nominal loop transfer function, then we can guarantee that the actual closed-loop system is stable. The idea is to utilize our set membership representation of the uncertainty to ensure that it is impossible to change the number of encirclements of the critical point for any loop transfer function $\tilde{g}(s)k(s)$ with $\tilde{g}(s) \in \mathcal{G}$. This will guarantee that the actual closed-loop system is stable, since our representation is such that $g_a(s) \in \mathcal{G}$. This is the type of sufficient condition for stability robustness that we seek.

To obtain this condition, we need a relationship between the Nyquist plots of $l(s)$ and $l_a(s)$ using our representation of the unstructured uncertainty. To start, we separate the Nyquist plot into three parts corresponding to the following three parts of the Nyqust contour D_r:

1. The nonnegative $j\omega$ axis
2. The negative $j\omega$ axis
3. The part of D_r that encircles the right half s-plane where $|s| \to \infty$

For the first part, we note from Equations 9.23 and 9.25 that $l_a(s)$ can be expressed as

$$l_a(s) = (1 + w(s)\Delta(s))l(s) \tag{9.27}$$

for some $\|\Delta\|_{\mathcal{H}_\infty} \leq 1$. Therefore, the multiplicative uncertainty representation developed earlier holds equally for $l(s) = g(s)k(s)$ as for $g(s)$. Extending the SISO interpretation of this uncertainty representation in Figure 9.4 through Figure 9.8, we note that for any nonnegative ω, the Nyquist plot of $l_a(s)$ is a point contained in the disk of radius $|w(j\omega)||l(j\omega)|$ centered at the point $l(j\omega)$, which is the Nyquist plot

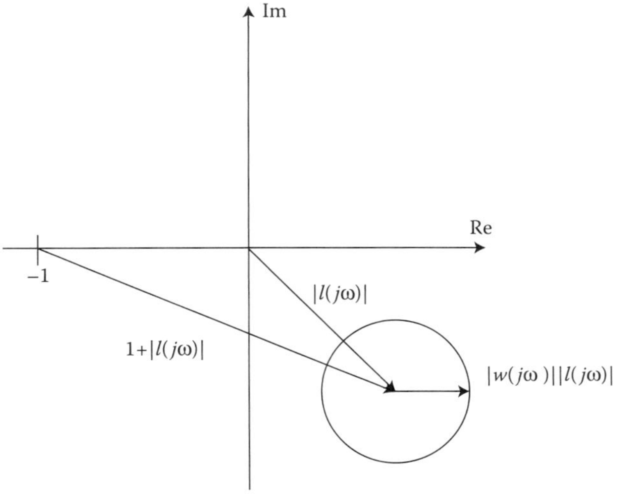

FIGURE 9.8 Multiplicative uncertainty in the Nyquist plot.

of $l(s)$ at that ω. Figure 9.8, then, gives a relationship between the Nyquist plot of $l(s)$ and $l_a(s)$ at a particular frequency for the part of the Nyquist contour along the positive $j\omega$ axis. For the second part of the Nyquist contour, we note that the Nyquist plot is simply a mirror image, with respect to the real axis, of the Nyquist plot for the first part, and thus this relationship will be identical. For the third part of the Nyquist contour, we note that since the model $g(s)$ represents a physical system, it should be strictly proper, and therefore, $|g(s)| \rightarrow 0$ as $|s| \rightarrow \infty$. In addition, the same holds for $g_a(s)$ since it is a physical system. Since the controller $k(s)$ is proper, it follows that the both the Nyquist plots for $l(s)$ and $l_a(s)$ are at the origin for the part of the Nyquist contour that encircles the right half s-plane.

For stability robustness, we only need to consider the relationship between $l(s)$ and $l_a(s)$ for the first part of the Nyquist contour. This is because for the second part the relationship is identical and thus the conclusions for stability robustness will be identical. Finally, for the third part, since the Nyquist plots for both $l(s)$ and $l_a(s)$ are at the origin, they are identical and cannot impact stability robustness. As a result, we only need to consider the relationship between $l(s)$ and $l_a(s)$ that is presented in Figure 9.8. We illustrate this relationship along a typical Nyquist plot of the nominal $l(s)$, for $\omega \geq 0$, in Figure 9.9 where, for clarity, we only illustrate the uncertainty disk at a few frequencies. Here, we note that the radius of the uncertainty disk changes as a function of frequency since both $w(j\omega)$ and $l(j\omega)$ varies with frequency.

From Figure 9.9, we note that it is impossible for the Nyquist plot of $l_a(s)$ to change the number of encirclements of the critical point $(-1, 0)$ if the disks representing the uncertainty in $l(s)$ do not intersect the critical point for all $\omega \geq 0$. To show this, let us prove the contrapositive. That is, if the Nyquist plot of $l_a(s)$ does change the number of encirclements of the critical point, then by continuity there exists a frequency ω^* where the line connecting $l(j\omega^*)$ and $l_a(j\omega^*)$ must intersect the critical point. Since the uncertainty disk at that frequency is convex, this uncertainty disk must also intersect the critical point.

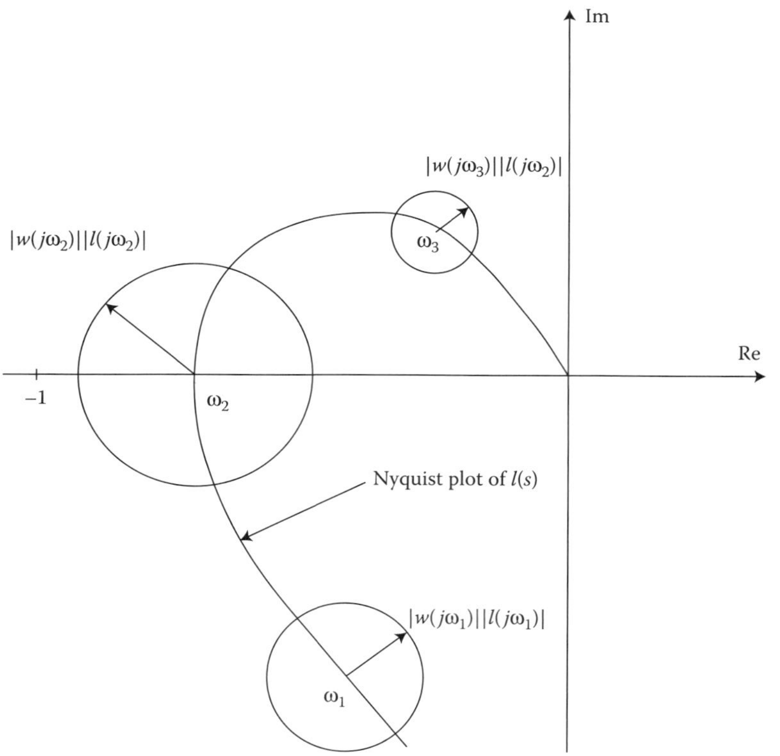

FIGURE 9.9 Stability robustness using the Nyquist stability criterion.

Graphically, then, we can guarantee stability robustness if the distance between the critical point and the Nyquist plot of $l(s)$ is strictly greater than the radius of the uncertainty disk for all $\omega \geq 0$. From Figure 9.8, the distance between the critical point and the Nyquist plot of the nominal loop transfer function, $l(s)$, is simply $|1 + l(j\omega)|$, which is the magnitude of the return difference. Therefore, a condition for stability robustness is

$$|w(j\omega)||l(j\omega)| < |1 + l(j\omega)| \quad \forall\, \omega \geq 0 \tag{9.28}$$

or equivalently,

$$|w(j\omega)||g(j\omega)k(j\omega)| < |1 + g(j\omega)k(j\omega)| \quad \forall\, \omega \geq 0 \tag{9.29}$$

We note that although the Nyquist stability criterion is both a necessary and sufficient condition for stability, the above condition for stability robustness is clearly only sufficient. That is, even if we violate this condition at a particular frequency or at a range of frequencies, the Nyquist plot of the actual system may not change the number of encirclements of the critical point and thus the actual closed-loop system may be stable. This is illustrated in Figure 9.10, where we note that if the Nyquist plot of $l_a(s)$ follows $l_{a2}(s)$, then the actual system is stable. The key here is that since we do not know $l_a(s)$, we cannot say whether or not the actual system is stable when the condition is violated. Therefore, to guarantee actual stability, we need to ensure a safe distance or margin, which may not be necessary, between the nominal Nyquist plot and the critical point. This margin is in terms of the uncertainty disk. We see that this conservatism stems from the fact that we admit total lack of knowledge concerning the phase of the actual plant, which led to the representation of the uncertainty as a disk in the complex plane.

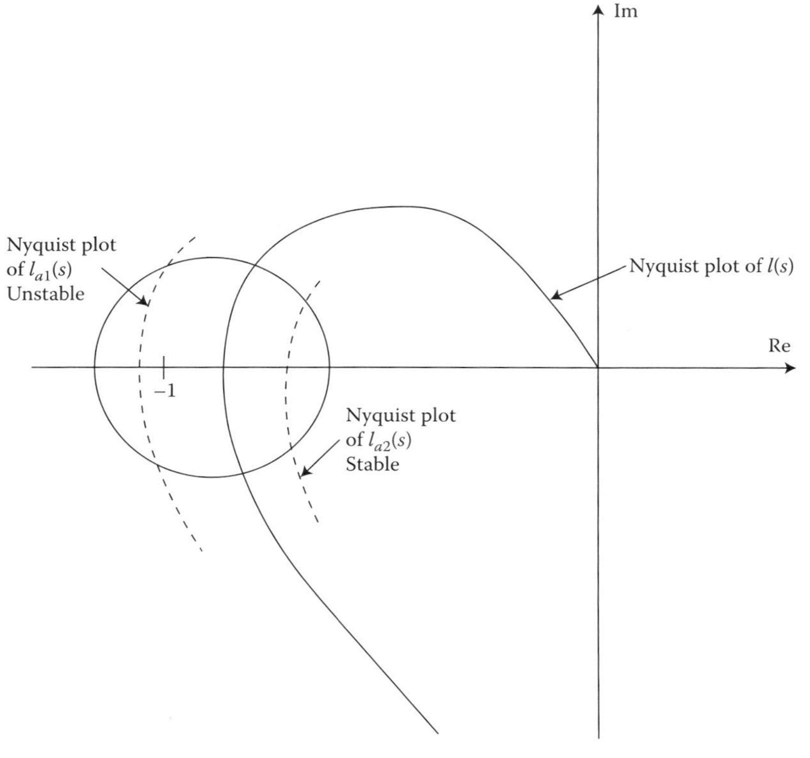

FIGURE 9.10 Sufficiency of the stability robustness condition.

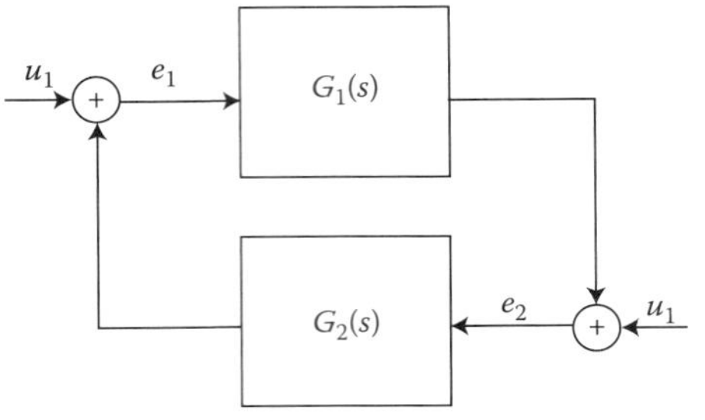

FIGURE 9.11 Standard feedback form.

9.3.1.2 Stability Robustness Using Small Gain Theorem

We now seek to derive an equivalent condition for stability robustness for SISO systems using the small gain theorem, see for example Dahleh and Diaz-Bobillo [1]. The goal here is to introduce another methodology whereby conditions for stability robustness can be derived. As we will see in the sequel, this methodology can be easily extended to MIMO systems. We begin with a statement of the small gain theorem, specialized to LTI systems, which addresses the stability of a closed-loop system in the standard feedback form given in Figure 9.11.

Theorem 9.1: Small Gain Theorem

Under the assumption that $G_1(s)$ and $G_2(s)$ are stable in the feedback system in Figure 9.11, the closed-loop transfer function matrix from (u_1, u_2) to (e_1, e_2) is stable if the small gain condition

$$\|G_1(j\omega)\|_{\mathcal{H}_\infty}\|G_2(j\omega)\|_{\mathcal{H}_\infty} < 1 \tag{9.30}$$

is satisfied.

Proof. We first show that the sensitivity transfer function matrix, $S(s) = (I - G_1(s)G_2(s))^{-1}$ is stable. For this, we need to show that if the small gain condition Equation 9.30 is satisfied, then $S(s) = (I - G_1(s)G_2(s))^{-1}$ is analytic in the closed right-half s-plane. An equivalent statement is that the return difference, $D(s) = I - G_1(s)G_2(s)$ is nonsingular for all s in the closed right-half plane.

For arbitrary input u and for all complex s,

$$\|D(s)u\| = \|(I - G_1(s)G_2(s))u\| = \|u - G_1(s)G_2(s)u\| \tag{9.31}$$

where $\|\cdot\|$ represents the standard Euclidean norm. From the triangle inequality,

$$\|u - G_1(s)G_2(s)u\| \geq \|u\| - \|G_1(s)G_2(s)u\| \tag{9.32}$$

from the definition of the maximum singular value,

$$\|G_1(s)G_2(s)u\| \leq \sigma_{\max}[G_1(s)G_2(s)]\|u\| \tag{9.33}$$

and from the submultiplicative property of induced norms,

$$\sigma_{\max}[G_1(S)G_2(s)] \leq \sigma_{\max}[G_1(s)]\sigma_{\max}[G_2(s)] \tag{9.34}$$

Substituting Equations 9.32 through 9.34 into Equation 9.31 gives for all complex s,

$$\begin{aligned}\|D(s)u\| &= \|u - G_1(s)G_2(s)u\| \\ &\geq \|u\| - \|G_1(s)G_2(s)u\| \\ &\geq \|u\| - \sigma_{\max}[G_1(s)G_2(s)]\|u\| \\ &\geq \|u\| - \sigma_{\max}[G_1(s)]\sigma_{\max}[G_2(s)]\|u\| \\ &= (1 - \sigma_{\max}[G_1(s)]\sigma_{\max}[G_2(s)])\|u\| \end{aligned} \tag{9.35}$$

Since G_1 and G_2 are stable, they are analytic in the closed right-half plane. Therefore, by the maximum modulus theorem [2],

$$\begin{aligned}\sigma_{\max}[G_1(s)] &\leq \sup_{\omega} \sigma_{\max}[G_1(j\omega)] = \|G_1(j\omega)\|_{\mathcal{H}_\infty} \\ \sigma_{\max}[G_2(s)] &\leq \sup_{\omega} \sigma_{\max}[G_2(j\omega)] = \|G_2(j\omega)\|_{\mathcal{H}_\infty}\end{aligned} \tag{9.36}$$

for all s in the closed right-half plane. Substituting Equation 9.36 into Equation 9.35 gives

$$\|D(s)u\| \geq (1 - \|G_1(j\omega)\|_{\mathcal{H}_\infty}\|G_2(j\omega)\|_{\mathcal{H}_\infty})\|u\| \tag{9.37}$$

for all s in the closed right-half plane. From the small gain condition, there exists an $\epsilon > 0$ such that

$$\|G_1(j\omega)\|_{\mathcal{H}_\infty}\|G_2(j\omega)\|_{\mathcal{H}_\infty} < 1 - \epsilon \tag{9.38}$$

Therefore, for all s in the closed right-half plane,

$$\|D(s)u\| \geq \epsilon\|u\| > 0 \tag{9.39}$$

for any arbitrary u, which implies that $D(s)$ is nonsingular in the closed right-half s-plane.

From a similar argument, we can show that $(I - G_2(s)G_1(s))$ is also stable. Therefore, the transfer function matrix relating (u_1, u_2) to (e_1, e_2), which is given by

$$\begin{bmatrix} (I - G_2(s)G_1(s))^{-1} & -(I - G_2(s)G_1(s))^{-1}G_2(s) \\ (I - G_1(s)G_2(s)^{-1}G_1(s) & (I - G_1(s)G_2(s))^{-1} \end{bmatrix} \tag{9.40}$$

is stable.

We note that the small gain theorem is only a sufficient condition for stability. For example, in the SISO case, the small gain condition (Equation 9.30) can be expressed as

$$|g_1(j\omega)g_2(j\omega)| < 1 \quad \forall\, \omega \geq 0 \tag{9.41}$$

which implies that the Nyquist plot of the loop transfer function $g_1(s)g_2(s)$ lies strictly inside the unit circle centered at the origin, as shown in Figure 9.12. This is sufficient for stability because $g_1(s)$ and $g_2(s)$ are stable; however, we note that it is clearly not necessary. In addition, we note that the small gain theorem applies equally well to MIMO systems, since the proof was actually done for the MIMO case. In fact, the general form of the small gain theorem applies to nonlinear, time-varying operators over any normed signal space. For a treatment of the general small gain theorem and its proof, the reader is referred to [3].

For stability robustness, we are, as before, interested in the stability of the two feedback loops given in Figures 9.1 and 9.2. From our representation of unstructured uncertainty, we can express the actual plant as

$$g_a(s) = (1 + w(s)\Delta(s))g(s) \tag{9.42}$$

for some $\|\Delta(j\omega)\|_{\mathcal{H}_\infty} \leq 1$. Therefore, the actual feedback loop can also be represented by the block diagram in Figure 9.13. In the figure, we note that we choose, merely by convention, to reflect the

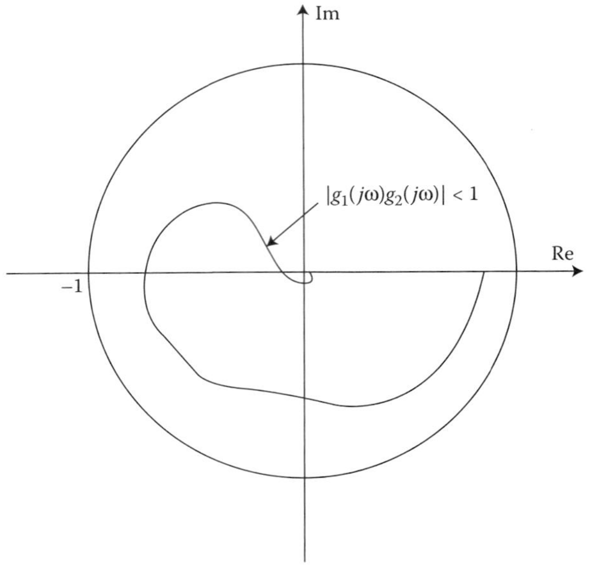

FIGURE 9.12 Sufficiency of SISO small gain theorem.

unstructured uncertainty to the output of the plant instead of to the input since for the SISO case the two are equivalent. In addition, we note that Figure 9.13 can also be interpreted as being the feedback loop for all perturbed plants, $\tilde{g}(s)$, belonging to the set $\mathcal{G}$. The key for stability robustness, then, is to show stability for this feedback loop using the small gain theorem for all $\|\Delta(j\omega)\|_{\mathcal{H}_\infty} \leq 1$ and, therefore, for all $\tilde{g}(s) \in \mathcal{G}$.

To apply the small gain theorem, we first reduce the feedback loop in Figure 9.13 to the standard feedback form in Figure 9.11. To do this, we isolate $\Delta(s)$ and calculate the transfer function from the output of Δ, v, to its input, z. From Figure 9.13,

$$\begin{aligned} z(s) &= -w(s)g(s)k(s)y(s) \\ y(s) &= v(s) - g(s)k(s)y(s) \end{aligned} \tag{9.43}$$

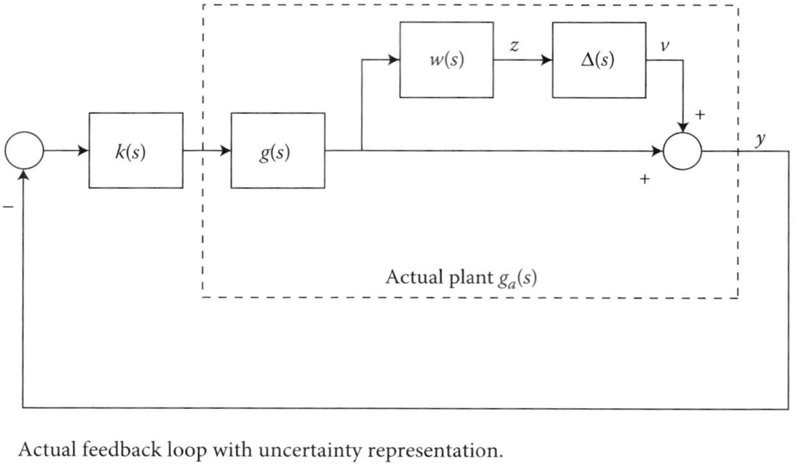

FIGURE 9.13 Actual feedback loop with uncertainty representation.

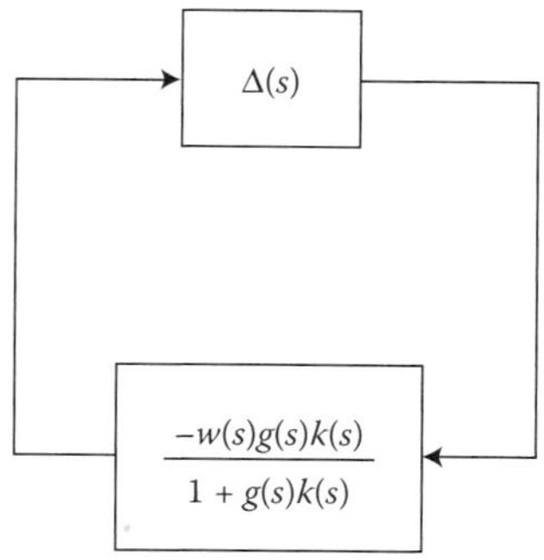

FIGURE 9.14 Actual feedback loop in standard form.

As a result, the transfer function seen by Δ is given by $m(s)$ where

$$m(s) = \frac{-w(s)g(s)k(s)}{1+g(s)k(s)} \tag{9.44}$$

and the reduced block diagram is given in Figure 9.14. We note that $m(s)$ is simply the product of the complementary sensitivity transfer function for the nominal feedback loop and the weight $w(s)$. Since we assumed that the nominal closed loop is stable and $w(s)$ is stable, $m(s)$ is stable. Furthermore, since we also assumed that $\Delta(s)$ is stable, the assumption for the small gain theorem is satisfied for the closed-loop system in Figure 9.14. Applying the small gain theorem, this closed-loop system is stable if the small gain condition

$$\left\| \frac{-w(j\omega)g(j\omega)k(j\omega)}{1+g(j\omega)k(j\omega)} \right\|_{\mathcal{H}\infty} \left\| \Delta(j\omega) \right\| < 1 \tag{9.45}$$

is satisfied. Since $\left\| \Delta(j\omega) \right\|_{\mathcal{H}_\infty} \leq 1$, an equivalent condition for stability to all $\left\| \Delta(j\omega) \right\|_{\mathcal{H}_\infty} \leq 1$ is

$$\left\| \frac{w(j\omega)g(j\omega)k(j\omega)}{1+g(j\omega)k(j\omega)} \right\|_{\mathcal{H}_\infty} < 1 \tag{9.46}$$

which is a sufficient condition for the stability of the feedback loop in Figure 9.13 for all $\tilde{g}(s) \in \mathcal{G}$. Since $g_a(s) \in \mathcal{G}$, this is a sufficient condition for stability robustness.

We now proceed to show that the stability robustness condition in Equation 9.46 is equivalent to that derived using the Nyquist stability criterion. From the definition of the $\mathcal{H}_\infty$ norm, an equivalent condition to Equation 9.46 is given by

$$\left| \frac{w(j\omega)g(j\omega)k(j\omega)}{1+g(j\omega)k(j\omega)} \right| < 1 \quad \forall\, \omega \geq 0 \tag{9.47}$$

Since everything is scalar, this condition is equivalent to

$$|w(j\omega)||g(j\omega)k(j\omega)| < |1+g(j\omega)k(j\omega)| \quad \forall\, \omega \geq 0 \tag{9.48}$$

which is exactly the condition for stability robustness derived using the Nyquist stability criterion. This equivalence is not due to the equivalence of the Nyquist stability criterion and the small gain theorem, since the former is a necessary and sufficient condition for stability while the latter is only sufficient. Rather, this equivalence is due to our particular approach in applying the small gain theorem and to our representation of the uncertainty. What this equivalence gives us is an alternative to the more familiar Nyquist stability criterion in analyzing the stability robustness problem. Unlike the Nyquist stability criterion, the small gain theorem applies equally well to MIMO systems. Therefore, our analysis is easily extended to the MIMO case, as we will do in Section 9.3.2.

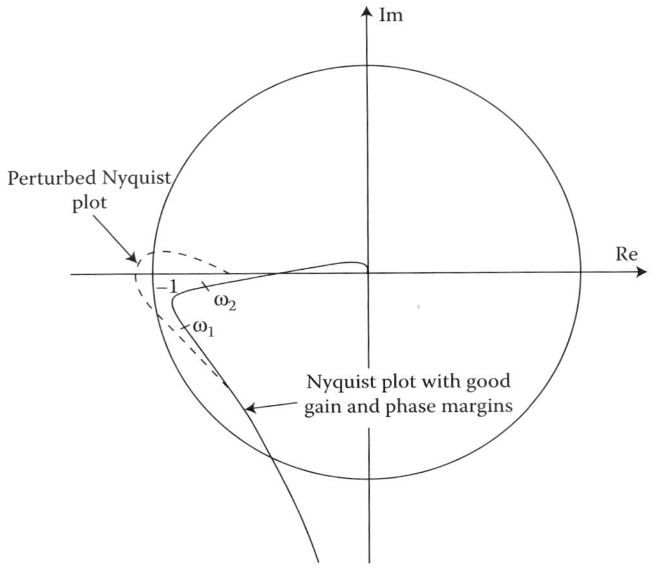

FIGURE 9.15 Insufficiency of gain and phase margins.

9.3.1.3 Comparison to Gain and Phase Margins

In closing this section on stability robustness for SISO systems, we would like to compare our notion of robustness to more traditional notions such as gain and phase margins. As we recall, gain margin is the amount of additional gain that the SISO open-loop transfer function can withstand before the closed loop goes unstable, and phase margin is the amount of additional phase shift or pure delay that the loop transfer function can withstand before the closed loop goes unstable. To be sure, gain and phase margins are measures of robustness for SISO systems, but they are, in general, insufficient in guaranteeing stability in the face of dynamic uncertainty such as those due to unmodeled dynamics. This is because gain and phase margins only deal with uncertainty in terms of pure gain variations or pure phase variations but not a combination of both. That is, the open loop can exhibit large gain and phase margins and yet be close to instability as shown in the Nyquist plot in Figure 9.15. In the figure, we note that for frequencies between ω_1 and ω_2 the Nyquist plot is close to the critical point so that a combination of gain and phase variation along these frequencies such as that in the perturbed Nyquist plot will destabilize the closed loop. This combination of gain and phase variations can be the result of unmodeled dynamics. In such a case, gain and phase margins will give a false sense of stability robustness. In contrast, we could get a true sense of stability robustness by explicitly accounting for the dynamic uncertainty in terms of unstructured uncertainty.

In addition, gain and phase margins are largely SISO measures of stability robustness since they are inadequate in capturing the cross coupling between inputs and outputs of the dynamics of MIMO systems. For MIMO systems, we usually think of gain and phase margins as being independent gain and phase variations that are allowed at each input channel. These variations clearly cannot cover the combined gain, phase, and directional variations due to MIMO dynamic uncertainty. As a result, the utility of our notion of stability robustness over traditional gain and phase margin concepts becomes even more clear in the MIMO case.

9.3.2 Stability Robustness for MIMO Systems

In this section, we analyze the stability robustness of MIMO feedback systems to multiplicative unstructured uncertainty. As mentioned beforehand, we will use the small gain theorem since it offers a natural extension from the SISO case to the MIMO case. As shown in the SISO case, the general procedure for analyzing stability robustness using the small gain theorem is as follows:

1. Start with the block diagram of the actual feedback loop with the actual plant represented by the nominal model perturbed by the uncertainty. Note that this block diagram also represents the feedback loop for all perturbed plants belonging to the set $\mathcal{G}$, which contains the actual plant.
2. Reduce the feedback loop to the standard feedback form by isolating the $\Delta(s)$ block and calculating the TFM from the output of Δ to the input of Δ. Denote this TFM as $M(s)$.
3. Apply the small gain theorem. In particular, since $\|\Delta(j\omega)\|_{\mathcal{H}_\infty} \leq 1$, the small gain theorem guarantees stability for the feedback loop for all perturbed plants in the set $\mathcal{G}$ and therefore guarantees robust stability if $\|M(j\omega)\|_{\mathcal{H}_\infty} < 1$.

We will follow this procedure in our analysis of MIMO stability robustness. For MIMO systems, as shown in Section 9.2.3, there is a difference between reflecting the modeling errors to the input and the output of the plant. As a consequence, we separate the two cases and derive a different stability robustness condition for each case. In the end, we will relate these two stability robustness tests and discuss their differences.

9.3.2.1 Uncertainty at Plant Output

We start with the case where the modeling errors are reflected to the plant output. In this case, the actual plant is of the form

$$G_a(s) = (I + w_o(s)\Delta(s))G(s) \tag{9.49}$$

for some $\|\Delta(j\omega)\|_{\mathcal{H}_\infty} \leq 1$. Following step 1 of the procedure, the block diagram of the actual feedback loop given this representation of the uncertainty is depicted in Figure 9.16a. For step 2, we calculate the TFM from the output of Δ, v, to its input, z. From Figure 9.16a,

$$\begin{aligned} z(s) &= -w_o(s)G(s)K(s)y(s) \\ y(s) &= v(s) - G(s)K(s)y(s) \end{aligned} \tag{9.50}$$

As a result, $M(s)$, the TFM seen by $\Delta(s)$, is

$$M(s) = -w_o(s)G(s)K(s)(I + G(s)K(s))^{-1} \tag{9.51}$$

and the reduced block diagram is given in Figure 9.16b. We note that $M(s)$ is simply the product of the complementary sensitivity TFM, $C(s) = G(s)K(s)(I + G(s)K(s))^{-1}$, for the nominal feedback loop and the scalar weight $w_o(s)$. Since we assumed that the nominal closed loop and $w_o(s)$ are both stable, $M(s)$ is stable. Furthermore, since we also assumed that $\Delta(s)$ is stable, the assumption for the small gain theorem is satisfied for the closed loop system in Figure 9.16b. For step 3, we apply the small gain theorem, which gives

$$\|w_o(j\omega)G(j\omega)K(j\omega)(I + G(j\omega)K(j\omega))^{-1}\|_{\mathcal{H}_\infty} < 1 \tag{9.52}$$

as a sufficient condition for stability robustness. Using the definition of the $\mathcal{H}_\infty$ norm, an equivalent condition for stability robustness is

$$\sigma_{\max}[w_o(j\omega)G(j\omega)K(j\omega)(I + G(j\omega)K(j\omega))^{-1}] < 1 \quad \forall\, \omega \geq 0 \tag{9.53}$$

Since $w_o(s)$ is scalar, another sufficient condition for stability robustness to multiplicative uncertainty at the plant output is

$$\sigma_{\max}[G(j\omega)K(j\omega)(I + G(j\omega)K(j\omega))^{-1}] < \frac{1}{|w_o(j\omega)|} \quad \forall\, \omega \geq 0 \tag{9.54}$$

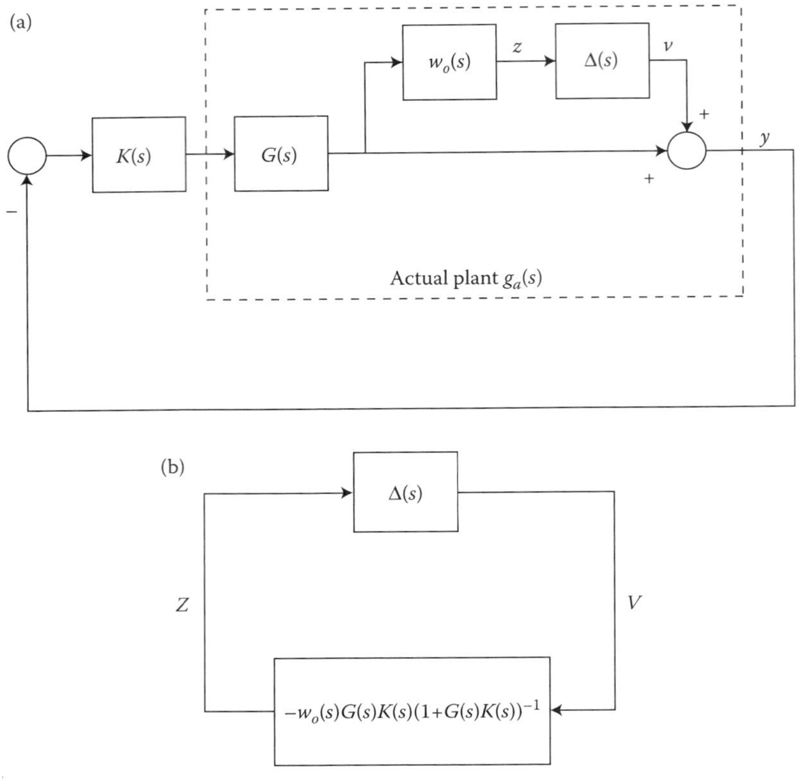

FIGURE 9.16 Stability robustness for multiplicative uncertainty at plant output.

9.3.2.2 Uncertainty at Plant Input

In the case where the modeling errors are reflected to the plant input, the actual plant is of the form

$$G_a(s) = G(s)(I + w_i(s)\Delta(s)) \tag{9.55}$$

for some $\|\Delta(j\omega)\|_{\mathcal{H}_\infty} \leq 1$. Following the procedure in Section 9.3.2, a sufficient condition for stability robustness is given by

$$\left\| w_i(j\omega)(I + K(j\omega)G(j\omega))^{-1}K(j\omega)G(j\omega) \right\|_{\mathcal{H}_\infty} < 1 \tag{9.56}$$

Using the definition of the $\mathcal{H}_\infty$ norm and the fact that $w_i(s)$ is scalar, an equivalent condition for stability robustness is

$$\sigma_{\max}[(I + K(j\omega)G(j\omega))^{-1}K(j\omega)G(j\omega)] < \frac{1}{|w_i(j\omega)|} \quad \forall\, \omega \geq 0 \tag{9.57}$$

Comparison between the sufficient conditions Equations 9.54 and 9.57 reveals that, although the two conditions both address the same stability robustness problem, they are indeed different. When we reflect the modeling errors to the plant output, our stability robustness condition is based on the nominal complementary sensitivity TFM, $C(s)$. This TFM is the closed loop TFM formed from the loop TFM with the loop broken at the plant output. On the other hand, when we reflect the modeling errors to the plant input, our stability robustness condition is based on the nominal input complementary sensitivity TFM, $C_i(s)$. In general, $C_i(s)$ is different from $C(s)$. As mentioned in Section 9.2.3, this difference stems from the fact that, although the sets $\mathcal{G}$ for the two cases both contain the actual plant, they are different because matrix multiplication does not commute. As a result, the two different conditions can give different results. Since the two conditions are both only sufficient, only one of them needs to be satisfied in order to conclude stability robustness. The fact that we only need to satisfy one of the conditions to achieve stability robustness means that one condition is less conservative than the other. This is indeed true since the sets $\mathcal{G}$ are different, and thus, invariably, one is larger than the other, resulting in a more conservative measure of modeling error. Finally, we note that if $G(s)$ and $K(s)$ are both SISO, then both Equations 9.54 and 9.57 reduce to the sufficient condition derived for the SISO case.

9.4 Impact of Stability Robustness on Closed-Loop Performance

In the previous section, we have established, for both the SISO and MIMO cases, sufficient conditions for stability robustness under the multiplicative representation for unstructured uncertainty. To the control engineer, these conditions will have to be satisfied by design in order to ensure the stability of the actual closed-loop system. These conditions, then, become additional constraints on the design that will invariably impact the performance of the closed-loop system, such as command following and disturbance rejection. As shown in the chapter on MIMO frequency response analysis in this handbook, these performance specifications can be defined in the frequency domain in terms of the maximum singular value of the sensitivity TFM, $S(s) = (I + G(s)K(s))^{-1}$. In particular, we require $\sigma_{\max}[S(j\omega)]$ to be small in the frequency range of the command and/or disturbance signals. In this section, we seek an interpretation of our stability robustness conditions in the frequency domain in order to discuss its impact on command following and disturbance rejection.

We start with the SISO case and the stability robustness condition given in Equation 9.48. Since everything is scalar and positive, this condition is equivalent to

$$\left|\frac{g(j\omega)k(j\omega)}{1+g(j\omega)k(j\omega)}\right| < \frac{1}{|w(j\omega)|} \quad \forall\, \omega \geq 0 \tag{9.58}$$

where the left side is the magnitude of the closed-loop or complementary sensitivity transfer function, $c(j\omega)$, of the nominal design. This condition is illustrated in the frequency domain in Figure 9.17. Interpreting from the figure, the stability robustness condition states that the magnitude plot of the nominal closed-loop transfer function must lie strictly below the plot of the inverse of $|w(j\omega)|$. As is typically the case, the inverse of the magnitude of $w(j\omega)$ will be large at low frequencies and small at high frequencies, since modeling errors increase with increasing frequency. Specifically, the modeling errors become significant near and above the frequency ω_m defined by

$$|w(j\omega_m)| = 0 \text{ dB} \tag{9.59}$$

Therefore, the stability robustness condition limits the bandwidth of the nominal closed-loop design. That is, the bandwidth of the nominal closed-loop design, ω_b, is constrained to be less than ω_m, as shown in the figure. Indeed, as indicated in Figure 9.17, the presence of significant modeling errors at frequencies beyond ω_m forces the rapid roll-off of the designed closed-loop transfer function at high frequencies. Physically, this roll-off prevents energy at these frequencies from exciting the unmodeled dynamics and, therefore, prevents the possible loss of stability. In terms of closed-loop performance, we recall from the chapter on MIMO frequency response analysis that a necessary condition for $\sigma_{\max}[S(j\omega)]$ to be small at a certain frequency is that the singular values of the closed-loop or complementary sensitivity TFM, $C(s) = G(s)K(s)(I + G(s)K(s))^{-1}$, must be close to unity (0 dB) at that frequency. For SISO systems, this simply translates to requiring that the magnitude of the closed-loop transfer function be close to unity (0 dB). From Figure 9.17, it is clear that the stability robustness condition limits the range of frequencies over which we can expect to achieve good command following and/or output disturbance rejection

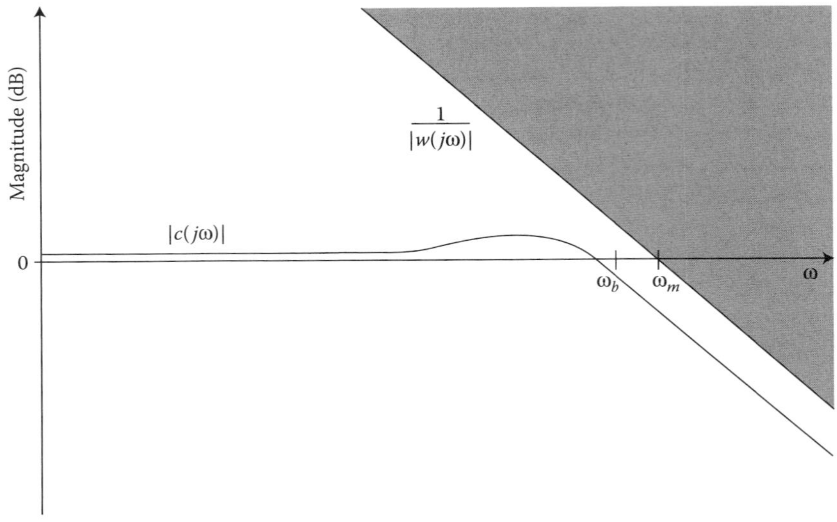

FIGURE 9.17 Interpretation of stability robustness for SISO systems.

to below ω_m. In other words, we should not expect good performance at frequencies where there are significant modeling errors.

For the MIMO case where the modeling errors are reflected to the plant output, the stability robustness condition given in Equation 9.54 can be illustrated in terms of singular value plots, as shown in Figure 9.18. This figure gives a similar interpretation concerning the impact of stability robustness on the nominal closed-loop performance as that for the SISO case. From the figure, the output bandwidth of the nominal closed-loop design, ω_{bo}, is constrained by the stability robustness condition to be less than ω_{mo}, where ω_{mo} is defined as the frequency at and beyond which the modeling error reflected to the plant output

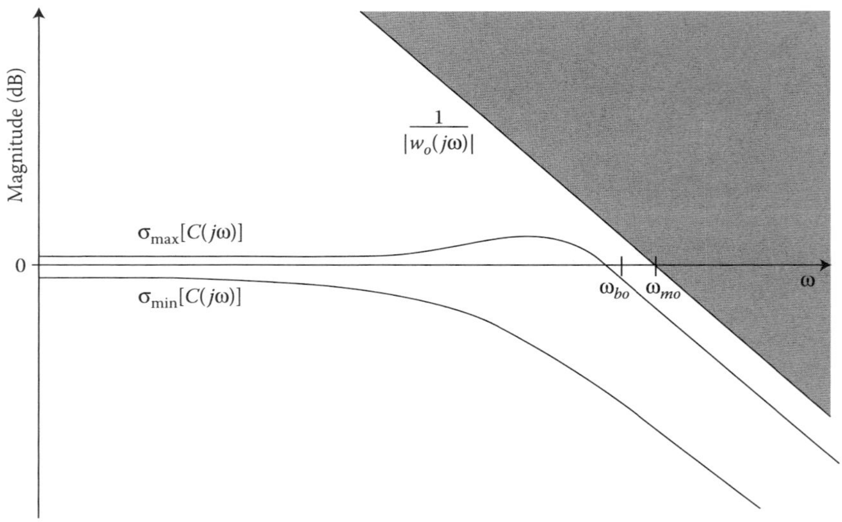

FIGURE 9.18 Interpretation of stability robustness for MIMO systems with uncertainty reflected to plant output.

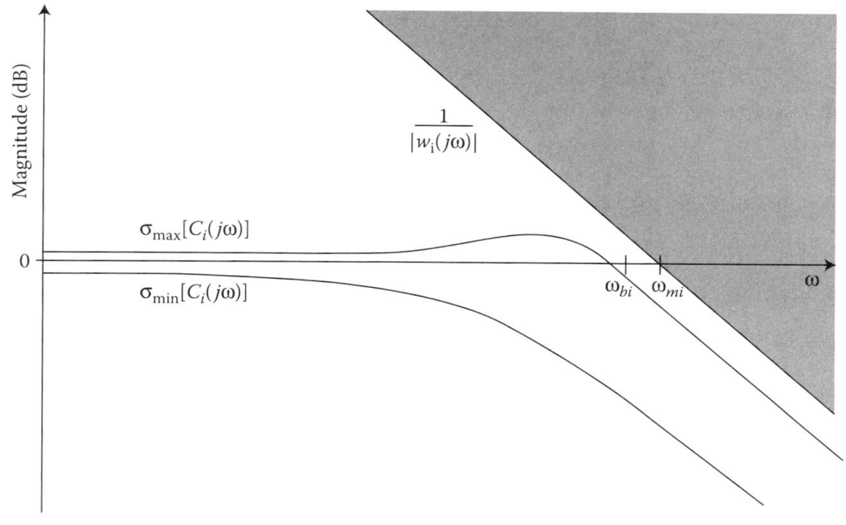

FIGURE 9.19 Interpretation of stability robustness for MIMO systems with uncertainty reflected to plant input.

becomes significant. In other words, ω_{mo} is defined as

$$|w_o(j\omega_{mo})| = 0 \text{ dB} \tag{9.60}$$

Therefore, as in the SISO case, the stability robustness condition limits the range of frequencies over which we can expect to achieve good command following and/or output disturbance rejection to below ω_{mo}. On the other hand, for the MIMO case where the modeling errors are reflected to the plant input, we do not get a similar interpretation. In this case, the stability robustness condition given in Equation 9.57 can be illustrated in terms of singular value plots, as shown in Figure 9.19. The difference lies in the fact that this condition depends on the input complementary sensitivity TFM, $C_i(s) = (I + K(s)G(s))^{-1}K(s)G(s)$, instead of $C(s)$. Under this stability robustness condition, the input bandwidth of the nominal closed-loop design, ω_{bi}, is constrained to be less than ω_{mi}, where ω_{mi} is defined as the frequency at and beyond which the modeling error reflected to the plant input becomes significant. Here, ω_{mi} is defined as

$$|w_i(j\omega_{mi})| = 0 \text{ dB} \tag{9.61}$$

which may be different than ω_{mo} since $w_i(s)$ is, in general, not the same as $w_o(s)$. Since command following and output disturbance rejection depends on the sensitivity TFM, $S(s) = (I + G(s)K(s))^{-1}$, instead of the input sensitivity TFM, $S_i(s) = (I + K(s)G(s))^{-1}$, we cannot directly ascertain the impact of the stability robustness condition in Equation 9.57 on these performance measures. However, we can obtain insight on the impact of this stability robustness condition on input disturbance rejection because the TFM from the disturbance at the plant input to the plant output is equal to $G(s)(I + K(s)G(s))^{-1}$. Therefore, if the input sensitivity TFM is small then we have good input disturbance rejection. As before, a necessary condition for $\sigma_{\max}[S_i(j\omega)]$ to be small at a certain frequency is that the singular values of $C_i(j\omega)$ are close to unity (0 dB). From Figure 9.19, this stability robustness condition clearly limits the range of frequencies at which we can expect to achieve good input disturbance rejection to below ω_{mi}. This is because we must roll-off $C_i(j\omega)$ above ω_{mi} to satisfy the stability robustness condition.

To illustrate the concept of stability robustness to multiplicative unstructured uncertainty and its impact on closed-loop performance, we present the following example.

Example 9.2: Integrator with Time Delay

Suppose that we wish to control the plant from Example 1 using a simple proportional controller. Our objective is to push the bandwidth of the nominal closed-loop system as high as possible. We use the simple integrator model

$$g(s) = \frac{1}{s} \tag{9.62}$$

for our design. In this case, the nominal loop transfer function is simply $g(s)k$ where k is the proportional gain of our controller, and the nominal closed loop transfer function is simply

$$c(s) = \frac{k}{s+k} \tag{9.63}$$

Since we require stability of the actual closed loop, we need to satisfy the stability robustness condition given in Equation 9.58 where, from Example 1, the weight $w(s)$ of the uncertainty for this model is

$$w(s) = \frac{0.21s}{0.05s+1} \tag{9.64}$$

As shown in Figure 9.20, the maximum value for k at which the stability robustness condition is satisfied is $k = 5.9$. With this value of k, the bandwidth achieved for the nominal system based on this model is 5.9 rad/s.

Now, consider the case where the time delay can vary up to 0.4 s instead of 0.2 s. In this case, we use the same model as above, but our uncertainty representation must change in order to cover this additional uncertainty. Specifically, we must choose a new weight $w(s)$. As in Example 1, we need to choose $w(s)$

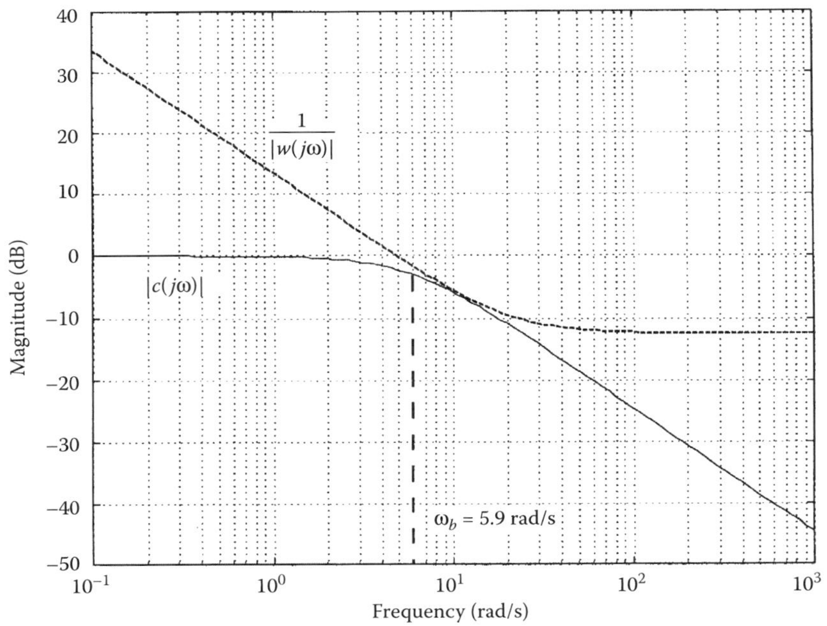

FIGURE 9.20 Stability robustness condition for 0.2 s maximum delay uncertainty.

such that

$$|w(j\omega)| \geq \max_{\tau} \left|2 \sin\left(\frac{\omega\tau}{2}\right)\right| \quad \forall\, \omega \geq 0 \tag{9.65}$$

where τ now ranges from 0 to 0.4 s. A simple $w(s)$ that will satisfy Equation 9.65 is

$$w(s) = \frac{0.41s}{0.1s + 1} \tag{9.66}$$

This is shown in Figure 9.21, where $|w(j\omega)|$ and $2\left|\sin(\frac{\omega\tau}{2})\right|$ are plotted together on a magnitude Bode plot for $\tau = 0.4$, which is now the worst case value. We note that $w(s)$ is proper and strictly stable as required.

Since we require stability of the actual closed loop, we again need to satisfy the stability robustness condition given in Equation 9.58. As shown in Figure 9.22, the maximum value for k at which the stability robustness condition is satisfied is now $k = 2.9$. With this value of k, the bandwidth achieved for the nominal system based on this model is 2.9 rad/s.

Comparing the two cases, it is clear that the uncertainty set for the second case is larger because the variation in the unknown time delay is larger. Physically, the larger time delay means larger unmodeled dynamics that we must cover in our uncertainty representation. This results in a larger magnitude for $w(s)$ at low frequencies, as shown in Figure 9.23. In particular, the frequency at which the modeling errors become significant, ω_m, is lower for the second case than the first, and therefore, further limiting the achievable bandwidth of the nominal closed loop. As a result, we are able to push the bandwidth higher for the first case. This results in better achievable performance, such as command following and output disturbance rejection, as shown by the sensitivity bode plots in Figure 9.24.

In summary, the requirement of stability robustness has a definite impact on the achievable nominal performance of the closed loop. This impact is typically a limitation on the nominal closed-loop bandwidth which, in turn, constrains the frequency range at which good command following and output disturbance rejection is achievable. As is evident in the above example, there is a trade-off between stability robustness

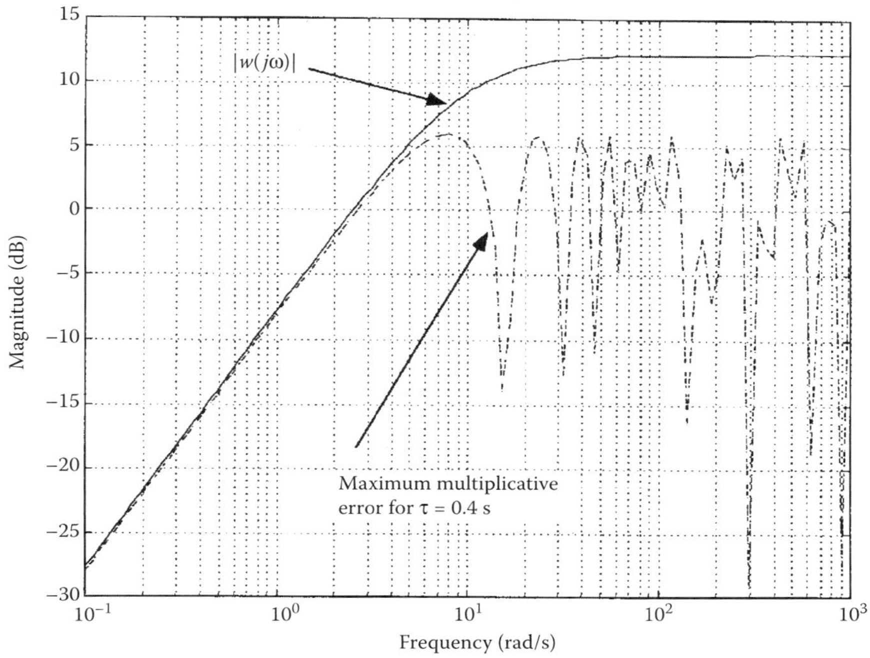

FIGURE 9.21 $w(s)$ for 0.4 s maximum delay uncertainty.

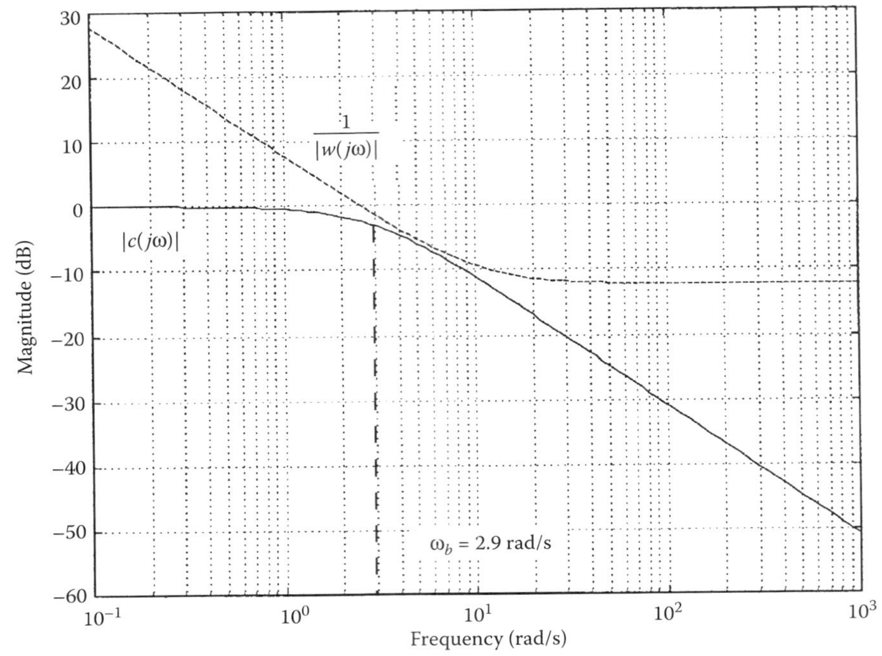

FIGURE 9.22 Stability robustness condition for 0.4 s maximum delay uncertainty.

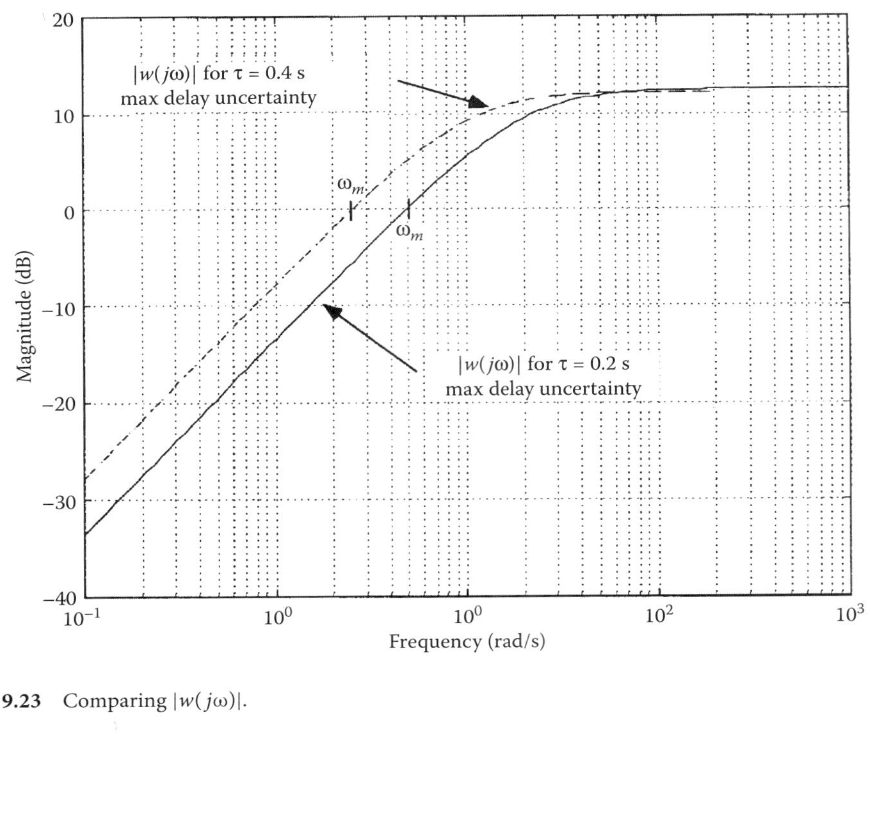

FIGURE 9.23 Comparing $|w(j\omega)|$.

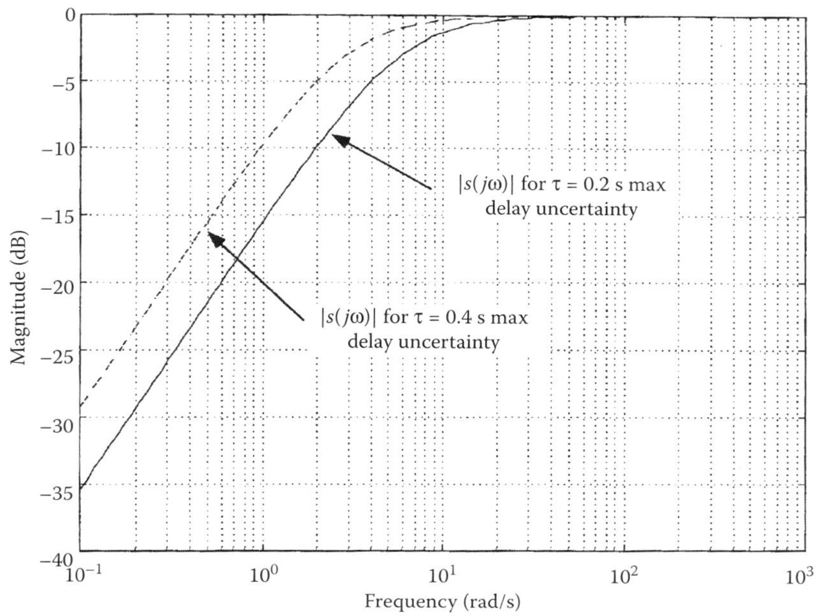

FIGURE 9.24 Comparing sensitivity magnitude plots.

and performance. That is, a controller that is designed to be robust to a larger uncertainty set will result in poorer performance due to the fact that its bandwidth will be further limited. In addition, there is a connection between this trade-off and the conservatism of our stability robustness conditions. In particular, our representation of the uncertainty is one of overbounding the set of TFMs from our modeling process with a larger set. If we choose $w(s)$ where the overbounding is not tight, then this will further limit the amount of achievable performance that we can obtain for the closed loop. Therefore, we see that both the concept of unstructured uncertainty and the way we represent this uncertainty in the stability robustness problem have an impact on achievable nominal performance. As a result, to improve performance we may have to reduce the level of uncertainty by having a better model and/or by obtaining a better representation for the uncertainty. Finally, we wish to point out that only nominal performance or the closed-loop performance of the model without the uncertainty is discussed. The actual performance of the loop will be different. However, for low frequencies where we care about command following and output disturbance rejection, the modeling errors will be typically small. Therefore, the difference between actual performance and nominal performance at these frequencies will be small.

9.5 Other Representations for Model Uncertainty

In this section, we present some alternative set membership representations for unstructured uncertainty. For each representation, we will start with the definition of the set $\mathcal{G}$. In defining $\mathcal{G}$, we assume throughout that

1. The weight is a fixed, proper, and strictly stable scalar transfer function
2. $\Delta(s)$ is a strictly stable TFM
3. No unstable or imaginary axis poles of $G(s)$ are cancelled in forming any element $\tilde{G}(s)$ of the set $\mathcal{G}$

With this definition, the appropriate stability robustness condition is obtained using the procedure outlined in Section 9.3.2. Like those derived for the multiplicative uncertainty representation, these conditions are all conservative since they are derived using the small gain theorem. The focus here will be primarily on the presentation of results rather than detailed treatments of their derivation. MIMO systems are treated as SISO systems that are simply special cases. Comments on the usefulness of these representations and interpretations of their results are given as appropriate.

9.5.1 Additive Uncertainty

An alternative representation for unstructured uncertainty is additive uncertainty. In this case, the set $\mathcal{G}$ is defined as

$$\mathcal{G} = \{\tilde{G}(s) \mid \tilde{G}(s) = G(s) + w_a(s)\Delta(s), \|\Delta(j\omega)\|_{\mathcal{H}_\infty} \leq 1\} \tag{9.67}$$

Following the procedure in Section 9.3.2, the stability robustness condition for additive uncertainty is

$$\sigma_{\max}[(I + K(j\omega)G(j\omega))^{-1}K(j\omega)] < \frac{1}{|w_a(j\omega)|} \quad \forall\, \omega \geq 0 \tag{9.68}$$

From Equation 9.67, the additive error is defined as

$$E_a(s) = w_a(s)\Delta(s) = \tilde{G}(s) - G(s) \tag{9.69}$$

which is simply the difference between the perturbed plant and the model. As a result, the additive representation is a more natural representation for differences in the internal dynamics between the actual plant and the model. In particular, it is no longer necessary to reflect these modeling errors to the plant input or output. However, we note that the resulting stability robustness condition is explicitly dependent on the controller TFM, $K(s)$, and the loop TFM, $K(s)G(s)$, instead of simply on $K(s)G(s)$. This is because, unlike multiplicative uncertainty, the uncertainty representation here does not apply equally to the loop TFM as to the model, $G(s)$. The result is that there will be added complexity in designing a $K(s)$ that will satisfy the condition because shaping the TFM, $(I + K(j\omega)G(j\omega))^{-1}K(j\omega)$, to satisfy Equation 9.68 will require shaping both $K(j\omega)$ and $K(j\omega)G(j\omega)$. Due to this complication, the multiplicative uncertainty or another form of cascaded uncertainty representation, where the representation applies equally to $G(s)$ as to $K(s)G(s)$, is often used instead in practice.

9.5.2 Division Uncertainty

Another representation for unstructured uncertainty is the division uncertainty. Like the multiplicative uncertainty representation, the division uncertainty represents the modeling error in a cascade form, and therefore, the modeling error can be reflected either to the plant output or the plant input. In the case where the modeling errors are reflected to the plant output, the set $\mathcal{G}$ is defined as

$$\mathcal{G} = \{\tilde{G}(s) \mid \tilde{G}(s) = (I + w_{do}(s)\Delta(s))^{-1}G(s), \|\Delta(j\omega)\|_{\mathcal{H}_\infty} \leq 1\} \tag{9.70}$$

Following the procedure in Section 9.3.2, the stability robustness condition for division uncertainty at the plant output is given by

$$\sigma_{\max}[(I + G(j\omega)K(j\omega))^{-1}] < \frac{1}{|w_{do}(j\omega)|} \quad \forall\, \omega \geq 0 \tag{9.71}$$

Similarly, for the case where the modeling errors are reflected to the plant input, the set $\mathcal{G}$ is defined as

$$\mathcal{G} = \{\tilde{G}(s) \mid \tilde{G}(s) = G(s)(I + w_{di}(s)\Delta(s))^{-1}, \|\Delta(j\omega)\|_{\mathcal{H}_\infty} \leq 1\} \tag{9.72}$$

and the resulting stability robustness condition is given by

$$\sigma_{\max}[(I + K(j\omega)G(j\omega))^{-1}] < \frac{1}{|w_{di}(j\omega)|} \quad \forall\, \omega \geq 0 \tag{9.73}$$

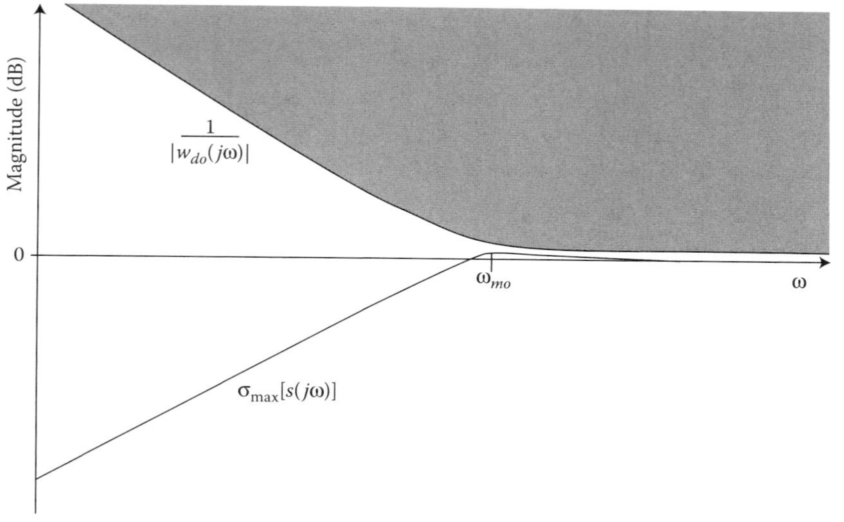

FIGURE 9.25 Interpretation of stability robustness for division uncertainty at plant output.

As shown in Figure 9.25, the stability robustness condition for division uncertainty at the plant output requires the maximum singular value plot for the sensitivity TFM of the nominal system, $\sigma_{\max}[S(j\omega)]$, to lie strictly below the plot of the inverse of $|w_{do}(j\omega)|$. Similarly, for division uncertainty at the plant input, the stability robustness condition requires that the maximum singular value plot for the input sensitivity TFM of the nominal system, $\sigma_{\max}[S_i(j\omega)]$, must lie strictly below the plot of the inverse of $|w_{di}(j\omega)|$. As shown in the figure, the inverse of the magnitude of $w_{do}(s)$ is typically large at low frequencies and approaches unity (0 dB) at high frequencies where the modeling errors become significant. Since the stability robustness condition for division uncertainty at the plant output depends on the nominal sensitivity TFM, its impact on command following and output disturbance rejection is clear from Figure 9.25. In particular, if the modeling error becomes significant at a particular frequency, ω_{mo} such that the plot of the inverse of $|w_{do}(j\omega)|$ is close to 0 dB at that frequency and beyond, the control must be designed such that $\sigma_{\max}[S(j\omega)]$ is below this barrier. That is, $\sigma_{\max}[S(j\omega)]$ is not allowed to be much greater than 0 dB for $\omega \geq \omega_{mo}$. However, we know from the Bode integral theorem [4] that if we suppress the sensitivity at high frequencies then we cannot also suppress it at low frequencies. Therefore, we again see how modeling errors can place a limitation on the range of frequencies over which we can expect to achieve good command following and output disturbance rejection. The same can be said for input disturbance rejection with respect to the stability robustness condition for division uncertainty at the plant input.

9.5.3 Representation for Parametric Uncertainty

Finally, we can also define the set $\mathcal{G}$ as

$$\mathcal{G} = \{\tilde{G}(s) \mid \tilde{G}(s) = (I + w_p(s)G(s)\Delta(s))^{-1}G(s), \|\Delta(j\omega)\|_{\mathcal{H}_\infty} \leq 1\} \tag{9.74}$$

In this case, the stability robustness condition is given by

$$\sigma_{\max}[(I + G(j\omega)K(j\omega))^{-1}G(j\omega)] < \frac{1}{|w_p(j\omega)|} \quad \forall\, \omega \geq 0 \tag{9.75}$$

As shown in the following example, we can use this representation to handle parametric uncertainty in the A matrix of the nominal model $G(s)$.

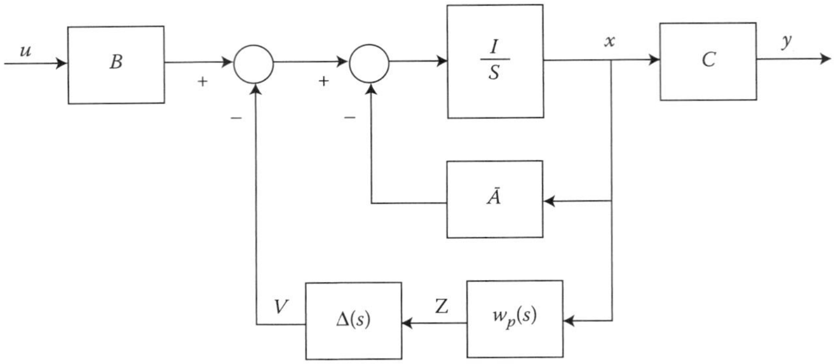

FIGURE 9.26 Representation for parametric uncertainty in the A matrix.

Example 9.3: Parametric Uncertainty in the *A* Matrix

Consider the dynamics of the actual plant in state-space form

$$\begin{aligned} \dot{x} &= Ax + Bu \\ y &= Cx \end{aligned} \tag{9.76}$$

where the parameters of the A matrix are uncertain but are known to be constant and contained in a certain interval. In particular, consider the case where the A matrix is given as

$$A = \bar{A} + \delta A \tag{9.77}$$

where the elements of $\bar{A}$ contain the midpoint values for each corresponding element in A and each element of δA is known to exist in the interval

$$-1 \leq \delta A_{ij} \leq 1 \tag{9.78}$$

We note that since

$$\|\delta A\| \leq n \tag{9.79}$$

for all possible δA satisfying Equation 9.78 where n is the dimension of A, it is clear that the set of possible δA is contained in the set $\{w_p(s)\Delta(s) | \, \|\Delta(j\omega)\|_{\mathcal{H}_\infty} \leq 1, w_p(s) = n\}$. Therefore, we can express the A matrix as

$$A = \bar{A} + w_p(s)\Delta(s) \tag{9.80}$$

for $w_p(s) = n$ and for some $\|\Delta(j\omega)\|_{\mathcal{H}_\infty} \leq 1$. In block diagram form, the state-space equations in Equation 9.76 can then be represented as in Figure 9.26, where the model $G(s)$ is taken to be $(sI - \bar{A})^{-1}$. We note that this is equivalent to the uncertainty representation given in Equation 9.74.

We note that the above representation for parametric uncertainty is, in general, very conservative. This is because the representation allows for complex perturbations of the matrix A since $\Delta(s)$ is complex. Since the parameters are real, the perturbed set of plants is much larger than the set that will actually be realized by the plant, which results in the conservatism.

Notes

The majority of the material in this article is adopted from Doyle and Stein [5], Dahleh and Diaz-Bobillo [1], Maciejowski [6], and Doyle et al. [7]. Other excellent references include Green and Limebeer [8] and Morari and Zafiriou [9].

References

1. Dahleh, M. A. and Diaz-Bobillo, I. J., *Control of Uncertain Systems: A Linear Programming Approach*, Englewood Cliffs, NJ: Prentice Hall, 1995.
2. Hildebrand, F. B., *Advanced Calculus for Applications*, 2nd Ed., Englewood Cliffs, NJ: Prentice Hall, 1976.
3. Desoer, C. A. and Vidyasagar, M., *Feedback Systems: Input–Output Properties*, New York: Academic Press, 1975.
4. Freudenburg, J. S. and Looze, D. P., *Frequency Domain Properties of Scalar and Multivariable Feedback Systems*, Berlin: Springer-Verlag, 1987.
5. Doyle, J.C. and Stein, G., Multivariable Feedback Design: Concepts for a Classical/Modern Synthesis, *IEEE Trans. Autom. Control*, 26(1), 4–16, 1981.
6. Maciejowski, J. M., *Multivariable Feedback Design*, Reading, MA: Addison-Wesley, 1989.
7. Doyle J. C., Francis, B. A., and Tannenbaum, A. R., *Feedback Control Theory*, New York: Macmillan, 1992.
8. Green, M. and Limebeer, D. J. N., *Linear Robust Control*, Englewood Cliffs, NJ: Prentice-Hall, 1995.
9. Morari, M. and Zafiriou, E., *Robust Process Control*, Englewood Cliffs, NJ: Prentice Hall, 1989.

10

Trade-Offs and Limitations in Feedback Systems

Douglas P. Looze
University of Massachusetts Amherst

James S. Freudenberg
University of Michigan

Julio H. Braslavsky
The University of Newcastle

Richard H. Middleton
National University of Ireland, Maynooth

10.1 Introduction and Motivation

It is well known that feedback may be used in a control design to obtain desirable properties that are not achievable with open-loop control. Among these are the ability to stabilize an unstable system and to reduce the effects of plant disturbances and modeling errors upon the system output. On the other hand, use of feedback control can also have undesirable consequences: feedback may destabilize a system, introduce measurement noise, amplify the effects of disturbances and modeling errors, and generate large control signals. A satisfactory control design will, if possible, achieve the potential benefits of feedback without incurring excessive costs.

Unfortunately, feedback systems possess limitations that manifest themselves as *design trade-offs* between the benefits and costs of feedback. For example, there exists a well-known trade-off between the response of a feedback system to plant disturbances and to measurement noise. This trade-off is an inherent consequence of feedback system topology; indeed, a plant disturbance cannot be attenuated without a measurement of its effect upon the system output. Other design trade-offs are a consequence of system properties such as unstable poles, nonminimum phase zeros, time delays, and bandwidth limitations.

The study of feedback design limitations and trade-offs dates back at least to the seminal work of Bode [7]. In his classic work, Bode stated the famous gain-phase relation and analyzed its implications for the classical loop-shaping problem. He also derived the Bode sensitivity integral. The importance of the sensitivity integral to feedback control design was emphasized by Horowitz [36]. Connections between the classical loop-shaping problem, including the gain-phase relation, and modern control techniques were developed by Doyle and Stein [22]. Further results pertaining to design limitations were derived by Freudenberg and Looze [28,31], who studied inherent design limitations present when the system to be controlled has unstable poles, nonminimum phase zeros, time delays, and/or bandwidth constraints. A result dual to the Bode sensitivity integral, and applicable to the complementary sensitivity function, was obtained by Middleton and Goodwin [43,45]. In all these works, emphasis is placed upon frequency response properties and, in particular, the insight into feedback design that may be obtained from a Bode plot.

In this chapter, we describe several trade-offs that are present in feedback system design due to the structure of a feedback loop, bandwidth limitations, and plant properties such as unstable poles, nonminimum phase zeros, and time delays. We focus primarily on linear time-invariant single input single output systems (SISOs) and later give a brief description and references to a number of extensions of these results. In Section 10.2, we show how the closed-loop transfer functions of a feedback system may be used to describe signal response, differential sensitivity, and stability robustness properties of the system. In Section 10.3, we describe design trade-offs imposed by the topology of a feedback loop and the relation between feedback properties and open-loop gain and phase. Next, in Section 10.4, we describe design limitations imposed by the Bode gain-phase relation, Bode sensitivity integral, and the dual complementary sensitivity integral. These trade-offs are present due to the fact that the closed-loop system is stable, linear, and time invariant. When the plant has nonminimum phase zeros, unstable poles, and/or time delays, additional design trade-offs are present. As discussed in Section 10.5, these may be described using the Poisson integral. An alternate view of some of the performance limitations using time responses is described in Section 10.6. In Section 10.7, we give a brief summary and references to various extensions of the theory of feedback performance limitations, including discrete-time systems, systems with feedback over a communication channel, limits on optimal cheap control and H_2 performance, multivariable systems, and nonlinear and time-varying systems. We conclude the section with a brief discussion of cases in which performance limitations may be alleviated in tracking problems. A general summary and suggestions for further reading are presented in Section 10.8.

10.2 Quantification of Performance for Feedback Systems

10.2.1 Overview

Throughout this chapter, we consider the single-degree-of-freedom unity feedback system depicted in Figure 10.1. In this figure, $P(s)$ and $C(s)$ denote the transfer functions of the plant and the compensator,

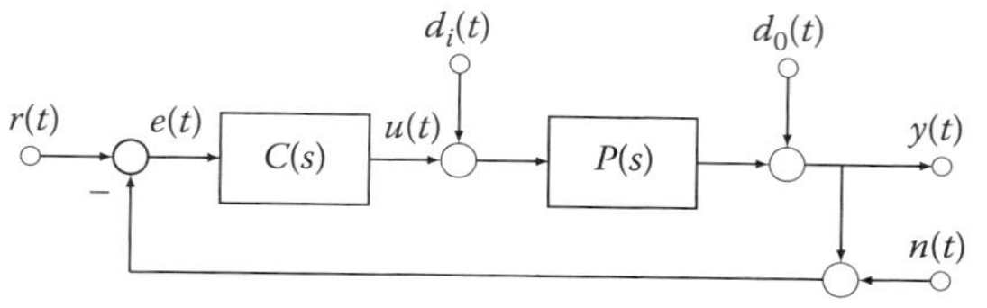

FIGURE 10.1 Linear time-invariant feedback system.

respectively. The system output is denoted $y(t)$, the control signal by $u(t)$, and the command input by $r(t)$. An important goal in many feedback control problems is that of forcing $y(t)$ to track $r(t)$ despite the presence of input and output disturbances $d_i(t)$ and $d_o(t)$. To do so, a measurement of the system output is compared to the command input to form an error signal $e(t)$, and used as an input to the controller. The effects of measurement noise are denoted $n(t)$.

Other feedback architectures are possible, including a two-degree-of-freedom architecture, in which the command input and measured output are processed separately. For example, the command input may be shaped by a precompensator before the error signal is formed. Similarly, any disturbances that are measurable may also be processed separately. Throughout this chapter, we discuss the feedback architecture in Figure 10.1, and assume that any precompensation is absorbed into the definition of the command input $r(t)$.

The ability of the plant output to follow the command input is referred to as tracking. In order that tracking be feasible, the closed-loop system must, at a minimum, be stable. As Figure 10.1 illustrates, exogenous signals can act at various points in the feedback loop to affect the response of the system. Signals that enter the loop between the control action and the output to be controlled are referred to as disturbance signals. The ability of the control system to eliminate the effects of the disturbances is referred to as disturbance rejection. Signals that enter the loop between the system output and the comparison of the measurement with the command input are referred to as measurement noise signals. Measurement noise only affects the output through the action of the compensator.

In addition to the uncertainty caused by the disturbance signals and measurement noise, the behavior of the plant as described by the model will differ from that of the true plant. A feedback system that is stable for the plant model is called nominally stable. A feedback loop that maintains stability when subject to variations in the plant model is robustly stable. Nominal performance is achieved when tracking and disturbance rejection objectives are met for the plant model. Robust performance is achieved if the feedback loop maintains performance (tracking or disturbance rejection) when subject to variations in the plant model.

To proceed with our analysis, we define closed-loop transfer functions from the exogenous input signals in Figure 10.1 to the system output, measured error, and the control signal, and shall quantify performance in terms of the peak in a Bode gain plot of each transfer function. We shall also see that these closed-loop transfer functions describe the sensitivity and robustness properties of a feedback system.

10.2.2 Nominal Signal Response

Consider the linear time-invariant feedback system shown in Figure 10.1. Three transfer functions are essential in studying properties of this system. Define the open-loop transfer function

$$L(s) = P(s)C(s), \tag{10.1}$$

the sensitivity function

$$S(s) = \frac{1}{1 + L(s)}, \tag{10.2}$$

and the complementary sensitivity function

$$T(s) = \frac{L(s)}{1 + L(s)}. \tag{10.3}$$

The response of the system of Figure 10.1 to the exogenous inputs (i.e., the command input, the input disturbance, the output disturbance, and the sensor noise) is governed by the closed-loop transfer functions (Equations 10.2 and 10.3). The output of the system is given by

$$Y(s) = S(s)\left(D_o(s) + P(s)D_i(s)\right) + T(s)\left(R(s) - N(s)\right). \tag{10.4}$$

The feedback loop error between the measured output and the command input is given by

$$E(s) = S(s)\left(R(s) - D_o(s) - P(s)D_i(s) - N(s)\right). \tag{10.5}$$

The performance error between the output and the command input is given by

$$E_p(s) = R(s) - Y(s) = S(s)\left(R(s) - D_o(s) - P(s)D_i(s)\right) - T(s)N(s). \tag{10.6}$$

The control input generated by the compensator is given by

$$U(s) = -T(s)D_i(s) + S(s)C(s)\left(R(s) - N(s) - D_o(s)\right). \tag{10.7}$$

Because the principal objective for control system performance is to have the plant output track the command input without using too much control energy, it is desirable that the performance error (Equation 10.6) and the plant input (Equation 10.7) be small. It is apparent from Equations 10.6 and 10.7 that each of these signals can be made small by making the sensitivity and complementary sensitivity transfer functions small. In particular, the performance error (Equation 10.6) can be made small at any frequency $s = j\omega$ by making the sensitivity function small relative to the command response and disturbances, and by making the complementary sensitivity small relative to the measurement noise. The plant input can also be kept within a desired range by bounding the magnitude of the complementary sensitivity function relative to the external signals. The presence of the plant transfer function in the denominator of the second term in Equation 10.7 implies that the magnitude of the control signal will increase as the plant magnitude decreases, unless there is a corresponding decrease in the magnitudes of either the complementary sensitivity function or the external command, measurement noise, and output disturbance signals. The presence of high-frequency measurement noise, or disturbances together with a bound on the desired control magnitude thus imposes a bandwidth constraint on the closed-loop system.

10.2.3 Differential Sensitivity

Because the true plant behavior will be somewhat different from the behavior described by the plant model $P(s)$, it is important that performance be maintained for variations in the plant model that correspond to possible differences with reality. Although the precise characterization of robust performance is beyond the scope of this chapter (see [23]), incremental changes in the system response due to model variations can be characterized by the sensitivity function. Assume that the true plant model is given by

$$P'(s) = P(s) + \Delta P(s), \tag{10.8}$$

where $\Delta P(s)$ represents the difference between the true and nominal values of the plant transfer function. Assuming that the disturbance and measurement noise signals are zero, the nominal closed-loop response (i.e., the response to the command reference for the plant model $P(s)$) is

$$Y(s) = T(s)R(s). \tag{10.9}$$

The response for the feedback system using the true plant is

$$Y'(s) = T'(s)R(s), \tag{10.10}$$

where

$$Y'(s) = Y(s) + \Delta Y(s), \tag{10.11}$$

$$T'(s) = T(s) + \Delta T(s). \tag{10.12}$$

Thus the variation in the command response is proportional to the variation in the true complementary sensitivity function.

Let $S'(s)$ denote the sensitivity function for the feedback system using the true plant transfer function. Then, it can be shown that [20]

$$\frac{\Delta T(s)}{T(s)} = S'(s)\frac{\Delta P(s)}{P(s)}. \tag{10.13}$$

Equation 10.13 shows that the true sensitivity function determines the relative deviation in the command response due to a relative deviation in the plant transfer function. If the true sensitivity function is small, the variation in command response will be small relative to the variation in the plant model.

Although Equation 10.13 provides insight into how the command response varies, it is not useful as a practical analysis and design tool because the relationship is expressed in terms of the unknown true plant transfer function. However, as the plant variation $\Delta P(s)$ becomes small, the true sensitivity function $S'(s)$ approaches the nominal sensitivity function $S(s)$. Thus, for $\Delta P(s) = dP(s)$, Equation 10.13 becomes

$$\frac{dT(s)}{T(s)} = S(s)\frac{dP(s)}{P(s)}. \tag{10.14}$$

Equation 10.14 shows that the sensitivity function $S(s)$ governs the sensitivity of the system output to small variations in the plant transfer function. Hence, the variation in the command response will be small if the sensitivity function is small relative to the plant variation.

10.2.4 Robust Stability

The feedback system in Figure 10.1 will be internally stable if each transfer function $S(s)$, $T(s)$, $S(s)P(s)$, and $C(s)S(s)$ is proper and have no poles in the closed right half-plane (CRHP) [34]. If there are no CRHP pole–zero cancellations between the plant and the controller, then the feedback system will be stable if the sensitivity function is proper and has no CRHP poles. A number of techniques, such as the Routh–Hurwitz and Nyquist criteria [21,27], are available for evaluating whether the system is stable for the nominal plant model. A more challenging problem is to determine whether the feedback system will be robustly stable for the true but unknown plant.

The sensitivity and complementary sensitivity functions each characterize stability robustness of the system against particular classes of plant variations (e.g., see [70], Table 9.1). One of the most significant types of plant modeling error is the high-frequency uncertainty that is present in any model of a physical system. All finite-order transfer function models of physical systems deteriorate at high-frequencies due to neglected dynamics (such as actuator and measurement lags, and flexible modes), the effects of approximating distributed systems with lumped models, and time delays. The result of these neglected physical phenomena is that the uncertainty in the gain of the true plant increases (eventually differing from the assumed plant model by more than 100%) and the phase of the true plant becomes completely uncorrelated with that of the model (i.e., there is $\pm 180°$ uncertainty in the phase).

This type of uncertainty can be represented as a relative deviation from the nominal plant. The true plant transfer function is represented by a multiplicative error model

$$P'(s) = P(s)(1 + \Delta(s)). \tag{10.15}$$

Given a true plant $P'(s)$, the multiplicative error $\Delta(s)$ is the relative plant error

$$\Delta(s) = \frac{P'(s) - P(s)}{P(s)}. \tag{10.16}$$

It will be assumed that the multiplicative error is stable. From Equation 10.16, it is apparent that this assumption will be satisfied if the number and location of any unstable poles of the true plant are the same as those of the plant model.

The characteristics of increasing gain and phase uncertainty can be represented by assuming that the multiplicative error is unknown except for an increasing upper bound on its magnitude. That is, it will be assumed that $\Delta(s)$ is any stable transfer function that satisfies

$$|\Delta(j\omega)| < M_\Delta(\omega), \tag{10.17}$$

where (typically) $M_\Delta(\omega) \to \infty$ as $\omega \to \infty$.

If the only information available about the uncertainty is Equation 10.17, then $\Delta(s)$ is referred to as unstructured uncertainty. In that case, a necessary and sufficient condition for the feedback system to be stable for all plants described by Equations 10.15 through 10.17 is that the system be stable when $\Delta(s) = 0$ and that the complementary sensitivity function satisfy the bound [22]

$$|T(j\omega)| < \frac{1}{M_\Delta(\omega)}, \quad \forall\omega. \tag{10.18}$$

Equation 10.18 demonstrates that the presence of high-frequency uncertainty forces the complementary sensitivity function to become small to insure that the system is robustly stable. This in turn implies that an upper limit on bandwidth is imposed on the closed-loop system.

10.2.5 Summary of Feedback System Performance Specification

In the previous section we saw that the sensitivity and complementary sensitivity functions each characterize important properties of a feedback system. This fact motivates stating design specifications directly as frequency-dependent bounds upon the magnitudes of these functions. Hence, we typically require that

$$|S(j\omega)| \le M_S(\omega), \quad \forall\omega, \tag{10.19}$$

and

$$|T(j\omega)| \le M_T(\omega), \quad \forall\omega. \tag{10.20}$$

The bounds $M_S(\omega)$ and $M_T(\omega)$ will generally depend on the size of the disturbance and noise signals, the level of plant uncertainty, and the extent to which the effect of these phenomena upon the system output must be diminished (see Equations 10.6, 10.7, 10.14, and 10.18).

It is interesting to note that the two design specifications represent two different aspects of control system design. The bound (Equation 10.19) on the sensitivity function typically represents the potential benefits of feedback, such as disturbance rejection, tracking and robust performance. The bound (Equation 10.20) on the complementary sensitivity function represents desired limits on the cost of feedback: amplification of measurement noise, increased control signal requirements, and possible introduction of instability into the feedback loop.

10.3 Design Trade-Offs at a Single Frequency

10.3.1 Introduction

Often the requirements imposed by various design objectives are mutually incompatible. It is therefore important to understand when a given design specification is achievable and when it must be relaxed by making trade-offs between conflicting design goals. The objective of this section is to explore the limitations that are imposed by the structure of the feedback loop in Figure 10.1.

The structural trade-offs presented in this section express the limitation that there is only one degree of design freedom in the feedback loop. This degree-of-freedom can be exercised by specifying any one of the associated transfer functions. Because the sensitivity and complementary sensitivity functions are both uniquely determined by the loop transfer function, these transfer functions cannot be specified independently. Hence, the properties of the feedback loop are completely determined once any one of $L(s), S(s)$, or $T(s)$ are specified, and these properties can be analyzed in terms of the chosen transfer function. In particular, the performance bounds (Equations 10.19 and 10.20) are not independent, and may lead to conflicts.

This section presents explicit relationships between the sensitivity function, the complementary sensitivity function, and the loop transfer function of the feedback system of Figure 10.1. These relationships can be used to explore whether design specifications in the form of (Equation 10.19 and 10.20) are consistent, and if not, how the specifications might be modified to be achievable.

10.3.2 Relationship between Closed-Loop Transfer Functions and Design Specifications

One important design trade-off may be quantified by noting that the sensitivity and complementary sensitivity functions satisfy the identity

$$S(j\omega) + T(j\omega) = 1. \tag{10.21}$$

Equation 10.21 is a structural identity in the sense that it is a consequence of the topology of the feedback loop.

It follows from Equation 10.21 that $|S(j\omega)|$ and $|T(j\omega)|$ both cannot be very small at the same frequency. Hence, at each frequency, there exists a trade-off between those feedback properties such as sensitivity reduction and disturbance response that are quantified by $|S(j\omega)|$ and those properties such as measurement noise response and robustness to high-frequency uncertainty that are quantified by $|T(j\omega)|$.

In applications it often happens that levels of uncertainty and sensor noise become significant at high frequencies, while disturbance rejection and sensitivity reduction are generally desired over a lower frequency range. Hence, the trade-off imposed by Equation 10.21 is generally performed by requiring $M_S(\omega)$ to be small at low frequencies, $M_T(\omega)$ to be small at high frequencies, and neither $M_S(\omega)$ nor $M_T(\omega)$ to be excessively large at any frequency.

This situation is illustrated in Figure 10.2. The bound $M_S(\omega)$ is small at low frequencies, representing the frequencies over which disturbance rejection and tracking are important. It increases with increasing frequency until it becomes greater than one at high frequencies. Conversely, the bound $M_T(\omega)$ begins somewhat greater than one at low frequencies, and rolls off (decreases in magnitude) at higher frequencies where the plant model error and measurement noise become significant. The sensitivity and complementary sensitivity functions shown in Figure 10.2 satisfy the design objectives (Equations 10.19 and 10.20).

Equation 10.21 can be used to derive conditions that the design objective functions $M_S(\omega)$ and $M_T(\omega)$ must satisfy. Taking the absolute value of the right-hand side of Equation 10.21, using the triangle inequality and applying the bounds on the sensitivity and complementary sensitivity functions (Equations 10.19 and 10.20), the following inequality is obtained:

$$M_S(\omega) + M_T(\omega) \geq 1. \tag{10.22}$$

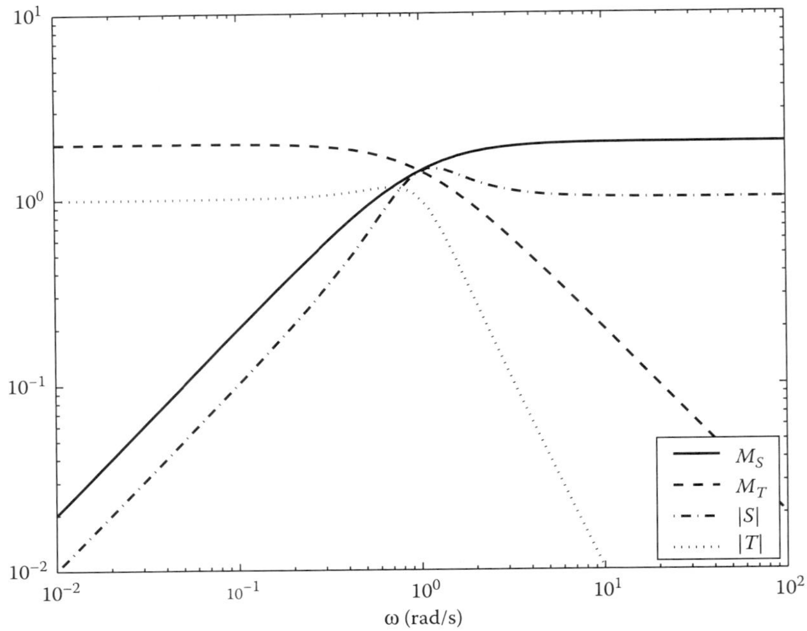

FIGURE 10.2 Typical sensitivity function, complementary sensitivity function, and specification bounds.

Equation 10.22 reinforces the observations made previously: the design objectives cannot be specified to require both the sensitivity function and the complementary sensitivity function to be small at the same frequency.

The structural identity can also be used to explore limits on more detailed specifications of control system objectives. For example, suppose that at a given frequency ω the dominant objective that specifies the weighting on the complementary sensitivity function is that the control signal should remain bounded when subject to an output disturbance which is unknown but bounded:

$$|U(j\omega)| < M_u(\omega), \quad \forall\, |D_o(j\omega)| \leq M_{d_o}(\omega). \tag{10.23}$$

Then, combining Equations 10.7 and 10.23, the design specification on the complementary sensitivity Equation 10.20 becomes

$$|T(j\omega)| \leq |P(j\omega)|\frac{M_u(\omega)}{M_{d_o}(\omega)} = M_T(\omega). \tag{10.24}$$

Substituting this value for the design specification bound on the complementary sensitivity into the inequality (Equation 10.22) yields

$$M_S(\omega) + |P(j\omega)|\frac{M_u(\omega)}{M_{d_o}(\omega)} \geq 1. \tag{10.25}$$

Inequality (Equation 10.25) requires the design specification on the sensitivity function to increase as the plant magnitude decreases relative to the available control authority for a given disturbance. Thus, the desire to reject disturbances (reflected by smaller $M_S(\omega)$) is limited. The trade-off between disturbance rejection and control amplitude is quantified by Equation 10.25.

10.3.3 Relation between Open and Closed-Loop Specifications

Because specifying one of the transfer functions $S(s)$, $T(s)$, or $L(s)$ completely determines the others, any one of them can be used as a basis for analysis and design. Classical "loop-shaping" design methods proceeded by directly manipulating the loop transfer function $L(s)$ (using, e.g., lead and lag filters) to alter the feedback properties of the system. As is well known, these methods are quite effective in coping with the type of design problems for which they were developed. One reason for the success of these methods is that, for a scalar system, open-loop gain and phase can be readily related to feedback properties. Indeed, the following relations are well known (e.g., [22,36]) and can readily be deduced from Equations 10.2 and 10.3:

$$|L(j\omega)| \gg 1 \quad \Leftrightarrow \quad |S(j\omega)| \ll 1 \text{ and } T(j\omega) \approx 1 \tag{10.26}$$

and

$$|L(j\omega)| \ll 1 \quad \Leftrightarrow \quad S(j\omega) \approx 1 \text{ and } |T(j\omega)| \ll 1. \tag{10.27}$$

At frequencies for which open-loop gain is approximately unity, feedback properties depend critically upon the value of open-loop phase:

$$|L(j\omega)| \approx 1 \text{ and } \angle L(j\omega) \approx \pm 180^\circ \quad \Leftrightarrow \quad |S(j\omega)| \gg 1 \text{ and } |T(j\omega)| \gg 1. \tag{10.28}$$

These approximations (Equations 10.26 through 10.28) yield the following rules of thumb useful in design. First, large loop gain yields small sensitivity and good disturbance rejection properties, although noise appears directly in the system output. Second, small loop gain is required for small noise response and for robustness against large multiplicative uncertainty. Finally, at frequencies near gain crossover ($|L(j\omega)| \approx 1$), the phase of the system must remain bounded sufficiently far away from $\pm 180^\circ$ to provide an adequate stability margin and to prevent amplifying disturbances and noise.

It is also possible to relate open-loop gain to the magnitude of the plant input. From Equations 10.26 and 10.2, it follows that

$$|L(j\omega)| \gg 1 \quad \Leftrightarrow \quad |S(j\omega)C(j\omega)| \approx |P^{-1}(j\omega)|. \tag{10.29}$$

Hence, requiring loop gain to be large at frequencies for which the plant gain is small may lead to unacceptably large response of the plant input to noise and disturbances (see also Equation 10.25).

Recall the discussion of the requirements imposed on the magnitudes of the sensitivity and complementary sensitivity by the performance objectives (Equations 10.19 and 10.20) (see Figure 10.2). From that discussion and the approximations (Equations 10.26 through 10.28), it follows (*cf.*, [22]) that corresponding specifications upon open-loop gain and phase might appear as in Figure 10.3. These specifications reflect the fact that loop gains must be large at low frequencies for disturbance rejection and sensitivity reduction and must be small at high frequencies to provide stability robustness. At intermediate frequencies the phase of the system must remain bounded away from $\pm 180^\circ$ to prevent excessively large values of $|S(j\omega)|$ and $|T(j\omega)|$. To obtain the benefits of feedback over as large a frequency range as possible, it is also desirable that ω_L be close to ω_H. Of course, gain and phase must also satisfy the encirclement condition dictated by the Nyquist stability criterion.

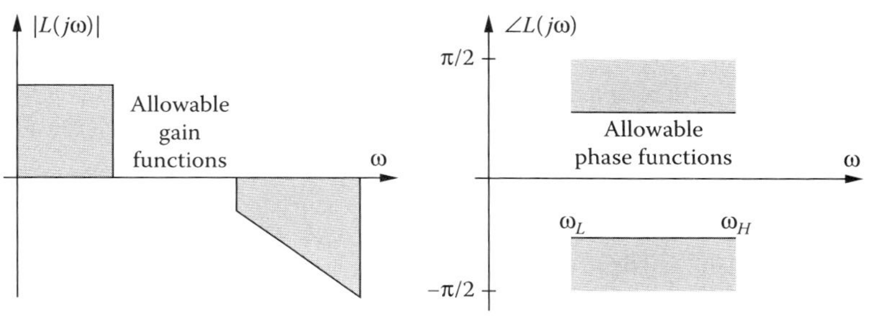

FIGURE 10.3 Gain and phase specifications.

10.4 Constraints Imposed by Stability

10.4.1 Introduction

Implicit in our construction of design specifications such as those illustrated by Figures 10.2 and 10.3 is the assumption that the transfer functions can be prescribed independently in different frequency regions or that open-loop gain and phase can be independently manipulated. However, the requirement that the closed-loop system be stable imposes additional constraints on the transfer functions that relate the values of the functions in different frequency regions. These constraints will be referred to as analytic constraints because they result from the requirement that the transfer functions have certain analytic properties.

As has been noted in Section 10.3, any of the transfer functions $S(s)$, $T(s)$, or $L(s)$ can be used to characterize the performance and properties of the feedback system. This section will present the constraints imposed by stability for each of these transfer functions.

10.4.2 Bode Gain-Phase Relation and Interpretations

We have shown in the previous section how classical control approaches view design specifications in terms of limits on the gain and the phase of the open loop transfer function. However, the gain and phase are not mutually independent: the value of one is generally determined once the other is specified. There are many ways to state this relationship precisely; the one most useful for our purposes was derived by Bode [7]. This relation has been used by many authors (*cf.* [22,36]) to analyze the implications that the gain-phase relation has upon feedback design.

Theorem 10.1: Bode Gain-Phase Relation

*Assume that $L(s)$ is a rational function with real coefficients and with no poles or zeros in the CRHP. Then, at each frequency $s = j\omega_0$, the phase of $L(s)$ must satisfy the integral relation**

$$\angle L(j\omega_0) - \angle L(0) = \frac{1}{\pi}\int_{-\infty}^{\infty} \frac{d\log|L|}{d\nu} \log\coth\frac{|\nu|}{2}\,d\nu, \tag{10.30}$$

where

$$\nu = \log\left(\frac{\omega}{\omega_0}\right). \tag{10.31}$$

Theorem 10.1 states conditions (i.e., $L(s)$ rational, stable, and minimum phase) under which knowledge of open-loop gain along the $j\omega$-axis suffices to determine open-loop phase to within a factor of $\pm\pi$. Constraints analogous to Equation 10.30 hold for unstable or nonminimum phase systems. Hence, gain and phase cannot be manipulated independently in design.

Equation 10.30 shows that the phase of a transfer function is related to the slope (on a log–log scale) of the magnitude of the transfer function. The presence of the weighting function $\log\coth(|\nu|/2) = \log|(\omega+\omega_0)/(\omega-\omega_0)|$ shows that the dependence of $\angle L(j\omega_0)$ upon the rate of gain decrease at frequency ω diminishes rapidly as the distance between ω and ω_0 increases (see Figure 10.4). Hence this integral supports a rule of thumb stating that a $20N$ dB/decade rate of gain decrease in the vicinity of frequency ω_0 implies that $\angle L(j\omega_0) \approx -90N°$. In many practical situations, the open loop transfer function is sufficiently well behaved that the value of phase at a frequency is largely determined by that of the gain over a decade-wide interval centered at the frequency of interest [7].

The Bode gain-phase relation may be used to assess whether a design specification of the type shown in Figure 10.3 is achievable. Since a $20N$ dB/decade rate of gain decrease in the vicinity of crossover implies

* Throughout this article, the notation $\log(\cdot)$ will be used to denote the natural logarithm.

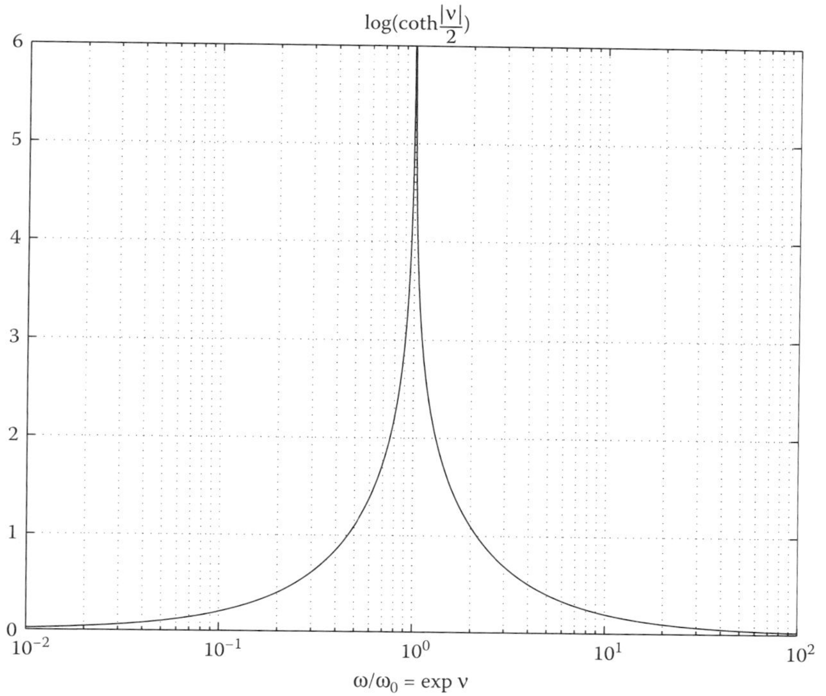

FIGURE 10.4 Weighting function in gain-phase integral.

that phase at crossover is roughly $-90N°$, it follows that the rate of gain decrease cannot be much greater than 20 dB/decade if the Nyquist stability criterion is to be satisfied and if an acceptable phase margin is to be maintained. One implication of this fact is that the frequency ω_L in Figure 10.3 cannot be too close to the frequency ω_H. Hence, the frequency range over which loop gain can be large to obtain sensitivity reduction is limited by the need to ensure stability robustness against uncertainty at higher frequencies, and to maintain reasonable feedback properties near crossover. As discussed in [22,36], relaxing the assumption that $L(s)$ has no right-half-plane poles or zeros does not lessen the severity of this trade-off. Indeed, the trade-off only becomes more difficult to accomplish.

10.4.3 The Bode Sensitivity Integral

The purpose of this section is to present and discuss the constraint imposed by stability on the sensitivity function. This constraint was first developed in the context of feedback systems in [7]. This integral quantifies a trade-off between sensitivity reduction and sensitivity increase that must be performed whenever the open-loop transfer function has at least two more poles than zeros.

The magnitude of the sensitivity function of a scalar feedback system can be obtained easily using a Nyquist plot of $L(j\omega)$. Indeed, since $S(j\omega) = 1/(1 + L(j\omega))$, the magnitude of the sensitivity function is equal to the reciprocal of the distance from the Nyquist plot to the critical point. In particular, sensitivity is less than one at frequencies for which $L(j\omega)$ is outside the unit circle centered at the critical point. Sensitivity is greater than one at frequencies for which $L(j\omega)$ is inside this unit circle.

To motivate existence of the integral constraint, consider the open-loop transfer function $L(s) = 2/(s + 1)^2$. As shown in Figure 10.5, there exists a frequency range over which the Nyquist plot of $L(j\omega)$ penetrates the unit circle centered at $-1 + j0$ and sensitivity is thus greater than one. In practice, the open-loop transfer function will generally have at least two more poles than zeros [36]. If $L(s)$ is stable, then,

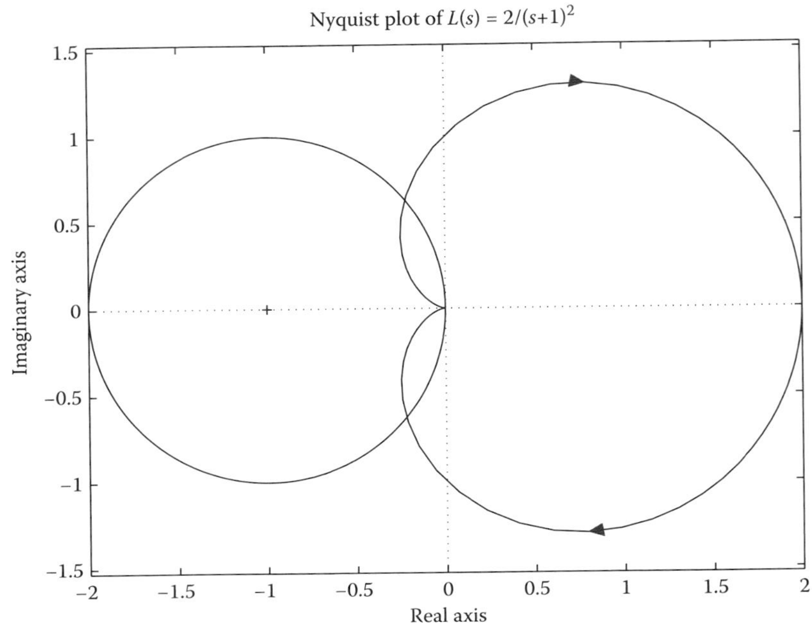

FIGURE 10.5 Effect of a two-pole rolloff upon the Nyquist plot of $L(s) = 2/(s+1)^2$.

using the gain-phase relation (Equation 10.30), it is straightforward to show that $L(j\omega)$ will asymptotically have a phase lag of at least $-180°$. Hence there will always exist a frequency range over which sensitivity is greater than one. This behavior may be quantified using a classical theorem due to Bode [7], which was extended in [31] to allow unstable poles in the open-loop transfer function. An excellent discussion of the implications that the Bode sensitivity integral has for open-loop unstable systems can be found in [62].

Theorem 10.2: Bode Sensitivity Integral

Suppose that the open-loop transfer function $L(s)$ is rational and has right-half-plane poles $\{p_i : i = 1, \ldots, N_p\}$, with multiple poles included according to their multiplicity. If $L(s)$ has at least two more poles than zeros, and if the associated feedback system is stable, then the sensitivity function must satisfy

$$\int_0^\infty \log|S(j\omega)|d\omega = \pi \sum_{i=1}^{N_p} \mathrm{Re}(p_i). \tag{10.32}$$

It is possible to extend Theorem 10.2 to systems for which $L(s)$ has relative degree one and a possible time delay. For details, see [56, Theorem 3.1.4].

Theorem 10.2 shows that a trade-off exists between sensitivity properties in different frequency ranges. Indeed, for stable open-loop systems, the area of sensitivity reduction must equal the area of sensitivity increase on a plot of the logarithm of the sensitivity versus linear frequency (see Figure 10.6). In this respect, the benefits and costs of feedback are balanced exactly.

The extension of Bode's theorem to open-loop unstable systems shows that the area of sensitivity increase exceeds that of sensitivity reduction by an amount proportional to the distance from the unstable poles to the left half-plane. A little reflection reveals that this additional sensitivity increase is plausible for

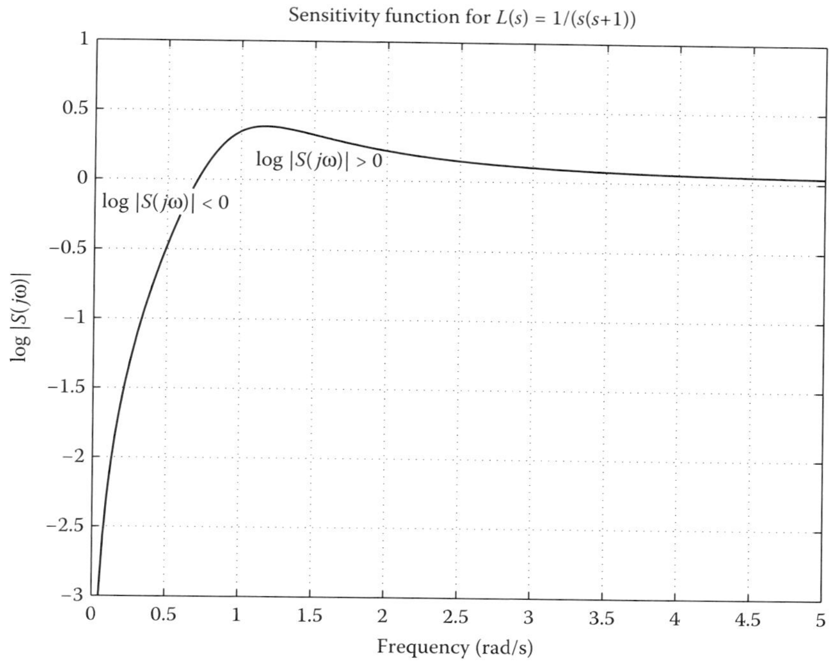

FIGURE 10.6 Areas of sensitivity reduction ($\log|S(j\omega)| < 0$) and sensitivity increase ($\log|S(j\omega)| > 0$) for $L(s) = 1/(s(s+1))$.

the following reason. When the system is open-loop unstable, then it is obviously necessary to use feedback to achieve closed-loop stability, as well as to obtain sensitivity reduction. One might expect that this additional benefit of feedback would be accompanied by a certain cost, and the integral (Equation 10.32) substantiates that hypothesis.

By itself, the trade-off quantified by Equation 10.32 does not impose a meaningful design limitation. It is true that requiring a large area of sensitivity reduction over a low-frequency interval implies that an equally large area of sensitivity increase must be present at higher frequencies. However, it is possible to achieve an arbitrary large area of sensitivity increase by requiring $|S(j\omega)| = 1 + \delta$, $\forall\omega \in (\omega_1, \omega_2)$, where δ can be chosen arbitrarily small and the interval (ω_1, ω_2) is adjusted to be sufficiently large.

The analysis in the preceding paragraph ignores the effect of limitations upon system bandwidth that are always present in a practical design. For example, it is almost always necessary to decrease open-loop gain at high frequencies to maintain stability robustness against large modeling errors due to unmodeled dynamics. Small open-loop gain is also required to prevent sensor noise from appearing at the system output. Finally, requiring open-loop gain to be large at a frequency for which plant gain is small may lead to unacceptably large response of the plant input to noise and disturbances. Hence the natural bandwidth of the plant also imposes a limitation upon open-loop bandwidth.

One or more of the bandwidth constraints just cited is usually present in any practical design. It is reasonable, therefore, to assume that open-loop gain must satisfy a frequency-dependent bound of the form

$$|L(j\omega)| \leq \epsilon \left(\frac{\omega_c}{\omega}\right)^{1+k}, \quad \forall\omega \geq \omega_c, \tag{10.33}$$

where $\epsilon < 1/2$ and $k > 0$. This bound imposes a constraint upon the rate at which loop gain rolls off, as well as the frequency at which rolloff commences and the level of gain at that frequency.

When a bandwidth constraint such as Equation 10.33 is imposed, it is obviously not possible to require the sensitivity function to exceed one over an arbitrarily large frequency interval. When Equation 10.33 is satisfied, there is an upper bound on the area of sensitivity increase which can be present at frequencies greater than ω_c. The corresponding limitation imposed by the sensitivity integral (Equation 10.32) and the rolloff constraint (Equation 10.33) is expressed by the following result [28].

Corollary 10.1:

Suppose, in addition to the assumptions of Theorem 10.2, that $L(s)$ satisfies the bandwidth constraint given by Equation 10.33. Then the tail of the sensitivity integral must satisfy

$$\left| \int_{\omega_c}^{\infty} \log |S(j\omega)| d\omega \right| \leq \frac{3\epsilon\omega_c}{2k}. \tag{10.34}$$

The bound in Equation 10.34 implies that the sensitivity trade-off imposed by the integral constraint in Equation 10.32 must be accomplished primarily over a finite frequency interval. As a consequence, the amount by which $|S(j\omega)|$ must exceed one cannot be arbitrarily small. Suppose that the sensitivity function is required to satisfy the upper bound

$$|S(j\omega)| \leq \alpha < 1, \quad \forall \omega \leq \omega_\ell < \omega_c. \tag{10.35}$$

If the bandwidth constraint (Equation 10.33) and the sensitivity bound (Equation 10.35) are both satisfied, then the integral constraint (Equation 10.32) may be manipulated to show that [28]

$$\sup_{\omega \in (\omega_\ell, \omega_c)} \log |S(j\omega)| \geq \frac{1}{\omega_c - \omega_\ell} \left(\pi \sum_{i=1}^{N_p} \mathrm{Re}(p_i) + \omega_\ell \log \left(\frac{1}{\alpha} \right) - \frac{3\epsilon\omega_c}{2k} \right). \tag{10.36}$$

The bound in Equation 10.36 shows that increasing the area of low-frequency sensitivity reduction by requiring α to be very small or ω_ℓ to be very close to ω_c, will necessarily cause a large peak in sensitivity at frequencies between ω_ℓ and ω_c. Hence, the integral constraint (Equation 10.32) together with the bandwidth constraint (Equation 10.33) imposes a trade-off between sensitivity reduction and sensitivity increase which must be accounted for in design.

It may be desirable to impose a bandwidth constraint such as Equation 10.33 directly on the complementary sensitivity function $T(s)$ since $T(s)$ directly expresses the feedback properties of sensor noise response and stability robustness, while $L(s)$ does so only indirectly. Analogous results to those stated in Corollary 10.1 can be obtained in this case.

10.4.4 Complementary Sensitivity Integral

The complementary sensitivity function is constrained by the stability requirement for the closed-loop system in a manner analogous to the sensitivity function. The following result is due to Middleton and Goodwin [45]. One difference between the two results is that the trade-off described by the Bode sensitivity integral (Equation 10.32) is worsened if the system is open-loop unstable. As we shall see, the analogous trade-off described by the complementary sensitivity integral becomes worse if the open-loop transfer function contains a time delay and/or zeros in the right half-plane.

Theorem 10.3: Complementary Sensitivity Integral

Suppose that the open loop transfer function $L(s)$ is given by the product of a rational transfer function and a delay element,

$$L(s) = \tilde{L}(s)e^{-s\tau}, \tag{10.37}$$

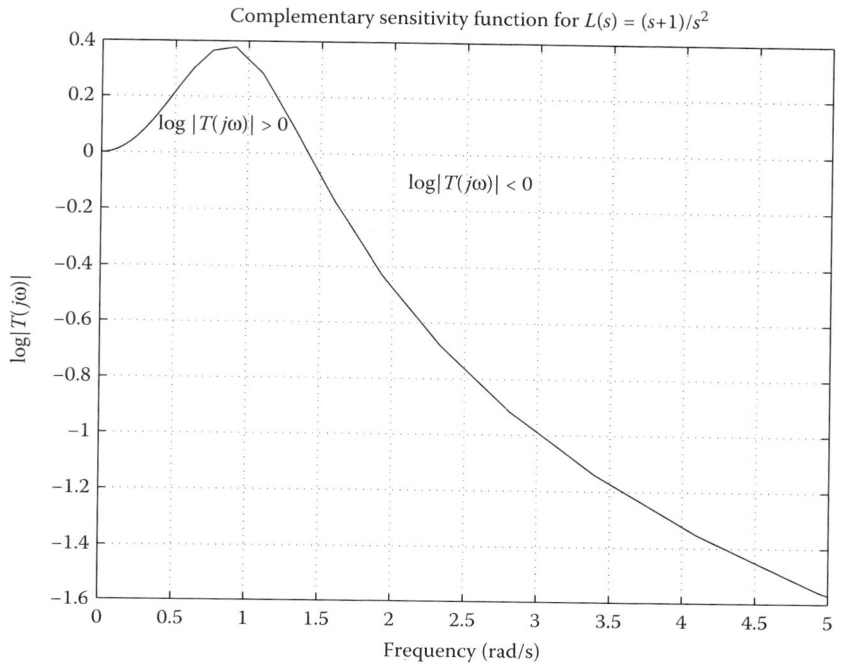

FIGURE 10.7 Areas of complementary sensitivity increase ($\log|T(j\omega)| < 0$) and complementary sensitivity decrease ($\log|T(j\omega)| > 0$) for $L(s) = (s+1)/s^2$.

where $\tilde{L}(s)$ is assumed to be rational with right-half-plane zeros $\{z_i : i = 1, \ldots, N_z\}$ (with multiple zeros included according to their multiplicity). If $L(s)$ has at least one pole at the origin (i.e., one integrator), and if the associated feedback system is stable, then the complementary sensitivity function must satisfy

$$\int_0^\infty \log|T(j\omega)|\frac{d\omega}{\omega^2} = \pi\sum_{i=1}^{N_z}\mathrm{Re}\left(\frac{1}{z_i}\right) + \frac{\pi}{2}\tau - \frac{\pi}{2}K_v^{-1}, \quad (10.38)$$

where K_v is the velocity constant of the system,

$$K_v = \lim_{s\to 0} sL(s). \quad (10.39)$$

The complementary sensitivity integral (Equation 10.38) has a similar interpretation to the Bode sensitivity integral. Recall that the complementary sensitivity function characterizes the response of the system to sensor noise and the robustness of the system to high-frequency model errors. Theorem 10.3 states that if the open-loop transfer function is minimum phase (i.e., it has no right-half-plane zeros, and no delay) and is a Type II system, then the area of amplified sensor noise response must equal the area of attenuated sensor noise response. In the case of the complementary sensitivity function, the areas are computed with respect to an inverse frequency scale (see Figure 10.7). The presence of nonminimum phase zeros or delays worsens the trade-off (i.e., increases the required area of noise amplification). For Type I systems, the trade-off is improved by the term involving the velocity constant on the right-hand side of Equation 10.39.

As for the sensitivity integral, the complementary sensitivity integral does not imply that the peak in the complementary sensitivity transfer function must be large. It is possible to accommodate the required increase in the magnitude of the complementary sensitivity function by allowing it to be only slightly greater than one for frequencies close to zero (since the area is computed with respect to inverse frequency). However, when combined with tracking requirements imposed at low frequencies (analogous to the rolloff requirement in Equation 10.33), the integral constraint in Equation 10.38 can be used to develop a lower bound on the peak of the complementary sensitivity function.

Assume that the open-loop transfer function satisfies

$$|L(j\omega)| \geq \delta \left(\frac{\omega_p}{\omega}\right)^{1+k}, \quad \forall \omega \leq \omega_p, \tag{10.40}$$

where $\delta > 2$ and $k > 0$. This bound imposes a constraint upon the tracking performance of the system.

When a performance constraint such as Equation 10.40 is imposed, it is obviously not possible to require the complementary sensitivity function to exceed one over an arbitrarily large inverse frequency interval. When Equation 10.40 is satisfied, there is an upper bound on the area of complementary sensitivity increase which can be present at frequencies less than ω_p. The corresponding limitation imposed by the complementary sensitivity integral (Equation 10.38) and the rolloff constraint (Equation 10.40) is expressed by the following result.

Corollary 10.2:

Suppose, in addition to the assumptions of Theorem 10.3, that $L(s)$ satisfies the performance constraint given by Equation 10.40. Then the low-frequency tail of the complementary sensitivity integral must satisfy

$$\left| \int_0^{\omega_p} \log |T(j\omega)| \frac{d\omega}{\omega^2} \right| \leq \frac{3}{2k\delta\omega_p}. \tag{10.41}$$

The bound in Equation 10.41 implies that the complementary sensitivity trade-off imposed by the integral constraint in Equation 10.38 must be accomplished primarily over a finite inverse frequency interval. As a consequence, the amount by which $|T(j\omega)|$ must exceed one cannot be arbitrarily small.

Suppose that the complementary sensitivity function is required to satisfy the upper bound

$$|T(j\omega)| \leq \alpha < 1, \quad \forall \omega \geq \omega_h > \omega_p. \tag{10.42}$$

If the performance constraint given by Equation 10.40 and the complementary sensitivity bound in Equation 10.42 are both satisfied, then the integral constraint (Equation 10.38) may be manipulated to show that

$$\sup_{\omega \in (\omega_p, \omega_h)} \log |T(j\omega)| \geq \frac{1}{\dfrac{1}{\omega_p} - \dfrac{1}{\omega_h}} \left\{ \pi \sum_{i=1}^{N_z} \text{Re}\left(\frac{1}{z_i}\right) + \frac{\log \frac{1}{\alpha}}{\omega_h} - \frac{3}{2k\delta\omega_p} + \frac{\pi}{2}\tau - \frac{\pi}{2} K_v^{-1} \right\}. \tag{10.43}$$

The bound in Equation 10.43 shows that increasing the area of high-frequency complementary sensitivity reduction by requiring α to be very small or ω_h to be very close to ω_p will necessarily cause a large peak in sensitivity at frequencies between ω_p and ω_h. Hence, the integral constraint (Equation 10.38) together with the performance constraint (Equation 10.40) imposes a trade-off between complementary sensitivity reduction and complementary sensitivity increase which must be accounted for in design.

10.5 Limitations Imposed by Right-Half-Plane Poles and Zeros

10.5.1 Introduction

As discussed in Section 10.2, design specifications are often stated in terms of frequency-dependent bounds on the magnitude of closed-loop transfer functions. It has long been known that control system design is more difficult for nonminimum phase or unstable systems. The sensitivity and complementary sensitivity integrals presented in Section 10.4 indicated that nonminimum phase zeros and unstable poles could worsen the individual design trade-offs. In fact, right-half-plane poles and zeros impose additional constraints upon the control system design. This section examines these limitations in detail.

10.5.2 Limitations for Nonminimum Phase Systems

Suppose that the plant possesses zeros in the open right-half-plane. Examples of such systems abound, including the inverted pendulum and cart [56], rear steering bicycle [4], fuel cells [63], acoustic ducts [54], and continuously variable transmission [41]. Then the internal stability requirement dictates that these zeros also appear, with at least the same multiplicity, in the open-loop transfer function $L(s) = P(s)C(s)$. Let the set of all open right-half-plane zeros of $L(s)$ (including any present in the compensator) be denoted by

$$\{z_i : i = 1, \ldots, N_z\}. \tag{10.44}$$

Defining the Blaschke product (all-pass filter)

$$B_z(s) = \prod_{i=1}^{N_z} \frac{z_i - s}{\bar{z}_i + s}, \tag{10.45}$$

we can factor the open-loop transfer function into the form

$$L(s) = L_m(s)B_z(s), \tag{10.46}$$

where $L_m(s)$ has no zeros in the open right-half-plane. Note that

$$|L(j\omega)| = |L_m(j\omega)|, \quad \forall\omega, \tag{10.47}$$

and, in the limit as $\omega \to \infty$,

$$\angle\left(\frac{z_i - j\omega}{\bar{z}_i + j\omega}\right) \to -180^\circ. \tag{10.48}$$

These facts show that open right-half-plane zeros contribute additional phase lag without changing the gain of the system (hence the term "nonminimum phase zero"). The effect that this additional lag has upon feedback properties can best be illustrated using a simple example.

Consider the nonminimum phase plant $P(s) = \frac{1}{(s+1)}\frac{(1-s)}{(1+s)}$ and its minimum phase counterpart $P_m(s) = 1/s+1$. Figure 10.8 shows that the additional phase lag contributed by the zero at $s = 1$ causes the Nyquist plot to penetrate the unit circle and the sensitivity to be larger than one. Experiments with various compensation schemes reveal that using large loop gain over some frequency range to obtain small sensitivity in that range tends to cause sensitivity to be large at other frequencies.

Assume that the open-loop transfer function can be factored as

$$L(s) = L_0(s)B_z(s)B_p^{-1}(s)e^{-s\tau}, \tag{10.49}$$

where $\tau \geq 0$ represents a possible time delay, $L_0(s)$ is a proper rational function with no poles or zeros in the open right plane, and $B_z(s)$ is the Blaschke product (Equation 10.45) containing the open right-half-plane zeros of the plant plus those of the compensator. The Blaschke product

$$B_p(s) = \prod_{i=1}^{N_p} \frac{p_i - s}{\bar{p}_i + s} \tag{10.50}$$

contains all the poles of both plant and compensator in the open right-half-plane, again counted according to multiplicity. We emphasize once more that internal stability requirements dictate that

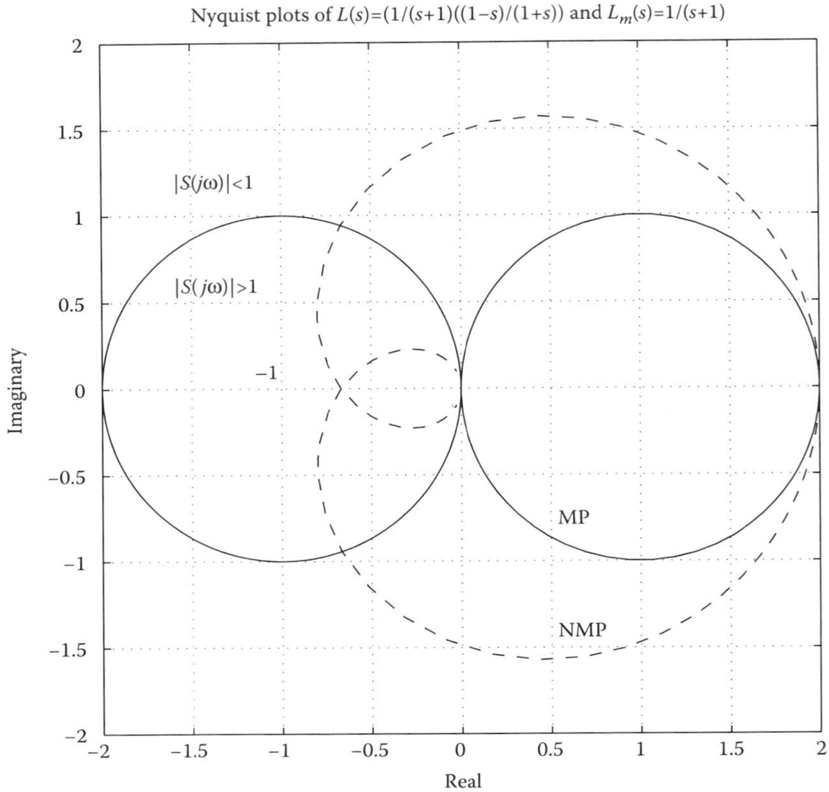

FIGURE 10.8 Additional phase lag contributed by a nonminimum phase zero.

all right-half-plane poles and zeros of the plant must appear with at least the same multiplicity in $L(s)$ and hence cannot be canceled by right-half-plane zeros and poles of the compensator.

One constraint that right-half-plane zeros impose upon the sensitivity function is immediately obvious from the definition $S(j\omega) = 1/(1 + L(j\omega))$. Suppose that $L(s)$ has a zero at $s = z$. It follows that

$$S(z) = 1. \tag{10.51}$$

Poles of $L(s)$ also constrain the sensitivity function. If $L(s)$ has a pole at $s = p$, then

$$S(p) = 0. \tag{10.52}$$

From Equations 10.51 and 10.52, it is clear that if the plant (and thus $L(s)$) has zeros or poles at points of the open right-half-plane, then the value of the sensitivity function is constrained at those points. Naturally, the value of sensitivity along the $j\omega$-axis, where the design specifications are imposed and the conjectured trade-off must take place, is of more concern.

Theorem 10.4:

Suppose that the open-loop transfer function $L(s)$ has a zero, $z = x + jy$, with $x > 0$. Assume that the associated feedback system is stable. Then the sensitivity function must satisfy

$$\int_0^{\infty} \log|S(j\omega)|\, W(z,\omega)\, d\omega = \pi \log\left|B_p^{-1}(z)\right|, \tag{10.53}$$

where $W(z, \omega)$ is a weighting function. For a real zero, $z = x$,

$$W(x, \omega) = \frac{2x}{x^2 + \omega^2} \tag{10.54}$$

and, for a complex zero, $z = x + jy$,

$$W(z, \omega) = \frac{x}{x^2 + (y - \omega)^2} + \frac{x}{x^2 + (y + \omega)^2}. \tag{10.55}$$

A number of remarks about this theorem are in order. First, as discussed in [28], the integral relations are valid even if $S(s)$ has zeros (or poles) on the $j\omega$-axis. Second, a zero z with multiplicity $m > 1$ imposes additional interpolation constraints on the first $m - 1$ derivatives of $\log S(s)$ evaluated at the zero. These interpolation constraints also have equivalent statements as integral relations [28].

We now show that Equation 10.53 imposes a sensitivity trade-off. To see this, note first that the weighting function satisfies $W(z, \omega) > 0, \forall \omega$, and that the Blaschke product satisfies $\log |B_p^{-1}(z)| \geq 0$. Using these facts, it follows easily from Equation 10.53 that requiring sensitivity reduction ($\log |S(j\omega)| < 0$) over some frequency range implies that there must be sensitivity amplification ($\log |S(j\omega)| > 0$) at other frequencies. Hence, if the plant is nonminimum phase, one cannot use feedback to obtain the benefits of sensitivity reduction over one frequency range unless one is willing to pay the attendant price in terms of increased sensitivity elsewhere.

Theorem 10.4 verifies the conjecture that a sensitivity trade-off is present whenever a system is nonminimum phase. Recall it was also conjectured that the severity of the trade-off is a function of the phase lag contributed by the zero at frequencies for which sensitivity reduction is desired. This conjecture can be verified by using the form of the weighting function $W(z, \omega)$ defined in Equations 10.54 and 10.55.

Consider first the case of a real zero $z = x$. Equation 10.46 shows that, as a function of frequency, the additional phase lag contributed by this zero is

$$\theta(x, \omega) = \angle \frac{x - j\omega}{x + j\omega}. \tag{10.56}$$

Noting that

$$\frac{d\theta(x, \omega)}{d\omega} = \frac{-2x}{x^2 + \omega^2}, \tag{10.57}$$

it follows that the weighting function in Equation 10.54 satisfies

$$W(x, \omega) = -\frac{d\theta(x, \omega)}{d\omega}. \tag{10.58}$$

Hence the weighting function appearing in the sensitivity constraint is equal to (minus) the rate at which the phase lag due to the zero increases with frequency.

One can use the weighting function (Equation 10.58) to compute the weighted length of a frequency interval. Note that sensitivity reduction is typically required over a low-frequency interval $\Omega = (0, \omega_1)$ and that the weighted length of such an interval equals

$$\begin{aligned} W(x, \Omega) &= \int_0^{\omega_1} W(x, \omega) d\omega \\ &= -\theta(x, \omega_1). \end{aligned} \tag{10.59}$$

Hence, the weighted length of the interval is equal to (minus) the phase lag contributed by the zero at the upper endpoint of the interval. It follows that, as $\omega_1 \to \infty$, the weighted length of the $j\omega$-axis equals π.

For a complex zero, the weighting function (Equation 10.55) is equal to minus the average of the additional phase lag contributed by the zero and by its complex conjugate:

$$W(z,\omega) = -\frac{1}{2}\left(\frac{d\theta(z,\omega)}{d\omega} + \frac{d\theta(\bar{z},\omega)}{d\omega}\right). \tag{10.60}$$

Hence, the weighted length of the frequency interval $\Omega = (0, \omega_1)$ is

$$W(x,\omega) = -\frac{1}{2}\left(\theta(z,\omega_1) + \theta(\bar{z},\omega_1)\right). \tag{10.61}$$

As we have already remarked, the integral constraint in Equation 10.53 implies that a trade-off exists between sensitivity reduction and sensitivity increase in different frequency ranges. An interesting interpretation of this trade-off is available using the weighting function. Suppose first that $L(s)$ has no poles in the open right-half-plane. Then the integral constraint is

$$\int_0^\infty \log|S(j\omega)|\,W(x,\omega)\,d\omega = 0. \tag{10.62}$$

Equation 10.62 states that the weighted area of sensitivity increase must equal the weighted area of sensitivity reduction. Since the weighted length of the $j\omega$-axis is finite, it follows that the amount by which sensitivity must exceed one at higher frequencies cannot be made arbitrarily small.

If the open-loop system has poles in the open-right-half-plane, then the weighted area of sensitivity increase must exceed that of sensitivity reduction. In particular,

$$\log\left|B_p^{-1}(z)\right| = \sum_{i=1}^{N_p} \log\left|\frac{\bar{p}_i + z}{p_i - z}\right|. \tag{10.63}$$

The right-hand side of Equation 10.63 is always greater than zero, and becomes large whenever the zero z approaches the value of one of the unstable poles p_i. It follows (unsurprisingly) that systems with approximate pole–zero cancellations in the open right half-plane will necessarily have poor sensitivity properties.

We can use the integral constraint in Equation 10.53 to obtain some simple lower bounds on the size of the peak in sensitivity accompanying a given level of sensitivity reduction over a low-frequency interval. Bounds of this type were first discussed by Francis and Zames [26, Theorem 3]. The results presented here will show how the relative location of the zero to the interval of sensitivity reduction influences the size of the peak in sensitivity outside that interval.

Suppose that the sensitivity function is required to satisfy the upper bound

$$|S(j\omega)| \leq \alpha < 1, \tag{10.64}$$

where $\Omega = (0, \omega_1)$ is a low-frequency interval of interest. Define the infinity norm of the sensitivity function:

$$\|S\|_\infty = \sup_{\omega \geq 0} |S(j\omega)|. \tag{10.65}$$

Assuming that the upper bound in Equation 10.64 is satisfied, the integral constraint in Equation 10.53 can be used to compute a lower bound on $\|S\|_\infty$ for each nonminimum phase zero of $L(s)$.

Corollary 10.3:

Suppose that the conditions in Theorem 10.4 are satisfied and that the sensitivity function is bounded as in Equation 10.64. Then the following lower bound must be satisfied at each nonminimum phase zero of $L(s)$:

$$\|S\|_\infty \geq \left(\frac{1}{\alpha}\right)^{\frac{W(z,\Omega)}{\pi - W(z,\Omega)}} \left|B_p^{-1}(z)\right|^{\frac{\pi}{\pi - W(z,\Omega)}}. \tag{10.66}$$

The bound in Equation 10.66 shows that if sensitivity is required to be very small over the interval $(0, \omega_1)$, then there necessarily exists a large peak in sensitivity outside this interval. Furthermore, the smallest possible size of this peak will become larger if the open-loop system has unstable poles near any zero.

The size of the sensitivity peak also depends upon the location of the interval $(0, \omega_1)$ relative to the zero. Assume for simplicity that the system is open-loop unstable and the zero is real. Then

$$\|S\|_\infty \geq \left(\frac{1}{\alpha}\right)^{\frac{W(x,\Omega)}{\pi - W(x,\Omega)}}. \tag{10.67}$$

Recall that the weighted length of the interval $\Omega = (0, \omega_1)$ is equal to (minus) the phase lag contributed by the zero at the upper endpoint of that interval. Since the zero eventually contributes 180° phase lag, it follows that as $\omega_1 \to \infty$, $W(x, \Omega) \to \pi$. Thus the exponent in Equation 10.67 becomes unbounded and, since $\alpha < 1$, so does the peak in sensitivity. To summarize, requiring sensitivity to be small throughout a frequency range extending into the region where the nonminimum phase zero contributes a significant amount of phase lag implies that there will necessarily exist a large peak in sensitivity at higher frequencies. On the other hand, if the zero is located so that it contributes only a negligible amount of phase lag at frequencies for which sensitivity reduction is desired, then it does not impose a serious limitation upon sensitivity properties of the system. Analogous results hold, with appropriate modifications, for a complex zero.

Suppose now that the open-loop system has poles in the open right-half-plane. It is interesting to note that, in this case, the bound (Equation 10.66) implies the existence of a peak in sensitivity even if no sensitivity reduction is present!

Recall next the approximation in Equation 10.26 which shows that small sensitivity can be obtained only by requiring open-loop gain to be large. It is easy to show that $|S(j\omega)| \leq \alpha < 1$ implies that $|L(j\omega)| \geq (1/\alpha) - 1$. The inequality in Equation 10.67 implies that, to prevent poor feedback properties, open-loop gain should not be large over a frequency interval extending into the region for which a nonminimum phase zero contributes significant phase lag. This observation substantiates a classical design rule of thumb: loop gain must be rolled off before the phase lag contributed by the zero becomes significant. However, if one is willing and able to adopt some nonstandard design strategies (such as having multiple gain crossover frequencies) then [37] it is possible to manipulate the design trade-off imposed by a nonminimum phase zero to obtain some benefits of large loop gain at higher frequencies. One drawback of these strategies is that loop gain must be small, and hence the benefits of feedback must be lost over an intermediate frequency range.

10.5.3 Limitations for Unstable Systems

We shall show in this section that unstable poles impose constraints upon the complementary sensitivity function which, loosely speaking, are dual to those imposed upon the sensitivity function by nonminimum phase zeros. That such constraints exist might be conjectured from the existence of the interpolation constraint in Equation 10.52 and the algebraic identity in Equation 10.21. Together, these equations show that if $L(s)$ has a pole $s = p$, then the complementary sensitivity function satisfies

$$T(p) = 1. \tag{10.68}$$

Furthermore, if $L(s)$ has a zero at $s = z$, then

$$T(z) = 0. \tag{10.69}$$

The previous results for the sensitivity function, together with the fact that $T(s)$ is constrained to equal one at open right-half-plane poles of $L(s)$, suggests that similar constraints might exist for $|T(j\omega)|$ due to the presence of such poles. It is also possible to motivate the presence of the integral constraint on $|T(j\omega)|$ using an argument based upon the inverse Nyquist plot [53] and the fact that $|T(j\omega)| > 1$ whenever $L^{-1}(j\omega)$ is inside the unit circle centered at the critical point.

As in Section 10.5.2, it is assumed that $L(s)$ has the form given in Equation 10.49. The following theorem states the integral constraint on the complementary sensitivity function due to unstable poles.

Theorem 10.5:

Suppose that the open-loop transfer function has a pole, $p = x + jy$, with $x > 0$. Assume that the associated feedback system is stable. Then the complementary sensitivity function must satisfy

$$\int_0^\infty \log|T(j\omega)|W(x,\omega)d\omega = \pi \log\left|B_z^{-1}(p)\right| + \pi x\tau, \qquad (10.70)$$

where $W(p,\omega)$ is a weighting function. For a real pole, $p = x$,

$$W(x,\omega) = \frac{2x}{x^2+\omega^2}, \qquad (10.71)$$

and, for a complex pole, $p = x + jy$,

$$W(p,\omega) = \frac{x}{x^2+(y-\omega)^2} + \frac{x}{x^2+(y+\omega)^2}. \qquad (10.72)$$

Remarks analogous to those following Theorem 10.4 apply to this result also. The integral relations are valid even if $T(s)$ has zeros on the $j\omega$-axis, and there are additional constraints on the derivative of $\log T(s)$ at poles with multiplicity greater than one.

The integral constraint in Equation 10.70 shows that there exists a trade-off between sensor noise response properties in different frequency ranges whenever the system is open-loop unstable. Since $|T(j\omega)|$ is the reciprocal of the stability margin against multiplicative uncertainty, it follows that a trade-off between stability robustness properties in different frequency ranges also exists. Using analysis methods similar to those in the preceding section, one can derive a lower bound on the peak in the complementary sensitivity function present whenever $|T(j\omega)|$ is required to be small over some frequency interval. One difference is that $|T(j\omega)|$ is generally required to be small over a high, rather than a low, frequency range.

It is interesting that time delays worsen the trade-off upon sensor noise reduction imposed by unstable poles. This is plausible for the following reason. Use of feedback around an open-loop unstable system is necessary to achieve stability. Time delays, as well as nonminimum phase zeros, impede the processing of information around a feedback loop. Hence, it is reasonable to expect that design trade-offs due to unstable poles are exacerbated when time delays or nonminimum phase zeros are present. This interpretation is substantiated by the fact that the term due to the time delay in Equation 10.70 is proportional to the product of the length of the time delay and the distance from the unstable pole to the left half-plane.

10.5.4 Summary

Nonminimum phase or unstable systems impose additional trade-offs for control system design. Nonminimum phase zeros limit the frequency range over which control system performance can be achieved, while unstable poles require active control over certain frequency ranges and reduce the overall performance that can be achieved. Quantitative expressions of these trade-offs are given by the integral constraints of Theorems 10.4 and 10.5. These constraints can be used together with bounds on the desired performance to compute approximations that provide useful insight into the design trade-offs.

10.6 Time-Domain Integral Constraints

10.6.1 Introduction

One class of limitations on feedback system performance can be described as time domain integral constraints. These date back at least as far as classical works such as [68], which analyze various error constants and their effects on performance. Such constraints, for single-degree-of-freedom unity feedback systems, can be described in a straightforward manner using classical control techniques and have been explored for example in [43,45]. They are based on the unity feedback system depicted in Figure 10.1.

10.6.2 Double Integrators

As an initial case, suppose that we have a feedback loop that incorporates two integrators. For example, this may arise from a situation where the plant includes an integrator and we also specify a controller that incorporates integral action. In this case, it follows that the open-loop transfer function can be written as $L(s) = \frac{1}{s^2}\bar{L}(s)$, where $\bar{L}(0) \neq 0$. Then consider the error response to a unit step change in the reference, $R(s) = \frac{1}{s}$,

$$E(s) = S(s)R(s) = \left(\frac{1}{1+L(s)}\right)\frac{1}{s} = \left(\frac{s^2}{s^2+\bar{L}(s)}\right)\frac{1}{s}. \tag{10.73}$$

Assuming closed-loop stability, it follows that $s = 0$ is in the region of convergence of $E(s)$ and $E(0) = S(0)R(0) = 0$, that is,

$$E(0) = \int_0^\infty e(t)\, dt = 0. \tag{10.74}$$

Equation 10.74 can be viewed as an equal area criterion on the step response. This in turn implies that either the error signal is zero for all time, or it must alternate signs. Therefore, since the error must be negative for some times, there must be overshoot in the step response, since $y(t) = 1 - e(t)$ will exceed 1 whenever $e(t)$ is negative. Figure 10.9 illustrates an example response for a system containing a double integrator.

We conclude from this motivating example that a stable unity feedback system with a double integrator displays unavoidable overshoot in the step response. More general versions of the integral constraint (Equation 10.74) can be derived to quantify the effect of open-loop CRHP poles and zeros on the closed-loop step response, as we discuss next.

10.6.3 Right-Half-Plane Poles and Zeros

We have seen in Section 10.5 that open-loop right-half-plane poles and zeros impose interpolation constraints on the sensitivity and complementary sensitivity functions. Namely, if $L(s)$ has a CRHP zero at $s = z$, then for any internally stable closed-loop system the interpolation constraints given in Equations 10.51 and 10.69 hold. On the other hand, if $L(s)$ has a CRHP pole at $s = p$, then Equations 10.52 and 10.68 hold.

Recall that $S(s)$ and $T(s)$ govern the output and error response of the system to exogenous inputs through Equations 10.4 and 10.5. Closed-loop stability implies that CRHP poles and zeros are in the region of convergence of the Laplace transform, and it thus follows from Equation 10.52 that $E(p) = S(p)R(p) = 0$, and from Equation 10.69 that $Y(z) = T(z)R(z) = 0$. These constraints are expressed in integral form in the following theorem.

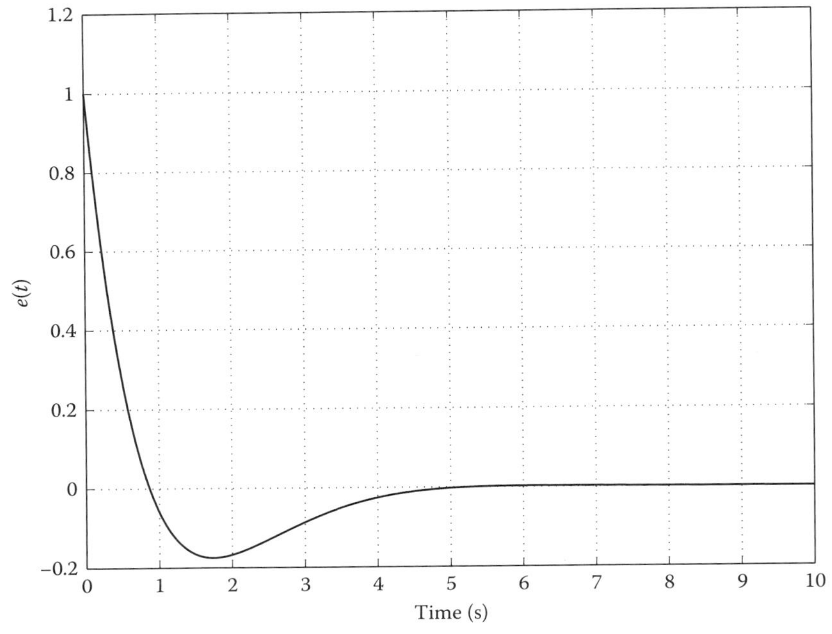

FIGURE 10.9 Example closed loop error step response for a double integrator open loop transfer function.

Theorem 10.6:

Consider any closed-loop stable linear time-invariant SISO system as illustrated in Figure 10.1 with zero initial conditions and a step reference signal. Then, for any CRHP pole p of P(s), the error signal e(t) satisfies

$$\int_0^\infty e^{-pt} e(t)\, dt = 0. \tag{10.75}$$

Furthermore, for any CRHP zero z of P(s), the output signal y(t) satisfies

$$\int_0^\infty e^{-zt} y(t)\, dt = 0. \tag{10.76}$$

The integral constraints in Equations 10.75 and 10.76 are weighted equal area criteria on the step responses $e(t)$ and $y(t)$. Some direct consequences of these results follow as trade-offs in the step response specifications (see Figure 10.10) in the case of *real* CRHP poles and zeros (see [43,45] for further details):

- The closed-loop unity feedback step response of a plant with a real RHP pole must overshoot, that is, the error signal, $e(t)$, must change sign. If we take a slightly unusual definition of rise time

$$t_r = \max\{t_r : y(t) < t/t_r \quad \forall t \leq t_r\},$$

  and if we define the overshoot as the maximum amount by which the output exceeds the reference, $y_{os} = \max_{t\geq 0}\{y(t) - 1\}$, then for a real open-loop pole at $s = p > 0$, the overshoot must satisfy the

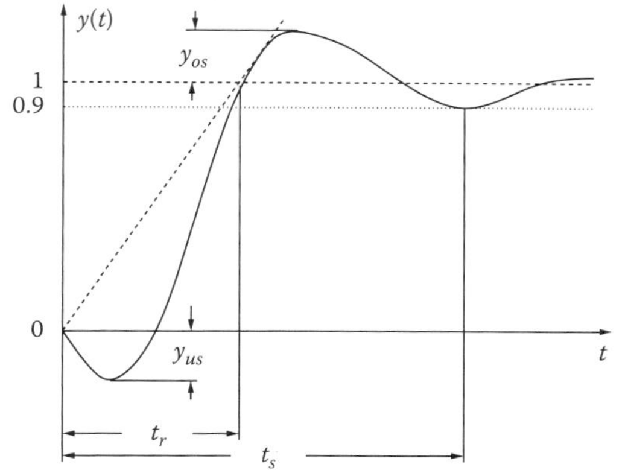

FIGURE 10.10 Step response specifications.

lower bound

$$y_{os} \geq \frac{pt_r}{2}. \tag{10.77}$$

- The closed-loop step response of a plant with a real RHP zero must undershoot, that is, there must exist times for which the output is negative. If we take the non standard definition of settling time

$$t_s = \min_{t_s \geq 0}\{t_s : y(t) \geq 0.9 \quad \forall t \geq t_s\},$$

then for any system with a real open-loop zero at $s = z > 0$, the undershoot must satisfy the lower bound

$$y_{us} = -\min_t\{-y(t)\} \geq \frac{0.9}{e^{zt_s} - 1}. \tag{10.78}$$

The above time domain integrals provide a number of insights into fundamental limits on the behavior of stable closed-loop systems with real RHP open-loop poles or zeros. The effects of slow OLHP open-loop poles on transient performance have been examined in [46]. Trade-offs in the step response due to $j\omega$-axis zeros have been studied in [33]. Extensions to multivariable systems has been pursued in [40], in which directional properties of multivariable systems may permit step response trade-offs that include possible transients in other loops.

10.7 Further Extensions and Comments

A range of extensions to the basic trade-offs and limitations discussed previously are available. In this section, we briefly review some of the main points and key references for a selection of these results. We conclude the section with a brief discussion of cases in which tracking performance limitations may be alleviated.

10.7.1 Limitations in Discrete-Time and Sampled-Data Control

Integral constraints imposed by feedback stability on the sensitivity and complementary sensitivity functions also apply to discrete-time systems. The first discrete-time extensions of the Bode integral for the sensitivity and complementary sensitivity functions were obtained by Sung and Hara [64,65]. A unified formulation of continuous and discrete-time results appeared in [43] and [45]. These results show that the design trade-offs arising from these integral constraints in continuous-time systems carry over to discrete-time systems with analogous interpretations.

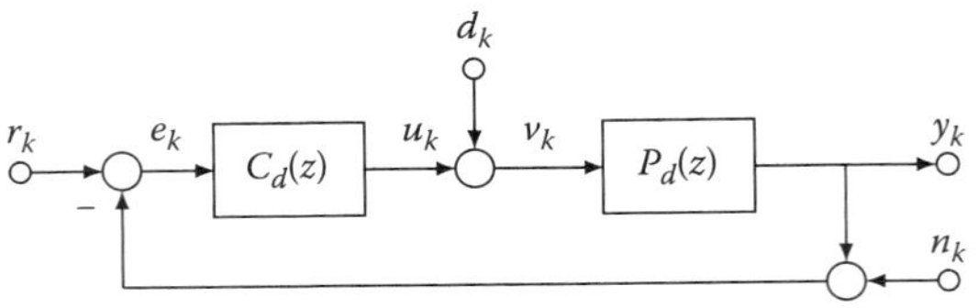

FIGURE 10.11 Linear time-invariant discrete-time feedback system.

We present here the discrete-time Bode sensitivity integral as given in [64]. Consider the discrete-time feedback system shown in Figure 10.11, where all signals are now functions of the discrete-time variable $k = 0, 1, 2, \ldots$, and the plant and controller are represented by the discrete-time rational transfer functions $P_d(z)$ and $C_d(z)$.

The discrete-time open-loop, sensitivity, and complementary sensitivity functions are defined analogously to their continuous-time counterparts:

$$L_d(z) = P_d(z)\,C_d(z), \quad S_d(z) = \frac{1}{1 + L_d(z)}, \quad T_d(z) = \frac{L_d(z)}{1 + L_d(z)}.$$

Suppose that the open-loop transfer function $L_d(s)$ has no pole–zero cancellations in $\bar{\mathbb{D}}^c = \{z : |z| \geq 1\}$. Then the discrete-time closed-loop system is stable if $S_d(z)$ and $T_d(z)$ have no poles in the open unit disk $\mathbb{D} = \{z : |z| < 1\}$.

Theorem 10.7: Bode Discrete-Time Sensitivity Integral

If the loop transfer function $L_d(z)$ is strictly proper and the closed-loop system shown in Figure 10.11 is stable, then

$$\frac{1}{\pi}\int_0^{\pi} \log|S_d(e^{j\theta})|\, d\theta = \sum_{i=1}^{N_p} \log|\phi_i|, \tag{10.79}$$

where $\{\phi_i \in \bar{\mathbb{D}}^c, i = 1, \ldots, N_p\}$ are the unstable poles of $L_d(z)$.

As in the continuous-time case, the Bode discrete-time sensitivity integral (Equation 10.79) shows that there also exists a balance of areas of sensitivity reduction and amplification for discrete-time systems. A significant difference with the continuous-time case is that in the Bode discrete-time sensitivity integral (Equation 10.79) integration is performed over a *finite* interval, and thus the balance of areas of sensitivity reduction and amplification directly implies design trade-offs, even if no bandwidth constraints are imposed on the discrete-time system.

One should note that if the discrete plant corresponds to the linear time-invariant discretization of an analog plant to be controlled digitally through periodic sampling and hold devices, the analysis of design trade-offs and limitations based on the study of discrete-time sensitivity functions may not capture the real (continuous-time) behavior. Indeed, it is known, for example, that the sampled zeros of a discretized plant can be manipulated by appropriate choice of the hold device [24, Chapter 10]. Hence, the discretization of a nonminimum phase analog plant may be rendered minimum phase, consequently relaxing design limitations in the *sampled* response. However, design limitations and design trade-offs remain in the intersample response independently of the type of hold device used, as shown in [29]. Intersample behavior must thus be taken into account to detect potential difficulties in the continuous-time response.

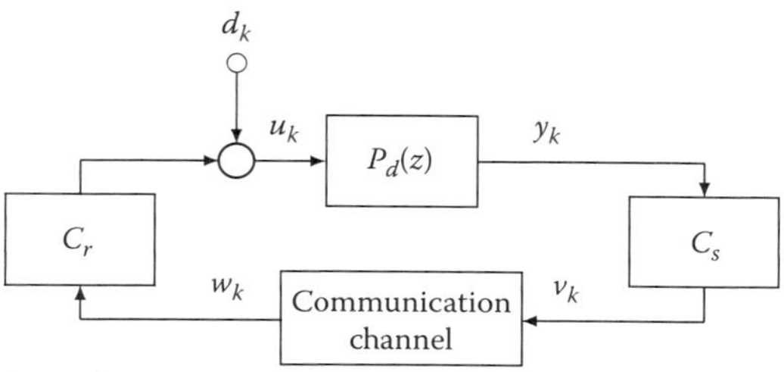

FIGURE 10.12 Feedback control over a communication channel.

10.7.2 Limitations in Control with Communication Constraints

The increased application of communications and networking technology in the new generation of control systems has prompted many recent studies of limitations in control schemes in which the feedback loop is closed over a communication channel (see, e.g., the papers in [2] and the references therein). While a comprehensive theory of limitations in control over communication channels is yet unavailable, several important questions have been answered for simplified channel models (see, e.g., [11,50]).

A simple scheme of a (discrete-time) control system with feedback over a communication channel is illustrated in Figure 10.12. Here, the block C_s encapsulates processes such as encoding, sensor filtering, and control computations to generate the signal v_k sent over the communication channel. The communication channel may include quantization effects, delay, noise, data loss, and data-rate contraints. The block C_r encapsulates processes such as decoding and actuator signal processing based on the received signal w_k.

This section presents a generalization of the discrete-time Bode Sensitivity Integral seen in Section 10.7.1 to systems with feedback over a noisy communication channel with limited capacity. This result links a fundamental limitation to feedback performance to a fundamental limitation in communications by using the information theoretic concepts developed by Shannon [60].

For an additive white Gaussian noise channnel (AWGN) model, the received signal w_k in Figure 10.12 is given as

$$w_k = v_k + n_k, \qquad (10.80)$$

where n_k is a white Gaussian signal with variance $\mathcal{N}$ independent of the transmitted signal v_k. The transmitted signal v_k is subject to a channel input power constraint $E\{v_k^2\} < \mathcal{P}$. The (information) capacity $\mathcal{C}$ of the channel is given by the famous Shannon formula [60]

$$\mathcal{C} = \frac{1}{2}\log_2\left(1 + \frac{\mathcal{P}}{\mathcal{N}}\right) \quad \text{bits/sample.} \qquad (10.81)$$

Given a source that produces information at rate $\mathcal{R}$, Shannon showed that it is possible to communicate this information reliably (i.e., with arbitrarily small probability of error) over a channel with capacity $\mathcal{C} \geq \mathcal{R}$. On the other hand, if $\mathcal{R} > \mathcal{C}$, then reliable communication is not possible.

Suppose that a feedback system uses information transmitted over a communication channel in the feedback loop, as depicted in Figure 10.12. Then it is natural to ask how the capacity of the channel limits properties of the feedback system. For example, it is of interest to determine the minimum channel capacity required to stabilize an unstable discrete-time plant. The authors of [50,66] show that stabilization is possible if and only if the data rate of the channel is greater than $\sum_{i=1}^{N_p} \log|\phi_i|$, where the ϕ_i are the unstable plant eigenvalues. Similarly, it is shown in [11] that an unstable discrete-time plant with relative degree equal to one and no zeros outside the open unit disk may be stabilized over a Gaussian communication channel if and only if the channel capacity (Equation 10.81) satisfies the same lower bound identified in [50]:

$$C \geq \sum_{i=1}^{N_p} \log |\phi_i|. \tag{10.82}$$

In addition to stabilization, feedback may also be used in a control design for the purpose of disturbance attenuation. We have seen that the Bode sensitivity integral limits the ability to attenuate disturbances for the discrete-time feedback system of Figure 10.11. A generalized version of the Bode sensitivity integral derived by Martins and Dahleh [42] shows that the disturbance attenuation properties of a feedback system with a communication channel in the feedback loop, as depicted in Figure 10.12, are limited by the capacity of the channel. To state this result, assume in Figure 10.12 that the blocks C_s and C_r are (possibly nonlinear and time-varying) causal operators, and let the exogenous disturbance d_k be a stationary Gaussian moving average process with a nontrivial power spectral density $\Phi_d(\omega)$. Assume also that the feedback system is mean-square stable, and that the control input sequence u_k is asymptotically stationary and has a well-defined power spectral density $\Phi_u(\omega)$. Then a sensitivity function can be defined as [42, Definition 3.2]

$$S_i(\omega) = \sqrt{\Phi_u(\omega)/\Phi_d(\omega)}. \tag{10.83}$$

The following result is found in [42, Theorem 6.3].

Theorem 10.8:

If the transfer function $P_d(z)$ is strictly proper and the closed-loop system shown in Figure 10.12 is stable, then

$$\frac{1}{\pi}\int_0^{\pi} \min\{0, \log S_i(\omega)\}\, d\omega \geq \sum_{k=1}^{N_p} \log |\phi_i| - C, \tag{10.84}$$

where $\{\phi_i \in \bar{\mathbb{D}}^c, i = 1, \ldots, N_p\}$ are the unstable poles of $P_d(z)$.

Theorem 10.8 shows that if the capacity of the Gaussian channel is equal to the minimum required for stabilization, then it is not possible to achieve $S_i(\omega) < 1$ at any frequency, and thus disturbance attenuation is not possible. If disturbance attenuation is required in a feedback design, it is thus necessary that the channel capacity exceed that required for stabilization.

Although we have presented Theorem 10.8 in the context of a Gaussian channel, in fact, the result holds for any channel, and the reader is referred to Martins and Dahleh [42] for further discussion.

10.7.3 Cheap Control and Achievable H_2 Performance

An alternate approach to considering some performance limitations has been to consider so-called *cheap control* problems, wherein an output performance is optimized, while (for the sake of analysis) ignoring any cost associated with the control signal. For example, it was shown in [52] that for the system as shown in Figure 10.1, and with a unit step reference, the minimum achievable quadratic error performance for any stabilizing control, even given full state measurement and a nonlinear controller, must satisfy the so-called *cheap control* performance limitation:

$$\int_0^{\infty} e^2(t)dt \geq 2 \sum_{z_i \in ORHP} z_i^{-1}. \tag{10.85}$$

A series of works such as [17] considered a similar problem, but in an output feedback situation where it was shown that in the case of linear time-invariant control, the inequality in Equation 10.85 is strict if and only if the plant includes both ORHP poles and zeros.

The quantification of performance limitations via cheap control (time domain) has been linked to the Bode sensitivity and complementary sensitivity integrals (frequency domain) by Seron et al. [57] and Middleton and Braslavsky [44], the former of which also extends the analysis to nonlinear systems. A key observation in that paper is that the best achievable performance, as the cost associated with the control signal tends to zero, is determined by the least amount of "energy" (in the L_2 sense) required to stabilize the unstable *zero dynamics* of the plant, which is nonzero if the plant is nonminimum phase.

Limitations to optimal cheap control performance have been studied also in linear filtering problems [9,39], and decoupling problems in multivariable systems [12].

10.7.4 MIMO Systems

Multivariable feedback systems incorporate a number of extra challenges compared to SISO loops. In particular, directional properties of transfer functions may play an important role. In addition, the range of transfer functions that are important to overall performance is larger, since the relevant open-loop transfer function matrices, $L_o(s) = P(s)C(s)$ and $L_i(s) = C(s)P(s)$, are identical only in special cases. In the multiple input multiple output (MIMO) case, the output sensitivity, $S_o(s)$, and complementary sensitivity, $T_o(s)$, (both of which take values in $\mathbb{C}^{m\times m}$ where m is the number of outputs) are defined as

$$\begin{aligned} S_o(s) &= (I + L_o(s))^{-1}, \\ T_o(s) &= (I + L_o(s))^{-1} L_o(s). \end{aligned} \tag{10.86}$$

Several classes of results have been obtained for multivariable systems, and we briefly review some of the key points below. For simplicity of exposition, we restrict attention to (1) loops referred to the output; (2) cases where the loop transfer function and its inverse are generically full rank; and (3) cases where the open-loop poles and zeros are disjoint. In particular, we assume:

$$\begin{aligned} \det(L_o(s)) = 0 &\quad \text{only for isolated values of } s \in \mathbb{C} \\ \det(L_o^{-1}(s)) = 0 &\quad \text{only for isolated values of } s \in \mathbb{C} \\ \{\det(L_o(z)) = 0\} &\ \Rightarrow\ \left\{\det(L_o^{-1}(z)) \neq 0\right\} \quad \text{for any } z \in \mathbb{C}. \end{aligned} \tag{10.87}$$

10.7.4.1 "Average" Sensitivity Integral

Consider the output sensitivity and output complementary sensitivity functions defined in Equation 10.86. There are various measures of interest for a transfer function matrix, but one that is particularly amenable to performance limitations analysis is the determinant. In particular, as shown in [28],

$$\int_0^{\infty} \log\left|\det(S_o(j\omega))\right| d\omega = \pi \sum_{i=1}^{N_p} \mathrm{Re}(p_i). \tag{10.88}$$

If we let $\sigma_i(\cdot)$ denote the ith singular value of a matrix, Equation 10.88 can be rewritten as

$$\int_0^{\infty} \sum_{j=1}^{m} \log\left(\sigma_i(S_o(j\omega))\right) d\omega = \pi \sum_{i=1}^{N_p} \mathrm{Re}(p_i).$$

Therefore, this generalization of the Bode Sensitivity Integral (Equation 10.32) gives a measure of the overall performance. However, it lacks detail about the worst-case or individual loop performances and therefore several extensions of this analysis have been considered.

10.7.4.2 Sensitivity Integral Inequalities

One form of generalizations that provides some alternate information to the "Average" performance results of Section 10.7.4.1 examines the behavior of the maximum singular value (i.e., the induced 2-norm) of the output sensitivity matrix. Indeed, it can be shown (see, e.g., [8,15]) that

$$\int_0^{\infty} \log \|S_o(j\omega)\|\, d\omega \geq \pi \max_{i=1\ldots N_p} \mathrm{Re}(p_i). \tag{10.89}$$

Inequality in Equation 10.89 arises from the directional properties of the multivariable system (see, e.g., Definition 10.1), which we now turn to study in more detail.

10.7.4.3 Direction-Based Analysis

Directional properties of vector signals and matrix transfer functions play an important role in multivariable systems. The appropriate MIMO generalization of a zero* of the loop transfer function matrix, $L_o(s)$, is as an isolated value of s for which $L_o(s)$ drops rank. Under Equation 10.87 this is equivalent to isolated values of s for which $\det(L_o(s)) = 0$. Similarly, a pole of $L_o(s)$ is an isolated value of s for which $\det(L_o^{-1}(s)) = 0$. This leads to the following definition of MIMO zeros and poles and their input directions†.

Definition 10.1: Multivariable Poles and Zeros

1. *Subject to Equation 10.87, we say $L_o(s)$ has a zero at $s = z$ with input direction d_z if*

$$L_o(z)d_z = 0.$$

2. *Subject to Equation 10.87, we say $L_o(s)$ has a pole at $s = p$ with input direction d_p if*

$$L_o^{-1}(p)d_p = 0.$$

It then follows that we have interpolation constraints on the output sensitivity and complementary sensitivity functions as follows.

Lemma 10.1: Multivariable Interpolation Constraints

Subject to Equation 10.87, we have the following:

1. *If $s = z$ is a zero of $L_o(s)$ with input direction d_z, then*

$$S_o(z)d_z = d_z,$$
$$T_o(z)d_z = 0.$$

---

* Note that in fact there are a number of different definitions of MIMO zeros. In this case, we are considering MIMO *transmission* zeros.

† Similar definitions and results apply to output directions, which we omit for brevity.

2. *If $s = p$ is a pole of $L_o(s)$ with input direction d_p, then*

$$S_o(p)d_p = 0,$$
$$T_o(p)d_p = d_p.$$

Using Lemma 10.1 we can generalize earlier SISO results such as Theorems 10.4 and 10.5 as follows:

Theorem 10.9:

Take any appropriately dimensioned vector d.

1. *If $s = z$ is a zero of $L_o(s)$ with input direction d_z, then*

$$\int_0^{\infty} \log \left| d^T S_o(j\omega) d_z \right| W(z,\omega) d\omega \geq \pi \log |d^T d_z|.$$

2. *If $s = p$ is a pole of $L_o(s)$ with input direction d_p, then*

$$\int_0^{\infty} \log \left| d^T T_o(j\omega) d_p \right| W(p,\omega) d\omega \geq \pi \log |d^T d_p|.$$

Note that the results in Theorem 10.9 are inequalities, since in general the expression for the relevant Blaschke products are much more complex in this case than in the SISO case (Equation 10.50). We also note that by taking d aligned with d_z or d_p as appropriate, and using unit length vectors, we can make the right-hand sides of the bounds in Theorem 10.9 equal to zero.

10.7.4.4 Other MIMO Results

There are a wide range of multivariable feedback results available, and this section has concentrated on a few specific forms of these results. A summary and analysis of many of the available results is contained in [56]. The results of [17] considered achievable H_2 performance of multivariable systems for example. The authors of [30] consider systems in which the performance variable is not identical to the measured variable and the disturbance and actuator affect the system through different dynamics. A range of other results are included in the special issue [16].

10.7.5 Time-Varying and Nonlinear Systems

The theory of fundamental feedback limitations for general time-varying and nonlinear systems is much less developed than that for linear time-invariant systems. One of the main challenges in extending notions such as the Bode integral is that, in contrast to the linear time-invariant case, it is in general impossible to obtain a complete characterization of the system action as an operator via a simple transformation to the frequency domain. A number of significant results, however, have been obtained in the last couple of decades showing that some fundamental feedback limitations remain in more general classes of systems.

One way of extending notions associated with transfer functions to nonlinear systems is to apply the theory of input/output (I/O) nonlinear operators on linear spaces. Using this approach, Shamma [59] shows that nonminimum phase dynamics impose an "area balance" constraint to the nonlinear sensitivity I/O operator analogous to that captured by the Bode integral for linear systems. The result shows that if the nonlinear plant is nonminimum phase (defined by conditions on the domain and range of the plant I/O operator [56, §12]), an arbitrarily small value of frequency-weighted sensitivity necessarily implies an arbitrarily large response to some admissible disturbance. A dual "area balance" constraint exists for the nonlinear complementary sensitivity I/O operator when the plant is open-loop unstable [56, §13.4].

Nonlinear equivalents of the interpolation constraints (Equations 10.51 and 10.69) can also be developed, and applied to quantify control design trade-offs [56, §13] and [58].

Another approach to characterize fundamental limitations in classes of nonlinear systems arises from the study of cheap control problems, as in Section 10.7.3. This idea has been applied to study limitations in nonminimum phase, and nonright invertible strict-feedback nonlinear systems [10,57].

An information theoretic approach has been pursued in Iglesias [38,69] to obtain a time-domain interpretation of the Bode sensitivity integral and extend fundamental sensitivity limitations results to classes of time-varying, and nonlinear systems. Following a similar approach, Martins and Dahleh [42] extended the Bode sensitivity integral to control systems with feedback over capacity-constrained communication channels (see Section 10.7.2) that may include arbitrary time-varying and nonlinear components in the loop, with the only restriction of being causal.

10.7.6 Alleviation of Tracking Performance Limitations

From the previous discussions, it can be seen that there are a number of inherent limitations on feedback system performance, including tracking performance. For plants with CRHP zeros, we have seen a number of fundamental limitations on achievable performance. For example, the time-domain integrals based on CRHP zeros in Theorem 10.6 constrain the output behavior, for any reference signal, for any stabilizing control. In particular, a simple generalization of the time-domain integral in Theorem 10.6 shows that with a nonminimum phase zero at $s = z$ and a given reference signal with Laplace transform, $R(s)$, then

$$\int_0^{\infty} e^{-zt} e(t)\, dt = R(z). \qquad (10.90)$$

Frequency-domain integrals such as the complementary sensitivity integral (see Section 10.4.4) and the Poisson sensitivity integral (Theorem 10.4) apply for any stabilizing linear time-invariant control scheme. In addition, there are various extensions of the cheap control results of Equation 10.85 to alternate reference signals. Here we wish to consider various problem settings in which it may be possible to alleviate the limitations on tracking performance imposed by plant CRHP zeros.

10.7.6.1 Preview Control

Preview control refers to a scenario wherein the reference trajectory, $r(t)$, may be prespecified, and therefore, at time t, the control may use advance knowledge of future or impending reference changes, $r(\tau){:}\tau \in [t, t + T_{pre}]$. In the case of infinite preview ($T_{pre} \to +\infty$), references such as [14] showed how nonlinear system inversion (and therefore perfect tracking) may be performed. A more detailed analysis of such systems from a performance limitations perspective is discussed in [18,47]. In particular, it is shown from a time-domain integral perspective; from a Poisson sensitivity integral perspective; and, from an achievable H_∞ performance perspective, that use of preview T_{pre} with $\mathrm{Re}\{zT_{pre}\}$ sufficiently large almost eliminates the tracking performance limitations due to a CRHP plant zero z. Of course, such preview alters only reference tracking performance and does not alter feedback properties such as noise performance, disturbance response, and stability robustness.

10.7.6.2 Path Tracking

Path tracking (see, e.g., [1,48]) is an alternate control paradigm in which a reference trajectory, $r(t)$, is not given as a prespecified function of time. Instead, the primary tracking objective may be to ensure that at all times, the output is close to a prespecified path ($r(\theta(t))$), that is, that $\|y(t) - r(\theta(t))\|$ is small for all time where $\theta(t)$ is, at least partially, free for the control designer to choose. This allows some additional degrees-of-freedom, and may be helpful in removing some of the limitations on tracking performance imposed by nonminimum phase zeros.

For example, if we have some freedom in the selection of $r(t)$, it may be possible to select $\theta(t)$ in such a way that $R(z)$ is small or even zero. In this case, the constraint (Equation 10.90) on the time-domain performance does not necessarily demand poor tracking performance. Clearly, the feedback noise, disturbance and robustness properties will still be affected by the usual performance trade-offs.

10.7.6.3 Reset Controllers and Other Hybrid Structures

Reset controllers have been shown to overcome some of the limitations inherent to linear time-invariant control, alleviating design trade-offs in the system time response, such as that between overshoot and rise time discussed in Section 10.6 [6,25]. A reset controller is a linear time-invariant system with states that reset to zero when the controller input is zero. This idea was first introduced by Clegg [19], who studied the effect of a reset integrator in a feedback loop. A reset control system may be viewed as a *hybrid* system, which incorporates both continuous dynamics and discrete events. It is possible that hybrid systems present a different set of performance limitations, and advantages in some problems as those discussed in [6,25]. However, a general theory of performance limitations for hybrid control systems remains to be developed.

10.8 Summary and Further Reading

In this chapter, we have discussed design limitations and trade-offs present in feedback design problems. Beginning with the pioneering work of Bode on fundamental feedback relationships, and their implications for feedback amplifier design, a rather complete theory has been developed for linear time-invariant feedback systems. The core of this chapter presented the main results of this theory, as quantified by fundamental relationships in the frequency domain for the sensitivity and complementary sensitivity functions. As discussed, these relationships arise from the basic structure of the feedback loop and the requirement of feedback stability, and are parameterized by the right-half-plane poles and zeros, and time delays of the open-loop system. These relationships have direct implications on the achievable performance and robustness properties of the feedback loop, and thus help a control engineer to achieve informed design trade-offs.

There have been many extensions to the theory of fundamental limitations since the first version of this article was published. We have included a brief account of several of these topics: alternative developments in the time-domain, extensions to multivariable systems, time-varying and nonlinear systems, and control systems with feedback over a communication channel. Finally, we have also included a section discussing the alleviation of performance limitations in some tracking problems.

Diverse applications of the theory of fundamental design limitations also abound. A few examples to illustrate the scope of applications include combustion [5], platooning [55], magnetic bearings [67], haptics [35], ship roll stabilization [51], and autocatalysis [13].

The study of performance limitations in feedback systems is at the core of feedback design, and can provide helpful knowledege in practical control engineering problems. On the one hand, the knowledge of these limitations and their associated trade-offs provides benchmarks against which different control structures and designs may be compared, and guides the deliberate selection of a design solution that achieves a reasonable compromise between conflicting design goals. On the other hand, if a reasonable compromise cannot be achieved for a particular system, then knowledge of design trade-offs may be used to modify the plant, to improve sensors and actuators, and to develop better models so that a tractable design problem is obtained.

The reader who wants to know more can find material in many places. Several textbooks include chapters on the theory of fundamental design limitations, including Franklin et al. [27], Doyle et al. [23] Zhou et al. [70], Goodwin et al. [34], Skogestad and Postlethwaite [61], Aström and Murray [3], and Glad and Ljung [32]. A comprehensive treatment of the theory up through 1997 is given by Seron et al. [56].

Bode's original book [7] is still instructive, and a recent monograph describes the context in which the limitations described by the gain-phase relation arose [49]. An excellent perspective on the importance of the theory of fundamental design limitations was given by Gunter Stein in the inaugural Bode Lecture in 1989, reprinted in [62].

References

1. A.P. Aguiar, J.P. Hespanha, and P.V. Kokotovic. Path-following for non-minimum phase systems removes performance limitations. *IEEE Transactions on Automatic Control*, 50(2):234–239, 2005.
2. P. Antsaklis and J. Baillieul (Eds). Special issue on networked control systems. *IEEE Transactions on Automatic Control*, 49(9), 2004.
3. K. J. Åström and R. M. Murray. *Feedback Systems: An Introduction for Scientists and Engineers*. Princeton University Press, Princeton, NJ, 2008.
4. K. J. Åström, R. E. Klein, and A. Lennartsson. Bicycle dynamics and control. *IEEE Control Systems Magazine*, 25(4):26–47, 2005.
5. A. Banaszuk, P. G. Mehta, C. A. Jacobson, and A. I. Khibnik. Limits of achievable performance of controlled combustion processes. *IEEE Transactions on Control Systems Technology*, 14(5):881–895, 2006.
6. O. Beker, C. V. Hollot, and Y. Chait. Plant with integrator: An example of reset control overcoming limitations of linear feedback. *IEEE Transactions on Automatic Control*, 46(11):1797–1799, 2001.
7. H. W. Bode. *Network Analysis and Feedback Amplifier Design*. D. van Nostrand, New York, 1945.
8. S. Boyd and C. Desoer. Subharmonic functions and performance bounds on linear time invariant feedback systems. *IMA Journal of Math. Control and Information*, 2(2): 153–170, 1995.
9. J. H. Braslavsky, M. M. Seron, D.Q. Mayne, and P.V. Kokotović. Limiting performance of optimal linear filters. *Automatica*, 35(2):189–199, 1999.
10. J. H. Braslavsky, R. H. Middleton, and J. S. Freudenberg. Cheap control performance of a class of non-right-invertible nonlinear systems. *IEEE Transactions on Automatic Control*, 47(8):1314–1319, 2002.
11. J. H. Braslavsky, R. H. Middleton, and J. S. Freudenberg. Feedback stabilization over signal to noise ratio constrained channels. *IEEE Transactions on Automatic Control*, 52(8):1391–1403, 2007.
12. T. S. Brinsmead and G. C. Goodwin. Cheap decoupled control. *Automatica*, 37(9):1465–1471, 2001.
13. F. A. Chandra, G. Buzi, and J. C. Doyle. Linear control analysis of the autocatalytic glycolysis system. In *Proceedings of the 2009 American Control Conference*, St. Louis, MI, 319–324, June 2009.
14. D. Chen and B. Paden. Stable inversion of nonlinear nonminimum phase systems. *International Journal of Control*, 64(1):81–97, 1996.
15. J. Chen. Sensitivity integral relations and design tradeoffs in linear multivariable feedback systems. *IEEE Transactions on Automatic Control*, 40(10):1700–1716, 1995.
16. J. Chen and R.H. Middleton (Eds). Special issue on 'new developments and applications in performance limitation of feedback control'. *IEEE Transactions on Automatic Control*, 48(8):1297–1393, 2003.
17. J. Chen, L. Qiu, and O. Toker. Limitations on maximal tracking accuracy. *IEEE Transactions on Automatic Control*, 45(2):326–331, 2000.
18. J. Chen, R. Zhang, S. Hara, and L. Qiu. Optimal tracking performance: Preview control and exponential signals. *IEEE Transactions on Automatic Control*, 46(10):1647–1653, 2001.
19. J. C. Clegg. A nonlinear integrator for servomechanisms. *Transactions of the AIEE (Part II)*, 77:41–42, 1958.
20. J. B. Cruz and W. R. Perkins. A new approach to the sensitivity problem in multivariable feedback design. *IEEE Transactions on Automatic Control*, 9:216–223, 1964.
21. R. C. Dorf. *Modern Control Systems*. Addison-Wesley, Reading, MA, 1991.
22. J. C. Doyle and G. Stein. Multivariable feedback design: Concepts for a classical/modern synthesis. *IEEE Transactions on Automatic Control*, 26(1):4–16, 1981.
23. J. C. Doyle, B. A. Francis, and A. R. Tannenbaum. *Feedback Control Theory*. Macmillan Publishing Company, New York, 1992.
24. A. Feuer and G.C. Goodwin. *Sampling in Digital Signal Processing and Control*. Birkhauser, Boston, MA, 1996.
25. A. Feuer, G. C. Goodwin, and M. Salgado. Potential benefits of hybrid control for linear time invariant plants. In *Proceedings of the American Control Conference*, Albuquerque, NM, 5:2790–2794, 1997.

26. B.A. Francis and G. Zames. On H_∞-optimal sensitivity theory for SISO feedback systems. *IEEE Transactions on Automatic Control*, 29(1):9–16, 1984.
27. G.F. Franklin, J.D. Powell, and A. Emami-Naeini. *Feedback Control of Dynamic Systems*. Prentice-Hall, (4th ed.) NJ, Englewood Cliffs, 2002.
28. J. S. Freudenberg and D. P. Looze. *Frequency Domain Properties of Scalar and Multivariable Feedback Systems*, Vol. 104, Lecture Notes in Control and Information Sciences. Springer-Verlag, Berlin, 1988.
29. J. S. Freudenberg, R. H. Middleton, and J. H. Braslavsky. Robustness of zero-shifting via generalized sampled-data hold functions. *IEEE Transactions on Automatic Control*, 42(12):1681–1692, 1997.
30. J. S. Freudenberg, C. V. Hollot, R. H. Middleton, and V. Toochinda. Fundamental design limitations of the general control configuration. *IEEE Transactions on Automatic Control*, 48(8):1355–1370, 2003.
31. J.S. Freudenberg and D.P. Looze. Right half plane poles and zeros and design tradeoffs in feedback systems. *IEEE Transactions on Automatic Control*, 30(6):555–565, 1985.
32. T. Glad and L. Ljung. *Control Theory, Multivariable and Nonlinear Methods*. Taylor & Francis, New York, 2000.
33. G. C. Goodwin, A. R. Woodyatt, R. H. Middleton, and J. Shim. Fundamental limitations due to $j\omega$-axis zeros in SISO systems. *Automatica*, 35(5):857–863, 1999.
34. G. C. Goodwin, S. F. Graebe, and M. E. Salgado. *Control System Design*. Prentice-Hall, Englewood Cliffs, NJ, 2001.
35. P. G. Griffiths, R. B. Gillespie, and J. S. Freudenberg. A fundamental tradeoff between performance and sensitivity within haptic rendering. *IEEE Transactions on Robotics*, 24(3):537–548, 2008.
36. I. M. Horowitz. *Synthesis of Feedback Systems*. Academic Press, New York, 1963.
37. I. M. Horowitz and Y.-K. Liao. Limitations of non-minimum-phase feedback systems. *International Journal of Control*, 40(5):1003–1013, 1984.
38. P. A. Iglesias. Tradeoffs in linear time-varying systems: An analogue of Bode's sensitivity integral. *Automatica*, 37(10):1541–1550, 2001.
39. L. B. Jemaa and E. J. Davison. Limiting performance of optimal linear discrete filters. *Automatica*, 39(7): 1221–1226, 2003.
40. K.H. Johansson. Interaction bounds in multivariable control systems. *Automatica*, 38(6):1045–1051, 2002.
41. S. Liu and A. G. Stefanopoulou. Effects of control structure on performance for an automotive powertrain with a continuously variable transmission. *IEEE Transactions on Control Systems Technology*, 10(5): 701–708, 2002.
42. N. C. Martins and M. A. Dahleh. Feedback control in the presence of noisy channels: "Bode-like" fundamental limitations of performance. *IEEE Transactions on Automatic Control*, 53(7):1604–1615, 2008.
43. R. H. Middleton. Tradeoffs in linear control system design. *Automatica*, 27(2):281–292, 1991.
44. R. H. Middleton and J. H. Braslavsky. On the relationship between logarithmic sensitivity integrals and limiting optimal control problems. In *Proceedings of the IEEE Conference on Decision and Control*, Sydney, Australia, 5:4990–4995, 2000.
45. R.H. Middleton and G.C. Goodwin. *Digital Control and Estimation. A Unified Approach*. Prentice-Hall, Inc., Englewood Cliffs, NJ, 1990.
46. R.H. Middleton and S.F. Graebe. Slow stable open loop poles: To cancel or not to cancel. *Automatica*, 35(5):877–886, 1999.
47. R.H. Middleton, J. Chen, and J.S. Freudenberg. Tracking sensitivity and achievable hinfinity performance in preview control. *Automatica*, 40(8):1297–1306, 2004.
48. D.E. Miller and R.H. Middleton. On limitations to the achievable path following performance for linear multivariable plants. *IEEE Transactions on Automatic Control*, 53(11):2586–2601, 2008.
49. D. A. Mindell. *Between Human and Machine: Feedback, Control, and Computing before Cybernetics*. Princeton University Press, Princeton, NJ, 2002.
50. G.N. Nair and R.J. Evans. Exponential stabilisability of finite-dimensional linear systems with limited data rates. *Automatica*, 39(4):585–593, 2003.
51. T. Perez. *Ship Motion Control: Course Keeping and Roll Stabilisation Using Rudder and Fins*. Springer-Verlag, London, UK, 2005.
52. L. Qui and E.J. Davison. Performance limitations of nonminimum phase systems in the servomechanism problem. *Automatica*, 29(2):337–349, 1993.
53. H. H. Rosenbrock. *Computer-Aided Control System Design*. Academic Press, London, 1974.
54. R. S. Sánchez Peña, M. A. Cugueró, A. Masip, J. Quevedo, and V. Puig. Robust identification and feedback design: An active noise control case study. *Control Engineering Practice*, 16(11):1265–1274, 2008.

55. P. Seiler, A. Pant, and K. Hedrick. Disturbance propagation in vehicle strings. *IEEE Transactions on Automatic Control*, 49(10):1835–1841, 2004.
56. M. M. Seron, J. H. Braslavsky, and G. C. Goodwin. *Fundamental Limitations in Filtering and Control.* Springer, Berlin, 1997.
57. M. M. Seron, J. H. Braslavsky, P.V. Kokotović, and D.Q. Mayne. Feedback limitations in nonlinear systems: From Bode integrals to cheap control. *IEEE Transactions on Automatic Control*, 44(4):829–833, 1999.
58. M.M. Seron and G.C. Goodwin. Sensitivity limitations in nonlinear feedback control. *Systems and Control Letters*, 27(4):249–254, 1996.
59. J.S. Shamma. Performance limitations in sensitivity reduction for nonlinear plants. *Systems and Control Letters*, 17(1):43–47, 1991.
60. C. E. Shannon. A mathematical theory of communication. *Bell Systems Technical Journal*, 27:379–423, 623–656, 1948.
61. S. Skogestad and I. Postlethwaite. *Multivariable Feedback Control: Analysis and Design.* Wiley, New York, 1996.
62. G. Stein. Respect the unstable. *Control Systems Magazine*, 23(4):12–25, 2003.
63. K.-W. Suh and A. G. Stefanopoulou. Performance limitations of air flow control in power-autonomous fuel cell systems. *IEEE Transactions on Control Systems Technology*, 15(3):465–473, 2007.
64. H.-K. Sung and S. Hara. Properties of sensitivity and complementary sensitivity functions in single-input single-output digital control systems. *International Journal of Control*, 48(6):2429–2439, 1988.
65. H.-K. Sung and S. Hara. Properties of complementary sensitivity function in SISO digital control systems. *International Journal of Control*, 50(4):1283–1295, 1989.
66. S. Tatikonda and S. M. Mitter. Control under communication constraints. *IEEE Transactions on Automatic Control*, 49(7):1056–1068, 2004.
67. N. M. Thibeault and R. S. Smith. Magnetic bearing measurement configurations and associated robustness and performance limitations. *ASME Journal of Dynamic Systems, Measurement, and Control*, 124 (4):589–597, 2002.
68. J.G. Truxal. *Automatic Feedback Control System Synthesis.* McGraw-Hill, New York, 1955.
69. G. Zang and P.A. Iglesias. Nonlinear extension of bode's integral based on an information-theoretic interpretation. *Systems and Control Letters*, 50(1):11–19, 2003.
70. K. Zhou, J. C. Doyle, and K. Glover. *Robust and Optimal Control.* Prentice-Hall, Englewood Cliffs, NJ, 1996.

11

Modeling Deterministic Uncertainty

Jörg Raisch
Technical University of Berlin

Bruce Francis
University of Toronto

11.1 Introduction

At first glance, the notion of deterministic uncertainty may seem to be a contradiction in terms—after all, the word "deterministic" is often used to signify the absence of any form of uncertainty. We will explain later on that, properly interpreted, this choice of words does indeed make sense. For the moment, we concentrate on the notion of uncertainty.

Uncertainty in the control context comes in two basic versions—uncertain signals and uncertainty in the way the plant maps input signals into output signals ("plant uncertainty").

Most processes we wish to control are subject to influences from their environment—some of them known (measured disturbances, reference inputs), others uncertain signals (unmeasured disturbances, and noise corrupting measurements). All these signals are labeled "external," because they originate in the "outside world."*

A plant model, by definition, is a simplified representation of a real system, and is usually geared toward a specific purpose. Models that are meant to be used for feedback controller design tend to be especially crude. This is because (1) most popular design techniques can handle only very restricted classes of models and (2) in a feedback configuration, one can potentially get away with more inaccurate models than in applications that are based on a pure feedforward structure.

Meaningful analysis and design, however, are possible only if signal and plant uncertainty is, in some sense, "limited." In other words, we have to assume that there exists some, however incomplete, knowledge about signal and plant uncertainty—we need an *uncertainty model*. Such an uncertainty model defines an admissible *set of plant models*, $\mathcal{G}$, and an admissible *set of uncertain external input signals*, $\mathcal{W}$ (Figure 11.1). The adjective in "deterministic uncertainty model" points to the fact that we do not attempt to assign probabilities (or probability densities) to the elements of the sets $\mathcal{G}$ and $\mathcal{W}$—every element is considered to be as likely as any other one. Based on this, one can ask the key (robustness) questions in control

* Control inputs, on the other hand, are generated within the control loop, and are called "internal input signals."

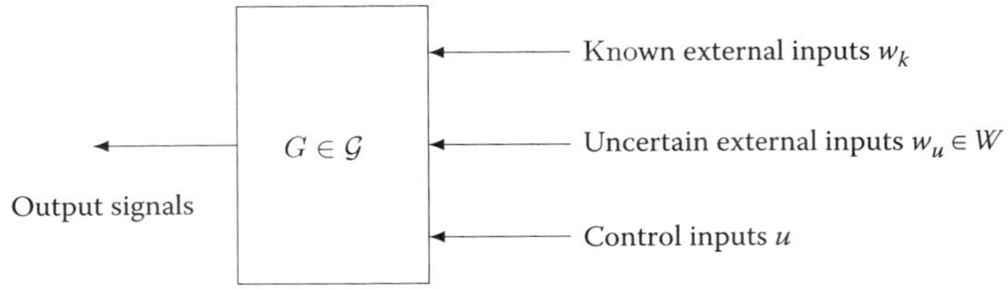

FIGURE 11.1 Signal and plant uncertainty models.

systems analysis: Is closed-loop stability guaranteed for every plant model in $\mathcal{G}$? Do desired closed-loop performance properties hold for every external input in $\mathcal{W}$ and every $G \in \mathcal{G}$?

One has to keep in mind, however, that no class of mathematical models is able to describe every detail of reality, that is, the physical plant is not a mathematical model and therefore cannot be an element in $\mathcal{G}$. Robustness of a desired closed-loop property with respect to any specific plant uncertainty model does not therefore guarantee that the *real* control system will also have this property. If the uncertainty model is chosen in a sensible way (that's what this chapter is about), it will, however, increase the likelihood for the real system to "function properly."

We will first discuss signal uncertainty. Then, we will summarize the most common plant uncertainty models. Finally, we will briefly mention the topic of model validation, that is, whether a given set of experimental data is compatible with given uncertainty models $\mathcal{G}$ and $\mathcal{W}$. We will work in a continuous-time framework, and all signals will be real and (except for some of the examples) vector valued.

11.2 Characterization of Uncertain Signals

Formulating a signal uncertainty model almost always involves a trade-off between conflicting principles: One wants the model to be "tight," that is, contain only signals that make physical sense; tightening uncertainty, however, typically implies imposing additional restrictions on the model—it gets unwieldy and more difficult to use for analysis and design purposes.

The following two widely used uncertainty models are on the extreme (simple) end of the spectrum. The first is

$$\mathcal{W}_2(c) := \{w_u(t) \mid \|w_u\|_2 \leq c\}, \tag{11.1}$$

where the norm is

$$\|w_u\|_2 := \left(\int_{-\infty}^{\infty} w_u(t)^T w_u(t)\, \mathrm{d}t\right)^{1/2},$$

that is, the admissible input set consists of all signals with $\mathcal{L}_2$-norm (energy) less than or equal to a given constant c; the second is

$$\mathcal{W}_\infty(c) := \{w_u(t) \mid \|w_u\|_\infty \leq c\}, \tag{11.2}$$

where the norm is

$$\|w_u\|_\infty := \sup_t \max_i |w_{u_i}(t)|,$$

that is, the admissible input set consists of all signals with $\mathcal{L}_\infty$-norm (maximum magnitude) less than or equal to a given constant c. If necessary, these models can be refined by introducing suitable weights or filters: In this case, admissible input signals are

$$\bar{w}_u(t) := \int_{-\infty}^{t} W(t-\tau) w_u(\tau)\, \mathrm{d}\tau, \tag{11.3}$$

where $w_u(t)$ ranges over the sets (Equation 11.1 or 11.2). Even such a modified uncertainty description remains pretty crude. Whether it is adequate depends on the control problem at hand. If the answer turns out to be "no," one has to cut back the "size" of the admissible signal classes by bringing in additional *a priori* information. This is illustrated in the following example.

Example 11.1:

Suppose we want to control room temperature. Clearly, outdoor temperature, T_o, is a disturbance signal for our control problem. Assume that, for one reason or another, we cannot measure T_o. In Toronto, T_o can go up to $+30°$C in summer and down to $-30°$C in winter. In this case, a simple uncertainty model is given by

$$T_o(t) \in \mathcal{W}_\infty(30°) = \{w_u(t) \mid \|w_u\|_\infty \le 30°\text{C}\}.$$

Clearly, this set contains many signals that do not make physical sense; it admits, for example, temperatures of $+30°$ during a winter night and $-30°$ at noon in summer. A tighter uncertainty set is obtained if we write $T_o(t) = T_a(t) + T_\Delta(t)$, where $T_a(t)$ represents the average seasonal and daily variation of temperature, and $T_\Delta(t)$ is a deviation term. The signal $T_a(t)$ could, for example, be modeled as the output of an autonomous dynamic system (a so-called *exosystem*) with a pair of poles at $s = \pm j(365\text{ days})^{-1}$ (accounting for seasonal variations) and $s = \pm j(24\text{ h})^{-1}$ (accounting for daily variations). For the deviation term, we can now assume a much stricter $\mathcal{L}_\infty$-norm bound, for example, $10°$. Furthermore, it makes sense to restrict the maximal rate of change of $T_\Delta(t)$ to, say, $5°$/h. This is achieved if we define the admissible set of deviations via

$$T_\Delta(t) = \int_{-\infty}^{t} W(t-\tau)w_u(\tau)\,d\tau,$$

where the weighting function $W(t)$ is given by

$$W(t) = \begin{cases} 5e^{-t/2}, & t \ge 0 \\ 0, & t < 0 \end{cases}$$

(with t in hours) and $w_u(t)$ lives in the set $\mathcal{W}_\infty(1°)$.

11.3 Characterization of Uncertain Plant Models

Sometimes the plant can be modeled by a finite set, say by linearizing at a finite number of operating points. Then a controller can be designed for each plant in the set. Implementation involves a table lookup for which controller to apply, together with a way of smoothly transitioning when a switch is required.

More commonly in robust control design, one tries to characterize the plant family $\mathcal{G}$ by specifying a nominal plant model together with a family of perturbations, denoted $\mathcal{D}$, away from the nominal. As for the set of admissible input signals, we have two requirements for $\mathcal{D}$:

1. $\mathcal{D}$ should be *tight* (i.e., contain only perturbations that somehow "mirror" the difference between our nominal model and the real system). If we include perturbations that make no physical sense, we will end up with a controller that tries to accommodate "too many" models and might therefore be "too conservative."
2. $\mathcal{D}$ should be easy to use for analysis and design purposes.

Again, both requirements rarely go together. By imposing *structure* on a perturbation model, we will in general be able to capture uncertainty more precisely; *unstructured* perturbation models, on the other hand, are in general easier to use in design procedures.

For the lack of space, we can only present the most common plant uncertainty models—the ones that are widely used for controller design and synthesis purposes. It is not surprising that, in a sense, this purpose reflects back onto the model itself: Some models we will be dealing with involve restrictive assumptions that can only be justified through a utilitarian argument—without these assumptions it would be a lot harder (or even impossible) to give robustness tests. We will look at unstructured perturbation models first. Such models are typically specified by giving, at every frequency, an upper bound on the maximum singular value of the transfer function matrix, this being a matrix generalization of the magnitude of a complex number. Then, we will deal with a specific class of structured perturbations that can be graphically characterized by Nyquist arrays. Finally, we will discuss a fairly general class of structured perturbations that includes the others.

11.3.1 Unstructured Plant Uncertainty Models

In what follows, $G(s)$ will denote the nominal plant model, $G_r(s)$ the transfer function matrix of a perturbed plant model, and $\Delta(s)$ the transfer function matrix of a perturbation. The dimensions of $G(s)$ are $p \times q$. It is assumed that the elements of $G(s)$, $G_r(s)$, and $\Delta(s)$ are proper real-rational transfer functions. The number of *unstable* poles of G and G_r, that is, those in the closed right half-plane, are denoted m_G and m_{G_r}.

11.3.1.1 Additive Modeling Perturbations

Additive perturbations are defined by (see Figure 11.2)

$$G_r(s) := G(s) + \Delta_A(s). \tag{11.4}$$

The class of unstructured additive perturbations we will be looking at is given by

$$\mathcal{D}_A := \left\{ \Delta_A \mid \bar{\sigma}[\Delta_A(j\omega)] < l_A(\omega);\ m_{G_r} = m_G \right\}. \tag{11.5}$$

Thus, the set of admissible Δ_A is characterized by two assumptions: (1) for each frequency ω, we know a (finite) upper bound $l_A(\omega)$ for the size of Δ_A (in the sense of the maximal singular value); (2) Δ_A cannot change the number of unstable poles of the model. Thus l_A is an *envelope function* on the size of the perturbation. The regularity of the function l_A is not too important; it can be assumed to be, for example, piecewise continuous. The second assumption seems pretty artificial, but is needed if we want to give simple robustness tests.* It trivially holds for all stable perturbation matrices. Obviously, $l_A(\omega)$ being finite for every ω implies that Δ_A does not have any poles on the imaginary axis. A typical perturbation bound $l_A(\omega)$ is shown in Figure 11.3: The DC gain is usually known precisely ($\Delta_A(0) = 0$); at high frequencies, we often have $G_r \to 0$, $G \to 0$ (G_r and G are strictly proper), and therefore $\Delta_A \to 0$.

The following example illustrates how an additive perturbation model typically can be obtained from frequency-response data.

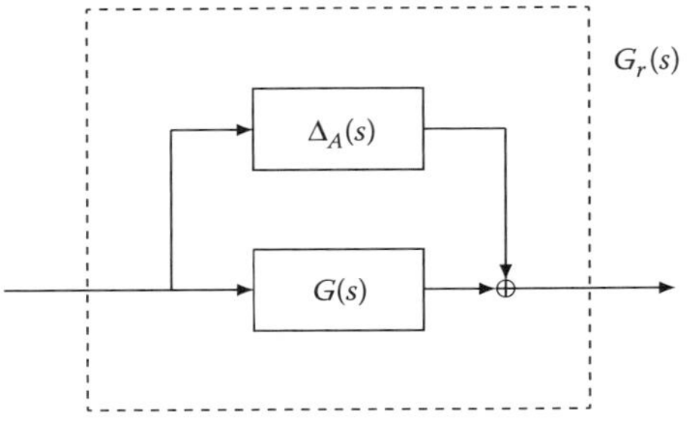

FIGURE 11.2 Additive model perturbation.

* A stability robustness test was first reported in Cruz et al. (1981).

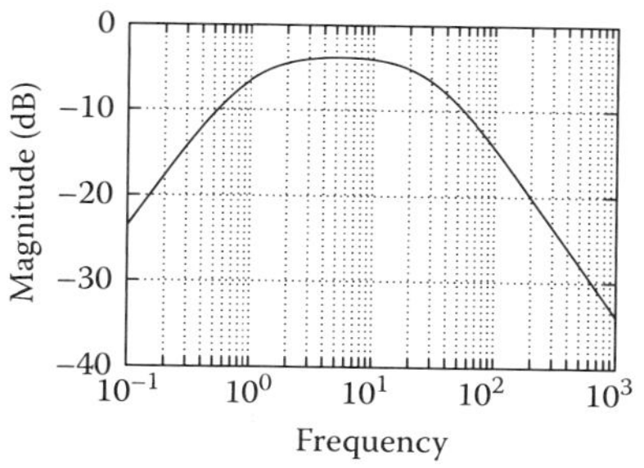

FIGURE 11.3 Typical bound for additive modeling perturbation.

Example 11.2:

For simplicity, assume the plant is single-input/single-output (SISO). Suppose that the plant is stable and its transfer function is arrived at by means of frequency-response experiments: Magnitude and phase are measured at a number of frequencies, $\omega_i, i = 1, \ldots, M$, and this experiment is repeated several, say N, times. Let the magnitude-phase measurement for frequency ω_i and experiment k be denoted (G_{ik}, ϕ_{ik}). Based on these data, select nominal magnitude-phase pairs (G_i, ϕ_i) for each frequency ω_i, and fit a nominal transfer function $G(s)$ to these data. Then fit a weighting function $l_A(s)$ so that

$$\left|G_{ik}e^{j\phi_{ik}} - G_i e^{j\phi_i}\right| < |l_A(\omega_i)|, \quad i = 1, \ldots, M; \ k = 1, \ldots, N.$$

The next example shows how to "cover" a parameter-uncertainty model by an additive perturbation model.

Example 11.3:

Consider the plant model

$$\frac{k}{s+1}, \quad 5 < k < 10.$$

Thus, the gain k is uncertain. Let us take the midpoint for the nominal plant:

$$G(s) = \frac{7.5}{s+1}.$$

Then the envelope function is determined via

$$\left|\frac{k}{j\omega + 1} - G(j\omega)\right| < l_A(\omega), \quad 5 < k < 10,$$

that is, $l_A(\omega) = |2.5/(j\omega + 1)|$.

11.3.1.2 Multiplicative Perturbations

Multiplicative (or proportional) perturbations at the plant output are defined by the following relation between G and G_r (see Figure 11.4):

$$G_r(s) := (I_p + \Delta_M(s))G(s). \tag{11.6}$$

Such proportional perturbations are invariant with respect to multiplication (from the right) by a known transfer function matrix (e.g., a compensator).

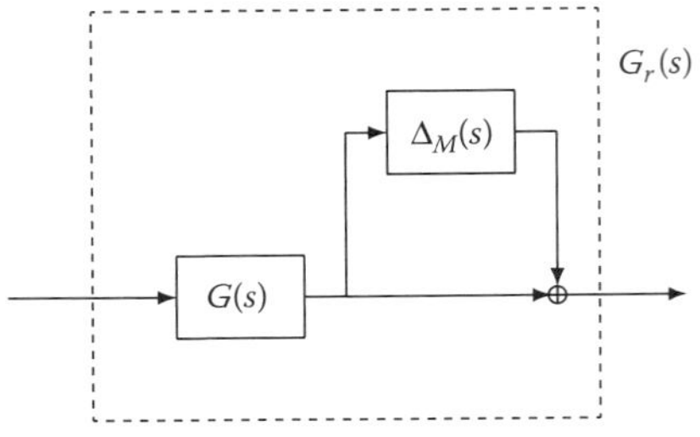

FIGURE 11.4 Multiplicative (proportional) model perturbation.

We consider the following class of (proportional perturbation) models:

$$\mathcal{D}_M := \mathcal{D}_{M1} \cup \mathcal{D}_{M2}, \tag{11.7}$$

$$\mathcal{D}_{M1} := \left\{ \Delta_M \mid \bar{\sigma}[\Delta_M(j\omega)] < l_M(\omega);\ \Delta_M \text{ stable} \right\}, \tag{11.8}$$

$$\mathcal{D}_{M2} := \left\{ \Delta_M \mid \bar{\sigma}[\Delta_M(j\omega)] < l_M(\omega);\ m_{G_r} = m_G \right\}. \tag{11.9}$$

Hence, admissible perturbations are either stable or do not change the number of unstable poles of the plant model. Again, in both cases, we assume that we know an upper bound for the perturbation frequency response $\Delta_M(j\omega)$. A typical perturbation bound l_M is shown in Figure 11.5: Exact knowledge of DC gain implies $\Delta_M(0) = 0$; neglecting high-order dynamics often causes Δ_M to be greater than 1 at high frequency.

The perturbation class $\mathcal{D}_A$ is of a simpler form than Equations 11.7 through 11.9, because an additive stable perturbation transfer function matrix does not affect the number of unstable poles. As the following example shows, this is not always true for multiplicative perturbations: A stable proportional perturbation *can* cancel unstable model poles. For example,

$$G(s) = \frac{-4}{s-1}, \quad \Delta_M(s) = \frac{-2}{s+1},$$

$$G_r(s) = [1 + \Delta_M(s)]G(s) = \frac{-4}{s+1}.$$

Stability robustness results for this class of pertubations have been given in Doyle and Stein (1981).

11.3.1.3 Coprime Factor Perturbations

For this uncertainty description, we need the concept of a left coprime factorization over $\mathcal{RH}_\infty$* (Vidyasagar, 1985). Here, we will give only a very brief introduction to this subject. Details can be

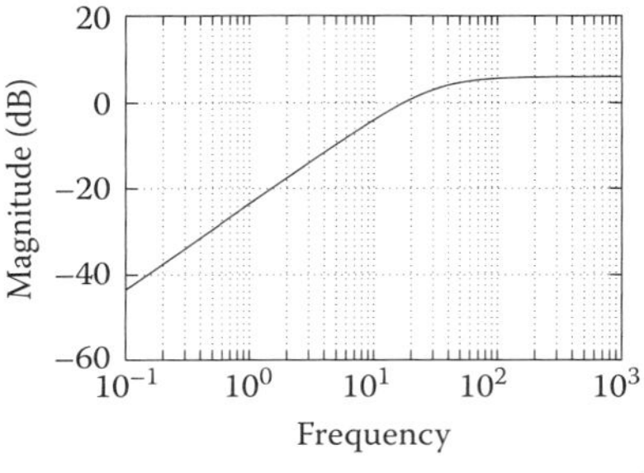

FIGURE 11.5 Typical bound for multiplicative model perturbation.

* $\mathcal{RH}_\infty$ denotes the set of proper stable real-rational transfer functions.

found elsewhere in this book. Consider a finite-dimensional linear time-invariant (FDLTI) SISO system. Clearly, its transfer function, $G(s)$, can be represented by a pair of polynomials with real coefficients—$N_P(s)$, a numerator, and $M_P(s)$, a denominator polynomial: $G(s) = N_P(s)/M_P(s)$. Furthermore, $N_P(s)$ and $M_P(s)$ can always be chosen to be coprime, meaning that all common divisors of $N_P(s)$, $M_P(s)$ are invertible in the set of polynomials (i.e., are real constants). Thus, there is no pole–zero cancellation when forming $N_P(s)/M_P(s)$.

For several reasons, it proves to be an advantage to use a straightforward generalization, and to replace polynomials by proper stable transfer functions (i.e., elements from the set $\mathcal{RH}_\infty$): We write $G(s) = N(s)/M(s)$, where $N(s), M(s) \in \mathcal{RH}_\infty$ can always be chosen to be coprime. This is called a *coprime factorization over* $\mathcal{RH}_\infty$. Coprimeness in the $\mathcal{RH}_\infty$-context means that all common divisors are invertible in $\mathcal{RH}_\infty$ (i.e., stable, minimum-phase, biproper transfer functions). Hence, when forming $N(s)/M(s)$, no cancellation of poles and zeros in the closed right half-plane can occur. Obviously, such a factorization is nonunique. However, it can be made unique up to sign by requiring $|N(j\omega)|^2 + |M(j\omega)|^2 = 1$, that is, the 1×2 matrix $\begin{bmatrix} N(s) & M(s) \end{bmatrix}$ is allpass. This is called a *normalized* coprime factorization (Vidyasagar, 1988).

These concepts can be easily extended to the multivariable case. However, as matrix multiplication is not commutative, we have to distinguish between *left* and *right coprime factorizations*: A pair $\widetilde{N}(s)$, $\widetilde{M}(s)$ of stable transfer function matrices with appropriate dimensions is called a left coprime factorization of $G(s)$ if $G(s) = \widetilde{M}^{-1}(s)\widetilde{N}(s)$, and all common left divisors of $\widetilde{N}(s)$ and $\widetilde{M}(s)$ are invertible in $\mathcal{RH}_\infty$. Similar for the concept of right coprime factorization.

Now we are in a position to define coprime factor perturbations. These are additive stable perturbation terms in both the numerator and the denominator of a left (or right) coprime factorization (see Figure 11.6):

$$\begin{aligned} G &= \widetilde{M}^{-1}\widetilde{N}; \quad \widetilde{M},\, \widetilde{N} \ldots \text{left coprime} \\ G_r &= \underbrace{(\widetilde{M} + \Delta_M)^{-1}}_{\widetilde{M}_r^{-1}} \underbrace{(\widetilde{N} + \Delta_N)}_{\widetilde{N}_r}; \quad \widetilde{M}_r,\, \widetilde{N}_r \ldots \text{left coprime.} \end{aligned} \tag{11.10}$$

Specifically, we will consider the following class of unstructured coprime factor perturbations:

$$\mathcal{D}_{MN} := \left\{ \begin{bmatrix} \Delta_M & \Delta_N \end{bmatrix} \mid \bar{\sigma}[\Delta_M(j\omega)\ \Delta_N(j\omega)] < l_{MN}(\omega);\ \Delta_M, \Delta_N \text{ stable} \right\}. \tag{11.11}$$

A typical perturbation bound $l_{MN}(\omega)$ may look as shown in Figure 11.3. Restricting the class $\mathcal{D}_{MN}$ to stable transfer function matrices does not imply any loss of generality, as both $\widetilde{M}, \widetilde{N}$ (denominator, numerator of the nominal model) and $\widetilde{M} + \Delta_M, \widetilde{N} + \Delta_N$ (denominator, numerator of any perturbed model) are stable by definition. Note that we do not have any restrictive assumptions regarding the number of unstable poles of admissible models $G_r(s)$: Perturbations from the set $\mathcal{D}_{MN}$ can both increase or decrease the

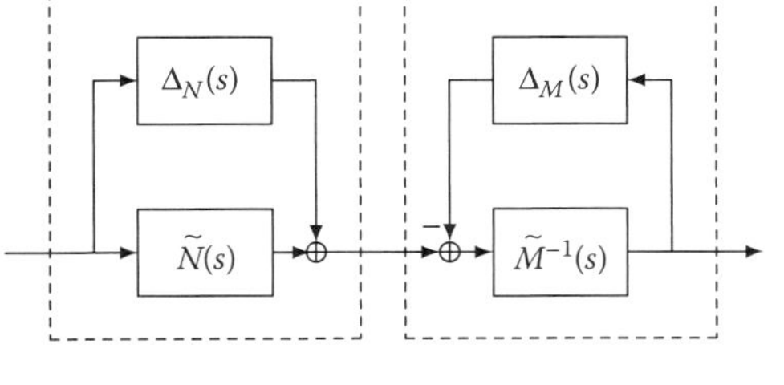

FIGURE 11.6 Coprime factor perturbation.

number of unstable model poles. For example,

$$G(s) = \frac{1}{s+\varepsilon} = \underbrace{\left[\frac{s+\varepsilon}{s+1}\right]^{-1}}_{\widetilde{M}(s)^{-1}} \underbrace{\left[\frac{1}{s+1}\right]}_{\widetilde{N}(s)},$$

$$\Delta_N = 0, \quad \Delta_M(s) = \frac{-2\varepsilon}{s+1},$$

$$G_r(s) = \frac{1}{s-\varepsilon} = \underbrace{\left[\frac{s-\varepsilon}{s+1}\right]^{-1}}_{\widetilde{M}_r(s)^{-1}} \underbrace{\left[\frac{1}{s+1}\right]}_{\widetilde{N}_r(s)}.$$

Coprime factor perturbations are especially useful for describing uncertainty in flexible structures. They allow covering a family of transfer function matrices with slightly damped uncertain pole-pairs by a (relatively speaking) small perturbation set.

Example 11.4:

Consider a nominal transfer function

$$G(s) = \frac{10}{(s+0.05-5j)(s+0.05+5j)}.$$

Suppose a perturbed model has a slightly different pair of poles:

$$G_r(s) = \frac{10}{(s+0.05-6j)(s+0.05+6j)}.$$

We first determine the magnitude of the smallest additive perturbation that, centered around $G(s)$, covers $G_r(s)$, and scale it (at each frequency) by $1/|G(j\omega)|$ (this is of course equivalent to computing the magnitude of the multiplicative perturbation connecting $G(s)$ and $G_r(s)$). The result is shown in the left part of Figure 11.7.

We now compute normalized coprime factorizations, $\{N(s), M(s)\}$ and $\{N_r(s), M_r(s)\}$, for $G(s)$ and $G_r(s)$, and plot the size (maximal singular value) of the perturbation matrix

$$\left[N_r(j\omega) - N(j\omega) \quad M_r(j\omega) - M(j\omega)\right]$$

over frequency. This gives a *relative (scaled)* measure of perturbation size, as both $\left[N(s) \quad M(s)\right]$ and $\left[N_r(s) \quad M_r(s)\right]$ are allpass. It is shown in the right half of Figure 11.7.

Clearly, there is an order of magnitude difference between the peak values of both perturbations.

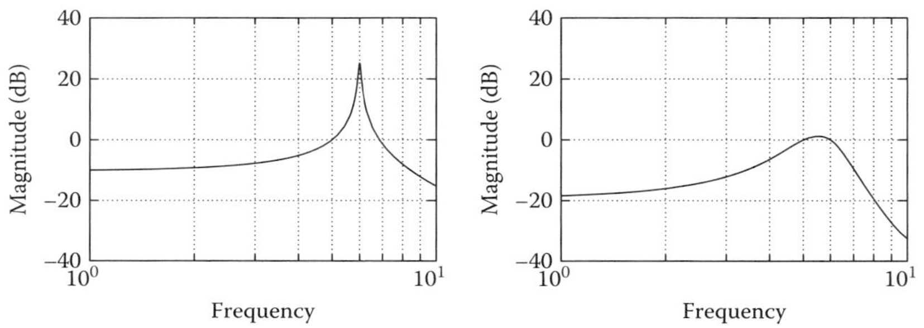

FIGURE 11.7 Perturbation sizes (normalized coprime factorization case shown on the right).

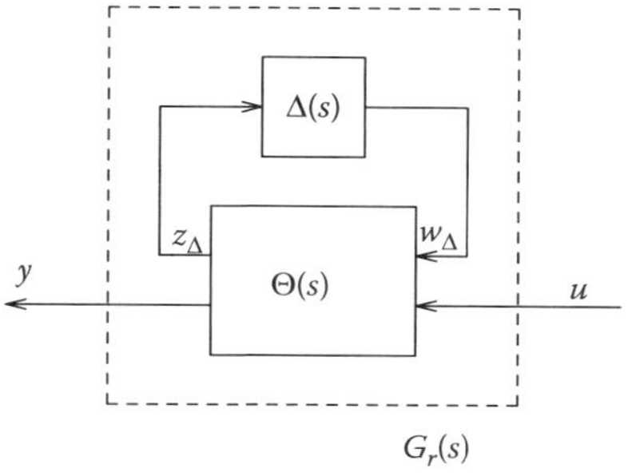

FIGURE 11.8 Generalized unstructured perturbation.

(Normalized) coprime factor perturbations and their use in H_∞-design methods have been extensively investigated in McFarlane and Glover (1989).

11.3.1.4 Generalized Unstructured Perturbations

Additive, multiplicative, and coprime factor uncertainty can be conveniently represented in a unified way—via *generalized unstructured perturbations* (McFarlane and Glover, 1989). For this purpose, we introduce a "new" (proper) transfer function matrix

$$\Theta(s) = \begin{bmatrix} \Theta_{11}(s) & \Theta_{12}(s) \\ \Theta_{21}(s) & \Theta_{22}(s) \end{bmatrix}. \tag{11.12}$$

Its partitioning implies a partitioning of its input and output vector. The lower parts of both vectors represent the "usual" plant input and output signals. The upper part of the output signal, z_Δ, is fed back into the upper part of the input signal, w_Δ, via a perturbation $\Delta(s)$ (Figure 11.8). Denote the number of rows and columns of Δ by m_Δ and l_Δ, respectively. Then, the transfer function matrix G_r of the perturbed model is given as an Upper *Linear Fractional Transformation* of Δ with respect to Θ:

$$G_r = \Theta_{22} + \Theta_{21}\Delta(I_{l_\Delta} - \Theta_{11}\Delta)^{-1}\Theta_{12} \tag{11.13}$$

$$= \Theta_{22} + \Theta_{21}(I_{m_\Delta} - \Delta\Theta_{11})^{-1}\Delta\Theta_{12}. \tag{11.14}$$

For $\Delta = 0$ (zero perturbation), we expect to recover the nominal plant model as the transfer function matrix between the plant input and the output vector. Hence, we must have $\Theta_{22} = G$. For Equations 11.13 and 11.14 to make sense, we also have to assume that

1. $(I_{l_\Delta} - \Theta_{11}(\infty)\Delta(\infty))$ (or, equivalently, $(I_{m_\Delta} - \Delta(\infty)\Theta_{11}(\infty))$) is nonsingular (i.e., G_r is uniquely defined and proper) for all admissible perturbations.
2. The transfer function matrix $\Theta(s)$ is stabilizable from the plant input and detectable from the plant output.

Specifically, we consider the following class of perturbations:

$$\mathcal{D} := \mathcal{D}_1 \cup \mathcal{D}_2, \tag{11.15}$$

$$\mathcal{D}_1 := \left\{\Delta \mid \bar{\sigma}[\Delta(j\omega)] < l(\omega),\ \ \Delta \text{ stable}\right\}, \tag{11.16}$$

$$\mathcal{D}_2 := \left\{\Delta \mid \bar{\sigma}[\Delta(j\omega)] < l(\omega),\ m_{G_r} = m_G\right\} \tag{11.17}$$

We admit all bounded (on $s = j\omega$) perturbation transfer function matrices that are either stable or do not change the number of unstable poles of the plant model.

We have not specified yet how to choose Θ if we want to translate additive, multiplicative, and coprime factor uncertainty into this general framework. By substituting into Equation 11.13, it is easy to check that

- for additive perturbations,

$$\Theta = \begin{bmatrix} 0 & I_q \\ I_p & G \end{bmatrix}, \quad \Delta = \Delta_A, \tag{11.18}$$

- for multiplicative perturbations,

$$\Theta = \begin{bmatrix} 0 & G \\ I_p & G \end{bmatrix}, \quad \Delta = \Delta_M, \tag{11.19}$$

- and for coprime factor perturbations,

$$\Theta = \begin{bmatrix} \begin{bmatrix} -\widetilde{M}^{-1} \\ 0 \end{bmatrix} & \begin{bmatrix} -G \\ I_q \end{bmatrix} \\ \widetilde{M}^{-1} & G \end{bmatrix}, \quad \Delta = \begin{bmatrix} \Delta_M & \Delta_N \end{bmatrix}. \tag{11.20}$$

For additive uncertainty, $\mathcal{D}_1 \subset \mathcal{D}_2$ (stable perturbations do not change the number of unstable model poles). Hence, in this case, $\mathcal{D}$ and $\mathcal{D}_A$ are the same. In the case of coprime factor uncertainty, all perturbation terms are stable by definition. We can therefore write $\mathcal{D} = \mathcal{D}_{MN}$ without restricting generality.

This framework is especially useful for theoretical investigations. Instead of proving, say, robust stability for the three sets of perturbations $\mathcal{D}_A$, $\mathcal{D}_M$, and $\mathcal{D}_{MN}$, it suffices to establish this result for the class $\mathcal{D}$ of generalized perturbations.

11.3.1.5 Which Unstructured Perturbation Model?

All of the above perturbation models are reasonably simple and easy to use in design procedures. Hence, the decision on which model to choose hinges primarily on their potential to cover a given set $\mathcal{G}$ without including too many "physically impossible" plant models. In other words: we want the least conservative perturbation set that "does the job." Whether this is a set of additive, multiplicative, or coprime factor perturbations depends on the problem at hand. A useful rule of thumb is: Whenever one deals with low-damped mechanical systems, it is a good idea to try coprime factor uncertainty first (compare the example in Section 11.3.1.3). For stable nonoscillatory processes, one might try an additive perturbation model first—it is more intuitive, and, if the nominal model is stable, excludes any unstable $G_r(s)$.

11.3.2 Structured Plant Uncertainty

11.3.2.1 Uncertain Parameters

Physical (or theoretical), as opposed to experimental, model building usually results in a state model, and model uncertainty is often in the form of parameter uncertainty: The (real-valued) parameters of the state model have physical meaning and are assumed to lie within given intervals. When converting to a transfer function matrix, however, one usually gets a fairly complicated set of admissible parameters. This is due to the fact that, in general, each parameter of the resulting transfer function matrix depends on several parameters of the underlying state model. Only in exceptionally simple cases (as in the following example) can we expect the set of admissible parameters to form a parallelepiped.

Example 11.5:

Consider the spring–damper system in Figure 11.9. Force is denoted by u, position by y. Newton's law gives the transfer function

$$G(s) = \frac{1}{ms^2 + ds + c}.$$

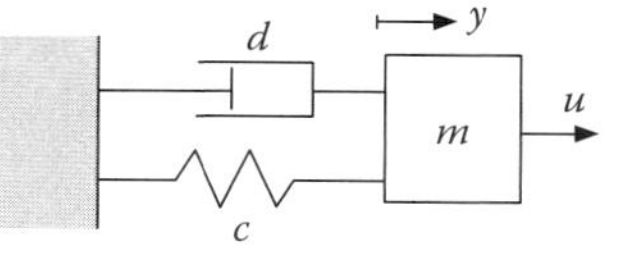

FIGURE 11.9 Spring–damper system.

In most cases, one cannot expect to know the precise values of mass, damping, and spring constant. Hence, it makes sense to define a class of models by stating admissible parameter intervals:

$$\begin{aligned}\Delta_m &\in [m_u - m,\ m_o - m] := \mathcal{D}_m\\ \Delta_d &\in [d_u - d,\ d_o - d] := \mathcal{D}_d\\ \Delta_c &\in [c_u - c,\ c_o - c] := \mathcal{D}_c\end{aligned}$$

$$\mathcal{G}_{mdc} := \left\{ \frac{1}{(m+\Delta_m)s^2 + (d+\Delta_d)s + (c+\Delta_c)} \,\middle|\, \Delta_m \in \mathcal{D}_m,\ \Delta_d \in \mathcal{D}_d,\ \Delta_c \in \mathcal{D}_c \right\}.$$

11.3.2.2 Independent Additive Perturbations in the Elements of a Transfer Function Matrix

Another straightforward (and often very useful) structured uncertainty description is the following: Consider an additive perturbation matrix $\Delta_A(s) = G_r(s) - G(s)$ and give frequency-dependent bounds $l_{ik}(\omega)$ for the magnitude of each of its elements:

$$\underbrace{\begin{bmatrix} |\Delta_{A_{11}}(j\omega)| & \cdots & |\Delta_{A_{1q}}(j\omega)| \\ \vdots & \ddots & \vdots \\ |\Delta_{A_{p1}}(j\omega)| & \cdots & |\Delta_{A_{pq}}(j\omega)| \end{bmatrix}}_{:=|\Delta_A|_e} \leq_e \underbrace{\begin{bmatrix} l_{11}(\omega) & \cdots & l_{1q}(\omega) \\ \vdots & \ddots & \vdots \\ l_{p1}(\omega) & \cdots & l_{pq}(\omega) \end{bmatrix}}_{:=L(\omega)} \tag{11.21}$$

($\leq_e$ denotes "elementwise less or equal"; $|\Delta_A|_e$ is a matrix with entries $|\Delta_{A_{ik}}|$). Again, we make the additional assumption that no admissible perturbation shall change the number of unstable poles. Hence, we get the following classes of perturbations and plant models:

$$\mathcal{D}_{Ae} := \left\{\Delta_A \mid |\Delta_A(j\omega)|_e \leq_e L(\omega);\ m_{G_r} = m_G\right\} \tag{11.22}$$

$$\mathcal{G}_{Ae} := \{G + \Delta_A \mid \Delta_A \in \mathcal{D}_{Ae}\}. \tag{11.23}$$

Such a set of models can be easily represented in a graphical way (Figure 11.10): Consider first the Nyquist plots of each element $g_{ik}(j\omega)$ of the nominal model ("Nyquist array"). For each frequency, draw a circle with radius $l_{ik}(\omega)$ around $g_{ik}(j\omega)$. This gives a set of bands covering the nominal Nyquist plots. Then, a transfer function matrix G_r is a member of the model set (Equation 11.23) if and only if the Nyquist plot of each element $g_{r_{ik}}(j\omega)$ is contained in the appropriate band, and G_r and G have the same number of unstable poles. The usefulness of this uncertainty model for controller design purposes stems mainly from the fact that it fits "naturally" into the framework of Nyquist-array methods. Stability robustness results can be found in Owens and Chotai (1984) and Lunze (1984).

11.3.2.3 Generalized Structured Plant Uncertainty

A more general structured perturbation model has become very popular, as the so-called μ-*theory* (Doyle, 1982, 1985) provides analysis and synthesis tools to deal with such uncertainty sets. As a motivation for this general perturbation model, consider the following example from Maciejowski (1989).

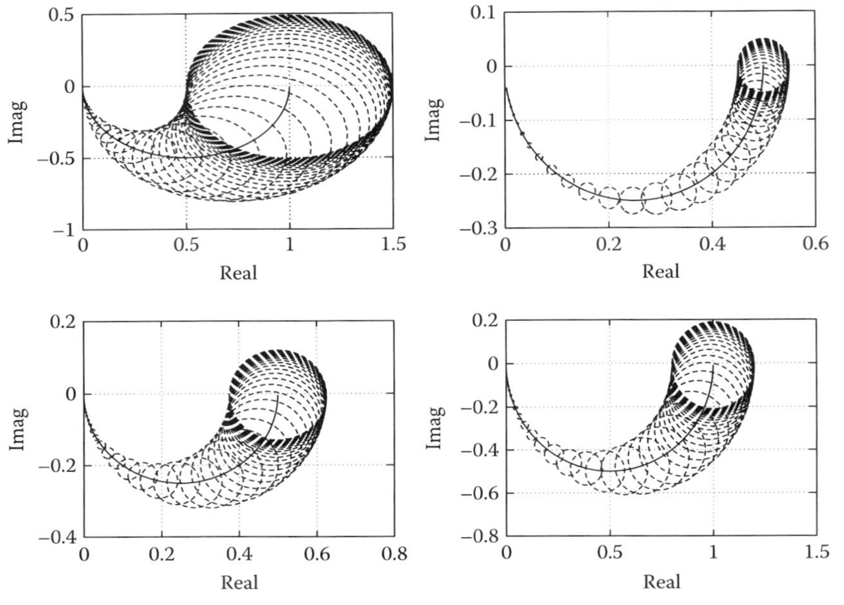

FIGURE 11.10 Graphical representation of the model class $\mathcal{G}_{Ae}$.

Example 11.6:

Let $G(s)$ be a 2×2 plant model with additive unstructured uncertainty $\Delta_A(s)$ and independent multiplicative perturbations $\tilde{\delta}_1(s), \tilde{\delta}_2(s)$ in each input channel (Figure 11.11). Assume that $\Delta_A(s), \tilde{\delta}_1(s)$, and $\tilde{\delta}_2(s)$ are stable, proper, and bounded by

$$|\tilde{\delta}_1(j\omega)| < l_1(\omega),$$
$$|\tilde{\delta}_2(j\omega)| < l_2(\omega),$$
$$\bar{\sigma}[\Delta_A(j\omega)] < l_3(\omega).$$

It is easy to see that we could subsume all three uncertainty terms in a single (additive) perturbation matrix $\overline{\Delta_A}(s)$:

$$\begin{aligned} G_r &= (G + \Delta_A)\left(I_2 + \begin{bmatrix} \tilde{\delta}_1 & 0 \\ 0 & \tilde{\delta}_2 \end{bmatrix}\right) \\ &= G + \underbrace{\left(\Delta_A \begin{bmatrix} 1+\tilde{\delta}_1 & 0 \\ 0 & 1+\tilde{\delta}_2 \end{bmatrix} + G \begin{bmatrix} \tilde{\delta}_1 & 0 \\ 0 & \tilde{\delta}_2 \end{bmatrix}\right)}_{:=\overline{\Delta}_A}. \end{aligned}$$

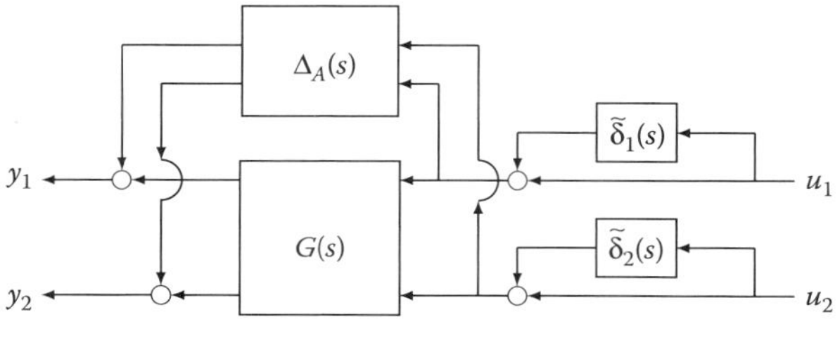

FIGURE 11.11 Example for a model with independent perturbations.

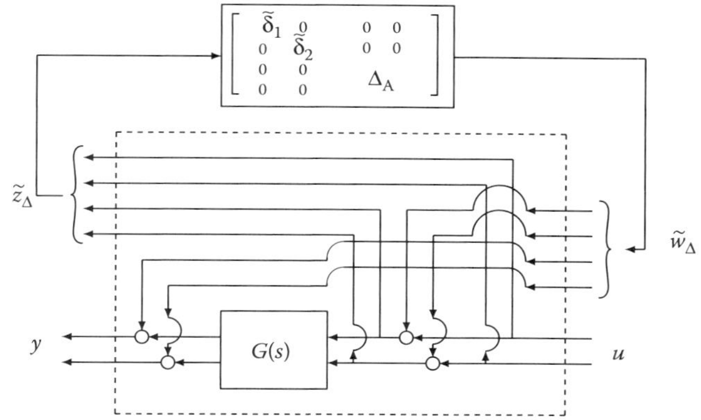

FIGURE 11.12 Generalized structured perturbation model.

Using the properties of singular values, we can derive both structured and unstructured bounds for the overall perturbation $\overline{\Delta}_A$:

$$\bar{\sigma}[\overline{\Delta}_A(j\omega)] \leq \bar{\sigma}[\Delta_A(j\omega)]\,\bar{\sigma}\begin{bmatrix}1+\tilde{\delta}_1(j\omega) & 0\\ 0 & 1+\tilde{\delta}_2(j\omega)\end{bmatrix} + \bar{\sigma}[G(j\omega)]\,\bar{\sigma}\begin{bmatrix}\tilde{\delta}_1(j\omega) & 0\\ 0 & \tilde{\delta}_2(j\omega)\end{bmatrix}$$
$$< l_3(\omega)\ \left(1+\max(l_1(\omega), l_2(\omega))\right) + \bar{\sigma}[G(j\omega)]\max(l_1(\omega), l_2(\omega))$$

or

$$\left|\overline{\Delta}_{Aik}(j\omega)\right| < l_3(\omega)(1+l_k(\omega)) + |g_{ik}(j\omega)|l_k(\omega).$$

In both cases, the perturbation bounds will probably be very *conservative*, that is, the resulting uncertainty class will cover more than the perturbations $\Delta_A(s)$, $\tilde{\delta}_1(s)$, and $\tilde{\delta}_2(s)$ we started off with. It is therefore a good idea to preserve perturbation structure when combining different "sources" of model uncertainty. This can be accomplished in the following way: As in Section 11.3.1.4, we define a "new" transfer function matrix $\Theta(s)$ that—apart from the plant input and output—contains another pair of input and output vectors, $\tilde{w}_\Delta$ and $\tilde{z}_\Delta$. $\tilde{z}_\Delta$ is fed back into $\tilde{w}_\Delta$ via a *blockdiagonal* perturbation matrix. Graphically, this corresponds to "pulling out" all perturbations from Figure 11.11 and rearranging them in a blockdiagonal structure (Figure 11.12).

Note that Figure 11.12 looks like Figure 11.8—the only difference being the blockdiagonal structure of the perturbation matrix in the feedback loop—that mirrors the structure of the underlying perturbation model.

In the general case, one proceeds in exactly the same way: The class of admissible models is represented by an Upper Linear Fractional Transformation of a blockdiagonal perturbation transfer function matrix $\Delta_s(s)$ with respect to a suitably defined transfer function matrix $\Theta(s)$:

$$G_r = \mathcal{F}_U(\Theta, \Delta_s). \tag{11.24}$$

Structure and dimension of Δ_s depend of course on the number of independent perturbation terms and their dimensions. Without loss of generality, we can assume that scalar uncertainty is always listed first in Δ_s. In this general framework, we can also restrict Δ_s to be stable: If unstable perturbations have to be considered, we can always circumvent this problem by introducing a coprime factorization with stable perturbations. Often, notation is simplified by normalizing the size of the perturbation blocks. This can be easily done, if each perturbation bound $l_i(\omega)$ can be written as magnitude of a frequency response $w_i(j\omega)$: In this case, we just have to multiply Θ_{11} and Θ_{12} from the left (or Θ_{11} and Θ_{21} from the right) by a suitably dimensioned diagonal matrix containing the $w_i(s)$. As only magnitude is important, we can always choose the transfer functions $w_i(s)$ to be stable and minimum-phase. Hence, we get the following class of perturbations:

$$\mathcal{D}_s := \left\{ \Delta_s \;\middle|\; \Delta_s = \begin{bmatrix} \delta_1 I_{l_1} & 0 & \dots & & & 0 \\ 0 & \ddots & & & & \vdots \\ \vdots & & \delta_k I_{l_k} & & & \\ & & & \Delta_{k+1} & & \\ & & & & \ddots & 0 \\ 0 & \dots & & & 0 & \Delta_v \end{bmatrix}, \ \Delta_s \text{ stable}, \ \bar{\sigma}[\Delta_s(j\omega)] < 1 \right\}. \tag{11.25}$$

Example 11.7:

Let us look at the previous example again: Suppose $w_1(s)$, $w_2(s)$, and $w_3(s)$ are stable minimum-phase transfer functions with

$$|w_1(j\omega)| = l_1(\omega),$$
$$|w_2(j\omega)| = l_2(\omega),$$
$$|w_3(j\omega)| = l_3(\omega).$$

Then we get

$$\Theta(s) = \begin{bmatrix} \begin{bmatrix} 0 & 0 \\ w_3 I_2 & 0 \end{bmatrix} & \begin{bmatrix} \text{diag}\{w_1, w_2\} \\ w_3 I_2 \end{bmatrix} \\ \begin{bmatrix} G & I_2 \end{bmatrix} & G \end{bmatrix}$$

$$\Delta_s = \begin{bmatrix} \delta_1 & 0 & 0 & 0 \\ 0 & \delta_2 & 0 & 0 \\ 0 & 0 & \multicolumn{2}{c}{\multirow{2}{*}{Δ_3}} \\ 0 & 0 & & \end{bmatrix}$$

$$= \begin{bmatrix} \frac{\tilde{\delta}_1}{w_1} & 0 & 0 & 0 \\ 0 & \frac{\tilde{\delta}_2}{w_2} & 0 & 0 \\ 0 & 0 & \multicolumn{2}{c}{\multirow{2}{*}{$\frac{1}{w_3}\Delta_A$}} \\ 0 & 0 & & \end{bmatrix}$$

and—as expected –

$$G_r = \mathcal{F}_U(\Theta, \Delta_s)$$
$$= G + \left(\Delta_A \begin{bmatrix} 1+\tilde{\delta}_1 & 0 \\ 0 & 1+\tilde{\delta}_2 \end{bmatrix} + G \begin{bmatrix} \tilde{\delta}_1 & 0 \\ 0 & \tilde{\delta}_2 \end{bmatrix} \right).$$

All structured perturbation models in this section can be written as in Equations 11.24 and 11.25. However, when "translating" parameter perturbations into this framework, one has to take into account that $\mathcal{D}_s$ admits complex perturbations, whereas parameters (and therefore parameter perturbations) in a state or transfer function model are real.

11.4 Model Validation

Model validation is understood to be the procedure of establishing whether a set of experimental data is compatible with given signal and plant uncertainty models, $\mathcal{W}$ and $\mathcal{G}$. It is *not* the (futile) attempt to show that an uncertainty model can *always* explain the true plant's input/output behavior—future experiments might well provide data that are inconsistent with $\mathcal{W}$ and $\mathcal{G}$.

Model validation is a rapidly expanding area, and—for the lack of space—we can only hope to give a flavor of the subject by looking at a comparatively simple version of the problem: We assume that

1. $\mathcal{W}$ is a singleton (i.e., there is no signal uncertainty).
2. Plant uncertainty is in the form of stable generalized unstructured perturbation (i.e., $G_r(s) = \mathcal{F}_U(\Theta(s), \Delta(s))$, where $\Delta(s) \in \mathcal{D}_1$—see Section 11.3.1.4), and $\Theta_{21}(s)$ is invertible.
3. Experimental data are given in the form of M frequency-response measurements $G_r(j\omega_1), \ldots, G_r(j\omega_M)$.

The plant uncertainty model of item 2 contains the cases of coprime factor perturbations and of stable additive and multiplicative perturbations.

With the invertibility condition for Θ_{21} in force, we can rewrite the perturbed plant model as follows:

$$G_r = \mathcal{F}_U(\Theta, \Delta) \tag{11.26}$$

$$= \Theta_{22} + \Theta_{21}(I_{m_\Delta} - \Delta\Theta_{11})^{-1}\Delta\Theta_{12} \tag{11.27}$$

$$= \left(\Delta\Gamma_{12} + \Gamma_{22}\right)^{-1}\left(\Delta\Gamma_{11} + \Gamma_{21}\right), \tag{11.28}$$

where

$$\begin{bmatrix} \Gamma_{11} & \Gamma_{12} \\ \Gamma_{21} & \Gamma_{22} \end{bmatrix} := \begin{bmatrix} \Theta_{12} - \Theta_{11}\Theta_{21}^{-1}\Theta_{22} & -\Theta_{11}\Theta_{21}^{-1} \\ \Theta_{21}^{-1}\Theta_{22} & \Theta_{21}^{-1} \end{bmatrix}. \tag{11.29}$$

This can be easily checked by substituting Equation 11.29 into Equation 11.28. Multiplying Equation 11.28 by $(\Delta\Gamma_{12} + \Gamma_{22})$ from the left gives

$$\Delta\underbrace{(\Gamma_{12}G_r - \Gamma_{11})}_{:=W} = \underbrace{\Gamma_{21} - \Gamma_{22}G_r}_{:=U}. \tag{11.30}$$

Consistency of experimental data $G_r(j\omega_1), \ldots, G_r(j\omega_M)$ and uncertainty model is equivalent to the existence of a transfer function matrix $\Delta(s) \in \mathcal{D}_1$ that solves Equation 11.30 for $\omega_1, \ldots, \omega_M$. Clearly, a necessary condition for this is that the system of linear equations over $\mathbb{C}$

$$\Delta_i W(j\omega_i) = U(j\omega_i) \tag{11.31}$$

has a solution Δ_i with $\bar{\sigma}[\Delta_i] < l(\omega_i)$ for each $i \in \{1, \ldots, M\}$. Using interpolation theory, it has been shown in Boulet and Francis (1994) that this condition is also sufficient*. If suitable Δ_i exist, one can always find a stable transfer function matrix $\Delta(s)$ such that $\Delta(j\omega_i) = \Delta_i$, $i = 1, \ldots, M$, and $\bar{\sigma}[\Delta(j\omega)] < l(\omega)$ for all $\omega \in \mathbb{R}$.

11.5 Further Reading

Any book on robust control contains information on plant model uncertainty. We especially recommend Vidyasagar (1985), McFarlane and Glover (1989), Lunze (1989), Glad and Ljung (2000), and Skogestad and Postlethwaite (2005). Parts of this chapter are based on Raisch (1994). See also Doyle et al. (1992).

* Their proof is for coprime factor perturbations. It carries over to the slightly more general case considered here.

On the problem of fitting a model to frequency-response data, see Hindi et al. (2002) and Balas et al. (2009), and from time-domain data see Volker and Engell (2005).

Several papers related to model validation can be found in a special issue of the *IEEE Transactions on Automatic Control* dealing with "System identification for robust control design" (IEEE, 1992).

References

Balas, G. J., A. K. Packard, and P. J. Seiler. Uncertain model set calculation from frequency domain data. In P. M. J. Hof, C. Scherer, and P. S. C. Heuberger (Eds). *Model-Based Control*. Springer, Berlin, 2009.

Boulet, B. and B. Francis. Consistency of open-loop experimental frequency–response data with coprime factor plant models. *IEEE Transactions on Automatic Control*, 43, 1680–1691, 1998.

Cruz, J., J. S. Freudenberg, and D. P. Looze. A relationship between sensitivity and stability of multivariable feedback systems. *IEEE Transactions Automatic Control*, 26, 66–74, 1981.

Doyle, J. C. Analysis of feedback systems with structured uncertainties. *Proceedings of IEE Part D*, 129, 242–250, 1982.

Doyle, J. C. Structured uncertainty in control system design. In 24th *IEEE Conference on Decision and Control*, 260–265, 1985.

Doyle, J. C., B. A. Francis, and A. R. Tannenbaum. *Feedback Control Theory*. Macmillan, 1992. Republished by Dover, New York, 2008.

Doyle, J. C., and G. Stein. Multivariable feedback design: Concepts for a classical/modern synthesis. *IEEE Transactions Automatic Control*, 26, 4–16, 1981.

Glad, T. and L. Ljung. *Control Theory: Multivariable and Nonlinear Methods*. Taylor & Francis, New York, 2000.

Hindi, H., C.-Y. Seong, and S. Boyd. Computing optimal uncertainty models from frequency domain data. In 41st *IEEE Conference on Decision and Control*, 2898–2905, 2002.

Special issue on system identification for robust control design. *IEEE Transactions on Automatic Control*, Vol. 37, July 1992.

Lunze, J. Robustness tests for feedback control systems using multidimensional uncertainty bounds. *System and Control Letters*, 4, 85–89, 1984.

Lunze, J. *Robust Multivariable Control*. Prentice-Hall, Englewood Cliffs, NJ, 1989.

Maciejowski, J. M. *Multivariable Feedback Design*. Addison-Wesley, Reading, MA, 1989.

McFarlane, D. and K. Glover. *Robust Controller Design using Normalized Coprime Factor Plant Descriptions*, Vol. 138 Lecture Notes in Control and Information Sciences. Springer-Verlag, Berlin, 1990.

Owens, D. H. and A. Chotai. On eigenvalues, eigenvectors, and singular values in robust stability analysis. *International Journal Control*, 40, 285–296, 1984.

Raisch, J. *Mehrgrößenregelung im Frequenzbereich*. R. Oldenbourg-Verlag, München/Wien, 1994.

Skogestad, S. and I. Postlethwaite. *Multivariable Feedback Control: Analysis and Design* (2nd edition), Wiley, New York, 2005.

Vidyasagar, M. *Control System Synthesis*. MIT Press, Cambridge, Mass, 1985.

Vidyasagar, M. Normalized comprime factorizations for non-strictly proper systems. *IEEE Trans. Automatic Control*, 33, 300–301, 1988.

Volker, M. and S. Engell. Computation of tight uncertainty bounds from time-domain data with application to a reactive distillation column. In 41st *IEEE Conference on Decision and Control*, 2939–2944, 2002.

II

Kalman Filter and Observers

12

Linear Systems and White Noise

William S. Levine
University of Maryland

12.1 Introduction

A linear system with white noise added to the input and, often but not always, white noise added to the output is the most common model for randomness in control systems. It is the basis for Kalman filtering and the linear quadratic Gaussian (LQG) or H_2 optimal regulator. This chapter presents the basic facts about linear systems and white noise and the intuition and ideas underlying them. Results are emphasized, not mathematically rigorous proofs. It is assumed that the reader is familiar with the elementary aspects of probability at, for example, the level of Leon-Garcia [1].

There are many reasons why people so commonly use a linear system driven by white noise as a model of randomness despite the fact that no system is truly linear and no noise is truly white. One of the reasons is that such a model is both elementary and tractable. The calculations are easy. The results can be understood without a deep knowledge of the mathematics of stochastic processes.

A second reason is that a large class of stochastic processes can be represented as the output of a linear system driven by white noise. The precise result can be found later in this chapter. The generality and tractability of this model can be quite dangerous because they often lead people to use it inappropriately. Some of the limitations of the model will also be discussed.

Scalar discrete-time stochastic processes are described first, because this is the simplest case. This is followed by a discussion of the ways single-input single-output (SISO) discrete-time linear systems operate on scalar discrete-time stochastic processes. Vector discrete-time stochastic processes and multiple-input multiple-output (MIMO) linear systems, a notationally more difficult but conceptually identical situation, are then briefly covered.

The second half of this chapter describes continuous-time stochastic processes and linear systems in the same order as was used for discrete time.

12.2 Discrete Time

The only difference between an n-dimensional vector random variable and a scalar discrete-time stochastic process over n time steps is in how they are interpreted. Mathematically, they are identical. As an aid to understanding discrete-time stochastic processes, this equivalence will be emphasized in the following.

12.2.1 Basics

The usual precise mathematical definition of a random variable [2,3,9] is unnecessary here. It is sufficient to define an n-dimensional random variable by means of its probability density function.

Definition 12.1:

An n-dimensional random variable, denoted $\underline{\mathbf{x}} = [\mathbf{x}_1\ \mathbf{x}_2 \ldots \mathbf{x}_n]'$, *takes values* $\underline{x} \in R^n$ *according to a probability that can be determined from the probability density function (pdf)* $p_{\underline{\mathbf{x}}}(\underline{x})$.

Unfortunately, the notation needed to describe stochastic processes precisely is very complicated. Thus, it is important to recognize that bold letters always denote random variables while the values that may be taken by a random variable are denoted by standard letters. For example, $\mathbf{x}$ is a random variable that may take values x. Underlined lower case letters will always denote vectors or discrete-time stochastic processes. Underlined capital letters will denote matrices. A list of all the notation used in this chapter appears at the end of the chapter. The reader is assumed to know the basic properties of pdfs.

Definition 12.2:

A scalar discrete-time stochastic process over n time steps, $\mathbf{x}(1), \mathbf{x}(2), \ldots, \mathbf{x}(n)$, *denoted* $\underline{\mathbf{x}}$, *takes values* $x(k) \in R$, *where R denotes the real numbers and* $k = 1, 2, \ldots, n$, *according to a probability that can be determined from the pdf* $p_{\underline{\mathbf{x}}}(\underline{x})$.

The equivalence of the two definitions is obvious once $\mathbf{x}(1), \mathbf{x}(2), \ldots, \mathbf{x}(n)$ is written as a vector $[\mathbf{x}(1)\ \mathbf{x}(2) \ldots \mathbf{x}(n)]' = [\mathbf{x}_1\ \mathbf{x}_2 \ldots \mathbf{x}_n]'$, where $'$ denotes the transpose. The equivalence is emphasized by using the same notation for both. The context will make it clear which is meant whenever it matters.

It is important to be able to specify the relationships among a collection of random variables. A simple special case is that they are completely unrelated.

Definition 12.3:

The n-vector random variable (equivalently, discrete-time stochastic process) $\underline{\mathbf{x}}$, *is composed of n independent random variables* $\mathbf{x}_1, \mathbf{x}_2, \ldots, \mathbf{x}_n$ *if and only if their joint pdf has the form*

$$p_{\underline{\mathbf{x}}}(\underline{x}) = \prod_{k=1}^{n} p_{\mathbf{x}_k}(x_k) = p_{\mathbf{x}_1}(x_1)p_{\mathbf{x}_2}(x_2)\ldots p_{\mathbf{x}_n}(x_n). \tag{12.1}$$

Independence is an extreme case. More typically, and more interestingly, the individual elements of a vector random variable or a scalar discrete-time stochastic process are related. One way to characterize these relationships is by means of the conditional pdf, $p_{\underline{\mathbf{x}}_1|\underline{\mathbf{x}}_2}(\underline{x}_1|\underline{\mathbf{x}}_2)$. When $p_{\underline{\mathbf{x}}_2}(\underline{x}_2) \neq 0$, the conditional pdf is given by

$$p_{\underline{\mathbf{x}}_1|\underline{\mathbf{x}}_2}(\underline{x}_1|\underline{\mathbf{x}}_2) = \frac{p_{\underline{\mathbf{x}}}(\underline{x})}{p_{\underline{\mathbf{x}}_2}(\underline{x}_2)}, \tag{12.2}$$

where $\underline{\mathbf{x}}_1$ is the $m-$vector $[\mathbf{x}_1\ \mathbf{x}_2 \ldots \mathbf{x}_m]'$; $\underline{\mathbf{x}}_2$ is the $(n-m)-$vector $[\mathbf{x}_{m+1}\ \mathbf{x}_{m+2} \ldots \mathbf{x}_n]'$; $\underline{\mathbf{x}}$ is the $n-$vector $[\mathbf{x}_1\ \mathbf{x}_2 \ldots \mathbf{x}_n]'$; and the $\underline{x}_i$s have the same dimensions as the corresponding $\underline{\mathbf{x}}_i$s. It is often possible to avoid working directly with pdfs, especially in the study of linear systems. Instead, one uses expectations.

12.2.2 Expectation

Definition 12.4:

The expected value, expectation, or mean, of a scalar random variable, $\mathbf{x}$, *is denoted by* $E(\mathbf{x})$ *or* m *and given by*

$$m \triangleq E(\mathbf{x}) \triangleq \int_{-\infty}^{\infty} x p_{\mathbf{x}}(x)\, dx. \tag{12.3}$$

Applying Definition 12.4 to the scalar random variable $\mathbf{x}(k)$, the kth element of the scalar discrete-time stochastic process, $\underline{\mathbf{x}}$, gives

$$m(k) \triangleq E(\mathbf{x}(k)) = \int_{-\infty}^{\infty} \cdots \int_{-\infty}^{\infty} x(k) p_{\underline{\mathbf{x}}}(\underline{x}) dx_1 dx_2 \ldots dx_n. \tag{12.4}$$

Note that the integrations over $x_\ell, \ell \neq k$, simply give the necessary marginal density $p_{\mathbf{x}(k)}(x(k))$.

The computation in Equation 12.4 can be repeated for all $k = 1, 2, \ldots, n$ and the result is organized as an n-vector

$$E(\underline{\mathbf{x}}) = [E(\mathbf{x}(1))\ E(\mathbf{x}(2)) \ldots E(\mathbf{x}(n))]'. \tag{12.5}$$

One can also take the expectation with respect to functions of $\underline{\mathbf{x}}$. Two of these are particularly important.

Definition 12.5:

The covariance of the scalar random variables, $\mathbf{x}_k$ *and* $\mathbf{x}_\ell$, *is denoted by* $r_{k\ell}$, *and given by*

$$r_{k\ell} \triangleq E((\mathbf{x}_k - m(k))(\mathbf{x}_\ell - m(\ell))) = \int_{-\infty}^{\infty}\int_{-\infty}^{\infty} (x_k - m(k))(x_\ell - m(\ell)) p_{\mathbf{x}_k\mathbf{x}_\ell}(x_k, x_\ell) dx_k dx_\ell. \tag{12.6}$$

Definition 12.5 can be applied to each pair $\mathbf{x}(k)\mathbf{x}(\ell), k, \ell = 1, 2, \ldots, n$ of elements of the discrete-time scalar stochastic process $\underline{\mathbf{x}}$. The result is a collection of n^2 elements $r_{k\ell}, k, \ell = 1, 2 \ldots n$. It is conventional to emphasize the time dependence by defining

$$r(k, \ell) \triangleq r_{k\ell}. \tag{12.7}$$

It should be obvious from Equation 12.6 that

$$r(k,\ell) = r(\ell,k) \quad \text{for all} \quad k,\ell \tag{12.8}$$

and that (because $p_{\underline{\mathbf{x}}}(\underline{x}) \geq 0$ for all $\underline{x}$),

$$r(k,k) \geq 0 \quad \text{for all } k. \tag{12.9}$$

In the context of discrete-time stochastic processes, $r(k,\ell)$ is known as the autocovariance function of the stochastic process $\underline{\mathbf{x}}$. It can be helpful, in trying to understand the properties of the autocovariance function, to write it as a matrix.

$$\underline{R} = \begin{bmatrix} r(1,1) & r(1,2) & \dots & r(1,n) \\ r(2,1) & r(2,2) & & \\ \vdots & & \ddots & \vdots \\ r(n,1) & \dots & & r(n,n) \end{bmatrix}. \tag{12.10}$$

In the context of n-vector random variables, the matrix $\underline{R}$ is known as the covariance matrix. The autocovariance will be discussed further subsequently. Another important expectation will be defined first.

Definition 12.6:

The characteristic function of a scalar random variable, $\mathbf{x}$, is denoted by $f_{\mathbf{x}}(\omega)$ and given by

$$f_{\mathbf{x}}(\omega) \triangleq E(e^{j\omega\mathbf{x}}) = \int_{-\infty}^{\infty} e^{j\omega x} p_{\mathbf{x}}(x)\, dx, \tag{12.11}$$

where

$$j \triangleq \sqrt{-1}.$$

Note that $f_{\mathbf{x}}(\omega)$ is a deterministic function of ω, not a random variable. Note also that $f_{\mathbf{x}}(-\omega)$ is the Fourier transform of $p_{\mathbf{x}}(x)$ and is thus equivalent to $p_{\mathbf{x}}(x)$ in the same way that Fourier transform pairs are usually equivalent; there is a unique correspondence between a function and its transform.

The generalization to the case of n-vector random variables or discrete-time stochastic processes over n time steps is as follows.

Definition 12.7:

The characteristic function of an n-vector random variable (or discrete-time stochastic process), $\underline{\mathbf{x}}$, is denoted $f_{\underline{\mathbf{x}}}(\underline{\omega})$ and given by

$$f_{\underline{\mathbf{x}}}(\underline{\omega}) = E\left(e^{j\sum_{k=1}^{n} \mathbf{x}_k \omega_k}\right) = E(e^{j\underline{\mathbf{x}}'\underline{\omega}}). \tag{12.12}$$

The characteristic function is particularly useful for studying the effect of linear mappings on a stochastic process. As an example, consider the operation of an $m \times n$ real matrix $\underline{L}$ on the n-vector random variable $\underline{\mathbf{x}}$.

Let $\underline{\mathbf{y}} = \underline{L}\underline{\mathbf{x}}$;

$$f_{\underline{\mathbf{y}}}(\underline{\omega}) = E(e^{j\underline{\omega}'\underline{\mathbf{y}}}) = E(e^{j\underline{\omega}'\underline{L}\underline{\mathbf{x}}}) = E(e^{j(\underline{L}'\underline{\omega})'\underline{\mathbf{x}}}) = f_{\underline{\mathbf{x}}}(\underline{L}'\underline{\omega}). \tag{12.13}$$

12.2.3 Example: Discrete-Time Gaussian Stochastic Processes

Definition 12.8:

An n-vector random variable (or discrete-time stochastic process), $\underline{\mathbf{x}}$, *is a Gaussian (normal) random variable (discrete-time stochastic process) if and only if it has the n-dimensional Gaussian (normal) pdf*

$$p_{\underline{\mathbf{x}}}(\underline{x}) = \frac{1}{(2\pi)^{n/2}(det\underline{R})^{1/2}} e^{-\frac{1}{2}(\underline{x}-\underline{m})'\underline{R}^{-1}(\underline{x}-\underline{m})}, \tag{12.14}$$

where

$$\underline{m} = E(\underline{\mathbf{x}}),$$
$$\underline{R} = E((\underline{\mathbf{x}} - \underline{m})(\underline{\mathbf{x}} - \underline{m})'), \quad \textit{the covariance matrix of } \underline{\mathbf{x}}.$$

The definition implies that $\underline{R}$ is symmetric ($\underline{R} = \underline{R}'$, see Equations 12.8 and 12.10 and that $\underline{R}$ must be positive definite ($\underline{y}'\underline{R}\underline{y} > 0$ for all n-vectors $\underline{y} \neq 0$ and often indicated by $\underline{R} > 0$).

It is easy to demonstrate that the characteristic function of the n-vector Gaussian random variable, $\underline{\mathbf{x}}$, with mean $\underline{m}$ and covariance matrix $\underline{R}$ is

$$f_{\underline{\mathbf{x}}}(\underline{\omega}) = e^{j\underline{\omega}'\underline{m} - \frac{1}{2}\underline{\omega}'\underline{R}\underline{\omega}}. \tag{12.15}$$

Any n-vector random variable or stochastic process that has a characteristic function in the form of Equation 12.15 is Gaussian. In fact, some authors [2] define a scalar Gaussian random variable by Equation 12.15 with m and R scalars, rather than by Equation 12.14. The reason is that Equation 12.15 is well defined when $R = 0$, whereas Equation 12.14 blows up. A scalar Gaussian random variable $\mathbf{x}$ with variance $R = 0$ and mean m makes perfectly good sense. It is the deterministic equality $\mathbf{x} = m$.

One other special case of the n-vector Gaussian random variable is particularly important. When $r_{k\ell} = 0$ for all $k \neq \ell$, $\underline{R}$ is a diagonal matrix. The pdf becomes

$$\begin{aligned} p_{\underline{\mathbf{x}}}(\underline{x}) &= \frac{1}{(2\pi)^{n/2} r_{11}^{1/2} r_{22}^{1/2} \ldots r_{nn}^{1/2}} e^{-\frac{1}{2}\left(\frac{(x_1-m_1)^2}{r_{11}} + \frac{(x_2-m_2)^2}{r_{22}} + \cdots + \frac{(x_n-m_n)^2}{r_{nn}}\right)} \\ &= \left(\frac{1}{\sqrt{2\pi} r_{11}^{1/2}} e^{-\frac{1}{2}\frac{(x_1-m_1)^2}{r_{11}}}\right)\left(\frac{1}{\sqrt{2\pi} r_{22}^{1/2}} e^{-\frac{1}{2}\frac{(x_2-m_2)^2}{r_{22}}}\right) \cdots \left(\frac{1}{\sqrt{2\pi} r_{nn}^{1/2}} e^{\frac{1}{2}\frac{(x_n-m_n)^2}{r_{nn}}}\right) \\ &= p_{\mathbf{x}_1}(x_1) p_{\mathbf{x}_2}(x_2) \ldots p_{\mathbf{x}_n}(x_n). \end{aligned} \tag{12.16}$$

In other words, the $\mathbf{x}_k$ are independent random variables. It is an important property of Gaussian random variables that they are independent if and only if their covariance matrix is diagonal, as has just been proven.

Finally, the characteristic function will be used to prove that if $\underline{y} = \underline{L}\underline{\mathbf{x}}$, where $\underline{L}$ is an $m \times n$ real matrix and $\underline{\mathbf{x}}$ is an n-vector Gaussian random variable with mean $\underline{m}$ and covariance $\underline{R}$, then $\underline{y}$ is an m-vector Gaussian random vector with mean $\underline{Lm}$ and covariance $\underline{LRL'}$. The proof is as follows:

$$\underline{\mathbf{y}} = \underline{L}\underline{\mathbf{x}},$$
$$f_{\underline{\mathbf{y}}}(\underline{\omega}) = f_{\underline{\mathbf{x}}}(\underline{L}'\underline{\omega})$$

by Equation 12.15

$$= e^{j(\underline{L}'\underline{\omega})'\underline{m} - \frac{1}{2}(\underline{L}'\underline{\omega})'\underline{R}(\underline{L}\underline{\omega})} = e^{j\underline{\omega}'\underline{L}\underline{m} - \frac{1}{2}\underline{\omega}'\underline{L}\underline{R}\underline{L}'\underline{\omega}} = e^{j\underline{\omega}'(\underline{L}\underline{m}) - \frac{1}{2}\underline{\omega}'(\underline{L}\underline{R}\underline{L}')\underline{\omega}}. \tag{12.17}$$

Finally, the uniqueness of Fourier transforms (characteristic functions), and the fact that Equation 12.17 is the characteristic function of an m-vector Gaussian random variable with mean $\underline{Lm}$ and covariance matrix $\underline{LRL'}$, completes the proof.

Define

$$\mathbf{m}_\ell = \frac{1}{2\ell} \sum_{k=-\ell}^{\ell} \mathbf{x}_k, \tag{12.26}$$

$$\mathbf{r}_\ell(k) = \frac{1}{2\ell} \sum_{i=-\ell}^{\ell} \mathbf{x}_{i+k}\mathbf{x}_i. \tag{12.27}$$

Note that $\mathbf{m}_\ell$ and $\mathbf{r}_\ell(k)$ are denoted by bold letters in Equations 12.26 and 12.27. This indicates that they are random variables, unlike the expectations m and $r(k)$, which are deterministic quantities. It can then be proved that

$$\lim_{\ell \to \infty} E[(\mathbf{m}_\ell - m)^2] = 0, \tag{12.28}$$

provided

$$\lim_{|k| \to \infty} r(k) = 0. \tag{12.29}$$

A similar result holds for $\mathbf{r}_\ell(k)$ (see [2, pp. 77–80]).

In order to apply this result to a real problem one must know, *a priori*, that Equation 12.29 holds. Again, this is often known from the physics.

The convergence result in Equation 12.28 is fairly weak. It would be preferable to prove that $\mathbf{m}_\ell$ converges to m almost surely (with probability 1). It is possible to prove this stronger result if the process is ergodic as well as stationary. This subject is both complicated and technical. See [2,4] for engineering-oriented discussions and [3] for a more mathematical introduction.

12.2.5 Transforms

In principle, one can apply any transform that is useful in the analysis of discrete-time signals to discrete-time stochastic processes. The two obvious candidates are the Z-transform and the discrete Fourier transform [5]. There is a slight theoretical complication. To see this, consider the discrete Fourier transform of the discrete-time stochastic process $\mathbf{x}_{(-\infty,\infty)}$

$$\mathbf{x}_f(\Omega) \triangleq \sum_{k=-\infty}^{\infty} \mathbf{x}(k) e^{-j\Omega k}. \tag{12.30}$$

Note that $\mathbf{x}_f(\Omega)$ is an infinite-dimensional random vector for each fixed value of Ω (it is a function of the random variables $\mathbf{x}(k)$; so it is a random variable) and $\mathbf{x}_f(\Omega)$ is defined for all Ω, $-\infty < \Omega < \infty$, not just integer values of Ω. In other words, $\mathbf{x}_f(\Omega)$ is a stochastic process in the continuous variable Ω. As is usual with the discrete Fourier transform, $\mathbf{x}_f(\Omega)$ is periodic in Ω with period 2π.

A very important use of transforms in the study of discrete-time stochastic processes is the spectral density function.

Definition 12.13:

Let $\mathbf{x}_{(-\infty,\infty)}$ be a wide-sense stationary scalar discrete-time stochastic process with mean m and autocovariance function $r(k)$. Assume $\sum_{k=-\infty}^{\infty} |r(k)| < \infty$. Then the discrete Fourier transform of the autocovariance function $r(k)$,

$$s(\Omega) \triangleq \sum_{k=-\infty}^{\infty} r(k) e^{-j\Omega k}, \tag{12.31}$$

is well defined and known as the spectral density function of $\mathbf{x}_{(-\infty,\infty)}$.

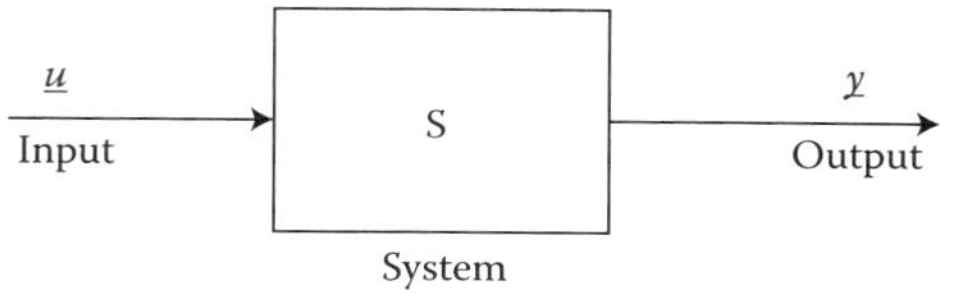

FIGURE 12.1 A symbolic representation of a discrete-time linear system, S, with input $\underline{u} = [\ldots u_{-2}\ u_{-1}\ u_0\ u_1 \ldots]'$ and output $\underline{y} = [\ldots y_{-1}\ y_0\ y_1 \ldots]'$.

Note, as the notation emphasizes, that no part of Equation 12.31 is random. Both the spectral density and the autocovariance are deterministic descriptors of the stochastic process, $\mathbf{x}_{(-\infty,\infty)}$.

The inverse of Equation 12.31 is the usual inverse of the discrete Fourier transform. That is,

$$r(k) = \frac{1}{2\pi} \int_{2\pi} s(\Omega) e^{j\Omega k} d\Omega, \tag{12.32}$$

where $\int_{2\pi}$ means the integral is taken over any interval of duration 2π.

When $m = 0$,

$$r(0) = E(\mathbf{x}_k^2) = \int_{2\pi} s(\Omega) \frac{d\Omega}{2\pi}. \tag{12.33}$$

If $\mathbf{x}_{(-\infty,\infty)}$ is a voltage, current, or velocity, then $r(0)$ can be interpreted as average power, at least to within a constant of proportionality. With this interpretation of $r(0)$, $s(\Omega_0)$ (where Ω_0 is any fixed frequency) must be the average power per unit frequency in $\mathbf{x}_{(-\infty,\infty)}$ at frequency Ω_0 [2]. This is why it is called the spectral density.

Example 12.2:

Suppose $\mathbf{x}_{(-\infty,\infty)}$ is a white-noise process (see Definition 12.12) with $r(k) = \delta(k)$. The spectral density function for this process is

$$s(\Omega) = \sum_{k=-\infty}^{\infty} \delta(k) e^{-j\Omega k} = 1. \tag{12.34}$$

Hence it is called "white noise." Like white light, all frequencies are equally present in a white noise process. A reasonable conjecture is that it is called "noise" because, in the early days of radio and telephone, such a stochastic process was heard as a recognizable and unwanted sound.

12.2.6 Single-Input Single-Output Discrete-Time Linear Systems

It is convenient, both pedagogically and notationally, to begin with SISO discrete-time linear systems described by their impulse response, $h(k, \ell)$. The notation is shown in Figure 12.1.

Definition 12.14:

The impulse response of a SISO discrete-time linear system is denoted by $h(k, \ell)$, were $h(k, \ell)$ is the value of the output at instant k, when the input is a unit impulse at instant ℓ.

The response of a linear system S to an arbitrary input, $u_{(-\infty,\infty)}$, is then given by a convolution sum

$$y(k) = \sum_{\ell=-\infty}^{\infty} h(k, \ell) u(\ell); \quad -\infty < k < \infty. \tag{12.35}$$

The crucial point is that S is a linear map, as is easily proved from Equation 12.35. Linear maps take Gaussian stochastic processes (random variables) into Gaussian stochastic processes. A special case of this, when S can be written as an $n \times m$ matrix, was proved earlier (Equation 12.17).

Example 12.3:

Consider a SISO discrete-time linear system, S, with input, $\mathbf{u}_{(-\infty,\infty)}$, a discrete-time Gaussian stochastic process having mean $m_u(k)$, and autocovariance $r_u(k,\ell)$. Knowing that the system is linear and that the input is Gaussian, the output, $\mathbf{y}_{(-\infty,\infty)}$, must be a Gaussian stochastic process; what are its mean and autocovariance? They can be calculated element by element:

$$\begin{aligned} m_y(k) = E(\mathbf{y}(k)) &= E\left(\sum_{\ell=-\infty}^{\infty} h(k,\ell)\mathbf{u}(\ell)\right) = \sum_{\ell=-\infty}^{\infty} h(k,\ell)E(\mathbf{u}(\ell)) \\ &= \sum_{\ell=-\infty}^{\infty} h(k,\ell)m_u(\ell); \quad -\infty < k < \infty, \end{aligned} \tag{12.36}$$

$$\begin{aligned} r_y(k,\ell) &= E((\mathbf{y}(k) - m_u(k))(\mathbf{y}(\ell) - m_u(\ell))) \\ &= E\left(\left(\sum_{i=-\infty}^{\infty} h(k,i)(\mathbf{u}(i) - m_u(i))\right)\left(\sum_{j=-\infty}^{\infty} h(\ell,j)(\mathbf{u}(j) - m_u(j))\right)\right) \\ &= \sum_{i=-\infty}^{\infty}\sum_{j=-\infty}^{\infty} h(k,i)h(\ell,j)E\left(\left(\mathbf{u}(i) - m_u(i)\right)\left(\mathbf{u}(j) - m_u(j)\right)\right) \\ &= \sum_{i=-\infty}^{\infty}\sum_{j=-\infty}^{\infty} h(k,i)h(\ell,j)r_u(i,j). \end{aligned} \tag{12.37}$$

Of course, not every stochastic process is Gaussian. However, careful review of the previous example shows that Equations 12.36 and 12.37 are valid formulas for the mean and autocovariance of $y_{(-\infty,\infty)}$ whenever $\mathbf{u}_{(-\infty,\infty)}$ is a second-order scalar discrete-time stochastic process with mean $m_u(k)$ and autocovariance $r_u(i,j)$ and the system S is asymptotically stable. This is a very important point. The second-order properties of a stochastic process are easily computed. The calculations are even simpler if the input process is wide-sense stationary and the linear system is time-invariant.

Example 12.4:

Consider a SISO discrete-time linear time-invariant (LTI) system, S, with input, $\mathbf{u}_{(-\infty,\infty)}$, a wide-sense stationary second-order discrete-time stochastic process having mean m_u and autocovariance $r_u(k)$. Denote the impulse response of S by $h(k)$, where $h(k) \triangleq h(k+\ell,\ell)$ and assume that S is asymptotically stable.

$$\begin{aligned} m_y(k) = E(\mathbf{y}(k)) &= E\left(\sum_{\ell=-\infty}^{\infty} h(k-\ell)\mathbf{u}(\ell)\right) = \sum_{\ell=-\infty}^{\infty} h(k-\ell)E(\mathbf{u}(\ell)) \\ &= \sum_{\hat{\ell}=-\infty}^{\infty} h(\hat{\ell})E(\mathbf{u}(k-\hat{\ell})) = \alpha m_u, \end{aligned} \tag{12.38}$$

where

$$\alpha = \sum_{\hat{\ell}=-\infty}^{\infty} h(\hat{\ell}).$$

Similarly, using Equation 12.37, and defining $\hat{\ell} = k - i$ and $\hat{k} = \ell - j$

$$r_y(k,\ell) = \sum_{i=-\infty}^{\infty} \sum_{j=-\infty}^{\infty} h(k-i)h(\ell-j)r_u(i-j) = \sum_{\hat{\ell}=-\infty}^{\infty} \sum_{\hat{k}=-\infty}^{\infty} h(\hat{\ell})h(\hat{k})r_u(k-\ell-\hat{\ell}+\hat{k}). \tag{12.39}$$

Note that the convolution sum in Equation 12.39 depends only on the difference $k-\ell$. Thus, Equations 12.38 and 12.39 prove that $m_y(k) = m_y$, a constant, and that $r_y(k,\ell) = r_y(k-\ell) = r_y(\bar{k})$, where $\bar{k} \triangleq k - \ell$. This proves that $\mathbf{y}_{(-\infty,\infty)}$ is also wide-sense stationary.

Although the computations in Equation 12.39 still appear to be difficult, they clearly involve deterministic convolution. It is well known that the Fourier transform can be used to simplify such calculations. The impulse responses of many discrete-time LTI systems, $h(k), -\infty < k < \infty$, have discrete Fourier transforms, $h_f(\Omega), 0 \le \Omega < 2\pi$. The exceptions include unstable systems. Assume $h_f(\Omega)$ exists in the preceding example and that $\mathbf{u}_{(-\infty,\infty)}$ is wide-sense stationary with mean m_u and spectral density function $s(\Omega)$. From the definition of the discrete Fourier transform

$$h_f(\Omega) = \sum_{k=-\infty}^{\infty} h(k)e^{-j\Omega k}, \tag{12.40}$$

it is evident that Equation 12.38 becomes

$$m_y = h_f(0)m_u. \tag{12.41}$$

Using the inversion formula (Equation 12.32) in Equation 12.39 gives

$$\begin{aligned} r_y(k) &= \sum_{\ell=-\infty}^{\infty} \sum_{i=-\infty}^{\infty} h(\ell)h(i) \left(\frac{1}{2\pi} \int_{2\pi} s_u(\Omega)e^{j\Omega(k-\ell+i)} d\Omega \right) \\ &= \int_{2\pi} \sum_{i=-\infty}^{\infty} h(i)s_u(\Omega)e^{j\Omega(k+i)} \left(\sum_{\ell=-\infty}^{\infty} h(\ell)e^{-j\Omega\ell} \right) \frac{d\Omega}{2\pi} \\ &= \int_{2\pi} \left(\sum_{i=-\infty}^{\infty} h(i)e^{j\Omega i} \right) h_f(\Omega)s_u(\Omega)e^{j\Omega k} \frac{d\Omega}{2\pi} \\ &= \int_{2\pi} h_f(-\Omega)h_f(\Omega)s_u(\Omega)e^{j\Omega k} \frac{d\Omega}{2\pi} \\ &= \int_{2\pi} |h_f(\Omega)|^2 s_u(\Omega)e^{j\Omega k} \frac{d\Omega}{2\pi}. \end{aligned} \tag{12.42}$$

By the uniqueness of Fourier transforms, Equation 12.42 implies

$$s_y(\Omega) = |h_f(\Omega)|^2 s_u(\Omega). \tag{12.43}$$

SISO linear systems in state-space form generally have an n-vector state. When either the initial state is random or the input is a stochastic process, this state is a vector stochastic process. The ideas and notation for such processes are described in the following section. This is followed by a description of linear systems in state-space form.

12.2.7 Vector Discrete-Time Stochastic Processes and LTI Systems

The vector case involves much more complicated notation but no new concepts.

Definition 12.15:

An m-vector discrete-time stochastic process over n time steps, denoted $\underline{\mathbf{X}} \triangleq \{[\mathbf{x}_1(k)\ \mathbf{x}_2(k) \ldots \mathbf{x}_m(k)]'; k = 1, 2, \ldots n\}$, *takes values* $\underline{X} = \{[x_1(h)\ x_2(k) \ldots x_m(k)]'; k = 1, 2, \ldots, n\}$ *according to a probability that can be determined from the pdf* $p_{\underline{\mathbf{X}}}(\underline{X})$.

It can be helpful to visualize such a process as an $m \times n$ matrix

$$\underline{\mathbf{X}} = \begin{bmatrix} \mathbf{x}_1(1) & \mathbf{x}_1(2) & \ldots & \mathbf{x}_1(n) \\ \mathbf{x}_2(1) & \mathbf{x}_2(2) & \ldots & \mathbf{x}_2(n) \\ \vdots & \vdots & \ddots & \vdots \\ \mathbf{x}_m(1) & \mathbf{x}_m(2) & \ldots & \mathbf{x}_m(n) \end{bmatrix}. \tag{12.44}$$

The notation is that bold capital underlined letters denote vector stochastic processes.

All of the scalar results apply, with obvious modifications, to the vector case. For example, $E(\mathbf{X})$ can be computed element by element from Definition 12.4.

$$E(x_\ell(k)) = \int_{-\infty}^{\infty} x p_{\mathbf{x}_\ell(k)}(x)\, dx \tag{12.45}$$

for all $\ell = 1, 2, \ldots, m, k = 1, 2, \ldots n$ (see Equation 12.4).

Then, the nm results of Equation 12.45 can be organized as an m-vector over n time steps, $\underline{m}(k) = [m_1(k)\ m_2(k) \ldots m_m(k)]'$, where

$$m_\ell(k) = E(\mathbf{x}_\ell(k)). \tag{12.46}$$

Similarly, the autocovariance of $\underline{\mathbf{X}}$ in Equation 12.44 can be computed element by element using Definition 12.5 and Equation 12.7. The results are conventionally organized as an autocovariance matrix at each pair of times. That is

$$\underline{R}(k, \ell) = \begin{bmatrix} r_{11}(k, \ell) & r_{12}(k, \ell) & \ldots & r_{1m}(k, \ell) \\ r_{21}(k, \ell) & r_{22}(l, \ell) & \ldots & r_{2m}(k, \ell) \\ \vdots & \vdots & \ddots & \vdots \\ r_{m1}(k, \ell) & r_{m2}(k, \ell) & \ldots & r_{mm}(k, \ell) \end{bmatrix}, \tag{12.47}$$

where

$$r_{ij}(k, \ell) \triangleq E((x_i(k) - m_i(k))(x_j(\ell) - m_j(\ell))).$$

Example 12.5:

Discrete-time m-vector white noise: Any white noise must be wide-sense stationary and have zero mean. Thus, letting $\underline{\xi}(k)$ denote an m-vector white noise at the instant k,

$$\begin{aligned} \underline{m}_\xi(k) &= E(\underline{\xi}(k)) = \underline{0} \quad \text{for all} \quad k, -\infty < k < \infty, \\ \underline{R}_\xi(k, \ell) &= E(\underline{\xi}(k)\underline{\xi}'(\ell)) = \underline{0} \quad \text{for all} \quad k \neq \ell, -\infty < k,\ \ell < \infty, \end{aligned} \tag{12.48}$$

because white noise must have autocovariance equal to zero except when $k = \ell$. When $k = \ell$, the autocovariance can be any positive-semidefinite matrix. Thus,

$$\underline{R}_\xi(k, \ell) = \underline{R}_\xi \delta(k - \ell), \tag{12.49}$$

where $\underline{R}_\xi$ can be any positive-semidefinite matrix ($\underline{y}'\underline{R}_\xi\underline{y} \geq 0$ for all m-vectors $\underline{y}$).

Finally, as is customary with stationary processes,

$$\underline{R}_\xi(k) \triangleq \underline{R}_\xi(k+\ell,\ell) = \underline{R}_\xi\delta(k). \tag{12.50}$$

Reading vector versions of all the previous results would be awfully tedious. Thus, the vector versions will not be written out here. Instead, an example of the way linear systems operate on vector discrete-time stochastic processes will be presented.

The previous discussion of linear systems dealt only with impulse responses and, under the added assumption of time invariance, Fourier transforms. Initial conditions were assumed to be zero. The following example includes nonzero initial conditions and a state-space description of the linear system. The system is assumed to be LTI. The time-varying case is not harder, but the notation is messy.

Example 12.6:

Let $\underline{\mathbf{x}}_0$ be an n-dimensional second-order random variable with mean $\underline{m}_{x_0}$ and covariance $\underline{R}_{x_0}$. Let $\underline{\boldsymbol{\xi}}$ and $\underline{\boldsymbol{\theta}}$ be, respectively, m-dimensional and p-dimensional second-order discrete-time stochastic processes on $0 \le k \le k_f$. Let $E(\underline{\boldsymbol{\xi}}) = \underline{0}, E(\underline{\boldsymbol{\theta}}) = \underline{0}, E(\underline{\boldsymbol{\xi}}(k)\underline{\boldsymbol{\xi}}'(\ell)) = \underline{\Xi}\delta(k-\ell), E(\underline{\boldsymbol{\theta}}(k)\underline{\boldsymbol{\theta}}'(\ell)) = \underline{\Theta}\delta(k-\ell)$, and $E(\underline{\boldsymbol{\xi}}(k)\underline{\boldsymbol{\theta}}'(\ell)) = \underline{0}$ for all $0 \le k, \ell \le k_f$. Strictly speaking, $\underline{\Xi}$ and $\underline{\Theta}$ are not white-noise processes because they are defined only on a finite interval. However, in the context of state-space analysis it is standard to call them white noise processes. Finally, assume $E((\underline{\mathbf{x}}_0 - \underline{m}_{x_0})\underline{\boldsymbol{\xi}}'(k)) = 0$ and $E((\underline{\mathbf{x}}_0 - \underline{m}_{x_0})\underline{\boldsymbol{\theta}}'(k)) = 0$ for all $0 \le k \le k_f$.

Suppose now that the n-vector random variable $\underline{\mathbf{x}}(k+1)$ and the p-vector random variable $\underline{y}(t)$ are defined recursively for $k = 0, 1, 2, \dots, k_f$ by

$$\underline{\mathbf{x}}(k+1) = \underline{A}\underline{\mathbf{x}}(k) + \underline{L}\underline{\boldsymbol{\xi}}(k); \quad \underline{\mathbf{x}}(0) = \underline{\mathbf{x}}_0, \tag{12.51}$$

$$\underline{\mathbf{y}}(k) = \underline{C}\underline{\mathbf{x}}(k) + \underline{\boldsymbol{\theta}}(k), \tag{12.52}$$

where $\underline{A}$ is a deterministic $n \times n$ matrix, $\underline{L}$ is a deterministic $n \times m$ matrix, and $\underline{C}$ is a deterministic $p \times n$ matrix. What can be said about the stochastic processes $\underline{\mathbf{X}} = \{\underline{\mathbf{x}}(k); \quad k = 0, 1, 2, \dots, k_f + 1\}$ and $\underline{\mathbf{Y}} = \{\underline{\mathbf{y}}(k); k = 0, 1, 2, \dots, k_f\}$?

The means and autocovariances of $\underline{\mathbf{X}}$ and $\underline{\mathbf{Y}}$ are easily computed. For example,

$$E(\underline{\mathbf{x}}(k+1)) = E(\underline{A}\underline{\mathbf{x}}(k) + \underline{L}\underline{\boldsymbol{\xi}}(k)) = \underline{A}E(\underline{\mathbf{x}}(k)) + \underline{L}E(\underline{\boldsymbol{\xi}}(k)) = \underline{A}E(\underline{\mathbf{x}}(k)). \tag{12.53}$$

Similarly,

$$E(\underline{\mathbf{y}}(k)) = \underline{C}E(\underline{\mathbf{x}}(k)). \tag{12.54}$$

Note that Equation 12.53 is a simple deterministic recursion for $E(\underline{\mathbf{x}}(k+1))$ starting from $E(\underline{\mathbf{x}}(0)) = \underline{m}_{x_0}$. Thus,

$$E(\underline{\mathbf{x}}(k)) = \underline{m}_x(k) = \underline{A}^k\underline{m}_{x_0}, \tag{12.55}$$

$$E(\underline{\mathbf{y}}(k)) = \underline{C}\underline{A}^k\underline{m}_{x_0}. \tag{12.56}$$

Similarly, recursive equations for the autocovariances of $\underline{\mathbf{X}}$ and $\underline{\mathbf{Y}}$ can be derived. Because the expressions for $\underline{\mathbf{Y}}$ are just deterministic algebraic transformations of those for $\underline{\mathbf{X}}$, only those for $\underline{\mathbf{X}}$ are given here.

$$\begin{aligned}\underline{R}_x(k+1,k+1) &\triangleq E((\underline{\mathbf{x}}(k+1) - \underline{m}_x(k+1))(\underline{\mathbf{x}}(k+1) - \underline{m}_x(k+1))') \\ &= E((\underline{A}(\underline{\mathbf{x}}(k) - \underline{m}_x(k)) + \underline{L}\underline{\xi}(k))(\underline{A}(\underline{\mathbf{x}}(k) - \underline{m}_x(k)) + \underline{L}\underline{\xi}(k))') \\ &= \underline{A}E((\underline{\mathbf{x}}(k) - \underline{m}_x(k))(\underline{\mathbf{x}}(k) - \underline{m}(k))')\underline{A}' + \underline{L}\underline{\Xi}\underline{L}' \\ &= \underline{A}\underline{R}_x(k,k)\underline{A}' + \underline{L}\underline{\Xi}\underline{L}',\end{aligned} \tag{12.57}$$

$$\underline{R}_x(k+1,k) \triangleq E((\underline{\mathbf{x}}(k+1) - \underline{m}_x(k+1))(\underline{\mathbf{x}}(k) - \underline{m}_x(k))') = \underline{A}\underline{R}_x(k,k). \tag{12.58}$$

Note that Equations 12.57 and 12.58 use the assumption that $E((\underline{\mathbf{x}}(k) - \underline{m}_x(k))\underline{\boldsymbol{\xi}}(\ell)) = 0$ for all $0 \leq k, \ell \leq k_f$. Furthermore, the recursion in Equation 12.57 begins with $\underline{R}_x(0,0) = \underline{R}_{x0}$.

12.3 Continuous Time

Continuous-time stochastic processes are technically much more complicated than discrete-time stochastic processes. Fortunately, most of the complications can be avoided by the restriction to second-order processes and linear systems. This is what will be done in this article because linear systems and second-order processes are by far the most common analytical basis for the design of control systems involving randomness.

A good understanding of the discrete-time case is helpful because the continuous-time results are often analogous to those in discrete time.

12.3.1 Basics

It is very difficult to give a definition of continuous-time stochastic process that is both elementary and mathematically correct. For discrete-time stochastic processes every question involving probability can be answered in terms of the finite-dimensional pdfs. For continuous-time stochastic processes the finite-dimensional pdfs are not sufficient to answer every question involving probability. See [2, pp. 59–62] for a readable discussion of the difficulty.

In the interest of simplicity and because the questions of interest here are all answerable in terms of the finite-dimensional pdfs, the working definition of a continuous-time stochastic process will be as follows.

Definition 12.16:

A scalar continuous-time stochastic process, denoted by $\mathbf{x}_{[t_s,t_f]}$, takes values $x(t) \in R$ for all real $t, t_s \leq t \leq t_f$, according to a probability that can be determined from the family of finite-dimensional pdfs. $p_{\mathbf{x}(t_1),\mathbf{x}(t_2),\ldots,\mathbf{x}(t_n)}(x(t_1), x(t_2), \ldots, x(t_n))$ for all finite collections of t_k, t_k real and all possible positive integers n, $t_s \leq t_k \leq t_f$ $k = 1, 2, \ldots, n$.

The essential feature of this definition is the idea of a signal that is a function of the continuous time, t, and also random. The definition itself will not be used directly. Its main purpose is to allow the definitions of expectation and second-order process.

12.3.2 Expectations

As for discrete-time stochastic processes, the expectation, expected value, or mean of a scalar continuous-time stochastic process follows directly from Definitions 12.16 and 12.4.

$$m(t) \triangleq E(\mathbf{x}(t)) = \int_{-\infty}^{\infty} x p_{\mathbf{x}(t)}(x)dx. \tag{12.59}$$

Note that $m(t)$ is defined where $\mathbf{x}_{[t_s,t_f]}$ is, that is, on the interval $t_s \leq t \leq t_f$.

The covariance function is defined and computed for continuous-time stochastic processes in exactly the same way as for discrete-time stochastic processes. See Equations 12.6 and 12.7.

Definition 12.17:

The covariance function of a scalar continuous-time stochastic process $\mathbf{x}_{[t_s,t_f]}$ *is denoted* $r(t,\tau)$ *and is given by*

$$\begin{aligned} r(t,\tau) &= E((\mathbf{x}(t)-m(t))(\mathbf{x}(\tau)-m(\tau)) \\ &= \int_{-\infty}^{\infty}\int_{-\infty}^{\infty}(x-m(t))(y-m(\tau))p_{\mathbf{x}(t),\mathbf{x}(\tau)}(x,y)\,dx\,dy. \end{aligned} \tag{12.60}$$

As in the discrete-time case, it should be obvious from Equation 12.60 that

$$r(t,\tau)=r(\tau,t) \quad \text{for all } t,\tau, \quad t_s \le t,\ \ \tau \le t_f \tag{12.61}$$

and

$$r(t,t) \ge 0 \quad \text{for all } t, \ \ t_s \le t \le t_s. \tag{12.62}$$

Because $r(t,\tau)$ is a function of two variables and not a matrix, it is necessary to extend the idea of nonnegative definiteness to such functions in order to define and demonstrate this aspect of the "shape" of the covariance function. The idea behind the following definition is to form every possible symmetric matrix from time samples of $r(t,\tau)$ and then to require that all those matrices be positive semidefinite in the usual sense.

Definition 12.18:

A real-valued function $g(t,\tau)$; $t,\tau \in R, t_s \le t,\tau \le t_f$; *is positive semidefinite if, for every finite collection* $t_1,t_2,\ldots,t_n$; $t_s \le t_i \le t_f$ *for* $i=1,2,\ldots,n$ *and every real n-vector* $\underline{\alpha}=[\alpha_1,\ \alpha_2,\ldots,\alpha_n]'$

$$\sum_{k-1}^{n}\sum_{\ell-1}^{n}\alpha_k\alpha_\ell g(t_k,t_\ell) \ge 0. \tag{12.63}$$

The function $g(t,\tau)$ is positive definite if strict inequality holds in Equation 12.63 whenever $t_1,t_2,\ldots,t_n$ are distinct and $\underline{\alpha} \ne \underline{0}$.

It is easy to prove that an autocovariance function, $r(t,\tau)$, must be positive semidefinite. First,

$$E\left(\sum_{k=1}^{n}\alpha_k(\mathbf{x}(t_k)-m(t_k))\right)^2 \ge 0 \tag{12.64}$$

because the expectation of a perfect square must be ≥ 0. Then,

$$\begin{aligned} 0 &\le E\left(\sum_{k=1}^{n}\alpha_k(\mathbf{x}(t_k)-m(t_k))\right)^2 \\ &= E\left(\sum_{k=1}^{n}\sum_{\ell=1}^{n}\alpha_k\alpha_\ell(\mathbf{x}(t_k)-m(t_k))(\mathbf{x}(t_\ell)-m(t_\ell))\right) \\ &= \sum_{k=1}^{n}\sum_{\ell=1}^{n}\alpha_k\alpha_\ell E((\mathbf{x}(t_k)-m(t_k))(\mathbf{x}(t_\ell)-m(t_\ell))) \\ &= \sum_{k=1}^{n}\sum_{\ell=1}^{n}\alpha_k\alpha_\ell r(t_k,t_\ell). \end{aligned}$$

The definition of the characteristic function in the case of a continuous-time stochastic process is obvious from Equations 12.11, 12.59, and 12.60. Because it is generally a function of time as well as ω it is not as useful in the continuous-time case. For this reason, it is not given here.

12.3.3 Example: Continuous-Time Gaussian Stochastic Processes

Definition 12.19:

A scalar continuous-time stochastic process $\mathbf{x}_{[t_s,t_f]}$ is a Gaussian process if the collection of random variables $\{\mathbf{x}(t_1), \mathbf{x}(t_2), \ldots, \mathbf{x}(t_n)\}$ is an n-vector Gaussian random variable for every finite set $\{t_1, t_2, \ldots, t_n; t_s \leq t_i \leq t_f, i = 1, 2, \ldots, n\}$.

It follows from Definition 12.8 in Section 12.2.3 that a Gaussian process is completely specified by its mean

$$m(t) = E(\mathbf{x}(t)) \tag{12.65}$$

and autocovariance

$$r(t, \tau) = E((\mathbf{x}(t) - m(t))(\mathbf{x}(\tau) - m(\tau))). \tag{12.66}$$

Example 12.7:

Wiener process (also known as Brownian motion): Let $\mathbf{x}_{[0,\infty)}$ be a Gaussian process with

$$\begin{aligned} m(t) &= 0, \\ r(t, \tau) &= \min(t, \tau). \end{aligned} \tag{12.67}$$

Note that the interval of definition is infinite and open at the right. This is a very minor expansion of Definition 12.16.

The Wiener process plays a fundamental role in the theory of stochastic differential equations (see Chapter 59); so it is worthwhile to derive one of its properties. Consider any ordered set of times $t_0 < t_1 < t_2 < \cdots < t_n$.

Define

$$\mathbf{y}_k = \mathbf{x}_{t_k} - \mathbf{x}_{t_k-1} \quad k = 1, 2, \ldots, n.$$

Then $\mathbf{y}_k$ is an increment in the process $\mathbf{x}_{[0,\infty)}$. Consider the collection of increments

$$\underline{\mathbf{y}} = \{\mathbf{y}_k, k = 1, 2, \ldots, n\}.$$

Because $\mathbf{x}_{[0,\infty)}$ is a Wiener process, each of the $\mathbf{x}_{t_k}$ is a Gaussian random variable (see Definition 12.19). Therefore $\mathbf{y}_k$, the difference of Gaussian random variables, is also a Gaussian random variable for every k and $\underline{\mathbf{y}}$ is an n-vector Gaussian random variable. Therefore, $\underline{\mathbf{y}}$ is completely characterized by its mean and covariance. The mean is zero. The covariance is diagonal, as can be seen from the following calculation.

Consider, for $j \leq k - 1$ (the means are zero),

$$\begin{aligned} E(\mathbf{y}_k \mathbf{y}_j) &= E((\mathbf{x}_{t_k} - \mathbf{x}_{t_{k-1}})(\mathbf{x}_{t_j} - \mathbf{x}_{t_{j-1}})) \\ &= E(\mathbf{x}_{t_k}\mathbf{x}_{t_j}) - E(\mathbf{x}_{t_{k-1}}\mathbf{x}_{t_j}) - E(\mathbf{x}_{t_k}\mathbf{x}_{t_{j-1}}) + E(\mathbf{x}_{t_{k-1}}\mathbf{x}_{t_{j-1}}) \\ &= t_j - t_j - t_{j-1} + t_{j-1} = 0. \end{aligned}$$

A similar calculation for $j \geq k + 1$ completes the demonstration that the covariance of $\mathbf{y}$ is diagonal. Because $\mathbf{y}$ is Gaussian, the fact that its covariance is diagonal proves that the $\underline{\mathbf{y}}_k$ $k = 1, 2, \ldots, n$ are independent. This is a very important property of the Wiener process. It is usually described thus: the Wiener process has independent increments.

12.3.4 Stationarity, Ergodicity, and White Noise

The concepts of stationarity and wide-sense stationarity for continuous-time stochastic processes are virtually identical to those for discrete-time stochastic processes. All that is needed is to define the continuous-time stochastic process on the infinite interval

$$\mathbf{x}_{(-\infty,\infty)} \triangleq \mathbf{x}(t) \quad \text{for all } t, -\infty < t < \infty$$

and adjust the notation in Definitions 12.9 and 12.11. Only the latter is included here so as to emphasize the focus on second-order processes.

Definition 12.20:

A scalar continuous-time stochastic process, $\mathbf{x}_{(-\infty,\infty)}$, is a second-order process if and only if $E(x^2(t)) < \infty$ for all t, $-\infty < t < \infty$.

Definition 12.21:

A second-order, continuous-time stochastic process $\mathbf{x}_{(-\infty,\infty)}$ is wide-sense stationary if and only if its mean $m(t)$ and covariance function $r(t,\tau)$ satisfy

$$m(t) = m, \text{ a constant} \quad \text{for all } t, \quad -\infty < t < \infty, \tag{12.68}$$

$$r(t,\tau) = r(t+\sigma,\tau+\sigma) \quad \text{for all } t,\tau,\sigma, \quad -\infty < t,\tau,\sigma < \infty. \tag{12.69}$$

It is customary to define

$$r(t) = r(t+\tau,\tau) \quad \text{for all } t,\tau, \quad -\infty < t < \infty. \tag{12.70}$$

The apparatus necessary to define continuous-time white noise is now in place.

Definition 12.22:

A scalar second-order continuous-time stochastic process, $\mathbf{x}_{(-\infty,\infty)}$ is a white-noise process if and only if

$$m(t) = E(\mathbf{x}(t)) = 0 \quad \text{for all } t, \quad -\infty < t < \infty, \tag{12.71}$$

$$r(t) = r\delta(t), \quad r > 0, \quad -\infty < t < \infty, \tag{12.72}$$

where $\delta(t)$ is the unit impulse (equivalently, the Dirac delta) function [5]. *As in discrete time, the process is Gaussian white noise if, in addition, $\mathbf{x}(t)$ is Gaussian for all t.*

As in discrete time, it is often necessary to determine the properties of a stochastic process from a combination of prior knowledge and observations of one sample function. In order for this to be possible it is necessary that the process be stationary and that, in some form, averages over time of $\mathbf{x}(t)$, $-\infty < t < \infty$, converge to expectations. Mathematically, this can be expressed as follows. Given a stationary continuous-time stochastic process $\mathbf{x}_{(-\infty<t<\infty)}$, one wants

$$\lim_{t\to\infty}\frac{1}{2t}\int_{-t}^{t} f(\mathbf{x}(t))dt = E(f(\mathbf{x}(0))) \tag{12.73}$$

to hold for any reasonable function $f(\cdot)$. This is true, with some minor technical conditions for processes that are stationary and ergodic. See [4] for an introduction to this important but difficult subject.

12.3.5 Transforms

Again, in parallel to the discrete-time case, any transform that is useful in the analysis of continuous-time signals can be applied to continuous-time stochastic processes. The two obvious candidates are the Laplace and Fourier transforms [5]. The most important application is the completely deterministic Fourier transform of the autocovariance of a wide-sense stationary continuous-time stochastic process.

Definition 12.23:

Let $\mathbf{x}_{(-\infty,\infty)}$ be a wide-sense stationary scalar continuous-time stochastic process with mean m and autocovariance $r(t)$. Assume $\int_{-\infty}^{\infty}|r(t)|dt < \infty$. Then the Fourier transform of the autocovariance function $r(t)$,

$$s(\omega) = \int_{-\infty}^{\infty} r(t)e^{-j\omega t}\,dt \tag{12.74}$$

is well defined and known as the spectral density of $\mathbf{x}_{(-\infty,\infty)}$.

The inverse of Equation 12.74 is the usual inverse of the Fourier transform.

$$r(t) = \frac{1}{2\pi}\int_{-\infty}^{\infty} s(\omega)e^{j\omega t}d\omega. \tag{12.75}$$

When $m = 0$,

$$r(0) = E(\mathbf{x}^2(t)) = \int_{-\infty}^{\infty} s(\omega)\frac{d\omega}{2\pi}. \tag{12.76}$$

If $\mathbf{x}_{(-\infty,\infty)}$ is a voltage, current, or velocity, then $r(0)$ can be interpreted as average power, at least to within a constant of proportionality. Then $s(\omega_0)$, where ω_0 is any fixed frequency, must be the average power per unit frequency in $\mathbf{x}_{(-\infty,\infty)}$ at the frequency ω_0 [2]. This is why $s(\omega)$ is called the spectral density.

Example 12.8:

Suppose $\mathbf{x}_{(-\infty,\infty)}$ is a white noise process (see Definition 12.24) with $r = 1$. The spectral density for this process is

$$s(\omega) = \int_{-\infty}^{\infty} \delta(t)e^{-j\omega t}dt = 1.$$

Because all frequencies are equally present, as in white light, a process having $s(\omega) = 1$ is called a white-noise process.

12.3.6 SISO Continuous-Time Linear Systems

It is convenient to begin with linear systems described by their impulse response, $h(t, \tau)$. The notation is shown in Figure 12.2.

Definition 12.24:

The impulse response of a SISO continuous-time system is denoted by $h(t,\tau)$, where $h(t,\tau)$ is the value of the output at instant t when the input is a unit impulse, $\delta(t-\tau)$, at instant τ.

The response of linear system, S, to an arbitrary input, $\boldsymbol{u}_{(-\infty,\infty)}$, is given by

$$y(t)=\int_{-\infty}^{\infty} h(t,\tau)u(\tau)d\tau, \quad -\infty<t<\infty. \tag{12.77}$$

Because the second-order properties of $\mathbf{y}_{(-\infty,\infty)}$ are completely described by its mean and autocovariance, it is worthwhile to know how to compute them from $h(t,\tau)$ and the second-order properties of $\mathbf{u}_{(-\infty,\infty)}$. The trick is to compute them at each time, t. That is,

$$m_y(t)\triangleq E(\mathbf{y}(t))=E\left(\int_{-\infty}^{\infty} h(t,\tau)\mathbf{u}(\tau)d\tau\right).$$

By the linearity of integration

$$=\int_{-\infty}^{\infty} E(h(t,\tau)\mathbf{u}(\tau))d\tau.$$

Because $h(t,\tau)$ is deterministic,

$$=\int_{-\infty}^{\infty} h(t,\tau)E(\mathbf{u}(\tau))d\tau.$$

That is,

$$m_y(t)=\int_{-\infty}^{\infty} h(t,\tau)m_u(\tau)d\tau, \tag{12.78}$$

where

$$m_u(t)\triangleq E(\mathbf{u}(t)).$$

Similarly,

$$\begin{aligned}
r_y(t,\tau)&=E((y(t)-m_y(t))(y(\tau)-m_y(\tau)))\\
&=E\left(\int_{-\infty}^{\infty} h(t,\sigma_1)(\mathbf{u}(\sigma_1)-m_u(\sigma_1))d\sigma_1\int_{-\infty}^{\infty} h(\tau,\sigma_2)(\mathbf{u}(\sigma_2)-m_u(\sigma_2))d\sigma_2\right)\\
&=E\left(\int_{-\infty}^{\infty}\int_{-\infty}^{\infty} h(t,\sigma_1)h(\tau,\sigma_2)(\mathbf{u}(\sigma_1)-m_u(\sigma_1))(\mathbf{u}(\sigma_2)-m_u(\sigma_2))d\sigma_1 d\sigma_2\right)\\
&=\int_{-\infty}^{\infty}\int_{-\infty}^{\infty} h(t,\sigma_1)h(\tau,\sigma_2)E((\mathbf{u}(\sigma_1)-m_u(\sigma_1))\mathbf{u}(\sigma_2)-m_u(\sigma_2)))d\sigma_1 d\sigma_2\\
&=\int_{-\infty}^{\infty}\int_{-\infty}^{\infty} h(t,\sigma_1,\sigma_2)r_u(\sigma_1,\sigma_2)d\sigma_1 d\sigma_2,
\end{aligned} \tag{12.79}$$

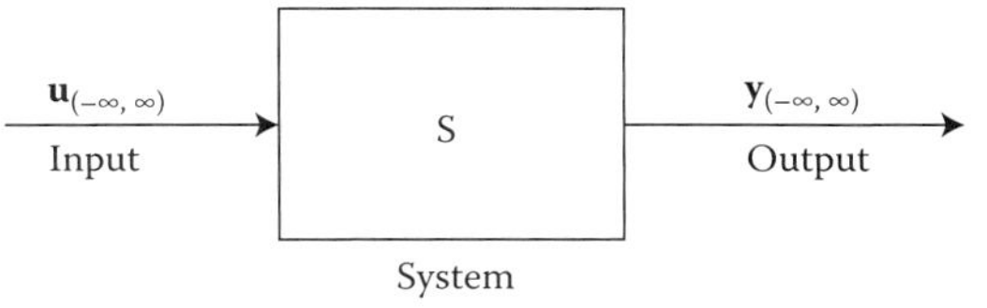

FIGURE 12.2 Representation of a system with input $\mathbf{u}_{(-\infty,\infty)}$ and output $\mathbf{y}_{(-\infty,\infty)}$.

where

$$r_u(t,\tau) \triangleq E((\mathbf{u}(t) - m_u(t))(\mathbf{u}(\tau) - m_u(\tau))).$$

As in the discrete-time case, the Fourier transform can be used to simplify the calculations when the linear system is also time-invariant and $\mathbf{u}_{(-\infty,\infty)}$ is wide-sense stationary. Denote the impulse response by $h(t)$, where $h(t) = h(t+\tau,\tau)$. Let $m_u = E(\mathbf{u}(t))$, and $r_u(t) = E((\mathbf{u}(t+\tau) - m_u)(\mathbf{u}(\tau) - m_u))$. It is then easy to show, by paralleling the argument leading to Equation 12.41, that

$$m_y = h_f(0)m_u, \tag{12.80}$$

where

$$h_f(\omega) = \int_{-\infty}^{\infty} h(t)e^{-j\omega t}dt, \text{ the Fourier transform of } h(t).$$

Similarly,

$$r_y(t) = \int_{-\infty}^{\infty} |h_f(\omega)|^2 s_u(\omega)e^{j\omega t}\frac{d\omega}{2\pi}, \tag{12.81}$$

where $s_u(\omega)$ denotes the spectral density of $\mathbf{u}_{(-\infty,\infty)}$ (see Equation 12.74). This follows from virtually the same argument as Equation 12.42. By the uniqueness of Fourier transforms, Equation 12.81 implies

$$s_y(\omega) = |h_f(\omega)|^2 s_u(\omega). \tag{12.82}$$

It is necessary to define the notation for n-vector continuous-time stochastic processes before describing continuous-time systems in state-space form. Both of these are done in the following section.

12.3.7 Vector Continuous-Time Stochastic Process and LTI Systems

The vector case involves no new concepts but requires a more complicated notation. Because only second-order stochastic processes are discussed here all that is needed is a notation for the process, its mean, and its autocovariance.

An n-vector continuous-time stochastic process, $\underline{\mathbf{x}}_{[t_1,t_2]}$, is an n-vector random variable, $\underline{\mathbf{x}}(t) = [\mathbf{x}_1(t)\,\mathbf{x}_2(t)\ldots\mathbf{x}_n(t)]'$ at each instant of time t, $t_1 \le t \le t_2$. As throughout this chapter, vectors are denoted by lower-case underlined letters. Random variables and stochastic processes are bold letters.

The mean of a vector stochastic process is defined at each instant of time; the autocovariance is defined for paired times. Thus, when $\underline{\mathbf{x}}(t)$ is an n-vector random variable defined for all t, $t_1 \le t \le t_2$ (i.e., $\mathbf{x}_{[t_1,t_2]}$ is an n-vector stochastic process)

$$\underline{m}_x(t) = E(\underline{\mathbf{x}}(t)) \quad t_1 \le t \le t_2 \tag{12.83}$$

will denote its mean and

$$\underline{R}_x(t,\tau) = E((\underline{\mathbf{x}}(t) - \underline{m}_x(t))(\underline{\mathbf{x}}(\tau) - \underline{m}_x(\tau))') \quad t_1 \le t, \quad \tau \le t_2 \tag{12.84}$$

will denote its autocovariance. Note the transpose in Equation 12.84 and the capital $\underline{R}_x$. This emphasizes that the autocovariance is an $n \times n$ matrix for each pair t and τ. As in the discrete-time case, Equations 12.83 and 12.84 are evaluated element by element. Thus,

$$m_{x_i}(t) = E(\mathbf{x}_i(t)) \quad i = 1,2,\ldots,n, \tag{12.85}$$

$$r_{x_{ij}}(t,\tau) = E((\mathbf{x}_i(t) - m_{x_i}(t))(\mathbf{x}_j(\tau) - m_{\mathbf{x}_j}(\tau))). \tag{12.86}$$

Example 12.9:

Continuous-time n-vector white noise: Any white noise must be wide-sense stationary and have zero mean. Letting $\underline{\xi}(t)$ denote an n-vector continuous-time white noise at the instant t,

$$\underline{m}_{\xi}(t) = E(\underline{\boldsymbol{\xi}}(t)) = \underline{0} \quad \text{for all } t, \quad -\infty < t < \infty, \tag{12.87}$$

$$\underline{R}_{\xi}(t,\tau) = E(\underline{\boldsymbol{\xi}}(t)\underline{\boldsymbol{\xi}}'(\tau)) = \underline{0} \quad \text{for all } t \neq \tau \quad -\infty < t, \quad \tau < \infty \tag{12.88}$$

because the autocovariance of white noise must be equal to 0 except when $t = \tau$. At $t = \tau$, the autocovariance reduces to a covariance matrix (denoted by $\underline{\Xi}$, which must be positive semidefinite because it is a covariance matrix) times a unit impulse. A rigorous derivation of this is hard but the analogy to Equation 12.49 should be convincing. Thus, for all $-\infty < t, \tau < \infty$

$$\underline{R}_{\xi}(t,\tau) = \underline{\Xi}\delta(t-\tau), \tag{12.89}$$

where $\underline{\Xi}$ is any positive semidefinite matrix.

Now that vector continuous-time stochastic processes and vector white noise have been introduced it is possible to develop the state-space description of linear continuous-time systems driven by continuous-time stochastic processes. For convenience, only the case of LTI systems is described.

Example 12.10:

LTI system and white noise: Let $\underline{\mathbf{x}}_0$ be an n-dimensional second-order random variable with mean $\underline{m}_{x_0}$ and covariance $\underline{R}_{x_0}$. Let $\underline{\boldsymbol{\xi}}_{[0,\infty)}$ and $\underline{\boldsymbol{\theta}}_{[0,\infty)}$ be q- and p-dimensional wide-sense stationary second-order continuous-time stochastic processes satisfying, for all $t \in [0,\infty)$

$$E(\underline{\boldsymbol{\xi}}(t)) = \underline{0} \text{ and } E(\underline{\boldsymbol{\theta}}(t)) = \underline{0}, \tag{12.90}$$

$$\underline{R}_{\xi}(t) = E(\underline{\boldsymbol{\xi}}(t+\tau)\underline{\boldsymbol{\xi}}'(\tau)) = \underline{\Xi}\delta(t), \tag{12.91}$$

$$\underline{R}_{\theta}(t) = E(\underline{\boldsymbol{\theta}}(t+\tau)\underline{\boldsymbol{\theta}}'(\tau)) = \underline{\Theta}\delta(t), \tag{12.92}$$

where $\underline{\Xi}$ and $\underline{\Theta}$ are symmetric positive semidefinite matrices of appropriate dimension.

Strictly speaking $\underline{\boldsymbol{\xi}}_{[0,\infty)}$ and $\underline{\boldsymbol{\theta}}_{[0,\infty)}$ are not white noise processes because they are defined only in $[0,\infty)$. However, in the context of state-space analysis it is standard to call them white-noise processes. Obviously, what matters is not the name, but Equations 12.90 through 12.92.

For convenience, assume that

$$E((\underline{\mathbf{x}}_0 - \underline{m}_{x_0})\underline{\boldsymbol{\xi}}'(t)) = \underline{0}, \tag{12.93}$$

$$E((\underline{\mathbf{x}}_0 - \underline{m}_{x_0})\underline{\boldsymbol{\theta}}'(t)) = \underline{0}, \tag{12.94}$$

and

$$E(\underline{\boldsymbol{\xi}}(t)\underline{\boldsymbol{\theta}}'(\tau)) = \underline{0} \quad \text{for all } t, \quad \tau \in [0,\infty). \tag{12.95}$$

The assumptions in Equations 12.93 through 12.95 remove some terms from the subsequent expressions, thereby saving space and the effort to compute them, but are otherwise inessential.

Define the n- and p-vector random variables $\underline{\mathbf{x}}(t)$ and $\underline{\mathbf{y}}(t)$ by

$$\frac{d\underline{\mathbf{x}}}{dt}(t) = \underline{A}\underline{\mathbf{x}}(t) + \underline{B}\underline{u}(t) + \underline{L}\underline{\boldsymbol{\xi}}(t); \quad \underline{\mathbf{x}}(0) = \underline{\mathbf{x}}_0, \tag{12.96}$$

$$\underline{\mathbf{y}}(t) = \underline{C}\underline{\mathbf{x}}(t) + \underline{\boldsymbol{\theta}}(t), \tag{12.97}$$

where $\underline{u}(t)$ is a deterministic m-vector, the control, and all the matrices are real, deterministic, and have the appropriate dimensions (that is, $\underline{A}$ is $n \times n$, $\underline{B}$ is $n \times m$, $\underline{C}$ is $p \times n$, and $\underline{L}$ is $n \times q$). The structure of this system is shown in Figure 12.3.

It follows from Equations 12.96 and 12.97 that $\underline{\mathbf{x}}_{[0,\infty)}$ and $\underline{\mathbf{y}}_{[0,\infty)}$ are vector continuous-time stochastic processes. Assuming that $\underline{u}(t)$ is bounded for all t guarantees that $\underline{\mathbf{x}}_{[0,t_f)}$ and $\underline{\mathbf{y}}_{[0,t_f)}$ are second-order for all finite t_f. What are their second-order statistics?

To compute the mean of $\underline{\mathbf{x}}(t)$ take the expected value of both sides of Equation 12.96. That is,

$$E(\dot{\underline{\mathbf{x}}}(t)) = E(\underline{A}\underline{\mathbf{x}}(t) + \underline{B}\underline{u}(t) + \underline{L}\underline{\boldsymbol{\xi}}(t)).$$

Interchanging the order of expectation and d/dt (both are linear; so their order can be reversed) and using linearity on the right-hand side of the equation gives

$$\frac{d}{dt}E(\underline{\mathbf{x}}(t)) = \underline{A}E(\underline{\mathbf{x}}(t)) + \underline{B}\underline{u}(t); \quad E(\underline{\mathbf{x}}(0)) = \underline{m}_{x_0} \tag{12.98}$$

or

$$\dot{\underline{m}}_x(t) = \underline{A}\underline{m}_x(t) + \underline{B}\underline{u}(t); \quad \underline{m}_x(0) = \underline{m}_{x_0}.$$

This is just a deterministic ordinary differential equation in state-space form for $E(\underline{\mathbf{x}}(t))$. Once $E(\underline{\mathbf{x}}(t))$ has been computed, taking expectation of both sides of Equation 12.97 gives

$$E(\underline{\mathbf{y}}(t)) = \underline{C}E(\underline{\mathbf{x}}(t)). \tag{12.99}$$

This is a deterministic algebraic equation.

A derivation of the autocovariance of $\underline{\mathbf{x}}_{[0,t_f)}$ is beyond the scope of this chapter. A derivation can be found in [6, pp. 111–113].

The result is that

$$\underline{R}_x(t,t) = E((\underline{\mathbf{x}}(t) - \underline{m}_x(t))(\underline{\mathbf{x}}(t) - \underline{m}_x(t))')$$

satisfies the differential equation

$$\dot{\underline{R}}_x(t,t) = \underline{A}\underline{R}_x(t,t) + \underline{R}_x(t,t)\underline{A}' + \underline{L}\underline{\Xi}\underline{L}' \tag{12.100}$$

with initial condition $\underline{R}_x(0,0) = \underline{R}_{x_0}$ and $\underline{R}_x(t,\tau)$, with $t \geq \tau$, satisfies the differential equation

$$\frac{\partial \underline{R}_x(t,\tau)}{\partial t} = \underline{A}\underline{R}_x(t,\tau) \tag{12.101}$$

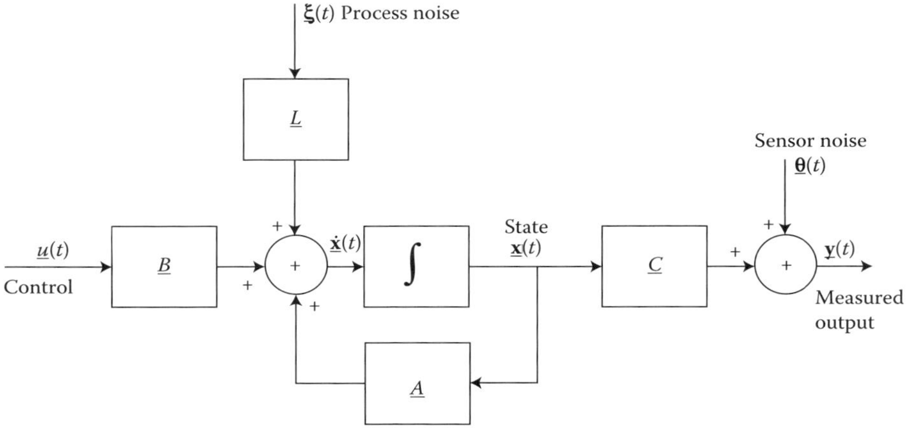

FIGURE 12.3 Block diagram of the LTI system described by Equations 12.96 and 12.97.

while $\underline{R}_x(\tau, t)$, again with $t \geq \tau$, satisfies

$$\frac{\partial R_x(\tau, t)}{\partial t} = \underline{R}_x(\tau, t)\underline{A}'. \tag{12.102}$$

The initial condition for both Equations 12.102 and 12.103 is $\underline{R}_x(\tau, \tau)$, computed from Equation 12.100.

12.4 Conclusions

The second-order properties of many zero-mean stochastic processes can be reproduced by an LTI system with white-noise input. The precise result for scalar continuous-time systems is the following [2]. Let $s(\omega)$ be a real nonnegative absolutely integrable function such that

$$\int_{-\infty}^{\infty} \frac{|\ln s(\omega)|}{1+\omega^2} d\omega < \infty. \tag{12.103}$$

Then there exists a square integrable function $h_f(\omega)$, $-\infty < \omega < \infty$, such that

$$|h_f(\omega)|^2 = s(\omega), \tag{12.104}$$

$$h(t) = \int_{-\infty}^{\infty} h_f(\omega) e^{j\omega t} \frac{d\omega}{2\pi} = 0 \quad \text{for } t < 0, \tag{12.105}$$

$$h_f(\sigma + j\omega) = \int_{-\infty}^{\infty} e^{j(\sigma+j\omega)t} h(t) dt \neq 0, \quad \omega > 0, \tag{12.106}$$

Because the spectral density of white noise is one for all ω, Equation 12.104 means that the spectral density, $s(\omega)$, is identical to the spectral density of the output of an LTI system with impulse response $h(t)$ driven by a white-noise input. Equation 12.105 means that $h(t)$ is causal. Equation 12.106 means that the transfer function $h_f(\omega)$ is minimum phase. Similar results hold in discrete time and for vector signals [7,8].

The restriction to second-order processes with zero mean is unimportant because the mean is deterministic. Thus, the only significant limitation on the use of a linear system driven by white noise as a model for second-order stochastic processes is Equation 12.103. The major limitation of the model is that the second-order statistics are not always a good description of a stochastic process. The second-order statistics are most useful when the process is approximately Gaussian. Processes that are very different from Gaussian, especially in the vicinity of the mean, are not usually well modeled by their second-order statistics.

12.5 Notation

Lower case letters such as x, y, and r denote real, deterministic scalars. The same letters when underlined denote vectors.

Lower case letters i, j, k, ℓ, m, and n denote integers (j is also used to indicate $\sqrt{-1}$ and m for mean—the meaning should be evident from the context).

Upper case, underlined, capital letters such as $\underline{A}, \underline{B}, \underline{C}, \underline{L}, \underline{\Theta}$, and $\underline{\Xi}$ denote real deterministic matrices.

Lower case, bold letters such as $\mathbf{x}$ and $\mathbf{y}$ denote scalar random variables. The same letters when underlined denote vector random variables (or discrete-time stochastic processes).

Upper case, underlined bold letters such as $\underline{\mathbf{X}}, \underline{\mathbf{Y}}, \underline{\boldsymbol{\Theta}}$, and $\underline{\boldsymbol{\Xi}}$ denote vector discrete-time stochastic processes.

Lower case, bold letters with interval subscripts such as $\mathbf{x}_{(-\infty,\infty)}$ denote stochastic processes.

References

1. Leon-Garcia, A., *Probability and Random Processes for Electrical Engineering*, Addison-Wesley, Reading, MA, 1989.
2. Wong, E., *Introduction to Random Processes*, Springer-Verlag, Amsterdam, 1983.
3. Breiman, L., *Probability*, SIAM (republication of a 1968 book published by Addison-Wesley), Philadelphia, PA, 1992.
4. Wong, E., *Stochastic Processes in Information and Dynamical Systems*, McGraw-Hill, New York, 1971.
5. Oppenheim, A.V. and Willsky, A.S., with S. H. Nawab, *Signals and Systems*. 2nd edition, Prentice-Hall, Upper Saddle River, NJ, 1997.
6. Davis, M.H.A., *Linear Estimation and Stochastic Control*, Chapman & Hall, London, 1977.
7. Astrom, K.J., *Introduction to Stochastic Control Theory*, Academic Press, New York, 1970.
8. Caines, P.E., *Linear Stochastic Systems*, John Wiley & Sons, New York, 1988.
9. Grimmett, G. and Stirzaker, D., *Probability and Random Processes*, 3rd Ed., Oxford University Press, New York, 2001.

13

Kalman Filtering

Michael Athans
Massachusetts Institute of Technology

13.1 Introduction

The purpose of this chapter is to provide an overview of Kalman filtering concepts. We cover only a small portion of the material associated with Kalman filters. Our choice of material is primarily motivated by the material needed in the design and analysis of multivariable feedback control systems for linear time-invariant (LTI) systems and, more specifically, by the design philosophy commonly called the linear quadratic Gaussian (LQG) method. Thus, from a technical perspective, we cover, without proofs, the so-called steady-state, constant-gain LTI Kalman filters.

More details regarding different formulations and solutions of Kalman filtering problems (continuous-time linear time-varying, discrete-time linear time-varying, and time-invariant) can be found in several classic and standard textbooks on the subject of estimation theory. Please see Further Reading at the end of this chapter.

The steady-state constant-gain Kalman filter is an algorithm that is used to estimate the state variables of a continuous-time LTI system, subject to stochastic disturbances, on the basis of noisy measurements of certain output variables. Thus, the Kalman filtering algorithm combines the information regarding the plant dynamics, the probabilistic information regarding the stochastic disturbances that influence the plant state variables, as well as that regarding the measurement noise that corrupts the sensor measurements, and the deterministic controls.

Credit for "inventing" this algorithm is usually given to Dr. Rudolph E. Kalman, who presented the key ideas in the late 1950s and early 1960s. Although the Kalman filter is an alternative representation of the Wiener filter [1], Kalman's contribution was to tie the state estimation problem to the state-space models, and the (then new) concepts of controllability and observability [2–5]. Thousands of papers have been written about the Kalman filter and its numerous applications to navigation, tracking, and estimator and controller design in defense and industrial applications.

13.2 Problem Definition

In this section we present the definition of the basic stochastic estimation problem for which the Kalman filter (KF) yields an "optimal" solution. First we present the plant state dynamics. We deal with a finite-dimensional LTI system whose state vector $\underline{\mathbf{x}}(t), \underline{\mathbf{x}}(t) \in R^n$ (a real n-vector), obeys the stochastic differential equation

$$\dot{\underline{\mathbf{x}}}(t) = \underline{A}\underline{\mathbf{x}}(t) + \underline{B}\underline{u}(t) + \underline{L}\underline{\xi}(t) \tag{13.1}$$

where $\underline{u}(t), \underline{u}(t) \in R^m$ is the deterministic control vector (assumed known), $\underline{\xi}(t), \underline{\xi}(t) \in R^q$ is a vector-valued stochastic process, often called the *process noise*, that acts as a disturbance to the plant dynamics. Vectors are underlined, lower case letters and matrices are underlined, upper case letters. The process noise $\underline{\xi}(t)$ is assumed to have certain statistical properties, corresponding to a stationary (time-invariant) continuous-time *white* Gaussian noise with zero mean, i.e.,

$$E\{\underline{\xi}(t)\} = 0, \quad \text{for all } t \tag{13.2}$$

and its covariance matrix is defined by

$$\text{cov}\,[\underline{\xi}(t); \underline{\xi}(\tau)] = E\{\underline{\xi}(t)\underline{\xi}'(\tau)\} = \underline{\Xi}\delta(t-\tau) \tag{13.3}$$

with $\delta(t-\tau)$ being the Dirac delta function (impulse at $t = \tau$). The matrix $\underline{\Xi}$ is called the intensity matrix of $\underline{\xi}(t)$ and it is a symmetric positive definite matrix, i.e.,

$$\underline{\Xi} = \underline{\Xi}' > 0 \tag{13.4}$$

Remark 13.1

Continuous-time white noise does not exist in nature; it is the limit of a broadband stochastic process. In the frequency domain, continuous-time white noise corresponds to a stochastic process with constant power spectral density as a function of frequency. This implies that continuous-time white noise has constant power at all frequencies, and therefore has infinite energy! White noise is completely unpredictable, as can be seen from Equation 13.3 because it is uncorrelated for any $t \neq \tau$, while it has finite variance and standard deviation at $t = \tau$. This is obviously an approximation to reality. As with the Dirac delta function, $\delta(t)$, white noise creates some subtle mathematical issues, but is nonetheless extremely useful in engineering.

Remark 13.2

The state $\underline{\mathbf{x}}(t)$, the solution of Equation 13.1, is a well-defined physical stochastic process, a so-called colored Gaussian random process, and it has finite energy. Its power spectral density rolls off at high frequencies.

Next, we turn our attention to the measurement equation. We assume that our sensors cannot directly measure all of the physical state variables, the components of the vector $\underline{\mathbf{x}}(t)$ of the plant given in Equation 13.1. Rather, in the classical Kalman filter formulation we assume that we can measure only certain output variables (linear combinations of the state variables) in the presence of additive continuous-time white noise.

The mathematical model of the measurement process is as follows:

$$\underline{\mathbf{y}}(t) = \underline{C}\underline{\mathbf{x}}(t) + \underline{\theta}(t) \tag{13.5}$$

The vector $\underline{\mathbf{y}}(t) \in R^p$ represents the sensor measurement. The measurement or sensor noise $\underline{\theta}(t) \in R^p$ is assumed to be a continuous-time white Gaussian random process, independent of $\underline{\xi}(t)$, with zero mean, i.e.,

$$E\{\underline{\theta}(t)\} = 0, \quad \text{for all } t \tag{13.6}$$

and covariance matrix

$$\text{cov}\,[\underline{\theta}(t); \underline{\theta}(\tau)] = E\{\underline{\theta}(t)\underline{\theta}'(\tau)\} = \underline{\Theta}\delta(t-\tau) \tag{13.7}$$

where the sensor noise intensity matrix is symmetric and positive definite, i.e.,

$$\underline{\Theta} = \underline{\Theta}' > 0 \tag{13.8}$$

Figure 13.1 shows a visualization, in block diagram form, of Equations 13.1 and 13.5.

13.2.1 The State Estimation Problem

Imagine that we have been observing the control $\underline{u}(\tau)$ and the output $\underline{\mathbf{y}}(\tau)$ over the infinite past up to the present time t. Let

$$U(t) = \{\underline{u}(\tau); -\infty < \tau \le t\} \tag{13.9}$$

$$Y(t) = \{\underline{\mathbf{y}}(\tau); -\infty < \tau \le t\} \tag{13.10}$$

denote the past histories of the control and output, respectively.

The state estimation problem is as follows: given $U(t)$ and $Y(t)$, find a vector $\hat{\underline{\mathbf{x}}}(t)$, at time t, which is an "optimal" estimate of the present state $\underline{\mathbf{x}}(t)$ of the system defined by Equation 13.1.

Under the stated assumptions regarding the Gaussian nature of $\underline{\xi}(t)$ and $\underline{\theta}(t)$ the "optimal" state estimate is the same for an extremely large class of optimality criteria [6]. This generally optimal estimate

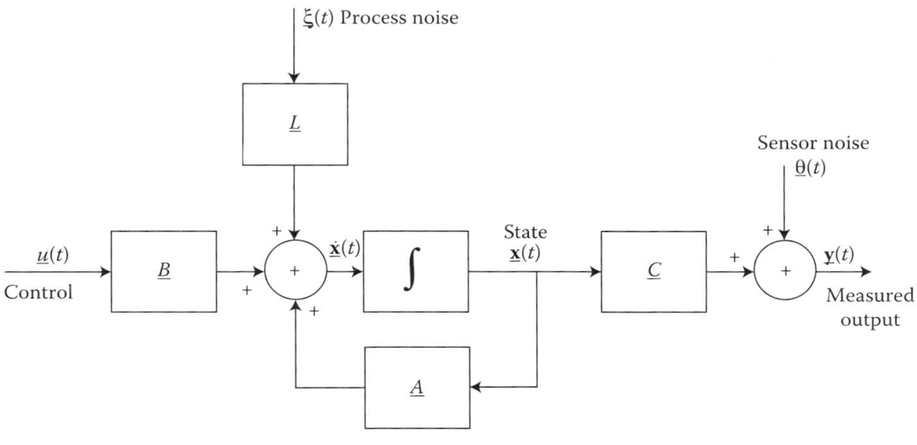

FIGURE 13.1 A stochastic linear dynamic system.

is the conditional mean of the state, i.e.,

$$\underline{\hat{\mathbf{x}}}(t) \triangleq E\{\underline{\mathbf{x}}(t)|U(t), Y(t)\} \tag{13.11}$$

One can relax the Gaussian assumption and define the optimality of the state estimate $\underline{\hat{\mathbf{x}}}(t)$ in different ways. One popular way is to demand that $\underline{\hat{\mathbf{x}}}(t)$ be generated by a *linear* transformation on the past "data" $U(t)$ and $Y(t)$, such that the state estimation error $\underline{\tilde{\mathbf{x}}}(t)$

$$\underline{\tilde{\mathbf{x}}} \triangleq \underline{\mathbf{x}}(t) - \underline{\hat{\mathbf{x}}}(t) \tag{13.12}$$

has zero mean, i.e.,

$$E\{\underline{\tilde{\mathbf{x}}}(t)\} = \underline{0} \tag{13.13}$$

and the cost functional

$$J = E\left\{\sum_{i=1}^{n} \tilde{\mathbf{x}}_i^2(t)\right\} = E\left\{\underline{\tilde{\mathbf{x}}}'(t)\underline{\tilde{\mathbf{x}}}(t)\right\} = tr[E\{\underline{\tilde{\mathbf{x}}}(t)\underline{\tilde{\mathbf{x}}}'(t)\}] \tag{13.14}$$

is minimized.

The cost functional J has the physical interpretation that it is the sum of the error variances $E\{\underline{\tilde{\mathbf{x}}}_i^2(t)\}$ for each state variable. If we let $\underline{\Sigma}$ denote the covariance matrix (stationary) of the state estimation error

$$\underline{\Sigma} \triangleq E\{\underline{\tilde{\mathbf{x}}}(t)\underline{\tilde{\mathbf{x}}}'(t)\} \tag{13.15}$$

then the cost, J, of Equation 13.14 can also be written as

$$J = tr[\underline{\Sigma}] \tag{13.16}$$

Bottom Line: We need an algorithm that translates the signals we can observe, $\underline{u}(t)$ and $\underline{\mathbf{y}}(t)$, into a state estimate $\underline{\hat{\mathbf{x}}}(t)$, such that the state estimation error $\underline{\tilde{\mathbf{x}}}$ is "small" in some well-defined sense. The KF is the algorithm that does just that!

13.3 Summary of the Kalman Filter Equations

13.3.1 Additional Assumptions

In this section we summarize the on-line and off-line equations that define the Kalman filter. Before we do that we make two additional "mild" assumptions

$$[\underline{A}, \underline{L}] \text{ is stabilizable (or controllable)} \tag{13.17}$$

$$[\underline{A}, \underline{C}] \text{ is detectable (or observable)} \tag{13.18}$$

$[\underline{A}, \underline{L}]$ is controllable, means that the process noise $\underline{\xi}(t)$ excites all modes of the system defined by Equation 13.1; $[\underline{A}, \underline{C}]$ is observable means that the "noiseless" output $\underline{\mathbf{y}}(t) = \underline{C}\underline{\mathbf{x}}(t)$ contains information about all state variables. If $[A, L]$ is stabilizable, the modes of the system that are not excited by $\underline{\xi}(t)$ are asymptotically stable; if $[A, C]$ is detectable, the unobserved modes are asymptotically stable.

13.3.2 The Kalman Filter Dynamics

The function of the KF is to generate in real time the state estimate $\underline{\hat{\mathbf{x}}}(t)$ of the state $\underline{\mathbf{x}}(t)$. The KF is actually an LTI dynamic system, of identical order (n) to the plant Equation 13.1, and is driven by (1) the deterministic control input $\underline{u}(t)$, and (2) the measured output vector $\underline{\mathbf{y}}(t)$. The Kalman filter dynamics are given as follows:

$$\frac{d\hat{\underline{\mathbf{x}}}(t)}{dt} = \underline{A}\hat{\underline{\mathbf{x}}}(t) + \underline{B}\underline{u}(t) + \underline{H}[\underline{\mathbf{y}}(t) - \underline{C}\hat{\underline{\mathbf{x}}}(t)] \tag{13.19}$$

A block diagram visualization of Equation 13.19 is shown in Figure 13.2. Note that in Equation 13.19 all variables have been defined previously, except for the KF gain matrix $\underline{H}$, whose calculation is carried out off-line and is discussed shortly.

The filter gain matrix $\underline{H}$ multiplies the so-called *residual* or *innovations* vector

$$\underline{\mathbf{r}}(t) \overset{\Delta}{=} \underline{\mathbf{y}}(t) - \underline{C}\hat{\underline{\mathbf{x}}}(t) \tag{13.20}$$

and updates the time rate of change, $d\hat{\underline{\mathbf{x}}}(t)/dt$, of the state estimate $\hat{\underline{\mathbf{x}}}(t)$. The residual $\underline{\mathbf{r}}(t)$ is like an "error" between the *measured* output $\mathbf{y}(t)$, and the *predicted* output $\underline{C}\hat{\underline{\mathbf{x}}}(t)$.

Remark 13.3

From an intuitive point of view the KF, defined by Equation 13.19 and illustrated in Figure 13.2, can be thought as a model-based observer or state reconstructor. The reader should carefully compare the structures depicted in Figures 13.1 and 13.2. The plant/sensor properties, reflected by the matrices $\underline{A}$, $\underline{B}$, and $\underline{C}$, are duplicated in the KF*. The state estimate $\hat{\mathbf{x}}(t)$ is continuously updated by the actual sensor measurements, through the formation of the residual $\underline{r}(t)$ and the "closing" of the loop with the filter gain matrix $\underline{H}$.

The KF dynamics of Equation 13.19 can also be written in the form

$$\frac{d\hat{\underline{\mathbf{x}}}(t)}{dt} = [\underline{A} - \underline{HC}]\hat{\underline{\mathbf{x}}}(t) + \underline{B}\underline{u}(t) + \underline{H}\underline{\mathbf{y}}(t) \tag{13.21}$$

From the structure of Equation 13.21 we can immediately see that the stability of the KF is governed by the matrix $\underline{A} - \underline{HC}$. At this point of our development we remark that the assumption in Equation 13.18,

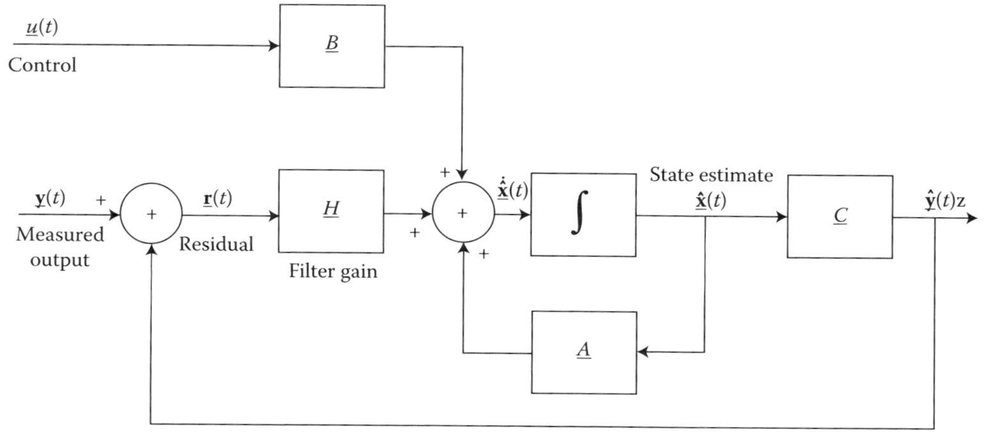

FIGURE 13.2 The structure of the Kalman Filter. The control, $\underline{u}(t)$, and measured output, $\underline{\mathbf{y}}(t)$, are those associated with the stochastic system of Figure 13.1. The filter gains matrix $\underline{H}$ is computed in a special way.

* No signal corresponding to $\underline{L}\underline{\xi}(t)$ shows up in Figure 13.2. This is because we assumed that $\underline{\xi}(t)$ had zero-mean and was completely unpredictable. Thus, the best estimate for $\underline{\xi}(t)$ given data up to time t is 0.

i.e., the detectability of $[\underline{A}, \underline{C}]$, guarantees the existence of at least one filter gain matrix $\underline{H}$ such that the KF is stable, i.e.,

$$Re\lambda_i[\underline{A} - \underline{HC}] < 0; \quad i = 1, 2, \ldots, n \tag{13.22}$$

where λ_i is the ith eigenvalue of $[\underline{A} - \underline{HC}]$.

13.3.3 Properties of Model-Based Observers

We have remarked that the KF gain matrix $\underline{H}$ is calculated in a very special way. However, it is extremely useful to examine the structure of Equations 13.19 or 13.21 and Figure 13.2 with a filter gain matrix $\underline{H}$ that is arbitrary except for the requirement that Equation 13.22 holds. Thus, for the development that follows in this subsection think of $\underline{H}$ as being a fixed matrix.

As before let $\underline{\tilde{\mathbf{x}}}(t)$ denote the state estimation error vector

$$\underline{\tilde{\mathbf{x}}}(t) \triangleq \underline{\mathbf{x}}(t) - \underline{\hat{\mathbf{x}}}(t) \tag{13.23}$$

It follows that

$$\frac{d\underline{\tilde{\mathbf{x}}}(t)}{dt} = \frac{d\underline{\mathbf{x}}(t)}{dt} - \frac{d\underline{\hat{\mathbf{x}}}(t)}{dt} \tag{13.24}$$

Next, we substitute Equations 13.1, 13.5, and 13.21 into Equation 13.24 and use Equation 13.23 as appropriate. After some easy algebraic manipulations we obtain the following stochastic vector differential equation for the state estimation error $\underline{\tilde{x}}(t)$:

$$\frac{d\underline{\tilde{\mathbf{x}}}(t)}{dt} = [\underline{A} - \underline{HC}]\underline{\tilde{\mathbf{x}}}(t) + \underline{L}\underline{\xi}(t) - \underline{H}\underline{\theta}(t) \tag{13.25}$$

Note that, in view of Equation 13.22, the estimation error dynamic system is stable. Also note that the deterministic signal $\underline{Bu}(t)$ does not appear in the error Equation 13.25.

Under our assumptions that the system is stable and was started at the indefinite past ($t_0 \to -\infty$), it is easy to verify that

$$E\{\underline{\tilde{\mathbf{x}}}(t)\} = \underline{0} \tag{13.26}$$

This implies that any stable model-based estimator of the form shown in Figure 13.2, with any filter gain matrix $\underline{H}$, gives us *unbiased* (that is, $E(\underline{\hat{\mathbf{x}}}(t)) = E(\underline{\mathbf{x}}(t))$ estimates.

Using next elementary facts from stochastic linear system theory one can calculate the error covariance matrix $\underline{\Sigma}$ of the state estimation error $\underline{\tilde{\mathbf{x}}}(t)$

$$\underline{\Sigma} \triangleq \text{cov}[\underline{\tilde{\mathbf{x}}}(t); \underline{\tilde{\mathbf{x}}}(t)] = E\{\underline{\tilde{\mathbf{x}}}(t)\underline{\tilde{\mathbf{x}}}'(t)\} \tag{13.27}$$

The matrix $\underline{\Sigma}$ is the solution of the so-called *Lyapunov matrix equation* (linear in $\underline{\Sigma}$)

$$[\underline{A} - \underline{HC}]\,\underline{\Sigma} + \underline{\Sigma}(\underline{A} - \underline{HC})' + \underline{L\Xi L'} + \underline{H\Theta H'} = \underline{0} \tag{13.28}$$

with

$$\underline{\Sigma} = \underline{\Sigma}' \geq \underline{0} \tag{13.29}$$

Thus, for any given filter gain matrix $\underline{H}$ we can calculate* the associated error covariance matrix $\underline{\Sigma}$ from Equation 13.28. Recalling Equation 13.16, we can evaluate, for a given $\underline{H}$, the quality of the estimator by calculating

$$J = tr[\underline{\Sigma}] \tag{13.30}$$

The specific way that the KF gain is calculated is by solving a constrained static deterministic optimization problem. Minimize Equation 13.30 with respect to the elements h_{ij} of the matrix $\underline{H}$ subject to the algebraic constraints given in Equations 13.28 and 13.29.

---

* The MATLAB® and MATRIXx™ software packages can solve Lyapunov equations.

13.3.4 The Kalman Filter Gain and Associated Filter Algebraic Riccati Equation (FARE)

We now summarize the off-line calculations that define fully the Kalman filter (Equation 13.19 or 13.21).

The KF gain matrix $\underline{H}$ is computed by

$$\underline{H} = \underline{\Sigma}\underline{C}'\underline{\Theta}^{-1} \tag{13.31}$$

where $\underline{\Sigma}$ is the unique, symmetric, and at least positive semidefinite solution matrix of the so-called filter algebraic Riccati equation (FARE)

$$\underline{0} = \underline{A}\underline{\Sigma} + \underline{\Sigma}\underline{A}' + \underline{L}\underline{\Xi}\underline{L}' - \underline{\Sigma}\underline{C}'\underline{\Theta}^{-1}\underline{C}\underline{\Sigma} \tag{13.32}$$

with

$$\underline{\Sigma} = \underline{\Sigma}' \geq \underline{0} \tag{13.33}$$

Remark 13.4

The formula for the KF gain can be obtained by setting

$$\frac{\partial}{\partial h_{ij}} tr[\underline{\Sigma}] = 0 \tag{13.34}$$

where $\underline{\Sigma}$ is given by Equation 13.28. The result is Equation 13.31. Substituting Equation 13.31 into Equation 13.28 one deduces the FARE (Equation 13.32).

13.3.5 Duality between the KF and LQ Problems

The mathematical problems associated with the solution of the LQ and KF are dual. This duality was recognized by R.E. Kalman as early as 1960 [2].

The duality can be used to deduce several properties of the KF simply by "dualizing" the results of the LQ problem. A summary of the KF properties is given in Section 13.4.

13.4 Kalman Filter Properties

13.4.1 Introduction

In this section we summarize the key properties of the Kalman filter. These properties are the "dual" of those for the LQ controller.

13.4.2 Guaranteed Stability

Recall that the KF algorithm is

$$\frac{d\hat{\underline{\mathbf{x}}}(t)}{dt} = [\underline{A} - \underline{H}\underline{C}]\hat{\mathbf{x}}(t) + \underline{B}u(t) + \underline{H}\underline{\mathbf{y}}(t) \tag{13.35}$$

Then, under the assumptions of Section 35.3, the matrix $[\underline{A} - \underline{H}\underline{C}]$ is strictly stable, i.e.,

$$Re\lambda_i[\underline{A} - \underline{H}\underline{C}] < 0; \quad i = 1, 2, \ldots, n \tag{13.36}$$

13.4.3 Frequency Domain Equality

One can readily derive a frequency domain equality for the KF. In the development that follows, let

$$\underline{\Xi} = \underline{I} \tag{13.37}$$

Let us make the following definitions: let $\underline{G}_{KF}(s)$ denote the KF loop-transfer matrix

$$\underline{G}_{KF}(s) \triangleq \underline{C}(s\underline{I} - \underline{A})^{-1}\underline{H} \tag{13.38}$$

$$\underline{G}_{KF}^{H}(s) \triangleq \underline{H}'(-s\underline{I} - \underline{A}')^{-1}\underline{C}' \tag{13.39}$$

where $[A]^H$ denotes the complex conjugate of the transpose of an arbitrary complex matrix A. Let $\underline{G}_{FOL}(s)$ denote the filter open-loop transfer matrix (from $\underline{\xi}(t)$ to $\underline{\mathbf{y}}(t)$)

$$\underline{G}_{FOL}(s) \triangleq \underline{C}(s\underline{I} - \underline{A})^{-1}\underline{L} \tag{13.40}$$

$$\underline{G}_{FOL}^{H}(s) \triangleq \underline{L}'(-s\underline{I} - \underline{A}')^{-1}\underline{C}' \tag{13.41}$$

Then the following equality holds

$$[\underline{I} + \underline{G}_{KF}(s)]\underline{\Theta}[\underline{I} + \underline{G}_{KF}(s)]^H = \underline{\Theta} + \underline{G}_{FOL}(s)\underline{G}_{FOL}^{H}(s) \tag{13.42}$$

If

$$\underline{\Theta} = \mu\underline{I} \quad \mu > 0 \tag{13.43}$$

then Equation 13.42 reduces to

$$[\underline{I} + \underline{G}_{KF}(s)][\underline{I} + \underline{G}_{KF}(s)]^H = \underline{I} + \frac{1}{\mu}\underline{G}_{FOL}(s)\underline{G}_{FOL}^{H}(s) \tag{13.44}$$

13.4.4 Guaranteed Robust Properties

The KF enjoys the same type of robustness properties as the LQ regulator. The following properties are valid if

$$\underline{\Theta} = \text{ diagonal matrix} \tag{13.45}$$

From the frequency domain equality (Equation 13.42) we deduce the inequality

$$[\underline{I} + \underline{G}_{KF}(s)][\underline{I} + \underline{G}_{KF}(s)]^H \geq \underline{I} \tag{13.46}$$

From the definition of singular values we then deduce that

$$\sigma_{\min}[\underline{I} + \underline{G}_{KF}(s)] \geq 1 \quad \text{or} \quad \sigma_{\max}[\underline{I} + \underline{G}_{KF}(s)]^{-1} \leq 1 \tag{13.47}$$

$$\sigma_{\min}[\underline{I} + \underline{G}_{KF}^{-1}(s)] \geq \frac{1}{2} \quad \text{or} \quad \sigma_{\max}[\underline{I} + \underline{G}_{KF}(s)]^{-1}\underline{G}_{KF} \leq 2 \tag{13.48}$$

13.5 The Accurate Measurement Kalman Filter

We summarize the properties of the Kalman Filter (KF) problem when the intensity of the sensor noise approaches zero. In a mathematical sense this is the "dual" of the so-called "cheap-control" LQR problem. The results are fundamental to the loop transfer recovery (LTR) method applied at the plant input.

13.5.1 Problem Definition

Consider as before, the stochastic LTI system

$$\dot{\underline{\mathbf{x}}}(t) = \underline{A}\mathbf{x}(t) + \underline{L}\underline{\xi}(t) \tag{13.49}$$

$$\underline{\mathbf{y}}(t) = \underline{C}\mathbf{x}(t) + \underline{\theta}(t) \tag{13.50}$$

We assume that the process noise $\underline{\xi}(t)$ is white, zero-mean, and with unit intensity, i.e.,

$$E\{\underline{\xi}(t)\underline{\xi}'(\tau) = \underline{I}\delta(t-\tau) \tag{13.51}$$

We also assume that the measurement noise $\underline{\theta}(t)$ is white, zero-mean, and with intensity indexed by μ, that is,

$$E\{\underline{\theta}(t)\underline{\theta}'(\tau) = \mu\underline{I}\delta(t-\tau) \tag{13.52}$$

Definition 13.1:

The accurate measurement KF problem is defined by the limiting case

$$\mu \rightarrow 0 \tag{13.53}$$

corresponding to essentially noiseless measurements.

Under the assumptions that $[\underline{A}, \underline{L}]$ is stabilizable and that $[\underline{A}, \underline{C}]$ is detectable we know that the KF is a stable system and generates the state estimates $\hat{\underline{\mathbf{x}}}(t)$ by

$$\frac{d\hat{\underline{\mathbf{x}}}(t)}{dt} = [\underline{A} - \underline{H}_\mu\underline{C}]\hat{\underline{\mathbf{x}}}(t) + \underline{H}_\mu\underline{\mathbf{y}}(t) \tag{13.54}$$

where we use the subscript μ to stress the dependence of the KF gain matrix $\underline{H}_\mu$ upon the parameter μ.

We recall that $\underline{H}_\mu$ is computed by

$$\underline{H}_\mu = \frac{1}{\mu}\underline{\Sigma}_\mu\underline{C}' \tag{13.55}$$

where the error covariance matrix $\underline{\Sigma}_\mu$, also dependent upon μ, is calculated by the solution of the FARE:

$$\underline{0} = \underline{A}\underline{\Sigma}_\mu + \underline{\Sigma}_\mu\underline{A}' + \underline{L}\underline{L}' - \frac{1}{\mu}\underline{\Sigma}_\mu\underline{C}'\underline{C}\underline{\Sigma}_\mu \tag{13.56}$$

We seek insight about the limiting behavior of both $\underline{\Sigma}_\mu$ and $\underline{H}_\mu$ as $\mu \rightarrow 0$.

13.5.2 The Main Result

In this section we summarize the main result in terms of a theorem.

Theorem 13.1:

Suppose that the transfer function matrix from the white noise $\underline{\xi}(t)$ to the output $\underline{\mathbf{y}}(t)$ for the system defined by Equations 13.49 and 13.50, i.e., the transfer function matrix

$$\underline{W}(s) \triangleq \underline{C}(s\underline{I} - \underline{A})^{-1}\underline{L} \tag{13.57}$$

is minimum phase. Then,

$$\lim_{\mu \to 0} \underline{\Sigma}_\mu = \underline{0} \tag{13.58}$$

and

$$\lim_{\mu \to 0} \sqrt{\mu}\underline{H}_\mu = \underline{LW}; \quad \underline{W}'\underline{W} = \underline{I} \tag{13.59}$$

Proof. This is theorem 4.14 in Kwakernaak and Sivan [7], pp. 370–371.

Remark 13.5

It can be shown that the requirement that $\underline{W}(s)$, given by Equation 13.57, be minimum phase is both a necessary and sufficient condition for the limiting properties given by Equations 13.58 and 13.59.

Remark 13.6

The implication of Equation 13.58 is that, in the case of exact measurements upon a minimum phase plant, the KF yields exact state estimates, since the error covariance matrix is zero. This assumes that the KF has been operating upon the data for a sufficiently long time so that initial transient errors have died out.

Remark 13.7

For a non-minimum phase plant

$$\lim_{\mu \to 0} \underline{\Sigma}_\mu \neq \underline{0} \tag{13.60}$$

Hence, perfect state estimation is impossible for non-minimum phase plants.

Remark 13.8

The limiting behavior (with $\underline{L} = \underline{B}$) of the Kalman Filter gain

$$\lim_{\mu \to 0} \sqrt{\mu}\underline{H}_\mu = \underline{BW}; \quad \underline{W}'\underline{W} = \underline{I} \tag{13.61}$$

is the precise dual of the limiting behavior of the LQ control gain

$$\lim_{\rho \to 0} \sqrt{\rho}\underline{G}_\rho = \underline{WC}; \quad \underline{W}'\underline{W} = \underline{I} \tag{13.62}$$

for the minimum phase plant

$$\underline{G}(s) = \underline{C}(s\underline{I} - \underline{A})^{-1}\underline{B} \tag{13.63}$$

The relation (Equation 13.61) has been used by Doyle and Stein [8] to apply the LTR method at the plant input, while Equation 13.62 has been used by Kwakernaak [9] to apply the LTR method at the plant output (see also Kwakernaak and Sivan [7], pp. 419–427).

References

1. Wiener, N., *Extrapolation, Interpolation and Smoothing of Stationary Time Series*, MIT Press and John Wiley & Sons, New York, 1950 (reprinted from a publication restricted for security reasons in 1942).
2. Kalman, R.E., A new approach to linear filtering and prediction problems, *ASME J. Basic Eng., Ser. D*, 82, 34–45, 1960.

3. Kalman, R.E., New methods and results in linear prediction and filtering theory, *Proc. Symp. Engineering Applications of Random Function Theory and Probability*, John Wiley & Sons, New York, 1961.
4. Kalman, R.E. and Bucy, R.S., New results in linear filtering and prediction theory, *ASME J. Basic Eng., Ser. D*, 83, 95–108, 1961.
5. Kalman, R.E., New methods in Wiener filtering, *Proc. First Symp. Engineering Applications of Random Function Theory and Probability*, John Wiley & Sons, New York, 1963, chap. 9.
6. Van Trees, H.L., *Detection Estimation, and Modulation Theory*, J. Wiley & Sons, New York, 1968.
7. Kwakernaak, H. and Sivan, R., *Linear Optimal Control Systems*, John Wiley & Sons, New York, 1972.
8. Doyle, J.C. and Stein, G., Multivariable feedback design: concepts for a classical/modern synthesis, *IEEE Trans. Autom. Control*, 1981.
9. Kwakernaak, H., Optimal low sensitivity linear feedback systems, *Automatica*, 5, 279–286, 1969.

Further Reading

There is a vast literature on Kalman filtering and its applications. A reasonable starting point is the recent textbook *Kalman Filtering: Theory and Practice* by M.S. Grewal and A.P. Andrews, Prentice Hall, Englewood Cliffs, NJ, 1993.

The books by Gelb et al., Jazwinski, and Maybeck have been standard references for those interested primarily in applications for some time.

1. Gelb, A., Kasper, J.F., Jr., Nash, R.A., Jr., Price, C.F., and Sutherland, A.A., Jr., *Applied Optimal Estimation*, MIT Press, Cambridge, MA, 1974.
2. Jazwinski, A.H., *Stochastic Processes and Filtering*, Academic Press, New York, 1970.
3. Maybeck, P.S., *Stochastic Models, Estimation, and Control*, Vol. 1, Academic Press, San Diego, 1979.
4. Maybeck, P.S., *Stochastic Models, Estimation, and Control*, Vol. 2, Academic Press, San Diego, 1982.

The reprint volume edited by H. Sorenson, *Kalman Filtering: Theory and Application*, IEEE Press, New York, 1985 contains reprints of many of the early theoretical papers on Kalman filtering, as well as a collection of applications.

A very readable introduction to the theory is

5. Davis, M.H.A., *Linear Estimation and Stochastic Control*, Halsted Press, New York, 1977.

14

Riccati Equations and Their Solution

Vladimír Kučera
Czech Technical University and Academy of Sciences

14.1 Introduction

An ordinary differential equation of the form

$$\dot{x}(t) + f(t)x(t) - b(t)x^2(t) + c(t) = 0 \tag{14.1}$$

is known as a *Riccati equation,* deriving its name from Jacopo Francesco, Count Riccati (1676–1754) [1,2], who studied a particular case of this equation from 1719 to 1724.

For several reasons, a differential equation of the form of Equation 14.1, and generalizations thereof comprise a highly significant class of nonlinear ordinary differential equations. First, they are intimately related to ordinary linear homogeneous differential equations of the second order. Second, the solutions of Equation 14.1 possess a very particular structure in that the general solution is a fractional linear function in the constant of integration. In applications, Riccati differential equations appear in the classical problems of the calculus of variations and in the associated disciplines of optimal control and filtering.

The *matrix* Riccati differential equation refers to the equation [3–5]

$$\dot{X}(t) + X(t)A(t) - D(t)X(t) - X(t)B(t)X(t) + C(t) = 0 \tag{14.2}$$

defined on the vector space of real $m \times n$ matrices. Here, $A, B, C,$ and D are real matrix functions of the appropriate dimensions. Of particular interest are the matrix Riccati equations that arise in optimal control and filtering problems and that enjoy certain symmetry properties. This chapter is concerned with these symmetric matrix Riccati differential equations and concentrates on the following four major topics:

- Basic properties of the solutions
- Existence and properties of constant solutions
- Asymptotic behavior of the solutions
- Methods for the numerical solution of the Riccati equations

14.2 Optimal Control and Filtering: Motivation

The following problems of optimal control and filtering are of great engineering importance and serve to motivate our study of the Riccati equations.

A linear-quadratic *optimal control* problem [6] consists of the following. Given a linear system

$$\dot{x}(t) = Fx(t) + Gu(t), \quad x(t_0) = c, \quad y(t) = Hx(t), \tag{14.3}$$

where x is the n-vector state, u is the q-vector control input, y is the p-vector of regulated variables, and F, G, H are constant real matrices of the appropriate dimensions. One seeks to determine a control input function u over some fixed time interval $[t_1, t_2]$ such that a given quadratic cost functional of the form

$$\eta(t_1, t_2, T) = \int_{t_1}^{t_2} [y'(t)y(t) + u'(t)u(t)]\,dt + x'(t_2)Tx(t_2), \tag{14.4}$$

with T being a constant real symmetric ($T = T'$) and nonnegative definite ($T \geq 0$) matrix, is afforded a minimum in the class of all solutions of Equation 14.3, for any initial state c.

A unique optimal control exists for all finite $t_2 - t_1 > 0$ and has the form

$$u(t) = -G'P(t, t_2, T)x(t),$$

where $P(t, t_2, T)$ is the solution of the matrix Riccati differential equation

$$-\dot{P}(t) = P(t)F + F'P(t) - P(t)GG'P(t) + H'H \tag{14.5}$$

subject to the terminal condition

$$P(t_2) = T.$$

The optimal control is a linear state feedback, which gives rise to the closed-loop system

$$\dot{x}(t) = [F - GG'P(t, t_2, T)]x(t)$$

and yields the minimum cost

$$\eta^*(t_1, t_2, T) = c'P(t_1, t_2, T)c. \tag{14.6}$$

A Gaussian *optimal filtering* problem [7] consists of the following. Given the p-vector random process z modeled by the equations

$$\begin{aligned} \dot{x}(t) &= Fx(t) + Gv(t), \\ z(t) &= Hx(t) + w(t), \end{aligned} \tag{14.7}$$

where x is the n-vector state and v, w are independent Gaussian white random processes (respectively, q-vector and p-vector) with zero means and identity covariance matrices. The matrices F, G, and H are constant real ones of the appropriate dimensions.

Given known values of z over some fixed time interval $[t_1, t_2]$ and assuming that $x(t_1)$ is a Gaussian random vector, independent of v and w, with zero mean and covariance matrix S, one seeks to determine an estimate $\hat{x}(t_2)$ of $x(t_2)$ such that the variance

$$\sigma(S, t_1, t_2) = Ef'[x(t_2) - \hat{x}(t_2)][x(t_2) - \hat{x}(t_2)]'f \tag{14.8}$$

of the error encountered in estimating any real-valued linear function f of $x(t_2)$ is minimized.

A unique optimal estimate exists for all finite $t_2 - t_1 > 0$ and is generated by a linear system of the form

$$\dot{\hat{x}}(t) = F\hat{x}(t) + Q(S, t_1, t)H'e(t), \quad \hat{x}(t_0) = 0, \quad e(t) = z(t) - H\hat{x}(t),$$

where $Q(S, t_1, t)$ is the solution of the matrix Riccati differential equation

$$\dot{Q}(t) = Q(t)F' + FQ(t) - Q(t)H'HQ(t) + GG' \tag{14.9}$$

subject to the initial condition

$$Q(t_1) = S.$$

The minimum error variance is given by

$$\sigma^*(S, t_1, t_2) = f'Q(S, t_1, t_2)f. \tag{14.10}$$

Equations 14.5 and 14.9 are special cases of the matrix Riccati differential equation 14.2 in that $A, B, C,$ and D are constant real $n \times n$ matrices such that

$$B = B', \quad C = C', \quad D = -A'.$$

Therefore, symmetric solutions $X(t)$ are obtained in the optimal control and filtering problems.

We observe that the control Equation 14.5 is solved *backward* in time, while the filtering Equation 14.9 is solved *forward* in time. We also observe that the two equations are *dual* to each other in the sense that

$$P(t, t_2, T) = Q(S, t_1, t)$$

on replacing $F, G, H, T,$ and $t_2 - t$ in Equation 14.5 respectively, by $F', H', G', S,$ and $t - t_1$ or, vice versa, on replacing $F, G, H, S,$ and $t - t_1$ in Equation 14.9 respectively, by $F', H', G', T,$ and $t_2 - t$. This makes it possible to dispense with both cases by considering only one prototype equation.

14.3 Riccati Differential Equation

This section is concerned with the basic properties of the prototype matrix Riccati differential equation

$$\dot{X}(t) + X(t)A + A'X(t) - X(t)BX(t) + C = 0, \tag{14.11}$$

where $A, B,$ and C are constant real $n \times n$ matrices with B and C being symmetric and nonnegative definite,

$$B = B', \ B \geq 0 \quad \text{and} \quad C = C', \ C \geq 0. \tag{14.12}$$

By definition, a *solution* of Equation 14.11 is a real $n \times n$ matrix function $X(t)$ that is absolutely continuous and satisfies Equation 14.11 for t on an interval on the real line R.

Generally, solutions of Riccati differential equations exist only locally. There is a phenomenon called finite escape time: the equation

$$\dot{x}(t) = x^2(t) + 1$$

has a solution $x(t) = \tan t$ in the interval $(-\frac{\pi}{2}, 0)$ that cannot be extended to include the point $t = -\frac{\pi}{2}$. However, Equation 14.11 with the sign-definite coefficients as shown in Equation14.12 does have global solutions.

Let $X(t, t_2, T)$ denote the solution of Equation 14.11 that passes through a constant real $n \times n$ matrix T at time t_2. We shall assume that

$$T = T' \quad \text{and} \quad T \geq 0. \tag{14.13}$$

Then the solution exists on every finite subinterval of R, is symmetric, nonnegative definite and enjoys certain monotone properties.

Theorem 14.1:

Under the assumptions of Equations 14.12 and 14.13, Equation 14.11 has a unique solution $X(t, t_2, T)$ satisfying

$$X(t, t_2, T) = X'(t, t_2, T), \quad X(t, t_2, T) \geq 0$$

for every T and every finite t, t_2, such that $t \geq t_2$.

This can most easily be seen by associating Equation 14.11 with the optimal control problem described in Equations 14.3 through 14.6. Indeed, using Equation 14.12, one can write $B = GG'$ and $C = H'H$ for some real matrices G and H. The quadratic cost functional η of Equation 14.4 exists and is nonnegative for every T satisfying Equation 14.13 and for every finite $t_2 - t$. Using Equation 14.6, the quadratic form $c'X(t, t_2, T)c$ can be interpreted as a particular value of η for every real vector c.

A further consequence of Equations 14.4 and 14.6 follows.

Theorem 14.2:

For every finite t_1, t_2 and τ_1, τ_2 such that $t_1 \leq \tau_1 \leq \tau_2 \leq t_2$,

$$X(t_1, \tau_1, 0) \leq X(t_1, \tau_2, 0)$$
$$X(\tau_2, t_2, 0) \leq X(\tau_1, t_2, 0)$$

and for every $T_1 \leq T_2$,

$$X(t_1, t_2, T_1) \leq X(t_1, t_2, T_2).$$

Thus, the solution of Equation 14.11 passing through $T = 0$ does not decrease as the length of the interval increases, and the solution passing through a larger T dominates that passing through a smaller T.

The Riccati Equation 14.11 is related in a very particular manner with linear Hamiltonian systems of differential equations.

Theorem 14.3:

Let

$$\Phi(t, t_2) = \begin{bmatrix} \Phi_{11} & \Phi_{12} \\ \Phi_{21} & \Phi_{22} \end{bmatrix}$$

be the fundamental matrix solution of the linear Hamiltonian matrix differential system

$$\begin{bmatrix} \dot{U}(t) \\ \dot{V}(t) \end{bmatrix} = \begin{bmatrix} A & -B \\ -C & -A' \end{bmatrix} \begin{bmatrix} U(t) \\ V(t) \end{bmatrix}$$

that satisfies the transversality condition

$$V(t_2) = TU(t_2).$$

If the matrix $\Phi_{11} + \Phi_{12}T$ *is nonsingular on an interval* $[t, t_2]$*, then*

$$X(t, t_2, T) = (\Phi_{21} + \Phi_{22}T)(\Phi_{11} + \Phi_{12}T)^{-1} \tag{14.14}$$

is a solution of the Riccati Equation 14.11.

Thus, if $V(t_2) = TU(t_2)$, then $V(t) = X(t, t_2, T)U(t)$ and the formula of Equation 14.14 follows.

Let us illustrate this with a simple example. The Riccati equation

$$\dot{x}(t) = x^2(t) - 1, \quad x(0) = T$$

satisfies the hypotheses of Equations 14.12 and 14.13. The associated linear Hamiltonian system of equations reads

$$\begin{bmatrix} \dot{u}(t) \\ \dot{v}(t) \end{bmatrix} = \begin{bmatrix} 0 & -1 \\ -1 & 0 \end{bmatrix} \begin{bmatrix} u(t) \\ v(t) \end{bmatrix}$$

and has the solution

$$\begin{bmatrix} u(t) \\ v(t) \end{bmatrix} = \begin{bmatrix} \cosh t & -\sinh t \\ -\sinh t & \cosh t \end{bmatrix} \begin{bmatrix} u(0) \\ v(0) \end{bmatrix},$$

where $v(0) = Tu(0)$. Then the Riccati equation has the solution

$$x(t, 0, T) = \frac{-\sinh t + T\cosh t}{\cosh t - T\sinh t}$$

for all $t \leq 0$. The monotone properties of the solution are best seen in Figure 14.1.

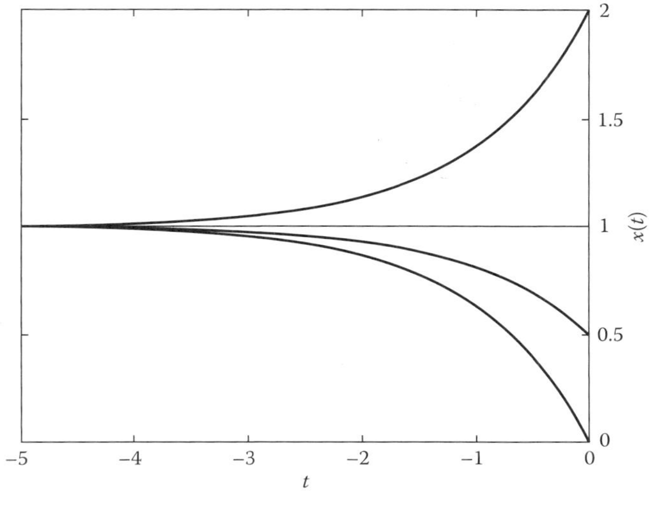

FIGURE 14.1 Graph of solutions.

14.4 Riccati Algebraic Equation

The constant solutions of Equation 14.11 are just the solutions of the quadratic equation

$$XA + A'X - XBX + C = 0, \tag{14.15}$$

called the *algebraic Riccati equation*. This equation can have real $n \times n$ matrix solutions X that are symmetric or nonsymmetric, sign definite or indefinite, and the set of solutions can be either finite or infinite. These solutions will be studied under the standing assumption of Equation 14.12, namely

$$B = B', \;\; B \geq 0 \quad \text{and} \quad C = C', \;\; C \geq 0.$$

14.4.1 General Solutions

The solution set of Equation 14.15 corresponds to a certain class of n-dimensional invariant subspaces of the associated $2n \times 2n$ matrix

$$H = \begin{bmatrix} A & -B \\ -C & -A' \end{bmatrix}. \tag{14.16}$$

This matrix has the *Hamiltonian* property

$$\begin{bmatrix} 0 & I \\ -I & 0 \end{bmatrix} H = -H' \begin{bmatrix} 0 & I \\ -I & 0 \end{bmatrix}.$$

It follows that H is similar to $-H'$ and therefore, the spectrum of H is symmetrical with respect to the imaginary axis.

Now suppose that X is a solution of Equation 14.15. Then

$$H \begin{bmatrix} I \\ X \end{bmatrix} = \begin{bmatrix} I \\ X \end{bmatrix} (A - BX).$$

Denote $J = U^{-1}(A - BX)U$, the Jordan form of $A - BX$ and put $V = XU$. Then

$$H \begin{bmatrix} U \\ V \end{bmatrix} = \begin{bmatrix} U \\ V \end{bmatrix} J,$$

which shows that the columns of

$$\begin{bmatrix} U \\ V \end{bmatrix}$$

are Jordan chains for H, that is, sets of vectors $x_1, x_2, \ldots, x_r$ such that $x_1 \neq 0$ and for some eigenvalue λ of H

$$\begin{aligned} Hx_1 &= \lambda x_1 \\ Hx_j &= \lambda x_j + x_{j+1}, \quad j = 2, 3, \ldots, r. \end{aligned}$$

In particular, x_1 is an eigenvector of H. Thus, we have the following result.

Theorem 14.4:

Equation 14.15 has a solution X if and only if there is a set of vectors $x_1, x_2, \ldots, x_n$ forming a set of Jordan chains for H and if

$$x_i = \begin{bmatrix} u_i \\ v_i \end{bmatrix},$$

where u_i is an n-vector, then $u_1, u_2, \ldots, u_n$ are linearly independent.

Furthermore, if

$$U = [u_1 \ldots u_n], \quad V = [v_1 \ldots v_n],$$

every solution of Equation 14.15 has the form $X = VU^{-1}$ *for some set of Jordan chains* $x_1, x_2, \ldots, x_n$ *for* H.

To illustrate, consider the scalar equation

$$X^2 + pX + q = 0,$$

where p, q are real numbers and $q \leq 0$. The Hamiltonian matrix

$$H = \begin{bmatrix} -\frac{p}{2} & -1 \\ q & \frac{p}{2} \end{bmatrix}$$

has eigenvalues λ and $-\lambda$, where

$$\lambda^2 = \left(\frac{p}{2}\right)^2 - q.$$

If $\lambda \neq 0$ there are two eigenvectors of H, namely

$$x_1 = \begin{bmatrix} 1 \\ -\frac{p}{2} + \lambda \end{bmatrix}, \quad x_2 = \begin{bmatrix} 1 \\ -\frac{p}{2} - \lambda \end{bmatrix},$$

which correspond to the solutions

$$X_1 = -\frac{p}{2} + \lambda, \quad X_2 = -\frac{p}{2} - \lambda.$$

If $\lambda = 0$ there exists one Jordan chain,

$$x_1 = \begin{bmatrix} 1 \\ -\frac{p}{2} \end{bmatrix}, \quad x_2 = \begin{bmatrix} 0 \\ 1 \end{bmatrix},$$

which yields the unique solution

$$X_1 = -\frac{p}{2}.$$

Theorem 14.4 suggests that, generically, the number of solutions of Equation 14.15 to be expected will not exceed the binomial coefficient $\binom{2n}{n}$, the number of ways in which the vectors $x_1, x_2, \ldots, x_n$ can be chosen from a basis of $2n$ eigenvectors for H. The solution set is infinite if there is a continuous family of Jordan chains. To illustrate this point consider Equation 14.15 with

$$A = \begin{bmatrix} 0 & 0 \\ 0 & 0 \end{bmatrix}, \quad B = \begin{bmatrix} 1 & 0 \\ 0 & 1 \end{bmatrix}, \quad C = \begin{bmatrix} 0 & 0 \\ 0 & 0 \end{bmatrix}.$$

The Hamiltonian matrix

$$H = \begin{bmatrix} 0 & 0 & -1 & 0 \\ 0 & 0 & 0 & -1 \\ 0 & 0 & 0 & 0 \\ 0 & 0 & 0 & 0 \end{bmatrix}$$

has the eigenvalue 0, associated with two Jordan chains

$$x_1 = \begin{bmatrix} a \\ b \\ 0 \\ 0 \end{bmatrix}, \quad x_2 = \begin{bmatrix} c \\ d \\ -a \\ -b \end{bmatrix}, \quad \text{and} \quad x_3 = \begin{bmatrix} c \\ d \\ 0 \\ 0 \end{bmatrix}, \quad x_4 = \begin{bmatrix} a \\ b \\ -c \\ -d \end{bmatrix},$$

where a, b and c, d are real numbers such that $ad - bc = 1$. The solution set of Equation 14.15 consists of the matrix

$$X_{13} = \begin{bmatrix} 0 & 0 \\ 0 & 0 \end{bmatrix}$$

and two continuous families of matrices

$$X_{12}(a, b) = \begin{bmatrix} ab & -a^2 \\ b^2 & -ab \end{bmatrix} \quad \text{and} \quad X_{34}(c, d) = \begin{bmatrix} -cd & c^2 \\ -d^2 & cd \end{bmatrix}.$$

Having in mind the applications in optimal control and filtering, we shall be concerned with the solutions of Equation 14.15 that are symmetric and nonnegative definite.

14.4.2 Symmetric Solutions

In view of Theorem 14.4, each solution X of Equation 14.15 gives rise to a factorization of the characteristic polynomial χ_H of H as

$$\chi_H(s) = (-1)^n q(s) q_1(s),$$

where $q = \chi_{A-BX}$. If the solution is *symmetric*, $X = X'$, then $q_1(s) = q(-s)$. This follows from

$$\begin{bmatrix} I & 0 \\ X & I \end{bmatrix}^{-1} \begin{bmatrix} A & -B \\ -C & -A' \end{bmatrix} \begin{bmatrix} I & 0 \\ X & I \end{bmatrix} = \begin{bmatrix} A - BX & -B \\ 0 & -(A - BX)' \end{bmatrix}.$$

There are two symmetric solutions that are of particular importance. They correspond to a factorization

$$\chi_H(s) = (-1)^n q(s) q(-s)$$

in which q has all its roots with nonpositive real part; it follows that $q(-s)$ has all its roots with nonnegative real part. We shall designate these solutions X_+ and X_-.

One of the basic results concerns the existence of these particular solutions [8]. To state the result, we recall some terminology. A pair of real $n \times n$ matrices (A, B) is said to be *controllable* (*stabilizable*) if the $n \times 2n$ matrix $[\lambda I - A \quad B]$ has linearly independent rows for every complex λ (respectively, for every complex λ such that $Re\ \lambda \geq 0$). The numbers λ for which $[\lambda I - A \quad B]$ loses rank are the eigenvalues of A that are not controllable (stabilizable) from B. A pair of real $n \times n$ matrices (A, C) is said to be *observable* (*detectable*) if the $2n \times n$ matrix $\begin{bmatrix} \lambda I - A \\ C \end{bmatrix}$ has linearly independent columns for every complex λ (respectively, for every complex λ such that $Re\ \lambda \geq 0$). The numbers λ for which $\begin{bmatrix} \lambda I - A \\ C \end{bmatrix}$ loses rank are the eigenvalues of A that are not observable (detectable) in C. Finally, let dim V denote the dimension of a linear space V and Im M, Ker M the image and the kernel of a matrix M, respectively.

Theorem 14.5:

There exists a unique symmetric solution X_+ of Equation 14.15 such that all eigenvalues of $A - BX_+$ have nonpositive real part if and only if (A, B) is stabilizable.

Theorem 14.6:

There exists a unique symmetric solution X_- of Equation 14.15 such that all eigenvalues of $A - BX_-$ have nonnegative real part if and only if $(-A, B)$ is stabilizable.

We observe that both (A, B) and $(-A, B)$ are stabilizable if and only if (A, B) is controllable. It follows that both solutions X_+ and X_- exist if and only if (A, B) is controllable.

For two real symmetric matrices X_1 and X_2, the notation $X_1 \geq X_2$ means that $X_1 - X_2$ is nonnegative definite. Since $A - BX_+$ has no eigenvalues with positive real part, neither has $X_+ - X$. Hence, $X_+ - X \geq 0$. Similarly, one can show that $X - X_- \geq 0$, thus introducing a partial order among the set of symmetric solutions of Equation 14.15.

Theorem 14.7:

Suppose that X_+ and X_- exist. If X is any symmetric solution of Equation 14.15, then

$$X_+ \geq X \geq X_-.$$

That is why X_+ and X_- are called the *extreme* solutions of Equation 14.15; X_+ is the maximal symmetric solution, while X_- is the minimal symmetric solution. The set of all symmetric solutions of Equation 14.15 can be related to a certain subset of the set of invariant subspaces of the matrix $A - BX_+$ or the matrix $A - BX_-$. Denote $\mathcal{V}_0$ and $\mathcal{V}_+$ the invariant subspaces of $A - BX_+$ that correspond, respectively, to the pure imaginary eigenvalues and to the eigenvalues having negative real part. Denote $\mathcal{W}_0$ and $\mathcal{W}_-$ the invariant subspaces of $A - BX_-$ that correspond, respectively, to the pure imaginary eigenvalues and to the eigenvalues having positive real part. Then it can be shown that $\mathcal{V}_0 = \mathcal{W}_0$ is the kernel of $X_+ - X_-$ and the symmetric solution set corresponds to the set of all invariant subspaces of $A - BX_+$ contained in $\mathcal{V}_+$ or, equivalently, to the set of all invariant subspaces of $A - BX_-$ contained in $\mathcal{W}_-$.

Theorem 14.8:

Suppose that X_+ and X_- exist. Let X_1, X_2 be symmetric solutions of Equation 14.15 corresponding to the invariant subspaces $\mathcal{V}_1$, $\mathcal{V}_2$ of $\mathcal{V}_+$ (or $\mathcal{W}_1$, $\mathcal{W}_2$ of $\mathcal{W}_-$). Then $X_1 \geq X_2$ if and only if $\mathcal{V}_1 \supset \mathcal{V}_2$ (or if and only if $\mathcal{W}_1 \subset \mathcal{W}_2$).

This means that the symmetric solution set of Equation 14.15 is a complete *lattice* with respect to the usual ordering of symmetric matrices. The maximal solution X_+ corresponds to the invariant subspace $\mathcal{V}_+$ of $A - BX_+$ or to the invariant subspace $\mathcal{W} = 0$ of $A - BX_-$, whereas the minimal solution X_- corresponds to the invariant subspace $\mathcal{V} = 0$ of $A - BX_+$ or to the invariant subspace $\mathcal{W}_-$ of $A - BX_-$.

This result allows one to count the distinct symmetric solutions of Equation 14.15 in some cases. Thus, let α be the number of distinct eigenvalues of $A - BX_+$ having negative real part and let $m_1, m_2, \ldots, m_\alpha$ be the multiplicities of these eigenvalues. Owing to the symmetries in H, the matrix $A - BX_-$ exhibits the same structure of eigenvalues with positive real part.

Theorem 14.9:

Suppose that X_+ and X_- exist. Then the symmetric solution set of Equation 14.15 has finite cardinality if and only if $A - BX_+$ is cyclic on $\mathcal{V}_+$ (or if and only if $A - BX_-$ is cyclic on $\mathcal{W}_-$). In this case, the set contains exactly $(m_1 + 1) \ldots (m_\alpha + 1)$ solutions.

Simple examples are most illustrative. Consider Equation 14.15 with

$$A = \begin{bmatrix} 0 & 0 \\ 0 & 1 \end{bmatrix}, \quad B = \begin{bmatrix} 1 & 0 \\ 0 & 1 \end{bmatrix}, \quad C = \begin{bmatrix} 1 & 0 \\ 0 & 3 \end{bmatrix}$$

and determine the lattice of symmetric solutions. We have

$$\chi_H(s) = s^4 - 5s^2 + 4$$

and the following eigenvectors of H:

$$x_1 = \begin{bmatrix} 1 \\ 0 \\ -1 \\ 0 \end{bmatrix}, \quad x_2 = \begin{bmatrix} 1 \\ 0 \\ 1 \\ 0 \end{bmatrix}, \quad x_3 = \begin{bmatrix} 0 \\ 1 \\ 0 \\ -1 \end{bmatrix}, \quad x_4 = \begin{bmatrix} 0 \\ 1 \\ 0 \\ 3 \end{bmatrix}$$

are associated with the eigenvalues $1, -1, 2$, and -2, respectively. Hence, the pair of solutions

$$X_+ = \begin{bmatrix} 1 & 0 \\ 0 & 3 \end{bmatrix}, \quad X_- = \begin{bmatrix} -1 & 0 \\ 0 & -1 \end{bmatrix}$$

corresponds to the factorization

$$\chi_H(s) = (s^2 - 3s + 2)(s^2 + 3s + 2)$$

and the solutions

$$X_{2,3} = \begin{bmatrix} 1 & 0 \\ 0 & -1 \end{bmatrix}, \quad X_{1,4} = \begin{bmatrix} -1 & 0 \\ 0 & 3 \end{bmatrix}$$

correspond to the factorization

$$\chi_H(s) = (s^2 - s - 2)(s^2 + s - 2).$$

There are four subspaces invariant under the matrices

$$A - BX_+ = \begin{bmatrix} -1 & 0 \\ 0 & -2 \end{bmatrix}, \quad A - BX_- = \begin{bmatrix} 1 & 0 \\ 0 & 2 \end{bmatrix}$$

each corresponding to one of the four solutions above. The partial ordering

$$X_+ \geq X_{2,3} \geq X_-, \quad X_+ \geq X_{1,4} \geq X_-$$

defines the lattice visualized in Figure 14.2.

As another example, we consider Equation 14.15 where

$$A = \begin{bmatrix} 0 & 0 \\ 0 & 0 \end{bmatrix}, \quad B = \begin{bmatrix} 1 & 0 \\ 0 & 1 \end{bmatrix}, \quad C = \begin{bmatrix} 1 & 0 \\ 0 & 1 \end{bmatrix}$$

and classify the symmetric solution set.

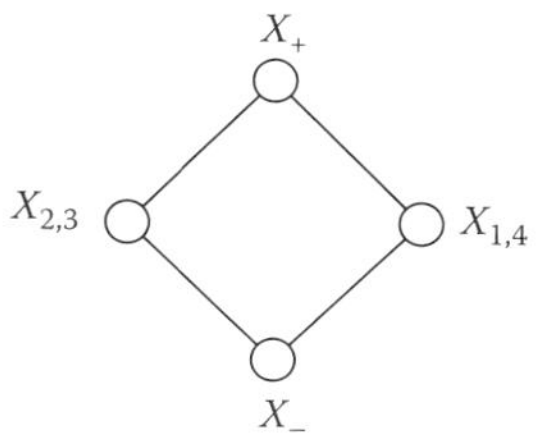

FIGURE 14.2 Lattice of solutions.

We have

$$\chi_H(s) = (s-1)^2(s+1)^2$$

and a choice of eigenvectors corresponding to the eigenvalues $1, -1$ of H is

$$x_1 = \begin{bmatrix} 1 \\ 0 \\ -1 \\ 0 \end{bmatrix}, \quad x_2 = \begin{bmatrix} 0 \\ 1 \\ 0 \\ -1 \end{bmatrix}, \quad x_3 = \begin{bmatrix} 1 \\ 0 \\ 1 \\ 0 \end{bmatrix}, \quad x_4 = \begin{bmatrix} 0 \\ 1 \\ 0 \\ 1 \end{bmatrix}.$$

Hence,

$$X_+ = \begin{bmatrix} 1 & 0 \\ 0 & 1 \end{bmatrix}, \quad X_- = \begin{bmatrix} -1 & 0 \\ 0 & -1 \end{bmatrix}$$

are the extreme solutions.

We calculate

$$A - BX_+ = \begin{bmatrix} -1 & 0 \\ 0 & -1 \end{bmatrix}, \quad A - BX_- = \begin{bmatrix} 1 & 0 \\ 0 & 1 \end{bmatrix}$$

and observe that the set of subspaces invariant under $A - BX_+$ or $A - BX_-$ (other than the zero and the whole space, which correspond to X_+ and X_-) is the family of one-dimensional subspaces parameterized by their azimuth angle θ. These correspond to the solutions

$$X_\vartheta = \begin{bmatrix} \cos\vartheta & \sin\vartheta \\ \sin\vartheta & -\cos\vartheta \end{bmatrix}.$$

Therefore, the solution set consists of X_+, X_- and the continuous family of solutions X_θ. It is a complete lattice and $X_+ \geq X_\theta \geq X_-$ for every θ.

14.4.3 Definite Solutions

Under the standing assumption (Equation 14.12), namely

$$B = B', \;\; B \geq 0 \quad \text{and} \quad C = C', \;\; C \geq 0,$$

one can prove that $X_+ \geq 0$ and $X_- \leq 0$. The existence of X_+, however, excludes the existence of X_- and vice versa, unless (A, B) is controllable.

If X_+ does exist, any other solution $X \geq 0$ of Equation 14.15 corresponds to a subspace $\mathcal{W}$ of $\mathcal{W}_-$ that is invariant under $A - BX$. From Equation 14.15,

$$X(A - BX) + (A - BX)'X = -XBX - C.$$

The restriction of $A - BX$ to $\mathcal{W}$ has eigenvalues with positive real part. Since $-XBX - C \leq 0$, it follows from the Lyapunov theory that X restricted to $\mathcal{W}$ is nonpositive definite and hence zero. We conclude that the solutions $X \geq 0$ of Equation 14.15 correspond to those subspaces $\mathcal{W}$ of $\mathcal{W}_-$ that are invariant under A and contained in Ker C.

The set of symmetric nonnegative definite solutions of Equation 14.15 is a sublattice of the lattice of all symmetric solutions [9,10]. Clearly X_+ is the largest solution and it corresponds to the invariant subspace $\mathcal{W} = 0$ of A. The smallest nonnegative definite solution will be denoted by X_* and it corresponds to $\mathcal{W}_*$, the largest invariant subspace of A contained in Ker C and associated with eigenvalues having positive real part.

The nonnegative definite solution set of Equation 14.15 has finite cardinality if and only if A is cyclic on W_*. In this case, the set contains exactly $(p_1 + 1) \ldots (_\rho+1)$ solutions, where ρ is the number of distinct eigenvalues of A associated with $\mathcal{W}_*$ and $p_1, p_2, \ldots, p_\rho$ are the multiplicities of these eigenvalues.

Analogous results hold for the set of symmetric solutions of Equation 14.15 that are nonpositive definite. In particular, if X_- exists, then any other solution $X \leq 0$ of Equation 14.15 corresponds to a subspace V of $\mathcal{V}_+$ that is invariant under A and contained in Ker C. Clearly X_- is the smallest solution and it corresponds to the invariant subspace $\mathcal{V} = 0$ of A. The largest nonpositive definite solution is denoted by $X_\times$ and it corresponds to $\mathcal{W}_\times$, the largest invariant subspace of A contained in Ker C and associated with eigenvalues having negative real part.

Let us illustrate this with a simple example. Consider Equation 14.15 where

$$A = \begin{bmatrix} 1 & 1 \\ 0 & 1 \end{bmatrix}, \quad B = \begin{bmatrix} 0 & 0 \\ 0 & 1 \end{bmatrix}, \quad C = \begin{bmatrix} 0 & 0 \\ 0 & 0 \end{bmatrix}$$

and classify the two sign-definite solution sets. We have

$$X_+ = \begin{bmatrix} 8 & 4 \\ 4 & 4 \end{bmatrix}, \quad X_- = \begin{bmatrix} 0 & 0 \\ 0 & 0 \end{bmatrix}.$$

The matrix A has one eigenvalue with positive real part, namely 1, and a basis for $\mathcal{W}_*$ is

$$x_1 = \begin{bmatrix} 1 \\ 0 \end{bmatrix}, \quad x_2 = \begin{bmatrix} 0 \\ -1 \end{bmatrix}.$$

Thus, there are three invariant subspaces of $\mathcal{W}_*$ corresponding to the three nonnegative definite solutions of Equation 14.15

$$X_+ = \begin{bmatrix} 8 & 4 \\ 4 & 4 \end{bmatrix}, \quad X_1 = \begin{bmatrix} 0 & 0 \\ 0 & 2 \end{bmatrix}, \quad X_* = \begin{bmatrix} 0 & 0 \\ 0 & 0 \end{bmatrix}.$$

These solutions make a lattice and

$$X_+ \geq X_1 \geq X_*.$$

The matrix A has no eigenvalues with negative real part. Therefore, $\mathcal{V}_* = 0$ and X_- is the only nonpositive definite solution of Equation 14.15.

Another example for Equation 14.15 is provided by

$$A = \begin{bmatrix} 0 & 0 \\ 0 & 0 \end{bmatrix}, \quad B = \begin{bmatrix} 1 & 0 \\ 0 & 0 \end{bmatrix}, \quad C = \begin{bmatrix} 0 & 0 \\ 0 & 0 \end{bmatrix}.$$

It is seen that neither (A, B) nor $(-A, B)$ is stabilizable; hence, neither X_+ nor X_- exists. The symmetric solution set consists of one continuous family of solutions

$$X_\alpha = \begin{bmatrix} 0 & 0 \\ 0 & \alpha \end{bmatrix}$$

for any real α. Therefore, both sign-definite solution sets are infinite; the nonnegative solution set is unbounded from above while the nonpositive solution set is unbounded from below.

14.5 Limiting Behavior of Solutions

The length of the time interval $t_2 - t_1$ in the optimal control and filtering problems is rather artificial. For this reason, an infinite time interval is often considered. This brings in the question of the limiting behavior of the solution $X(t, t_2, T)$ for the Riccati differential equation 14.11.

In applications to optimal control, it is customary to fix t and let t_2 approach $+\infty$. Since the coefficient matrices of Equation 14.11 are constant, the same result is obtained if t_2 is held fixed and t approaches $-\infty$. The limiting behavior of $X(t, t_2, T)$ strongly depends on the terminal matrix $T \geq 0$. For a suitable choice of T, the solution of Equation 14.11 may converge to a constant matrix $X \geq 0$, a solution of Equation 14.15. For some matrices T, however, the solution of Equation 14.11 may fail to converge to a constant matrix, but it may converge to a periodic matrix function.

Theorem 14.10:

Let (A, B) be stabilizable. If t and T are held fixed and $t_2 \to \infty$, then the solution $X(t, t_2, T)$ of Equation 14.11 is bounded on the interval $[t, \infty)$.

This result can be proved by associating an optimal control problem with Equation 14.11. Then stabilizability of (A, B) implies the existence of a stabilizing (not necessarily optimal) control. The consequent cost functional of Equation 14.4 is finite and dominates the optimal one.

If (A, B) is stabilizable, then X_+ exists and each real symmetric nonnegative definite solution X of Equation 14.15 corresponds to a subset $\mathcal{W}$ of $\mathcal{W}_*$, the set of A-invariant subspaces contained in Ker C and associated with eigenvalues having positive real part. The convergence of the solution $X(t, t_2, T)$ of Equation 14.11 to X depends on the properties of the image of $\mathcal{W}_*$ under T, see [11].

For simplicity, it is assumed that the eigenvalues $\lambda_1, \lambda_2, \ldots, \lambda_\rho$ of A associated with $\mathcal{W}_*$ are simple and, except for pairs of complex conjugate eigenvalues, have different real parts. Let the corresponding eigenvectors be ordered according to decreasing real parts of the eigenvalue

$$v_1, v_2, \ldots, v_\rho,$$

and denote $\mathcal{W}_i$ the A-invariant subspace of $\mathcal{W}_*$ spanned by $v_1, v_2, \ldots, v_i$.

Theorem 14.11:

Let (A, B) be stabilizable and the subspaces $\mathcal{W}_i$ of $\mathcal{W}_*$ satisfy the above assumptions. Then, for all fixed t and a given terminal condition $T \geq 0$, the solution $X(t, t_2, T)$ of Equation 14.11 converges to a constant solution of Equation 14.15 as $t_2 \to \infty$ if and only if the subspace $\mathcal{W}_{k+1}$ corresponding to any pair λ_k, λ_{k+1} of complex conjugate eigenvalues is such that dim $T\mathcal{W}_{k+1}$ equals either dim $T\mathcal{W}_{k-1}$ or dim $T\mathcal{W}_{k-1} + 2$.

Here is a simple example. Let

$$A = \begin{bmatrix} 1 & -1 \\ 1 & 1 \end{bmatrix}, \quad B = \begin{bmatrix} 1 & 0 \\ 0 & 1 \end{bmatrix}, \quad C = \begin{bmatrix} 0 & 0 \\ 0 & 0 \end{bmatrix}.$$

The pair (A, B) is stabilizable and A has two eigenvalues $1 + j$ and $1 - j$. The corresponding eigenvectors

$$v_1 = \begin{bmatrix} j \\ 1 \end{bmatrix}, \quad v_2 = \begin{bmatrix} -j \\ 1 \end{bmatrix}$$

span $\mathcal{W}_*$. Now consider the terminal condition

$$T = \begin{bmatrix} 1 & 0 \\ 0 & 0 \end{bmatrix}.$$

Then,

$$T\mathcal{W}_0 = 0, \quad T\mathcal{W}_2 = \text{Im} \begin{bmatrix} 1 \\ 0 \end{bmatrix}.$$

Theorem 14.11 shows that $X(t, t_2, T)$ does not converge to a constant matrix; in fact,

$$X(t, t_2, T) = \frac{1}{1 + e^{2(t-t_2)}} \begin{bmatrix} 2\cos^2(t - t_2) & -\sin 2(t - t_2) \\ -\sin 2(t - t_2) & 2\sin^2(t - t_2) \end{bmatrix}$$

tends to a periodic solution if $t_2 \to \infty$. On the other hand, if we select

$$T_0 = \begin{bmatrix} 0 & 0 \\ 0 & 0 \end{bmatrix}$$

we have

$$T_0\mathcal{W}_0 = 0, \quad T_0\mathcal{W}_2 = 0$$

and $X(t, t_2, T_0)$ does converge. Also, if we take

$$T_1 = \begin{bmatrix} 1 & 0 \\ 0 & 1 \end{bmatrix}$$

we have

$$T_1\mathcal{W}_0 = 0, \quad T_1\mathcal{W}_2 = R^2$$

and $X(t, t_2, T_1)$ converges as well.

If the solution $X(t, t_2, T)$ of Equation 14.11 converges to a constant matrix X_T as $t_2 \to \infty$, then X_T is a real symmetric nonnegative definite solution of Equation 14.15. Which solution is attained for a particular terminal condition?

Theorem 14.12:

Let (A, B) be stabilizable. Let

$$X_T = \lim_{t_2 \to \infty} X(t, t_2, T)$$

for a fixed $T \geq 0$. Then $X_T \geq 0$ is the solution of Equation 14.15 corresponding to the subspace $\mathcal{W}_T$ of $\mathcal{W}_*$, defined as the span of the real vectors v_i such that $T\mathcal{W}_i = T\mathcal{W}_{i-1}$ and of the complex conjugate pairs v_k, v_{k+1} such that $T\mathcal{W}_{k+1} = T\mathcal{W}_{k-1}$.

The cases of special interest are the extreme solutions X_+ and X_*. The solution $X(t, t_2, T)$ of Equation 14.12 tends to X_+ if and only if the intersection of $\mathcal{W}_*$ with Ker T is zero, and to X_* if and only if $\mathcal{W}_*$ is contained in Ker T.

This is best illustrated in the previous example, where

$$A = \begin{bmatrix} 1 & -1 \\ 1 & 1 \end{bmatrix}, \quad B = \begin{bmatrix} 1 & 0 \\ 0 & 1 \end{bmatrix}, \quad C = \begin{bmatrix} 0 & 0 \\ 0 & 0 \end{bmatrix}$$

and $\mathcal{W}_* = R^2$. Then $X(t, t_2, T)$ converges to X_+ if and only if T is positive definite; for instance, the identity matrix T yields the solution

$$X(t, t_2, I) = \frac{2}{1 + e^{2(t-t_2)}} \begin{bmatrix} 1 & 0 \\ 0 & 1 \end{bmatrix},$$

which tends to

$$X_+ = \begin{bmatrix} 2 & 0 \\ 0 & 2 \end{bmatrix}.$$

On the other hand, $X(t, t_2, T)$ converges to X_* if and only if T=0; then

$$X(t, t_2, 0) = 0$$

and X_*=0 is a fixed point of Equation 14.11.

14.6 Optimal Control and Filtering: Application

The problems of optimal control and filtering introduced in Section 14.2 are related to the matrix Riccati differential equations 14.5 and 14.9, respectively. These problems are defined over a finite horizon $t_2 - t_1$. We now apply the convergence properties of the solutions to study the two optimal problems in case the horizon becomes large.

To fix ideas, we concentrate on the optimal control problem. The results can easily be interpreted in the filtering context owing to the duality between Equations 14.5 and 14.9.

We recall that the finite horizon optimal control problem is that of minimizing the cost functional of Equation 14.4,

$$\eta(t_2) = \int_{t_1}^{t_2} [y'(t)y(t) + u'(t)u(t)]\, dt + x'(t_2)Tx(t_2)$$

along the solutions of Equation 14.3,

$$\dot{x}(t) = Fx(t) + Gu(t)$$
$$y(t) = Hx(t).$$

The optimal control has the form

$$u_0(t) = -G'X(t, t_2, T)x(t),$$

where $X(t, t_2, T)$ is the solution of Equation 14.11,

$$\dot{X}(t) + X(t)A + A'X(t) - X(t)BX(t) + C = 0$$

subject to the terminal condition $X(t_2) = T$, and where

$$A = F, \quad B = GG', \quad C = H'H.$$

The optimal control can be implemented as a state feedback and the resulting closed-loop system is

$$\begin{aligned} \dot{x}(t) &= [F - GG'X(t, t_2, T)]x(t) \\ &= [A - BX(t, t_2, T)]x(t). \end{aligned}$$

Hence, the relevance of the matrix $A - BX$, which plays a key role in the theory of the Riccati equation.

The *infinite horizon* optimal control problem then amounts to finding

$$\eta_* = \inf_{u(t)} \lim_{t_2 \to \infty} \eta(t_2) \tag{14.17}$$

and the corresponding optimal control $u_*(t)$, $t \geq t_1$ achieving this minimum cost.

The *receding horizon* optimal control problem is that of finding

$$\eta_{**} = \lim_{t_2 \to \infty} \inf_{u(t)} \eta(t_2) \tag{14.18}$$

and the limiting behavior $u_{**}(t)$, $t \geq t_1$ of the optimal control $u_o(t)$.

The question is whether η_* is equal to η_{**} and whether u_* coincides with u_{**}. If so, the optimal control for the infinite horizon can be approximated by the optimal control of the finite horizon problem for a sufficiently large time interval.

It turns out [12] that these two control problems have different solutions corresponding to different solutions of the matrix Riccati algebraic Equation 14.15,

$$XA + A'X - XBX + C = 0.$$

Theorem 14.13:

Let (A, B) be stabilizable. Then the infinite horizon optimal control problem of Equation 14.17 has a solution

$$\eta_* = x'(t_1)X_o x(t_1), \quad u_*(t) = -G'X_o x(t)$$

where $X_o \geq 0$ is the solution of Equation 14.15 corresponding to $\mathcal{W}_o$, the largest A-invariant subspace contained in $\mathcal{W}_* \cap$ Ker T.

Theorem 14.14:

Let (A, B) be stabilizable. Then the receding horizon optimal control problem of Equation 14.18 has a solution if and only if the criterion of Theorem 14.5 is satisfied and, in this case,

$$\eta_{**} = x'(t_1)X_T x(t_1), \quad u_{**}(t) = -G'X_T x(t)$$

where $X_T \geq 0$ is the solution of Equation 14.16 corresponding to $\mathcal{W}_T$ and defined in Theorem 14.5.

The equivalence result follows.

Theorem 14.15:

The solution of the infinite horizon optimal control problem is exactly the limiting case of the receding horizon optimal control problem if and only if the subspace $\mathcal{W}_* \cap$ KerT is invariant under A.

A simple example illustrates these points. Consider the finite horizon problem defined by

$$\dot{x}_1(t) = 2x_1(t) + u_1(t),$$
$$\dot{x}_2(t) = x_2(t) + u_2(t)$$

and

$$\eta(t_2) = [x_1(t_2) + x_2(t_2)]^2 + \int_{t_2}^{f} [u_1^2(\tau) + u_2^2(\tau)]\, d\tau,$$

which corresponds to the data

$$A = \begin{bmatrix} 2 & 0 \\ 0 & 1 \end{bmatrix}, \quad B = \begin{bmatrix} 1 & 0 \\ 0 & 1 \end{bmatrix}, \quad C = \begin{bmatrix} 0 & 0 \\ 0 & 0 \end{bmatrix}$$

and

$$T = \begin{bmatrix} 1 & 1 \\ 1 & 1 \end{bmatrix}.$$

Clearly $\mathcal{W}_* = R^2$ and the subspace

$$\mathcal{W}_* \cap \text{Ker}\, T = \text{Im} \begin{bmatrix} 1 \\ -1 \end{bmatrix}$$

is not invariant under A. Hence, the infinite and receding horizon problems are not equivalent.

The lattice of symmetric nonnegative definite solutions of Equation 14.11 has the four elements

$$X_+ = \begin{bmatrix} 4 & 0 \\ 0 & 2 \end{bmatrix}, \quad X_1 = \begin{bmatrix} 0 & 0 \\ 0 & 2 \end{bmatrix}, \quad X_2 = \begin{bmatrix} 4 & 0 \\ 0 & 0 \end{bmatrix}, \quad X_* = \begin{bmatrix} 0 & 0 \\ 0 & 0 \end{bmatrix}$$

depicted in Figure 14.3.

Since the largest A-invariant subspace of $\mathcal{W}_* \cap \text{Ker}T$ is zero, the optimal solution X_o of Equation 14.11 is the maximal element X_+. The infinite horizon optimal control reads

$$u_{1*}(t) = -4x_1(t),$$
$$u_{2*}(t) = -2x_2(t),$$

and affords the minimum cost

$$\eta_* = 4x_1^2(t_1) + 2x_2^2(t_1).$$

Now the eigenvectors of A spanning $\mathcal{W}_*$ are

$$v_1 = \begin{bmatrix} 1 \\ 0 \end{bmatrix}, \quad v_2 = \begin{bmatrix} 0 \\ 1 \end{bmatrix}$$

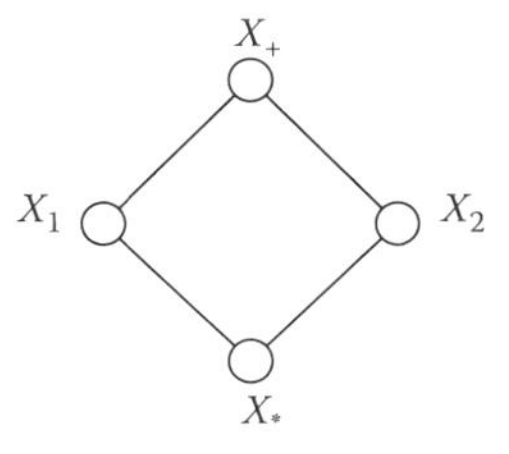

FIGURE 14.3 The four elements of the lattice of solutions.

and their T-images

$$Tv_1 = \begin{bmatrix} 1 \\ 1 \end{bmatrix}, \quad Tv_2 = \begin{bmatrix} 1 \\ 1 \end{bmatrix}$$

are linearly dependent. Hence, $\mathcal{W}_T$ is spanned by v_2 only,

$$\mathcal{W}_T = \text{Im}\begin{bmatrix} 0 \\ 1 \end{bmatrix}$$

and the optimal limiting solution X_T of Equation 14.11 equals X_2. The receding horizon optimal control reads

$$u_{1**}(t) = -4x_1(t)$$
$$u_{2**}(t) = 0$$

and affords the minimum cost

$$\eta_{**}(t) = 4x_1^2(t_1).$$

The optimal control problems with large horizon are practically relevant if the optimal closed-loop system

$$\dot{x}(t) = (A - BX)x(t)$$

is stable. A real symmetric nonnegative definite solution X of Equation 14.15 is said to be *stabilizing* if the eigenvalues of $A - BX$ all have negative real part. It is clear that the stabilizing solution, if it exists, is the maximal solution X_+. Thus, the existence of a stabilizing solution depends on $A - BX_+$. having eigenvalues with only negative real part.

Theorem 14.16:

Equation 14.15 has a stabilizing solution if and only if (A, B) is stabilizable and the Hamiltonian matrix H of Equation 14.16 has no pure imaginary eigenvalue.

The optimal controls over large horizons have a certain stabilizing effect. Indeed, if $X \geq 0$ is a solution of Equation 14.15 that corresponds to an A-invariant subspace $\mathcal{W}$ of $\mathcal{W}_*$, then the control $u(t) = -G'Xx(t)$ leaves unstable in $A - BX$ just the eigenvalues of A associated with $\mathcal{W}$; all the remaining eigenvalues of A with positive real part are stabilized. Of course, the pure imaginary eigenvalues of A, if any, cannot be stabilized; they remain intact in $A - BX$ for any solution X of Equation 14.15.

In particular, the infinite horizon optimal control problem leaves unstable the eigenvalues of A associated with $\mathcal{W}_o$, which are those not detectable either in C or in T, plus the pure imaginary eigenvalues. It follows that the infinite horizon optimal control results in a stable system if and only if X_o is the stabilizing solution of Equation 14.15. This is the case if and only if the hypotheses of Theorem 14.6 hold and $\mathcal{W}_o$, the largest A-invariant subspace contained in $\mathcal{W}_* \cap \text{Ker}T$, is zero. Equivalently, this corresponds to the pair

$$\left(\begin{bmatrix} C \\ T \end{bmatrix}, A\right)$$

being detectable.

The allocation of the closed-loop eigenvalues for the receding horizon optimal control problem is different, however. This control leaves unstable all eigenvalues of A associated with $\mathcal{W}_T$, where $\mathcal{W}_T$ is a subspace of $\mathcal{W}_*$ defined in Theorem 14.5. Therefore, the number of stabilized eigenvalues may be lower, equal to the dimension of $T\mathcal{W}_*$, whenever Ker T is not invariant under A. It follows that the receding horizon optimal control results in a stable system if and only if X_T is the stabilizing solution of Equation 14.15. This is the case if and only if the hypotheses of Theorem 14.6 hold and $\mathcal{W}_T$ is zero. Equivalently, this corresponds to $\mathcal{W}_* \cap \text{Ker}T = 0$. Note that this case occurs in particular if $T \geq X_+$.

It further follows that under the standard assumption, namely that

$$(A, B) \text{ stabilizable}$$
$$(A, C) \text{ detectable,}$$

both infinite and receding horizon control problems have solutions; these solutions are equivalent for any terminal condition T; and the resulting optimal system is stable.

14.7 Numerical Solution

The matrix Riccati *differential* equation 14.11 admits an analytic solution only in rare cases. A numerical integration is needed and the Runge–Kutta methods can be applied.

A number of techniques are available for the solution of the matrix Riccati *algebraic* Equation 14.15. These include invariant subspace methods [13] and the matrix sign function iteration [14]. We briefly outline these methods here with an eye on the calculation of the stabilizing solution to Equation 14.15.

14.7.1 Invariant Subspace Method

In view of Theorem 14.4, any solution X of Equation 14.15 can be computed from a Jordan form reduction of the associated $2n \times 2n$ Hamiltonian matrix

$$H = \begin{bmatrix} A & -B \\ -C & -A' \end{bmatrix}.$$

Specifically, compute a matrix of eigenvectors V to perform the following reduction:

$$V^{-1}HV = \begin{bmatrix} -J & 0 \\ 0 & J \end{bmatrix},$$

where $-J$ is composed of Jordan blocks corresponding to eigenvalues with negative real part only. If the stabilizing solution X exists, then H has no eigenvalues on the imaginary axis and J is indeed $n \times n$. Writing

$$V = \begin{bmatrix} V_{11} & V_{12} \\ V_{21} & V_{22} \end{bmatrix},$$

where each V_{ij} is $n \times n$, the solution sought is found by solving a system of linear equations,

$$X = V_{21}V_{11}^{-1}.$$

However, there are numerical difficulties with this approach when H has multiple or near-multiple eigenvalues. To ameliorate these difficulties, a method has been proposed in which a nonsingular matrix V of eigenvectors is replaced by an orthogonal matrix U of Schur vectors so that

$$U'HU = \begin{bmatrix} S_{11} & S_{12} \\ 0 & S_{22} \end{bmatrix},$$

where now S_{11} is a quasi-upper triangular matrix with eigenvalues having negative real part and S_{22} is a quasi-upper triangular matrix with eigenvalues having positive real part. When

$$U = \begin{bmatrix} U_{11} & U_{12} \\ U_{21} & U_{22} \end{bmatrix},$$

we observe that

$$\begin{bmatrix} V_{11} \\ V_{21} \end{bmatrix}, \quad \begin{bmatrix} U_{11} \\ U_{21} \end{bmatrix}$$

span the same invariant subspace and X can again be computed from

$$X = U_{21}U_{11}^{-1}.$$

14.7.2 Matrix Sign Function Iteration

Let M be a real $n \times n$ matrix with no pure imaginary eigenvalues. Let M have a Jordan decomposition $M = V\,J\,V^{-1}$ and let $\lambda_1, \lambda_2, \ldots, \lambda_n$ be the diagonal entries of J (the eigenvalues of M repeated according to their multiplicities). Then the *matrix sign function* of M is given by

$$\operatorname{sgn} M = V \begin{bmatrix} \operatorname{sgn} Re\lambda_1 & & \\ & \ddots & \\ & & \operatorname{sgn} Re\lambda_n \end{bmatrix} V^{-1}$$

It follows that the matrix $Z = \operatorname{sgn}\, M$ is diagonalizable with eigenvalues ± 1 and $Z^2 = I$. The key observation is that the image of $Z + I$ is the M-invariant subspace of R^n corresponding to the eigenvalues of M with negative real part.

This property clearly provides the link to Riccati equations, and what we need is a reliable computation of the matrix sign. Let $Z_0 = M$ be an $n \times n$ matrix whose sign is desired. For $k = 0, 1,$ perform the iteration

$$Z_{k+1} = \frac{1}{2c}(Z_k + c^2 Z_k^{-1}),$$

where $c = |\det Z_k|^{1/n}$. Then

$$\lim_{k\to\infty} Z_k = Z = \operatorname{sgn} M.$$

The constant c is chosen to enhance convergence of this iterative process. If $c = 1$, the iteration amounts to Newton's method for solving the equation

$$Z^2 - I = 0.$$

Naturally, it can be shown that the iteration is ultimately quadratically convergent.

Thus, to obtain the stabilizing solution X of Equation 14.15, provided it exists, we compute $Z = \operatorname{sgn}$ H, where H is the Hamiltonian matrix of Equation 14.16. The existence of X guarantees that H has no eigenvalues on the imaginary axis.

Writing

$$Z = \begin{bmatrix} Z_{11} & Z_{12} \\ Z_{21} & Z_{22} \end{bmatrix},$$

where each Z_{ij} is $n \times n$, the solution sought is found by solving a system of linear equations

$$\begin{bmatrix} Z_{12} \\ Z_{22} + I \end{bmatrix} X = -\begin{bmatrix} Z_{11} + I \\ Z_{21} \end{bmatrix}.$$

14.7.3 Concluding Remarks

We have discussed two numerical methods for obtaining the stabilizing solution of the matrix Riccati algebraic equation 14.15. They are both based on the intimate connection between the Riccati equation solutions and invariant subspaces of the associated Hamiltonian matrix. The method based on Schur vectors is a direct one, while the method based on the matrix sign function is iterative.

The Schur method is now considered one of the more reliable for Riccati equations and has the virtues of being simultaneously efficient and numerically robust. It is particularly suitable for Riccati equations with relatively small dense coefficient matrices, say, of the order of a few hundreds or less. The matrix sign function method is based on the Newton iteration and features global convergence, with ultimately quadratic order. Iteration formulas can be chosen to be of arbitrary order convergence in exchange for, naturally, an increased computational burden. The effect of this increased computation can, however, be ameliorated by parallelization.

The two methods are not limited to computing the stabilizing solution only. The matrix sign iteration can also be used to calculate X_-, the antistabilizing solution of Equation 14.15, by considering the matrix sgn $H - I$ instead of sgn $H + I$. The Schur approach can be used to calculate any, not necessarily symmetric, solution of Equation 14.15, by ordering the eigenvalues on the diagonal of S accordingly.

Acknowledgments

Acknowledgment to Project 1M0567 Ministry of Education of the Czech Republic.

References

Historical documents:

1. Riccati, J. F., Animadversationes in aequationes differentiales secundi gradus, *Acta Eruditorum Lipsiae*, 8, 67–73, 1724.
2. Boyer, C. B., *The History of Mathematics*, Wiley, New York, NY, 1974.

Tutorial textbooks:

3. Reid, W. T., *Riccati Differential Equations*, Academic Press, New York, NY, 1972.
4. Bittanti, S., Laub, A. J., and Willems, J. C., Eds., *The Riccati Equation*, Springer-Verlag, Berlin, 1991.

Survey paper:

5. Kučera, V., A review of the matrix Riccati equation, *Kybernetika*, 9, 42–61, 1973.

Original sources on optimal control and filtering:

6. Kalman, R. E., Contributions to the theory of optimal control, *Bol. Soc. Mat. Mexicana*, 5, 102–119, 1960.
7. Kalman, R. E. and Bucy, R. S., New results in linear filtering and prediction theory, *J. Basic Eng. (ASME Trans.)*, 83D, 95–108, 1961.

Original sources on the algebraic equation:

8. Willems, J. C., Least squares stationary optimal control and the algebraic Riccati equation, *IEEE Trans. Autom. Control*, 16, 612–634, 1971.
9. Kučera, V., A contribution to matrix quadratic equations, *IEEE Trans. Autom. Control*, 17, 344–347, 1972.
10. Kučera, V., On nonnegative definite solutions to matrix quadratic solutions, *Automatica*, 8, 413–423, 1972.

Original sources on the limiting behavior:

11. Callier, F. M. and Willems, J. L., Criterion for the convergence of the solution of the Riccati differential equation, *IEEE Trans. Autom. Control*, 26, 1232–1242, 1981.
12. Willems, J. L. and Callier, F. M., Large finite horizon and infinite horizon LQ-optimal control problems, *Optimal Control Appl. Methods*, 4, 31–45, 1983.

Original sources on the numerical methods:

13. Laub, A. J., A Schur method for solving algebraic Riccati equations, *IEEE Trans. Autom. Control*, 24, 913–921, 1979.
14. Roberts, J. D., Linear model reduction and solution of the algebraic Riccati equation by use of the sign function, *Int. J. Control*, 32, 677–687, 1980.

15

Observers

Bernard Friedland
New Jersey Institute of Technology

15.1 Introduction

An observer for a dynamic system $\mathcal{S}(x, y, u)$ with state x, output y, and input u is another dynamic system $\hat{\mathcal{S}}(\hat{x}, y, u)$ having the property that the state $\hat{x}$ of the observer $\hat{\mathcal{S}}$ converges to the state x of the process $\mathcal{S}$, independent of the input u or the state x.

Among the various applications for observers, perhaps the most important is for the implementation of closed-loop control algorithms designed by state-space methods. The control algorithm is designed in two parts: a "full-state feedback" part based on the assumption that all the state variables can be measured; and an observer to estimate the state of the process based on the observed output. The concept of separating the feedback control design into these two parts is known as the *separation principle,* which has rigorous validity in linear systems and in a limited class of nonlinear systems. Even when its validity cannot be rigorously established, the separation principle is often a practical solution to many design problems.

The concept of an observer for a dynamic process was introduced in 1966 by Luenberger [1]. The generic "Luenberger observer," however, appeared several years after the Kalman filter, which is in fact an important special case of a Luenberger observer—an observer optimized for the noise present in the observations and in the input to the process.

15.2 Linear Full-Order Observers

15.2.1 Continuous-Time Systems

Consider a linear, continuous-time dynamic system

$$\dot{x} = Ax + Bu \tag{15.1}$$

$$y = Cx \tag{15.2}$$

The more generic output

$$y = Cx + Du$$

can be treated by defining a modified output

$$\bar{y} = y - Du$$

and working with $\bar{y}$ instead of y. The direct coupling Du from the input to the output is absent in most physical plants.

A full-order observer for the linear process defined by Equations 15.1 and 15.2 has the generic form

$$\dot{\hat{x}} = \hat{A}\hat{x} + Ky + Hu \tag{15.3}$$

where the dimension of state $\hat{x}$ of the observer is equal to the dimension of process state x.

The matrices $\hat{A}$, K, and H appearing in Equation 15.3 must be chosen to conform with the required property of an observer: that the observer state must converge to the process state independent of the state x and the input u. To determine these matrices, let

$$e := x - \hat{x} \tag{15.4}$$

be the estimation error. From Equations 15.1 through 15.3

$$\begin{aligned}\dot{e} &= Ax + Bu - \hat{A}(x - e) - GCx - Hu,\\ &= \hat{A}e + (-\hat{A} + A - KC)x + (B - H)u,\end{aligned} \tag{15.5}$$

From Equation 15.5 it is seen that for the error to converge to zero, independent of x and u, the following conditions must be satisfied:

$$\hat{A} = A - KC \tag{15.6}$$

$$H = B \tag{15.7}$$

When these conditions are satisfied, the estimation error is governed by

$$\dot{e} = \hat{A}e \tag{15.8}$$

which converges to zero if $\hat{A}$ is a "stability matrix." When $\hat{A}$ is constant, this means that its eigenvalues must lie in the (open) left half-plane.

Since the matrices A, B, and C are defined by the plant, the only freedom in the design of the observer is in the selection of the gain matrix K.

To emphasize the role of the observer gain matrix, and accounting for requirements of Equations 15.6 and 15.7, the observer can be written as

$$\dot{\hat{x}} = A\hat{x} + Bu + K(y - C\hat{x}) \tag{15.9}$$

A block diagram representation of Equation 15.9, as given in Figure 15.1, aids in the interpretation of the observer. Note that the observer comprises a model of the process with an added input:

$$K(y - C\hat{x}) = Kr$$

The quantity

$$r := y - C\hat{x} = y - \hat{y} \tag{15.10}$$

often called the *residual*, is the difference between the actual observation y and the "synthesized" observation

$$\hat{y} = C\hat{x}$$

produced by the observer. The observer can be viewed as a feedback system designed to drive the residual to zero: as the residual is driven to zero, the input to Equation 15.9 due to the residual vanishes and the state of Equation 15.9 looks like the state of the original process.

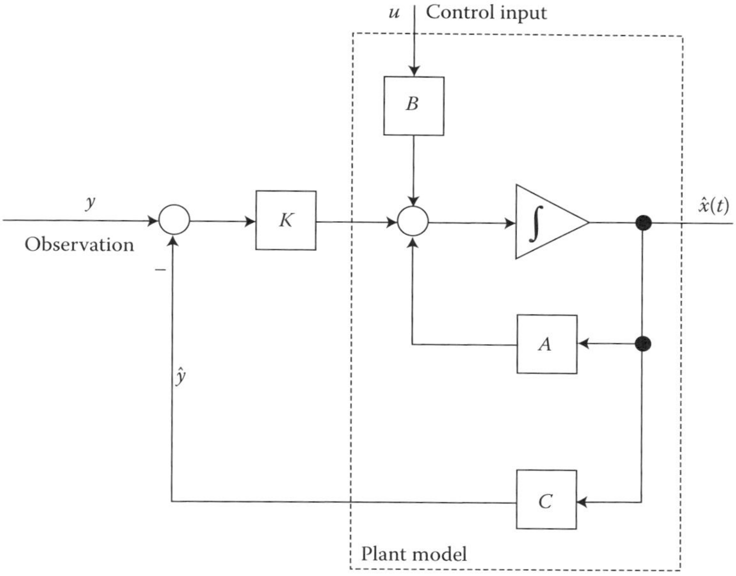

FIGURE 15.1 Full-order observer for linear process.

The fundamental problem in the design of an observer is the determination of the observer gain matrix K such that the closed-loop observer matrix

$$\hat{A} = A - KC \tag{15.11}$$

is a stability matrix.

There is considerable flexibility in the selection of the observer gain matrix. Two methods are standard: optimization and pole placement.

15.2.1.1 Optimization

Since the observer given by Equation 15.9 has the structure of a Kalman filter (see Chapter 13), its gain matrix can be chosen as a Kalman filter gain matrix, that is,

$$K = PC'R^{-1} \tag{15.12}$$

where P is the covariance matrix of the estimation error and satisfies the matrix Riccati equation

$$\dot{P} = AP + PA' - PC'R^{-1}CP + Q, \tag{15.13}$$

where R is a positive-definite matrix and Q is a positive-semidefinite matrix.

In most applications the steady-state covariance matrix is used in Equation 15.12. This matrix is given by setting $\dot{P}$ in Equation 15.13 to zero. The resulting equation is known as the *algebraic Riccati equation*. Algorithms to solve the algebraic Riccati equation are included in popular control system software packages such as MATLAB® or SciLab.

In order for the gain matrix given by Equations 15.12 and 15.13 to be genuinely optimum, the process noise and the observation noise must be white with the matrices Q and R being their spectral densities. It is rarely possible to determine these spectral density matrices in practical applications. Hence, the matrices Q and R are best treated as design parameters that can be varied to achieve overall system design objectives.

If the observer is to be used as a state estimator in a closed-loop control system, an appropriate form for the matrix Q is

$$Q = q^2 BB' \tag{15.14}$$

As has been shown by Doyle and Stein [2], as $q \to \infty$, this observer tends to "recover" the stability margins assured by a full-state feedback control law obtained by quadratic optimization.

15.2.1.2 Pole Placement

An alternative to solving the algebraic Riccati equation to obtain the observer gain matrix is to select K to place the poles of the observer, that is, the eigenvalues of $\hat{A}$ in Equation 15.11.

When there is a single observation, K is a column vector with exactly as many elements as eigenvalues of $\hat{A}$. Hence, specification of the eigenvalues of $\hat{A}$ uniquely determines the gain matrix K. A number of algorithms can be used to determine the gain matrix, some of which are incorporated into the popular control system design software packages. Some of the algorithms have been found to be numerically ill-conditioned; so caution should be exercised in using the results.

The present author has found the Bass–Gura [3] formula to be effective in most applications. This formula gives the gain matrix as

$$K = (OW)'^{-1}(\hat{a} - a) \tag{15.15}$$

where

$$a = \begin{bmatrix} a_1 & a_2 & \cdots & a_n \end{bmatrix}' \tag{15.16}$$

is the vector formed from the coefficients of the characteristic polynomial of the process matrix A

$$|sI - A| = s^n + a_1 s^{n-1} + \cdots + a_{n-1}s + a_n \tag{15.17}$$

and $\hat{a}$ is the vector formed from the coefficients of the desired characteristic polynomial

$$|sI - \hat{A}| = s^n + \hat{a}_1 s^{n-1} + \cdots + \hat{a}_{n-1}s + \hat{a}_n \tag{15.18}$$

The other matrices in Equation 15.15 are given by

$$O = \begin{bmatrix} C' & A'C' & \cdots & A'^{n-1}C' \end{bmatrix} \tag{15.19}$$

which is the *observability matrix* of the process, and

$$W = \begin{bmatrix} 1 & a_1 & \cdots & a_n \\ 0 & 1 & \cdots & a_{n-1} \\ \vdots & \vdots & \vdots & \vdots \\ 0 & 0 & \cdots & 1 \end{bmatrix} \tag{15.20}$$

The determinant of W is 1; hence it is not singular. If the observability matrix O is not singular, the inverse matrix required in Equation 15.15 exists. Hence, the gain matrix K can be found that places the observer poles at arbitrary locations if (and only if) the process for which an observer is sought is observable. Numerical problems occur, however, when the observability matrix is nearly singular. Other numerical problems can arise in determination of the characteristic polynomial $|sI - A|$ for high-order systems and in the determination of $sI - \hat{A}$ when the individual poles, and not the characteristic polynomial, are specified. In such instances, it may be necessary to use an algorithm designed to handle difficult numerical calculations.

When two or more quantities are observed, there are more elements in the gain matrix than eigenvalues of $\hat{A}$; hence specification of the eigenvalues of $\hat{A}$ does not uniquely specify the gain matrix K. In addition to placing the eigenvalues, more of the "eigenstructure" of $\hat{A}$ can be specified. This method of selecting the gain matrix is fraught with difficulty, however, and the use of the algebraic Riccati equation is usually preferable.

The Matlab algorithm place can be used to place the poles of a system with multiple observations, and uses the additional freedom to improve the robustness of the resulting observer.

15.2.2 Delayed Data

In some applications, the observation data may be delayed:

$$y_d(t) = y(t-T), \quad \text{with } y(t) = Cx(t) \tag{15.21}$$

in which T is the known time delay.

If the observation vector $y(t)$ has more than one component, then it is assumed that all the components are delayed by the same amount of time T.

On recognizing that the process is time-invariant, that is,

$$\dot{x}(t-T) = Ax(t-T) + Bu(t-T) \tag{15.22}$$

an observer for the delayed state is given by

$$\dot{\hat{x}}(t-T) = A\hat{x}(t-T) + Bu(t-T) + K(y_d(t) - C\hat{x}(t-T)) \tag{15.23}$$

where K is the observer gain matrix, which can be obtained by the methods discussed above.

Having thus obtained an estimate $\hat{x}(t-T)$ of the delayed state, we extrapolate the estimate to the present time. Since there are no data in the interval $[t-T, t]$, the extrapolation simply makes use of the state transition matrix e^{A_cT}, with

$$A_c = A - BG$$

of the closed-loop system over the interval $[t-T, T]$. Thus the estimate of the state at time t is given by

$$\hat{x}(t) = e^{A_cT}\hat{x}(t-T) \tag{15.24}$$

where $\hat{x}(t)$ is given by Equation 15.23.

The observer requires the delayed control signal $u(t-T)$, which can be implemented by any of the "standard" methods for implementing a delay line.

15.3 Linear Reduced-Order Observers

The observer described in the previous section has the same order as the plant, irrespective of the number of independent observations. A reduced-order observer of the order $n-m$, where n is the dimension of the state vector and m is the number of observations, can also be specified. When the number of observations is comparable to the dimension of the state vector, the reduced-order observer may represent a considerable simplification.

The description of the reduced-order observer is simplified if the state vector can be partitioned into two substates:

$$x = \begin{bmatrix} x_1 \\ \cdots \\ x_2 \end{bmatrix} \tag{15.25}$$

such that

$$x_1 = y = Cx \tag{15.26}$$

is the observation vector (of dimension m) and x_2 (of dimension $n-m$) comprises the components of the state vector that cannot be measured directly.

In terms of x_1, and x_2, the plant dynamics are written as

$$\dot{x}_1 = A_{11}x_1 + A_{12}x_2 + B_1u \tag{15.27}$$

$$\dot{x}_2 = A_{21}x_1 + A_{22}x_2 + B_2u \tag{15.28}$$

Since x_1 is directly measured, no observer is required for that substate, that is,

$$\hat{x}_1 = x_1 = y \tag{15.29}$$

For the remaining substate, we define the reduced-order observer by

$$\hat{x}_2 = Ky + z \tag{15.30}$$

where z is the state of a system of order $n - m$:

$$\dot{z} = \hat{A}z + Ly + Hu \tag{15.31}$$

A block-diagram representation of the reduced-order observer is given in Figure 15.2a.

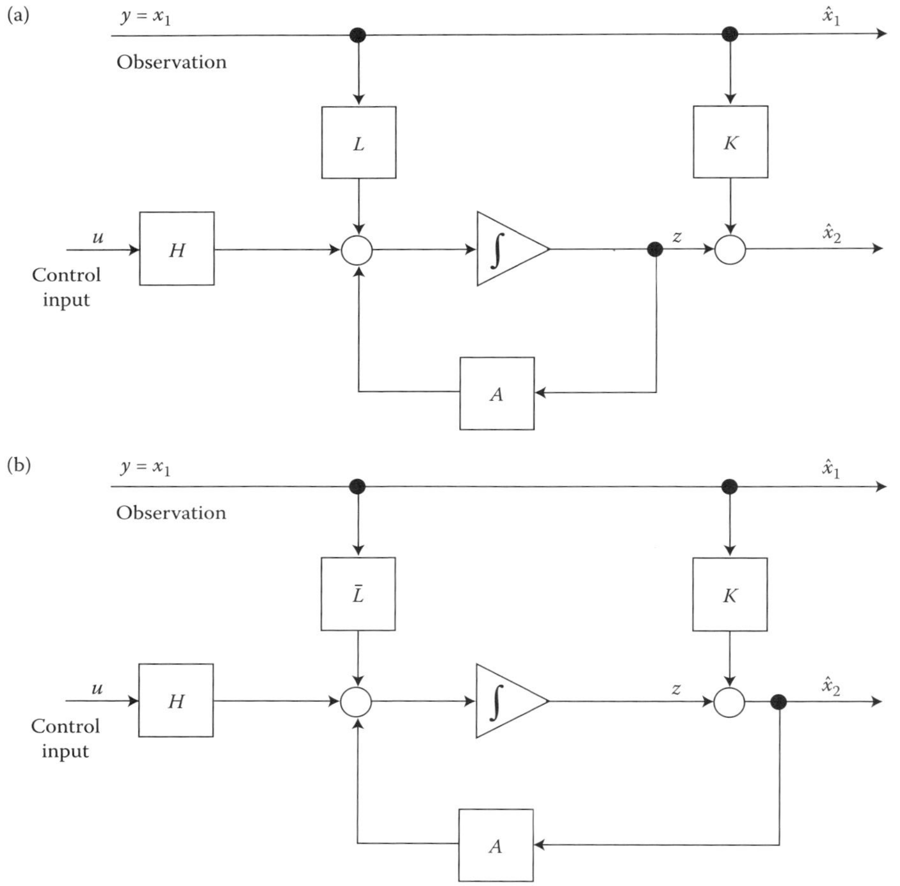

FIGURE 15.2 Reduced-order observer for linear process. (a) Feedback from z and (b) feedback from $\hat{x}_2$.

The matrices $\hat{A}, L, H$, and K are chosen, as in the case of the full-order observer, to ensure that the error in the estimation of the state converges to zero, independent of x, y, and u.

Since there is no error in the estimation of x_1, that is,

$$e_1 = x_1 - \hat{x}_1 = 0 \tag{15.32}$$

by virtue of Equation 15.29, it is necessary to ensure only the convergence of

$$e_2 = x_2 - \hat{x}_2 \tag{15.33}$$

to zero.

From Equations 15.28 through 15.31

$$\dot{e}_2 = (A_{21} - KA_{11} + \hat{A}K - L)x_1 + (A_{22} - KA_{12} - \hat{A})x_2 + \hat{A}e_2 + (B_2 - KB_1 - H)u \tag{15.34}$$

As in the case of the full-order observer, to make the coefficients of x_1, x_2, and u vanish it is necessary that the matrices in Equations 15.29 and 15.31 satisfy

$$\hat{A} = A_{22} - KA_{12}, \tag{15.35}$$

$$L = A_{21} - KA_{11} + \hat{A}K, \tag{15.36}$$

$$H = B_2 - KB_1 \tag{15.37}$$

Two of these conditions (Equations 15.35 and 15.37) are analogous to Equations 15.6 and 15.7 for the full-order observer; Equation 15.36 is a new requirement for the additional matrix L that is required by the reduced-order observer.

When these conditions are satisfied, the error in estimation of x_2 is given by

$$e_2 = \hat{A}e_2.$$

Hence, the gain matrix K must be chosen such that the eigenvalues of $\hat{A} = A_{22} - KA_{12}$ lie in the (open) left half-plane; A_{22} and A_{12} in the reduced-order observer take the roles of A and C in the full-order observer; once the gain matrix K is chosen, then there is no further freedom in the choice of L and H.

The specific form of the new matrix L in Equation 15.36 suggests another option for implementation of the dynamics of the reduced-order observer, namely

$$\dot{z} = \hat{A}\hat{x}_2 + \bar{L}y + Hu, \tag{15.38}$$

where

$$\bar{L} = A_{21} - KA_{11} \tag{15.39}$$

A block-diagram representation of this option is given in Figure 15.2b.

The selection of the gain matrix K of the reduced-order observer may be accomplished by any of the methods that can be used to select the gains of the full-order observer. In particular, pole placement, using any convenient algorithm, is feasible. Or the gain matrix can be obtained as the solution of a reduced-order Kalman filtering problem. For this purpose, one can use Equations 15.12 and 15.13, with A and C therein replaced by A_{22} and A_{12} of the reduced-order problem.

A more rigorous solution, taking into account the cross-correlation between the observation noise and the process noise [4], is available. Suppose the dynamic process is governed by

$$\dot{x}_1 = A_{11}x_1 + A_{12}x_2 + B_1u + F_1v, \tag{15.40}$$

$$\dot{x}_2 = A_{21}x_1 + A_{22}x_2 + B_2u + F_2v, \tag{15.41}$$

with the observation being noise free:

$$y = x_1 \tag{15.42}$$

In this case, the gain matrix is given by

$$K = (PA'_{12} + F_2QF'_1)R^{-1}, \tag{15.43}$$

where

$$R = F_1QF'_1,$$

and P is the covariance matrix of the estimation error e_2, as given by

$$\dot{P} = \tilde{A}P + P\tilde{A}' - PA_{12}R^{-1}A'_{12}P + \tilde{Q}, \tag{15.44}$$

where

$$\tilde{A} = A_{22} - F_2QF'_1R^{-1}A_{12}, \tag{15.45}$$

$$\tilde{Q} = F_2QF'_2 - F_2QF'_1R^{-1}F_1QF'_2 \tag{15.46}$$

Note that Equation 15.44 becomes homogeneous when

$$\tilde{Q} = 0 \tag{15.47}$$

One of the solutions of Equation 15.44 could be

$$P = 0. \tag{15.48}$$

which would imply that the error in estimating x_2 converges to zero! We cannot expect to achieve anything better than this. Unfortunately, $P = 0$ is not the only possible solution to Equation 15.47. To test whether it is, it is necessary to check whether the resulting observer dynamics matrix

$$\hat{A} = A_{22} - F_2F_1^{-1}A_{12} \tag{15.49}$$

is a stability matrix. If not, Equation 15.47 is not the correct solution to Equation 15.44.

The eigenvalues of the "zero steady-state variance" observer dynamics matrix Equation 15.49 have an interesting interpretation: as shown in [4], these eigenvalues are the transmission zeros of the plant with respect to the noise input to the process. Hence, the variance of the estimation error converges to zero if the plant is "minimum phase" with respect to the noise input.

For purposes of robustness, as discussed in Section 4, suggest that the noise distribution matrix F includes a term proportional to the control distribution matrix B, that is,

$$F = \bar{F} + q^2BB'$$

In this case, the zero-variance observer gain would satisfy

$$H = B_2 - KB_1 = 0, \tag{15.50}$$

as $q \to \infty$.

If Equation 15.50 is satisfied, the observer poles are located at the transmission zeros of the plant. Thus, in order to use the gain given by Equation 15.50, it is necessary for the plant to be minimum phase with respect to the input. Rynaski [5] has defined observers meeting this requirement as *robust observers*.

15.4 Discrete-Time Systems

Observers for discrete-time systems can be defined in a manner analogous to continuous-time systems.

Consider a discrete-time linear system

$$x_{n+1} = \Phi x_n + \Gamma u_n \tag{15.51}$$

with observations defined by

$$y_n = Cx_n \tag{15.52}$$

A full-order observer for Equation 15.51 is a dynamic system of the same order as the process whose state is to be estimated, excited by the inputs and outputs of that process and having the property that the estimation error (i.e., the difference between the state x_n of the process and the state $\hat{x}_n$ of the observer) converges to zero as $n \to \infty$, independent of the state of the process or its inputs and outputs.

Let the observer be defined by the general linear difference equation

$$\hat{x}_{n+1} = \hat{\Phi}\hat{x}_n + Ky_n + Hu_n \tag{15.53}$$

The goal is to find conditions on the matrices Φ, K, and H such that the requirements stated above are met. To find these conditions subtract Equation 15.53 from Equation 15.51

$$x_{n+1} - \hat{x}_{n+1} = \Phi x_n + \Gamma u_n - \hat{\Phi}\hat{x}_n - Ky_n - Hu_n \tag{15.54}$$

Letting

$$e_n = x_n - \hat{x}_n$$

and using Equation 15.52 we obtain from Equation 15.54

$$e_{n+1} = \Phi e_n + (\Phi - KC - \hat{\Phi})x_n + (\Gamma - H)u_n \tag{15.55}$$

Thus, in order to meet the requirements stated above, the transition matrix $\hat{\Phi}$ of the observer must be stable (i.e., the eigenvalues of Φ must lie within the unit circle) and, moreover,

$$\hat{\Phi} = \Phi - KC, \tag{15.56}$$

$$H = \Gamma \tag{15.57}$$

By virtue of these relations the observer can be expressed as

$$\hat{x}_{n+1} = \Phi x_n + \Gamma u_n + K(y_n - C\hat{x}_n) \tag{15.58}$$

It is seen from Equation 15.58 that the observer has the same dynamics as the underlying process, except that it has an additional input

$$K(y_n - C\hat{x}_n),$$

that is, a gain matrix K multiplying the *residual*

$$r_n = y_n - C\hat{x}_n$$

As in the continuous-time case, the observer can be interpreted as a feedback system, the role of which is that of driving residual r_n to zero.

The observer design thus reduces to the selection of the gain matrix K that makes the eigenvalues of $\hat{\Phi} = \Phi - KC$ lie at suitable locations within the unit circle.

If the discrete-time system is observable, the eigenvalues of $\Phi_c = \Phi - KC$ can be put anywhere. For a single-output plant, the Bass–Gura formula or other well-known algorithm can be used. For both single and multiple output processes, the observer gain matrix can be selected to make the observer a Kalman filter (i.e., a minimum variance estimator).

The gain matrix of the discrete-time Kalman filter is given by

$$K = \Phi PC'(PCP' + R)^{-1}, \tag{15.59}$$

where P is the covariance matrix of the estimation error, given (in the steady state) by the *discrete-time algebraic Riccati equation*

$$P = \Phi[P - PC'(CPC' + R)^{-1}CP]\Phi' + Q \tag{15.60}$$

The matrices Q and R are the covariance matrices of the excitation noise and the observation noise, respectively. As in the case of continuous-time processes, it is rarely possible to determine these matrices with any degree of accuracy. Hence, these matrices can be regarded as design parameters that can be adjusted by the user to provide desirable observer characteristics.

15.4.1 Delayed Data

Consider the discrete-time system defined by Equation 15.51 and with delayed observations defined by

$$z_n = Cx_n,$$

$$y_n = z_{n-N},$$

where N is the time delay, assumed to be an integer. The method described in Section 15.2.2 can readily be extended to this case. As an alternative the following method can be used. Define the metastate as the $\nu + kN$ dimensioned vector (where ν is the dimension of the state vector and k is the dimension of the original observation vector)

$$\mathbf{x}_n = \begin{bmatrix} x_n \\ z_n \end{bmatrix} \quad \text{with } z_n = \begin{bmatrix} z_{1n} & z_{2n} & \cdots & z_{Tn} \end{bmatrix}'$$

Then the metatstate evolves according to

$$\mathbf{x}_{n+1} = \mathbf{\Phi}\mathbf{x}_n + \mathbf{\Gamma}u_n,$$

where

$$\mathbf{\Phi} = \begin{bmatrix} \Phi & 0 & 0 & \cdots & 0 & 0 \\ C & 0 & 0 & \cdots & 0 & 0 \\ 0 & I & 0 & \cdots & 0 & 0 \\ \cdots & \cdots & \cdots & \cdots & \cdots\cdots & \\ 0 & 0 & 0 & \cdots & I & 0 \end{bmatrix}$$

and

$$\mathbf{\Gamma} = \begin{bmatrix} \Gamma' & 0 & \cdots & 0 & 0 \end{bmatrix}'$$

For all components of the observation vector concurrent, the observation equation is

$$y_n = \mathbf{Cx_n} = \begin{bmatrix} 0 & 0 & \cdots & 0 & I \end{bmatrix} x_n$$

For m time delays (each an integer multiple of the sampling time), the observation matrix is given by

$$\mathbf{C} = \begin{bmatrix} 0 & \cdots & U_1 & \cdots & U_2 & \cdots & U_m \end{bmatrix},$$

where U_i is a column vector with ones in the positions of the observation components present at the ith observation and zeros otherwise. Applying the theory developed earlier to the metasystem gives the observer the presence of delayed data.

It is noted that implementation of this method requires the storage of past observations, but permits the treatment of multiple observations with different time delays.

15.5 The Separation Principle

The predominant use of an observer is to estimate the state for purposes of feedback control. In particular, in a linear system with a control designed on the assumption of full-state feedback

$$u = -Gx, \tag{15.61}$$

when the state x is not directly measured, the state $\hat{x}$ of the observer is used in place of the actual state x in Equation 15.61. Thus, the control is implemented using

$$u = -G\hat{x}, \tag{15.62}$$

where

$$\hat{x} = x - e \tag{15.63}$$

Hence, when an observer is used, the closed-loop dynamics are given in part by

$$\dot{x} = Ax - BG(x - e) = (A - BG)x + BGe \tag{15.64}$$

This equation, together with the equation for the propagation of the error, defines the complete dynamics of the closed-loop system.

When a full-order observer is used

$$\dot{e} = \hat{A}e = (A - KC)e \tag{15.65}$$

Thus, the complete closed-loop dynamics are

$$\begin{bmatrix} \dot{x} \\ \dot{e} \end{bmatrix} = \begin{bmatrix} A - BG & BG \\ 0 & A - KC \end{bmatrix} \begin{bmatrix} x \\ e \end{bmatrix} \tag{15.66}$$

The closed-loop dynamics are governed by the upper triangular matrix

$$\mathbf{A} = \begin{bmatrix} A - BG & BG \\ 0 & A - KC \end{bmatrix}, \tag{15.67}$$

the eigenvalues of which are given by

$$|sI - \mathbf{A}| = |sI - A + BG||sI - A + KC| = 0, \tag{15.68}$$

that is, the closed-loop eigenvalues are the eigenvalues of $A - BG$, the full-state feedback system; and the eigenvalues of $A - KC$, the dynamics matrix of the observer. This is a statement of the well-known *separation principle,* which permits one to design the observer and the full-state feedback control independently, with the assurance that the poles of the closed-loop dynamic system will be the poles selected for the full-state feedback system and those selected for the observer.

When a reduced-order observer is used, it is readily established that the closed-loop dynamics are given by

$$\begin{bmatrix} \dot{x} \\ \dot{e}_2 \end{bmatrix} = \begin{bmatrix} A - BG & BG_2 \\ 0 & A_{22} - KA_{12} \end{bmatrix} \begin{bmatrix} x \\ e_2 \end{bmatrix} \tag{15.69}$$

and hence that the eigenvalues of the closed-loop system are given by

$$|sI - A + BG||sI - A_{22} + KA_{12}| = 0 \tag{15.70}$$

Thus, the separation principle also holds when a reduced-order observer is used.

It is important to recognize, however, that the separation principle applies only when the model of the process used in the observer agrees exactly with the actual dynamics of the physical process. It is not possible to meet this requirement in practice and, hence, the separation principle is an approximation at best. To assess the effect of a model discrepancy on the closed-loop dynamics, consider the following possibilities:

Case 1: Error in dynamics matrix

$$A = \bar{A} + \delta A$$

Case 2: Error in control distribution matrix

$$B = \bar{B} + \delta B$$

Case 3: Error in observation matrix

$$C = \bar{C} + \delta C$$

Using the "metastate"

$$\mathbf{x} = \begin{bmatrix} x \\ e \end{bmatrix}$$

it is readily determined [6] that the characteristic polynomial of the complete, closed-loop system for cases 1 and 3 is given by

$$|sI - \mathbf{A}| = \begin{vmatrix} sI - A_c & -BG \\ \delta A + K\delta C & sI - \hat{A} \end{vmatrix}, \tag{15.71}$$

where

$$A_c = \bar{A} - BG, \quad \hat{A} = \bar{A} - K\bar{C}$$

Similarly, using the metastate

$$\mathbf{x} = \begin{bmatrix} \hat{x} \\ e \end{bmatrix},$$

it is found that the characteristic polynomial for case 2 is given by

$$|sI - \mathbf{A}| = \begin{vmatrix} sI - \hat{A} & -\delta BG \\ KC & sI - A_c \end{vmatrix}, \tag{15.72}$$

where

$$A_c = A - \bar{B}G, \quad \hat{A} = A - KC$$

To assess the effect of perturbations of the dynamics matrices on the characteristic polynomial, the following determinantal identity can be used:

$$\begin{vmatrix} \mathcal{A} & \mathcal{B} \\ \mathcal{C} & \mathcal{D} \end{vmatrix} = |\mathcal{D}||\mathcal{A} - \mathcal{C}\mathcal{D}^{-1}\mathcal{B}| \tag{15.73}$$

Apply Equation 15.73 to Equation 15.72 to obtain

$$\begin{aligned} |sI - \mathbf{A}| &= |sI - A_c||sI - \hat{A} + \delta BG(sI - A_c)^{-1}KC| \\ &= |sI - A_c||sI - \hat{A}||I + \delta BG(sI - A_c)^{-1}KC(sI - \hat{A})^{-1}| \end{aligned} \tag{15.74}$$

upon use of

$$|\mathcal{A}\mathcal{B}| = |\mathcal{A}||\mathcal{B}|$$

The separation principle would continue to hold if the coefficient of δB in Equation 15.74 were to vanish. It does vanish if observer matrix K satisfies the *Doyle–Stein condition* [2]

$$K(I + C\Phi K)^{-1} = B(C\Phi B)^{-1}, \tag{15.75}$$

where

$$\Phi = (sI - A)^{-1}$$

is the plant resolvent.

To verify this, note that

$$\begin{aligned}(sI - \hat{A})^{-1} &= (sI - A + KC)^{-1} = (\Phi^{-1} + KC)^{-1} \\ &= \Phi - \Phi K(I + C\Phi K)^{-1}C\Phi\end{aligned} \tag{15.76}$$

When the Doyle–Stein condition (Equation 15.75) holds, Equation 15.76 becomes

$$(sI - \hat{A})^{-1} = \Phi - \Phi B(C\Phi B)^{-1}C\Phi,$$

and so

$$C(sI - \hat{A})^{-1} = C\Phi - C\Phi B(C\Phi B)^{-1}C\Phi = 0,$$

which ensures that the coefficient of δB in Equation 15.74 vanishes and, hence, that the separation principle applies.

Regarding the Doyle–Stein condition the following remarks are in order:

- The Doyle–Stein condition can rarely be satisfied exactly. But, as shown [2], it can be satisfied approximately by making the observer a Kalman filter with a noise matrix of the form given by Equation 15.14.
- The Doyle–Stein condition is not the only way the coefficient of δB can vanish. However, the Doyle–Stein condition ensures other robustness properties.
- An analogous condition for δA and δC can be specified.

In carrying out a similar analysis for a reduced-order observer it is found that the characteristic polynomial for the closed-loop control system, when a reduced-order observer is used and the actual control distribution matrix $B = \bar{B} + \delta B$ differs from the nominal (design) value $\bar{B}$, is given by

$$|sI - \mathbf{A}| = \begin{vmatrix} sI - F + \Delta G_2 & \Delta G \\ -BG_2 & sI - A_c \end{vmatrix}, \tag{15.77}$$

where

$$\Delta = K\delta B_1 - B_2 \tag{15.78}$$

It is seen that the characteristic polynomial of the closed-loop system reduces to that of Equation 15.70 when

$$\Delta = 0 \tag{15.79}$$

It is noted that Equation 15.79 can hold in a single-input system in which the loop gain is the only variable parameter. In this case

$$\delta B_1 = \rho B_1, \quad \delta B_2 = \rho B_2 \tag{15.80}$$

and thus

$$\Delta = \rho(KB_1 - B_2) = -\rho H$$

Hence, if the observer is designed with

$$H = B_2 - KB_1 = 0,$$

the separation principle holds for arbitrary changes in the loop gain.

If Equation 15.79 cannot be satisfied, then, as shown in [7], condition analogous to the Doyle–Stein condition can be derived from Equation 15.77 in the case of a scalar control input (Equation 15.80):

$$[I - K(I + A_{12}\Phi_{22}K)^{-1}A_{12}\Phi_{22}](B_2 - KB_1) = 0, \tag{15.81}$$

where

$$\Phi_{22} = (sI - A_{22})^{-1}$$

15.6 Nonlinear Observers

The concept of an observer carries over to nonlinear systems. However, for a nonlinear system, the structure of the observer is not nearly as obvious as it is for a linear system. The design of observers for nonlinear systems has been addressed by several authors, such as Thau [8] and Kou et. al. [9].

An observer for a plant, consisting of a dynamic system

$$\dot{x} = f(x, u) \tag{15.82}$$

with observations given by

$$y = g(x, u) \tag{15.83}$$

is another dynamic system, the state of which is denoted by $\hat{x}$, excited by the output y of the plant, having the property that the error

$$e = x - \hat{x} \tag{15.84}$$

converges to zero in the steady state.

One way of obtaining an observer is to imitate the procedure used in a linear system, namely to construct a model of the original system (Equation 15.1) and force it with the "residual":

$$r = y - \hat{y} = y - g(\hat{y}, u) \tag{15.85}$$

The equation of the observer thus becomes

$$\dot{\hat{x}} = f(\hat{x}, u) + \kappa(y - g(\hat{x}, u)), \tag{15.86}$$

where $\kappa(\,)$ is a suitably chosen nonlinear function. (How to choose this function will be discussed later.) A block diagram representation of a general nonlinear observer is shown in Figure 15.3.

The differential equation for the error e can be used to study its behavior. This equation is given by

$$\begin{aligned} \dot{e} &= \dot{x} - \dot{\hat{x}} \\ &= f(x, u) - f(\hat{x}, u) - \kappa(g(x, u) - g(\hat{x}, u)) \\ &= f(x, u) - f(x - e, u) + \kappa(g(x - e, u) - g(x, u)) \end{aligned} \tag{15.87}$$

Suppose that by the proper choice of $\kappa(\,)$ the error Equation 15.87 can be made asymptotically stable, so that an equilibrium state is reached for which

$$\dot{e} = 0$$

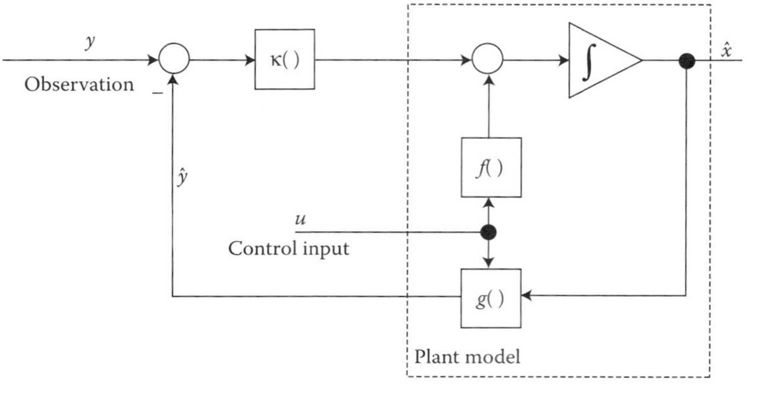

FIGURE 15.3 Structure of the nonlinear observer.

Then, in equilibrium, Equation 15.87 becomes

$$0 = f(x, u) - f(x - e, u) + \kappa(g(x - e, u) - g(x, u)) \tag{15.88}$$

Since the right-hand side of Equation 15.88 becomes zero when $e = 0$, *independent* of x and u, it is apparent that $e = 0$ is an equilibrium state of Equation 15.87. This implies that if $\kappa(\,)$ can be chosen to achieve asymptotic stability, the estimation error e converges to zero.

It is very important to appreciate that the right-hand side of Equation 15.88 becomes zero independent of x and u only when the nonlinear functions $f(\cdot,\cdot)$ and $g(\cdot,\cdot)$ used in the observer are exactly the same as in Equations 15.82 and 15.83, which define the plant dynamics and observations, respectively. Any discrepancy between the corresponding functions generally prevents the right-hand side of Equation 15.88 from vanishing and hence leads to a steady-state estimation error. Since the mathematical model of a physical process is always an approximation, in practice the steady-state estimation error generally does not tend to zero. But, by careful modeling, it is usually possible to minimize the discrepancies between the f and g functions of the true plant and the model used in the observer. This usually keeps the steady-state estimation error acceptably small.

For the same reason that the model of the plant and the observation that is used in the observer must be accurate, it is important that the control u that goes into the plant is the very same control used in the observer. If the control to the plant is subject to saturation, for example, then the nonlinear function that models the saturation must be included in the observer. Failure to observe this precaution can cause difficulties.

Including control saturation in the observer is particularly important as a means for avoiding the phenomenon known as *integrator windup*: the compensator, which has a pole at the origin, provides integral action. Imagine the transfer function of the compensator represented by an integrator in parallel with a second-order subsystem. The control signal to the integrator is oblivious to the fact that the input to the plant has saturated and hence keeps the integrator "winding up"; the error signal changes sign when the desired output reaches the set point, but the control signal does not drop from its maximum value. When the saturation is included in the observer, on the other hand, the control signal drops from its maximum value even before the error changes sign, thus correctly taking the dynamics (i.e., the lag) of the process into account.

The function $\kappa(\,)$ in the observer must be selected to ensure asymptotic stability of the origin ($e = 0$ in Equation 15.88). By the theorem of Lyapunov's first method (see Chapter 43), the origin is asymptotically stable if the Jacobian matrix of the dynamics, evaluated at the equilibrium state, corresponds to an asymptotically stable linear system. For the dynamics of the error Equation 15.87 the Jacobian matrix with respect to the error e evaluated at $e = 0$ is given by

$$A_c(x) = \left(\frac{\partial f}{\partial x}\right) - K\left(\frac{\partial g}{\partial x}\right). \tag{15.89}$$

This is the nonlinear equivalent of the closed-loop observer equation of a linear system

$$A_c = A - KC,$$

where A and C are the plant dynamics and observation matrices, respectively. The problem of selecting the gain matrix for a nonlinear observer is analogous to that of a linear observer, but somewhat more complicated by the presence of the nonlinearities that make the Jacobian matrices in Equation 15.89 dependent on the state x of the plant. Nevertheless, the techniques used for selecting the gain for a linear observer can typically be adapted for a nonlinear observer. Pole placement is one method and the other is to make the observer an extended Kalman filter which, as explained later, entails online computation of the gains via the linearized variance equation.

It should be noted that the observer closed-loop dynamics matrix depends on the actual state of the system and hence is time varying. The stability of the observer thus cannot be rigorously determined by the locations of the eigenvalues of A_c.

The choice of $\kappa(\ )$ may be aided through the use of Lyapunov's second method. Using this method, Thau [8] considered a "mildly nonlinear" process

$$\dot{x} = f(x) = Ax + \mu\phi(x), \tag{15.90}$$

where μ is a small parameter, with linear observations

$$y = Cx$$

For this case, κ can be simply a gain matrix chosen to stabilize the linear portion of the system

$$\kappa(r) = Kr,$$

where K is chosen to stabilize

$$\hat{A} = A - KC$$

This choice of K ensures asymptotic stability of the observer if $\phi(\ )$ satisfies a Lipschitz condition

$$||\phi(u) - \phi(v)|| \leq k||u - v||$$

and when

$$P\hat{A} + \hat{A}'P = -Q \leq -c_0 I$$

In this case, asymptotic stability of the observer

$$\dot{\hat{x}} = A\hat{x} + \mu\phi(\hat{x}) + K(y - C\hat{x})$$

is assured for

$$\mu < \frac{c_0}{2k||P||}$$

This analysis was substantially extended by Kou et al. [9].

Suggestions for the choice of the nonlinear function $\kappa(\ )$ have appeared in the technical literature. One suggestion [10], for example, is to use

$$\kappa(r) = \Psi(\hat{x})^{-1} Kr,$$

where K is a constant, possibly diagonal matrix, and $\Psi(x)$ is the Jacobian matrix defined by

$$\Psi(x) = \frac{\partial \Phi(x)}{\partial x}$$

with

$$\Phi(x) = \begin{bmatrix} L_f g(x) \\ L_f^2 g(x) \\ \vdots \\ L_f^n g(x) \end{bmatrix},$$

in which $L_f^n g(x)$ is the nth Lie derivative of $g(x)$ with respect to $f(x)$. (See Chapter 38.) This approach can possibly be viewed as a nonlinear generalization of the Bass–Gura formula given previously, since the matrix Ψ is akin to the observability matrix O defined earlier.

15.6.1 Using Zero-Crossing or Quantized Observations

The ability of an observer to utilize nonlinear observations is not limited to observations that exhibit only moderate nonlinearity; even highly nonlinear observations can be accommodated. Perhaps the most nonlinear observation is that of a zero-crossing detector, in which

$$y = \text{sgn}(x) = \begin{cases} 1, & x > 0 \\ -1, & x < 0 \end{cases}$$

This is the extreme special case of a quantizer, in which the output is quantized to only two levels.

Suppose, for example, that the only observation is of the zerocrossing of x_1. The observer for this process is then given by

$$\dot{\hat{x}} = f(\hat{x}, u) - k[y - \text{sgn}(\hat{x}_1)], \tag{15.91}$$

as illustrated in Figure 15.4.

Provided that a gain k, in this case a scalar parameter, can be found that stabilizes the observer at $e = 0$, the estimation error will be reduced to zero. The partial derivative of the nonlinear function with respect to the observation does not exist in this case because the observation is discontinuous with respect to the state. The stability of the observer cannot be established by linearizing about the origin. You have to use some other methods, such as Lyapunov's second method, or determine the appropriate range of k by simulation.

Some insight into how the observer operates can be gained by considering that both y and $\text{sgn}(x_1)$ are signals that take on the values of ± 1; their difference, which is the residual that appears in Equation 15.91, is either 0 or ± 2. Suppose the observer is working well; most of the time y and $\text{sgn}(x_1)$ will have the same sign and the residual will be zero. The residual will be nonzero for the short time interval in which y and $\text{sgn}(x_1)$ have different signs. The residual will thus consist of a train of narrow pulses, each of height ± 2 and of width proportional to the phase difference between y and $\text{sgn}(x_1)$, as shown in Figure 15.5. The effect of each pulse is to nudge the state of the observer to agree with the state of the plant.

The same idea extends to quantized observations, since a quantizer can be regarded as a multiple-level crossing detector.

Of course, if there are other observations in addition to the zero-crossing observation, they can be combined with the latter and, with an appropriate choice of gains, can provide enhanced performance.

15.6.2 Reduced-Order Observers

Nonlinear reduced-order observers can be developed by the same methods that one uses for linear, reduced-order observers.

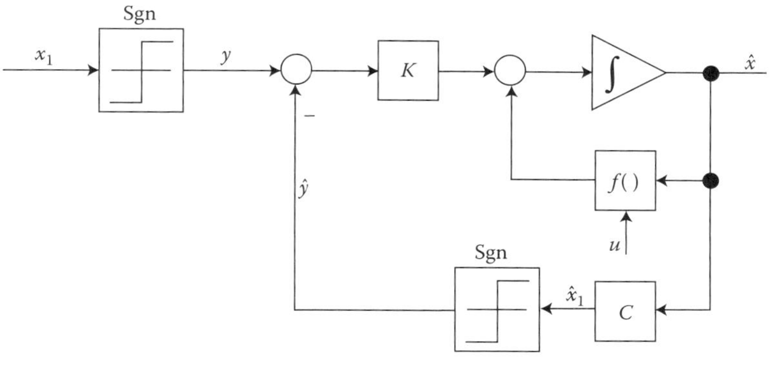

FIGURE 15.4 Observer using zero-crossing data.

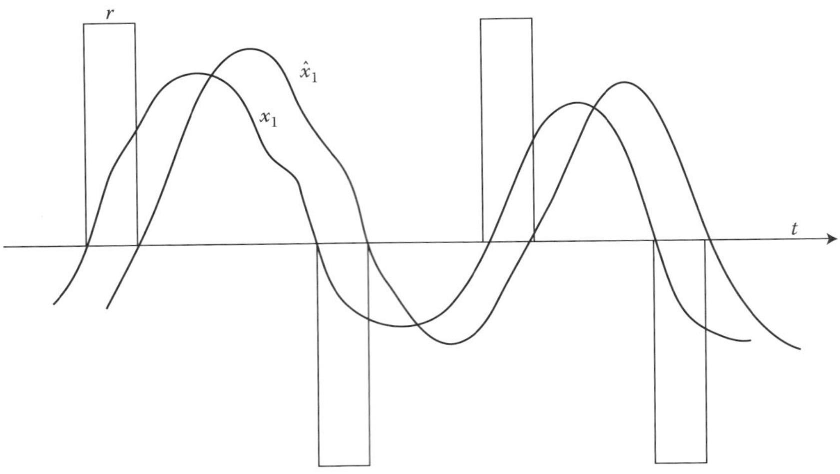

FIGURE 15.5 Residual for zero-crossing observations consists of pulses.

Suppose that the state is partitioned into two substates as given by Equation 15.25 with the observation given by

$$y = g(x_1, u)$$

Provided that this expression can be solved for x_1 as a function of y and u,

$$x_1 = \psi(y, u) = \bar{y},$$

we can use $\bar{y}$ as the observation. The completely general case in which y contains more state variables than its dimension can probably be handled in a manner similar to that used for linear systems, as discussed by Friedland [6]. The derivation is very tortuous in linear systems and is likely to be even more so in nonlinear systems and is, hence, omitted.

Corresponding to the partitioning of the state vector x as in Equation 15.25, the dynamic equations are written as

$$\dot{x}_1 = f_1(x_1, x_2, u), \tag{15.92}$$

$$\dot{x}_2 = f_2(x_1, x_2, u), \tag{15.93}$$

The nonlinear reduced-order observer is assumed to have the same structure as the corresponding linear observer. For the estimate of the substate x_1, we use the observation itself:

$$\hat{x}_1 = y; \tag{15.94}$$

while the substate x_2 is estimated using an observer of the form

$$\hat{x}_2 = Ky + z, \tag{15.95}$$

where z is the state of a dynamic system of the same order as the dimension of the subvector x_2 and is given by

$$\dot{z} = \phi(y, \hat{x}_2, u) \tag{15.96}$$

A block-diagram representation of the observer having the structure of Equations 15.94 through 15.96 is given in Figure 15.6.

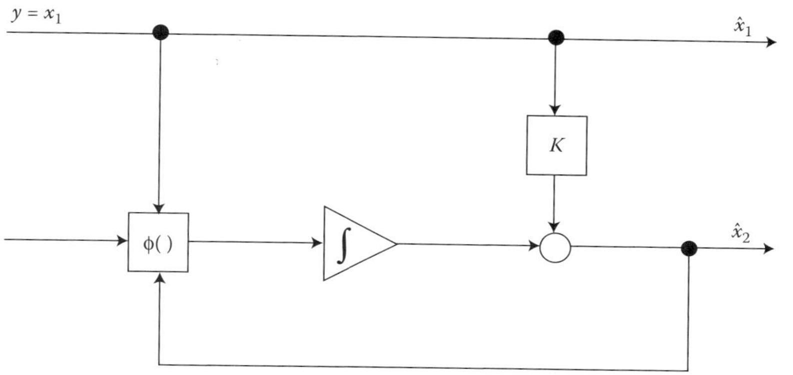

FIGURE 15.6 Reduced-order nonlinear observer.

The object of the observer design is the determination of the gain matrix K and the nonlinear function ϕ. As for the full-order observer, these are to be selected such that

- The steady-state error in estimating x_2 converges to zero, independent of x and u. (The error in estimating x_1 is already zero when $\hat{x}_1 = y$.)
- The observer is asymptotically stable.

As in the case of the full-order observer, we proceed by deriving the differential equation for the estimation error

$$e = x_2 - \hat{x}_2 \tag{15.97}$$

Using Equations 15.93, 15.95, and 15.92, we obtain

$$\dot{e} = \dot{x}_2 - \dot{\hat{x}}_2 = f_2(y, x_2, u) - Kf_1(y, x_2, u) - \phi(y, x_2 - e, u) \tag{15.98}$$

In order for the right-hand side of Equation 15.98 to vanish when $e = 0$, it is necessary that the function $\phi(\cdot, \cdot, \cdot)$ satisfy

$$\phi(y, x_2, u) = f_2(y, x_2, u) - Kf_1(y, x_2, u) \tag{15.99}$$

for all values of y, x_2, and u.

To achieve asymptotic stability, the linearized system

$$\dot{e} = A(x_2)e \tag{15.100}$$

with

$$A(x_2) = \left(\frac{\partial \phi}{\partial x_2}\right) = \left(\frac{\partial f_2}{\partial x_2}\right) - K\left(\frac{\partial f_1}{\partial x_2}\right) \tag{15.101}$$

must be asymptotically stable.

As in the case of the full-order observer, there are several techniques for selecting an appropriate gain matrix.

The reduced-order nonlinear observer can be further generalized by replacing the linear function Ky in Equation 15.95 by a nonlinear function $\kappa(y)$.

15.6.3 Extended Separation Principle

When an observer having the structure described above is used to estimate the state of a linear system, and the estimate is used in place of the actual state, the poles of the closed-loop system comprise the poles of the observer and the poles that would be present if full-state feedback were implemented. This is the separation principle of linear systems. (But remember that this result holds only when the model of the plant used in implementing the observer is a faithful model of the physical plant.)

The separation principle of linear systems can be extended to nonlinear systems. Consider, in particular, the nonlinear system

$$\dot{x} = f(x, u), \tag{15.102}$$

for which a control law

$$u = \gamma(x) \tag{15.103}$$

has been designed. Use of Equation 15.103 in Equation 15.102 gives the closed-loop dynamics

$$\dot{x} = f(x, \gamma(x)) = F(x) \tag{15.104}$$

Assume that the closed-loop dynamics of the system with full-state feedback, as represented by $F(x)$, has been designed—by whatever method might be appropriate—to achieve satisfactory behavior. How will the process behave when the state $\hat{x}$ of an observer is used in place of the true process state x [i.e., when $u = \gamma(\hat{x})$] ? As earlier, let

$$\hat{x} = x - e,$$

where e is the error in estimating the state. Then Equation 15.103 becomes

$$u = \gamma(\hat{x}) = \gamma(x - e)$$

Then Equation 15.104 becomes

$$\dot{x} = f(x, \gamma(x - e)), \tag{15.105}$$

which together with Equation 15.87, becomes

$$\dot{e} = f(x, \gamma(\hat{x})) - f(x - e, \gamma(\hat{x})) + K[g(x - e, \gamma(\hat{x})) - g(x, \gamma(\hat{x}))], \tag{15.106}$$

which define the closed-loop dynamics.

There is not much that can be done with Equations 15.105 and 15.106 when f, g, and γ are general functions. However suppose these functions are sufficiently smooth to permit the use of Taylor's theorem, that is,

$$f(x, \gamma(x - e)) = f(x, \gamma(\hat{x})) - (\partial f/\partial \gamma)(\partial \gamma/\partial e)e + O(e^2), \tag{15.107}$$

$$f(x - e, \gamma(\hat{x})) = f(x, \gamma(\hat{x})) - (\partial f/\partial x)e + O(e^2), \tag{15.108}$$

$$g(x - e, \gamma(\hat{x})) = g(x, \gamma(\hat{x})) - (\partial g/\partial x)e + O(e^2), \tag{15.109}$$

where $O(e^2)$ represents terms that go to zero as $\|e\|^2$. Then Equations 15.105 and 15.106 become

$$\dot{x} = F(x) - (\partial f/\partial \gamma)(\partial \gamma/\partial e)e + O(e^2), \tag{15.110}$$

$$\dot{e} = [\partial f/\partial x + K(\partial g/\partial x)]e + O(e^2) \tag{15.111}$$

With the terms of $O(e^2)$ omitted, a block-diagram representation of Equations 15.110 and 15.111 is shown in Figure 15.7. Note that the equation for the error has no input from the state estimation. The error thus converges asymptotically to zero as in the linear case. Since the error is the input to the full-state feedback control system, if the latter is asymptotically stable, the effect of the estimation error vanishes.

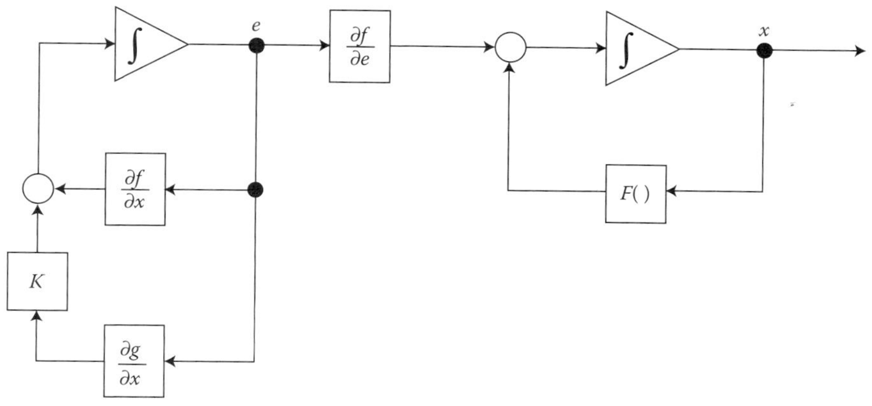

FIGURE 15.7 Illustration of the extended separation principle.

15.6.4 Extended Kalman Filter

Sometimes the nonlinear function $\kappa(\,)$ can simply be a constant gain. Often, however, there is no obvious method of choosing this gain and a more systematic method is required. It is often appropriate to make the observer an "extended Kalman filter," that is, to calculate the gain matrix online from the solution of the variance equation of the Kalman filter.

Few of the many applications for which Kalman filters have been used have met the linearity requirements of the theory. Nevertheless, the theory has been successfully, if not rigorously, applied. This is done by using a nonlinear observer of the form of Equation 15.88, but with the gain matrix K therein being computed, along with the state estimate, using Equations 15.12 and 15.13. The matrices A and C in these equations are the Jacobian matrices of the dynamics and observations

$$A = [\partial f/\partial x]_{x=\hat{x}}, \tag{15.112}$$

$$C = [\partial g/\partial x]_{x=\hat{x}}, \tag{15.113}$$

for the nonlinear process defined by

$$\dot{x} = f(x, u), \tag{15.114}$$

$$y = g(x, u) \tag{15.115}$$

In the linear case, the error covariance matrix, and through it the gain matrix K, does not depend on the estimated state. In principle, these matrices can be computed before the filter is implemented and stored in the filter's memory. In the nonlinear case, however, the matrices A and C that are used in computing P and K depend on the state estimate. Hence, in the extended Kalman filter, the observer and the Kalman filter gain matrix computation are coupled. This means that the equations for both the variance equation and the observer must be implemented online as shown schematically in Figure 15.8.

The requirement for online computation of the extended Kalman filter gain matrix can be a computational burden. Even considering that the covariance equation P is symmetric, there still are $k(k-1)/2$ scalar differential equations in Equation 15.13 that must be integrated numerically in addition to the k scalar observer equations for a kth-order dynamic process. In a 10-order process, for example, a total of 55 equations must be integrated. It does not take a process of much higher order to overwhelm even a supercomputer. Moreover, the matrix Riccati equation (Equation 15.13) for P is notorious for being poorly behaved. Unless special measures are taken, the numerical solution to Equation 15.13 is likely to lose its positive-definite character as the theory requires. If this happens, the resulting state estimate $\hat{x}$ will probably be useless.

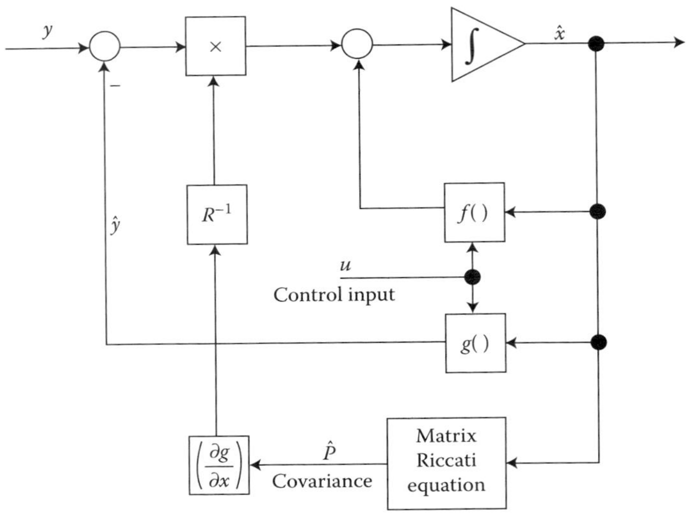

FIGURE 15.8 Schematic of the extended Kalman filter, showing coupling between state estimation and covariance computation.

Fortunately, it is rarely necessary to be a stickler for accuracy in the implementation of Equation 15.13. In the first place, the entire theory of the extended Kalman filter is only approximate. Moreover, the spectral density matrices Q and R that appear in Equation 15.13 are hardly ever known to be better than an order of magnitude. Hence, any computational method that gives a reasonable approximation to P and K is usually acceptable. Some of the approximations that have been considered include the following:

- Regard P as being piecewise constant and compute it relatively infrequently, using the discrete-time version of the Kalman filter.
- Simulate the observer and Equation 15.13 offline; examine the results and use appropriate approximations. It may be possible, for example, to approximate some of the gains by constants. The effect of the approximations must be evaluated by further simulation.
- Use a simpler model to represent the process in Equation 15.13 than is used as the process model in the observer; but be careful not to use an overly simple model in the implementation of the process dynamics.

Acknowledgment

The author is grateful to Dr. Sergey Levkov formerly of the Institute of Cybernetics in Kiev, Ukraine, for reviewing the manuscript and bringing to his attention some new references on nonlinear observers.

References

1. Luenberger, D., Observers for multivariable systems, *IEEE Trans. Automat. Control,* 11, 190–197, 1966.
2. Doyle, J.C. and Stein, G., Robustness with observers, *IEEE Trans. Automat. Control,* 24, 607–611, 1979.
3. Bass, R.W. and Gura, I., High-order system design via state-space considerations, *Proc. Joint. Automat. Control Conf.,* Troy, NY, 311–318, 1965.
4. Friedland, B., On the properties of reduced-order Kalman filters, *IEEE Trans. Automat. Control,* 34 (3), 321–324, 1989.

5. Rynaski, E.G., Flight control synthesis using robust observers, *Proc. AIAA Guidance Control Conf.*, San Diego, CA, 825–831, 1982.
6. Friedland, B., *Control System Design: An Introduction to State-Space Methods*, McGraw-Hill, New York, 1986.
7. Madiwale, A.N. and Williams, D.E., Some extensions of loop transfer recovery, *Proc. Am. Control Conf.*, Boston, MA, 790–795, 1985.
8. Thau, F.E., Observing the state of nonlinear dynamic systems, *Int. J. Control*, 17, 471-479, 1973.
9. Kou, S.R., Elliot, D.L., and Tarn, T.J., Exponential observers for nonlinear dynamic systems, *Inf. Control*, 29 (3), 204–216.
10. Ciccarella, G., DallaMora, M., and Germani, A., A Luenberger-like observer for nonlinear systems, *Int. J. Control*, 57 (3), 537–556, 1993.

III

Design Methods for MIMO LTI Systems

16

Eigenstructure Assignment*

Kenneth M. Sobel
The City College of New York

Eliezer Y. Shapiro
HR Textron

Albert N. Andry, Jr.
Teledyne Reynolds Group

16.1 Introduction

Eigenstructure assignment is a useful tool that allows the designer to satisfy damping, settling time, and *mode decoupling* specifications directly by choosing eigenvalues and eigenvectors. Andry et al. [1] have applied eigenstructure assignment to design a *stability augmentation system* for the lateral dynamics of the L-1011 aircraft. Both constant gain output feedback and *gain suppression* designs are proposed. Sobel and Shapiro [30] used eigenstructure assignment to design dynamic compensators for the L-1011 aircraft. First- and second-order compensators were proposed for the case in which sideslip angle could not be measured. Later, Sobel et al. [33] proposed a systematic method for choosing the elements of the feedback gain matrix which can be suppressed to zero with minimal effect on the eigenvalue and eigenvector

* Reprinted from Sobel, K.M., Shapiro, E.Y., and Andry, A.N., Jr., *Int. J. Control*, 59(1), 13–37, 1994. With permission [27].

assignment. A design of an eigenstructure assignment gain-suppression flight controller is shown for F-18 High Angle of Attack Research Vehicle (HARV) aircraft.

Specialized task tailored modes for highly maneuverable fighter aircraft have been designed by using eigenstructure assignment. Sobel and Shapiro [29] designed an eigenstructure assignment pitch pointing and vertical translation controller for the AFTI F-16 aircraft. Pilot command tracking was achieved by using a special case of O'Brien and Broussard's [21] command generator tracker. Sobel and Shapiro [28] designed a yaw pointing and lateral translation controller for the linearized lateral dynamics of the Flight Propulsion Control Coupling (FPCC) aircraft. This conceptual control-configured vehicle has a vertical canard in addition to the more conventional control surfaces.

This chapter is organized as follows. Section 16.2 describes eigenstructure assignment using constant gain output feedback. Section 16.3 extends eigenstructure assignment to allow the designer to suppress chosen elements of the feedback gain matrix to zero. A systematic method for choosing the entries to be suppressed is discussed. Section 16.4 describes eigenstructure assignment using dynamic compensation. Each section includes an application of eigenstructure assignment to the design of a stability augmentation system for the linearized lateral dynamics of the F-18 HARV. Finally, in Section 16.5 we present a robust, sampled data, eigenstructure assignment control law design for the yaw pointing/lateral translation maneuver of the FPCC aircraft.

16.2 Eigenstructure Assignment Using Output Feedback

Consider a system modeled by the linear time-invariant matrix differential equation described by

$$\dot{x} = Ax + Bu \tag{16.1}$$

$$y = Cx \tag{16.2}$$

where x is the state vector ($n \times 1$), u is the control vector ($m \times 1$), and y is the output vector ($r \times 1$). It is assumed that the m inputs and the r outputs are independent. Also, as is usually the case in aircraft problems, it is assumed that $m < r < n$. If there are no exogenous inputs such as pilot commands, the feedback control vector u equals a matrix times the output vector y :

$$u = -Fy \tag{16.3}$$

The feedback problem can be stated as follows: Given a set of desired eigenvalues, $(\lambda_i^d), i = 1, 2, \ldots, r$ and a corresponding set of desired eigenvectors, $(v_i^d), i = 1, 2, \ldots, r$, find the real $m \times r$ matrix F such that the eigenvalues of $A - BFC$ contain (λ_i^d) as a subset, and the corresponding eigenvectors of $A - BFC$ are close to the respective members of the set (v_i^d).

Srinathkumar [34] has shown that if (A, B) is a controllable pair, then the feedback gain matrix F will exactly assign r eigenvalues. It will also assign the corresponding eigenvectors, provided that v_i^d is chosen to be in the subspace spanned by the columns of $(\lambda_i I - A)^{-1}B$ for $i = 1, 2, \ldots, r$. This subspace is of dimension m, which is the number of independent control variables. In general, a chosen or desired eigenvector v_i^d will not reside in the prescribed subspace and, hence, cannot be achieved. Instead, a "best possible" choice for an achievable eigenvector is made. Andry et al. [1] showed that the best possible eigenvector is the projection of v_i^d onto the subspace spanned by the columns of $(\lambda_i I - A)^{-1}B$. An alternative representation, described by Kautsky et al. [11], showed that the subspace in which the eigenvector v_i must reside is also given by the null space of $U_1^{\mathrm{T}}(\lambda_i I - A)$. The matrix U_1 is obtained from the singular-value decomposition of B, given by

$$B = [U_0, U_1]\begin{bmatrix} \Sigma V^{\mathrm{T}} \\ 0 \end{bmatrix}. \tag{16.4}$$

The method of Kautsky et al. [11] for computing the subspaces is the preferred method for numerical computation.

Finally, the complete controllability assumption may be removed by using results derived by Liebst and Garrard [14] and by Liebst et al. [15]. These results allow the designer to alter eigenvectors that correspond to uncontrollable eigenvalues.

In many practical situations, complete specification of v_i^d is neither required or known, but rather the designer is interested only in certain elements of the eigenvector. Thus, assume that v_i^d has the following structure:

$$v_i^d = \left[v_{i1}, x, x, x, x, v_{ij}, x, x, v_{in}\right]^{\mathrm{T}},$$

where v_{ij} are designer-specified components and x is an unspecified component. Define, as shown by Andry et al. [1], a reordering operation $\{\ \}^{R_i}$ so that

$$\left\{v_i^d\right\}^{R_i} = \begin{bmatrix} \ell_i \\ d_i \end{bmatrix}, \tag{16.5}$$

where ℓ_i is a vector of specified components of v_i^d and d_i is a vector of unspecified components of v_i^d. The rows of the matrix $(\lambda_i I - A)^{-1}B$ are also reordered to conform with the reordered components of v_i^d. Thus,

$$\left\{(\lambda_i I - A)^{-1}B\right\}^{R_i} = \begin{bmatrix} \tilde{L}_i \\ D_i \end{bmatrix}. \tag{16.6}$$

Then, as shown by Andry et al. [1], the achievable eigenvector v_i^a is given by

$$v_i^a = (\lambda_i I - A)^{-1}B\tilde{L}_i^{\dagger}\ell_i, \tag{16.7}$$

where $(\cdot)^{\dagger}$ denotes the appropriate pseudoinverse of $(\cdot)$.

The output feedback gain matrix using eigenstructure assignment [1] is described by

$$F = -(Z - A_1 V)(CV)^{-1}, \tag{16.8}$$

where A_1 is the first m rows of the matrix A in Equation 16.1, V is the matrix whose columns are the r achievable eigenvectors, Z is a matrix whose columns are $\lambda_i z_i$, where the ith eigenvector v_i is partitioned as $v_i = \begin{bmatrix} z_i \\ w_i \end{bmatrix}$ with z_i an $m \times 1$ vector, and C is the output matrix in Equation 16.2. The result of Equation 16.8 assumes that the system described by Equations 16.1 and 16.2 has been transformed into a system in which the control distribution matrix B is a lead block identity matrix.

An alternative representation for the feedback gain matrix F developed by Sobel et al. [32] is

$$F = -V_b\Sigma_b^{-1}U_{b0}^{\mathrm{T}}(V\Lambda - AV)V_r\Sigma_r^{-1}U_{r0}^{\mathrm{T}}, \tag{16.9}$$

where the singular-value decompositions of the matrices B and CV are given by

$$B = [U_{b0}U_{b1}]\begin{bmatrix} \Sigma_b V_b^{\mathrm{T}} \\ 0 \end{bmatrix}, \tag{16.10}$$

$$CV = [U_{r0}U_{r1}]\begin{bmatrix} \Sigma_r V_r^{\mathrm{T}} \\ 0 \end{bmatrix} \tag{16.11}$$

and where Λ is an $r \times r$ diagonal matrix with entries $\lambda_i, i = 1, 2, \ldots, r$. The method described by Equation 16.9 is the preferred method for numerical computation.

We conclude the discussion with a comment about the closed-loop system stability. Unfortunately, it is not yet possible to ensure that stable open-loop eigenvalues do not move into the right half of the complex plane when an eigenstructure assignment output feedback controller is utilized. This is still an open area for further research. However, for aircraft flight control systems, the closed-loop stability requirement is neither necessary nor sufficient. For example, some modes, such as the dutch roll mode, are required to meet minimum frequency and damping specifications, as described in MIL-F-8785C [18]. For these modes, stability alone is not sufficient. Other modes, such as the spiral mode, may be unstable provided that the time to double amplitude is sufficiently large. For these modes, stability may not be necessary.

16.2.1 F-18 HARV Linearized Lateral Dynamics Design Example

Consider the lateral directional dynamics of the F-18 HARV aircraft linearized at a Mach number of 0.38, an altitude of 5000 ft, and an angle of attack of 5°. The aerodynamic model is augmented with first-order actuators and a yaw rate washout filter. The eight state variables are aileron deflection δ_a, stabilator deflection δ_s, rudder deflection δ_r, sideslip angle β, roll rate p, yaw rate r, bank angle ϕ, and washout filter state x_8. The three control variables are aileron command δ_{ac}, stabilator command δ_{sc}, and rudder command δ_{rc}. The four measurements are r_{wo}, p, β, and ϕ, where r_{wo} is the washed out yaw rate. All quantities are in the body axis frame of reference with units of degrees or degrees/second. The state-space matrices A, B, and C that completely describe the model are shown in the Appendix.

An output feedback gain matrix is now computed by using eigenstructure assignment. The desired dutch roll eigenvalues are chosen with a damping ratio of 0.707 and a natural frequency in the vicinity of 3 rad/s. The roll subsidence and spiral modes are chosen to be merged into a complex mode, as suggested by [1]. The desired eigenvalues are

Dutch roll mode:

$$\lambda_{dr}^{d} = -2 \pm j2$$

Roll mode:

$$\lambda_{roll}^{d} = -3 \pm j2$$

The desired eigenvectors are chosen to keep the quantity $|\phi/\beta|$ small. Therefore, the desired dutch roll eigenvectors will have zero entries in the rows corresponding to bank angle and roll rate. The desired roll mode eigenvectors will have zero entries in the rows corresponding to yaw rate, sideslip, and x_8 (which is filtered yaw rate). The desired eigenvectors are

$$v_{dr}^{d} = \begin{bmatrix} x \\ x \\ x \\ 1 \\ 0 \\ x \\ 0 \\ x \end{bmatrix} \pm j \begin{bmatrix} x \\ x \\ x \\ x \\ 0 \\ 1 \\ 0 \\ x \end{bmatrix} \qquad v_{roll}^{d} = \begin{bmatrix} x \\ x \\ x \\ 0 \\ 1 \\ 0 \\ x \\ 0 \end{bmatrix} \pm j \begin{bmatrix} x \\ x \\ x \\ 0 \\ x \\ 0 \\ 1 \\ 0 \end{bmatrix} \begin{matrix} \delta_a \\ \delta_s \\ \delta_r \\ \beta \\ p \\ r \\ \phi \\ x_8 \end{matrix}.$$

The achievable eigenvectors are computed by taking the orthogonal projection onto the null space of $U_1^{\mathrm{T}}(\lambda_i I - A)$. However, care must be taken when computing the pseudoinverse of $\tilde{L}_i$ because this matrix is ill-conditioned. Press et al. [24] suggest computing the pseudoinverse by using a singular-value decomposition in which the singular values, which are significantly smaller than the largest singular value, are treated as zero. The achievable eigenvectors given in this chapter were computed by using MATLAB® function PINV with TOL = 0.01. The achievable eigenvectors are shown in Table 16.1, where the underlined numbers indicate the small couplings between p, ϕ, and the dutch roll mode and between β, r, x_8, and the roll mode. Hence, the ratio $|\phi/\beta|$ can be expected to be small.

The feedback gain matrix is computed by using Equation 16.9 and is shown in Table 16.2. From the open-loop state responses to a 1° initial sideslip, shown in Figure 16.1, we conclude that the aircraft is poorly damped with strong coupling between the dutch roll mode and the roll mode. The closed-loop state response is shown in Figure 16.2. Observe that the maximum absolute values of the bank angle and roll rate are 0.0532 and 0.2819°/s., respectively.

Finally, we consider the multivariable gain and phase margins for our design. Suppose that the modeling errors may be described by the matrix L given by

$$L = Diag\left(\ell_1 e^{j\phi_1}, \ell_2 e^{j\phi_2}, \ldots, \ell_m e^{j\phi_m}\right).$$

Then, as shown by Lehtomaki [13], multivariable gain and phase margins at the inputs may be defined.

TABLE 16.1 Achievable Eigenvectors for Constant Gain Output Feedback

Dutch Roll Mode			Roll Mode			
1.1781		0.2929	0.3212		0.2785	δ_a
−2.7860		1.9039	0.3259		0.2058	δ_s
10.1406		−4.1580	−0.1617		0.0693	δ_r
0.9777	$\pm j$	1.7260	−0.0558	$\pm j$	0.0791	β
−0.0559		−0.0094	0.9983		−4.9603	p
5.7061		0.9932	−0.0016		−0.0246	r
−0.0915		−0.1302	−0.9938		0.9916	ϕ
−0.5258		−1.0322	−0.0022		0.0032	x_8

TABLE 16.2 Constant Gain Output Feedback Control Law

Feedback Gain Matrix (Degree/Degree)				Gain and Phase Margins	max $\lvert\phi\rvert$
r_{wo}	p	β	ϕ	(at inputs $\delta_{ac}, \delta_{sc}, \delta_{rc}$)	max $\lvert p\rvert$
−0.1704	0.1380	0.0277	0.4092	[−5.50 dB, 18.65 dB]	0.0532°
0.7164	0.1075	−1.7252	0.4867	±52.41°	0.2819°/s.
−2.2741	0.0173	4.4961	−0.3877		

Let $\sigma_{min}[I + FG(s)] > \alpha$, where the plant transfer matrix is given by $G(s) = C(sI - A)^{-1}B$. Then, the upward gain margin is at least as large as $1/(1-\gamma)$ and the gain reduction margin is at least as small as $1/(1+\gamma)$. The phase margins are at least $\pm SIN^{-1}(\gamma/2)$. The multivariable gain and phase margins for the constant gain output feedback design are shown in Table 16.2. We conclude that the margins are acceptable, especially because these margins are considered conservative.

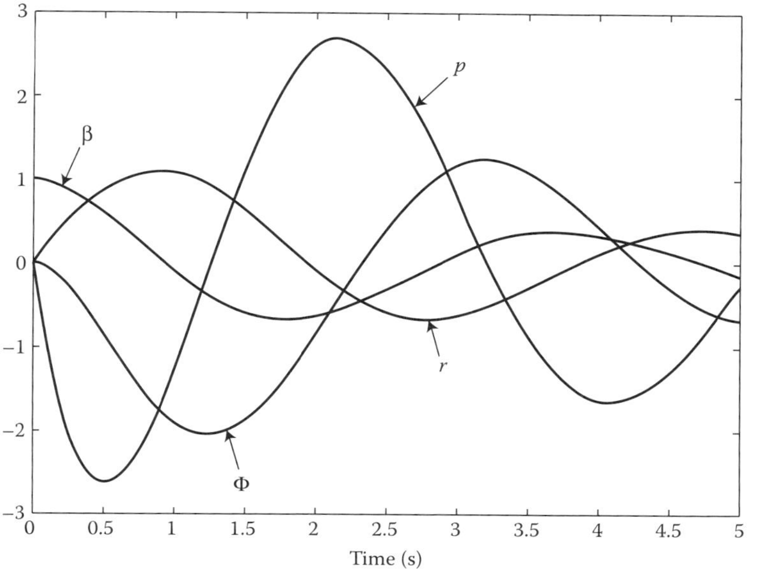

FIGURE 16.1 F-18 open-loop state responses.

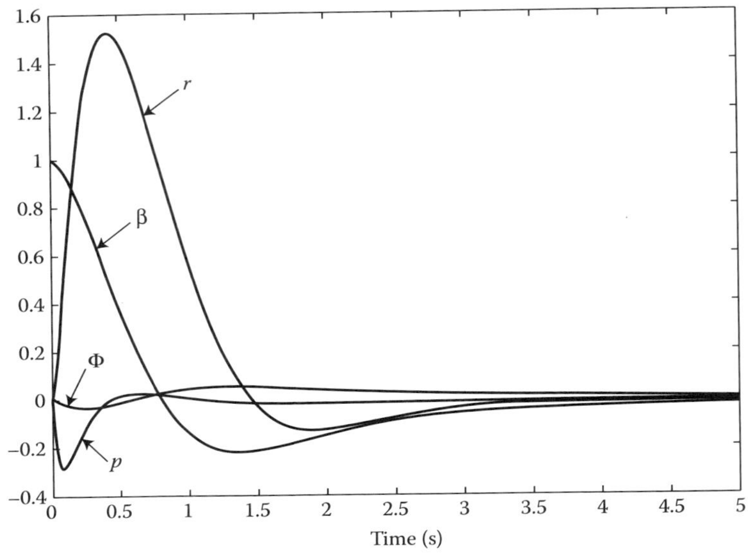

FIGURE 16.2 F-18 closed-loop state responses for output feedback.

16.3 Eigenstructure Assignment Using Constrained Output Feedback

The feedback gain matrix given by Equation 16.8 feeds back every output to every input. We now consider the problem of constraining certain elements of F to be zero. By suppressing certain gains to zero, the designer reduces controller complexity and increases reliability.

Using the development of [1], define

$$\Omega = CV \tag{16.12}$$

$$\Psi = Z - A_1 V \tag{16.13}$$

Then the expression for the feedback gain matrix F is given by (see Equation 16.8)

$$F = -\Psi\Omega^{-1}. \tag{16.14}$$

By using the Kronecker product and the lead block identity structure of the matrix B, each row of the feedback gain matrix can be computed independently of all the other rows [1]. Let ψ_i be the ith row of the matrix Ψ. Then the solution for f_i, which is the ith row of the feedback gain matrix F, is given by

$$f_i = -\psi_i\Omega^{-1}. \tag{16.15}$$

If f_{ij} is chosen to be constrained to zero, then Andry et al. [1] show that f_{ij} should be deleted from f_i and that the jth row of Ω should be deleted. Let $\widetilde{\Omega}$ be the matrix Ω with its jth row deleted and $\tilde{f}_i$ be the row vector f_i with its jth entry deleted. Then, by using a pseudoinverse, the solution for $\tilde{f}_i$, whose entries are the remaining active gains in the ith row of the matrix F, is given by

$$f_i = -\psi_i\widetilde{\Omega}^{\dagger}, \tag{16.16}$$

where $(\cdot)^{\dagger}$ denotes the appropriate pseudoinverse of $(\cdot)$. If more than one gain in a row of F is to be set to zero, the $\tilde{f}_i$ and $\widetilde{\Omega}$ must be appropriately modified.

16.3.1 Eigenvalue/Eigenvector Derivative Method for Choosing Gain-Suppression Structure

Calvo-Ramon [5] has proposed a method for choosing *a priori* which gains should be set to zero based on the sensitivities of the eigenvalues to changes in the feedback gains. The first-order sensitivity of the hth eigenvalue to changes in the ijth entry of the matrix F is denoted by $\partial\lambda_h/\partial f_{ij}$. The expected shift in the eigenvalue λ_h when constraining feedback gain f_{ij} to zero is given by

$$S_{ij}^h = (f_{ij})\frac{\partial\lambda_h}{\partial f_{ij}}. \tag{16.17}$$

Next, combine all the eigenvalue shifts that are related to the same feedback gain f_{ij} to form a decision matrix $D^\lambda = \{d_{ij}\}, D^\lambda \varepsilon R^{m\times r}$, where

$$d_{ij}^\lambda = \frac{1}{n}\left[\sum_{h=1}^{n}\left(\bar{S}_{ij}^h\right)\left(S_{ij}^h\right)\right]^{1/2} \tag{16.18}$$

and $(\bar{\cdot})$ denotes the complex conjugate of $(\cdot)$.

The decision matrix D^λ is used to determine which feedback gains f_{ij} should be set to zero. If d_{ij}^λ is "small," then setting f_{ij} to zero will have a small effect on the closed-loop eigenvalues. Conversely, if d_{ij}^λ is "large," then setting f_{ij} to zero will have a significant effect on the closed-loop eigenvalues. The control system designer must determine which d_{ij}^λ are "small" and which are "large" for a particular problem. In this regard, it is assumed that the states, inputs, and outputs are scaled so that these variables are expressed in the same or equivalent units.

The decision matrix D^λ was used by Calvo-Ramon [5] to design a constrained output feedback controller using eigenstructure assignment. However, the sensitivities of the eigenvectors with respect to the gains were not considered when deciding which feedback gains should be set to zero. Recall that the eigenvalues determine transient response characteristics such as overshoot and settling time, whereas the eigenvectors determine mode decoupling. This mode decoupling is related, for example, to the $|\phi/\beta|$ ratio in an aircraft lateral dynamics problem. This ratio must be small, as specified in [18], which implies that the closed-loop aircraft should exhibit a significant degree of decoupling between the dutch roll mode and the roll mode. The approach of Calvo-Ramon [5] may yield a constrained controller with acceptable overshoot and settling time, but the mode decoupling may be unacceptable. Thus, consideration of both eigenvalue and eigenvector sensitivities is important when choosing which feedback gains should be constrained to zero.

Sobel et al. [33] extended the results of Calvo-Ramon [5] to include both eigenvalue and eigenvector sensitivities to the feedback gains. The eigenvector derivatives are used to compute the expected shift in eigenvector v_h when constraining feedback gain f_{ij} to be zero. This expected shift in eigenvector v_h is given by

$$\tilde{s}_{ij}^h = (f_{ij})(\partial v_h/\partial f_{ij}). \tag{16.19}$$

Then, all the eigenvector shifts related to the same feedback gain f_{ij} are combined to form an eigenvector decision matrix D^v,

$$D^v = \frac{1}{n}\left[\sum_{h=1}^{n}\left(\tilde{s}_{ij}^h\right)^*\left(\tilde{s}_{ij}^h\right)\right]^{1/2}, \tag{16.20}$$

where $(\cdot)^*$ denotes the complex-conjugate transpose of $(\cdot)$. The gains that should be set to zero are determined by first eliminating those f_{ij} corresponding to entries of D^λ that are considered to be small. Then, those entries of D^v corresponding to those f_{ij} that were chosen to be set to zero based on D^λ are reviewed. In this way, the designer can determine whether some of the f_{ij} that may be set to zero based on eigenvalue considerations should not be constrained based on eigenvector considerations.

16.3.2 F-18 HARV Linearized Lateral Dynamics Design Example

We return to the example which was first considered in Section 16.2.1 and now we seek a constrained-output feedback controller. The eigenvalue decision matrix D^{λ} and the eigenvector decision matrix D^{v} are shown in Table 16.3. The entries of D^{λ} considered large are underlined in Table 16.3. Observe that only seven of the 12 feedback gains are needed when only eigenvalue sensitivities are considered. The constrained-output feedback gain matrix based on using only the information available from the eigenvalue decision matrix D^{λ} is shown in Table 16.4. The state responses to a 6° initial sideslip are shown in Figure 16.3. Observe the significantly increased coupling between sideslip and bank angle as compared to the unconstrained design of Section 16.2.1. The maximum absolute values of the bank angle and roll rate are now 0.5102° and 1.7881°/s compared with 0.0532° and 0.2819°/s obtained with the unconstrained feedback gain matrix. The increased coupling is due to ignoring the eigenvector sensitivities and illustrates the importance of the eigenvectors in achieving adequate mode decoupling.

Next, consider the entries of D^{v} that correspond to those f_{ij} that were chosen to be set to zero based on the eigenvalue decision matrix. The two largest d^{v}_{ij} that belong to this class are d^{v}_{11} and d^{v}_{23}. A new constrained-output feedback gain matrix is computed in which f_{11} and f_{23} are not set to zero. Nine gains are now needed to be unconstrained when using both eigenvalue and eigenvector information. The new feedback gain matrix is given in Table 16.4 and the state responses for a 1° initial sideslip are shown in Figure 16.4. Observe that these time responses are almost identical to the responses in Figure 16.2, which were obtained by using all 12 feedback gains. Thus, a simpler controller is obtained with a negligible change in the aircraft time responses. Finally, the multivariable gain and phase margins are shown in Table 16.4. The values of these margins are considered acceptable because singular-value-based multivariable stability margin computation is conservative.

TABLE 16.3 Eigenvalue and Eigenvector Decision Matrices

Eigenvalue Decision D^{λ}				Eigenvector Decision Matrix D^{v}				
r_{wo}	p	β	ϕ	r_{wo}	p	β	ϕ	
0.0710	$\underline{0.5117}$	0.0019	$\underline{0.2379}$	$\underline{0.2379}$	0.4675	0.0107	0.1835	δ_a
$\underline{0.2286}$	$\underline{0.3150}$	0.0843	$\underline{0.3362}$	0.7927	0.2882	$\underline{0.5270}$	0.1693	δ_s
$\underline{0.6754}$	0.0162	$\underline{0.3347}$	0.0482	0.7780	0.0259	0.3194	0.0871	δ_r

TABLE 16.4 Comparison of Constant Gain Control Laws

Feedback Gain Matrix Degree/Degree				Gain and Phase Margins	max $\|\phi\|$
r_{wo}	p	β	ϕ	(at inputs $\delta_{ac}, \delta_{sc}, \delta_{rc}$)	max $\|p\|$
Unconstrained					
−0.1704	0.1380	0.0277	0.4092	[−5.50 dB, 18.65 dB]	0.0532°
0.7164	0.1075	−1.7252	0.4867	±52.41°	0.2819°/s
−2.2741	0.0173	4.4961	−0.3877		
Constrained D^{λ} only					
0.0	0.1926	0.0	0.6417	[−5.38 dB, 16.90 dB]	0.5102°
0.3371	0.2010	0.0	0.7554	±50.75°	1.7881°/s
−2.2648	0.0	4.4891	0.0		
Constrained D^{λ} and D^{v}					
−0.1643	0.1365	0.0	0.4049	[−5.32 dB, 16.22 dB]	0.0505°
0.7164	0.1075	−1.7252	0.4867	±50.01°	0.2662°/s
−2.2648	0.0	4.4891	0.0		

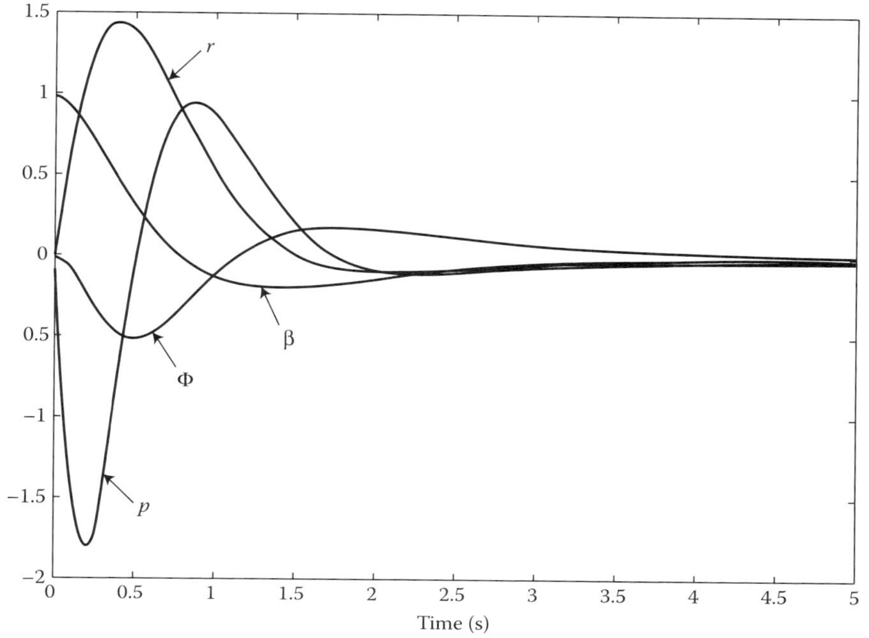

FIGURE 16.3 F-18 closed-loop state responses for constrained-output feedback (using only D^{λ}).

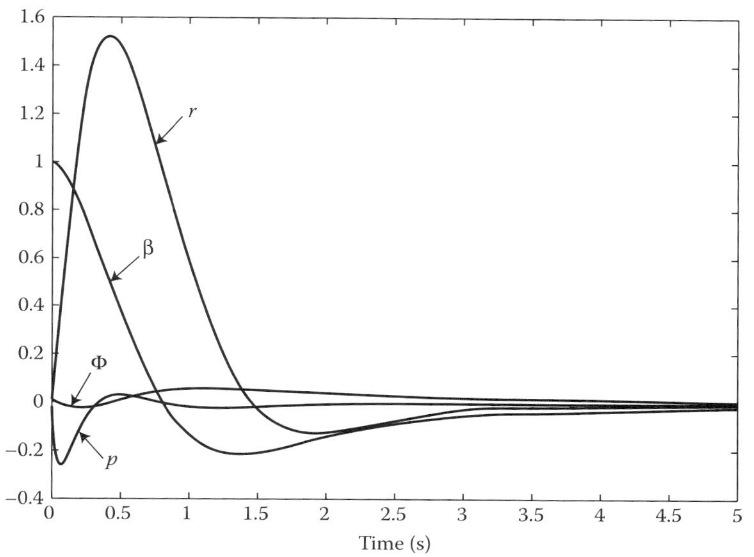

FIGURE 16.4 F-18 closed-loop state responses for constrained-output feedback (using both D^{λ} and D^{ν}).

16.4 Eigenstructure Assignment Using Dynamic Compensation

We now generalize the eigenstructure assignment flight control design methodology to include the design of low-order dynamic compensators of any given order $\ell, 0 \leq \ell \leq n-r$. Recall that n and r are the dimensions of the aircraft state and output vectors, respectively. Consider the linear time-invariant aircraft described by Equations 16.1 and 16.2 with a linear time-invariant dynamic controller specified by

$$\dot{z}(t) = Dz(t) + Ey(t), \tag{16.21}$$

$$u(t) = F(z) + Gy(t), \tag{16.22}$$

where the controller state vector $z(t)$ is of dimension $\ell, 0 \leq \ell \leq n-r$.

It is convenient to model the aircraft and compensator by the composite system originally proposed by Johnson and Athans [10]. Thus, define

$$\dot{\bar{x}} = \bar{A}\bar{x} + \bar{B}\bar{u}, \tag{16.23}$$

$$\bar{y} = \bar{C}\bar{x}, \tag{16.24}$$

$$\bar{u} = \bar{F}\bar{y}, \tag{16.25}$$

where

$$\bar{x} = \left[\frac{x}{z}\right], \quad \bar{A} = \left[\frac{A|0}{0|0}\right], \quad \bar{B} = \left[\frac{B|0}{0|I}\right], \quad \bar{C} = \left[\frac{C|0}{0|I}\right], \quad \bar{F} = \left[\frac{G|F}{E|D}\right].$$

Furthermore, the eigenvectors of the composite system may be described by

$$v_i = \left[\frac{v_i(x)}{v_i(z)}\right], \tag{16.26}$$

where $v_i(x)$ is the ith subeigenvector corresponding to the aircraft and $v_i(z)$ the ith subeigenvector corresponding to the compensator. Once the compensator dimension is chosen, the problem is solved as previously shown for the constant gain case.

The dynamic compensator design problem may be stated as follows. Given a set of desired aircraft eigenvalues $\{\lambda_i^d\}, i = 1, 2, \ldots, r+\ell$ and a corresponding set of desired aircraft subeigenvectors $v_i^d(x), i = 1, 2, \ldots, r+\ell$, find real matrices $D(\ell \times \ell), E(\ell \times r), F(m \times \ell)$, and $G(m \times r)$ so that the eigenvalues of $A + BFC$ contain $\{\lambda_i^d\}$ as a subset and corresponding subeigenvectors $\{v_i(x)\}$ are close to the respective members of the set $\{v_i^d(x)\}$.

16.4.1 F-18 HARV Linearized Lateral Dynamics Design Example

We again return to the example first considered in Section 16.2.1. If $y = [r_{wo}, p, \beta, \phi]^T$, as was the case in Sections 16.2.1 and 16.3.1, then the designer can specify both the dutch roll mode and roll mode eigenvalues. The designer might also specify three entries in the real and imaginary parts of the dutch roll eigenvectors and four entries in the real and imaginary parts of the roll mode eigenvectors as was done in Section 16.2.1. Then achievable eigenvectors are computed.

Now suppose that the measurement is given by $y = [r_{wo}, p, \phi]^T$, but the designer is still required to assign both the dutch roll and roll mode eigenvalues. Using the results of Section 16.4, we might utilize a first-order dynamic compensator with state z_1. The composite system has state vector $\bar{x} = [\delta_a, \delta_s, \delta_r, \beta, p, r, \phi, x_8, z_1]^T$, and the measurement vector is given by $\bar{y} = [r_{wo}, p, \phi, z_1]^T$. Thus, as before, the designer might choose to specify three or four entries of the real and imaginary parts of $v_i(x), i = 1,2,3,4$, which are the entries of the dutch roll and roll mode eigenvectors corresponding to the original aircraft state variables. Again, achievable eigenvectors will be computed.

We might ask whether the eigenvalue/eigenvector specifications are identical for both proposed problems. Certainly, the eigenvalue specifications and the corresponding $v_i^d(x)$ subeigenvectors may be identical. However, the $v_i^d(z)$ subeigenvector specifications must now be properly chosen. Otherwise, the modal matrix for the composite system may become numerically singular.

Finally, we remark that if the first-order compensator does not perform acceptably, then the designer might try a higher-order compensator. In the case of a second-order compensator, the composite system has state vector $\bar{x} = [\delta_a, \delta_s, \delta_r, \beta, p, r, \phi, x_8, z_1, z_2]^{\mathrm{T}}$, and the measurement vector is given by $\bar{y} = [r_{wo}, p, \phi, z_1, z_2]^{\mathrm{T}}$. In this case, the designer can also specify one of the compensator eigenvalues and some entries of its corresponding eigenvector.

Now consider the case when only the washed-out yaw rate, roll rate, and bank angle are measured. We form the composite system described by Equations 16.23 through 16.25 by appending a first-order compensator to the aircraft dynamics. We specify that the roll mode and dutch roll mode eigenvalues are the same as in the constant gain feedback problem in Section 16.2.1. The compensator pole is not specified, but it is chosen by the eigenstructure assignment algorithm to obtain eigenvalue and eigenvector assignment for the aircraft modes. Furthermore, because we have three sensors plus one compensator state, the eigenstructure assignment algorithm will allow us to specify four closed-loop eigenvalues. We specify that the desired aircraft subeigenvectors $v_i^d(x)$ are the same as the desired eigenvectors in the constant gain output feedback problem. The desired compensator subeigenvectors are chosen so that the dutch roll and roll modes participate in the compensator state solution. The control law is described by

$$\begin{bmatrix} \delta_a \\ \delta_s \\ \delta_r \end{bmatrix} = \begin{bmatrix} 0.1708 & -0.1383 & -0.4079 \\ -0.2723 & -0.3328 & -0.2408 \\ 1.0897 & 0.5836 & -1.5548 \end{bmatrix} \begin{bmatrix} r_{wo} \\ p \\ \phi \end{bmatrix} + \begin{bmatrix} -0.0040 \\ -1.4112 \\ 3.7731 \end{bmatrix} z_1, \tag{16.27}$$

$$\dot{z}_1 = -2.4784 z_1 + 0.5906 r_{wo} + 0.9817 p + 3.0796 \phi. \tag{16.28}$$

The desired eigenvalues, achievable eigenvalues, and desired eigenvectors are shown in Table 16.5. The separation principle does not apply to the dynamic compensator described by Equations 16.21 and 16.22. Thus, the composite system has eigenvalues at -1.7646 and -0.7549 which are due to the compensator and the yaw rate washout filter, respectively. The time responses of the aircraft states are shown in Figure 16.5 from which we observe that the responses with the dynamic compensator are slower than the responses with constant gain output feedback. However, the dynamic compensator controller is implemented without the need for a sideslip sensor. In addition, the ratio $|\phi/\beta| \approx 0.8$ as compared to $|\phi/\beta| \approx 4$ for the open-loop aircraft.

TABLE 16.5 Eigenvalues and Eigenvectors for Dynamic Compensator

Desired Eigenvalues	Achievable Eigenvalues	Desired Eigenvectors: Dutch Roll Mode Re	Dutch Roll Mode Im	Roll Mode Re	Roll Mode Im	
		x	x	x	x	δ_a
	$\lambda_{dr} = -2 \pm j2$	x	x	x	x	δ_s
	$\lambda_{roll} = -3 \pm j2$	x	x	x	x	δ_r
$\lambda_{dr} = -2 \pm j2$	$\lambda_{act} = -30.0000$	x	1	0	0	β
$\lambda_{roll} = -3 \pm j2$	$\lambda_{act} = -27.6993$	0	0	x	1	p
	$\lambda_{act} = -25.2634$	1	x	0	0	r
	$\lambda_{filt} = -0.7549$	0	0	1	x	ϕ
	$\lambda_{comp} = -1.7646$	x	x	0	0	x_8
		1	x	1	x	z_1

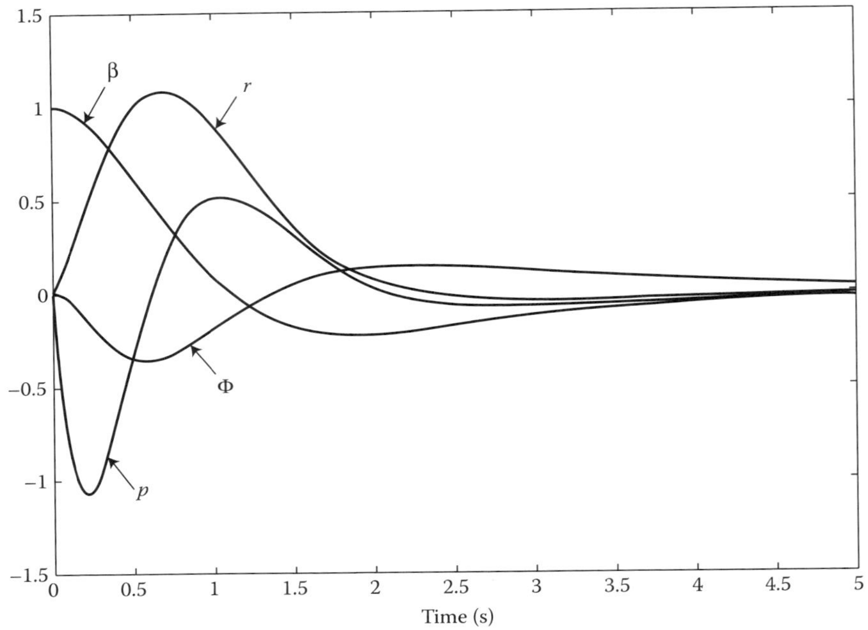

FIGURE 16.5 F-18 closed-loop state responses for first-order dynamic compensator.

The multivariable stability margins are computed using $\sigma_{\min}[I + KG(s)]$ where $G(s) = C(sI - A)^{-1}B$ and $K = -[F(sI - D)^{-1}E + G]$. The multivariable gain margins are $GM\varepsilon[-2.47$ dB, 3.46 dB $]$ and the multivariable phase margin is $\pm 18.89°$. If the designer considers these margins inadequate, then an optimization may be used to compute an eigenstructure assignment controller with a constraint on the minimum of the smallest singular value of the return difference matrix. In the next section, we present a robust, sampled data eigenstructure controller for the yaw pointing/lateral translation maneuver of the FPCC aircraft.

16.5 Robust, Sampled Data Eigenstructure Assignment

Sobel and Shapiro [28] used eigenstructure assignment to design a continuous-time controller for the yaw pointing/lateral translation maneuver of the FPCC aircraft. This conceptual control-configured aircraft has a vertical canard and is difficult to control because the control distribution matrix has a minimum singular value of 0.0546. The design of Sobel and Shapiro [28] is characterized by perfect decoupling, but the minimum of the smallest singular value of the return difference matrix at the aircraft inputs was only 0.18.

Sobel and Lallman [32] proposed a pseudocontrol strategy for reducing the dimension of the control space by using the singular-value decomposition. The FPCC yaw pointing/lateral translation design of Sobel and Lallman [32] yields a minimum of the smallest singular value of the return difference matrix at the aircraft inputs of 0.9835, but the lateral translation transient response has significant coupling to the heading angle.

Sobel and Shapiro [31] have proposed an extended pseudocontrol strategy. Piou and Sobel [22] extended eigenstructure assignment to linear time-invariant plants which are represented by Middleton and Goodwin's [17] unified delta model which is valid both for continuous-time and sampled data operation of the plant. Piou et al. [23] have extended Yedavalli's [35] Lyapunov approach for stability robustness of a linear time-invariant system to the unified delta system. In this section, we design a robust, sampled data, extended pseudocontrol, eigenstructure assignment flight control law for the yaw pointing/lateral translation maneuver of the FPCC aircraft. The main goal of this section is to describe a design methodology which incorporates robustness into the eigenstructure assignment method. However, a sampled data design is used for illustration.

16.5.1 Problem Formulation

Consider a nominal linear time-invariant system described by (A, B, C). The corresponding sampled data system is described by (A_δ, B_δ, C) and the unified delta model is described by (A_ρ, B_ρ, C). Suppose that the nominal delta system is subject to linear time-invariant uncertainties in the entries of A_ρ, B_ρ described by dA_ρ and dB_ρ, respectively. Then, the delta system with uncertainty is given by $(A_\rho + dA_\rho, B_\rho + dB_\rho, C)$. Here $dA_\rho = dA, \quad dB_\rho = dB$ in continuous time and $dA_\rho = dA_\delta, dB_\rho = dB_\delta$ in discrete time. Furthermore, suppose that bounds are available on the maximum absolute values of the elements of dA and dB so that $\{dA : dA^+ \leq A_{\max}\}$ and $\{dB : dB^+ \leq B_{\max}\}$ and where "$\leq$" is applied element by element to matrices.

Consider the constant gain output feedback control law described by $u(t) = F_\rho y(t)$, where $F_\rho = F$ in continuous time and $F_\rho = F_\delta$ in discrete time. Then, the nominal closed-loop unified delta system is given by $\rho x(t) = A_{\rho c}x(t)$, where $A_{\rho c} = A + BFC$ in continuous time and $A_\delta + B_\delta F_\delta C$ in discrete time. The uncertain closed-loop unified delta system is given by $\rho x(t) = A_{\rho c}x(t) + dA_{\rho c}x(t)$, where $dA_{\rho c} = dA + dB(FC)$ in continuous time and $dA_\delta + dB_\delta(F_\delta C)$ in discrete time. The reader is referred to Middleton and Goodwin [17] for a more detailed description.

16.5.2 Pseudocontrol and Robustness Results

The purpose of a pseudocontrol is to reduce the dimension of the control space. This reduction is needed for systems whose control distribution matrix B has a minimum singular value that is very small. After the eigenstructure assignment design is complete, the controller is mapped back into the original control space. Consider the singular-value decomposition of the matrix B_ρ given by

$$B_\rho = [U_1 U_2 U_0] \begin{bmatrix} \Sigma_1 & & \\ & \Sigma_2 & \\ & & 0 \end{bmatrix} \begin{bmatrix} V_1^T \\ V_2^T \\ V_0^T \end{bmatrix}, \tag{16.29}$$

where $\Sigma_1 = \text{diag}\,[\sigma_1, \ldots, \sigma_a]$ and $\Sigma_2 = \text{diag}\,[\sigma_{a+1}, \ldots, \sigma_b]$ and where $\sigma_b \leq \sigma_{b-1} \leq \cdots \leq \sigma_{a+1} \leq \varepsilon$ with ε small.

Lemma 16.1:

Let the system with the pseudocontrol $\tilde{u}(t)$ be described by

$$\rho x(t) = A\rho x(t) + \tilde{B}_\rho \tilde{u}(t), \tag{16.30}$$

$$y(t) = Cx(t), \tag{16.31}$$

where

$$\tilde{B}_\rho = U_1 + U_2[\alpha_1, \alpha_2]. \tag{16.32}$$

We design a feedback pseudocontrol for the system described by Equations 16.30 through 16.32. Then, the true control $u(t)$ for the system described by (A_ρ, B_ρ, C) is given by

$$u(t) = \left[V_1\Sigma_1^{-1} + V_2\Sigma_2^{-1}\alpha\right]\tilde{u}(t). \tag{16.33}$$

Furthermore, when $\alpha = [0,0]^T$, the control law $u(t)$ given by Equation 16.33 reduces to the control law given by Equation 20 in Sobel and Lallman [32].

Theorem 16.1:

The system matrix $A_{\rho c} + dA_{\rho c}$ is stable if

$$\sigma_{\max}\left(E_{2\max}^T P_\rho^+ E_{1\max}\right)_s < 1, \tag{16.34}$$

where

$$E_{1\max} = A_{\rho\max} + B_{\rho\max}(F_\rho C)^+$$

and

$$E_{2\max} = \left\{I_n + \Delta\left[A_\rho + B_\rho(F_\rho C)\right]\right\}^+ + (\Delta/2)E_{1\max}$$

and where P_ρ satisfies the Lyapunov equation given by

$$A_{\rho c}^T P_\rho + P_\rho A_{\rho c} + \Delta A_{\rho c} P_\rho A_{\rho c}^T = -2I_n$$

and where P_ρ^+ is the matrix formed by the modulus of the entries of the matrix P_ρ, and $(\cdot)_s$ denotes the symmetric part of a matrix.

16.5.3 FPCC Yaw Pointing/Lateral Translation Controller Design Using Robust, Sampled Data, Eigenstructure Assignment

We consider the FPCC aircraft linearized lateral dynamics which is described by Sobel and Lallman [32]. The state-space matrices A and B are shown in the Appendix. The state variables are sideslip angle β, bank angle ϕ, roll rate p, and lateral directional flight path angle $(\gamma = \Psi + \beta)$, where Ψ is the heading angle. The control variables are rudder δ_r, ailerons δ_a, and vertical canard δ_c. The angles and surface deflections are in degrees, and the angular rates are in degrees/second. The five measurements are β, ϕ, p, r, and γ.

First, we design an eigenstructure assignment control law by using an orthogonal projection. The delta state-space matrices A_δ and B_δ are computed by using the MATLAB Delta Toolbox. The sampling period Δ is chosen to be 0.02 s for illustrative purposes. The desired dutch roll, roll mode, and flight path mode eigenvalues are achieved exactly because five measurements are available for feedback. The achievable eigenvectors are computed by using the orthogonal projection of the ith desired eigenvector onto the subspace which is spanned by the columns of $(\gamma_i I - A_\delta)^{-1}B_\delta$. The closed-loop delta eigenvalues $\gamma_i, i = 1, \ldots, n$, and the feedback gain matrix F_δ are shown in Table 16.6. The desired closed-loop eigenvectors are shown in Table 16.7.

The orthogonal projection solution is characterized by excellent decoupling with the minimum of the smallest singular value of $(I + FG)$ equal to 0.18. Here the transfer function matrix of the delta plant is given by $G([e^{j\omega\Delta} - 1]/\Delta)$, where $0 < \omega < \pi/T$. Furthermore, the Lyapunov robust stability condition of Equation 16.34 is not satisfied.

To improve the minimum singular value of $(I + FG)$, we design a controller by using an orthogonal projection with the pseudocontrol of Sobel and Lallman [32]. This pseudocontrol mapping is given

TABLE 16.6 Comparison of FPCC Designs ($\Delta = 0.02$ s)

	Closed-Loop Eigenvalues	Feedback Gain Matrix					
		β	ϕ	p	r	γ	
Orthogonal projection design	$\gamma_{dr} = -1.9990 \pm j1.921$	1.4688	−0.2866	−0.0022	0.3799	−1.5332	δ_r
	$\gamma_{roll} = -2.95 \pm j1.883$	0.5652	−2.2990	−0.9044	−0.6691	−0.8890	δ_a
	$\gamma_{fp} = 0.4975$	9.2964	−1.2143	−0.514	−1.4219	−15.57	δ_c
Orthogonal projection design with pseudocontrol	$\gamma_{dr} = -1.999 \pm j1.921$	−0.4929	−0.0332	0.0076	0.6883	2.7584	δ_r
	$\gamma_{roll} = -2.95 \pm j1.883$	0.9517	−2.353	−0.9093	−0.7565	−2.5147	δ_a
	$\gamma_{fp} = -0.4975$	0.1203	−0.0530	−0.0245	−0.1535	−0.6023	δ_c
Robust pseudocontrol design	$\gamma_{1,2} = -3.37 \pm j1.56$	−0.2833	−0.0340	−0.0062	0.2377	0.5336	δ_r
	$\gamma_{3,4} = -3.16 \pm j1.473$	1.5501	−2.4760	−1.0370	−0.9812	−1.7788	δ_a
	$\gamma_{fp} = -0.4310$	5.6151	−0.0012	−0.1486	−4.6301	−10.285	δ_c
Robust pseudocontrol design with singular-value constraint	$\gamma_{1,2} = -2.55 \pm j1.19$	−0.1658	−0.1967	−0.0808	0.4895	0.8616	δ_r
	$\gamma_{3,4} = -2.60 \pm j1.22$	0.8390	−1.5479	−0.7663	−0.8565	−1.2679	δ_a
	$\gamma_{fp} = -0.3248$	1.1525	−0.8794	−0.4764	−1.9712	−3.2578	δ_c

Note: Eigenvalues are computed by using feedback gains with significant digits to machine precision.

by Equation 16.33 with $\alpha = [0, 0]$. This design is characterized by a lateral translation response with significant coupling between γ and Ψ with the minimum of the smallest singular value of $(I + FG)$ equal to 0.9835. The Lyapunov sufficient robust stability condition of Equation 16.34 is not satisfied. We note that the yaw pointing responses exhibit excellent decoupling for both orthogonal projection designs.

Next, to obtain a robust design with excellent decoupling, we design a robust pseudocontrol law by using the design method proposed by Piou et al. [23]. This new design method minimizes an objective function which weights the heading angle due to a lateral flight path angle command and the lateral flight path angle due to a heading command. Constraints are placed on the time constants of the dutch roll, roll, and flight path modes, the damping ratios of the dutch roll and roll modes, and the new sufficient condition for robust stability. Mathematically, the objective function to be minimized is given by

$$J = \sum_{k=1}^{100} \left[(1-\alpha)\left(\Psi_k^2\right)_{\gamma c} + \alpha\left(\gamma_k^2\right)_{\Psi c} \right]. \tag{16.35}$$

The upper limit on the index k is chosen to include the time interval $k\Delta\varepsilon[0, 2]$ during which most of the transient response occurs. Of course, computation of Equation 16.35 requires that two linear simulations be performed during each function evaluation of the optimization. The constraints for continuous time and the corresponding constraints for discrete time are shown in Table 16.8 where ζ is the damping ratio.

TABLE 16.7 FPCC Desired Closed-Loop Eigenvectors

Dutch Roll Mode	Roll Mode	Flight Path Mode	
$\begin{bmatrix} x \\ 0 \\ 0 \\ 1 \\ 0 \end{bmatrix} \pm j \begin{bmatrix} 1 \\ 0 \\ 0 \\ x \\ 0 \end{bmatrix}$	$\begin{bmatrix} 0 \\ x \\ 1 \\ 0 \\ 0 \end{bmatrix} \pm j \begin{bmatrix} 0 \\ 1 \\ x \\ 0 \\ 0 \end{bmatrix}$	$\begin{bmatrix} x \\ 0 \\ 0 \\ 0 \\ 1 \end{bmatrix}$	$\begin{matrix} \beta \\ \phi \\ p \\ r \\ \gamma \end{matrix}$

TABLE 16.8 Constraints for the FPCC Designs

Continuous Time	Discrete Time
For Complex Eigenvalues:	
$\text{Re}\lambda \in [-4, -1.5]$	$\lvert 1+\Delta\gamma\rvert \in \left[e^{-4\Delta}, e^{-1.5\Delta}\right]$
$\zeta \in [0.4, 0.9]$	$\lvert 1+\Delta\gamma\rvert \in \left[\exp\left(\frac{-0.9\phi}{[1-(0.9)^2]^{1/2}}\right), \exp\left(\frac{-0.4\phi}{[1-(0.4)^2]^{1/2}}\right),\right]$, where $\phi = \arg(1+\Delta\gamma)$
For the Real Eigenvalue:	
$\lambda \in [-1, -0.05]$	$\lvert 1+\Delta\gamma\rvert \in [e^{-\Delta}, e^{-0.05\Delta}]$
For Lyapunov Robustness:	
	$\sigma_{\max}\left(E_{2\max}^T P_\rho^+ E_{1\max}\right) < 0.999$
For Multivariable Stability Margins (Final Design only):	
	$\min \sigma_{\min}(I+FG) \geq 0.55; 0 < \omega < \pi/T$

For illustrative purposes we have chosen $A_{\max} = 0.085A^+$ and $B_{\max} = 0$. After many trials, we found that a good value for the weight α in Equation 16.35 is $\alpha = 0.0075$.

The parameter vector contains the quantities which may be varied by the optimization. This 17-dimensional vector includes Re γ_{dr}, Im γ_{dr}, Re γ_{roll}, Im γ_{roll}, γ_{fp}, Re $z_1(1)$, Re $z_1(2)$, Im $z_1(1)$, Im $z_1(2)$, Re $z_3(1)$, Re $z_3(2)$, Im $z_3(1)$, Im $z_3(2)$, $z_5(1)$, $z_5(2)$, and the two-dimensional pseudocontrol vector α of Equation 16.32. Here, the two-dimensional complex vectors z_i contain the free eigenvector parameters, that is, the ith eigenvector v_i may be written as

$$v_i = L_i z_i, \tag{16.36}$$

where the columns of $L_i = (\gamma_i I - A_\delta)^{-1}\tilde{B}_\delta$ are a basis for the subspace in which the ith eigenvector must reside. Thus, the free parameters are the vectors z_i rather than the eigenvectors v_i. The vectors z_i are two dimensional because the optimization is performed in the two-dimensional pseudocontrol space.

The optimization uses subroutine *constr* from the MATLAB Optimization Toolbox and subroutine *delsim* from the MATLAB Delta Toolbox. The optimization is initialized with the orthogonal projection pseudocontrol design which yields an initial value of 20.6 for the objective function of Equation 16.35 and a value of 1.66 for the right-hand-side (RHS) of the robustness condition of Equation 16.34. Unfortunately, the minimum of the smallest singular value of $(I + FG)$ is only 0.2607 which is less than desired.

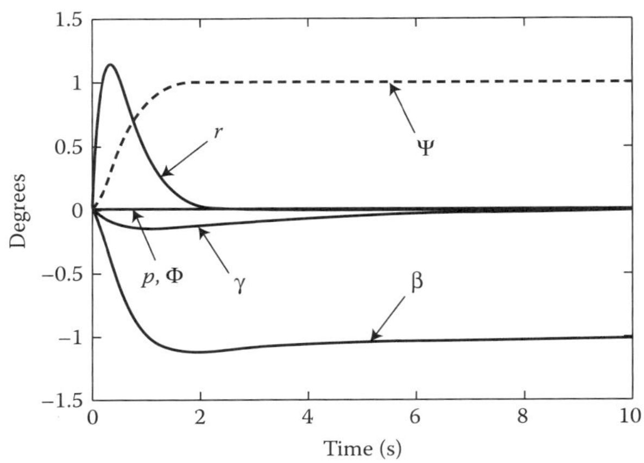

FIGURE 16.6 FPCC yaw pointing; robust pseudocontrol with singular-value constraint.

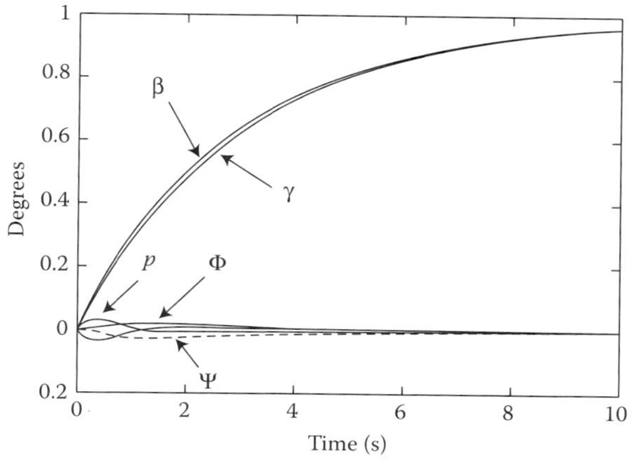

FIGURE 16.7 FPCC lateral translation; robust pseudocontrol with singular value constraint.

To achieve a design with excellent time responses, Lyapunov robustness, and an acceptable minimum of the smallest singular value of $(I + FG)$, we repeat the optimization with the additional constraint that $\sigma_{\min}(I + FG) \geq 0.55$. Once again we initialize the optimization at the orthogonal projection pseudocontrol design. The optimization yields an optimal objective function of 0.0556 and a value of 0.999 for the RHS of the robustness condition. The time responses for yaw pointing and lateral translation are shown in Figures 16.6 and 16.7, respectively. The lateral translation response is deemed excellent even though it has some small increase in coupling compared to the design without the additional singular-value constraint. Thus, we have obtained a controller which simultaneously achieves excellent time responses, Lyapunov robustness, and an acceptable minimum of the smallest singular-value of the return difference matrix at the aircraft inputs.

16.6 Defining Terms

Stability augmentation system: A feedback control designed to modify the inherent aerodynamic stability of the airframe.

Gain suppression: A feedback control in which one or more entries of the gain matrix are constrained to be zero.

Specialized task tailored modes: A feedback control system designed for a specific maneuver such as bombing, strafing, air-to-air combat, and so on.

Mode decoupling: The elimination of interactions between modes, for example, in an aircraft mode decoupling is important so that a sideslip angle disturbance will not cause rolling motion.

Eigenstructure assignment: A feedback control designed to modify both the system's eigenvalues and eigenvectors.

References

1. Andry, A.N., Shapiro, E.Y., and Chung, J.C., Eigenstructure assignment for linear systems, *IEEE Trans. Aerospace Electron. Systems*, 19, 711–729, 1983.
2. Ashari, A.E., Sedigh, A.K., and Yazdanpanah, M.J., Reconfigurable control system design using eigenstructure assignment: static, dynamic, and robust approaches, *Int. J. Control*, 78, 1005-1016, 2005.

3. Bruyere, L., Tsourdos, A., and White, B.A., Quasilinear parameter varying autopilot design using polynomial eigenstructure assignment with actuator constraints, *J. Guidance, Control, Dyn.*, 29, 1282–1294, 2006.
4. Burrows, S.P., Patton, R.J., and Szymanski, J.E., Robust eigenstructure assignment with a control design package, *IEEE Control Systems Mag.*, 9, 29–32, 1989.
5. Calvo-Ramon, J.R., Eigenstructure assignment by output feedback and residue analysis, *IEEE Trans. Automat. Control*, 31, 247–249, 1986.
6. Clarke, T., Griffin, S.J., and Ensor, J., A polynomial approach to eigenstructure assignment using projection with eigenvalue tradeoff, *Int. J. Control*, 76, 403–413,2003.
7. Doyle, J.C. and Stein, G., Multivariable feedback design: Concepts for a classical/modern synthesis, *IEEE Trans. Automat. Control*, 26, 4–16, 1981.
8. Gavito, V.F. and Collins, D.J., Application of eigenstructure assignment to design of robust decoupling controllers in mimo systems, *Proc. AIAA Guidance, Navig, and Control Conf.*, 1986, AIAA Paper No. 86-2246, 828–834.
9. Jiang, J., Design of reconfigurable control systems using eigenstructure assignments, *Int. J. Control*, 59, 395–410, 1994.
10. Johnson, T.L. and Athans, M., On the design of optimal constrained dynamic compensation for linear constant systems, *IEEE Trans. Automat. Control*, 15, 658–660, 1970.
11. Kautsky, J., Nichols, N.K., and Van Dooren, P., Robust pole assignment in linear state feedback, *Int. J. Control*, 41, 1129–1155, 1985.
12. Konstantopoulos, I., Optimal controller design and reconfigurable control under stability robustness, PhD Thesis, University of Notre Dame, Notre Dame, IN, 1996.
13. Lehtomaki, N.A., Practical robustness measures in multivariable control system analysis, PhD Thesis, Massachusetts Institute of Technology, Cambridge, MA, 1981.
14. Liebst, B.S. and Garrard, W.L., Application of eigenspace techniques to design of aircraft control systems, *Proc. Am. Control Conf.*, 475–480, 1985.
15. Liebst, B.S., Garrard, W.L., and Adams, W.M., Design of an active flutter suppression system, *J. Guidance, Control, Dyn.*, 9, 64–71, 1986.
16. Lu, J., Chiang, H.D., and Thorp, J.S., Eigenstructure assignment by decentralized feedback control, *IEEE Trans. Automat. Control*, 38(4), 587–594, 1993.
17. Middleton, R.H. and Goodwin, *Digital Control and Estimation: A Unified Approach*. Prentice-Hall, Englewood Cliffs, NJ, 1990.
18. Anon., *Military Specification—Flying Qualities of Piloted Airplanes*, US Department of Defense, Washington, DC, 1980, MIL-F8785C.
19. Nieto-Wire, C. and Sobel, K., Eigenstructure assignment for a tailless aircraft, *Proc. AIAA Guidance, Navig. Control Conf.*, AIAA Paper No. 2007-6417, 2007.
20. Nieto-Wire, C. and Sobel, K., Reconfigurable delta operator eigenstructure assignment for a tailless aircraft, *Proc. AIAA Guidance, Navig. Control Conf.*, AIAA Paper No. 2009-6306, 2009.
21. O'Brien, M.J. and Broussard, J.R., Feedforward control to track the output of a forced model, *Proc. 17th IEEE Conf. Decision Control*, San Diego, 1149–1155, 1978.
22. Piou, J.E. and Sobel, K.M., Robust sampled data eigenstructure assignment using the delta operator, *Proc. Guidance, Navig. Control Conf.*, Hilton Head, SC, 1992.
23. Piou, J.E., Sobel, K.M., and Shapiro, E.Y., Robust Lyapunov constrained sampled data eigenstructure assignment using the delta operator with application to flight control design, *Proc. First IEEE Conf. Control Appl.*, Dayton, Ohio, 1000–1005, 1992.
24. Press, W.H., Flannery, B.P., Teukolsky, S.A., and Vetterling, W.T., *Numerical Recipes: The Art of Scientific Computing*, Cambridge University Press, New York, 1987.
25. Satoh, A. and Sugimoto, K., Loose eigenstructure assignment via rank one LMI approach with application to transient response shaping in H-infinity control, *Int. J. Control*, 82, 497–507, 2009.
26. Schulz, M.J. and Inman, D.J., Eigenstructure assignment and controller optimization for mechanical systems, *IEEE Trans. Control Systems Tech.*, 2(2), 88–100, 1994.
27. Sobel, K.M., Shapiro, E.Y., and Andry, A.N., Jr., Eigenstructure Assignment, *Int. J. Control*, 59(1), 13–37, 1994.
28. Sobel, K.M. and Shapiro, E.Y., Application of eigensystem assignment to lateral translation and yaw pointing flight control, *Proc. 23rd IEEE Conf. Decision Control*, Las Vegas, 1423–1428, 1984.
29. Sobel, K.M. and Shapiro, E.Y., A design methodology for pitch pointing flight control systems, *J. Guidance, Control, Dyn.*, 8, 181–187, 1985.
30. Sobel, K.M. and Shapiro, E.Y., Application of eigenstructure assignment to flight control design: Some extensions, *J. Guidance, Control, Dyn.*, 10, 73–81, 1987.

31. Sobel, K.M. and Shapiro, E.Y., An extension to the pseudocontrol strategy with application to an eigenstructure assignment yaw pointing/lateral translation control law, *Proc. 30th IEEE Conf. Decision Control,* Brighton, UK, 515–516, 1991.
32. Sobel, K.M. and Lallman, F.J., Eigenstructure assignment for the control of highly augmented aircraft, *J. Guidance, Control, Dyn.,* 12, 318–324, 1989.
33. Sobel, K.M., Yu, W., and Lallman, F.J., Eigenstructure assignment with gain suppression using eigenvalue and eigenvector derivatives, *J. Guidance, Control, Dyn.,* 13, 1008–1013, 1990.
34. Srinathkumar, S., Eigenvalue/eigenvector assignment using output feedback, *IEEE Trans. Automat Control,* AC-23(1), 79–81, 1978.
35. Yedavalli, R.K., Perturbation bounds for robust stability in linear state space models, *Int. J. Control,* 42, 1507–1517, 1985.
36. Yu, W., Piou, J.E., and Sobel, K.M., Robust eigenstructure assignment for the extended medium range air to air missile, *Automatica,* 29(4), 889–898, 1993.

Further Reading

Kautsky et al. [11] suggested that eigenstructure assignment can be used to obtain a design with eigenvalues, which are least sensitive to parameter variation, by reducing one of several sensitivity measures. Among these measures are the quadratic norm condition number of the closed-loop modal matrix and the sum of the squares of the quadratic norms of the left eigenvectors. Burrows et al. [4] have proposed a stability augmentation system for a well-behaved light aircraft by using eigenstructure assignment with an optimization which minimizes the condition number of the closed-loop modal matrix. Such an approach may sometimes produce an acceptable controller. In contrast to eigensensitivity, Doyle and Stein [7] have characterized the stability robustness of a multi-input, multi-output system by the minimum of the smallest singular value of the return difference matrix at the plant input or output. Gavito and Collins [8] have proposed an eigenstructure assignment design in which a constraint is placed on the minimum of the smallest singular value of the return difference matrix at the inputs of both an L-1011 aircraft and a CH-47 helicopter. Several authors have proposed different approaches to eigenstructure assignment. Clarke et al. [6] present a method to trade exact closed-loop eigenvalue location against an improvement in the associated eigenvector match. Bruyere et al. [3] use eigenstructure assignment in a polynomial framework. Satoh and Sugimoto [25] present a regional eigenstructure assignment using a linear matrix inequalities (LMI) approach. Nieto-Wire and Sobel [19] applied eigenstructure assignment to the design of a flight control system for a tailless aircraft. Other applications of eigenstructure assignment include air-to-air missiles [36], mechanical systems [26], and power systems [16]. Jiang [9], Konstantopoulos [12], and Ashari et al. [2] have proposed methods for reconfiguration using eigenstructure assignment. Nieto-Wire and Sobel [20] used eigenstructure assignment for the accomodation of symmetric lock in place actuator failures for a tailless aircraft.

Appendix

Data for the F-18 HARV Lateral Directional Dynamics at M = 0.38, H = 5000 ft., and $\alpha = 5°$

$$A = \begin{bmatrix} -30.0000 & 0 & 0 & 0 & 0 & 0 & 0 & 0 \\ 0 & -30.0000 & 0 & 0 & 0 & 0 & 0 & 0 \\ 0 & 0 & -30.0000 & 0 & 0 & 0 & 0 & 0 \\ -0.0070 & -0.0140 & 0.0412 & -0.1727 & 0.0873 & -0.9946 & 0.0760 & 0 \\ 15.3225 & 12.0601 & 2.2022 & -11.0723 & -2.1912 & 0.7096 & 0 & 0 \\ -0.3264 & 0.2041 & -1.3524 & 2.1137 & -0.0086 & -0.1399 & 0 & 0 \\ 0 & 0 & 0 & 0 & 1.0000 & 0.0875 & 0 & 0 \\ 0 & 0 & 0 & 0 & 0 & 0.5000 & 0 & -0.5 \end{bmatrix},$$

$$B = \begin{bmatrix} 30.0000 & 0 & 0 \\ 0 & 30.0000 & 0 \\ 0 & 0 & 30.0000 \\ 0 & 0 & 0 \\ 0 & 0 & 0 \\ 0 & 0 & 0 \\ 0 & 0 & 0 \\ 0 & 0 & 0 \end{bmatrix}, \quad C = \begin{bmatrix} 0 & 0 & 0 & 0 & 0 & 1 & 0 & -1 \\ 0 & 0 & 0 & 0 & 1 & 0 & 0 & 0 \\ 0 & 0 & 0 & 1 & 0 & 0 & 0 & 0 \\ 0 & 0 & 0 & 0 & 0 & 0 & 1 & 0 \end{bmatrix},$$

Data for the FPCC Lateral Directional Dynamics

$$A = \begin{bmatrix} -0.340 & 0.0517 & 0.001 & -0.997 & 0 \\ 0 & 0 & 1 & 0 & 0 \\ -2.69 & 0 & -1.15 & 0.738 & 0 \\ 5.91 & 0 & 0.138 & -0.506 & 0 \\ -0.340 & 0.0517 & 0.001 & 0.0031 & 0 \end{bmatrix}, \quad B = \begin{bmatrix} 0.0755 & 0 & 0.0246 \\ 0 & 0 & 0 \\ 4.48 & 5.22 & -0.742 \\ -5.03 & 0.0998 & 0.984 \\ 0.0755 & 0 & 0.0246 \end{bmatrix}.$$

17

Linear Quadratic Regulator Control

Leonard Lublin
Massachusetts Institute of Technology

Michael Athans
Massachusetts Institute of Technology

17.1 Introduction

The linear quadratic regulator problem, commonly abbreviated as LQR, plays a key role in many control design methods. Not only is LQR a powerful design method, but in many respects it is also the mother of many current, systematic control design procedures for linear multiple-input, multiple output (MIMO) systems. Both the linear quadratic Gaussian, LQG or $\mathcal{H}_2$, and $\mathcal{H}_\infty$ controller design procedures have a usage and philosophy that are similar to the LQR methodology. As such, studying the proper usage and philosophy of LQR controllers is an excellent way to begin building an understanding of even more powerful design procedures.

From the time of its conception in the 1960s, the LQR problem has been the subject of volumes of research. Yet, as will be detailed, the LQR is nothing more than the solution to a convex, least squares optimization problem that has some very attractive properties. Namely the optimal controller automatically ensures a stable closed-loop system, achieves guaranteed levels of stability robustness, and is simple to compute. To provide a control systems engineer with the knowledge needed to take advantage of these attractive properties, this chapter is written in a tutorial fashion and from a user's point of view, without the complicated theoretical derivations of the results. A more in-depth reference that contains the theory for much of the material presented here is the text by Anderson and Moore [1], while the text by Kwakernaak and Sivan [2] is considered to be the classic text on the subject.

In addition to the standard LQR results, robustness properties and useful variations of the LQR are also discussed. In particular, we describe how sensitivity weighted LQR (SWLQR) can be used to make high-gain LQR controllers robust to parametric modeling errors, as well as how to include frequency domain specifications in the LQR framework through the use of frequency weighted cost functionals. Since this chapter can be viewed as an introduction to optimal control for MIMO, linear, time-invariant systems, we briefly discuss the full state feedback $\mathcal{H}_\infty$ controller and its relation to a mini-max quadratic optimal control problem for completeness.

17.2 The Time-Invariant LQR Problem

To begin, we simply state the LQR problem, its solution, and the assumptions used in obtaining the solution. The rest of this chapter is devoted to discussing the properties of the controller, how the mathematical optimization problem relates to the physics of control systems, and useful variations of the standard LQR.

Theorem 17.1: Steady State LQR

Given the system dynamics

$$\dot{x}(t) = Ax(t) + Bu(t)\,; \quad x(t=0) = x_0 \tag{17.1}$$

with $x(t) \in \mathbb{R}^n$ and $u(t) \in \mathbb{R}^m$ along with a linear combination of states to keep small

$$z(t) = Cx(t) \tag{17.2}$$

with $z(t) \in \mathbb{R}^p$. We define a quadratic cost functional

$$J = \int_0^\infty \left[z^T(t)z(t) + u^T(t)Ru(t)\right] \mathrm{d}t \tag{17.3}$$

in which the size of the states of interest, $z(t)$, is weighted relative to the amount of control action in $u(t)$ through the weighting matrix R.

If the following assumptions hold:

1. *The entire state vector $x(t)$ is available for feedback*
2. *$[A \;\; B]$ is stabilizable and $[A \;\; C]$ is detectable*
3. $R = R^T > 0$

Then

1. *The linear quadratic controller is the unique, optimal, full state feedback control law*

$$u(t) = -Kx(t) \quad \text{with } K = R^{-1}B^TS \tag{17.4}$$

   that minimizes the cost, J, subject to the dynamic constraints imposed by the open-loop dynamics in Equation 17.1.
2. *S is the unique, symmetric, positive semidefinite solution to the algebraic Riccati equation*

$$SA + A^TS + C^TC - SBR^{-1}B^TS = 0 \tag{17.5}$$

3. *The closed-loop dynamics arrived at by substituting Equation 17.4 into Equation 17.1*

$$\dot{x}(t) = [A - BK]x(t) \tag{17.6}$$

   are guaranteed to be asymptotically stable.
4. *The minimum value of the cost J in Equation 17.3 is $J = x_0^TSx_0$.*

While the theorem states the LQR result, it is still necessary to discuss how to take advantage of it. Before doing so, note that the control gains, K from Equation 17.4, can be readily computed for large-order systems using standard software packages such as MATLAB® and MATRIXx™. Hence, the remainder of this chapter is really about how the values of the design variables used in the optimization problem influence the behavior of the controllers. Here the design variables are z, the states to keep small, and R, the control weighting matrix. We begin the discussion by describing the physical motivation behind the optimization problem.

17.2.1 Physical Motivation for the LQR

The LQR problem statement and cost can be motivated in the following manner. Suppose that the system Equation 17.1 is initially excited, and that the net result of this excitation is reflected in the initial state vector x_0. This initial condition can be regarded as an undesirable deviation from the equilibrium position of the system, $x(t) = 0$. Given these deviations, the objective of the control can essentially be viewed as selecting a control vector $u(t)$ that regulates the state vector $x(t)$ back to its equilibrium position of $x(t) = 0$ as quickly as possible.

If the system Equation 17.1 is controllable, then it is possible to drive $x(t)$ to zero in an arbitrarily short period of time. This would require very large control signals which, from an engineering point of view, are unacceptable. Large control signals will saturate actuators and if implemented in a feedback form will require high-bandwidth designs that may excite unmodeled dynamics. Hence, it is clear that there must be a balance between the desire to regulate perturbations in the state to equilibrium and the size of the control signals needed to do so.

Minimizing the quadratic cost functional from Equation 17.3 is one way to quantify our desire to achieve this balance. Realize that the quadratic nature of both terms in the cost

$$u^T(t)Ru(t) > 0 \quad \text{for} \quad u(t) \neq 0 \tag{17.7}$$

$$z^T(t)z(t) = x^T(t)C^TCx(t) \geq 0 \quad \text{for} \quad x(t) \neq 0 \tag{17.8}$$

ensures that they will be non-negative for all t. Further since the goal of the control law is to make the value of the cost as small as possible, larger values of the terms (Equations 17.7 and 17.8) are penalized more heavily than smaller ones. More specifically, the term (Equation 17.7) represents a penalty that helps the designer keep the magnitude of $u(t)$ "small." Hence the matrix R, which is often called the control weighting matrix, is the designer's tool which influences how "small" $u(t)$ will be. Selecting large values of R leads to small values of $u(t)$, which is also evident from the control gain K given in Equation 17.4.

The other term, Equation 17.8, generates a penalty in the cost when the states that are to be kept small, $z(t)$, are different from their desired equilibrium value of zero. The selection of which states to keep small, that is the choice of C in Equation 17.2, is the means by which the control system designer communicates to the mathematics the relative importance of individual state variable deviations. That is, which errors are bothersome and to what degree they are so.

Example 17.1:

Consider the single degree of freedom, undamped, harmonic oscillator shown in Figure 17.1. Using Newton's laws, the equations of motion are

$$m\ddot{q}(t) + kq(t) = u(t) \tag{17.9}$$

Letting $x_1(t) = q(t)$ denote the position of the mass and $x_2(t) = \dot{q}(t)$ its velocity, the dynamics (Equation 17.9) can be expressed in the state space as

$$\begin{aligned} \dot{x}_1(t) &= x_2(t) \\ \dot{x}_2(t) &= -\omega^2 x_1(t) + \omega^2 u(t) \end{aligned} \tag{17.10}$$

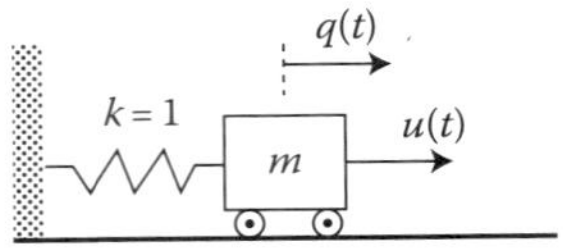

FIGURE 17.1 Single degree of freedom oscillator with mass m, spring stiffness $k = 1$, and control force input $u(t)$.

or

$$\dot{x}(t) = Ax(t) + Bu(t)$$

with

$$x(t) = \begin{bmatrix} x_1(t) \\ x_2(t) \end{bmatrix} \quad A = \begin{bmatrix} 0 & 1 \\ -\omega^2 & 0 \end{bmatrix} \quad B = \begin{bmatrix} 0 \\ \omega^2 \end{bmatrix}$$

where $\omega^2 = 1/m$ is the square of the natural frequency of the system with $k = 1$.

In the absence of control, initial conditions will produce a persistent sinusoidal motion of the mass at a frequency of ω rad/sec. As such, we seek a control law that regulates the position of the mass to its equilibrium value of $q(t) = 0$. Thus, we define the states of interest, z, as $x_1(t)$.

$$z(t) = Cx(t) \quad \text{with } C = \begin{bmatrix} 1 & 0 \end{bmatrix} \tag{17.11}$$

Since the control is scalar, the cost (Equation 17.3) takes the form

$$J = \int_0^\infty \left[z^2(t) + \rho u^2(t)\right] dt \quad \text{with } \rho > 0$$

where ρ is the control weighting parameter.

The optimal LQR control law takes the form $u(t) = -Kx(t)$ where the LQ gain K is a row vector $K = \begin{bmatrix} k_1 & k_2 \end{bmatrix}$. To determine the value of the gain, we must solve the algebraic Riccati equation 17.5. Recalling that we want a symmetric solution to the Riccati equation, letting

$$S = \begin{bmatrix} S_{11} & S_{12} \\ S_{12} & S_{22} \end{bmatrix} \tag{17.12}$$

and using the values of A, B, and C given above, the Riccati equation 17.5 becomes

$$0 = \begin{bmatrix} S_{11} & S_{12} \\ S_{12} & S_{22} \end{bmatrix} \begin{bmatrix} 0 & 1 \\ -\omega^2 & 0 \end{bmatrix} + \begin{bmatrix} 0 & -\omega^2 \\ 1 & 0 \end{bmatrix} \begin{bmatrix} S_{11} & S_{12} \\ S_{12} & S_{22} \end{bmatrix} + \begin{bmatrix} 1 \\ 0 \end{bmatrix} \begin{bmatrix} 1 & 0 \end{bmatrix}$$
$$- \frac{1}{\rho} \begin{bmatrix} S_{11} & S_{12} \\ S_{12} & S_{22} \end{bmatrix} \begin{bmatrix} 0 \\ \omega^2 \end{bmatrix} \begin{bmatrix} 0 & \omega^2 \end{bmatrix} \begin{bmatrix} S_{11} & S_{12} \\ S_{12} & S_{22} \end{bmatrix}$$

Carrying out the matrix multiplications leads to the following three equations in the three unknown S_{ij} from Equation 17.12.

$$0 = 2\omega^2 S_{12} - 1 + \frac{1}{\rho}\omega^4 S_{12}^2 \tag{17.13}$$

$$0 = -S_{11} + \omega^2 S_{22} + \frac{1}{\rho}\omega^4 S_{12} S_{22} \tag{17.14}$$

$$0 = -2S_{12} + \frac{1}{\rho}\omega^4 S_{22}^2 \tag{17.15}$$

Solving Equation 17.13 for S_{12} and simplifying yields

$$S_{12} = \frac{-\rho \pm \rho\sqrt{1 + 1/\rho}}{\omega^2}$$

Both the positive and negative choices for S_{12} are valid solutions of the Riccati equation. While we are only interested in the positive semidefinite solution, we still need more information to resolve which choice of S_{12} leads to the unique choice of $S \geq 0$. Rewriting Equation 17.15 as

$$2S_{12} = \frac{1}{\rho}\omega^4 S_{22}^2 \tag{17.16}$$

indicates that S_{12} must be positive to satisfy the equality of Equation 17.16, since the right-hand side of the equation must always be positive. Equation 17.16 indicates that there will also be a $\pm$ sign ambiguity in selecting the appropriate S_{22}. To resolve the ambiguity we use Sylvester's Test, which says that for $S \geq 0$ both

$$S_{11} \geq 0 \quad \text{and} \quad S_{11}S_{22} - S_{12}^2 \geq 0 \tag{17.17}$$

Solving Equation 17.15 and 17.13 using the relations in Equation 17.17, which clearly show that $S_{22} > 0$, gives the remaining elements of S

$$S_{11} = \frac{\rho}{\omega}\sqrt{2(1+1/\rho)\left(\sqrt{1+1/\rho}-1\right)}$$

$$S_{12} = \frac{-\rho + \rho\sqrt{1+1/\rho}}{\omega^2}$$

$$S_{22} = \frac{\rho}{\omega^3}\sqrt{2\left(\sqrt{1+1/\rho}-1v\right)}$$

The final step in computing the controller is to evaluate the control gains K from Equation 17.4. Doing so gives

$$k_1 = \sqrt{1+1/\rho}-1 \quad k_2 = \frac{1}{\omega}\sqrt{2\left(\sqrt{1+1/\rho}-1\right)}$$

Given the control gains as a function of the control weighting ρ it is useful to examine the locus of the closed-loop poles for the system as ρ varies over $0 < \rho < \infty$. Evaluating the eigenvalues of the closed-loop dynamics from Equation 17.6 leads to a pair of complex conjugate closed-loop poles

$$\lambda_{1,2} = -\frac{\omega}{2}\sqrt{2\left(\sqrt{1+1/\rho}-1\right)} \pm \jmath\frac{\omega}{2}\sqrt{2\left(\sqrt{1+1/\rho}+1\right)}.$$

A plot of the poles as ρ varies over $0 < \rho < \infty$ is shown in Figure 17.2. Notice that for large values of ρ, the poles are near their open-loop values at $\pm\jmath\omega$, and as ρ decreases the poles move farther out into the left half-plane. This is consistent with how ρ influences the cost. Large values of ρ place a heavy penalty on the control and lead to low gains with slow transients, while small values of ρ tell the mathematics that large control gains with fast transients are acceptable.

17.2.2 Designing LQR Controllers

While the above section presents all the formulas needed to start designing MIMO LQR controllers, the point of this section is to inform the potential user of the limitations of the methodology. The most restrictive aspects of LQR controllers is that they are full-state feedback controllers. This means that every state that appears in the model of the physical system Equation 17.1 must be measured by a sensor. In fact, the notation of $z(t) = Cx(t)$ for the linear combination of states that are to be regulated to zero is deliberate. We do not call $z(t)$ the outputs of the system because all the states $x(t)$ must be measured in real time to implement the control law Equation 17.4.

Full-state feedback is appropriate and can be applied to systems whose dynamics are described by a finite set of differential equations and whose states can readily be measured. An aircraft in steady, level flight is an example of a system whose entire state can be measured with sensors. In fact, LQR control has been used to design flight control systems for modern aircraft.

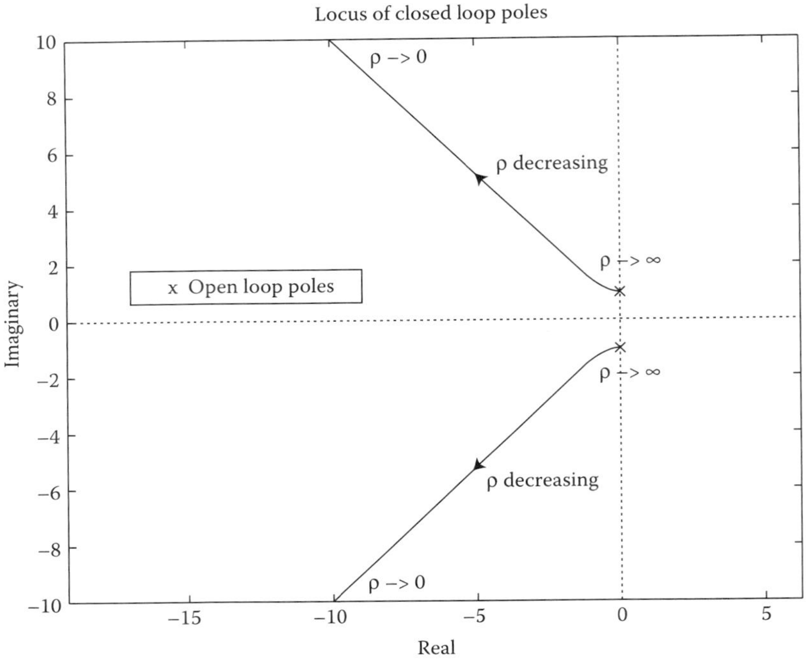

FIGURE 17.2 Locus of closed-loop pole locations for the single degree of freedom oscillator with $\omega = 1$ as ρ varies over $0 < \rho < \infty$.

On the other hand, full-state feedback is typically not appropriate for flexible systems. The dynamics of flexible systems are described by partial differential equations that often require very high-order, if not infinite-dimensional, state-space models. As such, it is not feasible to measure all the states of systems that possess flexible dynamics. Returning to the aircraft example, one could not use a LQR controller to regulate the aerodynamically induced vibrations of the aircraft's wing. The number of sensors that would be needed to measure the state of the vibrating wing prohibit this.

Having discussed the implications of full-state feedback, the next restrictive aspect of LQR to discuss is the gap between what the LQR controller achieves and the desired control system performance. Recall that the LQR is the control that minimizes the quadratic cost of Equation 17.3 subject to the constraints imposed by the system dynamics. This optimization problem and the resulting optimal controller have very little to do with more meaningful control system specifications like levels of disturbance rejection, overshoot in tracking, and stability margins. This gap must always be kept in mind when using LQR to design feedback controllers. The fact that LQR controllers are in some sense optimal is of no consequence if they do not meet the performance goals. Further since control system specifications are not given in terms of minimizing quadratic costs, it becomes the job of the designer to use the LQR tool wisely. To do so it helps to adopt a means-to-an-end design philosophy where the LQR is viewed as a tool, or the means, used to achieve the desired control system performance, or the end.

One last issue to point out is that LQR controller design is an iterative process even though the methodology systematically produces optimal, stabilizing controllers. Since the LQR formulation does not directly allow one to achieve standard control system specifications, trial and error iteration over the values of the weights in the cost is necessary to arrive at satisfactory controllers. Typically LQR designs are carried out by choosing values for the design weights, synthesizing the control law, evaluating how well the control law achieves the desired robustness and performance, and iterating through this process until a satisfactory controller is found. In the sequel, various properties and weight selection tools will be presented that provide good physical and mathematical guidance for selecting values for the LQR design variables. Yet these are only guides, and as such they will not eliminate the iterative nature of LQR controller design.

17.3 Properties of LQR

The LQR has several very important properties, which we summarize below. It is important to stress that the properties of LQR designs hinge upon the fact that full-state feedback is used and the specific way that the control gain matrix K is computed from the solution of the Riccati equation.

17.3.1 Robustness Properties

To visualize the robustness properties of LQR controllers, it is necessary to consider the loop transfer function matrix that results when implementing the control law of Equation 17.4. The LQR loop transfer function matrix, denoted by $G_{LQ}(s)$, induced by the control scheme of Equation 17.4 is given by

$$G_{LQ}(s) = K(sI - A)^{-1}B$$

and the closed-loop dynamics of Equation 17.6 using this representation are shown in Figure 17.3. An interesting fact is that $G_{LQ}(s)$ is always square and minimum phase. Note that as a consequence of this feedback architecture any unstructured modeling errors must be reflected to the inputs of the LQR loop, location 1 in Figure 17.3.

Under the *assumption that the control weight matrix $R = R^T > 0$ is diagonal,* the LQR loop transfer matrix is guaranteed to satisfy both of the inequalities

$$\begin{aligned}\sigma_{\min}\left[I + G_{LQ}(\jmath\omega)\right] &\geq 1 \quad \forall\omega \\ \sigma_{\min}\left[I + G_{LQ}(\jmath\omega)^{-1}\right] &\geq \frac{1}{2} \quad \forall\omega\end{aligned} \tag{17.18}$$

where $\sigma_{\min}$ denotes the minimum singular value. Since the multivariable robustness properties of any design depend on the size of $\sigma_{\min}\left[I + G(\jmath\omega)\right]$ and $\sigma_{\min}\left[I + G(\jmath\omega)^{-1}\right]$ the following guaranteed multivariable gain and phase margins are inherent to LQR controllers as a result of Equation 17.18.

17.3.1.1 LQR Stability Robustness Properties

1. Upward gain margin is infinite
2. Downward gain margin is at least 1/2
3. Phase margin is at least $\pm 60°$

These gain and phase margins can occur independently and simultaneously in all m control channels. To visualize this, consider Figure 17.4, where the $f_i(\cdot)$ can be viewed as perturbations to the individual inputs, $u_i = f_i(\cdot)\mu_i$. As a result of the gain margin properties (1) and (2), the LQR system shown in Figure 17.4

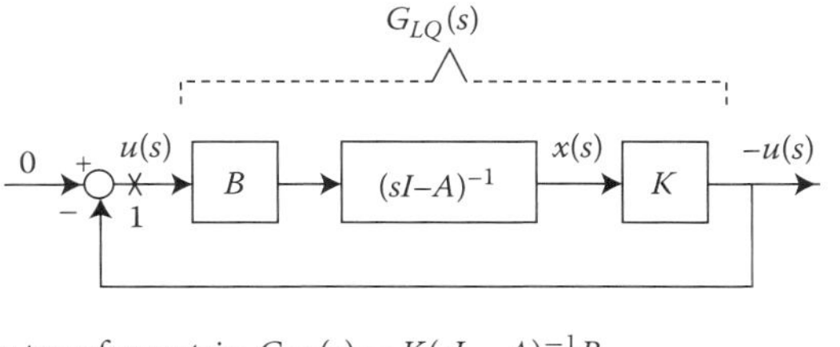

FIGURE 17.3 The LQR loop transfer matrix, $G_{LQ}(s) = K(sI - A)^{-1}B$.

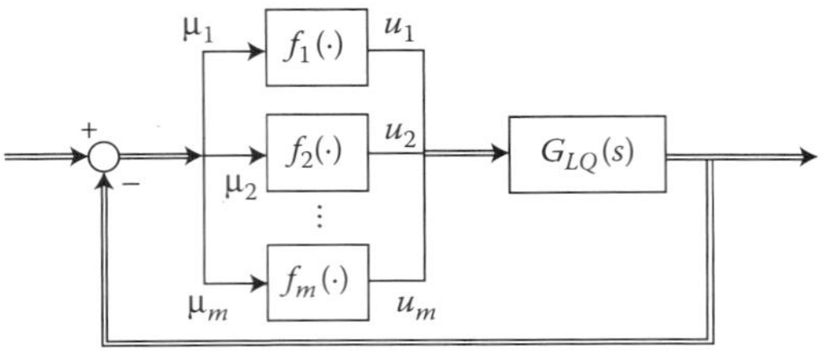

FIGURE 17.4 LQR loop used to visualize the guaranteed robustness properties.

is guaranteed to be stable for any set of scalar gains β_i with $f_i = \beta_i$ where the β_i lie anywhere in the range $1/2 < \beta_i < \infty$. The phase margin property (3) ensures the LQR system shown in Figure 17.4 is guaranteed to be stable for any set of scalar phases ϕ_i with $f_i = e^{\jmath\phi_i}$ where the ϕ_i can lie anywhere in the range $-60° < \phi_i < +60°$.

These inherent robustness properties of LQR designs are useful in many applications. To further appreciate what they mean, consider a single input system with a single variable we wish to keep small

$$\dot{x}(t) = Ax(t) + bu(t) \quad \text{with } z(t) = c^T x(t)$$

and let the control weight be a positive scalar, $R = \rho > 0$. Then the resulting LQR loop transfer function is scalar, and its robustness properties can be visualized by plotting the Nyquist diagram of $G_{LQ}(\jmath\omega)$ for all ω. Figure 17.5 contains a Nyquist plot for a scalar $G_{LQ}(\jmath\omega)$ that illustrates why the LQR obtains good gain and phase margins. Essentially, inequality Equation 17.18 guarantees that the Nyquist plot will not penetrate the unit circle centered at the critical point at $(-1, 0)$.

17.3.2 Asymptotic Properties of LQR Controllers

It is clear that the closed-loop poles of the LQR design will depend upon the values of the design parameters $z(t)$ and R. The exact numerical values of the closed-loop poles can only be determined using a digital computer, since we have to solve the Riccati equation 17.5. However, it is possible to qualitatively predict the asymptotic behavior of the closed-loop poles for the LQR as the size of the control gain is varied without solving the Riccati equation, evaluating the control gains, and computing the closed-loop poles.

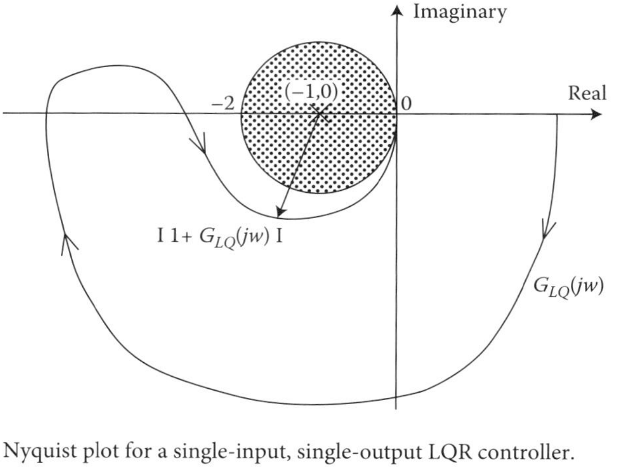

FIGURE 17.5 A typical Nyquist plot for a single-input, single-output LQR controller.

17.3.2.1 Assumptions for Asymptotic LQR Properties

1. The number of variables to keep small is equal to the number of controls. That is $\dim[z(t)] = \dim[u(t)] = m$.
2. The control weight is chosen such that $R = \rho\tilde{R}$ where ρ is a positive scalar and $\tilde{R} = \tilde{R}^T > 0$.
3. $G_z(s)$ is defined to be the square transfer function matrix between the variables we wish to keep small and the controls with the loop open, $z(s) = G_z(s)u(s)$ where $G_z(s) = C(sI - A)^{-1}B$, and q is the number of transmission zeros of $G_z(s)$.

Adjusting ρ directly influences the feedback control gain. When $\rho \to \infty$, we speak of the LQR controller as having low gain since under Assumption (2)

$$u(t) = -\frac{1}{\rho}\tilde{R}^{-1}B^T Sx(t) \to 0$$

as $\rho \to \infty$. Likewise, when $\rho \to 0$ we speak of the high gain controller since $u(t)$ will clearly become large. The asymptotic properties of the LQR closed-loop poles under the stated assumptions are as follows.

17.3.2.2 LQR Asymptotic Properties

1. *Low Gain:* As $\rho \to \infty$, the closed-loop poles start at
   a. The stable open-loop poles of $G_z(s)$
   b. The mirror image, about the $\jmath\omega$-axis, of any unstable open-loop poles of $G_z(s)$
   c. If any open-loop poles of $G_z(s)$ lie exactly on the $\jmath\omega$-axis, the closed-loop poles start just to the left of them
2. *High Gain:* As $\rho \to 0$, the closed-loop poles will
   a. Go to cancel any minimum phase zeros of $G_z(s)$
   b. Go to the mirror image, about the $\jmath\omega$-axis, of any non-minimum phase zeros of $G_z(s)$
   c. The rest will go off to infinity along stable Butterworth patterns

Realize that these rules are not sufficient for constructing the entire closed-loop pole root locus. However, they do provide good insight into the asymptotic behavior of LQR controllers. In particular, the second rule highlights how the choice of $z(t)$, which defines the zeros of $G_z(s)$, influences the closed-loop behavior of LQR controllers.

Extensions of these results exist for the case where Assumption (1) is not satisfied, that is when $\dim[z(t)] \neq \dim[u(t)]$. When $\dim[z(t)] \neq \dim[u(t)]$, the low gain asymptotic result will still hold, which is to be expected. The difficulty for non-square $G_z(s)$ arises in the high-gain case as $\rho \to 0$ because it is not a simple manner to evaluate the zeros that the closed-loop poles will go toward. For details of how to evaluate the zero locations that the closed-loop poles will go toward in the cheap control limit, one can refer to [3] and [2, *problem 3.14*].

17.4 LQR Gain Selection Tools

In this section we present some variations on the standard LQR problem given in Theorem 17.1. It is best to view the following variations on LQR as tools a designer can use to arrive at controllers that meet a set of desired control system specifications. While the tools are quite useful, the choice of which tool to use is very problem specific. Hence, we also present some physical motivation to highlight when the various tools might be useful. Realize that these tools do not remove the iterative nature of LQR control design. In fact, these tools are only helpful when used iteratively and when used with an understanding of the underlying physics of the design problem and the limitations of the design model. The variations on LQR presented here are by no means a complete list.

17.4.1 Cross-Weighted Costs

Consider the linear time-invariant system with dynamics described by the state equation in Equation 17.1 and the following set of variables we wish to keep small

$$z(t) = Cx(t) + Du(t) \tag{17.19}$$

Substituting Equation 17.19 into the quadratic cost from Equation 17.3 produces the cost function

$$J = \int_0^\infty \left\{ x^T(t)C^T Cx(t) + 2x^T(t)C^T Du(t) + u^T(t)\left[R + D^T D\right]u(t)\right\} \mathrm{d}t \tag{17.20}$$

which contains a cross penalty on the state and control, $2x^T(t)C^T Du(t)$. The solution to the LQR controller that minimizes the cost of Equation 17.20 will be presented here, but first we give some motivation for such a cost and choice of performance variables.

Cross-weighted cost functions and performance variables with control feed-through terms are common. The control feed-through term, $Du(t)$ in Equation 17.19, arises physically when the variables we wish to keep small are derivatives of the variables that appear in the state vector of the open-loop dynamics Equation 17.1. To see this, consider the single degree of freedom oscillator from Example 17.1 where the acceleration of the mass is the variable we wish to keep small, $z(t) = \ddot{q}(t)$. Using the definitions from the example and Equations 17.11, this can be expressed as

$$\begin{aligned} z(t) &= \ddot{q}(t) = \dot{x}_2(t) \\ &= -\omega^2 x_1(t) + \omega^2 u(t) = [-\omega^2 \;\; 0]x(t) + \omega^2 u(t) \end{aligned}$$

Clearly, penalizing acceleration requires the control feed-through term in the definition of the variables we wish to keep small. Control feed-through terms also arise when penalizing states of systems whose models contain neglected or unknown, higher-order modes. Such models often contain control feed-through terms to account for the low-frequency components of the dynamics of the modes that are not present in the design model. Further, as will be seen below, cross-coupled costs such as Equation 17.20 arise when frequency-dependent weighting terms are used to synthesize LQR controllers.

Theorem 17.2: Cross-Weighted LQR

Consider the system dynamics

$$\dot{x}(t) = Ax(t) + Bu(t) \quad x(t = 0) = x_0 \tag{17.21}$$

with $x(t) \in \mathbb{R}^n$ and $u(t) \in \mathbb{R}^m$ and the quadratic cost with a cross penalty on state and control

$$J = \int_0^\infty \left[x^T(t)R_{xx}x(t) + 2x^T(t)R_{xu}u(t) + u^T(t)R_{uu}u(t)\right] \mathrm{d}t \tag{17.22}$$

in which the size of the states is weighted relative to the amount of control action in $u(t)$ through the state weighting matrix R_{xx}, the control weighting matrix R_{uu}, and the cross weighting matrix R_{xu}.

If the following assumptions hold:

1. *The entire state vector $x(t)$ is available for feedback*
2. $R_{xx} = R_{xx}^T \geq 0$, $R_{uu} = R_{uu}^T > 0$, *and* $\begin{bmatrix} R_{xx} & R_{xu} \\ R_{xu}^T & R_{uu} \end{bmatrix} \geq 0$

3. $[A \quad B]$ *is stabilizable and* $[A \quad R_{xx}^{1/2}]$ *is detectable**

Then

1. *The linear quadratic controller is the unique, optimal, full-state feedback control law*

$$u(t) = -Kx(t) \quad \text{with } K = R_{uu}^{-1}\left(R_{xu}^{T} + B^{T}S\right) \tag{17.23}$$

that minimizes the cost, J of Equation 17.22, subject to the dynamic constraints imposed by the open-loop dynamics in Equation 17.21.

2. *Defining* $A_r = (A - BR_{uu}^{-1}R_{xu}^{T})$, *S is the unique, symmetric, positive semi-definite solution to the algebraic Riccati equation*

$$SA_r + A_r^{T}S + \left(R_{xx} - R_{xu}R_{uu}^{-1}R_{xu}^{T}\right) - SBR_{uu}^{-1}B^{T}S = 0 \tag{17.24}$$

3. *The closed-loop dynamics arrived at by substituting Equation 17.23 into Equation 17.21 are guaranteed to be asymptotically stable.*

Note that when the cost is defined by Equation 17.3 and the variables to be kept small are defined by Equation 17.19, $R_{xx} = C^{T}C, R_{xu} = C^{T}D$, and $R_{uu} = \left[R + D^{T}D\right]$ as can be seen from Equation 17.20. Also, if $R_{xu} = 0$ this theorem reduces to the standard LQR result presented in Theorem 17.1.

The assumptions used in Theorem 17.2 are essentially those used in the standard LQR result. One difference is that here we require $R_{xx} = R_{xx}^{T} \geq 0$, while in Theorem 17.1 this was automatically taken care of by our definition of the state cost as $x^{T}(t)C^{T}Cx(t)$ because $C^{T}C \geq 0$. Unfortunately, the guaranteed robustness properties presented in the previous section are not necessarily true for the LQR controllers that minimize the cost (Equation 17.22).

17.4.2 The Stochastic LQR Problem

In the stochastic LQR problem, we allow the inclusion of a white noise process in the state dynamics. Namely, the state vector $x(t)$ now satisfies the stochastic differential equation

$$\dot{x}(t) = Ax(t) + Bu(t) + L\xi(t) \tag{17.25}$$

where $\xi(t)$ is a vector valued, zero-mean, white noise process. We still assume that all the state variables can be measured exactly and used for feedback as necessary. Physically, the term $L\xi(t)$ in Equation 17.25 should be viewed as the impact of a persistent set of disturbances on a system that corrupt the desired equilibrium value of the state, $x(t) = 0$. The L matrix describes how the disturbances impact the system, and the stochastic process $\xi(t)$ captures the dynamics of how the disturbances evolve with time. In particular, since the disturbances are a white noise process, they are completely unpredictable from moment to moment.

As with the standard LQR problem, we would like to find the control law that minimizes some quadratic cost functional which captures a trade off between state cost and control cost. However, the cost functional for the standard LQR problem, Equation 17.3, cannot be used here as a result of the stochastic nature of the states. The value of the cost functional would be infinity because of the continuing excitation from $\xi(t)$. To accommodate the stochastic nature of the state, we seek the optimal control that minimizes the

* $R_{xx}^{1/2}$ is the matrix square root of R_{xx} which always exists for positive semidefinite matrices. For the cross-coupled cost in Equation 17.20, $R_{xx} = C^{T}C$ and $R_{xx}^{1/2} = C$.

expected value, or mean, of the standard LQR cost functional

$$J = \mathrm{E}\left\{\lim_{\tau\to\infty}\frac{1}{\tau}\int_0^{\tau}\left[z^T(t)z(t)+u^T(t)Ru(t)\right]\mathrm{d}t\right\} \tag{17.26}$$

where E is the expected value operator. The normalization by the integration time τ is necessary to ensure that the cost functional is finite. Surprisingly, the optimal control for this stochastic version of the LQR problem is identical to that of the corresponding deterministic LQR problem. In particular, Theorem 17.1 also describes the unique optimal full-state feedback control law that minimizes the cost Equation 17.26 subject to the constraints Equation 17.25. This is a result of the fact that $\xi(t)$ is a white noise process. Since the disturbances are completely unpredictable, the optimal thing to do is ignore them. As a consequence of all this, the properties of the deterministic LQR problem from Theorem 17.1 discussed in Section 17.3 are also true for this stochastic LQR problem.

Unfortunately, these results do not make it any easier to design LQR controllers to specifically reject disturbances that directly impact the state dynamics. This is because physical disturbances are typically not white in nature, and the stochastic LQR result only applies for white noise disturbances. However, these results do play a key role in synthesizing multivariable controllers when some of the states cannot be measured and fed back for control.

17.4.3 Sensitivity Weighted LQR

Synthesizing LQR controllers requires an accurate model of the system dynamics. Not only are the control gains, K, a function of the system matrices A and B, but in the high-gain limit we know that the closed-loop poles will go to cancel the zeros of $G_z(s)$. If the models upon which the control design is based are inaccurate, the closed-loop system at a minimum will not achieve the predicted performance and will, in the worst case, be unstable. This is particularly relevant for systems with lightly damped poles and zeros near the $j\omega$ axis. Hence, it is important to address the issue of stability robustness, since designers often must work with models that do not capture the dynamical behavior of their systems exactly.

Modeling errors can be classified into two main groups: unmodeled dynamics and parametric errors. Unmodeled dynamics are typically the result of unmodeled non-linear behavior and neglected high-frequency dynamics which lie beyond the control bandwidth of interest. How one can design LQR controllers that are robust to unmodeled dynamics will be discussed in the following section on frequency weighted LQR. Parametric uncertainty arises when there are errors in the values of the system parameters used to form the model. Sensitivity weighted LQR (SWLQR), is a variation on the standard LQR problem that can be used to increase the stability robustness of LQR controllers to parametric modeling errors. Before presenting the SWLQR results, we will discuss parametric modeling errors.

17.4.3.1 Parametric Modeling Errors

When parametric errors exist, the model will be able to capture the system's nominal dynamics, but the model cannot capture the exact behavior. Parametric errors occur often because it is difficult to know the exact properties of the physical devices that make up a system and because the physical properties may vary with time or the environment in which they are placed. For example, consider the single degree of freedom oscillator from Example 17.1 as the model of a car tire. While we might have some knowledge of the tire's stiffness, it is unrealistic to expect to know its exact value, as the tire is a complex physical device whose properties change with wear and temperature. Hence, the value of the spring stiffness k, which models the tire's elasticity, will be parametrically uncertain.

To deal with any sort of modeling error, one must formulate a model of the uncertainty in the nominal design model. Parametrically uncertain variables may appear directly as the elements of, or be highly nonlinear functions of, the elements of the A and B matrices. When the latter is the case, one often reduces the uncertainty model to parametric uncertainty in more fundamental system quantities such as the frequency and/or damping of the system's poles and zeros. In either case one must decide which elements of the model, be they pole frequencies or elements of the A matrix, are uncertain and by how much they are uncertain.

This digression on parametric modeling errors and the presentation of the SWLQR results that follows is vital. To achieve high levels of performance, which means using high-gain control, LQR requires an accurate plant model. If a model contains parametric modeling errors and high-gain LQR control ($\rho \to 0$) is used, then the closed-loop poles will go to cancel zeros of $G_z(s)$ which will be in different locations than the true zeros. Such errors often lead to instability, and this is why we worry about parametric modeling errors.

17.4.3.2 SWLQR

Designing feedback controllers that are robust to parametric uncertainty is a very difficult problem. Research on this problem is ongoing, and a great deal of research has gone into this problem in the past, see [4] for a survey on this topic. Here we present only one of the many methods, SWLQR, that can be used to design LQR controllers that are robust to parametric modeling errors. While the treatment is by no means complete, it should give one a sense of how to deal with parametric modeling errors when designing LQR controllers.

The philosophy behind SWLQR is that it aims, as the title implies, to desensitize the LQR controller to parametric uncertainty. The advantage of this philosophy is that it produces a technique that is nothing more than a special way to choose the weighting matrices of the cross-weighted LQR problem presented in Section 17.4.1. A shortcoming of the method is that it does not result in a controller that is guaranteed to be robust to the parametric uncertainty in the model.

In presenting the SWLQR results, the vector of uncertain parameters to which you wish to desensitize a standard LQR controller will be denoted by α, and α_i will denote the ith element of α. To desensitize LQR controllers to the uncertain variables in α, a quadratic penalty on the sensitivity of the state to each of the uncertain variables α_i,

$$\frac{\partial x^T(t)}{\partial \alpha_i} R_{\alpha\alpha_i} \frac{\partial x(t)}{\partial \alpha_i} \tag{17.27}$$

is added to the cross-weighted LQR cost functional (Equation 17.22). In this expression $\frac{\partial x(t)}{\partial \alpha_i}$ is known as the sensitivity state of the system and $R_{\alpha\alpha_i}$ is the sensitivity state weighting matrix, which must be positive semidefinite. Quadratic penalties for each α_i (Equation 17.27) are included in the LQR cost functional to tell the mathematics that we care not only about deviations in the state from its equilibrium position but that we also care about how sensitive the value of the state is to the known uncertain parameters of the system. The larger the value of $R_{\alpha\alpha_i}$, the more we tell the cost that we are uncertain about the influence of parameter α_i in our model on the trajectory of the state. In turn, the larger the uncertainty in the influence of parameter α_i on the state trajectory, the more robust the resulting SWLQR controller will be to the uncertain parameter α_i.

Making certain assumptions about the dynamics of the sensitivity states leads to the SWLQR result [6,7]. It is not necessary to understand or discuss these assumptions to design SWLQR controllers, and we thus simply present the SWLQR result.

The SWLQR Result

Given a stable open-loop system (Equation 17.21), let α denote the vector of uncertain variables to which you want to desensitize a LQR controller and define the sensitivity state weighting matrices $R_{\alpha\alpha_i}$ such that

$$R_{\alpha\alpha_i} = R_{\alpha\alpha_i}^T \geq 0 \quad \textit{where } R_{\alpha\alpha_i} \in \mathbb{R}^{n\times n}$$

Then the SWLQR controller results from synthesizing the cross-coupled cost LQR controller from Theorem 17.2 with the following set of design weights

$$R_{xx} = \tilde{R}_{xx} + \sum_{i=1}^{n_\alpha} \frac{\partial A^T}{\partial \alpha_i} A^{-T} R_{\alpha\alpha_i} A^{-1} \frac{\partial A}{\partial \alpha_i} \tag{17.28}$$

$$R_{xu} = \tilde{R}_{xu} + \sum_{i=1}^{n_\alpha} \frac{\partial A^T}{\partial \alpha_i} A^{-T} R_{\alpha\alpha_i} A^{-1} \frac{\partial B}{\partial \alpha_i} \tag{17.29}$$

$$R_{uu} = \tilde{R}_{uu} + \sum_{i=1}^{n_\alpha} \frac{\partial B^T}{\partial \alpha_i} A^{-T} R_{\alpha\alpha_i} A^{-1} \frac{\partial B}{\partial \alpha_i} \tag{17.30}$$

In these expressions n_α denotes the number of uncertain parameters in α, and $\tilde{R}_{xx}, \tilde{R}_{xu}$, and $\tilde{R}_{uu}$ are the nominal values of the weights from a standard, unrobust LQR design. The matrix derivatives $\partial A/\partial\alpha_i$ and $\partial B/\partial\alpha_i$ are taken term by term. For example, $\partial A/\partial\alpha_i$ is a matrix of the same size as A whose elements are the partial derivatives with respect to α_i of the corresponding elements in the A matrix. For

$$A = \begin{bmatrix} A_{11} & A_{12} & \cdots & A_{1n} \\ A_{21} & & & \\ \vdots & \ddots & & \\ A_{n1} & & & A_{nn} \end{bmatrix}$$

$$\frac{\partial A}{\partial \alpha_i} = \begin{bmatrix} \frac{\partial A_{11}}{\partial \alpha_i} & \frac{\partial A_{12}}{\partial \alpha_i} & \cdots & \frac{\partial A_{1n}}{\partial \alpha_i} \\ \frac{\partial A_{21}}{\partial \alpha_i} & & & \\ \vdots & \ddots & & \\ \frac{\partial A_{n1}}{\partial \alpha_i} & & & \frac{\partial A_{nn}}{\partial \alpha_i} \end{bmatrix}$$

While it might not be obvious that the modified cost expressions (Equations 17.28 through 17.30) lead to controllers that are more robust to the uncertain parameters in α, extensive empirical and experimental results verify that they do [7]. To see this mathematically, consider the state cost, $x^T(t)R_{xx}x(t)$ from Equation 17.28, when the variables we wish to keep small are defined by $z(t) = Cx(t)$,

$$\begin{aligned} x^T(t)R_{xx}x(t) &= x^T(t)C^TCx(t) + x^T(t)\frac{\partial A^T}{\partial \alpha_1} A^{-T} R_{\alpha\alpha_1} A^{-1} \frac{\partial A}{\partial \alpha_1} x(t) \\ &\quad + \cdots + x^T(t)\frac{\partial A^T}{\partial \alpha_{n_\alpha}} A^{-T} R_{\alpha\alpha_{n_\alpha}} A^{-1} \frac{\partial A}{\partial \alpha_{n_\alpha}} x(t) \end{aligned}$$

This expression can also be expressed as $z^T(t)z(t)$ with $z(t) = \bar{C}x(t)$ where $\bar{C}^T = [C^T \quad C_{\alpha_1}^T \quad \cdots \quad C_{\alpha_{n_\alpha}}^T]$ and C_{α_i} is the minimal order matrix square root of

$$\frac{\partial A^T}{\partial \alpha_i} A^{-T} R_{\alpha\alpha_i} A^{-1} \frac{\partial A}{\partial \alpha_i}.$$

From this we see that the SWLQR cost modifications essentially add new variables to the vector of variables we wish to keep small, $z(t)$. In addition to the standard variables we wish to keep small for performance defined by $Cx(t)$, we tell the cost to keep the variables $C_{\alpha_i}x(t)$ small, and these are related to the sensitivity of the state to parametrically uncertain variables. Furthermore with this new $z(t)$ vector, the zeros of $G_z(s)$

will change. As a result, in the high-gain limit as $\rho \to 0$ the closed-loop poles will not go to cancel the uncertain zeros defined by the original vector of variables we wish to keep small, $z(t) = Cx(t)$.

Selecting the values of the state sensitivity weights $R_{\alpha\alpha_i}$ is, as to be expected, an iterative process. Typically $R_{\alpha\alpha_i}$ is chosen to be either $\beta_i I$ or $\beta_i \tilde{R}_{xx}$ where β_i is a positive scalar constant. Both these choices simplify the selection of $R_{\alpha\alpha_i}$ by reducing the choice to a scalar variable. When using $R_{\alpha\alpha_i} = \beta_i \tilde{R}_{xx}$, one tends to directly trade off performance, reflected by the nominal state cost $x^T(t)\tilde{R}_{xx}x(t)$ with robustness to the parametric error now captured by the state cost

$$\beta_i x^T(t)\frac{\partial A^T}{\partial \alpha_i}A^{-T}\tilde{R}_{xx}A^{-1}\frac{\partial A}{\partial \alpha_i}x(t)$$

On the other hand, selecting $R_{\alpha\alpha_i} = \beta_i I$ is a way of telling the cost that you care about both performance and robustness since the various terms in the state cost are not directly related as when $R_{\alpha\alpha_i} = \beta_i \tilde{R}_{xx}$.

Example 17.2:

Consider the mass, spring, dashpot system shown in Figure 17.6. The four masses are coupled together by dampers with identical coefficients, $c_i = 0.05\ \forall i$, and springs with identical stiffnesses, $k_i = 1\ \forall i$. In this system, the mass m_3 is uncertain, but known to lie within the interval $0.25 \leq m_3 \leq 1.75$ while the values of m_1, m_2, and m_4 are all known to be 1. A unit intensity, zero-mean, white noise disturbance force $\xi(t)$ acts on m_4, and the control $u(t)$ is a force exerted on m_2. The performance variable of interest is the position of the tip mass m_4, so $z(t) = q_4(t)$. The design goals are to synthesize a controller which is robust to the uncertainty in m_3 and that rejects the effect of the disturbance source on the position of m_4. A combination of SWLQR and the stochastic LQR results from Section 17.4.2 will be used to meet these design goals.

Using Newton's laws, the dynamics of this system can be expressed as

$$\mathcal{M}\ddot{q}(t) + \mathcal{C}\dot{q}(t) + \mathcal{K}q(t) = T_1 u(t) + T_2\xi(t) \tag{17.31}$$

where $q^T(t) = [q_1(t)\ q_2(t)\ q_3(t)\ q_4(t)]^T$,

$$\mathcal{M} = \begin{bmatrix} m_1 & 0 & 0 & 0 \\ 0 & m_2 & 0 & 0 \\ 0 & 0 & m_3 & 0 \\ 0 & 0 & 0 & m_4 \end{bmatrix} \quad T_1 = \begin{bmatrix} 0 \\ 1 \\ 0 \\ 0 \end{bmatrix} \quad T_2 = \begin{bmatrix} 0 \\ 0 \\ 0 \\ 1 \end{bmatrix}$$

$$\mathcal{K} = \begin{bmatrix} k_1+k_2 & -k_2 & 0 & 0 \\ -k_2 & k_2+k_3 & -k_3 & 0 \\ 0 & -k_3 & k_3+k_4 & -k_4 \\ 0 & 0 & -k_4 & k_4 \end{bmatrix} \quad \mathcal{C} = \begin{bmatrix} c_1+c_2 & -c_2 & 0 & 0 \\ -c_2 & c_2+c_3 & -c_3 & 0 \\ 0 & -c_3 & c_3+c_4 & -c_4 \\ 0 & 0 & -c_4 & c_4 \end{bmatrix}$$

Then by defining the state vector as

$$x(t) = \begin{bmatrix} q(t) \\ \dot{q}(t) \end{bmatrix} \tag{17.32}$$

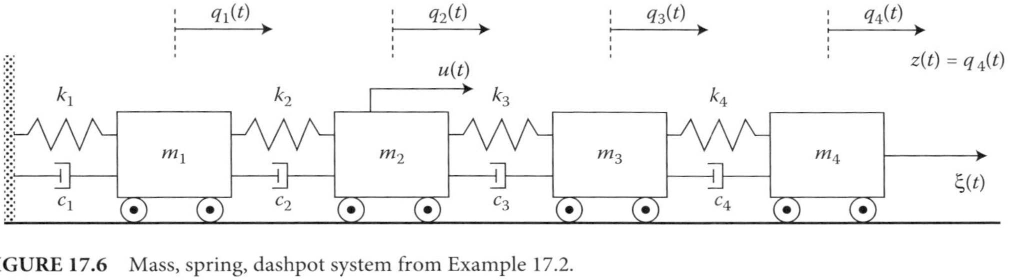

FIGURE 17.6 Mass, spring, dashpot system from Example 17.2.

the dynamics from Equation 17.31 can be represented in the state space as

$$\dot{x}(t) = Ax(t) + Bu(t) + L\xi(t)$$

with

$$A = \begin{bmatrix} 0 & I \\ -\mathcal{M}^{-1}\mathcal{K} & -\mathcal{M}^{-1}\mathcal{C} \end{bmatrix} \quad B = \begin{bmatrix} 0 \\ \mathcal{M}^{-1}T_1 \end{bmatrix} \quad L = \begin{bmatrix} 0 \\ \mathcal{M}^{-1}T_2 \end{bmatrix}$$

With the states defined by Equation 17.32, the performance variables we wish to keep small are

$$z(t) = Cx(t) \quad \text{where } C = \begin{bmatrix} 0 & 0 & 0 & 1 & 0 & 0 & 0 & 0 \end{bmatrix} \tag{17.33}$$

The first step in designing a SWLQR controller is to determine satisfactory values for the nominal design weights $\tilde{R}_{xx}$, $\tilde{R}_{xu}$, and $\tilde{R}_{uu}$. This can be done by ignoring the uncertainty and designing a LQR controller that achieves the desired system performance. Given the single-input, single-output (SISO) system with the desire to control the position of m_4, we find a nominal LQR controller by minimizing the stochastic LQR cost functional

$$J = \mathrm{E}\left\{ \lim_{\tau\to\infty} \frac{1}{\tau} \int_0^{\tau} \left[z^T(t)z(t) + \rho u(t)^2 \right] \mathrm{d}t \right\} \quad \rho > 0 \tag{17.34}$$

Substituting $z(t)$ from Equation 17.33 into the cost brings the integrand into the form of that in the cross-weighted cost from Equation 17.22 with $\tilde{R}_{xx} = C^T C$, $\tilde{R}_{xu} = 0$, and $\tilde{R}_{uu} = \rho$. Since m_3 is uncertain, we let it take on its median value of $m_3 = 1$ to synthesize actual controllers. After some iteration over the value of ρ, we decide that $\rho = .01$ leads to a nominal LQR controller that achieves good performance.

Figure 17.7 shows a set of transient responses for the mass spring system to an impulse applied to the disturbance source of the system. Responses for various values of the uncertain mass m_3 are shown, to illustrate how well the controllers behave in the presence of the parametric uncertainty. Notice that the LQR controller designed assuming $m_3 = 1$ is not stable over the entire range of possible values for m_3 and that even when the LQR controller is stable for values of $m_3 \neq 1$ the performance degrades. A better way to view the performance robustness of this controller to the uncertain mass in the presence of a white noise disturbance source is shown in Figure 17.8. This figure compares the root mean square (RMS) value of the tip mass position, $q_4(t)$, normalized by its open-loop RMS value as a function of the value of the uncertain mass m_3. Such plots are often called cost "buckets," for obvious reasons. When the closed-loop system is unstable, the cost blows up so that the wider the cost "bucket" the more robust the system will be to the parametric uncertainty. The cost bucket of Figure 17.8 clearly shows that the LQR controller is not robust to the parametric modeling error in m_3, since the cost "bucket" completely lies within the bounds on m_3.

Now to robustify this nominal LQR controller, we use the SWLQR modified costs from Equations 17.28 through 17.30. Here $\alpha = m_3$, $\frac{\partial B}{\partial m_3} = 0$ and

$$\frac{\partial A}{\partial m_3} = \begin{bmatrix} 0 & 0 \\ -\frac{\partial(\mathcal{M}^{-1})}{\partial m_3}\mathcal{K} & -\frac{\partial(\mathcal{M}^{-1})}{\partial m_3}\mathcal{C} \end{bmatrix}$$

where

$$\frac{\partial \mathcal{M}^{-1}}{\partial m_3} = \begin{bmatrix} 0 & 0 & 0 & 0 \\ 0 & 0 & 0 & 0 \\ 0 & 0 & -m_3^{-2} & 0 \\ 0 & 0 & 0 & 0 \end{bmatrix}$$

Using $R_{\alpha\alpha} = \beta I$ and $m_3 = 1$, we found that $\beta = 0.01$ led to a design that was significantly more robust than the LQR design. This can be seen in Figures 17.7 and 17.8 that compare the transient response and cost "buckets" of the SWLQR controller to the LQR controller. Both figures demonstrate that the SWLQR controller is stable over the entire range of possible values for m_3. As to be expected though, we do sacrifice some nominal performance to achieve the stability robustness.

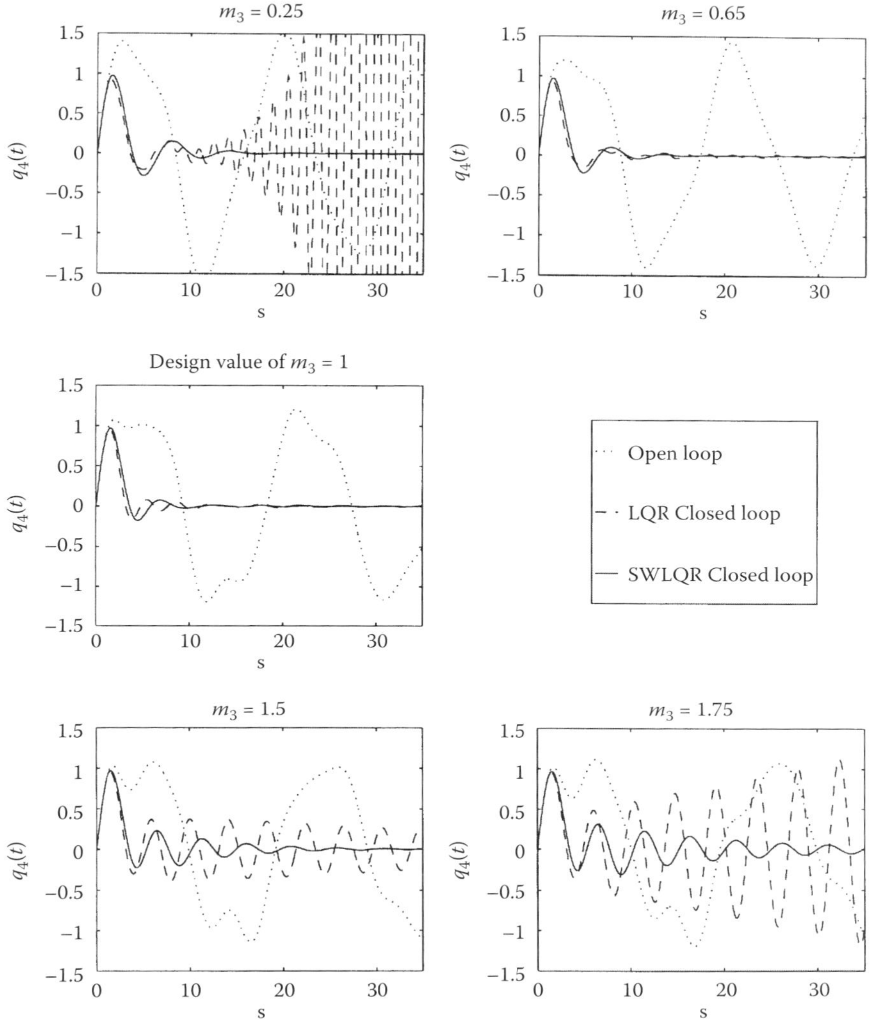

FIGURE 17.7 Transient response of the 4 mass, spring, dashpot system to an impulse applied at the disturbance source $\xi(t)$ for various values of m_3. The open-loop transients are shown with dots, the LQR transients are shown with a dashed line, and the SWLQR transients are shown with a solid line.

17.4.4 LQR with Frequency Weighted Cost Functionals

As discussed in Section 17.2, there is often a gap between what a LQR controller achieves and the desired control system performance. This gap can essentially be viewed as the discrepancy between the time domain optimization method, LQR, and the control system specifications. Often, both control system performance and robustness specifications can be expressed and analyzed in the frequency domain, but the time domain nature of the LQR problem makes it difficult to take advantage of these attributes. Using frequency weighted cost functionals to synthesize LQR controllers leads to a useful and practical variation on the standard LQR problem that narrows the gap between the time and frequency domains.

Recall that the quadratic cost functional is our means of communicating to the mathematics the desired control system performance. We've discussed the physical interpretation of the cost in the time domain, and now to understand it in the frequency domain we apply Parseval's Theorem to it.

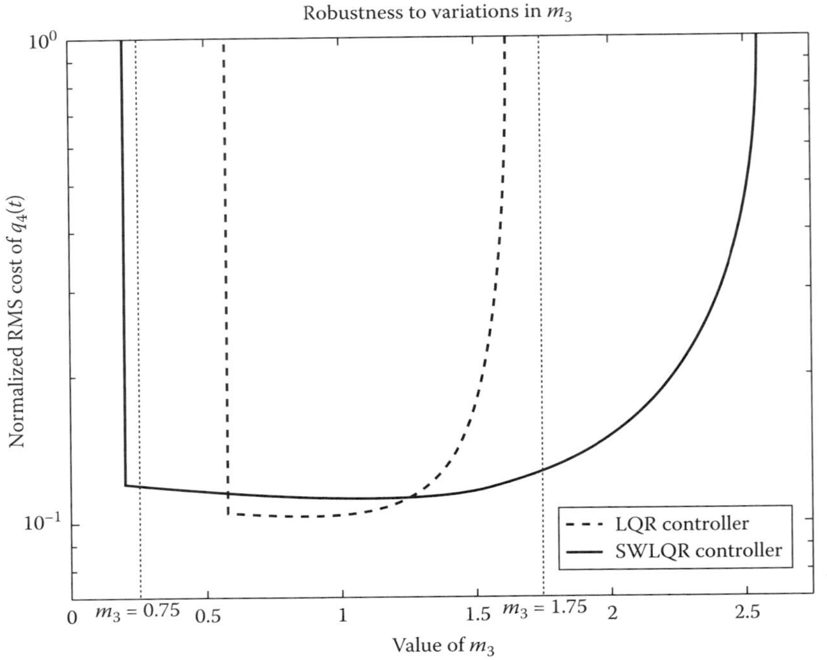

FIGURE 17.8 Comparison of cost "buckets" between the LQR and SWLQR controllers.

Theorem 17.3: Parseval's

For a vector valued signal $h(t)$ *defined on* $-\infty < t < +\infty$ *with*

$$\int_{-\infty}^{\infty} h^T(t)h(t)\,\mathrm{d}t < \infty$$

if we denote the Fourier transform of $h(t)$ *by* $h(j\omega)$ *then*

$$\int_{-\infty}^{\infty} h^T(t)h(t)\,\mathrm{d}t = \frac{1}{2\pi}\int_{-\infty}^{\infty} h^*(j\omega)h(j\omega)\,\mathrm{d}\omega$$

where $h^*(j\omega) = h^T(-j\omega)$ *is the complex conjugate transpose of* $h(j\omega)$.

Applying this theorem to the quadratic cost functional from Equation 17.3 with $R = \rho I$ and performance variables $z(t) = Cx(t)$ produces the cost

$$J = \frac{1}{2\pi}\int_{-\infty}^{\infty}\left[z^*(j\omega)z(j\omega) + \rho u^*(j\omega)u(j\omega)\right]\mathrm{d}\omega$$

Notice that the time domain weight, ρ, remains constant in the frequency domain. This illustrates that the standard LQR cost functional tells the mathematics to weight the control equally at all frequencies.

To influence the desired control system performance on a frequency by frequency basis, we employ the frequency weighted cost functional

$$J = \frac{1}{2\pi}\int_{-\infty}^{\infty}\left[z^*(\jmath\omega)W_1^*(\jmath\omega)W_1(\jmath\omega)z(\jmath\omega) + u^*(\jmath\omega)W_2^*(\jmath\omega)W_2(\jmath\omega)u(\jmath\omega)\right]\mathrm{d}\omega \tag{17.35}$$

in which both $W_1(\jmath\omega)$ and $W_2(\jmath\omega)$ are user-specified frequency weights [8]. Looking at the cost (Equation 17.35) for a SISO system in which the design weights will be scalar functions produces

$$J = \frac{1}{2\pi}\int_{-\infty}^{\infty}\left[|W_1(\jmath\omega)|^2|z(\jmath\omega)|^2 + |W_2(\jmath\omega)|^2|u(\jmath\omega)|^2\right]\mathrm{d}\omega$$

and helps to illustrate the benefit of using frequency weighted cost functionals. The frequency weights allow us to place distinct penalties on the state and control cost at various frequencies, which is not possible to do when constant weights are used. Their main advantage is that they facilitate the incorporation of known and desired control system behavior into the synthesis process. This is why frequency weighting plays a key role in modern control design [9]. If one knows over what frequency range it is important to achieve performance and where the control energy must be small, this knowledge can be reflected into the frequency weights. Since we seek to minimize the quadratic cost (Equation 17.35), large terms in the integrand incur greater penalties than small terms and more effort is exerted to make them small. For example, if there is a rigorous bandwidth constraint set by a region of high-frequency unmodeled dynamics and if the control weight $W_2(\jmath\omega)$ is chosen to have a large magnitude over this region then the resulting controller would not exert substantial energy in the region of unmodeled dynamics. This in turn would limit the controller's bandwidth. Similar arguments can be used to select $W_1(\jmath\omega)$ to tell the controller synthesis at what frequencies the size of the performance variables need to be small.

To synthesize LQR controllers that minimize the frequency weighted cost functional (Equation 17.35), it is necessary to transform the cost functional back into the time domain. Doing so requires state space realization of the frequency weights $W_1(\jmath\omega)$ and $W_2(\jmath\omega)$. Hence we let

$$z_1(s) = W_1(s)z(s) \quad \text{with } W_1(s) = C_1(sI - A_1)^{-1}B_1 + D_1 \tag{17.36}$$

$$z_2(s) = W_2(s)u(s) \quad \text{with } W_2(s) = C_2(sI - A_2)^{-1}B_2 + D_2 \tag{17.37}$$

Using these definitions and Parseval's Theorem, the cost (Equation 17.35) becomes

$$J = \frac{1}{2\pi}\int_{-\infty}^{\infty}\left[z_1^*(\jmath\omega)z_1(\jmath\omega) + z_2^*(\jmath\omega)z_2(\jmath\omega)\right]\mathrm{d}\omega = \int_{-\infty}^{\infty}\left[z_1^T(t)z_1(t) + z_2^T(t)z_2(t)\right]\mathrm{d}t \tag{17.38}$$

This cost can be brought into the form of an LQR problem that we know how to solve by augmenting the dynamics of the weights (Equations 17.36 and 17.37) to the open-loop dynamics (Equations 17.1 and 17.2). Using $x_1(t)$ and $x_2(t)$ to denote, respectively, the states of $W_1(s)$ and $W_2(s)$ to carry out the augmentation produces the system

$$\begin{aligned}\dot{\mathcal{X}}(t) &= \mathcal{A}\mathcal{X}(t) + \mathcal{B}u(t)\\ \mathcal{Z}(t) &= \mathcal{C}\mathcal{X}(t) + \mathcal{D}u(t)\end{aligned} \tag{17.39}$$

with

$$\mathcal{A} = \begin{bmatrix} A & 0 & 0\\ B_1C & A_1 & 0\\ 0 & 0 & A_2\end{bmatrix} \quad \mathcal{B} = \begin{bmatrix} B\\ 0\\ B_2\end{bmatrix} \quad \mathcal{C} = \begin{bmatrix} D_1C & C_1 & 0\\ 0 & 0 & C_2\end{bmatrix} \quad \mathcal{D} = \begin{bmatrix} 0\\ D_2\end{bmatrix}$$

where $\mathcal{X}^T(t) = [x^T(t)\ x_1^T(t)\ x_2^T(t)]$ and $\mathcal{Z}^T(t) = [z_1^T(t)\ z_2^T(t)]$ contains the outputs of $W_1(s)$ and $W_2(s)$. Notice that

$$\begin{aligned}\mathcal{Z}^T(t)\mathcal{Z}(t) &= z_1^T(t)z_1(t) + z_2^T(t)z_2(t)\\ &= \mathcal{X}^T(t)\mathcal{C}^T\mathcal{C}\mathcal{X}(t) + 2\mathcal{X}^T(t)\mathcal{C}^T\mathcal{D}u(t) + u^T(t)\mathcal{D}^T\mathcal{D}u(t)\end{aligned}$$

and thus the frequency weighted cost (Equation 17.38) becomes

$$J = \int_{-\infty}^{\infty} \left[\mathcal{X}^T(t)\mathcal{C}^T\mathcal{C}\mathcal{X}(t) + 2\mathcal{X}^T(t)\mathcal{C}^T\mathcal{D}u(t) + u^T(t)\mathcal{D}^T\mathcal{D}u(t)\right] \mathrm{d}t$$

Realize that minimizing this cost subject to the dynamic constraints (Equation 17.39) is not only equivalent to minimizing the frequency weighted cost functional (Equation 17.35), but it is also a simple manner to carry out the optimization by using the cross-weighted LQR results from Theorem 17.2 with

$$R_{xx} = \mathcal{C}^T\mathcal{C} \quad R_{xu} = \mathcal{C}^T\mathcal{D} \quad R_{uu} = \mathcal{D}^T\mathcal{D} \tag{17.40}$$

Applying Theorem 17.2 produces the control law

$$u(t) = -K\mathcal{X}(t) = -[K_x \ K_1 \ K_2] \begin{bmatrix} x(t) \\ x_1(t) \\ x_2(t) \end{bmatrix} \tag{17.41}$$

where $K = R_{uu}^{-1}\left(R_{xu}^T + \mathcal{B}^T S\right)$, S is the solution to the Riccati equation

$$S\left(\mathcal{A} - \mathcal{B}R_{uu}^{-1}R_{xu}^T\right) + \left(\mathcal{A}^T - R_{xu}R_{uu}^{-1}\mathcal{B}^T\right)S + \left(R_{xx} - R_{xu}R_{uu}^{-1}R_{xu}^T\right) - S\mathcal{B}R_{uu}^{-1}\mathcal{B}^T S = 0,$$

and the control weights are defined by Equation 17.40.

Clearly the augmented state vector $\mathcal{X}(t)$ must be fed back to implement the control law (Equation 17.41). Since $x_1(t)$ and $x_2(t)$ are the fictitious states of the weights $W_1(s)$ and $W_2(s)$, it is necessary to deliberately create them in real time to apply the feedback control law (Equation 17.41). That is, using frequency weighted cost functionals to synthesize LQR controllers leads to dynamic compensators. The control law is no longer a set of static gains multiplying the measurement of the state vector $x(t)$. It is a combination of the dynamics of the weighting matrices and the static gains K. A block diagram showing the feedback architecture for the dynamic LQR controller arrived at by minimizing the frequency weighted cost functional of Equation 17.35 is shown in Figure 17.9. Realize that having to implement a dynamic compensator is the price we pay for using frequency weights to incorporate known frequency domain information into the LQR controller synthesis. Though since the order of the compensator is only equal to the combined order of the design weights, this fact should not be troublesome.

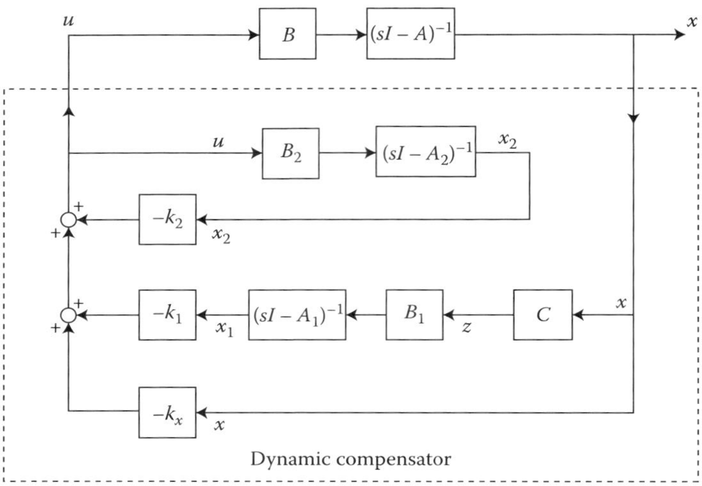

FIGURE 17.9 Block diagram for the LQR controller synthesized using the frequency weighted cost from Equation 17.35.

17.4.4.1 Selecting the Frequency Weights

Although we do not present the results here, it is possible to use any combination of frequency weights and constant weights when synthesizing LQR controllers with frequency weighted costs. For example, one could use a constant state penalty, $W_1(s) = I$, with a frequency weighted control. In any case, to satisfy the underlying LQR assumptions one must use stable design weights and ensure that $D_2^T D_2 > 0$ (so that R_{uu}^{-1} exists).

When synthesizing LQR controllers with frequency weighted cost functionals, the iteration over the values of the design weights involves selecting the dynamics of multivariable systems. Although it is not necessary, choosing the weighting matrices to be scalar functions multiplying the identity matrix,

$$W_1(s) = w_1(s)I \quad \text{and} \quad W_2(s) = w_2(s)I$$

simplifies the process of selecting useful weights. Then the process of selecting the weights reduces to selecting the transfer functions $w_1(s)$ and $w_2(s)$ so that their magnitudes have the desired effect on the cost.

To arrive at a useful set of weights, one must use one's knowledge of the physics of the system and the desired control system performance. Keeping this in mind, simply select the magnitude of $w_2(s)$ to be large relative to $|w_2(0)|$ over the frequency range where you want the control energy to be small and choose $w_1(s)$ to have a large magnitude relative to $|w_1(0)|$ over the frequency regions where you want the performance variables to be small. It is important to note that here large is relative to the values of the DC gains of the weights, $|w_1(0)|$ and $|w_2(0)|$. The DC gains specify the nominal penalties on the state and control cost in a manner similar to the constant weights of a standard LQR problem. As such, when choosing the frequency weights, it is beneficial to break the process up into two steps. The first step is to choose an appropriate DC gain to influence the overall characteristics of the controller, and the second step is to choose the dynamics so that the magnitudes of the weights reflect the relative importance of the variables over the various frequency ranges.

When choosing the frequency weights it is vital to keep in mind that the controls will have to be large over the frequency ranges where the performance variables are to be small. Likewise, if the control energy is specified to be small over a frequency range, it will not be possible to make the performance variables small there. Even though we could tell the cost to make both the performance variables and the control signals small in the same frequency range through the choice of the frequency weights, doing so will most likely result in a meaningless controller since we are asking the mathematical optimization problem to defy the underlying physics.

17.5 Mini-Max and $\mathcal{H}_\infty$ Full-State Feedback Control

Consider the problem of rejecting the effect of the disturbances, $d(t)$, on the performance variables of interest, $z(t)$, for the system

$$\begin{aligned} \dot{x}(t) &= Ax(t) + Bu(t) + Ld(t) \\ z(t) &= Cx(t) \end{aligned} \tag{17.42}$$

LQR control is not well suited to handle this problem because the optimal control that minimizes the quadratic cost (Equation 17.3) subject to the dynamic constraints (Equation 17.42) wants to know the future values of the disturbances, which is not realistic. The stochastic version of the LQR problem is also inappropriate unless $d(t)$ is white noise, which is rarely the case. To deal optimally with the disturbances using a full-state feedback controller, it is necessary to adopt a different philosophy than that of the LQR. Rather than treating the disturbances as known or white noise signals, they are assumed to behave in a "worst case" fashion. Treating the disturbances in this way leads to the so called $\mathcal{H}_\infty$ full-state feedback controller. $\mathcal{H}_\infty$ controllers have become as popular as LQR controllers in recent years as a result of their own attractive properties [10]. We introduce $\mathcal{H}_\infty$ controllers here using the quadratic cost functional optimization point of view.

The "worst case" philosophy for dealing with the disturbances arises by including them in the quadratic cost functional with their own weight, γ, much in the same way that the controls are included in the LQR cost functional. That is we seek to optimize the quadratic cost

$$J(u,d) = \frac{1}{2}\int_0^\infty \left[z^T(t)z(t) + \rho u^T(t)u(t) - \gamma^2 d^T(t)d(t)\right]\mathrm{d}t \quad \gamma, \rho > 0 \tag{17.43}$$

subject to the dynamic constraints of Equation 17.42. In this optimization problem both the controls and disturbances are the unknown quantities that the cost is optimized over. Note that since the disturbances enter the cost functional as a negative quadratic, they will seek to maximize $J(u,d)$. At the same time the controls seek to minimize $J(u,d)$ since they enter the cost functional as a positive quadratic. Hence, by using the cost (Equation 17.43), we are playing a mini-max differential game in which nature tries to maximize the cost through the choice of the disturbances, and we as control system designers seek to minimize the cost through the choice of the control $u(t)$. This mini-max optimization problem can be compactly stated as

$$\min_u \max_d J(u,d). \tag{17.44}$$

Since nature is allowed to pick the disturbances $d(t)$ which maximize the cost, this optimization problem deals with the disturbances by producing a control law that is capable of rejecting specific worst case disturbances.

The solution to the mini-max optimization problem (Equation 17.44) is not guaranteed to exist for all values of γ. When the solution does exist, it produces a full-state feedback control law similar in structure to the LQR controller. The solution to the mini-max differential game is summarized in the following theorem.

Theorem 17.4: Mini-Max Differential Game

Given the system dynamics

$$\dot{x}(t) = Ax(t) + Bu(t) + Ld(t) \tag{17.45}$$

with $x(t) \in \mathbb{R}^n$, $u(t) \in \mathbb{R}^m$, and $d(t) \in \mathbb{R}^q$ along with the performance variables we wish to keep small $z(t) = Cx(t)$ with $z(t) \in \mathbb{R}^p$, we define the mini-max quadratic cost functional

$$J(u,d) = \frac{1}{2}\int_0^\infty \left[z^T(t)z(t) + \rho u^T(t)u(t) - \gamma^2 d^T(t)d(t)\right]\mathrm{d}t \quad \gamma, \rho > 0$$

in which ρ and γ are user-specified design variables that weight the relative influence of the controls and disturbances. Under the following assumptions

1. *The entire state vector $x(t)$ is available for feedback*
2. *$d(t)$ is a deterministic, bounded energy signal with $\int_0^\infty d^T(t)d(t)\,\mathrm{d}t < \infty$*
3. *Both $[A \;\; B]$ and $[A \;\; L]$ are stabilizable, and $[A \;\; C]$ is detectable*

If the optimum value of the cost $J(u,d)$ constrained by the system dynamics (Equation 17.45) exists, it is a unique saddle point of $J(u,d)$ where

1. *The optimal mini-max control law is*

$$u(t) = -Kx(t) \quad \textit{with } K = \frac{1}{\rho}B^T S \tag{17.46}$$

2. *The optimal, worst case, disturbance is*

$$d(t) = \frac{1}{\gamma^2} L^T S x(t)$$

3. *S is the unique, symmetric, positive semidefinite solution of the matrix Riccati equation*

$$SA + A^T S + C^T C - S\left(\frac{1}{\rho} BB^T - \frac{1}{\gamma^2} LL^T\right) S = 0 \quad (17.47)$$

4. *The closed-loop dynamics for Equation 17.45 using Equation 17.46*

$$\dot{x}(t) = (A - BK)x(t) + Ld(t) \quad (17.48)$$

are guaranteed to be asymptotically stable.

If the solution to this optimization problem exists, it produces a stabilizing full-state feedback controller with the same structure as the LQR controller but with a different scheme for evaluating the feedback gains. Since the L matrix of Equation 17.45 appears in the Riccati equation 17.47, the mini-max control law directly incorporates the information of how the disturbances impact the system dynamics. The facts that the min-max controller guarantees a stable closed-loop and takes into consideration the nature of the disturbances make it an attractive alternative to LQR for synthesizing controllers.

17.5.1 Synthesizing Mini-Max Controllers

Mini-max controllers are not guaranteed to exist for arbitrary values of the design weight γ in the quadratic cost functional $J(u, d)$. Since γ influences the size of the $-\frac{1}{\gamma^2} LL^T$ term in the Riccati equation 17.47, there will exist values of $\gamma > 0$ for which there is either no solution to the Riccati equation or for which S will not be positive semidefinite. It turns out that there is a minimum value of γ, $\gamma_{\min}$, for which the mini-max optimization problem has a solution. Hence, useful values of γ will lie in the interval $\gamma_{\min} \leq \gamma < \infty$. Note that as $\gamma \to \infty$, the Riccati equation 17.47 becomes identical to the LQR one from Equation 17.5, and we recover the LQR controller. Likewise when $\gamma = \gamma_{\min}$, we have another special case which is known as the full-state feedback $\mathcal{H}_\infty$ controller*. For any other value of γ in $\gamma_{\min} \leq \gamma < \infty$ we still have an admissible stabilizing mini-max controller.

As with LQR, synthesizing mini-max controllers requires solving an algebraic Riccati equation. However, the presence of the $-\frac{1}{\gamma^2} LL^T$ term in the mini-max Riccati equation 17.47 makes the process of computing an $S = S^T \geq 0$ that satisfies Equation 17.47 more complicated than finding an $S = S^T \geq 0$ that satisfies the LQR Riccati equation 17.5. The reasons for this are directly related to the issue of whether or not a solution to the mini-max optimization problem exists.

To understand how one computes mini-max controllers, it is necessary to understand how current algebraic Riccati equation solvers work. While there is a rich theory for the topic, we summarize the key results in the following theorem.

Theorem 17.5: The Algebraic Riccati Equation

The Riccati equation

$$A^T S + SA + SVS - Q = 0 \quad (17.49)$$

* See Chapter 18 for details.

is solved by carrying out a spectral factorization of its associate Hamiltonian matrix*

$$H = \begin{bmatrix} A & V \\ Q & -A^T \end{bmatrix}$$

If $V = -BB^T, Q = -C^TC, [A \quad B]$ *is stabilizable, and* $[A \quad C]$ *is detectable then the Riccati equation solvers produce the unique, symmetric, positive semidefinite solution,* $S = S^T \geq 0$*, to the Riccati equation 17.49. Otherwise, as long as H has no* $j\omega$*-axis eigenvalues, the spectral factorization can be performed and a solution, S, which satisfies Equation 17.49 is produced.*

The Hamiltonian matrix for the mini-max Riccati equation, H_γ, is

$$H_\gamma = \begin{bmatrix} A & \frac{1}{\gamma^2}LL^T - \frac{1}{\rho}BB^T \\ -C^TC & -A^T \end{bmatrix} \tag{17.50}$$

The sign indefinite nature of the $1/\gamma^2 LL^T - 1/\rho BB^T$ term in H_γ makes it quite difficult to numerically test whether or not a solution to the mini-max optimization problem exists. Thus, a constructive algorithm based on the existing algebraic Riccati equation solvers is used to synthesize mini-max controllers.

Theorem 17.6: Algorithm for Computing Mini-Max Controllers

Pick a value of γ *and check to see if* H_γ *from Equation 17.50 has any* $j\omega$*-axis eigenvalues. If it does, increase* γ *and start over. If it does not, use an algebraic Riccati equation solver to produce a solution S to Equation 17.47. Test if* $S \geq 0$*. If it is not, increase gamma and start over. If* $S \geq 0$*, check to see if the closed-loop dynamics from Equation 17.48 are stable. If they are not, increase gamma and start over. If they are, you have constructed a mini-max controller, since you've found a* $S = S^T \geq 0$ *that satisfies the Riccati equation 17.47.*

Theoretically, the final step of the algorithm, which requires checking the closed-loop stability is not necessary. However, we highly recommend it since the numerical stability of the solution to the Riccati equations for values of γ near $\gamma_{\min}$ can be questionable.

From the algorithm for computing mini-max controllers, it can be seen that evaluating $\mathcal{H}_\infty$ controllers for a fixed value of ρ will require a trial-and-error search over γ. To compute $\gamma_{\min}$ it is best to use a bisection search over γ in the algorithm to find the smallest value of γ for which $S = S^T \geq 0$ and the closed-loop system (Equation 17.48) is stable. Current control system design packages such as MATLAB and MATRIXx employ such algorithms for computing the $\mathcal{H}_\infty$ controller that in turn determines $\gamma_{\min}$.

While the mini-max and $\mathcal{H}_\infty$ controllers are quite distinct from LQR controllers all the advice given for designing LQR controllers applies to mini-max controllers as well. Namely, it is necessary to iterate over the values of the design weights and independently check the robustness and performance characteristics for each design when synthesizing mini-max and $\mathcal{H}_\infty$ controllers. It can be shown through manipulating the Riccati equation 17.47 that the robustness properties of LQR controllers from Section 17.3 apply to $\mathcal{H}_\infty$ and mini-max controllers as well. Furthermore, modifications of the powerful sensitivity and frequency weighted LQR design tools from Section 17.4 do exist and can be used to incorporate stability robustness and known design specifications into the controller synthesis of mini-max and $\mathcal{H}_\infty$ controllers.

* Spectral factorizations are essentially eigenvalue decompositions.

References

1. Anderson, B.D.O. and Moore, J.B., *Optimal Control: Linear Quadratic Methods,* Prentice Hall, Englewood Cliffs, NJ, 1990.
2. Kwakernaak, H. and Sivan, R., *Linear Optimal Control Systems,* John Wiley & Sons, New York, 1972.
3. Emami-Naeini, A. and Rock, S.M., On asymptotic behavior of non-square linear optimal regulators, in *Proc. 23rd Conf. Decision and Control,* Las Vegas, NV, December 1984, pp.1762–1763.
4. Weinmann, A., *Uncertain Models and Robust Control,* Springer-Verlag, New York, 1991.
5. Grocott, S.C.O., Comparision of Control Techniques for Robust Performance on Uncertain Structural Systems, Master's thesis, Massachusetts Institute of Technology, 1994. MIT SERC report No. 2-94.
6. Sesak, J.R., Sensitivity Constrained Linear Optimal Control Analysis and Synthesis, Ph.D. thesis, University of Wisconsin, 1974.
7. Grocott, S.C.O., How, J.P., and Miller, D.W., A comparison of robust control techniques for uncertain structural systems, *Proc. AIAA Guidance Navigation and Control Conference,* Scottsdale, AZ, August 1994, pp. 261–271.
8. Gupta, N., Frequency-shaped cost functionals: extension of linear-quadratic-Gaussian methods, *J. Guidance Control Dynam.,* 3(6), 529–535, 1980.
9. Doyle, J., Francis, B., and Tannenbaum, A., *Feedback Control Theory,* Macmillan, New York, 1992.
10. Kwakernaak, H., Robust control and $\mathcal{H}_\infty$ optimization-tutorial paper, *Automatica,* 29(2), 255–273, 1993.

18

$\mathcal{H}_2$ (LQG) and $\mathcal{H}_\infty$ Control

Leonard Lublin
Massachusetts Institute of Technology

Simon Grocott
Massachusetts Institute of Technology

Michael Athans
Massachusetts Institute of Technology

18.1 Introduction

The fundamentals of output feedback $\mathcal{H}_2$ (linear quadratic Gaussian or LQG) and $\mathcal{H}_\infty$ controllers, which are the primary synthesis tools available for linear time-invariant systems, are presented in an analogous and tutorial fashion without rigorous mathematics. Since $\mathcal{H}_2$ and $\mathcal{H}_\infty$ syntheses are carried out in the modern control design paradigm, a review of the paradigm is presented, along with the definitions of the $\mathcal{H}_2$ and $\mathcal{H}_\infty$ norms and the methods used to compute them. The state-space formulae for the optimal controllers, under less restrictive assumptions than are usually found in the literature, are provided in an analogous fashion to emphasize the similarities between them. Rather than emphasizing the derivation of the controllers, we elaborate on the physical interpretation of the results and how one uses frequency weights to design $\mathcal{H}_\infty$ and $\mathcal{H}_2$ controllers. Finally, a simple disturbance rejection design for the longitudinal motion of an aircraft is provided to illustrate the similarities and differences between $\mathcal{H}_\infty$ and $\mathcal{H}_2$ controller synthesis.

18.2 The Modern Paradigm

$\mathcal{H}_2$ and $\mathcal{H}_\infty$ syntheses are carried out in the modern control paradigm. In this paradigm both performance and robustness specifications can be incorporated in a common framework along with the controller synthesis. In the modern paradigm, all of the information about a system is cast into the generalized block diagram shown in Figure 18.1 [1–3]. The generalized plant, P, which is assumed to be linear and time-invariant throughout this article contains all the information a designer would like to incorporate into the synthesis of the controller, K. System dynamics, models of the uncertainty in the system's dynamics, frequency weights to influence the controller synthesis, actuator dynamics, sensor dynamics, and implementation hardware dynamics from amplifiers, and analog-to-digital and digital-to-analog converters are all included in P. The inputs and outputs of P are, in general, vector valued signals. The sensor measurements that are used by the feedback controller are denoted y, and the inputs generated

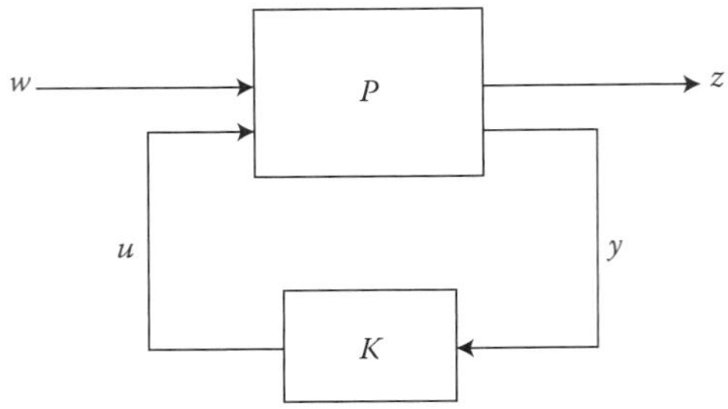

FIGURE 18.1 Generalized block diagram of the modern paradigm.

by the controller are denoted u. The components of w are all the exogenous inputs to the system. Typically these consist of disturbances, sensor noise, reference commands, and fictitious signals that drive frequency weights and models of the uncertainty in the dynamics of the system. The components of z are all the variables we wish to control. These include the performance variables of interest, tracking errors between reference signals and plant outputs, and the actuator signals which cannot be arbitrarily large and fast.

The general control problem in this framework is to synthesize a controller that will keep the size of the performance variables, z, small in the presence of the exogenous signals, w. For a classical disturbance rejection problem, z would contain the performance variables we wish to keep small in the presence of the disturbances contained in w that would tend to drive z away from zero. Hence, the disturbance rejection performance would depend on the "size" of the closed-loop transfer function from w to z, which we shall denote as $T_{zw}(s)$. This is also true for a command following control problem in which z would contain the tracking error that we would like to keep small in the presence of the commands in w that drive the tracking error away from zero.

Clearly then, the "size" of $T_{zw}(s)$ influences the effect that the exogenous signals in w have on z. Thus, in this framework, we seek controllers that minimize the "size" of the closed-loop transfer function $T_{zw}(s)$. Given that $T_{zw}(s)$ is a transfer function matrix, it is necessary to use appropriate norms to quantify its size. The two most common and physically meaningful norms that are used to classify the "size" of $T_{zw}(s)$ are the $\mathcal{H}_2$ and $\mathcal{H}_\infty$ norms. As such, we seek controllers that minimize either the $\mathcal{H}_2$ or $\mathcal{H}_\infty$ norm of $T_{zw}(s)$ in the modern control paradigm.

18.2.1 System Norms

Here we define and discuss the $\mathcal{H}_2$ and $\mathcal{H}_\infty$ norms of the linear, time-invariant, stable system with transfer function matrix

$$G(s) = C(sI - A)^{-1}B$$

This notation is meant to be general, and the reader should not think of $G(s)$ as only the actuator to sensor transfer function of a system. Realize that $G(s)$ is a system and thus requires an appropriate norm to classify its size. By a norm, we mean a positive, scalar number that is a measure of the size of $G(s)$ over all points in the complex s-plane. This is quite different from, for example, the maximum singular value of a matrix, $\sigma_{\max}[A]$, which is a norm that classifies the size of the matrix A.

The $\mathcal{H}_2$ Norm

Definition 18.1: $\mathcal{H}_2$ Norm

The $\mathcal{H}_2$ norm of $G(s)$, denoted $\|G\|_2$, is defined as

$$\|G\|_2 = \left(\frac{1}{2\pi}\int_{-\infty}^{\infty} \text{trace}\left[G(j\omega)G^*(j\omega)\right] d\omega\right)^{\frac{1}{2}} = \left(\frac{1}{2\pi}\int_{-\infty}^{\infty}\sum_{i=1}^{r}\sigma_i^2[G(j\omega)]d\omega\right)^{\frac{1}{2}}$$

where σ_i denotes the ith singular value, $G^*(j\omega)$ is the complex conjugate transpose of $G(j\omega)$, and r is the rank of $G(j\omega)$.

The $\mathcal{H}_2$ norm has an attractive, physically meaningful interpretation. If we consider $G(s)$ to be the transfer function matrix of a system driven by independent, zero mean, unit intensity white noise, u, then the sum of the variances of the outputs y is exactly the square of the $\mathcal{H}_2$ norm of $G(s)$. That is

$$\mathrm{E}\left[y^T(t)y(t)\right] = \|G(s)\|_2^2 \tag{18.1}$$

The $\mathcal{H}_2$ norm of $G(s)$ thus gives a precise measure of the "power" or signal strength of the output of a system driven with unit intensity white noise. Note that in the scalar case $\sqrt{\mathrm{E}[y^T(t)y(t)]}$ is the RMS or root mean squared value for $y(t)$ so the $\mathcal{H}_2$ norm specifies the RMS value of $y(t)$. A well-known fact for stochastic systems is that the mean squared value of the outputs can be computed by solving the appropriate Lyapunov equation [4]. As such, a state space procedure for computing the $\mathcal{H}_2$ norm of $G(s)$ is as follows [2].

Computing the $\mathcal{H}_2$ Norm

If L_c denotes the controllability Gramian of (A, B) and L_o the observability Gramian of (A, C), then

$$AL_c + L_cA^T + BB^T = 0 \quad A^TL_o + L_oA + C^TC = 0$$

and

$$\|G\|_2 = \left[\text{trace}(CL_cC^T)\right]^{\frac{1}{2}} = \left[\text{trace}(B^TL_oB)\right]^{\frac{1}{2}}$$

Note that this procedure for computing the $\mathcal{H}_2$ norm involves the solution of linear Lyapunov equations and can be done without iteration.

The $\mathcal{H}_\infty$ Norm

Definition 18.2: $\mathcal{H}_\infty$ Norm

The $\mathcal{H}_\infty$ norm of $G(s)$, denoted $\|G\|_\infty$, is defined as

$$\|G\|_\infty = \sup_\omega \sigma_{\max}[G(j\omega)]$$

In this definition "sup" denotes the supremum or least upper bound of the function $\sigma_{\max}[G(j\omega)]$, and thus the $\mathcal{H}_\infty$ norm of $G(s)$ is nothing more than the maximum value of $\sigma_{\max}[G(j\omega)]$ over all frequencies ω. The supremum must be used in the definition since, strictly speaking, the maximum of $\sigma_{\max}[G(j\omega)]$ may not exist even though $\sigma_{\max}[G(j\omega)]$ is bounded from above.

$\mathcal{H}_\infty$ norms also have a physically meaningful interpretation when considering the system $y(s) = G(s)u(s)$. Recall that when the system is driven with a unit magnitude sinusoidal input at a specific frequency, $\sigma_{\max}[G(\jmath\omega)]$ is the largest possible output size for the corresponding sinusoidal output. Thus, the $\mathcal{H}_\infty$ norm is the largest possible amplification over all frequencies of a unit sinusoidal input. That is, it classifies the greatest increase in energy that can occur between the input and output of a given system. A state space procedure for calculating the $\mathcal{H}_\infty$ norm is as follows.

Computing the $\mathcal{H}_\infty$ Norm

Let $\|G\|_\infty = \gamma_{\min}$. For the transfer function $G(s) = C(sI - A)^{-1}B$ with A stable and $\gamma > 0$, $\|G\|_\infty < \gamma$ if and only if the Hamiltonian matrix

$$H = \begin{bmatrix} A & \frac{1}{\gamma^2}BB^T \\ -C^TC & -A^T \end{bmatrix}$$

has no eigenvalues on the $\jmath\omega$-axis. This fact lets us compute a bound, γ, on $\|G\|_\infty$ such that $\|G\|_\infty < \gamma$. So to find $\gamma_{\min}$, select a $\gamma > 0$ and test if H has eigenvalues on the $\jmath\omega$-axis. If it does, increase γ. If it does not, decrease γ and recompute the eigenvalues of H. Continue until $\gamma_{\min}$ is calculated to within the desired tolerance.

The iterative computation of the $\mathcal{H}_\infty$ norm, which can be carried out efficiently using a bisection search over γ, is to be expected given that by definition we must search for the largest value of $\sigma_{\max}[G(\jmath\omega)]$ over all frequencies.

Note, the $\mathcal{H}_2$ norm is not an induced norm, whereas the $\mathcal{H}_\infty$ norm is. Thus, the $\mathcal{H}_2$ norm does not obey the submultiplicative property of induced norms. That is, the $\mathcal{H}_\infty$ norm satisfies

$$\|G_1G_2\|_\infty \leq \|G_1\|_\infty \|G_2\|_\infty$$

but the $\mathcal{H}_2$ norm does not have the analogous property. This fact makes synthesizing controllers that minimize $\|T_{zw}(s)\|_\infty$ attractive when one is interested in directly shaping loops to satisfy norm bounded robustness tests*. On the other hand, given the aforementioned properties of the $\mathcal{H}_2$ norm, synthesizing controllers that minimize $\|T_{zw}(s)\|_2$ is attractive when the disturbances, w, are stochastic in nature. In fact, $\mathcal{H}_2$ controllers are nothing more than linear quadratic Gaussian (LQG) controllers so the vast amount of insight into the well-understood LQG problem can be readily applied to $\mathcal{H}_2$ synthesis.

Example 18.1:

In this example the $\mathcal{H}_2$ and $\mathcal{H}_\infty$ norms are calculated for the simple four-spring, four-mass system shown in Figure 18.2. The equations of motion for this system can be found in Example 17.2 in Chapter 17. The system has force inputs on the second and fourth masses along with two sensors that provide a measure of the displacement of these masses. The singular values of the transfer function from the inputs to outputs, which we denote by $G(s)$, are shown in Figure 18.3. The $\mathcal{H}_\infty$ norm of the system is equal to the peak of $\sigma_1 = 260.4$, and the $\mathcal{H}_2$ norm of the system is equal to the square root of the sum of the areas under the square of each of the singular values, 14.5. Note that when considering the $\mathcal{H}_2$ norm, observing the log log plot of the transfer function can be very deceiving, since the integral is of σ_i, not $\log(\sigma_i)$, over ω, not $\log \omega$.

As pointed out in the example, the differences between the $\mathcal{H}_\infty$ and $\mathcal{H}_2$ norms for a system $G(s)$ are best viewed in the frequency domain from a plot of the singular values of $G(\jmath\omega)$. Specifically, the $\mathcal{H}_\infty$ norm is the peak value of $\sigma_{\max}[G(\jmath\omega)]$ while the $\mathcal{H}_2$ norm is related to the area underneath the singular values of $G(\jmath\omega)$. For a more in-depth treatment of these norms the reader is referred to [1,2,5,6].

* See Chapter 20 for a detailed exposition of this concept.

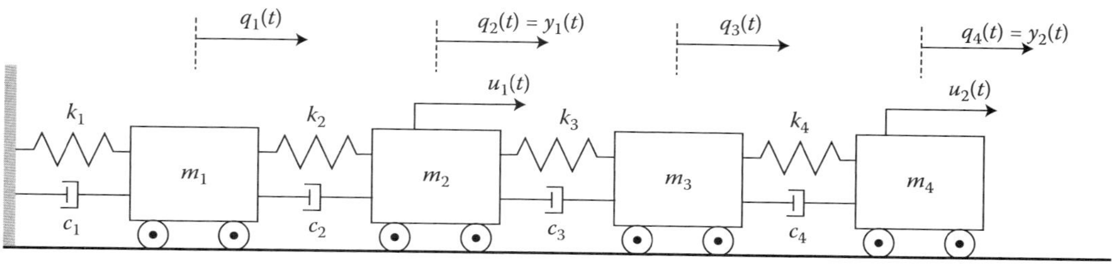

FIGURE 18.2 Mass, spring, dashpot system from Example 18.1. For the example $k_i = m_i = 1\ \forall i$, and $c_i = 0.05\ \forall i$.

18.3 Output Feedback $\mathcal{H}_\infty$ and $\mathcal{H}_2$ Controllers

Given that all the information a designer would like to include in the controller synthesis is incorporated into the system P, the synthesis of $\mathcal{H}_2$ and $\mathcal{H}_\infty$ controllers is quite straightforward. In this respect all of the design effort is focused on defining P. Below, we discuss how to define P using frequency weights to meet typical control system specifications. Here we simply present the formulas for the controllers.

All the formulas will be based on the following state-space realization of P,

$$P := \left[\begin{array}{c|cc} A & B_1 & B_2 \\ \hline C_1 & D_{11} & D_{12} \\ C_2 & D_{21} & D_{22} \end{array}\right]$$

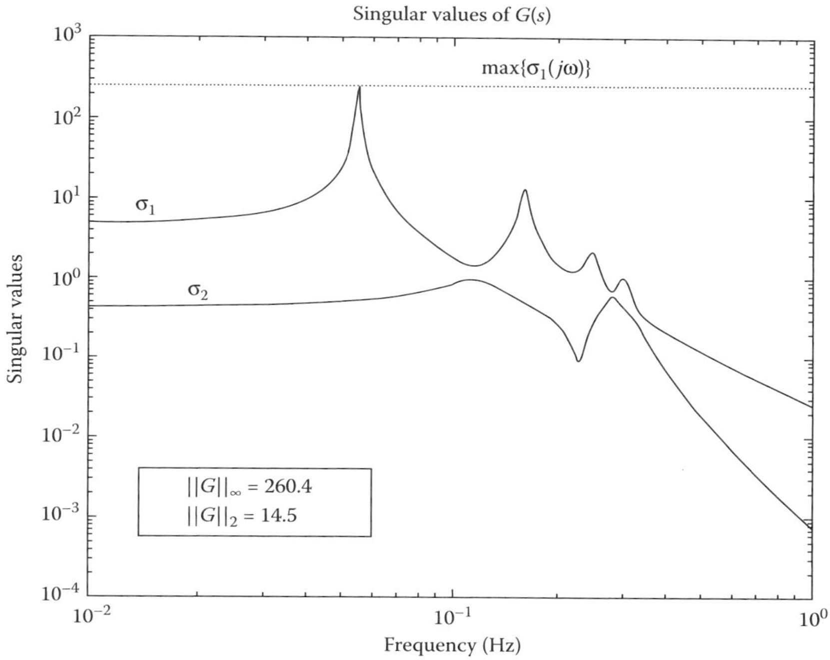

FIGURE 18.3 Singular values of the transfer function between the inputs and outputs of the mass, spring system shown in Figure 18.2.

This notation is a shorthand representation for the system of equations

$$\dot{x}(t) = Ax(t) + B_1 w(t) + B_2 u(t) \tag{18.2}$$

$$z(t) = C_1 x(t) + D_{11} w(t) + D_{12} u(t) \tag{18.3}$$

$$y(t) = C_2 x(t) + D_{21} w(t) + D_{22} u(t) \tag{18.4}$$

Additionally, the following assumptions concerning the allowable values for the elements of P are made.

Assumptions on P

1. $D_{11} = 0$ (A.1)
2. $[A \quad B_2]$ is stabilizable (A.2)
3. $[A \quad C_2]$ is detectable (A.3)
4. $V = \begin{bmatrix} B_1 \\ D_{21} \end{bmatrix} \begin{bmatrix} B_1^T & D_{21}^T \end{bmatrix} := \begin{bmatrix} V_{xx} & V_{xy} \\ V_{xy}^T & V_{yy} \end{bmatrix} \geq 0$ with $V_{yy} > 0$ (A.4)
5. $R = \begin{bmatrix} C_1^T \\ D_{12}^T \end{bmatrix} \begin{bmatrix} C_1 & D_{12} \end{bmatrix} := \begin{bmatrix} R_{xx} & R_{xu} \\ R_{xu}^T & R_{uu} \end{bmatrix} \geq 0$ with $R_{uu} > 0$ (A.5)

Assumption A.1 ensures that none of the disturbances feed through to the performance variables which is necessary for $\mathcal{H}_2$ synthesis but may be removed for $\mathcal{H}_\infty$ synthesis (see [7] for details.) Assumptions A.2 and A.3 are needed to guarantee the existence of a stabilizing controller while the remaining assumptions are needed to guarantee the existence of positive semidefinite solutions to the Riccati equations associated with the optimal controllers.

Theorem 18.1: $\mathcal{H}_2$ Output Feedback

Assuming that $w(t)$ is a unit intensity white noise signal, $E[w(t)w^T(\tau)] = I\delta(t-\tau)$, the unique, stabilizing, optimal controller which minimizes the $\mathcal{H}_2$ norm of $T_{zw}(s)$ is

$$K_2 := \left[\begin{array}{c|c} A + B_2F_2 + L_2C_2 + L_2D_{22}F_2 & -L_2 \\ \hline F_2 & 0 \end{array}\right] \tag{18.5}$$

where

$$\begin{aligned} F_2 &= -R_{uu}^{-1}\left(R_{xu}^T + B_2^T X_2\right) \\ L_2 &= -\left(Y_2 C_2^T + V_{xy}\right) V_{yy}^{-1} \end{aligned} \tag{18.6}$$

and X_2 and Y_2 are the unique, positive semidefinite solutions to the following Riccati equations

$$0 = X_2 A_r + A_r^T X_2 + R_{xx} - R_{xu} R_{uu}^{-1} R_{xu}^T - X_2 B_2 R_{uu}^{-1} B_2^T X_2 \tag{18.7}$$

$$0 = A_e Y_2 + Y_2 A_e^T + V_{xx} - V_{xy} V_{yy}^{-1} V_{xy}^T - Y_2 C_2^T V_{yy}^{-1} C_2 Y_2 \tag{18.8}$$

where

$$A_r = \left(A - B_2 R_{uu}^{-1} R_{xu}^T\right) \quad \text{and} \quad A_e = \left(A - V_{xy} V_{yy}^{-1} C_2\right)$$

Theorem 18.2: $\mathcal{H}_\infty$ Output Feedback [8]

Assuming that $w(t)$ is a bounded $\mathcal{L}_2$ signal, $\int_{-\infty}^{\infty} w^T(t)w(t)\,dt < \infty$, a stabilizing controller which satisfies $\|T_{zw}(\jmath\omega)\|_\infty < \gamma$ is

$$K_\infty := \left[\begin{array}{c|c} A & -Z_\infty L_\infty \\ \hline F_\infty & 0 \end{array}\right] \tag{18.9}$$

where

$$A_\infty = A + \left(B_1 + L_\infty D_{21}\right) W_\infty + B_2 F_\infty + Z_\infty L_\infty C_2 + Z_\infty L_\infty D_{22} F_\infty$$

where

$$F_\infty = -R_{uu}^{-1}\left(R_{xu}^T + B_2^T X_\infty\right) \quad W_\infty = \frac{1}{\gamma^2} B_1^T X_\infty$$

$$L_\infty = -\left(Y_\infty C_2^T + V_{xy}\right) V_{yy}^{-1} \quad Z_\infty = \left(I - \frac{1}{\gamma^2} Y_\infty X_\infty\right)^{-1}$$

and X_∞ and Y_∞ are the solutions to the following Riccati equations

$$0 = X_\infty A_r + A_r^T X_\infty + R_{xx} - R_{xu} R_{uu}^{-1} R_{xu}^T - X_\infty \left(B_2 R_{uu}^{-1} B_2^T - \frac{1}{\gamma^2} B_1 B_1^T\right) X_\infty \tag{18.10}$$

$$0 = A_e Y_\infty + Y_\infty A_e^T + V_{xx} - V_{xy} V_{yy}^{-1} V_{xy}^T - Y_\infty \left(C_2^T V_{yy}^{-1} C_2 - \frac{1}{\gamma^2} C_1^T C_1\right) Y_\infty \tag{18.11}$$

that satisfy the following conditions

1. $X_\infty \geq 0$
2. *The Hamiltonian matrix for Equation 18.10,*

$$\begin{bmatrix} A - B_2 R_{uu}^{-1} R_{xu}^T & -B_2 R_{uu}^{-1} B_2^T + \frac{1}{\gamma^2} B_1 B_1^T \\ -R_{xx} + R_{xu} R_{uu}^{-1} R_{xu}^T & -\left(A - B_2 R_{uu}^{-1} R_{xu}^T\right)^T \end{bmatrix}$$

   has no $\jmath\omega$-axis eigenvalues, or equivalently $A + B_1 W_\infty + B_2 F_\infty$ is stable
3. $Y_\infty \geq 0$
4. *The Hamiltonian matrix for Equation 18.11,*

$$\begin{bmatrix} \left(A - V_{xy} V_{yy}^{-1} C_2\right)^T & -C_2^T V_{yy}^{-1} C_2 + \frac{1}{\gamma^2} C_1^T C_1 \\ -V_{xx} + V_{xy} V_{yy}^{-1} V_{xy}^T & -A + V_{xy} V_{yy}^{-1} C_2 \end{bmatrix}$$

   has no $\jmath\omega$-axis eigenvalues, or equivalently $A + L_\infty C_2 + \frac{1}{\gamma^2} Y_\infty C_1^T C_1$ is stable
5. $\rho(Y_\infty X_\infty) < \gamma^2$, *where $\rho(\cdot) = \max_i |\lambda_i(\cdot)|$ is the spectral radius*

The (sub)optimal central $\mathcal{H}_\infty$ controller which minimizes $\|T_{zw}\|_\infty$ to within the desired tolerance is K_∞ with γ equal to the smallest value of $\gamma > 0$ that satisfies conditions 1 to 5.

Unlike the $\mathcal{H}_2$ controller, the $\mathcal{H}_\infty$ controller presented here is not truly optimal. Since there is no closed-form, state-space solution to the problem of minimizing the infinity norm of a multiple-input, multiple-output (MIMO) transfer function matrix $T_{zw}(s)$, the connections between the mini-max optimization problem

$$\inf_{u}\sup_{w}\int_{0}^{\infty}\left[z^{T}(t)z(t)-\gamma^{2}w^{T}(t)w(t)\right]dt \tag{18.12}$$

and $\mathcal{H}_\infty$ optimization are used to arrive at the constructive approach for synthesizing suboptimal $\mathcal{H}_\infty$ controllers given in Theorem 18.2. In fact, satisfying the conditions 1 to 5 of Theorem 18.2 is analogous to finding a saddle point of the optimization problem (Equation 18.12), and the search for $\gamma_{\min}$ is analogous to finding the global minimum over all the possible saddle points. As such, any value of $\gamma > \gamma_{\min}$ will also satisfy conditions 1 to 5 of Theorem 18.2, and thus produce a stabilizing controller. Such controllers are neither $\mathcal{H}_2$ nor $\mathcal{H}_\infty$ optimal. Since in the limit as $\gamma \to \infty$ the equations from Theorem 18.2 reduce to the equations for the $\mathcal{H}_2$ optimal controller, controllers with values of γ between $\gamma_{\min}$ and infinity provide a trade off between $\mathcal{H}_\infty$ and $\mathcal{H}_2$ performance. Along these lines, it is also worth noting that there is a rich theory for mixed $\mathcal{H}_2/\mathcal{H}_\infty$ controllers that minimize the $\mathcal{H}_2$ norm of $T_{zw}(s)$ subject to additional $\mathcal{H}_\infty$ constraints. See [9–11] for details.

The value of $w(t)$ that maximizes the cost in Equation 18.12 is known as the worst case disturbance, as it seeks to maximize the detrimental effect the disturbances have on the system. In this regard, $\mathcal{H}_\infty$ controllers provide optimal disturbance rejection to worst case disturbance, whereas the $\mathcal{H}_2$ controllers provide optimal disturbance rejection to stochastic disturbances.

Both $\mathcal{H}_2$ and $\mathcal{H}_\infty$ controllers are observer-based compensators [2], which can be seen from their block diagrams, shown in Figures 18.4 and 18.5. The regulator gains F_2 and F_∞ arise from synthesizing the full-state feedback controller, which minimizes the appropriate size of $z^T(t)z(t)$ constrained by the system dynamics Equation 18.2. Then the control law is formed by applying these regulator gains to an estimate of the states $x(t)$. The states, $x(t)$, are estimated using the noisy measurements of $y(t)$ from Equation 18.4, and L_2 and $Z_\infty L_\infty$ are the corresponding filter gains of the estimators.

In particular, F_2 is the full-state feedback LQR gain that minimizes the quadratic cost

$$J_{\mathrm{LQ}} = \mathrm{E}\left\{\lim_{\tau\to\infty}\frac{1}{\tau}\int_0^{\tau}\left[z^T(t)z(t)\right]dt\right\}$$

constrained by the dynamics of Equation 18.2, and L_2 is the Kalman filter gain from estimating the states x based on the measurements $y(t)$. Under the assumption that $z(t)$ is an ergodic process*

$$J_{\mathrm{LQ}} = \lim_{\tau\to\infty}\frac{1}{\tau}\int_0^{\tau} z^T(t)z(t)dt = \mathrm{E}\left[z^T(t)z(t)\right] = \|T_{zw}\|_2^2 \tag{18.13}$$

and this is exactly why $\mathcal{H}_2$ synthesis is nothing more than LQG control.

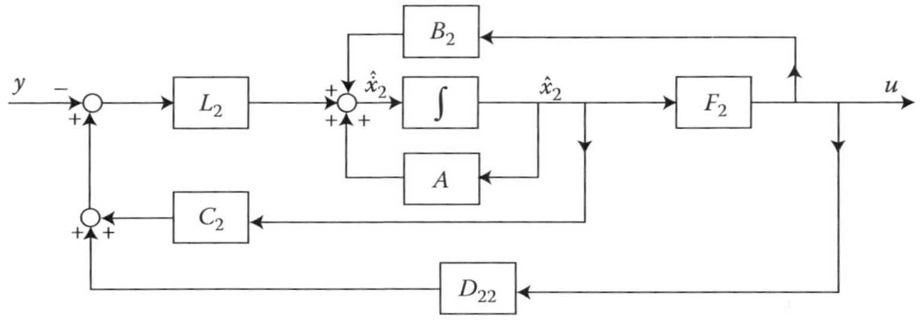

FIGURE 18.4 Block diagram of K_2 from Equation 18.9. Note, the Kalman Filter estimate of the states $x(t)$ from Equation 18.2, $\hat{x}_2(t)$, are the states of K_2.

* Assuming $z(t)$ is ergodic implies that its mean can be computed from the time average of a measurement of $z(t)$ as $t \to \infty$ [4].

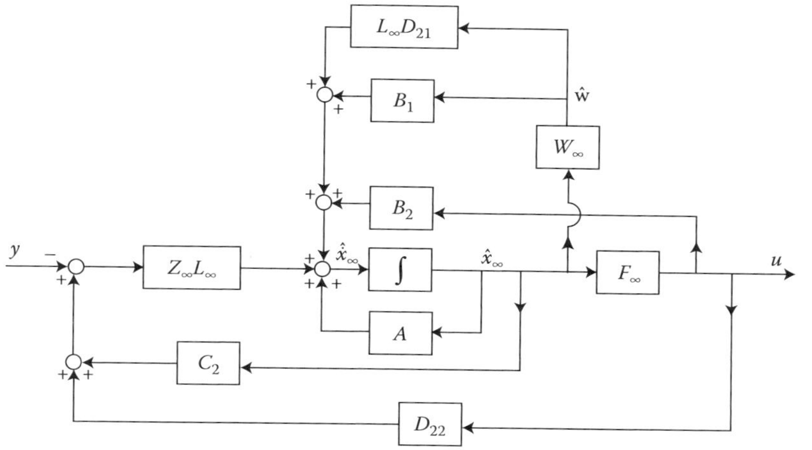

FIGURE 18.5 Block diagram of K_∞ from Equation 18.9. Note, the $\mathcal{H}_\infty$ optimal estimate of the states $x(t)$ from Equation 18.2, $\hat{x}_\infty(t)$, are the states of K_∞, and $\hat{w}(t)$ is an estimate of the worst case disturbance.

Analogously, F_∞ is the full-state feedback $\mathcal{H}_\infty$ control gain that results from optimizing the mini-max cost of Equation 18.12, and W_∞ is the full-state feedback gain that produces the worst case disturbance which maximizes the cost of Equation 18.12*. Unlike the Kalman filter in the $\mathcal{H}_2$ controller, the $\mathcal{H}_\infty$ optimal estimator must estimate the states of P in the presence of the worst case disturbance which is evident from the block diagram of K_∞ shown in Figure 18.5 [12]. This is why the filter gain of the $\mathcal{H}_\infty$ optimal estimator, $Z_\infty L_\infty$, is coupled to the regulator portion of the problem through X_∞ from Equation 18.10.

Since the $\mathcal{H}_2$ controller is an LQG controller, the closed-loop poles of $T_{zw}(s)$ separate into the closed-loop poles of the regulator, $\text{eig}(A - B_2F_2)$, and estimator, $\text{eig}(A - L_2C_2)$. A consequence of this separation property is that the $\mathcal{H}_2$ Riccati equations (Equations 18.7 and 18.8) can be solved directly without iteration. Since the worst case disturbance must be taken into consideration when synthesizing the $\mathcal{H}_\infty$ optimal estimator, the regulator and estimator problems in the $\mathcal{H}_\infty$ synthesis are coupled. Thus, the $\mathcal{H}_\infty$ controller does not have a separation structure that is analogous to that of the $\mathcal{H}_2$ controller. In addition, the $\mathcal{H}_\infty$ Riccati equations (Equations 18.10 and 18.11) are further coupled through the γ parameter, and we must iterate over the value of γ to find solutions of the $\mathcal{H}_\infty$ Riccati equations that satisfy conditions 1 to 5 of Theorem 18.2.

Note that in the literature the following set of additional, simplifying assumptions on the values of the elements of P are often made to arrive at less complicated sets of equations for the optimal $\mathcal{H}_\infty$ and $\mathcal{H}_2$ controllers [6,13,14].

Additional Assumptions on P	
1. $D_{22} = 0$	(No control feed-through term)
2. $C_1^T D_{12} = 0$	(No cross penalty on control and state)
3. $B_1 D_{21}^T = 0$	(Uncorrelated process and sensor noise)
4. $D_{12}^T D_{12} = I$	(Unity penalty on every control)
5. $D_{21}^T D_{21} = I$	(Unit intensity sensor noise on every measurement)

* See the section on $\mathcal{H}_\infty$ Full State Feedback in Chapter 17 for details.

18.4 Designing $\mathcal{H}_2$ and $\mathcal{H}_\infty$ Controllers

The results presented in the previous section are powerful because they provide the control system designer with a systematic means of designing controllers for systems whose entire state cannot be fed back. In order to take full advantage of these powerful tools, it is up to the designer to communicate to the optimization problems the desired control system performance and robustness. In the modern paradigm, this is done through the choice of the system matrix P. Since systems and their associated desired performance are diverse, there is no systematic procedure for defining P. However, by exploiting the rich mathematics of the $\mathcal{H}_2$ and $\mathcal{H}_\infty$ optimization problems along with their physical interpretations, it is possible to formulate guidelines for selecting appropriate systems P for a wide variety of problems.

Regardless of the synthesis employed, P will contain both the system model and the design weights used to communicate to the optimization the desired control system performance. Any linear interconnection of design weights and model can be selected so long as Assumptions A.1 through A.5 are satisfied. To satisfy the assumption that $R_{uu} > 0$, all of the control signals must appear explicitly in z. This is to be expected, since we cannot allow the synthesis to produce arbitrarily large control signals. Similarly, to ensure $V_{yy} > 0$, every measurement y must be corrupted by some sensor noise so as to avoid singular estimation problems.

Frequency-dependent weighting matrices are often included in P, since they provide greater freedom in telling the synthesis the desired control system performance. The synthesis of $\mathcal{H}_2$ and $\mathcal{H}_\infty$ controllers with frequency weights is just as straightforward as classical LQG synthesis with constant weights. Once the interconnection of the model with the defined performance variables, disturbances, and weights is specified, it is just a simple matter of state augmentation and block diagram manipulation to realize the state space form of P in Equations 18.2 through 18.4. Then given a state space representation of P, the formulas for the optimal controllers found in Theorems 18.1 and 18.2 can be used. The ability to use any admissible system interconnection with any combination of frequency weights is a direct consequence of the fact that we build an estimator for the entire state of P into the controllers. As such, the dynamics of any frequency weights will be reflected in the compensator whose order will be the same as that of P.

In either the $\mathcal{H}_2$ or $\mathcal{H}_\infty$ framework, arriving at a satisfactory design will involve iteration over the values of the frequency weights. Thus, it is vital to have an in-depth understanding of the dynamics of the system when choosing the system interconnection and the values for the design variables.

18.4.1 $\mathcal{H}_2$ Design

Given that the $\mathcal{H}_2$ optimal controller is an LQG controller, it is useful to adopt a stochastic framework and use the insights afforded by the well-known LQG problem when selecting P for $\mathcal{H}_2$ synthesis, see [13,15]. In this respect, $w(t)$ must contain both the process and sensor noises, while $z(t)$ must contain linear combinations of both the states and controls. Furthermore, the system P must be comprised of the system model and all the design weights such as the noise intensities and the state and control weighting matrices. For example, Figure 18.6 illustrates a possible system interconnection for the classical LQG problem of minimizing a weighted sum of state and control penalties given a system whose dynamics

$$\begin{aligned}\dot{x}(t) &= Ax(t) + Bu(t) + Ld(t) \\ y(t) &= Cx(t) + v(t)\end{aligned} \tag{18.14}$$

are driven by the uncorrelated stochastic disturbances, $d(t)$, and sensor noise, $v(t)$.

In the interconnection of Figure 18.6, W_i are weighting matrices, or design variables, that the designer selects. For the classical LQG problem, all the W_i are constant matrices. Since $w^T(t) = [w_1^T(t) \quad w_2^T(t)]$ must be a unit intensity, white noise process, W_1 and W_2 are the matrix square roots of the intensity

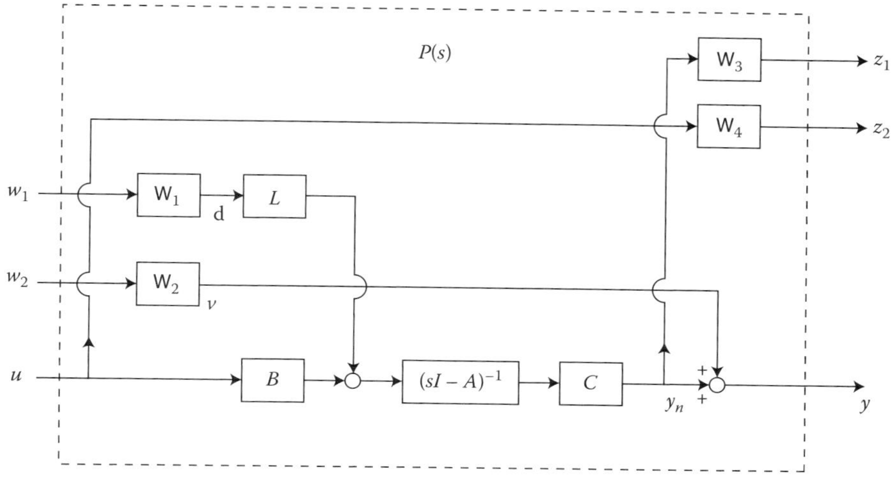

FIGURE 18.6 Block diagram interconnection for a typical $P(s)$.

matrices for the process and sensor noises d and v such that

$$\mathrm{E}\left\{\begin{bmatrix} d(t) \\ v(t) \end{bmatrix}\begin{bmatrix} d^T(\tau) & v^T(\tau) \end{bmatrix}\right\} = \begin{bmatrix} \mathsf{W}_1\mathsf{W}_1^T & 0 \\ 0 & \mathsf{W}_2\mathsf{W}_2^T \end{bmatrix}\delta(t-\tau)$$

As for the performance variable weights, W_3 is a weight on the outputs that produces a particular state weighting, and W_4 is the matrix square root of the control weighting matrix. These define the cost J_{LQ} from Equation 18.13 to be

$$J_{\mathrm{LQ}} = \mathrm{E}\left\{\lim_{\tau\to\infty}\frac{1}{\tau}\int_0^\tau \left[x^T(t)C^T\mathsf{W}_3^T\mathsf{W}_3Cx(t) + u^T(t)\mathsf{W}_4^T\mathsf{W}_4u(t)\right]dt\right\}$$

A drawback of classical LQG synthesis is that the weighting matrices are constant and thus limit our ability to place distinct penalties on the disturbances and performance variables at various frequencies. When synthesizing $\mathcal{H}_2$ controllers the weights W_i can, in general, be functions of frequency. Since performance and robustness specifications are readily visualized in the frequency domain, using frequency weights provides much more freedom in telling the optimization problem the desired control system behavior.

When choosing the values of the frequency weights, one should use the fact that $\mathcal{H}_2$ synthesis is equivalent to LQG synthesis. Any frequency weights that appear on the performance variables can be chosen using the insights afforded by the LQR problem with frequency weighted cost functionals as a result of Equation 18.13*. In brief, one uses Parseval's Theorem to arrive at a frequency domain representation of J_{LQ} from Equation 18.13. For the system interconnection shown in Figure 18.6 with scalar frequency weights $\mathsf{W}_3(s) = \mathsf{w}_3(s)I$ and $\mathsf{W}_4(s) = \mathsf{w}_4(s)I$

$$J_{\mathrm{LQ}} \approx \frac{1}{2\pi}\int_{-\infty}^{\infty}\left[|\mathsf{w}_3(\jmath\omega)|^2 y_n^*(\jmath\omega)y_n(\jmath\omega) + |\mathsf{w}_4(\jmath\omega)|^2 u^*(\jmath\omega)u(\jmath\omega)\right]d\omega$$

From this expression of the $\mathcal{H}_2$ cost, it is clear that the weights should be chosen to have a large magnitude over the frequencies where we want the outputs, y_n, and controls, u, to be small.

Frequency weights that appear on the disturbance signals should be viewed as shaping filters that specify the spectral content of the process and sensor noises. The values of the weights can then be chosen

* Section 4.4 in Chapter 17 has a detailed exposition of this.

to capture the true spectral content of the disturbances, as $w(t)$ must be unit intensity white noise to apply Theorem 18.2, or they can be chosen to influence the controller to produce some desired behavior. For example, in the system shown in Figure 18.6 if $\mathsf{W}_1(s) = \mathsf{w}_1(s)I$, then the control will work hard to reject the disturbances, d, over the frequencies where $|\mathsf{w}_1(j\omega)|$ is large. Likewise, if $\mathsf{W}_2(s) = \mathsf{w}_2(s)I$ and $|\mathsf{w}_2(j\omega)|$ is large over a particular frequency range, then the controller will not exert much effort there because we are telling the synthesis that the sensor measurements are very noisy there.

18.4.2 $\mathcal{H}_\infty$ Design

In the $\mathcal{H}_\infty$ framework it is possible to use loop shaping, see [1], to achieve performance and robustness specifications that can be expressed in the frequency domain. This is due to the fact that

$$\|T_{zw}\|_\infty < \gamma \;\Rightarrow\; \|\left(T_{zw}\right)_{ij}\|_\infty < \gamma \quad \forall i,j \tag{18.15}$$

where $\left(T_{zw}\right)_{ij}$ denotes the closed-loop transfer function matrix between exogenous disturbance w_j and performance variable z_i. To take advantage of Equation 18.15, it is necessary to define P so that the closed-loop transfer function matrices we wish to shape appear directly in $T_{zw}(s)$ and are multiplied by frequency-dependent design weights.

These concepts are best illustrated through an example. Consider the system Equation 18.14, which can be represented in the frequency domain as

$$\begin{aligned} y(s) &= G_1(s)d(s) + G_2(s)u(s) + v(s) \\ G_1(s) &= C(sI - A)^{-1}L \\ G_2(s) &= C(sI - A)^{-1}B \end{aligned}$$

where the disturbances are now considered to be unknown but bounded $\mathcal{L}_2$ signals. Suppose that we are interested in designing a controller that rejects the effect of the disturbances $d(t)$ on the outputs $y(t)$ and that is robust to an unstructured additive error in the input to output system model. Then it is necessary to independently shape the closed-loop transfer function between d and y, $S(s)G_1(s)$, and the closed-loop transfer function $K(s)S(s)$. In particular, we require $S(s)G_1(s)$ to have a desirable shape, and we need to satisfy the standard additive error stability robustness test

$$\sigma_{\max}[K(j\omega)S(j\omega)] < \frac{1}{|e_a(j\omega)|}$$

where $S(s) = [I - G_2(s)K(s)]^{-1}$ and $e_a(s)$ is a transfer function whose magnitude bounds the additive error *. If $\mathsf{W}_1 = I$ and $\mathsf{W}_2 = I$, then the system interconnection shown in Figure 18.6 is suitable for designing $\mathcal{H}_\infty$ controllers that achieve the loop-shaping objectives, because

$$T_{zw}(s) = \begin{bmatrix} \mathsf{W}_3 S(s)G_1(s) & \mathsf{W}_3 C(s) \\ \mathsf{W}_4 K(s)S(s)G_1(s) & \mathsf{W}_4 K(s)S(s) \end{bmatrix} \tag{18.16}$$

where $C(s) = S(s)G_2(s)K(s)$.

Notice that both of the loops of interest, $S(s)G_1(s)$ and $K(s)S(s)$, appear directly in Equation 18.16 multiplied by the design weights. By selecting scalar frequency-dependent weights, $\mathsf{W}_3 = \mathsf{w}_3(s)I$ and $\mathsf{W}_4 = \mathsf{w}_4(s)I$, an $\mathcal{H}_\infty$ controller that achieves a specific value of γ ensures that

$$\sigma_{\max}[S(j\omega)G_1(j\omega)] < \frac{\gamma}{|\mathsf{w}_3(j\omega)|} \quad \forall\omega \tag{18.17}$$

$$\sigma_{\max}[K(j\omega)S(j\omega)] < \frac{\gamma}{|\mathsf{w}_4(j\omega)|} \quad \forall\omega \tag{18.18}$$

as a result of Equation 18.15. Similar bounds will also hold for the other $\left(T_{zw}\right)_{ij}$ in Equation 18.16. To take advantage of Equations 18.17 and 18.18, set $\gamma = 1$ and select the values of $\mathsf{w}_3(s)$ and $\mathsf{w}_4(s)$ to provide

* See Chapter 9.

desirable bounds on $S(s)G_1(s)$ and $K(s)S(s)$. For example, let $\mathsf{w}_4(s) = e_a(s)$. Then if the $\mathcal{H}_\infty$ controller based on these values of the weights achieves $\|T_{zw}\|_\infty \approx 1$, the desired loops will in fact be shaped to satisfy Equations 18.17 and 18.18. This is how one should choose the values of the design variables to shape the loops of interest in an $\mathcal{H}_\infty$ design.

In using this method of weight selection there are a few issues the designer must keep in mind. First of all, realize that the bounds implied by Equation 18.15 and exemplified by Equation 18.17 are not necessarily tight over all frequencies. As a result it helps to graphically inspect all the constraints implicit in the choice of $T_{zw}(s)$ as one iterates through the values of the design variables. More importantly, simply assuming $\gamma = 1$ when the values of the weights are chosen does not ensure an $\mathcal{H}_\infty$ controller that achieves $\|T_{zw}\|_\infty \approx 1$. In fact, when $\|T_{zw}\|_\infty \gg 1$ it is a strong indication that the values of the design variables impose unrealistic constraints on the system's dynamics. One cannot choose $\mathsf{w}_3(s)$ and $\mathsf{w}_4(s)$ arbitrarily. They must complement each other. Another reason why the design variables cannot be chosen arbitrarily involves the fact that $\| \left(T_{zw}\right)_{ij} \|_\infty < \gamma \ \forall i,j$. Not only will $\mathsf{w}_3(s)$ shape the weighted sensitivity transfer function $S(s)G_1(s)$, it will also shape $C(s)$. Since $S(s) + C(s) = I$, there will clearly be restrictions on the choice of $\mathsf{w}_3(s)$. While loops such as $C(s)$ may not be of primary interest, they will influence the overall performance of the controller and should be kept in mind when selecting the values of the weights.

The choice of P in Figure 18.6 with $\mathsf{W}_1 = \mathsf{W}_2 = I$ could also have been made using structured singular value concepts*. In this context, the performance variables z_2 and disturbances w_2 can be viewed as the inputs and outputs to an unknown but norm bounded unstructured uncertainty that captures the additive error in the input to output model. Likewise, the performance variables z_1 and disturbances w_1 can be viewed as the inputs and outputs to a fictitious, unknown, norm bounded unstructured uncertainty that captures the desire to reject the disturbances d at the outputs y_n. Then selecting the values of the design weights is akin to scaling the system in the same way that the D-scales, used in the D-K iteration, scale the system.

18.4.3 Additional Notes for Selecting Design Weights

To ensure that Assumptions A.1 through A.5 are satisfied once the dynamics of the frequency weights are augmented to the system model, it is necessary to use proper, stable, minimum phase weights. For example, in the system shown in Figure 18.6, $\mathsf{W}_4(s)$ must contain an output feedthrough term to ensure $R_{uu} > 0$.

An important issue to be aware of when using frequency weights is that it is possible to define a set of weights with repetitive information. For example, in the system of Figure 18.6 with $\mathsf{W}_2(s) = \mathsf{w}_2(s)I$ and $\mathsf{W}_4(s) = \mathsf{w}_4(s)I$, specifying the magnitudes of $\mathsf{w}_2(s)$ and $\mathsf{w}_4(s)$ to be large over the same frequency region tells both optimization problems the same information, make the controls small there. Not only is such information redundant, it is also undesirable, since the order for the compensator is equal to the order of P.

18.5 Aircraft Design Example

To illustrate more clearly how one uses frequency weights to design $\mathcal{H}_2$ and $\mathcal{H}_\infty$ controllers, we shall discuss the design of a wind gust disturbance rejection controller for a linearized model of an F-8 aircraft. As you shall see, the modern paradigm allows us to incorporate frequency domain performance and robustness specifications naturally and directly into the controller synthesis.

The F-8 is an "old-fashioned" aircraft that has been used by NASA as part of their digital fly-by-wire research program. Assuming that the aircraft is flying at a constant altitude in equilibrium flight allows us to linearize the nonlinear equations of motion. In doing so, the longitudinal dynamics decouple from

* See Chapter 20 for more details.

the lateral dynamics. The variables needed to characterize the longitudinal motion, which are defined in the schematic drawing of the F-8 shown in Figure 18.7, are the horizontal velocity, $\nu(t)$, pitch angle, $\theta(t)$, pitch rate, $q(t) = \dot{\theta}(t)$, angle of attack, $\alpha(t)$, and flight path angle, $\beta(t) = \theta(t) - \alpha(t)$. To control the longitudinal motion, elevators, $\delta_e(t)$, and flaperons, $\delta_f(t)$, which are just like the elevators except that they move in the same direction, were used. While the thrust also influences the longitudinal motion of the aircraft, it is considered to be constant in our designs. The measurements are the pitch and flight path angles, $y^T(t) = [\theta(t) \quad \beta(t)]$.

The effect of wind gust disturbances, which primarily corrupt the angle of attack, is modeled as the output of a shaping filter driven with unit intensity white noise, $d(t)$. Using the state vector

$$x^T(t) = [\theta(t) \quad \beta(t) \quad q(t) \quad \nu(t) \quad x_d(t)]$$

in which $x_d(t)$ is the state of the first-order shaping filter of the wind gust disturbance model, the linearized, longitudinal equations of the F-8 aircraft are

$$\begin{aligned} \dot{x}(t) &= Ax(t) + Bu(t) + Ld(t) \\ y(t) &= Cx(t) + v(t) \end{aligned} \tag{18.19}$$

with $u^T(t) = [\delta_e(t) \quad \delta_f(t)]$ and

$$A = \begin{bmatrix} 0.0 & 0.0 & 1.0 & 0.0 & 0.0 \\ 1.50 & -1.50 & 0.0 & 0.0057 & 1.50 \\ -12.0 & 12.0 & -0.60 & -0.0344 & -12.0 \\ -0.8520 & 0.290 & 0.0 & -0.0140 & -0.290 \\ 0.0 & 0.0 & 0.0 & 0.0 & -0.730 \end{bmatrix} \quad B = \begin{bmatrix} 0.0 & 0.0 \\ 0.160 & 0.80 \\ -19.0 & -3.0 \\ -0.0115 & -0.0087 \\ 0.0 & 0.0 \end{bmatrix}$$

$$L^T = \begin{bmatrix} 0.0 & 0.0 & 0.0 & 0.0 & 1.1459 \end{bmatrix} \quad C = \begin{bmatrix} 1 & 0 & 0 & 0 & 0 \\ 0 & 1 & 0 & 0 & 0 \end{bmatrix}$$

The units for the angles and control signals are in degrees while the velocity has units of ft/s. The outputs are modeled as a nominal signal with additive white noise $v(t)$ that has an intensity of $\mu = 0.01$ deg^2/s, $\mathrm{E}\left\{v^T(t)v(\tau)\right\} = \mu I\delta(t - \tau)$, to capture the limited accuracy of the sensors.

The objective is to design controllers that reduce the effect of the wind disturbance on the system. Specifically we would like the magnitude of each output to be less than 0.25 degrees up to 1.0 rad/s as the aircraft passes through wind gusts. In addition, we require the control system to be robust to an unstructured multiplicative error reflected to the output of the plant whose magnitude is bounded by the function

$$e_m(s) = 0.1s^2$$

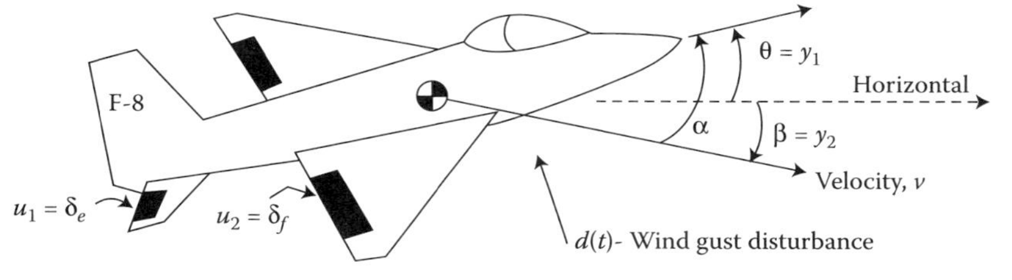

FIGURE 18.7 Definition of variables for the longitudinal dynamics of the F-8.

This multiplicative error captures the unmodeled dynamics associated with the flexibility of the aircraft's airframe. It will essentially constrain the bandwidth of the design to prevent these unmodeled modes from being excited.

Both of the design specifications can be represented in the frequency domain. To meet the performance specification, we require the closed-loop transfer function from d to y, $S(s)G_1(s)$, to satisfy

$$\sigma_{\max}[S(\jmath\omega)G_1(\jmath\omega)] < 0.25 \quad \text{for } 0 < \omega \leq 1.0 \text{ rad/s} \tag{18.20}$$

where

$$S(s) = [I - G_2(s)K(s)]^{-1}$$
$$G_2(s) = C(sI - A)^{-1}B$$
$$G_1(s) = C(sI - A)^{-1}L$$

To ensure stability robustness to the multiplicative error we require that

$$\sigma_{\max}[C(\jmath\omega)] < \frac{1}{|e_m(\jmath\omega)|} \quad \forall\omega \tag{18.21}$$

where $C(s) = S(s)G_2(s)K(s)$.

Given this representation of the design goals, we shall synthesize $\mathcal{H}_2$ and $\mathcal{H}_\infty$ controllers that shape the closed-loop transfer functions $S(s)G_1(s)$ and $C(s)$ to meet these constraints. Using the system interconnection for P shown in Figure 18.8 makes good mathematical and physical sense for this problem. Mathematically, $P(s)$ shown in Figure 18.8 leads to the following closed-loop transfer function matrix,

$$T_{zw}(s) = \begin{bmatrix} \mathrm{w}_1(s)S(s)G_1(s) & \sqrt{\mu}\mathrm{w}_2(s)C(s) \\ \rho\mathrm{w}_1(s)K(s)S(s)G_1(s) & \rho\sqrt{\mu}\mathrm{w}_2(s)K(s)S(s) \end{bmatrix} \tag{18.22}$$

Notice that the loops of interest, $S(s)G_1(s)$ and $C(s)$, appear directly in Equation 18.22 and are directly influenced by the scalar frequency weights $\mathrm{w}_1(s)$ and $\mathrm{w}_2(s)$. Realize that the coloring filter dynamics for the wind gust disturbance are already included in the system dynamics Equation (18.19), so that $\mathrm{w}_1(s)$ should not be viewed as a shaping filter for d. Rather $\mathrm{w}_1(s)$ is a design variable that overemphasizes the frequency range in which the impact of d is most vital, and it is chosen to reflect in the optimization problem our desire to appropriately shape $S(s)G_1(s)$. The scalar constant weight ρ, which is a penalty on the control that must be included in the synthesis to satisfy Assumption A.5, was allowed to vary, whereas μ was held fixed to capture the limited sensor accuracy. $P(s)$ also makes good sense in terms of the physics of the design objectives. It includes the effects of both the process and sensor noises, and its performance variables, z, contain the outputs we wish to keep small in the presence of the disturbances.

To illustrate that there is a strong connection between the physical, stochastic motivation used to select the values of the weights in the $\mathcal{H}_2$ framework and the more mathematical norm bound motivation used in the $\mathcal{H}_\infty$ framework, we compare the results of an $\mathcal{H}_\infty$ and an $\mathcal{H}_2$ design that both use the same values of the weights. The weights were chosen as described in the previous section. After some iteration we found that with

$$\rho = 0.01, \quad \mathrm{w}_1(s) = \frac{0.1(s+100)}{(s+1.25)}, \quad \text{and} \quad \mathrm{w}_2(s) = \frac{5000(s+3.5)}{3.5(s+1000)}$$

both the $\mathcal{H}_\infty$ and $\mathcal{H}_2$ designs met the desired performance and robustness specifications. Note that as a result of using these frequency weights, the controllers had eight states.

Figures 18.9 and 18.10 show that the loops of interest have in fact been shaped in accordance with the design goals. As seen in Figure 18.9, which compares the open and closed-loop disturbance to output transfer functions, both designs meet the performance goal from Equation 18.20. In this figure, $1/|\mathrm{w}_1(\jmath\omega)|$

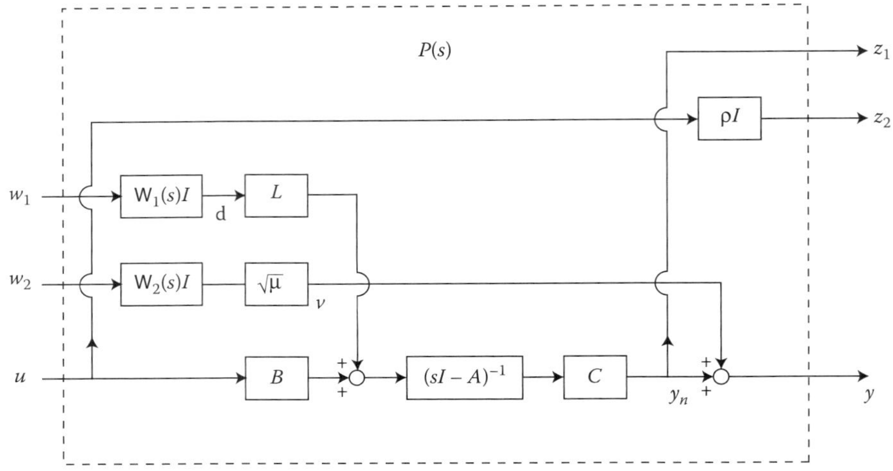

FIGURE 18.8 Generalized system $P(s)$ used in the F-8 designs.

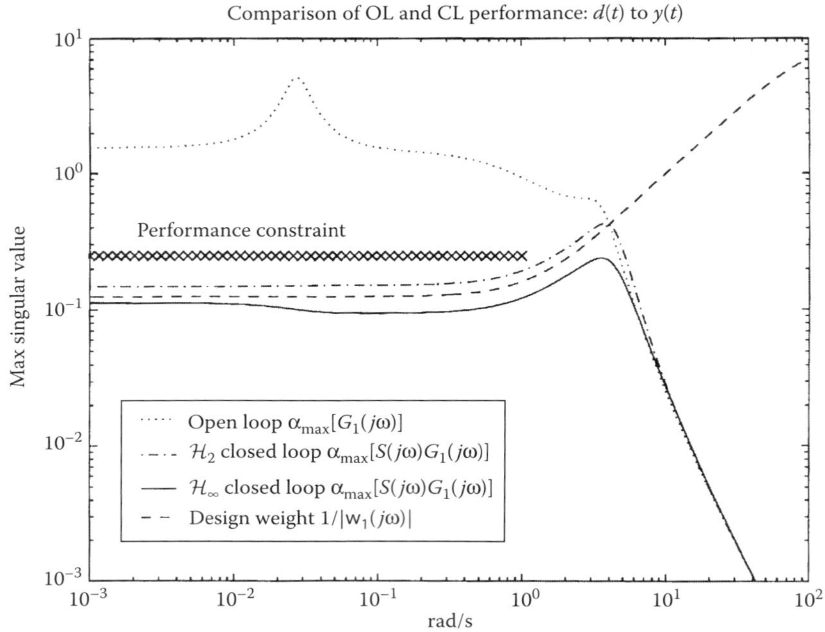

FIGURE 18.9 Comparison of the open- and closed-loop disturbance to output transfer functions for the $\mathcal{H}_\infty$ and $\mathcal{H}_2$ designs. The frequency weight $W_1(s)$ used to shape $S(s)G_1(s)$ is also shown.

is also shown to illustrate how the value of $W_1(s)$ was chosen. From an $\mathcal{H}_2$ perspective, the $|W_1(j\omega)|$ is large over $0 < \omega \leq 1.0$ rad/s and small elsewhere to tell the synthesis that the intensity of the disturbance is large where we desire good disturbance rejection. In the context of the $\mathcal{H}_\infty$ design, we assumed that $\gamma = 1$ and selected $1/|W_1(j\omega)|$ to appropriately bound $\sigma_{max}[S(j\omega)G_1(j\omega)]$ in accordance with the fact that

$$\sigma_{max}[S(j\omega)G_1(j\omega)] < \frac{\gamma}{|W_1(j\omega)|} \quad \forall\omega \tag{18.23}$$

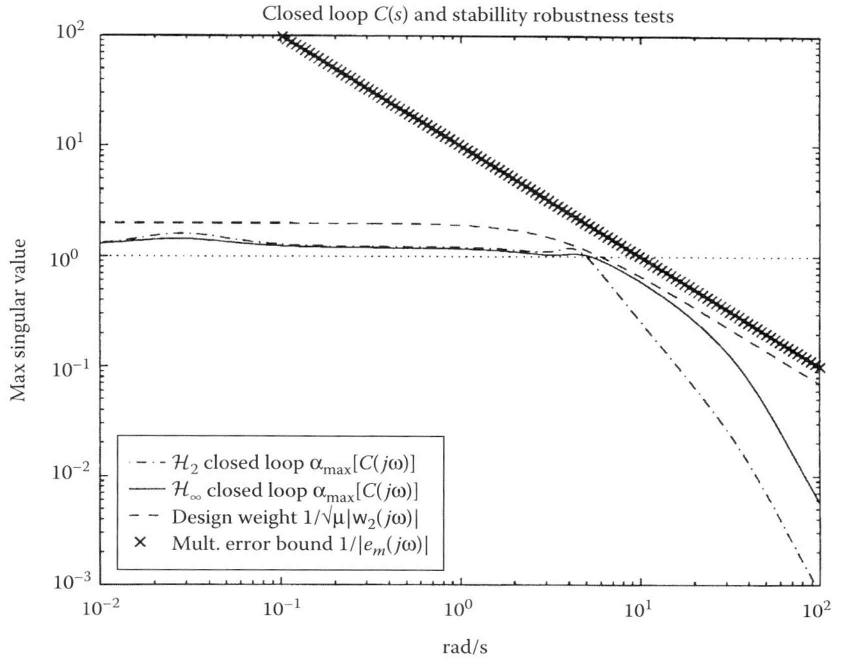

FIGURE 18.10 Comparison of $\sigma_{\max}[C(j\omega)]$ for the $\mathcal{H}_\infty$ and $\mathcal{H}_2$ designs. The bound on the multiplicative error, $1/|e_m(j\omega)|$, and the frequency weight $w_2(s)$ used to shape $C(s)$ are also shown.

for any $\mathcal{H}_\infty$ design which achieves $\|T_{zw}\|_\infty < \gamma$. For the values of the weights given above, $\|T_{zw}\|_\infty = 0.95$ for the $\mathcal{H}_\infty$ design, which ensures that Equation 18.23 was satisfied. A difference in the performance achieved by both designs is expected since the same values of the weights are used to minimize different measures of the size of $T_{zw}(s)$.

Figure 18.10, which compares $\sigma_{\max}[C(j\omega)]$ for the designs, illustrates that the stability robustness criterion from Equation 18.21 is also satisfied. Again, $1/\sqrt{\mu}|w_2(j\omega)|$ is shown to illustrate the manner in which the value of $w_2(s)$ was chosen. As seen in the figure, $w_2(s)$ forces the $\mathcal{H}_\infty$ design to roll off below the stability robustness bound, $1/|e_m(j\omega)|$, in accordance with the fact that

$$\sigma_{\max}[C(j\omega)] < \frac{\gamma}{\sqrt{\mu}|w_2(j\omega)|} \quad \forall\omega. \tag{18.24}$$

Since $|w_2(j\omega)|$ is large beyond 5.0 rad/s, it tells the $\mathcal{H}_2$ synthesis that the sensor noise is large there, which in turn limits the control energy beyond 5.0 rad/s.

The value of ρ also played an important role in the designs. In the $\mathcal{H}_2$ design, adjusting ρ directly influenced the amount of control effort used, just as a control weight would in LQG synthesis. For the $\mathcal{H}_\infty$ design, ρ minimized the constraints that the values of $w_1(s)$ and $w_2(s)$ placed on the closed-loop transfer functions $K(s)S(s)G_1(s)$ and $K(s)S(s)$ in Equation 18.22. Since these loops are not of primary interest, choosing a small value of ρ ensured that $w_1(s)$ and $w_2(s)$ would not overly constrain these since, for example,

$$\sigma_{\max}[K(j\omega)S(j\omega)G_1(j\omega)] < \frac{\gamma}{\rho|w_1(j\omega)|} \quad \forall\omega$$

for any $\mathcal{H}_\infty$ design.

The similarities in the achieved loop shapes are not coincidental. In fact, the dynamics of the $\mathcal{H}_\infty$ and $\mathcal{H}_2$ controllers presented here are quite similar. There is a clear reason why the similarity exists even though the optimization problems used are distinct. Once all of the desired control system performance is incorporated into $P(s)$ via the design variables, the task of minimizing the $\mathcal{H}_2$ norm of $T_{zw}(s)$ becomes nearly identical to the task of minimizing the $\mathcal{H}_\infty$ norm of $T_{zw}(s)$. This can be seen in Figure 18.11, which compares the values of $\sigma_{\max}[P_{zw}(j\omega)]$ and $\sigma_{\max}[T_{zw}(j\omega)]$ for the two designs. Here $P_{zw}(s)$ denotes the open-loop transfer function matrix between w and z of $P(s)$. As such, $\sigma_{\max}[P_{zw}(j\omega)]$ is an indication of the nominal cost that the controllers seek to minimize. Specifically, to minimize the $\mathcal{H}_\infty$ norm of $T_{zw}(s)$ the peak in $\sigma_{\max}[P_{zw}(j\omega)]$ must be flattened out so that it looks like a low pass filter. Then the DC gain of the filter must be reduced to further minimize the $\mathcal{H}_\infty$ norm of $T_{zw}(s)$. This is also the case for minimizing the $\mathcal{H}_2$ norm of $T_{zw}(s)$ which is dominated by the area under the spike in $\sigma_{\max}[P_{zw}(j\omega)]$ (recall that the area is evaluated linearly and that we are using a log log plot). While the optimization problems are distinct, the manner in which the cost is minimized is similar.

Figure 18.11 also provides a clear indication of how the optimization problems differ. Notice that the $\mathcal{H}_2$ design rolls off faster than the $\mathcal{H}_\infty$ design. This is because the $\mathcal{H}_2$ design minimizes energy, or the area under $\sigma_{\max}[T_{zw}(j\omega)]$, at the expense of its peak value, whereas the $\mathcal{H}_\infty$ design seeks to minimize the peak of $\sigma_{\max}[P_{zw}(j\omega)]$ at the expense of allowing there to be more energy at higher frequencies.

It would be improper to draw conclusions about which synthesis approach is better based on these designs, especially since the same values for the weights in $P(s)$ were used. Rather, our intent has been to illustrate the connections between the $\mathcal{H}_\infty$ and $\mathcal{H}_2$ frameworks and how one can go about synthesizing $\mathcal{H}_\infty$ and $\mathcal{H}_2$ controllers to meet frequency domain design specifications. We should also note that the methodology used in this example has been applied to and experimentally verified on a much more complex system [16].

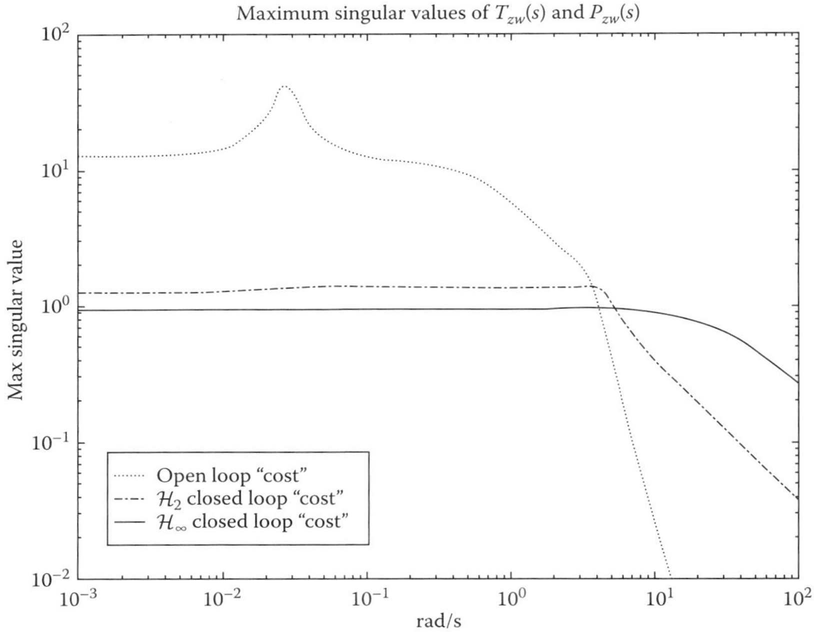

FIGURE 18.11 Comparison of the open-loop "cost," $\sigma_{\max}[P_{zw}(j\omega)]$, and the closed-loop "cost," $\sigma_{\max}[T_{zw}(j\omega)]$, for the $\mathcal{H}_\infty$ and $\mathcal{H}_2$ controllers.

References

1. Doyle, J., Francis, B., and Tannenbaum, A., *Feedback Control Theory*, Macmillan, New York, 1992.
2. Boyd, S. P. and Barrat, C. H., *Linear Controler Design: Limits of Performance*, Prentice Hall, Englewood Cliffs, NJ, 1991.
3. Zhou, K. Doyle, J., and Glover, K., *Robust and Optimal Control*, Prentice Hall, Englewood Cliffs, NJ, 1995.
4. Papoulis, A., *Probability, Random Variables, and Stochastic Processes*, 3rd ed., McGraw Hill, New York, 1991.
5. Francis, B. A., *A Course in $\mathcal{H}_\infty$ Control Theory*, Springer-Verlag, Berlin, 1987.
6. Doyle,J., Glover,K., Khargonekar, P., and Francis, B., State space solutions to standard $\mathcal{H}_2/\mathcal{H}_\infty$ control problems, *IEEE Trans. Autom. Control*, 34(8), 831–847, 1989.
7. Maciejowski, J., *Multivariable Feedback Design*, Addison-Wesley, Wokingham, England, 1989.
8. Glover, K. and Doyle, J., State-space formulae for all stabilizing controllers that satisfy an $\mathcal{H}_\infty$-norm bound and relations to risk sensitivity, *Syst. Control Lett.*, 11, 167–172, 1988.
9. Bernstein, D.S. and Haddad, W.M., LQG control with an $\mathcal{H}_\infty$ performance bound: A Riccati equation approach, *IEEE Trans. Autom. Control*, 34(3), 293–305, 1989.
10. Doyle, J., Zhou, K., Glover, K., and Bodenheimer, B., Mixed $\mathcal{H}_2$ and $\mathcal{H}_\infty$ performance objectives, II. Optimal control, *IEEE Trans. Autom. Control*, 39(8), 1575–1587, 1994.
11. Scherer, C.W., Multiobjective $\mathcal{H}_2/\mathcal{H}_\infty$ control, *IEEE Trans. Autom. Control*, 40(6), 1054–1062, 1995.
12. Nagpal, K. and Khargonekar, P., Filtering and smoothing in a $\mathcal{H}_\infty$ setting, *IEEE Trans. Autom. Control*, 36(2), 152–166, 1991.
13. Kwakernaak, H. and Sivan, R., *Linear Optimal Systems*, John Wiley & Sons, New York, 1972.
14. Kwakernaak, H., Robust control and $\mathcal{H}_\infty$ optimization—tutorial paper, *Automatica*, 29(2), 255–273, 1993.
15. Anderson, B.D.O. and Moore, J.B., *Optimal Control: Linear Quadratic Methods*, Prentice Hall, Englewood Cliffs, NJ, 1990.
16. Lublin, L. and Athans, M., An experimental comparison of $\mathcal{H}_2$ and $\mathcal{H}_\infty$ designs for an interferometer testbed, in *Lecture Notes in Control and Information Sciences: Feedback Control, Nonlinear Systems, and Complexity*, Francis, B. and Tannenbaum, A., Eds., Springer-Verlag, Amsterdam, 1995, 150–172.

19

ℓ_1 Robust Control: Theory, Computation, and Design

Munther A. Dahleh
Massachusetts Institute of Technology

19.1 Introduction*

Feedback controllers are designed to achieve certain performance specifications in the presence of both plant and signal uncertainty. Typical controller design formulations consider quadratic cost functions on the errors primarily for mathematical convenience. However, practical situations may dictate other kinds of measures. In particular, the peak values of error signals are often more appropriate for describing desired performance. This can be a consequence of uniform tracking problems, saturation constraints, rate limits, or simply a disturbance rejection problem. In addition, disturbance and noise are in general bounded and persistent because they continue to act on the system as long as the system is in operation. Such signals are better described by giving information about both the signals' frequency content and time-domain bounds on peak values.

The above class of problems motivated a formulation that involves the Peak-to-Peak gain of a system, which is mathematically given by the ℓ_1 norm of the pulse response. This formulation was first reported in [13], and the problem was completely solved in several subsequent articles [5–8,11]. The extension of the theory to incorporate plant uncertainty was reported in [3,4,9,10]. An extensive coverage of this theory with detailed references can be found in [2].

The need for developing this problem was further intensified by the failure of the frequency-domain techniques to address time-domain specifications. For instance, attempting to achieve an overshoot

* Research Supported by AFOSR under grant AFOSR-91-0368, by NSF under grant 9157306-ECS, and by Draper Laboratory under grant DL-H-467128.

constraint using $\mathcal{H}_\infty$ or $\mathcal{H}_2$ by appropriately adjusting the weighting matrices can be a very frustrating experience. On the other hand, solutions to such problems will no longer be in closed form due to the complexity of the performance objectives. Exact or approximate solutions will be obtained from solving equivalent yet simpler optimization problems. The derivation of such simpler problems and the computational properties are essential components of this research direction.

19.1.1 A Design Tool

The motivation behind research in ℓ_1 theory is developing a design tool for MIMO uncertain systems. A powerful design tool in general should have three ingredients:

1. *Practicality:* The ability to translate a large set of performance specifications into conditions or constraints readily acceptable by the design tool. It is evident that not all design specifications can be immediately translated into mathematical conditions. However, the mathematical formulation should well approximate these objectives.
2. *Computability:* It is in general straightforward to formulate controller design problems as constrained optimization problems. What is not so straightforward is formulating problems that can be solved efficiently and with acceptable complexity.
3. *Flexibility:* The ability to alter a design to achieve additional performance specifications with small marginal cost.

   It is evident that practicality and computability are conflicting ingredients. Computational complexity grows as a function of several parameters, which include the dimension of the system, the uncertainty description, and the performance specifications. Flexibility makes it possible to design a controller in stages, i.e., by altering a nominally good design to achieve additional specifications.

19.1.2 Motivation

We give some reasons behind the development of such a design tool, by quoting from the book: *Control of Uncertain Systems: A Linear Programming Approach.*

1. *Complex systems:* Many of today's systems, ranging from space structures to high purity distillation columns, are quite complex. The complexity comes from the very high order of the system as well as the large number of inputs and outputs. Modeling such systems accurately is a difficult task and may not be possible. A powerful methodology that deals systematically with multiple inputs and outputs and with various classes of structured uncertainty is essential.
2. *High performance requirement:* Systems are built to perform specific jobs with high accuracy. Robots are already used to perform accurate jobs such as placing components on integrated circuit boards. Aircraft are built with high maneuverability and are designed specifically for such tasks. Classical SISO design techniques cannot accommodate these problems resulting in designs that are conservative and perform poorly.
3. *Limits of performance:* In complex systems, it is time-consuming to establish, by trial and error, whether a system can meet certain performance objectives (even without uncertainty). Thus, it is necessary to develop systematic methods to quantify the fundamental limitations of systems and to highlight the trade-offs of a given design.
4. *A systematic design process:* It is inevitable that designing a controller for a system will involve iterations. Unless this procedure is made systematic, the design process can become very cumbersome and slow. The design procedure should not only exhibit a controller. It should also provide the designer with indicators for the next iteration, by showing which of the constraints are the limiting ones and which part of the uncertainty is causing the most difficulty. Note also that a general procedure should be able to accommodate a variety of constraints, both in the time and in the frequency domain.

5. *Computable methods:* It is quite straightforward to formulate important control problems, but it is not so easy to formulate solvable problems that can provide computable methods for analysis and synthesis. Much of the current research invokes high level mathematics to provide simple computable results. The computability of a methodology is the test of its success. By computability we do not mean necessarily closed form solutions. Problems that give rise to such solutions have limited applicability. However, computability means that we can synthesize controllers via computable algorithms and simultaneously obtain qualitative information about the solution.
6. *Technological advancement:* Many aspects of technological development will affect the design of control systems. The availability of "cheaper" sensors and actuators that are well-modeled allows for designing control systems with a large number of inputs and outputs. The availability of fast microprocessors, as well as memory, makes it possible to implement more complex feedback systems with high order. The limiting factor in controller implementation is no longer the order of the controller. Instead, it is the computational power and the memory availability.
7. *Available methods:* The available design techniques have concentrated on frequency-domain performance specifications by considering errors in terms of energy rather than peak values. These methods (such as $\mathcal{H}_2$ and $\mathcal{H}_\infty$) have elegant solutions. However, this elegance is lost if additional time-domain specifications (e.g., overshoot, undershoot, settling time ..) are added. This created a need for a time-domain based computational methodology that can accommodate additional frequency-domain constraints. This design tool aims at achieving this objective.

In the sequel we will summarize the ℓ_1 design tool by discussing the three ingredients, *practicality, computability,* and *flexibility.*

19.2 Practicality

19.2.1 Examples

To motivate the formulation of the ℓ_1 problem, we begin with two examples.

The first is the control design for an *Earth Observing System* (*EOS*). EOS is a spacecraft that orbits the earth and points in a specific location. It carries on its platform various sensory instruments, with the objective of collecting data from earth. An example of such an instrument is an array of cameras intended to provide images of various points and landscapes on earth. The spacecraft is subjected to various kinds of disturbances: external pressures, noise generated from the instruments on board, and the spacecraft itself. The objective of the control design is to point the spacecraft accurately in a specific direction, otherwise known as attitude control.

The second example is the control design of an active suspension of an automobile. A simplified one-dimensional problem is shown in Figure 19.1.

The objective of the controller design is to maximize ride comfort, while simultaneously maintaining handling (road holding ability) in the presence of bad road conditions.

These examples have several common features:

1. The objectives in both problems are to keep the maximum deviations of signals from set points bounded by some prescribed value to attain uniform tracking or disturbance rejection. In the EOS example, performance is measured in terms of maximum deviations of the attitude angles from a set point. In the active suspension problem, performance is measured in terms of the maximum force acting on the system and the maximum deviation of the distance of the wheel from the ground.

   In mathematical terms, both performance specifications are stated in terms of *peak* values of signals, i.e.,

$$\|w\|_\infty = \max_{i=1,\ldots n} \sup_t |w(t)|.$$

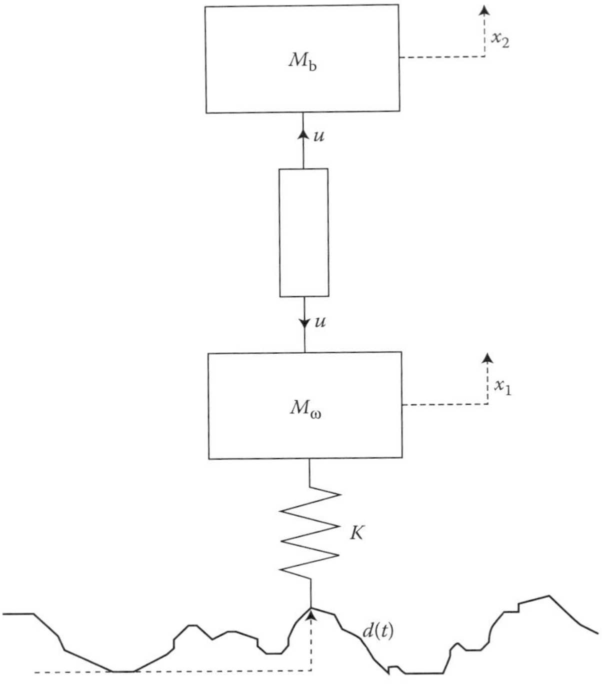

FIGURE 19.1 Active suspension.

This is known as the ℓ_∞-norm of the signal (or its peak value). For the active suspension problem, the performance can be stated as

$$\left\| \begin{pmatrix} x_1 - d \\ u \end{pmatrix} \right\|_\infty \leq \gamma,$$

for some performance bound γ. It is straightforward to incorporate additional scalings and weights into the performance.

2. The disturbances in both examples are unknown, persistent in time, but, bounded in magnitude (peak value). A good model for the disturbance in both cases is given by

$$d = Ww, \quad \|w\|_\infty \leq 1,$$

where W is a linear time-invariant filter that gives information about the frequency content of the disturbance. This model of disturbance accommodates persistent disturbances that are not necessarily periodic. It does not assume that the signal dies out asymptotically.

3. In both problems, saturation constraints are quite important and play a role in limiting the performance. In the active suspension problem, the saturation constraint is given by the maximum deflection of the hydraulic actuator, i.e.,

$$\|x_1 - x_2\|_\infty \leq \gamma_{sat}.$$

These constraints combined with the performance objectives have to be satisfied for all disturbances w such that $\|w\|_\infty \leq 1$.

4. Both examples are difficult to model precisely, and thus the control strategies have to accommodate unmodelled dynamics. In this article, we will not discuss in detail the robust performance problems. We refer the reader to [2] for details.

It is evident from the above discussion that peak values of signals are natural quantities in stating design specifications.

19.2.2 Formulation

Figure 19.2 shows a general setup for posing performance specifications. The variables as defined in the figure are

$$u = \text{control inputs}$$
$$y = \text{measured outputs}$$
$$w = \text{exogenous Inputs} = \begin{cases} \text{fixed commands} \\ \text{unknown commands} \\ \text{disturbances} \\ \text{noise} \\ \vdots \end{cases}$$
$$z = \text{regulated outputs} = \begin{cases} \text{tracking Errors} \\ \text{control Inputs} \\ \text{measured Outputs} \\ \text{states} \\ \vdots \end{cases}$$

The operator G is a 2×2 block matrix mapping the inputs w and u to the outputs z and y:

$$\begin{bmatrix} z \\ y \end{bmatrix} = \begin{bmatrix} G_{11} & G_{12} \\ G_{21} & G_{22} \end{bmatrix} \begin{bmatrix} w \\ u \end{bmatrix}.$$

The actual process or *plant* is the submatrix G_{22}. Both the exogenous inputs and the regulated outputs are auxiliary signals that need not be part of the closed loop system. The feedback controller is denoted by K.

From our discussion above, the set of exogenous inputs consists of unknown, persistent, but bounded, disturbances,

$$\mathcal{D} = \{w \in \ell_\infty \mid \|w\|_\infty \leq 1\}.$$

The performance measure (combined with constraints) is stated as

$$\|z\|_\infty \leq \gamma, \quad \forall\, w \in \mathcal{D}.$$

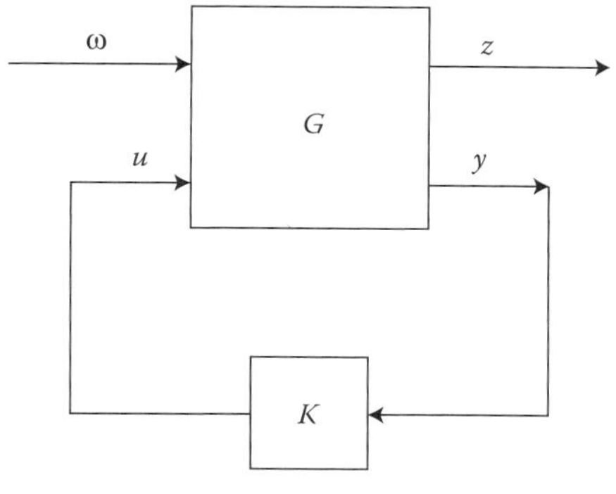

FIGURE 19.2 General setup.

If Φ is the linear time-invariant system mapping w to z, then

$$\|z\|_\infty \leq \gamma, \quad \text{for all w} \in \mathcal{D} \iff \|\Phi\|_1 \leq \gamma,$$

where

$$\|\Phi\|_1 = \max_{1\leq i\leq n_z} \sum_{j=1}^{n_w} \sum_{k=0}^{\infty} |\phi_{ij}(k)|.$$

The latter is the expression for the ℓ_1 norm of the system. In conclusion, the ℓ_1 norm of a system is the *peak-to-peak* gain of the system and can directly describe time-domain performance specifications. The nominal performance problem can be stated as

$$\inf_{K \text{ stabilizing}} (\sup_w \|\Phi\|_1).$$

19.2.3 Comparison to $\mathcal{H}_\infty$

Suppose the exogenous inputs are such that $\|w\|_2 \leq 1$ but are otherwise arbitrary ($\|w\|_2$ is the energy contained in the signal). If the objective is to minimize the energy of the regulated output, then the nominal performance problem is defined as

$$\inf_{K \text{ stabilizing}} (\sup_w \|\Phi w\|_2) = \inf_{K \text{ stabilizing}} \sup_\theta \sigma_{\max}[\hat{\Phi}(e^{i\theta})].$$

Both of these norm minimization problems fall under the same paradigm of minimax optimality. Minimizing the $\mathcal{H}_\infty$ norm results in attenuating the energy of the regulated signal but may still result in signals that have large amplitudes. Minimizing the ℓ_1 norm results in attenuating the amplitude of the regulated output, and overbounds the maximum energy due to bounded energy inputs because

$$\|\hat{\Phi}\|_{\mathcal{H}_\infty} \leq \sqrt{m}\|\Phi\|_1,$$

where m is the number of rows in Φ. On the other hand, the following inequality holds:

$$\|\Phi\|_1 \leq 2(N+1)\sqrt{n}\|\hat{\Phi}\|_{\mathcal{H}_\infty},$$

where N is the McMillan degree of the system, and n is the number of columns of Φ. The latter bound is the tightest possible bound (i.e., equality holds for certain classes of systems) and it shows that the gap between these measures can be large if the order of the system is high.

19.2.4 Prototypes

The following prototypes have been discussed in [2]. These are representative problems quite common in applications. We will use these prototypes to illustrate the significance of the ℓ_1 design methodology.

19.2.4.1 Disturbance Rejection

In the context of ℓ_∞ signals, the disturbance rejection problem is defined as follows: Find a feedback controller that minimizes the maximum amplitude of the regulated output over all possible disturbances of bounded magnitude. The two-input two-output system shown in Figure 19.3 depicts the particular case where the disturbance enters the system at the plant output. Its mathematical representation is given by

$$z = P_0 u + Ww,$$
$$y = P_0 u + Ww.$$

The disturbance rejection problem provides a general enough structure to represent a broad class of interesting control problems.

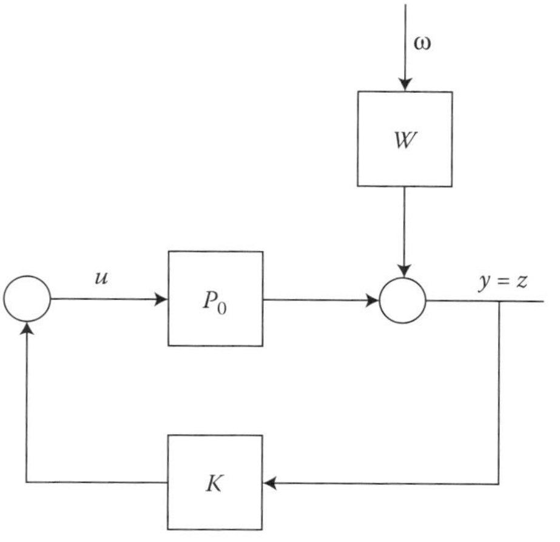

FIGURE 19.3 A disturbance rejection problem.

19.2.4.2 Command Following with Saturation

The command following problem, equivalent to a disturbance rejection problem, is shown in Figure 19.4. We will show how to pose this problem in the presence of saturation nonlinearities at the input of the plant, as an ℓ_1-optimal control problem.

Define the function

$$\text{Sat}(u) = \begin{cases} u & |u| \leq U_{\max} \\ U_{\max} sgn(u) & |u| \geq U_{\max} \end{cases}.$$

Let the plant be described as

$$Pu = P_0 \text{Sat}(u)$$

where P_0 is LTI. Let the commands be modeled as

$$r = Ww \quad \text{where } \|w\|_\infty \leq 1.$$

The objective is to find a controller K so that y follows r uniformly in time. Keeping in mind the saturation function, and in order to stay in the linear region of operation, the allowable control inputs must have $\|u\|_\infty \leq U_{\max}$. Let γ be the (tracking) performance level desired, and define

$$z = \begin{bmatrix} (y - r)/\gamma \\ u/U_{\max} \end{bmatrix}$$

with

$$y = P_0 u.$$

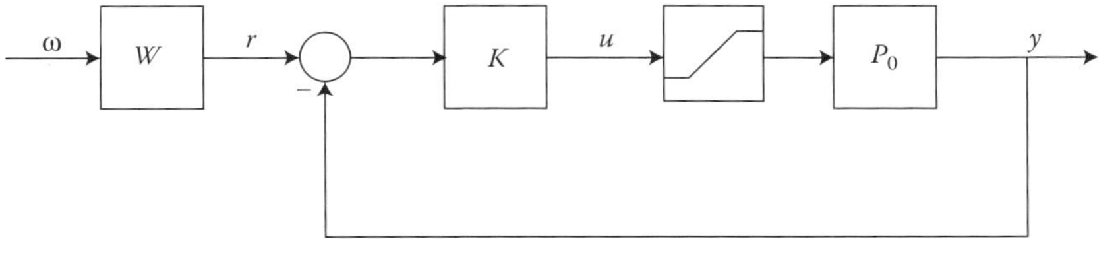

FIGURE 19.4 Command following with input saturation.

The problem is equivalent to finding a controller so that

$$\sup_{w} \|z\|_\infty < 1,$$

which is an ℓ_1-optimal control problem.

The above closed loop system will remain stable even if the input saturates, as long as it does so infrequently. The solution to the above problem will determine the limits of performance when the system is required to operate in the linear region. Also, the stability for such a system will mean the local stability of the nonlinear system.

19.2.4.3 Saturation and Rate Limits

In the previous example, actuator limitations may require that the rate of change of the control input be bounded. This is captured in the condition

$$\left| \frac{u(k) - u(k-1)}{T_s} \right| \leq U_{\text{der}}$$

where T_s is the sampling period. Let

$$W_{\text{der}} = \frac{1-\lambda}{T_s U_{\text{der}}}.$$

This condition can be easily incorporated in the objective function by defining z as

$$z = \begin{bmatrix} (y-r)/\gamma \\ u/U_{\max} \\ W_{\text{der}} u \end{bmatrix}.$$

The result is a standard ℓ_1-optimal control problem.

19.2.5 Stability and Performance Robustness

The power of any design methodology is in its ability to accommodate plant uncertainty. The ℓ_1 norm gives a good measure of performance. Because it is a gain over a class of signals, it will also provide a good measure for robustness. This is a consequence of the small gain theorem which is stated below [2].

Let M be a linear time-invariant system and Δ be a strictly proper ℓ_∞-stable perturbation. The closed-loop system shown in Figure 19.5 is ℓ_∞-stable for all Δ with $\|\Delta\|_{\ell_\infty-\text{ind}} \leq 1$ if and only if $\|M\|_1 < 1$.

The above result indicates that the ℓ_1 norm is the exact measure of robustness when the perturbations are known to be BIBO stable, bounded gain, and possibly nonlinear or time-varying. The result can be adapted to derive stability robustness conditions for a variety of plant uncertainty descriptions. We describe one such situation below.

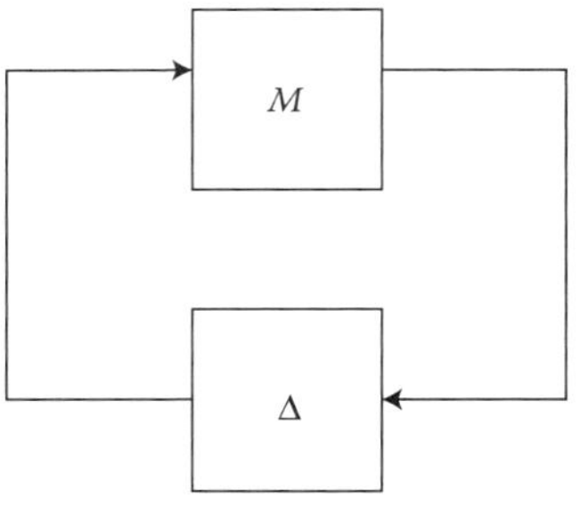

FIGURE 19.5 Stability robustness problem.

19.2.5.1 Unstructured Multiplicative Perturbations

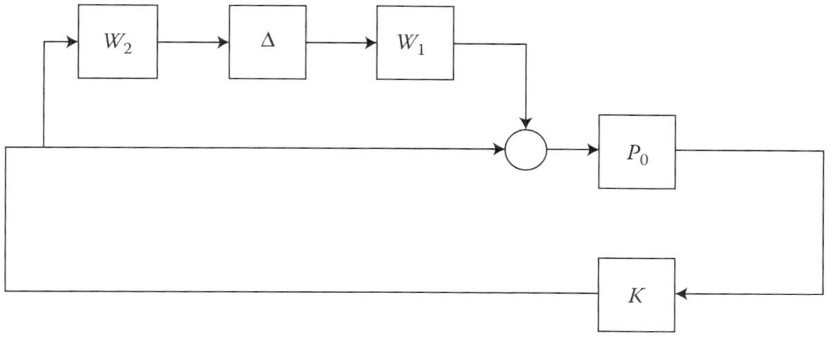

FIGURE 19.6 Multiplicative perturbations.

Consider the case where the system has input uncertainty in a multiplicative form as in Figure 19.6, i.e., let

$$\Omega = \{P \mid P = P_o(I + W_1 \Delta W_2) \quad \text{and} \quad \|\Delta\|_{\ell_\infty-\text{ind}} \leq 1\}.$$

If a controller is designed to stabilize P_0, under what conditions will it stabilize every system in the set Ω? By simple manipulations of the closed-loop system, the problem is equivalent to the stability robustness of the feedback system in Figure 19.5, with $M = W_2(I - KP_o)^{-1}KP_oW_1$. In general this manipulation is done in a systematic way: Cut the loop at the inputs and outputs of Δ, and then calculate the map from the output of Δ, w, to the input of Δ, z. A sufficient condition for robust stability is then given by $\|M\|_1 < 1$. The resulting two-input two-output description is given by

$$y = P_0 u + P_0 W_1 w,$$
$$z = W_2 u.$$

This is a standard ℓ_1 minimization problem.

19.2.5.2 Structured Uncertainty

In many applications, uncertainty is described in a structured fashion where independent perturbations are introduced in various locations of the closed loop system. It turns out that one can derive an exact necessary and sufficient condition in terms of a scaled ℓ_1 norm of the system to guarantee stability robustness in the presence of such structured perturbations. It can also be shown that the problem of achieving robust performance (where performance is measured in terms of the ℓ_1 norm) is equivalent to robustly stabilizing a plant with structured uncertainty. In this article, we will not discuss this problem. Instead, we refer the reader to [2] for more details.

19.3 Computability

Since it is quite hard, in general, to obtain closed form solutions to general optimization problems, we need to be precise about the meaning of a "solution." A closed form solution has two important features: the first is the ability to compute the optimal solution through efficient algorithms, and the second is to provide a qualitative insight into the properties of the optimal solution. A numerical solution should offer both of these ingredients. In particular, it should provide

1. The exact solution whenever it is possible

2. Upper and lower approximations of the objective function when exact solutions cannot be obtained and a methodology for synthesizing suboptimal controllers
3. Qualitative information about the controller, e.g., the order of the controller

Solutions based on general algorithms even for convex optimization problems offer only approximate upper bounds on the solution. To obtain more information, one needs to invoke duality theory. Duality theory provides a simple reformulation of the optimization problem from which lower bounds on the objective function and, possibly, exact solutions can be found.

19.3.1 Summary of Results

To minimize the ℓ_1 norm, first the Youla parameterization of all stabilizing controllers is invoked. The resulting optimization problem can be stated as an infinite-dimensional linear program in the free parameter. Two cases occur:

1. The infinite dimensional LP is exactly equivalent to a finite-dimensional LP. This happens if the dimension of the control input is at least as large as the dimension of the regulated variables and the dimension of the measured output is at least as large as the exogenous inputs. This means that the controller has a lot of degrees of freedom.
2. If any of the above conditions is violated, then the problem is inherently infinite-dimensional. However, duality theory can be used to provide approximate solutions to this problem with guaranteed performance levels.

The details of the computations for both of the above cases can be found in [2]. The most successful algorithm for computing solutions for the second case is not based on straightforward approximation, but rather on embedding the problem in another that falls under the first case. This procedure generates approximate solutions with converging upper and lower bounds, and also provides information about the order of the actual optimal controller. Other emerging techniques are based on dynamic programming and viability theory and can be found in [1,12].

19.4 Flexibility

Flexibility is the ability to use the design tool to alter a given nominal design so that additional specifications are met with minimal expense. Examples of additional specifications include fixed input constraints and frequency-domain constraints. The computational cost of alteration should be much less than the incorporation of the specification directly in the problem. In addition, it is desirable to maintain the qualitative information of the original solution.

Since the general synthesis problem is equivalent to an infinite dimensional LP, many additional specifications (not directly addressed by ℓ_1 norms) can be incorporated as additional linear constraints. Frequency-domain constraints can be well-approximated by linear constraints. Below we consider an example of adding fixed input constraints to the ℓ_1 problem.

19.4.1 ℓ_1 Performance Objective with Fixed Input Constraints

Consider the case where the specifications given are such that the control signal resulting from a step input must be constrained uniformly in time (e.g., to avoid actuator saturation). We want to bound the controller response to a step input and at the same time minimize the ℓ_1 norm of the transfer functions from the disturbance to both the control signal and plant output. In such a case we augment the basic ℓ_1

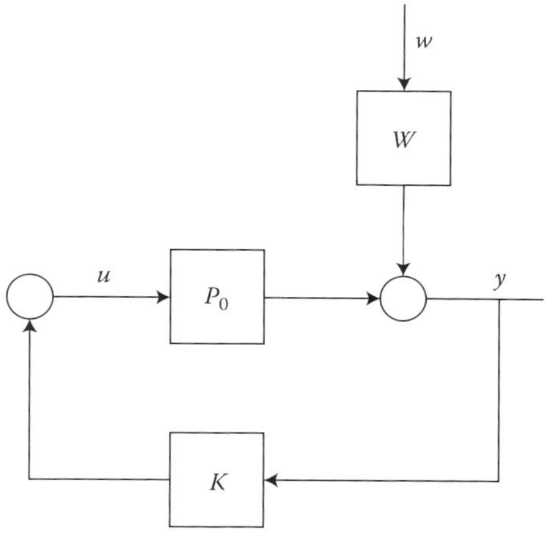

FIGURE 19.7 A disturbance rejection problem.

problem in the following way:

$$\inf_{K \text{ stab.}} \left\| \begin{matrix} K(I+PK)^{-1}W \\ (I+PK)^{-1}W \end{matrix} \right\|_1$$

subject to

$$\|K(I+PK)^{-1}w_f\|_\infty \leq U_{\max}$$

where w_f is a unit step input disturbance and $U_{\max}$ is the specified bound. The above modification results in adding infinitely many constraints on the sequence $K(I+PK)^{-1}$ (i.e., convolution of a unit step with $K(I+PK)^{-1}$). However, since the peak is typically achieved in early samples, only a finite number of constraints must be included (the rest being inactive). This is a particular case of nondecaying template constraints which arise frequently in control system design.

19.4.1.1 Trade-Offs in Design

We take these specifications a step further by asking the following questions: What are the trade-offs in the design? How does the bound on the control signal step response affect the overall performance? And, how does it affect the structure of the optimal solution?

These questions can be readily answered with the ℓ_1 machinery. It amounts to solving a family of mixed ℓ_1 problems parameterized in $U_{\max}$. Solutions for a range of values of $U_{\max}$ are presented in Figures 19.8 and 19.9 by showing the performance degradation and the controller order growth as $U_{\max}$ decreases. The numerical values are based on a model for the X29 aircraft (for details, see [2]). The following conclusions can be drawn from this analysis:

1. The results present the trade-offs clearly. It is possible to reduce $U_{\max}$ by 50% without losing too much performance in terms of the ℓ_1 norm of the system. This implies that the controller can be altered to satisfy stricter conditions on the step response without losing the nominal performance. The curve in Figure 19.8 also shows the smallest possible achievable $U_{\max}$.
2. The trade-offs in the order of the controller are valuable. The trade-off curve in Figure 19.9 shows that, by adding two additional states, $U_{\max}$ can be reduced to about 50% of its unconstrained value.
3. To compute such solutions, the unconstrained problem is solved first. The performance for a step input is then checked, and constraints are added only at the time instants where the peak value of

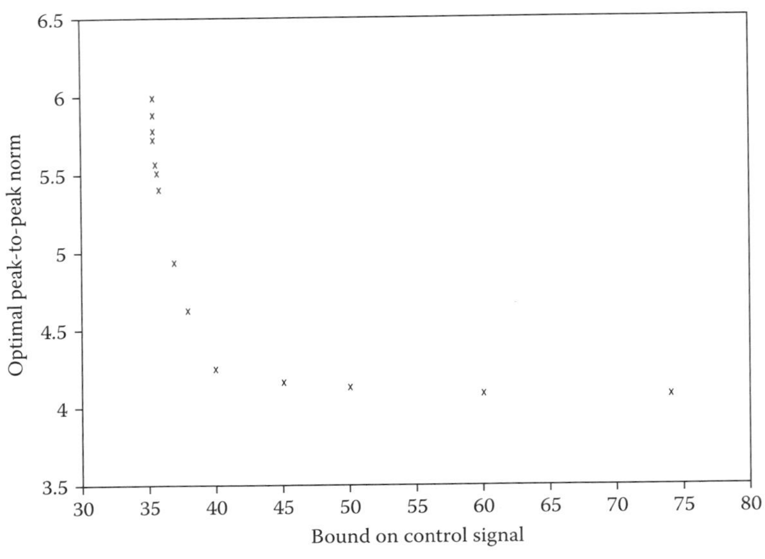

FIGURE 19.8 X29: Trade-offs in performance vs. control signal bound, $U_{\max}$.

the input u is larger than $U_{\max}$. This is a simpler problem than incorporating the infinite-horizon constraints at all the time instants of the step response.

Finally, such constraints are hard to deal with by selecting weights and solving an ℓ_1 problem or an $\mathcal{H}_\infty$ problem. The advantage that the ℓ_1 problem has over $\mathcal{H}_\infty$ is that such constraints can be incorporated in the problem, as described earlier, and then solved using the same solution techniques.

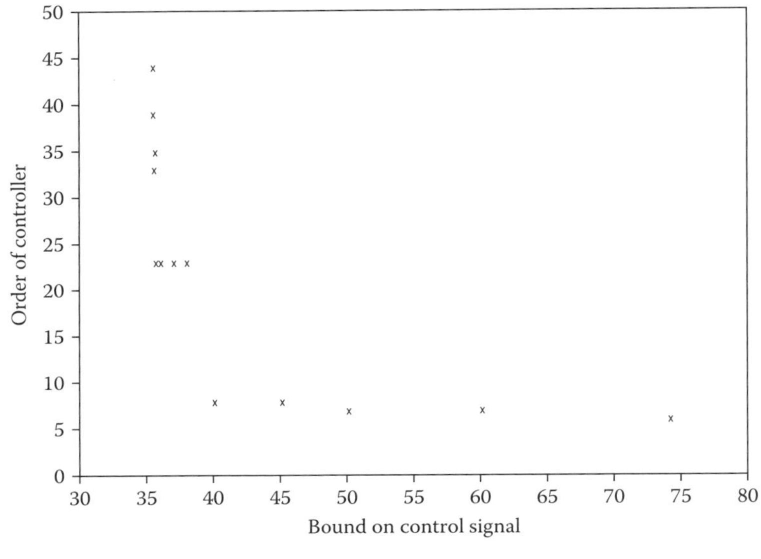

FIGURE 19.9 X29: Trade-offs in controller order vs. control signal bound, $U_{\max}$.

19.5 Conclusions

In this chapter, we gave an overview of the ℓ_1 theory for robust control. The presentation was not detailed. However, it was intended to serve as an introduction to a more detailed account of the theory that can be found in the book, *Control of Uncertain Systems: A Linear Programming Approach*, and references therein.

We highlighted three ingredients of the ℓ_1 design tool, practicality, computability and flexibility. These properties allow for implementing a computer-aided-design environment based on ℓ_1 nominal designs, in which the designer has the flexibility to incorporate frequency-domain and fixed-input constraints, without losing the qualitative information about the structure of the controllers obtained from the nominal designs. Such an environment has proven very powerful in designing controllers for real applications.

References

1. Barabanov, A. and Sokolov, A., The geometrical approach to l_1 optimal control, *Proc. 33rd IEEE Conf. Decision Control,* 1994.
2. Dahleh, M.A. and Diaz-Bobillo, I., *Control of Uncertain Systems: A Linear Programming Approach,* Prentice Hall, Englewood Cliffs, NJ, 1995.
3. Dahleh, M.A. and Khammash, M.H., Controller design for plants with structured uncertainty, *Automatica,* 29(1), 1993.
4. Dahleh, M.A. and Ohta, Y., A necessary and sufficient condition for robust BIBO stability, *Syst. Contr. Lett.,* 11, 1988.
5. Dahleh, M.A. and Pearson, J.B., $\mathcal{L}_1$ optimal feedback compensators for continuous time systems, *IEEE Trans. Automat. Control,* 32, October 1987.
6. Dahleh, M.A. and Pearson, J.B., ℓ_1 optimal feedback controllers for mimo discrete-time systems, *IEEE Trans. Automat. Control,* April 1987.
7. Dahleh, M.A. and Pearson, J.B., Optimal rejection of persistent disturbances, robust stability and mixed sensitivity minimization, *IEEE Trans. Automat. Control,* 33, August 1988.
8. Diaz-Bobillo, I.J. and Dahleh, M.A., Minimization of the maximum peak-to-peak gain: The general multiblock problem, *IEEE Trans. Automat. Control,* October 1993.
9. Khammash, M. and Pearson, J.B., Performance robustness of discrete-time systems with structured uncertainty, *IEEE Trans. Automat. Control,* 36, 1991.
10. Khammash, M. and Pearson, J.B., Robust disturbance rejection in ℓ_1-optimal control systems, *Syst. Control Lett.,* 14, 1990.
11. McDonald, J.S. and Pearson, J.B., ℓ_1-optimal control of multivariable systems with output norm constraints, *Automatica,* 27, 1991.
12. Shamma, J.S., Nonlinear state feedback for ℓ_1 optimal control, *Syst. Control Lett.,* to appear.
13. Vidyasagar, M., Optimal rejection of persistent bounded disturbances, *IEEE Trans. A-C,* 31(6), 1986.

20

The Structured Singular Value (μ) Framework

Gary J. Balas
University of Minnesota

Andy Packard
University of California, Berkeley

20.1 Introduction

This chapter gives a brief overview of the structured singular value (μ). The μ-based methods discussed are useful for analyzing the performance and robustness properties of linear feedback systems. Computational software for μ-based analysis and synthesis is available in commercial software products [2,4]. The interested reader might also consult the tutorials in references [8,14], and application-oriented papers, such as [1,6,13].

20.2 Shortcomings of Simple Robustness Analysis

Many of the theoretical robustness results for single-input, single-output (SISO) systems show that if a single-loop system has good robust stability characteristics, and good nominal performance characteristics, then, necessarily, it has reasonably good robust performance characteristics. Unfortunately, this "fact" is not, in general, true for multiloop systems. Also, for multiloop systems, checking the robustness via individual loop-at-a-time calculations can be misleading, because the interactions between the deviations are not accounted for in such an analysis. In this chapter, we illustrate these difficulties with examples and introduce the structured singular value as an analytical tool for uncertain, multivariable systems.

The first example concerns control of the angular velocity of a satellite spinning about one of its principal axes. Its mathematical origins are due to Doyle, and are alluded to in [14]. The closed-loop multivariable (MIMO) system is shown in Figure 20.1.

Set $a := 10$, and define

$$G := \frac{1}{s^2+a^2}\begin{bmatrix} s-a^2 & a(s+1) \\ -a(s+1) & s-a^2 \end{bmatrix}, \quad K_1 := \begin{bmatrix} 1 & 0 \\ 0 & 1 \end{bmatrix}, \quad K_2 := \frac{1}{1+a^2}\begin{bmatrix} 1 & -a \\ a & 1 \end{bmatrix}.$$

A minimal, state-space realization for the plant G is

$$G = \left[\begin{array}{c|c} A_G & B_G \\ \hline C_G & D_G \end{array}\right] = \left[\begin{array}{cc|cc} 0 & a & 1 & 0 \\ -a & 0 & 0 & 1 \\ \hline 1 & a & 0 & 0 \\ -a & 1 & 0 & 0 \end{array}\right] \tag{20.1}$$

In order to assess the robustness margins to perturbations in the input channels into the plant, consider the four-input, four-output system, denoted by M, in Figure 20.2. The lines from r_1 and r_2 which run above K_2 and G are included to define the tracking error, e_1 and e_2 explicitly.

Some important transfer functions are

$$M_{ry} = \frac{1}{s+1}I_2, \quad M_{w_1,z_1} = M_{w_2,z_2} = -\frac{1}{s+1}.$$

These imply that the nominal closed-loop system has decoupled command response, with a bandwidth of 1 rad/s, the crossover frequency in the first feedback loop is 1 rad/s, with phase margin of 90°, the gain margin in the first channel is infinite, the crossover frequency in the second loop is 1 rad/s, with phase margin of 90°, and the gain margin in the second channel is infinite. These suggest that the performance of the closed-loop system is excellent and that it is quite robust to perturbations in each input channel. Yet, consider a 5% variation in each channel at the input to the plant. Referring to Figure 20.3, let $\delta_1 = 0.05$, and $\delta_2 = -0.05$. The output response $y(t)$ to a unit-step reference input in channel 1 is shown in Figure 20.4 (along with the nominal responses). Note that the ideal behavior of the nominal system has degraded sharply despite the seemingly innocuous perturbations and excellent gain/phase margins in the closed-loop system. In fact, for a slightly larger perturbation, $\delta_1 = 0.11, \delta_2 = -0.11$, the closed-loop

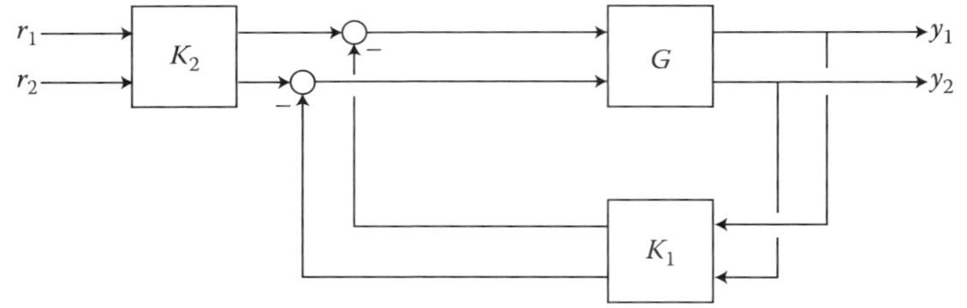

FIGURE 20.1 Nominal multiloop feedback system.

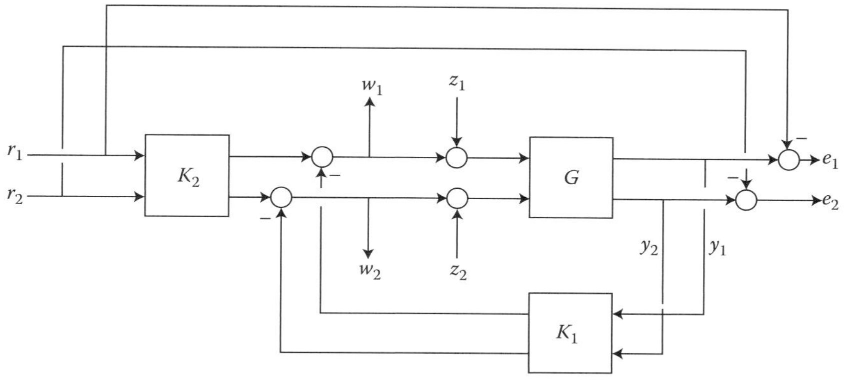

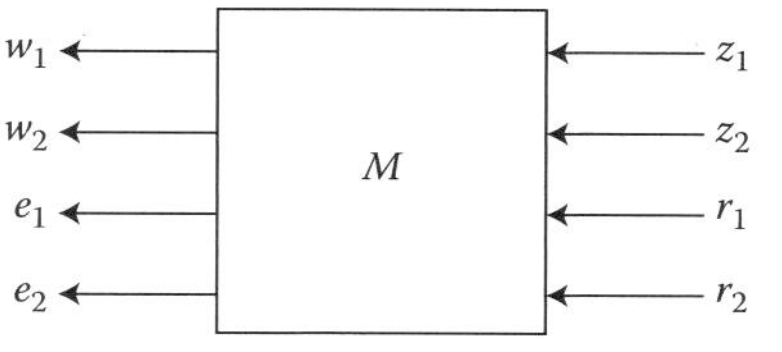

FIGURE 20.2 Closed-loop system with uncertainty model.

system is actually unstable. Why do these small perturbations cause such a significant degradation in performance? To answer this, calculate the 4×4 transfer matrix M represented in Figure 20.2, giving

$$\begin{bmatrix} w_1 \\ w_2 \\ e_1 \\ e_2 \end{bmatrix} = \begin{bmatrix} -\frac{1}{s+1} & -\frac{a}{s+1} & \frac{s-a^2}{s+1} & -a \\ \frac{a}{s+1} & -\frac{1}{s+1} & a & \frac{s-a^2}{s+1} \\ \frac{1}{s+1} & \frac{a}{s+1} & -\frac{s}{s+1} & 0 \\ -\frac{a}{s+1} & \frac{1}{s+1} & 0 & -\frac{s}{s+1} \end{bmatrix} \begin{bmatrix} z_1 \\ z_2 \\ r_1 \\ r_2 \end{bmatrix} =: M \begin{bmatrix} z_1 \\ z_2 \\ r_1 \\ r_2 \end{bmatrix}.$$

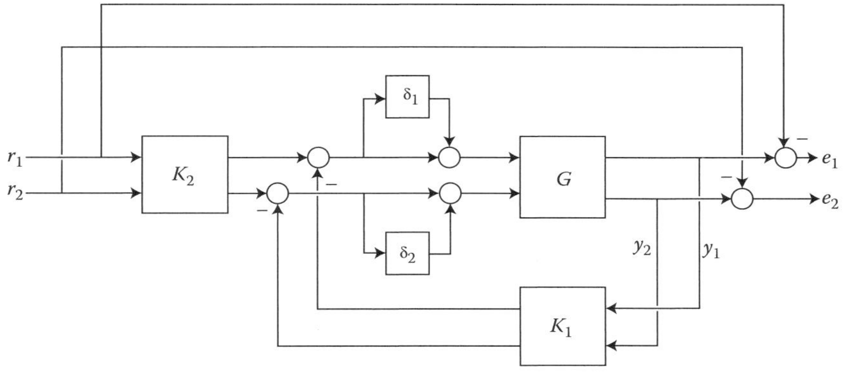

FIGURE 20.3 Satellite: Closed-loop system with uncertain elements, δ_1 and δ_2.

Note that the earlier calculations about the closed-loop system yielded information only about the $(1, 1)$, $(2,2)$, and $(3:4, 3:4)$ entries. The notation $(3:4, 3:4)$ denotes the 2×2 matrix formed by rows 3 to 4 and columns 3 to 4 of M. In particular, these entries are all small, in some sense. The neglected entries, $(1,2)$, $(2,1)$, $(1:2, 3:4)$, $(3:4, 1:2)$ are all quite large, because $a = 10$. It is these large off-diagonal entries, and the manner in which they enter, that causes the extreme sensitivity of the closed-loop system's performance to the perturbations δ_1 and δ_2. For instance, with $\delta_2 \equiv 0$, the perturbation δ_1 can only cause instability by making the transfer function $(1 - M_{w1,z1}\delta_1)^{-1}$ unstable. Similarly, with $\delta_1 \equiv 0$, the perturbation δ_2 can only cause instability by making the transfer function $(1 - M_{w2,z2}\delta_2)^{-1}$ unstable. Because both $M_{w1,z1}$ and $M_{w2,z2}$ are "small", this requires large perturbations, and the single-loop gain/phase margins reported earlier are accurate. However, acting together, the perturbations can cause instability by making

$$\left[\begin{bmatrix} 1 & 0 \\ 0 & 1 \end{bmatrix} - \begin{bmatrix} -\dfrac{1}{s+1} & -\dfrac{a}{s+1} \\ \dfrac{a}{s+1} & -\dfrac{1}{s+1} \end{bmatrix} \begin{bmatrix} \delta_1 & 0 \\ 0 & \delta_2 \end{bmatrix}\right]^{-1}$$

unstable. The denominator of this multivariable transfer function is

$$s^2 + (2 + \delta_1 + \delta_2)\, s + \left[1 + \delta_1 + \delta_2 + (a^2 + 1)\delta_1\delta_2\right].$$

By choosing $\delta_1 = \frac{1}{\sqrt{a^2+1}} \approx 0.1$, and $\delta_2 = -\delta_1$, the characteristic equation has a root at $s = 0$, indicating marginal stability. For slightly larger perturbations, a root moves into the right half-plane. The simultaneous nature of the perturbations has resulted in a much smaller destabilizing perturbation than predicted by the gain/phase margin calculations.

In terms of robust stability, the loop-at-a-time gain/phase margins only depended on the scalar transfer functions $M_{w1,z1}$ and $M_{w2,z2}$, but the robust stability properties of the closed-loop system to simultaneous perturbations actually depend on the 2×2 transfer function matrix $M_{w,z}$. Similarly, assessing the robust performance characteristics of the closed-loop system involves additional transfer functions ignored in the simple-minded analysis. Consider the perturbed closed-loop system in Figure 20.3. In terms of the

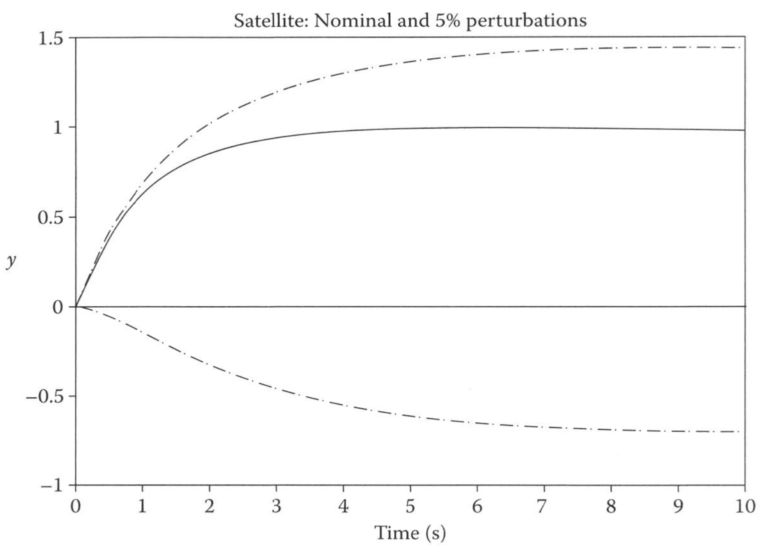

FIGURE 20.4 Satellite: Nominal (solid) and 5% perturbation responses (dashed).

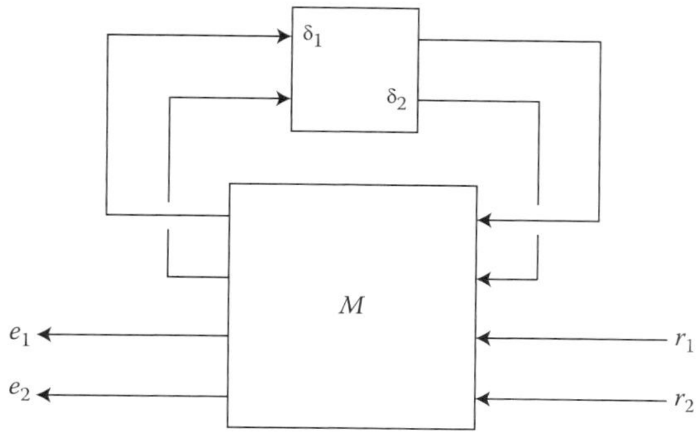

FIGURE 20.5 Perturbed system.

transfer function matrix M, the perturbed $r \to e$ transfer function can be drawn as shown in Figure 20.5. Partition the transfer function matrix M into four 2×2 blocks, as

$$M = \left[\begin{array}{c|c} M_{11} & M_{12} \\ \hline M_{21} & M_{22} \end{array}\right].$$

Then the perturbed closed-loop transfer function from r to e can be written as

$$e = \left[M_{22} + M_{21}\Delta\left(I - M_{11}\Delta\right)^{-1} M_{12}\right] d$$

where Δ is the structured matrix of perturbations, $\Delta = \text{diag}\,[\delta_1, \delta_2]$. Our initial information about the closed-loop system consisted of the diagonal entries of M_{11} and the entire matrix M_{22}. We have seen that the large off-diagonal entries of M_{11} created destabilizing interactions between the two perturbations. In the robust performance problem, there are additional relevant transfer functions, M_{21} and M_{12}, which were not analyzed in the loop-at-a-time robustness tests, or in the $r \to y$ nominal command response, though it is clear that these transfer functions may play a pivotal role in the robust performance characteristics of the closed-loop system.

Hence, by calculating the two single loop-at-a-time robustness tests and a nominal performance test, 10 of the 16 elements of the relevant 4×4 transfer function matrix are ignored. Any test which accounts for simultaneous perturbations along with the subsequent degradation of performance must be performed on the whole matrix. The point of this example is to show that there are some interesting issues in multivariable system analysis and standard SISO ideas that cannot be made into useful analytical tools simply by applying loop-at-a-time analysis. The structured singular value (μ), introduced in the next section, is a linear algebra tool useful for these types of MIMO system analyses.

20.3 Complex Structured Singular Value

This section is devoted to defining the structured singular value, a matrix function denoted by $\mu(\cdot)$ [5]. The notation we use is standard from linear algebra and systems theory. **R** denotes the set of real numbers, **C** denotes the set of complex numbers, $|\cdot|$ is the absolute value of elements in **R** or **C**, $\mathbf{R}^n$ is the set of real n vectors, $\mathbf{C}^n$ is the set of complex n vectors, $\|v\|$ is the Euclidean norm for $v \in \mathbf{C}^n$, $\mathbf{R}^{n\times m}$ is the set of $n \times m$ real matrices, $\mathbf{C}^{n\times m}$ is the set of $n \times m$ complex matrices, I_n is the $n \times n$ identity matrix, and $0_{n\times m}$ is an entirely zero matrix. For $M \in \mathbf{C}^{n\times m}$, M^T is the transpose of M, M^* is the complex-conjugate transpose of M, and $\bar{\sigma}(M)$ is the maximum singular value of M. For $M \in \mathbf{C}^{n\times n}$, $\lambda_i(M)$ is an eigenvalue of M, $\rho(M) := \max_i |\lambda_i(M)|$ is the *spectral radius of* M, and $\rho_R(M)$ is the real spectral radius of M, $\rho_R(M) := \max\{|\lambda| : \lambda \in \mathbf{R}, \det(\lambda I - M) = 0\}$, with $\rho_R(M) := 0$ if M has no real eigenvalues. If $M \in \mathbf{C}^{n\times n}$ satisfies $M = M^*$, then $M > 0$ denotes that M is positive definite, and $M^{\frac{1}{2}}$ means the unique positive definite Hermitian square root. For $M = M^*$, then $\lambda_{\max}(M)$ denotes the most positive eigenvalue.

We consider matrices $M \in \mathbf{C}^{n\times n}$. In the definition of $\mu(M)$, there is an underlying structure Δ, (a prescribed set of block diagonal matrices) on which everything following depends. This structure may be defined differently for each problem depending on the uncertainty and performance objectives. Defining the structure involves specifying three things: the total number of blocks, the type of each block, and their dimensions.

20.3.1 Purely Complex μ

Two types of blocks—*repeated scalar* and *full* blocks are considered. Two nonnegative integers, S and F, denote the number of *repeated scalar* blocks and the number of *full* blocks, respectively. To track the block dimensions, we introduce positive integers $r_1, \ldots, r_S$; $m_1, \ldots, m_F$. The ith repeated scalar block is $r_i \times r_i$, while the jth full block is $m_j \times m_j$. With those integers given, define $\Delta \subset \mathbf{C}^{n\times n}$ as

$$\Delta := \left\{\text{diag}\left[\delta_1 I_{r_1}, \ldots, \delta_S I_{r_S}, \Delta_{S+1}, \ldots, \Delta_{S+F}\right] : \ \delta_i \in \mathbf{C}, \Delta_{S+j} \in \mathbf{C}^{m_j\times m_j}, 1 \le i \le S, 1 \le j \le F\right\}. \quad (20.2)$$

For consistency among all the dimensions, $\sum_{i=1}^{S} r_i + \sum_{j=1}^{F} m_j$ must equal n. Often, we will need norm bounded subsets of Δ, and we introduce the notation $\mathbf{B}_\Delta := \{\Delta \in \Delta : \bar{\sigma}(\Delta) \le 1\}$. Note that in Equation 20.2 all of the repeated scalar blocks appear first, followed by the full blocks. This is done to simplify the notation and can easily be relaxed. The full blocks are also assumed to be square, but again, this is only to simplify notation.

Definition 20.1:

For $M \in \mathbf{C}^{n\times n}$, $\mu_\Delta(M)$ is defined

$$\mu_\Delta(M) := \frac{1}{\min\{\bar{\sigma}(\Delta) : \Delta \in \Delta, \det(I - M\Delta) = 0\}} \quad (20.3)$$

unless no $\Delta \in \Delta$ makes $I - M\Delta$ singular, in which case $\mu_\Delta(M) := 0$.

It is instructive to consider a "feedback" interpretation of $\mu_\Delta(M)$ at this point. Let $M \in \mathbf{C}^{n\times n}$ be given, and consider the loop shown in Figure 20.6. This picture represents the loop equations $u = Mv$, $v = \Delta u$. As long as $I - M\Delta$ is nonsingular, the only solutions of u, v to the loop equations are $u = v = 0$. However, if $I - M\Delta$ is singular, then there are infinitely many solutions to the equations, and the norms $\|u\|$, $\|v\|$ of the solutions can be arbitrarily large. Motivated by connections with stability of systems, we call this

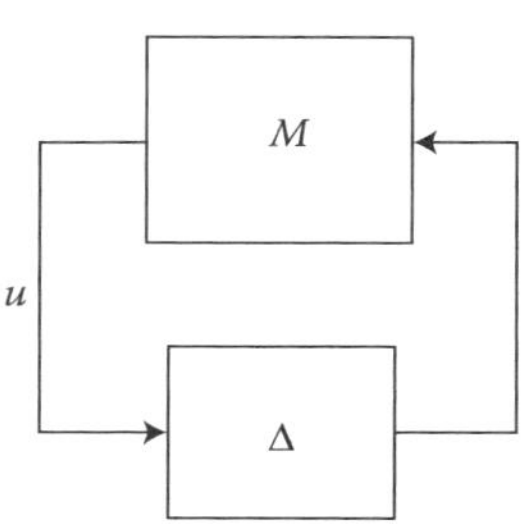

FIGURE 20.6 $M - \Delta$ interconnection.

constant matrix feedback system "unstable". Likewise, the term "stable" will describe the situation when the only solutions are identically zero. In this context then, $\mu_\Delta(M)$ provides a measure of the smallest structured Δ that causes "instability" of the constant matrix feedback loop in Figure 20.6. The norm of this "destabilizing" Δ is exactly $1/\mu_\Delta(M)$.

Consider $M \in \mathbf{C}^{5\times 4}$,

$$M := \begin{bmatrix} 0.100+0.070i & -0.154+0.162i & 0-0.560i & 0+42.000i \\ 0-0.273i & -0.300-0.280i & 2.860+0.546i & -26.000+72.800i \\ 0.100+0.175i & 0.077-0.108i & -0.400+0.210i & 5.000+3.500i \\ 0+0.002i & -0.004-0.002i & 0.006+0.011i & 0.200+0.420i \\ 0.024+0.028i & 0+0.027i & -0.066+0.042i & 0+0.700i \end{bmatrix} \tag{20.4}$$

to show the dependence of $\mu_\Delta(M)$ on the set Δ. Two different block structures compatible (in the sense of dimensions) with M are

$$\Delta_1 = \left\{\text{diag }[\delta_1, \delta_2, \Delta_3] : \delta_1, \delta_2 \in \mathbf{C}, \Delta_3 \in \mathbf{C}^{2\times 3}\right\}, \quad \text{and} \quad \Delta_2 = \left\{\text{diag }[\delta_1, \delta_2, \delta_3, \Delta_4] : \delta_i \in \mathbf{C}, \Delta_4 \in \mathbf{C}^{1\times 2}\right\}.$$

The definition yields $\mu_{\Delta_1}(M) \approx 8.32$, while $\mu_{\Delta_2}(M) \approx 2.42$.

An alternative expression for $\mu_\Delta(M)$ follows easily from the definition.

Lemma 20.1:

$$\mu_\Delta(M) = \max_{\Delta \in \mathbf{B}_\Delta} \rho(\Delta M)$$

Continuity of the function $\mu : \mathbf{C}^{n\times n} \to \mathbf{R}$ is apparent from this lemma. In general, though, the function $\mu : \mathbf{C}^{n\times n} \to \mathbf{R}$ is not a norm, because it doesn't satisfy the triangle inequality. However, for any $\alpha \in \mathbf{C}$, $\mu(\alpha M) = |\alpha|\mu(M)$, so it is related to how "big" the matrix is.

We can relate $\mu_\Delta(M)$ to familiar linear algebra quantities when Δ is one of two extreme sets:

- If $\Delta = \{\delta I : \delta \in \mathbf{C}\}$ $(S{=}1, F{=}0, r_1{=}n)$, then $\mu_\Delta(M) = \rho(M)$.
  Proof: The only Δ's in Δ which satisfy the $\det(I - M\Delta) = 0$ constraint are reciprocals of nonzero eigenvalues of M. The smallest one of these is associated with the largest (magnitude) eigenvalue, so, $\mu_\Delta(M) = \rho(M)$.
- If $\Delta = \mathbf{C}^{n\times n}$ $(S{=}0, F{=}1, m_1{=}n)$, then $\mu_\Delta(M) = \bar{\sigma}(M)$.
  Proof: If $\bar{\sigma}(\Delta) < \frac{1}{\bar{\sigma}(M)}$, then $\bar{\sigma}(M\Delta) < 1$, so $I - M\Delta$ is nonsingular. Applying Equation 20.3 implies $\mu_\Delta(M) \le \bar{\sigma}(M)$. On the other hand, let u and v be unit vectors satisfying $Mv = \bar{\sigma}(M)u$, and define $\Delta := \frac{1}{\bar{\sigma}(M)} vu^*$. Then $\bar{\sigma}(\Delta) = 1/\bar{\sigma}(M)$ and $I - M\Delta$ is obviously singular. Hence, $\mu_\Delta(M) \ge \bar{\sigma}(M)$.

Obviously, for a general Δ as in Equation 20.2, $\{\delta I_n : \delta \in \mathbf{C}\}$ must be included in $\Delta \subset \mathbf{C}^{n\times n}$. Hence directly from the definition of μ, and the two special cases above, we conclude that

$$\rho(M) \le \mu_\Delta(M) \le \bar{\sigma}(M). \tag{20.5}$$

These bounds by themselves may provide little information about the value of μ, because the gap between ρ and $\bar{\sigma}$ can be large. To see this, consider the matrix

$$M = \begin{bmatrix} 0 & \alpha \\ 0 & 0 \end{bmatrix}$$

and two different block structures, $\Delta_1 := \{\delta I_2 : \delta \in \mathbf{C}\}$ and $\Delta_2 := \left\{\Delta \in \mathbf{C}^{2\times 2}\right\}$. μ with respect to Δ_1, which corresponds to $\rho(M)$, is 0, independent of α. μ with respect to Δ_2, which corresponds to $\bar{\sigma}(M)$, is $|\alpha|$.

The bounds on μ can be refined with transformations on M that do not affect $\mu_\Delta(M)$, but do affect ρ and $\bar{\sigma}$. To do this, define two subsets, $\mathbf{Q}_\Delta$ and $\mathbf{D}_\Delta$ of $\mathbf{C}^{n\times n}$,

$$\mathbf{Q}_\Delta = \{Q \in \Delta : Q^*Q = I_n\} \tag{20.6}$$

$$\mathbf{D}_\Delta = \{\text{diag}[D_1, \ldots, D_S, d_{S+1}I_{m_1}, \ldots, d_{S+F}I_{m_F}] : D_i \in \mathbf{C}^{r_i \times r_i}, D_i = D_i^* > 0, d_{S+j} \in \mathbf{R}, d_{S+j} > 0\} \tag{20.7}$$

For any $\Delta \in \Delta, Q \in \mathbf{Q}_\Delta$, and $D \in \mathbf{D}_\Delta$,

$$Q^* \in \mathbf{Q}_\Delta, \quad Q\Delta \in \Delta, \quad \Delta Q \in \Delta \quad \bar{\sigma}(Q\Delta) = \bar{\sigma}(\Delta Q) = \bar{\sigma}(\Delta) \tag{20.8}$$

$$D^{\frac{1}{2}}\Delta = \Delta D^{\frac{1}{2}} \tag{20.9}$$

Theorem 20.1:

For all $Q \in \mathbf{Q}_\Delta$ and $D \in \mathbf{D}_\Delta$

$$\mu_\Delta(MQ) = \mu_\Delta(QM) = \mu_\Delta(M) = \mu_\Delta\left(D^{\frac{1}{2}}MD^{-\frac{1}{2}}\right) \tag{20.10}$$

Proof 20.1. For all $D \in \mathbf{D}_\Delta$ and $\Delta \in \Delta$,

$$\begin{aligned}\det(I - M\Delta) &= \det\left(I - MD^{-\frac{1}{2}}D^{\frac{1}{2}}\Delta\right)\\ &= \det\left(I - MD^{-\frac{1}{2}}\Delta D^{\frac{1}{2}}\right)\\ &= \det\left(D^{-\frac{1}{2}}D^{\frac{1}{2}} - D^{-\frac{1}{2}}MD^{-\frac{1}{2}}\Delta\right)\\ &= \det\left(I - D^{\frac{1}{2}}MD^{-\frac{1}{2}}\Delta\right)\end{aligned}$$

because D commutes with Δ. Therefore $\mu_\Delta(M) = \mu_\Delta\left(D^{\frac{1}{2}}MD^{-\frac{1}{2}}\right)$. Also, for each $Q \in \mathbf{Q}_\Delta$, $\det(I - M\Delta) = 0$ if and only if $\det(I - MQQ^*\Delta) = 0$. Because $Q^*\Delta = \Delta$ *and* $\bar{\sigma}(Q^*\Delta) = \bar{\sigma}(\Delta)$, $\mu_\Delta(MQ) = \mu_\Delta(M)$ as desired. The argument for QM is the same.

Therefore, the bounds in Equation 20.5 can be tightened to

$$\begin{aligned}\max_{Q\in\mathbf{Q}} \rho(QM) &\leq \max_{\Delta\in\mathbf{B}_\Delta} \rho(\Delta M)\\ &= \mu_\Delta(M) \leq \inf_{D\in\mathbf{D}} \bar{\sigma}\left(D^{\frac{1}{2}}MD^{-\frac{1}{2}}\right)\end{aligned} \tag{20.11}$$

where the equality comes from Lemma 20.1.

The lower bound, $\max_{Q\in\mathbf{Q}} \rho(QM)$, is always an equality [5]. Unfortunately, the quantity $\rho(QM)$ can have multiple local maxima which are not global. Thus local search cannot be guaranteed to obtain μ, but can only yield a lower bound. The upper bound can be reformulated as a convex optimization problem, so that the global minimum can, in principle, be found. Unfortunately, the upper bound is not always equal to μ. For block structures Δ satisfying $2S + F \leq 3$, the upper bound is always equal to $\mu_\Delta(M)$, and for block structures with $2S + F > 3$, matrices exist for which μ is less than the infimum [5,9].

Convexity properties make the upper bound computationally attractive. The simplest convexity property is given in the following theorem, which shows that the function $\bar{\sigma}\left(D^{\frac{1}{2}}MD^{-\frac{1}{2}}\right)$ has convex level sets.

Theorem 20.2:

Let $M \in \mathbf{C}^{n\times n}$ be given, along with a scaling set $\mathbf{D}_\Delta$, and $\beta > 0$. Then the set $\left\{D \in \mathbf{D}_\Delta : \bar{\sigma}\left(D^{\frac{1}{2}}MD^{-\frac{1}{2}}\right) < \beta\right\}$ is convex.

Proof 20.2. The following chain of equivalences comprises the proof:

$$\begin{aligned}\bar{\sigma}\left(D^{\frac{1}{2}}MD^{-\frac{1}{2}}\right) < \beta \quad &\Leftrightarrow \quad \lambda_{\max}\left(D^{-\frac{1}{2}}M^*D^{\frac{1}{2}}D^{\frac{1}{2}}MD^{-\frac{1}{2}}\right) < \beta^2 \\ &\Leftrightarrow \quad D^{-\frac{1}{2}}M^*D^{\frac{1}{2}}D^{\frac{1}{2}}MD^{-\frac{1}{2}} - \beta^2 I < 0 \\ &\Leftrightarrow \quad M^*DM - \beta^2 D < 0\end{aligned} \tag{20.12}$$

The latter is clearly a convex condition in D because it is linear.

The final condition in Equation 20.12 is called a *Linear Matrix Inequality* (LMI) in the variable D. In [3], a large number of control synthesis and analysis problems are cast as solutions of LMI's.

20.3.2 Mixed μ: Real and Complex Uncertainty

Until this point, this section has dealt with complex-valued perturbation sets. In specific instances, it may be more natural to describe modeling errors with real perturbations, for instance, when the real coefficients of a linear differential equation are uncertain. Although these perturbations can be treated simply as complex, proceeding with a complex μ analysis, the results may be conservative. Hence, theory and algorithms to test for robustness and performance degradation with mixed (real blocks and complex blocks) perturbation have been developed.

Definition 20.1 of μ can be used for more general sets Δ, such as those containing real and complex blocks. There are 3 types of blocks, *repeated real scalar, repeated complex scalar,* and *complex full* blocks. As before in Section 20.3, S and F, denote the number of *repeated, complex scalar* blocks and the number of *complex full* blocks, respectively. V denotes the number of *repeated, real scalar* blocks. The block dimensions of the real block are denoted by the positive integers $t_1, \ldots, t_V$. With these integers given, and r_i and m_j as defined in Section 20.3, define Δ as

$$\begin{aligned}\Delta = \Big\{&\text{diag}\left[\delta_1^r I_{t_1}, \ldots, \delta_V^r I_{t_V}, \delta_{V+1}^c I_{r_1}, \ldots, \delta_{V+S}^c I_{r_S}, \Delta_{V+S+1}, \ldots, \Delta_{V+S+F}\right] : \\ &\delta_k^r \in \mathbf{R}, \delta_{V+i}^c \in \mathbf{C}, \Delta_{V+S+j} \in \mathbf{C}^{m_j\times m_j}, 1 \le k \le V, 1 \le i \le S, 1 \le j \le F\Big\}.\end{aligned} \tag{20.13}$$

For consistency among all the dimensions, $\sum_{k=1}^{V} t_k + \sum_{i=1}^{S} r_i + \sum_{j=1}^{F} m_j$ must equal $= n$.

The mixed μ function inherits many of the properties of the purely complex μ function, [5,7]. However, in some aspects such as continuity, the mixed μ problem can be fundamentally different from the complex μ problem. It is now well-known that real μ problems can be discontinuous in the problem data. Beside adding computational difficulties to the problem, the utility of real μ is doubtful as a robustness measure in such cases, the system model is always a mathematical abstraction from the real world, computed to finite precision. It has been shown that, for many *mixed* μ problems, μ is continuous. The main idea is that, *if a mixed μ problem has complex uncertainty blocks that "count," then the function is continuous.* From an engineering viewpoint, "count" implies that the complex uncertainty affects the value of μ. This is the usual case because one is usually interested in robust performance problems (which therefore contain at least one complex block – see Section 20.5.2), or robust stability problems with some unmodeled dynamics, which are naturally covered with complex uncertainty. Purely real problems can be made continuous by adding a small amount of complex uncertainty to each real uncertainty (see [1] for an example). Consequently, a small amount of phase uncertainty is added to the gain uncertainty.

The theory for bounding (both lower and upper) mixed real/complex bounds is much more complicated to describe than the bounding theory for complex μ. The lower bound for the mixed case is a real eigenvalue maximization problem. Techniques to solve approximately for a mixed μ lower bound using power algorithms have been derived, and are similar to those used for a lower bound for complex μ [15]. The mixed μ upper bound takes the form of a more complicated version of the same problem, involving an additional "G scaling matrix" which scales only the real uncertainty blocks. This minimization, involving an LMI expression similar to Equation 20.12, is computed using convex optimization techniques similar to those for the purely complex upper bound. See references [7,15] for more computational details.

20.3.3 Frequency Domain, Robust Stability Tests with μ

The best-known use of μ as a robustness analysis tool is in the frequency domain [5,14]. Suppose $\hat{G}(s)$ is a stable, multi-input, multioutput transfer function of a linear system. For clarity, assume $\hat{G}$ has n_z inputs and n_w outputs. Let Δ be a block structure, as in Equation 20.2, and assume that the dimensions are such that $\Delta \subset \mathbf{C}^{n_z \times n_w}$. We want to consider feedback perturbations to $\hat{G}$, themselves dynamical systems, with the block-diagonal structure of the set Δ, in Figure 20.7.

Let $\mathbf{S}$ denote the set of real-rational, proper, stable, transfer matrices (of appropriate dimensions, which should be clear from context). Associated with any block structure Δ, let $\mathbf{S}_\Delta$ denote the set of all block diagonal, stable rational transfer functions, with diagonal block structure as in Δ.

$$\mathbf{S}_\Delta := \left\{\Delta \in \mathbf{S} : \Delta(s_o) \in \Delta \text{ for all } \ s_o \in \bar{\mathbf{C}}_+\right\}.$$

Theorem 20.3:

Let $\beta > 0$. The loop in Figure 20.7 is well-posed and internally stable for all $\Delta \in \mathbf{S}_\Delta$ with $\|\Delta\|_\infty < \frac{1}{\beta}$, where $\|\Delta(s)\|_\infty := \max_{\omega\in\mathcal{R}} \bar{\sigma}\left(\Delta(j\omega)\right)$ if, and only if,

$$\|G\|_\Delta := \sup_{\omega\in\mathbf{R}} \ \mu_\Delta\left(\hat{G}(j\omega)\right) \leq \beta$$

In summary, the peak value on the μ plot of the frequency response of the known linear system that the perturbation "sees" determines the size of perturbations against which the loop is robustly stable.

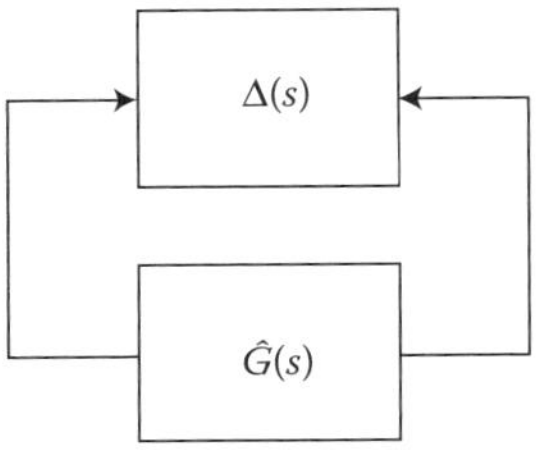

FIGURE 20.7 $\hat{G} - \Delta$ feedback loop block diagram.

Remark 20.1

If the peak occurs at $\omega = 0$, there are systems G where the theorem statement needs to be modified to be correct. In particular, it may be impossible to do what the theorem statement implies, that is, construct a real-rational perturbation $\Delta \in \mathbf{S}_\Delta$ with $\|\Delta\|_\infty = \left(\|G\|_\Delta\right)^{-1}$ and $(I - G\Delta)^{-1}$ unstable. Rather, for any $\epsilon > 0$, there will be a real-rational perturbation $\Delta \in \mathbf{S}_\Delta$ with $\|\Delta\|_\infty = \left(\|G\|_\Delta\right)^{-1} + \epsilon$ and $(I - G\Delta)^{-1}$ unstable. These facts can be ascertained from results in [9,10].

The overall implication of this modification can be viewed in two opposite ways. If the theorem is used for actual robustness analysis, the original viewpoint that $\left(\|G\|_\Delta\right)^{-1}$ is the size of the smallest real-rational perturbation causing instability is certainly the "right" mental model to use, because the small correction that may be needed is arbitrarily small, and hence of little engineering relevance. On the other hand, if Theorem 20.3 is being used to prove another theorem, then one needs to be very careful.

In the next section, the linear algebra results which extend μ from a robust stability tool to a robust performance tool are covered. These linear algebra results are then applied to give a frequency-domain robust performance test in Section 20.5.2.

20.4 Linear Fractional Transformations and μ

The use of μ in control theory depends to a great extent on its intimate relationship with a class of general linear feedback loops called *Linear Fractional Transformations* (LFTs) [11]. This section explores this relationship with some simple theorems that can be obtained almost immediately from the definition of μ. To introduce these, consider a complex matrix M partitioned as

$$M = \begin{bmatrix} M_{11} & M_{12} \\ M_{21} & M_{22} \end{bmatrix} \tag{20.14}$$

and suppose that there is a defined block structure $\boldsymbol{\Delta}_1$ compatible in size with M_{11} (for any $\Delta_1 \in \boldsymbol{\Delta}_1$, $M_{11}\Delta_1$ is square). For $\Delta_1 \in \boldsymbol{\Delta}_1$, consider the loop equations

$$z = M_{11}w + M_{12}d; \quad e = M_{21}w + M_{22}d; \quad w = \Delta_1 z \tag{20.15}$$

which correspond to the block diagram in Figure 20.8 (note the similarity to Figure 20.5 in Section 20.2).

The set of Equations 20.15 is called *well posed* if, for any vector d, unique vectors w, z, and e exist satisfying the loop equations. The set of equations is well-posed if, and only if, the inverse of $I - M_{11}\Delta_1$ exists. If not, then depending on d and M, there is either no solution to the loop equations, or there are an infinite number of solutions. When the inverse does exist, the vectors e and d must satisfy $e = F_u(M, \Delta_1)d$, where

$$F_u(M, \Delta_1) := M_{22} + M_{21}\Delta_1(I - M_{11}\Delta_1)^{-1}M_{12}. \tag{20.16}$$

$F_u(M, \Delta_1)$ is called a Linear Fractional Transformation on M by Δ_1 and, in a feedback diagram, appears as in Figure 20.8. $F_u(M, \Delta_1)$ denotes that the "upper" loop of M is closed by Δ_1. An analogous formula

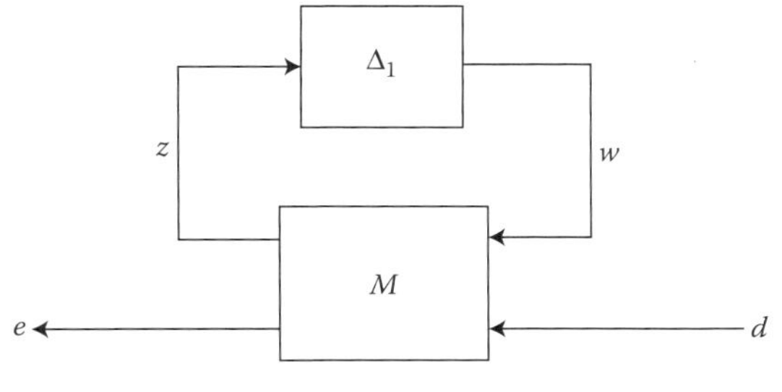

FIGURE 20.8 Linear fractional transformation.

describes $F_l(M, \Delta_2)$ which is the resulting matrix obtained by closing the "lower" loop of M with a matrix $\Delta_2 \in \mathbf{\Delta}_2$.

In this formulation, the matrix M_{22} is assumed to be something nominal, and $\Delta_1 \in \mathbf{B}_{\Delta_1}$ is viewed as a norm-bounded perturbation from an allowable perturbation class, $\mathbf{\Delta}_1$. The matrices M_{12}, M_{21}, and M_{22} and the formula $F_u(M, \Delta_1)$ reflect prior knowledge showing how the unknown perturbation affects the nominal map, M_{22}. This type of uncertainty, called *linear fractional*, is natural for many control problems and encompasses many other special cases considered by researchers.

20.4.1 Well-Posedness and Performance for Constant LFTs

Let M be a complex matrix partitioned as

$$M = \begin{bmatrix} M_{11} & M_{12} \\ M_{21} & M_{22} \end{bmatrix} \tag{20.17}$$

and suppose that there are two defined block structures, $\mathbf{\Delta}_1$ and $\mathbf{\Delta}_2$, compatible in size with M_{11} and M_{22} respectively. Define a third structure $\mathbf{\Delta}$ as

$$\mathbf{\Delta} = \left\{ \begin{bmatrix} \Delta_1 & 0 \\ 0 & \Delta_2 \end{bmatrix} : \Delta_1 \in \mathbf{\Delta}_1, \Delta_2 \in \mathbf{\Delta}_2 \right\}. \tag{20.18}$$

Now there are three structures for which we may compute μ. The notation we use to keep track of this is as follows: $\mu_1(\cdot)$ is with respect to $\mathbf{\Delta}_1$, $\mu_2(\cdot)$ is with respect to $\mathbf{\Delta}_2$, : $\mu_{\mathbf{\Delta}}(\cdot)$ is with respect to $\mathbf{\Delta}$. In view of this, $\mu_1(M_{11})$, $\mu_2(M_{22})$ and $\mu_{\mathbf{\Delta}}(M)$ all make sense, though, for instance, $\mu_1(M)$ does not. For notation, let $\mathbf{B}_1 := \{\Delta_1 \in \mathbf{\Delta}_1 : \bar{\sigma}(\Delta_1) \leq 1\}$.

Clearly, the linear fractional transformation $F_u(M, \Delta_1)$ is well-posed for all $\Delta_1 \in \mathbf{B}_{\Delta_1}$ if, and only if, $\mu_1(M_{11}) < 1$. As the "perturbation" Δ_1 deviates from zero, the matrix $F_u(M, \Delta_1)$ deviates from M_{22}. The range of values for $\mu_2(F_u(M, \Delta_1))$ is intimately related to $\mu_{\mathbf{\Delta}}(M)$, as follows:

Theorem 20.4: Main Loop Theorem

The following are equivalent:

$$\mu_{\mathbf{\Delta}}(M) < 1 \quad \Longleftrightarrow \quad \begin{cases} \mu_1(M_{11}) < 1, \text{ and} \\ \max\limits_{\Delta_1 \in \mathbf{B}_1} \mu_2(F_u(M, \Delta_1)) < 1 \end{cases}.$$

Proof 20.3. The proof of this theorem is based on the definition of μ and Schur formulae for determinants of block partitioned matrices as in [9]. The Main Loop Theorem forms the basis for all uses of μ in linear system robustness analysis, whether from a state-space, frequency-domain, or Lyapunov approach.

20.5 Robust Performance Tests Using μ and Main Loop Theorem

Often, stability is not the only property of a closed-loop system that must be robust to perturbations. Typically there are exogenous disturbances acting on the system (wind gusts, sensor noise) which result in tracking and regulation errors. Under perturbation, the effect of these disturbances on error signals can greatly increase. In most cases, long before the onset of instability, the closed-loop performance will be unacceptably degraded. Hence the need for a "robust performance" test to indicate the worst-case level of performance degradation for a given level of perturbations.

20.5.1 Characterization of Performance in μ Setting

Within the structured singular value setting, the most natural (mathematical) way to characterize acceptable performance is in terms of MIMO $\|\cdot\|_\infty$ ($\mathcal{H}_\infty$) norms, discussed in detail in other Chapters (29 and 40) of this Handbook. In this section, we quickly review the $\mathcal{H}_\infty$ norm, and interpretations.

Suppose T is a MIMO stable linear system, with transfer function matrix $T(s)$. For a given driving signal $\tilde{d}(t)$, define $\tilde{e}$ as the output, as in the left-hand diagram of Figure 20.9.

Assume that the dimensions of T are $n_e \times n_d$. Let $\beta > 0$ be defined as

$$\beta := \|T\|_\infty := \max_{\omega\in\mathbf{R}} \bar{\sigma}\left[T(j\omega)\right]. \tag{20.19}$$

A time-domain, sinusoidal, steady-state interpretation of this quantity is as follows:

Fact: For any frequency $\bar{\omega} \in \mathbf{R}$, any vector of amplitudes $a \in \mathbf{R}_{n_d}$, and any vector of phases $\phi \in \mathbf{R}^{n_d}$, with $\|a\|_2 \leq 1$, define a time signal

$$\tilde{d}(t) = \begin{bmatrix} a_1 \sin\left(\bar{\omega}t + \phi_1\right) \\ \vdots \\ a_{n_d} \sin\left(\bar{\omega}t + \phi_{n_d}\right) \end{bmatrix}.$$

Apply this input to the system T, resulting in a steady-state response $\tilde{e}_{ss}$ of the form

$$\tilde{e}_{ss}(t) = \begin{bmatrix} b_1 \sin\left(\bar{\omega}t + \psi_1\right) \\ \vdots \\ b_{n_e} \sin\left(\bar{\omega}t + \psi_{n_e}\right) \end{bmatrix}.$$

The vector $b \in \mathbf{R}^{n_e}$ will satisfy $\|b\|_2 \leq \beta$. Moreover, β, as defined in Equation 20.19, is the smallest number for which this fact is true for every $\|a\|_2 \leq 1$, $\bar{\omega}$, and ϕ.

Multivariable performance objectives are represented by a single, MIMO $\|\cdot\|_\infty$ objective on a closed-loop transfer function. Because many objectives are being lumped into one matrix and the associated cost is the norm of the matrix, it is important to use frequency-dependent weighting functions, so that different requirements can be meaningfully combined into a single cost function.

In the weighting function selection, diagonal weights are most easily interpreted. Consider the right-hand diagram of Figure 20.9. Assume that W_L and W_R are diagonal, stable transfer function matrices, with diagonal entries denoted L_i and R_i. Bounds on the quantity $\|W_L T W_R\|_\infty$ will imply bounds about the sinusoidal steady-state behavior of the signals $\tilde{d}$ and $\tilde{e}(= T\tilde{d})$. Specifically, for sinusoidal signal $\tilde{d}$,

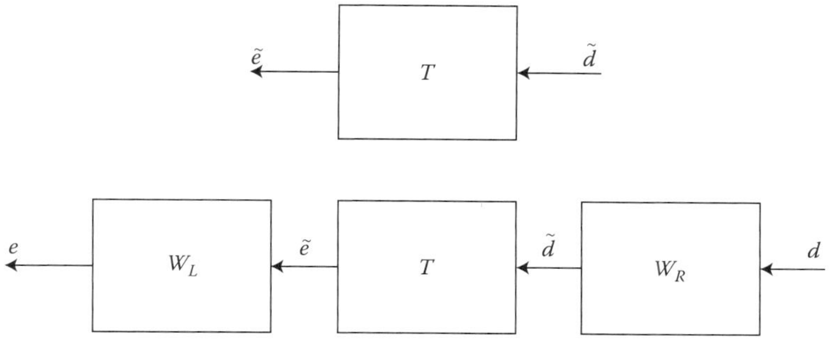

FIGURE 20.9 Unweighted and weighted MIMO systems.

the steady-state relationship between $\tilde{e}(= T\tilde{d})$, $\tilde{d}$ and $\|W_L T W_R\|_\infty$ follows. The steady-state solution $\tilde{e}_{ss}$, denoted as

$$\tilde{e}_{ss}(t) = \begin{bmatrix} \tilde{e}_1 \sin\left(\bar{\omega}t + \psi_1\right) \\ \vdots \\ \tilde{e}_{n_e} \sin\left(\bar{\omega}t + \psi_{n_d}\right) \end{bmatrix},$$

satisfies $\sum_{i=1}^{n_e} \left|L_i(j\bar{\omega})\tilde{e}_i\right|^2 \leq 1$ for all sinusoidal input signals $\tilde{d}$ of the form,

$$\tilde{d}(t) = \begin{bmatrix} \tilde{d}_1 \sin\left(\bar{\omega}t + \phi_i\right) \\ \vdots \\ \tilde{d}_{n_d} \sin\left(\bar{\omega}t + \phi_{n_d}\right) \end{bmatrix},$$

satisfying

$$\sum_{i=1}^{n_d} \frac{\left|\tilde{d}_i\right|^2}{\left|R_i(j\bar{\omega})\right|^2} \leq 1$$

if, and only if, $\|W_L T W_R\|_\infty \leq 1$.

20.5.2 Frequency-Domain Robust Performance Tests

Recall from Section 20.3.1, that if Δ is a single full complex block, then the function μ_Δ is simply the maximum singular value function. We can use this fact, along with the Main Loop Theorem (Theorem 20.4) and the $\mathcal{H}_\infty$ notion of performance, to obtain the central robust performance theorem for perturbed transfer functions.

Assume G is a stable linear system, with real-rational, proper transfer function $\hat{G}$. The dimension of G is $n_z + n_d$ inputs and $n_w + n_e$ outputs. Partition G in the obvious manner, so that G_{11} has n_z inputs and n_w outputs, and so on. Let $\boldsymbol{\Delta} \subset \mathbf{C}^{n_w \times n_z}$ be a block structure, as in Equation 20.2. For $\Delta \in \mathbf{S}_{\boldsymbol{\Delta}}$, consider the behavior of the perturbed system in Figure 20.10.

Define an augmented block structure

$$\boldsymbol{\Delta}_P := \left\{ \begin{bmatrix} \Delta & 0 \\ 0 & \Delta_F \end{bmatrix} : \Delta \in \boldsymbol{\Delta}, \Delta_F \in \mathbf{C}^{n_d \times n_e} \right\}.$$

Δ_F corresponds to the Δ_2 block of the Main Loop Theorem. It is used to compute bounds on $\bar{\sigma}\left(\cdot\right)$ of the perturbed transfer function $F_u\left(\hat{G}, \Delta\right)$ as Δ takes on values in $\mathbf{S}_{\boldsymbol{\Delta}}$.

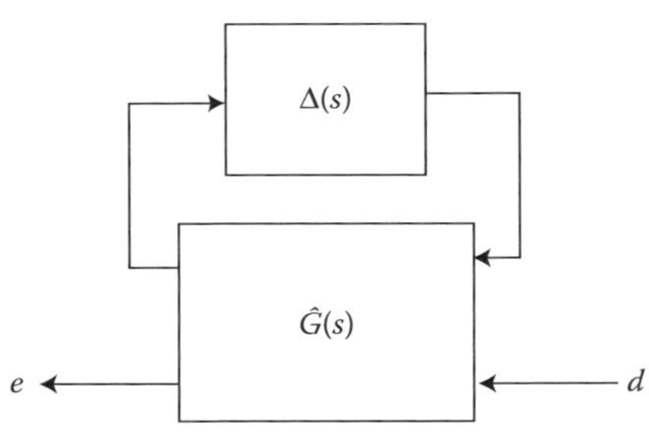

FIGURE 20.10 Robust performance LFT.

Theorem 20.5:

Let $\beta > 0$. For all $\Delta \in \mathbf{S}_\Delta$ with $\|\Delta\|_\infty < \frac{1}{\beta}$, the perturbed system in Figure 20.10 is well-posed, internally stable, and $\left\|F_u\left(\hat{G}, \Delta\right)\right\|_\infty \leq \beta$ if, and only if,

$$\|G\|_{\Delta_P} := \sup_{\omega \in \mathbf{R}} \mu_{\Delta_P}\left(\hat{G}(j\omega)\right) \leq \beta.$$

See [14] for details of the proof. The robust performance theorem provides a test to determine if the performance of the system $F_u\left(\hat{G}, \Delta\right)$ remains acceptable for all possible norm-bounded perturbations.

20.5.3 Robust Performance Example

It is instructive to carry out these steps on a simple example. Here, we analyze the robust stability of a simple single-loop feedback regulation system with two uncertainties. The plant is a lightly-damped, nominal two-state system with uncertainty in the (2, 1) entry of the A matrix (the frequency-squared coefficient) and unmodeled dynamics (in the form of multiplicative uncertainty) at the control input. The overall block diagram of the uncertain closed-loop system is shown in Figure 20.11.

The two-state system with uncertainty in the A matrix is represented as an upper linear fractional transformation about a two-input, two-output, two-state system H, whose realization is

$$A = \begin{bmatrix} 0 & 1 \\ -16 & -0.16 \end{bmatrix}, \quad B = \begin{bmatrix} 0 & 0 \\ 1 & 1 \end{bmatrix}, \quad C = \begin{bmatrix} 6.4 & 0 \\ 16 & 0 \end{bmatrix}, \quad \text{and} \quad D = \begin{bmatrix} 0 & 0 \\ 0 & 0 \end{bmatrix}$$

The resulting second order system takes the form

$$F_u\left(H, \delta_1\right) = \frac{16}{s^2 + 0.16s + 16(1 + 0.4\delta_1)}$$

If we assume that δ_1 is unknown, but satisfies $|\delta_1| \leq 1$, then we interpret the second-order system to have 40% uncertainty in the denominator entry of the natural frequency-squared coefficient.

The plant is also assumed to have unmodeled dynamics at the input. This could arise from an unmodeled, or unreliable, actuator, for instance. The uncertainty is assumed to be about 20% at low frequency, rising to 100% at 6.5 rad/s. We model it using the multiplicative uncertainty model, using a first-order weight, $W_u = (6.5s + 8)/(s + 42)$. In the block diagram, this is represented with the simple linear fractional transformation involving δ_2.

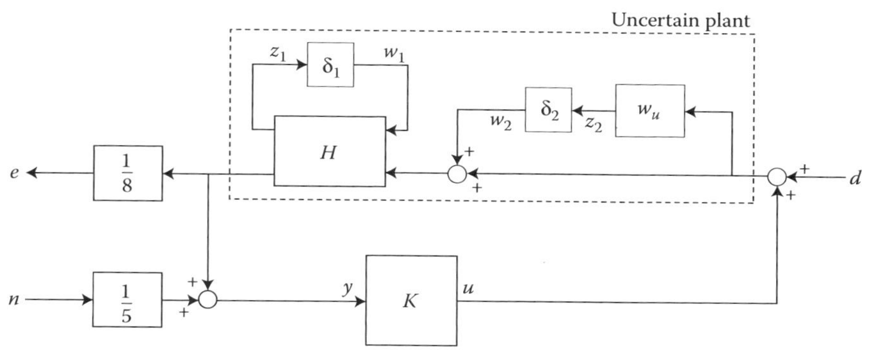

FIGURE 20.11 Robust stability/performance example.

The closed-loop performance objective is $\|T\|_\infty \leq 1$, where T is the transfer function from input disturbance d and sensor noise n to the output error e,

$$e = T\begin{bmatrix} d \\ n \end{bmatrix} = [T_1 \; T_2]\begin{bmatrix} d \\ n \end{bmatrix}$$

Note that the closed-loop T is a function of both δ_1 and δ_2. The scalar blocks which weight the error and the noise are used to normalize the two transfer functions that make up T. Finally, for comparison, the open-loop system has $\|T_1\|_\infty \approx 6$, and $\|T_2\|_\infty = 0$.

For this example, the controller is chosen as

$$K = \frac{-12.56s^2 + 17.32s + 67.28}{s^3 + 20.37s^2 + 136.74s + 179.46}.$$

Finally, $G(s)$ in Figure 20.12 denotes the closed-loop transfer function matrix from Figure 20.11. The dimensions of G are two states, four inputs and three outputs.

In terms of G, we have

$$T = F_u\left(G, \begin{bmatrix} \delta_1 & 0 \\ 0 & \delta_2 \end{bmatrix}\right)$$

Hence, using Theorems 20.3 and 20.5, the robust stability and robust performance of the closed-loop system can be ascertained by appropriate structured singular value calculations on G (or particular subblocks of G). In the next section, we analyze the robust stability and robust performance of the closed-loop system for a variety of assumptions on the uncertain elements, δ_1 and δ_2.

20.5.3.1 Analysis

For notational purposes, partition $G(s)$ into

$$G = \begin{bmatrix} G_{11} & G_{12} \\ G_{21} & G_{22} \end{bmatrix} \tag{20.20}$$

where $G_{11}(s)$ is 2×2, and $G_{22}(s)$ is 2×1. The first two inputs and outputs of G are associated with the perturbation structure and the third and fourth inputs and third output correspond to an exogenous multivariable disturbance signal, and associated error. For robust stability calculations we are only interested in the submatrix G_{11}, and for robust performance calculations the entire matrix G.

Robust stability calculations are performed with respect to two different block structures:

$$\Delta_1 := \left\{\text{diag}\,[\delta_1, \delta_2] : \delta_1, \delta_2 \in \mathbf{C}\right\},$$
$$\Delta_2 := \left\{\text{diag}\,[\delta_1, \delta_2] : \delta_1 \in \mathbf{R}, \delta_2 \in \mathbf{C}\right\}$$

For robust performance calculations, a 2×1 full block is appended to Δ_i for the performance calculation, yielding $\Delta_P \subset \mathbf{C}^{4\times 3}$. The two block structures used to evaluate robust performance are:

$$\Delta_{P_1} := \left\{\text{diag}\,[\delta_1, \delta_2, \Delta_F] : \delta_1, \delta_2 \in \mathbf{C}, \Delta_F \in \mathbf{C}^{2\times 1}\right\},$$
$$\Delta_{P_2} := \left\{\text{diag}\,[\delta_1, \delta_2, \Delta_F] : \delta_1 \in \mathbf{R}, \delta_2 \in \mathbf{C}, \Delta_F \in \mathbf{C}^{2\times 1}\right\}$$

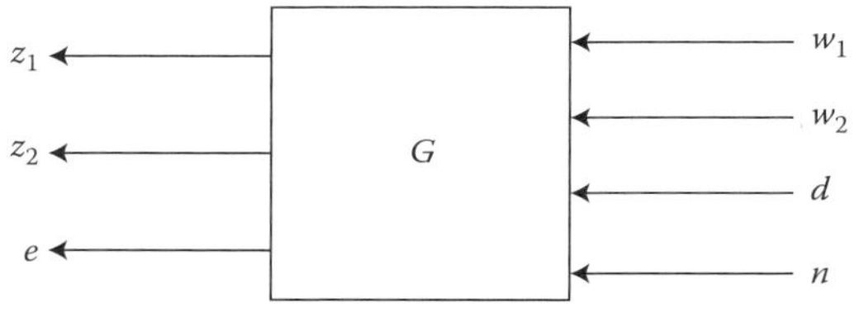

FIGURE 20.12 Closed-loop interconnection.

All the upper and lower bounds μ calculations are performed using the μ *Analysis and Synthesis Toolbox* [2].

20.5.3.2 Robust Stability

The robustness of the closed-loop system with respect to linear, complex time-invariant structured perturbations, Δ_1, is a μ test on $G_{11}(j\omega)$. The complex, robust stability bounds from the μ calculation are shown in Figure 20.13 (Note that the upper and lower bounds are identical.). The peak μ value is about 1.29, hence for any $\Delta(s) \in \mathbf{S}_{\Delta_1}$ stability is preserved as long as $\|\Delta(s)\|_\infty < \frac{1}{1.29}$, and there is a perturbation $\Delta_{dest}(s)$, of the correct structure, with $\|\Delta_{dest}\|_\infty = \frac{1}{1.29}$ that does cause instability.

The nominal performance of this system is defined by the $\mathcal{H}_\infty$ norm of the transfer function G_{22} is $\|G_{22}\|_\infty = 0.22$. The maximum singular value of G_{22} is plotted across frequency in Figure 20.14. The robustness and performance measures were originally scaled to be less than 1 when they were achieved. Therefore, the system *is not* robustly stabilized with respect to linear, time-invariant structured complex perturbations of size 1, but it achieves the performance objective on the nominal system.

Recall that the first uncertainty, δ_1, corresponds to uncertainty in the $A(2, 1)$ coefficient and the second uncertainty, δ_2, corresponds to input multiplicative modeling error. The $A(2, 1)$ coefficient uncertainty can be treated as a real uncertainty. This would imply that the magnitude $A(2, 1)$ varies between 9.6 and 22.4. In the initial robust stability analysis, both of these uncertainties were modeled as complex perturbations, a potentially more conservative representation of the uncertainty. Let us re-analyze the system with respect to Δ_2 where δ_1 is treated as a real perturbation.

We can analyze the robust stability of the system with respect to one of the uncertainties being real and the other uncertainty complex. This is shown in the mixed robust stability plot shown in Figure 20.13. Notice that when the $A(2, 1)$ uncertainty is treated as a real perturbation, and the input multiplicative uncertainty is complex, the mixed robust stability μ value is reduced from 1.29 to 0.84. Hence the

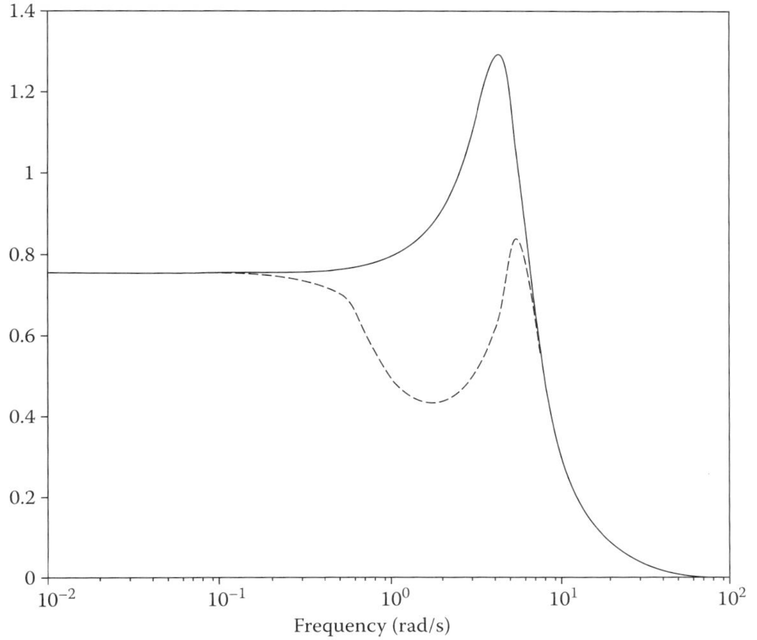

FIGURE 20.13 Complex robust stability (solid) and mixed robust stability (dashed) plots.

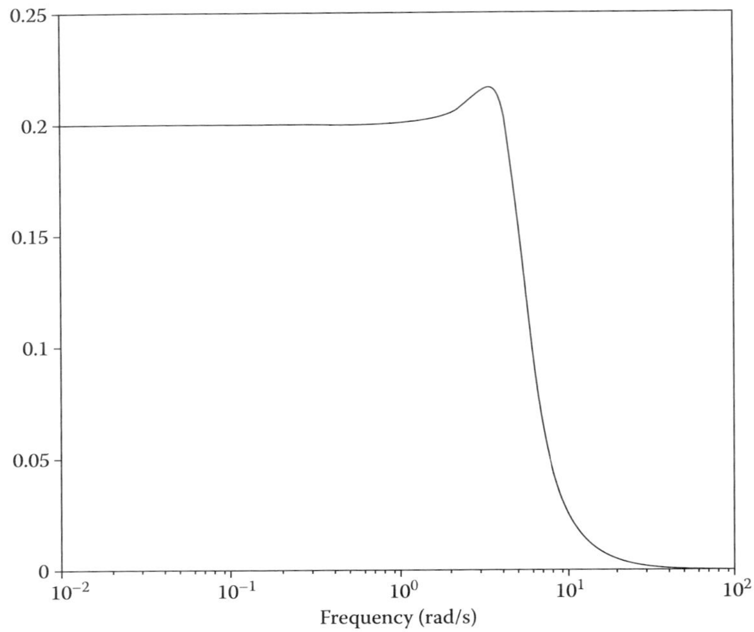

FIGURE 20.14 Nominal performance plot.

system *is* robustly stabilized with respect to real uncertainty in the A(2,1) coefficient and complex input multiplicative uncertainty. In this example, it is very conservative to treat the variation in the coefficient, $A(2, 1)$, as a complex uncertainty.

20.5.3.3 Robust Performance

The closed-loop system under perturbations becomes $F_u(G, \Delta)$. To analyze the degradation of performance due to the uncertainty, we use Theorem 20.5, and the augmented block structure Δ_{P_1}. The plot in Figure 20.15 of $\mu_{\Delta_{P_1}}(G(j\omega))$ is shown. The peak is approximately at 1.41. Applying Theorem 20.5 implies that for any structured $\Delta(s) \in \mathbf{S}_{\Delta_{P1}}$ with $\|\Delta(s)\|_\infty < \frac{1}{1.41}$, the perturbed loop remains stable *and*, the $\|\cdot\|_\infty$ norm of $F_u(G, \Delta)$ is guaranteed to be ≤ 1.41. Also, the converse of the theorem shows that there is a perturbation Δ, whose $\|\cdot\|_\infty$ is arbitrarily close to $\frac{1}{1.41}$ that causes $\left\|F_u\left(\Delta, \hat{G}\right)\right\|_\infty > 1.41$. Therefore robust performance was not achieved.

Figure 20.15 also shows the results of a mixed μ analysis on $G(\jmath\omega)$ with respect to Δ_{P_2}. The peak value of μ is 0.99. This implies that for a real perturbation δ_1 and a finite dimensional, linear time-invariant complex perturbation $\delta_2(s)$, stability is preserved and the performance objective achieved. Therefore the robust performance objective is achieved when the frequency-squared coefficient is treated as a real perturbation and the input multiplicative uncertainty is treated as a complex perturbation.

20.6 Spinning Satellite: Robust Performance Analysis with μ

Consider the 4×4 transfer function M shown (along with its with internal structure) in Figure 20.2. The perturbations δ_1 and δ_2 enter as shown in Figure 20.3. The appropriate block structure for the robust performance assessment of this example is $\left\{\text{diag}(\delta_1, \delta_2, \Delta_F) : \delta_i \in \mathbf{C}, \Delta_F \in \mathbf{C}^{2\times 2}\right\}$. This implies that there

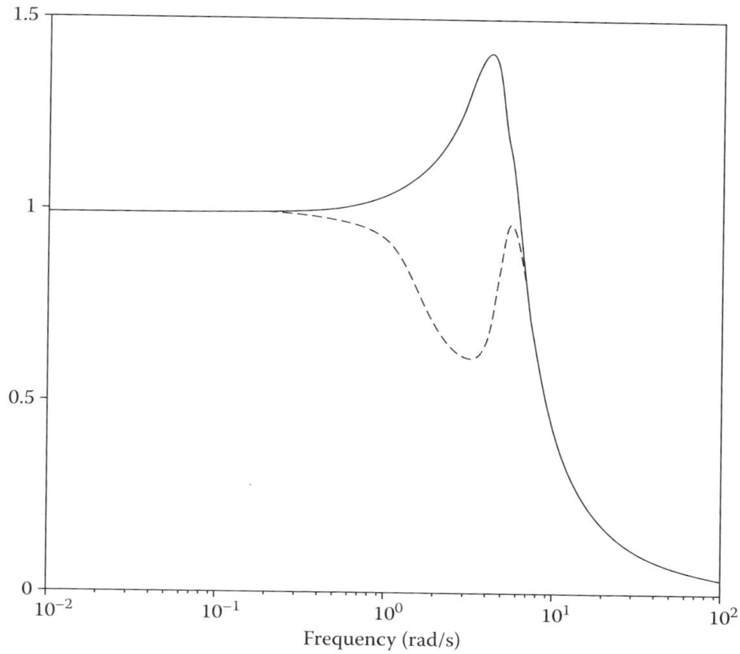

FIGURE 20.15 Complex (solid) and mixed (dashed) robust performance plots.

is independent uncertainty in each of the actuators but that the rest of the model is accurate. Computing the structured singular value of $\mu_\Delta(M)$ across frequency yields a peak of about 11. This implies that a diagonal perturbation *diag* $[\delta_1, \delta_2]$ of size $1/11$ exists so that the perturbed reference-to-error transfer function has a singular value peak of approximately 11. This μ analysis clearly detects the poor robust performance characteristics of the closed-loop system. In the next section, we turn our attention to design techniques which use the structured singular value as a design objective.

20.7 Control Design via μ Synthesis

Consider the standard linear fractional description of the control problem shown in Figure 20.16. The P block represents the open-loop interconnection and contains all of the known elements including the nominal plant model, uncertainty structure, and performance and uncertainty weighting functions. The Δ_{pert} block represents the structured set of norm-bounded uncertainty being considered and K represents the controller. Δ_{pert} parameterizes all of the assumed model uncertainty in the problem. Three groups of inputs enter P, perturbations z, disturbances d, and controls u, and three groups of outputs are generated, perturbations w, errors e, and measurements y. The set of systems to be controlled is described by the LFT

$$\left\{F_u\left(\Delta_{pert}, P\right) : \ \Delta_{pert} \in \mathbf{S}_{\Delta_{\text{pert}}}\right\}.$$

The design objective is to find a stabilizing controller K, so that, for all $\Delta_{pert} \in \mathbf{S}_{\Delta_{\text{pert}}}$, $\|\Delta_{pert}\|_\infty \leq 1$, the closed-loop system is stable and satisfies

$$\|F_u\left[F_l\left(P, K\right), \Delta_{pert}\right]\|_\infty \leq 1.$$

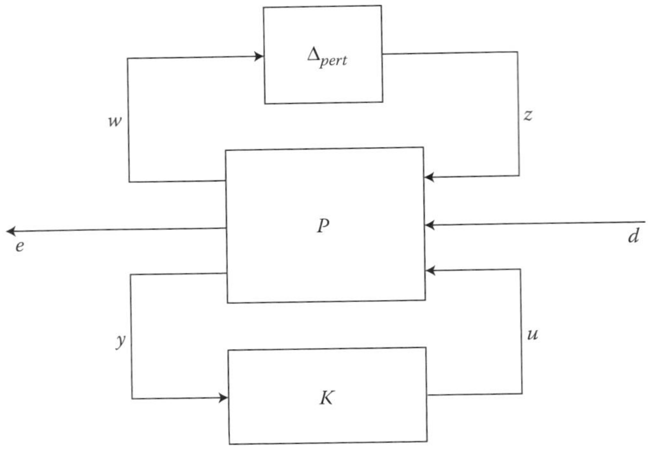

FIGURE 20.16 Linear fractional transformation description of control problem.

The performance objective involves a robust performance test on the linear fractional transformation $F_l\left(P, K\right)$. To assess the robust performance of the closed-loop system, define an augmented perturbation structure, Δ,

$$\Delta = \left\{ \begin{bmatrix} \Delta_{pert} & 0 \\ 0 & \Delta_F \end{bmatrix} : \Delta_{pert} \in \boldsymbol{\Delta}_{pert}, \Delta_F \in \mathbf{C}^{n_d \times n_e} \right\}.$$

The goal of μ synthesis is to minimize overall stabilizing controllers K, the peak value of $\mu_\Delta(\cdot)$ of the closed-loop transfer function $F_l(P, K)$. More formally,

$$\min_{\substack{K \\ \text{stabilizing}}} \max_{\omega} \mu_\Delta [F_l(P, K)(j\omega)] \tag{20.21}$$

For tractability of the μ synthesis problem, $\mu_\Delta[\cdot]$ is replaced by the upper bound for μ, $\bar{\sigma}\left[D(\cdot)D^{-1}\right]$. The scaling matrix D is a member of the appropriate set of scaling matrices $\mathbf{D}$ for the perturbation set Δ. One can reformulate this optimization problem as follows:

$$\min_{\substack{K \\ \text{stabilizing}}} \max_{\omega} \min_{D_\omega \in \mathbf{D}} \bar{\sigma}\left[D_\omega F_l(P, K)(j\omega) D_\omega^{-1}\right]. \tag{20.22}$$

Here, the D minimization is an approximation to the $\mu_\Delta[F_l(P, K)(j\omega)]$. D_ω is chosen from the set of scalings, $\mathbf{D}$, independently at every ω. Hence,

$$\min_{\substack{K \\ \text{stabilizing}}} \min_{\substack{D_{(\cdot)} \\ D_\omega \in \mathbf{D}}} \max_{\omega} \bar{\sigma}\left[D_\omega F_l(P, K)(j\omega) D_\omega^{-1}\right]. \tag{20.23}$$

The expression $\max_{\omega} \bar{\sigma}[\cdot]$ corresponds to $\|[\cdot]\|_\infty$, leaving

$$\min_{\substack{K \\ \text{stabilizing}}} \min_{\substack{D_{(\cdot)} \\ D_\omega \in \mathbf{D}}} \left\| \left[D_. F_l(P, K)(j\cdot) D_.^{-1}\right] \right\|_\infty. \tag{20.24}$$

Assume, for simplicity, that the uncertainty block Δ_{pert} has only full blocks. Then the set $\mathbf{D}_\Delta$ is of the form

$$\mathbf{D} = \left\{ \text{diag}\,[d_1 I, d_2 I, \ldots, d_{F-1} I, I] : d_i > 0 \right\}. \tag{20.25}$$

For any complex matrix M, the elements of $\mathbf{D}_\Delta$, which were originally defined as real and positive, can take on any nonzero complex values and without changing the value of the upper bound, $\inf_{D \in \mathcal{D}} \bar{\sigma}\left(DMD^{-1}\right)$.

Hence, we can restrict the scaling matrix to be a real-rational, stable, minimum-phase transfer function, $\hat{D}(s)$. The optimization is now

$$\min_{\substack{K \\ \text{stabilizing}}} \quad \min_{\substack{\hat{D}(s)\in\mathbf{D} \\ \text{stable,min-phase}}} \quad \|\hat{D}F_l(P,K)\hat{D}^{-1}\|_\infty. \tag{20.26}$$

This approximation to μ synthesis, is currently "solved" by an iterative approach, referred to as "$D-K$ iteration."

To solve Equation 20.26, first consider holding $\hat{D}(s)$ fixed. Given a stable, minimum-phase, real-rational $\hat{D}(s)$, solve the optimization $\min_{\substack{K \\ \text{stabilizing}}} \|\hat{D}F_l(P,K)\hat{D}^{-1}\|_\infty$. This equation is an $\mathcal{H}_\infty$ optimization control problem. The solution to the $\mathcal{H}_\infty$ problem is well-known, consisting of solving algebraic Riccati equations in terms of the state-space system.

Now suppose that a stabilizing controller, $K(s)$, is given, we then solve the following minimization corresponding to the upper bound for μ.

$$\min_{D_\omega\in\mathbf{D}} \bar{\sigma}\left[D_\omega F_l(P,K)(\jmath\omega)D_\omega^{-1}\right]$$

This minimization is done over the real, positive D_ω from the set $\mathbf{D}_\Delta$ defined in Equation 20.25. Recall that the addition of phase to each d_i does not affect the value of $\bar{\sigma}\left[D_\omega F_l(P,K)(\jmath\omega)D_\omega^{-1}\right]$. Hence, each discrete function, d_i, of frequency is fit (in magnitude) by a proper, stable, minimum-phase transfer function, $\hat{d}_{R_i}(s)$. These are collected together in a diagonal transfer function matrix $\hat{D}(s)$,

$$\hat{D}(s) = \text{diag}\left[\hat{d}_{R_1}(s)I, \hat{d}_{R_2}(s)I, \ldots \hat{d}_{R_{F-1}}(s)I, I\right],$$

and absorbed into the original open-loop generalized plant P. Iterating on these two steps comprises the current approach to $D-K$ iteration.

There are several problems with the $D-K$ iteration control design procedure. The first is that we have approximated $\mu_\Delta(\cdot)$ by its upper bound. This is not serious because the value of μ and its upper bound are often close. The most serious problem, that the $D-K$ iteration does not always converge to a global, or even, local minimum, [14] is a more severe limitation of the design procedure. However, in practice the $D-K$ iteration control design technique has been successfully applied to many engineering problems such as vibration suppression for flexible structures, flight control, chemical process control problems, and acoustic reverberation suppression in enclosures.

20.8 F-14 Lateral-Directional Control System Design

Consider the design of a lateral-directional axis controller for the F-14 aircraft during powered approach to landing. The linearized F-14 model is found at an angle-of-attack (α) of 10.5 *degs* and airspeed of 140 *knots*. The problem is posed as a robust performance problem with multiplicative plant uncertainty at the plant input and minimization of weighted-output transfer functions as the performance criterion. A diagram for the closed-loop system, which includes the feedback structure of the plant and controller and elements associated with the uncertainty models and performance objectives, is shown in Figure 20.17.

The overall performance objective is to have the "true" airplane, represented by the dashed box in Figure 20.17, respond effectively to the pilot's lateral stick and rudder pedal inputs. The performance objective includes

1. Decoupled response of the lateral stick, δ_{lstk}, to roll rate, p, and rudder pedals, δ_{rudp}, to side-slip angle, β. The lateral stick and rudder pedals have a maximum deflection of ±1 *inch*. Therefore they are represented as unweighted signals in Figure 20.17.

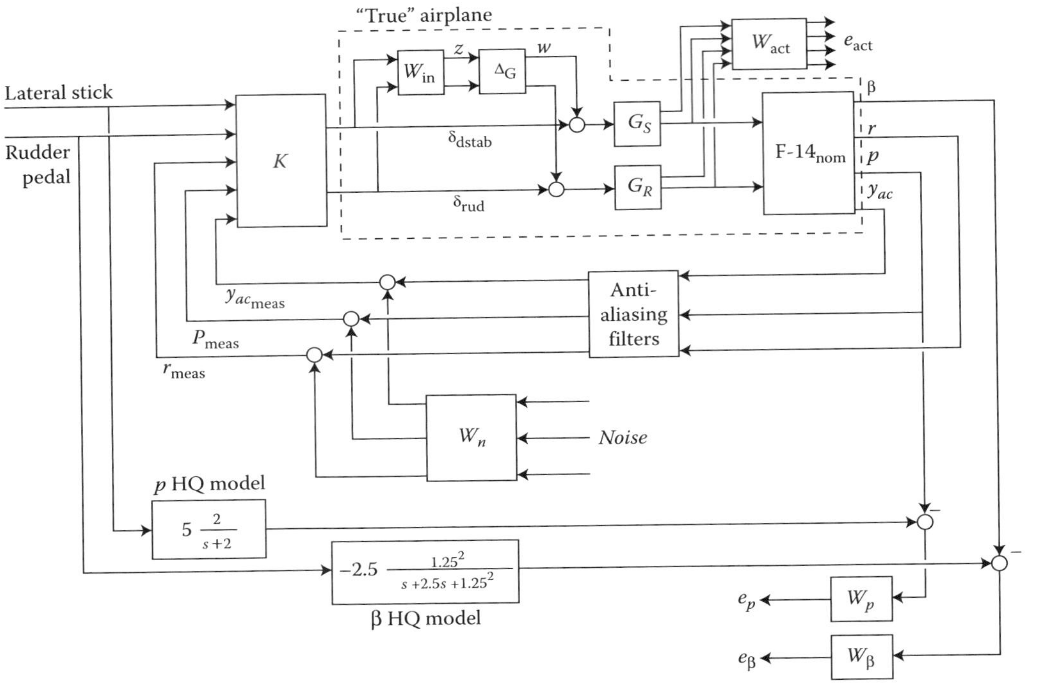

FIGURE 20.17 F-14 control block diagram.

2. The aircraft handling quality (HQ) response from the lateral stick to roll rate should be a first-order system, $5(2)/(s+2)\frac{degs/s}{inch}$. The aircraft handling quality response from the rudder pedals to side-slip angle should be $-2.5\frac{1.25^2}{s^2+2.5s+1.25^2}\frac{degs/s}{inch}$.
3. The stabilizer actuators have $\pm 20°$ and $\pm 90°/s$ deflection and deflection rate limits. The rudder actuators have $\pm 30°$ and $\pm 125°/s$ deflection and deflection rate limits.
4. The three measurement signals, roll rate, yaw rate and lateral acceleration, are passed through second-order antialiasing filters prior to being fed to the controller. The natural frequency and damping values for the yaw rate and lateral acceleration filters are 12.5Hz and 0.5, respectively and 4.1Hz and 0.7 for the roll rate filter. The antialiasing filters have unity gain at DC (see Figure 20.17). These signals are also corrupted by noise.

The performance objectives are accounted for in this framework via minimizing weight transfer function norms. Weighting functions serve two purposes in the $\mathcal{H}_\infty$ and μ framework: they allow the direct comparison of different performance objectives with the same norm and they allow incorporating frequency information into the analysis. The F-14 performance weighting functions include:

1. Limits on the actuator deflection magnitude and rates are included via the W_{act} weight. W_{act} is a 4×4 constant, diagonal scaling matrix described by $W_{act} = \text{diag}(1/90, 1/20, 1/125, 1/30)$. These weights correspond to the stabilizer and rudder deflection rate and deflection limits.
2. W_n is a 3×3 diagonal, frequency varying weight used to model the magnitude of the sensor noise. $W_n = \text{diag}(0.025, 0.025, 0.0125\frac{s+1}{s+100})$ which corresponds to the noise levels in the roll rate, yaw rate and lateral acceleration channels.
3. The desired δ_{lstk}-to-p and δ_{rudp}-to-β responses of the aircraft are formulated as a model matching problem in the μ framework. The difference between the ideal response of the transfer functions, δ_{lstk} filtered through the roll rate HQ model and δ_{rudp} filtered through the side-slip angle HQ

model, and the aircraft response, p and β, is used to generate an error that is to be minimized. The W_{p} transfer function, see Figure 20.17, weights the difference between the idealized roll rate response and the actual aircraft response, p.

$$W_{\mathrm{p}} = \frac{0.05s^4 + 2.90s^3 + 105.93s^2 + 6.17s + 0.16}{s^4 + 9.19s^3 + 30.80s^2 + 18.33s + 3.95}.$$

The magnitude of W_{p} emphasizes the frequency range between 0.06 and 30 rad/s. The desired performance frequency range is limited due to a right half-plane zero in the model at 0.002 rad/s, therefore, accurate tracking of sinusoids below 0.002 rad/s isn't required. Between 0.06 and 30 rad/s, a roll rate tracking error of less than 5% is desired. The performance weight on the β tracking error, W_β, is just $2 \times W_{\mathrm{p}}$. This also corresponds to a 5% tracking error objective.

All the weighted performance objectives are scaled for an $\mathcal{H}_\infty$ less than 1 when they are achieved. The performance of the closed-loop system is evaluated by calculating the maximum singular value of the weighted transfer functions from the disturbance and command inputs to the error outputs, as in Figure 20.18.

20.8.1 Nominal Model and Uncertainty Models

The pilot has the ability to command the lateral directional response of the aircraft with the lateral stick (δ_{lstk}) and rudder pedals (δ_{rped}). The aircraft has two control inputs, differential stabilizer deflection (δ_{dstab}, *degs*) and rudder deflection (δ_{rud}, *degs*), three measured outputs, roll rate (p, *degs/s*), yaw rate (r, *degs/s*) and lateral acceleration (y_{ac}, g's), and a calculated output side-slip angle (β). Note that β is not a measured variable but is used as a performance measure. The lateral directional F-14 model, F-14_{nom}, has four states, lateral velocity (v), yaw rate (r), roll rate (p) and roll angle (ϕ). These variables are related by the state-space equations

$$\begin{bmatrix} \dot{v} \\ \dot{r} \\ \dot{p} \\ \dot{\phi} \\ \beta \\ p \\ r \\ y_{ac} \end{bmatrix} = \left[\begin{array}{c|c} A & B \\ \hline C & D \end{array}\right] \begin{bmatrix} v \\ r \\ p \\ \phi \\ \delta_{\mathrm{dstab}} \\ \delta_{\mathrm{drud}} \end{bmatrix}$$

$$A = \begin{bmatrix} -1.16e-1 & -2.27e+2 & 4.30e+1 & 3.16e+1 \\ 2.65e-3 & -2.59e-1 & -1.45e-1 & 0.00e+0 \\ -2.11e-2 & 6.70e-1 & -1.36e+0 & 0.00e+0 \\ 0.00e+0 & 1.85e-1 & 1.00e+0 & 0.00e+0 \end{bmatrix}, B = \begin{bmatrix} 6.22e-02 & 1.01e-1 \\ -5.25e-03 & -1.12e-2 \\ -4.67e-02 & 3.64e-3 \\ 0.00e+00 & 0.00e+0 \end{bmatrix},$$

$$C = \begin{bmatrix} 2.47e-1 & 0.00e+0 & 0.00e+0 & 0.00e+0 \\ 0.00e+0 & 0.00e+0 & 5.73e+1 & 0.00e+0 \\ 0.00e+0 & 5.73e+1 & 0.00e+0 & 0.00e+0 \\ -2.83e-3 & -7.88e-3 & 5.11e-2 & 0.00e+0 \end{bmatrix}, D = \begin{bmatrix} 0.00e+00 & 0.00e+0 \\ 0.00e+00 & 0.00e+0 \\ 0.00e+00 & 0.00e+0 \\ 2.89e-03 & 2.27e-3 \end{bmatrix},$$

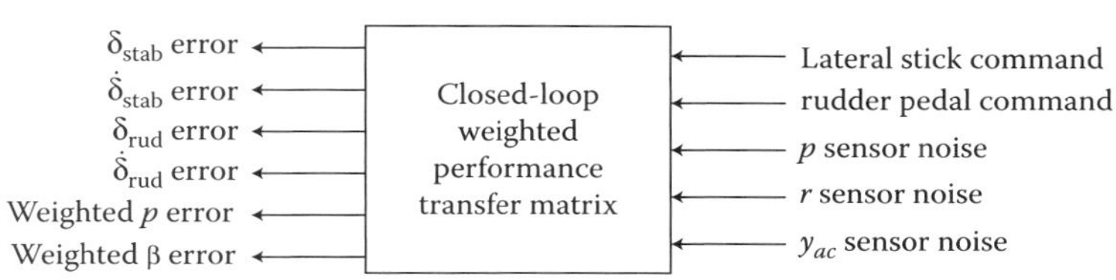

FIGURE 20.18 F-14 weighted performance objectives transfer matrix.

The dashed box represents the "true" airplane, corresponding to a set of F-14 plant models define by $\mathcal{G}$. Inside the box is the nominal model of the airplane dynamics, F-14$_{\text{nom}}$, models of the actuators, G_S and G_R, and two elements, W_{in} and Δ_G, which parameterize the uncertainty in the model. This type of uncertainty is called multiplicative plant input uncertainty. The transfer function W_{in} is assumed known and reflects the amount of uncertainty in the model. The transfer function Δ_G is assumed stable and unknown, except for the norm condition, $\|\Delta_G\|_\infty \leq 1$.

A "first principles" set of uncertainties in the aircraft model would include

1. Uncertainty in the stabilizers and the rudder actuators. The electrical signals that command deflections in these surfaces must be converted to actual mechanical deflections by the electronics and hydraulics of the actuators. Unlike the models, this is not done perfectly in the actual system.
2. Uncertainty in the forces and moments generated on the aircraft, due to specific deflections of the stabilizers and rudder. As a first approximation, this arises from the uncertainties in the aerodynamic coefficients, which vary with flight conditions, as well as uncertainty in the exact geometry of the airplane.
3. Uncertainty in the linear and angular accelerations produced by the aerodynamically generated forces and moments. This arises from the uncertainty in the various inertial parameters of the airplane, in addition to neglected dynamics, such as fuel slosh and airframe flexibility.
4. Other forms of uncertainty that are less well understood.

In this example, we choose not to model the uncertainty in this detailed manner but rather to lump all of these effects together into one, complex full-block, multiplicative uncertainty at the input of the rigid body aircraft nominal model.

The stabilizer and rudder actuators, G_S and G_R, are modeled as first order transfer functions, $25/(s+25)$. Given the actuator and aircraft nominal models (denoted by $G_{\text{nom}}(s)$), we also specify a stable, 2×2 transfer function matrix $W_{\text{in}}(s)$ called the uncertainty weight. These transfer matrices parameterize an entire set of plants, $\mathcal{G}$, which must be suitably controlled by the robust controller K.

$$\mathcal{G} := \left\{ G_{\text{nom}} \left(I + \Delta_G W_{\text{del}} \right) : \Delta_G \text{ stable}, \|\Delta_G\|_\infty \leq 1 \right\}.$$

All of the uncertainty in modeling the airplane is captured in the normalized, unknown transfer function Δ_G. The unknown transfer function $\Delta_G(s)$ is used to parameterize the potential differences between the nominal model $G_{\text{nom}}(s)$, and the actual behavior of the real airplane, denoted by $\mathcal{G}$.

In this example, the uncertainty weight W_{in} is of the form, $W_{\text{in}}(s) := \text{diag}(w_1(s), w_2(s))I_2$, for particular scalar valued functions $w_1(s)$ and $w_2(s)$. The $w_1(s)$ weight associated with the differential stabilizer input is selected to be $w_1(s) = \frac{2(s+4)}{s+160}$. The $w_2(s)$ weight associated with the differential rudder input is selected to be $w_2(s) = \frac{1.5(s+20)}{s+200}$. Hence the set of plants that are represented by this uncertainty weight

$$\mathcal{G} := \left\{ \text{F-14}_{\text{nom}} \begin{bmatrix} \frac{25}{s+25} & 0 \\ 0 & \frac{25}{s+25} \end{bmatrix} \left(I_2 + \begin{bmatrix} \frac{2(s+4)}{s+100} & 0 \\ 0 & \frac{1.5(s+20)}{s+200} \end{bmatrix} \Delta_G(s) \right) : \Delta_G(s) \text{ stable}, \|\Delta_G\|_\infty \leq 1 \right\}$$

Note that the weighting functions are used to normalize the size of the unknown perturbation Δ_G. At any frequency ω, the value of $|w_1(j\omega)|$ and $|w_2(j\omega)|$ can be interpreted as the percentage of uncertainty in the model at that frequency. The dependence of the uncertainty weight on frequency indicates that the level of uncertainty in the airplane's behavior depends on frequency.

The particular uncertainty weights chosen imply that, in the differential stabilizer channel at low frequency, there is potentially a 5% modeling error, and at a frequency of 93 rad/s, the uncertainty in channel 1 can be as much as 100%, and can get larger at higher frequencies. The rudder channel has more uncertainty at low frequency, up to 15% modeling error, and at a frequency of 177 rad/s, the uncertainty is at 100%. To illustrate the variety of plants represented by the set $\mathcal{G}$, some step responses of different systems from $\mathcal{G}$ are shown in Figure 20.19.

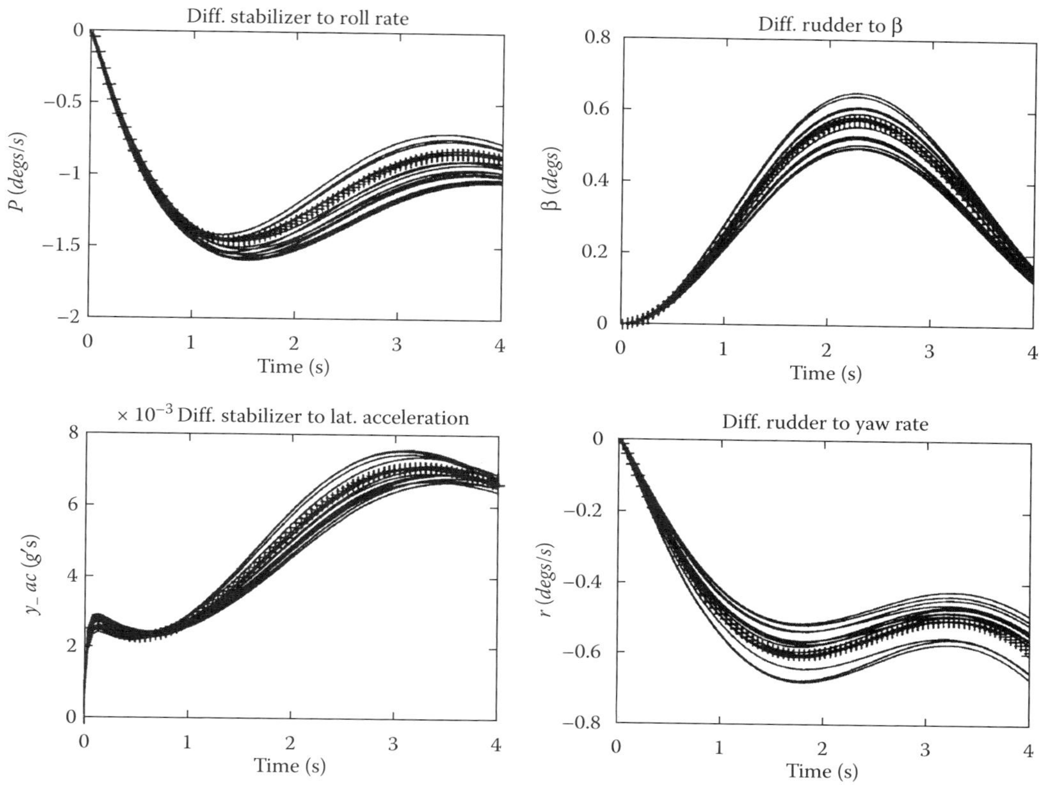

FIGURE 20.19 Unit-step responses of the nominal model (+) and 15 perturbed models from $\mathcal{G}$.

The control design objective is a stabilizing controller K so that for all stable perturbations $\Delta_G(s)$, with $\|\Delta_G\|_\infty \leq 1$, the perturbed closed-loop system remains stable, and the perturbed weighted performance transfer functions has an $\mathcal{H}_\infty$ norm less than 1 for all such perturbations. These mathematical objectives fit exactly into the structured singular value framework.

20.8.2 Controller Design

The control design block diagram shown in Figure 20.17 is redrawn as $P(s)$, shown in Figure 20.20. $P(s)$, the 25-state, six-input, six-output open-loop transfer matrix, corresponds to the P in the linear fractional block diagram in Figure 20.16.

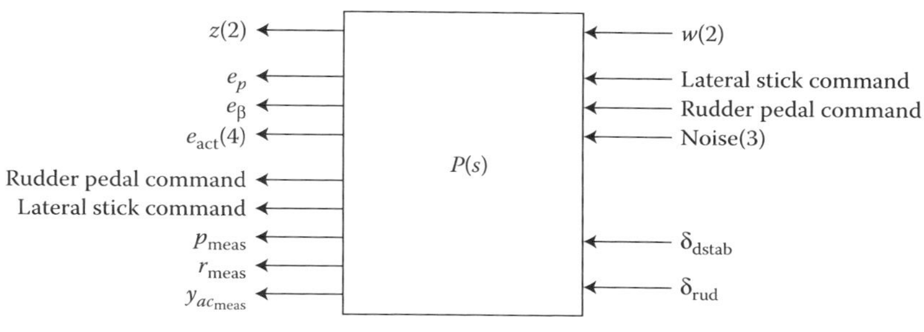

FIGURE 20.20 F-14 generalized plant.

TABLE 20.1 F-14 $D-K$ iteration information

Iteration number	1	2	3	4
Total D-scale order	0	4	4	4
Controller order	25	29	29	29
$\mathcal{H}_\infty$ norm achieved	1.562	1.079	1.025	1.017
Peak μ value	1.443	1.079	1.025	1.017

The first step in the $D-K$ iteration control design procedure is to design an $\mathcal{H}_\infty$ (sub)optimal controller for the open-loop interconnection, P. In terms of the $D-K$ iteration, this amounts to holding the d variable fixed (at 1) and minimizing the $\|\cdot\|_\infty$ norm of $F_l\left(P,K\right)$ over the controller variable K. The resulting controller is labeled K_1.

The second step in the $D-K$ iteration involves solving a μ analysis problem corresponding to the closed-loop system, $F_l\left(P,K_1\right)$. This calculation produces a frequency dependent scaling variable d_ω, the (1,1) entry in the scaling matrix. In a general problem (with more than two blocks), there would be several d variables, and the overall matrix is referred to as "the D-scales." The varying variables in the D-scales are fit (in magnitude) with proper, stable, minimum-phase rational functions and absorbed into the generalized plant for additional iterations. These scalings are used to "trick" the $\mathcal{H}_\infty$ minimization to concentrate more on minimizing μ rather than $\bar{\sigma}$ across frequency. For the first iteration in this example, the d scale data is fit with a first-order transfer function.

The new generalized plant used in the second iteration has 29 states, four more states than the original 25-state generalized plant, P. These extra states are due to the D-scale data being fitted with a rational function and absorbed into the generalized plant for the next iteration. Four $D-K$ iterations are performed until μ reaches a value of 1.02. Information about the $D-K$ iterations is shown in Table 20.1. All the analysis and synthesis results were obtained with *The* μ *Analysis and Synthesis Toolbox, Version 2.0* [2].

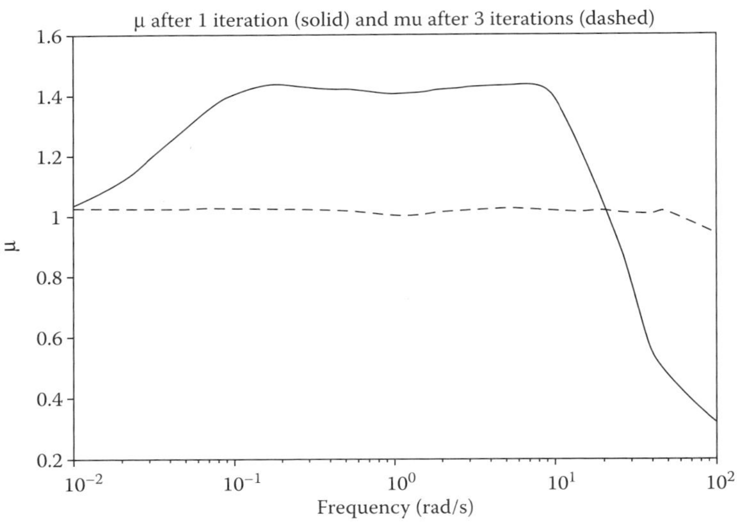

FIGURE 20.21 F-14 robust performance μ plots with K_1 and K_4 implemented.

20.8.2.1 Analysis of the Controllers

The robust performance properties of the controllers can be analyzed using μ analysis. Robust performance is achieved if, and only if, for every frequency, $\mu_\Delta\left(F_l(P,K)(j\omega)\right)$ of the closed-loop frequency response is less than 1. Plots of μ of the closed-loop system with K_1 and K_4 implemented are shown in Figure 20.21.

The controlled system with K_1 implemented *does not* achieve *robust performance*. This conclusion follows from the μ plot, which peaks to a value of 1.44, at a frequency of 7 rad/s. Because μ is 1.44, there is a perturbation matrix Δ_G, so that $\|\Delta_G\|_\infty = \frac{1}{1.44}$, and the perturbed weighted performance transfer functions gets "large." After four $D-K$ iterations the peak robust performance μ value is reduced to 1.02 (Figure 20.21), thereby, nearly achieving all of our robust performance objectives.

Illustrating the robustness of the closed-loop system in the time domain, time responses of the ideal model, the nominal closed-loop system and the "worst-case" closed-loop system from $\mathcal{G}$ (using perturbations of size 1) are shown in Figure 20.22. Controller K_4 is implemented in the closed-loop simulations.

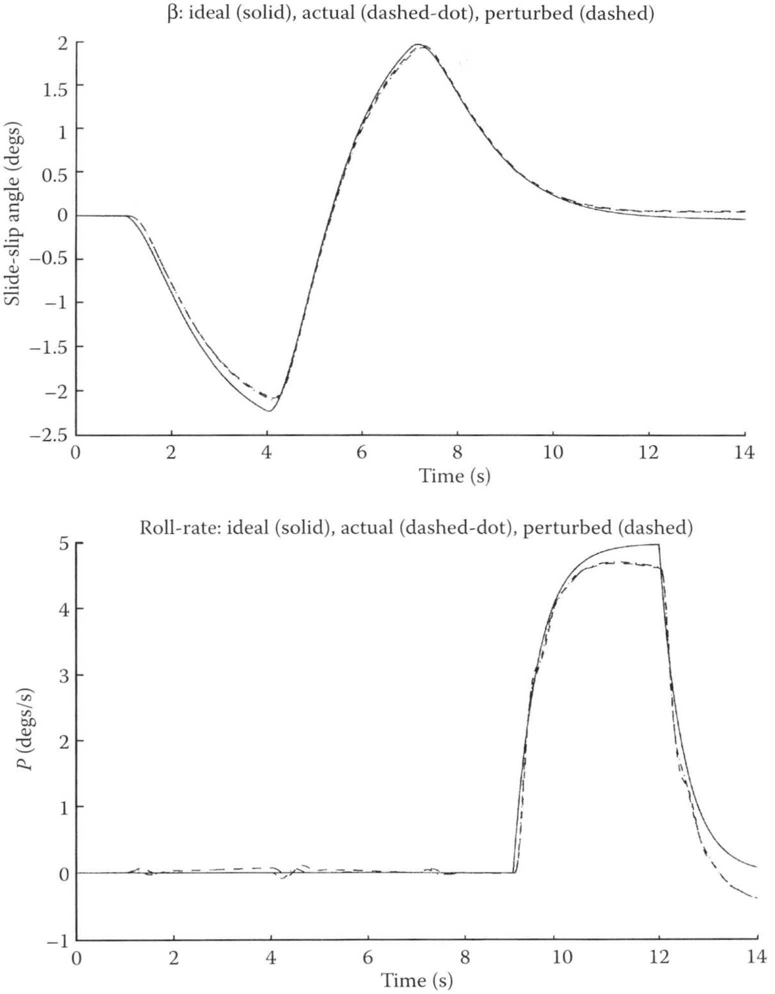

FIGURE 20.22 Time response plots of the F-14 lateral directional control system.

A 1-inch lateral stick command is given at 9 s, held at 1 inch until 12 s, and then returns to zero. The rudder is commanded at 1 s with a positive 1 inch pedal deflection and held at 1 inch until 4 s. At 4 s a −1-inch pedal deflection is commanded, held to 7 s, and then returned to zero. One can see from the time responses that the closed-loop response is nearly identical for the nominal closed-loop system and the "worst-case" closed-loop system. The ideal time response for β and p are plot for reference.

20.9 Conclusion

This chapter outlined the usefulness of the structured singular value (μ) analysis and synthesis techniques in designing and analyzing multiloop feedback control systems. Through examples, we have shown some pitfalls with simple-minded analytical techniques, and illustrated the usefulness of the analytic framework provided by the structured singular value. We outlined an approach to robust controller synthesis, the $D - K$ iteration. As an example, these techniques were applied to the design of a lateral directional control system for the F-14 aircraft.

References

1. Balas, G.J. and Doyle, J.C., Control of lightly damped, flexible modes in the controller crossover region, *AIAA J. Guidance, Dyn. Control*, 17(2), 370–377, 1994.
2. Balas, G.J., Doyle, J.C., Glover, K., Packard, A., and Smith, R., The μ analysis and synthesis toolbox, mathworks, Natick, MA, 1991.
3. Boyd, S., El Ghaoui, L., Feron, E., Balakrishnan, V., *Linear Matrix Inequalities in System and Control Theory*, SIAM Studies in Applied Mathematics, SIAM, Philadelphia, 1994.
4. Chiang, R. and Safonov, M., Robust control toolbox, MathWorks, Natick, MA, 1988.
5. Doyle, J.C., Analysis of feedback systems with structured uncertainties, *IEEE Proc.*, 129 Part D(6), 242–250, 1982.
6. Doyle, J.C., Lenz, K., and Packard, A., Design examples using μ synthesis: Space shuttle lateral axis FCS during reentry, *IEEE CDC*, 1986, 2218–2223; also *Modelling, Robustness and Sensitivity Reduction in Control Systems,* NATO ASI Series, Curtain, R.F., Ed., Springer, 1987, vol. F34, 128–154.
7. Fan, M., Tits, A., and Doyle, J.C., Robustness in the presence of joint parametric uncertainty and unmodeled dynamics, *IEEE Trans. Automat. Control*, 36, 25–38, 1991.
8. Packard, A., Doyle, J.C., and Balas, G.J., Linear, multivariable robust control with a μ perspective, *ASME J. Dyn., Meas. Control, Special Ed. on Control*, 115(2b), 426–438, 1993.
9. Packard, A. and Doyle, J.C., The complex structured singular value, *Automatica*, 29(1), 71–109, 1993.
10. Packard, A. and Pandey, P., Continuity properties of the real/complex structured singular value, *IEEE Trans. Automat. Control*, 38(3), 415–428, 1993.
11. Redheffer, On a certain linear fractional transformation, *J. Math. Phys.*, 39, 269–286, 1960.
12. Safonov, M.G., Stability margins of diagonally perturbed multivariable feedback systems, *Proc. IEEE*, 129 Part D(6), 251–256, 1982.
13. Skogestad, S., Morari, M., and Doyle, J.C., Robust control of ill-conditioned plants: high-purity distillation, *IEEE Trans. Automat. Control*, 33(12), 1092–1105, 1988.
14. Stein, G. and Doyle, J.C., Beyond singular values and loopshapes, *AIAA J. Guidance and Control*, 14(1), 5–16, 1991.
15. Young, P.M., Newlin, M.P., and Doyle, J.C., μ Analysis with real parametric uncertainty, In *Proc. 30th IEEE Conf. Decision and Control*, Hawaii, 1991, 1251–1256.

21

Algebraic Design Methods

Vladimír Kučera
Czech Technical University and Academy of Sciences

21.1 Introduction

One of the features of modern control theory is the growing presence of algebra. Algebraic formalism offers several useful tools for control system design, including the so-called "factorization" approach.

This approach is based on the input–output properties of linear systems. The central idea is that of "factoring" the transfer matrix of a (not necessarily stable) system as the "ratio" of two *stable* transfer matrices. This is a natural step for the linear systems whose transfer matrices are rational, that is, for the lumped-parameter systems. Under certain conditions, however, this approach is productive also for the distributed-parameter systems.

The starting point of the factorization approach is to obtain a simple parameterization of *all* controllers that stabilize a given plant. One could then, in principle, choose the best controller for various applications. The key point here is that the parameter appears in the closed-loop system transfer matrix in a linear manner, thus making it easier to meet additional design specifications.

The actual design of control systems is an engineering task that cannot be reduced to algebra. Design contains many additional aspects that have to be taken into account: sensor placement, computational constraints, actuator constraints, redundancy, performance robustness, among many others. There is a need for an understanding of the control process, a feeling for what kinds of performance objectives are unrealistic, or even dangerous, to ask for. The algebraic approach to be presented, nevertheless, is an elegant and useful tool for the mathematical part of the controller design.

21.2 Systems and Signals

The fundamentals of the factorization approach will be explained for linear systems with *rational* transfer functions whose input u and output y are scalar quantities. We suppose that u and y live in a space of functions mapping a time set into a value set. The time set is a subset of real numbers bounded on the left, say R_+ (the nonnegative reals) in the case of continuous-time systems and Z_+ (the nonnegative integers) for discrete-time systems. The value set is taken to be the set of real numbers R.

Let the input and output spaces of a continuous-time system be the spaces of locally (Lebesgue) integrable functions f from R_+ into R, and define a p-norm

$$||f||_{L_p} = \left[\int_0^\infty |f(t)|^p \, dt\right]^{1/p} \qquad \text{if } i = 1 \leq p < \infty,$$

$$||f||_{L_\infty} = \operatorname*{ess\,sup}_{t\geq 0} |f(t)| \quad \text{if } p = \infty.$$

The corresponding normed space is denoted by L_p.

The systems having the desirable property of preserving these functional spaces are called *stable*. More precisely, a system is said to be L_p stable if any input $u \in L_p$ gives rise to an output $y \in L_p$. The systems that are L_∞ stable are also termed to be bounded-input bounded-output (BIBO) stable.

The transfer function of a continuous-time system is the Laplace transform of its impulse response $g(t)$,

$$G(s) = \int_0^\infty g(t)e^{-st} \, dt.$$

It is well known that a system with a rational transfer function $G(s)$ is BIBO stable if and only if $G(s)$ is proper and Hurwitz stable, that is, bounded at infinity with all poles having negative real parts.

In the study of discrete-time systems, we let the input and output spaces be the spaces of infinite sequences $f = (f_0, f_1, \ldots)$ mapping Z_+ into R and define a p-norm as follows:

$$\|f\|_{l_p} = \left[\sum_{i=0}^\infty |f_i|^p\right]^{1/p} \qquad \text{if } 1 \leq p < \infty,$$

$$||f||_{l_\infty} = \sup_{i\geq 0} |f_i| \quad \text{if } p = \infty.$$

A discrete-time system is said to be l_p stable if it transforms any input $u \in l_p$ to an output $y \in l_p$. The systems that are l_∞ stable are also known as BIBO stable systems.

The transfer function of a discrete-time system is defined as the z-transform of its unit pulse response $(h_0, h_1, \ldots)$,

$$H(z) = \sum_{i=0}^\infty h_i z^{-i},$$

and it is always proper. A system with a proper rational transfer function $H(z)$ is BIBO stable if and only if $H(z)$ is Schur stable, that is, its all poles have modulus less than one.

Of particular interest are discrete-time systems that are finite-input finite-output (FIFO) stable. Such a system transforms finite-input sequences into finite-output sequences, its unit pulse response is finite, and its transfer function $H(z)$ has no poles outside the origin $z = 0$, that is, $H(z)$ is a polynomial in z^{-1}.

21.3 Fractional Descriptions

Consider a rational function $G(s)$. By definition, it can be expressed as the ratio

$$G(s) = \frac{B(s)}{A(s)}$$

of two qualified rational functions A and B.

A well-known example is the *polynomial* description, in which case A and B are coprime polynomials, that is, polynomials having no roots in common.

Another example is to take for A and B two coprime, *proper* and *Hurwitz-stable* rational functions. When $G(s)$ is, say,

$$G(s) = \frac{s+1}{s^2+1},$$

then one can take

$$A(s) = \frac{s^2+1}{(s+\lambda)^2}, \quad B(s) = \frac{s+1}{(s+\lambda)^2},$$

where $\lambda > 0$ is a real number. We recall that two proper and Hurwitz-stable rational functions are coprime if they have no infinite nor unstable zeros in common. Therefore, in the example above, the denominator of A and B can be any strictly Hurwitz polynomial of degree exactly 2; if its degree is lower, then A would not be proper and if it is higher, then A and B would have a common zero at infinity. The set of proper and Hurwitz-stable rational functions is denoted by $R_H(s)$.

The proper rational functions $H(z)$ arising in discrete-time systems can be treated in a similar manner. One can write

$$H(z) = \frac{B(z)}{A(z)},$$

where A and B are coprime, *Schur-stable* (hence proper) rational functions. Coprimeness means having no unstable zeros (i.e., in the closed disc $|z| \geq 1$) in common. For example, if

$$H(z) = \frac{1}{z-1},$$

then one can take

$$A(z) = \frac{z-1}{z-\lambda}, \quad B(z) = \frac{1}{z-\lambda}$$

for any real number λ such that $|\lambda| < 1$. The set of Schur-stable rational functions is denoted by $R_S(z)$.

The particular choice of λ=0 in the example above leads to

$$A(z) = \frac{z-1}{z} = 1 - z^{-1}, \quad B(z) = \frac{1}{z} = z^{-1}.$$

In this case, A and B are in fact polynomials in z^{-1}.

21.4 Feedback Systems

To control a system means to alter its dynamics so that a desired behavior is obtained. This can be done by feedback. A typical feedback system consists of two subsystems, S_1 and S_2, connected, as shown in Figure 21.1.

In most applications, it is desirable that the feedback system be BIBO stable in the sense that whenever the exegenous inputs u_1 and u_2 are bounded in magnitude, so too are the output signals y_1 and y_2.

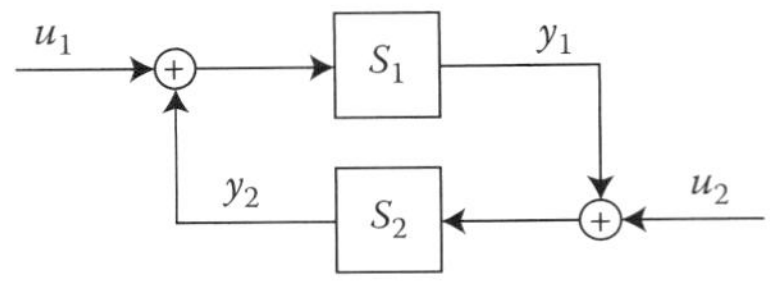

FIGURE 21.1 Feedback system.

In order to study this property, we express the transfer functions of S_1 and S_2 as ratios of proper stable rational functions and seek for conditions under which the transfer function of the feedback system is proper and stable.

To fix ideas, consider continuous-time systems and write

$$S_1 = \frac{B(s)}{A(s)}, \quad S_2 = -\frac{Y(s)}{X(s)},$$

where A, B and X, Y are two couples of coprime rational functions from $R_H(s)$. The transfer matrix of the feedback system

$$\begin{bmatrix} y_1 \\ y_2 \end{bmatrix} = \begin{bmatrix} \frac{S_1}{1-S_1S_2} & \frac{S_1S_2}{1-S_1S_2} \\ \frac{S_1S_2}{1-S_1S_2} & \frac{S_2}{1-S_1S_2} \end{bmatrix} \begin{bmatrix} u_1 \\ u_2 \end{bmatrix}$$

is then given by

$$\begin{bmatrix} y_1 \\ y_2 \end{bmatrix} = \frac{1}{AX+BY} \begin{bmatrix} BX & -BY \\ -BY & -AY \end{bmatrix} \begin{bmatrix} u_1 \\ u_2 \end{bmatrix}.$$

We observe that the numerator matrix has all its elements in $R_H(s)$ and that no infinite or unstable zeros of the denominator can be absorbed in all these elements. We therefore conclude that the transfer functions belong to $R_H(s)$ if and only if the inverse of $AX + BY$ is in $R_H(s)$.

We illustrate with the example where S_1 is a differentiator and S_2 is an invertor such that

$$S_1(s) = s, \quad S_2(s) = -1.$$

We take

$$A(s) = \frac{1}{s+\lambda}, \quad B(s) = \frac{s}{s+\lambda}$$

for any real $\lambda > 0$ and

$$X(s) = 1, \quad Y(s) = 1.$$

Then

$$(AX + BY)^{-1}(s) = \frac{s+\lambda}{s+1}$$

resides in $R_H(s)$ and hence the feedback system is BIBO stable.

The above analysis applies also to discrete-time systems; the set $R_H(s)$ is just replaced by $R_S(z)$. However, we note that any closed loop around a discrete-time system involves some information delay, no matter how small. Indeed, a control action applied to S_1 cannot affect the measurement from which it was calculated in S_2. Therefore, either $S_1(z)$ or $S_2(z)$ must be strictly proper; we shall assume that it is $S_1(z)$ that has this property.

To illustrate the analysis of discrete-time systems, consider a summator S_1 and an amplifier S_2,

$$S_1(z) = \frac{1}{z-1}, \quad S_2(z) = -k.$$

Taking

$$A(z) = \frac{z-1}{z-\lambda}, \quad B(z) = \frac{1}{z-\lambda}$$

for any real λ in magnitude less than 1 and

$$X(z) = 1, \quad Y(z) = k,$$

one obtains

$$(AX + BY)^{-1}(z) = \frac{z-\lambda}{z-(1-k)}.$$

Therefore, the closed-loop system is BIBO stable if and only if $|1-k| < 1$.

To summarize, the fractional representation used should be matched with the goal of the analysis. The denominators A, X and the numerators B, Y should be taken from the set of stable transfer functions, either $R_H(s)$ or $R_S(z)$, depending on the type of the stability studied. This choice makes the analysis more transparent and leads to a simple algebraic condition: the inverse of $AX + BY$ is stable. Any other type of stability can be handled in the same way, provided one can identify the set of the transfer functions that these stable systems will have.

21.5 Parameterization of Stabilizing Controllers

The design of feedback control systems consists of the following: given one subsystem, say S_1, we seek to determine the other subsystem, S_2, so that the resulting feedback system shown in Figure 21.1 meets the design specifications. We call S_1 the *plant* and S_2 the *controller*. Our focus is first on achieving BIBO stability. Any controller S_2 that BIBO stabilizes the plant S_1 is called a *stabilizing* controller for this plant.

Suppose S_1 is a continuous-time plant that gives rise to the transfer function

$$S_1(s) = \frac{B(s)}{A(s)}$$

for some coprime elements A and B of $R_H(s)$. It follows from the foregoing analysis that a stabilizing controller exists and that all controllers that stabilize the given plant are generated by all solution pairs X, Y with $X \neq 0$ of the Bézout equation

$$AX + BY = 1$$

over $R_H(s)$. There is no loss of generality in setting $AX + BY$ to the identity rather than to any rational function whose inverse is in $R_H(s)$: this inverse is absorbed by X and Y and therefore cancels in forming

$$S_2(s) = -\frac{Y(s)}{X(s)}.$$

The solution set of the equation $AX + BY = 1$ with A and B coprimes in $R_H(s)$ can be parameterized as

$$X = X' + BW, \quad Y = Y' - AW,$$

where X', Y' represent a particular solution of the equation, and W is a free parameter, which is an arbitrary function in $R_H(s)$.

The parameterization of the family of all stabilizing controllers S_2 for the plant S_1 now falls out almost routinely:

$$S_2(s) = -\frac{Y'(s) - A(s)W(s)}{X'(s) + B(s)W(s)},$$

where the parameter W varies over $R_H(s)$ while satisfying $X' + BW \neq 0$.

In order to determine the set of all controllers S_2 that stabilize the plant S_1, one needs to do two things: (1) express $S_1(s)$ as a ratio of two coprime elements from $R_H(s)$ and (2) find a particular solution in $R_H(s)$ of a Bézout equation, which is equivalent to finding one stabilizing controller for S_1. Once these two steps are completed, the formula above provides a parameterization of the set of all stabilizing controllers for S_1. The condition $X' + BW \neq 0$ is not very restrictive, as $X' + BW$ can identically vanish for at most one choice of W.

As an example, we shall stabilize an integrator plant S_1. Its transfer function can be expressed as

$$S_1(s) = \frac{1/s+1}{s/s+1},$$

where $s+1$ is an arbitrarily chosen Hurwitz polynomial of degree one. Suppose that using some design procedure we have found a stabilizing controller for S_1, namely

$$S_2(s) = -1.$$

This corresponds to a particular solution $X' = 1$, $Y' = 1$ of the Bézout equation

$$\frac{s}{s+1}X + \frac{1}{s+1}Y = 1.$$

The solution set in $R_H(s)$ of this equation is

$$X(s) = 1 + \frac{1}{s+1}W(s), \quad Y(s) = 1 - \frac{s}{s+1}W(s).$$

Hence, all controllers S_2 that BIBO stabilize S_1 have the transfer function

$$S_2(s) = -\frac{1-(s/s+1)W(s)}{1+(1/s+1)W(s)},$$

where W is any function in $R_H(s)$.

It is clear that the result is independent of the particular fraction taken to represent S_1. Indeed, if $s+1$ is replaced by another Hurwitz polynomial $s+\lambda$ in the above example, one obtains

$$S_2(s) = -\frac{\lambda-(s/s+\lambda)W'(s)}{1+(1/s+\lambda)W'(s)},$$

which is the same set when

$$W'(s) = \left(\frac{s+\lambda}{s+1}\right)^2 W(s) + \frac{s+\lambda}{s+1}(\lambda - 1).$$

21.6 Parameterization of Closed-Loop Transfer Functions

The utility of the fractional approach derives not merely from the fact that it provides a parameterization of all controllers that stabilize a given plant in terms of a free parameter W, but also from the simple manner in which this parameter enters the resulting (stable) closed-loop transfer matrix.

In fact,

$$\begin{bmatrix} y_1 \\ y_2 \end{bmatrix} = \begin{bmatrix} B(X'+BW) & -B(Y'-AW) \\ -B(Y'-AW) & -A(Y'-AW) \end{bmatrix} \begin{bmatrix} u_1 \\ u_2 \end{bmatrix},$$

and we observe that all four transfer functions are *affine* in the free parameter W.

This result serves to parameterize the performance specifications, and it is the starting point for the selection of the best controller for the application at hand. The search for S_2 is thus replaced by a search for W. The crucial point is that the resulting selection/optimization problem is linear in W, whereas it is nonlinear in S_2.

21.7 Optimal Performance

The performance specifications often involve a norm minimization.

Let us consider the problem of *disturbance attenuation*. We are given, say, a continuous-time plant S_1 having two inputs: the control input u and an unmeasurable disturbance d (see Figure 21.2). The objective is to determine a BIBO stabilizing controller S_2 for the plant S_1 such that the effect of d on the plant output y is minimized in some sense.

We describe the plant by two transfer functions

$$S_{1u}(s) = \frac{B(s)}{A(s)}, \quad S_{1d}(s) = \frac{C(s)}{A(s)},$$

where A, B, and C is a triple of coprime functions from $R_H(s)$. The set of stabilizing controllers for S_1 is given by the transfer function

$$S_2(s) = -\frac{Y'(s) - A'(s)W(s)}{X'(s) + B'(s)W(s)},$$

where A', B' represent a *coprime* fraction over $R_H(s)$ for S_{1u},

$$\frac{B(s)}{A(s)} = \frac{B'(s)}{A'(s)}$$

and X', Y' represent a particular solution over $R_H(s)$ of the equation

$$A'X + B'Y = 1,$$

such that $X' + B'W \neq 0$.

The transfer function, $G(s)$, between d and y in a stable feedback system is

$$G = \frac{S_{1d}}{1 - S_{1u}S_2} = C(X' + B'W)$$

and it is affine in the proper and Hurwitz-stable rational parameter W.

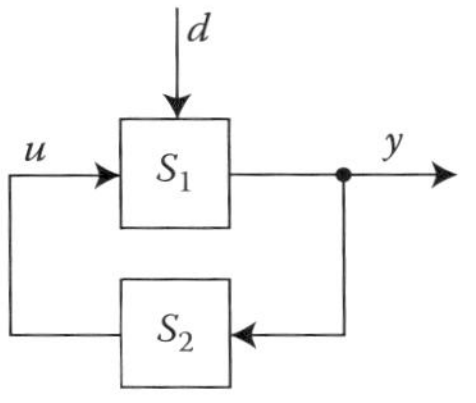

FIGURE 21.2 Disturbance attenuation.

Now suppose that the disturbance d is any function from L_∞, that is, any essentially bounded real function on R_+. Then

$$||y||_{L_\infty} \leq ||G||_1||d||_{L_\infty},$$

where

$$||G||_1 = \int_0^\infty |g(t)|\, dt$$

and $g(t)$ is the impulse response corresponding to $G(s)$. The parameter W can be used to minimize the norm $||G||_1$ and hence the maximum output amplitude.

If d is a stationary white noise, the steady-state output variance equals

$$Ey^2 = ||G||_2^2 Ed^2,$$

where

$$||G||_2^2 = \int_0^\infty |g(t)|^2\, dt = \frac{1}{2\pi j}\oint G(-s)G(s)\, ds.$$

The last integral is a contour integral up the imaginary axis and then around an infinite semicircle in the left half-plane. Again, W can be selected so as to minimize the norm $||G||_2$, thus minimizing the steady-state output variance.

Finally, suppose that d is any function from L_2, that is, any finite-energy real function on R_+. Then one obtains

$$||y||_{L_2} \leq ||G||_\infty ||d||_{L_2},$$

where

$$||G||_\infty = \sup_{Re\, s>0} |G(s)|.$$

Therefore, choosing W to make the norm $||G||_\infty$ minimal, one minimizes the maximum output energy.

The above system norms provide several examples showing how the effect of the disturbance on the plant output can be measured. The optimal attenuation is achieved by minimizing these norms.

Minimizing the 1-norm involves a linear program while minimizing the ∞-norm requires a search. The 2-norm minimization has a closed-form solution, which will be now described.

We recall that

$$G(s) = P(s) + Q(s)W(s),$$

where $P = CX'$ and $Q = CB'$. The norm $||G||_2$ is finite if and only if G is strictly proper and has no poles on the imaginary axis; hence we assume that Q has no zeros on the imaginary axis. We factorize

$$Q = Q_{ap}Q_{mp},$$

where Q_{ap} satisfies $Q_{ap}(-s)Q_{ap}(s) = 1$ (the so-called all-pass function) and Q_{mp} is such that Q_{mp}^{-1} is in $R_H(s)$ (the so-called minimum-phase function); this factorization is unique up to the sign. Let Q_{ap}^* denote the function $Q^*_{ap}(s) = Q_{ap}(-s)$. Then

$$\begin{aligned}||G||_2^2 &= ||P + QW||_2^2 \\ &= ||Q^*_{ap}P + Q_{mp}W||_2^2.\end{aligned}$$

Decompose $Q^*_{ap}P$ as

$$Q^*_{ap}P = (Q^*_{ap}P)_{st} + (Q^*_{ap}P)_{un}$$

where $(Q^*_{ap}P)_{st}$ is in $R_H(s)$ and $(Q^*_{ap}P)_{un}$ is unstable but strictly proper; this decomposition is unique. Then the cross-terms contribute nothing to the norm and

$$||G||_2^2 = ||(Q^*_{ap}P)_{un}||_2^2 + ||(Q^*_{ap}P)_{st} + Q_{mp}W||_2^2.$$

Since the first term is independent of W,

$$\min_W ||G||_2 = ||(Q^*_{ap}P)_{un}||_2$$

and this minimum is attained by

$$W = -\frac{(Q^*_{ap}P)_{st}}{Q_{mp}}.$$

Here is an illustrative example. The plant is given by

$$S_{1u}(s) = \frac{s-2}{s+1}, \quad S_{1d}(s) = 1$$

and we seek to find a stabilizing controller S_2 such that

$$G(s) = \frac{S_{1d}(s)}{1 - S_{1u}(s)S_2(s)}$$

has minimum 2-norm.

We write

$$A(s) = 1, \quad B(s) = \frac{s-2}{s+1}, \quad C(s) = 1$$

and find all stabilizing controllers first. Since the plant is already stable, these are given by

$$S_2(s) = -\frac{-W(s)}{1 + (s-2)/(s+1)W(s)},$$

where W is a free parameter in $R_H(s)$.

Then

$$G(s) = 1 + \frac{s-2}{s+1}W(s),$$

so that

$$P(s) = 1, \quad Q(s) = \frac{s-2}{s+1}.$$

Clearly,

$$Q_{ap}(s) = \frac{s-2}{s+2}, \quad Q_{mp}(s) = \frac{s+2}{s+1}$$

and

$$Q^*_{ap}(s)P(s) = \frac{s+2}{s-2} = 1 + \frac{4}{s-2}.$$

Therefore,

$$\|G\|_2^2 = \|\frac{4}{s-2}\|_2^2 + \|1 + \frac{s+2}{s+1}W\|_2^2,$$

so that the least norm

$$\min_W \|G\|_2 = \|\frac{4}{s-2}\|_2 = \|\frac{4}{s+2}\|_2 = 2$$

is attained by

$$W(s) = -\frac{s+1}{s+2}.$$

21.8 Robust Stabilization

The actual plant can differ from its nominal model. We suppose that a nominal plant description is available together with a description of the plant uncertainty. The objective is to design a controller that stabilizes all plants lying within the specified domain of uncertainty. Such a controller is said to *robustly* stabilize the family of plants.

The plant uncertainty can be modeled conveniently in terms of its fractional description. To fix ideas, we shall consider discrete-time plants factorized over $R_S(z)$ and endow $R_S(z)$ with the ∞-norm: for any function $H(z)$ from $R_S(z)$,

$$\|H\|_\infty = \sup_{|z|>1} |H(z)|.$$

For any two such functions, $H_1(z)$ and $H_2(z)$, we define

$$\begin{aligned}\|[H_1 H_2]\|_\infty &= \|\begin{bmatrix} H_1 \\ H_2 \end{bmatrix}\|_\infty \\ &= \sup_{|z|>1} (|H_1(z)|^2 + |H_2(z)|^2)^{1/2}.\end{aligned}$$

Let S_{10} be a nominal plant giving rise to a strictly proper transfer function

$$S_{10}(z) = \frac{B(z)}{A(z)},$$

where A and B are coprime functions from $R_S(z)$. We denote $S_1(A, B, \mu)$ the family of plants having strictly proper transfer functions

$$S_1(z) = \frac{B(z) + \Delta B(z)}{A(z) + \Delta A(z)},$$

where ΔA and ΔB are functions from $R_S(z)$ such that

$$\|[\Delta A\ \Delta B]\|_\infty < \mu$$

for some nonnegative real number μ.

Now, let S_2 be a BIBO stabilizing controller for S_{10}. Therefore,

$$S_2 = -\frac{Y' - AW}{X' + BW},$$

where $AX' + BY' = 1$ and W is an element of $R_S(z)$. Then S_2 will BIBO stabilize all plants from $S_1(A, B, \mu)$ if and only if the inverse of

$$(A + \Delta A)(X' + BW) + (B + \Delta B)(Y' - AW) = 1 + [\Delta A \quad \Delta B]\begin{bmatrix} X' + BW \\ Y' - AW \end{bmatrix}$$

is in $R_S(z)$. This is the case whenever

$$\|[\Delta A \; \Delta B]\begin{bmatrix} X' + BW \\ Y' - AW \end{bmatrix}\|_\infty < 1;$$

thus, we have the following condition of robust stability:

$$\mu\|\begin{bmatrix} X' + BW \\ Y' - AW \end{bmatrix}\|_\infty \leq 1.$$

The best controller that robustly stabilizes the plant corresponds to the parameter W that minimizes the ∞-norm above. This requires a search; closed-form solutions exist only in special cases. One such case is presented next.

Suppose the nominal model

$$S_{10}(z) = \frac{1}{z-1}$$

has resulted from

$$S_1(z) = \frac{z+\delta}{(z-1)(z-\varepsilon)}$$

by neglecting the second-order dynamics, where $\delta \geq 0$ and $0 \leq \varepsilon < 1$. Rearranging,

$$S_1(z) = \frac{(1/z) + (1/z)\,(\delta + \varepsilon/z - \varepsilon)}{(z - 1/z)}$$

and one identifies

$$\Delta A = 0, \quad \Delta B = \frac{1}{z}\frac{\delta+\varepsilon}{z-\varepsilon}.$$

Hence,

$$\|[\Delta A \; \Delta B]\|_\infty = \frac{\delta+\varepsilon}{1-\varepsilon}$$

and the true plant belongs to the family

$$S_1\left(\frac{z-1}{z}, \quad \frac{1}{z}, \quad \frac{\delta+\varepsilon}{1-\varepsilon}\right).$$

All controllers that BIBO stabilize the nominal plant S_{10} are given by

$$S_2(z) = -\frac{1-(z-1/z)W(z)}{1+(1/z)W(z)},$$

where W is a free parameter in $R_S(z)$. Which controller yields the best stability margin against δ and ε? The one that minimizes the ∞-norm in

$$\frac{\delta+\varepsilon}{1-\varepsilon}\|\begin{bmatrix} 1 + \dfrac{1}{z}W \\ 1 - \dfrac{z-1}{z}W \end{bmatrix}\|_\infty < 1.$$

Suppose we wish to obtain a controller of McMillan degree zero, $S_2(z) = -K$. Then

$$W(z) = (1-K)\frac{z}{z-(1-K)}$$

and $|1-K| < 1$. The norm

$$\left\| \begin{bmatrix} 1 + \frac{1}{z} W \\ 1 - \frac{z-1}{z} W \end{bmatrix} \right\|_\infty = \sqrt{(1+K^2)} \| \frac{z}{z-(1-K)} \|_\infty$$

attains the least value of $\sqrt{2}$ by $K = 1$, which corresponds to $W(z) = 0$. It follows that the controller

$$S_2(z) = -1$$

stabilizes all plants $S_1(z)$ for which

$$\frac{\delta + \varepsilon}{1-\varepsilon} < \frac{1}{\sqrt{2}}.$$

21.9 Robust Performance

The performance specifications often result in divisibility conditions. A typical example is the problem of *reference tracking.*

Suppose we are given a discrete-time plant S_1, with transfer function

$$S_1(z) = \frac{B(z)}{A(z)}$$

in coprime fractional form over $R_S(z)$, together with a reference r whose z-transform is of the form

$$r = \frac{E(z)}{D(z)},$$

where only D is specified. We recall that $S_1(z)$ is strictly proper. The objective is to design a BIBO stabilizing controller S_2 such that the plant output y asymptotically tracks the reference r (see Figure 21.3). The controller can operate on both r (feedforward) and y (feedback), so it is described by two transfer functions

$$S_{2y}(z) = -\frac{Y(z)}{X(z)}, \quad S_{2r}(z) = \frac{Z(z)}{X(z)},$$

where X, Y, and Z are from $R_S(z)$.

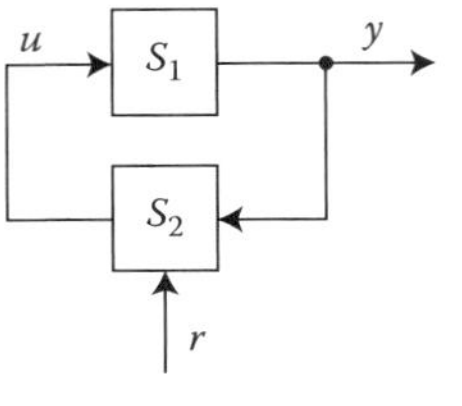

FIGURE 21.3 Reference tracking.

The requirement of tracking imposes that the tracking error

$$e = r - y = \left(1 - \frac{BZ}{AX + BY}\right)\frac{E}{D}$$

belong to $R_S(z)$. Since $AX + BY$ has inverse in $R_S(z)$ for every stabilizing controller and E is unspecified, D must divide $1 - BZ$ in $R_S(z)$. Hence, there must exist a function V in $R_S(z)$ such that $1 - BZ = DV$. Therefore, S_2 exists if and only if B and D are coprime in $R_S(z)$, and the two controller transfer functions evolve from solving the two Bézout equations

$$AX + BY = 1,$$
$$DV + BZ = 1,$$

where the function V serves to express the tracking error as

$$e = VE.$$

The reference tracking is said to be *robust* if the specifications are met even as the plant is slightly perturbed. We call S_{10} the nominal plant and $S_1(A, B, \mu)$ the neighborhood of S_{10} defined by

$$S_1 = \frac{B + \Delta B}{A + \Delta A},$$

where ΔA and ΔB are functions of $R_S(z)$, such that

$$\|[\Delta A \ \Delta B]\|_\infty < \mu$$

for some nonnegative real number μ. We recall that all $S_1(z)$ are strictly proper.

Now $A + \Delta A$ and $B + \Delta B$ are not specified, but $(A + \Delta A)X + (B + \Delta B)Y$ still has inverse in $R_S(z)$; call it U. We have

$$e = \left(\frac{A + \Delta A}{U}X + \frac{B + \Delta B}{U}(Y - Z)\right)\frac{E}{D}.$$

Hence, for robust tracking, D must divide both X and $Y - Z$ in $R_S(z)$. But, it is sufficient that D divides X; this condition already implies the other one as can be seen on subtracting the two Bézout equations above.

We illustrate on a discrete-time plant S_1 given by

$$S_1(z) = \frac{1}{z - 2}$$

whose output is to track every sinusoidal sequence of the form

$$r = \frac{az + b}{z^2 - z + 1},$$

where a, b are unspecified real numbers. Taking

$$A(z) = \frac{z - 2}{z}, \quad B(z) = \frac{1}{z}, \quad D(z) = \frac{z^2 - z + 1}{z^2}$$

and solving the pair of Bézout equations

$$\frac{z - 2}{z}X(z) + \frac{1}{z}Y(z) = 1,$$
$$\frac{z - z + 1}{z^2}V(z) + \frac{1}{z}Z(z) = 1$$

yields the tracking controllers in parametric form

$$S_{2y}(z) = -\frac{2-(z-2/z)W_1(z)}{1+(1/z)W_1(z)},$$

$$S_{2r}(z) = \frac{(z-1/z)-(z^2-z+1/z^2)W_2(z)}{1+(1/z)W_1(z)}$$

for any elements W_1, W_2 of $R_S(z)$. The resulting error is

$$e = \left[1+\frac{1}{z}W_2(z)\right]\frac{az+b}{z^2}.$$

Not all of these controllers, however, achieve a robust tracking of the reference. The divisibility condition is fulfilled if and only if W_1 is restricted to

$$W_1(z) = -\frac{z-1}{z} + \frac{z^2-z+1}{z^2}W(z),$$

where W is free in $R_S(z)$.

It is to be noted that the requirement of asymptotic tracking leaves enough degrees of freedom to meet additional design specifications.

21.10 Finite Impulse Response

Transients in discrete-time systems can settle in finite time. Systems having the property that any input sequence with a finite number of nonzero elements produces an output sequence with a finite number of nonzero elements have been called FIFO stable. We recall that a system with proper rational transfer function $H(z)$ is FIFO stable if and only if $H(z)$ is a polynomial in z^{-1}.

Let us consider the feedback system shown in Figure 21.1 and focus on achieving FIFO stability. To this end, we write the transfer function of the plant as

$$S_1(z) = \frac{B(z)}{A(z)},$$

where this time A and B are coprime polynomials in z^{-1}. We recall that the plant incorporates the necessary delay so that $S_1(z)$ is strictly proper. Repeating the arguments used to design a BIBO stable system, we conclude that all controllers S_2 that FIFO stabilize the plant S_1 have the transfer function

$$S_2(z) = -\frac{Y(z)}{X(z)},$$

where X, Y represent the solution class of the polynomial Bézout equation

$$AX + BY = 1.$$

In particular, if X' and Y' define any FIFO stabilizing controller for S_1, the set of all such controllers can be parameterized as

$$S_2(z) = -\frac{Y' - AW}{X' + BW},$$

where $W(z)$ is a free polynomial in z^{-1}.

It is a noteworthy fact that the parametric expressions for the sets of BIBO stable and FIFO stable controllers are the same; the only difference is that the free parameter of FIFO stabilizing controllers is permitted to range over only the smaller set of polynomials in z^{-1}, whereas in BIBO stabilizing controllers it is permitted to range over the larger set of Schur-stable rational functions in z. Indeed, FIFO stability is more restrictive than BIBO stability.

The design options offered by FIFO stability are remarkable. The parameter W can be selected so as to minimize the McMillan degree of S_2, or to achieve the shortest impulse response of the closed-loop system. Various norm minimizations can also be performed.

A well-known example is the deadbeat controller. We consider a double-summator plant with transfer function

$$S_1(z) = \frac{1}{(z-1)^2}$$

and interpret the exogenous inputs u_1 and u_2 as accounting for the effect of the initial conditions of S_1 and S_2. The requirement of FIFO stability is then equivalent to achieving finite responses y_1 and y_2 for all initial conditions. Since in this case

$$A(z) = (1-z^{-1})^2, \quad B(z) = z^{-2}$$

and the Bézout equation

$$(1-z^{-1})^2 X(z) + z^{-2} Y(z) = 1$$

has a particular solution

$$X'(z) = 1 + z^{-1}, \quad Y'(z) = 3 - 2z^{-1},$$

we obtain all deadbeat (or FIFO stabilizing) controllers as

$$S_2(z) = -\frac{3 - 2z^{-1} - (1-z^{-1})^2 W(z)}{1 + 2z^{-1} + z^{-2} W(z)}.$$

The deadbeat controller of least McMillan degree (=1) is obtained for $W(z) = 0$. The choice $W(z) = -3$ leads to a deadbeat controller that rejects step disturbances u_1 (hence, persistent) at the plant output y_1 in finite time. And when u_1 is a stationary white noise, then $W(z) = 2.5$ minimizes the steady-state variance of y_1 among all deadbeat controllers of McMillan degree 2.

21.11 Multivariable Systems

Up until now we have considered only single-input single-output (SISO) plants and controllers. In the case of multiple inputs and/or outputs, the input–output properties of linear systems are represented by a *matrix* of transfer functions. The additional intricacies introduced by these systems stem mainly from the fact that the matrix multiplication is not commutative.

Consider a rational transfer matrix $G(s)$ whose dimensions are, say, $m \times n$. Then it is always possible to factorize G as follows:

$$\begin{aligned} G(s) &= B_R(s) A_R^{-1}(s) \\ &= A_L^{-1}(s) B_L(s), \end{aligned}$$

where the factors B_R, A_R and A_L, B_L are, respectively, $m \times n$, $n \times n$ and $m \times m$, $m \times n$ matrices of qualified rational functions, say from $R_H(s)$, such that

$$\begin{aligned} &A_R, \quad B_R \text{ are right coprime,} \\ &A_L, \quad B_L \text{ are left coprime.} \end{aligned}$$

These "matrix fractions" are unique except for the possibility of multiplying the "numerator" and the "denominator" matrices by a matrix whose determinant has inverse in $R_H(s)$. That is, if $G(s)$ can also be expressed as

$$G(s) = B'_R(s)A'^{-1}_R(s)$$
$$= A'^{-1}_L(s)B'_L(s),$$

where the factors are matrices of functions from $R_H(s)$, such that

$$A'_R, B'_R \text{ are right coprime,}$$
$$A'_L, B'_L \text{ are left coprime,}$$

then

$$A'_R(s) = A_R(s)U_R(s), \quad B'_R(s) = B_R(s)U_R(s),$$
$$A'_L(s) = U_L(s)A_L(s), \quad B'_L(s) = U_L(s)B_L(s)$$

for some matrices U_R and U_L over $R_H(s)$, whose determinants have stable inverses in $R_H(s)$.

Analogous results hold for discrete-time systems. To illustrate, consider the transfer matrix

$$G(z) = \begin{bmatrix} \dfrac{1}{z-1} & \dfrac{2-z}{z^2-z} \\ \dfrac{1}{z-1} & \dfrac{1}{z-1} \end{bmatrix}$$

and determine its left and right coprime factorizations over $R_S(z)$. One obtains, for instance,

$$G(z) = \begin{bmatrix} \dfrac{1}{z} & \dfrac{2-z}{z^2-\lambda z} \\ 0 & \dfrac{1}{z-\lambda} \end{bmatrix} \begin{bmatrix} 1 & 0 \\ -1 & \dfrac{z-1}{z-\lambda} \end{bmatrix}^{-1}$$
$$= \begin{bmatrix} 1 & -1 \\ 0 & \dfrac{z-1}{z-\mu} \end{bmatrix}^{-1} \begin{bmatrix} 0 & -\dfrac{2}{z} \\ \dfrac{1}{z-\mu} & \dfrac{1}{z-\mu} \end{bmatrix}$$

for any real λ and μ with modulus less than one.

Let us now consider the feedback system shown in Figure 21.1 where S_1 and S_2 are multivariable systems and analyze its BIBO stability. We therefore factorize the two transfer matrices over $R_H(s)$,

$$S_1(s) = B_R(s)A_R^{-1}(s) = A_L^{-1}(s)B_L(s),$$
$$S_2(s) = -X_L^{-1}(s)Y_L(s) = -Y_R(s)X_R^{-1}(s),$$

where the two pairs A_R, B_R and X_R, Y_R are right coprimes while the two pairs A_L, B_L and X_L, Y_L are left coprimes. The transfer matrix of the feedback system

$$\begin{bmatrix} y_1 \\ y_2 \end{bmatrix} = \begin{bmatrix} S_1(I-S_2S_1)^{-1} & S_1(I-S_2S_1)^{-1}S_2 \\ S_2(I-S_1S_2)^{-1}S_1 & S_2(I-S_1S_2)^{-1} \end{bmatrix} \begin{bmatrix} u_1 \\ u_2 \end{bmatrix}$$

then reads

$$\begin{bmatrix} y_1 \\ y_2 \end{bmatrix} = \begin{bmatrix} B_R(X_LA_R+Y_LB_R)^{-1}X_L & -B_R(X_LA_R+Y_LB_R)^{-1}Y_L \\ I-A_R(X_LA_R+Y_LB_R)^{-1}X_L & -A_R(X_LA_R+Y_LB_R)^{-1}Y_L \end{bmatrix} \begin{bmatrix} u_1 \\ u_2 \end{bmatrix}$$

or alternatively

$$\begin{bmatrix} y_1 \\ y_2 \end{bmatrix} = \begin{bmatrix} X_R(A_LX_R+B_LY_R)^{-1}B_L & X_R(A_LX_R+B_LY_R)^{-1}A_L-I \\ -Y_R(A_LX_R+B_LY_R)^{-1}B_L & -Y_R(A_LX_R+B_LY_R)^{-1}A_L \end{bmatrix} \begin{bmatrix} u_1 \\ u_2 \end{bmatrix}.$$

The feedback system is BIBO stable if and only if this transfer matrix has entries in $R_H(s)$. We therefore conclude that the feedback system is BIBO stable if and only if the common denominator $X_LA_R + Y_LB_R$, or alternatively $A_LX_R + B_LY_R$, has inverse with entries in $R_H(s)$.

A parameterization of all controllers S_2 that BIBO stabilize the plant S_1 is now at hand. Given left and right coprime factorizations over $R_H(s)$ of the plant transfer matrix

$$S_1 = B_R A_R^{-1} = A_L^{-1} B_L,$$

we select matrices X'_L, Y'_L and X'_R, Y'_R with entries in $R_H(s)$, such that

$$X'_L A_R + Y'_L B_R = I, \quad A_L X'_R + B_L Y'_R = I.$$

Then the family of all stabilizing controllers has the transfer matrix

$$\begin{aligned} S_2 &= -(X'_L + W_L B_L)^{-1}(Y'_L - W_L A_L) \\ &= -(Y'_R - A_R W_R)(X'_R + B_R W_R)^{-1}, \end{aligned}$$

where W_L is a matrix parameter whose entries vary over $R_H(s)$ such that $X'_L + W_L B_L$ is nonsingular, and W_R is a matrix parameter whose entries vary over $R_H(s)$ such that $X'_R + B_R W_R$ is nonsingular.

As an example, determine all BIBO stabilizing controllers for the discrete-time plant considered earlier, with the transfer matrix

$$S_1(z) = \begin{bmatrix} \dfrac{1}{z-1} & \dfrac{2-z}{z^2-z} \\ \dfrac{1}{z-1} & \dfrac{1}{z-1} \end{bmatrix}.$$

The left and right coprime factors over $R_S(z)$ can be taken as

$$A_R(z) = \begin{bmatrix} 1 & 0 \\ -1 & \dfrac{z-1}{z} \end{bmatrix}, \quad B_R(z) = \begin{bmatrix} \dfrac{1}{z} & \dfrac{2-z}{z^2} \\ 0 & \dfrac{1}{z} \end{bmatrix}$$

and

$$A_L(z) = \begin{bmatrix} 1 & -1 \\ 0 & \dfrac{z-1}{z} \end{bmatrix}, \quad B_L(z) = \begin{bmatrix} 0 & -\dfrac{2}{z} \\ \dfrac{1}{z} & \dfrac{1}{z} \end{bmatrix}.$$

The Bézout equations

$$X'_L A_R + Y'_L B_R = I, \quad A_L X'_R + B_L Y'_R = I$$

have particular solutions

$$X'_L(s) = \begin{bmatrix} 1 & 0 \\ 1 & 1 \end{bmatrix}, \quad Y'_L(s) \begin{bmatrix} 0 & 0 \\ 0 & 1 \end{bmatrix}$$

and

$$X'_R(s) = \begin{bmatrix} 1 & 1 \\ 0 & 1 \end{bmatrix}, \quad Y'_R(s) \begin{bmatrix} 0 & 1 \\ 0 & 0 \end{bmatrix}.$$

The set of stabilizing controllers is given by

$$S_2(s) = -\left(\begin{bmatrix} 1 & 0 \\ 1 & 1 \end{bmatrix} + W_L \begin{bmatrix} 0 & -2z^{-1} \\ z^{-1} & z^{-1} \end{bmatrix}\right)^{-1} \left(\begin{bmatrix} 0 & 0 \\ 0 & 1 \end{bmatrix} - W_L \begin{bmatrix} 1 & -1 \\ 0 & 1-z^{-1} \end{bmatrix}\right),$$

where W_L varies over $R_S(z)$, or by

$$S_2(s) = -\left(\begin{bmatrix} 0 & 1 \\ 0 & 0 \end{bmatrix} - \begin{bmatrix} 1 & 0 \\ -1 & 1-z^{-1} \end{bmatrix} W_R\right) \left(\begin{bmatrix} 1 & 1 \\ 0 & 1 \end{bmatrix} + \begin{bmatrix} z^{-1} & -z^{-1}+2z^{-2} \\ 0 & z^{-1} \end{bmatrix} W_R\right)^{-1},$$

where W_R varies over $R_S(z)$ as well.

It is clear that the two parameterizations of S_2 are equivalent. To each controller S_2 there is a unique parameter W_L such that $S_2 = -(X_L' + W_L B_L)^{-1}(Y_L' - W_L A_L)$ as well as a unique parameter W_R such that $S_2 = -(Y_R' - A_R W_R)(X_R' + B_R W_R)^{-1}$, and these two are related by

$$W_R - W_L = X_L' Y_R' - Y_L' X_R'.$$

It is easy to see that the transfer matrix of the closed-loop system is *affine* in the free parameter W_L or W_R. Indeed,

$$\begin{bmatrix} y_1 \\ y_2 \end{bmatrix} = \begin{bmatrix} B_R(X_L' + W_L B_L) & -B_R(Y_L' - W_L A_L) \\ I - A_R(X_L' + W_L B_L) & -A_R(Y_L' - W_L A_L) \end{bmatrix} \begin{bmatrix} u_1 \\ u_2 \end{bmatrix}$$

or alternatively

$$\begin{bmatrix} y_1 \\ y_2 \end{bmatrix} = \begin{bmatrix} (X_R' + B_R W_R)B_L & (X_R' + B_R W_R)A_L - I \\ -(Y_R' - A_R W_R)B_L & -(Y_R' - A_R W_R)A_L \end{bmatrix} \begin{bmatrix} u_1 \\ u_2 \end{bmatrix}.$$

Thus, control synthesis problems beyond stabilization can be handled by determining the parameters W_L or W_R as described for SISO systems.

Let us consider the disturbance attenuation problem for the discrete-time plant

$$S_{1u}(z) = \begin{bmatrix} \frac{1}{z-1} & \frac{2-z}{z^2-z} \\ \frac{1}{z-1} & \frac{1}{z-1} \end{bmatrix}, \quad S_{1d}(z) = \begin{bmatrix} \frac{1}{z-1} \\ \frac{1}{z-1} \end{bmatrix},$$

where the disturbance d is assumed to be an arbitrary l_∞ sequence. We seek to find a BIBO stabilizing controller that minimizes the maximum amplitude of the plant output y.

We write

$$S_{1u}(z) = A_L^{-1}(z)B_L(z) = B_R(z)A_R^{-1}(z),$$
$$S_{1d}(z) = A_L^{-1}(z)C_L(z),$$

where

$$A_L(z) = \begin{bmatrix} 1 & -1 \\ 0 & 1-z^{-1} \end{bmatrix}, \quad B_L(z) = \begin{bmatrix} 1 & -2z^{-1} \\ z^{-1} & z^{-1} \end{bmatrix}, \quad C_L(z) = \begin{bmatrix} 0 \\ z^{-1} \end{bmatrix}$$

$$A_R(z) = \begin{bmatrix} 1 & 0 \\ -1 & 1-z^{-1} \end{bmatrix}, \quad B_R(z) = \begin{bmatrix} z^{-1} & -z^{-1}+2z^{-1} \\ 0 & z^{-1} \end{bmatrix}.$$

The set of BIBO stabilizing controllers has been found to have the transfer matrix

$$S_2(z) = -\left(\begin{bmatrix} 0 & 1 \\ 0 & 0 \end{bmatrix} - \begin{bmatrix} 1 & 0 \\ -1 & 1-z^{-1} \end{bmatrix} W_R\right) \left(\begin{bmatrix} 1 & 1 \\ 0 & 1 \end{bmatrix} + \begin{bmatrix} z^{-1} & -z^{-1}+2z^{-2} \\ 0 & z^{-1} \end{bmatrix} W_R\right)^{-1},$$

where W_R varies over $R_S(z)$.

The disturbance-output transfer matrix equals

$$G(z) = (I - S_{1u}S_2)^{-1}S_{1d}$$
$$= (X_R' + B_R W_R)^{-1} C_L.$$

When

$$W_R = \begin{bmatrix} W_{11} & W_{12} \\ W_{21} & W_{22} \end{bmatrix},$$

one obtains the expression

$$G(z) = \begin{bmatrix} z^{-1} + z^{-2}W_{12} - (z^{-2} - 2z^{-3})W_{22} \\ z^{-1} + z^{-2}W_{22} \end{bmatrix}.$$

The 1-norm of an $m \times n$ matrix $G(z)$ with entries

$$G_{ij}(z) = \sum_{k=0}^{\infty} g_{ij,k}\, z^{-k}$$

is defined by

$$\|G\|_1 = \max_{i=1,..,m} \sum_{j=1}^{n} \|G_{ij}\|_1,$$

where

$$\|G_{ij}\|_1 = \sum_{k=0}^{n} |g_{ij,k}|.$$

In our case

$$\|G\|_1 = \max\left(\|z^{-1} + z^{-2}W_{12} - (z^{-2} - 2z^{-3})\, W_{22}\|_1, \quad \|z^{-1} + z^{-2}W_{22}\|_1\right),$$

and it is clear by inspection that $\|G\|_1$ attains its minimum for $W_{12}(z) = 0$, $W_{22}(z) = 0$ and

$$\min_{W_R} \|G\|_1 = \max(1, 1) = 1.$$

The corresponding optimal BIBO stabilizing controllers are

$$S_2(z) = -\begin{bmatrix} -W_{11} & 1 \\ W_{11} - (1 - z^{-1})W_{21} & 0 \end{bmatrix} \begin{bmatrix} 1 + z^{-1}W_{11} - (z^{-1} - 2z^{-2})W_{22} & 1 \\ z^{-1}W_{21} & 1 \end{bmatrix}^{-1}$$

for any functions W_{11} and W_{21} in $R_S(z)$. The one of least McMillan degree reads

$$S_2(z) = \begin{bmatrix} 0 & -1 \\ 0 & 0 \end{bmatrix}.$$

21.12 Extensions

The factorization approach presented here for linear time-invariant systems with rational transfer matrices can be generalized to extend the scope of the theory to include distributed-parameter systems, time-varying systems, and even nonlinear systems.

The transfer matrices of distributed-parameter systems are no longer rational and coprime factorizations cannot be assumed *a priori* to exist. The coefficients of time-varying systems are functions of time, and the operations of multiplication and differentiation do not commute. In nonlinear systems, transfer matrices are replaced by input–output maps. Suitable factorizations of these maps may not exist and, if they do, they are not commutative in general.

For many systems of physical and engineering interest, these difficulties can be circumvented and the algebraic factorization approach carries over with suitable modifications.

Acknowledgment

The author acknowledges Project 1M0567, Ministry of Education of the Czech Republic.

Further Reading

Tutorial textbooks

1. Desoer, C. A. and Vidyasagar, M., *Feedback Systems: Input–Output Properties,* Academic Press, New York, 1975.
2. Kučera, V., *Discrete Linear Control: The Polynomial Equation Approach,* Wiley & Sons, Chichester, UK, 1979.
3. Vidyasagar, M., *Control System Synthesis: A Factorization Approach,* MIT Press, Cambridge, MA, 1985.
4. Doyle, J. C., Francis, B. A., and Tannenbaum, A. R., *Feedback Control Theory,* Macmillan, New York, 1992.

Survey papers

5. Kučera, V., Diophantine equations in control—A survey, *Automatica,* 29, 1361–1375, 1993.
6. Kučera, V., Polynomial control: Past, present, and future, *Int. J. Robust Nonlinear Control,* 17, 682–705, 2007.

Original sources on the parameterization of all stabilizing controllers

7. Kučera, V., Stability of discrete linear feedback systems, Paper 44.1, *Prepr. 6th IFAC World Congr.,* Vol. 1, Boston, 1975.
8. Youla, D. C., Jabr, H. A., and Bongiorno, J. J., Modern Wiener–Hopf design of optimal controllers, II. The multivariable case, *IEEE Trans. Autom. Control,* 21, 319–338, 1976.
9. Desoer, C. A., Liu, R. W., Murray, J., and Saeks, R., Feedback system design: The fractional representation approach to analysis and synthesis, *IEEE Trans. Autom. Control,* 25, 399–412, 1980.

Original sources on norm minimization

10. Francis, B. A., On the Wiener–Hopf approach to optimal feedback design, *System Control Lett.,* 2, 197–201, 1982.
11. Chang, B. C. and Pearson, J. B., Optimal disturbance reduction in linear multivariable systems, *IEEE Trans. Autom. Control,* 29, 880–888, 1984.
12. Dahleh, M. A. and Pearson, J. B., l^1-optimal feedback controllers for MIMO discrete-time systems, *IEEE Trans. Autom. Control,* 32, 314–322, 1987.

Original sources on robust stabilization

13. Vidyasagar, M. and Kimura, H., Robust controllers for uncertain linear multivariable systems, *Automatica,* 22, 85–94, 1986.
14. Dahleh, M. A., BIBO stability robustness in the presence of coprime factor perturbations, *IEEE Trans. Autom. Control,* 37, 352–355, 1992.

Original source on FIFO stability and related designs

15. Kučera, V. and Kraus, F. J., FIFO stable control systems, *Automatica,* 31, 605–609, 1995.

Advanced use of the parameterization of all stabilizing controllers

16. Kučera, V., Parametrization of stabilizing controllers with applications, in *Advances in Automatic Control*, M. Voicu (Ed.), pp. 173–192, Kluwer, Boston, 2003.
17. Henrion, D., Tarbouriech, S., and Kučera V., Control of linear systems subject to input constraints: A polynomial aproach, *Automatica*, 37, 597–604, 2001.
18. Henrion, D., Šebek, M., and Kučera, V., Positive polynomials and robust stabilization with fixed-order controllers. *IEEE Trans. Autom. Control*, AC-48, 178–186, 2003.
19. Henrion, D., Tarbouriech, S., and Kučera, V., Control of linear systems subject to time-domain constraints with polynomial pole placement and LMIs. *IEEE Trans. Autom. Control*, 50, 1360–1364, 2005.
20. Henrion, D., Kučera, V., and Molina-Cristobal A., Optimizing simultaneously over the numerator and denominator polynomials in the Youla–Kuèera parametrization, *IEEE Trans. Autom. Control*, 50, 1369–1374, 2005.

22

Quantitative Feedback Theory (QFT) Technique

Constantine H. Houpis
Air Force Institute of Technology

22.1 Introduction

22.1.1 Quantitative Feedback Theory

Quantitative feedback theory (QFT)* is a very powerful design technique for the achievement of assigned performance tolerances over specified ranges of structured plant parameter uncertainties without and with

* The original version of this material was first published by the Advisory Group for Aerospace Research and Development, North Atlantic Treaty Organization (AGARD/NATO) in Lecture Series LS-191 "on Linear Dynamics and Chaos" in June 1993.

control effector failures [9]. It is a frequency domain design technique utilizing the Nichols chart (NC) to achieve a desired robust design over the specified region of plant parameter uncertainty. This chapter presents an introduction to QFT analog and discrete design techniques for both multiple-input single-output (MISO) [1,5,13] and multiple-input multiple-output (MIMO) [3,4,6,7,10–12] control systems. QFT computer-aided design (CAD) packages are readily available to expedite the design process. The purposes of this chapter are (1) to provide a basic understanding of QFT; (2) to provide the minimum amount of mathematics necessary to achieve this understanding; (3) to discuss the basic design steps; and (4) to present a practical example.

22.1.2 Why Feedback?

For the answer to the question of "Why do you need QFT?" consider the following system. The plant P responds to the input $r(t)$ with the output $y(t)$ in the face of disturbances $d_1(t)$ and $d_2(t)$ (see Figure 22.1). If it is desired to achieve a specified system transfer function $T(s)[= Y(s)/R(s)]$, then it is necessary to insert a prefilter whose transfer function is $T(s)/P(s)$, as shown in Figure 22.2. This compensated system produces the desired output as long as the plant does not change and there are no disturbances. This type of system is sensitive to changes in the plant (or uncertainty in the plant), and the disturbances are reflected directly into the output. Thus, it is necessary to feed back the information in the output in order to reduce the output sensitivity to parameter variation and to attenuate the effect of disturbances on the plant output.

In designing a feedback control system, it is desired to utilize a technique that:

- Addresses all known plant variations up front
- Incorporates information on the desired output tolerances
- Maintains reasonably low loop gain (reduce the "cost of feedback")

This last item is important in order to avoid the problems associated with high loop gains such as sensor noise amplification, saturation, and high-frequency uncertainties.

22.1.3 What Can QFT Do?

Assume that the characteristics of a plant that is to be controlled over a specified region of operation vary; that is, a plant with structured parameter uncertainty. This plant parameter uncertainty may be described by the Bode plots of Figure 22.3. This figure represents the range of variation of plant magnitude (dB) and phase over a specified frequency range. The bounds of this variation, for this example, can be described by six linear time-invariant (LTI) plant transfer functions. By the application of QFT for a MISO control system containing this plant, a single compensator and a prefilter may be designed to achieve a specified robust design.

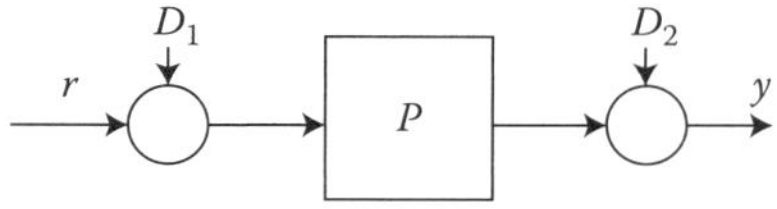

FIGURE 22.1 An open-loop system (basic plant).

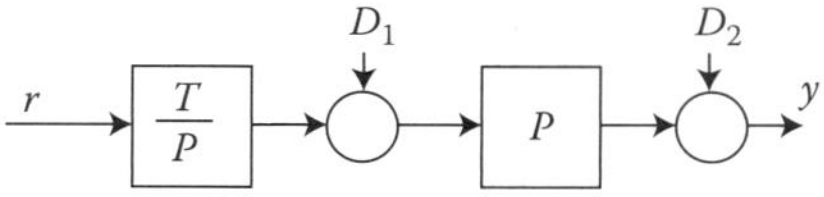

FIGURE 22.2 A compensated open-loop system.

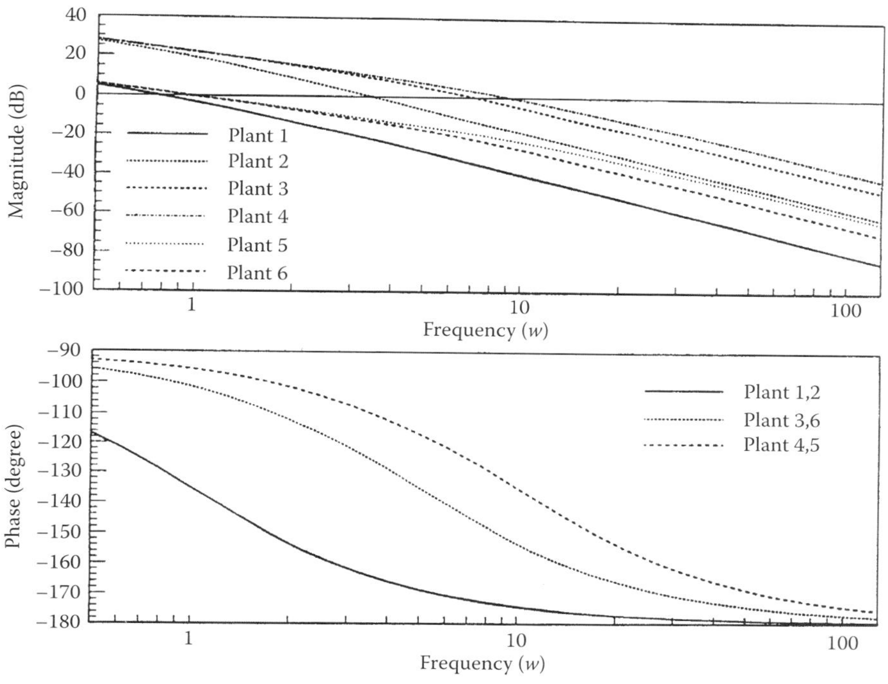

FIGURE 22.3 The Bode plots of six LTI plants that represent the range of the plant's parameter uncertainty.

22.1.4 Benefits of QFT

The benefits of QFT may be summarized as follows:

- The result is a robust design that is insensitive to plant variation
- There is one design for the full envelope (no need to verify plants inside templates)
- Any design limitations are apparent up front
- In comparison to other multivariable design techniques there is less development time for a full envelope design
- One can determine what specifications are achievable early in the design process
- One can redesign quickly for changes in the specifications
- The structure of compensator (controller) is determined up front

22.2 The MISO Analog Control System [1]

22.2.1 Introduction

The mathematical proof that an $m \times m$ feedback control system can be represented by m^2 equivalent MISO feedback control systems is given in Section 22.4.2. A 3×3 MIMO control system can be represented by the m^2 MISO equivalent loops shown in Figure 22.4. Thus, this and the next section present an introduction to the QFT technique by considering only a MISO feedback control system.

22.2.2 MISO System

The overview of the MISO QFT design technique is presented in terms of the minimum-phase (m.p.) LTI MISO system of Figure 22.5. The control ratios for tracking ($D = 0$) and for disturbance rejection ($R = 0$)

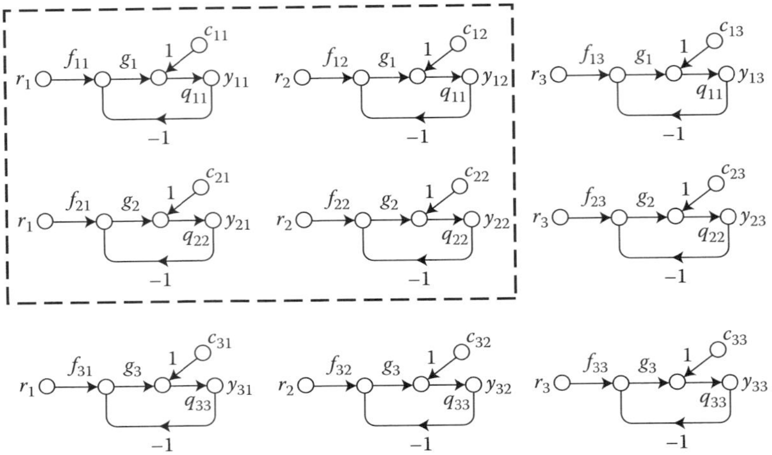

FIGURE 22.4 m^2 MISO equivalent of a 3×3 MIMO feedback control system.

are, respectively,

$$T_R = \frac{F(s)G(s)P(s)}{1+G(s)P(s)} = \frac{F(s)L(s)}{1+L(s)} \tag{22.1}$$

$$T_D = \frac{P(s)}{1+G(s)P(s)} = \frac{P(s)}{1+L(s)} \tag{22.2}$$

The design objective is to design the prefilter $F(s)$ and the compensator $G(s)$ so the specified robust design is achieved for the given region of plant parameter uncertainty. The design procedure to accomplish this objective is as follows:

Step 1: Synthesize the desired tracking model.
Step 2: Synthesize the desired disturbance model.
Step 3: Specify the J LTI plant models that define the boundary of the region of plant parameter uncertainty.
Step 4: Obtain plant templates, at specified frequencies, that pictorially describe the region of plant parameter uncertainty on the NC.
Step 5: Select the nominal plant transfer function $P_o(s)$.
Step 6: Determine the stability contour (U-contour) on the NC.
Steps 7–9: Determine the disturbance, tracking, and optimal bounds on the NC.
Step 10: Synthesize the nominal loop transmission function $L_o(s) = G(s)P_o(s)$ that satisfies all the bounds and the stability contour.
Step 11: Based upon Steps 1 to 10 synthesize the prefilter $F(s)$.
Step 12: Simulate the system in order to obtain the time response data for each of the J plants.

The following sections illustrate this design procedure.

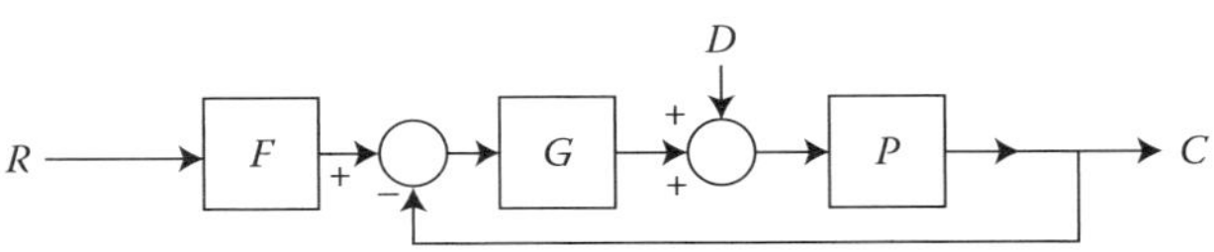

FIGURE 22.5 A MISO plant.

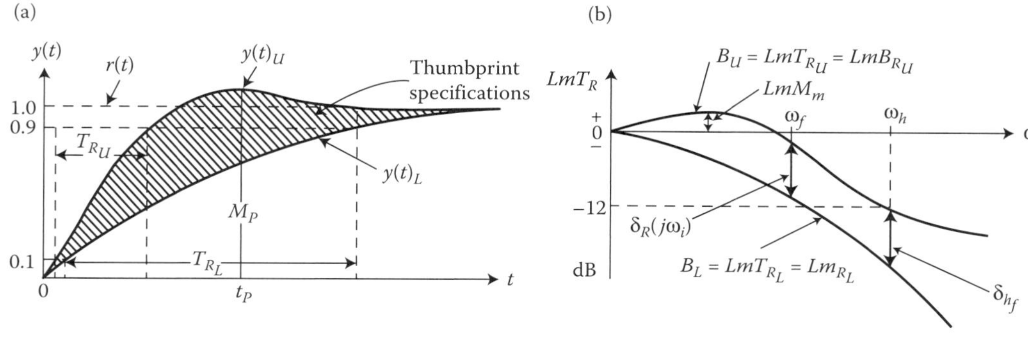

FIGURE 22.6 Desired response characteristic: (a) thumbprint specifications; (b) Bode plots of T_R.

22.2.3 Synthesize Tracking Models

The tracking thumbprint specifications, based upon satisfying some or all of the step-forcing function figures of merit for underdamped $(M_p, t_p, t_s, t_r, K_m)$ and overdamped (t_s, t_r, K_m) responses, respectively, for a simple-second order system, are depicted in Figure 22.6a. The Bode plots corresponding to the time responses $y(t)_U$ (Equation 22.3) and $y(t)_L$ (Equation 22.4) in Figure 22.6b represent the upper bound B_U and lower bound B_L, respectively, of the thumbprint specifications; i.e., an acceptable response $y(t)$ must lie between these bounds. Note that for *m.p.* plants, only the tolerance on $|T_R(j\omega_i)|$ need be satisfied for a satisfactory design. For nonminimum-phase (*n.m.p.*) plants, tolerances on $\angle T_R(j\omega_i)$ must also be specified and satisfied in the design process [4,5]. It is desirable to synthesize the tracking control ratios

$$T_{R_U} = \frac{(\omega_n^2/a)(s+a)}{s^2 + 2\zeta\omega_n s + \omega_n^2} \tag{22.3}$$

$$T_{R_L} = \frac{K}{(s-\sigma_1)(s-\sigma_2)(s-\sigma_3)} \tag{22.4}$$

corresponding to the upper and lower bounds T_{R_U} and T_{R_L}, respectively, so that $\delta_R(j\omega_i) = B_U - B_L$ increases as ω_i increases above the 0-dB crossing frequency of T_{R_U}. This characteristic of δ_R simplifies the process of synthesizing $L_o(s) = G(s)P_o(s)$. This synthesis process requires the determination of the tracking bounds $B_R(j\omega_i)$ that are obtained based upon $\delta_R(j\omega_i)$. The achievement of the desired performance specification is based upon the frequency bandwidth (BW), $0 < \omega \leq \omega_h$, which is determined by the intersection of the -12-dB line and the B_U curve in Figure 22.6b.

22.2.4 Disturbance Model

The simplest disturbance control ratio model specification is $|T_D(j\omega)| = |Y(j\omega)/D(j\omega)| \leq \alpha_p$ a constant [the maximum magnitude of the output based upon a unit-step disturbance input (d_1 of Figure 22.1)]. Thus, the frequency domain disturbance specification is log magnitude $(Lm)\ T_D(j\omega) \leq Lm\ \alpha_p$ over the desired specified BW $0 \leq \omega \leq \omega_h$ as defined in Figure 22.6b. Thus, the disturbance specification is represented by only an upper bound on the NC over the specified BW.

22.2.5 J LTI Plant Models

The simple plant

$$P_j(s) = \frac{Ka}{s(s+a)} \tag{22.5}$$

where $K \epsilon \{1, 10\}$ and $a \epsilon \{1, 10\}$, is used to illustrate the MISO QFT design procedure. The region of plant parameter uncertainty is illustrated by Figure 22.7. This region of uncertainty may be described by J LTI

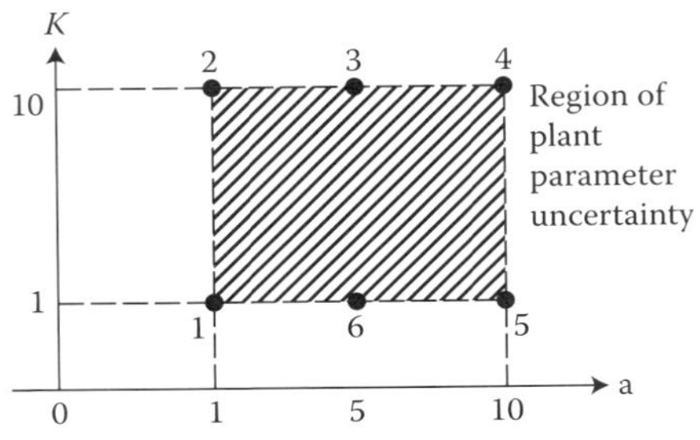

FIGURE 22.7 Region of plant uncertainty characterizing Equation 22.5.

plants, where $j = 1, 2, \ldots, J$. These plants lie on the boundary of this region of uncertainty; that is, the boundary points 1 to 6 are utilized to obtain six LTI plant models that adequately define the region of plant parameter uncertainty.

22.2.6 Plant Templates of $P_j(s)$, $\Im P(j\omega_i)$

With $L = GP$, Equation 22.1 yields

$$LmT_R = LmF - Lm\left[\frac{L}{1+L}\right] \tag{22.6}$$

The change in T_R due to the uncertainty in P, since F is LTI, is

$$\Delta(LmT_R) = LmT_R - LmF = Lm\left[\frac{L}{1+L}\right] \tag{22.7}$$

By the proper design of $L_o = GP_o$ and F, this change in T_R is restricted so that the actual value of Lm T_R always lies between B_U and B_L of Figure 22.6b. The first step in synthesizing an L_o is to make NC templates that characterize the variation of the plant uncertainty (see Figure 22.8), as described by $j = 1, 2, \ldots, J$ plant transfer functions, for various values of ω_i over a specified frequency range. The boundary of the plant template can be obtained by mapping the boundary of the plant parameter uncertainty region, Lm $P_j(j\omega_i)$ vs. $\angle P_j(j\omega_i)$, as shown on the NC in Figure 22.8. A curve is drawn through the points 1, 2, 3, 4, 5, and 6 where the shaded area is labeled $\Im P(j1)$, which can be represented by a plastic template. Templates for other values of ω_i are obtained in a similar manner. A characteristic of these templates is that, starting

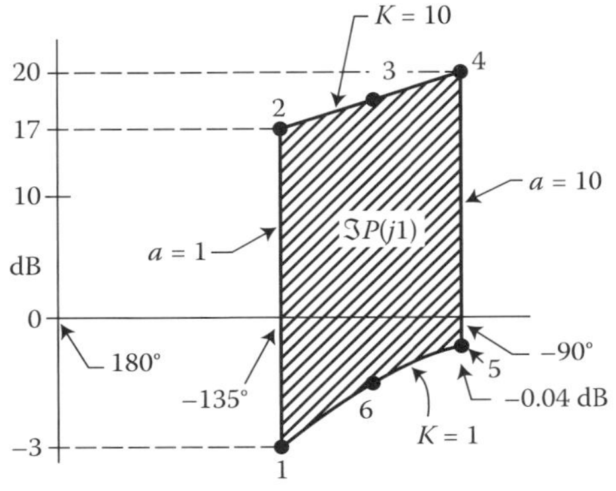

FIGURE 22.8 The template $\Im P(j1)$ characterizing Equation 22.5.

from a "low value" of ω_i, the templates widen (angular width becomes larger) for increasing values of ω_i, then, as ω_i takes on larger values and approaches infinity, they become narrower and eventually approach a straight line of height V dB (see Equation 22.9).

22.2.7 Nominal Plant

While any plant case can be chosen, select whenever possible a plant whose NC template point is always at the lower left corner for all frequencies for which the templates are obtained.

22.2.8 U-Contour (Stability Bound)

The specifications on system performance in the frequency domain (see Figure 22.6b) identify a minimum damping ratio ζ for the dominant roots of the closed-loop system, which becomes a bound on the value $M_p \approx M_m$. On the NC this bound on $M_m = M_L$ (see Figures 22.6b and 22.9) establishes a region that must not be penetrated by the templates and the loop transmission function $L(j\omega)$ for all ω. The boundary of this region is referred to as the universal high-frequency boundary (UHFB) or stability bound—the U-contour, because this becomes the dominating constraint on $L(j\omega)$. Therefore, in Figure 22.9 the top portion, *efa*, of the M_L contour becomes part of the U-contour. For a large problem class, as $\omega \to \infty$, the limiting value of the plant transfer function approaches

$$\lim_{\omega\to\infty} [P(j\omega)] = \frac{K}{\omega^{\lambda}} \tag{22.8}$$

where λ represents the excess of poles over zeros of $P(s)$. The plant template for this problem class approaches a vertical line of length equal to

$$\Delta \lim_{(\omega\to\infty)} [LmP_{\text{max}} - LmP_{\text{min}}] = LmK_{\text{max}} - LmK_{\text{min}} = VdB \tag{22.9}$$

If the nominal plant is chosen at $K = K_{\text{min}}$, then measuring VdB down from the bottom portion *age* of the constraint M_L gives the *bcd* portion of the U-contour *abcdefa* of Figure 22.9. The remaining portions, *ab* and *de*, of the stability contour are determined during the process of determining the tracking bounds.

22.2.9 Optimal Bounds $B_o(j\omega_i)$ on $L_o(j\omega_i)$

The determination of the tracking $B_R(j\omega_i)$ and the disturbance $B_D(j\omega_i)$ bounds are required in order to yield the optimal bounds $B_o(j\omega_i)$ on $L_o(j\omega_i)$.

22.2.9.1 Tracking Bounds

The solution for $B_R(j\omega_i)$ requires that the condition (actual)$\Delta T_R(j\omega_i) \leq \delta_R(j\omega_i)$ dB (see Figure 22.6b) must be satisfied. Thus, it is necessary to determine the resulting constraint, or bound $B_R(j\omega_i)$, on $L(j\omega_i)$.

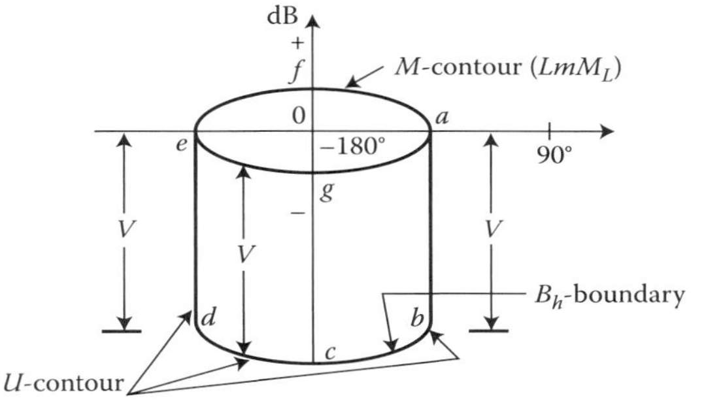

FIGURE 22.9 U-contour construction.

The procedure is to pick a nominal plant $P_o(s)$ and to derive tracking bounds on the NC, at specified values of frequency, by use of templates or a CAD package. That is, along a phase angle grid line on the NC, move the nominal point on the template $\Im P(j\omega_i)$ up or down, without rotating the template, until it is tangent to two M-contours whose difference in M values is essentially equal to δ_R. When this condition has been achieved, the location of the nominal point on the template becomes a point on the tracking bound $B_R(j\omega_i)$ on the NC. This procedure is repeated on sufficient angle grid lines on the NC to provide sufficient points to draw $B_R(j\omega_i)$ and for all values of frequency for which templates have been obtained. In general, the templates are moved from right to left starting from a phase angle grid line to the right of the M_L contour. When the templates become tangent to the M_L contour, the nominal point on the templates yields points on the *ab* and *de* portions of the stability contour. For m.p. systems, the condition $\Delta T_R(j\omega_i) \leq \delta_R(j\omega_i)$ requires that the synthesized loop transmission must satisfy the requirement that Lm $L(j\omega_i)$ is on or above the corresponding tracking bound Lm $B_R(j\omega_i)$.

22.2.9.2 Disturbance Bounds

The general procedure for determining disturbance bounds for the MISO control system of Figure 22.5 is outlined as follows, but more details are given in [1]. From Equation 22.2 the following equation is obtained:

$$T_D = \frac{P_o}{(P_o/P) + L_o} = \frac{P_o}{W} \tag{22.10}$$

where $W = (P_o/P) + L_o$. From Equation 22.10, setting $\text{Lm}T_D = \delta_D = \text{Lm}\alpha_p$, the following relationship is obtained:

$$LmW = LmP_o - \delta_D \tag{22.11}$$

For each value of frequency for which the NC templates are obtained, the magnitude of $|W(j\omega_i)|$ is obtained from Equation 22.11. This magnitude, in conjunction with the equation $W(j\omega_i) = [P_o(j\omega_i)/P(j\omega_i)]$, is utilized to obtain a graphical solution for $B_D(j\omega_i)$ as shown in Figure 22.10. Note that in this figure the template is plotted in rectangular or polar coordinates.

22.2.9.3 Optimal Bounds

For the case shown in Figure 22.11, $B_o(j\omega_i)$ is composed of those portions of each respective bound $B_R(j\omega_i)$ and $B_D(j\omega_i)$ that have the largest dB values. The synthesized $L_o(j\omega_i)$ must lie on or just above the bound $B_o(j\omega_i)$ of Figure 22.11.

22.2.10 Synthesizing (or Loop-Shaping) $L_o(s)$ and $F(s)$

The shaping of $L_o(j\omega)$ is shown by the dashed curve in Figure 22.11. A point such as Lm $L_o(j2)$ must be on or above $B_o(j2)$. Further, in order to satisfy the specifications, $L_o(j\omega)$ cannot violate the U-contour. In this example, a reasonable $L_o(j\omega)$ closely follows the U-contour up to $\omega = 40$ rad/s and must stay below it above $\omega = 40$, as shown in Figure 22.11. It also must be at least a Type 1 $L_o(s)$ transfer function (one or

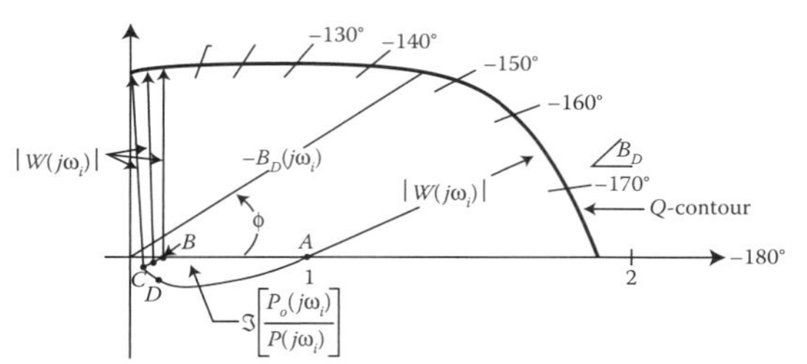

FIGURE 22.10 Graphical evaluation of $B_D(j\omega_i)$.

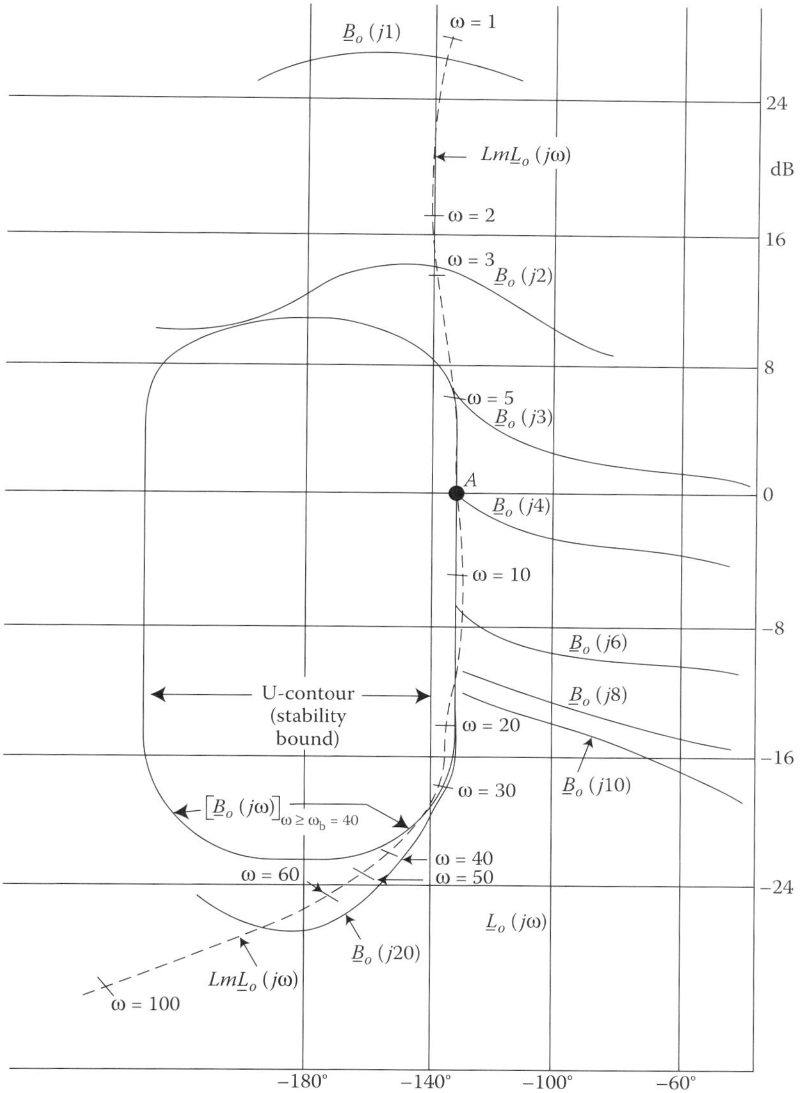

FIGURE 22.11 Bounds $B_o(j\omega_i)$ and loop-shaping.

more poles at the origin) for tracking a step-forcing function with zero steady-state error [1]. Synthesizing a rational function $L_o(s)$ that satisfies the above specification involves building up the function

$$L_o(j\omega) = L_{ok}(j\omega) = P_o(j\omega) \sqcap_{k=0}^{w} [K_k G_k(j\omega)] \tag{22.12}$$

where for $k = 0$, $G_0 = 1\angle 0°$ and $K = \prod_{k=0}^{w} K_k$. In order to minimize the order of the compensator, a good starting point for "building up" the loop transmission function is to assume initially that $L_{o0}(j\omega) = P_o(j\omega)$ as indicated in Equation 22.13. $L_o(j\omega)$ is built up term-by-term or by a CAD loop-shaping routine [8], in order (1) that the point $L_o(j\omega_i)$ lies on or above the corresponding optimal bound $B_o(j\omega_i)$, (2) that it passes close to the trough of the low frequency bounds, for achieving minimal gain, and (3) to stay just outside the U-contour in the NC of Figure 22.11. The design of a proper $L_o(s)$ guarantees only that the variation in $|T_R(j\omega_i)|$ is less than or equal to that allowed; i.e., $\delta_R(j\omega_i)$. The purpose of the prefilter $F(s)$

is to position Lm $[T(j\omega)]$ within the frequency domain specifications; i.e., that it always lies between B_U and B_L (see Figure 22.6b) for all J plants. The method for determining $F(s)$ is discussed in the next section. Once a satisfactory $L_o(s)$ is achieved, then the compensator is given by $G(s) = L_o(s)/P_o(s)$. Note that for this example $L_o(j\omega)$ slightly intersects the U-contour at frequencies above ω_h. Because of the inherent overdesign feature of the QFT technique, as a first trial design no effort is made to fine-tune the synthesis of $L_o(s)$. If the simulation results are not satisfactory, then a fine tuning of the design can be made. The available CAD packages simplify and expedite this fine tuning.

22.2.11 Prefilter Design [1,2,4,5]

Design of a proper $L_o(s)$ guarantees only that the variation in $|T_R(j\omega)|$ is less than or equal to that allowed; i.e., $[LmT_R(j\omega)] \leq \delta_R(j\omega)$ The purpose of the prefilter $F(s)$ is to position

$$LmT(j\omega) = Lm\frac{L(j\omega)}{1+L(j\omega)} \tag{22.13}$$

within the frequency domain specifications. A method for determining the bounds on $F(s)$ is as follows:

Step 1: Place the nominal point of the ω_i plant template on the $L_o(j\omega_i)$ point on the $L_o(j\omega)$ curve on the NC (see Figure 22.12).

Step 2: Traversing the template, determine the M-contours that yield the maximum LmT_{max} and the minimum LmT_{min} values of Equation 22.13.

Step 3: Based upon obtaining sufficient data points, by repeating Step 2 within the desired frequency bandwidth for various values of ω_i, and in conjunction with the data used to obtain Figure 22.6b the plots of Figure 22.13 are obtained.

Step 4: Utilizing Figure 22.13, the straight-line Bode technique, and the condition

$$\lim_{s\to 0} F(s) = 1 \tag{22.14}$$

for a step-forcing function, an $F(s)$ is synthesized that lies within the upper and lower plots in Figure 22.13.

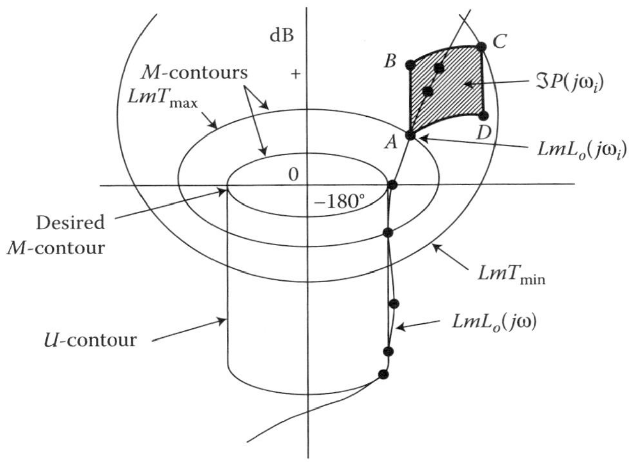

FIGURE 22.12 Prefilter determination.

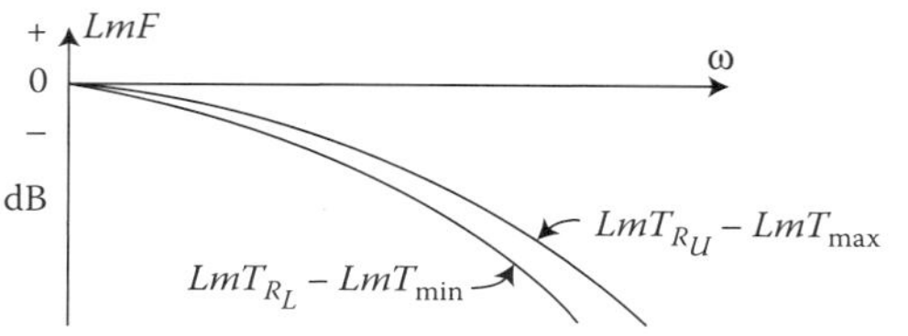

FIGURE 22.13 Frequency bounds on $F(s)$.

22.2.12 Simulation

The "goodness" of the synthesized $L_o(s)$ and $F(s)$ is determined by simulating the QFT-designed control system for all J plants. MISO QFT CAD packages, as discussed in Section 22.2.4, are available to expedite this simulation phase of the complete design process.

22.2.13 MISO QFT CAD Packages

The first usable MISO QFT CAD package was developed in 1986 for the analog design and in 1991 for the discrete design at the Air Force Institute of Technology (AFIT). These CAD packages have been a catalyst in assisting the newcomer to QFT to understand the fundamentals of this powerful design technique. The QFT CAD package illustrated in Appendix B can be used for both MISO and MIMO control system designs.

22.2.13.1 MISO QFT CAD

The flowchart of the MISO QFT CAD options in the AFIT package called TOTAL-P.C is shown in Appendix A. Those desiring a copy of this package can contact Professor C. H. Houpis, AFIT/ENG, Wright-Patterson AFB, OH 45433. This package has been designed as an educational tool. The QFT CAD package of Appendix B can also be used for the design of a MISO control system.

22.2.13.2 MISO QFT PC CAD

Dr. Yossi Chait, University of Massachusetts, and Dr. Oded Yaniv, Tel-Aviv University, Israel, have developed a MISO QFT PC CAD package for both analog and discrete system design available in MATLAB®.

22.3 The MISO Discrete Control System [13]

22.3.1 Introduction

The bilinear transformation, z-domain to the w'-domain and vice-versa, is utilized in order to accomplish the QFT design for both MISO and MIMO sampled-data (discrete) control system design in the w'-domain. This transformation enables the MISO QFT analog design technique to be readily used, with minor exceptions, to perform the QFT design for the controller $G(w')$. If the w'-domain simulations satisfy the desired performance specification, then by use of the bilinear transformation the z-domain controller $G(z)$ is obtained. With this z-domain controller, a discrete-time domain simulation is obtained to verify the "goodness" of the design. The QFT technique requires the determination of the minimum sampling frequency $(\omega_s)_{min}$ BW that is needed for a satisfactory design [13,14]. The larger the plant uncertainty and the narrower the system performance tolerances, the larger must be the value of $(\omega_s)_{min}$. *Henceforth, the prime is omitted from w'; whenever the symbol w is used it is to be interpreted as w'.*

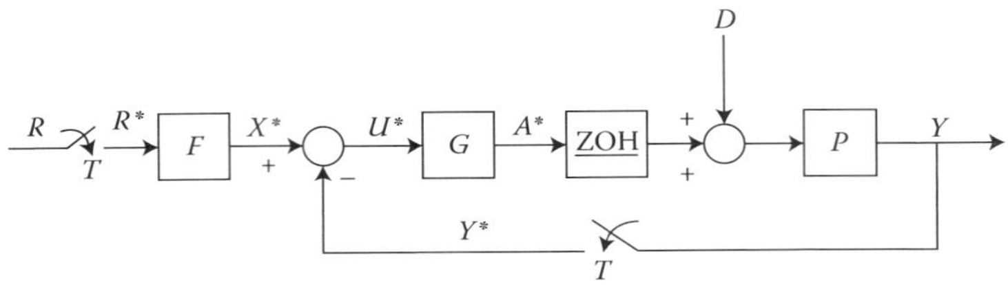

FIGURE 22.14 A MISO sampled-data control system.

22.3.2 The MISO Sampled-Data Control System

Figure 22.14 represents the MISO discrete control system, having plant uncertainty, that is to be designed by the QFT technique. The equations that describe this system are as follows:

$$P_z(z) = G_{zo}P(z) = (1 - z^{-1})Z\left[\frac{P(s)}{s}\right] = (1 - z^{-1})P_e(z) \tag{22.15}$$

$$L(z) = G_{zo}P(z)G_1(z), \quad P_e \equiv \frac{P(s)}{s}, \quad P_e(z) = Z\left[\frac{P(s)}{s}\right] = Z[P_e] \tag{22.16}$$

$$D(s) = \frac{1}{s} \quad P_e(s) = P(s)D(s)$$

$$P_e(z) = Z[P(s)D(s)] = PD(z) \tag{22.17}$$

$$T_R = \frac{F(z)L(z)}{1 + L(z)} \quad Y_D = \frac{PD(z)}{1 + L(z)} \tag{22.18}$$

$$\begin{aligned} Y(z) &= \left[\frac{L(z)F(z)}{1 + L(z)}\right]R(z) + \frac{PD(z)}{1 + L(z)} = Y_R(z) + Y_D(z) \\ &= T_R(z)R(z) + Y_D(z) \end{aligned} \tag{22.19}$$

22.3.3 *w*-Domain

The pertinent *s*-, *z*-, and *w*-plane relationships are as follows:

$$\alpha^2 = \left(\frac{\sigma T}{2}\right)^2 < 2, \quad \frac{\omega T}{2} \leq 0.297 \tag{22.20}$$

$$s = \sigma + j\omega, \quad (a)$$

$$w = u + jv = \left(\frac{2}{T}\right)\left[\frac{z - 1}{z + 1}\right] \quad (b) \tag{22.21}$$

$$z = \frac{Tw + 2}{-Tw + 2}, \quad v = \frac{2}{T}\tan\left(\frac{\omega T}{2}\right) = \left(\frac{\pi}{\omega_s}\right)\tan\left(\frac{\omega\pi}{\omega_s}\right) \tag{22.22}$$

$$\omega_s = 2\pi/T, \quad z = \epsilon^{\sigma T}\angle\omega T = |z|\angle\omega T \tag{22.23}$$

22.3.4 Assumptions

For this chapter, the following assumptions are made:

- Minimum-phase (m.p.) stable plants
- The analog design models, Equations 22.3 and 22.4, yield the desired time response characteristics for the discrete-time system

- The sampling time T is small enough so that over the BW, $0 < \omega < \omega_h$, Equation 22.23 is valid permit- ting the approximation $s \approx w$ and, in turn,

$$T_R(w) \approx [T_R(s)]_{s=w} \tag{22.24}$$

Both the upper and lower bound w-domain tracking models are obtained in this manner. The disturbance specification is the same as for the analog case.

22.3.5 Nonminimum Phase $L_o(w)$

It is important to note that in the w domain any practical $L(w)$ is n.m.p. containing a zero at $2/T$ (the sampling zero). This result is due to the fact that any practical $L(z)$ has an excess of at least one pole over zeros. Thus, the design technique *for a stable uncertain plant* is modified [14] to incorporate the allpass filter (apf)

$$A(w) = \frac{w - (2/T)}{w + (2/T)} = -A'(w) = -\left[\frac{(2/T) - w}{(2/T) + w}\right] \tag{22.25}$$

as follows. Let the nominal loop transmission be defined as

$$L_o \equiv -L_{mo}(w)A(w) = L_{mo}(w)A'(w) \tag{22.26}$$

From Equation 22.26 it is seen that

$$\angle L_{mo}(jv) = \angle L_o(jv) - \angle A'(jv) \tag{22.27}$$

where

$$-\angle A'(jv) = 2\tan^{-1}\frac{vT}{2} > 0 \tag{22.28}$$

An analysis of Equations 22.26 through 22.28 reveals that the bounds $B'_o(jv_i)$ on $L_o(jv)$ become the bounds $B_{mo}(jv_i)$ on $L_{mo}(jv)$ by shifting, over the desired $BW, B'_o(jv_i)$ positively (to the right on the NC) by the angle $\angle A'(jv_i)$, as shown in Figure 22.15. The U-contour (B'_h) must also be shifted to the right by the same amount, at the specified frequencies v_i, to obtain the shifted U-contour $B_h(jv_i)$. The contour B'_h is shifted to the right until it reaches the vertical line $\angle L_{mo_1}(jv_K) = 0°$. The value of v_K, which is a function of ω_s and the phase margin angle γ as shown in Figure 22.15, [13] is given by

$$2\tan^{-1}\left(\frac{v_k T}{2}\right) = 180° - \gamma \tag{22.29}$$

It should be mentioned that loop-shaping or synthesizing $L_o(w)$ can be done directly without the use of an apf.

22.3.6 Plant Templates $\Im$ $P(jv_1)$

The plant templates in the w-domain have the same characteristic as those for the analog case (see Section 22.2.4) for the frequency range $0 < \omega_i \leq \omega_s/2$ as shown in Figure 22.16a. In the frequency range $\omega_s/2 < \omega_i < \infty$, the w-domain templates widen once again, then eventually approach a vertical line as shown in Figure 22.16b.

22.3.7 Synthesizing $L_{mo}(w)$

The frequency spectrum can be divided into four general regions for the purpose of synthesizing (loop-shaping) an $L_{mo}(w)$ that satisfies the desired system performance specifications for the plant having plant parameter uncertainty. These four regions are

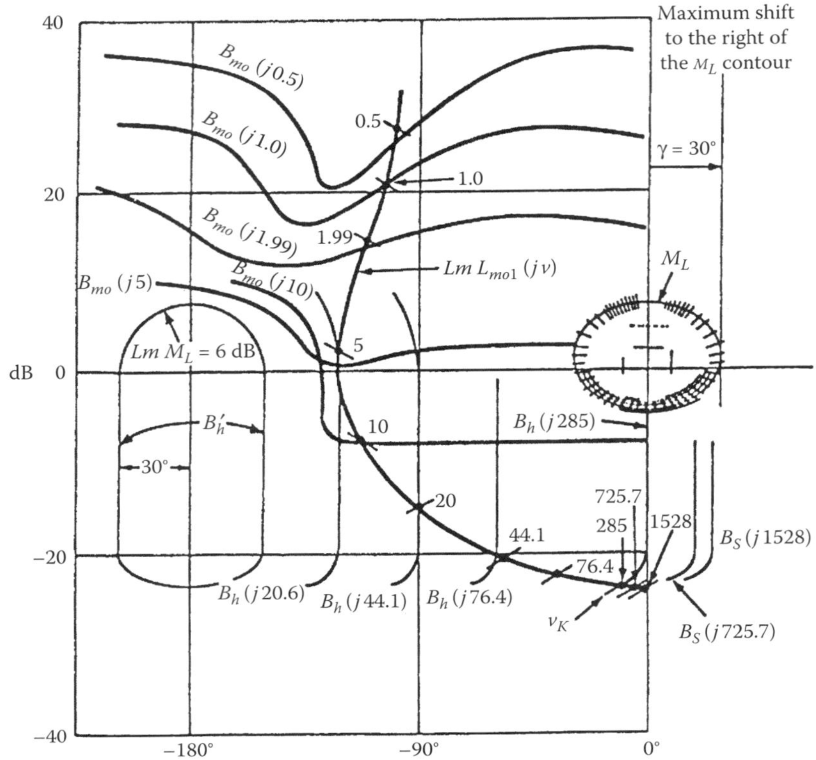

FIGURE 22.15 The shifted bounds on the NC. *Note:* Curves drawn approximately to scale.

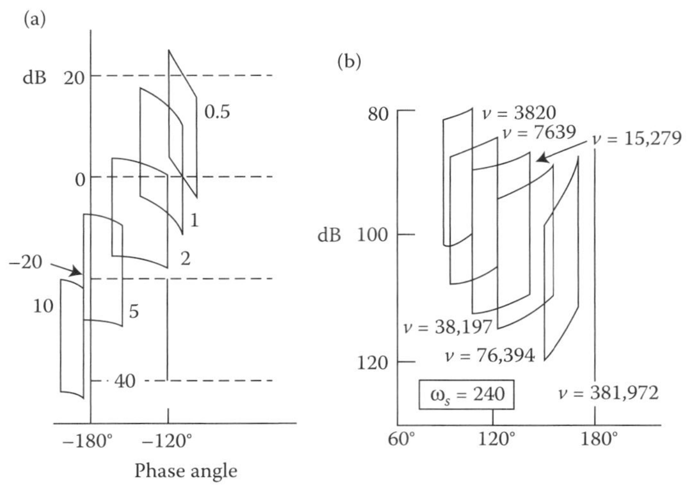

FIGURE 22.16 *w*-domain plant templates.

Region 1: For the frequency range where Equation 22.20 is essentially satisfied, use the analog templates; i.e., $\Im P(j\omega_i) \approx \Im P(jv_i)$. The w-domain tracking, disturbance, and optimal bounds and the U-contour are essentially the same as those for the analog system. The templates are used to obtain these bounds on the NC in the same manner as for the analog system.

Region 2: For the frequency range $v_{0.25} < v_i \leq v_h$, where $\omega_i \leq 0.25\omega_s$, use the w-domain templates. These templates are used to obtain all three types of bounds, in the same manner as for the analog system, in this region; and the corresponding $B'_h(jv_i)$ contours are also obtained. Depending on the value of T, v_h may be less than $v_{0.25}$.

Region 3: For the frequency range $v_h < v_i \leq v_K$, for the specified value of ω_s, only the B'_h contours are plotted.

Region 4: For the frequency range $v_h > v_K$, use the w-domain templates. Since the templates $\Im P_e(jv_i)$ broaden out again for $v_i > v_K$, as shown in Figure 22.16, it is necessary to obtain the more stringent (stability) bounds B_S shown in Figure 22.17. The templates are used only to determine the stability bounds B_S.

The synthesis (or loop-shaping) of $L_{mo}(w)$ involves the synthesis of the following function:

$$L_{mo}(jv) = P_{eo}(jv) \prod_{k=0}^{w} [K_k G_k(jv)] \tag{22.30}$$

where the nominal plant $P_{eo}(w)$ is the plant from the J plants that has the smallest dB value and the largest (most negative) phase lag characteristic. The final synthesized $L_{mo}(w)$ function must be one that satisfies the following conditions:

1. In Regions 1 and 2 the point on the NC that represents the dB value and phase angle of $L_{mo}(jv_i)$ must be such that it lies on or above the corresponding $B_{mo}(jv_i)$ bound (see Figure 22.15).
2. The values of Equation 22.30 for the frequency range of region 3 must lie to the right of or just below the corresponding B'_h contour (see Figure 22.15).
3. The value of Equation 22.30 for the frequency range of region 4 must lie below the B_S contour for negative phase angles on the NC (see condition 4 next).
4. In utilizing the bilinear transformation of Equation 22.21, the w-domain transfer functions are all equal order over equal order.
5. The Nyquist stability criterion dictates that the $L_{mo}(jv)$ plot is on the "right side" or the "bottom right side" of the $B_h(jv_i)$ contours for the frequency range of $0 \leq v_i \leq v_K$. It has been shown that [3]
   a. $L_{mo}(jv)$ must reach the *right-hand bottom* of $B_h(jv_K)$, (i.e., approximately point K in Figure 22.17) at a value of $v \leq v_K$.
   b. $\angle L_{mo}(jv_K) < 0°$ in order that there exists a practical L_{mo} that satisfies the bounds $B(jv)$ and provides the required stability.
6. For the situation where one or more of the J LTI plants that represent the uncertain plant parameter characteristics represent unstable plants and one of these unstable plants is selected as the nominal plant, *then the apf to be used in the QFT design must include all right half-plane (RHP) zeros of* P_{zo}. This situation is not discussed in this chapter. **Note:** For experienced QFT control system designers, $L_o(v)$ can be synthesized without the use of apf. This approach also is not covered in this chapter.

The synthesized $L_{mo}(w)$ obtained following the guidelines of this section is shown in Figure 22.17.

22.3.8 Prefilter Design

The procedure for synthesizing $F(w)$ is the same as for the analog case (see Section 22.2.11) over the frequency range $0 < v_i \leq v_h$. In order to satisfy condition 4 of Section 22.3.7, a nondominating zero or zeros ("far left" in the w-plane) are inserted so that the final synthesized $F(w)$ is equal order over equal order.

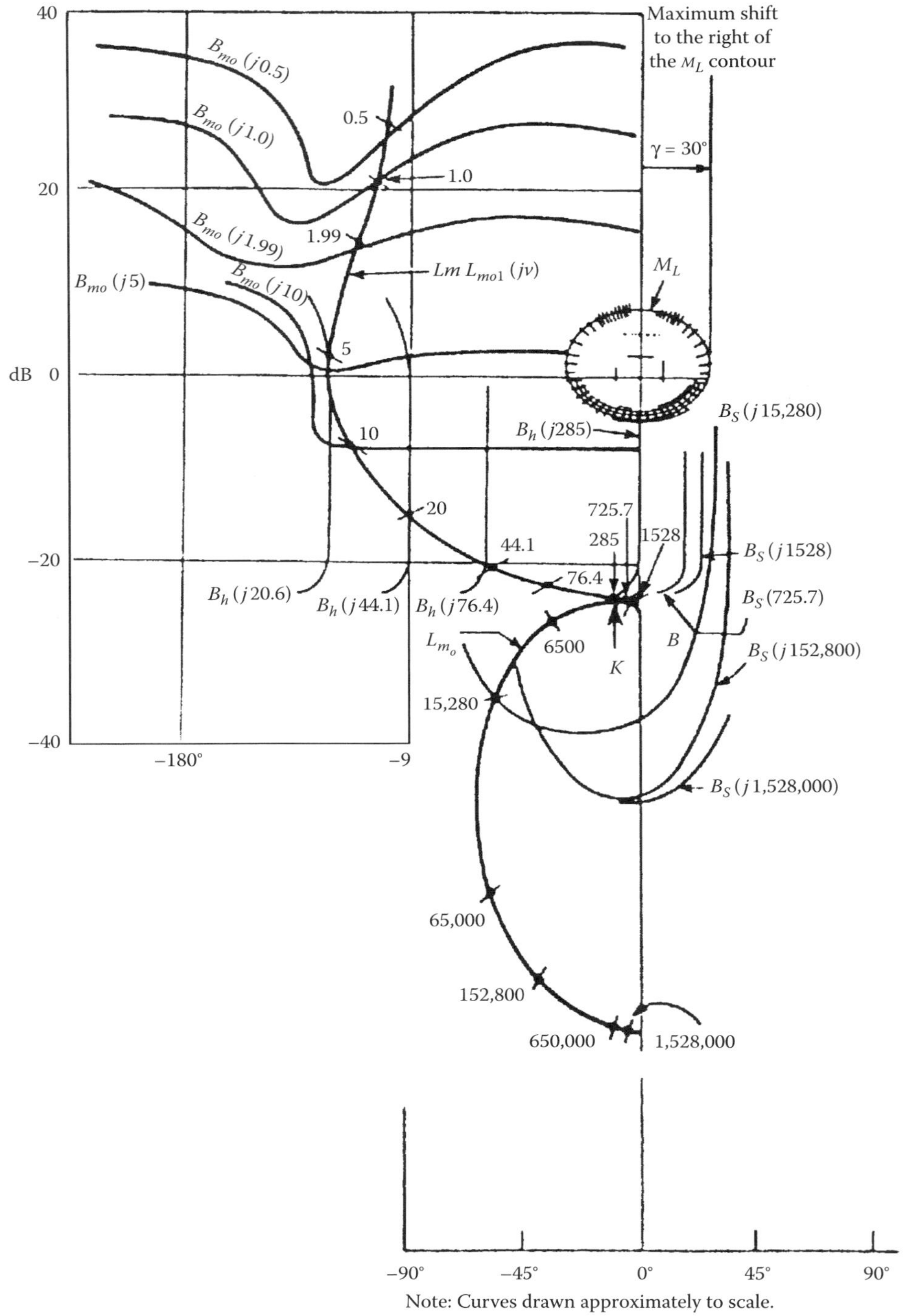

FIGURE 22.17 A satisfactory design: $L_{mo}(jv)$ at $\omega_s = 240$.

22.3.9 w-Domain Simulation

The "goodness" of the synthesized $L_{mo}(w)$ [or $L_o(w)$] and $F(w)$ is determined by first simulating the QFT w-domain designed control system for all J plants in the w-domain (an "analog" time domain simulation). See Section 22.2.13 for MISO QFT CAD packages that expedite this simulation.

22.3.10 z-Domain

The two tests of the "goodness" of the w-domain QFT-designed system is a discrete-time domain simulation of the system shown in Figure 22.14. To accomplish this simulation, the w-domain transfer functions $G(w)$ and $F(w)$ are transformed to the z-domain by use of the bilinear transformation of Equation 22.21. This transformation is utilized since the degree of the numerator and denominator polynomials of these functions are equal and the controller and prefilter do not contain a zero-order-hold device.

22.3.10.1 Comparison of the Controller's w- and z-Domain Bode Plots

Depending on the value of the sampling time T, warping may be sufficient to alter the loop-shaping characteristics of the controller when it is transformed from the w-domain into the z-domain. For the warping effect to be minimal, the Bode plots (magnitude and angle) of the w- and z-domain controllers must essentially lie on top of one another within the frequency range $0 < \omega \leq [(2/3)(\omega_s/2)]$. If the warping is negligible, then a discrete-time simulation can proceed. If not, a smaller value of sampling time needs to be selected.

22.3.10.2 Accuracy

The CAD packages that are available to the designer determines the degree of accuracy of the calculations and simulations. The smaller the value of T, the greater the degree of accuracy that is required to be maintained. The accuracy can be enhanced by implementing $G(z)$ and $F(z)$ as a set of g and f cascaded transfer functions, respectively; that is,

$$\begin{aligned} G(z) &= G_1(z)G_2(z)\cdots G_g(z), \\ F(z) &= F_1(z)F_2(z)\cdots F_f(z) \end{aligned} \tag{22.31}$$

22.3.10.3 Analysis of Characteristic Equation $Q_j(z)$

Depending on the value of T and the plant parameter uncertainty, the pole-zero configuration in the vicinity of the $-1+j0$ point in the z-plane for one or more of the J LTI plants can result in an unstable discrete-time response. Thus, before proceeding with a discrete-time domain simulation, an analysis of the characteristic equation $Q_j(z)$ for all J LTI plants must be made. If an unstable system exists, an analysis of $Q_j(z)$ and the corresponding root locus may reveal that a slight relocation of one or more controller poles in the vicinity of the $-1+j0$ point toward the origin may ensure a stable system for all J plants without essentially affecting the desired loop-shaping characteristic of $G(z)$.

22.3.10.4 Simulation and CAD Packages

With the "design checks" of Sections 22.3.10.1 through 22.3.10.3 satisfied, then a discrete-time simulation is performed to verify that the desired performance specifications have been achieved. In order to enhance the MISO QFT discrete control system design procedure that is presented in this chapter, the CAD flow chart of Section 22.2.13.1 is shown in Appendix B.

22.4 MIMO Systems

22.4.1 Introduction

Figure 22.18 represents an $m \times m$ MIMO closed-loop system in which F, G, and P are each $m \times m$ matrices, and $\mathcal{P} = \{P\}$ is a set of J matrices due to plant parameter uncertainty. There are m^2 closed-loop system transfer functions (transmissions) t_{ij} contained within its system transmission matrix (i.e., $T = \{t_{ij}\}$) relating the outputs y_i to the inputs r_j (e.g., $y_i = t_{ij}r_j$). These relationships hold for both the s- and w-domain analysis of a MIMO system. In a quantitative problem statement there are tolerance bounds on each t_{ij}, giving a set of m^2 acceptable regions τ_{ij} that are to be specified in the design; thus, $t_{ij} \epsilon \tau_{ij}$ and $\Im = \{\tau_{ij}\}$. From Figure 22.18 the system control ratio relating r to y is:

$$T = [I + PG]^{-1} PGF \tag{22.32}$$

The t_{ij} expressions derived from this expression are very complex and not suitable for analysis. The QFT design procedure systematizes and simplifies the manner of achieving a satisfactory system design for the entire range of plant uncertainty. In order to readily apply the QFT technique, another mathematical system description is presented in the next section. The material presented in this chapter pertains to both the s- and w-domain analysis of MIMO systems [2,12,16,17].

22.4.2 Derivation of m^2 MISO System Equivalents

The G, F, P and P^{-1} matrices are defined as follows:

$$G = \begin{bmatrix} g_1 & 0 & \cdots & 0 \\ 0 & g_2 & \cdots & 0 \\ \vdots & \vdots & \vdots & \vdots \\ 0 & 0 & \cdots & g_m \end{bmatrix} \quad F = \begin{bmatrix} f_{11} & f_{12} & \cdots & f_{1m} \\ f_{21} & f_{22} & \cdots & f_{2m} \\ \vdots & \vdots & \vdots & \vdots \\ f_{m1} & f_{m2} & \cdots & f_{mm} \end{bmatrix} \tag{22.33}$$

$$P = \begin{bmatrix} p_{11} & p_{12} & \cdots & p_{1m} \\ p_{21} & p_{22} & \cdots & p_{2m} \\ \vdots & \vdots & \vdots & \vdots \\ p_{m1} & p_{m2} & \cdots & p_{mm} \end{bmatrix} \quad P^{-1} = \begin{bmatrix} p^*_{11} & p^*_{12} & \cdots & p^*_{1m} \\ p^*_{21} & p^*_{22} & \cdots & p^*_{2m} \\ \vdots & \vdots & \vdots & \vdots \\ p^*_{m1} & p^*_{m2} & \cdots & p^*_{mm} \end{bmatrix} \tag{22.34}$$

Although only a diagonal G matrix is considered, the use of a nondiagonal G matrix may allow the designer more design flexibility [2]. The m^2 effective plant transfer functions are based upon defining:

$$q_{ij} \equiv \frac{1}{p^*_{ij}} = \frac{detP}{adjP_{ij}} \tag{22.35}$$

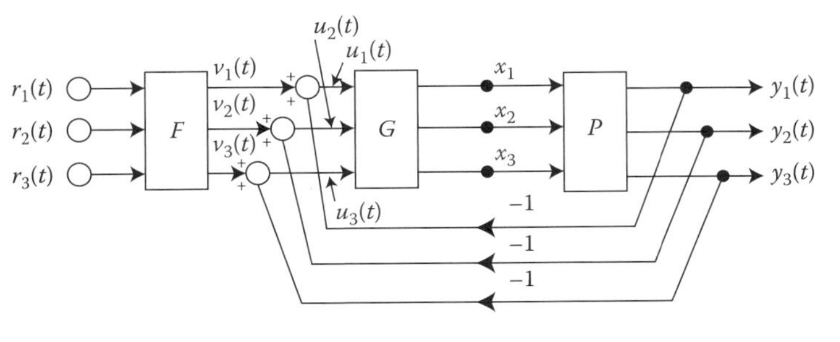

FIGURE 22.18 A 3 × 3 MIMO feedback control system.

There is a requirement that det.P be m.p. The Q matrix is then formed as:

$$Q = \begin{bmatrix} q_{11} & q_{12} & \cdots & q_{1m} \\ q_{21} & q_{22} & \cdots & q_{2m} \\ \vdots & \vdots & \vdots & \vdots \\ q_{m1} & q_{m2} & \cdots & q_{mm} \end{bmatrix} = \begin{bmatrix} \frac{1}{p_{11}^*} & \frac{1}{p_{12}^*} & \cdots & \frac{1}{p_{1m}^*} \\ \frac{1}{p_{21}^*} & \frac{1}{p_{22}^*} & \cdots & \frac{1}{p_{2m}^*} \\ \vdots & \vdots & \vdots & \vdots \\ \frac{1}{p_{m1}^*} & \frac{1}{p_{m2}^*} & \cdots & \frac{1}{p_{mm}^*} \end{bmatrix} \tag{22.36}$$

The matrix P^{-1} is partitioned to the form:

$$P^{-1} = [P_{ij}^*] = \left[\frac{1}{q_{ij}}\right] = \Lambda + B \tag{22.37}$$

where Λ is the diagonal part of P^{-1} and B is the balance of P^{-1}; thus $\lambda_{ij} = 1/q_{ii} = p_{ii}^*$, $b_{ii} = 0$, and $b_{ij} = 1/q_{ij} = p_{ij}^*$ for $i \neq j$. Premultiplying Equation 22.32 by $[I + PG]$ yields

$$[I + PG]T = PGF \rightarrow [P^{-1} + G]T = GF \tag{22.38}$$

where P is nonsingular. Using Equation 22.37 with G diagonal, Equation 22.38 can be rearranged to the form:

$$T = [\Lambda + G]^{-1}[GF - BT] \tag{22.39}$$

Equation 22.39 is used to define the desired fixed-point mapping where each of the m^2 matrix elements on the right-hand side of this equation can be interpreted mathematically as representing a MISO system. Proof of the fact that design of each MISO system yields a satisfactory MIMO design is based on the Schauder fixed-point theorem [7]. This theorem is described, based upon a *unit impulse input,* by defining a mapping

$$Y(T_i) = [\Lambda + G]^{-1}[GF - BT_i] \equiv T_j \tag{22.40}$$

where each member of T is from the acceptable set $\Im$. If this mapping has a fixed point [i.e., $T \epsilon \Im$ such that $Y(T_i) = T_j$], then this T is a solution of Equation 22.39. The y_{11} output obtained from Equation 22.40, for the 3×3 case, is given by:

$$y_{11} = \frac{q_{11}}{1 + g_1 q_{11}} \left[g_1 f_{11} - \left(\frac{t_{21}}{q_{12}} + \frac{t_{31}}{q_{13}} \right) \right] \tag{22.41}$$

Based upon the derivation of all the y_{ij} expressions from Equation 22.40 yields the four effective MISO loops (in the boxed area) in Figure 22.4, resulting from a 2×2 system and the nine effective MISO loops resulting from a 3×3 system [4]. The control ratios for the desired tracking inputs r_i by the corresponding outputs y_i for each feedback loop of Equation 22.40 have the form

$$y_{ii} = w_{ij}(v_{ij} + c_{ij}) \tag{22.42}$$

where $w_{ii} = q_{ii}/(1 + g_i q_{ii})$ and $v_{ij} = g_i f_{ij}$. The interaction between the loops has the form

$$c_{ij} = -\sum_{k \neq i} \left[\frac{t_{kj}}{q_{ik}}\right] \quad k = 1, 2, 3, \ldots, m \tag{22.43}$$

and appears as a "cross-coupling effect" input in each of the feedback loops. *Thus, Equation 22.42 represents the control ratio of the* i*th MISO system.* The transfer function $w_{ii}v_{ij}$ relates the "desired" ith output to the jth input r_j, and the transfer function $w_{ii}c_{ij}$ relates the ith output to the jth cross-coupling effect input c_{ij}. The outputs given in Equation 22.42 can thus be expressed as

$$y_{ij} = (y_{ij})_{r_i} + (y_{ij})_{c_{ij}} = y_{r_i} + y_{c_{ij}} \tag{22.44}$$

or, based on a *unit impulse input*,

$$t_{ij} = t_{r_{ij}} + t_{c_{ij}} \tag{22.45}$$

where

$$t_{r_{ij}} = y_{r_i} = w_{ii}v_{ij} \quad t_{c_{ij}} = y_{c_{ij}} = w_{ii}c_{ij} \tag{22.46}$$

and where now the upper bound, in the low-frequency range ($0 < \omega \leq \omega_h$), is expressed as b'_{ij}. Thus,

$$\tau_{c_{ij}} = b_{ij} - b'_{ij} \tag{22.47}$$

represents the maximum portion of b_{ij} allocated toward the cross-coupling effect rejection, and b'_{ij} represents the upper bound for the tracking portion, respectively, of t_{ij}. For each MISO system there is a cross-coupling effect input that is a function of all the other loop outputs. The object of the design is to have each loop track its desired input while minimizing the outputs due to the disturbance (cross-coupling effects) inputs.

In each of the nine structures of Figure 22.4 it is necessary that the control ratio t_{ij} must be a member of the acceptable $t_{ij} \epsilon \tau_{ij}$. All the g_i and f_{ij} must be chosen to ensure that this condition is satisfied, thus constituting nine MISO design problems. If all of these MISO problems are solved, there exists a fixed point; then y_{ij} on the left-hand side of Equation 22.40 may be replaced by t_{ij}, and all the elements of T on the right-hand side by t_{kj}. This means that there exist nine t_{ij} and t_{kj}, each in its acceptable set, which is a solution to Figure 22.18. If each element is 1:1, then this solution must be unique. A more formal and detailed treatment is given in [7].

22.4.3 Tracking and Cross-Coupling Specifications

The presentation for the remaining portion of this chapter is based upon not only a diagonal G matrix but also a diagonal F matrix. Thus, in Figure 22.4 the t_{ij} terms, for $i \neq j$, represent disturbance responses due to the cross-coupling effect whereas the t_{ij} terms, for $i = j$ (see Equation 22.45) is composed of a desired tracking term t_r and of an unwanted cross-coupling term t_c. Therefore, the desired tracking specifications for the diagonal MISO systems of Figure 22.4 contain an upper-bound and a lower bound as shown in Figure 22.6. The disturbance specification for all MISO loops is given by only an upper bound. These performance specifications are shown in Figure 22.19 for a 2×2 (in the boxed area) and for a 3×3 MIMO feedback control system.

22.4.3.1 Tracking Specifications

Based upon the analysis of Equations 22.45 through 22.47, the specifications for the t_{ii} responses shown in Figure 22.19 need to be modified as shown in Figure 22.20. As shown in this figure, a portion of $\delta_R(j\omega_i)$ (see Figure 22.6) has been allocated [2,7] for the disturbance (cross-coupling effect) specification. Thus, based upon this modification and given an uncertain plant $\mathcal{P} = \{P_j\}(j = 1, 2, \ldots, J)$ and the BW ω_h above which output sensitivity is ignored, it is desired to synthesize G and F such that for all $P \epsilon \mathcal{P}$

$$a'_{ii} \leq \left|t_{ii}(j\omega)\right| \leq b'_{ii} \quad \text{for } \omega \leq \omega_h \tag{22.48}$$

A finite ω_h is recommended because, in strictly proper systems, feedback is not effective in the high-frequency range.

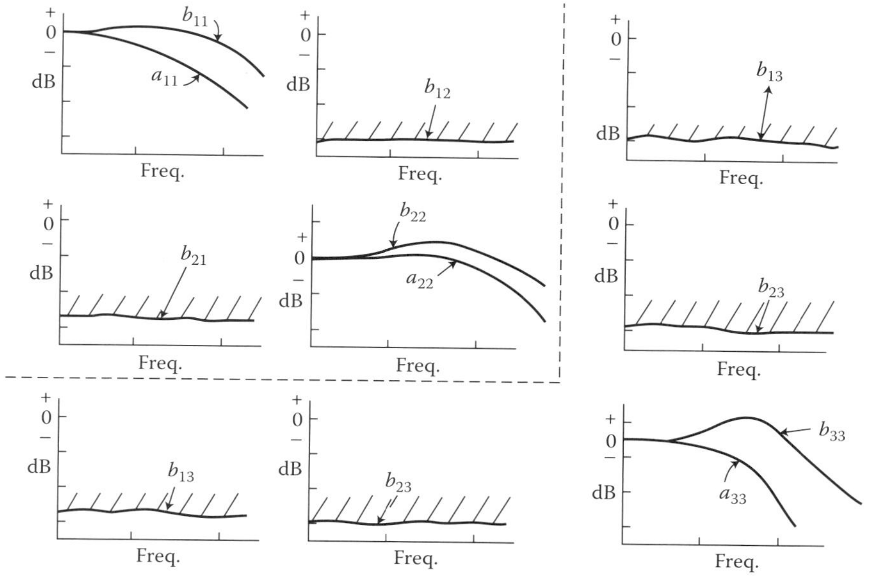

FIGURE 22.19 Tracking and cross-coupling specifications for a 2 × 2 (in boxed area) and for a 3 × 3 MIMO system.

22.4.3.2 Disturbance Specification (Cross-Coupling Effect)

Based upon the previous discussion the disturbance specification, an upper bound, is expressed as

$$|t_{c_{ik}}| \leq |b_{ij}| \tag{22.49}$$

Thus, the synthesis of G must satisfy both Equations 22.48 and 22.49.

22.4.4 Determination of Tracking, Cross-Coupling, and Optimal Bounds

The remaining portion of the MIMO QFT approach is confined to a 2 × 2 system. The reader can refer to the references for higher-order systems ($m > 2$). From Equation 22.39 the following equations are obtained:

$$t_{11} = \frac{L_1 f_{11} + c_{11} q_{11}}{1 + L_1}$$

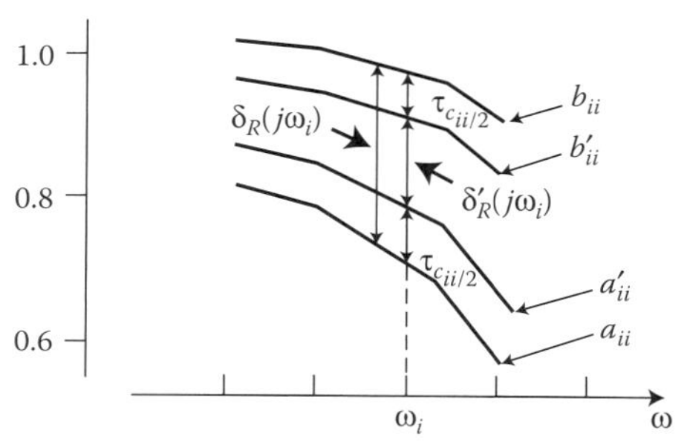

FIGURE 22.20 Allocation for tracking and cross-coupling specifications for the t_{ii} responses.

where

$$c_{11} = -\frac{t_{21}}{q_{12}}, \quad L_1 = q_{11}g_1$$
$$t_{12} = \frac{c_{12}q_{11}}{1+L_1} \tag{22.50}$$

where

$$c_{12} = -\frac{t_{22}}{q_{12}}, \quad f_{12} = 0$$
$$t_{21} = \frac{c_{21}q_{22}}{1+L_2} \tag{22.51}$$

where

$$c_{21} = -\frac{t_{11}}{q_{21}}, \quad L_2 = q_{22}g_2, \quad f_{21} = 0,$$
$$t_{22} = \frac{L_2 f_{22} + c_{22}q_{22}}{1+L_2} \tag{22.52}$$

where

$$c_{22} = -\frac{t_{12}}{q_{21}} \tag{22.53}$$

Equations 22.50 and 22.51 correspond to the MISO systems for the first row of loops in Figure 22.4, and Equations 22.52 and 22.53 correspond to the MISO loops for the second row.

22.4.4.1 Tracking Bounds

The tracking bounds for the *ii* MISO system is determined in the same manner as for the MISO system of PART II (see Section 22.2.9.1). By use of the templates for the *ii* loop plant, the value of $\delta_R(j\omega_i)'$, shown in Figure 22.20, is used to satisfy the constraint of Equation 22.48.

22.4.4.2 Cross-Coupling Bounds

From Equations 22.51 and 22.53, considering only the first row of MISO loops in Figure 22.4, the following cross-coupling transfer functions are obtained (see Figures 22.20 and 22.19, respectively):

$$|t_{c_{11}}| = \left|\frac{c_{11}q_{11}}{1+L_1}\right| \le \tau_{c_{ij}} \equiv \tau_{c_{11}} \quad \cdots \tag{22.54}$$

$$|t_{c_{12}}| = \left|\frac{c_{12}q_{11}}{1+L_1}\right| \le b_{ij} \equiv b_{12} \quad \cdots \tag{22.55}$$

Substituting for c_{11} and c_{12} into Equations 22.54 and 22.55, respectively, and replacing t_{21} and t_{22} by their respective upper bound values b_{21} and b_{22}, and rearranging these equations yields:

$$\left|\frac{1}{1+L_1}\right| =\le \left|\frac{q_{12}}{q_{11}}\right| \frac{\tau_{c_{11}}}{b_{21}} = M_{m_{11}} \tag{22.56}$$

$$\left|\frac{1}{1+L_1}\right| \le \left|\frac{q_{12}}{q_{11}}\right| \frac{b_{12}}{b_{22}} = M_{m_{12}} \tag{22.57}$$

Substituting into these equations $L_1 = 1/l_1$ yields:

$$\left|\frac{l_1}{1+l_1}\right| \le M_{m_{11}} \quad \cdots \quad \left|\frac{l_1}{1+l_1}\right| \le M_{m_{12}} \tag{22.58}$$

By analyzing these equations for each of the J plants over the desired BW, the maximum value M_m that each of these equations can have, for each value of ω_i(or ν_i), is readily determined by use of a CAD package. Thus, since $L_1 = 1/l_1$, the reciprocals of these values yield the value of the corresponding M-contours or cross-coupling bounds, for $\omega = \omega_i$(or ν_i), on the NC.

22.4.4.3 Optimal Bounds

The points on the optimal bound for a given value of frequency and for a given row of MISO loops of Figure 22.4 are determined by selecting the largest dB value, for a given NC phase angle, from all the tracking and cross-coupling bounds for these loops at this frequency. The MIMO QFT CAD package is designed to perform this determination of the optimal bounds.

22.4.5 QFT Methods of Designing MIMO Systems

There are two methods of achieving a QFT MIMO design. Method 1 involves synthesizing the loop transmission function L_i and the prefilter f_{ii} independent of the previous synthesized loop transmission functions and prefilters. Method 2 substitutes the synthesized g_i and f_{ii} of the first (or prior) MISO loop(s) that is (are) designed into the equations that describe the remaining loops to be designed. For Method 2, it is necessary to make the decision as to the order that the L_i functions are to be synthesized. Generally, the loops are chosen on the basis of the phase margin frequency ω_ϕ requirements. That is, the loop having the smallest value of ω_ϕ is chosen as the first loop to be designed; the loop having the next smallest value of ω_ϕ is selected as the second loop to be designed; etc. This is an important requirement for Method 2.

22.4.5.1 Method 1

This method involves overdesign (worst-case scenario), i.e., in getting the M_m values of Equations 22.56 and 22.57, for the 2×2 case, the maximum magnitude that q_{12} and the minimum magnitude that q_{11} can have, for each value of ω_i, over the entire J LTI plants are utilized. This method requires that the diagonal dominance condition [2,7] be satisfied. When this condition is not satisfied, then Method 2 needs to be utilized.

22.4.5.2 Method 2

Once the order in which the loops are to be designed is designated accordingly (loop 1, loop 2, etc.), then the compensator g_1 and the prefilter f_{11} are synthsized. These are now known LTI functions, which are utilized to define the loop 2 effective plant transfer function. That is, substitute Equation 22.51 into Equation 22.53, then rearrange the result to obtain a new expression for t_{12} in terms of g_1 and f_{11} as follows:

$$t_{21} = -\left[\frac{f_{11}L_1 q_{22_e}/q_{21}(1+L_1)}{1+g_2 q_{22_e}}\right] \tag{22.59}$$

where the effective loop 2 transfer function is

$$q_{22_e} = \frac{q_{22}(1+L_1)}{(1+L_1)-\gamma_{12}} \quad \text{where } \gamma_{12} = \frac{q_{11}q_{22}}{q_{12}q_{21}} \tag{22.60}$$

Repeating a similar procedure, the expression for t_{22} is

$$t_{22} = \frac{f_{22}g_2 q_{22_e}}{1+g_2 q_{22_e}} \tag{22.61}$$

Remember that a diagonal prefilter matrix has been specified. Note that Equations 22.59 through 22.61 involve the known f_{11} and g_1, which reduces the overdesign of loop 2.

22.4.6 Synthesizing the Loop Transmission and Prefilter Functions

Once the optimal bound has been determined for each L_i loop, then the synthesis procedures for determining the loop transmission and prefilter functions are the same as for the MISO analog and discrete systems as discussed in Sections 22.2 and 22.3, respectively.

22.4.7 Overview of the MIMO QFT CAD Package

The MIMO QFT CAD package [15], implemented using Mathematica, is capable of carrying an analog or a discrete MISO or MIMO QFT design from problem setup through the design process to a frequency domain analysis of the compensated system. For analog control problems, the design process is performed in the s-plane. The design process is performed in either the w-plane or the s-plane using the pseudo-continuous-time (PCT) representation of the sampled-data system. A flowchart of the CAD package is given in Appendix B.

22.5 QFT Application

A MIMO QFT example, [18] a 2×2 analog flight control system (see the boxed-in loops of Figure 22.4), is presented to illustrate the power of this design technique. Also, this example illustrates the increased accuracy and efficiency achieved by the MIMO QFT CAD package [15] and the straightforward method for designing an analog MIMO control system. The specifications require a robust analog design for an aircraft that provides stability and meets time domain performance requirements for the specified four flight conditions (Table 22.1) and the six aircraft failure modes (Table 22.2). Table 22.3 lists the resulting set of 24 plant cases that incorporate these flight conditions and failure modes. For stability, a 45° phase

TABLE 22.1 Flight Conditions

	Aircraft Parameters	
Flight Condition	Mach	Altitude
1	0.2	30
2	0.6	30,000
3	0.9	20,000
4	1.6	30,000

TABLE 22.2 Failure Modes

Failure Mode	Failure Condition
1	Healthy aircraft
2	One horizontal tail fails
3	One flaperon fails
4	One horizontal tail and one flaperon fail, same side
5	One horizontal tail and one flaperon fail, opposite side
6	Both flaperons fail

TABLE 22.3 Plant Models

	Flight Condition			
Failure Mode	1	2	3	4
1	#1	#7	#13	#19
2	#2	#8	#14	#20
3	#3	#9	#15	#21
4	#4	#10	#16	#22
5	#5	#11	#17	#23
6	#6	#12	#18	#24

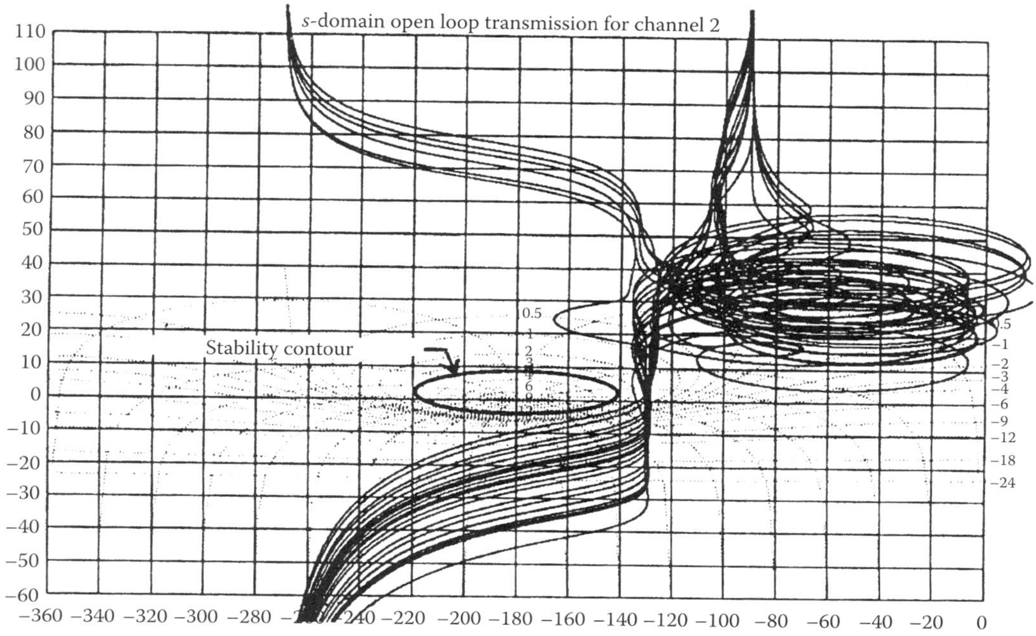

FIGURE 22.21 Open-loop transmissions on NC.

margin is required for each of the two feedback loops. Frequency domain performance specifications, when met, result in the desired closed-loop system performance in the time domain. The frequency domain specifications are shown as dashed lines on the Bode plots of Figure 22.22.

The specifications, the plant models [18] for the 24 cases, and the weighting matrix are entered into the QFT CAD package. The automated features accessed through the designer interface of the CAD package result in synthesizing the compensators $g_1(s)$ and $g_2(s)$ in the manner described in Section 22.2.10. That is, the nominal loop transmission functions $L_{1_0}(s) = g_1(s)q_{11_0}(s)$ and $L_{2_0}(s) = g_2(s)q_{22_0}(s)$ are synthesized (or shaped) so that they satisfy their respective stability bounds and their respective optimal bounds $B_{1_0}(j\omega_i)$ and $B_{2_0}(j\omega_i)$. Note that q_{11_0} and q_{22_0} are the nominal plant transfer functions. The first step in a validation check is to plot the loop transmission functions $L_{2_\iota}(s)$, where $\iota = 1, \ldots, 24$, for all 24 cases on the NC. This is accomplished by a CAD routine, as shown in Figure 22.21 for the purpose of a stability check (m.p. plants). As is seen, none of the cases violate the M_L stability contour (the dark ellipse). In this design, when synthesizing $L_{2_0}(s)$ a trade-off exists between performance and bandwidth. In this example, the designer chooses to accept the consequences of violating the disturbance bound for $\omega = 2$ rad/s. With $L_{1_0}(s)$ and $L_{2_0}(s)$ synthesized, the automated features of the CAD package expedite the design of the prefilters $f_{11}(s)$ and $f_{22}(s)$.

For the second step in the design validation process, the 2×2 array of Bode plots shown in Figure 22.22 is generated, showing on each plot the 24 possible closed-loop transmissions from an input to an output of the completed system. The consequence of violating the channel 2 disturbance bound for $\omega = 2$ rad/s is seen where the closed-loop transmissions violate b_{21}, denoted by dashed line, beginning at $\omega = 2$ rad/s. Violation of performance bounds during loop-shaping may result in violation of the performance specifications for the closed-loop system.

As seen in Figure 22.22, a robust design has been achieved for this 2×2 MIMO analog flight control system. The time domain results, although not drawn, meet all specifications.

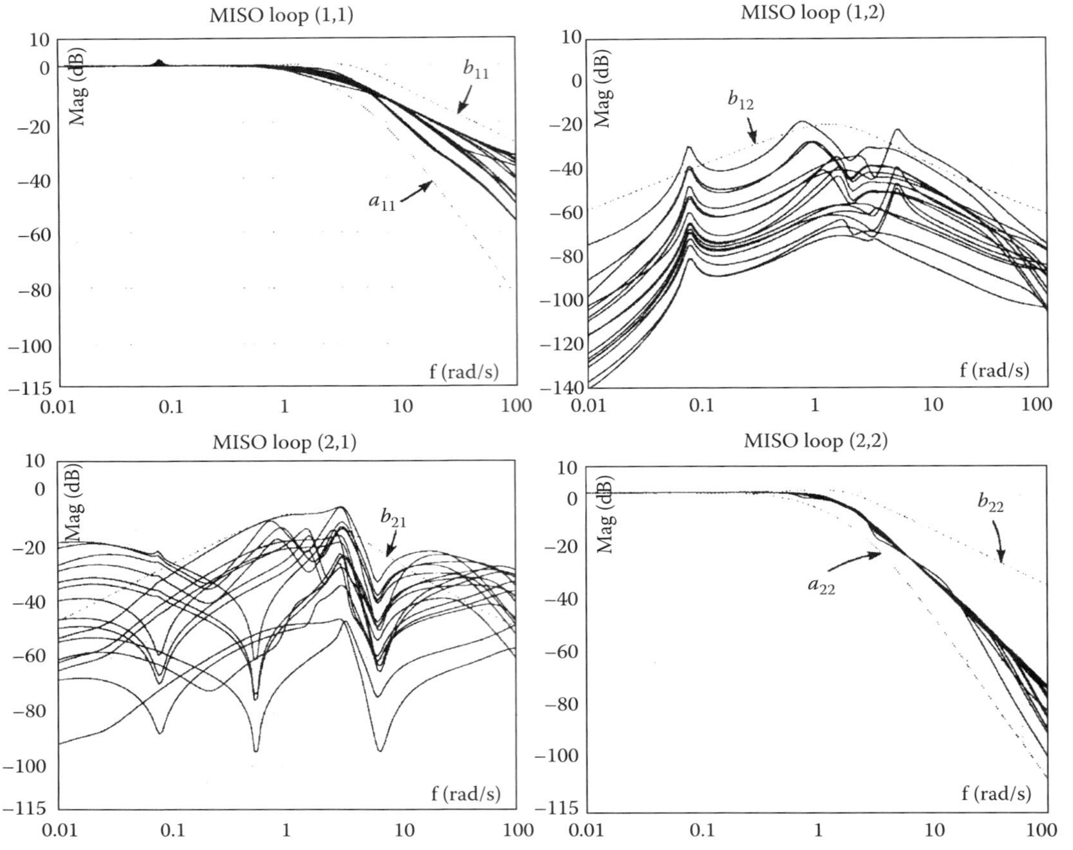

FIGURE 22.22 Closed-loop transmissions for an analog design system.

References

1. D'Azzo, J.J., and Houpis, C.H., *Linear Control System Analysis and Design,* 4th Ed., McGraw-Hill, New York, 1995.
2. Houpis, C. H., Quantitative feedback theory (QFT) for the engineer: A paradigm for the design of control systems for uncertain nonlinear plants, WL-TR-95-3061, AF Wright Aeronautical Laboratory, Wright-Patterson AFB, OH, 1987. (Available from National Technical Information Service, 5285 Port Royal Road, Springfield, VA 22151, document number AD-A297574.)
3. Horowitz, I. M. and Sidi, M., Synthesis of feedback systems with large plant ignorance for prescribed time domain tolerances, *Int. J. Control,* 16, 287–309, 1972.
4. Horowitz, I. M. and Loecher, C., Design of a 3 × 3 multivariable feedback system with large plant uncertainty, *Int. J. Control,* 33, 677–699, 1981.
5. Horowitz, I. M., Optimum loop transfer function in single-loop minimum phase feedback systems, *Int. J. Control,* 22, 97–113, 1973.
6. Horowitz, I. M., Synthesis of feedback systems with non-linear time uncertain plants to satisfy quantitative performance specifications, *IEEE Proc.,* 64, 123–130, 1976.
7. Horowitz, I. M., Quantitative synthesis of uncertain multiple-input multiple-output feedback systems, *Int. J. Control,* 30, 81–106, 1979.
8. Thompson, D. F. and Nwokah, O.D.I., Optimal loop synthesis in quantitative feedback theory, *Proc. Am. Control Conf.,* San Diego, CA, 626–631, 1990.

9. Houpis, C. H. and Chandler, P.R., Eds., *Quantitative Feedback Theory Symposium Proceedings*, WL-TR-92-3063, Wright Laboratories, Wright-Patterson AFB, OH, 1992.
10. Keating, M. S., Pachter, M., and Houpis, C.H., Damaged aircraft control system design using QFT, *Proc. Nat. Aerosp. Electron. Conf.* (NAECON), Vol. 1, Ohio, 621–628, May 1994.
11. Reynolds, O.R., Pachter, M., and Houpis, C.H., Design of a subsonic flight control system for the Vista F-16 using quantitative feedback theory, *Proc. Am. Control Conf.*, 350–354, 1994.
12. Trosen, D. W., Pachter, M., and Houpis, C.H., Formation flight control automation, *Proc. Am. Inst. Aeronaut. Astronaut. (AIAA) Conf.*, Scottsdale, AZ, 1379–1404, 1994.
13. Houpis, C. H. and Lamont, G., *Digital Control Systems: Theory, Hardware, Software,* 2nd ed., McGraw-Hill, NY, 1992.
14. Horowitz, I.M. and Liao, Y.K., Quantitative feedback design for sampled-data system, *Int. J. Control,* 44, 665–675, 1986.
15. Sating, R.R., Horowitz, I.M., and Houpis, C.H., Development of a MIMO QFT CAD package (Version 2), paper presented at the 1993 *Am. Control Conf.*, Air Force Institute of Technolgy, Wright-Patterson AFB, Ohio.
16. Boyum, K. E., Pachter, M., and Houpis, C.H., High angle of attack velocity rolls, *13th IFAC Symp. Autom. Control Aerosp.,* 51–57, Palo Alto, CA, September 1994.
17. Schneider, D. L., QFT digital flight control design as applied to the AFTI/F-16, M.S. thesis, AFIT/GE/ENG/86D-4, School of Engineering, Air Force Institute of Technolgy, Ohio, 1986.
18. Arnold, P. B., Horowitz, I.M., and Houpis, C.H., YF-16CCV flight control system reconfiguration design using quantitative feedback theory, *Proc. Nat. Aerosp. Electron. Conf.* (NAECON), Vol. 1, 578–585, 1985.

22.6 Appendix A

CAD flowchart for MISO QFT design.

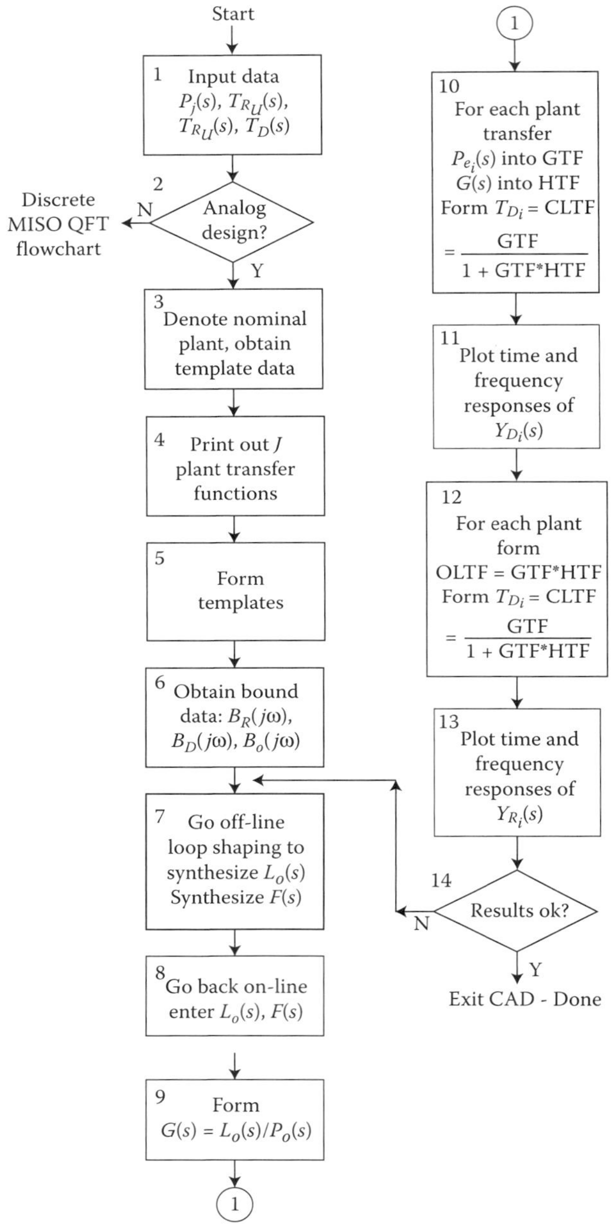

22.7 Appendix B

MIMO QFT flowchart for analog and discrete control systems [15].

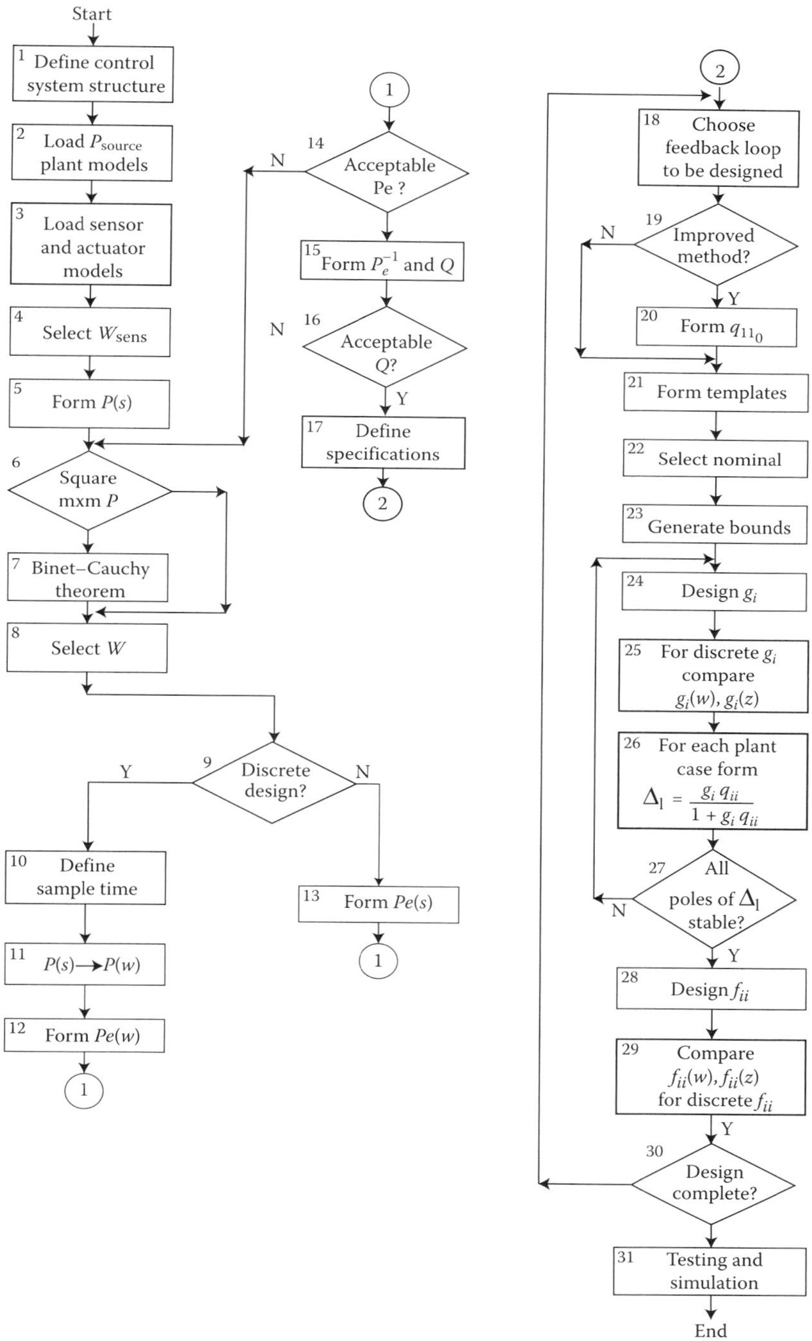

23

Robust Servomechanism Problem

Edward J. Davison
University of Toronto

23.1 Introduction

The so-called *servomechanism problem* is one of the most basic problems to occur in the field of automatic control, and it arises in almost all application problems of the aerospace and process industries. In the servomechanism problem, it is desired to design a controller for a plant (or "system") so that the outputs of the plant are independent, as much as possible, from disturbances which may affect the system (i.e., *regulation* occurs), and also such that the outputs asymptotically track any specified reference input signals applied to the system (i.e., *tracking* occurs), subject to the requirements of maintaining closed-loop system stability.

This chapter examines some aspects of controller synthesis for the multivariable servomechanism problem when the plant to be controlled is subject to uncertainty. In this case, a controller is to be designed so that desired regulation and tracking takes place in spite of the fact that the plant dynamics or/and parameters may vary by arbitrary, large amounts, subject only to the condition that the resultant closed-loop perturbed system remains stable. This problem is called the *robust servomechanism problem* (RSP).

23.2 Preliminary Results

23.2.1 Plant Model

The plant to be controlled is assumed to be described by the following linear time-invariant (LTI) model:

$$\dot{x} = Ax + Bu + E\omega$$

$$
\begin{aligned}
y &= Cx + Du + F\omega \\
y_m &= C_m x + D_m u + F_m\omega \\
e &= y - y_{\text{ref}}
\end{aligned}
\tag{23.1}
$$

where $x \in \mathbb{R}^n$ is the state, $u \in \mathbb{R}^m$ are the inputs that can be manipulated, $y \in \mathbb{R}^r$ are the outputs that are to be regulated and $y_m \in \mathbb{R}^{r_m}$ are the outputs which can be measured. Here $\omega \in \mathbb{R}^{\Omega}$ correspond to the disturbances in the system, which in general cannot necessarily be measured, and $e \in \mathbb{R}^r$ is the error in the system, which is the difference between the output y and the reference input signal y_{ref}, in which it is desired that the outputs y should track.

23.2.2 Class of Tracking/Disturbance Signals

It is assumed that the disturbances ω arise from the following class of systems:

$$
\dot{\eta}_1 = \mathcal{A}_1\eta_1, \quad \omega = \mathcal{C}_1\eta_1; \quad \eta_1 \in \mathbb{R}^{n_1}
\tag{23.2}
$$

and that the reference input signals y_{ref} arise from the following class of systems:

$$
\dot{\eta}_2 = \mathcal{A}_2\eta_2, \quad \rho = \mathcal{C}_2\eta_2, \quad y_{\text{ref}} = G\rho; \quad \eta_2 \in \mathbb{R}^{n_2}
\tag{23.3}
$$

It is assumed for nontriviality that $\text{sp}(\mathcal{A}_1) \subset \mathbb{C}^+$, $\text{sp}(\mathcal{A}_2) \subset \mathbb{C}^+$, where $\text{sp}(\cdot)$ denotes the eigenvalues of $(\cdot)$ and $\mathbb{C}^+$ denotes the closed right complex half-plane. It is also assumed with no loss of generality that $(\mathcal{C}_1, \mathcal{A}_1)$, $(\mathcal{C}_2, \mathcal{A}_2)$ are observable and that $\text{rank}\left(\begin{smallmatrix} E \\ F \end{smallmatrix}\right) = \text{rank}\,\mathcal{C}_1 = \dim(\omega)$, $\text{rank}\, G = \text{rank}\,\mathcal{C}_2 = \dim(\rho)$.

This class of signals is quite broad and includes most classes of signals that occur in application problems, e.g., constant, polynomial, sinusoidal, polynomial-sinusoidal, etc.

The following definitions will be used in the development to follow.

Definition 23.1:

Given the systems represented by Equations 23.2, and 23.3, let $\{\lambda_1, \lambda_2, \ldots, \lambda_p\}$ be the zeros of the least common multiple of the nominal polynomial of $\mathcal{A}_1$ and minimal polynomial of $\mathcal{A}_2$ (multiplicities repeated), and call

$$
\Lambda := \{\lambda_1, \lambda_2, \ldots, \lambda_p\}
\tag{23.4}
$$

the disturbances/tracking poles of Equations 23.2 and 23.3.

Definition 23.2:

Given the model represented by Equation 23.1, consider the system

$$
\begin{aligned}
\dot{x} &= Ax + Bu; \quad u \in R^m, \ y \in R^r, \ x \in R^n \\
y &= Cx + Du
\end{aligned}
$$

Then $\lambda \in \mathbb{C}$ is said to be a transmission zero (TZ) [3] of (C, A, B, D) if

$$
\text{rank}\begin{bmatrix} A - \lambda I & B \\ C & D \end{bmatrix} < n + \min(r, m)
$$

In particular, the transmission zeros are the zeros (multiplicities included) of the greatest common divisor of all $[n + \min(r, m)] \times [n + \min(r, m)]$ minors of $\begin{bmatrix} A - \lambda I & B \\ C & D \end{bmatrix}$.

Definition 23.3:

Given the system (C, A, B, D), assume that one or more of the transmission zeros of (C, A, B, D) are contained in the closed right complex half-plane; then (C, A, B, D) is said to be a nonminimum-phase system. If (C, A, B, D) is not nonminimum phase, then it is said to be a minimum-phase system.

23.2.3 Robust Servomechanism Problem

The *robust servomechanism problem* (RSP) for Equation 23.1 consists in finding an LTI controller that has inputs y_m, y_{ref} and outputs u for the plant so that:

1. The resultant closed-loop system is asymptotically stable.
2. Asymptotic tracking occurs; that is,

$$\lim_{t\to\infty} e(t) = 0, \quad \forall x(0) \in \mathbb{R}^n, \quad \forall \eta_1(0) \in \mathbb{R}^{n_1}, \quad \forall \eta_2(0) \in \mathbb{R}^{n_2}$$

   for all controller initial conditions.
3. Condition 2 holds for any arbitrary perturbations in the plant model Equation 23.1 (e.g., plant parameters or plant dynamics, including changes in model order) that do not cause the resultant closed-loop system to become unstable.

In this problem statement, there is no requirement made regarding the transient behavior of the closed-loop system; thus, the following problem statement is now made.

Perfect RSP

Given the plant represented by Equation 23.1, it is desired to find a controller such that:

1. It solves the RSP for the class of disturbances/tracking signals given by Equations 23.2 and 23.3.
2. The controller gives *perfect error regulation* when applied to the nominal plant model Equation 23.1, i.e., given $x(0)$, $z_1(0)$, $z_2(0)$, located on the unit sphere, with $\eta(0) = 0$, where $\eta(0)$ is the initial condition of the servo-compensator (see Equation 23.9), then $\forall\epsilon > 0$, there exists a controller (parameterized by ϵ) that satisfies property 1 and has the property that $\int_0^\infty e'(\tau)e(\tau)\,d\tau < \epsilon$, with no unbounded peaking occurring in the response of e; i.e., there exists a constant ρ, independent of ϵ, such that $\sup_{t\geq 0} |e(t)| < \rho$.

Thus, in the perfect RSP, arbitrarily good transient error, with no unbounded peaking in the error response of the system, can be obtained for any initial condition of the plant and for any disturbance/tracking signals that belong to Equations 23.2 and 23.3.

23.3 Main Results

The following results are obtained concerning the existence of a solution to the RSP [5,6].

Theorem 23.1:

There exists a solution to the RSP for Equation 23.1, if and only if the following conditions are all satisfied:

1. *(C_m, A, B) is stabilizable and detectable.*
2. *$m \geq r$.*

3. *The transmission zeros of* (C, A, B, D) *exclude the disturbance/tracking poles* λ_i, $i = 1, 2, \ldots, p$.
4. $y \subset y_m$; *i.e., the outputs* y *are measurable.*

Remark 23.1

The conditions 2 and 3 are equivalent to the condition:

$$\text{rank}\begin{bmatrix} A - \lambda_i I & B \\ C & D \end{bmatrix} = n + r, \quad i = 1, 2, \ldots, p \tag{23.5}$$

The following existence results are obtained concerning the existence of a solution to the perfect RSP [12]:

Theorem 23.2:

There exists a solution to the perfect RSP for Equation 23.1, if and only if the following conditions are all satisfied:

1. (C_m, A) *is detectable*
2. $m \geq r$
3. (C, A, B, D) *is minimum phase*
4. $y \subset y_m$

Remark 23.2

If $m = r$, the above conditions simplify to just conditions 3 and 4.

The following definitions of a stabilizing compensator and servo-compensator are required in the development to follow.

Definition 23.4:

Given the stabilizable, detectable system (C_m, A, B, D) *obtained from Equation 23.1, an LTI stabilizing compensator*

$$\begin{aligned} \dot{\xi} &= \Lambda_1 \xi + \Lambda_2 y_m \\ u &= K_1 \xi + K_2 y_m \end{aligned} \tag{23.6}$$

is defined to be a controller that asymptotically stabilizes the resultant closed-loop system, such that "desired" transient behavior occurs.

This compensator is not a unique device and may be designed by using a number of different techniques.

Definition 23.5:

Given the disturbance/reference input poles λ_i, $i = 1, 2, \ldots, p$, *the matrix* $C \in R^{p \times p}$ *and vector* $\gamma \in R^p$ *are defined by*

$$\mathcal{C} := \begin{bmatrix} 0 & 1 & 0 & \dots & 0 \\ 0 & 0 & 1 & \dots & 0 \\ \vdots & \vdots & \vdots & & \vdots \\ -\delta_1 & -\delta_2 & -\delta_3 & \dots & -\delta_p \end{bmatrix}, \quad \gamma := \begin{bmatrix} 0 \\ 0 \\ \vdots \\ 0 \\ 1 \end{bmatrix} \tag{23.7}$$

where the coefficients δ_i, $i = 1, 2, \dots, p$ *are given by the coefficients of the polynomial* $\prod_{i=1}^{p}(\lambda - \lambda_i)$; *i.e.,*

$$\lambda^p + \delta_p\lambda^{p-1} + \dots + \delta_2\lambda + \delta_1 := \prod_{i=1}^{p}(\lambda - \lambda_i) \tag{23.8}$$

The following compensator, called a *servo-compensator*, is of fundamental importance in the design of controllers to solve the RSP [5].

Definition 23.6:

Consider the class of disturbance/tracking signals given by Equations 23.2 and 23.3, and consider the system of Equation 23.1; then a servo-compensator for Equation 23.1 is a controller with input $e \in R^r$ *and output* $\eta \in R^{rp}$ *given by*

$$\dot{\eta} = \mathcal{C}^*\eta + \mathcal{B}^*e \tag{23.9}$$

where

$$\mathcal{C}^* := \text{block diag}\,(\underbrace{\mathcal{C}, \mathcal{C}, \dots, \mathcal{C}}_{r}) \tag{23.10}$$

$$\mathcal{B}^* := \text{block diag}\,(\underbrace{\gamma, \gamma, \dots, \gamma}_{r}) \tag{23.11}$$

where $\mathcal{C}$, γ *are given by Equation 23.7.*

The servo-compensator is unique within the class of coordinate transformations and nonsingular input transformations.

Given the servo-compensator of Equation 23.9, now let $\mathcal{D} \in R^{r \times rp}$ be defined by

$$\mathcal{D} := \text{block diag}\,(\underbrace{\delta, \delta, \dots, \delta}_{r}) \tag{23.12}$$

where $\delta \in R^{1 \times p}$ is given by:

$$\delta := (1\ 0\ 0 \dots 0) \tag{23.13}$$

The servo-compensator has the following properties:

Lemma 23.1: [12]

Given the plant represented by Equation 23.1, assume that the existence conditions of Theorem 23.1 all hold; then

1. *The system*

$$\left\{\begin{pmatrix} C_m & 0 \\ 0 & \mathcal{D} \end{pmatrix}, \begin{pmatrix} A & 0 \\ \mathcal{B}^*C & \mathcal{C}^* \end{pmatrix}, \begin{pmatrix} B \\ \mathcal{B}D \end{pmatrix}\right\}$$

is stabilizable and detectable and has centralized fixed modes [8] (i.e., those modes of the system that are not both simultaneously controllable and observable) equal to the centralized fixed modes of (C_m, A, B, D_m).

2. *The transmission zeros of*

$$\left\{(0\ \mathcal{D}), \begin{pmatrix} A & 0 \\ \mathcal{B}^*C & \mathcal{C}^* \end{pmatrix}, \begin{pmatrix} B \\ \mathcal{B}D \end{pmatrix}\right\}$$

are equal to the transmission zeros of (C, A, B, D).

23.3.1 Robust Servomechanism Controller

Consider the system of Equation 23.1, and assume that the existence conditions of Theorem 23.1 hold; then any LTI controller that solves the RSP for Equation 23.1 consists of the following structure [6] (see Figure 23.1):

$$u = \xi + K\eta \tag{23.14}$$

where $\eta \in R^{rp}$ is the output of the servo-compensator (Equation 23.9), and ξ is the output of a stabilizing compensator $\mathcal{S}$ with inputs $y_m, y_{\text{ref}}, \eta, u$, where $\mathcal{S}, K$ are found to stabilize and give "desired behavior" to the following stabilizable and detectable system:

$$\begin{aligned} \begin{pmatrix} \dot{x} \\ \dot{\eta} \end{pmatrix} &= \begin{bmatrix} A & 0 \\ \mathcal{B}^*C & \mathcal{C}^* \end{bmatrix} \begin{pmatrix} x \\ \eta \end{pmatrix} + \begin{pmatrix} B \\ \mathcal{B}^*D \end{pmatrix} y_{\text{ref}} \\ \tilde{y}_m &= \begin{bmatrix} C_m & 0 \\ 0 & I \\ 0 & 0 \end{bmatrix} \begin{pmatrix} x \\ \eta \end{pmatrix} + \begin{pmatrix} D_m \\ 0 \\ I \end{pmatrix} y_{\text{ref}} \end{aligned} \tag{23.15}$$

where, from Lemma 23.1, the centralized fixed modes (if any) of

$$\left\{\begin{pmatrix} C_m & 0 \\ 0 & I \end{pmatrix}, \begin{bmatrix} A & 0 \\ \mathcal{B}^*C & \mathcal{C}^* \end{bmatrix}, \begin{pmatrix} B \\ \mathcal{B}^*D \end{pmatrix}\right\}$$

are equal to the centralized fixed modes of (C_m, A, B); i.e., there always exists a coordinate transformation and nonsingular input transformation, by which any controller that solves the RSP for Equation 23.1 can be described by Equation 23.14. It is to be noted that this controller always has order $\geq rp$.

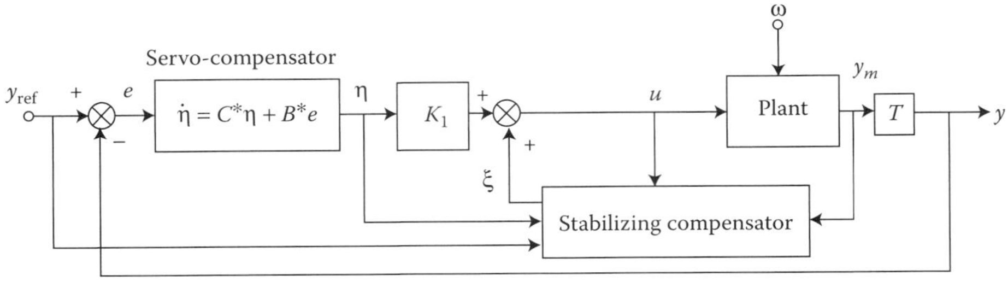

FIGURE 23.1 General controller to solve robust servomechanism problem.

Properties of Robust Servomechanism Controller

Some properties of the robust servomechanism controller (RSC) represented by Equation 23.14 are as follows [5]:

1. In the above controller, it is required to know only the disturbance/reference input poles $\{\lambda_1, \lambda_2, \ldots, \lambda_p\}$; i.e., it is not necessary to know E, F of Equation 23.1 nor $\mathcal{A}_1$, $\mathcal{A}_2$, $\mathcal{C}_1$, $\mathcal{C}_2$, G of Equations 23.2 and 23.3.
2. A controller exists generically [2] for almost all plants described by Equation 23.1, provided that (a) $m \geq r$, and (b) the outputs y can be measured; if either condition (a) or (b) fails to hold, there is no solution to the RSP.

23.3.2 Various Classes of Stabilizing Compensators

Various special classes of stabilizing compensators $\mathcal{S}$ that can be used in the RSC (Equation 23.14) are as follows. It is assumed that the existence conditions of Theorem 23.1 are all satisfied in order to implement these proposed controllers.

Multivariable Three-Term Controller

(See Figure 23.2). In order to use this controller, it is assumed that:

1. The plant of Equation 23.1 is open loop asymptotically stable.
2. The disturbance/tracking poles $\{\lambda_1, \lambda_2, \ldots, \lambda_p\}$ are of the polynomial-sinusoidal type; i.e., it is assumed that $\mathrm{Re}(\lambda_i) = 0$, $i = 1, 2, \ldots, p$.

If these assumptions hold, then the following generalized three-term controller solves the RSP [9]:

$$\begin{aligned} u &= (K_0 + K_1 s)e + K_2\eta \\ \dot{\eta} &= \mathcal{C}^*\eta + \mathcal{B}^* e \end{aligned} \tag{23.16}$$

An algorithm is given [7], that shows that a stabilizing K_2, with $K_0 = 0$, $K_1 = 0$, can always be found for this controller.

23.3.3 Complementary Controller

(See Figure 23.3) In order to use this controller, it is assumed that the plant of Equation 23.1 is open loop asymptotically stable. If this assumption holds, then the following controller, called a complementary

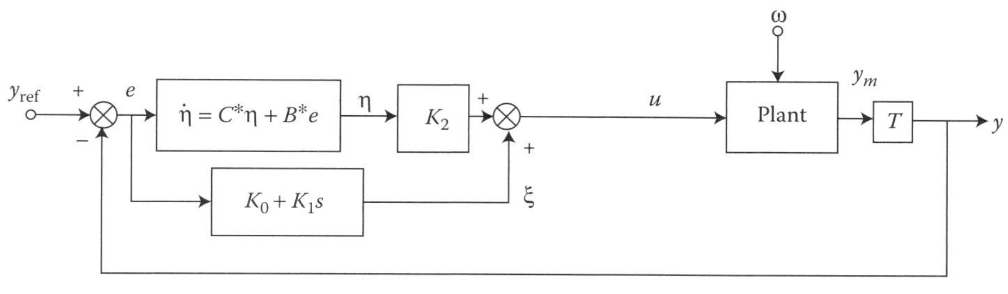

FIGURE 23.2 Generalized three term controller to solve the robust servomechanism problem.

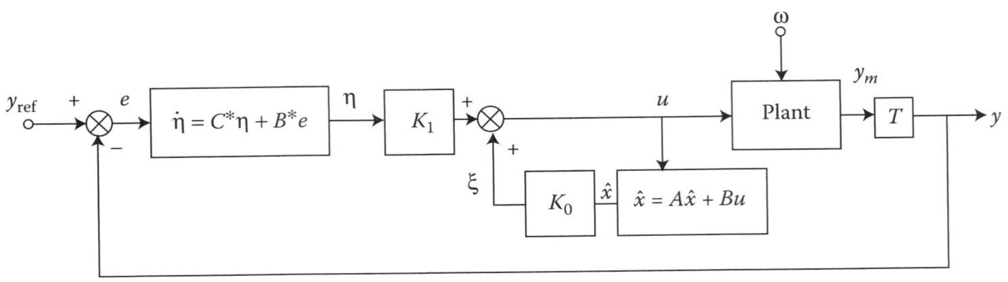

FIGURE 23.3 Complementary controller to solve the robust servomechanism problem.

controller, will solve the RSP [5]:

$$\begin{aligned} u &= K_0\hat{x} + K_1\eta \\ \dot{\eta} &= \mathcal{C}^*\eta + \mathcal{B}^*e \\ \dot{\hat{x}} &= A\hat{x} + Bu \end{aligned} \tag{23.17}$$

where (K_0, K_1) are found to stabilize the stabilizable system

$$\dot{\tilde{x}} = \begin{bmatrix} A & 0 \\ \mathcal{B}^*C & \mathcal{C}^* \end{bmatrix}\tilde{x} + \begin{pmatrix} B \\ \mathcal{B}^*D \end{pmatrix}u \tag{23.18}$$

using state feedback; i.e., $u = (K_0 \;\; K_1)\tilde{x}$.

23.3.4 Observer-Based Stabilizing Controller

(See Figure 23.4) No additional assumptions are required in order to implement this controller. The controller is given as follows [9]:

$$\begin{aligned} u &= K_0\hat{x} + K_1\eta \\ \dot{\eta} &= \mathcal{C}^*\eta + \mathcal{B}^*e \\ \dot{\hat{x}} &= \{A + (B - \Lambda D_m)K_0 - \Lambda C_m\}\hat{x} + (B - \Lambda D_m)K_1\eta + \Lambda y_m \end{aligned} \tag{23.19}$$

where (K_0, K_1) are found to stabilize the system of Equation 23.18, and Λ is an observer gain matrix found to stabilize the system matrix $(A - \Lambda C_m)$ where (C_m, A) is detectable. (A reduced-order observer could also be used to replace the full-order observer in Equation 23.19.)

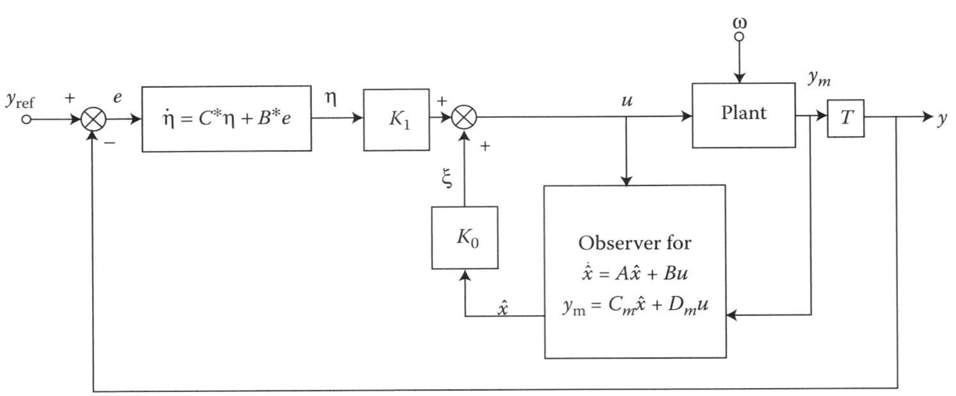

FIGURE 23.4 Observer-based stabilizing compensator to solve the robust servomechanism problem.

23.4 Applications and Example Calculations

We initially demonstrate the theory by considering the design of a controller for a nonminimum-phase plant (Example 23.1), and a minimum-phase plant (Example 23.2), and thence conclude with a case study on the control of a nontrivial distillation column system.

Example 23.1: Nonminimum-Phase System

Consider the following system

$$\dot{x} = \begin{pmatrix} 2 & -1 \\ 0 & 0 \end{pmatrix} x + \begin{pmatrix} 1 \\ 1 \end{pmatrix} u$$
$$y = (1 \quad 0)x + \omega; \quad y_m = y \tag{23.20}$$

which is open loop unstable and nonminimum phase (with a transmission zero at 1), and in which it is desired to design a controller to solve the RSP for the case of constant disturbances and constant reference input signals. In this case, the disturbance/tracking poles = {0}, and it can be directly verified that the existence conditions for a solution to the problem are satisfied from Theorem 23.1. In the controller design, it is initially assumed that the control input should not be "excessively large."

Controller Development

On applying the servo-compensator of Equation 23.9 for constant disturbances/reference input signals to Equation 23.20, the following system is obtained:

$$\begin{pmatrix} \dot{x} \\ \dot{\eta} \end{pmatrix} = \begin{pmatrix} 2 & -1 & 0 \\ 0 & 0 & 0 \\ 1 & 0 & 0 \end{pmatrix} \begin{pmatrix} x \\ \eta \end{pmatrix} + \begin{pmatrix} 1 \\ 1 \\ 0 \end{pmatrix} \omega + \begin{pmatrix} 0 \\ 0 \\ -1 \end{pmatrix} y_{\text{ref}}$$
$$y = (1 \quad 0 \quad 0) \begin{pmatrix} x \\ \eta \end{pmatrix} + 1\omega \tag{23.21}$$

and, on minimizing the performance index [12] for Equation 23.21

$$J_\epsilon = \int_0^\infty (e'e + \epsilon \dot{u}'\dot{u})\, d\tau, \quad \epsilon = 1 \tag{23.22}$$

the following controller is obtained:

$$u = (k_1 \ k_2)\hat{x} + k \int_0^t (y - y_{\text{ref}}) \tag{23.23}$$

where

$$k_1 = -13.77, \quad k_2 = 8.721, \quad k = 1.000$$

and where $\hat{x}$ is the output of an observer for Equation 23.20. On using a reduced-order observer with observer pole = −1, the following controller is thence obtained.

Robust Servomechanism Controller

$$
\begin{aligned}
u &= -22.498y + \eta + 8.7214\sigma \\
\dot{\eta} &= y - y_{\text{ref}} \\
\dot{\sigma} &= 16.443\sigma + 2\eta - 41.996y
\end{aligned}
\tag{23.24}
$$

which, when applied to Equation 23.20, gives the following closed-loop poles:

$$
\begin{pmatrix} -0.552 \pm j0.456 \\ -1.00 \\ -1.95 \end{pmatrix}
$$

Properties of Controller

On applying the controller of Equation 23.24 to the plant of Equation 23.20, the following closed-loop system is obtained:

$$
\begin{pmatrix} \dot{x} \\ \dot{\eta} \\ \dot{\sigma} \end{pmatrix} = \begin{bmatrix} -20.498 & -1 & 1 & 8.7214 \\ -22.498 & 0 & 1 & 8.7214 \\ 1 & 0 & 0 & 0 \\ -41.996 & 0 & 2 & 16.443 \end{bmatrix} \begin{pmatrix} x \\ \eta \\ \sigma \end{pmatrix} + \begin{pmatrix} 0 \\ 0 \\ -1 \\ 0 \end{pmatrix} y_{\text{ref}} + \begin{pmatrix} -22.49 \\ -22.49 \\ 1 \\ -41.99 \end{pmatrix} \omega
$$

$$
y = [1 \quad 0 \quad 0 \quad 0] \begin{pmatrix} x \\ \eta \\ \sigma \end{pmatrix} + \omega
$$

$$
u = [-22.498 \quad 0 \quad 1 \quad 8.7214] \begin{pmatrix} x \\ \eta \\ \sigma \end{pmatrix} + (-22.498)\omega
$$

which gives the Bode magnitude response between e, u and input y_{ref} in Figure 23.5, e, u and input ω in Figure 23.6, and unit-step responses in y_{ref} in Figure 23.7 and in ω in Figure 23.8, with zero initial conditions. It is seen that satisfactory tracking/regulation occurs using the controller.

Response of Closed-Loop System to Unbounded Reference Input/Disturbance Signals

According to Theorem 3 in [12], the RSC of Equation 23.14 has the property of not only achieving asymptotic tracking/regulation for the class of constant reference input/disturbance signals, but also bringing about *exact* asymptotic tracking/regulation for unbounded signals that have the property that:

$$
\begin{aligned}
\lim_{t\to\infty} \dot{y}_{\text{ref}}(t) &= 0 \\
\lim_{t\to\infty} \dot{\omega}(t) &= 0
\end{aligned}
\tag{23.25}
$$

To illustrate this result, Figure 23.9 gives the response of the closed-loop system for the case when the tracking input signal is given by

$$
y_{\text{ref}}(t) = t^{1/4}, \quad t \geq 0 \tag{23.26}
$$

with zero initial conditions. It is seen that exact asymptotic tracking indeed does take place using this controller.

Optimum Response of Nonminimum-Phase System

The RSC of Equation 23.24 was obtained so that the magnitude of the control input signal is constrained by letting $\epsilon = 1$ in the performance index of Equation 23.22. It is of interest to determine what type of

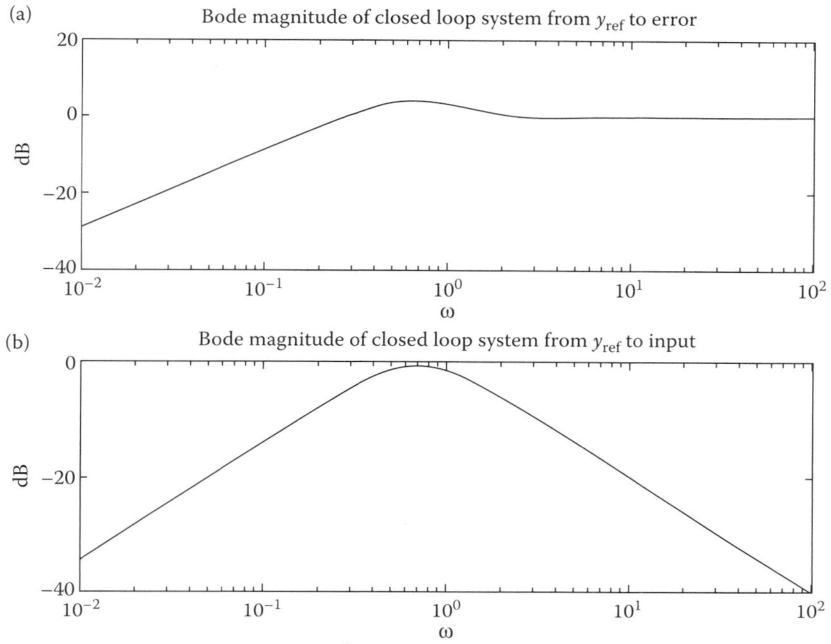

FIGURE 23.5 Bode plot magnitude of closed-loop system using robust servomechanism controller from reference input signal: (a) to error; (b) to control input.

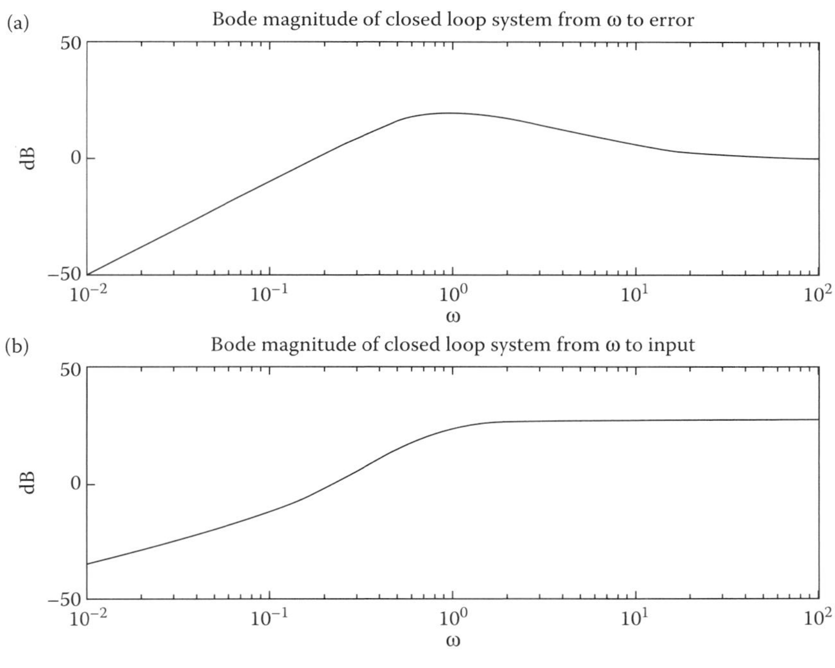

FIGURE 23.6 Bode plot magnitude of closed-loop system using robust servomechanism controller from disturbance signal: (a) to error; (b) to control input.

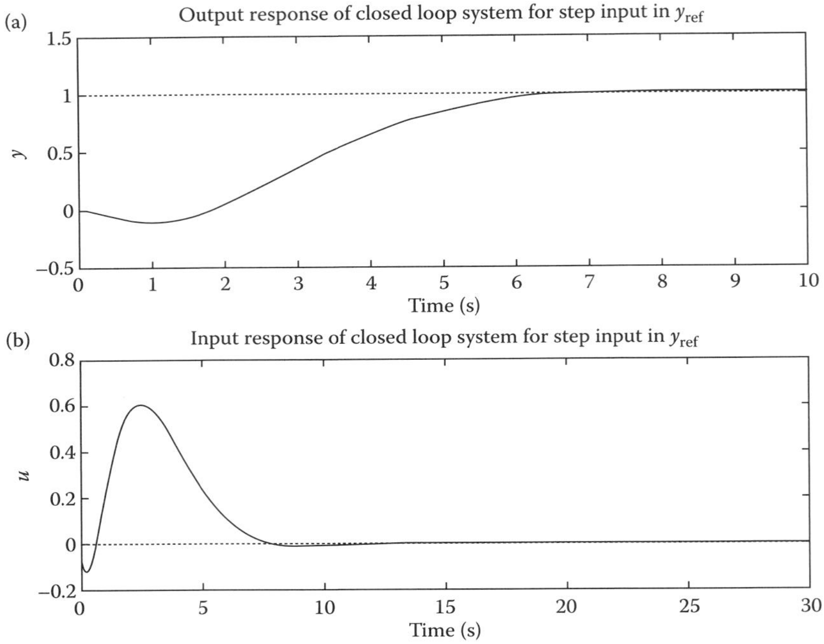

FIGURE 23.7 Response of closed-loop system using robust servomechanism controller for unit step in reference input signal: (a) output; (b) input.

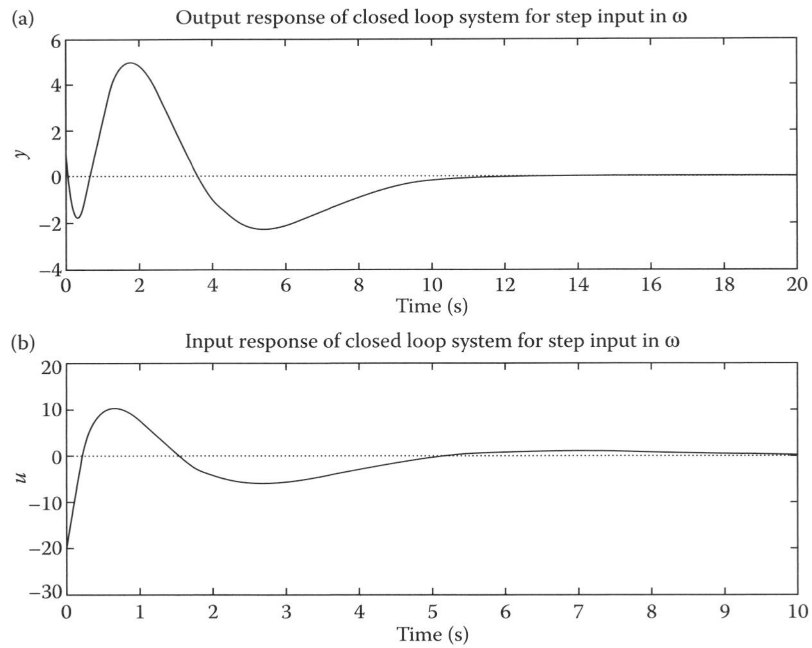

FIGURE 23.8 Response of closed-loop system using robust servomechanism controller for unit step in disturbance input signal: (a) output; (b) input.

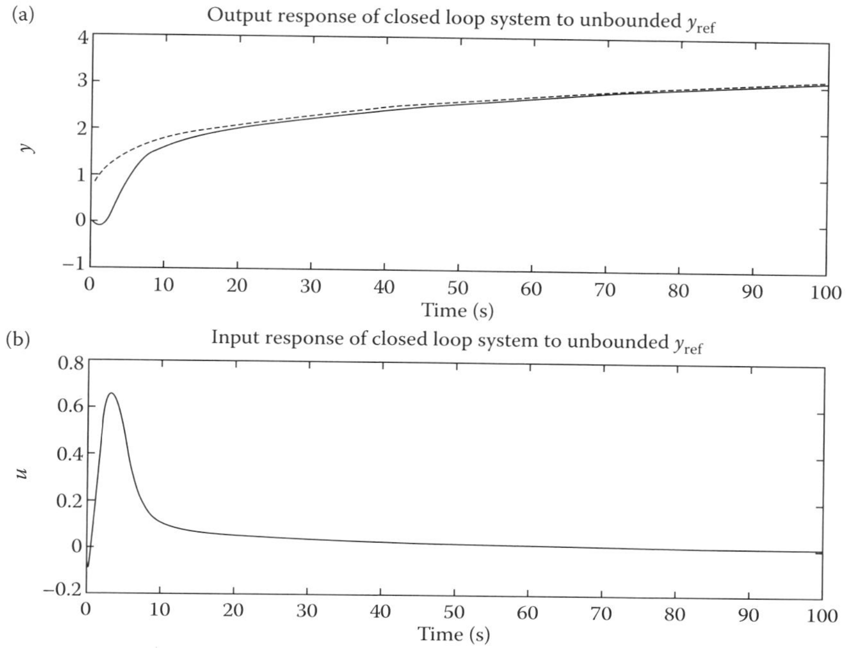

FIGURE 23.9 Response of closed-loop system using robust servomechanism controller for unbounded reference input signal $= t^{1/4}$: (a) output; (b) input.

transient performance can be obtained for the system if the controller is designed without any regard for the magnitude of the control input signal; i.e., by letting $\epsilon = 10^{-8}$, say, in the performance index (Equation 23.22). According to Theorem 23.2, "perfect control" cannot be achieved for the system; in particular, if the plant's and controller's initial conditions are equal to zero and $\omega = 0$, then the limiting performance that can be achieved is given by [13]

$$\lim_{\epsilon \to 0} J_\epsilon = 2 \sum_{i=1}^{l} \frac{1}{\lambda_i^*} y_{\text{ref}}^2 \tag{23.27}$$

where $\{\lambda_i^*,\ i = 1, 2, \ldots, l\}$ are the nonminimum TZ of the system, and J_ϵ is given by Equation 23.22. In this case, $l = 1$, $\lambda_1^* = 1$, so that the limiting performance index is given by $\lim_{\epsilon \to 0} J_\epsilon = 2y_{\text{ref}}^2$.

The following optimal controller is now obtained on putting $\epsilon = 10^{-8}$ in the performance index of Equation 23.22:

$$\begin{aligned} u &= (-2.043 \times 10^4 \quad 2.029 \times 10^4)x + 10^4 \eta \\ \dot{\eta} &= y - y_{\text{ref}} \end{aligned} \tag{23.28}$$

This controller results in the following closed-loop system poles:

$$\begin{pmatrix} -70.7 \pm j70.7 \\ -1.00 \end{pmatrix}$$

with the following optimal performance index, for $x(0) = 0$, $\omega = 0$:

$$J = 2.014 y_{\text{ref}}^2 \tag{23.29}$$

which is "close" to the optimal limiting performance index of $2y_{\text{ref}}^2$.

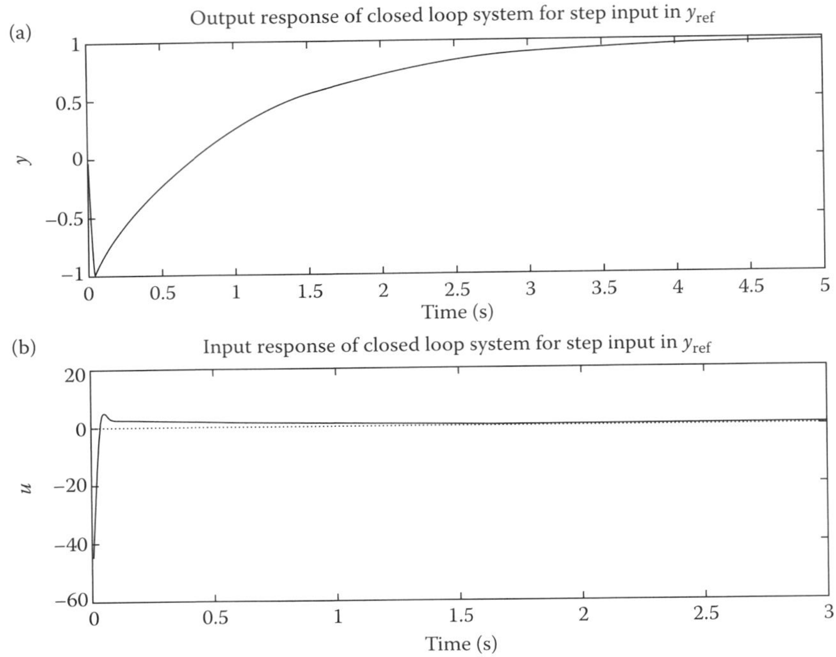

FIGURE 23.10 Response of closed-loop system using robust servomechanism controller with high gain for unit step in reference input signal: (a) output; (b) input.

A response of the closed-loop system using the controller of Equation 23.28 for a unit step in y_{ref}, with zero initial conditions for the plant, is given in Figure 23.10. It is seen that the response of the closed-loop system is only about twice as fast as that obtained using the controller of Equation 23.24, in spite of the fact that the control signal is now some 40 times larger than that obtained using Equation 23.24, which confirms the fact that nonminimum-phase systems are fundamentally "difficult to control."

High-Gain Servomechanism Control

The same example as considered in the previous sections will now be considered, except that a high-gain servomechanism controller (HGSC) [12], which is simpler than an RSC, will now be applied and compared with the RSC of Equation 23.24.

Plant Model

$$\begin{aligned} \dot{x} &= \begin{pmatrix} 2 & -1 \\ 0 & 0 \end{pmatrix} x + \begin{pmatrix} 1 \\ 1 \end{pmatrix} u \\ y &= (1 \quad 0)x + \omega \end{aligned} \tag{23.30}$$

Controller Development

Given the cheap control performance index

$$\tilde{J}_\epsilon = \int_0^\infty (y'y + \epsilon u'u)\, d\tau \tag{23.31}$$

let $\epsilon = 1$; in this case, the optimal control that minimizes $\tilde{J}_\epsilon$ for the system of Equation 23.30 with $\omega = 0$ is given by

$$u = kx, \quad k = (-10.292 \quad 5.6458) \tag{23.32}$$

which results in the closed-loop system of Equations 23.30 and 23.33 having poles given by $(-0.457, -2.19)$.

On letting $\mathcal{K} := -[c(A + Bk)^{-1}B]^{-1} = -1$, the HGSC is now given by [12]:

$$u = \mathcal{K}(y_{\text{ref}} - y) + (k + \mathcal{K}c)\hat{x} \tag{23.33}$$

where $\hat{x}$ is the output of an observer for Equation 23.30. On choosing a reduced-order observer with observer gain $= -1$, and simplifying, the following controller is finally obtained:

High-Gain Servomechanism Controller

$$\begin{aligned} u &= -y_{\text{ref}} - 15.94y - 5.646\xi \\ \dot{\xi} &= 10.29\xi + 28.87y + 2y_{\text{ref}} \end{aligned} \tag{23.34}$$

Properties of HGSC

On applying Equation 23.34 to Equation 23.30, the following closed-loop system is obtained:

$$\begin{pmatrix} \dot{x} \\ \dot{\xi} \end{pmatrix} = \begin{bmatrix} -13.94 & -1 & -5.646 \\ -15.94 & 0 & -5.646 \\ 28.87 & 0 & 10.29 \end{bmatrix} \begin{pmatrix} x \\ \xi \end{pmatrix} + \begin{pmatrix} -1 \\ -1 \\ 2 \end{pmatrix} y_{\text{ref}} + \begin{pmatrix} 0 \\ 0 \\ 28.87 \end{pmatrix} \omega$$

$$e = [1 \quad 0 \quad 0] \begin{pmatrix} x \\ \xi \end{pmatrix} - 1y_{\text{ref}} + 1\omega$$

$$y = [1 \quad 0 \quad 0] \begin{pmatrix} x \\ \xi \end{pmatrix} + 1\omega$$

$$u = [-15.94 \quad 0 \quad -5.646] \begin{pmatrix} x \\ \xi \end{pmatrix} - 1y_{\text{ref}} - 15.94\omega$$

which has closed-loop poles given by $(-0.457, -1.00, -2.19)$. The Bode magnitude response of this system with respect to output e, u and input y_{ref} is given in Figure 23.11. The unit-step function response of this system for an increase in y_{ref} with zero initial conditions is given in Figure 23.12. It is seen that satisfactory tracking occurs using this controller.

Comparison of Robustness Properties of HGSC to RSP

Although the HGSC of Equation 23.34 is "simpler" than the RSC of Equation 23.24, the outstanding advantage of the RSP is that it is robust; i.e., it provides *exact* asymptotic tracking/regulation for any perturbations of the plant model, that do not destabilize the perturbed closed-loop system. As an example of this behavior, consider the HGSC (Equation 23.34) and RSC (Equation 23.24) controlling the following (slightly) perturbed model of Equation 23.20:

Perturbed Model

$$\begin{aligned} \dot{x} &= \begin{pmatrix} 2 & -1 \\ 0 & -0.1 \end{pmatrix} x + \begin{pmatrix} 1 \\ 1 \end{pmatrix} u \\ y &= (1 \quad 0)x + \omega \end{aligned} \tag{23.35}$$

In this case, the resultant perturbed closed-loop system remains stable for each of the controllers, but the HGSC no longer provides tracking; e.g., $\lim_{t\to\infty} y(t) = 0.3$ when $y_{\text{ref}} = 1$. In contrast, it may be verified that the RSC (Equation 23.24) still provides exact tracking when applied to Equation 23.35.

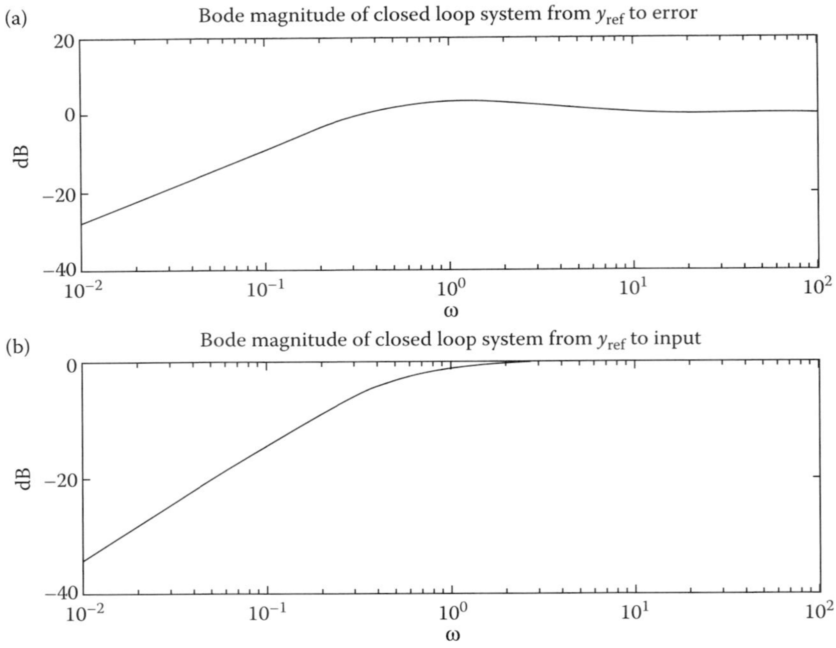

FIGURE 23.11 Bode plot magnitude of closed-loop system using high-gain servomechanism controller from reference input signal: (a) to error; (b) to control input.

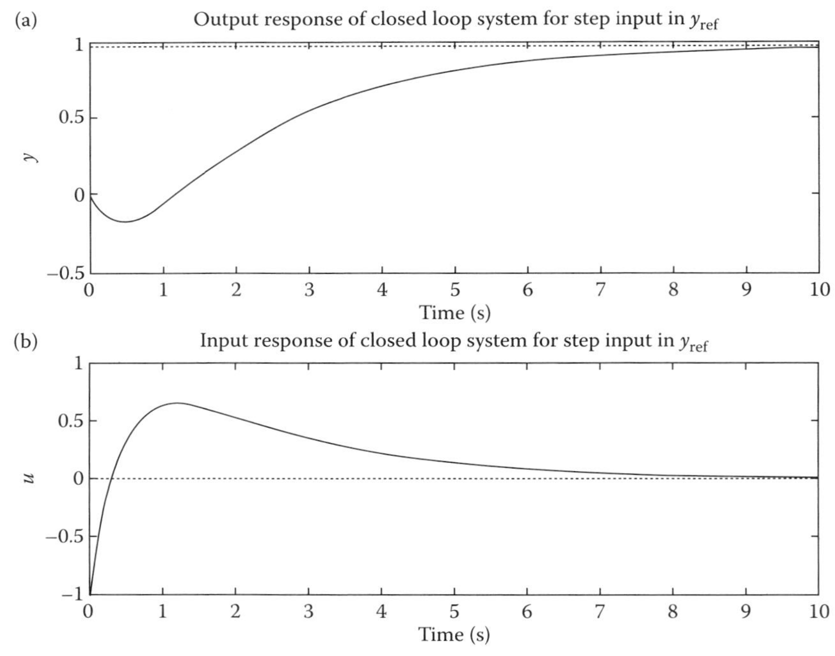

FIGURE 23.12 Response of closed-loop system using high-gain servomechanism controller for unit step in reference input signal: (a) output; (b) input.

Example 23.2: Minimum-Phase System

Consider the following system, which is a model of a pressurized head box in paper manufacturing:

$$\dot{x} = \begin{pmatrix} 0.395 & 0.01145 \\ -0.011 & 0 \end{pmatrix} x + \begin{pmatrix} 0.03362 & 1.038 \\ 0.000966 & 0 \end{pmatrix} u$$
$$y = \begin{pmatrix} 1 & 0 \\ 0 & 1 \end{pmatrix} x; \quad y_m = y \tag{23.36}$$

which has open-loop eigenvalues $= (-3.19 \times 10^{-4}, -3.95 \times 10^{-1})$. In this case, the plant represented by Equation 23.36 has no transmission zeros, which implies that it is minimum phase, and so there exists a solution to the RSP for the class of constant disturbances/reference input signals such that perfect control occurs (see Theorem 23.2).

The following development illustrates how "perfect control" in a minimum-phase system results in *arbitrarily good approximate error regulation* (AGAER) [12] occurring; i.e., although the servo-compensator is designed to give exact asymptotic error regulation for constant disturbances/reference input signals, AGAER will occur for other classes of disturbances/reference input signals, e.g., for the class of sinusoidal signals $y_{\text{ref}} = \bar{y}\sin(\omega t)$.

The following perfect controller is now obtained:

$$u = K_0 y + K \int_0^t (y - y_{\text{ref}})\, d\tau \tag{23.37}$$

which, on minimizing the performance index

$$J_\epsilon = \int_0^\infty (e'e + \epsilon \dot{u}'\dot{u})\, d\tau \tag{23.38}$$

gives

$$K_0 = \begin{pmatrix} -4.33 & -4549 \\ -138.3 & 158.7 \end{pmatrix}, \quad K = \begin{pmatrix} -371.9 & -10^4 \\ -10^4 & 371.9 \end{pmatrix} \quad \text{for } \epsilon = 10^{-8}$$
$$K_0 = \begin{pmatrix} -13.8 & -1.439 \times 10^4 \\ -438.2 & 477.3 \end{pmatrix}, \quad K = \begin{pmatrix} -338.8 & -9.994 \times 10^4 \\ -9.994 \times 10^4 & 3388 \end{pmatrix} \quad \text{for } \epsilon = 10^{-10} \tag{23.39}$$

It may now be demonstrated that AGAER occurs; e.g., when the reference input signal $y_{\text{ref}} = \bar{y}\sin(\omega t)$, $\omega = 0.1$ is applied, we obtain the following *error coefficient* [12]:

$${}_0K_t^0(j\omega) = \begin{pmatrix} 1.4 \times 10^{-3} & 1.0 \times 10^{-4} \\ 1.0 \times 10^{-4} & 4.6 \times 10^{-2} \end{pmatrix} \quad \text{for } \epsilon = 10^{-8}$$
$$= \begin{pmatrix} 4.4 \times 10^{-4} & 1.0 \times 10^{-5} \\ 1.0 \times 10^{-5} & 1.4 \times 10^{-2} \end{pmatrix} \quad \text{for } \epsilon = 10^{-10} \tag{23.40}$$

which implies that excellent (but approximate) tracking occurs as $\epsilon \to 0$ for the tracking signal $y_{\text{ref}} = \bar{y}\sin(\omega t)$, $\omega = 0.1$.

Stability Robustness Concerns

The previous example studies have ignored "stability robust concerns"; i.e., the recognition that any plant model is really only an approximation to a model of the actual physical system, and hence that a given physical system may become unstable under feedback if the model used is sufficiently inaccurate.

The following shows that for some classes of systems, a system may be relatively insensitive to plant perturbations, whereas for other classes of systems, a system may be highly sensitive to plant perturbations. This implies that any controller that is designed to solve the RSP must *always* take into account stability robustness considerations for the particular problem being studied.

Consider the head box example (Equation 23.36) and the controller

$$u = K_0 y + K \int_0^t (y - y_{\text{ref}})\, d\tau \tag{23.41}$$

which has been designed for the class of constant disturbances/reference input signals. It is now desired to determine the optimal controller gain matrices K_0, K that minimize the performance index

$$J_\epsilon = \int_0^\infty (e'e + \epsilon \dot{u}'\dot{u})\, d\tau \quad \epsilon = 10^{-8} \tag{23.42}$$

with and without a *gain margin* (GM) constraint [10] imposed on the system. This constraint can be imposed by using the computer-aided design (CAD) approach of Davison and Ferguson [9]. Let Γ denote the optimal cost matrix of Equation 23.44, and define $J_{opt} = \text{trace}\ (\Gamma)$. The following results are now obtained (see Table 23.1)

In this case, the optimal controller that minimizes J_ϵ (Equation 23.44), subject to a fairly demanding GM constraint of (0.2,2) is only about "two times slower" than the optimal controller, which does not take gain margin into account; i.e., the system is relatively insensitive to plant perturbations.

Consider now, however, the following system:

$$\begin{aligned} \dot{x} &= \begin{pmatrix} -1 & 0 \\ 0 & -2 \end{pmatrix} x + \begin{pmatrix} 7 & 8 \\ 12 & 14 \end{pmatrix} u \\ y &= \begin{pmatrix} 7 & -8 \\ -6 & 7 \end{pmatrix} x; \quad y_m = y \end{aligned} \tag{23.43}$$

and the controller of Equation 23.43. It is desired now to find optimal controller parameters K_0, K so as to minimize the performance index J_ϵ (Equation 23.44) with and without a modest GM of (0.9,1.1) applied. The following results are obtained from Davison and Copeland [10] (see Table 23.2).

In this case, it can be seen that a very modest demand that the controller should have a GM of only 10% has produced a dramatic effect in terms of the controller. The controller obtained with the 10% GM constraint being imposed has a performance index that is some 10^5 times "worse" than the case when no GM constraint is imposed; i.e., the system is extremely sensitive to plant perturbations. Thus, this example emphasizes the need to always apply some type of stability robustness constraint when solving "high-performance controller" problems.

TABLE 23.1 Optimal Controller Parameters Obtained

GM Constraint	J_{opt}	(K_0, K)	Closed-Loop Poles
None	0.469	$\begin{bmatrix} -4.33 & -4550 & -371.9 & -9993 \\ -138 & 158.7 & -9993 & 371.9 \end{bmatrix}$	$\begin{pmatrix} -2.20 \pm j2.20 \\ -72.1 \pm j72.1 \end{pmatrix}$
(0.2,2)	0.700	$\begin{bmatrix} 2330 & 245 & 6700 & 281 \\ -504 & -820 & -1960 & -1580 \end{bmatrix}$	$\begin{pmatrix} -1.43 \pm j1.51 \\ -5.31 \\ -437 \end{pmatrix}$

TABLE 23.2 Optimal Controller Parameters Obtained

GM Constraint	J_{opt}	(K_0, K)	Closed-Loop Poles
None	0.100	$\begin{bmatrix} -58.1 & -15.2 & -326 & -9995 \\ 220 & -185 & 9995 & -326 \end{bmatrix}$	$\begin{pmatrix} -10.1 \pm j10.1 \\ -700 \pm j700 \end{pmatrix}$
(0.9,1.1)	24.4	$\begin{bmatrix} -0.600 & -0.962 & -0.636 & 0.109 \\ 0.740 & 0.759 & 0.620 & -0.380 \end{bmatrix}$	$\begin{pmatrix} -0.0538 \\ -0.350 \\ -0.897 \pm j4.21 \end{pmatrix}$

More Complex Tracking/Disturbance Signal Requirements

The following example illustrates the utilization of more complex servo-compensator construction. In this case, the head box problem modeled by Equation 23.36 is to be controlled for the class of disturbance/reference input signals that have the structure:

$$y_{\text{ref}} = \bar{y}_1 \sin(\omega_1 t) + \bar{y}_2 \sin(\omega_2 t) + \bar{y}_3 \sin(\omega_3 t) \tag{23.44}$$

where $\omega_1 = \pi$, $\omega_2 = 3\pi$, $\omega_3 = 5\pi$, and $\bar{y}_1, \bar{y}_2, \bar{y}_3$ are arbitrary real two-dimensional vectors.

From Theorem 23.1, there exists a solution to the problem, and the servo-compensator is now given from Equation 23.9 by:

$$\dot{\eta} = \text{block diag}\left[\begin{pmatrix} 0 & I \\ -\omega_1^2 I & 0 \end{pmatrix}, \begin{pmatrix} 0 & I \\ -\omega_2^2 I & 0 \end{pmatrix}, \begin{pmatrix} 0 & I \\ -\omega_3^2 I & 0 \end{pmatrix}\right]\eta + \begin{bmatrix} 0 \\ I \\ 0 \\ I \\ 0 \\ I \end{bmatrix}(y - y_{\text{ref}}) \tag{23.45}$$

and the following controller is now obtained:

$$u = K_0 y + (K_1 \ \ K_2 \ \ K_3)\eta \tag{23.46}$$

In this case, the following performance index

$$\bar{J}_\epsilon = \int_0^\infty \left\{(x' \ \eta')Q\begin{pmatrix} x \\ \eta \end{pmatrix} + \epsilon u' u\right\} d\tau \tag{23.47}$$

where

$$\begin{aligned} Q &:= (0 \ \ I \ \ 0 \ \ I \ \ 0 \ \ I \ \ 0)'(0 \ \ I \ \ 0 \ \ I \ \ 0 \ \ I \ \ 0) \\ \epsilon &= 10^{-8} \end{aligned} \tag{23.48}$$

is to be minimized in order to determine the optimal controller parameters (K_0, K_1, K_2, K_3). As before, let Γ denote the optimal cost matrix of Equation 23.47, corresponding to a minimization of $\bar{J}_\epsilon$, and define $\bar{J}_{opt} = \text{trace}(\Gamma)$. The following results are now obtained:

$$\bar{J}_{opt} = 36.3$$

and the optimal K_0, K_1, K_2, K_3 are given in Table 23.3.

The eigenvalues of the resultant closed-loop system using this controller are given in Table 23.4.

In this case, the response of the resultant closed-loop system for a triangular tracking signal of period 4 seconds is given in Figure 23.13. It is seen that the tracking performance of the system is excellent; it is to be noted that the dominant harmonics of this triangular wave are given by π, 3π, 5π rad/s, and thus the servo-compensator is approximating the tracking of a triangular wave, in this case, by tracking/regulating the main harmonic terms of the periodic signal.

TABLE 23.3 Optimal Value of K_0, K_1, K_2, K_3 Obtained

$K_0 =$	−1.8047e+00	−2.1315e+03		
	−5.8028e+01	8.0400e+01		
$K_1 =$	−3.3062e+02	8.0559e+03	−4.0912e+01	−1.8825e+03
	−9.7821e+03	−2.4767e+02	−6.5109e+02	8.0833e+01
$K_2 =$	−2.6297e+02	9.7588e+03	−2.1678e+01	−2.2892e+02
	−8.2362e+03	−3.2042e+02	−6.0072e+02	9.6246e+00
$K_3 =$	−1.8240e+02	9.9091e+03	−1.7523e+01	−8.3041e+01
	−5.7704e+03	−3.2592e+02	−5.1951e+02	3.3729e+00

23.4.1 CAD Approach

In designing controllers for actual physical systems, it is often important to impose on the controller construction constraints that are related to the specific system being considered. Such constraints can be incorporated directly using a CAD approach (e.g., see [9]). The following example illustrates the type of results that may be obtained.

Consider the head box problem modeled by Equation 23.36, controlled by the controller

$$u = K_0 y + K \int_0^t (y - y_{\text{ref}})\, d\tau \tag{23.49}$$

and assume that it is desired to determine the controller gain matrices K_0, K_1, so as to minimize the performance index:

$$J_\epsilon = \int_0^\infty (e'e + \epsilon \dot{u}'\dot{u})\, d\tau, \quad \epsilon = 10^{-8} \tag{23.50}$$

such that all elements of K_0, K satisfy the constraint:

$$|k_{ij}| \leq 100$$

This constraint could arise, for example, in terms of attempting to regulate the control signal magnitude level for a system. The following results are obtained in this case [9] (see Table 23.5).

TABLE 23.4 Closed-Loop Eigenvalues

−1.5016e+01	+2.8300e+01i
−1.5016e+01	−2.8300e+01i
−3.0344e+01	
−1.2005e−01	+1.3596e+01i
−1.2005e−01	−1.3596e+01i
−1.9565e−02	+1.5708e+01i
−1.9565e−02	−1.5708e+01i
−5.4357e−02	+9.4249e+00i
−5.4357e−02	−9.4249e+00i
−3.5863e−02	+6.7460e+00i
−3.5863e−02	−6.7460e+00i
−1.0298e+00	
−4.4094e−01	+3.2555e+00i
−4.4094e−01	−3.2555e+00i

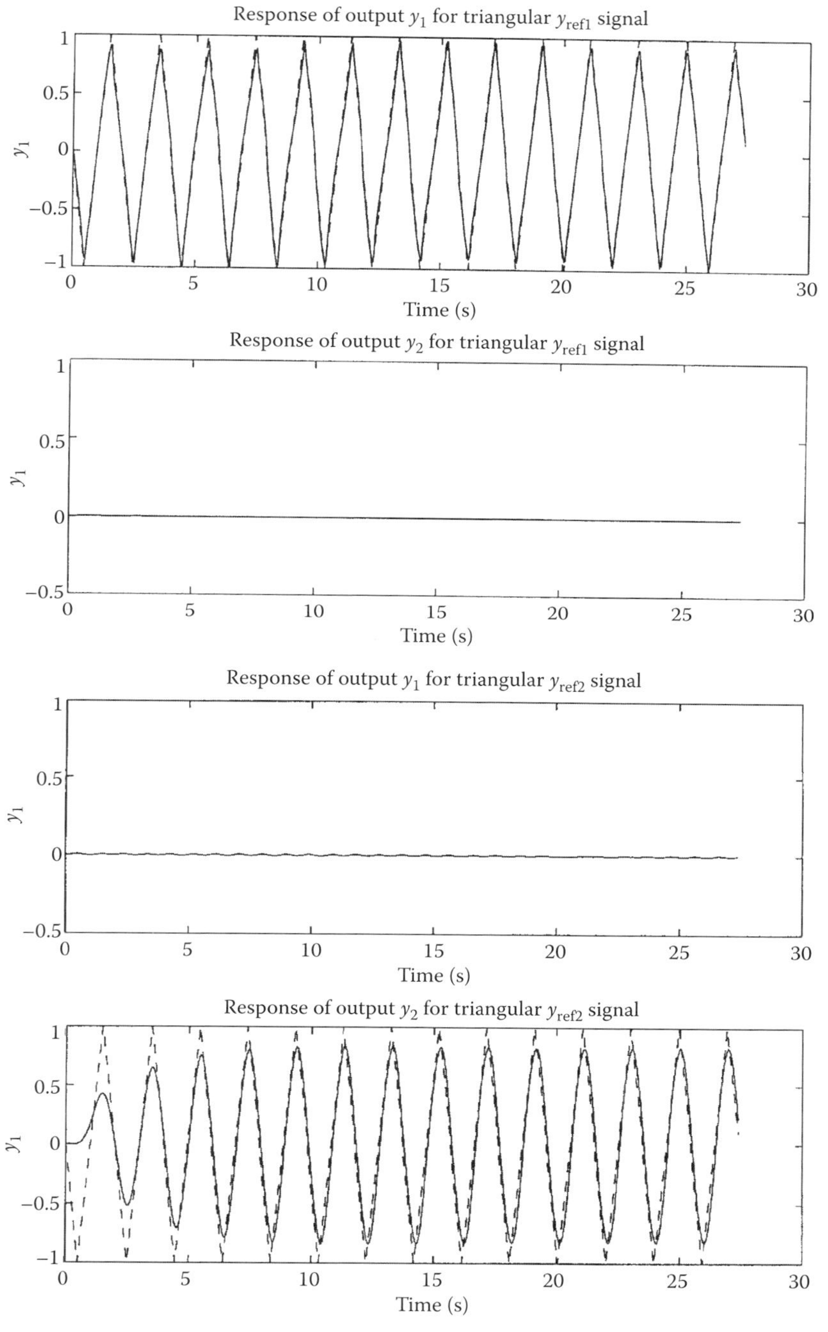

FIGURE 23.13 Response of closed-loop system for head box example using robust servomechanism controller for triangular reference input signal.

TABLE 23.5 Results Obtained

Constraint	J_{opt}	(K_0, K)	Closed-Loop Poles
None	0.516	$\begin{bmatrix} -4.33 & -4550 & -372 & -9993 \\ -138 & 159 & -9993 & 372 \end{bmatrix}$	$\begin{pmatrix} -2.20 \pm j2.20 \\ -72.1 \pm j72.1 \end{pmatrix}$
$\lvert k_{ij} \rvert \leq 100$	22.2	$\begin{bmatrix} -70 & -100 & -100 & -100 \\ -2.6 & 78 & -29 & 22 \end{bmatrix}$	$\begin{pmatrix} -0.157 \pm j0.330 \\ -2.61 \pm j5.62 \end{pmatrix}$

Consider now the following model, which approximately describes the behavior of a DC motor:

$$\dot{x} = \begin{bmatrix} -7.535 \times 10^{-2} & 5.163 \\ -209.4 & -198.1 \end{bmatrix} x + \begin{bmatrix} 0 \\ 188.7 \end{bmatrix} u + \begin{bmatrix} -4.651 \\ 0 \end{bmatrix} \omega$$
$$y = [1 \quad 0]x; \quad y_m = x \tag{23.51}$$

and consider the following controller:

$$u = K_0 y_m + K \int_0^t (y - y_{\text{ref}})\, d\tau \tag{23.52}$$

where (K_0, K) are to be obtained so as to minimize the performance index

$$J_\epsilon = \int_0^\infty (e'e + \dot{u}'\dot{u})\, d\tau, \quad \epsilon = 10^{-8} \tag{23.53}$$

subject to the constraint that the damping factor of the closed-loop system should have the property that $\zeta \geq 1$, in order to prevent an excessively oscillatory response, say. The following results are obtained [9] (see Table 23.6).

Discrete Systems

The previous results have considered a continuous system. For discrete-time systems, equivalent conditions for the existence of a solution to the RSP and the necessary controller structure can be obtained (e.g., see [4]). The following example illustrates this point.

Consider the following discrete system:

$$x_{k+1} = \begin{pmatrix} 0.9512 & 0 \\ 0 & 0.09048 \end{pmatrix} x_k + \begin{pmatrix} 4.877 & 4.877 \\ -1.1895 & 3.369 \end{pmatrix} u_k$$
$$y_k = \begin{pmatrix} 1 & 0 \\ 0 & 1 \end{pmatrix} x_k; \quad y^m = y \tag{23.54}$$

in which it is desired to solve the RSP for the class of constant disturbances/constant tracking signals. In this case, the above system is controllable and observable and has no transmission zeros, so that there

TABLE 23.6 Results Obtained

Constraint	J_{opt}	(K_0, K)	Closed-Loop Poles
None	9.93_{10}^{-3}	$[-98.1 \quad -1.49 \quad -10000]$	$\begin{pmatrix} -116 \pm j161 \\ -249 \end{pmatrix}$
$\zeta \geq 1$	1.00_{10}^{-2}	$[-109 \quad -1.62 \quad -9480]$	$\begin{pmatrix} -153 \pm j153 \\ -197 \end{pmatrix}$

exists a solution to the RSP; the servomechanism controller becomes in this case:

$$\begin{aligned} u_k &= K_0 y_k + K\eta_k \\ \eta_{k+1} &= \eta_k + y_k - y_{\text{ref}} \end{aligned} \tag{23.55}$$

and, on minimizing the performance index

$$J_\epsilon^* = \sum_{k=1}^{\infty} [e_k' e_k + \epsilon (u_{k+1} - u_k)'(u_{k+1} - u_k)], \quad \epsilon = 10^{-8} \tag{23.56}$$

the following controller gain matrices are obtained:

$$\begin{aligned} J_{opt}^* &= 6.00 \\ (K_0, K)_{opt} &= \begin{bmatrix} -0.296 & 0.239 & -0.152 & 0.219 \\ -0.104 & -0.239 & -0.0535 & -0.219 \end{bmatrix} \end{aligned} \tag{23.57}$$

The closed-loop poles obtained by applying the above controller to the plant modeled by Equation 23.54 are, in this case, given by $-8.4 \times 10^{-10} \pm j1.6 \times 10^{-5}$, $7.2 \times 10^{-11} \pm j8.1 \times 10^{-6}$); i.e., a "dead-beat" closed-loop system time response is obtained.

23.4.2 Case Study Problem—Distillation Column

The following model of a binary distillation column with pressure variation is considered:

$$\begin{aligned} \dot{x} &= Ax + Bu + E\omega \\ y &= Cx \end{aligned} \tag{23.58}$$

where (C, A, B, E) are given in Table 23.7. Here y_1 is the composition of the more volatile component in the bottom of the column, y_2 is the composition of the more volatile component in the top of the column, and y_3 is the pressure in the column; ω_1 is the input feed disturbance in the column; and u_1 the reheater input, u_2 the condensor input, and u_3 the reflux in the system.

Eigenvalues and Transmission Zeros of Distillation Column

The open-loop eigenvalues and transmission zeros of the distillation column are given in Table 23.8, which implies that the system is minimum phase.

It is desired now to find a controller that solves the RSP problem for this system for the case of constant disturbances and constant reference input signals. In this case, the existence conditions of Theorem 23.1 hold, so that a solution to the problem exists; in particular, there exists a solution to the "perfect control robust servomechanism" problem (see Theorem 23.2) for the system.

Perfect Robust Controller

The following controller is obtained from Zhang and Davison [14], and it can be shown to produce "perfect control" (i.e., the transient error in the system can be made arbitrarily small) in the system as $\epsilon \to 0$:

$$u = \frac{1}{\epsilon} \frac{(s+1)^2}{(\epsilon^2 s + 1)^2} \frac{\Theta}{s} (y - y_{\text{ref}}) \tag{23.59}$$

where

$$\Theta := \begin{bmatrix} 1.7599 \times 10^0 & -3.4710 \times 10^6 & -1.0869 \times 10^3 \\ -1.7599 \times 10^0 & 3.4710 \times 10^6 & -1.0870 \times 10^3 \\ -3.9998 \times 10^2 & -3.0545 \times 10^4 & -7.8258 \times 10^0 \end{bmatrix} \tag{23.60}$$

TABLE 23.7 Data Matrices for Distillation Column Model

$A =$

	x1	x2	x3	x4	x5
x1	−0.01400	0.00430	0	0	0
x2	0.00950	−0.01380	0.00460	0	0
x3	0	0.00950	−0.01410	0.00630	0
x4	0	0	0.00950	−0.01580	0.01100
x5	0	0	0	0.00950	−0.03120
x6	0	0	0	0	0.02020
x7	0	0	0	0	0
x8	0	0	0	0	0
x9	0	0	0	0	0
x10	0	0	0	0	0
x11	0.02550	0	0	0	0

	x6	x7	x8	x9	x10
x1	0	0	0	0	0
x2	0	0	0	0	0
x3	0	0	0	0	0
x4	0	0	0	0	0
x5	0.01500	0	0	0	0
x6	−0.03520	0.02200	0	0	0
x7	0.02020	−0.04220	0.02800	0	0
x8	0	0.02020	−0.04820	0.03700	0
x9	0	0	0.02020	−0.05720	0.04200
x10	0	0	0	0.02020	−0.04830
x11	0	0	0	0	0.02550

	x11
x1	0
x2	0.00050
x3	0.00020
x4	0
x5	0
x6	0
x7	0
x8	0.00020
x9	0.00050
x10	0.00050
x11	−0.01850

$B =$

	u1	u2	u3
x1	0	0	0
x2	5.00000e−06	−4.00000e−05	0.00250
x3	2.00000e−06	−2.00000e−05	0.00500
x4	1.00000e−06	−1.00000e−05	0.00500
x5	0	0	0.00500
x6	0	0	0.00500
x7	−5.00000e−06	1.00000e−05	0.00500
x8	−1.00000e−05	3.00000e−05	0.00500
x9	−4.00000e−05	5.00000e−06	0.00250
x10	−2.00000e−05	2.00000e−06	0.00250
x11	0.00046	0.00046	0

TABLE 23.7 (continued) Data Matrices for Distillation Column Model

$C =$

	x1	x2	x3	x4	x5
y1	0	0	0	0	0
y2	1.00000	0	0	0	0
y3	0	0	0	0	0

	x6	x7	x8	x9	x10
y1	0	0	0	0	1.00000
y2	0	0	0	0	0
y3	0	0	0	0	0

	x11
y1	0
y2	0
y3	1.00000

$E =$

	w1
x1	0
x2	0
x3	0
x4	0
x5	0.01
x6	0
x7	0
x8	0
x9	0
x10	0
x11	0

Properties of Closed-Loop System

In order to determine the potential "speed of response" that may be obtained by the controller modeled by Equation 23.59, the following closed-loop eigenvalues of the system are obtained with $\epsilon = 0.1$ (see Table 23.9).

TABLE 23.8 Properties of Distillation Column Model

Open-Loop Eigenvalues	Transmission Zeros
−9.6031e−02	−9.1024e−02
−7.0083e−02	−6.6830e−02
−5.0545e−02	−5.0627e−02
−1.2152e−04	−2.9083e−02
−3.2355e−03	−2.2078e−02
−3.3900e−02	−9.5168e−03
−7.7594e−03	−6.4089e−03
−1.9887e−02	
−1.8154e−02	
−2.4587e−02	
−1.4196e−02	

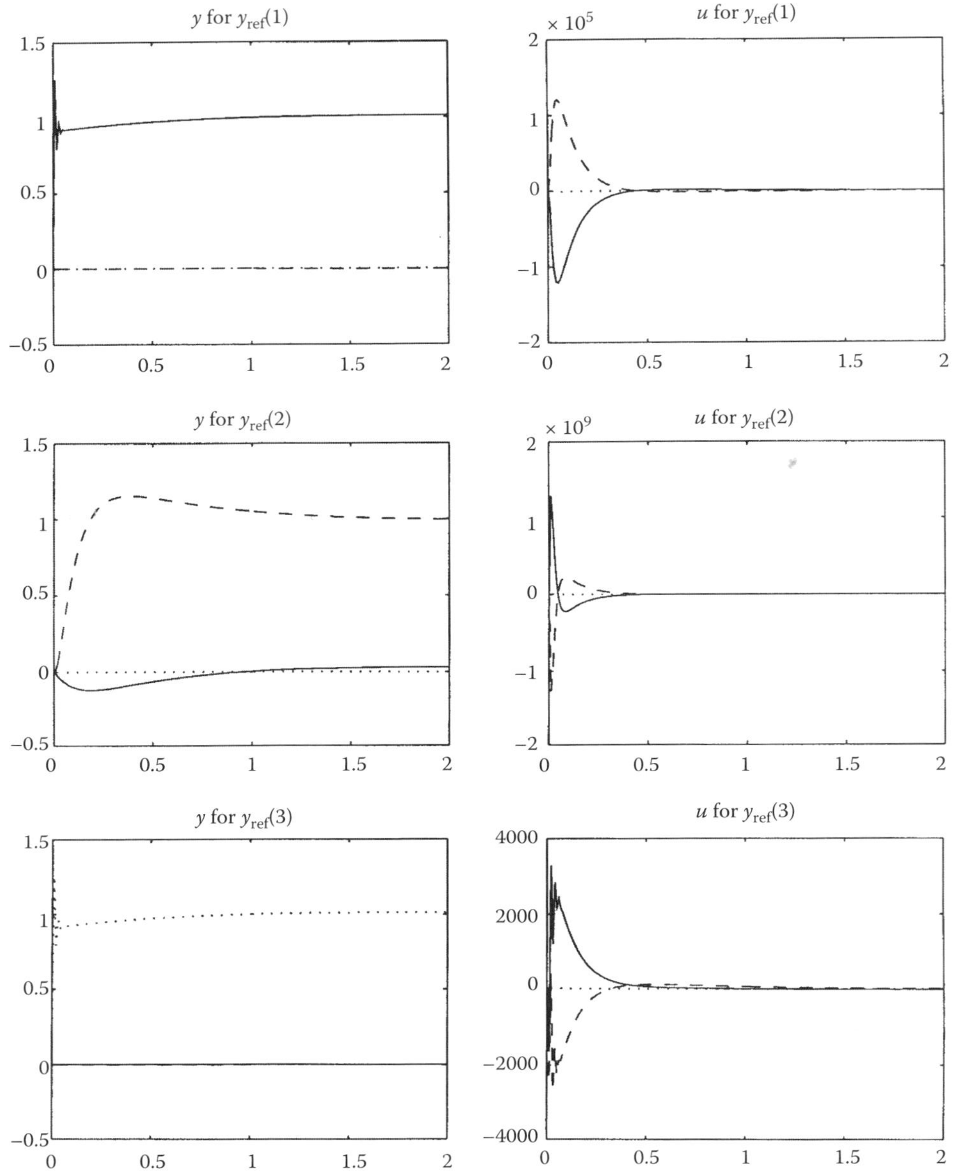

FIGURE 23.14 Response of closed-loop system for distillation column example using perfect robust servomechanism controller for unit step in y_{ref} given by $y_{ref}(1) = (1\,0\,0)'$, $y_{ref}(2) = (0\,1\,0)'$, $y_{ref}(3) = (0\,0\,1)'$.

Using the controller of Equation 23.59 with $\epsilon = 0.1$, the response of Figure 23.14 is then obtained for the case of a unit-step increase in $y_{ref} = (1\,0\,0)', (0\,1\,0)', (0\,0\,1)'$, respectively, with zero initial conditions. It is seen that "perfect control" indeed does take place; i.e., all transients have died down in less than 1 second, the system displays "low interaction," and no "excessive peaking occurs." In real life, however, this controller would not be used because the control inputs are excessively large (see Figure 23.14).

The following decentralized controller, obtained by the procedure given in [11], would be quite realistic to implement, however, since the control input signals do not "peak" now as they did in the previous controller.

TABLE 23.9 Closed-Loop Eigenvalues

−9.9102e+01	+3.1595e+02i
−9.9102e+01	−3.1595e+02i
−9.9097e+01	+3.1594e+02i
−9.9097e+01	−3.1594e+02i
−1.2776e+02	
−5.9778e+01	
−1.0030e+01	
−1.6633e+00	
−7.8470e−01	
−9.1245e−01	+2.8213e−01i
−9.1245e−01	−2.8213e−01i
−9.1207e−01	+2.8314e−01i
−9.1207e−01	−2.8314e−01i
−9.1027e−02	
−6.6832e−02	
−6.4090e−03	
−9.5169e−03	
−2.9083e−02	
−5.0626e−02	
−2.2078e−02	

23.4.3 Decentralized Robust Controller

$$\begin{aligned} u &= \mathcal{K}_1 y + \mathcal{K}_2 \eta + \mathcal{K}_3 \rho \\ \dot{\eta} &= y - y_{\text{ref}} \\ \dot{\rho} &= \mathcal{K}_5 \rho + \mathcal{K}_4 y \end{aligned} \tag{23.61}$$

where $\mathcal{K}_1, \mathcal{K}_2, \mathcal{K}_3, \mathcal{K}_4, \mathcal{K}_5$ are given in Table 23.10.

TABLE 23.10 Decentralized Controller Gains Obtained

$\mathcal{K}_1 =$	1.7409e+05	0	0
	0	1.9140e+04	0
	0	0	−7.7460e+03
$\mathcal{K}_2 =$	4.5284e+02	0	0
	0	1.2143e+01	0
	0	0	−3.7682e+00
$\mathcal{K}_3 =$	5.7841e+05	0	0
	0	3.0466e+05	0
	0	0	−1.0636e+07
$\mathcal{K}_4 =$	1.0000e+00	0	0
	0	1.0000e+00	0
	0	0	1.0000e+00
$\mathcal{K}_5 =$	−1.1024e+05	0	0
	0	−8.1558e+07	0
	0	0	−1.3506e+04

TABLE 23.11 Closed-Loop Eigenvalues

−8.1558e+07	
−1.3506e+04	
−1.1023e+03	
−1.7595e+00	+4.1365e+01i
−1.7595e+00	−4.1365e+01i
−9.5681e−03	+7.5529e−02i
−9.5681e−03	−7.5529e−02i
−8.9273e−02	
−6.5720e−02	
−5.0556e−02	
−2.9058e−02	
−2.2081e−02	
−9.5034e−03	
−6.4076e−03	
−2.5926e−03	
−6.2808e−04	
−4.4157e−04	

Properties of Closed-Loop System

The following closed-loop eigenvalues are obtained by applying the controller of Equation 23.61 to Equation 23.58 (see Table 23.11).

Using the controller of Equation 23.61, the response of Figure 23.15 is then obtained for the case of a unit step in $y_{\text{ref}} = (1\ 0\ 0)', (0\ 1\ 0)', (0\ 0\ 1)'$, respectively, and for the case of a unit step in the disturbance term ω with zero initial conditions. It is seen that excellent tracking/regulation takes place; i.e., the output responses obtained show little interaction effects with no peaking occurring, and the control input signals are now quite realistic to implement. In addition, the controller has the additional advantage of being decentralized; i.e., the controller is particularly simple to implement. The time response of the closed-loop system, however, is now much slower compared to the case when perfect control is applied.

23.5 Concluding Remarks

In this overview on the RSP, the emphasis has been placed on the control of LTI continuous systems, and existence conditions and corresponding required controller construction to solve the RSP have been reviewed. To demonstrate the principles involved, various simple nonminimum- and minimum-phase examples were initially considered, and then a case study of a nontrivial system example was studied.

23.6 Defining Terms

Arbitrarily good approximate error regulation (AGAER): The property of a closed-loop system that permits arbitrarily good regulation to occur for arbitrary disturbance/tracking signals of a specified class.

Centralized fixed modes: Those modes of an LTI system that are not both simultaneously controllable and observable.

Decentralized control: Refers to a controller in which the information flow between the inputs and outputs is constrained to be block diagonal.

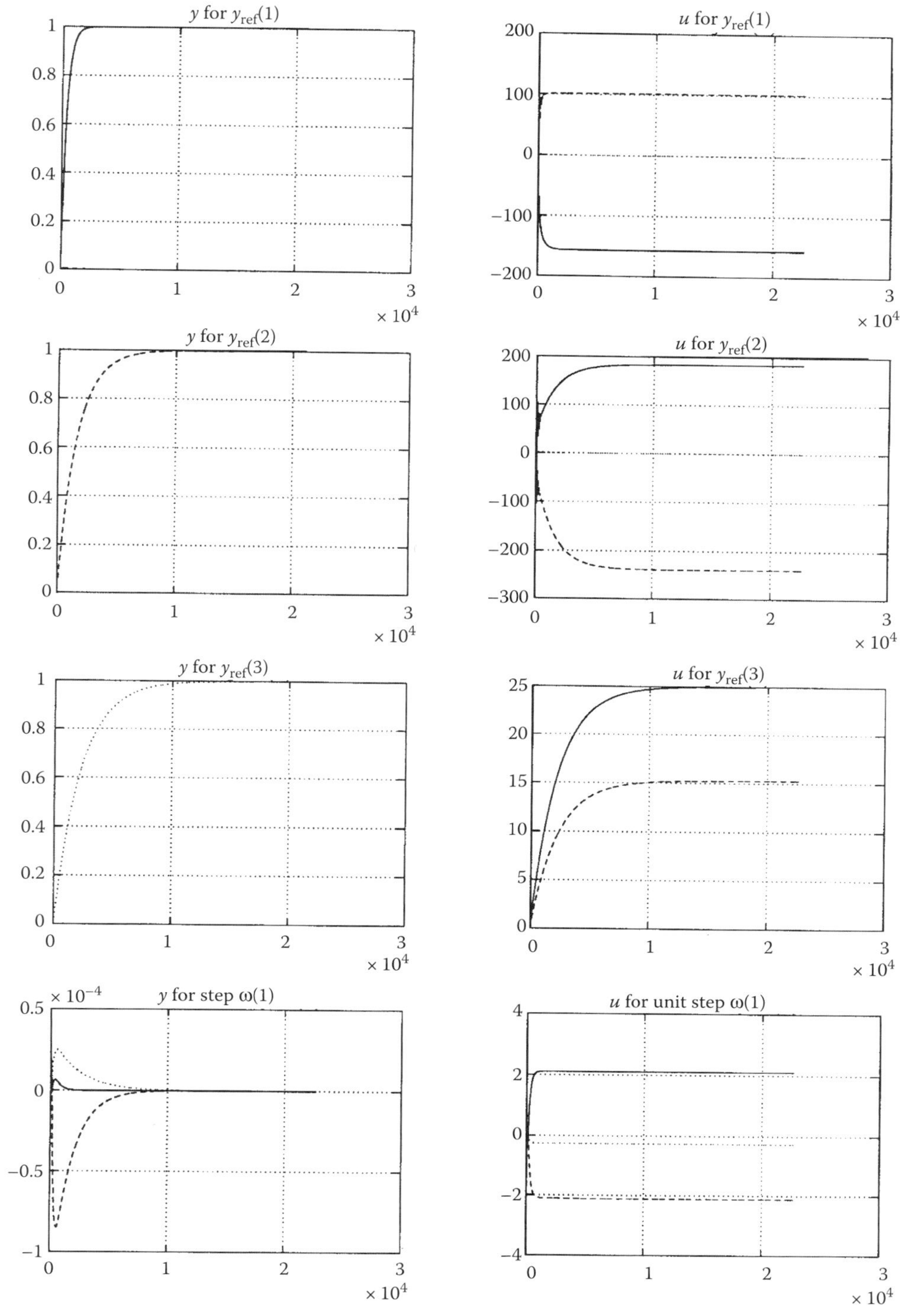

FIGURE 23.15 Response of closed-loop system for distillation column example using decentralized robust servomechanism controller for unit-step in y_{ref} given by $y_{ref}(1) = (1\,0\,0)'$, $y_{ref}(2) = (0\,1\,0)'$, $y_{ref}(3) = (0\,0\,1)'$ and for a unit-step disturbance.

Error coefficient: The steady-state error coefficient matrix associated with a closed-loop system for a given class of disturbance or reference input signals.

Gain margin (GM): Given a stable closed-loop system, the GM (θ, β) refers to the largest perturbation of gain in the system's transfer function matrix that may occur before instability occurs.

High-gain servomechanism controller (HGSC): A controller that gives perfect tracking for continuous minimum-phase systems.

Minimum phase: A system whose transmission zeros are all contained in the open left complex half-plane.

Nonminimum phase: A system that is not minimum phase.

Perfect control: The ability of a controller to provide arbitrarily good transient response in the system.

Perfect robust controller: A controller that solves the RSP such that perfect control occurs.

Robust servomechanism problem (RSP): The problem of finding a controller to solve the servomechanism problem that has the property of providing exact asymptotic error regulation, independent of any perturbations in the plant that do not destabilize the system.

Servomechanism problem: The problem of finding a controller to provide asymptotic error regulation and tracking for a system, subject to a specified class of disturbances and tracking signals.

Servo-compensator: A compensator that is used in the construction of a controller to solve the RSP.

Stabilizing compensator: A controller that stabilizes a system.

Transmission zero: A generalization of the notion of a zero of a single-input/single-output system to multivariable systems.

References

1. Davison, E.J., The feedforward control of linear time invariant multivariable systems, *Automatica,* 9(5), 561–573, 1973.
2. Davison, E.J. and Wang, S.H., Properties of linear time invariant multivariable systems subject to arbitrary output and state feedback, *IEEE Trans. Autom. Control,* 18, 24–32, 1973.
3. Davison, E.J. and Wang, S.H., Properties and calculation of transmission zeros of linear multivariable time-invariant systems, *Automatica,* 10, 643–658, 1974.
4. Goldenberg, A. and Davison, E.J., The feedforward and robust control of a general servomechanism problem with time lag, *8th Annu. Princeton Conf. Inf. Sci. Syst.,* 80–84, 1974.
5. Davison, E.J. and Goldenberg, A., The robust control of a general servomechanism problem: The servo compensator, *Automatica,* 11, 461–471, 1975.
6. Davison, E.J., The robust control of a servomechanism problem for linear time-invariant multivariable systems, *IEEE Trans. Autom. Control,* 21, 25–34, 1976.
7. Davison, E.J., Multivariable tuning regulators: The feedforward and robust control of a general servomechanism problem, *IEEE Trans. Autom. Control,* 21, 35–47, 1976.
8. Davison, E.J., The robust decentralized control of a general servomechanism problem, *IEEE Trans. Autom. Control,* 21, 14–24, 1976.
9. Davison, E.J. and Ferguson, I.J., The design of controllers for the multivariable robust servomechanism problem using parameter optimization methods, *IEEE Trans. Autom. Control,* 26, 93–110, 1981.
10. Davison, E.J. and Copeland, B., Gain margin and time lag tolerance constraints applied to the stabilization problem and robust servomechanism problem, *IEEE Trans. Autom. Control,* 30, 229–239, 1985.
11. Davison, E.J. and Chang, T., Decentralized controller design using parameter optimization methods, *Control: Theor. Adv. Technol.* 2, 131–154, 1986.
12. Davison, E.J. and Scherzinger, B., Perfect control of the robust servomechanism problem, *IEEE Trans. Autom. Control,* 32(8), 689–702, 1987.
13. Qiu, Li and Davison, E.J., Performance limitations of non-minimum phase systems in the servomechanism problem, *Automatica,* 29(2), 337–349, 1993.
14. Zhang, H. and Davison, E.J., A uniform high gain compensator for multivariable systems, *1994 IEEE Control Decision Conf.,* 892–897, 1994.

Further Reading

There are a number of important issues that have not yet been considered in this chapter. For example, when disturbances are measurable, so-called *feedforward control* [1,7] can be highly effective in minimizing the effects of disturbances in the servomechanismproblem.

In many classes of problems, the controller must often be constrained to be decentralized, e.g., in process control systems, power system problems, transportation system problems. A treatment of the so-called *decentralized robust servomechanism problem*, which arises in this case, is given in [8] and [11].

The effect of transportation delay in a system is often of critical importance in the design of controllers to solve the servomechanism problem. A treatment of systems that have *time lag* is given in [4].

Finally, it is often the case that no mathematical model is actually available to describe the plant that is to be controlled. In this case, if the plant is open-loop asymptotically stable, so-called *tuning regulator theory* [7] can be applied to obtain existence conditions and to design a controller, which can then be applied to the plant to solve the servomechanism problem.

The above treatment of the servomechanism problem has been carried out in a state-space setting; alternative treatments may be found using other settings such as geometric methods, frequency-domain methods, polynomial matrix representation methods, coprime matrix factorization methods, etc.

24

Linear Matrix Inequalities in Control

Carsten Scherer
University of Stuttgart

Siep Weiland
Eindhoven University of Technology

24.1 Introduction

Optimization problems and control problems are highly intertwined. If a control configuration has been decided upon, controller parameters or control input signals can be interpreted as decision variables of an optimization problem that reflects the desired specifications and constraints of the controlled system.

In recent years, linear matrix inequalities (LMIs) have emerged as a powerful tool for approaching control problems that appear difficult if not impossible to solve in an analytic fashion. Although the history of LMIs goes back to the 1940s, with a major emphasis on their role in control in the 1960s through the work of Kalman, Yakubovich, Popov, and Willems, only during the last decades have powerful numerical interior point techniques been developed to solve LMIs in a practical and efficient manner (Nesterov, Nemirovskii). Today, several commercial and noncommercial software packages are available that allow for simple coding of general LMI problems. For example, Yalmip [12] is a very flexible and noncommercial toolbox for defining and solving advanced optimization problems.

Boosted by the availability of fast LMI solvers, research in robust control theory has experienced a significant paradigm shift. Instead of arriving at an analytical solution of an optimal control problem and implementing such a solution in software so as to synthesize optimal controllers, today a substantial body of research is devoted to reformulating a control problem to the question of whether a specific LMI is solvable or, alternatively, to optimizing functionals over LMI constraints. It is the purpose of this chapter to highlight the main role and use of LMIs in solving a large variety of control and estimation problems.

24.1.1 Notation

$\mathbb{R}$ and $\mathbb{C}$ denote the fields of real and complex numbers. Sets of real and complex matrices of dimension $m \times n$ are denoted by $\mathbb{R}^{m\times n}$ and $\mathbb{C}^{m\times n}$. A matrix $A \in \mathbb{C}^{m\times n}$ is Hermitian if it is square ($m = n$) and if $A = A^*$, where the star denotes taking complex conjugate transpose. For real matrices, this amounts to saying that $A = A^\top$ in which case A is said to be symmetric. The vector spaces of all $n \times n$ Hermitian and symmetric matrices will be denoted by $\mathbb{H}^n$ and $\mathbb{S}^n$, respectively, and we will omit superscript n if the dimension is not relevant for the context. A Hermitian or symmetric matrix A is called negative definite, negative semidefinite, positive definite, or positive semidefinite if $x^*Ax < 0, \leq 0, > 0$ or ≥ 0 for all nonzero complex vectors $x \in \mathbb{C}^n$. We will denote this by $A \prec 0$, $A \preccurlyeq 0$, $A \succ 0$, and $A \succcurlyeq 0$, and extend this notation to expressions $A \preccurlyeq B$ to mean that $A - B \preccurlyeq 0$ for any $A, B \in \mathbb{H}^n$. A congruence transformation of a square matrix M is a mapping $M \mapsto T^\top MT$ with some nonsingular T. The operator col$(\cdot)$ stacks its arguments in a vector, as in $\mathrm{col}(a, b) = \left(\begin{smallmatrix} a \\ b \end{smallmatrix}\right)$, where a and b are vectors or matrices with the same number of columns. $\mathscr{L}_2{}^n$ denotes the space of all signals $x : [0, \infty) \to \mathbb{R}^n$ with finite energy $\|x\|_{\mathscr{L}_2} := \sqrt{\int_0^\infty \|x(t)\|^2\,\mathrm{d}t}$, where $\|.\|$ is the Euclidean vector norm. We refer to the appendix for a brief summary on notions of convex sets and convex functions.

24.2 LMIs and Convexity

An LMI is a constraint of the form

$$F(x) := F_0 + x_1F_1 + \cdots + x_nF_n \prec 0, \tag{24.1}$$

where $F_0, F_1, \ldots, F_n$ are given real symmetric matrices and where $x = \mathrm{col}(x_1, \ldots, x_n)$ is a vector of unknown real scalar *decision variables.*

The inequality $F(x) \prec 0$ means that x should render the symmetric matrix $F(x)$ negative definite, that is, the maximum eigenvalue of $F(x)$ should be negative. The LMI (Equation 24.1) gives rise to two types of questions:

1. The *LMI feasibility problem* amounts to testing whether there exist real variables $x_1, \ldots, x_n$ such that Equation 24.1 holds.
2. The *LMI optimization problem* amounts to minimizing a cost function $c(x) = c_1x_1 + \cdots + c_nx_n$ over all $x_1, \ldots, x_n$ that satisfy the constraint (Equation 24.1).

Classical linear programs easily fit in this formalism, but also quadratic programs and some instances of convex quadratically constrained quadratic programs can be reformulated in this setting. In fact, the LMI optimization problem is a natural generalization of a linear program in which inequalities are defined by the cone of positive definite matrices.

In most control applications, LMIs arise with matrix variables rather than vector variables. This means that we consider inequalities of the more general form

$$F(X) \prec 0,$$

in which X is a matrix that belongs to an arbitrary finite-dimensional vector space $\mathcal{X}$ of matrices and where $F : \mathcal{X} \to \mathbb{S}^m$ is an *affine function*. We recall that affine functions assume the form $F(x) = F_0 + T(x)$, where F_0 is fixed and where T is a linear map. Affine functions are, therefore, linear mappings plus some offset. For this reason, an LMI is actually better called an affine matrix inequality, but the world has decided to accept this erroneous nomenclature.

Example 24.1

A simple example of an LMI in the matrix valued unknown $X = X^\top$ is

$$F(X) = A^\top X + XA + Q \prec 0,$$

where A and $Q = Q^\top$ are given square real matrices. This is a special case of Equation 24.1 by expanding X as $X = \sum_{k=1}^n x_k X_k$, where $X_k = X_k^\top$, $k = 1, \ldots, n$, is a basis of the set of symmetric matrices $\mathcal{X}$. Indeed, F is affine and $F(X) = F(\sum_{k=1}^n x_k X_k) = Q + \sum_{k=1}^n x_k(A^\top X_k + X_k A)$ which is of the form as in Equation 24.1 with $F_0 = Q$ and $F_k = A^\top X_k + X_k A$.

Of course, one is led to wonder what the practical interest in studying constraints of the special form as in Equation 24.1 might be. There are a number of answers to this question. First, Equation 24.1 defines a *convex constraint* on the decision variable. This means that the solution set $\mathcal{S} := \{x \mid F(x) \prec 0\}$ of the LMI $F(x) \prec 0$ is convex. We refer the reader to the appendix for a brief overview of notions and results on convex sets and convex functions. Although the convex constraint $F(x) \prec 0$ on x may seem rather special, many convex sets can be represented in this way and, in fact, have more attractive properties than general convex sets. Second, the solution set of every finite set of LMIs

$$F_1(x) \prec 0, \ldots, F_k(x) \prec 0$$

can again be represented as one single LMI $F(x) \prec 0$ by setting $F(x) = \text{diag}(F_1(x), \ldots, F_k(x))$. Hence, multiple LMI constraints on a decision variable x is again an LMI constraint. Third, and very importantly, the partitioned LMI

$$F(x) = \begin{pmatrix} F_{11}(x) & F_{12}(x) \\ F_{21}(x) & F_{22}(x) \end{pmatrix} \prec 0$$

is equivalent to

$$\begin{cases} F_{11}(x) \prec 0 \\ S_{22}(x) := F_{22}(x) - F_{21}(x)F_{11}(x)^{-1}F_{12}(x) \prec 0 \end{cases}$$

and at the same time equivalent to

$$\begin{cases} S_{11}(x) := F_{11}(x) - F_{12}(x)F_{22}(x)^{-1}F_{21}(x) \prec 0. \\ F_{22}(x) \prec 0 \end{cases}$$

Since F is affine, this equivalence means that also specific types of quadratic and rational inequalities can be reformulated as Equation 24.1. The expressions $S_{11}(x)$ and $S_{22}(x)$ are called *Schur complements* of $F_{22}(x)$ and $F_{11}(x)$ in $F(x)$. The above equivalences follow from the fact that congruence transformations of symmetric matrices leave the number of positive and negative eigenvalues invariant. In particular, $M \prec 0$, if and only if $T^\top MT \prec 0$ for any nonsingular matrix T. For example, the first follows by taking $M = F(x)$ and $T = \left(\begin{smallmatrix} I & -F_{11}(x)^{-1}F_{12}(x) \\ 0 & I \end{smallmatrix}\right)$.

24.3 Numerical Solutions of LMIs

Optimization problems over symmetric semidefinite matrix constraints belong to the realm of semidefinite programming or semidefinite optimization. Although we mainly focus on control problems here, semidefinite programs occur in combinatorial optimization, polynomial optimization, topology optimization, and so on. In the last few decades, this research field has witnessed incredible breakthroughs in numerical tools, commercial and noncommercial software developments, and fast solution algorithms. In particular, the introduction of powerful interior point methods allow us to effectively decide about the feasibility of semidefinite programs and to determine their solutions.

This section aims to indicate briefly the main ideas behind interior point solvers of convex optimization programs. Let the solution set $\mathcal{S} := \{x \in \mathbb{R}^n \mid F(x) \prec 0\}$ of the LMI (Equation 24.1) be the domain of a convex function $f : \mathcal{S} \to \mathbb{R}$ which we wish to minimize. That is, we consider the convex optimization problem

$$v_{\text{opt}} = \inf_{x \in \mathcal{S}} f(x). \tag{24.2}$$

The idea behind interior point methods is to solve this constrained optimization problem by a sequence of unconstrained optimization problems. For this purpose, a *barrier function* ϕ is introduced. This is a function $\phi : \mathbb{R}^n \to \mathbb{R}$ which is required to

1. be *strictly convex* on the interior of $\mathcal{S}$ and
2. approach $+\infty$ along any sequence of points $\{x_n\}_{n=1}^{\infty}$ in the interior of $\mathcal{S}$ that converges to a boundary point of $\mathcal{S}$.

Given a barrier function ϕ, the constraint optimization problem is replaced by the *unconstrained optimization problem* to minimize the functional

$$f_t(x) := f(x) + t\phi(x), \tag{24.3}$$

where $t > 0$ is a *penalty parameter*. The main idea is to determine a curve $t \mapsto x(t)$ that associates with any $t > 0$ a minimizer $x(t)$ of f_t. The minimum of f_t is attained in the interior of $\mathcal{S}$. Subsequently, the behavior of this mapping is considered as the penalty parameter t decreases to zero. In almost all interior point methods, the unconstrained optimization problem is solved with the classical Newton–Raphson iteration technique to approximately determine a local minimizer of f_t. Since f_t is strictly convex on $\mathbb{R}^n$, every local minimizer of f_t is guaranteed to be the unique global minimizer. Under mild assumptions and for a suitably defined sequence of penalty parameters t_n, $t_n \to 0$ as $n \to \infty$, the sequence $x(t_n)$ will converge to a point x^*. That is, the limit $x^* := \lim_{n\to\infty} x(t_n)$ exists and $v_{\text{opt}} = f(x^*)$. If, in addition, x^* belongs to the interior of $\mathcal{S}$, then it is an optimal solution to Equation 24.2; otherwise an almost optimal solution of Equation 24.2 can be deduced from the sequence $x(t_n)$.

Interior point methods can be applied to either of the two LMI problems defined in the previous section. If we consider the *feasibility problem* associated with the LMI $F(x) \prec 0$, then (f does not play a role and) one candidate barrier function is the logarithmic function

$$\phi(x) := \begin{cases} \log\det(-F(x)^{-1}) & \text{if } x \in \mathcal{S}, \\ \infty & \text{otherwise.} \end{cases}$$

If $\mathcal{S}$ is bounded and nonempty, ϕ will be strictly convex. This implies the existence of a unique x_{opt} such that $\phi(x_{\text{opt}})$ is the global minimum of ϕ. The point x_{opt} belongs to $\mathcal{S}$ and is called the *analytic center* of the feasibility set $\mathcal{S}$.

The *LMI optimization problem* to minimize $c(x)$ subject to the LMI $F(x) \prec 0$ can be viewed as a feasibility problem for the LMI

$$G_t(x) := \begin{pmatrix} c(x) - t & 0 \\ 0 & F(x) \end{pmatrix} \prec 0,$$

where $t > t_0 := \inf_{x \in \mathcal{S}} c(x)$ is a penalty parameter. Using the same barrier function yields the unconstrained optimization problem to minimize

$$g_t(x) := \log\det(-G_t(x)^{-1}) = \log \frac{1}{t - c(x)} + \log\det(-F(x)^{-1})$$

for a sequence of decreasing positive values of t. Due to the strict convexity of g_t the minimizer $x(t)$ of g_t is unique for all $t > t_0$. Since closed-form expressions for the gradient and Hessian of g_t can be obtained, a Newton iteration is an efficient numerical method to find minimizers of g_t. Currently, much research is devoted to further exploiting the structure of LMIs in order to tailor dedicated solvers for specific semidefinite programs.

24.4 Stability Characterizations with LMIs

Around 1890, Aleksandr Mikhailovich Lyapunov made a systematic study of the local expansion and contraction properties of motions of dynamical systems around an attractor. He worked out the idea that an invariant set of a differential equation attracts all nearby solutions if one can find a function that is bounded from below and decreases along all solutions outside the invariant set. Such functions are called *Lyapunov functions* and they have been used to prove stability of equilibria of differential equations ever since.

Consider the differential equation

$$\dot{x}(t) = f(x(t), t) \tag{24.4}$$

with finite-dimensional state space $X = \mathbb{R}^n$ and where $f : X \times \mathbb{R} \to X$ is assumed to be sufficiently smooth so as to guarantee existence and uniqueness of the solution $x(t, t_0, x_0)$ that satisfies the initial condition $x(t_0, t_0, x_0) = x_0 \in X$. The differential equation (24.4) is *time-invariant*, if solutions satisfy

$$x(t + \tau, t_0 + \tau, x_0) = x(t, t_0, x_0)$$

for any $\tau \in \mathbb{R}$ with t and $t + \tau$ in the interval of existence. A point $x^* \in X$ is called an *equilibrium* or *fixed point* of the differential equation if $x(t, t_0, x^*) = x^*$ satisfies Equation 24.4 (which implies that $f(x^*, t) = 0$ for all $t \geq t_0$). A wealth of stability concepts associated with fixed points of differential equations exists. Here, we focus on just one.

The equilibrium point x^* of Equation 24.4 is called *exponentially stable* if for all $t_0 \in \mathbb{R}$, positive numbers α, β, and δ exist (all possibly depending on t_0) such that

$$\|x_0 - x^*\| \leq \delta \implies \|x(t, t_0, x_0) - x^*\| \leq \beta \|x_0 - x^*\| e^{-\alpha(t - t_0)} \quad \text{for all } t \geq t_0. \tag{24.5}$$

If α, β, and δ do not depend on t_0, then x^* is said to be *uniformly exponentially stable.* If δ is arbitrary, x^* is called *globally exponentially stable.*

Hence, a fixed point is exponentially stable if all solutions of the differential equation that initiate nearby x^* converge to x^* with an exponential rate $\alpha > 0$. The following result is standard and relates exponential stability of linear time-invariant differential equations to LMI feasibility.

Theorem 24.1:

Let $A \in \mathbb{R}^{n \times n}$. The following statements are equivalent.

1. The origin is an exponentially stable equilibrium point of $\dot{x} = Ax$.
2. All eigenvalues $\lambda(A)$ of A belong to $\mathbb{C}^- := \{s \in \mathbb{C} \mid \Re(s) < 0\}$ (i.e., A is *Hurwitz*).
3. The LMIs $A^\top X + XA \prec 0$ and $X \succ 0$ are feasible.

Any solution X of the LMIs in item (3) defines the quadratic function $V(x) := x^\top X x$ that serves as a Lyapunov function for the equilibrium point $x^* = 0$ of the differential equation $\dot{x} = Ax$. Indeed, V achieves its minimum at $x^* = 0$ and its derivative in the direction of the vector field Ax is

$$\frac{\mathrm{d}}{\mathrm{d}t} x(t)^\top X x(t) = \dot{x}(t)^\top X x(t) + x(t)^\top X \dot{x}(t) = x(t)^\top [A^\top X + XA] x(t),$$

and hence nonincreasing by item (3). Rather straightforward arguments lead to Equation 24.5, where δ is arbitrary, $\beta = \sqrt{\lambda_{\max}(X)/\lambda_{\min}(X)}$, and $\alpha > 0$ is any number for which $A^\top X + XA + 2\alpha X \prec 0$.

For many applications in control and engineering one may be interested in characterizing eigenvalue locations of A in more general stability regions than $\mathbb{C}^-$.

For a real symmetric matrix $P \in \mathbb{S}^{2m}$, the set of complex numbers

$$\mathcal{L}_P := \left\{ s \in \mathbb{C} \mid \begin{pmatrix} I \\ sI \end{pmatrix}^* P \begin{pmatrix} I \\ sI \end{pmatrix} \prec 0 \right\}$$

is called an *LMI region*.

Important stability regions such as half-planes $\mathbb{C}_{\text{stab }1} := \{s \mid \Re(s) < \alpha\}$, circles $\mathbb{C}_{\text{stab }2} := \{s \mid |s| < r\}$, or conic sectors $\mathbb{C}_{\text{stab }3} := \{s \mid \Re(s)\tan(\theta) < |\Im(s)|\}$ can be represented by LMI regions $\mathcal{L}_{P_1}, \mathcal{L}_{P_2}$, and $\mathcal{L}_{P_3}$, respectively, by taking

$$P_1 = \begin{pmatrix} -2\alpha & 1 \\ 1 & 0 \end{pmatrix}, \quad P_2 = \begin{pmatrix} -r^2 & 0 \\ 0 & 1 \end{pmatrix}, \quad P_3 = \begin{pmatrix} 0 & 0 & \sin(\theta) & \cos(\theta) \\ 0 & 0 & -\cos(\theta) & \sin(\theta) \\ \sin(\theta) & -\cos(\theta) & 0 & 0 \\ \cos(\theta) & \sin(\theta) & 0 & 0 \end{pmatrix}.$$

LMI regions include sets bounded by circles, ellipses, strips, parabolas, and hyperbolas. Since any finite intersection of LMI regions is again an LMI region, one can virtually approximate any convex region in $\mathbb{C}$ as long as it is symmetric with respect to the real axis.

To present the main result of this section, we recall the definition of the *Kronecker product* $A \otimes B$ of two matrices $A \in \mathbb{C}^{m\times n}$, $B \in \mathbb{C}^{k\times \ell}$, which is the $mk \times n\ell$ matrix

$$A \otimes B = \begin{pmatrix} A_{11}B & \dots & A_{1n}B \\ \vdots & & \vdots \\ A_{m1}B & \dots & A_{mn}B \end{pmatrix}.$$

The following result as originating from [4] is an interesting and elegant generalization of the stability characterization in Theorem 24.1.

Theorem 24.2:

Let $A \in \mathbb{R}^{n\times n}$. The following statements are equivalent.

1. *All eigenvalues of A are contained in the LMI region*

$$\left\{ s \in \mathbb{C} \mid \begin{pmatrix} I \\ sI \end{pmatrix}^* \begin{pmatrix} Q & S \\ S^\top & R \end{pmatrix} \begin{pmatrix} I \\ sI \end{pmatrix} \prec 0 \right\}.$$

2. *There exists* $X = X^\top$ *such that*

$$X \succ 0 \text{ and } \begin{pmatrix} I \\ A \otimes I \end{pmatrix}^* \begin{pmatrix} X \otimes Q & X \otimes S \\ X \otimes S^\top & X \otimes R \end{pmatrix} \begin{pmatrix} I \\ A \otimes I \end{pmatrix} \prec 0.$$

The condition in item (2) is an LMI in X. The result of Theorem 24.1 is recovered by taking $Q = 0, S = 1$, and $R = 0$. With $Q = -1, S = 0$, and $R = 1$, the LMI region corresponds to the open unit disc; hence A has all eigenvalues within the open unit disc iff there exists $X \succ 0$ such that $A^\top XA - X \prec 0$. In turn, this LMI test is equivalent to saying that the discrete-time system $x(k+1) = Ax(k)$ is exponentially stable.

Example 24.2

Consider the problem to find a stabilizing feedback law $u = Fx$ that simultaneously stabilizes the systems $\dot{x} = A_k x + B_k u$, where $k = 1, \ldots, 4$ and

$$A_1 = \begin{pmatrix} 1 & 0 \\ 1 & 1 \end{pmatrix}, \quad B_1 = \begin{pmatrix} 1 \\ 0 \end{pmatrix}, \quad A_2 = \begin{pmatrix} -1 & 2 \\ 1 & 2 \end{pmatrix}, \quad B_2 = \begin{pmatrix} 1 \\ 1 \end{pmatrix},$$
$$A_3 = \begin{pmatrix} 0 & 1 \\ 1 & 0 \end{pmatrix}, \quad B_3 = \begin{pmatrix} 0 \\ 1 \end{pmatrix}, \quad A_4 = \begin{pmatrix} 0 & 2 \\ 1 & -1 \end{pmatrix}, \quad B_4 = \begin{pmatrix} 2 \\ 1 \end{pmatrix}.$$

By Theorem 24.1, the equivalent problem is to find $X_k \succ 0$ and F such that $(A_k + B_k F)^\top X_k + X_k(A_k + B_k F) \prec 0$ for $k = 1, \ldots, 4$. Since both X_k and F are unknown, this is *not* an LMI constraint. However, assuming $X_1 = \cdots = X_4 = X$, a congruence transformation with the matrix $Y := X^{-1}$ transforms the five matrix inequalities to

$$Y \succ 0, \quad A_k Y + YA_k^\top + B_k M + M^\top B_k \prec 0, \quad k = 1, \ldots, 4,$$

where we set $M = FY$. These are LMIs in Y and M. When implemented with the given matrices, this set of LMIs turns out to be feasible and any solution defines a feedback $F = MY^{-1}$ that solves the stabilization problem. One of these feedbacks is computed to be $F = (-4.3874 \quad -10.6332)$. An analogous synthesis strategy can be applied for the *simultaneous pole-placement problem* which amounts to finding F such that eigenvalues of $A_k + B_k F$ belong to an LMI region $\mathcal{L}_P$ for all k. Its solution is then an application of Theorem 24.2.

24.5 Performance Characterizations with LMIs

In this section, we consider a linear system

$$\dot{x} = Ax + Bd, \quad e = Cx + Dd, \quad x(0) = 0, \tag{24.6}$$

in which d is viewed as an undesired external disturbance and e is an error output. Many control synthesis problems can be translated into a question of disturbance attenuation: the controller should reduce the effect of the disturbance d onto the error e as much as possible. In this section we quantify or analyze the effect of d onto e for the system (Equation 24.6) whose transfer function matrix is given by

$$T(s) = C(sI - A)^{-1}B + D.$$

24.5.1 $\mathcal{H}_2$-Performance

Let us assume for Equation 24.6 that A is Hurwitz and $D = 0$. If the number of inputs is m then $B = (b_1, \ldots, b_m)$ has m columns. If the system is excited with a unit impulse in the νth input, it responds with the output trajectory $z_\nu(t) = Ce^{At}b_\nu$. The energy of this output trajectory equals

$$\int_0^\infty [Ce^{At}b_\nu]^\top [Ce^{At}b_\nu]\,dt = b_\nu^\top \left(\int_0^\infty e^{A^\top t}C^\top Ce^{At}\,dt\right) b_\nu = b_\nu^\top Y_0 b_\nu \tag{24.7}$$

with Y_0 being the observability Gramian, the unique solution of the Lyapunov equation

$$A^\top Y_0 + Y_0 A + C^\top C = 0. \tag{24.8}$$

If we add the output energies for impulsive inputs in all input components we obtain

$$\sum_{\nu=1}^{m}\int_0^\infty z_\nu(t)^\top z_\nu(t)\,dt = \sum_{\nu=1}^{m}\mathrm{Trace}(b_\nu b_\nu^\top Y_0) = \mathrm{Trace}(BB^\top Y_0) = \mathrm{Trace}(B^\top Y_0 B),$$

where Trace denotes the sum of the diagonal elements of any matrix. In view of the explicit formula for the observability Gramian as used in Equation 24.7 combined with Parseval's theorem, we infer that $\mathrm{Trace}(B^\top Y_0 B)$ actually equals

$$\|T\|_{\mathcal{H}_2}^2 := \frac{1}{2\pi}\,\mathrm{Trace}\int_{-\infty}^{\infty}[C(i\omega I - A)^{-1}B]^*[C(i\omega I - A)^{-1}B]\,d\omega. \tag{24.9}$$

Note that $\|T\|_{\mathcal{H}_2}$ is the so-called $\mathcal{H}_2$-norm of the transfer matrix T, the name of which is motivated by the theory of Hardy-spaces in pure mathematics; this relation is not relevant for our purposes.

We have actually derived the *impulse-response performance interpretation* of the $\mathcal{H}_2$-norm. Moreover, this interpretation also provides a concise link to classical linear quadratic control, since the impulse response $z_\nu(\cdot)$ is identical to the output response if the system's state is initialized as $x(0) = b_\nu$.

Let us briefly touch upon the stochastic interpretation of the $\mathcal{H}_2$-norm. If w is white noise with unity covariance, the asymptotic variance of the output process of Equation 24.6 satisfies

$$\lim_{t\to\infty} E[z(t)^\top z(t)] = \mathrm{Trace}\left(\lim_{t\to\infty} CE[x(t)x(t)^\top]C^\top\right) = \mathrm{Trace}(CX_0C^\top),$$

where X_0 is the system's controllability Gramian, the unique solution of

$$AX_0 + X_0A^\top + BB^\top = 0. \tag{24.10}$$

Since $T(s)^\top = B^\top(sI - A^\top)^{-1}C^\top$, we can conclude that the asymptotic output variance is equal to $\|T^\top\|_{\mathcal{H}_2}$. Due to Equation 24.9 this is the same as $\|T\|_{\mathcal{H}_2}$, which leads to a dual version for computing $\|T\|_{\mathcal{H}_2}$. Let us summarize our findings as follows: *If $D = 0$ and X_0 and Y_0 are the system's controllability and observability Gramians satisfying Equations 24.10 and 24.8, respectively, then $\|T\|_{\mathcal{H}_2}^2 = \mathrm{Trace}(CX_0C^\top) = \mathrm{Trace}(B^\top Y_0 B)$ is the sum of the energies of the output of Equation 24.6 for impulsive inputs in each input channel, and it also equals the asymptotic output variance if the input is white noise with unity covariance.*

Note that $\mathcal{H}_2$-norms can be easily determined by solving linear equations and computing traces. Since these explicit formulas are inadequate for applying the synthesis procedure as developed in the next section, let us provide a genuine characterization of a bound on the $\mathcal{H}_2$-norm by LMIs. It is important to stress that this formulation actually combines an LMI characterization of system stability with a bound on system performance.

Theorem 24.3:

A is Hurwitz and $\|T\|_{\mathcal{H}_2} < \gamma$ iff $D = 0$ and there exist $X = X^\top$ and $W = W^\top$ with

$$\begin{pmatrix} A^\top X + XA & XB \\ B^\top X & -\gamma I\end{pmatrix} \prec 0, \quad \begin{pmatrix} X & C^\top \\ C & W\end{pmatrix} \succ 0 \quad \text{and} \quad \mathrm{Trace}(W) < \gamma. \tag{24.11}$$

Sketch of proof of "if"

For the general manipulation of performance specifications it is an instructive illustration of how to show that the feasibility of the system of LMIs (Equation 24.11) implies $\|T\|_{\mathscr{H}_2} < \gamma$. Indeed, by taking Schur complements we infer that Equation 24.11 leads to

$$A^\top X + XA + \frac{1}{\gamma} XBB^\top X \prec 0, \quad X \succ 0, \quad CX^{-1}C^\top \prec W, \quad \mathrm{Trace}(W) < \gamma.$$

The first inequality implies $A^\top X + XA \prec 0$. Together with $X \succ 0$ this implies (Theorem 24.1) that A is Hurwitz. On the other hand, $\hat{X}_0 = (\gamma X)^{-1}$ satisfies

$$A\hat{X}_0 + \hat{X}_0 A^\top + BB^\top \prec 0 \quad \text{and} \quad \mathrm{Trace}(C\hat{X}_0 C^\top) < \mathrm{Trace}(\gamma W) < \gamma^2.$$

Combining the latter inequalities with Equation 24.10, we get $A(\hat{X}_0 - X_0) + (\hat{X}_0 - X_0)A^\top \prec 0$, which implies, using the stability of A, that $\hat{X}_0 - X_0 \succ 0$. Consequently, also $\mathrm{Trace}(CX_0C^\top) < \gamma^2$, which yields that $\|T\|_{\mathscr{H}_2} < \gamma$.

24.5.2 $\mathscr{H}_\infty$-Performance

A different way of quantifying the effect of the disturbance d on the output e in Equation 24.6 is in terms of the so-called energy gain

$$\|T\|_{\mathscr{H}_\infty} := \sup_{0<\|d\|_{\mathscr{L}_2}<\infty} \frac{\|e\|_{\mathscr{L}_2}}{\|d\|_{\mathscr{L}_2}}.$$

The norm reflects the worst amplification of disturbances on outputs if measuring the sizes of the input and output signals in terms of their $\mathscr{L}_2$-norm or energy. Dissipativity theory [19] provides a direct path toward an LMI characterization of an upper bound $\|T\|_{\mathscr{H}_\infty} < \gamma$.

Theorem 24.4:

A is Hurwitz and $\|T\|_{\mathscr{H}_\infty} < \gamma$ iff there exists $X \succ 0$ with

$$\begin{pmatrix} I & 0 \\ A & B \end{pmatrix}^\top \begin{pmatrix} 0 & X \\ X & 0 \end{pmatrix} \begin{pmatrix} I & 0 \\ A & B \end{pmatrix} + \begin{pmatrix} 0 & I \\ C & D \end{pmatrix}^\top \begin{pmatrix} -\gamma^2 I & 0 \\ 0 & I \end{pmatrix} \begin{pmatrix} 0 & I \\ C & D \end{pmatrix} \prec 0. \tag{24.12}$$

Sketch of proof of "if"

Let us assume that Equation 24.12 holds. Then the left-upper block of Equation 24.12 just reads as $A^\top X + XA + C^\top C \prec 0$, which implies $A^\top X + XA \prec 0$. Therefore $X \succ 0$ guarantees that A is Hurwitz. Since the inequality Equation 24.12 is strict, it continues to hold if we replace $-\gamma^2$ by $-(\gamma - \epsilon)^2$ for some suitably small $\epsilon > 0$. Let us choose any d with $0 < \|d\|_{\mathscr{L}_2} < \infty$ and let e be the output of Equation 24.6. If we right-multiply the perturbed version of Equation 24.12 with $\mathrm{col}(x(t), d(t))$ and left-multiply with its transpose, and if we exploit the relations in Equation 24.6, we obtain

$$\begin{aligned} &\begin{pmatrix} x(t) \\ \dot{x}(t) \end{pmatrix}^\top \begin{pmatrix} 0 & X \\ X & 0 \end{pmatrix} \begin{pmatrix} x(t) \\ \dot{x}(t) \end{pmatrix} + \begin{pmatrix} d(t) \\ e(t) \end{pmatrix}^\top \begin{pmatrix} -(\gamma-\epsilon)^2 I & 0 \\ 0 & I \end{pmatrix} \begin{pmatrix} d(t) \\ e(t) \end{pmatrix} \\ &= \frac{\mathrm{d}}{\mathrm{d}t} x(t)^\top X x(t) + e(t)^\top e(t) - (\gamma-\epsilon)^2 d(t)^\top d(t) \le 0. \end{aligned}$$

By integration on $[0, T]$ and with $x(0) = 0$ we infer

$$x(T)^\top X x(T) + \int_0^T e(t)^\top e(t)\,\mathrm{d}t \le (\gamma-\epsilon)^2 \int_0^T d(t)^\top d(t)\,\mathrm{d}t \quad \text{for all } T \ge 0.$$

Since both $x(\cdot)$ and $\dot{x}(\cdot)$ are of finite energy, $x(T) \to 0$ for $T \to \infty$. After taking the limit $T \to \infty$ we hence obtain $\|e\|^2_{\mathscr{L}_2} \le (\gamma - \epsilon)^2 \|d\|^2_{\mathscr{L}_2}$ or $\|e\|_{\mathscr{L}_2}/\|d\|_{\mathscr{L}_2} \le \gamma - \epsilon$. Since this holds for all $0 < \|d\|_{\mathscr{L}_2} < \infty$ we finally arrive at $\|T\|_{\mathscr{H}_\infty} < \gamma$.

We started from the formulation of the LMIs (Equation 24.12), which gives the most insight for a dissipation-based proof of these results. In the literature, more common equivalent representations of the LMI (Equation 24.12) are

$$\begin{pmatrix} A^\top X + XA + C^\top C & XB + C^\top D \\ B^\top X + D^\top C & D^\top D - \gamma^2 I \end{pmatrix} \prec 0 \quad \text{or (Schur)} \quad \begin{pmatrix} A^\top X + XA & XB & C^\top \\ B^\top X & -\gamma^2 I & D^\top \\ C & D & -I \end{pmatrix} \prec 0. \tag{24.13}$$

Equivalently, $A^\top X + XA + C^\top C + (XB + C^\top D)(\gamma^2 I - D^\top D)^{-1}(B^\top X + D^\top C) \prec 0$, and $D^\top D - \gamma^2 I \prec 0$, which touches upon the relation to Riccati inequalities and equations.

24.5.3 The Kalman–Yakubovich–Popov (KYP) Lemma

We have established the link of Equation 24.12 to the time-domain energy-gain by dissipation arguments. The relation of this LMI to the frequency-domain is the subject of the celebrated KYP lemma. Algebraic arguments proceed as follows. If Equation 24.12 holds with $X = X^\top$, then $A^\top X + XA \prec 0$ implies that A has no eigenvalues on the imaginary axis. For $\omega \in \mathbb{R}$ let us observe that

$$\begin{aligned}
&\begin{pmatrix} (i\omega I - A)^{-1}B \\ I \end{pmatrix}^* \begin{pmatrix} I & 0 \\ A & B \end{pmatrix} \begin{pmatrix} 0 & X \\ X & 0 \end{pmatrix} \begin{pmatrix} I & 0 \\ A & B \end{pmatrix} \begin{pmatrix} (i\omega I - A)^{-1}B \\ I \end{pmatrix} \\
&= [(i\omega I - A)^{-1}B]^* \begin{pmatrix} I \\ i\omega I \end{pmatrix}^* \begin{pmatrix} 0 & X \\ X & 0 \end{pmatrix} \begin{pmatrix} I \\ i\omega I \end{pmatrix} (i\omega I - A)^{-1}B \\
&= [(i\omega I - A)^{-1}B]^* [i\omega X - i\omega X](i\omega I - A)^{-1}B = 0.
\end{aligned}$$

Therefore, feasibility of Equation 24.12 implies the validity of the frequency-domain inequality

$$\begin{pmatrix} I \\ T(i\omega) \end{pmatrix}^* \begin{pmatrix} -\gamma^2 I & 0 \\ 0 & I \end{pmatrix} \begin{pmatrix} I \\ T(i\omega) \end{pmatrix} \prec 0 \quad \text{for all } \omega \in \mathbb{R} \cup \{\infty\},$$

which follows for $\omega = \infty$ from the right-lower block of Equation 24.12. Note that this inequality translates into $T(i\omega)^* T(i\omega) - \gamma^2 I \prec 0$ or $\sigma_{\max}(T(i\omega)) < \gamma$ for all $\omega \in \mathbb{R} \cup \{\infty\}$. In turn, this reveals the relation of the $\mathscr{L}_2$-gain bound to the peak of the largest singular value of the system's frequency response, which is simply the classical $\mathscr{L}_\infty$-norm for transfer matrices without poles on the imaginary axis, or the $\mathscr{H}_\infty$-norm for transfer matrices whose poles are all contained in the open left-half-plane.

The following result captures a generalization of the strict version of the KYP lemma for an arbitrary symmetric matrix P and without involving any sign-constraint on X.

Theorem 24.5:

In the case that A has no eigenvalues on the imaginary axis then

$$\begin{pmatrix} I \\ T(i\omega) \end{pmatrix}^* P \begin{pmatrix} I \\ T(i\omega) \end{pmatrix} \prec 0 \quad \textit{for all } \omega \in \mathbb{R} \cup \{\infty\}, \tag{24.14}$$

iff there exists some $X = X^\top$ *with*

$$\begin{pmatrix} I & 0 \\ A & B \end{pmatrix}^\top \begin{pmatrix} 0 & X \\ X & 0 \end{pmatrix} \begin{pmatrix} I & 0 \\ A & B \end{pmatrix} + \begin{pmatrix} 0 & I \\ C & D \end{pmatrix}^\top P \begin{pmatrix} 0 & I \\ C & D \end{pmatrix} \prec 0. \tag{24.15}$$

The survey article [10] provides a nice historical account of the development around the KYP lemma with many references, also to the Russian literature, and discusses further versions of this result even without any *a priori* hypotheses. In particular [2] is a rich source for the link of the KYP lemma to semidefinite programming duality and the related control theoretic interpretations.

24.5.4 Variants

The books [3,7] contain a whole variety of concrete variations on the theme of formulating performance specifications with LMIs. Let us provide a sample.

24.5.4.1 Generalized $\mathscr{H}_2$-Performance

If replacing $\text{Trace}(W) < \gamma$ by $W \prec \gamma I$ in Theorem 24.3, the formulated LMIs characterize that the system gain from finite energy input signals to the peak value of the output signal is bounded by γ:

$$\sup_{0<\|d\|_{\mathscr{L}_2}<\infty} \frac{\|e\|_{\mathscr{L}_\infty}}{\|d\|_{\mathscr{L}_2}} < \gamma \quad \text{with } \|e\|_{\mathscr{L}_\infty} := \sup_{t\geq 0} \|e(t)\|.$$

This criterion allows to impose time-uniform bounds on the error variable under the assumption that the disturbance has bounded energy.

24.5.4.2 Quadratic Performance

Our discussion of $\mathscr{L}_2$-gain performance opens the path toward the following generalization. With a symmetric weighting matrix P, quadratic performance with index P is achieved if an $\epsilon > 0$ exists, such that the following integral quadratic constraint (IQC) on the performance channel $d \to e$ of the stable system in Equation 24.6 is satisfied:

$$\int_0^\infty \begin{pmatrix} d(t) \\ e(t) \end{pmatrix}^\top P \begin{pmatrix} d(t) \\ e(t) \end{pmatrix} \, dt \leq -\epsilon \|d\|^2_{\mathscr{L}_2} \quad \text{for all } d \in \mathscr{L}_2.$$

This time-domain specification directly translates into the frequency-domain inequality in Equation 24.14 and, due to Theorem 24.5, into feasibility of the LMI in Equation 24.15. If the right-lower block of P is positive semidefinite, it is elementary to see that stability of A is guaranteed by imposing the additional positivity constraint $X \succ 0$.

24.5.4.3 Discrete-Time

All described results have counterparts for discrete-time systems

$$x(t+1) = Ax(t) + Bd(t), \quad e(t) = Cx(t) + Dd(t), \quad x(0) = 0, \ \ t = 0, 1, 2, \ldots.$$

If A has all its eigenvalues in the unit disc (Schur stability), then quadratic performance holds, by definition, if there exists some $\epsilon > 0$ with

$$\sum_{t=0}^{\infty} \begin{pmatrix} d(t) \\ e(t) \end{pmatrix}^\top P \begin{pmatrix} d(t) \\ e(t) \end{pmatrix} \leq -\epsilon \sum_{t=0}^{\infty} d(t)^\top d(t).$$

This is equivalent to the frequency-domain inequality

$$\begin{pmatrix} I \\ T(z) \end{pmatrix}^* P \begin{pmatrix} I \\ T(z) \end{pmatrix} \prec 0 \quad \text{for all } z \in \mathbb{C}, \ |z| = 1$$

on the unit circle, which is, in turn, equivalent to the existence of $X = X^\top$ satisfying the LMI

$$\begin{pmatrix} I & 0 \\ A & B \end{pmatrix}^\top \begin{pmatrix} -X & 0 \\ 0 & X \end{pmatrix} \begin{pmatrix} I & 0 \\ A & B \end{pmatrix} + \begin{pmatrix} 0 & I \\ C & D \end{pmatrix}^\top P \begin{pmatrix} 0 & I \\ C & D \end{pmatrix} \prec 0.$$

If the right-lower block of P is positive semidefinite, Schur stability of A is once again guaranteed by $X \succ 0$. We stress the pleasing parallel structure of continuous-time and discrete-time performance formulations, which is even more striking for synthesis since the proposed general procedure applies to both domains without the need for any adaptation.

24.6 Optimal Performance Synthesis

Let us now consider a system

$$\begin{aligned} \dot{x} &= Ax + B_0 d + Bu, \\ e &= C_0 x + D_0 d + Eu, \\ y &= Cx + Fd, \end{aligned} \tag{24.16}$$

where, in addition to the disturbance d and the error e, the signal u is a control input and y is a measured output. In feedback synthesis, the goal is to determine a controller

$$\begin{aligned} \dot{x}_c &= A_c x_c + B_c y, \\ u &= C_c x_c + D_c y, \end{aligned} \tag{24.17}$$

which feeds the measurements y back to u such that the controlled system, the interconnection of Equations 24.16 and 24.17, is internally stable and satisfies a desired performance property (Figure 24.1).

Note that the controlled system with state $\xi = \text{col}(x, x_c)$ is easily seen to be described by

$$\begin{aligned} \dot{\xi} &= \mathcal{A}\xi + \mathcal{B}d \\ e &= \mathcal{C}\xi + \mathcal{D}d \end{aligned} \quad \text{with} \quad \left(\begin{array}{c|c} \mathcal{A} & \mathcal{B} \\ \hline \mathcal{C} & \mathcal{D} \end{array}\right) := \left(\begin{array}{cc|c} A + BD_cC & BC_c & B_0 + BD_cF \\ B_cC & A_c & B_cF \\ \hline C_0 + ED_cC & EC_c & D_0 + ED_cF \end{array}\right). \tag{24.18}$$

In view of the generalized plant framework [18,20], it is essential to understand that this innocent problem formulation comprises surprisingly many specific configurations as they are needed in one- or two-degrees of freedom controller synthesis for reference tracking and disturbance attenuation. In this section, emphasis is put on an $\mathcal{H}_\infty$-norm bound as a measure of performance.

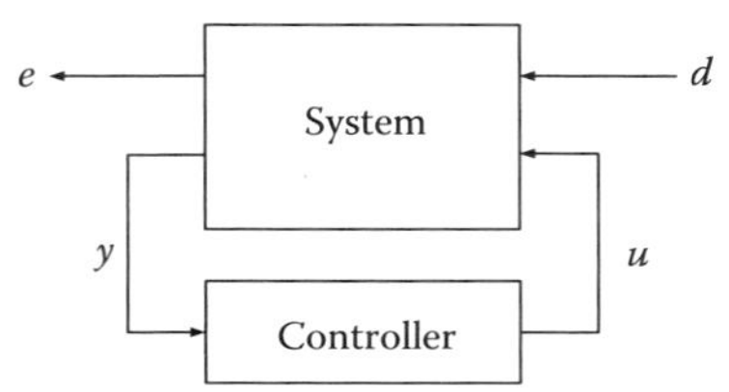

FIGURE 24.1 Generalized plant configuration.

24.6.1 State-Feedback Synthesis

The simplest control law is state-feedback $u = D_c x$ with a gain D_c, which leads to the controlled system

$$\begin{aligned} \dot{x} &= (A + BD_c)x + B_0 d, \\ e &= (C_0 + ED_c)x + D_0 d. \end{aligned} \tag{24.19}$$

Using Theorem 24.4 and with the LMI expressed as in Equation 24.13, this controller stabilizes the system in Equation 24.19 and renders the $\mathscr{H}_\infty$-norm of the transfer matrix $d \to e$ smaller than γ iff there exists some X with

$$X \succ 0 \quad \text{and} \quad \begin{pmatrix} (A+BD_c)^\top X + X(A+BD_c) & XB_0 & (C_0+ED_c)^\top \\ B_0^\top X & -\gamma^2 I & D_0^\top \\ (C_0+ED_c) & D_0 & -I \end{pmatrix} \prec 0. \tag{24.20}$$

Recall that the first inequality guarantees stability while the second captures performance. We observe that, in synthesis, we need to search for *both* D_c and X in order to satisfy Equation 24.20. The performance inequality is, with some abuse of notation, a so-called bilinear matrix inequality problem since the left-hand side is affine in X (for fixed D_c) and affine in D_c (for fixed X). Actually, a large variety of design problems in control can be easily seen to admit this structure. Unfortunately, however, bilinear matrix inequalities are as hard to handle as general nonlinear programs. Fortunately, for the problem at hand there is a surprisingly simple remedy.

There exists a celebrated procedure that actually turns the nonconvex bilinear matrix inequality (Equation 24.20) into a convex problem of LMIs. This is achieved by applying a nonlinear change of variables $(D_c, X) \to (M, Y)$ and a congruence transformation to Equation 24.20. Indeed if we transform Equation 24.20 by congruence with X^{-1} and $\text{diag}(X^{-1}, I, I)$, respectively, and if we introduce the new variables

$$Y := X^{-1} \quad \text{and} \quad M := D_c X^{-1}, \tag{24.21}$$

we arrive at

$$Y \succ 0 \quad \text{and} \quad \begin{pmatrix} (AY+BM)^\top + (AY+BM) & B_0 & (C_0Y+EM)^\top \\ B_0^\top & -\gamma^2 I & D_0^\top \\ (C_0Y+EM) & D_0 & -I \end{pmatrix} \prec 0. \tag{24.22}$$

Obviously the constraints in Equation 24.22 constitute LMIs in (M, Y) whose feasibility can be verified. If (M, Y) is a solution of Equation 24.22, we can solve Equation 24.21 for (D_c, X) as $X = Y^{-1}$ and $D_c = MY^{-1}$ and perform a congruence transformation of Equation 24.22 with Y^{-1} and $\text{diag}(Y^{-1}, I, I)$ which leads back to Equation 24.20. This proves, with X being a certificate, that D_c is indeed stabilizing and achieves the desired performance specification. On the other hand, if the inequalities in Equation 24.22 are not feasible, it is assured that no state-feedback gain can exist for which both these properties are satisfied.

24.6.2 Output-Feedback Synthesis

A general output-feedback controller leads to the controlled system in Equation 24.18. It achieves stability of $\mathcal{A}$ and $\|\mathcal{C}(sI - \mathcal{A})^{-1}\mathcal{B} + \mathcal{D}\|_{\mathscr{H}_\infty} < \gamma$ iff there exists some $\mathcal{X}$ with

$$\mathcal{X} \succ 0 \quad \text{and} \quad \begin{pmatrix} \mathcal{A}^\top \mathcal{X} + \mathcal{X}\mathcal{A} & \mathcal{X}\mathcal{B} & \mathcal{C}^\top \\ \mathcal{B}^\top \mathcal{X} & -\gamma^2 I & \mathcal{D}^\top \\ \mathcal{C} & \mathcal{D} & -I \end{pmatrix} \prec 0. \tag{24.23}$$

Due to the affine dependence of $(\mathcal{A}, \mathcal{B}, \mathcal{C}, \mathcal{D})$ on the controller matrices (A_c, B_c, C_c, D_c), this involves again a bilinear matrix inequality.

It is pleasing that one can identify, again, a convexifying controller parameter transformation [13,17]. If we partition

$$\mathcal{X} = \begin{pmatrix} X & U \\ U^\top & * \end{pmatrix} \quad \text{and} \quad \mathcal{X}^{-1} = \begin{pmatrix} Y & V \\ V^\top & * \end{pmatrix} \tag{24.24}$$

according to $\mathcal{A}$ (where X and Y share their dimension with A and the $*$'s denote matrices that are irrelevant for our purposes) it reads as

$$\begin{pmatrix} K & L \\ M & N \end{pmatrix} = \begin{pmatrix} XAY & 0 \\ 0 & 0 \end{pmatrix} + \begin{pmatrix} U & XB \\ 0 & I \end{pmatrix} \begin{pmatrix} A_c & B_c \\ C_c & D_c \end{pmatrix} \begin{pmatrix} V^\top & 0 \\ CY & I \end{pmatrix}. \tag{24.25}$$

We view this as a transformation $(\mathcal{X}, A_c, B_c, C_c, D_c) \to (X, Y, K, L, M, N) =: v$. This definition is motivated by the following easily verified relations:

$$\text{with } \mathcal{Y} := \begin{pmatrix} Y & I \\ V^\top & 0 \end{pmatrix} \text{ we have } \mathcal{Y}^\top \mathcal{X} \mathcal{Y} = \begin{pmatrix} Y & I \\ I & X \end{pmatrix} =: \boldsymbol{X}(v) \quad \text{and} \tag{24.26}$$

$$\left(\begin{array}{c|c} \mathcal{Y}^\top(\mathcal{X}\mathcal{A})\mathcal{Y} & \mathcal{Y}^\top(\mathcal{X}\mathcal{B}) \\ \hline \mathcal{C}\mathcal{Y} & \mathcal{D} \end{array}\right) = \left(\begin{array}{cc|c} AY + BM & A + BNC & B_0 + BNF \\ K & XA + LC & XB_0 + LF \\ \hline C_0 Y + EM & C_0 + ENC & D_0 + ENF \end{array}\right) =: \left(\begin{array}{c|c} \boldsymbol{A}(v) & \boldsymbol{B}(v) \\ \hline \boldsymbol{C}(v) & \boldsymbol{D}(v) \end{array}\right). \tag{24.27}$$

Although looking intricate, the key is the affine dependence of the blocks $\boldsymbol{X}(v), \boldsymbol{A}(v), \boldsymbol{B}(v), \boldsymbol{C}(v)$, and $\boldsymbol{D}(v)$ on the new variables v. If $\mathcal{Y}$ is nonsingular, a congruence transformation with $\mathcal{Y}$ and $\mathrm{diag}(\mathcal{Y}, I, I)$ on Equation 24.23 leads to

$$\mathcal{Y}^\top \mathcal{X} \mathcal{Y} \succ 0 \quad \text{and} \quad \begin{pmatrix} \mathcal{Y}^\top(\mathcal{A}^\top \mathcal{X})\mathcal{Y} + \mathcal{Y}^\top(\mathcal{X}\mathcal{A})\mathcal{Y} & \mathcal{Y}^\top(\mathcal{X}\mathcal{B}) & \mathcal{Y}^\top \mathcal{C}^\top \\ (\mathcal{B}^\top \mathcal{X})\mathcal{Y} & -\gamma^2 I & \mathcal{D}^\top \\ \mathcal{C}\mathcal{Y} & \mathcal{D} & -I \end{pmatrix} \prec 0, \tag{24.28}$$

which is, in turn, nothing but the following LMI in v:

$$\boldsymbol{X}(v) \succ 0 \quad \text{and} \quad \begin{pmatrix} \boldsymbol{A}(v)^\top + \boldsymbol{A}(v) & \boldsymbol{B}(v) & \boldsymbol{C}(v)^\top \\ \boldsymbol{B}(v)^\top & -\gamma^2 I & \boldsymbol{D}(v)^\top \\ \boldsymbol{C}(v) & \boldsymbol{D}(v) & -I \end{pmatrix} \prec 0. \tag{24.29}$$

This brings us to the following result which is true without any hypothesis on $\mathcal{Y}$.

Theorem 24.6:

There exists a controller as in Equation 24.17 which stabilizes the system in Equation 24.18 and renders the $\mathscr{H}_\infty$-norm of the transfer matrix $d \to e$ smaller than γ iff the LMI's in Equation 24.29 are feasible.

The actual design of a controller proceeds as follows. Find a solution of the LMIs in Equation 24.29; due to the first inequality in Equation 24.29 the matrix $I - XY$ is nonsingular (Schur); therefore, one can find square and nonsingular matrices U and V with $I - XY = UV^\top$; then we can solve Equation 24.25 for (A_c, B_c, C_c, D_c). This controller does the job since we can define the nonsingular matrix $\mathcal{Y}$ and solve for $\mathcal{X}$ in Equation 24.26, which implies that Equation 24.27 is valid; then Equation 24.29 is nothing but Equation 24.28; since $\mathcal{Y}$ is nonsingular, this transforms into Equation 24.23 by congruence.

In summary, we have reduced the design problem to find a stabilizing controller that establishes a bound on the $\mathscr{H}_\infty$-norm of the closed-loop system to an equivalent problem that amounts to checking the feasibility of the LMIs in Equation 24.29. Since γ^2 enters these constraints in an affine fashion, one can minimize γ^2 subject to the feasibility of these LMIs in order to directly compute the optimal $\mathscr{H}_\infty$-attenuation level that is achievable by stabilizing controllers.

24.6.3 General Synthesis Procedure

Although we discussed $\mathscr{H}_\infty$-synthesis in quite some detail, we can extract the following generic procedure for moving from analysis to synthesis inequalities for a whole variety of other performance specifications that can be expressed by LMIs.

- Rewrite the analysis inequalities in terms of the blocks $\mathcal{X}, \mathcal{XA}, \mathcal{XB}, \mathcal{C}$, and $\mathcal{D}$.
- Find a formal congruence transformation involving $\mathcal{Y}$ which leads to inequalities in terms of the blocks $\mathcal{Y}^\top\mathcal{XY}, \mathcal{Y}^\top(\mathcal{XA})\mathcal{Y}, \mathcal{Y}^\top(\mathcal{XB}), \mathcal{CY}$, and $\mathcal{D}$.
- Then the synthesis inequalities are obtained by the substitution

$$\mathcal{Y}^\top\mathcal{XY} \to \mathbf{X}(v), \quad \begin{pmatrix} \mathcal{Y}^\top(\mathcal{XA})\mathcal{Y} & \mathcal{Y}^\top(\mathcal{XB}) \\ \mathcal{C} & \mathcal{D} \end{pmatrix} \to \begin{pmatrix} \mathbf{A}(v) & \mathbf{B}(v) \\ \mathbf{C}(v) & \mathbf{D}(v) \end{pmatrix}$$

  with Equation 24.27 for $v := (X, Y, K, L, M, N)$.
- For state-feedback synthesis one can apply the very same procedure with the formulas $\mathbf{X}(v) = Y, \mathbf{A}(v) = AY + BM, \mathbf{B}(v) = B_0, \mathbf{C}(v) = C_0Y + EM, \mathbf{D}(v) = D_0$ for $v = (M, Y)$.
- The controller construction is independent from the particular analysis inequalities and remains, both for state-feedback and output-feedback synthesis, unaltered.

This procedure applies to $\mathscr{H}_2$-synthesis as well as to the variants of the LMI analysis specification discussed in Section 24.5.4. As an illustration, the discrete-time system $x(t+1) = \mathcal{A}x(t) + \mathcal{B}d(t), e(t) = \mathcal{C}x(t) + \mathcal{D}d(t)$ is Schur-stable and its l_2-gain is bounded by γ iff there exists some $\mathcal{X}$ with

$$\mathcal{X} \succ 0, \quad \begin{pmatrix} I & 0 \\ \mathcal{A} & \mathcal{B} \end{pmatrix}^\top \begin{pmatrix} -\mathcal{X} & 0 \\ 0 & \mathcal{X} \end{pmatrix} \begin{pmatrix} I & 0 \\ \mathcal{A} & \mathcal{B} \end{pmatrix} + \begin{pmatrix} 0 & I \\ \mathcal{C} & \mathcal{D} \end{pmatrix}^\top \begin{pmatrix} -\gamma^2 I & 0 \\ 0 & I \end{pmatrix} \begin{pmatrix} 0 & I \\ \mathcal{C} & \mathcal{D} \end{pmatrix} \prec 0.$$

Since these can also be expressed as

$$\mathcal{X} \succ 0, \quad \begin{pmatrix} \mathcal{X} & 0 \\ 0 & \gamma^2 I \end{pmatrix} - \begin{pmatrix} \mathcal{A} & \mathcal{B} \\ \mathcal{C} & \mathcal{D} \end{pmatrix}^\top \begin{pmatrix} \mathcal{X} & 0 \\ 0 & I \end{pmatrix} \begin{pmatrix} \mathcal{A} & \mathcal{B} \\ \mathcal{C} & \mathcal{D} \end{pmatrix} \succ 0,$$

a Schur complement argument reveals that these analysis inequalities can be written as

$$\left(\begin{array}{cc|cc} \mathcal{X} & 0 & \mathcal{A}^\top\mathcal{X} & \mathcal{C}^\top \\ 0 & \gamma^2 I & \mathcal{B}^\top\mathcal{X} & \mathcal{D}^\top \\ \hline \mathcal{XA} & \mathcal{XB} & \mathcal{X} & 0 \\ \mathcal{C} & \mathcal{D} & 0 & I \end{array}\right) \succ 0$$

and thus admit the precise format in order to apply the general synthesis procedure.

Based on the generic dualization and elimination results described in [16], it is often possible to eliminate matrix variables that only appear in one of these synthesis inequalities, with the benefit of reducing computational complexity. For example, one can eliminate all matrices K, L, M, and N from the $\mathscr{H}_\infty$-synthesis inequalities in order to arrive at the inequalities as proposed in [8,11].

Finally, in many practical problems the disturbance and error signals are partitioned as $d = \text{col}(d_1, \ldots, d_p)$ and $e = \text{col}(e_1, \ldots, e_q)$ in order to impose multiple individual performance requirements on some of the channels $d_\nu \to e_\mu$, possibly with different norms. The general controller synthesis procedure described in this section is applicable for this kind of multiobjective control problems provided that one is willing to allow some level of conservatism by expressing the combined desired specifications with one and the same matrix $\mathcal{X}$ in the closed-loop system. This variant of multiobjective control, which is addressed as a Lyapunov-shaping paradigm in [17] in more detail, even allows the incorporation of pole-placement requirements on $\mathcal{A}$ in terms of general LMI regions. We refer to [17] for an instructive controller design example. The introduction of slack-variables offers a possibility to reduce conservatism as shown for discrete-time systems in [5], while an LMI solution of the general multiobjective control problem based on the Youla–Kucera parameterization of all stabilizing controllers is discussed in [15] and its references.

24.6.4 Observer and Estimator Synthesis

Estimation problems are related to the configuration in Figure 24.2. Based on measurements y, the goal is to determine an estimator whose output approximates the system output z (which can be equal to the state or comprise components of d) as closely as possible, despite the corruption of the system by the disturbance d. If d is white noise with unity covariance, minimization of the asymptotic variance of e amounts to minimizing the $\mathscr{H}_2$-norm of $d \to e$ which results in the classical Kalman filter. Alternatively, using the energy-gain of $d \to e$ as a performance indicator amounts to considering the $\mathscr{H}_\infty$-estimation problem. These and many other variants of estimation problems can be rephrased as an output-feedback synthesis problem for a system as described by Equation 24.16 with $B = 0$ and $E = -I$ (which means that u does not excite the system dynamics). Let us stress that in this reformulation u admits the interpretation of the estimator output and e is the estimation error.

If the estimator admits the structure of an observer with a to-be-designed gain D_c,

$$\dot{x}_c = Ax_c + D_c(Cx_c - y), \quad u = C_0 x_c, \tag{24.30}$$

then the dynamics of the state-error $\xi = x - x_c$ is described by

$$\dot{\xi} = (A + D_c C)\xi + (B_0 + D_c F)d, \quad e = C_0\xi + D_0 d$$

and defines the transfer matrix $d \to e$. This representation is dual (transposed) to what we considered for state-feedback synthesis, and only slight modifications are required in order to determine the LMIs for synthesizing observer gains D_c that stabilize the error dynamics and achieve, for example, an $\mathscr{H}_\infty$- or $\mathscr{H}_2$-norm bound γ on $d \to e$.

Let us now assume that A in Equation 24.16 is stable. Instead of assuming a particular structure, we can then try to find a general estimator Equation 24.17 with stable A_c such that a desired performance level for $d \to e$ is achieved. In view of Equation 24.18 and the fact that $B = 0$, stability of A_c now boils down to stability of $\mathcal{A}$, and the very same output-feedback synthesis procedure can be followed in order to design optimal estimators by LMIs.

However, it is interesting to observe that the structural property $B = 0$ allows a simplification of the convexifying controller parameter transformation which will be essential for robust estimator synthesis [9]. Indeed, starting from Equation 24.24 and with $Z = Y^{-1}$ let us define

$$\begin{pmatrix} K & L \\ M & N \end{pmatrix} := \begin{pmatrix} U & 0 \\ 0 & I \end{pmatrix}\begin{pmatrix} A_c & B_c \\ C_c & D_c \end{pmatrix}\begin{pmatrix} V^\top Z & 0 \\ 0 & I \end{pmatrix} \quad \text{and} \quad \mathcal{Y} := \begin{pmatrix} I & I \\ V^\top Z & 0 \end{pmatrix}. \tag{24.31}$$

After a direct computation we obtain

$$\mathcal{Y}^\top \mathcal{X} \mathcal{Y} = \begin{pmatrix} Z & Z \\ Z & X \end{pmatrix} =: \mathbf{X}(v) \quad \text{and} \tag{24.32}$$

$$\left(\begin{array}{c|c} \mathcal{Y}^\top(\mathcal{X}\mathcal{A})\mathcal{Y} & \mathcal{Y}^\top(\mathcal{X}\mathcal{B}) \\ \hline \mathcal{C}\mathcal{Y} & \mathcal{D} \end{array}\right) = \left(\begin{array}{cc|c} ZA & ZA & ZB_0 \\ XA + LC + K & XA + LC & XB_0 + LF \\ \hline C_0 + ENC + EM & C_0 + ENC & D_0 + ENF \end{array}\right) =: \left(\begin{array}{c|c} \mathbf{A}(v) & \mathbf{B}(v) \\ \hline \mathbf{C}(v) & \mathbf{D}(v) \end{array}\right)$$

with an affine dependence on X, Z, K, L, M, and N. Therefore the general synthesis procedure in Section 24.6.3 does apply, with the only modification being the use of these new substitution formulas and by recalling the relation $Z = Y^{-1}$ for the actual design of the estimator.

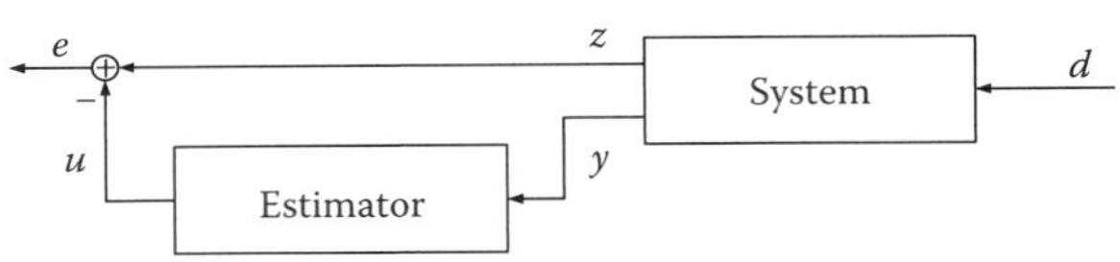

FIGURE 24.2 Estimator synthesis.

24.7 Polytopic Uncertainties and Robustness Analysis

First principle models of physical systems are often represented by state-space descriptions in which the various components of the state represent well-defined physical quantities. Variations, perturbations, or uncertainties in physical parameters lead to uncertainty in the model. Often, this uncertainty is reflected by variations in well-distinguished parameters or coefficients in the model, while, in addition, the nature and/or range of the uncertain parameters may be known, or partially known. Since very small parameter variations may have a major impact on the dynamics of a system, it is of evident importance to analyze parametric uncertainties of dynamical systems. Suppose that $\delta = (\delta_1, \ldots, \delta_p)$ is the vector that expresses the ensemble of all uncertain quantities in a given dynamical system. Then there are at least two distinct cases that are of independent interest:

a. *Time-invariant parametric uncertainties:* the vector δ is a fixed but unknown element of an uncertainty set $\boldsymbol{\delta} \subseteq \mathbb{R}^p$.
b. *Time-varying parametric uncertainties:* the vector δ is an unknown time-varying function $\delta : \mathbb{R} \to \mathbb{R}^p$ whose values $\delta(t)$ belong to an uncertainty set $\boldsymbol{\delta} \subseteq \mathbb{R}^p$, and possibly satisfy additional constraints.

24.7.1 Time-Invariant Parametric Uncertainty

Consider the uncertain time-invariant system defined by

$$\dot{x} = A(\delta)x, \tag{24.33}$$

where $A(\cdot)$ is a continuous function of the real-valued parameter vector $\delta = \mathrm{col}(\delta_1, \ldots, \delta_p)$ which is only known to be contained in an uncertainty set $\boldsymbol{\delta} \subseteq \mathbb{R}^p$. The problem of *robust stability* amounts to characterizing whether the equilibrium point $x^* = 0$ of Equation 24.33 is exponentially stable for all parameters $\delta \in \boldsymbol{\delta}$. With time-invariant uncertainties, Equation 24.33 is robustly stable iff $A(\delta)$ is Hurwitz for all $\delta \in \boldsymbol{\delta}$. Since $\boldsymbol{\delta}$ generally consists of infinitely many points, the verification of this condition is rather troublesome from a computational point of view.

The uncertain system (Equation 24.33) is called *quadratically stable* if there exists $X = X^\top$ such that

$$X \succ 0, \quad A(\delta)^\top X + XA(\delta) \prec 0 \quad \text{for all} \quad \delta \in \boldsymbol{\delta}. \tag{24.34}$$

The importance of this definition becomes apparent after observing that $V(x) := x^\top Xx$ is a quadratic Lyapunov function for Equation 24.33 which, by Theorem 24.1, implies that $A(\delta)$ is Hurwitz for all $\delta \in \boldsymbol{\delta}$. Hence, quadratic stability implies the origin of Equation 24.33 to be robust exponentially stable against time-invariant uncertainties $\delta \in \boldsymbol{\delta}$. Unless $\boldsymbol{\delta}$ has a finite number of points, Equation 24.34 cannot be verified easily. Therefore, the following result is of considerable interest.

Theorem 24.7:

If $A(\delta)$ is affine in δ and $\boldsymbol{\delta} = co\{\delta^1, \ldots, \delta^N\}$ then Equation 24.33 is quadratically stable if and only if there exists some X such that

$$X \succ 0, \quad A(\delta^k)^\top X + XA(\delta^k) \prec 0, \quad k = 1, \ldots, N.$$

Hence, for polytopic uncertainty sets *and* affine parametric dependence in Equation 24.33, quadratic stability can be numerically verified by a feasibility test for a *finite number* of LMIs only. The proof is an illustrative example of the use of convexity. It requires showing that $F(\delta) := A(\delta)^\top X + XA(\delta) \prec 0$ for all $\delta \in \boldsymbol{\delta}$ if $F(\delta^k) \prec 0$ for $k = 1, \ldots, N$. To see this, first observe that $F(\cdot)$ is a convex function on $\boldsymbol{\delta}$ whenever $A(\cdot)$ is affine. Second, any $\delta \in \boldsymbol{\delta}$ can be written as a convex combination of the points $\delta^1, \ldots, \delta^N$, say $\delta = \sum_{k=1}^{N} \alpha_k \delta^k$ with nonnegative coefficients α_k that sum up to 1; hence $F(\delta) = F(\sum_{k=1}^{N} \alpha_k \delta^k) \preccurlyeq \sum_{k=1}^{N} \alpha_k F(\delta^k) \prec 0$, which gives the result.

24.7.2 Time-Dependent Parametric Uncertainty

Robust stability against time-varying uncertainties is generally a more demanding requirement than robust stability against time-invariant uncertainties. Consider the system

$$\dot{x}(t) = A(\delta(t))x(t), \tag{24.35}$$

which is affected by an uncertain parameter curve $\delta : \mathbb{R} \to \boldsymbol{\delta}$. Unlike the case with time-invariant uncertainties, robust stability of the origin is now *not implied* by the condition that $A(\delta)$ is Hurwitz for all $\delta \in \boldsymbol{\delta}$. However, the uncertain system with time-varying parametric uncertainties is exponentially stable if there exists $X \succ 0$ such that Equation 24.34 holds. Therefore, quadratic stability does, in fact, imply robust stability against arbitrary fast time-varying parametric uncertainties. This is a nice, but in general conservative, test if additional *a priori* information on the uncertainty is available. For example, the parameter curves $\delta(\cdot)$ are often known to be continuously differentiable and constrained in terms of their values and their rate-of-variation as

$$\delta(t) \in \boldsymbol{\delta}, \quad \dot{\delta}(t) \in \boldsymbol{\rho} \quad \text{for all } t \in \mathbb{R}. \tag{24.36}$$

Less conservative robust stability tests can be inferred by postulating the existence of *parameter-dependent* Lyapunov functions. A popular instance of such functions takes the form $V(x, \delta) := x^\top X(\delta)x$ and requires a search over matrix functions $X(\delta) = X(\delta)^\top$ with $\delta \in \boldsymbol{\delta}$. For notational convenience, let us introduce, for a continuously differentiable matrix function $X(\delta)$, the "derivative"

$$\partial X(\delta, \rho) := \sum_{k=1}^{p} \partial_k X(\delta)\rho_k, \quad (\delta, \rho) \in \boldsymbol{\delta} \times \boldsymbol{\rho}, \tag{24.37}$$

where $\partial_k X(\cdot)$ denotes the partial derivative of the function $X(\cdot)$ with respect to the kth entry of δ and where ρ_k is the kth component of the vector ρ. (We stress that $\partial X(\delta, \rho)$ is purely a symbolic notation which is not to be confused with the partial derivative of $X(\cdot)$ itself.) The following result provides a sufficient condition for robust stability and actually covers many tests in the literature. The proof provides much insight into the understanding of stability arguments based on parameter-dependent Lyapunov functions.

Theorem 24.8:

Suppose that $\boldsymbol{\delta}$ and $\boldsymbol{\rho}$ are compact subsets of $\mathbb{R}^p$ and suppose that $X(\delta) = X(\delta)^\top$ is a continuously differentiable matrix function that satisfies

$$X(\delta) \succ 0, \quad \partial X(\delta, \rho) + A(\delta)^\top X(\delta) + X(\delta)A(\delta) \prec 0 \tag{24.38}$$

for all $\delta \in \boldsymbol{\delta}$ and $\rho \in \boldsymbol{\rho}$. Then the origin of Equation 24.35 is exponentially stable for all time-varying parametric uncertainties that satisfy Equation 24.36.

Proof. Suppose that $X(\delta)$ satisfies Equation 24.38. Continuity of $X(\cdot)$ and compactness of $\boldsymbol{\delta}$ and $\boldsymbol{\rho}$ guarantee the existence of constants $a, b, c > 0$ such that, for all $\delta \in \boldsymbol{\delta}$ and $\rho \in \boldsymbol{\rho}$,

$$aI \preccurlyeq X(\delta) \preccurlyeq bI, \quad \partial X(\delta, \rho) + A(\delta)^\top X(\delta) + X(\delta)A(\delta) \preccurlyeq -cI. \tag{24.39}$$

Let $\delta(\cdot)$ and $x(\cdot)$ be a parameter curve and a state trajectory that satisfy Equations 24.36 and 24.35, respectively. With $\xi(t) := x(t)^\top X(\delta(t))x(t)$ we clearly have

$$\dot{\xi}(t) = x(t)^\top \left[\sum_{k=1}^{p} \partial_k X(\delta(t))\dot{\delta}_k(t) \right] x(t) + x(t)^\top [A(\delta(t))^\top X(\delta(t)) + X(\delta(t))A(\delta(t))]x(t).$$

If we exploit Equation 24.39 we obtain $a\|x(t)\|^2 \leq \xi(t) \leq b\|x(t)\|^2$ and $\dot{\xi}(t) \leq -c\|x(t)\|^2$. This implies that $\dot{\xi}(t) \leq -\frac{c}{b}\xi(t)$ and hence $\xi(t) \leq \xi(t_0)\exp(-\frac{c}{b}(t-t_0))$ for all $t \geq t_0$. In turn, this leads to $\|x(t)\|^2 \leq \frac{b}{a}\|x(t_0)\|^2 e^{-\frac{c}{b}(t-t_0)}$, which is Equation 24.5 for $\alpha = c/(2b)$ and $\beta = \sqrt{b/a}$.

The constraints in Equation 24.38 on $X(\cdot)$ define a purely algebraic test that do not involve the system- or parameter-trajectories. The test is not easy to apply directly because the matrix function $X(\cdot)$ needs to satisfy a partial differential LMI. By considering specific classes of matrix functions $X(\cdot)$, Equation 24.38 can be converted and implemented with LMI solvers. One of these classes is the set of affine symmetric matrix functions $X : \boldsymbol{\delta} \to \mathbb{S}^n$.

A few special instances are worth mentioning. If the parameters are *time-invariant*, we have $\boldsymbol{\rho} = \{0\}$ and Equation 24.38 simplifies to the conditions $X(\delta) \succ 0$ and $A(\delta)^\top X(\delta) + X(\delta)A(\delta) \prec 0$. In that case, Equation 24.38 is also *necessary* for robust stability. If parameters vary *arbitrarily fast*, $\boldsymbol{\rho}$ is unbounded and Equation 24.38 is feasible only if the partial derivatives $\partial_k X(\delta)$ vanish identically. This means that $X(\delta) = X$ is not depending on δ and we recover the quadratic stability test.

24.7.3 Robust Performance

In the previous subsections we have shown how tests for robust stability against parametric uncertainties can be inferred from characterizations of nominal stability. The same generalization applies in order to obtain conditions for verifying robust performance. Consider the uncertain parameter depending system

$$\begin{aligned} \dot{x}(t) &= A(\delta(t))x(t) + B(\delta(t))d(t), \\ e(t) &= C(\delta(t))x(t) + D(\delta(t))d(t), \end{aligned} \tag{24.40}$$

where $\delta(\cdot)$ is a continuously differentiable rate-bounded uncertainty that satisfies Equation 24.36. In the case of $\mathscr{H}_\infty$-performance, we quantified the effect of d on e in terms of the $\mathscr{L}_2$ gain of the system. However, the output e of Equation 24.40 not only depends on the input d but also on the uncertainty $\delta(\cdot)$. Hence we say that the *robust $\mathscr{L}_2$-gain is smaller than* γ if

- for $d = 0$ and for all parameter curves $\delta(\cdot)$ satisfying Equation 24.36, $x^* = 0$ is an exponentially stable equilibrium of the system in Equation 24.40.
- for $x(0) = 0$ it holds that

$$\sup_{\delta(\cdot) \text{ satisfies Equation 24.36}} \ \sup_{0<\|d\|_{\mathscr{L}_2}<\infty} \frac{\|e\|_{\mathscr{L}_2}}{\|d\|_{\mathscr{L}_2}} < \gamma.$$

With some abuse of terminology, robust $\mathscr{L}_2$-gain performance is often referred to as *robust $\mathscr{H}_\infty$-performance*, but it is important to realize that frequency-domain characterizations do not make sense when considering time-dependent parametric uncertainties in Equation 24.40. Only with time-invariant parameter uncertainties ($\boldsymbol{\rho} = \{0\}$) does a robust $\mathscr{L}_2$-gain that is smaller than γ imply that $\|T_\delta\|_{\mathscr{H}_\infty} < \gamma$ for all $\delta \in \boldsymbol{\delta}$, where T_δ is the transfer function associated with Equation 24.40 for time-invariant parameters $\delta(t) = \delta$. The following result generalizes Theorem 24.4 to robust $\mathscr{L}_2$-gain performance.

Theorem 24.9:

Suppose there exists a continuously differentiable matrix function $X(\delta) = X(\delta)^\top$ *such that* $X(\delta) \succ 0$ *and*

$$\begin{pmatrix} \partial X(\delta, \rho) + A(\delta)^\top X(\delta) + X(\delta)A(\delta) & X(\delta)B(\delta) \\ B(\delta)^\top X(\delta) & 0 \end{pmatrix} + \begin{pmatrix} 0 & I \\ C(\delta) & D(\delta) \end{pmatrix}^\top \begin{pmatrix} -\gamma^2 I & 0 \\ 0 & I \end{pmatrix} \begin{pmatrix} 0 & I \\ C(\delta) & D(\delta) \end{pmatrix} \prec 0 \tag{24.41}$$

for all $\delta \in \boldsymbol{\delta}$ *and* $\rho \in \boldsymbol{\rho}$. *Then the uncertain system in Equation 24.40 has a robust* $\mathscr{L}_2$*-gain smaller than* γ.

The proof of this result is analogous to that of Theorem 24.4. The main merit of Theorem 24.9 is that it converts robust $\mathscr{L}_2$-gain performance to an algebraic property. As in Theorem 24.8, the condition in Equation 24.41 requires the numerical search for a matrix function $X(\cdot)$ that needs to satisfy a partial differential LMI. Moreover, the discussion about the extreme cases concerning time-invariant or arbitrary fast time-varying parameters in Section 24.7.2 remains valid for Equation 24.41. Finally, let us remark that Theorem 24.9 extends, mutatis mutandis, to all other performance specifications that have been mentioned in Section 24.5.

24.8 Robust State-Feedback and Estimator Synthesis

Consider the system

$$\begin{aligned} \dot{x}(t) &= A(\delta(t))x(t) + B_0(\delta(t))d(t) + B(\delta(t))u(t), \\ e(t) &= C_0(\delta(t))x(t) + D_0(\delta(t))d(t) + E(\delta(t))u(t), \\ y(t) &= C(\delta(t))x(t) + F(\delta(t))d(t), \end{aligned} \tag{24.42}$$

whose describing matrices are affected by a time-dependent parametric uncertainty $\delta(t) \in \mathbb{R}^p$. Let us also assume that the dependence of the system matrices is actually affine, and that $\delta(t)$ takes values that are, for $t \geq 0$, confined to the polytope

$$\boldsymbol{\delta} := \mathrm{co}\{\delta^1, \ldots, \delta^N\} \subset \mathbb{R}^p. \tag{24.43}$$

Robust controller synthesis deals with the problem of determining a feedback controller that processes measurements y to control inputs u, so as to guarantee robust stability and a desired robust performance specification on the mapping from the disturbance d to the output e of the controlled system.

The robust state-feedback synthesis problem is a special case in which the whole state is assumed to be measurable ($y = x$ in Equation 24.42) and the controller is a static feedback law $u = D_c x$ with some gain D_c. In that case, the resulting closed-loop system is described by

$$\begin{aligned} \dot{x}(t) &= [A(\delta(t)) + B(\delta(t))D_c]x(t) + B_0(\delta(t))d(t), \\ e(t) &= [C_0(\delta(t)) + E(\delta(t))D_c]x(t) + D_0(\delta(t))d(t). \end{aligned}$$

If applying Theorem 24.9 for a parameter-independent $X(\delta) = X$ and exploiting affine parameter-dependence as for Theorem 24.7, we infer that the robust $\mathscr{L}_2$-gain of the controlled system is smaller than γ if there exists $X = X^\top$ with

$$X \succ 0, \quad \begin{pmatrix} [A(\delta^k) + B(\delta^k)D_c]^\top X + X[A(\delta^k) + B(\delta^k)D_c] & XB_0(\delta^k) & [C_0(\delta^k) + E(\delta^k)D_c]^\top \\ B_0(\delta^k)^\top X & -\gamma^2 I & D_0(\delta^k)^\top \\ C_0(\delta^k) + E(\delta^k)D_c & D_0(\delta^k) & -I \end{pmatrix} \prec 0$$

$$\text{for all } k = 1, \ldots, N. \tag{24.44}$$

Literally following the nominal synthesis procedure of Section 24.6.1 with the convexifying transformation in Equation 24.21 we infer that there exist (D_c, X) satisfying Equation 24.44 iff there exist (M, Y) satisfying

$$Y \succ 0, \quad \begin{pmatrix} [A(\delta^k)Y + B(\delta^k)M]^\top + [A(\delta^k)Y + B(\delta^k)M] & B_0(\delta^k) & [C_0(\delta^k)Y + E(\delta^k)M]^\top \\ B_0(\delta^k)^\top & -\gamma^2 I & D_0(\delta^k)^\top \\ C_0(\delta^k)Y + E(\delta^k)M & D_0(\delta^k) & -I \end{pmatrix} \prec 0$$

for all $k = 1, \ldots, N$.

This LMI feasibility problem in (M, Y) can be readily solved. If (M, Y) is a solution, the state-feedback gain $D_c = MY^{-1}$ guarantees a robust $\mathcal{L}_2$-gain that is smaller than γ for the closed-loop system.

This procedure works smoothly because the transformation (Equation 24.21) does not involve data matrices that describe the open-loop system (Equation 24.42) and because X is assumed independent of δ. The reduction of conservatism by employing parameter-dependent functions $X(\delta)$ is possible, to a certain extent, by introducing slack variables as discussed in [5].

Unfortunately, robust output-feedback controllers cannot be designed by following the general synthesis procedure of Section 24.6.2 for the parameter-dependent system (Equation 24.42). As the main reason, the convexifying transformation (Equation 24.25) involves the system data while, in addition, the matrix blocks in the corresponding synthesis inequalities are not affine in the parameter vector δ. Similarly, robust observers with the structure appearing in Equation 24.30 cannot be designed, just because the observer is defined in terms of open-loop system data.

In contrast, for the estimation problem we have considered the convexifying transformation in Equation 24.31 of the estimator parameters that does not involve system matrices. Therefore it is possible to design robust estimators in complete analogy to what was discussed for state-feedback synthesis. Specifically, recall from Section 24.6.4 that estimator design involves a system description (Equation 24.42) with $B(\delta) = 0$ and $E(\delta) = E$. Then there exists an estimator (Equation 24.17) that achieves for $d \to e$ a robust $\mathcal{L}_2$-gain performance level smaller than γ (according to the definition in Section 24.7.3) if there exists $v = (X, Z, K, L, M, N)$ such that

$$\boldsymbol{X}(v) \succ 0, \quad \begin{pmatrix} \boldsymbol{A}(v, \delta^k)^\top + \boldsymbol{A}(v, \delta^k) & \boldsymbol{B}(v, \delta^k) & \boldsymbol{C}(v, \delta^k)^\top \\ \boldsymbol{B}(v, \delta^k)^\top & -\gamma^2 I & \boldsymbol{D}(v, \delta^k)^\top \\ \boldsymbol{C}(v, \delta^k) & \boldsymbol{D}(v, \delta^k) & -I \end{pmatrix} \prec 0 \quad \text{for all } k = 1, \ldots, N,$$

where $\boldsymbol{X}(v)$ is given in Equation 24.32 and

$$\left(\begin{array}{c|c} \boldsymbol{A}(v, \delta^k) & \boldsymbol{B}(v, \delta^k) \\ \hline \boldsymbol{C}(v, \delta^k) & \boldsymbol{D}(v, \delta^k) \end{array}\right) = \left(\begin{array}{cc|c} ZA(\delta^k) & ZA(\delta^k) & ZB_0(\delta^k) \\ XA(\delta^k) + LC(\delta^k) + K & XA(\delta^k) + LC(\delta^k) & XB_0(\delta^k) + LF(\delta^k) \\ \hline C_0(\delta^k) + ENC(\delta^k) + EM & C_0(\delta^k) + ENC(\delta^k) & D_0(\delta^k) + ENF(\delta^k) \end{array}\right)$$

are affine in the design parameters v. This set of $(N + 1)$ LMIs can be readily implemented. If v is a solution, the desired robust estimator is then given by the state-space representation matrices (A_c, B_c, C_c, D_c) that are resolved from Equation 24.31 with nonsingular U, V satisfying $Z - X = UV^T Z$.

24.9 Gain-Scheduling Synthesis

An interesting generalization of the robust controller synthesis problem treated in the previous section amounts to allowing the controller to be dependent on the parameter vector $\delta = \text{col}(\delta_1, \ldots, \delta_p)$. This means that the controller has online access to the time-varying parameter $\delta(t)$ through an additional measurement. Hence, the actual parameter value, although not known *a priori*, is used as extra information to control the system.

The classical *gain-scheduling* approach towards control system design fits in this line of reasoning and typically amounts to associating with $\delta(t)$ a specific operating condition of the plant. A (robust) controller is designed for each operating condition $\delta \in \boldsymbol{\delta}$ and these are *scheduled* by means of switching or interpolation rules inferred from measurement information. This design methodology has found widespread applications, but the assessment of guarantees on robust stability and robust performance in view of time-dependent changes of δ is usually difficult if not impossible.

For the linear parameter-varying (LPV) system in Equation 24.42, an LPV controller is a system of the form

$$\begin{aligned} \dot{x}_c(t) &= A_c(\delta(t))x_c(t) + B_c(\delta(t))y(t), \\ u(t) &= C_c(\delta(t))x_c(t) + D_c(\delta(t))y(t), \end{aligned} \tag{24.45}$$

where $\delta(t)$ satisfies the bound and rate constraints in Equation 24.36. As before, the aim will be to design an LPV controller (Equation 24.45) that renders the controlled system robustly stable and establishes a desired robust performance specification. The controlled system is given by

$$\begin{aligned} \dot{\xi}(t) &= \mathcal{A}(\delta(t))\xi(t) + \mathcal{B}(\delta(t))d(t), \\ e(t) &= \mathcal{C}(\delta(t))\xi(t) + \mathcal{D}(\delta(t))d(t), \end{aligned}$$

where the closed-loop matrix functions are structured as in Equation 24.18.

The synthesis techniques for the construction of LPV controllers closely resemble those for nominal controller design as we have investigated in Section 24.6.2. The LPV synthesis problem to achieve robust stability and a robust $\mathscr{L}_2$-gain smaller than γ is then solved if one can find a controller and a smooth function $\mathcal{X}(\delta) = \mathcal{X}(\delta)^\top$ such that for all $(\delta, \rho) \in \boldsymbol{\delta} \times \boldsymbol{\rho}$

$$\mathcal{X}(\delta) \succ 0 \quad \text{and} \quad \begin{pmatrix} \partial\mathcal{X}(\delta,\rho) + \mathcal{A}(\delta)^\top\mathcal{X}(\delta) + \mathcal{X}(\delta)\mathcal{A}(\delta) & \mathcal{X}(\delta)\mathcal{B}(\delta) & \mathcal{C}(\delta)^\top \\ \mathcal{B}(\delta)^\top\mathcal{X}(\delta) & -\gamma^2 I & \mathcal{D}(\delta)^\top \\ \mathcal{C}(\delta) & \mathcal{D}(\delta) & -I \end{pmatrix} \prec 0, \tag{24.46}$$

where $\partial X(\delta, \rho)$ is defined in Equation 24.37. Mutatis mutandis, the convexifying controller transformation $(\mathcal{X}, A_c, B_c, C_c, D_c) \to (X, Y, K, L, M, N) =: v$ from Section 24.6 still works to arrive at a synthesis procedure for LPV output-feedback controllers. However, note that all matrices in the transformed variable v have become functions of $\delta \in \boldsymbol{\delta}$ or of $(\delta, \rho) \in \boldsymbol{\delta} \times \boldsymbol{\rho}$ now. In fact, finding a controller and a continuously differentiable $\mathcal{X}(\cdot)$ satisfying Equation 24.46 on $\boldsymbol{\delta} \times \boldsymbol{\rho}$ is equivalent to finding $v(\cdot)$ such that the synthesis inequalities

$$\boldsymbol{X}(v) \succ 0, \quad \begin{pmatrix} \boldsymbol{Z}(v) + \boldsymbol{A}(v)^\top + \boldsymbol{A}(v) & \boldsymbol{B}(v) & \boldsymbol{C}(v)^\top \\ \boldsymbol{B}(v)^\top & -\gamma^2 I & \boldsymbol{D}(v)^\top \\ \boldsymbol{C}(v) & \boldsymbol{D}(v) & -I \end{pmatrix} \prec 0 \tag{24.47}$$

hold on all of $\boldsymbol{\delta} \times \boldsymbol{\rho}$, where $\boldsymbol{Z}(v(\delta, \rho)) := \text{diag}(-\partial Y(\delta, \rho), \partial X(\delta, \rho))$. This turns the LPV controller synthesis problem into a linear matrix-function inequality in the decision variable $v(\cdot)$. A description of the related design algorithm can be found in [1].

Let us consider a particular popular scenario in more detail. Suppose that the matrices in Equation 24.42 are affine functions of δ and that B, E, C, and F are actually *independent* of δ. Also, suppose that the time-varying parameters $\delta(t)$ assume their values in the polytope (Equation 24.43) without any constraints on their rate-of-variation. We will search for an LPV controller in which the matrix functions in Equation 24.45 are also *affine* in δ. Finally, we let $\mathcal{X}$ in Equation 24.46 be constant, which implies that $\partial\mathcal{X}(\delta, \rho) = 0$.

These assumptions immediately imply that the closed-loop system matrices become affine in δ. Therefore, Equation 24.46 is satisfied for all $\delta \in \boldsymbol{\delta}$ if and only if Equation 24.46 is satisfied for the generators $\delta = \delta^k, k = 1, \ldots, N$, of the set $\boldsymbol{\delta}$. Hence we achieve a robust $\mathscr{L}_2$-gain smaller than γ for the controlled system if there exists $\mathcal{X}$ with

$$\mathcal{X} \succ 0 \quad \text{and} \quad \begin{pmatrix} \mathcal{A}(\delta^k)^\top\mathcal{X} + \mathcal{X}\mathcal{A}(\delta^k) & \mathcal{X}\mathcal{B}(\delta^k) & \mathcal{C}(\delta^k)^\top \\ \mathcal{B}(\delta^k)^\top\mathcal{X} & -\gamma^2 I & \mathcal{D}(\delta^k)^\top \\ \mathcal{C}(\delta^k) & \mathcal{D}(\delta^k) & -I \end{pmatrix} \prec 0 \quad \text{for } k = 1, \ldots, N. \tag{24.48}$$

Under the present structural assumptions, the transformed variables (K, L, M, N) also become affine functions in δ and $\boldsymbol{Z}(v) = 0$. Now apply the general convexifying transformation

$$(\mathcal{X}, A_c(\delta^k), B_c(\delta^k), C_c(\delta^k), D_c(\delta^k)) \to (X, Y, K_k, L_k, M_k, N_k) := v_k$$

with $k = 1, \ldots, N$ as in Section 24.6.2. This transforms Equation 24.48 into

$$\begin{pmatrix} Y & I \\ I & X \end{pmatrix} \succ 0, \quad \begin{pmatrix} \boldsymbol{A}(v_k, \delta^k)^\top + \boldsymbol{A}(v_k, \delta^k) & \boldsymbol{B}(v_k, \delta^k) & \boldsymbol{C}(v_k, \delta^k)^\top \\ \boldsymbol{B}(v_k, \delta^k)^\top & -\gamma^2 I & \boldsymbol{D}(v_k, \delta^k)^\top \\ \boldsymbol{C}(v_k, \delta^k) & \boldsymbol{D}(v_k, \delta^k) & -I \end{pmatrix} \prec 0 \quad \text{for } k = 1, \ldots, N. \tag{24.49}$$

We end up with a system of genuine LMIs in the parameters $v_1, \ldots, v_N$ that can be readily solved. In summary, this leads to the following LPV controller design procedure:

- Verify feasibility of (or minimize γ over) the synthesis inequalities (Equation 24.49) in the variables $v_k = (X, Y, K_k, L_k, M_k, N_k)$ for $k = 1, \ldots, N$.
- If a solution has been found, construct $\mathcal{X}$ as in the standard output-feedback synthesis procedure in Section 24.6.2.
- For this fixed $\mathcal{X}$, find controller parameters $\left(\begin{smallmatrix} A_{c,k} & B_{c,k} \\ C_{c,k} & D_{c,k} \end{smallmatrix}\right)$ that render Equation 24.48 satisfied for each $k = 1, \ldots, N$. These controller parameters can be obtained by solving Equation 24.25.
- If $\delta \in \boldsymbol{\delta}$ is represented by $\delta = \sum_{k=1}^N \alpha_k \delta^k$ with $\alpha_k \geq 0$, $\sum_{k=1}^N \alpha_k = 1$, then Equation 24.46 holds for

$$\begin{pmatrix} A_c(\delta) & B_c(\delta) \\ C_c(\delta) & D_c(\delta) \end{pmatrix} = \sum_{k=1}^N \alpha_k \begin{pmatrix} A_{c,k} & B_{c,k} \\ C_{c,k} & D_{c,k} \end{pmatrix}$$

  with $\mathcal{X}$ independent of δ and, consequently, $\partial\mathcal{X}(\delta, \rho) = 0$.

The actual LPV controller is now obtained by taking time-varying convex combinations of the N controllers defined by the quadruples $(A_{c,k}, B_{c,k}, C_{c,k}, D_{c,k})$. It is given by Equation 24.45 with

$$\begin{pmatrix} A_c(\delta(t)) & B_c(\delta(t)) \\ C_c(\delta(t)) & D_c(\delta(t)) \end{pmatrix} = \sum_{k=1}^N \alpha_k(t) \begin{pmatrix} A_{c,k} & B_{c,k} \\ C_{c,k} & D_{c,k} \end{pmatrix},$$

where the parameter $\delta(t)$ at time t is represented as $\delta(t) = \sum_{k=1}^N \alpha_k(t)\delta^k$ with $\alpha_k(t) \geq 0$, $\sum_{k=1}^N \alpha_k(t) = 1$. It achieves robust stability and a robust $\mathcal{L}_2$-gain performance smaller than γ.

24.10 Robustness Analysis and Synthesis with Multipliers

24.10.1 Robustness Analysis with IQCs

Uncertain systems are often described by a feedback interconnection with a stable LTI system in the forward path and some uncertainty in the feedback path as depicted in Figure 24.3. Mathematically, this amounts to

$$\begin{aligned} \dot{x} &= Ax + B_1 w + B_2 d, \\ z &= C_1 x + D_1 w + D_{12} d \quad \text{interconnected with } w = \Delta(z), \\ e &= C_2 x + D_{21} w + D_2 d, \end{aligned} \tag{24.50}$$

where A is Hurwitz and Δ denotes the uncertainty, a map that is supposed to be contained in some class $\boldsymbol{\Delta}$. We do not intend to include the theory around general systems and their abstract properties as, for example, developed in [6]. Instead we just assume that the interconnection is well-posed: For all $\Delta \in \boldsymbol{\Delta}$

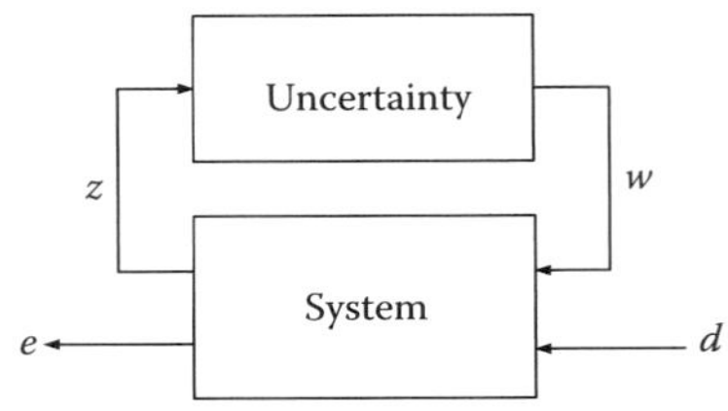

FIGURE 24.3 Uncertainty system.

and for all initial conditions $x(0)$ as well as arbitrary external signals d that are of finite energy on $[0, T]$ for all $T \geq 0$, the interconnection has unique responses x, z, w, and e that are of finite energy on finite subintervals of $[0, \infty)$.

The interconnection is said to be robustly stable if, for any $\Delta \in \mathbf{\Delta}$, arbitrary finite energy disturbance signals lead to responses of finite energy as well. The uncertain system has robust $\mathscr{L}_2$-gain smaller than γ if the interconnection is robustly stable and if there exists some $\epsilon > 0$ such that the $\mathscr{L}_2$-gain of $d \to e$ in Equation 24.50 for zero initial condition is bounded by $\gamma - \epsilon$ for all uncertainties $\Delta \in \mathbf{\Delta}$.

In order to computationally verify robust performance, we need to suitably capture information about the uncertainty set. Suppose, for example, that for all $\Delta \in \mathbf{\Delta}$ and all $z \in \mathscr{L}_2$ the signal $w = \Delta(z)$ satisfies

$$\int_0^T w(t)^\top w(t)\,dt \leq \int_0^T z(t)^\top z(t)\,dt \quad \text{for all } T \geq 0.$$

This just means that all uncertainties have an $\mathscr{L}_2$-gain bounded by 1. If, in addition, we also know that all uncertainties are passive, we have

$$\int_0^T w(t)^\top z(t)\,dt \geq 0 \quad \text{for all } T \geq 0.$$

These properties of the uncertainties can be expressed as IQCs on their input and output signals with the help of so-called *multipliers P*:

$$\int_0^T \begin{pmatrix} w(t) \\ z(t) \end{pmatrix}^\top P \begin{pmatrix} w(t) \\ z(t) \end{pmatrix} dt \geq 0 \quad \text{for all } T \geq 0 \quad \text{and} \quad P \in \{P_{\text{gain}}, P_{\text{passive}}\},$$

where the multipliers for the gain and passivity constraints are defined, respectively, as

$$P_{\text{gain}} = \begin{pmatrix} -I & 0 \\ 0 & I \end{pmatrix} \quad \text{and} \quad P_{\text{passive}} = \begin{pmatrix} 0 & I \\ I & 0 \end{pmatrix}. \tag{24.51}$$

With any set of such multipliers one can computationally verify robust performance on the basis of the following result.

Theorem 24.10:

*Let **P** be a family of symmetric matrices P which satisfy*

$$\int_0^T \begin{pmatrix} \Delta(z)(t) \\ z(t) \end{pmatrix}^\top P \begin{pmatrix} \Delta(z)(t) \\ z(t) \end{pmatrix} dt \geq 0 \quad \textit{for all } z \in \mathscr{L}_2[0, T],\ T \geq 0 \quad \textit{and} \quad \textit{all } \Delta \in \mathbf{\Delta}. \tag{24.52}$$

Then Equation 24.50 is robustly stable and has robust $\mathcal{L}_2$-gain performance smaller than γ if there exists $X = X^\top$ and $P = \begin{pmatrix} Q & S \\ S^\top & R \end{pmatrix} \in \boldsymbol{P}$ that satisfy the LMIs

$$X \succ 0 \quad \textit{and} \quad \left(\begin{array}{ccc} I & 0 & 0 \\ A & B_1 & B_2 \\ \hline 0 & I & 0 \\ C_1 & D_1 & D_{12} \\ \hline 0 & 0 & I \\ C_2 & D_{21} & D_2 \end{array}\right)^\top \left(\begin{array}{cc|cc|cc} 0 & X & 0 & 0 & 0 & 0 \\ X & 0 & 0 & 0 & 0 & 0 \\ \hline 0 & 0 & Q & S & 0 & 0 \\ 0 & 0 & S^\top & R & 0 & 0 \\ \hline 0 & 0 & 0 & 0 & Q_p & S_p \\ 0 & 0 & 0 & 0 & S_p^\top & R_p \end{array}\right) \left(\begin{array}{ccc} I & 0 & 0 \\ A & B_1 & B_2 \\ \hline 0 & I & 0 \\ C_1 & D_1 & D_{12} \\ \hline 0 & 0 & I \\ C_2 & D_{21} & D_2 \end{array}\right) \prec 0 \tag{24.53}$$

for the performance index $P_p = \begin{pmatrix} Q_p & S_p \\ S_p^\top & R_p \end{pmatrix} = \begin{pmatrix} -\gamma^2 I & 0 \\ 0 & I \end{pmatrix}$.

Proof. For some sufficiently small $\epsilon > 0$ we can add ϵI to the left-hand side of Equation 24.53 while keeping the inequality valid. Let us then choose any trajectory of the interconnection Equation 24.50 for an arbitrary uncertainty. Then right-multiply Equation 24.53 with $\text{col}(x(t), w(t), d(t))$ and left-multiply it with its transpose in order to obtain

$$\begin{pmatrix} x(t) \\ \dot{x}(t) \end{pmatrix}^\top \begin{pmatrix} 0 & X \\ X & 0 \end{pmatrix} \begin{pmatrix} x(t) \\ \dot{x}(t) \end{pmatrix} + \begin{pmatrix} w(t) \\ z(t) \end{pmatrix}^\top P \begin{pmatrix} w(t) \\ z(t) \end{pmatrix} + \begin{pmatrix} d(t) \\ e(t) \end{pmatrix}^\top P_p \begin{pmatrix} d(t) \\ e(t) \end{pmatrix}$$
$$+\epsilon \left(\|x(t)\|^2 + \|w(t)\|^2 + \|d(t)\|^2\right) \leq 0.$$

This inequality continues to hold after integration on $[0, T]$. As the key observation, we can exploit Equation 24.52 in order to conclude that one can drop the middle term in the first row without violating the inequality. Since the integral of the left-most term is $\int_0^T (\mathrm{d}/\mathrm{d}t)x(t)^T Xx(t)\,\mathrm{d}t = x(T)^\top Xx(T) - x(0)^\top Xx(0)$ and since $x(T)^\top Xx(T) \geq 0$, we obtain

$$\epsilon \int_0^T \|x(t)\|^2 + \|w(t)\|^2\,\mathrm{d}t + \int_0^T \|e(t)\|^2 - \gamma^2\|d(t)\|^2\,\mathrm{d}t \leq x(0)^\top Xx(0) - \epsilon \int_0^T \|d(t)\|^2\,\mathrm{d}t.$$

If d is of finite energy, we infer that $\int_0^T \|x(t)\|^2\,\mathrm{d}t$ and $\int_0^T \|w(t)\|^2\,\mathrm{d}t$ remain bounded for $T \to \infty$, which implies that x, w, and hence also z and e (due to Equation 24.50) are of finite energy. This proves stability. If $x(0) = 0$, we conclude that the $\mathcal{L}_2$-gain is strictly smaller than γ as for Theorem 24.4.

Clearly the very same result holds (with identical proof) for general robust quadratic performance (Section 24.5.4) with index

$$P_p = \begin{pmatrix} Q_p & S_p \\ S_p^\top & R_p \end{pmatrix} \quad \text{satisfying } R_p \succcurlyeq 0.$$

In order to verify robust stability and performance one only needs to check feasibility of the LMI in Equation 24.53. This is certainly possible if $\boldsymbol{P}$ is described by computationally tractable LMI constraints (such as in our example in which $\boldsymbol{P}$ just consists of two elements).

If Equation 24.52 holds for $\boldsymbol{P}$, it persists to hold for all P in the larger convex conic hull

$$\text{cc}(\boldsymbol{P}) := \left\{ \sum_{\nu=1}^{N} \tau_\nu P_\nu : P_\nu \in \boldsymbol{P},\ \tau_\nu \geq 0,\ \nu = 1, \ldots, N,\ N \in \mathbb{N} \right\},$$

which can lead to (sometimes substantially) improved computational results. Note that $\text{cc}(\boldsymbol{P})$ is a convex cone, since with any two of its elements P_1, P_2 it also contains their positive linear combination $\tau_1 P_1 + \tau_2 P_2$, $\tau_1, \tau_2 \geq 0$. Therefore, in applications, the set $\boldsymbol{P}$ in Theorem 24.10 is typically assumed to be a convex cone with an LMI representation.

24.10.2 Examples and Extensions

Many classical results on robust performance can be obtained as special cases of Theorem 24.10. Its real power, however, manifests itself in the flexibility to handle mixtures of structured uncertainties beyond such classical cases with ease. Here is a highly nonexhaustive sample.

24.10.2.1 Small-Gain and Passivity Theorems

With $\boldsymbol{P} = \{P_{\text{gain}}\}$ and $P_p = P_{\text{gain}}$ as defined in Equation 24.51, feasibility of Equation 24.53 just means that the $\mathscr{L}_2$-gain of the LTI system in Equation 24.50 is strictly smaller than one, while Equation 24.52 translates into $\Delta \in \boldsymbol{\Delta}$ having $\mathscr{L}_2$-gain smaller than one. This is simply the classical small-gain theorem. Similarly, for $\boldsymbol{P} = \{P_{\text{passive}}\}$ and $P_p = P_{\text{passive}}$, we obtain a passivity theorem which says that a passive interconnection in negative feedback with a passive system stays passive.

24.10.2.2 Sector Nonlinearities

As another variant let $\Delta(z)(t) = \phi(z(t))$ with any (Lipschitz-continuous) $\phi : \mathbb{R} \to \mathbb{R}$ whose graph is located in the conic sector between the lines $\{(x, y) \in \mathbb{R}^2 : \ y = \alpha x\}$ and $\{(x, y) \in \mathbb{R}^2 : \ y = \beta x\}$ for some real $\alpha \leq \beta$. For $D_1 = 0$, standard results on the existence of solutions of differential equations imply that Equation 24.50 is well-posed. Moreover, all these uncertainties satisfy the IQC in Equation 24.52 for all $P \in \boldsymbol{P}$ with $\boldsymbol{P}$ being equal to

$$\left\{ \begin{pmatrix} -2 & \alpha+\beta \\ \alpha+\beta & -2\alpha\beta \end{pmatrix} \right\} \text{ or its conic hull } \left\{ \tau \begin{pmatrix} -2 & \alpha+\beta \\ \alpha+\beta & -2\alpha\beta \end{pmatrix} : \ \tau \geq 0 \right\}. \tag{24.54}$$

This holds for example, with $\alpha = 0$ and $\beta > 0$ for the saturation nonlinearity

$$\phi(x) = \begin{cases} bx & \text{for } |x| \leq 1, \\ \beta \operatorname{sign}(x) & \text{for } |x| > 1. \end{cases}$$

For the interconnection $e = z = Gw + Gd$, $w = \Delta(z)$ with $G(s) = -(12(s+1)(s+2)(s+3))^{-1}$, a plot of the guaranteed $\mathscr{L}_2$-gain levels over the saturation slope β is shown in Figure 24.4. It indicates a well-known fundamental trade-off for obtaining less conservative results for larger sets of multipliers at the expense of higher computational complexity.

24.10.2.3 General Structured Uncertainties

Let $w = \text{col}(w_1, \ldots, w_q)$, $z = \text{col}(z_1, \ldots, z_q)$, and suppose that Δ is diagonally structured and defined by $w_\nu = \Delta_\nu(z_\nu)$, $\nu = 1, \ldots, q$. If Δ_ν satisfies an IQC with multiplier class $\boldsymbol{P}_\nu$, then Δ satisfies an IQC for all

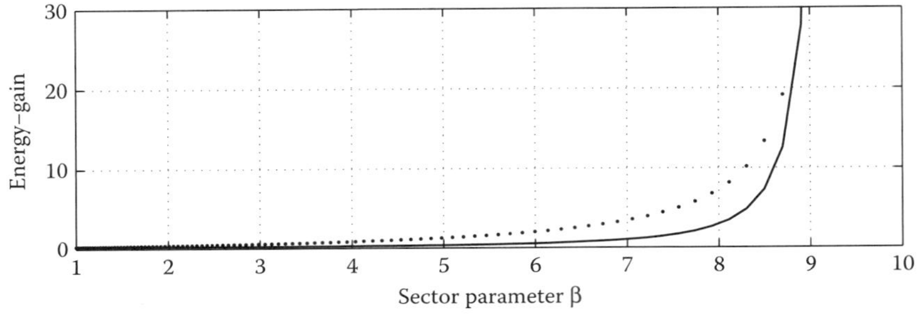

FIGURE 24.4 Results for multipliers in Equation 24.54 with $\alpha = 0$: Fixed (dotted); Cone (solid).

$P \in \boldsymbol{P}$ with

$$\boldsymbol{P} := \left\{ \begin{pmatrix} \text{diag}(Q_1, \ldots, Q_q) & \text{diag}(S_1, \ldots, S_q) \\ \text{diag}(S_1^\top, \ldots, S_q^\top) & \text{diag}(R_1, \ldots, R_q) \end{pmatrix} : \begin{pmatrix} Q_\nu & S_\nu \\ S_\nu^\top & P_\nu \end{pmatrix} \in \boldsymbol{P}_\nu, \;\; \nu = 1, \ldots, q \right\}.$$

If the sets $\boldsymbol{P}_\nu$, $\nu = 1, \ldots, q$, are represented by LMIs, the same is true for $\boldsymbol{P}$. Similarly if $\boldsymbol{P}_\nu$ are convex cones for $\nu = 1, \ldots, q$, so is the set $\boldsymbol{P}$. This diagonal augmentation procedure allows the construction of multipliers for diagonally structured uncertainties of an arbitrary nature. In this fashion, we can handle uncertainties whose diagonals are combinations of $\mathscr{L}_2$-gain, passive, sector-bounded static nonlinear, or time-varying parametric elements as discussed below.

24.10.2.4 Time-Varying Parametric Uncertainties

If $\Delta_\nu(z_\nu)(t) = \delta_\nu(t) z_\nu(t)$ for some $\delta_\nu(t) \in \mathbb{R}$ with $|\delta_\nu(t)| \leq 1$ for all $t \geq 0$, it is trivial to check that it satisfies an IQC for the convex cone

$$\boldsymbol{P}_\nu = \left\{ \begin{pmatrix} Q_\nu & S_\nu \\ S_\nu^\top & -Q_\nu \end{pmatrix} : \; Q_\nu \preccurlyeq 0, \; S_\nu + S_\nu^\top = 0 \right\}.$$

Diagonal augmentation leads to a set of diagonally structured multipliers for $\Delta(z)(t) = \Delta(t) z(t)$ with $\Delta(t) = \text{diag}(\delta_1(t) I, \ldots, \delta_q(t) I)$ as they are used in (frequency-by-frequency) structured singular value upper bound computations [20].

24.10.2.5 Full Block Multipliers

If $\Delta(z)(t) = \Delta(t) z(t)$ with some time-varying matrix $\Delta(t)$ which satisfies $\Delta(t) \in \text{co}\{\Delta_1, \ldots, \Delta_N\}$, a simple convexity argument reveals that Equation 24.52 holds for

$$P \in \left\{ P = P^\top : \begin{pmatrix} I \\ 0 \end{pmatrix}^\top P \begin{pmatrix} I \\ 0 \end{pmatrix} \preccurlyeq 0, \;\; \begin{pmatrix} \Delta_\nu \\ I \end{pmatrix}^\top P \begin{pmatrix} \Delta_\nu \\ I \end{pmatrix} \succcurlyeq 0, \;\; \nu = 1, \ldots, N \right\}.$$

Clearly this set of unstructured multipliers has an LMI description and can be readily implemented. If $\Delta(t)$ admits a diagonal structure as in the previous example, one can compare diagonally structured with full multipliers; the latter often lead to less conservative analysis results at a higher computational cost.

For time-varying parametric uncertainties as considered in the latter two examples, it is remarkable to note that the feasibility of the LMI in Equation 24.53 does actually imply nonsingularity of $I - D_1 \Delta(t)$ and hence well-posedness of the interconnection (Equation 24.50); moreover one can even prove exponential stability of Equation 24.50.

We have only hinted at the power of static IQCs, while the extension to allow for dynamics in the multipliers even further expands the wealth of the applications and their computational power [14].

24.10.3 Robust Synthesis

For the purpose of synthesis, the system (Equation 24.50) is extended with a control channel as

$$\begin{aligned} \dot{x} &= Ax + B_1 w + B_2 d + Bu \\ z &= C_1 x + D_1 w + D_{12} d + E_1 u \\ e &= C_2 x + D_2 w + D_{22} d + E_2 u \\ y &= Cx + F_1 w + F_2 d \end{aligned} \qquad \text{interconnected with } w = \Delta(z), \;\; \Delta \in \boldsymbol{\Delta}. \qquad (24.55)$$

The goal is to design a controller (Equation 24.17) which achieves robust performance for the closed-loop system. With a set $\boldsymbol{P}$ of IQC multipliers satisfying Equation 24.52, we need to enforce the LMI in Equation 24.53 for the closed-loop system description (after having verified well-posedness). This opens

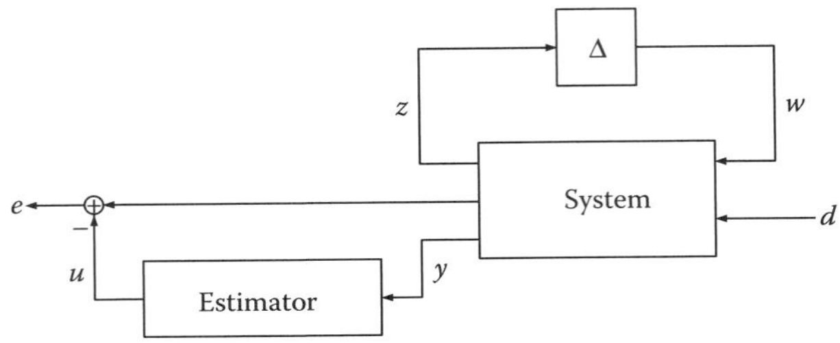

FIGURE 24.5 Robust estimator design.

up the opportunity to apply the general synthesis procedure from Section 24.6.3. Convex constraints result if $\boldsymbol{P}$ just consists of one element P whose right-lower block is positive semidefinite, since the problem then boils down to one of quadratic performance synthesis as discussed in Section 24.5.4. Unfortunately, convexification is impossible if $\boldsymbol{P}$ consists of a whole family of multipliers. This has led to the suggestion of various heuristic algorithms in order to approach the solution of the bilinear matrix inequalities for robust performance synthesis.

A particularly lucky case is the robust estimator synthesis problem for a configuration as depicted in Figure 24.5. Then Equation 24.55 specializes to $B = 0, E_1 = 0$, and $E_2 = -I$. After having checked well-posedness (which does not involve the to-be-designed estimator), we can apply the convexifying estimator parameter transformation from Section 24.6.4. The general procedure results in the synthesis inequalities

$$\begin{pmatrix} Z & Z \\ Z & X \end{pmatrix} \succ 0,$$

$$(\star)^\top \left(\begin{array}{cccc|cc|cc} 0 & 0 & I & 0 & 0 & 0 & 0 & 0 \\ 0 & 0 & 0 & I & 0 & 0 & 0 & 0 \\ I & 0 & 0 & 0 & 0 & 0 & 0 & 0 \\ 0 & I & 0 & 0 & 0 & 0 & 0 & 0 \\ \hline 0 & 0 & 0 & 0 & Q & S & 0 & 0 \\ 0 & 0 & 0 & 0 & S^\top & R & 0 & 0 \\ \hline 0 & 0 & 0 & 0 & 0 & 0 & Q_p & S_p \\ 0 & 0 & 0 & 0 & 0 & 0 & S_p^\top & R_p \end{array}\right) \left(\begin{array}{cc|cc} I & 0 & 0 & 0 \\ 0 & I & 0 & 0 \\ ZA & ZA & ZB_1 & ZB_2 \\ XA + LC + K & XA + LC & XB_1 + LF_1 & XB_2 + LF_2 \\ \hline 0 & 0 & I & 0 \\ C_1 & C_1 & D_1 & D_{12} \\ \hline 0 & 0 & 0 & I \\ C_2 - NC - M & C_2 - NC & D_{21} - NF_1 & D_2 - NF_2 \end{array}\right) \prec 0,$$

which are actually convex in X, Z and Q, S, R. Since $R_p \succcurlyeq 0$ can be factorized as $T_p^\top T_p$ with a full row-rank matrix T_p, one can easily turn the second constraint into a genuine LMI by taking the Schur complement.

Finally, let us stress that dualization arguments allow a similar convexification procedure for the design of robust state-feedback controller gains, as shown for parametric uncertainties in [16].

24.11 Conclusions

We have provided a selective overview on the application of semidefinite programming techniques in control in order to cover the most basic but essential ideas in the analysis and synthesis of nominal and robust controllers. During the last decades this field has grown dramatically and the scope of applications of LMIs in solving complex control problems has widened considerably. Notable activities have been devoted to broadening the LMI stability and performance specifications and to exploring their applicability to time-varying, time-delay, fuzzy and nonlinear systems. Considerable advances have been made in constructing systematic relaxation schemes for handling robust LMIs. Recently emerging applications of convex optimization to the synthesis of distributed controllers for large-scale systems offer exiting new avenues for progress.

Appendix: Convex Sets and Convex Functions

A set $\mathcal{S}$ in a linear vector space is said to be *convex* if

$$x_1, x_2 \in \mathcal{S} \text{ implies } x = \alpha x_1 + (1-\alpha)x_2 \in \mathcal{S} \quad \text{for all } \alpha \in (0,1).$$

Geometrically, this states that the line segment connecting any two points of the set belongs to the set as well. For all $\alpha \in (0,1)$, the point x defined in the above expression is called a *convex combination* of x_1 and x_2. More generally, the point $x = \sum_{k=1}^{n} \alpha_k x_k$ is a *convex combination* of $x_1, \ldots, x_n \in \mathcal{S}$ if $\alpha_k \geq 0$ for all k and $\sum_{k=1}^{n} \alpha_k = 1$. Many operations preserve convexity of sets. As the most important one, the intersection of an arbitrary collection of convex sets is convex. Simple examples of convex sets are *hyperplanes* $\{x \in \mathbb{R}^n \mid a^\top x = b\}$ and *half-spaces* $\{x \in \mathbb{R}^n \mid a^\top x \leq b\}$, where $a \in \mathbb{R}^n, b \in \mathbb{R}$. A *polyhedron* is the intersection of finitely many hyperplanes and half-spaces and is hence convex. A *polytope* is a compact polyhedron. It is easy to see that the set of all convex combinations of n points $x_1, \ldots, x_n \in \mathcal{S}$ is itself convex. For any subset $\mathcal{S}$ of a linear vector space the *convex hull* $\text{co}(\mathcal{S})$ is the set of all convex combinations of the elements of $\mathcal{S}$. The convex hull of a finite set of points is always a polytope and, conversely, any polytope is the convex hull of a finite set.

A (Hermitian-valued) function $F : \mathcal{S} \to \mathbb{H}^m$ is *convex* if its domain $\mathcal{S}$ is convex and

$$F\left(\sum_{k=1}^{n} \alpha_k x_k\right) \preccurlyeq \sum_{k=1}^{n} \alpha_k F(x_k) \quad \text{for all } x_1, \ldots, x_n \in \mathcal{S} \quad \text{and} \quad \alpha_k \geq 0 \text{ with } \sum_{k=1}^{n} \alpha_k = 1.$$

This is referred to as Jensen's inequality. F is *strictly convex* if Jensen's inequality holds with $\prec$ in the case that neither of the $\alpha_k{}'s$ equals one. If F is real valued, the inequalities are the same as the usual $\leq$ and $<$ for real numbers. Generally it is not easy to verify whether a function is convex. Twice continuously differentiable functions $F : \mathcal{S} \to \mathbb{R}$ on convex sets $\mathcal{S}$ with interior points are convex iff their Hessian satisfies $\partial^2 F(x) \succcurlyeq 0$ for all $x \in \mathcal{S}$. If F defined on $\mathcal{S}$ is convex, then the *sublevel sets* $\{x \in \mathcal{S} \mid F(x) \prec H\}$ are convex for any $H \in \mathbb{H}^m$. The most important reason for considering convex functions in optimization is the fact that local minimal points of convex functions are actually global minimal points. Precisely, $x_0 \in \mathcal{S}$ is a local minimal point of F if there exists an open neighborhood $\mathcal{N}(x_0)$ of x_0 such that $F(x_0) \preccurlyeq F(x)$ for all $x \in \mathcal{N}(x_0) \cap \mathcal{S}$. If F is convex one can conclude that $F(x_0) \preccurlyeq F(x)$ for all $x \in \mathcal{S}$, showing that x_0 is a global minimal point of F. This property is of great interest in numerical optimization. Indeed, many efficient algorithms exist for the numerical computation of local minima of real-valued functions. If these are convex, such algorithms actually determine global minimal points.

References

1. P. Apkarian and R. J. Adams. Advanced gain-scheduling techniques for uncertain systems. *IEEE Control System Magazine*, 6(1):21–32, 1998.
2. V. Balakrishan and L. Vandenberghe. Semidefinite programming duality and linear time-invariant systems. *IEEE Transactions on Autmatic Control*, 48(1):30–41, 2003.
3. S.P. Boyd, L. El Ghaoui, E. Feron, and V. Balakrishan. *Linear Matrix Inequalities in System and Control Theory*. SIAM Studies in Applied Mathematics 15. SIAM, Philadelphia, 1994.
4. M. Chilali and P. Gahinet. H_∞ design with pole placement constraints: An LMI approach. *IEEE Transactions on Automatic Control*, 41(3):358–367, 1996.
5. M.C. De Oliveira, J.C. Geromel, and J. Bernussou. Extended H_2 and H_∞-norm characterizations and controller parametrizations for discrete-time systems. *International Journal of Control*, 75(9):666–679, 2002.
6. C.A. Desoer and M. Vidyasagar. *Feedback Systems: Input–Output Approach*. Academic Press, London, 1975.
7. L. El Ghaoui and S.I. Niculescu, Eds. *Advances in Linear Matrix Inequality Methods in Control*. SIAM, Philadelphia, 2000.

8. P. Gahinet and P. Apkarian. A linear matrix inequality approach to H_∞ control. *International Journal of Robust and Nonlinear Control*, 4:421–448, 1994.
9. J.C. Geromel. Optimal linear filtering under parametric uncertainty. *IEEE Transactions on Signal Processing*, 47(1):168–175, 1999.
10. S.V. Gusev and A.L. Likhtarnikov. Kalman–Popov–Yakubovich lemma and the S-procedure: A historical essay. *Automation and Remote Control*, 67(11):1768–1810, 2006.
11. T. Iwasaki and R.E. Skelton. All controllers for the general $\mathcal{H}_\infty$ control problem: LMI existence conditions and state space formulas. *Automatica*, 30:1307–1317, 1994.
12. J. Löfberg. YALMIP: A toolbox for modeling and optimization in MATLAB. In *Proceedings of the CACSD Conference*, Taipei, Taiwan, 2004.
13. I. Masubuchi, A. Ohara, and N. Suda. LMI-based controller synthesis: A unified formulation and solution. *International Journal of Robust and Nonlinear Control*, 8:669–686, 1998.
14. A. Megretski and A. Rantzer. System analysis via integral quadratic constraints. *IEEE Transactions on Automatic Control*, 42:819–830, 1997.
15. C.W. Scherer. An efficient solution to multi-objective control problems with LMI objectives. *System & Control Letters*, 40(1):43–57, 2000.
16. C.W. Scherer. Robust mixed control and LPV control with full block scalings. In L. El Ghaoui and S.I. Niculescu, Eds, *Advances in Linear Matrix Inequality Methods in Control*, pp. 187–207. SIAM, Philadelphia, 2000.
17. C.W. Scherer, P. Gahinet, and M. Chilali. Multi-objective output-feedback control via LMI optimization. *IEEE Transactions on Automatic Control*, 42:896–911, 1997.
18. S. Skogestad and I. Postlethwaite. *Multivariable Feedback Control, Analysis and Design*. John Wiley & Sons, New York, 1996.
19. J.C. Willems. Dissipative dynamical systems, part II: Linear systems with quadratic supply rates. *Archive for Rational Mechanics and Analysis*, 45:352–393, 1972.
20. K. Zhou, J. Doyle, and K. Glover. *Robust and Optimal Control*. Prentice-Hall, Upper Saddle River, NJ, 1996.

25

Optimal Control

Frank L. Lewis
The University of Texas at Arlington

25.1 Introduction to Optimal Control Design

Many systems occurring naturally in fields such as biology and sociology use feedback control to achieve homeostasis, or equilibrium conducive to existence. Because the bounds within which life can continue are small (e.g., temperature changes of a few degrees can eliminate populations) and the *resources available* are limited, it is remarkable yet not expected that most of these feedback control systems have evolved into *optimal* systems where performance objectives are achieved efficiently with a minimum of control effort. Since naturally occurring systems are optimal, it makes sense to design man-made controllers from the view point of optimality.

25.1.1 The Philosophy of Classical Control

Classical control theory, developed in the mid-1900s, imparts a great deal of engineering insight. It was best developed for linear systems. Since computers were not available to solve complex design equations, the design algorithms are heuristic in terms of Bode plots, Nyquist plots, the root locus, and other single-input/single-output (SISO) graphical techniques that offer intuition and rely on the design engineer's expertise. Since most design was in the frequency domain, robustness to unknown disturbances, modeling errors, and noise was automatically built in.

Complex modern systems have multiple inputs and outputs. Examples include aircraft, satellites, and automobile engines, which, though nonlinear, can often be linearized about a desired operating point or trajectory. In such applications, classical design relies on successive loop closures based on *one-loop-at-a-time SISO design*. Unfortunately, using this approach, neither stability nor robustness of the overall system can be guaranteed, since one loop closure can destroy what has been gained in the design of previous loops.

25.1.2 The Philosophy of Optimal Control Design

About 1960, modern optimal control theory began developing for complex multivariable systems. It developed coincidentally with the space age (Sputnik was launched in 1957), the computer age, and the age of robotics. Optimal control is a branch of modern control theory that deals with designing controls for dynamical systems by minimizing a *performance index* that depends on the system variables. The performance index might include, for instance, a measure of operating error, a measure of control "effort," or any other characteristic important to the user of the control system. Under some mild assumptions, making the performance index small also guarantees that the system variables will be small, thus insuring *closed-loop stability.*

Classical design is concerned with directly selecting the feedback gains K in the inner *real-time control loops*. On the other hand, modern control design offers standard algorithms for implementing an *outer design loop* that automatically selects the inner loop feedback gains in such a fashion that *closed-loop stability and performance for MIMO systems is guaranteed.* In contrast to classical one-loop-at-a-time design, in modern control, *all of the feedback loops are closed simultaneously* by computing the feedback gains by solving standard *matrix design equations*. Special purpose software [9,10] is commercially available to solve these equations, so that control design for complex systems is straightforward with a personal computer (PC).

In this chapter, optimal control design is discussed for deterministic systems with a well-known mathematical model. The major emphasis is on linear systems in the state-space form. It will be assumed that *full state-variable feedback* is available; in the event that only partial information is available on the system states, the chapter on "Output Feedback" or the references should be consulted as well. The discussion in this chapter will center around continuous-time systems, with a follow-up discussion on discrete-time (DT) systems. Several design techniques will be covered, including the regulator problem, the tracker problem, minimum-time control, and polynomial design. Robustness of the linear quadratic regulator (LQR) will be discussed using multivariable frequency-domain design techniques, which allow one to draw close connections with classical control theory. This is a condensed version of presentations available in [7,8,11] where derivations and computer software appear. Important key foundation references are [1–6].

There are two basic approaches to optimal control. The Calculus of Variations leads to a formulation of the optimal control solution in terms of partial derivatives of the Hamiltonian, which is an energy-based construct that captures the prescribed optimality criteria in terms of motion along the system trajectories.

This approach leads to optimal control solutions in terms of the state equation, costate equation, and stationarity condition. Bellman's Optimality Principle, on the other hand, leads to Dynamic Programming (DP) solutions, which develop backwards in time. From this, one obtains minimizations with respect to the Hamiltonian, which leads to the Hamilton–Jacobi–Bellman (HJB) equations. First we present the Calculus of Variations results, then the DP results.

25.2 Optimal Control of Continuous-Time Systems

In this section nonlinear control design will be covered for general nonlinear systems. Then, the LQR will be developed for linear systems. The LQR is a cornerstone of modern control theory design.

25.2.1 The General Continuous-Time Optimal Control Problem

A state-variable model for a nonlinear time-varying dynamical system is given by Equation 25.1 in Table 25.1, where $x(t) \in \mathcal{R}^n$ is the vector of internal states and $u(t) \in \mathcal{R}^m$ is the vector of control inputs. This is the plant to be controlled. A broad range of performance objectives may be achieved by selecting the control $u(t)$ to minimize a *performance index* (PI) or a *cost* given by Equation 25.2 with t_0 the initial time and T the final time of interest. The *final-state weighting function* $\phi(x(T), T)$ and weighting function $L(x, u, t)$ are selected depending on the performance objectives.

The *optimal control problem* is to determine a control input $u(t)$ for the system that minimizes the PI and also insures that the *final state constraint* (Equation 25.3) is satisfied for a given function $\psi \in \mathcal{R}^p$. The roles of the final weighting function ϕ and the final constraint ψ should not be confused. The former is a function one would like to minimize, such as the final energy $x^T(T)S(T)x(T)$, with $S(T)$ a specified weighting matrix. On the other hand, $\psi(x(T), T)$ must be exactly equal to zero. A sample problem might be to find the control input $u(t)$ that drives a satellite, with dynamics described by Equation 25.1, from a given initial position $x(t_0)$ to an specified orbit, described by Equation 25.3, while minimizing the expended energy, as described by the PI (Equation 25.2).

25.2.1.1 Solution of the Nonlinear Optimal Control Problem

Using the Calculus of Variations approach to solve the optimal control problem, Lagrange multipliers are used to adjoin the constraints (Equations 25.1 and 25.3) to the performance index (Equation 25.2). Since the system Equation 25.1 is an equality constraint which must hold at each time, an associated multiplier $\lambda(t) \in \mathcal{R}^n$ is required that is a function of time. Thus, the *Hamiltonian function* is defined as Equation 25.4. Using the theory of Lagrange multipliers and the calculus of variations, the solution to the optimal control problem given in Table 25.1 is determined. It is assumed that the initial time t_0 and the initial state $x(t_0)$ are both known and fixed. In the boundary conditions, partial derivatives are denoted by subscripts (e.g., ψ_x represents $\frac{\partial \psi}{\partial x}$).

The equations in the table may be used as *design equations* for determining the control $u(t)$ that minimizes the PI. They are *necessary conditions* for the solution of the nonlinear optimal control problem. Any control $u(t)$ that results in a minimum value of the PI, when it is applied to the system, must satisfy the equations given there. Conditions under which these equations are sufficient as well are addressed in [3,7].

The structure of the equations is worth discussing. According to the table, the Lagrange multiplier $\lambda(t)$ is a dynamical variable that satisfies its own dynamical Equation 25.6; it is called the *costate*. The optimal control $u(t)$ is then generally determined in terms of $x(t)$ and $\lambda(t)$ by using the *stationarity condition* (Equation 25.7) (so named because this is the condition that guarantees a minimum or stationary point with respect to changes in $u(t)$). The value of $\lambda(t)$ is usually of no concern ultimately, but it is an *intermediate variable* which must be determined to solve for the optimal control $u(t)$, that minimizes the PI $J(t_0)$ while insuring that constraints (Equations 25.1 and 25.3) are satisfied. The appearance of

TABLE 25.1 Continuous Nonlinear Optimal Controller

System model:

$$\dot{x} = f(x, u, t), \quad t \geq t_0, \ t_0 \ \textit{fixed}. \tag{25.1}$$

Performance index:

$$J(t_0) = \phi(x(T), T) + \int_{t_0}^{T} L(x, u, t)\, dt. \tag{25.2}$$

Final state constraint:

$$\psi(x(T), T) = 0. \tag{25.3}$$

Optimal Controller:

Hamiltonian:

$$H(x, u, t) = L(x, u, t) + \lambda^T f(x, u, t). \tag{25.4}$$

State equation:

$$\dot{x} = \frac{\partial H}{\partial \lambda} = f, \quad t \geq t_0. \tag{25.5}$$

Costate equation:

$$-\dot{\lambda} = \frac{\partial H}{\partial x} = \frac{\partial f^T}{\partial x}\lambda + \frac{\partial L}{\partial x}, \quad t \leq T. \tag{25.6}$$

Stationarity condition:

$$0 = \frac{\partial H}{\partial u} = \frac{\partial L}{\partial u} + \frac{\partial f^T}{\partial u}\lambda. \tag{25.7}$$

Boundary conditions:

$$x(t_0) \ \text{given, initial condition,} \tag{25.8}$$

$$(\phi_x + \psi_x^T \nu - \lambda)^T |_T \ dx(T) + (\phi_t + \psi_t^T \nu + H) |_T \ dT = 0, \quad \text{final condition.} \tag{25.9}$$

intermediate variables, that are required to solve for the variables of interest, is typical of optimal control design.

The dynamical state and costate equations, along with the control specified by the stationarity condition, are called the *Hamiltonian system*. These equations may be used to derive Lagrange's and Hamilton's equations of motion in physics (see Example 25.1). The costate equation and stationarity condition are called *Euler's equations*. In the time-invariant case, f and L are not explicit functions of t, so that neither is H. In this situation

$$\dot{H} = 0. \tag{25.10}$$

Thus for time-invariant systems and cost functions, the Hamiltonian is a *constant* on the optimal trajectory. This is a general statement of the principle of conservation of energy.

25.2.1.2 Two-Point Boundary-Value Problems

The solution for the optimal control $u(t)$ in Table 25.1 depends on solving two coupled differential equations, the state Equation 25.5 and the costate Equation 25.6, each of which is of order n. These two dynamical equations comprise the Hamiltonian system once the stationarity condition has been used to eliminate $u(t)$. The costate equation develops *backward in time* (by defining a backward time variable $\tau = T - t, d\tau = -dt$), with the final condition $\lambda(T)$ determined by Equation 25.9. The boundary conditions consist of the initial conditions on the state

$$n \text{ conditions:} \quad x(t_0) \ \textit{given} \tag{25.11}$$

and the final conditions on the costate

$$p \text{ conditions:} \quad \psi(x(T), T) = 0 \tag{25.12}$$

$$n - p \text{ conditions:} \quad (\phi_x + \psi_x^T \nu - \lambda)^T \mid_T \ dx(T) = 0, \tag{25.13}$$

where it has been assumed for simplicity that the final time T is specified and hence fixed, so that $dT = 0$ in condition (Equation 25.9).

Since n boundary conditions are specified at the initial time t_0 and n conditions are specified at the final time T, this is a *two-point boundary-value problem*. There are many methods available for solving such problems, including the *shooting point method* and the *unit solution method*; good software is available for this purpose. The solution of the optimal control problem for nonlinear systems is often difficult, though, for some nonlinear plants, the design equations can be explicitly solved for the optimal control $u(t)$, yielding a great deal of insight. This includes the *Thrust Angle Programming* and *Intercept and Rendezvous* problems. In the special case that the plant is linear and the PI is quadratic, a solution is available, given subsequently.

Example 25.1: Hamilton's Principle

Both Lagrange's and Hamilton's equations of motion may be derived from Table 25.1.

LAGRANGE'S EQUATIONS OF MOTION

Define the generalized coordinate state vector q and the "control input" as the generalized velocities $u = \dot{q}$. Define the Lagrangian $L(q, u) \equiv T(q, u) - U(q)$ as the difference between the kinetic and potential energies. Then the "plant" (Equation 25.1) is

$$\dot{q} = u \equiv f(q, u)$$

where the function $f(\cdot)$ is given by the physics of the problem.

To find the trajectories of the motion, Hamilton's principle says we may minimize the performance index

$$J = \int_0^T L(q, u)\, dt,$$

so that the Hamiltonian (Equation 25.4) is $H = L + \lambda^T u$. According to Table 25.1, for a minimum

$$-\dot{\lambda} = \frac{\partial H}{\partial q} = \frac{\partial L}{\partial q}$$
$$0 = \frac{\partial H}{\partial u} = \frac{\partial L}{\partial u} + \lambda.$$

Combining these equations yields Lagrange's equations of motion,

$$\frac{\partial L}{\partial q} - \frac{d}{dt}\frac{\partial L}{\partial \dot{q}} = 0.$$

TABLE 25.2 Continuous-Time Linear Quadratic Regulator

System model:

$$\dot{x} = Ax + Bu, \quad t \geq t_0, \quad x(t_0) = x_0 \text{ given.} \tag{25.14}$$

Performance index:

$$J(t_0) = \frac{1}{2}x^T(T)S(T)x(T) + \frac{1}{2}\int_{t_0}^{T}(x^TQx + u^TRu)\,dt, \tag{25.15}$$

with

$$S(T) \geq 0, \quad Q \geq 0, \quad R > 0.$$

Optimal feedback control:

Riccati equation:

$$-\dot{S} = A^TS + SA - SBR^{-1}B^TS + Q, \quad t \leq T, \quad S(T) \text{ given.} \tag{25.16}$$

Optimal feedback gain:

$$K = R^{-1}B^TS. \tag{25.17}$$

Time-varying feedback:

$$u = -K(t)x. \tag{25.18}$$

Optimal cost:

$$J(t_0) = \frac{1}{2}x_0^TS(t_0)x_0. \tag{25.19}$$

HAMILTON'S EQUATIONS OF MOTION

Defining the generalized momentum vector by $\lambda = -\partial L/\partial\dot{q}$, the equations of motion may be expressed in Hamilton's form as

$$\dot{q} = \frac{\partial H}{\partial\lambda},$$

and

$$-\dot{\lambda} = \frac{\partial H}{\partial q}.$$

25.2.2 Continuous-Time Linear Quadratic Regulator

The nonlinear optimal control design equations in Table 25.1 are not easy to solve, and there is no design algorithm for doing so. In this section the design of optimal controllers for linear systems with quadratic performance indices will be discussed, the so-called LQR *problem*. The LQR is a cornerstone of modern optimal control design consisting of *explicit matrix design equations easily solved on a digital computer.* It has a wide range of relevance, because many systems are linear to begin with, and many nonlinear systems may be considered linear when operating near an equilibrium point. The LQR solution is given as a *closed-loop feedback control.*

Consider the multivariable linear system (Equation 25.14) in Table 25.2 with state $x \in \mathcal{R}^n$ and control input $u \in \mathcal{R}^m$. The plant matrices may be time-varying (e.g., A(t), B(t)), though for notational convenience this dependence will not be shown explicitly.

Choose the control that minimizes the quadratic PI (Equation 25.15). The *control weighting R, state weighting Q*, and *final state weighting S(T)* are symmetric matrices of design parameters chosen by the designer depending on the control objectives. For instance, if the elements of $S(T)$ are selected larger, then the control will force the final state $x(T)$ to be smaller to keep the PI small. Weight matrices Q and $S(T)$ are assumed positive-semidefinite ($Q \geq 0, S(T) \geq 0$). Thus Q and $S(T)$ have nonnegative eigenvalues so that $x^T Qx$ and $x^T(T)S(T)x(T)$ are nonnegative for all $x(t)$. Likewise, it will be assumed that R is positive definite ($R > 0$), that is, R has positive eigenvalues so that $u^T Ru > 0$ for all $u(t)$. In this case, J is always bounded below by zero, so that a sensible minimization problem results. Since the squares of the states and control inputs occur in Equation 25.15, the PI is a form of generalized energy (consider the case when some of the state components are velocities, or currents and voltages) and minimizing it will keep the states and controls small.

Using the equations in Table 25.1, the solution to the LQR optimal control problem may be derived, as given in Table 25.2. The state and costate equations are

$$\dot{x} = Ax + Bu \tag{25.20}$$

and

$$-\dot{\lambda} = Qx + A^T\lambda, \tag{25.21}$$

where the negative sign indicates that the costate equation must be solved backward in time. The stationarity condition gives the control in terms of the costate as $u(t) = -R^{-1}B^T\lambda(t)$. To find the optimal control, the two-point boundary-value problem associated with the state and costate dynamics is analytically solved using the *sweep method* [3] where it is assumed that $\lambda(t) = S(t)x(t)$ for some unknown auxiliary matrix $S(t)$. The optimal control is given in terms of this intermediate matrix as follows. First, it can be determined that the auxiliary matrix $S(t)$ satisfies the bilinear *matrix Riccati equation* 25.16. In terms of the Riccati solution S(t), the optimal control is given by $u(t) = -R^{-1}B^T S(t)x(t)$. Thus, defining the *optimal feedback gain* by Equation 25.17, one may write the optimal control as the *state-feedback control law* (Equation 25.18).

A block diagram of the LQR is shown in Figure 25.1. It is a *feedback control system* with time-varying feedback gains $K(t)$, and a formal *design outer loop*. Even if the system (A, B) is time-invariant, the optimal control $u(t)$ is a *time-varying state feedback*. This is why the optimal LQ controller may not be determined using classical frequency-domain techniques. If the system model (Equation 25.14) is not an exact description of the plant, the LQR still performs well if it is fairly close. In fact, it will be seen in Section 25.2.4.2 that the LQR has important *guaranteed robustness properties*.

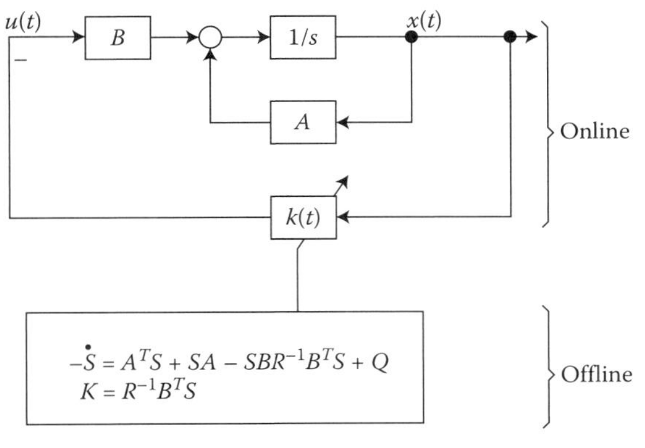

FIGURE 25.1 Linear quadratic regulator.

25.2.2.1 LQR Design

To design the optimal LQR, the design engineer first selects the design parameter weight matrices Q, R, $S(T)$. Then, the Riccati equation (RE) is solved for the auxiliary matrix function $S(t)$, which is used to compute $K(t)$. The RE is solved *backward in time* using, as a final condition, the value of the weighting matrix $S(T)$ selected for the PI. This equation may be solved *off-line* for $S(t)$, and the optimal feedback gain $K(t)$ computed and stored. Next, a *computer simulation* is generally performed to verify the closed-loop performance. If the performance is not suitable, new design matrices $Q, R, S(T)$ are selected and the entire procedure is repeated. With commercially available software, the entire process is fast and convenient. Finally, during the implementation or control run, the states are measured and the feedback control $u(t) = -K(t)x$ applied to the plant.

Thus, by the sweep method, the LQR problem has been decomposed into two stages: offline computation of the optimal gains using a backward differential equation, followed by the actual control of the plant using feedback. Such *hierarchical control schemes*, consisting of an inner linear feedback loop whose gain is computed by an outer quadratic design equation, are typical of modern control schemes. Optimal control is fundamentally a *noncausal design algorithm* requiring future information about the plant and performance objectives.

The LQR design procedure is in stark contrast to classical control design, where the gain matrix K is selected directly. In modern optimal control design, some parameter matrices Q, R, and $S(T)$ are selected by the engineer. Then, the feedback gain K is automatically given by *matrix design equations*. This has the significant advantages of allowing *all the control loops in a multiloop system to be closed simultaneously, while guaranteeing closed-loop stability*.

The optimal cost of using this controller is given in terms of the initial state by Equation 25.19. The initial state of the plant is known. Therefore, this expression allows computation of the optimal cost *before* the control is actually applied to the plant, or even before the optimal gain $K(t)$ is computed and it is simulated on a computer. If the cost is too high, the engineer can select different weighting matrices Q, R, and $S(T)$ in the performance index and try another design. This *preview feature* is typical of optimal control design using state feedback.

Example 25.2: LQR for Armature-Controlled DC Motor

The system equations of an armature-controlled DC motor are

$$\begin{aligned} \dot{i} &= -ai - k'\omega + bu \\ \dot{\omega} &= -\alpha\omega + ki \end{aligned} \tag{25.22}$$

with $i(t)$ the armature current, $\omega(t)$ the motor speed, control input $u(t)$ the armature voltage, $1/a$ the electrical time constant, $1/\alpha$ the mechanical time constant, and the remaining variables other motor parameters.

Defining the state as $x = [i \quad \omega]^T$,

$$\dot{x} = \begin{bmatrix} -a & -k' \\ k & -\alpha \end{bmatrix} x + \begin{bmatrix} b \\ 0 \end{bmatrix} u \equiv Ax + Bu. \tag{25.23}$$

It is required to determine $u(t)$ to minimize the PI

$$J = \frac{1}{2} x^T(T) \begin{bmatrix} s_i & 0 \\ o & s_\omega \end{bmatrix} x(T) + \frac{1}{2}\int_0^T \left[x^T \begin{bmatrix} q_i & 0 \\ 0 & q_\omega \end{bmatrix} x + ru^2 \right] dt \tag{25.24}$$

with s_i, s_ω the final state weights, q_i, q_ω the (intermediate) state weights, and r the control weight. These are design parameters that may be adjusted or *tuned* using computer simulations to yield suitable closed-loop behavior, as will be seen in this example. The minimization of J corresponds to the regulation objective of driving the motor to a speed of zero from any initial speed, while keeping the control energy small. If the model represents a linearization of a nonlinear motor about a set point, the control will regulate the motor speed to that set point.

Since the RE solution $S(t)$ is symmetric, one may assume that

$$S = \begin{bmatrix} s_1 & s_2 \\ s_2 & s_3 \end{bmatrix} \tag{25.25}$$

where the scalars $s_i(t)$ are to be determined. Substituting A, B from the state equation and $S(T), Q, R$ from the PI in the RE in Table 25.2 yields the three nonlinear scalar coupled differential equations

$$\begin{aligned} -\dot{s}_1 &= -2as_1 + 2ks_2 - \beta s_1^2 + q_i, \\ -\dot{s}_2 &= -(a+\alpha)s_2 - k's_1 + ks_3 - \beta s_1 s_2, \\ -\dot{s}_3 &= -2\alpha s_3 - 2k's_2 - \beta s_2^2 + q_\omega, \end{aligned} \tag{25.26}$$

where $\beta \equiv b^2/r$.

Writing the feedback gain as $K(t) = [k_i \quad k_\omega]$, the table shows that $K = R^{-1}B^T S$, so that

$$k_i = \frac{b}{r}s_1, \quad k_\omega = \frac{b}{r}s_2. \tag{25.27}$$

Then, the optimal control is given by the time-varying feedback $u = -k_i i - k_\omega \omega$.

Although the equations for $s_i(t)$ are difficult to solve analytically, it is easy to use computer software to solve the RE. Using such software, the optimal state trajectories and control voltage are plotted for $r = 1$ and several values of $q = q_i = q_\omega$. The results are displayed in Figure 25.2. Note that the states go to zero more quickly as q increases, while the controls become larger. Final weights of $s_i = s_\omega = 0$ were used. Based on the simulation results, suitable values for the PI weights can be selected. Then, the associated $K(t)$ may be stored in memory and applied to the actual motor during the control implementation run.

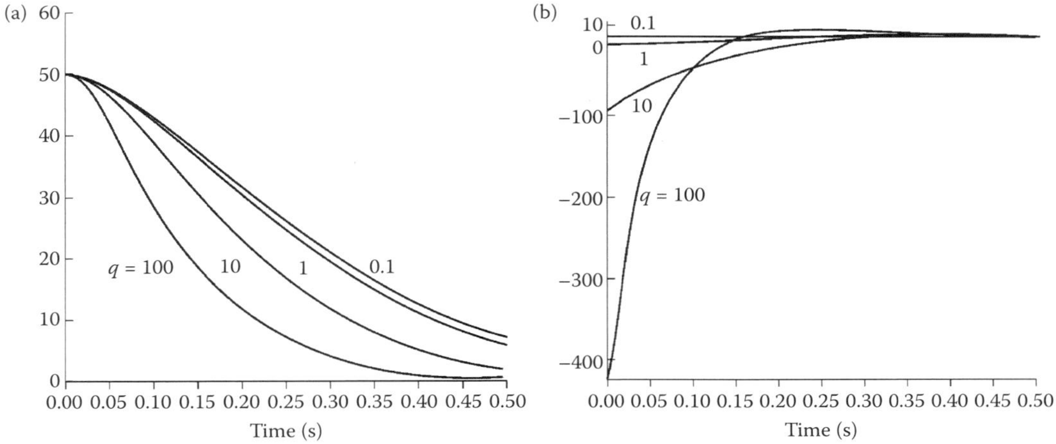

FIGURE 25.2 Results of DC motor simulation. (a) Motor speed. (b) Optimal control voltage.

25.2.3 Steady-State and Suboptimal Control

Even for time-invariant plants, the optimal LQ control is a *time-varying* state-variable feedback. Such feedbacks are inconvenient to implement, because they require the storage in computer memory of time-varying gains. An alternative control scheme will now be given in which the time-varying optimal gain $K(t)$ is replaced by its constant steady-state (e.g., $t \to \infty$) value. In most practical applications, this use of the steady-state feedback gain is adequate.

The results of this section are important. Showing how to find *multi-loop feedback gains for multi-input systems that are guaranteed to stabilize the closed-loop system*. The gains are determined simply by solving a *matrix design equation* using computer routines available in standard software packages such as MATLAB® and MATRIXx™. This goes far beyond what can be achieved with classical design techniques, which revolve around one-loop-at-a-time procedures offering no stability guarantees.

25.2.3.1 Steady-State Control—Guaranteed Stability of the LQR

Suppose the plant to be controlled has the linear description

$$\dot{x} = Ax + Bu, \tag{25.28}$$

with $x \in \mathcal{R}^n$ and control input $u \in \mathcal{R}^m$. For this section it will be necessary to assume that the plant is time-invariant.

Now, the control should be selected to minimize the quadratic PI

$$J(t_0) = \frac{1}{2}\int_0^{\infty} (x^TQx + u^TRu)\, dt, \tag{25.29}$$

with $Q \geq 0$ and $R > 0$. Since the integration interval is infinite, this is called an *infinite horizon* performance index; the performance objectives are referred to an infinite control interval $[0, \infty)$.

The control law of Table 25.2 still applies; however, because the control horizon is infinite, the RE may reach a steady-state solution where $\dot{S} = 0$. In this case, the RE may be replaced by the *algebraic Riccati equation (ARE)*

$$0 = A^TS + SA - SBR^{-1}B^TS + Q. \tag{25.30}$$

This is a symmetric matrix quadratic equation. $S(T)$ no longer appears in this steady-state formulation.

The ARE can have multiple solutions. However, if certain mild assumptions on the system and PI matrices hold, then there is a single *positive definite* solution S_∞, namely, the limiting solution to the time-varying RE for any $S(T)$. Then, the optimal infinite-horizon gain is the *constant* matrix given by

$$K_\infty = R^{-1}B^TS_\infty. \tag{25.31}$$

Thus, the optimal steady-state control is the *constant state-variable feedback*

$$u(t) = -K_\infty x(t). \tag{25.32}$$

Moreover, the optimal cost is given in terms of the initial state by

$$J = \frac{1}{2}x^T(0)S_\infty x(0). \tag{25.33}$$

Under the influence of the steady-state control the closed-loop plant has the *time-invariant* dynamics

$$\dot{x} = (A - BK_\infty)x \equiv A_cx. \tag{25.34}$$

The advantages of this simplified control that uses a constant feedback are clear. The next theorem is vital to modern control theory, and shows that the steady-state LQR is *guaranteed to stabilize the system*, even if it has multiple inputs, as long as the plant satisfies some basic properties. The system (Equation 25.28) is said to be *stabilizable* if the control input $u(t)$ can be selected to stabilize all the modes in the closed-loop. This is a weaker property than *reachability*, which requires that there is a $u(t)$ that drives any given initial state to any desired final state. These are both controllability properties of the plant. The system is said to be *observable* through the output $y = Cx$ if measurements of only $y(t)$ can be used to reconstruct the entire initial state $x(0)$. This is a stronger condition than *detectability* of (A, C), which says that if $y(t) \to 0$ then $x(t) \to 0$. These are both observability properties of the plant.

Reachability, stabilizability, observability, and detectability are basic *open-loop properties* that hold for any well-behaved system. There are simple tests for reachability and observability. In fact, if n is equal to the number of states, the system is reachable if, and only if, the reachability matrix

$$U = \begin{bmatrix} B & AB & A^2B & \dots & A^{n-1}B \end{bmatrix} \tag{25.35}$$

has rank n. This implies stabilizability as well. The system is observable if, and only if, the observability matrix

$$V = \begin{bmatrix} C \\ CA \\ \vdots \\ CA^{n-1} \end{bmatrix} \tag{25.36}$$

has rank n. This implies detectability as well. Standard software packages such as MATLAB and $MATRIX_X$ offer routines to compute these block matrices.

The next result is a cornerstone of LQR theory.

Theorem 25.1: Stability of Closed-Loop System

Let C be any square root of Q so that $Q = C^TC$. Suppose (C, A) is detectable and (A, B) is stabilizable. Then:

1. *There is a unique symmetric positive-semidefinite limiting solution S_∞ to the Riccati equation 25.16 independent of the choice of $S(T)$. Furthermore, S_∞ is the unique positive-definite solution of the ARE.*
2. *The closed-loop plant A_c is asymptotically stable.*

This result means that, as long as the system and PI satisfy certain basic controllability and observability requirements, *the steady-state LQ regulator will yield gains that stabilize the system*. Considering the difficulty encountered by classical control techniques in stabilizing multi-input systems, this is a remarkable property. Exactly as in classical control theory, the theorem predicts the *closed-loop* stability properties of the system in terms of *open-loop* system properties that are easily tested using matrix rank techniques.

The detectability of $(\sqrt{Q}, A)$ is needed for the stability result. This property plays out as follows. If the infinite integral cost (Equation 25.29) has been minimized, then it has a finite value so that the integrand goes to zero with time. Then $x^TQx = \|\sqrt{Q}\,x\|^2 \to 0$. Then detectability of $(\sqrt{Q}, A)$ guarantees that $x(t) \to 0$. Detectability means in fact that all the unstable plant modes should be weighted in the PI, which imposes a design requirement on the engineer as he selects the weighting matrix Q.

The closed-loop poles will depend on the selection of the design matrices Q and R; however, the poles will always be stable as long as the engineer selects $R > 0$ and $Q \geq 0$ with $(\sqrt{Q}, A)$ observable. Thus, the elements of Q and R may be varied during an interactive computer-aided design procedure to obtain suitable closed-loop performance. The optimal gain K is found for given values of Q and R, and the closed-loop time responses are found by simulation. If these responses are unsuitable, new values for Q and R are selected and the design is repeated. Given good software to solve for K, this procedure is quite convenient. Such software is available, for instance, in MATLAB and $MATRIX_X$.

25.2.3.2 Suboptimal Control—Constant Feedback Gains

Even if the control interval $[0, T]$ is not infinite, the engineer may decide to use the steady-state gain K_∞ instead of the optimal time-varying gain $K(t)$ given in Table 25.2. The theorems guarantee stability of the closed-loop system using the steady-state LQR. On a finite interval $[0, T]$, the constant gain K_∞ is *suboptimal*, but the convenience gained by not having to implement a time-varying gain can more than compensate for the loss of optimality. Moreover, as T becomes large, the optimal gain $K(t)$ tends to K_∞ so that the decision to use the steady-state gain makes more and more sense. In addition to the ease of implementation of constant feedback gains, this suboptimal controller has other important advantages: (1) *it guarantees stability even for complex multi-loop systems*, and (2) there are efficient numerical routines available for the solution of the ARE (e.g., MATLAB and MATRIXx).

Example 25.3: Inverted Pendulum

Figure 25.3 shows a rod attached to a cart through a pivot. A force $u(t)$ is applied to the cart through a motor attached to an axle. The control objective is to use $u(t)$ to balance the pendulum upright while simultaneously keeping the horizontal movement $p(t)$ of the cart small. This is known as the *inverted pendulum problem*.

The state is $x = [\theta\ \dot{\theta}\ p\ \dot{p}]^T$. A force/moment balance approach or a Lagrangian approach may be used to obtain the dynamics. Assuming that $M = 5\ kg, m = 0.5\ kg, L = 1\ m$, we thus obtain the state equations by linearizing the dynamics about $\theta = \dot{\theta} = p = \dot{p} = 0$:

$$\dot{x} = \begin{bmatrix} 0 & 1 & 0 & 0 \\ 10.78 & 0 & 0 & 0 \\ 0 & 0 & 0 & 1 \\ -0.98 & 0 & 0 & 0 \end{bmatrix} x + \begin{bmatrix} 0 \\ -0.2 \\ 0 \\ 0.2 \end{bmatrix} u = Ax + Bu. \tag{25.37}$$

The open-loop poles are at $s = 0, 0, \pm 3.283$, so that, with no control input, the rod will clearly fall over due to the unstable pole at $s = 3.283$.

It is desired to select K in

$$u = -Kx = -(k_\theta \theta + k_{\dot{\theta}} \dot{\theta} + k_p p + k_{\dot{p}} \dot{p}) \tag{25.38}$$

to regulate the state x to zero. For this purpose, select the PI (Equation 25.29); then, K is determined by using the matrix design Equations 25.30 and 25.31. Values of $R = 1$ and $Q = diag\{100, 100, 10, 10\}$ were selected. The motivation for choosing this Q was to place heavy emphasis on keeping the angle $\theta(t)$ small; the cart position control does not matter if the rod falls over. Using MATLAB subroutines from the

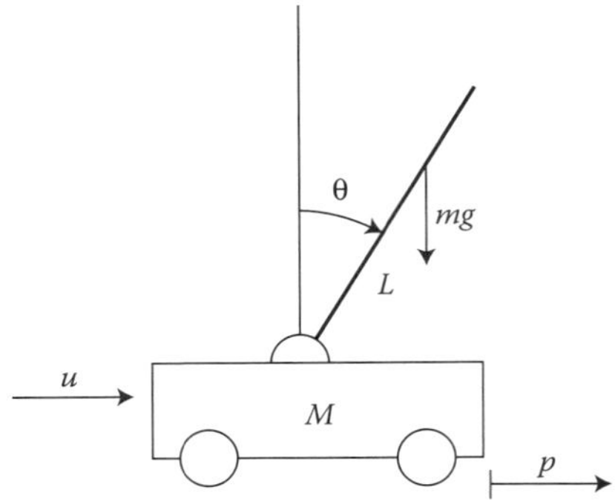

FIGURE 25.3 Inverted pendulum on a cart.

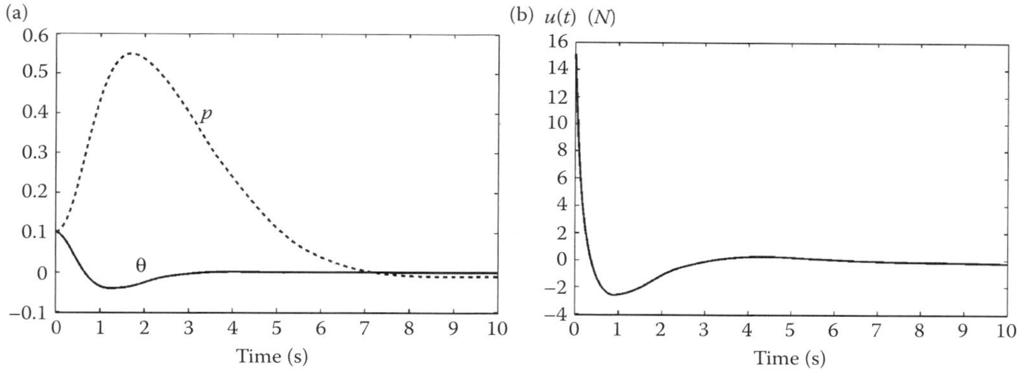

FIGURE 25.4 Inverted pendulum response. (a) Angle $\theta(t)$ and position $p(t)$. (b) Control input $u(t)$.

Control System Toolbox, the optimal gain was easily found to be $K = [-156.16 \quad -49.21 \quad -3.16 \quad -8.72]$, which yields closed-loop poles at $s = -0.60 \pm j0.45, -2.48, -4.41$.

Using MATLAB routines, the closed-loop response was *simulated* and plotted. The angle $\theta(t)$ and position $p(t)$ in response to an initial condition offset of $\theta(0) = 0.1 rad \approx 6°, p(0) = 0.1 m$ are shown in Figure 25.4a. The required control force $u(t)$ is shown in Figure 25.4b. These plots are quite interesting and bear discussion. Due to the initial offset of 6° in angle, a large control must be applied immediately to push the cart under the rod to catch it so it does not fall. Subsequent smaller control motions begin to move the cart slowly back to the desired horizontal position of $p = 0$ while balancing the rod. Thus, the fast closed-loop poles correspond to the rod motion. The slow complex pole pair is associated with the cart position. This *two-time scale behavior was induced* by the widely disparate weightings selected in the design matrix Q.

In fact, the value for Q was selected by performing *several design iterations* with different Q until computer simulation finally showed a good time response. Such design iterations, coupled with computer simulation, are common in modern control and are very easy using software like MATLAB.

25.2.3.3 Eigenstructure LQR Design

In most computer software routines for solving for the LQR gains, the ARE solution is determined from the eigenstructure of the Hamiltonian matrix H in the *Hamiltonian system*

$$\begin{bmatrix} \dot{x} \\ \dot{\lambda} \end{bmatrix} = \begin{bmatrix} A & -BR^{-1}B^T \\ -Q & -A^T \end{bmatrix} \begin{bmatrix} x \\ \lambda \end{bmatrix} \equiv H \begin{bmatrix} x \\ \lambda \end{bmatrix} \tag{25.39}$$

which consists of the state Equation 25.20 and the costate Equation 25.21, with $u(t)$ replaced by $-R^{-1}B^T\lambda(t)$.

The Hamiltonian matrix H enjoys the special property of having n stable poles and n unstable poles (their images in the $j\omega$ axis), where n is the dimension of the state vector x. If (A, B) is stabilizable, and $(\sqrt{Q}, A)$ is observable, so that the conditions of Theorem 25.1 hold, then the stable eigenvalues of H are also the poles of the optimal closed-loop system

$$\dot{x} = (A - BK_\infty)x \equiv A_c x, \tag{25.40}$$

with K_∞ the steady-state feedback gain. This provides an alternative proof of the stability of the optimal closed-loop system.

Select the eigenvectors of the stable eigenvalues of H, and partition them as $[X_i^T \quad \Lambda_i^T]^T$. Let X be an $n \times n$ matrix whose columns are X_i, and Λ be an $n \times n$ matrix whose columns are Λ_i. Then the solution to the ARE is given in terms of the eigenstructure of H by

$$S_\infty = \Lambda X^{-1}, \tag{25.41}$$

and the steady-state gain is given by

$$K_\infty = R^{-1} B^T \Lambda X^{-1}. \tag{25.42}$$

(If the eigenvalues are complex, then, in the definitions of X and Λ, it is necessary to use the real and imaginary parts of the associated vectors X_i and Λ_i instead of the complex conjugate vectors themselves.)

25.2.4 Frequency-Domain Results and Robustness of the LQR

It is possible to discuss the LQR from the view point of the frequency domain. Classical frequency-domain results are given in terms of scalar SISO systems and involve notions like the loop gain, return difference, sensitivity, and so on. Such ideas are extended in modern optimal design using multivariable transfer functions and the notions of the *singular value* and the *multivariable Bode plot*. For more details, see Chapter 8 or the references. An important property of any closed-loop system is *robustness* to uncertainties, including modeling errors, disturbances, and noise. The LQR has some important robustness properties that are detailed here.

25.2.4.1 LQR Frequency-Domain Relationships

Suppose that the plant is time-invariant and, in Figure 25.1, K is the constant optimal LQ state-feedback gain determined using the LQR ARE as in Section 25.2.3.1. Define the plant transfer function as $G(s) \equiv (sI - A)^{-1}B$. Then, the *loop gain* referred to the input is

$$KG(s) = K(sI - A)^{-1}B = KG(s) \tag{25.43}$$

and the *closed-loop return difference* is $[I + K(sI - A)^{-1}B] = I + KG(s)$.

25.2.4.1.1 Optimal Return Difference Relationship

Two key results are the following. The *optimal characteristic polynomial relationship* is

$$\Delta_c(s) = |I + K(sI - A)^{-1}B|\Delta(s) \tag{25.44}$$

where the open-loop characteristic polynomial is $\Delta(s) = |sI - A|$ and the closed-loop characteristic polynomial is $\Delta_c(s) = |sI - (A - BK)|$. The *optimal return difference relationship* is

$$[I + K(-sI - A)^{-1}B]^T R\,[I + K(sI - A)^{-1}B] = R + B^T(-sI - A)^{-T}Q(sI - A)^{-1}B. \tag{25.45}$$

These are extremely important because, exactly as in classical control theory, they express *closed-loop properties* in terms of *open-loop properties* that can be computed before the optimal controller is designed. They allow, for instance, the development of the *Chang-Letov design approach for LQR* which is an extension of root locus design to MIMO systems.

25.2.4.1.2 Optimal Singular Value Relationships

Select the control weighting matrix as $R = \rho I$, with ρ a positive design parameter. Denoting the ith singular value of a matrix M as $\sigma_i(M)$, Equation 25.45 yields the *optimal singular value relationship* of the LQR

$$\sigma_i[I + KG(j\omega)] = \left[1 + \frac{1}{\rho}\sigma_i^2[H(j\omega)]\right]^{\frac{1}{2}} \tag{25.46}$$

with

$$H(s) \equiv C(sI - A)^{-1}B \tag{25.47}$$

and matrix C defined by $Q = C^TC$. This is important because the right-hand side is known in terms of *open-loop* quantities before the optimal feedback gain is found by solution of the ARE, while the left-hand side is the closed-loop return difference.

According to this relationship, for all ω, the minimum singular value, denoted $\underline{\sigma}$, satisfies the *LQ optimal singular value constraint*

$$\underline{\sigma}[I + KG(j\omega)] \geq 1. \tag{25.48}$$

Thus, the LQ regulator always results in a *decreased sensitivity*.

25.2.4.2 Guaranteed Robustness of the Linear-Quadratic Regulator

The linear-quadratic regulator using full state feedback has many useful properties, including guaranteed closed-loop stability and ease of design by solving matrix design equations. It will now be shown that the steady-state LQR has certain *guaranteed robustness properties* that make it even more useful. These conclusions may be discovered using the *multivariable Nyquist criterion*, which shall be referred to the polar plot of the return difference $I + KG(s)$, where the origin is the critical point. A typical polar plot of $\underline{\sigma}[I + KG(j\omega)]$ is shown in Figure 25.5, where the optimal singular value constraint appears as the condition that *all the singular values remain outside the unit disc.*

25.2.4.2.1 Guaranteed Stability

The multivariable Nyquist criterion says that (as long as the open-loop system $G(s)$ is stable) the closed-loop system is stable if none of the singular value plots of $I + KG(j\omega)$ encircle the origin in the figure. Due to the optimal singular value constraint, no encirclements are possible. This constitutes a proof of the *guaranteed stability* of the LQR discussed in Section 25.2.3.1.

25.2.4.2.2 Gain Margin

Multiplying the optimal feedback K by any positive scalar gain $k > 1$ results in a loop gain of $kKG(s)$, which has a minimum singular value plot identical to the one in Figure 25.5 except that it is scaled outward; that is, the $\omega \to 0$ limit (i.e., the DC gain) will be larger, but the $\omega \to \infty$ limit will still be 1.

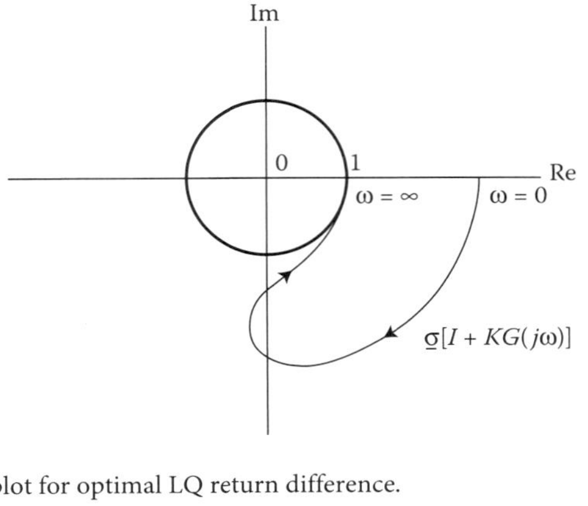

FIGURE 25.5 Typical polar plot for optimal LQ return difference.

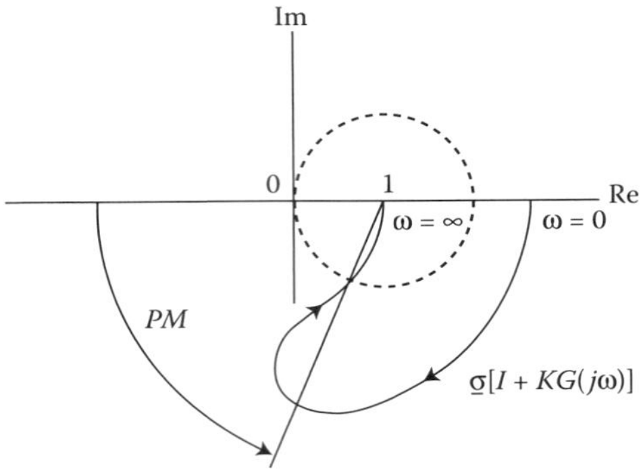

FIGURE 25.6 Definition of multivariable phase margin.

Thus, the closed-loop system will still be stable. In classical terms, the LQ regulator with full state feedback has *an infinite gain margin*.

25.2.4.2.3 Phase Margin

For multivariable systems the *phase margin* may be defined as the angle marked "PM" in Figure 25.6. As in the classical case, it is the angle through which the polar plot of $\underline{\sigma}[I + KG(j\omega)]$ must be rotated (about the point 1) clockwise to make the plot go through the critical point. By combining Figure 25.5 with Figure 25.6, it can be determined that, due to the LQ singular value constraint, the plot of $\underline{\sigma}[I + KG(j\omega)]$ must be rotated through at least 60° to make it pass through the origin. The LQR with full state feedback thus has a *guaranteed phase margin of at least* 60°. This means that a phase shift of up to 60° may be introduced in any of the m paths in Figure 25.1, or in all paths simultaneously, as long as the paths are not coupled to each other in the process. (Here, m is the number of control inputs, i.e., the number of control loops.)

This phase margin is excessive; it is higher than that normally required in classical control system design. This overdesign means that, in other performance aspects, the LQ regulator may have some deficiencies. One of these is that, at the *crossover frequency* (loop gain = 1), the slope of the *multivariable Bode plot* is -20 dB/decade, a relatively slow attenuation rate. (By allowing a Q weighting matrix in the PI that is not positive semidefinite, it is possible to obtain better LQ designs that have higher roll-off rates at high frequencies.)

25.2.4.2.4 Stability with Multiplicative Uncertainty

It can be shown that Equation 25.48 implies that

$$\underline{\sigma}[I + (KG(j\omega))^{-1}] \geq \frac{1}{2}. \tag{25.49}$$

This corresponds to the fact that the LQR with state feedback remains stable for all *multiplicative uncertainties* in the plant transfer function which satisfies $m(\omega) < \frac{1}{2}$.

25.3 The Tracker Problem

The function of the LQ regulator is to hold the states near zero, that is, to guarantee closed-loop stability. Another fundamental design problem in systems engineering is to control a system so that a specified output follows a given nonzero *reference trajectory* $r(t)$. An example is controlling an aircraft to follow a desired step input command (e.g., change in altitude). This is called the *tracking or servodesign* problem. For this purpose, the regulator control law must be modified. The fundamental issue here is that for optimal tracking some additional *feedforward terms* must be added to the control input besides the basic LQR feedback loop that gives closed-loop stability.

Consider the linear system given by Equation 25.51, with $z(t)$ in Equation 25.52 a *performance output* that should track the given reference input $r(t)$. In contrast to classical control, it is easy to include here the case where both $z(t)$ and $r(t)$ are vectors, that is, the case of multivariable tracking. The control input is given by

$$u = -Kx + v, \tag{25.50}$$

where $v(t)$ is a feedforward signal required for good tracking performance. The feedback gain K and feedforward signal $v(t)$ are determined so as to keep the *tracking error* (Equation 25.53) small.

The solution to the tracking problem is significantly more involved than the regulator problem. It is now discussed from three points of view. In this section, full state-variable feedback is assumed. In actual applications, only *output feedback* is allowed. In that case refer to Chapter 5.1 or to [8,11], where these techniques are extended.

25.3.1 Optimal LQ Tracker

The optimal LQ tracker can be derived along the same lines as the Hamiltonian approach in Section 25.2.2. The result is given in Table 25.3. The feedback structure is basically the same as the LQ regulator in

TABLE 25.3 Optimal Continuous-Time Linear Quadratic Tracker

System model:

$$\dot{x} = Ax + Bu, \quad t \geq t_0, \quad x(t_0) = x_0 \textit{ given}. \tag{25.51}$$

Performance output and tracking error:

$$z = Hx, \tag{25.52}$$

$$e = r - z. \tag{25.53}$$

Performance index:

$$J(t_0) = \frac{1}{2}e^T(T)Pe(T) + \frac{1}{2}\int_{t_0}^{T}(e^TQe + u^TRu)\,dt, \tag{25.54}$$

with

$$P \geq 0, \quad Q \geq 0, \quad R > 0.$$

Optimal tracking controller:

Riccati equation:

$$-\dot{S} = A^TS + SA - SBR^{-1}B^TS + H^TQH, \quad t \leq T, \quad S(T) = H^TPH. \tag{25.55}$$

Optimal feedback gain:

$$K = R^{-1}B^TS. \tag{25.56}$$

Feedforward system:

$$-\dot{w} = (A - BK)^Tw + H^TQr, \quad t \leq T, \quad w(T) = H^TPr(T). \tag{25.57}$$

Feedback plus feedforward control:

$$u(t) = -K(t)x(t) + R^{-1}B^Tw(t). \tag{25.58}$$

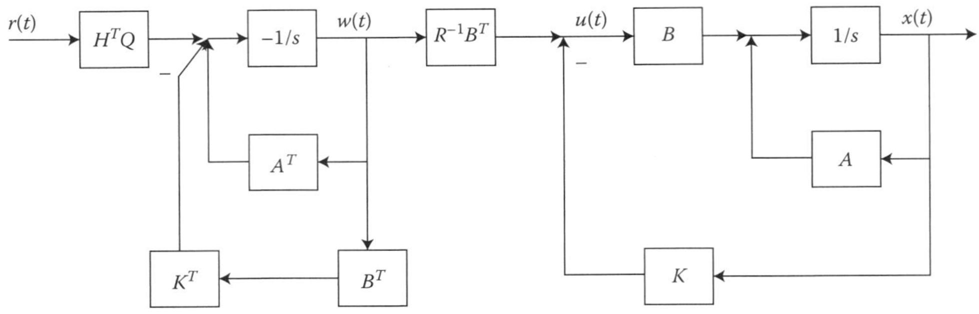

FIGURE 25.7 Optimal LQ tracker.

Table 25.2, but now the PI contains the tracking error, not the state, because e(t) is to be small in this application. The presence of $e(t)$ in the PI has the effect of adding a feedforward term $v(t)$ to the control signal. The tracker gain K and signal $v(t)$ determined with the LQ optimal approach are given in the table. The feedback gain K is found in the same manner as in the LQR case. The structure of the LQ optimal tracker is given in Figure 25.7.

The feedforward signal is computed using the dynamical system (Equation 25.57), which is called the *adjoint system*; like the RE, it is integrated *backward* in time. Thus, the RE and the adjoint system must be integrated off-line before the control run. In fact, the optimal LQ tracker is *noncausal*, because future values of the reference input $r(t)$ are needed to compute $w(t)$. The ramifications of this noncausal nature of the optimal tracker are illustrated in Figure 25.8, which shows the optimal tracker response for a scalar system using control weighting $R = 1$ and different values of the error weighting Q. The system output begins to change *before* the reference $r(t)$ does, so that the system anticipates the changes in $r(t)$. This *anticipatory behavior* is an important feature of the optimal tracker.

The noncausal nature of the optimal LQ tracker means that it cannot be implemented in practice when $r(t)$ is not predetermined. Therefore, two suboptimal strategies are now outlined that yield implementable tracking systems.

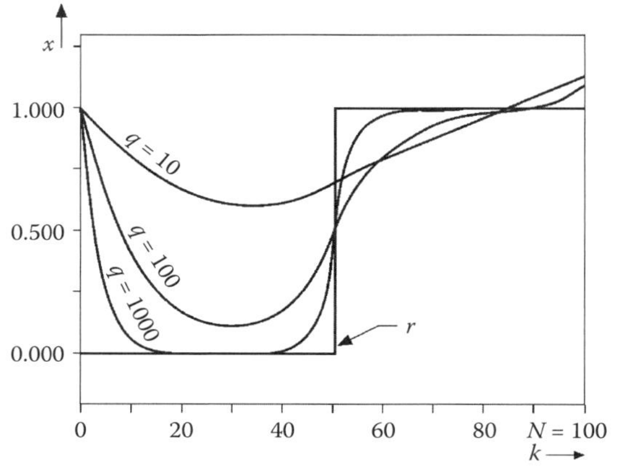

FIGURE 25.8 Anticipatory response of the optimal LQ tracker.

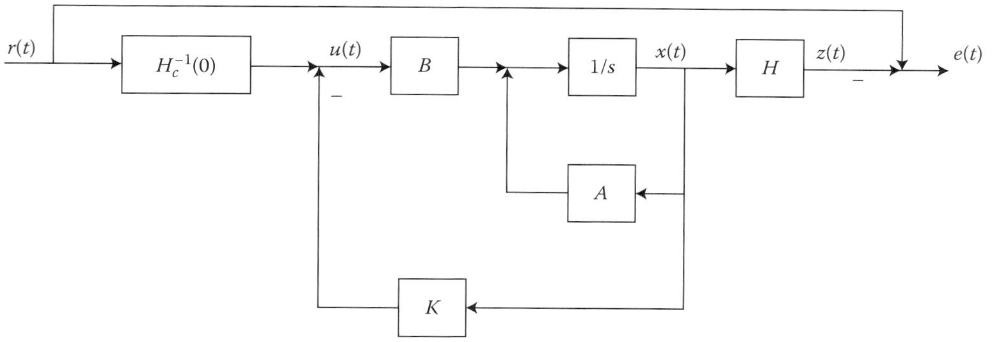

FIGURE 25.9 Tracker based on DC gain.

25.3.2 Conversion of an LQR to an LQ Tracker

As an alternative to the optimal tracking solution just presented, a causal tracker can be obtained as follows. First, the LQ regulator is designed using Table 25.2. Then, it is converted to an LQ tracker by adding a feedforward term. In the case where the reference signal $r(t)$ is a constant (i.e., step function) with magnitude r_0, the tracking control with state feedback is given by

$$u = -Kx + H_c^{-1}(0)r_0, \tag{25.59}$$

where the closed-loop transfer function is

$$H_c(s) = H(sI - (A - BK))^{-1}B \tag{25.60}$$

and $H_c(0)$ is the *DC gain* of the closed-loop system $(A - BK)$. The control gain K is found using the LQR design equations in Table 25.2. The structure of this suboptimal tracker is shown in Figure 25.9. Unfortunately, if the DC gain is not well known this tracker structure does not perform well, that is, this tracker is not *robust*.

25.3.3 A Practical Suboptimal Tracker

This section shows how to design a suboptimal tracker that works well for practical applications and is robust to uncertainties and disturbances. The key is in the use of engineering design insight and common sense to formulate the problem. One uses a *unity feedback gain outer loop*, which has proven effective in classical control approaches. This technique also relies on converting an LQR to a tracker, but differs from the work in the previous section.

25.3.3.1 Problem Formulation

A general class of systems is described by the equations

$$\begin{aligned} \dot{x} &= Ax + Bu + Er \\ z &= Hx \end{aligned} \tag{25.61}$$

which can contain both the plant plus some desirable *compensator dynamics*. The control input is allowed to have the form

$$u = -Kx - KFr, \tag{25.62}$$

which consists of state feedback plus a feedforward term of a special composition. Placing the control into the system yields the closed-loop system

$$\dot{x} = (A - BK)x + (E - BKF)r \equiv A_c x + B_c r. \tag{25.63}$$

Matrices E and F are chosen to have a structure that is sensible from a design point of view. Specifically, it is very desirable to incorporate a unity-gain outer tracking loop in the controller, as shown in Example 25.4.

25.3.3.2 Deviation System and LQR Design Step

Assume that the reference input is a unit step of magnitude r_0. Then, the *steady-state* system is

$$0 = A_c\bar{x} + B_c r_0,$$

where overbars denote steady-state values, so that the steady-state value of the state is $\bar{x} = -A_c^{-1}B_c r_0$. Though the reference input is assumed constant for design purposes, this is to allow good closed-loop rise time and overshoot qualities. Then, the designed controller works for *any reference input r(t), even though time-varying.*

Define the *deviations*

$$\begin{aligned} \tilde{x} &= x - \bar{x}, & \tilde{z} &= z - \bar{z}, \\ \tilde{u} &= u - \bar{u}, & \tilde{e} &= e - \bar{e}. \end{aligned} \tag{25.64}$$

Then, the deviations satisfy the dynamics of the *deviation system*

$$\dot{\tilde{x}} = A\tilde{x} + B\tilde{u} \tag{25.65}$$

$$\tilde{z} = H\tilde{x} \tag{25.66}$$

$$\tilde{u} = -K\tilde{x}. \tag{25.67}$$

Because $e = r - z$, the tracking error deviation is $\tilde{e} = -\tilde{z}$. To induce tracking behavior, define the performance index

$$J(t_0) = \frac{1}{2}\int_0^\infty (\tilde{x}^T Q\tilde{x} + \tilde{u}^T R\tilde{u})\, dt, \tag{25.68}$$

which makes the entire deviation state, and therefore $\tilde{e}$, small.

25.3.3.3 Tracker Design

The tracking problem may now be solved as follows. First, solve the LQ regulator problem for the deviation system using Table 25.2. Then, the tracking control input is given by Equation 25.62. This tracker has a much different structure than the DC-gain-based tracker in Figure 25.9. The next example shows that a sensible choice for matrices E and F based on classical control notions gives a robust tracker with a unity gain outer loop. Then, a sensible choice for the PI design matrices Q and R gives good control gains and *guaranteed stability*, even for complex multiloop tracking systems.

Note that $e = \tilde{e} + \bar{e}$, where $\bar{e}$ is the steady-state value of the tracking error. Because this technique only guarantees that $\tilde{e}$ is small, special steps must be taken to guarantee that $\bar{e}$ is also small. One way to do this is to include *integrators in all the feedforward loops*, as in the next example. As an alternative, a term involving $\bar{e}$ can be added to the PI (Equation 25.68). This gives more involved design equations, which are nevertheless still easily solved by digital computer. The details are in [8]. Finally, although the gain determined in this fashion is optimal for the deviation system, it is not optimal for the tracking problem in terms of the original dynamics (Equation 25.61). In practical applications, however, it is suitable provided that the design matrices are sensibly selected.

Example 25.4: Aircraft Pitch-Rate Control System

This example illustrates the tracker design procedure just presented. Good tracker system design relies on a sensible selection of the structure matrices E and F, and good feedback gains rely on a sensible selection of the design weighting matrices Q, R. *Compensator dynamics* can be accounted for using this procedure. Because this is an LQ-based approach, a reasonable formulation of the problem should result in *guaranteed closed-loop stability*. This is an important feature of modern control design techniques, and is in complete contrast to classical techniques where stability in multi-loop systems can be difficult to achieve.

AIRCRAFT AND CONTROL SYSTEM DYNAMICS

In a pitch-rate control system, the control input is elevator actuator voltage $u(t)$ and r is a reference step input corresponding to the desired pitch command. The performance output $z(t)$ is the pitch rate q. To ensure zero steady-state error, an integrator is added in the feedforward channel; this corresponds to *compensator dynamics*, and is easily dealt with in this approach. The integrator output is ϵ. It is assumed here that all states are available as measurements for feedback purposes; in practice, the output-feedback design technique in [11] will be required.

The design is based on a short period approximation to the F-16 dynamics linearized about a nominal flight condition of 502 ft/s, 0 ft altitude, level flight, with the center of gravity at 0.35. The basic aircraft states of interest are q and angle of attack α. An additional state is introduced by the elevator actuator, whose deflection is δ_e. The states of the plant plus compensator are $x = [\alpha \;\; q \;\; \delta_e \;\; \epsilon]^T$ and the system dynamics are described by Equation 25.61 with

$$A = \begin{bmatrix} -1.01887 & 0.90506 & -0.00215 & 0 \\ 0.82225 & -1.07741 & -0.17555 & 0 \\ 0 & 0 & -20.2 & 0 \\ 0 & -57.2958 & 0 & 0 \end{bmatrix}, \tag{25.69}$$

$$B = \begin{bmatrix} 0 \\ 0 \\ 20.2 \\ 0 \end{bmatrix}, \quad E = \begin{bmatrix} 0 \\ 0 \\ 0 \\ 1 \end{bmatrix}, \tag{25.70}$$

and

$$H = \begin{bmatrix} 0 & 57.2958 & 0 & 0 \end{bmatrix} \tag{25.71}$$

The factor of 57.2958 is added to convert angles from radians to degrees. The last line of the state equation using this A and E matrix describes the integrator, $\dot{\epsilon} = -57.2958q + r$.

CONTROL DESIGN

Select the control input $u(t)$ to yield good closed-loop response to a step input at r, which corresponds to a SISO tracker design problem. Since the integrator makes the system Type I, the steady-state error $\bar{e}$ is equal to zero and $e(t) = \tilde{e}(t)$. Thus, the design method just described is appropriate.

The control input is

$$u = -Kx = -\begin{bmatrix} k_\alpha & k_q & k_{\delta_e} & k_I \end{bmatrix} x = -k_\alpha \alpha - k_q q - k_{\delta_e} \delta_e - k_I \epsilon. \tag{25.72}$$

Therefore, referring to Equation 25.62 it is evident that $F = 0$; however, including the integrator output as a state variable in the dynamics (1) adds the feedforward path required for tracking behavior, that is, element k_I of the feedback matrix K is actually a "feedforward" gain.

To determine the gain matrix K, select the PI (Equation 25.68), and try weighting matrices $R = 1, Q = diag\{1, 10, 1, 1\}$. Now use the LQR routine from the MATLAB Control Systems Toolbox to determine the optimal gain $K = [-0.046 \;\; -1.072 \;\; 0 \;\; 3.381]$. Using MATLAB routines, the corresponding closed-loop poles are $s = -8.67 \pm j9.72, -9.85, -4.07, -1.04$. The resulting step response is shown in Figure 25.10, which displays good performance.

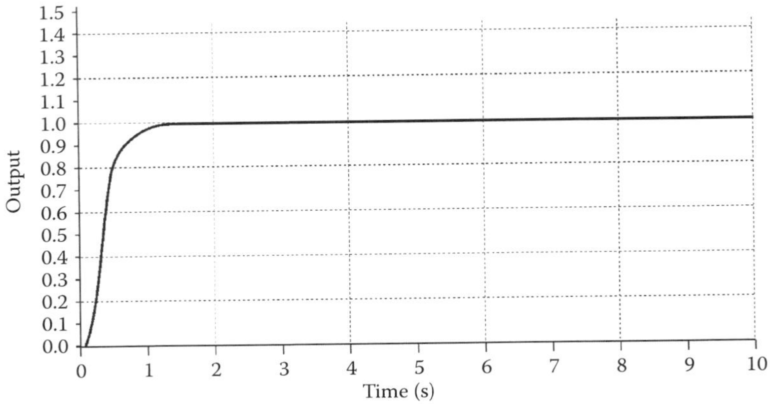

FIGURE 25.10 Pitch-rate step response.

25.4 Minimum-Time and Constrained-Input Design

An important class of control problems is concerned with achieving the performance objectives in *minimum time*. A suitable performance index for these problems is

$$J = \int_{t_0}^{T} 1\, dt = T - t_0. \tag{25.73}$$

Several sorts of minimum-time problems are now discussed.

25.4.1 Nonlinear Minimum-Time Problems

Suppose the objective is to drive the system

$$\dot{x} = f(x, u) \tag{25.74}$$

from a given initial state $x(t_0) \in \mathcal{R}^n$ to a specified final state $x(T)$ in minimum time. Then, from Table 25.1 the Hamiltonian is

$$H = 1 + \lambda^T f \tag{25.75}$$

and the Euler equations are the costate equation

$$-\dot{\lambda} = \frac{\partial f^T}{\partial x}\lambda \tag{25.76}$$

plus the stationarity condition

$$0 = \frac{\partial f^T}{\partial u}\lambda. \tag{25.77}$$

Since the final state is fixed (so that $dx(T) = 0$) but the final time is free, the final condition in Table 25.1 says that

$$0 = H(T) = 1 + \lambda^T(T) f[x(T), u(T)]. \tag{25.78}$$

If $f(x, u)$ is not an explicit function of time, then according to the conservation principle (Equation 25.10), $H(t)$ is zero for all time.

The stationarity condition (Equation 25.77) may often be used to solve for $u(t)$ in terms of $\lambda(t)$. Then, $u(t)$ may be eliminated in the state and costate equations to obtain the Hamiltonian system. To solve this, we require n initial conditions ($x(t_0)$ given) and n final conditions ($x(T)$ specified). However, the final time T is now unknown. The function of Equation 25.78 is to provide one more equation so that T can be solved for. Several nonlinear design problems can be explicitly solved, yielding great insight into the minimum-time control structure. Examples include *Zermelo's Problem* and the *Brachistochrone Problem*.

25.4.2 Linear Quadratic Minimum-Time Design

The general solution procedure given in the previous section for the nonlinear minimum-time problem is difficult to apply. Moreover, a reasonable solution may not exist. A general class of practical problems is covered by the case where it is required to find an optimal control for the linear system

$$\dot{x} = Ax + Bu \tag{25.79}$$

that minimizes the performance index

$$J = \frac{1}{2}x^T(T)S_Tx(T) + \frac{1}{2}\int_{t_0}^{T}(1 + x^TQx + u^TRu)\,dt \tag{25.80}$$

with $S_T \geq 0, Q \geq 0, R > 0$, and the final time T free. There is no constraint on the final state; thus, the control objective is to make the final state sufficiently small. Due to the term $\frac{1}{2}(T - t_0)$ arising from the integral, this must be accomplished in a short time period. This is a general sort of PI that allows for a trade-off between the minimum-time objective and a desire to keep the states and the controls small. Thus, if the engineer selects smaller Q and R, the term $\frac{1}{2}(T - t_0)$ in the PI dominates, and the control tries to make the transit time smaller. This is called the *LQ minimum-time problem*.

The solution for this problem is the same as in Table 25.2. The control is a linear time-varying state feedback given by

$$u = -K(t)x \tag{25.81}$$

with optimal gain

$$K = R^{-1}B^TS \tag{25.82}$$

and $S(t)$ the solution determined by integrating the RE backward from time T. Unfortunately, there is a problem in that the final time T is unknown.

To determine the value of T that minimizes the PI, an *extra condition* is needed, given by Equation 25.78, which yields

$$x^T(t_0)\dot{S}x(t_0) = 1, \tag{25.83}$$

with $x(t_0)$ the specified initial condition of the plant. The solution procedure for the LQ minimum-time problem is to integrate the RE

$$-\dot{S} = A^TS + SA + Q - SBR^{-1}B^TS \tag{25.84}$$

backward from some time τ using $S(\tau) = S_T$ as the final condition. At each time t, the left-hand side of Equation 25.83 is computed using the known initial state and $\dot{S}(t)$. Then, the minimum interval $(T - t_0)$ is equal to $(\tau - t)$ where t is the time for which Equation 25.83 first holds. This specifies the minimum final time T, and then allows the computation of the optimal feedback gain $K(t)$ on the interval $[t_0, T]$.

The Riccati derivative $\dot{S}$ is used to determine the optimal time interval, while S is used to determine the optimal feedback gain $K(t)$.

More details on this control scheme may be found in [12]. It is important to note that condition (Equation 25.83) may never hold. Then, the optimal solution is $T - t_0 = 0$, that is, the PI is minimized by using *no control*. Roughly speaking, if $x(t_0)$ and/or Q and $S(T)$ are selected large enough, then it makes sense to apply a nonzero control $u(t)$ to make $x(t)$ decrease. On the other hand, if Q and $S(T)$ are selected too small for the given initial state $x(t_0)$, then it is not worthwhile to apply any control to decrease $x(t)$, because a nonzero control and a nonzero time interval will increase the PI.

25.4.3 Constrained-Input Design and Bang-Bang Control

Up to this point minimum-time control has been presented based on the conditions of Table 25.1, which were derived using the calculus of variations. Under some smoothness assumptions on $f(x, u, t)$ and $L(x, u, t)$, the resulting controls are also smooth. Here, a fundamentally different sort of control strategy will be presented.

If the linear system

$$\dot{x} = Ax + Bu \tag{25.85}$$

with $x \in \mathcal{R}^n$, $u \in \mathcal{R}^m$ is prescribed, there are problems with using the pure minimum-time PI,

$$J(t_0) = \int_{t_0}^{T} 1\, dt, \tag{25.86}$$

where T is free. The way to minimize the time is to use infinite control energy! Since this optimal strategy is not acceptable, it is necessary to find a way to reformulate the minimum-time problem for linear systems.

Therefore, the control input now must satisfy the *magnitude constraint*

$$\|u(t)\| \leq 1 \tag{25.87}$$

for all $t \in [t_0, T]$. This constraint means that *each component* of the m-vector $u(t)$ must be no greater than 1. Thus, the control is constrained to an *admissible region* (in fact, a hypercube) of $\mathcal{R}^m$. If the constraints on the components of $u(t)$ have a value different from 1, then one may appropriately scale the corresponding columns of the B matrix to obtain the constraints in the form of Equation 25.87. A requirement like Equation 25.87 arises in many problems where the control magnitude is limited by physical considerations; for instance, the thrust of a rocket certainly has a maximum possible value, as has the armature voltage of a DC motor.

Referring to Table 25.1, the optimal control problem posed here is to find a control $u(t)$ that drives a given $x(t_0)$ to a final state $x(T)$ satisfying the final state constraint, minimizes the PI, and satisfies Equation 25.87 at all times. Intuitively, to minimize the time, the optimal control strategy appears to be to apply maximum effort (i.e., plus or minus 1) over the entire time interval. This idea will now be formalized. When a control component takes on a value at the boundary of its admissible region (i.e., ± 1), it is said to be *saturated*. Pontryagin and coworkers have shown that, in the case of constrained control, Table 25.1 still applies if the stationarity condition is replaced by the more general condition, known as *Pontryagin's Minimum Principle*,

$$H(x^*, u^*, \lambda^*, t) \leq H(x^*, u, \lambda^*, t), \quad \text{all admissible u.} \tag{25.88}$$

This is an extremely powerful result which can be employed to derive the following solution to the linear constrained-input minimum-time problem.

Define the *signum function* for scalar w as

$$sgn(w) = \begin{cases} 1, & w > 0 \\ indeterminate, & w = 0 \\ -1, & w < 0. \end{cases} \tag{25.89}$$

If w is a vector, define $v = sgn(w)$ as $v_i = sgn(w_i)$ for each i, where v_i, w_i are the components of v and w. Then, in terms of the costate, the optimal control is given by

$$u^*(t) = -sgn[B^T \lambda(t)]. \tag{25.90}$$

This may be interpreted as follows. For each column b_i of B, if $\lambda^T(t) b_i$ is positive, we should select $u_i(t) = -1$ to get the largest possible negative value of $\lambda^T(t) b_i u_i(t)$. On the other hand, if $\lambda^T(t) b_i$ is

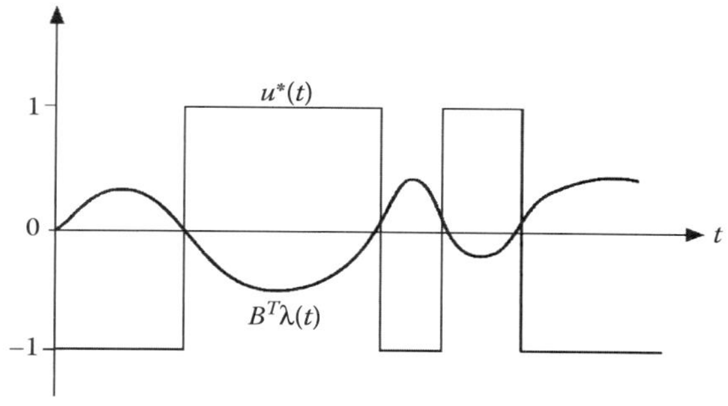

FIGURE 25.11 Sample switching function and associated optimal control.

negative, we should select $u_i(t)$ as its maximum admissible value of 1 to make $\lambda^T(t)b_iu_i(t)$ as negative as possible. If $\lambda^T(t)b_i$ is zero at a single point t in time, then $u_i(t)$ can be assigned any value at that time, because then $\lambda^T(t)b_iu_i(t)$ is zero for all values of $u(t)$.

The quantity $B^T\lambda(t)$ is called the *switching function.* A sample switching function and the optimal control it determines are shown in Figure 25.11. When the switching function changes sign, the control switches from one of its extreme values to another. The control in the figure switches four times. The optimal linear minimum-time control is always saturated since it switches back and forth between its extreme values, so it is called *bang-bang control.* In some problems, a component $b_i^T\lambda(t)$ of the switching function $B^T\lambda(t)$ can be zero over a finite time interval. If this happens, component $u_i(t)$ of the optimal control is not well-defined by Equation 25.90. This is called a *singular condition.* If this does not occur, the time-optimal problem is called *normal.*

The time-invariant plant (Equation 25.85) is reachable if, and only if, the reachability matrix (Equation 25.35) has full rank n. If b_i is the ith column of $B \in \mathcal{R}^{n\times m}$, then the plant is *normal* if

$$U_i = [b_i \;\; Ab_i \;\; \ldots \;\; A^{n-1}b_i] \tag{25.91}$$

has full rank n for each $i = 1, 2, \ldots, m$, that is, if the plant is reachable by each separate component u_i of $u \in \mathcal{R}^m$. Normality of the plant and normality of the minimum-time control problem are equivalent. Let the plant be normal (and hence reachable), and suppose it is desired to drive a given $x(t_0)$ to a desired fixed final state $x(T)$ in minimum time with a control satisfying Equation 25.87. Then, the following results have been achieved for time-invariant plants by Pontryagin and coworkers:

1. If the desired final state $x(T)$ is equal to zero, then a minimum-time control exists if the plant has no poles with positive real parts (i.e., no poles in the open right half plane).
2. For any fixed $x(T)$, if a solution to the minimum-time problem exists, then it is unique.
3. Finally, if the n plant poles are all real and if the minimum-time control exists, then each component $u_i(t)$ of the time-optimal control can switch at most $n-1$ times.

In both its computation and its final appearance, bang-bang control is fundamentally different from the smooth controls seen previously. The minimum principle leads to the expression (Equation 25.90) for $u^*(t)$, but it is difficult to solve explicitly for the optimal control. Instead, this condition specifies several different control laws, and it is necessary to select which among these is the optimal control. Thus, the minimum principle keeps one from having to examine all possible control laws for optimality, giving a small subset of potentially optimal controls to be investigated. In many cases, it is still possible to express $u^*(t)$ as a state-feedback control law.

Example 25.5: Bang-Bang Control

Any system obeying Newton's laws for point-mass motion is described by

$$\dot{y} = v,$$
$$\dot{v} = u, \tag{25.92}$$

with $y(t)$ the position, $v(t)$ the velocity, and $u(t)$ the input acceleration. The state is $x = [y \;\; v]^T$.

Let the acceleration input u be constrained in magnitude by

$$|u(t)| \leq 1. \tag{25.93}$$

The control objective is to bring the state from any initial point (y_0, v_0) to the origin in the minimum time T. The final state must be fixed at

$$\psi(x(T), T) = \begin{bmatrix} y(T) \\ v(T) \end{bmatrix} = 0. \tag{25.94}$$

Using Equation 25.88 the minimum-time control takes on only values of $u = \pm 1$. Moreover, there is at most one control switching because the maximum number of switchings is $n - 1$ when the plant poles are all real. The *phase plane* is a coordinate system whose axes are the state variables. Phase-plane plots of the state trajectories of (1) for $u = 1$ and for $u = -1$ are parabolas in the phase plane as shown in Figure 25.12. These parabolas represent minimum-time trajectories. The arrows indicate the direction of increasing time. For example, if the initial state (y_0, v_0) is as shown in Figure 25.12, then, under the influence of the control $u = -1$, the state will develop downward along the parabola, eventually passing through the point $(y = 0, v = -2)$. On the other hand, if a control of $u = 1$ is applied, the state will move upward and to the right.

It will now be argued that this figure represents a *state-feedback control law*, which brings any state to the origin in minimum time. Suppose the initial state is as shown in Figure 25.12. Then the only way to arrive at the origin, while satisfying the Pontryagin conditions, is to apply $u = -1$ to move the state

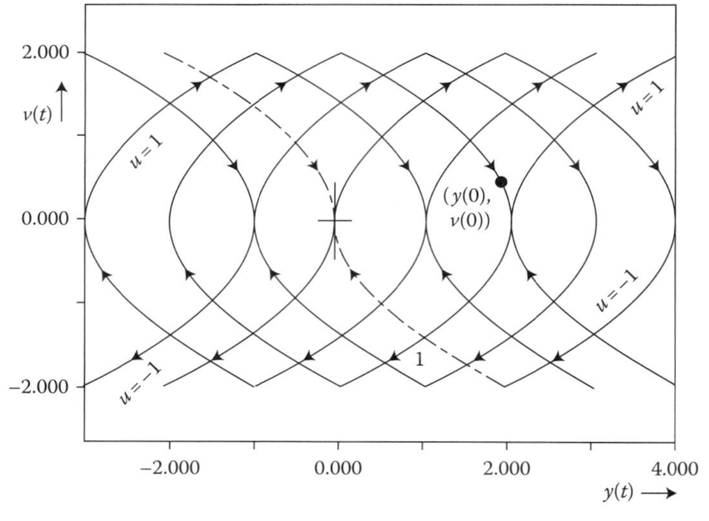

FIGURE 25.12 Phase plane trajectories for $u = 1$ and $u = -1$.

along a parabola to the dashed curve. At this point (labeled "a"), the control is switched to $u = 1$ to drive the state into the origin. Hence, the resulting seemingly roundabout trajectory is in fact a minimum-time path to the origin. The dashed curve is known as the *switching curve*. For initial states on this curve, a control of $u = 1$ (if $v_0 < 0$) or $u = -1$ (if $v_0 > 0$) for the entire control interval will bring the state to zero. For initial states off this curve, the state must first be driven onto the switching curve, and then the control must be switched to its other extreme value to bring the final state to zero. The switching curve is described by the equation $y = -\frac{1}{2}v\,|v|$.

Simply put, for initial states above the switching curve, the optimal control is $u = -1$, followed by $u = 1$, with the switching occurring when $y(t) = \frac{1}{2}v^2(t)$. For initial states below the switching curve, the optimal control is $u = 1$, followed by $u = -1$, with the switching occurring when $y(t) = -\frac{1}{2}v^2(t)$. Because the control at each time t is completely determined by the state (i.e., by the phase plane location), Figure 25.12 yields a feedback control law. This feedback law, represented graphically in the figure, can be stated as

$$u = \begin{cases} -1 & \text{if } y > -\frac{1}{2}v\,|v| \\ & \text{or if } y = -\frac{1}{2}v\,|v| \text{ and } y < 0 \\ 1 & \text{if } y < -\frac{1}{2}v\,|v| \\ & \text{or if } y = -\frac{1}{2}v\,|v| \text{ and } y > 0. \end{cases} \tag{25.95}$$

which makes it clear that the minimum-time control is indeed a state-feedback.

25.5 Optimal Control of Discrete-Time Systems

The discussion so far has applied to continuous-time (analog) systems. The discussion of the LQR problem for DT systems

$$x_{k+1} = Ax_k + Bu_k, \tag{25.96}$$

is identical in form, though more complicated in its details. The problem is to select the state-feedback matrix K in

$$u_k = -Kx_k \tag{25.97}$$

to minimize a performance index specified by the design engineer.

25.5.1 Discrete-Time LQR

In general the optimal DT linear quadratic regulator is a time-varying matrix gain sequence K_k. However, the practically useful solution is the optimal steady-state feedback gain obtained by using the infinite horizon PI (Equation 25.99). The design equations for the DT LQR are given in Table 25.4. The DT LQR equations are more complicated than the continuous-time equivalents; however, commercially available software (e.g., MATLAB) makes this irrelevant to the control designer. In practice, DT design is as straightforward as continuous-time design.

All the results discussed for continuous-time systems in Section 25.2 have their DT counterparts (see the references). Thus, as long as (A, B) is stabilizable and $(A, \sqrt{Q})$ observable, the discrete LQR has guaranteed properties of stability and robustness. Discrete versions of the tracker design problem are also given in the references.

25.5.2 Digital Control of Continuous-Time Systems

Using the DT LQR design equations in Table 25.4, optimal digital controllers may be designed for continuous-time systems. In fact, standard techniques are available for determining a DT description

TABLE 25.4 Discrete-Time Linear Quadratic Regulator.

System model:

$$x_{k+1} = Ax_k + Bu_k, \quad x_0 \quad given. \tag{25.98}$$

Performance index:

$$J = \frac{1}{2}\sum_0^{\infty}(x_k^T Qx_k + u_k^T Ru_k) \tag{25.99}$$

with

$$Q \geq 0, \quad R > 0.$$

Optimal feedback control:

Discrete-time algebraic Riccati equation:

$$0 = S - A^T SA + A^T SB(B^T SB + R)^{-1}B^T SA - Q. \tag{25.100}$$

Optimal feedback gain:

$$K = (B^T SB + R)^{-1}B^T SA. \tag{25.101}$$

Feedback control:

$$u_k = -Kx_k. \tag{25.102}$$

Optimal cost:

$$J = \frac{1}{2}x_0^T Sx_0. \tag{25.103}$$

given the continuous-time dynamics $\dot{x} = Ax + Bu$ and a specified sampling period T. Then, the table allows the design of digital controllers, because the feedback gain Equation 25.102 is expressed in DT, meaning that it can be directly programmed on a microprocessor or digital signal processor (DSP) and applied every T seconds to the plant. The next example shows some of the issues involved in digital control design, including selection of the sampling period and discretization of the plant.

Example 25.6: Digital Inverted Pendulum Controller

In Example 25.3 a continuous-time controller was designed for an inverted pendulum on a cart; it is now desired to design a digital controller.

DISCRETE INVERTED PENDULUM DYNAMICS

The continuous-time inverted pendulum dynamics are given in Example 25.3. Standard techniques for system discretization are covered in the chapter on Digital Control. The time histories and closed-loop poles in Example 25.3 reveal that a sampling period of $T = 0.1$ sec is very small compared to the speed of the plant response (e.g., about 1/10 of the smallest plant time constant). Therefore, this sampling period is selected.

Using the MATLAB Control System Toolbox to compute the zero-order-hold/step-invariant sampled dynamics yields the system

$$x_{k+1} = \begin{bmatrix} 1.054386 & 0.101806 & 0 & 0 \\ 1.097473 & 1.054386 & 0 & 0 \\ -0.004944 & -0.000164 & 1 & 0.1 \\ -0.099770 & -0.004944 & 0 & 1 \end{bmatrix} x_k + \begin{bmatrix} -0.001009 \\ -0.020361 \\ 0.001001 \\ 0.020033 \end{bmatrix} u_k = Ax_k + Bu_k \tag{25.104}$$

where the state is $x_k = [\theta_k \;\; \dot{\theta}_k \;\; p_k \;\; \dot{p}_k]^T$.

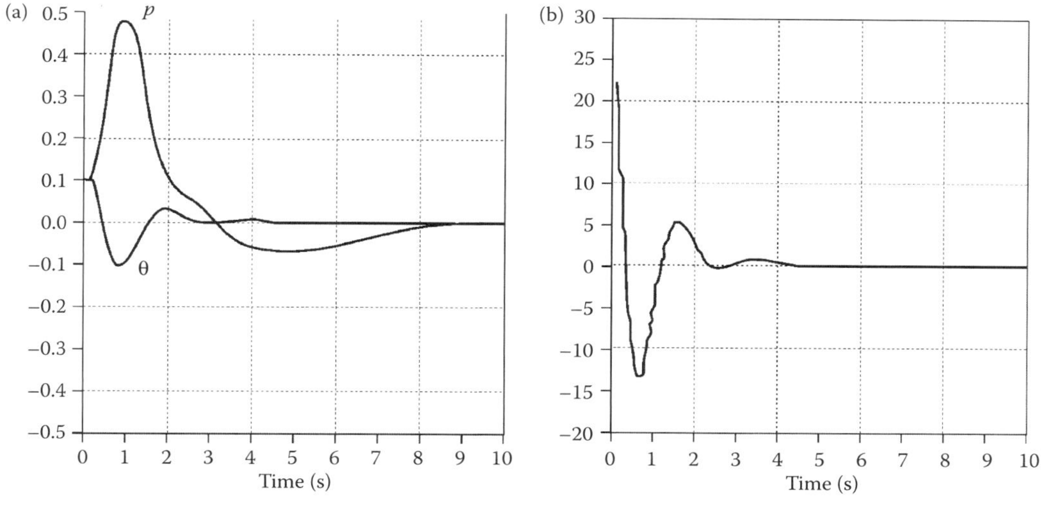

FIGURE 25.13 Response of inverted pendulum digital controller. (a) Rod angle $\theta(t)$ and cart position $p(t)$. (b) Control input $u(t)$.

The continuous system has poles at $s = 0, 0, 3.28, -3.28$. The discrete system has poles at $z = 1, 1, 1.3886, 0.7201$ which corresponds to the sampling transformation $z = e^{sT}$.

DIGITAL CONTROLLER DESIGN

To determine stabilizing control gains in

$$u_k = -Kx_k = -(k_\theta \theta_k + k_{\dot{\theta}} \dot{\theta}_k + k_p p_k + k_{\dot{p}} \dot{p}_k), \tag{25.105}$$

we may use the DT LQR in Table 25.4. Note that this is a *multiloop design problem*, yet the LQR approach easily deals with it. Trying weighting matrices of $R = 1, Q = diag\{10, 10, 1, 1\}$ and using the discrete ARE solver in MATLAB yields the gains $K = [-1.294 \quad -10.02 \quad 3.648 \quad 16.94]^T$ and corresponding closed-loop poles at $z = 0.37, 0.72, 0.82 \pm j.029$.

A simulation is easily performed to obtain the closed-loop response shown in Figure 25.13. It is very instructive to compare this with the response obtained in Example 25.3. The advantage of DT design is that the control input (2) may be computed every $T = 0.1$ *s* on a microprocessor and applied to the plant for real time control. The continuous-time feedback law needs to be applied using analog techniques or a very high sampling rate.

25.6 Optimal LQ Design for Polynomial Systems

The discussion thus far has focused on the state-space formulation. A dynamical system may be equally well described in transfer function or *polynomial form* as

$$A(z^{-1})y_k = z^{-d} B(z^{-1}) u_k, \tag{25.106}$$

with y_k the output and u_k the control input. The *system delay* is denoted d. This is a DT formulation with z^{-1} denoting the unit delay. For simplicity we discuss the SISO case; these notions may be extended to multivariable polynomial systems using the *matrix fraction descriptions* of the plant. The denominator polynomial

$$A(z^{-1}) = 1 + a_1 z^{-1} + a_2 z^{-2} + \cdots + a_n z^{-n} \tag{25.107}$$

has roots specifying the system poles, and the numerator

$$B(z^{-1}) = b_0 + b_1 z^{-1} + \cdots + b_m z^{-m} \tag{25.108}$$

has roots at the system zeros.

In contrast to the PI selected for state-space systems, which is a sum of squares, for polynomial systems, it is more convenient to select the *square of sums* PI

$$J_k = \left[\sum_{i=0}^{n_P} p_i y_{k+d-i} - \sum_{i=0}^{n_Q} q_i y_{k-i}\right]^2 + \left[\sum_{i=0}^{n_R} r_i u_{k-i}\right]^2. \tag{25.109}$$

The constants p_i, q_i, r_i are weighting coefficients (design parameters) selected by the engineer and w_k is a reference or command signal. Defining the *weighting polynomials*

$$\begin{aligned} P(z^{-1}) &= 1 + p_1 z^{-1} + \cdots + p_{n_P} z^{-n_P}, \\ Q(z^{-1}) &= q_0 + q_1 z^{-1} + \cdots + q_{n_Q} z^{-n_Q}, \\ R(z^{-1}) &= r_0 + r_1 z^{-1} + \cdots + r_{n_R} z^{-n_R}, \end{aligned} \tag{25.110}$$

the PI may be written in the streamlined form

$$J_k = (P y_{k+d} - Q w_k)^2 + (R u_k)^2. \tag{25.111}$$

This is a very general sort of PI. For instance, the tracking problem may be solved if we select $P = Q = 1$, $R = r_0$, for then

$$J_k = (y_{k+d} - w_k)^2 + (r_0 u_k)^2 \tag{25.112}$$

and a delayed version of the output y_k tries to follow a reference input w_k. The system delay d is explicitly accounted for. Thus, the polynomial tracker is very easy to compute and implement. In fact, it is *causal*, in contrast to the state-space LQR tracker where a noncausal feedforward signal was needed.

As another example, the regulator problem results if the weights are selected as $P = 1, Q = 0, R = r_0$, for then

$$J_k = (y_{k+d})^2 + (r_0 u_k)^2 \tag{25.113}$$

and the control tries to hold the output at zero without using too much energy.

The optimal control u_k that minimizes the PI is straightforward to determine. In the minimum-phase core (e.g., all roots of $B(z^{-1})$ stable), one solves the *Diophantine equation*

$$1 = AF + z^{-d} G \tag{25.114}$$

for the intermediate polynomials $F(z^{-1})$ and $G(z^{-1})$. Well-known routines are available for this. In fact, one may simply divide $A(z^{-1})$ into 1 until the remainder has a multiplier of z^{-d}. Then the quotient is $F(z^{-1})$ and the remainder yields $G(z^{-1})$. In terms of the Diophantine equation solution, the optimal control sequence is then given by the equation

$$\left(PBF + \frac{r_0}{b_0} R\right) u_k = -PG y_k + Q w_k. \tag{25.115}$$

This is nothing but a *difference equation* that gives the current control u_k in terms of y_k, w_k, and previous values of the control; it is easily implemented using a digital computer or microprocessor. Figure 25.14

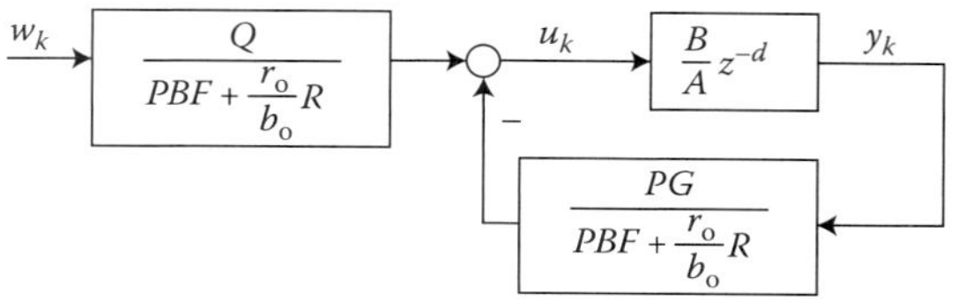

FIGURE 25.14 Optimal polynomial LQ regulator drawn as a two-degrees-of-freedom regulator.

shows the structure of the optimal LQ polynomial controller. Because it has a feedback and a feedforward component, it is called a *two-degrees-of-freedom regulator*. Such a controller can influence the closed-loop poles *as well as zeros*. Note that this controller actually requires full state feedback because the complete state is given by $y_k, y_{k-1}, \ldots, y_{k-n}, u_{k-d}, u_{k-d-1}, \ldots, u_{k-d-m}$.

Some fundamental points in polynomial LQ design, as contrasted to state-space design, are (1) the PI is a square of sums, (2) the role of the RE in state-space design is played by the Diophantine equation in polynomial design, and (3) the optimal tracker problem is easy to solve and implement since it is *causal*.

Example 25.7: Polynomial LQ Tracker

It is desired for the plant

$$y_k - 2y_{k-1} + \frac{3}{4}y_{k-2} = u_{k-1} - \frac{1}{2}u_{k-2} \tag{25.116}$$

to follow a given reference signal w_k using a fairly smooth control signal u_k. The control delay is $d = 1$. To accomplish the design, select the PI

$$J_k = (y_{k+1} - w_k)^2 + r^2(u_k - u_{k-1})^2. \tag{25.117}$$

This PI is motivated by Equation 25.112, but the *first difference* of the control is weighted to keep u_k smooth, as per the specifications. The scalar r is a *design parameter* used to tune the closed-loop performance at the end of the design (e.g., for suitable damping ratio, overshoot, etc.).

Inspecting the plant and PI, the polynomials defined in the discussion are

$$\begin{aligned} &A(z^{-1}) = 1 - 2z^{-1} + 0.75z^{-2}, \quad B(z^{-1}) = 1 - 0.5z^{-1}, \\ &P(z^{-1}) = Q(z^{-1}) = 1, \quad R(z^{-1}) = r(1 - z^{-1}). \end{aligned} \tag{25.118}$$

To find the required tracking controller, the Diophantine equation is easily solved (simply perform long division of $A(z^{-1})$ into 1 to obtain the quotient $F(z^{-1})$ and remainder $z^{-1}G(z^{-1})$), resulting in the intermediate quantities

$$\begin{aligned} F(z^{-1}) &= 1, \\ G(z^{-1}) &= 2 - 0.75z^{-1}. \end{aligned} \tag{25.119}$$

According to Equation 25.115, therefore, the control is given by

$$[(1 + r^2) - z^{-1}(0.5 + r^2)]u_k = -(2 - 0.75z^{-1})y_k + w_k. \tag{25.120}$$

The variable r is a *design parameter* that can be varied by the engineer as he performs computer simulations of the closed-loop system (1), (5). Then, based on the simulations, the best value of r is selected and the resulting controller is applied to the actual plant. Selecting, for instance, $r = \frac{1}{2}$, yields the difference equation

$$1.25u_k = 0.75u_{k-1} - 2y_k + 0.75y_{k-1} + w_k, \tag{25.121}$$

which is easily solved for the current control input u_k in terms of u_{k-1}, current and previous values of y_k, and the current w_k. The controller is of the form shown in Figure 25.14.

25.7 Dynamic Programming and the HJB Equation

The Calculus of Variations approach to optimal control design yields the state equation, costate equation, and stationarity condition in Table 25.1. A second approach to optimal control is through Bellman's Optimality Principle. This principle is a cornerstone of optimal control theory and states:

> "An optimal policy has the property that no matter what the previous decisions (i.e. controls) have been, the remaining decisions must constitute an optimal policy with regard to the state resulting from those previous decisions."

We shall apply this to solve the optimal control problem, first for DT systems, then continuous-time systems.

25.7.1 Dynamic Programming for DT Systems

Consider the nonlinear DT plant

$$x_{k+1} = f(x_k) + g(x_k)u_k \tag{25.122}$$

with performance index over the time interval $[k, N]$ given by

$$J(x_k) = \phi(x_N, N) + \sum_{i=k}^{N-1} L(x_i, u_i) \tag{25.123}$$

It is desired to find a state variable feedback policy

$$u_k = h(x_k) \tag{25.124}$$

that minimizes the PI. The optimal cost is

$$J^*(x_k) = \min_{\bar{u}(k:N-1)} \left(\phi(x_N, N) + \sum_{i=k}^{N-1} L(x_i, u_i) \right) \tag{25.125}$$

where $\bar{u}(k : N-1) = \{u_k, u_{k+1}, \cdots, u_{N-1}\}$. The optimal control is

$$\bar{u}^* = \arg\min_{\bar{u}(k:N-1)} \left(\phi(x_N, N) + \sum_{i=k}^{N-1} L(x_i, u_i) \right) \tag{25.126}$$

These equations are very difficult to solve.

One can write (Equation 25.123) as

$$J(x_k) = L(x_k, u_k) + \phi(x_N, N) + \sum_{i=k+1}^{N-1} L(x_i, u_i) \tag{25.127}$$

or

$$J(x_k) = L(x_k, u_k) + J(x_{k+1}) \tag{25.128}$$

which is a difference equation equivalent to (Equation 25.123). That is, given a stabilizing policy $u_k = h(x_k)$, one can find its associated cost either by computing the infinite sum (Equation 25.123) or by solving the difference equation 25.128. The latter is far easier.

Equation 25.128 is known as the Bellman Equation and is the basis for a host of reinforcement learning methods that learn the optimal cost and feedback policy online in real-time by observing data along the system trajectories. See the Chapter 6.

In terms of the Bellman equation, one can write the optimal cost as

$$J^*(x_k) = \min_{\bar{u}(k:N-1)} \big(L(x_k, u_k) + J(x_{k+1})\big) \tag{25.129}$$

The importance of Bellman's Optimality Principle is that one may write this is as

$$J^*(x_k) = \min_{u_k} \big(L(x_k, u_k) + J^*(x_{k+1})\big) \tag{25.130}$$

with $J^*(x_{k+1})$ the optimal cost from time $k+1$ on.

This is known as the Bellman optimality equation or the HJB equation. This equation yields the optimal cost at time k in terms of the optimal cost at time $k+1$. Therefore, it leads to a backwards-in-time solution procedure for the optimal control problem. This procedure leads to a host of methods for solution collected under the general name of Dynamic Programming. DP leads to tractable solutions in many situations, particularly if there are control constraints; See [7]. Since it proceeds backwards in time, DP is an off-line planning method for solving the optimal control problem.

To solve the DT optimal control problem, one first solves the HJB, then finds the optimal control using

$$u_k^* = \arg\min_{u_k} \big(L(x_k, u_k) + J^*(x_{k+1})\big) \tag{25.131}$$

which is far simpler to effect that Equation 25.126.

Let us show what this boils down to in the DT LQR case. Consider the linear time-invariant DT system (Equation 25.96) with associated cost (Equation 25.99). For the LQR, it is known that the cost for any stabilizing state variable feedback (Equation 25.97) is quadratic in the state so that

$$J(x_k) = \frac{1}{2}\sum_{i=k}^{\infty} x_i^T Q x_i + u_i^T R u_i = \frac{1}{2}\sum_{i=k}^{\infty} x_i^T (Q + K^T R K) x_i = \frac{1}{2} x_k^T S x_k \tag{25.132}$$

For some matrix $S = S^T > 0$ to be determined. The closed-loop system is

$$x_{k+1} = (A - BK)x_k \equiv A_c x_k \tag{25.133}$$

For the DT LQR, Bellman's equation 25.128 is

$$x_k^T S x_k = x_k^T Q x_k + u_k^T R u_k + x_{k+1}^T S x_{k+1} \tag{25.134}$$

or

$$x_k^T S x_k = x_k^T Q x_k + u_k^T R u_k + (A x_k + B u_k)^T S (A x_k + B u_k) \tag{25.135}$$

whence the minimization (Equation 25.131) is easily performed by differentiating with respect to u_k to obtain

$$Ru_k + B^T S(Ax_k + Bu_k) = 0 \tag{25.136}$$

or

$$u_k = -(R + B^T SB)^{-1} B^T SAx_k \tag{25.137}$$

so the optimal feedback gain is

$$K = (R + B^T SB)^{-1} B^T SA. \tag{25.138}$$

Substituting this into the Bellman equation 25.135 and simplifying yields the DT HJB equation

$$A^T SA - S + Q - A^T SB(R + B^T SB)^{-1} B^T SA = 0. \tag{25.139}$$

This is exactly the DT ARE in Equation 25.100!

25.7.2 Dynamic Programming for Continuous-Time Systems

Consider the nonlinear dynamical system

$$\dot{x} = f(x, u) \tag{25.140}$$

with associated cost

$$J(x(t)) = \int_t^{\infty} L(x(\tau), u(\tau))\, d\tau \tag{25.141}$$

To apply Bellman's optimality principle write this in the form

$$J(x(t)) = \int_t^{t+T} L(x(\tau), u(\tau))d\tau + J(x(t+T)) \tag{25.142}$$

For any $T > 0$. This is exactly in the form of the DT Bellman equation. According to Bellman's principle, the optimal value is given in terms of this construction as

$$J^*(x(t)) = \min_{\bar{u}(t:t+T)} \left(\int_t^{t+T} L(x(\tau), u(\tau))d\tau + J^*(x(t+T)) \right) \tag{25.143}$$

where $\bar{u}(t : t+T) = \{u(\tau) : t \le \tau < t+T\}$. The optimal control is

$$u^*(x(t)) = \arg\min_{\bar{u}(t:t+T)} \left(\int_t^{t+T} L(x(\tau), u(\tau))d\tau + J^*(x(t+T)) \right) \tag{25.144}$$

Let $J^*(x(t+T), t+T) = J^*(x(t) + \Delta x, t+T)$and perform a Taylor series expansion of Equation 25.143 and approximate the integral to obtain

$$J^*(x(t)) = \min_{\bar{u}(t:t+T)} \left(LT + J^*(x(t)) + \left(\frac{\partial J^*}{\partial x} \right)^T \Delta x + \left(\frac{\partial J^*}{\partial t} \right)^T T \right) \tag{25.145}$$

However, from Equation 25.140 $\Delta x = f(x,u)T$ so that, manipulating the terms that do not depend on $u(t)$ one has

$$-\left(\frac{\partial J^*}{\partial t}\right)^T T = \min_{\bar{u}(t:t+T)}\left(LT + \left(\frac{\partial J^*}{\partial x}\right)^T f(x,u)T\right) \tag{25.146}$$

whence, letting $T \to 0$ one obtains

$$-\left(\frac{\partial J^*}{\partial t}\right)^T = \min_{u(t)}\left(L + \left(\frac{\partial J^*}{\partial x}\right)^T f(x,u)\right) \tag{25.147}$$

This is the CT HJB equation. In terms of the Hamiltonian function (Equation 25.4) one may write it as

$$-\left(\frac{\partial J^*}{\partial t}\right)^T = \min_{u(t)}\left(H(x,u,\frac{\partial J^*}{\partial x},t)\right) \tag{25.148}$$

It is easily shown that, in the CT LQR case, this is exactly the CT RE (Equation 25.16).

References

1. Anderson, B.D.O, and Moore, J.B., *Optimal Control*, Prentice-Hall, Englewood Cliffs, NJ, 1990.
2. Athans, M, and Falb, P., *Optimal Control*, McGraw-Hill, New York, 1966.
3. Bryson, A.E., Jr. and Ho, Y.-C., *Applied Optimal Control*, Hemisphere, New York, 1975.
4. Grimble, M.J. and Johnson, M.A., *Optimal Control and Stochastic Estimation: Theory and Applications*, John Wiley & Sons, New York, Vol. 1, 1988.
5. Kirk, D.E., *Optimal Control Theory*, Prentice-Hall, Englewood Cliffs, NJ, 1970.
6. Kwakernaak, H. and Sivan, R., *Linear Optimal Control Systems*, John Wiley & Sons, New York, 1972.
7. Lewis, F.L. and Syrmos, V., *Optimal Control*, 2nd edition, John Wiley and Sons, New York, 1995.
8. Lewis, F.L., *Applied Optimal Control and Estimation*, Prentice-Hall, Englewood Cliffs, NJ, 1992.
9. MATRIXx, Integrated Systems, Inc., Santa Clara, CA, 1989.
10. Moler, C., Little, J., and Bangert, S., PC-MATLAB, The Mathworks, Inc., Sherborn, MA, 1987.
11. Stevens, B.L. and Lewis, F.L., *Aircraft Modelling, Dynamics, and Control*, John Wiley & Sons, New York, 1992.
12. Verriest, E.I. and Lewis, F.L., On the linear quadratic minimum-time problem, *IEEE Transactions on Automatic Control*, 859–863, 1991.

Further Reading

Further information may be obtained in the references, and in Chapter 62.

26

Decentralized Control

M. E. Sezer
Bilkent University

D. D. Šiljak
Santa Clara University

26.1 Introduction

The complexity and high performance requirements of present-day industrial processes place increasing demands on control technology. The orthodox concept of driving a large system by a central computer has become unattractive for either economic or reliability reasons. New emerging notions are subsystems, interconnections, distributed computing, parallel processing, and information constraints, to mention a few. In complex systems, where databases are developed around the plants with distributed sources of data, a need for fast control action in response to local inputs and perturbations dictates the use of distributed (that is, decentralized) information and control structures.

The accumulated experience in controlling complex industrial processes suggests three basic reasons for using decentralized control structures:

1. Dimensionality
2. Information structure constraints
3. Uncertainty

Because the amount of computation required to analyze and control a system of large dimension grows faster than its size, it is beneficial to decompose the system into subsystems, and design controls for each subsystem independently based on the local subsystem dynamics and its interconnections. In this way, special structural features of a system can be used to devise feasible and efficient decentralized strategies for solving large control problems previously impractical to solve by "one-shot" centralized methods.

A restriction on what and where the information is delivered in a system is a standard feature of interconnected systems. For example, the standard automatic generation control in power systems is decentralized because of the cost of excessive information requirements imposed by a centralized control strategy over distant geographic areas. The structural constraints on information make the centralized methods for control and estimation design difficult to apply, even to systems with small dimensions.

It is a common assumption that neither the internal nor the external nature of complex systems can be known precisely in deterministic or stochastic terms. The essential uncertainty resides in the interconnections between different parts of the system (subsystems). The local characteristics of each individual subsystem can be satisfactorily modeled in most practical situations. Decentralized control strategies are inherently robust with respect to a wide variety of structured and unstructured perturbations in the interconnections. The strategies can be made reliable to both interconnection and controller failures involving individual subsystems.

In decentralized control design, it is customary to use a wide variety of disparate methods and techniques that originated in system and control theory. Graph-theoretic methods have been devised to identify the special structural features of the system, which may help us cope with dimensionality problems and formulate a suitable decentralized control strategy. The concept of vector Lyapunov functions, each component of which determines the stability of a part of the system where others do not, is a powerful method for the stability analysis of large interconnected systems. Stochastic modeling and decentralized control have been used in a broad range of situations, involving LQG design, Kalman filtering, Markov processes, and stability analysis and design. Robustness considerations of decentralized control have been carried out since the early stages of its evolution, often preceding a similar development in the centralized control theory. Especially popular have been the adaptive decentralized schemes because of their flexibility and ability to cope efficiently with perturbations in both the interactions and the subsystems of a large system.

The objective of this chapter is to introduce the concept and methods of decentralized control. Due to a large number of results and techniques available, only the basic theory and practice of decentralized control will be reviewed. At the end of the chapter is a discussion of the larger background listing the books and survey papers on the subject. References related to more sophisticated treatment of decentralized control and the relevant applications are also discussed.

26.2 The Decentralized Control Problem

To introduce the decentralized control problem, consider two inverted penduli coupled by a spring as shown in Figure 26.1. The control objective is to keep the penduli in the upright position by applying feedback control via the inputs u_1 and u_2. The linearized equations of motion in the vicinity of $\theta_1 = \theta_2 = 0$ are

$$
\begin{aligned}
m\ell^2\ddot{\theta}_1 &= mg\ell\theta_1 - ka^2(\theta_1 - \theta_2) + u_1, \\
m\ell^2\ddot{\theta}_2 &= mg\ell\theta_2 - ka^2(\theta_2 - \theta_1) + u_2.
\end{aligned}
\tag{26.1}
$$

By choosing the state vector $x = (\theta_1, \dot{\theta}_1, \theta_2, \dot{\theta}_2)^T$ and the input vector $u = (u_1, u_2)^T$, the state space representation of the system is

$$
\mathbf{S}:\ \dot{x} = \left[\begin{array}{cc|cc}
0 & 1 & 0 & 0 \\
\dfrac{g}{\ell} - \dfrac{ka^2}{m\ell^2} & 0 & \dfrac{ka^2}{m\ell^2} & 0 \\
\hline
0 & 0 & 0 & 1 \\
\dfrac{ka^2}{m\ell^2} & 0 & \dfrac{g}{\ell} - \dfrac{ka^2}{m\ell^2} & 0
\end{array}\right] x + \left[\begin{array}{c|c}
0 & 0 \\
\dfrac{1}{m\ell^2} & 0 \\
\hline
0 & 0 \\
0 & \frac{1}{m\ell^2}
\end{array}\right] u\,. \tag{26.2}
$$

The fundamental restriction in choosing the feedback laws to control the system $\mathbf{S}$ is that each input u_1 and u_2 can depend only on the local states $x_1 = (\theta_1, \dot{\theta}_1)^T$ and $x_2 = (\theta_2, \dot{\theta}_2)^T$ of the corresponding penduli,

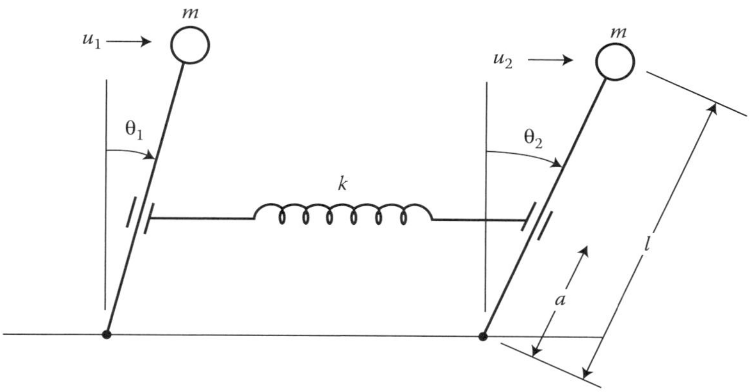

FIGURE 26.1 Inverted penduli.

that is, $u_1 = u_1(x_1)$ and $u_2 = u_2(x_2)$. This restriction is called the *decentralized information structure constraint.*

Since the system **S** is linear, a natural choice is the linear control laws

$$u_1 = k_1^T x_1 \,, \quad u_2 = k_2^T x_2 \tag{26.3}$$

where the feedback gain vectors $k_1 = (k_{11}, k_{12})^T$ and $k_2 = (k_{21}, k_{22})^T$ should be selected to *stabilize* the system **S**, that is, hold the penduli in the upright position.

In control design, it is fruitful to recognize the structure of the system **S** as an interconnection

$$\begin{aligned} \mathbf{S}: \dot{x}_1 &= \begin{bmatrix} 0 & 1 \\ \alpha & 0 \end{bmatrix} x_1 + \begin{bmatrix} 0 \\ \beta \end{bmatrix} u_1 + e \begin{bmatrix} 0 & 0 \\ -\gamma & 0 \end{bmatrix} x_1 + e \begin{bmatrix} 0 & 0 \\ \gamma & 0 \end{bmatrix} x_2, \\ \dot{x}_2 &= \begin{bmatrix} 0 & 1 \\ \alpha & 0 \end{bmatrix} x_2 + \begin{bmatrix} 0 \\ \beta \end{bmatrix} u_1 + e \begin{bmatrix} 0 & 0 \\ \gamma & 0 \end{bmatrix} x_1 + e \begin{bmatrix} 0 & 0 \\ -\gamma & 0 \end{bmatrix} x_2, \end{aligned} \tag{26.4}$$

of two subsystems

$$\begin{aligned} \mathbf{S}_1: \dot{x}_1 &= \begin{bmatrix} 0 & 1 \\ \alpha & 0 \end{bmatrix} x_1 + \begin{bmatrix} 0 \\ \beta \end{bmatrix} u_1, \\ \mathbf{S}_2: \dot{x}_2 &= \begin{bmatrix} 0 & 1 \\ \alpha & 0 \end{bmatrix} x_2 + \begin{bmatrix} 0 \\ \beta \end{bmatrix} u_2, \end{aligned} \tag{26.5}$$

where $\alpha = g/\ell$, $\beta = 1/m\ell^2$, $\gamma = \bar{a}^2 k/m\ell^2$, and $e = (a/\bar{a})^2$. One reason is that, in designing control for interconnected systems, the designer has to account for essential *uncertainty* in the interconnections among the subsystems. Though models of the subsystems are commonly available with sufficient accuracy, the shape and size of the interconnections cannot be predicted satisfactorily either for modeling or operational reasons. In the example, the interconnection parameter $e = a/\bar{a}$ is the uncertain height of the spring which is normalized by its nominal value $\bar{a}$.

An equally important reason for decomposition is present when controlling large dynamic systems. In complex systems with many variables, most of the variables are *weakly coupled*, if coupled at all, and the behavior of the overall system is dominated by strongly connected variables. Considerable conceptual and numerical simplification can be gained by controlling the strongly coupled variables with decentralized control.

26.3 Plant and Feedback Structures

Consider a linear constant system

$$\mathbf{S}: \begin{aligned} \dot{x} &= Ax + Bu, \\ y &= Cx, \end{aligned} \tag{26.6}$$

as an interconnected system

$$\mathbf{S}: \begin{aligned} \dot{x}_i &= A_i x_i + B_i u_i + \sum_{j \in \mathcal{N}} (A_{ij} x_j + B_{ij} u_j), \\ y_i &= C_i x_i + \sum_{j \in \mathcal{N}} C_{ij} x_j, \quad i \in \mathcal{N}, \end{aligned} \tag{26.7}$$

which is composed of N subsystems

$$\mathbf{S}_i: \begin{aligned} \dot{x}_i &= A_i x_i + B_i u_i, \\ y_i &= C_i x_i, \quad i \in \mathcal{N}, \end{aligned} \tag{26.8}$$

where $x_i(t) \in \mathbb{R}^{n_i}$, $u_i(t) \in \mathbb{R}^{m_i}$, $y_i(t) \in \mathbb{R}^{\ell_i}$ are the state, input, and output of the subsystem $\mathbf{S}_i$ at a fixed time $t \in \mathbb{R}$. All matrices have proper dimensions, and $\mathcal{N} = \{1, 2, \ldots, N\}$. At present we are interested in *disjoint* decompositions, that is,

$$\begin{aligned} x &= (x_1^T, x_2^T, \ldots, x_N^T)^T, \\ u &= (u_1^T, u_2^T, \ldots, u_N^T)^T, \\ y &= (y_1^T, y_2^T, \ldots, y_N^T)^T, \end{aligned} \tag{26.9}$$

and where $x(t) \in \mathbb{R}^n$, $u(t) \in \mathbb{R}^m$, and $y(t) \in \mathbb{R}^\ell$ are the state, input, and output of the overall system $\mathbf{S}$, so that

$$\begin{aligned} \mathbb{R}^n &= \mathbb{R}^{n_1} \times \mathbb{R}^{n_2} \times \cdots \times \mathbb{R}^{n_N}, \\ \mathbb{R}^m &= \mathbb{R}^{m_1} \times \mathbb{R}^{m_2} \times \cdots \times \mathbb{R}^{m_N}, \\ \mathbb{R}^\ell &= \mathbb{R}^{\ell_1} \times \mathbb{R}^{\ell_2} \times \cdots \times \mathbb{R}^{\ell_N}. \end{aligned} \tag{26.10}$$

A compact description of the interconnected system $\mathbf{S}$ is

$$\mathbf{S}: \begin{aligned} \dot{x} &= A_D x + B_D u + A_C x + B_C u \\ y &= C_D x + C_C x, \end{aligned} \tag{26.11}$$

where

$$\begin{aligned} A_D &= \text{diag}\{A_1, A_2, \ldots, A_N\}, \\ B_D &= \text{diag}\{B_1, B_2, \ldots, B_N\}, \\ C_D &= \text{diag}\{C_1, C_2, \ldots, C_N\}, \end{aligned} \tag{26.12}$$

and the coupling block matrices are

$$A_C = (A_{ij}), \quad B_C = (B_{ij}), \quad C_C = (C_{ij}). \tag{26.13}$$

The collection of N decoupled subsystems is described by

$$\mathbf{S}_D: \begin{aligned} \dot{x} &= A_D x + B_D u \\ y &= C_D x, \end{aligned} \tag{26.14}$$

obtained from (Equation 26.11) by setting the coupling matrices to zero.

Important special classes of interconnected systems are input ($B_C = 0$) and output ($C_C = 0$) decentralized systems, where inputs and outputs are not shared among the subsystems. Input–output decentralized systems are described as

$$\begin{aligned} \mathbf{S}: \ \dot{x} &= A_D x + B_D u + A_C x \\ y &= C_D x, \end{aligned} \tag{26.15}$$

where both B_C and C_C are zero. This structural feature helps to a great extent when decentralized controllers and estimators are designed for large plants.

A *static decentralized state feedback*,

$$u = -K_D x, \tag{26.16}$$

is characterized by a block-diagonal gain matrix,

$$K_D = \mathrm{diag}\{K_1, K_2, \ldots, K_N\}, \tag{26.17}$$

which implies that each subsystem $\mathbf{S}_i$ has its individual control law,

$$u_i = -K_i x_i, \quad i \in \mathcal{N}, \tag{26.18}$$

with a constant gain matrix K_i. The control law u of Equation 26.16, which is equivalent to the totality of subsystem control laws (Equation 26.18), obeys the decentralized information structure constraint requiring that each subsystem $\mathbf{S}_i$ is controlled on the basis of its locally available state x_i. The closed-loop system is described as

$$\hat{\mathbf{S}}: \ \dot{x} = (A_D - B_D K_D C_D)x + A_C x. \tag{26.19}$$

When *dynamic output feedback* is used under decentralized constraints, then controllers of the following type are considered:

$$\begin{aligned} \mathbf{C}_i: \ \dot{z}_i &= F_i z_i + G_i y_i, \\ u_i &= -H_i z_i - K_i y_i, \quad i \in \mathcal{N}, \end{aligned} \tag{26.20}$$

which can be written in a compact form as a single decentralized controller defined as

$$\begin{aligned} \mathbf{C}_D: \ \dot{z} &= F_D z + G_D y, \\ u &= -H_D z - K_D y, \end{aligned} \tag{26.21}$$

where

$$\begin{aligned} z &= (z_1^T, z_2^T, \ldots, z_N^T)^T, \ y = (y_1^T, y_2^T, \ldots, y_N^T)^T, \\ u &= (u_1^T, u_2^T, \ldots, u_N^T)^T, \end{aligned} \tag{26.22}$$

are the state $z \in \mathbb{R}^r$, input $y \in \mathbb{R}^\ell$, and output $u \in \mathbb{R}^m$ of the controller $\mathbf{C}_D$. By combining the system $\mathbf{S}$ and the decentralized dynamic controller $\mathbf{C}_D$, we get the composite closed-loop system as

$$\mathbf{S\&C}_D: \begin{bmatrix} \dot{x} \\ \dot{z} \end{bmatrix} = \begin{bmatrix} A_D - B_D K_D C_D + A_C & -B_D H_D \\ G_D C_D & F_D \end{bmatrix} \begin{bmatrix} x \\ z \end{bmatrix}. \tag{26.23}$$

26.4 Decentralized Stabilization

The fundamental problem in decentralized control theory and practice is choosing individual subsystem inputs to stabilize the overall interconnected system. In the previous section, the plant structures have been described, where the plant, inputs and outputs are all decomposed with each local controller responsible for the corresponding subsystem. While this is the most common situation in practice, it is by no means all inclusive. It is often advantageous, and sometime necessary, to decentralize the inputs and outputs without decomposing the plant. This is the situation that we consider first.

26.4.1 Decentralized Inputs and Outputs

Suppose that only the inputs and outputs, but not states, of system **S** in Equation 26.6 are partitioned as in Equation 26.9, and **S** is described as

$$\mathbf{S}: \dot{x} = Ax + \sum_{i\in\mathcal{N}} \tilde{B}_i u_i, \qquad y_i = \tilde{C}_i x, \quad i \in \mathcal{N}. \tag{26.24}$$

Then, the controllers $\mathbf{C}_i$ of Equation 26.20 still operate on local measurements y_i to generate local controls u_i, but now they are collectively responsible for the whole system. In this case,

$$\mathbf{S}\&\mathbf{C}_D: \begin{bmatrix} \dot{x} \\ \dot{z} \end{bmatrix} = \begin{bmatrix} A - BK_DC & -BH_D \\ G_DC & F_D \end{bmatrix} \begin{bmatrix} x \\ z \end{bmatrix}. \tag{26.25}$$

It is well-known that without the decentralization constraint on the controller, the closed-loop system of Equation 26.25 can be stabilized if, and only if, the uncontrollable or unobservable modes of the open-loop system **S** are stable; or equivalently, the set of (centralized) fixed modes of **S**, which is defined as

$$\Lambda_C = \bigcap_K \sigma(A - BKC) \tag{26.26}$$

is included in the open left half plane, where $\sigma(.)$ denotes the set of eigenvalues of the indicated matrix. This basic result has been extended in [34] to decentralized control of **S**, where it was shown that the closed-loop system Equation 26.25 can be made stable with suitable choice of the decentralized controllers $\mathbf{C}_i$ if, and only if, the set of decentralized fixed modes

$$\Lambda_D = \bigcap_{K_D} \sigma(A - BK_DC) = \bigcap_{K_1,\ldots,K_N} \sigma(A - \sum_{i\in\mathcal{N}} \tilde{B}_i K_i \tilde{C}_i) \tag{26.27}$$

is included in the open left half plane.

The result of [34] has been followed by extensive research on the following topics:

- State space and frequency domain characterization of decentralized fixed modes
- Development of various techniques for designing decentralized controllers (e.g., using static output feedback in all but one channel, distributing the control effort among channels, sequential stabilization, etc.)
- Generalization of the concept of decentralized fixed modes to arbitrary feedback structure constraints
- Formulation of the concept of structurally fixed modes, and their algebraic and graph-theoretical characterization

A useful and simple characterization of decentralized fixed modes was provided in [1]. For any subset $\mathcal{I} = \{i_1, \ldots, i_P\}$ of the index set $\mathcal{N}$, let $\mathcal{I}^C = \{j_1, \ldots, j_{N-P}\}$ denote the complement of $\mathcal{I}$ in $\mathcal{N}$, and define

$$\tilde{B}_{\mathcal{I}} = [\tilde{B}_{i_1}, \tilde{B}_{i_2}, \ldots, \tilde{B}_{i_P}], \qquad \tilde{C}_{\mathcal{I}^C} = \begin{bmatrix} \tilde{C}_{j_1} \\ \tilde{C}_{j_2} \\ \vdots \\ \tilde{C}_{j_{N-P}} \end{bmatrix}. \tag{26.28}$$

Then a complex number $\lambda \in \mathbb{C}$ is a decentralized fixed mode of **S** if, and only if,

$$\text{rank} \begin{bmatrix} A - \lambda I & \tilde{B}_{\mathcal{I}} \\ \tilde{C}_{\mathcal{I}^C} & 0 \end{bmatrix} < n \tag{26.29}$$

for some $\mathcal{I} \subset \mathcal{N}$. This result relates decentralized fixed modes to transmission zeros of the systems $(A, \tilde{B}_{\mathcal{I}}, \tilde{C}_{\mathcal{I}^C})$, called the complementary subsystems. Thus, appearance of a fixed mode corresponds to a special pole–zero cancellation, which can not be removed by constant decentralized feedback. However, under mild conditions, such fixed modes can be eliminated by time-varying decentralized feedback.

The characterization of decentralized fixed modes above prompts a generalization of the concept to arbitrary feedback structures. Let $\bar{K} = (\bar{k}_{ij})$ be an $m \times l$ binary matrix such that $\bar{k}_{ij} = 1$ if, and only if, a feedback link from output y_j to input u_i is allowed. Thus $\bar{K}$ specifies a constraint on the feedback structure, a special case of which is decentralized feedback. In this case, permissible controllers have the structure

$$\begin{aligned} \mathbf{C}_{\bar{K}} : \dot{z}_i &= F_i z_i + \sum_{j \in \mathcal{J}_i} g_{ij} y_j \\ u_i &= -h_i^T z_i - \sum_{j \in \mathcal{J}_i} k_{ij} y_i \end{aligned} \tag{26.30}$$

where $\mathcal{J}_i = \{j : \bar{k}_{ij} = 1\}$.

Let K denote any feedback matrix conforming to the structure of $\bar{K}$, that is, one with $k_{ij} = 0$ whenever $\bar{k}_{ij} = 0$. Then, the set

$$\Lambda_{\bar{K}} = \bigcap_{K} \sigma(A - BKC) \tag{26.31}$$

can conveniently be defined as the set of fixed modes with respect to the decentralized feedback structure constraint specified by $\bar{K}$. Then the closed-loop system consisting of **S** and the constrained controller $\mathbf{C}_{\bar{K}}$ can be stabilized if, and only if, $\Lambda_{\bar{K}}$ is included in the open left half-plane. Finally, it remains to characterize $\Lambda_{\bar{K}}$ as in Equation 26.29. This, however, is quite automatic; consider the index sets $\mathcal{I} \subset \mathcal{M} = \{1, 2, \ldots, M\}$ and replace $\mathcal{I}^C$ by $\mathcal{J} = \cup_{i \in \mathcal{I}^C} \mathcal{J}_i$, where now $\mathcal{I}^C$ refers to the complement of $\mathcal{I}$ in $\mathcal{M}$.

26.4.2 Structural Analysis

Structural analysis of large scale systems via graph-theoretic concepts and methods offers an appealing alternative to quantitative analysis which often faces difficulties due to high dimensionality and lack of exact knowledge of system parameters. Equipped with the powerful tools of graph theory, structural analysis provides valuable information concerning certain qualitative properties of the system under study by practical tests and algorithms [30].

One of the earliest problems of structural analysis is the graph-theoretic formulation of controllability [20]. Consider an uncontrollable pair (A, B). Loss of controllability is either due to a perfect matching of system parameters or due to an insufficient number of nonzero parameters, indicating a lack of sufficient linkage among system variables. In the latter case, the pair (A, B) is structurally uncontrollable in the sense that all pairs having the same structure as (A, B) are uncontrollable. Since the structure of (A, B) can be described by a directed graph (as explained below for a more general case), structural controllability can be checked by graph-theoretic means. Indeed, (A, B) is structurally controllable if, and only if, the system graph is input reachable (that is, each state variable is affected directly or indirectly by at least one input variable), and contains no dilations (that is, no subset of state variables exists whose number exceeds the total number of all state and input variables directly affecting these variables). These two conditions are equivalent to the spanning of the system graph by a minimal subgraph, called a cactus, which has a special structure.

The idea of treating controllability in a structural framework has led to formulation and graph-theoretic characterization of *structurally fixed modes* under constrained feedback [26]. Let $\mathbf{D} = (\mathcal{V}, \mathcal{E})$

be a directed graph associated with the system **S** of Equation 26.6, where $\mathcal{V} = \mathcal{U} \cup \mathcal{X} \cup \mathcal{Y}$ is a set of vertices corresponding to inputs, states, and outputs of **S**, and $\mathcal{E}$ is a set of directed edges corresponding to nonzero parameters of the system matrices A, B, and C. To every nonzero a_{ij}, there corresponds an edge from vertex x_j to vertex x_i, to every nonzero b_{ij}, an edge from u_j to x_i, and to every nonzero c_{ij}, one from x_j to y_i. Given a feedback pattern $\bar{K}$ and adding to **D** a feedback edge from y_j to u_i for every $\bar{k}_{ij} = 1$, one gets a digraph $\mathbf{D}_{\bar{K}} = (\mathcal{V}, \mathcal{E} \cup \mathcal{E}_{\bar{K}})$ completely describing the structure of both the system **S** and the feedback constraint specified by $\bar{K}$.

Two systems are said to be structurally equivalent if they have the same system graphs. A system **S** is said to have structurally fixed modes with respect to a given $\bar{K}$ if every system structurally equivalent to **S** has fixed modes with respect to $\bar{K}$. Having structurally fixed modes is a common property of a class of systems described by the same system graph; if a system has no structurally fixed modes, then either it has no fixed modes, or if it does, arbitrarily small perturbations of system parameters can eliminate the fixed modes. As a result, if a system has no structurally fixed modes with respect to $\bar{K}$, then generically it can be stabilized by a constrained controller of the form defined in Equation 26.30.

It was shown in [26] that a system **S** has no structurally fixed modes with respect to a feedback pattern $\bar{K}$ if, and only if

1. all state vertices of $\mathbf{D}_{\bar{K}}$ are covered by vertex disjoint cycles, and
2. no strong component of $\mathbf{D}_{\bar{K}}$ contains only state vertices, where a strong component is a maximal subgraph whose vertices are reachable from each other.

This simple graph-theoretic criterion has been used in an algorithmic way in problems such as choosing a minimum number of feedback links (or, if each feedback link is associated with a cost, choosing the cheapest feedback pattern) that avoid structurally fixed modes. As an example, consider a system with a system graph as in Figure 26.2. Let the costs of setting up feedback links (dotted lines) from each output to each input be given by a matrix

$$\begin{bmatrix} 6 & 2 \\ 3 & 7 \end{bmatrix}.$$

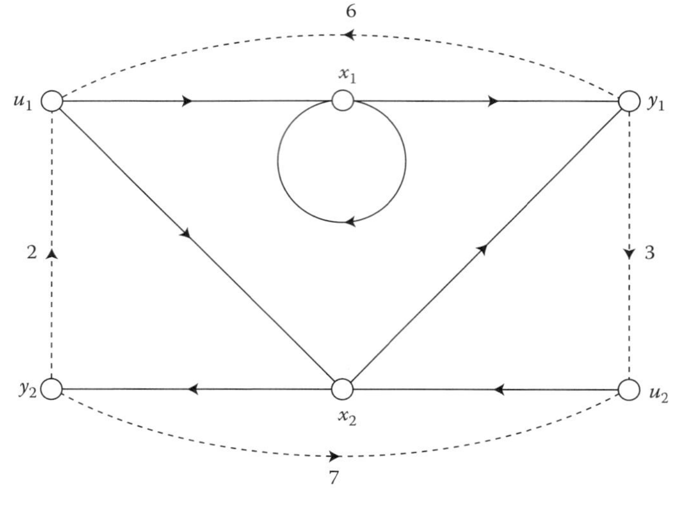

FIGURE 26.2 System graph.

It can easily be verified that any feedback pattern of the form

$$\begin{bmatrix} 1 & * \\ * & * \end{bmatrix} \quad \text{or} \quad \begin{bmatrix} * & 1 \\ 1 & * \end{bmatrix},$$

where $*$ stands for either a 0 or a 1, avoids structurally fixed modes. Clearly, the feedback patterns which contain the least number of links and which cost the least are, respectively,

$$\bar{K}_1 = \begin{bmatrix} 1 & 0 \\ 0 & 0 \end{bmatrix} \quad \text{or} \quad \bar{K}_2 = \begin{bmatrix} 0 & 1 \\ 1 & 0 \end{bmatrix}.$$

26.4.3 Decentrally Stabilizable Structures

Consider an interconnected system

$$\begin{aligned} \mathbf{S}: \dot{x}_i &= A_i x_i + B_i(u_i + \sum_{j \in \mathcal{N}} D_{ij} x_j) \\ y_i &= x_i \quad i \in \mathcal{N} \end{aligned} \tag{26.32}$$

which is a special case of the system **S** in Equation 26.7 in that $A_{ij} = B_i D_{ij}$, $B_{ij} = 0$, $C_i = I$, and $C_{ij} = 0$. Assuming that the decoupled subsystems described by the pairs (A_i, B_i) are controllable, it is easy to verify that **S** has no decentralized fixed modes. Thus **S** can be stabilized using a decentralized dynamic feedback controller of the form (Equation 26.21). However, because the subsystem outputs are the states, there should be no need to use dynamic controllers.

Choose the decentralized constant state feedbacks in Equation 26.18 to place the subsystem poles at $-\mu_{il}\rho, i \in \mathcal{N}, l = 1, 2, \ldots, n_i$, where $-\mu_{il}$ are distinct negative real numbers, and ρ is a parameter. Then a suitable change of coordinate frame transforms the closed-loop system of Equation 26.19 into the form

$$\hat{\mathbf{S}}: \dot{x} = (-\rho M + \hat{A}_C)x, \tag{26.33}$$

where $M = \text{diag}\{M_1, M_2, \ldots, M_N\}$, with $M_i = \text{diag}\{\mu_1, \mu_2, \ldots, \mu_{in_i}\}$, and $\hat{A}_C$ is independent of the parameter ρ. Clearly, $\hat{\mathbf{S}}$ is stable for a sufficiently large ρ.

The success of this high-gain decentralized stabilization technique results from the special structure of the interconnections among the subsystems. The interconnections from other subsystems affect a particular subsystem in the same way its local input does. This makes it possible to neutralize potentially destabilizing effects of the interconnections by a local state feedback and provide a high degree of stability to the decoupled subsystems. This special interconnection structure is termed the "matching conditions" [18].

Decentralized stabilizability of interconnected systems satisfying the matching conditions has motivated research in characterizing other decentrally stabilizable interconnection structures. Below, another such interconnection structure is described, where single-input subsystems are considered for convenience.

Let the interconnected system be described as

$$\mathbf{S}: \dot{x}_i = A_i x_i + b_i u_i + \sum_{j \in \mathcal{N}} A_{ij} x_j, \quad i \in \mathcal{N} \tag{26.34}$$

where, without loss of generality, the subsystem pairs (A_i, b_i) are assumed to be in controllable canonical form. For each interconnection matrix A_{ij}, define an integer m_{ij} as

$$m_{ij} = \begin{cases} \max\{q - p : a_{pq}^{ij} \neq 0\}, & A_{ij} \neq 0, \\ -n, & A_{ij} = 0, \end{cases} \tag{26.35}$$

Thus, m_{ij} is the distance between the main diagonal and a line parallel to the main diagonal which borders all nonzero elements of A_{ij}.

For an index set $\mathcal{I} \subset \mathcal{N}$, let $\mathcal{I}_P$ denote any permutation of $\mathcal{I}$. Then, the system **S** in Equation 26.34 is stabilizable by decentralized constant state feedback if

$$\sum_{\substack{i\in\mathcal{I}\\ j\in\mathcal{I}_P}} (m_{ij} - 1) < 0 \tag{26.36}$$

for all $\mathcal{I}$ and all permutations $\mathcal{I}_P$ [14,30]. In the case of matching interconnections, $m_{ij} = n_j - n_i$, so that Equation 26.36 guarantees decentralized stabilizability even when the elements of the interconnection matrices A_{ij} are bounded nonlinear, time-varying functions of the state variables. Therefore, the condition (Equation 26.36) and, thus, the matching conditions, are indeed structural conditions.

26.4.4 Vector Lyapunov Functions

A general way to establish the stability of nonlinear interconnected systems is to apply the Matrosov–Bellman concept of vector Lyapunov functions [17]. The concept has been developed to provide an efficient method of checking the stability of linear interconnected systems controlled by decentralized feedback [30]. First, each subsystem is stabilized using local state or output feedback. Then, for each stable closed-loop (but decoupled) subsystem, a Lyapunov function is chosen using standard methods. These functions are stacked to form a vector of functions, which can then be used to form a single scalar Lyapunov function for the overall system. The function establishes stability if we show positivity of the leading principal minors of a constant aggregate matrix whose dimension equals the number of subsystems.

Consider the linear interconnected system of Equation 26.7,

$$\mathbf{S}: \dot{x}_i = A_i x_i + B_i u_i + \sum_{j\in\mathcal{N}} e_{ij} A_{ij} x_j, \quad i \in \mathcal{N}, \tag{26.37}$$

where the output y_i is not included and $B_{ij} = 0$. We inserted the elements of $e_{ij} \in [0, 1]$ of the $N \times N$ interconnection matrix $E = (e_{ij})$ to capture the presence of uncertainty in coupling between the subsystems

$$\mathbf{S}_i: \ \dot{x}_i = A_i x_i + B_i u_i, \tag{26.38}$$

as illustrated by the example of the two penduli above.

We assume that each pair (A_i, B_i) is controllable and assign the eigenvalues $-\sigma_1^i \pm j\omega_1^i, \ldots, -\sigma_{p_i}^i \pm j\omega_{p_i}^i, \ldots, -\sigma_{2p_i+1}^i, \ldots, -\sigma_{n_i}^i$ to each closed-loop subsystem

$$\hat{\mathbf{S}}_i: \ \dot{x}_i = (A_i - B_i K_i) x_i \tag{26.39}$$

by applying decentralized feedback

$$u_i = -K_i x_i. \tag{26.40}$$

Using a nonsingular transformation,

$$x_i = T_i \tilde{x}_i, \tag{26.41}$$

we can obtain the closed-loop subsystems as

$$\tilde{\mathbf{S}}_i: \ \dot{\tilde{x}}_i = \Lambda_i \tilde{x}_i, \tag{26.42}$$

where the matrix $\Lambda_i = T_i^{-1}(A_i - B_i K_i) T_i$ has the diagonal form

$$\Lambda_i = \text{diag}\left\{ \begin{bmatrix} -\sigma_1^i & \omega_1^i \\ -\omega_1^i & -\sigma_1^i \end{bmatrix}, \ldots, \begin{bmatrix} -\sigma_{p_i}^i & \omega_{p_i}^i \\ -\omega_{p_i}^i & -\sigma_{p_i}^i \end{bmatrix}, -\sigma_{2p_i+1}, \ldots, -\sigma_{n_i} \right\}. \tag{26.43}$$

For each transformed subsystem, there exists a suitable Lyapunov function $v : \mathbb{R}^{n_i} \rightarrow \mathbb{R}_+$ of the form

$$v_i(\tilde{x}_i) = (\tilde{x}_i H_i \tilde{x}_i)^{\frac{1}{2}}, \tag{26.44}$$

where $H_i = I_i$ is the solution of the Lyapunov matrix equation

$$\Lambda_i H_i + H_i \Lambda_i = -G_i \tag{26.45}$$

for $G_i = \text{diag}\{\sigma_1^i, \sigma_1^i, \ldots, \sigma_{p_i}^i, \sigma_{2p_i+1}^i \cdots, \sigma_{n_i}^i\}$.

To determine the stability of the overall interconnected closed-loop system

$$\tilde{\mathbf{S}} : \dot{\tilde{x}}_i = \Lambda_i \tilde{x}_i + \sum_{j \in \mathcal{N}} e_{ij} \Delta_{ij} \tilde{x}_j \tag{26.46}$$

from the stability of the decoupled closed-loop subsystems $\tilde{\mathbf{S}}_i$, we consider subsystem functions v_i as components of a *vector Lyapunov function* $v = (v_1, v_2, \ldots, v_N)^T$, and form a candidate Lyapunov function $V : \mathbb{R}^n \rightarrow \mathbb{R}_+$ for the overall system $\tilde{\mathbf{S}}$ as

$$V(\tilde{x}) = \sum_{i \in \mathcal{N}} d_i v_i(\tilde{x}_i), \tag{26.47}$$

where the existence of positive numbers d_i for stability of $\tilde{\mathbf{S}}$ has yet to be established, and $\Delta_{ij} = T_i^{-1} A_{ij} T_j$.

Taking the total time derivative of $V(\tilde{x})$ with respect to $\tilde{\mathbf{S}}$, after lengthy but straightforward computations [30],

$$\dot{V}(\tilde{x}) \leq -d^T \bar{W} z\,, \tag{26.48}$$

with $d = (d_1, d_2, \ldots, d_N)^T$, $z = (\|\tilde{x}_1\|,\ \|\tilde{x}_2\|, \ldots, \|\tilde{x}_N\|)^T$, and $\bar{W} = (\bar{w}_{ij})$ is the $N \times N$ aggregate matrix defined as

$$\bar{w}_{ij} = \begin{cases} \frac{1}{2}\sigma_m^i - \bar{e}_{ii}\lambda_M^{1/2}(\Delta_{ii}^T \Delta_{ii}), & i = j \\ -\bar{e}_{ij}\lambda_M^{1/2}(\Delta_{ij}^T \Delta_{ij}), & i \neq j \end{cases} \tag{26.49}$$

where σ_m^i is the minimal value of all σ_k^i, and $\lambda_M(\,\cdot\,)$ is the maximal eigenvalue of the indicated matrix.

The elements $\bar{e}_{ij}$ of the fundamental interconnection matrix $\bar{E} = (\bar{e}_{ij})$ are binary numbers defined as

$$\bar{e}_{ij} = \begin{cases} 1, & \mathbf{S}_j \text{ acts on } \mathbf{S}_i \\ 0, & \mathbf{S}_j \text{ does not act on } \mathbf{S}_i. \end{cases} \tag{26.50}$$

In this way, the binary matrix describes the basic interconnection structure of the system $\mathbf{S}$. In the case of two penduli,

$$\bar{E} = \begin{bmatrix} 1 & 1 \\ 1 & 1 \end{bmatrix}. \tag{26.51}$$

It has been shown in [30] that stability of $-\bar{W}$ (all eigenvalues of $-\bar{W}$ have negative real parts) implies stability of the closed-loop system $\tilde{\mathbf{S}}$ and, hence, $\hat{\mathbf{S}}$. To explain this fact, we note first that $w_{ii} > 0$, $w_{ij} \leq 0$ $(i \neq j)$, which makes $\bar{W}$ an M-matrix (e.g., [30]) if, and only if, there exists a positive vector d $(d_i > 0,\ \ i \in \mathcal{N})$, so that the vector

$$c^T = d^T \bar{W} \tag{26.52}$$

is a positive vector as well. Positivity of c and d imply $V(\tilde{x}) > 0$ and $\dot{V}(\tilde{x}) < 0$ and, therefore, stability of $\hat{\mathbf{S}}$ by the standard Lyapunov argument. Finally, the M-matrix property of $\bar{W}$ is equivalent to stability of $-\bar{W}$.

Several comments are in order. First, we note that the M-matrix property of $\bar{W}$ can be tested by a simple determinantal condition

$$\begin{vmatrix} \bar{w}_{11} & \bar{w}_{12} & \cdots & \bar{w}_{1k} \\ \bar{w}_{21} & \bar{w}_{22} & \cdots & \bar{w}_{2k} \\ \cdots & \cdots & \cdots & \cdots \\ \bar{w}_{k1} & \bar{w}_{k2} & \cdots & \bar{w}_{kk} \end{vmatrix} > 0, \quad k \in \mathcal{N}. \tag{26.53}$$

Another important feature of the concept of vector Lyapunov functions is the *robustness* information about decentrally stabilized interconnected system $\hat{\mathbf{S}}$. The determinantal condition (Equation 26.53) is equivalent to the quasidominant diagonal property of $\bar{W}$,

$$\bar{w}_{ii} > d_i^{-1} \sum_{j \neq i}^{N} d_j |\bar{w}_{ij}|, \quad i \in \mathcal{N}. \tag{26.54}$$

where the d_i's are positive numbers. From Equation 26.54, it is obvious that, if $\bar{W}$ is an M-matrix, so is W for any $E \leq \bar{E}$, where the inequality is taken element by element; the system $\hat{\mathbf{S}}$ is *connectively stable* [30]. When a system is connectively stabilized by decentralized feedback, stability is robust and can tolerate variations in coupling among the subsystems. When the two penduli are stabilized for any given position $\bar{a}$ of the spring, including the entire length ℓ of the penduli, the penduli are stable for any position $a \leq \bar{a}$. In other words, if the penduli are stabilized for the fundamental interconnection matrix $\bar{E}$ of (51), they are stabilized for any interconnection matrix

$$E = \begin{bmatrix} e & e \\ e & e \end{bmatrix}, \tag{26.55}$$

whenever $e \in [0, 1]$.

Finally, the decentrally stabilized system can tolerate nonlinearities in the interconnections among the subsystems. The nonlinear interconnections need not be known since only their size is required to be limited. Once the closed-loop system $\hat{\mathbf{S}}$ is shown to be stable, it follows [30] that a nonlinear time-varying version

$$\hat{\mathbf{S}}_N : \dot{\tilde{x}}_i = (A_i - B_i K_i)\tilde{x}_i + h_i(t, \tilde{x}), \quad i \in \mathcal{N} \tag{26.56}$$

of $\hat{\mathbf{S}}$ is connectively stable, provided the conical constraints

$$\|h_i(t, \tilde{x})\| \leq \sum_{j=1}^{N} \bar{e}_{ij} \xi_{ij} \|\tilde{x}_j\|, \quad i \in \mathcal{N} \tag{26.57}$$

on interconnection functions $h_i : \mathbb{R} \times \mathbb{R}^n \to \mathbb{R}^{n_i}$ hold, where the nonnegative numbers ξ_{ij} do not exceed $\lambda_M^{1/2}(\Delta_{ij}^T \Delta_{ij})$. This robustness result is useful in practice because, typically, interconnections are poorly known, or they are changing during operation of the controlled system.

26.5 Optimization

There is no general method for designing optimal decentralized controls for interconnected systems, even if they are linear and time invariant. For this reason, standard design practice is to optimize each decoupled subsystem using linear quadratic (LQ) control laws. Then, suboptimality of the interconnected closed–loop system, which is driven by the union of the locally optimal LQ control laws, is determined with respect to the sum of the quadratic costs chosen for the subsystems. The suboptimal decentralized control design is attractive because, under relatively mild conditions, suboptimality implies stability.

Furthermore, the degree of suboptimality can serve as a measure of robustness with respect to a wide spectrum of uncertainties residing in both the subsystems and their interactions.

Consider again the interconnected system

$$\mathbf{S}: \dot{x}_i = A_i x_i + B_i u_i + \sum_{i \in \mathcal{N}} A_{ij} x_j, \quad i \in \mathcal{N} \tag{26.58}$$

in the compact form

$$\mathbf{S}: \dot{x} = A_D x + B_D u + A_C x. \tag{26.59}$$

We assume that the subsystems

$$\mathbf{S}_i: \dot{x}_i = A_i x_i + B_i u_i \tag{26.60}$$

or, equivalently, their union

$$\mathbf{S}: \dot{x} = A_D x + B_D u, \tag{26.61}$$

is controllable, that is, all pairs (A_i, B_i) are controllable.

With $\mathbf{S}_D$ we associate a quadratic cost

$$J_D(x_0, u) = \int_0^\infty (x^T Q_D x + u^T R_D u)\, dt, \tag{26.62}$$

where $Q_D = \text{diag}\{Q_1, Q_2, \ldots, Q_N\}$ is a symmetric nonnegative definite matrix, $R_D = \text{diag}\{R_1, R_2, \ldots, R_N\}$ is a symmetric positive definite matrix, and the pair $(A_D, Q_D^{1/2})$ is observable. The cost J_D can be considered as a sum of subsystem costs

$$J_i(x_{i0}, u_i) = \int_0^\infty (x_i^T Q_i x_i + u_i^T R u_i)\, dt. \tag{26.63}$$

In order to satisfy the decentralized constraints on the control law, we solve the standard LQ optimal control problem $(\mathbf{S}_D, J_D)$ to get

$$u_D^{\odot} = -K_D x, \tag{26.64}$$

where $K_D = \text{diag}\{K_1, K_2, \ldots, K_N\}$ is given as

$$K_D = R_D^{-1} B_D^T P_D,$$

and $P_D = \text{diag}\{P_1, P_2, \ldots, P_N\}$ is the unique symmetric positive definite solution of the algebraic Riccati equation

$$A_D^T P_D + P_D A_D - P_D B_D R_D^{-1} B_D^T P_D + Q_D = 0. \tag{26.65}$$

The control $u_D^{\odot}$, when applied to $\mathbf{S}_D$, results in the closed-loop system

$$\hat{\mathbf{S}}_D^{\odot}: \dot{x} = (A_D - B_D K_D)\, x, \tag{26.66}$$

which is optimal and produces the optimal cost

$$J_D^{\odot}(x_0) = x_0^T P_D x_0. \tag{26.67}$$

The important fact about the locally optimal control $u_D^{\odot}$ is that it is decentralized. Each component

$$u_i^{\odot} = -K_i x_i \tag{26.68}$$

of $u_D^{\odot}$ uses only the local state x_i. Generally, the proposed control strategy is not globally optimal, but we can proceed to determine if the cost $J_D^{\oplus}(x_0)$ corresponding to the closed-loop interconnected system

$$\hat{\mathbf{S}}^{\oplus}: \dot{x} = (A_D - B_D K_D + A_C)\, x \tag{26.69}$$

is finite. If it is, then $\mathbf{S}^{\oplus}$ is suboptimal and a positive number μ exists such that

$$J_D^{\oplus}(x_0) \le \mu^{-1} J_D^{\odot}(x_0) \tag{26.70}$$

for all $x_0 \in \mathbb{R}^n$. The number μ is called the degree of suboptimality of $u_D^{\odot}$.

We can determine the index μ by first computing the performance index

$$J_D^{\oplus}(x_0) = x_0^T H x_0, \tag{26.71}$$

where

$$H = \int_0^{\infty} \exp(\hat{A}^T t) G_D \exp(\hat{A} t)\, dt,$$
$$G_D = Q_D + P_D B_D R_D^{-1} B_D^T P_D, \tag{26.72}$$

and the closed-loop matrix is

$$\hat{A} = A_D - B_D K_D + A_C. \tag{26.73}$$

It is important to note that $u_D^{\odot}$ is suboptimal if, and only if, the symmetric matrix H exists. The existence of H is guaranteed by the stability of $\hat{\mathbf{S}}$, in which case we can compute H as the unique solution of the Lyapunov matrix equation

$$\hat{A}^T H + H\hat{A} = -G_D. \tag{26.74}$$

The degree of suboptimality, which is the largest we can obtain in this context, is given as

$$\mu^* = \lambda_M^{1/2}(HP_D^{-1}). \tag{26.75}$$

Details of this development, as well as the broad scope of suboptimality, were described in [30], where special attention was devoted to the robustness implications of suboptimality. First, we can explicitly characterize suboptimality in terms the of interconnection matrix A_C. The system $\hat{\mathbf{S}}^{\oplus}$ is suboptimal with degree μ if the matrix

$$F(\mu) = A_C^T P_D + P_D A_C - (1-\mu)(Q_D + P_D B_D R_D^{-1} B_D^T P_D) \tag{26.76}$$

is nonpositive definite. This is a sufficient condition for suboptimality, but one that implies stability if the pair $\{A_D + A_C, Q_D^{1/2}\}$ is detectable.

Another important aspect of nonpositivity of $F(\mu)$ is that it implies stability even if each control $u_i^{\odot}$ is replaced by a nonlinearity $\phi_i(u_i^{\odot})$, which is contained in a sector, or by a linear time-invariant dynamic element. Furthermore, if the subsystems are single-input systems, then each subsystem feedback loop has infinite gain margin, at least $\pm\cos^{-1}(1-\frac{1}{2}\mu)$ phase margin, and at least $50\mu\%$ gain reduction tolerance. These are the standard robustness characteristics of an optimal LQ control law, which are modified by the degree of suboptimality. It is interesting to note that the optimal robustness characteristics can be recovered by solving the inverse problem of optimal decentralized control. The matching conditions are one of the conditions that guarantee the solution of the problem.

The concept of suboptimality extends to the case of *overlapping subsystems*, when subsystems share common parts, and control is required to conform with the *overlapping information structure constraints*. By expanding the underlying state space, the subsystems become disjoint and decentralized control can be designed for the expanded system by standard techniques. Finally, the control laws obtained are contracted for implementation in the original system. This expansion–contraction framework is known as the *inclusion principle*. For a comprehensive presentation of this principle, see [30].

26.6 Adaptive Decentralized Control

As mentioned in the section on decentrally stabilizable structures, many large scale interconnected systems with a good interconnection structure can be stabilized by a high-gain type decentralized control. How high the gain should be depends on how strong the interconnections are. If a bound on the interconnections is known, then stability can be guaranteed by a fixed high-gain controller. However, if such a bound is not available, then one has to use an adaptive controller which adjusts the gain to a value needed for overall stability.

Consider an interconnected system consisting of single-input subsystems

$$\mathbf{S}: \dot{x}_i(t) = A_i x_i(t) + b_i[u_i(t) + h_i(t, x(t))], \quad i \in \mathcal{N} \tag{26.77}$$

where, without loss of generality, the pairs (A_i, b_i) are assumed to be in controllable canonical form, and the nonlinear matching interconnections $h_i : \mathbb{R} \times \mathbb{R}^n \to \mathbb{R}$ are assumed to satisfy

$$|h_i(t, x)| \le \sum_{j \in \mathcal{N}} \alpha_{ij} \|x_j\| \tag{26.78}$$

for some *unknown* constants $\alpha_{ij} \ge 0$. Let a decentralized state feedback

$$u_i(t) = -\rho(t) k_i^T R_i(\rho(t)), \quad i \in \mathcal{N} \tag{26.79}$$

be applied to $\mathbf{S}$, where $R_i(\rho) = \text{diag}\{\rho^{n_i - 1}, \ldots, \rho, 1\}$, with $\rho(t)$ being a time-varying gain, and k_i^T are such that the matrices $\hat{A}_i = A_i - b_i k_i^T$ have distinct eigenvalues λ_{il}, $i \in \mathcal{N}, l = 1, 2, \ldots, n_i$. Let T_i denote the modal matrices of $\hat{A}_i$, i.e., $T_i \hat{A}_i T_i^{-1} = M_i = \text{diag}\{\lambda_{i1}, \lambda_{i2}, \ldots, \lambda_{in_i}\}$. Then a time-varying coordinate transformation $z_i(t) = T_i R_i(\rho(t)) x_i(t)$ transforms the closed-loop system $\hat{\mathbf{S}}$ into

$$\hat{\mathbf{S}}: \dot{z}_i(t) = \rho(t) M_i z_i(t) + g_i(t, z(t)), \quad i \in \mathcal{N}, \tag{26.80}$$

where, provided $0 \le \dot{\rho}(t) \le 1 \le \rho(t)$,

$$\|g_i(t, z)\| \le \sum_{j \in \mathcal{N}} \beta_{ij} \|z_j\| \tag{26.81}$$

for some *unknown* constants $\beta_{ij} \ge 0$. From Equations 26.80 and 26.81 it follows that there exists a $\rho^* > 0$ so that $\hat{\mathbf{S}}$ is stable for all $\rho(t)$ satisfying $0 \le \dot{\rho}(t) \le 1 \le \rho^* \le \rho(t)$, as can be shown by the vector Lyapunov approach. However, the crucial point is that ρ^* depends on the unknown bounds β_{ij}. Fortunately, the difficulty can be overcome by increasing $\rho(t)$ adaptively until it is high enough to guarantee stability of $\hat{\mathbf{S}}$. A simple adaptation rule that serves the purpose is

$$\dot{\rho}(t) = \min\{1, \gamma \|x(t)\|\} \tag{26.82}$$

where $\gamma > 0$ is arbitrary. Although the control law is decentralized, $\rho(t)$ is adjusted based on complete state information.

The same idea can also be used in constructing adaptive decentralized dynamic output feedback controllers for various classes of large scale systems with structured nonlinear, time-varying interconnections. A typical example is a system described by

$$\begin{aligned} \mathbf{S}: \dot{x}_i(t) &= A_i x_i(t) + b_i u_i(t) + h_i(t, x(t)), \\ y_i(t) &= c_i^T x_i(t), \quad i \in \mathcal{N} \end{aligned} \tag{26.83}$$

where

1. The decoupled subsystems described by the triples (A_i, b_i, c_i^T) are controllable and observable.
2. The transfer functions $G_i(s) = c_i^T (sI - A_i)^{-1} b_i$ of the decoupled systems are minimum phase, have *known* relative degree q_i and *known* high frequency gain $\kappa_i = \lim_{s \to \infty} s^{q_i} G_i(s)$.

3. The nonlinear interconnections $h_i : \mathbb{R} \times \mathbb{R}^n \rightarrow \mathbb{R}^{n_i}$ are of the form $h_i(t,x) = b_i f_i(t,x) + g_i(t,y)$ where $f_i : \mathbb{R} \times \mathbb{R}^n \rightarrow \mathbb{R}$ and $g_i : \mathbb{R} \times \mathbb{R}^n \rightarrow \mathbb{R}^{n_i}$ satisfy

$$
\begin{aligned}
|f_i(t,x)| &\leq \sum_{j \in \mathcal{N}} \alpha^f_{ij} \|x_j\| \\
\|g_i(t,x)\| &\leq \sum_{j \in \mathcal{N}} \alpha^g_{ij} \|y_j\|
\end{aligned}
\tag{26.84}
$$

for some *unknown* constants α^f_{ij}, α^g_{ij} where $x(t) = [x_1^T(t), x_2^T(t), \ldots x_N^T(t)]^T$ and $y(t) = [y_1(t), y_2^T(t), \ldots y_N(t)]^T$ are the state and the output of the overall system.

Finally, suitable adaptive decentralized control schemes can be developed by forcing an interconnected system of the form (Equation 26.83) to track a decoupled stable linear reference model described as

$$
\begin{aligned}
\mathbf{S}_M : \dot{x}_{Mi}(t) &= A_{Mi} x_{Mi}(t) + b_{Mi} r_i(t), \\
y_{Mi}(t) &= c^T_{Mi} x_{Mi}(t), \quad i \in \mathcal{N},
\end{aligned}
\tag{26.85}
$$

under reasonable assumptions on $\mathbf{S}$ and $\mathbf{S}_M$.

26.7 Discrete and Sampled-Data Systems

Most of the results concerning the stability and stabilization of continuous-time interconnected systems can be carried over to the discrete case with suitable modifications. Yet, there is a distinct approach to the stability analysis of discrete systems, which is to translate the problem in to that of a continuous system for which abundant results are available. For an idea of this approach, consider a system

$$
\mathbf{S}_{SD} : x(t+1) = (A_0 + \sum_{k \in \mathcal{K}} p_k A_k) x(t) \tag{26.86}
$$

where A_0 is a stable matrix additively perturbed by $p_k A_k$, $k \in \mathcal{K} = \{1, 2, \ldots, K\}$ with p_k standing for one of K perturbation parameters. The purpose is to find the largest region in the parameter space within which $\mathbf{S}_{SD}$ remains stable. By choosing a Lyapunov function $v(x) = x^T P x$, where P is the positive definite solution of the discrete Lyapunov equation,

$$
A_0^T P A_0 - P = -I, \tag{26.87}
$$

it can be shown that $\mathbf{S}_{SD}$ is stable, provided $I - W(p)$ is positive definite, where

$$
W(p) = \sum_{k \in \mathcal{K}} p_k (A_k^T P A_0 + A_0^T P A_k) + \sum_{k,l \in \mathcal{K}} p_k p_l A_k^T P A_l. \tag{26.88}
$$

Since the perturbation parameters appear nonlinearly in $W(p)$, characterization of a stability region in the parameter space is not easy. However, $I - W(p)$ is positive definite if the continuous system

$$
\dot{\xi}(t) = (-I + \sum_{k \in \mathcal{K}} p_k E_k) \xi(t) \tag{26.89}
$$

is stable, where

$$
E_k = \begin{bmatrix} 0 & P^{1/2} A_k \\ E_k^T P^{1/2} & A_k^T P A_0 + A_0^T P A_k \end{bmatrix}. \tag{26.90}
$$

An analysis of the stability of the perturbed continuous system in Equation 26.89 provides a sufficient condition for the stability of the discrete system in Equation 26.86. This idea can be generalized to the stability analysis of discrete interconnected systems by treating the interconnections as perturbations to nominal stable decoupled subsystems.

A major difference between discrete and continuous systems is that characterizing decentrally stabilizable interconnections for discrete systems is not as easy as for continuous systems. For example, there is no discrete counterpart to the matching conditions. On the other hand, most existing control schemes for continuous systems seem applicable to sampled-data systems provided the sampling rate is sufficiently high. To illustrate this observation, consider the decentralized control of an interconnected system,

$$\mathbf{S}: \dot{x}_i(t) = A_i x_i(t) + b_i[u_i(t) + \sum_{j \in \mathcal{N}} d_{ij}^T x_j(t)], \quad i \in \mathcal{N}, \tag{26.91}$$

using sampled-data feedback of the form

$$u_i(t) = -k_i^T(t - t_m)x_i(t_m), t_m \leq t < t_{m+1}, \tag{26.92}$$

where t_m are the sampling instants, and $k_i(t)$ are time-varying local feedback gains. With $T_m = t_{m+1} - t_m$ denoting the mth sampling period, it can be shown that the choice of

$$k_i^T(t) = [\delta^{n_i}(t) \ldots \delta'(t)\ \delta(t)] \tag{26.93}$$

or similar feedback gains having impulsive behavior, stabilize **S** provided T_m are sufficiently small. How small the sampling periods should be requires knowledge of the bounds on the interconnections. If these bounds are not available, then a simple centralized adaptation scheme, such as

$$T_{m+1}^{-1} = T_m^{-1} + \sum_{j \in \mathcal{N}} \gamma_i \|x_i(t_{m-m_i})\|, \tag{26.94}$$

with $\gamma_i > 0$, decreases T_m to the value needed for stability. Clearly, this is a high-gain stabilization scheme coupled with fast sampling, owing its success to the matching structure of the interconnections [36]. Similar adaptive sampled-data control schemes are available for more general classes of interconnected systems.

26.8 Graph-Theoretic Decompositions

Decomposition of large scale systems and their associated problems is often desirable for computational reasons. In such cases, decentralization or any other structural constraints on the controllers, estimators, or the design process itself, is preferred rather than necessary. Depending on the particular problem in hand, one may be interested in obtaining lower block triangular (LBT) decompositions, input and/or output reachable acyclic decompositions, ϵ-decompositions, overlapping decompositions, etc. [30]. In all of these decomposition schemes, the problem is to find a suitable partitioning and reordering of the input, state, or output variables so that the resulting decomposed system has some desirable structural properties. As expected, the system graph plays the key role, with graph-theory providing the tools.

26.8.1 LBT Decompositions

LBT decompositions are used to reorder the states of system **S** in Equation 26.6, so that the subsystems have a hierarchical interconnection pattern as

$$\begin{aligned}\mathbf{S}: \dot{x}_i &= \sum_{j=1}^{i} A_{ij} x_j + B_i u, \quad i \in \mathcal{N}, \\ y &= \sum_{i \in \mathcal{N}} C_i x_i.\end{aligned} \tag{26.95}$$

Such a decomposition corresponds to transforming the A matrix into a Lower Block–Triangular form by symmetric row and column permutations (hence the name LBT decomposition). In terms of system graph, LBT decomposition is the almost trivial problem of identifying the strong components of the truncated digraph $\mathbf{D}_x = (\mathcal{X}, \mathcal{E}_x)$, where $\mathcal{E}_x \subset \mathcal{E}$ contains only the edges connecting state vertices.

LBT decompositions offer computational simplification in the standard state feedback or observer design problems. For example, the problem of designing a state feedback

$$u = -Kx = -\sum_{i\in\mathcal{N}} K_i x_i \tag{26.96}$$

for arbitrary pole placement, can be reduced to computation of the individual blocks K_i of K in a recursive scheme involving the subsystems only.

26.8.2 Acyclic IO Reachable Decompositions

In acyclic input–output (IO) reachable decompositions, the purpose is to decompose $\mathbf{S}$ into the form

$$\begin{aligned} \mathbf{S}: \dot{x}_i &= \sum_{j=1}^{i} A_{ij}x_j + \sum_{j=1}^{i} B_{ij}u_j, \\ y_i &= \sum_{j=1}^{i} C_{ij}x_j, \quad i \in \mathcal{N}. \end{aligned} \tag{26.97}$$

That is, in addition to the A matrix, the B and C matrices must have LBT structure. In addition to the desired structure of the system matrices, it is also necessary that the decoupled subsystems represented by (A_{ii}, B_{ii}, C_{ii}) are at least structurally controllable and observable, and that none is further decomposable.

Because the LBT structure is concerned with the reachability properties of the system, both this structure and input and/or output reachability requirements for the subsystems, which are necessary for structural controllability and/or observability, can be taken care of by a suitable decomposition scheme based on binary operations on the reachability matrix of the system digraph. The requirement that the subsystems be dilation free, which is the second condition for structural controllability and/or observability, is of a different nature, however, and should be checked separately after the input–output reachability decomposition has been obtained.

When outputs are of no concern, it is easy to identify all possible acyclic, irreducible, input reachable decompositions of a given system. If some of the resulting decoupled subsystems turn out to contain dilations (destroying structural controllability), then they can suitably be combined with one or more subsystems at a higher level of hierarchy to eliminate the dilations without destroying the LBT structure. Provided that the overall system is structurally controllable, this process eventually gives an acyclic, irreducible decomposition in which all subsystems are structurally controllable. Of course, dual statements are valid for acyclic output reachable decompositions.

Once an acyclic decomposition into controllable subsystems is obtained, many design problems can be decomposed accordingly. An obvious example is the state feedback structure in Equation 26.96. A more complicated problem is the suboptimal state feedback design discussed in the section on optimization. For the system in Equation 26.97, the test matrix $F(\mu)$, with the inclusion of the input coupling terms B_{ij}, becomes

$$F(M_D) = F_D(M_D) + F_C(M_D) + F_C^T(M_D), \tag{26.98}$$

where $M_D = \text{diag}\{\mu_1, \mu_2, \ldots, \mu_N\}$, allowing different μ_i's for $\mathbf{S}_i$'s, $F_D(M_D) = [(1-\mu_i^{-1})(Q_i + K_i^T R_i K_i)]$, and $F_C(M_D) = [F_{ij}(\mu_i)]$ with

$$F_{ij}(\mu_i) = \begin{cases} \mu_i^{-1} P_i (A_{ij} - B_{ij}K_j), & i > j \\ 0, & i \le j. \end{cases} \tag{26.99}$$

From the structure of $F(M_D)$ it is clear that the choice $\mu_i = \epsilon^{N+1-i}$, $i \in \mathcal{N}$, results in a negative definite $F(M_D)$ for sufficiently small ϵ. This guarantees existence of a suboptimal state feedback control law with the degree of suboptimality $\mu = \epsilon^N$. In practice, it is possible to achieve a much better μ by a careful choice of the weight matrices Q_i and R_i.

In a similar way, acyclic, structurally observable decompositions can be used to design suboptimal state estimators, which are discussed below in the context of sequential optimization for acyclic IO decompositions.

To illustrate the use of acyclic IO decompositions in a standard LQG optimization problem, it suffices to consider decomposition of a discrete-time system into only two subsystems as

$$\begin{aligned} \mathbf{S}_1 : x_1(t+1) &= A_{11}x_1(t) + B_{11}u_1(t) + w_1(t), \\ y_1(t) &= C_{11}x_1(t) + v_1(t), \\ \mathbf{S}_2 : x_2(t+1) &= A_{21}x_1(t) + A_{22}x_2(t) + B_{21}u_1(t) + B_{22}u_2(t) + w_2(t), \\ y_2(t) &= C_{21}x_1(t) + C_{22}x_2(t) + v_2(t), \end{aligned} \tag{26.100}$$

with the usual assumptions on the input and measurement noises ω_i and $v_i, i = 1, 2$. Let each subsystem be associated with a performance criterion

$$\mathcal{E}J_i = \mathcal{E}\left\{\lim_{T\to\infty} T^{-1} \sum_{t=0}^{T-1} \left[x_i^T(t)Q_ix_i(t) + u_i^T(t)R_iu_i(t)\right]\right\}, \quad i = 1, 2 \tag{26.101}$$

where $\mathcal{E}$ denotes expectation.

The sequential optimization procedure consists of minimizing $\mathcal{E}J_1$ and $\mathcal{E}J_2$ subject to the dynamic equations for the systems $\mathbf{S}_1$ and $(\mathbf{S}_1, \mathbf{S}_2)$, respectively. The first problem has the standard solution $u_1^*(t) = -K_1\hat{x}_1(t)$, where K_1 is the optimal control gain found from the solution of the associated Riccati equation, and $\hat{x}_1(t)$ is the best estimate of $x_1(t)$ given the output information $\mathcal{Y}_1^{t-1} = \{y_1(0), \ldots, y_1(t-1)\}$. The estimate $\hat{x}_1(t)$ is generated by the Kalman filter

$$\hat{x}_1(t+1) = A_{11}\hat{x}_1(t) + B_{11}u_1^*(t) + L_1[y_1(t) - c_{11}\hat{x}_1(t)] \tag{26.102}$$

where L_1 is the steady-state estimator gain. With the control u_1^* applied to $\mathbf{S}_1$, the overall system becomes

$$\mathbf{S}: \begin{bmatrix} \hat{x}_1(t+1) \\ x_1(t+1) \\ x_2(t+1) \end{bmatrix} = \begin{bmatrix} A_{11} - B_{11}K_1 - L_1C_{11} & L_1C_{11} & 0 \\ -B_{11}K_1 & A_{11} & 0 \\ -B_{21}K_1 & A_{21} & A_{22} \end{bmatrix} \begin{bmatrix} \hat{x}_1(t) \\ x_1(t) \\ x_2(t) \end{bmatrix} + \begin{bmatrix} 0 \\ 0 \\ B_{22} \end{bmatrix} u_2(t) + \begin{bmatrix} L_1v_1(t) \\ w_1(t) \\ w_2(t) \end{bmatrix} \tag{26.103}$$

which preserves the LBT structure of the original system. Assuming that both $\mathcal{Y}_1^{t-1}$ and $\mathcal{Y}_2^{t-1} = \{y_2(0), \ldots, y_2(t-1)\}$ are available for constructing the control u_2^* (which is consistent with the idea of sequential optimization), the problem reduces to minimization of $\mathcal{E}J_2$ subject to (103). An analysis of the standard solution procedure reveals that the optimal control law can be expressed as

$$u_2^*(t) = -K_{21}\hat{x}_1(t) - K_{22}\hat{\xi}(t) \tag{26.104}$$

where $K = [K_{21}\ K_{22}]$ is the optimal control gain, and $\hat{\xi}(t)$ is the optimal estimate of $x(t) = [x_1^T(t)\ \ x_2^T(t)]^T$, given $\mathcal{Y}_1^{t-1}$ and $\mathcal{Y}_2^{t-1}$. Furthermore, the $2n_1 + n_2$-dimensional Riccati equation, from which K is constructed, can be decomposed into an n_2-dimensional Riccati equation involving the parameters of the second isolated subsystem and a Lyapunov equation corresponding to an $n_2 \times 2n_1$ dimensional matrix. This results in considerably simplifying the solution of the optimal control gain. However, the Kalman filter for $\hat{\xi}(t)$ still requires the solution of an $(n_1 + n_2)$-dimensional Riccati equation.

Other sequential optimization schemes based on various information structure constraints can be analyzed similarly; for details, see [30].

26.8.3 Nested Epsilon Decompositions

Epsilon decomposition of a square matrix M is concerned with transforming M by symmetric row and column permutations into a form

$$P^T MP = M_D + \epsilon M_C \tag{26.105}$$

where M_D is block diagonal, and ϵ is a prescribed small number [27]. The problem is equivalent to identifying the connected components of a subgraph $\mathbf{D}^\epsilon$ of the digraph $\mathbf{D}$ associated with M, which is obtained by deleting all edges of $\mathbf{D}$ corresponding to those elements of M with magnitude smaller than ϵ. All of the vertices of a connected component of $\mathbf{D}^{\epsilon_1}$ appear in the same connected component of $\mathbf{D}^{\epsilon_2}$ for any $\epsilon_2 < e_1$. Thus one can identify a number of distinct values $\epsilon_1 > \epsilon_2 > \cdots > \epsilon_K$ such that

$$P^T MP = (\ldots((M_0 + \epsilon_1 M_1) + \epsilon_2 M_2) + \cdots + \epsilon_K M_K), \tag{26.106}$$

which is a *nested epsilon decomposition* of M as illustrated in Figure 26.3.

As seen from the figure, a large ϵ results in a finer decomposition than a small ϵ does. Thus the choice of ϵ provides a compromise between the size and the number of components and the strength of the interconnections among them. A nice property of nested epsilon decompositions is that once the decomposition corresponding to some ϵ_k is obtained, the decomposition corresponding to ϵ_{k+1} can be found by working with a smaller digraph obtained by condensing $\mathbf{D}^{\epsilon_k}$ with respect to its components.

An immediate application of the nested epsilon decompositions is the stability analysis of a large scale system via vector Lyapunov functions, where the matrix M is identified with the matrix A of the system in Equation 26.6. Provided the subsystems resulting from the decomposition are stable, the stability of the overall system can easily be established by means of the aggregate matrix W in Equation 26.49, whose off-diagonal elements are of the order of ϵ.

The nested epsilon decomposition algorithm can also be applied with some modifications to decompose a system with inputs as

$$\dot{x}_i = A_{ii}x_i + B_{ii}u_i + \epsilon \sum_{j \neq i}^{N} (A_{ij}x_j + B_{ij}u_j), \quad i \in \mathcal{N}. \tag{26.107}$$

If each decoupled subsystem identified by a pair (A_{ii}, B_{ii}) is stabilized by a local state feedback of the form $u_i = -K_i x_i$, $i \in \mathcal{N}$, with the local gains not excessively high, then the closed-loop system preserves the

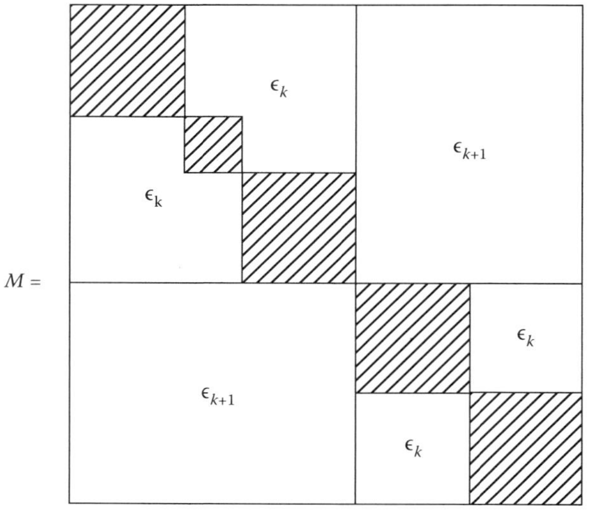

FIGURE 26.3 Nested epsilon decompositions.

weak-coupling property of the open-loop system, providing an easy way to stabilize the overall system. The same idea can also be employed in designing decentralized estimators [30] based on a suitable epsilon decomposition of the pair (A, C).

26.8.4 Overlapping Decompositions

Consider a system

$$\tilde{\mathbf{S}}: \dot{\tilde{x}}(t) = \tilde{A}\tilde{x}(t) \tag{26.108}$$

with an $\tilde{n}$-dimensional state vector $\tilde{x}$. Let columns of the matrix $V \in \mathbb{R}^{\tilde{n}\times n}$ form a basis for an n-dimensional A-invariant subspace of $\mathbb{R}^{\tilde{n}}$, and let A be the restriction of $\tilde{A}$ to $\operatorname{Im} V \simeq \mathbb{R}^n$, that is, $\tilde{A}V = VA$. Then the smaller order system

$$\mathbf{S}: \dot{x}(t) = Ax(t) \tag{26.109}$$

is called a restriction of $\tilde{\mathbf{S}}$. Conversely, starting with the system $\mathbf{S}$, one can obtain an expansion $\tilde{\mathbf{S}}$ of $\mathbf{S}$ by defining $\tilde{A} = VAV^L + M$, where V^L is any left inverse of V, and M is any complementary matrix satisfying $MV = 0$. The very definition of a restriction implies that $\mathbf{S}$ is stable if $\tilde{\mathbf{S}}$ is.

In many problems associated with large scale systems, it may be desirable to expand a system $\mathbf{S}$ to a larger dimensional one which possess some nice structural properties. The increase in dimensionality of the problem may very well be offset by the nice structure of the expansion. As an example, consider a system $\mathbf{S}$ with

$$A = \left[\begin{array}{cc|c} A_{11} & A_{12} & \epsilon A_{13} \\ \epsilon A_{21} & \multicolumn{1}{|c|}{A_{22}} & \epsilon A_{23} \\ \hline \epsilon A_{31} & \multicolumn{1}{|c}{A_{32}} & A_{33} \end{array}\right] \tag{26.110}$$

where ϵ is a small parameter. Letting

$$V = \begin{bmatrix} I_1 & 0 & 0 \\ 0 & I_2 & 0 \\ 0 & I_2 & 0 \\ 0 & 0 & I_3 \end{bmatrix} \tag{26.111}$$

where I_k denotes an identity matrix of order n_k, one obtains an expansion $\tilde{\mathbf{S}}$ with

$$\tilde{A} = \left[\begin{array}{cc|cc} A_{11} & A_{12} & 0 & \epsilon A_{13} \\ \epsilon A_{21} & A_{22} & 0 & \epsilon A_{23} \\ \hline \epsilon A_{21} & 0 & A_{22} & \epsilon A_{23} \\ \epsilon A_{31} & 0 & A_{32} & A_{33} \end{array}\right]. \tag{26.112}$$

Since $\tilde{\mathbf{S}}$ has an obvious decomposition into two weakly coupled subsystems, one can take advantage of this structural property in stability analysis, which is not available for the original system $\mathbf{S}$.

One can easily notice from the structure of V in Equation 26.114 that the expansion $\tilde{\mathbf{S}}$ of $\mathbf{S}$ is obtained simply by repeating the equation for the middle part x_2 of the state vector $x = [x_1^T\ x_2^T\ x_3^T]^T$. In some sense, x_2 is treated as common to two overlapping components $\tilde{x}_1 = [x_1^T\ x_2^T]^T$ and $\tilde{x}_2 = [x_2^T\ x_3^T]^T$ of x. Thus the partitioning of the A matrix in Equation 26.113 is termed the *overlapping decomposition*.

Although the expansion matrix V can be any matrix with full column rank, if it is restricted to contain one and only one unity element in each row (which corresponds, as in the case above, to repeating some of the state equations in the expanded domain), then one can develop a suitable graph-theoretic algorithm to find the smallest expansion which has a disjoint decomposition (into decoupled or ϵ-coupled components) with the property that no component is further decomposable.

The idea of overlapping decompositions via expansions can be extended to systems with inputs. A system

$$\tilde{\mathbf{S}}: \dot{\tilde{x}}(t) = \tilde{A}\tilde{x}(t) + \tilde{B}u(t) \tag{26.113}$$

is said to be an expansion of

$$\mathbf{S}:\ \dot{x}(t) = Ax(t) + Bu(t) \tag{26.114}$$

if $\tilde{B} = VB$ in addition to $\tilde{A}V = VA$. Consider the optimal control problems of minimizing the performance criteria

$$\begin{aligned} J &= \int_0^\infty [x^T(t)Qx(t) + u^T(t)Ru(t)]\, dt \\ \tilde{J} &= \int_0^\infty [\tilde{x}^T(t)\tilde{Q}\tilde{x}(t) + u^T(t)Ru(t)]\, dt \end{aligned} \tag{26.115}$$

associated with $\mathbf{S}$ and $\tilde{\mathbf{S}}$. The optimal solutions are

$$u(t) = -Kx(t), \quad \text{and} \quad u(t) = -\tilde{K}\tilde{x}(t), \tag{26.116}$$

respectively, resulting in closed-loop systems

$$\begin{aligned} \hat{\mathbf{S}}&:\ \dot{x}(t) = (A - BK)x(t), \\ \hat{\tilde{\mathbf{S}}}&: \dot{\tilde{x}}(t) = (\tilde{A} - \tilde{B}\tilde{K})\tilde{x}(t). \end{aligned} \tag{26.117}$$

Thus, $\hat{\mathbf{S}}$ is a restriction of $\hat{\tilde{\mathbf{S}}}$ if $(\tilde{A} - \tilde{B}\tilde{K})V = V(A - BK)$, or equivalently, if $\tilde{K} = KV$. The last condition is satisfied if $\tilde{Q}$ and Q are related as $Q = V^T\tilde{Q}V$, in which case the optimal cost matrices are also related as $P = V^T\tilde{P}V$. This analysis shows that, if the cost matrices $\tilde{Q}$ and R of the expanded system are chosen to be block diagonal with diagonal blocks associated with the decoupled expanded subsystems, then its optimal (in case of complete decoupling) or suboptimal (in case of weak decoupling) solution can be contracted back to an optimal or suboptimal solution of the original system with respect to a suitably chosen performance criterion.

References

1. Anderson, B. D. O. and Clements, D. J., Algebraic characterization of fixed modes in decentralized control, *Automatica*, 17, 703–712, 1981.
2. Brusin, V. A. and Ugrinovskaya, E. Ya., Decentralized adaptive control with a reference model, *Avtomatika i Telemekhanika*, 10, 29–36, 1992.
3. Chae, S. and Bien, Z., Techniques for decentralized control for interconnected systems, in *Control and Dynamic Systems*, C. T. Leondes, Ed., Academic Press, Boston, 41, 273–315, 1991.
4. Chen, Y. H., Decentralized robust control design for large-scale uncertain systems: The uncertainty is time-varying, *J. Franklin Institute*, 329, 25–36, 1992.
5. Chen, Y. H. and Han, M. C., Decentralized control design for interconnected uncertain systems, in *Control and Dynamic Systems*, C. T. Leondes, Ed., Academic Press, Orlando, FL, 56, 219–266, 1993.
6. Cheng, C. F., Wang, W. J., and Lin, Y. P., Decentralized robust control of decomposed uncertain interconnected systems, *Trans. ASME*, 115, 592–599, 1993.
7. Cho, Y. J. and Bien, Z., Reliable control via an additive redundant controller, *Int. J. Control*, 50, 385–398, 1989.
8. Date, R. A. and Chow, J. H., A parametrization approach to optimal H_2 and H_∞ decentralized control problems, *Automatica*, 29, 457–463, 1993.
9. Datta, A., Performance improvement in decentralized adaptive control: A modified model reference scheme, *IEEE Trans. Automatic Control*, 38, 1717–1722, 1993.
10. Gajić, Z. and Shen, X., *Parallel Algorithms for Optimal Control of Large Scale Linear Systems*, Springer–Verlag, Berlin, Germany, 1993.

11. Geromel, J. C., Bernussou, J., and Peres, P. L. D., Decentralized control through parameter space optimization, *Automatica*, 30, 1565–1578, 1994.
12. Gündes, A. N. and Kabuli, M. G., Reliable decentralized control, *Proc. Am. Control Conf.*, Baltimore, MD, pp. 3359–3363, 1994.
13. Iftar, A., Decentralized estimation and control with overlapping input, state, and output decomposition, *Automatica*, 29, 511–516, 1993.
14. Ikeda, M., Decentralized control of large scale systems, in *Three Decades of Mathematical System Theory*, H. Nijmeiyer and J. M. Schumacher, Eds., Springer-Verlag, New York, 219–242, 1989.
15. Jamshidi, M., *Large-Scale Systems. Modeling and Control*, North-Holland, New York, 1983.
16. Lakshmikantham, V. and Liu, X. Z., *Stability Analysis in Terms of Two Measures*, World Scientific, Singapore, 1993.
17. Lakshmikantham, V., Matrosov, V. M., and Sivasundaram, S., *Vector Lyapunov Functions and Stability Analysis of Nonlinear Systems*, Kluwer, The Netherlands, 1991.
18. Leitmann, G., One approach to the control of uncertain systems, *ASME J. Dyn. Syst., Meas., and Control*, 115, 373–380, 1993.
19. Leondes, C. T., Ed., *Control and Dynamic Systems*, Vols. 22–24, *Decentralized/Distributed Control and Dynamic Systems*, Academic Press, Orlando, FL, 1985.
20. Lin, C. T., Structural controllability, *IEEE Trans. Auto. Control*, AC–19, 201–208, 1974.
21. Lyon, J., Note on decentralized adaptive controller design, *IEEE Trans. Auto. Control*, 40, 89–91, 1995.
22. Michel, A. N., On the status of stability of interconnected systems, *IEEE Trans. Circuits Syst.*, CAS–30, 326–340, 1983.
23. Mills, J. K., Stability of robotic manipulators during transition to and from compliant motion, *Automatica*, 26, 861–874, 1990.
24. Sandell, N. R., Jr., Varaiya, P., Athans, M., and Safonov, M. G., Survey of decentralized control methods for large scale systems, *IEEE Trans. Auto. Control*, AC–23, 108–128, 1978.
25. Savastuk, S. V. and Šiljak, D. D., Optimal decentralized control, *Proc. Am. Control Conf.*, Baltimore, MD, pp. 3369–3373, 1994.
26. Sezer, M.E. and Šiljak, D. D., Structurally fixed modes, *Syst. Control Lett.*, 1, 60–64, 1981.
27. Sezer, M. E. and Šiljak, D. D., Nested ϵ-decomposition and clustering of complex systems, *Automatica*, 22, 321–331, 1986.
28. Shi, L. and Singh, S. K., Decentralized adaptive controller design for large-scale systems with higher order interconnections, *IEEE Trans. Auto. Control*, 37, 1106–1118, 1992.
29. Šiljak, D. D., *Large-Scale Dynamic Systems: Stability and Structure*, North-Holland, New York, 1978.
30. Šiljak, D. D., *Decentralized Control of Complex Systems*, Academic Press, Cambridge, MA, 1991.
31. Tamura, H. and Yoshikawa, T., *Large-Scale Systems Control and Decision Making*, Marcel Dekker, New York, 1990.
32. Ünyelioğlu, K. A. and Özgüler, A. B., Reliable decentralized stabilization of feed-forward and feedback interconnected systems, *IEEE Trans. Auto. Control*, 37, 1119–1132, 1992.
33. Voronov, A. A., Present state and problems of stability theory, *Automatika i Telemekhanika*, 5, 5–28, 1982.
34. Wang, S. H. and Davison, E. J., On the stabilization of decentralized control systems, *IEEE Trans. Auto. Control*, AC–18, 473–478, 1973.
35. Wu, W. and Lin, C., Optimal reliable control system design for steam generators in pressurized water reactors, *Nuclear Technology*, 106, 216–224, 1994.
36. Yu, R., Ocali, O., and Sezer, M. E., Adaptive robust sampled-data control of a class of systems under structured perturbations, *IEEE Trans. Auto. Control*, 38, 1707–1713, 1993.

Further Reading

There is a number of survey papers on decentralized control and large scale systems [3,14,24]. The books on the subject are [10,15,19,29,31]. For a comprehensive treatment of decentralized control theory, methods, and applications, with a large number of references, see [30].

For further information on vector Lyapunov functions and stability analysis of large scale interconnected systems, see the survey papers [22,33], and books [16,17].

Adaptive decentralized control has been of widespread recent interest, see [2,9,21,23,28,30,36].

Robustness of decentralized control to both structured and unstructured perturbations has been one of the central issues in the control of large scale systems. For the background of robustness issues in control, which are relevant to decentralized control, see [18,30]. For new and interesting results on the subject, see [4–6,8].

There is a number of papers devoted to design of decentralized control via parameter space optimization, which rely on powerful convex optimization methods. For recent results and references, see [11].

Overlapping decentralized control and the inclusion principle are surveyed in [30]. Useful extensions were presented in [13]. The concept of overlapping is basic to reliable control under controller failures using multiple decentralized controllers [30]. For more information about this area, see [7,12,32,35].

In a recent development [25], it has been shown how optimal decentralized control of large scale interconnected systems can be obtained in the classical optimization framework of Lagrange. Both sufficient and necessary conditions for optimality are derived in the context of Hamilton–Jacobi equations and Pontryagin's maximum principle.

27

Decoupling

Trevor Williams
University of Cincinnati

Panos J. Antsaklis
University of Notre Dame

27.1 Introduction

Multi-input/multi-output systems are usually difficult for human operators to control directly, since changing any one input generally affects many, if not all, outputs of the system. As an example, consider the vertical landing of a vertical take off and landing jet or of a lunar landing rocket. Moving to a desired landing point to the side of the current position requires tilting the thrust vector to the side; but this reduces the vertical thrust component, which was balancing the weight of the craft. The aircraft therefore starts to descend, which is not desired. To move to the side at a constant height thus requires smooth, simultaneous use of both attitude control and throttle. It would be simpler for the pilot if a single control existed to do this maneuver; hence the interest in control methods that make the original system behave in a way that is easier to control manually. One example of such technique is when a compensator is sought that makes the compensated system diagonally dominant. If this can be achieved, it is then possible to regard the system as, to first order, a set of independent single-input/single-output systems, which is far easier to control than the original plant. Another approach is that of decoupling, where the system transfer matrix is made to be exactly diagonal. Each output variable is therefore affected by only one input signal, and each input/output pair can then be controlled by an easier-to-design single-input/single-output controller or manually by a human operator.

This chapter studies the problem of making the transfer function matrix of the system diagonal using feedback control and, in particular, state feedback, state feedback with dynamic precompensation, and constant output feedback control laws. This problem is referred to as the *dynamical decoupling* problem, as it leads to a compensated system where the input actions are decoupled; it is also called a noninteracting control problem for similar reasons. Stability is an important issue and it also is examined here. Conditions for decoupling with stability and algorithms to determine such control laws are described. The problems of block diagonal or triangular decoupling are also addressed. They are of interest when full diagonal decoupling using a particular form of feedback control, typically state feedback, is not possible. Note that the approach taken in this chapter follows the development in [14]. Static decoupling is also briefly discussed; references for approximate diagonal decoupling are provided in "Further Reading."

27.1.1 Diagonal Decoupling

Diagonal decoupling of a system with equal numbers of inputs and outputs is the most straightforward type of problem in the field of noninteracting control. The goal is to apply some form of control law to the system so as to make the ith output of the closed-loop system independent of all but the ith closed-loop input signal. Each output can then be controlled by a dedicated simpler single-input/single-output controller, or by a human operator. The main questions to be answered when investigating diagonal decoupling of a given system are

- Can it be decoupled at all?
- If so, what form of controller is required to achieve this?

Three classes of controllers that are customarily considered are

1. Constant output feedback $u = Hy + Gr$, where the output y of the system is simply multiplied by a constant gain matrix H and this is fed back as the control signal u, with r the new external input to the system and G a constant gain matrix
2. Linear state feedback $u = Fx + Gr$, where the control signal consists of a constant matrix F multiplying the internal state variable vector x of the system
3. State feedback plus precompensation, where a feedforward dynamic control system is added to the state feedback controller

Note that the compensator in class 3 corresponds to dynamic output feedback, where the input and output signal vectors r and y are multiplied by dynamic transfer function gain matrices rather than constant ones.

The problem of diagonally decoupling a square system was the first decoupling question to be studied, and it can be answered in a fairly straightforward fashion. First of all, diagonal decoupling by state feedback plus precompensation, or by dynamic output feedback, amounts to finding a transfer matrix that, when the open-loop transfer matrix is multiplied by it, produces a diagonal closed-loop transfer matrix. This problem is therefore closely related to the problem of finding an inverse for the open-loop plant. As a result of this, any square plant that has a full rank transfer matrix can be diagonally decoupled by this type of control. This result was proved by Rekasius [10]. A system that does not satisfy this condition does not have linearly independent outputs, so it follows that it is impossible to decouple by any form of controller. It is of great practical interest to establish whether a given plant can actually be decoupled by a simpler type of controller than this. Falb and Wolovich [3] established the necessary and sufficient condition under which diagonal decoupling by state feedback alone is possible. This condition, which can be easily tested from either a state-space or a transfer matrix model of the plant, can be expressed as follows.

A square system can be diagonalized by state feedback alone if and only if the constant matrix $B^\star$ is nonsingular, where this matrix is defined below first from the state-space and then from the transfer matrix description of the system.

State-space representation. Let the given system be $\dot{x} = Ax + Bu, y = Cx + Du$ in the continuous-time case, or $x(k+1) = Ax(k) + Bu(k), y(k) = Cx(k) + Du(k)$ in the discrete-time case; let A, B, C, D be $n \times n$, $n \times p, p \times n, p \times p$ real matrices, respectively; and assume for simplicity that the system is controllable and observable. Then the $p \times p$ matrix $B^\star$ is constructed as follows: If the ith row of the direct feedthrough matrix D is nonzero, this becomes the ith row of the constant matrix $B^\star$. Otherwise, find the lowest integer, f_i, for which the ith row of $CA^{f_i-1}B$ is nonzero. This then becomes the ith row of the constant matrix $B^\star$.

Transfer matrix representation. Let $T(s)$, with s the Laplace transform variable, be the $p \times p$ transfer function matrix of the continuous-time system; or $T(z)$, with z the Z-transform variable, be the transfer function matrix of the discrete-time system. Let $D(s)$ [or $D(z)$] be the diagonal matrix $D(s) = diag(s^{f_i})$ where the nonnegative integers $\{f_i\}$ are so that all rows of $\lim_{s\to\infty} D(s)T(s)$ are constant and nonzero.

This limit matrix is $B^\star$; that is,

$$\lim_{s \to \infty} D(s)T(s) = B^\star \tag{27.1}$$

The integers $\{f_i\}$ are known as the decoupling indices of the system. They can be determined from either the state-space or the transfer function descriptions as described above; note that $f_i = 0$ corresponds to the ith row of D being nonzero. In either case, of course, the resulting matrix $B^\star$ is the same. It should be noted that systems will generically satisfy the decoupling condition; that is, if all entries of the A, B, C (and D) matrices are chosen at random, the resulting $B^\star$ will have full rank. Diagonal decoupling by state feedback is therefore likely to be feasible for a wide variety of systems.

Example 27.1:

$$A = \begin{pmatrix} 0 & 1 & 0 \\ 0 & 0 & 1 \\ -6 & -11 & -6 \end{pmatrix}, \quad B = \begin{pmatrix} 0 & 0 \\ 0 & 0 \\ -1 & 2 \end{pmatrix}, \quad C = \begin{pmatrix} 3 & 6 & 1 \\ 2 & 0 & 0 \end{pmatrix}, \quad D = \begin{pmatrix} 0 & 0 \\ 0 & 0 \end{pmatrix}$$

This gives $f_1 = 1$, $f_2 = 3$, and $B^\star = \begin{pmatrix} -1 & 2 \\ -2 & 4 \end{pmatrix}$. This matrix is clearly singular; therefore, the system cannot be decoupled by state feedback.

Example 27.2:

$$T(s) = \begin{pmatrix} \dfrac{1}{s} & \dfrac{2}{s+1} \\ \dfrac{4}{s+3} & \dfrac{8s}{s+4} \end{pmatrix}$$

This gives $B^\star = \begin{pmatrix} 1 & 2 \\ 0 & 8 \end{pmatrix}$, with decoupling indices $f_1 = 1$, $f_2 = 0$. This system can therefore be diagonally decoupled by state feedback.

Example 27.3:

$$T(s) = \begin{pmatrix} \dfrac{1}{s} & \dfrac{2}{s+1} \\ \dfrac{4}{s+3} & \dfrac{8}{s+4} \end{pmatrix}$$

The same as previously, but with the (2, 2) entry divided by s. We now obtain $B^\star = \begin{pmatrix} 1 & 2 \\ 4 & 8 \end{pmatrix}$, with decoupling indices $f_1 = 1$, $f_2 = 1$. $B^\star$ is now singular, so this system cannot be diagonally decoupled by state feedback alone.

27.1.2 Diagonal Decoupling with Internal Stability

A question of great practical interest is whether the closed-loop system that is obtained after decoupling can be made stable. It can be shown constructively (for instance, by use of the algorithm given below) that all of the poles that are evident from the diagonal closed-loop transfer matrix can be assigned any desired values. The question therefore becomes: Can the closed-loop system be made internally stable, where there are no "hidden" cancellations between unstable poles and zeros? Such unstable modes are particularly dangerous in practice, as they will not be revealed by an examination of the transfer matrix. However, the hidden unstable state behavior they represent will very likely cause problems, such as burnout of internal electronic components of the system. It was shown by Gilbert [5] that a given plant may indeed have hidden fixed modes when it is diagonally decoupled by state feedback, with or without precompensation. Wolovich and Falb [15] then showed that these modes are the same for both cases; furthermore, they are a subset of the transmission zeros of the plant. In fact, they are those transmission zeros z_i that do not make any of the rows of the transfer matrix $T(s)$ equal to zero when evaluating $T(z_i)$; they are called *diagonal coupling zeros*. Thus, any plant with square, full-rank transfer matrix for which all the diagonal coupling zeros are in the left half-plane can be diagonally decoupled with internal stability by state feedback plus precompensation; or by state feedback alone if $B^\star$ is nonsingular. Therefore, there will never be any problems with internal stability when decoupling a minimum-phase system, as all of its transmission zeros are in the left half-plane.

An algorithm to diagonally decouple a system, when $B^\star$ has full rank, using state feedback is now presented. This algorithm is based on a procedure to obtain a stable inverse of a system that is described below. This procedure is applied to the system $D(s)T(s) = \hat{T}(s)$, where $D(s) = diag(s^{f_i})$ as in Equation 27.1, that can be shown to have a state-space realization $\{A, B, \hat{C}, \hat{D}\}$. In fact $\hat{D} = B^\star$, which is assumed to have full rank p; and this implies that a proper right inverse of the system $\hat{T}(s)$ exists. Here it is assumed that the system has the same number of inputs and outputs, and this simplifies the selection of F, G as in this case they are unique; see the algorithm for the inverse below for the nonsquare case. In particular, if the state feedback $u = Fx + Gr$ with

$$F = -(B^\star)^{-1}\hat{C}, \quad G = (B^\star)^{-1} \tag{27.2}$$

is applied to the system $\dot{x} = Ax + Bu$, $y = \hat{C}x + B^\star u$, then it can be shown that $\hat{T}_{F,G}(s) = D(s)T_{F,G}(s) = I_p$. This implies that if the state feedback $u = Fx + Gr$ with F, G as in Equation 27.2 is applied to the given system $\dot{x} = Ax + Bu$, $y = Cx + Du$ with transfer matrix $T(s)$, then

$$T_{F,G}(s) = D^{-1}(s) \tag{27.3}$$

which is diagonal with entries s^{-f_i}. Note that here the state feedback matrix F assigns all the n closed-loop eigenvalues at the locations of the n zeros of $\hat{T}(s)$; that is, at the zeros of $T(s)$ and of $D(s)$. The closed-loop eigenvectors are also appropriately assigned so the eigenvalues cancel all the zeros to give $D(s)T_{F,G}(s) = I_p$. This explains the control mechanism at work here and also makes quite apparent the changes necessary to ensure internal stability. Simply instead of $D(s)$ use $\hat{D}(s) = diag\{p_i(s)\}$ with $p_i(s)$ stable polynomials of degree s^{f_i}; that is, $p_i(s) = s^{f_i}$ + lower-degree terms. Then it can be shown that $\lim_{s\to\infty} \hat{D}(s)T(s) = B^\star$ and that $\{A, B, \tilde{C}, B^\star\}$ is a realization of $\hat{D}(s)T(s) = \tilde{T}(s)$. State feedback with

$$F = -(B^\star)^{-1}\tilde{C}, \quad G = (B^\star)^{-1} \tag{27.4}$$

gives

$$T_{F,G}(s) = \hat{D}^{-1}(s) = diag\{p_i^{-1}(s)\} \tag{27.5}$$

which is stable. Note that in this case the closed-loop eigenvalues are at the assumed stable zeros of $T(s)$ and at the selected stable zeros of the polynomials $p_i(s)$, $i = 1, \ldots, p$.

Example 27.4:

Let $T(s) = \begin{pmatrix} \dfrac{s+1}{s^2} & 0 \\ \dfrac{1}{s(s-1)} & \dfrac{-1}{s-1} \end{pmatrix}$.

Here

$$\lim_{s\to\infty} D(s)T(s) = \lim_{s\to\infty} diag(s,s)T(s) = \begin{pmatrix} 1 & 0 \\ 0 & -1 \end{pmatrix} = B^\star.$$

Since $B^\star$ has full rank, the system can be decoupled using state feedback $u = Fx + Gr$. The system has one transmission zero at -1 and there are no diagonal coupling zeros, so it can be decoupled with internal stability. Let $\hat{D}(s) = \begin{pmatrix} s+1 & 0 \\ 0 & s+2 \end{pmatrix}$. A minimal (controllable and observable) realization of $\tilde{T}(s) = \hat{D}(s)T(s)$ is $\{A, B, \tilde{C}, B^\star\}$ where

$$A = \begin{pmatrix} 0 & 1 & 0 \\ 0 & 0 & 0 \\ -1 & 0 & 1 \end{pmatrix}, \quad B = \begin{pmatrix} 0 & 0 \\ 1 & 0 \\ 0 & 1 \end{pmatrix}, \quad \tilde{C} = \begin{pmatrix} 1 & 2 & 0 \\ 3 & 1 & -3 \end{pmatrix}.$$

In view now of Equations 27.4 and 27.5, for

$$F = -(B^\star)^{-1}\tilde{C} = \begin{pmatrix} -1 & -2 & 0 \\ 3 & 1 & -3 \end{pmatrix}$$

and

$$G = (B^\star)^{-1} = \begin{pmatrix} 1 & 0 \\ 0 & -1 \end{pmatrix},$$

$$T_{F,G}(s) = \hat{D}(s)^{-1} = \begin{pmatrix} \dfrac{1}{s+1} & 0 \\ 0 & \dfrac{1}{s+2} \end{pmatrix}.$$

The closed-loop eigenvalues are in this case located at the transmission zero of the plant at -1 and at the selected locations -1 and -2, the poles of $\hat{D}(s)^{-1}$. Note that it is not necessary to cancel the transmission zero at -1 in order to decouple the system since it is not a coupling zero; it could appear as a zero in the decoupled system instead. To illustrate this, consider Example 27.5 where $T(s)$ is the same except that the zero is now unstable at $+1$.

Example 27.5:

Let $T(s) = \begin{pmatrix} \dfrac{s-1}{s^2} & 0 \\ \dfrac{1}{s(s-1)} & \dfrac{-1}{s-1} \end{pmatrix}$ where again

$$\lim_{s\to\infty} D(s)T(s) = \lim_{s\to\infty} diag(s,s)T(s) = \begin{pmatrix} 1 & 0 \\ 0 & -1 \end{pmatrix} = B^\star.$$

Since $B^\star$ has full rank, the system can be decoupled using state feedback. Since there are no diagonal coupling zeros, the system can be decoupled with internal stability. Write $T(s) = \begin{pmatrix} s-1 & 0 \\ 0 & 1 \end{pmatrix} T_N(s)$

and apply the algorithm to diagonally decouple $T_N(s)$. Now $D_N(s) = \begin{pmatrix} s^2 & 0 \\ 0 & s \end{pmatrix}$ and take

$$\hat{D}_N(s) = \begin{pmatrix} (s+2)(s+3) & 0 \\ 0 & s+1 \end{pmatrix}.$$

A minimal (controllable and observable) realization of $\tilde{T}_N(s) = \hat{D}_N(s)T_N(s)$ is $\{A, B, \tilde{C}_N, B_N^\star\}$ where

$$A = \begin{pmatrix} 0 & 1 & 0 \\ 0 & 0 & 0 \\ -1 & 0 & 1 \end{pmatrix}, \quad B = \begin{pmatrix} 0 & 0 \\ 1 & 0 \\ 0 & 1 \end{pmatrix}, \quad \tilde{C}_N = \begin{pmatrix} 6 & 5 & 0; \\ 2 & 1 & -2 \end{pmatrix}$$

and

$$B_N^\star = \begin{pmatrix} 1 & 0 \\ 0 & -1 \end{pmatrix} = B^\star.$$

In view now of Equations 27.4 and 27.5, for $F = -(B^\star)^{-1}\tilde{C}_N = \begin{pmatrix} -6 & -5 & 0 \\ 2 & 1 & -2 \end{pmatrix}$ and

$$G = (B_N^\star)^{-1} = \begin{pmatrix} 1 & 0 \\ 0 & -1 \end{pmatrix},$$

$$(T_N)_{F,G}(s) = \hat{D}_N(s)^{-1} = \begin{pmatrix} \frac{1}{(s+2)(s+3)} & 0 \\ 0 & \frac{1}{s+1} \end{pmatrix}.$$

If now this state feeback is applied to the minimal realization $\{A, B, C\}$ of $T(s)$—note that A, B are the same as above—then $T_{F,G}(s) = \begin{pmatrix} s-1 & 0 \\ 0 & 1 \end{pmatrix} \hat{D}_N^{-1} = \begin{pmatrix} \frac{s-1}{(s+2)(s+3)} & 0 \\ 0 & \frac{1}{s+1} \end{pmatrix}$. Note that the unstable noncoupling transmission zero at +1 appears on the diagonal of the compensated system; the closed-loop eigenvalues are at the arbitrarily chosen stable locations −1,−2 and −3.

Algorithm to Obtain a Proper Stable Right Inverse Using State Feedback

Let $\dot{x} = Ax + Bu$, $y = Cx + Du$ with A, B, C, D $n \times n, n \times m, p \times n, p \times m$ real matrices, respectively, and assume that the system is controllable and observable. Let $T(s)$ be its transfer function matrix. It is known that there exists a proper right inverse $T_R(s)$, such that $T(s)T_R(s) = I_p$, if and only if $rankD = p$. If, in addition, all the zeros of $T(s)$ (that is, the transmission zeros of the system) are stable, then a stable right inverse of order n can be constructed with $k(< n)$ of its poles equal to the k stable zeros of $T(s)$ with the remaining $n-k$ poles arbitrarily assignable. This can be accomplished as follows:

Let $T_{eq} = F[sI - (A+BF)]^{-1}BG + G$ where F, G are $n \times m, m \times p$, respectively, and note that

$$\begin{aligned} T(s)T_{eq}(s) &= [C(sI-A)^{-1}B + D][F[sI-(A+BF)]^{-1}BG + G] \\ &= (C+DF)[sI-(A+BF)]^{-1}BG + DG = T_{F,G}(s) \end{aligned} \tag{27.6}$$

which is the transfer matrix one obtains when the state feedback control law $u = Fx + Gr$ is applied to the given system. Note that the second line of Equation 27.6 results from application of a well-known formula for the matrix inverse. If now F, G are such that

$$C + DF = 0, \quad DG = I_p \tag{27.7}$$

then $T_{F,G}(s) = I_p$ and T_{eq} is a proper right inverse $T_R(s)$. The additional freedom in the choice of F when $p < m$ is now used to derive a stable inverse; when $p = m$, F, G are uniquely determined from $F = -D^{-1}C$, $G = D^{-1}$.

If the nonsingular $m \times m$ matrix M is such that $DM = (I_p \ \ 0)$, then $C + DF = C + DMM^{-1}F = C + (I_p \ \ 0)\begin{pmatrix}\hat{F}_1 \\ \hat{F}_2\end{pmatrix} = 0$ from which $F = M\begin{pmatrix}-C \\ \hat{F}_2\end{pmatrix}$ with $\hat{F}_2$ arbitrary. Also, from $DG = DMM^{-1}G = I_p$, $G = M\begin{pmatrix}I_p \\ \hat{G}_2\end{pmatrix}$ with $\hat{G}_2$ arbitrary. The eigenvalues of $A + BF = A + BM\begin{pmatrix}-C \\ \hat{F}_2\end{pmatrix} = A + (\hat{B}_1 \ \ \hat{B}_2)\begin{pmatrix}-C \\ \hat{F}_2\end{pmatrix} = A - \hat{B}_1 C + \hat{B}_2\hat{F}_2$ are the poles of $T_R(s)$. It can be shown that the uncontrollable eigenvalues of $(A - \hat{B}_1 C \ \ \hat{B}_2)$ are exactly the (k) zeros of the system; they cannot be changed via $\hat{F}_2$. The remaining $n - k$ controllable eigenvalues can be arbitrarily assigned using $\hat{F}_2$. In summary, the steps to derive a stable proper inverse are

Step 1. Find nonsingular $m \times m$ matrix M such that $DM = (I_p \ \ 0)$.
Step 2. Calculate $(\hat{B}_1 \ \ \hat{B}_2) = BM$, and $A - \hat{B}_1 C$.
Step 3. Find $\hat{F}_2$ that assigns the controllable eigenvalues of $(A - \hat{B}_1 C \ \ \hat{B}_2)$ to the desired locations. The remaining uncontrollable eigenvalues are the stable zeros of the system.
Step 4.

$$\left\{A + BM\begin{pmatrix}-C \\ \hat{F}_2\end{pmatrix},\ BM\begin{pmatrix}I_p \\ \hat{G}_2\end{pmatrix},\ M\begin{pmatrix}-C \\ \hat{F}_2\end{pmatrix},\ M\begin{pmatrix}I_p \\ \hat{G}_2\end{pmatrix}\right\} \tag{27.8}$$

where $\hat{G}_2$, a $(m - p) \times p$ arbitrary real matrix, is a stable right inverse.

$T_{eq}(s)$ above is the open-loop equivalent to the state feedback control law. In view of Equations 27.6 and 27.7 the above algorithm selects F, G in a state feedback control law $u = Fx + Gr$ so that the closed-loop transfer matrix $T_{F,G}(s) = I_p$ and the closed-loop system is internally stable; that is, all the n eigenvalues of $A + BF$ are stable. Note that when $p = m$, then F, G are uniquely given by $F = -D^{-1}C$, $G = D^{-1}$; the eigenvalues of $A + BF$ are then the n zeros of the system. In this development of stable inverses via state feedback, the approach in [1] was taken; see also [7,12].

In order to implement decoupling by state feedback in practice, it is often necessary to estimate the internal state variables by means of an observer. Certain plants can be decoupled by *constant output feedback*, avoiding the need for an observer. The necessary and sufficient conditions under which this is possible were proved by Wolovich [18]: it is that $B^\star$ not only be nonsingular, but also that the modified inverse transfer matrix $B^\star T^{-1}(s)$ have only constant off-diagonal elements. This appears to be a very stringent condition, so diagonal decoupling by means of constant output feedback is not likely to be possible for any but a relatively small class of plants. This is in clear contrast with the state feedback case, as mentioned previously. If diagonal decoupling by output feedback is possible, any gain matrix H that achieves it must give all off-diagonal entries of $B^\star H$ equal to those of $B^\star T^{-1}(s)$. It can therefore be seen that any constant matrix of the form $(B^\star)^{-1}Z$ can be added to H, where Z is an arbitrary diagonal matrix, and still give a gain matrix that satisfies the required condition. There is thus a small amount of controller design freedom available, which can be used, for instance, to assign closed-loop poles to some extent. However, it does not appear possible to quantify this pole-placement freedom in any straightforward manner.

27.1.3 Block Decoupling

If diagonal decoupling by linear state feedback is not possible, an alternative to applying precompensation may still exist. It may be possible to use state feedback, or perhaps even output feedback, to reduce the system to a set of smaller subsystems that are independent; that is, decoupled. Controlling each of these small systems can then be performed in isolation from all the others, thus reducing the original plant control problem to several simpler ones. This is the idea behind block decoupling, where the goal is to transform the plant transfer matrix to one that is block diagonal rather than strictly diagonal. For square plants, each of these k diagonal blocks will also be square: the ith will be taken to have p_i inputs and p_i outputs, with $\sum p_i = p$.

One question associated with block decoupling can be answered immediately: namely, any plant with nonsingular transfer matrix can be block decoupled by linear state feedback plus precompensation. This follows from the fact that any such system can be diagonally decoupled by this form of compensation and so is trivially of any desired block diagonal form. The two types of compensation that have to be addressed here are therefore state feedback and constant output feedback.

If we are interested in block decoupling a given system by state feedback, this implies that it was not fully diagonalizable by state feedback. Hence, the matrix $B^\star$ must have been singular. As the inverse of this matrix played a significant role in the development of diagonal decoupling compensators, it seems likely that overcoming this singularity may lead toward designing block decoupling compensators for systems that cannot be diagonalized by state feedback. An equivalent way of stating that $B^\star$ is singular is to note that, although all rows of $\lim_{s\to\infty} D(s)T(s)$ are certainly finite and nonzero, some of these rows must have been linearly dependent on the preceding ones. Suppose the ith row is one such. It is then possible to add multiples of rows $1, \ldots, i-1$ to row i in order to zero out the ith row in $B^\star$; that is, to make what had been the leading coefficient vector of this row of $D(s)T(s)$ zero. If the new leading term in this row is now of order s^{-k}, multiplying the row by s^k yields a new finite and nonzero limit as s goes to infinity. If this row vector is independent of the preceding ones, we have now increased the rank of the modified $B^\star$-like matrix; if not, the same process can be repeated until successful. This basic procedure leads to the following definition, which has proved to be very useful for studying block decoupling problems.

The interactor $X_T(s)$ of $T(s)$ is the unique polynomial matrix of the form $X_T(s) = H(s)\Delta(s)$, where $\Delta(s) = diag(s^{\hat{f}_i})$ and $H(s)$ is a lower triangular polynomial matrix with 1s on the diagonal and the nonzero off-diagonal elements divisible by s, for which

$$\lim_{s\to\infty} X_T(s)T(s) = K_T \tag{27.9}$$

is finite and full rank. The interactor can be found from the transfer matrix of the system [16]; from a state-space representation [4]; or from a polynomial matrix fraction description for it [13]. The basic procedure can be illustrated by applying it to two examples discussed previously.

Example 27.6:

$$T(s) = \begin{pmatrix} \dfrac{1}{s} & \dfrac{2}{s+1} \\ \dfrac{4}{s+3} & \dfrac{8s}{s+4} \end{pmatrix}.$$

This gives $B^\star = \begin{pmatrix} 1 & 2 \\ 0 & 8 \end{pmatrix}$, with decoupling indices $f_1 = 1, f_2 = 0$. $B^\star$ is nonsingular, so it satisfies the definition of the desired matrix K_T. Thus, $K_T = B^\star = \begin{pmatrix} 1 & 2 \\ 0 & 8 \end{pmatrix}$ here, and $X_T(s) = diag(s^{f_1}, s^{f_2}) = \begin{pmatrix} s & 0 \\ 0 & 1 \end{pmatrix}$.

Example 27.7:

$$T(s) = \begin{pmatrix} \dfrac{1}{s} & \dfrac{2}{s+1} \\ \dfrac{4}{s+3} & \dfrac{8}{s+4} \end{pmatrix}$$

$B^{\star} = \begin{pmatrix} 1 & 2 \\ 4 & 8 \end{pmatrix}$, which is singular, with decoupling indices $f_1 = 1, f_2 = 1$. Subtracting 4 times row 1 of $diag(s^{f_i})T(s)$ from row 2 eliminates the linearly dependent leading coefficient vector. The resulting lower-degree polynomial row vector can then be multiplied by s, so as to again obtain a finite limit as s goes to infinity. We then have

$$\hat{T}_1(s) = \begin{pmatrix} 1 & 0 \\ 0 & s \end{pmatrix} \begin{pmatrix} 1 & 0 \\ -4 & 1 \end{pmatrix} \begin{pmatrix} s & 0 \\ 0 & s \end{pmatrix}$$

$$T(s) = \begin{pmatrix} 1 & \dfrac{2s}{s+1} \\ \dfrac{-12s}{s+3} & \dfrac{-24s^2}{(s+1)(s+4)} \end{pmatrix}.$$

Unfortunately, $\hat{T}_1(s)$ has limit as s goes to infinity of

$$\begin{pmatrix} 1 & 2 \\ -12 & -24 \end{pmatrix},$$

which is still singular. We therefore have to repeat the procedure, this time adding 12 times row 1 to row 2 to eliminate the leading coefficients and multiplying the resulting row by s to give it a finite limit. We then obtain $\begin{pmatrix} 1 & 0 \\ 0 & s \end{pmatrix} \begin{pmatrix} 1 & 0 \\ 12 & 1 \end{pmatrix} \hat{T}_1(s) = \begin{pmatrix} 1 & \dfrac{2s}{s+1} \\ \dfrac{36s}{s+3} & \dfrac{96s^2}{(s+1)(s+4)} \end{pmatrix}$, which has limit as s goes to infinity of $\begin{pmatrix} 1 & 2 \\ 36 & 96 \end{pmatrix}$. This is clearly nonsingular, so $K_T = \begin{pmatrix} 1 & 2 \\ 36 & 96 \end{pmatrix}$ for this system. The interactor is then

$$\begin{aligned} X_T(s) &= \begin{pmatrix} 1 & 0 \\ 0 & s \end{pmatrix} \begin{pmatrix} 1 & 0 \\ 12 & 1 \end{pmatrix} \begin{pmatrix} 1 & 0 \\ 0 & s \end{pmatrix} \begin{pmatrix} 1 & 0 \\ -4 & 1 \end{pmatrix} \begin{pmatrix} s & 0 \\ 0 & s \end{pmatrix} \\ &= \begin{pmatrix} s & 0 \\ -4s^3 + 12s^2 & s^3 \end{pmatrix} \\ &= \begin{pmatrix} 1 & 0 \\ -4s^2 + 12s & 1 \end{pmatrix} \begin{pmatrix} s & 0 \\ 0 & s^3 \end{pmatrix} \end{aligned}$$

which is of the desired form $H(s)\Delta(s)$.

It can be seen that, if $B^{\star}$ is nonsingular, no additional row operations are needed to modify it to give the nonsingular K_T. Thus, in this case $B^{\star} = K_T$ and $D(s) = X_T(s)$. But we already know that diagonal decoupling by state feedback is possible in this case; that is, diagonalization by state feedback is possible if and only if the interactor of the system is diagonal. This suggests the following general result.

A square system can be block decoupled by state feedback if and only if its interactor is of this same block diagonal structure.

A proof of this result is based on the fact that state feedback matrices F, G can always be found that make the closed-loop transfer matrix equal to the inverse of its interactor; see the algorithms discussed previously and [2,6]. Thus, if this matrix is block diagonal, so is the closed-loop transfer matrix. The state feedback that achieves this form can be found in an analogous manner to the state feedback matrices determined above that diagonally decouple the system.

Note that the structure algorithm of Silverman [11] is quite closely related to the interactor. This method determines a polynomial matrix $X(s)$ such that $\lim_{s\to\infty} X(s)T(s)$ is finite and nonsingular; however, $X(s)$ is not of any particular structure, unlike the interactor. This makes $X_T(s)$ better suited to obtaining clear block decoupling results.

Another question that generalizes naturally from the diagonal case is that of stability. The only fixed modes when diagonalizing were the diagonal coupling zeros, which were all zeros of the original plant that were not also zeros of any of the rows of the plant transfer matrix. In the case of block decoupling, the only fixed poles are the block coupling zeros, which are all zeros of the plant that are not also zeros of one of the $(p_i \times m)$ row blocks of $T(s)$. As in the diagonal case, these zeros must be cancelled by closed-loop poles in the decoupled transfer matrix, so creating unobservable modes; all other poles can be assigned arbitrarily.

Finally, it may be possible to achieve block decoupling by the simpler constant output feedback compensation. It can be shown that the interactor also allows a simple test for this question. In fact, block decoupling by constant output feedback is possible if and only if the interactor of the system is block diagonal and the modified inverse system $K_T T^{-1}(s)$ has only constant entries outside the diagonal blocks. The output feedback gain matrix H that achieves block decoupling is such that $K_T H$ is equal to the constant term in $K_T T^{-1}(s)$ outside the diagonal blocks. This is very similar to the diagonal decoupling result. As there, a certain degree of flexibility exists in the design of H, due to the fact that the diagonal blocks of $K_T T^{-1}(s)$ are essentially arbitrary; this can be used to provide a small amount of pole assignment flexibility when decoupling.

27.1.4 Decoupling Nonsquare Systems

The previous development has been primarily for plants with equal numbers of inputs and outputs. Plants that are not square present additional complications when studying decoupling. For instance, if there are more outputs than inputs, it is clearly impossible to assign a separate input to control each output individually; diagonal decoupling in its standard form is therefore not feasible. Similarly, decoupling the system into several independent square subsystems is also impossible. On the other hand, plants with more inputs than outputs present the opposite difficulty: there are now more input variables than are required to control each output individually.

Fortunately, the classical decoupling problem can be generalized in a straightforward fashion to cover nonsquare plants as well as square ones. In view of the preceding remarks, it is clear that systems with more outputs than inputs $(p > m)$ must be analyzed separately from those with more inputs than outputs $(p < m)$. The former case leads to decoupling results that are barely more complicated than those for the square case; the additional design freedom available in the latter case means that conditions that were necessary and sufficient for $p = m$ become only sufficient for $p < m$.

Taking the case of more inputs than outputs $(p < m)$, the following results can be shown to hold for diagonal decoupling. First, any such plant that is right-invertible (that is, for which the transfer matrix is of full rank, p) can be decoupled by state feedback plus precompensation; this follows from the close connections between this type of decoupling control law and finding a right inverse of the system. If we restrict ourselves to state feedback, two sufficient conditions for diagonal decoupling can be stated. First, the plant can be diagonalized by state feedback if its matrix $B^\star$ is of full row rank, p. This is extremely easy to test, but can be somewhat conservative. A tighter sufficient condition is as follows: The plant can be diagonalized by state feedback if a constant $(m \times p)$ matrix G can be found for which the $B^\star$ matrix of the square-modified transfer matrix $T(s)G$ is nonsingular.

It may be thought that these two sufficient conditions are identical. To see that they are not, consider the following simple example: $T(s) = \begin{pmatrix} 1/s & 1/s^2 & 1/s \\ (s+2)/s^2 & 1/s^2 & 1/s \end{pmatrix}$ has $B^\star = \begin{pmatrix} 1 & 0 & 1 \\ 1 & 0 & 1 \end{pmatrix}$, which has only rank 1. The first sufficient condition for diagonal decoupling is therefore violated. However, post-multiplying $T(s)$ by the matrix

$$G = \begin{pmatrix} 1 & 0 \\ 0 & 1 \\ -1 & 0 \end{pmatrix}$$

gives

$$T(s)G = \frac{1}{s^2}\begin{pmatrix} 0 & 1 \\ 2 & 1 \end{pmatrix},$$

which clearly has nonsingular $B^{\star}$ matrix of $\begin{pmatrix} 0 & 1 \\ 2 & 1 \end{pmatrix}$. The role of the G matrix is basically to cancel those higher-power terms in s in $T(s)$ that give rise to linearly dependent rows in $B^{\star}$; in the example, the first column of G, $(1\ 0\ -1)^T$, can be seen to be orthogonal to the repeated row vector $[1\ 0\ 1]$ in the original $B^{\star}$. Lower-power terms in $T(s)$ then become the leading terms, so their coefficients contribute to the new $B^{\star}$; these terms may well be independent of the first ones. An algorithm that goes into the details of constructing such a G, if it exists, for any right-invertible $T(s)$ is given by Williams [13].

Very similar results apply to the problem of block decoupling a system with more inputs than outputs ($p < m$) by means of state feedback. The more conservative sufficient condition states that the plant can be block decoupled if its interactor matrix has the desired block diagonal structure. This can then be tightened somewhat by proving that the plant $T(s)$ can be block decoupled if there exists some $(m \times p)$ constant matrix G that has interactor of the desired block diagonal form. Furthermore, the algorithm described previously for block decoupling of square plants can be applied equally in this case, either to $T(s)$ or $T(s)G$. The only distinction of significance between the square case and $p < m$ is that the algorithm was proved to use decoupling precompensation of lowest possible order in the square case; for nonsquare plants, minimality of this order cannot be proven.

In the case of plants with more outputs than inputs ($p > m$), the main complication is in modifying the definition of a "decoupled" closed-loop structure. Once this is done, the actual technical results are rather straightforward. As already noted, it is no longer possible to assign a single input to each individual output, as is required in the classical diagonal decoupling problem. The closest analog to this problem is one where the closed-loop system is decoupled into a set of m independent single-input/multi-output subsystems; each closed-loop control input influences a set of outputs, but does not affect any of the others. Similarly, it is not possible to assign equal numbers of independent inputs and outputs to each decoupled subsystem, as holds for square block decoupling. What we must do instead is to define decoupled subsystems that generally have more outputs than inputs; that is, they are of dimensions $p_i \times m_i$, where $p_i \leq m_i$; of course, $\sum p_i = p \leq \sum m_i = m$.

It can be shown that a very simple rank condition on the plant transfer matrix determines whether or not these decoupling problems have a solution. The simplest question to answer is whether the desired decoupled structure is achievable by means of a combination of state feedback and precompensation. The test is as follows:

Take the p_i rows of the plant transfer function corresponding to the outputs that are to be assigned to the ith decoupled subsystem. If this $p_i \times m$ transfer matrix has rank m_i, and this holds for each i, then the plant can be decoupled into $p_i \times m_i$ subsystems by means of state feedback plus precompensation.

The significance of this result is easiest to see for the special case where $m_i = 1$ for each i, the closest analog to diagonal decoupling for systems with $p > m$. If decoupling is to be possible, we must have that each $p_i \times m$ transfer matrix of the ith subsystem is of rank 1. This implies that the rows of this transfer matrix are all polynomial multiples of some common factor row vector. In other words, the p_i outputs of this subsystem are all made up of combinations of derivatives of a single "underlying" output variable. Similarly, decoupling into $p_i \times m_i$ subsystems is possible if and only if the p_i outputs making up the ith subsystem are actually made up of some combinations of m_i "underlying" output variables.

In practice, applying these rank conditions to the plant transfer matrix dictates what block dimensions are possible as closed-loop block decoupled structure. They also show which outputs must be taken as members of the same decoupled subsystem. For instance, if we wish to achieve $p_i \times 1$ decoupling and two rows of the plant transfer matrix are linearly dependent, the corresponding outputs must clearly be placed in the same subsystem.

But this approach also has one further very important implication. Consider a system that satisfies these submatrix rank conditions. If we take the m_i "underlying" output variables for each of the r subsystems, write down the corresponding $m_i \times m$ transfer matrix, and then concatenate these, we obtain a new $m \times m$ transfer matrix, denoted by $T_m(s)$. It can then be shown (see [13]) that a controller will decouple $T(s)$ into $p_i \times m_i$ blocks if and only if it also decouples $T_m(s)$ into square $m_i \times m_i$ blocks. We can therefore take all of the decoupling results derived previously for square plants and use them to solve the problem of decoupling systems with more outputs than inputs. In particular, $T(s)$ can be decoupled into $p_i \times m_i$ blocks by state feedback if and only if it satisfies the submatrix rank conditions and the interactor matrix of $T_m(s)$ is $m_i \times m_i$ block diagonal. Also, it can be shown that $T_m(s)$ has precisely the same zeros as $T(s)$. The two systems therefore clearly also have the same coupling zeros, so the fixed poles when decoupling $T(s)$ are the same as the fixed poles when decoupling $T_m(s)$. Finally, decoupling by means of output feedback can also be studied by applying the existing results for square systems to the associated $T_m(s)$.

Example 27.8:

The state-space model

$$A = \begin{pmatrix} 0 & 0 & 1 \\ 1 & 0 & 1 \\ 1 & 1 & 0 \end{pmatrix}, \quad B = \begin{pmatrix} 1 & 0 \\ 1 & 0 \\ 1 & 1 \end{pmatrix}, \quad C = I_3$$

has transfer matrix

$$T(s) = \frac{1}{(s+1)(s^2-s-1)} \begin{pmatrix} s(s+1) & s \\ (s+1)^2 & s+1 \\ (s+1)^2 & s^2 \end{pmatrix}.$$

Clearly, the first two rows are linearly dependent, so this system can be decoupled into the block diagonal form $\begin{pmatrix} \star & 0 \\ \star & 0 \\ 0 & \star \end{pmatrix}$ by state feedback plus precompensation. In fact, the associated invertible transfer function for this system is

$$T_m(s) = \frac{1}{(s+1)(s^2-s-1)} \begin{pmatrix} s+1 & 1 \\ (s+1)^2 & s^2 \end{pmatrix},$$

which has interactor $\begin{pmatrix} s^2 & 0 \\ 0 & s \end{pmatrix}$ diagonal $\left[\text{with } K_T = \begin{pmatrix} 1 & 0 \\ 1 & 1 \end{pmatrix}\right]$. Thus, block decoupling is actually possible for this system using state feedback alone.

As a final point on general block decoupling, note that this problem can also be studied using the geometric approach; see [19]. This state-space technique is based on considering the supremal (A, B)-invariant subspaces contained in the kernels of the various subsystems formed by deleting the outputs corresponding to each desired block in turn. The ranges of these subspaces determine whether decoupling is possible by state feedback. If it is not, the related "efficient extension" approach allows a precompensator of relatively low order to be found that will produce the desired block diagonal structure. This approach is somewhat involved, and the interested reader is referred to Morse and Wonham [8] for further details.

27.1.5 Triangular Decoupling

There is a form of "partially decoupled" system that can be of particular value for certain plants. This is the triangularized form, where all entries of the closed-loop transfer matrix above its leading diagonal are made zero. The first closed-loop output, y_1, is therefore affected only by the first input r_1; the second, y_2, is influenced only by inputs r_1 and r_2; etc. This type of transfer matrix can be used in the following sequential control scheme. First, input r_1 is adjusted until output y_1 is as desired, and the control is then frozen. Output variable y_2 is then affected only by r_2 and the fixed r_1, so r_2 can be adjusted until this output is also as desired. The third input, r_3, can then be used to set output y_3, etc. This scheme can be seen to be less powerful than diagonal decoupling, as the outputs must be adjusted sequentially rather than fully independently. However, it has one strong point in its favor: *any right-invertible plant can be triangularized by state feedback alone,* regardless of whether additional precompensation is required to make it diagonally decoupled. Proof of this follows directly from the fact that there always exists some state feedback gains F, G for which $T_{F,G}(s) = X_T^{-1}(s)$, and the interactor is, by definition, lower triangular. Of course, similar results apply for generalized rather than standard interactors also. Therefore, it can be shown, as originally proved by Morse and Wonham [9], that all closed-loop poles of the triangularly decoupled system can be arbitrarily assigned.

Finally, it may also be possible to triangularize a system by means of the simpler constant output feedback. If the original plant is square and strictly proper ($D = 0$), it can be shown that this is possible if and only if all entries of the modified inverse transfer matrix $K_T T^{-1}(s)$ that lie above the leading diagonal are constant. This is quite a simple condition to test and is very similar to the test for diagonal decoupling by output feedback. The required gain matrix H is given from the fact that the upper triangular part of $K_T H$ is precisely the upper triangular constant part of $K_T T^{-1}(s)$. It can be noted that there is therefore some non-uniqueness in the choice of the gain H: in particular, we can add a term of the form $K_T^{-1}Z$ to H, where Z is any lower triangular constant matrix, and still get a suitable output gain matrix. If it is possible to triangularize a given system by output feedback, there is consequently some freedom to assign closed-loop poles also. However, it is difficult to quantify this freedom in any concrete way.

27.1.6 Static Decoupling

Static decoupling, as opposed to dynamic decoupling already described, is much easier to achieve. A system is statically decoupled if a step change in the (static) steady-state level of the ith input is reflected by a change in the steady-state level of the ith output and only that output. To derive the conditions for static decoupling, assume that the system is described by a $p \times p$ transfer matrix $T(s)$ that is bounded-input/bounded-output stable; that is, all of its poles are in the open left half of the s-plane and none is on the imaginary axis. Note that stability is necessary for the steady-state values of the outputs to be well defined. Assume now that the p inputs are step functions described by $u_i(s) = \frac{k_i}{s}$, $i = 1, \ldots, p$. The steady-state value of the output vector y, y_{ss}, can then be found using the final value theorem, as follows:

$$y_{ss} = \lim_{s\to\infty} y(t) = \lim_{s\to 0} sT(s)\frac{1}{s}\begin{pmatrix} k_1 \\ k_2 \\ \vdots \\ k_p \end{pmatrix} = T(0)\begin{pmatrix} k_1 \\ k_2 \\ \vdots \\ k_p \end{pmatrix} \tag{27.10}$$

It is now clear that $T(s)$ is statically decoupled if and only if $T(0)$ is a diagonal nonsingular matrix; that is, all the off-diagonal entries of $T(s)$ must be divisible by s, while the entries on the diagonal should not be divisible by s. It can be shown easily that a system described by a $p \times p$ transfer matrix $T(s)$ that is bounded-input/bounded-output stable can be statically decoupled, via $u = Gr$, if and only if

$$rank\ T(0) = p \tag{27.11}$$

that is, if and only if there is no transmission zero at $s = 0$. Note that this condition, if a controllable and observable state-space description is given, is

$$rank\begin{pmatrix} A & B \\ C & D \end{pmatrix} = n + p \tag{27.12}$$

If this is the case, any feedforward constant gain G, in $u = Gr$, such that $T(0)G$ is a diagonal and nonsingular matrix will statically decouple the system. To illustrate, consider the following example:

Example 27.9:

$$T(s) = \begin{pmatrix} \dfrac{s+2}{s+1} & \dfrac{2}{s+3} \\ \dfrac{s(s+1)}{(s+3)^2} & \dfrac{1}{s+1} \end{pmatrix}$$

Here $T(0) = \begin{pmatrix} 2 & 2/3 \\ 0 & 1 \end{pmatrix}$, which has full rank; therefore, it can be statically decoupled. Let $T(0)G = \begin{pmatrix} 2 & 0 \\ 0 & 1 \end{pmatrix}$; then $G = \begin{pmatrix} 1 & -1/3 \\ 0 & 1 \end{pmatrix}$. Note that

$$T(s)G = \begin{pmatrix} \dfrac{s+2}{s+1} & \dfrac{-s(s-1)}{3(s+1)(s+3)} \\ \dfrac{s(s+1)}{(s+3)^2} & \dfrac{-s^3+s^2+17s+27}{3(s+1)(s+3)^2} \end{pmatrix}$$

where all the off-diagonal entries of $T(s)$ are divisible by s, while the entries on the diagonal are not divisible by s. If now the input $\frac{1}{s}\begin{pmatrix} k_1 \\ k_2 \end{pmatrix}$ is applied to $T(s)G$, the steady-state output is $T(0)G\begin{pmatrix} k_1 \\ k_2 \end{pmatrix} = \begin{pmatrix} 2k_1 \\ k_2 \end{pmatrix}$.

27.2 Defining Terms

Decoupling: Separating the system into a number of independent subsystems.

Non-interacting control: The control inputs and the outputs can be partitioned into disjoint subsets; each subset of outputs is controlled by only one subset of inputs, and each subset of inputs affects only one subset of outputs. From an input/output viewpoint, the system is split into independent subsystems; it is called decoupled.

References

1. Antsaklis, P.J., Stable proper n-th order inverses, *IEEE Trans. Autom. Control,* 23, 1104–1106, 1978.
2. Antsaklis, P.J., Maximal order reduction and supremal (A,B)-invariant and controllability subspaces, *IEEE Trans. Autom. Control,* 25, 44–49, 1980.
3. Falb, P.L. and Wolovich, W.A., Decoupling in the design and synthesis of multivariable control systems, *IEEE Trans. Autom. Control,* 12, 651–659, 1967.
4. Furuta, K. and Kamiyama, S., State feedback and inverse system, *Intern. J. Control,* 25(2), 229–241, 1977.
5. Gilbert, E.G., The decoupling of multivariable systems by state feedback, *SIAM J. Control,* 50–63, 1969.
6. Kamiyama, S. and Furuta, K., Integral invertibility of linear time-invariant systems, *Intern. J. Control,* 25(3), 403–412, 1977.
7. Moore, B.C. and Silverman, L.M., A new characterization of feedforward, delay-free inverses, *IEEE Trans. Inf. Theory,* 19, 126–129, 1973.
8. Morse, A.S. and Wonham, W.M., Decoupling and pole assignment by dynamic compensation, *SIAM J. Control,* 317–337, 1970.
9. Morse, A.S. and Wonham, W.M., Triangular decoupling of linear multivariable systems, *IEEE Trans. Autom. Control,* 447–449, 1970.

10. Rekasius, Z.V., Decoupling of multivariable systems by means of state variable feedback, *Proc. 3rd Allerton Conf.*, 439–448, 1965.
11. Silverman, L.M., Decoupling with state feedback and precompensation, *IEEE Trans. Autom. Control*, 15, 487–489, 1970.
12. Silverman, L.M. and Payne, H.J., Input–output structure of linear systems with application to the decoupling problem, *SIAM J. Control*, 9, 188–233, 1971.
13. Williams, T., Inverse and Decoupling Problems in Linear Systems, Ph.D. thesis, Imperial College, London, 1981.
14. Williams, T. and Antsaklis, P.J., A unifying approach to the decoupling of linear multivariable systems, *Intern. J. Control*, 44(1), 181–201, 1986.
15. Wolovich, W.A. and Falb, P.L., On the structure of multivariable systems, *SIAM J. Control*, 437–451, 1969.
16. Wolovich, W.A. and Falb, P.L., Invariants and canonical forms under dynamic compensation, *SIAM J. Control*, 996–1008, 1976.
17. Wolovich, W.A., *Linear Multivariable Systems*, Springer-Verlag, New York, 1974.
18. Wolovich, W.A., Output feedback decoupling, *IEEE Trans. Autom. Control*, 148–149, 1975.
19. Wonham, W.M., *Linear Multivariable Control: A Geometric Approach*, Springer-Verlag, New York, 1979.

Further Reading

Making the system diagonally dominant is a powerful design approach. Details on how to achieve diagonal dominance using Rosenbrock's Inverse Nyquist Array method can be found in Rosenbrock, H.H. 1974. *Computer-Aided Control System Design*, Academic Press, New York.

A good introduction to the geometric approach and to the decoupling problem using that approach can be found in Wohnam, W.M. 1985. *Linear Multivariable Control: A Geometric Approach*, Springer-Verlag, New York. The problem of disturbance decoupling or disturbance rejection, where a disturbance in the state equations must become unobservable from the output, is also studied there using the geometric approach.

A geometric approach has also been used to study non-interacting control in nonlinear systems; see, for example, Battilotti, S. 1994. *Noninteracting Control with Stability for Nonlinear Systems*, Springer-Verlag, New York.

For the decoupling of singular systems see, for example Paraskevopoulos, P.N. and Koumboulis, F.N. 1992. The decoupling of generalized state-space systems with state feedback, *IEEE Trans. Autom. Control*, pp. 148–152, vol.37.

The following journals report advances in all areas of decoupling including diagonal, block and triangular decoupling: *IEEE Transactions on Automatic Control*, *International Journal of Control* and *Automatica*.

28

Linear Model Predictive Control in the Process Industries

Jay H. Lee
Korea Advanced Institute of Science and Technology

Manfred Morari
Swiss Federal Institute of Technology

28.1 Introduction

Model predictive control (MPC) refers to a class of control algorithms that compute a sequence of control moves based on an explicit prediction of outputs within some future horizon. The computed control moves are typically implemented in a receding horizon fashion, meaning only the moves for the current time are implemented and the whole calculation is repeated at the next sample time. In essence, MPC is a *feedback* control strategy based on *repeated* calculation of *open-loop* control trajectories.

It is difficult to attribute MPC to any single individual, since the idea of MPC has appeared in many different forms and in the context of a variety of applications. In the process industries, serious applications and research on the subject began in the late 1970s, fueled by seminal papers by several industrialists, who outlined the basic algorithm and pointed out their potential for providing effective solutions to difficult process control problems [1,6]. Owing to its unique ability to handle process interactions and constraints in a unified manner, MPC progressed rapidly, establishing an impressive track record along the way. The initial applications were mainly in the petrochemical industries, but it has been applied to a variety of industries including chemicals, food, and pulp and paper. Now MPC has become a standard tool for process control and there are several vendors that market general-purpose MPC software and commissioning services.

The objective of this chapter is to introduce the *linear MPC* technique as viewed from the process industries. After introducing a prototypical algorithm in Section 28.2, we discuss some implementation issues in Section 28.3 and introduce several notable idiosyncratic features of various other commercial MPC algorithms. Finally, we point out some future research needs.

28.2 Industrial MPC Algorithm

Dynamic matrix control (DMC) was one of the first commercial implementations of MPC. In this section, we describe the basic ideas of the algorithm.

28.2.1 The Idea of Moving Horizon Control

Consider the diagram in Figure 28.1. At the present time k the behavior of the process over a horizon p is considered. Using the model, the response of the process output to changes in the manipulated variable (MV) is predicted. Current and future moves of the MVs are selected such that the predicted response has certain desirable (or *optimal*) characteristics. For instance, a commonly used objective is to minimize the sum of squares of the future errors, that is, the deviations of the controlled variable (CV) from a desired target (setpoint). This minimization can also take into account constraints, that may be present on the MVs and the outputs.

The idea is appealing but would not work very well in practice if the moves of the MV determined at time k were applied blindly over the future horizon. Disturbances and modeling errors may lead to deviations between the predicted behavior and the actual observed behavior, so that the computed MV moves may not be appropriate any more. Therefore only the first one of the computed moves is actually implemented. At the next time step $k + 1$, a measurement is taken, the horizon is shifted forward by one

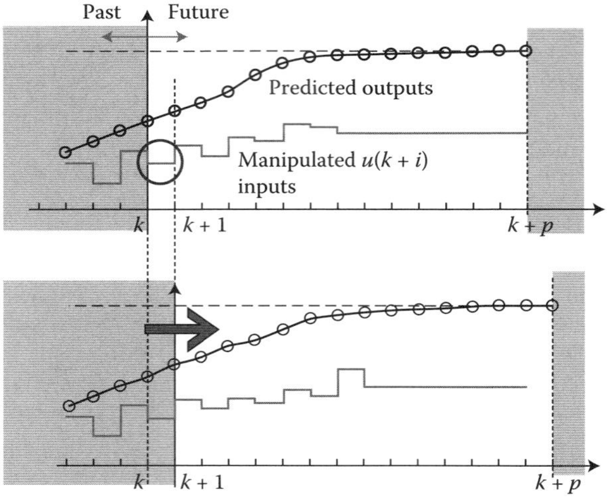

FIGURE 28.1 Moving horizon control.

step, and the optimization is done again over this shifted horizon based on the updated prediction of the system behavior. Therefore, this control strategy is also referred to as *moving horizon control*.

A similar strategy is used in many other nontechnical situations. One example is computer chess where the computer moves after evaluating all possible moves over a specified "depth" (the horizon). At the next turn the evaluation is repeated based on the current board situation. Another example would be investment planning. A five-year plan is established to maximize the return. Periodically, a new five-year plan is put together over a shifted horizon to take into account changes that have occurred in the economy.

The DMC algorithm includes as one of its major components a technique to predict the future output of the system as a function of the inputs and disturbances. This prediction capability is necessary to determine the optimal future control inputs and will be discussed next. Afterward we will state the objective function, and formulate the optimization problem.

28.2.2 Multistep Prediction

We consider the setup depicted in Figure 28.2 where we have three different types of external inputs: the MV u, whose effect on the output, usually a CV, is described by P_u; the measured disturbance variable (DV) d whose effect on the output is described by P_d; and finally the unmeasured and unmodeled disturbances w_y, which add a bias to the system output. The overall system can be described by

$$y(k) = \begin{bmatrix} P^u & P^d \end{bmatrix} \begin{bmatrix} u(k) \\ d(k) \end{bmatrix} + w_y(k). \tag{28.1}$$

We assume that step response models S^u, S^d are available for the system dynamics P_u and P_d, respectively. We can define the overall multivariable step response model

$$S = \begin{bmatrix} S^u & S^d \end{bmatrix}, \tag{28.2}$$

which is driven by the known overall input

$$\Delta v(k) = \begin{bmatrix} \Delta u(k) \\ \Delta d(k) \end{bmatrix}. \tag{28.3}$$

Let us adopt as the system state

$$\tilde{Y}(k) = \left[\tilde{y}_0^T(k), \quad \tilde{y}_1^T(k), \quad \ldots, \quad \tilde{y}_{n-1}^T(k)\right]^T, \tag{28.4}$$

where n is the number of sample steps it takes for the system to settle down after a step change is made to any of the inputs. The elements of the state represent the future system outputs

$$\tilde{Y}(k) = \begin{bmatrix} y(k) \\ y(k+1) \\ \vdots \\ y(k+n-1) \end{bmatrix} \tag{28.5}$$

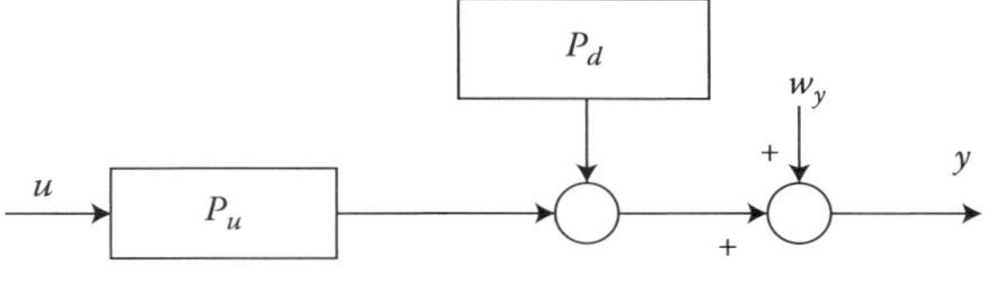

FIGURE 28.2 Basic problem setup.

obtained under the assumption that the system inputs do not change from the previous values, that is,

$$\begin{aligned}\Delta u(k) &= \Delta u(k+1) = \cdots = 0,\\ \Delta d(k) &= \Delta d(k+1) = \cdots = 0.\end{aligned} \tag{28.6}$$

Also, the state does not include any unmeasureed disturbance information and hence it is assumed in the definition that

$$w_y(k) = w_y(k+1) = \cdots = 0. \tag{28.7}$$

The state is updated according to

$$\tilde{Y}(k) = M \cdot \tilde{Y}(k-1) + S\Delta v(k-1). \tag{28.8}$$

where M is a shift operator expressed as

$$M = \left.\begin{bmatrix} 0 & I & 0 & \dots & \dots & 0 & 0 \\ 0 & 0 & I & 0 & \dots & 0 & 0 \\ \vdots & & & & & & \vdots \\ 0 & 0 & \dots & \dots & \dots & 0 & I \\ 0 & 0 & \dots & \dots & \dots & 0 & I \end{bmatrix}\right\} n \tag{28.9}$$

and

$$S = \begin{bmatrix} S_1 \\ \vdots \\ S_n \end{bmatrix}, \tag{28.10}$$

where S_i is the ith step response coefficient matrix. The equation reflects the effect of the input change $\Delta v(k-1)$ on the future evolution of the system assuming that there are no further input changes. The influence of the input change manifests itself through the step response matrix S. The effect of any future input changes is described as well by the appropriate step response matrix. Let us consider the predicted output over the next p time steps

$$\begin{aligned}\begin{bmatrix} y(k+1|k) \\ y(k+2|k) \\ \vdots \\ \vdots \\ y(k+p|k) \end{bmatrix} &= \begin{bmatrix} \tilde{y}_1(k) \\ \tilde{y}_2(k) \\ \vdots \\ \vdots \\ \tilde{y}_p(k) \end{bmatrix} + \begin{bmatrix} S_1^u \\ S_2^u \\ \vdots \\ \vdots \\ S_p^u \end{bmatrix} \Delta u(k|k) + \begin{bmatrix} 0 \\ S_1^u \\ S_2^u \\ \vdots \\ S_{p-1}^u \end{bmatrix} \Delta u(k+1|k) + \cdots + \cdots \\ &+ \begin{bmatrix} 0 \\ 0 \\ \vdots \\ 0 \\ S_1^u \end{bmatrix} \Delta u(k+p-1|k) + \begin{bmatrix} S_1^d \\ S_2^d \\ \vdots \\ \vdots \\ S_p^d \end{bmatrix} \Delta d(k) + \begin{bmatrix} 0 \\ S_1^d \\ S_2^d \\ \vdots \\ S_{p-1}^d \end{bmatrix} \Delta d(k+1|k) + \cdots \\ &+ \begin{bmatrix} 0 \\ 0 \\ \vdots \\ 0 \\ S_1^d \end{bmatrix} \Delta d(k+p-1|k) + \begin{bmatrix} w_y(k+1|k) \\ w_y(k+2|k) \\ \vdots \\ \vdots \\ w_y(k+p|k) \end{bmatrix}.\end{aligned} \tag{28.11}$$

Here the first term on the right-hand side, the first p elements of the state, describes the future evolution of the system when all the future input changes are zero. The remaining terms describe the effect of the present and future changes of the manipulated inputs $\Delta u(k+i|k)$, the measured disturbances $\Delta d(k+i|k)$, and the unmeasured and unmodeled disturbances $w_y(k+i|k)$. The notation $y(k+i|k)$ represents the prediction of $y(k+i)$ made based on the information available at time k. The same notation applies to Δd and w_y.

The values of most of these variables are not available at time k and have to be *predicted* in a rational fashion. From the measurement at time k $d(k)$ is known and therefore $\Delta d(k) = d(k) - d(k-1)$. Unless some additional process information or "upstream" measurements are available to conclude about the future disturbance behavior, the disturbances are assumed not to change in the future for the derivation of the DMC algorithm.

$$\Delta d(k+1|k) = \Delta d(k+2|k) = \cdots = \Delta d(k+p-1|k) = 0. \tag{28.12}$$

This assumption is reasonable when the disturbances are varying only infrequently. Similarly, we will assume that the future unmodeled disturbances $w_y(k+i|k)$ do not change.

$$w_y(k|k) = w_y(k+1|k) = w_y(k+2|k) = \cdots = w_y(k+p|k). \tag{28.13}$$

We can obtain an estimate of the present unmodeled disturbance from Equation 28.1

$$w_y(k|k) \approx y_m(k) - \tilde{y}_0(k), \tag{28.14}$$

where $y_m(k)$ represents the value of the output as actually measured in the plant. Here $\tilde{y}_0(k)$, the first component of the state $\tilde{Y}(k)$, is the model prediction of the output at time k (assuming $w_y(k) = 0$) based on the information up to this time. The difference between this predicted output and the measurement provides a good estimate of the unmodeled disturbance.

For generality, we want to consider the case where the manipulated inputs are not varied over the whole horizon p but only over the next m steps $(\Delta u(k|k), \Delta u(k+1|k), \ldots, \Delta u(k+m-1|k))$ and that the input changes are set to zero after that.

$$\Delta u(k+m|k) = \Delta u(k+m+1|k) = \cdots = \Delta u(k+p-1)|k) = 0. \tag{28.15}$$

With these assumptions Equation 28.11 becomes

$$\mathcal{Y}(k+1|k) = \underbrace{\begin{bmatrix} \tilde{y}_1(k) \\ \tilde{y}_2(k) \\ \vdots \\ \vdots \\ \tilde{y}_p(k) \end{bmatrix}}_{\substack{\mathcal{M}\tilde{Y}(k) \\ \text{from the memory}}} + \underbrace{\begin{bmatrix} S_1^d \\ S_2^d \\ \vdots \\ \vdots \\ S_p^d \end{bmatrix}}_{\substack{\mathcal{S}^d \Delta d(k) \\ \text{feedforward term}}} \Delta d(k) + \underbrace{\begin{bmatrix} y_m(k) - \tilde{y}_0(k) \\ y_m(k) - \tilde{y}_0(k) \\ \vdots \\ \vdots \\ y_m(k) - \tilde{y}_0(k) \end{bmatrix}}_{\substack{\mathcal{I}_p(y_m(k) - \tilde{y}_0(k)) \\ \text{feedback term}}}$$

$$+ \underbrace{\begin{bmatrix} S_1^u & 0 & \cdots & \cdots & 0 \\ S_2^u & S_1^u & 0 & \cdots & 0 \\ \vdots & \vdots & \ddots & \ddots & \vdots \\ S_m^u & S_{m-1}^u & \cdots & \cdots & S_1^u \\ \vdots & \ddots & \ddots & \ddots & \vdots \\ S_p^u & S_{p-1}^u & \cdots & \cdots & S_{p-m+1}^u \end{bmatrix}}_{\substack{\mathcal{S}^u \\ \text{dynamic matrix}}} \underbrace{\begin{bmatrix} \Delta u(k|k) \\ \Delta u(k+1|k) \\ \vdots \\ \vdots \\ \Delta u(k+m-1|k) \end{bmatrix}}_{\substack{\Delta\mathcal{U}(k) \\ \text{future input moves}}}. \tag{28.16}$$

Here we have introduced the new symbols

$$\mathcal{Y}(k+1|k) = \begin{bmatrix} y(k+1|k) \\ y(k+2|k) \\ \vdots \\ y(k+p|k) \end{bmatrix}, \tag{28.17}$$

$$\mathcal{S}^u = \begin{bmatrix} S_1^u & 0 & \cdots & 0 \\ S_2^u & S_1^u & \cdots & 0 \\ \vdots & \vdots & & \vdots \\ S_m^u & S_{m-1}^u & \cdots & S_1^u \\ \vdots & \vdots & & \vdots \\ S_p^u & S_{p-1}^u & \cdots & S_{p-m+1}^u \end{bmatrix}, \quad S^d := \begin{bmatrix} S_1^d \\ S_2^d \\ \vdots \\ S_p^d \end{bmatrix}, \tag{28.18}$$

$$\mathcal{I}_p = \left.\begin{bmatrix} I \\ I \\ \vdots \\ I \end{bmatrix}\right\} p, \tag{28.19}$$

$$\Delta\mathcal{U}(k) = \begin{bmatrix} \Delta u(k|k) \\ \Delta u(k+1|k) \\ \vdots \\ \Delta u(k+m-1|k) \end{bmatrix}, \tag{28.20}$$

$$\mathcal{M} = \begin{cases} \left.\begin{bmatrix} 0 & I & 0 & \cdots & \cdots & \cdots & \cdots & 0 \\ 0 & 0 & I & 0 & \cdots & \cdots & \cdots & 0 \\ k\vdots & \vdots & \ddots & \ddots & \ddots & \ddots & \vdots & \vdots \\ 0 & \cdots & \cdots & 0 & I & 0 & \cdots & 0 \end{bmatrix}\right\} p \quad \text{for } p < n, \\ \\ \left.\begin{bmatrix} 0 & I & 0 & \cdots & 0 \\ 0 & 0 & I & \cdots & 0 \\ \vdots & \vdots & \ddots & \ddots & \vdots \\ 0 & \cdots & \cdots & 0 & I \\ \vdots & \vdots & & & \vdots \\ 0 & \cdots & \cdots & 0 & I \end{bmatrix}\right\} p \quad \text{for } p \geq n, \end{cases} \tag{28.21}$$

With this new notation the p-step ahead prediction becomes

$$\mathcal{Y}(k+1|k) = \mathcal{M}\tilde{Y}(k) + S^d\Delta d(k) + \mathcal{I}_p(y_m(k) - \tilde{y}_0(k)) + \mathcal{S}^u\Delta\mathcal{U}(k), \tag{28.22}$$

where the first three terms are completely defined by past control actions ($\tilde{Y}(k)$, $\tilde{y}_0(k)$) and present measurements ($y_m(k)$, $\Delta d(k)$) and the last term describes the effect of future MV moves $\Delta\mathcal{U}(k)$.

This prediction equation can be easily adjusted if different assumptions are made on the future behavior of the measured and unmeasured disturbances. For instance, if the disturbances are expected to evolve in a ramp-like fashion then we would set

$$\Delta d(k) = \Delta d(k+1|k) = \cdots = \Delta d(k+p-1|k) \tag{28.23}$$

and

$$w_y(k+\ell|k) = w_y(k|k) + \ell(w_y(k|k) - w_y(k-1|k-1)). \tag{28.24}$$

28.2.3 State-Space Formulation

Although the first generation of industrial algorithms adopted a finite impulse response or a step response model, given their intuitive appeal to the practitioners, its limitations were quickly pointed out by the academics who went on to propose more general formulations based on state-space models. These formulations were later adopted by the vendor companies to develop a second generation of commercial MPC algorithms. These formulations also incorporated state estimation techniques into MPC for enhanced disturbance estimation and noise filtering.

The state-space model form used was

$$\begin{aligned} x(k+1) &= Ax(k) + B_u u(k) + B_w w_x(k) + B_d d(k), \\ y(k) &= Cx(k) + w_y(k). \end{aligned} \tag{28.25}$$

Here w_x and w_y are state and output disturbances. Since these disturbances tend to be "persistent," integrating states are created to reflect their nature:

$$\begin{aligned} w_x(k+1) &= w_x(k) + \varepsilon_1(k), \\ w_y(k+1) &= w_y(k) + \varepsilon_2(k), \end{aligned} \tag{28.26}$$

where ε_1 and ε_2 are white-noise sequences. The above can be expressed as

$$\begin{aligned} \begin{bmatrix} x(k+1) \\ w_x(k+1) \\ w_y(k+1) \end{bmatrix} &= \begin{bmatrix} A & B_w & 0 \\ 0 & I & 0 \\ 0 & 0 & I \end{bmatrix} \begin{bmatrix} x(k) \\ w_x(k) \\ w_y(k) \end{bmatrix} + \begin{bmatrix} B_u \\ 0 \\ 0 \end{bmatrix} u(k) + \begin{bmatrix} B_d \\ 0 \\ 0 \end{bmatrix} d_k + \begin{bmatrix} 0 & 0 \\ I & 0 \\ 0 & I \end{bmatrix} \begin{bmatrix} \varepsilon_1(k) \\ \varepsilon_2(k) \end{bmatrix}, \\ y(k) &= \begin{bmatrix} C & 0 & I \end{bmatrix} \begin{bmatrix} x(k) \\ w_x(k) \\ w_y(k) \end{bmatrix}, \end{aligned} \tag{28.27}$$

which will be denoted using the notation

$$\begin{aligned} z(k+1) &= \Phi z(k) + \Gamma_u u(k) + \Gamma_d d_k + \Gamma_\varepsilon \varepsilon(k), \\ y(k) &= \Xi z(k). \end{aligned} \tag{28.28}$$

The above is a standard state-space model with white state noise input and Kalman filtering can be used to design a state estimator of the following form:

$$\begin{aligned} \hat{z}(k|k-1) &= \Phi \hat{z}(k-1|k-1) + \Gamma_u u(k) + \Gamma_d d_k, \\ \hat{z}(k|k) &= \hat{z}(k-1|k-1) + K\left(y(k) - \Xi \hat{z}(k|k-1)\right), \end{aligned} \tag{28.29}$$

where K is the Kalman filter gain.

One potential problem for using the above model is the lack of detectability: One cannot have more disturbance states (w_x and w_y), which have integrating dynamics, than the number of outputs for detectability to hold. Hence, to use the above model, one would have to choose between input and output disturbances. Alternatively, it may be convenient to adopt the following differenced model form:

$$\begin{aligned} \begin{bmatrix} \Delta x(k+1) \\ y(k+1) \end{bmatrix} &= \begin{bmatrix} A & 0 \\ CA & I \end{bmatrix} \begin{bmatrix} x(k) \\ y(k) \end{bmatrix} + \begin{bmatrix} B_u \\ CB_u \end{bmatrix} \Delta u(k) + \begin{bmatrix} B_d \\ CB_d \end{bmatrix} \Delta d_k + \begin{bmatrix} B_w & 0 \\ CB_w & I \end{bmatrix} \begin{bmatrix} \varepsilon_1(k) \\ \varepsilon_2(k) \end{bmatrix}, \\ y(k) &= \begin{bmatrix} 0 & I \end{bmatrix} \begin{bmatrix} \Delta x(k) \\ y(k) \end{bmatrix}. \end{aligned} \tag{28.30}$$

The above system is detectable as long as the original (C, A) was a detectable pair. Once again, the above system is in the standard state-space form of Equation 28.28 and the Kalman filter of Equation 28.29 can be designed for state estimation.

The multistep prediction equation can be constructed and has the following structure:

$$\mathcal{Y}(k+1|k) = \mathcal{S}^z\hat{z}(k|k) + \mathcal{S}^d\Delta d(k) + \mathcal{S}^u\Delta\mathcal{U}(k). \tag{28.31}$$

The readers are referred to [4] for a detailed derivation of the matrices in the above equation. The above equation has the same linear structure of $\mathcal{Y}(k+1|k) = b(k) + \mathcal{S}^u\Delta\mathcal{U}(k)$ as before and the subsequent derivations presented hereafter hold for both the step response model-based formulation and the state-space model-based formulation (with some obvious modifications).

28.2.4 Objective Function

Plant operation requirements determine the performance criteria of the control system. These criteria must be expressed in mathematical terms so that a control law can be obtained in algorithmic form. In DMC, a *quadratic* objective function is used, which can be stated in its simplest form as*

$$\min_{\Delta u(k|k)\ldots\Delta u(k+m-1|k)} \sum_{\ell=1}^{p} \|y(k+\ell|k) - r(k+\ell)\|^2. \tag{28.32}$$

This criterion minimizes the sum of squared deviations of the predicted CV values from a time–varying reference trajectory or setpoint $r(k+\ell)$ over p future time steps. The quadratic criterion penalizes large deviations proportionally more than smaller ones so that on the average the output remains close to its reference trajectory and large excursions are avoided.

Note that the MVs are assumed to be constant after m intervals of time into the future, or equivalently,

$$\Delta u(k+m|k) = \Delta u(k+m+1|k) = \cdots = \Delta u(k+p-1|k) = 0,$$

where $m \leq p$ always. This means that DMC determines the next m moves, only. The choices of m and p affect the closedly–loop behavior. Moreover, m, the number of degrees of freedom, has a dominant influence on the computational effort. Also, it does not make sense to make the horizon longer than $m+n$ ($p \leq m+n$), because for an FIR system of order n the system reaches a steady state after $m+n$ steps. Increasing the horizon beyond $m+n$ would simply add identical constant terms to the objective function (Equation 28.32).

Due to inherent process interactions, it is generally not possible to keep all outputs close to their corresponding reference trajectories simultaneously. Therefore, in practice only a subset of the outputs is controlled well at the expense of larger excursions in others. This can be influenced transparently by including weights in the objective function as follows:

$$\min_{\Delta u(k|k)\ldots\Delta u(k+m-1|k)} \sum_{\ell=1}^{p} \|\Gamma_\ell^y[y(k+\ell|k) - r(k+\ell)]\|^2. \tag{28.33}$$

For example, for a system with two outputs y_1 and y_2, and constant diagonal weight matrices of the form

$$\Gamma_\ell^y = \begin{bmatrix} \gamma_1 & 0 \\ 0 & \gamma_2 \end{bmatrix}; \quad \forall \ell \tag{28.34}$$

the objective becomes

$$\min_{\Delta u(k|k)\ldots\Delta u(k+m-1|k)} \left\{ {\gamma_1}^2 \sum_{\ell=1}^{p} [y_1(k+\ell|k) - r_1(k+\ell)]^2 + {\gamma_2}^2 \sum_{\ell=1}^{p} [y_2(k+\ell|k) - r_2(k+\ell)]^2 \right\}. \tag{28.35}$$

Thus, the larger the weight is for a particular output, the larger is the contribution of its sum of squared deviations to the objective. This will make the controller bring the corresponding output closer to its reference trajectory.

* $\|x\|$ denotes the norm $(x^Tx)^{1/2}$ of the vector x.

Finally, the MV moves that make the output follow a given trajectory could be too severe to be acceptable in practice. This can be corrected by adding a penalty term for the MV moves to the objective as follows:

$$\min_{\Delta\mathcal{U}(k)} \sum_{\ell=1}^{p} \|\Gamma_\ell^y[y(k+\ell|k) - r(k+\ell)]\|^2 + \sum_{\ell=1}^{m} \|\Gamma_\ell^u[\Delta u(k+\ell-1)]\|^2. \qquad (28.36)$$

Note that the larger the elements of the matrix Γ_ℓ^u, the smaller the resulting moves, and consequently, the output trajectories will not be followed as closely. Thus, the relative magnitudes of Γ_ℓ^y and Γ_ℓ^u will determine the trade-off between following the trajectory closely and reducing the action of the MVs.

Of course, not every practical performance criterion is faithfully represented by this quadratic objective. However, many control problems can be formulated as trajectory tracking problems and therefore this formulation is very useful. Most importantly this formulation leads to an optimization problem for which there exist effective solution techniques.

28.2.5 Constraints

In many control applications the desired performance cannot be expressed solely as a trajectory following the problem. Many practical requirements are more naturally expressed as constraints on process variables.

There are three types of process constraints:

MV constraints: these are hard limits on inputs $u(k)$ to take care of, for example, valve saturation constraints.

MV rate constraints: these are hard limits on the size of the MV moves $\Delta u(k)$ to directly influence the rate of change of the MVs.

Output variable constraints: hard or soft limits on the outputs of the system are imposed to, for example, avoid overshoots and undershoots. These can be of two kinds:

- *CV:* limits for these variables are specified even though deviations from their setpoints are minimized in the objective function
- *Associated variables:* no setpoints exist for these output variables but they must be kept within bounds (i.e., corresponding rows of Γ_ℓ^y are zero for the projections of these variables in the objective function given in Equation 28.36.

The three types of constraints in DMC are enforced by formulating them as linear inequalities. In the following, we explicitly formulate these inequalities.

28.2.5.1 MV Constraints

The solution vector of DMC contains not only the current moves to be implemented but also the moves for the future m intervals of time. Although violations can be avoided by constraining only the move to be implemented, constraints on future moves can be used to allow the algorithm to anticipate and prevent future violations, thus producing a better overall response. The MV value at a future time $k+\ell$ is constrained to be

$$u_{low}(\ell) \le \sum_{j=0}^{\ell} \Delta u(k+j|k) + u(k-1) \le u_{high}(\ell); \quad \ell = 0, 1, \ldots m-1,$$

where $u(k-1)$ is the implemented previous value of the MV. For generality, we allowed the limits $u_{low}(\ell), u_{high}(\ell)$ to vary over the horizon. These constraints are expressed in matrix form for all projections as

$$\begin{bmatrix} -I_L \\ I_L \end{bmatrix} \Delta\mathcal{U}(k) \geq \begin{bmatrix} u(k-1) - u_{high}(0) \\ \vdots \\ u(k-1) - u_{high}(m-1) \\ u_{low}(0) - u(k-1) \\ \vdots \\ u_{low}(m-1) - u(k-1) \end{bmatrix}, \tag{28.37}$$

where

$$I_L = \begin{bmatrix} I & 0 & \cdots & 0 \\ I & I & \cdots & 0 \\ \vdots & \vdots & \ddots & \vdots \\ I & I & \cdots & I \end{bmatrix}. \tag{28.38}$$

28.2.5.2 MV Rate Constraints

Often MPC is used in a supervisory mode where there are limitations on the rate at which lower-level controller setpoints are moved. These are enforced by adding constraints on the MV move sizes:

$$\begin{bmatrix} -I \\ I \end{bmatrix} \Delta\mathcal{U}(k) \geq \begin{bmatrix} -\Delta u_{max}(0) \\ \vdots \\ -\Delta u_{max}(m-1) \\ -\Delta u_{max}(0) \\ \vdots \\ -\Delta u_{max}(m-1) \end{bmatrix}, \tag{28.39}$$

where $\Delta u_{max}(\ell) > 0$ is the possibly time-varying bound on the magnitude of the moves.

28.2.5.3 Output Variable Constraints

The algorithm can make use of the output predictions (Equation 28.22 or 28.31) to anticipate future constraint violations:

$$\mathcal{Y}_{low} \leq \mathcal{Y}(k+1|k) \leq \mathcal{Y}_{high}. \tag{28.40}$$

Substituting from Equation 28.22, we obtain constraints on $\Delta\mathcal{U}(k)$:

$$\begin{bmatrix} -\mathcal{S}^u \\ \mathcal{S}^u \end{bmatrix} \Delta\mathcal{U}(k) \geq \begin{bmatrix} \mathcal{M}\tilde{Y}(k) + S^d \Delta d(k) + \mathcal{I}_p(y_m(k) - \tilde{y}_0(k)) - \mathcal{Y}_{high} \\ -(\mathcal{M}\tilde{Y}(k) + S^d \Delta d(k) + \mathcal{I}_p(y_m(k) - \tilde{y}_0(k))) + \mathcal{Y}_{low} \end{bmatrix}. \tag{28.41}$$

where

$$\mathcal{Y}_{low} = \begin{bmatrix} y_{low}(1) \\ y_{low}(2) \\ \vdots \\ y_{low}(p) \end{bmatrix}; \quad \mathcal{Y}_{high} = \begin{bmatrix} y_{high}(1) \\ y_{high}(2) \\ \vdots \\ y_{high}(p) \end{bmatrix}$$

are vectors of output constraint trajectories $y_{low}(\ell), y_{high}(\ell)$ over the horizon length p.

28.2.5.4 Combined Constraints

The MV constraints (Equation 28.37), MV rate constraints (Equation 28.39) and output variable constraints (Equation 28.41) can be combined into one convenient expression

$$\mathcal{C}^u \Delta\mathcal{U}(k) \geq \mathcal{C}(k+1|k), \tag{28.42}$$

where $\mathcal{C}^u$ combines all the matrices on the left-hand side of the inequalities as follows:

$$\mathcal{C}^u = \begin{bmatrix} -I_L \\ I_L \\ -I \\ I \\ -\mathcal{S}^u \\ \mathcal{S}^u \end{bmatrix}. \tag{28.43}$$

The vector $\mathcal{C}(k+1|k)$ on the right-hand side collects all the "error" vectors on the constraint equations as follows:

$$\mathcal{C}(k+1|k) = \begin{bmatrix} u(k-1) - u_{high}(0) \\ \vdots \\ u(k-1) - u_{high}(m-1) \\ u_{low}(0) - u(k-1) \\ \vdots \\ u_{low}(m-1) - u(k-1) \\ -\Delta u_{max}(0) \\ \vdots \\ -\Delta u_{max}(m-1) \\ -\Delta u_{max}(0) \\ \vdots \\ -\Delta u_{max}(m-1) \\ \mathcal{M}\tilde{Y}(k) + S^d \Delta d(k) + \mathcal{I}_p(y_m(k) - \tilde{y}_0(k)) - \mathcal{Y}_{high} \\ -(\mathcal{M}\tilde{Y}(k) + S^d \Delta d(k) + \mathcal{I}_p(y_m(k) - \tilde{y}_0(k))) + \mathcal{Y}_{low} \end{bmatrix}. \tag{28.44}$$

28.2.6 Formulation of Control Problem as a Quadratic Program

We make use of the prediction equation 28.22 to rewrite the objective

$$\min_{\Delta\mathcal{U}(k)} \sum_{\ell=1}^{p} \|\Gamma_\ell^y [y(k+\ell|k) - r(k+\ell)]\|^2 + \sum_{\ell=1}^{m} \|\Gamma_\ell^u [\Delta u(k+\ell-1)]\|^2 \tag{28.45}$$

and add the constraints (Equation 28.42) to obtain the optimization problem

$$\begin{aligned} &\min_{\Delta\mathcal{U}(k)} \left\{ \|\Gamma^y [\mathcal{Y}(k+1|k) - \mathcal{R}(k+1)]\|^2 + \|\Gamma^u \Delta\mathcal{U}(k)\|^2 \right\} \\ &\text{s.t.}\ \ \mathcal{Y}(k+1|k) = \mathcal{M}\tilde{Y}(k) + S^d \Delta d(k) + \mathcal{I}_p(y_m(k) - \tilde{y}_0(k)) + \mathcal{S}^u \Delta\mathcal{U}(k) \mathcal{C}^u \Delta\mathcal{U}(k) \geq \mathcal{C}(k+1|k), \end{aligned} \tag{28.46}$$

where

$$\Gamma^u = \text{diag}\left\{\Gamma_1^u, \ldots, \Gamma_m^u\right\} \tag{28.47}$$

and

$$\Gamma^y = \text{diag}\left\{\Gamma^y_1, \cdots, \Gamma^y_p\right\} \tag{28.48}$$

are the weight matrices in block diagonal form, and

$$\mathcal{R}(k+1) = \begin{bmatrix} r(k+1) \\ r(k+2) \\ \vdots \\ r(k+p) \end{bmatrix} \tag{28.49}$$

is the vector of reference trajectories.

We can substitute the prediction equation into the objective function to obtain

$$\begin{aligned}
&\|\Gamma^y\left[\mathcal{Y}(k+1|k) - \mathcal{R}(k+1)\right]\|^2 + \|\Gamma^u \Delta\mathcal{U}(k)\|^2 && (28.50)\\
&= \|\Gamma^y\left[\mathcal{S}^u \Delta\mathcal{U}(k) - E_p(k+1|k)\right]\|^2 + \|\Gamma^u \Delta\mathcal{U}(k)\|^2 && (28.51)\\
&= \Delta\mathcal{U}^T(k)(\mathcal{S}^{u^T}\Gamma^{y^T}\Gamma^y\mathcal{S}^u + \Gamma^{u^T}\Gamma^u)\Delta\mathcal{U}(k) && (28.52)\\
&\quad - 2E_p(k+1|k)^T\Gamma^{y^T}\Gamma^y\mathcal{S}^u\Delta\mathcal{U}(k) + E_p^T(k+1|k)\Gamma^{y^T}\Gamma^y E_p(k+1|k).
\end{aligned}$$

Here we have defined

$$\begin{aligned}
E_p(k+1|k) &= \begin{bmatrix} e(k+1|k) \\ e(k+2|k) \\ \vdots \\ e(k+p|k) \end{bmatrix} && (28.53)\\
&\triangleq \mathcal{R}(k+1) - \left[\mathcal{M}\tilde{Y}(k) + S^d\Delta d(k) + \mathcal{I}_p(y_m(k) - \tilde{y}_0(k))\right],
\end{aligned}$$

which is the measurement corrected vector of future output deviations from the reference trajectory (i.e., errors), assuming that all future control moves are zero. Note that this vector includes the effect of the measurable disturbances ($\mathcal{S}^d\Delta d(k)$) on the prediction. For the state-space model of Equation 28.31, E_p changes to

$$E_p \triangleq \mathcal{R}(k+1) - \left(S^z\hat{z}(k|k) + S^d\Delta d(k)\right). \tag{28.54}$$

The optimization problem with a quadratic objective and linear inequalities, which we have defined, is a quadratic program (QP). By converting to the standard QP formulation the DMC problem becomes*

$$\begin{aligned}
\min_{\Delta\mathcal{U}(k)} \quad & \frac{1}{2}\Delta\mathcal{U}(k)^T\mathcal{H}^u\Delta\mathcal{U}(k) - \mathcal{G}(k+1|k)^T\Delta\mathcal{U}(k)\\
\text{s.t.} \quad & \mathcal{C}^u\Delta\mathcal{U}(k) \geq \mathcal{C}(k+1|k),
\end{aligned} \tag{28.55}$$

where the Hessian of the QP is

$$\mathcal{H}^u = \mathcal{S}^{u^T}\Gamma^{y^T}\Gamma^y\mathcal{S}^u + \Gamma^{u^T}\Gamma^u \tag{28.56}$$

* The term $E_p^T(k+1|k)E_p(k+1|k)$ is independent of $\Delta\mathcal{U}(k)$ and can be removed from the objective function.

and the gradient vector is

$$\mathcal{G}(k+1|k) = \mathcal{S}^{u^T} \Gamma^{y^T} \Gamma^y E_p(k+1|k). \tag{28.57}$$

28.3 Implementation Issues

As explained in the introduction of this chapter the implementation of DMC is done in a *moving horizon* fashion. This implies that the QP derived above will be solved at each controller execution time. Because of this feature, the algorithm can be configured online as required to take care of unexpected situations. For example, in case an actuator is lost during the implementation, the high and low constraint limits on that particular MV can be set to be equal. Then the MPC problem with the remaining MVs is solved. Similarly, the weight parameters in the objective function can also be adjusted online, giving the user the ability to tune the control law. In this section we discuss the different implementation issues associated with the DMC.

28.3.1 Moving Horizon Algorithm

The constrained MPC algorithm is implemented online as follows.

1. *Preparation.* Do not vary the MVs for at least n time intervals ($\Delta u(-1) = \Delta u(-2) = \cdots = \Delta u(-n) = 0$) and assume the measured disturbances are zero ($\Delta d(-1) = \Delta d(-2) = \cdots = \Delta d(-n) = 0$) during that time. Then the system will be at rest at $k = 0$.
2. *Initialization* ($k = 0$). Measure the output $y_m(0)$ and initialize the model prediction vector as*

$$\tilde{Y}(k) = \Bigg[\underbrace{y_m(0)^T, y_m(0)^T, \ldots, y_m(0)^T}_{n}\Bigg]^T. \tag{28.58}$$

3. *State update:* Set $k = k+1$. Then, update the state according to

$$\tilde{Y}(k) = M \cdot \tilde{Y}(k-1) + S^u \Delta u(k-1) + S^d \Delta d(k-1), \tag{28.59}$$

   where the first element of $\tilde{Y}(k)$, $\tilde{y}(k|k)$, is the model prediction of the output $y_m(k)$ at time k.
4. *Obtain measurements:* Obtain measurements ($y_m(k), \Delta d(k)$).
5. Compute the reference trajectory error vector

$$E_p(k+1|k) = \mathcal{R}(k+1) - \mathcal{M}\tilde{Y}(k) + S^d \Delta d(k) + \mathcal{I}_p(y_m(k) - \tilde{y}_0(k)). \tag{28.60}$$

6. Compute the QP gradient vector

$$\mathcal{G}(k+1|k) = \mathcal{S}^{u^T} (\Gamma^y)^T \Gamma^y E_p(k+1|k). \tag{28.61}$$

* If Equation 28.58 is used for intialization and changes in the past n inputs did actually occur, then the initial operation of the algorithm will not be smooth. The transfer from *manual* to *automatic* will introduce a disturbance; it will not be "bumpless."

7. Compute the constraint equation's right-hand side vector

$$\mathcal{C}(k+1|k) = \begin{bmatrix} u(k-1) - u_{high}(0) \\ \vdots \\ u(k-1) - u_{high}(m-1) \\ u_{low}(0) - u(k-1) \\ \vdots \\ u_{low}(m-1) - u(k-1) \\ -\Delta u_{max}(0) \\ \vdots \\ -\Delta u_{max}(m-1) \\ -\Delta u_{max}(0) \\ \vdots \\ -\Delta u_{max}(m-1) \\ -E_p(k+1|k) + \mathcal{R}(k+1) - \mathcal{Y}_{high} \\ E_p(k+1|k) - \mathcal{R}(k+1) + \mathcal{Y}_{low} \end{bmatrix}. \tag{28.62}$$

8. Solve the QP

$$\begin{array}{ll} \min_{\Delta\mathcal{U}(k)} & \frac{1}{2}\Delta\mathcal{U}(k)^T\mathcal{H}^u\Delta\mathcal{U}(k) - \mathcal{G}(k+1|k)^T\Delta\mathcal{U}(k) \\ \text{s.t.} & \mathcal{C}^u\Delta\mathcal{U}(k) \geq \mathcal{C}(k+1|k) \end{array} \tag{28.63}$$

   and implement $\Delta u(k|k)$ as $\Delta u(k)$ on the plant.
9. Go to 3.

Note that the sequence of moves produced by the moving horizon implementation of the QP will be different from the sequence of moves $\Delta\mathcal{U}(k)$.

28.3.2 Solving the QP

In a moving horizon framework the QP in Equation 28.63 is solved at each controller execution time after a new prediction is obtained. The only time-varying elements in this problem are the vectors $E_p(k+1|k)$ (or equivalently $\mathcal{G}(k+1|k)$) and $\mathcal{C}(k+1|k)$. That is, the Hessian $\mathcal{H}^u$ of the QP remains constant for all executions. In that case, as explained above, a parametric QP algorithm which employs the preinverted Hessian in its computations, is preferable in order to reduce online computation effort. Of course, in case either Γ^y or Γ^u (or the step response coefficients) need to be updated, or the model's step response coefficients have changed, the Hessian must be recomputed and inverted in background mode in order not to increase the online computational requirements.

QP is a *convex* program and therefore, is fundamentally tractable, meaning a global optimal solution within a specified tolerance can be assured. Although not as extensively as linear programs (LPs), QPs have been well studied and reliable algorithms have been developed and coded. General-purpose QP solvers like *QPSOL* are readily available but use of tailored algorithms that take advantage of specific problem structures can offer significant computational savings.

The conventional approach for solving QPs is the so-called *active set* method. In this method, one initiates the search by assuming a set of active constraints. For an assumed active set, one can easily solve the resulting least-squares problem (where the active constraints are treated as equality constraints) through the use of Lagrange multiplier. In general, in the active set one starts out with that it will not be the correct one. Through the use of the Karush–Kuhn–Tucker (KKT) condition,* one can modify the active set iteratively until the correction is found. Most active set algorithms are *feasible path* algorithms, in which the constraints must be met at all times. Hence, the number of constraints can have a significant effect on the computational time.

More recently, a promising new approach called the *interior point (IP)* method has been getting a lot of attention. The idea of the IP method is to "trap" the solution within the feasible region by including a so-called "barrier" function in the objective function. With the modified objective function, the Newton iteration is applied to find the solution. Though originally developed for LPs, the IP method can be readily generalized to QPs and other more general constrained optimization problems. Even though not formally proven, it has been observed empirically that the Newton iteration converges within 5–50 steps. Significant work has been carried out in using this solution approach for solving QPs that arise in MPC, but details are out of the scope of this chapter.

Computational properties of QPs vary with problems. As the number of constraints increases, more iterations are generally required to find the QP solution, and therefore the solution time increases. This may have an impact on the minimum control execution time possible. Also, note that the dimension of the QP (i.e., the number of degrees of freedom $m \cdot n_u$) influences the execution time proportionately.

Storage requirements are also affected directly by the number of degrees of freedom and the number of projections $n \cdot n_y$. For example, the Hessian size increases quadratically with the number of degrees of freedom. Also, because of the prediction algorithm, $\tilde{Y}(k)$ must be stored for use in the next controller execution (both $E_p(k+1|k)$ and $\mathcal{C}(k+1|k)$ can be computed from $\tilde{Y}(k)$).

28.3.3 Proper Constraint Formulation

Many engineering control objectives are stated in the form of constraints. Therefore, it is very tempting to translate them into linear inequalities and to include them in the QP control problem formulation. In this section we want to demonstrate that constraints make it very difficult to predict the behavior of the control algorithm under real operating conditions. Therefore, they should be used only when necessary and then only with great caution.

First of all constraints tend to greatly increase the time needed to solve the QP. Thus, we should introduce them sparingly. For example, if we wish an output constraint to be satisfied over the whole future horizon, we may want to state it as a linear inequality only at *selected* future sampling times rather than at *all* future sampling times. Unless we are dealing with a highly oscillatory system, a few output constraints at the beginning and one at the end of the horizon should keep the output more or less inside the constraints throughout the horizon. Note that even when constraint violations occur in the prediction, this does not imply constraint violations in the actual implementation because of the moving horizon policy. The future constraints serve only to prevent the present control move from being short-sighted.

Output constraints can also lead to an "infeasibility." A QP is *infeasible* if there does not exist any value of the vector of independent variables (the future MV move $\Delta\mathcal{U}(k)$), which satisfies all the constraints — regardless of the value of the objective function. Physically, this situation can arise when there are output constraints to be met but the MVs are not sufficiently effective—either because they are constrained or because there is deadtime in the system that delays their effect. Needless to say, provisions must be built into the online algorithm such that an infeasibility never occurs.

Mathematically an infeasibility can only occur when the right-hand side of the output constraint equations is positive. This implies that a nonzero move *must* be made in order to satisfy the constraint equations. Otherwise, infeasibility is not an issue since $\Delta\mathcal{U}(k) = 0$ is feasible.

A simple example of infeasibility arises in the case of deadtimes in the response. For illustration, assume a single-input single-output (SISO) system with θ units of deadtime. The output constraint equations for

* The KKT condition is a necessary condition for the solution to a general constrained optimization problem. For QP, it is a necessary and sufficient condition.

this system will look like

$$\begin{bmatrix} 0 & 0 & \cdots & 0 \\ \vdots & & & \vdots \\ 0 & 0 & \cdots & 0 \\ -S^u_{\theta+1} & 0 & \cdots & 0 \\ -S^u_{\theta+2} & -S^u_{\theta+1} & \cdots & 0 \\ \vdots & & & \vdots \end{bmatrix} \Delta\mathcal{U}(k) \geq \begin{bmatrix} c(k+1|k) \\ \vdots \\ c(k+\theta|k) \\ c(k+\theta+1|k) \\ c(k+\theta+2|k) \\ \vdots \end{bmatrix}.$$

Positive elements $c(k+1|k),\ldots,c(k+\theta|k)$ indicate that a violation is projected unless the MVs are changed ($\Delta\mathcal{U}(k) \neq 0$). Since the corresponding coefficients in the left-hand side matrix are zero, the inequalities cannot be satisfied and the QP is infeasible. Of course, this problem can be removed by simply not including these initial θ inequalities in the QP.

Because inequalities are dealt with *exactly* by the QP, the corrective action against a projected violation is equivalent to that generated by a very tightly tuned controller. As a result, the moves produced by the QP to correct for violations may be undesirably severe (even when feasible). Both infeasibilities and severe moves can be dealt with in various ways.

One way is to include a *constraint window* on the output constraints similar to what we suggested above for computational savings. For each output a time $k + H_c$ in the future is chosen at which constraint violations will start to be checked (Figure 28.3).

For the above illustration, this time should be picked to be at least equal to $\theta + 1$. This allows the algorithm to check for violations after the effects of deadtimes and inverse responses have passed. For each situation there is a minimal value of H_c necessary for feasibility. If this minimal value is chosen large, constraint violations may occur over a significant period of time. In many cases, if a larger value of H_c is chosen, smaller constraint violations may occur over a longer time interval. Thus, there is a trade-off between magnitude and duration of constraint violation.

In general, it is difficult to select a value of H_c for each constrained output such that the proper compromise is achieved. Furthermore, in multivariable cases, constraints may need to be relaxed according to the priorities of the constrained variables. The selection of constraint windows is greatly complicated by the fact that appropriate amount and location for relaxation are usually time dependent due to varying disturbances and occurrences of actuator and sensor failures. Therefore, it is usually preferred to "soften" the constraint by adding a slack variable ϵ and penalizing this violation through an additional term in the objective function.

$$\min_{\epsilon,\Delta\mathcal{U}(k)} [\text{Usual Objective}] + \lambda\epsilon^2$$

$$y_{min} - \epsilon \leq y(k+\ell|k) \leq y_{max} + \epsilon$$
$$\text{plus other constraints.}$$

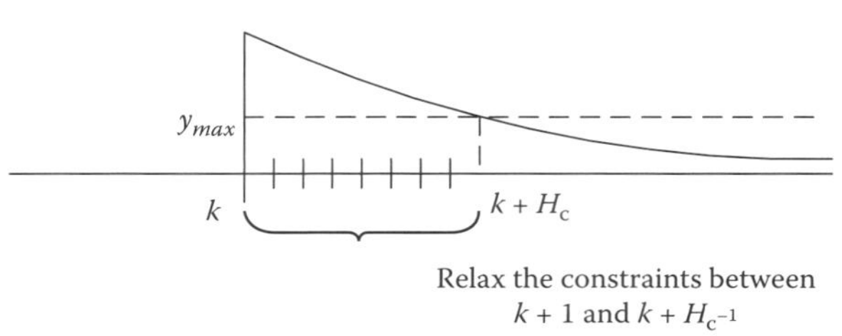

FIGURE 28.3 Relaxing the constraints.

The optimization seeks a compromise between minimizing the original performance objective and minimizing the constraint violations expressed by ϵ^2. The parameter λ determines the relative importance of the two terms. The degree of constraint violation can be fine-tuned arbitrarily by introducing a separate slack variable ϵ for each output and time step, and associating with it a separate penalty parameter λ.

Finally, we must realize that while unconstrained MPC is a form of *linear* feedback control, constrained MPC is a *nonlinear* control algorithm. Thus, its behavior for small deviations can be drastically different from that for large deviations. This may be surprising and undesirable and is usually very difficult to analyze *a priori*.

28.3.4 Choice of Horizon Length

On the one hand, the prediction horizon p and the control horizon m should be kept short to reduce the computational effort; on the other hand, they should be made long to prevent short-sighted control policies. Making m short is generally conservative because we are imposing constraints (forcing the control to be constant after m steps) that do not exist in the actual implementation because of the moving horizon policy. Therefore, a small m will tend to give rise to a cautious control action.

Choosing p small is "short-sighted" and will generally lead to an aggressive control action. If constraint violations are checked only over a small control horizon p, this policy may lead the system into a "dead alley" from which it can escape only with difficulty, that is, only with large constraint violations and/or large MV moves.

When p and m are infinity and when there are no disturbance changes and unknown inputs, the sequence of control moves determined at time k is the same sequence that is realized through the moving horizon policy. In this sense our control actions are truly optimal. When the horizon lengths are shortened, then the sequence of moves determined by the optimizer and the sequence of moves actually implemented on the system will become increasingly different. Thus the short-time objective, which is optimized, will have less and less to do with the actual value of the objective realized when the moving horizon control is implemented. This may be undesirable.

In general, we should try to choose a small m to keep the computational effort manageable, but large enough to give us a sufficient number of degrees of freedom. We should choose p as large as possible, possibly ∞, to completely capture the consequences of the control actions. This is possible in several ways. Because an FIR system will settle after $m+n$ steps, choosing a horizon $p = m+n$ is a sensible choice used in many commercial systems (Figure 28.4).

Instead or in addition we can impose a large output penalty at the end of the prediction horizon forcing the system effectively to settle to zero at the end of the horizon. Then, with $p = m+n$, the error after $m+n$ is essentially zero and there is little difference between the finite and the infinite horizon objective.

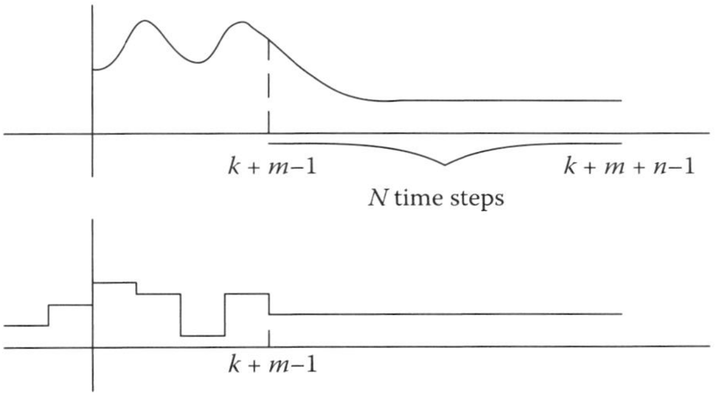

FIGURE 28.4 Choosing the horizon.

28.3.5 Input Blocking

As said, use of a large control horizon is generally preferred from the viewpoint of performance, but available computational resource may limit its size. One way to relax this limit is through a procedure called *blocking*, which allows to the user to "block out" the input moves at selected locations from the calculation by setting them to zero *a priori*. The result is a reduction in the number of input moves that need to be computed through the optimization, hopefully without a significant sacrifice in the solution quality. Obviously, judicious selection of blocking locations is critical for achieving the intended effect. The selection is done mostly on an *ad hoc* basis, although there are some qualitative rules like blocking less of the immediate moves and more of the distant ones.

At a more general level, blocking can be expressed as follows:

$$\Delta \mathcal{U} = \mathbf{B} \Delta \mathcal{U}^b, \tag{28.64}$$

where $\Delta \mathcal{U}^b$ represents the reduced input parameters to be calculated through the optimization. $\mathbf{B}$ is the blocking matrix that needs to be designed for a good performance. Typically, the rows of $\mathbf{B}$ corresponding to the blocked moves would contain all zeros. In general, columns of $\mathbf{B}$ can be designed to represent different basis in the input space. Note that dimension of $\mathcal{U}^b$, which is less than that of $\mathcal{U}$, must also be determined in the design.

28.3.6 Filtering of the Feedback Signal

In practice, feedback measurements can contain significant noise and other fast-varying disturbances. Since in DMC the effect of unmeasured disturbances is projected as a constant bias in the prediction, the high-frequency contents of a feedback signal must be filtered out in order to obtain a meaningful long-term prediction. For this, one can pass the feedback signal through a low-pass filter of some sort, perhaps a first- or second-order filter, before putting it into the prediction equation. Use of state estimation, discussed in [3], for example, allows one to model the statistical characteristics of disturbances and noise and perform the filtering in an optimal manner.

28.4 Features Found in Other MPC Algorithms

What we just covered is the basic form of a multivariable control algorithm called DMC, which was one of the first MPC algorithms applied to industrial processes with success. The original DMC algorithm did not use QP to handle constraints; instead, it added an extra output to the prediction to drive the input back to the feasible region whenever a predicted future input came close to a constraint. This was somewhat *ad hoc* and it was not until the 1980s that engineers at Shell Oil proposed the use of QP to handle input and output constraints explicitly and rigorously. They called this modified version QDMC. Currently, this basic form of DMC is still used in a commercial package called *DMC-PLUS*, which is marketed by Aspen Technology.

Besides DMC and QDMC, there are several other MPC algorithms that have seen, and are still seeing, extensive use in practice. These include model predictive heuristic control (MPHC), which led to popular commercial algorithms such as IDCOM and SMC-IDCOM marketed by Setpoint (now Aspen Technology) and Hierarchical Constraint Control (HIECON) and Predictive Functional Control (PFC) marketed by Adersa; Predictive Control Technology (PCT), which was marketed by Profimatics (now Honeywell); and more recent Robust Model Predictive Control Technology (RMPCT), which is currently being marketed by Honeywell. These algorithms share same fundamentals but differ in details of implementation. Rather than elaborating on the details of each algorithm, we will touch upon some popular features not seen in the basic DMC/QDMC method.

28.4.1 Reference Trajectories

In DMC, output deviation from the desired setpoint is penalized in the optimization. Other algorithms such as IDCOM, HIECON, and PFC let the user specify not only *where* the output should go but also *how*. For this, a *reference trajectory* is introduced for each CV, which is typically defined as a first-order path from the current output value to the desired setpoint. The time constant of the path can be adjusted according to the speed of the desired closed-loop response. This is displayed in Figure 28.5.

Reference trajectories provide an intuitive way to control the aggressiveness of control, which is adjusted through the weighting matrix for the input move penalty term in DMC. One could argue that the controller's aggressiveness is more conveniently tuned by specifying the speed of output response rather than through input weight parameters, whose effects on the speed of response is highly system dependent.

28.4.2 Coincidence Points

Some commercial algorithms such as IDCOM and PFC allowed the option of penalizing the output error only at a few chosen points in the prediction horizon called *coincidence points*. This is motivated primarily by reduction in computation it brings. When the number of input moves has to be kept small (in order to keep the computational burden low), use of a large prediction horizon, which is sometimes necessary due to large inverse responses, and long dynamics, results in a sluggish control behavior. This problem can be obviated by penalizing output deviation only at a few carefully selected points. At the extreme, one could ask the output to match the reference trajectory value at a single time point, which can be achieved with a single control move. Such formulation was used, for example, in IDCOM-M, an offspring of the original IDCOM algorithm, marketed by setpoint.

Clearly, the choice of coincidence points is critical for performance, especially when the number of points used is small. Alhough some guidelines exist on choosing these points, there is no systematic method for the selection. Because the response time of different outputs can vary significantly, coincidence points are usually defined separately for each output.

28.4.3 The Funnel Approach

The RMPCT algorithm differs from other MPC algorithms in that it attempts to keep each controlled output within a user-specified zone called *funnel*, rather than to keep it on a specific reference trajectory. The typical shape of a funnel is displayed in Figure 28.5. The user sets the maximum and minimum limits and also the slope of the funnel through a parameter called "performance ratio," which is the desired time to return to the limit zone divided by the open-loop response time. The gap between the maximum and minimum can be closed for exact setpoint control, or left open for range control.

The algorithm solves the following QP at each time:

$$\min_{y^r,u} \sum_{i=1}^{p} \|y(k+i|k) - y^r(k+i|k)\|_Q^2 + \sum_{j=0}^{m-1} \|\Delta u(k+j|k)\|_R^2 \tag{28.65}$$

or

$$\min_{y^r,u} \sum_{i=1}^{p} \|y(k+i|k) - y^r(k+i|k)\|_Q^2 + \sum_{j=0}^{m-1} \|u(k+j|k) - u^r\|_R^2 \tag{28.66}$$

subject to usual constraints plus the funnel constraint

$$y^f_{min}(k+i|k) \le y^r(k+i|k) \le y^f_{max}(k+i|k), \quad 1 \le i \le p, \tag{28.67}$$

where $y^f_{min}(k+i|k)$ and $y^f_{min}(k+i|k)$ represent the upper and lower limit values of the funnel for $k+i$ in the prediction horizon as specified at time k. u^r is the desired settling value for the input. Note that the

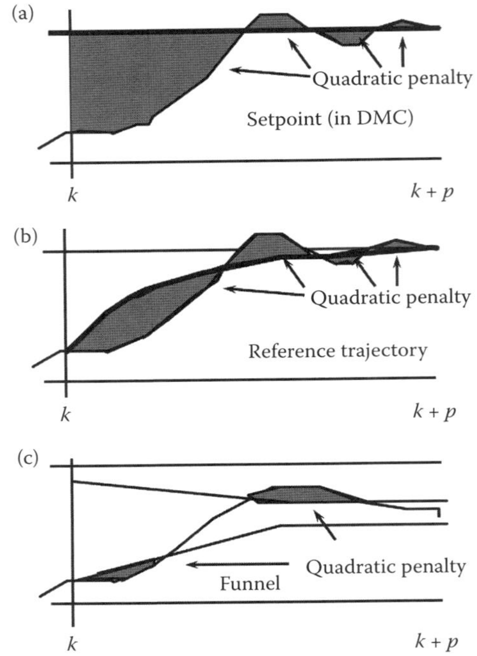

FIGURE 28.5 Output penalties used in various formulations.

reference trajectory y^r is a free parameter, which is optimized to lie within the funnel. Typically, $Q \gg R$ in order to keep the outputs within the funnel as much as possible. Then one can think of the above as a multiobjective optimization, in which the primary objective is to minimize the funnel constraint violation by the output and the secondary objective is to minimize the size of input movement (or input deviation from the desired settling value in the case of Equation 28.66). In this case, as long as there exists an input trajectory that keeps the output within the funnel, the first penalty term will be made exactly zero. Typically, there will be an infinite number of solutions that achieve this, leading to a "degenerate" QP. The algorithm thus finds the minimum norm solution, which corresponds to the least amount of input adjustment—hence the name "Robust" MPCT . However, if there is no input that can keep the output within the funnel, the first term will be the primary factor that determines the input.

The use of funnel is motivated by the fact that, in multivariable systems, the shape of desirable trajectories for outputs is not always clear due to system interaction. Thus, it is argued that an attractive formulation is to let the user specify an acceptable dynamic zone for each output as a funnel and then find the minimum size input moves (or inputs with minimum deviation from their desired values) that keep the outputs within the zone—or, if not possible, minimize the extent of violation.

28.4.4 Use of Other Norms

In defining the objective function, use of norms other than 2-norm is certainly possible. For example, the possibility of using 1-norm (sum of absolute values) has been explored to a great extent. Use of infinity norm has also been investigated with the aim of minimizing worst-case deviation over time. In both cases, one gets an LP, for which a plethora of theories and software exist due to its significance in economics. However, one difficulty with these formulations is in tuning. This is because the solution of an LP lies at the intersection of binding constraints and it can switch abruptly from one vertex to another as one varies tuning parameters (such as the input weight parameters). The solution behavior of a QP is much smoother and therefore, it is a preferred formulation for control.

28.4.5 Input Parameterization

In some algorithms such as PFC, the input trajectory can be parameterized using continuous basis functions like polynomials. This can be useful if the objective is to follow *smooth* setpoint trajectories precisely, such as in mechanical servo applications, and the sampling time cannot be made sufficiently small to allow this with piecewise constant inputs.

In other commercial algorithms such as HIECON and IDCOM-M, only a single control move is calculated,which would correspond to $m = 1$ in DMC. With this setting, the calculation is greatly simplified. On the other hand, use of $m = 1$ in DMC would limit the closed-loop performance in general. These algorithms get around this problem by using a single coincidence point, at which the output is asked to match the reference value exactly.

28.4.6 Model Conditioning

In multivariable plants, two or more outputs can behave very similarly in response to all the inputs. This phenomenon is referred to as "ill-conditioning" and is reflected by a gain matrix that is nearly singular. An implication is that it can be very difficult to control these outputs independently with the inputs, as it will require an excessive amount of input movement in order to move the outputs in certain directions. Using an ill-conditioned process model for control calculation is not recommended as it can lead to numerical problems (e.g., inversion of a nearly singular matrix) and also excessive input movements and/or even an instability.

Even though one would check the conditioning of the model at the design stage, because control structure can change due to constraints and failures of sensors and actuators, one must make sure at each execution time that an ill-conditioned process model is not directly inverted in the input calculation.

In DMC, direct inversion of an ill-conditioned process model can be circumvented by including a substantive input move penalty term, which effectively increases the magnitudes of the diagonal elements of the dynamic matrix that is inverted during the least-squares calculation.

In other algorithms that do not include an input move penalty in the objective function, ill-conditioning must be checked at each execution time. In RMPCT, this is done through a method called *Singular Value Threshholding*, where a procedure called "singular value decomposition" is performed on the gain matrix to determine those CV directions for which the gain is too low for any effective control at all. Those directions with singular values lower than a threshhold value are given up for control and only the remaining "high-gain" directions are controlled. SMC-IDCOM addresses this based on the user-defined ranking of CVs. Here, whenever an ill-conditioning is detected, CVs are dropped from the control calculation in the order of their ranks, starting from the one with the least assigned priority, until the condition number improves to an acceptable level. When two CVs are seen to behave very similarly, the user can rank the less important CV with a very low priority. Even though the control on the dropped CV is given up, it is hoped that it would be controlled indirectly since it behaves similarly to the other high-ranked CV.

28.4.7 Prioritization of CVs and MVs

In most practical control problems, it is not possible to satisfy all constraints and also drive all outputs and inputs to their desired resting values. Hence, priorities need to be assigned to express their relative importance. In DMC, these priorities are determined through weight parameters, which enter into the various quadratic penalty terms in the objective function. For large, complex problems, determining proper weights that lead to an intended behavior can be a daunting task. Even if a set of weights consistent with the control specification is found, the weights can differ vastly in magnitude from one to another, causing a numerical conditioning problem.

Algorithms such as HIECON and SMC-IDCOM attempt to address this difficulty by letting the user *rank* various objectives in the order of their importance. For example, constraint satisfaction may be the most critical aspect, which must be taken care of before satisfying other objectives. Also driving the CVs to their desired setpoints may be more important than driving the MVs to their most economic values. In these algorithms, an optimization would be solved with the most important objective first and then remaining degrees of freedom would be used to address the other objectives in the order of priority. These algorithms also allow the user to rank each CV and MV according to its priority. Hence, for constraint softening, one may specify the order in which constraints for various CVs must be relaxed. Also, in setpoint tracking, one can prioritize the CVs so that CVs with higher ranks are driven to their setpoints before those with lower ranks are considered.

28.4.8 Bi-Level Optimization

The MPC calculation can be split into two parts for an added flexibility. First, a local *steady-state* optimization can be performed to obtain target values for each input and output. This can be followed by a *dynamic* optimization to determine the most desirable dynamic trajectory to these target values. Even though the local steady-state optimization can be based on an economic index, it does not replace the more comprehensive nonlinear optimization that often runs above the MPC layer—at a much slower rate—in order to provide an optimal range of inputs and outputs for the plant condition experienced during a particular optimization cycle. The local optimization performed in MPC is based on a linear steady-state model, which may be obtained by linearizing a nonlinear model or simply the steady-state version of the step response model used in the dynamic optimization.

The reasons for running the local optimization may vary. For example, one may want to perform an economic optimization at a higher frequency to account for local disturbances. Even if there is no economic objective in the given control problem, the steady-state optimization can be helpful to determine best feasible target values for CVs and the corresponding MV settling values.

The two-stage optimization can be formulated as below:

- Step 1: *Steady-State Optimization* The general form of a steady-state prediction equation is

$$y(\infty|k) = K_s \underbrace{(u(\infty|k) - u(k-1))}_{\Delta u_s(k)} + b(k), \tag{28.68}$$

where $y(\infty|k)$ and $u(\infty|k)$ are the steady-state values of the output and input projected at time k. With only m input moves considered,

$$\Delta u_s(k) = \Delta u(k) + \Delta u(k+1) + \cdots + \Delta u(k+m-1). \tag{28.69}$$

Note that, for the step response model,

$$y(\infty|k) = y(k+m+n-1|k) \tag{28.70}$$

and $K_s = S_n$. Also,

$$b(k) = \tilde{y}_{n-1}(k) + S_n^d \Delta d(k) + (y_m(k) - \tilde{y}_0(k)). \tag{28.71}$$

This steady-state prediction model can be used to optimize a given economic objective function subject to various input and output constraints:

$$\min_{\Delta u_s(k)} \ell(u(\infty|k), y(\infty|k)). \tag{28.72}$$

Since an economic objective function is typically linear and the prediction equation is also linear, an LP results. Alternatively, one can also solve

$$\min_{\Delta u_s(k)} \|\Delta u_s(k)\|, \tag{28.73}$$

$$\min_{\Delta u_s(k)} \|r - y(\infty|k))\|_Q. \tag{28.74}$$

In the first case, we would be looking for a minimum input change such that all the constraints are satisfied. In the second case, we would be seek a minimum deviation from the setpoint values that are achievable within the given constraints. The solution sets the target settling values for the inputs and outputs.

- Step 2: *Dynamic Optimization* The dynamic prediction equation is the same as before. A quadratic regulation objective of the following is minimized subject to the given constraints through QP:

$$\left[\sum_{i=1}^{m+n-2} (y(k+i|k) - y^*(\infty|k))^T Q(y(k+i|k) - y^*(\infty|k)) + \sum_{j=0}^{m-1} \Delta u^T(k+j|k) R \Delta u(k+j|k)\right], \tag{28.75}$$

where $y^*(\infty|k)$ is the solution from the steady-state optimization. An additional constraint may be added to match the settling values of the optimized input trajectories to those computed from the steady-state optimization:

$$\Delta u(k|k) + \Delta u(k+1|k) + \cdots + \Delta u(k+m-1|k) = \Delta u_s^*(k). \tag{28.76}$$

This also forces $y(k+m+n-1|k)$ to be at the optimal steady-state value $y^*(k+\infty|k)$.

Note that, this steady-state optimization may be performed as often as at every sample time, that is at the same execution rate as the dynamic optimization. However, it is critical to filter the noise and other high-frequency variations from the feedback signal. Otherwise, the solution from the steady-state solution can fluctuate wildly from sample time to sample time, especially in the case of an LP.

28.5 Future Needs

28.5.1 Better Identification

Conventionally, step response models used in DMC (or other industrial MPC algorithms) are identified through a series of step tests. In some cases, pseudorandom binary sequence (PRBS) tests instead of step tests are used and the step response coefficients are fitted through least squares. In almost all cases, input channels are perturbed *one at a time*, leading to SISO identification. While this practice is simple and easy to implement, it does not always yield a multivariable model of required accuracy. This is usually dealt with in practice by detuning or dropping a part of control space, not a desirable remedy from a performance standpoint.

Independent testing of input channels emphasizes the accuracy of individual transfer function elements, and can result in a poor fit of control-relevant multivariable system characteristics (e.g., the gain directionality) [5]. For highly interactive processes, test signals for different input channels need to be correlated in order to yield an accurate model for multivariable control. In general, this points to the need for a systematic method to design identification experiments with control requirements directly considered. In addition, a nonconservative way to quantify the model quality (e.g., uncertainty bounds) would greatly aid the much needed integration between identification and control.

In terms of identification algorithms, development and use of multi-input multi-output (MIMO) algorithms that are capable of capturing output correlations can be helpful. They will not only improve prediction in conventional control problems, but will also enable construction of a model useful for inferential control. Traditional polynomial-model-based identification algorithms are poorly suited to MIMO identification. Recently proposed subspace identification algorithms [8] may fill this need, but they need to be tested further and perhaps tailored to suit the process control problems.

28.5.2 Robust MPC

Model uncertainty is an important aspect of every process control problem, as most processes are very difficult and time consuming to model accurately. Hence, it is desirable to incorporate the model quality information (e.g., uncertainty bounds, parameter probability distribution) directly into the control computation, instead of achieving robustness margins indirectly through various tuning parameters. In terms of the robust MPC controller synthesis, the most popular approach has been the so-called *min–max predictive control* that aims at minimizing the worst-case error with respect to the model set. Other methods include a stochastic approach where the expectation of the error is minimized, given a certain probability distribution of model parameters [7]. None of these methods have been tested in real problems. See [2] for a review and references. In order for these methods to find practical applications, what is most needed is a method that reliably computes the uncertainty bounds or probability distribution from identification data.

28.5.3 Performance Monitoring, Diagnosis, and Adaptation

It has been reported that many MPCs perform well when commissioned, but their performances deteriorate over time and they eventually have to be taken offline. There are a wide variety of causes for this, including instrumentation problems, process nonlinearity, and parameter variations. In order to sustain the benefits of these controllers over a long period of time, a mechanism to detect and diagnose the cause of significant performance deterioration is needed. The results can be communicated to engineers and can also be used to adapt control parameters. Vigorous research has already begun in this area and some vendors (such as Honeywell) have expressed interests in including such features in the next generation of commercial software.

28.6 Conclusion

This chapter presented a brief review and outlook of MPC in the process industries. The original industrial algorithms, albeit simple and restrictive, are continuing to be used quite successfully in process control applications. Some generalizations and extensions introduced mostly by academics to enhance the original industrial technology have found their way into the next generation of commercial algorithms being marketed currently. This indicates a healthy synergism between academia and industry in this particular area. This chapter also provided some needs (in both research and practice) to make MPC a more appealing and broadly applicable technique in the future.

References

1. C. R. Cutler and B. L. Ramaker. Dynamic matrix control—A computer control algorithm. In *Proceedings of the Joint Automatic Control Conference*, San Francisco, CA, 1980.
2. J. H. Lee and B. Cooley. Recent advances in model predictive control and related areas. In *5th International Conference on Chemical Process Control*, AIChE Symposium Series, Vol. 93, pp. 201–216, New York, NY, 1997.
3. J. H. Lee, M. Morari, and C. E. Garcia. State space interpretation of model predictive control. *Automatica*, 30:707–717, 1994.

4. J. H. Lee and Z. H. Yu. Tuning of model predictive control for robust performance. *Computers & Chemical Engineering*, 18:15–37, 1994.
5. W. Li and J. H. Lee. Control relevant identification of ill-conditioned systems. *Computers & Chemical Engineering*, 20:1023–1042, 1996.
6. J. Richalet, A. Rault, Testud J. L., and J. Papon. Model predictive heuristic control: Applications to industrial processes. *Automatica*, 14(5):413–428, 1978.
7. E. Y. Tse and M. Athans. Adaptive stochastic control for a class of systems. *IEEE Transactions on Automatic Control*, 17:38–52, 1972.
8. P. Van Overschee and B. De Moor. N4SID: Subspace algorithms for the identification of combined deterministic-stochastic systems. *Automatica*, 30(1):75–94, 1994.

IV

Analysis and Design of Hybrid Systems

29

Computation of Reach Sets for Dynamical Systems

Alex A. Kurzhanskiy
University of California, Berkeley

Pravin Varaiya
University of California, Berkeley

29.1 Introduction

Traditional control theory is concerned with the design of linear feedback control with desirable asymptotic behavior, such as stability and small steady-state and tracking errors, while properties of transient behavior are expressed in terms of overshoot and speed of response. External disturbances can be handled by modeling these as random processes, leading to the Linear Quadratic Gaussian (LQG) problem formulation. This theory has some limitations.

Because the feedback law is specified to be linear, it is not possible for design methods to explicitly incorporate hard bounds on the control values, for example, the requirement that the applied force should not exceed a specified limit. Second, it is not possible to express finite time requirements, for example, the requirement that the system state reach a prespecified value at a prespecified time. Third, it is not possible to demand guaranteed performance in the face of disturbances, for example, the requirement that a certain target state be reached, no matter what the disturbance. Thus control problems with hard bounds on the control values, restrictions on the state trajectory over a finite time horizon, and guaranteed behavior despite the disturbances, are difficult to solve using frequency-based design methods.

In order to address these problems one needs to study system evolution in the time domain. The central concept that emerges in such studies is that of the *reach* set, which is the set of states that can be reached by using all possible controls. This chapter is devoted to the formulation and computation of the reach set of a linear system with disturbances. The concept of reachability was introduced in [28]; [29] shows the reach set can be computed by solving the forward Hamilton–Jacobi–Bellman–Isaacs (HJBI) partial differential equation; and the notion of backward reachability with its application to aiming at a specified target set is described in [20]. Reachability of hybrid systems is addressed in [30,35]. Over the last decade, significant advances were made in the characterization of reach sets and their computation for linear systems. These advances are described in this chapter.

Section 29.2 introduces the forward and backward reach sets, the classes of open- and closed-loop controls, and different kinds of reach sets that are appropriate in dealing with disturbances. Although some of the discussion applies to nonlinear systems, explicit formulas for reach sets are available only for linear systems. These formulas lead to explicit algorithms for computing reach sets (of linear systems) and Section 29.3 critically reviews the most promising algorithms. Section 29.4 is devoted to a set of algorithms based on the ellipsoidal calculus. These algorithms have a lower computational complexity, greater accuracy, and can work with systems of a larger size, compared with those reviewed in Section 29.3. Finally, Section 29.5 presents three examples to illustrate the ellipsoidal-based approach.

29.2 Basics of Reachability Analysis

29.2.1 Systems without Disturbances

Consider a general continuous-time

$$\dot{x}(t) = f(t, x, u), \tag{29.1}$$

or discrete-time dynamical system

$$x(t+1) = f(t, x, u), \tag{29.1d}$$

wherein t is time*, $x \in \mathbf{R}^n$ is the state, $u \in \mathbf{R}^m$ is the control, and f is a measurable vector function taking values in $\mathbf{R}^n$.† The control values $u(t, x(t))$ are restricted to a closed compact control set $\mathcal{U}(t) \subset \mathbf{R}^m$. An *open-loop* control does not depend on the state, $u = u(t)$; for a *closed-loop* control, $u = u(t, x(t))$.

Definition 29.1: Reach Set

The (forward) reach set $\mathcal{X}(t, t_0, x_0)$ at time $t > t_0$ from the initial position (t_0, x_0) is the set of all states $x(t)$ reachable at time t by system (Equation 29.1, or 29.1d), with $x(t_0) = x_0$ through all possible controls $u(\tau, x(\tau)) \in \mathcal{U}(\tau)$, $t_0 \le \tau < t$. For a given set of initial states $\mathcal{X}_0$, the reach set $\mathcal{X}(t, t_0, \mathcal{X}_0)$ is

$$\mathcal{X}(t, t_0, \mathcal{X}_0) = \bigcup_{x_0 \in \mathcal{X}_0} \mathcal{X}(t, t_0, x_0).$$

Here are two facts about forward reach sets.

1. $\mathcal{X}(t, t_0, \mathcal{X}_0)$ is the same for open-loop and closed-loop control.
2. $\mathcal{X}(t, t_0, \mathcal{X}_0)$ satisfies the semigroup property,

$$\mathcal{X}(t, t_0, \mathcal{X}_0) = \mathcal{X}(t, \tau, \mathcal{X}(\tau, t_0, \mathcal{X}_0)), \quad t_0 \le \tau < t. \tag{29.2}$$

* In discrete-time case, t assumes integer values.

† We are being general when giving the basic definitions. However, it is important to understand that for any specific *continuous-time* dynamical system it must be determined whether the solution exists and is unique, and in which class of solutions these conditions are met. we shall assume that function f is such that the solution of the differential equation 29.1 exists and is unique in Fillipov sense. This allows the right-hand side to be discontinuous. For discrete-time systems, this problem does not exist.

For linear systems

$$f(t,x,u) = A(t)x(t) + B(t)u, \tag{29.3}$$

with matrices $A(t)$ in $\mathbf{R}^{n\times n}$ and $B(t)$ in $\mathbf{R}^{m\times n}$. For a continuous-time linear system, the state transition matrix is

$$\dot{\Phi}(t,t_0) = A(t)\Phi(t,t_0), \quad \Phi(t,t) = I,$$

which for constant $A(t) \equiv A$ simplifies as

$$\Phi(t,t_0) = e^{A(t-t_0)}.$$

For a discrete-time linear system, the state transition matrix is

$$\Phi(t+1,t_0) = A(t)\Phi(t,t_0), \quad \Phi(t,t) = I,$$

which for constant $A(t) \equiv A$ simplifies as

$$\Phi(t,t_0) = A^{t-t_0}.$$

If the state transition matrix is invertible, $\Phi^{-1}(t,t_0) = \Phi(t_0,t)$. The transition matrix is always invertible for continuous-time and for sampled discrete-time systems. However, if for some $\tau, t_0 \leq \tau < t, A(\tau)$ is degenerate (singular), $\Phi(t,t_0) = \prod_{\tau=t_0}^{t-1} A(\tau)$ is also degenerate and cannot be inverted.

Following Cauchy's formula, the reach set $\mathcal{X}(t,t_0,\mathcal{X}_0)$ for a linear system can be expressed as

$$\mathcal{X}(t,t_0,\mathcal{X}_0) = \Phi(t,t_0)\mathcal{X}_0 \oplus \int_{t_0}^{t} \Phi(t,\tau)B(\tau)\mathcal{U}(\tau)\,d\tau \tag{29.4}$$

in continuous-time, and as

$$\mathcal{X}(t,t_0,\mathcal{X}_0) = \Phi(t,t_0)\mathcal{X}_0 \oplus \sum_{\tau=t_0}^{t-1} \Phi(t,\tau+1)B(\tau)\mathcal{U}(\tau) \tag{29.4d}$$

in discrete-time case.

The operation "$\oplus$" is the *geometric sum*, also known as *Minkowski sum*.* The geometric sum and linear (or affine) transformations preserve compactness and convexity. Hence, if the initial set $\mathcal{X}_0$ and the control sets $\mathcal{U}(\tau)$, $t_0 \leq \tau < t$, are compact and convex, so is the reach set $\mathcal{X}(t,t_0,\mathcal{X}_0)$.

Definition 29.2: Backward Reach Set

The backward reach set $\mathcal{Y}(t_1,t,y_1)$ for the target position (t_1,y_1) is the set of all states $y(t)$ for which there exists some control $u(\tau,x(\tau)) \in \mathcal{U}(\tau)$, $t \leq \tau < t_1$, that steers system (Equation 29.1 or 29.1d) to the state y_1 at time t_1. For the target set $\mathcal{Y}_1$ at time t_1, the backward reach set $\mathcal{Y}(t_1,t,\mathcal{Y}_1)$ is

$$\mathcal{Y}(t_1,t,\mathcal{Y}_1) = \bigcup_{y_1\in\mathcal{Y}_1} \mathcal{Y}(t_1,t,y_1).$$

The backward reach set $\mathcal{Y}(t_1,t,\mathcal{Y}_1)$ is the largest *weakly invariant* set with respect to the target set $\mathcal{Y}_1$ and time values t and t_1.†

---

* Minkowski sum of sets $\mathcal{W},\mathcal{Z} \subseteq \mathbf{R}^n$ is defined as $\mathcal{W}\oplus\mathcal{Z} = \{w+z \mid w\in\mathcal{W},\ z\in\mathcal{Z}\}$. Set $\mathcal{W}\oplus\mathcal{Z}$ is nonempty if and only if both $\mathcal{W}$ and $\mathcal{Z}$ are nonempty. If $\mathcal{W}$ and $\mathcal{Z}$ are convex, set $\mathcal{W}\oplus\mathcal{Z}$ is convex.

† $\mathcal{M}$ is weakly invariant with respect to the target set $\mathcal{Y}_1$ and times t_0 and t, if for every state $x_0 \in \mathcal{M}$ there exists a control $u(\tau,x(\tau)) \in \mathcal{U}(\tau)$, $t_0 \leq \tau < t$, that steers the system from x_0 at time t_0 to some state in $\mathcal{Y}_1$ at time t. If *all* controls in $\mathcal{U}(\tau)$, $t_0 \leq \tau < t$ steer the system from every $x_0 \in \mathcal{M}$ at time t_0 to $\mathcal{Y}_1$ at time t, set $\mathcal{M}$ is said to be *strongly* invariant with respect to $\mathcal{Y}_1$, t_0, and t.

Remark 29.1

Backward reach set can be computed for continuous-time system only if the solution of Equation 29.1 exists for $t < t_1$; and for discrete-time system only if the right-hand side of Equation 29.1d is invertible*.

These two facts about the backward reach set $\mathcal{Y}$ are similar to those for forward reach sets.

1. $\mathcal{Y}(t_1, t, \mathcal{Y}_1)$ is the same for open-loop and closed-loop control.
2. $\mathcal{Y}(t_1, t, \mathcal{Y}_1)$ satisfies the semigroup property,

$$\mathcal{Y}(t_1, t, \mathcal{Y}_1) = \mathcal{Y}(\tau, t, \mathcal{Y}(t_1, \tau, \mathcal{Y}_1)), \quad t \leq \tau < t_1. \tag{29.5}$$

For the linear system (Equation 29.3) the backward reach set can be expressed as

$$\mathcal{Y}(t_1, t, \mathcal{Y}_1) = \Phi(t, t_1)\mathcal{Y}_1 \oplus \int_{t_1}^{t} \Phi(t, \tau)B(\tau)\mathcal{U}(\tau)\, d\tau \tag{29.6}$$

in the continuous-time case, and as

$$\mathcal{Y}(t_1, t, \mathcal{Y}_1) = \Phi(t, t_1)\mathcal{Y}_1 \oplus \sum_{\tau=t}^{t_1-1} -\Phi(t, \tau)B(\tau)\mathcal{U}(\tau) \tag{29.6d}$$

in the discrete-time case. The last formula makes sense only for discrete-time linear systems with invertible state transition matrix. Degenerate discrete-time linear systems have unbounded backward reach sets and such sets cannot be computed with available software tools.

Just as in the case of forward reach set, the backward reach set of a linear system $\mathcal{Y}(t_1, t, \mathcal{Y}_1)$ is compact and convex if the target set $\mathcal{Y}_1$ and the control sets $\mathcal{U}(\tau)$, $t \leq \tau < t_1$ are compact and convex.

Remark 29.2

In the computer science literature, the reach set is said to be the result of operator *post*, and the backward reach set is the result of operator *pre*. In the control literature, the backward reach set is also called the *solvability set*.

29.2.2 Systems with Disturbances

Consider the continuous-time dynamical system with disturbance

$$\dot{x}(t) = f(t, x, u, v), \tag{29.7}$$

or the discrete-time dynamical system with disturbance

$$x(t+1) = f(t, x, u, v), \tag{29.7d}$$

in which we also have the disturbance input $v \in \mathbf{R}^d$ with values $v(t)$ restricted to a closed compact set $\mathcal{V}(t) \subset \mathbf{R}^d$.

In the presence of disturbances, the open-loop reach set (OLRS) is different from the closed-loop reach set (CLRS).

Given the initial time t_0, the set of initial states $\mathcal{X}_0$, and terminal time t, there are two types of OLRS.

Definition 29.3: OLRS of Maxmin Type

The maxmin open-loop reach set $\overline{\mathcal{X}}_{OL}(t, t_0, \mathcal{X}_0)$ *is the set of all states* x*, such that for any disturbance* $v(\tau) \in \mathcal{V}(\tau)$*, there exists an initial state* $x_0 \in \mathcal{X}_0$ *and a control* $u(\tau) \in \mathcal{U}(\tau)$, $t_0 \leq \tau < t$*, that steers system (Equation 29.7 or 29.7d) from* $x(t_0) = x_0$ *to* $x(t) = x$.

* There exists $f^{-1}(t, x, u)$ such that $x(t) = f^{-1}(t, x(t+1), u, v)$.

Definition 29.4: OLRS of Minmax Type

The minmax open-loop reach set $\underline{\mathcal{X}}_{OL}(t, t_0, \mathcal{X}_0)$ *is the set of all states* x*, such that there exists a control* $u(\tau) \in \mathcal{U}(\tau)$ *that for all disturbances* $v(\tau) \in \mathcal{V}(\tau)$*,* $t_0 \leq \tau < t$*, assigns an initial state* $x_0 \in \mathcal{X}_0$ *and steers system (Equation 29.7 or 29.7d), from* $x(t_0) = x_0$ *to* $x(t) = x$*.*

In the maxmin case, the control is chosen *after* knowing the disturbance over the entire time interval $[t_0, t]$, whereas in the minmax case, the control is chosen *before* any knowledge of the disturbance. Consequently, the OLRS do not satisfy the semigroup property.

The terms "maxmin" and "minmax" come from the fact that $\overline{\mathcal{X}}_{OL}(t, t_0, \mathcal{X}_0)$ is the subzero level set of the value function

$$\underline{V}(t,x) = \max_{v} \min_{u} \{\mathbf{dist}(x(t_0), \mathcal{X}_0) \mid x(t) = x,\ u(\tau) \in \mathcal{U}(\tau),\ v(\tau) \in \mathcal{V}(\tau),\ t_0 \leq \tau < t\}, \tag{29.8}$$

that is., $\overline{\mathcal{X}}_{OL}(t, t_0, \mathcal{X}_0) = \{x \mid \underline{V}(t,x) \leq 0\}$, and $\underline{\mathcal{X}}_{OL}(t, t_0, \mathcal{X}_0)$ is the subzero level set of the value function

$$\overline{V}(t,x) = \min_{u} \max_{v} \{\mathbf{dist}(x(t_0), \mathcal{X}_0) \mid x(t) = x,\ u(\tau) \in \mathcal{U}(\tau),\ v(\tau) \in \mathcal{V}(\tau),\ t_0 \leq \tau < t\}, \tag{29.9}$$

in which $\mathbf{dist}(\cdot,\cdot)$ denotes Hausdorff semidistance.* Since $\underline{V}(t,x) \leq \overline{V}(t,x)$, $\underline{\mathcal{X}}_{OL}(t, t_0, \mathcal{X}_0) \subseteq \overline{\mathcal{X}}_{OL}(t, t_0, \mathcal{X}_0)$.

Note that maxmin and minmax OLRS imply *guarantees*: these are states that can be reached no matter what the disturbance is, whether it is known in advance (maxmin case) or not (minmax case). The OLRS may be empty.

Fixing time instant τ_1, $t_0 < \tau_1 < t$, define the *piecewise maxmin open-loop reach set with one correction,*

$$\overline{\mathcal{X}}^1_{OL}(t, t_0, \mathcal{X}_0) = \overline{\mathcal{X}}_{OL}(t, \tau_1, \overline{\mathcal{X}}_{OL}(\tau_1, t_0, \mathcal{X}_0)), \tag{29.10}$$

and the *piecewise minmax open-loop reach set with one correction,*

$$\underline{\mathcal{X}}^1_{OL}(t, t_0, \mathcal{X}_0) = \underline{\mathcal{X}}_{OL}(t, \tau_1, \underline{\mathcal{X}}_{OL}(\tau_1, t_0, \mathcal{X}_0)). \tag{29.11}$$

The piecewise maxmin OLRS $\overline{\mathcal{X}}^1_{OL}(t, t_0, \mathcal{X}_0)$ is the subzero level set of the value function

$$\underline{V}^1(t,x) = \max_{v} \min_{u} \{\underline{V}(\tau_1, x(\tau_1)) \mid x(t) = x,\ u(\tau) \in \mathcal{U}(\tau),\ v(\tau) \in \mathcal{V}(\tau),\ \tau_1 \leq \tau < t\}, \tag{29.12}$$

with $V(\tau_1, x(\tau_1))$ given by Equation 29.8, which yields

$$\underline{V}^1(t,x) \geq \underline{V}(t,x),$$

and thus,

$$\overline{\mathcal{X}}^1_{OL}(t, t_0 \mathcal{X}_0) \subseteq \overline{\mathcal{X}}_{OL}(t, t_0, \mathcal{X}_0).$$

On the other hand, the piecewise minmax OLRS $\underline{\mathcal{X}}^1_{OL}(t, t_0, \mathcal{X}_0)$ is the subzero level set of the value function

$$\overline{V}^1(t,x) = \min_{u} \max_{v} \{\overline{V}(\tau_1, x(\tau_1)) \mid x(t) = x,\ u(\tau) \in \mathcal{U}(\tau),\ v(\tau) \in \mathcal{V}(\tau),\ \tau_1 \leq \tau < t\}, \tag{29.13}$$

* Hausdorff semidistance between compact sets $\mathcal{W}, \mathcal{Z} \subseteq \mathbf{R}^n$ is defined as

$$\mathbf{dist}(\mathcal{W}, \mathcal{Z}) = \min\{\langle w - z, w - z\rangle^{1/2} \mid w \in \mathcal{W},\ z \in \mathcal{Z}\},$$

where $\langle \cdot,\cdot \rangle$ denotes the inner product.

with $V(\tau_1, x(\tau_1))$ given by Equation 29.9, which yields

$$\overline{V}(t,x) \geq \overline{V}^1(t,x),$$

and thus,

$$\underline{\mathcal{X}}_{OL}(t, t_0\mathcal{X}_0) \subseteq \underline{\mathcal{X}}^1_{OL}(t, t_0, \mathcal{X}_0).$$

We can now recursively define piecewise maxmin and minmax OLRS with k corrections for $t_0 < \tau_1 < \cdots < \tau_k < t$. The maxmin piecewise OLRS with k corrections is

$$\overline{\mathcal{X}}^k_{OL}(t, t_0, \mathcal{X}_0) = \overline{\mathcal{X}}_{OL}(t, \tau_k, \overline{\mathcal{X}}^{k-1}_{OL}(\tau_k, t_0, \mathcal{X}_0)), \tag{29.14}$$

which is the subzero level set of the corresponding value function

$$\underline{V}^k(t,x) = \max_v \min_u \{\underline{V}^{k-1}(\tau_k, x(\tau_k)) \mid x(t) = x,\ u(\tau) \in \mathcal{U}(\tau),\ v(\tau) \in \mathcal{V}(\tau),\ \tau_k \leq \tau < t\}. \tag{29.15}$$

The minmax piecewise OLRS with k corrections is

$$\underline{\mathcal{X}}^k_{OL}(t, t_0, \mathcal{X}_0) = \underline{\mathcal{X}}_{OL}(t, \tau_k, \underline{\mathcal{X}}^{k-1}_{OL}(\tau_k, t_0, \mathcal{X}_0)), \tag{29.16}$$

which is the subzero level set of the corresponding value function

$$\overline{V}^k(t,x) = \min_u \max_v \{\overline{V}^{k-1}(\tau_k, x(\tau_k)) \mid x(t) = x,\ u(\tau) \in \mathcal{U}(\tau),\ v(\tau) \in \mathcal{V}(\tau),\ \tau_k \leq \tau < t\}. \tag{29.17}$$

From Equations 29.12, 29.13, 29.15, and 29.17 it follows that,

$$\underline{V}(t,x) \leq \underline{V}^1(t,x) \leq \cdots \leq \underline{V}^k(t,x) \leq \overline{V}^k(t,x) \leq \cdots \leq \overline{V}^1(t,x) \leq \overline{V}(t,x).$$

Hence,

$$\begin{aligned} \underline{\mathcal{X}}_{OL}(t, t_0, \mathcal{X}_0) &\subseteq \underline{\mathcal{X}}^1_{OL}(t, t_0, \mathcal{X}_0) \subseteq \cdots \subseteq \underline{\mathcal{X}}^k_{OL}(t, t_0, \mathcal{X}_0) \\ &\subseteq \overline{\mathcal{X}}^k_{OL}(t, t_0, \mathcal{X}_0) \subseteq \cdots \subseteq \overline{\mathcal{X}}^1_{OL}(t, t_0, \mathcal{X}_0) \subseteq \overline{\mathcal{X}}_{OL}(t, t_0, \mathcal{X}_0). \end{aligned} \tag{29.18}$$

We call

$$\overline{\mathcal{X}}_{CL}(t, t_0, \mathcal{X}_0) = \overline{\mathcal{X}}^k_{OL}(t, t_0, \mathcal{X}_0), \quad k = \begin{cases} \infty & \text{for continuous-time system,} \\ t - t_0 - 1 & \text{for discrete-time system,} \end{cases} \tag{29.19}$$

the *maxmin closed-loop reach set* of system (Equation 29.7 or 29.7d) at time t, and we call

$$\underline{\mathcal{X}}_{CL}(t, t_0, \mathcal{X}_0) = \underline{\mathcal{X}}^k_{OL}(t, t_0, \mathcal{X}_0), \quad k = \begin{cases} \infty & \text{for continuous-time system,} \\ t - t_0 - 1 & \text{for discrete-time system,} \end{cases} \tag{29.20}$$

the *minmax closed-loop reach set* of system (Equation 29.7 or 29.7d) at time t.

Definition 29.5: CLRS of Maxmin Type

Given initial time t_0 and the set of initial states $\mathcal{X}_0$, the maxmin CLRS $\overline{\mathcal{X}}_{CL}(t, t_0, \mathcal{X}_0)$ of system (Equation 29.7 or 29.7d) at time $t > t_0$, is the set of all states x, for each of which and for every disturbance $v(\tau) \in \mathcal{V}(\tau)$, there exist an initial state $x_0 \in \mathcal{X}_0$ and a control $u(\tau, x(\tau)) \in \mathcal{U}(\tau)$, such that the trajectory $x(\tau|v(\tau), u(\tau, x(\tau)))$ satisfying $x(t_0) = x_0$ and

$$\dot{x}(\tau|v(\tau), u(\tau, x(\tau))) \in f(\tau, x(\tau), u(\tau, x(\tau)), v(\tau))$$

in the continuous-time case, or

$$x(\tau + 1|v(\tau), u(\tau, x(\tau))) \in f(\tau, x(\tau), u(\tau, x(\tau)), v(\tau))$$

in the discrete-time case, with $t_0 \leq \tau < t$, is such that $x(t) = x$.

Definition 29.6: CLRS of Minmax Type

Given initial time t_0 and the set of initial states $\mathcal{X}_0$, the maxmin CLRS $\underline{\mathcal{X}}_{CL}(t, t_0, \mathcal{X}_0)$ of system (Equation 29.7 or 29.7d), at time $t > t_0$ is the set of all states x, for each of which there exists a control $u(\tau, x(\tau)) \in \mathcal{U}(\tau)$, and for every disturbance $v(\tau) \in \mathcal{V}(\tau)$ there exists an initial state $x_0 \in \mathcal{X}_0$, such that the trajectory $x(\tau, v(\tau)|u(\tau, x(\tau)))$ satisfying $x(t_0) = x_0$ and

$$\dot{x}(\tau, v(\tau)|u(\tau, x(\tau))) \in f(\tau, x(\tau), u(\tau, x(\tau)), v(\tau))$$

in the continuous-time case, or

$$x(\tau + 1, v(\tau)|u(\tau, x(\tau))) \in f(\tau, x(\tau), u(\tau, x(\tau)), v(\tau))$$

in the discrete-time case, with $t_0 \leq \tau < t$, is such that $x(t) = x$.

By construction, both maxmin and minmax CLRS satisfy the semigroup property (Equation 29.2).

For some classes of dynamical systems and some types of constraints on initial conditions, controls and disturbances, the maxmin and minmax CLRS may coincide. This is the case for continuous-time linear systems with convex compact bounds on the initial set, controls and disturbances under the condition that the initial set $\mathcal{X}_0$ is large enough to ensure that $\mathcal{X}(t_0 + \epsilon, t_0, \mathcal{X}_0)$ is nonempty for some small $\epsilon > 0$.

Consider the linear system case,

$$f(t, x, u) = A(t)x(t) + B(t)u + G(t)v, \tag{29.21}$$

where $A(t)$ and $B(t)$ are as in Equation 29.3, and $G(t)$ takes its values in $\mathbf{R}^d$.

The maxmin OLRS for the continuous-time linear system can be expressed through set valued integrals,

$$\overline{\mathcal{X}}_{OL}(t, t_0, \mathcal{X}_0) = \left(\Phi(t, t_0)\mathcal{X}_0 \oplus \int_{t_0}^{t} \Phi(t, \tau)B(\tau)\mathcal{U}(\tau)\, d\tau\right) \dot{-} \int_{t_0}^{t} \Phi(t, \tau)(-G(\tau))\mathcal{V}(\tau)\, d\tau, \tag{29.22}$$

and for discrete-time linear system through set-valued sums,

$$\overline{\mathcal{X}}_{OL}(t, t_0, \mathcal{X}_0) = \left(\Phi(t, t_0)\mathcal{X}_0 \oplus \sum_{\tau=t_0}^{t-1} \Phi(t, \tau + 1)B(\tau)\mathcal{U}(\tau)\right) \dot{-} \sum_{\tau=t_0}^{t-1} \Phi(t, \tau + 1)(-G(\tau))\mathcal{V}(\tau). \tag{29.22d}$$

Similarly, the minmax OLRS for the continuous-time linear system is

$$\underline{\mathcal{X}}_{OL}(t, t_0, \mathcal{X}_0) = \left(\Phi(t, t_0)\mathcal{X}_0 \dot{-} \int_{t_0}^{t} \Phi(t, \tau)(-G(\tau))\mathcal{V}(\tau)\, d\tau\right) \oplus \int_{t_0}^{t} \Phi(t, \tau)B(\tau)\mathcal{U}(\tau)\, d\tau, \tag{29.23}$$

and for the discrete-time linear system, it is

$$\underline{\mathcal{X}}_{OL}(t, t_0, \mathcal{X}_0) = \left(\Phi(t, t_0)\mathcal{X}_0 \dot{-} \sum_{\tau=t_0}^{t-1} \Phi(t, \tau + 1)(-G(\tau))\mathcal{V}(\tau)\right) \oplus \sum_{\tau=t_0}^{t-1} \Phi(t, \tau + 1)B(\tau)\mathcal{U}(\tau). \tag{29.23d}$$

The operation '$\dot{-}$' is *geometric difference*, also known as *Minkowski difference*.*

* The Minkowski difference of sets $\mathcal{W}, \mathcal{Z} \in \mathbf{R}^n$ is defined as $\mathcal{W} \dot{-} \mathcal{Z} = \{\xi \in \mathbf{R}^n \mid \xi \oplus \mathcal{Z} \subseteq \mathcal{W}\}$. If $\mathcal{W}$ and $\mathcal{Z}$ are convex, $\mathcal{W} \dot{-} \mathcal{Z}$ is convex if it is nonempty.

Now consider the piecewise OLRS with k corrections. Expression 29.14 translates into

$$\overline{\mathcal{X}}^k_{OL}(t,t_0,\mathcal{X}_0)=\left(\Phi(t,\tau_k)\overline{\mathcal{X}}^{k-1}_{OL}(\tau_k,t_0,\mathcal{X}_0)\oplus\int_{\tau_k}^{t}\Phi(t,\tau)B(\tau)\mathcal{U}(\tau)\,d\tau\right)\dot{-}\int_{\tau_k}^{t}\Phi(t,\tau)(-G(\tau))\mathcal{V}(\tau)\,d\tau, \tag{29.24}$$

in the continuous-time case, and for the discrete-time case into

$$\overline{\mathcal{X}}^k_{OL}(t,t_0,\mathcal{X}_0)=\left(\Phi(t,\tau_k)\overline{\mathcal{X}}^{k-1}_{OL}(\tau_k,t_0,\mathcal{X}_0)\oplus\sum_{\tau=\tau_k}^{t-1}\Phi(t,\tau+1)B(\tau)\mathcal{U}(\tau)\right)\dot{-}\sum_{\tau=\tau_k}^{t-1}\Phi(t,\tau+1)(-G(\tau))\mathcal{V}(\tau). \tag{29.24d}$$

Expression 29.16 translates into

$$\underline{\mathcal{X}}^k_{OL}(t,t_0,\mathcal{X}_0)=\left(\Phi(t,\tau_k)\underline{\mathcal{X}}^{k-1}_{OL}(t,t_0,\mathcal{X}_0)\dot{-}\int_{\tau_k}^{t}\Phi(t,\tau)(-G(\tau))\mathcal{V}(\tau)\,d\tau\right)\oplus\int_{\tau_k}^{t}\Phi(t,\tau)B(\tau)\mathcal{U}(\tau)\,d\tau, \tag{29.25}$$

in the continuous-time case, and for the discrete-time case into

$$\underline{\mathcal{X}}^k_{OL}(t,t_0,\mathcal{X}_0)=\left(\Phi(t,\tau_k)\underline{\mathcal{X}}^{k-1}_{OL}(\tau_k,t_0,\mathcal{X}_0)\dot{-}\sum_{\tau=\tau_k}^{t-1}\Phi(t,\tau+1)(-G(\tau))\mathcal{V}(\tau)\right)\oplus\sum_{\tau=\tau_k}^{t-1}\Phi(t,\tau+1)B(\tau)\mathcal{U}(\tau). \tag{29.25d}$$

Since for any $\mathcal{W}_1,\mathcal{W}_2,\mathcal{W}_3\subseteq\mathbf{R}^n$ it is true that

$$(\mathcal{W}_1\dot{-}\mathcal{W}_2)\oplus\mathcal{W}_3=(\mathcal{W}_1\oplus\mathcal{W}_3)\dot{-}(\mathcal{W}_2\oplus\mathcal{W}_3)\subseteq(\mathcal{W}_1\oplus\mathcal{W}_3)\dot{-}\mathcal{W}_2,$$

from Equations 29.24 and 29.25 and from Equations 29.24d and 29.25d, it is clear that Equation 29.18 is true.

For linear systems, if the initial set $\mathcal{X}_0$, control bounds $\mathcal{U}(\tau)$ and disturbance bounds $\mathcal{V}(\tau)$, $t_0\le\tau<t$, are compact and convex, the CLRS $\overline{\mathcal{X}}_{CL}(t,t_0,\mathcal{X}_0)$ and $\underline{\mathcal{X}}_{CL}(t,t_0,\mathcal{X}_0)$ are compact and convex, provided they are nonempty. For continuous-time linear systems, $\overline{\mathcal{X}}_{CL}(t,t_0,\mathcal{X}_0)=\underline{\mathcal{X}}_{CL}(t,t_0,\mathcal{X}_0)=\mathcal{X}_{CL}(t,t_0,\mathcal{X}_0)$.

Just as for forward reachability case, the backward reach sets can be open-loop (OLBRS) or closed-loop (CLBRS).

Definition 29.7: OLBRS of Maxmin Type

Given the terminal time t_1 and target set $\mathcal{Y}_1$, the maxmin open-loop backward reach set $\overline{\mathcal{Y}}_{OL}(t_1,t,\mathcal{Y}_1)$ of system (Equation 29.7 or 29.7d) at time $t<t_1$ is the set of all y, such that for any disturbance $v(\tau)\in\mathcal{V}(\tau)$ there exists a terminal state $y_1\in\mathcal{Y}_1$ and control $u(\tau)\in\mathcal{U}(\tau)$, $t\le\tau<t_1$, which steers the system from $y(t)=y$ to $y(t_1)=y_1$.

$\overline{\mathcal{Y}}_{OL}(t_1,t,\mathcal{Y}_1)$ is the subzero level set of the value function

$$\underline{V}_b(t,y)=\max_v\min_u\{\mathbf{dist}(y(t_1),\mathcal{Y}_1)\mid y(t)=y,\ u(\tau)\in\mathcal{U}(\tau),\ v(\tau)\in\mathcal{V}(\tau),\ t\le\tau<t_1\}, \tag{29.26}$$

Definition 29.8: OLBRS of Minmax Type

Given the terminal time t_1 and target set $\mathcal{Y}_1$, the minmax open-loop backward reach set $\underline{\mathcal{Y}}_{OL}(t_1,t,\mathcal{Y}_1)$ of system (Equation 29.7 or 29.7d) at time $t<t_1$ is the set of all y, such that there exists a control $u(\tau)\in\mathcal{U}(\tau)$ that for all disturbances $v(\tau\in\mathcal{V}(\tau)$, $t\le\tau<t_1$, assigns a terminal state $y_1\in\mathcal{Y}_1$ and steers the system from $y(t)=y$ to $y(t_1)=y_1$.

$\underline{\mathcal{Y}}_{OL}(t_1, t, \mathcal{Y}_1)$ is the subzero level set of the value function

$$\overline{V}_b(t, y) = \min_u \max_v \{\mathbf{dist}(y(t_1), \mathcal{Y}_1) \mid y(t) = y,\ u(\tau) \in \mathcal{U}(\tau),\ v(\tau) \in \mathcal{V}(\tau),\ t \leq \tau < t_1\}, \tag{29.27}$$

Remark 29.3

The backward reach set can be computed for a continuous-time system only if the solution of Equation 29.7 exists for $t < t_1$, and for a discrete-time system, only if the right-hand side of Equation 29.7d is invertible.

Similar to the forward reachability case, we construct piecewise OLBRS with one correction at time τ_1, $t < \tau_1 < t_1$. The piecewise maxmin OLBRS with one correction is

$$\overline{\mathcal{Y}}^1_{OL}(t_1, t, \mathcal{Y}_1) = \overline{\mathcal{Y}}_{OL}(\tau_1, t, \overline{\mathcal{Y}}_{OL}(t_1, \tau_1, \mathcal{Y}_1)), \tag{29.28}$$

and it is the subzero level set of the function

$$\underline{V}^1_b(t, y) = \max_v \min_u \{\underline{V}_b(\tau_1, y(\tau_1)) \mid y(t) = y,\ u(\tau) \in \mathcal{U}(\tau),\ v(\tau) \in \mathcal{V}(\tau),\ t \leq \tau < \tau_1\}. \tag{29.29}$$

The piecewise minmax OLBRS with one correction is

$$\underline{\mathcal{Y}}^1_{OL}(t_1, t, \mathcal{Y}_1) = \underline{\mathcal{Y}}_{OL}(\tau_1, t, \underline{\mathcal{Y}}_{OL}(t_1, \tau_1, \mathcal{Y}_1)), \tag{29.30}$$

and it is the subzero level set of the function

$$\overline{V}^1_b(t, y) = \min_u \max_v \{\overline{V}_b(\tau_1, y(\tau_1)) \mid y(t) = y,\ u(\tau) \in \mathcal{U}(\tau),\ v(\tau) \in \mathcal{V}(\tau),\ t \leq \tau < \tau_1\}, \tag{29.31}$$

Recursively define maxmin and minmax OLBRS with k corrections for $t < \tau_k < \cdots < \tau_1 < t_1$. The maxmin OLBRS with k corrections is

$$\overline{\mathcal{Y}}^k_{OL}(t_1, t, \mathcal{Y}_1) = \overline{\mathcal{Y}}_{OL}(\tau_k, t, \overline{\mathcal{Y}}^{k-1}_{OL}(t_1, \tau_k, \mathcal{Y}_1)), \tag{29.32}$$

which is the subzero level set of function

$$\underline{V}^k_b(t, y) = \max_v \min_u \{\underline{V}^{k-1}_b(\tau_k, y(\tau_k)) \mid y(t) = y,\ u(\tau) \in \mathcal{U}(\tau),\ v(\tau) \in \mathcal{V}(\tau),\ t \leq \tau < \tau_k\}. \tag{29.33}$$

The minmax OLBRS with k corrections is

$$\underline{\mathcal{Y}}^k_{OL}(t_1, t, \mathcal{Y}_1) = \underline{\mathcal{Y}}_{OL}(\tau_k, t, \underline{\mathcal{Y}}^{k-1}_{OL}(t_1, \tau_k, \mathcal{Y}_1)), \tag{29.34}$$

which is the subzero level set of the function

$$\overline{V}^k_b(t, y) = \min_u \max_v \{\overline{V}^{k-1}_b(\tau_k, y(\tau_k)) \mid y(t) = y,\ u(\tau) \in \mathcal{U}(\tau),\ v(\tau) \in \mathcal{V}(\tau),\ t \leq \tau < \tau_k\}, \tag{29.35}$$

From Equations 29.29, 29.31, 29.33, and 29.35 it follows that

$$\underline{V}_b(t, y) \leq \underline{V}^1_b(t, y) \leq \cdots \leq \underline{V}^k_b(t, y) \leq \overline{V}^k_b(t, y) \leq \cdots \leq \overline{V}^1_b(t, y) \leq \overline{V}_b(t, y).$$

Hence,

$$\begin{aligned} \underline{\mathcal{Y}}_{OL}(t_1, t, \mathcal{Y}_1) &\subseteq \underline{\mathcal{Y}}^1_{OL}(t_1, t, \mathcal{Y}_1) \subseteq \cdots \subseteq \underline{\mathcal{Y}}^k_{OL}(t_1, t, \mathcal{Y}_1) \\ &\subseteq \overline{\mathcal{Y}}^k_{OL}(t_1, t, \mathcal{Y}_1) \subseteq \cdots \subseteq \overline{\mathcal{Y}}^1_{OL}(t_1, t, \mathcal{Y}_1) \subseteq \overline{\mathcal{Y}}_{OL}(t_1, t, \mathcal{Y}_1). \end{aligned} \tag{29.36}$$

We say that

$$\overline{\mathcal{Y}}_{CL}(t_1, t, \mathcal{Y}_1) = \overline{\mathcal{Y}}^k_{OL}(t_1, t, \mathcal{Y}_1), \quad k = \begin{cases} \infty & \text{for continuous-time system} \\ t_1 - t - 1 & \text{for discrete-time system} \end{cases} \tag{29.37}$$

is the *maxmin closed-loop backward reach set* of system (Equation 29.7 or 29.7d) at time t. We say that

$$\underline{\mathcal{Y}}_{CL}(t_1, t, \mathcal{Y}_1) = \underline{\mathcal{Y}}^k_{OL}(t_1, t, \mathcal{Y}_1), \quad k = \begin{cases} \infty & \text{for continuous-time system} \\ t_1 - t - 1 & \text{for discrete-time system} \end{cases} \tag{29.38}$$

is the *minmax closed-loop backward reach set* of system (Equation 29.7 or 29.7d) at time t.

Definition 29.9: CLBRS of Maxmin Type

Given the terminal time t_1 and target set $\mathcal{Y}_1$, the maxmin CLBRS $\overline{\mathcal{Y}}_{CL}(t_1, t, \mathcal{Y}_1)$ of system (Equation 29.7 or 29.7d) at time $t < t_1$ is the set of all states y, for each of which for every disturbance $v(\tau) \in \mathcal{V}(\tau)$ there exists terminal state $y_1 \in \mathcal{Y}_1$ and control $u(\tau, y(\tau)) \in \mathcal{U}(\tau)$ that assigns trajectory $y(\tau, |v(\tau), u(\tau, y(\tau)))$ satisfying

$$\dot{y}(\tau|v(\tau), u(\tau, y(\tau))) \in f(\tau, y(\tau), u(\tau, y(\tau)), v(\tau))$$

in continuous-time case, or

$$y(\tau+1|v(\tau), u(\tau, y(\tau))) \in f(\tau, y(\tau), u(\tau, y(\tau)), v(\tau))$$

in discrete-time case, with $t \leq \tau < t_1$, such that $y(t) = y$ and $y(t_1) = y_1$.

Definition 29.10: CLBRS of Minmax Type

Given the terminal time t_1 and target set $\mathcal{Y}_1$, the minmax CLBRS $\underline{\mathcal{Y}}_{CL}(t_1, t, \mathcal{Y}_1)$ of system (Equation 29.7 or 29.7d) at time $t < t_1$ is the set of all states y, for each of which there exists control $u(\tau, y(\tau)) \in \mathcal{U}(\tau)$ that for every disturbance $v(\tau) \in \mathcal{V}(\tau)$ assigns terminal state $y_1 \in \mathcal{Y}_1$ and trajectory $y(\tau, v(\tau)|u(\tau, y(\tau)))$ satisfying

$$\dot{y}(\tau, v(\tau)|u(\tau, y(\tau))) \in f(\tau, y(\tau), u(\tau, y(\tau)), v(\tau))$$

in the continuous-time case, or

$$y(\tau+1, v(\tau)|u(\tau, y(\tau))) \in f(\tau, y(\tau), u(\tau, y(\tau)), v(\tau))$$

in the discrete-time case, with $t \leq \tau < t_1$, such that $y(t) = y$ and $y(t_1) = y_1$.

Both maxmin and minmax CLBRS satisfy the semigroup property (Equation 29.5).

The maxmin OLBRS for the continuous-time linear system can be expressed through set valued integrals,

$$\overline{\mathcal{Y}}_{OL}(t_1, t, \mathcal{Y}_1) = \left(\Phi(t, t_1)\mathcal{Y}_1 \oplus \int_{t_1}^{t} \Phi(t, \tau)B(\tau)\mathcal{U}(\tau)\, d\tau\right) \dot{-} \int_{t}^{t_1} \Phi(t, \tau)G(\tau)\mathcal{V}(\tau)\, d\tau, \tag{29.39}$$

and for the discrete-time linear system through set-valued sums,

$$\overline{\mathcal{Y}}_{OL}(t_1, t, \mathcal{Y}_1) = \left(\Phi(t, t_1)\mathcal{Y}_1 \oplus \sum_{\tau=t}^{t_1-1} -\Phi(t, \tau+1)B(\tau)\mathcal{U}(\tau)\right) \dot{-} \sum_{\tau=t}^{t_1-1} \Phi(t, \tau+1)G(\tau)\mathcal{V}(\tau). \tag{29.39d}$$

Similarly, the minmax OLBRS for the continuous-time linear system is

$$\underline{\mathcal{Y}}_{OL}(t_1, t, \mathcal{Y}_1) = \left(\Phi(t, t_1)\mathcal{Y}_1 \dot{-} \int_{t}^{t_1} \Phi(t, \tau)G(\tau)\mathcal{V}(\tau)\, d\tau\right) \oplus \int_{t_1}^{t} \Phi(t, \tau)B(\tau)\mathcal{U}(\tau)\, d\tau, \tag{29.40}$$

and for the discrete-time linear system it is

$$\underline{\mathcal{Y}}_{OL}(t_1, t, \mathcal{Y}_1) = \left(\Phi(t, t_1)\mathcal{Y}_1 \dot{-} \sum_{\tau=t}^{t_1-1} \Phi(t, \tau+1)G(\tau)\mathcal{V}(\tau)\right) \oplus \sum_{\tau=t}^{t_1-1} -\Phi(t, \tau+1)B(\tau)\mathcal{U}(\tau). \tag{29.40d}$$

Now consider piecewise OLBRS with k corrections. Expression 29.32 translates into

$$\overline{\mathcal{Y}}^k_{OL}(t_1, t, \mathcal{Y}_1) = \left(\Phi(t, \tau_k)\overline{\mathcal{Y}}^{k-1}_{OL}(t_1, \tau_k, \mathcal{Y}_1) \oplus \int_{\tau_k}^{t} \Phi(t, \tau)B(\tau)\mathcal{U}(\tau)\, d\tau \right) \dot{-} \int_{t}^{\tau_k} \Phi(t, \tau)G(\tau)\mathcal{V}(\tau)\, d\tau, \tag{29.41}$$

in the continuous-time case, and for the discrete-time case into

$$\overline{\mathcal{Y}}^k_{OL}(t_1, t, \mathcal{Y}_1) = \left(\Phi(t, \tau_k)\overline{\mathcal{Y}}^{k-1}_{OL}(t_1, \tau_k, \mathcal{Y}_1) \oplus \sum_{\tau=t}^{\tau_k - 1} -\Phi(t, \tau+1)B(\tau)\mathcal{U}(\tau) \right) \dot{-} \sum_{\tau=t}^{\tau_k - 1} \Phi(t, \tau+1)G(\tau)\mathcal{V}(\tau). \tag{29.41d}$$

Expression 29.34 translates into

$$\underline{\mathcal{Y}}^k_{OL}(t_1, t, \mathcal{Y}_1) = \left(\Phi(t, \tau_k)\overline{\mathcal{Y}}^{k-1}_{OL}(t_1, \tau_k, \mathcal{Y}_1) \dot{-} \int_{t}^{\tau_k} \Phi(t, \tau)G(\tau)\mathcal{V}(\tau)\, d\tau \right) \oplus \int_{\tau_k}^{t} \Phi(t, \tau)B(\tau)\mathcal{U}(\tau)\, d\tau, \tag{29.42}$$

in the continuous-time case, and for the discrete-time case into

$$\underline{\mathcal{Y}}^k_{OL}(t_1, t, \mathcal{Y}_1) = \left(\Phi(t, \tau_k)\overline{\mathcal{Y}}^{k-1}_{OL}(t_1, \tau_k, \mathcal{Y}_1) \dot{-} \sum_{\tau=t}^{\tau_k - 1} \Phi(t, \tau+1)G(\tau)\mathcal{V}(\tau) \right) \oplus \sum_{\tau=t}^{\tau_k - 1} -\Phi(t, \tau+1)B(\tau)\mathcal{U}(\tau). \tag{25.42d}$$

For continuous-time linear systems $\overline{\mathcal{Y}}_{CL}(t_1, t, \mathcal{Y}_1) = \underline{\mathcal{Y}}_{CL}(t_1, t, \mathcal{Y}_1) = \mathcal{Y}_{CL}(t_1, t, \mathcal{Y}_1)$ under the condition that the target set $\mathcal{Y}_1$ is large enough to ensure that $\underline{\mathcal{Y}}_{CL}(t_1, t_1 - \epsilon, \mathcal{Y}_1)$ is nonempty for some small $\epsilon > 0$.

Computation of backward reach sets for discrete-time linear systems makes sense only if the state transition matrix $\Phi(t_1, t)$ is invertible.

If the target set $\mathcal{Y}_1$, control sets $\mathcal{U}(\tau)$, and disturbance sets $\mathcal{V}(\tau)$, $t \leq \tau < t_1$, are compact and convex, then CLBRS $\overline{\mathcal{Y}}_{CL}(t_1, t, \mathcal{Y}_1)$ and $\underline{\mathcal{Y}}_{CL}(t_1, t, \mathcal{Y}_1)$ are compact and convex, if they are nonempty.

29.2.3 Reachability Problem

Reachability analysis is concerned with the computation of the forward $\mathcal{X}(t, t_0, \mathcal{X}_0)$ and backward $\mathcal{Y}(t_1, t, \mathcal{Y}_1)$ reach sets (the reach sets may be maxmin or minmax) in a way that can effectively meet requests like the following:

1. For the given time interval $[t_0, t]$, determine whether the system can be steered into the given target set $\mathcal{Y}_1$. In other words, is the set $\mathcal{Y}_1 \cap \bigcup_{t_0 \leq \tau \leq t} \mathcal{X}(\tau, t_0, \mathcal{X}_0)$ nonempty? And if the answer is "yes," find a control that steers the system to the target set (or avoids the target set).*
2. If the target set $\mathcal{Y}_1$ is reachable from the given initial condition $\{t_0, \mathcal{X}_0\}$ in the time interval $[t_0, t]$, find the shortest time to reach $\mathcal{Y}_1$,

$$\arg\min_{\tau}\{\mathcal{X}(\tau, t_0, \mathcal{X}_0) \cap \mathcal{Y}_1 \neq \emptyset \mid t_0 \leq \tau \leq t\}.$$

3. Given the terminal time t_1, target set $\mathcal{Y}_1$, and time $t < t_1$, find the set of states starting at time t from which the system can reach $\mathcal{Y}_1$ within time interval $[t, t_1]$. In other words, find $\bigcup_{t \leq \tau < t_1} \mathcal{Y}(t_1, \tau, \mathcal{Y}_1)$.
4. Find a closed-loop control that steers a system with disturbances to the given target set in given time.
5. Graphically display the projection of the reach set along any specified two- or three-dimensional subspace.

---

* So-called verification problems often consist in ensuring that the system is unable to reach an "unsafe" target set within a given time interval.

For linear systems, if the initial set $\mathcal{X}_0$, target set $\mathcal{Y}_1$, control bounds $\mathcal{U}(\cdot)$, and disturbance bounds $\mathcal{V}(\cdot)$ are compact and convex, so are the forward $\mathcal{X}(t, t_0, \mathcal{X}_0)$ and backward $\mathcal{Y}(t_1, t, \mathcal{Y}_1)$ reach sets. Hence reachability analysis requires the computationally effective manipulation of convex sets, and performing the set-valued operations of unions, intersections, geometric sums, and differences.

Existing reach set computation tools can deal reliably only with linear systems with convex constraints. A claim that a certain tool or method can be used *effectively* for nonlinear systems must be treated with caution, and the first question to ask is for what class of nonlinear systems and with what limit on the state space dimension does this tool work? Some "reachability methods for nonlinear systems" reduce to the local linearization of a system followed by the use of well-tested techniques for linear system reach set computation. Thus these approaches in fact use reachability methods for linear systems.

29.3 Overview of Computational Methods and Tools

Before choosing a method and a tool for reachability analysis, one should answer the following questions to specify the requirements.

1. Do you really need to compute reach sets, or it is enough to perform a safety check, for example, to ensure that trajectories of a system never enter a given target set, or never leave a given initial set? Barrier functions or invariant sets, described in the end of this section, may be sufficient for safety checking.
2. Do you need to compute reach sets exactly, or will approximations, external and internal, be enough? Except for very specific classes of systems, exact reach set computation is not possible, and approximation techniques are required. Unless a reach set has simple structure, its exact representation is possible only for low state-space dimension. Hence, the next question.
3. What is the dimension of your system? The higher the system dimension, the rougher the reach set approximation.

Another important quality of a computational method for reach sets is the preservation of the semigroup property. It is highly desirable that the semigroup property is maintained by the algorithm as well as by its software implementation.

29.3.1 Level Set Method

We start the overview of computational techniques for reach sets with the *level set* method as it points out the essence of the reachability problem, and has been used in practice for specific nonlinear systems. The idea is to solve the HJBI partial differential equation

$$\frac{\partial V}{\partial t} + \max_{u} \left\langle \frac{\partial V}{\partial x}, f(t, x, u) \right\rangle = 0,$$

with initial condition

$$V(t_0, x) = \mathbf{dist}(x, \mathcal{X}_0),$$

for $t > t_0$, and then to find the reach set $\mathcal{X}(t, t_0, \mathcal{X}_0)$ as the subzero level set of the solution $V(t, x)$,

$$\mathcal{X}(t, t_0, \mathcal{X}_0) = \left\{x \in \mathbf{R}^n \mid V(t, x) \leq 0\right\}.$$

This *forward* HJBI equation was introduced in [29]. For systems with disturbances, the HJBI equation is

$$\frac{\partial V}{\partial t} + \min_{v} \max_{u} \left\langle \frac{\partial V}{\partial x}, f(t, x, u, v) \right\rangle = 0$$

in the maxmin case, and

$$\frac{\partial \overline{V}}{\partial t} + \max_u \min_v \left\langle \frac{\partial \overline{V}}{\partial x}, f(t,x,u,v) \right\rangle = 0$$

in the minmax case.

For backward reach sets, the HJBI equation is solved in backward time,

$$\frac{\partial V_b}{\partial t} + \min_u \left\langle \frac{\partial V_b}{\partial x}, f(t,x,u) \right\rangle,$$

with boundary condition

$$V_b(t_1, x) = \mathbf{dist}(x, \mathcal{Y}_1),$$

for $t < t_1$. For systems with disturbances, the backward HJBI is

$$\frac{\partial \underline{V}_b}{\partial t} + \max_v \min_u \left\langle \frac{\partial \underline{V}_b}{\partial x}, f(t,x,u,v) \right\rangle = 0$$

in the maxmin case, and

$$\frac{\partial \overline{V}_b}{\partial t} + \min_u \max_v \left\langle \frac{\partial \overline{V}_b}{\partial x}, f(t,x,u,v) \right\rangle = 0$$

in the minmax case.

Computation of reach sets as level sets of HJBI solutions was introduced in [21,22,24] with special emphasis on linear systems. In [31], the authors applied the level set method to reachability analysis of hybrid systems. The level set method is implemented in the *Level Set Toolbox* [5], which uses numerical algorithms for time-dependent HJBI equations and structured grids. Work is under way to implement fast marching methods. These are effective numerical schemes that work for time-independent HJBI, but whose major restriction is the need for the control to have the same dimension as the state. Level Set Toolbox tries to compute the surface of the reach set exactly with accuracy dependent on the choice of the grid. This plus the exponential growth of computational complexity with the system dimension makes the level set method impractical for systems with dimension larger than three.

Level Set Toolbox deals with continuous-time systems. To use the level set method in discrete-time case, one has to solve the Bellman equation under the condition that the right-hand side of system (Equation 29.1d) is invertible,

$$V(t+1, x) = \min_u (V(t, f^{-1}(t,x,u)))$$

with initial condition

$$V(t_0, x) = \mathbf{dist}(x, \mathcal{X}_0)$$

for $t > t_0$, and then find the forward reach set

$$\mathcal{X}(t, t_0, \mathcal{X}_0) = \{x \in \mathbf{R}^n \mid V(t,x) \leq 0\}.$$

The backward reach set is the subzero level set of the value function $V_b(t,x)$ obtained from the backward Bellman equation,

$$V_b(t-1, x) = \min_u (V_b(t, f(t-1, x, u))),$$

with boundary condition

$$V_b(t_1, x) = \mathbf{dist}(x, \mathcal{Y}_1),$$

for $t < t_1$. For systems with disturbances $\min_u$ is substituted with $\max_v \min_u$ or with $\min_u \max_v$ in both forward and backward Bellman equations, and the functions f and f^{-1}, whose existence is required, depend on an additional parameter v.

Even though in the discrete-time case the computation of the value function does not involve solving a partial differential equation (PDE), it is still very burdensome, especially for nonlinear systems whose reach sets are nonconvex. Computing the distance function for such sets and minimizing it over u may be difficult. Even more difficult it is to search for maxmin or minmax.

The conclusion is that although the level set method handles nonlinear systems, it is computationally costly, and the need to maintain a grid with the value function values, which must be rather dense to ensure proper accuracy, makes it practical only for low-dimensional dynamical systems.

29.3.2 Quantifier Elimination

For some classes of systems the reach sets can be computed symbolically using *quantifier elimination*. Quantifier elimination is the removal of all quantifiers (the universal quantifier $\forall$ and the existential quantifier $\exists$) from a quantified system. Each quantified formula is substituted with quantifier-free expression with operations $+$, $\times$, $=$ and $<$. For example, consider the discrete-time linear system (Equations 29.1d, 29.3), with $A(t) = A = \begin{bmatrix} 0 & 1 \\ 0 & 0 \end{bmatrix}$ and $B(t) = B = \begin{bmatrix} 0 \\ 1 \end{bmatrix}$. For initial conditions $x_0 \in \{x \in \mathbf{R}^2 \mid \|x\|_\infty \leq 1\}$ and controls $u(t) \in \{u \in \mathbf{R} \mid -1 \leq u \leq 1\}$, the reach set for $t \geq 0$ is given by the quantified formula

$$\left\{ x \in \mathbf{R}^2 \mid \exists x_0, \exists t \geq 0, \exists u(\tau), 0 \leq \tau < t : x = A^t x_0 + \sum_{\tau=0}^{t-1} A^{t-\tau-1} Bu(\tau) \right\},$$

which is equivalent to the quantifier-free expression

$$-1 \leq [1 \ \ 0]x \leq 1 \ \wedge \ -1 \leq [0 \ \ 1]x \leq 1.$$

It is proved in [27] that if A is constant and nilpotent or is diagonalizable with rational real or purely imaginary eigenvalues, the quantifier elimination package returns a quantifier free formula describing the reach set. This class of systems is evidently rather limited.

Requiem [10] is a Mathematica notebook which, given a linear system, the set of initial conditions and control bounds, symbolically computes the exact reach set, using the experimental quantifier elimination package.

29.3.3 Polytope Method

Definition 29.11: Hyperplane

The hyperplane $H(c, \gamma)$ in $\mathbf{R}^n$ is the set

$$H = \{x \in \mathbf{R}^n \mid \langle c, x \rangle = \gamma\}$$

with fixed $c \in \mathbf{R}^n$ and $\gamma \in \mathbf{R}$.

A hyperplane defines two (closed) *halfspaces*,

$$\mathbf{S}_1 = \{x \in \mathbf{R}^n \mid \langle c, x \rangle \leq \gamma\}, \quad \text{and} \quad \mathbf{S}_2 = \{x \in \mathbf{R}^n \mid \langle c, x \rangle \geq \gamma\}.$$

Definition 29.12: Polytope

The polytope $P(C, g)$ is the intersection of a finite number of closed halfspaces:

$$P = \{x \in \mathbf{R}^n \mid Cx \leq g\},$$

with fixed $C = [c_1 \ \cdots \ c_m]^T \in \mathbf{R}^{m \times n}$ and $g = [\gamma_1 \ \cdots \ \gamma_m]^T \in \mathbf{R}^m$.

For linear discrete-time systems (Equations 29.7d and 29.21), with $\mathcal{X}_0, \mathcal{U}(t)$ and $\mathcal{V}(t), t \geq t_0$, being polytopes in $\mathbf{R}^n$, $\mathbf{R}^m$, and $\mathbf{R}^d$ respectively, the reach sets $\overline{\mathcal{X}}_{CL}(t, t_0, \mathcal{X}_0)$ and $\underline{\mathcal{X}}_{CL}(t, t_0, \mathcal{X}_0)$ are also polytopes, because the polytope structure is closed under the operations of affine transformation, geometric sum, and geometric difference. (For continuous-time systems the reach sets need not be polytopes.) Starting with initial condition

$$\overline{\mathcal{X}}_{CL}(t_0, t_0, \mathcal{X}_0) = \underline{\mathcal{X}}_{CL}(t_0, t_0, \mathcal{X}_0) = \mathcal{X}_0,$$

for time step $t > t_0$ these reach sets are computed *exactly* as

$$\overline{\mathcal{X}}_{CL}(t, t_0, \mathcal{X}_0) = A(t-1)\mathcal{X}(t-1, t_0, \mathcal{X}_0) \oplus B(t-1)\mathcal{U}(t-1) \dot{-} G(t-1)\mathcal{V}(t-1),$$

and

$$\underline{\mathcal{X}}_{CL}(t, t_0, \mathcal{X}_0) = A(t-1)\mathcal{X}(t-1, t_0, \mathcal{X}_0) \dot{-} G(t-1)\mathcal{V}(t-1) \oplus B(t-1)\mathcal{U}(t-1).$$

A similar procedure works for backward reach sets if the matrices $A(t)$, $t < t_1$ are nondegenerate.

As we see, the polytope method consists in sequential computation of affine transformations of polytopes, geometric sum of two polytopes, and geometric difference of two polytopes (or, geometric difference first, then geometric sum for minmax CLRS). In the absence of disturbances the term $G(t-1)\mathcal{V}(t-1)$ vanishes and no geometric difference operation is performed. Each operation of geometric sum or geometric difference for two polytopes consists in finding the vertices of the resulting polytope and calculating its convex hull.

This method is implemented in the *MultiParametric Toolbox* (MPT) for MATLAB® [7,26]. Among its advantages are its simplicity, the fact that the reach sets are computed exactly, and that it is easy to compute the distance between two polytopes and to check whether two or more polytopes intersect, or whether a polytope intersects a hyperplane or a halfspace.

However, the polytope method has its limitations. The convex hull algorithm employed by MPT is based on the Double description method [32] and implemented in the *CDD/CDD+* package [1]. Its complexity is K^n, where K is the number of vertices and n is the state-space dimension. Hence, the use of MPT in general is practical only for low-dimensional systems, or for systems with very special structure of matrices A, B, and G that ensure that the number of polytope vertices does not grow too much with each time step. But even in low-dimensional systems the number of vertices in the reach set polytope can grow very large with the number of time steps. For example, consider the discrete-time linear time-invariant system with $A(t) = A = \begin{bmatrix} \cos 1 & -\sin 1 \\ \sin 1 & \cos 1 \end{bmatrix}$, $B(t) = I$, $G(t) = 0$ $u_k \in \{u \in \mathbf{R}^2 \mid \|u\|_\infty \leq 1\}$, and $x_0 \in \{x \in \mathbf{R}^2 \mid \|x\|_\infty \leq 1\}$. Starting with a rectangular initial set, the number of vertices of the reach set polytope is $4t + 4$ at the tth step.

29.3.4 *d/dt*

In *d/dt* [3], the reach set of a discrete-time linear system is approximated by unions of rectangular polytopes [13]. The algorithm works as follows. First, given the set of initial conditions $\mathcal{X}_0$ defined as a polytope at time t_0, the evolution in time of the polytope's extreme points is computed $\mathcal{X}(\tau_1, t_0, \mathcal{X}_0)$. Second, the algorithm computes the convex hull of vertices of both, the initial polytope $\mathcal{X}_0$ and $\mathcal{X}(\tau_1, t_0, \mathcal{X}_0)$. The resulting polytope is then bloated (magnified) to include $\bigcup_{t_0 \leq \tau \leq \tau_1} \mathcal{X}(\tau, t_0, \mathcal{X}_0)$. Finally, this overapproximating polytope is in its turn overapproximated by the union of rectangles. The same procedure is repeated for the next time interval $[\tau_1, \tau_2]$, and the union of both rectangular approximations is taken, and so on.

Rectangular polytopes are easy to represent and the number of facets grows linearly with dimension, but a large number of rectangles must be used to assure that the approximation is not overly conservative. Besides, the important part of this method is again the convex hull calculation whose implementation relies on the same CDD/CDD+ library. This limits the dimension of the system and time interval for which it is feasible to calculate the reach set.

d/dt is implemented in C.

29.3.5 Zonotope Method

Polytopes can give arbitrarily close approximations to any convex set, but the number of vertices can grow prohibitively large and, as shown in [14], the computation of a polytope by its convex hull becomes intractable for large number of vertices in high dimensions. Symmetric polytopes, called zonotopes [11], could be a solution.

Definition 29.13: Zonotope

A zonotope is a special class of polytopes of the form

$$Z = \left\{ x \in \mathbf{R}^n \mid x = c + \sum_{i=1}^{k} \alpha_i g_i, \quad -1 \leq \alpha_i \leq 1 \right\},$$

wherein c *and* $g_1, \ldots, g_k$ *are vectors in* $\mathbf{R}^n$.

Thus, a zonotope Z is compactly represented by its center c and generator vectors $g_1, \ldots, g_k$. The value k/n is called the order of the zonotope.

The zonotope method for external approximation of reach sets of discrete-time linear systems was introduced in [17], implemented in the *MATISSE* package for MATLAB [6], and further discussed in [18]. In [18], the authors introduce computational tricks that work only for *time-invariant* linear systems. The advantage of zonotopes is that they are closed under the operations of affine transformation and geometric sum, hence, the reach set of a discrete-time linear system (Equations 29.1d and 29.3), with $\mathcal{X}_0$ and $\mathcal{U}(t)$, $t \geq t_0$, being zonotopes, is also an zonotope. Similar properties hold for the backward reach set.

The problem with using zonotopes is that with every time step the order of the approximating zonotope increases by k/n. This difficulty can be averted by limiting the number of generator vectors, and overapproximating zonotopes whose number of generator vectors exceeds this limit by lower-order zonotopes. This may affect the accuracy of the reach set approximation and potentially destroy the semigroup property that is inherently present in the zonotope method.

Further limitations of zonotopes are that geometric difference of two zonotopes, intersections of zonotopes or zonotopes with hyperplanes or halfspaces, are not zonotopes. That presents a difficulty for the computation of reach sets for systems with disturbances and hybrid systems. Effective zonotope approximation algorithms for the geometric difference and intersections are needed. Currently, MATISSE does not provide a zonotope library in which these operations are implemented.

29.3.6 CheckMate

CheckMate [2] is a MATLAB toolbox that can evaluate specifications for trajectories starting from the set of initial (continuous) states corresponding to the parameter values at the vertices of the parameter set. This provides preliminary insight into whether the specifications will be true for all parameter values. The method of oriented rectangular polytopes for external approximation of reach sets is introduced in [36]. The basic idea is to construct an oriented rectangular hull of the reach set for every time step, whose orientation is determined by the singular value decomposition of the sample covariance matrix for the states reachable from the vertices of the initial polytope. The limitation of CheckMate and the method of oriented rectangles is that only autonomous (i.e., there is no control) systems are allowed, and only an external approximation of the reach set is provided.

Currently, the development of CheckMate is discontinued. Therefore, we refer the reader to *PHAVer* [8], the newly developed verification tool that uses *Parma Polyhedra Library* (PPL) [9] for its polyhedral computations.

29.3.7 Ellipsoidal Method

All the geometric methods for reach set computation described above, namely polytopes, zonotopes, rectangular hulls, and oriented rectangles employ the notion of time step. At every time step a certain algorithm runs producing a new reach set for that time step. This can work only for discrete-time systems. The ellipsoidal method offers a different approach that works for continuous- and discrete-time linear systems with disturbances, with ellipsoidal constraints on the initial or target set, controls, and disturbances.

Definition 29.14: Ellipsoid

The ellipsoid $\mathcal{E}(q, Q)$ *in* $\mathbf{R}^n$ *with center* q *and shape matrix* Q *is the set*

$$\mathcal{E}(q, Q) = \left\{x \in \mathbf{R}^n \mid \langle (x-q), Q^{-1}(x-q)\rangle \le 1\right\},$$

wherein Q *is positive definite* ($Q = Q^T$ *and* $\langle x, Qx\rangle > 0$ *for all nonzero* $x \in \mathbf{R}^n$).

Definition 29.15: Support Function

The support function of a set $\mathcal{X} \subseteq \mathbf{R}^n$ *is*

$$\rho(l \mid \mathcal{X}) = \sup_{x\in\mathcal{X}} \langle l, x\rangle.$$

In particular, the support function of an ellipsoid is

$$\rho(l \mid \mathcal{E}(q, Q)) = \langle l, q\rangle + \langle l, Ql\rangle^{1/2}. \tag{29.43}$$

We say that the ellipsoid $\mathcal{E}$ *tightly overapproximates* a given convex set $\mathcal{X}$ if there exist $l \in \mathbf{R}^n$ such that

$$\rho(\pm l \mid \mathcal{E}) = \rho(\pm l \mid \mathcal{X}) \quad \text{and} \quad \mathcal{X} \subseteq \mathcal{E}.$$

We say that ellipsoid $\mathcal{E}$ *tightly underapproximates* given convex set $\mathcal{X}$ if there exist $l \in \mathbf{R}^n$ such that

$$\rho(\pm l \mid \mathcal{E}) = \rho(\pm l \mid \mathcal{X}) \quad \text{and} \quad \mathcal{E} \subseteq \mathcal{X}.$$

The equality $\rho(\pm l \mid \mathcal{E}) = \rho(\pm l \mid \mathcal{X})$ means that the boundaries of $\mathcal{E}$ and $\mathcal{X}$ touch in directions l and $-l$.

In [23], the authors introduce parametrized families of external and internal ellipsoids that tightly overapproximate and underapproximate the reach set and derive the differential equations that govern the evolution in time of the center and the shape matrices of these ellipsoids. The reach set is represented as the intersection of tight external and as the union of tight internal ellipsoids. In [25], this result is extended to the discrete-time case with special emphasis on systems with degenerate matrices $A(t)$. In the next section we present the equations that describe ellipsoidal overapproximation and underapproximation of reach sets.

The ellipsoidal method provides the following benefits:

- Approximating the reach set of an n-dimensional discrete-time linear system by L ellipsoids over t time steps requires $t[L(8n^3 + 4n^2 + 2n) + 2n^2]$ scalar multiplications. The computational complexity grows polynomially with the system dimension, in contrast with the exponential growth of the polytope method complexity.

- It is possible to refine the reach set approximation as much as needed by adding more ellipsoids to the parameterized family. Theoretically, it is possible to exactly represent the reach set of linear system through both external and internal ellipsoids.
- It is possible to single out individual external and internal approximating ellipsoids that are optimal for a given criterion (e.g., trace, volume, diameter), or a combination of such criteria.
- For systems with no disturbance, there are simple analytical expressions for control sequences that steer the state to a desired target.

Ellipsoidal Toolbox (ET) for MATLAB [4] implements the reach set computations described here.

29.3.8 Parallelotope Method

The parallelotope* method [19] employs the idea of the ellipsoidal method to compute the reach sets of linear systems. The reach set is represented as the intersection of a parametrized family of tight external, and the union of a parametrized family of tight internal parallelotopes. The evolution equations for the centers and orientation matrices of both external and internal parallelotopes are provided. This method also finds controls that can drive the system to the boundary points of the reach set, similar to [23,37]. The computation to solve the evolution equations for tight approximating parallelotopes, however, is more involved than the one for ellipsoids, and in the case of discrete-time systems, this method does not deal with singular state transition matrices.

29.3.9 Other Methods

As was mentioned above, for certain verification problems, computation of reach sets can be avoided. For example, it may be enough to ensure that for given set of initial conditions $\mathcal{X}_0$, the trajectories of system (Equation 29.1) never enter a given target set $\mathcal{Y}_1$. In this case, the method of *barrier certificates* [34] may help. The idea as well as the main difficulty is to find a Lyapunov-like function $C(x)$ such that

1. $C(x) > 0$ in $\mathcal{Y}_1$
2. $C(x) \leq 0$ in $\mathcal{X}_0$
3. $\langle D_x C(x), f(t,x,u)\rangle \leq 0$ where $C(x) = 0$

If such a function exists, system (Equation 29.1) is "safe" with respect to the initial set $\mathcal{X}_0$ and the target set $\mathcal{Y}_1$, that is, system trajectories emanating from $\mathcal{X}_0$ never reach $\mathcal{Y}_1$.

Another example for which reach sets need not be computed exactly occurs when it is possible to ensure that for given initial set $\mathcal{X}_0$ there exist system trajectories that never leave $\mathcal{X}_0$. The set $\mathcal{X}_0$ is said to be *invariant* with respect to those trajectories.

In [12], the authors show that for certain classes of discrete-time dynamical systems with disturbances (Equation 29.7d) and certain initial sets $\mathcal{X}_0$, convex constraints on controls, and disturbances, for every disturbance there exist closed-loop control strategies that keep the state of the system inside $\mathcal{X}_0$.

For more information about invariant sets, we refer the reader to the survey paper [15] and references therein.

29.4 Ellipsoidal Method

Consider a continuous-time linear system

$$\dot{x}(t) = A(t)x(t) + B(t)u + G(t)v, \tag{29.44}$$

in which $x \in \mathbf{R}^n$ is the state, $u \in \mathbf{R}^m$ is the control, and $v \in \mathbf{R}^d$ is the disturbance. $A(t)$, $B(t)$, and $G(t)$ are continuous and take their values in $\mathbf{R}^{n\times n}$, $\mathbf{R}^{n\times m}$, and $\mathbf{R}^{n\times d}$, respectively. Control $u(t, x(t))$ and

* Parallelotope is a zonotope with n generator vectors in $\mathbf{R}^n$.

disturbance $v(t)$ are measurable functions restricted by ellipsoidal constraints: $u(t, x(t)) \in \mathcal{E}(p(t), P(t))$ and $v(t) \in \mathcal{E}(q(t), Q(t))$. The set of initial states at initial time t_0 is assumed to be the ellipsoid $\mathcal{E}(x_0, X_0)$.

The reach sets for systems with disturbances computed by the ET are CLRS. Henceforth, when describing backward reachability, reach sets refer to CLRS or CLBRS. Recall that for continuous-time linear systems maxmin and minmax CLRS coincide, and the same is true for maxmin and minmax CLBRS.

If the matrix $Q(\cdot) = 0$, the system (Equation 29.44) becomes an ordinary affine system with known $v(\cdot) = q(\cdot)$. If $G(\cdot) = 0$, the system becomes linear. For these two cases, ($Q(\cdot) = 0$ or $G(\cdot) = 0$), the reach set is as given in Definition 29.1, and so the reach set will be denoted as $\mathcal{X}_{CL}(t, t_0, \mathcal{E}(x_0, X_0)) = \mathcal{X}(t, t_0, \mathcal{E}(x_0, X_0))$.

The reach set $\mathcal{X}(t, t_0, \mathcal{E}(x_0, X_0))$ is a symmetric compact convex set, whose center evolves in time according to

$$\dot{x}_c(t) = A(t)x_c(t) + B(t)p(t) + G(t)q(t), \quad x_c(t_0) = x_0. \tag{29.45}$$

Fix a vector $l_0 \in \mathbf{R}^n$, and consider the solution $l(t)$ of the adjoint equation

$$\dot{l}(t) = -A^T(t)l(t), \quad l(t_0) = l_0, \tag{29.46}$$

which is equivalent to

$$l(t) = \Phi^T(t_0, t)l_0.$$

If the reach set $\mathcal{X}(t, t_0, \mathcal{E}(x_0, X_0))$ is nonempty, there exist tight external and tight internal approximating ellipsoids $\mathcal{E}(x_c(t), X_l^+(t))$ and $\mathcal{E}(x_c(t), X_l^-(t))$, respectively, such that

$$\mathcal{E}(x_c(t), X_l^-(t)) \subseteq \mathcal{X}(t, t_0, \mathcal{E}(x_0, X_0)) \subseteq \mathcal{E}(x_c(t), X_l^+(t)), \tag{29.47}$$

and

$$\rho(l(t) \mid \mathcal{E}(x_c(t), X_l^-(t))) = \rho(l(t) \mid \mathcal{X}(t, t_0, \mathcal{E}(x_0, X_0))) = \rho(l(t) \mid \mathcal{E}(x_c(t), X_l^+(t))). \tag{29.48}$$

The equation for the shape matrix of the external ellipsoid is

$$\begin{aligned}\dot{X}_l^+(t) = {} & A(t)X_l^+(t) + X_l^+(t)A^T(t) + \pi_l(t)X_l^+(t) + \frac{1}{\pi_l(t)}B(t)P(t)B^T(t) \\ & - (X_l^+(t))^{1/2}S_l(t)(G(t)Q(t)G^T(t))^{1/2} - (G(t)Q(t)G^T(t))^{1/2}S_l^T(t)(X_l^+(t))^{1/2}, \qquad (29.49)\\ X_l^+(t_0) = {} & X_0, \qquad (29.50)\end{aligned}$$

in which

$$\pi_l(t) = \frac{\langle l(t), B(t)P(t)B^T(t)l(t)\rangle^{1/2}}{\langle l(t), X_l^+(t)l(t)\rangle^{1/2}},$$

and the orthogonal matrix $S_l(t)$ ($S_l(t)S_l^T(t) = I$) is determined by the equation

$$S_l(t)(G(t)Q(t)G^T(t))^{1/2}l(t) = \frac{\langle l(t), G(t)Q(t)G^T(t)l(t)\rangle^{1/2}}{\langle l(t), X_l^+(t)l(t)\rangle^{1/2}}(X_l^+(t))^{1/2}l(t).$$

In the presence of disturbance, if the reach set is empty, the matrix $X_l^+(t)$ becomes sign indefinite. For a system without disturbance, the terms containing $G(t)$ and $Q(t)$ vanish from Equation 29.49.

The equation for the shape matrix of the internal ellipsoid is

$$\dot{X}_l^-(t) = A(t)X_l^-(t) + X_l^-(t)A^T(t) + (X_l^-(t))^{1/2}T_l(t)(B(t)P(t)B^T(t))^{1/2} \\ + (B(t)P(t)B^T(t))^{1/2}T_l^T(t)(X_l^-(t))^{1/2} - \eta_l(t)X_l^-(t) - \frac{1}{\eta_l(t)}G(t)Q(t)G^T(t), \quad (29.51)$$

$$X_l^-(t_0) = X_0, \quad (29.52)$$

in which

$$\eta_l(t) = \frac{\langle l(t), G(t)Q(t)G^T(t)l(t)\rangle^{1/2}}{\langle l(t), X_l^+(t)l(t)\rangle^{1/2}},$$

and the orthogonal matrix $T_l(t)$ is determined by the equation

$$T_l(t)(B(t)P(t)B^T(t))^{1/2}l(t) = \frac{\langle l(t), B(t)P(t)B^T(t)l(t)\rangle^{1/2}}{\langle l(t), X_l^-(t)l(t)\rangle^{1/2}}(X_l^-(t))^{1/2}l(t).$$

Similar to the external case, the terms containing $G(t)$ and $Q(t)$ vanish from Equation 29.51 for a system without disturbance.

The point where the external and internal ellipsoids touch the boundary of the reach set is given by

$$x_l^*(t) = x_c(t) + \frac{X_l^+(t)l(t)}{\langle l(t), X_l^+(t)l(t)\rangle^{1/2}}.$$

The boundary points $x_l^*(t)$ form trajectories, which we call *extremal trajectories*. Due to the nonsingular nature of the state transition matrix $\Phi(t, t_0)$, every boundary point of the reach set belongs to an extremal trajectory. To follow an extremal trajectory specified by parameter l_0, the system has to start at time t_0 at initial state

$$x_l^0 = x_0 + \frac{X_0 l_0}{\langle l_0, X_0 l_0\rangle^{1/2}}. \quad (29.53)$$

In the absence of disturbances, the open-loop control

$$u_l(t) = p(t) + \frac{P(t)B^T(t)l(t)}{\langle l(t), B(t)P(t)B^T(t)l(t)\rangle^{1/2}}. \quad (29.54)$$

steers the system along the extremal trajectory defined by the vector l_0. When a disturbance is present, this control keeps the system on an extremal trajectory if and only if the disturbance plays against the control always taking its extreme values.

Expressions 29.47 and 29.48 lead to the following fact:

$$\bigcup_{\langle l_0, l_0\rangle = 1} \mathcal{E}(x_c(t), X_l^-(t)) = \mathcal{X}(t, t_0, \mathcal{E}(x_0, X_0)) = \bigcap_{\langle l_0, l_0\rangle = 1} \mathcal{E}(x_c(t), X_l^+(t)).$$

In practice, this means that the more values of l_0 we use to compute $X_l^+(t)$ and $X_l^-(t)$, the better will be our approximation.

Remark about Discrete-Time Systems

For discrete-time linear system

$$x(t+1) = A(t)x(t) + B(t)u(t, x(t)) + G(t)v(t), \tag{29.44d}$$

the equivalent of Equation 29.46 is

$$l(t+1) = \left(A^T\right)^{-1}(t)l(t), \quad l(t_0) = l_0, \tag{29.46d}$$

which implies nonsingular $A(t)$.*

For discrete-time systems, maxmin and minmax CLRS do not coincide and are computed separately. For maxmin CLRS, the ellipsoidal approximation (external or internal) $\mathcal{E}(x_c(t+1), \overline{X}_l(t+1))$ defined by parameter $l_0 \in \mathbf{R}^n$ is computed as tight external or internal approximating ellipsoid of

$$\mathcal{E}(A(t)x_c(t), A(t)\overline{X}_l(t)A^T(t)) \oplus \mathcal{E}(B(t)p(t), B(t)P(t)B^T(t)) \dot{-} \mathcal{E}(G(t)q(t), G(t)Q(t)G^T(t)),$$

and for minmax CLRS, the ellipsoidal approximation $\mathcal{E}(x_c(t+1), \underline{X}_l(t+1))$ is computed as tight external or internal approximating ellipsoid of

$$\mathcal{E}(A(t)x_c(t), A(t)\underline{X}_l(t)A^T(t)) \dot{-} \mathcal{E}(G(t)q(t), G(t)Q(t)G^T(t)) \oplus \mathcal{E}(B(t)p(t), B(t)P(t)B^T(t))$$

specified by direction $l(t)$ that satisfies Equation 29.46d.

For details and equations related to the discrete-time case, we refer the reader to the manual of the ET [4].

Analogous results hold for the backward reach set.

Given the terminal time t_1 and ellipsoidal target set $\mathcal{E}(y_1, Y_1)$, the CLBRS $\mathcal{Y}_{CL}(t_1, t, \mathcal{Y}_1) = \mathcal{Y}(t_1, t, \mathcal{Y}_1)$, $t < t_1$, if it is nonempty, is a symmetric compact convex set whose center is governed by

$$y_c(t) = Ay_c(t) + B(t)p(t) + G(t)q(t), \quad y_c(t_1) = y_1. \tag{29.55}$$

Fix a vector $l_1 \in \mathbf{R}^n$, and consider

$$l(t) = \Phi(t_1, t)^T l_1. \tag{29.56}$$

If the backward reach set $\mathcal{Y}(t_1, t, \mathcal{E}(y_1, Y_1))$ is nonempty, there exist tight external and tight internal approximating ellipsoids $\mathcal{E}(y_c(t), Y_l^+(t))$ and $\mathcal{E}(y_c(t), Y_l^-(t))$ respectively, such that

$$\mathcal{E}(y_c(t), Y_l^-(t)) \subseteq \mathcal{Y}(t_1, t, \mathcal{E}(y_1, Y_1)) \subseteq \mathcal{E}(y_c(t), Y_l^+(t)), \tag{29.57}$$

and

$$\rho(l(t) \mid \mathcal{E}(y_c(t), Y_l^-(t))) = \rho(l(t) \mid \mathcal{Y}(t_1, t, \mathcal{E}(y_0, Y_0))) = \rho(l(t) \mid \mathcal{E}(y_c(t), Y_l^+(t))). \tag{29.58}$$

The equation for the shape matrix of the external ellipsoid is

$$\begin{aligned} \dot{Y}_l^+(t) &= A(t)Y_l^+(t) + Y_l^+(t)A^T(t) - \pi_l(t)Y_l^+(t) - \frac{1}{\pi_l(t)}B(t)P(t)B^T(t) \\ &\quad + (Y_l^+(t))^{1/2}S_l(t)(G(t)Q(t)G^T(t))^{1/2} + (G(t)Q(t)G^T(t))^{1/2}S_l^T(t)(Y_l^+(t))^{1/2}, \qquad (29.59) \\ Y_l^+(t_1) &= Y_1, \qquad (29.60) \end{aligned}$$

* The case when $A(t)$ is singular is described in [25]. The idea is to substitute $A(t)$ with the nonsingular $A_\delta(t) = A(t) + \delta U(t)W(t)$, in which $U(t)$ and $W(t)$ are obtained from the singular-value decomposition

$$A(t) = U(t)\Sigma(t)W(t).$$

The parameter δ can be chosen based on the number of time steps for which the reach set must be computed and the required accuracy. The issue of inverting ill-conditioned matrices is also addressed in [25].

in which

$$\pi_l(t) = \frac{\langle l(t), B(t)P(t)B^T(t)l(t)\rangle^{1/2}}{\langle l(t), Y_l^+(t)l(t)\rangle^{1/2}},$$

and the orthogonal matrix $S_l(t)$ satisfies the equation

$$S_l(t)(G(t)Q(t)G^T(t))^{1/2}l(t) = \frac{\langle l(t), G(t)Q(t)G^T(t)l(t)\rangle^{1/2}}{\langle l(t), Y_l^+(t)l(t)\rangle^{1/2}}(Y_l^+(t))^{1/2}l(t).$$

The equation for the shape matrix of the internal ellipsoid is

$$\begin{aligned}\dot{Y}_l^-(t) &= A(t)Y_l^-(t) + Y_l^-(t)A^T(t) - (Y_l^-(t))^{1/2}T_l(t)(B(t)P(t)B^T(t))^{1/2}\\ &\quad - (B(t)P(t)B^T(t))^{1/2}T_l^T(t)(Y_l^-(t))^{1/2} + \eta_l(t)Y_l^-(t) + \frac{1}{\eta_l(t)}G(t)Q(t)G^T(t),\end{aligned} \tag{29.61}$$

$$Y_l^-(t_1) = Y_1, \tag{29.62}$$

in which

$$\eta_l(t) = \frac{\langle l(t), G(t)Q(t)G^T(t)l(t)\rangle^{1/2}}{\langle l(t), Y_l^+(t)l(t)\rangle^{1/2}},$$

and the orthogonal matrix $T_l(t)$ is determined by the equation

$$T_l(t)(B(t)P(t)B^T(t))^{1/2}l(t) = \frac{\langle l(t), B(t)P(t)B^T(t)l(t)\rangle^{1/2}}{\langle l(t), Y_l^-(t)l(t)\rangle^{1/2}}(Y_l^-(t))^{1/2}l(t).$$

Just as in the forward reachability case, the terms containing $G(t)$ and $Q(t)$ vanish from Equations 29.59 and 29.61 in the absence of disturbances. The boundary value problems (Equations 29.55, 29.59 and 29.61) are converted to the initial value problems by the change of variables $s = -t$.

Owing to Equations 29.57 and 29.58,

$$\bigcup_{\langle l_1, l_1\rangle = 1} \mathcal{E}(y_c(t), Y_l^-(t)) = \mathcal{Y}(t_1, t, \mathcal{E}(y_1, Y_1)) = \bigcap_{\langle l_1, l_1\rangle = 1} \mathcal{E}(y_c(t), Y_l^+(t)).$$

Remark 29.3

In expressions 29.49, 29.51, 29.59, and 29.61, the terms $1/\pi_l(t)$ and $1/\eta_l(t)$ may not be well defined for some vectors l, because matrices $B(t)P(t)B^T(t)$ and $G(t)Q(t)G^T(t)$ may be singular. In such cases, we set these entire expressions to zero.

29.5 Applications

We illustrate the ellipsoidal approach with three applications.

29.5.1 Steering the System to a Target

Given system Equation 29.44, target set defined by ellipsoid $\mathcal{E}(y_1, Y_1)$, and terminal time t_1, we want to find a closed-loop control that steers the system from some state y_0 at time $t_0 < t_1$ to $\mathcal{E}(y_1, Y_1)$ at t_1.

First we compute external and internal ellipsoidal approximations $\mathcal{E}(y_c(t), Y_l^+(t))$ using Equations 29.59 and 29.60, and $\mathcal{E}(y_c(t), Y_l^-(t))$ using Equations 29.61 and 29.62, $t_0 \leq t < t_1$, for different values of the parameter $l_1 \in \mathbf{R}^n$. If there exists an external ellipsoid $\mathcal{E}(y_c(t_0), Y_l^+(t_0))$ such that $y_0 \notin \mathcal{E}(y_c(t_0), Y_l^+(t_0))$, there is no closed-loop control that can guarantee taking the system from y_0 at t_0 to a state within $\mathcal{E}(y_1, Y_1)$ at t_1. On the other hand, if there exists an internal ellipsoid $\mathcal{E}(y_c(t_0), Y_l^-(t_0))$ defined by the choice of l_1, such that $y_0 \in \mathcal{E}(y_c(t_0), Y_l^-(t_0))$, such a control does exist.

We build the closed-loop control $u(t, y(t))$ so as to keep the system state $y(t)$ inside, if possible, or as close as we can, if not, to the internal approximating ellipsoid $\mathcal{E}(y_c(t), Y_l^-(t))$ for $t_0 \leq t < t_1$. The steps below describe control synthesis at time t.

1. Compute
$$\gamma(t) = \langle y(t) - y_c(t), (Y_l^-(t))^{-1}(y(t) - y_c(t))\rangle.$$
If $\gamma(t) \leq 1$, then $y(t) \in \mathcal{E}(y_c(t), Y_l^-(t))$, and the control $u(t, y(t))$ can be chosen arbitrarily in $\mathcal{E}(p(t), P(t))$. For example, set $u(t, y(t)) = p(t)$.
2. Otherwise, if $\gamma(t) > 1$, $y(t)$ is a boundary point of ellipsoid $\mathcal{E}(y_c(t), \gamma(t)Y_l^-(t))$ corresponding to the direction $m(t) \in \mathbf{R}^n$,
$$m(t) = (Y_l^-(t))^{-1}(y(t) - y_c(t)).$$
In order to steer the system closer to the internal approximating ellipsoid, control $u(t, y(t))$ must act in the direction $-m(t)$.
3. Choose $u(t, y(t))$ so that the vector $B(t)u(t, y(t))$ is a boundary point of the ellipsoid $\mathcal{E}(B(t)p(t), B(t)P(t)B^T(t)) \subset \mathbf{R}^n$ in the direction $-m(t)$,
$$u(t, y(t)) = p(t) - \frac{P(t)B^T(t)m(t)}{\langle m(t), B(t)P(t)B^T(t)m(t)\rangle^{1/2}}.$$

To summarize,

$$u(t, y(t)) = \begin{cases} p(t), & \text{if } \langle y(t) - y_c(t), (Y_l^-(t))^{-1}(y(t) - y_c(t))\rangle \leq 1, \\ p(t) - \dfrac{P(t)B^T(t)(Y_l^-(t))^{-1}(y(t) - y_c(t))}{\langle (Y_l^-(t))^{-1}(y(t) - y_c(t)), B(t)P(t)B^T(t)(Y_l^-(t))^{-1}(y(t) - y_c(t))\rangle^{1/2}}, & \text{otherwise.} \end{cases} \tag{29.63}$$

The rigorous proof that this closed-loop control works can be found in [22]. In [16], the authors apply this technique to stop a high-dimensional oscillating system using the ET for backward reach set computation.

Formula (Equation 29.63) holds for discrete-time linear systems, except that instead of $Y_l^-(t)$, shape matrices of internal approximating ellipsoids for maxmin or minmax CLBRS must be used.

29.5.2 Switching System

A *switching system* is a system whose dynamics changes at known times. Consider the RLC circuit shown in Figure 29.1. It has two inputs, the voltage v and current i. Define

- x_1, the voltage across capacitor C_1, so $C_1\dot{x}_1$ is the corresponding current.
- x_2, the voltage across capacitor C_2, so the corresponding current is $C_2\dot{x}_2$.
- x_3, the current through the inductor L, so the voltage across the inductor is $L\dot{x}_3$.

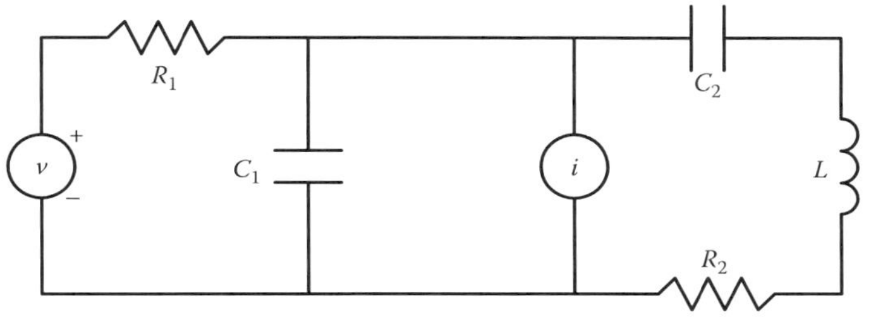

FIGURE 29.1 RLC circuit with two inputs.

Applying Kirchoff's laws we arrive at the linear system,

$$\begin{bmatrix} \dot{x}_1 \\ \dot{x}_2 \\ \dot{x}_3 \end{bmatrix} = \begin{bmatrix} -\frac{1}{R_1C_1} & 0 & -\frac{1}{C_1} \\ 0 & 0 & \frac{1}{C_2} \\ \frac{1}{L} & -\frac{1}{L} & -\frac{R_2}{L} \end{bmatrix} \begin{bmatrix} x_1 \\ x_2 \\ x_3 \end{bmatrix} + \begin{bmatrix} \frac{1}{R_1C_1} & \frac{1}{C_1} \\ 0 & 0 \\ 0 & 0 \end{bmatrix} \begin{bmatrix} v \\ i \end{bmatrix}. \tag{29.64}$$

The parameters R_1, R_2, C_1, C_2, and L, as well as the inputs, may depend on time. Suppose, for time $0 \le t < 2$, $R_1 = 2\Omega$, $R_2 = 1\Omega$, $C_1 = 3$ F, $C_2 = 7$ F, $L = 2$ H, and both inputs, v and i, are present and bounded by ellipsoid $\mathcal{E}(0, I)$; and for time $t \ge 2$, $R_1 = R_2 = 2\Omega$, $C_1 = C_2 = 3$ F, $L = 6$ H, the current source is turned off, and $|v| \le 1$. Then, system (Equation 29.64) can be rewritten as

$$\begin{bmatrix} \dot{x}_1 \\ \dot{x}_2 \\ \dot{x}_3 \end{bmatrix} = \begin{cases} \begin{bmatrix} -\frac{1}{6} & 0 & -\frac{1}{3} \\ 0 & 0 & \frac{1}{7} \\ \frac{1}{2} & -\frac{1}{2} & -\frac{1}{2} \end{bmatrix} \begin{bmatrix} x_1 \\ x_2 \\ x_3 \end{bmatrix} + \begin{bmatrix} \frac{1}{6} & \frac{1}{3} \\ 0 & 0 \\ 0 & 0 \end{bmatrix} \begin{bmatrix} v \\ i \end{bmatrix}, & 0 \le t < 2, \\ \begin{bmatrix} -\frac{1}{6} & 0 & -\frac{1}{3} \\ 0 & 0 & \frac{1}{3} \\ \frac{1}{6} & -\frac{1}{6} & -\frac{1}{3} \end{bmatrix} \begin{bmatrix} x_1 \\ x_2 \\ x_3 \end{bmatrix} + \begin{bmatrix} \frac{1}{6} \\ 0 \\ 0 \end{bmatrix} v, & 2 \le t. \end{cases} \tag{29.65}$$

We can use the ET to compute the reach set of Equation 29.65 for some time $t > 2$, say, $t = 3$.

```
>> % define system 1:
>> A1 = [-1/6 0 -1/3; 0 0 1/7; 1/2 -1/2 -1/2];
>> B1 = [1/6 1/3; 0 0; 0 0];
>> U1 = ellipsoid(eye(2));
>> s1 = linsys(A1, B1, U1);
>>
>> % define system 2:
>> A2 = [-1/6 0 -1/3; 0 0 1/3; 1/6 -1/6 -1/3];
>> B2 = [1/6; 0; 0];
>> U2 = ellipsoid(1);
>> s2 = linsys(A2, B2, U2);
>>
>> X0 = ellipsoid(0.01*eye(3));  % set of initial states
>> L0 = eye(3);  % 3 initial directions
>> TS = 2;  % time of switch
>> T = 3;  % terminal time
>>
>> % compute the reach set:
>> rs1 = reach(s1, X0, L0, TS);  % reach set of the first system
>> % computation of the second reach set starts
>> % where the first left off
>> rs2 = evolve(rs1, T, s2);
>>
>> % obtain projections onto (x1, x2) subspace:
```

```
>> BB = [1 0 0; 0 1 0]';  % (x1, x2) subspace basis
>> ps1 = projection(rs1, BB);
>> ps2 = projection(rs2, BB);
>>
>> % plot the results:
>> subplot(2, 2, 1);
>> plot_ea(ps1, 'r');  % external apprx. of reach set 1 (red)
>> hold on;
>> plot_ia(ps1, 'g');  % internal apprx. of reach set 1 (green)
>> plot_ea(ps2, 'y');  % external apprx. of reach set 2 (yellow)
>> plot_ia(ps2, 'b');  % internal apprx. of reach set 2 (blue)
>>
>> % plot the 3-dimensional reach set at time t = 3:
>> subplot(2, 2, 2);
>> plot_ea(cut(rs2, 3), 'y');
>> hold on;
>> plot_ia(cut(rs2, 3), 'b');
```

Figure 29.2a shows how the reach set of system (Equation 29.65) projected onto the (x_1, x_2) plane evolves in time from $t = 0$ to $t = 3$. The external reach set approximation for the first dynamics is in light gray, the internal approximation is in dark gray. The dynamics switches at $t = 2$. The external reach set approximation for the second dynamics is in very light gray, its internal approximation is in dark gray. The full three-dimensional external (very light gray) and internal (dark gray) approximations of the reach set at $t = 3$ are shown in Figure 29.2b.

To find out where the system should start at time $t = 0$ in order to reach a neighborhood `M` of the origin at time $t = 3$, we compute the backward reach set from $t = 3$ to $t = 0$.

```
>> M  = ellipsoid(0.01*eye(3));  % terminal set
>> TT = 3;  % terminal time
>>
>> % compute backward reach set:
>> % compute the reach set:
>> brs2 = reach(s2, M, L0, [TT TS]);  % second system comes first
>> brs1 = evolve(brs2, 0, s1);  % then the first system
>>
>> % obtain projections onto (x1, x2) subspace:
>> bps1 = projection(brs1, BB);
>> bps2 = projection(brs2, BB);
>>
>> % plot the results:
>> subplot(2, 2, 3);
>> plot_ea(bps1, 'r');  % external apprx. of backward reach set 1
       (red)
>> hold on;
>> plot_ia(bps1, 'g');  % internal apprx. of backward reach set 1
       (green)
>> plot_ea(bps2, 'y');  % external apprx. of backward reach set 2
       (yellow)
>> plot_ia(bps2, 'b');  % internal apprx. of backward reach set 2
       (blue)
>>
```

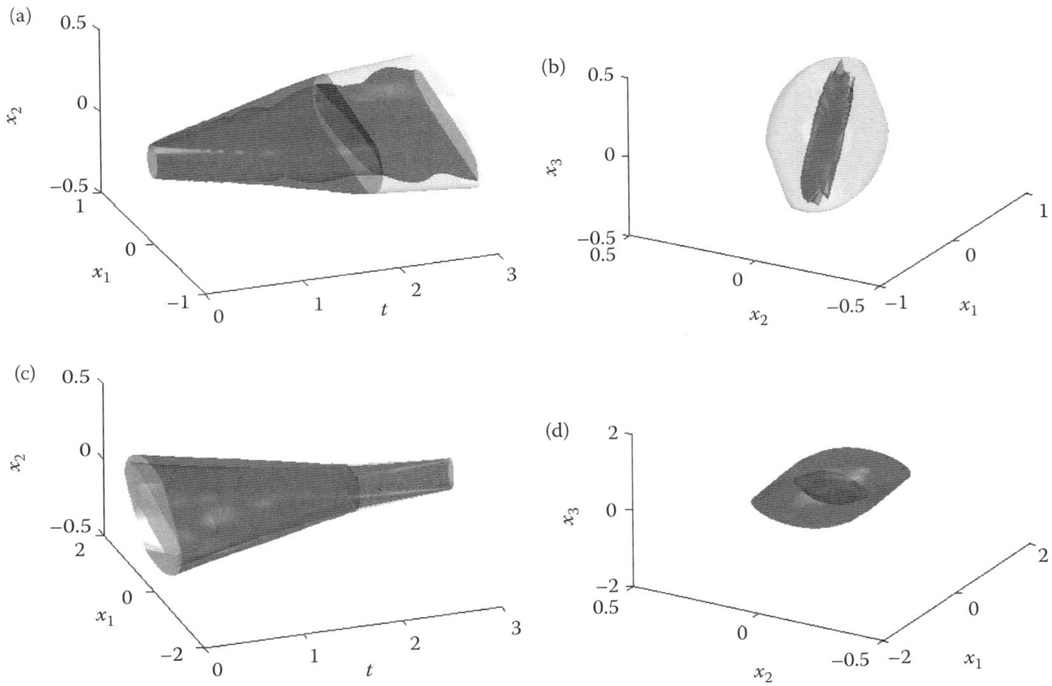

FIGURE 29.2 Forward and backward reach sets of the switched system (external and internal approximations). The dynamics switches at $t = 2$. (a) Forward reach set for the time interval $0 \leq t \leq 3$ projected onto (x_1, x_2) subspace. (b) Forward reach set at $t = 3$ in $\mathbf{R}^3$. (c) Backward reach set evolving from $t = 3$ to $t = 0$ projected onto (x_1, x_2) subspace. (d) Backward reach set at $t = 0$ in $\mathbf{R}^3$.

```
>> % plot the 3-dimensional backward reach set at time t = 0:
>> subplot(2, 2, 4);
>> plot_ea(cut(brs1, 0), 'r');
>> hold on;
>> plot_ia(cut(brs1, 0), 'g');
```

Figure 29.2c presents the evolution of the reach set projected onto the (x_1, x_2) plane in backward time. Again, external and internal approximations corresponding to the first dynamics are shown in dark and light gray, and these corresponding to the second dynamics in very light and dark gray. The full dimensional backward reach set external and internal approximations of system (Equation 29.65) at time $t = 0$ is shown in Figure 29.2d.

29.5.3 Hybrid System

There is no explicit implementation of the reachability analysis for hybrid systems in the ET. Nonetheless, the operations of intersection available in the toolbox allow us to work with certain class of hybrid systems, namely, hybrid systems with affine continuous dynamics whose guards are ellipsoids, hyperplanes, halfspaces, or polytopes.

We consider the *switching-mode model* of highway traffic presented in [33]. The highway segment is divided into N cells as shown in Figure 29.3. In this particular case, $N = 4$. The traffic density in cell i is x_i vehicles per mile, $i = 1, 2, 3, 4$.

Define

- v_i, the average speed in mph, in the ith cell, $i = 1, 2, 3, 4$.
- w_i, the backward congestion wave propagation speed in mph, in the ith highway cell, $i = 1, 2, 3, 4$.

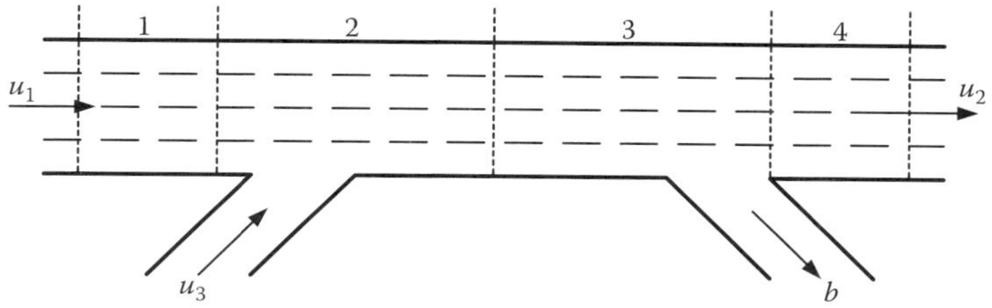

FIGURE 29.3 Highway model. (Adapted from L. Muñoz, et al. In *American Control Conference*, pp. 3750–3755, 2003.)

- x_{Mi}, the maximum allowed density in the ith cell; when this value is reached, there is a traffic jam, $i = 1, 2, 3, 4$.
- d_i, the length of ith cell in miles, $i = 1, 2, 3, 4$.
- T_s, the sampling time in hours.
- b, the split ratio for the off-ramp.
- u_1, the traffic flow coming into the highway segment, in vehicles per hour (vph).
- u_2, the traffic flow coming out of the highway segment (vph).
- u_3, the on-ramp traffic flow (vph).

Highway traffic operates in two modes: *free-flow* in normal operation; and *congested* mode, when there is a jam. Traffic flow in free-flow mode is described by

$$\begin{bmatrix} x_1[t+1] \\ x_2[t+1] \\ x_3[t+1] \\ x_4[t+1] \end{bmatrix} = \begin{bmatrix} 1-\dfrac{v_1T_s}{d_1} & 0 & 0 & 0 \\ \dfrac{v_1T_s}{d_2} & 1-\dfrac{v_2T_s}{d_2} & 0 & 0 \\ 0 & \dfrac{v_2T_s}{d_3} & 1-\dfrac{v_3T_s}{d_3} & 0 \\ 0 & 0 & (1-b)\dfrac{v_3T_s}{d_4} & 1-\dfrac{v_4T_s}{d_4} \end{bmatrix} \begin{bmatrix} x_1[t] \\ x_2[t] \\ x_3[t] \\ x_4[t] \end{bmatrix} + \begin{bmatrix} \dfrac{v_1T_s}{d_1} & 0 & 0 \\ 0 & 0 & \dfrac{v_2T_s}{d_2} \\ 0 & 0 & 0 \\ 0 & 0 & 0 \end{bmatrix} \begin{bmatrix} u_1 \\ u_2 \\ u_3 \end{bmatrix}. \tag{29.66}$$

The equation for the congested mode is

$$\begin{aligned} \begin{bmatrix} x_1[t+1] \\ x_2[t+1] \\ x_3[t+1] \\ x_4[t+1] \end{bmatrix} = {} & \begin{bmatrix} 1-\dfrac{w_1T_s}{d_1} & \dfrac{w_2T_s}{d_1} & 0 & 0 \\ 0 & 1-\dfrac{w_2T_s}{d_2} & \dfrac{w_3T_s}{d_2} & 0 \\ 0 & 0 & 1-\dfrac{w_3T_s}{d_3} & \dfrac{1}{1-b}\dfrac{w_4T_s}{d_3} \\ 0 & 0 & 0 & 1-\dfrac{w_4T_s}{d_4} \end{bmatrix} \begin{bmatrix} x_1[t] \\ x_2[t] \\ x_3[t] \\ x_4[t] \end{bmatrix} + \begin{bmatrix} 0 & 0 & \dfrac{w_1T_s}{d_1} \\ 0 & 0 & 0 \\ 0 & 0 & 0 \\ 0 & -\dfrac{w_4T_s}{d_4} & 0 \end{bmatrix} \begin{bmatrix} u_1 \\ u_2 \\ u_3 \end{bmatrix} \\ & + \begin{bmatrix} \dfrac{w_1T_s}{d_1} & -\dfrac{w_2T_s}{d_1} & 0 & 0 \\ 0 & \dfrac{w_2T_s}{d_2} & -\dfrac{w_3T_s}{d_2} & 0 \\ 0 & 0 & \dfrac{w_3T_s}{d_3} & -\dfrac{1}{1-b}\dfrac{w_4T_s}{d_3} \\ 0 & 0 & 0 & \dfrac{w_4T_s}{d_4} \end{bmatrix} \begin{bmatrix} x_{M1} \\ x_{M2} \\ x_{M3} \\ x_{M4} \end{bmatrix}. \end{aligned} \tag{29.67}$$

The switch from the free-flow to the congested mode occurs when the density x_2 reaches x_{M2}. In other words, the hyperplane $H([0\ 1\ 0\ 0]^T, x_{M2})$ is the guard. (When the state enters the guard, the system equation switches.)

We indicate how to implement the reach set computation of this hybrid system using the ET. We first define the two linear systems and the guard.

```
>> % assign parameter values:
>> v1 = 65; v2 = 60; v3 = 63; v4 = 65;  % mph
>> w1 = 10; w2 = 10; w3 = 10; w4 = 10;  % mph
>> d1 = 2; d2 = 3; d3 = 4; d4 = 2;  % miles
>> Ts = 2/3600;  % sampling time in hours
>> xM1 = 200; xM2 = 200; xM3 = 200; xM4 = 200;  % vehicles per lane
>> b = 0.4;
>>
>> A1 = [(1-(v1*Ts/d1)) 0 0 0
        (v1*Ts/d2) (1-(v2*Ts/d2)) 0 0
        0 (v2*Ts/d3) (1-(v3*Ts/d3)) 0
        0 0 ((1-b)*(v3*Ts/d4)) (1-(v4*Ts/d4))];
>> B1 = [v1*Ts/d1 0 0; 0 0 v2*Ts/d2; 0 0 0; 0 0 0];
>> U1 = ellipsoid([180; 150; 50], [100 0 0; 0 100 0; 0 0 25]);
>>
>> A2 = [(1-(w1*Ts/d1)) (w2*Ts/d1) 0 0
        0 (1-(w2*Ts/d2)) (w3*Ts/d2) 0
        0 0 (1-(w3*Ts/d3)) ((1/(1-b))*(w4*Ts/d3))
        0 0 0 (1-(w4*Ts/d4))];
>> B2 = [0 0 w1*Ts/d1; 0 0 0; 0 0 0; 0 -w4*Ts/d4 0];
>> U2 = U1;
>> G2 = [(w1*Ts/d1) (-w2*Ts/d1) 0 0
        0 (w2*Ts/d2) (-w3*Ts/d2) 0
        0 0 (w3*Ts/d3) ((-1/(1-b))*(w4*Ts/d3))
        0 0 0 (w4*Ts/d4)];
>> V2 = [xM1; xM2; xM3; xM4];
>>
>> % define linear systems:
>> s1 = linsys(A1, B1, U1, [], [], [], [], 'd');  % free-flow mode
>> s2 = linsys(A2, B2, U2, G2, V2, [], [], 'd');  % congestion mode
>>
>> % define guard:
>> GRD = hyperplane([0; 1; 0; 0], xM2);
```

We assume that initially the system is in free-flow mode. Given a set of initial conditions, we compute the reach set according to dynamics (Equation 29.66) for certain number of time steps. We consider an external approximation of the reach set by a single ellipsoid.

```
>> initial conditions:
>> X0 = [170; 180; 175; 170] + 10*ell_unitball(4);
>
>> L0 = [1; 0; 0; 0];  % single initial direction
>> N = 100;  % number of time steps
>>
>> ffrs = reach(s1, X0, L0, N);  % free-flow reach set
>> EA = get_ea(ffrs);  % 101x1 array of external ellipsoids
```

Having obtained the ellipsoidal array EA representing the reach set evolving in time, we determine the ellipsoids in the array that intersect the guard.

```
>> I = hpintersection(EA, GRD); % some of the intersections are empty
>> D = find(~isempty(I));   % determine nonempty intersections
>> min(D)

ans =

     19

>> max(D)

ans =

     69
```

Analyzing the values in array D, we conclude that the free-flow reach set has nonempty intersection with hyperplane GRD at $t = 18$ for the first time, and at $t = 68$ for the last time. Between $t = 18$ and $t = 68$ the reach set crosses the guard. Figure 29.4a shows the free-flow reach set projected onto the (x_1, x_2, x_3) subspace for $t = 10$, before the guard crossing; Figure 29.4b for $t = 50$, during the guard crossing; and Figure 29.4c for $t = 80$, after the guard was crossed.

For each time step that the intersection of the free-flow reach set and the guard is nonempty, we establish a new initial time and a set of initial conditions for the reach set computation according to dynamics (Equation 29.67). The initial time is the array index minus one, and the set of initial conditions is the intersection of the free-flow reach set with the guard.

```
>> crs = [];
>> for i = 1:size(D, 2)
     rs = reach(s2, I(D(i)), L0, [(D(i)-1) N]);
     crs = [crs rs];
   end
```

The union of reach sets in array crs forms the reach set for the congested dynamics.

A summary of the reach set computation of the linear hybrid system (Equations 29.66 and 29.67) for $N = 100$ time steps with one guard crossing is given in Figure 29.4d, which shows the projection of the reach set trace onto the (x_1, x_2) subspace. The system starts evolving in time in free-flow mode from a set of initial conditions at $t = 0$, whose boundary is shown in magenta. The free-flow reach set evolving from $t = 0$ to $t = 100$ is shown in blue. Between $t = 18$ and $t = 68$, the free-flow reach set crosses the guard. The guard is shown in red. For each nonempty intersection of the free-flow reach set and the guard, the congested mode reach set starts evolving in time until $t = 100$. All the congested mode reach sets are shown in green. Observe that in the congested mode the density x_2 in the congested part decreases slightly, while the density x_1 upstream of the congested part increases. The blue set above the guard is not actually reached, because the state evolves according to the green region.

29.6 Conclusion

Control problems with hard bounds on the control set and for which finite time behavior has to meet guarantees (reaching or avoiding a target set) despite disturbances cannot be addressed by traditional methods of design. Central to recent approaches to solving these design problems are the concept and calculation of the OLRS or CLRS. Effective reach set computational tools have been developed over the

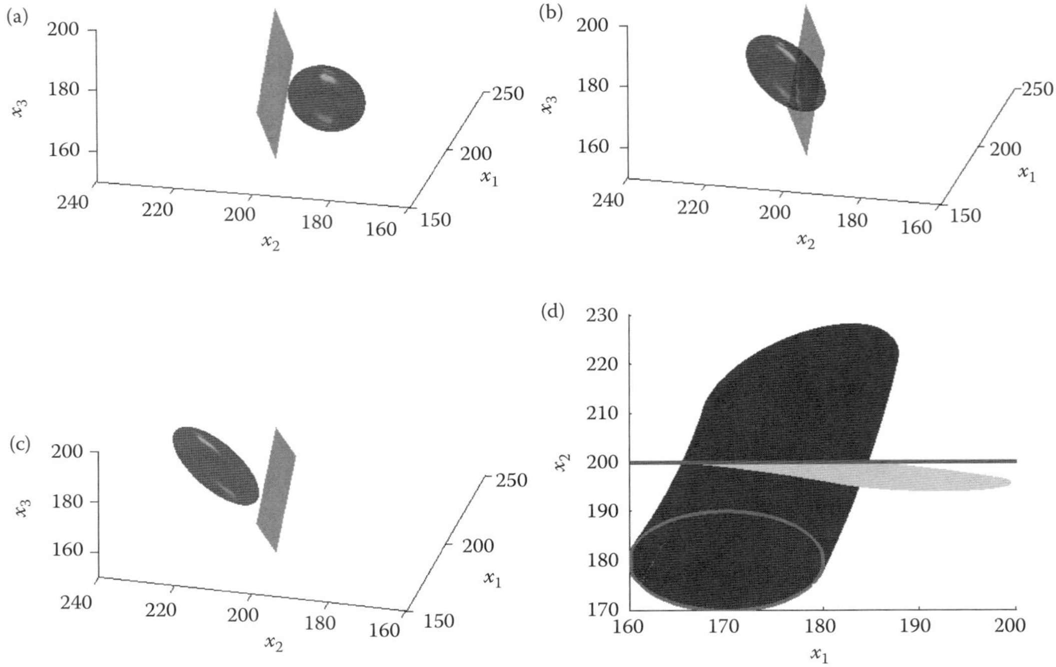

FIGURE 29.4 Reach set of the free-flow system is dark gray, reach set of the congested system is very light gray, the guard is gray straight line. (a) Reach set of the free-flow system at $t = 10$, before reaching the guard (projection onto (x_1, x_2, x_3)). (b) Reach set of the free-flow system at $t = 50$, crossing the guard (projection onto (x_1, x_2, x_3)). (c) Reach set of the free-flow system at $t = 80$, after the guard is crossed (projection onto (x_1, x_2, x_3)). (d) Reach set trace from $t = 0$ to $t = 100$, free-flow system in dark gray, congested system in green; bounds of initial conditions are marked with light gray (projection onto (x_1, x_2)).

past decade. These tools have been described. Among these tools, the ellipsoidal approach is the most promising. That approach, embodied in the ET, is illustrated by three examples.

References

1. CDD/CDD+ homepage. www.cs.mcgill.ca/~fukuda/soft/cdd_home/cdd.html.
2. CheckMate homepage. www.ece.cmu.edu/~webk/checkmate.
3. *d/dt* homepage. www-verimag.imag.fr/~tdang/ddt.html.
4. Ellipsoidal Toolbox homepage. code.google.com/p/ellipsoids.
5. Level Set Toolbox homepage. www.cs.ubc.ca/~mitchell/ToolboxLS.
6. MATISSE homepage. www.seas.upenn.edu/~agirard/Software/MATISSE.
7. MPT homepage. control.ee.ethz.ch/~mpt.
8. PHAVer homepage. www-verimag.imag.fr/~frehse/phaver_web.
9. PPL homepage. www.cs.unipr.it/ppl.
10. Requiem homepage. www.seas.upenn.edu/~hybrid/requiem/requiem.html.
11. Zonotope methods on Wolfgang Kühn homepage. www.decatur.de.
12. Z. Artstein and S. Raković. Feedback and invariance under uncertainty via set-iterates. *Automatica*, 44(2):520–525, 2008.
13. E. Asarin, O. Bournez, T. Dang, and O. Maler. Approximate reachability analysis of piecewise linear dynamical systems. In N. Lynch and B. H. Krogh, Eds, *Hybrid Systems: Computation and Control*, Lecture Notes in Computer Science, Vol. 1790, pp. 21–31. Springer, Berlin, 2000.

14. D. Avis, D. Bremner, and R. Seidel. How good are convex hull algorithms? *Computational Geometry: Theory and Applications*, 7:265–301, 1997.
15. F. Blanchini. Set-invariance in control. *Automatica*, 35(11):1747–1767, 1999.
16. A. N. Daryin, A. B. Kurzhanskiy, and I. V. Vostrikov. Reachability approaches and ellipsoidal techniques for closed-loop control of oscillating systems under uncertainty. In *IEEE Conference on Decision and Control*, Seville, Spain, 2006.
17. A. Girard. Reachability of uncertain linear systems using zonotopes. In M. Morari, L. Thiele, and F. Rossi, editors, *Hybrid Systems: Computation and Control*, Lecture Notes in Computer Science, Vol. 3414, pp. 291–305. Springer, Berlin, 2005.
18. A. Girard, C. Le Guernic, and O. Maler. Efficient computation of reachable sets of linear time-invariant systems with inputs. In J. Hespanha and A. Tiwari, Eds, *Hybrid Systems: Computation and Control*, Lecture Notes in Computer Science, Vol. 3927, pp. 257–271. Springer, Berlin, 2006.
19. E. K. Kostousova. Control synthesis via parallelotopes: Optimization and parallel computations. *Optimization Methods and Software*, 14(4):267–310, 2001.
20. N. N. Krasovski and A. I. Subbotin. *Positional Differential Games*. Springer-Verlag, New York, 1988.
21. A. B. Kurzhanski. Set-valued calculus and dynamic programming in problems of feedback control. Vol. 124, *International Series of Numerical Mathematics (ISNM)*, pp. 163–174. Birkhäuser, Basel, 1998.
22. A. B. Kurzhanski and I. Vályi. *Ellipsoidal Calculus for Estimation and Control*. Ser. SCFA. Birkhäuser, Boston, 1997.
23. A. B. Kurzhanski and P. Varaiya. On ellipsoidal techniques for reachability analysis. *Optimization Methods and Software*, 17:177–237, 2000.
24. A. B. Kurzhanski and P. Varaiya. Dynamic optimization for reachability problems. *Optimization Theory and Applications*, 108(2):227–251, 2001.
25. A. A. Kurzhanskiy and P. Varaiya. Ellipsoidal techniques for reachability analysis of discrete-time linear systems. *IEEE Transactions on Automatic Control*, 52(1):26–38, 2007.
26. M. Kvasnica, P. Grieder, M. Baotić, and M. Morari. Multi-parametric toolbox (MPT). In R. Alur and G. J. Pappas, Eds, *Hybrid Systems: Computation and Control*, Lecture Notes in Computer Science, Vol. 2993, pp. 448–462. Springer, Berlin, 2004.
27. G. Lafferriere, G. J. Pappas, and S. Yovine. Symbolic reachability computation for families of linear vector fields. *Journal of Symbolic Computation*, 32:231–253, 2001.
28. E. B. Lee and L. Markus. *Foundations of Optimal Control Theory*. Wiley & Sons, New York, 1961.
29. G. Leitmann. Optimality and reachability with feedback controls. In A. Blaquiere and G. Leitmann, Eds, *Dynamical Systems and Microphysics*, Vol. 1790, pp. 119–141. Academic Press, New York, NY, 1982.
30. J. Lygeros, C. Tomlin, and S. Sastry. Controllers for reachability specifications for hybrid systems. *Automatica*, 35:349–370, 1999.
31. I. Mitchell and C. Tomlin. Level set methods for computation in hybrid systems. In N. Lynch and B. H. Krogh, Eds, *Hybrid Systems: Computation and Control*, Lecture Notes in Computer Science, Vol. 1790, pp. 21–31. Springer, Berlin, 2000.
32. T. S. Motzkin, H. Raiffa, G. L. Thompson, and R. M. Thrall. The double description method. In H. W. Kuhn and A. W. Tucker, Eds, *Contributions to Theory of Games*, Vol. 2. Princeton University Press, Princeton, NJ, 1953.
33. L. Muñoz, X. Sun, R. Horowitz, and L. Alvarez. Traffic density estimation with the cell transmission model. In *American Control Conference*, pp. 3750–3755, 2003.
34. S. Prajna. Barrier certificates for nonlinear model validation. *Automatica*, 42(1):117–126, 2006.
35. A. Puri and P. Varaiya. Decidability of hybrid systems with rectangular differential inclusion. In *Proceedings of the 6th International Conference on Computer Aided Verification*, Lecture Notes in Computer Science, Vol. 818, pp. 95–104. Springer, Berlin, 1994.
36. O. Stursberg and B. H. Krogh. Efficient representation and computation of reachable sets for hybrid systems. In O. Maler and A. Pnueli, Eds, *Hybrid Systems: Computation and Control*, Lecture Notes in Computer Science, Vol. 2623, pp. 482–497. Springer, Berlin, 2003.
37. P. Varaiya. Reach set computation using optimal control. *Proceedings of KIT Workshop on Verification on Hybrid Systems*. Verimag, Grenoble., 1998.

30

Hybrid Dynamical Systems: Stability and Stabilization

Hai Lin
National University of Singapore

Panos J. Antsaklis
University of Notre Dame

30.1 Introduction

Hybrid systems are heterogeneous dynamical systems, the behaviors of which are determined by interacting continuous variables and discrete switching logics [1,2]. By heterogeneity, we mean hybrid systems containing two different kinds of dynamics. One is time-driven continuous variable dynamics, usually described by differential or difference equations; the other is event-driven discrete logic dynamics, whose evolutions depend on "if-then-else" type of rules and may be described by automata or Petri nets. In addition, these two kinds of dynamics interact with each other and generate complex dynamical behaviors, such as switching once certain continuous variable passes through a threshold, or state jumping when certain discrete event occurs. As a simple example, the temperature regulation in an air-conditioned room can be considered as a hybrid system; the room temperature evolution forms the continuous variable dynamics following thermophysical laws, whereas the on–off evolution of the air conditioner can be modeled as a discrete event process.

Hybrid systems have been identified in a wide variety of applications; in the control of mechanical systems, in process control, in automotive industry, power systems, aircraft, and traffic control, among many other fields. Specifically, hybrid systems have a central role in embedded control systems, where program codes interact with the physical world. In particular, the logic rules programmed in the embedded devices, which can be modeled as discrete event systems, are affected and influenced by the continuous variable physical processes, such as spatial location, temperature, and pressure evolutions. Studies in hybrid systems could provide a unified modeling framework for embedded systems, and systematic methods for performance analysis, verification, and embedded microcontroller design. Therefore, hybrid systems have attracted the attention of researchers not only from control engineering, but also from computer science and mathematics. Topics, such as modeling, verification, stability, controllability, optimal control, and supervisory control, have been extensively studied in the hybrid system literature, and the interested readers may refer to [1,2,4,5,13] and the references therein. In this chapter, we focus on the stability issues of hybrid systems.

It is known that the stability of hybrid systems includes several interesting phenomena due to the interaction of continuous variable dynamics and discrete switching logics [3,6,8]. For example, even when all the continuous variable subsystems are exponentially stable, the hybrid system may have divergent trajectories under certain discrete switching logic. On the other hand, one may carefully switch between unstable continuous variable subsystems to make the overall hybrid system exponentially stable. As these examples suggest, the stability of hybrid systems depends not only on the continuous variable dynamics of each subsystem, but also on the properties of discrete switching logics. Therefore, the stability study of hybrid systems can be roughly divided into two kinds of problems. One is the stability analysis of hybrid systems under given discrete switching logics; the other is the synthesis of stabilizing switching logics for a given collection of continuous variable dynamical systems.

We mainly focus on a subclass of hybrid systems that consists of a finite number of continuous-variable subsystems and a discrete logical rule, which orchestrates switching between these subsystems. The systems are usually called switched systems in the literature [6,8]. In this chapter, we use the terms "hybrid systems" and "switched systems" interchangeably. One convenient way to classify hybrid/switched systems is based on the dynamics of their subsystems, for example, continuous-time or discrete-time, linear or nonlinear and so on. In this chapter, we focus our attention to hybrid/switched systems where all subsystems are linear time-invariant (LTI) systems. The generalization of these results to nonlinear switched systems or more general cases are well documented in the literature; see, for example, survey papers [3,6,8] for further references.

The rest of this chapter is organized as follows. First, we focus on the stability analysis of hybrid systems under given discrete switching logics. In particular, some results on the stability analysis for hybrid systems under arbitrary switching are introduced in Section 30.2, while the stability under slow switching (like dwell time and average dwell time) is studied in Section 30.3.1. The general case of hybrid system stability under restricted switching is investigated in Section 30.3 through multiple Lyapunov functions. Then, we turn to the synthesis of stabilizing switching logic for a given collection of continuous variable dynamical systems in Section 30.4, where several stabilization conditions and design methods are described. Finally, the chapter concludes with remarks and a list of references.

30.2 Arbitrary Switching

In this section, we first consider the stability analysis problem where there are no restrictions on the discrete event dynamics of the hybrid system. This may be due to our lack of knowledge of the discrete event logic, or of the partitions of the state space, or of the constraints in the hybrid system of concern. Under these circumstances, one usually tends to be more conservative and assumes that all discrete switchings are possible; this is called *arbitrary switching* in the literature. On the other hand, when the stability under arbitrary switching is guaranteed, this could provide us with flexibility in the discrete logic design, where one may focus on improving the performances, since the closed-loop system stability is not a problem any longer.

30.2.1 Common Lyapunov Function

We know that a hybrid system may become unstable even when all subsystems are exponentially stable. Therefore, to identify conditions under which a hybrid system is stable under arbitrary switching is nontrivial and interesting. For this, it is necessary that all the subsystems are asymptotically stable, since if one subsystem were unstable, one switching strategy would have been to always select that subsystem all the time, which is a valid "switching logic" as well, and that would make the system unstable. In general, the above subsystems' stability assumption is not sufficient to ensure stability for the hybrid systems under arbitrary switching. However, if there exists a *common Lyapunov function* for all the subsystems, that is, a continuously differentiable, radially unbounded, positive definite function $V : \mathbb{R} \times \mathbb{R}^n \to \mathbb{R}$, for which the derivative $\dot{V}(x,t)$ is negative definite along all subsystems' trajectories, then the stability of the hybrid systems is guaranteed under arbitrary switchings. This provides us with a possible way to solve this problem, and a lot of research efforts have been focused on finding common quadratic Lyapunov functions (CQLFs).

30.2.1.1 Common Quadratic Lyapunov Functions

First, we consider a collection of continuous-time LTI systems

$$\dot{x}(t) = A_i x(t), \quad t \in \mathbb{R}^+, \quad i \in \mathfrak{I}, \tag{30.1}$$

where $\mathfrak{I}$ stands for a finite index set. For all $i \in \mathfrak{I}$, the state matrices $A_i \in \mathbb{R}^{n\times n}$. Note that the origin $x_e = 0$ is a common equilibrium for the systems described in Equation 30.1. The hybrid system of interest is built by allowing arbitrary switching among these LTI systems (Equation 30.1).

A CQLF for Equation 30.1 is a special class of Lyapunov functions of the form

$$V(x) = x^T P x, \tag{30.2}$$

where $P = P^T$ (symmetric) and $P > 0$ (positive definite). In addition, its time derivative along any trajectory of systems (Equation 30.1) is negative definite, or alternatively

$$A_i^T P + P A_i = -Q_i, \quad i \in \mathfrak{I}, \tag{30.3}$$

where Q_i are symmetric and positive definite for all $i \in \mathfrak{I}$. The existence of a CQLF for all its subsystems assures the quadratic stability of the hybrid system. Quadratic stability is a special class of exponential stability, which implies asymptotic stability, and has attracted a lot of research efforts due to its importance in practice.

A CQLF for Equation 30.1 can be obtained by solving a set of linear matrix inequalities (LMIs). for example, there exists a positive definite symmetric matrix P, $P \in \mathbb{R}^{n\times n}$, such that

$$PA_i + A_i^T P < 0, \ \forall i \in \mathcal{I}, \tag{30.4}$$

hold simultaneously. However, the standard interior point methods for LMIs may become ineffective as the number of subsystems increases. This motivates researchers to identify easily verifiable conditions that guarantee the existence of a CQLF for Equation 30.1. Here, we take a look at a well-studied special case in the literature; interested readers may refer to [6,8] for further references.

30.2.1.2 Commutative Systems

Let us first look at a special case, where the subsystems' state matrices are pairwise commutative, that is, $A_i A_j = A_j A_i$ for all $i, j \in \mathfrak{I}$. Because of the commutativity, it is easy to show that

$$A_i^{k_1} A_j^{k_2} = A_j^{k_2} A_i^{k_1}$$

for any nonnegative integer k_1 and k_2, and

$$e^{A_i t_1} e^{A_j t_2} = e^{A_j t_2} e^{A_i t_1}$$

for any nonnegative real number t_1 and t_2. By direct computation, it is straightforward to verify that in this case the arbitrary switching system is stable if and only if all its subsystems are stable.

Theorem 30.1:

For a collection of LTI systems (Equation 30.1) with the index set $\Im = \{1, \ldots, N\}$, if all subsystem matrices are stable (i.e., all eigenvalues of A_i have negative real part) and commute pairwise ($A_iA_j = A_jA_i, \forall i,j \in \Im$), then the hybrid system with subsystems given by Equation 30.1 is asymptotically stable under arbitrary switching.

A CQLF exists in this case, and can be determined by solving a collection of chained Lyapunov equations as shown in the following:

Theorem 30.2:

Assume that the index set $\Im = \{1, \ldots, N\}$. Let $P_1, \ldots, P_N$ be the unique symmetric positive definite matrices that satisfy the Lyapunov equations

$$A_1^T P_1 + P_1 A_1 = -I, \tag{30.5}$$

$$A_i^T P_i + P_i A_i = -P_{i-1}, \quad i = 2, \ldots, N. \tag{30.6}$$

Then the function $V(x) = x^T P_N x$ is a CQLF for systems $\dot{x}(t) = A_i x(t)$, $i = 1, \ldots, N$.

In addition, the matrix P_N can be expressed in integral form as

$$P_N = \int_0^\infty e^{A_N^T t_N} \cdots \left(\int_0^\infty e^{A_1^T t_1} e^{A_1 t_1} \, dt_1 \right) \cdots e^{A_N t_N} \, dt_N.$$

It is not difficult to extend this result to the discrete-time case.

Theorem 30.3:

Let $P_1, \ldots, P_N$ be the unique symmetric positive definite matrices that satisfy the Lyapunov equations

$$A_1^T P_1 A_1 + P_1 = -I, \tag{30.7}$$

$$A_i^T P_i A_i + P_i = -P_{i-1}, \quad i = 2, \ldots, N. \tag{30.8}$$

Then the function $V(x) = x^T P_N x$ is a common Lyapunov function for the systems $x[k+1] = A_i x[k]$, $i = 1, \ldots, N$.

In the literature, there exist several interesting necessary and also sufficient algebraic conditions for the existence of a CQLF for more general cases but usually for low-dimensional systems, and interested readers may consult [6,8] for further references. Note that Lie algebraic conditions were proposed in the literature; see, for example, [6], for arbitrary switching systems, which are based on the solvability of the Lie algebra generated by the subsystems' state matrices. It was shown that if the Lie algebra generated by the set of state matrices A_i is solvable, then there exists a CQLF, and the hybrid system is stable under arbitrary switching.

30.2.2 Switched Quadratic Lyapunov Functions

It should be pointed out that the existence of a CQLF is only sufficient for the stability of arbitrary switching systems. Therefore, in general, the existence of a CQLF is only sufficient for the asymptotic or exponential stability of hybrid systems under arbitrary switching, and could be rather conservative. Hence, some attention has been paid to a less conservative class of Lyapunov functions, called *switched quadratic Lyapunov functions*.

In this subsection, we investigate the stability of the following discrete-time arbitrary switching LTI systems:

$$x[k+1] = A_i x[k], \quad t \in \mathbb{Z}^+, \tag{30.9}$$

where $x \in \mathbb{R}^n$, and $i \in \mathfrak{I}$. Basically, since every subsystem is stable, there exists a positive definite symmetric matrix P_i that solves the Lyapunov equation for each ith subsystem:

$$A_i^T P_i A_i - P_i < 0,$$

for all $i \in \mathcal{I}$. Next, these matrices, P_i, are patched together based on the switching signals to construct a global Lyapunov function as

$$V(k, x[k]) = x^T[k] P_{\sigma(k)} x[k], \tag{30.10}$$

where $\sigma(k) : k \rightarrow \mathcal{I}$ stands for the switching signal at step k. Since all P_i are positive definite, it is clear that the function $V(k, x[k]) = x^T[k]P_{\sigma(k)}x[k]$ is also positive definite. If it further holds that $\Delta V(k, x[k]) = V(k+1, x[k+1]) - V(k, x[k])$ is negative definite along the solution of Equation 30.9, then the origin of the system (Equation 30.9) is globally asymptotically stable. In particular, a sufficient condition for the stability of the arbitrary switching system (Equation 30.9) is given as follows.

Theorem 30.4:

If there exist positive definite symmetric matrices $P_i \in \mathbb{R}^{n\times n}$ $(P_i = P_i^T)$ for $i \in \mathcal{I}$, satisfying

$$\begin{bmatrix} P_i & A_i^T P_j \\ P_j A_i & P_j \end{bmatrix} > 0 \tag{30.11}$$

for all $i, j \in \mathcal{I}$, then the linear system (Equation 30.9) with arbitrary switching is asymptotically stable.

The stability checking for arbitrary switching linear systems can be performed by solving LMIs.

It is clear that when $P_i = P_j$ for all $i, j \in \mathcal{I}$, the switched quadratic Lyapunov function becomes the CQLF. Therefore, the stability criteria based on the switched quadratic Lyapunov function generalize the CQLF approach and usually give us less conservative results. However, it is worth pointing out that the switched quadratic Lyapunov function method is still only a sufficient condition.

30.2.3 Necessary and Sufficient Conditions

In the sequel, we will provide some necessary and sufficient conditions for the asymptotic stability of arbitrary switching linear systems. It is shown that the asymptotic stability problem for hybrid linear systems with arbitrary switching is equivalent to the robust asymptotic stability problem for polytopic uncertain linear time-variant systems, for which several strong stability conditions exist.

Theorem 30.5: [8,10]

The following statements are equivalent:

1. *The arbitrary switching system*
$$\dot{x}(t) = A_{\sigma(t)}x(t),$$
where $A_{\sigma(t)} \in \{A_1, A_2, \ldots, A_N\}$, *is asymptotically stable;*
2. *The linear time-variant system*
$$\dot{x}(t) = A(t)x(t),$$
where $A(t) \in \mathcal{A} \hat{=} \mathrm{Conv}\{A_1, A_2, \ldots, A_N\}$, *where* $\mathrm{Conv}\{\cdot\}$ *stands for the convex combination, is asymptotically stable;*
3. *There exists a full column rank matrix* $L \in \mathbb{R}^{m\times n}$, $m \geq n$, *and a family of matrices* $\{\bar{A}_i \in \mathbb{R}^{m\times n} : i \in \mathcal{I}\}$ *with strictly negative row dominating diagonal, that is, for each* $\bar{A}_i, i \in \mathcal{I}$ *its elements satisfying*
$$\bar{a}_{kk} + \sum_{k\neq l} |\bar{a}_{kl}| < 0, \quad k = 1, \ldots, m,$$
such that the matrix relations
$$LA_i = \bar{A}_i L$$
are satisfied.

It is interesting to note that the nice property of $\bar{A}_i$ $(i \in \mathcal{I})$ implies the existence of a CQLF for the higher-dimensional arbitrary switching system. Unfortunately, applying the above Theorem is still difficult, because, in general, the numerical search for the matrix L is not simple. However, this equivalence bridges two research fields, namely the fields of hybrid system and robust stability. Therefore, existing results in the robust stability area, which has been extensively studied for over two decades, can be directly introduced to study the arbitrarily switching systems and vice versa. For example, it is known in the robust stability literature that the global attractiveness, (global) asymptotic stability, and (global) exponential stability are all equivalent for the polytopic uncertain linear time-variant systems [10]. Hence, these stability concepts are also equivalent for arbitrary switching systems. Similar results can be developed for the discrete-time case as shown below.

Theorem 30.6: [8,10]

The following statements are equivalent:

1. *The arbitrary switching system* $x[k+1] = A_{\sigma(k)}x[k]$, *where* $A_{\sigma(k)} \in \{A_1, A_2, \ldots, A_N\}$, *is asymptotically stable.*
2. *The linear time-variant system* $x[k+1] = A(k)x[k]$, *where* $A(k) \in \mathcal{A} \hat{=} \mathrm{Conv}\{A_1, A_2, \ldots, A_N\}$, *is robustly asymptotically stable.*
3. *There exists an integer* $m \geq n$ *and* $L \in \mathbb{R}^{n\times m}$, $rank(L) = n$ *such that for all* $A_i, i \in \mathcal{I}$, *there exists* $\bar{A}_i \in \mathbb{R}^{m\times m}$ *with the following properties:*
   a. $A_i^T L = L\bar{A}_i^T$.
   b. *Each column of* $\bar{A}_i$ *has no more than n nonzero elements and*
$$\|\bar{A}_i\|_\infty = \max_{1\leq k\leq m} \sum_{l=1}^{m} |\bar{a}_{kl}| < 1.$$

Based on the equivalence between the asymptotic stability of arbitrary switching linear systems and the robust stability of polytopic uncertain linear time-variant systems, some well-established converse Lyapunov theorems can be introduced for arbitrary switching linear systems. For example, the following results were taken from [10].

Theorem 30.7:

If the arbitrary switching system is exponentially stable, then it has a strictly convex, homogenous (of second order) common Lyapunov function of a quasi-quadratic form $V(x) = x^T L(x)x$, where $L(x) = L^T(x) = L(\tau x)$ for all nonzero $x \in \mathbb{R}^n$ and $\tau \in \mathbb{R}$.

Furthermore, we may restrict our search to include only polyhedral Lyapunov functions (also known as piecewise linear Lyapunov functions) as the following result pointed out.

Theorem 30.8:

If an arbitrary switching linear system is asymptotically stable, then there exists a polyhedral Lyapunov function, which is monotonically decreasing along the switched system's trajectories.

This converse Lyapunov theorem holds for both discrete-time and continuous-time cases, which suggests that the existence of a common Lyapunov function (not necessarily quadratic) is not only sufficient, but also necessary for the stability of a hybrid system under arbitrary switching.

Before we move on to another topic, let us take a look at the following example, which is taken from the robust stability literature.

Example 30.1:

Consider an arbitrary switching system, $\dot{x} = A_i x, i \in \{1, 2\}$, where

$$A_1 = \begin{bmatrix} 0 & 1 \\ -0.06 & -1 \end{bmatrix}; \quad A_2 = \begin{bmatrix} 0 & 1 \\ -1.94 & -1 \end{bmatrix}.$$

It is known that no CQLF exits. However, the arbitrary switching system is asymptotically stable, which is assured by the existence of a piecewise quadratic Lyapunov function; a particular piecewise linear Lyapunov function is also suggested in the robust literature.

30.3 Restricted Switching

Hybrid systems may fail to preserve stability under arbitrary switching. On the other hand, one may have some knowledge about the occurrence of possible discrete event dynamics in the hybrid systems and this knowledge can be translated into restrictions on the switching signals. For example, there must exist certain bounds on the time interval between two successive switchings, which may be due to the fact that the state trajectories have to spend some finite period of time in traveling from the initial set to certain boundary sets before switching, if these two sets are separated. With such kind of prior knowledge about the switching signals, we may derive stronger results on the stability for a given hybrid system instead of just using the worst-case arguments of the previous section.

30.3.1 Slow Switching

By studying the cases where divergent trajectories are generated through switching between two stable systems, one may note that the unboundedness is caused by the failure to absorb the energy increase caused by frequent switchings. In addition, when there is an unstable subsystem present (e.g., controller failure or sensor fault), if one either stays too long on it or switches too frequently to it, this may cause instability. Therefore, a natural question is what if we restrict the switching signals to some constrained subclasses. Intuitively, if one stays at stable subsystems long enough and switches less frequently, that is, slow switching, one may trade off the energy increase caused by switching or unstable modes, and it should perhaps become possible to attain stability. These ideas are proved to be reasonable and are captured by concepts such as *dwell time* and *average dwell time* [4] between switchings that are introduced below.

The simplest way to characterize the concept of slow switching is perhaps to request a lower bound on two consecutive switching times.

Definition 30.1:

A positive scalar τ_d is called the dwell time if the time interval between any two consecutive switchings is no smaller than τ_d.

Assume that all subsystems of the hybrid system are exponentially stable. Then, it can be shown that there exists a scalar $\tau_d > 0$ such that the hybrid system remains exponentially stable if the dwell time is larger than τ_d. In addition, one may give an estimate on the bound of the dwell time and decay rate.

In fact, it really does not matter if one occasionally has a smaller dwell time between switching, provided this does not occur too frequently. This concept is captured by the concept of "average dwell-time."

Definition 30.2:

A positive constant τ_a is called the average dwell time if $N_\sigma(t) \le N_0 + t/\tau_a$ holds for all $t > 0$ and some scalar $N_0 \ge 0$, where $N_\sigma(t)$ denotes the number of discontinuities of a given switching signal σ over $[0, t)$.

Here the constant τ_a is called the *average dwell time* and N_0 the *chatter bound.* The reason for calling a class of switching signals that satisfy

$$N_\sigma(t) \le N_0 + \frac{t}{\tau_a}$$

having an average dwell no less than τ_a is because

$$N_\sigma(t) \le N_0 + \frac{t}{\tau_a} \iff \frac{t}{N_\sigma(t) - N_0} \ge \tau_a.$$

This means that on an average the "dwell time" between any two consecutive switchings is no smaller than τ_a. The idea is that there may exist consecutive switching separated by less than τ_a, but the average time interval between consecutive switching is not less than τ_a.

Theorem 30.9:

Assume that all subsystems, $\dot{x} = A_i x$ for $i \in \mathfrak{I}$, in the hybrid system are exponentially stable. Then, there exists a scalar $\tau_a > 0$ such that the hybrid system is exponentially stable if the average dwell time is larger than τ_a.

Moreover, we can also obtain a bound on the decay rate.

Theorem 30.10:

Given a positive scalar λ_0 such that $A_i + \lambda_0 I$ is stable for all $i \in \mathfrak{I}$. Then, for any given $\lambda \in (0, \lambda_0)$, there exists a finite constant τ_a such that the hybrid system is exponentially stable with decay rate λ provided that the average dwell time is no less than τ_a.

The stability results for slow switching can be extended to discrete-time case, where the dwell time τ_d or average dwell time τ_a are counted as the number of sampling periods. In particular,

---

Definition 30.3:

A positive constant τ_a is called the average dwell time if $N_\sigma(k) \leq N_0 + k/\tau_a$ holds for all $k > 0$ and some scalar $N_0 \geq 0$, where $N_\sigma(k)$ denotes the number of switchings of a given switching signal σ over $[0, k)$.

Theorem 30.11:

Given a positive scalar λ_0 such that A_i/λ_0 is stable for all $i \in \mathfrak{I}$. Then, for any given $\lambda \in (\lambda_0, 1)$, there exists a finite constant τ_a such that the hybrid system, consisting of $x[k+1] = A_i x[k]$ as its subsystems, is exponentially stable with decay rate λ provided that the average dwell time is no less than τ_a.

Interested readers may refer to the survey papers [4,6,8] for further references on the stability of hybrid systems under slow switchings.

We continue our study of the stability of hybrid systems under restricted switchings in this section. It should be pointed out that not all restrictions on switching signals can be captured by the dwell time or average dwell time. For example, it is difficult to transform the invariant set constraints, guard set constraints, and so on, which determine the switching signals, into only dwell-time or average dwell time restrictions on switching signals. The main difficulty comes from the fact that most constraints in hybrid systems are state dependent and in the form of partitions of the state space, and so it is hard to transform them into pure time-dependent constraints such as dwell time, and so on. This calls for a more general tool to study hybrid system stability, and we introduce a powerful tool, *multiple Lyapunov functions* (MLFs).

30.3.2 Multiple Lyapunov Functions

The stability analysis under constrained switching has usually been pursued in the framework of MLFs. The basic idea is to use multiple Lyapunov or Lyapunov-like functions, each of which may correspond to a single subsystem or certain regions in the state space, concatenated together to produce a nontraditional Lyapunov function. The nontraditionality is in the sense that the MLF may not be monotonically decreasing along the state trajectories, may have discontinuities, and be piecewise differentiable. The reason for considering nontraditional Lyapunov functions is that the traditional Lyapunov function may not exist for hybrid systems with restricted switching signals. For such cases, one may still construct a collection of Lyapunov-like functions, which only requires nonpositive Lie-derivative for certain subsystems in a certain region of the state space instead of globally negativity conditions.

Lyapunov-like functions are defined as a family of real-valued functions $\{V_i, i = 1, \ldots, N\}$ with certain properties, each associated with the vector field $\dot{x} = f_i(x)$ that represents the continuous dynamics for the hybrid system under the ith discrete mode.

Definition 30.4: Lyapunov-Like Function

By saying that a subsystem has an associated Lyapunov-like function V_i in region $\Omega_i \subseteq \mathbb{R}^n$, we mean that

1. *There exist constant scalars $\beta_i \geq \alpha_i > 0$, such that*

$$\alpha_i \|x\|^2 \leq V_i(x) \leq \beta_i \|x\|^2$$

   holds for any $x \in \Omega_i$.
2. *For all $x \in \Omega_i$ and $x \neq 0$, $\dot{V}_i(x) < 0$.*

Here $\dot{V}_i(x) = (\partial V_i(x)/\partial x) f_i(x)$. The first condition implies positiveness and radius unboundedness for $V_i(x)$ when $x \in \Omega_i$, while the second condition guarantees the decrease of the abstracted energy, value of function $V_i(x)$, along the trajectories of subsystem i inside Ω_i.

Suppose that these regions Ω_i cover the whole state space, and so a cluster of Lyapunov-like functions is obtained. By concatenating these Lyapunov-like functions together, we obtain a nontraditional Lyapunov function, called MLF, which can be used to study the global stability of hybrid systems. MLFs are proved to be a powerful tool for studying the stability of switched systems and hybrid systems; see, for example, [3,6,8,9]. There are several versions of MLF results in the literature. A very intuitive MLF result [3] is illustrated in Figure 30.1, where the Lyapunov-like function decreases when the corresponding mode is active and does not increase its value at each switching instant. Formally, this result can be stated by the following theorem [3].

Theorem 30.12:

Suppose that each subsystem has an associated Lyapunov-like function V_i in its active region Ω_i, each with equilibrium point $x = 0$. Also, suppose that $\bigcup_i \Omega_i = \mathbb{R}^n$. Let $\sigma(t)$ be a class of piecewise-constant switching

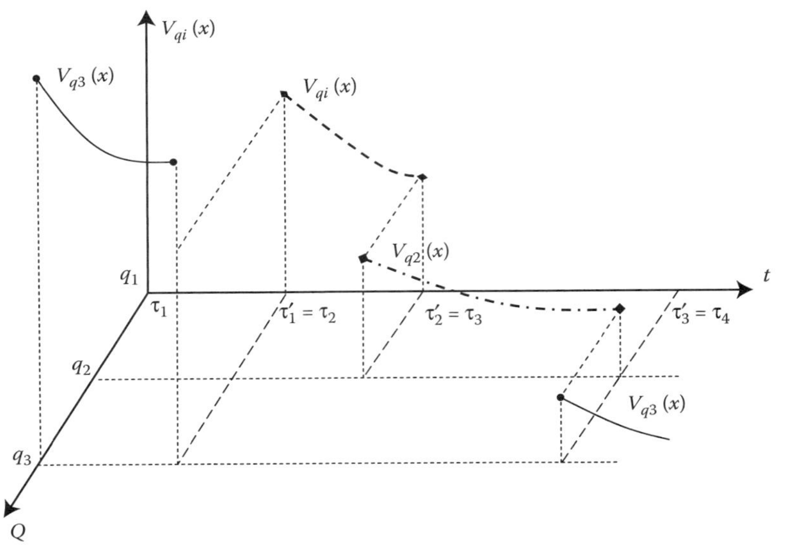

FIGURE 30.1 The hybrid system is asymptotically stable if the Lyapunov-like functions' values at the switching instants form a decreasing sequence.

sequences such that $\sigma(t)$ *can take value i only if* $x(t) \in \Omega_i$*, and in addition*

$$V_j(x(t_{i,j})) \leq V_i(x(t_{i,j})),$$

where $t_{i,j}$ *denotes the time when switched system switches from subsystem i to subsystems j, that is,* $x(t_{i,j}^-) \in \Omega_i$ *while* $x(t_{i,j}) \in \Omega_j$*. Then, the switched linear system (Equation 30.1) is exponentially stable under the switching signals* $\sigma(t)$*.*

The above MLF theorem requires that at each switching instant the Lyapunov-like function does not increase its value, which is quite conservative. Actually, one may obtain less conservative results. For example, the switching signals may be restricted in such a way that at every time when we exit (switch from) a certain subsystem its corresponding Lyapunov-like function value is smaller than its value at the previous exiting time. Then the switched system is asymptotically stable. In other words, for each subsystem, the corresponding Lyapunov-like function values at every exiting instant form a monotonically decreasing sequence. Alternatively, the decreasing tendency is captured by the Lyapunov-like function's value at the entering instant instead. This case is illustrated in Figure 30.2. This result can be presented as follows.

Theorem 30.13: [3]

Assume that there exists a family of Lyapunov functions $\{V_i : i \in \mathfrak{I}\}$ *for each stable subsystem. If for any two switching instants* t_i *and* t_j *such that* $i < j$ *and* $\sigma(t_i) = \sigma(t_j)$ *we have*

$$V_{\sigma(t_j)}(x(t_{j+1})) - V_{\sigma(t_i)}(x(t_{i+1})) \leq -\rho\|x(t_{i+1})\|^2,$$

for some constant $\rho > 0$*, then the switched system is asymptotically stable.*

Furthermore, as shown in [9], the Lyapunov-like function may increase its value during a time interval, only if the increment is bounded by certain kind of continuous functions as illustrated in Figure 30.3.

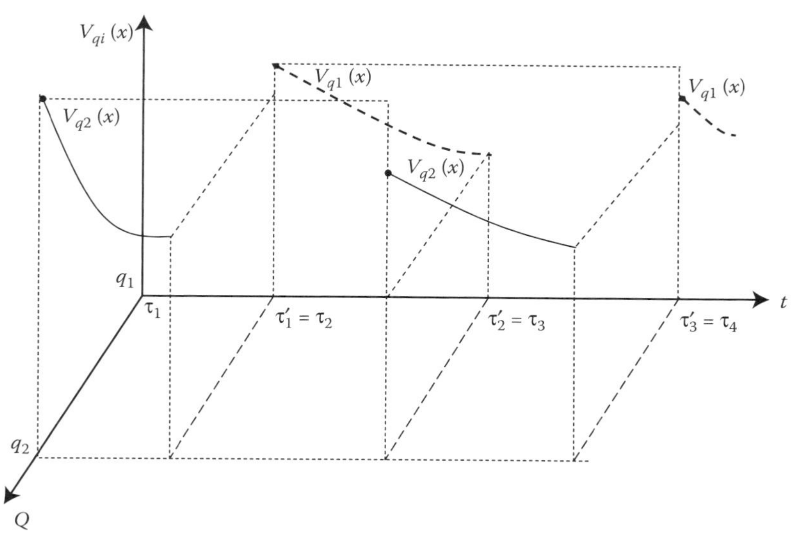

FIGURE 30.2 For every subsystem, its Lyapunov-like function's value V_i at the start point of each interval exceeds the value at the start point of the next interval on which the ith subsystem is activated, then the hybrid system is asymptotically stable.

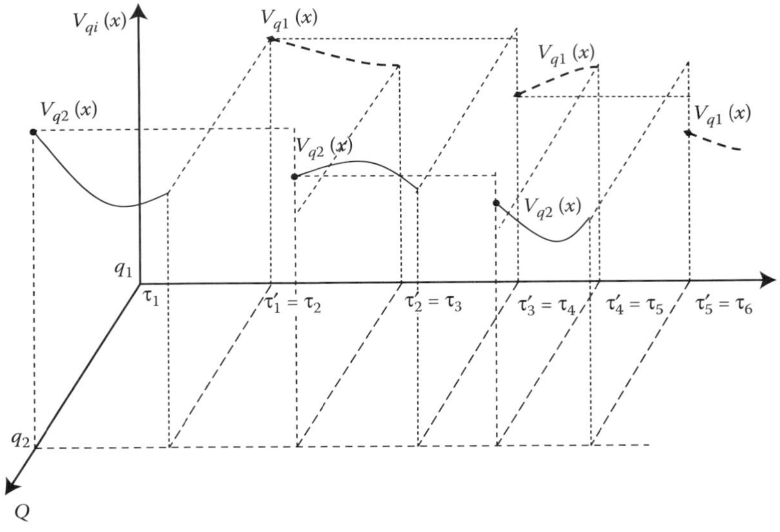

FIGURE 30.3 The hybrid system can remain stable even when the Lyapunov-like function increases its value during a certain period.

Interested readers may refer to the survey papers [3,6,9] and references therein. Note that all the arguments for continuous-time hybrid/switched systems can be extended to the discrete-time case without essential differences.

30.3.3 Piecewise Quadratic Lyapunov Functions

The critical challenge of applying the MLF theorems to practical switched/hybrid systems is how to construct a proper family of Lyapunov-like functions. Usually this is a hard problem. However, if one focuses on the linear case, piecewise quadratic Lyapunov-like functions could be attractive candidates, since the stability conditions in the MLF theorems can be formulated as LMIs [3,5], for which efficient software solution packages exist.

Considering the hybrid system with an LTI subsystem, $\dot{x}(t) = A_i x(t)$, since we do not assume that the subsystem is stable, there may not exist a quadratic Lyapunov function in a classical sense. However, it is still possible to restrict our search to certain regions of the state space, say $\Omega_i \subset \mathbb{R}^n$, and the energy of the ith subsystem could be decreasing along the trajectories inside this region (there is no decreasing requirements outside Ω_i). Suppose that the union of all these regions Ω_i covers the whole state space; then we obtain a cluster of Lyapunov-like functions. Broadly speaking, the problem entails searching for Lyapunov-like functions whose associated Ω-region covers the state space.

Assume that the state space $\mathbb{R}^n$ has a partition given by $\{\Omega_1, \ldots, \Omega_N\}$, and these regions Ω_i are defined *a priori* as a restriction of the possible switching signals (state-dependent). In this subsection, we present LMI conditions for the existence of quadratic Lyapunov-like functions of the form of $V_i(x) = x^T P_i x$, assigned to each region Ω_i. A Lyapunov-like function $V_i(x) = x^T P_i x$ needs to satisfy the following two conditions:

Condition 30.1:

There exist constant scalars $\beta_i \geq \alpha_i > 0$ *such that*

$$\alpha_i \|x\|^2 \leq V_i(x) \leq \beta_i \|x\|^2$$

hold for all $x \in \Omega_i$.

Consider a quadratic Lyapunov-like function candidate, $V_i(x) = x^T P_i x$, and require that

$$\alpha_i x^T I x \leq x^T P_i x \leq \beta_i x^T I x$$

holds for any $x \in \Omega_i$. That is

$$\begin{cases} x^T(\alpha_i I - P_i)x \leq 0 \\ x^T(P_i - \beta_i I)x \leq 0 \end{cases}$$

holds for all $x \in \Omega_i$.

Condition 30.2:

For all $x \in \Omega_i$ and $x \neq 0$, $\dot{V}_i(x) < 0$.

This negativeness of the Lyapunov-like function's derivative along the trajectories of a subsystem can be represented as $\exists P_i, (P_i = P_i^T)$ such that

$$x^T[A_i^T P_i + P_i A_i]x < 0 \tag{30.12}$$

for $x \in \Omega_i$.

30.3.3.1 Switching Condition

In addition, based on the MLF theorem of [3], for stability it is also required that the Lyapunov-like functions' values at switching instant are nonincreasing, which can be expressed by

$$x^T P_j x \leq x^T P_i x$$

for $x \in \Omega_{i,j} \subseteq \Omega_i \bigcap \Omega_j$. The region $\Omega_{i,j}$ stands for the states where the trajectory passes from region Ω_i to Ω_j.

Note that all the above matrix inequalities are constrained in a local region, such as $x \in \Omega_i$ or $\Omega_{i,j}$. A technique called $\mathcal{S}$-procedure can be applied to replace a constrained matrix inequality condition by a condition without constraints. To employ the $\mathcal{S}$-procedure, the regions Ω_i and $\Omega_{i,j}$ need to be expressed by or be contained in regions characterized by quadratic forms. For simplicity, we assume here that each region Ω_i has a quadratic representation or approximation, that is

$$\Omega_i = \{x|\ x^T Q_i x \geq 0\},$$

and regions $\Omega_{i,j}$ can be expressed or approximated by

$$\Omega_{i,j} = \{x|\ x^T Q_{i,j} x \geq 0\}.$$

Then the above matrix inequalities can be transformed into unconstrained ones based also on the $\mathcal{S}$-procedure, namely

---

Theorem 30.14:

The system (Equation 30.1) is (exponentially) stable if there exist matrices P_i $(P_i = P_i^T)$ and scalars $\alpha > 0, \beta > 0, \mu_i \geq 0, \nu_i \geq 0, \vartheta_i \geq 0$, and $\eta_{i,j} \geq 0$, such that

$$\begin{cases} \alpha I + \mu_i Q_i \leq P_i \leq \beta I - \nu_i Q_i \\ A_i^T P_i + P_i A_i + \vartheta_i Q_i \leq -I \\ P_j + \eta_{i,j} Q_{i,j} \leq P_i \end{cases} \tag{30.13}$$

are satisfied.

The above theorem is an adaptation of a result in [3]. If there is a solution to the above LMI problem, the exponential stability is verified. In addition, a bound on the convergence rate can be estimated:

$$\|x(t)\| \leq \sqrt{\frac{\beta}{\alpha}} e^{-\frac{1}{2\beta}t} \|x_0\|,$$

where $x(t)$ is the continuous trajectory with an initial state x_0, and the constants α, β are solutions of the LMI (Equation 30.13). Based on similar arguments, LMI- based sufficient conditions for the discrete-time case can be derived; see, for example, [8].

An example is now presented to illustrate Theorem 30.14.

Example 30.2:

Consider a hybrid system

$$\begin{cases} \dot{x}(t) = \begin{bmatrix} 0 & 10 \\ 0 & 0 \end{bmatrix} x(t), & \text{if } x(t) \in \Omega_1 = \{x | x^T Q_1 x \geq 0\}, \\ \dot{x}(t) = \begin{bmatrix} 1.5 & 2 \\ -2 & -0.5 \end{bmatrix} x(t), & \text{if } x(t) \in \Omega_2 = \{x | x^T Q_2 x \geq 0\}, \end{cases} \tag{30.14}$$

where $Q_1 = \begin{bmatrix} -0.25 & -0.25 \\ -0.25 & 2 \end{bmatrix}$ and $Q_2 = \begin{bmatrix} 0.25 & 0.25 \\ 0.25 & -2 \end{bmatrix}$. Since $Q_1 = -Q_2$, it is straightforward to verify that $\Omega_1 \bigcup \Omega_1 = \mathbb{R}^2$.

Solving the LMI problem in Theorem 30.14 results in a solution

$$P_1 = \begin{bmatrix} 0.1000 & -0.4500 \\ -0.4500 & 41.1167 \end{bmatrix}, \quad P_2 = \begin{bmatrix} 4.3792 & 3.8292 \\ 3.8292 & 6.8833 \end{bmatrix}$$

with a value of $\beta = 41.12$. Hence, the hybrid system is exponentially stable. Interested readers may refer to [3] for details and illustration of trajectories and Lyapunov-level curves.

Note that the above conditions are all based on MLF theorems; hence the results developed in this subsection are only sufficient. To reduce the possible conservativeness, a new kind of polynomial Lyapunov functions has been introduced and investigated for the stability analysis of hybrid systems. The computation of such polynomial Lyapunov functions can be efficiently performed using convex optimization, based on the sum of squares (SOS) decomposition of multivariate polynomials. It is also possible to use SOS techniques together with the $\mathcal{S}$-procedure to construct piecewise polynomial Lyapunov functions, with each polynomial as a SOS while incorporating the state constraints, and hence to generalize piecewise quadratic Lyapunov functions. Interested readers may refer to the survey paper [8] for further references.

30.4 Switching Stabilization

Implicitly, the above MLF results provide methodologies for the design of switching logics between vector fields so as to achieve a stable trajectory, since MLF results characterize the conditions on switching signals, under which the hybrid system is stable. In this section, we explicitly consider the design of stabilizing switching logics for hybrid systems. The formulation of the problem can be stated as follows.

- Given a collection of LTI systems $\dot{x} = A_i x$, design switching logics so that the induced hybrid system is stable.

This is usually called the *switching stabilization problem* in the literature. It is known that even when all subsystems are unstable, there still may exist stabilizing switching signals.

30.4.1 Quadratic Switching Stabilization

In the switching stabilization literature, most of the work has focused on quadratic stabilization for certain classes of systems. A hybrid system is called quadratically stabilizable when there exist switching signals that stabilize the system along a quadratic Lyapunov function, $V(x) = x^T Px$.

It is known that a necessary and sufficient condition for a pair of LTI systems to be quadratically stabilizable is the existence of a stable convex combination of the two subsystems' matrices. Specifically,

---

Theorem 30.15:

A hybrid system that contains two LTI subsystems, $\dot{x}(t) = A_i x(t), i = 1, 2$, is quadratically stabilizable if and only if the matrix pencil $\gamma_\alpha(A_1, A_2) = \{A_\alpha \mid A_\alpha = \alpha A_1 + (1-\alpha)A_2,\ 0 \leq \alpha \leq 1\}$ contains a stable matrix.

A generalization to more than two LTI subsystems was suggested by using a "min-projection strategy," that is,

$$\sigma(t) = \arg\min_{i \in \mathcal{I}} x(t)^T PA_i x(t). \tag{30.15}$$

Theorem 30.16:

If there exist constants $\alpha_i \in [0, 1]$, and $\sum_{i \in \mathcal{I}} \alpha_i = 1$ such that

$$A_\alpha = \sum_{i \in \mathcal{I}} \alpha_i A_i,$$

is stable, then the min-projection strategy (Equation 30.15) quadratically stabilizes the switched system.

However, the existence of a stable convex combination matrix A_α is only sufficient for switched LTI systems with more than two subsystems. There are example systems for which no stable convex combination state matrix exists, yet the system is quadratically stabilizable using certain switching signals. A necessary and sufficient condition for the quadratic stabilizability of switched controller systems is as follows.

Theorem 30.17: [12]

The switched system is quadratically stabilizable if and only if there exists a positive definite real symmetric matrix $P = P^T > 0$ such that the set of matrices $\{A_i P + PA_i^T\}$ is strictly complete, that is, for any $x \in \mathbb{R}^n/\{0\}$, there exists $i \in \mathcal{I}$ such that $x^T(A_i P + PA_i^T)x < 0$. In addition, a stabilizing switching signal can be selected as $\sigma(t) = \min_i\{x^T(t)(A_i P + PA_i^T)x(t)\}$.

Analogously, for the discrete-time case, it is necessary and sufficient for quadratic stabilizability to check whether there exists a positive symmetric matrix P such that the set of matrices $\{A_i^T PA_i - P\}$ is strictly complete. Obviously, the existence of a convex combination of state matrices A_α automatically satisfies the above strict completeness conditions due to convexity, while the inverse statement is not true in general. Unfortunately, checking the strict completeness of a set of matrices is NP hard [12]. Interested readers may refer to survey papers [3,6,8] for further references.

Quadratic stability means that there exists a positive constant ϵ such that $\dot{V}(x) \leq -\epsilon x^T x$. All of these methods guarantee stability by using a common quadratic Lyapunov function, which is conservative in the sense that there are switched systems that can be asymptotically (or exponentially) stabilized in case when a CQLF function does not exist. Therefore, we will turn our attention to multiple Lyapunov functions, and describe constructive synthesis methods based on the piecewise quadratic Lyapunov function in the next section, which are mainly based on [11].

30.4.2 Piecewise Quadratic Switching Stabilization

According to Theorem 30.14, if there exist real matrices P_i ($P_i = P_i^T$) and scalars $\alpha > 0, \beta > 0, \mu_i \geq 0, \nu_i \geq 0, \vartheta_i \geq 0$, and $\eta_{i,j} \geq 0$, satisfying

$$\begin{cases} \alpha I + \mu_i Q_i \leq P_i \leq \beta I - \nu_i Q_i \\ A^T P_i + P_i A + \vartheta_i Q_i \leq -I \\ P_j + \eta_{i,j} Q_{i,j} \leq P_i \end{cases},$$

then the switched linear system (Equation 30.1) is exponentially stable.

In contrast to the stability analysis problem, here the state-space partitions Ω_i are not given *a priori* any more. Actually, the state partitions Ω_i, which induce the state-dependent switching signals, are to be designed. Moreover, the state space cannot be partitioned in an arbitrary way. The partition of the state space should facilitate the search of proper quadratic Lyapunov-like functions, and satisfy the nonincreasing conditions when switching occurs. This will be discussed in detail in the following.

30.4.2.1 State-Space Partition

Once again, the purpose of dividing the whole state space $\mathbb{R}^n$ into pieces, denoted as Ω_i, is to facilitate the search for Lyapunov-like functions for one of these subsystems. After successfully obtaining these Lyapunov-like functions associated with each region Ω_i, one may patch them together, following the conditions of the above MLF theorem, so as to guarantee global stability.

For this purpose, it is necessary that these regions Ω_i cover the whole state space, that is, the following *covering property* holds:

$$\Omega_1 \bigcup \Omega_2 \bigcup \cdots \bigcup \Omega_N = \mathbb{R}^n.$$

This condition merely says that there are no regions in the state space where none of the subsystems are activated.

Since we will restrict our attention to quadratic Lyapunov-like functions for the purpose of computational efficiency, we will consider regions given (or approximated) by quadratic forms

$$\Omega_i = \{x \in \mathbb{R}^n |\, x^T Q_i x \geq 0\},$$

where $Q_i \in \mathbb{R}^{n\times n}$ are symmetric matrices, and $i \in \{1, \ldots, N\}$.

The following lemma gives a sufficient condition for the covering property.

Lemma 30.1: [11]

If for every $x \in \mathbb{R}^n$

$$\sum_{i=1}^{N} \theta_i x^T Q_i x \geq 0, \tag{30.16}$$

where $\theta_i \geq 0$, $i \in \mathcal{I}$, *then* $\bigcup_{i=1}^{N} \Omega_i = \mathbb{R}^n$.

30.4.2.2 Switching Condition

In order to guarantee exponential stability we also need to make sure that

1. Subsystem i is active only when $x(t) \in \Omega_i$.
2. When switching occurs, it is required to guarantee that the Lyapunov-like function values do not increase.

To verify the first requirement, we consider the *largest region function strategy*, that is,

$$\sigma(x(t)) = \arg\left(\max_{i\in\mathcal{I}} x(t)^T Q_i x(t)\right). \tag{30.17}$$

This is due to the selection of subsystems (at state $x(t)$) corresponding to the largest value of the region function $x(t)^T Q_i x(t)$.

Suppose that the covering condition (Equation 30.16) holds, that is,

$$\sum_{i=1}^{N} \theta_i x^T Q_i x \geq 0$$

for some $\theta_i \geq 0, i \in \mathcal{I}$. Then, based on the largest region function strategy, the state x with the current active mode i satisfies $x^T Q_i x \geq 0$. This implies that $x \in \Omega_i$. So the first condition holds for the largest region function strategy (Equation 30.17).

To satisfy the second energy decreasing condition at switching instants, we need to know in which direction the state trajectory $x(t)$ passes through the switching surfaces. However, the switching surface is to be designed, and so such information is lacking in general. Then we make a compromise and require that

$$x^T P_i x = x^T P_j x$$

for states at the switching plane, that is, $x \in \Omega_i \cap \Omega_j$. Assume that the set $\Omega_i \cap \Omega_j$ can be represented by the following quadratic form:

$$\Omega_i \cap \Omega_j = \{x | x^T (Q_i - Q_j) x = 0\}.$$

Again, applying the $\mathcal{S}$-procedure, we obtain

$$P_i - P_j + \eta_{i,j}(Q_i - Q_j) = 0,$$

for an unknown scalar $\eta_{i,j}$, as the switching condition.

30.4.2.3 Synthesis Condition

The above discussion can be summarized by the following sufficient conditions for the collection of continuous-time systems (Equation 30.1) to be exponentially stabilized.

Theorem 30.18: [11]

If there exist real matrices P_i ($P_i = P_i^T$) and scalars $\alpha > 0, \beta > 0, \mu_i \geq 0, \nu_i \geq 0, \theta_i \geq 0, \vartheta_i \geq 0$, and $\eta_{i,j}$, solving the optimization problem

$$\min \beta$$
$$\text{s.t.} \begin{cases} \alpha I + \mu_i Q_i \leq P_i \leq \beta I - \nu_i Q_i \\ A^T P_i + P_i A + \vartheta_i Q_i \leq -I \\ P_j = P_i + \eta_{i,j}(Q_i - Q_j) \\ \theta_1 Q_1 + \cdots + \theta_N Q_N \geq 0 \end{cases}$$

for all $i, j \in \{1, \ldots, N\}$, then the switched linear system (Equation 30.1) can be exponentially stabilized (with decay rate $\frac{1}{2\beta}$) by the largest region function strategy (Equation 30.17).

The extension of the synthesis method for continuous-time switched linear systems to discrete-time counterpart is not obvious. The main difficulty is that, unlike the continuous-time case, discrete-time switched systems do not have the nice property that the switching occurs exactly on the switching surface. Instead, the switching happens in a region around the switching surface. As a result, we cannot simply capture the switching instants for discrete-time switched systems at the time instants when the state trajectories cross the switching surfaces. Therefore, in order to guarantee the nonincreasing requirement at the switching instants for the discrete-time case, we need to include more constraints involving state transitions for the discrete-time switched systems around the switching surfaces. This makes the switching stabilization problem for discrete-time switched systems more challenging.

Some remarks are in order. First, for both the continuous-time and discrete-time cases, the optimization problem above is a Bilinear Matrix Inequality (BMI) problem, due to the product of unknown scalars and matrices. BMI problems are non-convex, and not computationally efficient. However, practical algorithms for optimization problems over BMIs exist and typically involve approximations, heuristics, branch-and-bound, or local search. One possible way to solve the BMI problem is to grid up the unknown scalars, and then solve a set of LMIs for fixed values of these parameters. It is argued in [11] that the gridding of the unknown scalars can be made quite sparsely.

Example 30.3: [11]

To illustrate the synthesis procedure, consider the case of two unstable subsystems given by

$$A_1 = \begin{bmatrix} 1 & -5 \\ 0 & 1 \end{bmatrix}, \quad A_2 = \begin{bmatrix} 1 & 0 \\ 5 & 1 \end{bmatrix}.$$

It can be shown that there is no stable convex combination of these two matrices, which means that the system cannot be quadratically stabilized. However, solving the BMI in Theorem 30.18 through gridding up the unknown parameters results in a solution

$$\beta = 3.7941, \quad \alpha = 0.2101$$

and

$$Q_1 = -Q_2 = \begin{bmatrix} -0.08242 & 0.8648 \\ 0.8648 & 0.8053 \end{bmatrix}, \tag{30.18}$$

$$P_1 = \begin{bmatrix} 1.1896 & 1.1440 \\ 1.1440 & 3.2447 \end{bmatrix}, \quad P_2 = \begin{bmatrix} 3.3325 & -1.1044 \\ -1.1044 & 1.1509 \end{bmatrix}. \tag{30.19}$$

Hence, the switched linear system can be exponentially stabilized by the largest region function strategy (Equation 30.17), and the estimate of the exponential convergence becomes $\|x(t)\| \leq 4.2495e^{-0.1318t}\|x_0\|$.

So far, we have only derived sufficient conditions for the existence of stabilizing switching signals for a given collection of linear systems. A more difficult problem has been the necessity part of the switching stabilizability problem, and a particularly challenging part has been the problem of finding necessary and sufficient conditions for switching stabilizability. In [7], a necessary and sufficient condition was proposed for the existence of a switching control law (in static state-feedback form) for asymptotic stabilization of continuous-time switched linear systems.

30.5 Conclusion

In this chapter, we discussed, by necessity, a brief introduction to the basic concepts and results of the field of stability and stabilizability of hybrid systems. For further references, we would suggest several survey papers on the stability of hybrid and switched systems, for example [3,6,8,9].

References

1. P. J. Antsaklis and A. Nerode. Hybrid control systems: An introductory discussion to the special issue. *IEEE Trans. Automat. Control*, 43(4):457–460, 1998.
2. P. J. Antsaklis, Ed., *Proceedings of the IEEE, Special Issue on Hybrid Systems: Theory and Applications*, Vol. 88. IEEE Press, Piscataway, NJ, July 2000.
3. R. A. Decarlo, M. S. Branicky, S. Pettersson, and B. Lennartson. Perspectives and results on the stability and stabilizability of hybrid systems. In P. J. Antsaklis, Ed., *Proceedings of the IEEE: Special Issue on Hybrid Systems*, Vol. 88, pp. 1069–1082. IEEE Press, Piscataway, NJ, July 2000.
4. J. P. Hespanha. Stabilization through hybrid control. In H. Unbehauen, Ed., *Encyclopedia of Life Support Systems (EOLSS)*, Vol. *Control Systems, Robotics, and Automation*. Developed under the auspices of the UNESCO, Eolss Publishers, Oxford, UK, 2004.
5. M. Johansson. *Piecewise Linear Control Systems—A Computational Approach*, Lecture Notes in Control and Information Sciences, Vol. 284, Springer-Verlag, Berlin, 2002.
6. D. Liberzon and A. S. Morse. Basic problems in stability and design of switched systems. *IEEE Control System Magazine*, 19(5):59–70, 1999.
7. H. Lin and P. J. Antsaklis. Switching stabilizability for continuous-time uncertain switched linear systems. *IEEE Transactions on Automatic Control*, 52(4):633–646, 2007.
8. H. Lin and P. J. Antsaklis. Stability and stabilizability of switched linear systems: A survey of recent results. *IEEE Transactions on Automatic Control*, 54(2):308–322, 2009.
9. A. N. Michel. Recent trends in the stability analysis of hybrid dynamical systems. *IEEE Transactions on Circuits and System I*, 46(1):120–134, 1999.
10. A. P. Molchanov and E. Pyatnitskiy. Criteria of asymptotic stability of differential and difference inclusions encountered in control theory. *Systems & Control Letters*, 13:59–64, 1989.
11. S. Pettersson. Synthesis of switched linear systems. In *Proceedings of 42nd IEEE Conference Decision Control*, pp. 5283–5288, Maui, HI, 9-12, December 2003.
12. E. Skafidas, R. J. Evans, A. V. Savkin, and I. R. Petersen. Stability results for switched controller systems. *Automatica*, 35(4):553–564, 1999.
13. Z. Sun and S. S. Ge. *Switched Linear Systems: Control and Design*. Springer-Verlag, Berlin, 2005.

31

Optimal Control of Switching Systems via Embedding into Continuous Optimal Control Problem

Sorin Bengea
United Technologies Research Center

Kasemsak Uthaichana
Chiang Mai University

Milos Žefran
University of Illinois at Chicago

Raymond A. DeCarlo
Purdue University

31.1 Introduction

People, engineers, and scientists encounter a variety of switched systems everyday. Gear selection in automatic transmissions [1], control of robots subject to constraints [2], power management in hybrid electric vehicles [3], load balancing in a computer cluster [4], coordination of flexible AC transmission systems (FACTS) devices [5], and output voltage regulation in DC/DC converters [6–10] are but a few examples. Minimizing energy/power usage and/or tracking errors are typical objectives in these control systems. These objectives are achieved by formulating an appropriate cost function to be optimized over some switching functions in combination with the usual continuous control input. Switched systems are a subclass of hybrid systems in which both discrete (switching) and the continuous control inputs are present. Additionally, there may be internal or uncontrolled switches that result when the system trajectory and/or continuous inputs enter certain regions of the state and input spaces, respectively. These switches are called autonomous.

There are a variety of techniques for solving hybrid optimal control problems. In general, the problem can be subdivided into three tasks: (1) finding the optimal sequence of switching instants, (2) finding the optimal sequence of discrete input modes, and (3) finding the optimal value for the continuous control input. For general cases, completing all three tasks is difficult. For example, Giua et al. [11], Xu and Antsaklis [12], and Loxton et al. [7] fix the switching sequence, and then compute the continuous control input. Bemporad and Morari [13] suggest using mixed integer programming to find the optimal solution whose computational complexity is NP-hard and increases exponentially with the number of modes. The minimum principle [14–16] has also been applied to solve for solutions. The dynamic programming approach adopted in [1,17,18] has the curse of dimensionality as a drawback. In [19], the Heaviside function (calculus of variations) is used to re-create a continuous system from the hybrid system with state jumps.

Based on the result in [20], we show, in this chapter, that for quite a general class of hybrid optimal control problems, the computational complexity of the problem is no greater than that of smooth optimal control problems. In Appendix A, we also describe how to extend the embedding methodology of [20] to incorporate hybrid behavior stemming from memoryless autonomous switches that results in plant equations with piecewise smooth vector fields. Further we point out that the approach from [20] can be readily extended to systems with an arbitrary number of modes with only a linear increase in complexity [21]. A direct collocation approach to the solution of the hybrid optimal control problems is summarized in Appendix B.

The switched systems studied herein exhibit two types of switching behavior: autonomous (uncontrolled) switches and controlled switches. They both result in discontinuous jumps in the vector fields governing the evolution of the continuous state of the system. In the case of *autonomous* switches, the vector fields of the system undergo discontinuous jumps as a result of the state and the input entering different regions in the combined state and input space. Such switches are *uncontrolled,* meaning that the switches cannot be affected directly through a separate switching mechanism. An example of a system with autonomous switches is the one subject to continuous state-dependent constraints, where the autonomous switches correspond to different combinations of constraints that are active in a particular continuous state. A practical example is a mobile robot that encounters a patch of ice in which rolling motion changes abruptly to sliding motion. The second type of switches involves discontinuous jumps in the vector fields that can be directly controlled, and thus are called *controlled* switches. An example of a system with such switches is a continuous control system whose control mechanism consists of a finite number of different continuous controllers, and the controller to be used is determined at a supervisory or decision-making level of the overall control system. Such would be the case for power management in a hybrid electric vehicle, or, for example, a large ship's propulsion/electrical system. We assume that the set of different operating regimes of the system defined through the autonomous and controlled switches is finite.

As mentioned, in hybrid electric vehicles, the energy or power management problem is naturally modeled as an optimal switched system in which the electric drive operates either as a propelling machine or as a generating machine for recharging the battery. A suboptimal model predictive control (MPC) approach was adopted to solve for the energy management solutions in [3,22–24]. A nonlinear MPC (NMPC) version provides more accurate hybrid electric vehicle (HEV) solution but at the expense of computational requirement [3]. To improve the computational time, a few optimization subproblems can be solved offline and the solutions stored as maps, thereby decreasing the computational load [25,26]. Real-time MPC is detailed for boost converters in [8–10].

The formal definition of the switched optimal control problem (SOCP) is provided in Section 31.2. This section also illustrates mathematically how to parameterize the family of problems as the embedded optimal control problem (EOCP). This section suggests solving the SOCP via first solving the EOCP. The benefits of the embedding technique are also provided.

Section 31.3 summarizes sufficient and the necessary conditions for solvability of the EOCP. It will be seen therein that sufficiency conditions can be met rather easily by many systems. The necessary conditions can be seen as a generalized Maximum Principles without explicit assumptions on fixing the number or sequence of switchings. This is followed by a description of the extension to three modes (and implicitly to n) in Section 31.4.

The application of the necessary conditions for solving the EOCP is illustrated in Section 31.5. Therein a simplified vehicle with two gears is investigated [20]. Through the usage of the generalized Hamiltonian one can apply the EOCP to characterize the solutions, including the singular cases.

Since applying the necessary condition requires solving for the state and the adjoint equations simultaneously, the sensitivity of the solution on the initial state or adjoint state is known to be large, precluding the use of single shooting methods. Section 31.6 applies the direct collocation method, a more robust technique to solve an example of MPC of a unicycle. A summary of the direct collocation method is provided in Appendix B. Concluding remarks and further pointers to the literature are given in Section 31.7.

31.2 The Switched and EOCPs

This section formalizes a model and problem formulation for a two-switched system. Generalization to 3-modes is described in Section 31.4 and the extension to n-modes is straightforward [21]. The well-known SOCP is developed first. Then as per Bengea and DeCarlo [20], the embedded formulation is presented. The relationship between the two formulations is then set forth. One can then show that the embedded problem is a viable approach for solving the SOCP and in contrast to other approaches has linearly increasing complexity with the number of modes rather than combinatorial complexity.

31.2.1 Model of a Two-Switched System

The state dynamics of a two-switched system is

$$\dot{x}_S(t) = f_{v_S(t)}(t, x_S(t), u_S(t)), \quad x_S(t_0) = x_0 \in \mathbb{R}^n, \quad t \geq t_0 \tag{31.1}$$

where (1) the continuously differentiable vector fields, $f_0, f_1 : R \times \mathbb{R}^n \times \mathbb{R}^m \to \mathbb{R}^n$, specify the dynamics of each of two possible system configurations; (2) the classical control input, $u_S(t) \in \Omega \subset \mathbb{R}^m$, is constrained to the bounded and convex set, Ω, at each time instant; and (3) a mode switching mechanism, represented through the subscript $v_S(t) \in \{0, 1\}$, identifies which of the two possible system configurations, f_0 or f_1, is operational. These "controls" ensure that the system vector field can be controlled both through a selection of the control and switching inputs. We also assume that the initial time, t_0, initial state, $x_S(t_0)$, final time, t_f, and final state, $x_S(t_f)$, are restricted to a boundary set B as follows: $(t_0, x_S(t_0), t_f, x_S(t_f)) \in B \triangleq T_0 \times B_0 \times T_f \times B_f \subset \mathbb{R}^{2n+2}$.

In addition to controlled switches, there are often autonomous switches and even switches that are only possible in certain regions of the state space. Appendix A describes how autonomous switches can be easily incorporated into the framework set forth in this chapter.

In general, the behavior of switched systems can be quite complex and might lead to anomalies such as Zeno behavior or deadlock states. Furthermore, the systems of Equation 31.1 (or Equations 31.42 and 31.44) belong to the class of systems with discontinuous right-hand sides [27]; hence, the questions of existence and uniqueness of solutions have to be carefully studied. Although certainly important, these issues are outside the scope of this chapter and we will assume that the existence and uniqueness (in the appropriate sense) are guaranteed. We refer the interested reader to the conference series [28,29], special issues [30,31], and [32,33] for further reading.

31.2.2 Performance Index and the SOCP

The extent to which it is preferable (when the same trajectories can be generated in both modes) or feasible (when only one mode can generate desired trajectories) to modulate the continuous inputs or choose the switching control input value depends on the input and state constraint set and a performance index (PI). For measuring the degree of optimality of triplets $(u_S(\cdot), v_S(\cdot), x_S(\cdot))$, we introduce the following cost PI:

$$J_S(x_0, u_S, v_S) = g(t_0, x_0, t_f, x_f) + \int_{t_0}^{t_f} F_{v_S(t)}(t, x_S(t), u_S(t))\, dt \tag{31.2}$$

where the function g penalizes the endpoints and is defined on a neighborhood of B, and the integrands F_0 and F_1 are real-valued continuously differentiable functions that penalize the running cost in each mode, respectively. Depending on the engineering applications, the functions F_0 and F_1 represent the cost of operation of various subsystems that are active only during individual modes of operations. For a hybrid electric vehicle example, during acceleration we want to minimize fuel consumption, while during braking one tries to maximize the regenerative power [3,24].

The previously introduced dynamics, input and state constraints, and performance measure define the SOCP:

$$\min_{u_S \in \Omega, v_S \in \{0,1\}} J_S(x_0, u_S, v_S)$$

constrained by

1. Dynamical state equation 31.1.
2. Endpoint constraints $(t_0, x_S(t_0), t_f, x_S(t_f)) \in B$.

In the selection of the above switching system dynamics and performance measure, we make several transparent assumptions that limit the application of the proposed approach:

1. The dynamical equation 31.1 models the switching control input $v_S(t)$ as an independent control input; therefore, the mode switching can occur independently of the state values and control input values. In practical examples, this may not be the case as would be for the speed-dependent gear-switching mechanism of a vehicle. The state-dependent switching can still be studied via the formulated SOCP by penalizing the switching (a formulation known as a "soft" constraint).
2. Mode-switching is assumed to occur instantaneously. In engineering application, switching among the vector fields is implemented by generating new actuator values that interconnect and/or disconnect various subsystems. In power electronics, such switching takes place over a time interval that is negligible in comparison with other component dynamics. When this switching action absorbs large amounts of energy or affects the dynamics of the physical system beyond the connection/disconnection process, the switching dynamics may need to be modeled.

31.2.3 The Embedded Optimal Control Problem

To display the interaction between the discrete and continuous control inputs Equation 31.1 can be rewritten as a convex combination:

$$\dot{x}_E(t) = [1 - v_E(t)] \cdot f_0(t, x_E(t), u_{E0}(t)) + v_E(t) \cdot f_1(t, x_E(t), u_{E1}(t)) \tag{31.3}$$

which reduces to Equation 31.1 under the conditions: $v_E(t) = v_S(t)$ and $u_{E0}(t) = u_{E1}(t) = u_S(t)$. However, although Equation 31.3 is equivalent to Equation 31.1 under the indicated conditions, this new equation invites exploration of additional trajectories generated with an enlarged domain of the switching control denoted here by $v_E \in [0, 1]$ and independent continuous-time controls $u_{E0}(t), u_{E1}(t) \in \Omega$.

For illustrating the possible trajectories generated with $v_E \in [0, 1]$, we sketch in Figure 31.1 vector fields and trajectories for a hypothetical switching system and the corresponding switching controls v_E. With

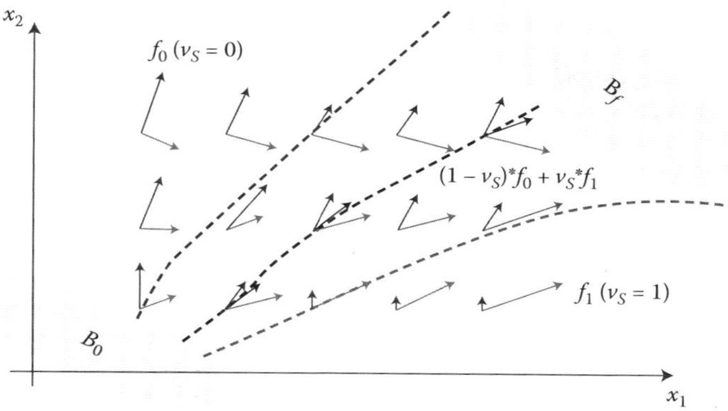

FIGURE 31.1 Trajectories generated with various control inputs v_E for a hypothetical switch system: $v_E = 0$(solid upward arrow), $v_E = 1$(solid rightward arrow); some $v_E \in (0, 1)$ (solid dark arrow).

this enlarged control input set the original switching control becomes a special case; it would appear that the generated trajectories are no longer feasible for the original switched system.

This expansion of the control input domains follows relaxation techniques from optimal control employed when certain sets that guarantee solution existence need to be rendered convex [34]. We call this relaxation of the switching system an "embedding." The PI of Equation 31.2 has a similar embedding given by

$$J_E(x_0, u_{E0}, u_{E1}, v_E) = g(t_0, x_E(t_0), t_f, x_E(t_f)) + \int_{t_0}^{t_f} \{[1 - v_E(t)] \cdot F_0(t, x_E(t), u_{E0}(t)) + v_E(t) \cdot F_1(t, x_E(t), u_{E1}(t))\}\, dt \tag{31.4}$$

which leads to the EOCP:

$$\min_{u_{E0}, u_{E1} \in \Omega, v_E \in [0,1]} J_E(x_0, u_{E0}, u_{E1}, v_E)$$

subject to

1. Dynamical state equation 31.3.
2. Endpoint constraints $(t_0, x_E(t_0), t_f, x_E(t_f)) \in B$.

The EOCP becomes now a classical optimization problem with continuous control inputs. Although application of established techniques is now possible for solving this problem, the EOCP's solutions $(x_E^*, u_{E0}^*, u_{E1}^*, v_E^*)$ may have $v_E^* \in (0, 1)$ for almost all time instants $t \in [t_0, t_f]$. One wonders how this might help solve the original SOCP where the switching control $v_S^* \in \{0, 1\}$. If one selects a signal v_S that switches between 0 and 1 with an appropriate duty cycle per (small) unit time, then a trajectory of Equation 31.1, so generated, ought to approximate an embedded trajectory x_E as is illustrated in Figure 31.2. The idea of approximating trajectories generated with v_E by appropriately selecting an on–off signal $v_S \in \{0, 1\}$ is made more rigorous in Theorem 31.1.

Theorem 31.1:

Let $u_{E0}, u_{E1} \in \Omega$ and $v_E \in [0, 1]$ be a control triplet for the embedded system Equation 31.3 and x_E the generated trajectory on the interval $[t_0, t_f]$. Let both the switched and embedded systems have the same

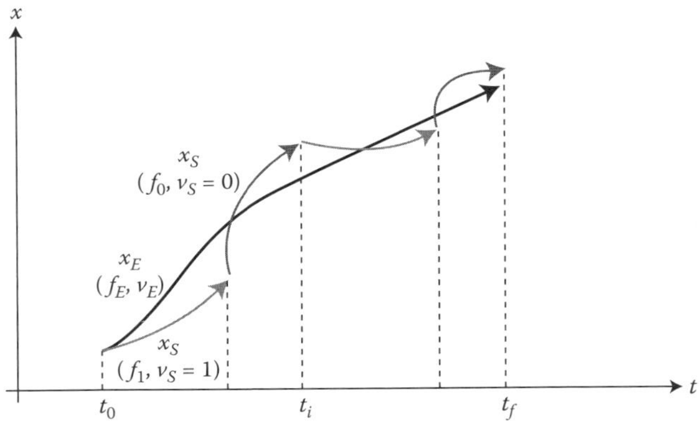

FIGURE 31.2 Embedded system trajectory x_E(black), approximating switching system trajectory x_S, and corresponding switching signal.

initial condition, $x_S(t_0) = x_E(t_0)$. Then for any desired trajectory-approximation error $\varepsilon > 0$, there are control inputs $v_{S,\varepsilon}(t) \in \{0, 1\}$ and $u_{S,\varepsilon}(t) \in \Omega$ defined on $[t_0, t_f]$ such that the generating switching trajectory has the property $\|x_{S,\varepsilon}(t) - x_E(t)\| < \varepsilon$ for all $t \in [t_0, t_f]$.

The proof of the theorem and details on the approximation can be found in [20]. By establishing that the set of embedded system trajectories is not significantly larger than the set of the trajectory set of Equation 31.1, the theorem validates the approach of generating optimal solution for the SOCP by solving the EOCP, a classical optimization problem. Indeed, the set of trajectories of Equation 31.1 is dense in the set of trajectories of the embedded system. After application of standard techniques for generating a solution $(x_E^*, u_{E0}^*, u_{E1}^*, v_E^*)$ for the EOCP, at least a suboptimal solution of the SOCP (x_S, u_S, v_S) is guaranteed to exist by Theorem 31.1. How best to achieve the approximation remains an area of open research.

Based on Theorem 31.1 and on the properties of the EOCP, it can be shown that the relationships of Table 31.1 hold (see Propositions 3 and 5 in [20]). An alternate proof of Theorem 31.1 based on the Lyapunov (integral) theorem is given in [35].

The embedding approach enables the study of the cases when the SOCP does and does not have solutions. The presence of singular solutions, with $v_E^* \in (0, 1)$, can reveal situations when the SOCP does not have solutions. The only case, 2a in Table 31.1, when the EOCP does not solve the SOCP is the case when the end-state of the embedded system trajectory $x_E^*(t_f)$ is on the boundary of the closed set B_f. And this happens because the end-states of approximating trajectories $x_{S,\varepsilon}(t_f)$ are not guaranteed to meet this final constraint. This case has not been completely studied, and the current analysis does not

TABLE 31.1 Relationship between SOCP's and EOCP's Solutions

	EOCP Solutions	SOCP Solutions	Remarks
1.	Bang-bang solutions with $v_E^* \in \{0, 1\}$	The same as ECOP bang-bang solution with $v_S^* = v_E^*$ and $x_S^* = x_E^*$	SOCP is solved
2a.	Singular solutions only $v_E^* \in (0, 1)$	May have solutions	EOCP does not solve SOCP
2b.		Does not have solutions	Suboptimal solutions can be constructed via Theorem 31.1

exclude its existence. From an engineering perspective, however, this case has little relevance. By using engineering approximations for a particular application, the terminal constraint set can in most cases be slightly enlarged making Case 2a unlikely. Hence, approximate feasible solutions of the SOCP can always be constructed, motivating a continued focus on the embedding approach.

31.3 Sufficient and Necessary Conditions for Solvability of the EOCP

This section presents conditions that guarantee the existence of EOCP solutions and properties of these solutions. Combined with the EOCP and SOCP relationships of Table 31.1, the existence conditions for the EOCP solutions enable the study of sufficient conditions for the SOCP. In deriving existence conditions for the EOCP solutions, we employ the generic theorem of Berkovitz [34, Theorem 51, p. 61]. With appropriate notational adaptations to the EOCP formulation, this theorem states that a solution exists for a generic optimal control problem, if the following conditions are met:

1. The set of admissible pairs states-control inputs, $(x_E, u_{E0}, u_{E1}, v_E)$, is not empty.
2. There is a compact set that includes all the points $(t, x_E(t))$ for all $t \in [t_0, t_f]$.
3. The terminal constraint set B is compact.
4. The input constraint set $\Omega \times \Omega \times [0, 1]$ is compact.
5. The set

$$Q_E^+ \triangleq \big\{(y^0, y) : y^0 > (1-\mu)F_0(t, x_E, u_{E0}) + \mu F_1(t, x_E, u_{E1}), \quad y = (1-\mu)f_0(t, x_E, u_{E0}) + \mu f_1(t, x_E, u_{E1}), \mu \in [0, 1], u_{E0}, u_{E1} \in \Omega)\big\} \tag{31.5}$$

is convex.

Among all the above conditions, meeting the convexity requirement of the set Q_E^+ is the most challenging. From [34], this condition is met when a system is affine and the penalty cost is convex in the continuous control input. The particular form of the EOCP, where both the vector field and the penalty costs are affine in the additional switching control input, makes possible the extension of the mentioned results to the EOCP.

Proposition 31.1:

The set Q_E^+, defined in Equation 31.5, is convex for the following class of EOCP vector fields:

(S1) The vector fields for the two modes of operation are linear in their control inputs: $f_0(t, x_E, u_{E0}) = A_0(t, x_E) + B_0(t, x_E) \cdot u_{E0}$, $f_1(t, x_E, u_{E1}) = A_1(t, x_E) + B_1(t, x_E) \cdot u_{E1}$, with A_0, B_0, A_1, B_1 continuously differentiable functions.

(S2) The cost functional integrands are convex in their continuous control inputs: for every (t, x_E); the functions $F_0(t, x_E, u_{E0})$ and $F_1(t, x_E, u_{E1})$ are convex in the inputs u_{E0} and u_{E1}, respectively.

Proposition 31.1 and the previously presented relationships of Table 31.1 between the SOCP's and EOCP's solutions are the main implements one can use for analyzing the existence of SOCP's solutions. Before understanding what particular case of Table 31.1 is applicable to a specific problem, the EOCP solutions must be generated. Assuming that the EOCP has a solution, a characterization of these solutions can be made by using classical results of optimal control theory. In the following, we use these results and the particular form of EOCP vector fields and cost functional to derive conditions corresponding to the cases presented in Table 31.1.

The Hamiltonian associated with the EOCP is a function $H_E : \mathbb{R} \times \mathbb{R}^n \times \mathbb{R}^m \times \mathbb{R}^m \times [0,1] \times \mathbb{R} \times \mathbb{R}^n \to \mathbb{R}$ defined as

$$\begin{aligned} H_E(t, x_E, u_{E0}, u_{E1}, v_E, \lambda_E^0, \lambda_E) = \lambda_E^0 \left[(1 - v_E) F_0(t, x_E, u_{E0}) + v_E F_1(t, x_E, u_{E1})\right] \\ + \lambda_E^T \left[(1 - v_E) f_0(t, x_E, u_{E0}) + v_E f_1(t, x_E, u_{E1})\right] \end{aligned} \tag{31.6}$$

To emphasize the affine dependency on the generalized switching input v_E, the Hamiltonian is factored as

$$H_E(t, x_E, u_{E0}, u_{E1}, v_E, \lambda_E^0, \lambda_E) = E_0(t, x_E, u_{E0}, u_{E1}, \lambda_E^0, \lambda_E) + v_E \cdot E_1(t, x_E, u_{E0}, u_{E1}, \lambda_E^0, \lambda_E) \tag{31.7}$$

with the obvious definitions of E_0 and E_1. Assuming that an optimal solution of the EOCP exists, denoted hereafter as $(t, x_E^*, u_{E0}^*, u_{E1}^*, v_E^*)$, a characterization of its properties is possible via application of Maximum Principle [34, Theorem 31, p. 185, and Corollary 3.1, p. 186]. These conditions guarantee the existence of a constant λ_E^{0*} and an absolutely continuous function $\lambda_E^*(t)$ on $[t_0, t_f]$ such that for almost all $t \in [t_0, t_f]$ the following hold (the argument t is not included for simplicity):

$$\begin{cases} \dot{x}_E^* = \left.\dfrac{\partial H_E}{\partial \lambda_E}\right]_{(x_E^*, u_{E0}^*, u_{E1}^*, v_E^*, \lambda_E^{0*}, \lambda_E^*)} \\ \dot{\lambda}_E^* = -\left.\dfrac{\partial H_E}{\partial x_E}\right]_{(x_E^*, u_{E0}^*, u_{E1}^*, v_E^*, \lambda_E^{0*}, \lambda_E^*)} \end{cases} \tag{NC1}$$

where $(\partial H_E / \partial \lambda_E)$ and $(\partial H_E / \partial x_E)$ denote appropriate partial derivatives;

$$H_E(t, x_E^*, u_{E0}^*, u_{E1}^*, v_E^*, \lambda_E^{0*}, \lambda_E^*) \geq H_E(t, x_E^*, u_{E0}, u_{E1}, v_E, \lambda_E^{0*}, \lambda_E^*) \tag{NC2}$$

for all $u_{E0}, u_{E1} \in \Omega$ and $v_E \in [0,1]$. Using simple manipulations detailed in [20], condition NC2 can be shown to be equivalent to

$$\begin{aligned} &H_E(t, x_E^*, u_{E0}^*, u_{E1}^*, v_E^*, \lambda_E^{0*}, \lambda_E^*) \\ &= \max\left\{ \max_{u_{E0}, u_{E1}} \{H_E(t, x_E^*, u_{E0}, u_{E1}, 0, \lambda_E^{0*}, \lambda_E^*)\}, \max_{u_{E0}, u_{E1}} \{H_E(t, x_E^*, u_{E0}, u_{E1}, 1, \lambda_E^{0*}, \lambda_E^*)\} \right\} \end{aligned} \tag{31.8}$$

where the two inner Hamiltonian expressions are directly associated with the two modes of the switching system. We observe this is the first step in the derivation where the SOCP's solutions emerge as solution of the EOCP. However, Equation 31.8 alone does not yet guarantee that EOCP solutions are of the bang-bang type, that is, $v_E^* \in \{0, 1\}$. In analyzing the cases when the EOCP has bang-bang-type solutions, the expression

$$\begin{aligned} &E_1(t, x_E, u_{E0}, u_{E1}, \lambda_E^0, \lambda_E) \\ &\quad = \lambda_E^0 \left[F_1(t, x_E, u_{E1}) - F_0(t, x_E, u_{E0})\right] + \lambda_E^T \left[f_1(t, x_E, u_{E1}) - f_0(t, x_E, u_{E0})\right] \end{aligned} \tag{31.9}$$

of Equation 31.7 plays a critical role, specifically when it becomes zero. This role motivates the introduction of the following set of time instants:

$$T \triangleq \left\{ t \in [t_0, t_f] : E_1(t, x_E^*, u_{E0}^*, u_{E1}^*, v_E, \lambda_E^{0*}, \lambda_E^*) = 0 \right\} \tag{31.10}$$

As an intuitive explanation, suppose the expression E_1 of Equation 31.9 is nonzero almost everywhere along the optimal solution then its sign would indicate which of the two maxima of Equation 31.8 is attained. This in turn specifies the optimal value of the control input v_E^*, which in this case would be restricted to the set {0,1}. These statements are formalized in the following theorem:

Theorem 31.2:

For almost all $t \in [t_0, t_f] - T$ the following hold:

1. *(Mode 0) If* $\max_{u_{E0},u_{E1}}\{H_E(t, x_E^*, u_{E0}, u_{E1}, 0, \lambda_E^{0*}, \lambda_E^*)\} > \max_{u_{E0},u_{E1}}\{H_E(t, x_E^*, u_{E0}, u_{E1}, 1, \lambda_E^{0*}, \lambda_E^*)\}$, *then* $v_E^*(t) = 0$ *and the equations of NC1 become*

$$\begin{cases} \dot{x}_E^* = f_0\left(t, x_E^*, u_{E0}^*\right) \\ \dot{\lambda}_E^* = -\lambda_E^0 \left[\dfrac{\partial F_0}{\partial x}\right]_{(t,x_E^*,u_{E0}^*)} - \lambda_E^T \left[\dfrac{\partial f_0}{\partial x}\right]_{(t,x_E^*,u_{E0}^*)} \end{cases} \quad (31.11)$$

The optimal control input u_{E1}^* *has an indeterminate value and*

$$u_{E0}^* = \arg\min_{u_{E0}\in\Omega}\{H_E(t, x_E^*, u_{E0}, u_{E1}, 0, \lambda_E^{0*}, \lambda_E^*)\} \quad (31.12)$$

(here H_E does not depend on u_{E1}).

2. *(Mode 1) If* $\max_{u_{E0},u_{E1}}\{H_E(t, x_E^*, u_{E0}, u_{E1}, 0, \lambda_E^{0*}, \lambda_E^*)\} < \max_{u_{E0},u_{E1}}\{H_E(t, x_E^*, u_{E0}, u_{E1}, 1, \lambda_E^{0*}, \lambda_E^*)\}$, *then* $v_E^*(t) = 1$ *and the equations of NC1 become*

$$\begin{cases} \dot{x}_E^* = f_1\left(t, x_E^*, u_{E1}^*\right) \\ \dot{\lambda}_E^* = -\lambda_E^0 \left[\dfrac{\partial F_1}{\partial x}\right]_{(t,x_E^*,u_{E1}^*)} - \lambda_E^T \left[\dfrac{\partial f_1}{\partial x}\right]_{(t,x_E^*,u_{E1}^*)} \end{cases} \quad (31.13)$$

The optimal control input u_{E0}^* *has an indeterminate value and*

$$u_{E1}^* = \arg\min_{u_{E1}\in\Omega}\{H_E(t, x_E^*, u_{E0}, u_{E1}, 1, \lambda_E^{0*}, \lambda_E^*)\} \quad (31.14)$$

(here H_E does not depend on u_{E0}).

3. *(Nonsingularity) If* $\max_{u_{E0},u_{E1}}\{H_E(t, x_E^*, u_{E0}, u_{E1}, 0, \lambda_E^{0*}, \lambda_E^*)\} = \max_{u_{E0},u_{E1}}\{H_E(t, x_E^*, u_{E0}, u_{E1}, 1, \lambda_E^{0*}, \lambda_E^*)\}$, *then either* $v_E^* \in \{0, 1\}$ *or the corresponding . Equations 31.11 or 31.13 hold, but additional constraints need to be used to determine exactly which of the two modes are optimal at time t.*

Theorem 31.2 characterizes all the situations of Case 1 of Table 31.1. The singular solutions, summarized in Case 2 of Table 31.1, may be generated when the expression E_1 of Equation 31.9 equals zero for some time interval of nonzero measure; in this case

$$\max_{u_{E0},u_{E1}}\{H_E(t, x_E^*, u_{E0}, u_{E1}, 0, \lambda_E^{0*}, \lambda_E^*)\} = \max_{u_{E0},u_{E1}}\{H_E(t, x_E^*, u_{E0}, u_{E1}, 1, \lambda_E^{0*}, \lambda_E^*)\} \quad (31.15)$$

As such, the optimal solution x_E^* is generated with some $v_E^* \notin \{0, 1\}$ on the same nonzero measure time interval. In summary, all these cases occur when the set T of Equation 31.10 is positive time-invariant, on some time interval, for the dynamical system of NC1, and Equations 31.12, 31.14, and 31.15 hold simultaneously for some $v_E^* \notin \{0, 1\}$.

31.4 Optimal Control of a Switched System with Three Modes of Operation

The results of Section 31.3 can be extended to a switched system with multiple modes of operation, and we illustrate here the extension to three modes only for simplifying the notational complexity. Let the vector fields of these modes be denoted as $f_{00}(t, x_S, u_S)$, $f_{01}(t, x_S, u_S)$, and $f_1(t, x_S, u_S)$. The subscripts are selected to correspond to different combinations of two discrete signals $(v_{S0}, v_{S1}) \in \{0, 1\} \times \{0, 1\}$ as described below by the dynamics of the switched system (where we drop the time dependency to simplify the exposition):

$$\dot{x}_S = (1 - v_{S0}) \cdot \{(1 - v_{S1}) \cdot f_{00}(x_S, u_S) + v_{S1} \cdot f_{01}(x_S, u_S)\} + v_{S0} \cdot f_1(x_S, u_S) \tag{31.16}$$

One observes that, for example, the mode-switching input combination $(v_{S0} = 0, v_{S1} = 1)$ selects the dynamics $\dot{x}_S = f_{01}(x_s, u_S)$, and therefore three modes are possible: 00, 01, and 1. Other vector field combinations can be selected to describe the same switched system dynamics:

$$\dot{x}_S = v_{S0} \cdot f_{00}(x_S, u_S) + v_{S1} \cdot f_{01}(x_S, u_S) + (1 - v_{S0} - v_{S1}) \cdot f_1(x_S, u_S) \tag{31.17}$$

Although the results are the same for the two approaches, the one described by Equation 31.16 has two advantages:

1. A parsimonious selection of switching inputs. For n modes, Equation 31.16 uses at most $\lceil \log_2 n \rceil$ inputs (the smallest integer larger than $\log_2 n$), whereas Equation 31.17 uses $(n - 1)$ inputs. Although in our case, for $n = 3$, $\lceil \log_2 n \rceil = n - 1$, the number of discrete inputs becomes larger for Equation 31.17 as n increases.
2. The results of Sections 31.3 and 31.4 can be directly applied in a few nested steps.

We continue to use Equation 31.16, and refer the reader to [21] for development of similar results derived based on Equation 31.17. The embedding of switched system of Equation 31.16 follows the same approach as presented in Section 31.4 with the switching control inputs redefined as $(v_{E0}, v_{E1}) \in [0, 1] \times [0, 1]$:

$$\dot{x}_E = (1 - v_{E0}) \cdot \{(1 - v_{E1}) \cdot f_{00}(x_E, u_{E00}) + v_{E1} \cdot f_{01}(x_E, u_{E01})\} + v_{E0} \cdot f_1(x_E, u_{E1}) \tag{31.18}$$

Intuitively, and this is based on the construction theorem presented in [20], Theorem 31.1 holds for systems (Equations 31.16 and 31.17). This motivates the study of the EOCP and we focus on the necessary conditions for the embedded system solution $(x_E^*, u_{E00}^*, u_{E01}^*, u_{E1}^*, v_{E0}^*, v_{E1}^*)$ to be a solution of the SOCP, with $(v_{E0}^*, v_{E1}^*) \in \{0, 1\} \times \{0, 1\}$ and $x_S^* = x_E^*$.

With the optimization cost integrands, similarly to Equation 31.4, denoted by $F_{00}(x_E, u_{E00})$, $F_{01}(x_E, u_{E01})$, and $F_1(x_E, u_{E1})$, the Hamiltonian of the embedded system (Equation 31.18) can be written as

$$\begin{aligned} H_E(x_E, u_{E00}, u_{E01}, u_{E1}, v_{E0}, v_{E1}, \lambda_E^0, \lambda_E) = \lambda_E^0 \cdot \{(1 - v_{E0}) \cdot [(1 - v_{E1}) \cdot F_{00}(x_E, u_{E00}) + v_{E1} \cdot F_{01}(x_E, u_{E01})] \\ + v_{E0} \cdot F_1(x_E, u_{E1})\} + \lambda_E^T \cdot \{(1 - v_{E0}) \cdot [(1 - v_{E1}) \cdot f_{00}(x_E, u_{E00}) \\ + v_{E1} \cdot f_{01}(x_E, u_{E01})] + v_{E0} \cdot f_1(x_E, u_{E1})\} \end{aligned} \tag{31.19}$$

Similarly to Equation 31.7, the three-mode switched system Hamiltonian of Equation 31.19 can be rewritten more compactly as

$$H_E = v_{E0} \cdot v_{E1} \cdot E_1 + v_{E1} \cdot E_2 + v_{E0} \cdot E_3 + E_4 \tag{31.20}$$

where the arguments $(x_E, u_{E00}, u_{E01}, u_{E1}, v_{E0}, v_{E1}, \lambda_E^0, \lambda_E)$ are dropped for H_E and the new expressions E_1, E_2, E_3, and E_4. These expressions are very similar to those of Equation 31.7: they are weighted summations of differences between vector fields and penalties corresponding to pairwise switching modes. These expressions are

$$E_1 = \lambda_E^0 \cdot [F_{00} - F_{01}] + \lambda_E^T \cdot [f_{00} - f_{01}] \tag{31.21}$$

$$E_2 = \lambda_E^0 \cdot [F_1 - F_{00}] + \lambda_E^T \cdot [f_1 - f_{00}] \tag{31.22}$$

$$E_3 = \lambda_E^0 \cdot [F_{01} - F_{00}] + \lambda_E^T \cdot [f_{01} - f_{00}] \tag{31.23}$$

$$E_4 = \lambda_E^0 \cdot F_{00} + \lambda_E^T \cdot f_{00} \tag{31.24}$$

where again the arguments are dropped.

The compact form of the switched system dynamics that resulted in the Hamiltonian of Equation 31.20 can now be used to generate the final result of this section. The following necessary condition for the optimal solution $(x_E^*, u_{E00}^*, u_{E01}^*, u_{E1}^*, v_{E0}^*, v_{E1}^*, \lambda_E^{0*}, \lambda_E^*)$ is derived by applying in two steps the results of Lemma 10 of [20] in two steps (again, the arguments are not included).

$$\begin{aligned}
H_E\left(\text{Optimal Sol}\right) &= \max_{\substack{(v_{E0},v_{E1})\in[0,1]\times[0,1]\\(u_{E00},u_{E01},u_{E1})\in\Omega\times\Omega\times\Omega}} H_E\left(x_E^*, u_{E00}, u_{E01}, u_{E1}, v_{E0}, v_{E1}, \lambda_E^{0*}, \lambda_E^*\right)\\
&= \max_{\substack{(v_{E0},v_{E1})\in[0,1]\times[0,1]\\(u_{E00},u_{E01},u_{E1})\in\Omega\times\Omega\times\Omega}} \{v_{E0}\cdot v_{E1}\cdot E_1 + v_{E1}\cdot E_2 + v_{E0}\cdot E_3 + E_4\}\\
&= \max\left\{\begin{array}{l}\max\limits_{\substack{v_{E0}=0,v_{E1}\in[0,1]\\(u_{E00},u_{E01},u_{E1})\in\Omega\times\Omega\times\Omega}} \{v_{E1}\cdot E_2 + E_4\},\\ \max\limits_{\substack{v_{E0}=1,v_{E1}\in[0,1]\\(u_{E00},u_{E01},u_{E1})\in\Omega\times\Omega\times\Omega}} \{v_{E1}\cdot E_1 + v_{E1}\cdot E_2 + E_3 + E_4\}\end{array}\right\}\\
&= \max\left\{\begin{array}{l}\max\left\{\max\limits_{\substack{v_{E0}=0,v_{E1}=0\\(u_{E00},u_{E01},u_{E1})\in\Omega\times\Omega\times\Omega}} \{E_4\},\ \max\limits_{\substack{v_{E0}=0,v_{E1}=1\\(u_{E00},u_{E01},u_{E1})\in\Omega\times\Omega\times\Omega}} \{E_2 + E_4\}\right\},\\ \max\left\{\max\limits_{\substack{v_{E0}=1,v_{E1}=0\\(u_{E00},u_{E01},u_{E1})\in\Omega\times\Omega\times\Omega}} \{E_3 + E_4\},\ \max\limits_{\substack{v_{E0}=1,v_{E1}=1\\(u_{E00},u_{E01},u_{E1})\in\Omega\times\Omega\times\Omega}} \{E_1 + E_2 + E_3 + E_4\}\right\}\end{array}\right\}\\
&= \max\left\{\begin{array}{l}\max\limits_{(u_{E00},u_{E01},u_{E1})\in\Omega\times\Omega\times\Omega} \{E_4\},\ \max\limits_{(u_{E00},u_{E01},u_{E1})\in\Omega\times\Omega\times\Omega} \{E_2 + E_4\},\\ \max\limits_{(u_{E00},u_{E01},u_{E1})\in\Omega\times\Omega\times\Omega} \{E_3 + E_4\}\end{array}\right\} \quad (\text{we used } E_1 = -E_3)
\end{aligned}$$

This is an equivalent way of saying that the embedded-system Hamiltonian calculated at the optimal solution is equal to the maximum of the Hamiltonians associated with all modes, similarly to Equation 31.8.

$$H_E\left(\text{Optimal Sol}\right) = \max\left\{\begin{array}{l}\max\limits_{(u_{E00},u_{E01},u_{E1})\in\Omega\times\Omega\times\Omega} H_E\left(x_E^*, u_{E00}, u_{E01}, u_{E1}, v_{E0}=0, v_{E1}=0, \lambda_E^{0*}, \lambda_E^*\right),\\ \max\limits_{(u_{E00},u_{E01},u_{E1})\in\Omega\times\Omega\times\Omega} H_E\left(x_E^*, u_{E00}, u_{E01}, u_{E1}, v_{E0}=0, v_{E1}=1, \lambda_E^{0*}, \lambda_E^*\right),\\ \max\limits_{(u_{E00},u_{E01},u_{E1})\in\Omega\times\Omega\times\Omega} H_E\left(x_E^*, u_{E00}, u_{E01}, u_{E1}, v_{E0}=1, v_{E1}=0, \lambda_E^{0*}, \lambda_E^*\right)\end{array}\right\} \tag{31.25}$$

Therefore, a result similar to Theorem 31.2 can be derived providing necessary conditions for optimality for the embedded system of Equation 31.18.

31.5 Two-Gear Car Example with Application of Necessary Conditions

This section uses the embedded methodology developed in Sections 31.2 and 31.3 to analyze a crude two-dimensional model of a car with two gears [1, Example 1, p. 3975] having speed-dependent efficiencies

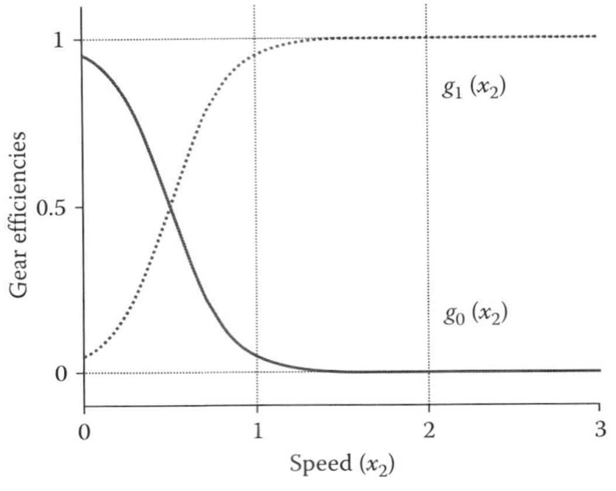

FIGURE 31.3 Gear efficiencies as a function of normalized speed.

$g_0(\varsigma)$ and $g_1(\varsigma)$ as plotted in Figure 31.3. Let x_1 denote the car's position and x_2 its velocity with respect to some coordinate system.

The embedded system has the form ($\dot{x} = (1-v)f_0(x, u_0) + vf_1(x, u_1)$):

$$\begin{bmatrix} \dot{x}_1(t) \\ \dot{x}_2(t) \end{bmatrix} = \left(1 - v(t)\right) \begin{bmatrix} x_2(t) \\ g_0\left(x_2(t)\right) \cdot u_0(t) \end{bmatrix} + v(t) \begin{bmatrix} x_2(t) \\ g_1\left(x_2(t)\right) \cdot u_1(t) \end{bmatrix} \tag{31.26}$$

The inputs $u_0(t), u_1(t) \in \Omega = [-1, 1]$ represent control of the brake or throttle. The terminal state constraints are $x(0) = [-5, 0]^T$ and $x(t_f) = [0, 0]^T$ in which case the terminal constraint set is $\mathrm{B} = \left\{(t_0, x_0, t_f, x_f) = \left(0, [-5, 0]^T, t_f, [0, 0]^T\right) | t_f \in \mathbb{R}\right\}$. Additionally, we require that the vehicle begins in Mode 0 and returns to Mode 0 at t_f, but for the moment we do not impose these constraints; it turns out that they follow directly from the solution of the "mode unconstrained" scenario. Finally, we define the embedded PI to be

$$J_E(x, v, u_0, u_1) = \int_0^{t_f} \left[(1 - v(t))F_0(t, x, u) + v(t)F_1(t, x, u)\right] \mathrm{d}t = \int_0^{t_f} \mathrm{d}t = t_f$$

where we have set $F_0 = F_1 = 1$; this is a minimum time problem. One observes that the sufficiency conditions for the existence of an optimal solution are satisfied. We denote the optimal solution by $\left(x^*(t), u_0^*(t), u_1^*(t), v^*(t)\right)$.

The Hamiltonian associated with system Equation 31.26 is with $\lambda = [\lambda_1, \lambda_2]^T$ (suppressing t-dependencies)

$$\begin{aligned} H_E\left(t, x, u_0, u_1, v, \lambda^0, \lambda\right) &= v \cdot \lambda_2 \left[g_1(x_2)u_1 - g_0(x_2)u_0\right] + \left[\lambda^0 + \lambda_1 x_2 + \lambda_2 g_0(x_2) u_0\right] \\ &\triangleq v \cdot E_1(t, x, u_0, u_1, \lambda^0, \lambda) + E_0(t, x, u_0, u_1, \lambda^0, \lambda) \end{aligned} \tag{31.27}$$

We like to argue that the Hamiltonian of Equation 31.27 implies the existence of bang-bang solutions as well as other information about the optimal solution. This requires that $E_1(t, x^*(t), u_0^*(t), u_1^*(t), \lambda^0, \lambda(t)) \neq 0$ almost everywhere on $[0, t_f]$, which is shown in Proposition 4.2.1 [36], that is, there are only bang-bang solutions to the hybrid optimal control problem. Let us now determine the values for the optimal solution.

From the material of Section 31.3, there exists a constant $\lambda^0 \leq 0$ and an absolutely continuous function $\lambda(\cdot) : [0, t_f] \to \mathbb{R}^2$ such that $\left(\lambda^0, \lambda(t)\right) \neq 0$ on $[0, t_f]$, and for almost all $t \in [0, t_f]$, the state equations

$$\begin{bmatrix} \dot{x}_1^*(t) \\ \dot{x}_2^*(t) \end{bmatrix} = \begin{bmatrix} x_2^*(t) \\ \left(1 - v^*(t)\right) g_0\left(x_2^*(t)\right) \cdot u_0^*(t) + v^*(t) g_1\left(x_2^*(t)\right) \cdot u_1^*(t) \end{bmatrix} \tag{31.28}$$

and costate equations (suppressing t-dependencies)

$$\begin{bmatrix} \dot{\lambda}_1 \\ \dot{\lambda}_2 \end{bmatrix} = \begin{bmatrix} 0 \\ -\lambda_1 - \lambda_2(1-v^*)u_0^* \left[\dfrac{dg_0}{dx_2}\right]_{x_2^*} - \lambda_2 v^* u_1^* \left[\dfrac{dg_1}{dx_2}\right]_{x_2^*} \end{bmatrix} \tag{31.29}$$

hold.

We can draw two conclusions: (1) from Equation 31.29, $\lambda_1(t) = \lambda_1$, a constant that can be shown to be greater than zero; and from the transversality condition,

$$H_E\left(t_f, x^*(t_f), u_0^*(t_f), u_1^*(t_f), v^*(t_f), \lambda^0, \lambda(t_f)\right) = 0 \tag{31.30}$$

Thus, Equations 31.27 and 31.30, with $x_2^*(t_f) = 0$, imply that

$$\lambda^0 + \lambda_2(t_f)\left[v^*(t_f)g_1(0)u_1^*(t_f) + \left(1 - v^*(t_f)\right)g_0(0)u_0^*(t_f)\right] = 0 \tag{31.31}$$

With Equation 31.31 and a "proof by contradiction," one can also show that $\lambda_2(\cdot) \neq 0$ (not identically zero) on any nonzero subinterval of $[t_0, t_f]$. Hence, using an equivalent form of Equation 31.8, we conclude that

$$\begin{aligned} &\max_{u_0 \in \Omega} H_E\left(t, x^*(t), u_0, u_1, 0, \lambda^0, \lambda(t)\right) - \max_{u_1 \in \Omega} H_E\left(t, x^*(t), u_0, u_1, 1, \lambda^0, \lambda(t)\right) \\ &= \left\{\lambda^0 + \lambda_1 x_2^*(t) + \max_{u_0 \in [-1,1]} \left[\lambda_2(t) g_0\left(x_2^*(t)\right) u_0\right]\right\} - \left\{\lambda^0 + \lambda_1 x_2^*(t) + \max_{u_1 \in [-1,1]} \left[\lambda_2(t) g_1\left(x_2^*(t)\right) u_1\right]\right\} \\ &= \left[g_0\left(x_2^*(t)\right) - g_1\left(x_2^*(t)\right)\right] \max_{u \in [-1,1]} \left[\lambda_2(t) u\right] \end{aligned} \tag{31.32}$$

From Equation 31.32, we observe that the optimal mode of operation is the mode with the largest gear ratio, and, excepting the time instants when (possibly) $\lambda_2(t) = 0$, the optimal throttle/braking control is either $+1$ or -1, depending on the sign of $\lambda_2(t)$. This brings us to the following optimal operation:

1. Similar to the discussion in [1], the optimal mode $v(t)$ is given by the mode with the largest efficiency at time instant t.
2. For Mode 0 ($E_1(\cdot) > 0$). If $g_0\left(x_2^*(t)\right) > g_1\left(x_2^*(t)\right)$, that is, $x_2^*(t) < 0.5$ for some $t \in [0, t_f]$, then $v(t) = 0$ and
$$u_0(t) = \begin{cases} \text{sgn}\left(\lambda_2(t)\right), & \lambda_2(t) \neq 0 \\ \text{indeterminate in } [-1, 1], & \lambda_2(t) = 0 \end{cases} \tag{31.33}$$
3. For Mode 1 ($E_1(\cdot) < 0$). If $g_0\left(x_2^*(t)\right) < g_1\left(x_2^*(t)\right)$, that is, $x_2^*(t) > 0.5$ for some $t \in [0, t_f]$, then $v(t) = 1$ and
$$u_1(t) = \begin{cases} \text{sgn}\left(\lambda_2(t)\right) & \lambda_2(t) \neq 0, \\ \text{indeterminate in } [-1, 1] & \lambda_2(t) = 0 \end{cases} \tag{31.34}$$
4. From the endpoint constraints, $x_2^*(0) = x_2^*(t_f) = 0$, it follows that $v^*(0) = v^*(t_f) = 0$ since $g_0\left(x_2^*(t)\right) > g_1\left(x_2^*(t)\right)$ for $t = 0, t_f$; thus optimality enforces the physically meaningful mode constraint at the endpoints. Finally, we mention again that the constant $\lambda_1 > 0$, and that as per [36] there exists $0 < t_1 < t_f$ such that $\lambda_2(t) > 0$ for all $t \in [0, t_1)$ and $\lambda_2(t) < 0$ for all $t \in (t_1, t_f]$. We make the following further conclusions:
   1. If $t < t_1$ and $x_2^*(t) < 0.5$, then $v^*(t) = 0$ and $u_0^*(t) = 1$.
   2. If $t < t_1$ and $x_2^*(t) > 0.5$, then $v^*(t) = 1$ and $u_1^*(t) = 1$.
   3. If $t > t_1$ and $x_2^*(t) < 0.5$, then $v^*(t) = 0$ and $u_0^*(t) = -1$.
   4. If $t > t_1$ and $x_2^*(t) > 0.5$, then $v^*(t) = 1$ and $u_1^*(t) = -1$.

   It turns out that this solution is also optimal for a penalty on switching. See Figure 31.4 for simulation results. A variation of this example for control through a communication network with fixed delay is considered in [37].

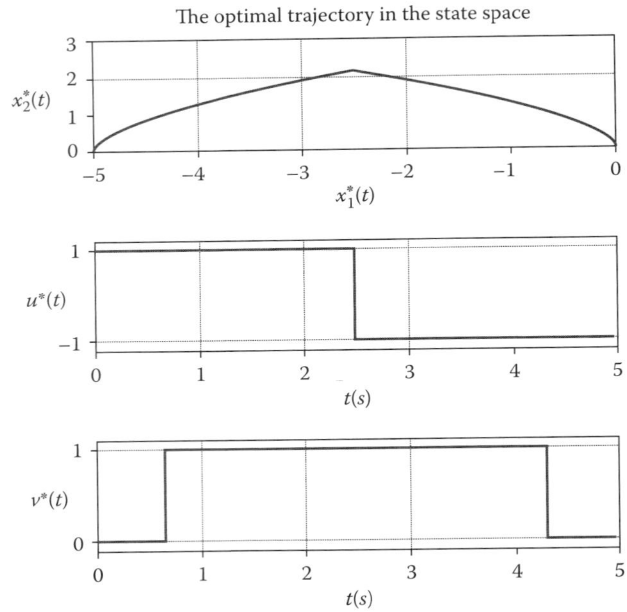

FIGURE 31.4 Optimal $x_1^*(t)$, $x_2^*(t)$, $u^*(t)$, and $v^*(t)$.

31.6 Unicycle Example with Direct Collocation

To demonstrate the full power of the embedding approach, we show how to numerically compute optimal trajectories for a system exhibiting both controlled and autonomous switches. Further details on how the embedding technique can be applied to such systems can be found in Appendix B and [21]. The example considers a unicycle driving on a horizontal plane (Figure 31.5). The wheel of the unicycle can either roll or slide, resulting in autonomous switches. In addition, we assume that the unicycle has a regenerative brake that can be turned on or off. These switches are controlled. We assume that the unicycle contains a separate motor and a generator, both connected to a battery pack. This implies that the system can brake either by applying a negative torque on the wheels or by using the regenerative braking.

Referring to Figure 31.5, the forward velocity of the wheel is controlled by the torque u_1 applied to the wheel's axle, while its heading is controlled by the torque u_2. Our objective is to drive the unicycle to the origin within an allotted time while minimizing weighted power usage. We use MPC [38] to compute the control inputs for the system.

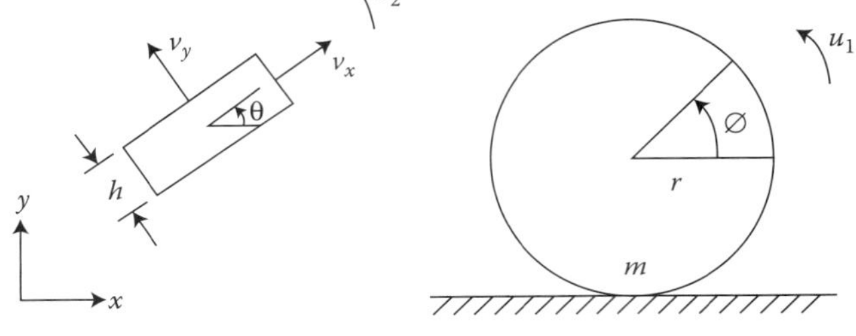

FIGURE 31.5 A top view and a side view of the unicycle.

31.6.1 Unicycle Model

The generalized coordinates for the unicycle are its center of mass position x and y, body orientation θ (heading) relative to the x-axis, and the angular position of the wheel φ. Since we are not interested in φ itself, the state variables for the system are $z^T = [x, y, \theta, v_x, v_y, \dot{\theta}, \dot{\varphi}]^T \in R^7$, where $[v_x, v_y]$ is the velocity of the center of mass of the unicycle, expressed in the body frame, $\dot{\theta}$ is the turning velocity of the unicycle, and $\dot{\varphi}$ is the angular velocity of the wheel as it spins on its axle. The equations of motion for the unicycle take the form

$$\dot{x} = v_x \cos(\theta) - v_y \sin(\theta) \tag{31.35a}$$

$$\dot{y} = v_x \sin(\theta) + v_y \cos(\theta) \tag{31.35b}$$

$$\dot{\theta} = \omega \tag{31.35c}$$

$$\dot{v}_x = \frac{F_x(z)}{m} + \dot{\theta} v_y \tag{31.35d}$$

$$\dot{v}_x = \frac{F_x(z)}{m} + \dot{\theta} v_y \tag{31.35e}$$

$$\ddot{\theta} = \frac{1}{I_2} u_2 \tag{31.35f}$$

$$\ddot{\varphi} = \frac{F_x r}{I_1} + \frac{1}{I_1} u_1 \tag{31.35g}$$

where (1) m is the mass of the unicycle, (2) r is the radius of the wheel, (3) I_1 is the moment of inertia of the wheel around its axis, (4) I_2 is the moment of inertia of the unicycle about the vertical axis through the center of mass, and (5) F_x and F_y are the forces between the ground and the wheel in the forward and lateral directions, respectively.

The autonomous switched behavior of the unicycle occurs because $F_x(z)$ and $F_y(z)$ depend on whether the unicycle is rolling (F_x and F_y are ground reaction forces that oppose slipping) or sliding (where F_x and F_y are frictional forces). We note that when rolling, the relative velocity, $v_r = [v_{rx}, v_{ry}]^T = [v_x + \dot{\varphi} r, v_y]^T$, between the ground and the wheel's point of contact is zero, that is,

$$[v_{rx}, v_{ry}] = [v_x + \dot{\varphi} r, v_y] = [0, 0] \tag{31.36}$$

Thus from Equations 31.35a and 31.36, with μ_d the coefficient of dynamic friction and g the gravitational constant

$$[F_x(z), F_y(z)] = \begin{cases} -mr\begin{bmatrix} \dfrac{u_1}{mr^2 + I_1} & \dot{\theta}\dot{\varphi} \end{bmatrix} & \text{Rolling} \\ -\dfrac{\mu_d mg}{\|v_r\|}\begin{bmatrix} v_x + \dot{\varphi} r & v_y \end{bmatrix} & \text{Sliding} \end{cases} \tag{31.37}$$

The autonomous switch from rolling to sliding occurs when the magnitude of the constraint force $F = [F_x, F_y]^T$ exceeds the maximum possible magnitude of the static friction, $\mu_s mg$ (μ_s being the coefficient of static friction), that is, $\|F\| > \mu_s mg \Rightarrow$ "*rolling* $\rightarrow$ *sliding*". On the other hand, the switch from sliding to rolling occurs when (1) $v_r = [v_{rx}, v_{ry}]^T = 0$ and (2) the maximum magnitude of the frictional force exceeds that of the constraint force F, that is, $\|v_r\| = 0$ and $\|F\| < F_{s,\max} = \mu_s mg \Rightarrow$ "*sliding* $\rightarrow$ *rolling*".

In contrast to the autonomous switches, described above, the regenerative brake can be switched off (Mode 0) or switched on (Mode 1) arbitrarily. For Mode 0,

$$u_1 = u_{1A}^0 \in [-20, \ 20] \tag{31.38a}$$

denotes an actuating torque that can be either propelling or braking. Mode 1 denotes the use of regenerative braking alone in which case

$$u_1 = u_1^1 = \begin{Bmatrix} -K_b\dot{\varphi}, & |\dot{\varphi}| \le 2 \\ -20\text{sgn}(\dot{\varphi}), & |\dot{\varphi}| > 2 \end{Bmatrix} \tag{31.38b}$$

where $K_b = 10$ is a fixed regenerative braking coefficient; note that the magnitude of u_1 saturates at 20 Nm. Note that both control modes are possible in either sliding or rolling, an example of decoupled switches. The resulting torque control has the form

$$u_1(t) = \left(1 - v(t)\right) u_1^0(t) + v(t)u_1^1(t) \tag{31.39}$$

where $v(t) \in \{0, 1\}$ denotes the original problem whereas $v(t) \in [0, 1]$ the embedded problem which is solved in this investigation. Because of the autonomous switches determined by Equation 31.37, the right-hand side of Equation 31.35 is piecewise continuous, provided the system does not chatter about the boundary implicitly defined in Equation 31.37; if there is such chattering then we must interpret the solution of the equations in the sense of Filippov [27].

31.6.2 Control Objective, PI, and MPC

The objective of the control design is to drive the unicycle from a given starting initial state, $z_0^T = [x(0), y(0), v_x(0), v_y(0), \theta(0), \omega(0), \phi(0)]^T = [0, 4, 0, 0, 1, 0, 1]^T$ back to the origin while minimizing the energy usage. In addition, we like to limit the undesirable sliding motion of the wheel as it implies a loss of controllability. As such, the PI takes the form

$$J = c_0 \left\|z(T)\right\|^2 + \int_0^T \left[c_1 \left(1 - v\right) \left(u_1\right)^2 + c_2 \left(u_2\right)^2 + c_3 \left\|v_r\right\|^2\right] \mathrm{d}t \tag{31.40}$$

where the constant weights $c_i > 0$. The term (1) $c_0 \left(z^T(T)z(T)\right)$ drives the final position of the unicycle toward the origin; (2) $c_1 \left(1 - \alpha\right) \left(u_1\right)^2$ penalizes the actuating power usage; (3) $c_2 \left(u_2\right)^2$ penalizes the heading power usage; and (4) $c_3 \left(\left\|v_r\right\|\right)^2$ limits the sliding motion. The terminal constraints are enforced through the cost functional (as soft constraints) rather than imposed as hard constraints because the system is stabilizable but not controllable in the sliding regime. Therefore, using hard constraints could make the optimal control problem infeasible. Note also that there is no penalty for regenerative braking.

The control objective is to minimize the PI of Equation 31.40 subject to the embedded state dynamics given by Equations 31.35 and the initial state z_0. However, when applying the computed controls to the actual model that differs from the nominal model due to the presence of disturbances or modeling uncertainties, the state trajectory might deviate from the desired trajectory, and fail to reach the desired final state within the allotted time interval. To cope with such disturbances and uncertainties, an MPC-type controller that is well known for its robustness is utilized. The MPC approach can be summarized as follows:

1. Given z_0, partition the time interval T into N equal subintervals of length $h = T/N$, for the purpose of computing a (backward) piecewise constant control sequence $\left\{\hat{u}_1, \ldots, \hat{u}_N\right\}$, where $\hat{u}_i = \left[u_1(ih) \quad u_2(ih)\right]^T$, and the state values $\{z_1, \ldots, z_n\}$.
2. For $k = 1, \ldots, N$, solve the embedded problem of the unicycle over the receding horizon $[k, N]$ by minimizing the PI given by Equation 31.40 subject to the *nominal* model with the initial state z_{k-1} and obtain the (look ahead) control sequence $\left\{\hat{u}_k, \ldots, \hat{u}_N\right\}$.
3. Apply the control input $\hat{u}_k$ for the time interval $t_{k-1} \le t < t_k$ to the real model. The value of the state of the real model at the end of the interval becomes , the initial condition for the next iteration.
4. Repeat steps 2 and 3 until $k = N$.

31.6.3 Simulation Results and Discussion

A variation of the direct collocation method [39] is used to numerically solve the EOCP at each step of the MPC algorithm. The number of points for the discretization was $N = 20$. Nominal trajectories are for the model without any disturbances. Comparisons are made to the MPC-controlled process with a frictional disturbance. In all cases the embedded problem has a bang-bang solution, meaning that it is also a solution to the original hybrid optimal control problem.

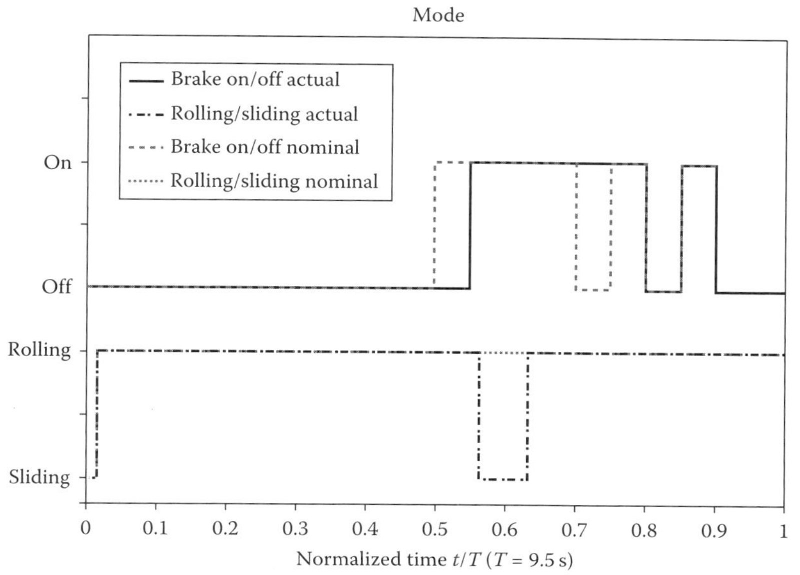

FIGURE 31.6 Modes of operation.

31.6.3.1 Unicycle Trajectories for the PI of Equation 31.40

The plots show trajectories of the unicycle that minimize the minimum energy PI of Equation 31.40, and its behavior under MPC control. The computed cost of the nominal optimal trajectory is 972.2 versus 1036.3 for the disturbed MPC-controlled trajectory. The nominal response is for the model without disturbances, whereas the MPC control is for the frictional disturbance where in an annulus $0.9 \leq r \leq 1.4$ around the origin, the static coefficient of friction μ_s drops from 0.7 to 0.002, and the dynamic coefficient of friction μ_d drops from 0.6 to 0.001. Moreover, the unicycle's nominal parameters $m_{\text{nominal}} = 1$ and $r_{\text{nominal}} = 4$ during the simulation were perturbed so that $m_{\text{actual}} = 1.05$ and $r_{\text{actual}} = 3.9$. After starting in the sliding mode, the unicycle is driven to the rolling mode after about 0.2 s. At around 5 s the unicycle encounters the frictional disturbances and switches from rolling to sliding as indicated in Figure 31.6. However, after leaving the slippery area at about 6 s, in conjunction with the corrective action of the MPC controller, the unicycle starts to roll again. Figures 31.7 and 31.8 show the unicycle's trajectories and the evolutions of two states V_x and V_y. It clearly shows that the unicycle can still reach the origin in the required time despite disturbances and model errors. Figure 31.9 displays the control inputs that again adapt in accordance with state and model changes. The results thus confirm that the influence of disturbances on system performance is small and the MPC scheme achieves good performance and robustness. Recall that the electric motor can apply both a propelling torque and a braking torque u_1^0 in Mode 0, while a regenerative braking torque u_1^1 is applied in Mode 1. One observes that during the final 4 s of the simulation, the switches of the braking torque in Mode 0 and the regenerative braking torque in Mode 1 are coordinated to reduce cost and reach the origin on time.

31.7 Concluding Remarks

This study has developed an approach to the solution of optimal control of switching systems that converts the nonconvex SOCP into a convex EOCP, which allows for direct solution to the SOCP or for arbitrarily close approximations to the SOCP except in rare circumstances. This chapter illustrated the use of the

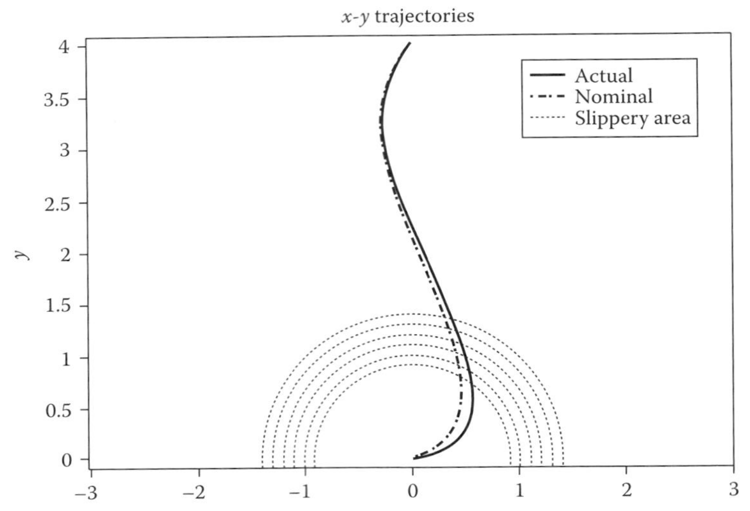

FIGURE 31.7 Wheel's position on the x–y plane.

approach in two examples while directing the reader to the references for additional examples and for further developments. What was not discussed in this chapter is the development of discrete-time optimal control problems and the related ongoing work. To this end we offer the following recent work.

A general framework based on the method of approximate dynamic programming is developed in [40] for the controller synthesis of discrete-time switched linear systems (SLS) as well as discrete-time nonlinear systems, particularly for their optimal control and stabilization. As pointed out earlier in this chapter, direct solution of switched system problems is challenged by their combinatorial nature, the size of

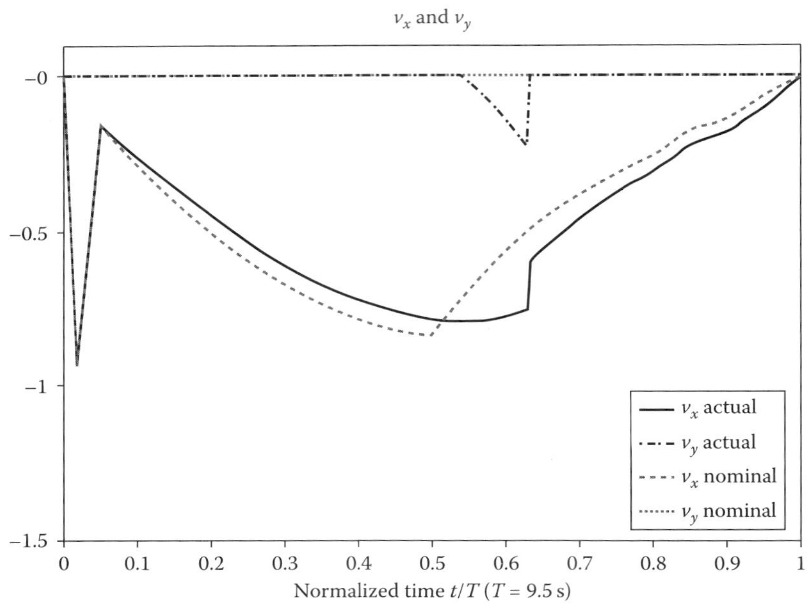

FIGURE 31.8 Forward and lateral velocities.

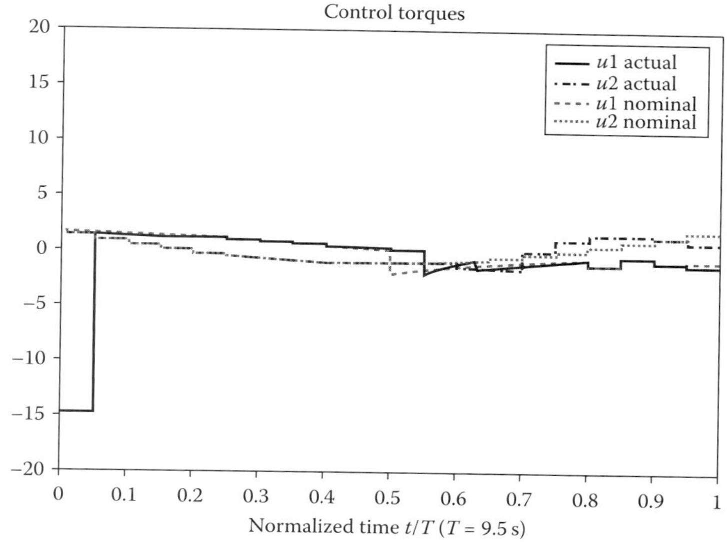

FIGURE 31.9 Control inputs.

control policy space, which consists of traditional controls and discrete mode sequences that become prohibitively large as the time horizon increases. Using an iterative implementation of approximate dynamic programming, and by permitting a small compromise in optimality, the computational complexity of finding a (sub)optimal solution may be significantly reduced. For SLS, efficient solution algorithms (the so-called *relaxed switched Riccati iterations*) exist that find the optimal controllers for fairly large dimensional discrete-time systems. The reduction in computational complexity suggests that the (iterative) algorithms are amenable to practical implementation [41,42]. A relationship between stabilization and the optimal control of an SLS is established in [43].

31.8 Appendix A: Modeling with Autonomous Switches

Four quantities are essential to describe the evolution of a system subject to autonomous and controlled switches: (1) the usual continuous state $x(t) \in \mathbb{R}^n$; (2) the usual continuous control input $u(t) \in \mathbb{R}^m$; (3) a switching (discrete) control input, $v(t) \in D_v = \{1, 2, \ldots, d_v\}$; and (4) a discrete state $\xi \in D_\xi = \{1, 2, \ldots, d_\xi\}$ that identifies autonomous switches. This chapter only considers autonomous switches that depend on the continuous state $x(t)$ and the continuous input $u(t)$ but do not depend on the current values of the discrete states, $\xi(t)$ or $v(t)$. Such systems are usually called memoryless systems. The evolution of the discrete state, $\xi(t)$, of the memoryless system is defined by a piecewise continuous* function $\eta : \mathbb{R}^n \times \mathbb{R}^m \to D_\xi$, such that

$$\xi(t^+) = \eta(x(t), u(t)) \in D_\xi \tag{31.41}$$

Thus, for each $i \in D_\xi$, we define

$$M_i = \{(x, u) \in \mathbb{R}^n \times \mathbb{R}^m | \eta(x, u) = i\} \subseteq \mathbb{R}^n \times \mathbb{R}^m$$

* By piecewise continuous we mean a function that is continuous everywhere except on a finite union of switching surfaces that are smooth submanifolds of $\mathbb{R}^n \times \mathbb{R}^m$ with measure 0 where it undergoes discontinuous jumps, but has well-defined limits in all directions.

$M_i \subseteq \mathbb{R}^n \times \mathbb{R}^m$ is the set of pairs (x, u) corresponding to the discrete state $i \in D_\xi$. Let $f_{(i,j)} : M_i \to \mathbb{R}^n$, $i \in D_\xi, j \in D_v$, be a collection of C^1 vector fields associated with a system. The evolution of the continuous state $x(t)$ is then described by

$$\dot{x}(t) = f_{(\eta(x(t),u(t)),v(t))}\left(x(t), u(t)\right), \quad x(t_0) = x_0. \tag{31.42}$$

At each $t \geq t_0$ and for each discrete state $\xi(t) \in D_\xi$, the switching control input $v(t) \in D_v$ thus selects the particular vector field that governs the evolution of the continuous state.

As mentioned in Section 31.3, we assume that the continuous control input $u(t) \in \Omega$, a convex and compact set in $\mathbb{R}^m$, and that the switching control input $v(t)$ and the continuous control input $u(t)$ are both measurable functions. Note that we restrict our attention to time-invariant systems, but the results can be easily generalized to time-varying systems.

Given that the discrete state $\xi(t)$ is completely determined by $x(t)$ and $u(t)$ through Equation 31.41, we can define for each $j \in D_v$ a piecewise C^1 vector field:*

$$f_j(x(t), u(t)) \triangleq f_{(\eta(x(t),u(t)),j)}(x(t), u(t)) \tag{31.43}$$

and rewrite Equation 31.42 in a more convenient form:

$$\dot{x}(t) = f_{\upsilon(t)}(x(t), u(t)), \quad x(t_0) = x_0 \tag{31.44}$$

This means the vector fields f_j *all have the same set of points of discontinuity*. We thus refer to the systems described by Equations 31.42 and 31.44 as *systems with decoupled switches.*

We are interested in computing optimal control laws for the system described by Equation 31.42 or Equation 31.44. If the system only undergoes autonomous switches ($D_v = \{1\}$), only the continuous input $u(t)$ needs to be computed. This suggests that the complexity of the optimal control problem might not be any different than in the traditional case. In contrast, for systems with controlled switches we need to compute the sequence of switching times $t_1, \ldots, t_n$ (including n), the sequence of discrete inputs $v_1, \ldots, v_n$, as well as the continuous input $u(t)$ on each interval $[t_i, t_i + 1]$ for $i = 0, \ldots, n-1$. It would therefore appear that for systems with controlled switches the optimal control problem has combinatorial complexity. As discussed in this chapter, *both these cases have the same complexity and are amenable to traditional nonlinear programming techniques such as sequential quadratic programming (SQP).* This further implies that for the systems with memoryless autonomous switches and controlled switches the optimal control problem is no more complex than the traditional smooth problem.

31.9 Appendix B: Numerical Solution Using Direct Collocation

This section overviews the direct numerical solution algorithm and strategy used in the solution of the unicycle example without first having to apply the necessary conditions for optimality. Specifically, we discuss the collocation method for solving hybrid optimal control problems [21,44–47].

Given the embedded PI and the state equation and constraints mentioned in Section 31.3, one discretizes these equations using the collocation method. These discretized equations convert the EOCP into a finite-dimensional nonlinear programming problem (NLP), where states and inputs are treated as unknown variables. The NLP can be solved using an SQP solver, such as *fmincon* in the optimization toolbox of MATLAB®. The discretization-and-collocation technique consists of several steps that have two main stages: (1) time discretization, and state and input function approximations by a finite number

* Similarly as before, by piecewise C^1, we mean a function that is C^1 everywhere, except on a finite union of switching surfaces that are smooth submanifolds $\mathbb{R}^n \times \mathbb{R}^m$ with measure 0 where the function is not differentiable and undergoes discontinuous jumps, but has well defined limits in all directions.

of polynomial basis functions; (2) approximation of the continuous state dynamics and PI integrand by discrete-state and discrete-input-dependent counterparts.

Without going through a lengthy derivation, the continuous-time interval $[t_0, \ t_f]$ is discretized into a sequence of points $t_0 < t_1 < t_2 < \cdots < t_{N-1} < t_N = T$, where, for simplicity, we take $t_j - t_{j-1} = h$, for $j = 1, \ldots, N$. A "hat" notation is also used to distinguish the numerically estimated state and control values from their actual counterparts that are "hatless," for example, $\hat{x}_j = \hat{x}(t_j)$, $\hat{u}_{0,j} = \hat{u}_0(t_j)$, $\hat{u}_{1,j} = \hat{u}_1(t_j)$, and $\hat{v}_j = \hat{v}(t_j)$. The collocation method used here assumes triangular basis functions for the state and piecewise constant basis functions (derivatives of triangular functions) for the controls. Specifically, the estimated state is given by

$$\hat{x}(t) = \sum_{j=0}^{N} \hat{x}_j \varphi_j(t) \tag{31.45}$$

where the $\hat{x}_j$'s are to be determined and the triangular basis functions are given by

$$\varphi_j(t) = \begin{cases} \dfrac{t - t_{j-1}}{h}, & t_{j-1} < t \leq t_j \\ \dfrac{t_{j+1} - t}{h}, & t_j < t \leq t_{j+1} \\ 0, & \text{elsewhere} \end{cases} \tag{31.46}$$

We note two points: the method is not restricted to using triangular basis functions and each of the $\varphi_j(t)$'s is a time shift of the previous one.

As summarized in [45], the theoretical approach for computing the controls is to extend the state space with new state variables, $x_{\text{ext}} \in R^{m+1}$, whose derivative are the desired controls, $u(t) \in R^m$ and $v(t) \in [0, 1] \subset R$, to be computed. However, our choice of triangular basis functions for the states renders the control inputs piecewise constant and we simply solve directly for these (constant) control values. Specifically, the estimates of the control inputs are given by

$$\begin{bmatrix} \hat{u}(t) \\ \hat{v}(t) \end{bmatrix} = \sum_{j=1}^{N} \begin{bmatrix} \hat{u}_j \\ \hat{v}_j \end{bmatrix} \psi_j(t) \tag{31.47}$$

where the piecewise constant basis functions are given by

$$\psi_j(t) = \begin{cases} 1 & t_{j-1} < t \leq t_j \\ 0 & \text{elsewhere} \end{cases} \tag{31.48}$$

Here we note that by the definition of the basis functions in Equation 31.48, the control values computed at t_j are enforced over the interval $t_{j-1} < t \leq t_j$.

The essence of the midpoint rule in the collocation method is to enforce the constraints at the midpoints of each interval $[t_{j-1}, t_j]$ for $j = 1, \ldots, N$. There results the discretized embedded state dynamics

$$\hat{x}_j = \hat{x}_{j-1} + h \cdot (1 - \hat{v}_j) \cdot f_0 \left(\frac{\hat{x}_{j-1} + \hat{x}_j}{2}, \hat{u}_{0j} \right) + h \cdot \hat{v}_j \cdot f_1 \left(\frac{\hat{x}_{j-1} + \hat{x}_j}{2}, \hat{u}_{1j} \right) \tag{31.49}$$

for $j = 1, \ldots, N$, with $f_0(\cdot)$ and $f_1(\cdot)$ the discretized state dynamics in modes -0 and -1, respectively. Thus the solution to the EOCP is given by the following NLP: Minimize

$$\hat{J} = g_N \left(t_N, x_N\right) + \sum_{j=1}^{N} \frac{1}{2} h \left\{ F_E \left(t_j, \hat{x}_j, \hat{u}_{0j}, \hat{u}_{1j}, \hat{v}_j, \hat{p}_j\right) + F_E \left(t_{j-1}, \hat{x}_{j-1}, \hat{u}_{0j}, \hat{u}_{1j}, \hat{v}_j, \hat{p}_j\right) \right\} \tag{31.50}$$

over the controls $(\hat{u}_j, \hat{v}_j) \in \Omega \times [0, 1]$, subject to Equation 31.49 and all other equality constraints represented as $g(\hat{x}_{j-1}, \hat{x}_j, \hat{u}_j, \hat{v}_j, \hat{p}_j) = 0$. Here $F_E(\cdot)$ is the integrand of the PI properly discretized and $\hat{p}_j$ represents various intermediate constraints.

References

1. Hedlund, S. and Rantzer, A. Optimal control of hybrid systems. *Proceedings of the IEEE Conference on Decision and Control*, Phoenix, AZ, USA, 3972–3977, 1999.
2. Wei, S., Žefran, M., Uthaichana, K., and DeCarlo, R.A. Hybrid model predictive control for stabilization of wheeled mobile robots subject to wheel slippage. *Proceedings of the IEEE International Conference on Robotics and Automation*, Rome, 2373–2378, 2007.
3. Uthaichana, K., Bengea, S., DeCarlo, R., Pekarek, S., and Žefran, M. Hybrid model predictive control tracking of a sawtooth driving profile for an HEV. *Proceedings of the American Control Conference*, Seattle, WA, 967–974, 2008.
4. Moerdyk, B., De Carlo, R., Birdwell, D., Žefran, M., and Chiasson, J. Hybrid optimal control for load balancing in a cluster of computer nodes. *Proceedings of the IEEE International Conference on Control Applications*, Munich, 1713–1718, 2008.
5. Domiìnguez-Navarro, J.A., Bernal-Agustiìn, J.L., Diìaz, A., Requena, D., and Vargas, E.P. Optimal parameters of FACTS devices in electric power systems applying evolutionary strategies, *International Journal of Electrical Power and Energy Systems*, 29(1): 83–90, 2007.
6. Ho, C.Y.F., Ling, B.W.K., Liu, Y.Q., Tam, P.K.S., and Teo, K.L., Optimal PWM control of switched-capacitor DC–DC power converters via model transformation and enhancing control techniques, *IEEE Transactions on Circuits and Systems I: Regular Papers*, 55(5): 1382–1391, 2008.
7. Loxton, R.C., Teo, K.L., Rehbock, V., and Ling, W.K. Optimal switching instants for a switched-capacitor DC/DC power converter, *Automatica*, 45(4): 973–980, 2009.
8. Oettmeier, F.M., Neely, J., Pekarek, S., DeCarlo, R., and Uthaichana, K. MPC of switching in a boost converter using a hybrid state model with a sliding mode observer, *IEEE Transactions on Industrial Electronics*, 56(9): 3453–3466, 2009.
9. Neely, J., Pekarek, S., and DeCarlo, R. Hybrid optimal-based control of a boost converter. *IEEE Applied Power Electronics Conference and Exposition*, Washington, DC, pp. 1129–1137, February 15–19, 2009.
10. Neely, J., Pekarek, S., DeCarlo, R., and Vaks, N. Real-time hybrid model predictive control of a boost converter with constant power load. *25th Annual IEEE Applied Power Electronics Conference and Exposition*, Palm Springs, CA, pp. 480–490, February 23–28, 2010.
11. Giua, A., Seatzu, C., and Van Der Mee, C. Optimal control of switched autonomous linear systems. *Proceedings of the IEEE Conference on Decision and Control*, Orlando, FL, 3: 2472–2477, 2001.
12. Xu, X. and Antsaklis, P.J. Optimal control of switched systems: New results and open problems. *Proceedings of the American Control Conference*, Chicago, IL, 4: 2683–2687, June 28–30, 2000.
13. Bemporad, A. and Morari, M. Control of systems integrating logic, dynamics, and constraints, *Automatica*, 35(3): 407–427, 1999.
14. Sussmann, H.J. Maximum principle for hybrid optimal control problems. *Proceedings of the IEEE Conference on Decision and Control*, Phoenix, AZ, 1: 425–430, 1999.
15. Shaikh, M.S. and Caines, P.E. On the hybrid optimal control problem: Theory and algorithms, *IEEE Transactions on Automatic Control*, 52(9): 1587–1603, 2007.
16. Riedinger, P., Kratz, F., Iung, C., and Zanne, C. Linear quadratic optimization for hybrid systems. *Proceedings of the IEEE Conference on Decision and Control*, Phoenix, AZ, USA, 3: 3059–3064, 1999.
17. Xu, X. and Antsaklis, P.J. Results and perspectives on computational methods for optimal control of switched systems. *Proceedings of the 6th International Conference on Hybrid Systems: Computation and Control*, Lecture Notes in Computer Science, 2623: 540–555, 2003.
18. Borrelli, F., Baotic, M., Bemporad, A., and Morari, M. Dynamic programming for constrained optimal control of discrete-time linear hybrid systems, *Automatica*, 41(10): 1709–1721, 2005.
19. Gapaillard, M. Continuous representation and control of hybrid systems, *International Journal of Control*, 81(1): 1–20, 2008.
20. Bengea, S.C. and DeCarlo, R.A. Optimal control of switching systems, *Automatica*, 41(1): 11–27, 2005.
21. Wei, S., Uthaichana, K., Žefran, M., DeCarlo, R.A., and Bengea, S. Applications of numerical optimal control to nonlinear hybrid systems, *Nonlinear Analysis: Hybrid Systems*, 1(2): 264–279, 2007.
22. Ripaccioli, G., Bemporad, A., Assadian, F., Dextreit, C., Cairano, S.D., and Kolmanovsky, I.V. Hybrid modeling, identification, and predictive control: An application to hybrid electric vehicle energy management, *Hybrid Systems: Computation and Control*, Lecture Notes in Computer Science, 5469: 321–335, 2009.
23. Borhan, H.A., Vahidi, A., Phillips, A.M., Kuang, M.L., and Kolmanovsky, I.V. Predictive energy management of a power-split hybrid electric vehicle. *Proceedings of the American Control Conference*, St. Louis, MO, 3970–3976, 2009.

24. Pekarek, S., Uthaichana, K., Bengea, S., DeCarlo, R., and Žefran, M. Modeling of an electric drive for a HEV supervisory level power flow control problem. *IEEE Vehicle Power and Propulsion Conference,* Chicago, IL, 396–401, 2005.
25. Arce, A., Del Real, A.J., and Bordons, C. MPC for battery/fuel cell hybrid vehicles including fuel cell dynamics and battery performance improvement, *Journal of Process Control*, 19(8): 1289–1304, 2009.
26. Tate, E.D., Grizzle, J.W., and Peng, H. SP-SDP for fuel consumption and tailpipe emissions minimization in an EVT hybrid, *IEEE Transactions on Control Systems Technology*, 18(3): 673–687, May 2010.
27. Filippov, A.F. Differential Equations with discontinuous right-hand sides, *American Mathematical Society Translations*, Series 2, 42: 199–231, 1964.
28. Antsaklis, P., Kohn, W., Nerode, A., and Sastry, S. (Eds) *Hybrid Systems IV*, Lecture Notes in Computer Science, Vol. 1273, Springer, Berlin, 1997.
29. Vaandrager, F.W. and van Schuppen, J.H. (Eds) *Hybrid Systems: Computation and Control*, Lecture Notes in Computer Science, Springer-Verlag, Berlin, 1999.
30. Antsaklis, P.J. (Ed) Hybrid systems: Theory and applications a brief introduction to the theory and applications of hybrid systems, *Proceedings of the IEEE*, 88, 879–887, 2000.
31. Morse, A., Pantelides, C., Sastry, S., and Schumacher, J. (Eds) Hybrid Systems, *Automatica*, 35(3): 347–348, 1999.
32. Imura, J.I. and Van Der Schaft, A. Characterization of well-posedness of piecewise-linear systems, *IEEE Transactions on Automatic Control*, 45(9): 1600–1619, 2000.
33. Lygeros, J., Johansson, K.H., Simić, S.N., Zhang, J., and Sastry, S.S. Dynamical properties of hybrid automata, *IEEE Transactions on Automatic Control*, 48(1): 2–17, 2003.
34. Berkovitz, L.D. *Optimal Control Theory*, Springer, Berlin, 1974.
35. Bai, X. and Yang, X.S. A new proof of a theorem on optimal control of switched systems, *Journal of Mathematical Analysis and Applications*, 331(2): 895–901, 2007.
36. Bengea, S.C. Optimal control of switched hybrid systems with applications to the control of hybrid electric vehicles. PhD Thesis, School of ECE, Purdue University, 2005.
37. Bengea, S.C., DeCarlo, R.A., Hokayem, P.F., and Abdallah, C.T. Suboptimal control techniques for hybrid systems operating via networks, in *Advanced in Communication Control Networks*, Tarbouriech, S., Abdallah, C.T. and Chiasson, J. (Eds), Springer-Verlag, Berlin, 281–302, 2005.
38. Camacho, E.F. and Bordons, C. *Model Predictive Control*, Springer-Verlag, Berlin, 2004.
39. Von Stryk, O. Numerical solution of optimal control problems by direct collocation in optimal control—Calculus of variations, optimal control theory, and numerical methods, *International Series of Numerical Mathematics*, 111: 129–143, 1993.
40. Zhang, W. Controller synthesis for switched systems using approximate dynamic programming, PhD dissertation, School of Electrical and Computer Engineering, Purdue University, December 2009.
41. Zhang, W. and J. Hu, On the value functions of the discrete-time switched LQR problem, *IEEE Transactions on Automatic Control*, 54(11): 2669–2674, 2009.
42. Zhang, W., Abate, A., and Hu, J. Efficient suboptimal solutions of switched LQR problems, *Proceedings of American Control Conference*, St Louis, MO, 1084–1091, 2009.
43. Zhang, A. Abate, Hu, J., and Vitus, M. P. Exponential stabilization of discrete-time switched linear systems, *Automatica*, 45(11): 2526–2536, 2009.
44. Gregory, J. and Lin, C. *Constrained Optimization in the Calculus of Variations and Optimal Control Theory*, 1st edn. Van Nostrand Reinhold, London, 1992.
45. Žefran, M. Continuous methods for motion planning. PhD Thesis, University of Pennsylvania, 1996.
46. Pytlak, R. and Vinter, R.B. Second-order method for optimal control problems with state constraints and piecewise-constant controls. *Proceedings of the IEEE Conference on Decision and Control*, New Orleans, LA, 1995.
47. Neuman, C.P. and Sen, A. A suboptimal control algorithm for constrained problems using cubic splines, *Automatica*, 9: 601–613, 1973.

V

Adaptive Control

32

Automatic Tuning of PID Controllers

Tore Hägglund
Lund Institute of Technology

Karl J. Åström
Lund Institute of Technology

32.1 Introduction

Methods for automatic tuning of PID controllers have been one of the results of the active research on adaptive control. PID controllers are defined in Section 9.5. A result of this development is that the design of PID controllers is going through a very interesting phase. Practically, all PID controllers that are designed now have at least some features for automatic tuning. Automatic tuning has also made it possible to generate automatically gain schedules. Many controllers also have adaptation of feedback and feedforward gains. Overviews of the development are given in [1,2].

The most important component of the adaptive controllers and automatic tuning procedures is the design method. The next section presents some of the most common design methods for PID controllers. These design methods are divided into three categories: (1) future-based techniques, (2) analytical methods, and (3) methods that are based on optimization.

Section 32.3 treats adaptive techniques. An overview of different uses of these techniques is first presented, followed by a more detailed treatment of automatic tuning, gain scheduling, and adaptive control. Section 32.4 gives an overview of how the adaptive techniques have been used in commercial controllers. References are at the end of the chapter.

32.2 Design Methods

To obtain rational methods for designing controllers it is necessary to deal with specifications and models. In the classical Ziegler–Nichols methods, the process dynamics are characterized by two parameters, a gain and a time. Another approach is used in the analytical design methods, where the controller parameters are obtained from the specifications and the process transfer function by a direct calculation. Optimization methods allow for compromise between several different criteria. These approaches are discussed here.

32.2.1 Specifications

When solving a control problem, it is necessary to understand the primary goal of control. Two common control objectives are to follow the setpoint and to reject disturbances. It is also important to have an assessment of the major limitations, which can be system dynamics, nonlinearities, disturbances, or process uncertainty. Typical specifications on a control system may include attenuation of load disturbances, setpoint following, robustness to model uncertainty, and lack of sensitivity to measurement noise.

32.2.1.1 Attenuation of Load Disturbances

Attenuation of load disturbances is of primary concern for process control. The disturbances may enter the system in many different ways, but it is often assumed that they enter at the process input. Let e be the error caused by a unit step load disturbance at the process input. Typical quantities used to characterize the error are maximum error, time to reach maximum, settling time, decay ratio, and the integrated absolute error (IAE) which is defined by

$$IAE = \int_0^{\infty} |e(t)|dt. \tag{32.1}$$

32.2.1.2 Setpoint Following

Setpoint following is of primary interest in motion control, but of less importance for process control because production rates are not changed so often. Furthermore, the response to setpoint changes can be improved by setpoint weighting or by prefiltering of the command signal. Specifications on setpoint following may include requirements on rise time, settling time, decay ratio, overshoot, and steady-state offset for step and ramp changes in setpoint.

32.2.1.3 Robustness to Model Uncertainty

It is important that the controller parameters are chosen in such a way that the closed-loop system is not too sensitive to changes in process dynamics. There are many ways to specify the sensitivity. Many different criteria are conveniently expressed in terms of the Nyquist plot of the loop transfer function, and its distance to the critical point -1. The gain and phase margins are classical robustness measures, see Section 9.1. A drawback is that even if both are specified the Nyquist curve can still be close to the critical point. Maximum sensitivity M_s is a better robustness measure since $1/M_s$ is the shortest distance between the Nyquist plot and the critical point. A requirement on M_s simultaneously captures requirements on both gain and phase margins because of the following inequalities:

$$g_m \geq \frac{M_s}{M_s - 1},$$

$$\varphi_m \geq 2 \arcsin\left(\frac{1}{2M_s}\right).$$

32.2.1.4 Sensitivity to Measurement Noise

Care should always be taken to reduce measurement noise by appropriate filtering, since it will be fed into the system through the feedback. It will generate control actions and control errors. Measurement noise is typically of high frequency. The high-frequency gain of a PID controller is

$$K_{hf} = K(1 + N),$$

where K is the controller gain and N is the derivative gain limitation factor. See Section 9.5. Note that $N = 0$ corresponds to PI control, and $N \to \infty$ corresponds to PID control without filtering and with

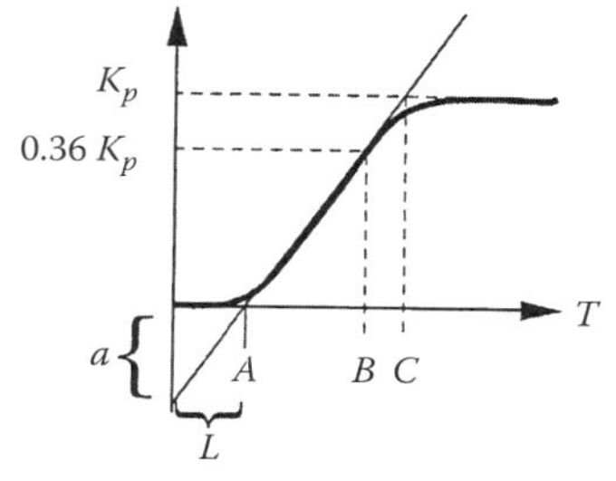

FIGURE 32.1 Determining a first-order plus dead-time model from a step response. Time constant T can be obtained either as the distance AB or the distance AC.

infinite high-frequency gain. Multiplication of the measurement noise by K_{hf} gives the fluctuations in the control signal that are caused by the measurement noise. Also note that there may be a significant difference in K_{hf} for PI and PID control. It is typically an order of magnitude larger for a PID controller, since the gain normally is higher for a PID controller than for a PI controller, and N is typically around 10. It could also be advantageous to use a second-order filter as discussed in Section 9.1.

32.2.2 Feature-Based Techniques

The simplest design methods are based on a few features of the process dynamics that are easy to obtain experimentally. Typical time-domain features are static gain K_p, dominant time constant T, and dominant dead time L. They can all be determined from a step response of the process, see Figure 32.1. Static gain K_p, dominant time constant T, and dominant dead time L can be used to obtain an approximate first-order plus dead-time model for the process as given in Equation 32.2.

$$G_p(s) = \frac{K_p}{1+sT}e^{-sL} \tag{32.2}$$

Typical frequency-domain features are static gain K_p, ultimate gain K_u, and ultimate period T_u. They are defined in Figure 32.2.

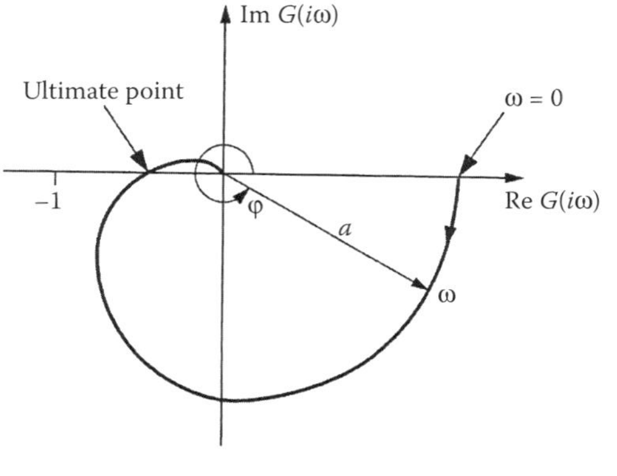

FIGURE 32.2 Static gain K_p, ultimate gain K_u, and ultimate period T_u defined in the Nyquist diagram. Static gain K_p is the point on the Nyquist plot at $\omega = 0$. Ultimate gain K_u is -1 divided by the ultimate point. Ultimate period T_u is 2π divided by the frequency corresponding to the ultimate point.

TABLE 32.1 Controller Parameters Obtained from the Ziegler–Nichols Step Response Method

Controller	K	T_i	T_d
P	$1/a$		
PI	$0.9/a$	$3L$	
PID	$1.2/a$	$2L$	$L/2$

32.2.2.1 Ziegler–Nichols Methods

In 1942, Ziegler and Nichols presented two design methods for PID controllers: time-domain method and frequency-domain method [3]. The methods are based on determination of process dynamics in terms of only two parameters, gain and time. The controller parameters are then expressed in terms of these parameters by simple formulas. In both methods, the design specification of quarter amplitude decay ratio was used. The decay ratio is the ratio between two consecutive maxima of the control error after a step change in setpoint or load.

The time-domain method is based on a registration of the open-loop step response of the process. Ziegler and Nichols have given PID parameters directly as functions of a and L, defined in Figure 32.1. These are given in Table 32.1.

The second method presented by Ziegler and Nichols is based on the frequency response of the process. They have given simple formulas for the parameters of the controller in terms of ultimate gain K_u and ultimate period T_u. These parameters can be determined in the following way. Connect a controller to the process, set the parameters so that control action is proportional, that is, $T_i = \infty$ and $T_d = 0$. Increase the gain slowly until the process starts to oscillate. The gain when this occurs is K_u and the period of the oscillation is T_u. The parameters can also be determined approximately by relay feedback as is discussed in Section 32.3. The controller parameters are given in Table 32.2.

32.2.2.2 Modifications of the Ziegler–Nichols Methods

The Ziegler–Nichols methods do not give satisfactory control. Therefore, there have been many modifications of the method [2,4–6]. The reason is that they give closed-loop systems with very poor damping. The design criterion “quarter amplitude decay ratio” corresponds to a relative damping of $\zeta \approx 0.2$ which is much too small for most applications. The maximum sensitivity is also much too large, which means that the closed-loop systems obtained are too sensitive to parameter variations.

The Ziegler–Nichols methods do, however, have the advantage of being very easy to use. Many efforts have therefore been made to obtain tuning methods that retain the simplicity of the Ziegler–Nichols methods but give improved robustness.

32.2.2.3 AMIGO Tuning

Significantly better tuning rules can be obtained if the process dynamics are described in terms of three parameters instead of two. An early step in this direction was made by Cohen and Coon, who assumed

TABLE 32.2 Controller Parameters Obtained from the Ziegler–Nichols Frequency Response Method

Controller	K	T_i	T_d
P	$0.5K_u$		
PI	$0.4K_u$	$0.8T_u$	
PID	$0.6K_u$	$0.5T_u$	$0.12T_u$

that the process was given by Equation 32.2, which has three parameters [5]. Their design did, however, also give very sensitive systems.

The AMIGO (Approximate M-constraint Integral Gain Optimization) tuning rules [2] where derived in the following way. PI and PID controllers were designed for a large test batch consisting of 134 process models. The design goal was to maximize the controller integral gain subject to robustness constraints expressed by maximum sensitivities (MIGO). The process models were then approximated by simple models and relations between model parameters and controller parameters were derived. These relations form the AMIGO (Approximate MIGO) tuning rules. Two versions are available, one step response method and one frequency response method.

PID controllers were designed by maximizing integral gain subject to the constrains that the sensitivity functions M_s and M_t should be smaller than 1.4. In the step response method it was attempted to correlate controller gains to parameters of the simple FOTD (First order plus time delay) model (Equation 32.2). The parameters describing the process were obtained by approximating the step responses by Equation 32.2. Processes with integration are approximated by

$$G_p(s) = \frac{K_v}{s} e^{-sL}, \tag{32.3}$$

where K_v is the velocity gain and L the time delay.

Figure 32.3 illustrates the relations between the PI controller parameters and the process parameters for all processes in the test batch. The controller gain is normalized by multiplying it either with the static process gain K_p or with the parameter $a = K_p L/T = K_v L$. The integral time is normalized by dividing it by T or by L. The controller parameters in Figure 32.3 are plotted versus the normalized dead time $\tau = L/(L+T)$. The figure shows that there is a good correlation between the normalized controller parameters and normalized dead time.

The solid lines in Figure 32.3 correspond to the AMIGO tuning formula,

$$\begin{aligned} K &= \frac{0.15}{K_p} + \left(0.35 - \frac{LT}{(L+T)^2}\right)\frac{T}{K_p L}, \\ T_i &= 0.35L + \frac{13LT^2}{T^2 + 12LT + 7L^2}, \end{aligned} \tag{32.4}$$

and the dotted lines show the limits for 15 % variations in the controller parameters. Almost all processes included in the test batch fall within these limits.

For integrating processes, K_p and T go to infinity and $K_p/T = K_v$. Therefore, the AMIGO tuning rules (Equation 32.4) can be simplified to

$$\begin{aligned} K &= \frac{0.35}{K_v L}, \\ T_i &= 13.4L. \end{aligned} \tag{32.5}$$

for integrating processes.

The suggested AMIGO tuning rules for PID controllers are

$$\begin{aligned} K &= \frac{1}{K_p}\left(0.2 + 0.45\frac{T}{L}\right), \\ T_i &= \frac{0.4L + 0.8T}{L + 0.1T}L, \\ T_d &= \frac{0.5LT}{0.3L + T}. \end{aligned} \tag{32.6}$$

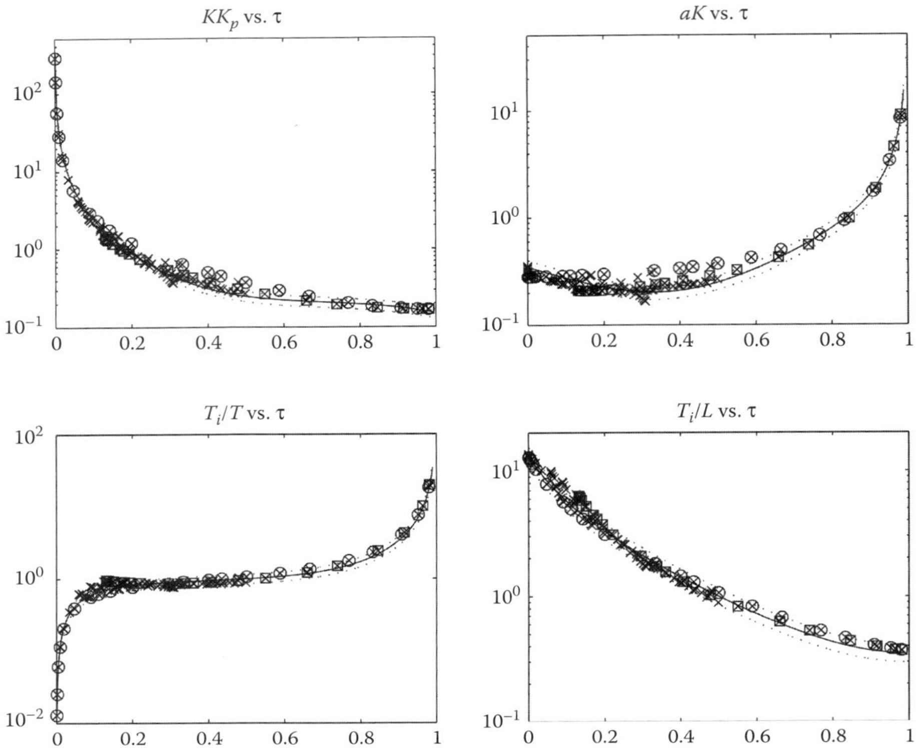

FIGURE 32.3 Normalized PI controller parameters plotted versus normalized time delay τ. The solid lines correspond to the AMIGO design rule (Equation 32.4), and the dotted lines indicate 15 % parameter variations.

For integrating processes, Equation 32.6 can be written as

$$\begin{aligned} K &= 0.45/(K_v L), \\ T_i &= 8L, \\ T_d &= 0.5L. \end{aligned} \tag{32.7}$$

In the AMIGO frequency-response method, the processes in the test batch are characterized by the three parameters static gain K_p, ultimate gain K_u, and ultimate period T_u. The processes are classified according to the gain ratio $\kappa = 1/(K_p K_u)$. The AMIGO tuning rules are appropriate for processes where $\kappa > 0.2$. The PI tuning rules are

$$\begin{aligned} K &= 0.16K_u, \\ T_i &= \frac{1}{1+4.5\kappa} T_u \end{aligned} \tag{32.8}$$

and the PID rules are

$$\begin{aligned} K &= (0.3 - 0.1\kappa^4)K_u, \\ T_i &= \frac{0.6}{1+2\kappa} T_u, \\ T_d &= \frac{0.15(1-\kappa)}{1-0.95\kappa} T_u. \end{aligned} \tag{32.9}$$

32.2.3 Tuning Based on Gain and Phase Margins

Figure 32.3 shows that integral time is close to the process time constant for a reasonably wide range of τ. We can use this insight to obtain a Ziegler–Nichols like formula which contains specifications on gain and phase margins. A PI controller that cancels the process pole is,

$$C(s) = \frac{K(1+sT)}{sT}.$$

The corresponding loop transfer function is

$$G_l = \frac{KK_p}{sT}e^{-sL},$$

and the gain crossover frequency is $\omega_{gc} = K_pK/T$. The phase at the gain crossover frequency is

$$\arg G_l(i\omega_{gc}) = \frac{\pi}{2} - \omega_{gc}L = \frac{\pi}{2} - \frac{K_pKL}{T}.$$

Requiring a phase margin φ_m gives

$$\omega_{gc} = \frac{\alpha_m}{L}, \qquad \alpha_m = \frac{\pi}{2} - \varphi_m.$$

With $\varphi_m = \pi/3$ we obtain $\alpha_m = \pi/6 = 0.52 \approx 0.5$. The tuning rule becomes

$$K = \frac{\alpha_m T}{K_pL}, \qquad k_i = \frac{K}{T} = \frac{\alpha_m}{K_pL}. \tag{32.10}$$

Note that $k_iK_pL = \alpha_m$.

The control law (Equation 32.10) gives good control when $L \approx T$, but not when $L >> T$ or $L << T$. The proportional gain goes to zero with L/T and the gain is too low for delay-dominated processes. To explore how proportional gain should be increased we approximate the transfer function by

$$P(s) \approx K_pe^{-sL},$$

and we find that the proportional gain can be increased to

$$K < \frac{1}{g_mK_p},$$

where g_m is the gain margin. A gain margin $g_m = 5$ gives $K < 0.2/K_p$. The formula (Equation 32.10) gives very poor damping for systems with lag-dominated dynamics, which can be approximated by

$$P(s) \approx \frac{K_p}{sT}e^{-sL}.$$

The gain crossover frequency for a proportional controller that gives the phase margin φ_m is then

$$\omega_{gc} = \frac{\alpha_m}{L},$$

and the controller has the gain

$$K = \frac{\alpha_m T}{K_pL}.$$

Note that this is the same gain as was obtained for systems with balanced dynamics, compare with Equation 32.10. Adding integral action reduces the phase margin. To avoid reducing it too much we require that $\omega_{gc} T_i > \beta$, which corresponds to an increase of the phase lag of β, where β typically is in the range of 0.1–0.5. The condition can also be written as $\alpha_m T_i \geq \beta L$. The corresponding condition on integral gain is

$$k_i = \frac{K}{T_i} \leq \frac{\alpha_m T}{\beta K_p L^2}.$$

Summarizing, we find the following tuning rule:

$$K = \begin{cases} \dfrac{\alpha_m T}{K_p L} & \text{for } \dfrac{L}{T} < \alpha_m g_m, \\ \dfrac{1}{g_m} & \text{for } \dfrac{L}{T} \geq \alpha_m g_m, \end{cases}$$
$$k_i = \begin{cases} \dfrac{\alpha_m T}{\beta K_p L^2} & \text{for } \dfrac{L}{T} < \dfrac{1}{\beta}, \\ \dfrac{\alpha_m}{K_p L} & \text{for } \dfrac{L}{T} \geq \dfrac{1}{\beta}, \end{cases} \tag{32.11}$$

where $\alpha_m = \pi/2 - \varphi_m$, and φ_m is the phase margin in radians and g_m is the gain margin.

32.2.4 Analytical Methods

If the process can be described well by a simple model, the controller parameters can be obtained by a direct calculation. This approach is treated in this section.

32.2.4.1 Pole Placement

If the process is described by a low-order transfer function, a complete pole-placement design can be performed. A PI controller has two parameters and the two poles can be placed and a PID controller can place three poles. The zeros can be influenced by using setpoint weighting. Equations for the controller parameters are given in [2]. To have a robust system the desired closed loop poles should be chosen with care [7].

32.2.4.2 λ-Tuning

Let G_p and G_c be the transfer functions of the process and the controller. The closed-loop transfer function obtained with error feedback is then

$$G_0 = \frac{G_p G_c}{1 + G_p G_c}.$$

Solving this equation for G_c gives

$$G_c = \frac{1}{G_p} \cdot \frac{G_0}{1 - G_0}. \tag{32.12}$$

If the closed-loop transfer function G_0 is specified and G_p is known, it is thus easy to compute G_c.

The method, called λ-tuning, was developed for processes with long dead time L [8]. Consider a process with the transfer function

$$G_p = \frac{K_p}{1 + sT} e^{-sL}. \tag{32.13}$$

Assume that the desired closed-loop transfer function is specified as

$$G_0 = \frac{e^{-sL}}{1 + s\lambda T}, \tag{32.14}$$

where λ is a tuning parameter. The time constants of the open- and closed-loop systems are the same when $\lambda = 1$. The closed-loop system responds faster than the open-loop system if $\lambda < 1$. It is slower when $\lambda > 1$.

It follows from Equation 32.12 that the controller transfer function becomes

$$G_c = \frac{1 + sT}{K_p(1 + \lambda sT - e^{-sL})}.$$

When $L = 0$ this becomes a PI controller with gain $K = 1/(\lambda K_p)$ and integral time $T_i = T$. The sensitivity function obtained with λ-tuning is given by

$$S(s) = 1 - \frac{e^{-sL}}{1 + s\lambda T} = \frac{1 + s\lambda T - e^{-sL}}{1 + s\lambda T}.$$

The maximum sensitivity M_s is always less than 2 if the model is correct. With unmodeled dynamics, the sensitivity may be larger. The parameter λ should be small to give a low IAE, but a small value of λ increases the sensitivity.

32.2.4.3 Internal Model Control

The internal model principle is a general method for design of control systems that can be applied to PID control [9]. A block diagram of such a system is shown in Figure 32.4. It is assumed that all disturbances acting on the process are reduced to an equivalent disturbance d at the process output. In this figure, G_m denotes a model of the process, $G_m^{\dagger}$ is an approximate inverse of G_m, and G_f is a low-pass filter. The name internal model controller derives from the fact that the controller contains a model of the process. This model is connected in parallel with the process.

If the model matches the process, that is, $G_m = G_p$, the signal e is equal to the disturbance d for all control signals u. If $G_f = 1$ and $G_m^{\dagger}$ is an exact inverse of the process, then the disturbance d will be canceled perfectly. The filter G_f is introduced to obtain a system that is less sensitive to modeling errors, and to ensure that the system $G_f G_m^{\dagger}$ is realizable. A common choice is $G_f(s) = 1/(1 + sT_f)$, where T_f is a design parameter.

The controller obtained by the internal model principle can be represented as an ordinary series controller with the transfer function

$$G_c = \frac{G_f G_m^{\dagger}}{1 - G_f G_m^{\dagger} G_m}. \tag{32.15}$$

From this expression it follows that controllers of this type cancel process poles and zeros. The controller is normally of high order. Using simple models it is, however, possible to obtain PI or PID controllers. To see this, consider a process with the transfer function

$$G_p(s) = \frac{K_p}{1 + sT} e^{-sL}.$$

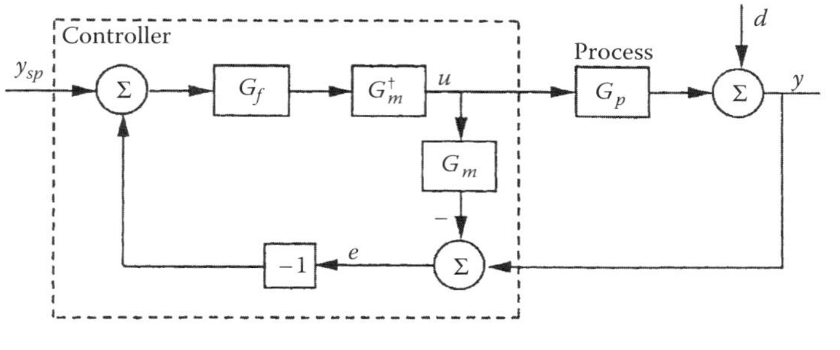

FIGURE 32.4 Block diagram of a closed-loop system with a controller based on the internal model principle.

An approximate inverse, where no attempt is made to find an inverse of the time delay, is given by

$$G_m^{\dagger}(s) = \frac{1+sT}{K_p}.$$

Choosing the filter

$$G_f(s) = \frac{1}{1+sT_f}$$

and approximating the time delay by

$$e^{-sL} \approx 1 - sL$$

Equation 32.15 now gives

$$G_c(s) = \frac{1+sT}{K_p s(L+T_f)},$$

which is a PI controller. If the time delay is approximated instead by a first-order Padé approximation

$$e^{-sL} \approx \frac{1-sL/2}{1+sL/2}.$$

Equation 32.15 gives instead the PID controller

$$G_c(s) = \frac{(1+sL/2)(1+sT)}{K_p s(L+T_f+sT_f L/2)} \approx \frac{(1+sL/2)(1+sT)}{K_p s(L+T_f)}.$$

An interesting feature of the internal model controller is that robustness is considered explicitly in the design. Robustness can be adjusted by selecting the filter G_f properly. A trade-off between performance and robustness can be made by using the filter constant as a design parameter.

The internal model control (IMC) method can be designed to give excellent responses to setpoint changes. Since the design method inherently implies that poles and zeros of the plant are canceled, the response to load disturbances may be poor if the canceled poles are slow in comparison with the dominant poles. This is discussed in the next section.

32.2.4.4 Skogestad's Internal Model Controller

Skogestad has developed a simple tuning method based on internal model control for FOTD systems. The closed loop transfer function is specified as

$$G_{yy_{sp}} = \frac{1}{1+sT_{cl}} e^{-sL}.$$

For an FOTD system it then follows from Equation 32.13 that the controller transfer function is

$$G_c(s) = \frac{1+sT}{K_p(1+sT_{cl} - e^{-sL})} \approx \frac{1+sT}{sK_p(T_{cl}+L)},$$

where the exponential function is approximated using a Taylor-series expansion. The closed-loop response time is specified to be proportional to the time delay L and the integral time is modified for lag-dominated processes leading to the following tuning rule for PI control:

$$K = \frac{T}{2K_pL}, \qquad T_i = \min\,(T, 8L). \tag{32.16}$$

Note that the proportional gain has the same form as for the Ziegler–Nicols rule but that the gain is smaller. Also note the similarity with Equation 32.11.

The same parameters are used for a PID controller in series form, and the derivative time is chosen as the shortest time constant.

32.2.5 Loop-Shaping

Most traditional control design techniques can also be applied to PID control. Loop-shaping tries to achieve a desired loop transfer function by suitable choice of the controller. The following procedures are minor modifications of techniques proposed in [10].

We start by choosing a desired gain crossover frequency ω_{gc}. The choice could be governed by the requirement of attenuation of load disturbances. Using a controller with integral action the attenuation of a sinusoidal load disturbances with frequency ω_d is approximately ω_{gc}/ω_d. This approximation is particularly good when the phase margin is 60° because the sensitivity crossover frequency are then equal to the gain crossover frequency. To have the phase margin ϕ_m we obtain the following condition:

$$\angle C(i\omega_{gc}) + \angle P(i\omega_{gc}) = -180^\circ + \phi_m, \tag{32.17}$$

where ϕ_m is the required phase margin.

Any stable process can be controlled by an integrating controller. Since an integrating controller has a phase lag of 90° we find that an integral controller can be used if the phase lag of the process is in the range of 0–90° $-\ \phi_m$ (0–30° with a phase margin $\phi_m = 60^\circ$). The integral gain is then given by

$$k_i = \frac{\omega_{gc}}{|P(i\omega_{gc})|}.$$

PI control can be used if higher gain crossover frequencies are desired. Since a PI controller has a phase lag between 0° and 90° it follows from Equation 32.17 that PI control can be used if the phase lag of the process is in the range of $90^\circ - \phi_m$ (30–120° with a phase margin $\phi_m = 60^\circ$). The phase lag of a PI controller at ω_{gc} is $90 - \arctan \omega_{gc} T_i$. Equation 32.17 then gives

$$\omega_{gc} T_i = -\angle P(i\omega_{gc}) - 90^\circ + \phi_m.$$

The integral gain is then given by

$$k_i = \frac{\omega_{gc}}{|G(i\omega_{gc})|\sqrt{1 + \omega_{gc}^2 T_i^2}}.$$

A PID controller can provide lead and it is then possible to choose even higher gain crossover frequencies. The phase lead is at most 90° which corresponds to pure derivative control, and the crossover frequency can be such that the process phase lag approaches $270^\circ - \phi_m$.

The limits of $270^\circ - \phi_m$ phase lag for PID control and $180^\circ - \phi_m$ for PI control are too optimistic because they correspond to pure derivative and pure proportional control. In practice, the allowable phase lags are smaller because integral action is needed.

To find a PID controller of the form

$$C(s) = k_i \frac{1 + sT_i + s^2 T_i T_d}{s(1 + sT_f)}, \tag{32.18}$$

we first allocate a phase lag to the filter by picking the filter time constant T_f so that the phase $\phi_f = \arctan \omega_{gc} T_f$ has a reasonable value, typically about 10°. The required phase advance is then

$$\phi = -\angle P(i\omega_{gc}) + \phi_f + \phi_m - 90^\circ.$$

This phase lag has to be provided by the numerator $1 + sT_i + s^2 T_d$ of the transfer function 32.18, hence

$$\tan \phi = \frac{\omega_{gc} T_i}{1 - \omega_{gc}^2 T_i T_d} = \frac{\alpha \omega_{gc} T_i}{\alpha - \omega_{gc}^2 T_i^2}.$$

Since there are two parameters T_i and T_d, one of them has been fixed by choosing $\alpha = T_i/T_d$ as a tuning parameter which will be adjusted later (typical values are in the range of 0.1–1). Solving the above equation for T_i gives

$$T_i = \begin{cases} \dfrac{-\alpha + \sqrt{\alpha^2 + 4\alpha \tan^2 \phi}}{2\omega_{gc} \tan \phi} & \text{for } \phi < 90^\circ, \\ \dfrac{-\alpha - \sqrt{\alpha^2 + 4\alpha \tan^2 \phi}}{2\omega_{gc} \tan \phi} & \text{for } 90^\circ \le \phi < 180^\circ. \end{cases} \tag{32.19}$$

Having determined T_i integral gain is then adjusted to give unit gain at the crossover frequency; hence

$$k_i = \frac{\omega_{gc} \sqrt{1 + \omega_{gc}^2 T_f^2}}{|P(i\omega_{gc})(1 - \alpha \omega_{gc}^2 T_i^2 + i\omega_{gc} T_i)|} \tag{32.20}$$

The design parameter α can be adjusted, for example, to maximize the integral gain. An interactive learning module for loop-shaping can be downloaded from http://www.calerga.com/contrib/index.html.

32.2.6 Optimization Methods

Optimization is a powerful tool for design of controllers. The method is conceptually simple. A controller structure with a few parameters is specified. Specifications are expressed as inequalities of functions of the parameters. The specification that is most important is chosen as the function to optimize. The method is well suited for PID controllers where the controller structure and the parameterization are given. There are several pitfalls when using optimization. Care must be exercised when formulating criteria and constraints; otherwise, a criterion will indeed be optimal, but the controller may still be unsuitable because of a neglected constraint. Another difficulty is that the loss function may have many local minima. A third is that the computations required may be excessive. Numerical problems may also arise. Nevertheless, optimization is a good tool that has successfully been used to design PID controllers.

Popular optimization criteria are the IAE, the integrated time absolute error (ITAE), and the integrated square error (ISE). They are mostly done for the first-order plus dead-time model as given in Equation 32.2. Many tables that provide controller parameters based on optimization have been published [6]. A very general optimization algorithm which considers many different constrains is presented in [11].

32.2.6.1 Modulus Optimum and Symmetrical Optimum

Modulus Optimum (BO) and Symmetrical Optimum (SO) are two design methods that are based on optimization. The acronyms BO and SO are derived from the German words Betrags Optimum and Symmetrische Optimum. These methods are based on the idea of finding a controller that makes the frequency response from setpoint to plant output as close to one as possible for low frequencies. If $G(s)$ is the transfer function from the setpoint to the output, the controller is determined in such a way that $G(0) = 1$ and that $d^n|G(i\omega)|/d\omega^n = 0$ at $\omega = 0$ for as many n as possible.

If the closed-loop system is given by

$$G(s) = \frac{\omega_0^2}{s^2 + \sqrt{2}\omega_0 s + \omega_0^2},$$

the first three derivatives of $|G(i\omega)|$ will vanish at the origin. If the transfer function G in the example is obtained by error feedback of a system with the loop transfer function G_{BO}, the loop transfer function is

$$G_{\mathrm{BO}}(s) = \frac{G(s)}{1 - G(s)} = \frac{\omega_0^2}{s(s + \sqrt{2}\omega_0)},$$

which is the desired loop transfer function for the method called modulus optimum.

If the closed-loop transfer function is given by

$$G(s) = \frac{\omega_0^3}{(s + \omega_0)(s^2 + \omega_0 s + \omega_0^2)} \tag{32.21}$$

the first five derivatives of $|G(i\omega)|$ will vanish at the origin. A system with this closed-loop transfer function can be obtained with a system having error feedback and the loop transfer function

$$G_\ell(s) = \frac{\omega_0^3}{s(s^2 + 2\omega_0 s + 2\omega_0^2)}.$$

The closed-loop transfer function (Equation 32.21) can also be obtained from other loop transfer functions if a two-degree of freedom controller is used. For example, if a process with the transfer function

$$G_p(s) = \frac{\omega_0^2}{s(s + 2\omega_0)}$$

is controlled by a PI controller having parameters $K = 2$, $T_i = 2/\omega_0$, and $b = 0$, the loop transfer function becomes

$$G_{\mathrm{SO}} = \frac{\omega_0^2(2s + \omega_0)}{s^2(s + 2\omega_0)}. \tag{32.22}$$

The symmetric optimum aims at obtaining the loop transfer function given by Equation 32.22. Note that the Bode diagram of this transfer function is symmetrical around the frequency $\omega = \omega_0$. This is the motivation for the name symmetrical optimum.

The methods BO and SO can be called loop-shaping methods since both methods try to obtain a specific loop transfer function. The design methods can be described as follows. It is first established which of the transfer functions, G_{BO} or G_{SO}, is most appropriate. The transfer function of the controller $G_c(s)$ is then chosen so that $G_\ell(s) = G_c(s)G_p(s)$, where G_ℓ is the chosen loop transfer function.

32.2.6.2 Cancellation of Process Poles

The PID controller has two zeros. Many design methods choose these zeros so that they cancel one or two of the dominant process poles. This often results in a simple design as in IMC or SO optimization. The response to setpoint changes is generally good. The responses to load disturbances may, however, be poor if the canceled poles are slower than the dominating closed-loop poles. Figure 32.5 illustrates the problem. A process with a time constant $T = 10$ is controlled with a PI controller with integral time $T_i = 10$ and a suitable controller gain. The design is made so that the closed-loop time constant is significantly shorter than the open-loop time constant. This is seen in the response to the setpoint change. However, the load disturbance response is very sluggish since the open-loop time constant is present and dominating. The figure also shows a retuned controller, where pole cancellation is avoided.

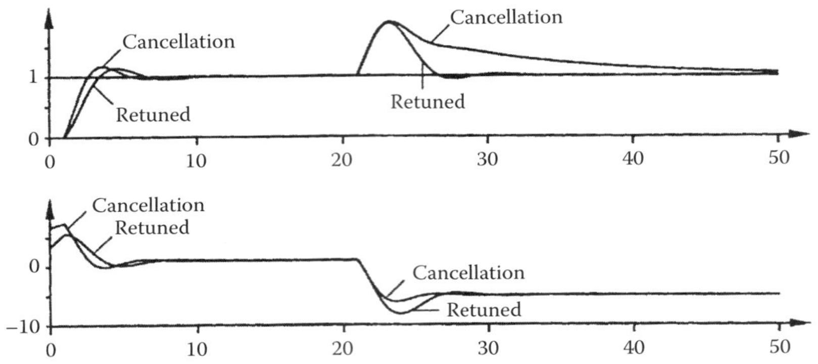

FIGURE 32.5 Simulation of a closed-loop system obtained by pole cancelation. The process transfer function is $G(s) = e^{-s}/(10s + 1)$, and the controller parameters are $K = 6.67$ and $T_i = 10$. The upper diagram shows setpcint $y_{sp} = 1$ and process output y, and the lower diagram shows control signal u. The figure also shows the responses to a retuned controller with $K = 6.67$, $T_i = 3$ and $b = 0.5$.

32.3 Adaptive Techniques

This section gives an overview of adaptive techniques. It starts with a discussion of uses of the different techniques, followed by a more detailed presentation of automatic tuning, gain scheduling, and adaptive control. The section ends with a presentation of how the adaptive techniques have been used in industrial controllers.

32.3.1 Use of the Adaptive Techniques

The word *adaptive techniques* is used to cover autotuning, gain scheduling, and adaptation. Although research on adaptive techniques has almost exclusively focused on adaptation, experience has shown that autotuning and gain scheduling have much wider industrial applicability. Figure 32.6 illustrates the appropriate use of the different techniques.

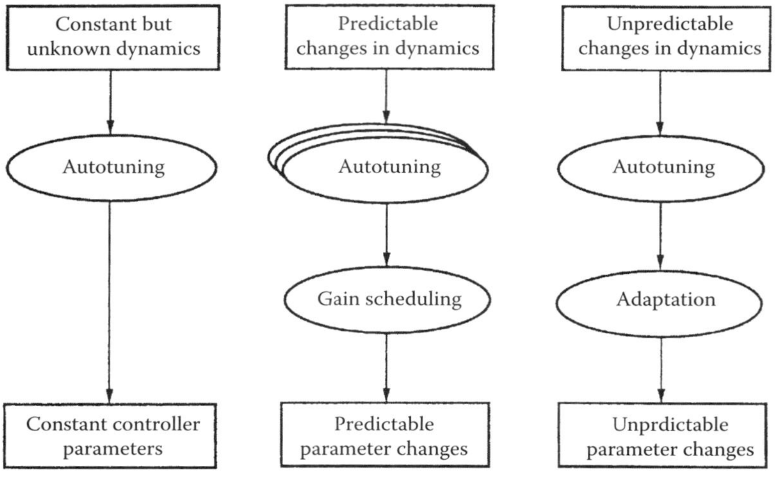

FIGURE 32.6 When to use different adaptive techniques.

Controller performance is the first issue to consider. If requirements are modest, a controller with constant parameters and conservative tuning can be used. Other solutions should be considered when higher performance is required.

If the process dynamics are constant, a controller with constant parameters should be used. The parameters of the controller can be obtained by autotuning.

If the process dynamics or the character of the disturbances are changing it is useful to compensate for these changes by changing the controller. If the variations can be predicted from measured signals, gain scheduling should be used since it is simpler and gives superior and more robust performance than continuous adaptation. Typical examples are variations caused by nonlinearities in the control loop. Autotuning can be used to build up the gain schedules automatically.

There are also cases where the variations in process dynamics are not predictable. Typical examples are changes due to unmeasurable variations in raw material, wear, fouling, and so on. These variations cannot be handled by gain scheduling but must be dealt with by adaptation. An autotuning procedure is often used to initialize the adaptive controller. It is then sometimes called pretuning or initial tuning.

32.3.2 Automatic Tuning

Automatic tuning (or autotuning) is a method where a controller is tuned automatically on demand from the user. Typically, the user will either push a button or send a command to the controller. Automatic tuning is sometimes called tuning on demand or one-shot tuning.

Automatic tuning can also be performed by external devices that are connected to the control loop during the tuning phase. Since these devices are supposed to work together with controllers from different manufacturers, they must be provided with quite a lot of information about the controller structure and parameterization in order to provide appropriate controller parameters. Such information includes controller structure (series or parallel form), sampling rate, filter time constants, and units of the different controller parameters (gain or proportional band, minutes or seconds, time or repeats/time).

The automatic tuning procedures can be divided into methods that are based on step response experiments, and methods based on frequency response experiments.

32.3.2.1 Step Response Methods

Most methods for automatic tuning of PID controllers are based on step response experiments. When the operator wishes to tune the controller, an open-loop step response experiment is performed. A process model is then obtained from the step response, and controller parameters are determined. This is usually done using simple look-up tables as in the Ziegler–Nichols method.

The greatest difficulty in carrying out the tuning automatically is in selecting the amplitude of the step. The user naturally wants the disturbance to be as small as possible so that the process is not disturbed more than necessary. On the other hand, it is easier to determine the process model if the disturbance is large. The result of this dilemma is usually that the operator himself has to decide how large the step in the control signal should be.

Controllers with automatic tuning which are based on this technique have become very common in the last few years. This is especially true of temperature controllers.

32.3.2.2 The Relay Autotuner

Frequency-domain characteristics of the process can be determined from experiments with relay feedback in the following way [12]. When the controller is to be tuned, a relay with hysteresis is introduced in the loop, and the PID controller is temporarily disconnected; see Figure 32.7. For large classes of processes, relay feedback gives an oscillation with period close to the ultimate frequency ω_u, as shown in Figure 32.8. The control signal is a square wave and the process output is close to a sinusoid. The gain of the transfer function at this frequency is also easy to obtain from amplitude measurements. Describing function analysis can be used to determine the process characteristics. The describing function of a relay with

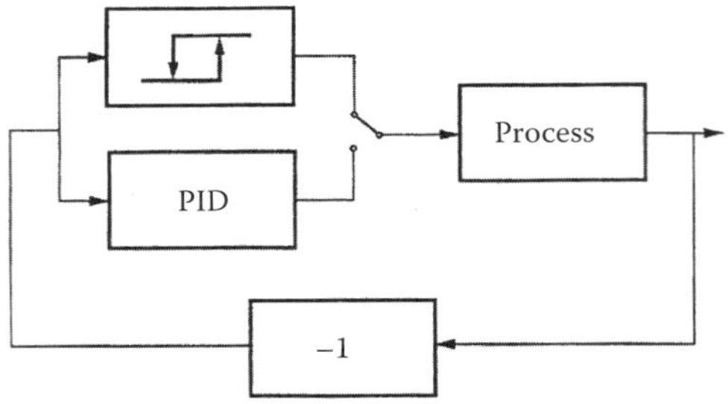

FIGURE 32.7 The relay autotuner. In the tuning mode, the process, is connected to relay feedback.

hysteresis is

$$N(a) = \frac{4d}{\pi a}\left(\sqrt{1 - \left(\frac{\epsilon}{a}\right)^2} - i\,\frac{\epsilon}{a}\right),$$

where d is the relay amplitude, ϵ is the relay hysteresis, and a is the amplitude of the input signal. The negative inverse of this describing function is a straight line parallel to the real axis; see Figure 32.9. The oscillation corresponds to the point where the negative inverse describing function crosses the Nyquist curve of the process, that is, where

$$G(i\omega) = -\frac{1}{N(a)}.$$

Since $N(a)$ is known, $G(i\omega)$ can be determined from the amplitude a and the frequency ω of the oscillation.

Note that the relay experiment is easily automated. There is often an initialization phase where the noise level in the process output is determined during a short period of time. The noise level is used to determine the relay hysteresis and a desired oscillation amplitude in the process output. After this initialization phase, the relay function is introduced. Since the amplitude of the oscillation is proportional to the relay output, it is easy to control it by adjusting the relay output. Also note in Figure 32.8 that a stable oscillation is established very quickly. The amplitude and the period can be determined after about 10 s only. The average residence time of the system is 12 s, which means that it would take about 40 s for an open-loop response to reach steady state.

32.3.3 Gain Scheduling

By gain scheduling we mean a system where controller parameters are changed depending on measured operating conditions. The scheduling variable can, for instance, be the measurement signal, controller

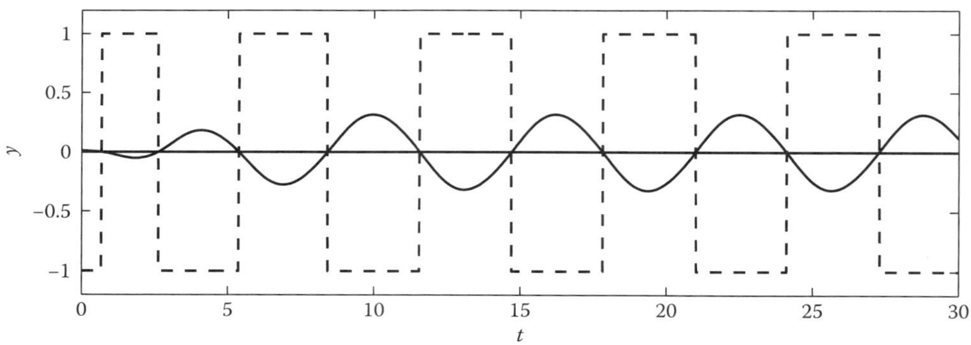

FIGURE 32.8 Relay output u (dashed) and process output y (solid) for a system under relay feedback.

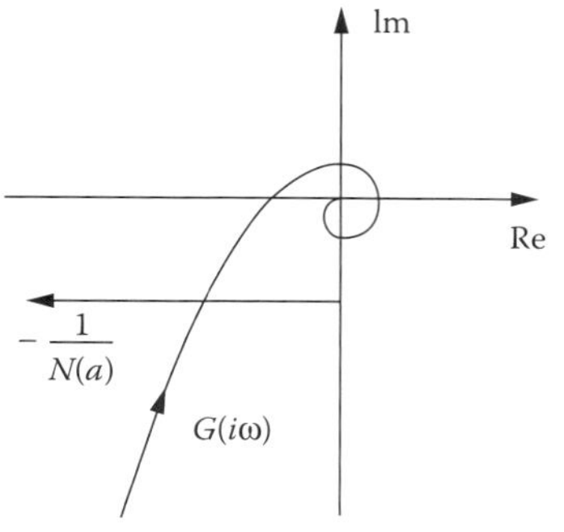

FIGURE 32.9 The negative inverse describing function of a relay with hysteresis $-1/N(a)$ and a Nyquist curve $G(i\omega)$.

output, or an external signal. For historical reasons the word "gain scheduling" is used even if other parameters such as derivative time or integral time are changed. Gain scheduling is a very effective way of controlling systems whose dynamics change with the operating conditions.

The notion of gain scheduling was originally used for flight control systems, but it is being used increasingly in process control. It is, in fact, a standard ingredient in some single-loop PID controllers. For process control applications, significant improvements can be obtained by using just a few sets of controller parameters.

32.3.4 Adaptive Control

An adaptive controller is a controller whose parameters are continuously adjusted to accommodate changes in process dynamics and disturbances. Parameters can be adjusted directly or indirectly via estimation of process parameters. There is a large number of both direct and indirect methods available. Adaptation can be applied both to feedback and feedforward control parameters. Adaptive controls can be described conveniently in terms of the methods used for modeling and control design.

32.3.4.1 Model-Based Methods

All indirect systems can be represented by the block diagram in Figure 32.10. There is a parameter estimator that determines the parameters of the model based on observations of process inputs and outputs. There is also a design block that computes controller parameters from the model parameters. The parameters can either be estimated recursively or batchwise.

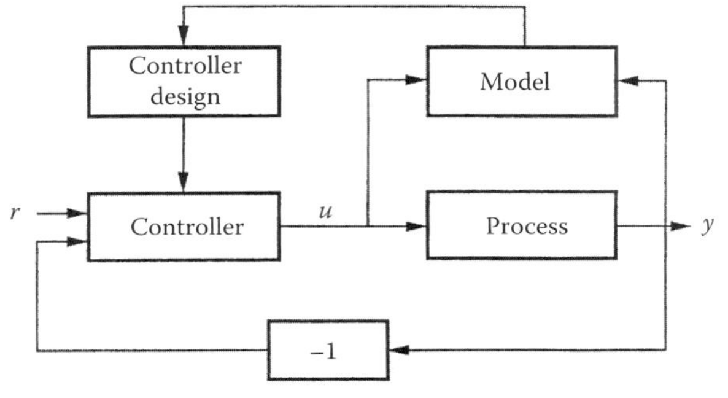

FIGURE 32.10 Block diagram of indirect systems.

32.3.4.2 Rule-Based Methods

In the direct methods, the key issues are to find suitable features that characterize relevant properties of the closed-loop system and appropriate ways of changing the controller parameters so that the desired properties are obtained.

The majority of the PID controllers in industry are tuned manually by instrument engineers. The tuning is done based on past experience and heuristics. By observing the pattern of the closed-loop response to a setpoint change, the instrument engineers use heuristics to directly adjust the controller parameters. The heuristics have been captured in tuning charts that show the responses of the system for different parameter values. A considerable insight into controller tuning can be developed by studying such charts and performing simulations. The heuristic rules have also been captured in knowledge bases in the form of crisp or fuzzy rules. Rules of this type are used in several commercial adaptive controllers. Most products will wait for setpoint changes or major load disturbances. When these occur, properties such as damping, overshoot, period of oscillations and static gains are estimated. Based on these properties, rules for changing the controller parameters to meet desired specifications are executed.

32.3.5 Adaptive Feed Forward

Feedforward control deserves special mention. It is a very powerful method for dealing with measurable disturbances. Use of feedforward control requires, however, good models of process dynamics. It is difficult to tune feedforward control loops automatically on demand, since the operator often cannot manipulate the disturbance used for the feedforward control. To tune the feedforward controller it is therefore necessary to wait for an appropriate disturbance. Adaptation is therefore particularly useful for the feedforward controller.

32.4 Some Commercial Products

Commercial PID controllers with adaptive techniques have been available since the beginning of the 1980s [2,13].

There is a distinction between *temperature controllers* and *process controllers*. Temperature controllers are primarily designed for temperature control, whereas process controllers are supposed to work for a wide range of control loops in the process industry such as flow, pressure, level, and pH control loops. Automatic tuning and adaptation are easier to implement in temperature controllers, since most temperature control loops have several common features. This is the main reason why automatic tuning has been introduced more rapidly in these controllers.

The process controllers can be separate hardware boxes for single loops, or distributed control systems (DCS) where many loops are handled by one system.

Since the processes that are controlled with process controllers may have large differences in their dynamics, tuning and adaptation becomes more difficult compared to the pure temperature control loops. In Table 32.3, a collection of process controllers is presented together with information about their adaptive techniques.

Automatic tuning is the most common adaptive technique in the industrial products. The usefulness of this technique is also obvious from Figure 32.6, where it is shown that the autotuning procedures are used not only to tune the controller, but also to obtain a comfortable operation of the other adaptive techniques. Most auto-tuning procedures are based on step response analysis.

Gain scheduling is often not available in the controllers. This is surprising, since gain scheduling is found to be more useful than continuous adaptation in most situations. Furthermore, the technique is much simpler to implement than automatic tuning or adaptation.

It is interesting to see that many industrial adaptive controllers are rule based instead of model based. The research on adaptive control at universities has been almost exclusively focused on model-based adaptive control.

TABLE 32.3 Industrial Adaptive Process Controllers

Manufacturer	Controller	Automatic Tuning	Gain Scheduling	Adaptive Feedback	Adaptive Feedforward
Bailey Controls	CLC04	Step	Yes	Model based	–
Control Techniques	Expert controller	Ramps	–	Model based	–
Fisher Controls	DPR900	Relay	Yes	–	–
	DPR910	Relay	Yes	Model based	Model based
Foxboro	Exact	Step	–	Rule based	–
Fuji	CC-S:PNA 3	Steps	Yes	–	–
Hartmann & Braun	Protronic P	Step	–	–	–
	Digitric P	Step	–	–	–
Honeywell	UDC 6000	Step	Yes	Rule based	–
Alfa Laval Automation	ECA40	Relay	Yes	–	–
	ECA400	Relay	Yes	Model based	Model based
Siemens	SIPART DR22	Step	Yes	–	–
Toshiba	TOSDIC-215D	PRBS	Yes	Model based	–
	EC300	PRBS	Yes	Model based	–
Turnbull Control Systems	TCS 6355	Steps	–	Model based	–
Yokogawa	SLPC-171,271	Step	Yes	Rule based	–
	SLPC-181,281	Step	Yes	Model based	–

One of the earliest rule-based adaptive controllers is the Foxboro EXACT [14]. It was released in 1984. In this controller, the user specifies a maximum damping and a maximum overshoot. Whenever the control loop is subjected to a larger load disturbance or setpoint change, the response is investigated and the controller parameters are adjusted according to certain rules to meet the specifications.

Adaptive feedforward control is seldom provided in the industrial controllers. This is surprising, since adaptive feedforward control is known to be of great value. Furthermore, it is easier to develop robust adaptive feedforward control than adaptive feedback control.

Alfa Laval Automation's ECA400 and Fisher Controls DPR910 are identical. This controller has automatic tuning, gain scheduling, adaptive feedback, and adaptive feedforward. The automatic tuning procedure is based on relay feedback. The automatic tuning procedure is also used to build the gain schedule automatically, and to initiate the adaptive feedback and feedforward. In this way, there is no need for the user to supply any information about the process dynamics. All adaptive features can be used automatically.

References

1. O'Dwyer, A., *Handbook of PI and PID Controller Tuning Rules*. Imperial College Press, London, 2009.
2. Åström, K.J. and Hägglund, T., *Advanced PID Control,* Instrumentation, Systems, and Automation Society, Research Triangle Park, NC, 2005.
3. Ziegler, J.G. and Nichols, N.B., Optimum settings for automatic controllers, *Trans. ASME,* 64, 759, 1942.
4. Chien, K.L., Hrones, J.A., and Reswick, J.B., On the automatic control of generalized passive systems, *Trans. ASME,* 74, 175, 1952.
5. Cohen, G.H. and Coon, G.A., Theoretical consideration of retarded control, *Trans. ASME,* 75, 827, 1953.
6. Kaya, A. and Scheib, T.J., Tuning of PID controls of different structures, *Control Eng.,* July, 62, 1988.
7. Åström, K.J. and Murray, R.M., *Feedback Systems and Introduction for Scientists and Engineers*. Princeton University Press, Princeton, NJ, 2008.
8. Dahlin, E.B., Designing and tuning digital controllers, *Instr. Control Syst.,* 42, 77, 1968.
9. Morari, M. and Zafiriou, E., *Robust Process Control,* Prentice-Hall, Englewood Cliffs, NJ, 1989.

10. Messner, W.C., Classical control revisited: Variations on a theme. *Proc. 10th Int. Workshop on Advanced Motion Control, AMC'08*. Trento, 2008, pp. 15–20.
11. Garpinger, O. and Hägglund, T. A., software tool for robust PID design. *Proc. 17th IFAC World Congress*, Seoul, Korea, 2008.
12. Hägglund, T. and Åström, K.J., Industrial adaptive controllers based on frequency response techniques, *Automatica*, 27, 599, 1991.
13. Åström, K.J., Hägglund, T., Hang, C.C., and Ho, W.K., Automatic tuning and adaptation for PID controllers—A survey, *Control Eng. Prac.*, 1(4), 699, 1993.
14. Kraus, T.W. and Myron, T.J., Self-tuning PID controller uses pattern recognition approach, *Control Eng.*, June, 106, 1984.

33

Self-Tuning Control

David W. Clarke
University of Oxford

33.1 Introduction

Most control theory is concerned with the design of feedback systems for time-invariant plants with *known* mathematical models, e.g., in the form of *given* transfer functions. The assumption of constant, known models is not valid for many modern technical or nontechnical systems, such as:

Robotics: Inertias as seen by the drive motors vary with the end-effector position and the load mass so that the dynamic model varies with the robot's attitude.

Chemical reactors: Transfer functions vary according to the mix of reagents and catalyst in the vessel and change as the reaction progresses.

People: The gain of the function {injected anaesthetic → unconsciousness} depends on the patient's metabolism.

So how can controllers be designed to cope with these initial model uncertainties and time variations in dynamics? One approach is to design a *robust* fixed controller that ensures stability for all possible plant dynamics, but this is often at the expense of "detuned" behavior. Instead we can embed algorithms inside computers that "learn from experience" and *self-tune* the controllers so as to improve closed-loop performance. Often this learning process builds up a mathematical model based on experimental input/output data; this operation is known as *system identification* or *parameter estimation*. The model

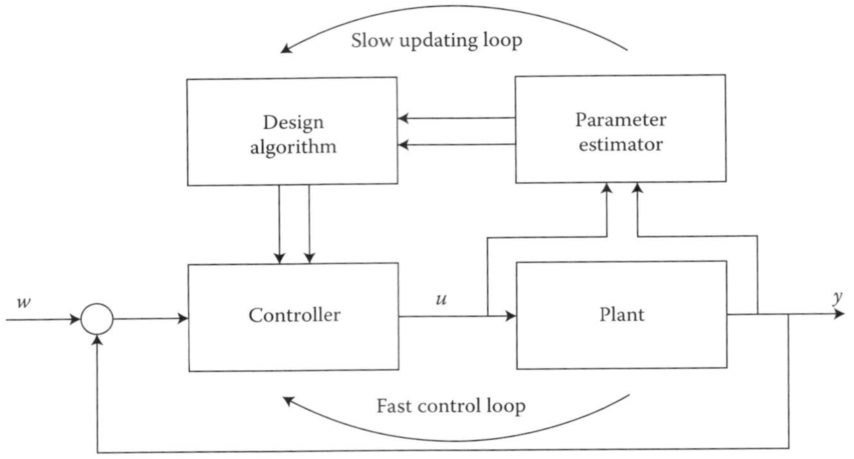

FIGURE 33.1 Structure of a self-tuning controller.

could be a complete transfer function or simply the gain and phase of the plant at a given input frequency. In the design of "learning" controllers there are two main themes:

- For controlling systems that have *unknown but constant* dynamics, using single-shot methods when the controller is first commissioned: If the target control law is the "industry-standard" PID (proportional-integral-derivative, or three-term) form, the approach is often called *autotuning*. If the controller is more complex, for example, when the plant has significant dead time for which PID control is not "tight," a full process model is estimated using system identification methods, and an analytic design procedure uses the model to *self-tune* the coefficients of a fixed control law.
- For *time-varying* systems: Here the identification algorithm operates all the time to "update" the model, and the coefficients of the control law are then automatically adjusted, as shown in Figure 33.1. This scheme must be "alert" so as to track *variations* in the plant's transfer function. This is the full *adaptive* control approach.

Figure 33.1 shows the basic structure of a self-tuning controller: the parameter estimator builds up the plant model from input/output data $\{u(t), y(t)\}$; a control design algorithm takes the estimated model parameters $\hat{\theta}$ and determines the "best" set of controller parameter θ_c (such as gains and time constants); and a controller applies these gains in a feedback loop. Such self-tuning designs have two time scales: a fast inner control loop and a slow outer "updating" loop.

33.1.1 Examples of Unknown and Time-Varying Systems

An annealing furnace thermally soaks metal over some prescribed temperature/time profile to attain desired material properties. The equation describing the furnace temperature T as a function of heat input u is

$$mc_p\frac{dT}{dt} = u - \beta T,$$

where m is the mass of the "burden," which differs from load to load, and β gives the heat loss from Newton's law of cooling. Hence the time-constant of the system, mc_p/β, varies with the burden, so that a fixed controller will lead to transients that differ accordingly. For example, a controller tuned for a large load would heat up a small mass too quickly. A self-tuned controller that deduces m (or an equivalent parameter) during warm-up provides more consistent results.

A hydraulically powered shaking table is used to test models of buildings to evaluate their earthquake resistance. It is required that the spectral density of the shaking should correspond to that of typical earthquakes at the design site. However, the model has mass and other dynamics that affect the behavior of the table (just as an inertial load added to a position control affects the loop dynamics). A dynamically complex (i.e., more than just PID) self-tuning controller is required to ensure that in closed loop the spectral density of the shaking is accurate. A similar problem occurs when testing suspension systems for racing cars using hydraulic actuators, where the objective is to replay the surface deviations of a given racing track in order to "tune" the suspension for best performance.

An exothermic chemical reactor generates heat $Q = kxe^{\alpha T}$, which varies exponentially with temperature T and depends on the proportion $x(t)$ of reagents left in the mixture. A linear model for small excursions around some nominal operating point T_0 can be deduced, so that the transfer function between u (control) and ΔT (deviations) is

$$G(s) = \frac{1}{sc + \beta - \gamma\alpha},$$

where $\gamma = kx\exp(\alpha T_0)$. Consider the behavior during a batch with $x(0) = 1$ initially and where the objective is to end up with $x = 0$ as quickly as possible. First T is small and $G(s)$ is stable. Heat is then added via u so the temperature and hence γ increase. This might cause the pole of G to become unstable. As the reagent "strength" $x(t)$ decreases during the reaction, the plant becomes stable again. In practice, many reactors are used for producing a range of chemicals for which variations *between* batches are highly significant. In such cases, a self-tuned controller that adjusts parameters according to the individual batch might suffice. Even better performance is possible if an adaptive controller is used, adjusting its parameters as the reaction progresses.

A materials testing machine has dynamics that depend on the stiffness of the specimen under test. If it is testing a manufactured item such as a bump-stop for a car suspension, the stiffness changes radically during the cycle. With a fixed controller, the response is either "sluggish" during one phase and acceptable in the other, or it can be successively well tuned initially and then oscillatory. The output of a "specimen stiffness estimator" can be used to "gain schedule" a PI (proportional-integrator) regulator to get good response over both parts of the cycle. This is full adaptive control, as rapid variations in parameters are seen.

The above examples indicate the range of possible controller tuning problems: PI settings for a simple fixed plant; a more general controller for a plant with rich dynamics; regulating a slowly changing plant; a rapidly time-varying plant for which the dynamics depend on some changing but measurable parameter.

Successful applications of self-tuning control have been for those cases where engineering knowledge leads to a simple model of the underlying dynamics for which bounds on parameter variation can be deduced. Early hopes for an effective "general-purpose" self-tuner have not been realized in practice. In particular, this is because of the requirement for "persistency of excitation" in the plant's input/output data, which cannot in general be guaranteed. Hence, in the following we concentrate on the fundamental ideas; the section "References" contains more details of, for example, multivariable self-tuning designs.

33.2 Some Simple Methods

One of the most basic problems in self-tuning is to find a "good" dynamic model of a plant. Suppose we take the classical first-order system $G(s) = K/(1 + sT)$, where K, T are unknown numbers that characterize its behavior. The simplest test is to inject a step of amplitude U and inspect the response and its derivative:

$$y(t) = KU(1 - e^{-t/T}) \rightarrow \dot{y}(t) = \frac{KU}{T}e^{-t/T}. \tag{33.1}$$

We note two things: (1) $y(t) \rightarrow KU$ as $t \rightarrow \infty$, and (2) the tangent at $t = 0$ has slope KU/T and meets the line $y = KU$ when $t = T$. Hence the "final value" gives K, and the meet of the tangent at the origin

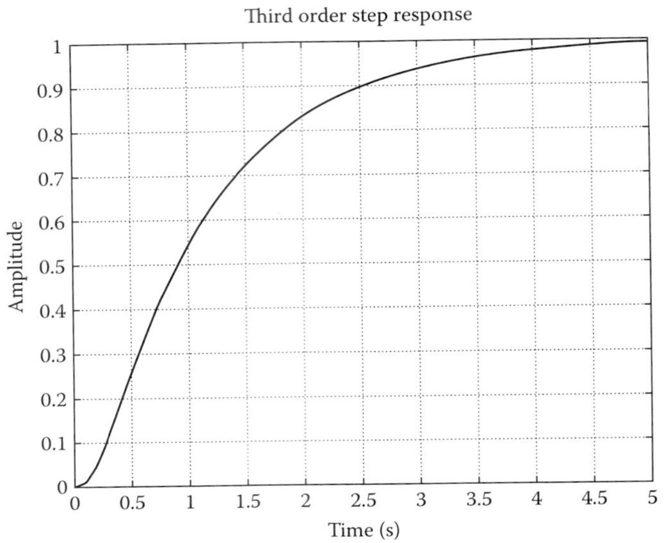

FIGURE 33.2 A third-order system with a dominant pole.

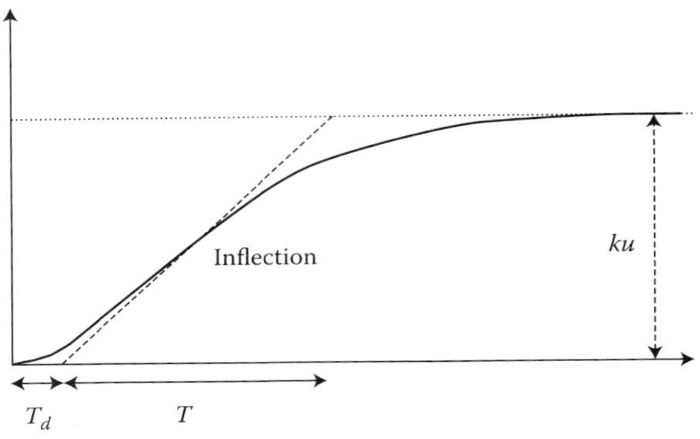

FIGURE 33.3 Finding K, T, T_d from a step-test.

with the final value gives T. Now many real plant responses can be *approximated* by a first-order system together with a dead time, giving:

$$G(s) \approx e^{-sT_d}\frac{K}{1+sT}. \tag{33.2}$$

The response of this model is simply the above first-order response shifted by the dead time T_d, so that T_d is found just by inspecting when the output starts to move after the input step U is injected. To give a sense of the accuracy of this approximation, Figure 33.2 shows the step response of $1/(s+1)(1+0.1s)^2$, having a pole at $s=-1$ and two "fast" poles at $s=-10$. It is seen that the initial dynamics look closely like those of a dead-time of about 0.2 seconds.

How can we use the above simple idea for finding K, T, T_d? What we can look for is the point of inflection of the output curve and extrapolate, as shown in Figure 33.3.

This is easily done by inspection, but such a procedure is actually quite hard to do in a computer. Even more tricky is to derive a method that is reliable with real data where the output response is corrupted by noise. We might be able to get "reasonable" values of the model parameters "by eye," but a computer algorithm will be completely confused. What is necessary is a method that considers the *overall* response without getting locked into local features. Hence, looking for points of inflection is not the answer. Revert to the original first-order system, and write it as a differential equation:

$$T\frac{dy}{dt} + y = Ku(t).$$

Integrating with respect to t from $t = t1$ to $t = t2$ gives:

$$T[y(t2) - y(t1)] + \int_{t1}^{t2} y\,dt = K\int_{t1}^{t2} u\,dt. \tag{33.3}$$

Given a series of output samples $y(0), y(h), y(2h), \ldots, y(ih), \ldots,$ where h is the sample interval, we can *approximate* the integrals in Equation 33.3 by, say, using the rectangular approximation:

$$\int_{jh}^{kh} y\,dt \approx h\sum_{i=j}^{i=k-1} y(ih),$$

leading to:

$$Ta_1 + a_2 = Kb_1, \tag{33.4}$$

where a_1, a_2, b_1 are available numerically. By repeating the procedure over another time period, a second such equation results, allowing the unknowns K, T to be deduced. We note that the use of the integrated equation to a certain extent "smoothes out" the noise.

Suppose (as with the annealing furnace) it is desired that the closed-loop transfer function is to be first-order with unit *dc* gain and a *fixed* time constant T_c *irrespective* of the values of K, T. With a PI controller of the form $C(s) = K_c(1 + 1/sT_i)$, the choices $T_i = \hat{T}$ and $K_c = T_i/(\hat{K}T_c)$ are appropriate. This completes the self-tuning design.

33.2.1 A Plant with an Unknown Time-Varying Gain

Consider now controlling an unknown-gain plant using discrete-time methods. At the sample instants, the output measurement $y(t)$ is made via an analog-to-digital converter, and the control $u(t)$ is calculated and applied via a digital-to-analog converter. We treat the general self-tuning design problem in two stages:

1. The design of a controller assuming *known* transfer function parameters
2. The estimation of the plant's dynamic parameters from the input/output data sequences

For discrete-time systems we will use a mixed Z-transform/shift-operator notation, where the z is considered to be the forward-shift operator: $zx(t) \to x(t+1)$. The current sample of a variable x is $x(t)$, the previous sample is $x(t-1)$, and so on. In particular, Δ is the backward-difference operator $\Delta = 1 - z^{-1}$, giving $\Delta x(t) \to x(t) - x(t-1)$.

A plant with known gain K has a sampled-data model $y(t+1) = Ku(t)$, as the control asserted at sample t affects the output measurement one sample later, so given the set point $w(t)$ the "best" *open-loop* control for attaining this value one sample later is clearly $u(t) = w(t)/K$. But we generally want to have *closed-loop* control so we compute an error signal:

$$e(t) = w(t) - y(t) \quad \text{or} \quad e(t) = w(t) - Ku(t-1),$$

in the usual way. But the "best" control is such that $w(t) = Ku(t)$, so replace $w(t)$ by $Ku(t)$ and rearrange to get the feedback control:

$$u(t) = u(t-1) + e(t)/K, \quad \text{or} \quad \Delta u(t) = u(t) - u(t-1) = e(t)/K,$$

i.e.,

$$u(t) = \frac{1}{K\Delta} e(t), \tag{33.5}$$

where $e(t)$ is the system error. Hence, the controller has the transfer function $1/K\Delta$: an integrator. We can modify the controller gain by a factor μ to give $\mu/K\Delta$ for which the "gain" $\mu = 1$ gives "ideal" single-step response to a change in w. This is the first step in a self-tuning design.

Now we design an estimator for the unknown plant gain parameter K. The basic idea is that if the *output* = *Gain*input*, then the *Gain estimate* = *output/input*. The way we will do it, however, will generalize to larger problems. The idea is to use the model to "predict the present" and to compare the prediction with the actual measured plant response.

A *prediction model* gives a forecast $\hat{y}(t+1|t)$ of the future output $y(t+1)$ depending on available input/output data and values of the model parameters. In our case, the forecast of the *current* output depends on the existing value of the gain estimate $\hat{K}(t-1)$ calculated at the previous sample:

$$\hat{y}(t|t-1) = \hat{K}(t-1)u(t-1),$$

and we define the *prediction error* $\epsilon(t)$ to be

$$\epsilon(t) = y(t) - \hat{y}(t|t-1). \tag{33.6}$$

The estimator updates at each t its "best guess" of plant gain. An "open-loop" best estimator would give $\hat{K}(t) = y(t)/u(t-1)$; but, as with control design, we prefer a *closed-loop* algorithm of the form:

$$\hat{K}(\text{new}) = \hat{K}(\text{old}) + f(\text{data}).\epsilon(t).$$

Hence we write, as for the controller design:

$$\hat{K}(t) = \frac{y(t)}{u(t-1)} = \frac{[\hat{K}(t-1)u(t-1) + \epsilon(t)]}{u(t-1)},$$

noting that

$$y(t) = \hat{y}(t) + \epsilon(t) = \hat{K}(t-1) + \frac{\lambda}{u(t-1)}\epsilon(t), \tag{33.7}$$

where the "estimator gain" λ is 1 for "optimal" performance. The sequence of operations is first to estimate $\epsilon(t)$ by comparing the old model's prediction $\hat{y}(t)$ of the newly acquired output data $y(t)$, then to update the model parameters depending on the chosen value of "adaptive gain" λ.

To get the full self-tuner we just couple the estimator of plant gain of Equation 33.7 to the feedback controller of Equation 33.5 by passing $\hat{K}$ from one to the other. Note the resemblance between controller and estimator as shown in Figure 33.4: both are integrators. The use of feedback in control is for good performance *despite* "bad" models and disturbances; similarly, with the estimator we can get good results despite noise added to y. Just as with control where loop gain μ is reduced to improve stability, we reduce λ to improve the estimator's robustness. With "optimal" values of μ and λ, the algorithm takes just two successive sample times to get perfect control: step 1 gives an accurate $\hat{K}$ and step 2 attains the set-point. As an example, consider a plant that at $t = 0$ has $K = 2$ and our model $\hat{K}(0) = 1$. At $t = 1$ the set point w is made 4, and at $t = 3$ the plant gain changes to 1. Iterating the equations for the first five samples gives Table 33.1.

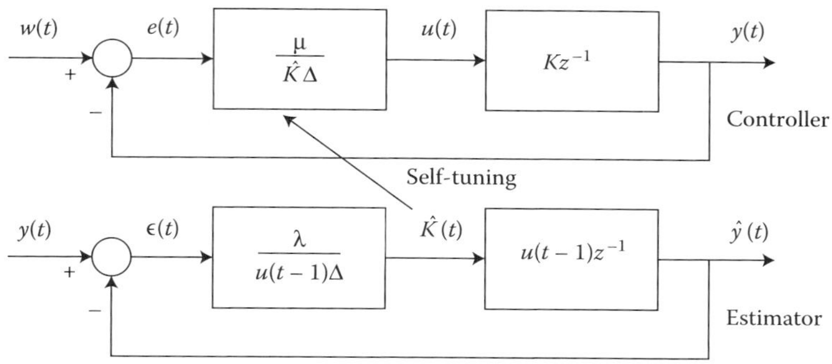

FIGURE 33.4 The controller and estimator as feedback loops.

33.3 Plant Models

A mathematical model of a plant is required in order to design a controller. The design procedure needs to be analytic so that a unique control comes from a given model and performance criterion. In most cases, we consider a model as a *predictor* of the behavior of the system, e.g., its output response $y(t+i)$ to inputs $u(t-j)$. The model will have a set of parameters θ; *identification* will provide values for these, which can be fed into the control design. It is to be hoped that the model can in fact represent the "real" behavior of the plant. (Aside: A distillation column might be described by four hundred simultaneous nonlinear differential equations—far too complex for our needs. Fortunately, it is often well represented by a second-order transfer function with dead time.)

We adopt a *model structure* $\mathcal{M}(\uparrow)$ that is *parameterized* by the set θ; for example, we could choose:

$$\mathcal{M}(\theta) = \frac{b_0}{a_1 s + a_0}, \quad \text{i.e., } \theta = \{a_0, a_1, b_0\},$$

but this would not be sensible as an infinite number of values θ would give the same transfer function. Clearly it is better to divide through by a_1 and assume:

$$\mathcal{M}(\theta) = \frac{b_0}{s + a_0}, \quad \text{i.e., } \theta = \{a_0, b_0\},$$

with one fewer parameter to estimate and a unique answer obtainable.

The choice of model structure is therefore important in the design of general-purpose system identification or self-tuning schemes:

- Can $\mathcal{M}$ represent a general class of plants (e.g., unstable, lightly damped, high order, ...) so that the self-tuner does not need to be redesigned for each case?

TABLE 33.1 The Signals in the Simple Self-Tuner

t	w	y	e	K	$\hat{K}$	$\hat{y}$	ϵ	u
0	0	0	0	2	1	0	0	0
1	4	0	4	2	1	0	0	4
2	4	8	−4	2	2	4	4	2
3	4	4	0	2	2	4	0	2
4	4	2	2	1	1	4	−2	4
5	4	4	0	1	1	4	0	4

- Does $\mathcal{M}$ use a *minimal* set of parameters so that the "true" input/output behavior is given by unique values of θ?
- Is the structure of $\mathcal{M}$ such that estimation is simple? So that it is *robust* to bad assumptions (e.g., about the number of poles and zeros in the real plant)?

We shall concentrate on dynamic models in which the underlying plant is continuous time and assumed to be in the locally linearized form:

$$y(t) = \frac{B(s)}{A(s)} u(t - T_d) + d(t), \tag{33.8}$$

where T_d is the dead time and $d(t)$ is an unmeasured disturbance. The inclusion of a dead time is for two reasons: (1) many industrial processes involve mass transport with corresponding delays, but also (2) plants with complex dynamics (e.g., many poles) can be *approximated* by a low-order model with dead time. The disturbance reflects reality: it corresponds to any factor that affects the output not included in the $G(s)$ such as measurement noise, "unmodeled dynamics," nonlinear effects, real disturbances (such as load torques) acting on the system. It will be found that estimation is a *compromise* between "alertness" to changes in the plant's $G(s)$ and susceptibility of the estimator to noise (fast estimators adapt rapidly to new output data, which sadly could be noise rather than real signals).

As most identification and self-tuning algorithms are implemented digitally, we shall assume *discrete-time* models obtained from $G(s)$ via sampling. These models can come in many forms, though all are usable as predictors. The simplest general-purpose form is the *pulse response* or weighting sequence. Consider injecting a unit pulse (i.e., of height 1 and over the sample interval h) into the plant and sampling the output to get a sequence $\{h_i\}$. *For a general input* $u(t)$, the sampled output $y(t)$ is then given by the *convolution sum*:

$$y(t) = \sum_{i=1}^{\infty} h_i u(t - i). \tag{33.9}$$

The only assumptions here are superposition and open-loop stability: the arbitrary linear plant has a parameter set $\theta = \{h_i\}$ being the sampled points on the pulse response.

Theoretically there are an *infinite* number of parameters in this model, but in practice we *truncate* after some point N, assuming $[h_i = 0, i > N]$ and where h_0 is zero, as the plant cannot respond instantaneously. If there is a dead time of k samples between a control and its initial effect on the output, then $[h_i = 0,\ \text{for}\ i = 1 \ldots k]$ also, so that the model handles dead time simply by having leading coefficients of zero.

One problem with FIR models is that they require a very large number of parameters accurately to represent "stiff" dynamic systems (i.e., with fast and slow modes in the same plant), or even a simple lightly damped pole pair. The sample interval h must be less than the smallest time constant of interest, and the model "length" must be such that Nh exceeds the plant settling time, which is typically five times the largest time constant. Hence, even with only a 1:10 range of time constants, at least 50 parameters may be necessary. Nevertheless, the associated computations are very simple (easily embedded in VLSI), and FIRs are commonly used in signal processing and some process control designs. It sometimes proves to be useful to consider the input to a plant as a series of *increments* $\Delta u(t) = u(t) - u(t-1)$ (i.e., steps or control "moves" as in a stepper-motor) and the response (by superposition) to be

$$y(t) = s_1 \Delta u(t-1) + s_2 \Delta u(t-2) + \cdots + s_i \Delta u(t-i) + \cdots,$$

where s_i is the ith point on the plant's *unit-step response*. It is easy to show that the predictor that gives the next plant output is

$$y(t+1) = y(t) + s_1 \Delta u(t) + \sum_{i=1}^{N} h_i \Delta u(t-i). \tag{33.10}$$

The predicted output $y(t+1)$ is the sum of three components: the current output $y(t)$, the *forced response* due to the current control "move" $\Delta u(t)$, plus the *free response* due to previous control moves.

33.3.1 Transfer-Function or Difference-Equation Models

By far the most popular model in self-tuning control is the DARMA (deterministic autoregressive and moving average) or general difference-equation form:

$$y(t) + a_1 y(t-1) + a_2 y(t-2) + \cdots + a_{na} y(t-na) = b_1 u(t-1) + b_2 u(t-2) + \cdots + b_{nb} u(t-nb). \tag{33.11}$$

Defining $A(z^{-1})$ and $B(z^{-1})$ to be polynomials in the backward-shift operator:

$$A(z^{-1}) = 1 + a_1 z^{-1} + \cdots + a_{na} z^{-na} \tag{33.12}$$

$$B(z^{-1}) = b_1 z^{-1} + \cdots + b_{nb} z^{-nb}, \tag{33.13}$$

this can be written

$$A(z^{-1}) y(t) = B(z^{-1}) u(t),$$

i.e., a transfer function

$$G(z^{-1}) = \frac{B(z^{-1})}{A(z^{-1})}. \tag{33.14}$$

The values of the parameters are obtained by taking Z-transforms of $G(s)$ (+ZOH) as previously shown. Note that:

- There is a unique correspondence [via the mapping $z = \exp(sh)$] between the continuous- and discrete-time poles; the degree na is the same as that of $G(s)$. A pole $s = -\alpha$ in the left half-plane (LHP) in s (stable) maps into $z = \exp(-\alpha h)$; i.e., within the unit circle.
- There is *no* simple mapping of zeros. Indeed, even if $G(s)$ has all its zeros in the stability region (LHP), this does *not* mean that $G(z^{-1})$ will be likewise (in unit circle). It is, in fact, extremely common for so-called *nonminimum-phase* discrete-time models to appear (zeros outside unit circle), e.g., when there is *fractional dead time* or when controlling a high-order plant with a small sample interval h.

33.3.2 Incorporating Disturbances

In continuous-time random processes it is useful to define "white noise"—a signal with a constant spectral power at all frequencies. In discrete time, the corresponding signal is a sequence of random independent (uncorrelated) variables with zero mean and common variance σ^2 [hence called a $(0, \sigma^2)$ uncorrelated random sequence (URS)]. Here we do not have an "infinite variance" signal, but instead something easily produced by a random signal generator such as RAND in MATLAB®. Such a signal has the following properties:

$$\mathcal{E}e(t) = 0; \quad \mathcal{E}e^2(t) = \sigma_e^2; \quad \mathcal{E}e(i)e(j) = 0, \quad \text{for } i \neq j, \tag{33.15}$$

where $\mathcal{E}$ denotes the expectation operation. As in continuous time, a *general* stationary discrete-time random process is modeled by *white noise passing through a transfer function* in z.

Suppose the controlled part of a plant is $G = B_1/A_1$ and the output y is affected by additive disturbances $C_1/D_1 e(t)$. Then we have:

$$y(t) = \frac{B_1}{A_1} u(t) + \frac{C_1}{D_1} e(t),$$

which, when multiplied up, gives:

$$A_1 D_1 y(t) = B_1 D_1 u(t) + C_1 A_1 e(t).$$

Hence, we make the common overall plant assumption of the CARMA (controlled autoregressive and moving average) model:

$$A(z^{-1}) y(t) = B(z^{-1}) u(t) + C(z^{-1}) e(t), \tag{33.16}$$

where $e(t)$ is a URS sequence of independent $(0, \sigma^2)$ random variables. The corresponding difference equation is then generated using the polynomials A, B, C.

However, the CARMA form is insufficient to characterize *offsets* for which $u = 0$ is not accompanied by $y = 0$. A model could include *deviations* from the mean levels u_0, y_0, say, but in control loops these mean levels would not, in general, be constant or known. Indeed, it is often found that disturbances are "steps"; for example, when passengers enter an elevator, there are steps in the load torque acting on the motor: these correspond to shifts in mean levels. One way to deal with this problem is to add an offset parameter d to the output. In a full model used in self-tuning, the value of d would need to be estimated along with the other (dynamic) parameters in the A, B, C polynomials.

In practice, the disturbance is likely to be a combination of factors such that no consistent values of the parameters $C(z^{-1})$ can be estimated. But it is known from the internal model principle that *disturbance elimination is best achieved by having an internal model of the disturbance in the control law*. Inspection provides the following "theorem": *the ubiquitous nature of PID regulators in industry implies that an integrator is the internal model of most practical disturbances*. Hence, the best *general* assumption is of a CAR*I*MA (integrated) model in which random disturbances are integrated:

$$A(z^{-1})y(t) = B(z^{-1})u(t) + C(z^{-1})x(t), \tag{33.17}$$

where $x(t)$ is of the form a/Δ for deterministic step disturbances, or $\Delta x(t) = e(t)$ [with $e(t)$ a URS] for random disturbances. Incorporating into the model gives an *incremental* form:

$$A(z^{-1})\Delta y(t) = B(z^{-1})\Delta u(t) + C(z^{-1})e(t), \tag{33.18}$$

or

$$\begin{aligned} y(t) = y(t-1) - a_1\Delta y(t-1) - \cdots a_{na}\Delta y(t-na) + b_1\Delta u(t-1) + \cdots b_{nb}\Delta u(t-nb) \\ + e(t) + c_1 e(t-1) + \cdots + c_{nc}e(t-nc). \end{aligned} \tag{33.19}$$

Note that the model deals with *increments* of the input/output data such as $\Delta y(t-1) = y(t-1) - y(t-2)$ and hence no *dc* term is involved as Δ.constant $= 0$. It is found that injecting a *pulse* into x (or e) gives a step change in y, and injecting a "white" random sequence or URS gives "Brownian motion" for which y "drifts" (Brownian motion is quite a good model of stock exchange prices).

33.4 Recursive Prediction Error (RPE) Estimators

The job of an estimator is to provide values for the model parameters based on fitting the model's responses to the measured plant input/output data. A *recursive* estimator *updates* the estimates at each sample instant based on the newly available information. One important point about recursive estimators is that the computational load must *not* increase with time. Note that potentially the amount of data is always increasing with more available from each sample. *Prediction error* methods consider the model to be a forecaster $\hat{y}(t|t-1)$ of the actual outcome $y(t)$, the difference or residual $\epsilon = y - \hat{y}$ being used to correct the estimates. Now define the following n vectors:

1. $\theta = [\theta_1, \ldots, \theta_n]'$ are the n unknown plant parameters.
2. $\hat{\theta}(t) = [\hat{\theta}_1(t), \ldots, \hat{\theta}_n(t)]'$ are the corresponding estimates at time t.
3. $\mathbf{x}(t) = [x_1(t), \ldots, x_n(t)]'$ are *known* data associated with the parameters.

A *prediction model* generates a forecast $\hat{y}(t)$ depending on $\hat{\theta}(t-1)$ and $\mathbf{x}(t)$:

$$\hat{y}(t) == \hat{y}(t|t-1) = f(\hat{\theta}(t-1), \mathbf{x}(t)). \tag{33.20}$$

The models we will consider are all in *linear-in-the-parameters* (LITP) form with:

$$\hat{y}(t) = \sum_{i=1}^{n} \hat{\theta}_i(t-1)x_i(t), \tag{33.21}$$

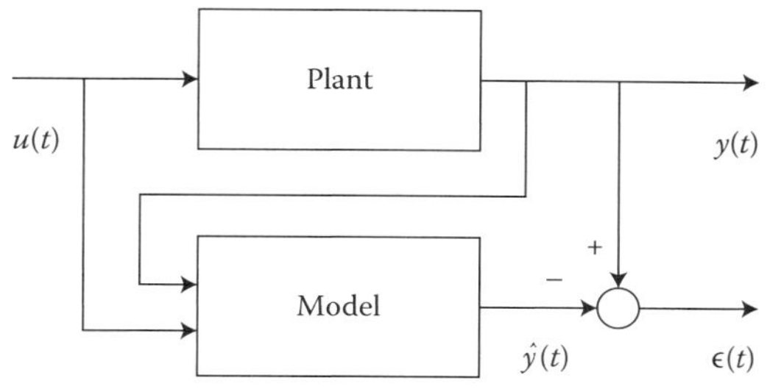

FIGURE 33.5 General RPE structure.

and where the measured plant output is assumed to satisfy the LITP equation:

$$y(t) = \sum_{i=1}^{n} \theta_i x_i(t) + e(t), \tag{33.22}$$

with $e(t)$ corresponding to "noise" or "disturbances" and assumed to be independent of the data $\mathbf{x}(t)$. The scalar *prediction error* is defined to be

$$\epsilon(t) = y(t) - \hat{y}(t) = y(t) - f(\hat{\theta}(t-1), \mathbf{x}(t)), \tag{33.23}$$

where $y(t)$ is the new output measurement. The general structure is shown in Figure 33.5.

Typical models that come under the general description of Equation 33.21 are:

Pulse response

$$y(t) = \sum_{i=1}^{n} h_i u(t-i)$$

where:

$$\theta = [h_1, h_2, \ldots, h_n]' \quad \text{and} \quad \mathbf{x}(t) = [u(t-1), u(t-2), \ldots, u(t-n)]'.$$

In this model, the parameter vector are points on the pulse response, and the data vector contains n past values of the control signal.

DARMA

$$y(t) = -a_1 y(t-1) - \cdots + b_1 u(t-1) + \cdots + b_{nb} u(t-nb)$$

where:

$$\theta = [a_1, \ldots a_{na}, b_1, \ldots, b_{nb}]'$$

and:

$$\mathbf{x}(t) = [-y(t-1), \ldots, -y(t-na),\ u(t-1), \ldots, u(t-nb)]'.$$

In this model, the parameters are the difference-equation constants, and the data vector contains past values of inputs *and* outputs.

Note that the LITP equations can be written concisely as:

$$\hat{y}(t) = \mathbf{x}'(t)\hat{\theta}(t-1),$$

and equivalently

$$y(t) = \mathbf{x}'(t)\theta + e(t),$$

so that if the *parameter error vector* $\tilde{\theta}(t)$ is defined to be $\theta - \hat{\theta}(t)$, then:

$$\epsilon(t) = y(t) - \hat{y}(t) = \mathbf{x}'(t)\tilde{\theta}(t-1) + e(t). \tag{33.24}$$

This important equation shows that the prediction error, has two components: the model error and the unknown disturbance. Our use of this result depends on whether we want to identify a time-invariant system (in which case we average over a lot of data to eliminate the effects of noise) or whether we want to track a time-varying plant (which is possible only if the signal to noise ratio is large and the variations are not too fast). We can associate with the prediction error a cost function J such as $J(t) = 0.5\epsilon^2(t)$. If there is no noise, this cost depends only on $\tilde{\theta}$ and so we can imagine a set of equal-cost contours in the n-space of the parameters $\hat{\theta}$ with the minimum value at the "true" parameters. If there is noise, these contours will correspond to average or expected values.

An RPE algorithm updates the estimates using:

$$\hat{\theta}(t) = \hat{\theta}(t-1) + a(t)\mathbf{M}(t)\mathbf{x}(t)\epsilon(t), \tag{33.25}$$

where:

- $a(t)$ is a scalar "gain factor," giving the step length.
- $\mathbf{x}(t)\epsilon(t)$ is along the gradient vector $-\nabla J$ of the cost-function surface, pointing down the slope of local steepest descent.
- $\mathbf{M}(t)$ is a rotation matrix that modifies the parameter update direction away from the steepest descent route.

We can reduce the parameter error using a "large" value of the gain $a(t)$, but this increases the effect of the disturbance $e(t)$. The compromise between rapid model error reduction and insensitivity to noise is a fundamental design issue: in practice, we want rapid initial convergence, followed by good noise immunity for steady-state tracking. In view of its long history of success in many applications, the most common estimator used in self-tuning is based on variants of the *least-squares* (LS) method, for which the current estimate $\theta(\hat{t})$ minimizes the "loss function":

$$J(t) = (\hat{\theta}(t) - \hat{\theta}(0))'\mathbf{S}(0)(\hat{\theta}(t) - 1\hat{\theta}(0)) + \sum_{i=1}^{t}(y(i) - \mathbf{x}'(i)\hat{\theta}(t))^2 \tag{33.26}$$

It can be shown (e.g., Isermann et al. [8]) that $J(t)$ is minimized by the recursions:

$$\mathbf{S}(t) = \mathbf{S}(t-1) + \mathbf{x}(t)\mathbf{x}'(t) \tag{33.27}$$

$$\hat{\theta}(t) = \hat{\theta}(t-1) + \mathbf{S}(t)^{-1}\mathbf{x}(t)\epsilon(t), \tag{33.28}$$

with $\hat{\theta}(0)$ being the initial "guess" of the parameters and $\mathbf{S}(0)$ an assertion about their likely accuracy (i.e., small entries in $\mathbf{S}$ implies low accuracy). It can be shown that:

$$\mathbf{S}(t)\hat{\theta}(t) = \mathbf{S}(0)\hat{\theta}(0) + [\mathbf{S}(t) - \mathbf{S}(0)]\theta, \tag{33.29}$$

so that as $\mathbf{S}(t)$ increases, the effect of the initial assumptions declines, and a solution is possible only if $\mathbf{S}$ has full rank: a "persistency of excitation" condition.

In practice, we prefer to propagate $\mathbf{P}(t) = \mathbf{S}^{-1}(t)$ to avoid the inversion in Equation 33.28; using $\mathbf{P}$, the important recursive least-squares (RLS) equations become:

Kalman gain vector:

$$\mathbf{k}(t) = \frac{\mathbf{P}(t-1)\mathbf{x}(t)}{1 + \mathbf{x}'(t)\mathbf{P}(t-1)\mathbf{x}(t)} \tag{33.30}$$

parameter update:

$$\hat{\theta}(t) = \hat{\theta}(t-1) + \mathbf{k}(t)\epsilon(t) \tag{33.31}$$

covariance update:

$$\mathbf{P}(t) = [\mathbf{I} - \mathbf{k}(t)\mathbf{x}'(t)]\mathbf{P}(t-1) \tag{33.32}$$

The algorithm needs to be initialized. Note that $\mathbf{P}$ is proportional to the "errors in the parameters," as $\mathbf{P} = \mathbf{S}^{-1}$, so that at $t = 0$ where nothing (or little) is known, the appropriate choice is $\gamma\mathbf{I}$ where γ is large. It can then be shown that the estimates obtained after n samples using RLS are the same as with "off-line" LS. For smaller γ, though, slower parameter movement is seen, for that choice indicates the chosen $\hat{\theta}(0)$ are fairly accurate. In this way it is possible to start recursive estimation with a reasonable "guessed" model rather than simply $(0, 0, 0 \ldots)'$.

What happens for large t? Assuming that there are good persistently exciting data, then $\mathbf{S} = \sum_0^t \mathbf{x}\mathbf{x}'$ increases all the time and hence $||\mathbf{k}|| \to 0$. This means that asymptotically RLS loses its "alertness" (though by that time there should be convergence to the "true" values despite noise). This is a natural consequence of the built-in assumptions: a fixed parameter plant where accurate estimates are wanted. Hence, the method is possibly suitable for self-tuning but not for adaptive control. Figure 33.6 shows RLS in action with data from a FIR model in which the parameters $\{h_1(t), h_2(t)\}$ each change by square waves. The input is white noise in "bursts" so it is not exciting during quiescent periods. The estimates do not move at all when there is zero excitation and change at an increasingly slow rate even when the data are "rich." Hence, RLS concentrates on accuracy (assuming fixed plant parameters) and not on tracking parameter changes.

33.4.1 Forgetting Factors

The cost function in LS weights all residuals $\epsilon(t-j)$ equally, no matter how far back in the past the data were acquired. If it is expected that the model dynamics varies with time, then recent data are more significant than old data. Thus, the idea is to "forget" so that the effect of data on the estimates decays in time. It is convenient to consider a simple single-parameter continuous-time estimator to explore the

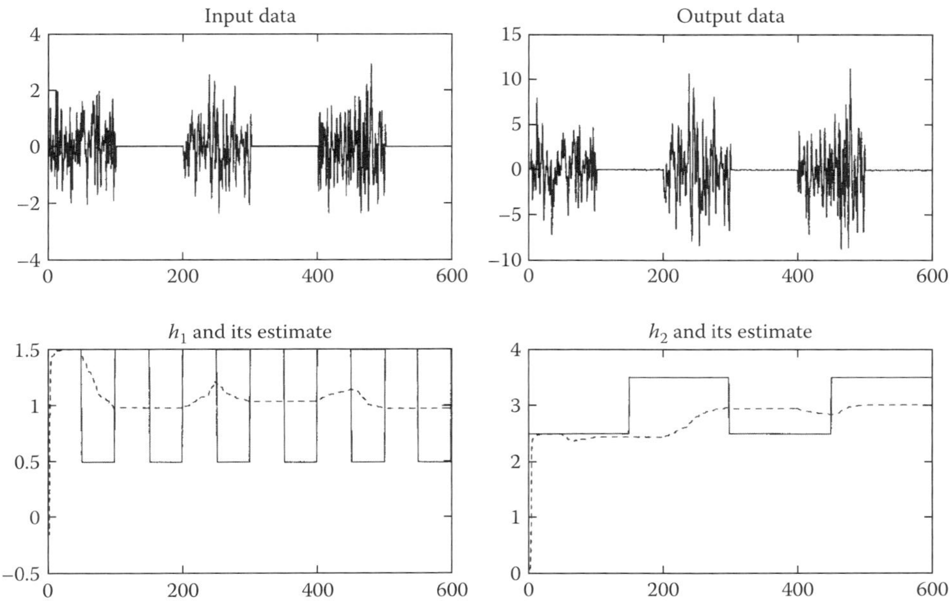

FIGURE 33.6 RLS with no forgetting.

ideas. As a preliminary, consider the integral:

$$I = \int_0^t e^{-\alpha(t-\lambda)} f(\lambda)\, d\lambda. \tag{33.33}$$

Then, from first principles:

$$\frac{dI}{dt} = f(t) - \alpha I. \tag{33.34}$$

A fixed forgetting factor exponentially weights past data at a rate α, so that using the continuous-time exponentially weighted version of Equation 33.26, the estimate $\hat{\theta}(t)$ minimizes:

$$J(t) = e^{-\alpha t} s(0)(\hat{\theta}(t) - \hat{\theta}(0))^2 + \int_0^t e^{-\alpha(t-\lambda)} (y(\lambda) - \hat{\theta}(t)x(\lambda))^2\, d\lambda, \tag{33.35}$$

giving, by setting the partial differential with respect to $\hat{\theta}$ to zero:

$$s(t)\hat{\theta}(t) = e^{-\alpha t} s(0)\hat{\theta}(0) + \int_0^t e^{-\alpha(t-\lambda)} y(\lambda)x(\lambda)\, d\lambda \tag{33.36}$$

where:

$$s(t) = e^{-\alpha t} s(0) + \int_0^t e^{-\alpha(t-\lambda)} x^2(\lambda)\, d\lambda. \tag{33.37}$$

Differentiating with respect to t and using the preliminary result gives the recursive equations of continuous-time RLS for a single parameter:

$$\frac{ds}{dt} + \alpha s = x^2 \tag{33.38}$$

$$\frac{d\hat{\theta}}{dt} + \frac{x^2}{s}\hat{\theta} = \frac{xy}{s}, \tag{33.39}$$

with initial conditions $\{s(0), \hat{\theta}(0)\}$. The Equation 33.39 shows that, for constant excitation $x = X$, $s(t)$ rises exponentially to X^2/α; for zero "forgetting" $\alpha = 0$, $s(t) \to \infty$ if the data are persistently exciting. In practice, the convention is to use the current prediction error $\epsilon = y - \hat{\theta}x$, so substituting we get as the update equation:

$$\frac{d\hat{\theta}}{dt} = \frac{x}{s}\epsilon. \tag{33.40}$$

It is instructive to find how the modeling error $\tilde{\theta} = \theta - \hat{\theta}$ changes with time. Substituting for y using $y = \theta x + e$ and assuming that θ varies in time we get:

$$\frac{d\tilde{\theta}}{dt} + \frac{x^2}{s}\tilde{\theta} = \frac{d\theta}{dt} - \frac{x}{s}e.$$

The left-hand side tries to drive $\tilde{\theta}$ to zero at a rate depending on x^2/s while the right-hand drives it away from zero. In RLS it is conventional to use the inverse $p = 1/s$ in the calculations. Substituting in the update for s we get:

$$\frac{dp}{dt} - \alpha p = -p^2x^2, \quad \text{or} \quad \frac{dp}{dt} + (x^2p - \alpha)p = 0. \tag{33.41}$$

The behavior of forgetting factors such as α is clearly shown in this formulation: the equation has an unstable mode when $x^2p < \alpha$. If there is no excitation $[x(t) = 0]$, $p(t)$ increases exponentially at a rate determined by α. If x is constant p increases until $p = \alpha/x^2$ and then becomes constant. Hence, clearly one method to avoid "blow up" is to add a small quantity x_0^2 to x^2 in the updating equation for p.

33.4.2 Variable Forgetting Factors

The loss function J can be written in terms of ϵ:

$$J(t) = J_0(t) + \int_0^t e^{-\alpha(t-\lambda)}\epsilon(\lambda)^2\, d\lambda.$$

Suppose now that estimation is perfect (i.e., $\hat{\theta} = \theta$) for all t so that $\epsilon(t) = e(t)$ always and the variance of the noise is σ_0^2. Then the expected value of the irreducible or "ideal loss" is given by:

$$\mathcal{E}\{J_{opt}\} = \int_0^t e^{-\alpha(t-\lambda)}\sigma_0^2\, d\lambda = \frac{\sigma_0^2}{\alpha}(1 - \exp^{-\alpha t}).$$

For large t, this approaches the value $\sigma_0^2/\alpha = T_c\sigma_0^2$, say, where T_c corresponds to the "asymptotic averaging time"; i.e., the period of past time that contains the data "most influential" in providing the estimate.

Consider the differential equation for J, which is:

$$\frac{dJ}{dt} + \alpha J = \epsilon^2(t).$$

We would expect the average value for ϵ^2 to be greater than σ_0^2, given that $\epsilon(t) = x\tilde{\theta}(t) + e(t)$. In particular, we expect that $\epsilon^2 \gg \sigma_0^2$ when the estimate is far from the true value of θ. One way to proceed is to assert that the value J (or $1/J$) contains the "information" about the parameter and that this should be constant in the steady state. This implies that a good value for α is obtained by making $\alpha J = \epsilon^2 = \alpha J_{opt}$, giving:

$$\alpha = \frac{1}{T_c}\left(\frac{\epsilon(t)}{\sigma_0}\right)^2,$$

and the update equation for s now reads:

$$\frac{ds}{dt} + \frac{1}{T_c}\left(\frac{\epsilon(t)}{\sigma_0}\right)^2 s = x^2. \tag{33.42}$$

Hence, adaptation is faster with a prediction error that is large compared with the assumed SD of the underlying noise.

33.4.3 Forgetting with Multiparameter Models

For discrete-time estimation we write $\beta = \exp(-\alpha h)$, where h is the sample interval (small $\alpha \rightarrow \beta \approx 1 - \alpha h$,) and then the RLS equations become:

$$\mathbf{k}(t) = \frac{\mathbf{P}(t-1)\mathbf{x}}{\beta + \mathbf{x}'\mathbf{P}(t-1)\mathbf{x}} \tag{33.43}$$

$$\hat{\theta}(t) = \hat{\theta}(t-1) + \mathbf{k}(t)\epsilon(t) \tag{33.44}$$

$$\mathbf{P}(t) = [\mathbf{I} - \mathbf{k}(t)\mathbf{x}'(t)]\mathbf{P}(t-1)/\beta \tag{33.45}$$

A useful measure of the amount of data effectively contributing to the current estimate is the *asymptotic sample length* (ASL) or time constant, given by $1/(1-\beta)$. A rapidly varying system might need an ASL of 20 samples ($\beta = 0.95$) to more than 1000 ($\beta = 0.999$). Figure 33.7 repeats the experiment of Figure 33.6 but using $\beta = 0.95$. Again, when the input is not exciting, the parameters freeze (and **P** "blows up"), but otherwise track changes rapidly. Moral: Use forgetting, but then make sure the input perturbs the plant.

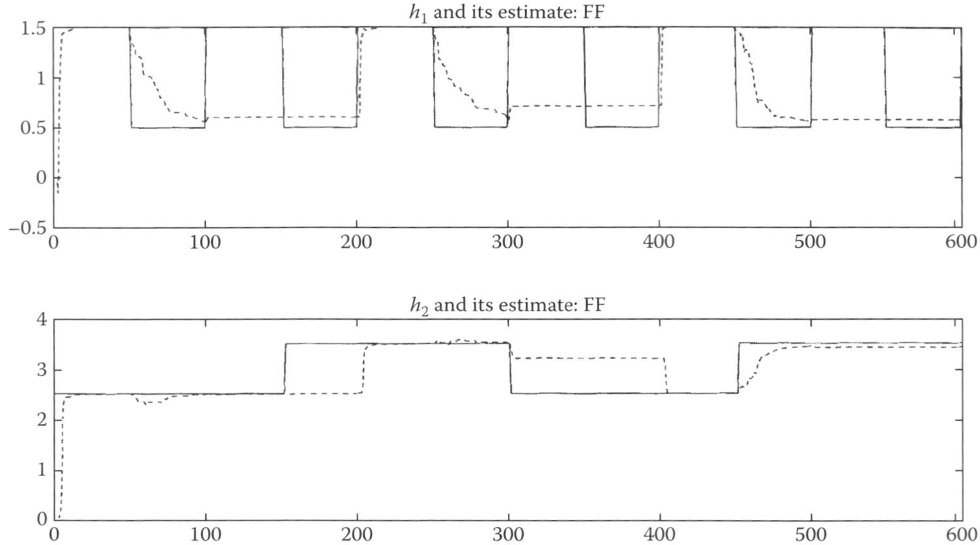

FIGURE 33.7 RLS with forgetting.

33.5 Predictive Models

Industrial processes have characteristics that make advanced control useful:

Disturbances: Fluctuations in the raw materials and the operating environment; sensor noise.
Dead-time: The effect of the current control is not seen in the measured response for a number of samples, because of material transport times.

Typical examples of processes like this are:

Steel rolling: Controls are via screws on the stands; gauge responses are after the end of the mill; x-ray gauge measurements are noisy.
Papermaking: The controls are on the head box at entry to the Fourdrinier wire; the basis weight of the dried paper is measured after the heating rolls.
Strip annealing: Inductive heating causes temperature changes, later measured by a pyrometer with large random fluctuations in its signal.
Distillation: The multiple lags (several hundred) arising from the thermal behavior of each tray appear like dead time; composition is measured by a chromatograph with sampling delays; ambient thermal variations induce disturbances.

Delays cause phase lag, which means that PID control gains must be reduced. Predictive control can give "perfect" control provided the delay is known; even better performance is obtainable if the disturbance process can be predicted also. Consider the problem of predicting the plant output with its two components:

Deterministic effects: Old inputs and outputs give initial conditions, from which the "free response" can be determined via the model. The "forced response" is the additional effect due to current and future controls.
Disturbances: Old URS values $e(t-i)$ can be reconstructed from the model and known data. The free response can then be computed; as nothing is known about future white noise, the best approach is simply to assume its mean value of 0.

Consider an *autoregressive* (AR) process driven by white noise giving the measured output $y(t)$:

$$y(t) = 0.9y(t-1) + e(t), \quad \text{or} \quad y(t) = \frac{1}{(1-0.9z^{-1})}e(t).$$

By expanding as a power series, the model can be written in pulse form:

$$y(t) = e(t) + 0.9e(t-1) + 0.9^2e(t-1) + 0.9^3e(t-2) + \cdots.$$

Consider its value two steps into the future:

$$y(t+2) = e(t+2) + 0.9e(t+1) + 0.9^2e(t) + 0.9^3e(t-1) + \cdots.$$

As $e(t+1), e(t+2)$ are not known at time t, the best prediction of $y(t+2)$ is

$$\hat{y}(t+2|t) = 0.9^2e(t) + 0.9^3e(t-1) + \cdots = \frac{0.9^2}{(1-0.9z^{-1})}e(t).$$

But the noise model gives $e(t) = (1-0.9z^{-1})y(t)$, giving the simple result that $\hat{y}(t+2|t) = 0.9^2y(t)$ (the free response of the noise model). The error in the prediction is

$$\tilde{y}(t+2|t) = e(t+2) + 0.9e(t+1).$$

To find the variance or mean square of this signal, simply square and take averages, taking $\mathcal{E}e(t+i)^2 = \sigma^2; \mathcal{E}e(t+i)e(t+j) = 0$, and the following should be noted:

1. The prediction error (p.e.) is *independent* of the "known" data [in this case, $e(t), e(t-1), \ldots$]; i.e., the maximum possible information has been extracted.
2. The variance of the p.e. is $(1+0.9^2)\sigma^2$, where σ^2 is the variance of the URS e.
3. The p.e. variance increases with the prediction *horizon*; here, two-steps ahead.
4. "Sluggish" disturbances (pole near 1) are predicted with more accuracy than rapidly moving disturbances.

To generalize this example, consider a noise model in *moving-average* (MA) form:

$$y(t) = N(z^{-1})e(t),$$

so that:

$$\begin{aligned} y(t+k) &= (1 + n_1z^{-1} + \cdots + n_{k-1}z^{-k+1})e(t+k) + (n_kz^{-k} + n_{k+1}z^{-k-1} + \cdots)e(t+k) \\ &= N_k^*(z^{-1})e(t+k) + N_k(z^{-1})e(t). \end{aligned} \tag{33.46}$$

The disturbance is split into "future" and "past" components, and the prediction uses known data:

$$\hat{y}(t+k|t) = N_k(z^{-1})e(t) = n_ke(t) + n_{k+1}e(t-1) + \cdots.$$

How does this procedure work with a transfer function (ARMA) structure:

$$y(t) = \frac{C(z^{-1})}{A(z^{-1})}e(t)?$$

Performing long division by A and stopping after k terms gives:

$$\frac{C(z^{-1})}{A(z^{-1})} = E(z^{-1}) + z^{-k}\frac{F(z^{-1})}{A(z^{-1})} = N_k^* + z^{-k}N_k.$$

In fact, instead of doing long division we multiply each side by A to get:

$$C(z^{-1}) = E(z^{-1})A(z^{-1}) + z^{-k}F(z^{-1}). \tag{33.47}$$

This key equation is known as a *Diophantine* (or *Bezoutian*) identity from which E and F can be obtained for given A, C, k by equating powers of z^{-1}. For example, consider:

$$(1-0.9z^{-1})y(t) = (1+0.7z^{-1})e(t) \rightarrow A = (1-0.9z^{-1}); \quad C = (1+0.7z^{-1}).$$

Hence, the Diophantine identity of Equation 33.47 for $k = 2$ becomes:

$$(1+0.7z^{-1}) = E(z^{-1})(1-0.9z^{-1}) + z^{-2}F(z^{-1}),$$

or

$$(1+0.7z^{-1}) = (e_0 + e_1z^{-1})(1-0.9z^{-1}) + z^{-2}f_0. \tag{33.48}$$

Equating coefficients of increasing powers of z^{-1} in Equation 33.48:

$$z^0:\ 1 = e_0;\ z^{-1}:\ 0.7 = -0.9 + e_1;$$
$$z^{-2}:\ 0 = -0.9e_1 + f_0.$$

Hence we have:

$$e_0 = 1; \quad e_1 = 1.6; \quad f_0 = 1.44,$$

i.e.,

$$E(z^{-1}) = 1 + 1.6z^{-1} \quad \text{and} \quad F(z^{-1}) = 1.44, \tag{33.49}$$

and so the two-step-ahead prediction becomes:

$$\hat{y}(t+2|t) = N_k(z^{-1})e(t) = \frac{F}{A}e(t) = \frac{1.44}{(1-0.9z^{-1})}e(t).$$

But we can reconstruct $e(t)$ from the measured value of $y(t)$ and the inverted model:

$$e(t) = \frac{A(z^{-1})}{C(z^{-1})}y(t),$$

giving the predictor:

$$\hat{y}(t+2|t) = \frac{1.44}{1+0.7z^{-1}}y(t). \tag{33.50}$$

The prediction error $\tilde{y}$ is given by $E(z^{-1})e(t+2)$, or:

$$\tilde{y}(t+2|t) = e(t+2) + 1.6e(t+1),$$

which has a variance of $(1+1.6^2)\sigma^2 = 3.56\sigma^2$. It is interesting to note that the actual variance of $y(t+2)$ is $14.47\sigma^2$, meaning that our predictor "explains" roughly three fourths of the output variance.

33.6 Minimum-Variance (MV) Control

A growing requirement in manufacturing is guaranteed and quantified quality, as measured, for example, by the proportion of a product lying outside some prespecified limit. In continuous processes, such as papermaking, it is important that the output (at worst) exceeds some lower quality limit (e.g., thickness of paper). To ensure that this is so, the *average* thickness must be set greater than the minimum by an amount dependent on the variance of the controlled output. Hence, if this variance is minimized, the manufacturer can reduce the average, as shown in Figure 33.8.

[Aside: the worst manufacturers sometimes make the best product. If they have a large *spread* in quality, they have to test everything and reject out-of-spec items; i.e., those below the lower statistical limit (LSL) of the figure. Hence, the majority of sales is at a higher quality than really needed. It is best to be "just good enough" = low spread of quality, and hence be profitable.]

Consider the plant with dead time k samples and with model:

$$A(z^{-1})y(t) = B(z^{-1})u(t-k) + C(z^{-1})e(t), \tag{33.51}$$

or, dividing:

$$y(t+k) = \frac{B(z^{-1})}{A(z^{-1})}u(t) + \frac{C(z^{-1})}{A(z^{-1})}e(t+k). \tag{33.52}$$

The second right-hand side term is the effect of the disturbances on the output, which can be predicted using the ideas of the previous section; the first term is the effect of the control (which by assumption can affect the output only after k samples). The idea of MV control, in essence, is to choose the control $u(t)$ that will counteract the *predicted* disturbance at time $t+k$. The development first solves the Diophantine identity Equation 33.47 to provide $E(z^{-1})$ and multiplies:

$$A(z^{-1})y(t+k) = B(z^{-1})u(t) + C(z^{-1})e(t+k)$$

(each side) by $E(z^{-1})$ to give:

$$EAy(t+k) = EBu(t) + ECe(t+k)$$

or, as $EA = C - z^{-k}F$ from Equation 33.47:

$$Cy(t+k) - Fy(t) = Gu(t) + ECe(t+k),$$

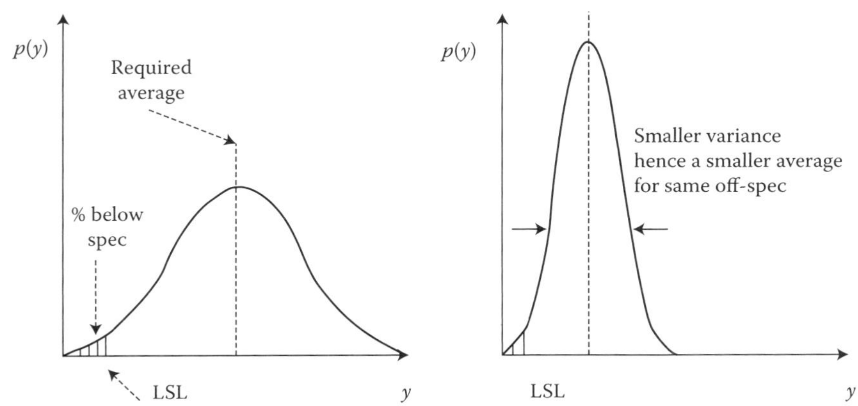

FIGURE 33.8 Using minimum-variance control.

where $G(z^{-1}) = E(z^{-1})B(z^{-1})$. Hence, we have the equation:

$$y(t+k) = \frac{Fy(t) + Gu(t)}{C} + Ee(t+k). \tag{33.53}$$

The first term on the right-hand side is the k-step-ahead predictor and the second is the prediction error. Hence, we can write the key prediction equations:

k step prediction:

$$\hat{y}(t+k|t) = \frac{F(z^{-1})y(t) + G(z^{-1})}{C(z^{-1})}u(t) \tag{33.54}$$

k step error:

$$\tilde{y}(t+k|t) = E(z^{-1})e(t+k). \tag{33.55}$$

As an example take the plant:

$$(1 - 0.9z^{-1})y(t) = 0.5u(t-2) + (1 + 0.7z^{-1})e(t),$$

where we have already solved the Diophantine identity of Equation 33.47 giving Equation 33.49 and so:

$$y(t+2) = \frac{1.44y(t) + (0.5 + 0.8z^{-1})u(t)}{(1 + 0.7z^{-1})} + (1 + 1.6z^{-1})e(t+2).$$

The MV control is then easy to determine: Simply choose $u(t)$ so that the first right-hand side term becomes 0; all that remains on the controlled output is the prediction error, which cannot be minimized further as it is comprised of only future white noise components $e(t+1)$ and $e(t+2)$. The feedback control law for our example is then:

$$1.44y(t) + (0.5 + 0.8z^{-1})u(t) = 0,$$

or:

$$u(t) = -1.6u(t-1) - 2.88y(t).$$

In the general case, the control becomes:

$$u(t) = -\frac{F}{G}y(t) = -\frac{F(z^{-1})}{B(z^{-1})E(z^{-1})}y(t); \tag{33.56}$$

this controller *cancels* the zeros of the plant transfer function. In closed loop, the characteristic equation is

$$1 + \frac{F}{BE}z^{-k}BA = 0, \quad \text{or} \quad B(EA + z^{-k}F) = 0,$$

so that, using the Diophantine identity of Equation 33.47, this reduces to:

$$B(z^{-1})C(z^{-1}) = 0. \tag{33.57}$$

The closed-loop modes are defined by those of C (which, in fact, are stable) and of $B(z^{-1})$. There is, therefore, a potential instability problem with MV control in cases where B has roots outside the unit-circle stability region (so-called *nonminimum-phase* zeros) as these appear as unstable poles of the closed loop; such nonminimum-phase zeros occur much more frequently in discrete systems than in continuous-time control.

33.7 Minimum-Variance Self-Tuning

All the machinery is now available for self-tuning: connect a parameter estimator to an MV controller by solving the Diophantine identity of Equation 33.47. Note that MV controller design requires knowledge of k, A, B, *and* C. In difference equation terms the CARMA plant model is

$$y(t) = -a_1 y(t-1) - \cdots + b_0 u(t-k) + \cdots + e(t) + c_1 e(t-1) + \cdots,$$

but the standard estimators can estimate only A, B; the driving noise $e(t)$ is not measurable and, hence, cannot be placed into the $\mathbf{x}$-vector to estimate C. There are methods (such as extended least squares) for estimating C, but these tend to be unreliable in practice. However, it transpires we can obtain self-tuned MV (giving a *self-tuning regulator*) by using a standard LS estimator without needing knowledge of C (in effect, assuming $C = 1$)!

There is potentially a further problem: the effect of feedback control on the parameter estimates. Suppose, for example, that a plant:

$$y(t) = ay(t-1) + bu(t-1) + e(t) \tag{33.58}$$

has a simple proportional controller (with zero set point):

$$u(t) = -\alpha y(t), \quad \text{or} \quad \alpha y(t-1) + u(t-1) = 0. \tag{33.59}$$

Then adding a fraction μ of Equation 33.59 to Equation 33.58 gives:

$$y(t) = (a + \mu\alpha)y(t-1) + (b + \mu)u(t-1) + e(t).$$

If we now use an estimator based on the two-parameter model:

$$y(t) = \theta_1 y(t-1) + \theta_2 u(t-1) + e(t) = \mathbf{x}'\theta + e,$$

then $\hat{\theta}_1 = (a + \mu\alpha), \hat{\theta}_2 = (b + \mu)$ will be obtained, where μ is arbitrary. Hence, if we use LS estimation in a closed-loop mode, the estimated $\hat{\theta}$ does not converge to a unique point but to a line where the estimated parameters can wander up and down in unison. This is a problem of using closed-loop data with only internal signals such as $e(t)$ stimulating the loop; to get an consistent estimate we must do one of the following:

1. Use externally generated test signals, such as step changes in set point.
2. Have a controller that is more complex (higher order) than the plant.
3. Have a time-varying controller.

This third solution is appropriate for self-tuning, though it is still best to make the data "rich" by exciting the plant with external signals (e.g., a PRBS added to the set point).

How is it that we can use LS? The key idea is not to go *estimate* → *design* → *controller* (giving what is called the *indirect* approach), but instead to proceed *estimate* → *controller* (the *direct* approach). What are estimated are the *controller* (in fact, the k-step-ahead predictor) rather than the *plant* parameters. How this is done is seen below.

Recall the prediction Equation 33.54 when multiplied up by $C(z^{-1})$:

$$(1 + c_1 z^{-1} + c_2 z^{-2} + \cdots)\hat{y}(t+k|t) = Fy(t) + Gu(t),$$

giving at time t:

$$\hat{y}(t|t-k) = Fy(t-k) + Gu(t-k) - \sum c_i z^{-i}\hat{y}(t|t-k).$$

But the point about MV control is that it makes the prediction zero by correct choice of u. Hence all the terms in the sum on the right-hand side are set to zero by *previous* controls, so that from:

$$y(t) = \hat{y}(t|t-k) + \tilde{y}(t|t-k), \tag{33.60}$$

we have:

$$y(t) = F(z^{-1})y(t-k) + G(z^{-1})u(t-k) + E(z^{-1})e(t). \tag{33.61}$$

This is the crucial equation: it obeys the LS rules of an LITP model with:

$$\mathbf{x}(t) = [\,y(t-k), y(t-k-1), \ldots, u(t-k), u(t-k-1), \ldots]' \tag{33.62}$$

$$\theta = [f_0, f_1, \ldots, g_0, g_1, \ldots]', \tag{33.63}$$

and, most importantly, the data $\mathbf{x}(t)$ are independent of the error term as the data are from $t-k$ backwards, whereas $E(z^{-1})e(t)$ finishes at $e_{k-1}e(t-k+1)$. Hence, LS leads directly to the required $\hat{F}$ and $\hat{G}$ parameters, so we get a self-tuner with feedback law:

$$\hat{F}(z^{-1})y(t) + \hat{G}(z^{-1})u(t) = 0. \tag{33.64}$$

The procedure, then, is as follows:

1. Assemble old data into the $\mathbf{x}$-vector as in Equation 33.63.
2. Use RLS to get $\hat{\theta} = \hat{F}, \hat{G}$.
3. Use the estimated parameters in the feedback law of Equation 33.64.

Of course, the above is simply a *plausibility* argument; in fact, the algorithm can by lengthy algebra be shown to converge to give the required control signals; i.e., satisfying the *self-tuning* property. The speed of convergence is found to depend on the roots of $C(z^{-1})$. As an example, consider the first-order system:

$$(1 - 0.9z^{-1})y(t) = 0.2u(t-2) + (1 + 0.9z^{-1})e(t),$$

which has the two-step-ahead prediction equation:

$$(1 + 0.9z^{-1})\hat{y}(t+2|t) = 1.62y(t) + 0.2(1 + 1.8z^{-1})u(t),$$

for which the MV controller with *known* parameters is

$$1.62y(t) + 0.2u(t) + 0.36u(t-1) = 0.$$

The corresponding model to estimate in self-tuning is

$$y(t) = f_0 y(t-2) + g_0 u(t-2) + g_1 u(t-3) + \epsilon(t).$$

The estimator for a self-tuner will have data and parameter vectors:

$$\mathbf{x} = [y(t-2), u(t-2), u(t-3)]'$$

$$\theta = [f_0, g_0, g_1]' \;\; \rightarrow \;\; [1.62, 0.2, 0.36]'.$$

The system was simulated for 1000 samples, the first 500 being "open loop." At $t = 500$ the self-tuner was switched on, giving the results seen in Figure 33.9. Observe how the variance has been reduced by STMV and how the estimated parameters "wander about."

As discussed above, the estimates of the parameters are not unique, as the control Equation 33.64 can be multiplied by an arbitrary factor μ without affecting $u(t)$. In principle, this is not a problem, but to avoid excessively large or small estimates we can "fix a parameter" to a guessed value and estimate the others. Typically, the fixed parameter is the value of g_0: the multiplier of the current control $u(t)$, whose nominal value is b_0. Suppose $\bar{g}_0$ is the choice. In our example, it means that the model becomes:

$$y(t) - \bar{g}_0 u(t-2) = y_1(t) = f_0 y(t-2) + g_1 u(t-3) + \epsilon(t).$$

Then the model to use in RLS has data and parameter vectors:

$$\mathbf{x}(t) = [y(t-k), y(t-k-1), \ldots, u(t-k-1), \ldots]' \tag{33.65}$$

$$= [y(t-2), u(t-3)]' \tag{33.66}$$

in our example;

$$\theta = [f_0, f_1, \ldots, g_1, \ldots]', \tag{33.67}$$

and with "output" $y_1(t) = y(t) - \bar{g}_0 u(t-2)$. The control to use is like Equation 33.64 but is based on the *chosen* fixed $\bar{g}_0$ and the remaining estimates:

$$u(t) = -[\sum \hat{f}_i y(t-i) + \sum \hat{g}_i u(t-i)]/\bar{g}_0.$$

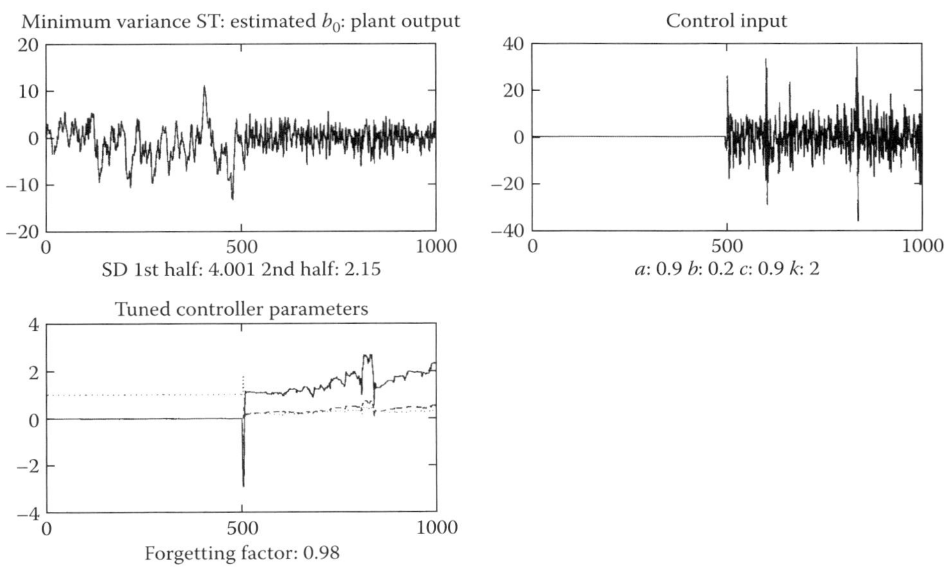

FIGURE 33.9 MV self-tuning, all parameters estimated.

A self-tuner based on this idea is coded in MATLAB as:

```
% M-file for the simple first-order A-W
% minimum-variance self-tuner.

nt = 1000; na = 1; nb = k-1; np = na+nb;
th=zeros(nt,np); P = 100*eye(np);
II = eye(np);

for i=k+1:nt,
 e(i) = rand;
 y(i) = a*y(i-1) + b*u(i-k) + e(i) + c*e(i-1); % plant
  if i > nt/2,
   x  = [y(i-k) u(i-2*k+1:i-k-1)]; % data vector
   ep = y(i)-g0*u(i-k) - x*th(i-1,:)'; % pred error
   kk = x*P/(beta+x*P*x');        % RLS
   th(i,:) = th(i-1,:) + kk*ep; % ..
   P=(II - kk'*x)*P/beta;         % update
   u(i) = - (th(i,1)*y(i) + u(i-nb:i-1)*th(i,2:np)')/g0;
  end;
end;
```

Does it matter if the wrong value of $\bar{g}_0$ is chosen? No, provided that:

$$\frac{1}{2} < \frac{\bar{g}_0}{b_0} < \infty. \tag{33.68}$$

This means that a "large" value of the fixed parameter is safe.

Figure 33.10 and Figure 33.11 show the behavior of a self-tuner with the parameter $\bar{g}_0$ set to 0.4 and then 2. Both are perfectly well behaved, though with slower convergence for $\bar{g}_0 = 2$. Note that in the original example with no "fixing" there is a point where the parameter estimates "jump." It is found that without fixing and with no external signals a self-tuner tends to "burst" like this every now and then.

Generalized minimum-variance (GMV) control was developed to overcome the problems of (1) MV's instability when B is nonminimum-phase and (2) the large control variance produced by MV, particularly when using "fast sampling." An *auxiliary* output $\phi(t)$ is defined:

$$\phi(t) = P(z^{-1})y(t) - R(z^{-1})w(t-k) + Q(z^{-1})u(t-k), \tag{33.69}$$

where P, Q, R are *design polynomials* whose choice gives a range of possible closed-loop objectives; see Harris and Billings [7] or Wellstead and Zarrop [11] for more details. GMV self-tuning simply uses the same approach as developed for MV:

1. Estimate polynomials $\hat{F}, \hat{G}, \hat{H}$ in the predictor model:

$$\phi(t) = \hat{F}(z^{-1})y(t-k) + \hat{G}(z^{-1})u(t-k) + \hat{H}(z^{-1})w(t-k) + \epsilon(t). \tag{33.70}$$

2. Using the estimates from Equation 33.70, compute the control:

$$\hat{G}(z^{-1})u(t) = -[\hat{F}(z^{-1})y(t) + \hat{H}(z^{-1})w(t)]. \tag{33.71}$$

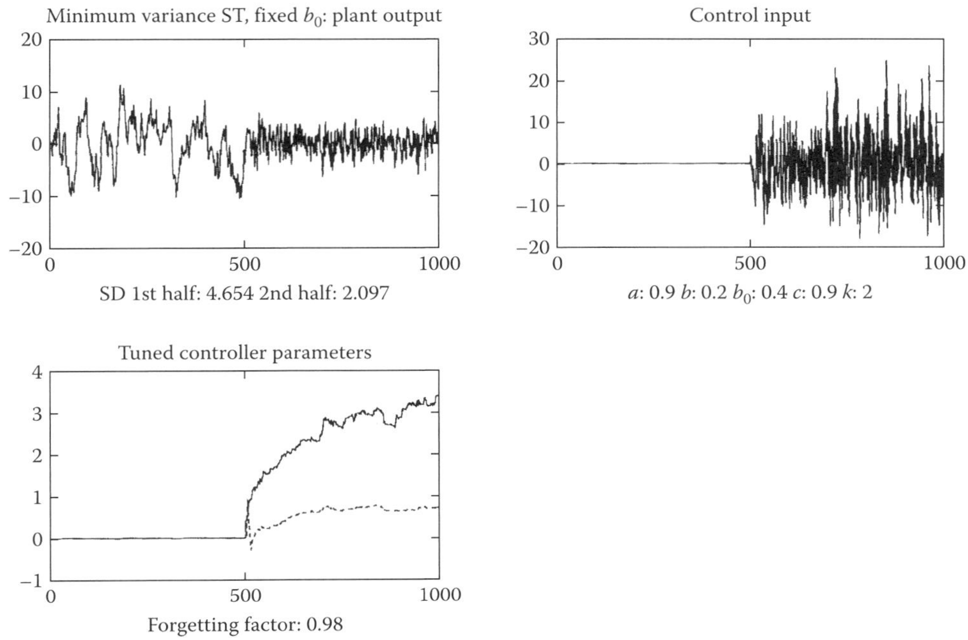

FIGURE 33.10 Simulation of a simple self-tuner with a fixed parameter $\bar{g}_0$.

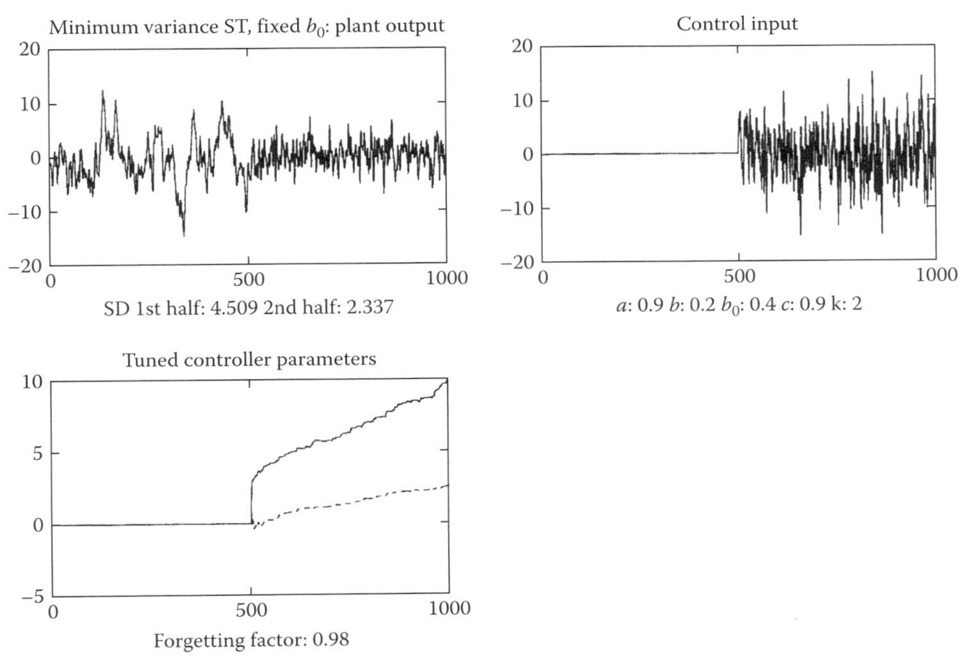

FIGURE 33.11 Simulation of STMV with a larger fixed parameter.

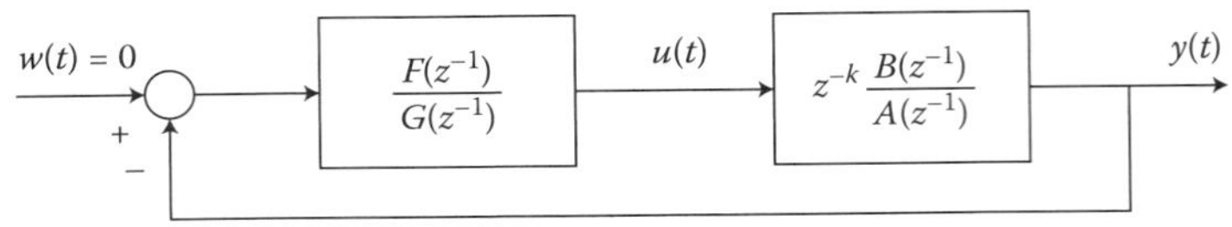

FIGURE 33.12 Feedback controller for pole placement.

A simple case of GMV is when $P = R = 1$ and $Q = \lambda$ for which it can be shown that the cost:

$$J_{GMV} = \mathcal{E}[(y(t+k) - w(t))^2 + \frac{\lambda}{b_0}u^2(t)|t] \tag{33.72}$$

is minimized, where the expectation is conditional on data available up to time t. It can further be shown that the characteristic polynomial now has a factor:

$$B(z^{-1}) + \frac{\lambda}{b_0}A(z^{-1}),$$

so that λ can be considered a root-locus parameter; for plants with a stable A polynomial, a large enough value of λ ensures loop stability even for nonminimum-phase plants. Unfortunately for self-tuning applications, a prior "good" value of λ needs to be known; nevertheless, λ can be used to trade off output variance against control activity.

33.8 Pole-Placement (PP) Self-Tuning

The problem with MV regulators if that they are unstable if the plant has a zero outside the unit circle, as the controller attempts to *cancel* the zero by having an unstable pole. Hence, there is a mode in the control signal that grows without limit. Therefore, alternative strategies such as GMV must be used for more complex plants; one such appeals to classical control design.

Recall the root-locus approach: it is a method of *analysis* showing how the poles of the closed-loop transfer function vary as some parameter, such as the controller gain, is changed. Design considers some pole-zero structure of a controller and then fixes the gain to give "nice" closed-loop pole positions, such as large ω_n for a desired value of ζ such as $1/\sqrt{2}$. An alternative procedure that is often used in discrete-time control is *first to choose* the desired pole positions and then back-calculate the appropriate controller.

Suppose that the feedback controller of Figure 33.12 is:

$$F(z^{-1})y(t) + G(z^{-1})u(t) = 0, \quad \text{where } g_0 = 1, \tag{33.73}$$

which gives the control action as:

$$u(t) = -g_1u(t-1) - g_2u(t-1) - \cdots - f_0y(t) - f_1y(t-1) - \cdots.$$

If the plant has the CARMA model:

$$A(z^{-1})y(t) = z^{-k}B(z^{-1})u(t) + C(z^{-1})e(t),$$

where the dead time k is explicitly factored from the zero polynomial, the closed loop is given by:

$$[A(z^{-1})G(z^{-1}) + z^{-k}B(z^{-1})F(z^{-1})]y(t) = C(z^{-1})G(z^{-1})e(t). \tag{33.74}$$

Suppose that the polynomials F and G are obtained by solving the Diophantine identity:

$$AG + z^{-k}BF = CT(z^{-1}), \tag{33.75}$$

where $T(z^{-1})$ is a polynomial *chosen by the designer* of the form:

$$T(z^{-1}) = 1 + t_1 z^{-1} + t_2 z^{-2} + \cdots = \prod_i (1 - \alpha_i z^{-1}),$$

with the α_i corresponding to the *desired closed-loop pole positions.* (For hand calculation, we compare powers of z^{-1} as before; in a computer, we use Euclid's algorithm.) By substituting Equation 33.75 into Equation 33.74, the closed loop is given by:

$$C(z^{-1})T(z^{-1})y(t) = C(z^{-1})G(z^{-1})e(t),$$

or:

$$y(t) = \frac{G(z^{-1})}{T(z^{-1})}e(t), \quad \text{and} \quad u(t) = -\frac{F(z^{-1})}{T(z^{-1})}e(t). \tag{33.76}$$

Hence, the closed-loop poles are given by the user-chosen polynomial T; as the B polynomial is *not* cancelled by this law, there is no longer any unstable mode in the control signal even if the plant is nonminimum-phase. As an example, consider again the system:

$$(1 - 0.9z^{-1})y(t) = 0.5u(t-2) + (1 + 0.7z^{-1})e(t),$$

and choose $T(z^{-1}) = 1 - 0.5z^{-1}$, giving a relatively fast closed-loop pole at $\alpha = 0.5$. Then the Diophantine identity of Equation 33.75 becomes:

$$(1 - 0.9z^{-1})(1 + g_1 z^{-1} + \cdots) + z^{-2}0.5(f_0 + f_1 z^{-1} \cdots) = (1 + 0.7z^{-1})(1 - 0.5z^{-1}).$$

Comparing coefficients of increasing powers of z^{-1} gives $g_1 = 1.1$ and $f_0 = 1.28$, all other coefficients being zero. Hence, the control law is given by

$$u(t) = -[1.1u(t-1) + 1.28y(t)],$$

and in closed loop we have from Equation 33.76:

$$y(t) = \frac{1 + 1.1z^{-1}}{1 - 0.5z^{-1}}e(t), \quad \text{and} \quad u(t) = -\frac{1.28}{1 - 0.5z^{-1}}e(t).$$

It can be shown that the output variance due to this control law is $5.6\sigma^2$, roughly double the MV result: *you don't get something for nothing*.

It is interesting to see that pole assignment (for this regulator case) can also be self-tuned using RLS *without* knowing or estimating the polynomial C. To see why this is so, let $\mathcal{A}(z^{-1})$ and $\mathcal{B}(z^{-1})$ be the solutions to a new Diophantine identity *that does not include the* C *polynomial* (compare with Equation 33.75):

$$\mathcal{A}G(z^{-1}) + z^{-k}\mathcal{B}F(z^{-1}) = T(z^{-1}). \tag{33.77}$$

The polynomials F and G provided as "input" to this identity are those obtained by using the previous identity of Equation 33.75 with the "true" plant polynomials A, B, C. Let us take our example, giving the new Diophantine identity of Equation 33.77:

$$\mathcal{A}(1 + 1.1z^{-1}) + z^{-2}\mathcal{B}1.28 = (1 - 0.5z^{-1}),$$

giving $\mathcal{A} = 1 - 1.6z^{-1}, \mathcal{B} = 1.6 \times 1.1/1.28 = 1.375$.

Consider operating the plant in closed loop using the "correct" controller $-F/G$ as computed by Equation 33.75; then the sequence:

$$\mathcal{A}(z^{-1})y(t) - z^{-k}\mathcal{B}(z^{-1})u(t) = \frac{\mathcal{A}G + z^{-k}\mathcal{B}F}{T(z^{-1})}e(t) = e(t), \quad (33.78)$$

using the closed-loop behavior from Equation 33.76 and then the second Diophantine identity of Equation 33.77. This result gives the key idea: when in closed loop with the right PP controller, there exists a relationship between the input and output signals given by the new polynomials $\mathcal{A}$ and $\mathcal{B}$ such that the error term is just the white driving noise.

Equation 33.78 shows that RLS can be used to get unbiased estimates of $\mathcal{A}$ and $\mathcal{B}$, as the right-hand side noise term is uncorrelated. To estimate the parameters, simply choose the data vector to be

$$\mathbf{x}(t) = [-y(t-1), -y(t-2), \ldots; u(t-1), u(t-2), \ldots]'$$

and with model output $y(t)$. Hence, if the data sequences follow those of the pole-placed closed loop of Equation 33.76, then LS simply gives $\hat{\theta} \to [\mathcal{A}, \mathcal{B}]$. If these estimates are placed as *input* to the *second* Diophantine identity of Equation 33.77, we can use the identity to recompute the controller F and G polynomials; i.e., the procedure of going $[A, B, C, T] \to [F, G]$ via 33.75 $\to [\mathcal{A}, \mathcal{B}]$ via RLS $\to [F, G]$ via 33.77 is self-consistent.

Hence, the self-tuned version of PP goes through the following steps:

1. Use RLS to obtain estimates $\hat{\mathcal{A}}$ and $\hat{\mathcal{B}}$ from the model $\mathcal{A}y(t) = z^{-k}\mathcal{B}u(t) + e(t)$ (i.e., no C estimated).
2. Resolve the Diophantine identity of Equation 33.77 at each sample for user-chosen T (using as input the estimates $\hat{\mathcal{A}}$ and $\hat{\mathcal{B}}$) to obtain $\hat{F}$ and $\hat{G}$, by equating powers of z^{-1} as usual.
3. Assert the control $\hat{F}y(t) + \hat{G}u(t) = 0$, or $u(t) = -[\hat{F}/\hat{G}]y(t)$.

As with self-tuned MV, the above argument is simply to give plausibility to the approach; in practice, the procedure converges provided that there is a solution to the Diophantine identity of Equation 33.77 (which is not possible if there are common roots in the estimated model polynomials).

33.9 Long-Range Predictive Control

For practical applications, an adaptive controller must be *robust* against the prior assumptions made about the plant to be controlled. For example, we must choose sometimes arbitrary values for the degrees of the estimated polynomials and have no assurance that, for all t, our model is of neither too high nor too low an order compared with the true plant model within the bandwidth of our closed loop. The estimates may be affected over periods of time by disturbances not fully captured by our assumptions about $C(z^{-1})$. There may be occasions when there are common factors in the estimated TFs. The dead time of the plant may vary so that the k assumed in k-step-ahead prediction for MV control is not correct (so that the "true" value of g_0 is zero). Fractional dead time and fast sampling might cause the model to become nonminimum phase. Hence:

Minimum-variance: We might get instability by assuming too small a value of k, or if the plant is nonminimum phase.

Pole placement: There is no solution to the Diophantine identity if there are common factors in the estimated $\mathcal{A}, \mathcal{B}$ polynomials.

Results of controlling a time-varying plant with an adaptive MV algorithm with a forgetting factor of 0.98 are shown in Figure 33.13. The plant changes its order n and delay k at various stages during the run. When the actual plant delay exceeds the assumed delay of the algorithm, MV goes unstable. Similarly,

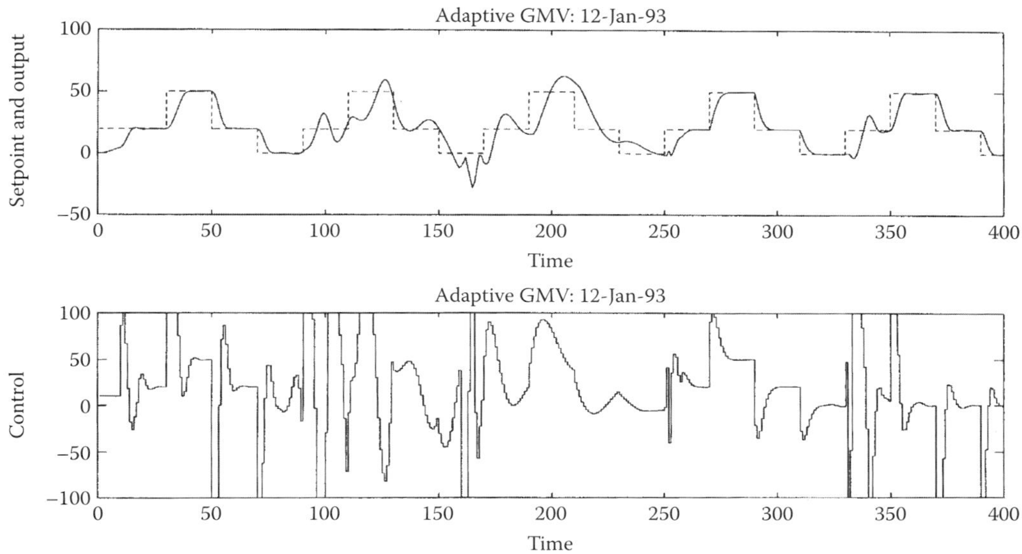

FIGURE 33.13 Simulation of the adaptive MV control of a dynamically time-varying plant.

it is found that with adaptive PP, the control goes unstable if the assumed model order is greater than the actual model order. With a fixed PID control the initial results are good, but instability sets in when the plant changes its dynamics. In practice, good engineering design using prior knowledge and proper choice of sample interval (e.g., at most 0.1 times the open loop rise time) can give good results with MV and PP. On the other hand, you can be unlucky.

Long-range predictive control (LRPC) is a more modern approach that overcomes many of the above problems and has many extra features, which makes it useful in applications. The basic idea is to predict the future output as the sum of a *free response* (based on past known data) and a *forced response* depending on *current and future control* actions, as shown in Figure 33.14. Consider the noise-free incremental model:

$$A(z^{-1})\Delta y(t) = B(z^{-1})\Delta u(t-1), \tag{33.79}$$

or:

$$y(t) = y(t-1) - a_1\Delta y(t-1) - \cdots + b_1\Delta u(t-1) + \cdots. \tag{33.80}$$

Consider using this model to give the prediction $p(t+1)$ of the free output response based on maintaining the control signal equal to the previous value $u(t-1)$:

$$p(t+1) = y(t) - a_1\Delta y(t) - \cdots + b_2\Delta u(t-1) + \cdots,$$

as by assumption $\Delta u(t) = 0$. Continuing the iteration further:

$$p(t+2) = p(t+1) - a_1\Delta p(t+1) - \cdots + b_3\Delta u(t-1) + \cdots,$$

where the term $\Delta p(t+1) = p(t+1) - y(t)$, the difference between the prediction and the available data $y(t)$. After some stage all the $\Delta u(t-1)$ terms drop out as the polynomial B has been exhausted, leaving the iterations:

$$p(t+i) = p(t+i-1) - a_1\Delta p(t+i-1) - a_2\Delta p(t+i-2) - \cdots. \tag{33.81}$$

Hence, the predictions can be expressed verbally as "*iterate the plant equations forward in time, assuming current and future control increments (moves) are zero, and using existing old $\Delta u, \Delta y$ to initialize the data.*"

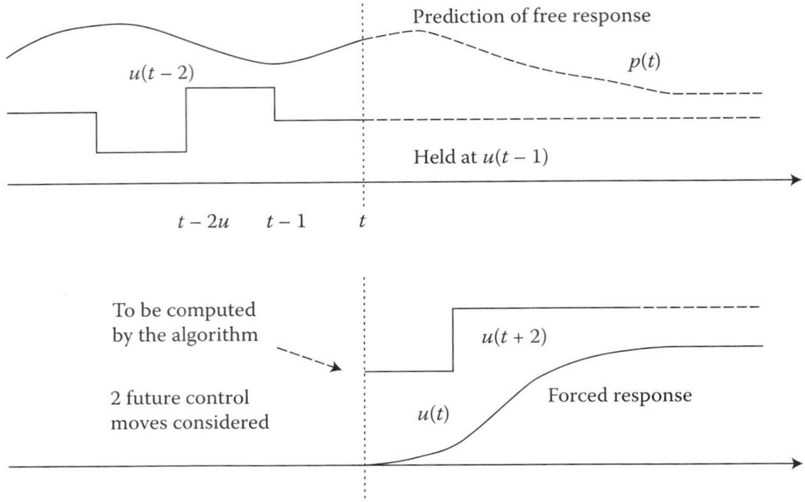

FIGURE 33.14 Long-range predictive control.

Now consider the forced component of response, with the control input being a set of moves $\Delta u(t), \Delta u(t+1), \ldots$, which are to be determined by the algorithm. This is simply a series of "steps" so that the output is just the superposition of a set of step responses $\{s_i\}$, giving the total response as:

$$
\begin{aligned}
y(t+1) &= s_1 \Delta u(t) + p(t+1) \\
y(t+2) &= s_1 \Delta u(t+1) + s_2 \Delta u(t) + p(t+2) \\
&\cdots \\
y(t+j) &= s_1 \Delta u(t+j-1) + s_2 \Delta u(t+j-2) + \cdots + s_j \Delta u(t) + p(t+j) \\
&\cdots
\end{aligned}
$$

Hence, we can collect N such equations into vector-matrix form:

$$
\mathbf{y} = \mathbf{G}\Delta\mathbf{u} + \mathbf{p}, \tag{33.82}
$$

where:

$$
\begin{aligned}
\mathbf{y} &= [y(t+1), y(t+2), \ldots, y(t+N)]' \\
\Delta\mathbf{u} &= [\Delta u(t), \Delta u(t+1), \ldots, \Delta u(t+N-1)]' \\
\mathbf{p} &= [p(t+1), p(t+2), \ldots, p(t+N)]'
\end{aligned}
$$

and:

$$
\mathbf{G} = \begin{bmatrix} s_1 & 0 & \cdots & 0 \\ s_2 & s_1 & \cdots & 0 \\ \cdots & & & \\ s_N & s_{N-1} & \cdots & s_1 \end{bmatrix}.
$$

Suppose that we had a "future set point" sequence $\{w(t+1)\cdots w(t+N)\}$ available at time t. In robotics, this could be the required future trajectory, whereas in process control we would normally assume the future set point to equal the current value. Either way, we can collect the sequence into a vector and hence define a vector of "future system errors":

$$\mathbf{e} = [w(t+1) - y(t+1), w(t+2) - y(t+2), \dots, w(t+N) - y(t+N)]',$$

giving:

$$\mathbf{e} = \mathbf{w} - (\mathbf{G}\Delta u + \mathbf{p}).$$

The only unknowns in this set of equations are the future controls, so we minimize $S = \sum e(t+j)^2$ over the predictions by the set of controls:

$$\Delta\mathbf{u} = (\mathbf{G}'\mathbf{G})^{-1}\mathbf{G}'(\mathbf{w} - \mathbf{p}). \tag{33.83}$$

This gives a set of "best" future control actions; we then use a *receding-horizon* strategy, which simply asserts the first of this sequence and repeats the whole calculation at each sample instant. Note, however, that **G** is an $N \times N$ matrix and we get a solution only if it is invertible, giving:

$$\Delta\mathbf{u} = \mathbf{G}^{-1}(\mathbf{w} - \mathbf{p}). \tag{33.84}$$

This solution is exact such that the control sequence would drive *all* the future system errors to zero. Good in theory; bad in practice as it would require excessive control signals. Moreover, what happens if the plant delay is, say, 2 so that $s_1 = 0$? The result is failure, as we cannot then invert **G**.

How can we derive more equations than unknowns to let LS "smooth" out our future errors? We can make *assumptions* about what controls will be exerted in the future. Consider controlling a simple Type 0 plant. We could inject a large initial signal to get it moving and then maintain a constant control of sufficient size to get it to the final set point. This would mean that at the initial time only two control increments are considered. Hence, we take:

$$\Delta\mathbf{u} = [\Delta u(t), \Delta u(t+1), 0, 0, \dots, 0]',$$

so that now we have fewer unknowns than equations. In general we can allow only *NU* control increments to be nonzero (called the "control horizon"; see Figure 33.14) and will define the "future control increment vector" to be

$$\Delta\mathbf{u} = [\Delta u(t), \Delta u(t+1), \dots, \Delta u(t+NU-1)]'.$$

This means that our set of equations for **e** involve a nonsquare matrix **G**.

One special case is where only *one* control increment is considered at time t; i.e., $NU = 1$. In essence we are asking, "What step change in control will minimize the sum-of-squares of the future system errors?" Suppose we make the prediction horizon N very large. Then if there were any steady-state error, the corresponding sum-of-squares would be large compared with errors accruing during the transient. The outcome is that the control step will be just the right size to make the steady-state error zero, and there will be just the same dynamics in closed loop as in open loop. This simple approach is called *mean-level control* (see Figure 33.15). Given that $NU = 1$, the matrix inversion is simple, as $\mathbf{G}'\mathbf{G}$ is just $\sum_1^N s_i^2$: a scalar.

Consider the case where N is not large. Then the initial transient errors become increasingly important in comparison with the steady-state error so that the initial control is made larger to reduce them; i.e., we obtain a faster response as N reduces.

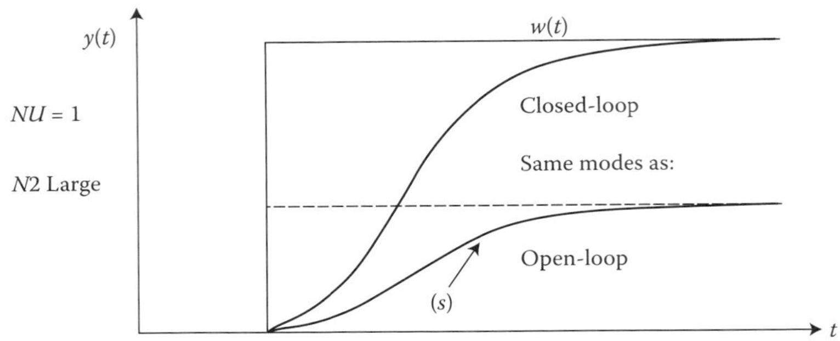

FIGURE 33.15 Mean-level control.

33.10 The Generalized Predictive Control (GPC) Cost Function

We can include other ideas into the LRPC algorithm. For example, if it is known that the plant has a dead time, then it is clear that the current control $u(t)$ *cannot affect* the future errors until the dead time is cleared. Hence, there is no point in putting $e(t+1)\cdots e(t+k-1)$ into the LS cost function. We might also want to have a mechanism for "trading" the cost of control (e.g., control variance) against output performance (i.e., error variance). Combining all these we get the generalized predictive control (GPC) cost-function:

$$J_{GPC}(N1, N2, NU, \lambda) = \sum_{i=N1}^{N2} e(t+i)^2 + \lambda \sum_{i=1}^{NU} \Delta u(t+i-1)^2,$$

where:

N1 is the *lower costing horizon*
N2 is the *upper costing horizon*
N1 → *N2* is the *costing range*
NU is the *control horizon;* i.e., the "degrees of freedom" in the control
λ is the *control weighting*

subject to the condition that assumed future control-increment sequence is zero after the control horizon.

Using GPC we have the solution

$$\Delta \mathbf{u} = (\mathbf{G}'\mathbf{G} + \lambda \mathbf{I})^{-1}\mathbf{G}'(\mathbf{w} - \mathbf{p}), \tag{33.85}$$

where the matrix **G** is now $N2 - N1 + 1$ by NU:

$$\mathbf{G} = \begin{bmatrix} s_{N1} & s_{N1-1} & \cdots & 0 \\ s_{N1+1} & s_{N1} & s_{N1-1} & \cdots \\ \vdots & & & \\ s_{N2} & s_{N2-1} & \cdots & s_{N2-NU+1} \end{bmatrix}.$$

There are many possible combinations of the four "design parameters" $(N1, N2, NU, \lambda)$ but in practice, two main choices are made. The first is mean-level control as above: $[k, \text{large}, 1, 0]$, where "large" means about 10. The other is based on "dead beat" control where the idea is to attain the set point as rapidly as possible such that the error *and all its derivatives* become zero simultaneously. This can be shown to be achieved by the following choices of horizon:

$$N1 = n, \quad N2 \geq 2n - 1, \quad NU = n \quad \text{and} \quad \lambda = 0,$$

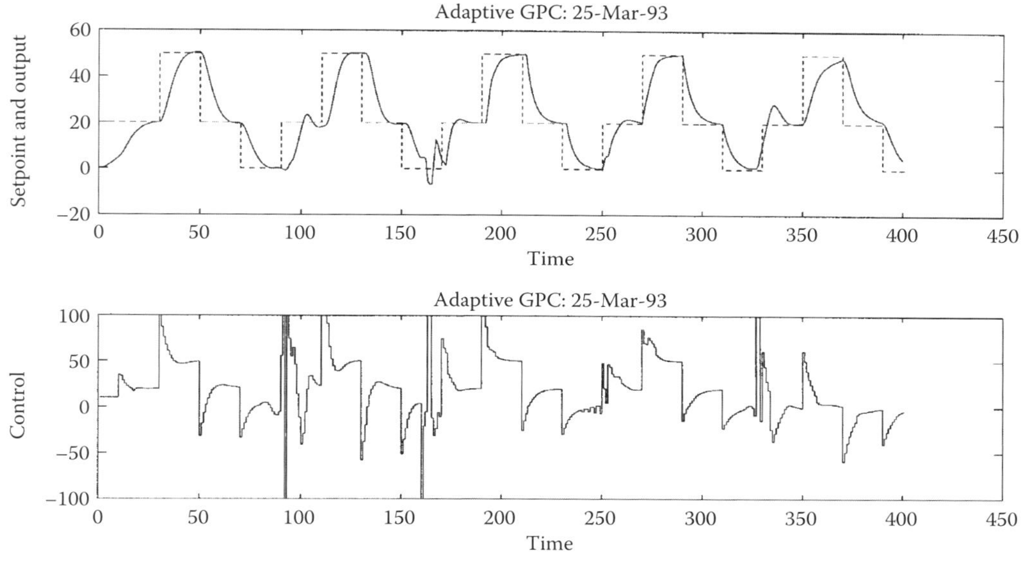

FIGURE 33.16 Adaptive GPC control of a time-varying plant.

where n is the largest power of z^{-1} found in the model (including the extra because of the $\Delta = 1 - z^{-1}$). Suppose we choose $N2 = 2n - 1$. Then $\mathbf{G}$ is $n \times n$, and, provided it is invertible the GPC solution gives $J = 0$ (as the equations are solved exactly) and hence $e(t+n) \cdots e(t+2n-1)$ are all 0. This means that the control is such as to make n successive future system errors zero, a prerequisite for state deadbeat control.

It is not just the flexibility of GPC (see for example Soeterboek [10]) that gives it the power. It can handle realistic control problems that cannot be treated by other designs. Suppose that we know the control signal to be *constrained* (as with torque saturation in motors) or that some internal variable (such as temperature in a catalytic cracker) must not exceed some limit. In principle, we can use the prediction idea to test if any of our variables is likely to hit constraints and, hence, modify the control signal suitably. By running a plant nearer to constraints, we enhance quality and profitability, so LRPC is increasingly popular in industry.

By connecting an RLS estimator to GPC on the plant as in Figure 33.13, using a prediction horizon of 10 and a control horizon of 1 (approximating to mean-level control) we obtain the results of Figure 33.16: much better!

The problem with "simple" GPC as presented above is the lack of stability proofs, except for some limiting cases such as mean-level control; with practical applications, mean-level control has been used most often and this has not been found to cause difficulties. However, in order to get stability *guarantees,* we can use infinite-horizon LQ (e.g., Bitmead et al. [2]) or adopt *terminal constraints* where the objective now is to minimize a cost function *subject* to the output $y(t+j)$ exactly matching the set point over some future constraint range (see Clarke [3], and Mosca [9], for details).

33.11 Robustness of Self-Tuning Controllers

Applications of self-tuning control have to take into account the following practical problems:

- Disturbances acting on a process are likely to be nonstationary and to have inconsistent behavior, e.g., during plant start-up.

- Certain patterns of disturbance can lead to the estimation of a poor plant model.
- The dynamic order of the process is likely to be significantly greater than that assumed by the self-tuning control design.
- Actuator nonlinearities (e.g., stiction) give unrepresentative small-signal behavior.

Some of these difficulties (in particular, the problem of *unmodeled dynamics*) have been treated in some detail; others (e.g., nonlinear actuation) require careful attention to engineering detail. In general, a "good" self-tuner requires (1) a robust estimator; (2) a robust control design; and (3) "jacketing" software that takes into account practical process features.

For improving the estimator, it is possible to add *normalization* and a *dead zone*. Normalization simply takes the regressor vector and produces a factor $m(t-1) = \max\{m, ||\mathbf{x}(t-1)||\}$ and divides the prediction error Equation 33.23 by $m(t-1)$. A dead zone takes into account that the "true" prediction is of the form:

$$\hat{y}(t) = \sum_{1}^{n} \hat{\theta}_i^m x_m(t-i) + \sum_{1}^{n} \hat{\theta}_i^u x_u(t-i),$$

where x_m is the "modeled" data input and x_u, the "unmodeled" input. If, after normalization, the prediction error is smaller than a certain amount, then this is deemed to be due to the (assumed small) effect of unmodeled dynamics and the estimation is temporarily frozen. While there is a good theoretical background for this approach (see, for example, Mosca [9]), there is not much practical experience in how to choose an appropriate value for the dead zone parameter.

Perhaps the most important advance for ensuring robustness is the reconsideration of the CARIMA model in the form:

$$A(z^{-1})y(t) = B(z^{-1})u(t) + \frac{T(z^{-1})}{\Delta}e(t), \tag{33.86}$$

where $T(z^{-1})$ is now considered to be an *assigned design* rather than an estimated polynomial for enhancing estimation and control robustness. Multiplying up in Equation 33.86 gives:

$$A(z^{-1})y^f(t) = B(z^{-1})u^f(t) + e(t), \tag{33.87}$$

where the signals are given by $y^f = \Delta y/T, u^f = \Delta u/T$; i.e., they are band-pass filtered by Δ/T. The effect of Δ is to remove *dc*-offset or constant-load disturbances; the effect of $1/T$ is to filter out high-frequency effects so that the estimation concentrates on low-frequency behavior.

The above assumption about the disturbance also affects predictive controller design, as optimal predictions take into account the assumed model for the disturbance. For example, it can be shown that a predictive controller in closed loop satisfies:

$$\alpha T(z^{-1})w(t) = R(z^{-1})\Delta u(t) + S(z^{-1})y(t), \tag{33.88}$$

where the polynomials R, S satisfy the Diophantine identity:

$$R(z^{-1})A(z^{-1})\Delta + B(z^{-1})S(z^{-1}) = P_c(z^{-1})T(z^{-1}), \tag{33.89}$$

where P_c gives the closed-loop poles (e.g., for mean-level control $P_c = A$). Then the closed loop is stable for *fixed* estimated polynomials $\hat{A}, \hat{B}$ provided that for all frequencies up to Nyquist:

$$\left|\frac{B}{A} - \frac{\hat{B}}{\hat{A}}\right| < \left|\frac{P_c}{\hat{A}}\right| \cdot \left|\frac{T}{S}\right|, \tag{33.90}$$

where S is deduced from Equation 33.89 using the estimated parameters. Hence, the right-hand side of Equation 33.90 is known from the estimated model, and in particular, T can be chosen to ensure the bound is satisfied, particularly at high frequencies where the undermodeling problem arises; see [3] for more details. It has been found by "benchmark" studies that good choice of T is highly significant. It is possible to consider *different* designs of T for the estimator and controller, but in practice it is convenient and near optimal to use the same polynomial. In general, the choice of $T = \hat{A}(1 - \gamma z^{-1})^m$, where γ is in the neighborhood of a dominant plant pole, is fairly effective.

References

1. Åström, K.J. and Wittenmark, B., *Adaptive Control,* Addison-Wesley, Reading, MA, 1989.
2. Bitmead, R.R., Gevers, M., and Wertz, V., *Adaptive Optimal Control,* Prentice Hall, Englewood Cliffs, NJ, 1990.
3. Clarke, D.W., Ed., *Advances in Model-Based Predictive Control,* Oxford University Press, UK, 1994.
4. Gawthrop, P.J., *Continuous-Time Self-Tuning Control,* Research Studies Press, Letchworth, UK, 1987.
5. Goodwin, G.C. and Sin, K.S., *Adaptive Filtering, Prediction and Control,* Prentice Hall, Englewood Cliffs, NJ, 1984.
6. Hang, C.C., Lee, T.H., and Ho, W.K., *Adaptive Control,* Instrument Society of America, Research Triangle Park, NC, 1993.
7. Harris, C.J. and Billings, S.A., Eds., *Self-Tuning and Adaptive Control: Theory and Applications,* Peter Perigrinus Ltd., Stevenage, UK, 1981.
8. Isermann, R., Lachmann, K.-H., and Drago, D., *Adaptive Control Systems,* Prentice Hall, Englewood Cliffs, NJ, 1992.
9. Mosca, E., *Optimal Predictive and Adaptive Control,* Prentice Hall, Englewood Cliffs, NJ, 1995.
10. Soeterboek, R., *Predictive Control: A Unified Approach,* Prentice Hall, Englewood Cliffs, NJ, 1992.
11. Wellstead, P.E. and Zarrop, M.B., *Self-Tuning Systems,* Wiley, New York, 1991.

34

Model Reference Adaptive Control

Petros Ioannou
University of Southern California

34.1 Introduction

Research in adaptive control has a long history of intense activity involving debates about the precise definition of adaptive control, examples of instabilities, stability and robustness proofs, and applications. Starting in the early 1950s, the design of autopilots for high-performance aircraft motivated an intense research activity in adaptive control. High-performance aircraft undergo drastic changes in their dynamics when they fly from one operating point to another. These changes cannot be handled by constant gain feedback control. A sophisticated controller, such as an adaptive controller, that would be able to learn and accommodate changes in the aircraft dynamics was needed. Model reference adaptive control (MRAC) was suggested by Whitaker et al. [9] to solve the autopilot control problem. The sensitivity method and the MIT rule [18] were used to design the adjustment or adaptive laws for estimating the unknown parameters for the various proposed MRAC schemes.

The work on adaptive flight control was characterized by "a lot of enthusiasm, bad hardware and nonexisting theory" [1]. The lack of stability proofs and the lack of understanding of the properties of the proposed adaptive control schemes, coupled with a disaster in a flight test [8] caused the interest in adaptive control in the late 1950s and early 1960s to diminish.

The 1960s became the most important period for the development of control theory and adaptive control in particular. State-space techniques and stability theory based on Lyapunov were introduced. Developments in dynamic programming [11], dual control [13], and stochastic control in general and in system identification and parameter estimation [10] played a crucial role in the reformulation and redesign of adaptive control. By 1966 Parks [6] and others found a way of redesigning the MIT rule-based adaptive laws used in the MRAC schemes of the 1950s by applying the Lyapunov design approach. Their work, even though applicable to a special class of linear, time-invariant (LTI) plants, set the stage for further rigorous stability proofs in MRAC for more general classes of plant models.

The advances in stability theory and the progress in control theory in the 1960s improved the understanding of adaptive control and contributed to a strong renewed interest in the field in the 1970s. On the other hand, the simultaneous development and progress in computers and electronics that made the implementation of complex controllers, such as the adaptive ones, feasible, contributed to an increased interest in applications of adaptive control. The 1970s witnessed several breakthrough results in the design of adaptive control. MRAC schemes using the Lyapunov design approach were developed and analyzed and the concepts of positivity and hyperstability were used to develop a wide class of MRAC schemes with well-established stability properties [4,5,12,18]. At the same time, parallel efforts for discrete-time plants in a deterministic and stochastic environment produced several classes of adaptive control schemes with rigorous stability proofs [14]. The excitement of the 1970s and the development of a wide class of adaptive control schemes with well-established stability properties was accompanied by a number of successful applications [15].

The successes of the 1970s, however, were soon followed by controversies over the practicality of adaptive control. As early as 1979 it was pointed out that the MRAC schemes of the 1970s could easily go unstable in the presence of small disturbances [12]. The nonrobust behavior of adaptive control became very controversial in the early 1980s when more examples of instabilities were published, demonstrating lack of robustness in the presence of unmodeled dynamics and or bounded disturbances [7,16]. These examples stimulated many researchers, whose objective was to understand the mechanisms of instabilities and find ways to counteract them. By the mid 1980s, a number of new redesigns and modifications were proposed and analyzed, leading to a body of work known as robust adaptive control. An adaptive controller is defined to be robust if it guarantees signal boundedness in the presence of "reasonable" classes of unmodeled dynamics and bounded disturbances as well as performance error bounds that are of the order of the modeling error. The work on robust adaptive control continued throughout the 1980s and involved the understanding of the various robustness modifications and their unification under a more general framework [2,12].

The solution of the robustness problem in adaptive control led to the solution of the long-standing problem of controlling a linear plant whose parameters are unknown and changing with time. By the end of the 1980s several breakthrough results were published in the area of adaptive control and in particular MRAC for linear time-varying plants [19] .

The focus of adaptive control research in the late 1980s and early 1990s was on performance properties and on extending the results of the 1980s to certain classes of nonlinear plants with unknown parameters. These efforts led to new classes of MRAC-type schemes motivated from nonlinear system theory [3], as well as to MRAC schemes with improved transient and steady-state performance [2].

Adaptive control has been traditionally divided into two classes, the MRAC-type schemes and adaptive pole placement control (APPC) schemes. In MRAC both the poles and zeros of the plant are changed so that the closed-loop plant has the same input–output properties as those of a given reference model. In APPC only the poles of the plant are changed. In this chapter we concentrate on MRAC for continuous-time plants that attracted considerable interest in the literature of adaptive control. For information on APPC and discrete-time adaptive control, the reader is referred to [17] and [14].

34.2 MRAC Schemes

Model reference adaptive control (MRAC) is derived from the model-following problem or model reference control (MRC) problem. In MRC, a good understanding of the plant and the performance requirements it has to meet allows the designer to come up with a model, referred to as the reference model, that describes the desired input–output properties of the closed-loop plant. The objective of MRC is to find the feedback control law that changes the structure and dynamics of the plant so that its input–output

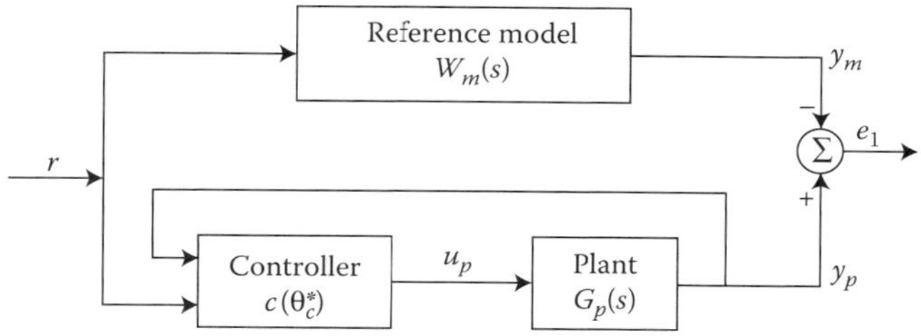

FIGURE 34.1 The diagram shows the basic structure of model reference control (MRC).

properties are exactly the same as those of the reference model. The structure of an MRC scheme for a LTI, single-input, single-output (SISO) plant is shown in Figure 34.1. The transfer function $W_m(s)$ of the reference model is designed so that, for a given reference input signal $r(t)$, the output $y_m(t)$ of the reference model represents the desired response the plant output $y_p(t)$ has to follow. The feedback controller, denoted by $C(\theta_c^*)$, is designed so that all signals are bounded and the closed-loop plant transfer function from r to y_p is equal to $W_m(s)$. This transfer function matching guarantees that for any given reference input $r(t)$, the tracking error $e_1(t)$, which represents the deviation of the plant output y_p from the desired trajectory y_m, converges to zero with time. The transfer function matching is achieved by cancelling the zeros of the plant transfer function $G_p(s)$ and replacing them with those of $W_m(s)$ through the use of the feedback controller $C(\theta_c^*)$. The cancellation of the plant zeros puts a restriction on the plant to be minimum phase, i.e., have stable zeros. If any plant zero is unstable, its cancellation may easily lead to unbounded signals.

The design of $C(\theta_c^*)$ requires knowledge of the coefficients of the plant transfer function $G_p(s)$. If θ^* is a vector containing all the coefficients of $G_p(s) = G_p(s, \theta^*)$, then the controller parameter vector θ_c^* may be computed by solving an algebraic equation of the form

$$\theta_c^* = F(\theta^*) \tag{34.1}$$

When θ^* is unknown the MRC scheme of Figure 34.1 cannot be implemented, since θ_c^* cannot be calculated using Equation 34.1 and is therefore unknown. One way of dealing with the unknown parameter case is to use the certainty equivalence approach [14,17]. In this context, the certainty equivalence approach is to replace the unknown θ_c^* in the control law with its estimate $\theta_c(t)$ obtained using the direct or the indirect approach. The resulting control schemes are known as MRAC and can be classified as indirect MRAC, shown in Figure 34.2, and direct MRAC, shown in Figure 34.3. In indirect MRAC the controller parameter vector θ_c is calculated at each time using the estimate of the plant parameter vector θ^* and the mapping defined by Equation 34.1. In direct MRAC the vector θ_c is adjusted directly without any intermediate calculations that involve estimates of θ^*. In this case the plant transfer function $G_p(s, \theta^*)$ is parameterized with respect to θ_c^* to obtain $G_p(s, \theta_c^*)$, whose form is used to estimate θ_c^* directly.

Different choices of on-line parameter estimators lead to further classifications of MRAC.

34.2.1 Model Reference Control

Consider the SISO, LTI plant described by the vector differential equation

$$\begin{aligned} \dot{x}_p &= A_p x_p + B_p u_p, \quad x_p(0) = x_0 \\ y_p &= C_p^T x_p \end{aligned} \tag{34.2}$$

where $x_p \in R^n; y_p, u_p \in R^1$ and A_p, B_p, C_p have the appropriate dimensions. The transfer function of the plant is given by

$$y_p = G_p(s) u_p = k_p \frac{Z_p(s)}{R_p(s)} u_p \tag{34.3}$$

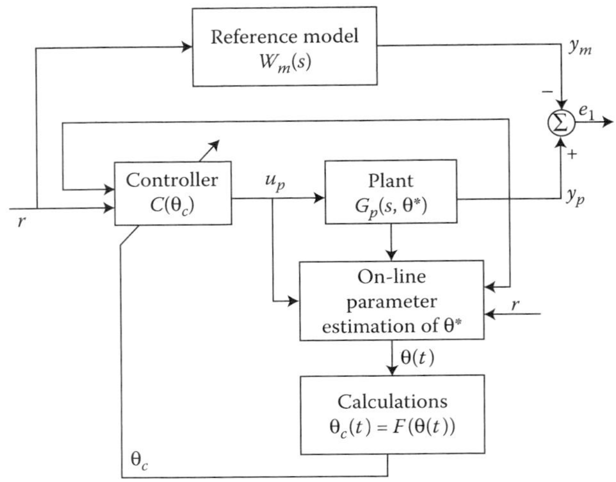

FIGURE 34.2 The diagram shows the basic structure of indirect MRAC.

where $Z_p(s)$, $R_p(s)$ are monic polynomials and k_p is a constant referred to as the high-frequency gain. The reference model, selected by the designer to describe the desired characteristics of the closed-loop system, is given by

$$y_m = W_m(s)r = k_m \frac{Z_m(s)}{R_m(s)} r \tag{34.4}$$

where $Z_m(s)$, $R_m(s)$ are monic polynomials of degree q_m, p_m, respectively, and k_m is a constant.

The MRC objective is to determine the plant input u_p so that all signals are bounded and the plant output y_p tracks the reference model output y_m as closely as possible for any given reference input $r(t)$ that is bounded and continuous. We refer to the problem of finding the desired u_p to meet the control objective as the MRC problem.

In order to meet the MRC objective with a control law that uses signals that are available for measurement, we assume that the plant and reference models satisfy the following assumptions:

Plant Assumptions:

P1. $Z_p(s)$ is a monic Hurwitz polynomial of degree m_p, i.e., $Z_p(s)$ is a monic polynomial of degree m_p that has all roots in the open left half s-plane

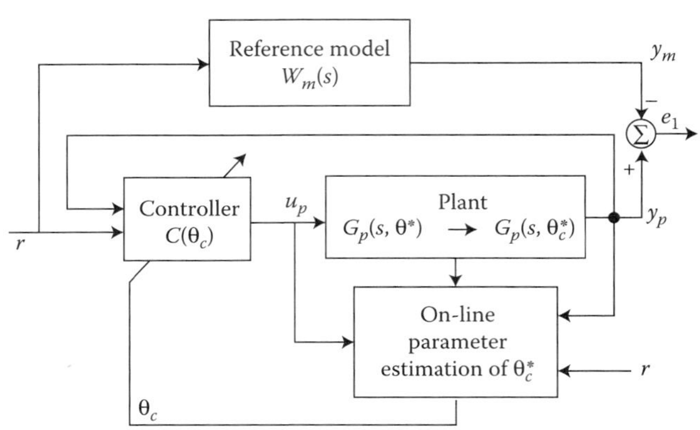

FIGURE 34.3 The diagram shows the basic structure of direct MRAC.

P2. An upper bound n of the degree n_p of $R_p(s)$
P3. the relative degree $n^* = n_p - m_p$ of $G_p(s)$ and
P4. the sign of the high-frequency gain k_p

are known.

Reference Model Assumptions:

M1. $Z_m(s), R_m(s)$ are monic Hurwitz polynomials of degree q_m, p_m, respectively, where $p_m \leq n$
M2. The relative degree $n_m^* = p_m - q_m$ of $W_m(s)$ is the same as that of $G_p(s)$, i.e., $n_m^* = n^*$

In addition to assumptions P1 to P4 and M1, M2, let us also assume that the plant parameters, i.e., the coefficients of $G_p(s)$, are known exactly. Since the plant is LTI and known, the design of the MRC scheme is achieved using linear system theory.

The MRC objective is met if u_p is chosen so that the closed-loop transfer function from r to y_p has stable poles and is equal to $W_m(s)$, the transfer function of the reference model. Such transfer function matching guarantees that for any reference input signal $r(t)$ the plant output y_p converges to y_m exponentially fast.

We consider the feedback control law

$$u_p = \theta_1^{*T}\omega_1 + \theta_2^{*T}\omega_2 + \theta_3^* y_p + c_0^* r = \theta_c^{*T}\omega \tag{34.5}$$

where $\theta_c^* = [\theta_1^{*T}, \theta_2^{*T}, \theta_3^*, c_0^*]^T$, $\omega = [\omega_1^T, \omega_2^T, y_p, r]^T$, $\alpha(s) = [s^{n-2}, s^{n-3}, \ldots, s, 1]^T$ $\omega_1 = \frac{\alpha(s)}{\Lambda(s)} u_p$, $\omega_2 = \frac{\alpha(s)}{\Lambda(s)} y_p$, $c_0^*, \theta_3^* \in R^1; \theta_1^*, \theta_2^* \in R^{n-1}$ are constant parameters to be designed and $\Lambda(s)$ is an arbitrary monic Hurwitz polynomial of degree $n-1$ that contains $Z_m(s)$ as a factor, i.e., $\Lambda(s) = \Lambda_0(s)Z_m(s)$, which implies that $\Lambda_0(s)$ is monic, Hurwitz, and of degree $n_0 = n - 1 - q_m$. The controller parameter vector $\theta_c^* \in R^{2n}$ is to be chosen so that the transfer function from r to y_p i.e., $y_p = G_c(s)r$ given by

$$G_c(s) = \frac{c_0^* k_p Z_p \Lambda^2}{\Lambda[(\Lambda - \theta_1^{*T}\alpha)R_p - k_p Z_p(\theta_2^{*T}\alpha + \theta_3^*\Lambda)]} \tag{34.6}$$

is stable and is equal to $W_m(s) = k_m \frac{Z_m}{R_m}$ for all s.

Since the degree of the denominator of $G_c(s)$ is $n_p + 2n - 2$ and that of $R_m(s)$ is $p_m \leq n$, for the matching equation

$$\frac{c_0^* k_p Z_p \Lambda^2}{\Lambda[(\Lambda - \theta_1^{*T}\alpha)R_p - k_p Z_p(\theta_2^{*T}\alpha + \theta_3^*\Lambda)]} = k_m \frac{Z_m}{R_m} \tag{34.7}$$

to hold, an additional $n_p + 2n - 2 - p_m$ zero-pole cancellations must occur in $G_c(s)$. Now since $Z_p(s)$ is Hurwitz by assumption, and $\Lambda(s) = \Lambda_0(s)Z_m(s)$ is designed to be Hurwitz, it follows that all the zeros of $G_c(s)$ are stable and therefore any zero-pole cancellation can only occur in C^-, the open left half of the complex plane. Choosing

$$c_0^* = \frac{k_m}{k_p} \tag{34.8}$$

and using $\Lambda(s) = \Lambda_0(s)Z_m(s)$ the matching Equation 34.7 becomes

$$(\Lambda - \theta_1^{*T}\alpha)R_p - k_p Z_p(\theta_2^{*T}\alpha + \theta_3^*\Lambda) = Z_p \Lambda_0 R_m \tag{34.9}$$

Dividing both sides of Equation 34.9 by $R_p(s)$, we obtain

$$\Lambda - \theta_1^{*T}\alpha - k_p \frac{Z_p}{R_p}(\theta_2^{*T}\alpha + \theta_3^*\Lambda) = Z_p\left(Q + k_p \frac{\Delta^*}{R_p}\right)$$

where $Q(s)$ (of degree $n - 1 - m_p$) is the quotient and $k_p\Delta^*$ (of degree at most $n_p - 1$) is the remainder of $\Lambda_0 R_m / R_p$, respectively. Then the solution for $\theta_i^*, i = 1, 2, 3$ can be found by inspection, i.e.,

$$\theta_1^{*T}\alpha(s) = \Lambda(s) - Z_p(s)Q(s)$$

$$\theta_2^{*T}\alpha(s)+\theta_3^*\Lambda(s)=\frac{Q(s)R_p(s)-\Lambda_0(s)R_m(s)}{k_p} \tag{34.10}$$

where the equality in the second equation is obtained by substituting for $\Delta^*(s)$ using the identity

$$\frac{\Lambda_0 R_m}{R_p}=Q+\frac{k_p\Delta^*}{R_p}$$

The parameters $\theta_i^*, i=1,2,3$ can now be obtained directly by equating the coefficients of the powers of s on both sides of Equation 34.10. Equation 34.10 indicates that, in general, the controller parameters $\theta_i^*, i=1,2,3$ are nonlinear functions of the coefficients of the plant polynomials $Z_p(s), R_p(s)$ due to the dependence of $Q(s)$ on the coefficients of $R_p(s)$. When $n=n_p$ and $n^*=1$, however, $Q(s)=1$ and the θ_i^*s are linear functions of the coefficients of $Z_p(s), R_p(s)$.

Lemma 34.1:

i. Let the degrees of $R_p, Z_p, \Lambda, \Lambda_0$ and R_m be as specified in Equation 34.5. Then the solution $\overline{\theta}_c^*=[\theta_1^{*T},\theta_2^{*T},\theta_3^{*T}]$ of Equation 34.9 or Equation 34.10 always exists.
ii. In addition, if R_p, Z_p are coprime and $n=n_p$, then the solution $\overline{\theta}_c^*$ is unique.

The proof is based on the solution of certain Diophantine equations and is given in [17].

It can be shown that the control law (Equation 34.5) with θ_c^* calculated from Equations 34.8 and 34.10 guarantees that the closed-loop plant is stable and the tracking error $e_1=y_p-y_m$ converges exponentially to zero for any given bounded reference input r.

34.2.2 Direct MRAC

A direct MRAC scheme is formed by combining the control law (Equation 34.5), with θ_c^* replaced by its estimate $\theta_c(t)$ at time t, i.e.,

$$u_p=\theta_c^T(t)\omega \tag{34.11}$$

with an adaptive law that generates $\theta_c(t)$ at each time t.

The estimate $\theta_c(t)$ of θ_c^* is generated by first obtaining an appropriate parameterization of the plant in terms of θ_c^* and then using parameter estimation techniques to form the adaptive law for $\theta_c(t)$. Such a parameterization is developed by using the plant and matching equations to obtain

$$e_1=W_m(s)\rho^*(u_p-\theta_c^{*T}\omega) \tag{34.12}$$

where $e_1=y_p-y_m$, and $\rho^*=1/c_0^*$. Using Equation 34.12 a wide class of adaptive laws may be developed to estimate θ_c^*, ρ^*. The adaptive laws may be split into two major classes; those with unnormalized signals and those with normalized signals leading to direct MRAC without normalization and direct MRAC with normalization.

34.2.2.1 Direct MRAC without Normalization

The derivation and complexity of the MRAC scheme depends on the relative degree n^* of the plant. For $n^*=1$ we choose the transfer function $W_m(s)$ of the reference model to be strictly positive real

(SPR) [17,18]. Substituting for $u_p = \theta_c^T(t)\omega$ in Equation 34.12 we obtain the error equation

$$e_1 = W_m(s)\rho^*\tilde{\theta}_c^T\omega \tag{34.13}$$

where $\tilde{\theta}_c = \theta_c - \theta_c^*$ is the parameter error. Since $W_m(s)$ is SPR we can use the SPR-Lyapunov design approach [17] to generate the adaptive law

$$\dot{\theta}_c = -\Gamma e_1 \omega sgn(\rho^*) \tag{34.14}$$

where

$$\Gamma = \Gamma^{\mathrm{T}} > 0$$

which, together with Equation 34.11 forms the MRAC scheme. The adaptive law (Equation 34.14) is chosen so that a certain positive definite function V of the error states of Equation 34.13 and Equation 34.14 is a Lyapunov function with the property

$$\frac{dV}{dt} \leq 0$$

which implies uniform stability and with additional arguments, $e_1(t) \to 0$ as $t \to \infty$.

When $n^* = 2$, $W_m(s)$ cannot be designed to be SPR because of assumption M2 and the fact that a transfer function of relative degree 2 cannot be SPR. In this case we rewrite Equation 34.12 as

$$e_1 = W_m(s)(s+p)\rho^*[u_f - \theta_c^{*T}\omega_f]$$

where $u_f = \frac{1}{s+p}u_p$, $\omega_f = \frac{1}{s+p}\omega$, and $p > 0$ is chosen so that $\overline{W}_m(s) = W_m(s)(s+p)$ is SPR. If we choose $u_f = \theta_c^T\omega_f$ then

$$e_1 = \overline{W}_m(s)\rho^*(\tilde{\theta}_c^T\omega_f)$$

which has the same form as Equation 34.13 and leads to the adaptive law

$$\dot{\theta}_c = -\Gamma e_1 \omega_f sgn(\rho^*) \tag{34.15}$$

The adaptive law (Equation 34.15) is generated by using $u_f = \theta_c^T\omega_f$. Since $u_f = \frac{1}{s+p}u_p$ the control input u_p has to be chosen so that $u_f = \theta_c^T\omega_f$. We have

$$u_p = (s+p)\theta_c^T\omega_f = \theta_c^T\omega + \dot{\theta}_c^T\omega_f$$

where the second equality is obtained by treating s as a differential operator. Since $\dot{\theta}_c$ is given by Equation 34.15, the MRAC scheme when $n^* = 2$ is given by

$$\begin{aligned} u_p &= \theta_c^T\omega + \dot{\theta}_c^T\omega_f \\ \dot{\theta}_c &= -\Gamma e_1\omega_f sgn(\rho^*) \end{aligned} \tag{34.16}$$

When $n^* = 3$ we can use the same procedure as in the case of $n^* = 2$ to obtain the error equation

$$e_1 = W_m(s)(s+p_1)(s+p_2)\rho^*[u_f - \theta_c^{*T}\phi]$$

where

$$u_f = \frac{1}{(s+p_1)(s+p_2)}u_p, \quad \phi = \frac{1}{(s+p_1)(s+p_2)}\omega$$

and $p_1, p_2 > 0$ are chosen so that $\overline{W}_m(s) = W_m(s)(s+p_1)(s+p_2)$ is SPR. In this case we cannot choose $u_f = \theta_c^T\phi$ since such a choice will have to include second derivatives of θ_c in the expression for u_p that are not available for measurement. We go around this difficulty by using the nonlinear tools "backstepping" and "nonlinear damping" as explained in [3,17] to obtain the MRAC scheme

$$\begin{aligned} u_p &= \theta_c^T \omega + \dot{\theta}_c^T \phi_1 - (s+p_2)\alpha_0(\phi^T \Gamma \phi)^2 r_0 \\ \dot{r}_0 &= -[p_1 + \alpha_0(\phi^T \Gamma \phi)^2] r_0 + \phi^T \Gamma \phi \epsilon_1 sgn(\rho^*) \\ \dot{\theta}_c &= -\Gamma \epsilon_1 \phi sgn(\rho^*), \quad \dot{\rho} = \gamma \epsilon_1 r_0 \\ \phi_1 &= \frac{1}{s+p_2}\omega, \quad \epsilon_1 = e_1 - \frac{1}{s+q_0}\rho r_0 \end{aligned} \tag{34.17}$$

where $\alpha_0 > 0$ is a design constant and $\overline{W}_m(s) = 1/(s+q_0)$. For $n^* > 3$ the procedure is the same, but it leads to much more complex MRAC schemes.

The above MRAC schemes guarantee that all signals are bounded and the tracking error converges to zero. If r is sufficiently rich then the MRAC scheme for $n^* = 1, 2$ guarantees exponential convergence of the parameter error $\tilde{\theta}_c$ and that the tracking error e_1 goes to zero. For $n^* \geq 3$ the convergence of $\tilde{\theta}_c$ to zero is asymptotic [17].

34.2.2.2 Direct MRAC with Normalized Adaptive Laws

This class of MRAC schemes dominated the literature of adaptive control due to the simplicity of their design as well as their robustness properties in the presence of modeling errors. The adaptive laws of these schemes are driven by a normalized error signal that "slows" down adaptation and improves robustness with respect to plant uncertainties. For this reason they are referred to as normalized adaptive laws. The MRAC law $u_p = \theta_c^T \omega$ in Equation 34.11 remains unchanged and the parametric model (Equation 34.12) is used to generate the adaptive law for θ_c. The parametric model (Equation 34.12) may be rewritten in various other forms, giving rise to a wide class of adaptive laws. For example, we can rewrite Equation 34.12 as

$$z = \theta_c^{*T} \phi_p \tag{34.18}$$

where $z = W_m(s)u_p$, $\phi_p = [W_m(s)\omega_1^T, W_m(s)\omega_2^T, W_m(s)y_p, y_p]^T$ or

$$e_1 = \rho^*(u_f - \theta_c^{*T}\phi) \tag{34.19}$$

where $u_f = W_m(s)u_p$, $\phi = W_m(s)\omega$. Equation 34.18 is obtained by first rewriting Equation 34.12 as $y_p - y_m = \rho^*[z - \theta_c^{*T}\phi_p - c_o^* y_m + c_o^* y_p]$ and then using the identity $\rho^* c_o^* = 1$.

Using Equation 34.12 and the SPR-Lyapunov approach [17], we have the adaptive law

$$\dot{\theta}_c = -\Gamma \epsilon \phi sgn(\rho^*) \quad \dot{\rho} = \gamma \epsilon \xi \tag{34.20}$$

where

$$\begin{aligned} \epsilon &= e_1 - \hat{e}_1 - W_m L(\epsilon n_s^2) \\ \hat{e}_1 &= W_m(s)L(s)[\rho(u_f - \theta_c^T \phi)] \\ \xi &= u_f - \theta_c^T \phi, \quad \phi = L^{-1}(s)\omega \\ u_f &= L^{-1}(s)u_p, \quad n_s^2 = \phi^T \phi + u_f^2 \end{aligned}$$

$L(s)$ is chosen so that $W_m L$ is SPR and proper, and $L^{-1}(s)$ is proper and stable.

Using Equation 34.19 and the gradient method [17] we have

$$\dot{\theta}_c = -\Gamma \epsilon \phi sgn(\rho^*) \quad \dot{\rho} = \gamma \epsilon \xi \tag{34.21}$$

where ε, ϕ, ξ are the same as in Equation 34.20 with $L^{-1}(s) = W_m(s)$.

Using Equation 34.18 we can generate a wide class of adaptive laws using the gradient method with different cost functions as well as least squares [17]. The gradient algorithm is given by

$$\dot{\theta}_c = Pr[\Gamma\epsilon\phi_p] \tag{34.22}$$

and the least squares is given by

$$\begin{aligned}\dot{\theta} &= Pr[P\epsilon\phi_p]\\ \dot{P} &= \overline{Pr}\left[-\frac{P\phi_p\phi_p^T P}{m^2}\right]\end{aligned} \tag{34.23}$$

where $\epsilon = \frac{z-\hat{z}}{m^2}$, $\hat{z} = \theta_c^T\phi_p$, $m^2 = 1 + \phi_p^T\phi_p$ and $Pr[.]$ is the projection operator that constrains $c_0(t)$, the estimate of c_0^*, to satisfy $|c_0(t)| \geq c_m, \forall t \geq 0$, where $c_m > 0$ is a lower bound for $|c_0^*|$. The $\overline{Pr}(.)$ operator sets $\dot{P} = 0$, when $|c_0(t)| = c_m$ and $\dot{c}_0 < 0$. The projection is used to guarantee that $1/(c_0(t))$ is bounded for all $t \geq 0$, a property that is used in the stability analysis of the MRAC scheme (Equation 34.11) with θ_c generated by Equation 34.22 or 34.23. For the implementation of projection we require the knowledge of c_m, a lower bound for $|c_0^*|$ and the sign of c_0^*.

The control law (Equation 34.11) with any one of the adaptive laws (Equation 34.20, 34.21, 34.22, or 34.23) forms a direct MRAC scheme. As shown in [17] these schemes guarantee signal boundedness and convergence of the tracking error to zero. If, in addition, the reference input $r(t)$ is sufficiently rich of order $2n$ then both the parameter and tracking errors converge to zero. The rate of convergence in the case of Equation 34.11 with Equation 34.22 or 34.23 is exponential, whereas for the case of the MRAC scheme (Equations 34.11 and 34.20, or 34.11 and 34.21), the convergence is asymptotic.

34.2.3 Indirect MRAC

In indirect MRAC the controller parameter vector $\theta_c(t)$ in the control law (Equation 34.11) is calculated at each time t using the estimates of k_p and of the coefficients of $Z_p(s)$, $R_p(s)$ that are generated using an adaptive law. The calculation of $\theta_c(t)$ is achieved by using the mapping defined by the matching Equations 34.8 and 34.10.

As in the direct MRAC case the adaptive laws for the estimated coefficients of $Z_p(s)$, $R_p(s)$ could be normalized or unnormalized. We concentrate on the normalized adaptive laws and refer the reader to [17] for results using unnormalized adaptive laws.

The adaptive law for estimating k_p and the coefficients of $Z_p(s)$, $R_p(s)$ is generated using the parametric plant model

$$z = \theta_p^{*T}\phi \tag{34.24}$$

where

$$z = \frac{s^n}{\Lambda_p(s)}y_p, \quad \phi = \left[\frac{\alpha_{n-1}^T(s)}{\Lambda_p(s)}u_p, \frac{-\alpha_{n-1}^T(s)}{\Lambda_p(s)}y_p\right]^T$$

$\theta_p^* = [0, \ldots, 0, b_m, p_1^T, p_2^T]^T \in R^{2n}$, $p_1 = [b_{m-1}, \ldots, b_0]^T$ and $p_2 = [a_{n-1}, \ldots, a_0]^T$ are the coefficient vectors of $k_p[Z_p(s) - s^m]$, $R_p(s) - s^n$, respectively, $\Lambda_p(s)$ is an nth order monic Hurwitz polynomial, $a_{n-1}(s) = [s^{n-1}, \ldots, s, 1]^T$ and $b_m = k_p$.

Using Equation 34.24 the estimate $\theta_p(t)$ of θ_p^* may be generated using adaptive laws that are based on the gradient or the least squares methods. The controller parameter vector $\theta_c(t) = [\theta_1^T(t), \theta_2^T(t), \theta_3(t), c_0(t)]^T$

is calculated from $\theta_p(t) = [0, \ldots 0, \hat{k}_p, \hat{p}_1^T, \hat{p}_2^T]^T$, the estimate of θ_p^* at each time t, as follows:

$$
\begin{aligned}
c_0(t) &= \frac{k_m}{\hat{k}_p} \\
\theta_1^T \alpha_{n-2}(s) &= \Lambda(s) - \hat{Z}_p(s,t) \bullet \hat{Q}(s,t) \\
\theta_2^T \alpha_{n-2}(s) + \theta_3 \Lambda(s) &= \frac{1}{\hat{k}_p}[\hat{Q}(s,t) \bullet \hat{R}_p(s,t) - \Lambda_0(s) R_m(s)]
\end{aligned}
\tag{34.25}
$$

where

$$
\begin{aligned}
\hat{Q}(s,t) &= \text{quotient of } \frac{\Lambda_0(s) R_m(s)}{\hat{R}_p(s,t)}, \\
\hat{Z}_p(s,t) &= \hat{k}_p s^m + \hat{p}_1^T \alpha_{m-1}(s) \\
\hat{R}_p(s,t) &= s^n + \hat{p}_2^T \alpha_{n-1}(s), \ \alpha_i(s) = [s^i, s^{i-1}, \ldots, s, 1]^T
\end{aligned}
$$

and $A \bullet B$ denotes pointwise in time multiplication. From Equation 34.25 it is clear that the adaptive law for $\hat{k}_p$ has to be modified using projection so that $|\hat{k}_p(t)| \geq k_m \geq 0$, where $k_m > 0$ is a lower bound for k_p. As an example of an adaptive law consider the gradient algorithm

$$
\begin{aligned}
\dot{\theta}_p &= Pr[\Gamma \epsilon \phi], \\
\epsilon &= \frac{z - \hat{z}}{m^2}, \quad \hat{z} = \theta_p^T \phi, \quad m^2 = 1 + \phi^T \phi
\end{aligned}
\tag{34.26}
$$

where $Pr[\bullet]$ is the projection operator that guarantees $|\hat{k}_p| \geq k_m > 0$ for all $t \geq 0$. The projection operator requires the knowledge of the sign of k_p and the lower bound k_m of k_p. The indirect MRAC scheme (Equations 34.11, 34.25, and 34.26) guarantees that $\theta_c(t)$ given by Equation 34.25 exists and is bounded for any bounded estimate θ_p, all signals in the closed-loop plant are bounded and the tracking error converges to zero with time. If, in addition, the reference signal r is sufficiently rich of order $2n$ then the parameter and tracking errors converge to zero exponentially fast [17].

34.2.4 Robust MRAC

The MRAC schemes presented above are designed for the plant model (Equation 34.2) that is free of disturbances and unmodeled dynamics. In the presence of disturbances and/or unmodeled dynamics the above schemes may be driven unstable, as shown by several examples in [16]. These schemes can be made robust by modifying the adaptive laws using leakage, dead-zone, projections and their by-products [17]. For the MRAC schemes without normalization these modifications guarantee the existence of a region of attraction in which all signals are bounded and the tracking error converges to a smaller residual set. For the MRAC schemes with normalization the region of attraction becomes the whole space provided a special normalizing signal is used to bound from above all the modeling error terms that are required to be small in the low-frequency range.

As an example, let us modify the direct MRAC scheme (Equations 34.11 and 34.20) for robustness using a leakage type of modification known as σ-modification. We have

$$
\begin{aligned}
u_p &= \theta_c^T \omega \\
\dot{\theta}_c &= -\Gamma \epsilon \phi sgn(\rho^*) - \sigma \Gamma \theta_c, \quad \dot{\rho} = -\gamma \epsilon \xi - \sigma_2 \gamma \rho \\
n_s^2 &= \phi^T \phi + u_f^2 + m_s \\
\dot{m}_s &= -\delta_0 m_s + u_p^2 + y_p^2, \quad m_s(0) = 0
\end{aligned}
$$

where $\sigma_1, \sigma_2 > 0$ are small positive constants, m_s is the dynamic normalizing signal, and $\delta_0 > 0$ is chosen so that the m_s bounds from above any modeling error term in the plant. The rest of the signals are as defined in Equation 34.20. If the above robust MRAC scheme is applied to the plant

$$y_p = G_p(s)(1 + \Delta_m(s))u_p$$

where $\Delta_m(s)$ is a multiplicative plant uncertainty with the property that $\Delta_m(s - \delta_{0/2})$ has stable poles, then for small $\Delta_\infty \triangleq \|W(s - \delta_{0/2})\Delta_m(s - \delta_{0/2})\|_\infty$, $\Delta_2 \triangleq \|W(s - \delta_{0/2})\Delta_m(s - \delta_{0/2})\|_2$ where $W(s - \delta_{0/2})$ is an arbitrary stable transfer function with stable $W^{-1}(s - \delta_{0/2})$ and $W(s)\Delta_m(s)$ is strictly proper, we have signal boundedness for any finite initial condition. Furthermore, the tracking error has a mean square value of the order of Δ_∞, Δ_2. The details of the design and analysis of robust MRAC schemes are given in [17].

34.3 Examples

In this section, we present several examples that illustrate the design and analysis of the MRAC schemes described in the previous sections.

34.3.1 Scalar Example: Adaptive Regulation

Consider the following scalar plant:

$$\dot{x} = ax + u, \quad x(0) = x_0 \tag{34.27}$$

where a is a constant but unknown. The control objective is to determine a bounded function $u = f(t, x)$ such that the state $x(t)$ is bounded and converges to zero as $t \to \infty$ for any given initial condition x_0. Let $-a_m$ be the desired closed-loop pole where $a_m > 0$ is chosen by the designer. In this case the reference model is

$$\dot{x}_m = -a_m x_m, \quad x_m(0) = x_0 \tag{34.28}$$

Control law: If the plant parameter a is known the control law

$$u = -k^* x \tag{34.29}$$

with $k^* = a + a_m$ could be used to meet the control objective, i.e., with Equation 34.29, the closed-loop plant is

$$\dot{x} = -a_m x, \quad x(0) = x_0$$

whose equilibrium $x_e = 0$ is exponentially stable in the large.

Since a is unknown, k^* cannot be calculated and therefore Equation 34.29 cannot be implemented. A possible procedure to follow in the unknown parameter case is to use the same control law as given in Equation 34.29, but with k^* replaced by its estimate $k(t)$, i.e., we use

$$u = -k(t)x \tag{34.30}$$

and search for an adaptive law to update $k(t)$ continuously with time.

Adaptive law: The adaptive law for generating $k(t)$ is developed by viewing the problem as an on-line identification problem for k^*. This is accomplished by first obtaining an appropriate parameterization for the plant (Equation 34.27) in terms of the unknown k^*, as follows.

We add and subtract the desired control input $-k^*x$ in the plant equation 34.27 to obtain

$$\dot{x} = ax - k^*x + k^*x + u$$

Since $a - k^* = -a_m$ we have

$$\dot{x} = -a_m x + k^*x + u$$

or

$$x = \frac{1}{s + a_m}(u + k^*x) \tag{34.31}$$

Equation 34.31 is a parameterization of the plant equation 34.27 in terms of the unknown controller parameter k^*. Since x, u are measured and $a_m > 0$ is known, many adaptive laws may be generated using Equation 34.31 as shown in [17].

Substituting for the control $u = -k(t)x$ in Equation 34.31, we obtain the error equation that relates the parameter error $\tilde{k} = k - k^*$ with the estimation error $\epsilon_1 = x$, i.e.,

$$\dot{\epsilon}_1 = -a_m\epsilon_1 - \tilde{k}x, \quad \epsilon_1 = x \tag{34.32}$$

Due to Equation 34.31 the estimation error ϵ_1, which is defined as the error that reflects the parameter error $\tilde{k}$, is equal to the regulation error x. The error equation 34.32 is in a convenient form for choosing an appropriate Lyapunov function to design the adaptive law for $k(t)$. We assume that the adaptive law is of the form

$$\dot{\tilde{k}} = \dot{k} = f_1(\epsilon_1, x, u) \tag{34.33}$$

where f_1 is some function to be selected, and propose

$$V(\epsilon_1, \tilde{k}) = \frac{\epsilon_1^2}{2} + \frac{\tilde{k}^2}{2\gamma}$$

for some $\gamma > 0$ as a potential Lyapunov function for the system defined by Equations 34.32 and 34.33. The time derivative of V along the trajectory of this system is given by

$$\dot{V} = -a_m\epsilon_1^2 - \tilde{k}\epsilon_1 x + \frac{\tilde{k}f_1}{\gamma}$$

Choosing $f_1 = \gamma\epsilon_1 x$, i.e.,

$$\dot{k} = \gamma\epsilon_1 x = \gamma x^2, \qquad k(0) = k_0 \tag{34.34}$$

we have

$$\dot{V} = -a_m\epsilon_1^2 \leq 0$$

Analysis: Since V is a positive definite function and $\dot{V} \leq 0$, we have $V \in \mathcal{L}_\infty$, which implies that $\epsilon_1, \tilde{k} \in \mathcal{L}_\infty$. Since $\epsilon_1 = x$, we also have that $x \in \mathcal{L}_\infty$ and therefore all signals in the closed-loop plant are bounded. Furthermore, $\epsilon_1 = x \in \mathcal{L}_2$, and $\dot{\epsilon}_1 = \dot{x} \in \mathcal{L}_\infty$ which imply that $\epsilon_1(t) = x(t) \to 0$ as $t \to \infty$. From $x(t) \to 0$ and the boundedness of k, we establish that $\dot{k}(t) \to 0, u(t) \to 0$ as $t \to \infty$.

We have shown that the combination of the control law (Equation 34.30) with the adaptive law (Equation 34.34) meets the control objective, in the sense that it forces the plant state to converge to zero while guaranteeing signal boundedness.

It is worth mentioning that we cannot establish that $k(t)$ converges to k^*, i.e., that the pole of the closed-loop plant converges to that of the reference model given by $-a_m$. The lack of parameter convergence is less crucial in adaptive control than in parameter identification, since in most cases the control objective can be achieved without requiring the parameters to converge to their true values. The simplicity of this scalar example, however, allows us to solve for $\varepsilon_1 = x$ explicitly, and study the properties of $k(t), x(t)$ as they evolve with time. We can verify that

$$\epsilon_1(t) = \frac{2ce^{-ct}}{c + k_0 - a + (c - k_0 + a)e^{-2ct}}\epsilon_1(0), \quad \epsilon_1 = x$$
$$k(t) = \alpha + \frac{c[(c + k_0 - \alpha)e^{2ct} - (c - k_0 + \alpha)]}{(c + k_0 - \alpha)e^{2ct} + (c - k_0 + \alpha)} \tag{34.35}$$

where $c^2 = \gamma x_0^2 + (k_0 - a)^2$, satisfy the differential equations 34.32 and 34.34 of the closed-loop plant. Equation 34.35 can be used to investigate the effects of initial conditions and adaptive gain γ on the transient and asymptotic behavior of $x(t)$, $k(t)$. We have $\lim_{t\to\infty} k(t) = a + c$, if $c > 0$, and $\lim_{t\to\infty} k(t) = a - c$ if $c < 0$, i.e.,

$$\lim_{t\to\infty} k(t) = k_\infty = a + [\gamma x_0^2 + (k_0 - a)^2]^{1/2}$$

Therefore for $x_0 \neq 0$, $k(t)$ converges to a stabilizing gain whose value depends on γ and the initial condition x_0, k_0. It is clear from Equation 34.35 that the value of k_∞ is independent of whether k_0 is a destabilizing gain, i.e., $0 < k_0 < a$, or a stabilizing one, i.e., $k_0 > a$, as long as $(k_0 - a)^2$ is the same. The use of different k_0, however, will affect the transient behavior as it is obvious from Equation 34.35. In the limit as $t \to \infty$, the closed-loop pole converges to $-(k_\infty - a)$ which may be different from $-a_m$. Since the control objective is to achieve signal boundedness and regulation of the state $x(t)$ to zero, the convergence of $k(t)$ to k^* is not crucial.

34.3.2 Scalar Example: Adaptive Tracking

Consider the following first-order plant:

$$\dot{x} = ax + bu \tag{34.36}$$

where a, b are unknown parameters but the sign of b is known. The control objective is to choose an appropriate control law u such that all signals in the closed-loop plant are bounded and x tracks the state x_m of the reference model given by

$$x_m = \frac{b_m}{s + a_m} r$$

for any bounded piecewise continuous signal $r(t)$, where $a_m > 0$, b_m are known and $x_m(t), r(t)$ are measured at each time t. It is assumed that a_m, b_m and r are chosen so that x_m represents the desired state response of the plant.

Control law: In order for x to track x_m for any reference input signal r, the control law should be chosen so that the closed-loop plant transfer function from the input r to the output x is equal to that of the reference model. We propose the control law

$$u = -k^* x + l^* r \tag{34.37}$$

where k^*, l^* are calculated so that

$$\frac{x(s)}{r(s)} = \frac{bl^*}{s - a + bk^*} = \frac{b_m}{s + a_m} = \frac{x_m(s)}{r(s)} \tag{34.38}$$

Equation 34.38 is satisfied if we choose

$$l^* = \frac{b_m}{b} \quad k^* = \frac{a_m + a}{b} \tag{34.39}$$

provided of course that $b \neq 0$, i.e., the plant is controllable. The control law (Equations 34.37 and 34.39) guarantees that the transfer function of the closed-loop plant, i.e., $x(s)/r(s)$ is equal to that of the reference model. Such a transfer function matching guarantees that $x(t) = x_m(t), \forall t \geq 0$ when $x(0) = x_m(0)$ or $|x(t) - x_m(t)| \to 0$ exponentially fast when $x(0) \neq x_m(0)$, for any bounded reference signal $r(t)$.

When the plant parameters a, b are unknown, Equation 34.37 cannot be implemented. Therefore, instead of Equation 34.37, we propose the control law

$$u = -k(t)x + l(t)r \tag{34.40}$$

where $k(t), l(t)$ is the estimate of k^*, l^*, respectively, at time t, and search for an adaptive law to generate $k(t), l(t)$ on-line.

Adaptive law: As before, we can view the problem as an on-line identification problem of the unknown constants k^*, l^*. We start with the plant equation, which we express in terms of k^*, l^* by adding and subtracting the desired input term $-bk^*x + bl^*r$ to obtain

$$x = \frac{b_m}{s + a_m} r + \frac{b}{s + a_m}(k^*x - l^*r + u) \tag{34.41}$$

Since $x_m = \frac{b_m}{s+a_m} r$ is a known bounded signal, we express Equation 34.41 in terms of the tracking error defined as $\epsilon_1 = x - x_m$, i.e.,

$$\epsilon_1 = \frac{b}{s + a_m}(k^*x - l^*r + u) \tag{34.42}$$

Substituting for $u = -k(t)x + l(t)r$ in Equation 34.42 and defining the parameter errors $\tilde{k} \triangleq k - k^*, \tilde{l} \triangleq l - l^*$, we have

$$\begin{aligned} \dot{\epsilon}_1 &= -a_m \epsilon_1 + b(-\tilde{k}x + \tilde{l}r) \\ \epsilon_1 &= x - x_m \end{aligned} \tag{34.43}$$

The development of the differential equation 34.43 relating the estimation error with the parameter error is a significant step in deriving the adaptive laws for updating $k(t), l(t)$. We assume that the structure of the adaptive law is given by

$$\dot{k} = f_1(\epsilon_1, x, r, u) \quad \dot{l} = f_2(\epsilon_1, x, r, u) \tag{34.44}$$

where the functions f_1, f_2 are to be designed.

Consider the function

$$V(\epsilon_1, \tilde{k}, \tilde{l}) = \frac{\epsilon_1^2}{2} + \frac{\tilde{k}^2}{2\gamma_1}|b| + \frac{\tilde{l}^2}{2\gamma_2}|b|$$

where $\gamma_1, \gamma_2 > 0$, as a Lyapunov candidate for the system (Equations 34.43 and 34.44). The time derivative $\dot{V}$ along any trajectory of the system is given by

$$\dot{V} = -a_m \epsilon_1^2 - b\tilde{k}\epsilon_1 x + b\tilde{l}\epsilon_1 r + \frac{|b|\tilde{k}}{\gamma_1} f_1 + \frac{|b|\tilde{l}}{\gamma_2} f_2 \tag{34.45}$$

Since $|b| = b sgn(b)$, the indefinite terms in Equation 34.43 disappear if we choose $f_1 = \gamma_1 \epsilon_1 x sgn(b)$, $f_2 = -\gamma_2 \epsilon_1 r sgn(b)$. Therefore, for the adaptive law

$$\dot{k} = \gamma_1 \epsilon_1 x sgn(b), \quad \dot{l} = -\gamma_2 \epsilon_1 r sgn(b) \tag{34.46}$$

we have

$$\dot{V} = -a_m \epsilon_1^2$$

Analysis: Treating $x_m(t), r(t)$ in Equation 34.43 as bounded arbitrary functions of time, it follows that V is a Lyapunov function for the third-order differential equations 34.43 and 34.46 and the equilibrium $\epsilon_{1e} = 0, \quad \tilde{k}_e = 0, \quad \tilde{l}_e = 0$ is uniformly stable. Furthermore, $\epsilon_1, \tilde{k}, \tilde{l} \in \mathcal{L}_\infty$ and $\epsilon_1 \in \mathcal{L}_2$. Since $\epsilon_1 = x - x_m$,

$x_m \in \mathcal{L}_\infty$, we also have $x \in \mathcal{L}_\infty$ and $u \in \mathcal{L}_\infty$ and therefore all signals in the closed-loop plant are bounded. Now from Equation 34.43 we have $\dot{\epsilon}_1 \in \mathcal{L}_\infty$, which together with $\epsilon_1 \in \mathcal{L}_2$, implies that $\epsilon_1(t) \to 0$, as $t \to \infty$. We have established that the control law (Equation 34.40), together with the adaptive law (Equation 34.46) guarantees boundedness for all signals in the closed-loop system. In addition, the plant state $x(t)$ tracks the state of the reference model x_m asymptotically with time for any reference input signal r which is bounded and piecewise continuous. These results do not imply that $k(t) \to k^*, l(t) \to l^*$ as $t \to \infty$, i.e., that the transfer function of the closed-loop plant approaches that of the reference model as $t \to \infty$. In order to achieve such a result, the reference input r has to be sufficiently rich of order 2. A sufficiently rich input is one that excites all the modes of the system [14,17]. For example $r(t) = \sin\omega t$ for some $\omega \neq 0$ is sufficiently rich of order 2 and guarantees the exponential convergence of $x(t)$ to $x_m(t)$ and of $k(t), l(t)$ to k^*, l^*, respectively. In general, a sufficiently rich reference input $r(t)$ is not desirable in cases where the control objective involves tracking of signals that are not rich in frequencies.

34.3.3 Example: Direct MRAC without Normalization ($n^* = 1$)

Let us consider the second-order plant

$$y_p = \frac{k_p(s+b_0)}{s^2 + a_1 s + a_0} u_p$$

where $k_p > 0, b_0 > 0, k_p, b_0, a_1, a_0$ are unknown constants. The desired performance of the plant is specified by the reference model

$$y_m = \frac{1}{s+1} r$$

Using the results of Section 34.2.2 the control law is designed as

$$\begin{aligned} \dot{\omega}_1 &= -2\omega_1 + u_p, \quad \omega_1(0) = 0 \\ \dot{\omega}_2 &= -2\omega_2 + y_p, \quad \omega_2(0) = 0 \\ u_p &= \theta_1\omega_1 + \theta_2\omega_2 + \theta_3 y_p + c_0 r \end{aligned}$$

by choosing $\Lambda(s) = s + 2$ in the general control law. The adaptive law is given by

$$\dot{\theta}_c = -\Gamma e_1 \omega, \quad \theta_c(0) = \theta_0$$

where $e_1 = y_p - y_m, \theta_c = [\theta_1, \theta_2, \theta_3, c_0]^T$, $\omega = [\omega_1, \omega_2, y_p, r]^T$ and $\Gamma = \Gamma^T$ is any positive definite matrix.

Analysis: From Equation 34.12 we have that the tracking error e_1 satisfies

$$e_1 = \frac{1}{s+1} \rho^* \tilde{\theta}_c^T \omega$$

where $\rho^* = k_p, \tilde{\theta}_c = \theta_c - \theta_c^*$, i.e., $\dot{e}_1 = -e_1 + k_p \tilde{\theta}_c^T \omega$.

We choose the positive definite function

$$V = \frac{e_1^2}{2} + k_p \frac{\tilde{\theta}_c^T \Gamma^{-1} \tilde{\theta}_c}{2}$$

then

$$\dot{V} = -e_1^2 + k_p \tilde{\theta}_c^T e_1 \omega - k_p \tilde{\theta}_c^T e_1 \omega = -e_1^2 \le 0$$

Therefore, e_1, θ_c are bounded, i.e., $e_1, \theta_c \in \mathcal{L}_\infty$ and e_1 is square integrable, i.e., $e_1 \in \mathcal{L}_2$. Since $y_m, e_1 \in \mathcal{L}_\infty$, we have $y_p \in \mathcal{L}_\infty$ and therefore $\omega_2 \in \mathcal{L}_\infty$. Now

$$\omega_1 = \frac{1}{s+2} u_p = \frac{(s^2 + a_1 s + a_0)}{(s+2)k_p(s+b_0)} y_p$$

Since $b_0 > 0$, i.e., the plant is minimum phase and $y_p \in \mathcal{L}_\infty$,we have $\omega_1 \in \mathcal{L}_\infty$. Hence, $\omega \in \mathcal{L}_\infty$, which implies that $u_p \in \mathcal{L}_\infty$. Since $e_1, \tilde{\theta}_c^T \omega \in \mathcal{L}_\infty$, we have $\dot{e}_1 \in \mathcal{L}_\infty$, which together with $e_1 \in \mathcal{L}_2$, implies that $e_1(t) \to 0$ as $t \to \infty$.

For parameter convergence, we choose r to be sufficiently rich of order 4. As an example, we select $r(t) = A_1 \sin \omega_1 t + A_2 \sin \omega_2 t$ for some nonzero constants $A_1, A_2, \omega_1, \omega_2$ with $\omega_1 \neq \omega_2$.

34.3.4 Example: Direct MRAC without Normalization ($n^\star = 2$)

Let us consider the second-order plant

$$y_p = \frac{k_p}{s^2 + a_1 s + a_0} u_p$$

where $k_p > 0$, a_1, a_0 are constants. The reference model is chosen as

$$y_m = \frac{5}{(s+5)^2} r$$

Using the results of Section 34.2.2 the control law is chosen as

$$\begin{aligned}
\dot{\omega}_1 &= -2\omega_1 + u_p, \quad \dot{\omega}_2 = -2\omega_2 + y_p \\
\dot{\phi} &= -\phi + \omega \\
u_p &= \theta_c^T \omega - \phi^T \Gamma \phi e_1
\end{aligned}$$

where $\omega = [\omega_1, \omega_2, y_p, r]^T$, $e_1 = y_p - y_m$, $p = 1$, $\Lambda(s) = s + 2$ and $5(s+1)/(s+5)^2$ is SPR. The adaptive law is given by

$$\dot{\theta}_c = -\Gamma e_1 \phi$$

where $\Gamma = \Gamma^T > 0$ is arbitrary and $\theta_c = [\theta_1, \theta_2, \theta_3, c_0]^T$.

Analysis: From Equation 34.12 by substituting for u_p we have that

$$e_1 = W_m(s) k_p (\tilde{\theta}_c^T \omega + \dot{\theta}_c^T \omega)$$

or

$$e_1 = W_m(s)(s+1) k_p \tilde{\theta}_c^T \phi \tag{34.47}$$

Since $W_m(s)(s+1)$ is SPR and $k_p > 0$ we can establish using the Lyapunov-like function

$$V = \frac{e^T P e}{2} + \frac{\tilde{\theta}_c^T \Gamma^{-1} \tilde{\theta}_c}{2} k_p$$

where e is the state of a state-space representation of Equation 34.47 and $P = P^T > 0$ satisfies the Lefschetz-Kalman-Yakubovich lemma [17] that

$$\dot{V} \leq -c e_1^2$$

for some constant $c > 0$. This implies that $e_1, \tilde{\theta}_c \in \mathcal{L}_\infty$ and $e_1 \in \mathcal{L}_2$. Proceeding as in the case of $n^* = 1$ we can establish that all signals are bounded and $e_1(t) \to 0$ as $t \to \infty$. For parameter convergence, the input r is chosen as $r(t) = A_1 \sin \omega_1 t + A_2 \sin \omega_2 t$ for some $A_1, A_2 \neq 0$, $\omega_1 \neq \omega_2$.

34.3.5 Example: Direct MRAC with Normalization

In contrast to the direct MRAC schemes without normalization the complexity of the design and analysis of direct MRAC schemes with normalization does not change with the relative degree of the plant. We demonstrate the design and analysis of MRAC with normalization using the first-order plant

$$\dot{x} = ax + bu$$

where a, b are unknown and $b > 0$. The closed-loop plant is required to be stable and the state x is required to track the state x_m of the reference model

$$x_m = \frac{b_m}{s + a_m} r$$

for any given bounded reference input signal r. If a, b were known the control law

$$u = -k^* x + l^* r \tag{34.48}$$

with $k^* = (a_m + a)/b, l^* = b_m/b$ could be used to meet the control objective exactly. Since a, b are unknown, we replace Equation 34.48 with

$$u = -k(t)x + l(t)r \tag{34.49}$$

where $k(t), l(t)$ are the on-line estimates of k^*, l^*, respectively. We design the adaptive laws for updating $k(t), l(t)$ by first developing appropriate parametric models for k^*, l^*. We can show that the tracking error e_1 satisfies

$$e_1 = \frac{b}{s + a_m}[u - (-k^* x + l^* r)]$$

which may be written in the form of Equation 34.19 in Section 34.2.2, i.e.,

$$e_1 = b(u_f - \theta_c^{*T} \phi)$$

where $\theta_c^* = [k^*, l^*]^T, \phi = W_m(s)[-x, r]^T, u_f = \frac{1}{s+a_m} u$. Using the gradient method and the fact that $b > 0$ we have

$$\dot{\theta}_c = -\Gamma \epsilon \phi, \quad \dot{\hat{b}} = \gamma \epsilon \xi \tag{34.50}$$

where $\theta_c, \hat{b}$ are the estimates of θ_c^*, b, respectively,

$$\epsilon = \frac{e_1 - \hat{e}_1}{m^2}, \quad \hat{e}_1 = \hat{b}[u_f - \theta_c^T \phi]$$

$$\xi = u_f - \theta_c^T \phi, \quad m^2 = 1 + \phi^T \phi + u_f{}^2$$

The stability analysis of the MRAC examples is accomplished as follows. First we show that Equation 34.50 guarantees that $\theta_c, \hat{b}, \epsilon, \epsilon m \in \mathcal{L}_\infty$ and $\epsilon, \epsilon m, \dot{\theta}_c, \dot{\hat{b}} \in \mathcal{L}_2$ independent of the choice of u and the boundedness of ϕ, u, e_1. These properties are then used to establish the boundedness of all signals in the control loop and the convergence of the tracking error e_1 to zero. The details of the analysis are given in [17].

34.3.6 Example: Indirect MRAC

Consider the following third-order plant:

$$y_p = \frac{1}{s^2(s+a)} u_p \tag{34.51}$$

where a is the only unknown parameter. The output y_p is required to track the output of y_m of the reference model

$$y_m = \frac{1}{(s+2)^3} r$$

The control law is given by

$$u_p = \theta_{11} \frac{s}{(s+\lambda_1)^2} u_p + \theta_{12} \frac{1}{(s+\lambda_1)^2} u_p + \theta_{21} \frac{s}{(s+\lambda_1)^2} y_p + \theta_{22} \frac{1}{(s+\lambda_1)^2} y_p + \theta_3 \dot{y}_p + c_0 r$$

where $\theta_c = [\theta_{11}, \theta_{12}, \theta_{21}, \theta_{22}, c_0]^T \in R^6$. In direct MRAC, θ_c is generated by a sixth-order adaptive law. In indirect MRAC, θ_c is calculated from the adaptive law as follows. Using the results of Section 34.2.2 the estimate $\hat{a}$ of the only unknown plant parameter a is given by

$$\dot{\hat{a}} = \gamma_a \phi_a \epsilon$$

$$\epsilon = \frac{z - \hat{z}}{1 + \phi^T \phi}, \quad \hat{z} = \theta_p^T \phi, \quad z = y_p + \lambda_p^T \phi_2$$

$$\phi = [\phi_1^T, \phi_2^T]^T, \quad \phi_1 = \frac{[s^2, s, 1]^T}{(s+\lambda_1)^3} u_p,$$

$$\phi_2 = -\frac{[s^2, s, 1]^T}{(s+\lambda_1)^3} y_p$$

where $\theta_p = [0, 0, 1, \hat{a}, 0, 0]^T$, $\Lambda(s)$ is chosen as $\Lambda(s) = (s+\lambda_1)^3, \lambda_p = [3\lambda_1, 3\lambda_1^2, 3\lambda_1^3]^T$, $\phi_a = [0, 0, 0, 1, 0, 0]\phi = -\frac{s^2}{(s+\lambda_1)^3} y_p$ and $\gamma_a > 0$ is a constant. The controller parameter vector is calculated as $c_0 = 1$,

$$\theta_1^T [s, 1]^T = (s+\lambda_1)^2 - \hat{Q}(s, t)$$

$$\theta_2^T [s, 1]^T + \theta_3 (s+\lambda_1)^2 = \hat{Q}(s, t) \bullet [s^3 + \hat{a} s^2] - (s+\lambda_1)^2 (s+2)^3$$

where $\hat{Q}(s, t)$ is the quotient of $(s+\lambda_1)^2 (s+2)^3 / (s^3 + \hat{a} s^2)$.

The example demonstrates that for the plant equation 34.50, the indirect scheme requires a first-order adaptive law, whereas the direct scheme requires a sixth-order one.

References

1. Åström, K.J., Theory and applications of adaptive control—a survey, *Automatica,* 19(5), 471–486, 1983.
2. Ioannou, P.A. and Datta, A., Robust adaptive control: a unified approach, *Proc. IEEE,* 79(12), 1735–1768, 1991.
3. Kanellakopoulos, I., Kokotovic, P.V., and Morse, A.S., Systematic design of adaptive controllers for feedback linearizable systems, *IEEE Trans. Autom. Control,* 36, 1241–1253, 1991.
4. Morse, A.S., Global stability of parameter adaptive control systems, *IEEE Trans. Autom. Control,* 25, 433–439, 1980.
5. Narendra, K.S., Lin, Y.H., and Valavani, L.S., Stable adaptive controller design, II. Proof of stability, *IEEE Trans. Autom. Control,* 25(3), 440–448, 1980.
6. Parks, P.C., Lyapunov redesign of model reference adaptive control systems, *IEEE Trans. Autom. Control,* 11, 362–367, 1966.

7. Rohrs, C.E., Valavani, L., Athans, M., and Stein, G., Robustness of continuous-time adaptive control algorithms in the presence of unmodeled dynamics, *IEEE Trans. Autom. Control,* 30(9), 881–889, 1985.
8. Taylor, L.W. and Adkins, E.J., Adaptive control and the X-15, *Proc. Princeton University Conference on Aircraft Flying Qualities,* Princeton University, 1965.
9. Whitaker, H.P., Yamron, J., and Kezer, A., Design of Model Reference Adaptive Control Systems for Aircraft, Report R-164, Instrumentation Laboratory, Massachusetts Institute of Technology, Cambridge, 1958.
10. Åström, K.J. and Eykhoff, P., System identification - a survey, *Automatica,* 20(1), 123, 1971.
11. Bellman, R.E., *Dynamic Control Processes—A Guided Tour,* Princeton University Press, Princeton, NJ, 1961.
12. Egardt, B., *Stability of Adaptive Controllers,* Lecture Notes in Control and Information Sciences, Vol. 20, Springer-Verlag, Berlin, 1979.
13. Fel'dbaum, A.A., *Optimal Control of Systems,* Academic Press, New York, 1965.
14. Goodwin, G.C, and Sin, K.C., *Adaptive Filtering Prediction and Control,* Prentice Hall, Englewood Cliffs, NJ, 1984.
15. Harris, C.J. and Billings, S.A., Eds., *Self-Tuning and Adaptive Control: Theory and Applications,* Peter Peregrinus, London, 1981.
16. Ioannou, P.A. and Kokotovic, P.V., *Adaptive Systems with Reduced Models,* Lecture Notes in Control and Information Sciences, Vol. 47, Springer-Verlag, New York, 1983.
17. Ioannou, P.A. and Sun, J., *Robust Adaptive Control,* Prentice Hall, Englewood Cliffs, NJ, 1996.
18. Landau, I.D., *Adaptive Control: The Model Reference Approach,* Marcel Dekker, New York, 1979.
19. Tsakalis, K.S. and Ioannou, P.A., *Linear Time Varying Systems: Control and Adaptation*, Prentice Hall, Englewood Cliffs, NJ, 1993.

Further Reading

Further details on MRAC for continuous-time plants can be found in the following textbooks:

1. *Robust Adaptive Control,* by P. Ioannou and J. Sun, Prentice Hall, Englewood Cliffs, NJ, 1996.
2. *Stable Adaptive Systems,* by K.S. Narendra and A. M. Annaswamy, Prentice Hall, Englewood Cliffs, NJ, 1989.
3. *Adaptive Control: Stability, Convergence and Robustness,* by S. Sastry and M. Bodson, Prentice Hall, Englewood Cliffs, NJ, 1989.
4. *Adaptive Control: The Model Reference Approach,* by I. D. Landau, Marcel Dekker, New York, 1979.
5. *Adaptive Control,* by K. J. Åström and B. Wittenmark, Addison-Wesley, Reading, MA, 1989.

More information on the design and analysis of MRAC for linear, time-varying plants can be found in the following monograph:

6. *Linear Time-Varying Systems: Control and Adaptation,* by K. Tsakalis and P. Ioannou, Prentice Hall, Englewood Cliffs, NJ, 1993.

More information on the design and analysis of robust MRAC can be found in the books 1, 2, 3, and 6 given above.

Details on MRAC for discrete-time plants can be found in the book 4 by Landau given above and in:

7. *Adaptive Filtering Prediction and Control* by G. Goodwin and K. Sin, Prentice Hall, Englewood Cliffs, NJ, 1984.

For applications of MRAC the reader is referred to the following books.

8. *Self-Tuning and Adaptive Control: Theory and Applications,* Edited by C. J. Harris and S. A. Billings, Peter Peregrinus, London, 1981.
9. *Adaptive and Learning Systems: Theory and Applications,* Edited by K. S. Narendra, Plenum Press, New York, 1986.

35 Robust Adaptive Control

Petros Ioannou
University of Southern California

Simone Baldi
University of Florence

35.1 Introduction

The design of autopilots for high-performance aircraft was one of the primary motivations for active research in adaptive control in the early 1950s. Aircrafts operate over a wide range of speeds and altitudes, and their dynamics are nonlinear and conceptually time-varying. For a given operating point, specified by the aircraft speed (Mach number) and altitude, the longitudinal nonlinear aircraft dynamics can be approximated by a linear model. As the aircraft goes through different flight conditions, the operating point changes. These changes cannot be handled by constant gain feedback control. Since the output response $y(t)$ carries information about the state as well as the parameters, one may argue that in principle, a sophisticated feedback controller should be able to learn about the plant changes by processing the input/output (I/O) measurements (u, y) and choosing the appropriate controller from a list or design a new one in real-time. The real-time or on-the-fly selection or design of the controller is what distinguishes adaptive from nonadaptive schemes. Figure 35.1 illustrates this general adaptive control structure. The structure covers almost all classes of adaptive control. The idea is to process the I/O and possibly auxiliary measurements and decide what controller to use in real-time. Under this generic structure one can include gain scheduling where the real time controller design block is just a look-up table with a scheduler logic. In identifier-based schemes, this block includes a parameter estimator and the online calculation of the controller whereas in nonidentifier-based schemes, the block may consist of multiple models, stored controllers, and so on and an appropriate logic for selecting the right controller in real-time. Structures such as direct and indirect adaptive control also fall into this general feedback structure.

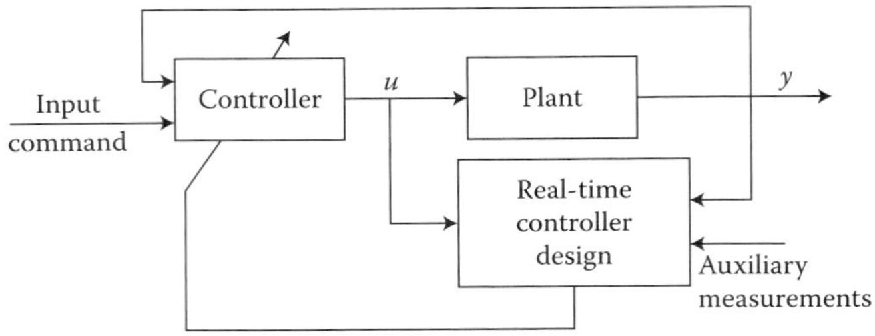

FIGURE 35.1 General adaptive control structure.

The assumption that the plant is linear time-invariant with constant parameters, some or all of which could be unknown, was used in almost all the early work on adaptive control. It was not till robustness issues were resolved that controlling linear plants with unknown time-varying parameters became possible [1]. An important part of most adaptive control schemes is the online parameter identifier or adaptive law, which generates estimates of the unknown parameters to be used for calculating or updating the controller parameters in real-time. The way the adaptive law is combined with the control law gives rise to different adaptive control structures. The class of adaptive controllers that employ online parameter estimators are labeled as *identifier-based* adaptive control schemes. If instead of a realtime, parameter identifier one could partition the parameter space into a finite set of regions for which a controller could be designed *a priori*, the problem then becomes that of identifying in real-time which of the controllers is the right one to be switched on. Similarly, a bank of possible models for the plant can be formed. For each model a controller is designed *a priori*. A switching logic then can identify which plant model is valid based on input output measurements, and therefore choose the appropriate controller from a given list. These schemes that do not involve a parameter estimator or adaptive law are referred to as *nonidentifier-based*. There is also the class of adaptive controllers which could be a combination of identifier-based and nonidentifier-based schemes. In the following sections we discuss these different structures and associated stability and performance results, after we present a brief history how these different adaptive control structures and schemes evolved during the past 50 years.

35.1.1 Brief History

Research in adaptive control has a long history of intense activities that involved debates about the precise definition of adaptive control, designs based on heuristics, stability-based schemes, examples of instabilities, stability and robustness proofs, and applications.

Starting in the early 1950s, the design of autopilots for high-performance aircraft motivated an intense research activity in adaptive control. High-performance aircraft undergo drastic changes in their dynamics when they fly from one operating point to another that cannot be handled by constant-gain feedback control. A sophisticated controller, such as an adaptive controller, that could learn and accommodate changes in the aircraft dynamics was needed. Model reference adaptive control (MRAC) was suggested by Whitaker and coworkers in [2] to solve the autopilot control problem. The sensitivity method and the MIT rule was used to design the adaptive laws of the various proposed adaptive control schemes. An adaptive pole placement scheme based on the optimal linear quadratic problem was suggested by Kalman in [3]. The lack of stability proofs and the lack of understanding of the properties of the proposed adaptive control schemes coupled with a disaster in a flight test caused the interest in adaptive control to diminish.

The 1960s became the most important period for the development of control theory and adaptive control in particular. State-space techniques and stability theory based on Lyapunov were introduced. Developments in dynamic programming [4], dual control [5], and stochastic control in general, and in system identification and parameter estimation played a crucial role in the reformulation and redesign of adaptive control. By 1966 Parks and others [6] found a way of redesigning the MIT rule-based adaptive laws used in the MRAC schemes of the 1950s by applying the Lyapunov design approach. Their work, even though applicable to a special class of LTI plants, set the stage for further rigorous stability proofs in adaptive control for more general classes of plant models in subsequent years.

The advances in stability theory and the progress in control theory in the 1960s improved the understanding of adaptive control and contributed to a strong renewed interest in the field in the 1970s. On the other hand, the simultaneous development and progress in computers and electronics that made the implementation of complex controllers feasible, contributed to an increased interest in applications of adaptive control. The 1970s witnessed several breakthrough results in the design of adaptive control. A Lyapunov stability-based design approach was used to design and analyze MRAC schemes in [7,8]. The concepts of positivity and hyperstability were used in [9] to develop a wide class of MRAC schemes with well-established stability properties. At the same time, parallel efforts for discrete-time plants in a deterministic and stochastic environment produced several classes of adaptive control schemes with rigorous stability proofs [10]. The excitement of the 1970s and the development of a wide class of adaptive control schemes with well-established stability properties was accompanied by several successful applications. The successes of the 1970s, however, were soon followed by controversies over the practicality of adaptive control. As early as 1979 it was pointed out that the adaptive schemes of the 1970s could easily go unstable in the presence of small disturbances [11]. The nonrobust behavior of adaptive control became very controversial in the early 1980s when more examples of instabilities were published demonstrating lack of robustness in the presence of unmodeled dynamics and/or bounded disturbances [12,13]. Understanding the mechanisms of instabilities and finding ways to counteract them became the focus of subsequent research studies. By the mid-1980s, several new redesigns and modifications were proposed and analyzed, leading to a body of work known as robust adaptive control. An adaptive controller is defined to be robust if it guarantees signal boundedness in the presence of "reasonable" classes of unmodeled dynamics and bounded disturbances as well as performance error bounds that are of the order of the modeling error. The work on robust adaptive control continued throughout the 1980s and involved the understanding of the various robustness modifications and their unification under a more general framework [14]. Global stability in the presence of unmodeled dynamics using various fixes and a dynamic normalizing signal was established in [15] for discrete-time systems. The use of the normalizing signal together with the switching σ-modification led to the proof of global stability in the presence of unmodeled dynamics for continuous-time plants in [16].

The solution of the robustness problem in adaptive control led to the solution of the long-standing problem of controlling a linear plant whose parameters are unknown and changing with time. By the end of the 1980s several breakthrough results were published in the area of adaptive control for linear time-varying plants [1]. The focus of adaptive control research in the late 1980s to early 1990s was on performance properties and on extending the results of the 1980s to certain classes of nonlinear plants with unknown parameters. These efforts led to new classes of adaptive schemes, motivated from nonlinear system theory [17] as well as to adaptive control schemes with improved transient and steady-state performance [18,19]. New concepts such as adaptive backstepping, nonlinear damping, and tuning functions are used to address the more complex problem of dealing with parametric uncertainty in classes of nonlinear systems.

In the late 1980s to early 1990s, the use of neural networks as universal approximators of unknown nonlinear functions led to the use of on-line parameter estimators to "train" or update the weights of the neural networks. Difficulties in establishing global convergence results soon arose since in multilayer neural networks the weights appear in a nonlinear fashion, leading to "nonlinear in the parameters" parameterizations for which globally stable on-line parameter estimators cannot be developed. This led to the consideration of single-layer neural networks where the weights can be expressed in certain parametric models that are convenient for estimation [20,21].

In the mid-1980s to recent years, several groups of researchers started looking at alternative methods of controlling plants with unknown parameters [22–27]. These methods avoid the use of online parameter estimators in general and use search methods to identify unknown parameters, multiple models to characterize parametric uncertainty, switching logic to identify the stabilizing controller, and so on. They were motivated from the fact that in identifier-based schemes the estimated plant has to be stabilizable at each instant of time in order for a controller to exist. Since there is no guarantee that the online estimator will generate estimates that correspond to a stabilizable plant at all times, it raises theoretical and implementation issues that need to be addressed. Designing all the stabilizing controllers to cover all possible plant parameter changes *a priori* eliminates this problem and transfers it to the ability of identifying which one of the *a priori* designed controllers is the right one to use. One distinct advantage of these efforts which are currently continuing is that well-established techniques from robust control for LTI systems can be employed. In the following sections we will elaborate further on some of the most popular adaptive control methodologies.

35.2 Identifier-Based Adaptive Control

35.2.1 Direct and Indirect Adaptive Control

An adaptive controller of the identifier-based class is formed by combining an *online parameter estimator*, which provides estimates of unknown parameters at each time instant, with a *control law* that is motivated from the known parameter case. The way the parameter estimator, also referred to as *adaptive law*, is combined with the control law gives rise to two different approaches. In the first approach, referred to as *indirect adaptive control*, the plant parameters are estimated online and used to calculate the controller parameters at each instant of time. This approach has also been referred to as *explicit adaptive control*, because the design is based on an explicit plant model. In the second approach, referred to as *direct adaptive control*, the plant model is parameterized in terms of the controller parameters that are estimated directly without intermediate calculations involving plant parameter estimates. This approach has also been referred to as *implicit adaptive control* because the design is based on the estimation of an implicit plant model.

The principle behind the design of direct and indirect adaptive control is conceptually simple. It treats the parameter estimates at each instant of time as if they are the true ones. In indirect adaptive control, the parameter estimates are associated with a plant parameterization, such as the coefficients of its transfer function in the case of an LTI plant, and so on. Hence at each time, the estimated parameters can be used to generate an estimated plant. The estimated plant is then treated as the true one and is used to calculate the controller by following the same techniques as in the known parameter case. In direct adaptive control, the plant is parametrized with respect to the desired controller parameters and the estimator generates the estimated controller parameters directly. Again the estimated parameters at each instant of time are treated as the true ones. This design approach of treating estimated parameters at each time instant as the true parameters and using them to generate the controller parameters is called *certainty equivalence* and has been used to generate a wide class of adaptive control schemes by combining different online parameter estimators with different control laws. The idea behind the certainty equivalence approach is that as the parameter estimates converge to the true ones, the performance of the adaptive controller tends to that achieved by the desired controller in the case of known parameters. The parameter estimator or adaptive law has a multiplicative nonlinearity that makes the closed-loop system nonlinear and time-varying. Because of this, the analysis and understanding of the stability and robustness of adaptive control schemes are more challenging, since most of the practical control design tools incorporating robustness and performance specifications used for LTI systems are not applicable to time-varying and nonlinear systems and therefore to adaptive control. For example, we can no longer use pole location and gain or phase margins considerations to specify or analyze stability and performance. Instead nonlinear techniques need to be developed specifically for adaptive control.

35.2.2 Online Parameter Estimation

The first step in the design of online parameter estimation algorithms is to lump the unknown parameters in a vector and separate them from known signals, transfer functions, and other known parameters in an equation that is convenient for parameter estimation. In the general case, the class of parameterizations of the form

$$z = \theta^{*T}\phi, \tag{35.1}$$

where $\theta^* \in \mathbb{R}^n$ is the vector with all the unknown parameters and $z \in \mathbb{R}$, $\phi \in \mathbb{R}^n$ are signals available for measurement, are referred to as the linear *static parametric model* (SPM). The SPM may represent a dynamic, static, linear, or nonlinear system. Any linear or nonlinear dynamics in the original system are hidden in the signals z, ϕ that usually consist of the I/O measurements of the system and their filtered values.

Another type of parameterization is of the form

$$z = W(q)\left(\theta^{*T}\phi\right), \tag{35.2}$$

where $z \in \mathbb{R}$, $\phi \in \mathbb{R}^n$ are signals available for measurement and $W(q)$ is a known stable proper transfer function; q is either the shift operator in discrete time (i.e., $q = z$) or the differential operator ($q = s$) in continuous time. We refer to Equation 35.2 as the linear *dynamic parametric model* (DPM). The importance of the SPM and DPM as compared to other possible parameterizations is that the unknown parameter vector θ^* appears linearly. For this reason we refer to Equations 35.1 and 35.2 as linear in the parameters parameterizations. This property is significant in designing online parameter estimators whose global convergence properties can be established analytically.

In some cases, the unknown parameters cannot be expressed in the form of the linear in the parameters models. In such cases, the parameter estimators algorithms based on such models cannot be shown to converge globally. Special cases of nonlinear in the parameters models for which convergence results exist is when the unknown parameters appear in the special bilinear form

$$z = \rho^*\left(\theta^{*T}\phi + z_1\right), \tag{35.3}$$

$$z = W(q)\rho^*\left(\theta^{*T}\phi + z_1\right), \tag{35.4}$$

where $z \in \mathbb{R}$, $\phi \in \mathbb{R}^n$, and $z_1 \in \mathbb{R}$ are signals available for measurement at each time t, and $\rho^* \in \mathbb{R}$, and $\theta^* \in \mathbb{R}^n$ are the unknown parameters. The transfer function $W(q)$ is a known stable transfer function. We refer to Equations 35.3 and 35.4 as the *bilinear* SPM (B-SPM) and *bilinear* DPM (B-DPM), respectively.

In some applications of parameter identification or adaptive control of plants in state-space form whose state x is available for measurement, the following parametric model may be used:

$$\dot{x} = A_m x + \Theta^{*T}\Phi, \tag{35.5}$$

where A_m is a stable design matrix; Θ^* is an unknown matrix; $\Phi = [x^T u^T]^T$, and x, u are signal vectors available for measurement. We refer to this class of parametric models as *state-space parametric models* (SSPMs). It is clear that SSPMs can be expressed in the form of DPMs and SPMs. Another class of state-space models that appear in adaptive control is

$$\dot{x} = A_m x + B\Theta^{*T}\Phi, \tag{35.6}$$

where B is also unknown but is positive definite, negative definite, or the sign of each of its elements is known. We refer to this class of parametric models as B-SSPMs. The B-SSPM model can be easily expressed as a set of scalar B-SPMs or B-DPMs.

The parameter estimation problem can now be stated as follows:

- Given the available measurements, generate $\theta(t)$, the estimate of the unknown vector θ^*, at each time t. The parameter estimator or adaptive law updates $\theta(t)$ with time so that as time evolves, $\theta(t)$ approaches or converges to θ^*. Since we are dealing with online parameter estimation, we would also expect that if θ^* changes, then the estimator will react to such change and update the estimate $\theta(t)$ to match the new value of θ^*.

The online parameter estimators generate estimates at each time t, by using the past and current measurements of signals. Convergence is achieved asymptotically as time evolves. For this reason they are referred to as recursive parameter estimators to be distinguished from the nonrecursive ones, in which all the measurements are collected *a priori* over large intervals of time and are processed off-line to generate the estimates of the unknown parameters. Generating the parametric models (Equations 35.1 through 35.6) is the first step in the design of the appropriate parameter estimators. The essential idea behind online estimation is the comparison of the observed system response $z(t)$, with the output of a parameterized model $\hat{z}(\theta; t)$ whose structure is the same as that of the plant model. The parameter vector $\theta(t)$ is adjusted continuously so that $\hat{z}(\theta; t)$ approaches $z(t)$ as t increases. Under certain input conditions, $\hat{z}(\theta; t)$ being close to $z(t)$ implies that $\theta(t)$ is close to the unknown parameter vector θ^* of the plant model. The adaptive law is usually a differential equation whose state is $\theta(t)$ and is designed using stability considerations or simple optimization techniques to minimize the difference between $z(t)$ and $\hat{z}(\theta; t)$ with respect to $\theta(t)$ at each time t.

For the SPM case, the estimation model has the same form as the SPM with the exception that the unknown parameter θ^* is replaced with its estimate at time t, denoted by $\theta(t)$, that is,

$$\hat{z} = \theta^T(t)\phi, \tag{35.7}$$

where $\hat{z}$ is the estimate of z based on the parameter estimate $\theta(t)$ at time t. Since θ^* is unknown, the difference $\tilde{\theta}(t) = \theta(t) - \theta^*$ is not available for measurement. Therefore, the only signal that we can generate, using available measurements, that reflects the difference between $\theta(t)$ and θ^* is the error signal

$$\epsilon = \frac{\hat{z} - z}{m_s^2}, \tag{35.8}$$

which we refer to as the *estimation error*. $m_s^2 > 1$ is a normalization signal designed to guarantee that ϕ/m_s is bounded. This property of m_s is used to establish the boundedness of the estimated parameters even when ϕ is not guaranteed to be bounded. A straightforward choice for m_s is $m_s^2 = 1 + a\phi^T\phi$, $a > 0$. If ϕ is bounded, we can take $m_s^2 = 1$. Using Equation 35.7 in Equation 35.8, we can express the estimation error as a function of the parameter error $\tilde{\theta}$, that is,

$$\epsilon = -\frac{\tilde{\theta}^T\phi}{m_s^2}. \tag{35.9}$$

We update $\theta(t)$ in a direction that minimizes a certain cost of the estimation error ϵ. As an example, consider the cost criterion

$$J(\theta) = \frac{(z - \theta^T\phi)^2}{2m_s^2}, \tag{35.10}$$

which we minimize with respect to θ using the gradient method to obtain the adaptive law

$$\dot{\theta} = -\Gamma\nabla J(\theta) = \Gamma\epsilon\phi. \tag{35.11}$$

The gradient algorithm (Equation 35.11) guarantees ϵ, ϵm_s, $\dot{\theta} \in \mathcal{L}_2 \cap \mathcal{L}_\infty$, and $\theta \in \mathcal{L}_\infty$.

The procedure for estimating θ^* in a linear model extends to the other parametric models: the details of which can be found in [28,29].

35.2.3 Model Reference Adaptive Control

MRAC has been one of the most popular approaches to adaptive control. The basic structure of an MRAC scheme is shown in Figure 35.2 for the indirect scheme and Figure 35.3 for the direct scheme. The reference model is chosen to generate the desired trajectory, y_m, that the plant output y_p has to follow. The tracking error $e_1 = y_p - y_m$ represents the deviation of the plant output from the desired trajectory. The closed-loop plant is made up of an ordinary feedback control law that contains the plant and a controller $C(\theta)$ and an adjustment mechanism that generates the controller parameter estimates $\theta(t)$ online.

MRAC schemes can be characterized as direct or indirect and with normalized or unnormalized adaptive laws (Figure 35.4). In direct MRAC, the parameter vector θ of the controller $C(\theta)$ is updated directly by an adaptive law, whereas in indirect MRAC, θ is calculated at each time t by solving a certain algebraic equation that relates θ with the online estimates of the plant parameters through some mapping $F(\cdot)$. In both direct and indirect MRAC with normalized adaptive laws, the form of $C(\theta)$, motivated from the known parameter case, is kept unchanged. The controller $C(\theta)$ is combined with an adaptive law (or an adaptive law and an algebraic equation in the indirect case) that is developed independently by following the techniques of Section 35.2.2. This design procedure allows the use of a wide class of adaptive laws that includes gradient, least-squares, and those based on the SPR–Lyapunov design approach [28,29].

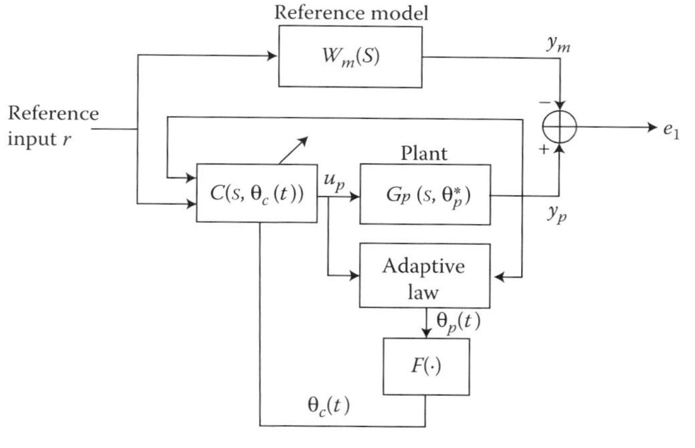

FIGURE 35.2 Indirect MRAC.

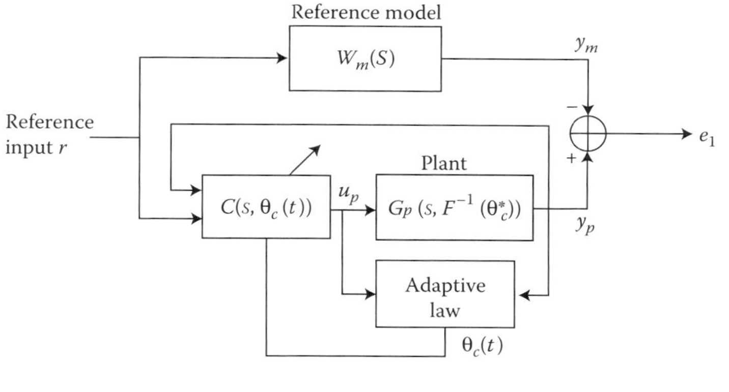

FIGURE 35.3 Direct MRAC.

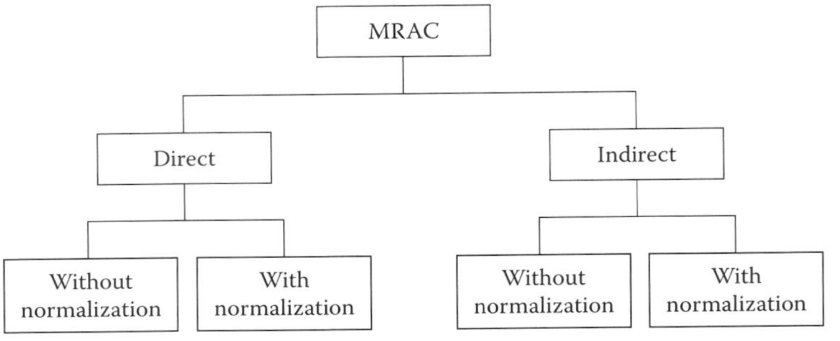

FIGURE 35.4 Classes of MRAC.

For most classes of MRAC, the controller and the parameter estimator are of the form

$$u(t) = \theta^T(t)\phi(t), \tag{35.12}$$

$$\dot{\theta}(t) = \Gamma\epsilon(t)\phi(t), \tag{35.13}$$

for the direct scheme and

$$u(t) = \theta^T(t)\phi(t), \tag{35.14}$$

$$\dot{\theta}_p(t) = \Gamma\epsilon(t)\phi_p(t), \tag{35.15}$$

$$\theta(t) = F(\theta_p(t)), \tag{35.16}$$

for the indirect scheme. These schemes guarantee that all signals are bounded and the plant output tracks the output of the reference model asymptotically with time for any input command. Furthermore, if the input command is sufficiently rich in frequencies, then $\theta(t)$ converges to the true parameter value θ^*.

In order to meet the model reference control (MRC) objective we must assume that the plant transfer function is minimum phase. The minimum-phase assumption is a consequence of the control objective which is met by designing an MRC control law that cancels the zeros of the plant and replaces them with those of the reference model in an effort to force the closed-loop plant transfer function from r to y_p to be equal to $W_m(s)$. Other assumptions such as the knowledge of an upper bound for the plant order or the knowledge of the relative degree may be relaxed at the expense of additional complexity in the control and adaptive laws.

35.2.4 Adaptive Pole Placement Control

The assumption that the plant is minimum phase, that is, it has stable zeros, is rather restrictive in many applications. For example, the approximation of time delays leads to plant models with unstable zeros. Furthermore, the desired properties of the plant to be controlled are often expressed in terms of desired pole locations to be placed by the controller. Consequently, there is no need to alter the zeros of the plant. If the plant satisfies the properties of controllability and observability, then a controller always exists to place the poles in the desired locations. These schemes are referred to as *pole placement* schemes and are applicable to both minimum- and nonminimum-phase LTI plants. The combination of a pole placement control law with a parameter estimator or an adaptive law leads to an *adaptive pole placement control* (APPC) scheme that can be used to control a wide class of LTI plants with unknown parameters.

As in the MRAC case, the APPC schemes may be divided into two classes: the *indirect* APPC schemes where the adaptive law generates online estimates of the coefficients of the plant transfer function that are then used to calculate the parameters of the pole placement control law by solving a certain algebraic equation; and the *direct* APPC where the parameters of the pole placement control law are generated directly by an adaptive law without any intermediate calculations that involve estimates of the plant parameters. The direct APPC schemes are restricted to scalar plants and to special classes of plants where the desired parameters of the pole placement controller can be expressed in the form of the linear or bilinear parametric model. The indirect APPC schemes, on the other hand, are easy to design and are applicable to a wide class of LTI plants that are not required to be minimum phase or stable. Because of this flexibility in choosing the controller design methodology (observer-based state feedback, linear quadratic, etc.) and adaptive law (least-squares, gradient, or SPR–Lyapunov type), indirect APPC is the most general class of adaptive control schemes. This class also includes indirect MRAC as a special case where some of the poles of the plant are assigned to be equal to the zeros of the plant to facilitate the required zero-pole cancellation for transfer function matching.

In the indirect case, the APPC scheme is of the form

$$u(t) = \theta^T(t)\phi(t), \tag{35.17}$$

$$\dot{\theta}_p(t) = \Gamma\epsilon(t)\phi_p(t), \tag{35.18}$$

$$F(\theta(t), \theta_p(t)) = 0, \tag{35.19}$$

where the last equation must be solved for the controller parameters $\theta(t)$ at each time t. It is a pole placement equation to be solved pointwise in time. Such an equation has a unique solution only if the estimated plant polynomials are strongly coprime at each time t. Such a strong condition cannot be guaranteed by the adaptive law without any additional modifications, giving rise to the so-called "stabilizability" or "admissibility" problem, which is the main drawback of indirect APPC. This drawback can be eliminated by modifying the indirect APPC schemes at the expense of adding more complexity. If the estimated plant polynomials are strongly coprime at each time t then all signals are bounded and the tracking error converges to zero asymptotically with time. The same result holds if we replace the gradient algorithm with any other adaptive law. Furthermore, if the reference trajectory is sufficiently rich in frequencies then convergence is exponential and we have parameter convergence of $\theta(t)$ to the true parameter value θ^*. The proof involves the manipulation of the estimation error and control law equations to express the plant input u_p and output y_p in terms of the estimation error in an equation of the form

$$\dot{x}(t) = A(t)x(t) + b\tilde{\theta}^T\phi + d, \tag{35.20}$$

where the state x is formed by the plant input u_p and output y_p and their derivatives, d is a bounded vector, and $\tilde{\theta}$ is the parameter error. The matrix $A(t)$ has stable eigenvalues at each frozen time t by design as a result of the pole placement approach. Furthermore, the adaptive law guarantees that $\|\dot{A}(t)\| \in \mathcal{L}_2$. These two properties are used to show that the homogeneous part of Equation 35.20 is uniformly asymptotically stable (u.a.s.). Using the properties of the $\mathcal{L}_2$ norm and Bellman–Gronwall lemma, we can establish boundedness of u_p and y_p and also $\tilde{\theta}^T\phi \in \mathcal{L}_2 \cap \mathcal{L}_\infty$. The convergence of e_1 to zero follows by using the control and estimation error equations to express e_1 as the output of proper stable LTI systems whose inputs are in $\mathcal{L}_2 \cap \mathcal{L}_\infty$. In the direct case and MRAC, the form of Equation 35.20 is the same with the exception that the matrix A is time-invariant with stable eigenvalues, implying that the homogeneous part is u.a.s. The rest of the analysis follows the same way as in the indirect case using the same tools. Similar results and techniques can be established for discrete time systems. A wide class of MRAC and APPC can be designed to meet the performance requirements of signal boundedness and convergence of the regulation or tracking error to zero. This speed of convergence is asymptotic unless the reference input has a sufficient number of frequencies in which case we also have parameter convergence. In most applications the use of sufficiently rich signals is not possible as it violates the tracking performance requirements. In such case, the estimated parameters may not converge to any constant values but continuously move on some manifolds that guarantee signal boudedness and convergence of the tracking error to zero. This ability of the adaptive control schemes to guarantee tracking performance without having parameter convergence to the true values is remarkable. The price paid, however, is that in the presence of uncertainties the schemes may lose stability as discussed in the next section.

35.2.5 Instability Phenomena in Adaptive Systems

The adaptive laws and control schemes discussed in previous section are based on a plant model that is free of noise, disturbances, and unmodeled dynamics. These schemes are to be implemented on actual plants that most likely deviate from the plant models on which their design is based. An actual plant may be infinite dimensional, nonlinear and its measured input and output may be corrupted by noise and external disturbances. The effect of these discrepancies on the stability and performance of the schemes designed using idealized models is therefore of paramount importance from the practical point of view. Let us examine the effect of modeling errors using the following simple plant example:

$$y(t) = \theta^* u(t) + d(t), \tag{35.21}$$

where d is an external bounded disturbance. The simple estimator

$$\dot{\theta}(t) = \gamma\epsilon(t)u(t), \quad \epsilon(t) = y(t) - \theta(t)u(t) \tag{35.22}$$

is designed assuming that the external disturbance does not exist, that is, $d = 0$. In the ideal case ($d = 0$) we can establish that $\epsilon, \theta \in \mathcal{L}_\infty$, and $\epsilon \to 0$ as $t \to \infty$. When d is different than zero these results cannot be guaranteed. More important is that the boundedness of the estimated parameters cannot be guaranteed either. In fact, we can find an input u such that $\epsilon \to 0$ and $|\theta| \to \infty$ as $t \to \infty$. This rather strange phenomenon is better understood by analyzing the parameter error equation

$$\dot{\tilde{\theta}}(t) = -\gamma u^2(t)\tilde{\theta}(t) + \gamma d(t)u(t) \tag{35.23}$$

obtained from Equation 35.22.

The homogeneous part of Equation 35.23 is not guaranteed to be u.a.s for all possible bounded inputs u. Consequently, bounded input does not imply bounded output. This instability phenomenon where the estimated parameters drift to infinity is known as *parameter drift*. It is mainly due to the pure integral action of the adaptive law, which, in addition to integrating the "good" signals, integrates the disturbance term as well, leading to the parameter drift phenomenon.

Another kind of instability is the so-called *high-gain instability*. Consider the plant transfer function

$$\frac{1-\mu s}{(s-a)(1+\mu s)} = \frac{1}{s-a}\left[1 - \frac{2\mu s}{1+\mu s}\right], \tag{35.24}$$

where μ is a small positive number which may be due to a small time constant in the plant. Let us now design an adaptive controller for the simplified first-order plant ($\mu = 0$) and use it to control the actual second-order plant, where $\mu > 0$. The adaptive control law

$$u(t) = -k(t)y(t), \quad \dot{k}(t) = \gamma y^2(t) \tag{35.25}$$

guarantees that when $\mu = 0$, $u, k, y \in \mathcal{L}_\infty$, and $y \to 0$ as $t \to \infty$ for all initial conditions. But when $\mu \neq 0$, for $k(0) > 1/\mu - a$, we have u, k, $y \to \infty$ as $t \to \infty$, indicating that the presence of modeling errors reduced the stability of the closed-loop system from global to local.

The instability examples presented demonstrate that the adaptive schemes designed for ideal plants, that is, plants with no modeling errors may easily go unstable in the presence of disturbances or unmodeled dynamics. The lack of robustness is primarily due to the adaptive law which is nonlinear in general and therefore more susceptible to modeling error effects. The lack of robustness of adaptive schemes in the presence of bounded disturbances was demonstrated as early as 1979 [11] and became a hot issue in the early 1980s when several adaptive control examples are used to show instability in the presence of unmodeled dynamics and bounded disturbances [12,13]. It was clear from these examples that new approaches and adaptive laws were needed to assure boundedness of all signals in the presence of plant uncertainties. These activities led to a new body of work referred to as *robust adaptive control.*

35.2.6 Robust Adaptive Laws

It turns out that the destabilizing effects of bounded disturbances and dynamic uncertainties can be counteracted by modifying the adaptive laws developed for the ideal plants using techniques such as leakage, dead zones, projection, signal normalization, and so on. One unique characteristics of these modifications is that they remove the pure integral action of the adaptive laws at all or some of the time. In this section we extend the results of Section 35.2.2 to a general class of parametric models with modeling errors that may arise in the online parameter estimation problem of a wide class of plants. If the objective is parameter convergence, then parameter drift can be prevented by making sure that the regressor vector is persistently exciting (PE) with a level of excitation higher than the level of the modeling error. In many applications, such as in adaptive control, the plant input is the result of feedback and cannot be designed to be sufficiently rich. In such situations, the objective is to drive the plant output to zero or force it to follow a desired trajectory rather than convergence of the online parameter estimates to their true values. It is therefore of interest to guarantee stability and robustness even in the absence of persistence of excitation. This can be achieved by modifying the adaptive laws of the previous sections to guarantee stability and robustness in the presence of modeling errors independent of the properties of the regressor vector ϕ.

A class of robust modifications involves the use of a small feedback around the "pure" integrator in the adaptive law, leading to the adaptive law structure

$$\dot{\theta}(t) = \Gamma\epsilon(t)\phi(t) - \sigma_l(t)\Gamma(t)\theta(t), \tag{35.26}$$

where $\sigma_l > 0$ is a small design parameter and $\Gamma = \Gamma^T > 0$ is the adaptive gain. The above modification is referred to as the σ-*modification* or as *leakage* [12,28,30]. Different choices of $\sigma_l(t)$ lead to different robust adaptive laws. The simpler choice is the fixed σ-modification:

$$\sigma_l(t) = \sigma > 0, \quad \forall t \geq 0. \tag{35.27}$$

Another possible choice of $\sigma_l(t)$ leads to the switching σ-modification: and M_0 is known:

$$\sigma_l = \begin{cases} 0 & \text{if } |\theta(t)| \leq M_0 \\ \left(\frac{|\theta(t)|}{M_0} - 1\right)^{q_0} \sigma_0 & \text{if } M_0 < |\theta(t)| \leq 2M_0, \\ \sigma_0 & \text{if } |\theta(t)| > 2M_0 \end{cases} \tag{35.28}$$

where $q_0 \geq 1$ is any finite integer, $\sigma_0 > 0$ is a design constant, and M_0 is a known upper bound of $|\theta^*|$.

Another effective way to guarantee bounded parameter estimates is to use projection to constrain the parameter estimates to lie inside a bounded convex set in the parameter space that contains the unknown θ^*. By requiring the parameter set to be bounded, projection can be used to guarantee that the estimated parameters are bounded by forcing them to lie within the bounded set. As an example, consider the set

$$\mathcal{P} = \left\{\theta \mid \theta^T\theta - M_0^2 \leq 0\right\}, \tag{35.29}$$

where M_0 is chosen so that $M_0 > |\theta^*|$. The adaptive law with projection becomes

$$\dot{\theta} = \begin{cases} \Gamma\epsilon\phi & \begin{array}{l}\text{if } \theta^T\theta < M_0^2 \text{ or} \\ \text{if } \theta^T\theta = M_0^2 \\ \text{and } \quad (\Gamma\epsilon\phi)^T\theta \leq 0\end{array} \\ \left(I - \frac{\Gamma\theta\theta^T}{\theta^T\Gamma\theta}\right)\Gamma\epsilon\phi & \text{otherwise.} \end{cases} \tag{35.30}$$

In the presence of modeling errors the static linear parametric model is of the form

$$z = \theta^{*T}\phi + \eta, \tag{35.31}$$

where η is the modeling error term. Let us now consider the estimation error

$$\epsilon = \frac{z - \theta^T\phi}{m_s^2} = \frac{-\tilde{\theta}^T\phi + \eta}{m_s^2}. \tag{35.32}$$

The signal ϵ that drives the adaptive law is a function of the good signal that depends on the parameter error $\tilde{\theta}$, and of the modeling error η. Large ϵm_s^2 implies that $\tilde{\theta}$ is large and that the effect of the modeling error η is small, and therefore the parameter estimates driven by ϵ move in a direction which reduces $\tilde{\theta}$. When ϵm_s is small, however, the effect of η may be more dominant than that of the signal $\tilde{\theta}^T\phi$, and the parameter estimates may be driven in a direction dictated mostly by η. Thus, it seems reasonable to update the parameter estimate θ only when the signal $\tilde{\theta}^T\phi$ is large relative to the disturbance η and switch-on adaptation when $\tilde{\theta}^T\phi$ is small relative to the size of η. This method of adaptation is referred to as *dead zone* because of the presence of a zone or interval where θ is constant, that is, no updating occurs. The adaptive law in this case takes the form:

$$\dot{\theta}(t) = \begin{cases} \Gamma\epsilon\phi & \text{if } |\epsilon m_s| > g_0 \\ 0 & \text{otherwise,} \end{cases} \tag{35.33}$$

where g_0 is a known upper bound of the normalized modeling error η/m_s. In other words, the parameters are updated in the direction of the steepest descent only when the estimation error is large relative to the modeling error, that is, when $|\epsilon m_s| > g_0$.

Another important modification that helps improve robustness and performance is the use of a *dynamic normalizing signal* in the adaptive law. The dynamic normalizing signal is designed to bound from above the modelling error signals and in some sense make sure that the speed of adaptation is slower than the speed with which the modeling error terms change. We demonstrate the design of the dynamic normalizing signal for the SPM with modeling error

$$z(t) = \theta^*\phi(t) + \eta(t), \tag{35.34}$$

where $\eta(t) = \Delta_1(s)u(t) + \Delta_2(s)y(t)$ in the case of an LTI plant with additive and multiplicative uncertainties. $\Delta_1(s)$, $\Delta_2(s)$ are strictly proper transfer functions with stable poles. Our objective is to design a dynamic normalizing signal m_s so that $\eta/m_s \in \mathcal{L}_\infty$. We assume that $\Delta_1(s)$, $\Delta_2(s)$ are analytic in $\Re[s] > -\delta_0$ for some known $\delta_0 > 0$. If we define $n_d = \|u_t\|_{2\delta_0} + \|y_t\|_{2\delta_0}$ which can be generated by the differential equation

$$\dot{n}_d(t) = -\delta_0 n_d(t) + u^2(t) + y^2(t). \tag{35.35}$$

then it follows that $m_s^2 = 1 + n_d$ bounds $|\eta(t)|$ from above and

$$\frac{|\eta(t)|}{m_s} \leq \|\Delta_1\|_{2\delta_0} + \|\Delta_2\|_{2\delta_0}, \tag{35.36}$$

where $\|H\|_{2\delta} \triangleq \frac{1}{\sqrt{2\pi}}\left\{\int_0^{2\pi}\left|H(j\omega - \frac{\delta}{2})\right|^2 d\omega\right\}^{1/2}$. The normalizing signal m_s, used in the adaptive laws for parametric models free of modeling errors, is required to bound the regressor ϕ from above. In the presence of modeling errors, m_s is chosen to bound both ϕ and the modeling error η from above for improved robustness properties. An example of such normalizing signal is

$$m_s^2 = 1 + n_s^2 + n_d, \tag{35.37}$$

where n_s is the static part and n_d the dynamic one. Examples of static and dynamic normalizing signals are $n_s^2 = \phi^T\phi$ or $n_s^2 = \phi^T P\phi$, where $P = P^T > 0$, and

$$\dot{n}_d(t) = -\delta_0 n_d(t) + \delta_1(u^2(t) + y^2(t)) \tag{35.38}$$

or

$$n_d = n_1^2, \quad \dot{n}_1(t) = -\delta_0 n_1(t) + \delta_1(|u(t)| + |y(t)|) \tag{35.39}$$

or

$$n_d = n_\infty^2, \quad n_\infty = \delta_1 \max\left(\sup_{\tau \le t} |u(\tau)|, \sup_{\tau \le t} |y(\tau)|\right). \tag{35.40}$$

Any one of the choices (Equations 35.38 through 35.40) can be shown to guarantee that $\eta/m_s \in \mathcal{L}_\infty$. Since $\phi = H(s)[u\ y]'$, for an appropriate transfer function $H(s)$, the dynamic normalizing signal can be chosen to bound ϕ from above, provided that $H(s)$ is analytic in $\mathcal{R}[s] \ge -\frac{\delta}{2}$, in which case $m_s^2 = 1 + n_d$ bounds both ϕ and η from above.

The above adaptive laws are shown to be robust with respect to a wide class of plant model uncertainties and can be combined with control laws to generate robust adaptive control schemes. Let us consider the plant

$$y(t) = G_0(s)(1 + \Delta_m(s))u(t) + d(t), \tag{35.41}$$

where $G_0(s)$ represents the dominant or modeled part of the plant transfer function, $\Delta_m(s)$ is a multiplicative perturbation, and $d(t)$ is a bounded disturbance, that is, $|d(t)| < d_0$. It is important to note that we can rewrite Equation 35.41 in the form of Equation 35.34. Suppose there exists a strictly proper transfer function $W(s)$ analytic in $\mathcal{R}[s] \ge -\delta_0/2$ and such that $W(s)\Delta_m(s)$ is strictly proper. The result is that there exists a $\delta^* > 0$ such that if $c\Delta_*^2 < \delta^*$, where $\Delta_* = \|W(s)\Delta_m(s)\|_{2\delta_0}$ and $c > 0$ is a finite constant, then all the signals are bounded and the tracking error e_1 satisfies

$$\frac{1}{T}\int_t^{t+T} e_1^2(\tau)\,d\tau \le c_0(\Delta_*^2 + d_0^2 + f_0) + \frac{c_1}{T}, \quad \forall t > 0 \tag{35.42}$$

for any $T > 0$ and some positive constants c_0, c_1, where $f_0 = 0$ in the case of switching σ-modification and projection, and $f_0 > 0$ in the case of fixed σ-modification ($f_0 = \sigma$) and dead zone ($f_0 = g_0$). Due to the presence of modeling errors we no longer have exact convergence of the tracking error to zero. Rather the tracking error is of the order of the modeling error in the mean square sense. This, however, does not guarantee that the tracking error will be bounded from above by the modeling error bound at all times. In fact, simulations show that a phenomenon known as "bursting" could occur. That is the tracking error converges to a small value where it stays for some time before it bursts to large values and then again converges back to the small steady-state value. Condition (Equation 35.42) does not exclude such phenomena. One explanation of the "bursting" phenomenon is that when the signals are small the modeling error becomes dominant and could force the estimated parameters to drift to the unstable region. The plant output then grows causing the good signals to be come dominant which then force the estimated parameters to move to a region that corresponds to a smaller tracking error. This phenomenon can keep repeating itself and is a major drawback. One way to get rid of it is to use PE signals but this is not practical as such signals will interfere with the performance objective. A more practical approach is to use a dead zone to switch-off adaptation when convergence to steady-state values is achieved [1].

35.3 Nonidentifier-Based Adaptive Control

The stabilizability issues encountered in the case of indirect adaptive control as well as the nonlinearities introduced by the adaptive law in the feedback control system do not allow the use of the powerful control design techniques developed for LTI plants. Because of these two main reasons a number of researchers explored the possibility of controlling plants with unknown parameters without necessarily following the traditional adaptive control procedure that relies on the certainty equivalence principle. This let to a wide class of adaptive schemes with and without parameter estimation, with switching and supervisory logic in an effort to utilize as much as possible tools from robust control of known LTI systems. In the following sections we discuss some of these main approaches.

35.3.1 Switching and Multiple Models

In multiple model adaptive control (MMAC) schemes [22,31], since the plant parameters are unknown, the parameter space is parameterized with respect to a set of plant models, which is used to design a finite set of controllers, so that each plant model from the set can be stabilized by at least one controller from the controller set. A switching approach is then developed so that the stabilizing controller is selected online based on the I/O data measurements. The general structure of this MMAC with switching, as it is often called, is shown in Figure 35.5.

In Figure 35.5, N controllers are used to control a plant whose parameters θ^* are unknown or could change with time. The *a priori* knowledge of where the elements of θ^* are located, such as lower and upper bounds, is used to parameterize the plant and generate a finite set of nominal models $M_1, \dots, M_N$, as well as a finite set of candidate controllers $C_1, \dots, C_N$ so that for each possible plant there exists at least one stabilizing controller from the set of the N controllers. This by itself could be a difficult task in some practical situations where the plant parameter regions are uncertain or change in an unpredictable manner. Furthermore, since there is an infinite number of plants within any given bound of parametric uncertainty, finding controllers to cover all possible parametric uncertainties may also be challenging. It is usually assumed that the set of controllers with the property that at least one of them is stabilizing is available. Once the set of controllers with the stabilizing property is available, the problem of finding the stabilizing one using I/O data has to be resolved. This is achieved by the use of a switching logic that differs in detail from one approach to another.

An example of such logic is described as follows: each nominal model M_i is used to construct an output-predictor E_i, a dynamical system whose inputs are the output y and the input u of the unknown plant, and whose output is the prediction output $\hat{y}_i$. Each $\hat{y}_i$ will converge to y asymptotically if the transfer

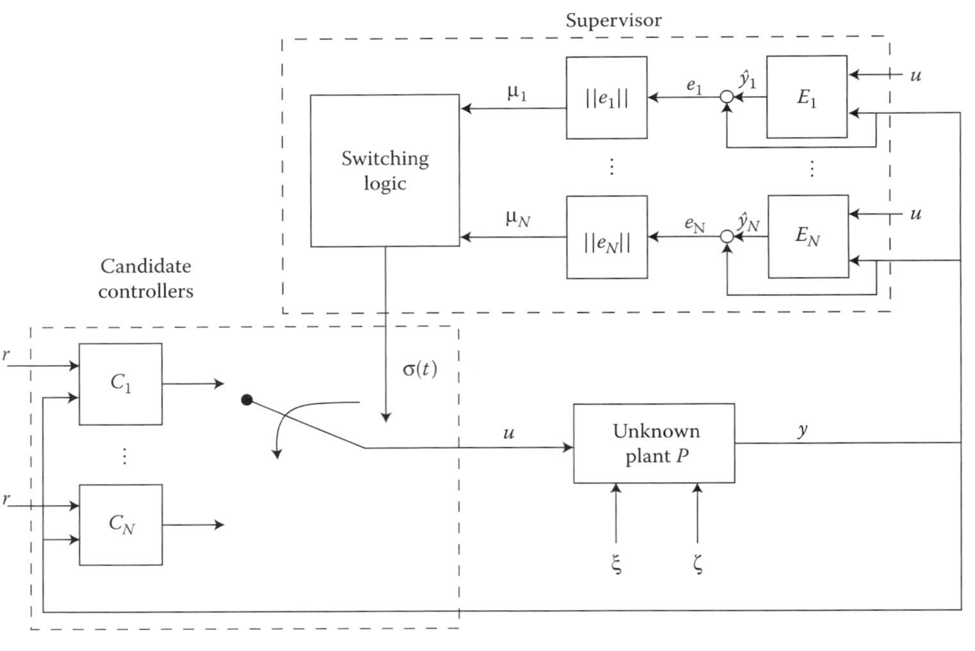

FIGURE 35.5 The MMAC architecture.

function of the unknown plant is equal to the nominal process model transfer function M_i and there is no noise or disturbances. Controller selection is decided by comparing in real-time suitably defined norm-squared prediction errors, also referred to as *performance signal*. Typically, the performance signal is an exponentially weighted integral of prediction errors

$$\mu_i(t) = \int_0^t e^{-2\lambda(t-\tau)} e_i^2(\tau)\, d\tau. \tag{35.43}$$

where $e_i(t) = y(t) - \hat{y}_i(t)$ is the prediction error associated with the ith nominal model.

In Figure 35.5, $\sigma(t)$ is a piecewise continuous switching signal that takes on values from the candidate controller index set. The candidate controller associated with the smallest performance signal is placed in the loop according to an appropriate switching logic. The stability results associated with the MMAC schemes state that there exists a number $\bar{\delta}$ such that if the unmodeled dynamics are smaller than $\bar{\delta}$, then all the closed-loop signals remain bounded. Furthermore, switching stops after some finite time [32].

Following the idea of supervisory control, logic-based switching and multiple models are combined with conventional adaptive control [22] with the objective of improving the sometimes poor transient performance of conventional adaptive schemes.

Another MMAC approach is the so-called robust MMAC (RMMAC), whose basic structure is described in Figure 35.6. RMMAC provides guidelines for designing both the candidate controller set and the supervisor [33,34]. The candidate controller set is designed by using robust control techniques such as the mixed-μ synthesis, in order to account for robust stability and performance requirements. The supervisor identifies the nearest probabilistic model via a dynamic hypothesis testing, and uses the probabilities obtained to weight the output of each candidate controller.

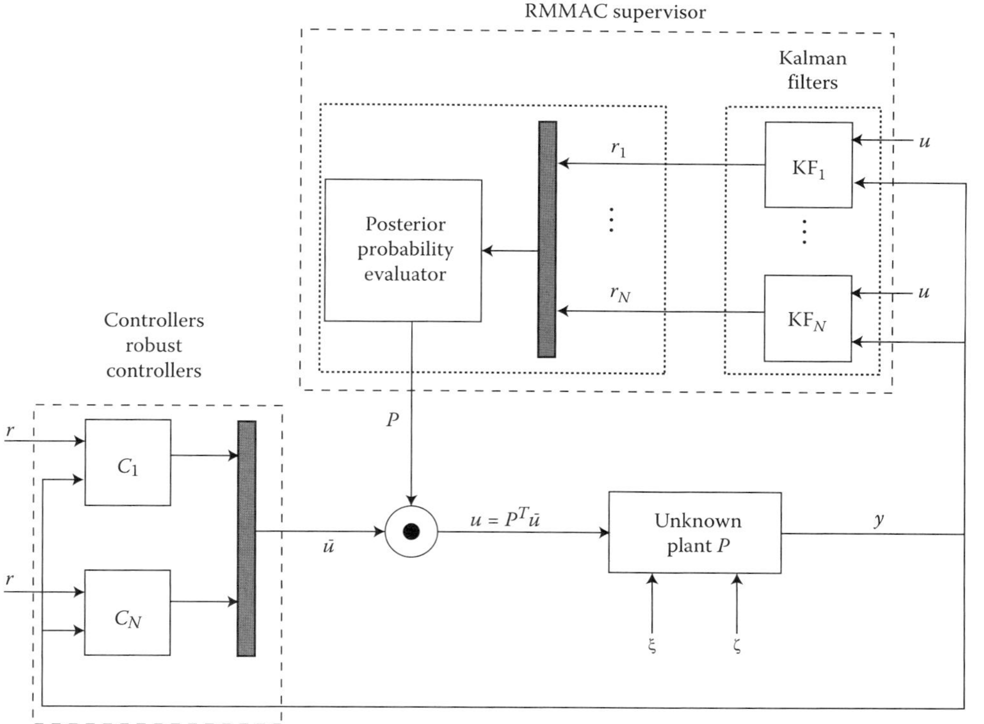

FIGURE 35.6 The RMMAC architecture.

Given a state-space description of the plant, where the system matrices depend upon a real-valued parameter vector $\theta \in \Omega$, the space Ω is partitioned into N small subsets Ω_i. For each subset Ω_i a candidate controller C_i is designed and a stochastic LTI model M_i is developed, including disturbance covariances, which are used to design a set of N Kalman filters (KFs), one for each model M_i.

The supervisor comprises the set of KFs plus a posterior probability evaluator. For each KF, the residual $r_i(t) = y(t) - \hat{y}_i(t|t-1)$ is calculated. The posterior probability $P_i(t)$, that is the probability that the model corresponding to the ith KF is the true one, is computed using the recursive formula

$$P_i(t) = \frac{\beta_i e^{r_i^T(t) S_i^{-1} r_i(t)}}{\sum_{j=1}^{N} \beta_j e^{r_j^T(t) S_j^{-1} r_j(t)}} P_i(t-1) \tag{35.44}$$

with $\beta_i = 1/\sqrt{2\pi \det(S_i)}$, and S_i is the residual covariance matrix, which can be precalculated using the steady-state covariance equations for the KF. The control input is the sum of the output of each candidate controller, weighted with the probabilities obtained:

$$u(t) = \sum_{i=1}^{N} P_i(t) u_i(t). \tag{35.45}$$

The RMMAC scheme requires the disturbances to be Gaussian random variables with known covariance. Furthermore, the KFs must satisfy a Baram proximity measure (BPM) requirement (see [33] for details). When all the assumptions are satisfied, extensive simulations show, at least empirically, the goodness of the method under various scenarios. A drawback of the scheme is that the performance of the RMMAC scheme may be sensitive to model assumptions (disturbance model, initial conditions, etc.). For this reason, the same authors developed a variant architecture, called RMMAC/XI, in order to account for possible performance degradation of the standard RMMAC when the plant disturbances have wide variability, at the expense of doubling the number of KFs. Another drawback of the method is that currently, there are no stability results. Despite these drawbacks, the scheme has been demonstrated to perform very well compared to other schemes in many situations. Recently, the RMMAC scheme was integrated with the Stability Overlay (SO), an algorithm that can be integrated with virtually any MMAC architecture, guaranteeing the stability of the closed-loop system, while preserving the high levels of performance observed in the simulation of the standard RMMAC, whenever the model assumptions are not violated [35]. The SO is based on a falsification philosophy: a controller receives a "reward" if a stability test is satisfied and the controller is disqualified or not, based on its rewards. However, unlike [36], where the rewards are also used to achieve performance, the SO is only responsible for the I/O stability of the plant, constraining the controllers that can be selected at each sampling time. The RMMAC algorithm is run in parallel in order to satisfy the posed performance requirements.

35.3.2 Unfalsified Adaptive Control

Another class of supervisory-based adaptive control schemes proposed in the literature aimed to relax the assumptions on the plant model required in the multiple model adaptive schemes. This family of nonidentifier-based schemes, referred to as unfalsified adaptive switching control (UASC), shown in Figure 35.7, compares at every time instant certain performance-related test functionals generated using a virtual reference input signal [36,37]. The currently operating controller is substituted whenever its tested performance turns out to be worse than the inferred performance of another candidate controller. Unfalsified control requires no *a priori* knowledge of the plant [26], and relies solely on I/O data to choose the stabilizing controller from a given set of controllers.

Given an unknown plant P and a family of N LTI controllers $C = \{C_i(s), i \in \overleftarrow{N}\}$, the switching supervisor aims at controlling the uncertain plant by switching at any time a selected controller $C_{\sigma(t)}$ from a set of candidate controllers according to a specified switching logic. The so-called *fictitious reference*

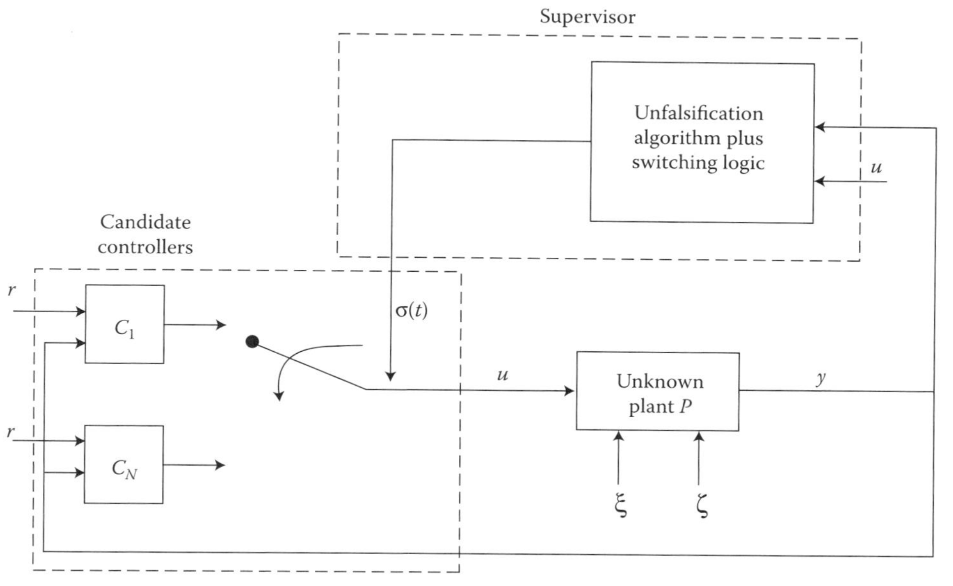

FIGURE 35.7 The UASC architecture.

signals or *virtual reference* v_i, are constructed using I/O data. Given at time t a set of past measured data $(u(\cdot), y(\cdot))$ obtained over $[0, t]$ from the feedback system $(P, C_{\sigma(\cdot)})$, v_i equals the fictitious reference that, if injected into the feedback system (P, C_i) would reproduce the unknown plant I/O pairs $(u(\cdot), y(\cdot))$.

For a candidate controller C_i, the performance index, or test functional $\mathcal{J}_i$ is calculated from past I/O data. The functional is a measure of performance of the closed-loop system over the interval $[0, t]$, had the controller C_i been in the closed loop with the corresponding v_i as the reference signal and the past I/O pairs $(u(\cdot), y(\cdot))$ as the measured data. A typical test functional used in UASC is

$$\mathcal{J}_i(t) = \max_{0 \leq \tau < t} \mathcal{L}_i(\tau), \quad \mathcal{L}_i(\tau) = \frac{\|\epsilon_i^\tau\|_2^2 + \rho\|u^\tau\|_2^2}{m^2 + \|v_i^\tau\|_2^2}, \tag{35.46}$$

where $\rho > 0$ is a design constant, $\epsilon_i(\tau) \triangleq v_i(\tau) - y(\tau)$, and $m^2 > 0$ prevents the denominator from assuming values too close to zero. The notation $\|x^\tau\|_2$ stands for the l_2 norm of the sequence $x(\cdot)$ up to time τ. The selection of the controller index $\sigma(t)$ is made, at each time t, via a *hysteresis switching logic* [24]. The virtual reference is well defined, provided that the C_i's be causal and stably causally invertible (CSCI). While the results summed up hinge upon the assumption that all C_i's be CSCI, in [38], it is proved that the same conclusions hold true for possibly nonstably invertible C_i's provided that a modified virtual reference, be appropriately used.

The main positive feature of UASC schemes is the fact that, even in the absence of any prior information on the uncertain plant (the plant can be of any order, unstable, nonminimum phase, nonlinear or of infinite order), they can select in finite time a final controller yielding, a finite affine gain from the reference to the data, under the minimal conceivable requirement, namely the existence of a stabilizing candidate controller in the candidate controller set. One of course could argue that to design a stabilizing controller to start with you need to know something about the plant. One major drawback is that there is no stability result that guarantees that convergence to the stabilizing controller will be achieved in general. In fact, the system could converge to an unstable controller leading to an internally unstable closed-loop system when the adaptive procedure is switched off. Since the method is driven by data, the quality of the data is crucial. If the data do not carry information about the plant or are corrupted by initial conditions and/or disturbances and other plant uncertainties the wrong controllers could be selected leading to

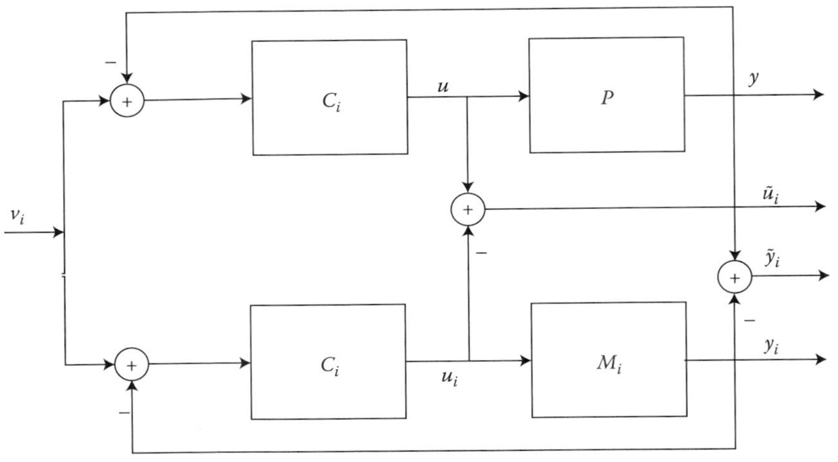

FIGURE 35.8 Detail of multi-model UASC.

large transients which from the practical point of view are unacceptable. Such bad transients have been observed by independent simulation studies [39,40].

When candidates of possible plant models are available, they can be incorporated into the design of the candidate controllers, and the construction of test functionals, to obtain a multimodel UASC (MUASC) scheme with improved performance [27]. The multimodel based UASC approach consists of embedding in the UASC scheme a family of nominal models M_i pairwise associated with the given candidate controllers C_i in such a way that the feedback interconnection (M_i, C_i) is stable.

Consider now, according to Figure 35.8, the feedback interconnection of each model M_i with the corresponding C_i

$$\begin{cases} y_i(t) = M_i(u_i)(t), \\ u_i(t) = C_i(v_i - y_i)(t), \end{cases}$$

where v_i is the virtual reference input for the ith controller. Define $z_i(t) = \left[u_i(t)\, y_i(t)\right]'$, and $\tilde{z}_i(t) = [\,\tilde{u}_i(t)\ \tilde{y}_i(t)\,]' = z(t) - z_i(t)$. Note that $\tilde{z}_i$ measures the discrepancy of (P, C_i) from the tuned-loop (M_i, C_i).

The MUASC algorithm replaces the test functional (Equation 35.46) with the *multimodel-based test functional*

$$\mathcal{J}_i(t) = \max_{\tau<t} \Lambda_i(\tau), \tag{35.47}$$

$$\Lambda_i^{1/2}(\tau) = \begin{cases} 0, & \text{if } \|z^\tau\|_2 = 0, \\ \dfrac{\|\tilde{z}_i^\tau\|_2}{\|z_i^\tau\|_2}, & \text{elsewhere.} \end{cases} \tag{35.48}$$

In [27], it is shown that the computation of v_i can be avoided in order to calculate Equations 35.47 and 35.48, so that the validity of Equations 35.47 through 35.48 holds true for general controllers C_i, even when, as in the case of nonstably invertible controllers, v_i is not well defined.

As in UASC, for the MUASC scheme, stability is guaranteed under the minimal conceivable assumption that a stabilizing candidate controller exists. Furthermore, with MUASC, guidelines are provided how to design the family of nominal models and the family of candidate controllers, so that the number of controller switching is reduced and the chance that destabilizing controllers be switched-on is moderated, which helps reduce transients. These properties cannot in general be guaranteed in schemes with no nominal models. Extensive simulation studies demonstrate the advantages of MUASC.

35.4 Mixed Identifier and Nonidentifier-Based Tools

35.4.1 Adaptive Control with Mixing

To reduce sensitivity to model assumptions that occurs in RMMAC, a design, called adaptive mixing control (AMC), shown in Figure 35.9, substitutes the posterior probability evaluator with a robust online parameter estimator plus a mixing strategy for the output of each candidate controller [41].

The supervisor consists of two subsystems: the online parameter estimators and the mixer. By monitoring the results of the robust parameter estimators at each time t the adaptive control mixer decides which outputs of controllers to combine and pass on to the plant. The candidate controllers are designed to handle overlapping parameter sets and therefore mixing their outputs up when the estimated parameters fall into these overlapping parameter sets is appropriate. Another advantage is that no switching takes place but rather a smooth transition from one controller or combination of controllers to another. We present below some of the technical details and stability results of the AMC approach. Consider the SISO LTI plant:

$$y = G_0(s,\theta^*)(1+\Delta_m(s))(u+\xi)+\zeta, \tag{35.49}$$

where $G_0(s,\theta^*)$ represents the nominal plant; the vector $\theta^* \in \Omega$ contains the unknown parameters of $G_0(s,\theta^*)$; $\Delta_m(s)$ is an unknown multiplicative perturbation; and ξ, ζ are bounded disturbances, that is, $|\xi(t)| \leq \xi_0$, $|\zeta(t)| \leq \zeta_0$, $\forall t \geq 0$.

As with the MMAC schemes, the parameter uncertainty set Ω is divided into n smaller subsets Ω_i, and a family of N candidate controllers $C_i(s), \ldots, C_N(s)$ is designed using the powerful tools from robust control for LTI plants. We use a mixing strategy to develop the multicontroller $\mathbb{C}(\beta,\theta)$, which is constructed from $C_i(s), \ldots, C_N(s)$ by taking into account linear combinations of these controllers. The multicontroller depends on the mixing signal $\beta \in \mathbb{R}^N$ which determines the participation level of the candidate controllers generating a mix of candidate control laws. For fixed values of $\beta \in [0,1]^N$, the multicontroller

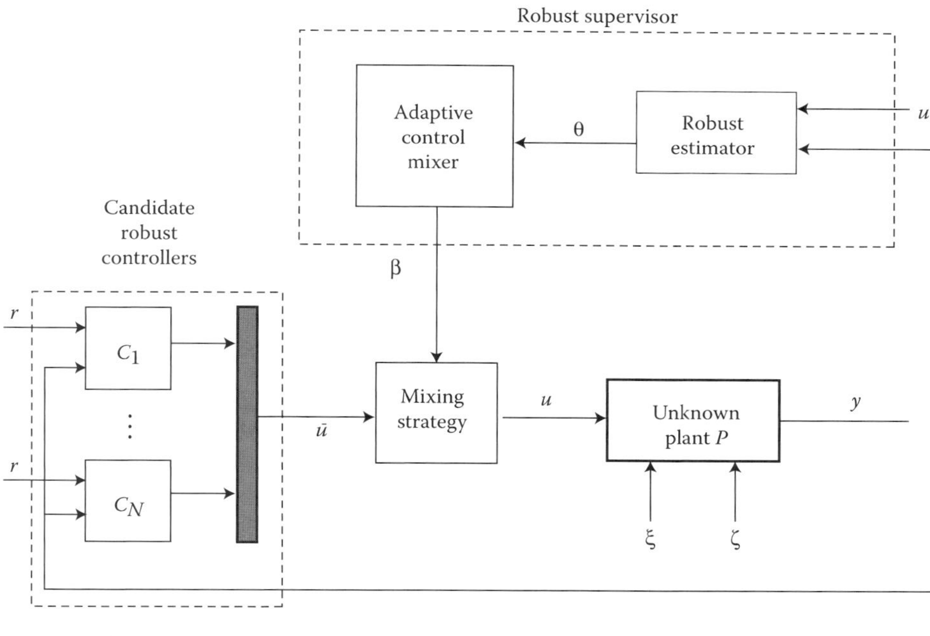

FIGURE 35.9 The AMC architecture.

$u = -C(s; \beta, \theta)y$ has the state-space form

$$\dot{x}_C = A_C(\beta, \theta)x_C + B_C(\beta, \theta)y,$$
$$u = -C_C(\beta, \theta)x_C.$$

The mixer determines the participation level β_i of each candidate controller, through the map $\beta : \Omega \mapsto [0, 1]^N$. Construction of the multicontroller involves interpolating the candidate controllers over the parameter overlaps. The mixing strategy must be designed to ensure stability on the overlapping regions. As in gain scheduling, interpolation methods may not satisfy pointwise stability over the set Ω; therefore the controllers should be designed by taking into account the interpolation used in order to guarantee stability.

The adaptive mixing law approach replaces θ^* with its estimate θ. Because of the presence of multiplicative uncertainty $\Delta_m(s)$ and disturbances ξ, ζ, a robust online parameter estimator, like those developed in Section 35.2.6 is used.

The main stability properties of the AMC scheme for the case of regulation are [41]

1. If $\Delta_m, \xi, \zeta = 0$, then $u, y \to 0$, $\dot{u}, \dot{y} \to 0$ as $t \to \infty$.
2. If $\Delta_m, \xi, \zeta \neq 0$, There exists $\mu^* > 0$ such that, if $c\Delta_1^2 < \mu^*$, where Δ_1 depends on Δ_m, and $c > 0$ is a finite constant, then the AMC scheme guarantees $u, y, \dot{u}, \dot{y} \in \mathcal{L}_\infty$, and

$$\int_0^t |y(\tau)|^2 \, d\tau \leq c_0\mu^2 t + c_1, \tag{35.50}$$

where $\mu^2 = c(\Delta_1^2 + \xi_0^2 + \zeta_0^2)$.

The stability properties and performance in simulations of different schemes have been studied using a benchmark example in [40] demonstrate the advantages and disadvantages of the different approaches.

35.5 Conclusions

The area of robust adaptive control has grown to include classes of different techniques and algorithms based on different assumptions with the goal to control plants with large parametric uncertainties which cannot be handled by nonadaptive schemes. A significant achievement of most of these efforts is that robust adaptive control in general can effectively deal with parametric uncertainties as no other schemes. In addition, it can also handle modeling errors, bounded disturbances, noise, and so on. While some schemes perform better than others, in general, all schemes suffer from the same fundamental limitations. Since they all rely on online learning to design or choose the appropriate controller, the plant performance relies on the quality of learning which in turn relies on the quality of the I/O data it processes. Corruption of the I/O data by initial conditions, disturbances, modeling errors, and so on could influence learning and lead to wrong selection of controllers for some intervals of time especially initially. This will lead to large transients which in turn improve the situation as the signal-to-noise ratio increases aiding the learning process and leading to the selection of a better controller. The scheme that causes the least transients is obviously better. Expecting an adaptive scheme to have similar transient behavior as in the nonadaptive case is very unrealistic and possibly counter intuitive in general. Another fundamental limitation of robust adaptive control is that the closed-loop system is nonlinear. Therefore, poles and zeros no longer make sense and measures such as gain and phase margins, bandwidth, and so on are not applicable as they are only defined for LTI systems. The nonidentifier-based schemes aim to reduce this drawback by switching from one LTI controller to another so that at least within each time interval the closed loop system behaves as an LTI system. Even in this case, however, convergence to the desired stabilizing controller cannot be guaranteed unless there is persistence of excitation which is unrealistic in most practical situations. Convergence to a nonstabilizing controller which in the closed loop guarantees zero regulation or tracking error is possible and this is what is usually achieved. As long as the scheme is switched-on, then every thing will stay as it is till the plant parameters change and that is when switching will start again. The library of adaptive control tools developed together with a good understanding of the advantages and limitations of different tools as well as a good understanding of the limitations associated with trying to control a plant while at the same time trying to learn its parameters from I/O data are very important in every application of robust adaptive control.

References

1. K.S. Tsakalis and P.A. Ioannou. *Linear Time Varying Systems: Control and Adaptation*. Prentice-Hall, Englewood Cliffs, NJ, 1993.
2. H.P. Whitaker, J. Yamron, and A. Kezer. *Design of Model Reference Adaptive Control Systems for Aircraft*. Report R-164, Instrumentation Laboratory, MIT, Cambridge, MA, 1958.
3. R.E. Kalman. Design of a self-optimizing control system. *Trans. ASME*, 80:468–478, 1958.
4. R.E. Bellman. *Dynamic Programming*. Princeton University Press, Princeton, NJ, 1957.
5. A.A. Feldbaum. *Optimal Control Systems*. Academic Press, New York, 1965.
6. P.C. Parks. Lyapunov redesign of model reference adaptive control systems. *IEEE Trans. Automat. Control*, 11:362–367, 1966.
7. A.S. Morse. Global stability of parameter adaptive control systems. *IEEE Trans. Automat. Control*, 25:433–439, 1980.
8. K.S. Narendra, Y.H. Lin, and L.S. Valavani. Stable adaptive controller design, Part II: Proof of stability. *IEEE Trans. Automat. Control*, 25:440–448, 1980.
9. I.D. Landau. *Adaptive Control: The Model Reference Approach*. Marcel Dekker, New York, 1979.
10. G.C. Goodwin, P.J. Ramadge, and P.E. Caines. Discrete-time multi-variable adaptive control. *IEEE Trans. Automat. Control*, 25:449–456, 1980.
11. B. Egardt. *Stability of Adaptive Controllers*, Lecture Notes in Control and Information Sciences, Springer-Verlag, Berlin, 1979.
12. P.A. Ioannou and P.V. Kokotovic. Instability analysis and improvement of robustness of adaptive control. *Automatica*, 20:583–594, 1984.
13. C.E. Rohrs, L. Valavani, M. Athans, and G. Stein. Robustness of continuous-time adaptive control algorithms in the presence of un- modeled dynamics. *IEEE Trans. Automat. Control*, 30:881–889, 1985.
14. P.A. Ioannou and J. Sun. Theory and design of robust direct and indirect adaptive control schemes. *Int. J. Control*, 47:775–813, 1988.
15. L. Praly. Robust model reference adaptive controllers, Part I: Stability analysis. *Proc. of the 22nd IEEE Conference on Decision and Control*, San Antonio, Texas, 1984.
16. P.A. Ioannou and K.S. Tsakalis. A robust direct adaptive controller. *IEEE Trans. Automat. Control*, 31:1033–1043, 1986.
17. M. Krstic, I. Kanellakopoulos, and P. Kokotovic. *Nonlinear and Adaptive Control Design*. John Willey & Sons Inc, New York, 1995.
18. Y. Zhang, B. Fidan, and P.A. Ioannou. Backstepping control of linear time-varying systems with known and unknown parameters. *IEEE Trans. Automat. Control*, 48:1908–1925, 2003.
19. B. Fidan, Y. Zhang, and P.A. Ioannou. Adaptive control of a class of slowly time-varying systems with modeling uncertainties. *IEEE Trans. Automat. Control*, 50:915–920, 2005.
20. M.M. Polycarpou. Stable adaptive neural control scheme for nonlinear systems. *IEEE Trans. Automat. Control*, 41:447–451, 1996.
21. H. Xu and P.A. Ioannou. Robust adaptive control of linearizable nonlinear single input systems with guaranteed error bounds. *Automatica*, 40:1905–1911, 2003.
22. K.S. Narendra and J. Balakrishnan. Adaptive control using multiple models. *IEEE Trans. Automat. Control*, 42:171–187, 1997.
23. B.D.O. Anderson, T.S. Brinsmead, D. Liberzon, and A.S. Morse. Multiple model adaptive control with safe switching. *Int. J. Adapt. Control Signal Process*, 15:445–470, 2001.
24. A.S. Morse, D.Q. Mayne, and G.C. Goodwin. Applications of hysteresis switching in parameter adaptive control. *IEEE Trans. Automat. Control*, 37:1343–1354, 1992.
25. P.V. Zhivoglyadov, R.H. Middleton, and M. Fu. Localization based switching adaptive control for time-varying discrete-time systems. *IEEE Trans. Automat. Control*, 45:752–755, 2000.
26. M. Stefanovic and M.G. Safonov. Safe adaptive switching control: Stability and convergence. *IEEE Trans. Automat. Control*, 53:2012–2021, 2008.

27. S. Baldi, G. Battistelli, E. Mosca, and P. Tesi. Multi-model unfalsified adaptive switching supervisory control. *Automatica*, 46:249–259, 2010.
28. P.A. Ioannou and J. Sun. *Robust Adaptive Control*. Prentice-Hall, Englewood cliffs, NJ, 1996.
29. P.A. Ioannou and B. Fidan. *Adaptive Control Tutorial*. Advances in design and control, SIAM, Philadelphia, PA, 2006.
30. P.A. Ioannou and A. Datta. Robust adaptive control: Design, analysis and robustness bounds, in P.V. Kokotovic Ed., *Grainger Lectures: Foundations of Adaptive Control*, Springer-Verlag, New York, 1991.
31. J.P. Hespanha, D. Liberzon, and A.S. Morse. Hysteresis-based switching algorithms for supervisory control of uncertain systems. *Automatica*, 39:263–272, 2003.
32. A.S. Morse. Supervisory control of families of linear set-point controllers, Part 1: Exact matching. *IEEE Trans. Automat. Control*, 41:1413–1431, 1996.
33. S. Fekri, M. Athans, and A. Pascoal. Issues, progress and new results in robust adaptive control. *Int. J. Adapt. Control Signal Process.*, 20:519–579, 2006.
34. S. Fekri, M. Athans, and A. Pascoal. Robust multiple model adaptive control (rmmac): A case study. *Int. J. Adapt. Control Signal Process.*, 21:1–30, 2007.
35. P. Rosa, J.S. Shamma, C.J. Silvestre, and M. Athans. Integration of the stability overlay (so) with the robust multiple-model adaptive control (rmmac). *Proc. of the 17th Mediterranean Conference on Automation and Control*, Thessaloniki, Greece, 2009.
36. M. Stefanovic, R. Wang, A. Paul, and M.G. Safonov. Cost detectability and stability of adaptive control systems. *Int. J. Robust Nonlinear Control*, 17:549–561, 2007.
37. M.G. Safonov and T.C. Tsao. The unfalsified control concept and learning. *IEEE Trans. Automat. Control*, 42:843–847, 1997.
38. C. Manuelli, S.G. Cheong, E. Mosca, and M.G. Safonov. Stability of unfalsified adaptive control with non-scli controllers and related performance under different prior knowledge. *Proc. of the European Control Conference*, Kos, Greece, 2007.
39. A. Dehghani, B.D.O. Anderson, and A. Lanzon. Unfalsified adaptive control: A new controller implementation and some remarks. *Proc. of the European Control Conference*, Kos, Greece, 2007.
40. S. Baldi, P. Ioannou, and E. Mosca. Evaluation of identifier based and non-identifier based adaptive supervisory control using a benchmark example. *4th International Symposium on Communications Control and Signal Processing*, Limassol, Cyprus, 2010.
41. M. Kuipers and P.A. Ioannou. Multiple model adaptive control with mixing. *IEEE Trans. Automat. Control*, to appear in 2010.

36

Iterative Learning Control

Douglas A. Bristow
Missouri University of Science and Technology

Kira L. Barton
University of Illinois at Urbana–Champaign

Andrew G. Alleyne
University of Illinois at Urbana–Champaign

36.1 Introduction

Iterative learning control (ILC) is a performance-enhancing feedforward control scheme for systems that repeat the same trajectory or task. Before the start of each iteration of the trajectory, the designed ILC algorithm uses the error signal from the previous iteration(s) to generate an updated feedforward control signal. The learning process converges after anywhere from a few to tens of iterations, depending on the algorithm. In the literature it is commonly reported that ILC improves the performance of physical systems by several orders of magnitude, measured by root mean square (RMS) or maximum error, as compared to those systems' feedback controllers.

The essential caveat in ILC is system repeatability, meaning that multiple repetitions of the trajectory or task yield nearly identical error signals. Although this may appear to be very restrictive, many practical systems are highly repetitive. This is particularly true in manufacturing systems, as will be discussed shortly. When the system is repeatable, one expects that for a given trajectory there exists a control signal that yields zero or nearly zero error. In practice, it is rare that this control signal is known *a priori*. However, in applications where the trajectory repeats ILC is used to search for, or *learn*, the optimal control signal. The process is illustrated in Figure 36.1.

ILC was originally developed in the mid-1980s for robotics [2] and, although robotics remains an important application area, it has expanded to a much broader class of manufacturing and chemical processing systems. Examples (see references in [3,5,6]) include computer numerically controlled (CNC) machine tools, wafer stage motion systems, injection molding machines, aluminum extruders, cold rolling

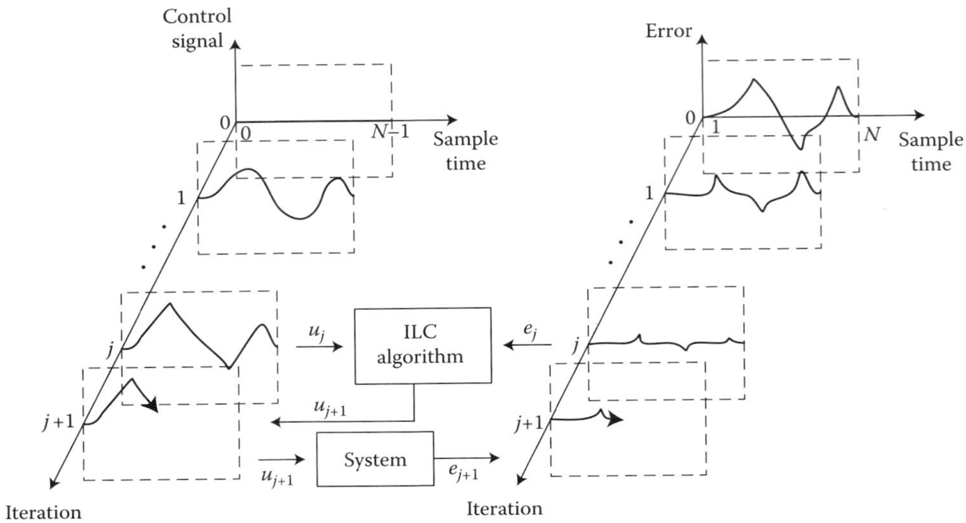

FIGURE 36.1 Time- and iteration-domain evolution of signals in ILC.

mills, induction motors, chain conveyor systems, rapid thermal processing, and semibatch chemical reactors. There are many books and surveys on ILC [3–6,9,12,13,18]. This chapter focuses on summarizing the main ideas in ILC and presenting established design techniques for linear systems. The reference list to this chapter is a good starting place for the reader interested in continuing beyond the information contained in the following sections.

36.1.1 Implementation Hardware

Before delving into the algorithms of ILC, the practicing engineer may be interested in the hardware needs of such an advanced control algorithm. Most advantageously, ILC is usually implemented as an offline algorithm, in between iterations. In an offline mode, the ILC algorithm can usually be computed using the control system's existing digital signal processor (DSP) or on another computational platform such as a PC. In high-throughput manufacturing where offline time is expensive, a simple ILC algorithm is preferable to minimize downtime. Using a modern DSP, offline time should be on the order of milliseconds for all but the most complex algorithms or longest trajectories. Additionally, offline time can be eliminated by "freezing" the control signal after convergence. That is, learning is used only on the first few iterations of the process until the control signal converges. The remaining process runs skip the learning step, eliminating downtime, and simply reuse the already converged feedforward control signal.

Because ILC requires the storage of error signals and control signals for offline computation, memory capacity (usually RAM) must be considered. The amount of memory required is calculated as follows:

$$\text{Memory (bits)} = 2 \times \begin{array}{c}\text{No. of time}\\ \text{samples/iteration}\end{array} \times \begin{array}{c}\text{Bits of}\\ \text{resolution}\end{array} \times \text{No. of axes} \times \begin{array}{c}\text{Order of}\\ \text{algorithm}\end{array},$$

where "2" is used to account for the error signal *and* the control signal.

36.1.2 Outline

The remainder of this chapter is organized as follows. A basic ILC algorithm is introduced and tuned for a servo system example problem in Section 36.2. Two design frameworks are presented in this chapter. The first one, frequency-domain design, is presented in Section 36.3. This is a simpler approach that is less general than the second one. The second technique, time-domain norm-optimal design, requires additional preliminary setup and results before it can be presented. As such, Section 36.4 presents the class of systems and Section 36.5 presents some analytical convergence and robustness results. The norm-optimal design is presented in Section 36.6. A brief discussion of advanced algorithms and concluding remarks are given in Section 36.7.

36.2 Example: A Servo System

Consider the discrete-time system,

$$G(z) = \frac{z - 0.5}{(z - 1)(z - 0.925)}, \tag{36.1}$$

whose Bode plot is shown in Figure 36.2. Discrete-time is assumed because ILC requires the storage of signals, which is typically done digitally. Here, G represents a servo-positioning system with viscous friction. Assume that the system is stabilized with a proportional feedback controller,

$$C(z) = 0.425. \tag{36.2}$$

The desired output trajectory is the triangle wave (shown in Figure 36.3a),

$$y_d(k) = \begin{cases} k/200, & 0 \le k \le 200, \\ 2 - k/200, & 201 \le k \le 400, \end{cases} \tag{36.3}$$

where $k = 0, 1, \ldots, 400$ is the time index. The triangle wave is used, for example, in scanning operations where the forward and backward motions are scanning at the same velocity. Figure 36.3b shows the error signal from the feedback controller. The goal is to reduce the tracking error by adding an ILC to the control system.

ILC is a plug-in-type controller that can be added to an existing control loop, as shown in Figure 36.4, by adding the ILC signal to the feedback controller's signal. The reader may note that the ILC signal is injected in the same location in the loop that is often used for feedforward control in servo systems. This does not, however, exclude the use of ILC on systems that already employ a feedforward controller [5]. Alternatively, ILC can be added to the reference signal, which is most commonly used on closed-architecture control systems [5,11].

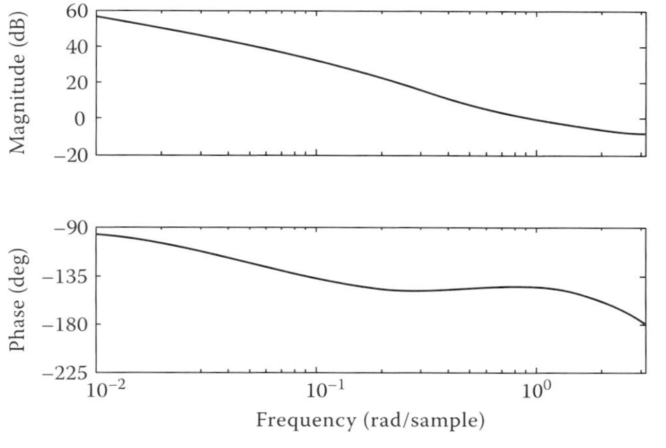

FIGURE 36.2 Bode plot of the open-loop servo plant in Equation 36.1.

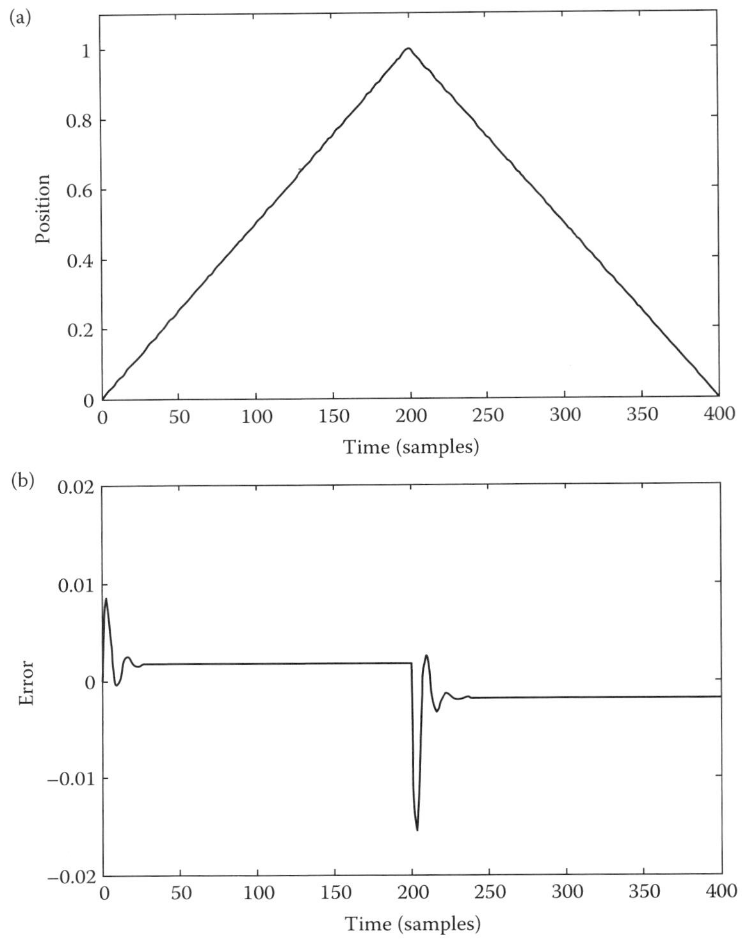

FIGURE 36.3 (a) Triangle reference signal. (b) Error signal using feedback controller, $e_0(k)$.

For ILC design, several assumptions are made:

A1. The reference signal repeats. Each repetition is referred to as a trial or iteration, and all iterations have the same length.
A2. Any external disturbance repeats identically on each iteration.
A3. Initial conditions (ICs) on each iteration are identical.

Assumptions A1–A3 are standard in ILC analysis and design. Although A2 and A3 are rarely strictly true, in practice iteration-to-iteration variations are often negligible in precision servo systems. The topic of iteration-to-iteration variation is discussed in more detail in Section 36.5.

Possibly the simplest approach is a first-order P-type ILC algorithm as given by

$$u_{j+1}(k) = u_j(k) + \gamma e_j(k+1), \tag{36.4}$$

where γ is the learning gain. One can think of Equation 36.4 as an integrator in the iteration domain, as follows. On each iteration that the error is nonzero, the control signal grows until the error finally reaches zero. The reader may note that the error signal is forward shifted by one time step in Equation 36.4. This

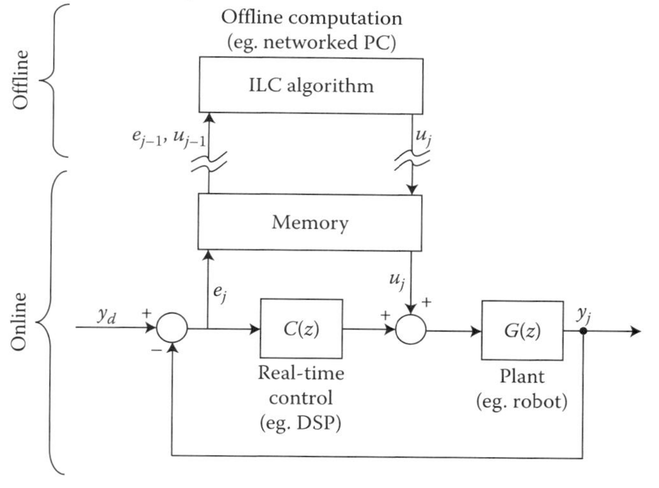

FIGURE 36.4 ILC added to a feedback control loop as a "plug-in" controller.

is standard in ILC and used to compensate for the usual one-step delay (or equivalently relative degree of one) in the plant, Equation 36.1. That is, $u_j(k)$ does not affect $e_j(k)$ because of the plant's one-step delay; hence, $e_j(k+1)$ is used instead.

The algorithm parameter γ can be tuned to achieve convergence with satisfactory performance. One suitable choice, obtained through trial-and-error tuning, for the previous example is $\gamma = 0.1$. The RMS error iteration series for this choice is shown in Figure 36.5. As shown in the figure, the error is asymptotically approaching zero at a rate of approximately one order of magnitude for every 50 iterations. Figure 36.6 shows the error, feedback control, and ILC time series for iterations 0, 1, 10, and 100. As the ILC converges, the control effort transfers from the feedback controller to the ILC, as shown by the signals in Figure 36.6. One measure of effectiveness of a feedforward control, such as ILC, is the magnitude of feedback control effort.

The P-type algorithm is the simplest algorithm that can be used. As a cautionary note, the P-type algorithm may not be sufficient for many systems. In particular, high-frequency resonances, common in servo systems, may be problematic for P-type learning [11]. In practice, it is more common to use an

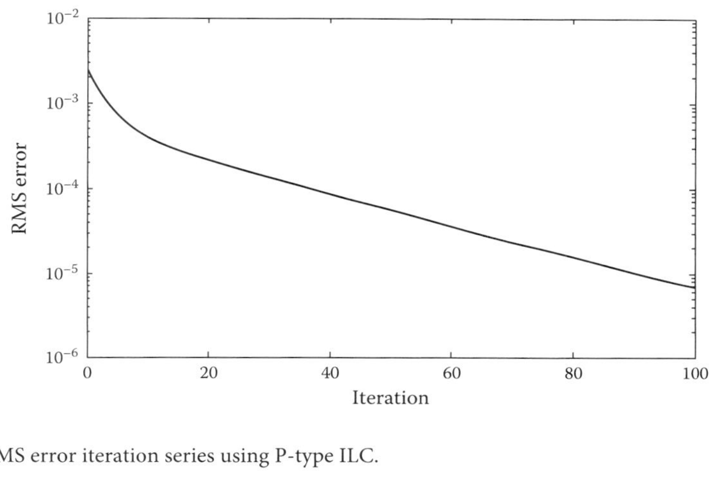

FIGURE 36.5 RMS error iteration series using P-type ILC.

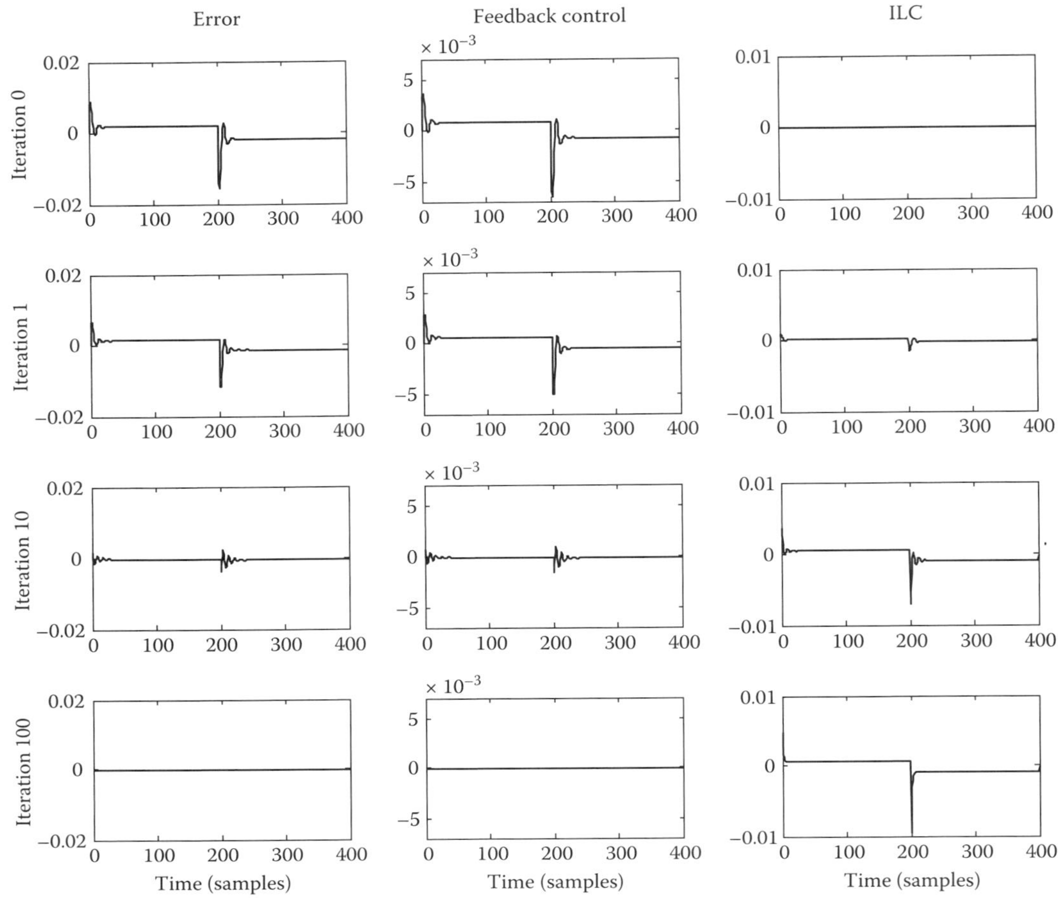

FIGURE 36.6 Error, feedback control, and ILC time series using P-type ILC.

algorithm of the form

$$u_{j+1}(k) = Q(z)(u_j(k) + L(z)e_j(k)), \tag{36.5}$$

where the mixture of time-domain signals and frequency-domain filters is a (standard) abuse of notation. The Q-filter is a low-pass filter used to limit the learning bandwidth to provide robustness and is particularly useful in servo systems with lightly damped system resonances. The L-filter, also called the learning filter, is designed to maximize the learnable bandwidth and convergence rate. Q and L designs are discussed in more detail in the following sections.

36.3 Frequency-Domain Design

The frequency-domain design framework uses classical control design tools such as z-transform and Nyquist plots [11,14]. A key additional assumption in this framework is that the iteration length (or reference length) is infinitely long. This is, of course, an approximation of reality because the iteration length is always finite in practice. However, this assumption permits the use of the z-transform, which greatly simplifies the design.

Assuming zero ICs, the output of the system shown in Figure 36.4 on the jth iteration can be calculated as

$$Y_j(z) = \frac{C(z)G(z)}{1+C(z)G(z)}Y_d(z) + \frac{G(z)}{1+C(z)G(z)}U_j(z). \quad (36.6)$$

The error on the jth iteration is $E_j(z) = Y_d(z) - Y_j(z)$, or,

$$E_j(z) = E_0(z) - H(z)U_j(z), \quad (36.7)$$

where

$$H(z) \equiv \frac{G(z)}{1+C(z)G(z)} \quad (36.8)$$

is the transfer function from the ILC input to the error and

$$E_0(z) \equiv \frac{1}{1+C(z)G(z)}Y_d(z) \quad (36.9)$$

is the initial error, or the error from the feedback control. It is easy to show that the addition of nonzero ICs and exogenous disturbance signals appear as additional terms in $E_0(z)$. Furthermore, changing the location of where the ILC enters into the control loop, such as the reference signal [5,11], changes only $H(z)$. Thus, one can conclude that the form of Equation 36.7 is general, although $E_0(z)$ and $H(z)$ depend on the specific control system architecture and disturbances.

36.3.1 Frequency-Domain Analysis

The frequency-domain description of the first-order ILC algorithm (Equation 36.5) is

$$U_{j+1}(z) = Q(z)(U_j(z) + L(z)E_j(z)). \quad (36.10)$$

Substituting Equation 36.7 into Equation 36.10, the iteration dynamics can be obtained as

$$U_{j+1}(z) = Q(z)(1 - L(z)H(z))U_j(z) + Q(z)L(z)E_0(z). \quad (36.11)$$

The goal in designing $L(z)$ and $Q(z)$ is to force a contractive mapping from $U_j(z)$ to $U_{j+1}(z)$, which is done by making

$$\|Q(z)(1 - L(z)H(z))\|_\infty < 1, \quad (36.12)$$

where $\|\bullet(z)\|_\infty \equiv \max_{-\pi\le\omega<\pi} |\bullet(e^{i\omega})|$. The contractive mapping ensures convergence of $U_j(z)$, which can be calculated from Equation 36.11 as

$$U_\infty(z) \equiv \lim_{j\to\infty} U_j(z) = \frac{Q(z)L(z)}{1 - Q(z)(1 - L(z)H(z))}E_0(z). \quad (36.13)$$

Substituting Equation 36.13 into Equation 36.7 yields the converged tracking error as

$$E_\infty(z) \equiv E_0(z) - H(z)U_\infty(z) = \left(\frac{1 - Q(z)}{1 - Q(z)(1 - L(z)H(z))}\right)E_0(z). \quad (36.14)$$

36.3.1.1 The Role of the Q-Filter

Fundamental to frequency-domain design is the low-pass Q-filter, which ideally has the frequency-domain characteristic

$$Q_{ideal}\left(e^{i\omega}\right) = \begin{cases} 1, & 0 \leq \omega \leq \omega_c \\ 0, & \omega_c < \omega \leq \pi \end{cases} \tag{36.15}$$

where ω_c is the cutoff frequency. For $Q_{ideal}(e^{i\omega})$, the stability condition (Equation 36.12) simplifies to

$$\left|1 - L(e^{i\omega})H(e^{i\omega})\right| < 1 \quad \text{for } 0 \leq \omega \leq \omega_c, \tag{36.16}$$

and the asymptotic error is given by

$$E_{\infty}^{Qideal}(e^{i\omega}) = \begin{cases} 0, & 0 \leq \omega \leq \omega_c \\ E_0(e^{i\omega}), & \omega_c < \omega \leq \pi. \end{cases} \tag{36.17}$$

Thus, the low-pass Q-filter simplifies the design problem by limiting the frequency range over which Equation 36.16 must be satisfied, but at the expense of performance. In fact, in the ideal Q-filter analysis, Q acts as a learning switch. When Q is unity at a particular frequency, that frequency is turned on for learning. When Q is zero at a particular frequency, that frequency is turned off. The goal in the Nyquist tuning approach presented in the following subsection is to design L to maximize the range $0 \leq \omega \leq \omega_c$ over which Equation 36.16 is satisfied. In doing so, Q can be "turned on" for the largest range of frequencies.

36.3.2 Tuning Based on Nyquist Analysis

The Nyquist-based tuning method, referred to more commonly in the literature simply as frequency-domain design [11], is a two-step design process as illustrated in Figure 36.7. The first step is to select a filter type for $L(z)$ and then to tune the filter parameters so that the Nyquist plot of $L(e^{i\omega})H(e^{i\omega})$ fits inside the unit circle centered at one for as high a frequency range $0 \leq \omega \leq \omega_c$ as the designer is able to achieve. This is equivalent to satisfying Equation 36.16 over the same frequency range. The second step is to construct a low-pass $Q(z)$ (e.g., Butterworth, Chebyshev Type I and II, etc.) that satisfies Equation 36.12. For an ideal low-pass filter, the bandwidth is ω_c, although for practical filters the actual bandwidth may vary. The result will give near-zero tracking error up to ω_c frequency.

Any type of filter may be used for $L(z)$, although some popular choices are listed in Table36.1. The phase lead (and PD-type to a lesser extent) adds positive phase to partially compensate for the plant's phase lag. The model inversion type is intended to fully compensate for the plant's dynamics, but may be challenging to implement in practice. First, the inverse of the plant may not be stable, although stable, approximate inversion [17] can be used. Second, because many plants have high-frequency rolloff, the inverse results in high-frequency gain, which may amplify noise. Third, the designer may be unaware of the model's inaccuracy and the inverse may create a false perception of attainable bandwidth. However, this may be true of any filter choice and the designer should take appropriate care in implementation and

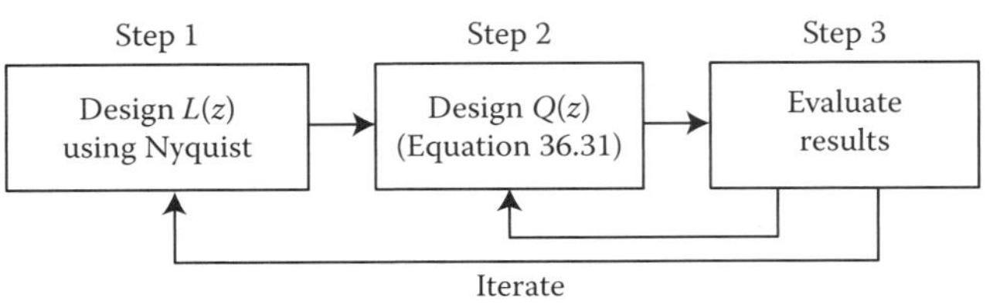

FIGURE 36.7 Nyquist tuning two-step design procedure.

TABLE 36.1 Common Filter Types for $L(z)$

Filter Name	Equation
P-type	$L(z) = \gamma z$
PD-type	$L(z) = \gamma z + \lambda(z-1)$
Phase lead	$L(z) = \gamma z^{s}$
Model inversion	$L(z) \approx H^{-1}(z)$

testing. When possible, it is safest to "tune up" the Q-filter bandwidth online by starting low and slowly increasing ω_c through repeated experiments.

36.3.2.1 Servo System Revisited: Robustness

Figure 36.8 shows the Nyquist plots of $H(z)$ and $L(z)H(z)$ for the servo system ILC design presented in Section 36.2. Recall that for this example, $L(z)$=0.1z. The 0.1 gain reduces the magnitude of $H(z)$ and the z adds enough phase lead to rotate the high frequencies of $H(z)$ into the unit circle centered at one. To illustrate robust design, consider now a high-frequency perturbation to $H(z)$,

$$H'(z) = H(z) \times \frac{1}{z+0.01}.$$

The Nyquist plots for the original system $L(z)H(z)$ and the perturbed system $L(z)H'(z)$ using the same learning filter are shown in Figure 36.9. The perturbed system leaves the unit circle at $\omega_c = 0.55$ rad/sample and therefore is not stable with the same learning filter as the nominal system. The designer has two options to robustly stabilize the nominal and perturbed system. The designer can look for a learning filter that stabilizes both systems, or the designer can use a Q-filter to limit the learning bandwidth to the frequency range that is stable for both systems, $\omega \in [0, 0.55]$. The latter option is explored next.

The stabilizing effect of the Q-filter can also be seen using other frequency-domain tools with which the user is comfortable. Here we use a Bode plot to demonstrate that the same tools from classical control apply. In Figure 36.10, a Bode plot of $Q(z)(1 - L(z)H(z))$ is shown for several scenarios. The goal is to keep the magnitude below one for all frequencies, which equates to stability (Equation 36.12). The solid

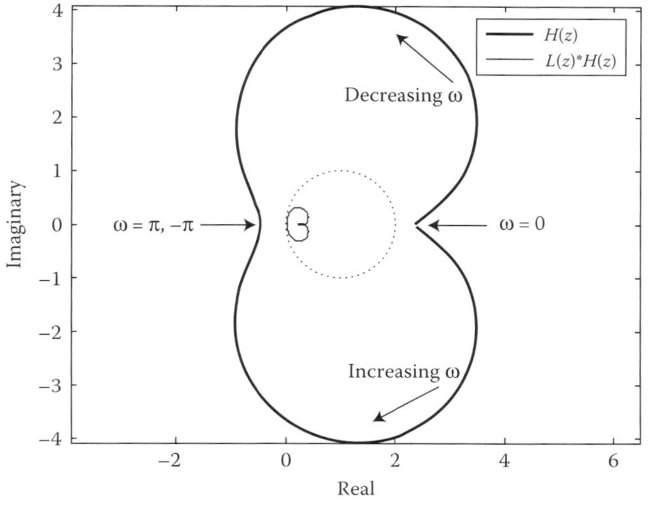

FIGURE 36.8 Nyquist plot of the servo system example.

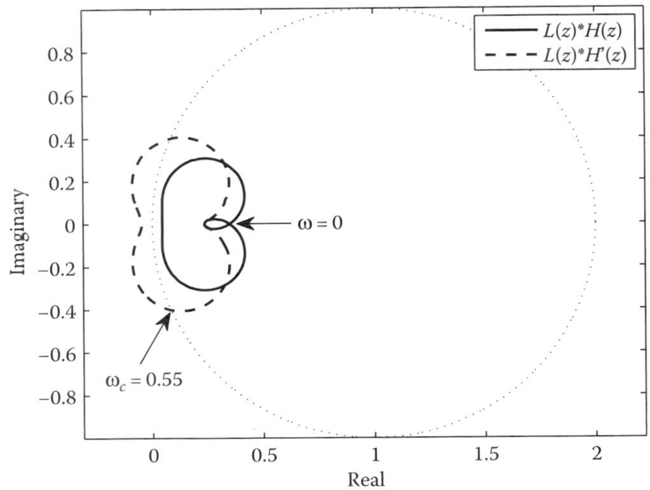

FIGURE 36.9 Nyquist plot of the perturbed servo system.

line shows the unperturbed plant and the P-type learning function (Equation 36.4) presented in Section 36.2 (in this case, $Q(z) = 1$). As noted previously, this learning algorithm is stable, which is confirmed on the Bode plot with magnitude less than 0 dB. The dashed line shows the Bode plot for the perturbed plant, $H'(z)$. In this case, the magnitude crosses 0 dB at $\omega = 0.55$ rad/sample in agreement with the Nyquist plot of Figure 36.9. The dash-dot line shows the perturbed plant again, but this time with a second-order Butterworth Q-filter with cutoff frequency $\omega_c = 0.55$ rad/sample. The Butterworth Q-filter successfully reduces the magnitude below 0 dB to achieve a stable learning algorithm.

As discussed previously, the Q-filter provides robust stability at the cost of asymptotic performance. The performance loss can be minimized through careful selection of the cutoff frequency so as not to be overly conservative. Filter type and order will also affect performance. Generally, sharp filter "rolloff" is

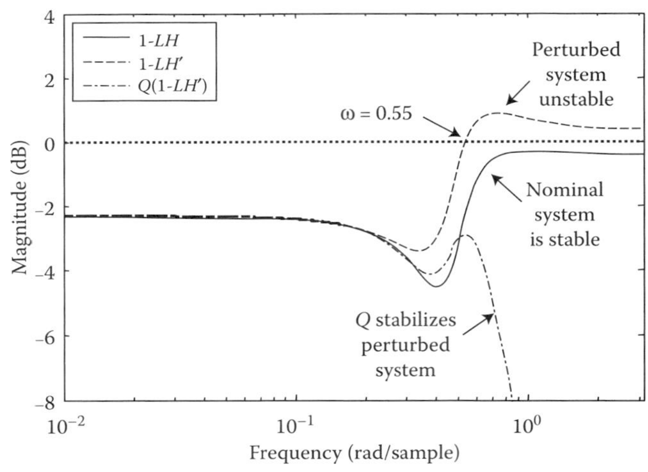

FIGURE 36.10 Bode plot of the perturbed servo system with second-order Butterworth Q-filter.

preferred to slow "rolloff" because it better approximates the ideal low-pass filter. Zero-phase Q-filters, or filters without lag, are also preferable (and popular) in ILC. These filters are discussed in the following subsection.

36.3.3 Zero-Phase Q-Filter

Phase lag in the Q-filter adds delay to the control signal, which degrades the asymptotic performance, particularly for high acceleration motions such as steps. The phase lag can be eliminated by using a zero-phase Q-filter. The zero-phase Q-filter is noncausal, although this is not problematic in ILC because the filtering is done offline after the entire signal has been collected. There are a number of methods for creating a zero-phase filter. Two methods, an infinite-impulse response (IIR) method and a finite-impulse response (FIR) method, are developed below using basic signal processing techniques.

36.3.3.1 Filtfilt (IIR Method)

This approach uses any nonzero-phase low-pass filter $F(z)$ in a modified filtering procedure to produce zero-phase shift. The procedure is as follows.

1. Filter the input signal; $s_1(k) = F(z)s_{in}(k)$, where $s_{in}(k)$ is the signal to be filtered.
2. Reverse the resultant signal in time; $s_2(k) = s_1(N-k)$.
3. Filter the signal again; $s_3(k) = F(z)s_2(k)$.
4. Reverse the resultant signal in time; $s_{out}(k) = s_3(N-k)$.

The phase lag introduced in step 1 is canceled in step 3 resulting in zero total phase shift. Note that the effective gain of the process is $|F(e^{i\omega})|^2$, owing to the double-filtering.

36.3.3.2 FIR Convolution (FIR Method)

In this approach, a zero-phase filter is constructed using an FIR approximation of the filter. Let $f(k)$, $k = 0, 1, \ldots, N_{FIR} - 1$ be the N_{FIR}-step truncation of the impulse response of $F(z)$. Then,

$$f_{NC}(k) = \frac{h(k)}{\sum_{l=-N_{FIR}+1}^{N_{FIR}-1} h(l)},$$

where $h(k) = f(k) * f(-k)$ with* defined as the convolution symbol is an FIR moving average filter with zero phase.

36.3.4 H_∞ Methods

The H_∞ method [7] is the reverse of the Nyquist tuning method in that $Q(z)$ is selected first, and $L(z)$ is designed second. $Q(z)$ can be a low-pass filter of any type, and $L(z)$ is optimally obtained as the solution to the model matching problem

$$L^*(z) = \arg\min_L \|Q(z)(I - L(z)H(z))\|_\infty.$$

This can be written equivalently as a lower linear fractional transform

$$Q(z)(I - L(z)H(z)) = G_{11}(z) + G_{12}(z)L(z)(I - G_{22}(z)L(z))^{-1}G_{21}(z),$$

where

$$G(z) = \begin{bmatrix} G_{11}(z) & G_{12}(z) \\ G_{21}(z) & G_{22}(z) \end{bmatrix} = \begin{bmatrix} Q(z) & Q(z) \\ -H(z) & 0 \end{bmatrix}.$$

In this form, standard H_∞ synthesis tools can be used to solve the problem. If no solution is found, the bandwidth of $Q(z)$ is lowered. If a solution is found, the bandwidth can be increased and a new $L(z)$ obtained.

While the frequency domain provides a simple design approach, the next three sections present a more rigorous and general design technique using a time-domain system description.

36.4 Generalized Time-Domain System Description

Consider the discrete-time (DT), linear time-varying (LTV) system

$$\begin{aligned} x_j(k+1) &= A(k)x_j(k) + B(k)u_j(k), \\ y_j(k) &= C(k)x_j(k) + d(k), \end{aligned} \tag{36.18}$$

where $x \in \mathrm{R}^n$, $u \in \mathrm{R}^r$ is the control input, $y \in \mathrm{R}^r$ is the output, $d \in \mathrm{R}^r$ is an iteration-invariant exogenous signal, $k = 0, 1, \ldots, N$ is the time index, and $j = 1, 2, \ldots$ is the iteration index. If $x_j(0) = x_0$ for all j, the input–output solution for (36.18) is given by

$$y_j(k) = C(k)\Phi(k,0)x_0 + \sum_{l=0}^{k-1} H(k,l)u_j(l) + d(k), \tag{36.19}$$

where

$$H(k,l) = C(k)\Phi(k,l+1)B(l), \tag{36.20}$$

$$\Phi(k,l) = A(k-1)A(k-2)\ldots A(l), \quad \Phi(k,k) = I. \tag{36.21}$$

Let $y_d(k)$ be the desired output for $k = 1, \ldots, N$ and define the error as

$$e_j(k) \equiv y_d(k) - y_j(k). \tag{36.22}$$

Then,

$$e_j(k) = e_0(k) - \sum_{l=0}^{k-1} H(k,l)u_j(l), \tag{36.23}$$

where

$$e_0(k) = y_d(k) - C(k)\Phi(k,0)x_0 - d(k). \tag{36.24}$$

The reader should note that (36.23) is the time-domain equivalent to (36.7).

The lifted-system (or supervector) representation is given by

$$\mathbf{e}_j = \mathbf{e}_0 - \mathbf{H}\mathbf{u}_j, \tag{36.25}$$

where

$$\mathbf{e}_j = \begin{bmatrix} e_j^{\mathrm{T}}(1) & e_j^{\mathrm{T}}(2) & \cdots & e_j^{\mathrm{T}}(N) \end{bmatrix}^{\mathrm{T}}, \tag{36.26}$$

$$\mathbf{u}_j = \begin{bmatrix} u_j^{\mathrm{T}}(0) & u_j^{\mathrm{T}}(1) & \cdots & u_j^{\mathrm{T}}(N-1) \end{bmatrix}^{\mathrm{T}}, \tag{36.27}$$

$$\mathbf{H} = \begin{bmatrix} H(1,0) & 0 & 0 & \cdots & 0 \\ H(2,0) & H(2,1) & 0 & & 0 \\ H(3,0) & H(3,1) & H(3,2) & \ddots & \vdots \\ \vdots & & & \ddots & 0 \\ H(N,0) & H(N,1) & H(N,2) & \cdots & H(N,N-1) \end{bmatrix}. \tag{36.28}$$

The lifted-system representation transforms the r-input, r-output, two-dimensional (time and iteration) system into an Nr-input, Nr-output, one-dimensional (iteration) system.

The most general first-order linear ILC algorithm for (36.25) is

$$\mathbf{u}_{j+1} = \mathbf{L_u}\mathbf{u}_j + \mathbf{L_e}\mathbf{e}_j, \tag{36.29}$$

where $\mathbf{L_u}, \mathbf{L_e} \in \mathrm{R}^{Nr\times Nr}$. Note that (36.29) has a slightly different form than (36.5). The convention in the norm-optimal design presented in Section 36.6 is to use the more general $\mathbf{L_u}, \mathbf{L_e}$ form, while the convention in frequency-domain design is $Q(z), L(z)$. For the purpose of exposition, (36.29) can be rewritten as

$$\mathbf{u}_{j+1} = \mathbf{L_u}\mathbf{u}_j + \mathbf{L_e}\mathbf{e}_j = \mathbf{Q}(\mathbf{u}_j + \mathbf{L}\mathbf{e}_j), \quad \text{where } \mathbf{L_u} = \mathbf{Q},\ \ \mathbf{L_e} = \mathbf{QL}. \tag{36.30}$$

Although the lifted representation is very useful in design and analysis, it is not necessary, or oftentimes desirable, to use the lifted representation in the implementation because $\mathbf{L_u}$ and $\mathbf{L_e}$ are large matrices when the iteration length is long. For example, consider the choice $\mathbf{L_u} = \mathbf{I}$ and $\mathbf{L_e} = \mathrm{diag}\{\gamma(0), \ldots, \gamma(N-1)\}$, $\gamma \in \mathrm{R}^{r\times r}$. This is the P-type algorithm, written in time domain as

$$u_{j+1}(k) = u_j(k) + \gamma(k)e_j(k+1), \tag{36.31}$$

and is the discrete-time analog to the P-type algorithm for continuous time presented in Section 36.2. Although both are equivalent, it is much more efficient to apply the filter description (36.31) in practice than the matrix description (36.29) for this choice of $\mathbf{L_u}$ and $\mathbf{L_e}$.

36.5 Time-Domain Analysis

36.5.1 Asymptotic Stability

Substituting (36.25) into (36.29) yields the closed-loop iteration domain dynamics

$$\mathbf{u}_{j+1} = (\mathbf{L_u} - \mathbf{L_e}\mathbf{H})\mathbf{u}_j + \mathbf{L_e}\mathbf{e}_0. \tag{36.32}$$

The iterative learning controller is asymptotic stability (AS) if and only if $|\lambda_i(\mathbf{L_u} - \mathbf{L_e}\mathbf{H})| < 1$ for $i = 1, 2, \ldots, Nr$, where $\lambda_i(\bullet)$ is the ith eigenvalue of $(\bullet)$. Using the P-type algorithm (36.31), stability simplifies to $|\lambda_i(\mathbf{I} - \mathit{diag}\{\gamma(0), \ldots, \gamma(N-1)\}\mathbf{H})| < 1$, or the following results:

$$|1 - \gamma(k)C(k+1)B(k)| < 1, \quad \text{for } k = 0, 1, \ldots, N. \tag{36.33}$$

This is the discrete-time analog to Arimoto et al.'s [2] original result for continuous-time systems.

If the controller is AS, and $(\mathbf{I} - \mathbf{L_u} + \mathbf{L_e}\mathbf{H})$ is nonsingular, then the asymptotic control can be found as

$$\lim_{j\to\infty} \mathbf{u}_j \equiv \mathbf{u}_\infty = (\mathbf{I} - \mathbf{L_u} + \mathbf{L_e}\mathbf{H})^{-1}\mathbf{L_e}\mathbf{e}_0. \tag{36.34}$$

The asymptotic error is

$$\begin{aligned}\lim_{j\to\infty} \mathbf{e}_j \equiv \mathbf{e}_\infty &= \mathbf{e}_0 - \mathbf{H}\mathbf{u}_\infty, \\ &= \left[\mathbf{I} - \mathbf{H}(\mathbf{I} - \mathbf{L_u} + \mathbf{L_e}\mathbf{H})^{-1}\mathbf{L_e}\right]\mathbf{e}_0.\end{aligned} \tag{36.35}$$

Interestingly, $\mathbf{e}_\infty = \mathbf{0}$ for an AS P-type control. The ease by which perfect tracking can be achieved with the simplest type of iterative algorithm may be surprising. The practicality of this result, however, is limited in actual practice by two challenges: convergence behavior and trial-varying disturbances.

36.5.2 Monotonic Convergence

For most practical systems, AS is not a strong enough condition. ILC systems that are AS may experience large transient learning growth, a problem studied extensively by Longman [11]. The transient learning growth appears as a rapid growth in the error over many iterations before convergence. Growth may be many orders of magnitude and convergence very slow [11], well outside the realm of practicality for physical implementations.

Monotonic convergence is a stronger stability requirement than AS and is not susceptible to transient learning growth. Monotonic convergence of the error requires that $\|\mathbf{e}_\infty - \mathbf{e}_j\|$ decreases at each iteration, where $\|\bullet\|$ is a suitable norm. For ILC design with monotonic convergence, the most commonly used norm is the induced 2-norm. Monotonic convergence of the control, $\left\|\mathbf{u}_\infty - \mathbf{u}_j\right\|$, may also be used. (36.29) is monotonically convergent if

$$\eta \equiv \|(\mathbf{L_u} - \mathbf{L_e H})\| < 1. \tag{36.36}$$

Furthermore, $\left\|\mathbf{u}_\infty - \mathbf{u}_{j+1}\right\| \le \eta \left\|\mathbf{u}_\infty - \mathbf{u}_j\right\|$ gives the rate of convergence.

36.5.3 Trial-Varying Disturbances

One of the key assumptions in most ILC literature is that disturbance signals repeat identically from one iteration to the next. In practice this is rarely the case. As will be shown, trial-varying disturbances do not affect stability. They do, however, affect the performance. Because ILC is a feedforward technique, it cannot compensate for random trial-varying disturbances. In fact, ILC will tend to learn from these disturbances and incorrectly attempt to compensate for them on subsequent iterations, amplifying their affect. Design parameters in Section 36.6 can be selected to minimize the amplification.

Trial-varying disturbances can be treated by adding the signal $\boldsymbol{\delta}_j$ to the system description

$$\mathbf{e}_j = \mathbf{e}_0 - \mathbf{H}\mathbf{u}_j - \boldsymbol{\delta}_j.$$

Substituting into (36.29) yields the iteration domain closed loop

$$\mathbf{u}_{j+1} = (\mathbf{L_u} - \mathbf{L_e H})\mathbf{u}_j + \mathbf{L_e}\mathbf{e}_0 - \mathbf{L_e}\boldsymbol{\delta}_j. \tag{36.37}$$

If $\boldsymbol{\delta}_j$ is bounded, then the AS from Section 36.5.1 becomes the bounded-input, bounded-output stability condition for trial-varying disturbances. Note that $\boldsymbol{\delta}_j$ is filtered by $\mathbf{L_e}$ in (36.37), and thus one expects trial-varying disturbance sensitivity to decrease when the gain $\mathbf{L_e}$ is reduced.

36.5.3.1 IC Variation

IC variation is a problem in many systems because feedback control system limitations and nonlinearities such as friction make perfect resetting challenging. IC variation can be treated as a trial-varying disturbance, by setting $\delta_j(k) = C(k)\Phi(k,0)x_{0,j}$, where $x_{0,j}$ is the IC on the jth iteration.

36.5.4 Robustness

As demonstrated in Section 36.3, small perturbations to the system dynamics may destabilize an ILC. Analytical results depend on the type of model uncertainty considered, which is an active area of research. In the following section, analysis is presented demonstrating that the norm-optimal design is robustly monotonically convergent to additive system perturbations $\boldsymbol{\Delta}$ of the form, $\mathbf{H}_t = \mathbf{H} + \boldsymbol{\Delta}$. The largest system perturbation $\bar{\sigma}(\Delta)$ for which the ILC is robustly monotonically convergent depends on the choice of design parameters.

36.6 Time-Domain Norm-Optimal Design

The time-domain or norm-optimal algorithm is designed to minimize the quadratic optimization problem [1,8,10],

$$\mathcal{J} = \mathbf{e}_{j+1}^T \mathbf{Q} \mathbf{e}_{j+1} + \mathbf{u}_{j+1}^T \mathbf{S} \mathbf{u}_{j+1} + \left(\mathbf{u}_{j+1} - \mathbf{u}_j\right)^T \mathbf{R} \left(\mathbf{u}_{j+1} - \mathbf{u}_j\right), \tag{36.38}$$

where $\{\mathbf{Q}, \mathbf{S}, \mathbf{R}\}$ are symmetric positive (semi)-definite real-valued matrices of appropriate dimension and $\mathbf{H}^T\mathbf{Q}\mathbf{H} + \mathbf{S}$ is positive definite. Often $\{\mathbf{Q}, \mathbf{S}, \mathbf{R}\} \equiv \{q\mathbf{I}, s\mathbf{I}, r\mathbf{I}\}$ with q, s, r real-valued positive scalars. Applying the substitution $\mathbf{e}_{j+1} = \mathbf{e}_j - \mathbf{H}(\mathbf{u}_{j+1} - \mathbf{u}_j)$, differentiating $\mathcal{J}$ with respect to $\mathbf{u}_{j+1}$, setting the result equal to zero, and rearranging the solution yields the norm-optimal ILC controller

$$\mathbf{u}_{j+1} = \mathbf{L}_\mathbf{u} \mathbf{u}_j + \mathbf{L}_\mathbf{e} \mathbf{e}_j, \tag{36.39}$$

$$\mathbf{L}_\mathbf{u} = (\mathbf{H}^\mathrm{T}\mathbf{Q}\mathbf{H} + \mathbf{S} + \mathbf{R})^{-1}(\mathbf{H}^\mathrm{T}\mathbf{Q}\mathbf{H} + \mathbf{R}), \tag{36.40}$$

$$\mathbf{L}_\mathbf{e} = (\mathbf{H}^\mathrm{T}\mathbf{Q}\mathbf{H} + \mathbf{S} + \mathbf{R})^{-1}\mathbf{H}^\mathrm{T}\mathbf{Q}. \tag{36.41}$$

As an essential part of the design process, we discuss here some guidelines [16] for designing the weighting matrices by studying the properties of the ILC system with respect to convergence, performance, robust convergence, and performance in the presence of trial-varying disturbances.

36.6.1 Weighting Matrix Design

It is straightforward to verify that the norm-optimal controller is monotonically convergent in the control, under the 2-norm, by substituting (36.39) into (36.36). The convergence rate is calculated as $\eta = \|\left(\mathbf{H}^T\mathbf{Q}\mathbf{H} + \mathbf{S} + \mathbf{R}\right)^{-1}\mathbf{R}\|_{i2}$. Note that the iteration-domain convergence speed depends strongly on $\mathbf{R}$: for $\|\mathbf{R}\|_{i2} = 0$ deadbeat control is achieved, as $\|\mathbf{R}\|_{i2} \to \infty$ the convergence speed approaches zero.

The asymptotic error for norm-optimal ILC, obtained by substitution of (36.39) into (36.35), is $\mathbf{e}_\infty = \left(\mathbf{I} - \mathbf{H}\left(\mathbf{H}^T\mathbf{Q}\mathbf{H} + \mathbf{S}\right)^{-1}\mathbf{H}^T\mathbf{Q}\right)\mathbf{e}_0$. For zero error, $\|\mathbf{S}\|_{i2} = 0$ is necessary, and hence $\mathbf{H}^\mathrm{T}\mathbf{Q}\mathbf{H}$ must be positive definite. An important result is that $\mathbf{e}_\infty$ is not a function of $\mathbf{R}$, and hence performance is not a function of convergence speed.

For robust convergence, the true system $\mathbf{H}_\mathbf{t} = \mathbf{H} + \mathbf{\Delta}$, with additive uncertainty $\mathbf{\Delta}$, is considered. As a result, the requirement for robust convergence is

$$\|\mathbf{L}_\mathbf{u} - \mathbf{L}_\mathbf{e}\mathbf{H}_\mathbf{t}\|_{i2} < 1 \longrightarrow \max_{\Delta} \left\|(\mathbf{H}^T\mathbf{Q}\mathbf{H} + \mathbf{S} + \mathbf{R})^{-1}(\mathbf{R} + \mathbf{H}^T\mathbf{Q}(\mathbf{H} - \mathbf{H}_\mathbf{t}))\right\|_{i2} < 1. \tag{36.42}$$

One can deduce that increasing $\|\mathbf{S}\|_{i2}$ allows for more robustness to model uncertainty.

Finally, for trial-varying disturbances $\boldsymbol{\delta}_j$, which affect steady-state error fluctuations, one should minimize $\|\mathbf{L}_\mathbf{e}\|_{i2}$. From (36.39), $\|\mathbf{L}_\mathbf{e}\|_{i2}$ is affected by $\mathbf{S}$ and $\mathbf{R}$. Given that $\mathbf{R}$ does not affect the performance, it is the natural candidate for reducing trial-varying disturbance sensitivity.

Based on the above information, the following tuning guidelines [16] for norm-optimal ILC control can be given. These guidelines are most easily implemented using common $\{q\mathbf{I}, s\mathbf{I}, r\mathbf{I}\}$ diagonal-type real-valued scalar gains.

1. Design $\mathbf{Q}$ to correspond to the desired weighting of the error. Usually, $\mathbf{Q} = \mathbf{I}$ for uniform weighting of the error.
2. The actual system dynamics will not usually be perfectly captured by the system model. Thus, $\mathbf{S}$ must be designed such that the system is robustly monotonically convergent. Start with an $\mathbf{S}$ yielding $\|\mathbf{S}\|_{i2} \approx 0.01\,\|\mathbf{H}\|_{i2}$. Note that the critical design parameter is the size of $\|\mathbf{S}\|_{i2}$ relative to the size of $\|\mathbf{H}\|_{i2}$, where the magnitude of $\|\mathbf{H}\|_{i2}$ is related to system uncertainty. Subsequently, reduce $\|\mathbf{S}\|_{i2}$ until the system diverges. Set $\|\mathbf{S}\|_{i2} = 2 \cdot \|\mathbf{S}\|_{i2}^{\min}$ to allow for a safety factor of 2.

3. When trial-varying disturbances are present, steady-state error fluctuations will occur. Start with $\|\mathbf{R}\|_{i2} = 0$ and increase $\|\mathbf{R}\|_{i2}$ until the fluctuations are within desired bounds, or the RMS error does not decrease anymore.

36.6.1.1 Servo System Revisited: Norm-Optimal Design

To illustrate the effects of varying $\{\mathbf{Q}, \mathbf{S}, \mathbf{R}\}$, this subsection provides performance results for various combinations of the weighting matrices, where $\{\mathbf{Q} = q\mathbf{I}, \mathbf{S} = s\mathbf{I}, \mathbf{R} = r\mathbf{I}\}$. The different weighting matrices are applied to the servo system example in Section 36.2.

Figure 36.11 shows the error, feedback control signal, and feedforward control signal for $\{\mathbf{Q}, \mathbf{S}, \mathbf{R}\} = \{\mathbf{I}, \mathbf{I}, \mathbf{I}\}$. Similar to the results shown in Figure 36.6 the feedback signal is slowly replaced by the ILC feedforward signal with increasing iteration.

To determine the effect of q on the error, s and r were held constant at $\{s, r\} = \{1, 1\}$, while the value of q was varied as $\{1, 0.5, 0.1, 0.001\}$. Figure 36.12 illustrates that the smaller the value of q, the larger the value of the RMS converged tracking error. Varying s affects both the performance and convergence. While a smaller s results in smaller converged RMS error, robustness to model uncertainty requires a larger s value. Holding $\{q, r\}$ constant at $\{1, 1\}$, respectively, Figure 36.13 illustrates how decreasing s results in a decrease in the RMS-converged error.

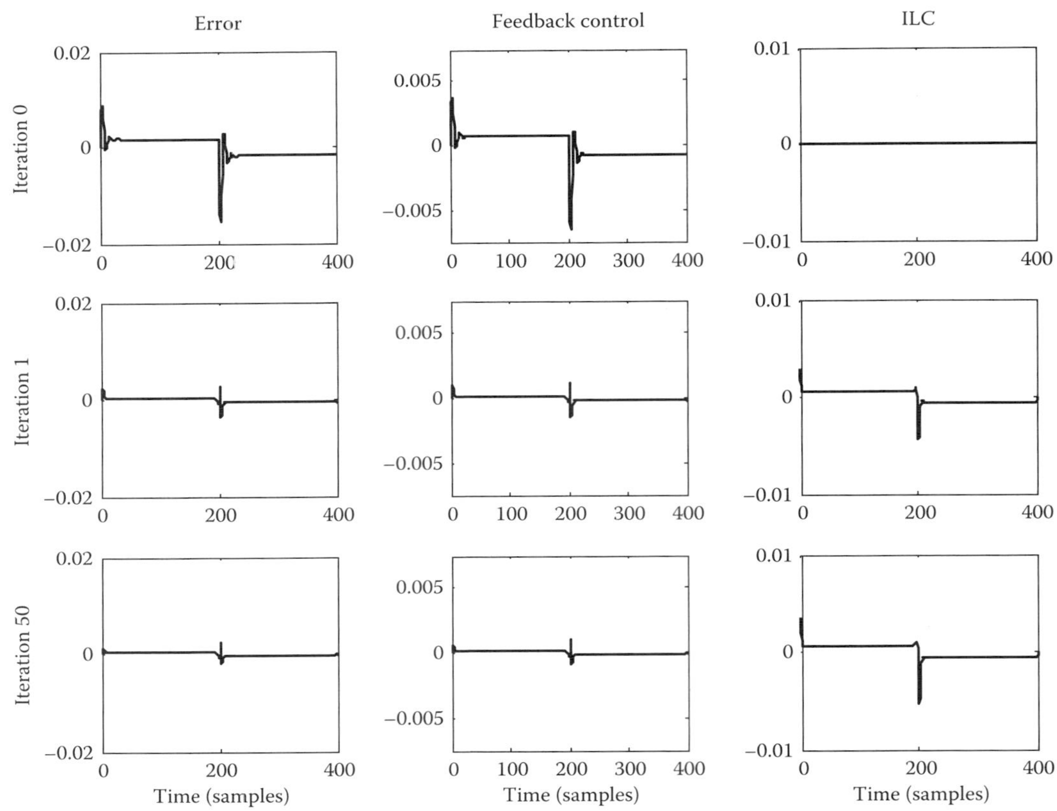

FIGURE 36.11 Error, feedback control, and ILC time series using norm-optimal ILC.

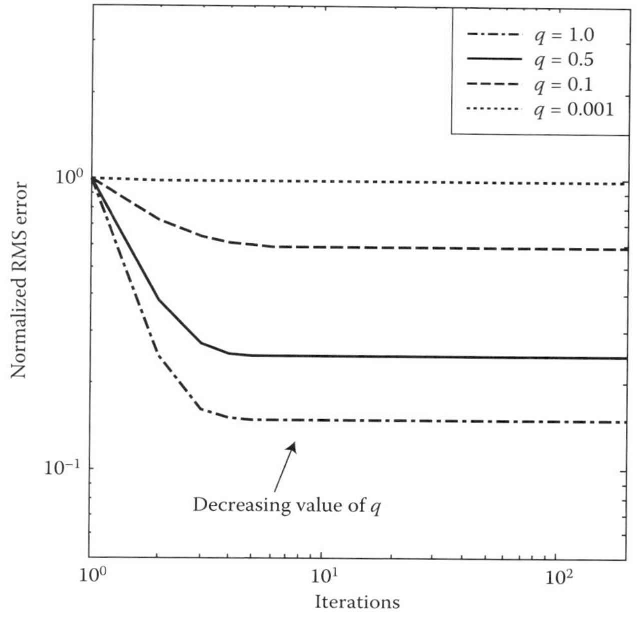

FIGURE 36.12 RMS converged error values for various Q weighting matrices.

Lastly, r can be increased to minimize trial-varying disturbances, at the expense of convergence rate. Holding $\{q, s\}$ constant at $\{1, 1\}$, respectively, Figure 36.14 shows how the convergence rate increases as r increases. Figure 36.15 illustrates the effects on the RMS error of adding white noise {(mean,var) = (0,1e-6)} to the system. As r increases, the effect of the noise on the system is decreased, as demonstrated by a decrease in converged RMS error.

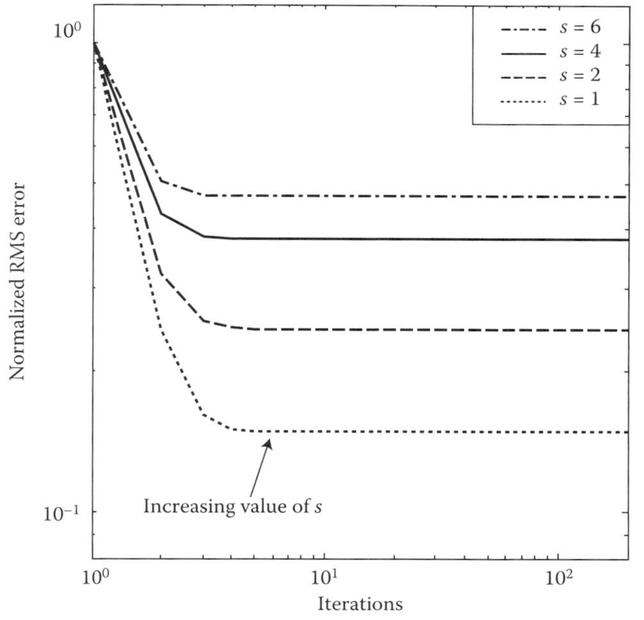

FIGURE 36.13 RMS error and monotonic convergence values for varying S weighting matrices.

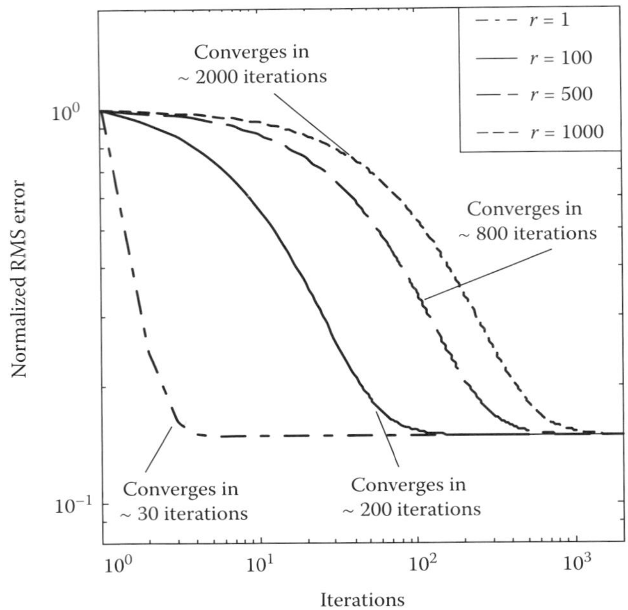

FIGURE 36.14 RMS error values for varying R weighting matrices. Note the longer iteration range.

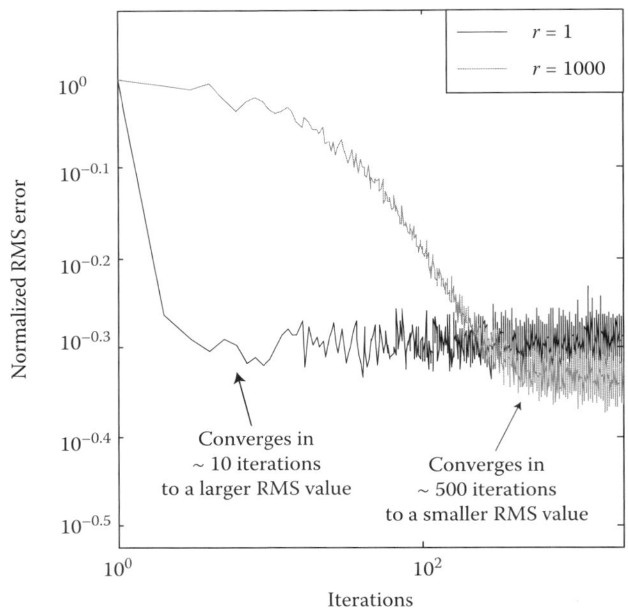

FIGURE 36.15 RMS error values for varying R weighting matrices with trial-varying noise. Note the longer iteration range.

36.7 Concluding Remarks

This chapter has introduced the basic ILC design and analysis tools for linear systems. The results and design details presented here only touch the surface of the field of ILC. In particular, a number of advanced algorithms have been developed, which may be of interest to the reader. Current-iteration-tracking-error (CITE) [6] ILC algorithms use the current iteration's error signal in the learning algorithm, effectively adding feedback control in ILC design. Most notably, CITE algorithms can use the feedback component to compensate for trial-varying disturbances. For stochastic measurement noise, one might consider the use of optimal trial-varying algorithms [15]. For faster convergence rate and better robustness, high-order algorithms [6] use several iterations of past control and error signals in the update algorithm.

References

1. Amann, N., Owens, D. H., and Rogers, E., Iterative learning control for discrete-time systems with exponential rate of convergence, *IEE Proceedings: Control Theory and Applications*, 143(2), 217–224, 1996.
2. Arimoto, S., Kawamura, S., and Miyazaki, F., Bettering pperation of robots by learning, *Journal of Robotic Systems*, 1, 123–140, 1984.
3. Ayn, H.-S., Chen, Y.Q., and Moore, K. L., Iterative learning control: Brief survey and categorization, *IEEE Transactions on Systems, Man, and Cybernetics—Part C: Applications and Reviews*, 37(6), 1099–1121, 2007.
4. Bien, Z. and Xu, J.-X., *Iterative Learning Control: Analysis, Design, Integration and Applications*, Boston: Kluwer Academic Publishers, 1998.
5. Bristow, D. A., Tharayil, M., and Alleyne, A. G., A survey of iterative learning control, *Control Systems Magazine*, 26(3), 96–114, 2006.
6. Chen, Y. and Wen, C., *Iterative Learning Control: Convergence, Robustness, and Applications,* London: Springer, 1999.
7. de Roover, D. and Bosgra, O. H., Synthesis of robust multivariable iterative learning controllers with application to a wafer stage motion system, *International Journal of Control*, 73(10), 968–979, 2000.
8. Gunnarsson, S. and Norrlof, M., On the design of ILC algorithms using optimization, *Automatica*, 37(12), 2011–2016, 2001.
9. Horowitz, R., Learning control of robot manipulators, *Transactions of the ASME Journal of Dynamic Systems, Measurement and Control*, 115(2B), 402–411, 1993.
10. Lee, J. H., Lee, K. S., and Kim, W. C., Model-based iterative learning control with a quadratic criterion for time-varying linear systems, *Automatica*, 36(5), 641–657, 2000.
11. Longman, R. W., Iterative learning control and repetitive control for engineering practice, *International Journal of Control*, 73(10), 930–954, 2000.
12. Moore, K. L., *Iterative Learning Control for Deterministic Systems,* London: Springer-Verlag, 1993.
13. Moore, K. L., Dahleh, M., and Bhattacharyya, S. P., Iterative learning control: A survey and new results, *Journal of Robotic Systems*, 9(5), 563–594, 1992.
14. Norrlof, M. and Gunnarsson, S., Time and frequency domain convergence properties in iterative learning control, *International Journal of Control*, 75(14), 1114–1126, 2002.
15. Saab, S. S., On a discrete-time stochastic learning control algorithm, *IEEE Transactions on Automatic Control*, 46(8), 1333–1336, 2001.
16. Steinbuch, M., Bosgra, O., and Dynamics and Control Technology Group, Technical University of Eindhoven, Eindhoven, The Netherlands, personal communication, 2007.
17. Tomizuka, M., Tsao, T.-C., and Chew, K.-K., Analysis and synthesis of discrete-time repetitive controllers, *Journal of Dynamic Systems, Measurement and Control, Transactions ASME*, 111(3), 353–358, 1989.
18. Xu, J.-X. and Tan, Y., *Linear and Nonlinear Iterative Learning Control*, Berlin: Springer, 2003.

VI

Analysis and Design of Nonlinear Systems

37

Nonlinear Zero Dynamics

Alberto Isidori
Sapienza University of Rome

Christopher I. Byrnes
Washington University

37.1 Input–Output Feedback Linearization

Consider a nonlinear single-input single-output system, described by equations of the form

$$\begin{aligned} \dot{x} &= f(x) + g(x)u \\ y &= h(x) \end{aligned} \tag{37.1}$$

and suppose $x = 0$ is an equilibrium of the vector field $f(x)$; that is, $f(0) = 0$, and $h(0) = 0$. Assume also that this system has relative degree $r < n$ at $x = 0$. Then there is a neighborhood U of $x = 0$ in $\mathbb{R}^n$ and a local change of coordinates $z = \Phi(x)$ defined on U [and satisfying $\Phi(0) = 0$] such that, in the new coordinates, the system is described by equations of the form (see Chapter 46 for details)

$$\begin{aligned} \dot{z}_1 &= z_2 \\ \dot{z}_2 &= z_3 \\ &\cdots \\ \dot{z}_{r-1} &= z_r \\ \dot{z}_r &= v \\ \dot{z}_{r+1} &= b(z) + a(z)u \\ \dot{z}_{r+1} &= q_{r+1}(z) \\ &\cdots \\ \dot{z}_n &= q_n(z) \\ y &= z_1. \end{aligned} \tag{37.2}$$

Equation 37.2, which describes the system in the new coordinates, can be more conveniently represented as follows. Set

$$\xi = \begin{pmatrix} z_1 \\ z_2 \\ \cdots \\ z_r \end{pmatrix} \qquad \eta = \begin{pmatrix} z_{r+1} \\ z_{r+2} \\ \cdots \\ z_n \end{pmatrix},$$

and recall that, in particular,

$$\xi = \begin{pmatrix} h(x) \\ L_f h(x) \\ \cdots \\ L_f^{r-1} h(x) \end{pmatrix}.$$

Moreover, define

$$A = \begin{pmatrix} 0 & 1 & 0 & \cdots & 0 & 0 \\ 0 & 0 & 1 & \cdots & 0 & 0 \\ \cdot & \cdot & \cdot & \cdots & \cdot & \cdot \\ 0 & 0 & 0 & \cdots & 0 & 1 \\ 0 & 0 & 0 & \cdots & 0 & 0 \end{pmatrix}, \quad B = \begin{pmatrix} 0 \\ 0 \\ \cdot \\ 0 \\ 1 \end{pmatrix},$$

$$C = \begin{pmatrix} 1 & 0 & 0 & \cdots & 0 & 0 \end{pmatrix}$$

and set

$$q(z) = \begin{pmatrix} q_{r+1}(z) \\ q_{r+2}(z) \\ \cdots \\ q_n(z) \end{pmatrix}.$$

Then, Equation 37.2 reduces to equations of the form

$$\begin{aligned} \dot{\xi} &= A\xi + B(b(\xi, \eta) + a(\xi, \eta)u) \\ \dot{\eta} &= q(\xi, \eta) \\ y &= C\xi. \end{aligned} \tag{37.3}$$

Suppose now the input u to the system of Equation 37.3 is chosen as

$$u = \frac{1}{a(\xi, \eta)}(-b(\xi, \eta) + v). \tag{37.4}$$

This feedback law yields a closed-loop system that is described by equations of the form

$$\begin{aligned} \dot{\xi} &= A\xi + Bv \\ \dot{\eta} &= q(\xi, \eta). \end{aligned} \tag{37.5}$$

This system clearly appears decomposed into a *linear subsystem*, of dimension r, which is the only one responsible for the input–output behavior, and a possibly nonlinear subsystem, of dimension $n - r$, whose behavior does not affect the output. In other words, this feedback law has changed the original system so as to obtain a new system whose input–output behavior coincides with that of a linear (controllable and observable) system of dimension r having transfer function

$$H(s) = \frac{1}{s^r}.$$

Remark 37.1

To interpret the role played by the feedback law of Equation 37.4, it is instructive to examine the effect produced by a feedback of this kind on a *linear system*. In this case, the system of Equations 37.3 is modeled by equations of the form

$$\begin{aligned}\dot{\xi} &= A\xi + B(R\xi + S\eta + Ku)\\ \dot{\eta} &= P\xi + Q\eta\\ y &= C\xi\end{aligned}$$

in which R and S are row vectors, of suitable dimensions, of real numbers; K is a nonzero real number; and P and Q are matrices, of suitable dimensions, of real numbers. The feedback of Equation 37.4 is a feedback of the form

$$u = -\frac{R}{K}\xi - \frac{S}{K}\eta + \frac{1}{K}v\,. \tag{37.6}$$

A feedback of this type indeed modifies the eigenvalues of the system on which it is imposed. Since, from the previous analysis, it is known that the transfer function of the resulting closed-loop system has no zeros and r poles at $s = 0$, it can be concluded that the effect of the feedback of Equation 37.6 is such as to place r eigenvalues at $s = 0$ and the remaining $n - r$ eigenvalues exactly where the $n - r$ zeros of the transfer function of the open-loop system are located. The corresponding closed-loop system, having $n - r$ eigenvalues coinciding with its $n - r$ zeros, is unobservable and its minimal realization has a transfer function that has no zeros and r poles at $s = 0$.

37.2 The Zero Dynamics

In this section we discuss an important concept that in many instances plays a role exactly similar to that of the "zeros" of the transfer function in a linear system.

Given a single-input single-output system, having relative degree $r < n$ at $x = 0$ represented by equations of the form of Equation 37.3, consider the following problem, which is sometimes called the *Problem of Zeroing the Output*. Find, if any, pairs consisting of an initial state x° and of an input function $u^\circ(\cdot)$, defined for all t in a neighborhood of $t = 0$, such that the corresponding output $y(t)$ of the system is identically zero for all t in a neighborhood of $t = 0$. Of course, the interest is to find *all* such pairs (x°, u°) and not simply the trivial pair $x^\circ = 0, u^\circ = 0$ (corresponding to the situation in which the system is initially at rest and no input is applied).

Recalling that, in the normal form of Equation 37.3

$$y(t) = \xi_1(t),$$

we observe that the constraint $y(t) = 0$ for all t implies

$$\dot{\xi}_1(t) = \dot{\xi}_2(t) = \ldots = \dot{\xi}_r(t) = 0\,,$$

that is, $\xi(t) = 0$ for all t. In other words, if the output of the system is identically zero, its state necessarily respects the constraint $\xi(t) = 0$ for all t. In addition, the input $u(t)$ must necessarily be the unique solution of the equation

$$0 = b(0, \eta(t)) + a(0, \eta(t))u(t)$$

[recall that $a(0, \eta(t)) \neq 0$ if $\eta(t)$ is close to 0]. As far as the variable $\eta(t)$ is concerned, it is clear that, $\xi(t)$ being identically zero, its behavior is governed by the differential equation

$$\dot{\eta}(t) = q(0, \eta(t)). \tag{37.7}$$

From this analysis it is possible to conclude the following. In order to have the output $y(t)$ of the system identically zero, necessarily the initial state must be such that $\xi(0) = 0$, whereas $\eta(0) = \eta^\circ$ can be chosen arbitrarily. According to the value of η°, the input must be set equal to the following function

$$u(t) = -\frac{b(0, \eta(t))}{a(0, \eta(t))}$$

where $\eta(t)$ denotes the solution of the differential equation

$$\dot{\eta}(t) = q(0, \eta(t))$$

with initial condition $\eta(0) = \eta^\circ$. Note also that for each set of initial conditions $(\xi, \eta) = (0, \eta^\circ)$ the input thus defined is the *unique* input capable of keeping $y(t)$ identically zero.

The dynamics of Equation 37.7 correspond to the dynamics describing the "internal" behavior of the system when input and initial conditions have been chosen in such a way as to constrain the output to remain identically zero. These dynamics, which are rather important in many instances, are called the *zero dynamics* of the system.

The previous analysis interprets the trajectories of the $(n - r)$-dimensional system

$$\dot{\eta} = q(0, \eta) \tag{37.8}$$

as "open-loop" trajectories of the system, when the latter is *forced* (by appropriate choice of input and initial condition) to constrain the output to be identically zero. However, the trajectories of Equation 37.8 can also be interpreted as *autonomous* trajectories of an appropriate "closed-loop system." In fact, consider again a system in the normal form of Equation 37.3 and suppose the feedback control law of Equation 37.4 is imposed, under which the input–output behavior becomes identical with that of a linear system. The corresponding closed-loop system thus obtained is described by Equations 37.5. If the linear subsystem is initially at rest and no input is applied, then $y(t) = 0$ for all values of t, and the corresponding internal dynamics of the whole (closed-loop) system are exactly those of Equation 37.8, namely, the zero dynamics of the open-loop system.

Remark 37.2

In a linear system, the dynamics of Equation 37.8 are determined by the *zeros* of the transfer function of the system itself. In fact, consider a linear system having relative degree r and let

$$H(s) = K\frac{b_0 + b_1 s + \cdots + b_{n-r-1}s^{n-r-1} + s^{n-r}}{a_0 + a_1 s + \cdots + a_{n-1}s^{n-1} + s^n}$$

denote its transfer function. Suppose the numerator and denominator polynomials are relatively prime and consider a minimal realization of $H(s)$

$$\begin{aligned} x &= Ax + Bu \\ y &= Cx \end{aligned}$$

with

$$A = \begin{pmatrix} 0 & 1 & 0 & \cdots & 0 \\ 0 & 0 & 1 & \cdots & 0 \\ \cdot & \cdot & \cdot & \cdots & \cdot \\ 0 & 0 & 0 & \cdots & 1 \\ -a_0 & -a_1 & -a_2 & \cdots & -a_{n-1} \end{pmatrix} \quad B = \begin{pmatrix} 0 \\ 0 \\ \cdots \\ 0 \\ K \end{pmatrix}$$

$$C = \begin{pmatrix} b_0 & b_1 & \cdots & b_{n-r-1} & 1 & 0 & \cdots & 0 \end{pmatrix}.$$

The realization in question can easily be reduced to the form of Equation 37.2. For the ξ coordinates one has to take

$$\begin{aligned}
\xi_1 &= Cx = b_0x_1 + b_1x_2 + \cdots + b_{n-r-1}x_{n-r} + x_{n-r+1} \\
\xi_2 &= CAx = b_0x_2 + b_1x_3 + \cdots + b_{n-r-1}x_{n-r+1} + x_{n-r+2} \\
&\cdots \\
\xi_r &= CA^{r-1}x = b_0x_r + b_1x_{r+1} + \cdots + b_{n-r-1}x_{n-1} + x_n.
\end{aligned}$$

while for the η coordinates it is possible to choose

$$\begin{aligned}
\eta_1 &= x_1 \\
\eta_2 &= x_2 \\
&\cdots \\
\eta_{n-r} &= x_{n-r}.
\end{aligned}$$

In the new coordinates we obtain equations in normal form, which, because of the linearity of the system, have the following structure

$$\begin{aligned}
\dot{\xi} &= A\xi + B(R\xi + S\eta + Ku) \\
\dot{\eta} &= P\xi + Q\eta
\end{aligned}$$

where R and S are row vectors and P and Q are matrices of suitable dimensions. The zero dynamics of this system, according to our previous definition, are those of

$$\dot{\eta} = Q\eta.$$

The particular choice of the last $n - r$ new coordinates (i.e., of the elements of η) entails a particularly simple structure for the matrices P and Q. As a matter of fact, it is easily checked that

$$P = \begin{pmatrix} 0 \\ 0 \\ \cdot \\ 0 \\ 1 \end{pmatrix}, \quad Q = \begin{pmatrix} 0 & 1 & 0 & \cdots & 0 \\ 0 & 0 & 1 & \cdots & 0 \\ \cdot & \cdot & \cdot & \cdots & \cdot \\ 0 & 0 & 0 & \cdots & 1 \\ -b_0 & -b_1 & -b_2 & \cdots & -b_{n-r-1} \end{pmatrix}.$$

From the particular form of this matrix, it is clear that the eigenvalues of Q coincide with the zeros of the numerator polynomial of $H(s)$, that is, with the zeros of the transfer function. Thus, it is concluded that in a linear system the zero dynamics are linear dynamics with eigenvalues coinciding with the zeros of the transfer function of the system.

These arguments also show that the linear approximation, at $\eta = 0$, of the zero dynamics of a system coincides with the zero dynamics of the linear approximation of the system at $x = 0$. In order to see this, consider for $f(x)$, $g(x)$ and $h(x)$ expansions of the form

$$\begin{aligned}
f(x) &= Ax + f_2(x) \\
g(x) &= B + g_1(x) \\
h(x) &= Cx + h_2(x)
\end{aligned}$$

where

$$A = \left[\frac{\partial f}{\partial x}\right]_{x=0}, \quad B = g(0), \quad C = \left[\frac{\partial h}{\partial x}\right]_{x=0}.$$

An easy calculation shows, by induction, that

$$L_f^k h(x) = CA^k x + d_k(x)$$

where $d_k(x)$ is a function such that

$$\left[\frac{\partial d_k}{\partial x}\right]_{x=0} = 0\,.$$

From this, one deduces that

$$\begin{aligned} CA^k B &= L_g L_f^k h(0) = 0 \quad \text{for all } k < r-1 \\ CA^{r-1} B &= L_g L_f^{r-1} h(0) \neq 0 \end{aligned}$$

that is, the relative degree of the linear approximation of the system at $x = 0$ is exactly r.

From this fact, it is concluded that taking the linear approximation of equations in normal form, based on expansions of the form

$$\begin{aligned} b(\xi,\eta) &= R\xi + S\eta + b_2(\xi,\eta) \\ a(\xi,\eta) &= K + a_1(\xi,\eta) \\ q(\xi,\eta) &= P\xi + Q\eta + q_2(\xi,\eta) \end{aligned}$$

yields a linear system in normal form. Thus, the Jacobian matrix

$$Q = \left[\frac{\partial q}{\partial \eta}\right]_{(\xi,\eta)=0}$$

which describes the linear approximation at $\eta = 0$ of the zero dynamics of the original nonlinear system, has eigenvalues that coincide with the zeros of the transfer function of the linear approximation of the system at $x = 0$.

37.3 Local Stabilization of Nonlinear Minimum-Phase Systems

In analogy with the case of linear systems, which are traditionally said to be "minimum phase" when all their transmission zeros have negative real part, nonlinear systems (of the form of Equation 37.1) whose zero dynamics (Equation 37.8) have a locally (globally) asymptotically stable equilibrium at $z = 0$ are also called locally (globally) *minimum-phase* systems. As in the case of linear systems, minimum-phase nonlinear systems can be asymptotically stabilized via state feedback. We discuss first the case of local stabilization.

Consider again a system in normal form of Equation 37.3 and impose a feedback of the form

$$u = \frac{1}{a(\xi,\eta)}(-b(\xi,\eta) - c_0\xi_1 - c_1\xi_2 - \cdots - c_{r-1}\xi_r) \tag{37.9}$$

where $c_0, c_1, \cdots, c_{r-1}$ are real numbers.

This choice of feedback yields a closed-loop system of the form

$$\begin{aligned} \dot{\xi} &= (A + BK)\xi \\ \dot{\eta} &= q(\xi,\eta) \end{aligned} \tag{37.10}$$

with

$$A + BK = \begin{pmatrix} 0 & 1 & 0 & \cdots & 0 \\ 0 & 0 & 1 & \cdots & 0 \\ \cdot & \cdot & \cdot & \cdots & \cdot \\ 0 & 0 & 0 & \cdots & 1 \\ -c_0 & -c_1 & -c_2 & \cdots & -c_{r-1} \end{pmatrix}.$$

In particular, the matrix $A+BK$ has a characteristic polynomial

$$p(s) = c_0 + c_1 s + \cdots + c_{r-1}s^{r-1} + s^r.$$

From this form of the equations describing the closed-loop system we deduce the following interesting property.

Proposition 37.1:

Suppose the equilibrium $\eta = 0$ of the zero dynamics of the system is locally asymptotically stable and all the roots of the polynomial $p(s)$ have negative real part. Then the feedback law of Equation 37.9 locally asymptotically stabilizes the equilibrium $(\xi, \eta) = (0, 0)$.

This is a consequence of the fact that the closed-loop system has a triangular form. According to a well-known property of systems in triangular form, since by assumption the subsystem

$$\dot{\eta} = q(0, \eta)$$

has a locally asymptotically stable equilibrium at $\eta = 0$ and the subsystem

$$\dot{\xi} = (A + BK)\xi$$

has a (globally) asymptotically stable equilibrium at $\xi = 0$, the equilibrium $(\xi, \eta) = (0, 0)$ of the entire system is locally asymptotically stable.

Note that the matrix

$$Q = \left[\frac{\partial q(\xi, \eta)}{\partial \eta}\right]_{(\xi,\eta)=(0,0)}$$

characterizes the linear approximation of the zero dynamics at $\eta = 0$. If this matrix had all its eigenvalues in the left complex half-plane, then the result stated in Proposition 37.1 would have been a trivial consequence of the Principle of Stability in the First Approximation, because the linear approximation of Equation 37.10 has the form

$$\begin{pmatrix} \dot{\xi} \\ \dot{\eta} \end{pmatrix} = \begin{pmatrix} A & 0 \\ \star & Q \end{pmatrix} \begin{pmatrix} \xi \\ \eta \end{pmatrix}.$$

However, Proposition 37.1 establishes a stronger result, because it relies only upon the assumption that $\eta = 0$ is *simply* an asymptotically stable equilibrium of the zero dynamics of the system, and this (as is well known) does not necessarily require, for a nonlinear dynamics, asymptotic stability of the linear approximation (i.e., all eigenvalues of Q having negative real part). In other words, the result in question may also hold in the presence of some eigenvalue of Q with zero real part.

In order to design the stabilizing control law there is no need to know explicitly the expression of the system in normal form, but only to know *the fact* that the system has a zero dynamics with a locally asymptotically stable equilibrium at $\eta = 0$. Recalling how the ξ coordinates and the functions $a(\xi, \eta)$ and $b(\xi, \eta)$ are related to the original description of the system, it is easily seen that, in the original coordinates, the stabilizing control law assumes the form

$$u = \frac{1}{L_g L_f^{r-1} h(x)} \left(-L_f^r h(x) - c_0 h(x) - c_1 L_f h(x) - \cdots - c_{r-1} L_f^{r-1} h(x)\right)$$

which is particularly interesting because expressed in terms of quantities that can be immediately calculated from the original data.

If an output function is not defined, the zero dynamics are not defined as well. However, it may happen that one is able to *design* a suitable dummy output whose associated zero dynamics have an asymptotically stable equilibrium. In this case, a control law of the form discussed before will guarantee asymptotic stability. This procedure is illustrated in the following simple example, taken from [5].

Example 37.1:

Consider the system

$$\dot{x}_1 = x_1^2 x_2^3$$
$$\dot{x}_2 = x_2 + u$$

whose linear approximation at $x = 0$ has an uncontrollable mode corresponding to the eigenvalue $\lambda = 0$. Suppose one is able to find a function $\gamma(x_1)$ such that

$$\dot{x}_1 = x_1^2 [\gamma(x_1)]^3$$

is asymptotically stable at $x_1 = 0$. Then, setting

$$y = h(x) = \gamma(x_1) - x_2$$

a system with an asymptotically stable zero dynamics is obtained. As a matter of fact, we know that the zero dynamics are those induced by the constraint $y(t) = 0$ for all t. This, in the present case, requires that the x_1 and x_2 respect the constraint

$$\gamma(x_1) - x_2 = 0\,.$$

Thus, the zero dynamics evolve exactly according to

$$\dot{x}_1 = x_1^2 [\gamma(x_1)]^3$$

and the system can be locally stabilized by means of the procedure discussed above. A suitable choice of $\gamma(x_1)$ will be, e.g.,

$$\gamma(x_1) = -x_1\,.$$

Accordingly, a locally stabilizing feedback is the one given by

$$\alpha(x) = \frac{1}{L_g h(x)}\left(-L_f h(x) - c h(x)\right) = -c x_1 - (1+c) x_2 - x_1^2 x_2^3$$

with $c > 0$.

37.4 Global Stabilization of Nonlinear Minimum-Phase Systems

In this section we consider a special class of nonlinear system that can be *globally* asymptotically stabilized via state feedback. The systems in question are systems that can be transformed, by means of a globally defined change of coordinates and/or feedback, into a system having this special normal form

$$\begin{aligned}
\dot{z} &= f_0(z, \xi_1)\\
\dot{\xi}_1 &= \xi_2\\
&\cdots\\
\dot{\xi}_{r-1} &= \xi_r\\
\dot{\xi}_r &= u.
\end{aligned} \tag{37.11}$$

Remark 37.3

Note that a system in the normal form of Equation 37.2, considered in the previous sections, can indeed be changed, via feedback, into a system of the form

$$\begin{aligned}\dot{\xi}_1 &= \xi_2\\ &\cdots\\ \dot{\xi}_{r-1} &= \xi_r\\ \dot{\xi}_r &= u\\ \dot{\eta} &= q(\xi,\eta).\end{aligned} \tag{37.12}$$

Moreover, if the normal form of Equation 37.2 is globally defined, so also is the feedback yielding the (globally defined) normal form of Equation 37.12. The form of Equation 37.11 is a special case of Equation 37.12, the one in which the function $q(\xi,\eta)$ depends only on the component ξ_1 of the vector ξ. In Equation 37.11, for consistency with the notations more frequently used in the literature on global stabilization, the vector z replaces the vector η of Equation 37.12 and the places of z and ξ are interchanged.

In order to describe how systems of the form of Equation 37.11 can be globally stabilized, we begin with the analysis of the (very simple) case in which $r = 1$. For convenience of the reader, we recall that a smooth function $V : \mathbb{R}^n \to \mathbb{R}$ is said to be *positive definite* if $V(0) = 0$ and $V(x) > 0$ for $x \neq 0$, and *proper* if, for any $a \in \mathbb{R}$, the set $V^{-1}([0,a]) = \{x \in \mathbb{R}^n : 0 \leq V(x) \leq a\}$ is compact.

Consider a system described by equations of the form

$$\begin{aligned}\dot{z} &= f(z,\xi)\\ \dot{\xi} &= u\end{aligned} \tag{37.13}$$

in which $(z,\xi) \in \mathbb{R}^n \times \mathbb{R}$, and $f(0,0) = 0$. Suppose the *subsystem*

$$\dot{z} = f(z,0)$$

has a globally asymptotically stable equilibrium at $z = 0$. Then, in view of a converse Lyapunov theorem, there exists a smooth positive definite and proper function $V(z)$ such that $\frac{\partial V}{\partial z} f(z,0)$ is negative for each nonzero z. Using this property, it is easy to show that the system of Equation 37.13 can be globally asymptotically stabilized. In fact, observe that the function $f(z,\xi)$ can be put in the form

$$f(z,\xi) = f(z,0) + p(z,\xi)\xi \tag{37.14}$$

where $p(z,\xi)$ is a smooth function. For it suffices to observe that the difference

$$\bar{f}(z,\xi) = f(z,\xi) - f(z,0)$$

is a smooth function vanishing at $\xi = 0$, and to express $\bar{f}(z,\xi)$ as

$$\bar{f}(z,\xi) = \int_0^1 \frac{\partial \bar{f}(z,s\xi)}{\partial s} ds = \int_0^1 \left[\frac{\partial \bar{f}(z,\zeta)}{\partial \zeta}\right]_{\zeta = s\xi} \xi ds\,.$$

Now consider the positive definite and proper function

$$W(z,\xi) = V(z) + \frac{1}{2}\xi^2, \tag{37.15}$$

and observe that

$$\begin{pmatrix}\frac{\partial W}{\partial z} & \frac{\partial W}{\partial \xi}\end{pmatrix}\begin{pmatrix}f(z,\xi)\\ u\end{pmatrix} = \frac{\partial V}{\partial z} f(z,0) + \frac{\partial V}{\partial z} p(z,\xi)\xi + \xi u\,.$$

Choosing

$$u = u(z,\xi) = -\xi - \frac{\partial V}{\partial z} p(z,\xi) \tag{37.16}$$

yields

$$\begin{pmatrix} \frac{\partial W}{\partial z} & \frac{\partial W}{\partial \xi} \end{pmatrix} \begin{pmatrix} f(z,\xi) \\ u(z,\xi) \end{pmatrix} < 0$$

for all nonzero (z, ξ). By the direct Lyapunov theorem, it is concluded that the system

$$\begin{aligned} \dot{z} &= f(z,\xi) \\ \dot{\xi} &= u(z,\xi) \end{aligned}$$

has a globally asymptotically stable equilibrium at $(z,\xi) = (0,0)$.

In other words, it has been shown that, if $\dot{z} = f(z,0)$ has a globally asymptotically stable equilibrium at $z = 0$, then the equilibrium $(z,\xi) = (0,0)$ of the system of Equation 37.13 can be rendered globally asymptotically stable by means of a smooth feedback law $u = u(z,\xi)$.

The result thus proven can be easily extended by showing that, for the purpose of stabilizing the equilibrium $(z,\xi) = (0,0)$ of Equation 37.13, it suffices to assume that the equilibrium $z = 0$ of

$$\dot{z} = f(z,\xi)$$

is *stabilizable* by means of a smooth law $\xi = v^\star(z)$.

Lemma 37.1:

Consider a system described by equations of the form of Equation 37.13. Suppose there exists a smooth real-valued function

$$\xi = v^\star(z)\,,$$

with $v^\star(0) = 0$, *and a smooth real-valued function* $V(z)$, *which is positive definite and proper, such that*

$$\frac{\partial V}{\partial z} f(z, v^\star(z)) < 0$$

for all nonzero z. *Then, there exists a smooth static feedback law* $u = u(z,\xi)$ *with* $u(0,0) = 0$, *and a smooth real-valued function* $W(z,\xi)$, *which is positive definite and proper, such that*

$$\begin{pmatrix} \frac{\partial W}{\partial z} & \frac{\partial W}{\partial \xi} \end{pmatrix} \begin{pmatrix} f(z,\xi) \\ u(z,\xi) \end{pmatrix} < 0$$

for all nonzero (z,ξ).

In fact, it suffices to consider the (globally defined) change of variables

$$y = \xi - v^\star(z)\,,$$

which transforms Equation 37.13 into

$$\begin{aligned} \dot{z} &= f(z, v^\star(z) + y) \\ \dot{y} &= -\frac{\partial v^\star}{\partial z} f(z, v^\star(z) + y) + u\,, \end{aligned} \tag{37.17}$$

and observe that the feedback law

$$u = \frac{\partial v^\star}{\partial z} f(z, v^\star(z) + y) + u'$$

changes the latter into a system satisfying the hypotheses that are at the basis of the previous construction.

Using repeatedly the property indicated in Lemma 37.1, it is straightforward to derive the following stabilization result about a system in the form of Equation 37.11.

Theorem 37.1:

Consider a system of the form of Equation 37.11. Suppose there exists a smooth real-valued function

$$\xi_1 = v^\star(z),$$

with $v^\star(0) = 0$, and a smooth real-valued function $V(z)$, which is positive definite and proper, such that

$$\frac{\partial V}{\partial z} f_0(z, v^\star(z)) < 0$$

for all nonzero z. Then, there exists a smooth static feedback law

$$u = u(z, \xi_1, \ldots, \xi_r)$$

with $u(0, 0, \ldots, 0) = 0$, which globally asymptotically stabilizes the equilibrium $(z, \xi_1, \ldots, \xi_r) = (0, 0, \ldots, 0)$ of the corresponding closed-loop system.

Of course, a special case in which the result of Theorem 37.1 holds is when $v^\star(z) = 0$, that is, when $\dot{z} = f_0(z, 0)$ has a globally asymptotically stable equilibrium at $z = 0$. This is the case of a system whose *zero dynamics* have a globally asymptotically stable equilibrium at $z = 0$, that is, the case of a globally minimum-phase system.

The stabilization procedure outlined above is illustrated in the following example, taken from [1].

Example 37.2:

Consider the problem of globally asymptotically stabilizing the equilibrium $(x_1, x_2, x_3) = (0, 0, 0)$ of the nonlinear system

$$\begin{aligned} \dot{x}_1 &= x_2^3 \\ \dot{x}_2 &= x_3^3 \\ \dot{x}_3 &= u. \end{aligned} \tag{37.18}$$

To this end, observe that a "dummy output" of the form

$$y = x_3 - v^\star(x_1, x_2)$$

yields a system having relative degree $r = 1$ at each $x \in \mathbb{R}^3$ and two-dimensional zero dynamics. The latter, that is, the dynamics obtained by imposing on Equation 37.18 the constraint $y = 0$, are those of the autonomous system

$$\begin{aligned} \dot{x}_1 &= x_2^3 \\ \dot{x}_2 &= (v^\star(x_1, x_2))^3. \end{aligned} \tag{37.19}$$

From the discussion above we know that, if it is possible to find a function $v^\star(x_1, x_2)$ that globally asymptotically stabilizes the equilibrium $(x_1, x_2) = (0, 0)$ of Equation 37.19, then there exists an input $u(x_1, x_2, x_3)$ that globally asymptotically stabilizes the equilibrium $(x_1, x_2, x_3) = (0, 0, 0)$ of Equation 37.18.

It is easy to check that the function

$$v^\star(x_1,x_2) = -x_1\exp(x_1x_2)$$

accomplishes this task. In fact, consider the system

$$\begin{aligned}\dot{x}_1 &= x_2^3\\ \dot{x}_2 &= -x_1^3\exp(3x_1x_2),\end{aligned} \tag{37.20}$$

and choose a candidate Lyapunov function

$$V(x_1,x_2) = x_1^4 + x_2^4,$$

which yields

$$\dot{V} = 4(x_1x_2)^3(1-\exp(3x_1x_2)).$$

This function is nonpositive for all (x_1,x_2) and zero only at $x_1 = 0$ or $x_2 = 0$. Since no nontrivial trajectory of Equation 37.20 is contained in the set

$$M = \{(x_1,x_2) : \dot{V} = 0\},$$

by Lasalle's invariance principle it is concluded that the equilibrium $(x_1,x_2) = (0,0)$ of Equation 37.20 is globally asymptotically stable.

In order to obtain the input function that globally stabilizes the equilibrium $(x_1,x_2,x_3) = (0,0,0)$ of Equation 37.18, it is necessary to use the construction indicated in the proof of Lemma 37.1. In fact, consider the change of variables

$$y = x_3 - v^\star(x_1,x_2)$$

which transforms Equation 37.18 into

$$\begin{aligned}\dot{x}_1 &= x_2^3\\ \dot{x}_2 &= (y+v^\star(x_1,x_2))^3\\ \dot{y} &= u - \frac{\partial v^\star}{\partial x_1}x_2^3 - \frac{\partial v^\star}{\partial x_2}(y+v^\star(x_1,x_2))^3.\end{aligned} \tag{37.21}$$

Choosing a preliminary feedback

$$u = \frac{\partial v^\star}{\partial x_1}x_2^3 + \frac{\partial v^\star}{\partial x_2}(y+v^\star(x_1,x_2))^3 + u'$$

yields

$$\begin{aligned}\dot{x}_1 &= x_2^3\\ \dot{x}_2 &= (y+v^\star(x_1,x_2))^3\\ \dot{y} &= u'.\end{aligned} \tag{37.22}$$

which has exactly the form of Equation 37.13, namely,

$$\begin{aligned}\dot{z} &= f(z,\xi)\\ \dot{\xi} &= u',\end{aligned}$$

with

$$z = \begin{pmatrix} x_1\\ x_2\end{pmatrix}$$

$$f(z,\xi) = \begin{pmatrix} x_2^3\\ (v^\star(x_1,x_2))^3\end{pmatrix} + \xi\begin{pmatrix} 0\\ 3(v^\star(x_1,x_2))^2 + 3v^\star(x_1,x_2)\xi + \xi^2\end{pmatrix}$$

and $\dot{z} = f(z,0)$ has a globally asymptotically stable equilibrium at $z = 0$. As a consequence, this system can be globally asymptotically stabilized by means of a feedback law $u' = u'(z,\xi)$ of the form of Equation 37.16.

Bibliography

1. Byrnes, C.I. and Isidori, A., New results and examples in nonlinear feedback stabilization, *Syst. Control Lett.*, 12, 437–442, 1989.
2. Tsinias, J., Sufficient Lyapunov-like conditions for stabilization, *Math. Control Signals Syst.*, 2, 424–440, 1989.
3. Byrnes, C.I. and Isidori, A., Asymptotic stabilization of minimum phase nonlinear systems, *IEEE Trans. Autom. Control,* 36, 1122–1137, 1991.
4. Byrnes, C.I. and Isidori, A., On the attitude stabilization of a rigid spacecraft, *Automatica*, 27, 87–96, 1991.
5. Isidori, A., *Nonlinear Control Systems,* 2nd ed., Springer-Verlag, Berlin, 1989.

38

The Lie Bracket and Control

V. Jurdjevic
University of Toronto

38.1 Introduction

Time-dependent events $a(t)$ and $b(t)$ are said to commute if the occurrence of $a(t)$ during a time interval T followed by the occurrence of $b(t)$ during an interval S leads to an outcome that does not change when the order of events is reversed. Denoting by $b(S)a(T)$ the occurrence of a followed by b, then $a(T)b(S)$ denotes the reversed order, and a and b commute if $a(T)b(S) = b(S)a(T)$. Otherwise, a and b are said to be noncommutative events.

Most events do not commute, as is evident from common experience. For instance, filling the swimming pool with water and then diving into the pool results in a state different from one in which the order is reversed. Driving an automobile relies on a more subtle use of noncommutativity (based on nonholonomy). Any automobile driver knows that the rotations of the steering wheel do not commute with either forward or backward motions of the automobile. The ability to park the automobile in a tight spot, a most demanding challenge for an inexperienced driver, is only possible because of the noncommuting nature of these events. An experienced driver knows that it is possible to execute a parallel displacement of an automobile in any space which is large enough to allow some forward and backward movement, although the number of maneuvers may be so large that it might very well be advisable to look for another parking spot with more space to spare.

Control of many dynamic systems, like the control of an automobile, consists of a time-sequential application of noncommuting events. Knowing which event to apply at a given time is a basic requirement of successful control. The recognition of noncommutativity as a fundamental issue of control is a starting point for geometric control; this chapter describes the main mathematical tools required for its understanding.

38.2 Notations and Basic Assumptions

The subsequent discussion is confined to control systems described by a differential equation $\frac{dx}{dt} = F(x, u)$ in which the control functions $u(t) = (u_1(t), \ldots, u_m(t))$ take values in a fixed set U in R^m. Most of the theory presented in this chapter is extracted through F and its derivatives, and for that reason it is expedient to assume that the state variable $x(t)$ belongs to an analytic manifold M and that $F(x, u)$ is an analytic vector field for each u in U. The reader not familiar with these notions may assume at the beginning that M is a finite dimensional vector space and that for each $u \in U, F(x, u)$ can be represented by its Taylor series at each point x in M. The extensions to arbitrary manifolds will be defined as needed.

Throughout this chapter $\mathcal{A}(x, T)$ will denote the set of states reachable from an initial state x in exactly T units of time. Then, $\mathcal{A}(x, \leq T) = \bigcup_{0 \leq t \leq T} \mathcal{A}(x, t)$, and $\mathcal{A}(x) = \bigcup_{t \geq 0} \mathcal{A}(x, t)$. As will be explained later, a differential control system may also be viewed as a family of vector fields $\mathcal{F}$, in which case the reachable sets will be subscripted $\mathcal{A}_{\mathcal{F}}(x, T), \mathcal{A}_{\mathcal{F}}(x, \leq T)$, and $\mathcal{A}_{\mathcal{F}}(x)$ in order to emphasize their dependence on a given system $\mathcal{F}$. These notations will be needed particularly when discussing several systems simultaneously.

38.3 Vector Fields

Geometric control theory begins with a distinctive view of a differential equation, initiated by H. Poincaré in the latter part of the 19th century, that the solutions of

$$\frac{dx}{dt} = F(x(t)) \quad x(t) \text{ in } M \tag{38.1}$$

can be analyzed in simple mathematical terms without ever having to solve the differential equation. The corresponding theory is based on a single assumption that for each initial state x_0 differential Equation 38.1 admits a unique solution through that state defined for all times t. Assuming that this is the case, let $x(x_0, t)$ denote the solution of Equation 38.1 for which $x(x_0, 0) = x_0$. The mapping $(x_0, t) \to x(x_0, t)$ is called the flow, or dynamic system induced by the vector field F. Its essential properties are

1. $x(x_0, 0) = x_0$ for each x_0 in M
2. $x(x_0, t + s) = x(x(x_0, t), s)) = x(x(x_0, s), t)$ for all x_0 in M and s, t in R
3. $\frac{\partial}{\partial t} x(x_0, t) = F(x(x_0, t))$

At each instant of time, t, the flow of F induces a transformation on M which will be denoted by $\exp tF$: $\exp tF$ maps each initial point x_0 onto $x(x_0, t)$. It follows from (1) and (2) that $\exp 0F =$ Identity and that $\exp(t + s)F = (\exp tF)(\exp sF) = (\exp sF)(\exp tF)$ for any s and t, with $(\exp sF)(\exp tF)$ denoting the composition of mappings. Since $\exp 0F =$ Identity, it follows that $(\exp tF)^{-1} = \exp -tF$, and therefore, the mappings $\{\exp tF : t \in R\}$ form a commutative group. This group is called the one-parameter group of transformations induced by F.

Example 38.1:

If F is a linear vector field, i.e., $F(x) = Ax$ for some linear mapping on M then $x(x_0, t) = e^{tA} x_0$ with $e^{tA} = \sum_{k=0}^{\infty} \frac{t^k}{k!} A^k$. Thus, in this case $\exp tF$ is a linear transformation on M equal to the exponential of a linear mapping.

Example 38.2:

If $F(x) = b$ is a constant vector field, then $x(x_0, t) = x_0 + tb$, and therefore each transformation expt$F$ is a translation in the direction of b.

Vector fields, no matter how nonlinear, admit linear interpretations on an infinite dimensional vector space in which expt$F$ becomes the exponential of a linear mapping. In this interpretation vector fields act linearly on functions on M as follows: let

$$(\text{expt}F)(f)(x_0) = f(x(x_0, t))$$

for any real-valued function, f. Then Ff is the function on M defined by

$$Ff = \frac{d}{dt}(\text{expt}F)(f)|_{t=0}$$

Ff is called the derivation of f along the vector field F.

Ff admits a simple description in any system of coordinates on M. Assuming that M is a linear vector space, then any basis $a_1, \ldots, a_n$ in M induces coordinates $x_1, \ldots, x_n$ on M by the formula $x = \sum_{i=1}^{n} x_i a_i$. Then $F(x) = \sum_{i=1}^{n} F_i(x_1, \ldots, x_n)a_i$ and Equation 38.1 is written as a system of differential equations in R^n

$$\frac{dx_i}{dt} = F_i(x_1, \ldots, x_n) \quad i = 1, \ldots, n. \tag{38.2}$$

Any real-valued function on M becomes a function on R^n by the correspondence $f(x) = f(x_1, \ldots x_n)$. Then,

$$\frac{d}{dt} f(x_1(t), \ldots, x_n(t))|_{t=0} = \sum_{i=1}^{n} \frac{\partial f}{\partial x_i}(x_1, \ldots, x_n)F_i(x_1, \ldots, x_n)$$

and therefore

$$(Ff)(x_1, \ldots, x_n) = \sum_{i=1}^{n} \frac{\partial f}{\partial x_1} F_i(x_1, \ldots, x_n)$$

It follows that $(Ff)(x)$ is equal to the directional derivative of f at x in the direction of $F(x)$. In particular, when $f = x_i$, then $Fx_i = F_i, i = 1, \ldots, n$. Evidently F acts linearly on functions (as a directional derivative) and therefore satisfies further properties:

$Ff = 0$ for constant functions f, and $F(fg) = f(Fg) + g(Ff)$ with respect to the products of functions. (fg denotes the product of f and g, i.e., $(fg)(x) = f(x)g(x)$ and $f(Fg)$ is the product of f with Fg. Note that $f(Fg) \neq (Ff)g$.)

As a mapping on the space of functions expt$F$ satisfies

$$\frac{d}{dt}\text{expt}F = F(\text{expt}F) = (\text{expt}F)F \quad \text{and} \quad \frac{d^n}{dt^n}\text{expt}F = F^n\text{expt}F \tag{38.3}$$

with each product denoting the composition of linear mappings. Analytic vector fields can be represented by their Taylor series and therefore $(\text{expt}F)(f) = \sum_{k=0}^{\infty} \frac{t^k}{k!} F^k(f)$ for any analytic function f. The reader can easily verify that in each system of coordinates, $F^2 f = F(Ff) = F\left(\sum_{i=1}^{n} \frac{\partial f}{\partial x_i} F_i\right) = \sum_{i,j=1}^{n} \left(\frac{\partial^2 f}{\partial x_i \partial x_j} F_i F_j + \frac{\partial f}{\partial x_i}\frac{\partial F_i}{\partial x_j}\right)$. Then $F^3 f$ is the directional derivative of $F^2 f$ in the direction F and so on for each derivate $F^n f$.

38.4 The Lie Bracket

With these notions and notations at our disposal, let us return to the commutativity issue raised in the introduction, and consider commutativity properties of exptF and expsG corresponding to vector fields F and G. This question may be further motivated through the following control theoretic context:

Consider a control system

$$\frac{dx}{dt} = u(t)F(x(t)) + (1 - u(t))G(x(t))$$

with switching control u which can only take values 0 or 1 (on-off controls). During any time interval that the control is turned on the system follows F and during the time intervals that the control is switched off the system follows G.

Let $u_1(t)$ and $u_2(t)$ denote the following control functions defined in an interval $[0, T + S]$:

$$u_1(t) = \begin{cases} 1, & t \in [0, T) \\ 0, & t \in [T, T + S] \end{cases}$$

$$u_2(t) = \begin{cases} 0, & t \in [0, S) \\ 1, & t \in [S, T + S] \end{cases}$$

The corresponding trajectories $x_1(t)$ and $x_2(t)$, both initiating from the same initial point x_0, are given by:

$$x_1(t) = (\text{expt}F)(x_0), \quad t \in [0, T)$$
$$x_1(t) = (\exp(t - T)G)(\text{expt}TF)(x_0), \quad t \in [T, T + S]$$

and

$$x_2(t) = (\text{expt}G)(x_0), \quad t \in [0, S)$$
$$x_2(t) = (\exp(t - S)F)(\exp SG)(x_0) \text{ for } t \in [S, S + T]$$

At the terminal time $t = T + S, x_1(t) = (\exp SG)(\exp TF)(x_0)$, and $x_2(t) = (\exp TF)(\exp SG)(x_0)$.

Assuming that the control actions of $u_1(t)$ and $u_2(t)$ lead to the same terminal state independently of the initial point x_0 and of the switching times S and T, then $(\exp SG)(\exp TF) = (\exp TF)(\exp SG)$. This equality remains unaltered when the domain is extended to the space of functions and therefore, $\frac{\partial}{\partial t}\frac{\partial}{\partial s}(\text{exps}G)(\text{expt}F)(f) = \frac{\partial}{\partial s}\frac{\partial}{\partial t}(\text{expt}F)(\text{exps}G)(f)$ for any function f.

Taking advantage of formulas (Equation 38.3)

$$\frac{\partial}{\partial t}\frac{\partial}{\partial s}(\text{exps}G)(\text{expt}F)f = \frac{\partial}{\partial t}G(\text{exps}G)(\text{expt}F)(f) = G(\text{exps}G)(F\text{expt}F)f,$$

and

$$\frac{\partial}{\partial s}\frac{\partial}{\partial t}(\text{expt}F)(\text{exps}G)(f) = F(\text{expt}F)(G\text{exps}G)f$$

Evaluating these derivatives at $t = s = 0$ gives

$$G(Ff) = F(Gf)$$

Definition 38.1:

Let F and G be any vector fields on M. Then $[F, G](f) = G(Ff) - F(Gf)$ for any real-valued function f on M. $[F, G]$ is called the Lie bracket of F with G.

It follows that $[F, G]$ is a vector field for any vector fields F and G. Note that $[F, G] = -[G, F]$. The ith coordinate of the Lie bracket $[F, G]$ is given by $[F, G](x_i)$. Therefore,

$$[F, G](x_i) = G(F_i) - F(G_i) = \sum_{j=1}^{n} \frac{\partial F_i}{\partial x_j} G_j - \frac{\partial G_i}{\partial x_j} F_j \tag{38.4}$$

The calculations above show that if $(\exp sG)(\exp tF) = (\exp tF)(\exp sG)$ for all s and t then $[F, G] = 0$. Somewhat remarkably, the converse is also true; i.e., if $[F, G] = 0$, then $(\exp tF)(\exp sG) = (\exp sG)(\exp tF)$.

Example 38.3:

If $F(x) = b$ and $G(x) = c$ are any constant vector fields, then their coordinates are constant functions, and therefore $[F, G] = 0$ (as can be seen from Equation 38.4). Then, $(\exp tF)(x) = x + tb$ and $(\exp sG)(x) = x + sc$, and therefore $(\exp tF)(\exp sG)(x) = (x + sc) + tb = (x + tb) + sc = (\exp sG)(\exp tF)(x)$, confirming that the flows commute.

Example 38.4:

Let $F(x) = Ax$ be a linear vector field and $G(x) = b$ a constant vector field. In terms of any linear coordinates $x_1, \ldots, x_n, F_i(x_1, \ldots, x_n) = \sum_{j=1}^{n} A_{ij} \cdot x_j$. (A_{ij}) is the matrix of A relative to this basis. Then

$$[F, G](x_i) = \sum_{j=1}^{n} \frac{\partial F_i}{\partial x_j} G_j - \frac{\partial G_i}{\partial x_j} F_j = \sum_{j=1}^{n} A_{ij} b_j - O \cdot F_j = \sum_{j=1}^{n} A_{ij} b_j$$

Thus $[G, F]$ is a constant vector field equal to Ab. $[G, F] = 0$ if and only if $b \in \ker A$.

For instance, the rotation $A = \begin{pmatrix} 0 & -1 & 0 \\ 1 & 0 & 0 \\ 0 & 0 & 0 \end{pmatrix}$ and $b = \begin{pmatrix} 0 \\ 0 \\ 1 \end{pmatrix}$ commute, while A does not commute with $b = \begin{pmatrix} 1 \\ 0 \\ 0 \end{pmatrix}$ (Figure 38.1).

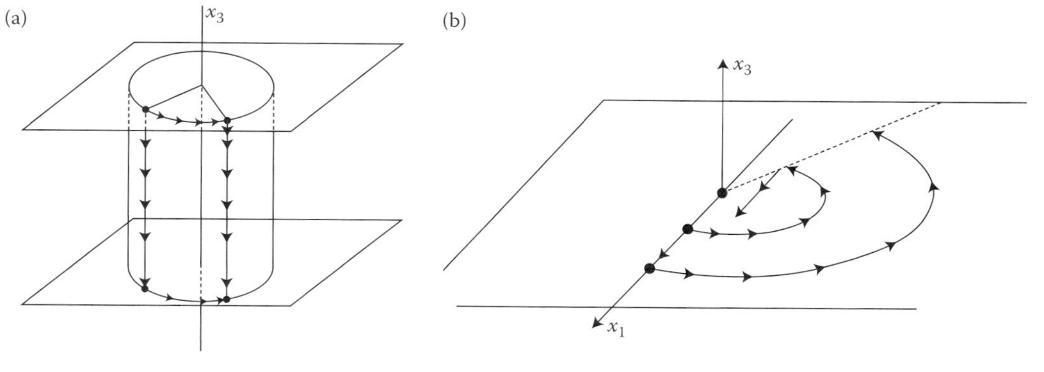

FIGURE 38.1 Commutativity: (a) Commuting case and (b) Noncommuting case.

Example 38.5:

Let $F(x) = Ax$ and $G(x) = Bx$ be linear vector fields on M. Then $[F, G]$ is a linear vector field given by $[F, G](x) = (AB - BA)(x)$.

For example, if $A = \begin{pmatrix} 1 & 0 \\ 0 & -1 \end{pmatrix}$ and $B = \begin{pmatrix} 0 & 1 \\ 1 & 0 \end{pmatrix}$ are the matrices corresponding to F and G (relative to a linear system of coordinates) then $C = 2\begin{pmatrix} 0 & 1 \\ -1 & 0 \end{pmatrix}$ is the matrix that corresponds to $[F, G]$. The flows of these fields are shown in Figure 38.2.

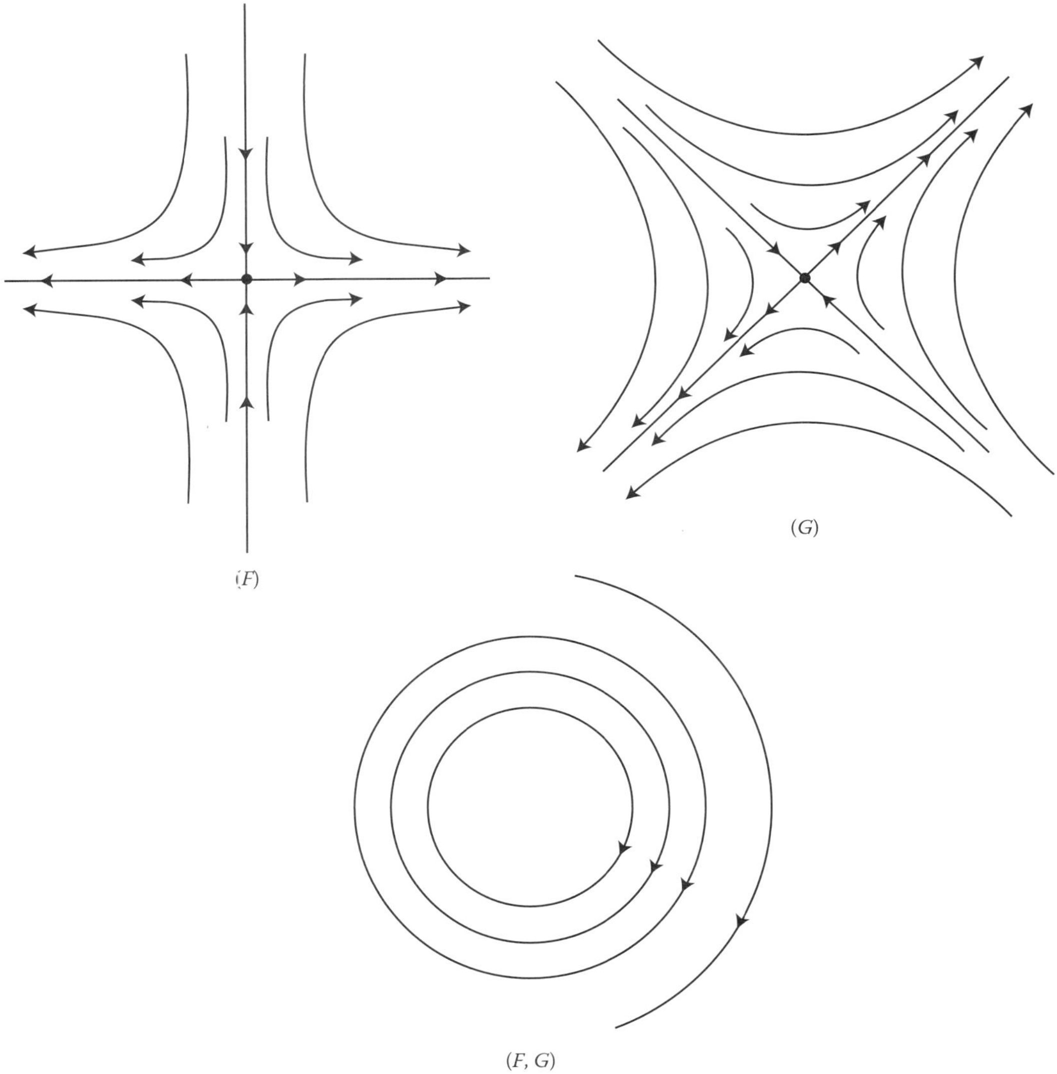

FIGURE 38.2 Lie brackets of vector fields.

Return now to the general discussion, and consider the expression

$$(\exp -tG)(\exp -tF)(\exp tG)(\exp tF) \tag{38.5}$$

As shown earlier, this expression reduces to the identity when $[F, G] = 0$. Begin with a simple case with F and G linear vector fields defined by linear mappings A and B. Then $\exp tF = e^{tA} = \sum_{k=0}^{\infty} \frac{t^k}{k!} A^k$, and $\exp tG = \sum_{k=0}^{\infty} \frac{t^k}{k!} B^k$. We shall be interested only in the terms up to and including the second-order terms in t. Then,

$$e^{tB}e^{tA} = I + t(A+B) + t^2 BA + \frac{t^2}{2!}(A^2 + B^2)$$

and

$$e^{-tB}e^{-tA} = I - t(A+B) + t^2 BA + \frac{t^2}{2!}(A^2 + B^2)$$

Therefore,

$$\begin{aligned} e^{-tB}e^{-tA}e^{tB}e^{tA} &= \left(I - t(A+B) + t^2BA + \frac{t^2}{2!}(A^2+B^2)\right)\left(I + t(A+B) + t^2BA + \frac{t^2}{2!}(A^2+B^2)\right) \\ &= I + t(A+B) + t^2BA + \frac{t^2}{2}(A^2+B^2) - t(A+B) - t^2(A+B)^2 + t^2BA + \frac{t^2}{2}(A^2+B^2) \\ &= I - t^2(AB+BA) + 2t^2BA = I + t^2(BA - AB) \end{aligned}$$

These calculations show that up to second-order terms in t,

$$(\exp -tG)(\exp -tF)(\exp tG)(\exp tF) = I + t^2[G, F]$$

This result extends to arbitrary vector fields, as can be demonstrated through the formulas $\exp tF = \sum_{k=0}^{\infty} \frac{t^k}{k!} F^k(f)$ and $(\exp tG)(f) = \sum_{k=0}^{\infty} \frac{t^k}{k!} G^k(f)$ described earlier. These arguments show that the proof for the general flows is the same as it is for the linear flows.

The asymptotic formula obtained above shows that the curve $\sigma(t) = (\exp -\sqrt{t}G)(\exp -\sqrt{t}F)$ $(\exp t\sqrt{t}G)(\exp t\sqrt{t}F)(x_0)$ satisfies $\sigma(0) = x_0$ and $\frac{d\sigma}{dt}(0) = [G, F](x_0)$.

The preceding developments may be stated in a control theoretic context, as follows.

Theorem 38.1:

Suppose that $\frac{dx}{dt} = F(x, u)$ is any system in M with the control functions taking place in a set $U \subset R^m$. Suppose that $F_1(x) = F(x, u_1)$ and $F_2(x) = F(x, u_2)$ for some choices u_1 and u_2 of control values. Assume that there exist control values u_3 and u_4 in U such that $-F_1(x) = F(x, u_3)$ and $-G(x) = F(x, u_4)$ for all x. Then there is a curve $\sigma(t)$ contained in the reachable set $A(x, \leq \epsilon)$ for any $\epsilon > 0$ such that $\frac{d\sigma}{dt}(0) = [F_1, F_2](x)$. That is, the system can move infinitesimally in the direction of the Lie bracket.

Example 38.6:

Let $A = \begin{pmatrix} 0 & 1 \\ 1 & 0 \end{pmatrix}$ and $B = \begin{pmatrix} 1 & 0 \\ 0 & -1 \end{pmatrix}$ define the bilinear system $\frac{dx}{dt} = (1-u)Ax + uBx$ in $M = R^2$, with $0 \leq u(t) \leq 1$. Then, $u = 0$ follows $F(x) = Ax$ and $u = 1$ follows $G(x) = Bx$. The Lie bracket $[F, G]$ is a

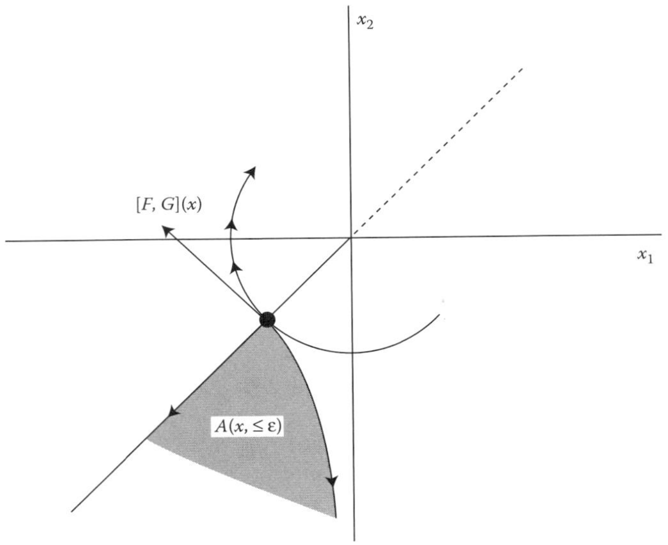

FIGURE 38.3 Noncontrollability of the Lie bracket.

linear field given by the rotation matrix $2\begin{pmatrix} 0 & 1 \\ -1 & 0 \end{pmatrix}$. Since

$$\frac{dx_1}{dt} = ux_1 + (1-u)x_2 \quad \text{and} \quad \frac{dx_2}{dt} = -x_2u + x_1(1-u),$$

$$\frac{d}{dt}x_1(t)x_2(t) = (ux_1 + (1-u)x_2)x_2 + (-x_2u + x_1(1-u))x_1 = (1-u)(x_1^2 + x_2^2) \ge 0.$$

Thus, $x_1(t)x_2(t)$ is increasing for any admissible control function. The system cannot move infinitesimally in the direction of $[F, G]$ because neither $-F$ nor $-G$ can be traced by the admissible controls (Figure 38.3).

38.5 The Lie Algebras

From a geometric point of view, a control system is a family of vector fields parameterized by controls. Each control value determines a vector field, and the corresponding trajectory is a solution curve of this field. As the control switches to a new value so does the vector field and the trajectory begins to follow the direction of the new field. From this view a trajectory generated by a piecewise constant control is a continuous curve in M having discontinuous derivatives with the breaks in the derivative corresponding to the changes in vector fields caused by the switches in the control. Such curves are conveniently called continuous broken trajectories. Between any consecutive breaks, the trajectory is a solution curve of a vector field in the family.

From this perspective, a linear control system $\frac{dx}{dt} = Ax + Bu$, $u \in U$ is a collection of affine vector fields $\mathcal{F}$ of the form $F_u(x) = Ax + Bu$ with $u \in U$. An affine vector field F is any vector field in M which satisfies $F(\sum \lambda_i x_i) = \sum \lambda_i F(x_i)$ for any affine combination $\sum \lambda_i x_i$ with $\sum \lambda_i = 1$. It can be shown that any affine field is the sum of a linear vector field and a constant vector field. The exponential map expt$F$ of any affine vector field $F(x) = Ax + b$ is given by $(\text{expt}F)(x) = e^{At}(x + \int_0^t e^{-As} b\, ds)$. For instance, if $b \in \ker A$, then $e^{-As}b = b$, and $(\text{expt}F)(x) = e^{At}x + bt$.

Vector fields form a vector space under pointwise addition and multiplication by scalars. The solution curves of F and λF, λ a real number, differ only by reparameterization of time; i.e., if $\frac{dx}{dt}(t) = F(x(t))$ then $\frac{d}{dt}x(\lambda t) = \lambda F(x(\lambda t))$, and consequently $y(t) = x(\lambda t)$ is a solution curve of λF.

For sums of vector fields the situation is different. Thus, $F(x) = \begin{pmatrix} 0 & 1 \\ 0 & 0 \end{pmatrix}\begin{pmatrix} x_1 \\ x_2 \end{pmatrix}$, $G(x) = \begin{pmatrix} 0 & 0 \\ 1 & 0 \end{pmatrix}\begin{pmatrix} x_1 \\ x_2 \end{pmatrix}$ add to $H(x) = \begin{pmatrix} 0 & 1 \\ 1 & 0 \end{pmatrix}\begin{pmatrix} x_1 \\ x_2 \end{pmatrix}$.

The phase portraits of these flows are shown in Figure 38.4.

It is known that every solution curve of the sum $F + G$ can be approximated by a curve that oscillates between the solution curves of F and the solution curves of G, as shown in Figure 38.5.

The vector space structure, together with the Lie bracket operation, turns the set of all vector fields into an algebra called the Lie algebra.

Any set of vector fields $\mathcal{F}$ generates the smallest sub-algebra that contains $\mathcal{F}$. We will use Lie($\mathcal{F}$) to denote the Lie algebra generated by a family of vector fields $\mathcal{F}$.

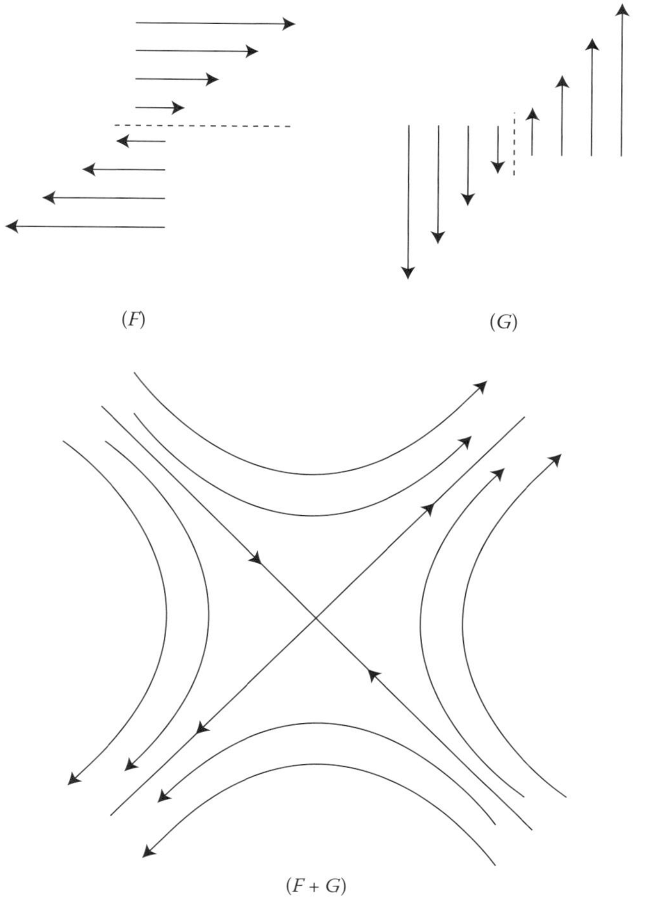

FIGURE 38.4 Sums of vector fields.

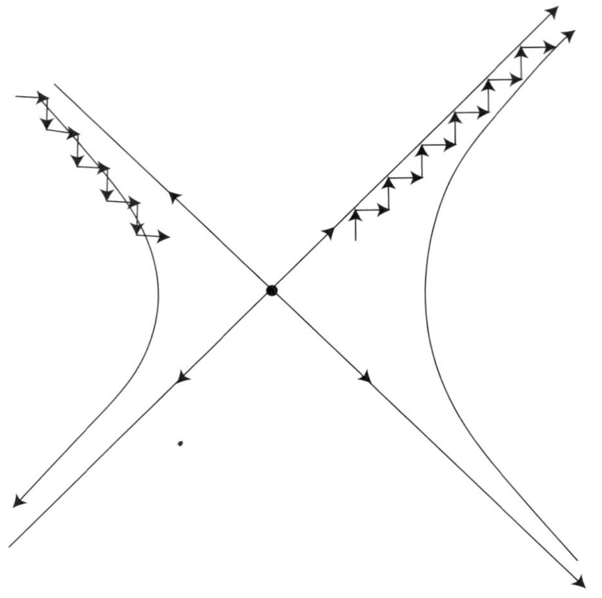

FIGURE 38.5 Chattering along the sum of vector fields.

Example 38.7:

Let $\mathcal{F}$ consist of two vector fields F and G in R^2 given by $Ff = x_1^2 \frac{\partial f}{\partial x_1}$ and $Gf = x_1^2 \frac{\partial f}{\partial x_2}$. Then, $[F,G]f = -2x_1^3 \frac{\partial f}{\partial x_2}$. Furthermore, $[F,[F,G]] = 6x_1^4 \frac{\partial}{\partial x_2}$. It then follows by induction that each vector field $x_1^n \frac{\partial f}{\partial x_2}$ is contained in Lie($\mathcal{F}$), and therefore Lie($\mathcal{F}$) is an infinite dimensional algebra of vector fields.

Example 38.8:

Consider now the family of affine vector fields $\mathcal{F} = \{F_u(x) = Ax + Bu : u \in U\}$ defined by a linear system with a constraint set U. When the linear span of points in U spans R^m, Lie($\mathcal{F}$) coincides with the Lie algebra generated by $F_0(x) = Ax$ and the constant vector fields with values in the range of B.

As we have already shown in Example 38.2, $[F_0, G] = Ab$ for any constant vector field $G = b$. Let $\mathcal{B}$ denote the range space of B. Then, it follows by the preceding observation that the space of constant vector fields in $\mathcal{B} + A\mathcal{B}$ is contained in Lie($\mathcal{F}$). Denote by $\mathcal{L}_k$ the vector space of all constant vector fields with values in $\mathcal{B} + A\mathcal{B} + \cdots + A^{k-1}\mathcal{B}$. It follows that $\mathcal{L}_{k+1} = [F_0, \mathcal{L}_k] + \mathcal{L}_k$ and therefore each space $\mathcal{L}_k$ is contained in Lie($\mathcal{F}$). Since each power A^{n+k} is a linear combination of $\{A^k : 0 \le k \le n-1\}$, as a consequence of the Cayley–Hamilton theorem, it follows that Lie($\mathcal{F}$) is equal to the linear span of F_0 and $\mathcal{L}_n$, and is therefore a finite dimensional Lie algebra. The space $\mathcal{L}_n$ is usually described as the range space of the controllability matrix

$$(B\,AB \ldots A^{n-1}B)$$

An arbitrary family of vector fields $\mathcal{F}$ in Lie($\mathcal{F}$), when evaluated at each point x, defines a linear space of tangent vectors at x. For instance, in Example 38.7, each vector in Lie($\mathcal{F}$) evaluated at $x_1 = 0$ is equal to zero. Therefore, each element in Lie($\mathcal{F}$) is equal to zero at such points, and consequently the corresponding space of tangent vectors is zero-dimensional. At any other point of the plane G and F are linearly independent, and their span is two-dimensional.

The Lie algebra in Example 38.8 evaluated at the origin is equal to the range space of the controllability matrix. The Lie algebra evaluated at any other point is the vector sum of Ax and the range of the controllability matrix.

We shall use $\mathrm{Lie}_x(\mathcal{F})$ to denote the vector space of all tangent vectors $F(x)$ with F in $\mathrm{Lie}(\mathcal{F})$. Then $\dim \mathrm{Lie}_x(\mathcal{F})$ will denote the dimension of this vector space.

The following theorem is of fundamental importance for geometric control theory.

Theorem 38.2:

Let $\mathcal{F}$ be any family of vector fields, and let $A_{\mathcal{F}}(x, \leq T)$ denote its reachable set from x in T units of time (in accordance with the conventions outlined earlier). Then $\dim \mathrm{Lie}_x(\mathcal{F}) = \dim M$ *is a necessary and sufficient condition that $\mathcal{A}_{\mathcal{F}}(x, \leq T)$ has a non-empty interior in M. Furthermore, when* $\dim \mathrm{Lie}_x(M) = \dim M$, *then*

$$cl\mathcal{A}_{\mathcal{F}}(x, \leq T) \subseteq \text{ int } \mathcal{A}_{\mathcal{F}}(x, \leq T+\epsilon) \subseteq \mathcal{A}_{\mathcal{F}}(x, \leq T+\epsilon) \quad \textit{for any } T > 0 \quad \textit{and} \quad \textit{any } \epsilon > 0.$$

In the preceding notation, $\mathrm{cl}(A)$ denotes the topological closure of a set A, which means that $\mathrm{cl}(A)$ consists of A along with all points of M which are limit points of elements in A. The interior of any set A, denoted by int (A), consists of all points a in A contained in an open ball in M centered at a which is entirely contained in A.

Typically, the initial point x belongs to the boundary of the reachable set $A(x, \leq T)$, as in Example 38.6. x is said to be small-time controllable by $\mathcal{F}$ whenever x belongs to the interior of $\mathcal{A}_{\mathcal{F}}(x, \leq T)$ for any $T > 0$. The following example shows that x can be small-time locally controllable even for families of two vector fields neither of which vanishes at x.

Example 38.9:

Let $\mathcal{F}$ be the family consisting of linear vector fields F and G in R^2 described by $A = \begin{pmatrix} 0 & 1 \\ 1 & 0 \end{pmatrix}$ and $B = \begin{pmatrix} -1 & 0 \\ 0 & -2 \end{pmatrix}$. The solution curves of F are hyperbolas $x_2^2 - x_1^2 = \text{const}$ while the integral curves of G are parabolas $x_2 = cx_1^2$. The curves are tangent to each other along the lines $x_1 = \pm\sqrt{2}x_2$. Any initial point x along such a line is in the interior of $\mathcal{A}_{\mathcal{F}}(x, \leq T)$, as Figure 38.6 shows.

For linear systems, any trajectory $x(t)$ which originates at $x_0 = 0$ is of the form $x(t) = e^{At} \int_0^t e^{-As} Bu(s)\, ds$, and is therefore necessarily contained in the range space of the controllability matrix, as can be seen from the expression

$$e^{A(t-s)} Bu(s) = \sum_{k=0}^{\infty} \frac{(t-s)^k}{k!} A^k Bu(s)\, ds$$

confirming the necessity of the Lie algebraic criterion stated in Theorem 38.2.

38.6 The Lie Saturate

We now shift to the invariance properties of control systems using the closure of the reachable sets as the basic criterion for invariance. Let $\mathcal{F}_1$ and $\mathcal{F}_2$ be any families of vector fields. Then $\mathcal{F}_1$ is said to be strongly equivalent to $\mathcal{F}_2$ if

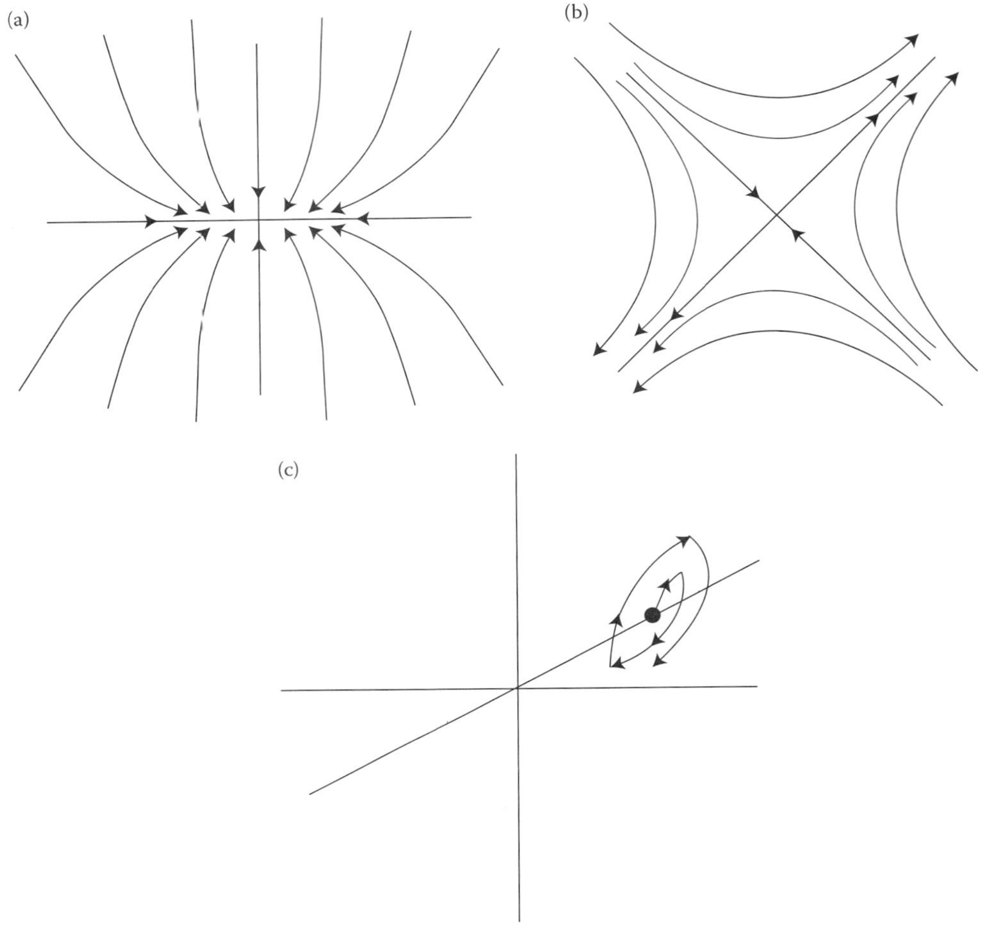

FIGURE 38.6 Small-time local controllability: (a) solution curves of G; (b) solution curves of F; and (c) closed cycles.

1. $\text{Lie}_x(\mathcal{F}_1) = \text{Lie}_x(\mathcal{F}_2)$ for all x in M
2. $\text{cl}\mathcal{A}_{\mathcal{F}_1}(x, \leq T) = \text{cl}\mathcal{A}_{\mathcal{F}_2}(x, \leq T)$ for all $T > 0$ and all x in M

$\mathcal{F}_1$ and $\mathcal{F}_2$ are said to be equivalent if (2) is replaced by $\text{cl}\mathcal{A}_{\mathcal{F}_1}(x) = \text{cl}\mathcal{A}_{\mathcal{F}_2}(x)$.

Definition 38.2:

The (strong) Lie saturate of a given control system is the largest family of vector fields (strongly) equivalent to $\mathcal{F}$. The (strong) Lie saturate will be denoted by $\mathcal{LS}_s(\mathcal{F})$, and $\mathcal{LS}(\mathcal{F})$ will denote the Lie saturate of $\mathcal{F}$.

Definition 38.3:

A control system $\mathcal{F}$ is said to be strongly controllable if $\mathcal{A}_{\mathcal{F}}(x, \leq T) = M$ for each x in M and each $T > 0$. It is said to be controllable if $\mathcal{A}_{\mathcal{F}}(x) = M$ for each x in M.

Theorem 38.3:

i. *A control system $\mathcal{F}$ is strongly controllable if and only if the strong Lie saturate of $\mathcal{F}$ is equal to Lie($\mathcal{F}$).*
ii. *$\mathcal{F}$ is controllable if and only if the Lie saturate of $\mathcal{F}$ is equal to Lie($\mathcal{F}$).*

The above theorem is called the Lie saturate criterion of controllability. As elegant as the criterion seems, its practical value rests on the constructive means of calculating the Lie saturate. The next theorems describe permissible system enlargements, prolongations, which respect the reachable sets and may be used to generate a procedure for calculating the Lie saturate. We say that $\mathcal{F}_2$ is a prolongation of $\mathcal{F}_1$ if $\mathcal{F}_1 \subset \mathcal{F}_2$ and if $\mathcal{F}_1$ and $\mathcal{F}_2$ have the same Lie saturates.

We begin by noting that the convex hull of any family of vector fields is contained in the strong Lie saturate of the family. Recall that the convex hull of $\mathcal{F}$ consists of all convex combinations $\sum_{i=1}^{m} \lambda_i F_i$, with $\sum_{i=1}^{m} \lambda_i = 1, \lambda_i \geq 0$, and each F_i in $\mathcal{F}$. It is known that

$$\left(\exp\left(t \sum \lambda_i F_i\right)\right)(x) \in \mathrm{cl}\mathcal{A}_{\mathcal{F}}(x, t)$$

for each $t > 0$, because any trajectory of a convex combination can be approximated by a trajectory of $\mathcal{F}$. The approximation is achieved by switching sufficiently fast along the trajectories of $F_1, \ldots, F_m$ around the trajectory of the convex sum (as illustrated in Figure 38.5). It may happen that the terminal point of the trajectory of the convex sum is not reachable by the original system, as the example below shows.

Example 38.10:

The point $x = 1, y = 0$ cannot be reached in time $t = 1$ from the origin by the trajectories of

$$\frac{dx}{dt} = -y^2 + 1$$
$$\frac{dy}{dt} = u \quad \text{with } u(t) = \pm 1$$

The convex hull of $U = \{-1, 1\}$ is the closed interval $-1 \leq u \leq 1$. Therefore, $u(t) = 0$ is in the convex hull of $U = \{-1, 1\}$. The corresponding trajectory which originates at 0 is given by $x(t) = t, y(t) = 0$ and reaches $x = 1, y = 0$ at $t = 1$.

Having taken the closure of the reachable sets as the criterion for equivalence, it becomes natural to pass to topologically closed families of vector fields. The choice of topology for the space of vector fields is not particularly important. In this context it is convenient to topologize the space of all vector fields by the C^{∞} topology on compact subsets of M. Rather than going into the mathematical details of this topology, let us illustrate the use with an example.

Suppose that $X_{\lambda}(x) = \lambda(Ax + \frac{1}{\lambda}b)$ is a family of affine vector fields parameterized by λ. For each $\lambda \neq 0$, $(\exp tX_{\lambda})x = e^{t\lambda A}x + \int_0^t e^{\lambda(t-s)A}b\, ds$, $\lim_{\lambda \to 0}(\exp tX_{\lambda})(x) = x + bt$ because $\lim_{\lambda \to 0} e^{t\lambda A} = I$ uniformly in t. Thus the limiting curve $x + bt$ is equal to $(\exp tX_0)(x)$ with $X_0 = \lim_{\lambda \to 0} X_{\lambda}$.

It can be shown in general that if a sequence of vector fields converges to a vector field F then each curve $\sigma_n(t) = (\exp tF_n)(x_0)$ converges uniformly in t to $\sigma(t) = (\exp tF)(x_0)$. Therefore, each family $\mathcal{F}$ may be prolonged to its topological closure.

In addition to the convexification and the topological closure, there is another means of prolonging a given family of vector fields based on reparameterizations of trajectories.

Note that $y(t) = x(\lambda t)$ remains in the reachable set $\mathcal{A}_{\mathcal{F}}(x_0, \leq T)$ for any trajectory $x(t)$ of $\mathcal{F}$ for which $x(0) = x_0$ provided that $0 \leq \lambda \leq 1$. $y(t) \in \mathcal{A}_{\mathcal{F}}(x_0)$ for any $\lambda \geq 0$. Thus, $\lambda F \in \mathcal{L}S_s(\mathcal{F})$ for any $0 \leq \lambda \leq 1$ and any F in $\mathcal{L}S_s(\mathcal{F})$. It will be useful for further references to assemble these prolongations into a theorem.

Theorem 38.4:

i. *The Lie saturate of any system is a closed convex cone, i.e.,* $\sum_{i=1}^{m} \lambda_i F_i \in \mathcal{L}S(\mathcal{F})$ *for any vector fields* $F_1, \ldots, F_m$ *in* $\mathcal{L}S(\mathcal{F})$ *and any numbers* $\lambda_1 \geq 0, \ldots, \lambda_m \geq 0$.
ii. *The strong Lie saturate of any family of vector fields is a closed convex body, i.e.,* $\sum_{i=1}^{m} \lambda_i F_i \in \mathcal{L}S_s(\mathcal{F})$ *for any elements* $F_1, \ldots, F_m$ *in* $\mathcal{L}S_s(\mathcal{F})$ *and any non-negative numbers* $\lambda_1, \ldots, \lambda_n$ *such that* $\sum_{i=1}^{m} \lambda_i \leq 1$.

We now describe another operation which may be used to prolong the system without altering its reachable sets. This operation is called the normalization of the system.

An invertible map $\Phi : M \to M$ is called a strong normalizer for $\mathcal{F}$ if $\Phi(\mathcal{A}_{\mathcal{F}}(\Phi^{-1}(x), \leq T)) \subset \mathrm{cl}\mathcal{A}_{\mathcal{F}}(x, \leq T)$ for all x in M and $T > 0$. Φ is called a normalizer for $\mathcal{F}$ if $\Phi\mathcal{A}_{\mathcal{F}}(\Phi^{-1}(x)) \subset \mathrm{cl}\mathcal{A}_{\mathcal{F}}(x)$. It may be also said that Φ is a strong normalizer if both $\Phi(x)$ and $\Phi^{-1}(x)$ are contained in $\mathrm{cl}\mathcal{A}_{\mathcal{F}}(x, \leq T)$ and that Φ is a normalizer if both $\Phi(x)$ and $\Phi^{-1}(x)$ belong to $\mathrm{cl}\mathcal{A}_{\mathcal{F}}(x)$. In this notation $\Phi(\mathcal{A}_{\mathcal{F}}(\Phi^{-1}(x), \leq T))$ is equal to the set of points $\Phi(y)$ with y belonging to $\mathcal{A}_{\mathcal{F}}(\Phi^{-1}(x), \leq T)$. If Φ is any invertible transformation, and if F is any vector field then $(\Phi)(\mathrm{expt}F)\Phi^{-1}$ is a one-parameter group of transformations and is itself generated by a vector field. That is, there is a vector field G such that

$$(\mathrm{expt}G) = \Phi(\mathrm{expt}F)\Phi^{-1}$$

It can be shown that $G = (d\Phi)F(\Phi^{-1})$ where $d\Phi$ denotes the derivative of Φ. We shall use $\Phi_{\#}(F)$ to denote the vector field $(d\Phi)(F\Phi^{-1})$.

Example 38.11:

i. Let Φ be a transformation $\Phi(x) = x + b$, and F a linear vector field $F(x) = Ax$. Then,

$$\Phi\mathrm{expt}F\Phi^{-1}(x) = e^{At}(x - b) + b$$

Therefore, $\frac{d}{dt}e^{At}(x - b) + b|_{t=0} = A(x - b) = Ax - Ab$. Thus, $\Phi_{\#}F$ is an affine vector field.
ii. If Φ is a linear transformation, then $d\Phi$ is also linear, and therefore, $\Phi_{\#}F = \Phi A\Phi^{-1}$, i.e., $\Phi_{\#}F$ is a linear vector field for any linear field F.

Theorem 38.5:

i. If Φ is a strong normalizer for a family of vector fields F then,

$$\Phi_{\#}(\mathcal{L}S_s(\mathcal{F})) \cap \mathrm{Lie}(\mathcal{F}) \subset \mathcal{L}S_s(\mathcal{F})$$

ii. If Φ is a normalizer for $\mathcal{F}$, then

$$\Phi_{\#}(\mathcal{L}S(\mathcal{F})) \cap \mathrm{Lie}(\mathcal{F}) \subset \mathcal{L}S(\mathcal{F})$$

38.7 Applications to Controllability

The geometric ideas that led to the Lie saturate criterion of controllability provide a beautiful proof of controllability of linear systems, demonstrating at the same time that linearity plays an inessential role. This proof goes as follows.

We use the induction on k to show that each controllability space $\mathcal{L}_k = \mathcal{B} + A\mathcal{B} + \cdots + A^{k-1}\mathcal{B}$ defined in Example 38.8 is contained in the strong Lie saturate of the system.

Let $\mathcal{F}$ denote the family of affine vector fields $F_u(x) = Ax + Bu$ defined by the linear system $\frac{dx}{dt} = Ax + Bu$. For each real number $\lambda, 0 \le \lambda \le 1$ and each F_u in $\mathcal{F}$

$$F_{\lambda,u}(x) = \lambda(Ax + \frac{1}{\lambda}Bu)$$

belongs to $\mathcal{LS}_s(\mathcal{F})$ by Theorem 38.4 (ii). Its limit as $\lambda \to 0$ also belongs to $\mathcal{LS}_s(\mathcal{F})$ since the latter is closed. It follows that $\lim_{\lambda\to 0} F_{\lambda,u} = Bu$ and therefore $\mathcal{L}_1 = \mathcal{B}$ is contained in $\mathcal{LS}_s(\mathcal{F})$.

Now assume that $\mathcal{L}_{k-1} \subset \mathcal{LS}_s(\mathcal{F})$. Let b be any element of $\mathcal{L}_{k-1}$ and let α be any real number. The constant vector field $F_\alpha = \alpha b$ is in $\mathcal{L}_{k-1}$ for each α. Let $\Phi_\alpha = \exp F_\alpha$. Then $(\Phi_\alpha)^{-1} = \exp F_{-\alpha}$ and therefore both $\Phi_\alpha(x)$ and $\Phi_\alpha^{-1}(x)$ remain in $\mathrm{cl}\mathcal{A}_{\mathcal{F}}(x, \le T)$ for any $x \in M$ and $T > 0$. Therefore, Φ_α is a strong normalizer for $\mathcal{F}$. According to Theorem 38.5, $(\Phi_\alpha)_\#(Fu) \subset \mathcal{LS}_s(\mathcal{F})$ provided that $(\Phi_\alpha)_\#(F_u) \in \mathrm{Lie}(\mathcal{F})$. Then $((\Phi_\alpha)_\#(F_0))(x) = \alpha\Phi_\alpha A\Phi_{-\alpha}(x) = A(x - \alpha b)$ because the derivative map of a translation is equal to the identity map. Thus, $(\Phi_\alpha)_\#(Fu)$ belongs to $\mathrm{Lie}(\mathcal{F})$. An analogous argument used in the first step of the induction procedure applied to the limit of $\lambda(\Phi_{\frac{\alpha}{\lambda}})_\#(F_0)$ as λ tends to 0 shows that the constant vector field $-\alpha Ab$ is contained in $\mathcal{LS}_s(\mathcal{F})$ for each real number α. But then $\mathcal{L}_k \subset \mathcal{LS}_s(\mathcal{F})$ because the convex hull of two vector spaces is the vector space spanned by their sum, i.e., $\mathcal{L}_k = \mathcal{L}_{k-1} + A\mathcal{L}_{k-1}$. Therefore, each $\mathcal{L}_k$ is in $\mathcal{LS}_s(\mathcal{F})$.

When $\mathcal{L}_{n-1} = M$, the space of all constant vector fields is in the strong Lie saturate and hence $\mathrm{cl}(\mathcal{A}_{\mathcal{F}}(x, \le T)) = M$ for each $x \in M$ and $T > 0$. But then it follows from Theorem 38.2 that $M = \mathrm{cl}\mathcal{A}_{\mathcal{F}}(x, \le T) \subset \mathrm{int}\,\mathcal{A}_{\mathcal{F}}(x, \le T + \epsilon) \subset \mathcal{A}_{\mathcal{F}}(x, \le T + \epsilon)$. Therefore, the system is strongly controllable.

The inductive procedure can also be described pictorially as follows:

Step 1: Prolong the original system to its closed convex body $\mathcal{F}_1$. Geometrically $\mathcal{F}(x)$ is the translate of $\mathcal{B}$ to Ax. For each u, $Ax + \lambda Bu$ is the line through Ax parallel to Bu, as shown in Figure 38.7a. $\mathcal{F}_1(x)$ is the union of all translates of $\mathcal{B}$ to points $\lambda Ax, 0 \le \lambda \le 1$, as shown in Figure 38.7b.

Step 2: $\mathcal{F}_1$ contains the vector space $\mathcal{B}$ as its edge. Conjugate the original family $\mathcal{F}$ by $\mathcal{B}$ to obtain a prolonged family $\mathcal{F}_2$ given by $\frac{dx}{dt} = Ax + Bu + ABv$ with both u and v as controls. $\mathcal{F}_2(x)$ is the translate of $\mathcal{B} + A\mathcal{B}$ to Ax, while the convex body $\mathcal{F}_3$ generated by $\mathcal{F}_2$ at each point x is the union

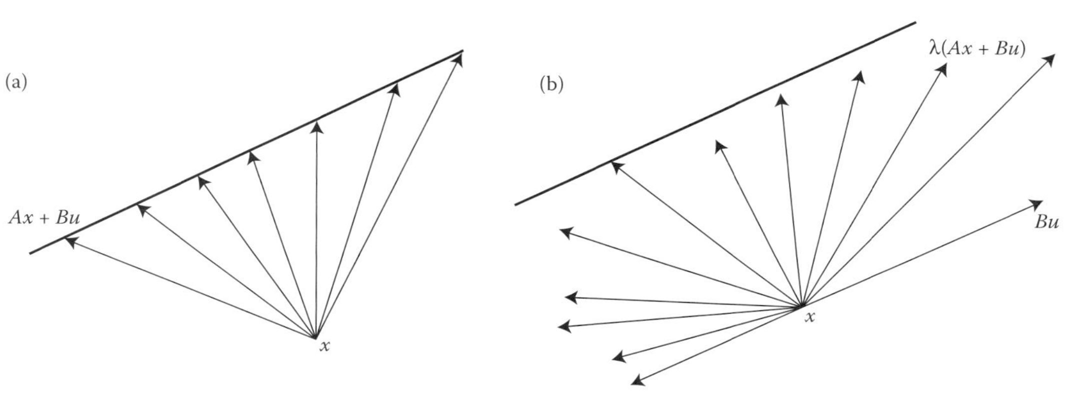

FIGURE 38.7 llustration for step 1: (a) $\mathcal{F}(x)$ and (b) $\mathcal{F}_1(x)$.

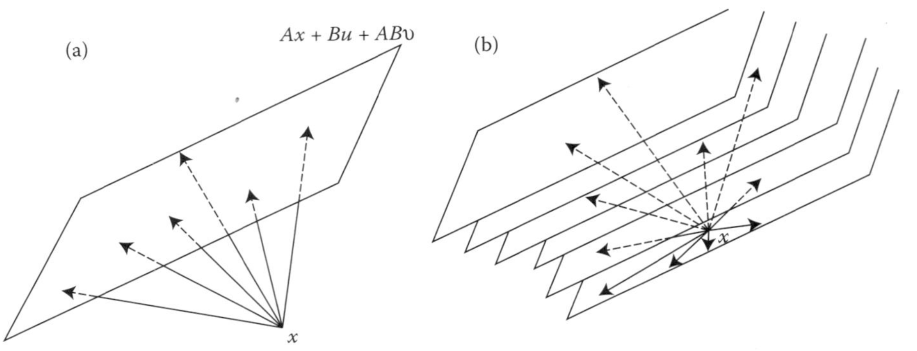

FIGURE 38.8 Illustration for step 2: (a) $\mathcal{F}_2(x)$ and (b) $\mathcal{F}_3(x)$.

of all translates of $\mathcal{B} + A\mathcal{B}$ to λAx as λ ranges in the interval $[0, 1]$. Figure 38.8 illustrates their differences.

Step 3: Conjugate the original family by the edge $\mathcal{B} + A\mathcal{B}$ of $\mathcal{F}_3$. The prolonged family is given by $\frac{dx}{dt} = Ax + Bu + ABv + A^2Bw$.

A repetition of these steps embodied in the induction scheme leads to the saturated system from which the controllability properties are evident.

We now illustrate the importance of the Lie saturate by considering controllability of linear systems with bounded controls. Strong controllability is not possible when the constraint set U is compact, because each set $\mathcal{A}_{\mathcal{F}}(x, \le T)$ is compact. It is also known that controllability is not possible whenever the drift vector field Ax has an eigenvalue with non-zero real parts. We will now use the geometric framework provided by the Lie saturate criterion to obtain affirmative controllability results when the real part of the spectrum of A is zero. For simplicity the proof will be given for a particular case only when all the eigenvalues of A are zero, i.e., when A is nilpotent.

Theorem 38.6:

Suppose that U is a compact neighborhood of the origin in R^m, and suppose further that the linear drift is nilpotent, i.e., suppose that there is a positive integer p such that $A^p \neq 0$ but $A^{p+1} = 0$. Then $\frac{dx}{dt} = Ax + Bu$, $u \in U$ is controllable, provided that the rank of $(B\ AB\ \ldots\ A^{n-1}B)$ is equal to $\dim M$.

Proof. There is no loss of generality in assuming that U is the cube $|u_i| \le \epsilon\ i = 1, \ldots, m$. Then the reachable set $\mathcal{A}(0)$ is a convex neighborhood of the origin in M. Any trajectory $x(t)$ which originates at $x(0) = 0$ is of the form

$$x(t) = \int_0^t e^{A(t-s)}Bu(s)\,ds = \sum_{k=0}^{p} \frac{A^kB}{k!} \int_0^t (t-s)^k u(s)\,ds$$

For any $u \in R^m$, and any real number λ, there exists $T > 0$ such that $\frac{|\lambda u_i|}{T^p} < \epsilon$ for all $i = 1, \ldots, m$. Let $u(T) = \frac{\lambda u}{T^p}$. The corresponding response $x(T)$ is equal to

$$\lambda \left(\frac{Bu}{T^p} + \frac{ABu}{T^{p-1}} + \cdots + \frac{A^{p-1}Bu}{p!T} + \frac{A^pBu}{(p+1)!} \right),$$

and therefore, $\lim_{T\to\infty} x(T) = \frac{\lambda A^pBu}{(p+1)!}$. Therefore, the line through A^pBu is contained in the closure of $\mathcal{A}(0)$. The convex hull of these lines as u ranges over R^m is equal to the vector space $A^p\mathcal{B}$.

Take now $u(T) = \frac{\lambda u}{T^{p-1}}$. The corresponding trajectory $x(T)$ is given by $\lambda(\frac{Bu}{T^{p-1}} + \cdots + \frac{A^{p-1}Bu}{p!} + \frac{TA^pBu}{(p+1)!})$. Then, $\frac{1}{2}(x(T) - \frac{\lambda TA^pBu}{(p+1)!}) \in \text{cl}\mathcal{A}(0)$, since the latter is convex. But then $\lim_{T\to\infty} \frac{1}{2}(x(T) - \frac{\lambda TA^pBu}{(p+1)!}) = \frac{\lambda A^{p-1}Bu}{p!}$. A repetition of the previous argument shows that the sum of $A^{p-1}\mathcal{B}$ and $A^p\mathcal{B}$ is contained in $\text{cl}\mathcal{A}(0)$. Further repetitions of the same argument show that $\text{cl}\mathcal{A}(0) = \mathcal{B} + A\mathcal{B} + \cdots + A^p\mathcal{B}$. The latter is equal to M by the rank assumption.

Since $-A$ is also nilpotent, the above proof is applicable to the time reversed system $\frac{dx}{dt} = -Ax - Bu$, with $u \in U$, to show that its reachable set from the origin is the entire space M. Therefore, any initial point x_0 can be steered to the origin in some finite time T_1 using the time-reversed system. But then the origin can be steered to any terminal state x_1 as a consequence of the fact proved above that $\mathcal{A}(0) = M$. Thus, $\mathcal{A}(x_0) = M$ for any x_0 in M and our proof is finished.

Remark 38.1

We have implicitly used the Lie saturate criterion to conclude that $\mathcal{A}(x) = M$ for all $x \in M$ whenever $\text{cl}\mathcal{A}(x) = M$ for all x in M.

38.8 Rotations

The group of rotations in R^3 is a natural state space for many mechanical control problems, because the kinematics of a rigid body can be described by the movements of an orthonormal frame fixed on the body relative to an orthonormal frame fixed in the ambient space. Recall that the rotation group consists of all linear transformations R which leave the Euclidean metric $\langle\ ,\ \rangle$ in R^3 invariant. A Euclidean metric in R^3 is any positive definite scalar product. So if e_1, e_2, e_3 is any orthonormal basis in R^3 and if $x = \sum_{i=1}^3 x_i e_i$, and $y = \sum_{i=1}^3 y_i e_i$ then $\langle x, y\rangle = \sum_{i=1}^3 x_i y_i$. R is a rotation if $\langle Rx, Ry\rangle = \langle x, y\rangle$ for all x and y in R^3.

Denoting by a_1, a_2, a_3 an orthonormal frame fixed on the body, then any motion of the body is monitored by the rotation through which the moving frame a_1, a_2, a_3 undergoes relative to the fixed frame e_1, e_2, e_3. This rotation, when expressed relative to the basis e_1, e_2, e_3, becomes a 3×3 matrix whose columns consist of the coordinates of a_1, a_2, a_3 relative to the fixed basis e_1, e_2, e_3.

The group of all such matrices whose determinant is equal to 1 is called the special orthogonal group and is denoted by $SO_3(R)$. $SO_3(R)$ is a three-dimensional manifold, which, together with its group structure, accounts for a rich geometric base, which needs to be properly understood as a prerequisite for effective control of mechanical systems.

Let us first outline the manifold structure of $SO_3(R)$. To begin with, the tangent space of $SO_3(R)$ at any point R_0 consists of all tangent vectors $\frac{d}{d\epsilon}R(\epsilon)|_{\epsilon=0}$ for curves $R(\epsilon)$ in $SO_3(R)$ which satisfy $R(0) = R_0$. The tangent space at the group identity I plays a special role and is called the Lie algebra of $SO_3(R)$. It consists of all matrices A for which $e^{Ae} \in SO_3(R)$. Each such matrix A is antisymmetric because the rotations satisfy $R^{-1} = R^*$ with R^* equal to the transpose, and $e^{-A\epsilon} = (e^{A\epsilon})^* = e^{A^*\epsilon}$. Consequently, $A^* = -A$.

The space of 3×3 antisymmetric matrices is a three-dimensional vector space and is denoted by $so_3(R)$. Each rotation, consisting of orthonormal column vectors, is defined by six orthonormality relations in a nine-dimensional group of all 3×3 matrices. Therefore, $SO_3(R)$ is a three-dimensional manifold, and consequently, each tangent space is three-dimensional. But then, the tangent space at I is equal to $so_3(R)$.

Consider now the tangent space at an arbitrary point R_0. For any antisymmetric matrix A each of the curves $R_1(\epsilon) = R_0 e^{A\epsilon}$ and $R_2(\epsilon) = e^{A\epsilon}R_0$ is a curve in $SO_3(R)$ which passes through R_0 at $\epsilon = 0$. Therefore, both $\frac{dR_1}{d\epsilon}(0) = R_0 A$ and $\frac{dR_2}{d\epsilon}(0) = AR_0$ are tangent vectors at R_0. These vectors are different from each other because of noncommutativity of R_0 with A. The first vector is called the left-translation of A by R_0, and the second is called the right-translation of A by R_0. It follows that the tangent space at R_0 can be described by either left- or right-translations of $so_3(R)$.

Denote by A_1, A_2, A_3 the standard basis of $so_3(R)$,

$$A_1 = \begin{pmatrix} 0 & 0 & 0 \\ 0 & 0 & -1 \\ 0 & 1 & 0 \end{pmatrix} \quad A_2 = \begin{pmatrix} 0 & 0 & 1 \\ 0 & 0 & 0 \\ -1 & 0 & 0 \end{pmatrix} \quad A_3 = \begin{pmatrix} 0 & -1 & 0 \\ 1 & 0 & 0 \\ 0 & 0 & 0 \end{pmatrix}$$

Since $A_i e_i = 0$, it follows that each $e^{A_i\epsilon}$ is a rotation about the axis containing $e_i, i = 1, 2, 3$. For any antisymmetric matrix $A = \begin{pmatrix} 0 & -a_2 & a_2 \\ a_2 & 0 & -a_1 \\ -a_2 & a_1 & 0 \end{pmatrix}$ we will use $\hat{A}$ to denote the column vector $\begin{pmatrix} a_1 \\ a_2 \\ a_3 \end{pmatrix}$. $\hat{A}$ is the coordinate vector of A relative to the standard basis.

Any antisymmetric matrix A induces vector fields on $SO_3(R)$. The first vector field is given by $F_l(R) = RA$, and the second is $F_r(R) = AR$. F_l is called the left-invariant vector field induced by A because its tangent vector at R is a left translation by R of its tangent vector at the group identity. Similar explanations apply to right-invariant vector fields F_r. We will use $\overrightarrow{A_l}$ to denote the left-invariant vector field whose tangent at I is equal to A, i.e., $\overrightarrow{A_l}(R) = RA$. Similarly, $\overrightarrow{A_r}$ denotes the right-invariant field $\overrightarrow{A_r}(R) = AR$.

Then $(\overrightarrow{A_1})_l, (\overrightarrow{A_2})_l$ and $(\overrightarrow{A_3})_l$ is a basis of left-invariant vector fields which span each tangent space and $(\overrightarrow{A_1})_r, (\overrightarrow{A_2})_r$ and $(\overrightarrow{A_3})_r$ is a basis of right-invariant vector fields with the same property.

Any differentiable curve $R(t)$ in $SO_3(R)$ defines a curve of tangent vectors $\frac{dR}{dt}$ at $R(t)$, which can be expressed by either right or left basis. Let $\frac{dR}{dt} = \sum_{i=1}^{3} \omega_i(t)(\overrightarrow{A_i})_r(R) = \sum_{i=1}^{3} \Omega_i(t)(\overrightarrow{A_i})_l(R(t))$ denote the corresponding coordinates of $\frac{dR}{dt}$. Vectors $\omega(t) = \begin{pmatrix} \omega_1(t) \\ \omega_2(t) \\ \omega_3(t) \end{pmatrix}$, and $\Omega(t) = \begin{pmatrix} \Omega_1(t) \\ \Omega_2(t) \\ \Omega_3(t) \end{pmatrix}$ are called the angular velocities of $R(t)$. In analogy with the kinematics of a rigid body, the first angular velocity is called the (absolute) angular velocity, while the second is called the body angular velocity. The above differential equations can be rewritten as

$$\begin{pmatrix} 0 & -\omega_3(t) & \omega_2(t) \\ \omega_3(t) & 0 & -\omega_1(t) \\ -\omega_2(t) & \omega_1(t) & 0 \end{pmatrix} R(t) = \frac{dR(t)}{dt} = R(t) \begin{pmatrix} 0 & -\Omega_3(t) & \Omega_2(t) \\ \Omega_3(t) & 0 & -\Omega_1(t) \\ -\Omega_2(t) & \Omega_1(t) & 0 \end{pmatrix}$$

It can be shown that $\Omega(t) = R^{-1}(t)\omega(t)$.

Any differentiable curve $R(t)$ whose angular velocity is constant is a solution curve of an invariant vector field. If $\omega(t)$ is constant, then $\frac{dR}{dt} = AR$ with $\hat{A} = \omega$, and if $\Omega(t)$ is constant then $\frac{dR}{dt} = RA$ with $\hat{A} = \Omega$. In the first case, $R(t) = e^{At}R_0$ while in the second case $R(t) = R_0 e^{At}$.

It can be shown that the Lie bracket of a right (respectively, left) invariant vector fields is a right (respectively, left) invariant vector field, with $[\overrightarrow{A_l}, \overrightarrow{B_l}](R) = R(BA - AB)$ and $[\overrightarrow{A_r}, \overrightarrow{B_r}](R) = (AB - BA)R$.

It is easy to verify that the commutator $AB - BA$ can also be expressed in terms of the cross-product of $\hat{A}$ and $\hat{B}$ in R^3 as follows: let $[A, B]_r = AB - BA$ and $[A, B]_l = BA - AB$. Then, $\widehat{[A, B]_r} = \hat{A} \times \hat{B}$, while $\widehat{[A, B]_l} = \hat{B} \times \hat{A}$.

Except for the cross-product correspondence, all of these concepts extend to the rotation group $SO_n(R)$ of R^n, and its $\frac{n(n-1)}{2}$-dimensional Lie algebra $so_n(R)$ of $n \times n$ antisymmetric matrices.

38.9 Controllability in $SO_n(R)$

A unit sphere which rolls on a horizontal plane without slipping and without spinning along the axis perpendicular to the point of contact, can be described by the following equations:

$$\frac{dx_1}{dt} = u_1(t)$$

$$\frac{dx_2}{dt} = u_2(t)$$

$$\frac{dR(t)}{dt} = \begin{pmatrix} 0 & 0 & u_1 \\ 0 & 0 & u_2 \\ -uo & -u_2 & 0 \end{pmatrix} R(t)$$

$x_1(t)$ and $x_2(t)$ are the coordinates of the center of the sphere ($x_3 = 1$), and $R(t)$ is the orientation of the sphere relative to an absolute frame e_1, e_2, e_3. The angular velocity $\omega(t) = \begin{pmatrix} -u_2 \\ u_1 \\ 0 \end{pmatrix}$ of the sphere is always orthogonal to the velocity of its center.

The rotational kinematics of the sphere may be viewed as a left-invariant control system on $SO_3(R)$ with two controls u_1 and u_2. This control system has no drift, and therefore, according to a well-known theorem of geometric control theory, the system is strongly controllable whenever the Lie algebra generated by the controlling vector fields is equal to the Lie algebra of the group (in this case $SO_3(R)$). It follows that the controlling vector fields are $F_1(R) = A_2R$ and $F_2(R) = -A_1R$ corresponding to $u_1 = 1, u_2 = 0$ and $u_2 = 0, u_1 = 1$. The rotational part is strongly controllable since $[F_1, F_2](R) = A_3R$. It can also be shown that the overall system in $R^3 \times SO_3(R)$ is strongly controllable because the Lie algebra generated by the controlling vector fields is equal to $R^2 \times so_3(R)$.

There is a simple argument showing that any states in $R^2 \times SO_3(R)$ can be transferred to each other by two switches in controls. Note first that for any angular velocity $\hat{A}$ the corresponding rotation e^A is the rotation about $\hat{A}$ through the angle $\|\hat{A}\|$. Figure 38.9 shows that any rotation can be achieved by one switch in controls (two angular velocities ω_1 and ω_2).

The proof begins with the observation that each unit circle in the e_1, e_2 plane centered at the origin has a line ω in common with the circle in the a_1, a_2 plane also centered at the origin. ω is in the plane $\omega_3 = 0$ as shown in the picture. The first move consists of rotating about ω so that a_3 coincides with $-e_1$. Then rotate through π radians along the midpoint of the arc between a_2 and e_2. These two moves rotate any frame a_1, a_2, a_3 into the standard frame. The remaining moves are used to roll for the position of the point of contact along a line segment whose length is an integral multiple of 2π. Such moves do not alter the orientation of the ball. Any two points in the plane can be joined along the sides of an isosceles triangle with equal sides equal to $2\pi m$, as shown in Figure 38.10.

The reader may note the similarity of this argument with the one used to show that any rotation in R^3 may be achieved by the rotations through the Euler angles ϕ, θ, and ψ.

This exposition ends with a controllability theorem whose proof also relies on the Lie saturate.

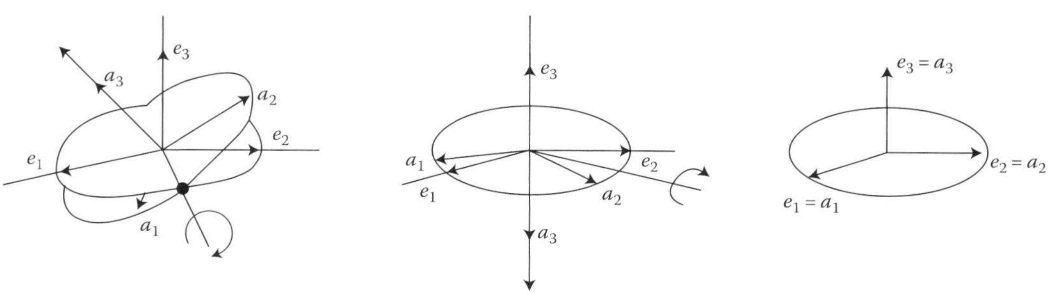

FIGURE 38.9 Rotational kinematics.

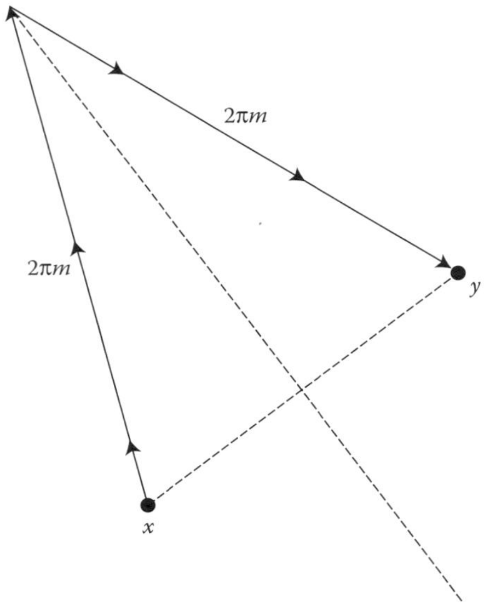

FIGURE 38.10 Translational kinematics.

Theorem 38.7:

Suppose that $\mathcal{F}$ is any family of right (or left) invariant vector fields on $SO_n(R)$ (or any other compact Lie group G). Then $\mathcal{F}$ is controllable if and only if Lie($\mathcal{F}$), evaluated at I, is equal to the Lie algebra of $SO_n(R)$ (or G).

The proof consists in showing that $-\mathcal{F}$ is contained in the Lie saturate of $\mathcal{F}$. Therefore, the vector span of $\mathcal{F}$ is contained in $\mathcal{L}\ S(\mathcal{F})$ by the convexity property of $\mathcal{L}S(\mathcal{F})$. But then the Lie algebra of $\mathcal{F}$ is contained in the Lie saturate and, hence, must be equal to it. The controllability result then follows from the Lie saturate criterion. So, the proof will be complete once we showed that $(\exp -tF)(R) \in \text{cl}\mathcal{A}_{\mathcal{F}}(R)$ for any $t > 0$ and any $F \in \mathcal{F}$. Let $F(R) = AR$. Then, $(\exp -tF)(R) = e^{-tA}R$, and therefore $(\exp -tF)(R)$ belongs to $\text{cl}\mathcal{A}_{\mathcal{F}}(R)$ if and only if e^{-tA} belongs to the closure of the reachable set from the group identity.

$S0_n(R)$ is a compact group and therefore there exists a sequence of times t_n tending to ∞ such that $\lim e^{t_nA}$ exists. Let $\lim_{t\to\infty} e^{t_nA} = R_0$. Then, $R_0^{-1} = \lim_{n\to\infty} e^{-At_n}$. If necessary, choose a subsequence so that $t_{n+1} - t_n$ also tends to ∞. Then, $I = R_0^{-1}R_0 = (\lim_{n\to\infty} e^{-t_nA})(\lim e^{t_{n+1}A}) = \lim_{n\to\infty} e^{(t_{n+1}-t_n)A}$.

The preceding argument shows that e^{tA} comes arbitrarily close to the identity for large values of time. Then,

$$e^{-tA} = e^{-tA}(\lim_{n\to\infty} e^{(t_{n+1}-t_n)A}) = \lim_{n\to\infty} e^{((t_{n+1}-t_n)-t)A}$$

Since $t_{n+1} - t_n \to \infty$, $(t_{n+1} - t_n) - t > 0$ for large values of n and therefore $e^{-tA} \in \text{cl}\mathcal{A}_{\mathcal{F}}(I)$. The proof is now finished.

Theorem 38.7 might be used to show that the orientation of a rigid body may be controlled by any number of gyros situated on the body as long as the Lie algebra generated by their angular velocities has full rank.

Further Reading

The proofs of all theorems quoted in this paper can be found in the forthcoming book titled *Geometric Control Theory* by V. Jurdjevic, (to appear in *Studies in Advanced Mathematics*, Cambridge University Press.) The material for this article is taken out of the first part of the book dealing with the reachable sets of Lie determined systems (which includes analytic systems).
The reader may also find some of this material in the following publications:

1. Jurdjevic, V. and Kupka, I.A., Polynomial control system, *Math. Ann.*, 361–368, 1985.
2. Jurdjevic, V. and Sussmann, H.J., Control systems on Lie groups, *J. Diff. Eqs.*, 12, 313–329, 1972.
3. Sussmann, H.J. and Jurdjevic, V., Controllability of non-linear systems, *J. Diff. Eqs.*, 12, 95–116, 1972.

Convexification of control systems is also known as the relaxation of controls in the early literature on control. See for instance,

4. Hermes, H. and LaSalle, J.P., *Functional Analysis and Time Optimal Control*, Academic Press, New York, 1969.
5. Warga, T., *Optimal Control of Differential and Functional Equations*, Academic Press, New York, 1972.

For other applications of Lie theory to control systems the reader may consult *Geometric Methods in Systems Theory*, Proceedings of the NATO Advanced Study Series, Editors: R. Brockett and D. Q. Mayne, Reidel Publishing, 1973.

39

Two Timescale and Averaging Methods

Hassan K. Khalil
Michigan State University

39.1 Introduction

In this chapter we present the asymptotic methods of averaging and singular perturbation. Suppose we are given the state equation $\dot{x} = f(t, x, \varepsilon)$, where ε is a "small" positive parameter, and, under certain conditions, the equation has an exact solution $x(t, \varepsilon)$. Equations of this type are encountered in many applications. The goal of an asymptotic method is to obtain an approximate solution $\tilde{x}(t, \varepsilon)$ such that the approximation error $x(t, \varepsilon) - \tilde{x}(t, \varepsilon)$ is small, in some norm, for small ε and the approximate solution $\tilde{x}(t, \varepsilon)$ is expressed in terms of equations simpler than the original equation. The practical significance of asymptotic methods is in revealing underlying multiple timescale structures inherent in many practical problems. Quite often the solution of the state equation exhibits the phenomenon that some variables move in time faster than other variables, leading to the classification of variables as "slow" and "fast." Both the averaging and singular perturbation methods deal with the interaction of slow and fast variables.

39.2 Asymptotic Methods

We start by a brief description of the perturbation method that seeks an approximate solution as a finite Taylor expansion of the exact solution. Then, we introduce the averaging method in its simplest form, which is sometimes called "periodic averaging" since the right-hand side function is periodic in time. Next, we introduce the singular perturbation model and give its two timescale properties. Finally, we show how to improve the accuracy of the reduced model of a singularly perturbed system.

39.2.1 The Perturbation Method

Consider the system

$$\dot{x} = f(t, x, \varepsilon), \quad x(t_0) = \eta \tag{39.1}$$

where f is sufficiently smooth in its arguments in the domain of interest, and ε is a positive parameter. The solution of Equation 39.1 depends on the parameter ε, a point that we shall emphasize by writing the solution as $x(t,\varepsilon)$. The goal of the perturbation method is to exploit the "smallness" of the perturbation parameter ε to construct approximate solutions that will be valid for sufficiently small ε. The simplest approximation results by setting $\varepsilon = 0$ in Equation 39.1 to obtain the nominal or unperturbed problem:

$$\dot{x} = f(t, x, 0), \quad x(t_0) = \eta \tag{39.2}$$

Suppose this problem has a unique solution $x_0(t)$ defined on $[t_0, t_1]$. By continuous dependence of the solutions of differential equations on parameters, we know that, for sufficiently small ε, the system equation 39.1 has a unique solution $x(t,\varepsilon)$, defined on $[t_0, t_1]$, such that

$$\|x(t,\varepsilon) - x_0(t)\| \leq k\varepsilon, \quad \forall\, \varepsilon < \varepsilon_1,\ \forall\, t \in [t_0, t_1]$$

for some $k > 0$ and $\varepsilon_1 > 0$. In this case, we say that the error is of the order $O(\varepsilon)$ and write $x(t,\varepsilon) - x_0(t) = O(\varepsilon)$. This order of magnitude notation will be used frequently. It is defined as follows.

Definition 39.1:

$\delta_1(\varepsilon) = O(\delta_2(\varepsilon))$ *if there exist positive constants k and c such that*

$$|\delta_1(\varepsilon)| \leq k|\delta_2(\varepsilon)|, \quad \forall\, |\varepsilon| < c$$

Higher-order approximations for the solution of Equation 39.1 can be obtained in a straightforward manner. We construct a finite Taylor series

$$x(t,\varepsilon) = x_0(t) + \sum_{k=1}^{N-1} x_k(t)\varepsilon^k + \varepsilon^N r(t,\varepsilon) \tag{39.3}$$

Substitution of Equation 39.3 into Equation 39.1 yields

$$\sum_{k=0}^{N-1} \dot{x}_k(t)\varepsilon^k + \varepsilon^N \dot{r}(t,\varepsilon) = f(t, x(t,\varepsilon), \varepsilon) \stackrel{\text{def}}{=} h(t,\varepsilon) \tag{39.4}$$

where the coefficients of the Taylor series of $h(t,\varepsilon)$ are functions of the coefficients of the Taylor series of $x(t,\varepsilon)$. Since the equation holds for all sufficiently small ε, it must hold as an identity in ε. Hence, coefficients of like powers of ε must be equal. Matching those coefficients we can derive the equations that must be satisfied by x_0, x_1, and so on. The zeroth-order term $h_0(t)$ is given by $h_0(t) = f(t, x_0(t), 0)$. Hence, matching coefficients of ε^0 in Equation 39.4, we determine that $x_0(t)$ satisfies

$$\dot{x}_0 = f(t, x_0, 0), \quad x_0(t_0) = \eta$$

which, not surprisingly, is the unperturbed problem of Equation 39.2. The first-order term $h_1(t)$ is given by

$$h_1(t) = \frac{\partial f}{\partial x}(t, x_0(t), 0)\, x_1(t) + \frac{\partial f}{\partial \varepsilon}(t, x_0(t), 0)$$

Matching coefficients of ε in Equation 39.4 we find that $x_1(t)$ satisfies

$$\dot{x}_1 = A(t)x_1 + g_1(t, x_0(t)), \quad x_1(t_0) = 0 \tag{39.5}$$

where

$$A(t) = \frac{\partial f}{\partial x}(t, x_0(t), 0), \quad g_1(t, x_0(t)) = \frac{\partial f}{\partial \varepsilon}(t, x_0(t), 0)$$

This linear equation has a unique solution defined on $[t_0, t_1]$. This process can be continued to derive the equations satisfied by x_2, x_3, and so on. By straightforward error analysis, it can be established that

$$x(t, \varepsilon) - \sum_{k=0}^{N-1} x_k(t)\varepsilon^k = O(\varepsilon^N) \tag{39.6}$$

The $O(\varepsilon^N)$ error bound in Equation 39.6 is valid only on finite time intervals $[t_0, t_1]$. It does not hold on intervals like $[t_0, T/\varepsilon]$ nor on the infinite time interval $[t_0, \infty)$. The reason is that the constant k in the bound $k\varepsilon^N$ depends on t_1 and may grow unbounded as t_1 increases. The error bound in Equation 39.6 can be extended to the infinite time interval $[t_0, \infty)$ if some additional conditions are added to ensure stability of the solution of the nominal system of Equation 39.2. In particular, suppose that Equation 39.2 has an *exponentially stable equilibrium point* p^*, then Equation 39.6 holds on the infinite time interval $[t_0, \infty)$, provided η is sufficiently close to p^*. We recall that a solution $\bar{x}(t)$ of a state equation is exponentially stable if for $x(0)$ sufficiently close to $\bar{x}(0)$, the inequality

$$\|x(t) - \bar{x}(t)\| \le k\|x(0) - \bar{x}(0)\| \exp(-\gamma t), \quad \forall\, t \ge 0$$

is satisfied with some positive constants k and γ. This definition applies whether $\bar{x}$ is an equilibrium point or a *periodic solution*. For a *time-invariant system* $\dot{x} = f(x)$, an equilibrium point p^* is exponentially stable if the Jacobian matrix $[\partial f/\partial x]$, evaluated at $x = p^*$, has eigenvalues with negative real parts.

39.2.2 Averaging

The averaging method applies to a system of the form

$$\dot{x} = \varepsilon f(t, x, \varepsilon) \tag{39.7}$$

where ε is a small positive constant and $f(t, x, \varepsilon)$ is T-periodic in t, uniformly in (x, ε); that is,

$$f(t, x, \varepsilon) = f(t + T, x, \varepsilon)$$

for all x and ε. We assume that f is sufficiently smooth in its arguments over the domain of interest. The method approximates the solution of Equation 39.7 by the solution of the time-invariant *average system*

$$\dot{x} = \varepsilon f_{av}(x) \tag{39.8}$$

where

$$f_{av}(x) = \frac{1}{T}\int_0^T f(\tau, x, 0)\, d\tau \tag{39.9}$$

The intuition behind this approximation can be seen as follows. When ε is small, the solution x will vary "slowly" with t relative to the periodic fluctuation of $f(t, x, \varepsilon)$. Therefore, x will be determined predominantly by the average of f. This intuition has its roots in linear system theory where we know that if the bandwidth of the system is much smaller than the bandwidth of the input, then the system will act as a low-pass filter that rejects the high-frequency component of the input.

The basic problem in the averaging method is to determine in what sense the behavior of the time-invariant system of Equation 39.8 approximates the behavior of the time-varying system of Equation 39.7.

We shall address this problem by showing, via a change of variables, that the system of Equation 39.7 can be represented as a perturbation of the system of Equation 39.8. Define

$$u(t,x) = \int_0^t [f(t,x,0) - f_{av}(x)]\, d\tau \tag{39.10}$$

Since $f(t,x,0) - f_{av}(x)$ is T-periodic in t with zero mean, the function $u(t,x)$ is T-periodic in t. It can be also shown that $\partial u/\partial t$ and $\partial u/\partial x$ are T-periodic in t. The change of variables

$$x = y + \varepsilon u(t,y) \tag{39.11}$$

transforms Equation 39.7 into the form

$$\dot{y} = \varepsilon f_{av}(y) + \varepsilon^2 q(t,y,\varepsilon) \tag{39.12}$$

where $q(t,y,\varepsilon)$ is T-periodic in t. This equation is a perturbation of the average system of Equation 39.8. It can be represented as a standard perturbation problem by changing the time variable from t to $s = \varepsilon t$. In the s timescale the equation is given by

$$\frac{dy}{ds} = f_{av}(y) + \varepsilon q\left(\frac{s}{\varepsilon}, y, \varepsilon\right) \tag{39.13}$$

where $q(s/\varepsilon, y, \varepsilon)$ is εT-periodic in s.

The problem has now been reduced to the perturbation problem we studied in the previous section. If, for a given initial state $x(0) = \eta$, the average system

$$\frac{dy}{ds} = f_{av}(y), \quad y(0) = \eta$$

has a unique solution $\bar{y}(s)$ defined on $[0, b]$, then for sufficiently small ε the perturbed system of Equation 39.13 will have a unique solution defined for all $s \in [0, b]$ and the two solutions will be $O(\varepsilon)$ close. Since $t = s/\varepsilon$ and $x - y = O(\varepsilon)$, by Equation 39.11, *the solution of the average system of Equation 39.8 provides an $O(\varepsilon)$ approximation for the solution of Equation 39.7 over the time interval $[0, b/\varepsilon]$ in the t timescale. If the average system of Equation 39.8 has an exponentially stable equilibrium point p^* and Ω is a compact subset of its* **region of attraction**, *then for all initial states in Ω, the $O(\varepsilon)$ approximation will be valid for all $t \geq 0$.*

Investigation of Equation 39.13 reveals another interesting relationship between Equations 39.7 and 39.8. *If Equation 39.8 has an exponentially stable equilibrium point p^*, then Equation 39.7 has a unique exponentially stable T-periodic solution in an $O(\varepsilon)$ neighborhood of p^*.*

The averaging method can be extended to systems where the right-hand side of Equation 39.7 is not periodic in t, if an average of $f(t,x,0)$ can be defined by the limit

$$f_{av}(x) = \lim_{T\to\infty} \frac{1}{T} \int_t^{t+T} f(\tau, x, 0)\, d\tau$$

A simple example is the case when $f(t,x,0) = f_1(t,x) + f_2(t,x)$, where f_1 is periodic in t while f_2 satisfies $\|f_2(t,x)\| \leq ke^{-\gamma t}$ over the domain of interest.

39.2.3 Singular Perturbation

While the perturbation method of Section 39.2.1 applies to state equations that depend smoothly on a small parameter ε, in this section we face a perturbation problem characterized by discontinuous dependence of system properties on the perturbation parameter ε. We shall study the singular perturbation model

$$\dot{x} = f(t,x,z,\varepsilon) \tag{39.14}$$

$$\varepsilon\dot{z} = g(t,x,z,\varepsilon) \tag{39.15}$$

where setting $\varepsilon = 0$ causes a fundamental and abrupt change in the dynamic properties of the system, as the differential equation $\varepsilon\dot{z} = g$ degenerates into the algebraic or transcendental equation $0 = g(t,x,z,0)$.

The essence of the theory described in this section is that the discontinuity of solutions caused by singular perturbations can be avoided if analyzed in separate timescales. This multi timescale approach is a fundamental characteristic of the singular perturbation method.

Consider the singularly perturbed system of Equations 39.14 and 39.15 where $x \in R^n$ and $z \in R^m$. We assume that the functions f and g are sufficiently smooth in the domain of interest. When we set $\varepsilon = 0$ in Equation 39.15, the dimension of the state equation reduces from $n+m$ to n because the differential equation 39.15 degenerates into the equation

$$0 = g(t, x, z, 0) \tag{39.16}$$

We shall say that the model of Equations 39.14 and 39.15 is in the *standard form* if and only if Equation 39.16 has $k \geq 1$ isolated real solutions

$$z = h_i(t, x), \quad i = 1, 2, \ldots, k \tag{39.17}$$

for each (t, x) in the domain of interest. This assumption ensures that a well-defined n-dimensional *reduced model* will correspond to each solution of Equation 39.16. To obtain the ith reduced model, we substitute Equation 39.17 into Equation 39.14, at $\varepsilon = 0$, to obtain

$$\dot{x} = f(t, x, h(t, x), 0) \tag{39.18}$$

where we have dropped the subscript i from h. It is usually clear from the context in which the solution of Equation 39.17 is being used. This model is called quasi-steady-state model because z, whose velocity $\dot{z} = g/\varepsilon$ can be large when ε is small and $g \neq 0$, may rapidly converge to a solution of Equation 39.16 which is the equilibrium of Equation 39.15. The model of Equation 39.18 is also known as the slow model or the reduced model.

Singular perturbations cause a two timescale behavior characterized by the presence of slow and fast transients in the system's response. Loosely speaking, the slow response is approximated by the reduced model of Equation 39.18, while the discrepancy between the response of the reduced model (Equation 39.18) and that of the full model of Equations 39.14 and 39.15 is the fast transient. To see this point, let us consider the problem of solving the state equation

$$\dot{x} = f(t, x, z, \varepsilon), \quad x(t_0) = \xi \tag{39.19}$$

$$\varepsilon\dot{z} = g(t, x, z, \varepsilon), \quad z(t_0) = \eta \tag{39.20}$$

Let $x(t, \varepsilon)$ and $z(t, \varepsilon)$ denote the solution of the full problem of Equations 39.19 and 39.20. When we define the corresponding problem for the reduced model of Equation 39.18, we can only specify n initial conditions since the model is of nth order. Naturally we retain the initial state for x, to obtain the reduced problem

$$\dot{x} = f(t, x, h(t, x), 0), \quad x(t_0) = \xi \tag{39.21}$$

Denote the solution of Equation 39.21 by $\bar{x}(t)$. Since the variable z has been excluded from the reduced model and substituted by its "quasi-steady-state" $h(t, x)$, the only information we can obtain about z by solving Equation 39.21 is to compute $\bar{z}(t) \stackrel{\text{def}}{=} h(t, \bar{x}(t))$, which describes the quasi-steady-state behavior of z when $x = \bar{x}$. By contrast to the original variable z, starting at t_0 from a prescribed η, the quasi-steady-state $\bar{z}$ is not free to start from a prescribed value, and there may be a large discrepancy between its initial value $\bar{z}(t_0) = h(t_0, \xi)$ and the prescribed initial state η. Thus $\bar{z}(t)$ cannot be a uniform approximation of $z(t, \varepsilon)$. The best we can expect is that the estimate $z(t, \varepsilon) - \bar{z}(t) = O(\varepsilon)$ will hold on an interval excluding t_0, that is, for $t \in [t_b, t_1]$ where $t_b > t_0$. On the other hand, it is reasonable to expect the estimate $x(t, \varepsilon) - \bar{x}(t) = O(\varepsilon)$ to hold uniformly for all $t \in [t_0, t_1]$ since $x(t_0, \varepsilon) = \bar{x}(t_0)$.

If the error $z(t, \varepsilon) - \bar{z}(t)$ is indeed $O(\varepsilon)$ over $[t_b, t_1]$, then it must be true that during the initial ("boundary layer") interval $[t_0, t_b]$ the variable z approaches $\bar{z}$. Let us remember that the speed of z

can be large since $\dot{z} = g/\varepsilon$. In fact, having set $\varepsilon = 0$ in Equation 39.15, we have made the transient of z instantaneous whenever $g \neq 0$. To analyze the behavior of z in the boundary layer, we set $y = z - h(t, x)$, to shift the quasi-steady-state of z to the origin, and change the timescale from t to $\tau = (t - t_0)/\varepsilon$. The new time variable τ is "stretched"; that is, if ε tends to zero, τ tends to infinity even for finite t only slightly larger than t_0 by a fixed (independent of ε) difference. In the τ timescale, y satisfies the equation

$$\frac{dy}{d\tau} = g(t, x, y + h(t, x), \varepsilon) - \varepsilon\frac{\partial h}{\partial t}(t, x, y + h(t, x), \varepsilon) - \varepsilon\frac{\partial h}{\partial x}(t, x, y + h(t, x), \varepsilon)f(t, x, y + h(t, x), \varepsilon) \tag{39.22}$$

with $y(0) = \eta - h(t_0, \xi)$. The variables t and x in the foregoing equation will be slowly varying since, in the τ timescale, they are given by

$$t = t_0 + \varepsilon\tau, \quad x = x(t_0 + \varepsilon\tau, \varepsilon)$$

Setting $\varepsilon = 0$ freezes these variables at their initial values and reduces Equation 39.22 to the time-invariant system

$$\frac{dy}{d\tau} = g(t_0, \xi, y + h(t_0, \xi), 0), \quad y(0) = \eta(0) - h(t_0, \xi) \tag{39.23}$$

which has equilibrium at $y = 0$. The frozen parameters (t_0, ξ_0) in Equation 39.23 depend on the given initial time and initial state. In our investigation of the stability of the origin of Equation 39.23 we should allow the frozen parameters to take any values in the region of the slowly varying parameters (t, x). We rewrite Equation 39.23 as

$$\frac{dy}{d\tau} = g(t, x, y + h(t, x), 0) \tag{39.24}$$

where (t, x) are treated as fixed parameters. We shall refer to Equation 39.24 as the *boundary-layer model* or the boundary-layer system. The crucial stability property we need for the boundary-layer system is exponential stability of its origin, uniformly in the frozen parameters. The following definition states this property precisely.

Definition 39.2:

The equilibrium $y = 0$ of the boundary-layer system of Equation 39.24 is exponentially stable uniformly in (t, x) if there exist positive constants k and γ and a compact set Ω, containing the origin, such that for each $y(0) \in \Omega$ the solution of Equation 39.24 satisfies the inequality

$$\|y(\tau)\| \leq k\|y(0)\| \exp(-\gamma\tau), \quad \forall\, \tau \geq 0 \tag{39.25}$$

for all (t, x) in the domain of interest.

If the Jacobian matrix $[\partial g/\partial y]$ satisfies the eigenvalue condition

$$Re\left[\lambda\left\{\frac{\partial g}{\partial y}(t, x, h(t, x), 0)\right\}\right] \leq -c < 0 \tag{39.26}$$

for all (t, x) in the domain of interest, then there exist k, γ, and Ω for which the inequality of Equation 39.25 is satisfied.

Under the boundary-layer stability condition, the fundamental result of singular perturbation, known as Tikhonov's theorem, states that *if the reduced problem of Equation 39.21 has a unique solution $\bar{x}(t)$,* *defined on* $[t_0, t_1]$*, and* $\eta - h(t_0, \xi) \in \Omega$*, then for sufficiently small* ε*, the full problem of Equations 39.19 and 39.20 has a unique solution* $(x(t,\varepsilon), z(t,\varepsilon))$ *defined on* $[t_0, t_1]$*, and*

$$x(t,\varepsilon) - \bar{x}(t) = O(\varepsilon) \tag{39.27}$$

$$z(t,\varepsilon) - h(t,\bar{x}(t)) - \hat{y}((t-t_0)/\varepsilon) = O(\varepsilon) \tag{39.28}$$

hold uniformly for $t \in [t_0, t_1]$*, where* $\hat{y}(\tau)$ *is the solution of the boundary-layer model of Equation 39.23. Moreover, given any* $t_b > t_0$*,*

$$z(t,\varepsilon) - h(t,\bar{x}(t)) = O(\varepsilon) \tag{39.29}$$

holds uniformly for $t \in [t_b, t_1]$. An infinite time version of this result holds when the reduced system of Equation 39.18 has an exponentially stable equilibrium point.

39.2.4 Model Order Reduction

The singular perturbation method provides a systematic procedure to obtain a reduced-order model of a two timescale system by neglecting its fast dynamics. In this section we show how we can improve the accuracy of the reduced model. We consider a time-invariant special case of Equations 39.14 and 39.15, namely,

$$\dot{x} = f(x,z) \tag{39.30}$$

$$\varepsilon\dot{z} = g(x,z) \tag{39.31}$$

in which f and g do not depend on ε. Let $z = h(x)$ be an isolated solution of the equation $0 = g(x,z)$ and suppose the assumptions of Tikhonov's theorem are satisfied with a nonsingular Jacobian matrix $[\partial g/\partial z]$. The equation $z = h(x)$ describes an n-dimensional manifold in the the $(n+m)$-dimensional state space of x and z. It is an invariant manifold for the system

$$\dot{x} = f(x,z)$$

$$0 = g(x,z)$$

since a trajectory starting in the manifold $z = h(x)$ stays in the manifold for all future (or past) time. The existence of an invariant manifold will carry over to the case $\varepsilon > 0$ when ε is sufficiently small. We seek the manifold equation in the form

$$z = \phi(x,\varepsilon) \tag{39.32}$$

where ϕ is sufficiently smooth. Equation 39.32 defines an ε-dependent n-dimensional manifold. For $z = \phi(x,\varepsilon)$ to be an invariant manifold for the system of Equations 39.30 and 39.31, it must be true that

$$\varepsilon\frac{\partial\phi}{\partial x} f(x,\phi(x,\varepsilon)) = g(x,\phi(x,\varepsilon)) \tag{39.33}$$

for all x in the domain of interest and all $\varepsilon \in [0,\varepsilon_0]$. It can be shown that, for sufficiently small ε, there is a function ϕ that satisfies Equation 39.33 and $\phi(x,0) = h(x)$. The reduced-order model

$$\dot{x} = f(x,\phi(x,\varepsilon)) \tag{39.34}$$

is an *exact slow model* that describes the motion of the system on the invariant manifold $z = \phi(x,\varepsilon)$. The reduced model of the previous section, namely,

$$\dot{x} = f(x,h(x))$$

is an approximation of the model of Equation 39.34. We can improve the approximation of the model by seeking the solution of Equation 39.33 in the power-series form

$$\phi(x,\varepsilon) = \phi_0(x) + \varepsilon\phi_1(x) + \varepsilon^2\phi_2(x) + \cdots \tag{39.35}$$

where $\phi_0(x) = h(x)$. The series of Equation 39.35 is substituted into Equation 39.33 and the terms ϕ_0, $\phi_1, \cdots$ are calculated by matching the coefficients of like powers of ε. The more terms we include in the approximation of ϕ, the more accurate the reduced model is.

39.3 Examples

We give four examples to illustrate the foregoing discussion. Example 39.1 illustrates the averaging method. Example 39.2 shows how the averaging method can be used to detect the existence of limit cycles in weakly nonlinear second-order systems. Example 39.3 illustrates the singular perturbation method. Example 39.4 shows how the accuracy of the reduced model can be improved.

Example 39.1:

Consider the scalar system

$$\dot{x} = \varepsilon(x\sin^2 t - 0.5x^2) = \varepsilon f(t,x)$$

The function $f(t,x)$ is π-periodic in t. The average function $f_{av}(x)$ is given by

$$f_{av}(x) = \frac{1}{\pi}\int_0^{\pi}(x\sin^2 t - 0.5x^2)\,dt = 0.5(x - x^2)$$

The average system

$$\dot{x} = 0.5\varepsilon(x - x^2)$$

has an exponentially stable equilibrium point at $x = 1$ since $[df_{av}/dx](1) = -0.5$. Hence, for sufficiently small ε, the original system has an exponentially stable π-periodic solution in an $O(\varepsilon)$ neighborhood of $x = 1$. Moreover, for initial states sufficiently near $x = 1$, solving the average system with the same initial state as the original system yields the approximation

$$x(t,\varepsilon) - x_{av}(\varepsilon t) = O(\varepsilon), \quad \forall\, t \geq 0$$

Figure 39.1 shows the solutions of the original and average systems for the initial state $x(0) = 0.5$ and $\varepsilon = 0.2$.

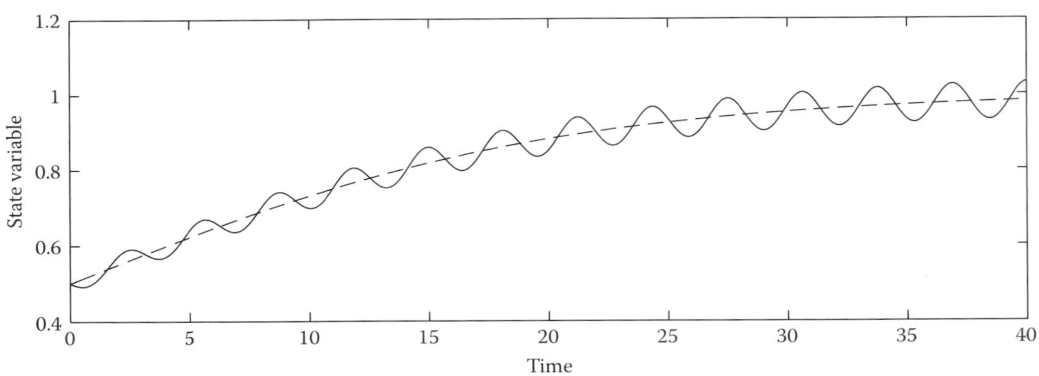

FIGURE 39.1 The exact (solid) and average (dashed) solutions of Example 39.1 with $\varepsilon = 0.2$.

Example 39.2:

Consider the Van der Pol equation

$$\dot{x}_1 = x_2$$
$$\dot{x}_2 = -x_1 + \varepsilon x_2(1 - x_1^2)$$

Representing the system in the polar coordinates

$$x_1 = r \sin \phi, \quad x_2 = r \cos \phi$$

it can be shown that

$$\dot{r} = \varepsilon r \cos^2 \phi \,(1 - r^2 \sin^2 \phi) \tag{39.36}$$
$$\dot{\phi} = 1 - \varepsilon \sin \phi \cos \phi \,(1 - r^2 \sin^2 \phi) \tag{39.37}$$

Divide Equation 39.36 by Equation 39.37 to obtain

$$\frac{dr}{d\phi} = \frac{\varepsilon r \cos^2 \phi \,(1 - r^2 \sin^2 \phi)}{1 - \varepsilon \sin \phi \cos \phi \,(1 - r^2 \sin^2 \phi)} \stackrel{\text{def}}{=} \varepsilon f(\phi, r, \varepsilon) \tag{39.38}$$

If we view ϕ as the independent variable, then Equation 39.38 takes the form of Equation 39.7, where $f(\phi, r, \varepsilon)$ is 2π-periodic in ϕ. The average system

$$\frac{dr}{d\phi} = \varepsilon \left(\frac{1}{2} r - \frac{1}{8} r^3 \right) \tag{39.39}$$

has an exponentially stable equilibrium point at $r = 2$ since $[df_{av}/dr](2) = -1$. Therefore, for sufficiently small ε, Equation 39.38 has an exponentially stable 2π-periodic solution $r = R(\phi, \varepsilon)$ in an $O(\varepsilon)$ neighborhood of $r = 2$. Substitution of $r = R(\phi, \varepsilon)$ into Equation 39.37 yields

$$\dot{\phi} = 1 - \varepsilon \sin \phi \cos \phi \,(1 - R^2(\phi, \varepsilon) \sin^2 \phi) \tag{39.40}$$

Let $\phi^*(t, \varepsilon)$ be the solution of Equation 39.40 that starts at $\phi^*(0, \varepsilon) = 0$. It can be argued that there is $T(\varepsilon) = 2\pi + O(\varepsilon)$ such that

$$\phi^*(t + T(\varepsilon), \varepsilon) = 2\pi + \phi^*(t, \varepsilon), \quad \forall\, t \geq 0$$

Hence,

$$R(\phi^*(t + T(\varepsilon), \varepsilon), \varepsilon) = R(2\pi + \phi^*(t, \varepsilon), \varepsilon) = R(\phi^*(t, \varepsilon), \varepsilon)$$

which shows that the Van der Pol equation has a *stable limit cycle* in an $O(\varepsilon)$ neighborhood of $r = 2$. The period of oscillation is $O(\varepsilon)$ close to 2π.

Example 39.3:

Consider the singular perturbation problem

$$\dot{x} = x^2(1 + t)/z, \qquad x(0) = 1$$
$$\varepsilon \dot{z} = -[z + (1 + t)x]\, z\, [z - (1 + t)], \quad z(0) = \eta$$

Setting $\varepsilon = 0$ yields the equation

$$0 = -[z + (1+t)x]\, z\, [z - (1+t)]$$

which has three isolated solutions

$$z = -(1+t)x; \quad z = 0; \quad z = 1 + t$$

in the region $\{t \geq 0 \text{ and } x \geq k\}$, where $0 < k < 1$. Consider first the solution $z = -(1+t)x$. The boundary-layer model of Equation 39.24 is

$$\frac{dy}{d\tau} = -y[y - (1+t)x][y - (1+t)x - (1+t)]$$

Using the Lyapunov function $V(y) = \frac{1}{2}y^2$, it can be verified that the boundary-layer stability condition is satisfied for $|y(0)| \leq \rho < (1+t)x$. The reduced problem

$$\dot{x} = -x, \quad x(0) = 1$$

has the unique solution $\bar{x}(t) = \exp(-t)$ for all $t \geq 0$. The boundary-layer problem with $t = 0$ and $x = 1$ is

$$\frac{dy}{d\tau} = -y(y-1)(y-2), \quad y(0) = \eta + 1$$

and has a unique exponentially decaying solution $\hat{y}(\tau)$ for $\eta < 0$. Consider next the solution $z = 0$. The boundary-layer model of Equation 39.24 is

$$\frac{dy}{d\tau} = -[y + (1+t)x]\, y\, [y - (1+t)]$$

By sketching the right-hand side function, it can be shown that the origin is unstable. Hence, Tikhonov's theorem does not apply to this case. Finally, the boundary-layer model for the solution $z = 1 + t$ is

$$\frac{dy}{d\tau} = -[y + (1+t) + (1+t)x][y + (1+t)]y$$

Similar to the first case, it can be shown that the origin is exponentially stable uniformly in (t, x). The reduced problem

$$\dot{x} = x^2, \quad x(0) = 1$$

has the unique solution $\bar{x}(t) = 1/(1-t)$ for all $t \in [0, 1)$. Notice that $\bar{x}(t)$ has a finite escape time at $t = 1$. However, Tikhonov's theorem still holds for $t \in [0, t_1]$ with $t_1 < 1$. The boundary-layer problem, with $t = 0$ and $x = 1$,

$$\frac{dy}{d\tau} = -(y+2)(y+1)y, \quad y(0) = \eta - 1$$

has a unique exponentially decaying solution $\hat{y}(\tau)$ for $\eta > 0$. Among the three solutions of Equation 39.16, only two solutions, $h = -(1+t)x$ and $h = 1 + t$, give rise to valid reduced models. Tikhonov's theorem applies to the solution $h = -(1+t)x$ if $\eta < 0$ and to the solution $h = 1 + t$ if $\eta > 0$. Figure 39.2 shows the exact and approximate solutions of x and z for $\eta = -2$ and $\varepsilon = 0.3$. The z solution exhibits a two timescale behavior. It starts with a fast transient from η to the reduced solution $\bar{z}(t)$. After the decay of this transient, it remains close to $\bar{z}(t)$. The approximation $\bar{z}(t) + \hat{y}(t/\varepsilon)$ is valid for all $t \in [0, 1]$, while the approximation $\bar{z}(t)$ is valid only after the boundary-layer period. As for the slow variable x, the approximation $\bar{x}(t)$ is valid for all $t \in [0, 1]$.

Example 39.4:

Consider the singularly perturbed system

$$\dot{x}_1 = x_2$$

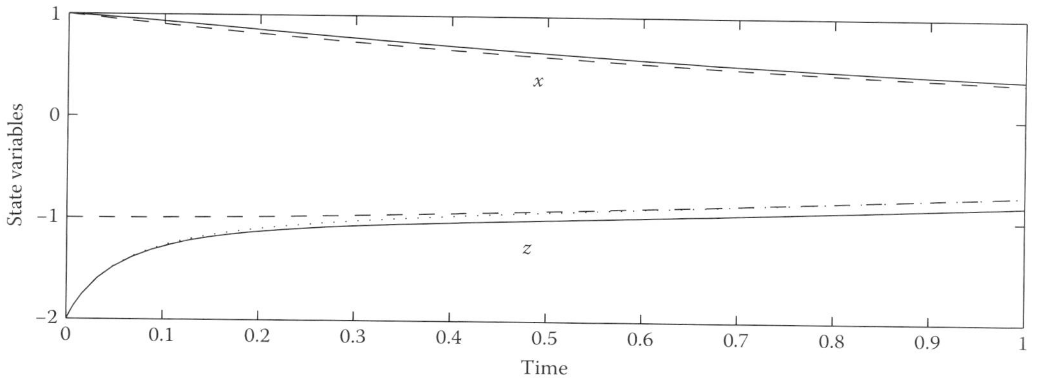

FIGURE 39.2 The exact (solid), reduced (dashed), and approximate (dotted) solutions of Example 39.3 with $\varepsilon = 0.3$.

$$\dot{x}_2 = -x_1 + (1 - x_1^2)z$$
$$\varepsilon\dot{z} = x_2 - z$$

The manifold condition of Equation 39.33 takes the form

$$\varepsilon\left[\frac{\partial\phi}{\partial x_1}x_2 + \frac{\partial\phi}{\partial x_2}(-x_1 + (1 - x_1^2)\phi)\right] = x_2 - \phi$$

Seeking the solution ϕ in the power series of Equation 39.35 and matching the coefficients of like powers of ε, we see that $\phi_0 = x_2$, which results in the reduced model

$$\dot{x}_1 = x_2$$
$$\dot{x}_2 = -x_1 + (1 - x_1^2)x_2$$

The next term ϕ_1 is given by $\phi_1 = x_1 - (1 - x_1^2)x_2$. Approximating ϕ by $\phi_0 + \varepsilon\phi_1$ results in the corrected reduced model

$$\dot{x}_1 = x_2$$
$$\dot{x}_2 = -x_1 + (1 - x_1^2)[x_2 + \varepsilon(x_1 - (1 - x_1^2)x_2)]$$

Figure 39.3 shows the improvement in approximating $x(t)$ as we go from the reduced to the corrected reduced models. The calculations are done with the initial conditions $x_1(0) = x_2(0) = 1$ and $z(0) = 0$.

39.4 Defining Terms

Average system: A time-invariant system obtained by averaging the fast periodic right-hand side function of a time-varying system.

Boundary-layer model: A reduced order model that describes the motion of the fast variables of a singularly perturbed system in a fast timescale where the slow variables are treated as constant parameters.

Equilibrium point: A constant solution of $\dot{x} = f(t, x)$. For the time-invariant system $\dot{x} = f(x)$, the equilibrium points are the real solutions of the equation $0 = f(x)$.

Exponentially stable solution: A solution (e.g., equilibrium point or periodic solution) is exponentially stable if other solutions in its neighborhood converge to it faster than an exponentially decaying function.

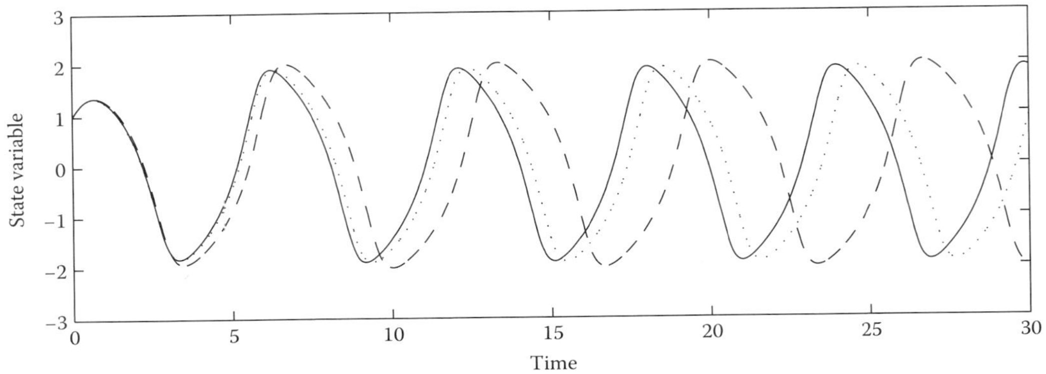

FIGURE 39.3 The exact (solid), reduced (dashed), and corrected reduced (dotted) solutions of Example 39.4 with $\varepsilon = 0.1$.

Linearization: Approximation of the nonlinear state equation in the vicinity of a nominal solution by a linear state equation, obtained by dropping second- and higher-order terms of the Taylor expansion (about the nominal solution) of the right-hand side function.

Periodic solution (orbit): A periodic solution $x(t)$ of $\dot{x} = f(t, x)$ satisfies the condition $x(t + T) = x(t)$, for all $t \geq 0$ for some positive constant T. The image of a periodic solution in the state space is a closed orbit.

Reduced Model: A reduced-order model that describes the motion of the slow variables of a singularly perturbed system. The model is obtained by setting $\varepsilon = 0$ and eliminating the fast variables.

Region of attraction: A domain containing an asymptotically stable equilibrium point such that all trajectories starting in the domain converge to the point.

Stable limit cycle: An isolated periodic orbit such that all trajectories in its neighborhood asymptotically converge to it.

Standard singularly perturbed model: A singularly perturbed model where upon setting $\varepsilon = 0$, the degenerate equation has one or more isolated solutions.

Time-invariant system: A state equation where the right-hand side function is independent of the time variable.

References

1. Khalil, H.K. 2002. *Nonlinear Systems*, 3rd Edn., Prentice-Hall, Upper Saddle River, NJ.
2. Kokotovic, P., Khalil, H.K. and O'Reilly, J. 1999. *Singular Perturbation Methods in Control: Analysis and Design*, Classic Edition, SIAM, Pennsylvania, PA.

For Further Information

Our presentation of the asymptotic methods is based on the textbook by Khalil [1]. For further information on this topic, the reader is referred to Chapters 10 and 11 of Khalil's book. Chapter 10 covers the perturbation method and averaging, and Chapter 11 covers the singular perturbation method. Proofs of the results stated here are given in the book.

The discussion of model order reduction is based on Chapter 1 of Kokotovic et al. [2]. This book gives a broader view of the use of singular perturbation methods in systems and control. Modeling two timescale systems in the singularly perturbed form is discussed in Chapters 1 and 2 of Kokotovic [2] and Chapter 11 of Khalil [1].

For a broader view of the averaging method, the reader may consult Sanders, J.A., Verhulst, F. and Murdock, J. 2007. *Averaging Methods in Nonlinear Dynamical Systems*, 2nd Edn., Springer, Berlin. A unified treatment of singular perturbation and averaging is given in Teel, A.R., Moreau, L. and Nesic, D. 2003. A unified framework for input-to-state stability in systems with two timescales, *IEEE Transactions on Automatic Control*, vol. 48, no. 9, 1526–1544.

40

Volterra and Fliess Series Expansions for Nonlinear Systems

Françoise Lamnabhi-Lagarrigue
Supélec

40.1 Motivation

40.1.1 Some Simple Examples

Consider the linear system

$$\begin{cases} \dot{x}(t) = Ax(t) + Bu(t), & x \in \mathbf{R}^n, \ u \in \mathbf{R}, \ x(0) = x_0 \\ y(t) = Cx(t) \end{cases}$$

Its solution may be written in the form

$$y(t) = Ce^{At}x_0 + \int_0^t Ce^{A(t-\tau)}Bu(\tau)\,d\tau. \tag{40.1}$$

On the other hand, the scalar time-varying linear system

$$\begin{cases} \dot{x}(t) = a(t)x + b(t)u(t), & x, y, u \in \mathbf{R}, \ x(0) = x_0 \\ y(t) = c(t)x(t) \end{cases},$$

has a solution

$$y(t) = c(t)e^{(\int_0^t a(\tau)d\tau)}x_0 + \int_0^t c(t)e^{(\int_\tau^t a(\sigma)d\sigma)}b(\tau)u(\tau)\,d\tau. \tag{40.2}$$

Now consider the system

$$\begin{cases}\dot{x}(t) = (Dx(t) + B)u(t), & x \in \mathbf{R}^2,\ y, u \in \mathbf{R},\ x(0) = (0,0)^T \\ y(t) = Cx(t)\end{cases}$$

where

$$D = \begin{bmatrix} 0 & 0 \\ 1 & 0 \end{bmatrix}, \quad B = \begin{bmatrix} 1 \\ 0 \end{bmatrix}, \quad \text{and} \quad C = \begin{bmatrix} 0 & 1 \end{bmatrix}.$$

The solution of this system may be written in the form

$$x(t) = C \int_0^t e^{(D \int_\tau^t u(\sigma) d\sigma)} Bu(\tau)\, d\tau.$$

Since $D^2 = 0$, the series definition of the exponential gives

$$e^{(D \int_\tau^t u(\sigma) d\sigma)} = \begin{bmatrix} 1 & 0 \\ \int_\tau^t u(\sigma)\, d\sigma & 1 \end{bmatrix}$$

and therefore

$$y(t) = \int_0^t \int_\tau^t u(\sigma) u(\tau) d\sigma d\tau$$

or

$$y(t) = \int_0^t \int_0^t \mu(\sigma - \tau) u(\sigma) u(\tau) d\sigma d\tau \tag{40.3}$$

where μ is the step function $\mu(t) = \begin{cases} 1 & t \geq 0 \\ 0 & t < 0 \end{cases}$.

Before introducing various types of expansions for the response of nonlinear control systems, let us summarize some classical results for the solution of linear differential equations.

40.1.2 Linear Differential Equations

Let us consider the linear time-varying differential equation

$$\dot{x}(t) = \sum_{i=1}^{m} \alpha_i(t) A_i x(t), \quad x \in \mathbf{R}^n,\ x(0) = x_0 \tag{40.4}$$

where for $i = 1, \ldots, m$, $\alpha_i : \mathbf{R} \to \mathbf{R}$ are locally Lebesgue integrable functions and A_i are constant $n \times n$ matrices. We may also write

$$x(t) = x_0 + \sum_{i=1}^{m} \int_0^t \alpha_i(\sigma) A_i x(\sigma)\, d\sigma$$

From the classical Peano–Baker scheme, there exists a series solution of Equation 40.4 of the form [1]

$$\begin{aligned} x(t) = x_0 &+ \sum_{i=1}^{m} \left(\int_0^t \alpha_i(\sigma_1) d\sigma_1 \right) A_i x_0 + \sum_{i,j=1}^{m} \left(\int_0^t \int_0^{\sigma_1} \alpha_i(\sigma_1) \alpha_j(\sigma_2) d\sigma_1 d\sigma_2 \right) A_i A_j x_0 + \cdots \\ &+ \sum_{i_1,\ldots,i_k=1}^{m} \left(\int_0^t \int_0^{\sigma_1} \cdots \int_0^{\sigma_{k-1}} \alpha_{i_1}(\sigma_1) \ldots \alpha_{i_k}(\sigma_k) d\sigma_1 \ldots d\sigma_k \right) A_{i_1} \ldots A_{i_k} x_0 + \cdots . \end{aligned} \tag{40.5}$$

This series expansion was used in quantum electrodynamics [11].

Under certain unspecified conditions of convergence, the solution of Equation 40.4 may also be written [37]

$$x(t) = e^{\Omega(t)}x_0 \tag{40.6}$$

with

$$\begin{aligned}\Omega(t) = {} & \sum_{i=1}^{m}\left(\int_0^t \alpha_i(\sigma_1)d\sigma_1\right)A_i + \frac{1}{2}\sum_{i,j=1}^{m}\left(\int_0^t\int_0^{\sigma_1}\alpha_i(\sigma_1)\alpha_j(\sigma_2)d\sigma_1 d\sigma_2\right)[A_i, A_j] \\ & + \frac{1}{4}\sum_{i,j,k=1}^{m}\left(\int_0^t\int_0^{\sigma_1}\int_0^{\sigma_2}\alpha_i(\sigma_1)\alpha_j(\sigma_2)\alpha_k(\sigma_3)d\sigma_1 d\sigma_2 d\sigma_3\right)[A_i,[A_j,A_k]] \\ & + \frac{1}{12}\sum_{i,j,k=1}^{m}\left(\int_0^t\int_0^{\sigma_1}\int_0^{\sigma_2}\alpha_i(\sigma_1)\alpha_j(\sigma_2)\alpha_k(\sigma_3)d\sigma_1 d\sigma_2 d\sigma_3\right)[[A_i,A_j],A_k] + \cdots,\end{aligned} \tag{40.7}$$

where the commutator product or *Lie-product* is defined by

$$[A_i, A_j] = A_iA_j - A_iA_j.$$

Indeed, for instance, the first terms of the expansion 40.6 are given by

$$\begin{aligned}&\sum_{i=1}^{m}\left(\int_0^t \alpha_i(\sigma_1)d\sigma_1\right)A_i + \frac{1}{2}\sum_{i,j=1}^{m}\left(\int_0^t\int_0^{\sigma_1}\alpha_i(\sigma_1)\alpha_j(\sigma_2)d\sigma_1 d\sigma_2\right) \\ &\quad\times[A_i,A_j] + \frac{1}{2!}\left(\sum_{i=1}^{m}\left(\int_0^t \alpha_i(\sigma_1)d\sigma_1\right)A_i\right)^2\end{aligned}$$

Using an integration by parts, this leads to

$$\sum_{i=1}^{m}\left(\int_0^t \alpha_i(\sigma_1)d\sigma_1\right)A_i + \sum_{i,j=1}^{m}\left(\int_0^t\int_0^{\sigma_1}\alpha_i(\sigma_1)\alpha_j(\sigma_2)d\sigma_1 d\sigma_2\right)A_iA_j$$

which corresponds to the first two terms of the expansion 40.5.

40.2 Functional Expansions for Nonlinear Control Systems

The general problem we consider here is how to generalize Equations 40.1 through 40.3, and 40.5 or 40.6 to nonlinear control systems of the form

$$\frac{dx}{dt} = f_0(x(t)) + \sum_{i=1}^{m}u_i(t)f_i(x(t)) \tag{40.8}$$

where $f_0, f_1, \ldots, f_m$ are C^∞ vector fields on a n-dimensional manifold M, x takes values in M and $u_i : \mathbf{R}^+ \to \mathbf{R}$, $i = 1, \ldots, m$ are piecewise continuous. In a local coordinate chart, $x = (x_1, \ldots, x_n)^T$, Equation 40.8

can be written

$$\dot{x}^k(t) = f_0^k(x(t)) + \sum_{i=1}^{m} u_i(t) f_i^k(x(t)), \quad 1 \le k \le n \tag{40.9}$$

where the functions $f_i^k : \mathbf{R}^n \to \mathbf{R}$ are C^∞. Let $h \in C^\infty(M)$. Then

$$\frac{d}{dt}h(x) = dh(x)\dot{x} = L_{f_0}h(x(t)) + \sum_{i=1}^{m} u_i(t) L_{f_i} h(x(t)),$$

where $L_{f_i}h$ is the *Lie derivative of h* along the vector field f_i, $i = 0, \ldots n$,

$$L_{f_i}h = \sum_{j=1}^{n} f_i^j(x) \frac{\partial}{\partial x_j} h(x).$$

Thus,

$$h(x(t)) = h(x(0)) + \sum_{i=0}^{m} \int_0^t u_i(\sigma) L_{f_i} h(x(\sigma))\, d\sigma$$

where $u_0(t) = 1,\ t \ge 0$. Also,

$$L_{f_i}h(x(t)) = L_{f_i}h(x(0)) + \sum_{j=0}^{m} \int_0^t u_j(\sigma) L_{f_j} L_{f_i} h(x(\sigma))\, d\sigma,$$

so that

$$\begin{aligned} h(x(t)) = h(x(0)) &+ \sum_{i=0}^{m} \left(\int_0^t u_i(\sigma) d\sigma \right) L_{f_i} h(x(0)) \\ &+ \sum_{i,j=0}^{m} \int_0^t \int_0^{\sigma_2} u_i(\sigma_2) u_j(\sigma_1) L_{f_j} L_{f_i} h(x(\sigma_1)) d\sigma_1 d\sigma_2. \end{aligned}$$

Iterating this procedure yields

$$\begin{aligned} h(x(t)) = h(x(0)) &- \sum_{i=0}^{m} \left(\int_0^t u_i(\sigma) d\sigma \right) L_{f_i} h(x(0)) + \sum_{\nu \ge 2}^{N} \sum_{j_1,\ldots,j_\nu=0}^{m} \\ &\times \left(\int_0^t \int_0^{\sigma_\nu} \cdots \int_0^{\sigma_2} u_{j_\nu}(\sigma_\nu) \ldots u_{j_2}(\sigma_2) u_{j_1}(\sigma_1) d\sigma_1 d\sigma_2 \ldots d\sigma_\nu \right) \times L_{f_{j_\nu}} \ldots L_{f_{j_2}} L_{f_{j_1}} h(x(0)) \\ &+ \sum_{j_1,\ldots,j_{N+1}=0}^{m} \int_0^t \int_0^{\sigma_{N+1}} \cdots \int_0^{\sigma_2} u_{j_{N+1}}(\sigma_{N+1}) \ldots u_{j_2}(\sigma_2) u_{j_1}(\sigma_1) \\ &\times L_{f_{j_{N+1}}} \ldots L_{f_{j_2}} L_{f_{j_1}} h(x(\sigma_1))(\sigma_1) d\sigma_1 d\sigma_2 \ldots d\sigma_{N+1} \end{aligned} \tag{40.10}$$

It is not difficult to see that the remainder [45]

$$\begin{aligned} R_N = &\sum_{j_1,\ldots,j_{N+1}=0}^{m} \int_0^t \int_0^{\sigma_{N+1}} \cdots \int_0^{\sigma_2} u_{j_{N+1}}(\sigma_{N+1}) \ldots u_{j_2}(\sigma_2) u_{j_1}(\sigma_1) \\ &\times L_{f_{j_\nu}} \ldots L_{f_{j_2}} L_{f_{j_1}} h(x(\sigma_1))(\sigma_1) d\sigma_1 d\sigma_2 \ldots d\sigma_{N+1} \end{aligned}$$

is such that

$$\|R_N\| \le \frac{A_t^{N+1} t^{N+1}}{(N+1)!} C_N,$$

where C_N is such that

$$|L_{f_{j_{N+1}}} \dots L_{f_{j_2}} L_{f_{j_1}} h(x)| \leq C_N$$

on some compact set and

$$A_t = \sup\left(1, \max_{0\leq\tau\leq t, 1\leq i\leq n} |u_i(\tau)|\right).$$

If the vector fields f_i are analytic, and the function h is also analytic, then the previous result can be strengthened. One can actually prove [12,45] that the series

$$h(x(0)) + \sum_{i=0}^{m}\left(\int_0^t u_i(\sigma)d\sigma\right) L_{f_i} h(x(0))$$
$$+ \sum_{\nu\geq 2}^{N} \sum_{j_1,\dots,j_\nu=0}^{m} \left(\int_0^t \int_0^{\sigma_\nu} \dots \int_0^{\sigma_2} u_{j_\nu}(\sigma_\nu)\dots u_{j_1}(\sigma_1)d\sigma_1 \dots d\sigma_\nu\right) L_{f_{j_\nu}} \dots L_{f_{j_2}} L_{f_{j_1}} h(x(0)) \tag{40.11}$$

converges to $h(x(t))$. Indeed in this case, there exists a constant $C > 0$ such that

$$|L_{f_{j_\nu}} \dots L_{f_{j_2}} L_{f_{j_1}} h(x)| \leq (\nu)!C^\nu.$$

for all $\nu \geq 1$.

40.2.1 Volterra Series

In the following, we consider scalar input, scalar output nonlinear systems on $\mathbf{R}^n$ called *linear-analytic*,

$$\begin{cases} \dot{x}(t) = f(x(t)) + u(t)g(x(t)), & x(0) = x_0 \\ y(t) = h(x(t)) \end{cases} \tag{40.12}$$

We always assume that f, g, and h are analytic functions in x, in some neighborhood of the free response (when $u(t) = 0$, $\forall t \geq 0$). Analyticity is important but the restriction to scalar inputs and outputs can be easily removed. We say that a linear–analytic system admits a Volterra series representation if there exist locally bounded, piecewise continuous functions

$$w_n : \mathbf{R}^n \to \mathbf{R}, \quad n = 0, 1, 2, \dots,$$

such that for each $T > 0$ there exists $\epsilon(T) > 0$ with the property that for all piecewise continuous functions $u(.)$ with $| u(T) | \leq \epsilon$ on $[0, T]$ we have

$$y(t) = w_0(t) + \sum_{n=1}^{\infty} \int_0^t \dots \int_0^t w_n(t, \sigma_1, \dots, \sigma_n)u(\sigma_1)\dots u(\sigma_n)d\sigma_1 \dots d\sigma_n \tag{40.13}$$

with the series converging absolutely and uniformly on $[0, T]$.

40.2.2 Bilinear Systems

It is not difficult to show that the following class of nonlinear systems called *bilinear systems*

$$\begin{cases} \dot{x}(t) = [A(t) + u(t)B(t)]\, x(t), & x(0) = x_0 \\ y(t) = c(t)x(t) \end{cases} \tag{40.14}$$

admits a Volterra series representation.

Let Φ_A denote the transition matrix for $A(t)$. Make the change of variable $z(t) = \Phi_A(0,t)x(t)$ in order to eliminate A. This gives

$$\begin{cases} \dot{z}(t) = u(t)\tilde{B}(t)z(t), \\ y(t) = \tilde{c}(t)z(t) \end{cases}$$

where $\tilde{B}(t) = \Phi_A(0,t)B(t)\Phi_A(t,0)$ and $\tilde{c}(t) = c(t)\Phi_A(0,t)$. Applying the Peano-Baker formula (Equation 40.5) construction, we have

$$z(t) = \left(I + \int_0^t u(\sigma_1)\tilde{B}(\sigma_1)d\sigma_1 + \int_0^t \int_0^{\sigma_2} u(\sigma_2)\tilde{B}(\sigma_2)u(\sigma_1)\tilde{B}(\sigma_1)d\sigma_1 d\sigma_2 + \cdots\right) z(0)$$

Thus, the Volterra kernels for $y(t)$ are given in triangular form by

$$w_n(t,\sigma_1,\sigma_2,\ldots,\sigma_n) \begin{cases} c(t)\Phi_A(t,\sigma_n)B(\sigma_n)\Phi_A(\sigma_n,\sigma_{n-1})B(\sigma_{n-1})\ldots B(\sigma_1)\Phi_A(\sigma_1,0)x(0) \\ 0 \qquad \text{if } \sigma_{i+p} < \sigma_p, \ \ i,p = 1,2,3,\ldots \end{cases}$$

For A, B, and u bounded this series converges uniformly on any compact interval.

The existence and the computation of the Volterra series for more general nonlinear systems is less straightforward. Several authors [6,15,25,36] gave the main results at about the same time.

For the existence of the Volterra series, let us recall for instance Brockett's result: Suppose that

$$f(.,.) : \mathbf{R} \times \mathbf{R}^n \to \mathbf{R}^n \text{ and } g(.,.) : \mathbf{R} \times \mathbf{R}^n \to \mathbf{R}^n$$

are continuous with respect to their first argument and analytic with respect to their second. Given any interval $[0, T]$ such that the solution of

$$\dot{x}(t) = f(t, x(t)), \quad x(0) = 0,$$

exists on $[0, T]$, there exists an $\epsilon > 0$ and a Volterra series for

$$\dot{x}(t) = f(t, x(t)) + u(t)g(t, x(t)), \quad x(0) = 0, \tag{40.15}$$

with the Volterra series converging uniformly on $[0, T]$ to the solution of Equation 40.15 provided $|u(t)| < \epsilon$.

Although the computation of the Volterra kernels is given in the previous referenced papers, their expressions may also be obtained from the Fliess algebraic framework [13] summarized in the next section.

40.2.3 Fliess Series

Let us recall some definitions and results from the Fliess algebraic approach [12]. Let $u_1(t), u_2(t), \ldots, u_m(t)$ be some piecewise continuous inputs and $Z = \{z_0, z_1, \ldots, z_m\}$ be a finite set called the alphabet. We denote by Z^* the set of words generated by Z. The algebraic approach introduced by Fliess may be sketched as follows. Let us consider the letter z_0 as an operator which codes the integration with respect to time and the letter $z_i, i = 1, \ldots, m$, as an operator which codes the integration with respect to time after multiplying by the input $u_i(t)$. In this way, any word $w \in Z^*$ gives rise to an iterated integral, denoted by $\mathbf{I}^t\{w\}$, which can be defined recursively as follows:

$$\mathbf{I}^t\{\emptyset\} = 1$$

$$\mathbf{I}^t\{w\} = \begin{cases} \int_0^t d\tau \mathbf{I}^\tau\{v\} & \text{if } w = z_0 v \\ \int_0^t u_i(\tau) d\tau \mathbf{I}^\tau\{v\} & \text{if } w = z_1 v \end{cases}, \; v \in Z^*. \tag{40.16}$$

Using the previous formalism and an iterative scheme like Equation 40.10, the solution $y(t)$ of the nonlinear control system

$$\begin{cases} \dot{x}(t) = f(x(t)) + \sum_{i=1}^{m} u_i(t) g_i(x(t)), \; x(0) = x_0 \\ y(t) = h(x(t)) \end{cases} \tag{40.17}$$

may be written [12]

$$y(t) = h(x_0) + \sum_{\nu \geq 0} \sum_{j_0, j_1, \ldots, j_\nu = 0}^{m} L_{f_{j_\nu}} \ldots L_{f_{j_2}} L_{f_{j_1}} h(x_0) \mathbf{I}^t\{z_{j_0} z_{j_1} \ldots z_{j_\nu}\} \tag{40.18}$$

with the series converging uniformly for small t and small $|u_i(\tau)|$, $0 \leq \tau \leq t$, $1 \leq i \leq m$. This functional expansion is called the *Fliess fundamental formula* or *Fliess expansion* of the solution. To this expansion can be also associated [12] an absoluting converging power series for small t and small $|u_i(\tau)|$, $0 \leq \tau \leq t$, $1 \leq i \leq m$, called the *Fliess generating power series* or *Fliess series* denoted by $\mathbf{g}$ of the following form

$$\mathbf{g} = h(x_0) + \sum_{\nu \geq 0} \sum_{j_0, j_1, \ldots, j_\nu = 0}^{m} L_{f_{j_\nu}} \ldots L_{f_{j_2}} L_{f_{j_1}} h(x_0) z_{j_0} z_{j_1} \ldots z_{j_\nu}. \tag{40.19}$$

This algebraic setting allows us to generalize to the nonlinear domain the Heaviside calculus for linear systems. This will appear clearly in the next section devoted to the effective computation of the Volterra series.

A lot of work uses this formalism, see for instance the work on bilinear realizability [44], some analytic aspects and local realizability of generating power series, on realization and input–output relations [49], or works establishing links with other algebras [10,17].

40.2.4 Links Between Volterra and Fliess Series

The following result [6,13,31,36] gives the expression of the Volterra kernels of the response of the nonlinear control system (40.12) in terms of the vector fields and the output function defining the system,

$$y(t) = w_0(t) + \sum_{n=1}^{\infty} \int_0^t \int_0^{\tau_2} \cdots \int_0^{\tau_n} w_n(t, \tau_n, \ldots, \tau_1) u(\tau_n) \ldots u(\tau_1) d\tau_n \cdots d\tau_1, \tag{40.20}$$

where the kernels are analytic functions of the form

$$\begin{aligned} w_0(t) &= \sum_{\nu \geq 0} L_f^\nu h(x_0) \frac{t^\nu}{\nu!} = e^{tL_f} h(x_0), \\ w_1(t, \tau_1) &= \sum_{\nu_0, \nu_1 \geq 0} L_f^{\nu_0} L_g L_f^{\nu_1} h(x_0) \frac{(t - \tau_1)^{\nu_1} \tau_1^{\nu_0}}{\nu_1! \nu_0!} \\ &= e^{\tau_1 L_f} L_g e^{(t-\tau_1) L_f} h(x_0), \\ &\vdots \\ w_n(t, \tau_n, \tau_{n-1}, \ldots, \tau_1) &= \sum_{\nu_0, \nu_1, \cdots, \nu_n \geq 0} L_f^{\nu_0} L_g L_f^{\nu_1} \ldots L_g L_f^{\nu_n} h(x_0) \frac{(t - \tau_n)^{\nu_n} \ldots \tau_1^{\nu_0}}{\nu_n! \ldots \nu_0!} \\ &= e^{\tau_1 L_f} L_g e^{(\tau_2 - \tau_1) L_f} \ldots L_g e^{(t - \tau_n) L_f} h(x_0) \end{aligned} \tag{40.21}$$

In order to show this, let us use the fundamental formula (Equation 40.18). The zero order kernel is the free response of the system. Indeed, from Equation 40.18 we have

$$w_0(t) = h(x_0) + \sum_{\nu \geq 0} \sum_{j_0, \cdots, j_\nu = 0} L_{f_{j_\nu}} \ldots L_{f_{j_2}} L_{f_{j_1}} h(x_0) \int_0^t d\xi_{j_\nu} d\xi_{j_{\nu-1}} \cdots d\xi_{j_0},$$

which can also be written as

$$y(t) = \sum_{l \geq 0} L_f^l h(x_0) \frac{t^l}{l!},$$

or using a formal notation,

$$y(t) = e^{tL_f} h(x_0).$$

This formula is nothing other than the classical formula given in [16].

For the computation of the first order kernel, let us consider the terms of Equation 40.18 which contain only one contribution of the input u; therefore,

$$\int_0^t w_1(t, \tau_1) u(\tau_1)\, d\tau_1 = \sum_{\nu_0, \nu_1 \geq 0} L_f^{\nu_0} L_g L_f^{\nu_1} h(x_0) \int_0^t \underbrace{d\xi_0 \ldots d\xi_0}_{\nu_1\text{- times}} d\xi_1 \underbrace{d\xi_0 \ldots d\xi_0}_{\nu_0\text{- times}}.$$

But the iterated integral inside can be proven to be equal to

$$\int_0^t \frac{(t-\tau_1)^{\nu_1} \tau_1^{\nu_0}}{\nu_1! \nu_0!} u(\tau_1)\, d\tau_1.$$

So, the first order kernel may be written as

$$\begin{aligned} w_1(t, \tau_1) &= \sum_{\nu_0, \nu_1 \geq 0} L_f^{\nu_0} L_g L_f^{\nu_1} h(x_0) \frac{(t-\tau_1)^{\nu_1} \tau_1^{\nu_0}}{\nu_1! \nu_0!} \\ &= e^{\tau_1 L_f} L_g e^{(t-\tau_1) L_f} h(x_0). \end{aligned}$$

For the computation of the second order kernel, let us regroup the terms of Equation 40.18, which contain exactly two contributions of the input u; therefore,

$$\begin{aligned} \int_0^t \int_0^{\tau_2} w_2(t, \tau_1, \tau_2) u(\tau_1) u(\tau_2) d\tau_1 d\tau_2 = &\sum_{\nu_0, \nu_1, \nu_2 \geq 0} L_f^{\nu_0} L_g L_f^{\nu_1} L_g L_f^{\nu_2} h(x_0) \\ &\times \int_0^t \int_0^{\tau_2} \underbrace{d\xi_0 \ldots d\xi_0}_{\nu_2\text{- times}} d\xi_1 \underbrace{d\xi_0 \ldots d\xi_0}_{\nu_1\text{- times}} d\xi_1 \underbrace{d\xi_0 \ldots d\xi_0}_{\nu_0\text{- times}}. \end{aligned}$$

The iterated integral inside this expression can be proven to be equal to

$$\int_0^t \int_0^{\tau_2} \frac{(t-\tau_2)^{\nu_2} (\tau_2 - \tau_1)^{\nu_1} \tau_1^{\nu_0}}{\nu_2! \nu_1! \nu_0!} u(\tau_1) u(\tau_2) d\tau_1 d\tau_2.$$

Thus, the second order kernel may be written as

$$\begin{aligned} w_2(t, \tau_1, \tau_2) &= \sum_{\nu_0, \nu_1, \nu_2 \geq 0} L_f^{\nu_0} L_g L_f^{\nu_1} L_g L_f^{\nu_2} h(x_0) \frac{(t-\tau_2)^{\nu_2} (\tau_2 - \tau_1)^{\nu_1} \tau_1^{\nu_0}}{\nu_2! \nu_1! \nu_0!} \\ &= e^{\tau_2 L_f} L_g e^{(\tau_1 - \tau_2) L_f} L_g e^{(t-\tau_1) L_f} h(x_0). \end{aligned}$$

The higher order is obtained in the same way.

Using the Campbell-Baker-Hausdorff formula

$$e^{\sigma L_f} L_g e^{-\sigma L_f} h(x_0) = \sum_{i=1}^{\infty} \frac{\sigma^i}{i!} ad^i_{L_f} L_g,$$

the expressions for the kernels (Equation 40.21) may be written,

$$\begin{aligned}
w_0(t) &= e^{tL_f} h(x_0), \\
w_1(t, \tau_1) &= e^{\tau_1 L_f} L_g e^{(t-\tau_1)L_f} h(x_0) \\
&= \sum_{i=1}^{\infty} \frac{\tau_1^i}{i!} ad^i_{L_f} L_g e^{tL_f} h(x_0) \\
w_2(t, \tau_n, \tau_{n-1}, \dots, \tau_1) &= e^{\tau_1 L_f} L_g e^{(\tau_2-\tau_1)L_f} L_g e^{(t-\tau_2)L_f} h(x_0) \\
&= \sum_{i,j=1}^{\infty} \frac{\tau_1^i}{i!} \frac{\tau_2^j}{j!} ad^i_{L_f} L_g ad^j_{L_f} L_g e^{tL_f} h(x_0) \\
&\vdots
\end{aligned} \tag{40.22}$$

These kernel expressions lead to techniques which may, for example, be used in singular optimal control problems [32]. This will be sketched in the next section.

40.3 Effective Computation of Volterra Kernels

Example 40.1

Let us consider the system [39],

$$\ddot{y}(t) + (\omega^2 + u(t))y(t) = 0, \quad t \geq 0, \ \ y(0) = 0, \ \ \dot{y}(0) = 1,$$

After two integrations, we obtain

$$y(t) + \omega^2 \int_0^t \int_0^\tau y(\sigma) d\sigma d\tau + \int_0^t \int_0^\tau u(\sigma) y(\sigma) d\sigma d\tau - t = 0$$

The associated algebraic equation for the Fliess series (see Section 40.2.3) $\mathbf{g}$ is given by

$$(1 + \omega^2 z_0^2)\mathbf{g} + z_0 z_1 \mathbf{g} - z_0 = 0.$$

In order to solve this equation, let us use the following iterative scheme

$$\mathbf{g} = \mathbf{g}_0 + \mathbf{g}_1 + \mathbf{g}_2 + \cdots + \mathbf{g}_i + \cdots$$

where $\mathbf{g}_i$ contains all the terms of the solution $\mathbf{g}$ having exactly i occurrences in the variable z_1,

$$\begin{aligned}
\mathbf{g}_0 &= (1 + \omega^2 z_0^2)^{-1} z_0, \\
\mathbf{g}_1 &= -(1 + \omega^2 z_0^2)^{-1} z_0 z_1 \mathbf{g}_0 \\
&= -(1 + \omega^2 z_0^2)^{-1} z_0 z_1 (1 + \omega^2 z_0^2)^{-1} z_0, \\
\mathbf{g}_2 &= -(1 + \omega^2 z_0^2)^{-1} z_0 z_1 \mathbf{g}_1 \\
&= (1 + \omega^2 z_0^2)^{-1} z_0 z_1 (1 + \omega^2 z_0^2)^{-1} z_0 z_1 (1 + \omega^2 z_0^2)^{-1} z_0, \\
&\vdots
\end{aligned}$$

Each $\mathbf{g}_i, i \geq 0$ is a (rational) generating power series of functionals $y_i, i \geq 0$ which represents the ith order term of the Volterra series associated with the solution $y(t)$. Let us now compute $y_i(t)$ associated with $\mathbf{g}_i$, $i \geq 0$. First, note that

$$(1 - az_0)^{-1} = \sum_{n\geq 0} a^n z_0^n, \quad a \in \mathbf{C},$$

represents in the algebraic domain, the function e^{-at}. Indeed,

$$\mathbf{I}^t(z_0^n) = \frac{t^n}{n!}.$$

Consider now

$$\mathbf{g}_0 = -\frac{1}{2j\omega}(1 + j\omega z_0)^{-1} + \frac{1}{2j\omega}(1 - j\omega z_0)^{-1}.$$

Hence,

$$y_0(t) = w_0(t) = -\frac{1}{2j\omega}e^{-j\omega t} + \frac{1}{2j\omega}e^{j\omega t} = \frac{1}{\omega}\sin(wt).$$

The power series

$$\mathbf{g}_1 = -(1 + \omega^2 z_0^2)^{-1} z_0 z_1 (1 + \omega^2 z_0^2)^{-1} z_0,$$

after decomposing into partial fractions the term on the right and on the left of z_1,

$$\left[\frac{1}{2j\omega}(1 + j\omega z_0)^{-1} - \frac{1}{2j\omega}(1 - j\omega z_0)^{-1}\right] z_1 \left[-\frac{1}{2j\omega}(1 + j\omega z_0)^{-1} + \frac{1}{2j\omega}(1 - j\omega z_0)^{-1}\right]$$

or

$$\begin{aligned}\mathbf{g}_1 = \frac{1}{4\omega^2}&\left[(1 + j\omega z_0)^{-1} z_1 (1 + j\omega z_0)^{-1} - (1 + j\omega z_0)^{-1} z_1 (1 - j\omega z_0)^{-1}\right.\\ &\left.- (1 - j\omega z_0)^{-1} z_1 (1 + j\omega z_0)^{-1} + (1 - j\omega z_0)^{-1} z_1 (1 - j\omega z_0)^{-1}\right]\end{aligned}$$

In order to obtain the equivalent expression in the time domain, we need the following result [27,31].

The rational power series

$$(1 - a_0 z_0)^{-p_0} z_1 (1 - a_1 z_0)^{-p_1} z_1 \ldots z_1 (1 - a_l z_0)^{-p_l}, \tag{40.23}$$

where $a_0, a_1, \ldots, a_l \in \mathbf{C}, p_0, p_1, \ldots, p_l \in \mathbf{N}$, in the symbolic representation of

$$\int_0^t \int_0^{\tau_l} \cdots \int_0^{\tau_2} f_{a_0}^{p_0}(t - \tau_l) \ldots f_{a_{l-1}}^{p_{l-1}}(\tau_2 - \tau_1) f_{a_l}^{p_l}(\tau_1) u(\tau_l) \ldots u(\tau_1) d\tau_l \ldots d\tau_1,$$

where $f_a^p(t)$ denotes the exponential polynomial

$$\left(\sum_{j=0}^{p-1} \frac{\binom{j}{p-1}}{j!} a^j t^j\right) e^{at}.$$

For the previous example

$$y_1(t) = \int_0^t \frac{1}{2j\omega} e^{-j\omega(t-\tau)} u(\tau) \left[\frac{-1}{2j\omega} e^{-j\omega\tau} + \frac{1}{2j\omega} e^{j\omega\tau}\right] d\tau - \int_0^t \frac{1}{2j\omega} e^{j\omega(t-\tau)} u(\tau) \left[\frac{-1}{2j\omega} e^{-j\omega\tau} + \frac{1}{2j\omega} e^{j\omega\tau}\right] d\tau.$$

Therefore,

$$y_1(t) = \int_0^t w_1(t,\tau)u(\tau)\,d\tau,$$

with $w_1(t,\tau) = -\frac{1}{\omega^2}\sin[\omega(t-\tau)]\sin\omega t$.

The higher-order kernels can be computed in the same way after decomposing into partial fraction each rational power series.

40.3.1 Noncommutative Padé-Type Approximants

Assume that the functions $f_i^k : \mathbf{R}^n \to \mathbf{R}$ of Equation 40.9 are C^ω, with

$$f_i^k(x_1,\ldots,x^n) = \sum_{j_1,\ldots,j_n \geq 0} a^{k,i}_{j_1,\ldots,j_n}(x_1)^{j_1}\ldots(x_n)^{j_n}$$
$$h(x_1,\ldots,x^n) = \sum_{j_1,\ldots,j_n \geq 0} h_{j_1,\ldots,j_n}(x_1)^{j_1}\ldots(x_n)^{j_n}.$$

Let γ denote an equilibrium point of the system (Equation 40.9) and let

$$x^{<p>}_{j_1,\ldots,j_n}$$

denote the monomial or new state

$$(x_1)^{j_1}\ldots(x_n)^{j_n}, \quad j_1+\cdots j_n \leq p.$$

Then the Brockett bilinear system

$$\begin{cases} \dot{x}^{<p>}_{j_1,\ldots,j_n} &= \sum_{k=1}^n j_k\left(\sum_{i=0}^m u_i \sum_{i_1,\ldots,i_n} a^{k,i}_{j_1,\ldots,j_n} x^{<p>}_{j_1+i_1,\ldots,j_k+i_k-1,\ldots,j_n+i_n}\right), \\ y^{<p>} &= \sum_{i_1,\ldots,i_n} h_{j_1,\ldots,j_n} x^{<p>}_{j_1,\ldots,j_n} \end{cases} \tag{40.24}$$

with initial conditions

$$x^{<p>}_{j_1,\cdots,j_n}(0) = (\gamma_1)^{j_1}\ldots(\gamma_n)^{j_n}$$

where

$$x^{<p>}_{j_1,\ldots,-1,\ldots,j_n} = 0$$

for all $j_1,\ldots,j_n \geq 0$ and

$$x^{<p>}_{j_1,\ldots,j_n} = 0$$

if $j_1+\ldots+j_n > p$, has the same Volterra series up to order p as the Volterra series of the nonlinear system (Equation 40.9).

This system may be interpreted in the algebraic context by defining the generating power series $\mathbf{g}^{<p>}_{j_1,\ldots,j_n}$ associated with $x^{<p>}_{j_1,\ldots,j_n}$ and $\mathbf{g}^{<p>}$ associated with $y^{<p>}$,

$$\begin{cases} \mathbf{g}^{<p>}_{j_1,\ldots,j_n} &= \sum_{k=1}^n j_k\left(\sum_{i=0}^m z_i \sum_{i_1,\ldots,i_n} a^{k,i}_{j_1,\ldots,j_n} \mathbf{g}^{<p>}_{j_1+i_1,\ldots,j_k+i_k-1,\ldots,j_n+i_n}\right), \\ \mathbf{g}^{<p>} &= \sum_{i_1,\ldots,i_n} h_{j_1,\ldots,j_n} \mathbf{g}^{<p>}_{j_1,\ldots,j_n} \end{cases} \tag{40.25}$$

The rational power series $\mathbf{g}^{<p>}$ may be seen as a *noncommutative Padé-type approximant* for the forced differential system (Equation 40.9) which generalizes the notion of Padé-type approximant obtained in [4]. Using algebraic computing, these approximants may be derived explicitly [38]. These algebraic tools for the first time enable one to derive the Volterra kernels.

Other techniques have recently been introduced in order to compute approximate solutions to the response of nonlinear control systems. The method in [34] is based on the combinatorial notion of L-species. Links between this combinatorial method and the Fliess algebraic setting have been studied in [33,35]. Another approach using automata representations [41] is proposed in [23].

The Volterra series (Equation 40.13) terminating with the term involving the pth kernel is called a Volterra series *of length* p. In the following we will summarize some properties of an input–output map having a finite Volterra series. The main question is how to characterize a state space representation (Equation 40.12) that admits a finite Volterra series representation [8]. This study lead in particular to the introduction of a large class of approximating systems, having a solvable but not necessarily nilpotent Lie algebra [9] (see also [24]).

40.4 Approximation Abilities of Volterra Series

40.4.1 Analysis of Responses of Systems

In this part we show how to compute the response of nonlinear systems to typical inputs. We assume here $m = 1$. This method, based on the use of the formal representation of the Volterra kernels (Equation 40.25), is also easily implementable on a computer using formal languages [38]. These algebraic tools allow us to derive exponential polynomial expressions depending explicitly on time for the truncated Volterra series associated with the response [27,31] and therefore lead to a finer analysis than pure numerical results.

To continue our use of algebraic tools, let us introduce the Laplace–Borel transform associated with a given analytic function input

$$u(t) = \sum_{n\geq 0} a_n \frac{t^n}{n!}.$$

Its Laplace–Borel transform is

$$\mathbf{g}_u = \sum_{n\geq 0} a_n z_0^n.$$

For example, the Borel transformation of

$$\cos\omega t = \frac{1}{2}e^{j\omega t} + \frac{1}{2}e^{-j\omega t}.$$

is given by

$$\mathbf{g}_u = \frac{1}{2}(1 - j\omega t z_0)^{-1} + \frac{1}{2}(1 + j\omega t z_0)^{-1} = (1 + \omega^2 z_0^2)^{-1}.$$

Before seeing the algebraic computation itself in order to compute the first terms of the response to typical inputs, let us introduce a new operation on formal power series, *the shuffle product.*

Given two formal power series,

$$\mathbf{g}_1 = \sum_{w\in Z^*} (\mathbf{g}_1, w)w \quad \text{and} \quad \mathbf{g}_2 = \sum_{w\in Z^*} (\mathbf{g}_2, w)w.$$

The *shuffle product* of two formal power series $\mathbf{g}_1$ and $\mathbf{g}_2$ is given by

$$\mathbf{g}_1 \diamond \mathbf{g}_2 = \sum_{w_1,w_2\in Z^*} (\mathbf{g}_1, w_1)(\mathbf{g}_2, w_2) w_1 \diamond w_2,$$

where shuffle product of two words is defined as follows,

- $1\diamond 1 = 1$
- $\forall z \in Z,\ 1\diamond z = z\diamond 1 = z$

- $\forall z, z' \in Z,\ \forall w, w' \in Z^*$
  $zw \diamond z'w' = z[w \diamond z'w'] + z'[zw \diamond w']$

This operation consists of shuffling all the letters of the two words by keeping the order of the letters in the two words. For instance,

$$z_0 z_1 \diamond z_1 z_0 = 2 z_0 z_1^2 z_0 + z_0 z_1 z_0 z_1 + z_1 z_0 z_1 z_0 + z_1 z_0^2 z_1.$$

It has been shown that the Laplace–Borel transform of Equation 40.23, for a given input $u(t)$ with the Laplace-Borel transform $\mathbf{g}_u$, is obtained by substituting from the right, each variable z_1 by the operator $z_0[\mathbf{g}_u \diamond \cdot]$.

Therefore, in order to apply this result, we need to know how to compute shuffle product of algebraic expressions of the form

$$\mathbf{g}_n = (1 + a_0 z_0)^{-1} z_{i_1} (1 + a_1 z_0)^{-1} z_{i_2} \ldots (1 + a_{n-1} z_0)^{-1} z_{i_n} (1 + a_n z_0)^{-1}$$

where $i_1, i_2, \cdots, i_n \in \{0, 1\}$. This computation is very simple as it amounts to adding some singularities. For instance

$$(1 + a z_0)^{-1} \diamond (1 + b z_0)^{-1} = (1 + (a + b) z_0)^{-1}.$$

Consider two generating power series of the form (Section 40.4.1),

$$\mathbf{g}_p = (1 + a_0 z_0)^{-1} z_{i_1} (1 + a_1 z_0)^{-1} z_{i_2} \ldots (1 + a_{p-1} z_0)^{-1} z_{i_p} (1 + a_p z_0)^{-1}$$

and

$$\mathbf{g}_q = (1 + b_0 z_0)^{-1} z_{j_1} (1 + b_1 z_0)^{-1} z_{j_2} \ldots (1 + b_{q-1} z_0)^{-1} z_{j_q} (1 + b_q z_0)^{-1}$$

where p and $q \in \mathbf{N}$, the indices $i_1, i_2, \ldots, i_p \in \{0, 1\}, j_1, j_2, \ldots, j_q \in \{0, 1\}$ and $a_i, b_j \in \mathbf{C}$. The shuffle product of these expressions is given by induction on the length

$$\mathbf{g}_p \diamond \mathbf{g}_q = \mathbf{g}_p \diamond \mathbf{g}_{q-1} z_{j_q} (1 + (a_p + b_q) z_0)^{-1} + \mathbf{g}_{p-1} \diamond \mathbf{g}_q z_{i_p} (1 + (a_p + b_q) z_0)^{-1}.$$

See [30] for case-study examples and some other rules for computing directly the stationary response to harmonic inputs or the response of a Dirac function and see [14] for the algebraic computation of the response to white noise inputs. This previous effective computation of the rational power series $\mathbf{g}$ and of the response to typical entries has been applied to the analysis of nonlinear electronics circuits [2] and to the study of laser semi-conductors [18].

40.4.2 Optimality

Volterra series expansions have been used in order to study control variations for the output of nonlinear systems combined with some multiple integral identities [10]. This analysis [32,42,43], demonstrates links between the classical Hamiltonian formalism and the Lie algebra associated with the nonlinear control problem. To be more precise, let us consider the control system

$$\sum \begin{cases} \dot{x}(t) = f(x(t), u(t)), \\ x(0) = x_0 \end{cases}, \tag{40.26}$$

where $f : \mathbf{R}^n \times \mathbf{R}^m \to \mathbf{R}^n$ is a smooth mapping. Let $h : \mathbf{R}^n \to \mathbf{R}$ be a smooth function, let

$$\gamma(t, x_0) = e^{tL_f} x_0$$

be the free response of the system and let $x(t, x_0, u)$ be the solution relative to the control u, u being an integrable function taking values in some given bounded open set $U \in \mathbf{R}^m$. An example of control problem is the following: find necessary conditions such that

$$h(\gamma(T)) = \min_u h(x(T, x_0, u)). \tag{40.27}$$

Let $w_0 : [0, T] \times \mathbf{R}^n \to \mathbf{R}$ be defined as follows,

$$w_0(t, x) = h(e^{(T-t)L_f}x).$$

It is easy to see that the map $\lambda : [0, T] \to (\mathbf{R})^*$ given by

$$\lambda(t) = \frac{\partial w_0}{\partial x}(t, \gamma(t))$$

is the solution of the adjoint equation

$$-\dot{\lambda}(t) = \lambda(t)\frac{\partial f(x, 0)}{\partial x}(t, \gamma(t))$$

and $\lambda(T) = dh(\gamma(t))$.

A first order necessary condition is provided by the application of the Maximum Principle: If $\gamma(t)$ satisfies (Equation 40.27) for $t \in [0, T]$, then

$$ad_{f_0} f_i w_0(t, \gamma(t)) = 0 \quad \text{for } t \in [0, T] \tag{40.28}$$

and the matrix

$$((f_{ij} w_0(t, \gamma(t)))) \tag{40.29}$$

is a non-negative matrix for $t \in [0, T]$ where

$$f_0(x) = f(0, x), \quad f_i(x) = \frac{\partial f}{\partial u_i}(x, 0),$$

and

$$f_{ij}(x) = \frac{\partial^2 f}{\partial u_i \partial u_j}(x, 0), \quad i, j = 1, \ldots, m.$$

The reference trajectory γ is said *extremal* if it satisfies Equation 40.28 and is said *singular* if it is extremal and if all the terms in the matrix (Equation 40.29) vanish. If γ is singular and it satisfies Equation 40.27, then it can be shown for instance (see [32]) that if there exists $s \geq 1$ such that for $t \in [0, T]$ and $i, j = 1, \ldots, m$,

$$[ad_{f_0}^{k+1} f_i, ad_{f_0}^k f_j] w_0(t, \gamma(t)) = 0 \quad \text{for } k = 0, 1, s-1,$$

then

$$[ad_{f_0}^{k_1} f_i, ad_{f_0}^{k_2} f_j] w_0(t, \gamma(t)) = 0 \quad \text{for } k_1, k_2 \geq 0 \quad \text{with } k_1 + k_2 = 0, \ldots, 2s$$

and the matrix

$$\left(\left([ad_{f_0}^{s+1} f_i, ad_{f_0}^s f_j] w_0(t, \gamma(t))\right)\right)$$

is a symmetric non-negative matrix for $t \in [0, T]$.

As a dual problem, sufficient conditions for local controllability have been derived using Volterra series expansions (see for instance [3]).

40.4.3 Search for Limit Cycles and Bifurcation Analysis

The Hopf bifurcation theorem deals with the appearance and growth of a limit cycle as a parameter is varied in a nonlinear system. Several authors have given a rigorous proof using various mathematical tools, series expansions, central manifold theorem, harmonic balance, Floquet theory, or Lyapunov methods. Using Volterra series [47] did provide a conceptual simplification of the Hopf proof. In many problems, the calculations involved are simplified leading to practical advantages as well as theoretical ones.

40.5 Other Approximations: Application to Motion Planning

Let us consider a control system

$$\dot{x} = \sum_{i=1}^{m} u_i(t) f_i(x). \tag{40.30}$$

The dynamical exact motion planning problem is the following: given two state vectors p and q, find an input function $u(t) = (u_1(t), u_2(t), \ldots, u_m(t))$ that drives exactly the state vector from p to q. In order to solve this problem, several expansions for the solutions which are intimately linked with the Fliess series are used.

When the vector fields f_i, $i = 1, \ldots, m$ are real analytic and complete and such that the generated control Lie algebra is everywhere of full rank and nilpotent, then in [26,46], the authors described a complete solution of the previous control problem using P. Hall basis.

Let $A(z_1, z_2, \ldots, z_m)$ denote the algebra of noncommutative polynomials in $(z_1, z_2, \ldots, z_m)$ and let $L(z_1, z_2, \ldots, z_m)$ denote the Lie subalgebra of $A(z_1, z_2, \ldots, z_m)$ generated by $z_1, z_2, \ldots, z_m$ with the Lie bracket defined by $[P, Q] = PQ - QP$. The elements of $L(z_1, z_2, \ldots, z_m)$ are known as *Lie polynomials* in $z_1, z_2, \ldots, z_m$. Let $\mathcal{F}_m$ be the set of *formal Lie monomials* in $z_1, z_2, \ldots, z_m$. A *P. Hall* basis of $L(z_1, z_2, \ldots, z_m)$ is a totally ordered subset $(\mathcal{B}, <)$ of $\mathcal{F}_m$ such that,

The z_i belong to $\mathcal{B}$.
If $A, B \in \mathcal{B}$, and degree$(A) <$ degree(B), then $A < B$.
If $P \in \mathcal{F}_m$ and P is not one of the z_i, then $P \in \mathcal{B}$ if and only if $P = [A, B]$ with $A, B \in \mathcal{B}$, $A < B$, and either $B = z_i$ for some i or $B = [C, D]$ with $C, D \in \mathcal{B}$, $C \leq A$.

If $L_k(z_1, z_2, \ldots, z_m)$ denote the nilpotent version of $L(z_1, z_2, \ldots, z_m)$ obtained by killing all the monomials of degree $k+1$, it is not difficult to see that $\{M \in \hat{\mathcal{B}} : degree(\mathrm{M}) \leq k\}$ is a basis of $L_k(z_1, z_2, \ldots, z_m)$, where $\hat{\mathcal{B}}$ is the set of all elements of $L(z_1, z_2, \ldots, z_m)$ obtained by actually evaluating the Lie brackets.

Now, let $z_1, \ldots, z_m, z_{m+1}, \ldots, z_r$ a P. Hall basis of $L_k(z_1, z_2, \ldots, z_m)$ and let E_f be the evaluation map that assigns to each P the vector field obtained by plugging in the f_i, $i = 1, \ldots, m$, for the corresponding z_i. We assume that the vector fields $f_{m+1}, \ldots, f_r$ are given by $f_j = E_f(z_j)$ for $j = m+1, \ldots r$.

An expansion for Equation 40.30, and consequently a solution to the exact motion planning problem, is then obtained [26] from the solution of

$$\dot{S}(t) = S(t)\left(u_1(t)z_1 + u_2(t)z_2 + \cdots + u_r(t)z_r\right), \quad S(0) = 1,$$

written as a product

$$S(t) = e^{h_r(t)z_r} e^{h_{r-1}(t)z_{r-1}} \cdots e^{h_2(t)z_2} e^{h_1(t)z_1}.$$

For example, in the case $k=3$, $m=2$, we may choose $z_3=[z_1,z_2]$, $z_4=[z_1,[z_1,z_2]]$ and $z_5=[z_2,[z_1,z_2]]$. For this choice, the functions $h_j, j=1,\ldots,r$ are computed by solving

$$\begin{aligned}
\dot{h}_1 &= u_1\\
\dot{h}_2 &= u_2\\
\dot{h}_3 &= h_1u_2+u_3\\
\dot{h}_4 &= \frac{1}{2}h_1^2u_2+h_1u_3+u_4\\
\dot{h}_5 &= h_2u_3+h_1h_2u_2
\end{aligned}$$

with $h_j(0)=0, j=1,\ldots,5$.

In order to take into account systems with drift, for the system (Equation 40.30) where $u_1(t)$ may be identically equal to 1, a different basis, called the Lyndon basis, has been used in [21]. Without going into the details, for the previous case, it is shown that the solution is given by the following exponential product expansion

$$x(t)=e^{\xi_1(t)f_1}e^{\xi_2(t)f_2}e^{\xi_3(t)f_3}e^{\xi_4(t)f_4}e^{\xi_5(t)f_5}\cdot Id(x)|_{x(0)}$$

with here $z_3=[z_2,z_1]$, $z_4=[z_2,[z_1,z_1]]$, $z_5=[z_2,[z_2,z_1]]$, $f_j=E_f(z_j)$ for $j=3,\ldots,5$ and

$$\begin{aligned}
\xi_1 &= \int_0^t u_1(\tau)\,d\tau\\
\xi_2 &= \int_0^t u_2(\tau)\,d\tau\\
\xi_3 &= \int_0^t \xi_2\,d\xi_1\\
\xi_4 &= \int_0^t \xi_3\,d\xi_1\\
\xi_5 &= \frac{1}{2}\int_0^t \xi_2^2\,d\xi_1.
\end{aligned}$$

References

1. d'Alessandro, P., Isidori, A., and Ruberti, A., Realizations and structure theory of bilinear dynamical systems, *SIAM J. Control,* 12, 517–535, 1974.
2. Baccar, S., Lamnabhi-Lagarrigue, F., and Salembier, G., Utilisation du calcul formel pour la modélisation et la simulation des circuits électroniques faiblement non linéaires, *Annales des Télécommunications,* 46, 282–288, 1991.
3. Bianchini, R.M. and Stefani, G., A high order maximum principle and controllability, 1991.
4. Brezinski, C., *Padé-type approximants and general orthogonal polynomials,* INSM, 50, Birkhaüser, 1980.
5. Brockett, R.W., Nonlinear systems and differential geometry, *Proceedings IEEE,* 64, 61–72, 1976.
6. Brockett, R.W., Volterra series and geometric control theory, *Automatica,* 12, 167–176, 1976; Brockett, R.W. and Gilbert, E.G., An addendum to Volterra series and geometric control theory, *Automatica,* 12, 635, 1976.
7. Chen, K.T., Integration of paths, geometric invariants and a generalized Baker-Hausdorff formula, *Ann. Math.,* 65, 163–178, 1957.
8. Crouch, P.E., Dynamical realizations of finite Volterra series, *SIAM J. Control and Optimization,* 19, 177–202, 1981.
9. Crouch, P.E., Solvable approximations to control systems, *SIAM J. Control and Optimization,* 22(1), 40–54, 1984.
10. Crouch, P.E. and Lamnabhi-Lagarrigue, F., Algebraic and multiple integral identities, *Acta Applicanda Mathematicae,* 15, 235–274, 1989.

11. Feynman, R.P., An operator calculus having applications in quantum electrodynamics, *Physical Review,* 84, 108–128, 1951.
12. Fliess, M., Fonctionnelles causales non linéaires et indéterminées non commutatives, *Bull. Soc. Math. France,* 109, 3–40, 1981.
13. Fliess, M., Lamnabhi, M., and Lamnabhi-Lagarrigue, F., Algebraic approach to nonlinear functional expansions, *IEEE CS,* 30, 550–570, 1983.
14. Fliess, M., and Lamnabhi-Lagarrigue, F., Application of a new functional expansion to the cubic anharmonic oscillator, *J. Math. Physics,* 23, 495–502, 1982.
15. Gilbert, E.G., Functional expansions for the response of nonlinear differential systems, *IEEE AC,* 22, 909–921, 1977.
16. Gröbner, W., *Die Lie-Reihen und ihre Anwendungen* (2e édition), VEB Deutscher Verlag der Wissenschaften, Berlin, 1967.
17. Grunenfelder, L., Algebraic aspects of control systems and realizations, *J. Algebra,* 165, 446–464, 1994.
18. Hassine, L., Toffano, Z., Lamnabhi-Lagarrigue, F. Destrez, A., and Birocheau, C., Volterra functional series expansions for semiconductor lasers under modulation, *IEEE J. Quantum Electron.,* 30, 918–928, 1994; Hassine, L., Toffano, Z., Lamnabhi-Lagarrigue, F., Joindot, I., and Destrez, A., Volterra functional series for noise in semiconductor lasers, *IEEE J. Quantum Electron.,* 30, 2534–2546, 1994.
19. Hoang Ngoc Minh, V. and Jacob, G., Evaluation transform and its implementation in MACSYMA, in *New Trends in Systems Theory,* G. Conte, A.M. Perdon, B. Wyman, Eds., Progress in Systems and Control Theory, Birkhäuser, Boston, 1991.
20. Jacob, G., Algebraic methods and computer algebra for nonlinear systems' study, *Proc. IMACS Symposium MCTS,* 1991.
21. Jacob, G., Motion planning by piecewise constant or polynomial inputs, *Proc. NOLCOS'92,* 1992.
22. Jacob, G. and Lamnabhi-Lagarrigue, F., *Algebraic Computing in Control,* Lect. Notes Contr. Inform. Sc., Springer-Verlag, 165, 1991.
23. Hespel, C. and Jacob, G., Truncated bilinear approximants: Carleman, finite Volterra, Padé-type, geometric and structural automata, in *Algebraic Computing in Control,* Lect. Note Contr. Inform. Sc., G. Jacob and F. Lamnabhi-Lagarrigue, Eds., Springer-Verlag, 165, 1991.
24. Jakubczyk, B. and Kaskosz, B., Realizability of Volterra series with constant kernels, *Nonlinear Analysis, Theory, Methods and Applications,* 5, 167–183, 1981.
25. Krener, A.J., Local approximations of control systems, *J. Differential Equations,* 19, 125–133, 1975.
26. Lafferriere, G. and Sussmann, H.J., Motion planning for controllable systems without drift: a preliminary report, *Report SYCON-90-04,* Rutgers Center for Systems and Control, June 1990.
27. Lamnabhi, M., A new symbolic response of nonlinear systems, *Systems and Control Letters,* 2, 154–162, 1982.
28. Lamnabhi, M., Functional analysis of nonlinear circuits: a generating power series approach, *IEE Proceedings,* 133, 375–384, 1986.
29. Lamnabhi–Lagarrigue, F., *Séries de Volterra et commande optimale singulière,* Thèse d'état, Université Paris–Sud, Orsay, 1985.
30. Lamnabhi–Lagarrigue, F., *Analyse des systèmes non linéaires,* Hermes Ed., 1994.
31. Lamnabhi–Lagarrigue F. and Lamnabhi, M., Détermination algébrique des noyaux de Volterra associés à certains systèmes non linéaires. *Ricerche di Automatica,* 10, 17–26, 1979.
32. Lamnabhi-Lagarrigue, F. and Stefani, G., Singular optimal control problems: on the necessary conditions for optimality, *SIAM J. Control Optim.,* 28, 823–840, 1990.
33. Lamnabhi–Lagarrigue, F., Leroux, P., and Viennot, X.G., Combinatorial approximations of Volterra series by bilinear systems, in *Analyse des Systèmes Contrôlés,* B. Bonnard, B. Bride, J.P. Gautier, and I. Kupka, Eds., Progress in Systems and Control Theory, Birkhäuser, 304–315, 1991.
34. Leroux, P. and Viennot, X.G., A combinatorial approach to nonlinear functional expansions, *Theoritical Computer Science,* 79, 179–183, 1991.
35. Leroux, P., Martin A., and Viennot, X.G., Computing iterated derivatives along trajectories of nonlinear systems, *NOLCOS'92,* M. Fliess, Ed., Bordeaux, 1992.
36. Lesiak, C. and Krener, A.J., The existence and uniqueness of Volterra series for nonlinear systems, *IEEE AC,* 23, 1090–1095, 1978.
37. Magnus, W., On the exponential solution of differential equations for a linear operator, *Comm. Pure Appl. Math.,* 7, 649–673, 1954.
38. Martin, A., Calcul d'approximations de la solution d'un système non linéaire utilisant le logiciel SCRATCHPAD, in *Algebraic Computing in Control,* Lect. Note Contr. Inform. Sc., G. Jacob and F. Lamnabhi-Lagarrigue, Eds., Springer-Verlag, 165, 1991.
39. Rugh, W.J., *Nonlinear System Theory,* The John Hopkins University Press, Baltimore, MD, 1981.

40. Schetzen, I.W., *The Volterra and Wiener Theories of Nonlinear Systems*, John Wiley & Sons, New York, 1980.
41. Schutzenberger, M.P., On the definition of a family of automata, *Information and Control*, 4, 245–270, 1961.
42. Stefani, G., Volterra approximations, *NOLCOS'90*, Nantes, 1990.
43. Stefani, G. and Zezza, P., A new type of sufficient optimality conditions for a nonlinear constrained optimal control problem, *NOLCOS'92*, Bordeaux, 1992.
44. Sontag, E.D., Bilinear realizability is equivalent to existence of a singular affine differential input/output equation, *Syst. Control Lett.*, 11, 181–187, 1988.
45. Sussmann, H.J., Lie brackets and local controllability: A sufficient condition for scalar-input systems, *SIAM J. Control Optimiz.*, 21, 686–713, 1983.
46. Sussmann, H.J., Local controllability and motion planning for some classes of systems with drift, *Proc. 30th IEEE CDC*, 1110–1113, 1991.
47. Tang, Y.-S., Mees, A.I., and Chua, L.O., Hopf bifurcation via Volterra series, *IEEE AC*, 28(1), 42–53, 1983.
48. Volterra, V., *Theory of Functionals* (translated from Spanish), Blackie, London, 1930 (reprinted by Dover, New York, 1959).
49. Wang, Y. and Sontag, E.D., Generating series and nonlinear systems: Analytic aspects, local realizability, and input–output representations, *Forum Math.*, 4, 299–322, 1992; Algebraic differential equations and rational control systems, *SIAM J. Control Optimiz.*,
50. Wei, J. and Norman, E., On global representations of the solutions of linear differential equations as a product of exponentials, *Proc. Am. Math. Soc.*, 15, 327–334, 1964.
51. Wiener, N., *Nonlinear Problems in Random Theory*, John Wiley & Sons, New York, 1958.

41

Integral Quadratic Constraints

Alexandre Megretski
Massachusetts Institute of Technology

Ulf T. Jönsson
Royal Institute of Technology

Chung-Yao Kao
National Sun Yat-Sen University

Anders Rantzer
Lund University

41.1 Introduction

Integral quadratic constraints (IQC*) are inequalities used to describe (partially) possible signal combinations within a given dynamical system. IQC offer a framework for abstracting "challenging" (e.g., nonlinear, time-varying, uncertain, or distributed) elements of dynamical system models to aid in rigorous analysis of robust stability and performance (more specifically, to establish L_2 gain bounds, passivity, and other system properties, which can be expressed, exactly or approximately, in terms of generalized dissipativity). While the technique can be employed to prove general theorems, it is most powerful when used to derive optimization-based algorithms for certification of stability and robustness of specific feedback systems.

IQC can be viewed as implicit generalized dissipation inequalities with known quadratic supply rates and unspecified storage functions, not necessarily quadratic or sign definite. Alternatively, they have a frequency domain interpretation as bounds on the degree of harmonic distortion produced by a specific element of the complete model. The past research in nonlinear systems and robust control can be harvested to extract rich IQC descriptions of commonly used components of feedback systems. These IQC can then be reused in a modular approach to system analysis.

The IQC framework is closely related to multiplier-based passivity, upper bounding of structured singular values, quadratic relaxations in nonconvex optimization (including the sums of squares approach to positivity of multivariable polynomials), and other constructive techniques for handling nonlinearity and uncertainty. The IQC approach could be viewed as a unifying framework where ideas from several research directions have been combined and developed in order to obtain a flexible and powerful

* We will use "IQC" for both singular and plural forms.

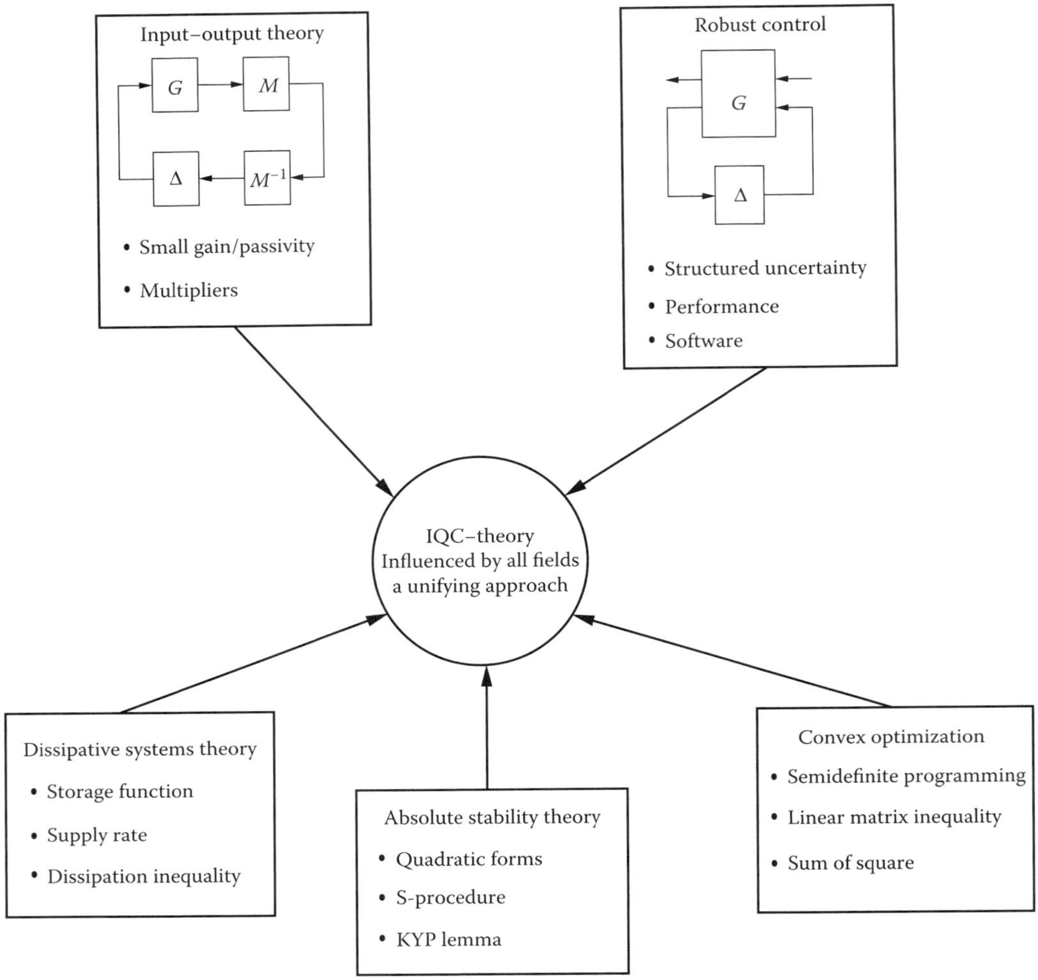

FIGURE 41.1 The IQC theory unifies powerful ideas from several important research fields: (1) The input–output theory; (2) the dissipative systems theory that was developed primarily in the west during 1960–1975; (3) The absolute stability theory that was developed in the Soviet Union during 1960–1975; (4) The robust control field from 1980–1995, and finally; and (5) tools from convex optimization that have been developed since the late 1980s.

framework that can be easily implemented as a MATLAB®-based software package. This is illustrated in Figure 41.1.

41.1.1 Notation

The real numbers are denoted by $\mathbb{R}$ and the complex numbers by $\mathbb{C}$. The subset of nonnegative real numbers is denoted $\mathbb{R}_+$. We treat elements of $\mathbb{R}^d$ (or $\mathbb{C}^d$) as d-by-1 column matrices. The notation $[x; y]$ means the column vector $\begin{bmatrix} x' & y' \end{bmatrix}'$ obtained by stacking two column vectors x and y on top of each other, and $'$ is the operation of Hermitian conjugation (transposition for real matrices).

Central to the development in this article is the use of quadratic forms.

41.1.1.1 Quadratic Forms

A *quadratic form* $\sigma : \mathbb{R}^d \mapsto \mathbb{R}$ is a function defined by $\sigma(x) = x'Px$ where P is a real d-by-d matrix (the set of all such matrices to be denoted by $\mathbb{R}^{d\times d}$). Without loss of generality, P can be assumed to be symmetric, that is, such that $P = P'$. The *Hermitian extension* $\sigma^H : \mathbb{C}^d \mapsto \mathbb{R}$ of σ is then the function defined by $\sigma(x) = x'Px$ for all $x \in \mathbb{C}^d$. Any function that has the same form as that of σ^H is called a *Hermitian form*.

The notation $\sigma \geq 0$ means that the quadratic form is positive semidefinite, that is, $\sigma(x) \geq 0$ for all $x \in \mathbb{R}^d$. This is equivalent to P being positive semidefinite, which is denoted as $P \geq 0$.

41.2 Getting Started with IQC

In this section we introduce basic elements of the IQC framework by applying it to the classical problem of global analysis of a feedback interconnection of a single-input single-output (SISO) linear time-invariant system (LTI) and a memoryless "rate limited" nonlinearity. The overall model can be described either by the block diagram

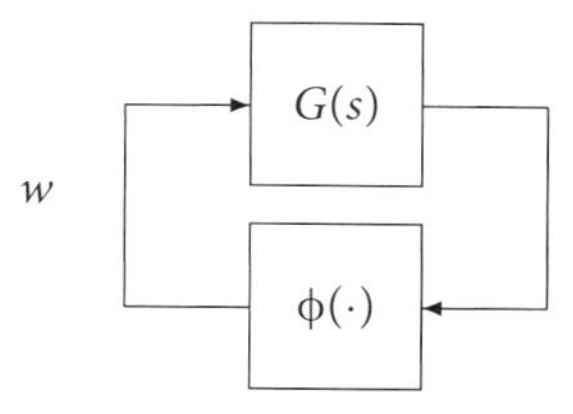

or by the n-dimensional ordinary differential equation

$$\dot{x} = Ax + Bw, \quad w = \phi(Cx), \tag{41.1}$$

where A, B, C are given real matrices with A Hurwitz (i.e., $\det(sI - A) \neq 0$ for $\mathrm{Re}(s) \geq 0$), $G(s) = C(sI - A)^{-1}B$ is the transfer function of the LTI subsystem, and $\phi : \mathbb{R} \mapsto \mathbb{R}$ is such that $\phi(0) = 0$ and $\dot{\phi} \in [0, 1]$. The analysis objective is to establish global asymptotic stability of the equilibrium $x = 0$.

41.2.1 IQC Modeling

IQC analysis begins with *IQC modeling*, which includes (1) recognizing the *exact model* $\mathcal{S}$ of the dynamical system under consideration; (2) specifying the analysis objective as an IQC to be established for $\mathcal{S}$; and (3) finding a sufficiently rich family of IQC readily known to be satisfied on $\mathcal{S}$.

To apply IQC analysis to the system in Equation 41.1, let $\mathcal{S}$ be its *behavioral model* in terms of signals w and x:

$$\mathcal{S} = \{[w; x] : \dot{x} = Ax + Bw, \; w = \phi(Cx)\}. \tag{41.2}$$

41.2.1.1 Complete and Conditional IQC

For a set $\mathcal{S} = \{q\}$ of d-dimensional signals $q = q(t)$, an IQC is defined by a quadratic form* $\sigma : \mathbb{R}^d \mapsto \mathbb{R}$ (referred to as the *supply rate* of the IQC), and a statement of the *existence* of a lower bound $\kappa = \kappa(q(0))$

* While it is possible to construct a similar *theory* utilizing general supply rates σ, requiring σ to be *quadratic* is important for practical feasibility of the approach: eventually, checking positive semidefiniteness of linear combinations of different σ will be required to reach an analysis conclusion, and quadratic functions form the only generic class for which this can be done efficiently.

for the integrals of $\sigma(q(t))$ over long intervals of time: a *complete IQC* $\sigma \succ 0$ on $\mathcal{S}$ means existence of a continuous function $\kappa : \mathbb{R}^d \mapsto \mathbb{R}_+$ such that $\kappa(0) = 0$ and

$$\int_0^T \sigma(q(t))\, dt \geq -\kappa(q(0)) \tag{41.3}$$

for all $q \in \mathcal{S}$ and $T \geq 0$, while a *conditional IQC* $\sigma \rhd 0$ on $\mathcal{S}$ states that

$$\int_0^\infty \sigma(q(t))\, dt \geq -\kappa(q(0)) \tag{41.4}$$

for all signals $q \in \mathcal{S}$ of *finite energy*.

Informally, one can think of a complete IQC $\sigma \succ 0$ as an *implicit dissipation inequality* $\sigma(q(t)) \geq dV(x_h(t))/dt$ where $x_h = x_h(t)$ is the *hidden state* of the system, and $V = V(x_h)$ is an *unknown* nonnegative storage function satisfying an *unknown** upper bound $V(x_h(0)) \leq \kappa(q(0))$. In a similar way, a conditional IQC $\sigma \rhd 0$ corresponds to the case where the storage function is not necessarily nonnegative, but satisfies $V(0) \geq 0$.

41.2.1.2 IQC as Analysis Objective and Background Information

IQC can be used to define an *objective of system analysis*. For example, since A is a Hurwitz matrix, global asymptotic stability of the equilibrium $x(0) = 0$ in Equation 41.1 is *implied* by the energy bound

$$\int_0^T |w(t)|^2 dt \leq \kappa(x(0)), \quad \forall T \geq 0, \tag{41.5}$$

where $\kappa : \mathbb{R}^n \mapsto \mathbb{R}_+$ is continuous and such that $\kappa(0) = 0$. By definition, Equation 41.5 is a complete IQC $\sigma_* \succ 0$ on $\mathcal{S}$, where

$$\sigma_*([w; x]) = -|w|^2 \quad (w \in \mathbb{R},\ x \in \mathbb{R}^n). \tag{41.6}$$

In the standard IQC analysis flow, the main effort in proving $\sigma_* \succ 0$ is devoted to finding a quadratic form σ_0 such that $\sigma_0 \rhd 0$ and $\sigma_* \geq \sigma_0$. This is accomplished by compiling a large set $\Lambda_0 = \{\sigma\}$ of quadratic forms for which the IQC $\sigma \rhd 0$ are either trivial or readily established, and then optimizing over the convex hull Λ of Λ_0 to satisfy $\sigma_* \geq \sigma$ for some $\sigma \in \Lambda$. Thus, the IQC $\sigma \rhd 0$ for $\sigma \in \Lambda_0$ become elementary pieces of "background information" about $\mathcal{S}$, and success of IQC analysis depends on one's ability to generate a set Λ_0 which is rich enough.

41.2.1.3 Sector IQC

Since $\phi(0) = 0$ and $\dot{\phi} \in [0, 1]$, $\phi(v)$ is between 0 and v for all $v \in \mathbb{R}$, that is, the points $[v; w] \in \mathbb{R}^2$ such that $w = \phi(v)$ lies in the sector $w(v - w) \geq 0$.

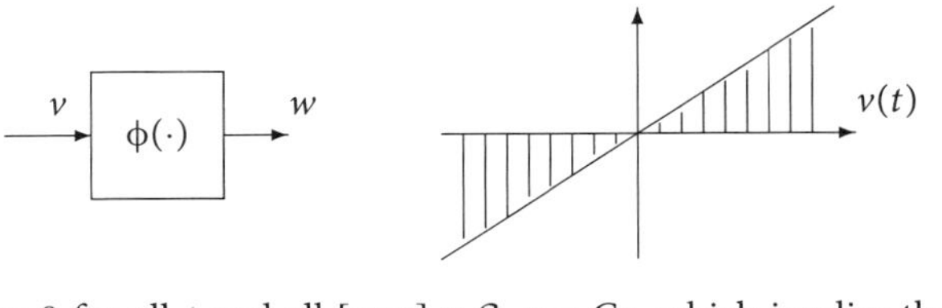

Hence, $w(t)[v(t) - w(t)] \geq 0$ for all t and all $[w; x] \in \mathcal{S}$, $v = Cx$, which implies that the complete IQC $\sigma_s \succ 0$ is satisfied on $\mathcal{S}$ for

$$\sigma_s([w; x]) = 2w(Cx - w) \quad (w \in \mathbb{R},\ x \in \mathbb{R}^n). \tag{41.7}$$

The corresponding lower bound $\kappa = \kappa_s$ can be chosen as $\kappa_s \equiv 0$.

* In most applications, it is possible to have *explicit* upper bounds $\kappa(\cdot)$, which can be useful in performing advanced analysis tasks.

41.2.1.4 Popov IQC

The sector IQC $\sigma_s \succ 0$ does not reflect the time-invariant nature of the relation between $v = Cx$ and w: it is still satisfied when the equation $w(t) = \phi(v(t))$ in the definition of $\mathcal{S}$ is replaced by $w(t) = \alpha(t)^{-1}\phi(\alpha(t)v(t))$, where $\alpha = \alpha(t)$ is an arbitrary time-varying nonzero coefficient. Two less obvious IQC can be derived from the observation that for every $[w; x] \in \mathcal{S}$ and $v = Cx$ we have

$$w\dot{v} = \frac{d}{dt}\psi(v), \quad \text{where } \psi(h) = \int_0^h \phi(\tau)\, d\tau.$$

Since, due to $\phi(0) = 0$ and $\dot{\phi} \in [0, 1]$, we have $0 \le \psi(v) \le v^2/2$ for all $v \in \mathbb{R}$, the IQC $\sigma_p \succ 0$ (with $\kappa([w; x]) = |Cx|^2$ for $w \in \mathbb{R}$, $x \in \mathbb{R}^n$) and $-\sigma_p \rhd 0$ (with $\kappa \equiv 0$) hold on $\mathcal{S}$ for

$$\sigma_p([w; x]) = 2wC(Ax + Bw) \quad (w \in \mathbb{R},\ x \in \mathbb{R}^n). \tag{41.8}$$

41.2.1.5 Pure Integrator IQC

Since, due to Equation 41.1, $Ax + Bw$ is the derivative of x, for every symmetric real matrix $Q = Q'$ the signal $2x'Q(Ax + Bw)$ is the derivative of $x'Qx$. Hence, both $\sigma_I \rhd 0$ on $\mathcal{S}$ for all $Q = Q'$ and $\sigma_I \succ 0$ on $\mathcal{S}$ for all $Q = Q' \ge 0$, where

$$\sigma_I([w; x]) = 2x'Q(Ax + Bw) \quad (w \in \mathbb{R},\ x \in \mathbb{R}^n), \tag{41.9}$$

with $\kappa([w; x]) = x'Qx$.

41.2.1.6 Zames–Falb IQC

To give a taste of less obvious IQC relations describing the memoryless nonlinearity of system in Equation 41.1, consider the following IQC from a (much larger) family established by a classical result by G. Zames and P. Falb (see Section 41.4.2 for details). Define an *extension* $\mathcal{S}_e$ of $\mathcal{S}$ by

$$\mathcal{S}_e = \{[w; x; \xi; \dot{\xi}] :\ [w; x] \in \mathcal{S},\ \ddot{\xi} + 3\dot{\xi} + 2\xi = 2w,\ \dot{\xi}(0) = \xi(0) = 0\}.$$

It turns out that, due to the rate limit $\dot{\phi} \in [0, 1]$, the complete IQC $\sigma_z \succ 0$ is satisfied on $\mathcal{S}_e$, where

$$\sigma_z([w; x; \xi_0; \xi_1]) = 2(Cx - w)(w - \xi_0) \quad (w, \xi_0, \xi_1 \in \mathbb{R}, x \in \mathbb{R}^n), \tag{41.10}$$

with $\kappa \equiv 0$.

This IQC has an interesting interpretation in the frequency domain. Assume that $v = Cx$ has finite energy, and let $y = v - w = v - \phi(v)$. It can then be shown that the IQC $\sigma_z \rhd 0$ claims that, due to the rate limit imposed on the memoryless nonlinearity ϕ, the integral of $\text{Re}[\tilde{w}(j\omega)'G_e(j\omega)\tilde{y}(j\omega)]$, where $G_e(s) = (s^2 + 3s)/(s^2 + 3s + 2)$ and $\tilde{w}$, $\tilde{y}$ denote the Fourier transforms, is also not negative. In other words, the IQC $\sigma_z \rhd 0$ sets a limit on the amount of harmonic distortion, which can be introduced by the nonlinear transformation $v \mapsto \phi(v)$. Note that this frequency domain interpretation is far from trivial since in general, the time domain relation $w(t)y(t) \ge 0$ does not imply the corresponding inequality $\text{Re}[\tilde{w}(j\omega)'\tilde{y}(j\omega)] \ge 0$ for the Fourier transforms.

While it can be shown that the IQC $\sigma_z \succ 0$ indicates *existence* of a nonnegative storage function $V_z = V_z(\dot{\xi}, \xi)$ such that $(Cx - w)(w - 2\xi) \ge dV_z/dt$, no explicit form for V_z is available.* Despite its "implicit" nature, the IQC $\sigma_z \succ 0$, can be very useful in analysis of specific nonlinear systems.

---

* To the best of authors knowledge, the only available proof is based on a converse storage function argument, and, as such, is not very constructive.

Note that applying the "extension" $\mathcal{S} \mapsto \mathcal{S}_e$ does not remove any of the original IQC: the sector and Popov IQC $\sigma_s \succ 0$, $\sigma_p \succ 0$, $-\sigma_p \rhd 0$ are still valid on $\mathcal{S}_e$*, the "objective" IQC $\sigma_* \succ 0$ has the same meaning on $\mathcal{S}$ and $\mathcal{S}_e$, and the family of "pure integration" IQC $\sigma_I \rhd 0$ generalizes to $\sigma_{Ie} \rhd 0$, where

$$\sigma_{Ie}([w; x; \xi_0; \xi_1]) = 2 \begin{bmatrix} x \\ \xi_0 \\ \xi_1 \end{bmatrix}' Q_e \begin{bmatrix} Ax + Bw \\ \xi_1 \\ -2\xi_0 - 3\xi_1 + 2w \end{bmatrix}, \quad Q_e = Q_e'. \tag{41.11}$$

41.2.1.7 Combining IQC

As long as *linear* operations are concerned, IQC can be handled as usual inequalities: if $\sigma_1 \rhd 0$ and $\sigma_2 \rhd 0$ on $\mathcal{S}$ then $c_1\sigma_1 + c_2\sigma_2 \rhd 0$ on $\mathcal{S}$ for arbitrary nonnegative real numbers c_i. Similarly, if $\sigma_1 \succ 0$ and $\sigma_2 \succ 0$ then $c_1\sigma_1 + c_2\sigma_2 \succ 0$. This allows the user of the IQC framework to convert individual IQC satisfied for subsystems of a larger model into one convex set $\Lambda = \{\sigma\}$ of quadratic forms σ defining valid conditional IQC $\sigma \rhd 0$ on $\mathcal{S}$. In addition, since many of the classical IQC are readily shown to be complete, a convex subset $\Lambda_+ \subset \Lambda$ such that $\sigma \succ 0$ for all $\sigma \in \Lambda_+$ can be constructed as well.

In particular, for the behavioral model $\mathcal{S}$ defined by Equation 41.1, the set Λ compiled so far is given by

$$\Lambda = \{c_s\sigma_s + c_p\sigma_p + \sigma_I : \ c_s \in \mathbb{R}_+, \ c_p \in \mathbb{R}, \ Q = Q'\}, \tag{41.12}$$

where $\sigma_s, \sigma_p, \sigma_I$ are defined in Equations 41.7 through 41.9. In contrast, the set Λ_+ of *recognized* complete IQC consists of $\sigma = c_s\sigma_s + c_p\sigma_p + \sigma_I$ with $c_s \geq 0$, $c_p \geq 0$, $Q = Q' \geq 0$. It is important to understand that, depending on the coefficient matrices A, B, C, there could be some $\sigma \in \Lambda$ such that $\sigma \succ 0$ on $\mathcal{S}$ but $\sigma \notin \Lambda_+$: those are the quadratic forms defining valid complete IQC which are not recognized as such.

Similarly, for the extended model $\mathcal{S}_e$,

$$\Lambda = \{c_s\sigma_s + c_p\sigma_p + \sigma_{Ie} + c_z\sigma_z : \ c_s, c_z \in \mathbb{R}_+, \ c_p \in \mathbb{R}, \ Q_e = Q_e'\}, \tag{41.13}$$

with σ_{Ie} defined in Equation 41.11, and the forms from Λ_+ satisfy the additional constraints $c_p \geq 0$ and $Q_e \geq 0$.

41.2.2 Feasibility Optimization and Postfeasibility Analysis

Once the exact model $\mathcal{S}$ is defined, the objective IQC σ_* is selected, and a convex set $\Lambda = \{\sigma\}$ of valid IQC $\sigma \rhd 0$ on $\mathcal{S}$ is compiled, the next step in the IQC framework is *feasibility analysis*, understood as the search for $\sigma_0 \in \Lambda$ satisfying $\sigma_* \geq \sigma_0$. In a typical application, there is also a "cost" parameter to be minimized (in the IQC model for Equation 41.1, this could be γ, subject to $\gamma I \geq |c_p|C'C + Q$, bounding κ in Equation 41.5), so there is a well-defined optimization criterion in addition to the feasibility task.

If there is no $\sigma_0 \in \Lambda$ such that $\sigma_* \geq \sigma_0$, the IQC analysis fails, and the only option is to return to the IQC modeling step to find a larger set Λ. Sometimes it helps to consider (as in the Zames–Falb IQC example) replacing $\mathcal{S}$ with an *extension* $\mathcal{S}_e$ of $\mathcal{S}$: a cleverly defined set $\mathcal{S}_e = \{q_e\}$ of signals of fixed dimension $d_e > d$ such that for every $q \in \mathcal{S}$ there is at least one signal g such that $[q; g] \in \mathcal{S}_e$.

Having $\sigma_* \geq \sigma_0$ for some $\sigma_0 \in \Lambda$ means that $\sigma_* \rhd 0$, that is, either $\sigma_* \succ 0$, as desired, or $\sigma_* \rhd 0$ but $\sigma_* \not\succ 0$. In most applications, the latter means that the system is strongly unstable: subject to mild assumptions, the conditional IQC $\sigma_* \rhd 0$ on $\mathcal{S}$ is a "certificate of dichotomy," that is, it implies that $\mathcal{S}$ is either quite stable or very unstable, no middle ground. In most situations, general theorems of *postfeasibility analysis* provide easy criteria to decide which outcome actually takes place.

---

* Here we allow some abuse of notation, using the same identifier for a quadratic form $\sigma : \mathbb{R}^d \mapsto \mathbb{R}$ and its "extension" $\tilde{\sigma} : \mathbb{R}^{d+2} \mapsto \mathbb{R}$ defined by $\tilde{\sigma}([w; x; \xi_0; \xi_1]) = \sigma([w; x])$.

41.2.2.1 Feasibility Optimization and Semidefinite Programming

When the set Λ is linearly parameterized by a vector variable which is in turn subject to a linear matrix inequality (LMI) constraint, the search for $\sigma_0 \in \Lambda$ satisfying $\sigma_* \geq \sigma_0$ becomes a *semidefinite program*.

For example, in the analysis of Equation 41.1 with Λ defined by Equation 41.12, one possible semidefinite program to solve is: min γ subject to $\gamma I \geq \pm c_p C'C + Q$, $c_s \geq 0$, and

$$\begin{bmatrix} 2c_s - 1 - 2c_p CB & -c_s C - c_p CA - B'Q \\ -c_s C' - c_p A'C' - QB & -QA - A'Q \end{bmatrix} \geq 0, \tag{41.14}$$

where the decision variables are $\gamma, c_p, c_s \in \mathbb{R}$ and $Q = Q' \in \mathbb{R}^{n\times n}$. While the LMI in Equation 41.14 looks quite cumbersome, the quadratic form notation is usually much more straightforward. For example, the functional version of Equation 41.14 is

$$-|w|^2 - 2c_s w(Cx - w) - 2c_p wC(Ax + Bw) - 2x'Q(Ax + Bw) \geq 0 \; \forall \; w, x.$$

Certain programming tools can be used to convert quadratic form notation into the format of the standard semidefinite program solvers [6].

41.2.2.2 Feasibility Optimization and the Kalman–Yakubovich–Popov Lemma

While conversion to a semidefinite program, to be solved numerically, appears to be the only practical approach to feasibility analysis of advanced IQC models, valuable theoretical insight can be gained by applying the following version of the classical *Kalman–Yakubovich–Popov (KYP) Lemma* (sometimes also called the *positive real Lemma*).

Theorem 41.1:

For real matrices A, B of dimensions n-by-n and n-by-m respectively, and a Hermitian form $\sigma : \mathbb{C}^{m+n} \mapsto \mathbb{R}$ with real coefficients, let Ω be the set of $\omega \in \mathbb{R}$ such that the matrix $A_\omega = j\omega I_n - A$ is invertible. For $\omega \in \Omega$ let $L_\omega = A_\omega^{-1}B$. Consider the following statements:

a. $\sigma([u; L_\omega u]) \geq 0$ *for all* $u \in \mathbb{C}^m$ *and* $\omega \in \Omega$
b. *There exists* $\omega \in \Omega$ *such that* $\sigma([u; L_\omega u]) > 0$ *for all* $u \in \mathbb{C}^m$, $u \neq 0$
c. *The quadratic form* $\sigma([u; x]) - 2x'Q(Ax + Bu)$ *(where* $x \in \mathbb{R}^n$ *and* $u \in \mathbb{R}^m$*) is positive semidefinite for some real matrix* $Q = Q'$

Then

i. *(c) implies (a)*
ii. *when the pair (A, B) is controllable, (a) implies (c)*
iii. *when the pair (A, B) is stabilizable, (a) and (b) together imply (c)*

Theorem 41.1 can be used to formulate frequency domain conditions of feasibility in IQC analysis when the "pure integration" term $2x'Q(Ax + Bu)$ is present in the description of Λ.

41.2.2.3 The Circle Criterion

In the special case where only the pure integrator and sector IQC are used to represent system Equation 41.1, the feasibility analysis calls for finding $c_s \in \mathbb{R}_+$ and $Q = Q'$ such that

$$-|w|^2 - 2c_s w(Cx - w) - 2x'Q(Ax + Bw) \geq 0 \quad \forall \; w \in \mathbb{R}, \; x \in \mathbb{R}^n. \tag{41.15}$$

Theorem 41.1 offers valuable insight into the feasibility of the associate semidefinite program: together with the identity $CL_\omega = G(j\omega)$, it allows one to conclude that $Q = Q'$ satisfying Equation 41.15 exists for some $c_s \in \mathbb{R}_+$ if and only if (assuming A is Hurwitz and (A, B) is controllable)

$$-|w|^2 - c_s\text{Re}[G(j\omega) - 1]|w|^2 \geq 0, \quad \forall w \in \mathbb{C}, \omega \in \mathbb{R}$$

Since c_s can be taken arbitrarily large, it can be seen that the frequency domain condition holds if $\rho < 1$, where $\rho = \sup_{\omega \in \mathbb{R}} \text{Re}[G(j\omega)]$ (the minimal upper bound of $\text{Re}[G(j\omega)]$ over $\omega \in \mathbb{R}$).

When $\rho \geq 1$, the IQC analysis proves nothing: system in Equation 41.1 may be stable or unstable depending on fine details not reflected in the IQC model used so far. When $\rho < 1$, we have $\sigma_* \geq \sigma_0$ where $\sigma_0 = c_s\sigma_s + \sigma_I$ for some $c_s > 0$ and $Q = Q'$. Since $\sigma_s \succ 0$ on $\mathcal{S}$ and $\sigma_I \succ 0$ on $\mathcal{S}$ for $Q \geq 0$, the desired complete IQC $\sigma_* \succ 0$ will be established as long as it is possible to guarantee that $Q = Q'$ is positive semidefinite. When Q is found numerically by solving a semidefinite program, the inequality $Q \geq 0$ can be checked directly. Otherwise, checking that $Q \geq 0$ becomes a typical *postfeasibility analysis* task.

Substituting $w = 0$ into the inequality in Equation 41.15 yields $QA + A'Q \leq 0$. Since A is a Hurwitz matrix, which in turn implies $Q \geq 0$, recovering the CIRCLE CRITERION: *system in Equation 41.1 is globally asymptotically stable when* $\text{Re}[G(j\omega)] < 1$ *for all* $\omega \in \mathbb{R} \cup \{\infty\}$.

41.2.2.4 The Popov Criterion

In the case where the pure integrator, sector, and Popov IQC are used together to represent system in Equation 41.1, the feasibility analysis calls for finding $c_s \in \mathbb{R}_+$, $c_p \in \mathbb{R}$, and $Q = Q'$ such that

$$-|w|^2 - 2c_s w(Cx - w) - 2c_p wC(Ax + Bw) - 2x'Q(Ax + Bw) \geq 0 \quad \forall\, w \in \mathbb{R}, x \in \mathbb{R}^n. \tag{41.16}$$

Theorem 41.1 can be used to show that existence of $c_s \in \mathbb{R}_+$, $c_p \in \mathbb{R}$, $Q = Q' \in \mathbb{R}^{n\times n}$ satisfying in Equation 41.16 is equivalent to existence of $p \in \mathbb{R}$ such that $\rho_p < 1$, where ρ_p is the minimal upper bound of $\text{Re}[(1 + j\omega p)G(j\omega)]$ over $\omega \in \mathbb{R}$ (when $\rho_p < 1$, one can use $c_p = pc_s$ with $c_s > 0$ large enough).

Since the storage function for the IQC $\sigma_s + p\sigma_p + \sigma_I \rhd 0$ is $V(x) = x'Qx + 2p\psi(Cx)$, the desired IQC $\sigma_* \succ 0$ is established when nonnegativity of V is assured. Since Equation 41.16 with $w = 0$ implies $QA + A'Q \leq 0$, we know that $Q \geq 0$ is positive semidefinite. Since $\psi(y) \geq 0$ for all $y \in \mathbb{R}$, this implies $V \geq 0$ when $p \geq 0$.

The postfeasibility analysis becomes trickier in the case $p \leq 0$. The upper bound $\psi(y) \leq y^2/2$ can be used to show that $V \geq 0$ whenever $Q + c_pC'C \geq 0$. However, this special treatment is not necessary, as, according to the general postfeasibility analysis theorems discussed in Section 41.3.2, the inequality $\sigma_* \geq c_s\sigma_s + c_p\sigma_p + \sigma_I$ implies the complete IQC $c_s\sigma_s + c_p\sigma_p + \sigma_I \succ 0$ whenever $c_s \geq 0$ and A is a Hurwitz matrix. This establishes the classical POPOV CRITERION: *system in Equation 41.1 is globally asymptotically stable when there exists* $p \in \mathbb{R}$ *such that* $\text{Re}[(1 + pj\omega)G(j\omega)] < 1$ *for all* $\omega \in \mathbb{R} \cup \{\infty\}$.

41.3 Theory of IQC Analysis

In general, IQC analysis follows the steps highlighted in Section 41.2 (IQC modeling, feasibility optimization, postfeasibility analysis). This section gives a formal presentation of the framework.

41.3.1 IQC Modeling

An IQC model consists of a *system* (understood as a set $\mathcal{S}$ of signals of fixed dimension d) and an *analysis objective* in terms of IQC, where each IQC is a property of $\mathcal{S}$ defined by a quadratic form σ on $\mathbb{R}^d$.

41.3.1.1 Signals

A *continuous time* (CT) signal of dimension d is an element of the set $\mathcal{L}^d$ of all measurable locally bounded functions $q: [0, \infty) \mapsto \mathbb{R}^d$ (we also use $\mathcal{L}$ for $\mathcal{L}^d$ with $d = 1$).

We define $\dot{q} = dq/dt$ for $q \in \mathcal{L}^d$ (in which case the constraint $dq/dt \in \mathcal{L}^d$ is imposed automatically). We use $\mathcal{I}[q]$ to denote the total integral of q:

$$\mathcal{I}[q] = \int_0^\infty q(t)\,dt \quad \text{for } q \in \mathcal{L}^d$$

($\mathcal{I}[q]$ is not defined for some q, and it is possible to have $\mathcal{I}[q] = \infty$ or $\mathcal{I}[q] = -\infty$). We use $\|q\|$ to denote the square root of the *energy* $\|q\|^2 = \mathcal{I}[|q|^2]$ of signal q, and denote by $\mathcal{L}_0^d$ the set of all finite energy signals in $\mathcal{L}^d$. Given $T \geq 0$ and $q \in \mathcal{L}^d$, $p = [q]_T \in \mathcal{L}_0^d$ denotes the *past history* of q: $p(t) = q(t)$ for $t \leq T$, $p(t) = 0$ for $t > T$.

When $\sigma : \mathbb{R}^d \mapsto \mathbb{R}$ is a quadratic form, we use $\sigma\langle q\rangle$ as a shortcut notation for $\mathcal{I}[\sigma(q)]$ when $q \in \mathcal{L}_0^d$. In the special case $\sigma(x) = |x|^2$ ($x \in \mathbb{R}^d$), we use $\|x\|_T^2$ in place of $\sigma\langle[x]_T\rangle$ for $x \in \mathcal{L}^d$ and $T \geq 0$.

41.3.1.2 Systems

Mathematically, we view a general system as a *behavioral model*, that is, simply a set $\mathcal{S}$ of signals of given dimension. The elements of $\mathcal{S}$ are assumed to represent all possible combinations of all signals of interest (the "outputs" may include components of the *hidden state* as well as auxiliary signals introduced specifically to simplify the analysis). Accordingly, a system is a subset of $\mathcal{L}^d$, where d is a positive integer.

We also consider the special case of *operator* models, which are functions $\Delta : \mathcal{L}^k \mapsto \mathcal{L}^m$ mapping signals to signals. An operator model Δ can be described by a behavioral model defined as the *graph* $\mathcal{G}_\Delta = \{[\Delta(v); v] : v \in \mathcal{L}^k\}$ of Δ. The operator model Δ is called *causal* when $[\Delta(v)]_T$ is completely defined by $[v]_T$ for all $T \geq 0$, that is, $[\Delta(v_1)]_T = [\Delta(v_2)]_T$ whenever $[v_1]_T = [v_2]_T$. The operator model Δ is called *stable* (in the "bounded input energy implies bounded output energy" sense) if $\|\Delta(v)\| < \infty$ whenever $\|v\| < \infty$.

In particular, a *finite order LTI* operator defined by real matrices A, B, C, D is a causal operator model mapping v to $y = Cx + Dv \in \mathcal{L}^k$ where x is defined by $\dot{x} = Ax + Bv$ and $x(0) = 0$. An LTI operator is completely defined by its *transfer matrix* $G(s) = D + C(sI - A)^{-1}B$, an element of the set $RL^{k\times m}$ of all real rational k-by-m matrix functions of scalar complex argument s which are *proper* (i.e., bounded as $|s| \to \infty$).

41.3.1.3 Definition of IQC

Let $\sigma : \mathbb{R}^d \mapsto \mathbb{R}$ be a quadratic form. We say that system $\mathcal{S} \subset \mathcal{L}^d$ *satisfies the conditional* IQC *defined by* σ (shortcut notation $\sigma \rhd 0$ or $\sigma(q) \rhd 0$ when it is more convenient) if there exists a continuous function $\kappa : \mathbb{R}^d \mapsto \mathbb{R}_+$, such that $\kappa(0) = 0$ and

$$\sigma\langle q\rangle \geq -\kappa(q(0)) \quad \text{whenever } q \in \mathcal{S}, \quad \|q\| < \infty. \tag{41.17}$$

We say that system $\mathcal{S}$ *satisfies the complete* IQC *defined by* σ (shortcut notation $\sigma \succ 0$ or $\sigma(q) \succ 0$) if there exists a continuous function $\kappa : \mathbb{R}^d \mapsto \mathbb{R}_+$, such that $\kappa(0) = 0$ and

$$\sigma\langle[q]_T\rangle \geq -\kappa(q(0)) \quad \text{whenever } q \in \mathcal{S}, \quad T \geq 0. \tag{41.18}$$

41.3.1.4 Extended Systems and IQC

Introducing new signals into consideration (an action referred to as *extension* here) frequently helps to expose IQC relations in the original model. In general, the behavioral model $\mathcal{S} \subset \mathcal{L}^{d+N}$ is called an extension of $\mathcal{S}_0 \subset \mathcal{L}^d$ if for every $q_0 \in \mathcal{S}_0$ there exists $g \in \mathcal{L}^N$ such that $[q_0; g] \in \mathcal{S}$.

Example 41.1:

Consider the behavioral model $\mathcal{S}_0 = \{[u^3; u^2] : u \in \mathcal{L}\}$ (it is the graph of the memoryless nonlinear operator $v \mapsto v^{2/3}$). It can be shown that every quadratic form $\sigma : \mathbb{R}^2 \mapsto \mathbb{R}$ such that $\sigma \rhd 0$ on $\mathcal{S}_0$ is positive semidefinite, that is, $\mathcal{S}_0$ does not satisfy any nontrivial IQC. The extension $\mathcal{S} = \{[u^3; u^2; u; 1] : u \in \mathcal{L}\}$, however, satisfies a set of useful IQC $\sigma \succ 0$, where

$$\sigma([x_3; x_2; x_1; x_0]) = c_1(x_0x_2 - x_1^2) + c_2(x_0x_3 - x_1x_2) + c_3(x_1x_3 - x_2^2),$$

and the coefficients $c_i \in \mathbb{R}$ are arbitrary.

41.3.1.5 Stable LTI Extensions and Frequency Weighted IQC

The situation in which an extension $\mathcal{S}$ of $\mathcal{S}_0$ is defined by a stable LTI operator E according to $\mathcal{S} = \{[q_0; Eq_0] : q_0 \in \mathcal{S}_0\}$ is of special importance in the IQC framework. Allowing some abuse of notation, let $E = E(s)$ also denote the transfer matrix of E. Assume that $\sigma \rhd 0$ on $\mathcal{S}$ for some quadratic form $\sigma : \mathbb{R}^{d+N} \mapsto \mathbb{R}$.

Let $\Pi = \mathcal{P}\{\sigma, E\} \in RL^{d\times d}$ be the transfer matrix defined (uniquely) by the identities

$$\sigma^H([u; E(j\omega)u]) = u'\Pi(j\omega)u, \quad \Pi(j\omega) = \Pi(j\omega)' \quad \forall\, u \in \mathbb{C}^d,\ \omega \in \mathbb{R},$$

where σ^H denotes the Hermitian extension of σ. The IQC $\sigma\langle q\rangle \rhd 0$ on the extended system $\mathcal{S}$ can be interpreted as a *frequency domain weighted* IQC on the imaginary axis (shortcut notation $\sigma^H([q_0; E(j\omega)q_0]) \rhd 0$ or $q_0'\Pi(j\omega)q_0 \rhd 0$). Indeed, due to the Parseval identity, $\sigma\langle q\rangle$ can be represented as the integral

$$\int_{-\infty}^{\infty} \tilde{q}_0(j\omega)'\Pi(j\omega)\tilde{q}_0(j\omega)\, d\omega$$

for all $q_0 \in \mathcal{L}_0^d$, $q = [q_0; Eq_0]$, where $\tilde{q}_0$ denotes the Fourier transform of $q_0 \in \mathcal{L}_0^d$. While Π is uniquely defined by E and σ, a single Π corresponds to a variety of pairs (E, σ). All such "realizations" are equivalent in the framework of conditional IQC.

41.3.1.6 Extended IQC for I/O System and State-Space Interpretation

In the most common scenario, extended IQC are derived for *input/output relations*, that is, for behavioral models of the form

$$\mathcal{S}_0 = \{q_0 = [w; v] : v \in \mathcal{L}^k,\ w \in \mathcal{L}^m,\ w = \Delta(v)\},$$

where $\Delta : \mathcal{L}^k \mapsto \mathcal{L}^m$ is a causal and stable operator. Suppose the transfer matrix E in Section 41.3.1.5 has a state-space realization $E(s) = C_e(sI - A_e)^{-1}\begin{bmatrix} B_{e1} & B_{e2}\end{bmatrix} + \begin{bmatrix} D_{e1} & D_{e2}\end{bmatrix}$, where $A_e, B_{e1}, B_{e2}, C_e, D_{e1}, D_{e2}$ are fixed real matrices, and A_e is a Hurwitz N-by-N matrix. Then the stable dynamical extension $\mathcal{S}$ can alternatively be represented as

$$\mathcal{S}_{ss} = \{q = [w; v; x_e] \in \mathcal{L}^{k+m+N} :\ w = \Delta(v),\ \dot{x}_e = A_ex_e + B_{e1}v + B_{e2}w,\ x_e(0) = 0\}, \qquad (41.19)$$

(see Figure 41.2 for an illustration) and the corresponding quadratic form is

$$\sigma_{ss}(q) := \sigma([w; v; C_ex_e + D_{e1}v + D_{e2}w])$$

Clearly $\sigma_{ss} \rhd 0$ on $\mathcal{S}_{ss}$ if and only if $\sigma_{ss}^H([q_0; E(j\omega)q_0]) \rhd 0$ on $\mathcal{S}_0$.

For example, in the special case of the Zames–Falb IQC in Section 41.2.1.6, the IQC $\sigma_z \rhd 0$ on $\mathcal{S}_e$ can be expressed as a frequency-weighted IQC $(v - w)(w - H(j\omega)w) \rhd 0$, where $H(s) = 2/(s^2 + 3s + 2)$, relating signals $v = Cx \in \mathcal{L}_0$ and $w = \phi(y) \in \mathcal{L}_0$. In this case, the explicit expression for Π is

$$\Pi(j\omega) = \begin{bmatrix} 0 & 1 - H(j\omega) \\ 1 - H(j\omega)' & 2\mathrm{Re}[1 - H(j\omega)] \end{bmatrix}.$$

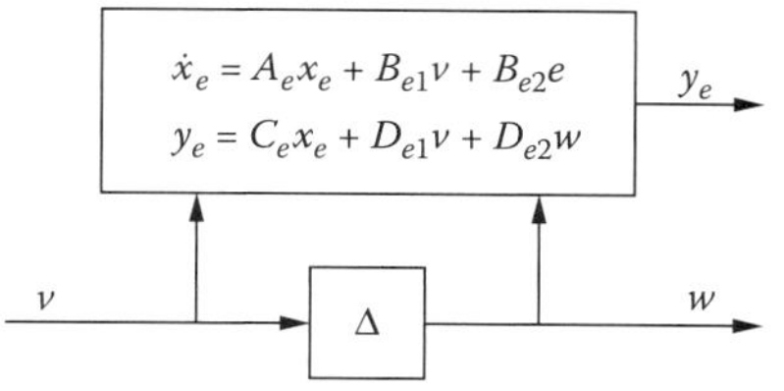

FIGURE 41.2 Extended IQC setup for I/O system Δ.

41.3.2 Feasibility Optimization

This section briefly discusses the problem of feasibility optimization. The attention is restricted to a special system structure that frequently appears in applications.

41.3.2.1 Stable Operator Feedback Setup

An important class of IQC analysis scenarios is given by the *stable operator feedback setup* which calls for certifying the complete IQC $r|f|^2 - |e|^2 \succ 0$ (with $r \geq 0$ being as small as possible) in the feedback interconnection of a stable causal operator Δ (input $v \in \mathcal{L}^k$, output $w \in \mathcal{L}^m$) and a stable LTI system M (input $[f; w] \in \mathcal{L}^{l+m}$, output $[e; v] \in \mathcal{L}^{d+k}$, state $x \in \mathcal{L}^n$), assuming the interconnection satisfies the IQC $\sigma_\Delta\langle[f; w; x]\rangle \rhd 0$.

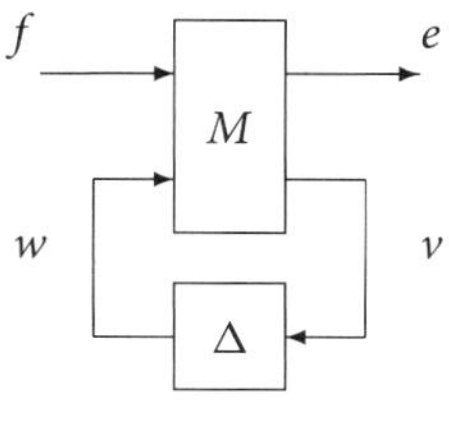

The IQC $\sigma_\Delta \rhd 0$ is interpreted as one describing Δ. While a typical IQC model will have multiple IQC for every subsystem, the single IQC setup is general enough to formulate abstract statements relevant to feasibility optimization and postfeasibility analysis.

The quadratic form $\sigma_\Delta : \mathbb{R}^{k+m+n} \mapsto \mathbb{R}$, as well as the real coefficient matrices A, B_i, C_i, D_{ij} (where the pair $(A, [B_1, B_2])$ is controllable) in the equations

$$M: \begin{bmatrix} e \\ v \\ \dot{x} \end{bmatrix} = \begin{bmatrix} D_{11} & D_{12} & C_1 \\ D_{21} & D_{22} & C_2 \\ B_1 & B_2 & A \end{bmatrix} \begin{bmatrix} f \\ w \\ x \end{bmatrix}, \quad x(0) \in \mathcal{X}_0 \tag{41.20}$$

are given, while the feedback operator Δ and the (nonempty) set of initial conditions $\mathcal{X}_0 \subset \mathbb{R}^n$ are not expected to be known in detail.

The associated IQC model is given by the triplet $(\mathcal{S}, \sigma_*, \Lambda)$, where $\mathcal{S} = \{q\}$ is the behavioral model consisting of all signals $q = [f; w; x]$, satisfying the Equation 41.20 and $w = \Delta(v)$ for some e, v,

$$\sigma_*([f; w; x]) = r|f|^2 - |C_1 x + D_{11} f + D_{12} w|^2 \tag{41.21}$$

is the parameterized "analysis objective" quadratic form, and

$$\Lambda = \{\sigma = c\sigma_\Delta + \sigma_I : \quad c \geq 0, \; Q = Q' \in \mathbb{R}^{n \times n}\}$$

where $\sigma_I = \sigma_I([f; w; x]) = 2x'Q(Ax + B_1 f + B_2 w)$ ("pure integrator" IQC). This is the set of quadratic forms σ for which the IQC $\sigma \rhd 0$ on $\mathcal{S}$ is readily established. In particular, as mentioned in Section 41.2.1.5,

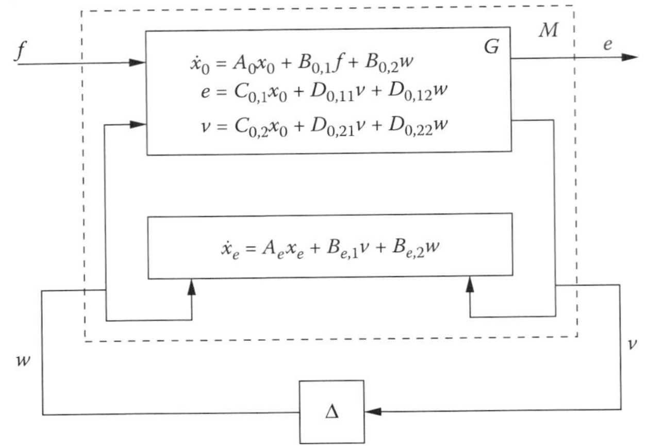

FIGURE 41.3 Illustration of how the state-space extension contributes to the dynamics of M. The extended state vector does not affect the signals in the feedback loop (e, v, w). It is only used in the IQC that describes Δ.

for continuous time signals of finite energy, the relation $\dot{x} = Ax + B_1 f + B_2 w$ implies $\sigma_I\langle[f; w; x]\rangle \rhd 0$ for every $Q = Q'$. Moreover, $\sigma_I\langle[f; w; x]\rangle \succ 0$ when $Q \geq 0$ is positive semidefinite.

The LTI system M in general includes a nominal LTI part of the system under investigation as well as a stable state-space extension used for the IQC modeling. As an example, consider Figure 41.3, where a linear system G is interconnected with Δ. By using an extended IQC description as in Figure 41.2 we obtain an LTI model M with state vector $x = [x_0; x_e]$. The IQC $\sigma_\Delta\langle[f; w; x]\rangle \rhd 0$ on $\mathcal{S}$ that is used to describe Δ then depends on x not only via the output $v = C_{0,2}x_0 + D_{0,21}v + D_{0,22}w$ but also via the extended state vector x_e. This IQC also has a frequency domain representation. To see this, let

$$\Gamma = \begin{bmatrix} \Gamma_{11} & \Gamma_{12} \\ \Gamma_{21} & \Gamma_{22} \end{bmatrix} = \mathcal{P}\{\sigma_\Delta, (sI - A)^{-1}[B_1, B_2]\}, \tag{41.22}$$

where the operation $\mathcal{P} = \mathcal{P}\{\sigma, E\}$ is defined in Section 41.3.1.5. Since M is assumed to be stable (i.e., A is Hurwitz), the IQC $\sigma_\Delta \rhd 0$ on $\mathcal{S}$ can be expressed in the frequency domain format as $[f; w]'\Gamma(j\omega)[f; w] \rhd 0$, as long as $\mathcal{X}_0 = \{0\}$.

41.3.2.2 Feasibility Optimization in Time and Frequency Domain

Consider again the operator feedback setup in Section 41.3.2.1. The task of finding $\sigma_0 \in \Lambda$ such that $\sigma_* \geq \sigma_0$ with r as small as possible can be delegated to an appropriate optimization engine as a semidefinite program with decision variables $c \in \mathbb{R}_+$, $Q = Q' \in \mathbb{R}^{n\times n}$, $r \in \mathbb{R}$ and objective $r \to \min$.

It is possible to find an equivalent frequency domain formulation. By the KYP Lemma (Theorem 41.1), the existence of $Q = Q' \in \mathbb{R}^{n\times n}$ such that the quadratic form

$$\sigma_*([f; w; x]) - c\sigma_\Delta(f; w; x]) - \sigma_I([f; w, x]) \geq 0,$$

is positive semidefinite is equivalent to the following condition*:

$$r|f|^2 - |M_1(j\omega)[f; w]|^2 - c[f; w]'\Gamma(j\omega)[f; w] \geq 0, \ \forall [f; w] \in \mathbb{C}^{l+m}, \quad \omega \in \mathbb{R} \tag{41.23}$$

where $M_1(s) = [M_{11}(s), M_{12}(s)] = C_1(sI - A)^{-1}[B_1, B_2] + [D_1, D_2]$, and $\Gamma(s)$ is as defined in Equation 41.22. One can verify that there exists $r, c \in \mathbb{R}_+$ such that Equation 41.23 holds if and only if the

* We use that A is Hurwitz and assume that $(A, [B_1, B_2])$ is controllable.

condition

$$\Gamma_{22}(j\omega) \leq -\epsilon(\Gamma_{12}(j\omega)'\Gamma_{12}(j\omega) + M_{12}(j\omega)'M_{12}(j\omega)) \quad \forall\, \omega \in \mathbb{R}, \tag{41.24}$$

is satisfied for some $\epsilon > 0$.

41.3.2.3 Need for Postfeasibility Analysis

In general, the existence of $c \geq 0$ and $Q = Q'$ such that $\sigma_* \geq c\sigma_\Delta + \sigma_I$ does not imply the complete IQC $\sigma_* \succ 0$. This could be due to the fact that $\sigma_\Delta \rhd 0$ but $\sigma_\Delta \not\succ 0$, or because $Q = Q'$ is not positive semidefinite. On the other hand, it is possible to have $\sigma_\Delta \not\succ 0$ and $Q \not\geq 0$ while $\sigma_\Delta + \sigma_I \succ 0$.

For example, consider the stable operator feedback setup from Section 41.3.2.1 given by state-space equations $\dot{x}_1 = -x_1 + f$, $\dot{x}_2 = -x_2 + w$, $e = v = x_1 + x_2$ with $\mathcal{X}_0 = \{0\}$:

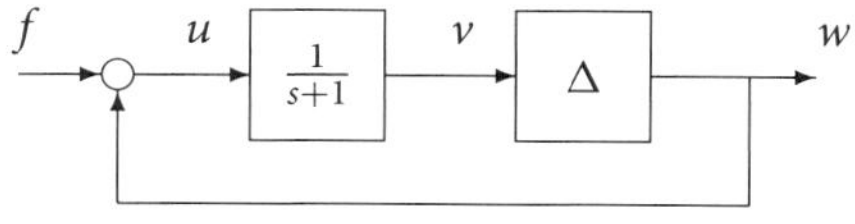

The conditional IQC $\sigma_\Delta \rhd 0$, where $\sigma_\Delta([f; w; x_1; x_2]) = |x_1|^2 - 0.25|w|^2$ (or, equivalently, $|v - \frac{w}{j\omega+1}|^2 - 0.25|w|^2 \rhd 0$) is satisfied for $\Delta = \Delta_0$ as well as for $\Delta = \Delta_1$, where $\Delta_0(v) \equiv 0$ and $\Delta_1(v) = 2v$. Also, existence of $Q = Q'$, $c \geq 0$, and $r \in \mathbb{R}$ such that $\sigma_* \geq c\sigma_\Delta + \sigma_I$ is guaranteed by the frequency domain condition (Equation 41.24), as in this case

$$\Gamma(s) = \begin{bmatrix} \frac{1}{1-s^2} & 0 \\ 0 & -0.25 \end{bmatrix}, \quad M_{ij}(s) = \frac{1}{1+s}.$$

Nevertheless, the complete IQC $\sigma_* \succ 0$ holds for $r \geq 1$ when $\Delta = \Delta_0$, but the feedback interconnection is unstable, and the power gain from f to e equals infinity, when $\Delta = \Delta_1$.

41.3.3 Postfeasibility Analysis

In this section, we briefly discuss conditions, to be imposed on a behavioral model $\mathcal{S} \subset \mathcal{L}^d$ and quadratic forms $\sigma_*, \sigma_0 : \mathbb{R}^d \mapsto \mathbb{R}$, which guarantee that conditional IQC $\sigma_0 \rhd 0$ and the inequality $\sigma_* \geq \sigma_0$ imply the complete IQC $\sigma_* \succ 0$.

Three approaches can be considered: minimal stability, the homotopy, and the minimax approaches. The later two are often convenient to apply in analysis of complex systems with many different components. This will be discussed in further detail in the next section.

The minimal stability is a simple condition* to be imposed on $\sigma_q = \sigma_* - c\sigma_\Delta$, which guarantees that $Q \geq 0$ whenever $\sigma_q \geq \sigma_I$. Then $\sigma_I \succ 0$ and $\sigma_* \succ 0$ follows as long as it is known that $\sigma_\Delta \succ 0$.

In terms of the stable operator feedback setup from Section 41.3.2.1, minimal stability means existence of real matrices K_1, K_2 such that $A_K = A + B_1K_1 + B_2K_2$ is stable (i.e., A_K is Hurwitz), and $\sigma_q(K_1x, K_2x, x) \leq 0$ for all $x \in \mathbb{R}^n$. Indeed, since $\sigma_q \geq \sigma_I$, minimal stability implies $QA_K + A_K'Q \leq 0$, hence $Q \geq 0$. Such K_1, K_2 are usually easy to find when σ_Δ has a simple structure, as in the classical theory of absolute stability.

For example, the derivation of the circle criterion in Section 41.2.2.3 uses the stable operator feedback setup with $B_1 = 0$, $B_2 = B$, $\Delta(y) = \phi(y)$, and $\sigma_\Delta([f; w; x]) = 2c_s w(Cx - w)$ satisfying $\sigma_\Delta \succ 0$. It employs a minimal stability argument, with $K_2 = 0$, to show that $Q \geq 0$. The derivation of the Popov criterion in Section 41.2.2.4 can use a similar argument, with $\sigma_\Delta([f; w; x]) = 2c_s w(Cx - w) + 2c_p wC(Ax + Bw)$,

* Invented and frequently used by V. Yakubovich.

when $c_p \geq 0$ (and hence $2c_p wC(Ax + Bw) \succ 0$). When, $c_p < 0$, σ_δ should be redefined as

$$\sigma_\delta([f; w; x]) = 2c_s w(Cx - w) + 2c_p wC(Ax + Bw) - 2c_p x'C'C(Ax + Bw).$$

Since $2c_p wC(Ax + Bw) - 2c_p x'C'C(Ax + Bw) \succ 0$ when $c_p \leq 0$, the minimal stability argument can be used, with $K_2 = C$ being the natural selection. Establishing that $A_K = A + BC$ is a Hurwitz matrix is the extra effort required in this case.

The homotopy approach does not utilize information about σ but requires the IQC $\sigma_* \rhd 0$ to be "strict" and the system to be parameter-dependent. It considers a family $\{\mathcal{S}(\tau)\}$ of behavioral models $\mathcal{S}(\tau) \subset \mathcal{L}^d$ depending continuously on parameter $\tau \in [0, 1]$ (so that $\mathcal{S}(0)$ is easy to analyze and $\mathcal{S}(1)$ is the true system of interest), and a quadratic form $\sigma_* : \mathbb{R}^d \mapsto \mathbb{R}$ such that the conditional IQC $\sigma_* \rhd 0$ is satisfied on $\mathcal{S}(\tau)$ for all $\tau \in [0, 1]$. Conditions can then be derived under which, the complete IQC $\sigma_* \succ 0$ on $\mathcal{S}(1)$ is implied by the complete IQC $\sigma_* \succ 0$ on $\mathcal{S}(0)$.

The homotopy approach is particularly easy to use in the stable operator setup in Section 41.3.2.1. Then the homotopy is easy to construct due to the fact that all operators in the loop are stable, see Section 41.4 and [3,8,9] for further details. The extension to more general situations is discussed in [11].

The "minimax" approach does not utilize information about σ and does not require one to construct a homotopy of systems. Instead, it assumes that $\mathcal{S} = \{q = [w; v; x]\}$ is a feedback connection of a marginally stable "nominal" LTI system with inputs v, w and state x, and a causal stable feedback $w = \Delta(v)$. This includes the setup in Section 41.3.2.1 with the exception that the system matrix A could be marginally stable. The quadratic form σ_0 is assumed to be such that, subject to the nominal LTI equations, the conditional IQC $\sigma_* \rhd 0$ is satisfied not only for the actual Δ, but also for $\Delta = 0$, and is a convex constraint on w. These assumptions allow one to use the minimax theorem to prove the complete IQC $\sigma_* \succ 0$. A more elaborate discussion can be found in [7].

41.4 Application of IQC Analysis

The framework discussed in the previous section is a flexible and versatile approach for systems analysis that can be used in many contexts. In this section we illustrate how IQC can be used in analysis of systems consisting of a nominal part interconnected with a number of more challenging components Δ_k, $k = 1, \ldots, N$; see Figure 41.4. We restrict our attention to the case where the components Δ_k are stable causal operators and the nominal part is LTI and stable with a transfer function representation $G(s)$. It is assumed that $G(s)$ has a state-space realization as in Figure 41.3. The purpose of the analysis is as before to verify an energy gain from input f to output e.

One of the main advantages of IQC is that components of very different nature can be characterized in a unified way. The analysis framework can therefore be implemented as a software package consisting of a library of component descriptions, a compiler that assembles the IQC descriptions, and finally, an optimization engine to solve the resulting feasibility optimization problem. One such prototype, IQCβ, is implemented in MATLAB; interested readers are referred to [6] for more details.

In this section we briefly describe the IQC analysis flow and apply it to a simple example. A brief "library" containing the two IQC descriptions used in the example is also included.

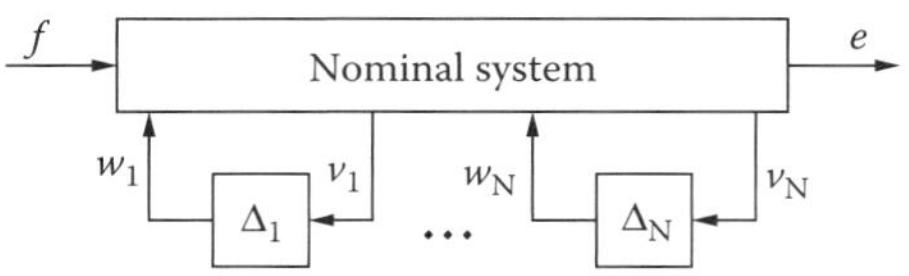

FIGURE 41.4 Illustration of a system consisting of a nominal part and complicating elements Δ_k, $k = 1, \ldots, N$.

41.4.1 The IQC Analysis Flow

The IQC framework applied to the system in Figure 41.4 involves the following three steps.

IQC modeling: The first step is to find IQC descriptions for the Δ_k. For this purpose, we consider behavioral models that usually contain a stable LTI extension

$$\mathcal{S}_{\Delta_k} = \{q_k = [w_k; v_k; x_{ek}] : w_k = \Delta_k(v_k); \dot{x}_{ek} = A_{ek}x_{ek} + B_{ek1}v_k + B_{ek2}w_k\}$$

and set of quadratic forms $\Lambda_k = \{\sigma_k\}$ such that $\sigma_k\langle q_k\rangle \rhd 0$ on $\mathcal{S}_{\Delta_k}$ for each $\sigma_k \in \Lambda_k$. These IQC descriptions can be assembled to obtain a complete IQC model of the form $(\mathcal{S}, \sigma_*, \Lambda)$, where $\mathcal{S} = \{q = [f; w; x]\}$ is a behavioral model for the whole interconnected system, consisting of signals satisfying a state-space equation of the form in Equation 41.20 with $x = [x_0; x_{e1}, \ldots, x_{eN}]$, $v = [v_1; \ldots; v_N]$, $w = [w_1; \ldots; w_N]$, and $w_k = \Delta_k(v_k)$. The quadratic form σ_* is the analysis objective of the form in Equation 41.21 and

$$\Lambda = \left\{\sigma = \sum c_k\sigma_k + \sigma_I : \sigma_k \in \Lambda_k;\ c_k \geq 0\right\}$$

is a set of quadratic forms for which $\sigma \rhd 0$ on $\mathcal{S}$.

Feasibility analysis: The feasibility analysis can be formulated as the semidefinite program

$$\text{minimize } r \text{ subject to } \sigma_* - \sum c_{k,l}\sigma_{k,l} - \sigma_I > 0; \quad c_{k,l} \geq 0 \tag{41.25}$$

for some fixed $\sigma_{k_l} \in \Lambda_k$, $k = 1, \ldots, n$. Since we are using a strict inequality feasibility of this optimization problem implies that the IQC $\sigma_* \rhd 0$ is strict, which is necessary in order to use the homotopy approach for postfeasibility analysis.

This optimization problem can sometimes be of a very large dimension, which makes solving the problem numerically a challenge. In such cases, in order to improve the computational efficiency, it becomes crucial to explore and exploit special structures of the optimization problem. One idea, proposed and developed in [12], is to solve the dual formulation of the semidefinite program in Equation 41.25. The number of decision variables to be optimized in the dual formulation is usually significantly smaller than those of Equation 41.25; thus the computational complexity of the dual problem is substantially reduced. Another idea is to reformulate Equation 41.25 into an equivalent semi-infinite form and solve the semi-infinite problem using specialized algorithms. Along this line of thought, several algorithms have been proposed, developed, and tested. See [4,5] for more details.

Postfeasibility analysis: By assembling the IQC descriptions for the Δ_k we arrive at the stable operator setup in Section 41.3.2.1 with $\Delta(v) = [\Delta_1(v_1); \ldots; \Delta_N(v_N)]$. Application of the homotopy argument to the stable operator feedback setup is generally straightforward. It is sufficient to use a parametrization of the form

$$\mathcal{S}(\tau) = \{[f; w; x] \in \mathcal{L}^{l+m+n} : w = \Delta_\tau(v);\ \text{Equation 41.20 is satisfied}\},$$

where Δ_τ is a continuous parametrization of Δ such that

1. There exists $\gamma > 0$ such that $\|\Delta_{\tau_1} - \Delta_{\tau_2}\| \leq \gamma|\tau_1 - \tau_2|$, for $\tau_1, \tau_2 \in [0, 1]$
2. $\Delta_1 = \Delta$
3. $\mathcal{S}(0)$ is a stable system, that is, $\sigma_* \succ 0$ on $\mathcal{S}(0)$
4. $\sigma_\Delta \rhd 0$ on $\mathcal{S}(\tau)$, for $\tau \in [0, 1]$

The simple parametrization $\Delta_\tau = \tau\Delta$ is often sufficient.

It is sometimes also possible to apply the minimax approach by exploiting convexity properties of the quadratic forms that define the IQC. The later approach is particularly easy to apply in the frequency domain formulation described next. This formulation is in line with how IQC analysis was presented in [8].

We have seen in the previous sections the IQC can be formulated in both time and frequency domain. Each of the IQC descriptions considered above is equivalent to a frequency domain IQC $[v_k; w_k]'\Pi_k(j\omega)[v_k; w_k] \triangleright 0$, where

$$\Pi_k = \begin{bmatrix} \Pi_{k,11} & \Pi_{k,12} \\ \Pi'_{k,12} & \Pi_{k,22} \end{bmatrix} = \mathcal{P}\{\sigma_{\Delta_k}, (sI - A_{ek})^{-1}[B_{ek1}, B_{ek2}]\}.$$

In fact, we may use a convex set of IQC obtained as

$$\Pi_{\Delta_k} = \left\{ \sum c_{k,l}\Pi_{k,l} : c_{k,l} \geq 0 \right\},$$

where $\mathcal{P}\{\sigma_{\Delta_{k,l}}, (sI - A_{ek})^{-1}[B_{ek1}, B_{ek2}]$ for some fixed $\sigma_{\Delta_{k,l}} \in \Lambda_k$. The three-step procedure of IQC analysis goes as follows.

IQC modelling: The model of the linear part of the system has the form $\mathcal{S}_{lin} = \{[f; w; e; v] : [e; v] = G[f; w]\}$, where G has block row structure $G = [G_0; G_1; \ldots; G_N]$ with dimensions consistent with the output signals $[e; v_1; \ldots; v_N]$. The complete system has the behavioral model

$$\mathcal{S} = \{q = [f; w; e; v] : q \in \mathcal{S}_{lin};\ w_k = \Delta_k(v_k)\}.$$

For this, we have the analysis objective

$$r|f|^2 - |e|^2 \succ 0$$

and the IQC descriptions

$$[v_k; w_k]'\Pi_k(j\omega)[v_k; w_k] \triangleright 0, \quad \Pi_k \in \Pi_{\Delta_k}.$$

Feasibility analysis: The feasibility criterion can be formulated as a frequency domain constraint

$$r|f|^2 - |G_0(j\omega)[f; w]|^2 > \sum_{k=1}^{N} [G_k(j\omega)[f; w]]'\Pi_k(j\omega)[G_k(j\omega)[f; w]]$$

for all $\omega \in \mathbb{R} \cup \{\infty\}$, $[f; w] \in \mathbb{C}$. This can equivalently be rewritten

$$\begin{bmatrix} G \\ I \end{bmatrix}' \left[\begin{array}{cc|cc} 1 & 0 & 0 & 0 \\ 0 & \Pi_{11} & 0 & \Pi_{12} \\ \hline 0 & 0 & -r & 0 \\ 0 & \Pi'_{12} & 0 & \Pi_{22} \end{array}\right] \begin{bmatrix} G \\ I \end{bmatrix}' (j\omega) < 0 \tag{41.26}$$

for all $w \in \mathbb{R} \cup \{\infty\}$, where $\Pi_{kl} = \text{diag}(\Pi_{1,kl}, \ldots, \Pi_{N,kl})$, $k, l = 1, 2$. The feasibility optimization can thus be formulated as

$$\text{minimize } r \text{ subject to Equation 41.26}; \quad \Pi_k \in \Pi_{\Delta_k}.$$

Post-feasibility analysis: The minimax approach applies to this setup if $\Pi_{11}(j\omega) \geq 0$ and $\Pi_{22}(j\omega) \leq 0$ for all $\omega \in \mathbb{R}$. Otherwise, the homotopy approach applies under the same assumptions as discussed above.

The above analysis expression can be arrived at more efficiently as we will see in the example presented in Section 41.4.3.

41.4.2 IQC Library

There exists a large number of IQC established for common nonlinear, time-varying, uncertain, and distributed elements of dynamical system models. Here we present two that will be used below.

41.4.2.1 Monotonic Odd Nonlinearity

Let $\phi : \mathbb{R} \to \mathbb{R}$ be an odd function such that $\phi(0) = 0$ and $\dot{\phi} \in [0, k]$ for some constant k. Let $\mathcal{S}_{\text{mon}} = \{[w; v] : v \in \mathcal{L}, w(t) = \phi(v(t))\}$ be the behavioral model describing the mapping $v(t) \mapsto \phi(v(t))$, and

$$\mathcal{S}_{\text{mon,e}} = \left\{[w; v; f] : [w; v] \in \mathcal{S}, f(t) = \int_{-\infty}^{\infty} h(t-\tau)w(\tau)\, d\tau\right\},$$

be an extended model for $\mathcal{S}_{\text{mon}}$, where h is any function which satisfies $\int_{-\infty}^{\infty} |h(t)|\, dt \leq 1$; that is, the L_1-norm of h is bounded by 1. It can be shown that $\mathcal{S}_{\text{mon,e}}$ satisfies IQC $\sigma_{\text{zf}} \succ 0$, where

$$\sigma_{\text{zf}}([w; v; f]) = \alpha(v - w/k)(w - f),$$

and $\alpha \geq 0$ is any nonnegative real number. This IQC is derived from a classical paper by G. Zames and P. Falb [16], and sometimes referred to as the "Zames–Falb" IQC. Note that, although any h whose L_1-norm is bounded by 1 can be chosen to form the extended model $\mathcal{S}_{\text{mon,e}}$, one would usually select an h which is the impulse response of some rational transfer function in order to take advantage of utilizing convex optimization tools in the feasibility analysis step.

In the case when the function ϕ is not odd but satisfies the other properties above, the IQC $\sigma_{\text{zf}} \succ 0$ is still valid as long as the additional condition $h(t) \geq 0$, $\forall t$ holds. For example, the IQC presented in Section 41.2.1.6 corresponds to selecting $h(t) = 2(\mathrm{e}^{-t} - \mathrm{e}^{-2t})$ for $t \geq 0$ and $h(t) = 0$ when $t < 0$. It can be readily verified that h is the impulse response of transfer function $H(s) = 2/(s^2 + 3s + 2)$, and the L_1-norm of h is equal to 1. The corresponding extended model $\mathcal{S}_{\text{mon,e}}$ can be equivalently expressed as

$$\mathcal{S}_{\text{mon,e}} = \left\{[w; v; f; \dot{f}] : [w; v] \in \mathcal{S}, \ddot{f} + 3\dot{f} + 2f = 2w, \dot{f}(0) = f(0) = 0\right\},$$

which satisfies IQC $2(v - w)(w - f) \succ 0$.

41.4.2.2 LTI Unmodeled Dynamics

Let $\Delta : \mathcal{L}^n \to \mathcal{L}^m$ be a LTI bounded operator with the L_2-induced gain being less than or equal to 1. Let $\mathcal{S}_{\text{ud}} = \{[w; v] : v \in \mathcal{L}^n, w = \Delta v\}$ be the behavioral model describing the LTI transformation $v \mapsto \Delta v$. It can be shown that $\mathcal{S}_{\text{ud}}$ satisfies the frequency weighted IQC $\sigma_{\text{ud}} \rhd 0$, where

$$\sigma_{\text{ud}}^H([w; v]) = \alpha(j\omega)(|v|^2 - |w|^2),$$

for every rational function $\alpha \in RL^{1\times 1}$ which is bounded and nonnegative on the imaginary axis. This IQC can be proven by observing that the Fourier transforms $\tilde{v}$, $\tilde{w}$ of finite energy $[w; v] \in \mathcal{S}_{\text{ud}}$ satisfy

$$\tilde{w}(j\omega) = \tilde{\Delta}(j\omega)\tilde{v}(j\omega), \quad \forall \omega \in \mathbb{R},$$

where $\tilde{\Delta}(s)$ is the representation of Δ in the Laplace domain. Since $\sup_\omega \bar{\sigma}\left(\tilde{\Delta}(j\omega)\right) \leq 1$, where $\bar{\sigma}(M)$ denotes the maximal singular value of M, we conclude that

$$\|\tilde{w}(j\omega)\|^2 \leq \|\tilde{v}(j\omega)\|^2, \quad \forall \omega \in \mathbb{R},$$

and hence the IQC.

41.4.3 Example

Figure 41.5 shows a model of a servo system consisting of a nominal plant P, a controller K, a standard saturation nonlinearity, and a dynamic uncertainty Δ, which represents unmodeled flexible modes. It is assumed that the energy gain of Δ is upper bounded by κ. The analysis objective is to verify whether the

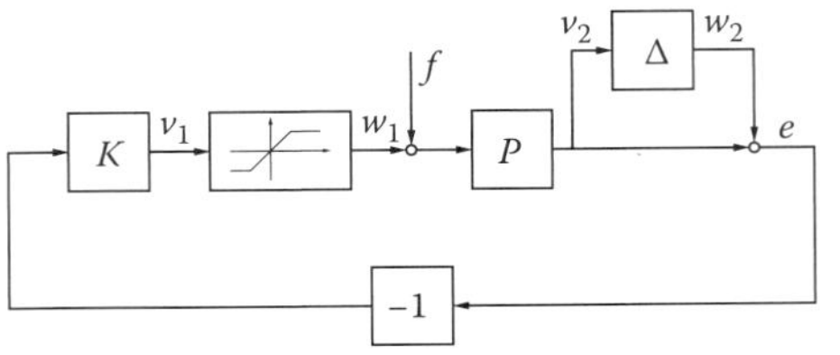

FIGURE 41.5 Feedback system with saturation and unmodeled dynamics.

energy gain from the input disturbance f to the output e is less than $\sqrt{r}$. In this case we use quadratic form $\sigma_* = r|e|^2 - |f|^2$ as the analysis objective, the IQC description

$$\alpha_1 \cdot 2\text{Re}[(v_1 - w_1)(1 - H(j\omega))w_1] \rhd 0 \tag{41.27}$$

for the saturation nonlinearity, and the IQC description

$$\alpha_2(j\omega)(\kappa^2|v_2|^2 - |w_2|^2) \rhd 0 \tag{41.28}$$

for the dynamic uncertainty, where $\alpha_1 \geq 0$, $\alpha_2(j\omega) \geq 0$, and H is a transfer function whose impulse response function has L_1-norm less than one. The frequency domain formulation of the analysis condition can directly be formulated as

$$\begin{aligned} r|P(j\omega)(f + w_1) + w_2|^2 - |f|^2 &> \alpha_2(j\omega)(\kappa^2|P(j\omega)(f + w_1)|^2 - |w_2|^2) \\ &+ 2\alpha_1 \text{Re}[\left(-K(j\omega)(w_2 + P(j\omega)(w_1 + f)) - w_1\right)(1 - H(j\omega))w_1] \end{aligned} \tag{41.29}$$

for all $\omega \in \mathbb{R} \cup \{\infty\}$ and $f, w_1, w_2 \in \mathbb{C}$. It is easy to see that the postfeasibilty conditions discussed above are satisfied.

In the previous section, we mentioned that the IQC analysis framework is implemented in MATLAB. The toolbox, named IQCβ, consists of a parser and a library of IQC descriptions for various kind of uncertain/nonlinear operators. The toolbox allows the users to perform IQC analysis in the MATLAB environment, using syntax similar to the language of MATLAB. The parser interprets the IQC models, descriptions, and the analysis objective that users input to the computer, defines the corresponding optimization problem for feasibility analysis, and calls for an optimization engine (an LMI solver if the problem is formulated as a set of finite dimensional LMIs) to solve the optimization problem. In the following, we briefly illustrate how to use IQCβ to perform the energy gain analysis described above. For the sake of illustration, let the plant P and the controller K be modelled by the following transfer functions:

$$P(s) = \frac{-s+1}{s^2 + 0.01s + 1}, \quad K(s) = \frac{1.92s - 3.471}{s^2 + 4.628s + 11.02}.$$

The MATLAB codes for calculating an upper bound of the energy gain are given below.

```
s = tf([1,0],1);                          %(1)
P = (-s+1)/(s*s+0.01*s+1);                %(2)
K = (1.92*s-3.471)/(s*s+4.628*s+11.02);   %(3)
abst_init_iqc                             %(4)
f = signal;                               %(5)
w1 = signal;                              %(6)
w2 = signal;                              %(7)
v2 = P*(f+w1);                            %(8)
```

```
e = w2+v2;                                %(9)
v1 = K*(-1)*e;                            %(10)
w1 == iqc_monotonic(v1,1,1,1);            %(11)
w2 == iqc_ltiunmod(v2,1,1,0.01);          %(12)
iqc_gain_tbx(f,e)                         %(13)
```

Command lines (1) through (3) define the transfer functions $P(s)$ and $K(s)$. The function/operations used in these command lines are native to MATLAB and not a part of the IQCβ toolbox. Command lines (4) through (13), on the other hand, utilize functions and operations which are parts of IQCβ and will not work if the toolbox is not installed. Command line (4) initiates the environment for IQC analysis under MATLAB. Command lines (5) through (10) define the linear part of the system in Figure 41.5; that is, the corresponding $\mathcal{S}_{lin}$ for this system. In this example, the linear part of the system has three "external inputs" f, w_1, and w_2, which are defined by command lines (5), (6), and (7), respectively. The "internal signals" v_2, e, and v_1, are related to f, w_1, and w_2 by command lines (8) through (10). Command line (11) defines the IQC model for the saturation nonlinearity, which is embedded as a monotonic odd nonlinearity. The IQC relation in Equation 41.27 for signals w_1 and v_1 is defined in the function `iqc_monotonic.m`. Here the corresponding $H(j\omega)$ is $1/(j\omega + 1)$. Likewise, Command line (12) defines the IQC model for the dynamic uncertainty Δ, and the IQC relation in Equation 41.28 for signals w_2 and v_2 is defined in the function `iqc_ltiunmod.m`. The corresponding $\alpha_2(j\omega)$ has the form $x/(j\omega + 1)$, where parameter x is a positive real number. We also note that for this illustration, we assume the energy gain of Δ to be less than 0.01. Finally, command line (13) executes the IQC parser, which collects the information defined by command lines (1) through (12), formulates the corresponding optimization problem in Equation 41.29 as LMIs, and call the generic LMI solver provided by MATLAB to find the least upper bound of energy gain. In this example, the bound found by the optimization engine is around 142.9.

41.5 Historical Remarks and References

The use of IQC in systems analysis has a long history. The term was, to our knowledge, introduced by V.A. Yakubovich in the 1960s in a sequence of works where advanced sector and Popov criteria were developed using IQC. Yakubovich continued to develop this approach and a recent survey can be found in [15].

Several related and equally important directions were initiated in the 1960s and 1970s. The theory of dissipative dynamical systems was introduced by J.C. Willems with a focus on the use of dissipation inequalities and positive definite storage functions as a natural way to quantify energy dissipation in physical systems. See, for example, the recent survey in [14]. Another approach to stability analysis was the input–output theory developed by Zames, Sandberg, and others. The most powerful results are obtained by using the multipliers, an idea, which to our knowledge was introduced by Brockett and Willems [1] and further developed to allow the use of noncausal multipliers by Zames and Falb [16]. Extensive accounts of the input–output theory can be found in the books [2,13].

Later in the 1980s and 1990s the attention turned to more complex systems with structured uncertainties. The development of efficient algorithms and software for convex optimization motivated the computational frameworks for robustness analysis in, for example, [10]. However, as more complex uncertainty structures were considered it became evident that the input–output theory required certain factorization conditions on the multipliers, which complicated the analysis. This was overcome by using IQC defined in the frequency domain as in presented [8]. This tutorial paper presents the IQC approach closer to the way it is implemented in the custom-made MATLAB toolbox IQCβ [6]. It provides a general tool for analysis of complex systems containing heterogeneous components. A previous tutorial on IQC can be found in [3] and further perspectives are given in [9,11].

References

1. R. Brockett and J. Willems. Frequency domain stability criteria — Part I. *IEEE Transactions on Automatic Control*, 10(3):255–261, 1965.
2. C.A. Desoer and M. Vidyasagar. *Feedback Systems: Input–Output Properties*. Academic Press, New York, 1975.
3. U. Jönsson. Lecture notes on integral quadratic constraints. Technical Report TRITA/MAT-00-OS12, Department of Mathematics, Royal Institute of Technology, September 2000. http://www.math.kth.se/ulfj/5B5744/Lecturenotes.ps.
4. C.-Y. Kao and A. Megretski. On the interior point method for a class of semi-definite programs. *SIAM Journal on Control and Optimization*, 46(2):468–495, 2007.
5. C.-Y. Kao, A. Megretski, and U. Jönsson. Specialized fast algorithms for IQC feasibility and optimization problems. *Automatica*, 40(2):239–252, 2004.
6. C.-Y. Kao, A. Megretski, U. Jönsson, and A. Rantzer. A MATLAB toolbox for robustness analysis. In *IEEE Conference on Computer Aided Control Systems Design*, pp. 297–302, Taipei, Taiwan, 2004.
7. A. Megretski. KYP lemma for non-strict inequalities and the associated minimax theorem. arXiv:1008.2552 [math,oc], 2010.
8. A. Megretski and A. Rantzer. System analysis via integral quadratic constraints. *IEEE Transactions on Automatic Control*, 42(6):819–830, 1997.
9. A. Megretski and A. Rantzer. System analysis via integral quadratic constraints. Part I. Technical Report ISRN LUTFD2/TFRT–7531–SE, Department of Automatic Control, Lund University, Sweden, April 1995.
10. A. Packard and J.C. Doyle. The complex structured singular value. *Automatica*, 29(1):71–109, 1993.
11. A. Rantzer and A. Megretski. System analysis via integral quadratic constraints. Part II. Technical Report ISRN LUTFD2/TFRT–7559–SE, Department of Automatic Control, Lund University, Sweden, September 1997.
12. R. Wallin, A. Hansson, and J.H. Johansson. A structure exploiting preprocessor for semidefinite programs derived from the Kalman–Yakubovich–Popov lemma. *IEEE Transactions on Automatic Control*, 54(4):697–704, April 2009.
13. J.C. Willems. *The Analysis of Feedback Systems*. MIT Press, Cambridge, MA, 1971.
14. J.C. Willems. Dissipative dynamical systems. *European Journal of Control*, 13:134–151, 2007.
15. V.A. Yakubovich. Popov's method and its subsequent development. *European Journal of Control*, 2:2000–2008, 2002.
16. G. Zames and P.L. Falb. Stability conditions for systems with monotone and slope-restricted nonlinearities. *SIAM Journal of Control*, 6(1):89–108, 1968.

42

Control of Nonholonomic and Underactuated Systems

Kevin M. Lynch
Northwestern University

Anthony M. Bloch
University of Michigan

Sergey V. Drakunov
Embry-Riddle Aeronautical University

Mahmut Reyhanoglu
Embry-Riddle Aeronautical University

Dmitry Zenkov
North Carolina State University

42.1 Introduction

In this chapter, we study motion planning and control for systems subject to *nonholonomic* and *underactuation* constraints. All such systems can be written in the form

$$\dot{x} = f(x) + \sum_{i=1}^{m} u_i g_i(x), \quad u_i \in \mathbb{R}, \tag{42.1}$$

where $f(x)$ is a drift vector field, $g_i(x)$ are linearly independent control vector fields, and $u = [u_1, \ldots, u_m]^T$ is the control. Unless otherwise specified, we work in local coordinates and treat the state as $x \in \mathcal{M} = \mathbb{R}^p$.

While many nonlinear systems can be written in the control-affine form (Equation 42.1), we are particularly interested in first-order kinematic systems and second-order dynamic mechanical systems. In the case of kinematic systems, the state x is simply the *configuration* $q \in \mathcal{Q} = \mathbb{R}^n$ (i.e., $p = n$), the control vector fields are velocity vector fields, and the controls u are speeds. For second-order mechanical systems, the state is $x = (q, \dot{q})$ in the tangent bundle $T\mathcal{Q} = \mathbb{R}^{2n}$, the control vector fields encode acceleration directions, and the controls u are generalized forces. For both first-order kinematic and second-order mechanical systems, we are interested in the case of fewer control inputs than configuration variables, $m < n$.

A classical nonholonomic constraint is a velocity constraint that cannot be integrated to yield an equivalent configuration constraint. A common kind of velocity constraint is a *Pfaffian* constraint of the form

$$a(q)\dot{q} = 0. \tag{42.2}$$

If there exists a function $h(q)$ such that $\partial h/\partial q = \mu(q)a(q)$ for a suitable function $\mu(q)$, then the Pfaffian constraint is not nonholonomic, but is the derivative of a *holonomic* constraint. A holonomic constraint reduces the dimension of configuration space. If there is no equivalent holonomic constraint, then the constraint is nonholonomic and reduces the dimension of the feasible velocities, but does not reduce the dimension of the reachable configuration space.

Our interest here is in nonintegrable Pfaffian constraints of the form $A(q)\dot{q} = 0$, where k velocity constraints are encoded in the $n \times k$ matrix $A(q)$. Such nonholonomic constraints arise from rolling without slip, and may be implicit in some conservation laws, such as conservation of angular momentum. The constraints define a *distribution* $\mathcal{D}$ in the tangent bundle $T\mathcal{Q}$, and $\mathcal{D}_q$ denotes the subspace of feasible velocities at each q.

While kinematic systems may be subject to nonholonomic velocity constraints, second-order systems may be subject to both velocity and acceleration constraints. Acceleration constraints arise due to the fact that the system is underactuated—it has fewer control inputs than configuration variables ($m < n$).* If acceleration constraints cannot be integrated to yield equivalent velocity constraints, we refer to them as *second-order nonholonomic constraints.*

Thus the systems of interest in this chapter are kinematic systems with velocity constraints, and second-order mechanical systems with velocity and/or acceleration constraints. All of the systems we consider can be expressed in the form (Equation 42.1).

Control of nonholonomic and underactuated systems is challenging, as we experience every time we try to parallel park a car. This chapter focuses on the following challenges:

- *Evaluating controllability.* Given a description of the system in the form (Equation 42.1), Lie algebraic tests can be used to study the reachable state space. Some controllability concepts are specific to second-order mechanical systems and allow simplified tests.
- *Feedback stabilization.* A classic result due to Brockett prompts the search for discontinuous or time-varying feedback control laws. Another strategy is to derive motion planners and to use feedback control to stabilize planned trajectories.
- *Motion planning.* The motion planning problem is to find an efficient algorithm that gives an open-loop control steering the system from a start state to a goal state.

In the next section we provide examples of nonholonomic and underactuated systems, and the remaining sections address the issues above. General references for further reading include [2,5,11,13,18].

42.2 Notation and Examples

We can classify example systems according to whether they are first-order kinematic systems or second-order mechanical systems. For first-order kinematic systems, the system state x is simply the configuration

* At least two definitions of "underactuated" are possible: (1) the number of control inputs m is less than the number of configuration variables n, as adopted in this chapter or (2) m is less than the dimension of $\mathcal{D}_q$, the number of linearly independent instantaneously feasible velocity directions. For example, a simple model of a car has $n = 3$ (the position and heading direction of the car in the plane) and $m = 2$ controls (the forward and turning speeds). The car is underactuated by the first definition, but not by the second—the no-slip constraint of the car means that the control vector fields span the space of feasible velocities. We could modify our definition of "underactuated" to be that m is less than the number of "degrees of freedom," but this also does not settle the matter: some authors equate "degrees of freedom" to n, while others equate it to the dimension of $\mathcal{D}_q$. In any case, once the system is converted to the form (Equation 42.1), there is no ambiguity, and $m < n$ for all systems we study.

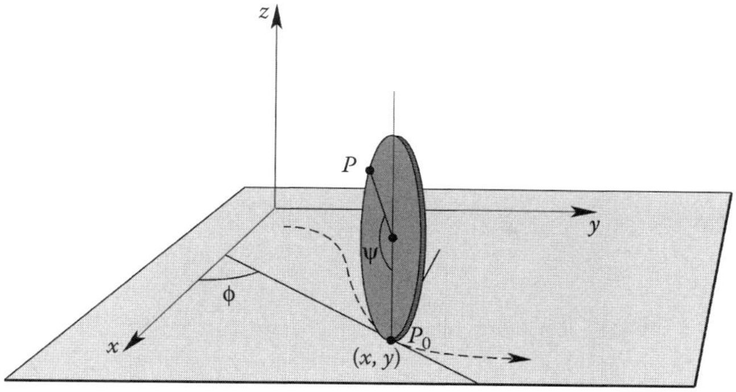

FIGURE 42.1 The geometry and configuration coordinates for the vertical rolling disk.

q, and $k = n - m \geq 1$, that is, the number of velocity constraints is equal to the dimension of the configuration space minus the dimension of the control space. For second-order mechanical systems, there are three possibilities: $k = 0$ (and $n - m \geq 1$), that is, there are no nonholonomic velocity constraints; $k = n - m \geq 1$, that is, the actuators can apply a generalized force in each unconstrained velocity direction; and $n - m > k \geq 1$, that is, there is one or more unconstrained velocity direction which has no associated actuator. We give examples of each of these cases below. More detail on these examples can be found in [2,11–13,21].

42.2.1 Kinematic Systems

Example 42.1: The Vertical Rolling Disk

The configuration of a disk of radius R rolling on a horizontal plane can be parameterized by $q = [x, y, \psi, \phi]^T$, as shown in Figure 42.1. The $k = 2$ nonholonomic rolling constraints can be written $A(q)\dot{q} = 0$, where

$$A(q) = \begin{bmatrix} 1 & 0 & -R\cos\phi & 0 \\ 0 & 1 & -R\sin\phi & 0 \end{bmatrix}. \tag{42.3}$$

The $m = 2$ controls are the wheel rolling velocity $u_1 = \dot{\psi}$ and the heading rate of change $u_2 = \dot{\phi}$, with corresponding control vector fields $g_1(q) = [R\cos\phi, R\sin\phi, 1, 0]^T$ and $g_2(q) = [0, 0, 0, 1]^T$. There is no drift field, $f(q) = 0$.

It is easy to confirm that the rolling disk can reach any configuration in $\mathcal{Q}$ (see Section 42.3). Therefore, the constraints $A(q)\dot{q} = 0$ are indeed nonholonomic—they do not restrict the reachable configuration space. If we add one more velocity constraint, $u_2 = \dot{\phi} = 0$, however, we see that the three constraints taken together can be integrated to obtain three holonomic constraints. The reachable configuration space has been reduced from four-dimensional to one-dimensional integral manifold of the vector field g_1; the disk can only roll back and forth on a line, and its position on the line, together with the initial configuration, determines ψ. While the reduction in the reachable space is obvious in this example, and indeed in any drift-free example with a single control vector field, in general the integrability of constraints can be evaluated by considering the Lie algebra of the system vector fields (Section 42.3).

Example 42.2: The Heisenberg System (Nonholonomic Integrator)

Consider the vertical disk example, but eliminating ψ from the representation of the configuration. The system can be written

$$\dot{x} = vR\cos\phi,$$

$$\dot{y} = vR \sin\phi,$$
$$\dot{\phi} = \omega,$$

where the forward velocity and heading velocity controls are v and ω, respectively. We can define a change of coordinates $F(\phi)$

$$\begin{bmatrix} x_1 \\ x_2 \\ z \end{bmatrix} = F(\phi) \begin{bmatrix} x \\ y \\ \phi \end{bmatrix},$$

where

$$F(\phi) = \begin{bmatrix} 0 & 0 & 1 \\ \cos\phi & \sin\phi & 0 \\ \phi\cos\phi - 2\sin\phi & \phi\sin\phi + 2\cos\phi & 0 \end{bmatrix},$$

and a nonsingular state-dependent transformation of the controls

$$u_1 = \omega,$$
$$u_2 = Rv + \left(\frac{z}{2} - \frac{x_1 x_2}{2}\right)\omega,$$

yielding the system

$$\dot{x}_1 = u_1, \quad (42.4)$$
$$\dot{x}_2 = u_2, \quad (42.5)$$
$$\dot{z} = x_1 u_2 - x_2 u_1 \quad (42.6)$$

defined by the control vector fields $g_1 = [1, 0, -x_2]^T$ and $g_2 = [0, 1, x_1]^T$. This system is called the *Heisenberg system* or *nonholonomic integrator* [8,9]. The importance of this system is that the associated three-dimensional Lie algebra is the Heisenberg algebra appearing in quantum mechanics. Generalizations can be used to model a number of other important systems.

Example 42.3: Wheeled Mobile Robots

Kinematic wheeled mobile robots are perhaps the best-known examples of nonholonomic systems. Some of the most common wheeled mobile robots can be described by the Heisenberg system. We denote the configuration of the robot in a horizontal plane as $q = [x, y, \phi]^T$, where (x, y) is the location of a reference point halfway between two wheels of the robot, and ϕ is the orientation of the robot (Figure 42.2). For a differential-drive robot, the two wheels are not steered but their speeds can be controlled independently. For a car-like robot, the reference point is between the unsteered rear wheels. In both cases, the system is described by the vector fields $g_1 = [\cos\phi, \sin\phi, 0]^T$ and $g_2 = [0, 0, 1]^T$ with controls $u_1 = v$, the forward velocity, and $u_2 = \omega$, the angular velocity.

A number of different control sets have been studied for wheeled mobile robots. Four sets are shown in Figure 42.2, corresponding to a car with a limited forward-reverse speed and limited turning radius; a differential-drive robot with limited wheel speeds; a "Reeds–Shepp" car, which is a car that can only take forward–reverse velocities in the set $\{-v_{max}, v_{max}\}$; and a forward-only "Dubins" car that has v fixed at v_{max}. For simplicity, in this chapter, we treat control sets as $\mathbb{R}^m$, but some results also apply to restricted control sets as in Figure 42.2.

42.2.2 Dynamic Mechanical Systems

42.2.2.1 Systems with Velocity Constraints

To describe a constrained mechanical system, we need the following elements: a symmetric positive-definite inertia tensor $M(q) \in \mathbb{R}^{n \times n}$ and a potential energy $U(q) \in \mathbb{R}$ defining the Lagrangian

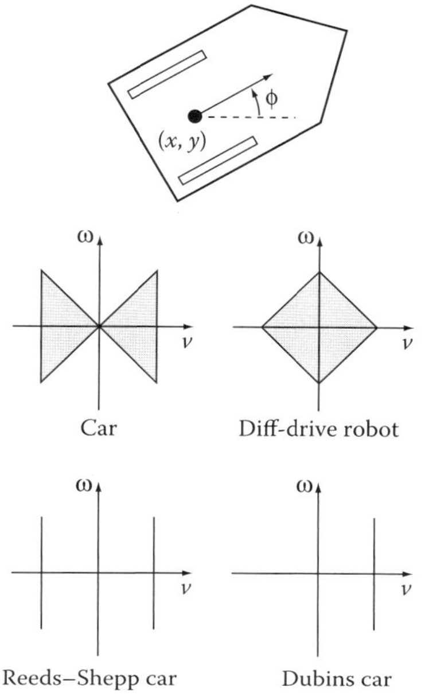

FIGURE 42.2 Configuration coordinates of a wheeled mobile robot, and four different control sets in the space of forward velocities v and turning rates ω.

$L(q,\dot{q}) = \frac{1}{2}\dot{q}^T M(q)\dot{q} - U(q)$; a set of controls $\mathcal{U}$, which we take to be $\mathbb{R}^m$ for simplicity; and an input tensor $T(q) \in \mathbb{R}^{n\times m}$ indicating how a control u acts on the generalized coordinates q. If the system is subject to Pfaffian velocity constraints, then the k constraints are written $A(q)\dot{q} = 0$, where $A(q) \in \mathbb{R}^{k\times n}$, and these constraints result in a set of constraint forces $A^T(q)\lambda$, where $\lambda \in \mathbb{R}^k$.

We define the n^3 Christoffel symbols of $M(q)$

$$\Gamma^i_{jk}(q) = \frac{1}{2}\left(\frac{\partial m_{ij}(q)}{\partial q_k} + \frac{\partial m_{ik}(q)}{\partial q_j} - \frac{\partial m_{kj}(q)}{\partial q_i}\right),$$

where $m_{ij}(q)$ is the (i,j)th element of $M(q)$. Then Lagrange's equations yield the dynamics

$$M(q)\ddot{q} + b(q,\dot{q}) = T(q)u + A^T(q)\lambda, \tag{42.7}$$

$$A(q)\dot{q} = 0, \tag{42.8}$$

where $b(q,\dot{q}) = \dot{q}^T\Gamma(q)\dot{q} + \partial U(q)/\partial q$, and the ith element of the n-vector $\dot{q}^T\Gamma(q)\dot{q}$ of Coriolis and centripetal terms is defined as $\dot{q}^T\Gamma^i(q)\dot{q}$.

For Equation 42.8 to be satisfied at all times, we must also have

$$\dot{A}(q)\dot{q} + A(q)\ddot{q} = 0, \tag{42.9}$$

allowing us to solve for the the Lagrange multipliers λ. Dropping the dependence on q,

$$\lambda = (AM^{-1}A^T)^{-1}(AM^{-1}(b - Tu) - \dot{A}\dot{q}). \tag{42.10}$$

Plugging Equation 42.10 into Equation 42.7 to eliminate the Lagrange multipliers, we obtain the n dynamic equations

$$\ddot{q} = M^{-1}\big(A^T(AM^{-1}A^T)^{-1}\dot{A}\dot{q} - P^{\perp}b + P^{\perp}Tu\big), \tag{42.11}$$

where

$$P^{\perp} = \mathcal{I}_n - A^T(AM^{-1}A^T)^{-1}AM^{-1} \tag{42.12}$$

is an $n \times n$ matrix of rank $n-k$ that orthogonally projects (by the kinetic energy metric) generalized forces to the components that do work on the system, and $\mathcal{I}_n$ is the $n \times n$ identity matrix. Equation 42.11 can be expressed more compactly as

$$P^{\perp}(q)\big(M(q)\ddot{q} + b(q,\dot{q})\big) = P^{\perp}(q)T(q)u$$

or

$$P(q)\big(\ddot{q} + M^{-1}(q)b(q,\dot{q})\big) = \underbrace{P(q)M^{-1}(q)T(q)}_{Y(q)}\,u, \tag{42.13}$$

where $P = M^{-1}P^{\perp}M$ is an $n \times n$ matrix of rank $n-k$ that orthogonally projects motions to those that satisfy the velocity constraints. P and $P^{\perp}$ are orthogonal by the kinetic energy metric. We call the columns of $Y(q)$ *input vector fields*, discussed in Section 42.3.2.2.

Using Equation 42.11, we could now write the $2n$-dimensional drift and control vector fields of Equation 42.1. If Equation 42.11 is integrated exactly, the system state evolves on the $(2n-k)$-dimensional distribution $\mathcal{D}$ defined by the k nonholonomic velocity constraints. Note, however, that there are only $n-k$ independent equations in Equations 42.11 and 42.13, and when we examine controllability of the system using Lie brackets, we use Taylor expansions of the system vector fields. Because these only approximate the vector fields, our controllability analysis may incorrectly conclude that the system can escape $\mathcal{D}$ to other points in $T\mathcal{Q}$. See Section 42.3.2.2 for controllability tests based on the dynamics (Equation 42.13).

As an alternative, we can use the nonholonomic Pfaffian constraints to eliminate k velocity state variables and arrive at a reduced set of equations on the $(2n-k)$-dimensional distribution $\mathcal{D}$. Reordering the configuration variables and labeling them $q = [q_1^T, q_2^T]^T$, we can write the Pfaffian constraints as

$$\underbrace{A(q)}_{k\times n}\,\dot{q} = \big[\underbrace{A_1(q)}_{k\times n-k}\;\;\underbrace{A_2(q)}_{k\times k}\big]\begin{bmatrix}\dot{q}_1\\ \dot{q}_2\end{bmatrix} = 0, \tag{42.14}$$

where $A_2(q)$ is invertible. The k-dependent velocities $\dot{q}_2$ can be calculated from the $n-k$-independent velocities $\dot{q}_1$ by

$$\dot{q}_2 = D(q)\dot{q}_1 = -A_2^{-1}(q)A_1(q)\dot{q}_1.$$

Defining the $n \times (n-k)$ matrix

$$C(q) = \begin{bmatrix}\mathcal{I}_{n-k}\\ D(q)\end{bmatrix},$$

we obtain

$$\dot{q} = C(q)\dot{q}_1, \tag{42.15}$$

$$\ddot{q} = \dot{C}(q)\dot{q}_1 + C(q)\ddot{q}_1. \tag{42.16}$$

Substituting Equations 42.15 and 42.16 into Equation 42.7 and premultiplying both sides by $C^T(q)$, we obtain

$$C^T(q)M(q)C(q)\ddot{q}_1 = C^T(q)\big[T(q)u - b(q, C(q)\dot{q}_1) - M(q)\dot{C}(q)\dot{q}_1\big].$$

Premultiplying each side by the inverse of the full rank $(n-k)\times(n-k)$ matrix $\widetilde{M}(q) = C^T(q)M(q)C(q)$, and dropping the dependence on q, we obtain

$$\ddot{q}_1 = \underbrace{\widetilde{M}^{-1}C^T T}_{H(q)}\,u - \underbrace{\widetilde{M}^{-1}C^T(b + M\dot{C}\dot{q}_1)}_{c(q,\dot{q}_1)}. \tag{42.17}$$

We have reduced the n dynamic equations with constraint forces in Equation 42.7 to $n-k$ equations for $\ddot{q}_1$. Defining the state variables $x_1 = q_1$, $x_2 = q_2$, $x_3 = \dot{q}_1$, we can write the $2n-k$ state equations in the form (Equation 42.1):

$$\dot{x} = \begin{bmatrix} \dot{x}_1 \\ \dot{x}_2 \\ \dot{x}_3 \end{bmatrix} = \underbrace{\begin{bmatrix} x_3 \\ D(x_1, x_2)x_3 \\ -c(x) \end{bmatrix}}_{f(x)} + \underbrace{\begin{bmatrix} 0 \\ 0 \\ H(x_1, x_2) \end{bmatrix}}_{G(x_1, x_2)} u. \tag{42.18}$$

The m columns of $G(x_1, x_2) \in \mathbb{R}^{(2n-k)\times m}$ are the control vector fields $g_i(x)$ in Equation 42.1.

42.2.2.2 Systems without Velocity Constraints

In the case of no velocity constraints ($k = 0$), we can simplify the equations by first partitioning the configuration variables into those that are acted upon by the inputs and those that are not, $q^T = [q_1^T, q_2^T]$, $q_1 \in \mathbb{R}^m$, $q_2 \in \mathbb{R}^{n-m}$.* Lagrange's equations can then be written

$$M_{11}(q)\ddot{q}_1 + M_{12}(q)\ddot{q}_2 + b_1(q, \dot{q}) = B(q)u, \tag{42.19}$$

$$M_{21}(q)\ddot{q}_1 + M_{22}(q)\ddot{q}_2 + b_2(q, \dot{q}) = 0, \tag{42.20}$$

where $B(q) \in \mathbb{R}^{m\times m}$ is invertible for all $q \in \mathcal{Q}$, $b_1(q, \dot{q}) \in \mathbb{R}^m$, $b_2(q, \dot{q}) \in \mathbb{R}^{n-m}$, and $M_{ij}(q)$, $i, j = 1, 2$, are components of $M(q)$. We may solve for $\ddot{q}_2$ as

$$\ddot{q}_2 = -M_{22}^{-1}(q)[M_{21}(q)\ddot{q}_1 + b_2(q, \dot{q})],$$

and substitute into Equation 42.19 to obtain

$$\bar{M}(q)\ddot{q}_1 + \bar{b}(q, \dot{q}) = B(q)u,$$

where

$$\bar{M}(q) = M_{11}(q) - M_{12}(q)M_{22}^{-1}(q)M_{21}(q),$$
$$\bar{b}(q, \dot{q}) = b_1(q, \dot{q}) - M_{12}(q)M_{22}^{-1}(q)b_2(q, \dot{q}).$$

Consequently, using the partial feedback linearizing controller

$$u = B^{-1}(q)[\bar{M}(q)v + \bar{b}(q, \dot{q})],$$

Equations 42.19 and 42.20 can be rewritten as

$$\ddot{q}_1 = v, \tag{42.21}$$

$$\ddot{q}_2 = J(q)\ddot{q}_1 + R(q, \dot{q}), \tag{42.22}$$

where

$$J(q) = -M_{22}^{-1}(q)M_{21}(q),$$
$$R(q, \dot{q}) = -M_{22}^{-1}(q)b_2(q, \dot{q}).$$

Equations 42.21 and 42.22 have a special triangular or cascade form that appropriately captures the important attributes of underactuated mechanical systems. Equation 42.21 defines the linearized dynamics of the m completely actuated degrees of freedom. Equation 42.22 defines the dynamics of the $n - m$ unactuated degrees of freedom; these are expressed in terms of equalities involving the generalized accelerations. If these latter relations do not admit any nontrivial integral, that is, any nonconstant smooth

* It may be necessary to perform a coordinate transformation to obtain this form.

function $h(q, \dot{q}, t)$ such that $dh/dt = 0$ along the solutions, then these relations may be interpreted as $n - m$ completely nonintegrable acceleration constraints (or second-order nonholonomic constraints). As will be seen in the subsequent development, controllability and stabilizability properties of underactuated mechanical systems are closely related to this property. Hence, it is crucial to identify underactuated mechanical systems where the acceleration relations defined by Equation 42.22 are completely nonintegrable.

Equations 42.21 and 42.22 can be expressed in the form (Equation 42.1) by defining the state variables

$$x_1 = q_1, \quad x_2 = q_2, \quad x_3 = \dot{q}_1, \quad x_4 = \dot{q}_2.$$

Then the state equations are given by

$$\dot{x}_1 = x_3, \tag{42.23}$$

$$\dot{x}_2 = x_4, \tag{42.24}$$

$$\dot{x}_3 = v, \tag{42.25}$$

$$\dot{x}_4 = J(x_1, x_2)v + R(x_1, x_2, x_3, x_4). \tag{42.26}$$

Equations 42.23 through 42.26 define a drift vector field $f(x) = [x_3^T, x_4^T, 0^T, R^T(x_1, x_2, x_3, x_4)]^T$ and m control vector fields $g_i(x) = [0^T, 0^T, e_i^T, J_i^T(x_1, x_2)]^T$, where e_i denotes the ith standard basis vector in $\mathbb{R}^m$ and $J_i(x_1, x_2)$ denotes the ith column of the matrix-valued function $J(x_1, x_2)$.

42.2.2.3 Examples

Example 42.4: The Dynamic Rolling Disk

Consider the rolling disk of Figure 42.1, but now with rolling and steering torques as inputs, instead of angular velocities. In this example, $n = 4$, $m = 2$, and $k = 2$, and therefore $k = n - m = 2$—there is an actuator for each unconstrained velocity direction. Following the development in Equations 42.14 through 42.18, we reorder the configuration variables from the kinematic case to $q = [q_1^T, q_2^T]^T$, where $q_1 = [\psi, \phi]^T$ and $q_2 = [x, y]^T$. The system is described by the reordered rolling constraints from Equation 42.3, $A_1(q)\dot{q}_1 + A_2(q)\dot{q}_2 = 0$, where

$$A_1(q) = \begin{bmatrix} -R\cos\phi & 0 \\ -R\sin\phi & 0 \end{bmatrix}, \quad A_2(q) = \begin{bmatrix} 1 & 0 \\ 0 & 1 \end{bmatrix},$$

and

$$M(q) = \begin{bmatrix} J & 0 & 0 & 0 \\ 0 & I & 0 & 0 \\ 0 & 0 & m & 0 \\ 0 & 0 & 0 & m \end{bmatrix}, \quad T(q) = \begin{bmatrix} 1 & 0 \\ 0 & 1 \\ 0 & 0 \\ 0 & 0 \end{bmatrix},$$

where m is the mass of the disk, J is the inertia of the disk about an axis perpendicular to the face of the disk and through the disk's center, and I is the inertia of the disk about a vertical axis through its center. The disk has zero potential energy, $U(q) = 0$. Setting $x_1 = q_1$, $x_2 = q_2$, $x_3 = \dot{q}_1$, the drift and control vector fields in Equation 42.18 on the six-dimensional distribution $\mathcal{D}$ are described by

$$D(x_1, x_2) = \begin{bmatrix} R\cos\phi & 0 \\ R\sin\phi & 0 \end{bmatrix}, \quad c(x) = \begin{bmatrix} 0 \\ 0 \end{bmatrix}, \quad H(x_1, x_2) = \begin{bmatrix} \dfrac{1}{J + mR^2} & 0 \\ 0 & \dfrac{1}{I} \end{bmatrix}.$$

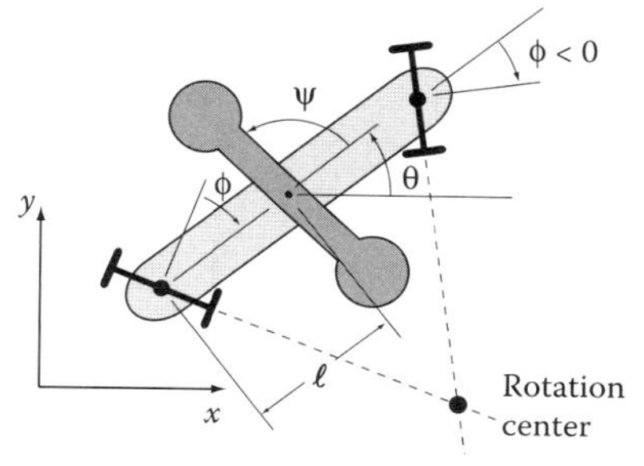

FIGURE 42.3 The snakeboard model.

Example 42.5: The Snakeboard

The *snakeboard* is a commercial toy whose concept is derived from the well-known skateboard. It is composed of two steerable wheeled platforms joined by a coupling bar, and the rider propels itself forward without touching the ground by steering the wheels and twisting it body back and forth. A simple model of the snakeboard is shown in Figure 42.3. Here a momentum rotor simulates the rider by spinning back and forth, and by conservation of angular momentum about the rotation center chosen by the wheels, the snakeboard body moves.

Our simple model of the snakeboard has $n = 5, m = 2$, and $k = 2$, and therefore $n - m > k = 2$—there is no actuator for each unconstrained velocity direction. Let the configuration of the snakeboard be represented by $q = [\theta, \psi, \phi, x, y]^T$, where (x, y) represents the Cartesian position of the center of the snakeboard coupler, θ is its angle, and ψ and ϕ are the angle of the rotor and the steering angle of the wheels, respectively, expressed in the body frame. The inertia matrix for the simplified model of the snakeboard is given by

$$M = \begin{bmatrix} I + I_r + I_w & I_r & 0 & 0 & 0 \\ I_r & I_r & 0 & 0 & 0 \\ 0 & 0 & I_w & 0 & 0 \\ 0 & 0 & 0 & m & 0 \\ 0 & 0 & 0 & 0 & m \end{bmatrix},$$

where m is the total mass of the snakeboard, I is the inertia of the coupler about its center of mass, I_r is the rotor inertia, and $\frac{1}{2}I_w$ is the inertia of each set of wheels about its pivot point. (Note that because the inertia matrix is invariant to the configuration, the Christoffel symbols are zero.) The system is subject to two control inputs: a torque u_1 that controls the rotor angle ψ, and a torque u_2 controlling the steering angle ϕ. Therefore, $T(q)$ can be written

$$T(q) = \begin{bmatrix} 0 & 0 \\ 1 & 0 \\ 0 & 1 \\ 0 & 0 \\ 0 & 0 \end{bmatrix}.$$

The wheels are assumed to roll without lateral slipping, and the wheel angle chooses a rotation center along a line perpendicular to the body of the snakeboard and through its center. The no-slip constraints can be written in the form $A(q)\dot{q} = 0$, where

$$A(q) = \begin{bmatrix} -\ell \cos\theta \cos\phi & 0 & 0 & \sin\phi & 0 \\ -\ell \sin\theta \cos\phi & 0 & 0 & 0 & \sin\phi \end{bmatrix}.$$

Following the development in Equations 42.14 through 42.18, we could choose $q_1 = [\theta, \psi, \phi]^T$, $q_2 = [x, y]^T$, and $A_2(q)$ to be the right 2×2 submatrix of $A(q)$. This yields the state equations 42.18 on the eight-dimensional distribution $\mathcal{D}$ away from states where $\phi \in \{0, \pi\}$, where $A_2(q)$ is noninvertible.

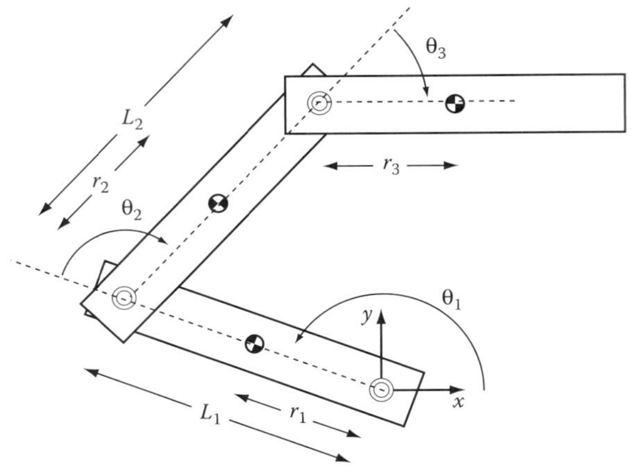

FIGURE 42.4 The 3R robot with three revolute joints.

Example 42.6: The 3R Robot

A 3R robot manipulator consists of three revolute joints (Figure 42.4). We consider such a robot operating in a horizontal plane with a passive joint—one joint lacks an actuator. This is an example of an underactuated system with no velocity constraints ($k = 0$).

This system is described by $q = [\theta_1, \theta_2, \theta_3]^T$, zero potential energy, and

$$M(q) = \begin{bmatrix} m_{11} & m_{12} & m_{13} \\ m_{21} & m_{22} & m_{23} \\ m_{31} & m_{32} & m_{33} \end{bmatrix},$$

where

$$\begin{aligned} m_{11} &= I_1 + I_2 + I_3 + m_1 r_1^2 + m_2(L_1^2 + r_2^2) + m_3(L_1^2 + L_2^2 + r_3^2) + 2m_2 L_1 r_2 c_2 \\ &\quad + 2m_3(L_1 L_2 c_2 + L_2 r_3 c_3 + L_1 r_3 c_{23}), \\ m_{12} &= m_{21} = I_2 + I_3 + m_2(r_2^2 + L_1 r_2 c_2) + m_3(L_2^2 + r_3^2 + L_1 L_2 c_2 + 2L_2 r_3 c_3 + L_1 r_3 c_{23}, \\ m_{13} &= m_{31} = I_3 + m_3(r_3^2 + L_2 r_3 c_3 + L_1 r_3 c_{23}), \\ m_{22} &= I_2 + I_3 + m_2 r_2^2 + m_3(L_2^2 + r_3^2 + 2L_2 r_3 c_3), \\ m_{23} &= m_{32} = I_3 + m_3(r_3^2 + L_2 r_3 c_3), \\ m_{33} &= I_3 + m_3 r_3^2, \end{aligned}$$

and m_i is the mass of link i, I_i is the inertia of link i about its center of mass, and $c_2 = \cos\theta_2, c_3 = \cos\theta_3$, $c_{23} = \cos(\theta_2 + \theta_3)$.

The 3×2 input matrix $T(q)$ consists of two of the three columns $[1, 0, 0]^T$, $[0, 1, 0]^T$, and $[0, 0, 1]^T$, depending on which actuator is missing. This system can be written in the form (Equations 42.21 and 42.22). Accordingly, this system is subject to an acceleration constraint. If the missing actuator is at the first joint, the acceleration constraint integrates to a velocity constraint—the angular momentum of the robot is conserved. The acceleration constraint does not integrate to a velocity constraint when the actuator is missing at the second or the third joint.

Example 42.7: Underactuated Surface Vessel

The underactuated surface vessel is another system with $k = 0$, but now with added damping. The problem is to control the Cartesian position and orientation of a vessel with two independent

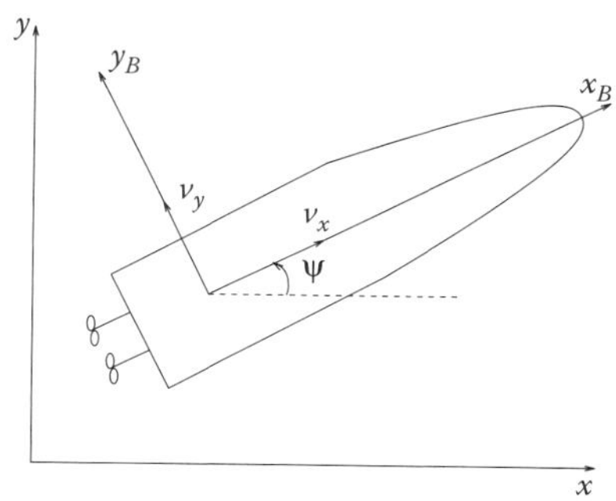

FIGURE 42.5 Underactuated surface vessel.

propellers as shown in Figure 42.5. The kinematic model describing the geometrical relationship between the earth-fixed (I-frame) and the vehicle-fixed (B-frame) motion is given as

$$\dot{x} = v_x \cos\psi - v_y \sin\psi, \tag{42.27}$$

$$\dot{y} = v_x \sin\psi + v_y \cos\psi, \tag{42.28}$$

$$\dot{\psi} = \omega_z, \tag{42.29}$$

where (x, y) denotes the I-frame position of the center of mass of the vehicle, ψ denotes the orientation, and (v_x, v_y) and ω_z are the linear and angular velocities of the vehicle in the B-frame. For simplicity, the origin of the B-frame is assumed to be located at the center of mass of the vehicle. It is also assumed that the vehicle is neutrally buoyant. Then the dynamic equations of motion of the vehicle can be expressed in the B-frame as

$$M\dot{\nu} + C(\nu)\nu + D(\nu)\nu = \tau, \tag{42.30}$$

where $\nu = [v_x, v_y, \omega_z]^T$ denotes the velocity vector; $\tau = [F_x, 0, T_z]^T$ denotes the vector of external force and torque generated by the two propellers; $M \in \mathbb{R}^{3\times3}$ is the inertia matrix, including hydrodynamic added mass; and $C(\nu) \in \mathbb{R}^{3\times3}$ and $D(\nu) \in \mathbb{R}^{3\times3}$ denote the Coriolis/centrifugal and damping matrices, respectively. The following simplified model can be obtained by assuming that both the inertia matrix M and the damping matrix D are constant and diagonal:

$$m_{11}\dot{v}_x - m_{22}v_y\omega_z + d_{11}v_x = F_x, \tag{42.31}$$

$$m_{22}\dot{v}_y + m_{11}v_x\omega_z + d_{22}v_y = 0, \tag{42.32}$$

$$m_{33}\dot{\omega}_z + (m_{22} - m_{11})v_xv_y + d_{33}\omega_z = T_z, \tag{42.33}$$

where m_{ii}, d_{ii}, $i = 1, 2, 3$, are positive constants.

The surface vessel considered here has no side thruster, that is, $F_y = 0$, but the controllability analysis and control synthesis can be easily extended to the cases where $\tau = [F_x, F_y, 0]^T$ or $\tau = [0, F_y, T_z]^T$.

Let (ξ, η) denote the position of the center of mass of the vehicle in the B-frame, which is given by

$$(\xi, \eta) = (x\cos\psi + y\sin\psi, -x\sin\psi + y\cos\psi).$$

Denote by $\mathcal{Q} = \mathbb{S}^1 \times \mathbb{R}^2$ the configuration space parameterized by $q = [\psi, \xi, \eta]^T$.

By defining control input transformations from (F_x, T_z) to new control inputs (u_1, u_2) the equations of motion can be written as

$$\ddot{\psi} = u_1, \tag{42.34}$$

$$\ddot{\xi} = u_2, \tag{42.35}$$

$$\ddot{\eta} = -\xi u_1 - \alpha(\dot{\eta} + \dot{\psi}\xi) - (1+\beta)\dot{\psi}\dot{\xi} + \beta\dot{\psi}^2\eta, \tag{42.36}$$

where $\alpha = d_{22}/m_{22}$ and $\beta = m_{11}/m_{22}$.

Equation 42.36 can be rewritten as

$$\ddot{\eta} = -\xi\ddot{\psi} - \alpha(\dot{\eta} + \dot{\psi}\xi) - (1+\beta)\dot{\psi}\dot{\xi} + \beta\dot{\psi}^2\eta.$$

This equation represents a nonintegrable relation involving not only the generalized coordinates and velocities but also the generalized accelerations and, hence, it can be viewed as a second-order nonholonomic constraint.

42.3 Controllability

In this section we briefly review some nonlinear controllability definitions. We then provide tests for these properties, including tests specific to mechanical systems. Further reading on definitions and tests for controllability can be found in [2,11,13,20,24].

42.3.1 Controllability Definitions

Let V be a neighborhood of a point $x \in \mathcal{M}$. Let $R^V(x, T)$ indicate the set of reachable points at time T by trajectories remaining inside V and satisfying Equation 42.1, and let

$$R^V(x, \le T) = \bigcup_{0<t\le T} R^V(x, t).$$

We define the following versions of nonlinear controllability:

- The system is *controllable* from x if, for any $x_f \in \mathcal{M}$, there exists a $T > 0$ such that $x_f \in R^{\mathcal{M}}(x, \le T)$. In other words, any goal state is reachable from x in finite time.
- The system is *accessible* from x if $R^{\mathcal{M}}(x, \le T)$ contains a full-dimensional subset of $\mathcal{M}$ for some $T > 0$.
- The system is *locally strongly accessible* from x if $R^V(x, T)$ contains a full n-dimensional subset of $\mathcal{M}$ for all neighborhoods V and all sufficiently small $T > 0$.
- The system is *small-time locally accessible* (*STLA*) from x if $R^V(x, \le T)$ contains a full n-dimensional subset of $\mathcal{M}$ for all neighborhoods V and all $T > 0$. This is a weaker property than strong accessibility.
- The system is *small-time locally controllable* (*STLC*) from x if $R^V(x, \le T)$ contains a neighborhood of x for all neighborhoods V and all $T > 0$.

The phrase "small-time" indicates that the property holds for any time $T > 0$, and "locally" indicates that the property holds for arbitrarily small (but full-dimensional) wiggle room around the initial state.

Small-time local controllability is of particular interest. STLC implies that the system can locally maneuver in any direction, and if the system is STLC at all $x \in \mathcal{M}$, then the system can follow any curve on $\mathcal{M}$ arbitrarily closely. This allows the system to maneuver through cluttered spaces, since any motion of a system with no motion constraints can be approximated by a system that is STLC everywhere.

For second-order mechanical systems, by definition STLC can only hold at zero velocity states. Nonzero velocity means that in small time, the system will necessarily have changed its configuration in the direction of the velocity, effectively placing a unilateral constraint on the small-time reachable configurations. Also, for mechanical systems with velocity constraints, controllability and accessibility questions only make sense on the $(2n-k)$-dimensional distribution $\mathcal{D} \subset T\mathcal{Q}$.

For mechanical systems, we are often particularly interested in understanding the reachable configurations irrespective of the velocity. This is a more restricted question than understanding the reachable states. Beginning from an equilibrium state $x = [q^T, 0^T]^T$, a system is *small-time locally configuration accessible* (*STLCA*) from q if the locally reachable set is full-dimensional on $\mathcal{Q}$, and *small-time locally configuration controllable* (*STLCC*) from q if the locally reachable set on $\mathcal{Q}$ contains q in the interior. A stronger condition than STLCC is *small-time local equilibrium controllability* (*STLEC*) from q, which holds if the locally reachable set contains zero velocity states forming a neighborhood of q on $\mathcal{Q}$. STLEC is stronger than STLCC, as STLEC demands that nearby configurations be reachable at zero velocity, while STLCC says nothing about the velocity. Finally, the system is *equilibrium controllable* if any equilibrium state can be reached from any other equilibrium state.

In some cases, second-order mechanical systems can be modeled by first-order kinematic systems. Consider a second-order mechanical system written as

$$\dot{x} = f(x) + \sum_{i=1}^{m} u_i g_i(x), \tag{42.37}$$

and a first-order kinematic system

$$\dot{q} = \sum_{j=1}^{\ell} w_j v_j(q), \quad \text{where } q, v_j(q) \in \mathbb{R}^n, \tag{42.38}$$

where the velocity controls w_j are continuous. The kinematic system (Equation 42.38) is a *kinematic reduction* of Equation 42.37 if all feasible trajectories of Equation 42.38 are feasible for Equation 42.37. We further say that a mechanical system (Equation 42.37) is *maximally reducible to a kinematic system* if there exists a kinematic reduction such that all feasible trajectories of the mechanical system, starting with an initial configuration q_0 and velocity in the column space of $Y(q_0)$ (defined in Equation 42.13), are also trajectories of the kinematic reduction. For example, all fully actuated mechanical systems are maximally reducible to kinematic systems—we can equivalently assume the controls are forces or continuous velocities.

If there exist kinematic reductions such that the kinematic model is STLC on configuration space, the second-order mechanical system is called *small-time locally kinematically controllable* (STLKC). STLKC implies STLEC.

42.3.2 Controllability Tests

42.3.2.1 General Controllability Concepts

STLA and STLC are local concepts that can be established by looking at the Lie algebra of the system vector fields in a neighborhood of a state (i.e., using Taylor expansions of the vector fields at x). Accessibility and controllability, on the other hand, are global concepts. As a result, they may depend on things such as the topology of the space and nonlocal behavior of the system vector fields. In this section we focus on local tests.

The *Lie bracket* of vector fields g_1 and g_2 is another vector field, given in local coordinates as

$$[g_1, g_2] = \frac{\partial g_2}{\partial x} g_1 - \frac{\partial g_1}{\partial x} g_2.$$

To order ϵ^2, this is the direction of the net motion obtained by flowing along the vector field g_1 for time ϵ, g_2 for time ϵ, $-g_1$ for time ϵ, and $-g_2$ for time ϵ, where $\epsilon \ll 1$. We define the *Lie algebra* of a set of vector fields $\mathcal{G}$, written $\overline{\text{Lie}}\{\mathcal{G}\}$, to be the linear span of all iterated Lie brackets of vector fields in $\mathcal{G}$. For example, for $\mathcal{G} = \{g_1, g_2, g_3\}$, the Lie algebra is spanned by $g_1, g_2, g_3, [g_1, g_2], [g_1, g_3], [g_1, [g_1, g_2]], [g_3, [g_1, [g_2, g_3]]]$, and so on. Each of these terms is called a Lie product, and the degree of a Lie product is the total number of original vector fields in the product. Lie products satisfy the Jacobi identity $[g_1, [g_2, g_3]] + [g_2, [g_3, g_1]] + [g_3, [g_1, g_2]] = 0$, and this fact can be used to find a *Philip Hall basis*, a subset all possible Lie products that spans the Lie algebra [18].

Theorem 42.1:

The system (Equation 42.1) is STLA from x if (and only if for analytic vector fields) it satisfies the Lie algebra rank condition (LARC)—the Lie algebra of the vector fields, evaluated at x, is the tangent space at x, that is, $\overline{\text{Lie}}(\{f, g_1, \ldots, g_m\})(x) = T_x\mathcal{M}$. *If* $f = 0$, *and by assumption the control set is* $\mathcal{U} = \mathbb{R}^m$, *then the system is symmetric and the LARC also implies small-time local controllability.*

An early version of this result is due to W.-L. Chow, and it is sometimes called *Chow's theorem.*

Example 42.8:

The control vector fields for the kinematic rolling disk are $g_1 = [R\cos\phi, R\sin\phi, 1, 0]^T$ and $g_2 = [0, 0, 0, 1]^T$. We define

$$g_3 = [g_1, g_2] = [R\sin\phi, -R\cos\phi, 0, 0]^T,$$
$$g_4 = [g_2, [g_1, g_2]] = [R\cos\phi, R\sin\phi, 0, 0]^T,$$

and note that $\mathcal{G} = \{g_1, \ldots, g_4\}$ spans $T_x\mathcal{M}$ at all x. Because $f = 0$, the rolling disk is STLC at all x.

Second-order systems with nonzero drift are not symmetric for any control set. The system may still be STLC at zero velocity states, however, as symmetry plus the LARC is sufficient but not necessary for STLC. Sussmann [25] provided a more general sufficient condition for STLC that includes the symmetric case ($f = 0$) as a special case. To understand it, we first define a Lie product term to be a *bad bracket* if the drift term f appears an odd number of times in the product and each control vector field g_i, $i = 1, \ldots, m$, appears an even number of times (including zero). A *good bracket* is any Lie product that is not bad. For example, $[g_1, [f, g_1]]$ is a bad bracket and $[g_2, [g_1, [f, g_1]]]$ and $[g_1, [g_2, [g_1, g_2]]]$ are good brackets. With these definitions, we can state a version of Sussmann's theorem:

Theorem 42.2:

The system (Equation 42.1) is STLC at x if

1. $f(x) = 0$
2. *The LARC is satisfied by good Lie bracket terms up to degree k*
3. *Any bad bracket of degree* $j \leq k$ *can be expressed as a linear combination of good brackets of degree less than j*

The intuition behind the theorem is the following. Bad brackets are called bad because, after generating the net motion obtained by following the Lie bracket motion prescription, we find that the controls u_i only appear in the net motion with even exponents, meaning that the vector field can only be followed in one direction. In this sense, a bad bracket is similar to a drift field, and we must be able to compensate for it. Since motions in Lie product directions of high degree are essentially "slower" than those in directions with a lower degree, we should only try to compensate for bad bracket motions by good bracket motions of lower degree. If a bad bracket of degree j can be expressed as a linear combination of good brackets of degree less than j, the good brackets are said to *neutralize* the bad bracket. For the bad bracket of degree 1 (the drift vector field f), there are no lower degree brackets that can be used to neutralize it; hence we require $f(x) = 0$. Therefore, this result only holds at states x where the drift vanishes, that is, equilibrium states.

Example 42.9:

For the dynamic rolling disk, the configuration variables are $[x_1, x_2, x_3, x_4]^T = [\psi, \phi, x, y]^T$, the drift vector field is $f = [x_5, x_6, x_5 R\cos x_2, x_5 R\sin x_2, 0, 0]^T$, and the control vector fields are $g_1 = [0, 0, 0, 0, 1/(J + mR^2), 0]^T$ and $g_2 = [0, 0, 0, 0, 0, 1/I]^T$, each defined on the six-dimensional distribution $\mathcal{D}$. We define the good Lie brackets $g_3 = [f, g_1]$, $g_4 = [f, g_2]$, $g_5 = [[f, g_1], [f, g_2]]$, $g_6 = [[[f, g_1], [f, g_2]], [f, g_2]]$, and observe that $\mathcal{G} = \{g_1, \ldots, g_6\}$ spans $T_x\mathcal{M}$ at all x. By Theorem 42.1, the system is STLA at all x. The maximum degree of good Lie brackets in $\mathcal{G}$ is six, and we find that all bad brackets of equal or lesser degree (degree 1, 3, and 5) are zero or can be neutralized by good brackets of lower degree. Therefore, at zero velocity states where $f(x) = 0$, the dynamic rolling disk is STLC.

42.3.2.2 Simple Mechanical Systems with Zero Potential

For the case of second-order mechanical systems (Equations 42.7 and 42.8) with no damping or potential terms ($U(q) = 0$), we can use the structure of the Lie brackets to derive simplified controllability tests. See [11,13] for details.

We define the m columns of $Y(q)$ (from Equation 42.13) as the input vector fields $y_1(q), \ldots, y_m(q)$. The (constrained) *covariant derivative* of $y_2(q)$ with respect to $y_1(q)$ is

$$\nabla_{y_1(q)} y_2(q) = P(q)\left(\frac{\partial y_2(q)}{\partial q} y_1(q) + M^{-1}(q) y_1^T(q)\Gamma(q) y_2(q)\right). \tag{42.39}$$

With this definition, we can define the *symmetric product* of y_1 and y_2 as the vector field

$$\langle y_1 : y_2 \rangle = \nabla_{y_1} y_2 + \nabla_{y_2} y_1.$$

We can use the symmetric product in simplified controllability calculations for mechanical systems with no potential. The symmetric product captures patterns that appear when taking Lie brackets of the full system vector fields $f(x)$ and $g_i(x)$. Advantages of using the symmetric product are that the vector fields have half the number of elements, and controllability properties may be established by products of lower degree than with Lie brackets. As a result, the number of controllability computations is reduced.

Let $\mathcal{Y} = \{y_1(q), \ldots, y_m(q)\}$ denote the set of input vector fields, and let $\overline{\text{Sym}}(\mathcal{Y})$ be the span of $\mathcal{Y}$ and its iterated symmetric products. A symmetric product is "bad" if each of the vector fields appears an even number of times, and is "good" otherwise.

Theorem 42.3:

Beginning from an equilibrium state $x = [q^T, 0^T]^T$, the system (Equation 42.13) with zero potential is

1. *STLA from x if and only if $\overline{\text{Sym}}(\mathcal{Y})(q) = \mathcal{D}_q$ and $\overline{\text{Lie}}(\mathcal{D})(q) = T_q\mathcal{Q}$.*
2. *STLC from x if $\overline{\text{Sym}}(\mathcal{Y})(q) = \mathcal{D}_q$ and $\overline{\text{Lie}}(\mathcal{D})(q) = T_q\mathcal{Q}$, and every bad symmetric product can be expressed as a linear combination of good symmetric products of lower degree.*
3. *STLCA from q if and only if $\overline{\text{Lie}}(\overline{\text{Sym}}(\mathcal{Y}))(q) = T_q\mathcal{Q}$.*
4. *Both STLCC and STLEC from q if $\overline{\text{Lie}}(\overline{\text{Sym}}(\mathcal{Y}))(q) = T_q\mathcal{Q}$ and if every bad symmetric product can be expressed as a linear combination of good symmetric products of lower degree. If these conditions are satisfied at all $q \in \mathcal{Q}$, then the system is equilibrium controllable.*

We can also use the symmetric product to determine if the mechanical system is reducible to an equivalent kinematic system.

Theorem 42.4:

A second-order mechanical system is maximally reducible to a kinematic system if and only if $\overline{\text{Sym}}(\mathcal{Y}) = \text{span}(\mathcal{Y})$.

Even when the underactuated mechanical system is not fully reducible to a kinematic system, it may admit one or more kinematic reductions consisting of a single vector field. Such a kinematic reduction is also known as a *decoupling vector field*—a velocity vector field that the full mechanical system can follow at any speed.

Theorem 42.5:

A velocity vector field $v(q)$ *is a decoupling vector field of the second-order mechanical system if and only if* $v \in \text{span}(\mathcal{Y})$ *and* $\nabla_v v \in \text{span}(\mathcal{Y})$.

Example 42.10:

The input vector fields for the dynamic rolling disk are computed to be $y_1(q) = (J + mR^2)^{-1}(\partial/\partial\psi + (R\cos\phi)\partial/\partial x + (R\sin\phi)\partial/\partial y)$ and $y_2(q) = (1/I)\partial/\partial\phi$. The disk is trivially shown to be STLC at zero velocity states by Theorem 42.3 as $\text{span}(\{y_1, y_2\})(q) = \mathcal{D}_q$ for all q, and we have already shown that $\overline{\text{Lie}}(\mathcal{D})(q) = T_C\mathcal{Q}$. Further, by Theorem 42.4, the disk is maximally reducible to the kinematic disk system, with velocity vector fields $v_1 = y_1$, $v_2 = y_2$, and the system is STLKC with these vector fields.

Example 42.11:

For the snakeboard example, the two input vector fields y_1, y_2 of $Y(q)$ in Equation 42.13 correspond to torque rotating the rotor and the wheels, respectively. The snakeboard is STLC at zero velocity states, and therefore STLCC and STLEC, by Theorem 42.3. The vector fields y_1, y_2 are also decoupling vector fields making the system STLKC. The snakeboard is not maximally reducible to a kinematic system, however.

Example 42.12:

For the 3R robot with a missing actuator at the first joint, the acceleration constraint integrates to a velocity constraint (conservation of angular momentum), and therefore the system is constrained to a lower-dimensional subspace of $T\mathcal{Q}$. This system is maximally reducible to a kinematic system.

For a missing actuator at the third joint, the system admits two decoupling vector fields: translation along the length of the third link and rotation of the third link about a point on the link a distance $r_3 + I_3/m_3r_3$ from the third joint. (This point is known as the *center of percussion* of the link with respect to the third joint.) The 3R robot is STLC at zero velocity states, STLCC, STLEC, and STLKC away from the singularity where $\theta_2 \in \{0, \pi\}$. It is not maximally reducible to a kinematic system, however. See [12] for details.

42.3.2.3 Mechanical Systems with Nonintegrable Acceleration Relations and No Velocity Constraints

This section develops controllability and stabilizability results for underactuated systems with completely nonintegrable acceleration relations.

Consider Equations 42.21 and 42.22, and let I_p denote the set $\{1, \ldots, p\}$ for any integer $p \geq 1$. Define the $n-m$ covector fields

$$\omega^i = \sum_{j=1}^{m} J_{ij}(q)d\dot{q}_{1,j} - d\dot{q}_{2,i} + R_i(q,\dot{q})dt, \quad i \in I_{n-m}, \tag{42.40}$$

on $\mathcal{M} \times \mathbb{R}$ so that the $n-m$ relations given by Equation 42.22 can be rewritten as $\omega^i = 0$, $i \in I_{n-m}$. Augment the covector fields (Equation 42.40) with

$$\tilde{\omega}^j = dq_{1,j} - \dot{q}_{1,j}dt, \quad j \in I_m, \tag{42.41}$$

$$\tilde{\omega}^{m+k} = dq_{2,k} - \dot{q}_{2,k}dt, \quad k \in I_{n-m}, \tag{42.42}$$

and let $\Omega \subset T^*(\mathcal{M} \times \mathbb{R})$ denote the codistribution

$$\Omega = \text{span}\{\omega^i, \tilde{\omega}^j, i \in I_{n-m}, j \in I_n\}. \tag{42.43}$$

The annihilator of Ω, denoted $\Omega^\perp$, is spanned by $m+1$ linearly independent smooth vector fields

$$\tau_0 = \sum_{j=1}^{m} \dot{q}_{1,j}\frac{\partial}{\partial q_{1,j}} + \sum_{k=1}^{n-m}\left(\dot{q}_{2,k}\frac{\partial}{\partial q_{2,k}} + R_k(q,\dot{q})\frac{\partial}{\partial \dot{q}_{2,k}}\right) + \frac{\partial}{\partial t}, \tag{42.44}$$

$$\tau_j = \frac{\partial}{\partial \dot{q}_{1,j}} + \sum_{i=1}^{n-m} J_{ij}(q)\frac{\partial}{\partial \dot{q}_{2,i}}, \ j \in I_m. \tag{42.45}$$

We present the following definition.

Definition 42.1: [21]

Consider the distribution $\Omega^\perp$ and let $\tilde{\mathcal{C}}$ denote its accessibility algebra, that is, the smallest subalgebra of $\mathbf{V}^\infty(\mathcal{M} \times \mathbb{R})$ that contains $\tau_0, \tau_1, \ldots, \tau_m$. Let $\tilde{C}$ denote the accessibility distribution generated by the accessibility algebra $\tilde{\mathcal{C}}$. Then the acceleration relations defined by Equation 42.22 are said to be completely nonintegrable if

$$\dim\ \tilde{C}(x,t) = 2n+1, \quad \forall (x,t) \in \mathcal{M} \times \mathbb{R}.$$

Note that the above definition gives a coordinate-free characterization of nonintegrability for any set of acceleration relations of the form Equation 42.22. Note also that this definition is analogous to the definition given in [2] for the nonintegrability of a set of kinematic or velocity relations.

Examples of underactuated systems with completely nonintegrable acceleration relations include underactuated robot manipulators (such as our 3R robot), underactuated marine vehicles, the planar vertical takeoff and landing (PVTOL) aircraft, the rotational translational actuator (RTAC) system, the acrobot system, the pendubot system, and examples in [21].

In the rest of this section, we assume that the acceleration constraints (Equation 42.22) are completely nonintegrable. With this assumption, the underactuated mechanical system (Equations 42.21 and 42.22) is strongly accessible [21].

A particularly important class of solutions of Equations 42.21 and 42.22 are the equilibrium solutions with $v(t) = 0$, $\forall t \geq 0$. A solution is an equilibrium solution if it is a constant solution. If $(q, \dot{q}) = (q_e, 0)$ is an equilibrium solution, we refer to q_e as an equilibrium configuration. Clearly, the set of equilibrium configurations of the system (Equations 42.21 and 42.22) is given by

$$\{q \in \mathcal{Q} \mid R(q, 0) = 0\}.$$

It is well-known that strong accessibility is far from being sufficient for the existence of a feedback control that asymptotically stabilizes the underactuated system at an equilibrium solution. In certain cases it is possible to prove STLC at equilibrium states, which guarantees the existence of a piecewise analytic feedback law for asymptotic stabilization in the real analytic case. Since an underactuated mechanical system satisfies $1 \leq m < n$, the dimension of the state is at least four. Hence, in the real-analytic case, the STLC property also guarantees the existence of asymptotically stabilizing continuous time-periodic feedback laws (see, for example, [21] and the references therein).

We now briefly summarize a result of Bianchini and Stefani [1] that generalizes Theorem 42.2 and is utilized to prove the subsequent controllability results. Let $Br(X)$ denote the smallest Lie algebra of vector fields containing $f, g_1, \ldots, g_m$ and let B denote any bracket in $Br(X)$. Let $\delta^0(B), \delta^1(B), \ldots, \delta^m(B)$ denote the number of times $f, g_1, \ldots, g_m$, respectively, occur in the bracket B. For an admissible weight vector $\mathbf{l} = (l_0, l_1, \ldots, l_m)$, $l_i \geq l_0 \geq 0$, $\forall i$, the $\mathbf{l}$-degree of B is equal to the value of $\sum_{i=0}^{m} l_i \delta^i(B)$. The Bianchini and Stefani condition for STLC for a strongly accessible system is essentially that the so-called bad brackets, the brackets with $\delta^0(B)$ odd and $\delta^i(B)$ even for each i, must be $\mathbf{l}$-neutralized, that is, must be a linear combination of good brackets of lower $\mathbf{l}$-degree at the equilibrium.

Consider the system Equations 42.21 and 42.22, and rewrite the drift and control vector fields as

$$f = [\dot{q}_1^T, \dot{q}_2^T, 0^T, R^T(q, \dot{q})]^T,$$
$$g_j = [0^T, 0^T, e_j^T, J_j^T(q)]^T, \quad j \in I_m.$$

The following Lie bracket calculations are straightforward:

$$[g_j, g_i] \equiv 0, \quad i, j \in I_m,$$
$$[f, g_i] = [-e_i^T, -J_i^T(q), 0^T, *]^T, \quad i \in I_m,$$
$$[g_j, [f, g_i]] = [0^T, 0^T, 0^T, H_{ij}^T(q)]^T, \quad i, j \in I_m,$$
$$[f, [g_j, [f, g_i]]] = [0^T, -H_{ij}^T(q), 0^T, *]^T, \quad i, j \in I_m,$$

where

$$H_{ij}(q) = \frac{\partial J_i(q)}{\partial q} h_j(q) + \frac{\partial J_j(q)}{\partial q} h_i(q) - \frac{\partial}{\partial \dot{q}} \left(\frac{\partial R(q, \dot{q})}{\partial \dot{q}} h_i(q) \right) h_j(q), \quad i, j \in I_m, \tag{42.46}$$

$$h_i(q) = \begin{bmatrix} e_i \\ J_i(q) \end{bmatrix}, \quad i \in I_m. \tag{42.47}$$

Note that the vertical lift of h_i (considered as a vector field on the configuration space $\mathcal{Q}$) is the control vector field g_i. Note also that $H_{ij}(q) = H_{ji}(q)$, $\forall q \in \mathcal{Q}$, $\forall i, j \in I_m$.

We now present the following result.

Theorem 42.6:

Let $n - m \geq 1$ and let $(q_e, 0)$ denote an equilibrium solution. The underactuated mechanical system, defined by Equations 42.21 and 42.22, is STLC at $(q_e, 0)$ if there exists a set of $n - m$ pairs of indices $(i_k, j_k) \in I_m^2$

such that

$$\dim \operatorname{span}\{H_{i_k j_k}(q_e),\ i_k \neq j_k,\ k \in I_{n-m}\} = n - m \tag{42.48}$$

and $H_{ii}(q_e)$, $i \in I_m$, *can be written as a linear combination of* $H_{i_k j_k}(q_e)$, $i_k \neq j_k$, $k \in I_{n-m}$, *such that* $l_0 + 2l_i > l_0 + l_{i_k} + l_{j_k}$ *for some admissible weight vector* $\mathbf{l} = (l_0, l_1, \ldots, l_m)$, $l_i \geq l_0 \geq 0$, $\forall i$.

Note that for the condition (Equation 42.48) to hold, the condition $m(m+1) \geq 2n$ must be satisfied. This condition arises due to the fact that in the above result we have considered Lie brackets up to degree four only. It is possible to develop a result that weakens or even removes this restriction by also taking into account higher-order Lie brackets.

We now restrict our consideration to underactuated mechanical systems with no potential or friction forces. For such systems, when evaluated at the equilibrium, the only nontrivial brackets are those satisfying $\sum_{i=1}^{m} \delta^i(B) - \delta^0(B) = 0$ or $\sum_{i=1}^{m} \delta^i(B) - \delta^0(B) = 1$. Clearly, the brackets with $\sum_{i=1}^{m} \delta^i(B) - \delta^0(B) = 0$ are all good, and the only bad brackets are those with $\sum_{i=1}^{m} \delta^i(B) - \delta^0(B) = 1$, $\delta^0(B)$ odd, and $\delta^i(B)$ even, $\forall i \in I_m$.

We define the following sequence of collections of vector fields:

$$\begin{aligned}
\mathcal{G}_1 &= \{g_i,\ i \in I_m\},\\
\mathcal{G}_k &= \{[X, [f, Y]],\ X \in \mathcal{G}_i,\ Y \in \mathcal{G}_j,\ k = i + j\}, \quad k \geq 2,\\
\mathcal{G} &= \bigcup_{i \geq 2} \mathcal{G}_i.
\end{aligned}$$

Let X denote a vector field in $\mathcal{G}$. It is easy to show that X has the form $X = [0^T, 0^T, 0^T, N^T(q)]^T$, where $N(q)$ is an $n - m$ vector function, and its Lie bracket with f can be written as $[f, X] = [0^T, -N^T(q), 0^T, *]^T$. Now let $(q_e, 0)$ denote an equilibrium solution. Clearly, if there exists an integer $k^* \geq 2$ such that

$$\dim \operatorname{span}\left\{X(q_e, 0),\ X \in \bigcup_{i=2}^{k^*} \mathcal{G}_i\right\} = n - m, \tag{42.49}$$

then the system is strongly accessible at $(q_e, 0)$, that is, the system satisfies a necessary condition for STLC at $(q_e, 0)$. As shown in [21], all the bad brackets can be written as linear combinations of the bad brackets contained in $\mathcal{G}$. Thus, a sufficient condition for STLC at $(q_e, 0)$ can be obtained by considering the bad brackets in $\mathcal{G}$ and applying the Bianchini and Stefani condition.

The following result can now be stated.

Theorem 42.7:

Let $n - m \geq 1$ *and let* $(q_e, 0)$ *denote an equilibrium solution. Consider the underactuated mechanical system, defined by Equations 42.21 and 42.22, and assume that the components of* $R(q, \dot{q})$ *are of second-order in* $\dot{q}$*-variables. Also assume that the condition (Equation 42.49) is satisfied. Then, the system is STLC at* $(q_e, 0)$ *if there exists an admissible weight vector* $\mathbf{l} = (l_0, l_1, \ldots, l_m)$, $l_i \geq l_0 \geq 0$, $\forall i$, *such that every bad bracket in* $\mathcal{G}_{2k}$, $k \in \mathbb{Z}^+$, *can be* $\mathbf{l}$*-neutralized.*

Example 42.13:

Consider the surface vessel dynamics described by Equations 42.34 through 42.36. It can be shown that the following hold:

1. The system is strongly accessible since the space spanned by the vectors

$$g_1,\ g_2,\ [f,g_1],\ [f,g_2],\ [g_2,[f,g_1]],[f,[g_2,[f,g_1]]]$$

   has dimension 6 at any $(q,\dot{q}) \in \mathcal{M} = T\mathcal{Q}$.
2. The system is STLC at any equilibrium $(q_e, 0)$ since the sufficient conditions for STLC of Theorem 42.6 are satisfied with the admissible weight vector $\mathbf{l} = (2, 4, 3)$.
3. There exist both time-invariant piecewise analytic feedback laws and time-periodic continuous feedback laws that asymptotically stabilize $(q_e, 0)$.

These controllability properties guarantee the existence of the solution to the problem of controlling the surface vessel to any desired equilibrium configuration. The time-invariant discontinuous feedback control laws developed in [22] represent such solutions.

42.4 Feedback Stabilization

A beautiful general theorem on necessary conditions for feedback stabilization of nonlinear systems was given in [9].

Theorem 42.8: Brockett

Consider the nonlinear system $\dot{x} = f(x,u)$ with $f(x_0, 0) = 0$ and $f(\cdot,\cdot)$ continuously differentiable in a neighborhood of $(x_0, 0)$. Necessary conditions for the existence of a continuously differentiable control law for asymptotically stabilizing $(x_0, 0)$ are:

1. *The linearized system has no uncontrollable modes associated with eigenvalues with positive real part.*
2. *There exists a neighborhood N of $(x_0, 0)$ such that for each $\xi \in N$ there exists a control $u_\xi(t)$ defined for all $t > 0$ that drives the solution of $\dot{x} = f(x, u_\xi)$ from the point $x = \xi$ at $t = 0$ to $x = x_0$ at $t = \infty$.*
3. *The mapping $\gamma : N \times \mathbb{R}^m \to \mathbb{R}^n$, N a neighborhood of the origin, defined by $\gamma : (x,u) \to f(x,u)$ should be onto an open set of the origin.*

Based on an extension of this result to continuous feedback, the key consequence for kinematic systems is the following.

Proposition 42.1:

Consider the system

$$\dot{x} = \sum_{k=1}^{m} u_k g_k, \tag{42.50}$$

where $x \in \mathbb{R}^n$, $m < n$, and

$$\text{rank}\{g_1(0), \ldots, g_m(0)\} = m.$$

There exists no continuous feedback that locally asymptotically stabilizes the origin.

One way to deal with this problem is to use time-varying feedback. A key result on stabilization by time-varying feedback is that of [15]. Another approach is nonsmooth feedback (see below).

42.4.1 Kinematic Example: The Heisenberg System

In this section we consider the problem of feedback stabilization of the Heisenberg system Equations 42.4 through 42.6 to an equilibrium or a trajectory. This serves as a model system for the kinematic vertical disk and many wheeled mobile robots, for example. We then extend the results to the generalized Heisenberg system, which can be shown to be equivalent to a broad class of first-order nonholonomic systems.

42.4.1.1 Stabilization of the Heisenberg System

One of the possibilities suggested in [3] is to utilize sliding modes to stabilize the Heisenberg system, by applying the following feedback control law:

$$u_1 = -\alpha x_1 + \beta x_2 \operatorname{sign}(z), \tag{42.51}$$

$$u_2 = -\alpha x_2 - \beta x_1 \operatorname{sign}(z), \tag{42.52}$$

where α and β are positive constants. It is shown that if an initial condition is outside of the parabolic region defined by the inequality

$$\frac{\beta}{2\alpha}(x_1^2 + x_2^2) \leq |z|, \tag{42.53}$$

then the control (Equations 42.51 and 42.52) stabilizes the state z to zero in finite time, and on the manifold $z = 0$ the two remaining states x_1 and x_2 converge to the origin too. If the initial data are such that Equation 42.53 is true, that is, the state is inside the paraboloid, we can use any control law which steers it outside, and then apply (Equations 42.51 and 42.52).

Let us note here that because of the switchings, the above strategy assumes that the control input is a discontinuous function of the state variables. In such a system the sliding mode may exist when the solution moves along the set of discontinuity of the right-hand side (as happens with Equations 42.51 and 42.52 when $z = 0$). Thus, the existence of the corresponding solution of the differential equations (off the z-axis only) should be understood in the sense of the Filippov definition.

A similar way to stabilize system (Equations 42.4 through 42.6) to the origin is the following variable structure control law:

$$u_1 = -\alpha x_1 + \beta z x_2, \tag{42.54}$$

$$u_2 = -\alpha x_2 - \beta z x_1, \tag{42.55}$$

where $\alpha = \alpha(x_1, x_2, z)$ and $\beta = \beta(x_1, x_2, z)$ are positive state-dependent switching functions.

With this choice, the system (Equations 42.4 through 42.6) becomes

$$\dot{x}_1 = -\alpha x_1 + \beta z x_2, \tag{42.56}$$

$$\dot{x}_2 = -\alpha x_2 - \beta z x_1, \tag{42.57}$$

$$\dot{z} = -\beta z(x_1^2 + x_2^2). \tag{42.58}$$

There are a number of strategies for choosing α and β to stabilize the system. Let $V = x_1^2 + x_2^2$, then $\dot{V} = -2\alpha V$. It is clear from Equation 42.58 that if we initially choose $\alpha = 0$ and $\beta > 0$, then for x_1 or x_2 not equal to zero, z will be driven asymptotically to 0, while V will remain fixed. On the other hand, for $\alpha > 0$ and $\beta = 0$, V will be driven to 0. Thus, the stabilizing control in this case should be the following. (1) unless $V = 0$, apply $\alpha = 0$ and $\beta > 0$. This guarantees $z \to 0$. Then (2) apply $\alpha > 0$ and $\beta = 0$. In this case, while z remains zero, $V \to 0$. If initially $V = 0$, apply any other nonzero control to make $V \neq 0$, then apply (1). The convergence in phase (1) is asymptotic, but it is clear that we can achieve stabilization with any desired accuracy by alternating steps (1) and (2), if necessary.

As we show below, this strategy can be extended to the generalized system (Equations 42.80 and 42.81).

42.4.1.2 Tracking in the Heisenberg System

The robustness property of the sliding mode algorithm (Equations 42.51 and 42.52) allows trajectory tracking for arbitrary smooth three-dimensional trajectories, even though the control dimension is equal to two.

Let $X^*(t) = [x_1^*(t), x_2^*(t), z^*(t)]^T$ be an arbitrary smooth curve in $\mathbb{R}^3$. Let $\hat{z}$ be defined as

$$\hat{z}(t) = z(t) - x_1^*(t)x_2(t) + x_2^*(t)x_1(t). \tag{42.59}$$

Using Equations 42.4 through 42.6 and 42.59, the derivative of $\hat{z}$ can be written as

$$\dot{\hat{z}} = (x_1 - x_1^*)(u_2 - \dot{x}_2^*) - (x_2 - x_2^*)(u_1 - \dot{x}_1^*) + g(t, x_1, x_2), \tag{42.60}$$

where

$$g(t, x_1, x_2) = 2x_1\dot{x}_2^* - \dot{x}_2^* x_1^* - 2x_2\dot{x}_1^* + \dot{x}_1^* x_2^*. \tag{42.61}$$

The problem of tracking x_1^*, x_2^*, z^* by the variables x_1, x_2, z is equivalent to the one of stabilizing $\bar{x}_1 = x_1 - x_1^*, \bar{x}_2 = x_2 - x_2^*, \bar{z} = \hat{z} - z^*$.

The system (Equations 42.4 through 42.6) in the new variables can be written as

$$\dot{\bar{x}}_1 = \bar{u}_1, \tag{42.62}$$

$$\dot{\bar{x}}_2 = \bar{u}_2, \tag{42.63}$$

$$\dot{\bar{z}} = \bar{x}_1\bar{u}_2 - \bar{x}_2\bar{u}_1 + \bar{g}, \tag{42.64}$$

where we used the notations $\bar{u}_1 = u_1 - \dot{x}_1^*$, $\bar{u}_2 = u_2 - \dot{x}_2^*, \bar{g} = g - \dot{z}^*$.

If $\bar{g} \equiv 0$, that is, the desired trajectory satisfies the constraint

$$\dot{z}^* = x_1^*\dot{x}_2^* - x_2^*\dot{x}_1^*, \tag{42.65}$$

then the system (Equations 42.62 through 42.64) corresponds exactly to the Heisenberg system, and we can use the control Equations 42.51 and 42.52 or Equations 42.54 and 42.55 described in the previous section.

If $\bar{g} \neq 0$, then unlike in the previous case, perfect tracking is not possible, but we can track with arbitrary accuracy. In this case we can still apply control of the type (Equations 42.51 and 42.52), but with state-dependent α and β:

$$\bar{u}_1 = -\alpha\bar{x}_1 + \beta\bar{x}_2 \operatorname{sign}(\bar{z}), \tag{42.66}$$

$$\bar{u}_2 = -\alpha\bar{x}_2 - \beta\bar{x}_1 \operatorname{sign}(\bar{z}). \tag{42.67}$$

Substituting into Equations 42.62 through 42.64, we obtain

$$\dot{\bar{x}}_1 = -\alpha\bar{x}_1 + \beta\bar{x}_2 \operatorname{sign}(\bar{z}), \tag{42.68}$$

$$\dot{\bar{x}}_2 = -\alpha\bar{x}_2 - \beta\bar{x}_1 \operatorname{sign}(\bar{z}), \tag{42.69}$$

$$\dot{\bar{z}} = -\beta(\bar{x}_1^2 + \bar{x}_2^2) \operatorname{sign}(\bar{z}) + \bar{g}. \tag{42.70}$$

Here we assume that α is the following function of $\bar{x}_1$ and $\bar{x}_2$:

$$\alpha = \begin{cases} \alpha_0 & \text{if } \bar{x}_1^2 + \bar{x}_2^2 \geq \varepsilon^2 \\ \alpha_1 & \text{if } \bar{x}_1^2 + \bar{x}_2^2 < \varepsilon^2, \end{cases} \tag{42.71}$$

where $\alpha_0 > 0$ and $\alpha_1 < 0$ are constants, and $\varepsilon > 0$ is a constant that defines the tracking accuracy. Let us consider the Lyapunov function $V = \bar{x}_1^2 + \bar{x}_2^2$. As in the previous case, its derivative along the trajectories of the system (Equations 42.68 through 42.70) is $\dot{V} = -2\alpha V$. Due to this choice of α, the function

V converges to $V = \varepsilon^2 = \text{const.}$ for any initial conditions except for the origin. This means that for sufficiently large β such that

$$\beta\varepsilon^2 > |\bar{g}| \tag{42.72}$$

and for sufficiently large t in Equation 42.70,

$$\dot{\bar{z}} = -\beta V \operatorname{sign}(\bar{z}) + \bar{g} \tag{42.73}$$

we obtain sliding mode on the manifold

$$\bar{z} \equiv 0 \quad \text{for } t \geq t_1, \tag{42.74}$$

where the time t_1 can be chosen arbitrarily close to the initial moment by increasing β. In general, β should be chosen to be a function of x_1, x_2, and $X^*, \dot{X}^*$ to satisfy (Equation 42.72), but due to the separate convergence of $\bar{x}_1, \bar{x}_2$ to the ε-neighborhood of the origin. For a bounded reference curve with bounded derivative β can be constant.

Thus, the state trajectory $X(t) = [x_1(t), x_2(t), z(t)]^T$ enters an ε-neighborhood of the curve $X^*(t)$ not later than in time t_1 and stays in that neighborhood for all subsequent moments of time. The generalization of the ε-solution of the tracking problem for the case of higher-order systems can be found in [16].

42.4.1.3 The General Setting

In Section 42.2.1, we used the rolling vertical disk example to demonstrate the equivalence of the Heisenberg system to a particular class of nonholonomic systems. The generalization of this equivalence, at least locally, was obtained by Brockett [8]. He showed that controllable systems of the form

$$\dot{x} = B(x)u, \tag{42.75}$$

where $u \in \mathbb{R}^n$ and $x \in \mathbb{R}^{n(n+1)/2}$ is such that the first derived algebra of control vector fields spans the tangent space $T\mathbb{R}^{n(n+1)/2}$ at any point, is locally equivalent to

$$\dot{x} = u, \tag{42.76}$$

$$\dot{Y} = xu^T - ux^T, \tag{42.77}$$

where x and u are column vectors in $\mathbb{R}^n$ and $Y \in so(n)$, $n \geq 2$. Here $so(n)$ is the Lie algebra of $n \times n$ skew symmetric matrices: $Y^T = -Y$.

The system (Equations 42.4 through 42.6) is a particular case of the $so(n)$ system (Equations 42.76 and 42.77). That can be easily seen if we identify the variable z with the skew-symmetric matrix

$$Y = \begin{bmatrix} 0 & z \\ -z & 0 \end{bmatrix}$$

and observe that

$$\begin{bmatrix} x_1 \\ x_2 \end{bmatrix} [u_1, u_2] - \begin{bmatrix} u_1 \\ u_2 \end{bmatrix} [x_1, x_2] = \begin{bmatrix} 0 & x_1u_2 - x_2u_1 \\ -x_1u_2 + x_2u_1 & 0 \end{bmatrix}.$$

A different generalization of the Heisenberg system (Equations 42.4 through 42.6) is obtained by identifying the vectors $[x_1, x_2]^T$ and $[u_1, u_2]^T$ with the matrices

$$X = \frac{1}{\sqrt{2}} \begin{bmatrix} x_1 & -x_2 \\ -x_2 & -x_1 \end{bmatrix} \quad \text{and} \quad U = \frac{1}{\sqrt{2}} \begin{bmatrix} u_1 & -u_2 \\ -u_2 & -u_1 \end{bmatrix},$$

respectively. Then we have

$$[U,X] = UX - XU = \begin{bmatrix} 0 & x_1u_2 - x_2u_1 \\ -x_1u_2 + x_2u_1 & 0 \end{bmatrix}.$$

This suggests the following matrix system occurring in the Lie algebra $sl(n,\mathbb{R})$ of $n \times n$ matrices with trace 0:

$$\dot{X} = U, \tag{42.78}$$

$$\dot{Y} = [U,X], \tag{42.79}$$

where $X, U \in \text{sym}_0(n,\mathbb{R})$ and $Y \in so(n)$. Here $\text{sym}_0(n,\mathbb{R})$ is the space of $n \times n$ real symmetric matrices with zero trace. Note that $sl(n,\mathbb{R}) = \text{sym}_0(n,\mathbb{R}) \oplus so(n)$.

The system that generalizes both the $so(n)$ system (Equations 42.76 and 42.77) and the $sl(n,\mathbb{R})$ system (Equations 42.78 and 42.79) is the following system on a Lie algebra:

$$\dot{x} = u, \tag{42.80}$$

$$\dot{Y} = [u,x], \tag{42.81}$$

where $x, u \in \mathfrak{m}$, $Y \in \mathfrak{h}$ such that $\mathfrak{h}$ is a Lie subalgebra of a real semisimple Lie algebra $\mathfrak{g}$, and $\mathfrak{m}$ is the orthogonal complement of $\mathfrak{h}$ relative to the Killing form $B : \mathfrak{g} \times \mathfrak{g} \to \mathbb{R}$. The algebra $\mathfrak{g}$ has the structure of a direct sum decomposition $\mathfrak{g} = \mathfrak{m} \oplus \mathfrak{h}$, $[\mathfrak{h},\mathfrak{m}] \subseteq \mathfrak{m}$, and $[\mathfrak{m},\mathfrak{m}] = \mathfrak{h}$. We note that every simple Lie algebra with a Cartan decomposition is of this type.

The problem we consider herein is that of finding a stabilizing control for the general system (Equations 42.80 and 42.81). Since the dimension of $\mathfrak{m}$, which is where the control u takes its values, is smaller than the dimension of the state space $\mathfrak{g}$, the system fails Brockett's necessary condition for the existence of a continuous feedback law (see [9]). A discontinuous feedback for the state stabilization of the generalized Heizenberg system can be found in [3,4]. It completely solves the stabilization problem for Equations 42.80 and 42.81.

We now consider the general setting for stabilization of the system (Equations 42.80 and 42.81) in $\mathfrak{g} = \mathfrak{m} \oplus \mathfrak{h}$. The main part of the variable structure control is given by

$$u = -\alpha x + \beta[Y,x] - \gamma[Y,[Y,x]] = -\alpha x + \beta[Y,x] + \gamma N(Y)x, \tag{42.82}$$

where $\alpha, \beta, \gamma : \mathfrak{g} \to \mathbb{R}$ are switching functions, and $N(Y) = -(\text{ad}\, Y)^2|_{\mathfrak{m}}$ is a nonnegative symmetric operator on $\mathfrak{m}$ relative to the inner product. We will also need $M(x) = \epsilon(\text{ad}\, x)^2|_{\mathfrak{h}}$, a nonnegative symmetric operator on $\mathfrak{h}$ relative to the inner product. Here

$$\epsilon = \begin{cases} 1, & \text{if } \mathfrak{g} \text{ is of noncompact type} \\ -1, & \text{if } \mathfrak{g} \text{ is of compact type} \end{cases}, \tag{42.83}$$

Note that $(\text{ad}\, x)^2$ is nonnegative if $\mathfrak{g}$ is noncompact, and nonpositive if $\mathfrak{g}$ is compact, and therefore, introducing ϵ into the definition of $M(x)$ guarantees that it is a nonnegative operator. We will assume that $\alpha, \gamma \geq 0$ and $\beta\epsilon \leq 0$.

With (Equation 42.82) as our choice of u, the system (Equations 42.80 and 42.81) becomes

$$\dot{x} = -\alpha x + \beta[Y,x] + \gamma N(Y)x, \tag{42.84}$$

$$\dot{Y} = \beta\epsilon M(x)Y + \gamma[N(Y)x, x] = \beta\epsilon M(x)Y - \gamma\epsilon[Y, M(x)Y]. \tag{42.85}$$

Using (Equations 42.84 and 42.85), we compute

$$\frac{\text{d}}{\text{d}t}\|x\|^2 = -2\alpha\|x\|^2 + 2\gamma\langle x, N(Y)x\rangle \tag{42.86}$$

and

$$\frac{d}{dt}\|Y\|^2 = 2\beta\epsilon\langle Y, M(x)Y\rangle. \tag{42.87}$$

Since $\beta\epsilon \leq 0$ and $M(x)$ is a nonnegative operator, it follows that the right-hand side of Equation 42.87 is nonpositive. Thus, $\|Y\|$ is nonincreasing in general, and is constant if $\beta = 0$.

Our stabilization algorithm will be (necessarily) discontinuous, and will require switching of the control (Equation 42.82) between the following three cases 1–3 and case 4:

1. $\alpha > 0, \beta = \gamma = 0$
2. $\alpha = \kappa\lambda_*, \gamma = \kappa$, and $\beta = 0$, where λ_* is the largest eigenvalue of $N(Y)$
3. $\alpha = \gamma = 0, \beta\epsilon < 0$
4. $u = -\alpha(x - z_*)$, where $\alpha > 0$, and z_* denotes a fixed λ_*-eigenvector of $N(Y)$

In case 1, x is driven to 0 radially while Y remains fixed. If Y was not already 0 in the first place, implementing the control (Equation 42.82) with these parameter values will render the system unstabilizable. Thus, this case will only be used if $Y \equiv 0$.

In case 2, $\|Y\|$ is a constant and the spectrum of ad Y is constant. Therefore, the spectrum of the operator $N(Y)$ is constant, as are the dimensions of its eigenspaces. Let λ_* denote those eigenvalues of $N(Y)$ in this case, we have that Y evolves isospectrally with constant norm and asymptotically vanishing velocity, while x is driven to x_*, its (constant) projection onto the λ_*-eigenspace of $N(Y)$.

In case 3, $\|x\|$ is a constant. If we let $Y_\#$ denote the projection of Y onto the nullspace of $M(x)$, then noting that $Y_\# \equiv Y_\#|_{t=0}$ is constant, we conclude that $Y \to Y_\#$ asymptotically. Summarizing, in this case, we have that x evolves isospectrally with a constant norm, Y is driven to $Y_\#$, its (constant) projection onto the nullspace of $M(x)$, and using the orthogonality of the eigenspaces, we can compute

$$\|Y_\#\|^2 \leq \|Y\|^2 - \lambda_*. \tag{42.88}$$

Case 4 is used to move x from the origin along the eigenvector corresponding to the biggest eigenvalue λ_* of $N(Y)$.

The idea of the feedback strategy is the following. Since for $Y = 0$ step (1) stabilizes the system, the main task of the algorithm is to move Y to the origin. As mentioned above, step (2) allows to decrease the magnitude of $\|Y\|^2$ by λ_*, the biggest eigenvalue of $N(Y)$. On the other hand, applying (2) guarantees that $\|Y\|$ and the spectrum of the operator $N(Y)$ remain constant. Thus, Y evolves isospectrally with a constant norm, while x converges to the constant x_*. If $x_* \neq 0$, then go to (3). Otherwise, use (4). Then x converges to z_* while Y remains constant.

As shown in [4], by alternating steps (1)–(4) the system can be stabilized in a finite number of steps. The resulting algorithm is a variable structure feedback.

42.4.2 Energy Methods for Nonholonomic Mechanical Systems with Symmetries

We discuss here stabilization of nonholonomic systems with symmetry. In the simplest setting, the configuration space $\mathcal{Q}$ is the direct product $\mathcal{S} \times \mathcal{G}$, where $\mathcal{S}$ is a smooth manifold and $\mathcal{G}$ is the symmetry group.* The Lagrangian and the constraints are invariant with respect to the action of $\mathcal{G}$ by left translations on the second factor of the decomposition of $\mathcal{Q}$. If the group $\mathcal{G}$ is Abelian, the group variables become cyclic. We assume here that the Lagrangian $L : \mathcal{Q} \to \mathbb{R}$ equals kinetic minus potential energy of the system, and that the kinetic energy is given by a quadratic form on the configuration space.

In the presence of symmetry, we write the configuration coordinates as $q = (r, g)$, where $r \in \mathcal{S}$ is the shape variable and $g \in \mathcal{G}$ is the group variable. The state coordinates are $x = (r, g, \dot{r}, \Omega)$, where $\Omega \in \mathfrak{g}$ is the

* In general, this direct product structure is observed only locally, while globally one sees a (principal) fiber bundle.

constrained group velocity relative to the so-called body frame (see [2] for details).* With these notations, Equations 42.11 are usually written as

$$\ddot{r} = F(r, \dot{r}, \Omega), \tag{42.89}$$

$$\dot{\Omega} = \dot{r}^T \Lambda(r)\Omega + \Omega^T \Gamma(r)\Omega + \dot{r}^T \Upsilon(r)\dot{r}, \tag{42.90}$$

$$\dot{g} = g\Omega, \tag{42.91}$$

as discussed in [2]. Note that controls have not been introduced yet. Equations 42.89 and 42.90 are called the *shape* and *momentum* equations, respectively. Equation 42.91 is called the *reconstruction equation.* Below, stability is understood in the orbital sense, and thus one can study system (Equations 42.89 and 42.90).

42.4.2.1 The Energy–Momentum Method

Recall that relative equilibria are solutions of the full system (Equations 42.89 through 42.91) such that the shape and momentum variables are kept constant. In other words, after reduction relative equilibria of Equations 42.89 through 42.91 become equilibria of the reduced system (Equations 42.89 and 42.90). Stability of these equilibria implies orbital stability of corresponding relative equilibria.

The energy momentum method is a stability analysis technique that uses the restriction of energy on the momentum levels as a Lyapunov function. The momentum is always conserved in holonomic systems with symmetry.

Unlike the holonomic case, the momentum equation in the nonholonomic setting generically does not define conservation laws. Examples such as the rattleback and Chaplygin sleigh are well known. Sometimes, however, the components of momentum relative to an appropriately selected moving frame are conserved. This is observed in examples like the balanced Chaplygin sleigh, where the angular momentum relative to the vertical line through the contact point of the body and the supporting plane and the projection of the linear momentum onto the blade direction are conserved. We remark that this kind of momentum conservation is significantly different from that of in holonomic systems, in which, according to Noether's theorem, the *spatial momentum* is conserved.

In the special case $\Upsilon(r) \equiv 0$, $\Gamma(r) \equiv 0$, the momentum equation can be rewritten as

$$d\Omega = dr^T \Lambda(r)\Omega. \tag{42.92}$$

Theorem 42.9: [26]

If the distribution (Equation 42.92) is integrable, the system has conservation laws

$$\Omega = \mathcal{F}(r, c), \tag{42.93}$$

where c are constants.

These conservation laws enable one to extend the energy–momentum method for stability analysis to nonholonomic setting. Let $E(r, \dot{r}, \Omega)$ be the energy of the system; since the system is $\mathcal{G}$-invariant, the energy is independent of the group variable g. The energy itself is often not positive-definite at an equilibrium (r_e, Ω_e) of the reduced system, and thus the energy cannot be used as a Lyapunov function. If, however, the conditions of Theorem 42.9 are satisfied, one can construct a family of Lyapunov functions, one for each level (Equation 42.93).

* Here and below, $\mathfrak{g}$ denotes the Lie algebra of the group $\mathcal{G}$.

Theorem 42.10: [2]

Assume that the energy restricted to the level of the conservation law (Equation 42.93) through the equilibrium (r_e, Ω_e) of Equations 42.89 and 42.90 is positive-definite. Then this equilibrium is Lyapunov stable, and the corresponding relative equilibrium is orbitally stable.

The statement of this theorem can be extended to a more general class of systems and used for establishing partial asymptotic stability of relative equilibria of nonholonomic systems. Details may be found in [2].

42.4.2.2 Energy-Based Feedback Stabilization

Following [6], we now apply the energy–momentum approach to the problem of stabilization of relative equilibria of nonholonomic systems with symmetry. Recall that the dynamics is given by Equations 42.89 through 42.91. We assume that control inputs are $\mathcal{G}$-invariant, are applied in the symmetry directions, and are consistent with constraints. Thus, the controlled dynamics are $\mathcal{G}$-invariant and the corresponding reduced controlled dynamics are given by

$$\begin{aligned} \ddot{r} &= F(r, \dot{r}, \Omega), \\ \dot{\Omega} &= \dot{r}^T \Lambda(r)\Omega + \Omega^T \Gamma(r)\Omega + \dot{r}^T \Upsilon(r)\dot{r} + T(r)u, \end{aligned}$$

where u are the controls. Our strategy is to assign feedback control inputs that have the same structure as the right-hand side of the momentum equation. That is, the control inputs are given by homogeneous quadratic polynomials in $\dot{r}$ and Ω whose coefficients are functions of r. The controls are selected in such a way that the controlled momentum equation satisfies the conditions of Theorem 42.9, and thus the controlled momentum equation defines *controlled conservation laws*. We then utilize the remaining freedom in the control selection and make the equilibria of the system's dynamics, reduced to the levels of the controlled conservation laws, stable. In other words, the controls are used to shape the momentum levels in such a way that the energy reduced to the level through the equilibrium of interest becomes positive-definite, thus letting us use the energy–momentum method to conclude stability.

The proposed strategy is only capable of nonasymptotic stabilization. If partial asymptotic stabilization is desirable, one should add dissipation-emulating terms to the control inputs. In this case, stability is checked by the energy–momentum method, although it may be necessary to use methods for stability analysis for *nonconservative systems* such as the Lyapunov–Malkin theorem. See [2] for details.

To simplify the exposition, the details are given for systems with one shape degree of freedom.

Consider a system with Lagrangian

$$L(r, \dot{r}, \Omega) = K(r, \dot{r}, \Omega) - U(r)$$

and assume that the momentum equation of the *controlled* system is

$$\dot{\Omega} = \big(\Lambda(r) + \widetilde{\Lambda}(r)\big)\Omega\,\dot{r}.$$

According to our general strategy, the distribution defined by this equation;

$$d\Omega = \big(\Lambda(r) + \widetilde{\Lambda}(r)\big)\Omega\,\mathrm{d}r$$

is integrable and defines controlled conservation laws

$$\Omega = \mathcal{F}^c(r, c). \tag{42.94}$$

Dynamics on the levels of these conservation laws reads

$$\ddot{r} = F(r, \dot{r}, \mathcal{F}^c(r, c)). \tag{42.95}$$

This defines a family of one degree of freedom Lagrangian (or Hamiltonian) systems. The orbital stability analysis of relative equilibria of the original system reduces to the stability analysis of equilibria of Equation 42.95. This analysis is somewhat simpler to carry out if the Euler–Lagrange form of the equations of motion is used instead of Equation 42.95.

Skipping some technical details, the Euler–Lagrange form of the dynamics on the levels of controlled conservation laws (Equation 42.94) is

$$\frac{\mathrm{d}}{\mathrm{d}t}(g(r)\dot{r}) + Q(\mathcal{F}^c(r, c)) + \frac{\partial U}{\partial r} = 0, \tag{42.96}$$

where $Q(\cdot)$ is a quadratic form. Therefore, the equilibrium (r_e, Ω_e) is stable if the following condition is satisfied:

$$\frac{\mathrm{d}}{\mathrm{d}r}\left(Q(\mathcal{F}^c(r, c_e)) + \frac{\partial U}{\partial r}\right)\bigg|_{r=r_e} > 0, \tag{42.97}$$

where c_e is defined by the condition

$$\mathcal{F}^c(r_e, c_e) = \Omega_e.$$

This stability condition is obtained by using the energy of dynamics (Equation 42.96) as a Lyapunov function, which is justified by the energy–momentum approach to stability analysis; see [2] for details.

We illustrate the above techniques by the problem of stabilization of slow motions of a falling rolling disk along a straight line.

42.4.2.3 Stabilization of a Falling Disk

Consider a uniform disk rolling without sliding on a horizontal plane. The disk can reach any configuration, therefore the constraints imposed on the disk are nonholonomic. It is well-known that some of the steady-state motions are the uniform motions of the disk along a straight line. Such motions are unstable if the angular velocity of the disk is small. Stability is observed if the angular velocity of the disk exceeds a certain critical value; see [2] and [19] for details. Below we use a steering torque for stabilization of slow unstable motions of the disk.

We assume that the disk has a unit mass and a unit radius. The moments of inertia of the disk relative to its diameter and to the line orthogonal to the disk and through its center are I and J, respectively. The configuration coordinates for the disk are $[\theta, \psi, \phi, x, y]^T$ as in Figure 42.6. Following [19], we select e_1 to

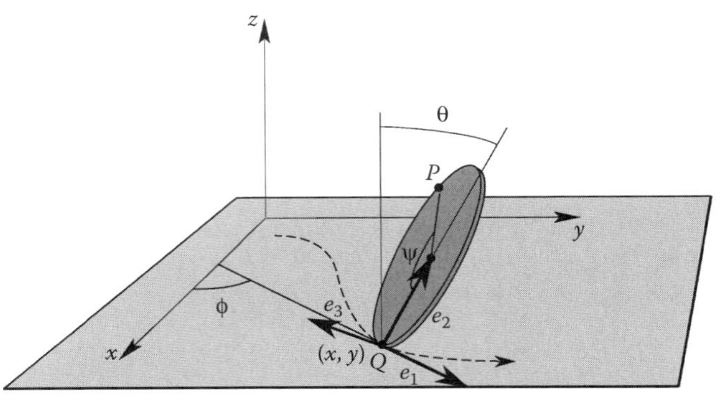

FIGURE 42.6 The geometry and configuration coordinates for the falling rolling disk.

be the vector in the xy-plane and tangent to the rim of the disk, e_2 to be the vector from the contact point to the center of the disk, and e_3 to be $e_1 \times e_2$, as shown in Figure 42.6.

At each $q \in \mathcal{Q}$, the fields e_1, e_2, and e_3 span the subspace $\mathcal{D}_q$ defined by the constraint distribution $\mathcal{D}$, the fields e_2 and e_3 span the constrained symmetry directions, and the dual of e_2 is the control direction. The component of disk's angular velocity along e_1 equals $\dot{\theta}$; the e_2 and e_3 components are denoted by ξ and η.

Using this frame, equations of motion are computed to be

$$(I+1)\ddot{\theta} + I\xi^2 \tan\theta - (J+1)\xi\eta - g\sin\theta = 0, \tag{42.98}$$

$$I\dot{\xi} - I\xi\dot{\theta}\tan\theta + J\eta\dot{\theta} = u, \tag{42.99}$$

$$(J+1)\dot{\eta} + \xi\dot{\theta} = 0, \tag{42.100}$$

where u is the steering torque and g is the acceleration of gravity. In the absence of the torque, the last two equations can be written as conservation laws of the form

$$\xi = \mathcal{F}_\xi(\theta, c_\xi, c_\eta), \quad \eta = \mathcal{F}_\eta(\theta, c_\xi, c_\eta); \tag{42.101}$$

here and below the parameters c_ξ and c_η label the levels of these conservation laws. These conservation laws are obtained by integrating the equations

$$I\frac{d\xi}{d\theta} = I\xi\tan\theta - J\eta, \quad (J+1)\frac{d\eta}{d\theta} = -\xi.$$

Now consider a steady-state motion $\theta = 0$, $\xi = 0$, $\eta = \eta_e$. This motion is unstable if η_e is small. Set

$$u = -f(\theta)\eta\dot{\theta}, \tag{42.102}$$

where $f(\theta)$ is a differentiable function. The motivation for the choice (Equation 42.102) for u is that it preserves the structure of Equations 42.99 and 42.100, and thus the controlled system will have conservation laws whose structure is similar to that of the uncontrolled system. Viewing θ as an independent variable, we replace Equations 42.99 and 42.100 with the linear system

$$I\frac{d\xi}{d\theta} = I\xi\tan\theta - (J + f(\theta))\eta, \quad (J+1)\frac{d\eta}{d\theta} = -\xi. \tag{42.103}$$

The general solution of system (Equation 42.103),

$$\xi = \mathcal{F}^c_\xi(\theta, c_\xi, c_\eta), \quad \eta = \mathcal{F}^c_\eta(\theta, c_\xi, c_\eta), \tag{42.104}$$

is interpreted as the *controlled conservation laws.* The functions that define these conservation laws are typically difficult or impossible to find explicitly.

Dynamics (Equation 42.96) on the level set of the conservation laws for the disk becomes

$$(I+1)\ddot{\theta} + I\tan\theta\,(\mathcal{F}^c_\xi(\theta, c_\xi, c_\eta))^2 - (J+1)\mathcal{F}^c_\xi(\theta, c_\xi, c_\eta)\mathcal{F}^c_\eta(\theta, c_\xi, c_\eta) - g\sin\theta = 0.$$

The condition for stability of the relative equilibrium $\theta = 0$, $\xi = 0$, $\eta = \eta_e$ is obtained using formula (Equation 42.97). Using Equation 42.103, the stability condition becomes

$$f(0) > \frac{Ig}{(J+1)\eta_e^2} - J. \tag{42.105}$$

That is, any function $f(\theta)$ whose value at $\theta = 0$ satisfies inequality (Equation 42.105) defines a stabilizing steering torque.

Let us reiterate that in the settings considered here the energy–momentum method gives conditions for nonlinear Lyapunov (nonasymptotic) stability. Hence stabilization by the torque (Equation 42.102) is nonlinear and nonasymptotic. Partial asymptotic stabilization can be achieved by adding dissipation-emulating terms to the control input.

42.5 Motion Planning

Motion planning problems for nonholonomic and underactuated systems can be characterized along a number of dimensions. Is the system drift-free or not? Are there configuration obstacles? What form do control constraints take? Is it more important to find a solution that minimizes time or energy, or is it more important to find a satisficing motion plan quickly?

Motion planning for nonholonomic and underactuated systems is an active research area, and in this section we only briefly describe a few approaches. More can be found in [2,11,13,17,18] and references therein.

42.5.1 Numerical Optimal Control

This approach typically involves choosing a finite-dimensional parameterization of the control history $u : [0, T] \to \mathcal{U}$ and solving for these design variables to minimize a cost function while satisfying equality and inequality constraints. Example constraints include control limits, terminal state conditions, and perhaps obstacle constraints, sampled at equally spaced points in time. The nonlinear optimization problem is typically solved by an algorithm that takes advantage of the gradient (and perhaps Hessian) of the cost function and constraints with respect to the design variables. A typical algorithm choice is Sequential Quadratic Programming. A variant of the approach is to dispense with the objective function and to simply cast the problem as a nonlinear root-finding problem. These approaches can be applied to general underactuated systems with drift. The success of these methods depends on problem-specific characteristics, such as the quality of the initial guess to the gradient-based algorithm, the number of local minima in the design space, and the method used to calculate gradients.

42.5.2 Optimal Control of the Heisenberg System

Some motion planning problems admit an analytical optimal solution. One such problem is to drive the Heisenberg system from an initial state $[x_1, x_2, z]^T = [0, 0, 0]^T$ to a goal state $[0, 0, a > 0]^T$ in time T while minimizing the cost functional

$$\frac{1}{2}\int_0^T (u_1^2 + u_2^2)\,\mathrm{d}t.$$

An equivalent formulation is the following: Minimize the integral

$$\frac{1}{2}\int_0^T (\dot{x}_1^2 + \dot{x}_2^2)\,\mathrm{d}t$$

among all curves $q(t)$ joining $q(0) = [0, 0, 0]^T$ to $q(T) = [0, 0, a]^T$ that satisfy the constraint

$$\dot{z} = x_2\dot{x}_1 - x_1\dot{x}_2.$$

Any solution must satisfy the Euler–Lagrange equations for the Lagrangian with a Lagrange multiplier inserted:

$$L(x_1, \dot{x}_1, x_2, \dot{x}_2, z, \dot{z}, \lambda, \dot{\lambda}) = \tfrac{1}{2}\left(\dot{x}_1^2 + \dot{x}_2^2\right) + \lambda\left(\dot{z} - x_2\dot{x}_1 + x_1\dot{x}_2\right).$$

The corresponding Euler–Lagrange equations are given by

$$\ddot{x}_1 - 2\lambda\dot{x}_2 = 0, \tag{42.106}$$

$$\ddot{x}_2 + 2\lambda\dot{x}_1 = 0, \tag{42.107}$$

$$\dot{\lambda} = 0. \tag{42.108}$$

We can show the solution of the optimal control problem is given by choosing initial conditions such that $\dot{x}_1(0)^2 + \dot{x}_2(0)^2 = 2\pi a/T^2$ and with the trajectory in the x_1x_2-plane given by the circle

$$\begin{bmatrix} x_1(t) \\ x_2(t) \end{bmatrix} = \frac{1}{2\lambda} \begin{bmatrix} \cos(2\pi t/T) - 1 & -\sin(2\pi t/T) \\ \sin(2\pi t/T) & \cos(2\pi t/T) - 1 \end{bmatrix} \begin{bmatrix} -\dot{x}_2(0) \\ \dot{x}_1(0) \end{bmatrix} \tag{42.109}$$

and with z given by

$$z(t) = \frac{ta}{T} - \frac{a}{2\pi} \sin\left(\frac{2\pi t}{T}\right).$$

A motion planner to an arbitrary final state $[x_{1d}, x_{2d}, z_d]^T$ could simply use the two controls to drive the first two configuration variables directly to the goal state, followed by the optimal trajectory to accomplish the remaining necessary motion in z. More details can be found in [2] and [9].

42.5.3 Motion Planning for the Generalized Heisenberg System

The approach above can be generalized to chained-form systems [13,18] and the generalized Heisenberg system (Equations 42.76 and 42.77).

As with the Heisenberg system, for the generalized Heisenberg system we first use the controls to drive the system directly to the desired x states, and then use sinusoidal controls to accomplish a net motion in Y while producing zero net motion in x. The idea is to proceed along loops in x-space, which gradually drives one through Y-space.

We choose the control law

$$u_i = \sum_k a_{ik} \sin kt + \sum_k b_{ik} \cos kt, \quad k = 1, \ldots, \tag{42.110}$$

where a_{ik} and b_{ik} are real numbers. Since $\dot{x}_i = u_i$, integration gives

$$x_i = -\sum_k \frac{a_{ik}}{k} \cos kt + \sum_k \frac{b_{ik}}{k} \sin kt + C_i, \tag{42.111}$$

where C_i is a constant depending on the initial value of x.

Substituting these equations for $x_i(t)$ and $u_i(t)$ into Equation 42.77 and integrating yields

$$Y_{ij}(2\pi) = \sum_k \frac{2\pi}{k} \left(b_{ik}a_{jk} - b_{jk}a_{ik}\right) + Y_{ij}(0), \tag{42.112}$$

since all integrals except those of the squares of cosine and sine vanish. Under this input, x has zero net motion. More details can be found in [2].

42.5.4 Search-Based Methods

Configuration obstacles can be accounted for by constraints in a numerical optimization or by using search algorithms from computer science [13,17]. Search-based methods for motion planning for underactuated systems usually involve choosing a small number of "representative" controls from the control set. One or more of these controls are integrated forward for a short time Δt from the initial state x_0 to create a new set of reached states $\mathcal{X}$. Trajectories that intersect obstacles are discarded, and the remaining states in $\mathcal{X}$ are put in a list sorted by the cost of reaching the state. The first state on the list is then removed, assigned to be x_0, and the process continues until x_0 is in a neighborhood of the goal state. To reduce the planning time or improve the plans, heuristics can be designed for choosing the controls, pruning states that are "close" to previously reached states, and assigning the cost of states. This approach is applied to parallel parking a car-like robot in Figure 42.7. The cost trades off the length of the path and the number of control changes (e.g., cusps).

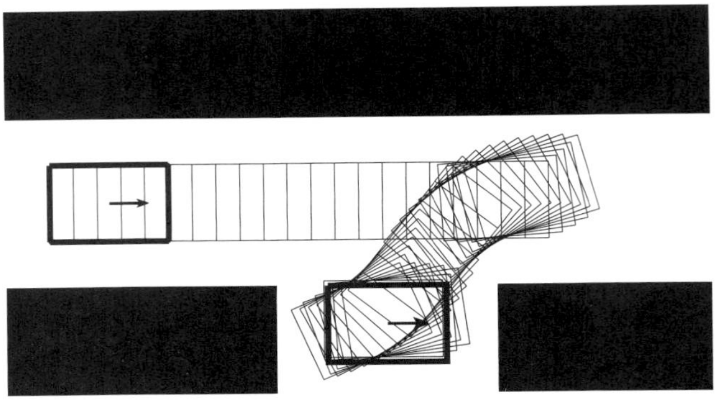

FIGURE 42.7 A car parallel parking.

42.5.5 Path Transformation Methods

If the system is kinematic and STLC, one can first plan a path among obstacles assuming the system has no nonholonomic constraints, using one of the many collision-free motion planning methods developed in the robotics literature. This plan can then be transformed into a motion plan that satisfies the velocity constraints [13]. Typically this transformation proceeds by recursively choosing a segment of the initial path and attempting to replace it with a path chosen by a local path planner. The local planner takes the initial and final configurations q_0 and q_f of the segment and chooses a path satisfying the nonholonomic constraints, but ignoring the obstacles. The replacement occurs if the new path segment is collision-free. If not, the initial segment is subdivided again. To guarantee success of the transformation process, the local path planner must have the property that the length of the chosen path goes to zero as $\| q_f - q_0 \|$ goes to zero. The final feasible path produced by the transformation could be further post-processed to reduce some measure of the cost of the path.

42.5.6 Kinematic Reductions

For dynamic underactuated systems with kinematic reductions yielding STLKC, trajectory planning from one equilibrium state to another can be reduced to the problem of path planning using the kinematic reductions, followed by time-scaling of the path to produce a trajectory [11–13]. This decoupling of the trajectory planning problem reduces the computational complexity by turning the constrained motion planning problem on the $2n$-dimensional state space into path planning on the n-dimensional configuration space. In addition, time-optimal time scaling algorithms can be used to turn the paths into fast feasible trajectories. Because the kinematic reductions take account of the underactuation constraints, the only constraints on path speed come from the actuator limits.

An example is shown in Figure 42.8 for the 3R robot of Figure 42.4. The robot arm has no actuator at the third link and is described by $L_{1,2} = 0.3$ m, $r_{1,2,3} = 0.15$ m, $m_i = \{2.0, 1.0, 0.5\}$ kg, $I_i = \{0.02, 0.01, 0.004125\}$ kgm^2 with actuator limits $|u_1| \leq 20$ Nm, $|u_2| \leq 10$ Nm. As mentioned earlier, the two decoupling vector fields are translation along the third link and rotation of the third link about its center of percussion with respect to the third joint. Using just these motions, we can use a path planner to plan point-to-point motions. Figure 42.8 shows an example path consisting of four segments along the decoupling vector fields. Also shown is the time-optimal time scaling, which allows the robot to follow the path as quickly as possible subject to actuator limits. Because the only common velocity between the two decoupling vector fields is zero velocity, the robot must come to a stop at the transitions between vector fields.

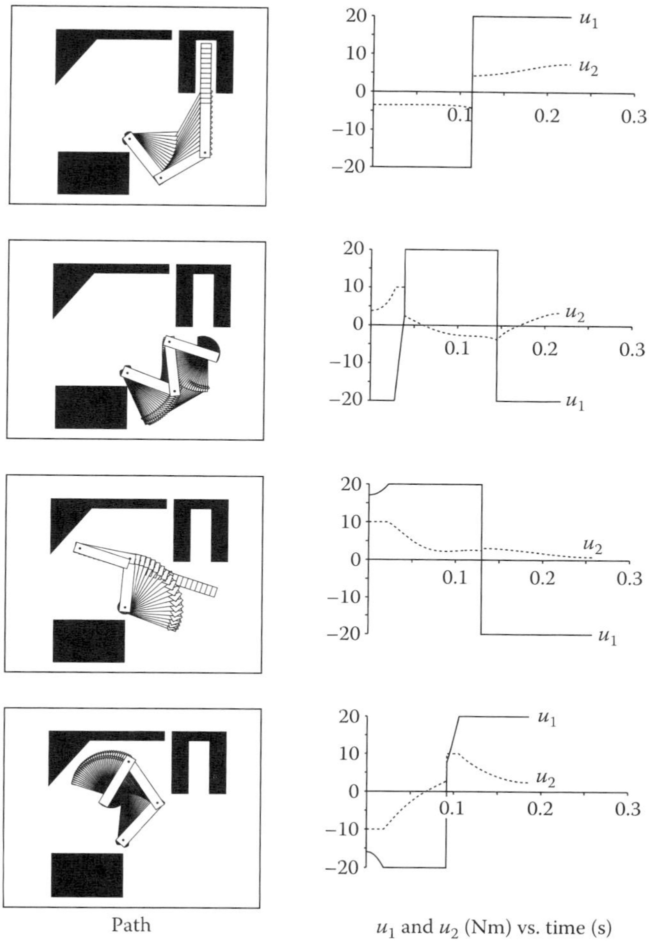

FIGURE 42.8 Four path segments and their time scalings for a 3R robot without an actuator at the third joint. Note that the time-optimal time scalings saturate one actuator at all times.

42.5.7 Differentially Flat Systems

Differentially flat and dynamically feedback linearizable systems have a structure that makes motion planning (in the absence of control and configuration constraints such as obstacles) particularly simple [23]. For a differentially flat system with a state x and $u \in \mathbb{R}^m$, there exists a set of m functions z_i, $i = 1, \ldots, m$, of the state, the control, and its derivatives,

$$z_i(x, u, \dot{u}, \ldots, u^{(r)}), \quad i = 1, \ldots, m,$$

such that the states and control inputs can be expressed as functions of z and its time-derivatives:

$$x = \phi(z, \dot{z}, \ldots, z^{(s)}),$$
$$u = \psi(z, \dot{z}, \ldots, z^{(s)}).$$

The functions z_i are known as the *flat outputs*. Armed with a set of flat outputs, the problem of finding a feasible trajectory $(x(t), u(t)), x(0) = x_0$, $x(t_f) = x_f$, $t \in [0, t_f]$ for the underactuated system is transformed to the problem of finding a curve $z(t)$ satisfying constraints on $z(0), \dot{z}(0), \ldots, z^{(s)}(0)$ and $z(t_f), \dot{z}(t_f), \ldots, z^{(s)}(t_f)$ specified by x_0 and x_f. In other words, the problem of finding a trajectory satisfying the underactuation constraints becomes the relatively simple algebraic problem of finding a curve to fit the start and end constraints on z. Any curve $z(t)$ maps directly to a consistent pair of state and control histories $x(t)$ and $u(t)$.

The flat outputs for mechanical systems are often a function of configuration variables only, and sometimes are just the location of particular points on the system. Unfortunately, there is no systematic way to determine if a system is differentially flat, or what the flat outputs for a system are. Many important systems have been shown to be differentially flat, however, such as chained-form systems, cars, and cars pulling trailers.

As an example, consider a vertical rolling disk, ignoring the configuration variable ψ. The configuration is $q = [x, y, \phi]^T$, and the controls are the forward velocity v and the turning rate ω (Figure 42.2). The flat outputs are simply $z_1 = x$ and $z_2 = y$. The state and controls can be derived from the flat outputs and their derivatives as follows:

$$[x, y, \phi]^T = \left[z_1, z_2, \tan^{-1} \frac{\dot{z}_2}{\dot{z}_1} \right]^T, \tag{42.113}$$

$$[v, \omega]^T = \left[\pm\sqrt{\dot{z}_1^2 + \dot{z}_2^2}, \frac{\dot{z}_1 \ddot{z}_2 - \ddot{z}_1 \dot{z}_2}{\dot{z}_1^2 + \dot{z}_2^2} \right]^T. \tag{42.114}$$

The orientation ϕ and the turning control ω are not well defined as a function of the flat outputs when the linear velocity of the disk is zero.

Now we would like to find a feasible trajectory from $q_0 = [0, 0, 0]^T$ to $q_f = [1, 1, 0]^T$. Since there are six state variables in the specification of the start and goal points, there are six constraints on the flat outputs z and their derivatives at the beginning and end of motion. These constraints can be written

$$\begin{aligned} z_1(0) = 0, \quad & z_2(0) = 0, \\ z_1(t_f) = 1, \quad & z_2(t_f) = 1, \\ & \dot{z}_2(0) = 0, \\ & \dot{z}_2(t_f) = 0, \end{aligned}$$

where the last two constraints indicate that the initial and final motion of the unicycle must be along the x-axis, indicating that the wheel is oriented with the x-axis. The simplest polynomial functions of time that have enough free coefficients to satisfy these constraints are

$$\begin{aligned} z_1(t) &= a_0 + a_1 t, \\ z_2(t) &= b_0 + b_1 t + b_2 t^2 + b_3 t^3. \end{aligned}$$

Setting the time of motion $t_f = 1$ and using the constraints to solve for the polynomial coefficients, we obtain

$$z_1(t) = t, \tag{42.115}$$

$$z_2(t) = 3t^2 - 2t^3. \tag{42.116}$$

The state and control can be obtained from Equations 42.113 and 42.114. The unicycle motion is shown in Figure 42.9.

In fitting a curve $z(t)$, we must choose a family of curves with enough degrees of freedom to satisfy the initial and terminal constraints. We may choose a family of curves with more degrees of freedom, however, and use the extra degrees of freedom to, individually or severally, (1) satisfy bounds on the control $u(t)$, (2) avoid obstacles in the configuration space, or (3) minimize a cost function. Incorporating these conditions in the calculation of $z(t)$ typically requires resorting to numerical optimization methods. A good way to generate an initial guess for the optimization is to solve exactly for a minimal number of coefficients to satisfy the initial and terminal constraints, setting the other coefficients to zero.

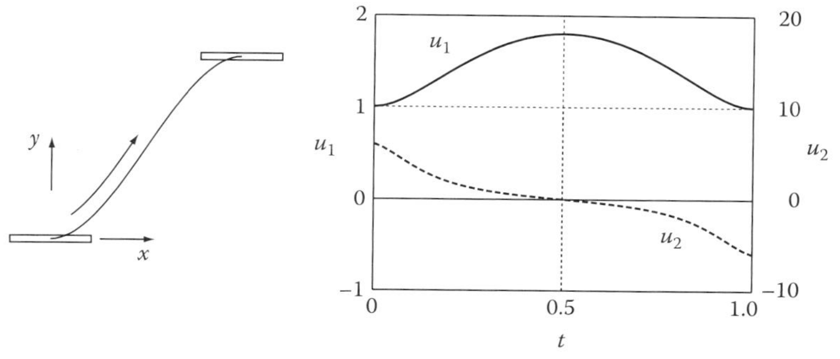

FIGURE 42.9 A feasible path for the unicycle from $[0, 0, 0]^T$ to $[1, 1, 0]^T$, and the controls.

References

1. Bianchini, R. M. and G. Stefani. Controllability along a trajectory: A variational approach, *SIAM Journal on Control and Optimization*, 31(4):900–927, 1993.
2. Bloch, A. M., with J. Baillieul, P. E. Crouch and J. E. Marsden. *Nonholonomic Mechanics and Control*, Springer-Verlag, New York, 2003.
3. Bloch, A. M. and S. V. Drakunov. Stabilization and tracking in the nonholonomic integrator via sliding modes, *Systems and Control Letters*, 29:91–99, 1996.
4. Bloch, A. M., S. V. Drakunov, and M. K. Kinyon. Stabilization of nonholonomic systems using isospectral flows, *SIAM Journal on Control and Optimization*, 38(3):855–874, 2000.
5. Bloch, A. M., P. S. Krishnaprasad, J. E. Marsden, and R. Murray. Nonholonomic mechanical systems with symmetry, *Archive for Rational Mechanical Analysis.*, 136:21–99, 1996.
6. Bloch, A. M., J. E. Marsden, and D. V. Zenkov. Quasivelocities and stabilization of relative equilibria of underactuated nonholonomic systems, *Proceedings of the IEEE Conference on Decision and Control*, Shanghai, China, vol. 48, pp. 3335–3340, 2009.
7. Bloch, A. M., M. Reyhanoglu, and H. McClamroch. Control and stabilization of nonholonomic dynamic systems, *IEEE Transations on Automatic Control*, 37(11):1746–1757, 1992.
8. Brockett, R. W. Control theory and singular Riemannian geometry, in *New Directions in Applied Mathematics*, P. J. Hilton and G. S. Young (Eds), Springer-Verlag, New York, pp. 11–27, 1981.
9. Brockett, R. W. Nonlinear control theory and differential geometry, in *Proceedings of International Congress of Mathematicians*, Warsaw, Poland, 1357–1368, 1983.
10. Brockett, R. W. Asymptotic stability and feedback stabilization, in *Differential Geometric Control Theory*, R. Brockett, R. Millman, and H. Sussmann (Eds), Birkhauser, Boston, pp. 181–191, 1983.
11. Bullo, F. and A. D. Lewis. *Geometric Control of Mechanical Systems*, Texts in Applied Mathematics, Vol. 49, Springer-Verlag, New York, 2005.
12. Bullo, F. and K. M. Lynch. Kinematic controllability for decoupled trajectory planning of underactuated mechanical systems, *IEEE Transactions on Robotics and Automation*, 17(4):402–412, 2001.
13. Choset, H., K. M. Lynch, S. Hutchinson, G. Kantor, W. Burgard, L. Kavraki, and S. Thrun, *Principles of Robot Motion*, MIT Press, Cambridge, MA, 2005.
14. Coron, J. M. A necessary condition for feedback stabilization, *Systems and Control Letters*, 14(3): 227–232, 1990.
15. Coron, J. M. Global asymptotic stabilization for controllable systems without drift, *Mathematics of Controls, Signals and Systems*, 5(3):295–312, 1992.
16. Drakunov, S. V., T. Floquet, and W. Perruquetti. Stabilization and tracking for an extended Heisenberg system with a drift, *Systems and Control Letters*, 54(5):435–445, 2005.
17. LaValle, S. M. *Planning Algorithms*, Cambridge University Press, Cambridge, UK, 2006.
18. Murray, R. M., Z. Li, and S. Sastry. *A Mathematical Introduction to Robotic Manipulation*, CRC Press, Boca Raton, FL, 1994.
19. Neimark J. I and N. A. Fufaev. *Dynamics of Nonholonomic Systems*, Series Translations of Mathematical Monographs. AMS, Providence, Rhode Island: vol. 33, 1972.
20. Nijmeijer, H. and A.J. van der Schaft. *Nonlinear Dynamical Control Systems*, Springer-Verlag, New York, 1990.

21. Reyhanoglu, M., A. J. van der Schaft, N.H. McClamroch, and I. Kolmanovsky. Dynamics and control of a class of underactuated mechanical systems, *IEEE Transactions on Automatic Control*, 44(9): 1663–1671, 1999.
22. Reyhanoglu, M., S. Cho, and N.H. McClamroch. Discontinuous feedback control of a class of underactuated mechanical systems, *International Journal of Robust and Nonlinear Control*, 10(4): 265–281, 2000.
23. Sira-Ramirez, H. and S. K. Agrawal. *Differentially Flat Systems*, CRC Press, Boca Raton, FL, 2004.
24. Sussmann, H. J. Lie brackets, real analyticity and geometric control, in *Differential Geometric Control Theory*, R. Brockett, R. Millman, and H. Sussmann (Eds), Birkhauser, Boston, 1983.
25. Sussmann, H. J. A general theorem on local controllability, *SIAM Journal on Control and Optimization*, 25(1):158–194, 1987.
26. Zenkov, D. V. Linear conservation laws of nonholonomic systems with symmetry, *Discrete and Continuous Dynamical Systems*, Extended volume: 963–972, 2003.

VII

Stability

43

Lyapunov Stability

Hassan K. Khalil
Michigan State University

43.1 Introduction

Stability theory plays a central role in systems theory and engineering. It deals with the system's behavior over a long time period. There are several ways to characterize stability. For example, we may characterize stability from an input–output viewpoint, by requiring the output of a system to be "well-behaved" in some sense, whenever the input is well behaved. Alternatively, we may characterize stability by studying the asymptotic behavior of the state of the system near steady-state solutions, like equilibrium points or periodic orbits.

In this chapter we introduce Lyapunov's method for determining the stability of equilibrium points. Lyapunov laid the foundation of this method over a century ago, but of course the method as we use it today is the result of intensive research efforts by many engineers and applied mathematicians. The attractive features of the method include a solid theoretical foundation, the ability to conclude stability without knowledge of the solution (no extensive simulation effort), and an analytical framework that makes it possible to study the effect of model perturbations and to design feedback control. Its main drawback is the need to search for an auxiliary function that satisfies certain conditions.

43.2 Stability of Equilibrium Points

We consider a nonlinear system represented by the state model

$$\dot{x} = f(x) \tag{43.1}$$

where the components of the n-dimensional vector $f(x)$ are *locally Lipschitz* functions of x, defined for all x in a domain $D \subset R^n$. A function $f(x)$ is locally Lipschitz at a point x_0 if it satisfies the *Lipschitz condition* $\|f(x) - f(y)\| \leq L\|x - y\|$ for all x, y in some neighborhood of x_0, where L is a positive constant and $\|x\| = \sqrt{x_1^2 + x_2^2 + \cdots + x_n^2}$. The Lipschitz condition guarantees that Equation 43.1 has a unique solution for a given initial state $x(0)$. Suppose $\bar{x} \in D$ is an *equilibrium point* of Equation 43.1; that is,

$f(\bar{x}) = 0$. Whenever the state of the system starts at $\bar{x}$ it will remain at $\bar{x}$ for all future time. Our goal is to characterize and study the stability of $\bar{x}$. For convenience, we take $\bar{x} = 0$. There is no loss of generality in doing so because any equilibrium point $\bar{x}$ can be shifted to the origin via the change of variables $y = x - \bar{x}$. Therefore, we shall always assume that $f(0) = 0$, and study stability of the origin $x = 0$.

The equilibrium point $x = 0$ of Equation 43.1 is *stable*, if for each $\varepsilon > 0$, there is $\delta = \delta(\varepsilon) > 0$ such that $\|x(0)\| < \delta$ implies that $\|x(t)\| < \varepsilon$, for all $t \geq 0$. It is *asymptotically stable*, if it is stable and δ can be chosen such that $\|x(0)\| < \delta$ implies that $x(t)$ converges to the origin as t tends to infinity. When the origin is asymptotically stable, the *region of attraction* (also called region of asymptotic stability, domain of attraction, or basin) is defined as the set of all points x such that the solution of Equation 43.1 that starts from x at time $t = 0$, approaches the origin as t tends to ∞. When the region of attraction is the whole space, we say that the origin is *globally asymptotically stable*. A stronger form of asymptotic stability arises when there exist positive constants c, k, and λ such that the solutions of Equation 43.1 satisfy the inequality

$$\|x(t)\| \leq k\|x(0)\|e^{-\lambda t}, \quad \forall\, t \geq 0 \tag{43.2}$$

for all $\|x(0)\| < c$. In this case, the equilibrium point $x = 0$ is said to be *exponentially stable*. It is said to be globally exponentially stable if the inequality is satisfied for any initial state $x(0)$.

43.2.1 Linear Systems

For the linear time-invariant system

$$\dot{x} = Ax \tag{43.3}$$

the stability properties of the origin can be characterized by the location of the eigenvalues of A. The origin is stable if and only if all the eigenvalues of A satisfy $\text{Re}[\lambda_i] \leq 0$ and for every eigenvalue with $\text{Re}[\lambda_i] = 0$ and algebraic multiplicity $q_i \geq 2$, $\text{rank}(A - \lambda_i I) = n - q_i$, where n is the dimension of x and q_i is the multiplicity of λ_i as a zero of $\det(\lambda I - A)$. The origin is globally exponentially stable if and only if all eigenvalues of A have negative real parts; that is, A is a *Hurwitz matrix*. For linear systems, the notions of asymptotic and exponential stability are equivalent because the solution is formed of exponential modes. Moreover, due to linearity, if the origin is exponentially stable, then the inequality of Equation 43.2 will hold for all initial states.

43.2.2 Linearization

Suppose the function $f(x)$ of Equation 43.1 is continuously differentiable in a domain D containing the origin. The Jacobian matrix $[\partial f/\partial x]$ is an $n \times n$ matrix whose (i, j) element is $\partial f_i/\partial x_j$. Let A be the Jacobian matrix evaluated at the origin $x = 0$. It can be shown that

$$f(x) = [A + G(x)]x, \quad \text{where } \lim_{x \to 0} G(x) = 0$$

This suggests that in a small neighborhood of the origin we can approximate the nonlinear system $\dot{x} = f(x)$ by its linearization about the origin $\dot{x} = Ax$. Indeed, we can draw conclusions about the stability of the origin as an equilibrium point for the nonlinear system by examining the eigenvalues of A. The origin of Equation 43.1 is exponentially stable if and only if A is Hurwitz. It is unstable if $\text{Re}[\lambda_i] > 0$ for one or more of the eigenvalues of A. Linearization fails when $\text{Re}[\lambda_i] \leq 0$ for all i, with $\text{Re}[\lambda_i] = 0$ for some i, for in this case we cannot draw a conclusion about the stability of the origin of Equation 43.1.

43.2.3 Lyapunov Method

Let $V(x)$ be a continuously differentiable scalar function defined in a domain $D \subset R^n$ that contains the origin. The function $V(x)$ is said to be *positive definite* if $V(0) = 0$ and $V(x) > 0$ for $x \neq 0$. It is said to be *positive semidefinite* if $V(x) \geq 0$ for all x. A function $V(x)$ is said to be *negative definite* or *negative semidefinite* if $-V(x)$ is positive definite or positive semidefinite, respectively. The derivative of V along the trajectories of Equation 43.1 is given by

$$\dot{V}(x) = \sum_{i=1}^{n} \frac{\partial V}{\partial x_i} \dot{x}_i = \frac{\partial V}{\partial x} f(x)$$

where $[\partial V/\partial x]$ is a row vector whose ith component is $\partial V/\partial x_i$.

Lyapunov's stability theorem states that *the origin is stable if there is a continuously differentiable positive-definite function* $V(x)$ *so that* $\dot{V}(x)$ *is negative semidefinite, and it is asymptotically stable if* $\dot{V}(x)$ *is negative definite.*

A function $V(x)$ satisfying the conditions for stability is called a *Lyapunov function.* The surface $V(x) = c$, for some $c > 0$, is called a *Lyapunov surface* or a level surface. Using Lyapunov surfaces, Figure 43.1 makes the theorem intuitively clear. It shows Lyapunov surfaces for decreasing constants $c_3 > c_2 > c_1 > 0$. The condition $\dot{V} \leq 0$ implies that when a trajectory crosses a Lyapunov surface $V(x) = c$, it moves inside the set $\Omega_c = \{V(x) \leq c\}$ and can never come out again, since $\dot{V} \leq 0$ on the boundary $V(x) = c$. When $\dot{V} < 0$, the trajectory moves from one Lyapunov surface to an inner Lyapunov surface with a smaller c. As c decreases, the Lyapunov surface $V(x) = c$ shrinks to the origin, showing that the trajectory approaches the origin as time progresses. If we only know that $\dot{V} \leq 0$, we cannot be sure that the trajectory will approach the origin, but we can conclude that the origin is stable since the trajectory can be contained inside any ε neighborhood of the origin by requiring the initial state $x(0)$ to lie inside a Lyapunov surface contained in that neighborhood.

When $\dot{V}(x)$ is only negative semidefinite, we may still conclude asymptotic stability of the origin if we can show that no solution can stay identically in the set $\{\dot{V}(x) = 0\}$, other than the zero solution $x(t) \equiv 0$. Under this condition, $V(x(t))$ must decrease toward 0, and consequently $x(t)$ converges to zero as t tends to infinity. This extension of the basic theorem is known as *the invariance principle.*

Lyapunov functions can be used to estimate the region of attraction of an asymptotically stable origin, that is, to find sets contained in the region of attraction. Let $V(x)$ be a Lyapunov function that satisfies the conditions of asymptotic stability over a domain D. For a positive constant c, let Ω_c be the component of $\{V(x) \leq c\}$ that contains the origin in its interior. The properties of V guarantee that, by choosing c small enough, Ω_c will be bounded and contained in D. Then, every trajectory starting in Ω_c remains in Ω_c, and approaches the origin as $t \to \infty$. Thus, Ω_c is an estimate of the region of attraction. If $D = R^n$ and $V(x)$ is radially unbounded, that is, $\|x\| \to \infty$ implies that $V(x) \to \infty$, then any point $x \in R^n$ can be included in a bounded set Ω_c by choosing c large enough. Therefore, *the origin is globally asymptotically stable if there is a continuously differentiable, radially unbounded function V(x) such that for all* $x \in R^n$,

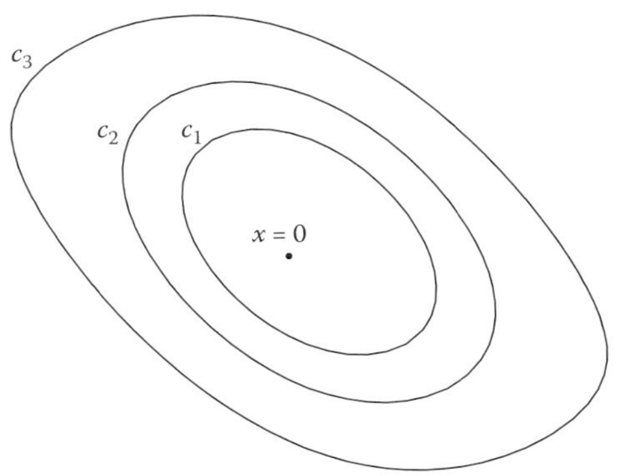

FIGURE 43.1 Lyapunov surfaces $V(x) = c_i$ with $c_3 > c_2 > c_1$.

$V(x)$ is positive definite and $\dot{V}(x)$ is either negative definite or negative semidefinite, but no solution can stay identically in the set $\{\dot{V}(x) = 0\}$ other than the zero solution $x(t) \equiv 0$.

Lyapunov's method is a powerful tool for studying the stability of equilibrium points. However, there are two drawbacks of the method that we should be aware of. First, there is no systematic method for finding a Lyapunov function for a given system. Second, the conditions of the method are only sufficient; they are not necessary. Failure of a Lyapunov function candidate to satisfy the conditions for stability or asymptotic stability does not mean that the origin is not stable or asymptotically stable.

43.2.4 Time-Varying Systems

Equation 43.1 is time-invariant because f does not depend on t. The more general time-varying system is represented by

$$\dot{x} = f(t, x) \tag{43.4}$$

In this case, we may allow the Lyapunov function candidate V to depend on t. Let $V(t,x)$ be a continuously differentiable function defined for all $t \geq 0$ and all $x \in D$. The derivative of V along the trajectories of Equation 43.4 is given by

$$\dot{V}(t,x) = \frac{\partial V}{\partial t} + \frac{\partial V}{\partial x} f(t,x)$$

If there are positive-definite functions $W_1(x)$, $W_2(x)$, and $W_3(x)$ such that

$$W_1(x) \leq V(t,x) \leq W_2(x) \tag{43.5}$$

$$\dot{V}(t,x) \leq -W_3(x) \tag{43.6}$$

for all $t \geq 0$ and all $x \in D$, then the origin is uniformly asymptotically stable, where "uniformly" indicates that the ε–δ definition of stability and the convergence of $x(t)$ to zero are independent of the initial time t_0. Such uniformity annotation is not needed with time-invariant systems since the solution of a time-invariant state equation starting at time t_0 depends only on the difference $t - t_0$, which is not the case for time-varying systems. If the inequalities of Equations 43.5 and 43.6 hold globally and $W_1(x)$ is radially unbounded, then the origin is globally uniformly asymptotically stable. If $W_1(x) = k_1\|x\|^a$, $W_2(x) = k_2\|x\|^a$, and $W_3(x) = k_3\|x\|^a$ for some positive constants k_1, k_2, k_3, and a, then the origin is exponentially stable.

43.2.5 Perturbed Systems

Consider the system

$$\dot{x} = f(t,x) + g(t,x) \tag{43.7}$$

where f and g are continuous in t and locally Lipschitz in x, for all $t \geq 0$ and $x \in D$, in which $D \subset R^n$ is a domain that contains the origin $x = 0$. Suppose $f(t,0) = 0$ and $g(t,0) = 0$ so that the origin is an equilibrium point of Equation 43.7. We think of the system of Equation 43.7 as a perturbation of the nominal system

$$\dot{x} = f(t,x) \tag{43.8}$$

The perturbation term $g(t,x)$ could result from modeling errors, uncertainties, or disturbances, which exist in any realistic problem. In a typical situation, we do not know $g(t,x)$, but we know some information about it, like knowing an upper bound on $\|g(t,x)\|$. Suppose the nominal system has an exponentially stable equilibrium point at the origin, what can we say about the stability of the origin as an equilibrium point of the perturbed system? A natural approach to address this question is to use a Lyapunov function for the nominal system as a Lyapunov function candidate for the perturbed system.

Let $V(t,x)$ be a Lyapunov function that satisfies

$$c_1\|x\|^2 \le V(t,x) \le c_2\|x\|^2 \tag{43.9}$$

$$\frac{\partial V}{\partial t} + \frac{\partial V}{\partial x} f(t,x) \le -c_3\|x\|^2 \tag{43.10}$$

$$\left\|\frac{\partial V}{\partial x}\right\| \le c_4\|x\| \tag{43.11}$$

for all $x \in D$ for some positive constants c_1, c_2, c_3, and c_4. Suppose the perturbation term $g(t,x)$ satisfies the linear growth bound

$$\|g(t,x)\| \le \gamma\|x\|, \quad \forall\, t \ge 0, \;\; \forall\, x \in D \tag{43.12}$$

where γ is a nonnegative constant. We use V as a Lyapunov function candidate to investigate the stability of the origin as an equilibrium point for the perturbed system. The derivative of V along the trajectories of Equation 43.7 is given by

$$\dot{V}(t,x) = \frac{\partial V}{\partial t} + \frac{\partial V}{\partial x} f(t,x) + \frac{\partial V}{\partial x} g(t,x)$$

The first two terms on the right-hand side are the derivative of $V(t,x)$ along the trajectories of the nominal system, which is negative definite and satisfies the inequality of Equation 43.10. The third term, $[\partial V/\partial x]g$, is the effect of the perturbation. Using Equations 43.10 through 43.12, we obtain

$$\dot{V}(t,x) \le -c_3\|x\|^2 + \left\|\frac{\partial V}{\partial x}\right\| \; \|g(t,x)\| \le -c_3\|x\|^2 + c_4\gamma\|x\|^2$$

If $\gamma < c_3/c_4$, then

$$\dot{V}(t,x) \le -(c_3 - \gamma c_4)\|x\|^2, \quad (c_3 - \gamma c_4) > 0$$

which shows that the origin is an exponentially stable equilibrium point of Equation 43.7.

43.3 Examples and Applications

Example 43.1:

A simple pendulum moving in a vertical plane can be modeled by the state equation

$$\dot{x}_1 = x_2$$
$$\dot{x}_2 = -a \sin x_1 - b x_2$$

where $a > 0$, $b \ge 0$, and x_1 is the angle between the pendulum and the vertical line through the pivot point. The case $b = 0$ is an idealized frictionless pendulum. To find the equilibrium points, we set $\dot{x}_1 = \dot{x}_2 = 0$ and solve for x_1 and x_2. The first equation gives $x_2 = 0$ and the second one gives $\sin x_1 = 0$. Thus, the equilibrium points are located at $(n\pi, 0)$, for $n = 0, \pm 1, \pm 2, \ldots$ The pendulum has only two equilibrium positions corresponding to the equilibrium points $(0,0)$ and $(\pi, 0)$. The other equilibrium points are repetitions of these two positions that correspond to the number of full swings the pendulum would make before it rests at one of the two equilibrium positions.

To start with, let us investigate the stability of the equilibrium points by linearization. The Jacobian matrix is given by

$$\frac{\partial f}{\partial x} = \begin{bmatrix} \dfrac{\partial f_1}{\partial x_1} & \dfrac{\partial f_1}{\partial x_2} \\ \dfrac{\partial f_2}{\partial x_1} & \dfrac{\partial f_2}{\partial x_2} \end{bmatrix} = \begin{bmatrix} 0 & 1 \\ -a\cos x_1 & -b \end{bmatrix}$$

To investigate the stability of the origin, we evaluate the Jacobian matrix at $x = 0$.

$$A = \left.\frac{\partial f}{\partial x}\right|_{x=0} = \begin{bmatrix} 0 & 1 \\ -a & -b \end{bmatrix}$$

The eigenvalues of A are $\lambda_{1,2} = \left(-b \pm \sqrt{b^2 - 4a}\right)/2$. For all positive values of a and b, the eigenvalues satisfy $\text{Re}[\lambda] < 0$. Hence, the equilibrium point at the origin is exponentially stable. In the absence of friction ($b = 0$), both eigenvalues are on the imaginary axis. In this case, we cannot determine the stability of the origin through linearization. To investigate the stability of $(\pi, 0)$, we evaluate the Jacobian at this point. This is equivalent to performing a change of variables $z_1 = x_1 - \pi$, $z_2 = x_2$ to shift the equilibrium point to the origin, and then evaluating the Jacobian $[\partial f/\partial z]$ at $z = 0$.

$$\tilde{A} = \left.\frac{\partial f}{\partial x}\right|_{x_1=\pi, x_2=0} = \begin{bmatrix} 0 & 1 \\ a & -b \end{bmatrix}$$

The eigenvalues of $\tilde{A}$ are $\lambda_{1,2} = \left(-b \pm \sqrt{b^2 + 4a}\right)/2$. For all $a > 0$ and $b \geq 0$, there is, at least, one eigenvalue with positive real part. Hence, $(\pi, 0)$ is unstable.

Let us now use Lyapunov's method to study the stability of the origin. As a Lyapunov function candidate, we use the energy of the pendulum, which is defined as the sum of its potential and kinetic energies, namely,

$$V(x) = \int_0^{x_1} a \sin y \, dy + \frac{1}{2}x_2^2 = a(1 - \cos x_1) + \frac{1}{2}x_2^2$$

The reference of the potential energy is chosen such that $V(0) = 0$. The function $V(x)$ is positive definite over the domain $-2\pi < x_1 < 2\pi$. The derivative of V along the trajectories of the system is given by

$$\dot{V}(x) = a\dot{x}_1 \sin x_1 + x_2\dot{x}_2 = -\, bx_2^2$$

When friction is neglected ($b = 0$), $\dot{V}(x) = 0$ and we can conclude that the origin is stable. Moreover, $V(x)$ is constant during the motion of the system. Since $V(x) = c$ forms a closed contour around $x = 0$ for small $c > 0$, we see that the trajectory will be confined to one such contour and will not approach the origin. Hence the origin is not asymptotically stable. On the other hand, in the case with friction ($b > 0$), $\dot{V}(x) = -bx_2^2 \leq 0$ is negative semidefinite and we can conclude that the origin is stable. Note that $\dot{V}(x)$ is only negative semidefinite and not negative definite because $\dot{V}(x) = 0$ for $x_2 = 0$ irrespective of the value of x_1. Therefore, we cannot conclude asymptotic stability using Lyapunov's stability theorem. Here comes the role of the invariance principle. Consider the set $\{\dot{V}(x) = 0\} = \{x_2 = 0\}$. Suppose that a solution of the state equation stays identically in this set. Then

$$x_2(t) \equiv 0 \;\Rightarrow\; \dot{x}_2(t) \equiv 0 \;\Rightarrow\; \sin x_1(t) \equiv 0$$

Hence, on the segment $-\pi < x_1 < \pi$ of the $x_2 = 0$ line, the system can maintain the $\dot{V}(x) = 0$ condition only at the origin $x = 0$. Noting that the solution is confined to Ω_c (the component $\{V(x) \leq c\}$ that contains the origin) and for sufficiently small c, $\Omega_c \subset \{-\pi < x_1 < \pi\}$, we conclude that no solution can stay forever in the set $\Omega_c \cap \{x_2 = 0\}$ other than the trivial solution $x(t) \equiv 0$. Hence, the origin is asymptotically stable. We can also estimate the region of attraction by Ω_c with $c < 2a$ to ensure that Ω_c is bounded and contained in the strip $\{-\pi < x_1 < \pi\}$.

Example 43.2:

Consider the system

$$\dot{x}_1 = x_2$$
$$\dot{x}_2 = -g_1(x_1) - g_2(x_2)$$

where g_1 and g_2 are locally Lipschitz and satisfy

$$g_i(0) = 0, \quad yg_i(y) > 0, \quad \forall\, y \neq 0, \ i = 1, 2$$

and $\int_0^y g_1(z)\, dz \to \infty$, as $|y| \to \infty$. The system has an isolated equilibrium point at the origin. It can be viewed as a generalized pendulum equation with $g_2(x_2)$ as the friction term. Therefore, a Lyapunov function candidate may be taken as the energy-like function

$$V(x) = \int_0^{x_1} g_1(y)\, dy + \frac{1}{2}x_2^2,$$

which is positive definite in R^2 and radially unbounded. The derivative of $V(x)$ along the trajectories of the system is given by

$$\dot{V}(x) = g_1(x_1)x_2 + x_2[-g_1(x_1) - g_2(x_2)] = -x_2 g_2(x_2) \leq 0$$

Thus, $\dot{V}(x)$ is negative semidefinite. Note that $\dot{V}(x) = 0$ implies $x_2 g_2(x_2) = 0$, which implies $x_2 = 0$. Therefore, the only solution that can stay identically in the set $\{x \in R^2 \mid x_2 = 0\}$ is the zero solution $x(t) \equiv 0$. Thus, the origin is globally asymptotically stable.

Example 43.3:

The second-order system

$$\dot{x}_1 = -x_2$$
$$\dot{x}_2 = x_1 + (x_1^2 - 1)x_2$$

has a unique equilibrium point at the origin. Linearization at the origin yields the matrix

$$A = \left.\frac{\partial f}{\partial x}\right|_{x=0} = \begin{bmatrix} 0 & -1 \\ 1 & -1 \end{bmatrix}$$

which is Hurwitz. Hence, the origin is exponentially stable. By viewing the nonlinear system $\dot{x} = f(x)$ as a perturbation of the linear system $\dot{x} = Ax$, we can find a Lyapunov function for the linear system and use it as a Lyapunov function candidate for the nonlinear system. For the linear system $\dot{x} = Ax$ with Hurwitz matrix A, a quadratic Lyapunov function is given by $V(x) = x^T Px$, where P is the solution of the Lyapunov equation

$$PA + A^T P = -Q$$

for any positive-definite symmetric matrix Q. This is so because the solution P of the Lyapunov equation is a positive-definite symmetric matrix and the derivative of $V(x) = x^T Px$ along the trajectories of $\dot{x} = Ax$ is $-x^T Qx$. For our example, taking $Q = I$ results in

$$P = \begin{bmatrix} 1.5 & -0.5 \\ -0.5 & 1 \end{bmatrix}$$

Now we use $V(x) = x^T Px$ as a Lyapunov function candidate for the nonlinear system. The perturbation term $f(x) - Ax$ satisfies the linear growth bound of Equation 43.12 and the constant γ can be made arbitrarily small by limiting the analysis to a sufficiently small neighborhood of the origin. This is so because $f(x) - Ax = G(x)x$, where $\lim_{x\to 0} G(x) = 0$. Therefore, from the earlier analysis of perturbed systems we know that $V(x)$ is a Lyapunov function for the nonlinear system in some neighborhood of the origin. Let us estimate the region of attraction. The function $V(x)$ is positive definite for all x. We need to determine a domain D about the origin where $\dot{V}(x)$ is negative definite and a set $\Omega_c = \{V(x) \le c\} \subset D$. The set Ω_c is bounded for any $c > 0$. We are interested in the largest set Ω_c that we can determine, that is, the largest value for the constant c. The derivative of $V(x)$ along the trajectories of the system is given by

$$\dot{V}(x) = -(x_1^2 + x_2^2) - x_1^2 x_2(x_1 - 2x_2)$$

The right-hand side of $\dot{V}(x)$ is written as the sum of two terms. The first term, $-\|x\|^2$, is the contribution of the linear part Ax, while the second term is the contribution of the nonlinear term $f(x) - Ax$. Using the inequalities $|x_1 - 2x_2| \le \sqrt{5}\|x\|$ and $|x_1x_2| \le \frac{1}{2}\|x\|^2$, we see that $\dot{V}(x)$ satisfies the inequality $\dot{V}(x) \le -\|x\|^2 + (\sqrt{5}/2)\|x\|^4$. Hence $\dot{V}(x)$ is negative definite in the region $\{\|x\| < r\}$, where $r^2 = 2/\sqrt{5}$. We would like to choose a positive constant c such that $\{V(x) \le c\}$ is a subset of this region. Since $x^T Px \ge \lambda_{min}(P)\|x\|^2$, we can choose $c < \lambda_{min}(P)r^2$. Using $\lambda_{min}(P) \ge 0.69$, we choose $c = 0.615 < 0.69(2/\sqrt{5}) = 0.617$. The set $\{V(x) \le 0.615\}$ is an estimate of the region of attraction.

Example 43.4:

Consider the time-varying system

$$\begin{aligned}\dot{x}_1 &= -x_1 - g(t)x_2\\ \dot{x}_2 &= x_1 - x_2\end{aligned}$$

where $g(t)$ is continuously differentiable and satisfies $0 \le g(t) \le k$ and $\dot{g} \le g(t)$ for all $t \ge 0$. The system has an equilibrium point at the origin. Consider a Lyapunov function candidate $V(t,x) = x_1^2 + [1 + g(t)]x_2$. The function V satisfies the inequalities

$$x_1^2 + x_2^2 \le V(t,x) \le x_1^2 + (1+k)x_2^2$$

The derivative of V along the trajectories of the system is given by

$$\dot{V} = -2x_1^2 + 2x_1x_2 - [2 + 2g(t) - \dot{g}(t)]x_2^2$$

Using the bound on $\dot{g}(t)$, we have $2 + 2g(t) - \dot{g}(t) \ge 2 + 2g(t) - g(t) \ge 2$. Therefore,

$$\dot{V} \le -2x_1^2 + 2x_1x_2 - 2x_2^2 = -x^T \begin{bmatrix} 2 & -1 \\ -1 & 2 \end{bmatrix} x = -x^T Qx$$

The matrix Q is positive definite. Hence, the origin is globally exponentially stable.

43.3.1 Feedback Stabilization

Consider the nonlinear system

$$\begin{aligned}\dot{x} &= f(x,u)\\ y &= h(x)\end{aligned}$$

where x is an n-dimensional state, u is an m-dimensional control input, and y is a p-dimensional measured output. Suppose f and h are continuously differentiable in the domain of interest, and $f(0,0) = 0$, $h(0) = 0$ so that the origin is an open-loop equilibrium point and the output y vanishes at the origin. Suppose we want to design an output feedback controller to stabilize the origin, that is, to make the origin an asymptotically stable equilibrium point of the closed-loop system. We can pursue the design via linearization. The linearization of the system about the point $(x = 0,\ u = 0)$ is given by

$$\dot{x} = Ax + Bu$$
$$y = Cx$$

$$A = \left.\frac{\partial f}{\partial x}\right|_{x=0,u=0}, \quad B = \left.\frac{\partial f}{\partial u}\right|_{x=0,u=0}, \quad C = \left.\frac{\partial h}{\partial x}\right|_{x=0}$$

Assuming that (A, B) is stabilizable and (A, C) is detectable, that is, uncontrollable and unobservable eigenvalues, if any, have negative real parts, we can design a dynamic output feedback controller

$$\dot{z} = Fz + Gy,$$
$$u = Hz + Ky,$$

where z is a q-dimensional vector, such that the matrix

$$\mathcal{A} = \begin{bmatrix} A + BKC & BH \\ GC & F \end{bmatrix}$$

is Hurwitz. When the feedback controller is applied to the nonlinear system it results in a system of order $(n + q)$, whose linearization at the origin is the matrix $\mathcal{A}$. Hence, the origin of the closed-loop system is exponentially stable. By solving the Lyapunov equation $\mathcal{P}\mathcal{A} + \mathcal{A}^T\mathcal{P} = -\mathcal{Q}$ for some positive-definite matrix $\mathcal{Q}$, we can use $V = \mathcal{X}^T\mathcal{P}\mathcal{X}$, where $\mathcal{X} = [x\ \ z]^T$, as a Lyapunov function for the closed-loop system, and we can estimate the region of attraction of the origin, as illustrated in Example 43.3.

43.4 Defining Terms

Asymptotically stable equilibrium point: A stable equilibrium point with the additional feature that all trajectories starting at nearby points approach the equilibrium point as time approaches infinity.

Equilibrium point: A constant solution of $\dot{x} = f(t, x)$. For the time-invariant system $\dot{x} = f(x)$, equilibrium points are the real solutions of the equation $0 = f(x)$.

Exponentially stable equilibrium point: An asymptotically stable equilibrium point with the additional feature that the norm of the state is bounded by an exponentially decaying function of time whose amplitude is proportional to the norm of the difference between the initial state and the equilibrium point.

Globally asymptotically stable equilibrium point: An asymptotically stable equilibrium point where the region of attraction is the whole space.

Hurwitz matrix: A square real matrix is Hurwitz if all its eigenvalues have negative real parts.

Linearization: Approximation of the nonlinear state equation in the vicinity of an equilibrium point by a linear state equation, obtained by calculating the Jacobian matrix of the right-hand side at the equilibrium point.

Lipschitz condition: A condition imposed on a function $f(x)$ to ensure that it has a finite slope. For a vector-valued function, it takes the form $\|f(x) - f(y)\| \leq L\|x - y\|$ for some positive constant L.

Locally Lipschitz function: A function $f(x)$ is locally Lipschitz at a point if it satisfies the Lipschitz condition in the neighborhood of that point.

Lyapunov equation: A linear algebraic matrix equation of the form $PA + A^TP = -Q$, where A and Q are real square matrices. When Q is symmetric and positive definite, the equation has a (unique) positive-definite solution P if and only if A is Hurwitz.

Lyapunov function: A scalar positive-definite function of the state whose derivative along the trajectories of the system is negative semidefinite.

Lyapunov surface: A set of the form $V(x) = c$ where $V(x)$ is a Lyapunov function and c is a positive constant.

Negative (semi-) definite function: A scalar function of a vector argument $V(x)$ is negative (semi-) definite if $V(0) = 0$ and $V(x) < 0$ (≤ 0) for all $x \neq 0$ in some neighborhood of $x = 0$.

Positive (semi-) definite function: A scalar function of a vector argument $V(x)$ is positive (semi-) definite if $V(0) = 0$ and $V(x) > 0$ (≥ 0) for all $x \neq 0$ in some neighborhood of $x = 0$.

Positive-definite matrix: A symmetric real square matrix P is positive definite if the quadratic form $V(x) = x^T Px$ is a positive definite function. Equivalently, P is positive definite if and only if all its eigenvalues are positive.

Region of attraction: For an asymptotically stable equilibrium point, the region of attraction is the set of all points with the property that the trajectories starting at these points asymptotically approach the equilibrium point. It is an open connected set that contains the equilibrium point in its interior.

Stable equilibrium point: An equilibrium point where all solutions can be confined to an ε-neighborhood of the point be constraining the initial states to belong to a δ-neighborhood.

Reference

1. Khalil, H.K. 2002. *Nonlinear Systems*, 3rd Edition, Prentice-Hall, Upper Saddle River, NJ.

Further Reading

The presentation of Lyapunov stability is based on the textbook by Khalil (see [1]). For further information on Lyapunov stability, the reader is referred to Chapters 4, 8, and 9 of Khalil's book. Chapter 4 covers the basic Lyapunov theory. Chapter 8 covers more advanced topics, including the use of the center manifold theorem when linearization fails. Chapter 9 covers the stability of perturbed systems.

Other engineering textbooks where Lyapunov stability is emphasized include Vidyasagar, M. 2002. *Nonlinear Systems Analysis*, classic Ed., SIAM Philadelphia, PA; Slotine, J-J. and Li, W. 1991. *Applied Nonlinear Control*, Prentice-Hall, Englewood Cliffs; Sastry, S. 1999. *Nonlinear Systems: Analysis, Stability, and Control*, Springer, New York.

For a deeper look into the theoretical foundation of Lyapunov stability, there are excellent references, including Rouche, N., Habets, P., and Laloy, M. 1977. *Stability Theory by Lyapunov's Direct Method*, Springer-Verlag, New York; Hahn, W. 1967. *Stability of Motion*, Springer-Verlag, New York; Krasovskii, N.N. 1963. *Stability of Motion*, Stanford University Press, Palo Alto, CA.

Control journals often include articles where Lyapunov's method is used in system analysis or control design. Examples are the *IEEE Transactions on Automatic Control* and the IFAC Journal *Automatica*.

44

Input–Output Stability

A.R. Teel
University of Minnesota

T.T. Georgiou
University of Minnesota

L. Praly
Mines Paris Institute of Technology

Eduardo D. Sontag
Rutgers University

44.1 Introduction

A common task for an engineer is to design a system that reacts to stimuli in some specific and desirable way. One way to characterize appropriate behavior is through the formalism of input–output stability. In this setting a notion of well-behaved input and output signals is made precise and the question is posed: do well-behaved stimuli (inputs) produce well-behaved responses (outputs)?

General input–output stability analysis has its roots in the development of the electronic feedback amplifier of H.S. Black in 1927 and the subsequent development of classical feedback design tools for linear systems by H. Nyquist and H.W. Bode in the 1930s and 1940s, all at Bell Telephone Laboratories. These latter tools focused on determining input–output stability of linear feedback systems from the characteristics of the feedback components. Generalizations to nonlinear systems were made by several researchers in the late 1950s and early 1960s. The most notable contributions were those of G. Zames, then at M.I.T., I.W. Sandberg at Bell Telephone Laboratories, and V.M. Popov. Indeed, much of this chapter is based on the foundational ideas found in [5,7,10], with additional insights drawn from [6]. A thorough understanding of nonlinear systems from an input–output point of view is still an area of ongoing and intensive research.

The strength of input–output stability theory is that it provides a method for anticipating the qualitative behavior of a feedback system with only rough information about the feedback components. This, in turn, leads to notions of robustness of feedback stability and motivates many of the recent developments in modern control theory.

44.2 Systems and Stability

Throughout our discussion of input–output stability, a *signal* is a "reasonable" (e.g., piecewise continuous) function defined on a finite or semi-infinite time interval, i.e., an interval of the form $[0, T)$ where T is either a strictly positive real number or infinity. In general, a signal is vector-valued; its components typically represent actuator and sensor values. A *dynamical system* is an object which produces an output signal for each input signal.

To discuss stability of dynamical systems, we introduce the concept of a *norm function*, denoted $||\cdot||$, which captures the "size" of signals defined on the semi-infinite time interval. The significant properties of a norm function are that 1) the norm of a signal is zero if the signal is identically zero, and is a strictly positive number otherwise, 2) scaling a signal results in a corresponding scaling of the norm, and 3) the triangle inequality holds, i.e., $||u_1+u_2|| \leq ||u_1||+||u_2||$. Examples of norm functions are the p-norms. For any positive real number $p \geq 1$, the p-norm is defined by

$$||u||_p := \left(\int_0^\infty |u(t)|^p\right)^{\frac{1}{p}} \tag{44.1}$$

where $|\cdot|$ represents the standard Euclidean norm, i.e., $|u| = \sqrt{\sum_{i=1}^{n} u_i^2}$. For $p=\infty$, we define

$$||u||_\infty := \sup_{t\geq 0} |u(t)|. \tag{44.2}$$

The ∞-norm is useful when amplitude constraints are imposed on a problem, and the 2-norm is of more interest in the context of energy constraints. The norm of a signal may very well be infinite. We will typically be interested in measuring signals which may only be defined on finite time intervals or measuring truncated versions of signals. To that end, given a signal u defined on $[0,T)$ and a strictly positive real number τ, we use u_τ to denote the *truncated signal* generated by extending u onto $[0,\infty)$ by defining $u(t)=0$ for $t\geq T$, if necessary, and then truncating, i.e., u_τ is equal to the (extended) signal on the interval $[0,\tau]$ and is equal to zero on the interval (τ,∞).

Informally, a system is *stable* in the input–output sense if small input signals produce correspondingly small output signals. To make this concept precise, we need a way to quantify the dependence of the norm of the output on the norm of the input applied to the system. To that end, we define a *gain function* as a function from the nonnegative real numbers to the nonnegative real numbers which is continuous, nondecreasing, and zero when its argument is zero. For notational convenience we will say that the "value" of a gain function at ∞ is ∞. A dynamical system is stable (with respect to the norm $||\cdot||$) if there is a gain function γ which gives a bound on the norm of truncated output signals as a function of the norm of truncated input signals, i.e.,

$$||y_\tau|| \leq \gamma(||u_\tau||), \quad \text{for all } \tau. \tag{44.3}$$

In the very special case when the gain function is linear, i.e., there is at most an amplification by a constant factor, the dynamical system is *finite gain stable*. The notions of finite gain stability and closely related variants are central to much of classical input–output stability theory, but in recent years much progress has been made in understanding the role of more general (nonlinear) gains in system analysis.

The focus of this chapter will be on the stability analysis of interconnected dynamical systems as described in Figure 44.1. The composite system in Figure 44.1 will be called a *well-defined interconnection*

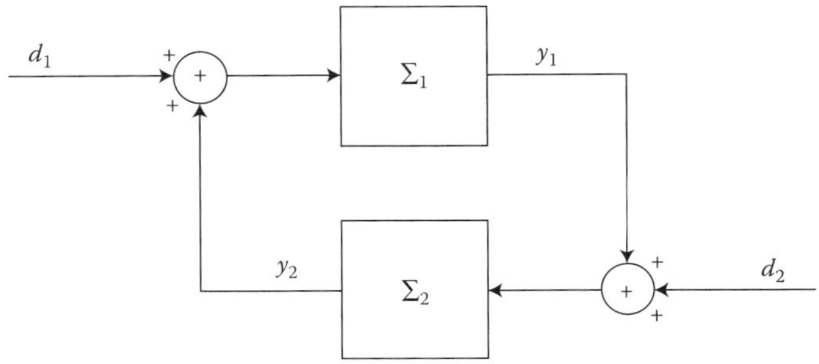

FIGURE 44.1 Standard feedback configuration.

if it is a dynamical system with $\begin{pmatrix} d_1 \\ d_2 \end{pmatrix}$ as input and $\begin{pmatrix} y_1 \\ y_2 \end{pmatrix}$ as output, i.e., given an arbitrary input signal $\begin{pmatrix} d_1 \\ d_2 \end{pmatrix}$, a signal $\begin{pmatrix} y_1 \\ y_2 \end{pmatrix}$ exists so that, for the dynamical system Σ_1, the input $d_1 + y_2$ produces the output y_1 and, for the dynamical system Σ_2, the input $d_2 + y_1$ produces the output y_2. To see that not every interconnection is well-defined, consider the case where both Σ_1 and Σ_2 are the identity mappings. In this case, the only input signals $\begin{pmatrix} d_1 \\ d_2 \end{pmatrix}$ for which an output $\begin{pmatrix} y_1 \\ y_2 \end{pmatrix}$ can be found are those for which $d_1 + d_2 = 0$. The dynamical systems which make up a well-defined interconnection will be called its *feedback components.*

For stability of well-defined interconnections, it is not necessary for either of the feedback components to be stable nor is it sufficient for both of the feedback components to be stable. On the other hand, necessary and sufficient conditions for stability of a well-defined interconnection can be expressed in terms of the set of all possible input–output pairs for each feedback component. To be explicit, following are some definitions. For a given dynamical system Σ with input signals u and output signals y, the set of its ordered input–output pairs (u, y) is referred to as the *graph* of the dynamical system and is denoted $\mathcal{G}_\Sigma$. When the input and output are exchanged in the ordered pair, i.e., (y, u), the set is referred to as the *inverse graph* of the system and is denoted $\mathcal{G}^I_\Sigma$. Note that, for the system in Figure 44.1, the inverse graph of Σ_2 and the graph of Σ_1 lie in the same Cartesian product space called the *ambient space.* We will use as norm on the ambient space the sum of the norms of the coordinates.

The basic observation regarding input–output stability for a well-defined interconnection says, in informal terms, that if a signal in the inverse graph of Σ_2 is near any signal in the graph of Σ_1 then it must be small. To formalize this notion, we need the concept of the *distance* to the graph of Σ_1 from signals x in the ambient space. This (truncated) distance is defined by

$$d_\tau(x, \mathcal{G}_{\Sigma_1}) := \inf_{z \in \mathcal{G}_{\Sigma_1}} ||(x - z)_\tau||. \tag{44.4}$$

Theorem 44.1: Graph Separation Theorem

A well-defined interconnection is stable if, and only if, a gain function γ exists which gives a bound on the norm of truncated signals in the inverse graph of Σ_2 as a function of the truncated distance from the signals to the graph of Σ_1, i.e.,

$$x \in \mathcal{G}^I_{\Sigma_2} \implies ||x_\tau|| \leq \gamma\left(d_\tau\left(x, \mathcal{G}_{\Sigma_1}\right)\right), \quad \textit{for all } \tau. \tag{44.5}$$

In the special case where γ is a linear function, the well-defined interconnection is finite gain stable.

The idea behind this observation can be understood by considering the signals that arise in the closed loop which belong to the inverse graph of Σ_2, i.e., the signals $(y_2, y_1 + d_2)$. (Stability with these signals taken as output is equivalent to stability with the original outputs.) Notice that, for the system in Figure 44.1, signals in the graph of Σ_1 have the form $(y_2 + d_1, y_1)$. Consequently, signals $x \in \mathcal{G}^I_{\Sigma_2}$ and $z \in \mathcal{G}_{\Sigma_1}$, which satisfy the feedback equations, also satisfy

$$(x - z)_\tau = (d_1, -d_2)_\tau \tag{44.6}$$

and

$$||(x - z)_\tau|| = \|(d_1, d_2)_\tau\| \tag{44.7}$$

for truncations within the interval of definition. If there are signals x in the inverse graph of Σ_2 with large truncated norm but small truncated distance to the graph of Σ_1, i.e., there exists some $z \in \mathcal{G}_{\Sigma_1}$ and $\tau > 0$ such that $||(x - z)_\tau||$ is small, then we can choose (d_1, d_2) to satisfy Equation 44.6 giving, according to

Equation 44.7, a small input which produces a large output. This contradicts our definition of stability. Conversely, if there is no z which is close to x, then only large inputs can produce large x signals and thus the system is stable.

The distance observation presented above is the unifying idea behind the input–output stability criteria applied in practice. However, the observation is rarely applied directly because of the difficulties involved in exactly characterizing the graph of a dynamical system and measuring distances. Instead, various simpler conditions have been developed which constrain the graphs of the feedback components to guarantee that the graph of Σ_1 and the inverse graph of Σ_2 are sufficiently separated. There are many such sufficient conditions, and, in the remainder of this chapter, we will describe a few of them.

44.3 Practical Conditions and Examples

44.3.1 The Classical Small Gain Theorem

One of the most commonly used sufficient conditions for graph separation constrains the graphs of the feedback components by assuming that each feedback component is finite gain stable. Then, the appropriate graphs will be separated if the product of the coefficients of the linear gain functions is sufficiently small. For this reason, the result based on this type of constraint has come to be known as the small gain theorem.

Theorem 44.2: Small Gain Theorem

If each feedback component is finite gain stable and the product of the gains (the coefficients of the linear gain functions) is less than one, then the well-defined interconnection is finite gain stable.

Figure 44.2 provides the intuition for the result. If we were to draw an analogy between a dynamical system and a static map whose graph is a set of points in the plane, the graph of Σ_1 would be constrained to the darkly shaded conic region by the finite gain stability assumption. Likewise, the inverse graph of Σ_2 would be constrained to the lightly shaded region. The fact that the product of the gains is less than one guarantees the positive aperture between the two regions and, in turn, that the graphs are separated sufficiently.

To apply the small gain theorem, we need a way to verify that the feedback components are finite gain stable (with respect to a particular norm) and to determine their gains. In particular, any linear dynamical system that can be represented with a real, rational transfer function $G(s)$ is finite gain stable in any of the p-norms if, and only if, all of the poles of the transfer function have negative real parts. A popular norm to work with is the 2-norm. It is associated with the energy of a signal. For a single-input, single-output (SISO) finite gain stable system modeled by a real, rational transfer function $G(s)$, the smallest possible coefficient for the stability gain function with respect to the 2-norm, is given by

$$\bar{\gamma} := \sup_{\omega} |G(j\omega)|. \tag{44.8}$$

For multi-input, multioutput systems, the magnitude in Equation 44.8 is replaced by the maximum singular value. In either case, this can be established using *Parseval's theorem*. For SISO systems, the quantity in Equation 44.8 can be obtained from a quick examination of the Bode plot or Nyquist plot for the transfer function. If the Nyquist plot of a stable SISO transfer function lies inside a circle of radius $\bar{\gamma}$ centered at the origin, then the coefficient of the 2-norm gain function for the system is less than or equal to $\bar{\gamma}$.

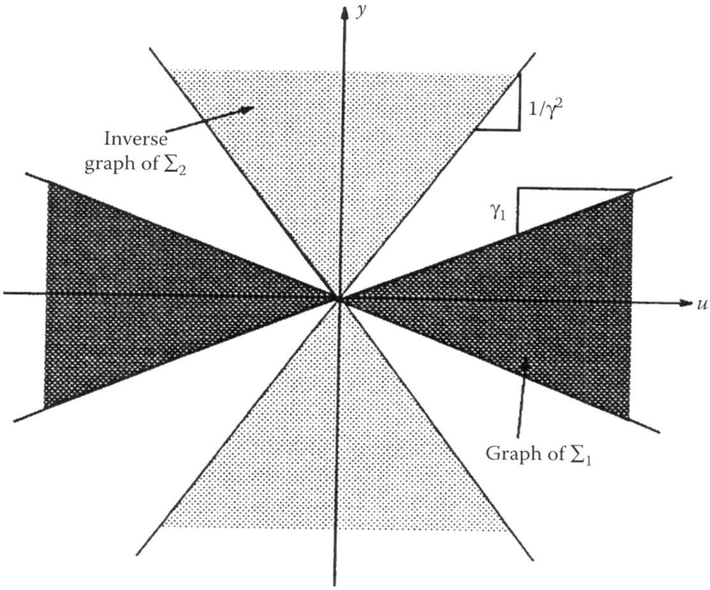

FIGURE 44.2 Classical small gain theorem.

More generally, consider a dynamical system that can be represented by a finite dimensional ordinary differential equation with zero initial state:

$$\dot{x} = f(x, u), \quad x(0) = 0 \quad \text{and} \quad y = h(x, u). \tag{44.9}$$

Suppose that f has globally bounded partial derivatives and that positive real numbers ℓ_1 and ℓ_2 exist so that

$$|h(x, u)| \leq \ell_1|x| + \ell_2|u|. \tag{44.10}$$

Under these conditions, if the trajectories of the unforced system with nonzero initial conditions,

$$\dot{x} = f(x, 0), \quad x(0) = x_\circ, \tag{44.11}$$

satisfy

$$|x(t)| \leq k \exp(-\lambda t)|x_\circ|, \tag{44.12}$$

for some positive real number k and λ and any $x_\circ \in \mathbb{R}^n$, then the system (Equation 44.9) is finite gain stable in any of the p-norms. This can be established using Lyapunov function arguments that apply to the system (Equation 44.11). The details can be found in the textbooks on nonlinear systems mentioned later.

Example 44.1:

Consider a nonlinear control system modeled by an ordinary differential equation with state $x \in \mathbb{R}^n$, input $v \in \mathbb{R}^m$ and disturbance $d_1 \in \mathbb{R}^m$:

$$\dot{x} = f(x, v + d_1). \tag{44.13}$$

Suppose that f has globally bounded partial derivatives and that a control $v = \alpha(x)$ can be found, also with a globally bounded partial derivative, so that the trajectories of the system

$$\dot{x} = f(x, \alpha(x)), \quad x(0) = x_\circ \tag{44.14}$$

satisfy the bound

$$|x(t)| \leq k \exp(-\lambda t)|x_\circ| \tag{44.15}$$

for some positive real numbers k and λ and for all $x_\circ \in \mathbb{R}^n$. As mentioned above, for any function h satisfying the type of bound in Equation 44.10, this implies that the system

$$\dot{x} = f(x, \alpha(x) + d_1), \quad x(0) = 0 \quad \text{and} \quad y = h(x, d_1) \tag{44.16}$$

has finite 2-norm gain from input d_1 to output y. We consider the output

$$y := \dot{\alpha} = \frac{\partial \alpha}{\partial x} f(x, \alpha(x) + d_1) \tag{44.17}$$

which satisfies the type of bound in Equation 44.10 because α and f both have globally bounded partial derivatives.

We will show, using the small gain theorem, that disturbances d_1 with finite 2-norm continue to produce outputs y with finite 2-norm even when the actual input v to the process is generated by the following fast dynamic version of the commanded input $\alpha(x)$:

$$\begin{aligned} \epsilon \dot{z} &= Az + B(\alpha(x)) + d_2, \quad z(0) = -A^{-1}B\alpha(x(0)) \\ v &= Cz. \end{aligned} \tag{44.18}$$

Here, ϵ is a small positive parameter, the eigenvalues of A all have strictly negative real part (thus A is invertible), and $-CA^{-1}B = I$. This system may represent unmodeled actuator dynamics.

To see the stability result, we will consider the composite system in the coordinates x and $\zeta = z + A^{-1}B\alpha(x)$. Using the notation from Figure 44.1,

$$\Sigma_1: \quad \begin{aligned} \dot{x} &= f(x, \alpha(x) + u_1), \quad x(0) = 0 \\ y_1 &= A^{-1}B\dot{\alpha}(x), \end{aligned} \tag{44.19}$$

and

$$\Sigma_2: \quad \begin{aligned} \dot{\zeta} &= \epsilon^{-1}A\zeta + u_2, \quad \zeta = 0 \\ y_2 &= C\zeta, \end{aligned} \tag{44.20}$$

with the interconnection conditions

$$u_1 = y_2 + d_1, \quad \text{and} \quad u_2 = y_1 + \epsilon^{-1}d_2. \tag{44.21}$$

Of course, if the system is finite gain stable with the inputs d_1 and $\epsilon^{-1}d_2$, then it is also finite gain stable with the inputs d_1 and d_2. We have already discussed that the system Σ_1 in Equation 44.19 has finite 2-norm gain, say γ_1. Now consider the system Σ_2 in Equation 44.20. It can be represented with the transfer function

$$\begin{aligned} G(s) &= C(sI - \epsilon^{-1}A)^{-1}, \\ &= \epsilon C(\epsilon sI - A)^{-1}, \\ &=: \epsilon\bar{G}(\epsilon s). \end{aligned} \tag{44.22}$$

Identifying $\bar{G}(s) = C(sI - A)^{-1}$, we see that, if

$$\gamma_2 := \sup_\omega \sigma(\bar{G}(j\omega)), \tag{44.23}$$

then

$$\sup_\omega \sigma(G(j\omega)) = \epsilon\gamma_2. \tag{44.24}$$

We conclude from the small gain theorem that, if $\epsilon < \dfrac{1}{\gamma_1\gamma_2}$, then the composite system (Equations 44.19 through 44.21), with inputs d_1 and d_2 and outputs $y_1 = A^{-1}B\dot{\alpha}(x)$ and $y_2 = C\zeta$, is finite gain stable.

44.3.2 The Classical Passivity Theorem

Another very popular condition used to guarantee graph separation is given in the *passivity theorem*. For the most straightforward passivity result, the number of input channels must equal the number of output channels for each feedback component. We then identify the relative location of the graphs of the feedback components using a condition involving the integral of the product of the input and the output signals. This operation is known as the *inner product*, denoted $\langle \cdot, \cdot \rangle$. In particular, for two signals u and y of the same dimension defined on the semi-infinite interval,

$$\langle u, y \rangle := \int_0^\infty u^T(t) y(t)\, dt. \tag{44.25}$$

Note that $\langle u, y \rangle = \langle y, u \rangle$ and $\langle u, u \rangle = ||u||_2^2$. A dynamical system is *passive* if, for each input–output pair (u, y) and each $\tau > 0$, $\langle u_\tau, y_\tau \rangle \geq 0$. The terminology used here comes from the special case where the input and output are a voltage and a current, respectively, and the energy absorbed by the dynamical system, which is the inner product of the input and output, is nonnegative.

Again by analogy to a static map whose graph lies in the plane, passivity of a dynamical system can be viewed as the condition that the graph is constrained to the darkly shaded region in Figure 44.3, i.e., the first and third quadrants of the plane. This graph and the inverse graph of a second system would be separated if, for example, the inverse graph of the second system were constrained to the lightly shaded region in Figure 44.3, i.e., the second and fourth quadrants but bounded away from the horizontal and vertical axes by an increasing and unbounded distance. But, this is the same as asking that the graph of the second system followed by the scaling "−1," i.e., all pairs $(u, -y)$, be constrained to the first and third quadrants, again bounded away from the axes by an increasing and unbounded distance, as in Figure 44.4a. For classical passivity theorems, this region is given a linear boundary as in Figure 44.4b. Notice that, for points $(u_\circ, y_\circ)$ in the plane, if $u_\circ \cdot y_\circ \geq \epsilon(u_\circ^2 + y_\circ^2)$ then $(u_\circ, y_\circ)$ is in the first or third quadrant, and $(\epsilon)^{-1}|u_\circ| \geq |y_\circ| \geq \epsilon|u_\circ|$ as in Figure 44.4b. This leads to

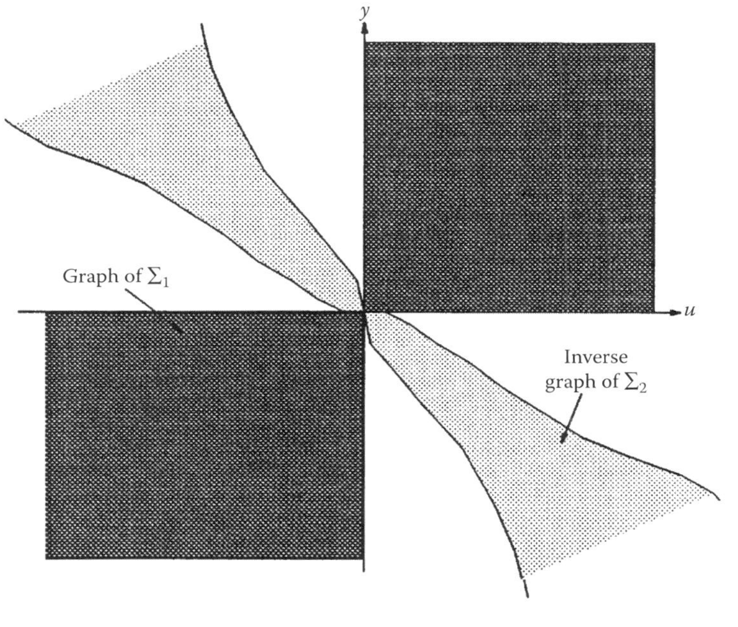

FIGURE 44.3 General passivity-based interconnection.

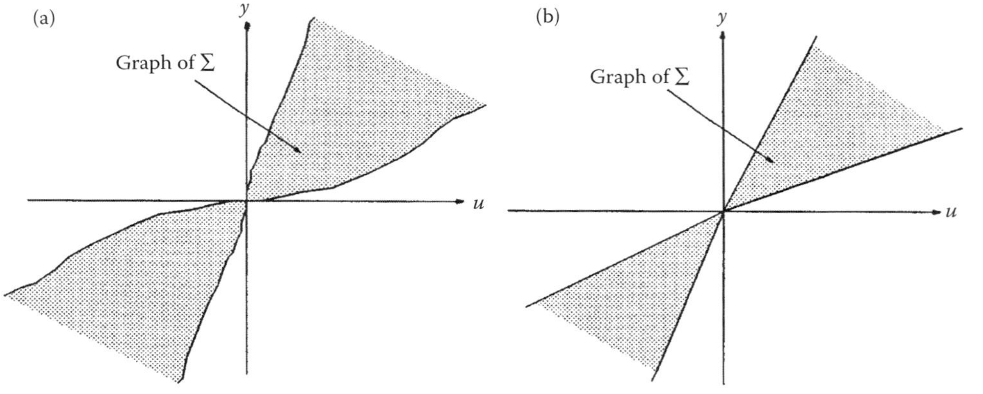

FIGURE 44.4 Different notions of input and output strict passivity.

the following stronger version of passivity. A dynamical system is *input and output strictly passive* if a strictly positive real number ϵ exists so that, for each input–output pair (u, y) and each $\tau > 0$, $\langle u_\tau, y_\tau \rangle \geq \epsilon \left(||u_\tau||_2^2 + ||y_\tau||_2^2\right)$.

There are intermediate versions of passivity which are also useful. These correspond to asking for an increasing and unbounded distance from either the horizontal axis or the vertical axis but not both. For example, a dynamical system is *input strictly passive* if a strictly positive real number ϵ exists so that, for each input–output pair (u, y) and each $\tau > 0$, $\langle u_\tau, y_\tau \rangle \geq \epsilon ||u_\tau||_2^2$. Similarly, a dynamical system is *output strictly passive* if a strictly positive real number ϵ exists so that, for each input–output pair (u, y) and each $\tau > 0$, $\langle u_\tau, y_\tau \rangle \geq \epsilon ||y_\tau||_2^2$. It is worth noting that input and output strict passivity is equivalent to input strict passive plus finite gain stability. This can be shown with standard manipulations of the inner product. Also, the reader is warned that all three types of strict passivity mentioned above are frequently called "strict passivity" in the literature.

Again by thinking of a graph of a system as a set of points in the plane, output strict passivity is the condition that the graph is constrained to the darkly shaded region in Figure 44.5, i.e., the first and third quadrants with an increasing and unbounded distance from the vertical axis. To complement such a graph, consider a second dynamical system which, when followed by the scaling "-1," is also output strictly passive. Such a system has a graph (without the "-1" scaling) constrained to the second and fourth quadrants with an increasing and unbounded distance from the vertical axis. In other words, its inverse graph is constrained to the lightly shaded region of Figure 44.5, i.e., to the second and fourth quadrants but with an increasing and unbounded distance from the *horizontal* axis. The conclusions that we can then draw, using the graph separation theorem, are summarized in the following passivity theorem.

Theorem 44.3: Passivity Theorem

If one dynamical system and the other dynamical system followed by the scaling "-1" are

- *both input strictly passive, OR*
- *both output strictly passive, OR*
- *respectively, passive* and *input and output strictly passive,*

then the well-defined interconnection is finite gain stable in the 2-norm.

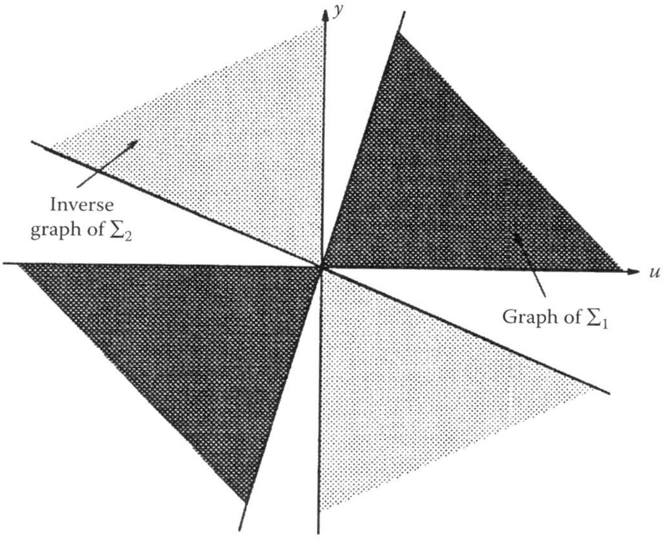

FIGURE 44.5 Interconnection of output strictly passive systems.

To apply this theorem, we need a way to verify that the (possibly scaled) feedback components are appropriately passive. For stable SISO systems with real, rational transfer function $G(s)$, it again follows from Parseval's theorem that, if

$$\text{Re } G(j\omega) \geq 0,$$

for all real values of ω, then the system is passive. If the quantity $\text{Re}\, G(j\omega)$ is positive and uniformly bounded away from zero for all real values of ω, then the linear system is input and output strictly passive. Similarly, if $\epsilon > 0$ exists so that, for all real values of ω,

$$\text{Re } G(j\omega - \epsilon) \geq 0, \tag{44.26}$$

then the linear system is output strictly passive. So, for SISO systems modeled with real, rational transfer functions, passivity and the various forms of strict passivity can again be easily checked by means of a graphical approach such as a Nyquist plot.

More generally, for any dynamical system that can be modeled with a smooth, finite dimensional ordinary differential equation,

$$\begin{aligned} \dot{x} &= f(x) + g(x)u, \quad x(0) = 0 \\ y &= h(x), \end{aligned} \tag{44.27}$$

if a strictly positive real number ϵ exists and a nonnegative function $V : \mathbb{R}^n \to \mathbb{R}_{\geq 0}$ with $V(0) = 0$ satisfying

$$\frac{\partial V}{\partial x}(x) f(x) \leq -\epsilon h^T(x) h(x), \tag{44.28}$$

and

$$\frac{\partial V}{\partial x}(x) g(x) = h^T(x), \tag{44.29}$$

then the system is output strictly passive. With $\epsilon = 0$, the system is passive. Both of these results are established by integrating $\dot{V}$ over the semi-infinite interval.

Example 44.2:

(This example is prompted by the work in Berghuis and Nijmeijer, *Syst. Control Lett.*, 1993, 21, 289–295.) Consider a "completely controlled dissipative Euler-Lagrange" system with generalized "forces" F, generalized coordinates q, uniformly positive definite "inertia" matrix $I(q)$, Rayleigh dissipation function $R(\dot{q})$ and, say positive, potential $V(q)$ starting from the position q_d. Let the dynamics of the system be given by the Euler-Lagrange-Rayleigh equations,

$$\overbrace{\frac{\partial L}{\partial \dot{q}}(q,\dot{q})}^{\cdot} = \frac{\partial L}{\partial q}(q,\dot{q}) + F^{\top} - \frac{\partial L}{\partial q}(\dot{q})$$
$$q(0) = qd, \quad \dot{q}(0) = 0, \tag{44.30}$$

where L is the Lagrangian

$$L(q,\dot{q}) = \frac{1}{2}\,\dot{q}^{\top} I(q)\,\dot{q} - V(q). \tag{44.31}$$

Along the solution of Equation 44.30,

$$\dot{L} = \frac{\partial L}{\partial \dot{q}}\ddot{q} + \frac{\partial L}{\partial q}\dot{q} = \frac{\partial L}{\partial \dot{q}}\ddot{q} + \left[\overbrace{\frac{\partial L}{\partial \dot{q}}}^{\cdot} - F^{\top} + \frac{\partial R}{\partial \dot{q}}\right]\dot{q}, \tag{44.32}$$

$$= \overbrace{\left(\frac{\partial L}{\partial \dot{q}}\dot{q}\right)}^{\cdot} - \left[F^{\top} - \frac{\partial R}{\partial \dot{q}}\right]\dot{q} = \overbrace{\left(\partial \dot{q}^{\top} I(q)\,\dot{q}\right)}^{\cdot} - \left[F^{\top} - \frac{\partial R}{\partial \dot{q}}\right]\dot{q}, \tag{44.33}$$

$$= 2\,\overbrace{L + V}^{\cdot} - \left[F^{\top} - \frac{\partial R}{\partial \dot{q}}\right]\dot{q} = -2\,\dot{V} + \left[F^{\top} - \frac{\partial R}{\partial \dot{q}}\right]\dot{q}. \tag{44.34}$$

We will suppose that $\epsilon > 0$ exists so that

$$\frac{\partial R}{\partial \dot{q}}(\dot{q})\,\dot{q} \geq \epsilon\,|\dot{q}|^2. \tag{44.35}$$

Now let V_d be a function so that the modified potential

$$V_m = V + V_d \tag{44.36}$$

has a global minimum at $q = q_d$, and let the generalized "force" be

$$F = -\frac{\partial V_d}{\partial q}(q) + F_m. \tag{44.37}$$

We can see that the system (Equation 44.30) combined with Equation 44.37, having input F_m and output $\dot{q}$, is output strictly passive by integrating the derivative of the defined Hamiltonian,

$$H = \frac{1}{2}\,\dot{q}^{\top} I(q)\,\dot{q} + V_m(q) = L + 2\,V + V_d. \tag{44.38}$$

Indeed the derivative is

$$\dot{H} = \left[F_m^{\top} - \frac{\partial R}{\partial \dot{q}}\right]\dot{q} \tag{44.39}$$

and, integrating, for each τ

$$\left\langle \left[F_{m_\tau} - \frac{\partial R}{\partial \dot{q}}(\dot{q}_\tau)^{\top}\right], \dot{q}_\tau \right\rangle = H(\tau) - H(0). \tag{44.40}$$

Since $H \geq 0$, $H(0) = 0$ and Equation 44.35 holds

$$\langle F_{m\tau}, \dot{q}_\tau \rangle \geq \epsilon ||\dot{q}_\tau||_2^2. \tag{44.41}$$

Using the notation from Figure 44.1, let Σ_1 be the system (Equations 44.30 and 44.37) with input F_m and output $\dot{q}$. Let Σ_2 be any system that, when followed by the scaling "-1," is output strictly passive. Then, according to the passivity theorem, the composite feedback system as given in Figure 44.1 is finite gain stable using the 2-norm. One possibility for Σ_2 is minus the identity mapping. However, there is interest in choosing Σ_2 followed by the scaling "-1" as a linear, output strictly passive compensator which, in addition, has no direct feed-through term. The reason is that if, d_2 in Figure 44.1 is identically zero, we can implement Σ_2 with measurement only of q and without $\dot{q}$. In general,

$$G(s)\dot{q} = G(s)s\left(\frac{1}{s}\dot{q}\right) = G(s)s(q - q_d), \tag{44.42}$$

and the system $G(s)s$ is implementable if $G(s)$ has no direct feed-through terms. To design an output strictly passive linear system without direct feed-through, let A be a matrix having all eigenvalues with strictly negative real parts so that, by a well-known result in linear systems theory, a positive definite matrix P exists satisfying

$$A^T P + PA = -I. \tag{44.43}$$

Then, for any B matrix of appropriate dimensions, the system modeled by the transfer function,

$$G(s) = -B^T P(sI - A)^{-1} B, \tag{44.44}$$

followed by the scaling "-1," is output strictly passive. To see this, consider a state-space realization

$$\begin{aligned} \dot{x} &= Ax + Bu \quad x(0) = 0 \\ y &= B^T Px, \end{aligned} \tag{44.45}$$

and note that

$$\overbrace{x^{\top} P x}^{\cdot} = -x^{\top} x + 2x^{\top} P B u \tag{44.46}$$

$$= -x^{\top} x + 2y^{\top} u. \tag{44.47}$$

But, with Equation 44.45, for some strictly positive real number c,

$$2c\,y^{\top} y \leq x^{\top} x. \tag{44.48}$$

So, integrating Equation 44.47 and with P positive definite, for all τ,

$$\langle y_\tau, u_\tau \rangle \geq c\|y_\tau\|_2^2. \tag{44.49}$$

As a point of interest, one could verify that

$$G(s)s = -B^T PA(sI - A)^{-1} B - B^T PB. \tag{44.50}$$

44.3.3 Simple Nonlinear Separation Theorems

In this section we illustrate how allowing regions with nonlinear boundaries in the small gain and passivity contexts may be useful. First we need a class of functions to describe nonlinear boundaries. A *proper separation function* is a function from the nonnegative real numbers to the nonnegative real numbers which is continuous, zero at zero, strictly increasing and unbounded. The main difference between a gain function and a proper separation function is that the latter is invertible, and the inverse is another proper separation function.

44.3.3.1 Nonlinear Passivity

We will briefly discuss a definition of nonlinear input and output strict passivity. To our knowledge, this idea has not been used much in the literature. The notion replaces the linear boundaries in the input and output strict passivity definition by nonlinear boundaries as in Figure 44.4a. A dynamical system is *nonlinearly input and output strictly passive* if a proper separation function ρ exists so that, for each input–output pair (u, y) and each $\tau > 0$, $\langle u_\tau, y_\tau \rangle \geq \|u_\tau\|_2 \rho(\|u_\tau\|_2) + \|y_\tau\|_2 \rho(\|y_\tau\|_2)$. (Note that in the classical definition of strict passivity, $\rho(\zeta) = \epsilon\zeta$ for all $\zeta \geq 0$.)

Theorem 44.4: Nonlinear Passivity Theorem

If one dynamical system is passive and the other dynamical system followed by the scaling "−1" is nonlinearly input and output strictly passive, then the well-defined interconnection is stable using the 2-norm.

Example 44.3:

Let Σ_1 be a single integrator system,

$$\begin{aligned} \dot{x}_1 &= u_1 \quad x_1(0) = 0 \\ y_1 &= x_1. \end{aligned} \tag{44.51}$$

This system is passive because

$$0 \leq \tfrac{1}{2}x_1(\tau)^2 = \int_0^{\tau} \frac{d}{dt}\tfrac{1}{2}x(t)^2\,dt = \int_0^{\tau} y_1(t)u_1(t)\,dt = \langle y_{1_\tau}, u_{1_\tau}\rangle. \tag{44.52}$$

Let Σ_2 be a system which scales the instantaneous value of the input according to the energy of the input:

$$\begin{aligned} \dot{x}_2 &= u_2^2 \quad x_2(0) = 0 \\ y_2 &= -u_2\left(\frac{1}{1+|x_2|^{0.25}}\right). \end{aligned} \tag{44.53}$$

This system followed by the scaling "−1" is nonlinearly strictly passive. To see this, first note that

$$x_2(t) = ||u_{2_t}||_2^2 \tag{44.54}$$

which is a nondecreasing function of t. So,

$$\begin{aligned} \langle -y_{2_\tau}, u_{2_\tau}\rangle &= \int_0^{\tau} u_2^2(t)\left(\frac{1}{1+|x_2(t)|^{0.25}}\right)dt, \\ &\geq \left(\frac{1}{1+|x_2(\tau)|^{0.25}}\right)\int_0^{\tau} u_2^2(t)\,dt, \\ &= \left(\frac{1}{1+||u_{2_\tau}||_2^{0.5}}\right)||u_{2_\tau}||_2^2. \end{aligned} \tag{44.55}$$

Now we can define

$$\rho(\zeta) := \frac{0.5\zeta}{1+\zeta^{0.5}}, \tag{44.56}$$

which is a proper separation function, so that

$$\langle -y_{2_\tau}, u_{2_\tau}\rangle \geq 2\rho(||u_{2_\tau}||_2)||u_{2_\tau}||_2. \tag{44.57}$$

Finally, note that

$$||y_{2_\tau}||_2^2 = \int_0^{\tau} u_2^2(t)\frac{1}{\left(1+x_2^{0.25}(t)\right)^2}\,dt \leq ||u_{2_\tau}||_2^2, \tag{44.58}$$

so that

$$\langle -y_{2_\tau}, u_{2_\tau}\rangle \geq \rho(||u_{2_\tau}||_2)||u_{2_\tau}||_2 + \rho(||y_{2_\tau}||_2)||y_{2_\tau}||_2. \tag{44.59}$$

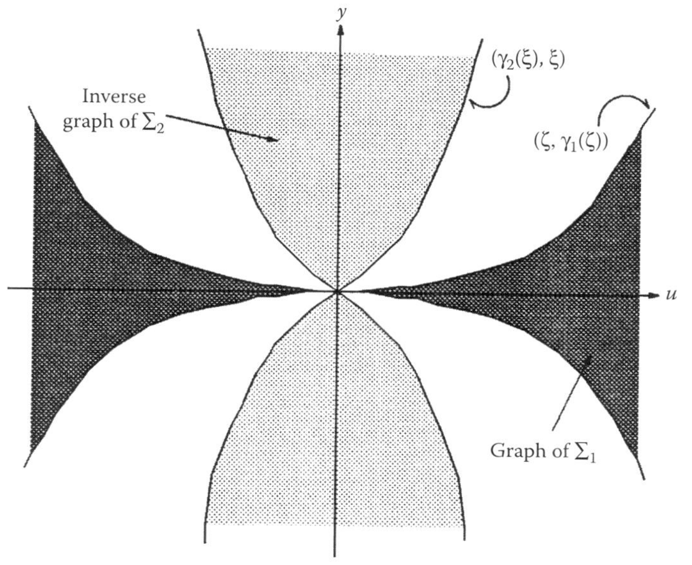

FIGURE 44.6 Nonlinear small gain theorem.

The conclusion that we can then draw from the nonlinear passivity theorem is that the interconnection of these two systems:

$$\begin{aligned} \dot{x}_1 &= -(x_1+d_2)\left(\frac{1}{1+|x_2|^{0.25}}\right)+d_1, \quad \dot{x}_2=(x_1+d_2)^2, \\ y_1 &= x_1, \quad \text{and} \quad y_2=-(x_1+d_2)\left(\frac{1}{1+|x_2|^{0.25}}\right) \end{aligned} \tag{44.60}$$

is stable when measuring input (d_1, d_2) and output (y_1, y_2) using the 2-norm.

44.3.3.2 Nonlinear Small Gain

Just as with passivity, the idea behind the small gain theorem does not require the use of linear boundaries. Consider a well-defined interconnection where each feedback component is stable but not necessarily finite gain stable. Let γ_1 be a stability gain function for Σ_1 and let γ_2 be a stability gain function for Σ_2. Then the graph separation condition will be satisfied if the distance between the curves $(\zeta, \gamma_1(\zeta))$ and $(\gamma_2(\xi), \xi)$ grows without bound as in Figure 44.6. This is equivalent to asking whether it is possible to add to the curve $(\zeta, \gamma_1(\zeta))$ in the vertical direction and to the curve $(\gamma_2(\xi), \xi)$ in the horizontal direction, by an increasing and unbounded amount, to obtain new curves $(\zeta, \gamma_1(\zeta)+\rho(\zeta))$ and $(\gamma_2(\xi)+\rho(\xi), \xi)$ where ρ is a proper separation function, so that the modified first curve is never above the modified second curve. If this is possible, we will say that the composition of the functions γ_1 and γ_2 is a *strict contraction*. To say that a curve $(\zeta, \tilde{\gamma}_1(\zeta))$ is never above a second curve $(\tilde{\gamma}_2(\xi), \xi)$ is equivalent to saying that $\tilde{\gamma}_1(\tilde{\gamma}_2(\zeta)) \le \zeta$ or $\tilde{\gamma}_2(\tilde{\gamma}_1(\zeta)) \le \zeta$ for all $\zeta \ge 0$. (Equivalently, we will write $\tilde{\gamma}_1 \circ \tilde{\gamma}_2 \le \text{Id}$ or $\tilde{\gamma}_2 \circ \tilde{\gamma}_1 \le \text{Id}$.) So, requiring that the composition of γ_1 and γ_2 is a strict contraction is equivalent to requiring that a strictly proper separation function ρ exists so that $(\gamma_1+\rho)\circ(\gamma_2+\rho) \le \text{Id}$ (equivalently $(\gamma_2+\rho)\circ(\gamma_1+\rho) \le \text{Id}$). This condition was made precise in [3]. (See also [2].) Note that it is not enough to add to just one curve because it is possible for the vertical or horizontal distance to grow without bound while the total distance remains bounded. Finally, note that, if the gain functions are linear, the condition is the same as the condition that the product of the gains is less than one.

Theorem 44.5: Nonlinear Small Gain Theorem

If each feedback component is stable (with gain functions γ_1 and γ_2) and the composition of the gains is a strict contraction, then the well-defined interconnection is stable.

To apply the nonlinear small gain theorem, we need a way to verify that the feedback components are stable. To date, the most common setting for using the nonlinear small gain theorem is when measuring the input and output using the ∞-norm. For a nonlinear system which can be represented by a smooth, ordinary differential equation,

$$\dot{x} = f(x,u), \quad x(0) = 0, \quad \text{and} \quad y = h(x,u), \tag{44.61}$$

where $h(0,0) = 0$, the system is stable (with respect to the ∞-norm) if there exist a positive definite and radially unbounded function $V : \mathbb{R}^n \to \mathbb{R}_{\geq} 0$, a proper separation function ψ, and a gain function $\tilde{\gamma}$ so that

$$\frac{\partial V}{\partial x} f(x,u) \leq -\psi(|x|) + \tilde{\gamma}(|u|). \tag{44.62}$$

Since V is positive definite and radially unbounded, additional proper separation functions $\underline{\alpha}$ and $\bar{\alpha}$ exist so that

$$\underline{\alpha}(|x|) \leq V(x) \leq \bar{\alpha}(|x|). \tag{44.63}$$

Also, because h is continuous and zero at zero, gain functions ϕ_x and ϕ_u exist so that

$$|h(x,u)| \leq \phi_x(|x|) + \phi_u(|u|). \tag{44.64}$$

Given all of these functions, a stability gain function can be computed as

$$\gamma = \phi_x \circ \underline{\alpha}^{-1} \circ \bar{\alpha} \circ \psi^{-1} \circ \tilde{\gamma} + \phi_u. \tag{44.65}$$

For more details, the reader is directed to [8].

Example 44.4:

Consider the composite system,

$$\begin{aligned} \dot{x} &= Ax + B\text{sat}(z + d_1), \quad x(0) = 0 \\ \dot{z} &= -z + \epsilon(\exp(|x| + d_2) - 1), \quad z(0) = 0, \end{aligned} \tag{44.66}$$

where $x \in \mathbb{R}^n, z \in \mathbb{R}$, the eigenvalues of A all have strictly negative real part, ϵ is a small parameter, and $\text{sat}(s) = \text{sgn}(s)\min\{|s|, 1\}$. This composite system is a well-defined interconnection of the subsystems

$$\Sigma_1 : \quad \begin{aligned} \dot{x} &= Ax + B\text{sat}(u_1), \quad x(0) = 0 \\ y_1 &= |x| \end{aligned} \tag{44.67}$$

and

$$\Sigma_2 : \quad \begin{aligned} \dot{z} &= -z + \epsilon\left(\exp(u_2) - 1\right), \quad z(0) = 0 \\ y_2 &= z. \end{aligned} \tag{44.68}$$

A gain function for the Σ_1 system is the product of the ∞-gain for the linear system

$$\begin{aligned} \dot{x} &= Ax + Bu, \quad x(0) = 0 \\ y &= x, \end{aligned} \tag{44.69}$$

which we will call $\bar{\gamma}_1$, with the function sat(s), i.e., for the system Σ_1,

$$||y||_\infty \leq \bar{\gamma}_1 \text{sat}(||u_1||_\infty). \tag{44.70}$$

For the system Σ_2,

$$||z||_\infty \leq |\epsilon| \left(\exp\left(||u_2||_\infty\right) - 1\right). \tag{44.71}$$

The distance between the curves $(\zeta, \bar{\gamma}_1 \text{sat}(\zeta))$ and $(|\epsilon| \left(\exp\left(\xi\right) - 1\right), \xi)$ must grow without bound. Graphically, one can see that a necessary and sufficient condition for this is that

$$|\epsilon| < \frac{1}{\exp(\bar{\gamma}_1) - 1}. \tag{44.72}$$

44.3.4 General Conic Regions

There are many different ways to partition the ambient space to establish the graph separation condition in Equation 44.5. So far we have looked at only two very specific sufficient conditions, the small gain theorem and the passivity theorem. The general idea in these theorems is to constrain signals in the graph of Σ_1 within some conic region, and signals in the inverse graph of Σ_2 outside of this conic region. Conic regions more general than those used for the small gain and passivity theorems can be generated by using operators on the input–output pairs of the feedback components.

Let **C** and **R** be operators on truncated ordered pairs in the ambient space, and let γ be a gain function. We say that the graph of Σ_1 is inside Cone$(\mathbf{C}, \mathbf{R}, \gamma)$ if, for each $(u, y) =: z$ belonging to the graph of Σ_1,

$$||\mathbf{C}(z_\tau)|| \leq \gamma(||\mathbf{R}(z_\tau)||), \quad \text{for all } \tau. \tag{44.73}$$

On the other hand, we say that the inverse graph of Σ_2 is strictly outside Cone$(\mathbf{C}, \mathbf{R}, \gamma)$ if a proper separation function ρ exists so that, for each $(y, u) =: x$ belonging to the inverse graph of Σ_2,

$$||\mathbf{C}(x_\tau)|| \geq \gamma \circ (\text{Id} + \rho)(||\mathbf{R}(x_\tau)||) + \rho(||x_\tau||), \quad \text{for all } \tau. \tag{44.74}$$

We will only consider the case where the maps **C** and **R** are incrementally stable, i.e., a gain function $\bar{\gamma}$ exists so that, for each x_1 and x_2 in the ambient space and all τ,

$$\begin{aligned} ||\mathbf{C}(x_{1_\tau}) - \mathbf{C}(x_{2_\tau})|| &\leq \bar{\gamma}(||x_{1_\tau} - x_{2_\tau}||) \\ ||\mathbf{R}(x_{1_\tau}) - \mathbf{R}(x_{2_\tau})|| &\leq \bar{\gamma}(||x_{1_\tau} - x_{2_\tau}||). \end{aligned} \tag{44.75}$$

In this case, the following result holds.

Theorem 44.6: Nonlinear Conic Sector Theorem

If the graph of Σ_1 is inside CONE$(\mathbf{C}, \mathbf{R}, \gamma)$ *and the inverse graph of Σ_2 is strictly outside* CONE$(\mathbf{C}, \mathbf{R}, \gamma)$, *then the well-defined interconnection is stable.*

When γ and ρ are linear functions, the well-defined interconnection is finite gain stable.

The small gain and passivity theorems we have discussed can be interpreted in the framework of the nonlinear conic sector theorem. For example, for the nonlinear small gain theorem, the operator **C** is a projection onto the second coordinate in the ambient space, and **R** is a projection onto the first coordinate;

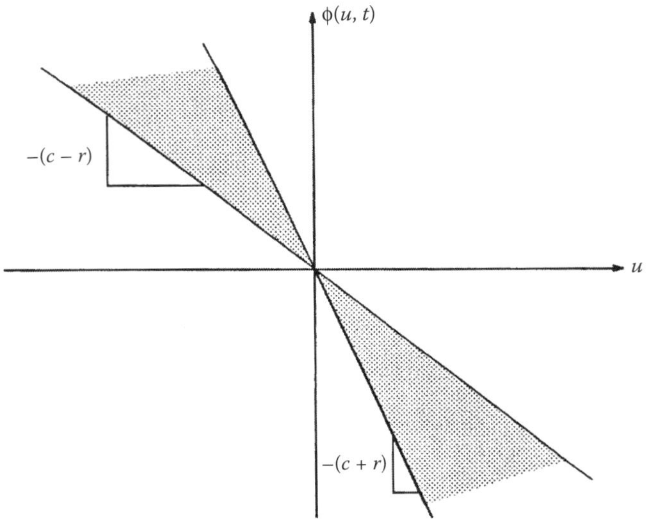

FIGURE 44.7 Instantaneous sector.

γ is the gain function γ_1, and the small gain condition guarantees that the inverse graph of Σ_2 is strictly outside of the cone specified by this **C**, **R** and γ.

In the remaining subsections, we will discuss other useful choices for the operators **C** and **R**.

44.3.4.1 The Classical Conic Sector (Circle) Theorem

For linear SISO systems connected to memoryless nonlinearities, there is an additional classical result, known as the circle theorem, which follows from the nonlinear conic sector theorem using the 2-norm and taking

$$\begin{aligned} \mathbf{C}\left(u, y\right) &= y + cu \\ \mathbf{R}\left(u, y\right) &= ru \quad r \geq 0 \\ \gamma(\zeta) &= \zeta. \end{aligned} \tag{44.76}$$

Suppose ϕ is a memoryless nonlinearity which satisfies

$$|\phi(u, t) + cu| \leq |ru| \quad \text{for all } t, u. \tag{44.77}$$

Graphically, the constraint on ϕ is shown in Figure 44.7. (In the case shown, $c > r > 0$.) We will use the notation Sector$[-(c+r), -(c-r)]$ for the memoryless nonlinearity. It is also clear that the graph of ϕ lies in the Cone(**C**, **R**, γ) with **C**, **R**, γ defined in Equation 44.76. For a linear, time invariant, finite dimensional SISO system, whether its inverse graph is strictly outside of this cone can be determined by examining the Nyquist plot of its transfer function. The condition on the Nyquist plot is expressed relative to a disk $\mathcal{D}_{c,r}$ in the complex plane centered on the real axis passing through the points on the real axis with real parts $-1/(c+r)$ and $-1/(c-r)$ as shown in Figure 44.8.

Theorem 44.7: Circle Theorem

Let $r \geq 0$, and consider a well-defined interconnection of a memoryless nonlinearity belonging to Sector$[-(c+r), -(c-r)]$ *with a SISO system having a real, rational transfer function $G(s)$. If*

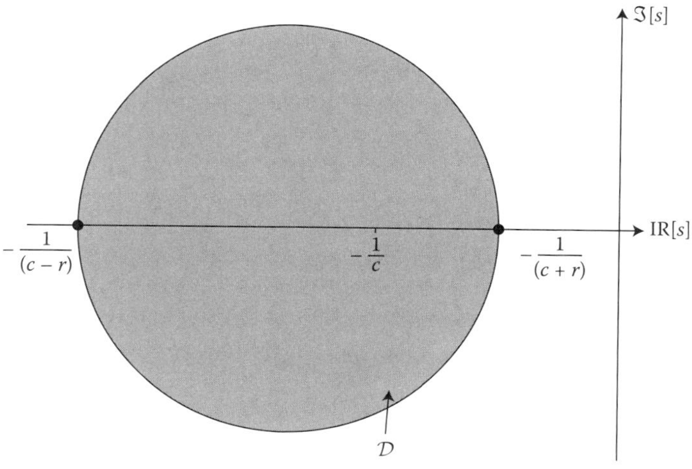

FIGURE 44.8 A disc in the complex plane.

- *$r > c$, $G(s)$ is stable and the Nyquist plot of $G(s)$ lies in the interior of the disc $\mathcal{D}_{c,r}$, or*
- *$r = c$, $G(s)$ is stable and the Nyquist plot of $G(s)$ is bounded away and to the right of the vertical line passing through the real axis at the value $-1/(c+r)$, or*
- *$r < c$, the Nyquist plot of $G(s)$ (with Nyquist path indented into the right-half plane) is outside of and bounded away from the disc $\mathcal{D}_{c,r}$, and the number of times the plot encircles this disc in the counterclockwise direction is equal to the number of poles of $G(s)$ with strictly positive real parts,*

then the interconnection is finite gain stable.

Case 1 is similar to the small gain theorem, and case 2 is similar to the passivity theorem. We will now explain case 3 in more detail. Let $n(s)$ and $d(s)$ represent, respectively, the numerator and denominator polynomials of $G(s)$. Since the point $(-1/c, 0)$ is inside the disc $\mathcal{D}_{c,r}$, it follows, from the assumption of the theorem together with the well-known Nyquist stability condition, that all of the roots of the polynomial $d(s) + cn(s)$ have negative real parts. Then $y = G(s)u = N(s)D(s)^{-1}u$ where

$$D(s) := \frac{d(s)}{d(s) + cn(s)}, \quad \text{and} \quad N(s) := \frac{n(s)}{d(s) + cn(s)}, \tag{44.78}$$

and, by taking $z = D(s)^{-1}u$, we can describe all of the possible input–output pairs as

$$(u, y) = \left(D(s)z, \quad N(s)z\right). \tag{44.79}$$

Notice that $D(s) + cN(s) = 1$, so that

$$||u + cy||_2 = ||z||_2. \tag{44.80}$$

To put a lower bound on this expression in terms of $||u||_2$ and $||y||_2$, to show that the graph is strictly outside of the cone defined in Equation 44.76, we will need the 2-norm gains for systems modeled by the transfer functions $N(s)$ and $D(s)$. We will use the symbols γ_N and γ_D for these gains. The condition of the circle theorem guarantees that $\gamma_N < r^{-1}$. To see this, note that

$$N(s) = \frac{G(s)}{1 + cG(s)} \tag{44.81}$$

implying

$$\gamma_N := \sup_{\omega \in \mathbb{R}} \left| \frac{G(j\omega)}{1 + cG(j\omega)} \right|. \tag{44.82}$$

But

$$\begin{aligned} &|1 + c\ G(j\omega)|^2 - r^2 |G(j\omega)|^2 \\ &= \left(c\ \text{Re}\left\{G(j\omega)\right\} + 1\right)^2 + c^2\ \text{Im}^2\left\{G(j\omega)\right\} - r^2\ \text{Re}^2\left\{G(j\omega)\right\} - r^2\ \text{Im}^2\left\{G(j\omega)\right\}, \\ &= (c^2 - r^2)\left(\text{Re}\left\{G(j\omega)\right\} + \frac{c}{c^2 - r^2}\right)^2 + (c^2 - r^2)\ \text{Im}^2\left\{G(j\omega)\right\} - \frac{r^2}{c^2 - r^2}. \end{aligned} \tag{44.83}$$

Setting the latter expression to zero defines the boundary of the disc $\mathcal{D}_{c,r}$. Since the expression is positive outside of this disc, it follows that $\gamma_N < r^{-1}$.

Returning to the calculation initiated in Equation 44.80, note that $\gamma_N < r^{-1}$ implies that a strictly positive real number ϵ exists so that

$$(1 - \epsilon\gamma_D)\gamma_N^{-1} \geq r + 2\epsilon. \tag{44.84}$$

So,

$$\begin{aligned} ||u + cy||_2 = ||z||_2 &= (1 - \epsilon\gamma_D)||z||_2 + \epsilon\gamma_D||z||_2, \\ &\geq (1 - \epsilon\gamma_D)\gamma_N^{-1}||y||_2 + \epsilon||u||_2, \\ &\geq (r + \epsilon)||y||_2 + \epsilon(||u||_2 + ||y||_2). \end{aligned} \tag{44.85}$$

We conclude that the inverse graph of the linear system is strictly outside of the CONE($\mathbf{C}, \mathbf{R}, \gamma$) as defined in Equation 44.76.

Note, incidentally, that $N(s)$ is the closed loop transfer function from d_1 to y_1 for the special case where the memoryless nonlinearity satisfies $\phi(u) = -cu$. This suggests another way of determining stability: first make a preliminary loop transformation with the feedback $-cu$, changing the original linear system into the system with transfer function $N(s)$ and changing the nonlinearity into a new nonlinearity $\tilde{\phi}$ satisfying $|\tilde{\phi}(u,t)| \leq r|u|$. Then apply the classical small gain theorem to the resulting feedback system.

Example 44.5:

Let

$$G(s) = \frac{175}{(s-1)(s+4)^2}. \tag{44.86}$$

The Nyquist plot of $G(s)$ is shown in Figure 44.9. Because G(s) has one pole with positive real part, only the third condition of the circle theorem can apply. A disc centered at −8.1 on the real axis and with radius 2.2 can be placed inside the left loop of the Nyquist plot. Such a disc corresponds to the values $c = 0.293$ and $r = 0.079$. Because the Nyquist plot encircles this disc once in the counterclockwise direction, it follows that the standard feedback connection with the feedback components $G(s)$ and a memoryless nonlinearity constrained to the SECTOR[−0.372,−0.214] is stable using the 2-norm.

44.3.4.2 Coprime Fractions

Typical input–output stability results based on stable coprime fractions are corollaries of the conic sector theorem. For example, suppose both Σ_1 and Σ_2 are modeled by transfer functions $G_1(s)$ and $G_2(s)$. Moreover, assume stable (in any p-norm) transfer functions $N_1, D_1, \tilde{N}_1, \tilde{D}_1, N_2$ and D_2 exist so that D_1,

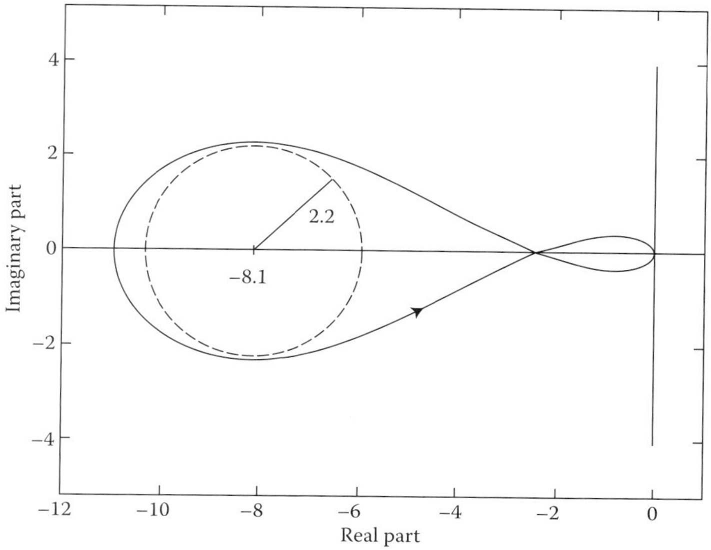

FIGURE 44.9 The Nyquist plot for $G(s)$ in Example 44.5.

D_2 and $\tilde{D}_1$ are invertible, and

$$\begin{aligned} G_1 &= N_1 D_1^{-1} = \tilde{D}_1^{-1}\tilde{N}_1 \\ G_2 &= N_2 D_2^{-1} \\ \mathrm{Id} &= \tilde{D}_1 D_2 - \tilde{N}_1 N_2. \end{aligned} \tag{44.87}$$

Let $\mathbf{C}(u,y) = \tilde{D}_1(s)y - \tilde{N}_1(s)u$, which is incrementally stable in any p-norm, let $\mathbf{R}(u,y) = 0$, and let $\gamma \equiv 0$. Then, the graph of Σ_1 is inside and the inverse graph of Σ_2 is strictly outside $\textsc{Cone}(\mathbf{C},\mathbf{R},\gamma)$ and thus the feedback loop is finite gain stable in any p-norm.

To verify these claims about the properties of the graphs, first recognize that the graph of Σ_i can be represented as

$$\mathcal{G}_{\Sigma_i} = \big(D_i(s)z, \quad N_i(s)z\big) \tag{44.88}$$

where z represents any reasonable signal. Then, for signals in the graph of Σ_1,

$$\mathbf{C}\left(D_1(s)z_\tau, N_1(s)z_\tau\right) = \tilde{D}_1(s)N_1(s)z_\tau - \tilde{N}_1(s)D_1(s)z_\tau \equiv 0. \tag{44.89}$$

Conversely, for signals in the inverse graph of Σ_2,

$$\begin{aligned} \left\|\mathbf{C}\left(N_2(s)z_\tau, D_2(s)z_\tau\right)\right\| &= \left\|\tilde{D}_1(s)D_2(s)z_\tau - \tilde{N}_1(s)N_2(s)z_\tau\right\| \\ &= \|z_\tau\| \geq \epsilon \left\|\left(N_2(s)z_\tau, D_2(s)z_\tau\right)\right\| \end{aligned} \tag{44.90}$$

for some strictly positive real number ϵ. The last inequality follows from the fact that D_2 and N_2 are finite gain stable.

Example 44.6:

(This example is drawn from the work in Potvin, M-J., Jeswiet, J., and Piedboeuf, J.-C. , *Trans. NAMRI/SME 1994, XXII,* pp 373–377.) Let Σ_1 represent the fractional Voigt–Kelvin model for the relation between stress and strain in structures displaying plasticity. For suitable values of Young's modulus, damping magnitude, and order of derivative for the strain, the transfer function of Σ_1 is

$$g_1(s) = \frac{1}{1+\sqrt{s}}.$$

Integral feedback control, $g_2(s) = -\frac{1}{s}$, may be used for asymptotic tracking. Here

$$\begin{aligned} N_1(s) &= \frac{1}{s+1}, & D_1(s) &= \frac{1+\sqrt{s}}{s+1}, \\ N_2(s) &= -\frac{s+1}{1+s(1+\sqrt{s})}, & D_2(s) &= \frac{s(s+1)}{1+s(1+\sqrt{s})}. \end{aligned} \tag{44.91}$$

It can be shown that these fractions are stable linear operators, and thereby incrementally stable in the 2-norm. (This fact is equivalent to proving nominal stability and can be shown using Nyquist theory.) Moreover, it is easy to see that $D_1 D_2 - N_1 N_2 = 1$ so that the feedback loop is stable and finite gain stable.

44.3.4.3 Robustness of Stability and the Gap Metric

It is clear from the original graph separation theorem that, if a well-defined interconnection is stable, i.e., the appropriate graphs are separated in distance, then modifications of the feedback components will not destroy stability if the modified graphs are close to the original graphs.

Given two systems Σ_1 and Σ, define $\vec{\delta}(\Sigma_1, \Sigma) = \alpha$ if α is the smallest number for which

$$x \in \mathcal{G}_\Sigma, \implies d_\tau(x, \mathcal{G}_{\Sigma_1}) \leq \alpha \|x\|_\tau \quad \text{for all } \tau.$$

The quantity $\vec{\delta}(\cdot,\cdot)$ is called the "directed gap" between the two systems and characterizes basic neighborhoods where stability as well as closed-loop properties are preserved under small perturbations from the nominal system Σ_1 to a nearby system Σ.

More specifically, if the interconnection of (Σ_1, Σ_2) is finite gain stable, we define the gain $\beta_{\Sigma_1,\Sigma_2}$ as the smallest real number so that

$$\left\| \begin{pmatrix} d_1 + y_2 \\ y_1 \end{pmatrix} \right\|_\tau \leq \beta_{\Sigma_1,\Sigma_2} \left\| \begin{pmatrix} d_1 \\ d_2 \end{pmatrix} \right\|_\tau, \quad \text{for all } \tau.$$

If Σ is such that

$$\vec{\delta}(\Sigma_1, \Sigma)\beta_{\Sigma_1,\Sigma_2} < 1,$$

then the interconnection of (Σ, Σ_2) is also finite gain stable.

As a special case, let $\Sigma, \Sigma_1, \Sigma_2$ represent linear systems acting on finite energy signals. Further, assume that stable transfer functions N, D exist where D is invertible, $G_1 = ND^{-1}$, and N and D are normalized so that $D^T(-s)D(s) + N^T(-s)N(s) = \text{Id}$. Then, the class of systems in a ball with radius $\gamma \geq 0$, measured in the directed gap, is given by $\textsc{Cone}(\mathbf{C}, \mathbf{R}, \gamma)$, where $\mathbf{R} = \text{Id}$ and

$$\mathbf{C} = \text{Id} - \begin{pmatrix} D(s) \\ N(s) \end{pmatrix} \mathbf{P}_+(D^T(-s), N^T(-s))$$

where $\mathbf{P}_+$ designates the truncation of the Laplace transform of finite energy signals to the part with poles in the left half plane. At the same time, if $\beta_{\Sigma_1,\Sigma_2} < 1/\gamma$, then it can be shown that Σ_2 is strictly outside the cone $\textsc{Cone}(\mathbf{C}, \mathbf{R}, \gamma)$ and, therefore, stability of the interconnection of Σ with Σ_2 is guaranteed for any Σ inside $\textsc{Cone}(\mathbf{C}, \mathbf{R}, \gamma)$.

Given Σ and Σ_1, the computation of the directed gap reduces to a standard $\mathcal{H}_\infty$-optimization problem (see [1]). Also, given Σ_1, the computation of a controller Σ_2, which stabilizes a maximal cone around

Σ_1, reduces to a standard $\mathcal{H}_\infty$-optimization problem [1] and forms the basis of the $\mathcal{H}_\infty$-loop shaping procedure for linear systems introduced in [4].

A second key result which prompted introducing the gap metric is the claim that the behavior of the feedback interconnection of Σ and Σ_2 is "similar" to that of the interconnection of Σ_1 and Σ_2 if, and only if, the distance between Σ and Σ_1, measured using the gap metric, is small (i.e., Σ lies within a "small aperture" cone around Σ_1). The "gap" function is defined as

$$\delta(\Sigma_1, \Sigma) = \max\{\vec{\delta}(\Sigma_1, \Sigma), \vec{\delta}(\Sigma, \Sigma_1)\}$$

to "symmetrize" the distance function $\vec{\delta}(\cdot, \cdot)$ with respect to the order of the arguments. Then, the above claim can be stated more precisely as follows: for each $\epsilon > 0$, a $\zeta(\epsilon) > 0$ exists so that

$$\delta(\Sigma_1, \Sigma) < \zeta(\epsilon) \implies \|x - x_1\|_\tau < \epsilon\|d\|_\tau$$

where $d = \begin{pmatrix} d_1 \\ d_2 \end{pmatrix}$ is an arbitrary signal in the ambient space and x (resp. x_1) represents the response $\begin{pmatrix} d_1 + y_2 \\ y_1 \end{pmatrix}$ of the feedback interconnection of (Σ, Σ_2) (resp. (Σ_1, Σ_2)). Conversely, if $\|x - x_1\|_\tau < \epsilon\|d\|_\tau$ for all d and τ, then $\delta(\Sigma_1, \Sigma) \leq \epsilon$.

Defining Terms

Ambient space: the Cartesian product space containing the inverse graph of Σ_2 and the graph of Σ_1.

Distance (from a signal to a set): measured using a norm function; the infimum, over all signals in the set, of the norm of the difference between the signal and a signal in the set; see Equation 44.4; used to characterize necessary and sufficient conditions for input–output stability; see Section 44.2.

Dynamical system: an object which produces an output signal for each input signal.

Feedback components: the dynamical systems which make up a well-defined interconnection.

Finite gain stable system: a dynamical system is finite gain stable if a nonnegative constant exists so that, for each input–output pair, the norm of the output is bounded by the norm of the input times the constant.

Gain function: a function from the nonnegative real numbers to the nonnegative real numbers which is continuous, nondecreasing and zero when its argument is zero; used to characterize stability; see Section 44.2; some form of the symbol γ is usually used.

Graph (of a dynamical system): the set of ordered input–output pairs (u, y).

Inner product: defined for signals of the same dimension defined on the semi-infinite interval; the integral from zero to infinity of the component-wise product of the two signals.

Inside (or strictly outside) Cone$(\mathbf{C}, \mathbf{R}, \gamma)$**:** used to characterize the graph or inverse graph of a system; determined by whether or not signals in the graph or inverse graph satisfy certain inequalities involving the operators **C** and **R** and the gain function γ; see Equations 44.73 and 44.74; used in the conic sector theorem.

Inverse graph (of a dynamical system): the set of ordered output-input pairs (y, u).

Norm function ($\|\cdot\|$): used to measure the size of signals defined on the semi-infinite interval; examples are the p-norms $p \in [1, \infty]$ (see Equations 44.1 and 44.2).

Parseval's theorem: used to make connections between properties of graphs for SISO systems modeled with real, rational transfer functions and frequency domain characteristics of their transfer functions; Parseval's theorem relates the inner product of signals to their Fourier transforms if they exist. For example, it states that, if two scalar signals u and y, assumed to be zero for negative values of time, have Fourier transforms $\hat{u}(j\omega)$ and $\hat{y}(j\omega)$ then

$$\langle u, y\rangle = \frac{1}{2\pi}\int_{-\infty}^{\infty} \hat{y}^*(j\omega)\hat{u}(j\omega)\, d\omega.$$

Passive: terminology resulting from electrical network theory; a dynamical system is passive if the inner product of each input–output pair is nonnegative.

Proper separation function: a function from the nonnegative real numbers to the nonnegative real numbers which is continuous, zero at zero, strictly increasing and unbounded; such functions are invertible on the nonnegative real numbers; used to characterize nonlinear separation theorems; some form of the symbol ρ is usually used.

Semi-infinite interval: the time interval $[0, \infty)$.

Signal: a "reasonable" vector-valued function defined on a finite or semi-infinite time interval; by "reasonable" we mean piecewise continuous or measurable.

SISO systems: an abbreviation for single input, single output systems.

Stable system: a dynamical system is stable if a gain function exists so that, for each input–output pair, the norm of the output is bounded by the gain function evaluated at the norm of the input.

Strict contraction: the composition of two gain functions γ_1 and γ_2 is a strict contraction if a proper separation function ρ exists so that $(\gamma_1 + \rho) \circ (\gamma_2 + \rho) \leq \text{Id}$, where $\text{Id}(\zeta) = \zeta$ and $\tilde{\gamma}_1 \circ \tilde{\gamma}_2(\zeta) = \tilde{\gamma}_1(\tilde{\gamma}_2(\zeta))$. Graphically, this is the equivalent to the curve $(\zeta, \gamma_1(\zeta) + \rho(\zeta))$ never being above the curve $(\gamma_2(\xi) + \rho(\xi), \xi)$. This concept is used to state the nonlinear small gain theorem.

Strictly passive: We have used various notions of strictly passive including input-, output-, input and output-, and nonlinear input and output-strictly passive. All notions strengthen the requirement that the inner product of the input–output pairs be positive by requiring a positive lower bound that depends on the 2-norm of the input and/or output.

Truncated signal: A signal defined on the semi-infinite interval which is derived from another signal (not necessarily defined on the semi-infinite interval) by first appending zeros to extend the signal onto the semi-infinite interval and then keeping the first part of the signal and setting the rest of the signal to zero. Used to measure the size of finite portions of signals.

Well-defined interconnection: An interconnection of two dynamical systems in the configuration of Figure 44.1 which results in another dynamical system, i.e., one in which an output signal is produced for each input signal.

References

1. Georgiou, T.T. and Smith, M.C., Optimal robustness in the gap metric, *IEEE Trans. Auto. Control*, 35, 673–686, 1990.
2. Jiang, Z.P., Teel, A.R., and Praly, L., Small-gain theorem for ISS systems and applications, *Math. Control, Sign., Syst.*, 7(2), 95–120, 1995.
3. Mareels, I.M.Y. and Hill, D.J., Monotone stability of nonlinear feedback systems, *J. Math. Syst., Est. Control*, 2(3), 275–291, 1992.
4. McFarlane, D.C. and Glover, K., *Robust Controller Design Using Normalized Coprime Factor Plant Descriptions,* Lecture Notes in Control and Information Sciences, Springer-Verlag, vol. 138, 1989.
5. Popov, V.M., Absolute stability of nonlinear systems of automatic control, *Auto. Remote Control*, 22, 857–875, 1961.
6. Safonov, M., *Stability and Robustness of Multivariable Feedback Systems,* The MIT Press, Cambridge, MA, 1980.
7. Sandberg, I.W., On the $\mathcal{L}_2$-boundedness of solutions of nonlinear functional equations, *Bell Sys. Tech. J.*, 43, 1581–1599, 1964.
8. Sontag, E., Smooth stabilization implies coprime factorization, *IEEE Trans. Auto. Control*, 34, 435–443, 1989.
9. Zames, G., On the input–output stability of time-varying nonlinear feedback systems, Part I: Conditions using concepts of loop gain, conicity, and positivity, *IEEE Trans. Auto. Control*, 11, 228–238, 1969.
10. Zames, G., On the input–output stability of time-varying nonlinear feedback systems, Part II: Conditions involving circles in the frequency plane and sector nonlinearities, *IEEE Trans. Auto. Control*, 11, 465–476, 1966.

Further Reading

As mentioned at the outset, the material presented in this chapter is based on the results in [6,7,9,10]. In the latter, a more general feedback interconnection structure is considered where nonzero initial conditions can also be consider as inputs.

Other excellent references on input–output stability include *The Analysis of Feedback Systems*, 1971, by J.C. Willems and *Feedback Systems: Input–Output Properties*, 1975, by C. Desoer and M. Vidyasagar. A nice text addressing the factorization method in linear systems control design is *Control Systems Synthesis: A Factorization Approach*, 1985, by M. Vidyasagar. A treatment of input–output stability for linear, infinite dimensional systems can be found in Chapter 6 of *Nonlinear Systems Analysis*, 1993, by M. Vidyasagar. That chapter also discusses many of the connections between input–output stability and state-space (Lyapunov) stability. Another excellent reference is *Nonlinear Systems*, 1992, by H. Khalil.

There are results similar to the circle theorem that we have not discussed. They go under the heading of "multiplier" results and apply to feedback loops with a linear element and a memoryless, nonlinear element with extra restrictions such as time invariance and constrained slope. Special cases are the well-known Popov and off-axis circle criterion. Some of these results can be recovered using the general conic sector theorem although we have not taken the time to do this. Other results, like the Popov criterion, impose extra smoothness conditions on the external inputs which are not found in the standard problem. References for these problems are *Hyperstability of Control Systems*, 1973, V.M. Popov, the English translation of a book originally published in 1966, and *Frequency Domain Criteria for Absolute Stability*, 1973, by K.S. Narendra and J.H. Taylor.

Another topic closely related to these multiplier results is the structured small gain theorem for linear systems which lends to much of the μ-synthesis control design methodology. This is described, for example, in *μ-Analysis and Synthesis Toolbox*, 1991, by G. Balas, J. Doyle, K. Glover, A. Packard and R. Smith.

There are many advanced topics concerning input–output stability that we have not addressed. These include the study of small-signal stability, well-posedness of feedback loops, and control design based on input–output stability principles. Many articles on these topics frequently appear in control and systems theory journals such as *IEEE Transactions on Automatic Control, Automatica, International Journal of Control, Systems and Control Letters, Mathematics of Control, Signals, and Systems*, to name a few.

45

Input-to-State Stability

Eduardo D. Sontag
Rutgers University

45.1 Introduction

The notion of input-to-state stability (ISS) was introduced in [21]. Together with several variants, also discussed in this article, it provides theoretical concepts that describe various desirable stability features of a mapping $u(\cdot) \mapsto (\cdot)$ from (time-dependent) inputs to outputs (or internal states). Prominent among these features are that inputs that are bounded, "eventually small," "integrally small," or convergent, should lead to outputs with the respective property. In addition, ISS and related notions quantify in what manner initial states affect transient behavior. The discussion in this article focuses on stability notions relative to globally attractive steady states, but a more general theory is also possible, that allows consideration of more arbitrary attractors, as well as robust and/or adaptive concepts. The reader is referred to the cited literature, as well as the textbooks [5,7,8,12,14,15,20], for extensions of the theory as well as applications, The paper [26] may also be consulted for further references and an exposition of many extensions of the concepts and results discussed in this chapter.

45.1.1 Operator and Lyapunov Stability

Broadly speaking, there are two main competing approaches to system stability: the *state-space approach* usually associated with the name of Lyapunov, and the *operator* approach, of which George Zames was one of the main proponents and developers and which was the subject of major contributions by Sandberg, Willems, Safonov, and others. In the operator approach, one studies the i/o mapping:

$$\left(x^0, u(\cdot)\right) \mapsto y(\cdot),$$
$$\mathbb{R}^n \times \left[\mathcal{L}_q(0, +\infty)\right]^m \to \left[\mathcal{L}_q(0, +\infty)\right]^p,$$

that sends initial states and input functions into output functions. This includes the special case when the output is the internal state of a system. The notation $\mathcal{L}_q$ refers to spaces of functions whose qth power is integrable; typical choices are $q = 2$ or $q = \infty$. This approach permits the use of Hilbert or Banach space techniques, and elegantly generalizes to nonlinear systems many properties of linear systems, especially in the context of robustness analysis. The state-space approach, in contrast, is geared to the study of systems without inputs, but is better suited to the study of nonlinear dynamics, and it allows the use of geometric and topological ideas. The ISS notion combines these dual views of stability.

45.1.2 The Class of Systems

This chapter considers systems with inputs and outputs in the usual sense of control theory [24]:

$$\dot{x}(t) = f(x(t), u(t)), \quad y(t) = h(x(t))$$

(the arguments "t" is often omitted). There are n state variables, m input channels, and p output channels. States $x(t)$ take values in Euclidean space $\mathbb{R}^n$, and the inputs (also called "controls" or "disturbances" depending on the context) are measurable locally essentially bounded maps $u(\cdot) : [0, \infty) \to \mathbb{R}^m$. Output values $y(t)$ take values in $\mathbb{R}^p$, The map $f : \mathbb{R}^n \times \mathbb{R}^m \to \mathbb{R}^n$ is assumed to be locally Lipschitz with $f(0, 0) = 0$, and $h : \mathbb{R}^n \to \mathbb{R}^p$ is continuous with $h(0) = 0$. These two properties mean that the state $x = 0$ is an equilibrium when the input is $u = 0$, and the corresponding output is $y = 0$. Many of these assumptions can be weakened considerably, and the cited references should be consulted for more details. The solution, defined on some maximal interval $[0, t_{\max}(x^0, u))$, for each initial state x^0 and input u, is denoted as $x(t, x^0, u)$, and in particular, for systems with no inputs

$$\dot{x}(t) = f(x(t)),$$

just as $x(t, x^0)$. The *zero-system* associated to $\dot{x} = f(x, u)$ is by definition the system with no inputs $\dot{x} = f(x, 0)$. Euclidean norm is written as $|x|$. For a function of time, typically an input or an output, $\|u\|$, or $\|u\|_\infty$ for emphasis, is the (essential) supremum or "sup" norm (possibly $+\infty$, if u is not bounded). The norm of the restriction of a signal to an interval I is denoted by $\|u_I\|_\infty$ (or just$\|u_I\|$).

45.1.3 Notions of Stability

It is convenient to introduce "comparison functions" to quantify stability. A *class $\mathcal{K}_\infty$ function* is a function $\alpha : \mathbb{R}_{\geq 0} \to \mathbb{R}_{\geq 0}$ which is continuous, strictly increasing, unbounded, and satisfies $\alpha(0) = 0$ and a *class $\mathcal{KL}$ function* is a function $\beta : \mathbb{R}_{\geq 0} \times \mathbb{R}_{\geq 0} \to \mathbb{R}_{\geq 0}$ such that $\beta(\cdot, t) \in \mathcal{K}_\infty$ for each t and $\beta(r, t)$ decreases to zero as $t \to \infty$, for each fixed r.

For a system with no inputs $\dot{x} = f(x)$, there is a well-known notion of global asymptotic stability (GAS), or "*0-GAS*" when referring to the zero-system $\dot{x} = f(x, 0)$ associated to a given system with inputs $\dot{x} = f(x, u)$) due to Lyapunov, and usually defined in "ε-δ" terms. It is an easy exercise to show that this standard definition is in fact equivalent to the following statement:

$$(\exists \beta \in \mathcal{KL}) \quad |x(t,x^0)| \leq \beta\left(|x^0|,t\right), \quad \forall x^0, \ \forall \geq 0.$$

Observe that, since β decreases on t, in particular:

$$|x(t,x^0)| \ \leq \ \beta\left(|x^0|,0\right), \quad \forall x^0, \ \forall t \geq 0,$$

which provides the Lyapunov-stability or "small overshoot" part of the GAS definition (because $\beta\left(|x^0|,0\right)$ is small whenever $\left|x^0\right|$ is small, by continuity of $\beta(\cdot,0)$ and $\beta(0,0)=0$), while the fact that $\beta \to 0$ as $t \to \infty$ gives

$$|x(t,x^0)| \leq \beta\left(|x^0|,t\right) \underset{t\to\infty}{\longrightarrow} 0, \quad \forall x^0,$$

which is the attractivity (convergence to steady state) part of the GAS definition.

In [23, Proposition 7], it is shown that for each $\beta \in \mathcal{KL}$ there exist two class $\mathcal{K}_\infty$ functions α_1, α_2 such that:

$$\beta(r,t) \ \leq \ \alpha_2\left(\alpha_1(r)e^{-t}\right), \quad \forall s,t \geq 0,$$

which means that the GAS estimate can be also written in the form:

$$|x(t,x^0)| \ \leq \ \alpha_2\left(\alpha_1(|x^0|)e^{-t}\right)$$

and thus suggests a close analogy between GAS and an exponential stability estimate $|x(t,x^0)| \leq \left|x^0\right| e^{-at}$.

In general, 0-GAS does not guarantee good behavior with respect to inputs. To explain why this is relevant, let us briefly recall the case of linear systems. A *linear* system in control theory is the one for which both f and h are linear mappings:

$$\dot{x} = Ax + Bu, \quad y = Cx$$

with $A \in \mathbb{R}^{n\times n}$, $B \in \mathbb{R}^{n\times m}$, and $C \in \mathbb{R}^{p\times n}$. It is well-known that a linear system is 0-GAS (or "internally stable") if and only if the matrix A is a Hurwitz matrix, that is to say, all the eigenvalues of A have negative real parts. Such a 0-GAS linear system automatically satisfies all reasonable i/o stability properties: bounded inputs result in bounded state trajectories as well as outputs, inputs converging to zero imply solutions (and outputs) converging to zero, and so forth [24]. But *the 0-GAS property is not equivalent*, in general, to i/o, or even input/state, stability of any sort. The implication that 0-GAS implies i/o stability is in general false for nonlinear systems. For a simple example, consider the following one-dimensional ($n=1$) system, with scalar ($m=1$) inputs:

$$\dot{x} = -x + (x^2+1)u.$$

This system is clearly 0-GAS, since it reduces to $\dot{x} = -x$ when $u \equiv 0$. On the other hand, solutions diverge even for some inputs that converge to zero. For example, take the control $u(t) = (2t+2)^{-1/2}$ and $x^0 = \sqrt{2}$, This results in the unbounded trajectory $x(t) = (2t+2)^{1/2}$. This is in spite of the fact that the unforced system is GAS. Thus, the converging-input converging-state property does not hold. Even worse, the bounded input $u \equiv 1$ results in a finite-time explosion. This example is not artificial, as it arises in feedback-linearization design, mentioned below.

45.1.4 Gains for Linear Systems

For linear systems, the three most typical ways of defining i/o stability in terms of operators

$$\left\{L^2, L^\infty\right\} \to \left\{L^2, L^\infty\right\}$$

are as follows. The statements below should be read, more precisely, as asking that there should exist positive c and λ such that the given estimates hold for all $t \geq 0$ and all solutions of $\dot{x} = Ax + Bu$ with

$x(0) = x^0$ and arbitrary inputs $u(\cdot)$. The estimates are written in terms of states $x(t)$, but similar notions can be defined for more general outputs $y = Cx$.

$$\text{“}L^\infty \to L^\infty\text{”}: \; c\left|x(t,x^0,u)\right| \leq \left|x^0\right| e^{-\lambda t} + \sup_{s\in[0,t]} |u(s)|$$

$$\text{“}L^2 \to L^\infty\text{”}: \; c\left|x(t,x^0,u)\right| \leq \left|x^0\right| e^{-\lambda t} + \int_0^t |u(s)|^2 \, ds$$

$$\text{“}L^2 \to L^2\text{”}: \; c\int_0^t \left|x(s,x^0,u)\right|^2 \, ds \leq \left|x^0\right| + \int_0^t |u(s)|^2 \, ds$$

(the missing case $L^\infty \to L^2$ is less interesting, being too restrictive).

For linear systems, these estimates are all equivalent in the following sense: if an estimate of one type exists, then the other two estimates exist too, although the actual numerical values of the constants c, λ appearing in the different estimates are not necessarily the same: they are associated to various types of norms on input spaces and spaces of solutions, such as "H_2" and "H_∞" gains, [4]. It is easy to see that existence of the above estimates is simply equivalent to the requirement that the A matrix be Hurwitz, that is to say, to 0-GAS, the asymptotic stability of the unforced system $\dot{x} = Ax$.

45.1.5 Nonlinear Coordinate Changes

A "geometric" view of nonlinear dynamics suggests that *notions of stability should be invariant under (nonlinear) changes of variables*: under a change of variables, a system which is stable in some technical sense should remain stable, in the same sense, when written in the new coordinates. This principle leads to the ISS notion when starting from the above linear notions, as elaborated next. In this article, a *change of coordinates* is any map

$$T : \mathbb{R}^n \to \mathbb{R}^n$$

such that the following properties hold: $T(0) = 0$ (this fixes the equilibrium at $x = 0$), T is continuous, and it admits an inverse map $T^{-1} : \mathbb{R}^n \to \mathbb{R}^n$, which is well-defined and continuous as well. In other words, T is a homeomorphism which fixes the origin. One could add the requirement that T should be differentiable, or that it be differentiable at least for $x \neq 0$, but the discussion to follow does not require this additional condition. Now suppose that a system $\dot{x} = f(x)$ is exponentially stable:

$$|x(t,x^0)| \leq c\,|x^0| e^{-\lambda t} \quad \forall\, t \geq 0 \quad (\text{some } c, \lambda > 0)$$

and that a change of variables is performed:

$$x(t) = T(z(t)).$$

Consider, for this transformation T, the following two functions:

$$\underline{\alpha}(r) := \min_{|x|\geq r} |T(x)| \quad \text{and} \quad \overline{\alpha}(r) := \max_{|x|\leq r} |T(x)|,$$

which are well-defined because T and its inverse are both continuous, and are both functions of class $\mathcal{K}_\infty$ (easy exercise). Then,

$$\underline{\alpha}(|x|) \leq |T(x)| \leq \overline{\alpha}(|x|), \quad \forall\, x \in \mathbb{R}^n$$

and therefore, substituting $x(t,x^0) = T(z(t,z^0))$ in the exponential stability estimate:

$$\underline{\alpha}(|z(t,z^0)|) \leq c\,\overline{\alpha}(|z^0|)\, e^{-\lambda t},$$

where $z^0 = T^{-1}(x^0)$. Thus, the estimate in z-coordinates takes the following form:

$$|z(t,z^0)| \leq \beta\left(|z^0|, t\right),$$

where $\beta(r,t) = \underline{\alpha}^{-1}\left(c\overline{\alpha}\left(re^{-\lambda t}\right)\right)$ is a function of class $\mathcal{KL}$. (As remarked earlier, any possible function of class $\mathcal{KL}$ can be written in this factored form, actually.)

In summary, the concept of GAS is rederived simply by making coordinate changes on globally exponentially stable systems. The same considerations, applied to systems with inputs, lead to ISS and related notions. In addition to the state transformation $x(t) = T(z(t))$, there is now also a transformation $u(t) = S(v(t))$, where S is a change of variables in the space of input values $\mathbb{R}^m$. Arguing analogously as for systems without inputs, one arrives to the following three concepts:

$$L^\infty \to L^\infty \rightsquigarrow \alpha\left(|x(t)|\right) \le \beta(|x^0|, t) + \sup_{s\in[0,t]} \gamma(|u(s)|),$$

$$L^2 \to L^\infty \rightsquigarrow \alpha\left(|x(t)|\right) \le \beta(|x^0|, t) + \int_0^t \gamma(|u(s)|)\, ds$$

$$L^2 \to L^2 \rightsquigarrow \int_0^t \alpha\left(|x(s)|\right)\, ds \le \alpha_0(|x^0|) + \int_0^t \gamma(|u(s)|)\, ds$$

($x(t)$ is written instead of the more cumbersome $x(t, x^0, u)$). If more general outputs $y = h(x)$ instead of states are the object of interest, these notions can be modified in several ways, as discussed later in the chapter. Unless otherwise stated, the convention when giving an estimate like the ones above is that there should exist comparison functions ($\alpha, \alpha_0 \in \mathcal{K}_\infty$, $\beta \in \mathcal{KL}$) such that the estimates hold for all inputs and initial states. These three notions will be studied one at a time.

45.2 ISS and Feedback Redesign

The "$L^\infty \to L^\infty$" estimate, under changes of variables, leads to the concept of ISS: there should exist some $\beta \in \mathcal{KL}$ and $\gamma \in \mathcal{K}_\infty$ such that

$$|x(t)| \le \beta(|x^0|, t) + \gamma\left(\|u\|_\infty\right) \qquad \text{(ISS)}$$

holds for all solutions (meaning that the estimate is valid for all inputs $u(\cdot)$, all initial conditions x^0, and all $t \ge 0$). Note that there is now no function "α" is the left-hand side because, redefining β and γ, one can assume, without loss of generality, that α is the identity: if $\alpha(r) \le \beta(s,t) + \gamma(t)$ holds, then also $r \le \alpha^{-1}(\beta(s,t) + \gamma(t)) \le \alpha^{-1}(2\beta(s,t)) + \alpha^{-1}(2\gamma(t))$; since $\alpha^{-1}(2\beta(\cdot,\cdot)) \in \mathcal{KL}$ and $\alpha^{-1}(2\gamma(\cdot)) \in \mathcal{K}_\infty$, an estimate of the same type, but now with no "α," is obtained. In addition, note that the supremum $\sup_{s\in[0,t]} \gamma(|u(s)|)$ over the interval $[0,t]$ is the same as $\gamma(\|u_{[0,t]}\|_\infty) = \gamma(\sup_{s\in[0,t]}(|u(s)|))$, because the function γ is increasing, so one may replace this term by $\gamma(\|u\|_\infty)$, where $\|u\|_\infty = \sup_{s\in[0,\infty)} \gamma(|u(s)|)$ is the sup norm of the input, because the solution $x(t)$ depends only on values $u(s), s \le t$ (so, one could equally well consider the input that has values $\equiv 0$ for all $s > t$).

Note that a potentially weaker definition might be that the ISS estimate should hold merely for all $t \in [0, t_{\max}(x^0, u))$. However, this potentially different definition turns out to be equivalent. Indeed, if the estimate holds *a priori* only on such a maximal interval of definition, then, since the right-hand is bounded on $[0, T]$, for any $T > 0$ (recall that inputs are by definition assumed to be bounded on any bounded interval), it follows that the maximal solution of $x(t, x^0, u)$ is bounded, and therefore that $t_{\max}(x^0, u) = +\infty$ (e.g., Proposition C.3.6 in [24]). In other words, the ISS estimate holds for all $t \ge 0$ automatically, if it is required to hold merely for maximal solutions.

Since, in general, $\max\{a, b\} \le a + b \le \max\{2a, 2b\}$, one can restate the ISS condition in a slightly different manner, namely, asking for the existence of some $\beta \in \mathcal{KL}$ and $\gamma \in \mathcal{K}_\infty$ (in general, different from the ones in the ISS definition) such that

$$|x(t)| \le \max\left\{\beta(|x^0|, t),\ \gamma\left(\|u\|_\infty\right)\right\}$$

holds for all solutions. Such redefinitions, using "max" instead of sum, are also possible for each of the other concepts to be introduced later.

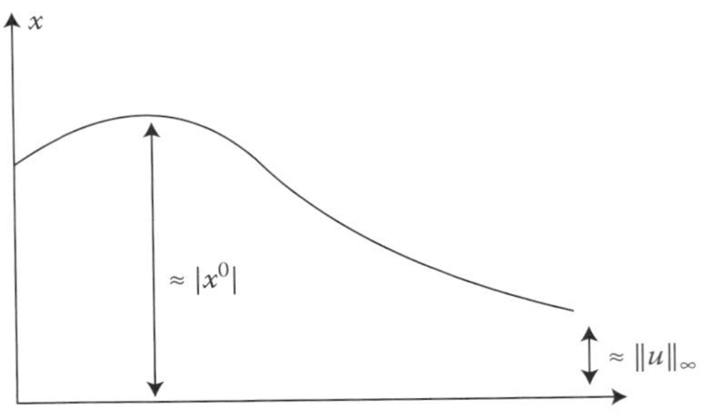

FIGURE 45.1 ISS combines overshoot and asymptotic behavior.

Intuitively, the definition of ISS requires that, for t large, the size of the state must be bounded by some function of the sup norm—that is to say, the amplitude—of inputs (because $\beta(|x^0|, t) \to 0$ as $t \to \infty$). On the other hand, the $\beta(|x^0|, 0)$ term may dominate for small t, and this serves to quantify the magnitude of the transient (overshoot) behavior as a function of the size of the initial state x^0 (Figure 45.1).

The *ISS superposition theorem*, discussed later, shows that ISS is, in a precise mathematical sense, the conjunction of two properties, one of them dealing with asymptotic bounds on $|x^0|$ as a function of the magnitude of the input, and the other one providing a transient term obtained when one ignores inputs.

45.2.1 Linear Case, for Comparison

For internally stable linear systems $\dot{x} = Ax + Bu$, the variation of parameters formula gives immediately the following inequality:

$$|x(t)| \leq \beta(t)\,|x^0| + \gamma\,\|u\|_\infty,$$

where

$$\beta(t) = \|e^{tA}\| \to 0 \quad \text{and} \quad \gamma = \|B\| \int_0^\infty \|e^{sA}\|\,ds < \infty.$$

This is a particular case of the ISS estimate, $|x(t)| \leq \beta(|x^0|, t) + \gamma\left(\|u\|_\infty\right)$, with linear comparison functions.

45.2.2 Feedback Redesign

The notion of ISS arose originally as a way to precisely formulate, and then answer the following question. Suppose that, as in many problems in control theory, a system $\dot{x} = f(x, u)$ has been stabilized by means of a feedback law $u = k(x)$ (Figure 45.2), that is to say, k was chosen such that the origin of the closed-loop system $\dot{x} = f(x, k(x))$ is globally asymptotically stable. (See e.g. [25] for a discussion of mathematical

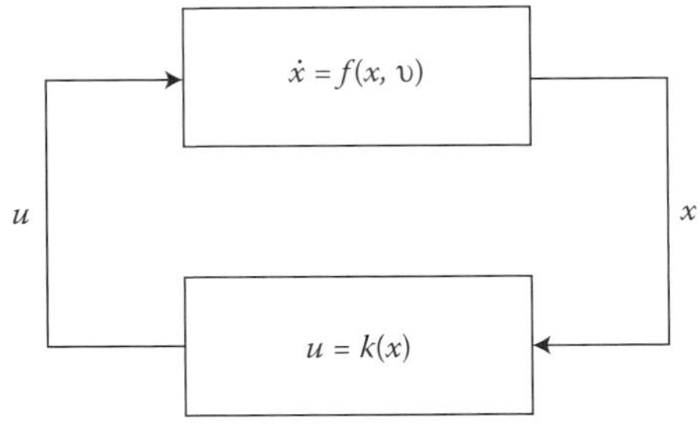

FIGURE 45.2 Feedback stabilization, closed-loop system $\dot{x} = f(x, k(x))$.

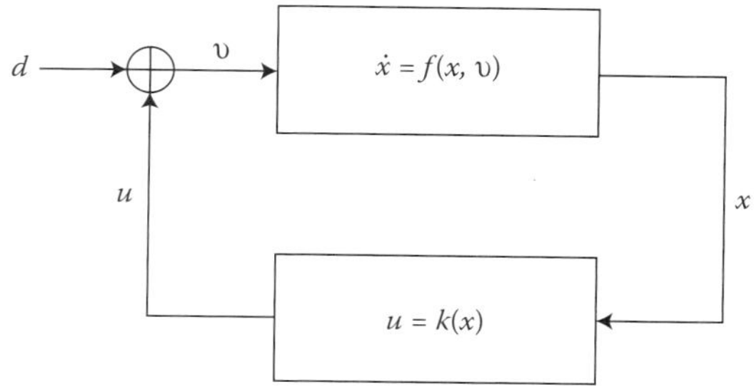

FIGURE 45.3 Actuator disturbances, closed-loop system $\dot{x} = f(x, k(x) + d)$.

aspects of state feedback stabilization.) Typically, the design of k was performed by ignoring the effect of possible *input disturbances* $d(\cdot)$ (also called actuator disturbances). These "disturbances" might represent true noise or perhaps errors in the calculation of the value $k(x)$ by a physical controller, or modeling uncertainty in the controller or the system itself. What is the effect of considering disturbances? In order to analyze the problem, d is incorporated into the model, and one studies the new system $\dot{x} = f(x, k(x) + d)$, where d is seen as an input (Figure 45.3). One may then ask what is the effect of d on the behavior of the system.

Disturbances d may well destabilize the system, and the problem may arise even when using a routine technique for control design, feedback linearization. To appreciate this issue, take the following very simple example. Given is the system

$$\dot{x} = f(x, u) = x + (x^2 + 1)u.$$

In order to stabilize it, substitute $u = \tilde{u}/(x^2 + 1)$ (a preliminary feedback transformation), rendering the system linear with respect to the new input $\tilde{u}$: $\dot{x} = x + \tilde{u}$, and then use $\tilde{u} = -2x$ in order to obtain the closed-loop system $\dot{x} = -x$. In other words, in terms of the original input u, the feedback law is

$$k(x) = \frac{-2x}{x^2 + 1},$$

so that $f(x, k(x)) = -x$. This is a GAS system. The effect of the disturbance input d is analyzed as follows. The system $\dot{x} = f(x, k(x) + d)$ is

$$\dot{x} = -x + (x^2 + 1)\, d.$$

As seen before, this system has solutions that diverge to infinity even for inputs d that converge to zero; moreover, the constant input $d \equiv 1$ results in solutions that explode in finite time. Thus $k(x) = -2x/(x^2 + 1)$ was not a good feedback law, in the sense that its performance degraded drastically once actuator disturbances were taken into account.

The key observation for what follows is that, if one adds a correction term "$-x$" to the above formula for $k(x)$, so that now

$$\tilde{k}(x) = \frac{-2x}{x^2 + 1} - x,$$

then the system $\dot{x} = f(x, \tilde{k}(x) + d)$ with disturbance d as input becomes, instead

$$\dot{x} \; = \; -2x - x^3 + (x^2 + 1)\, d$$

and this system is much better behaved: it is still GAS when there are no disturbances (it reduces to $\dot{x} = -2x - x^3$) but, in addition, it is ISS (easy to verify directly, or appealing to some of the characterizations mentioned later). Intuitively, for large x, the term $-x^3$ serves to dominate the term $(x^2 + 1)d$, for all bounded disturbances $d(\cdot)$, and this prevents the state from getting too large.

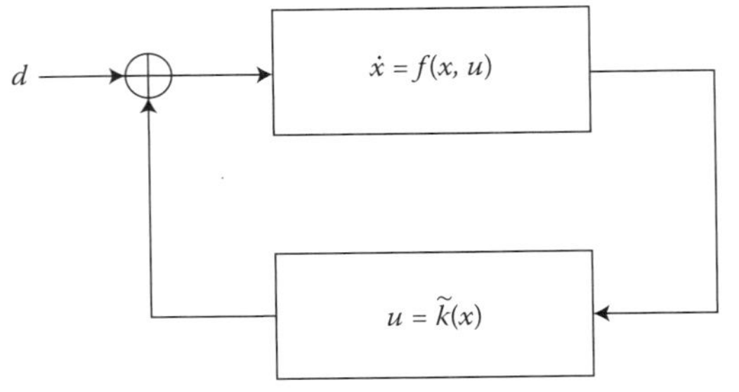

FIGURE 45.4 Different feedback ISS-stabilizes.

45.2.3 A Feedback Redesign Theorem for Actuator Disturbances

This example is an instance of a general result, which says that whenever there is some feedback law that stabilizes a system, there is also a (possibly different) feedback so that the system with external input d (Figure 45.4) is ISS.

Theorem 45.1: [21]

Consider a system affine in controls

$$\dot{x} = f(x, u) = g_0(x) + \sum_{i=1}^{m} u_i g_i(x) \quad (g_0(0) = 0)$$

and suppose that there is some differentiable feedback law $u = k(x)$ so that

$$\dot{x} = f(x, k(x))$$

has $x = 0$ as a GAS equilibrium. Then, there is a feedback law $u = \tilde{k}(x)$ such that

$$\dot{x} = f(x, \tilde{k}(x) + d)$$

is ISS with input $d(\cdot)$

The proof is very easy, once the appropriate technical machinery has been introduced: one starts by considering a smooth Lyapunov function V for GAS of the origin in the system $\dot{x} = f(x, k(x))$ (such a V always exists, by classical converse theorems); then $\hat{k}(x) := -(L_G V(x))^T = -(\nabla V(x) G(x))^T$, where G is the matrix function whose columns are the $g_i, i = 1, \ldots, m$ and T indicates transpose, provides the necessary correction term to add to k. This term has the same degree of smoothness as the vector fields making up the original system. Somewhat less than differentiability of the original k is enough for this argument: continuity is enough. However, if no continuous feedback stabilizer exists, then no smooth V can be found. (Continuous stabilization of nonlinear systems is basically equivalent to the existence of what are called smooth control-Lyapunov functions, see e.g. [25].) In that case, if only discontinuous stabilizers are available, the result can still be generalized, see [17], but the situation is harder to analyze, since even the notion of "solution" of the closed-loop system $\dot{x} = f(x, k(x))$ has to be carefully defined.

There is also a redefinition procedure for systems that are not affine on inputs, but the result as stated above is false in that generality, and is much less interesting; see [22] for a discussion.

The above feedback redesign theorem is merely the beginning of the story. The reader is referred to the book [15], and the references given later, for many further developments on the subjects of recursive feedback design, the "backstepping" approach, and other far-reaching extensions.

45.3 Equivalences for ISS

This section reviews results that show that ISS is equivalent to several other notions, including asymptotic gain (AG), existence of robustness margins, dissipativity, and an energy-like stability estimate.

45.3.1 Nonlinear Superposition Principle

Clearly, if a system is ISS, then the system with no inputs $\dot{x} = f(x, 0)$ is GAS: the term $\|u\|_\infty$ vanishes, leaving precisely the GAS property. In particular, then, the system $\dot{x} = f(x, u)$ is *0-stable*, meaning that the origin of the system without inputs $\dot{x} = f(x, 0)$ is stable in the sense of Lyapunov: for each $\varepsilon > 0$, there is some $\delta > 0$ such that $|x^0| < \delta$ implies $|x(t, x^0)| < \varepsilon$. (In comparison function language, one can restate 0-stability as: there is some $\gamma \in \mathcal{K}$ such that $|x(t, x^0)| \leq \gamma(|x^0|)$ holds for all small x^0.)

On the other hand, since $\beta(|x^0|, t) \to 0$ as $t \to \infty$, for t large one has that the first term in the ISS estimate $|x(t)| \leq \max\left\{\beta(|x^0|, t), \gamma\left(\|u\|_\infty\right)\right\}$ vanishes. Thus, an ISS system satisfies the following *"AG" property*: there is some $\gamma \in \mathcal{K}_\infty$ so that:

$$\overline{\lim_{t\to+\infty}} \left|x(t, x^0, u)\right| \leq \gamma\left(\|u\|_\infty\right), \quad \forall x^0, u(\cdot) \qquad \text{(AG)}$$

(see Figure 45.5). In words, for all large enough t, the trajectory exists, and it gets arbitrarily close to a sphere whose radius is proportional, in a possibly nonlinear way quantified by the function γ, to the amplitude of the input. In the language of robust control, the estimate (AG) would be called an "ultimate boundedness" condition; it is a generalization of attractivity (all trajectories converge to zero, for a system $\dot{x} = f(x)$ with no inputs) to the case of systems with inputs; the "lim sup" is required since the limit of $x(t)$ as $t \to \infty$ may well not exist. From now on (and analogously when defining other properties), we will just say "the system is AG" instead of the more cumbersome "satisfies the AG property."

Observe that, since only large values of t matter in the limsup, one can equally well consider merely tails of the input u when computing its sup norm. In other words, one may replace $\gamma(\|u\|_\infty)$ by $\gamma(\overline{\lim}_{t\to+\infty} |u(t)|)$, or (since γ is increasing), $\overline{\lim}_{t\to+\infty}\gamma(|u(t)|)$.

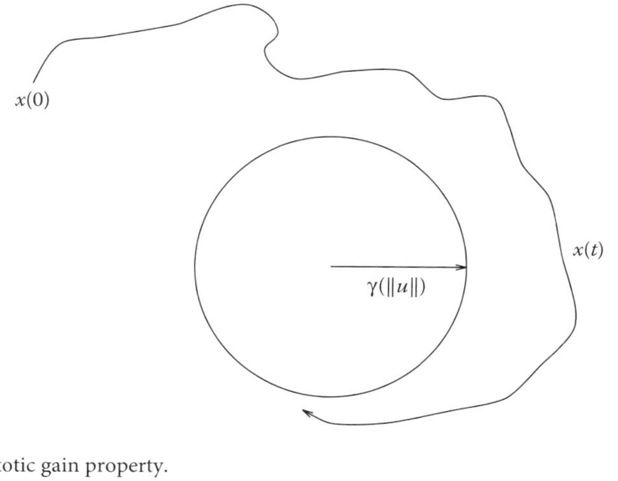

FIGURE 45.5 Asymptotic gain property.

The surprising fact is that these two necessary conditions are also sufficient. This is summarized by the *ISS superposition theorem*:

Theorem 45.2: [29]

A system is ISS if and only if it is 0-stable and AG.

The basic difficulty in the proof of this theorem is in establishing uniform convergence estimates for the states, that is, in constructing the β function in the ISS estimate, independently of the particular input. As in optimal control theory, one would like to appeal to compactness arguments (using weak topologies on inputs), but there is no convexity to allow this. The proof hinges upon a lemma given in [29], which may be interpreted [6] as a relaxation theorem for differential inclusions, relating GAS of an inclusion $\dot{x} \in F(x)$ to GAS of its convexification.

A minor variation of the above superposition theorem is as follows. Let us consider the *limit property (LIM)*:

$$\inf_{t\geq 0} |x(t,x^0,u)| \leq \gamma(\|u\|_\infty), \quad \forall x^0,\ u(\cdot) \qquad \text{(LIM)}$$

(for some $\gamma \in \mathcal{K}_\infty$).

Theorem 45.3: [29]

A system is ISS if and only if it is 0-stable and LIM.

45.3.2 Robust Stability

In this article, a system is said to be *robustly stable* if it admits a *margin of stability* ρ, that is, a smooth function $\rho \in \mathcal{K}_\infty$ so the system

$$\dot{x} = g(x,d) := f(x, d\rho(|x|))$$

is GAS uniformly in this sense: for some $\beta \in \mathcal{KL}$,

$$\left|x(t,x^0,d)\right| \leq \beta(\left|x^0\right|, t)$$

for all possible $d(\cdot) : [0,\infty) \to [-1,1]^m$. An alternative way to interpret this concept (*cf.* [28]) is as uniform GAS of the origin with respect to all possible time-varying feedback laws Δ bounded by ρ: $|\Delta(t,x)| \leq \rho(|x|)$. In other words, the system

$$\dot{x} = f(x, \Delta(t,x))$$

(Figure 45.6) is stable uniformly over all such perturbations Δ. In contrast to the ISS definition, which deals with all possible "open-loop" inputs, the present notion of robust stability asks about all possible closed-loop interconnections. One may think of Δ as representing uncertainty in the dynamics of the original system, for example.

Theorem 45.4: [28]

A system is ISS if and only if it is robustly stable.

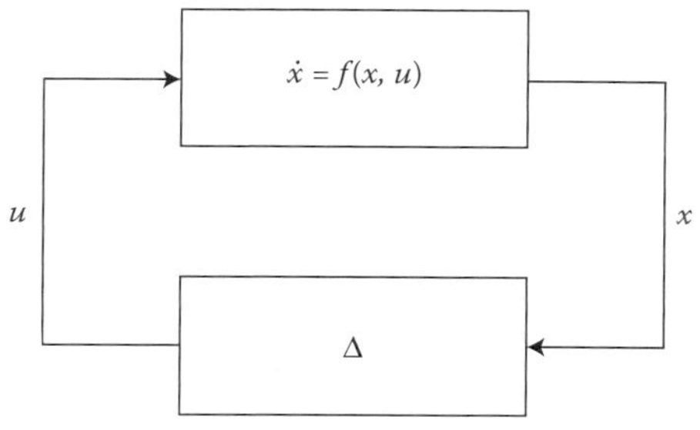

FIGURE 45.6 Margin of robustness.

Intuitively, the ISS estimate $|x(t)| \le \max\left\{\beta(|x^0|, t), \gamma\left(\|u\|_\infty\right)\right\}$ says that the β term dominates as long as $|u(t)| \ll |x(t)|$ for all t, but $|u(t)| \ll |x(t)|$ amounts to $u(t) = d(t).\rho(|x(t)|)$ with an appropriate function ρ. This is an instance of a "small gain" argument, see below. One analog for linear systems is as follows: if A is a Hurwitz matrix, then $A + Q$ is also Hurwitz, for all small enough perturbations Q; note that when Q is a nonsingular matrix, $|Qx|$ is a $\mathcal{K}_\infty$ function of $|x|$.

45.3.3 Dissipation

Another characterization of ISS is as a dissipation notion stated in terms of a Lyapunov-like function. A continuous function $V : \mathbb{R}^n \to \mathbb{R}$ is said to be a *storage function* if it is positive definite, that is, $V(0) = 0$ and $V(x) > 0$ for $x \neq 0$, and proper, that is, $V(x) \to \infty$ as $|x| \to \infty$. This last property is equivalent to the requirement that the sets $V^{-1}([0, A])$ should be compact subsets of $\mathbb{R}^n$, for each $A > 0$, and in the engineering literature it is usual to call such functions *radially unbounded*. It is an easy exercise to show that $V : \mathbb{R}^n \to \mathbb{R}$ is a storage function if and only if there exist $\underline{\alpha}, \overline{\alpha} \in \mathcal{K}_\infty$ such that

$$\underline{\alpha}(|x|) \le V(x) \le \overline{\alpha}(|x|) \quad \forall x \in \mathbb{R}^n$$

(the lower bound amounts to properness and $V(x) > 0$ for $x \neq 0$, while the upper bound guarantees $V(0) = 0$). For convenience, $\dot{V} : \mathbb{R}^n \times \mathbb{R}^m \to \mathbb{R}$ is the function

$$\dot{V}(x, u) := \nabla V(x) \cdot f(x, u),$$

which provides, when evaluated at $(x(t), u(t))$, the derivative $dV(x(t))/dt$ along solutions of $\dot{x} = f(x, u)$.

An *ISS-Lyapunov function* for $\dot{x} = f(x, u)$ is by definition a smooth storage function V for which there exist functions $\gamma, \alpha \in \mathcal{K}_\infty$ so that

$$\dot{V}(x, u) \le -\alpha(|x|) + \gamma(|u|), \quad \forall x, u. \qquad \text{(L-ISS)}$$

Integrating, an equivalent statement is that, along all trajectories of the system, there holds the following dissipation inequality:

$$V(x(t_2)) - V(x(t_1)) \le \int_{t_1}^{t_2} w(u(s), x(s))\, ds,$$

where, using the terminology of [31], the "supply" function is $w(u, x) = \gamma(|u|) - \alpha(|x|)$. For systems with no inputs, an ISS-Lyapunov function is precisely the same object as a Lyapunov function in the usual sense.

Theorem 45.5: [28]

A system is ISS if and only if it admits a smooth ISS-Lyapunov function.

Since $-\alpha(|x|) \leq -\alpha(\overline{\alpha}^{-1}(V(x)))$, the ISS-Lyapunov condition can be restated as

$$\dot{V}(x,u) \leq -\tilde{\alpha}(V(x)) + \gamma(|u|), \quad \forall x, u$$

for some $\tilde{\alpha} \in \mathcal{K}_\infty$. In fact, one may strengthen this a bit [19]: for any ISS system, there is a always a smooth ISS-Lyapunov function satisfying the "exponential" estimate $\dot{V}(x,u) \leq -V(x) + \gamma(|u|)$.

The sufficiency of the ISS-Lyapunov condition is easy to show, and was already in the original paper [21]. A sketch of the proof is as follows, assuming for simplicity a dissipation estimate in the form $\dot{V}(x,u) \leq -\alpha(V(x)) + \gamma(|u|)$. Given any x and u, either $\alpha(V(x)) \leq 2\gamma(|u|)$ or $\dot{V} \leq -\alpha(V)/2$. From here, one deduces by a comparison theorem that, along all solutions,

$$V(x(t)) \leq \max\left\{\beta(V(x^0), t),\ \alpha^{-1}(2\gamma(\|u\|_\infty))\right\},$$

where the $\mathcal{KL}$ function $\beta(s,t)$ is the solution $y(t)$ of the initial value problem

$$\dot{y} = -\frac{1}{2}\alpha(y) + \gamma(u), \quad y(0) = s.$$

Finally, an ISS estimate is obtained from $V(x^0) \leq \overline{\alpha}(x^0)$.

The proof of the converse part of the theorem is based upon first showing that ISS implies robust stability in the sense already discussed, and then obtaining a converse Lyapunov theorem for robust stability for the system $\dot{x} = f(x, d\rho(|x|)) = g(x,d)$, which is asymptotically stable uniformly on all Lebesgue-measurable functions $d(\cdot) : \mathbb{R}_{\geq 0} \to B(0,1)$. This last theorem was given in [16], and is basically a theorem on Lyapunov functions for differential inclusions. The classical result of Massera [18] for differential equations (with no inputs) becomes a special case.

45.3.4 Using "Energy" Estimates Instead of Amplitudes

In linear control theory, H_∞ theory studies $L^2 \to L^2$-induced norms, which under coordinate changes leads to the following type of estimate:

$$\int_0^t \alpha\left(|x(s)|\right) ds \leq \alpha_0(|x^0|) + \int_0^t \gamma(|u(s)|)\, ds$$

along all solutions, and for some $\alpha, \alpha_0, \gamma \in \mathcal{K}_\infty$. Just for the statement of the next result, a system is said to *satisfy an integral–integral estimate* if for every initial state x^0 and input u, the solution $x(t, x^0, u)$ is defined for all $t > 0$ and an estimate as above holds. (In contrast to ISS, this definition explicitly demands that $t_{\max} = \infty$.)

Theorem 45.6: [23]

A system is ISS if and only if it satisfies an integral–integral estimate.

This theorem is quite easy to prove, in view of previous results. A sketch of the proof is as follows. If the system is ISS, then there is an ISS-Lyapunov function satisfying $\dot{V}(x,u) \leq -V(x) + \gamma(|u|)$, so, integrating along any solution:

$$\int_0^t V(x(s))\, ds \leq \int_0^t V(x(s))\, ds + V(x(t)) \leq V(x(0)) + \int_0^t \gamma(|u(s)|)\, ds$$

and thus an integral–integral estimate holds. Conversely, if such an estimate holds, one can prove that $\dot{x} = f(x, 0)$ is stable and that an AG exists.

45.4 Cascade Interconnections

One of the main features of the ISS property is that it behaves well under composition: a cascade (Figure 45.7) of ISS systems is again ISS, see [21]. This section sketches how the cascade result can also be seen as a consequence of the dissipation characterization of ISS, and how this suggests a more general feedback result. For more details regarding the rich theory of ISS small-gain theorems, and their use in nonlinear feedback design, the references should be consulted. Consider a cascade as follows:

$$\dot{z} = f(z, x),$$
$$\dot{x} = g(x, u),$$

where each of the two subsystems is assumed to be ISS. Each system admits an ISS-Lyapunov function V_i. But, moreover, it is always possible (see [27]) to redefine the V_i's so that the comparison functions for both are matched in the following way:

$$\dot{V}_1(z, x) \leq \theta(|x|) - \alpha(|z|),$$
$$\dot{V}_2(x, u) \leq \tilde{\theta}(|u|) - 2\theta(|x|).$$

Now it is obvious why the full system is ISS: simply use $V := V_1 + V_2$ as an ISS-Lyapunov function for the cascade:

$$\dot{V}((x, z), u) \ \leq\ \tilde{\theta}(|u|) - \theta(|x|) - \alpha(|z|).$$

Of course, in the special case in which the x-subsystem has no inputs, this also proved that the cascade of a GAS and an ISS system is GAS.

More generally, one may allow a "small gain" feedback as well (Figure 45.8). That is, one allows inputs $u = k(z)$ as long as they are small enough:

$$|k(z)| \ \leq\ \tilde{\theta}^{-1}((1 - \varepsilon)\alpha(|z|)).$$

The claim is that the closed-loop system

$$\dot{z} = f(z, x),$$
$$\dot{x} = g(x, k(x))$$

is GAS. This follows because the same V is a Lyapunov function for the closed-loop system; for $(x, z) \neq 0$:

$$\tilde{\theta}(|u|) \leq (1 - \varepsilon)\alpha(|z|) \ \rightsquigarrow\ \dot{V}(x, z) \leq -\theta(|x|) - \varepsilon\alpha(|z|) \ < 0.$$

A far more interesting version of this result, resulting in a composite system with inputs being itself ISS, is the *ISS small-gain theorem* due to Jiang et al. [10].

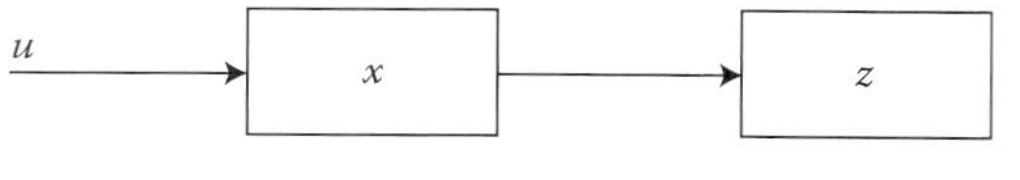

FIGURE 45.7 Cascade.

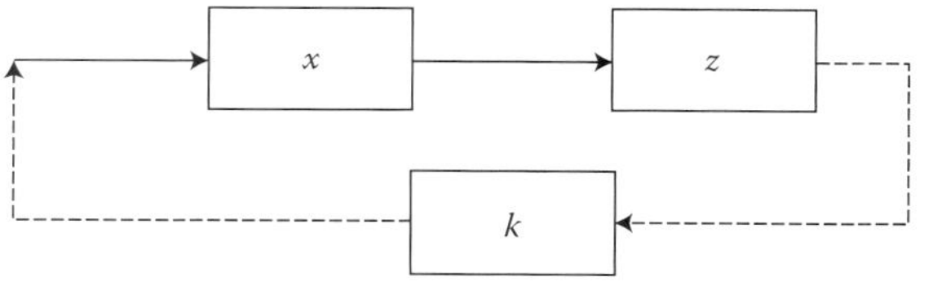

FIGURE 45.8 Adding a feedback to the cascade.

45.5 Integral ISS

Several different properties, including "integral-to-integral" stability, dissipation, robust stability margins, and AG properties, were all shown to be exactly equivalent to ISS. Thus, it would appear to be difficult to find a general and interesting concept of nonlinear stability that is truly distinct from ISS. One such concept, however, does arise when considering a mixed notion which combines the "energy" of the input with the amplitude of the state. It is obtained from the "$L^2 \to L^\infty$" gain estimate, under coordinate changes, and it provides a genuinely new concept [23].

A system is said to be *integral-input-to-state stable (iISS)* provided that there exist $\alpha, \gamma \in \mathcal{K}_\infty$ and $\beta \in \mathcal{KL}$ such that the estimate

$$\alpha\left(|x(t)|\right) \leq \beta(|x^0|, t) + \int_0^t \gamma(|u(s)|)\,\mathrm{d}s \qquad \text{(iISS)}$$

holds along all solutions. Just as with ISS, one could state this property merely for all times $t \in t_{\max}(x^0, u)$. Since the right-hand side is bounded on each interval $[0, t]$ (because, inputs are by definition assumed to be bounded on each finite interval), it is automatically true that $t_{\max}(x^0, u) = +\infty$ if such an estimate holds along maximal solutions. So forward-completeness (solution exists for all $t > 0$) can be assumed with no loss of generality.

45.5.1 Other Mixed Notions

A change of variables starting from a system that satisfies a finite operator gain condition from L^p to L^q, with $p \neq q$ both finite, leads naturally to the following type of "weak integral-to-integral" mixed estimate:

$$\int_0^t \underline{\alpha}(|x(s)|)\,\mathrm{d}s \leq \kappa(|x^0|) + \alpha\left(\int_0^t \gamma(|u(s)|)\,\mathrm{d}s\right)$$

for appropriate $\mathcal{K}_\infty$ functions (note the additional "α"). See [3] for more discussion on how this estimate is reached, as well as the following result:

Theorem 45.7: [3]

A system satisfies a weak integral-to-integral estimate if and only if it is iISS.

Another interesting variant is found when considering mixed *integral/supremum* estimates:

$$\underline{\alpha}(|x(t)| \leq \beta(|x^0|, t) + \int_0^t \gamma_1(|u(s)|)\,\mathrm{d}s + \gamma_2(\|u\|_\infty)$$

for suitable $\beta \in \mathcal{KL}$ and $\underline{\alpha}, \gamma_i \in \mathcal{K}_\infty$. One then has:

Theorem 45.8: [3]

A system satisfies a mixed estimate if and only if it is iISS.

45.5.2 Dissipation Characterization of iISS

Recall that a storage function is a continuous $V : \mathbb{R}^n \to \mathbb{R}$ which is positive definite and proper. A smooth storage function V is an *iISS-Lyapunov function* for the system $\dot{x} = f(x, u)$ if there are a $\gamma \in \mathcal{K}_\infty$ and an

$\alpha : [0, +\infty) \to [0, +\infty)$ which is merely *positive definite* (i.e., $\alpha(0) = 0$ and $\alpha(r) > 0$ for $r > 0$) such that the inequality

$$\dot{V}(x, u) \leq -\alpha(|x|) + \gamma(|u|) \qquad \text{(L-iISS)}$$

holds for all $(x, u) \in \mathbb{R}^n \times \mathbb{R}^m$. To compare, recall that an ISS-Lyapunov function is required to satisfy an estimate of the same form but where α is required to be of class $\mathcal{K}_\infty$; since every $\mathcal{K}_\infty$ function is positive definite, an ISS-Lyapunov function is also an iISS-Lyapunov function.

Theorem 45.9: [2]

A system is iISS if and only if it admits a smooth iISS-Lyapunov function.

Since an ISS-Lyapunov function is also an iISS one, ISS implies iISS. However, iISS is a strictly weaker property than ISS, because α may be bounded in the iISS-Lyapunov estimate, which means that V may increase, and the state become unbounded, even under bounded inputs, so long as $\gamma(|u(t)|)$ is larger than the range of α. This is also clear from the iISS definition, since a constant input with $|u(t)| = r$ results in a term in the right-hand side that grows like rt.

An interesting general class of examples is given by *bilinear* systems

$$\dot{x} = \left(A + \sum_{i=1}^{m} u_i A_i\right) x + Bu$$

for which the matrix A is Hurwitz. Such systems are always iISS (see [23]), but they are not in general ISS. For instance, in the case when $B = 0$, boundedness of trajectories for all constant inputs already implies that $A + \sum_{i=1}^{m} u_i A_i$ must have all eigenvalues with nonpositive real part, for all $u \in \mathbb{R}^m$, which is a condition involving the matrices A_i (e.g., $\dot{x} = -x + ux$ is iISS but it is not ISS).

The notion of iISS is useful in situations where an appropriate notion of detectability can be verified using LaSalle-type arguments. There follow two examples of theorems along these lines.

Theorem 45.10: [2]

A system is iISS if and only if it is 0-GAS and there is a smooth storage function V such that, for some $\sigma \in \mathcal{K}_\infty$*:*

$$\dot{V}(x, u) \leq \sigma(|u|)$$

for all (x, u).

The sufficiency part of this result follows from the observation that the 0-GAS property by itself already implies the existence of a smooth and positive-definite, but not necessarily proper, function V_0 such that $\dot{V}_0 \leq \gamma_0(|u|) - \alpha_0(|x|)$ for all (x, u), for some $\gamma_0 \in \mathcal{K}_\infty$ and positive-definite α_0 (if V_0 were proper, then it would be an iISS-Lyapunov function). Now one uses $V_0 + V$ as an iISS-Lyapunov function (V provides properness).

Theorem 45.11: [2]

A system is iISS if and only if there exists an output function $y = h(x)$ *(continuous and with* $h(0) = 0$*), which provides zero-detectability* ($u \equiv 0$ *and* $y \equiv 0 \Rightarrow x(t) \to 0$) *and dissipativity in the following* *sense: there exists a storage function V and $\sigma \in \mathcal{K}_\infty$, α positive definite, so that:*

$$\dot{V}(x,u) \leq \sigma(|u|) - \alpha(h(x))$$

holds for all (x,u).

The paper [3] contains several additional characterizations of iISS.

45.5.3 Superposition Principles for iISS

There are also AG characterizations for iISS. A system is *bounded energy weakly converging state (BEWCS)* if there exists some $\sigma \in \mathcal{K}_\infty$ so that the following implication holds:

$$\int_0^{+\infty} \sigma(|u(s)|)\,ds < +\infty \;\Rightarrow\; \liminf_{t\to+\infty} \left|x(t,x^0,u)\right| = 0 \qquad \text{(BEWCS)}$$

(more precisely: if the integral is finite, then $t_{\max}(x^0,u) = +\infty$ and the lim inf is zero). It is *bounded energy frequently bounded state (BEFBS)* if there exists some $\sigma \in \mathcal{K}_\infty$ so that the following implication holds:

$$\int_0^{+\infty} \sigma(|u(s)|)\,ds < +\infty \;\Rightarrow\; \liminf_{t\to+\infty} \left|x(t,x^0,u)\right| < +\infty \qquad \text{(BEFBS)}$$

(again, meaning that $t_{\max}(x^0,u) = +\infty$ and the lim inf is finite).

Theorem 45.12: [1]

The following three properties are equivalent for any given system $\dot{x} = f(x,u)$:

- *The system is iISS*
- *The system is BEWCS and 0-stable*
- *The system is BEFBS and 0-GAS*

45.6 Output Notions

Until now, the chapter discussed only stability of states with respect to inputs. For systems with outputs $\dot{x} = f(x,u)$, $y = h(x)$, several new notions can be introduced.

45.6.1 Input-to-Output Stability

If one simply replaces states by outputs in the left-hand side of the estimate defining ISS, there results the notion of *input-to-output stability (IOS)*: there exist some $\beta \in \mathcal{KL}$ and $\gamma \in \mathcal{K}_\infty$ such that

$$\left|y(t)\right| \leq \beta(|x^0|,t) + \gamma\left(\|u\|_\infty\right) \qquad \text{(IOS)}$$

holds for all solutions, where $y(t) = h(x(t,x^0,u))$. (Meaning that the estimate is valid for all inputs $u(\cdot)$, all initial conditions x^0, and all $t \geq 0$, and imposing as a requirement that the system be forward complete, that is, $t_{\max}(x^0,u) = \infty$ for all initial states x^0 and inputs u.) As earlier, $x(t)$, and hence $y(t) = h(x(t))$, depend only on past inputs ("causality"), so one could have used just as well simply used the supremum of $|u(s)|$ for $s \geq t$ in the estimate.

A system is *bounded-input bounded-state stable (BIBS)* if, for some $\sigma \in \mathcal{K}_\infty$, the following estimate

$$|x(t)| \leq \max\left\{\sigma(|x^0|),\, \sigma(\|u\|_\infty)\right\}$$

holds along all solutions. (Note that forward completeness is a consequence of this inequality, even if it is only required on maximal intervals, since the state is upper bounded by the right-hand side expression.)

An *IOS-Lyapunov function* is any smooth function $V : \mathbb{R}^n \to \mathbb{R}_{\geq 0}$ so that, for some $\alpha_i \in \mathcal{K}_\infty$:

$$\alpha_1(|h(x)|) \leq V(x) \leq \alpha_2(|x|), \quad \forall\, x \in \mathbb{R}^n, \;\; u \in \mathbb{R}^m$$

and, for all x, u:

$$V(x) > \alpha_3(|u|) \;\Rightarrow\; \nabla V(x) f(x,u) < 0.$$

One of the key results for IOS is as follows:

Theorem 45.13: [30]

A BIBS system is IOS if and only if it admits an IOS-Lyapunov function.

45.6.2 Detectability and Observability

Recall (see [24] for precise definitions) that an *observer* for a given system with inputs and outputs $\dot{x} = f(x,u)$, $y = h(x)$ is another system which, using only information provided by past input and output signals, provides an asymptotic (i.e., valid as $t \to \infty$) estimate $\hat{x}(t)$ of the state $x(t)$ of the system of interest (Figure 45.9).

One may think of the observer as a physical system or as an algorithm implemented by a digital computer. The general problem of building observers is closely related to "incremental" ISS-like notions, a subject not yet studied enough. This chapter will limit itself to an associated but easier question. When the ultimate goal is that of stabilization to an equilibrium, say $x = 0$ in Euclidean space, sometimes a weaker type of estimate suffices: it may be enough to obtain a *norm-estimator* which provides merely an *upper bound* on the norm $|x(t)|$ of the state $x(t)$; see [9,11,19].

Suppose that an observer exists for a given system. Since $x^0 = 0$ is an equilibrium for $\dot{x} = f(x,0)$, and also $h(0) = 0$, the solution $x(t) \equiv 0$ is consistent with $u \equiv 0$ and $y \equiv 0$. Thus, the estimation property $\hat{x}(t) - x(t) \to 0$ implies that $\hat{x}(t) \to 0$. Now consider *any* state x^0 for which $u \equiv 0$ and $y \equiv 0$, that is, so that $h(x(t,x^0,0)) \equiv 0$. The observer output, which can only depend on u and y, must be the same $\hat{x}$ as when $x^0 = 0$, so $\hat{x}(t) \to 0$; then, using once again the definition of observer $\hat{x}(t) - x(t,x^0,0) \to 0$, it follows that $x(t,x^0,0) \to 0$. In summary, a *necessary* condition for the existence of an observer is that the "subsystem" of $\dot{x} = f(x,u)$, $y = h(x)$, consisting of those states for which $u \equiv 0$ produces the output $y \equiv 0$, must have $x = 0$ as a GAS state (Figure 45.10); one says in that case that the system is *zero detectable*. (For *linear* systems, zero detectability is equivalent to detectability or "asymptotic observability" [24]: two trajectories than produce the same output must approach each other. But this equivalence need not hold for nonlinear systems.) In a nonlinear context, zero detectability is not "well-posed" enough: to get a well-behaved notion, one should add explicit requirements to ask that small inputs and outputs imply that internal states are small too (Figure 45.11), and that inputs and outputs converging to zero

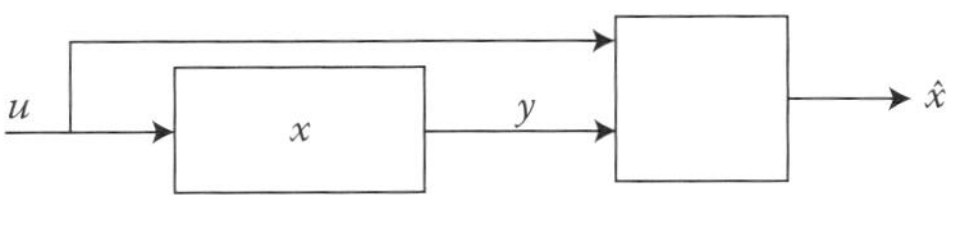

FIGURE 45.9 Observer provides estimate $\hat{x}$ of state x; $\hat{x}(t) - x(t) \to 0$ as $t \to \infty$.

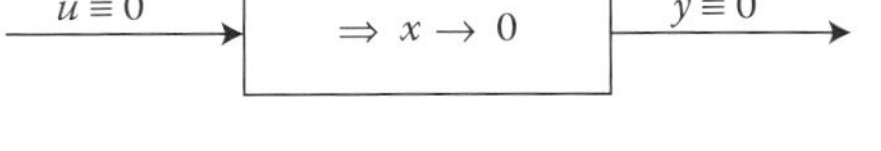

FIGURE 45.10 Zero detectability.

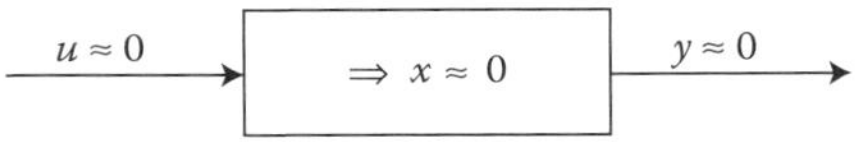

FIGURE 45.11 Small inputs and outputs imply small states.

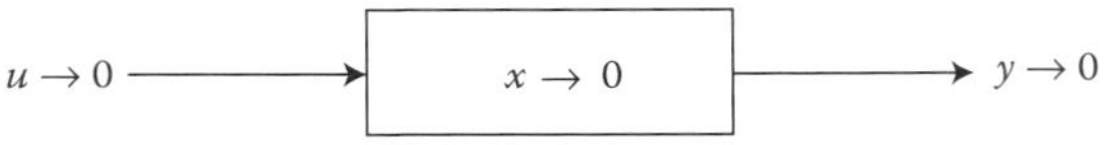

FIGURE 45.12 Converging inputs and outputs imply convergent states.

as $t \to \infty$ implies that states do, too (Figure 45.12). These properties are needed so that "small" errors in measurements of inputs and outputs processed by the observer give rise to small errors. Furthermore, one should impose asymptotic bounds on states as a function of input/output bounds, and it is desirable to quantify "overshoot" (transient behavior). This leads us to the following notion.

45.6.3 Dualizing ISS to OSS and IOSS

A system is *input/output to state stable (IOSS)* if, for some $\beta \in \mathcal{KL}$ and $\gamma_u, \gamma_y \in \mathcal{K}_\infty$,

$$x(t) \le \beta(|x^0|, t) + \gamma_1\left(\left\|u_{[0,t]}\right\|_\infty\right) + \gamma_2\left(\left\|y_{[0,t]}\right\|_\infty\right) \qquad \text{(IOSS)}$$

for all initial states and inputs, and all $t \in [0, T_{\xi,u})$. Just as ISS is stronger than 0-GAS, IOSS is stronger than zero detectability. A special case is when one has no inputs, *output to state stability:*

$$|x(t, x^0)| \le \beta(|x^0|, t) + \gamma\left(\left\|y|_{[0,t]}\right\|_\infty\right)$$

and this is formally "dual" to ISS, simply replacing inputs u by outputs in the ISS definition. This duality is only superficial, however, as there seems to be no useful way to obtain theorems for OSS by dualizing ISS results. (Note that the outputs y depend on the state, not vice versa.)

45.6.4 Lyapunov-Like Characterization of IOSS

To formulate a dissipation characterization, we define an *IOSS-Lyapunov function* as a smooth storage function so that

$$\nabla V(x) f(x, u) \le -\alpha_1(|x|) + \alpha_2(|u|) + \alpha_3(|y|)$$

for all $x \in \mathbb{R}^n, u \in \mathbb{R}^m, y \in \mathbb{R}^p$. The main result is

---

Theorem 45.14: [13]

A system is IOSS if and only if it admits an IOSS-Lyapunov function.

45.6.5 Norm-Estimators

A *state-norm-estimator* (or *state-norm-observer*) for a given system is another system

$$\dot{z} = g(z, u, y), \qquad \text{with output } k : \mathbb{R}^\ell \times \mathbb{R}^p \to \mathbb{R}_{\ge 0}$$

evolving in some Euclidean space $\mathbb{R}^\ell$, and driven by the inputs and outputs of the original system. It is required that the output k should be IOS with respect to the inputs u and y, and the true state should be asymptotically bounded in norm by some function of the norm of the estimator output, with a transient (overshoot) which depends on both initial states. Formally

- there are $\hat{\gamma}_1, \hat{\gamma}_2 \in \mathcal{K}$ and $\hat{\beta} \in \mathcal{KL}$ so that, for each initial state $z^0 \in \mathbb{R}^\ell$, and inputs $\mathbf{u}$ and $\mathbf{y}$, and every t in the interval of definition of the solution $z(\cdot, z^0, \mathbf{u}, \mathbf{y})$:

$$k\left(z(t, z^0, \mathbf{u}, \mathbf{y}), \mathbf{y}(t)\right) \leq \hat{\beta}(|z^0|, t) + \hat{\gamma}_1\left(\left\|\mathbf{u}|_{[0,t]}\right\|\right) + \hat{\gamma}_2\left(\left\|\mathbf{y}|_{[0,t]}\right\|\right);$$

- there are $\rho \in \mathcal{K}$, $\beta \in \mathcal{KL}$ so that, for all initial states x^0 and z^0 of the system and observer, and every input $\mathbf{u}$:

$$\left|x(t, x^0, \mathbf{u})\right| \leq \beta(|x^0| + |z^0|, t) + \rho\left(k\left(z(t, z^0, \mathbf{u}, \mathbf{y}_{x^0,\mathbf{u}}), \mathbf{y}_{x^0,\mathbf{u}}(t)\right)\right)$$

for all $t \in [0, t_{\max}(x^0, \mathbf{u}))$, where $\mathbf{y}_{x^0,\mathbf{u}}(t) = y(t, x^0, \mathbf{u})$.

Theorem 45.15: [13]

A system admits a state-norm-estimator if and only if it is IOSS.

45.7 The Fundamental Relationship among ISS, IOS, and IOSS

The definitions of the basic ISS-like concepts are consistent and related in an elegant conceptual manner, as follows:
A system is ISS if and only if it is both IOS and IOSS.
In informal terms:

External stability and detectability $\iff$ **Internal stability**

as it is the case for linear systems. Intuitively, consider the three possible signals in Figure 45.13

The basic idea of the proof is as follows. Suppose that external stability and detectability hold, and take an input so that $u \to 0$. Then $y \to 0$ (by external stability), and this then implies that $x \to 0$ (by detectability). Conversely, if the system is internally stable, then it is i/o stable and detectable. Suppose that $u \to 0$. By internal stability, $x \to 0$, and this gives $y(t) \to 0$ (i/o stability). Detectability is even easier: if both $u(t) \to 0$ and $y(t) \to 0$, then in particular $u \to 0$, so $x \to 0$ by internal stability. The proof that ISS is equivalent to the conjunction of IOS and IOSS must keep careful track of the estimates, but the idea is similar.

45.8 Response to Constant and Periodic Inputs

Systems $\dot{x} = f(x, u)$ that are ISS have certain noteworthy properties when subject to constant or, more generally periodic, inputs.

Let V be an ISS-Lyapunov function that satisfies the inequality $\dot{V}(x, u) \leq -V(x) + \gamma(|u|)$ for all x, u, for some $\gamma \in \mathcal{K}_\infty$. To start with, suppose that $\bar{u}$ is any fixed bounded input, and let $a := \gamma(\|\bar{u}\|_\infty)$, pick any initial state x^0, and consider the solution $x(t) = x(t, x^0, \bar{u})$ for this input. Letting $v(t) := V(x(t))$, it follows that $\dot{v}(t) + v(t) \leq a$ so, using e^t as an integrating factor, $v(t) \leq a + e^{-t}(v(0) - a)$ for all $t \geq 0$. In particular, if $v(0) \leq a$ it will follow that $v(t) \leq a$ for all $t \geq 0$, that is to say, the sublevel set $K := \{x \mid V(x) \leq a\}$ is a forward-invariant set for this input: if $x^0 \in K$ then $x(t) = x(t, x^0, \bar{u}) \in K$ for all $t \geq 0$. Therefore,

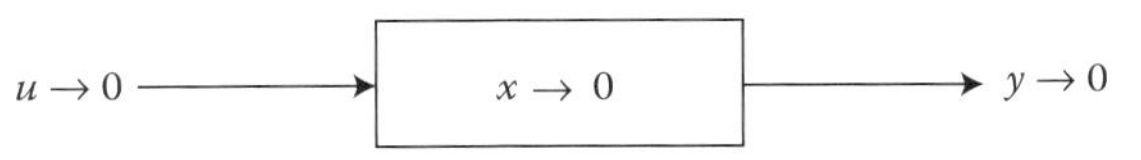

FIGURE 45.13 Convergent input, state, and/or output.

$M_T : x^0 \mapsto x(T, x^0, \bar{u})$ is a continuous mapping from K into K, for each fixed $T > 0$, and thus, provided that K has a fixed-point property (every continuous map $M : K \to K$ has some fixed point), then for each $T > 0$ there exists some state x^0 such that $x(T, x^0, \bar{u}) = x^0$. The set K indeed has the fixed-point property, as does any sublevel set of a Lyapunov function. To see this, note that V is a Lyapunov function for the zero-input system $\dot{x} = f(x, 0)$, and thus, if B is any ball which includes K in its interior, then the map $Q : B \to K$ which sends any $\xi \in B$ into $x(t_\xi, \xi)$, where t_ξ is the first time such that $x(t, \xi) \in K$ is continuous (because the vector field is transversal to the boundary of K since $\nabla V(x).f(x, 0) < 0$), and is the identity on K (that is, Q is a topological retraction). A fixed point of the composition $M \circ Q : B \to B$ is a fixed point of M.

Now suppose that $\bar{u}$ is periodic of period T, $\bar{u}(t + T) = \bar{u}(t)$ for all $t \geq 0$, and pick any x^0 which is a fixed point for M_T. Then the solution $x(t, x^0, \bar{u})$ is periodic of period T as well. In other words, *for each periodic input, there is a solution of the same period*. In particular, if $\bar{u}$ is constant, one may pick for each $h > 0$ a state x_h so that $x(h, x_h, \bar{u}) = x_h$, and therefore, picking a convergent subsequence $x_h \to \bar{x}$ gives that $0 = (1/h)(x(h, x_h, \bar{u}) - x_h) \to f(\bar{x}, \bar{u})$, so $f(\bar{x}, \bar{u}) = 0$. Thus one also has the conclusion that *for each constant input, there is a steady state.*

References

1. D. Angeli, B. Ingalls, E.D. Sontag, and Y. Wang. Separation principles for input–output and integral-input-to-state stability. *SIAM J. Control Optim.*, 43(1):256–276, 2004.
2. D. Angeli, E.D. Sontag, and Y. Wang. A characterization of integral input-to-state stability. *IEEE Trans. Automat. Control*, 45(6):1082–1097, 2000.
3. D. Angeli, E.D. Sontag, and Y. Wang. Further equivalences and semiglobal versions of integral input to state stability. *Dynam. Control*, 10(2):127–149, 2000.
4. J.C. Doyle, B. Francis, and A. Tannenbaum. *Feedback Control Theory*. MacMillan Publishing Co., London, 1990.
5. R.A. Freeman and P.V. Kokotović. *Robust Nonlinear Control Design, State-Space and Lyapunov Techniques*. Birkhauser, Boston, 1996.
6. B. Ingalls, E.D. Sontag, and Y. Wang. An infinite-time relaxation theorem for differential inclusions. *Proc. Am. Math. Soc.*, 131(2):487–499, 2003.
7. A. Isidori. *Nonlinear Control Systems II*. Springer-Verlag, London, 1999.
8. A. Isidori, L. Marconi, and A. Serrani. *Robust Autonomous Guidance: An Internal Model-Based Approach*. Springer-Verlag, London, 2003.
9. Z.-P. Jiang and L. Praly. Preliminary results about robust lagrange stability in adaptive nonlinear regulation. *Intern. J. Control*, 6:285–307, 1992.
10. Z.-P. Jiang, A. Teel, and L. Praly. Small-gain theorem for ISS systems and applications. *Math. Control, Signals, and Systems*, 7:95–120, 1994.
11. G. Kaliora, A. Astolfi, and L. Praly. Norm estimators and global output feedback stabilization of nonlinear systems with ISS inverse dynamics. In *Proc. 43rd IEEE Conference on Decision and Control, Paradise Island, Bahamas*, p. ThC07.2, 2004.
12. H.K. Khalil. *Nonlinear Systems*, 2nd edn. Prentice-Hall, Upper Saddle River, NJ, 1996.
13. M. Krichman, E.D. Sontag, and Y. Wang. Input-output-to-state stability. *SIAM J. Control Optim.*, 39(6):1874–1928, 2001.
14. M. Krstić and H. Deng. *Stabilization of Uncertain Nonlinear Systems*. Springer-Verlag, London, 1998.
15. M. Krstić, I. Kanellakopoulos, and P.V. Kokotović. *Nonlinear and Adaptive Control Design*. John Wiley & Sons, New York, 1995.
16. Y. Lin, E.D. Sontag, and Y. Wang. A smooth converse Lyapunov theorem for robust stability. *SIAM J. Control Optim.*, 34(1):124–160, 1996.
17. M. Malisoff, L. Rifford, and E.D. Sontag. Global asymptotic controllability implies input-to-state stabilization. *SIAM J. Control Optim.*, 42(6):2221–2238, 2004.
18. J.L. Massera. Contributions to stability theory. *Ann. Math.*, 64:182–206, 1956.
19. L. Praly and Y. Wang. Stabilization in spite of matched unmodelled dynamics and an equivalent definition of input-to-state stability. *Math. Control, Signals, and Systems*, 9:1–33, 1996.
20. R. Sepulchre, M. Jankovic, and P.V. Kokotović. *Constructive Nonlinear Control*. Springer-Verlag, New York, 1997.

21. E.D. Sontag. Smooth stabilization implies coprime factorization. *IEEE Trans. Automat. Control*, 34(4):435–443, 1989.
22. E.D. Sontag. Further facts about input to state stabilization. *IEEE Trans. Automat. Control*, 35(4): 473–476, 1990.
23. E.D. Sontag. Comments on integral variants of ISS. *Systems Control Lett.*, 34(1–2):93–100, 1998.
24. E.D. Sontag. *Mathematical Control Theory. Deterministic Finite-Dimensional Systems*, Vol. 6 *Texts in Applied Mathematics*. 2nd edn., Springer-Verlag, New York, 1998.
25. E.D. Sontag. Stability and stabilization: Discontinuities and the effect of disturbances. In *Nonlinear analysis, differential equations and control* (Montreal, QC, 1998), Vol. 528 *NATO Sci. Ser. C Math. Phys. Sci.*, pp. 551–598. Kluwer Acad. Publ., Dordrecht, 1999.
26. E.D. Sontag. Input to state stability: Basic concepts and results. In P. Nistri and G. Stefani, editors, *Nonlinear and Optimal Control Theory*, pp. 163–220. Springer-Verlag, Berlin, 2006.
27. E.D. Sontag and A.R. Teel. Changing supply functions in input/state stable systems. *IEEE Trans. Automat. Control*, 40(8):1476–1478, 1995.
28. E.D. Sontag and Y. Wang. On characterizations of the input-to-state stability property. *Systems Control Lett.*, 24(5):351–359, 1995.
29. E.D. Sontag and Y. Wang. New characterizations of input-to-state stability. *IEEE Trans. Automat. Control*, 41(9):1283–1294, 1996.
30. E.D. Sontag and Y. Wang. Lyapunov characterizations of input to output stability. *SIAM J. Control Optim.*, 39(1):226–249, 2000.
31. J.C. Willems. Mechanisms for the stability and instability in feedback systems. *Proc. IEEE*, 64:24–35, 1976.

VIII

Design

46

Feedback Linearization of Nonlinear Systems

Alberto Isidori
Sapienza University of Rome

Maria Domenica Di Benedetto
University of L'Aquila

46.1 The Problem of Feedback Linearization

A basic problem in control theory is how to use feedback in order to modify the original internal dynamics of a controlled plant so as to achieve some prescribed behavior. In particular, feedback may be used for the purpose of imposing, on the associated closed-loop system, the (unforced) behavior of some prescribed *autonomous linear system*. When the plant is modeled as a linear time-invariant system, this is known as the problem of pole placement, while, in the more general case of a nonlinear model, this is known as the problem of *feedback linearization* (see [1–4]).

The purpose of this chapter is to present some of the basic features of the theory of feedback linearization.

Consider a *plant* modeled by nonlinear differential equations of the form

$$\dot{x} = f(x) + g(x)u \tag{46.1}$$

$$y = h(x) \tag{46.2}$$

having *internal state* $x = (x_1, x_2, \cdots, x_n) \in \mathbb{R}^n$, *control input* $u \in \mathbb{R}$ and *measured output* $y \in \mathbb{R}$. The functions

$$f(x) = \begin{pmatrix} f_1(x_1, x_2, \cdots, x_n) \\ f_2(x_1, x_2, \cdots, x_n) \\ \cdots \\ f_n(x_1, x_2, \cdots, x_n) \end{pmatrix},$$

$$g(x) = \begin{pmatrix} g_1(x_1, x_2, \cdots, x_n) \\ g_2(x_1, x_2, \cdots, x_n) \\ \cdots \\ g_n(x_1, x_2, \cdots, x_n) \end{pmatrix},$$

$$h(x) = h(x_1, x_2, \cdots, x_n)$$

are nonlinear functions of their arguments that are assumed to be differentiable a sufficient number of times.

Changes in the description and in the behavior of this system will be investigated under two types of transformations: (1) changes of coordinates in the state space and (2) static state feedback control laws, i.e., *memoryless* state feedback laws.

In the case of a *linear* system,

$$\dot{x} = Ax + Bu \tag{46.3}$$

$$y = Cx \tag{46.4}$$

a static state feedback control law takes the form

$$u = Fx + Gv, \tag{46.5}$$

in which v represents a new control input and F and G are matrices of appropriate dimensions. Moreover, only linear changes of coordinates are usually considered. This corresponds to the substitution of the original state vector x with a new vector z related to x by a transformation of the form

$$z = Tx$$

where T is a nonsingular matrix. Accordingly, the original description of the system of Equations 46.3 and 46.4 is replaced by a new description

$$\dot{z} = \widetilde{A}z + \widetilde{B}u$$
$$y = \widetilde{C}z$$

in which

$$\widetilde{A} = TAT^{-1}, \quad \widetilde{B} = TB, \quad \widetilde{C} = CT^{-1}.$$

In the case of a nonlinear system, a static state feedback control law is a control law of the form

$$u = \alpha(x) + \beta(x)v, \tag{46.6}$$

where v represents a new control input and $\beta(x)$ is assumed to be nonzero for all x. Moreover, *nonlinear* changes of coordinates are considered, i.e., transformations of the form

$$z = \Phi(x) \tag{46.7}$$

where z is the new state vector and $\Phi(x)$ represents a (n-vector-valued) function of n variables,

$$\Phi(x) = \begin{pmatrix} \phi_1(x) \\ \phi_2(x) \\ \cdots \\ \phi_n(x) \end{pmatrix} = \begin{pmatrix} \phi_1(x_1, x_2, \cdots, x_n) \\ \phi_2(x_1, x_2, \cdots, x_n) \\ \cdots \\ \phi_n(x_1, x_2, \cdots, x_n) \end{pmatrix}$$

with the following properties:

1. $\Phi(x)$ is invertible; i.e., there exists a function $\Phi^{-1}(z)$ such that

   $$\Phi^{-1}(\Phi(x)) = x, \qquad \Phi(\Phi^{-1}(z)) = z$$

   for all $x \in \mathbb{R}^n$ and all $z \in \mathbb{R}^n$.
2. $\Phi(x)$ and $\Phi^{-1}(z)$ are both smooth mappings i.e., continuous partial derivatives of any order exist for both mappings.

A transformation of this type is called a *global diffeomorphism*. The first property is needed to guarantee the invertibility of the transformation to yield the original state vector as

$$x = \Phi^{-1}(z)$$

while the second one guarantees that the description of the system in the new coordinates is still a smooth one.

Sometimes a transformation possessing both these properties and defined for all x is hard to find and the properties in question are difficult to check. Thus, in most cases, transformations defined only in a neighborhood of a given point are of interest. A transformation of this type is called a *local diffeomorphism*. To check whether a given transformation is a local diffeomorphism, the following result is very useful.

Proposition 46.1:

Suppose $\Phi(x)$ is a smooth function defined on some subset U of $\mathbb{R}^n$. Suppose the Jacobian matrix

$$\frac{\partial \Phi}{\partial x} = \begin{pmatrix} \frac{\partial \phi_1}{\partial x_1} & \frac{\partial \phi_1}{\partial x_2} & \cdots & \frac{\partial \phi_1}{\partial x_n} \\ \frac{\partial \phi_2}{\partial x_1} & \frac{\partial \phi_2}{\partial x_2} & \cdots & \frac{\partial \phi_2}{\partial x_n} \\ \cdot & \cdot & \cdots & \cdot \\ \frac{\partial \phi_n}{\partial x_1} & \frac{\partial \phi_n}{\partial x_2} & \cdots & \frac{\partial \phi_n}{\partial x_n} \end{pmatrix}$$

is nonsingular at a point $x = x^0$. Then, for some suitable open subset U^0 of U, containing x^0, $\Phi(x)$ defines a local diffeomorphism between U^0 and its image $\Phi(U^0)$.

The effect of a change of coordinates on the description of a nonlinear system can be analyzed as follows. Set

$$z(t) = \Phi(x(t))$$

and differentiate both sides with respect to time to yield

$$\dot{z}(t) = \frac{dz}{dt} = \frac{\partial \Phi}{\partial x}\frac{dx}{dt} = \frac{\partial \Phi}{\partial x}(f(x(t)) + g(x(t))).$$

Then, expressing $x(t)$ as $\Phi^{-1}(z(t))$, one obtains

$$\dot{z}(t) = \tilde{f}(z(t)) + \tilde{g}(z(t))u(t)$$

$$y(t) = \tilde{h}(z(t))$$

where

$$\tilde{f}(z) = \left(\frac{\partial \Phi}{\partial x} f(x)\right)_{x=\Phi^{-1}(z)}$$

$$\tilde{g}(z) = \left(\frac{\partial \Phi}{\partial x} g(x)\right)_{x=\Phi^{-1}(z)}$$

$$\tilde{h}(z) = (h(x))_{x=\Phi^{-1}(z)}.$$

The latter are the formulas relating the new description of the system to the original one.

Given the nonlinear system of Equation 46.1, the problem of *feedback linearization* consists of finding, if possible, a change of coordinates of the form of Equation 46.7 and a static state feedback of the form of Equation 46.6 such that the composed dynamics of Equations 46.1 through 46.6, namely the system

$$\dot{x} = f(x) + g(x)\alpha(x) + g(x)\beta(x)v, \tag{46.8}$$

expressed in the new coordinates z, is the linear and controllable system

$$\begin{aligned}\dot{z}_1 &= z_2\\ \dot{z}_2 &= z_3\\ &\cdots\\ \dot{z}_{n-1} &= z_n\\ \dot{z}_n &= v.\end{aligned}$$

46.2 Normal Forms of Single-Input Single-Output Systems

Single-input single-output nonlinear systems can be locally given, by means of a suitable change of coordinates in the state space, a "normal form" of special interest, in which several important properties can be put in evidence and which is useful in solving the problem of feedback linearization. In this section, methods for obtaining this normal form are presented.

Given a real-valued function of $x = (x_1, \ldots, x_n)$

$$\lambda(x) = \lambda(x_1, \ldots, x_n)$$

and a (n-vector)-valued function of x

$$f(x) = \begin{pmatrix} f_1(x_1, \ldots, x_n) \\ \cdots \\ f_n(x_1, \ldots, x_n) \end{pmatrix},$$

we define a new real-valued function of x, denoted $L_f\lambda(x)$, in the following way

$$L_f\lambda(x) = L_f\lambda(x_1, \ldots, x_n) = \sum_{i=1}^{n} \frac{\partial\lambda}{\partial x_i} f_i(x_1, \ldots, x_n).$$

Setting

$$\frac{\partial\lambda}{\partial x} = \begin{pmatrix} \frac{\partial\lambda}{\partial x_1} & \cdots & \frac{\partial\lambda}{\partial x_n} \end{pmatrix}$$

the function $L_f\lambda(x)$ can be expressed in the simple form

$$L_f\lambda(x) = \frac{\partial\lambda}{\partial x} f(x).$$

The new function $L_f\lambda(x)$ thus defined is sometimes called the derivative of $\lambda(x)$ along $f(x)$. Repeated use of this operation is possible. Thus, for instance, by differentiating $\lambda(x)$ first along $f(x)$ and then along $g(x)$, we may construct the function

$$L_g L_f\lambda(x) = \frac{\partial L_f\lambda}{\partial x} g(x),$$

or, differentiating k times $\lambda(x)$ along $f(x)$, we may construct a function recursively defined as

$$L_f^k\lambda(x) = \frac{\partial L_f^{k-1}\lambda}{\partial x} f(x).$$

With the help of this operation, we introduce the notion of *relative degree* of a system.

Definition 46.1:

The single-input single-output nonlinear system

$$\dot{x} = f(x) + g(x)u$$
$$y = h(x)$$

has relative degree r *at* x^0 *if:*

1. $L_g L_f^k h(x) = 0$ *for all* x *in a neighborhood of* x^0 *and all* $k < r - 1$.
2. $L_g L_f^{r-1} h(x^0) \neq 0$.

Example 46.1:

Consider the system

$$\dot{x} = \begin{pmatrix} 0 \\ x_1^2 + \sin x_2 \\ -x_2 \end{pmatrix} + \begin{pmatrix} \exp(x_2) \\ 1 \\ 0 \end{pmatrix} u \tag{46.9}$$

$$y = h(x) = x_3\,. \tag{46.10}$$

For this system we have

$$\frac{\partial h}{\partial x} = (0 \quad 0 \quad 1), \quad L_g h(x) = 0, \quad L_f h(x) = -x_2$$
$$\frac{\partial (L_f h)}{\partial x} = (0 \quad -1 \quad 0), \quad L_g L_f h(x) = -1.$$

Thus, the system has relative degree 2 at any point x°. However, if the output function were, for instance

$$y = h(x) = x_2$$

then $L_g h(x) = 1$ and the system would have relative degree 1 at any point x°.

The notion of relative degree has the following interesting interpretation. Suppose the system at some time t^0 is in the state $x(t^0) = x^0$. Calculate the value of $y(t)$, the output of the system, and of its derivatives with respect to time, $y^{(k)}(t)$, for $k = 1, 2, \ldots$, at $t = t^0$, to obtain

$$y(t^0) = h(x(t^0)) = h(t^0)$$
$$y^{(1)}(t) = \frac{\partial h}{\partial x}\frac{dx}{dt} = \frac{\partial h}{\partial x}(f(x(t)) + g(x(t))u(t)) = L_f h(x(t)) + L_g h(x(t))u(t).$$

If the relative degree r is larger than 1, for all t such that $x(t)$ is near x^0, i.e., for all t near $t = t^0$, we have $L_g h(x(t)) = 0$ and therefore

$$y^{(1)}(t) = L_f h(x(t)).$$

This yields

$$y^{(2)}(t) = \frac{\partial L_f h}{\partial x}\frac{dx}{dt} = \frac{\partial L_f h}{\partial x}(f(x(t)) + g(x(t))u(t))$$
$$= L_f^2 h(x(t)) + L_g L_f h(x(t))u(t)$$

Again, if the relative degree is larger than 2, for all t near $t = t^0$, we have $L_g L_f h(x(t)) = 0$ and

$$y^{(2)}(t) = L_f^2 h(x(t)).$$

Continuing in this way, we get

$$y^{(k)}(t) = L_f^k h(x(t))$$

for all $k < r$ and all t near $t = t^0$, and

$$y^{(r)}(t^0) = L_f^r h(x^0) + L_g L_f^{r-1} h(x^0) u(t^0).$$

Thus, r is exactly the number of times $y(t)$ has to be differentiated at $t = t^0$ for $u(t^0)$ to appear explicitly.

The above calculations suggest that the functions $h(x), L_f h(x), \ldots, L_f^{r-1} h(x)$ have a special importance. As a matter of fact, it is possible to show that they can be used in order to define, at least partially, a local coordinate transformation near x^0, where x^0 is a point such that $L_g L_f^{r-1} h(x^0) \neq 0$. This is formally expressed in the following statement.

Proposition 46.2:

Let the system of Equations 46.1 and 46.2 be given and let r be its relative degree at $x = x^0$. Then $r \leq n$. Set

$$\begin{aligned} \phi_1(x) &= h(x) \\ \phi_2(x) &= L_f h(x) \\ &\cdots \\ \phi_r(x) &= L_f^{r-1} h(x). \end{aligned}$$

If r is strictly less than n, it is always possible to find $n - r$ additional functions $\phi_{r+1}(x), \ldots, \phi_n(x)$ such that the mapping

$$\Phi(x) = \begin{pmatrix} \phi_1(x) \\ \cdots \\ \phi_n(x) \end{pmatrix}$$

has a Jacobian matrix that is nonsingular at x^0 and therefore qualifies as a local coordinates transformation in a neighborhood of x^0. The value at x^0 of these additional functions can be fixed arbitrarily. Moreover, it is always possible to choose $\phi_{r+1}(x), \ldots, \phi_n(x)$ in such a way that

$$L_g \phi_i(x) = 0$$

for all $r + 1 \leq i \leq n$ and all x around x^0.

The description of the system in the new coordinates $z_i = \phi_i(x)$, $1 \leq i \leq n$ can be derived very easily. From the previous calculations, we obtain for $z_1, \ldots, z_r$

$$\frac{dz_1}{dt} = \frac{\partial \phi_1}{\partial x}\frac{dx}{dt} = \frac{\partial h}{\partial x}\frac{dx}{dt} = L_f h(x(t)) = \phi_2(x(t)) = z_2(t)$$

$$\cdots$$

$$\frac{dz_{r-1}}{dt} = \frac{\partial \phi_{r-1}}{\partial x}\frac{dx}{dt} = \frac{\partial L_f^{r-2} h}{\partial x}\frac{dx}{dt} = L_f^{r-1} h(x(t)) = \phi_r(x(t)) = z_r(t).$$

For z_r we obtain

$$\frac{dz_r}{dt} = L_f^r h(x(t)) + L_g L_f^{r-1} h(x(t)) u(t).$$

On the right-hand side of this equation we must now replace $x(t)$ by its expression as a function of $z(t)$, i.e., $x(t) = \Phi^{-1}(z(t))$. Thus, set

$$a(z) = L_g L_f^{r-1} h(\Phi^{-1}(z))$$
$$b(z) = L_f^r h(\Phi^{-1}(z)),$$

to obtain

$$\frac{dz_r}{dt} = b(z(t)) + a(z(t))u(t).$$

As far as the remaining coordinates are concerned, we cannot expect any special structure for the corresponding equations. If $\phi_{r+1}(x), \ldots, \phi_n(x)$ have been chosen so that $L_g\phi_i(x) = 0$, then

$$\frac{dz_i}{dt} = \frac{\partial\phi_i}{\partial x}(f(x(t)) + g(x(t))u(t)) = L_f\phi_i(x(t)) + L_g\phi_i(x(t))u(t) = L_f\phi_i(t).$$

Setting

$$q_i(z) = L_f\phi_i(\Phi^{-1}(z))$$

for $r+1 \leq i \leq n$, the latter can be rewritten as

$$\frac{dz_i}{dt} = q_i(z(t)).$$

Thus, in summary, the state space description of the system in the new coordinates is as follows

$$\begin{aligned}
\dot{z}_1 &= z_2 \\
\dot{z}_2 &= z_3 \\
&\cdots \\
\dot{z}_{r-1} &= z_r \\
\dot{z}_r &= b(z) + a(z)u \\
\dot{z}_{r+1} &= q_{r+1}(z) \\
&\cdots \\
\dot{z}_n &= q_n(z).
\end{aligned}$$

In addition to these equations, one has to specify how the output of the system is related to the new state variables. Since $y = h(x)$, it is immediately seen that

$$y = z_1$$

The equations thus defined are said to be in *normal form*. Note that at point $z^0 = \Phi(x^0)$, $a(z^0) \neq 0$ by definition. Thus, the coefficient $a(z)$ is nonzero for all z in a neighborhood of z^0.

Example 46.2:

Consider the system of Equations 46.9 and 46.10. In order to find the normal form, we set

$$z_1 = \phi_1(x) = h(x) = x_3$$
$$z_2 = \phi_2(x) = L_f h(x) = -x_2.$$

We now have to find a function $\phi_3(x)$ that completes the coordinate transformation and is such that $L_g\phi_3(x) = 0$, i.e.,

$$\frac{\partial \phi_3}{\partial x} g(x) = \frac{\partial \phi_3}{\partial x_1} \exp(x_2) + \frac{\partial \phi_3}{\partial x_2} = 0\,.$$

The function

$$\phi_3(x) = 1 + x_1 - \exp(x_2)$$

satisfies the equation above. The transformation $z = \Phi(x)$ defined by the functions $\phi_1(x)$, $\phi_2(x)$, and $\phi_3(x)$ has a Jacobian matrix

$$\frac{\partial \Phi}{\partial x} = \begin{pmatrix} 0 & 0 & 1 \\ 0 & -1 & 0 \\ 1 & -\exp(x_2) & 0 \end{pmatrix}$$

which is nonsingular for all x, and $\Phi(0) = 0$. Hence, $z = \Phi(x)$ defines a global change of coordinates. The inverse transformation is given by

$$\begin{aligned} x_1 &= z_3 - 1 + \exp(-z_2) \\ x_2 &= -z_2 \\ x_3 &= z_1\,. \end{aligned}$$

In the new coordinates the system is described by

$$\begin{aligned} \dot{z}_1 &= z_2 \\ \dot{z}_2 &= (1 - z_3 - \exp(-z_2))^2 + \sin z_2 - u \\ \dot{z}_3 &= \exp(-z_2)(\sin z_2 - (z_3 - 1 + \exp(-z_2))^2)\,. \end{aligned}$$

These equations describe the system in normal form and are globally valid because the coordinate transformation we considered is global.

46.3 Conditions for Exact Linearization via Feedback

In this section, conditions and constructive procedures are given for a single-input single-output nonlinear system to be transformed into a linear and controllable system via change of coordinates in the state space and static state feedback.

The discussion is based on the normal form developed in the previous section. Consider a nonlinear system having at some point $x = x^0$ relative degree equal to the dimension of the state space, i.e., $r = n$. In this case, the change of coordinates required to construct the normal form is given by the function $h(x)$ and its first $n - 1$ derivatives along $f(x)$, i.e.,

$$\Phi(x) = \begin{pmatrix} \phi_1(x) \\ \phi_2(x) \\ \cdots \\ \phi_n(x) \end{pmatrix} = \begin{pmatrix} h(x) \\ L_f h(x) \\ \cdots \\ L_f^{n-1} h(x) \end{pmatrix}. \tag{46.11}$$

No additional functions are needed to complete the transformation. In the new coordinates

$$z_i = \phi_i(x) = L_f^{i-1}(x) \qquad 1 \le i \le n$$

the system is described by equations of the form:

$$\begin{aligned}\dot{z}_1 &= z_2\\ \dot{z}_2 &= z_3\\ &\cdots\\ \dot{z}_{n-1} &= z_n\\ \dot{z}_n &= b(z) + a(z)u\end{aligned}$$

where $z = (z_1, z_2, \ldots, z_n)$. Recall that at the point $z^0 = \Phi(x^0)$, and hence at all z in a neighborhood of z^0, the function $a(z)$ is nonzero.

Choose now the following state feedback control law

$$u = \frac{1}{a(z)}(-b(z) + v), \tag{46.12}$$

which indeed exists and is well defined in a neighborhood of z^0. The resulting closed-loop system is governed by the equations

$$\dot{z}_1 = z_2 \tag{46.13}$$

$$\dot{z}_2 = z_3 \tag{46.14}$$

$$\cdots \tag{46.15}$$

$$\dot{z}_{n-1} = z_n \tag{46.16}$$

$$\dot{z}_n = v, \tag{46.17}$$

i.e., it is linear and controllable. Thus we conclude that any nonlinear system with relative degree n at some point x^0 can be transformed into a system that is linear and controllable by means of (1) a local change of coordinates, and (2) a local static state feedback.

The two transformations used in order to obtain the linear form can be interchanged: one can first apply a feedback and then change the coordinates in the state space, without altering the result. The feedback needed to achieve this purpose is exactly the feedback of Equation 46.12, but now expressed in the x coordinates as

$$u = \frac{1}{a(\Phi(x))}(-b(\Phi(x)) + v). \tag{46.18}$$

Comparing this with the expressions for $a(z)$ and $b(z)$ given in the previous section, one immediately realizes that this feedback, expressed in terms of the functions $f(x), g(x), h(x)$, which characterize the original system, has the form

$$u = \frac{1}{L_g L_f^{n-1}(x)}(-L_f^n(x) + v)$$

An easy calculation shows that the feedback of Equation 46.18, together with the same change of coordinates used so far (Equation 46.11), exactly yields the same linear and controllable system.

If x^0 is an equilibrium point for the original nonlinear system, i.e., if $f(x^0) = 0$, and if also $h(x^0) = 0$, then

$$\phi_1(x^0) = h(x^0) = 0$$

and

$$\phi_i(x^0) = \frac{\partial L_f^{i-1} h}{\partial x} f(x^0) = 0$$

for all $2 \le i \le n$, so that $z^0 = \Phi(x^0) = 0$. Note that a condition like $h(x^0) = 0$ can always be satisfied by means of a suitable translation of the origin of the output space. Thus, we conclude that, if x^0 is an equilibrium point for the original system, and this system has relative degree n at x^0, there is a feedback control law (defined in a neighborhood of x^0) and a coordinate transformation (also defined in a neighborhood of x^0) that change the system into a linear and controllable one, defined in a neighborhood of the origin.

New feedback controls can be imposed on the linear system thus obtained; for example,

$$v = Kz$$

where

$$K = \begin{pmatrix} k_1 & k_2 & \cdots & k_n \end{pmatrix}$$

can be chosen to meet some given control specifications, e.g., to assign a specific set of eigenvalues or to satisfy an optimality criterion. Recalling the expression of the z_is as functions of x, the feedback in question can be rewritten as

$$v = k_1 h(x) + k_2 L_f h(x) + \cdots + k_n L_f^{n-1} h(x)$$

i.e., in the form of a nonlinear feedback from the state x of the original description of the system.

Up to this point of the presentation, the existence of an "output" function $h(x)$ relative to which the system of Equations 46.1 and 46.2 has relative degree exactly equal to n (at x^0) has been key in making it possible to transform the system into a linear and controllable one. Now, if such a function $h(x)$ is not available beforehand, either because the actual output of the system does not satisfy the conditions required to have relative degree n or simply because no specific output is defined for the given system, the question arises whether it is possible to find an appropriate $h(x)$ that allows output linearization. This question is answered in the remaining part of this section.

Clearly, the problem consists of finding a function, $h(x)$, satisfying the conditions

$$L_g h(x) = L_g L_f h(x) = \cdots = L_g L_f^{n-2} h(x) = 0$$

for all x near x^0 and

$$L_g L_f^{n-2} h(x^0) \neq 0.$$

We shall see that these conditions can be transformed into a partial differential equation for $h(x)$, for which conditions for existence of solutions as well as constructive integration procedures are well known. In order to express this, we need to introduce another type of differential operation. Given two (n-vector)-valued functions of $x = (x_1, \ldots, x_n)$, $f(x)$ and $g(x)$, we define a new (n-vector)-valued function of x, denoted $[f, g](x)$, in the following way

$$[f, g](x) = \frac{\partial g}{\partial x} f(x) - \frac{\partial f}{\partial x} g(x)$$

where $\frac{\partial g}{\partial x}$ and $\frac{\partial f}{\partial x}$ are the Jacobian matrices of $g(x)$ and $f(x)$, respectively. The new function thus defined is called the *Lie product* or *Lie bracket* of $f(x)$ and $g(x)$. The Lie product can be used repeatedly. Whenever a function $g(x)$ is "Lie-multiplied" several times by a function $f(x)$, the following notation is frequently used

$$\begin{aligned} ad_f g(x) &= [f, g](x) \\ ad_f^2 g(x) &= [f, [f, g]](x) \\ &\cdots \\ ad_f^k g(x) &= [f, ad_f^{k-1} g](x). \end{aligned}$$

We shall see now that the conditions a function $h(x)$ must obey in order to be eligible as "output" of a system with relative degree n can be re-expressed in a form involving the gradient of $h(x)$ and a certain number of the repeated Lie products of $f(x)$ and $g(x)$. For, note that, since

$$L_{[f,g]}h(x) = L_f L_g h(x) - L_g L_f h(x),$$

if $L_g h(x) = 0$, the two conditions $L_{[f,g]}h(x) = 0$ and $L_g L_f h(x) = 0$ are equivalent. Using this property repeatedly, one can conclude that a system has relative degree n at x^0 if and only if

$$L_g h(x) = L_{ad_f g} h(x) = \cdots = L_{ad_f^{n-2} g} h(x) = 0$$

for all x near x^0, and

$$L_{ad_f^{n-1} g} h(x^0) \neq 0.$$

Keeping in mind the definition of derivative of $h(x)$ along a given (n-vector)-valued function, the first set of conditions can be rewritten in the following form

$$\frac{\partial h}{\partial x}\begin{pmatrix} g(x) & ad_f g(x) & \cdots & ad_f^{n-2} g(x) \end{pmatrix} = 0. \tag{46.19}$$

This partial differential equation for $h(x)$ has important properties. Indeed, if a function $h(x)$ exists such that

$$L_g h(x) = L_{ad_f g} h(x) = \cdots = L_{ad_f^{n-2} g} h(x) = 0$$

for all x near x^0, and

$$L_{ad_f^{n-1} g} h(x^0) \neq 0,$$

then necessarily the n vectors

$$g(x^0) \quad ad_f g(x^0) \quad \cdots \quad ad_f^{n-2} g(x^0) \quad ad_f^{n-1} g(x^0)$$

must be linearly independent. So, in particular, the matrix

$$\begin{pmatrix} g(x) & ad_f g(x) & \cdots & ad_f^{n-2} g(x) \end{pmatrix}$$

has rank $n - 1$. The conditions for the existence of solutions to a partial differential equation of the form of Equation 46.19 where the matrix

$$\begin{pmatrix} g(x) & ad_f g(x) & \cdots & ad_f^{n-2} g(x) \end{pmatrix}$$

has full rank are given by the well-known *Frobenius' theorem*.

Theorem 46.1:

Consider a partial differential equation of the form

$$\frac{\partial h}{\partial x}\begin{pmatrix} X_1(x) & X_2(x) & \cdots & X_k(x) \end{pmatrix} = 0,$$

in which $X_1(x), \cdots, X_k(x)$ are (n-vector)-valued functions of x. Suppose the matrix

$$\begin{pmatrix} X_1(x) & X_2(x) & \cdots & X_k(x) \end{pmatrix}$$

has rank k at the point $x = x^0$. There exist $n - k$ real-valued functions of x, say $h_1(x), \ldots, h_{n-k}(x)$, defined in a neighborhood of x^0, that are solutions of the given partial differential equation, and are such that the

Jacobian matrix

$$\begin{pmatrix} \dfrac{\partial h_1}{\partial x} \\ \cdots \\ \dfrac{\partial h_{n-k}}{\partial x} \end{pmatrix}$$

has rank $n-k$ *at* $x = x^0$ *if and only if, for each pair of integers* (i,j), $1 \le i,j \le k$, *the matrix*

$$\begin{pmatrix} X_1(x) & X_2(x) & \cdots & X_k(x) & [X_i, X_j](x) \end{pmatrix}$$

has rank k *for all* x *in a neighborhood of* x^0.

Remark 46.1

A set of k (n-vector)-valued functions $\{X_1(x), \ldots, X_k(x)\}$, such that the matrix:

$$\begin{pmatrix} X_1(x) & X_2(x) & \cdots & X_k(x) \end{pmatrix}$$

has rank k at the point $x = x^0$, is said to be *involutive* near x^0 if, for each pair of integers (i,j), $1 \le i,j \le k$, the matrix

$$\begin{pmatrix} X_1(x) & X_2(x) & \cdots & X_k(x) & [X_i, X_j](x) \end{pmatrix}$$

still has rank k for all x in a neighborhood of x^0. Using this terminology, the necessary and sufficient condition indicated in the previous theorem can be simply referred to as the *involutivity* of the set $\{X_1(x), \ldots, X_k(x)\}$.

The arguments developed thus far can be summarized formally as follows.

Proposition 46.3:

Consider a system:

$$\dot{x} = f(x) + g(x)u$$

There exists an "output" function $h(x)$ *for which the system has relative degree* n *at a point* x^0 *if and only if the following conditions are satisfied:*

1. *The matrix*

$$\begin{pmatrix} g(x^0) & ad_f g(x^0) & \cdots & ad_f^{n-2} g(x^0) & ad_f^{n-1} g(x^0) \end{pmatrix}$$

   has rank n.
2. *The set* $\{g(x), ad_f g(x), \ldots, ad_f^{n-2} g(x)\}$ *is involutive near* x^0.

In view of the results illustrated at the beginning of the section, it is now possible to conclude that conditions 1 and 2 listed in this statement are necessary and sufficient conditions for the existence of a state feedback and of a change of coordinates transforming, at least locally around the point x^0, a given nonlinear system of the form

$$\dot{x} = f(x) + g(x)u$$

into a linear and controllable one.

Remark 46.2

For a nonlinear system whose state space has dimension $n = 2$, condition 2 is always satisfied since $[g, g](x) = 0$. Hence, by the above result, any nonlinear system whose state space has dimension $n = 2$ can be transformed into a linear system, via state feedback and change of coordinates, around a point x^0 if and only if the matrix

$$(g(x^0) \quad ad_f g(x^0))$$

has rank 2. If this is the case, the vector $g(x^0)$ is nonzero and it is always possible to find a function $h(x) = h(x_1, x_2)$, defined locally around x^0, such that

$$\frac{\partial h}{\partial x} g(x) = \frac{\partial h}{\partial x_1} g_1(x_1, x_2) + \frac{\partial h}{\partial x_2} g_2(x_1, x_2) = 0.$$

If a nonlinear system of the form of Equations 46.1 and 46.2 having relative degree strictly less than n meets requirements 1 and 2 of the previous proposition, there exists a different "output" function, say $k(x)$, with respect to which the system has relative degree exactly n. Starting from this new function, it is possible to construct a feedback $u = \alpha(x) + \beta(x)v$ and a change of coordinates $z = \Phi(x)$, that transform the system

$$\dot{x} = f(x) + g(x)u$$

into a linear and controllable one. However, in general, the real output of the system expressed in the new coordinates

$$y = h(\Phi^{-1}(z))$$

is still a nonlinear function of the state z. Then the question arises whether there exist a feedback and a change of coordinates transforming the entire description of the system, output function included, into a linear and controllable one. The appropriate conditions should include the previous ones with some additional constraints arising from the need to linearize the output map. For the sake of completeness, a possible way of stating these conditions is given hereafter.

Proposition 46.4:

Let the system of Equations 46.1 and 46.2 be given and let r be its relative degree at $x = x^0$. There exist a static state feedback and a change of coordinates, defined locally around x^0, so that the system is transformed into a linear and controllable one

$$\dot{x} = Ax + Bu$$
$$y = Cx$$

if and only if the following conditions are satisfied:

1. *The matrix*

$$\left(g(x^0) \quad ad_f g(x^0) \quad \cdots \quad ad_f^{n-2} g(x^0) \quad ad_f^{n-1} g(x^0)\right)$$

   has rank n.
2. *The (n-vector)-valued functions defined as*

$$\tilde{f}(x) = f(x) - \frac{L_f^r h(x)}{L_g L_f^{r-1} h(x)} g(x) \; \tilde{g}(x) = \frac{1}{L_g L_f^{r-1} h(x)} g(x)$$

   are such that

$$\left[ad_{\tilde{f}}^i \tilde{g}, ad_{\tilde{f}}^j \tilde{g}\right] = 0$$

   for all pairs (i, j) such that $0 \le i, j \le n$.

Example 46.3:

Consider the system of Equations 46.9 and 46.10. In order to see if this system can be transformed into a linear and controllable system via static state feedback and change of coordinates, we have to check conditions 1 and 2 of Proposition 46.3. We first compute $ad_f g(x)$ and $ad_f^2 g(x)$:

$$ad_f g(x) = \begin{pmatrix} \exp(x_2)(x_1^2 + \sin x_2) \\ -2x_1 \exp(x_2) - \cos x_2 \\ 1 \end{pmatrix}$$

$$ad_f^2 g(x) = \begin{pmatrix} \exp(x_2)(x_1^2 + \sin x_2)(x_1^2 + \sin x_2 + \cos x_2) \\ x_1^2(\sin x_2 - 4x_1 \exp(x_2)) + 1 - 4x_1 \exp(x_2) \sin x_2 + 2x_1 \exp(x_2) \cos x_2 \\ -2x_1 \exp(x_2) - \cos x_2. \end{pmatrix}$$

The matrix

$$\begin{pmatrix} g(x) & ad_f g(x) & ad_f^2 g(x) \end{pmatrix}$$

has rank 3 at all points x where its determinant, an analytic function of x, is different from zero. Hence, condition 1 is satisfied almost everywhere. Note that at point $x = 0$ the matrix

$$\begin{pmatrix} g(x) & ad_f g(x) & ad_f^2 g(x) \end{pmatrix}$$

has rank 2, and this shows that condition 1 is not satisfied at the origin.

The product $[g, ad_f g](x)$ has the form

$$[g, ad_f g](x) = \begin{pmatrix} 4x_1 \exp(2x_2) + \exp(x_2)(x_1^2 + \sin x_2 + 2\cos x_2) \\ \sin x_2 - 2\exp(2x_2) - 2x_1 \exp(x_2) \\ 0 \end{pmatrix}.$$

Then one can see that the matrix

$$(g(x) \quad ad_f g(x) \quad [g, ad_f g](x))$$

has rank 2 at all points x for which its determinant

$$\exp(x_2)(2\exp(2x_2) + 6x_1 \exp(x_2) + 2\cos x_2 + x_1^2)$$

is zero. This set of points has measure zero. Hence, condition 2 is not satisfied at any point x of the state space.

In summary, the system of Equations 46.9 and 46.10 satisfies condition 1 almost everywhere but does not satisfy condition 2. Hence, it is not locally feedback linearizable.

Example 46.4:

Consider the system

$$\dot{x} = \begin{pmatrix} x_3 - x_2 \\ 0 \\ x_3 + x_1^2 \end{pmatrix} + \begin{pmatrix} 0 \\ \exp(x_1) \\ \exp(x_1) \end{pmatrix} u$$

$$y = x_2\,.$$

This system has relative degree 1 at all x since $L_g h(x) = \exp(x_1)$. It is easily checked that conditions 1 and 2 of Proposition 46.3 are satisfied. Hence, there exists a function $h(x)$ for which the system has relative degree 3. This function has to satisfy

$$\frac{\partial h}{\partial x}\left(g(x) \quad ad_f g(x)\right) = 0.$$

A solution to this equation is given by

$$h(x) = x_1.$$

The system can be transformed into a linear and controllable one by means of the static state feedback

$$u = \frac{-L_f^3 h(x) + v}{L_g L_f^2 h(x)} = \frac{2x_1x_2 - 2x_1x_3 - x_3 - x_1^2 + v}{\exp(x_1)}$$

and the coordinates transformation

$$\begin{aligned} z_1 &= h(x) = x_1 \\ z_2 &= L_f h(x) = x_3 - x_2 \\ z_3 &= L_f^2 h(x) = x_3 + x_1^2 . \end{aligned}$$

The original output of the system $y = x_2$ is a nonlinear function of z:

$$y = -z_2 + z_3 - z_1^2.$$

To determine whether the entire system, output function included, can be transformed into a linear and controllable one, condition 2 of Proposition 46.4 should be checked. Since $L_f h(x) = 0, \widetilde{f}(x) = f(x)$ and $\widetilde{g}(x) = \begin{pmatrix} 0 \\ 1 \\ 1 \end{pmatrix}$. Easy calculations yield

$$[ad_{\widetilde{f}}\widetilde{g}] = \begin{pmatrix} 0 \\ 0 \\ -1 \end{pmatrix}$$

$$[ad_{\widetilde{f}}^2\widetilde{g}] = \begin{pmatrix} 1 \\ 0 \\ 1 \end{pmatrix}$$

$$[ad_{\widetilde{f}}^3\widetilde{g}] = \begin{pmatrix} -1 \\ 0 \\ -2x_1 - 1 \end{pmatrix}$$

One can check that $[ad_{\widetilde{f}}^2\widetilde{g}, ad_{\widetilde{f}}^3\widetilde{g}] \neq 0$. Hence, condition 2 of Proposition 46.4 is not satisfied. Therefore, the system with its output cannot be transformed into a linear and controllable one.

References

1. Jakubczyk, B. and Respondek, W., On linearization of control systems, *Bull. Acad. Polonaise Sci. Ser. Sci. Math.*, 28, 517–522, 1980.
2. Su, R., On the linear equivalents of nonlinear systems, *Syst. Control Lett.*, 2, 48–52, 1982.
3. Isidori, A., Krener, A.J., Gori-Giorgi, C., and Monaco, S., Nonlinear decoupling via feedback: a differential geometric approach, *IEEE Trans. Autom. Control*, 26, 331–345, 1981.
4. Hunt, L.R., Su, R., and Meyer, G., Design for multi-input systems, in *Differential Geometric Control Theory*, Brockett, R.W., Millman, R.S., and Sussmann, H.J., Eds., Birkhauser, 268–298, 1983.
5. Isidori, A., *Nonlinear Control Systems*, 2nd Ed., Springer-Verlag, 1989.

47

The Steady-State Behavior of a Nonlinear System

Alberto Isidori
Sapienza University of Rome

Christopher I. Byrnes
Washington University

47.1 Limit Sets

In linear system theory, if the variables which characterize the behavior of a system are either constant or periodic functions of time, the system is said to be in *steady state*. In a stable linear system, the steady state can be seen as a *limit* behavior, approached either as the *actual* time t tends to $+\infty$ or, alternatively, as the *initial* time t_0 tends to $-\infty$ (the two viewpoints being in fact equivalent). For a general nonlinear dynamical system, concepts yielding to a notion of steady-state repose on certain fundamental ideas dating back to the works of H. Poincaré and G.D. Birkhoff.* In particular, a fundamental role is played by the concept of ω-limit set of a given point, which is defined as follows. Consider an *autonomous* ordinary differential equation

$$\dot{x} = f(x), \tag{47.1}$$

in which $x \in \mathbb{R}^n$. Suppose that $f(x)$ is locally Lipschitz. Then, it is well-known that, for any $x_0 \in \mathbb{R}^n$, the solution of Equation 47.1 with initial condition $x(0) = x_0$, denoted in what follows by $x(t, x_0)$, exists on some open interval of the point $t = 0$ and is unique.

Suppose that, for some x_0, the solution $x(t, x_0)$, is defined for all $t \geq 0$, that is, for all forward times. A point x is said to be an ω-limit *point* of the trajectory $x(t, x_0)$ if there exists a sequence of times $\{t_k\}$, with $\lim_{k\to\infty} t_k = \infty$, such that

$$\lim_{k\to\infty} x(t_k, x_0) = x\,.$$

The ω-limit *set* of a point x_0, denoted $\omega(x_0)$, is *the union* of all ω-limit points of the trajectory $x(t, x_0)$.

It is obvious from this definition that an ω-limit point *is not* necessarily a limit of $x(t, x_0)$ as $t \to \infty$, since the solution in question may not admit any limit as $t \to \infty$. However, it happens that if the motion $x(t, x_0)$ is *bounded*, then $x(t, x_0)$ asymptotically approaches *the set* $\omega(x_0)$. In fact, the following property holds [1, p. 198].

* Relevant, in this regard, are the concepts introduced by G.D. Birkhoff in his classical 1927 essay, where he asserts that "with an arbitrary dynamical system ... there is associated always a closed set of 'central motions' which do possess this property of regional recurrence, towards which all other motions of the system in general tend asymptotically" [1, p. 190].

Lemma 47.1:

*Suppose $x(t, x_0)$ is bounded in forward time, that is, there is a $M > 0$ such that $\|x(t, x_0)\| \leq M$ for all $t \geq 0$. Then, $\omega(x_0)$ is a nonempty compact connected set, invariant under the dynamics equation 47.1.** Moreover, the distance of $x(t, x_0)$ from $\omega(x_0)$ tends to 0 as $t \to \infty$.†

One of the features of the set $\omega(x_0)$, highlighted in this Lemma, is that this set is *invariant* for Equation 47.1. Hence, the set $\omega(x_0)$ is filled by trajectories of Equation 47.1 which are defined for all backward and forward times, and also bounded, because so is the set $\omega(x_0)$. The other relevant feature is that $x(t, x_0)$ asymptotically *approaches* the set $\omega(x_0)$ as $t \to \infty$, in the sense that the distance of the point $x(t, x_0)$ from the set $\omega(x_0)$ tends to 0 as $t \to \infty$.

Suppose now that the initial conditions of Equation 47.1 range over some set B and that $x(t, x_0)$ is bounded, in forward time, for any $x_0 \in B$. Since any trajectory which is bounded in forward time asymptotically approaches its own ω-limit set $\omega(x_0)$, it is concluded that any trajectory obtained by picking x_0 in B asymptotically approaches the set

$$\psi(B) = \bigcup_{x_0 \in B} \omega(x_0).$$

The set in question is filled by trajectories of Equation 47.1 which are bounded in forward and backward time. Since the set $\psi(B)$ is approached asymptotically by any trajectory with initial condition in B, it seems plausible to regard the set of all trajectories evolving in $\psi(B)$ as the set of steady-state "behaviors" of Equation 47.1. There is, however, a problem in taking this as the definition of steady-state behavior of a nonlinear system: the convergence of $x(t, x_0)$ to $\psi(B)$ is not guaranteed to be *uniform* in x_0, even if the set B is compact (see, e.g., [2]).

Uniform convergence to the steady state, which is automatically guaranteed in the case of linear systems, is an important feature to be kept in extending the notion of steady state from linear to nonlinear systems. In fact, the notion of steady-state would lose much of its practical relevance if the convergence were not uniform, that is, if the time needed to get within an ε-distance from the steady state could grow unbounded when the initial state is varied (even when the latter is picked within a fixed bounded set). Thus, the set $\psi(B)$ is not a good candidate for a definition of steady state in a nonlinear system. There is a larger set, however, which does have this property of uniform convergence. This set, known as the ω limit set *of a set*, is defined as follows. Suppose that, for all $x_0 \in B$, the trajectory $x(t, x_0)$ is defined for all $t \geq 0$. The ω-limit set of B, denoted $\omega(B)$, is the set of all points x for which there exists a sequence of pairs $\{x_k, t_k\}$, with $x_k \in B$ and $\lim_{k\to\infty} t_k = \infty$ such that

$$\lim_{k\to\infty} x(t_k, x_k) = x.$$

It is clear from this definition that if B consists of only one single point x_0, all x_k's in the definition above are necessarily equal to x_0 and the definition in question reduces to the definition of ω-limit set of a point. It is also clear from this definition that, if for some $x_0 \in B$ the set $\omega(x_0)$ is nonempty, all points of $\omega(x_0)$ are points of $\omega(B)$. In fact, all such points have the property indicated in the definition, if all the x_k's are taken equal to x_0. Thus, in particular, if all motions with $x_0 \in B$ are bounded in forward time, $\psi(B) \subset \omega(B)$. However, the converse inclusion is not true in general. The simplest example in which this

* A set S is invariant for Equation 47.1 if, for any $x_0 \in S$, the solution $x(t, x_0)$ exists for all $t \in \mathbb{R}$ and $x(t, x_0) \in S$ for all $t \in \mathbb{R}$.

† The distance of a point $x \in \mathbb{R}^n$ from a set $S \subset \mathbb{R}^n$, written as $\text{dist}(x, S)$, is defined as $\inf_{y\in S} \|y - x\|$.

fact can be checked is the the case of a stable Van der Pol oscillator

$$\dot{x} = y,$$
$$\dot{y} = -x - \epsilon(1 - x^2)y \quad \epsilon > 0.$$

If B is a disc of sufficiently large radius, centered at $(x, y) = (0, 0)$, the set $\psi(B)$ consists of the union of the (unstable) equilibrium at $(x, y) = (0, 0)$ and of the (stable) limit cycle. On the contrary, the set $\omega(B)$ consists of all points on the limit cycle and of all points inside this limit cycle.

The counterpart of Lemma 47.1 is the following result, which characterizes the relevant features of the concept of the ω-limit set of a set [3, p. 8].

Lemma 47.2:

Let B be a nonempty bounded subset of $\mathbb{R}^n$ and suppose there is a number M such that $\|x(t, x_0)\| \leq M$ for all $t \geq 0$ and all $x_0 \in B$. Then $\omega(B)$ is a nonempty compact set, invariant under Equation 47.1. Moreover, the distance of $x(t, x_0)$ from $\omega(B)$ tends to 0 as $t \to \infty$, uniformly in $x_0 \in B$. If B is connected, so is $\omega(B)$.

As in the case of the set $\psi(B)$, it is seen that the set $\omega(B)$ is filled with trajectories which are defined for all backward and forward times, and bounded. But, above all, it is seen that the set in question is *uniformly* approached by trajectories with initial state $x_0 \in B$, a property that the set $\psi(B)$ does not have. The set $\omega(B)$ asymptotically attracts, as $t \to \infty$, all motions that start in B. Since the convergence to $\omega(B)$ is uniform in x_0, it is also true that, whenever $\omega(B)$ is contained in the interior of B, the set $\omega(B)$ is *asymptotically stable,* in the sense of Lyapunov, that is, for every number $\varepsilon > 0$ there is a number $\delta > 0$ such that, if the distance of x_0 from $\omega(B)$ is less than δ, then the distance of $x(t, x_0)$ from $\omega(B)$ is less than ε for all $t \geq 0$. This property is very important in nonlinear feedback design, because it is a key property in establishing the existence of Lyapunov functions, as often required in most results concerning the design of stabilizing feedback laws.

47.2 The Steady-State Behavior of a System

Consider an autonomous finite-dimensional system

$$\dot{x} = f(x). \tag{47.2}$$

with initial conditions in a closed subset $X \subset \mathbb{R}^n$, and suppose the set X is *forward invariant*, that is, for any initial condition $x_0 \in X$, the solution $x(t, x_0)$ exists for all $t \geq 0$ and $x(t, x_0) \in X$ for all $t \geq 0$.

The motions of this system are said to be *ultimately bounded* if there is a bounded subset B with the property that, for every compact subset $\bar{X}$ of X, there is a time $\bar{T} > 0$ such that $x(t, x_0) \in B$ for all $t \geq \bar{T}$ and all $x_0 \in \bar{X}$. In other words, the motions are ultimately bounded if any admissible trajectory in finite time (this finite time possibly being dependent on the chosen initial condition) enters a bounded set B, and remains in this set for all future times.

If the motions of a system are ultimately bounded, in particular all trajectories with initial conditions in B are bounded. As a consequence, the limit set $\omega(B)$ is nonempty. Since all trajectories with initial conditions in X eventually enter the set B, the set $\omega(B)$—which by Lemma 47.2 attracts all trajectories with initial condition in B—also attracts all trajectories with initial conditions in X. It is therefore natural to regard $\omega(B)$ as a set to which all admissible trajectories of Equation 47.2 converge. While the set B, in the definition of ultimate boundedness, is not uniquely characterized, it is easy to see that the set $\omega(B)$ is, on the contrary, a uniquely defined object. In fact, it is possible to prove that if the motions of

Equation 47.2 are ultimately bounded and if $B' \neq B$ is any other bounded subset with the property that, for every compact subset $\bar{X}$ of X, there is a time $\bar{T} > 0$ such that $x(t, x_0) \in B'$ for all $t \geq \bar{T}$ and all $x_0 \in \bar{X}$, then $\omega(B') = \omega(B)$ (see, e.g., [2]).

In view of these properties, it is concluded that if the motions of a system are ultimately bounded, any trajectory asymptotically approaches a uniquely defined compact invariant set, the set $\omega(B)$. The latter is nonempty, compact, and invariant. In other words, any trajectory of the system approaches a nonempty set, which is in turn filled by other trajectories, which are defined and bounded in forward and backward time. Thus it is natural to look at any of such trajectories as a steady-state trajectory and to regard the set $\omega(B)$ as the set in which the steady-state behavior of the system takes place. This leads to the following definition (see [2]).

Definition 47.1:

Suppose the motions of system (Equation 47.2), with initial conditions in a closed and forward invariant set X, are ultimately bounded. A steady-state motion is any motion with initial condition $x(0) \in \omega(B)$. The set $\omega(B)$ is the steady-state locus of Equation 47.2 and the restriction of Equation 47.2 to $\omega(B)$ is the steady-state behavior of Equation 47.2.

47.3 The Steady-State Response

The definition given in the previous section recaptures the classical notion of steady state for linear systems and provides a powerful tool to deal with similar issues in the case of nonlinear systems. We discuss in what follows a number of relevant cases.

Consider an n-dimensional, single-input, *asymptotically stable* linear system

$$\dot{z} = Az + Bu, \tag{47.3}$$

forced by the harmonic input $u(t) = u_0 \sin(\Omega t + \phi_0)$. It is well-known that, regardless of what the initial condition is, the response $z(t)$ converges to a unique, well-defined, steady-state response, a periodic function of period $2\pi/\Omega$. A simple (geometric) method to determine such response consists in viewing the forcing input $u(t)$ as provided by an autonomous "signal generator" of the form

$$\dot{w} = Sw \quad u = Qw$$

in which

$$S = \begin{pmatrix} 0 & \Omega \\ -\Omega & 0 \end{pmatrix} \quad Q = (1 \quad 0)$$

and in analyzing the state behavior of the associated "augmented" system

$$\begin{aligned} \dot{w} &= Sw, \\ \dot{z} &= BQw + Az. \end{aligned} \tag{47.4}$$

Since the matrices S and A do not have common eigenvalues, the Sylvester equation $\Pi S = A\Pi + BQ$ has a unique solution. The augmented system possesses two complementary invariant subspaces: a *stable invariant subspace* and a *center invariant subspace*. The former is the set of all pairs (w, z) in which $w = 0$, while the latter is the graph of a linear map

$$\begin{aligned} \pi \colon \mathbb{R}^2 &\to \mathbb{R}^n \\ w &\mapsto \Pi w. \end{aligned}$$

All trajectories of Equation 47.4 approach the center invariant subspace as $t \to \infty$, and hence the limit behavior of Equation 47.4 is determined by the restriction of its motions to this invariant subspace. As a consequence, the steady-state response of Equation 47.3 to the periodic input $u(t) = u_0 \sin(\Omega t + \phi_0)$ is given by

$$x(t) = \Pi w(t) = \Pi \begin{pmatrix} u_0 \sin(\Omega t + \phi_0) \\ u_0 \cos(\Omega t + \phi_0) \end{pmatrix}.$$

Revisiting this analysis from the viewpoint of the more general notion of steady state introduced above, let $W \subset \mathbb{R}^2$ be a set of the form

$$W = \{w \in \mathbb{R}^2 : \|w\| \leq c\} \tag{47.5}$$

in which c is a fixed number, and suppose the set of initial conditions for Equation 47.4 is $W \times \mathbb{R}^n$. This is the case when the problem of evaluating the periodic response of Equation 47.3 to harmonic inputs whose amplitude does not exceed a fixed number c is addressed. Note that the set W is compact and invariant for the upper subsystem of Equation 47.4.

The set $W \times \mathbb{R}^n$ is closed and forward invariant for the full system Equation 47.4 and, moreover, since the lower subsystem of Equation 47.4 is a linear asymptotically stable system driven by a bounded input, it is immediate to check that the motions of system Equation 47.4, with initial conditions taken in $W \times \mathbb{R}^n$, are ultimately bounded. In particular,

$$\omega(B) = \{(w, z) \in \mathbb{R}^2 \times \mathbb{R}^n : w \in W, z = \Pi w\},$$

that is, $\omega(B)$ is the graph of the restriction of the map π to the set W. The restriction of Equation 47.4 to the invariant set $\omega(B)$ characterizes the steady-state behavior of Equation 47.3 under the family of all harmonic inputs of fixed angular frequency ω, and amplitude not exceeding c.

A similar result holds if $u(t)$ is provided by a nonlinear "signal generator" of the form

$$\dot{w} = s(w), \quad u = q(w). \tag{47.6}$$

In fact, consider an augmented system of the form

$$\begin{aligned} \dot{w} &= s(w), \\ \dot{z} &= Bq(w) + Az, \end{aligned} \tag{47.7}$$

in which $w \in W \subset \mathbb{R}^s$, with W a compact set, invariant for the the upper subsystem of Equation 47.7. Suppose, as before, that the matrix A has eigenvalues with negative real part.

As in the previous example, since the lower subsystem of Equation 47.7 is a linear asymptotically stable system driven by the bounded input $u(t) = q(w(t, w_0))$, the motions of system Equation 47.7, with initial conditions taken in $W \times \mathbb{R}^n$, are ultimately bounded. A simple calculation (see [2]) shows that the steady-state locus of Equation 47.7 is the graph of the map

$$\begin{aligned} \pi : W &\to \mathbb{R}^n \\ w &\mapsto \pi(w), \end{aligned}$$

defined by

$$\pi(w) = \lim_{T\to\infty} \int_{-T}^{0} e^{-A\tau} Bq(w(\tau, w))\, d\tau. \tag{47.8}$$

As a consequence, it is concluded that the steady-state response of Equation 47.3 to an input $u(t)$ produced by a signal generator of the form (Equation 47.6) can be expressed as $z(t) = \pi(w(t))$.

There are various ways in which the result discussed in the previous example can be generalized. For instance, it can be extended to describe the steady-state response of a nonlinear system

$$\dot{z} = f(z, u) \tag{47.9}$$

to an input provided by a nonlinear signal generator of the form (Equation 47.6), if system (Equation 47.9) is input-to-state stable.* In this case, in fact, the composition of Equations 47.9 and 47.6

$$\begin{aligned} \dot{w} &= s(w), \\ \dot{z} &= f(z, q(w)), \end{aligned} \tag{47.10}$$

with initial conditions $(w_0, z_0) \in W \times \mathbb{R}^n$ is a system whose motions are ultimately bounded and hence a well-defined steady-state locus exists.

A common feature of the two examples (Equations 47.4 and 47.7) is the fact that the steady-state locus of the system can be expressed as the *graph of a map*, defined on the set W. This means that, so long as this is the case, the system has a *unique* well-defined *steady-state response* to the input $u(t) = q(w(t))$, expressible as $z(t) = \pi(w(t))$. Of course, in general, this may not be the case and the global structure of the steady-state locus could be very complicated. In particular, the set $\omega(B)$ may fail to be the graph of a map defined on W and *multiple* steady-state responses to a given input may occur. This is the counterpart—in the context of forced motions—of the fact that, in general, a nonlinear system may possess multiple equilibria. In these cases, the steady-state response is determined not only by the forcing input, but also by the initial state of the system to which the input is applied.

Even though, in general, uniqueness of the steady-state response of a system (Equation 47.9) to inputs generated by a system of the form (Equation 47.6) cannot be guaranteed, if the set W is compact and invariant (as assumed above), for each $w \in W$ there is always at least one initial condition z of Equation 47.9 such that the pair (w, z) produces a steady-state response.

Lemma 47.3:

Let W be a compact set, invariant under the flow of Equation 47.6. Let Z be a closed set and suppose that the motions of Equation 47.10 with initial conditions in $W \times Z$ are ultimately bounded. Then, the steady state locus of Equation 47.10 is the graph of a set-valued map defined on the whole of W.

This result is particularly useful in establishing certain necessary conditions in the analysis of the problem of nonlinear output regulation.

References

1. Birkhoff, G.D., *Dynamical Systems*, American Mathematical Society, 1927.
2. Isidori, A. and Byrnes, C.I., Steady-state behaviors in nonlinear systems, with an application to robust disturbance rejection, *Annual Reviews in Control*, 32, 1–16, 2008.
3. Hale J.K., Magalhães L.T., and Oliva W.M., *Dynamics in Infinite Dimensions*, Springer-Verlag, New York, NY, 2002.

* For a definition of input-to-state stability, see Chapter 45.

48

Nonlinear Output Regulation

Alberto Isidori
Sapienza University of Rome

Lorenzo Marconi
University of Bologna

48.1 The Problem

A classical problem in control theory is to impose, via feedback, a prescribed steady-state response to every external command in a given family. This may include, for instance, the problem of having the output of a controlled plant asymptotically track any prescribed reference signal in a certain class of functions of time, as well as the problem of having this output asymptotically reject any undesired disturbance in a certain class of disturbances. In both cases, the issue is to force a suitably defined *tracking error* to zero, as time tends to infinity, for every reference output and every undesired disturbance ranging over prescribed families of functions of time.

Generally speaking, the problem can be cast as follows. Consider a finite-dimensional, time-invariant, nonlinear system modeled by equations of the form

$$\begin{aligned} \dot{x} &= F(w, x, u), \\ e &= H(w, x), \end{aligned} \tag{48.1}$$

in which $x \in \mathbb{R}^n$ is a vector of state variables, $u \in \mathbb{R}$ is a vector of inputs used for *control* purposes, $w \in \mathbb{R}^s$ is a vector of inputs that cannot be controlled and include *exogenous* commands, exogenous disturbances, and model uncertainties, and $e \in \mathbb{R}$ is a vector of *regulated* outputs that include tracking errors and any other variable that needs to be steered to 0. The problem is to design a controller, which receives $e(t)$ as input and produces $u(t)$ as output, able to guarantee that, in the resulting closed-loop system, $x(t)$ remains bounded and $e(t) \to 0$ as $t \to \infty$, regardless of what the exogenous input $w(t)$ actually is.

The ability to successfully address this problem very much depends on how much the controller is allowed to know about the exogenous disturbance $w(t)$. In the ideal situation in which $w(t)$ is available to the controller in real time, the design problem indeed looks much simpler. This is, however, only an extremely optimistic situation which does not represent, in any circumstance, a realistic scenario. The other extreme situation is the one in which nothing is known about $w(t)$. In this, pessimistic, scenario the best result one could hope for is the fulfillment of some prescribed ultimate bound for $|e(t)|$, but certainly not a sharp goal such as the convergence of $e(t)$ to 0. A more comfortable, intermediate, situation is the one in which $w(t)$ is only known *to belong to a fixed family* of functions of time, for instance, the family of all solutions obtained from a fixed ordinary differential equation of the form

$$\dot{w} = s(w) \tag{48.2}$$

as the corresponding initial condition $w(0)$ is allowed to vary on a prescribed set. This situation is in fact sufficiently distant from the ideal but unrealistic case of perfect knowledge of $w(t)$ and from the realistic but conservative case of totally unknown $w(t)$. But, above all, this way of thinking about the exogenous inputs covers a number of cases of major practical relevance. There is, in fact, an abundance of design problems in which parameter uncertainties, reference command, and/or exogenous disturbances can be modeled as functions of time that satisfy an ordinary differential equation.

The control law is to be provided by a system modeled by equations of the form

$$\begin{aligned} \dot{x}_c &= F_c(x_c, e), \\ u &= H_c(x_c, e), \end{aligned} \tag{48.3}$$

with state $x_c \in \mathbb{R}^\nu$. The initial conditions $x(0)$ of the *controlled plant* (Equation 48.1), $w(0)$ of the *exosystem* (Equation 48.2), and $x_c(0)$ of the *controller* (Equation 48.3) are allowed to range over a fixed *compact* sets $X \subset \mathbb{R}^n$, $W \subset \mathbb{R}^s$, and $X_c \subset \mathbb{R}^\nu$ respectively. All maps characterizing the model of the controlled plant, of the exosystem, and of the controller are assumed to be sufficiently differentiable.

The problem that is analyzed in this chapter, known as the *problem of output regulation* (or *generalized tracking problem* or also *generalized servomechanism problem*), is to design a feedback controller of the form (Equation 48.3) so as to obtain a closed-loop system in which all trajectories are bounded and the regulated output $e(t)$ asymptotically decays to 0 as $t \to \infty$. More precisely, it is required that the composition of Equations 48.1, 48.2, and 48.3, that is the *autonomous* system

$$\begin{aligned} \dot{w} &= s(w), \\ \dot{x} &= F(w, x, H_c(x_c, H(w, x)), \\ \dot{x}_c &= F_c(x_c, H(w, x)), \end{aligned} \tag{48.4}$$

with output

$$e = H(w, x),$$

be such that

- The positive orbit of $W \times X \times X_c$ is bounded, that is, there exists a bounded subset S of $\mathbb{R}^s \times \mathbb{R}^n \times \mathbb{R}^\nu$ such that, for any $(w_0, x_0, x_{c,0}) \in W \times X \times X_c$, the integral curve $(w(t), x(t), x_c(t))$ of Equation 48.4 passing through $(w_0, x_0, x_{c,0})$ at time $t = 0$ remains in S for all $t \geq 0$.
- $\lim_{t\to\infty} e(t) = 0$, uniformly in the initial condition, that is, for every $\varepsilon > 0$ there exists a time $\bar{t}$, depending only on ε and *not on* $(w_0, x_0, x_{c,0}) \in W \times X \times X_c$, such that the integral curve $(w(t), x(t), x_c(t))$ of Equation 48.4 passing through $(w_0, x_0, x_{c,0})$ at time $t = 0$ satisfies $|e(t)| \leq \varepsilon$ for all $t \geq \bar{t}$.

Cast in these terms, the problem is readily seen to be equivalent to a problem to design a controller yielding a closed-loop system that possesses a steady-state locus* entirely immersed in the set of all (w, x) at which the regulated output $e = H(w, x)$ is 0. This being the case, there is no loss of generality in assuming from the very beginning that *the exosystem is in steady state*, which is the case when the compact set W is *invariant* under the dynamics of Equation 48.2. This will be assumed throughout the entire chapter.

* See Chapter 47 for a definition of steady state in a nonlinear system.

Note also that an approach of this kind covers also the case in which the exosystem dynamics admit a decomposition of the form

$$\dot{w}_1 = s_1(w_1, w_2),$$
$$\dot{w}_2 = 0,$$

in which some of the components of the exogenous input have a trivial dynamics, that is, are constant. The elements of w_2 comprise any uncertain constant parameters (assumed to range on a compact set) affecting the model of the controlled plant (Equation 48.1), as well as the dynamics of the time-varying components of w. Thus, solving a design problem cast in these terms provides *robustness* with respect to structured parametric uncertainties in the model of the plant, as well as in the model of the exogenous inputs to be tracked and/or rejected.

48.2 The Case of Linear Systems as a Design Paradigm

As an introduction, we describe in this section how the problem of output regulation can be analyzed and solved for linear systems. This provides in fact an instructive design paradigm that can be successfully followed in handling the corresponding general problem. The first step is the analysis of the steady state, which in turn entails the derivation of certain *necessary conditions.*

Consider the case in which the composition of exosystem (Equation 48.2) and controlled plant (Equation 48.1) is modeled by equations of the form

$$\begin{aligned} \dot{w} &= Sw, \\ \dot{x} &= Pw + Ax + Bu, \\ e &= Qw + Cx, \end{aligned} \tag{48.5}$$

and the controller (Equation 48.3) is modeled by equations of the form

$$\begin{aligned} \dot{x}_c &= A_c x_c + B_c e, \\ u &= C_c x_c + D_c e. \end{aligned} \tag{48.6}$$

The associated closed-loop system is the autonomous linear system

$$\begin{pmatrix} \dot{w} \\ \dot{x} \\ \dot{x}_c \end{pmatrix} = \begin{pmatrix} S & 0 & 0 \\ P + BD_cQ & A + BD_cC & BC_c \\ B_cQ & B_cC & A_c \end{pmatrix} \begin{pmatrix} w \\ x \\ x_c \end{pmatrix}. \tag{48.7}$$

If the controller (Equation 48.6) solves the problem of output regulation, the trajectories of Equation 48.7 are bounded and, necessarily, all the eigenvalues of the matrix

$$\begin{pmatrix} A + BD_cC & BC_c \\ B_cC & A_c \end{pmatrix}$$

have negative real part. Since by assumption S has all eigenvalues on the imaginary axis, system (Equation 48.7) possesses two complementary invariant subspaces: a *stable invariant subspace* and a *center invariant subspace*. The latter, in particular, is the graph of a linear map

$$w \mapsto \begin{pmatrix} x \\ x_c \end{pmatrix} = \begin{pmatrix} \Pi \\ \Pi_c \end{pmatrix} w$$

in which Π and Π_c are solutions of the Sylvester equation

$$\begin{pmatrix} \Pi \\ \Pi_c \end{pmatrix} S = \begin{pmatrix} A + BD_cC & BC_c \\ B_cC & A_c \end{pmatrix} \begin{pmatrix} \Pi \\ \Pi_c \end{pmatrix} + \begin{pmatrix} P + BD_cQ \\ B_cQ \end{pmatrix}. \tag{48.8}$$

Any trajectory of Equation 48.7 has a unique decomposition into a component entirely contained in the stable invariant subspace and a component entirely contained in the center invariant subspace. The former, which asymptotically decays to 0 as $t \to \infty$ is the *transient component* of the trajectory. The latter, in which $x(t)$ and $x_c(t)$ have, respectively, the form

$$x(t) = \Pi w(t), \quad x_c(t) = \Pi_c w(t) \tag{48.9}$$

is the *steady-state component* of the trajectory.

If the controller (Equation 48.6) solves the problem of output regulation, the steady-state component of any trajectory must be contained in the kernel of the map $e = Qw + Cx$ and hence the solution (Π, Π_c) of the Sylvester equation 48.8 necessarily satisfies $Q + C\Pi = 0$. Entering this constraint into Equation 48.8 it is concluded that if the controller (Equation 48.6) solves the problem of output regulation, necessarily there exists a pair (Π, Π_c) satisfying

$$\begin{aligned} \Pi S &= A\Pi + BC_c\Pi_c + P, \\ \Pi_c S &= A_c\Pi_c, \\ 0 &= C\Pi + Q. \end{aligned}$$

Setting $\Psi = C_c\Pi_c$ the first and third equations are more conveniently rewritten in the (controller-independent) form

$$\begin{aligned} \Pi S &= A\Pi + B\Psi + P, \\ 0 &= C\Pi + Q, \end{aligned} \tag{48.10}$$

in which, of course, Ψ is a matrix satisfying

$$\begin{aligned} \Psi &= C_c\Pi_c, \\ \Pi_c S &= A_c\Pi_c, \end{aligned} \tag{48.11}$$

for some choice of Π_c, A_c, C_c. The linear equations 48.10 are known as Francis' equations and the existence of a solution pair (Π, Ψ) is—as shown—a *necessary condition* for the solution of the problem of output regulation [1,10,11].

Equations 48.11, from a general viewpoint, could be regarded as a constraint on the component Ψ of the solution of Francis' equations 48.10. However, as an easy calculation shows, this constraint is actually irrelevant. In fact, given any pair S, Ψ it is always possible to fulfill conditions like (Equation 48.11). Let

$$d(\lambda) = s_0 + s_1\lambda + \cdots + s_{d-1}\lambda^{d-1} + \lambda^d$$

denote the minimal polynomial of S and set

$$T = \begin{bmatrix} \Psi \\ \Psi S \\ \cdots \\ \Psi S^{d-2} \\ \Psi S^{d-1} \end{bmatrix}, \quad \Phi = \begin{pmatrix} 0 & 1 & 0 & \cdot & 0 \\ 0 & 0 & 1 & \cdots & 0 \\ \cdot & \cdot & \cdot & \cdots & \cdot \\ 0 & 0 & 0 & \cdots & 1 \\ -s_0 & -s_1 & -s_2 & \cdots & -s_{d-1} \end{pmatrix}, \quad \Gamma = (1 \quad 0 \quad 0 \quad \cdots \quad 0).$$

Then, it is immediate to check that

$$\begin{aligned} \Psi &= \Gamma T, \\ TS &= \Phi T, \end{aligned} \tag{48.12}$$

which is precisely a constraint of the form (Equation 48.11). Note, in particular, that the pair (Φ, Γ) thus defined is observable.

The relevant role, however, of Equations 48.11 is that they interpret the ability, of the controller, to generate the *feedforward input* necessary to keep the regulated variable identically zero in steady state. In steady state—as shown—the state $x(t)$ of the plant and $x_c(t)$ of the controller evolve as in Equation 48.9 and $e(t) = 0$. Consequently, in steady state the controller is driven by input which is identically zero and generates, as output, a control of the form

$$u_{ss}(t) = C_c x_c(t) = C_c \Pi_c w(t) = \Psi w(t).$$

The latter, as predicated by Francis' equations, is a control able to force a steady state trajectory of the form $x(t) = \Pi w(t)$ and consequently to keep $e(t)$ identically zero. The property thus described is usually referred to as the *internal model property*: any controller that solves the problem of output regulation necessarily embeds a model of the feedforward inputs needed to keep $e(t)$ identically zero [2].

We proceed now with the design of a control that ensures asymptotic convergence to the required steady state. In view of the above analysis, to solve the problem of output regulation it is natural to assume that Francis' equations 48.10 have a solution, and to consider a controller of the form

$$\begin{aligned} u &= \Gamma\eta + v, \\ \dot{\eta} &= \Phi\eta + v', \end{aligned} \tag{48.13}$$

where Γ and Φ satisfy (Equation 48.12) for some T (which, as shown, is always possibile) and where v, v' are additional controls. If these controls vanish in steady state, the graph of the linear map

$$w \mapsto \begin{pmatrix} x \\ \eta \end{pmatrix} = \begin{pmatrix} \Pi \\ T \end{pmatrix} w \tag{48.14}$$

is, by construction, an invariant subspace of the composite system

$$\begin{aligned} \dot{w} &= Sw, \\ \dot{x} &= Pw + Ax + B(\Gamma\eta + v), \\ \dot{\eta} &= \Phi\eta + v', \\ e &= Qw + Cx. \end{aligned} \tag{48.15}$$

The regulated variable e vanishes on the graph of Equation 48.14. Thus, if the additional controls v and v' are able to (robustly) steer all trajectories to this invariant subspace, the problem of output regulation is solved.

Let now the states x and η of Equation 48.15 be replaced by the differences

$$\tilde{x} = x - \Pi w, \quad \tilde{\eta} = \eta - Tw,$$

in which case the equations describing the system, by virtue of Equations 48.10 and 48.12, become

$$\begin{aligned} \dot{\tilde{x}} &= A\tilde{x} + B\Gamma\tilde{\eta} + Bv, \\ \dot{\tilde{\eta}} &= \Phi\tilde{\eta} + v', \\ e &= C\tilde{x}. \end{aligned} \tag{48.16}$$

To steer all trajectories of Equation 48.15 to the graph of Equation 48.14 is the same as to stabilize the equilibrium $(\tilde{x}, \tilde{\eta}) = (0, 0)$ of Equation 48.16.

A simple design option, at this point, is to set $v' = Gv$, and to seek a (possibly dynamic) controller, with input e and output v, which (robustly) stabilizes the equilibrium $(\tilde{x}, \tilde{\eta}) = (0, 0)$ of

$$\begin{aligned} \dot{\tilde{x}} &= A\tilde{x} + B\Gamma\tilde{\eta} + Bv, \\ \dot{\tilde{\eta}} &= \Phi\tilde{\eta} + Gv, \\ e &= C\tilde{x}. \end{aligned} \tag{48.17}$$

A sufficient condition under which such a robust stabilizer exists is that *all zeros of* Equation 48.17 *have negative real part.* The zeros of system (Equation 48.17), on the other hand, are the roots of the equation

$$\begin{aligned} 0 = \det\begin{pmatrix} A-\lambda I & B\Gamma & B \\ 0 & \Phi - \lambda I & G \\ C & 0 & 0 \end{pmatrix} &= \det\begin{pmatrix} A-\lambda I & 0 & B \\ 0 & \Phi - G\Gamma - \lambda I & G \\ C & 0 & 0 \end{pmatrix} \\ &= \det\begin{pmatrix} A-\lambda I & B \\ C & 0 \end{pmatrix} \det(\Phi - G\Gamma - \lambda I), \end{aligned}$$

and hence it is readily seen that a sufficient condition for the existence of the desired robust stabilizer is that all roots of the equation

$$0 = \det\begin{pmatrix} A-\lambda I & B \\ C & 0 \end{pmatrix} \tag{48.18}$$

and all eigenvalues of the matrix $\Phi - G\Gamma$ have negative real part. The latter condition can always be fulfilled. In fact, as observed earlier, it is always possible to find an *observable* pair (Φ, Γ) which render (Equation 48.12) fulfilled for some T. Hence, there always exists a vector G which makes $\Phi - G\Gamma$ a Hurwitz matrix. In view of this, it is concluded that a sufficient condition for the existence of a robust stabilizer for Equation 48.17, once the vector G has been chosen in this way, is simply that all roots of Equation 48.18, or—what is the same—all zeros of

$$\begin{aligned} \dot{x} &= Ax + Bu, \\ e &= Cx \end{aligned} \tag{48.19}$$

have negative real part. This condition, on the other hand, also guarantees that the Francis' equation 48.10 have a solution.

48.3 Steady-State Analysis

Controlling the nonlinear plant (Equation 48.1) by means of the nonlinear controller (Equation 48.3) yields a closed-loop system modeled by Equations 48.4. If the problem of output regulation is solved, the positive orbit of the set $W \times X \times X_c$ of initial conditions is bounded and hence all trajectories asymptotically approach a steady-state locus $\omega(W \times X \times X_c)$. This set is the graph of a (possibly set-valued) map defined on W.* To streamline the analysis, we assume that this map is *single-valued*, that is, that there exists a pair of maps $x = \pi(w)$ and $x_c = \pi_c(w)$, defined on W, such that

$$\omega(W \times X \times X_c) = \{(w, x, x_c) : w \in W, x = \pi(w), x_c = \pi_c(w)\}. \tag{48.20}$$

This is equivalent to assume that, in the closed-loop system, for each given exogenous input function $w(t)$, there exists a *unique* steady state response, which therefore can be expressed as $x(t) = \pi(w(t))$ and $x_c(t) = \pi_c(w(t))$. Moreover, for convenience, we also assume that the maps $\pi(w)$ and $\pi_c(w)$ are

* See Chapter 47 for details.

continuously differentiable. This enables us to characterize in simple terms the property that the steady state locus is invariant under the flow of the closed-loop system (Equation 48.4). If this is the case, in fact, to say that the locus (Equation 48.20) is invariant under the flow of Equation 48.4 is the same as to say that $\pi(w)$ and $\pi_c(w)$ satisfy

$$\begin{aligned} \frac{\partial \pi}{\partial w} s(w) &= F(w, \pi(w), H_c(\pi_c(w), H(w, \pi(w)))), \\ \frac{\partial \pi_c}{\partial w} s(w) &= F_c(\pi_c(w), H(w, \pi(w))), \end{aligned} \qquad \forall w \in W. \tag{48.21}$$

These equations are the nonlinear counterpart of the Sylvester equations 48.8. If the controller solves the problem of output regulation, the steady state locus, which is asymptotically approached by the trajectories of the closed-loop system, must be a subset of the set of all pairs (w, x) for which $H(w, x) = 0$ and hence the map $\pi(w)$ necessarily satisfies $H(w, \pi(w)) = 0$. Entering this constraint into Equation 48.21 it follows that

$$\begin{aligned} \frac{\partial \pi}{\partial w} s(w) &= F(w, \pi(w), H_c(\pi_c(w), 0)), \\ \frac{\partial \pi_c}{\partial w} s(w) &= F_c(\pi_c(w), 0), \\ 0 &= H(w, \pi(w)). \end{aligned} \tag{48.22}$$

Proceeding as in the case of linear systems and setting $\psi(w) = H_c(\pi_c(w), 0)$, the first and third Equations of 48.22 can be rewritten in controller-independent form as

$$\begin{aligned} \frac{\partial \pi}{\partial w} s(w) &= F(w, \pi(w), \psi(w)), \\ 0 &= H(w, \pi(w)), \end{aligned} \qquad \forall w \in W. \tag{48.23}$$

These equations, introduced in [3] and known as the *nonlinear regulator equations*, are the nonlinear counterpart of the Francis' equations 48.10.

Observe that the map $\psi(w)$ appearing in Equation 48.23 satisfies

$$\begin{aligned} \psi(w) &= H_c(\pi_c(w), 0), \\ \frac{\partial \pi_c}{\partial w} s(w) &= F_c(\pi_c(w), 0). \end{aligned} \tag{48.24}$$

These constraints also can be formally rewritten in controller-independent form. In fact, the constraint in question simply expresses the existence of an integer d, of an autonomous dynamical system

$$\dot{\eta} = \varphi(\eta), \quad \eta \in \mathbb{R}^d \tag{48.25}$$

with output

$$u = \gamma(\eta), \tag{48.26}$$

and of a map $\tau : W \to \mathbb{R}^d$ such that

$$\begin{aligned} \psi(w) &= \gamma(\tau(w)), \\ \frac{\partial \tau}{\partial w} s(w) &= \varphi(\tau(w)), \end{aligned} \qquad \forall w \in W. \tag{48.27}$$

These are nonlinear counterparts of the constraints (Equation 48.12). However, while in the case of linear systems constraints of this form are irrelevant (i.e., can always be satisfied), in the case of nonlinear systems some technical problems arise and the existence of a triplet $\{\varphi(\cdot), \gamma(\cdot), \tau(\cdot)\}$ satisfying Equation 48.27 may require extra hypotheses. Issues associated with the existence and the design of such a triplet will be discussed later in Section 48.4. For the time being, we observe that the constraints (Equation 48.24) still interpret the ability, of the controller, to generate the feedforward input necessary to keep $e(t) = 0$ in steady state. In steady state, in fact, a controller that solves the problem generates a control of the form

$$u_{ss}(t) = H_c(x_c(t), 0) = H_c(\pi_c(w(t)), 0) = \psi(w(t)),$$

which, as predicated by the nonlinear regulator equations, is a control able to force a steady state trajectory of the form $x(t) = \pi(w(t))$ and consequently to keep $e(t)$ identically zero.

The nonlinear regulator equations can be given a more tangible form if the model (Equation 48.1) of the plant is affine in the input u and, viewed as a system with input u and output e, has a globally defined *normal form*. * This means that, in suitable (globally defined) coordinates, the composition of plant (Equation 48.1) and exosystem (Equation 48.2) can be modeled by equations of the form

$$\begin{aligned} \dot{w} &= s(w), \\ \dot{z} &= f(w, z, \xi_1, \ldots, \xi_r), \\ \dot{\xi}_1 &= \xi_2 \\ &\cdots \\ \dot{\xi}_{r-1} &= \xi_r, \\ \dot{\xi}_r &= a(w, z, \xi_1, \ldots, \xi_r) + b(w, z, \xi_1, \ldots, \xi_r)u, \\ e &= \xi_1 \end{aligned} \tag{48.28}$$

in which r is the relative degree of the system, $z \in \mathbb{R}^{n-r}$ and $b(w, z, \xi_1, \ldots, \xi_r)$, the so-called *high-frequency gain*, is nowhere zero.

If the model of the plant is available in normal form, the nonlinear regulator equations 48.23 can be dealt with as follows. Let $\pi(w)$ be partitioned, consistently with the partition of the state $(z, \xi_1, \ldots, \xi_r)$ of Equation 48.28, into

$$\pi(w) = \text{col}(\pi_0(w), \pi_1(w), \ldots, \pi_r(w))$$

in which case the equations in question become

$$\begin{aligned} \frac{\partial \pi_0}{\partial w} s(w) &= f(w, \pi_0(w), \pi_1(w), \ldots, \pi_r(w)), \\ \frac{\partial \pi_i}{\partial w} s(w) &= \pi_{i+1}(w) \quad i = 1, \ldots, r-1, \\ \frac{\partial \pi_r}{\partial w} s(w) &= a(w, \pi_0(w), \pi_1(w), \ldots, \pi_r(w)) + b(w, \pi_0(w), \pi_1(w), \ldots, \pi_r(w))\psi(w), \\ 0 &= \pi_1(w). \end{aligned}$$

From these, we deduce that

$$\pi_1(w) = \cdots = \pi_r(w) = 0,$$

while $\pi_0(w)$ satisfies

$$\frac{\partial \pi_0}{\partial w} s(w) = f_0(w, \pi_0(w)), \tag{48.29}$$

in which

$$f_0(w, z) = f(w, z, 0, \ldots, 0).$$

Moreover,

$$\psi(w) = -\, q_0(w, \pi_0(w)), \tag{48.30}$$

* See Section 57.1 of Chapter 57 for the definition of *relative degree* and *normal form*.

in which

$$q_0(w,z) = \frac{a(w,z,0,\ldots,0)}{b(w,z,0,\ldots,0)}.$$

The autonomous system

$$\begin{aligned} \dot{w} &= s(w), \\ \dot{z} &= f_0(w,z), \end{aligned} \tag{48.31}$$

characterizes the so-called *zero dynamics* of Equation 48.28. Thus, to say that the nonlinear regulator equations 48.23 have a solution is to say that the zero dynamics of Equation 48.28 possess an invariant manifold expressible as the graph of a map $z = \pi_0(w)$ defined on W.

48.4 Convergence to the Required Steady State

Mimicking the design philosophy chosen in the case of linear systems, it is natural to look at a controller of the form

$$\begin{aligned} u &= \gamma(\eta) + v, \\ \dot{\eta} &= \varphi(\eta) + v', \end{aligned} \tag{48.32}$$

where $\gamma(\cdot)$ and $\varphi(\cdot)$ satisfy (Equation 48.27) for some $\tau(\cdot)$ and where v, v' are additional controls. If these controls vanish in steady state, the graph of the nonlinear map

$$w \in W \mapsto \begin{pmatrix} x \\ \eta \end{pmatrix} = \begin{pmatrix} \pi(w) \\ \tau(w) \end{pmatrix} \tag{48.33}$$

is by construction an invariant manifold in the composite system

$$\begin{aligned} \dot{w} &= s(w), \\ \dot{x} &= f(w,x,\gamma(\eta)+v), \\ \dot{\eta} &= \varphi(\eta) + v'. \end{aligned} \tag{48.34}$$

The regulated variable e vanishes on the graph of Equation 48.33. Thus, if the additional controls v and v' are able to steer all trajectories to this invariant manifold, the problem of output regulation is solved. In the case of linear systems, the success of a similar design philosophy was made possible by the additional assumption that all zeros of the controlled plant had negative real part. In what follows, we show how the approach in question can be extended to the case of nonlinear systems.

For convenience, we begin by addressing the special case in which $r = 1$ and $b = 1$, deferring to Section 48.6 the discussion of more general cases. Picking, as in the case of linear systems, $v' = Gv$, system (Equation 48.34) reduces to a system of the form

$$\begin{aligned} \dot{w} &= s(w), \\ \dot{z} &= f(w,z,\xi_1), \\ \dot{\xi}_1 &= a(w,z,\xi_1) + \gamma(\eta) + v, \\ \dot{\eta} &= \varphi(\eta) + Gv, \\ e &= \xi_1, \end{aligned} \tag{48.35}$$

which, in what follows, will be referred to as the *augmented system*. The (compact) set of admissible initial conditions of Equation 48.35 is a set of the form $W \times Z \times \Xi \times H$.

System (Equation 48.35) still has relative degree $r = 1$ between input v and output e, with a normal form which can be revealed by simply changing η into

$$\chi = \eta - G\xi_1,$$

which yields

$$\begin{aligned}
\dot{w} &= s(w),\\
\dot{z} &= f(w, z, \xi_1),\\
\dot{\chi} &= \varphi(\chi + G\xi_1) - G\gamma(\chi + G\xi_1) - Ga(w, z, \xi_1),\\
\dot{\xi}_1 &= a(w, z, \xi_1) + \gamma(\chi + G\xi_1) + v,\\
e &= \xi_1.
\end{aligned} \tag{48.36}$$

Observe, in this respect, that the zero dynamics of this system, which will be referred to as the *augmented zero-dynamics*, obtained by entering the constraint $e = 0$, are those of the autonomous system

$$\begin{aligned}
\dot{w} &= s(w),\\
\dot{z} &= f_0(w, z),\\
\dot{\chi} &= \varphi(\chi) - G\gamma(\chi) - Gq_0(w, z).
\end{aligned} \tag{48.37}$$

By virtue of Equation 48.27 through 48.30, it is readily seen that the manifold

$$\mathcal{M} = \{(w, z, \chi) : w \in W, z = \pi_0(w), \chi = \tau(w)\} \tag{48.38}$$

is an invariant manifold of Equation 48.37.

It is now convenient to regard system (Equation 48.36) as feedback interconnection of a system with input ξ_1 and state (w, z, χ) and of a system with inputs (w, z, χ) and v and state ξ_1. In particular, setting

$$p = \mathrm{col}(w, z, \chi)$$

the system in question can be regarded as a system of the form

$$\begin{aligned}
\dot{p} &= M(p) + N(p, \xi_1),\\
\dot{\xi}_1 &= H(p) + J(p, \xi_1) + b(p, \xi_1)v,
\end{aligned} \tag{48.39}$$

in which $M(p)$ and $H(p)$ are defined as

$$M(p) = \begin{pmatrix} s(w) \\ f_0(w, z) \\ \varphi(\chi) - G\gamma(\chi) - Gq_0(w, z) \end{pmatrix}$$

and

$$H(p) = q_0(w, z) + \gamma(\chi),$$

while $N(p, \xi_1)$ and $J(p, \xi_1)$ are residual functions satisfying $N(p, 0) = 0$ and $J(p, 0) = 0$, and $b(p, \xi_1) = 1$.

The advantage of seeing system (Equation 48.36) in this form is that, under appropriate hypotheses, the control

$$v = -k\xi_1 \tag{48.40}$$

keeps all admissible trajectories bounded and forces $\xi_1(t)$ to zero as $t \to \infty$. In fact, the following result holds (see, e.g., [4,9]).

Theorem 48.1:

Consider a system of the form (Equation 48.39) with v as in Equation 48.40. Suppose that $M(p)$, $N(p,\xi_1)$, $H(p)$, $J(p,\xi_1)$, and $b(p,\xi_1)$ are at least locally Lipschitz and $b(p,\xi_1) > 0$. Let the initial conditions of the system range in a compact set $P \times \Xi$. Suppose there exists a set $\mathcal{A}$ which is locally exponentially stable for $\dot{p} = M(p)$, with a domain of attraction that contains the set P. Suppose also that $H(p) = 0$ for all $p \in \mathcal{A}$. Then, there is a number k^* such that, for all $k > k^*$, the set $\mathcal{A} \times \{0\}$ is locally exponentially stable for the interconnection (Equations 48.39 through 48.40), with a domain of attraction that contains $P \times \Xi$.

Applying this result to system (Equation 48.36), it is observed that the system $\dot{p} = M(p)$ coincides with Equation 48.37, that is with the zero dynamics of the augmented system (Equation 48.36). The set (Equation 48.38) is an invariant manifold of these dynamics and, by construction, the map $H(p)$ vanishes on this set. Thus, it is concluded that if the set (Equation 48.38) is locally exponentially stable for Equation 48.37, with a domain of attraction that contains the set of all admissible initial conditions, the choice of a high-gain control as in Equation 48.40 suffices to steer ξ_1 to zero and hence to solve the problem of output regulation.

For convenience, we summarize the result obtained so far as follows.

Corollary 48.1:

Consider a system in normal form (Equation 48.28) with $r = 1$ and $b = 1$. Suppose (Equation 48.29) holds for some $\pi_0(w)$. Let $\psi(w)$ be defined as in Equation 48.30 and $\varphi(\eta)$ and $\gamma(\eta)$ be such that (Equation 48.27) hold for some $\tau(w)$. Consider a controller of the form

$$\begin{aligned} u &= \gamma(\eta) - ke, \\ \dot{\eta} &= \varphi(\eta) - Gke. \end{aligned} \tag{48.41}$$

If the manifold (Equation 48.38) is locally exponentially stable for Equation 48.37, with a domain of attraction that contains the set of all admissible initial conditions, there exists k^* such that, for all $k > k^*$, the positive orbit of the set of admissible initial conditions in bounded and $e(t) \to 0$ as $t \to \infty$.

The main issue that remains to be addressed, in this framework, is to determine whether or not the desired asymptotic properties of the invariant manifold (Equation 48.38) of Equation 48.37 can be obtained. In this respect, the asymptotic properties of the zero dynamics (Equation 48.31) of the controlled plant—on the one hand—and the choice of $\{\varphi(\cdot), \gamma(\cdot), G\}$—on the other hand—indeed play a major role. This issue is addressed in the next section. For the time being, we observe that, in the case of linear systems, the dynamics of Equation 48.37 reduce to linear dynamics, modeled by

$$\begin{aligned} \dot{w} &= Sw, \\ \dot{z} &= B_0 w + A_0 z, \\ \dot{\chi} &= (\Phi - G\Gamma)\chi - G(D_0 w + C_0 z), \end{aligned}$$

in which A_0 is a matrix whose eigenvalues coincide with the zeros of Equation 48.19. The maps $\pi_0(w)$ and $\tau(w)$ are linear maps, $\pi_0(w) = \Pi_0 w$ and $\tau(w) = Tw$, satisfying

$$\Pi_0 S = B_0 + A_0 \Pi_0, \quad TS = \Phi T, \quad \Gamma T = -(D_0 + C_0 \Pi_0).$$

Changing z, χ into $\tilde{z} = z - \Pi_0 w$ and $\tilde{\chi} = \chi - Tw$, respectively, these dynamics can be rewritten as

$$\dot{w} = Sw,$$

$$\dot{\tilde{z}} = A_0\tilde{z},$$
$$\dot{\tilde{\chi}} = (\Phi - G\Gamma)\tilde{\chi} - GC_0\tilde{z},$$

from which it is readily seen, as expected, that if all zeros of Equation 48.19 and all eigenvalues of $(\Phi - G\Gamma)$ have negative real part, the dynamics in question have the desired asymptotic properties.

48.5 The Design of the Internal Model

System (Equation 48.37) can be interpreted as the cascade of

$$\begin{aligned}\dot{w} &= s(w),\\ \dot{z} &= f_0(w, z),\\ y &= q_0(w, z),\end{aligned} \tag{48.42}$$

and of

$$\dot{\chi} = \varphi(\chi) - G\gamma(\chi) - Gy. \tag{48.43}$$

We are interested in finding hypotheses under which there exists a triplet $\{\varphi(\cdot), \gamma(\cdot), G\}$ such that Equation 48.27, with $\psi(w) = -q_0(w, \pi_0(w))$, holds for some $\tau(\cdot)$ and such that the resulting invariant manifold (Equation 48.38) is locally exponentially stable, with a domain of attraction that contains the set of all admissible initial conditions. The first obvious hypothesis is that the set $z = \pi_0(w)$ is a locally exponentially stable (invariant) manifold of Equation 48.42, with a domain of attraction that contains the set $W \times Z$. This assumption is the nonlinear analogue of the assumption that system (Equation 48.19) has all zeros with negative real part and is usually referred to, with some abuse of terminology, as the *minimum-phase* assumption.

Once this is assumed, the matter is to determine the existence of a triplet $\{\varphi(\cdot), \gamma(\cdot), G\}$ with the appropriate properties. As we have seen, this is always possible for a linear system. The basic argument behind the construction of the pair (Φ, Γ) that makes conditions (Equation 48.12) fulfilled (recall that these are the linear version of Equation 48.27) is that, by Cayley–Hamilton's theorem,

$$\Psi S^d = -(s_0\Psi + s_1\Psi S + \cdots + s_{d-1}\Psi S^{d-1}).$$

It is seen from this that the function $u_{ss}(t) = \Psi w(t)$, with $w(t)$ solution of $\dot{w} = Sw$, satisfies the linear differential equation

$$u_{ss}^{(d)}(t) = -s_0 u_{ss}(t) - s_1 u_{ss}^{(1)}(t) - \cdots - s_{d-1} u_{ss}^{(d-1)}(t).$$

Motivated by this interpretation, we may *assume*, in the nonlinear case, the existence of an integer d, of a (locally Lipschitz) function $\phi : \mathbb{R}^d \to \mathbb{R}$ with the property that, for any $w_0 \in W$, the solution $w(t)$ of $\dot{w} = s(w)$ passing through w_0 at time $t = 0$ is such that the function $u_{ss}(t) = \psi(w(t))$, in which $\psi(w) = -q_0(w, \pi_0(w))$, satisfies the *nonlinear* differential equation

$$u_{ss}^{(d)}(t) = \phi(u_{ss}(t), u_{ss}^{(1)}(t), \ldots, u_{ss}^{(d-1)}(t)). \tag{48.44}$$

If this property is assumed, it is easy to find a pair $\varphi(\cdot), \gamma(\cdot)$ such that Equation 48.27 holds for some $\tau(\cdot)$. In fact, it suffices to set

$$\tau(w) = \mathrm{col}(\psi(w), L_s\psi(w), \ldots, L_s^{d-1}\psi(w)),$$

to pick any function $\phi_c : \mathbb{R}^d \to \mathbb{R}$ which is globally Lipschitz and agrees with $\phi(\cdot)$ on $\tau(W)$, and set

$$\varphi(\eta) = \begin{pmatrix} \eta_2 \\ \cdots \\ \eta_d \\ \phi_c(\eta_1, \ldots, \eta_d), \end{pmatrix}, \quad \gamma(\eta) = \eta_1. \tag{48.45}$$

A simple calculation shows that conditions (Equation 48.27) hold.

Once the $\varphi(\cdot), \gamma(\cdot)$ have been determined, it remains to show that a vector G can be found yielding the desired asymptotic properties. To this end, set

$$\tilde{\chi} = \chi - \tau(w), \quad \tilde{y} = q_0(w, z) - q_0(w, \pi_0(w))$$

and observe that

$$\dot{\tilde{\chi}} = \varphi(\tilde{\chi} + \tau(w)) - \varphi(\tau(w)) - G\tilde{\chi}_1 - G\tilde{y}. \tag{48.46}$$

Since $z = \pi_0(w)$ is an asymptotically stable (invariant) manifold of Equation 48.42, the input $\tilde{y}$ of Equation 48.46 decays to zero as $t \to \infty$. Moreover, by construction, $\tilde{\chi} = 0$ is an equilibrium of Equation 48.46 when $\tilde{y} = 0$.

The pair $\{\varphi(\cdot), \gamma(\cdot)\}$ defined in Equation 48.45 is *uniformly observable* (see [5]). Therefore, it is likely that the desired asymptotic properties of Equation 48.46 could be achieved by picking G as in the design of *high-gain observers*. Proceeding in this way, choose

$$G = D_\kappa G_0,$$

in which

$$D_\kappa = \mathrm{diag}(\kappa, \kappa^2, \ldots, \kappa^d),$$
$$G_0 = \mathrm{col}(c_{d-1}, c_{d-2}, \ldots, c_0),$$

with $c_{d-1}, \ldots, c_0$ coefficients of a Hurwitz polynomial

$$p(\lambda) = c_0 + c_1\lambda + \cdots + c_{d-1}\lambda^{d-1} + \lambda^d.$$

It is possible to prove (see, e.g., [6]) that, if κ is large enough, system (Equation 48.46) is globally input-to-state stable, actually with a linear gain function, and that the equilibrium $\tilde{\chi} = 0$—achieved for $\tilde{y} = 0$—is globally exponentially stable. This, coupled with the assumption that $z = \pi_0(w)$ is a locally exponentially stable (invariant) manifold of Equation 48.42, proves that Equation 48.38 is a locally exponentially stable (invariant) manifold of Equation 48.37, with a domain of attraction that contains the set $W \times Z \times \mathbb{R}^d$. This makes the result of Corollary 48.1 applicable.

We have shown, in this way, that under the assumptions that the controlled plant is minimum phase and that the family of all "steady state inputs" $u_{ss}(t) = \psi(w(t))$ obeys a (possibly nonlinear) high-order differential equation of the form (Equation 48.44), the problem of output regulation can be solved. One may wonder whether these assumptions can be weakened, in particular the assumption of the existence of a differential equation of the form (Equation 48.44). There is, in fact, an alternative design strategy, in which the assumption in question is not needed. This strategy is based on seeking the fulfillment of Equation 48.27 with a $\varphi(\eta)$ of the form

$$\varphi(\eta) = F\eta + G\gamma(\eta),$$

which entails, for the system (Equation 48.43), a structure of the form

$$\dot{\chi} = F\chi - Gy, \tag{48.47}$$

with F a Hurwitz matrix. In this case, to say that Equation 48.27 are fulfilled is to say that there exists a pair (F, G), with F a Hurwitz matrix and a map $\gamma(\eta)$ such that

$$\begin{aligned} \psi(w) &= \gamma(\tau(w)), \\ \frac{\partial \tau}{\partial w}s(w) &= F\tau(w) + G\psi(w), \end{aligned} \qquad \forall w \in W \tag{48.48}$$

hold for some $\tau(w)$. The relevant result, in this respect, is that such a triplet always exists, if the dimension d of F is sufficiently large and $\gamma(\eta)$ is allowed to be only continuous (and, thus, possibly not locally

Lipschitz). Specifically, note that, regardless of what the dimension of the matrix F is, if the latter is Hurwitz, a map $\tau(w)$ fulfilling the second equation of Equation 48.48 always exists. In fact, if the controlled plant is minimum phase, and hence all trajectories of Equation 48.42 with initial conditions in $W \times Z$ are bounded, and if the matrix F is Hurwitz, also all trajectories of

$$\begin{aligned} \dot{w} &= s(w), \\ \dot{z} &= f_0(w,z), \\ \dot{\chi} &= F\chi - Gq_0(w,z) \end{aligned} \tag{48.49}$$

for any initial conditions in $W \times Z \times \mathbb{R}^d$ are bounded, and converge to a steady state locus. The latter is the graph of a map defined on W, in which $z = \pi_0(w)$ and $\chi = \tau(w)$, where $\tau(w)$, if continuously differentiable, satisfies

$$\frac{\partial \tau}{\partial w} s(w) = F\tau(w) - Gq_0(w, \pi_0(w)) = F\tau(w) + G\psi(w). \tag{48.50}$$

Thus, the real issue is simply when there exists a map $\gamma(\eta)$ such that the map $\tau(w)$ which characterizes the steady state locus of Equation 48.49 satisfies the first condition in Equation 48.48. This problem has been recently answered in [4], in the following terms.

Proposition 48.1:

There is an integer $\ell > 0$ such that, if the eigenvalues of F have real part which is less than $-\ell$, there exists a unique continuously differentiable $\tau(w)$ which satisfies (Equation 48.50). Suppose

$$d \geq 2\dim(w) + 2.$$

Then for almost all choices (see [4] for details) of a controllable pair (F, G), with F a Hurwitz matrix whose eigenvalues have real part which is less than $-\ell$, the map $\tau(w)$ satisfies

$$\tau(w_1) = \tau(w_2) \;\Rightarrow\; \psi(w_1) = \psi(w_2), \quad \forall (w_1, w_2) \in W \times W.$$

As a consequence, there exists a continuous map $\gamma : \mathbb{R}^d \to \mathbb{R}$ such that the first identity in Equation 48.48 holds.

This shows that the existence of a triplet $\{F, G, \gamma(\cdot)\}$ with the desired properties can always be achieved, so long as the integer d is large enough. As a consequence, the design procedure outlined earlier in Corollary 48.1 is always applicable, so long as the controlled plant satisfies the minimum-phase assumption. From the constructive viewpoint, however, it must be observed that the result indicated in Proposition 48.1 is only an existence result and that the function $\gamma(\cdot)$, whose existence is guaranteed, is only known to be continuous. Obtaining continuous differentiability of such $\gamma(\cdot)$ is likely to require further hypotheses.*

48.6 The Case of Higher Relative Degree

Consider now the case of a system having relative degree higher than 1, but still assume, for simplicity, that $b(w, z, \xi_1, \ldots, \xi_r) = 1$. In addition, *assume* that the function $f(w, z, \xi_1, \ldots, \xi_r)$ is independent of

* Closed-form expressions for $\gamma(\cdot)$ and other relevant constructive aspects are discussed in [7].

$\xi_2, \ldots, \xi_r$. Choose, as in Section 48.4, a control of the form (Equation 48.32) with $v' = Gv$. This yields an augmented system that can be written in the form

$$\begin{aligned} \dot{w} &= s(w), \\ \dot{z} &= f(w, z, C\bar{z}), \\ \dot{\bar{z}} &= A\bar{z} + B\xi_r, \\ \dot{\xi}_r &= a(w, z, \bar{z}, \xi_r) + \gamma(\eta) + v, \\ \dot{\eta} &= \varphi(\eta) + Gv, \\ e &= C\bar{z}, \end{aligned} \tag{48.51}$$

in which $\bar{z} = \text{col}(\xi_1, \ldots, \xi_{r-1})$ and

$$A = \begin{pmatrix} 0 & 1 & 0 & \cdot & 0 \\ 0 & 0 & 1 & \cdots & 0 \\ \cdot & \cdot & \cdot & \cdots & \cdot \\ 0 & 0 & 0 & \cdots & 1 \\ 0 & 0 & 0 & \cdots & 0 \end{pmatrix}, \quad B = \begin{pmatrix} 0 \\ 0 \\ \cdot \\ 0 \\ 1 \end{pmatrix}, \quad C = (1 \quad 0 \quad 0 \quad \cdots \quad 0).$$

The idea is to try to transform this system into a system of the form (Equation 48.39) and to try to use again, if possible, the result of Theorem 48.1. Transformation into a system of the form (Equation 48.39) can be achieved by changing ξ_r into

$$\theta = \xi_r - K\bar{z},$$

and η into

$$\chi = \eta - G\theta, \tag{48.52}$$

in which K is a vector of design parameters yet to be determined. In this way, the augmented system (Equation 48.51) can be written as

$$\begin{aligned} \dot{p} &= M(p) + N(p, \theta), \\ \dot{\theta} &= H(p) + J(p, \theta) + b(p, \theta)v, \end{aligned} \tag{48.53}$$

in which $p = \text{col}(w, z, \bar{z}, \chi)$. The subsystem $\dot{p} = M(p)$, which coincides with the zero dynamics of Equation 48.51 if the latter is viewed as a system with input v and output θ, is a system of the form

$$\begin{pmatrix} \dot{w} \\ \dot{z} \\ \dot{\bar{z}} \\ \dot{\chi} \end{pmatrix} = \begin{pmatrix} s(w) \\ f_0(w, z) + f_1(w, z, C\bar{z}) \\ (A + BK)\bar{z} \\ \varphi(\chi) - G\gamma(\chi) - Gq_0(w, z) - Gq_1(w, z, \bar{z}) \end{pmatrix}, \tag{48.54}$$

in which

$$\begin{aligned} f_1(w, z, C\bar{z}) &= f(w, z, C\bar{z}) - f_0(w, z), \\ q_1(w, z, \bar{z}) &= a(w, z, \bar{z}, K\bar{z}) - K(A + BK)\bar{z} - q_0(w, z) \end{aligned}$$

are functions vanishing at $\bar{z} = 0$. The map $H(p)$ is a map of the form

$$H(p) = q_0(w, z) + \gamma(\chi) + q_1(w, z, \bar{z}),$$

the residual functions $N(p, \theta)$ and $J(p, \theta)$ vanish at $\theta = 0$, and $b(p, \theta) = 1$. Observe that the manifold

$$\mathcal{M} = \{(w, z, \bar{z}, \chi) : w \in W, z = \pi_0(w), \bar{z} = 0, \chi = \tau(w)\} \tag{48.55}$$

is invariant for the dynamics of Equation 48.54 and that the map $N(p)$ vanishes on this set.

It follows from Theorem 48.1 that, if the manifold (Equation 48.55) is locally exponentially stable for Equation 48.54, with a domain of attraction that contains the set of all admissible initial conditions, the choice of a high-gain control

$$v = -k\theta = -k(\xi_r - K\bar{z})$$

suffices to steer θ to zero and p to Equation 48.55. Since ξ_1, a component of $\bar{z}$, is zero in Equation 48.55, then ξ_1 also is steered to zero and the problem of output regulation is solved.

The control law proposed in this way is a law which presupposes the availability of $\xi_1, \dots, \xi_r$, that is, of the regulated variable e and its derivatives $e^{(1)}, \dots, e^{(r-1)}$. This is not a problem, however, since appropriate substitutes for these variables can be generated, so long as the set of admissible initial conditions is compact, by means of an appropriate r-dimensional system driven by e, as suggested in [8]. The applicability of the methods depends therefore on the ability to choose the design parameters in such a way that the manifold (Equation 48.55) is locally exponentially stable for Equation 48.54, with a domain of attraction that contains the set of all admissible initial conditions. In this respect, it must be observed that the system in question can be seen as a cascade of

$$\begin{aligned}\dot{w} &= s(w),\\ \dot{z} &= f_0(w,z) + f_1(w,z,C\bar{z}),\\ \dot{\bar{z}} &= (A+BK)\bar{z},\\ y &= q_0(w,z) + q_1(w,z,\bar{z}),\end{aligned} \tag{48.56}$$

and of system (Equation 48.43).

It is known that, if the controlled plant is minimum phase, that is, if the set $z = \pi_0(w)$ is a locally exponentially stable (invariant) manifold for the dynamics of Equation 48.42, with a domain of attraction that contains the set $W \times Z$, then given any compact set $\bar{Z}$, there is a matrix K such that the set $(z, \bar{z}) = (\pi_0(w), 0)$ is a locally exponentially stable invariant manifold for the dynamics of Equation 48.56, with a domain of attraction that contains the set $W \times Z \times \bar{Z}$ (see, e.g., [9]). Thus, if K is chosen in this way, either one of the two methods for the design of $\{\varphi(\cdot), \gamma(\cdot), G\}$ suggested in Section 48.5 can be used to complete the design of the regulator.

We conclude by observing that, if the high-frequency gain on the system is not equal to 1, identical results hold, which can be proven using the change of variable

$$\chi = \eta - G\int_0^{\theta} \frac{1}{b(w,z,\xi_1,\dots,\xi_{r-1},s)}\,ds,$$

instead of Equation 48.52. On the contrary, the assumption that $f(w, z, \xi_1, \dots, \xi_r)$ is independent of $\xi_2, \dots, \xi_r$ can only be removed at the expense of other assumptions, such as the property, of system

$$\begin{aligned}\dot{w} &= s(w),\\ \dot{z} &= f(w,z,\xi_1,\dots,\xi_r),\end{aligned}$$

of being input-to-state stable, in the input $(\xi_1, \dots, \xi_r)$.

References

1. Francis, B.A., The linear multivariable regulator problem, *SIAM J. Control Optimiz.*, 14, 486–505, 1976.
2. Francis, B.A. and Wonham, W.M., The internal model principle of control theory, *Automatica*, 12, 457–465, 1977.
3. Isidori, A. and Byrnes, C.I., Output regulation of nonlinear systems, *IEEE Trans. Automat. Control*, 35, 131–140, 1990.

4. Marconi, L., Praly, L., and Isidori, A., Output stabilization via nonlinear Luenberger observers, *SIAM J. Control Optimiz.*, 45, 2277–2298, 2006.
5. Gauthier, J.P. and Kupka,I., *Deterministic Observation Theory and Applications*, Cambridge University Press, Cambridge, UK 2001.
6. Byrnes, C.I. and Isidori, A., Nonlinear internal models for output regulation, *IEEE Trans. Automat. Control*, 49, 2244–2247, 2004.
7. Marconi, L. and Praly, L., Uniform practical nonlinear output regulation, *IEEE Trans. Automat. Control*, 53, 1184–1202, 2008.
8. Esfandiari, F, and Khalil, H., Output feedback stabilization of fully linearizable systems, *Int. J. Control*, 56, 1007–1037, 1992.
9. Byrnes, C.I. and Isidori, A., Asymptotic stabilization of minimum phase nonlinear systems, *IEEE Trans. Automat. Control*, 36, 1122–1137, 1991.
10. Isidori, A., Marconi, L., and Serrani, A., *Robust Autonomous Guidance: An Internal-Model Approach*, Springer-Verlag, London, 2003.
11. Huang, J., *Nonlinear Output Regulation: Theory and Applications, PA*, SIAM, Philadephia, 2004.
12. Pavlov, A., van de Wouw, N., and Nijmeijer, H., *Uniform Output Regulation of Nonlinear Systems: A Convergent Dynamics Approach*, Birkhauser, Boston, 2006.

49

Lyapunov Design

Randy A. Freeman
University of California, Santa Barbara

Petar V. Kokotović
University of California, Santa Barbara

49.1 Introduction

Lyapunov functions represent the primary tool for the stability analysis of nonlinear systems. They verify the stability of a given trajectory, and they also provide an estimate of its region of attraction. The purpose of this text is to illustrate the utility of Lyapunov functions in the *synthesis* of nonlinear control systems. We will focus on recursive state-feedback design methods which guarantee robust stability for systems with uncertain nonlinearities. Lyapunov design is used in many other contexts, such as dynamic feedback, output feedback, gain assignment, estimation, and adaptive control, but such topics are beyond the scope of this chapter.

Given a state-space model of a plant, the Lyapunov design strategy is conceptually straightforward and consists of two main steps:

1. Construct a candidate Lyapunov function V for the closed-loop system.
2. Construct a controller which renders its derivative $\dot{V}$ negative for all admissible uncertainties.

Such a controller design guarantees, by standard Lyapunov theorems, the robust stability of the closed-loop system. The difficulty lies in the first step, because only carefully constructed Lyapunov functions can lead to success in the second step. In other words, for an arbitrary Lyapunov function candidate V, it is likely that no controller can render $\dot{V}$ negative in the entire region of interest. Those select candidates which do lead to success in the second step are called *control Lyapunov functions.* Our first design step should, therefore, be to construct a control Lyapunov function for the given system; this will then insure the existence of controllers in the second design step.

In Section 49.2 we review the *Lyapunov redesign* method, in which a Lyapunov function is known for the *nominal system* (the system without uncertainties) and is used as the control Lyapunov function for the uncertain system. We will see that this method is essentially limited to systems whose uncertainties satisfy a restrictive *matching condition.* In Section 49.3 we show how such limitations can be avoided by taking the uncertainty into account while building the control Lyapunov function. We then present a recursive robust control design procedure in Section 49.4 for a class of uncertain nonlinear systems. Flexibilities in this recursive design are discussed in Section 49.6.

49.2 Lyapunov Redesign

A standard method for achieving robustness to state-space uncertainty is *Lyapunov redesign;* see [12]. In this method, one begins with a Lyapunov function for a nominal closed-loop system and then uses this Lyapunov function to construct a controller which guarantees robustness to given uncertainties. To illustrate this method, we consider the system,

$$\dot{x} = F(x) + G(x)u + \Delta(x,t), \tag{49.1}$$

where F and G are known functions comprising the *nominal system* and Δ is an uncertain function known only to lie within some bounds. For example, we may know a function $\rho(x)$ so that $\mid \Delta(x,t) \mid \leq \rho(x)$. A more general uncertainty Δ would also depend on the control variable u, but for simplicity we do not consider such uncertainty here. We assume that the nominal system is stabilizable, that is, that some state feedback $u_{nom}(x)$ exists so that the nominal closed-loop system,

$$\dot{x} = F(x) + G(x)u_{nom}(x), \tag{49.2}$$

has a globally asymptotically stable equilibrium at $x = 0$. We also assume knowledge of a Lyapunov function V for this system so that

$$\nabla V(x)\,[F(x) + G(x)u_{nom}(x)] < 0 \tag{49.3}$$

whenever $x \neq 0$. Our task is to design an additional robustifying feedback $u_{rob}(x)$ so that the composite feedback $u = u_{nom} + u_{rob}$ robustly stabilizes the system (Equation 49.1), that is, guarantees stability for every admissible uncertainty Δ. It suffices that the derivative of V along closed-loop trajectories is negative for all such uncertainties. We compute this derivative as follows:

$$\dot{V} = \nabla V(x)\,[F(x) + G(x)u_{nom}(x)] + \nabla V(x)\,[G(x)u_{rob}(x) + \Delta(x,t)] \tag{49.4}$$

Can we make this derivative negative by some choice of $u_{rob}(x)$? Recall from Equation 49.3 that the first of the two terms in Equation 49.4 is negative; it remains to examine the second of these terms. For those values of x for which the coefficient $\nabla V(x) \cdot G(x)$ of the control $u_{rob}(x)$ is nonzero, we can always choose the value of $u_{rob}(x)$ large enough to overcome any finite bound on the uncertainty Δ and thus make the second term in Equation 49.4 negative. The only problems occur on the set where $\nabla V(x) \cdot G(x) = 0$, because on this set

$$\dot{V} = \nabla V(x) \cdot F(x) + \nabla V(x) \cdot \Delta(x,t) \tag{49.5}$$

regardless of our choice for the control. Thus to guarantee the negativity of $\dot{V}$, the uncertainty Δ must satisfy

$$\nabla V(x) \cdot F(x) + \nabla V(x) \cdot \Delta(x,t) \leq 0 \tag{49.6}$$

at all points where $\nabla V(x) \cdot G(x) = 0$. This inequality constraint on the uncertainty Δ is necessary for the Lyapunov redesign method to succeed. Unfortunately, there are two undesirable aspects of this necessary condition. First, the allowable size of the uncertainty Δ is dictated by F and V and can thus be severely restricted. Second, this inequality (Equation 49.6) cannot be checked a priori on the system (Equation 49.1) because it depends on the choice for V.

These considerations lead to the following question. Are there structural conditions that can be imposed on the uncertainty Δ so that the necessary condition (Equation 49.6) is automatically satisfied? One such structural condition is obvious. If we require that the uncertainty Δ is of the form,

$$\Delta(x,t) = G(x) \cdot \bar{\Delta}(x,t), \tag{49.7}$$

for some uncertain function $\bar{\Delta}$, then clearly $\nabla V \cdot \Delta = 0$ at all points where $\nabla V \cdot G = 0$, and thus the necessary condition (Equation 49.6) is satisfied. In the literature, Equation 49.7 is called the *matching condition* because it allows the system (Equation 49.1) to be written

$$\dot{x} = F(x) + G(x)[u + \bar{\Delta}(x,t)] \tag{49.8}$$

where now the uncertainty $\bar{\Delta}$ is *matched* with the control u, that is, it enters the system through the same channel as the control [2,4,12].

There are many methods available for the design of $u_{rob}(x)$ when the matching condition (Equation 49.7) is satisfied. For example, if the uncertainty $\bar{\Delta}$ is such that $|\bar{\Delta}(x,t)| \leq \bar{\rho}(x)$ for some known function $\bar{\rho}$, then the control,

$$u_{rob}(x) = -\bar{\rho}(x)\frac{\left(\nabla V(x)\cdot G(x)\right)^T}{|\nabla V(x)\cdot G(x)|}, \tag{49.9}$$

yields

$$\dot{V} \leq \nabla V(x)\,[F(x) + G(x)u_{nom}(x)] + |\nabla V(x)\cdot G(x)|\left[-\bar{\rho}(x) + |\bar{\Delta}(x,t)|\right] \tag{49.10}$$

The first term in Equation 49.10 is negative from the nominal design (Equation 49.3), and the second term is also negative because we know that $|\bar{\Delta}(x,t)| \leq \bar{\rho}(x)$. The composite control $u = u_{nom} + u_{rob}$ thus guarantees stability and robustness to the uncertainty $\bar{\Delta}$. This controller (Equation 49.9), proposed, for example, by [8], is likely to be discontinuous at points where $\nabla V(x)\cdot G(x) = 0$. Indeed, in the scalar input case, Equation 49.9 becomes

$$u_{rob}(x) = -\bar{\rho}(x)\,\mathrm{sgn}\left(\nabla V(x)\cdot G(x)\right) \tag{49.11}$$

which is discontinuous unless $\bar{\rho}(x) = 0$ whenever $\nabla V(x)\cdot G(x) = 0$. Corless and Leitmann [4] introduced a continuous approximation to this controller which guarantees convergence, not to the point $x = 0$, but to an arbitrarily small prescribed neighborhood of this point. We will return to this continuity issue in Section 49.5.

We have seen that, because of the necessary condition (Equation 49.6), the Lyapunov redesign method is essentially limited to systems whose uncertainties satisfy the restrictive matching condition. In the next sections, we will take a different look at Equation 49.6 and obtain much weaker structural conditions on the uncertainty, which still allow a systematic robust controller design.

49.3 Beyond Lyapunov Redesign

In the previous section, we have seen that, if a Lyapunov function V is to guarantee robustness to an uncertainty Δ, then the inequality

$$\nabla V(x)\cdot F(x) + \nabla V(x)\cdot \Delta(x,t) \leq 0 \tag{49.12}$$

must be satisfied at all points where $\nabla V(x)\cdot G(x) = 0$. In the Lyapunov redesign method, this inequality was viewed as a constraint on the uncertainty Δ. Now let us instead view this inequality as a constraint on the Lyapunov function V. This new look at Equation 49.12 will lead us beyond Lyapunov redesign: our construction of V will be based on Equation 49.12 rather than on the nominal system. In other words, we will take the uncertainty Δ into account during the construction of V itself.

To illustrate our departure from Lyapunov redesign, consider the second-order, single-input uncertain system,

$$\dot{x}_1 = x_2 + \Delta_1(x,t) \tag{49.13}$$

$$\dot{x}_2 = u + \Delta_2(x,t) \tag{49.14}$$

where Δ_1 and Δ_2 are uncertain functions which satisfy some known bounds. Let us try Lyapunov redesign. The first step would be to find a state feedback $u_{nom}(x)$ so that the nominal closed-loop system,

$$\dot{x}_1 = x_2 \tag{49.15}$$

$$\dot{x}_2 = u_{nom}(x) \tag{49.16}$$

has a globally asymptotically stable equilibrium at $x = 0$. Because the nominal system is linear, this step can be accomplished with a linear control law $u_{nom}(x) = Kx$, and we can obtain a quadratic Lyapunov function $V(x) = x^T Px$ for the stable nominal closed-loop system. In this case, the necessary condition Equation 49.12 becomes

$$2x^T P \begin{bmatrix} x_2 + \Delta_1(x,t) \\ \Delta_2(x,t) \end{bmatrix} \leq 0 \tag{49.17}$$

at all points where $2x^T P[0 \quad 1]^T = 0$, that is, where $x_2 = -cx_1$ for some constant $c > 0$. We substitute $x_2 = -cx_1$ in Equation 49.17, and, after some algebra, we obtain

$$x_1 \Delta_1(x_1, -cx_1, t) \leq cx_1^2 \tag{49.18}$$

for all $x_1, t \in R$. Now suppose our knowledge of the uncertainty $\Delta_1(x_1, -cx_1, t)$ consists of a bound $\rho_1(x_1)$ so that $| \Delta_1(x_1, -cx_1, t) | \leq \rho_1(x_1)$. Then Equation 49.18 implies that $\rho_1(x_1) \leq c \mid x_1 \mid$, that is, that the uncertainty Δ_1 is restricted to exhibit only *linear* growth in x_1 at a rate determined by the constant c. In other words, if the uncertainty Δ_1 does not satisfy this c-linear growth, then this particular Lyapunov redesign fails. This was to be expected because the uncertainty Δ_1 does not satisfy the matching condition.

The above Lyapunov redesign failed because it was based on the linear nominal system which suggested a *quadratic* Lyapunov function V. Let us now ignore the nominal system and base our search for V directly on the inequality (Equation 49.12). Let $\mu(x_1)$ be a smooth function so that $\mu(0) = 0$, and consider the Lyapunov function

$$V(x) = x_1^2 + [x_2 - \mu(x_1)]^2 . \tag{49.19}$$

This function V is smooth, positive definite, and radially unbounded and thus qualifies as a candidate Lyapunov function for our system (Equations 49.13 and 49.14). We will justify this choice for V in the next section; our goal here is to illustrate how we can use our freedom in the choice for the function μ to derive a necessary condition on the uncertainty Δ_1 which is much less restrictive than Equation 49.18.

For V in Equation 49.19, $\nabla V(x) \cdot G(x) = 0$ if, and only if, $x_2 = \mu(x_1)$, so that the necessary condition Equation 49.12 becomes

$$x_1 \mu(x_1) + x_1 \Delta_1(x_1, \mu(x_1), t) \leq 0 \tag{49.20}$$

for all $x_1, t \in R$. Because we have left the choice for μ open, this inequality can be viewed as a constraint on the choice of V (through μ) rather than a constraint on the uncertainty Δ_1. We need only impose a structural condition on Δ_1 which guarantees the *existence* of a suitable function μ. An example of such a condition would be the knowledge of a bound $\rho_1(x_1)$ so that $| \Delta_1(x_1, x_2, t) | \leq \rho_1(x_1)$; then Equation 49.20 becomes

$$x_1 \mu(x_1) + | x_1 | \rho_1(x_1) \leq 0 \tag{49.21}$$

for all $x_1 \in R$. It is then clear that we can satisfy Equation 49.21 by choosing, for example,

$$\mu(x_1) = -x_1 - \rho_1(x_1)\text{sgn}(x_1). \tag{49.22}$$

A technical detail is that this μ is not smooth at $x_1 = 0$ unless $\rho_1(0) = 0$, which means V in Equation 49.19 may not strictly qualify as a Lyapunov function. As we will show in Section 49.5, however, smooth approximations always exist that will end up guaranteeing convergence to a neighborhood of $x = 0$ in the final design. What is important is that this design succeeds for any function $\rho_1(x_1)$, regardless of its growth. Thus the c-linear growth condition on Δ_1 which appeared in the above Lyapunov redesign through Equation 49.18 is gone; this new design allows arbitrary growth (in x_1) of the uncertainty Δ_1.

We have not yet specified the controller design; rather, we have shown how the limitations of Lyapunov redesign can be overcome through a reinterpretation of the necessary condition (Equation 49.12) as a constraint on the choice of V. Let us now return to the controller design problem and motivate our choice of V in Equation 49.19.

49.4 Recursive Lyapunov Design

Let us consider again the system (Equations 49.13 and 49.14):

$$\dot{x}_1 = x_2 + \Delta_1(x,t) \tag{49.23}$$

$$\dot{x}_2 = u + \Delta_2(x,t) \tag{49.24}$$

We assume knowledge of two bounding functions $\rho_1(x_1)$ and $\rho_2(x)$ so that all admissible uncertainties are characterized by the inequalities,

$$|\ \Delta_1(x,t)\ | \leq \rho_1(x_1) \tag{49.25}$$

$$|\ \Delta_2(x,t)\ | \leq \rho_2(x) \tag{49.26}$$

for all $x \in R^2$ and all $t \in R$. Note that the bound ρ_1 on the uncertainty Δ_1 is allowed to depend only on the state x_1; this is the structural condition suggested in the previous section and will be characterized more completely below. We will take a recursive approach to the design of a robust controller for this system. This approach is based on the *integrator backstepping* technique developed by [11] for the adaptive control of nonlinear systems. The first step in this approach is to consider the scalar system,

$$\dot{x}_1 = \bar{u} + \Delta_1(x_1, \bar{u}, t), \tag{49.27}$$

which we obtain by treating the state variable x_2 in Equation 49.23 as a control variable $\bar{u}$. This new system (Equation 49.27) is only conceptual; its relationship to the actual system (Equations 49.23–49.24) will be explored later. Let us next design a robust controller $\bar{u} = \mu(x_1)$ for this conceptual system. By construction, this new system satisfies the matching condition, and so we may use the Lyapunov redesign method to construct the feedback $\bar{u} = \mu(x_1)$. The nominal system is simply $\dot{x}_1 = \bar{u}$ which can be stabilized by a nominal feedback $\bar{u}_{nom} = -x_1$. A suitable Lyapunov function for the nominal closed-loop system $\dot{x}_1 = -x_1$ would be $V_1(x_1) = x_1^2$. We then choose $\bar{u} = \bar{u}_{nom} + \bar{u}_{rob}$, where $\bar{u}_{rob}$ is given, for example, by Equation 49.9 with $\bar{\rho} = \rho_1$. The resulting feedback function for $\bar{u}$ is

$$\mu(x_1) = -x_1 - \rho_1(x_1)\mathrm{sgn}(x_1). \tag{49.28}$$

If we now apply the feedback $\bar{u} = \mu(x_1)$ to the conceptual system (Equation 49.27), we achieve $\dot{V}_1 \leq -2x_1^2$ and thus guarantee stability for every admissible uncertainty Δ_1. Let us assume for now that this function μ is (sufficiently) smooth; we will return to the question of smoothness in Section 49.5.

The idea behind the backstepping approach is to use the conceptual controller (Equation 49.28) in constructing a control Lyapunov function V for the actual system (Equations 49.23–49.24). If we treat the Lyapunov function as a penalty function, it is reasonable to include in V a term penalizing the difference between the state x_2 and the conceptual control $\bar{u}$ designed for the conceptual system (Equation 49.27):

$$V(x) = V_1(x_1) + [x_2 - \mu(x_1)]^2. \tag{49.29}$$

We have seen in the previous section that this choice satisfies the necessary condition (Equation 49.12) and is thus a good candidate for our system (Equations 49.23–49.24). It is no coincidence that we arrived at the same expression for μ in Equations 49.22 and 49.28 both from the viewpoint of the necessary condition (Equation 49.12) and from the viewpoint of the conceptual system (Equation 49.27).

Let us now verify that the choice of Equation 49.29 for V leads to a robust controller design for the system (Equations 49.23–49.24). Computing $\dot{V}$ we obtain

$$\dot{V} = 2x_1[x_2 + \Delta_1(x,t)] + 2[x_2 - \mu(x_1)]\left[u + \Delta_2(x,t) - \mu'(x_1)[x_2 + \Delta_1(x,t)]\right] \tag{49.30}$$

where $\mu' := d\mu/dx$. Rearranging terms and using Equation 49.28,

$$\dot{V} \leq -2x_1^2 + 2[x_2 - \mu(x_1)]\left[x_1 + u + \Delta_2(x,t) - \mu'(x_1)[x_2 + \Delta_1(x,t)]\right] \tag{49.31}$$

The effect of backstepping is that the uncertainties enter the expression (Equation 49.31) with the same coefficient as the control variable u; in other words, the uncertainties effectively satisfy the matching condition. As a result, we can again apply the Lyapunov redesign technique; the feedback control $u = u_{nom} + u_{rob}$ with a nominal control,

$$u_{nom}(x) = -\left[x_2 - \mu(x_1)\right] - x_1 + \mu'(x_1)x_2, \tag{49.32}$$

yields

$$\dot{V} \leq -2x_1^2 - 2\left[x_2 - \mu(x_1)\right]^2 + 2\left[x_2 - \mu(x_1)\right]\left[u_{rob} + \Delta_2(x,t) - \mu'(x_1)\Delta_1(x,t)\right]. \tag{49.33}$$

We may complete the design by choosing u_{rob} as in Equation 49.9:

$$u_{rob}(x) = -\bar{\rho}(x)\text{sgn}\left[x_2 - \mu(x_1)\right] \tag{49.34}$$

where $\bar{\rho}$ is some function satisfying $| \Delta_2(x,t) - \mu'(x_1)\Delta_1(x,t) | \leq \bar{\rho}(x)$. This yields

$$\dot{V} \leq -2x_1^2 - 2\left[x_2 - \mu(x_1)\right]^2 \tag{49.35}$$

for all admissible uncertainties Δ_1 and Δ_2, and thus robust stability is guaranteed.

Example 49.1:

The above second-order design applies to the following system:

$$\dot{x}_1 = x_2 + \Delta_1(x,t) \tag{49.36}$$

$$\dot{x}_2 = u \tag{49.37}$$

where we let the uncertainty Δ_1 be any function satisfying $| \Delta_1(x,t) | \leq |x_1|^3$. In this case, the function μ in Equation 49.28 would be $\mu(x_1) = -x_1 - x_1^3$ which is smooth as required. The nominal control u_{nom} is given by Equation 49.32, which, for this example, becomes

$$u_{nom}(x) = -\left[x_2 + x_1 + x_1^3\right] - x_1 - (1 + 3x_1^2)x_2 \tag{49.38}$$

Adding the robustifying term (Equation 49.34), we obtain the following robust controller:

$$u(x) = -\left[x_2 + x_1 + x_1^3\right] - x_1 - \left(1 + 3x_1^2\right)x_2 - \left(| x_1 |^3 + 3 | x_1 |^5\right)\text{sgn}\left(x_2 + x_1 + x_1^3\right). \tag{49.39}$$

This robust controller is not continuous at points where $x_2 + x_1 + x_1^3 = 0$; an alternate smooth design will be proposed in Section 49.5.

The above controller design for the system (Equations 49.23 and 49.24) is a two-step design. In the first step, we considered the scalar system (Equation 49.27) and designed a controller $\bar{u} = \mu(x_1)$ which guaranteed robust stability. In the second step, we used the Lyapunov function (Equation 49.29) to construct a controller (Equations 49.32 + 49.34) for the actual system (Equations 49.23 and 49.24). This step-by-step design can be repeated to obtain controllers for higher order systems. For example, suppose that instead of the system (Equations 49.23 and 49.24), we have the system

$$\dot{x}_1 = x_2 + \Delta_1(x,t) \tag{49.40}$$

$$\dot{x}_2 = z + \Delta_2(x,t)$$

$$\dot{z} = \upsilon + \Delta_3(x,z,t) \tag{49.41}$$

where Δ_1 and Δ_2 are as in Equations 49.25 and 49.26, Δ_3 is a new uncertainty, and υ is the control variable. We can use the controller $u(x) := u_{nom}(x) + u_{rob}(x)$ designed above in Equations 49.32 + 49.34 to obtain the following Lyapunov function W for our new system:

$$W(x,z) = V(x) + [z - u(x)]^2 \tag{49.42}$$

Here V is the Lyapunov function (Equation 49.29) used above, and we have simply added a term which penalizes the difference between the state variable z and the controller $u(x)$ designed above for the old system (Equations 49.23 and 49.24). If $u(x)$ is smooth, then Equation 49.42 qualifies as a candidate Lyapunov function for our new system; see Section 49.5 below for details on choosing a smooth function $u(x)$. We can now construct a controller $\upsilon = \upsilon(x, z)$ for our new system in the same manner as the construction of $\mu(x_1)$ and $u(x)$ above.

We have thus obtained a systematic method for constructing Lyapunov functions and robust controllers for systems of the form,

$$\dot{x}_1 = x_2 + \Delta_1(x, t), \tag{49.43}$$

$$\dot{x}_2 = x_3 + \Delta_2(x, t),$$

$$\vdots$$

$$\dot{x}_n = u + \Delta_n(x, t). \tag{49.44}$$

The Lyapunov function will be of the form

$$V(x) = x_1^2 + \sum_{i=1}^{n-1} \left[x_{i+1} - \mu_i(x_1, \ldots, x_i)\right]^2 \tag{49.45}$$

where the functions μ_i are constructed according to the recursive procedure described above. For this approach to succeed, it is sufficient that the uncertainties Δ_i satisfy the following *strict feedback condition*:

$$| \Delta_i(x, t) | \leq \rho_i(x_1, \ldots, x_i) \tag{49.46}$$

for known functions ρ_i. The restriction here is that the ith bound ρ_i can depend only on the first i states $(x_1, \ldots, x_i)$.

This recursive Lyapunov design technique applies to uncertain systems more general than Equations 49.43–49.44. Multi-input versions are possible, and the strict feedback condition (Equation 49.46) can be relaxed to allow the bound ρ_i to depend also on the state x_{i+1}. In particular, the bound ρ_n on the last uncertainty Δ_n can also depend on the control variable u. More details can be found in [5,14,16,17].

49.5 Smooth Control Laws

The control law μ_i designed in the ith step of the recursive design becomes part of the Lyapunov function (Equation 49.45) in the next step. It is imperative, therefore, that each such function μ_i be differentiable. To illustrate how smooth functions can be obtained at each step, let us return to the second-order design in Section 49.4. The first step was to design a robust controller $\bar{u} = \mu(x_1)$ for the conceptual system,

$$\dot{x}_1 = \bar{u} + \Delta_1(x_1, \bar{u}, t) \tag{49.47}$$

with $| \Delta_1 | \leq \rho_1(x_1)$. In general, when $\rho_1(0) \neq 0$, we cannot choose μ as in Equation 49.28 because of the discontinuity at $x_1 = 0$. One alternative is to approximate Equation 49.28 by smooth function as follows:

$$\mu(x_1) = -x_1 - \rho_1(x_1) \frac{x_1}{| x_1 | + \delta(x_1)} \tag{49.48}$$

where $\delta(x_1)$ is a smooth, strictly positive function. This choice for μ is once differentiable, and it reduces to the discontinuous function (Equation 49.28) when $\delta \equiv 0$. We compute the derivative of $V_1(x_1) = x_1^2$

for the system (Equation 49.47) with the smooth feedback (Equation 49.48):

$$\begin{aligned}\dot{V} &\leq -2x_1^2 - 2\rho_1(x_1)\frac{x_1^2}{|x_1| + \delta(x_1)} + 2\rho_1(x_1)\,|x_1| \\ &\leq -2x_1^2 + 2\rho_1(x_1)\frac{\delta(x_1)\,|x_1|}{|x_1| + \delta(x_1)} \\ &\leq -2x_1^2 + 2\rho_1(x_1)\delta(x_1)\end{aligned} \tag{49.49}$$

If we now choose $\delta(x_1)$ so that $\rho_1\delta$ is small, we see that $\dot{V}_1 < 0$ except in a small neighborhood of $x_1 = 0$.

In the second design step, we will choose u_{nom} as before in Equation 49.32, but instead of Equation 49.33 we obtain the Lyapunov derivative,

$$\dot{V} \leq -2x_1^2 + 2\rho_1(x_1)\delta(x_1) - 2\,[x_2 - \mu(x_1)]^2 + 2\,[x_2 - \mu(x_1)]\left[u_{rob} + \Delta_2(x,t) - \mu'(x_1)\Delta_1(x,t)\right] \tag{49.50}$$

where the extra term $2\rho_1\delta$ comes from Equation 49.49 and is a result of our smooth choice for $\mu(x_1)$. Our remaining task is to construct the robustifying term u_{rob}. Using the bound $|\Delta_2(x,t) - \mu'(x_1)\Delta_1(x,t)| \leq \bar{\rho}(x)$ as before, we obtain

$$\dot{V} \leq -2x_1^2 + 2\rho_1(x_1)\delta(x_1) - 2\,[x_2 - \mu(x_1)]^2 + 2\,[x_2 - \mu(x_1)]\,u_{rob} + 2\,|x_2 - \mu(x_1)|\,\bar{\rho}(x) \tag{49.51}$$

We cannot choose u_{rob} as before in Equation 49.34 because it is not continuous at points where $x_2 = \mu(x_1)$. We could choose a smooth approximation to Equation 49.34, as we did above for the function μ, but, to illustrate an alternative approach, we will instead make use of Young's inequality,

$$2ab \leq \frac{1}{\varepsilon}a^2 + \varepsilon b^2 \tag{49.52}$$

which holds for any $a, b \in R$ and any $\varepsilon > 0$. Applying this inequality to the last term in Equation 49.51, we obtain

$$\dot{V} \leq -2x_1^2 + 2\rho_1(x_1)\delta(x_1) - \left[2 - \frac{1}{\varepsilon(x)}\right][x_2 - \mu(x_1)]^2 + 2\,[x_2 - \mu(x_1)]\,u_{rob} + \varepsilon(x)\,[\bar{\rho}(x)]^2 \tag{49.53}$$

where $\varepsilon(x)$ is a smooth, strictly positive function to be chosen below. Thus

$$u_{rob}(x) = -\frac{1}{2\varepsilon(x)}\,[x_2 - \mu(x_1)] \tag{49.54}$$

is smooth and yields

$$\dot{V} \leq -2x_1^2 - 2\,[x_2 - \mu(x_1)]^2 + 2\rho_1(x_1)\delta(x_1) + \varepsilon(x)\,[\bar{\rho}(x)]^2\,. \tag{49.55}$$

It is always possible to choose $\delta(x_1)$ and $\varepsilon(x)$ so that the right-hand side of Equation 49.55 is negative, except possibly in a neighborhood of $x = 0$. Thus we have gained smoothness in the control law but lost exact convergence of the state to zero.

49.6 Design Flexibilities

The degrees of freedom in the recursive Lyapunov design procedure outlined above are numerous and allow for the careful shaping of the closed-loop performance. However, this procedure is new, and guidelines for exploiting design flexibilities are only beginning to appear. Our purpose in this section is to point out several of these degrees of freedom and discuss the consequences of various design choices.

We have already seen that the choices for the functions μ_i in the Lyapunov function (Equation 49.45) are by no means unique, nor is it the final choice for the control law. For example, when choosing a robustifying term u_{rob} in some step of the design, should we choose a smooth approximation to Equation 49.9 as in Equation 49.48, or should we use Young's inequality and choose Equation 49.54? Also, how should we choose the nominal term u_{nom}? Is it always good to cancel nonlinearities and apply linearlike feedback as in Equation 49.32, and if so, which gain should we use? Such questions are difficult in general, but there are some guidelines that apply in many situations. For example, consider the task of robustly stabilizing the point $x = 0$ of the simple scalar system,

$$\dot{x} = -x^3 + u + \Delta(x, t) \tag{49.56}$$

where Δ is an uncertain function with a known bound $| \Delta(x, t) | \leq | x |$. Because the matching condition is satisfied, we can apply the Lyapunov redesign method and choose $u = u_{nom} + u_{rob}$. One choice for the nominal control would be $u_{nom}(x) = x^3 - x$, which, together with a robustifying term as in Equation 49.9 yields a control law

$$u(x) = x^3 - 2x \tag{49.57}$$

This control law is valid from the viewpoint of Lyapunov redesign, and it indeed guarantees robustness to the uncertainty Δ. In such a choice, however, large positive feedback x^3 is used to cancel the nonlinearity $-x^3$ in Equation 49.56. This is absurd because the nonlinearity $-x^3$ is *beneficial* for stabilization, and positive feedback x^3 will lead to unnecessarily large control effort and will cause robustness problems more severe than those caused by the uncertainty Δ. Clearly, a much more reasonable choice is $u(x) = -2x$, but how can we identify better choices in a more general setting? One option would be to choose the control to minimize some meaningful cost functional. For example, the control

$$u(x) = x^3 - x - x\sqrt{x^4 - 2x^2 + 2} \tag{49.58}$$

minimizes the worst-case cost functional

$$J = \sup_{|\Delta| \leq |x|} \int_0^\infty \left[x^2 + u^2\right] dt \tag{49.59}$$

for this system Equation 49.56. The two control laws (Equations 49.57 and 49.58) are shown in Figure 49.1. We see that the optimal control (Equation 49.58) recognizes the benefit of the nonlinearity $-x^3$ in Equation 49.56 and accordingly produces little control effort for large x. Moreover, this optimal control is never positive feedback. Unfortunately, the computation of the optimal control (Equation 49.58) requires the solution of a Hamilton–Jacobi–Isaacs partial differential equation, and will be difficult and expensive for all but the simplest problems.

As a compromise between the benefits of optimality and its computational burden, we might consider the *inverse* optimal control problem, summarized for example by [7]. In this approach, we start with a Lyapunov function as in the Lyapunov redesign method above. We then show that this Lyapunov function is in fact the value function for some meaningful optimal stabilization problem, and we use this information to compute the corresponding optimal control. Freeman and Kokotović [6] have shown that the pointwise solutions of static minimization problems of the form,

$$\text{minimize} \quad u^T S u, \quad S = S^T > 0, \tag{49.60}$$

$$\text{subject to} \quad \sup_{\Delta} \left[\dot{V}(x, u, \Delta) + \sigma(x)\right] \leq 0, \tag{49.61}$$

yield optimal controllers (in this inverse sense) for meaningful cost functionals, where V is a given control Lyapunov function and σ is a positive function whose choice represents a degree of freedom. When the system is jointly affine in the control u and the uncertainty Δ, this optimization problem

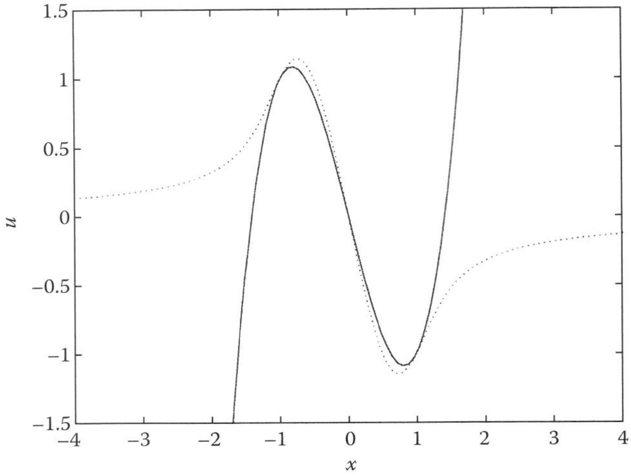

FIGURE 49.1 Comparison of the control laws Equation 49.57 (solid) and Equation 49.58 (dotted).

(Equations 49.60 and 49.61) is a quadratic program with linear constraints and thus has an *explicit* solution $u(x)$. For example, the solution to Equations 49.60 and 49.61 for the system (Equation 49.56) with $V = x^2$ and $\sigma = 2x^2$ yields the control law,

$$u(x) = \begin{cases} x^3 - 2x & \text{when} \quad x^2 \leq 2, \\ 0 & \text{when} \quad x^2 \geq 2. \end{cases} \tag{49.62}$$

As shown in Figure 49.2, this control (Equation 49.62) is qualitatively the same as the optimal control (Equation 49.58); both recognize the benefit of the nonlinearity $-x^3$ and neither one is ever positive feedback. The advantage of the inverse optimal approach is that the controller computation is simple once a control Lyapunov function is known.

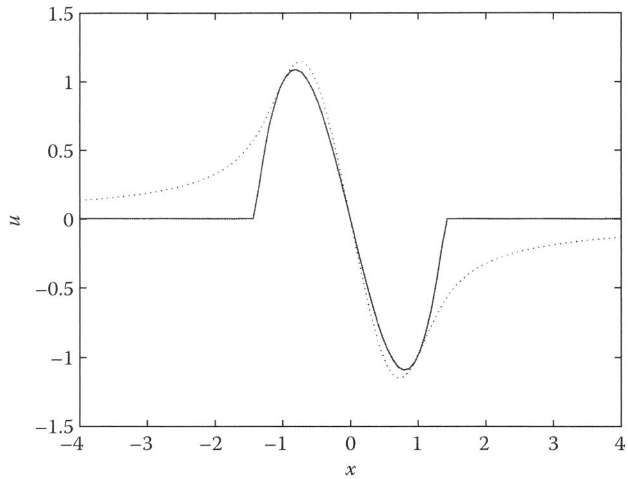

FIGURE 49.2 A comparison of the control laws Equation 49.62 (solid) and Equation 49.58 (dotted).

We thus have some guidelines for choosing the intermediate control laws μ_i at each step of the design, namely, we can avoid the wasteful cancellations of beneficial nonlinearities. However, these guidelines, when used at early steps of the recursive design, have not yet been proved beneficial for the final design.

We have shown that the choices for the functions μ_i in Equation 49.45 represent important degrees of freedom in the recursive Lyapunov design procedure. Perhaps even more important is the choice of the form of V itself. Recall that, given a Lyapunov function V_i and a control μ_i at the ith design step, we constructed the new Lyapunov function V_{i+1} as follows:

$$V_{i+1}(x_1,\dots,x_{i+1}) = V_i(x_1,\dots,x_i) + [x_{i+1} - \mu_i(x_1,\dots,x_i)]^2. \tag{49.63}$$

This choice for V_{i+1} is not the only choice that will lead to a successful design, and we are thus confronted with another degree of freedom in the design procedure. For example, instead of Equation 49.63, we can choose

$$V_{i+1}(x_1,\dots,x_{i+1}) = \kappa\,[V_i(x_1,\dots,x_i)] + [x_{i+1} - \mu_i(x_1,\dots,x_i)]^2. \tag{49.64}$$

where $\kappa : R_+ \to R_+$ is any smooth, positive-definite, unbounded function whose derivative is everywhere strictly positive. This function κ represents a nonlinear weighting on the term V_i and can have a large effect on the control laws obtainable in future steps.

The last degree of freedom we wish to discuss involves the quadratic-like term in Equations 49.63 and 49.64. Praly et al. [15] have shown that Equation 49.64 can be replaced by the more general expression

$$V_{i+1}(x_1,\dots,x_{i+1}) = \kappa\,[V_i(x_1,\dots,x_i)] + \int_{\mu_i(x_1,\dots,x_i)}^{x_{i+1}} \phi(x_1,\dots,x_i,s)\,ds \tag{49.65}$$

for a suitable choice of the function ϕ. Indeed, Equation 49.64 is a special case of Equation 49.65 for $\phi = 2[s - \mu_i(x_1,\dots,x_i)]$. This degree of freedom in the choice of ϕ can be significant; for example, it allowed extensions of the recursive design to the nonsmooth case by [15]. It can also be used to reduce the unnecessarily large control gains often caused by the quadratic term in Equation 49.63. To illustrate this last point, let us return to the second-order example (Equations 49.36 and 49.37) given by

$$\dot{x}_1 = x_2 + \Delta_1(x,t) \tag{49.66}$$

$$\dot{x}_2 = u \tag{49.67}$$

where the uncertainty Δ_1 is any function satisfying $|\,\Delta_1(x,t)\,| \le |\,x_1\,|^3$. Recall that using the Lyapunov function,

$$V(x) = x_1^2 + \left[x_2 + x_1 + x_1^3\right]^2 \tag{49.68}$$

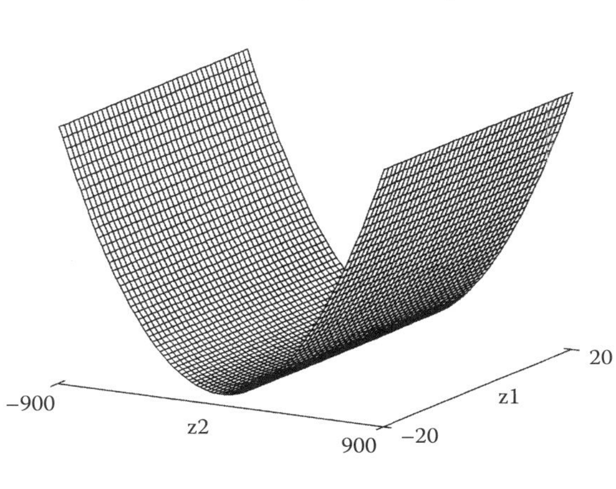

FIGURE 49.3 Quadratic-like Lyapunov function (Equation 49.68). (From Freeman, R. A. and Kokotović, P. V., *Automatica*, 29(6), 1425–1437, 1993. With permission.)

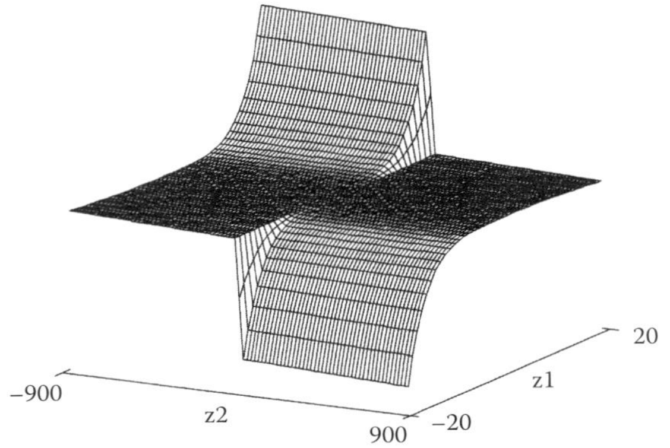

FIGURE 49.4 Controller from quadratic-like Lyapunov function (Equation 49.68). (From Freeman, R. A. and Kokotović, P. V., *Automatica* 29(6), 1425–1437, 1993. With permission.)

we designed the following robust controller (Equation 49.39) for this system:

$$u(x) = -\left[x_2 + x_1 + x_1^3\right] - x_1 - \left(1 + 3x_1^2\right) x_2 - \left(|x_1|^3 + 3|x_1|^5\right) \operatorname{sgn}\left(x_2 + x_1 + x_1^3\right). \tag{49.69}$$

This controller is not continuous at points where $x_2 + x_1 + x_1^3 = 0$. In other words, the local controller gain $\partial u/\partial x_2$ is *infinite* at such points. Such infinite gain will cause high-amplitude chattering along the manifold M described by $x_2 + x_1 + x_1^3 = 0$. As a result of such chattering, this control law may not be implementable because of the unreasonable demands on the actuator. However, as was shown in Section 49.5, we can use this same Lyapunov function (Equation 49.68) to design a *smooth* robust controller $\bar{u}$ for our system. Will such a smooth controller eliminate the chattering caused by the discontinuity in Equation 49.69? Surprisingly, the answer is no. One can show that the local controller gain $\partial \bar{u}/\partial x_2$, although finite because of the smoothness of $\bar{u}$, grows like x_1^6 along the manifold M. Thus for large signals, this local gain is extremely large and can cause chattering as if it were infinite. Figure 49.3 shows the Lyapunov function V in Equation 49.68, plotted as a function of the two variables, $z_1 := x_1$ and $z_2 := x_2 + x_1 + x_1^3$. A smooth control law $\bar{u}$ designed using this V is shown in Figure 49.4, again plotted as a function of z_1 and z_2. Note that the x_1^6 growth of the local gain of $\bar{u}$ along the manifold $z_2 = 0$ is clearly visible in this figure. One might conclude that such high gain is unavoidable for this particular system. This conclusion is false, however, because the x_1^6 growth of the local gain is an artifact of the *quadratic*

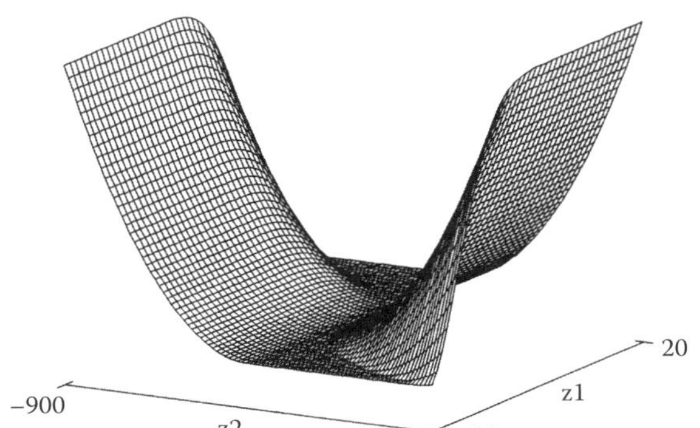

FIGURE 49.5 New flattened Lyapunov function. (From Freeman, R. A. and Kokotović, P. V., *Automatica*, 29(6), 1425–1437, 1993. With permission.)

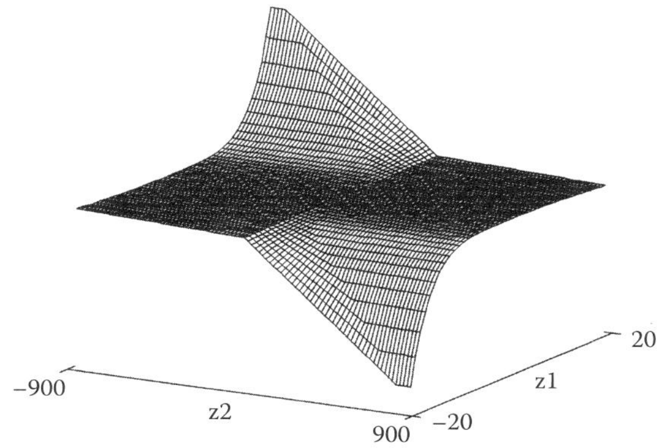

FIGURE 49.6 Controller from flattened Lyapunov function. (From Freeman, R. A. and Kokotović, P. V., *Automatica,* 29(6), 1425–1437, 1993. With permission.)

form of the Lyapunov function (Equation 49.68) and has nothing to do with robust stability requirements for this system. Freeman and Kokotović [6] have shown how to choose the function ϕ in Equation 49.65 to reduce greatly the growth of the local gain. For this example (Equations 49.66 and 49.67), they achieved a growth of x_1^2 as opposed to the x_1^6 growth caused by the quadratic V in Equation 49.68. Their new, *flattened* Lyapunov function is shown in Figure 49.5, and the corresponding control law is shown in Figure 49.6. The x_1^2 versus x_1^6 growth of the local gain is evident when comparing Figures 49.6 and 49.4. The control signals generated from a particular initial condition are compared in Figure 49.7. These controls produce virtually the same state trajectories, but the chattering caused by the old control law has been eliminated by the new one.

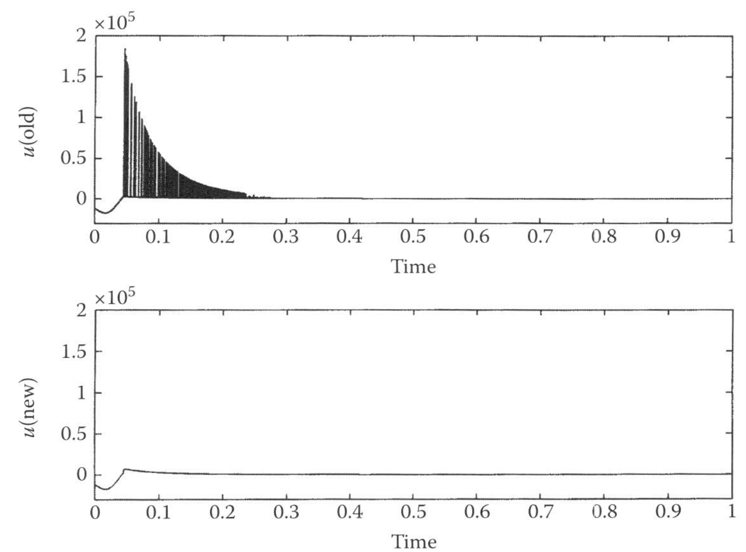

FIGURE 49.7 Comparison of control signals. (From Freeman, R. A. and Kokotović, P. V., *Automatica,* 29(6), 1425–1437, 1993. With permission.)

References

1. Artstein, Z., Stabilization with relaxed controls, *Nonlinear Anal.* 7(11), 1163–1173, 1983.
2. Barmish, B. R., Corless, M. J., and Leitmann, G., A new class of stabilizing controllers for uncertain dynamical systems, *SIAM J. Cont. Optimiz.*, 21, 246–255, 1983.
3. Corless, M. J., Robust stability analysis and controller design with quadratic Lyapunov functions, in *Variable Structure and Lyapunov Control,* Zinober, A., Ed., Springer, Berlin, 1993.
4. Corless, M. J. and Leitmann, G., Continuous state feedback guaranteeing uniform ultimate boundedness for uncertain dynamic systems, *IEEE Trans. Automat. Control,* 26(5), 1139–1144, 1981.
5. Freeman, R. A. and Kokotović, P. V., Design of 'softer' robust nonlinear control laws, *Automatica* 29(6), 1425–1437, 1993.
6. Freeman, R. A. and Kokotović, P. V., Inverse optimality in robust stabilization, *SIAM J. Control Optimiz.*, to appear revised November 1994.
7. Glad, S. T., Robustness of nonlinear state feedback — A survey, *Automatica,* 23(4), 425–435, 1987.
8. Gutman, S., Uncertain dynamical systems — Lyapunov min-max approach, *IEEE Trans. Automat. Control,* 24, 437–443, 1979.
9. Jacobson, D. H., *Extensions of Linear-Quadratic Control, Optimization and Matrix Theory,* Academic, London, 1977.
10. Jurdjevic, V. and Quinn, J. P., Controllability and stability, *J. Diff. Eq.*, 28, 381–389, 1978.
11. Kanellakopoulos, I., Kokotović, P. V., and Morse, A. S., Systematic design of adaptive controllers for feedback linearizable systems, *IEEE Trans. Automat. Control,* 36(11), 1241–1253, 1991.
12. Khalil, H. K., *Nonlinear Systems,* Third Edition, Prentice-Hall, Upper Saddle River, 2002.
13. Krasovsky, A. A., A new solution to the problem of a control system analytical design, *Automatica,* 7, 45–50, 1971.
14. Marino, R. and Tomei, P., Robust stabilization of feedback linearizable time-varying uncertain nonlinear systems, *Automatica,* 29(1), 181–189, 1993.
15. Praly, L., d'Andréa Novel, B., and Coron, J.-M., Lyapunov design of stabilizing controllers for cascaded systems, *IEEE Trans. Automat. Control,* 36(10), 1177–1181, 1991.
16. Qu, Z., Robust control of nonlinear uncertain systems under generalized matching conditions, *Automatica,* 29(4), 985–998, 1993.
17. Slotine, J. J. E. and Hedrick, K., Robust input–output feedback linearization, *Int. J. Control,* 57, 1133–1139, 1993.
18. Sontag, E. D., A Lyapunov-like characterization of asymptotic controllability, *SIAM J. Control Optimiz.*, 21(3), 462–471, 1983.

Further Reading

Contributions to the development of the Lyapunov design methodology for systems with no uncertainties include [1,9,10,13,18]. A good introduction to Lyapunov redesign can be found [12]. Corless [3] has recently surveyed various methods for the design of robust controllers for nonlinear systems using *quadratic* Lyapunov functions. Details of the recursive design presented in Section 49.4 can be found in [6, 14,16]. The state-space techniques discussed in this chapter can be combined with nonlinear input/output techniques to obtain more advanced designs (see Chapter 44). Finally, when the uncertain nonlinearities are given by *constant* parameters multiplying *known* nonlinearities, adaptive control techniques apply (see Chapter 53).

50

Variable Structure, Sliding-Mode Controller Design

Raymond A. DeCarlo
Purdue University

S.H. Żak
Purdue University

Sergey V. Drakunov
Embry-Riddle Aeronautical University

50.1 Introduction and Background

This chapter investigates variable structure control (VSC) as a high-speed switched feedback control resulting in a sliding mode. For example, the gains in each feedback path switch between two values according to a rule that depends on the value of the state at each time instant. The purpose of the switching control law is to drive the plant's state trajectory onto a prespecified (user-chosen) surface in the state space and to maintain the plant's state trajectory on this surface for all subsequent time. This surface is called a *switching surface* and the resulting motion of the state trajectory a *sliding mode*. When the plant state trajectory is "above" the surface, a feedback path has one gain and a different gain if the trajectory drops "below" the surface. This surface defines the rule for proper switching. The surface is also called a *sliding surface* (sliding manifold) because, ideally speaking, once intercepted, the switched control maintains the plant's state trajectory on the surface for all subsequent time and the plant's state trajectory then slides along this surface. The plant dynamics restricted to this surface represent the controlled system's behavior. The first critical phase of a VSC design is to properly construct a switching surface, so that the plant, restricted to the surface, has desired dynamics, such as stability to the origin, tracking, regulation, and so on.

In summary, a VSC control design generally breaks down into two phases. The first phase is to design or choose a sliding manifold/switching surface, so that the plant state restricted to the surface has desired dynamics. The second phase is to design a switched control that will drive the plant state to the switching surface and maintain it on the surface upon interception. A Lyapunov approach is used in this chapter to characterize this second design phase. Here, a generalized Lyapunov function, which characterizes the motion of the state trajectory to the surface, is specified in terms of the surface. For each chosen switched control structure, one chooses the "gains," so that the derivative of this Lyapunov function is negative definite with respect to the sliding surface, thus guaranteeing motion of the state trajectory to the surface.

As an introductory example, consider the first-order system $\dot{x}(t) = u(x,t)$ with control

$$u(x,t) = -\text{sgn}(x) = \begin{cases} -1, & \text{if } x > 0, \\ +1, & \text{if } x < 0. \end{cases}$$

Hence, the system with control satisfies $\dot{x} = -\text{sgn}(x)$ with trajectories plotted in Figure 50.1a. Here the control $u(x,t)$ switches, changing its value between ± 1 around the surface $\sigma(x,t) = x = 0$. Hence, for any initial condition x_0, a finite time t_1 exists for which $x(t) = 0$ for all $t \geq t_1$. Now, suppose $\dot{x}(t) = u(x,t) + v(t)$, where again $u(x,t) = -\text{sgn}(x)$ and $v(t)$ is a bounded disturbance for which $\sup_t |v(t)| < 1$. As before, the control $u(x,t)$ switches its value between ± 1 around the surface $\sigma(x,t) = x = 0$. It follows that if $x(t) > 0$, then $\dot{x}(t) = -\text{sgn}[x(t)] + v(t) < 0$, forcing motion toward the line $\sigma(x,t) = x = 0$, and if $x(t) < 0$, then $\dot{x}(t) = -\text{sgn}[x(t)] + v(t) > 0$, again forcing motion toward the line $\sigma(x,t) = x = 0$. For a positive initial condition, this is illustrated in Figure 50.1b. The rate of convergence to the line depends on the disturbance. Nevertheless, a finite time t_1 exists for which $x(t) = 0$ for all $t \geq t_1$. If the disturbance

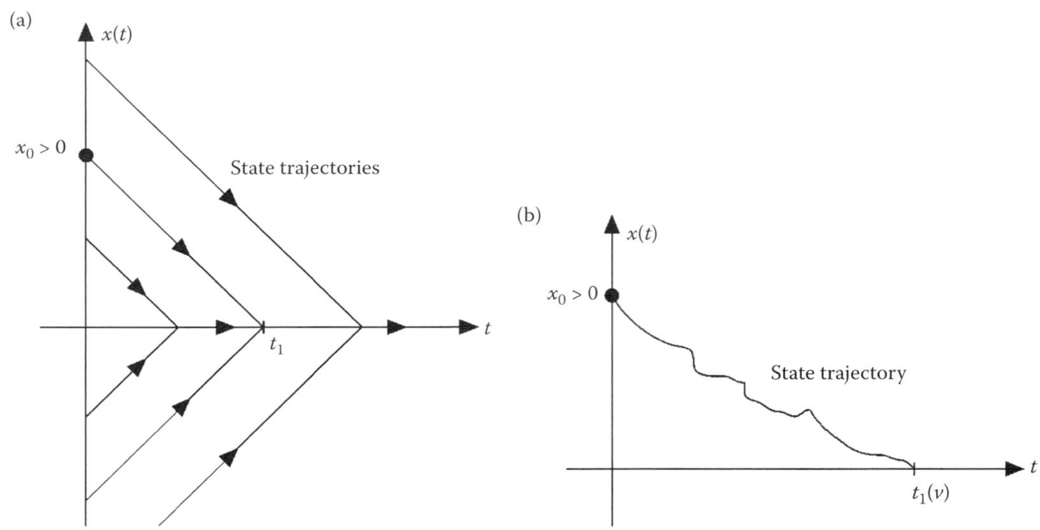

FIGURE 50.1 (a) State trajectories for the system $\dot{x} = -\text{sgn}(x)$; (b) State trajectory for the system $\dot{x}(t) = -\text{sgn}[x(t)] + v(t)$.

magnitude exceeds 1, then the gain can always be adjusted to compensate for the change. Hence, this VSC law is robust in the face of bounded disturbances, illustrating the simplicity and advantage of the VSC technique.

From the above example, one can see that VSC can provide a robust means of controlling (nonlinear) plants with disturbances and parameter uncertainties. Further, the advances in computer technology and high-speed switching circuitry have made the practical implementation of VSC quite feasible and of increasing interest. Indeed, pulse-width modulation control and switched dc-to-dc power converters [1] can be viewed in a VSC framework.

50.2 System Model, Control Structure, and Sliding Modes

50.2.1 System Model

The class of systems investigated herein has a state model nonlinear in the state vector $x(\cdot)$ and linear in the control vector $u(\cdot)$ of the form

$$\dot{x}(t) = F(x,t,u) = f(x,t) + B(x,t)u(x,t), \tag{50.1}$$

where $x(t) \in R^n$, $u(t) \in R^m$, and $B(x,t) \in R^{n\times m}$; further, each entry in $f(x,t)$ and $B(x,t)$ is assumed continuous with a bounded continuous derivative with respect to x. In the linear time-invariant case, Equation 50.1 becomes

$$\dot{x} = Ax + Bu \tag{50.2}$$

with A $n \times n$ and B $n \times m$ being constant matrices.

As mentioned in the previous section, the first phase of VSC or sliding mode control (SMC) design is to choose a manifold $S \subset R^n$, so that the control goal is reached once the state is maintained on S. As such we formally define the $(n-m)$-dimensional switching surface (also called a *discontinuity, sliding manifold*, or *equilibrium manifold*), as (the possibly time-varying)

$$S = \{x \in R^n | \sigma(x,t) = 0\} = \bigcap_{i=1}^{n} \{x \in R^n | \sigma_i(x,t) = 0\}, \tag{50.3}$$

where

$$\sigma(x,t) = [\sigma_1(x,t), \ldots, \sigma_m(x,t)]^T = 0. \tag{50.4}$$

(We will often refer to S as $\sigma(x,t) = 0$.) When there is no t-dependence, this $(n-m)$-dimensional manifold $S \subset R^n$ is determined as the intersection of m $(n-1)$-dimensional surfaces $\sigma_i(x,t) = 0$. These surfaces are designed in such a way that the system state trajectory, restricted to $\sigma(x,t) = 0$, has a desired behavior such as stability or tracking. Although general nonlinear time-varying surfaces as in Equation 50.3 are possible, linear ones are more prevalent in design [2–6]. Linear surface design is presented in Section 50.4.

50.2.2 Control Structure

After proper design of the surface, a controller $u(x,t) = [u_1(x,t), \ldots, u_m(x,t)]^{\mathrm{T}}$ is constructed, which generally has a switched form

$$u_i(x,t) = \begin{cases} u_i^+(x,t), & \text{when } \sigma_i(x,t) > 0, \\ u_i^-(x,t), & \text{when } \sigma_i(x,t) < 0. \end{cases} \tag{50.5}$$

Equation 50.5 indicates that the control changes its value depending on the sign of $\sigma(x,t)$. Here we can define that a discontinuity set, D, in the right-hand side is a union of the hypersurfaces defined by $\sigma_i(x,t) = 0$:

$$D = \bigcup_{i=1}^{n} \{x \in R^n | \sigma_i(x,t) = 0\}.$$

Thus, the (possibly t-dependent) hypersurfaces $\{x \in R^n | \sigma_i(x,t) = 0\}$ can be called switching surfaces and the functions $\sigma_i(x,t)$ switching functions. The goal of phase 2 is to stabilize the state to S. Off S, the control values $u_i^{\pm}$ are chosen so that the state trajectory converges to S in finite time, that is, the sliding mode exists on S, but the sliding mode may (or may not) also exist on some of the hypersurfaces $\{x \in R^n | \sigma_i(x,t) = 0\}$ while the state is converging to S.

50.2.3 Sliding Modes

The control $u(x,t)$ is designed in such a way that the system state trajectory is attracted to S and, once having intercepted S, remains there for all subsequent time; thus, the state trajectory can be viewed as sliding along S meaning that the system is in a sliding mode. A sliding mode exists if, in the vicinity of the switching surface, S, the tangent or velocity vectors of the state trajectory point toward the switching surface. If the state trajectory intersects the sliding surface, the value of the state trajectory or "representative point" remains within an ε-neighborhood of S. If a sliding mode exists on S, then S, or more commonly $\sigma(x,t) = 0$, is also termed a sliding surface. Note that interception of the surface $\sigma_i(x,t) = 0$ does not guarantee sliding on the surface for all subsequent time as illustrated in Figure 50.2, although this is possible.

An *ideal sliding mode* exists only when the state trajectory $x(t)$ of the controlled plant satisfies $\sigma(x(t),t) = 0$ at every $t \geq t_1$ for some t_1. This may require infinitely fast switching. In real systems, a switched controller has imperfections, such as delay, hysteresis, and so on, which limit switching to a finite frequency. The representative point then oscillates within a neighborhood of the switching surface. This oscillation, called *chattering* (discussed in a later section), is also illustrated in Figure 50.2. If the frequency of the switching is very high relative to the dynamic response of the system, the imperfections and the finite switching frequencies are often but not always negligible. The subsequent development focuses primarily on ideal sliding modes.

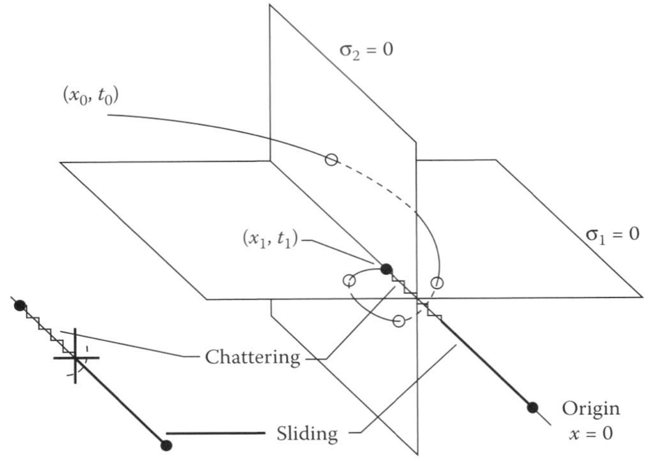

FIGURE 50.2 A situation in which a sliding mode exists on the intersection of the two indicated surfaces for $t \geq t_1$.

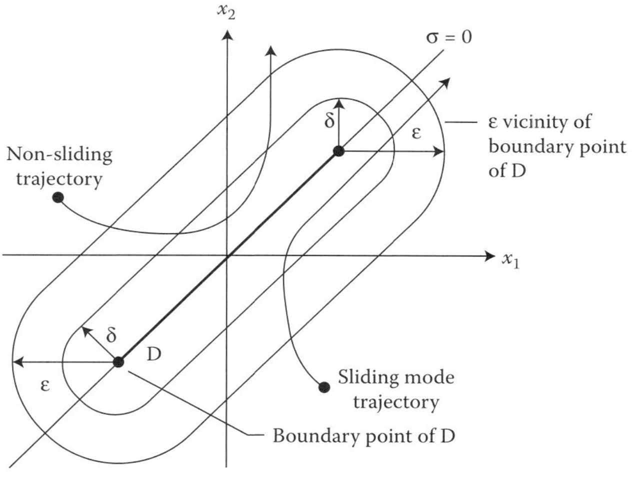

FIGURE 50.3 Two-dimensional illustration of the domain of a sliding mode.

50.2.4 Conditions for the Existence of a Sliding Mode

The existence of a sliding mode [2,5,6] requires stability of the state trajectory to the switching surface $\sigma(x,t) = 0$, that is, after some finite time t_1, the system representative point, $x(t)$, must be in some suitable neighborhood, $\{x|\ \|\sigma(x,t)\| < \varepsilon\}$, of S for suitable $\varepsilon > 0$. A domain, D, of dimension $n - m$ in the manifold, S, is a sliding-mode domain if, for each $\varepsilon>0$, there is a $\delta > 0$, so that any motion starting within an n-dimensional δ-vicinity of D may leave the n-dimensional ε-vicinity of D only through the n-dimensional ε-vicinity of the boundary of D as illustrated in Figure 50.3.

The *region of attraction* is the largest subset of the state space from which sliding is achievable. A sliding mode is globally reachable if the domain of attraction is the entire state space. The second method of Lyapunov provides the natural setting for a controller design leading to a sliding mode. In this effort one uses a generalized Lyapunov function, $V(t,x,\sigma)$, that is positive definite with a negative time derivative in the region of attraction.

Theorem 50.1: [5, p. 83]:

For the $(n-m)$-dimensional domain D to be the domain of a sliding mode, it is sufficient that in some n-dimensional domain $\Omega \supset D$, a function $V(t,x,\sigma)$ exists, continuously differentiable with respect to all of its arguments and satisfying the following conditions:

1. *$V(t,x,\sigma)$ is positive definite with respect to σ, that is, for arbitrary t and x, $V(t,x,\sigma) > 0$, when $\sigma \neq 0$ and $V(t,x,0) = 0$; on the sphere $||\sigma|| \leq \rho > 0$, for all $x \in \Omega$ and any t, the relationships*

$$\inf_{||\sigma||=\rho} V(t,x,\sigma) = h_\rho,\ h_\rho > 0 \quad \textit{and} \quad \sup_{||\sigma||=\rho} V(t,x,\sigma) = H_\rho,\ H_\rho > 0$$

hold, where h_ρ and H_ρ depend only on ρ with $h_\rho \neq 0$ if $\rho \neq 0$.

2. *The total time derivative of $V(t,x,\sigma)$ on the trajectories of the system of Equation 50.1 has a negative supremum for all $x \in \Omega$ except for x on the switching surface where the control inputs are undefined and the derivative of $V(t,x,\sigma)$ does not exist.*

In summary, two phases underlie VSC design. The first phase is to construct a switching surface $\sigma(x,t)=0$, so that the system restricted to the surface has a desired global behavior, such as stability, tracking, regulation, and so on. The second phase is to design a (switched) controller $u(x,t)$, so that away from the surface $\sigma(x,t)=0$, the tangent vectors of the state trajectories point toward the surface, that is, there is stability to the switching surface. This second phase is achieved by defining an appropriate Lyapunov function $V(t,x,\sigma)$, differentiating this function so that the control $u(x,t)$ becomes explicit, and adjusting controller gains so that the derivative is negative definite. The choice of $V(t,x,\sigma)$ determines the allowable controller structures. Conversely, a workable control structure has a set of possible Lyapunov functions to verify its viability. A later section discusses general control structures.

50.2.5 An Illustrative Example

To conclude this section, we present an illustrative example for a single pendulum system,

$$\dot{x}=A(x)x+Bu(x)=\begin{bmatrix}0 & 1\\ -\dfrac{\sin(x_1)}{x_1} & 0\end{bmatrix}\begin{bmatrix}x_1\\ x_2\end{bmatrix}+\begin{bmatrix}0\\ 1\end{bmatrix}u(x),$$

with a standard feedback control structure, $u(x)=k_1(x)x_1+k_2(x)x_2$, having nonlinear feedback gains switched according to the rule

$$k_i(x)=\begin{cases}\alpha_i(x), & \text{if } \sigma(x)x_i>0,\\ \beta_i(x), & \text{if } \sigma(x)x_i<0,\end{cases}$$

which depend on the linear switching surface $\left(\sigma(x)=[s_1\ s_2]x\right)$. Without loss of generality, assume $s_2>0$. For such single-input systems it is ordinarily convenient to choose a Lyapunov function of the form $V(t,x,\sigma)=0.5\sigma^2(x)$. To determine the gains necessary to drive the system state to the surface $\sigma(x)=0$, they may be chosen so that

$$\begin{aligned}\dot{V}(t,x,\sigma)&=0.5\frac{d\sigma^2}{dt}=\sigma(x)\frac{d\sigma(x)}{dt}=\sigma(x)[s_1\ s_2]\dot{x}\\ &=\sigma(x)x_1\left[s_2\left(k_1(x)-\frac{\sin(x_1)}{x_1}\right)\right]+\sigma(x)x_2\left[s_1+s_2k_2(x)\right]<0.\end{aligned}$$

This is satisfied whenever

$$\alpha_1(x)=\alpha_1<\min_{x_1}\left[\frac{\sin(x_1)}{x_1}\right]=-1,$$

$$\beta_1(x)=\beta_1>\max_{x_1}\left[\frac{\sin(x_1)}{x_1}\right]=1,$$

$\alpha_2<-(s_1/s_2)$ and $\beta_2>-(s_1/s_2)$. Thus, for properly chosen s_1 and s_2, the controller gains are readily computed.

This example proposed no methodology for choosing s_1 and s_2, that is, for designing the switching surface. Section 50.4 presents this topic. Further, this example was only single input. For the multi-input case, ease of computation of the control gains depends on a properly chosen Lyapunov function. For most cases, a quadratic Lyapunov function is adequate. This topic is discussed in Section 50.5.

50.3 Existence and Uniqueness of Solutions to VSC Systems

VSC produces system dynamics with discontinuous right-hand sides owing to the switching action of the controller. Thus they fail to satisfy conventional existence and uniqueness results of differential

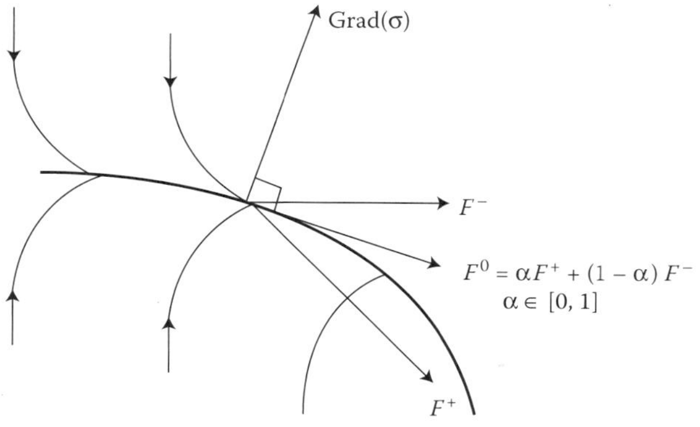

FIGURE 50.4 Illustration of the Filippov method of determining the desired velocity vector F^0 for the motion of the state trajectory on the sliding surface as per Equation 50.6.

equation theory. Nevertheless, an important aspect of VSC is the presumption that the plant behaves uniquely when restricted to $\sigma(x,t) = 0$. One of the earliest and conceptually straightforward approaches addressing existence and uniqueness is the method of Filippov [7]. The following briefly reviews this method in the two-dimensional, single-input case.

From Equation 50.1, $\dot{x}(t) = F(x,t,u)$ and the control $u(x,t)$ satisfy Equation 50.5. Filippov's work shows that the state trajectories of Equation 50.1 with control Equation 50.5 on the switching manifold Equation 50.3 solve the equation

$$\dot{x}(t) = \alpha F^+ + (1-\alpha)F^- = F^0, \quad 0 \leq \alpha \leq 1. \tag{50.6}$$

This is illustrated in Figure 50.4, where $F^+ = F(x,t,u^+)$, $F^- = F(x,t,u^-)$, and F^0 is the resulting velocity vector of the state trajectory in a sliding mode.

The problem is to determine α, which follows from solving the equation $\langle \text{grad}(\sigma), F^0 \rangle = 0$, where the notation $\langle a, b \rangle$ denotes the inner product of a and b, that is,

$$\alpha = \frac{\langle \text{grad}(\sigma), F^- \rangle}{\langle \text{grad}(\sigma), (F^- - F^+) \rangle},$$

provided:

1. $\langle \text{grad}(\sigma), (F^- - F^+) \rangle \neq 0$.
2. $\langle \text{grad}(\sigma), F^+ \rangle \leq 0$.
3. $\langle \text{grad}(\sigma), F^- \rangle \geq 0$.

Here, F^0 represents the "average" velocity, $\dot{x}(t)$ of the state trajectory restricted to $\sigma(x,t) = 0$. On average, the solution to Equation 50.1 with control Equation 50.5 exists and is uniquely defined on the switching surface S. This technique can also be used to determine the plant behavior in a sliding mode.

50.4 Switching-Surface Design

Switching-surface design is predicated based on the knowledge of the system behavior in a sliding mode. This behavior depends on the parameters of the switching surface. Nonlinear switching surfaces are nontrivial to design. In the linear case, the switching-surface design problem can be converted into an equivalent state feedback design problem. In any case, achieving a switching-surface design requires analytically specifying the motion of the state trajectory in a sliding mode. The so-called method of equivalent control is essential to this specification.

50.4.1 Equivalent Control

Equivalent control constitutes a control input which, when exciting the system, produces the motion of the system on the sliding surface whenever the initial state is on the surface. Suppose at t_1 the plant's state trajectory intercepts the switching surface and a sliding mode exists. The existence of a sliding mode implies that for all $t \geq t_1, \sigma(x(t), t) = 0$, and hence $\dot{\sigma}(x(t), t) = 0$. Using the chain rule, we define the equivalent control u_{eq} for systems of the form of Equation 50.1 as the input, satisfying

$$\dot{\sigma} = \frac{\partial \sigma}{\partial t} + \frac{\partial \sigma}{\partial x}\dot{x} = \frac{\partial \sigma}{\partial t} + \frac{\partial \sigma}{\partial x}f(x,t) + \frac{\partial \sigma}{\partial x}B(x,t)u_{eq} = 0.$$

Assuming that the matrix product $(\partial\sigma/\partial x)B(x,t)$ is nonsingular for all t and x, one can compute u_{eq} as

$$u_{eq} = -\left[\frac{\partial \sigma}{\partial x}B(x,t)\right]^{-1}\left(\frac{\partial \sigma}{\partial t} + \frac{\partial \sigma}{\partial x}f(x,t)\right). \tag{50.7}$$

Therefore, given that $\sigma(x(t_1), t_1) = 0$, then, for all $t \geq t_1$, the dynamics of the system on the switching surface will satisfy

$$\dot{x}(t) = \left[I - B(x,t)\left[\frac{\partial \sigma}{\partial x}B(x,t)\right]^{-1}\frac{\partial \sigma}{\partial x}\right]f(x,t) - B(x,t)\left[\frac{\partial \sigma}{\partial x}B(x,t)\right]^{-1}\frac{\partial \sigma}{\partial t}. \tag{50.8}$$

This equation represents the *equivalent system dynamics* on the sliding surface. The driving term is present when some form of tracking or regulation is required of the controlled system, for example, when

$$\sigma(x,t) = Sx + r(t) = 0$$

with $r(t)$ serving as a "reference" signal [4].

The $(n-m)$-dimensional switching surface, $\sigma(x,t) = 0$, imposes m constraints on the plant dynamics in a sliding mode. Hence, m of the state variables can be eliminated, resulting in an equivalent reduced-order system whose dynamics represent the motion of the state trajectory in a sliding mode. Unfortunately, the structure of Equation 50.8 does not allow convenient exploiting of this fact in switching-surface design. To set forth a clearer switching-surface design algorithm, we first convert the plant dynamics to the so-called regular form.

50.4.2 Regular Form of the Plant Dynamics

The *regular form* of the dynamics of Equation 50.1 is

$$\begin{aligned} \dot{z}_1 &= \hat{f}_1(z,t), \\ \dot{z}_2 &= \hat{f}_2(z,t) + \hat{B}_2(z,t)u(z,t), \end{aligned} \tag{50.9}$$

where $z_1 \in R^{n-m}$, $z_2 \in R^m$. This form can often be constructed through a linear state transformation, $z(t) = Tx(t)$, where T has the property

$$TB(x,t) = TB(T^{-1}z,t) = \begin{bmatrix} 0 \\ \hat{B}_2(z,t) \end{bmatrix},$$

and $\hat{B}_2(z,t)$ is an $(m \times m)$ nonsingular mapping for all t and z. In general, computing the regular form requires the nonlinear transformation,

$$z(t) = T(x,t) = \begin{bmatrix} T_1(x,t) \\ T_2(x,t) \end{bmatrix},$$

where

1. $T(x,t)$ is a diffeomorphic transformation, that is, a continuous differentiable inverse mapping $\tilde{T}(z,t) = x$ exists, satisfying $\tilde{T}(0,t) = 0$ for all t.
2. $T_1(\cdot,\cdot) : R^n \times R \to R^{n-m}$ and $T_2(\cdot,\cdot) : R^n \times R \to R^m$.
3. $T(x,t)$ has the property that

$$\frac{\partial T}{\partial x}B(x,t) = \begin{bmatrix} \frac{\partial T_1}{\partial x} \\ \frac{\partial T_2}{\partial x} \end{bmatrix} B(\tilde{T}(z,t),t) = \begin{bmatrix} 0 \\ \hat{B}_2(z,t) \end{bmatrix}.$$

This partial differential equation has a solution only if the so-called Frobenius condition is satisfied [8]. The resulting nonlinear regular form of the plant dynamics has the structure,

$$\begin{aligned} \dot{z}_1 &= \frac{\partial T_1}{\partial x} f(\tilde{T}(z,t),t) + \frac{\partial T_1}{\partial t} \triangleq \hat{f}_1(z,t), \\ \dot{z}_2 &= \frac{\partial T_2}{\partial x} f(\tilde{T}(z,t),t) + \frac{\partial T_2}{\partial t} + \frac{\partial T_2}{\partial x} B(\tilde{T}(z,t),t) \\ &\triangleq \hat{f}_2(z,t) + \hat{B}_2(z,t)u. \end{aligned} \tag{50.10}$$

Sometimes all nonlinearities in the plant model can be moved to $\hat{f}_2(z,t)$ so that

$$\dot{z}_1 = \hat{f}_1(z,t) = \begin{bmatrix} A_{11} & A_{12} \end{bmatrix} \begin{bmatrix} z_1 \\ z_2 \end{bmatrix}, \tag{50.11}$$

which solves the sliding-surface design problem with linear techniques (to be shown). If the original system model is linear, the regular form is given by

$$\begin{bmatrix} \dot{z}_1 \\ \dot{z}_2 \end{bmatrix} = \begin{bmatrix} A_{11} & A_{12} \\ A_{21} & A_{22} \end{bmatrix} \begin{bmatrix} z_1 \\ z_2 \end{bmatrix} + \begin{bmatrix} 0 \\ B_2 \end{bmatrix} u, \tag{50.12}$$

where $z_1 \in R^{n-m}$ and $z_2 \in R^m$ are as above.

50.4.3 Equivalent System Dynamics via Regular Form

The regular form of the equivalent state dynamics is convenient for analysis and switching-surface design. To simplify the development we make three assumptions: (1) the sliding surface is given in terms of the states of the regular form; (2) the surface has the linear time-varying structure,

$$\sigma(z,t) = Sz + r(t) = [S_1 \ \ S_2] \begin{bmatrix} z_1 \\ z_2 \end{bmatrix} + r(t) = 0,$$

where the matrix S_2 is chosen to be nonsingular; and (3) the system is in a sliding mode, that is, for some $t_1, \sigma(x(t),t) = 0$ for all $t \geq t_1$. With these three assumptions, one can solve for $z_2(t)$ as

$$z_2(t) = -S_2^{-1}S_1 z_1(t) - S_2^{-1}r(t). \tag{50.13}$$

Substituting Equation 50.13 into the nonlinear regular form of Equation 50.10 yields

$$\dot{z}_1 = \hat{f}_1(z_1,z_2,t) = \hat{f}_1\left(z_1, -S_2^{-1}S_1 z_1 - S_2^{-1}r(t), t\right).$$

The goal then is to choose S_1 and S_2 to achieve a desired behavior of this nonlinear system.

If this system is linear, that is, if Equation 50.11 is satisfied, then, using Equation 50.13, the reduced-order dynamics are

$$\dot{z}_1 = \left(A_{11} - A_{12}S_2^{-1}S_1\right) z_1 - A_{12}S_2^{-1}r(t). \tag{50.14}$$

50.4.4 Analysis of the State Feedback Structure of Reduced-Order Linear Dynamics

The equivalent reduced-order dynamics of Equation 50.14 exhibit a state feedback structure in which $F = S_2^{-1}S_1$ is a state feedback map and A_{12} represents an "input" matrix. Under the conditions that the original (linear) system is controllable, the following well-known theorem applies.

Theorem 50.2: [9]:

If the linear regular form of the state model (Equation 50.12) is controllable, then the pair (A_{11}, A_{12}) of the reduced-order equivalent system of Equation 50.14 is controllable.

This theorem leads to a wealth of switching-surface design mechanisms. First, it permits setting the poles of $A_{11} - A_{12}S_2^{-1}S_1$, for stabilizing the state trajectory to zero when $r(t) = 0$ or to a prescribed rate of tracking, otherwise. Alternatively, one can determine S_1 and S_2 to solve the LQR (linear quadratic regulator) problem when $r(t) = 0$.

As an example, suppose a system has the regular form of Equation 50.12 except that A_{21} and A_{22} are time-varying and nonlinear,

$$\begin{bmatrix} \dot{z}_1 \\ \dot{z}_2 \end{bmatrix} = \left[\begin{array}{ccc|cc} 0 & 1 & 0 & 0 & 0 \\ 0 & 0 & 0 & 1 & 0 \\ 0 & 0 & 0 & 0 & 1 \\ \hline a_{11} & a_{12} & a_{13} & a_{14} & a_{15} \\ a_{21} & a_{22} & a_{23} & a_{24} & a_{25} \end{array}\right] \begin{bmatrix} z_1 \\ z_2 \end{bmatrix} + \left[\begin{array}{cc} 0 & 0 \\ 0 & 0 \\ 0 & 0 \\ \hline 1 & 0 \\ 0 & 1 \end{array}\right] u,$$

where $a_{ij} = a_{ij}(t,x)$ and $a_{ij}^{\min} \leq a_{ij}(t,x) \leq a_{ij}^{\max}$. Let the switching surface be given by

$$\sigma(z) = [S_1 \;\; S_2] \begin{bmatrix} z_1 \\ z_2 \end{bmatrix} = 0.$$

The pertinent matrices of the reduced-order equivalent system matrices are

$$A_{11} = \begin{bmatrix} 0 & 1 & 0 \\ 0 & 0 & 0 \\ 0 & 0 & 0 \end{bmatrix} \quad \text{and} \quad A_{12} = \begin{bmatrix} 0 & 0 \\ 1 & 0 \\ 0 & 1 \end{bmatrix}.$$

To stabilize the system, suppose that the goal is to find F so that the equivalent system has eigenvalues at $\{-1,-2,-3\}$. Using MATLAB®'s Control System's Toolbox yields

$$F = \begin{bmatrix} 2 & 3 & 0 \\ 0 & 0 & 3 \end{bmatrix} = S_2^{-1}S_1.$$

Choosing $S_2 = I$ leaves $S_1 = F$. This then specifies the switching-surface matrix $S = [F \quad I]$.

Alternatively, suppose that the objective is to find the control that minimizes the performance index

$$J = \int_0^{\infty} (z_1^T Q z_1 + \hat{u}^T R \hat{u})\, dt,$$

where the lower limit of integration refers to the initiation of sliding. This is associated with the equivalent reduced-order system

$$\dot{z}_1 = A_{11}z_1 - A_{12}\hat{u},$$

where

$$\hat{u} = S_2^{-1} S_1 z_1 \equiv F z_1.$$

Suppose weighting matrices are taken as

$$Q = \begin{bmatrix} 1.0 & 0.5 & 1.0 \\ 0.5 & 2.0 & 1.0 \\ 1.0 & 1.0 & 3.0 \end{bmatrix}, \quad R = \begin{bmatrix} 2 & 0 \\ 0 & 1 \end{bmatrix}.$$

Using MATLAB Control Systems Toolbox, the optimal feedback is

$$F = \begin{bmatrix} 0.6420 & 1.4780 & 0.2230 \\ 0.4190 & 0.4461 & 1.7031 \end{bmatrix}.$$

Again, choosing $S_2 = I$, the switching-surface matrix is given by $S = [F \quad I]$. Here, the poles of the system in sliding are $\{-1.7742, -0.7034 \pm j\,0.2623\}$.

50.5 Controller Design

50.5.1 Stability to Equilibrium Manifold

As mentioned, in VSC a Lyapunov approach is used for deriving conditions on the control $u(x,t)$ that will drive the state trajectory to the equilibrium manifold. Ordinarily, it is sufficient to consider only quadratic Lyapunov function candidates of the form

$$V(t,x,\sigma) = \sigma^T(x,t) W \sigma(x,t), \tag{50.15}$$

where W is a symmetric positive-definite matrix. The control $u(x,t)$ must be chosen so that the time derivative of $V(t,x,\sigma)$ is negative definite for $\sigma \neq 0$. To this end, consider

$$\dot{V}(t,x,\sigma) = \dot{\sigma}^T W \sigma + \sigma^T W \dot{\sigma} = 2\sigma^T W \dot{\sigma}, \tag{50.16}$$

where we have suppressed specific x and t dependencies. Recalling Equation 50.1, it follows that

$$\dot{\sigma} = \frac{\partial \sigma}{\partial t} + \frac{\partial \sigma}{\partial x}\dot{x} = \frac{\partial \sigma}{\partial t} + \frac{\partial \sigma}{\partial x} f + \frac{\partial \sigma}{\partial x} B u. \tag{50.17}$$

Substituting Equation 50.17 into Equation 50.16 leads to a Lyapunov-like theorem.

Theorem 50.3:

A sufficient condition for the equilibrium manifold (Equation 50.3) to be globally attractive is that the control $u(x,t)$ be chosen so that

$$\dot{V} = 2\sigma^T W \frac{\partial \sigma}{\partial t} + 2\sigma^T W \frac{\partial \sigma}{\partial x} f + 2\sigma^T W \frac{\partial \sigma}{\partial x} B u < 0 \tag{50.18}$$

for $\sigma \neq 0$, that is, $\dot{V}(t,x,\sigma)$ is negative definite.

Observe that Equation 50.18 is linear in the control. Virtually all control structures for VSC are chosen so that this inequality is satisfied for appropriate W. Some control laws utilize an x- and t-dependent W requiring that the derivation above be generalized.

50.5.2 Various Control Structures

To make the needed control structures more transparent, recall the equivalent control of Equation 50.7,

$$u_{eq}(x,t) = -\left[\frac{\partial\sigma}{\partial x}B(x,t)\right]^{-1}\left(\frac{\partial\sigma}{\partial t} + \frac{\partial\sigma}{\partial x}f(x,t)\right)$$

computed assuming that the matrix product $\frac{\partial\sigma}{\partial x}B(x,t)$ is nonsingular for all t and x. We can now decompose the general control structure as

$$u(x,t) = u_{eq}(x,t) + u_N(x,t), \tag{50.19}$$

where $u_N(x,t)$ is as yet an unspecified substructure. Substituting the above into Equation 50.18 produces the following sufficient condition for stability to the switching surface: Choose $u_N(x,t)$ so that

$$\dot{V}(t,x,\sigma) = 2\sigma^T W\frac{\partial\sigma}{\partial x}B(x,t)u_N(x,t) < 0. \tag{50.20}$$

Because $\frac{\partial\sigma}{\partial x}B(x,t)$ is assumed to be nonsingular for all t and x, it is convenient to set

$$u_N(x,t) = \left[\frac{\partial\sigma}{\partial x}B(x,t)\right]^{-1}\hat{u}_N(x,t). \tag{50.21}$$

Often a switching surface $\sigma(x,t)$ can be designed to achieve a desired system behavior in sliding and, at the same time, to satisfy the constraint $\frac{\partial\sigma}{\partial x}B = I$ in which case $u_N = \hat{u}_N$. Without loss of generality, we make one last simplifying assumption, $W = I$, because $W > 0$, W is nonsingular, and can be compensated for in the control structure. Hence, stability on the surface reduces to finding $\hat{u}_N(x,t)$ such that

$$\dot{V} = 2\sigma^T W\left[\frac{\partial\sigma}{\partial x}B\right]\left[\frac{\partial\sigma}{\partial x}B\right]^{-1}\hat{u}_N = 2\sigma^T\hat{u}_N < 0. \tag{50.22}$$

These simplifications allow us to specify five common controller structures:

1. Relays with constant gains: $\hat{u}_N(x,t)$ is chosen so that

   $$\hat{u}_N = \alpha\,\text{sgn}(\sigma(x,t))$$

   with $\alpha = [\alpha_{ij}]$ an $m \times m$ matrix, and sgn $(\sigma(x,t))$ is defined componentwise. Stability to the surface is achieved if $\alpha = [\alpha_{ij}]$ is chosen diagonally dominant with negative diagonal entries [5]. Alternatively, if α is chosen to be diagonal with negative diagonal entries, then the control can be represented as

   $$\hat{u}_{Ni} = \alpha_{ii}\text{sgn}(\sigma_i(x,t))$$

   and, for $\sigma_i \neq 0$,

   $$2\sigma_i\hat{u}_{Ni} = 2\alpha_{ii}\sigma_i\text{sgn}(\sigma_i) = 2\alpha_{ii}|\sigma_i| < 0,$$

   which guarantees stability to the surface, given the Lyapunov function, $V(t,x,\sigma) = \sigma^T(x,t)\sigma(x,t)$.
2. Relays with state-dependent gains: Each entry of $\hat{u}_N(x,t)$ is chosen so that

   $$\hat{u}_{Ni} = \alpha_{ii}(x,t)\text{sgn}(\sigma_i(x,t)), \quad \alpha_{ii}(x,t) < 0.$$

   The condition for stability to the surface is that

   $$2\sigma_i\hat{u}_N = 2\alpha_{ii}(x,t)\sigma_i\text{sgn}(\sigma_i) = 2\alpha_{ii}(x,t)|\sigma_i| < 0 \quad \text{for } \sigma_i \neq 0.$$

   An adequate choice for $\alpha_{ii}(x,t)$ is to choose $\beta_i < 0$, $\gamma_i > 0$, and k a natural number with

   $$\alpha_{ii}(x) = \beta_i(\sigma_i^{2k}(x,t) + \gamma_i).$$

3. Linear state feedback with switched gains: Here $\hat{u}_N(x,t)$ is chosen so that

$$\hat{u}_N = \Psi x; \quad \Psi = \left[\Psi_{ij}\right]; \quad \Psi_{ij} = \begin{cases} \alpha_{ij} < 0, & \sigma_i x_j > 0, \\ \beta_{ij} > 0, & \sigma_i x_j < 0. \end{cases}$$

To guarantee stability, it is sufficient to choose α_{ij} and β_{ij} so that

$$\sigma_i \hat{u}_{Ni} = \sigma_i \left(\Psi_{i1}x_1 + \Psi_{i2}x_2 + \cdots + \Psi_{in}x_n\right) = \Psi_{i1}\sigma_i x_1 + \Psi_{i2}\sigma_i x_2 + \cdots + \Psi_{in}\sigma_i x_n < 0.$$

4. Linear continuous feedback: Choose

$$\hat{u}_N = -P\sigma(x,t), \quad P = P^T > 0,$$

that is, $P \in R^{m\times m}$ is a symmetric positive-definite constant matrix. Stability is achieved because

$$\sigma^T \hat{u}_N = -\sigma^T P \sigma < 0,$$

where P is often chosen as a diagonal matrix with positive diagonal entries.

5. Univector nonlinearity with scale factor: In this case, choose

$$\hat{u}_N = \begin{cases} \dfrac{\sigma(x,t)}{||\sigma(x,t)||}\rho, & \rho < 0 \quad \text{and} \quad \sigma \neq 0, \\ 0, & \sigma = 0. \end{cases}.$$

Stability to the surface is guaranteed because, for $\sigma \neq 0$,

$$\sigma^T \hat{u}_N = \frac{\sigma^T\sigma}{||\sigma||}\rho = ||\sigma||\rho < 0.$$

Of course, it is possible to make ρ time dependent, if necessary, for certain tracking problems. This concludes our discussion of control structures to achieve stability to the sliding surface.

50.6 Design Examples

This section presents two design examples illustrating typical VSC strategies.

Design Example 50.1:

In this example, we illustrate a constant gain relay control with nonlinear sliding surface design for a single-link robotic manipulator driven by a *dc* armature-controlled motor modeled by the normalized (i.e., scaled) simplified equations

$$\begin{bmatrix} \dot{x}_1 \\ \dot{x}_2 \\ \dot{x}_3 \end{bmatrix} = \begin{bmatrix} x_2 \\ \sin(x_1) + x_3 \\ x_2 + x_3 \end{bmatrix} + \begin{bmatrix} 0 \\ 0 \\ 1 \end{bmatrix} u \equiv f(x) + Bu$$

in the regular form.

To determine the structure of an appropriate sliding surface, recall the assumption that $\frac{\partial\sigma}{\partial x}B(x,t)$ is nonsingular. Because $B = [0\ 0\ 1]^T$, it follows that $\frac{\partial\sigma}{\partial x_3}$ must be nonzero. Without losing generality, we set $\frac{\partial\sigma}{\partial x_3} = 1$. Hence, it is sufficient to consider sliding surfaces of the form

$$\sigma(x) = \sigma(x_1, x_2, x_3) = \sigma_1(x_1, x_2) + x_3 = 0. \tag{50.23}$$

Our design presumes that the reduced-order dynamics have a second-order response represented by the reduced-order state dynamics,

$$\begin{bmatrix} \dot{x}_1 \\ \dot{x}_2 \end{bmatrix} = \begin{bmatrix} x_2 \\ -a_1x_1 - a_2x_2 \end{bmatrix} = \begin{bmatrix} 0 & 1 \\ -a_1 & -a_2 \end{bmatrix} \begin{bmatrix} x_1 \\ x_2 \end{bmatrix}.$$

This form allows us to specify the characteristic polynomial of the dynamics and thus the eigenvalues, that is, $\pi_A(\lambda) = \lambda^2 + a_2\lambda + a_1$. Proper choice of a_1 and a_2 leads to proper rise time, settling time, overshoot, gain margin, and so on.

The switching-surface structure of Equation 50.23 implies that, in a sliding mode,

$$x_3 = -\sigma_1(x_1, x_2).$$

Substituting the above equation into the given system model, the reduced-order dynam cs become

$$\begin{bmatrix} \dot{x}_1 \\ \dot{x}_2 \end{bmatrix} = \begin{bmatrix} x_2 \\ \sin(x_1) - \sigma_1(x_1, x_2) \end{bmatrix} = \begin{bmatrix} x_2 \\ -a_1x_1 - a_2x_2 \end{bmatrix}.$$

Hence the switching-surface design is completed by setting

$$\sigma_1(x_1, x_2) = \sin(x_1) + a_1x_1 + a_2x_2.$$

To complete the controller design, we first compute the equivalent control,

$$u_{eq} = -\begin{bmatrix} \frac{\partial\sigma_1}{\partial x_1} & \frac{\partial\sigma_1}{\partial x_2} & 1 \end{bmatrix} \begin{bmatrix} x_2 \\ \sin(x_1) + x_3 \\ x_2 + x_3 \end{bmatrix}.$$

For the constant gain relay control structure (Equation 50.19),

$$u_N = \alpha\, \text{sgn}(\sigma(x)).$$

Stability to the switching surface results whenever $\alpha < 0$ as

$$\sigma\dot{\sigma} = \alpha\sigma\, \text{sgn}(\sigma) = \alpha|\sigma| < 0.$$

Design Example 50.2:

Consider the fourth-order (linear) model of a mass–spring system that could represent a simplified model of a flexible structure in space with two-dimensional control (Figure 50.5).

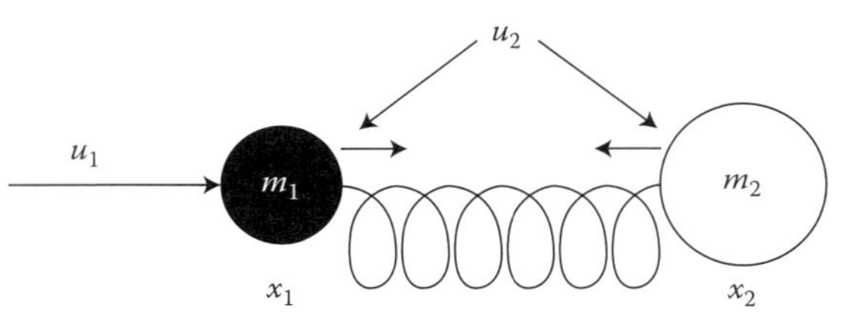

FIGURE 50.5 A mass spring system for design Example 50.2.

Here, x_1 is the position of m_1, x_2 the position of m_2, u_1 the force applied to m_1, and u_2 the force applied between m_1 and m_2. The differential equation model has the form

$$m_1\ddot{x}_1 + k(x_1 - x_2) = u_1 + u_2,$$
$$m_2\ddot{x}_2 + k(x_2 - x_1) = -u_2,$$

where k is the spring constant. Given that $x_3 = \dot{x}_1$ and $x_4 = \dot{x}_2$, the resulting state model in regular form is

$$\begin{bmatrix}\dot{x}_1\\ \dot{x}_2\\ \dot{x}_3\\ \dot{x}_4\end{bmatrix} = \begin{bmatrix} x_3\\ x_4\\ -\dfrac{k}{m_1}x_1 + \dfrac{k}{m_1}x_2\\ \dfrac{k}{m_2}x_1 - \dfrac{k}{m_2}x_2\end{bmatrix} + \begin{bmatrix} 0\\ 0\\ \dfrac{1}{m_1}u_1 + \dfrac{1}{m_1}u_2\\ -\dfrac{1}{m_2}u_2\end{bmatrix}.$$

There are two simultaneous control objectives:

1. Stabilize oscillations, that is, $x_1 = x_2$.
2. Track a desired trajectory, $x_2(t) = z_{ref}(t)$.

These goals are achieved if the following relationships are maintained for $c_1, c_2 > 0$:

$$\dot{x}_1 - \dot{x}_2 + c_1(x_1 - x_2) = 0 \implies x_1 - x_2 \longrightarrow 0$$

and

$$\dot{x}_2 - \dot{z}_{\text{ref}} + c_2(x_2 - z_{\text{ref}}) = 0 \implies x_2 - z_{\text{ref}} \longrightarrow 0.$$

The first step is to determine the appropriate sliding surface. To achieve the first control objective, set

$$\sigma_1(x,t) = x_3 - x_4 + c_1(x_1 - x_2) = 0,$$

and to achieve the desired tracking, set

$$\sigma_2(x,t) = x_4 - \dot{z}_{ref} + c_2(x_2 - z_{ref}) = 0.$$

The next step is to design a VSC law to drive the state trajectory to the intersection of these switching surfaces. In this effort, we illustrate two controller designs. The first is a hierarchical structure [2] so that, for $\sigma \neq 0$,

$$u_1 = \alpha_1 \text{sgn}(\sigma_1),$$
$$u_2 = \alpha_2 \text{sgn}(\sigma_2)$$

with the sign of $\alpha_1, \alpha_2 \neq 0$ to be determined.

For stability to the surface, it is sufficient to have $\sigma_1\dot{\sigma}_1 < 0$ and $\sigma_2\dot{\sigma}_2 < 0$, as can be seen from Equation 50.16, with $W = I$. Observe that

$$\dot{\sigma}_1 = \dot{x}_3 - \dot{x}_4 + c_1(\dot{x}_1 - \dot{x}_2) = \dot{x}_3 - \dot{x}_4 + c_1(x_3 - x_4)$$

and

$$\dot{\sigma}_2 = \dot{x}_4 - \ddot{z}_{ref} + c_2(\dot{x}_2 - \dot{z}_{ref}) = \dot{x}_4 - \ddot{z}_{ref} + c_2(x_4 - \dot{z}_{ref}).$$

Substituting for the derivatives of x_3 and x_4 leads to

$$\begin{bmatrix}\dot{\sigma}_1\\ \dot{\sigma}_2\end{bmatrix} = \begin{bmatrix}\dfrac{1}{m_1} & \dfrac{1}{m_1} + \dfrac{1}{m_2}\\ 0 & -\dfrac{1}{m_2}\end{bmatrix}\begin{bmatrix}u_1\\ u_2\end{bmatrix} + \begin{bmatrix}h_1\\ h_2\end{bmatrix} = \begin{bmatrix}\dfrac{1}{m_1} & \dfrac{1}{m_1} + \dfrac{1}{m_2}\\ 0 & -\dfrac{1}{m_2}\end{bmatrix}\begin{bmatrix}\alpha_1\text{sgn}(\sigma_1)\\ \alpha_2\text{sgn}(\sigma_2)\end{bmatrix} + \begin{bmatrix}h_1\\ h_2\end{bmatrix}, \quad (50.24)$$

where

$$h_1 = -\frac{k}{m_1}x_1 + \frac{k}{m_1}x_2 - \frac{k}{m_2}x_1 + \frac{k}{m_2}x_2 + c_1x_3 - c_1x_4$$

and

$$h_2 = \frac{k}{m_2}x_1 - \frac{k}{m_2}x_2 - \ddot{z}_{\text{ref}} + c_2x_4 - c_2\dot{z}_{\text{ref}}.$$

Taking a brute force approach to the computation of the control gains, stability to the switching surface is achieved, provided

$$\sigma_2\dot{\sigma}_2 = \frac{-\alpha_2}{m_2}\sigma_2\text{sgn}(\sigma_2) + \sigma_2h_2 < 0,$$

which is satisfied if

$$\alpha_2 > m_2|h_2|(> 0),$$

and provided

$$\sigma_1\dot{\sigma}_1 = \frac{\alpha_1}{m_1}\sigma_1\text{sgn}(\sigma_1) + \sigma_1\left[\left(\frac{1}{m_1} + \frac{1}{m_2}\right)\alpha_2\text{sgn}(\sigma_2) + h_1\right] < 0$$

which is satisfied if

$$\alpha_1 < -m_1|h_1| - \left(1 + \frac{m_1}{m_2}\right)\alpha_2.$$

In a second controller design, we recall Equation 50.24. For $\sigma_1 \neq 0$ and $\sigma_2 \neq 0$, it is convenient to define the controller as

$$\begin{bmatrix} u_1 \\ u_2 \end{bmatrix} = \begin{bmatrix} \frac{1}{m_1} & \frac{1}{m_1} + \frac{1}{m_2} \\ 0 & -\frac{1}{m_2} \end{bmatrix}^{-1} \begin{bmatrix} \beta_1\text{sgn}(\sigma_1) \\ \beta_2\text{sgn}(\sigma_2) \end{bmatrix},$$

where β_1 and β_2 are to be determined. It follows that

$$\begin{bmatrix} \dot{\sigma}_1 \\ \dot{\sigma}_2 \end{bmatrix} = \begin{bmatrix} \beta_1\text{sgn}(\sigma_1) \\ \beta_2\text{sgn}(\sigma_2) \end{bmatrix} + \begin{bmatrix} h_1 \\ h_2 \end{bmatrix}.$$

As in the first controller design, the state trajectory will intercept the sliding surface in finite time and sliding will occur for β_1 and β_2 sufficiently negative, thereby achieving the desired control objective.

A unifying characterization of sliding mode controllers that drive a trajectory to the sliding manifold is given in [10].

50.7 Chattering

The VSC controllers developed earlier assure the desired behavior of the closed-loop system. These controllers, however, require an infinitely (in the ideal case) fast switching mechanism. The phenomenon of nonideal but fast switching was labeled as chattering (actually, the word stems from the noise generated by the switching element). The high-frequency components of the chattering are undesirable because they may excite unmodeled high-frequency plant dynamics resulting in unforeseen instabilities. To reduce chatter, define a so-called boundary layer as

$$\{x|\ \|\sigma(x)\| \leq \varepsilon, \varepsilon > 0\}, \tag{50.25}$$

whose thickness is 2ε. Now, modify the control law of Equation 50.26 (suppressing t and x arguments) to

$$u = \begin{cases} u_{eq} + u_N, & \|\sigma\| \geq \varepsilon, \\ u_{eq} + p(\sigma, x), & \|\sigma\| \leq \varepsilon, \end{cases} \tag{50.26}$$

where $p(\sigma, x)$ is any continuous function satisfying $p(0, x) = 0$ and $p(\sigma, x) = u_N(x)$ when $\|\sigma(x)\| = \varepsilon$. This control guarantees that trajectories are attracted to the boundary layer. Inside the boundary layer, Equation 50.26 offers a continuous approximation to the usual discontinuous control action. Similar to Corless and Leitmann [11], asymptotic stability is not guaranteed but ultimate boundedness of trajectories to within an ε-dependent neighborhood of the origin is assured.

50.8 Robustness to Matched Disturbances and Parameter Variations

To explore the robustness of VSC to disturbances and parameter variations, one modifies Equation 50.1 to

$$\dot{x}(t) = [f(x,t) + \Delta f(x,t,q(t))] + [B(x,t) + \Delta B(x,t,q(t))]u(x,t) + d(t), \tag{50.27}$$

where $q(t)$ is a vector function representing parameter uncertainties, Δf and ΔB represent the cumulative effects of all plant uncertainties, and $d(t)$ denotes an external (deterministic) disturbance. The first critical assumption in our development is that all uncertainties and external disturbances satisfy the so-called *matching condition*, that is, Δf, ΔB, and $d(t)$ lie in the image of $B(x, t)$ for all x and t. As such they can all be lumped into a single vector function $\xi(x, t, q, d, u)$, so that

$$\dot{x}(t) = f(x,t) + B(x,t)u(x,t) + B(x,t)\xi(x,t,q,d,u). \tag{50.28}$$

The second critical assumption is that a positive continuous bounded function $\rho(x, t)$ exists, satisfying

$$||\xi(x,t,q,d,u)|| \leq \rho(x,t). \tag{50.29}$$

To incorporate robustness into a VSC design, we utilize the control structure of Equation 50.19, $u(x, t) = u_{eq}(x, t) + u_N(x, t)$, where $u_{eq}(x, t)$ is given by Equation 50.7. Given the plant and disturbance model of Equation 50.28, then, as per Equation 50.20, it is necessary to choose $u_N(x, t)$, so that

$$\dot{V}(t,x,\sigma) = 2\sigma^T W \frac{\partial\sigma}{\partial x} B(x,t)\left[u_N(x,t) + \xi(x,t,q,d,u)\right] < 0.$$

Choosing any one of the control structures outlined in Section 50.5, a choice of sufficiently "high" gains will produce a negative-definite $\dot{V}(t, x, \sigma)$. Alternatively, one can use a control structure [2],

$$u_N(x,t) = \begin{cases} -\dfrac{B^T\left[\frac{\partial\sigma}{\partial x}\right]^T\sigma}{\left\|B^T\left[\frac{\partial\sigma}{\partial x}\right]^T\sigma\right\|}[\rho(x,t) + \alpha(x,t)] & \text{for } \sigma(x,t) \neq 0, \\ 0 & \text{otherwise}, \end{cases} \tag{50.30}$$

where $\alpha(x, t)$ is to be determined. Assuming $W = I$, it follows that, for $\sigma \neq 0$,

$$\begin{aligned} \dot{V}(t,x,\sigma) &= -2\sigma^T \frac{\frac{\partial\sigma}{\partial x}BB^T\left[\frac{\partial\sigma}{\partial x}\right]^T\sigma}{\left\|B^T\left[\frac{\partial\sigma}{\partial x}\right]^T\sigma\right\|} \times [\rho(x,t) + \alpha(x,t)] + 2\sigma^T\frac{\partial\sigma}{\partial x}B\xi \\ &\leq -2\left\|B^T\left[\frac{\partial\sigma}{\partial x}\right]^T\sigma\right\|\alpha(x,t). \end{aligned}$$

Choosing $\alpha(x, t) = \alpha > 0$ leads to the stability of the state trajectory to the equilibrium manifold despite matched disturbances and parameter variations, demonstrating the robustness property of a VSC law.

50.9 Observer Design

"Observers" can be viewed as software algorithms that allow online estimation of the current state of a dynamic system when only the output and the input of the system can be measured. In the case of a linear system, we have

$$\begin{aligned} \dot{x} &= Ax + Bu + B\xi, \\ y &= Cx, \end{aligned} \tag{50.31}$$

where $C \in R^{p\times n}$, and we assume that the pair (C, A) is observable. The observer design problem is to construct a dynamic system that estimates the system state based on knowledge of the input and the output measurement. This results in the so-called Luenberger observer when $\xi = 0$,

$$\dot{\hat{x}} = A\hat{x} + Bu + L(y - C\hat{x}), \tag{50.32}$$

The estimation error $e(t) = x(t) - \hat{x}(t)$ satisfies: $\dot{e}(t) = (A - LC)e(t)$. Since (C, A) is observable, the eigenvalues of $A - LC$ can be assigned arbitrarily by a choice of the gain matrix L, although in practice this is limited by the bandwidth of the system.

The sliding mode concept can be used for designing an observer by replacing $L(y - C\hat{x})$ in Equation 50.32 with a discontinuous function $E_d(y, \hat{x})$ of and $\hat{x}$ yielding

$$\begin{aligned} \dot{\hat{x}} &= A\hat{x} + Bu + E_d(y, \hat{y}), \\ \hat{y} &= C\hat{x}, \end{aligned} \tag{50.33}$$

where E_d is a user-chosen function to insure convergence in the presence of uncertainties modeled by nonzero ξ in Equation 50.31.

One possibility is to choose $E_d(y, \hat{y}) = L(y - C\hat{x}) + BE(y, \hat{y})$, where L is chosen so that A-LC is a stability matrix (eigenvalues in the open left-half-complex plane) and

$$E(y, \hat{y}) = \eta \frac{F(y - \hat{y})}{\|F(y - \hat{y})\|}, \tag{50.34}$$

where η is a design parameter satisfying $\eta > ||\xi||$. Now, L, $F \in R^{m\times p}$, $p \geq$ m, and a matrix $P = P^T > 0$ must simultaneously satisfy:

1. $\text{eig}(A - LC) \subset C^-$
2. $FC = B^T P$, and
3. $(A - LC)^T P + P(A - LC) = -Q$

for an appropriate $Q = Q^T > 0$, if it exists. A solution for (L, F, P) exists if and only if

1. $\text{rank}(B) = \text{rank}(CB) = r$ and
2. $\text{rank}\begin{bmatrix} sI - A & B \\ C & 0 \end{bmatrix} = n + r, \text{Re}[s] \geq 0$

With the estimation error as $e(t) = x(t) - \hat{x}(t)$, the error dynamics become

$$\dot{e}(t) = (A - LC)e(t) - BE(y, \hat{y}) + B\xi. \tag{50.35}$$

It follows that

$$\begin{aligned} \frac{\mathrm{d}}{\mathrm{d}t}(e^T Pe) &= -e^T Qe - 2\eta \|FCe\| + 2e^T PB\xi \\ &\leq -e^T Qe - 2\eta \|FCe\| + 2\|FCe\| \|\xi\| \leq -e^T Qe, \end{aligned}$$

which implies $\lim_{t\to\infty} e(t) = 0$. For further analysis see [3,12,13]. For an alternate sliding mode observer structure, see [14–16].

50.9.1 Observer Design 2 [17,18]

Now consider $E_d(y, \hat{x}) = L\,\text{sign}(y - C\hat{x})$ resulting in the observer dynamics

$$\dot{\hat{x}} = A\hat{x} + Bu + L\,\text{sign}(y - C\hat{x}). \tag{50.36}$$

For the deterministic case ($\xi = 0$) the observation error satisfies $\dot{e} = Ae - L\,\text{sgn}(Ce)$. For such a system, a sliding mode is possible on the manifold $Ce = 0$. In order to describe the choice of the observer gain L and analyze the error dynamics let us consider a nonsingular transformation of the state x into a new set of coordinates such that the first p coordinates correspond to the observed vector y:

$$\begin{bmatrix} y \\ w \end{bmatrix} = \begin{bmatrix} C \\ M \end{bmatrix} x.$$

The transformed plant dynamics are

$$\begin{bmatrix} \dot{y} \\ \dot{w} \end{bmatrix} = \begin{bmatrix} A_{11} & A_{12} \\ A_{21} & A_{22} \end{bmatrix} \begin{bmatrix} y \\ w \end{bmatrix} + \begin{bmatrix} B_1 \\ B_2 \end{bmatrix} u. \tag{50.37}$$

The observer in the new coordinates is

$$\dot{\hat{y}} = A_{11}\hat{y} + A_{12}\hat{w} + B_1 u + L_1\,\text{sgn}(y - \hat{y}), \tag{50.38a}$$

$$\dot{\hat{w}} = A_{21}\hat{y} + A_{22}\hat{w} + B_2 u + L_2\,\text{sgn}(y - \hat{y}). \tag{50.38b}$$

Denoting $e_1(t) = y(t) - \hat{y}(t)$ and $e_2(t) = w(t) - \hat{w}(t)$, the error dynamics for the first subsystem is

$$\dot{e}_1 = A_{11}e_1 + A_{12}e_2 - L_1\,\text{sgn}(e_1), \tag{50.39a}$$

$$\dot{e}_2 = A_{21}e_1 + A_{22}e_2 - L_2\,\text{sgn}(e_1). \tag{50.39b}$$

By choosing an appropriate nonsingular gain matrix L_1 (it is a square matrix), we can enforce sliding regime in the first equation along the manifold $e_1(t) = 0$. Indeed, the equivalent control is obtained from Equation 50.39a under the condition, $\dot{e}_1(t) = 0$ as

$$[\text{sgn}(e_1)]_{eq} = L_1^{-1}A_{12}e_2. \tag{50.40}$$

The dynamics of the system in a sliding mode ($\dot{e}_1 = 0$) can be obtained by substituting this value into Equation 50.39b, to obtain the linear equation

$$\dot{e}_2 = (A_{22} - L_2L_1^{-1}A_{12})e_2. \tag{50.41}$$

Let us note that the observability of the original pair (C, A) implies observability of the pair (A_{12}, A_{22}) in the system (Equation 50.37). Using this fact it follows that we can assign any eigenvalues in this system by appropriate choice of L_2; thus, guaranteeing convergence $e_2(t) \to 0$ with any desired exponential rate. The dimension of the system or Equation 50.41 is n–p. The case when the output is corrupted by measurement noise was also considered in [18]. Similar observer structures and explanations can be found in [6,14–16]. An application of such an observer structure to state estimation of a magnetic bearing is considered in [19].

In [20], Drakunov proposed a sliding mode observer structure

$$\dot{\hat{x}} = \left[\frac{\partial H(\hat{x})}{\partial x}\right]^{-1} M(\hat{x}, t)\,\text{sgn}[V - H(\hat{x})] \tag{50.42}$$

that can be used for a nonlinear system of the form

$$\dot{x} = f(x),$$
$$y = h(x),$$

where the measurement map $h : R^n \to R$ is a scalar and where $H(x) = [h_1(x)\, h_2(x) \cdots h_n(x)]^T$ has entries defined using repeated Lie derivatives: $h_1(x) = h(x)$, $h_2(x) = L_f h(x)$, $h_3(x) =$

$L_f^2h(x), \ldots, h_n(x) = L_f^{n-1}h(x)$; $M(\hat{x}, t) = diag(m_1(\hat{x}, t), \ldots, m_n(\hat{x}, t))$ is a diagonal gain matrix and the vector $V = [v_1\ v_2\ \cdots\ v_n]^T$ has components defined recursively: $v_1(t) = y(t)$, $v_{i+1}(t) = \left[m_i(\hat{x}, t)\text{sgn}\left(v_i(t) - h_i(\hat{x})\right)\right]_{eq}$. The equivalent values can be obtained using an equivalent control filter such as a low-pass filter, although a first-order low-pass filter may not be sufficient; more complicated even nonlinear digital filters may need to be employed.

Example 50.3:

To illustrate the above nonlinear observer design, consider the nonlinear state model

$$\begin{aligned}\dot{x}_1 &= (1 - 2x_1 + 2x_2^2)x_2,\\ \dot{x}_2 &= -x_1 + x_2^2\end{aligned}$$

with the output $y = x_1 - x_2^2$. In this case, we have $h(x) = h_1(x_1, x_2) = x_1 - x_2^2$, and since $n = 2$ we need only the first Lie derivative: $h_2(x_1, x_2) = L_f h(x) = x_2$. Therefore, the corresponding map H and its Jacobian matrix are

$$H(x) = \begin{bmatrix} x_1 - x_2^2 \\ x_2 \end{bmatrix}, \quad \frac{\partial H}{\partial x} = \begin{bmatrix} 1 & -2x_2 \\ 0 & 1 \end{bmatrix} \Rightarrow \left(\frac{\partial H}{\partial x}\right)^{-1} = \begin{bmatrix} 1 & 2x_2 \\ 0 & 1 \end{bmatrix}.$$

The observer of Equation 50.42 is

$$\begin{aligned}\dot{\hat{x}}_1 &= m_1\,\text{sgn}(y - \hat{x}_1 + \hat{x}_2^2) + 2m_2\hat{x}_2\,\text{sgn}(v - \hat{x}_2),\\ \dot{\hat{x}}_2 &= m_2\,\text{sgn}(v - \hat{x}_2),\end{aligned}$$

where $v = \{m_1\,\text{sgn}(y - \hat{x}_1 + \hat{x}_2^2)\}_{eq}$. The second-order observer converges as long as the observer gains are sufficiently large, which means that $m_1 \geq |x_2|, m_2 \geq \left|x_1 - x_2^2\right|$. If the region of initial conditions and system trajectories are bounded, then the gains can be chosen to be constant. In general, the gains depend on $(\hat{x}_1, \hat{x}_2)$. The equivalent value operator $\{\cdots\}_{eq}$ can be implemented in different

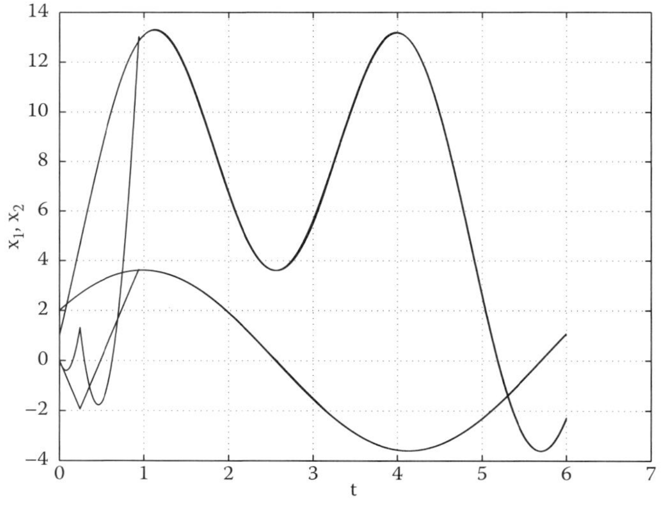

FIGURE 50.6 Nonlinear observer convergence.

ways, the easiest of which is a first-order low-pass filter $\dot{v} = -\lambda v + \lambda m_1 \text{sgn}(y - \hat{x}_1 + \hat{x}_2^2)$ for an appropriate value of $\lambda > 0$. The resulting solution to the filtering equation is $v = \{m_1 \text{sgn}(y - \hat{x}_1 + \hat{x}_2^2)\}_{eq}$. The simulation results are shown in Figure 50.6.

For further work on VSC systems and sliding mode observers we refer the reader to [16,20–22].

50.10 Concluding Remarks

This chapter has summarized the salient results of sliding mode control theory and illustrated the design procedures with various examples. A wealth of literature exists on the subject that cannot be included because of space limitations. In particular, the literature is replete with realistic applications [1,23], extensions to output feedback [24], extensions to decentralized control [25], and extensions to discrete-time systems. Additionally, there is some work, old and new, on higher-order sliding modes [16] and [26]. For extensions of the above methods to time delay systems, see [18,27,29]. The reader is encouraged to search the literature for many papers in this area.

50.11 Defining Terms

Chattering: The phenomenon of nonideal but fast switching. The term stems from the noise generated by a switching element.

Equilibrium (discontinuity) manifold: A specified, user-chosen manifold in the state space to which a system's trajectory is driven and maintained for all time subsequent to intersection of the manifold by a discontinuous control that is a function of the system's states, and hence, discontinuity manifold. Other terms commonly used are sliding surface and switching surface.

Equivalent control: The solution to the algebraic equation involving the derivative of the equation of the switching surface and the plant's dynamic model. The equivalent control is used to determine the system's dynamics on the sliding surface.

Equivalent system dynamics: The system dynamics obtained after substituting the equivalent control into the plant's dynamic model. It characterizes state motion parallel to the sliding surface if the system's initial state is off the surface and state motion is on the sliding surface if the initial state is on the surface.

Ideal sliding mode: Motion of a system's state trajectory along a switching surface when switching in the control law is infinitely fast.

Matching condition: The condition requiring the plant's uncertainties to lie in the image of the input matrix, that is, the uncertainties can affect the plant dynamics only through the same channels as the plant's input.

Region of attraction: A set of initial states in the state space from which sliding is achievable.

Regular form: A particular form of the state-space description of a dynamic system obtained by a suitable transformation of the system's state variables.

Sliding surface: See equilibrium manifold.

Switching surface: See equilibrium manifold.

References

1. Sira-Ramirez, H., Nonlinear P-I controller design for switchmode dc-to-dc power converters, *IEEE Trans. Circuits Systems*, 38(4), 410–417, 1991.

2. DeCarlo, R.A., Żak, S.H., and Matthews, G.P., Variable structure control of nonlinear multivariable systems: A tutorial, *Proc. IEEE*, 76(3), 212–232, 1988.
3. Hui, S. and Żak, S.H., Robust control synthesis for uncertain/nonlinear dynamical systems, *Automatica*, 28(2), 289–298, 1992.
4. Matthews, G.P. and DeCarlo, R.A., Decentralized tracking for a class of interconnected nonlinear systems using variable structure control, *Automatica*, 24(2), 187–193, 1988.
5. Utkin, V.I., *Sliding Modes and Their Application in Variable Structure Control*, Mir, Moscow, 1978.
6. Utkin, V.I., *Sliding Modes in Control and Optimization*, Springer, Berlin, 1992.
7. Filippov, A.F., *Differential Equations with Discontinuous Righthand Sides*, Kluwer Academic, Dordrecht, The Netherlands, 1988.
8. Hunt, L.R., Su, R., and Meyer, G., Global transformations of nonlinear systems, *IEEE Trans. Automat. Control*, AC-28(1), 24–31, 1983.
9. Young, K.-K.D., Kokotović, P.V., and Utkin, V.I., A singular perturbation analysis of high-gain feedback systems, *IEEE Trans. Automat. Control*, AC-22(6), 931–938, 1977.
10. DeCarlo, R.A., Drakunov, S., and Li, Xiaoqui, A unifying characterization of sliding mode control: A Lyapunov approach, *ASME J. Dynamic Systems, Measurement, Control,* special issue on Variable Structure Systems, 122(4), 708–718, 2000.
11. Corless, M.J. and Leitmann, G., Continuous state feedback guaranteeing uniform ultimate boundedness for uncertain dynamic systems, *IEEE Trans. Automat. Control*, AC-26(5), 1139–1144, 1981.
12. Edwards, C. and Spurgeon, S.K., On the development of discontinuous observers, *Int. J. Control*, 59(5), 1211–1229, 1994.
13. Hui, S. and Żak, S.H., Observer design for systems with unknown inputs, *Int. J. Appl. Math. Comput. Sci.*, 15(4), 431–446, 2005.
14. Perruquetti, W. and Barbot, J.P., *Sliding Mode Control in Engineering*, Marcel Dekker, New York, 2002.
15. Edwards, C. and Spurgeon, S.K., *Sliding Mode Control: Theory and Applications*, Taylor & Francis, London, UK, 1998.
16. Sabanovic, A., Fridman, L., and Spurgeon S. (Eds), *Variable Structure Systems: From Principles to Implementation*, IEE Control Engineering Series, The Institute of Electrical Engineers, UK, 2004.
17. Drakunov, S.V., Izosimov, D.B., Lukyanov, A. G., Utkin, V. A., and Utkin, V.I., The block control principle, Part II, *Autom. Remote Control*, 51(6), 737–746, 1990.
18. Drakunov, S. V., Adaptive quasioptimal filters with discontinuous parameters, *Autom. Remote Control*, 44(9), 1167–1175, 1984.
19. Rundell, A., Drakunov, S., and DeCarlo, R., A sliding mode observer and controller for stabilization of rotational motion of a vertical shaft magnetic bearing, *IEEE Trans. Control Systems Technol.*, 4(5), 598–608, 1996.
20. Drakunov, S.V., Sliding-mode observers based on equivalent control method, *Proc. 31st IEEE Conf. Decision Control*, Tucson, AZ, pp. 2368–2369, 1992.
21. Barbot, J.P., Boukhobza, T., and Djemai, M., Sliding mode observer for triangular input form. *Proc. 35th IEEE* CDC, Kobe, Japan, 1996.
22. Fridman, L., Levant, A., and Davila, J., Observation of linear systems with unknown inputs via high-order sliding modes, *Int. J. Syst. Sci.*, 38(10), 773–791, 2007.
23. Utkin, V., Guldner, J., and Shi, J., *Sliding Model Control in Electromechanical Systems*, Taylor & Francis, London, 1999.
24. El-Khazali, R. and DeCarlo, R.A., Output feedback variable structure control design, *Automatica*, 31(5), 805–816, 1995.
25. Matthews, G. and DeCarlo, R.A., Decentralized variable structure control of interconnected multi-input/multi-output nonlinear systems, *Circuits Syst. Signal Process.*, 6(3), 363–387, 1987.
26. Emelyanov, S.V., Korovin, S.K., and Levantovsky, L.V., Second-order sliding modes in controlling uncertain systems, *Sov. J. Comput. Syst. Sci.*, 24(4), 63–68, 1986.
27. Li, X. and DeCarlo, R.A., Robust sliding mode control of uncertain time-delay systems, *Int. J. Control*, 76(13), 1296–1305, 2003.
28. Walcott, B.L. and Żak, S.H., State observation of nonlinear/uncertain dynamical systems, *IEEE Trans. Automat. Control*, AC-32(2), 166–170, 1987.
29. Bengea, S., Li, X., and DeCarlo, R.A., Combined controller–observer design for time delay systems with application to engine idle speed control, *ASME J. Dyn. Sys., Meas. Control*, 126, 2004.

51

Control of Bifurcations and Chaos

Eyad H. Abed
University of Maryland

Hua O. Wang
United Technologies Research Center

Alberto Tesi
University of Florence

51.1 Introduction

This chapter deals with the control of bifurcations and chaos in nonlinear dynamical systems. This is a young subject area that is currently in a state of active development. Investigations of control system issues related to bifurcations and chaos began relatively recently, with most currently available results having been published within the past decade. Given this state of affairs, a unifying and comprehensive picture of control of bifurcations and chaos does not yet exist. Therefore, the chapter has a modest but, it is hoped, useful goal: to summarize some of the motivation, techniques, and results achieved to date on control of bifurcations and chaos. Background material on nonlinear dynamical behavior is also given, to make the chapter somewhat self-contained. However, interested readers unfamiliar with nonlinear dynamics will find it helpful to consult nonlinear dynamics texts to reinforce their understanding.

Despite its youth, the literature on control of bifurcations and chaos contains a large variety of approaches as well as interesting applications. Only a small number of approaches and applications can be touched upon here, and these reflect the background and interests of the authors. The Further Reading section provides references for those who would like to learn about alternative approaches or to learn more about the approaches discussed here.

Control system design is an enabling technology for a variety of application problems in which nonlinear dynamical behavior arises. The ability to manage this behavior can result in significant practical benefits. This might entail facilitating system operability in regimes where linear control methods break down; taking advantage of chaotic behavior to capture a desired oscillatory behavior without expending much control energy; or purposely introducing a chaotic signal in a communication system to mask a transmitted signal from an adversary while allowing perfect reception by the intended party.

The control problems addressed in this chapter are characterized by two main features:

1. Nonlinear dynamical phenomena impact system behavior
2. Control objectives can be met by altering nonlinear phenomena

Nonlinear dynamics concepts are clearly important in *understanding* the behavior of such systems (with or without control). Traditional linear control methods are, however, often effective in the *design* of control strategies for these systems. In other cases, such as for systems of the type discussed in Section 51.5.2, nonlinear methods are needed both for control design and performance assessment.

The chapter proceeds as follows. In Section 51.2, some basic nonlinear system terminology is recalled. This section also includes a new term, namely "candidate operating condition," which facilitates subsequent discussions on control goals and strategies. Section 51.3 contains a brief summary of basic bifurcation and chaos concepts that will be needed. Section 51.4 provides application examples for which bifurcations and/or chaotic behavior occur. Remarks on the control aspects of these applications are also given. Section 51.5 is largely a review of some basic concepts related to control of parameterized families of (nonlinear) systems. Section 51.5 also includes a discussion of what might be called "stressed operation" of a system. Section 51.6 is devoted to control problems for systems exhibiting bifurcation behavior. The subject of Section 51.7 is control of chaos. Conclusions are given in Section 51.8. The final section gives some suggestions for further reading.

51.2 Operating Conditions of Nonlinear Systems

In linear system analysis and control, a blanket assumption is made that the operating condition of interest is a particular equilibrium point, which is then taken as the origin in the state space. The topic addressed here relates to applying control to alter the dynamical behavior of a system possessing multiple possible operating conditions. The control might alter these conditions in terms of their location and amplitude, and/or stabilize certain possible operating conditions, permitting them to take the place of an undesirable open-loop behavior. For the purposes of this chapter, therefore, it is important to consider a variety of possible operating conditions, in addition to the single equilibrium point focused on in linear system theory. In this section, some basic terminology regarding operating conditions for nonlinear systems is established.

Consider a finite-dimensional continuous-time system

$$\dot{x}(t) = F(x(t)) \tag{51.1}$$

where $x \in \mathbb{R}^n$ is the system state and F is smooth in x. (The terminology recalled next extends straightforwardly to discrete-time systems $x^{k+1} = F(x^k)$.) An *equilibrium point* or *fixed point* of the system (Equation 51.1) is a constant steady-state solution, i.e., a vector x^0 for which $F(x^0) = 0$. A *periodic solution* of the system is a trajectory $x(t)$ for which there is a minimum $T > 0$ such that $x(t + T) = x(t)$ for all t. An *invariant set* is a set for which any trajectory starting from an initial condition within the set remains in the set for all future and past times. An *isolated invariant set* is a bounded invariant set a neighborhood of which contains no other invariant set. Equilibrium points and periodic solutions are examples of invariant sets.

A periodic solution is called a *limit cycle* if it is isolated. An *attractor* is a bounded invariant set to which trajectories starting from all sufficiently nearby initial conditions converge as $t \to \infty$. The *positive limit set* of Equation 51.1 is the ensemble of points that some system trajectory either approaches as $t \to \infty$ or makes passes nearer and nearer to as $t \to \infty$. For example, if a system is such that all trajectories converge either to an equilibrium point or a limit cycle, then the positive limit set would be the union of points on the limit cycle and the equilibrium point. The *negative limit set* is the positive limit set of the system run with time t replaced by $-t$. Thus, the positive limit set is the set where the system ends up at $t = +\infty$, while the negative limit set is the set where the system begins at $t = -\infty$ [30]. The *limit set* of the system is the union of its positive and negative limit sets.

It is now possible to introduce a term that will facilitate the discussions on control of bifurcations and chaos. A *candidate operating condition* of a dynamical system is an equilibrium point, periodic solution or other invariant subset of its limit set. Thus, a candidate operating condition is any possible steady-state solution of the system, without regard to its stability properties. This term, though not standard, is useful since it permits discussion of bifurcations, bifurcated solutions, and chaotic motion in general terms without having to specify a particular nonlinear phenomenon. The idea behind this term is that such a solution, if stable or stabilizable using available controls, might qualify as an operating condition of the system. Whether or not it would be *acceptable* as an actual operating condition would depend on the system and on characteristics of the candidate operating condition. As an extreme example, a steady spin of an airplane *is* a candidate operating condition, but certainly *is not* acceptable as an actual operating condition!

51.3 Bifurcations and Chaos

This section summarizes background material on bifurcations and chaos that is needed in the remainder of the chapter.

51.3.1 Bifurcations

A *bifurcation* is a change in the number of candidate operating conditions of a nonlinear system that occurs as a parameter is quasistatically varied. The parameter being varied is referred to as the bifurcation parameter. A value of the bifurcation parameter at which a bifurcation occurs is called a critical value of the bifurcation parameter. Bifurcations from a nominal operating condition can only occur at parameter values for which the condition (say, an equilibrium point or limit cycle) either loses stability or ceases to exist.

To fix ideas, consider a general one-parameter family of ordinary differential equation systems

$$\dot{x} = F^{\mu}(x) \tag{51.2}$$

where $x \in \mathbb{R}^n$ is the system state, $\mu \in \mathbb{R}$ denotes the bifurcation parameter, and F is smooth in x and μ. Equation 51.2 can be viewed as resulting from a particular choice of control law in a family of nonlinear control systems (in particular, as Equation 51.9 with the control u set to a fixed feedback function $u(x, \mu)$). For any value of μ, the equilibrium points of Equation 51.2 are given by the solutions for x of the algebraic equations $F^{\mu}(x) = 0$.

Local bifurcations are those that occur in the vicinity of an equilibrium point. For example, a small-amplitude limit cycle can emerge (bifurcate) from an equilibrium as the bifurcation parameter is varied. The bifurcation is said to occur regardless of the stability or instability of the "bifurcated" limit cycle. In another local bifurcation, a pair of equilibrium points can emerge from a nominal equilibrium point. In either case, the *bifurcated solutions* are close to the original equilibrium point—hence the name local bifurcation. *Global bifurcations* are bifurcations that are not local, i.e., those that involve a domain in phase space. Thus, if a limit cycle loses stability releasing a new limit cycle, a global bifurcation is said to take place.*

The nominal operating condition of Equation 51.2 can be an equilibrium point or a limit cycle. In fact, depending on the coordinates used, it is possible that a limit cycle in one mathematical model corresponds to an equilibrium point in another. This is the case, for example, when a truncated Fourier series is used to approximate a limit cycle, and the amplitudes of the harmonic terms are used as state variables in the approximate model. The original limit cycle is then represented as an equilibrium in the amplitude coordinates.

If the nominal operating condition of Equation 51.2 is an equilibrium point, then bifurcations from this condition can occur only when the linearized system loses stability. Suppose, for example, that the origin is the nominal operating condition for some range of parameter values. That is, let $F^{\mu}(0) = 0$ for all values of μ for which the nominal equilibrium exists. Denote the Jacobian matrix of Equation 51.2 evaluated at the origin by

$$A(\mu) := \frac{\partial F^{\mu}}{\partial x}(0).$$

Local bifurcations from the origin can only occur at parameter values μ where $A(\mu)$ loses stability.

The scalar differential equation

$$\dot{x} = \mu x - x^3 \tag{51.3}$$

provides a simple example of a bifurcation. The origin $x^0 = 0$ is an equilibrium point for all values of the real parameter μ. The Jacobian matrix is $A(\mu) = \mu$ (a scalar). It is easy to see that the origin loses stability as μ increases through $\mu = 0$. Indeed, a bifurcation from the origin takes place at $\mu = 0$. For $\mu \leq 0$, the only equilibrium point of Equation 51.3 is the origin. For $\mu > 0$, however, there are two additional equilibrium points at $x = \pm\sqrt{\mu}$. This pair of equilibrium points is said to *bifurcate* from the origin at the *critical parameter value* $\mu_c = 0$. This is an example of a pitchfork bifurcation, which will be discussed later.

51.3.1.1 Subcritical vs. Supercritical Bifurcations

In a very real sense, the fact that bifurcations occur when stability is lost is helpful from the perspective of control system design. To explain this, suppose that a system operating condition (the "nominal" operating condition) is not stabilizable beyond a critical parameter value. Suppose a bifurcation occurs at the critical parameter value. That is, suppose a new candidate operating condition emerges from the nominal one at the critical parameter value. Then it may be that the new operating condition is stable and occurs beyond the critical parameter value, providing an alternative operating condition near the nominal one. This is referred to as a *supercritical bifurcation*. In contrast, it may happen that the new operating condition is unstable and occurs prior to the critical parameter value. In this situation (called a *subcritical bifurcation*), the system state must leave the vicinity of the nominal operating condition for parameter values beyond the critical value. However, feedback offers the possibility of rendering such a bifurcation supercritical. This is true even if the nominal operating condition is not stabilizable. If such a feedback control can be found, then the system behavior beyond the stability boundary can remain close to its behavior at the nominal operating condition.

The foregoing discussion of bifurcations and their implications for system behavior can be gainfully viewed using graphical sketches called *bifurcation diagrams*. These are depictions of the equilibrium points and limit cycles of a system plotted against the bifurcation parameter. A bifurcation diagram is a schematic representation in which only a measure of the amplitude (or norm) of an equilibrium point or limit cycle need be plotted. In the bifurcation diagrams given in this chapter, a solid line indicates a stable solution, while a dashed line indicates an unstable solution.

* This use of the terms local bifurcation and global bifurcation is common. However, in some books, a bifurcation from a limit cycle is also referred to as a local bifurcation.

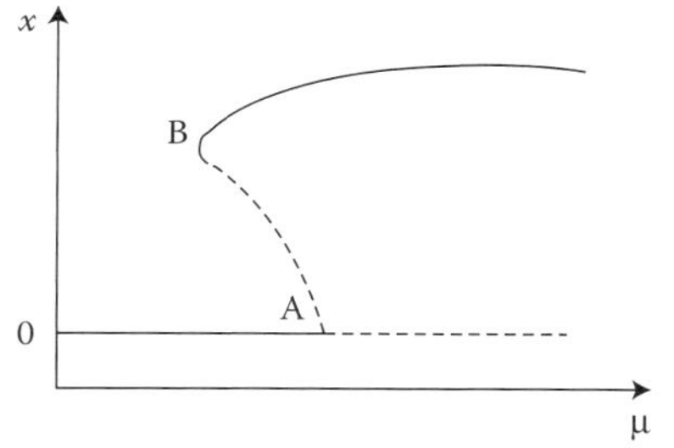

FIGURE 51.1 Subcritical bifurcation with hysteresis.

Several bifurcation diagrams will now be used to further explain the meanings of supercritical and subcritical bifurcation, and to introduce some common bifurcations. It should be noted that not all bifurcations are supercritical or subcritical. For example, bifurcation can also be transcritical. In such a bifurcation, bifurcated operating conditions occur both prior to and beyond the critical parameter value. Identifying a bifurcation as supercritical, subcritical, transcritical, or otherwise is the problem of determining the *direction of the bifurcation.* A book on nonlinear dynamics should be consulted for a more extensive treatment (e.g., [10,18,28,30,32,43,46]). In this chapter only the basic elements of bifurcation analysis can be touched upon.

Suppose that the origin of Equation 51.2 loses stability as μ *increases* through the critical parameter value $\mu = \mu_c$. Under mild assumptions, it can be shown that a bifurcation occurs at μ_c.

Figure 51.1 serves two purposes: it depicts a subcritical bifurcation from the origin, and it shows a common consequence of subcritical bifurcation, namely hysteresis. A subcritical bifurcation occurs from the origin at the point labeled A in the figure. It leads to the unstable candidate operating condition corresponding to points on the dashed curve connecting points A and B. As the parameter μ is decreased to its value at point B, the bifurcated solution merges with another (stable) candidate operating condition and disappears. A *saddle-node bifurcation* is said to occur at point B. This is because the unstable candidate operating condition (the "saddle" lying on the dashed curve) merges with a stable candidate operating condition (the "node" lying on the solid curve). These candidate operating conditions can be equilibrium points or limit cycles—both situations are captured in the figure. Indeed, the origin can also be reinterpreted as a limit cycle and the diagram would still be meaningful. Another common name for a saddle-node bifurcation point is a *turning point.*

The physical scenario implied by Figure 51.1 can be summarized as follows. Starting from operation at the origin for small μ, increasing μ until point A is reached does not alter system behavior. If μ is increased past point A, however, the origin loses stability. The system then transitions ("jumps") to the available stable operating condition on the upper solid curve. This large transition can be intolerable in many applications. As μ is then decreased, another transition back to the origin occurs but at a lower parameter value, namely that corresponding to point B. Thus, the combination of the subcritical bifurcation at A and the saddle-node bifurcation at B can lead to a hysteresis effect.

Figure 51.2 depicts a *supercritical bifurcation* from the origin. This bifurcation is distinguished by the fact that the solution bifurcating from the origin at point A is stable, and occurs locally for parameter values μ beyond the critical value (i.e., those for which the nominal equilibrium point is unstable). In marked difference with the situation depicted in Figure 51.1, here as the critical parameter value is crossed a smooth change is observed in the system operating condition. No sudden jump occurs.

Suppose closeness of the system's operation to the nominal equilibrium point (the origin, say) is a measure of the system's performance. Then supercritical bifurcations ensure close operation to the nominal equilibrium, while subcritical bifurcations may lead to large excursions away from the nominal equilibrium point. For this reason, a supercritical bifurcation is commonly also said to be *safe* or *soft*, while a subcritical bifurcation is said to be *dangerous* or *hard* [32,49,52].

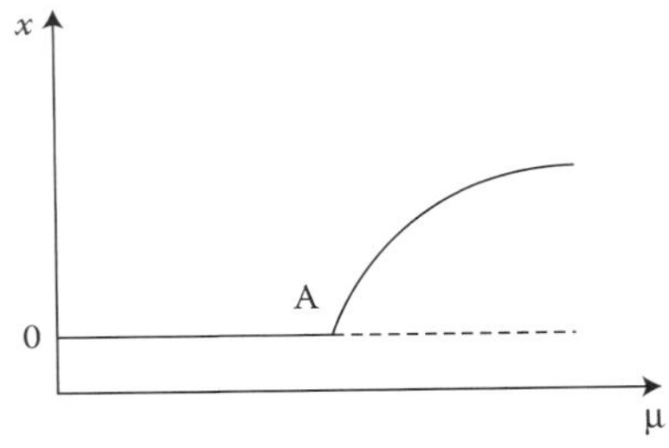

FIGURE 51.2 Supercritical bifurcation.

Given full information on the nominal equilibrium point, the occurrence of bifurcation is a consequence of the behavior of the linearized system at the equilibrium point. The manner in which the equilibrium point loses stability as the bifurcation parameter is varied determines the type of bifurcation that arises.

Three types of local bifurcation and a global bifurcation are discussed next. These are, respectively, the stationary bifurcation, the saddle-node bifurcation, the Andronov–Hopf bifurcation, and the period doubling bifurcation. All of these except the saddle-node bifurcation can be safe or dangerous. However, the saddle-node bifurcation is always dangerous.

There are analytical techniques for determining whether a stationary, Andronov-Hopf, or period doubling bifurcation is safe or dangerous. These techniques are not difficult to understand but involve calculations that are too lengthy to repeat here. The calculations yield formulas for so-called "bifurcation stability coefficients" [18], the meaning of which is addressed below. The references [1,2,4,18,28,32] can be consulted for details.

What is termed here as "stationary bifurcation" is actually a special case of the usual meaning of the term. In the bifurcation theory literature [10], stationary bifurcation is any bifurcation of one or more equilibrium points from a nominal equilibrium point. When the nominal equilibrium point exists both before and after the critical parameter value, a stationary bifurcation "from a known solution" is said to occur. If the nominal solution disappears beyond the critical parameter value, a stationary bifurcation "from an unknown solution" is said to occur. To simplify the terminology, here the former type of bifurcation is called a *stationary bifurcation*. The saddle-node bifurcation is the most common example of the latter type of bifurcation.

Andronov–Hopf bifurcation also goes by other names. "Hopf bifurcation" is the traditional name in the West, but this name neglects the fundamental early contributions of Andronov and his coworkers (see, e.g., [6]). The essence of this phenomenon was also known to Poincaré, who did not develop a detailed theory but used the concept in his study of lunar orbital dynamics [37, Secs. 51–52]. The same phenomenon is sometimes called flutter bifurcation in the engineering literature. This bifurcation of a limit cycle from an equilibrium point occurs when a complex conjugate pair of eigenvalues crosses the imaginary axis into the right half of the complex plane at $\mu = \mu_c$. A small-amplitude limit cycle then emerges from the nominal equilibrium point at μ_c.

51.3.1.2 Saddle-Node Bifurcation and Stationary Bifurcation

Saddle-node bifurcation occurs when the linearized system has a zero eigenvalue at $\mu = \mu_c$ but the origin does not persist as an equilibrium point beyond the critical parameter value. Saddle-node bifurcation was discussed briefly before, and will not be discussed in any detail in the following. Several basic remarks are, however, in order.

1. Saddle-node bifurcation of a nominal, stable equilibrium point entails the disappearance of the equilibrium upon its merging with an unstable equilibrium point at a critical parameter value.
2. The bifurcation occurring at point B in Figure 51.1 is representative of a saddle-node bifurcation.

3. The nominal equilibrium point possesses a zero eigenvalue at a saddle-node bifurcation.
4. An important feature of the saddle-node bifurcation is the *disappearance, locally, of any stable bounded solution of the system* (Equation 51.2).

Stationary bifurcation (according to the agreed upon terminology above) is guaranteed to occur when a single real eigenvalue goes from being negative to being positive as μ passes through the value μ_c. More precisely, the origin of Equation 51.2 undergoes a stationary bifurcation at the critical parameter value $\mu = 0$ if hypotheses (S1) and (S2) hold.

S1 F of system (Equation 51.2) is sufficiently smooth in x, μ, and $F^\mu(0) = 0$ for all μ in a neighborhood of 0. The Jacobian $A(\mu) := \frac{\partial F^\mu}{\partial x}(0)$ possesses a simple real eigenvalue $\lambda(\mu)$ such that $\lambda(0) = 0$ and $\lambda'(0) \neq 0$.

S2 All eigenvalues of the critical Jacobian $\frac{\partial F^\mu}{\partial x}(0)$ besides 0 have negative real parts.

Under (S1) and (S2), two new equilibrium points of Equation 51.2 emerge from the origin at $\mu = 0$. Bifurcation stability coefficients are quantities that determine the direction of bifurcation, and in particular the stability of the bifurcated solutions. Next, a brief discussion of the origin and meaning of these quantities is given.

Locally, the new equilibrium points occur for parameter values given by a smooth function of an auxiliary small parameter ϵ (ϵ can be positive or negative):

$$\mu(\epsilon) = \mu_1\epsilon + \mu_2\epsilon^2 + \mu_3\epsilon^3 + \cdots \tag{51.4}$$

where the ellipsis denotes higher order terms. One of the new equilibrium points occurs for $\epsilon > 0$ and the other for $\epsilon < 0$. Also, the stability of the new equilibrium points is determined by the sign of an eigenvalue $\beta(\epsilon)$ of the system linearization at the new equilibrium points. This eigenvalue is near 0 and is also given by a smooth function of the parameter ϵ:

$$\beta(\epsilon) = \beta_1\epsilon + \beta_2\epsilon^2 + \beta_3\epsilon^3 + \cdots \tag{51.5}$$

Stability of the bifurcated equilibrium points is determined by the sign of $\beta(\epsilon)$. If $\beta(\epsilon) < 0$ the corresponding equilibrium point is stable, while if $\beta(\epsilon) > 0$ the equilibrium point is unstable. The coefficients β_i, $i = 1, 2, \ldots$ in the expansion above are the bifurcation stability coefficients mentioned earlier, for the case of stationary bifurcation. The values of these coefficients determine the local nature of the bifurcation.

Since ϵ can be positive or negative, it follows that if $\beta_1 \neq 0$ the bifurcation is neither subcritical nor supercritical. (This is equivalent to the condition $\mu_1 \neq 0$.) The bifurcation is therefore generically transcritical. In applications, however, special structures of system dynamics and inherent symmetries often result in stationary bifurcations that are not transcritical. Also, it is sometimes possible to render a stationary bifurcation supercritical using feedback control. For these reasons, a brief discussion of subcritical and supercritical pitchfork bifurcations is given next.

If $\beta_1 = 0$ and $\beta_2 \neq 0$, a stationary bifurcation is known as a *pitchfork bifurcation*. The pitchfork bifurcation is subcritical if $\beta_2 > 0$; it is supercritical if $\beta_2 < 0$. The bifurcation diagram of a subcritical pitchfork bifurcation is depicted in Figure 51.3, and that of a supercritical pitchfork bifurcation is depicted in Figure 51.4. The bifurcation discussed previously for the example system (Equation 51.3) is a supercritical pitchfork bifurcation.

51.3.1.3 Andronov–Hopf Bifurcation

Suppose that the origin of Equation 51.2 loses stability as the result of a complex conjugate pair of eigenvalues of $A(\mu)$ crossing the imaginary axis. All other eigenvalues are assumed to remain stable, i.e., their real parts are negative for all values of μ. Under this simple condition on the linearization of a nonlinear system, the nonlinear system typically undergoes a bifurcation. The word "typically" is used because there is one more condition to satisfy, but it is a mild condition. The type of bifurcation

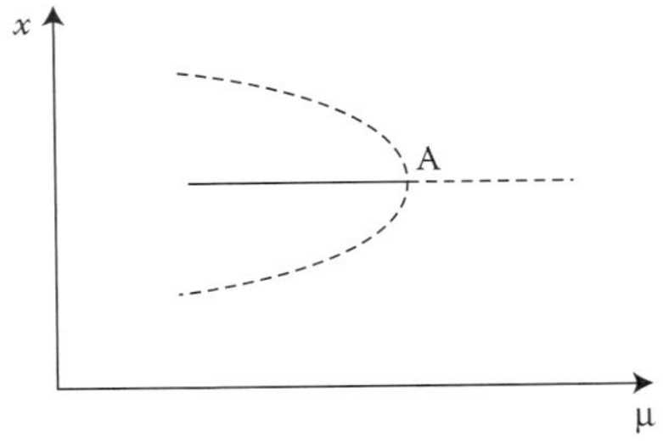

FIGURE 51.3 Subcritical pitchfork bifurcation.

that occurs under these circumstances involves the emergence of a limit cycle from the origin as μ is varied through μ_c. This is the Andronov–Hopf bifurcation, a more precise description of which is given next. The following hypotheses are invoked in the Andronov–Hopf Bifurcation Theorem. The critical parameter value is taken to be $\mu_c = 0$ without loss of generality.

AH1 F of system (Equation 51.2) is sufficiently smooth in x, μ, and $F^\mu(0) = 0$ for all μ in a neighborhood of 0. The Jacobian $\frac{\partial F^\mu}{\partial x}(0)$ possesses a complex-conjugate pair of (algebraically) simple eigenvalues $\lambda(\mu) = \alpha(\mu) + i\omega(\mu), \overline{\lambda(\mu)}$, such that $\alpha(0) = 0, \alpha'(0) \neq 0$ and $\omega_c := \omega(0) > 0$.

AH2 All eigenvalues of the critical Jacobian $\frac{\partial F^\mu}{\partial x}(0)$ besides $\pm i\omega_c$ have negative real parts.

The Andronov–Hopf Bifurcation Theorem asserts that, under (AH1) and (AH2), a small-amplitude nonconstant limit cycle (i.e., periodic solution) of Equation 51.2 emerges from the origin at $\mu = 0$. Locally, the limit cycles occur for parameter values given by a smooth and even function of the amplitude ϵ of the limit cycles:

$$\mu(\epsilon) = \mu_2\epsilon^2 + \mu_4\epsilon^4 + \cdots \tag{51.6}$$

where the ellipsis denotes higher order terms.

Stability of an equilibrium point of the system (Equation 51.2) can be studied using eigenvalues of the system linearization evaluated at the equilibrium point. The analogous quantities for consideration of limit cycle stability for Equation 51.2 are the *characteristic multipliers* of the limit cycle. (For a definition, see for example [10,18,28,30,32,43,46,48].) A limit cycle is stable (precisely: orbitally asymptotically stable) if its characteristic multipliers all have magnitude less than unity. This is analogous to the widely known fact that an equilibrium point is stable if the system eigenvalues evaluated there have negative real parts. The stability condition is sometimes stated in terms of the characteristic exponents of the limit cycle, quantities which are easily obtained from the characteristic multipliers. If the characteristic exponents have negative real parts, then the limit cycle is stable. Although it is not possible to discuss the basic theory of limit cycle stability in any detail here, the reader is referred to almost any text on differential equations, dynamical systems, or bifurcation theory for a detailed discussion (e.g., [10,18,28,30,32,43,46,48]).

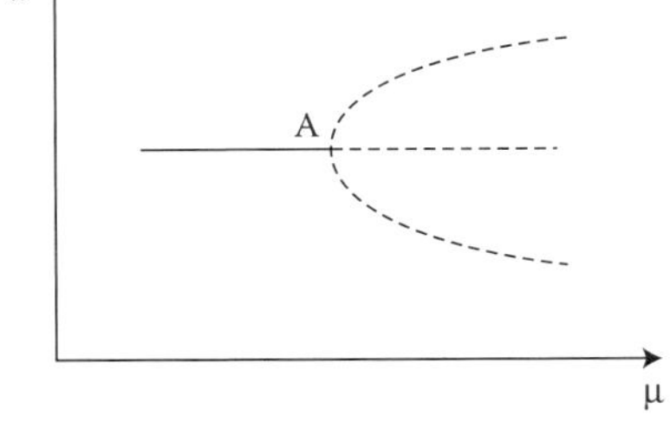

FIGURE 51.4 Supercritical pitchfork bifurcation.

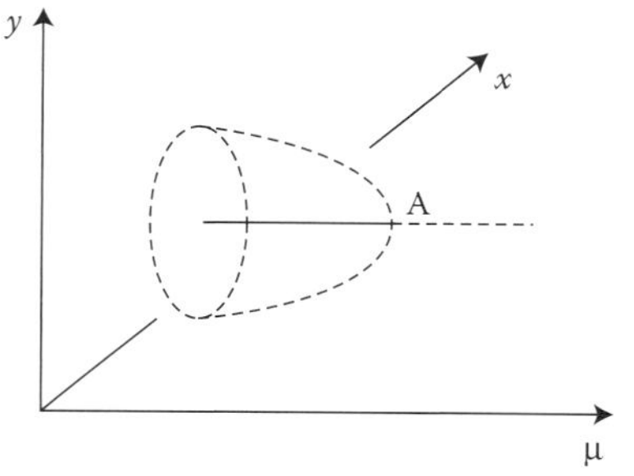

FIGURE 51.5 Subcritical Andronov–Hopf bifurcation.

The stability of the limit cycle resulting from an Andronov–Hopf bifurcation is determined by the sign of a particular characteristic exponent $\beta(\epsilon)$. This characteristic exponent is real and vanishes in the limit as the bifurcation point is approached. It is given by a smooth and even function of the amplitude ϵ of the limit cycles:

$$\beta(\epsilon) = \beta_2\epsilon^2 + \beta_4\epsilon^4 + \cdots \tag{51.7}$$

The coefficients μ_2 and β_2 in the expansions above are related by the exchange of stability formula

$$\beta_2 = -2\alpha'(0)\mu_2. \tag{51.8}$$

Generically, these coefficients do not vanish. Their signs determine the direction of bifurcation. The coefficients $\beta_2, \beta_4, \ldots$ in the expansion (Equation 51.7) are the bifurcation stability coefficients for the case of Andronov–Hopf bifurcation.

If $\beta_2 > 0$, then locally the bifurcated limit cycle is unstable and the bifurcation is subcritical. This case is depicted in Figure 51.5. If $\beta_2 < 0$, then locally the bifurcated limit cycle is stable (more precisely, one says that it is orbitally asymptotically stable [10]). This is the case of supercritical Andronov–Hopf bifurcation, depicted in Figure 51.6. If it happens that β_2 vanishes, then stability is determined by the first nonvanishing bifurcation stability coefficient (if one exists).

51.3.1.4 Period Doubling Bifurcation

The bifurcations considered above are all local bifurcations, i.e., bifurcations from an equilibrium point of the system (Equation 51.2). Solutions emerging at these bifurcation points can themselves undergo further bifurcations. A particularly important scenario involves a global bifurcation known as the *period doubling bifurcation*.

To describe the period doubling bifurcation, consider the one-parameter family of nonlinear systems (Equation 51.2). Suppose that Equation 51.2 has a limit cycle γ^μ for a range of values of the real parameter

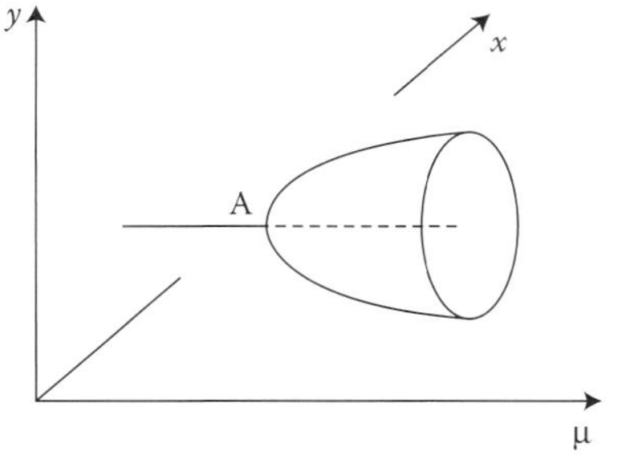

FIGURE 51.6 Supercritical Andronov–Hopf bifurcation.

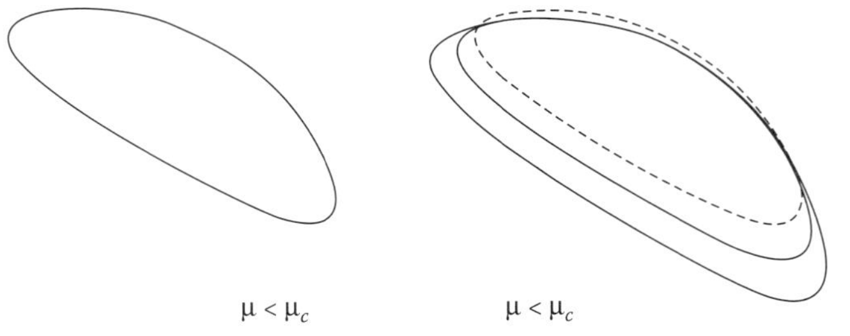

FIGURE 51.7 Period doubling bifurcation (supercritical case).

μ. Moreover, suppose that for all values of μ to one side (say, less than) a critical value μ_c, all the characteristic multipliers of γ^{μ} have magnitude less than unity. If exactly one characteristic multiplier exits the unit circle at $\mu = \mu_c$, and does so at the point $(-1, 0)$, and if this crossing occurs with a nonzero rate with respect to μ, then one can show that a period doubling bifurcation from γ^{μ} occurs at $\mu = \mu_c$. (See, e.g., [4] and references therein.)

This means that another limit cycle, initially of twice the period of γ^{μ_c}, emerges from γ^{μ} at $\mu = \mu_c$. Typically, the bifurcation is either *supercritical* or *subcritical*. In the supercritical case, the period doubled limit cycle is stable and occurs for parameter values on the side of μ_c for which the limit cycle γ^{μ} is unstable. In the subcritical case, the period doubled limit cycle is unstable and occurs on the side of μ_c for which the limit cycle γ^{μ} is stable. In either case, an *exchange of stability* is said to have occurred between the nominal limit cycle γ^{μ} and the bifurcated limit cycle. Figure 51.7 depicts a supercritical period doubling bifurcation. In this figure, a solid curve represents a stable limit cycle, while a dashed curve represents an unstable limit cycle. The figure assumes that the nominal limit cycle loses stability as μ increases through μ_c.

The direction of a period doubling bifurcation can easily be determined in discrete-time, using formulas that have been derived in the literature (see, e.g., [4]). Recently, an approximate test that applies in continuous-time has been derived using the harmonic balance approach [47].

51.3.2 Chaos

Bifurcations of equilibrium points and limit cycles are well understood and there is little room for alternative definitions of the main concepts. Although the notation, style, and emphasis may differ among various presentations, the main concepts and results stay the same. Unfortunately, the situation is not quite as tidy in regard to discussions of chaos. There are several distinct definitions of chaotic behavior of dynamical systems. There are also some aspects of chaotic motion that have been found to be true for many systems but have not been proved in general. The aim of this subsection is to summarize in a nonrigorous fashion some important aspects of chaos that are widely agreed upon.

The following working definition of chaos will suffice for the purposes of this chapter. The definition is motivated by [46, p. 323] and [11, p. 50]. It uses the notion of "attractor" defined in section two, and includes the definition of an additional notion, namely that of "strange attractor."

> A solution trajectory of a deterministic system (such as Equation 51.1) is *chaotic* if it converges to a strange attractor. A *strange attractor* is a bounded attractor that: (1) exhibits sensitive dependence on initial conditions, and (2) cannot be decomposed into two invariant subsets contained in disjoint open sets.

A few remarks on this working definition are in order. An aperiodic motion is one that is not periodic. Long-term behavior refers to steady-state behavior, i.e., system behavior that persists after

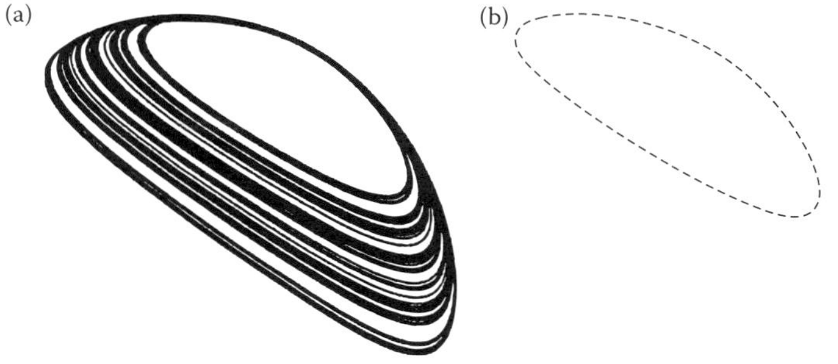

FIGURE 51.8 Strange attractor with embedded unstable limit cycle.

the transient decays. Sensitive dependence on initial conditions means that for almost any initial condition lying on the strange attractor, there exists another initial condition as close as desired to the given one such that the solution trajectories starting from these two initial conditions separate by at least some prespecified amount after some time. The requirement of nondecomposability simply ensures that strange attractors are considered as being distinct if they are not connected by any system trajectories.

Often, sensitive dependence on initial conditions is defined in terms of the presence of at least one positive "Lyapunov exponent" (e.g., [34,35]). Further discussion of this viewpoint would entail technicalities that are not needed in the sequel.

From a practical point of view, a chaotic motion can be defined as a bounded invariant motion of a deterministic system that is not an equilibrium solution or a periodic solution or a quasiperiodic solution [32, p. 277]. (A quasiperiodic function is one that is composed of finitely many periodic functions with incommensurate frequencies. See [32, p. 231].)

There are two more important aspects of strange attractors and chaos that should be noted, since they play an important role in a technique for control of chaos discussed in Section 51.7. These are:

1. A strange attractor generally has embedded within itself infinitely many unstable limit cycles. For example, Figure 51.8a depicts a strange attractor, and Figure 51.8b depicts an unstable limit cycle that is embedded in the strange attractor. Note that the shape of the limit cycle resembles that of the strange attractor. This is to be expected in general. The limit cycle chosen in the plot happens to be one of low period.
2. The trajectory starting from any point on a strange attractor will, after sufficient time, pass as close as desired to any other point of interest on the strange attractor. This follows from the indecomposability of strange attractors noted previously.

An important way in which chaotic behavior arises is through sequences of bifurcations. A well-known such mechanism is the *period doubling route to chaos*, which involves the following sequence of events:

1. A stable limit cycle loses stability, and a new stable limit cycle of double the period emerges. (The original stable limit cycle might have emerged from an equilibrium point via a supercritical Andronov–Hopf bifurcation.)
2. The new stable limit cycle loses stability, releasing another stable limit cycle of twice its period.
3. There is a cascade of such events, with the parameter separation between each two successive events decreasing geometrically.* This cascade culminates in a sustained chaotic motion (a strange attractor).

* This is true exactly in an asymptotic sense. The ratio in the geometric sequence is a universal constant discovered by Feigenbaum. See, e.g., [11,32,34,43,46,48].

51.4 Bifurcations and Chaos in Physical Systems

In this brief section, a list of representative physical systems that exhibit bifurcations and/or chaotic behavior is given. The purpose is to provide practical motivation for study of these phenomena and their control.

Examples of physical systems exhibiting bifurcations and/or chaotic behavior include the following.

- An aircraft stalls for flight under a critical speed or above a critical angle-of-attack (e.g., [8,13,39,51]).
- Aspects of laser operation can be viewed in terms of bifurcations. The simplest such observation is that a laser can only generate significant output if the pump energy exceeds a certain threshold (e.g., [19,46]). More interestingly, as the pump energy increases, the laser operating condition can exhibit bifurcations leading to chaos (e.g., [17]).
- The dynamics of ships at sea can exhibit bifurcations for wave frequencies close to a natural frequency of the ship. This can lead to large-amplitude oscillations, chaotic motion, and ship capsize (e.g., [24,40]).
- Lightweight, flexible, aircraft wings tend to experience flutter (structural oscillations) (e.g., [39]) (along with loss of control surface effectiveness (e.g., [15])).
- At peaks in customer demand for electric power (such as during extremes in weather), the stability margin of an electric power network may become negligible, and nonlinear oscillations or divergence ("voltage collapse") can occur (e.g., [12]).
- Operation of aeroengine compressors at maximum pressure rise implies reduced weight requirements but also increases the risk for compressor stall (e.g., [16,20,25]).
- A simple model for the weather consists of fluid in a container (the atmosphere) heated from below (the sun's rays reflected from the ground) and cooled from above (outer space). A mathematical description of this model used by Lorenz [27] exhibits bifurcations of convective and chaotic solutions. This has implications also for boiling channels relevant to heat-exchangers, refrigeration systems, and boiling water reactors [14].
- Bifurcations and chaos have been observed and studied in a variety of chemically reacting systems (e.g., [42]).
- Population models useful in the study and formulation of harvesting strategies exhibit bifurcations and chaos (e.g., [19,31,46]).

51.5 Control of Parametrized Families of Systems

Tracking of a desired trajectory (referred to as regulation when the trajectory is an equilibrium), is a standard goal in control system design [26]. In applying linear control system design to this standard problem, an evolving (nonstationary) system is modeled by a *parametrized family* of time-invariant (stationary) systems. This approach is at the heart of the gain scheduling method, for example [26].

In this section, some basic concepts in control of parameterized families of systems are reviewed, and a notion of stressed operation is introduced. Control laws for nonlinear systems usually consist of a feedforward control plus a feedback control. The feedforward part of the control is selected first, followed by the feedback part. This decomposition of control laws is discussed in the next subsection, and will be useful in the discussions of control of bifurcations and chaos. In the second subsection, a notion of "stressed operation" of a system is introduced. Stressed systems are not the only ones for which control of bifurcations and chaos are relevant, but they are an important class for which such control issues need to be evaluated.

51.5.1 Feedforward/Feedback Structure of Control Laws

Control designs for nonlinear system can usually be viewed as proceeding in two main steps [26, Chapters 2, 14], [45, Chapter 3]:

1. Feedforward control
2. Feedback control

In Step 1, the feedforward part of the control input is selected. Its purpose is to achieve a desired candidate operating condition for the system. If the system is considered as depending on one or more parameters, then the feedforward control will also depend on the parameters. The desired operating condition can often be viewed as an equilibrium point of the nonlinear system that varies as the system parameters are varied. There are many situations when this operating condition is better viewed as a limit cycle that varies with the parameters. In Step 2, an additional part of the control input is designed to achieve desired qualities of the transient response in a neighborhood of the nominal operating condition. Typically this second part of the control is selected in feedback form.

In the following, the feedforward part of the control input is taken to be designed already and reflected in the system dynamical equations. The discussion centers on design of the feedback part of the control input. Because of this, it is convenient to denote the feedback part of the control simply by u and to view the feedforward part as being determined *a priori*. It is also convenient to take u to be small (close to zero) near the nominal operating condition. (Any offset in u is considered part of the feedforward control.)

It is convenient to view the system as depending on a single exogeneous parameter, denoted by μ. For instance, μ can represent the set-point of an aircraft's angle-of-attack, or the power demanded of an electric utility by a customer. In the former example, the control u might denote actuation of the aircraft's elevator angles about the nominal settings. In the latter example, u can represent a control signal in the voltage regulator of a power generator.

Consider, then, a nonlinear system depending on a single parameter μ

$$\dot{x} = f^{\mu}(x, u). \tag{51.9}$$

Here $x \in \mathbb{R}^n$ is the n-dimensional state vector, u is the feedback part of the control input, and μ is a parameter. Both u and μ are taken to be scalar-valued for simplicity. The dependence of the system equations on x, u, and μ is assumed smooth; i.e., f is jointly continuous and several times differentiable in these variables. This system is thus actually a one-parameter *family* of nonlinear control systems. The parameter μ is viewed as being allowed to change so slowly that its variation can be taken as quasistatic.

Suppose that the nominal operating condition of the system is an equilibrium point $x^0(\mu)$ that depends on the parameter μ. For simplicity of notation, suppose that the state x has been chosen so that this nominal equilibrium point is the origin for the range of values of μ for which it exists ($x^0(\mu) \equiv 0$). Recall that the nominal equilibrium is achieved using feedforward control. Although the process of choosing a feedforward control is not elaborated on here, it is important to emphasize that in general this process aims at securing a particular desired candidate operating condition. The feedforward control thus is expected to result in an acceptable form for the nominal operating condition, but there is no reason to expect that other operating conditions will also behave in a desirable fashion. As the parameter varies, the nominal operating condition may interact with other candidate operating conditions in bifurcations. The design of the feedback part of the control should take into account the other candidate operating conditions in addition to the nominal one.

For simplicity, suppose the control u is not allowed to introduce additional dynamics. That is, suppose u is required to be a direct state feedback (or static feedback), and denote it by $u = u(x, \mu)$. Note that in general u can depend on the parameter μ. That is, it can be *scheduled*. Since in the discussion above it was assumed that u is small near the nominal operating condition, it is assumed that $u(0, \mu) \equiv 0$ (for the parameter range in which the origin is an equilibrium).

Denote the linearization of Equation 51.9 at $x^0 = 0$, $u = 0$ by

$$\dot{x} = A(\mu)x + b(\mu)u. \tag{51.10}$$

Here,

$$A(\mu) := \frac{\partial f^{\mu}}{\partial x}(0, 0)$$

and

$$b(\mu) := \frac{\partial f^{\mu}}{\partial u}(0, 0).$$

Consider how control design for the linearized system depends on the parameter μ. Recall the terminology from linear system theory that the pair $(A(\mu), b(\mu))$ is controllable if the system (Equation 51.10) is controllable. Recall also that there are several simple tests for controllability, one of which is that the controllability matrix

$$(b(\mu), A(\mu)b(\mu), (A(\mu))^2 b(\mu), \ldots, (A(\mu))^{(n-1)} b(\mu))$$

is of full rank. (In this case this is equivalent to the matrix being nonsingular, since here the controllability matrix is square.)

If μ is such that the pair $(A(\mu), b(\mu))$ is controllable, then a standard linear systems result asserts the existence of a linear feedback $u(x, \mu) = -k(\mu)x$ stabilizing the system. (The associated closed-loop system would be $\dot{x} = (A(\mu) - b(\mu)k(\mu))x$.) Stabilizability tests not requiring controllability also exist, and these are more relevant to the problem at hand. Even more interesting from a practical perspective is the issue of output feedback stabilizability, since not all state variables are accessible for real-time measurement in many systems. As μ is varied over the desired regime of operability, the system (Equation 51.10) may lose stability and stabilizability.

51.5.2 Stressed Operation and Break-Down of Linear Techniques

A main motivation for the study of control of bifurcations is the need in some situations to operate a system in a condition for which the stability margin is small and linear (state or output) feedback is ineffective as a means for increasing the stability margin. Such a system is sometimes referred to as being "pushed to its limits," or "heavily loaded." In such situations, the ability of the system to function in a state of increased loading is a measure of system performance. Thus, the link between increased loading and reduced stability margin can be viewed as a performance vs. stability trade-off. Systems operating with a reduced achievable margin of stability may be viewed as being "stressed." This trade-off is not a general fact that can be proved in a rigorous fashion, but has been found to occur in a variety of applications. Examples of this trade-off are given at the end of this subsection.

Consider a system that is weakly damped and for which the available controls cannot compensate with sufficient additional damping. Such a situation may arise for a system for some ranges of parameter values and not for others. Let the *operating envelope* of a system be the possible combinations of system parameters for which system operability is being considered. Linear control system methods lose their effectiveness on that part of the operating envelope for which the operating condition of interest of system (Equation 51.10) is either:

1. not stabilizable with linear feedback using available sensors and actuators
2. linearly stabilizable using available sensors and actuators but only with unacceptably high feedback gains
3. vulnerable in the sense that small parameter changes can destroy the operating condition completely (as in a saddle-node bifurcation)

Operation in this part of the desired operating envelope can be referred to by terms such as "stressed operation."

An example of the trade-off noted previously is provided by an electric power system under conditions of heavy loading. At peaks in customer demand for electric power (such as during extremes in weather), the stability margin may become negligible, and nonlinear dynamical behaviors or divergence may arise (e.g., [12]). The divergence, known as voltage collapse, can lead to system blackout. Another example arises in operation of an aeroengine compressor at its peak pressure rise. The increased pressure rise comes at the price of nearness to instability. The unstable modes that can arise are strongly related to flow asymmetry modes that are unstabilizable by linear feedback to the compression system throttle. However, bifurcation control techniques have yielded valuable nonlinear throttle actuation techniques that facilitate operation in these circumstances with reduced risk of stall [16,25].

51.6 Control of Systems Exhibiting Bifurcation Behavior

Most engineering systems are designed to operate with a comfortable margin of stability. This means that disturbances or moderate changes in system parameters are unlikely to result in loss of stability. For example, a jet airplane in straight level flight under autopilot control is designed to have a large stability margin. However, engineering systems with a usually comfortable stability margin may at times be operated at a reduced stability margin. A jet airplane being maneuvered at high angle-of-attack to gain an advantage over an enemy aircraft, for instance, may have a significantly reduced stability margin. If a system operating condition actually loses stability as a parameter (like angle-of-attack) is slowly varied, then generally it is the case that a bifurcation occurs. This means that at least one new candidate operating condition emerges from the nominal one at the point of loss of stability. The purpose of this section is to summarize some results on control of bifurcations, with an emphasis placed on control of local bifurcations. Control of a particular global bifurcation, the period doubling bifurcation, is considered in the next section on control of chaos. This is because control of a period doubling bifurcation can result in quenching of the period doubling route to chaos summarized at the end of section three.

Bifurcation control involves designing a control input for a system to result in a desired modification to the system's bifurcation behavior. In Section 51.5, the division of control into a feedforward component and a feedback component was discussed. Both components of a control law can be viewed in terms of bifurcation control. The feedforward part of the control sets the equilibrium points of the system, and may influence the stability margin as well. The feedback part of the control has many functions, one of which is to ensure adequate stability of the desired operating condition over the desired operating envelope. Linear feedback is used to ensure an adequate margin of stability over a desired parameter range. Use of linear feedback to "delay" the onset of instability to parameter ranges outside the desired operating range is a common practice in control system design. An example is the gain scheduling technique [26]. Delaying instability modifies the bifurcation diagram of a system. Often, the available control authority does not allow stabilization of the nominal operating condition beyond some critical parameter value. At this value, instability leads to bifurcations of new candidate operating conditions. For simplicity, suppose that a single candidate operating condition is born at the bifurcation point. Another important goal in bifurcation control is to ensure that the bifurcation is *supercritical* (i.e., *safe*) and that the resulting candidate operating condition remains stable and close to the original operating condition for a range of parameter values beyond the critical value. The need for control laws that soften (stabilize) a hard (unstable) bifurcation has been discussed earlier in this chapter. This need is greatest in stressed systems, since in such systems delay of the bifurcation by linear feedback is not viable. A soft bifurcation provides the possibility of an alternative operating condition beyond the regime of operability at the nominal condition.

Both of these goals (delaying and stabilization) basically involve local considerations and can be approached analytically (if a good system model is available). Another goal might entail a reduction in amplitude of any bifurcated solutions over some prespecified parameter range. This goal is generally impossible to work with on a completely analytical basis. It requires extensive numerical study in addition to local analysis near the bifurcation(s).

In the most severe local bifurcations (saddle-node bifurcations), neither the nominal equilibrium point nor any bifurcated solution exists past the bifurcation. Even in such cases, an understanding of bifurcations provides some insight into control design for safe operation. For example, it may be possible to use this understanding to determine (or introduce via added control) a warning signal that becomes more pronounced as the severe bifurcation is approached. This signal would alert the high-level control system (possibly a human operator) that action is necessary to avoid catastrophe.

In this section, generally it is assumed that the feedforward component of the control has been predetermined, and the goal is to design the feedback component. An exception is the following brief discussion of a real-world example of the use of feedforward control to modify a system's operating condition and its stability margin in the face of large parameter variations. In short, this is an example where feedforward controls are used to successfully *avoid* the possibility of bifurcation. During takeoff and landing of a commercial jet aircraft, one can observe the deployment of movable surfaces on the leading- and trailing-edges of the wings. These movable surfaces, called flaps and slats, or camber changers [13], result in a nonlinear change in the aerodynamics, and, in turn, in an increased lift coefficient [13,51]. This is needed to allow takeoff and landing at reduced speeds. Use of these surfaces has the drawback of reducing the critical angle-of-attack for stall, resulting in a reduced stability margin. A common method to alleviate this effect is to incorporate other devices, called vortex generators. These are small airfoil-shaped vanes, protruding upward from the wings [13]. The incorporation of vortex generators results in a further modification to the aeodynamics, moving the stall angle-of-attack to a higher value. Use of the camber changers and the vortex generators are examples of feedforward control to modify the operating condition within a part of the aircraft's operating envelope. References [13] and [51] provide further details, as well as diagrams showing how the lift coefficient curve is affected by these devices.

51.6.1 Local Direct State Feedback

To give a flavor of the analytical results available in the design of the feedback component in bifurcation control, consider the nonlinear control system (Equation 51.9), repeated here for convenience:

$$\dot{x} = f^{\mu}(x, u). \tag{51.11}$$

Here, u represents the feedback part of the control law; the feedforward part is assumed to be incorporated into the function f. The technique and results of [1,2] form the basis for the following discussion. Details are not provided since they would require introduction of considerable notation related to multivariate Taylor series. However, an illustrative example is given based on formulas available in [1,2].

Suppose for simplicity that Equation 51.11 with $u \equiv 0$ possesses an equilibrium at the origin for a parameter range of interest (including the value $\mu = 0$). Moreover, suppose that the origin of Equation 51.11 with the control set to zero undergoes either a subcritical stationary bifurcation or a subcritical Andronov–Hopf bifurcation at the critical parameter value $\mu = 0$. Feedback control laws of the form $u = u(x)$ ("static state feedbacks") are derived in [1,2] that render the bifurcation supercritical.

For the Andronov-Hopf bifurcation, this is achieved using a formula for the coefficient β_2 in the expansion (Equation 51.7) of the characteristic exponent for the bifurcated limit cycle. Smooth nonlinear controls rendering $\beta_2 < 0$ are derived. For the stationary bifurcation, the controlled system is desired to display a supercritical pitchfork bifurcation. This is achieved using formulas for the coefficients β_1 and β_2 in the expansion (Equation 51.5) for the eigenvalue of the bifurcated equilibrium determining stability. Supercriticality is insured by determining conditions on $u(x)$ such that $\beta_1 = 0$ and $\beta_2 < 0$.

The following example is meant to illustrate the technique of [2]. The calculations involve use of formulas from [2] for the bifurcation stability coefficients β_1 and β_2 in the analysis of stationary bifurcations. The general formulas are not repeated here.

Consider the one-parameter family of nonlinear control systems

$$\dot{x}_1 = \mu x_1 + x_2 + x_1 x_2^2 + x_1^3, \tag{51.12}$$

$$\dot{x}_2 = -x_2 - x_1 x_2^2 + \mu u + x_1 u. \tag{51.13}$$

Here x_1, x_2 are scalar state variables, and μ is a real-valued parameter. This is meant to represent a system after application of a feedforward control, so that u is to be designed in feedback form. The nominal operating condition is taken to be the origin $(x_1, x_2) = (0, 0)$, which is an equilibrium of (Equations 51.12 and 51.13) when the control $u = 0$ for all values of the parameter μ.

Local stability analysis at the origin proceeds in the standard manner. The Jacobian matrix $A(\mu)$ of the right side of (Equations 51.12 and 51.13) is given by

$$A(\mu) = \begin{pmatrix} \mu & 1 \\ 0 & -1 \end{pmatrix}. \tag{51.14}$$

The system eigenvalues are μ and -1. Thus, the origin is stable for $\mu < 0$ but is unstable for $\mu > 0$. The critical value of the bifurcation parameter is therefore $\mu_c = 0$. Since the origin persists as an equilibrium point past the bifurcation, and since the critical eigenvalue is 0 (not an imaginary pair), it is expected that a stationary bifurcation occurs. The stationary bifurcation that occurs for the open-loop system can be studied by solving the pair of algebraic equations

$$0 = \mu x_1 + x_2 + x_1 x_2^2 + x_1^3, \tag{51.15}$$

$$0 = -x_2 - x_1 x_2^2 \tag{51.16}$$

for a nontrivial (i.e., nonzero) equilibrium (x^1, x^2) near the origin for μ near 0. Adding Equation 51.15 to Equation 51.16 gives

$$0 = \mu x_1 + x_1^3. \tag{51.17}$$

Disregarding the origin, this gives two new equilibrium points that exist for $\mu < 0$, namely

$$x(\mu) = \begin{pmatrix} \pm\sqrt{-\mu} \\ 0 \end{pmatrix}. \tag{51.18}$$

Since these bifurcated equilibria occur for parameter values ($\mu < 0$) for which the nominal operating condition is unstable, the bifurcation is a *subcritical* pitchfork bifurcation.

The first issue addressed in the control design is the possibility of using linear feedback to delay the bifurcation to some positive value of μ. This would require stabilization of the origin at $\mu = 0$. Because of the way in which the control enters the system dynamics, however, the system eigenvalues are not affected by linear feedback at the parameter value $\mu = 0$. To see this, simply note that in Equation 51.13, the term μu vanishes when $\mu = 0$, and the remaining impact of the control is through the term $x_1 u$. This latter term would result only in the addition of *quadratic* terms to the right side of Equations 51.12 and 51.13 for any linear feedback u. Hence, the system is an example of a stressed nonlinear system for μ near 0.

Since the pitchfork bifurcation cannot be delayed by linear feedback, next consider the possibility of rendering the pitchfork bifurcation supercritical using nonlinear feedback. This rests on determining how feedback affects the bifurcation stability coefficients β_1 and β_2 for this stationary bifurcation. Once this is known, it is straightforward to seek a feedback that renders $\beta_1 = 0$ and $\beta_2 < 0$. The formulas for β_1 and β_2 derived in [2] simplify for systems with no quadratic terms in the state variables. For such systems, the coefficient β_1 always vanishes, and the calculation of β_2 also simplifies. Since the dynamical equations 51.12 and 51.13 in the example contain no quadratic terms in x, it follows that $\beta_1 = 0$ in the absence of control. Moreover, if the control contains no linear terms, then it will not introduce quadratic terms into the dynamics. Hence, for any smooth feedback $u(x)$ containing no linear terms in x, the bifurcation stability coefficient $\beta_1 = 0$.

As for the bifurcation stability coefficient β_2, the pertinent formula in [2] applied to the open-loop system yields $\beta_2 = 2$. Thus, a subcritical pitchfork bifurcation is predicted for the open-loop system, a fact that was established above using simple algebra. Now let the control consist of a quadratic function of the state and determine conditions under which $\beta_2 < 0$ for the closed-loop system.* Thus, consider u to be of the form

$$u(x_1, x_2) = -k_1x_1^2 - k_2x_1x_2 - k_3x_2^2. \tag{51.19}$$

The formula in [2] yields that β_2 for the closed-loop system is then given by

$$\beta_2 = 2(1 - k_1). \tag{51.20}$$

Thus, to render the pitchfork bifurcation in Equations 51.12 and 51.13 supercritical, it suffices to take u to be the simple quadratic function

$$u(x_1, x_2) = -k_1x_1^2, \tag{51.21}$$

with any gain $k_1 > 1$. In fact, other quadratic terms, as well as any other cubic or higher order terms, can be included along with this term without changing the local result that the bifurcation is rendered supercritical. Additional terms can be useful in improving system behavior as the parameter leaves a local neighborhood of its critical value.

51.6.2 Local Dynamic State Feedback

Use of a static state feedback control law $u = u(x)$ has potential disadvantages in nonlinear control of systems exhibiting bifurcation behavior. To explain this, consider the case of an equilibrium $x^0(\mu)$ as the nominal operating condition. The equilibrium is not translated to the origin to illustrate how it is affected by feedback. In general, a static state feedback

$$u = u(x - x^0(\mu)) \tag{51.22}$$

designed with reference to the nominal equilibrium path $x^0(\mu)$ of Equation 51.11 will affect not only the stability of this equilibrium but also the location and stability of other equilibria. Now suppose that Equation 51.11 is only an approximate model for the physical system of interest. Then the nominal equilibrium branch will also be altered by the feedback. A main disadvantage of such an effect is the wasted control energy that is associated with the forced alteration of the system equilibrium structure. Other disadvantages are that system performance is often degraded by operating at an equilibrium that differs from the one at which the system is designed to operate. Moreover, by influencing the locations of system equilibria, the feedback control is competing with the feedforward part of the control.

For these reasons, dynamic state feedback-type bifurcation control laws have been developed that do not affect the locations of system equilibria [22,23,50]. The method involves incorporation of filters called "washout filters" into the controller architecture. A washout filter-aided control law preserves all system equilibria, and does so without the need for an accurate system model.

A washout filter is a stable high pass filter with zero static gain [15, p. 474]. The typical transfer function for a washout filter is

$$G(s) = \frac{y_i(s)}{x_i(s)} = \frac{s}{s+d}. \tag{51.23}$$

where x_i is the input variable to the filter and y_i is the output of the filter. A washout filter produces a nonzero output only during the transient period. Thus, such a filter "washes out" sensed signals that have settled to constant steady-state values. Washout filters occur in control systems for power systems [5, p. 277], [41, Chapter 9] and aircraft [7,15,39,45].

* Cubic terms are not included because they would result in quartic terms on the right side of Equations 51.12 and 51.13, while the formula for β_2 in [2] involves only terms up to cubic order in the state.

Washout filters are positioned in a control system so that a sensed signal being fed back to an actuator first passes through the washout filter. If, due to parameter drift or intentional parameter variation, the sensed signal has a steady-state value that deviates from the assumed value, the washout filter will give a zero output and the deviation will not propagate. If a direct state feedback were used instead, the steady-state deviation in the sensed signal would result in the control modifying the steady-state values of the other controlled variables. As an example, washout filters are used in aircraft yaw damping control systems to prevent these dampers from "fighting" the pilot in steady turns [39, p. 947].

From a nonlinear systems perspective, the property of washout filters described above translates to achieving equilibrium preservation, i.e., zero steady-state tracking error, in the presence of system uncertainties.

Washout filters can be incorporated into bifurcation control laws for Equation 51.11. This should be done only after the feedforward control has been designed and incorporated *and* part of the feedback control ensuring satisfactory equilibrium point structure has also been designed and incorporated into the dynamics. Otherwise, since washout filters preserve equilibrium points, there will be no possibility of modifying the equilibria. It is assumed below that these two parts of the control have been chosen and implemented.

For each system state variable $x_i, i = 1, \ldots, n$, in Equation 51.11, introduce a washout filter governed by the dynamic equation

$$\dot{z}_i = x_i - d_i z_i \tag{51.24}$$

along with output equation

$$y_i = x_i - d_i z_i. \tag{51.25}$$

Here, the d_i are positive parameters (this corresponds to using stable washout filters). Finally, require the control u to depend only on the measured variables y, and that $u(y)$ satisfy $u(0) = 0$.

In this formulation, n washout filters, one for each system state, are present. In fact, the actual number of washout filters needed, and hence also the resulting increase in system order, can usually be taken less than n.

It is straightforward to see that washout filters result in equilibrium preservation and automatic equilibrium (operating point) following. Indeed, since $u(0) = 0$, it is clear that y vanishes at steady-state. Hence, the x subvector of a closed-loop equilibrium point (x, z) agrees exactly with the open-loop equilibrium value of x. Also, since y_i may be re-expressed as

$$y_i = x_i - d_i z_i = (x_i - x_i^0(\mu)) - d_i(z_i - z_i^0(\mu)), \tag{51.26}$$

the control function $u = u(y)$ is guaranteed to center at the correct operating point.

51.7 Control of Chaos

Chaotic behavior of a physical system can either be desirable or undesirable, depending on the application. It can be beneficial for many reasons, such as enhanced mixing of chemical reactants, or, as proposed recently [36], as a replacement for random noise as a masking signal in a communication system. Chaos can, on the other hand, entail large amplitude motions and oscillations that might lead to system failure. The control techniques discussed in this section have as their goal the replacement of chaotic behavior by a nonchaotic steady-state behavior. The first technique discussed is that proposed by Ott, Grebogi, and Yorke [33]. The Ott–Grebogi–Yorke (OGY) method sparked significant interest and activity in control of chaos. The second technique discussed is the use of bifurcation control to delay or extinguish the appearance of chaos in a family of systems.

51.7.1 Exploitation of Chaos for Control

Ott, Grebogi, and Yorke [33] proposed an approach to control of chaotic systems that involves use of small controls and exploitation of chaotic behavior. To explain this method, recall from section three that a strange attractor has embedded within itself a "dense" set (infinitely many) of unstable limit cycles. The strange attractor is a candidate operating condition, according to the definition in section two. In the absence of control, it is the actual system operating condition. Suppose that system performance would be significantly improved by operation at one of the unstable limit cycles embedded in the strange attractor. The OGY method replaces the originally chaotic system operation with operation along the selected unstable limit cycle.

Figure 51.8 depicts a strange attractor along with a particular unstable limit cycle embedded in it. It is helpful to keep such a figure in mind in contemplating the OGY method. The goal is to reach a limit cycle such as the one shown in Figure 51.8, or another of some other period or amplitude. Imagine a trajectory that lies on the strange attractor in Figure 51.8, and suppose the desire is to use control to force the trajectory to reach the unstable limit cycle depicted in the figure and to remain there for all subsequent time.

Control design to result in operation at the desired limit cycle is achieved by the following reasoning. First, note that the desired unstable limit cycle is also a candidate operating condition. Next, recall from section three that a trajectory on the strange attractor will, after sufficient time, pass as close as desired to any other point of interest on the attractor. Thus, the trajectory will eventually come close (indeed, arbitrarily close) to the desired limit cycle. Thus, no control effort whatsoever is needed in order for the trajectory to reach the desired limit cycle—chaos guarantees that it will. To maintain the system state on the desired limit cycle, a small stabilizing control signal is applied once the trajectory enters a small neighborhood of the limit cycle. Since the limit cycle is rendered stable by this control, the trajectory will converge to the limit cycle. If noise drives the trajectory out of the neighborhood where control is applied, the trajectory will wander through the strange attractor again until it once again enters the neighborhood and remains there by virtue of the control.

Note that the control obtained by this method is an example of a variable structure control, since it is "on" in one region in state space and "off" in the rest of the space. Also, the particular locally stabilizing control used in the neighborhood of the desired limit cycle has not been discussed in the foregoing. This is because several approaches are possible, among them pole placement [38]. See [34,35] for further discussion.

Two particularly significant strengths of the OGY technique are:

1. It requires only small controls
2. It can be applied to experimental systems for which no mathematical model is available

The first of these strengths is due to the assumption that operation at an unstable limit cycle embedded in the strange attractor is desirable. If none of the embedded limit cycles provides adequate performance, then a large control could possibly be used to introduce a new desirable candidate operating condition within the strange attractor. This could be followed by a small control possibly designed within the OGY framework. Note that the large control would be a feedforward control, in the terminology used previously in this chapter. For a discussion of the second main strength of the OGY control method mentioned above, see, e.g., [17,33,34]. It suffices to note here that a construction known as experimental delay coordinate embedding is one means to implement this control method without *a priori* knowledge of a reliable mathematical model.

Several interesting applications of the OGY control method have been performed, including control of cardiac chaos (see [35]). In [17], a multimode laser was controlled well into its usually unstable regime.

51.7.2 Bifurcation Control of Routes to Chaos

The bifurcation control techniques discussed in section six have direct relevance for issues of control of chaotic behavior of dynamical systems. This is because chaotic behavior often arises as a result of bifurcations, such as through the period doubling route to chaos. The bifurcation control technique discussed earlier in this chapter is model-based. Thus, the control of chaos applications of the technique also require availability of a reliable mathematical model.

Only a few comments are given here on bifurcation control of routes to chaos, since the main tool has already been discussed in section six. The cited references may be consulted for details and examples.

In [50], a thermal convection loop model is considered. The model, which is equivalent to the Lorenz equations mentioned in Section 51.4, displays a series of bifurcations leading to chaos. In [50], an Andronov–Hopf bifurcation that occurs in the model is rendered supercritical using *local* dynamic state feedback (of the type discussed in Section 51.6). The feedback is designed using local calculations at the equilibrium point of interest. It is found that this simple control law results in elimination of the chaotic behavior in the system. From a practical perspective, this allows operation of the convection loop in a *steady* convective state with a desired velocity and temperature profile.

Feedback control to render supercritical a previously subcritical period doubling bifurcation was studied in [4,47]. In [4], a discrete-time model is assumed, whereas a continuous-time model is assumed in [47]. The discrete-time model used in [4] takes the form (k is an integer)

$$x(k+1) = f^{\mu}(x(k), u(k)) \tag{51.27}$$

where $x(k) \in \mathbb{R}^n$ is the state, $u(k)$ is a scalar control input, $\mu \in \mathbb{R}$ is the bifurcation parameter, and the mapping f^{μ} is sufficiently smooth in x, u, and μ. The continuous-time model used in [47] is identical to Equation 51.9.

For discrete-time systems, a limit cycle must have an integer period. A period-1 limit cycle sheds a period-2 limit cycle upon period doubling bifurcation. The simplicity of the discrete-time setting results in explicit formulas for bifurcation stability coefficients and feedback controls [4]. By improving the stability characteristics of a bifurcated period-2 limit cycle, an existing period doubling route to chaos can be extinguished. Moreover, the period-2 limit cycle will then remain close to the period-1 limit cycle for an increased range of parameters.

For continuous-time systems, limit cycles cannot in general be obtained analytically. Thus, [47] employs an approximate analysis technique known as *harmonic balance*. Approximate bifurcation stability coefficients are obtained, and control to delay the onset of period doubling bifurcation or stabilize such a bifurcation is discussed.

The washout filter concept discussed in section six is extended in [4] to discrete-time systems. In [47], an extension of the washout filter concept is used that allows approximate preservation of limit cycles of a certain frequency.

51.8 Concluding Remarks

Control of bifurcations and chaos is a developing area with many interesting avenues for research and for application. Some of the tools and ideas that have been used in this area were discussed. Connections among these concepts, and relationships to traditional control ideas, have been emphasized.

Acknowledgment

During the preparation of this chapter, the authors were supported in part by the Air Force Office of Scientific Research (U.S.), the Electric Power Research Institute (U.S.), the National Science Foundation (U.S.), and the Ministero della Università e della Ricerca Scientifica e Tecnologica under the National Research Plan 40% (Italy).

References

1. Abed, E.H. and Fu, J.-H., Local feedback stabilization and bifurcation control, I. Hopf bifurcation, *Syst. Control Lett.*, 7, 11–17, 1986.
2. Abed E.H. and Fu J.-H., Local feedback stabilization and bifurcation control, II. Stationary bifurcation, *Syst. Control Lett.*, 8, 467–473, 1987.
3. Abed E.H. and Wang H.O., Feedback control of bifurcation and chaos in dynamical systems, in *Nonlinear Dynamics and Stochastic Mechanics*, W. Kliemann and N. Sri Namachchivaya, Eds., CRC Press, Boca Raton, 153–173, 1995.
4. Abed E.H., Wang H.O., and Chen R.C., Stabilization of period doubling bifurcations and implications for control of chaos, *Physica D*, 70(1–2), 154–164, 1994.
5. Anderson P.M. and Fouad A.A., *Power System Control and Stability*, Iowa State University Press, Ames, 1977.
6. Andronov A.A., Vitt, A.A., and Khaikin S.E., *Theory of Oscillators*, Pergamon Press, Oxford, 1966 (reprinted by Dover, New York, 1987), English translation of Second Russian Edition; Original Russian edition published in 1937.
7. Blakelock J.H., *Automatic Control of Aircraft and Missiles*, 2nd ed., John Wiley & Sons, New York, 1991.
8. Chapman, G.T., Yates, L.A., and Szady, M.J., Atmospheric flight dynamics and chaos: Some issues in modeling and dimensionality, in *Applied Chaos*, J.H. Kim and J. Stringer, Eds., John Wiley & Sons, New York, 87–141, 1992.
9. Chen, G. and Dong, X., From chaos to order—Perspectives and methodologies in controlling chaotic nonlinear dynamical systems, *Internat. J. Bifurcation Chaos*, 3, 1363–1409, 1993.
10. Chow, S.N. and Hale, J.K., *Methods of Bifurcation Theory*, Springer-Verlag, New York, 1982.
11. Devaney, R.L., *An Introduction to Chaotic Dynamical Systems*, 2nd ed., Addison-Wesley, Redwood City, CA, 1989.
12. Dobson, I., Glavitsch, H., Liu, C.-C., Tamura, Y., and Vu, K., Voltage collapse in power systems, *IEEE Circuits and Devices Magazine*, 8(3), 40–45, 1992.
13. Dole, C.E., *Flight Theory and Aerodynamics: A Practical Guide for Operational Safety*, John Wiley & Sons, New York, 1981.
14. Dorning, J.J. and Kim, J.H., Bridging the gap between the science of chaos and its technological applications, in *Applied Chaos*, J.H. Kim and J. Stringer, Eds., John Wiley & Sons, New York, 3–30, 1992.
15. Etkin, B., *Dynamics of Atmospheric Flight*, John Wiley & Sons, New York, 1972.
16. Eveker, K.M., Gysling, D.L., Nett, C.N., and Sharma, O.P., Integrated control of rotating stall and surge in aeroengines, in *Sensing, Actuation, and Control in Aeropropulsion*, J.D. Paduano, Ed., Proc. SPIE 2494, 21–35, 1995.
17. Gills, Z., Iwata, C., Roy, R., Schwartz, I.B., and Triandaf, I., Tracking unstable steady states: Extending the stability regime of a multimode laser system, *Phys. Rev. Lett.*, 69, 3169–3172, 1992.
18. Hassard, B.D., Kazarinoff, N.D., and Wan, Y.H., *Theory and Applications of Hopf Bifurcation*, Cambridge University Press, Cambridge, 1981.
19. Jackson, E.A., *Perspectives of Nonlinear Dynamics*, Vols. 1 and 2, Cambridge University Press, Cambridge, 1991.
20. Kerrebrock, J.L., *Aircraft Engines and Gas Turbines*, 2nd ed., MIT Press, Cambridge, MA, 1992.
21. Kim, J.H. and Stringer, J., Eds., *Applied Chaos*, John Wiley & Sons, New York, 1992.
22. Lee, H.C., *Robust Control of Bifurcating Nonlinear Systems with Applications*, Ph.D. Dissertation, Department of Electrical Engineering, University of Maryland, College Park, 1991.
23. Lee, H.-C. and Abed, E.H., Washout filters in the bifurcation control of high alpha flight dynamics, *Proc. 1991 Am. Control Conf.*, Boston, pp. 206–211, 1991.
24. Liaw, C.Y. and Bishop, S.R., Nonlinear heave-roll coupling and ship rolling, *Nonlinear Dynamics*, 8, 197–211, 1995.
25. Liaw, D.-C. and Abed, E.H., Analysis and control of rotating stall, *Proc. NOLCOS'92: Nonlinear Control System Design Symposium*, (M. Fliess, Ed.), June 1992, Bordeaux, France, pp. 88–93, Published by the International Federation of Automatic Control; See also: Active control of compressor stall inception: A bifurcation-theoretic approach, *Automatica*, 32, 1996 (to appear).
26. Lin, C.-F., *Advanced Control Systems Design*, Prentice Hall Series in Advanced Navigation, Guidance, and Control, and their Applications, Prentice Hall, Englewood Cliffs, NJ, 1994.
27. Lorenz, E.N., Deterministic nonperiodic flow, *J. Atmosph. Sci.*, 20, 130–141, 1963.
28. Marsden, J.E. and McCracken, M., *The Hopf Bifurcation and Its Applications*, Springer-Verlag, New York, 1976.

29. McRuer, D., Ashkenas, I., and Graham, D., *Aircraft Dynamics and Automatic Control*, Princeton University Press, Princeton, 1973.
30. Mees, A.I., *Dynamics of Feedback Systems*, John Wiley & Sons, New York, 1981.
31. Murray, J.D., *Mathematical Biology*, Springer-Verlag, New York, 1990.
32. Nayfeh, A.H. and Balachandran, B., *Applied Nonlinear Dynamics: Analytical, Computational, and Experimental Methods*, Wiley Series in Nonlinear Science, John Wiley & Sons, New York, 1995.
33. Ott, E., Grebogi, C., and Yorke, J.A., Controlling chaos, *Phys. Rev. Lett.*, 64, 1196–1199, 1990.
34. Ott, E., *Chaos in Dynamical Systems*, Cambridge University Press, Cambridge, 1993.
35. Ott, E., Sauer, T., and Yorke, J.A., Eds., *Coping with Chaos: Analysis of Chaotic Data and the Exploitation of Chaotic Systems*, Wiley Series in Nonlinear Science, John Wiley & Sons, New York, 1994.
36. Pecora, L.M. and T.L. Carroll, Synchronization in chaotic systems, *Phys. Rev. Lett.*, 64, 821–824, 1990.
37. Poincaré, H., *New Methods of Celestial Mechanics*, Parts 1, 2, and 3 (Edited and introduced by D.L. Goroff), Vol. 13 of the History of Modern Physics and Astronomy Series, American Institute of Physics, U.S., 1993; English translation of the French edition *Les Méthodes nouvelles de la Mécanique céleste*, originally published during 1892–1899.
38. Romeiras, F.J., Grebogi, C., Ott, E., and Dayawansa, W.P., Controlling chaotic dynamical systems, *Physica D*, 58, 165–192, 1992.
39. Roskam, J., *Airplane Flight Dynamics and Automatic Flight Controls (Part II)*, Roskam Aviation and Engineering Corp., Lawrence, Kansas, 1979.
40. Sanchez, N.E. and Nayfeh, A.H., Nonlinear rolling motions of ships in longitudinal waves, *Internat. Shipbuilding Progress*, 37, 247–272, 1990.
41. Sauer, P.W. and Pai, M.A., *Power System Dynamics and Stability*, Draft manuscript, Department of Electrical and Computer Engineering, University of Illinois at Urbana-Champaign, 1995.
42. Scott, S.K., *Chemical Chaos*, International Series of Monographs on Chemistry, Oxford University Press, Oxford, 1991.
43. Seydel, R., *Practical Bifurcation and Stability Analysis: From Equilibrium to Chaos*, 2nd ed., Springer-Verlag, Berlin, 1994.
44. Shinbrot, T., Grebogi, C., Ott, E., and Yorke, J.A., Using small perturbations to control chaos, *Nature*, 363, 411–417, 1993.
45. Stevens, B.L. and Lewis, F.L., *Aircraft Control and Simulation*, John Wiley & Sons, New York, 1992.
46. Strogatz, S.H., *Nonlinear Dynamics and Chaos: With Applications to Physics, Biology, Chemistry, and Engineering*, Addison-Wesley, Reading, MA, 1994.
47. Tesi, A., Abed, E.H., Genesio, R., and Wang, H.O., Harmonic balance analysis of period doubling bifurcations with implications for control of nonlinear dynamics, Report No. RT 11/95, Dipartimento di Sistemi e Informatica, Università di Firenze, Firenze, Italy, 1995 (submitted for publication).
48. Thompson, J.M.T. and Stewart, H.B., *Nonlinear Dynamics and Chaos*, John Wiley & Sons, Chichester, U.K., 1986.
49. Thompson, J.M.T., Stewart, H.B., and Ueda, Y., Safe, explosive, and dangerous bifurcations in dissipative dynamical systems, *Phys. Rev. E*, 49, 1019–1027, 1994.
50. Wang, H.O. and Abed, E.H., Bifurcation control of a chaotic system, *Automatica*, 31, 1995, in press.
51. Wegener, P.P., *What Makes Airplanes Fly?: History, Science, and Applications of Aerodynamics*, Springer-Verlag, New York, 1991.
52. Ziegler, F., *Mechanics of Solids and Fluids*, Springer-Verlag, New York, 1991.

Further Reading

Detailed discussions of bifurcation and chaos are available in many excellent books (e.g., [6,10,11,18,19, 28,30,32,34,35,43,46,48]). These books also discuss a variety of interesting applications. Many examples of bifurcations in mechanical systems are given in [52]. There are also several journals devoted to bifurcations and chaos. Of particular relevance to engineers are *Nonlinear Dynamics*, the *International Journal of Bifurcation and Chaos*, and the journal *Chaos, Solitons and Fractals*.

The book [45] discusses feedforward control in the context of "trimming" an aircraft using its nonlinear equations of motion and the available controls. Bifurcation and chaos in flight dynamics are discussed in [8]. Lucid explanations on specific uses of washout filters in aircraft control systems are given in [15, pp. 474–475], [7, pp. 144–146], [39, pp. 946–948 and pp. 1087–1095], and [45, pp. 243–246 and p. 276]. The book [31] discussed applications of nonlinear dynamics in biology and population dynamics.

Computational issues related to bifurcation analysis are addressed in [43]. Classification of bifurcations as safe or dangerous is discussed in [32,43,48,49].

The edited book [14] contains interesting articles on research needs in applications of bifurcations and chaos.

The article [3] contains a large number of references on bifurcation control, related work on stabilization, and applications of these techniques. The review papers [9,44] address control of chaos methods. In particular, [44] includes a discussion of use of sensitive dependence on initial conditions to direct trajectories to targets. The book [35] includes articles on control of chaos, detection of chaos in time series, chaotic data analysis, and potential applications of chaos in communication systems. The book [32] also contains discussions of control of bifurcations and chaos, and of analysis of chaotic data.

52

Open-Loop Control Using Oscillatory Inputs

J. Baillieul
Boston University

B. Lehman
Northeastern University

52.1 Introduction

The interesting discovery that the topmost equilibrium of a pendulum can be stabilized by oscillatory vertical movement of the suspension point has been attributed to Bogolyubov [11] and Kapitsa [26], who published papers on this subject in 1950 and 1951, respectively. In the intervening years, literature appeared analyzing the dynamics of systems with oscillatory forcing, e.g., [31]. Control designs based on oscillatory inputs have been proposed (for instance [8] and [9]) for a number of applications. Many classical results on the stability of operating points for systems with oscillatory inputs depend on the eigenvalues of the averaged system lying in the left half-plane. Recently, there has been interest in the stabilization of systems to which such classical results do not apply. Coron [20], for instance, has shown the existence of a time-varying feedback stabilizer for systems whose averaged versions have eigenvalues on the imaginary axis. This design is interesting because it provides smooth feedback stabilization for systems which Brockett [15] had previously shown were never stabilizable by smooth, time-invariant feedback. For conservative mechanical systems with oscillatory control inputs, Baillieul [7] has shown that stability of operating points may be assessed in terms of an energy-like quantity known as the *averaged potential*. Control designs with objectives beyond stabilization have been studied in path-planning for mobile robots [40] and in other applications where the models result in "drift-free" controlled differential equations. Work by Sussmann and Liu [36–38], extending earlier ideas of Haynes and Hermes, [21], has shown that, for drift-free systems satisfying a certain Lie algebra rank condition (LARC discussed in Section 52.2), arbitrary smooth trajectories may be interpolated to an arbitrary accuracy by appropriate choice of oscillatory controls. Leonard and Krishnaprasad [28] have reported algorithms for generating desired trajectories when certain "depth" conditions on the brackets of the defining vector fields are satisfied.

This chapter summarizes the current theory of open-loop control using oscillatory forcing. The recent literature has emphasized geometric aspects of the methods, and our discussion in Sections 52.2 and 52.3 will reflect this emphasis. Open-loop methods are quite appealing in applications in which the realtime sensor measurements needed for feedback designs are expensive or difficult to obtain. Because the methods work by virtue of the geometry of the motions in the systems, the observed effects may be quite robust. This is borne out by experiments described below. The organization of the article is as follows. In the present section we introduce oscillatory open-loop control laws in two very different ways. Example 52.1 illustrates the geometric mechanism through which oscillatory forcing produces nonlinear behavior in certain types of (drift-free) systems. Following this, the remainder of the section introduces a more classical analytical approach to control systems with oscillatory inputs. Section 52.2 provides a detailed exposition of open loop design methods for so-called "drift-free" systems. The principal applications are in kinematic motion control, and the section concludes with an application to grasp mechanics. Section 52.3 discusses some geometric results of oscillatory forcing for stabilization. Examples have been chosen to illustrate different aspects of the theory.

52.1.1 Noncommuting Vector Fields, Anholonomy, and the Effect of Oscillatory Inputs

We begin by describing a fundamental mathematical mechanism for synthesizing motions in a controlled dynamical system using oscillatory forcing. We shall distinguish among three *classes* of systems:

I. Drift-free systems with input entering linearly:

$$\dot{x} = \sum_{i=1}^{m} u_i g_i(x). \tag{52.1}$$

Here we assume each $g_i : \mathbf{R}^n \to \mathbf{R}^n$ is a smooth (i.e. analytic) vector field, and each "input" $u_i(\cdot)$ is a piecewise analytic function of time. Generally, we assume $m < n$.

II. Systems with drift and input entering affinely:

$$\dot{x} = f(x) + \sum_{i=1}^{m} u_i g_i(x). \tag{52.2}$$

The assumptions here are, as in the previous case, with $f : \mathbf{R}^n \to \mathbf{R}^n$ also assumed to be smooth.

III. Systems with no particular restriction on the way in which the control enters:

$$\dot{x} = f(x, u). \tag{52.3}$$

Here $u = (u_1, \ldots, u_m)^T$ is a vector of piecewise analytic inputs, as in the previous two cases, and $f : \mathbf{R}^n \times \mathbf{R}^m \to \mathbf{R}^n$ is analytic.

This is a hierarchy of types, each a special case of its successor in the list. More general systems could be considered, and indeed, in the Lagrangian models which are described in Section 52.3, we shall encounter systems in which the derivatives of inputs also enter the equations of motion. It will be shown that these systems can be reduced to Class III, however.

Remark 52.1

The extra term on the right hand side of Equation 52.2 is called a "drift" because, in the absence of control input, the differential equation "drifts" in the direction of the vector field f.

Example 52.1:

Even Class I systems possess the essential features of the general mechanism (anholonomy) by which oscillatory inputs may be used to synthesize desired motions robustly. (See remarks below on the robustness of open-loop methods.) Consider the simple and widely studied "Heisenberg" system (see [16]):

$$\begin{pmatrix} \dot{x}_1 \\ \dot{x}_2 \\ \dot{x}_3 \end{pmatrix} = \begin{pmatrix} u_1(t) \\ u_2(t) \\ u_2(t)x_1 - u_1(t)x_2 \end{pmatrix}. \tag{52.4}$$

This system is a special case of Equation 52.1 in which $m = 2$ and

$$g_1(x) = \begin{pmatrix} 1 \\ 0 \\ -x_2 \end{pmatrix}, \quad g_2(x) = \begin{pmatrix} 0 \\ 1 \\ x_1 \end{pmatrix}.$$

If we define the *Lie bracket* of these vector fields by $[g_1, g_2] = \frac{\partial g_1}{\partial x} g_2 - \frac{\partial g_2}{\partial x} g_1$, then a simple calculation reveals $[g_1, g_2] = (0, 0, -2)^T$. Another fairly straightforward calculation shows that, from general considerations, there is a choice of inputs $(u_1(\cdot), u_2(\cdot))$ which generates a trajectory pointing approximately in the direction $(0, 0, 1)^T$, and this approximation may be made arbitrarily accurate (see, Nijmeijer and Van der Schaft, [30], p. 77 or Bishop and Crittenden, [10], p. 18.). In the present case, we can be more precise and more explicit. Starting at the origin, $(x_1, x_2, x_3) = (0, 0, 0)$, motion in any of the three coordinate directions is possible. By choosing $u_1(t) \equiv 1,\ u_2(t) \equiv 0$, for instance, motion along the x_1-axis is produced, and motion along the x_2-axis may similarly be produced by reversing the role of $u_1(\cdot)$ and $u_2(\cdot)$. Motion along the x_3-axis is more subtle. If we let the inputs, $(u_1(\cdot), u_2(\cdot))$, trace a closed curve so that the states x_1 and x_2 end with the same values with which they began, the net motion of the system is along the x_3-axis. This is illustrated in Figure 52.1. Brockett [17] has observed that the precise shape of the input

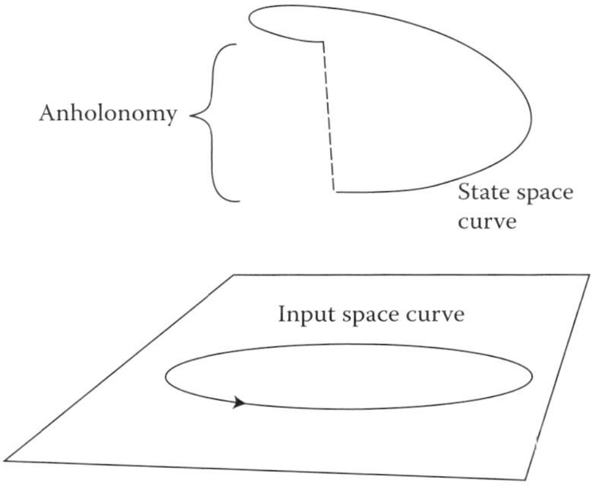

FIGURE 52.1 The anholonomy present in the Heisenberg system is depicted in a typical situation in which the input variables trace a closed curve and the state variables trace a curve which does not close. The distance between endpoints of the state space curve (measured as the length of the dashed vertical line) reflects the anholonomy in the system.

curve is unimportant, but the x_3-distance is twice the (signed) area circumscribed by the (x_1, x_2)-curve. For this simple system, we thus have a way to prescribe trajectories between any two points in $\mathbf{R}^3$. Taking the case of trajectories starting at the origin, for instance, we may specify a trajectory passing through any other point (x, y, z) at time $t = 1$ by finding a circular arc initiating at the origin in (x_1, x_2)-space with appropriate length, initial direction, and curvature. This construction of inputs may be carried out somewhat more generally, as discussed below in Section 52.2. Ideas along this line are also treated in more depth in [28].

The geometric mechanism by which motion is produced by oscillatory forcing is fairly transparent in the case of the Heisenberg system. For systems of the form of Equation 52.2 and especially of the form of Equation 52.3, there is no comparably complete geometric theory. Indeed, much of the literature on such systems makes no mention of geometry. A brief survey/overview of some of the classical literature on such systems is given next. We shall return to the geometric point of view in Sections 52.2 and 52.3.

52.1.2 Oscillatory Inputs to Control Physical Systems

The idea of using oscillatory forcing to control physical processes is not new. Prompted by work in the 1960s on the periodic optimal control of chemical processes, Speyer and Evans [33] derived a sufficiency condition for a periodic process to minimize a certain integral performance criterion. This approach also led to the observation that periodic paths could be used to improve aircraft fuel economy (see [32].). Previously cited work of Bogolyubov and Kapitsa [11] and [26], led Bellman et al. [8] and [9] to investigate the systematic use of *vibrational control* as an open loop control technique in which zero average oscillations are introduced into a system's parameters to achieve a dynamic response (such as stabilizing effects). For example, it has been shown in [8,9,24,25], that the oscillation of flow rates in a continuous stirred tank reactor allows operating exothermic reactions at (average) yields which were previously unstable. Similarly, [6] and [7] describe the general geometric mechanism by which oscillations along the vertical support of an inverted pendulum stabilize the upper equilibrium point.

This section treats the basic theory of vibrational control introduced in [8] and [9]. The techniques are primarily based on [8], with additional material taken from [24] and [25]. In particular, in [24] and [25], the technique of vibrational control has been extended to delay differential equations.

52.1.3 Problem Statement

Consider the nonlinear differential equation (Class III)

$$\frac{dx}{dt} = f(x, u), \tag{52.5}$$

where $f : \mathbf{R}^n \times \mathbf{R}^d \to \mathbf{R}^n$ is continuously differentiable, $x \in \mathbf{R}^n$ is the state, and $u = (u_1, \ldots, u_d)^T$ is a vector of control inputs assumed to be piecewise analytic functions of time. These are the quantities which we can directly cause to vibrate.

Introduce into Equation 52.5 oscillatory inputs according to the law $u(t) = \lambda_0 + \gamma(t)$ where λ_0 is a constant vector and $\gamma(t)$ is a *periodic average zero* (PAZ) vector. (For simplicity, $\gamma(t)$ has been assumed periodic. However, the following discussion can be extended to the case where $\gamma(t)$ is an almost periodic zero average vector [8,9,24,25].) Then Equation 52.5 becomes

$$\frac{dx}{dt} = f(x, \lambda_0 + \gamma(t)). \tag{52.6}$$

Assume that Equation 52.5 has a fixed equilibrium point $x_s = x_s(\lambda_0)$ for fixed $u(t) = \lambda_0$.

Definition 52.1:

An equilibrium point $x_s(\lambda_0)$ *of Equation 52.5 is said to be vibrationally stabilizable if, for any* $\delta > 0$*, there exists a PAZ vector* $\gamma(t)$ *such that Equation 52.6 has an asymptotically stable periodic solution,* $x^*(t)$*, characterized by*

$$\| \overline{x}^* - x_s(\lambda_0) \| \leq \delta, \quad \text{where } \overline{x}^* = \frac{1}{T}\int_0^T x^*(t)\,dt.$$

It is often preferable that Equations 52.5 and 52.6 have the same fixed equilibrium point, $x_s(\lambda_0)$. However, this is not usually the case because the right hand side of Equation 52.6 is time varying and periodic. Therefore, the technique of vibrational stabilization is to determine vibrations $\gamma(t)$ so that the (possibly unstable) equilibrium point $x_s(\lambda_0)$ bifurcates into a stable periodic solution whose average is close to $x_s(\lambda_0)$. The engineering aspects of the problem consist of 1) finding conditions for the existence of stabilizing vibrations, 2) determining which oscillatory inputs, $u(\cdot)$, are physically realizable, and 3) determining the shape (waveform type, amplitude, phase) of the oscillations which will insure the desired response. In Section 52.3, we shall present an example showing how oscillatory forcing induces interesting stable motion in neighborhoods of points which are not close to equilibria of the time-varying system of Equation 52.6.

52.1.4 Vibrational Stabilization

It is frequently assumed that Equation 52.6 can be decomposed as

$$\frac{dx}{dt} = f_1(x(t)) + f_2(x(t), \gamma(t)), \tag{52.7}$$

where λ_0 and $\gamma(\cdot)$ are as above and where $f_1(x(t)) = f_1(\lambda_0, x(t))$ and the function $f_2(x(t), \gamma(t))$ is linear with respect to its second argument. Systems for which this assumption does not hold are discussed in Section 52.3. For simplicity only, assume that f_1 and f_2 are analytic functions. Additionally, assume that $\gamma(t)$, the control, is periodic of period $T(0 < T \ll 1)$ and in the form, $\gamma(t) = \omega\hat{u}(\omega t)$, where $\omega = \frac{2\pi}{T}$, and $\hat{u}(\cdot)$ is some fixed period-2π function (e.g., sin or cos). We write $\gamma(\cdot)$ in this way because, although the theory is not heavily dependent on the exact shape of the waveform of the periodic input, there is a crucial dependence on the simultaneous scaling of the frequency and amplitude. Because we are usually interested in high frequency behavior, this usually implies that the amplitude of $\gamma(t)$ is large. It is possible, however, that $\hat{u}(\cdot)$ has small amplitude, making the amplitude of $\gamma(t)$ small also.

Under these assumptions, Equation 52.7 can be rewritten as

$$\frac{dx}{dt} = f_1(x(t)) + \omega f_2(x(t), \hat{u}(\omega t)). \tag{52.8}$$

To proceed with the stability analysis, Equation 52.8 will be transformed to an ordinary differential equation in "standard" form ($\frac{dx}{dt} = \epsilon f(x, t)$) so that the method of averaging can be applied (see [11] and [24]). This allows the stability properties of the time varying system Equation 52.8 to be related to the stability properties of a simpler autonomous differential equation (the averaged equation). To make this transformation, consider the so-called "generating equation"

$$\frac{dx}{dt} = f_2(x(t), \hat{u}(t)).$$

Suppose that this generating equation has a T-periodic general solution $h(t, c)$, for some $\hat{u}(\cdot)$ and $t \geq t_0$, where $h : \mathbf{R} \times \mathbf{R}^n \to \mathbf{R}^n$ and $c \in \mathbf{R}^n$ is uniquely defined for every initial condition $x(t_0) \in \Omega \subset \mathbf{R}^n$.

Introduce into Equation 52.8 the Lyapunov substitution $x(t) = h(\omega t, q(t))$ to obtain an equation for $q(\cdot)$:

$$\frac{dq}{dt} = \left[\frac{\partial h(\omega t, q(t))}{\partial q}\right]^{-1} f_1(h(\omega t, q(t))),$$

which, in slow time $\tau = \omega t$, with $z(\tau) = q(t)$ and $\epsilon = \frac{1}{\omega}$, becomes

$$\frac{dz}{d\tau} = \epsilon \left[\frac{\partial h[\tau, z(\tau)]}{\partial z}\right]^{-1} f_1(h(\tau, z(\tau))). \tag{52.9}$$

Equation 52.9 is a periodic differential equation in "standard" form and averaging can be applied. If T denotes the period of the right hand side of Equation 52.9, then the averaged equation (autonomous) corresponding to Equation 52.9 is given as

$$\frac{dy}{d\tau} = \epsilon \bar{Y}(y(\tau)); \quad \text{where}$$

$$\bar{Y}(c) = \frac{1}{T}\int_0^T \left[\frac{\partial h(\tau, c)}{\partial z}\right]^{-1} f_1(h(\tau, c))\, d\tau. \tag{52.10}$$

By the theory of averaging, it is known that an $\epsilon_0 > 0$ exists such that for $0 < \epsilon \leq \epsilon_0$, the hyperbolic stability properties of Equation 52.9 and Equation 52.10 are the same. Specifically, if y_s is an asymptotically stable equilibrium point of Equation 52.10, this implies that, for $0 < \epsilon \leq \epsilon_0$, a unique periodic solution, $z^*(\tau)$ of Equation 52.9 exists, in the vicinity of y_s that is asymptotically stable also. Since the transformation $x(t) = h(\omega t, q(t))$ is a homeomorphism, there will exist an asymptotically stable, periodic, solution to Equation 52.8 given by $x^*(t) = h(\omega t, z^*(\omega t))$ (converting back to fast time using the fact that $q(t) = z(\omega t)$, where $z(\cdot)$ is the solution to Equation 52.9). Using Definition 52.1, Equation 52.5 is said to be *vibrationally stabilized* provided that $\overline{x}^* = \frac{1}{T}\int_0^T x^*(t)dt$ remains in the vicinity of $x_s(\lambda_0)$. This can be formalized by the following theorem given in [8] and [24]:

Theorem 52.1:

Assume that Equation 52.6 with $\gamma(t) = \omega\hat{u}(\omega t)$ has the form of Equation 52.8, with f_1 and f_2 analytic. Assume, also, that $h(t, c)$ is periodic and that the function $\left[\frac{\partial h(\tau, z(\tau))}{\partial z}\right]^{-1} f_1(h(\tau, z(\tau)))$ in Equation 52.8 is continuously differentiable with respect to $z \in \Omega \subset \mathbf{R}^n$. Then the equilibrium point $x_s(\lambda_0)$ of Equation 52.5 is vibrationally stabilizable if a $\hat{u}(t)$ exists such that Equation 52.10 has an asymptotically stable equilibrium point characterized by $\frac{1}{T}\int_0^T h(\tau, y_s)\, d\tau = x_s$.

The technique of vibrational control now becomes clearer. Introduce open loop oscillatory forcing into Equation 52.5, $u(t) = \lambda_0 + \omega\hat{u}(\omega t)$, such that Equation 52.5 is in the form of Equation 52.8. Transform Equation 52.8 into Equation 52.9 and study the stability properties of the corresponding average of Equation 52.9, given by Equation 52.10. Then determine parameters of $\hat{u}$ (phase, amplitude and frequency) such that Equation 52.10 has an asymptotically stable equilibrium point y_s. If $\frac{1}{T}\int_0^T h(\tau, y_s)\, d\tau = x_s$, then the system is vibrationally stabilizable.

The procedure of vibrational stabilization described above is trial and error. Vibrations are inserted into a system until vibrations are found which give the desired response. However, if specific classes of vibrations are analyzed, explicit algorithms are known which explain the size and location of vibration needed to give specified responses (see [8,9,24]). For example, it is common to assume that the vibrations are in *linear multiplicative* form, $f_2(x(t), \gamma(t)) = B(t)x(t)$ or *vector additive* form, $f_2(x(t), \gamma(t)) = L(t)$. In each of these cases, sufficient conditions (sometimes necessary and sufficient) are known for vibrational

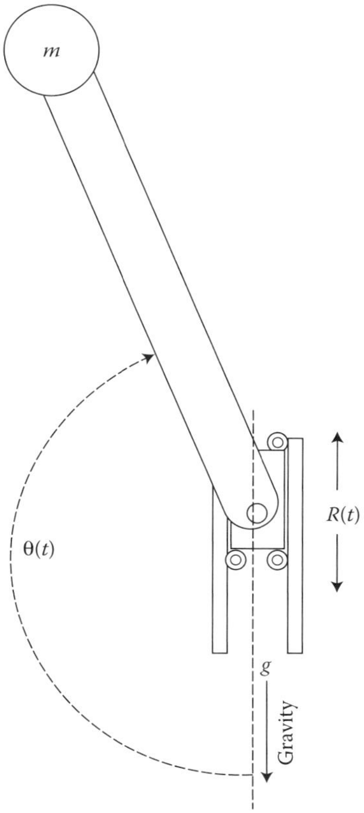

FIGURE 52.2 A simple pendulum whose hinge point undergoes vertical motion.

stabilization of systems. We return to linear multiplicative vibrations in Section 52.3, using a more geometric framework.

Finally, it should be noted that [8] and [24] can relate the transient behavior of Equation 52.10 to the "average" transient behavior of Equation 52.7 through the estimation $x(t) \approx h(\omega t, y(\omega t))$, where $y(\tau)$ is the solution of Equation 52.10 in slow time. Hence, the technique of vibrational control has the ability to both stabilize a system and control transient response issues.

Example 52.2: Oscillatory Stabilization of a Simple Pendulum

A classical example to which the theory applies involves stabilizing the upright equilibrium of a simple pendulum by the forced vertical oscillation of the pendulum's hinge point. Consider a simple pendulum consisting of a massless but rigid link of length ℓ to which a tip of mass m is attached. Suppose the hinge point of the pendulum undergoes vertical motion, and is located at time t at vertical position $R(t)$ with respect to some reference coordinate (see Figure 52.2.). Taking into account motion in this variable and the friction coefficient at the hinge ($b > 0$), the pendulum dynamics may be written as

$$\ell\ddot{\theta} + \left(\frac{b}{m}\right)\dot{\theta} + \ddot{R}\sin\theta + g\sin\theta = 0.$$

Consider the simple sinusoidal oscillation of the hinge-point, $R(t) = \alpha\sin\beta t$. Then $\ddot{R}(t) = -\eta\beta\sin\beta t$, where $\eta = \eta(\beta) = \alpha\beta$. Letting $x_1 = \theta$ and $x_2 = \dot{\theta}$, the system can be placed in the form of Equation 52.8. The corresponding generating equation is given as $\dot{x}_1 = 0$ and $\dot{x}_2 = (\eta/\ell)\sin t\sin x_1$, which has solution $x_1 = c_1 = h_1(t,c)$ and $x_2 = -(\eta/\ell)\cos t \sin c_1 + c_2 = h_2(t,c)$. Introducing the transformation $x_1 = z_1$ and $x_2 = -(\eta/\ell)\cos\beta t \sin z_1 + z_2$, letting $\tau = \beta t$, and letting $\epsilon = 1/\beta$, Equation 52.9 specializes to

$$\dot{z}_1 = \epsilon\left[-\left(\frac{\eta}{\ell}\right)\cos\tau \sin z_1 + z_2\right]$$

$$\dot{z}_2 = \epsilon\left[-\left(\frac{\eta^2}{\ell^2}\right)\cos^2\tau \cos z_1 \sin z_1 - \left(\frac{g}{\ell}\right)\sin z_1 + \left(\frac{\eta}{\ell}\right) z_2 \cos\tau \cos z_1 \right.$$
$$\left. + \left(\frac{\eta b}{m\ell^2}\right)\cos\tau \sin z_1 - \left(\frac{b}{m\ell}\right) z_2\right].$$

Therefore the averaged equations are $\dot{y}_1 = \epsilon y_2$ and $\dot{y}_2 = \epsilon\left[-\frac{\eta^2}{2\ell^2}\cos y_1 \sin y_1 - (g/\ell)\sin y_1 - (b/m\ell)y_2\right]$. The upper equilibrium point in the averaged equation has been preserved and $x_s = \frac{1}{T}\int_0^T h(\tau, y_s)\,d\tau$. Hence, by the above theorem, if the vertical equilibrium point is asymptotically stable for the averaged equation, then for sufficiently large β the inverted pendulum with oscillatory control has an asymptotically stable periodic orbit vibrating vertically about the point $\theta = \pi$. A simple linearization of the averaged equation reveals that its upper equilibrium point is asymptotically stable when $\alpha^2\beta^2 > 2g\ell$. Under these conditions, the upper equilibrium point of the inverted pendulum is said to be *vibrationally stabilized*.

52.2 The Constructive Controllability of Drift-Free (Class I) Systems

Class I control systems arise naturally as kinematic models of mechanical systems. In this section, we outline the current theory of motion control for such systems, emphasizing the geometric mechanism (anholonomy) through which oscillatory inputs to Equation 52.1 produce motions of the state variables. Explicit results along the lines given in Section 52.1 for the Heisenberg system have been obtained in a variety of settings, and some recent work will be discussed below. The question of when such explicit constructions are possible more generally for Class I systems does not yet have a complete answer. Thus we shall also discuss computational approaches that yield useful approximate solutions. After briefly outlining the state of current theory, we conclude the section with an example involving motion specification for a ball "grasped" between two plates.

The recent literature treating control problems for such systems suggests that it is useful distinguishing between two control design problems:

P1: The *prescribed endpoint steering problem* requires that, given any pair of points $x_0, x_1 \in \mathbf{R}^n$, a vector of piecewise analytic control inputs $u(\cdot) = (u_1(\cdot), \ldots, u_m(\cdot))$ is to be determined to steer the state of Equation 52.1 from x_0 at time $t = 0$ to x_1 at time $t = T > 0$.

P2: The *trajectory approximation steering problem* requires that, given any sufficiently "regular" curve $\gamma : [0, T] \to \mathbf{R}^n$, we determine a sequence $\{u^j(\cdot)\}$ of control input vectors such that the corresponding sequence of trajectories of Equation 52.1 converges (uniformly) to γ.

A general solution to either of these problems requires that a certain *Lie algebra rank condition* (LARC) be satisfied. More specifically, with the *Lie bracket* of vector fields defined as in Section 52.1, define a set of vector fields

$$\mathcal{C} = \{\xi : \xi = [\xi_j, [\xi_{j-1}, [\ldots, [\xi_1, \xi_0]\ldots]]];$$
$$\xi_i \in \{g_1, \ldots, g_m\}, i = 1, \ldots, j;\ j = 1, \ldots, \infty\}.$$

Then $\mathcal{L} = \mathrm{span}(\mathcal{C})$ (= the set of all linear combinations of elements of $\mathcal{C}$) is called the *Lie algebra* generated by $\{g_1, \ldots, g_m\}$. We say Equation 52.1 (or equivalently the set of vector fields $\{g_1, \ldots, g_m\}$)

satisfies the *Lie algebra rank condition* on $\mathbf{R}^n$ if $\mathcal{L}$ spans $\mathbf{R}^n$ at each point of $\mathbf{R}^n$. The following result is fundamental because it characterizes those systems for which the steering problems may, in principle, be solved.

Theorem 52.2:

A drift-free system Equation 52.1 is completely controllable in the sense that, given any $T > 0$ and any pair of points $x_0, x_1 \in \mathbf{R}^n$, there is a vector of inputs $u = (u_1, \ldots, u_m)$ which are piecewise analytic on $[0, T]$ and which steer the system from x_0 to x_1 in T units of time if, and only if, the system satisfies the Lie algebra rank condition.

As stated, this theorem is essentially due to W.L. Chow [19], but it has been refined and tailored to control theory by others ([13,35]). The various versions of this theorem in the literature have all been nonconstructive. Methods for the explicit determination of optimal (open-loop) control laws for steering Class I systems between prescribed endpoints have appeared in [1,2,16,18]. The common features in all this work are illustrated by the following:

A model nonlinear optimal control problem with three states and two inputs: Find controls $u_1(\cdot)$, $u_2(\cdot)$ which steer the system

$$\begin{pmatrix} \dot{x}_1 \\ \dot{x}_2 \\ \dot{x}_3 \end{pmatrix} = \begin{pmatrix} 0 & 0 & u_2 \\ 0 & 0 & -u_1 \\ -u_2 & u_1 & 0 \end{pmatrix} \begin{pmatrix} x_1 \\ x_2 \\ x_3 \end{pmatrix} \tag{52.11}$$

between prescribed endpoints to minimize the cost criterion

$$\eta = \int_0^1 u_1^2 + u_2^2 \, dt.$$

Several comments regarding the geometry of this problem are in order. First, an appropriate version of Chow's theorem shows that Equation 52.11 is controllable on any 2-sphere, $\mathcal{S} = \{x \in \mathbf{R}^3 : \|x\| = r\}$ for some $r > 0$, centered at the origin in $\mathbf{R}^3$. Hence, the optimization problem is well-posed precisely when the prescribed endpoints $x_0, x_1 \in \mathbf{R}^3$ satisfy $\|x_0\| = \|x_1\|$. Second, the problem may be interpreted physically as seeking minimum length paths on a sphere in which only motions composed of rotations about the x-axis (associated with input u_1) and y-axis (associated with u_2) are admissible. General methods for solving this type of problem appear in [1] and [2]. Specifically, in the first author's 1975 PhD thesis (see reference in [1] and [2]), it was shown that the optimal inputs have the form, $u_1(t) = \mu \sin(\omega t + \phi)$, $u_2(t) = \mu \cos(\omega t + \phi)$. The optimal inputs depend on three parameters reflecting the fact that the set (group) of rotations of the 2-sphere is three dimensional. The details for determining the values of the parameters μ, ω, and ϕ in terms of the end points x_0 and x_1 are given in the thesis cited in [1] and [2].

The general nonlinear quadratic optimal control problem of steering Equation 52.1 to minimize a cost of the form $\int_0^1 \|u\|^2 \, dt$ has not yet been solved in such an explicit fashion. The general classes of problems which have been discussed in [1,2,16,18] are associated with certain details of structure in the set of vector fields $\{g_1, \ldots, g_m\}$ and the corresponding Lie algebra $\mathcal{L}$. In [16] and [18], for example, Brockett discusses various higher dimensional versions of the Lie algebraic structure characterizing the above sphere problem and the Heisenberg system.

In addition to optimal control theory having intrinsic interest, it also points to a broader approach to synthesizing control inputs. Knowing the form of optimal trajectories, we may relax the requirement that inputs be optimal and address the simpler question of whether problems P1 and P2 may be solved using inputs with the given parametric dependence. Addressing the cases where we have noted the optimal inputs are phase-shifted sinusoids, we study the effect of varying each of the parameters. For instance, consider Equation 52.4 steered by the inputs $u_1(t) = \mu \sin(\omega t + \phi)$, $u_2(t) = \mu \cos(\omega t + \phi)$, with μ and ϕ

fixed and ω allowed to vary. As ω increases, the trajectories produced become increasingly tight spirals ascending about the z-axis. One consequence of this is that, although the vectorfields

$$\begin{pmatrix} 1 \\ 0 \\ -x_2 \end{pmatrix}, \quad \text{and} \quad \begin{pmatrix} 0 \\ 1 \\ x_1 \end{pmatrix}$$

are both perpendicular to the x_3-axis at the origin, pure motion in the x_3-coordinate direction may nevertheless be produced to an arbitrarily high degree of approximation. This basic example can be generalized, and extensive work on the use of oscillatory inputs for approximating arbitrary motions in Class I systems has been reported by Sussmann and Liu, [36]. The general idea is that, when a curve $\gamma(t)$ is specified, a sequence of appropriate oscillatory inputs $u^j(\cdot)$ is produced so that the corresponding trajectories of Equation 52.1 converge to $\gamma(\cdot)$ *uniformly*. The interested reader is referred to [37,38], and the earlier work of Haynes and Hermes, [21], for further details.

Progress on problem P1 has been less extensive. Although a general constructive procedure for generating motions of Equation 52.1 which begin and end exactly at specified points, x_0 and x_1, has not yet been developed, solutions in a number of special cases have been reported. For the case of arbitrary "nilpotent" systems, Lafferriere and Sussmann, [22] and [23], provide techniques for approximating solutions for general systems. Leonard and Krishnaprasad [28] have designed algorithms for synthesizing open-loop sinusoidal control inputs for point-to-point system maneuvers where up to depth-two brackets are required to satisfy the LARC.

Brockett and Dai [18] have studied a natural subclass of nilpotent systems within which the Heisenberg system Equation 52.4 is the simplest member. The underlying geometric mechanism through which a rich class of motions in $\mathbf{R}^3$ is produced by oscillatory inputs to Equation 52.4 is also present in systems with two vector fields but higher dimensional state spaces. These systems are constructed in terms of nonintegrable p-forms in the coordinate variables x_1 and x_2. We briefly describe the procedure, referring to [18] for more details.

The number of linearly independent p-forms in x_1 and x_2 is $p+1$. (Recall that a p-form in x_1 and x_2 is a monomial of the form $x_1^k x_2^{p-k}$. The linearly independent p-forms may be listed explicitly $\{x_1^p, x_1^{p-1}x_2, \ldots, x_2^p\}$.) Thus, there are $2(p+1)$ linearly independent expressions of the form

$$\eta = \phi(x_1, x_2)\dot{x}_1 + \psi(x_1, x_2)\dot{x}_2$$

where ϕ, ψ are homogeneous polynomials of degree p in x_1 and x_2. Within the set of such expressions, there is a set of $p+2$ linearly independent expressions of the form $\eta = d\gamma/dt$, where γ is a homogeneous polynomial in x_1, x_2 of degree $p+1$ (Such expressions are called exact differentials). There is a complementary p-dimensional family ($p = 2(p+1) - (p+2)$) of η's which are not integrable.

For example, if $p = 2$, there are 2 linearly independent nonintegrable forms η, and we may take these to be $\{x_1^2\dot{x}_2, x_2^2\dot{x}_1\}$. From these, we construct a completely controllable two-input system

$$\begin{pmatrix} \dot{x}_1 \\ \dot{x}_2 \\ \dot{x}_3 \\ \dot{x}_4 \\ \dot{x}_5 \end{pmatrix} = \begin{pmatrix} u \\ v \\ x_1 v - x_2 u \\ x_1^2 v \\ x_2^2 u \end{pmatrix}. \tag{52.12}$$

More generally, for each positive integer p we could write a completely controllable two-input system whose state space has dimension $2 + p(p+1)/2$. Brockett and Dai [18] consider the optimal control problem of steering Equation 52.12 between prescribed endpoints $x(0), x(T) \in \mathbf{R}^5$ to minimize the cost functional

$$\int_0^T u^2 + v^2 \, dt.$$

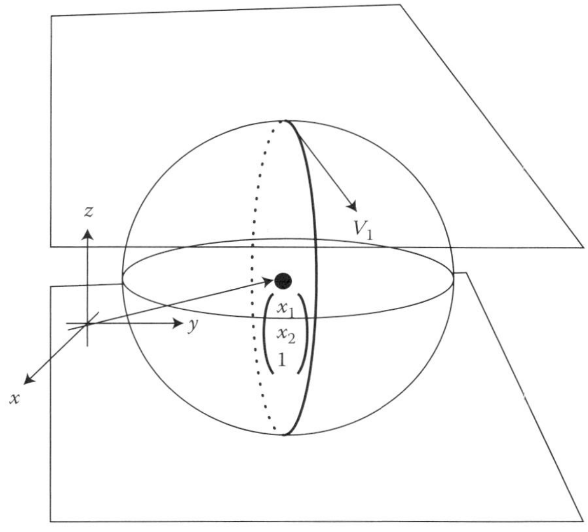

FIGURE 52.3 A ball rolling without slipping between two flat plates.

It is shown that explicit solutions may also be obtained in this case, and these are given in terms of elliptic functions.

Before treating an example problem in mechanics which makes use of these ideas, we summarize some other recent work on control synthesis for Class I systems. Whereas Sussmann and his coworkers [22,23,36–38] have used the concept of a *P. Hall basis* for free Lie algebras to develop techniques applicable in complete generality to Class I systems, an approach somewhat in the opposite direction has been pursued by S. Sastry and his coworkers [29,39,40]. This approach sought to characterize systems controlled by sinusoidal inputs. Motivated by problems in the kinematics of wheeled vehicles and robot grasping, they have defined a class of *chained systems* in which desired motions result from inputs which are sinusoids with integrally related frequencies. While the results are no less explicit than the Heisenberg example in Section 52.1, the systems themselves are very special. Conditions under which a Class I system may be converted to chained form are given in [29].

Example 52.3:

The example we discuss next (due to Brockett and Dai [18]) is prototypical of applications involving the kinematics of objects in the grasp of a robotic hand. Consider a ball that rolls without slipping between two flat horizontal plates. It is convenient to assume the bottom plate is fixed. Suppose that the ball has unit radius. Fix a coordinate system whose x- and y-axes lie in the fixed bottom plate with the positive z-axis perpendicular to the plate in the direction of the ball. Call this the (bottom) "plate frame." We keep track of the ball's motion by letting $(x_1, x_2, 1)$ denote the plate-frame coordinates of the ball's center. We also fix an orthonormal frame in the ball, and we denote the plate-frame directions of the three coordinate axes by a 3×3 orthogonal matrix

$$A = \begin{pmatrix} a_{11} & a_{12} & a_{13} \\ a_{21} & a_{22} & a_{23} \\ a_{31} & a_{32} & a_{33} \end{pmatrix}.$$

The ball's position and orientation are thus specified by the 4×4 matrix

$$H = \begin{pmatrix} A & \vec{x} \\ 0 & 1 \end{pmatrix} = \begin{pmatrix} a_{11} & a_{12} & a_{13} & x_1 \\ a_{21} & a_{22} & a_{23} & x_2 \\ a_{31} & a_{32} & a_{33} & 1 \\ 0 & 0 & 0 & 1 \end{pmatrix}.$$

As the top plate moves in the plate-frame x direction with velocity v_1, the ball's center also moves in the same direction with velocity $u_1 = v_1/2$. This motion imparts a counterclockwise rotation about the y-axis, and since the ball has unit radius, the angular velocity is also u_1. Similarly, if the top plate moves in the (plate-frame) y direction with velocity v_2, the ball's center moves in the y direction with velocity $u_2 = v_2/2$, and the angular velocity imparted about the x-axis is $-u_2$. The kinematic description of this problem is obtained by differentiating H with respect to time.

$$\dot{H} = \begin{pmatrix} \Omega A & \vec{u} \\ 0 & 0 \end{pmatrix}$$

where

$$\Omega = \begin{pmatrix} 0 & 0 & u_1 \\ 0 & 0 & u_2 \\ -u_1 & -u_2 & 0 \end{pmatrix} \quad \text{and} \quad \vec{u} = \begin{pmatrix} u_1 \\ u_2 \\ 0 \end{pmatrix}.$$

This velocity relationship, at the point $(x_1, x_2) = (0, 0)$, is

$$\dot{H} = u_1 U_1 H + u_2 U_2 H, \tag{52.13}$$

where

$$U_1 = \begin{pmatrix} 0 & 0 & 1 & 1 \\ 0 & 0 & 0 & 0 \\ -1 & 0 & 0 & 0 \\ 0 & 0 & 0 & 0 \end{pmatrix} \quad \text{and} \quad U_2 = \begin{pmatrix} 0 & 0 & 0 & 0 \\ 0 & 0 & 1 & 1 \\ 0 & -1 & 0 & 0 \\ 0 & 0 & 0 & 0 \end{pmatrix}.$$

Computing the Lie bracket of the vector fields U_1H and U_2H according to the formula given in Section 52.1, we obtain a new vector field $U_3H = [U_1H, U_2H]$, where

$$U_3 = \begin{pmatrix} 0 & -1 & 0 & 0 \\ 1 & 0 & 0 & 0 \\ 0 & 0 & 0 & 0 \\ 0 & 0 & 0 & 0 \end{pmatrix}.$$

Computing additional Lie brackets yields only quantities which may be expressed as linear combinations of U_1H, U_2H, and U_3H. Since the number of linearly independent vector fields obtained by taking Lie brackets is three, the system is completely controllable on a three dimensional space. Comparison with the problem of motion on a sphere discussed above shows that, by having the top plate execute a high-frequency oscillatory motion—$u_1(t) = \mu \sin(\omega t + \phi), u_2(t) = \mu \cos(\omega t + \phi)$, the ball may be made to rotate about its z-axis. The motion of the ball is "retrograde." If the top plate executes a small amplitude clockwise loop about the "plate-frame" z-axis, the motion of the ball is counterclockwise.

52.3 Systems with Drift—Stability Analysis Using Energy Methods and the Averaged Potential

Constructive design methods for the control of systems of Class II are less well developed than in the Class I case. Nevertheless, the classical results on stability described in Section 52.2 apply in many cases, and the special structure of Class II systems again reveals the geometric mechanisms underlying stability.

52.3.1 First and Second Order Stabilizing Effects of Oscillatory Inputs in Systems with Drift

Much of the published literature on systems of Class I and Class II is devoted to the case in which the vector fields are linear in the state. Consider the control system

$$\dot{x} = (A + \sum_{i=1}^{m} u_i(t)B_i)x, \tag{52.14}$$

where $A, B_1, \ldots, B_m$ are constant $n \times n$ matrices, $x(t) \in \mathbf{R}^n$ with control inputs of the type we have been considering. We shall be interested in the possibility of using high-frequency oscillatory inputs to create stable motions in a neighborhood of the origin $x = 0$.

Assume that $\hat{u}_1(\cdot), \ldots, \hat{u}_m(\cdot)$ are periodic functions and (for simplicity only) assume that each $\hat{u}_i(\cdot)$ has common fundamental period $T > 0$. To apply classical averaging theory to study the motion of Equation 52.14, we consider the effect of increasing the frequency of the forcing. Specifically, we study the dynamics of

$$\dot{x}(t) = (A + \sum \hat{u}_i(\omega t)B_i)x(t)$$

as ω becomes large. The analysis proceeds by scaling time and considering $\tau = \omega t$. Let $z(\tau) = x(t)$. This satisfies the differential equation

$$\frac{dz}{d\tau} = \frac{1}{\omega}(A + \sum \hat{u}_i(\tau)B_i)z. \tag{52.15}$$

Assuming $\omega > 0$ is large, and letting $\epsilon = \frac{1}{\omega}$, we see that Equation 52.15 is in a form to which classical averaging theory applies.

Theorem 52.3:

Consider the system of Equation 52.14 with $u_i(\cdot) = \hat{u}_i(\cdot)$ where, for $i = 1, \ldots, m$, $\hat{u}_i(\cdot)$ is continuous on $0 \le t < t_f \le \infty$ and periodic of period $T \ll t_f$. Let

$$\bar{u}_i = \frac{1}{T}\int_0^T \hat{u}_i(t)\,dt,$$

and let $y(\tau)$ be a solution of the constant coefficient linear system

$$\dot{y}(\tau) = \epsilon(A + \sum \bar{u}_i B_i)y(\tau). \tag{52.16}$$

If z_0 and y_0 are respective initial conditions associated with Equations 52.15 and 52.16 such that $|z_0 - y_0| = \mathcal{O}(\epsilon)$, then $|z(\tau) - y(\tau)| = \mathcal{O}(\epsilon)$ on a time scale $\tau \sim \frac{1}{\epsilon}$. If $A + \sum \bar{u}_i B_i$ has its eigenvalues in the left half plane, then $|z(\tau) - y(\tau)| = \mathcal{O}(\epsilon)$ as $\tau \to \infty$.

This theorem relies on classical averaging theory and is discussed in [7]. A surprising feature of systems of this form is that the stability characteristics can be modified differently if both the magnitude and frequency of the oscillatory forcing are increased. A contrast to Theorem 52.3 is the following:

Theorem 52.4:

Let $\hat{u}_1(\cdot), \ldots, \hat{u}_m(\cdot)$ be periodic functions of period $T > 0$ for $i = 1, \ldots, m$. Assume that each $\hat{u}_i(\cdot)$ has mean 0. Consider Equation 52.14, and assume that, for all $i, j = 1, \ldots, m$, $B_iB_j = 0$. Let $\epsilon = \frac{1}{\omega}$. Define for each

$i = 1, \ldots, m$, the periodic function $v_i(t) = \int_0^t \hat{u}_i(s)\,ds$, and let

$$\bar{v}_i = \frac{1}{T}\int_0^T v_i(s)\,ds, \quad i = 1, \ldots, m,$$

and

$$\sigma_{ij} = \frac{1}{T}\int_0^T v_i(s)v_j(s)\,ds, \quad i,j = 1, \ldots, m.$$

Let $y(t)$ be a solution of the constant coefficient linear system

$$\dot{y} = \left(A + \sum_{i,j}(\bar{v}_i\bar{v}_j - \sigma_{ij})B_iAB_j\right)y. \tag{52.17}$$

Suppose that the eigenvalues of Equation 52.17 have negative real parts. Then there is a $t_1 > 0$ such that for $\omega > 0$ sufficiently large (i.e., for $\epsilon > 0$ sufficiently small), if x_0 and y_0 are respective initial conditions for Equations 52.14 and 52.17 such that $|x_0 - y_0| = \mathcal{O}(\epsilon)$, then $|x(t) - y(t)| = \mathcal{O}(\epsilon)$ for all $t > t_1$.

The key distinction between these two theorems is that induced stability is a first order effect in Theorem 52.3 where we scale frequency alone, whereas in Theorem 52.4 where both frequency and magnitude are large, any induced stability is a second order effect (depending on the rms value of the integral of the forcing). Further details are provided in [7]. Rather than pursue these results, we describe a closely related theory which may be applied to mechanical and other physical systems.

52.3.2 A Stability Theory for Lagrangian and Hamiltonian Control Systems with Oscillatory Inputs

The geometric nature of emergent behavior in systems subject to oscillatory forcing is apparent in the case of conservative physical systems. In this subsection, we again discuss stabilizing effects of oscillatory forcing. The main analytical tool will be an energy-like quantity called the *averaged potential.* This is naturally defined in terms of certain types of Hamiltonian control systems of the type studied in [30]. Because we shall be principally interested in systems with symmetries most easily described by Lagrangian models, we shall start from a Lagrangian viewpoint and pass to the Hamiltonian description via the Legendre transformation. Following Brockett, [14] and Nijmeijer and Van der Schaft, [30], we define a *Lagrangian control system* on a differentiable manifold M as a dynamical system with inputs whose equations of motion are prescribed by applying the Euler-Lagrange operator to a function $L : TM \times U \to \mathbf{R}$, $L = L(q, \dot{q}; u)$, whose dependence on the configuration q, the velocity $\dot{q}$, and the control input u is smooth. U is a set of control inputs satisfying the general properties outlined in Section 52.1.

Lagrangian systems arising via reduction with respect to cyclic coordinates: Consider a Lagrangian control system with configuration variables $(q_1, q_2) \in \mathbf{R}^{n_1} \times \mathbf{R}^{n_2}$. The variable q_1 will be called *cyclic* if it does not enter into the Lagrangian when $u = 0$, i.e., if $\frac{\partial L}{\partial q_1}(q_1, q_2, \dot{q}_1, \dot{q}_2; 0) \equiv 0$. A symmetry is associated with the cyclic variables q_1 which manifests itself in the invariance of $L(q_1, q_2, \dot{q}_1, \dot{q}_2; 0)$ with respect to a change of coordinates $q_1 \mapsto q_1 + \alpha$ for any constant $\alpha \in \mathbf{R}^n$. We shall be interested in Lagrangian control systems with cyclic variables in which the cyclic variables may be directly controlled. In such systems, we shall show how the velocities associated with the cyclic coordinates may themselves be viewed as controls. Specifically, we shall consider systems of the form

$$L(q_1, q_2, \dot{q}_1, \dot{q}_2; u) = \mathcal{L}(q_2, \dot{q}_1, \dot{q}_2) + q_1^T u \tag{52.18}$$

where

$$\mathcal{L}(q_2, \dot{q}_1, \dot{q}_2) = \frac{1}{2}(\dot{q}_1^T, \dot{q}_2^T)\begin{pmatrix} m(q_2) & A^T(q_2) \\ A(q_2) & M(q_2) \end{pmatrix}\begin{pmatrix} \dot{q}_1 \\ \dot{q}_2 \end{pmatrix} - V(q_2).$$

The matrices $m(q_2)$ and $M(q_2)$ are symmetric and positive definite of dimension $n_1 \times n_1$ and $n_2 \times n_2$ respectively, where $\dim q_1 = n_1, \dim q_2 = n_2$, and the matrix $A(q_2)$ is arbitrary.

To emphasize the distinguished role to be played by the velocity associated with the cyclic variable q_1, we write $v = \dot{q}_1$. Applying the usual Euler–Lagrange operator to this function leads to the equations of motion,

$$\frac{d}{dt}\frac{\partial \mathcal{L}}{\partial v} = u$$

and

$$\left(\frac{d}{dt}\frac{\partial \mathcal{L}}{\partial \dot{q}_2} - \frac{\partial \mathcal{L}}{\partial q_2}\right)_{|(q_2,\dot{q}_2,v)} = 0. \tag{52.19}$$

The first of these equations may be written more explicitly as

$$\frac{d}{dt}(m(q_2)v + A(q_2)^T\dot{q}_2) = u. \tag{52.20}$$

Although, in the physical problem represented by the Lagrangian Equation 52.18, $u(\cdot)$ is clearly the input with $v(\cdot)$ determined via Equation 52.20, it is formally equivalent to take $v(\cdot)$ as the input with the corresponding $u(\cdot)$ determined by Equation 52.20. In actual practice, this may be done, provided we may command actuator inputs $u(\cdot)$ large enough to dominate the dynamics. The motion of q_2 is then determined by Equation 52.19 with the influence of the actual control input felt only through $v(\cdot)$ and $\dot{v}(\cdot)$. Viewing $v(\cdot)$ together with $\dot{v}(\cdot)$ as control input, Equation 52.19 is a Lagrangian control system in its own right. The defining Lagrangian is given by $\hat{L}(q_2,\dot{q}_2;v) = \frac{1}{2}\dot{q}_2^T M(q_2)\dot{q}_2 + \dot{q}_2^T A(q_2)v - V_a(q_2;v)$, where V_a is the *augmented potential* defined by $V_a(q;v) = V(q) - \frac{1}{2}v^T m(q_2)v$. In the remainder of our discussion, we shall confine our attention to controlled dynamical systems arising from such a *reduced Lagrangian*. Because the reduction process itself will typically not be central, we henceforth omit the subscript "2" on the generalized configuration and velocity variables which we wish to control. We write:

$$\hat{L}(q,\dot{q};v) = \frac{1}{2}\dot{q}^T M(q)\dot{q} + \dot{q}^T A(q)v - V_a(q;v). \tag{52.21}$$

Example 52.4: The Rotating Pendulum

As in [4], we consider a mechanism consisting of a solid uniform rectangular bar fixed at one end to a universal joint as depicted in Figure 52.4. The universal joint is comprised of two single degree of freedom revolute joints with mutually orthogonal intersecting axes (labeled x and y in Figure 52.2). These joints are assumed to be frictionless. Angular displacements about the x- and y-axes are denoted ϕ_1 and ϕ_2 respectively, with $(\phi_1,\phi_2) = (0,0)$ designating the configuration in which the pendulum hangs straight down. The pendulum also admits a controlled rotation about a spatially fixed vertical axis. Let θ denote the amount of this rotation relative to some chosen reference configuration.

To describe the dynamics of the forced pendulum, we choose a (principal axis) coordinate frame, fixed in the bar, consisting of the x- and y-axes of the universal joint together with the corresponding z-axis prescribed by the usual right-hand rule. When the pendulum is at rest, for some reference value of θ, the body frame $x-, y-$, and z-axes will coincide with corresponding axes x', y', and z' of an inertial frame, as in Figure 52.4, with respect to which we shall measure all motion. Let I_x, I_y, and I_z denote the principal moments of inertia with respect to the body (x,y,z)-coordinate system. Then the system has

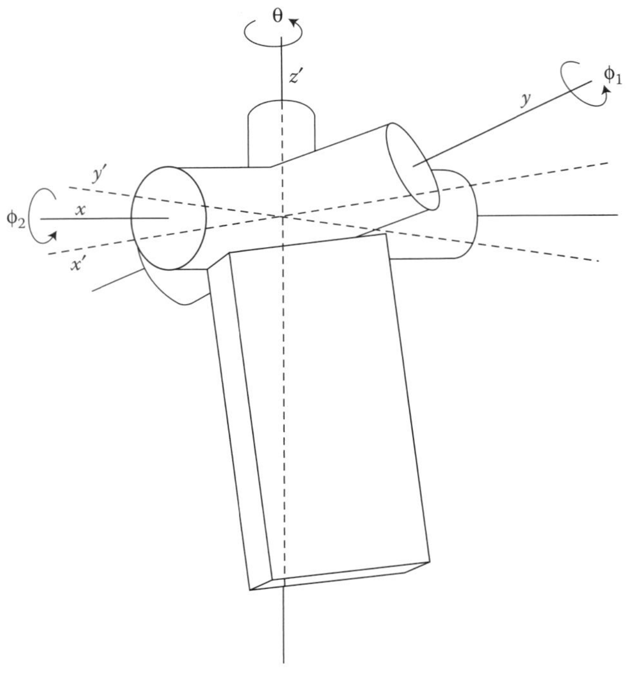

FIGURE 52.4 A rotating pendulum suspended from a universal joint.

the Lagrangian

$$L(\phi_1, \dot{\phi}_1; \phi_2, \dot{\phi}_2; \dot{\theta}) = \frac{1}{2}\left[I_x(\dot{\theta}s_2 + \dot{\phi}_1)^2 + I_y(\dot{\theta}c_2s_1 - \dot{\phi}_2c_1)^2 + I_z(\dot{\theta}c_1c_2 + \dot{\phi}_2s_1)^2\right] + c_1c_2, \quad (52.22)$$

where the last term is a normalized gravitational potential, and where $s_i = \sin\phi_i$, $c_i = \cos\phi_i$. We assume that there is an actuator capable of rotating the mechanism about the inertial z-axis with any prescribed angular velocity $\dot{\theta}$. We shall study the dynamics of this system when $I_x \geq I_y \gg I_z$, and as above, we shall view $v = \dot{\theta}(\cdot)$ as a control input. The *reduced Lagrangian* Equation 52.21 takes the form

$$\begin{aligned}\hat{L}(\phi, \dot{\phi}; v) = \frac{1}{2}\left[I_x\dot{\phi}_1^2 + (I_yc_1^2 + I_zs_1^2)\dot{\phi}_2^2\right] + v\left[I_xs_2\dot{\phi}_1 + (I_z - I_y)s_1c_1c_2\dot{\phi}_2\right] \\ + \left[(I_zc_1^2 + I_ys_1^2)c_2^2 + I_xs_2^2\right]v^2 + c_1c_2. \end{aligned} \quad (52.23)$$

The corresponding control system is given by applying the Euler-Lagrange operator to $\hat{L}$, and this is represented by a system of coupled second order differential equations,

$$\frac{d}{dt}\frac{\partial \hat{L}}{\partial \dot{\phi}} - \frac{\partial \hat{L}}{\partial \phi} = 0. \quad (52.24)$$

From the way in which v appears in the reduced Lagrangian, the equations of motion Equation 52.24 will have terms involving $\dot{v}$ as well as terms involving v. Though it is possible to analyze such a system directly, we shall not discuss this approach. The general analytical technique advocated here is to transform all Lagrangian models to Hamiltonian form.

The Hamiltonian viewpoint and the averaged potential: Recall that in Hamiltonian form the dynamics are represented in terms of the configuration variables q and conjugate momenta $p = \frac{\partial \hat{L}}{\partial \dot{q}}$ (see [30], p. 351).

Referring to the Lagrangian in Equation 52.21 and applying the Legendre transformation $H(q, p; v) = [p \cdot \dot{q} - \hat{L}(q, \dot{q}; v)]|_{(q,p)}$, we obtain the corresponding Hamiltonian

$$H(q, p; v) = \frac{1}{2}(p - vA)^T M^{-1}(p - vA) + V_a(v, q). \tag{52.25}$$

The equations of motion are then written in the usual way in position and momentum coordinates as

$$\dot{q} = M^{-1}(p - vA) \tag{52.26}$$

$$\dot{p} = -\frac{\partial}{\partial q}\left[\frac{1}{2}(p - vA)^T M^{-1}(p - vA) + V_a(v, q)\right]. \tag{52.27}$$

We may obtain an averaged version of this system by replacing all coefficients involving the input $v(\cdot)$ by their time-averages. Assuming $v(\cdot)$ is bounded, piecewise continuous, and periodic of period $T > 0$, $v(\cdot)$ has a Fourier series representation:

$$v(t) = \sum_{k=-\infty}^{\infty} a_k e^{\frac{2\pi k}{T} it}. \tag{52.28}$$

Equations 52.26 and 52.27 contain terms of order not greater than two in v. Averaging the coefficients we obtain

---

Proposition 52.1:

Suppose $v(\cdot)$ is given by Equation 52.28. Then, if all coefficients in Equations 52.26 and 52.27 are replaced by their time averages, the resulting averaged system is Hamiltonian with corresponding Hamiltonian function

$$\bar{H}(q, p) = \frac{1}{2}(p - A(q)\bar{v})^T M(q)^{-1}(p - A(q)\bar{v}) + V_A(q), \tag{52.29}$$

where

$$V_A(q) = V(q) + \frac{1}{2}\left(\Sigma(q) - \bar{v}^T A(q)^T M(q)^{-1} A(q)\bar{v}\right), \tag{52.30}$$

$$\bar{v} = \frac{1}{T}\int_0^T v\, dt,$$

and

$$\Sigma(q) = \frac{1}{T}\int_0^T v(t)^T \left(A(q)^T M(q)^{-1} A(q) - m(q)\right) v(t)\, dt.$$

Definition 52.2:

We refer to $\bar{H}(q, p)$ in Equation 52.29 as the averaged Hamiltonian associated with Equation 52.25. $V_A(q)$, defined in Equation 52.30, is called the averaged potential.

Remark 52.2

(Averaged kinetic and potential energies, the *averaged Lagrangian*.) Before describing the way in which the *averaged potential* may be used for stability analysis, we discuss its formal definition in more detail.

The Legendre transformation used to find the Hamiltonian corresponding to Equation 52.21 makes use of the conjugate momentum

$$p = p(q, \dot{q}, t) = \frac{\partial \hat{L}}{\partial \dot{q}} = M(q)\dot{q} + A(q)v(t).$$

This explicitly depends on the input $v(t)$. Given a point in the phase space, $(q, \dot{q})$, the corresponding *averaged* momentum is

$$p = M(q)\dot{q} + A(q)\bar{v}.$$

We may think of the first term $\frac{1}{2}(p - A(q)\bar{v})^T M(q)^{-1}(p - A(q)\bar{v})$ in Equation 52.29 as an "averaged kinetic energy." It is not difficult to see that there is an "averaged Lagrangian"

$$\bar{L}(q, \dot{q}) = \frac{1}{2}\dot{q}^T M(q)\dot{q} + \dot{q}^T A(q)\bar{v} - V_A(q)$$

from which the Hamiltonian $\bar{H}$ in Equation 52.29 is obtained by means of the Legendre transformation.

The averaged potential is useful in assessing the stability of motion in Hamiltonian (or Lagrangian) control systems. The idea behind this is that strict local minima of the averaged potential will correspond to stable equilibria of the averaged Hamiltonian system. The theory describing the relationship with the stability of the forced system is discussed in [6] and [7]. The connection with classical averaging is emphasized in [6], where Rayleigh dissipation is introduced to make the critical points hyperbolically stable. In [7], dissipation is not introduced, and a purely geometric analysis is applied within the Hamiltonian framework. We state the principal stability result.

Theorem 52.5:

Consider a Lagrangian control system prescribed by Equation 52.21 or its Hamiltonian equivalent (Equation 52.25). Suppose that the corresponding system of Equations 52.26 and 52.27 is forced by the oscillatory input given in Equation 52.28. Let q_0 be a critical point of the averaged potential which is independent of the period T (or frequency) of the forcing. Suppose, moreover, that, for all T sufficiently small (frequencies sufficiently large), q_0 is a strict local minimum of the averaged potential. Then $(q, \dot{q}) = (q_0, 0)$ is a stable equilibrium of the forced Lagrangian system, provided T is sufficiently small. If $(q, p) = (q_0, 0)$ is the corresponding equilibrium of the forced Hamiltonian system, then it is likewise stable, provided T is sufficiently small.

This theorem is proved in [7].

We end this section with two examples to which this theorem applies, followed by a simple example which does not satisfy the hypothesis and for which the theory is currently less complete.

Example 52.5: Oscillatory Stabilization of a Simple Pendulum

Consider, once again, the inverted pendulum discussed in Example 52.2. Assume now that no friction exists in the hinge and, therefore, $b = 0$. Using the classical theory of vibrational control in Example 52.6, it is not possible to draw conclusions on the stabilization of the upper equilibrium point by fast oscillating control, because the averaged equation will have purely imaginary eigenvalues when $b = 0$. The averaged potential provides a useful alternative in this case. For $b = 0$, the pendulum dynamics may be written

$$\ell\ddot{\theta} + \ddot{R}\sin\theta + g\sin\theta = 0,$$

where all the parameters have been previously defined in Example 52.2. Writing the pendulum's vertical velocity as $v(t) = \dot{R}(t)$, this is a system of the type we are considering with (reduced) Lagrangian

$\hat{L}(\theta,\dot{\theta};v)=(1/2)\ell\dot{\theta}^2+v\dot{\theta}\sin\theta+g\cos\theta$. To find stable motions using the theory we have presented, we construct the *averaged potential* by passing to the Hamiltonian description of the system. The momentum (conjugate to θ) is $p=\frac{\partial\hat{L}}{\partial\dot{\theta}}=\ell\dot{\theta}+v\sin\theta$. Applying the Legendre transformation, we obtain the corresponding Hamiltonian

$$H(\theta,p;v)=(p\,\dot{\theta}-\hat{L})_{|(\theta,p)}=\frac{1}{2\ell}(p-v\sin\theta)^2-g\cos\theta.$$

If we replace the coefficients involving $v(\cdot)$ with their time-averages over one period, we obtain the averaged Hamiltonian

$$\bar{H}(\theta,p)=\frac{1}{2\ell}(p-\bar{v}\sin\theta)^2+\frac{1}{2\ell}(\Sigma-\bar{v}^2)\sin^2\theta-g\cos\theta,$$

where $\bar{v}$ and Σ are the time averages over one period of $v(t)$ and $v(t)^2$ respectively. The averaged potential is just $V_A(\theta)=\frac{1}{2\ell}(\Sigma-\bar{v}^2)\sin^2\theta-g\cos\theta$. Consider the simple sinusoidal oscillation of the hinge-point, $R(t)=\alpha\sin\beta t$. Then $v(t)=\alpha\beta\cos\beta t$. Carrying out the construction we have outlined, the averaged potential is given more explicitly by

$$V_A(\theta)=\frac{\alpha^2\beta^2}{4\ell}\sin^2\theta-g\cos\theta. \tag{52.31}$$

Looking at the first derivative $V_A'(\theta)$, we find that $\theta=\pi$ is a critical point for all values of the parameters. Looking at the second derivative, we find that $V_A''(\pi)>0$ precisely when $\alpha^2\beta^2>2\ell g$. From Theorem 52.5 we conclude that, for sufficiently large values of the frequency β, the upright equilibrium is stable in the sense that motions of the forced system will remain nearby. This is of course completely consistent with classical results on this problem. (*cf.* Example 52.2.)

Example 52.6:

Example 52.4, reprise: oscillatory stabilization of a rotating pendulum.

Let us return to the mechanical system treated in Example 52.4. Omitting a few details, we proceed as follows. Starting from the Lagrangian in Equation 52.23, we obtain the corresponding Hamiltonian (the general formula for which is given by Equation 52.25). The averaged potential is given by the formula in Equation 52.30. Suppose the pendulum is forced to rotate at a constant rate, perturbed by a small-amplitude sinusoid, $v(t)=\omega+\alpha\sin\beta t$. Then the coefficients in Equation 52.30 are

$$\bar{v}=\frac{\beta}{2\pi}\int_0^{2\pi/\beta}v(t)\,dt=\omega,\ \text{and}$$

$$\Sigma=\frac{\beta}{2\pi}\int_0^{2\pi/\beta}v(t)^2\,dt=\omega^2+\frac{\alpha^2}{2},$$

and some algebraic manipulation shows that the averaged potential is given in this case by

$$V_A(\phi_1,\phi_2)=-c_1c_2-\frac{1}{4}\frac{I_yI_zc_2^2}{I_yc_1^2+I_zs_2^2}\alpha^2-\frac{1}{2}\left[I_xs_2^2+(I_ys_1^2+I_zc_1^2)c_2^2\right]\omega^2.$$

Stable modes of behavior under this type of forcing correspond to local minima of the averaged potential. A brief discussion of how this analysis proceeds will illustrate the utility of the approach.

When $\alpha=0$, the pendulum undergoes rotation at a constant rate about the vertical axis. For all rates ω, the pendulum is in equilibrium when it hangs straight down. There is a critical value, ω_{cr}, however, above which the vertical configuration is no longer stable. A critical point analysis of the averaged potential yields the relevant information and more. The partial derivatives of V_A with respect to ϕ_1 and ϕ_2 both vanish at $(\phi_1, \phi_2) = (0, 0)$ for all values of the parameters α, β, ω. To assess the stability of this critical point using Theorem 52.5, we compute the Hessian (matrix of second partial derivatives) of V_A evaluated at $(\phi_1, \phi_2) = (0, 0)$:

$$\begin{pmatrix} \frac{\partial^2 V_A}{\partial \phi_1^2}(0,0) & \frac{\partial^2 V_A}{\partial \phi_1 \partial \phi_2}(0,0) \\ \frac{\partial^2 V_A}{\partial \phi_1 \partial \phi_2}(0,0) & \frac{\partial^2 V_A}{\partial \phi_2^2}(0,0) \end{pmatrix} = \begin{pmatrix} 1 + \frac{1}{2}\frac{I_z}{I_y}(I_z - I_y)\alpha^2 - (I_y - I_z)\omega^2 & 0 \\ 0 & 1 + \frac{1}{2} I_z \alpha^2 - (I_x - I_z)\omega^2 \end{pmatrix}.$$

Let us treat the constant rotation case first. We have assumed $I_x \geq I_y \gg I_z$. When $\alpha = 0$, this means that the Hessian matrix above is positive definite for $0 \leq \omega^2 < 1/(I_x - I_z)$. This inequality gives the value of the critical rotation rate precisely as $\omega_{cr} = 1/\sqrt{I_x - I_z}$. We wish to answer the following question: Is it possible to provide a stabilizing effect by superimposing a small-amplitude, high-frequency sinusoidal oscillation on the constant-rate forced rotation? The answer emerges from Theorem 52.5 together with analysis of the Hessian. In the symmetric case, $I_x = I_y$, the answer is "no" because any nonzero value of α will decrease the $(1, 1)$-entry and hence the value of ω_{cr}. If $I_x > I_y$, however, there is the possibility of increasing ω_{cr} slightly, because, although the $(1, 1)$-entry is decreased, the more important $(2, 2)$-entry is increased.

Current research on oscillatory forcing to stabilize rotating systems (chains, shafts, turbines, etc.) is quite encouraging. Though only modest stabilization was possible for the rotating pendulum in the example above, more pronounced effects are generally possible with axial forcing. Because this approach to control appears to be quite robust (as seen in the next example), it merits attention in applications where feedback designs would be difficult to implement.

We conclude with an example to which Theorem 52.5 does not apply and for which the theory is currently less well developed. Methods of [6] can be used in this case.

Example 52.7:

Oscillation induced rest points in a pendulum on a cart. We consider a slight variation on Example 52.5 wherein we consider oscillating the hinge point of the pendulum along a line which is not vertical. More specifically, consider a cart to which there is a simple pendulum (as described in Example 52.5) attached so that the cart moves along a track inclined at an angle ψ to the horizontal. Suppose the position of the cart along its track at time t is prescribed by a variable $r(t)$. Then the pendulum dynamics are expressed

$$\ell\ddot{\theta} + \ddot{r}\cos(\theta - \psi) + g \sin\theta = 0.$$

Note that when $\psi = \pi/2$, the track is aligned vertically, and we recover the problem treated in Example 52.5. In the general case, let $v(t) = \dot{r}(t)$ and write the averaged potential

$$V_A(\theta) = -g\cos\theta + \frac{1}{2\ell}\cos^2(\theta - \alpha)\,(\Sigma - \bar{v}^2)$$

where

$$\bar{v} = \frac{1}{T}\int_0^T v(t)\,dt$$

and

$$\Sigma = \frac{1}{T}\int_0^T v(t)^2\,dt.$$

As in Example 52.5, we may take $v(t) = \alpha\beta \cos \beta t$. For sufficiently large frequencies β there are strict local minima of the averaged potential which are not equilibrium points of the forced system. Nevertheless, as noted in [6], the pendulum will execute motions in a neighborhood of such a point. To distinguish such emergent behavior from stable motions in neighborhoods of equilibria (of the nonautonomous system), we have called motions confined to neighborhoods of nonequilibrium critical points of the averaged potential *hovering motions*. For more information on such motions, the reader is referred to [41].

Remark on the Robustness of Open-Loop Methods

The last example suggests, and laboratory experiments bear out, that the stabilizing effects of oscillatory forcing of the type we have discussed are quite pronounced. Moreover, they are quite insensitive to the fine details of the mathematical models and to physical disturbances which may occur. Thus, the stabilizing effect observed in the inverted pendulum will be entirely preserved if the pendulum is perturbed or if the direction of the forcing isn't really vertical. Such robustness suggests that methods of this type are worth exploring in a wider variety of applications.

Remark on Oscillatory Control with Feedback

There are interesting applications (e.g., laser cooling) where useful designs arise through a combination of oscillatory forcing and certain types of feedback. For the theory of time-varying feedback designs, the reader is referred to [20] and the chapters on stability by Khalil, Teel, Sontag, Praly, and Georgiou appearing in this handbook.

52.4 Defining Terms

Anholonomy: Consider the controlled differential equation 52.1, and suppose that there is a non-zero function $\phi : \mathbf{R}^n \times \mathbf{R}^n \to \mathbf{R}$ such that $\phi(x, g_i(x)) \equiv 0$ for $i = 1, \ldots, m$. This represents a constraint on the state velocities which can be commanded. Despite such a constraint, it may happen that any two specified states can be joined by a trajectory $x(t)$ generated via Equation 52.1 by an appropriate choice of inputs $u_i(\cdot)$. Any state trajectory arising from Equation 52.1 constrained in this way is said to be determined from the inputs $u_i(\cdot)$ by *anholonomy*. In principle, the notation of *anholonomy* can be extended to systems given by Equation 52.2 or 52.3. Some authors who were consulted in preparation of this chapter objected to the use of the word in this more general context.

Averaged potential: An energy-like function that describes the steady-state behavior produced by high-frequency forcing of a physical system.

Completely controllable: A system of Equation 52.3 is said to be *completely controllable* if, given any $T > 0$ and any pair of points $x_0, x_1 \in \mathbf{R}^n$, there is a control input $u(\cdot)$ producing a motion $x(\cdot)$ of Equation 52.3 such that $x(0) = x_0$ and $x(T) = x_1$.

LARC: The *Lie algebra rank condition* is the condition that the defining vector fields in systems, such as Equation 52.1 or 52.2 together with their Lie brackets of all orders span $\mathbf{R}^n$.

Acknowledgment

The authors are indebted to many people for help in preparing this chapter. R.W. Brockett, in particular, provided useful guidance and criticism. The author J. Baillieul gratefully acknowledges support of the U.S. Air Force Office of Scientific Research under grant AFOSR-90-0226. Author B. Lehman gratefully acknowledges support of an NSF Presidential Faculty Fellow Award, NSF CMS-9453473.

References

1. Baillieul, J., Multilinear optimal control, *Proc. Conf. Geom. Control Eng.*, (NASA-Ames, Summer 1976), Brookline, MA: Math. Sci. Press, 337–359, 1977.
2. Baillieul, J., Geometric methods for nonlinear optimal control problems, *J. Optimiz. Theory Appl.*, 25(4), 519–548, 1978.
3. Baillieul, J., The behavior of super-articulated mechanisms subject to periodic forcing, in *Analysis of Controlled Dynamical Systems*, B. Bonnard, B. Bride, J.P. Gauthier, and I. Kupka, Eds., Birkhaüser, 35–50, 1991.
4. Baillieul, J. and Levi, M., Constrained relative motions in rotational mechanics, *Arch. Rational Mech. Anal.*, 115/2, 101–135, 1991.
5. Baillieul, J., The behavior of single-input super-articulated mechanisms, *Proc. 1991 Am. Control Conf.*, Boston, June 26–28, pp. 1622–1626, 1991.
6. Baillieul, J., Stable average motions of mechanical systems subject to periodic forcing, *Dynamics and Control of Mechanical Systems: The Falling Cat and Related Problems*, Fields Institute Communications, **1**, AMS, Providence, RI, 1–23, 1993.
7. Baillieul, J., Energy methods for stability of bilinear systems with oscillatory inputs, *Int. J. Robust Nonlinear Control*, Special Issue on the "Control of Nonlinear Mechanical Systems," H. Nijmeijer and A.J. van der Schaft, Guest Eds., July, 285–301, 1995.
8. Bellman, R., Bentsman, J., and Meerkov, S.M., Vibrational control of nonlinear systems: Vibrational stabilizability, *IEEE Trans. Automatic Control*, AC-31, 710–716, 1986.
9. Bellman, R., Bentsman, J., and Meerkov, S.M., Vibrational control of nonlinear systems: Vibrational controllability and transient behavior, *IEEE Trans. Automatic Control*, AC-31, 717–724, 1986.
10. Bishop, R.L. and Crittenden, R.J., *Geometry of Manifolds*, Academic Press, New York, 1964.
11. Bogolyubov, N.N., Perturbation theory in nonlinear mechanics, *Sb. Stroit. Mekh. Akad. Nauk Ukr.* SSR *14*, 9–34, 1950.
12. Bogolyubov, N.N. and Mitropolsky, Y.A., *Asymptotic Methods in the Theory of Nonlinear Oscillations*, 2nd ed., Gordon & Breach Publishers, New York, 1961.
13. Brockett, R.W., System theory on group manifolds and coset spaces, *SIAM J. Control*, 10(2), 265–284, 1972.
14. Brockett, R.W., Control Theory and Analytical Mechanics, in *Geometric Control Theory*, Vol. VII of Lie Groups: History, Frontiers, and Applications, C. Martin and R. Hermann, Eds., Math Sci Press, Brookline, MA, 1–46, 1977.
15. Brockett, R.W., Asymptotic stability and feedback stabilization, in *Differential Geometric Control Theory*, R.W. Brockett, R.S. Millman, and H.J. Sussmann, Eds., Birkhaüser, Basel, 1983.
16. Brockett, R.W., Control theory and singular Riemannian geometry, in *New Directions in Applied Mathematics*, Springer-Verlag, New York, 13–27, 1982.
17. Brockett, R.W., On the rectification of vibratory motion, *Sens. Actuat.*, 20, 91–96, 1989.
18. Brockett, R.W. and Dai, L., Nonholonomic kinematics and the role of elliptic functions in constructive controllability, in *Nonholonomic Motion Planning*, Kluwer Academic Publishers, 1–21, 1993.
19. Chow, W.L., Über Systeme von Linearen Partiellen Differentialgleichungen erster Ordnung, *Math. Ann.*, 117, 98–105, 1939.
20. Coron, J.M., Global asymptotic stabilization for controllable systems without drift, *Math. Control, Sig., Syst.*, 5, 295–312, 1992.
21. Haynes, G.W. and Hermes, H., Nonlinear controllability via Lie theory, *SIAM J. Control*, 8(4), 450–460, 1970.
22. Lafferriere, G. and Sussmann, H.J., Motion planning for controllable systems without drift, *Proc. IEEE Intl. Conf. Robot. Automat.*, 1148–1153, 1991.
23. Lafferriere, G. and Sussmann, H.J., A differential geometric approach to motion planning, in *Nonholonomic Motion Planning*, Kluwer Academic Publishers, 235–270, 1993.
24. Lehman, B., Bentsman, J., Lunel, S.V., and Verriest, E.I., Vibrational control of nonlinear time lag systems with bounded delay: Averaging theory, stabilizability, and transient behavior, *IEEE Trans. Auto. Control*, AC-39, 898–912, 1994.
25. Lehman, B., Vibrational control of time delay systems, in *Ordinary and Delay Equations*, J. Wiener and J. Hale, Eds., Pitman Research Notes in Mathematical Series (272), 111–115, 1992.
26. Kapitsa, P.L., Dynamic stability of a pendulum with a vibrating point of suspension, *Zh. Ehksp. Teor. Fiz.*, 21(5), 588–598, 1951.

27. Leonard, N.E. and Krishnaprasad, P.S., Control of switched electrical networks using averaging on Lie groups, *The 33rd IEEE Conference on Decision and Control*, Orlando, FL, Dec. 14–16, pp. 1919–1924, 1994.
28. Leonard, N.E. and Krishnaprasad, P.S., Motion control of drift-free, left-invariant systems on Lie groups, *IEEE Trans. Automat. Control*, AC40(9), 1539–1554, 1995.
29. Murray, R.M. and Sastry, S.S., Nonholonomic motion planning: steering using sinusoids, *IEEE Trans. Auto. Control*, 38(5), 700–716, 1993.
30. Nijmeijer, H. and van der Schaft, A.J., *Nonlinear Dynamical Control Systems*, Springer-Verlag, New York, 1990.
31. Sanders, J.A. and Verhulst, F., *Averaging Methods in Nonlinear Dynamical Systems*, Springer-Verlag, Applied Mathematical Sciences, 59, New York, 1985.
32. Speyer, J.L., Nonoptimality of steady-state cruise for aircraft, *AIAA J.*, 14(11), 1604–1610, 1976.
33. Speyer, J.L. and Evans, R.T., A second variational theory for optimal periodic processes, *IEEE Trans. Auto. Control*, AC-29(2), 138–148, 1984.
34. Stoker, J.J., *Nonlinear Vibrations in Mechanical and Electrical Systems*, J. Wiley & Sons, New York, 1950. Republished 1992 in Wiley Classics Library Edition.
35. Sussmann, H. and Jurdjevic, V., Controllability of nonlinear systems, *J. Diff. Eqs.*, 12, 95–116, 1972.
36. Sussmann, H.J. and Liu, W., Limits of highly oscillatory controls and approximation of general paths by admissible trajectories, *30th IEEE Conf. Decision Control*, Brighton, England, 1991.
37. Sussmann, H.J. and Liu, W., An approximation algorithm for nonholonomic systems, Rutgers University, Department of Mathematics preprint, SYCON-93-11. To appear in *SIAM J. Optimiz. Control.*
38. Sussmann, H.J. and Liu, W., Lie bracket extension and averaging: The single bracket case, in *Nonholonomic Motion Planning*, Kluwer Academic Publishers, 109–147, 1994.
39. Tilbury, D., Murray, R. and Sastry, S., Trajectory generation for the N-trailer problem using Goursat normal form, *30th IEEE Conf. Decision Control*, San Antonio, Texas, 971–977, 1993.
40. Tilbury, D., Murray, R. and Sastry, S., Trajectory generation for the N-trailer problem using Goursat normal form, *IEEE Trans. Auto. Control*, AC40(5), 802–819, 1995.
41. Weibel, S., Baillieul, J., and Kaper, T., Small-amplitude periodic motions of rapidly forced mechanical systems, *34th IEEE Conf. on Decision and Control,* New Orleans, 1995.

53

Adaptive Nonlinear Control

Miroslav Krstić
University of California, San Diego

Petar V. Kokotović
University of California, Santa Barbara

53.1 Introduction: Backstepping

Realistic models of physical systems are nonlinear and usually contain parameters (masses, inductances, aerodynamic coefficients, etc.) which are either poorly known or dependent on a slowly changing environment. If the parameters vary in a broad range, it is common to employ adaptation: a parameter estimator—*identifier*—continuously acquires knowledge about the plant and uses it to tune the controller "on-line."

Instabilities in nonlinear systems can be more explosive than in linear systems. During the parameter estimation transients, the state can "escape" to infinity in finite time. For this reason, adaptive nonlinear controllers cannot simply be the "adaptive versions" of standard nonlinear controllers.

Currently, the most systematic methodology for adaptive nonlinear control design is *backstepping*. We introduce the idea of backstepping by carrying out a *nonadaptive* design for the system

$$\dot{x}_1 = x_2 + \varphi(x_1)^{\mathrm{T}}\theta, \quad \varphi(0) = 0 \tag{53.1}$$

$$\dot{x}_2 = u, \tag{53.2}$$

where θ is a *known* parameter vector and $\varphi(x_1)$ is a smooth nonlinear function. Our goal is to stabilize the equilibrium $x_1 = 0, x_2 = -\varphi(0)^{\mathrm{T}}\theta = 0$. Backstepping design is recursive. First, the state x_2 is treated as a *virtual control* for the x_1-equation 53.1, and a *stabilizing function*

$$\alpha_1(x_1) = -c_1 x_1 - \varphi(x_1)^{\mathrm{T}}\theta, \quad c_1 > 0 \tag{53.3}$$

is designed to stabilize (Equation 53.1) assuming that $x_2 = \alpha_1(x_1)$ can be implemented. Since this is not the case, we define

$$z_1 = x_1, \tag{53.4}$$

$$z_2 = x_2 - \alpha_1(x_1), \tag{53.5}$$

where z_2 is an error variable expressing the fact that x_2 is not the true control. Differentiating z_1 and z_2 with respect to time, the complete system (Equations 53.1 and 53.2) is expressed in the error coordinates (Equations 53.4 and 53.5):

$$\dot{z}_1 = \dot{x}_1 = x_2 + \varphi^{\mathrm{T}}\theta = z_2 + \alpha_1 + \varphi^{\mathrm{T}}\theta = -c_1 z_1 + z_2, \tag{53.6}$$

$$\dot{z}_2 = \dot{x}_2 - \dot{\alpha}_1 = u - \frac{\partial \alpha_1}{\partial x_1}\dot{x}_1 = u - \frac{\partial \alpha_1}{\partial x_1}\left(x_2 + \varphi^{\mathrm{T}}\theta\right). \tag{53.7}$$

It is important to observe that the time derivative $\dot{\alpha}_1$ is implemented analytically, without a differentiator. For the system (Equations 53.6 and 53.7), we now design a control law $u = \alpha_2(x_1, x_2)$ to render the time derivative of a Lyapunov function negative definite. It turns out that the design can be completed with the simplest Lyapunov function

$$V(x_1, x_2) = \frac{1}{2}z_1^2 + \frac{1}{2}z_2^2. \tag{53.8}$$

Its derivative for Equations 53.6 and 53.7 is

$$\begin{aligned}\dot{V} &= z_1\left(-c_1 z_1 + z_2\right) + z_2\left[u - \frac{\partial \alpha_1}{\partial x_1}\left(x_2 + \varphi^{\mathrm{T}}\theta\right)\right] \\ &= -c_1 z_1^2 + z_2\left[u + z_1 - \frac{\partial \alpha_1}{\partial x_1}\left(x_2 + \varphi^{\mathrm{T}}\theta\right)\right].\end{aligned} \tag{53.9}$$

An obvious way to achieve negativity of $\dot{V}$ is to employ u to make the bracketed expression equal to $-c_2 z_2$ with $c_2 > 0$, namely,

$$u = \alpha_2(x_1, x_2) = -c_2 z_2 - z_1 + \frac{\partial \alpha_1}{\partial x_1}\left(x_2 + \varphi^{\mathrm{T}}\theta\right). \tag{53.10}$$

This control may not be the best choice because it cancels some terms which may contribute to the negativity of $\dot{V}$. Backstepping design offers enough flexibility to avoid cancellation. However, for the sake of clarity, we assume that none of the nonlinearities is useful, so that they all need to be cancelled as in the control law (Equation 53.10). This control law yields

$$\dot{V} = -c_1 z_1^2 - c_2 z_2^2, \tag{53.11}$$

which means that the equilibrium $z = 0$ is globally asymptotically stable. In view of Equations 53.4 and 53.5, the same is true about $x = 0$. The resulting closed-loop system in the z-coordinates is linear:

$$\begin{bmatrix}\dot{z}_1 \\ \dot{z}_2\end{bmatrix} = \begin{bmatrix}-c_1 & 1 \\ -1 & -c_2\end{bmatrix}\begin{bmatrix}z_1 \\ z_2\end{bmatrix}. \tag{53.12}$$

In the next four sections we present adaptive nonlinear designs through examples. Summaries of general design procedures are also provided but without technical details, for which the reader is referred to the text on nonlinear and adaptive control design by Krstić et al. [9]. Only elementary background on Lyapunov stability is assumed, while no previous familiarity with adaptive linear control is necessary. The two main methodologies for adaptive backstepping design are the *tuning functions design*, Section 53.2, and the *modular design*, Section 53.3. These sections assume that the full state is available for feedback.

Section 53.4 presents designs where only the output is measured. Section 53.5 discusses various extensions to more general classes of systems, followed by a brief literature review in Section 53.6.

53.2 Tuning Functions Design

In the tuning functions design, both the controller and the parameter update law are designed recursively. At each consecutive step a *tuning function* is designed as a potential update law. The tuning functions are not implemented as update laws. Instead, the stabilizing functions use them to compensate the effects of parameter estimation transients. Only the final tuning function is used as the parameter update law.

53.2.1 Introductory Examples

The tuning functions design will be introduced through examples with increasing complexity:

A	B	C
$\dot{x}_1 = u + \varphi(x_1)^{\mathrm{T}}\theta$	$\dot{x}_1 = x_2 + \varphi(x_1)^{\mathrm{T}}\theta$	$\dot{x}_1 = x_2 + \varphi(x_1)^{\mathrm{T}}\theta$
	$\dot{x}_2 = u$	$\dot{x}_2 = x_3$
		$\dot{x}_3 = u$

The adaptive problem arises because the parameter vector θ is *unknown*. The nonlinearity $\varphi(x_1)$ is known and for simplicity it is assumed that $\varphi(0) = 0$. The systems A, B, and C differ structurally: the number of integrators between the control u and the unknown parameter θ increases from zero at A, to two at C. Design A will be the simplest because the control u and the uncertainty $\varphi(x_1)^{\mathrm{T}}\theta$ are "matched," that is, the control does not have to overcome integrator transients in order to counteract the effects of the uncertainty. Design C will be the hardest because the control must act through two integrators before it reaches the uncertainty.

53.2.1.1 Design A

Let $\hat{\theta}$ be an estimate of the unknown parameter θ in the system

$$\dot{x}_1 = u + \varphi^{\mathrm{T}}\theta. \tag{53.13}$$

If this estimate were correct, $\hat{\theta} = \theta$, then the control law

$$u = -c_1 x_1 - \varphi(x_1)^{\mathrm{T}}\hat{\theta} \tag{53.14}$$

would achieve global asymptotic stability of $x = 0$. Because $\tilde{\theta} = \theta - \hat{\theta} \neq 0$, we have

$$\dot{x}_1 = -c_1 x_1 + \varphi(x_1)^{\mathrm{T}}\tilde{\theta}, \tag{53.15}$$

that is, the parameter estimation error $\tilde{\theta}$ continues to act as a disturbance which may destabilize the system. Our task is to find an update law for $\hat{\theta}(t)$ which preserves the boundedness of $x(t)$ and achieves its regulation to zero. To this end, we consider the Lyapunov function

$$V_1(x, \hat{\theta}) = \frac{1}{2}x_1^2 + \frac{1}{2}\tilde{\theta}^{\mathrm{T}}\Gamma^{-1}\tilde{\theta}, \tag{53.16}$$

where Γ is a positive definite symmetric matrix. The derivative of V_1 is

$$\dot{V}_1 = -c_1 x_1^2 + x_1\varphi(x_1)^{\mathrm{T}}\tilde{\theta} - \tilde{\theta}^{\mathrm{T}}\Gamma^{-1}\dot{\hat{\theta}} = -c_1 x_1^2 + \tilde{\theta}^{\mathrm{T}}\Gamma^{-1}\left(\Gamma\varphi(x_1)x_1 - \dot{\hat{\theta}}\right). \tag{53.17}$$

Our goal is to select an update law for $\dot{\hat{\theta}}$ to guarantee

$$\dot{V}_1 \leq 0. \tag{53.18}$$

The only way this can be achieved for any unknown $\tilde{\theta}$ is to choose

$$\dot{\hat{\theta}} = \Gamma\varphi(x_1)x_1. \tag{53.19}$$

This choice yields

$$\dot{V}_1 = -c_1 x_1^2, \tag{53.20}$$

which guarantees global stability of the equilibrium $x_1 = 0, \hat{\theta} = \theta$, and hence, the boundedness of $x_1(t)$ and $\hat{\theta}(t)$. By LaSalle's invariance theorem (see Chapter 39 by Khalil in this book), all the trajectories of the closed-loop adaptive system converge to the set where $\dot{V}_1 = 0$, that is, to the set where $c_1 x_1^2 = 0$, which implies that

$$\lim_{t\to\infty} x_1(t) = 0. \tag{53.21}$$

Alternatively, we can prove Equation 53.21 as follows. By integrating Equation 53.20 we obtain $\int_0^t c_1 x_1(\tau)^2 d\tau = V_1(x_1(0), \hat{\theta}(0)) - V_1(x_1(t), \hat{\theta}(t))$, which, thanks to the nonnegativity of V_1, implies that $\int_0^t c_1 x_1(\tau)^2 d\tau \le V_1(x_1(0), \hat{\theta}(0)) < \infty$. Hence, x_1 is square-integrable. Due to the boundedness of $x_1(t)$ and $\hat{\theta}(t)$, from Equations 53.15 and 53.19 it follows that $\dot{x}_1(t)$ and $\dot{\hat{\theta}}$ are also bounded. By Barbalat's lemma we conclude that $x_1(t) \to 0$.

The update law (Equation 53.19) is driven by the vector $\varphi(x_1)$, called the *regressor*, and the state x_1. This is a typical form of an update law in the tuning functions design: the speed of adaptation is dictated by the nonlinearity $\varphi(x_1)$ and the state x_1.

53.2.1.2 Design B

For the system

$$\begin{aligned} \dot{x}_1 &= x_2 + \varphi(x_1)^{\mathrm{T}}\theta, \\ \dot{x}_2 &= u, \end{aligned} \tag{53.22}$$

we have already designed a nonadaptive controller in Section 53.1. To design an adaptive controller, we replace the unknown θ by its estimate $\hat{\theta}$ both in the stabilizing function (Equation 53.3) and in the change of coordinate (Equation 53.5):

$$z_2 = x_2 - \alpha_1(x_1, \hat{\theta}), \quad \alpha_1(x_1, \hat{\theta}) = -c_1 z_1 - \varphi^{\mathrm{T}}\hat{\theta}. \tag{53.23}$$

Because in the system (Equation 53.22) the control input is separated from the unknown parameter by an integrator, the control law (Equation 53.10) will be strengthened by a term $\nu_2(x_1, x_2, \hat{\theta})$ which will compensate for the parameter estimation transients:

$$u = \alpha_2(x_1, x_2, \hat{\theta}) = -c_2 z_2 - z_1 + \frac{\partial \alpha_1}{\partial x_1}\left(x_2 + \varphi^{\mathrm{T}}\hat{\theta}\right) + \nu_2(x_1, x_2, \hat{\theta}). \tag{53.24}$$

The resulting system in the z coordinates is

$$\dot{z}_1 = z_2 + \alpha_1 + \varphi^{\mathrm{T}}\theta = -c_1 z_1 + z_2 + \varphi^{\mathrm{T}}\tilde{\theta}, \tag{53.25}$$

$$\begin{aligned} \dot{z}_2 = \dot{x}_2 - \dot{\alpha}_1 &= u - \frac{\partial \alpha_1}{\partial x_1}\left(x_2 + \varphi^{\mathrm{T}}\theta\right) - \frac{\partial \alpha_1}{\partial \hat{\theta}}\dot{\hat{\theta}} \\ &= -z_1 - c_2 z_2 - \frac{\partial \alpha_1}{\partial x_1}\varphi^{\mathrm{T}}\tilde{\theta} - \frac{\partial \alpha_1}{\partial \hat{\theta}}\dot{\hat{\theta}} + \nu_2(x_1, x_2, \hat{\theta}), \end{aligned} \tag{53.26}$$

or, in vector form,

$$\begin{bmatrix} \dot{z}_1 \\ \dot{z}_2 \end{bmatrix} = \begin{bmatrix} -c_1 & 1 \\ -1 & -c_2 \end{bmatrix}\begin{bmatrix} z_1 \\ z_2 \end{bmatrix} + \begin{bmatrix} \varphi^{\mathrm{T}} \\ -\dfrac{\partial \alpha_1}{\partial x_1}\varphi^{\mathrm{T}} \end{bmatrix}\tilde{\theta} + \begin{bmatrix} 0 \\ -\dfrac{\partial \alpha_1}{\partial \hat{\theta}}\dot{\hat{\theta}} + \nu_2(x_1, x_2, \hat{\theta}) \end{bmatrix}. \tag{53.27}$$

The term ν_2 can now be chosen to eliminate the last brackets:

$$\nu_2(x_1, x_2, \hat{\theta}) = \frac{\partial\alpha_1}{\partial\hat{\theta}}\dot{\hat{\theta}}. \tag{53.28}$$

This expression is implementable because $\dot{\hat{\theta}}$ will be available from the update law. Thus we obtain the *error system*

$$\begin{bmatrix}\dot{z}_1\\ \dot{z}_2\end{bmatrix} = \begin{bmatrix}-c_1 & 1\\ -1 & -c_2\end{bmatrix}\begin{bmatrix}z_1\\ z_2\end{bmatrix} + \begin{bmatrix}\varphi^{\mathrm{T}}\\ -\dfrac{\partial\alpha_1}{\partial x_1}\varphi^{\mathrm{T}}\end{bmatrix}\tilde{\theta}. \tag{53.29}$$

When the parameter error $\tilde{\theta}$ is zero, this system becomes the linear asymptotically stable system (Equation 53.12). Our remaining task is to select the update law $\dot{\hat{\theta}} = \Gamma\tau_2(x, \hat{\theta})$. Consider the Lyapunov function

$$V_2(x_1, x_2, \hat{\theta}) = V_1 + \frac{1}{2}z_2^2 = \frac{1}{2}z_1^2 + \frac{1}{2}z_2^2 + \frac{1}{2}\tilde{\theta}^{\mathrm{T}}\Gamma^{-1}\tilde{\theta}. \tag{53.30}$$

Because $\dot{\tilde{\theta}} = -\dot{\hat{\theta}}$, the derivative of V_2 is

$$\begin{aligned}\dot{V}_2 &= -c_1z_1^2 - c_2z_2^2 + [z_1,\ z_2]\begin{bmatrix}\varphi^{\mathrm{T}}\\ -\dfrac{\partial\alpha_1}{\partial x_1}\varphi^{\mathrm{T}}\end{bmatrix}\tilde{\theta} - \tilde{\theta}^{\mathrm{T}}\Gamma^{-1}\dot{\hat{\theta}}\\ &= -c_1z_1^2 - c_2z_2^2 + \tilde{\theta}^{\mathrm{T}}\Gamma^{-1}\left(\Gamma\left[\varphi, -\frac{\partial\alpha_1}{\partial x_1}\varphi\right]\begin{bmatrix}z_1\\ z_2\end{bmatrix} - \dot{\hat{\theta}}\right).\end{aligned} \tag{53.31}$$

The only way to eliminate the unknown parameter error $\tilde{\theta}$ is to select the update law

$$\dot{\hat{\theta}} = \Gamma\tau_2(x, \hat{\theta}) = \Gamma\left[\varphi,\ -\frac{\partial\alpha_1}{\partial x_1}\varphi\right]\begin{bmatrix}z_1\\ z_2\end{bmatrix} = \Gamma\left(\varphi z_1 - \frac{\partial\alpha_1}{\partial x_1}\varphi z_2\right). \tag{53.32}$$

Then $\dot{V}_2$ is nonpositive:

$$\dot{V}_2 = -c_1z_1^2 - c_2z_2^2, \tag{53.33}$$

which means that the global stability of $z = 0$, $\tilde{\theta} = 0$ is achieved. Moreover, by applying either the LaSalle or the Barbalat argument mentioned in Design A, we prove that $z(t) \to 0$ as $t \to \infty$. Finally, from Equation 53.23, it follows that the equilibrium $x = 0$, $\hat{\theta} = \theta$ is globally stable and $x(t) \to 0$ as $t \to \infty$.

The crucial property of the control law in Design B is that it incorporates the ν_2-term (Equation 53.28) which is proportional to $\dot{\hat{\theta}}$ and compensates for the effect of parameter estimation transients on the coordinate change (Equation 53.23). It is this departure from the certainty equivalence principle that makes the adaptive stabilization possible for systems with nonlinearities of arbitrary growth.

By comparing Equation 53.32 with Equation 53.19, we note that the first term φz_1 is the potential update law for the z_1-system. The functions

$$\tau_1(x_1) = \varphi z_1, \tag{53.34}$$

$$\tau_2(x_1, x_2, \hat{\theta}) = \tau_1(x_1) - \frac{\partial\alpha_1}{\partial x_1}\varphi z_2 \tag{53.35}$$

are referred to as the *tuning functions*, because of their role as potential update laws for intermediate systems in the backstepping procedure.

53.2.1.3 Design C

The system

$$\begin{aligned}\dot{x}_1 &= x_2 + \varphi(x_1)^{\mathrm{T}}\theta,\\ \dot{x}_2 &= x_3,\\ \dot{x}_3 &= u\end{aligned} \tag{53.36}$$

is obtained by augmenting system (Equation 53.22) with an integrator. The control law $\alpha_2(x_1, x_2, \hat{\theta})$ designed in (Equation 53.24) can no longer be directly applied because x_3 is a state and not a control input. We "step back" through the integrator $\dot{x}_3 = u$ and design the control law for the actual input u. However, we keep the stabilizing function α_2 and use it to define the third error coordinate

$$z_3 = x_3 - \alpha_2(x_1, x_2, \hat{\theta}). \tag{53.37}$$

The parameter update law (Equation 53.32) will have to be modified with an additional z_3-term. Instead of $\dot{\hat{\theta}}$ in Equation 53.28, the compensating term ν_2 will now use the potential update law (Equation 53.35) for the system (Equation 53.29):

$$\nu_2(x_1, x_2, \hat{\theta}) = \frac{\partial\alpha_1}{\partial\hat{\theta}}\Gamma\tau_2(x_1, x_2, \hat{\theta}). \tag{53.38}$$

Hence, the role of the tuning function τ_2 is to substitute for the actual update law in the compensation of the effects of parameter estimation transients.

With Equations 53.23, 53.35, 53.37, and 53.38, and the stabilizing function α_2 in Equation 53.24, we have

$$\begin{bmatrix}\dot{z}_1\\ \dot{z}_2\end{bmatrix} = \begin{bmatrix}-c_1 & 1\\ -1 & -c_2\end{bmatrix}\begin{bmatrix}z_1\\ z_2\end{bmatrix} + \begin{bmatrix}\varphi^{\mathrm{T}}\\ -\dfrac{\partial\alpha_1}{\partial x_1}\varphi^{\mathrm{T}}\end{bmatrix}\tilde{\theta} + \begin{bmatrix}0\\ z_3 + \dfrac{\partial\alpha_1}{\partial\hat{\theta}}(\Gamma\tau_2 - \dot{\hat{\theta}})\end{bmatrix}. \tag{53.39}$$

This system differs from the error system (Equation 53.29) only in its last term. Likewise, instead of the Lyapunov inequality (Equation 53.33) we have

$$\dot{V}_2 = -c_1 z_1^2 - c_2 z_2^2 + z_2 z_3 + z_2\frac{\partial\alpha_1}{\partial\hat{\theta}}(\Gamma\tau_2 - \dot{\hat{\theta}}) + \tilde{\theta}^{\mathrm{T}}(\tau_2 - \Gamma^{-1}\dot{\hat{\theta}}). \tag{53.40}$$

Differentiating Equation 53.37, we obtain

$$\begin{aligned}\dot{z}_3 &= u - \frac{\partial\alpha_2}{\partial x_1}\left(x_2 + \varphi^{\mathrm{T}}\theta\right) - \frac{\partial\alpha_2}{\partial x_2}x_3 - \frac{\partial\alpha_2}{\partial\hat{\theta}}\dot{\hat{\theta}}\\ &= u - \frac{\partial\alpha_2}{\partial x_1}\left(x_2 + \varphi^{\mathrm{T}}\hat{\theta}\right) - \frac{\partial\alpha_2}{\partial x_2}x_3 - \frac{\partial\alpha_2}{\partial\hat{\theta}}\dot{\hat{\theta}} - \frac{\partial\alpha_2}{\partial x_1}\varphi^{\mathrm{T}}\tilde{\theta}.\end{aligned} \tag{53.41}$$

We now stabilize the (z_1, z_2, z_3)-system (Equation 53.39 and 53.41 with respect to the Lyapunov function

$$V_3(x, \hat{\theta}) = V_2 + \frac{1}{2}z_3^2 = \frac{1}{2}z_1^2 + \frac{1}{2}z_2^2 + \frac{1}{2}z_3^2 + \frac{1}{2}\tilde{\theta}^{\mathrm{T}}\Gamma^{-1}\tilde{\theta}. \tag{53.42}$$

Its derivative along Equations 53.39 and 53.41 is

$$\begin{aligned}\dot{V}_3 = -c_1 z_1^2 - c_2 z_2^2 + z_2\frac{\partial\alpha_1}{\partial\hat{\theta}}(\Gamma\tau_2 - \dot{\hat{\theta}}) + z_3\left[z_2 + u - \frac{\partial\alpha_2}{\partial x_1}\left(x_2 + \varphi^{\mathrm{T}}\hat{\theta}\right) - \frac{\partial\alpha_2}{\partial x_2}x_3 - \frac{\partial\alpha_2}{\partial\hat{\theta}}\dot{\hat{\theta}}\right]\\ + \tilde{\theta}^{\mathrm{T}}\left(\tau_2 - \frac{\partial\alpha_2}{\partial x_1}\varphi z_3 - \Gamma^{-1}\dot{\hat{\theta}}\right).\end{aligned} \tag{53.43}$$

Again we must eliminate the unknown parameter error $\tilde{\theta}$ from $\dot{V}_3$. For this we must choose the update law as

$$\dot{\hat{\theta}} = \Gamma\tau_3(x_1, x_2, x_3, \hat{\theta}) = \Gamma\left(\tau_2 - \frac{\partial\alpha_2}{\partial x_1}\varphi z_3\right) = \Gamma\left[\varphi, \ -\frac{\partial\alpha_1}{\partial x_1}\varphi, \ \frac{\partial\alpha_2}{\partial x_1}\varphi\right]\begin{bmatrix} z_1 \\ z_2 \\ z_3 \end{bmatrix}. \tag{53.44}$$

Upon inspection of the bracketed terms in $\dot{V}_3$, we pick the control law:

$$u = \alpha_3(x_1, x_2, x_3, \hat{\theta}) = -z_2 - c_3 z_3 + \frac{\partial\alpha_2}{\partial x_1}\left(x_2 + \varphi^{\mathrm{T}}\hat{\theta}\right) + \frac{\partial\alpha_2}{\partial x_2}x_3 + \nu_3. \tag{53.45}$$

The compensation term ν_3 is yet to be chosen. Substituting Equation 53.45 into Equation 53.43, we obtain

$$\dot{V}_3 = -c_1 z_1^2 - c_2 z_2^2 - c_3 z_3^2 + z_2\frac{\partial\alpha_1}{\partial\hat{\theta}}(\Gamma\tau_2 - \dot{\hat{\theta}}) + z_3\left(\nu_3 - \frac{\partial\alpha_2}{\partial\hat{\theta}}\dot{\hat{\theta}}\right). \tag{53.46}$$

From this expression it is clear that ν_3 should cancel $\frac{\partial\alpha_2}{\partial\hat{\theta}}\dot{\hat{\theta}}$. In order to cancel the cross-term $z_2\frac{\partial\alpha_1}{\partial\hat{\theta}}(\Gamma\tau_2 - \dot{\hat{\theta}})$ with ν_3, it appears that we would need to divide by z_3. However, the variable z_3 might take a zero value during the transient, and should be regulated to zero to accomplish the control objective. We resolve this difficulty by noting that

$$\begin{aligned} \dot{\hat{\theta}} - \Gamma\tau_2 &= \dot{\hat{\theta}} - \Gamma\tau_3 + \Gamma\tau_3 - \Gamma\tau_2 \\ &= \dot{\hat{\theta}} - \Gamma\tau_3 - \Gamma\frac{\partial\alpha_2}{\partial x_1}\varphi z_3, \end{aligned} \tag{53.47}$$

so that $\dot{V}_3$ in Equation 53.46 is rewritten as

$$\dot{V}_3 = -c_1 z_1^2 - c_2 z_2^2 - c_3 z_3^2 + z_3\left(\nu_3 - \frac{\partial\alpha_2}{\partial\hat{\theta}}\Gamma\tau_3 + \frac{\partial\alpha_1}{\partial\hat{\theta}}\Gamma\frac{\partial\alpha_2}{\partial x_1}\varphi z_2\right). \tag{53.48}$$

From Equation 53.48, the choice of ν_3 is immediate:

$$\nu_3(x_1, x_2, x_3, \hat{\theta}) = \frac{\partial\alpha_2}{\partial\hat{\theta}}\Gamma\tau_3 - \frac{\partial\alpha_1}{\partial\hat{\theta}}\Gamma\frac{\partial\alpha_2}{\partial x_1}\varphi z_2. \tag{53.49}$$

The resulting $\dot{V}_3$ is

$$\dot{V}_3 = -c_1 z_1^2 - c_2 z_2^2 - c_3 z_3^2, \tag{53.50}$$

which guarantees that the equilibrium $x = 0$, $\hat{\theta} = \theta$ is globally stable, and $x(t) \to 0$ as $t \to \infty$. The Lyapunov design leading to Equation 53.48 is effective but does not reveal the stabilization mechanism. To provide further insight we write the (z_1, z_2, z_3)-system (Equations 53.39 and 53.41) with u given in Equation 53.45 but with ν_3 yet to be selected:

$$\begin{bmatrix} \dot{z}_1 \\ \dot{z}_2 \\ \dot{z}_3 \end{bmatrix} = \begin{bmatrix} -c_1 & 1 & 0 \\ -1 & -c_2 & 1 \\ 0 & -1 & -c_3 \end{bmatrix}\begin{bmatrix} z_1 \\ z_2 \\ z_3 \end{bmatrix} + \begin{bmatrix} \varphi^{\mathrm{T}} \\ -\frac{\partial\alpha_1}{\partial x_1}\varphi^{\mathrm{T}} \\ -\frac{\partial\alpha_2}{\partial x_1}\varphi^{\mathrm{T}} \end{bmatrix}\tilde{\theta} + \begin{bmatrix} 0 \\ \frac{\partial\alpha_1}{\partial\hat{\theta}}(\Gamma\tau_2 - \dot{\hat{\theta}}) \\ \nu_3 - \frac{\partial\alpha_2}{\partial\hat{\theta}}\Gamma\tau_3 \end{bmatrix}. \tag{53.51}$$

While ν_3 can cancel the matched term $\frac{\partial\alpha_2}{\partial\hat{\theta}}\dot{\hat{\theta}}$, it cannot cancel the term $\frac{\partial\alpha_1}{\partial\hat{\theta}}(\Gamma\tau_2 - \dot{\hat{\theta}})$ in the second equation. By substituting Equation 53.47, we note that $\frac{\partial\alpha_1}{\partial\hat{\theta}}(\Gamma\tau_2 - \dot{\hat{\theta}})$ has z_3 as a factor and absorb it into the "system matrix":

$$\begin{bmatrix} \dot{z}_1 \\ \dot{z}_2 \\ \dot{z}_3 \end{bmatrix} = \begin{bmatrix} -c_1 & 1 & 0 \\ -1 & -c_2 & 1 + \dfrac{\partial\alpha_1}{\partial\hat{\theta}}\Gamma\dfrac{\partial\alpha_2}{\partial x_1}\varphi \\ 0 & -1 & -c_3 \end{bmatrix} \begin{bmatrix} z_1 \\ z_2 \\ z_3 \end{bmatrix} + \begin{bmatrix} \varphi^{\mathrm{T}} \\ -\dfrac{\partial\alpha_1}{\partial x_1}\varphi^{\mathrm{T}} \\ -\dfrac{\partial\alpha_2}{\partial x_1}\varphi^{\mathrm{T}} \end{bmatrix} \tilde{\theta} + \begin{bmatrix} 0 \\ 0 \\ \nu_3 - \dfrac{\partial\alpha_2}{\partial\hat{\theta}}\Gamma\tau_3 \end{bmatrix}. \tag{53.52}$$

Now ν_3 in Equation 53.49 yields

$$\begin{bmatrix} \dot{z}_1 \\ \dot{z}_2 \\ \dot{z}_3 \end{bmatrix} = \begin{bmatrix} -c_1 & 1 & 0 \\ -1 & -c_2 & 1 + \dfrac{\partial\alpha_1}{\partial\hat{\theta}}\Gamma\dfrac{\partial\alpha_2}{\partial x_1}\varphi \\ 0 & -1 - \dfrac{\partial\alpha_1}{\partial\hat{\theta}}\Gamma\dfrac{\partial\alpha_2}{\partial x_1}\varphi & -c_3 \end{bmatrix} \begin{bmatrix} z_1 \\ z_2 \\ z_3 \end{bmatrix} + \begin{bmatrix} \varphi^{\mathrm{T}} \\ -\dfrac{\partial\alpha_1}{\partial x_1}\varphi^{\mathrm{T}} \\ -\dfrac{\partial\alpha_2}{\partial x_1}\varphi^{\mathrm{T}} \end{bmatrix} \tilde{\theta}. \tag{53.53}$$

This choice, which places the term $-\frac{\partial\alpha_1}{\partial\hat{\theta}}\Gamma\frac{\partial\alpha_2}{\partial x_1}\varphi$ at the (2,3) position in the system matrix, achieves skew-symmetry with its positive image above the diagonal. What could not be achieved by pursuing a linear-like form was achieved by designing a nonlinear system where the nonlinearities are "balanced" rather than cancelled.

53.2.2 General Recursive Design Procedure

A systematic backstepping design with tuning functions has been developed for the class of nonlinear systems transformable into the *parametric strict-feedback form*:

$$\begin{aligned} \dot{x}_1 &= x_2 + \varphi_1(x_1)^{\mathrm{T}}\theta, \\ \dot{x}_2 &= x_3 + \varphi_2(x_1, x_2)^{\mathrm{T}}\theta, \\ &\vdots \\ \dot{x}_{n-1} &= x_n + \varphi_{n-1}(x_1, \ldots, x_{n-1})^{\mathrm{T}}\theta, \\ \dot{x}_n &= \beta(x)u + \varphi_n(x)^{\mathrm{T}}\theta, \\ y &= x_1, \end{aligned} \tag{53.54}$$

where β and

$$F(x) = [\varphi_1(x_1),\ \varphi_2(x_1, x_2), \ldots, \varphi_n(x)] \tag{53.55}$$

are smooth nonlinear functions, and $\beta(x) \neq 0, \forall x \in \mathbb{R}^n$. (Broader classes of systems that can be controlled using adaptive backstepping are listed in Section 53.5.)

TABLE 53.1 Summary of the Tuning Functions Design for Tracking

$$z_i = x_i - y_r^{(i-1)} - \alpha_{i-1}, \tag{53.56}$$

$$\alpha_i(\bar{x}_i, \hat{\theta}, \bar{y}_r^{(i-1)}) = -z_{i-1} - c_i z_i - w_i^T \hat{\theta} + \sum_{k=1}^{i-1}\left(\frac{\partial \alpha_{i-1}}{\partial x_k} x_{k+1} + \frac{\partial \alpha_{i-1}}{\partial y_r^{(k-1)}} y_r^{(k)}\right) + \nu_i, \tag{53.57}$$

$$\nu_i(\bar{x}_i, \hat{\theta}, \bar{y}_r^{(i-1)}) = +\frac{\partial \alpha_{i-1}}{\partial \hat{\theta}} \Gamma \tau_i + \sum_{k=2}^{i-1} \frac{\partial \alpha_{k-1}}{\partial \hat{\theta}} \Gamma w_i z_k, \tag{53.58}$$

$$\tau_i(\bar{x}_i, \hat{\theta}, \bar{y}_r^{(i-1)}) = \tau_{i-1} + w_i z_i, \tag{53.59}$$

$$w_i(\bar{x}_i, \hat{\theta}, \bar{y}_r^{(i-2)}) = \varphi_i - \sum_{k=1}^{i-1} \frac{\partial \alpha_{i-1}}{\partial x_k} \varphi_k, \tag{53.60}$$

$$i = 1, \ldots, n, \quad \bar{x}_i = (x_1, \ldots, x_i), \quad \bar{y}_r^{(i)} = (y_r, \dot{y}_r, \ldots, y_r^{(i)}).$$

Adaptive Control Law:

$$u = \frac{1}{\beta(x)}\left[\alpha_n(x, \hat{\theta}, \bar{y}_r^{(n-1)}) + y_r^{(n)}\right]. \tag{53.61}$$

Parameter Update Law:

$$\dot{\hat{\theta}} = \Gamma \tau_n(x, \hat{\theta}, \bar{y}_r^{(n-1)}) = \Gamma W z. \tag{53.62}$$

Note: For notational convenience we define $z_0 \stackrel{\triangle}{=} 0, \alpha_0 \stackrel{\triangle}{=} 0, \tau_0 \stackrel{\triangle}{=} 0$.

The general design summarized in Table 53.1 achieves asymptotic tracking, that is, the output $y = x_1$ of the system (Equation 53.54) is forced to asymptotically track the reference output $y_r(t)$ whose first n derivatives are assumed to be known, bounded, and piecewise continuous.

The closed-loop system has the form

$$\dot{z} = A_z(z, \hat{\theta}, t) z + W(z, \hat{\theta}, t)^T \tilde{\theta}, \tag{53.63}$$

$$\dot{\hat{\theta}} = \Gamma W(z, \hat{\theta}, t) z, \tag{53.64}$$

where

$$A_z(z, \hat{\theta}, t) = \begin{bmatrix} -c_1 & 1 & 0 & \cdots & 0 \\ -1 & -c_2 & 1 + \sigma_{23} & \cdots & \sigma_{2n} \\ 0 & -1 - \sigma_{23} & \ddots & \ddots & \vdots \\ \vdots & \vdots & \ddots & \ddots & 1 + \sigma_{n-1,n} \\ 0 & -\sigma_{2n} & \cdots & -1 - \sigma_{n-1,n} & -c_n \end{bmatrix} \tag{53.65}$$

and

$$\sigma_{jk}(x, \hat{\theta}) = -\frac{\partial \alpha_{j-1}}{\partial \hat{\theta}} \Gamma w_k. \tag{53.66}$$

Because of the skew-symmetry of the off-diagonal part of the matrix A_z, it is easy to see that the Lyapunov function

$$V_n = \frac{1}{2} z^T z + \frac{1}{2} \tilde{\theta}^T \Gamma^{-1} \tilde{\theta} \tag{53.67}$$

has the derivative

$$\dot{V}_n = -\sum_{k=1}^{n} c_k z_k^2, \tag{53.68}$$

which guarantees that the equilibrium $z = 0, \hat{\theta} = \theta$ is globally stable and $z(t) \to 0$ as $t \to \infty$. This means, in particular, that the system state and the control input are bounded and asymptotic tracking is achieved: $\lim_{t\to\infty}\left[y(t) - y_r(t)\right] = 0$.

To help understand how the control design of Table 53.1 leads to the closed-loop system (Equations 53.63 through 53.66), we provide an interpretation of the matrix A_z for $n = 5$:

$$A_z = \begin{bmatrix} -c_1 & 1 & & & \\ -1 & -c_2 & 1 & & \\ & -1 & -c_3 & 1 & \\ & & -1 & -c_4 & 1 \\ & & & -1 & -c_5 \end{bmatrix} + \begin{bmatrix} & 0 & 0 & 0 & 0 \\ 0 & & \sigma_{23} & \sigma_{24} & \sigma_{25} \\ 0 & -\sigma_{23} & & \sigma_{34} & \sigma_{35} \\ 0 & -\sigma_{24} & -\sigma_{34} & & \sigma_{45} \\ 0 & -\sigma_{25} & -\sigma_{35} & -\sigma_{45} & \end{bmatrix}. \tag{53.69}$$

If the parameters were known, $\hat{\theta} \equiv \theta$, in which case we would not use adaptation, $\Gamma = 0$, the stabilizing functions (Equation 53.57) would be implemented with $\nu_i \equiv 0$, and hence $\sigma_{i,j} = 0$. Then A_z would be just the above constant tridiagonal asymptotically stable matrix. When the parameters are unknown, we use $\Gamma > 0$ and, owing to the change of variable $z_i = x_i - y_r^{(i-1)} - \alpha_{i-1}$, in each of the $\dot{z}_i$-equations a term $-\frac{\partial \alpha_{i-1}}{\partial \hat{\theta}}\dot{\hat{\theta}} = \sum_{k=1}^{n} \sigma_{ik} z_k$ appears. The term $\nu_i = -\sum_{k=1}^{i} \sigma_{ik} z_k - \sum_{k=2}^{i-1} \sigma_{ki} z_k$ in the stabilizing function (Equation 53.57) is crucial in compensating the effect of $\dot{\hat{\theta}}$. The σ_{ik}-terms above the diagonal in Equation 53.69 come from $\dot{\hat{\theta}}$. Their skew-symmetric negative images come from feedback ν_i.

It can be shown that the resulting closed-loop system (Equations 53.63 and 53.64), as well as each intermediate system, has a *strict passivity* property from $\tilde{\theta}$ as the input to τ_i as the output. The loop around this operator is closed (see Figure 53.1) with the vector integrator with gain Γ, which is a passive block. It follows from passivity theory that this feedback connection of one strictly passive and one passive block is globally stable.

53.3 Modular Design

In the tuning functions design, the controller and the identifier are derived in an interlaced fashion. This interlacing led to considerable controller complexity and inflexibility in the choice of the update law.

It is not hard to extend various standard identifiers for linear systems to nonlinear systems. It is therefore desirable to have adaptive designs where the controller can be combined with different identifiers (gradient, least-squares, passivity based, etc.). We refer to such adaptive designs as *modular*.

In nonlinear systems it is not a good idea to connect a good identifier with a controller which is good when the parameter is known (a "certainty equivalence" controller). To illustrate this, let us consider the

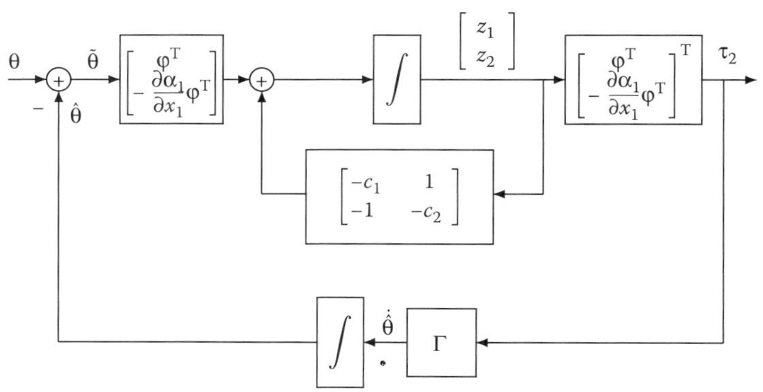

FIGURE 53.1 The closed-loop adaptive system (Equations 53.29 and 53.32).

error system

$$\dot{x} = -x + \varphi(x)\tilde{\theta} \tag{53.70}$$

obtained by applying a certainty equivalence controller $u = -x - \varphi(x)\hat{\theta}$ to the scalar system $\dot{x} = u + \varphi(x)\theta$. The parameter estimators commonly used in adaptive linear control generate bounded estimates $\hat{\theta}(t)$ with convergence rates not faster than exponential. Suppose that $\tilde{\theta}(t) = e^{-t}$ and $\varphi(x) = x^3$, which, upon substitution in Equation 53.70, gives

$$\dot{x} = -x + x^3 e^{-t}. \tag{53.71}$$

For initial conditions $|x_0| > \sqrt{\frac{3}{2}}$, the system (Equation 53.71) is not only unstable but its solution escapes to infinity in finite time:

$$x(t) \to \infty \quad \text{as } t \to \frac{1}{3}\ln\frac{x_0^2}{x_0^2 - 3/2}. \tag{53.72}$$

From this example we conclude that for nonlinear systems we need stronger controllers which prevent unbounded behavior caused by $\tilde{\theta}$.

53.3.1 Controller Design

We strengthen the controller for the preceding example, $u = -x - \varphi(x)\hat{\theta}$, with a *nonlinear damping* term $-\varphi(x)^2 x$, that is, $u = -x - \varphi(x)\hat{\theta} - \varphi(x)^2 x$. With this stronger controller, the closed-loop system is

$$\dot{x} = -x - \varphi(x)^2 x + \varphi(x)\tilde{\theta}. \tag{53.73}$$

To see that x is bounded whenever $\tilde{\theta}$ is, we consider the Lyapunov function $V = \frac{1}{2}x^2$. Its derivative along the solutions of Equation 53.73 is

$$\begin{aligned}\dot{V} &= -x^2 - \varphi(x)^2 x^2 + x\varphi(x)\tilde{\theta} = -x^2 - \left[\varphi(x)x - \frac{1}{2}\tilde{\theta}\right]^2 + \frac{1}{4}\tilde{\theta}^2 \\ &\le -x^2 + \frac{1}{4}\tilde{\theta}^2.\end{aligned} \tag{53.74}$$

From this inequality it is clear that $|x(t)|$ will not grow larger than $\frac{1}{2}|\tilde{\theta}(t)|$, because then $\dot{V}$ becomes negative and $V = \frac{1}{2}x^2$ decreases. Thanks to the nonlinear damping, the boundedness of $\tilde{\theta}(t)$ guarantees that $x(t)$ is bounded.

To show how nonlinear damping is incorporated into a higher-order backstepping design, we consider the system

$$\begin{aligned}\dot{x}_1 &= x_2 + \varphi(x_1)^{\mathrm{T}}\theta, \\ \dot{x}_2 &= u.\end{aligned} \tag{53.75}$$

Viewing x_2 as a control input, we first design a control law $\alpha_1(x_1, \hat{\theta})$ to guarantee that the state x_1 in $\dot{x}_1 = x_2 + \varphi(x_1)^{\mathrm{T}}\theta$ is bounded whenever $\tilde{\theta}$ is bounded. In the first stabilizing function we include a nonlinear damping term* $-\kappa_1|\varphi(x_1)|^2 x_1$:

$$\alpha_1(x_1, \hat{\theta}) = -c_1 x_1 - \varphi(x_1)^{\mathrm{T}}\hat{\theta} - \kappa_1|\varphi(x_1)|^2 x_1, \quad c_1, \kappa_1 > 0. \tag{53.76}$$

Then we define the error variable $z_2 = x_2 - \alpha_1(x_1, \hat{\theta})$, and for uniformity denote $z_1 = x_1$. The first equation is now

$$\dot{z}_1 = -c_1 z_1 - \kappa_1|\varphi|^2 z_1 + \varphi^{\mathrm{T}}\tilde{\theta} + z_2. \tag{53.77}$$

* The Euclidian norm of a vector v is denoted as $|v| = \sqrt{v^{\mathrm{T}}v}$.

If z_2 were zero, the Lyapunov function $V_1 = \frac{1}{2}z_1^2$ would have the derivative

$$\dot{V}_1 = -c_1 z_1^2 - \kappa_1 |\varphi|^2 z_1^2 + z_1 \varphi^{\mathrm{T}} \tilde{\theta} = -c_1 z_1^2 - \kappa_1 \left| \varphi z_1 - \frac{1}{2\kappa_1} \tilde{\theta} \right|^2 + \frac{1}{4\kappa_1} |\tilde{\theta}|^2 \leq -c_1 z_1^2 + \frac{1}{4\kappa_1} |\tilde{\theta}|^2, \quad (53.78)$$

so that z_1 would be bounded whenever $\tilde{\theta}$ is bounded. With $z_2 \neq 0$ we have

$$\dot{V}_1 \leq -c_1 z_1^2 + \frac{1}{4\kappa_1} |\tilde{\theta}|^2 + z_1 z_2. \quad (53.79)$$

Differentiating $x_2 = z_2 + \alpha_1(x_1, \hat{\theta})$, the second equation in Equation 53.75 yields

$$\dot{z}_2 = \dot{x}_2 - \dot{\alpha}_1 = u - \frac{\partial \alpha_1}{\partial x_1} \left(x_2 + \varphi^{\mathrm{T}} \theta \right) - \frac{\partial \alpha_1}{\partial \hat{\theta}} \dot{\hat{\theta}}. \quad (53.80)$$

The derivative of the Lyapunov function

$$V_2 = V_1 + \frac{1}{2} z_2^2 = \frac{1}{2} |z|^2 \quad (53.81)$$

along the solutions of Equations 53.77 and 53.80 is

$$\begin{aligned} \dot{V}_2 &\leq -c_1 z_1^2 + \frac{1}{4\kappa_1} |\tilde{\theta}|^2 + z_1 z_2 + z_2 \left[u - \frac{\partial \alpha_1}{\partial x_1} \left(x_2 + \varphi^{\mathrm{T}} \theta \right) - \frac{\partial \alpha_1}{\partial \hat{\theta}} \dot{\hat{\theta}} \right] \\ &\leq -c_1 z_1^2 + \frac{1}{4\kappa_1} |\tilde{\theta}|^2 + z_2 \left[u + z_1 - \frac{\partial \alpha_1}{\partial x_1} \left(x_2 + \varphi^{\mathrm{T}} \hat{\theta} \right) - \left(\frac{\partial \alpha_1}{\partial x_1} \varphi^{\mathrm{T}} \tilde{\theta} + \frac{\partial \alpha_1}{\partial \hat{\theta}} \dot{\hat{\theta}} \right) \right]. \end{aligned} \quad (53.82)$$

We note that now, in addition to the $\tilde{\theta}$-dependent disturbance term $\frac{\partial \alpha_1}{\partial x_1} \varphi^{\mathrm{T}} \tilde{\theta}$, we also have a $\dot{\hat{\theta}}$-dependent disturbance $\frac{\partial \alpha_1}{\partial \hat{\theta}} \dot{\hat{\theta}}$. No such term appeared in the scalar system (Equation 53.73). We now use nonlinear damping terms $-\kappa_2 \left| \frac{\partial \alpha_1}{\partial x_1} \varphi \right|^2 z_2$ and $-g_2 \left| \frac{\partial \alpha_1}{\partial \hat{\theta}}^{\mathrm{T}} \right|^2 z_2$ to counteract the effects of both $\tilde{\theta}$ and $\dot{\hat{\theta}}$:

$$u = -z_1 - c_2 z_2 - \kappa_2 \left| \frac{\partial \alpha_1}{\partial x_1} \varphi \right|^2 z_2 - g_2 \left| \frac{\partial \alpha_1}{\partial \hat{\theta}}^{\mathrm{T}} \right|^2 z_2 + \frac{\partial \alpha_1}{\partial x_1} \left(x_2 + \varphi^{\mathrm{T}} \hat{\theta} \right), \quad (53.83)$$

where $c_2, \kappa_2, g_2 > 0$. Upon completing the squares as in Equation 53.78, we obtain

$$\dot{V}_2 \leq -c_1 z_1^2 - c_2 z_2^2 + \left(\frac{1}{4\kappa_1} + \frac{1}{4\kappa_2} \right) |\tilde{\theta}|^2 + \frac{1}{4g_2} |\dot{\hat{\theta}}|^2, \quad (53.84)$$

which means that the state of the error system

$$\dot{z} = \begin{bmatrix} -c_1 - \kappa_2 |\varphi|^2 & 1 \\ -1 & -c_2 - \kappa_2 \left| \frac{\partial \alpha_1}{\partial x_1} \varphi \right|^2 - g_2 \left| \frac{\partial \alpha_1}{\partial \hat{\theta}}^{\mathrm{T}} \right|^2 \end{bmatrix} z + \begin{bmatrix} \varphi^{\mathrm{T}} \\ -\frac{\partial \alpha_1}{\partial x_1} \varphi^{\mathrm{T}} \end{bmatrix} \tilde{\theta} + \begin{bmatrix} 0 \\ -\frac{\partial \alpha_1}{\partial \hat{\theta}} \end{bmatrix} \dot{\hat{\theta}} \quad (53.85)$$

is bounded whenever the disturbance inputs $\tilde{\theta}$ and $\dot{\hat{\theta}}$ are bounded. Moreover, since V_2 is quadratic in z, see Equation 53.81, we can use Equation 53.84 to show that the boundedness of z is guaranteed also when $\dot{\hat{\theta}}$ is square-integrable but not bounded. This observation is crucial for the modular design with passive identifiers where $\dot{\hat{\theta}}$ cannot be *a priori* guaranteed to be bounded.

The recursive controller design for the parametric strict-feedback systems (Equation 53.54) is summarized in Table 53.2.

TABLE 53.2 Summary of the Controller Design in the Modular Approach

$$z_i = x_i - y_r^{(i-1)} - \alpha_{i-1}, \tag{53.86}$$

$$\alpha_i(\bar{x}_i, \hat{\theta}, \bar{y}_r^{(i-1)}) = -z_{i-1} - c_i z_i - w_i^T \hat{\theta} + \sum_{k=1}^{i-1}\left(\frac{\partial \alpha_{i-1}}{\partial x_k} x_{k+1} + \frac{\partial \alpha_{i-1}}{\partial y_r^{(k-1)}} y_r^{(k)}\right) - s_i z_i, \tag{53.87}$$

$$w_i(\bar{x}_i, \hat{\theta}, \bar{y}_r^{(i-2)}) = \varphi_i - \sum_{k=1}^{i-1} \frac{\partial \alpha_{i-1}}{\partial x_k}\varphi_k, \tag{53.88}$$

$$s_i(\bar{x}_i, \hat{\theta}, \bar{y}_r^{(i-2)}) = \kappa_i |w_i|^2 + g_i \left|\frac{\partial \alpha_{i-1}}{\partial \hat{\theta}}^T\right|^2, \tag{53.89}$$

$$i = 1, \ldots, n, \quad \bar{x}_i = (x_1, \ldots, x_i), \quad \bar{y}_r^{(i)} = (y_r, \dot{y}_r, \ldots, y_r^{(i)}).$$

Adaptive Control Law:

$$u = \frac{1}{\beta(x)}\left[\alpha_n(x, \hat{\theta}, \bar{y}_r^{(n-1)}) + y_r^{(n)}\right]. \tag{53.90}$$

Controller Module Guarantees:

$$\text{If} \quad \tilde{\theta} \in \mathcal{L}_\infty \quad \text{and} \quad \dot{\hat{\theta}} \in \mathcal{L}_2 \quad \text{or} \quad \mathcal{L}_\infty \ \text{then}\ x \in \mathcal{L}_\infty.$$

Note: For notational convenience we define $z_0 \triangleq 0, \alpha_0 \triangleq 0$.

Comparing the expression for the stabilizing function (Equation 53.87) in the modular design with the expression (Equation 53.57) for the tuning functions design we see that the difference is in the second lines. While the stabilization in the tuning functions design is achieved with the terms ν_i, in the modular design this is accomplished with the nonlinear damping term $-s_i z_i$, where

$$s_i(\bar{x}_i, \hat{\theta}, \bar{y}_r^{(i-2)}) = \kappa_i |w_i|^2 + g_i \left|\frac{\partial \alpha_{i-1}}{\partial \hat{\theta}}^T\right|^2. \tag{53.91}$$

The resulting error system is

$$\dot{z} = A_z(z, \hat{\theta}, t)z + W(z, \hat{\theta}, t)^T \tilde{\theta} + Q(z, \hat{\theta}, t)^T \dot{\hat{\theta}}, \tag{53.92}$$

where A_z, W, and Q are

$$A_z(z, \hat{\theta}, t) = \begin{bmatrix} -c_1 - s_1 & 1 & 0 & \cdots & 0 \\ -1 & -c_2 - s_2 & 1 & \ddots & \vdots \\ 0 & -1 & \ddots & \ddots & 0 \\ \vdots & \ddots & \ddots & \ddots & 1 \\ 0 & \cdots & 0 & -1 & -c_n - s_n \end{bmatrix},$$

$$W(z, \hat{\theta}, t)^T = \begin{bmatrix} w_1^T \\ w_2^T \\ \vdots \\ w_n^T \end{bmatrix}, \quad Q(z, \hat{\theta}, t)^T = \begin{bmatrix} 0 \\ -\frac{\partial \alpha_1}{\partial \hat{\theta}} \\ \vdots \\ -\frac{\partial \alpha_{n-1}}{\partial \hat{\theta}} \end{bmatrix}. \tag{53.93}$$

Since the controller module guarantees that x is bounded whenever $\tilde{\theta}$ is bounded and $\dot{\hat{\theta}}$ is either bounded or square-integrable, then we need identifiers which independently guarantee these properties. Both the boundedness and the square-integrability requirements for $\dot{\hat{\theta}}$ are essentially conditions which limit the speed of adaptation, and only one of them needs to be satisfied. The modular design needs *slow adaptation* because the controller does not cancel the effect of $\dot{\hat{\theta}}$, as was the case in the tuning functions design.

In addition to boundedness of $x(t)$, our goal is to achieve asymptotic tracking, that is, to regulate $z(t)$ to zero. With z and $\hat{\theta}$ bounded, it is not hard to prove that $z(t) \to 0$ provided

$$W(z(t), \hat{\theta}(t), t)^{\mathrm{T}}\tilde{\theta}(t) \to 0 \quad \text{and} \quad \dot{\hat{\theta}}(t) \to 0.$$

Let us factor the regressor matrix W, using Equations 53.93, 53.88, and 53.55 as

$$W(z, \hat{\theta}, t)^{\mathrm{T}} = \begin{bmatrix} 1 & 0 & \cdots & 0 \\ -\dfrac{\partial \alpha_1}{\partial x_1} & 1 & \ddots & \vdots \\ \vdots & \ddots & \ddots & 0 \\ -\dfrac{\partial \alpha_{n-1}}{\partial x_1} & \cdots & -\dfrac{\partial \alpha_{n-1}}{\partial x_{n-1}} & 1 \end{bmatrix} F(x)^{\mathrm{T}} \triangleq N(z, \hat{\theta}, t) F(x)^{\mathrm{T}}. \tag{53.94}$$

Since the matrix $N(z, \hat{\theta}, t)$ is invertible, the tracking condition $W(z(t), \hat{\theta}(t), t)^{\mathrm{T}}\tilde{\theta}(t) \to 0$ becomes $F(x(t))^{\mathrm{T}}\tilde{\theta}(t) \to 0$.

In the next two subsections we develop identifiers for the general parametric model

$$\dot{x} = f(x, u) + F(x, u)^{\mathrm{T}}\theta. \tag{53.95}$$

The parametric strict-feedback system (Equation 53.54) is a special case of this model with $F(x, u)$ given by Equation 53.55 and $f(x, u) = [x_2, \ldots, x_n, \beta_0(x)u]^{\mathrm{T}}$.

Before we present the design of identifiers, we summarize the properties required from the identifier module:

1. $\tilde{\theta} \in \mathcal{L}_\infty$ and $\dot{\hat{\theta}} \in \mathcal{L}_2$ or $\mathcal{L}_\infty$,
2. If $x \in \mathcal{L}_\infty$ then $F(x(t))^{\mathrm{T}}\tilde{\theta}(t) \to 0$ and $\dot{\hat{\theta}}(t) \to 0$.

We present two types of identifiers: the *passive* identifier and the *swapping* identifier.

53.3.2 Passive Identifier

For the parametric model (Equation 53.95), we implement, as shown in Figure 53.2, the "observer"

$$\dot{\hat{x}} = \left[A_0 - \lambda F(x, u)^{\mathrm{T}} F(x, u) P\right](\hat{x} - x) + f(x, u) + F(x, u)^{\mathrm{T}}\hat{\theta}, \tag{53.96}$$

where $\lambda > 0$ and A_0 is an arbitrary constant matrix such that

$$PA_0 + A_0^{\mathrm{T}}P = -I, \quad P = P^{\mathrm{T}} > 0. \tag{53.97}$$

By direct substitution it can be seen that the observer error

$$\epsilon = x - \hat{x} \tag{53.98}$$

is governed by

$$\dot{\epsilon} = \left[A_0 - \lambda F(x, u)^{\mathrm{T}} F(x, u) P\right]\epsilon + F(x, u)^{\mathrm{T}}\tilde{\theta}. \tag{53.99}$$

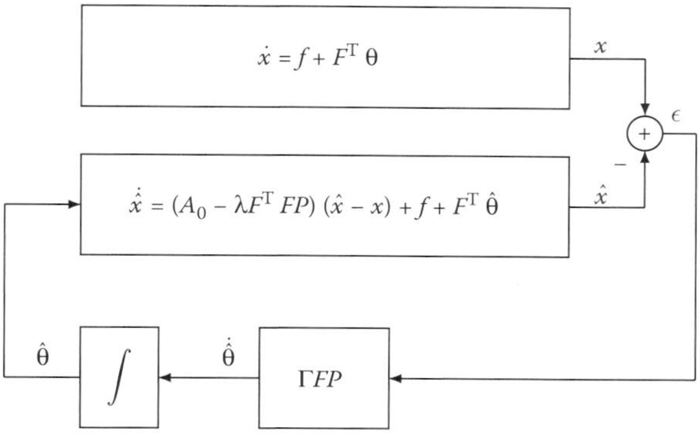

FIGURE 53.2 The passive identifier.

The observer error system (Equation 53.99) has a *strict passivity* property from the input $\tilde{\theta}$ to the output $F(x, u)P\epsilon$. A standard result of passivity theory is that the equilibrium $\tilde{\theta} = 0$, $\epsilon = 0$ of the negative feedback connection of one strictly passive and one passive system is globally stable. Using integral feedback such a connection as in Figure 53.3 can be formed. This suggests the use of the following update law:

$$\dot{\hat{\theta}} = \Gamma F(x, u)P\epsilon, \quad \Gamma = \Gamma^T > 0. \tag{53.100}$$

To analyze the stability properties of the passive identifier we use the Lyapunov function

$$V = \tilde{\theta}^T\Gamma^{-1}\tilde{\theta} + \epsilon^T P\epsilon. \tag{53.101}$$

After uncomplicated calculations, its derivative can be shown to satisfy

$$\dot{V} \leq -\epsilon^T\epsilon - \frac{\lambda}{\bar{\lambda}(\Gamma)^2}|\dot{\hat{\theta}}|^2. \tag{53.102}$$

This guarantees the boundedness of $\tilde{\theta}$ and ϵ, even when $\lambda = 0$. However, $\dot{\hat{\theta}}$ cannot be shown to be bounded (unless x and u are known to be bounded). Instead, for the passive identifier one can show that $\dot{\hat{\theta}}$ is square-integrable. For this we must use $\lambda > 0$, that is, we rely on the nonlinear damping term $-\lambda F(x, u)^T F(x, u)P$ in the observer. The boundedness of $\tilde{\theta}$ and the square-integrability of $\dot{\hat{\theta}}$ imply (*cf.* Table 53.2) that x is bounded.

To prove the tracking, we need to show that the identifier guarantees that, whenever x is bounded, $F(x(t))^T\tilde{\theta}(t) \to 0$ and $\dot{\hat{\theta}}(t) \to 0$. Both properties are established by Barbalat's lemma. The latter property can easily be shown to follow from the square-integrability of $\dot{\hat{\theta}}$. The regulation of $F(x)^T\tilde{\theta}$ to zero follows upon showing that both $\epsilon(t)$ and $\dot{\epsilon}(t)$ converge to zero. While the convergence of $\epsilon(t)$ follows by deducing its square-integrability from Equation 53.102, the convergence of $\dot{\epsilon}(t)$ follows from the fact that its integral, $\int_0^\infty \dot{\epsilon}(\tau)\,d\tau = \epsilon(\infty) - \epsilon(0) = -\epsilon(0)$, exists.

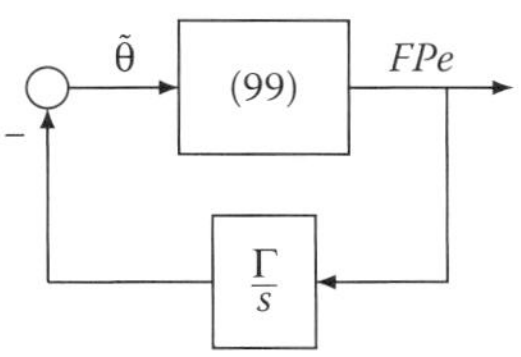

FIGURE 53.3 Negative feedback connection of the strictly passive system (Equation 53.99) with the passive system $\frac{\Gamma}{s}$.

53.3.3 Swapping Identifier

For the parametric model (Equation 53.95), we implement two filters,

$$\dot{\Omega}^{\mathrm{T}} = \left[A_0 - \lambda F(x,u)^{\mathrm{T}} F(x,u) P\right] \Omega^{\mathrm{T}} + F(x,u)^{\mathrm{T}}, \tag{53.103}$$

$$\dot{\Omega}_0 = \left[A_0 - \lambda F(x,u)^{\mathrm{T}} F(x,u) P\right] (\Omega_0 - x) - f(x,u), \tag{53.104}$$

where $\lambda \geq 0$ and A_0 is as defined in Equation 53.97. The *estimation error*,

$$\epsilon = x + \Omega_0 - \Omega^{\mathrm{T}}\hat{\theta}. \tag{53.105}$$

can be written in the form

$$\epsilon = \Omega^{\mathrm{T}}\tilde{\theta} + \tilde{\epsilon}, \tag{53.106}$$

where $\tilde{\epsilon} \stackrel{\triangle}{=} x + \Omega_0 - \Omega^{\mathrm{T}}\theta$ decays exponentially because it is governed by

$$\dot{\tilde{\epsilon}} = \left[A_0 - \lambda F(x,u)^{\mathrm{T}} F(x,u) P\right] \tilde{\epsilon}. \tag{53.107}$$

The filters (Equations 53.103 and 53.104) have converted the dynamic model (Equation 53.95) into the linear static parametric model (Equation 53.106) to which we can apply standard estimation algorithms. As our update law we will employ either the gradient

$$\dot{\hat{\theta}} = \Gamma \frac{\Omega\epsilon}{1 + \nu \mathrm{tr}\{\Omega^{\mathrm{T}}\Omega\}}, \quad \Gamma = \Gamma^{\mathrm{T}} > 0, \quad \nu \geq 0 \tag{53.108}$$

or the least-squares algorithm

$$\begin{aligned} \dot{\hat{\theta}} &= \Gamma \frac{\Omega\epsilon}{1 + \nu \mathrm{tr}\{\Omega^{\mathrm{T}}\Omega\}}, \\ \dot{\Gamma} &= -\Gamma \frac{\Omega\Omega^{\mathrm{T}}}{1 + \nu \mathrm{tr}\{\Omega^{\mathrm{T}}\Omega\}} \Gamma, \quad \Gamma(0) = \Gamma(0)^{\mathrm{T}} > 0, \quad \nu \geq 0. \end{aligned} \tag{53.109}$$

By allowing $\nu = 0$, we encompass unnormalized gradient and least squares. The complete swapping identifier is shown in Figure 53.4.

The update law normalization, $\nu > 0$, and the nonlinear damping, $\lambda > 0$, are two different means for slowing down the identifier in order to guarantee the boundedness and square-integrability of $\dot{\hat{\theta}}$.

For the gradient update law (Equation 53.108), the identifier properties (boundedness of $\tilde{\theta}$ and $\hat{\theta}$ and regulation of $F(x)\tilde{\theta}$ and $\dot{\hat{\theta}}$) are established via the Lyapunov function

$$V = \frac{1}{2}\tilde{\theta}^{\mathrm{T}}\Gamma^{-1}\tilde{\theta} + \tilde{\epsilon}P\tilde{\epsilon} \tag{53.110}$$

whose derivative is

$$\dot{V} \leq -\frac{3}{4}\frac{\epsilon^{\mathrm{T}}\epsilon}{1 + \nu \mathrm{tr}\{\Omega^{\mathrm{T}}\Omega\}}. \tag{53.111}$$

The Lyapunov function for the least-squares update law (Equation 53.109) is $V = \tilde{\theta}^{\mathrm{T}}\Gamma(t)^{-1}\tilde{\theta} + \tilde{\epsilon}P\tilde{\epsilon}$.

53.4 Output Feedback Designs

For linear systems, a common solution to the output-feedback problem is a stabilizing state-feedback controller employing the state estimates from an exponentially converging observer. Unfortunately,

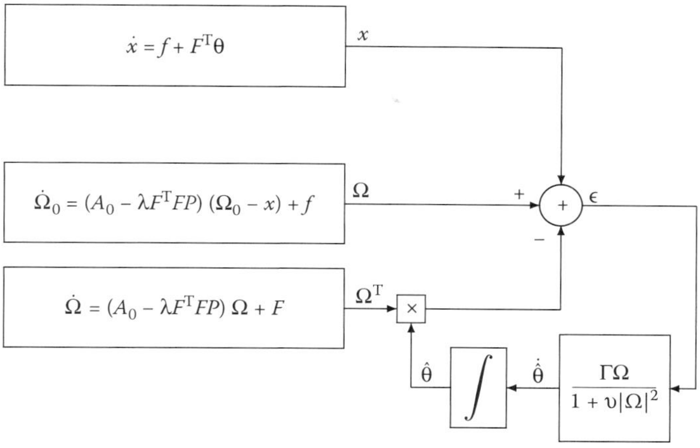

FIGURE 53.4 The swapping identifier.

this approach is not applicable to nonlinear systems. Additional difficulties arise when the nonlinear plant has unknown parameters because *adaptive* observers, in general, are not exponentially convergent.

These obstacles have been overcome for systems transformable into the *output-feedback form*:

$$\begin{aligned} \dot{x} &= Ax + \phi(y) + \Phi(y)a + \begin{bmatrix} 0 \\ b \end{bmatrix} \sigma(y)u, \quad x \in \mathbb{R}^n \\ y &= e_1^{\mathrm{T}} x, \end{aligned} \tag{53.112}$$

where only the output y is available for measurement,

$$A = \begin{bmatrix} 0 & \\ \vdots & I_{n-1} \\ 0 & \cdots\ 0 \end{bmatrix}, \tag{53.113}$$

$$\phi(y) = \begin{bmatrix} \varphi_{0,1}(y) \\ \vdots \\ \varphi_{0,n}(y) \end{bmatrix}, \quad \Phi(y) = \begin{bmatrix} \varphi_{1,1}(y) & \cdots & \varphi_{q,1}(y) \\ \vdots & & \vdots \\ \varphi_{1,n}(y) & \cdots & \varphi_{q,n}(y) \end{bmatrix}, \tag{53.114}$$

and the vectors of unknown constant parameters are

$$a = [a_1, \ldots, a_q]^{\mathrm{T}}, \quad b = [b_m, \ldots, b_0]^{\mathrm{T}}. \tag{53.115}$$

We make the following assumptions: the sign of b_m is known; the polynomial $B(s) = b_m s^m + \cdots + b_1 s + b_0$ is known to be Hurwitz; and $\sigma(y) \neq 0\ \forall y \in \mathbb{R}$. An important restriction is that the nonlinearities $\phi(y)$ and $\Phi(y)$ are allowed to depend only on the output y. Even when θ is known, this restriction is needed to achieve global stability.

TABLE 53.3 State Estimation Filters

Filters:	
$\dot{\xi} = A_0\xi + ky + \phi(y),$	(53.116)
$\dot{\Xi} = A_0\Xi + \Phi(y),$	(53.117)
$\dot{\lambda} = A_0\lambda + e_n\sigma(y)u.$	(53.118)
$v_j = A_0^j\lambda, \quad j = 0, \ldots, m,$	(53.119)
$\Omega^{\mathrm{T}} = [v_m, \ldots, v_1, v_0, \Xi].$	(53.120)

We define the parameter-dependent state estimate

$$\hat{x} = \xi + \Omega^{\mathrm{T}}\theta, \tag{53.121}$$

which employs the filters given in Table 53.3, with the vector $k = [k_1, \ldots, k_n]^{\mathrm{T}}$ chosen so that the matrix $A_0 = A - ke_1^{\mathrm{T}}$ is Hurwitz, that is,

$$PA_0 + A_0^{\mathrm{T}}P = -I, \quad P = P^{\mathrm{T}} > 0. \tag{53.122}$$

The state estimation error

$$\varepsilon = x - \hat{x} \tag{53.123}$$

is readily shown to satisfy

$$\dot{\varepsilon} = A_0\varepsilon. \tag{53.124}$$

The following two expressions for $\dot{y}$ are instrumental in the backstepping design:

$$\dot{y} = \omega_0 + \omega^{\mathrm{T}}\theta + \varepsilon_2 \tag{53.125}$$

$$= b_m v_{m,2} + \omega_0 + \bar{\omega}^{\mathrm{T}}\theta + \varepsilon_2, \tag{53.126}$$

where

$$\omega_0 = \varphi_{0,1} + \xi_2, \tag{53.127}$$

$$\omega = [v_{m,2}, v_{m-1,2}, \ldots, v_{0,2}, \Phi_{(1)} + \Xi_{(2)}]^{\mathrm{T}}, \tag{53.128}$$

$$\bar{\omega} = [0, v_{m-1,2}, \ldots, v_{0,2}, \Phi_{(1)} + \Xi_{(2)}]^{\mathrm{T}}. \tag{53.129}$$

Since the states $x_2, \ldots, x_n$ are not measured, the backstepping design is applied to the system

$$\dot{y} = b_m v_{m,2} + \omega_0 + \bar{\omega}^{\mathrm{T}}\theta + \varepsilon_2, \tag{53.130}$$

$$\dot{v}_{m,i} = v_{m,i+1} - k_i v_{m,1}, \quad i = 2, \ldots, \rho - 1, \tag{53.131}$$

$$\dot{v}_{m,\rho} = \sigma(y)u + v_{m,\rho+1} - k_\rho v_{m,1}. \tag{53.132}$$

The order of this system is equal to the relative degree of the plant (Equation 53.112).

53.4.1 Output-Feedback Design with Tuning Functions

The output-feedback design with tuning functions is summarized in Table 53.4. The resulting error system is

$$\dot{z} = A_z(z,t)z + W_\varepsilon(z,t)\varepsilon_2 + W_\theta(z,t)^{\mathrm{T}}\tilde{\theta} - b_m\left(\dot{y}_{\mathrm{r}} + \bar{\alpha}_1\right)e_1\tilde{\varrho}, \tag{53.133}$$

where

$$A_z = \begin{bmatrix} -c_1 - d_1 & \hat{b}_m & 0 & \cdots & \cdots & 0 \\ -\hat{b}_m & -c_2 - d_2\left(\frac{\partial \alpha_1}{\partial y}\right)^2 & 1+\sigma_{23} & \sigma_{24} & \cdots & \sigma_{2,\rho} \\ 0 & -1-\sigma_{23} & \ddots & \ddots & \ddots & \vdots \\ \vdots & -\sigma_{24} & \ddots & \ddots & \ddots & \sigma_{\rho-2,\rho} \\ \vdots & \vdots & \ddots & \ddots & \ddots & 1+\sigma_{\rho-1,\rho} \\ 0 & -\sigma_{2,\rho} & \cdots & -\sigma_{\rho-2,\rho} & -1-\sigma_{\rho-1,\rho} & -c_\rho - d_\rho\left(\frac{\partial \alpha_{\rho-1}}{\partial y}\right)^2 \end{bmatrix} \tag{53.134}$$

and

$$W_\varepsilon(z,t) = \begin{bmatrix} 1 \\ -\frac{\partial \alpha_1}{\partial y} \\ \vdots \\ -\frac{\partial \alpha_{\rho-1}}{\partial y} \end{bmatrix}, \quad W_\theta(z,t)^{\mathrm{T}} = W_\varepsilon(z,t)\omega^{\mathrm{T}} - \hat{\varrho}\left(\dot{y}_{\mathrm{r}} + \bar{\alpha}_1\right) e_1 e_1^{\mathrm{T}}. \tag{53.135}$$

The nonlinear damping terms $-d_i\left(\frac{\partial \alpha_{i-1}}{\partial y}\right)^2$ in Equation 53.134 are included to counteract the exponentially decaying state estimation error ε_2. The variable $\hat{\varrho}$ is an estimate of $\varrho = 1/b_m$.

TABLE 53.4 Output-Feedback Tuning Functions Design

$$z_1 = y - y_{\mathrm{r}}, \tag{53.136}$$

$$z_i = v_{m,i} - \hat{\varrho} y_{\mathrm{r}}^{(i-1)} - \alpha_{i-1}, \quad i = 2, \ldots, \rho. \tag{53.137}$$

$$\alpha_1 = \hat{\varrho}\bar{\alpha}_1, \quad \bar{\alpha}_1 = -\left(c_1 + d_1\right) z_1 - \omega_0 - \bar{\omega}^{\mathrm{T}}\hat{\theta}, \tag{53.138}$$

$$\alpha_2 = -\hat{b}_m z_1 - \left[c_2 + d_2\left(\frac{\partial \alpha_1}{\partial y}\right)^2\right] z_2 + \left(\dot{y}_{\mathrm{r}} + \frac{\partial \alpha_1}{\partial \hat{\varrho}}\right)\dot{\hat{\varrho}} + \frac{\partial \alpha_1}{\partial \hat{\theta}}\Gamma\tau_2 + \beta_2, \tag{53.139}$$

$$\begin{aligned} \alpha_i = {} & -z_{i-1} - \left[c_i + d_i\left(\frac{\partial \alpha_{i-1}}{\partial y}\right)^2\right] z_i + \left(y_{\mathrm{r}}^{(i-1)} + \frac{\partial \alpha_{i-1}}{\partial \hat{\varrho}}\right)\dot{\hat{\varrho}}, \\ & + \frac{\partial \alpha_{i-1}}{\partial \hat{\theta}}\Gamma\tau_i - \sum_{j=2}^{i-1}\frac{\partial \alpha_{j-1}}{\partial \hat{\theta}}\Gamma\frac{\partial \alpha_{i-1}}{\partial y} z_j + \beta_i, \quad i = 3, \ldots, \rho, \end{aligned} \tag{53.140}$$

$$\begin{aligned} \beta_i = {} & \frac{\partial \alpha_{i-1}}{\partial y}\left(\omega_0 + \omega^{\mathrm{T}}\hat{\theta}\right) + \frac{\partial \alpha_{i-1}}{\partial \xi}\left(A_0\xi + ky + \phi\right) + \frac{\partial \alpha_{i-1}}{\partial \Xi}\left(A_0\Xi + \Phi\right), \\ & + \sum_{j=1}^{i-1}\frac{\partial \alpha_{i-1}}{\partial y_{\mathrm{r}}^{(j-1)}} y_{\mathrm{r}}^{(j)} + k_i v_{m,1} + \sum_{j=1}^{m+i-1}\frac{\partial \alpha_{i-1}}{\partial \lambda_j}\left(-k_j\lambda_1 + \lambda_{j+1}\right). \end{aligned} \tag{53.141}$$

TABLE 53.4 (Continued) Output-Feedback Tuning Functions Design

$$\tau_1 = \left(\omega - \hat{\varrho}\left(\dot{y}_r + \bar{\alpha}_1\right) e_1\right) z_1, \tag{53.142}$$

$$\tau_i = \tau_{i-1} - \frac{\partial \alpha_{i-1}}{\partial y} \omega z_i, \quad i = 2, \ldots, \rho. \tag{53.143}$$

Adaptive Control Law:

$$u = \frac{1}{\sigma(y)} \left(\alpha_\rho - v_{m,\rho+1} + \hat{\varrho} y_r^{(\rho)}\right). \tag{53.144}$$

Parameter Update Laws:

$$\dot{\hat{\theta}} = \Gamma \tau_\rho \tag{53.145}$$

$$\dot{\hat{\varrho}} = -\gamma \operatorname{sgn}(b_m) \left(\dot{y}_r + \bar{\alpha}_1\right) z_1. \tag{53.146}$$

TABLE 53.5 Output-Feedback Controller in the Modular Design

$$z_1 = y - y_r, \tag{53.147}$$

$$z_i = v_{m,i} - \frac{1}{\hat{b}_m} y_r^{(i-1)} - \alpha_{i-1}, \quad i = 2, \ldots, \rho. \tag{53.148}$$

$$\alpha_1 = -\frac{\operatorname{sgn}(b_m)}{\varsigma_m} \left(c_1 + s_1\right) z_1 + \frac{1}{\hat{b}_m} \bar{\alpha}_1, \quad \bar{\alpha}_1 = -\omega_0 - \bar{\omega}^{\mathrm{T}} \hat{\theta}, \tag{53.149}$$

$$\alpha_2 = -\hat{b}_m z_1 - \left(c_2 + s_2\right) z_2 + \beta_2, \tag{53.150}$$

$$\alpha_i = -z_{i-1} - \left(c_i + s_i\right) z_i + \beta_i, \quad i = 3, \ldots, \rho, \tag{53.151}$$

$$\begin{aligned} \beta_i = {} & \frac{\partial \alpha_{i-1}}{\partial y} \left(\omega_0 + \omega^{\mathrm{T}} \hat{\theta}\right) + \frac{\partial \alpha_{i-1}}{\partial \xi} \left(A_0 \xi + k y + \phi\right) + \frac{\partial \alpha_{i-1}}{\partial \Xi} \left(A_0 \Xi + \Phi\right) \\ & + \sum_{j=1}^{i-1} \frac{\partial \alpha_{i-1}}{\partial y_r^{(j-1)}} y_r^{(j)} + k_i v_{m,1} + \sum_{j=1}^{m+i-1} \frac{\partial \alpha_{i-1}}{\partial \lambda_j} \left(-k_j \lambda_1 + \lambda_{j+1}\right). \end{aligned} \tag{53.152}$$

$$s_1 = d_1 + \kappa_1 \left| \bar{\omega} + \frac{1}{\hat{b}_m} \left(\dot{y}_r + \bar{\alpha}_1\right) e_1 \right|^2, \tag{53.153}$$

$$s_2 = d_2 \left(\frac{\partial \alpha_1}{\partial y}\right)^2 + \kappa_2 \left| \frac{\partial \alpha_1}{\partial y} \omega - z_1 e_1 \right|^2 + g_2 \left| \frac{\partial \alpha_1}{\partial \hat{\theta}}^{\mathrm{T}} - \frac{1}{\hat{b}_m^2} \dot{y}_r e_1 \right|^2, \tag{53.154}$$

$$s_i = d_i \left(\frac{\partial \alpha_{i-1}}{\partial y}\right)^2 + \kappa_i \left| \frac{\partial \alpha_{i-1}}{\partial y} \omega \right|^2 + g_i \left| \frac{\partial \alpha_{i-1}}{\partial \hat{\theta}}^{\mathrm{T}} - \frac{1}{\hat{b}_m^2} y_r^{(i-1)} e_1 \right|^2, \quad i = 3, \ldots, \rho. \tag{53.155}$$

Adaptive Control Law:

$$u = \frac{1}{\sigma(y)} \left(\alpha_\rho - v_{m,\rho+1} + \frac{1}{\hat{b}_m} y_r^{(\rho)}\right). \tag{53.156}$$

53.4.2 Output-Feedback Modular Design

In addition to $\text{sgn}(b_m)$, in the modular design we assume that a positive constant ς_m is known such that $|b_m| \geq \varsigma_m$.

The complete design of the control law is summarized in Table 53.5. The resulting error system is

$$\dot{z} = A_z^*(z,t)z + W_\varepsilon(z,t)\varepsilon_2 + W_\theta^*(z,t)^{\mathrm{T}}\tilde{\theta} + Q(z,t)^{\mathrm{T}}\dot{\hat{\theta}}, \tag{53.157}$$

where

$$A_z^*(z,t) = \begin{bmatrix} -\frac{|b_m|}{\varsigma_m}(c_1+s_1) & b_m & 0 & \cdots & 0 \\ -b_m & -(c_2+s_2) & 1 & \ddots & \vdots \\ 0 & -1 & \ddots & \ddots & 0 \\ \vdots & \ddots & \ddots & \ddots & 1 \\ 0 & \cdots & 0 & -1 & -(c_\rho+s_\rho) \end{bmatrix}, \tag{53.158}$$

$$W_\varepsilon(z,t) = \begin{bmatrix} 1 \\ -\frac{\partial\alpha_1}{\partial y} \\ \vdots \\ -\frac{\partial\alpha_{\rho-1}}{\partial y} \end{bmatrix} \qquad W_\theta^*(z,t)^{\mathrm{T}} = \begin{bmatrix} \bar{\omega}^{\mathrm{T}} + \frac{1}{\hat{b}_m}\left(\dot{y}_{\mathrm{r}} + \bar{\alpha}_1\right)e_1^{\mathrm{T}} \\ -\frac{\partial\alpha_1}{\partial y}\omega^{\mathrm{T}} + z_1 e_1^{\mathrm{T}} \\ -\frac{\partial\alpha_2}{\partial y}\omega^{\mathrm{T}} \\ \vdots \\ -\frac{\partial\alpha_{\rho-1}}{\partial y}\omega^{\mathrm{T}} \end{bmatrix}, \tag{53.159}$$

$$Q(z,t)^{\mathrm{T}} = \begin{bmatrix} 0 \\ -\frac{\partial\alpha_1}{\partial\hat{\theta}} + \frac{1}{\hat{b}_m^2}\dot{y}_{\mathrm{r}}e_1^{\mathrm{T}} \\ \vdots \\ -\frac{\partial\alpha_{\rho-1}}{\partial\hat{\theta}} + \frac{1}{\hat{b}_m^2}y_{\mathrm{r}}^{(\rho-1)}e_1^{\mathrm{T}} \end{bmatrix}. \tag{53.160}$$

53.4.2.1 Passive Identifier

For the parametric model (Equation 53.125), we introduce the scalar observer

$$\dot{\hat{y}} = -\left(c_0 + \kappa_0|\omega|^2\right)\left(\hat{y} - y\right) + \omega_0 + \omega^{\mathrm{T}}\hat{\theta}. \tag{53.161}$$

The observer error

$$\epsilon = y - \hat{y} \tag{53.162}$$

is governed by

$$\dot{\epsilon} = -\left(c_0 + \kappa_0|\omega|^2\right)\epsilon + \omega^{\mathrm{T}}\tilde{\theta} + \varepsilon_2. \tag{53.163}$$

The parameter update law is

$$\dot{\hat{\theta}} = \underset{\hat{b}_m}{\text{Proj}}\left\{\Gamma\omega\epsilon\right\}, \quad \begin{matrix} \hat{b}_m(0)\text{sgn}b_m > \varsigma_m \\ \Gamma = \Gamma^{\mathrm{T}} > 0 \end{matrix} \tag{53.164}$$

where the projection operator is employed to guarantee that $|\hat{b}_m(t)| \geq \varsigma_m > 0$, $\forall t \geq 0$.

53.4.2.2 Swapping Identifier

The estimation error

$$\epsilon = y - \xi_1 - \Omega_1^{\mathrm{T}}\hat{\theta} \tag{53.165}$$

satisfies the following equation linear in the parameter error:

$$\epsilon = \Omega_1^{\mathrm{T}}\tilde{\theta} + \varepsilon_1. \tag{53.166}$$

The update law for $\hat{\theta}$ is either the gradient:

$$\dot{\hat{\theta}} = \underset{\hat{b}_m}{\mathrm{Proj}}\left\{\Gamma\frac{\Omega_1\epsilon}{1+\nu|\Omega_1|^2}\right\}, \quad \begin{array}{l}\hat{b}_m(0)\mathrm{sgn}b_m > \varsigma_m \\ \Gamma = \Gamma^{\mathrm{T}} > 0 \\ \nu > 0\end{array} \tag{53.167}$$

or the least squares:

$$\begin{aligned} \dot{\hat{\theta}} &= \underset{\hat{b}_m}{\mathrm{Proj}}\left\{\Gamma\frac{\Omega_1\epsilon}{1+\nu|\Omega_1|^2}\right\}, \quad \hat{b}_m(0)\mathrm{sgn}b_m > \varsigma_m \\ \dot{\Gamma} &= -\Gamma\frac{\Omega_1\Omega_1^{\mathrm{T}}}{1+\nu|\Omega_1|^2}\Gamma, \quad \begin{array}{l}\Gamma(0) = \Gamma(0)^{\mathrm{T}} > 0 \\ \nu > 0.\end{array} \end{aligned} \tag{53.168}$$

53.5 Extensions

Adaptive nonlinear control designs presented in the preceding sections are applicable to classes of nonlinear systems broader than the parametric strict-feedback systems (Equation 53.54).

53.5.1 Pure-Feedback Systems

$$\begin{aligned} \dot{x}_i &= x_{i+1} + \varphi_i(x_1, \ldots, x_{i+1})^{\mathrm{T}}\theta, \quad i = 1, \ldots, n-1, \\ \dot{x}_n &= \left(\beta_0(x) + \beta(x)^{\mathrm{T}}\theta\right) u + \varphi_0(x) + \varphi_n(x)^{\mathrm{T}}\theta, \end{aligned} \tag{53.169}$$

where $\varphi_0(0) = 0,\ \varphi_1(0) = \cdots = \varphi_n(0) = 0$, and $\beta_0(0) \neq 0$. Because of the dependence of φ_i on x_{i+1}, the regulation or tracking for pure-feedback systems is, in general, not global, even when θ is known.

53.5.2 Unknown Virtual Control Coefficients

$$\begin{aligned} \dot{x}_i &= b_i x_{i+1} + \varphi_i(x_1, \ldots, x_i)^{\mathrm{T}}\theta, \quad i = 1, \ldots, n-1, \\ \dot{x}_n &= b_n\beta(x)u + \varphi_n(x_1, \ldots, x_n)^{\mathrm{T}}\theta, \end{aligned} \tag{53.170}$$

where, in addition to the unknown vector θ, the constant coefficients b_i are also unknown. The unknown b_i-coefficients are frequent in applications ranging from electric motors to flight dynamics. The signs of b_i, $i = 1, \ldots, n$, are assumed to be known. In the tuning functions design, in addition to estimating b_i, we also estimate its inverse $\varrho_i = 1/b_i$. In the modular design we assume that in addition to $\mathrm{sgn}b_i$, a positive constant ς_i is known such that $|b_i| \geq \varsigma_i$. Then, instead of estimating $\varrho_i = 1/b_i$, we use the inverse of the estimate $\hat{b}_i$, that is, $1/\hat{b}_i$, where $\hat{b}_i(t)$ is kept away from zero by using parameter projection.

53.5.3 Multi-Input Systems

$$\begin{aligned} \dot{X}_i &= B_i(\bar{X}_i)X_{i+1} + \Phi_i(\bar{X}_i)^{\mathrm{T}}\theta, \quad i = 1, \ldots, n-1, \\ \dot{X}_n &= B_n(X)u + \Phi_n(X)^{\mathrm{T}}\theta, \end{aligned} \tag{53.171}$$

where X_i is a ν_i-vector, $\nu_1 \leq \nu_2 \leq \cdots \leq \nu_n, \bar{X}_i = \left[X_1^{\mathrm{T}}, \ldots, X_i^{\mathrm{T}}\right]^{\mathrm{T}}, X = \bar{X}_n$. and the matrices $B_i(\bar{X}_i)$ have full rank for all $\bar{X}_i \in \mathbb{R}^{\sum_{j=1}^{i}\nu_j}$. The input u is a ν_n-vector. The matrices B_i can be allowed to be unknown, provided they are constant and positive definite.

53.5.4 Block Strict-Feedback Systems

$$\begin{aligned} \dot{x}_i &= x_{i+1} + \varphi_i(x_1, \ldots, x_i, \zeta_1, \ldots, \zeta_i)^{\mathrm{T}}\theta, \quad i = 1, \ldots, \rho-1, \\ \dot{x}_\rho &= \beta(x, \zeta)u + \varphi_\rho(x, \zeta)^{\mathrm{T}}\theta, \\ \dot{\zeta}_i &= \Phi_{i,0}(\bar{x}_i, \bar{\zeta}_i) + \Phi_i(\bar{x}_i, \bar{\zeta}_i)^{\mathrm{T}}\theta, \quad i = 1, \ldots, \rho, \end{aligned} \tag{53.172}$$

with the following notation: $\bar{x}_i = [x_1, \ldots, x_i]^{\mathrm{T}}, \bar{\zeta}_i = \left[\zeta_1^{\mathrm{T}}, \ldots, \zeta_i^{\mathrm{T}}\right]^{\mathrm{T}}, x = \bar{x}_\rho$, and $\zeta = \bar{\zeta}_\rho$. Each ζ_i-subsystem of Equation 53.172 is assumed to be bounded-input bounded-state (BIBS) stable with respect to the input $(\bar{x}_i, \bar{\zeta}_{i-1})$. For this class of systems it is quite simple to modify the procedure in Tables 53.1 and 53.2. Because of the dependence of φ_i on $\bar{\zeta}_i$, the stabilizing function α_i is augmented by the term $+\sum_{k=1}^{i-1} \frac{\partial \alpha_{i-1}}{\partial \zeta_k}\Phi_{k,0}$, and the regressor w_i is augmented by $-\sum_{k=1}^{i-1} \Phi_i \left(\frac{\partial \alpha_{i-1}}{\partial \zeta_k}\right)^{\mathrm{T}}$.

53.5.5 Partial State-Feedback Systems

In many physical systems there are unmeasured states as in the output-feedback form (Equation 53.112), but there are also states other than the output $y = x_1$ that are measured. An example of such a system is

$$\begin{aligned} \dot{x}_1 &= x_2 + \varphi_1(x_1)^{\mathrm{T}}\theta, \\ \dot{x}_2 &= \boldsymbol{x_3} + \varphi_2(x_1, x_2)^{\mathrm{T}}\theta, \\ \dot{\boldsymbol{x}}_3 &= \boldsymbol{x_4} + \varphi_3(x_1, x_2)^{\mathrm{T}}\theta, \\ \dot{\boldsymbol{x}}_4 &= x_5 + \varphi_4(x_1, x_2)^{\mathrm{T}}\theta, \\ \dot{x}_5 &= u + \varphi_5(x_1, x_2, x_5)^{\mathrm{T}}\theta. \end{aligned}$$

The states $\boldsymbol{x_3}$ and $\boldsymbol{x_4}$ are assumed not to be measured. To apply the adaptive backstepping designs presented in this chapter, we combine the state-feedback techniques with the output-feedback techniques. The subsystem $(x_2, \boldsymbol{x_3}, \boldsymbol{x_4})$ is in the output-feedback form with x_2 as a measured output; hence we employ a state estimator for $(x_2, \boldsymbol{x_3}, \boldsymbol{x_4})$ using the filters introduced in Section 53.4.

53.6 For Further Information

Here we have briefly surveyed representative results in adaptive nonlinear control, a research area that exhibited rapid growth in the 1990s and continues to be active today.

The first adaptive backstepping design was developed by Kanellakopoulos et al. [5]. Its overparametrization was removed by the tuning functions design of Krstić et al. [8]. Possibilities for extending the class of systems in [5] were studied by Seto et al. [17].

Among the early estimation-based results are Sastry and Isidori [16], Pomet and Praly [13], and so on. They were surveyed in Praly et al. [14]. All these designs involve some growth conditions. The modular approach of Krstić and Kokotović [10] removed the growth conditions and achieved a complete separation of the controller and the identifier.

One of the first output-feedback designs was proposed by Marino and Tomei [11]. Kanellakopoulos et al. [6] presented a solution to the partial state-feedback problem. Subsequent efforts in adaptive nonlinear control focused on broadening the class of nonlinear systems for which adaptive controllers are available. Jiang and Pomet [4] developed a design for nonholonomic systems using the tuning functions technique. Khalil [7] developed semiglobal output feedback designs for a class which includes some systems not transformable into the output feedback form.

For a complete and pedagogical presentation of adaptive nonlinear control the reader is referred to the text "Nonlinear and Adaptive Control Design" by Krstić et al. [9]. The book introduces backstepping and illustrates it with numerous applications (including jet engine, automotive suspension, aircraft wing rock, robotic manipulator, and magnetic levitation). It contains the details of methods surveyed here and their extensions. It also covers several important topics not mentioned here. Among them is the systematic improvement of *transient performance*. It also shows the advantages of applying adaptive backstepping to *linear* systems.

Hundreds of papers have been written on the subject of adaptive nonlinear control and backstepping design over the last decade and a half. A detailed citation survey is beyond the scope of this chapter, but we mention some of the key groups of results. Adaptive nonlinear controllers have been developed for the class of strict-feedforward systems, some classes of stochastic nonlinear systems, and some classes of nonlinear systems involving time delays. Robustness of adaptive backstepping controllers to disturbances and unmodeled dynamics has been studied, as well as parameter convergence and identifiability. Extensions to discrete-time nonlinear systems have been developed. Some classes of nonlinearly parametrized problems have also been considered. The fundamental problem of asymptotic behavior of parameter estimates in the absence of persistency of excitation has also been addressed, with surprising results—estimates may converge to destabilizing frozen values from large sets of initial conditions. Decentralized forms of adaptive backstepping have also been developed. Numerous control application results have been reported using adaptive backstepping methods, from automotive, aerospace, and underwater vehicles, to biochemical systems, magnetic levitation, HVAC, and so on.

A number of books have been inspired by the design frameworks introduced in [9]. We mention only a few of the books. Marino and Tomei [12] and Qu [15] presented additional or alternative recursive design techniques for adaptive and robust nonlinear control. Spooner et al. [18] combined the methods presented this chapter with neural and fuzzy approximation techniques. French et al. [3] studied the performance properties of adaptive backstepping controllers employing neural networks. Dawson et al. [2] presented a comprehensive methodology for adaptive nonlinear control of electric machines based on backstepping methods. Astolfi et al. [1] developed the idea of system immersion and manifold invariance, leading to modular rather than classical Lyapunov schemes.

References

1. A. Astolfi, D. Karagiannis, and R. Ortega, *Nonlinear and Adaptive Control with Applications*, Springer, Berlin, 2008.
2. D. M. Dawson, J. Hu, and T. C. Burg, *Nonlinear Control of Electric Machinery*, Marcel Dekker, 1998.
3. M. French, C. Szepesvari, and E. Rogers, *Performance of Nonlinear Approximate Adaptive Controllers*, Wiley, New York, 2003.
4. Z. P. Jiang and J. B. Pomet, Combining backstepping and time-varying techniques for a new set of adaptive controllers, *Proceedings of the 33rd IEEE Conference on Decision and Control*, Lake Buena Vista, FL, December 1994, pp. 2207–2212.
5. I. Kanellakopoulos, P. V. Kokotović, and A. S. Morse, Systematic design of adaptive controllers for feedback linearizable systems, *IEEE Transactions on Automatic Control*, 36, 1241–1253, 1991.
6. I. Kanellakopoulos, P. V. Kokotović, and A. S. Morse, Adaptive nonlinear control with incomplete state information, *International Journal of Adaptive Control and Signal Processing*, 6, 367–394, 1992.

7. H. Khalil, Adaptive output-feedback control of nonlinear systems represented by input–output models, *Proceedings of the 33rd IEEE Conference on Decision and Control*, Lake Buena Vista, FL, December 1994, pp. 199–204.
8. M. Krstić, I. Kanellakopoulos, and P. V. Kokotović, Adaptive nonlinear control without overparametrization, *Systems and Control Letters*, 19, 177–185, 1992.
9. M. Krstić, I. Kanellakopoulos, and P. V. Kokotović, *Nonlinear and Adaptive Control Design*, Wiley, New York, NY, 1995.
10. M. Krstić and P. V. Kokotović, Adaptive nonlinear design with controller-identifier separation and swapping, *IEEE Transactions on Automatic Control*, 40, 426–441, 1995.
11. R. Marino and P. Tomei, Global adaptive output-feedback control of nonlinear systems, Part I: Linear parametrization, *IEEE Transactions on Automatic Control*, 38, 17–32, 1993.
12. R. Marino and P. Tomei, *Nonlinear Control Design: Geometric, Adaptive, and Robust*, Prentice-Hall, Englewood Cliffs, NJ, 1995.
13. J. B. Pomet and L. Praly, Adaptive nonlinear regulation: Estimation from the Lyapunov equation, *IEEE Transactions on Automatic Control*, 37, 729–740, 1992.
14. L. Praly, G. Bastin, J.-B. Pomet, and Z. P. Jiang, Adaptive stabilization of nonlinear systems, in *Foundations of Adaptive Control*, P. V. Kokotović, Ed., Springer-Verlag, Berlin, pp. 347–434, 1991.
15. Z. Qu, *Robust Control of Nonlinear Uncertain Systems*, Wiley, New York, 1998.
16. S. S. Sastry and A. Isidori, Adaptive control of linearizable systems, *IEEE Transactions on Automatic Control*, 34, 1123–1131, 1989.
17. D. Seto, A. M. Annaswamy, and J. Baillieul, Adaptive control of a class of nonlinear systems with a triangular structure, *IEEE Transactions on Automatic Control*, 39, 1411–1428, 1994.
18. J. T. Spooner, M. Maggiore, R. Ordonez, and K. M. Passino, *Stable Adaptive Control and Estimation for Nonlinear Systems: Neural and Fuzzy Approximator Techniques*, Wiley, New York, 2002.

54 Intelligent Control

Kevin M. Passino
The Ohio State University

54.1 Introduction

Intelligent control, the discipline where control algorithms are developed by emulating certain characteristics of intelligent biological systems, is an emerging area of control that is being fueled by the advancements in computing technology (Antsaklis and Passino, 1993; Passino, 2005). For instance, software development and validation tools for *expert systems* (computer programs that emulate the actions of a human who is proficient at some task) are being used to construct "expert controllers" that seek to automate the actions of a human operator who controls a system. Other knowledge-based systems such as *fuzzy systems* (rule-based systems that use fuzzy logic for knowledge representation and inference) and *planning systems* (that emulate human planning activities) are being used in a similar manner to automate the perceptual, cognitive (deductive and inductive), and action-taking characteristics of humans who perform control tasks. Artificial *neural networks* emulate biological neural networks and have been used to (1) learn how to control systems by observing the way that a human performs a control task, and (2) learn in an online fashion how to best control a system by taking control actions, rating the quality of the responses achieved when these actions are used, then adjusting the recipe used for generating control actions so that the response of the system improves. *Genetic algorithms* are being used to evolve controllers via off-line computer-aided-design of control systems or in an online fashion by maintaining a population of controllers and using survival of the fittest principles where "fittest" is defined by the quality of the response achieved by the controller.

From these examples we see that computing technology is driving the development of the field of control by providing alternative strategies for the functionality and implementation of controllers for dynamical systems. In fact, there is a trend in the field of control to integrate the functions of intelligent systems, such as those listed above, with conventional control systems to form highly "autonomous" systems that have the capability to perform complex control tasks independently with a high degree of success. This trend toward the development of *intelligent autonomous control systems* is gaining momentum as control engineers have solved many problems and are naturally seeking control problems where broader issues must be taken into consideration and where the full range of capabilities of available computing technologies is used.

The development of such sophisticated controllers does, however, still fit within the conventional engineering methodology for the construction of control systems. Mathematical modeling using first principles or data from the system, along with heuristics are used. Some intelligent control strategies rely more on the use of heuristics (e.g., direct fuzzy control) but others utilize mathematical models in the same way that they are used in conventional control (e.g., see the chapter in this section on neural control of nonlinear systems), while still others use a combination of mathematical models and heuristics (see, e.g., the approaches to fuzzy adaptive control in Passino, 2005). There is a need for systematic methodologies for the construction of controllers. Some methodologies for the construction of intelligent controllers are quite *ad hoc* (e.g., for the fuzzy controller) yet often effective since they provide a method and formalism for incorporating and representing the nonlinearities that are needed to achieve high-performance control. Other methodologies for the construction of intelligent controllers are no more *ad hoc* than ones for conventional control (e.g., for neural and fuzzy adaptive controllers). There is a need for nonlinear analysis of stability, controllability, and observability properties. Although there has been significant progress recently in stability analysis of fuzzy, neural, and expert control systems, there is need for much more work in nonlinear analysis of intelligent control systems, especially in hybrid ones. Simulations and experimental evaluations of intelligent control systems are necessary. Comparative analysis of competing control strategies (conventional or intelligent) is, as always, important. Engineering cost–benefit analysis that involves issues of performance, stability, ease-of-design, lead-time to implementation, complexity of implementation, cost, and other issues must be used.

Overall, while the intelligent control paradigm focuses on biologically motivated (bioinspired) approaches (and uses what some call biomimicry), there are sometimes only small differences in the behavior of the resulting controllers that are finally implemented (intelligent are not mystical; they are simply nonlinear, often adaptive controllers). This is, however, not surprising since there seems to be an existing conventional control approach that is analogous to every new intelligent control approach that has been introduced. This is illustrated in Table 54.1. It is not surprising then that while there seem to be some new concepts growing from the field of intelligent control, there is a crucial role for the control engineer and control scientist to play in evaluating and developing the field of intelligent control. For more detailed discussions on the relationships between conventional and intelligent control; see Passino (2005).

TABLE 54.1 Analogies between Conventional and Intelligent Control

Intelligent Control Technique	Conventional Control Approach
Direct fuzzy control	Nonlinear control
Fuzzy adaptive/learning control	Adaptive control and identification
Fuzzy supervisory control	Gain-scheduled control, hierarchical control
Direct expert control	Controllers for automata, Petri nets, and other discrete event systems
Planning systems for control	Certain types of controllers for discrete event systems, receding horizon control of nonlinear systems, model predictive control
Neural control	Adaptive control and identification, optimal control
Genetic algorithms for computer-aided-design (CAD) of control systems, controller tuning, identification	CAD using heuristic optimization techniques, optimal control, receding horizon control, and stochastic adaptive control

In this chapter, we briefly examine the basic techniques of intelligent control, provide an overview of intelligent autonomous control, and discuss some advances that have focused on comparative analysis, modeling, and nonlinear analysis of intelligent control systems. The intent is only to provide a brief introduction to the field of intelligent control and to the next two chapters; the interested reader should consult the references provided at the end of the chapter, or the chapters on fuzzy and neural control for more details.

54.2 Intelligent Control Techniques

Major approaches to intelligent control are outlined below and references are provided in case the reader would like to learn more about any one of these approaches. It is important to note that while each of the approaches is presented separately, in practice there is a significant amount of work being done to determine the best ways to utilize various aspects of each of the approaches in "hybrid" intelligent control techniques. For instance, neural and fuzzy control approaches are often combined. In other cases, neural networks are trained with genetic algorithms. One can imagine justification for integration of just about any permutation of the presented techniques depending on the application at hand.

54.2.1 Fuzzy Control

A fuzzy controller can be designed to crudely emulate the human deductive process (i.e., the process whereby we successively infer conclusions from our knowledge). As shown in Figure 54.1 the fuzzy controller consists of four main components. The rule-base holds a set of "IF–THEN" rules that are quantified via fuzzy logic and used to represent the knowledge that human experts may have about how to solve a problem in their domain of expertise. The fuzzy inference mechanism successively decides what rules are most relevant to the current situation and applies the actions indicated by these rules. The fuzzification interface converts numeric inputs into a form that the fuzzy inference mechanism can use to determine which knowledge in the rule-base is most relevant at the current time. The defuzzification interface combines the conclusions reached by the fuzzy inference mechanism and provides a numeric value as an output. Overall, the fuzzy control design methodology, which primarily involves the specification of the rule-base, provides a heuristic technique to construct nonlinear controllers and this is one of its main advantages. For more details on direct fuzzy control see the next chapter or Passino and Yurkovich (1998).

Often it is the case that we have better knowledge about how to control a process such as, information on how to tune the controller while it is in operation or how to coordinate the application of different controllers based on the operating point of the system. For instance, in aircraft control certain key variables are used in the tuning (scheduling) of control laws and fuzzy control provides a unique approach to the construction and implementation of such a gain scheduler. In process control, engineers or process

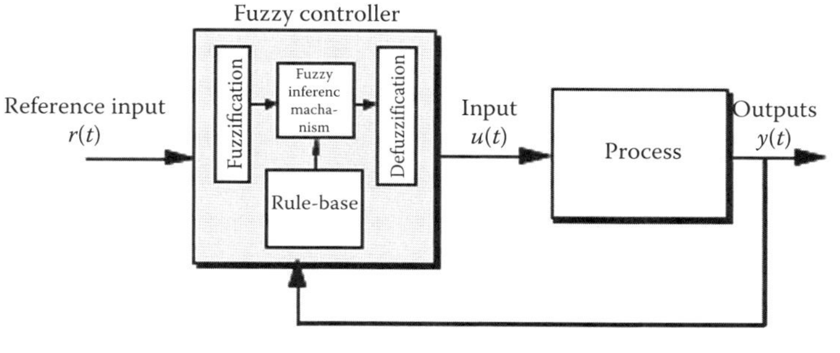

FIGURE 54.1 Fuzzy control system.

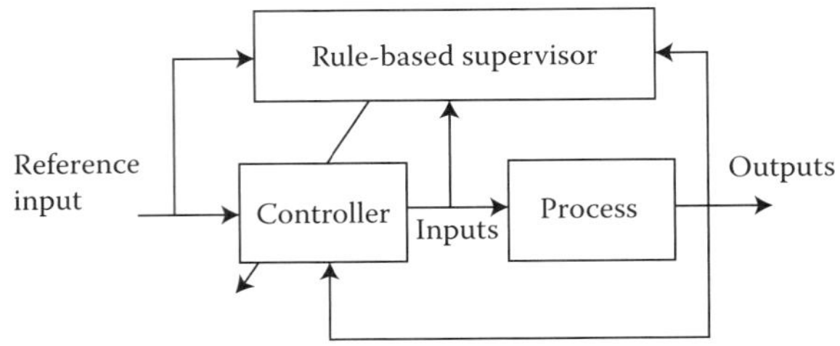

FIGURE 54.2 Rule-based supervisory control.

operators often have a significant amount of heuristic expertise on how to tune proportional-integral-derivative (PID) controllers while they are in operation and such information can be loaded into the rule-base of a "fuzzy PID tuner" and used to make sure that a PID controller is tuned properly at all times. In the more general case, we may have knowledge of how to tune and coordinate the application of conventional or fuzzy controllers and this can be used in a rule-based supervisor as it is shown in Figure 54.2. For more details on fuzzy supervisory control see the next chapter.

In other fuzzy control approaches, rather than implementing deductive systems, the goal is to implement an "inductive system," that is, one that can *learn* and generalize from particular examples (e.g., examples of how the system is behaving). Such approaches typically fall under the title of "fuzzy learning control" or "fuzzy adaptive control." In one approach, shown in Figure 54.3, called fuzzy model reference learning control (FMRLC) (Layne and Passino, 1993), there is a fuzzy controller with a rule-base that has no knowledge about how to control the system. A "reference model" with output $y_m(t)$ is used to characterize how you would like the closed-loop system to behave (i.e., it holds the performance specifications). Then, a learning mechanism compares $y(t)$ to $y_m(t)$ (i.e., the way that the system is currently performing to how you would like it to perform) and decides how to synthesize and tune the fuzzy controller so that the difference between $y(t)$ and $y_m(t)$ goes to zero and hence, the performance objectives are met.

Overall, our experiences with the FMRLC indicate that significant advantages may be obtained if one can implement a controller that can truly *learn* from its experiences (while forgetting appropriate information) so that when a similar situation is repeatedly encountered the controller already has a good idea of how to react. This seems to represent an advance over some adaptive controllers where parameters are adapted in a way such that each time the same situation is encountered, some amount of (often complete) readaptation must occur no matter how often this situation is encountered (for more details on this and other fuzzy learning/adaptive control approaches see the next chapter or Passino, 2005).

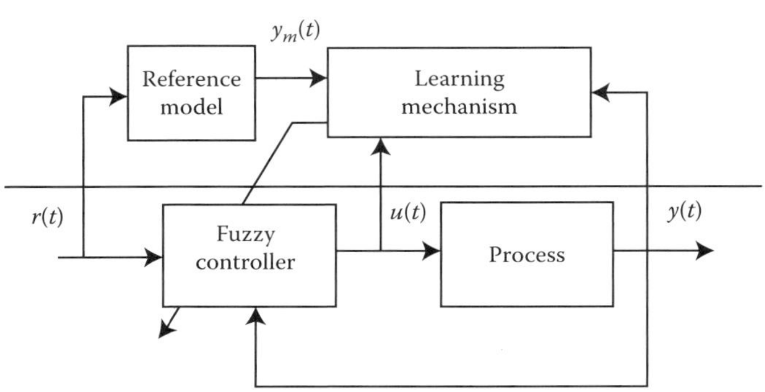

FIGURE 54.3 Fuzzy model reference learning control.

54.2.2 Expert and Planning Systems

While fuzzy control techniques similar to those described above have been employed in a variety of industrial control applications, more general "knowledge-based controllers" have also been used successfully. For instance, there are "expert controllers" that are being used to directly control complex processes or in a supervisory role similar to that shown in Figure 54.2 (Antsaklis and Passino, 1993; Passino and Lunardhi, 1995). Others are being used to supervise conventional control algorithms. For instance, the work in Astrom et al. (1986) describes the use of expert supervisory systems for conventional adaptive controllers. Expert systems are also being used as the basis for learning controllers.

In addition, there are planning systems (computer programs that emulate the way that experts plan) that have been used in path planning and high-level decisions about control tasks for robots (Antsaklis and Passino, 1993; Valavanis and Saridis, 1992; Dean and Wellman, 1991). A generic planning system, configured in the architecture of a standard control system, is shown in Figure 54.4. Here, the "problem domain" is the environment that the planner operates in, that is, the plant. There are measured outputs y_i at step i (variables of the problem domain that can be sensed in real-time), control actions u_i (the ways in which we can affect the problem domain), disturbances d_i (that represent random events that can affect the problem domain and hence, the measured variable y_i), and goals g_i (what we would like to achieve in the problem domain). There are closed-loop specifications that quantify performance specifications and stability requirements.

It is the task of the planner, shown in Figure 54.4, to monitor the measured outputs and goals and generate control actions that will counteract the effects of the disturbances and result in the goals and the closed-loop specifications to be achieved. To do this, the planner performs "plan generation" where it projects into the future (a finite number of steps using a model of the problem domain) and tries to determine a set of candidate plans. Next, this set of plans is pruned to one plan that is the best one to apply at the current time. The plan is executed and during execution the performance resulting from the plan is monitored and evaluated. Often, due to disturbances, plans will fail and hence the planner must generate a new set of candidate plans, select one, and then execute that one. While not pictured in Figure 54.4, some planning systems use "situation assessment" to try to estimate the state of the problem domain (this can be useful in execution monitoring and in plan generation), others perform "world modeling" where a model of the problem domain is developed in an online fashion (similar to online system identification), and "planner design" where information from the world modeler is used to tune the planner (so that it makes the right plans for the current problem domain). The reader will, perhaps, think of such a planning system as a general "self-tuning regulator." For more details on the use of planning systems for control see Antsaklis and Passino (1993), Dean and Wellman (1991), and Passino (2005).

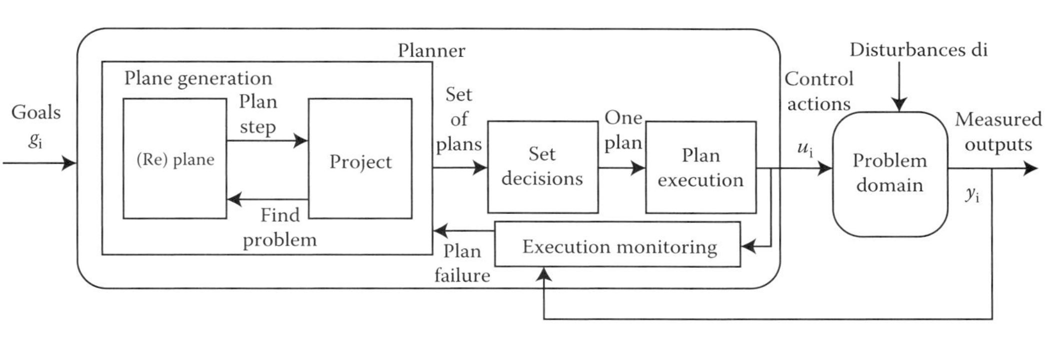

FIGURE 54.4 Closed-loop planning systems.

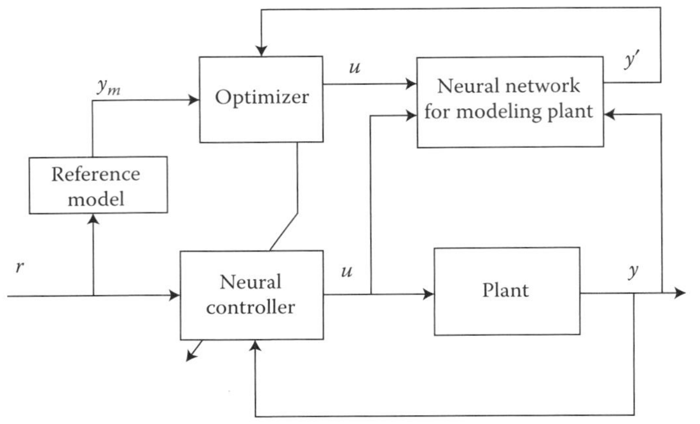

FIGURE 54.5 Neural predictive control.

54.2.3 Neural Networks for Control

There has been significant activity in the use of artificial neural networks for control (Hunt et al., 1994; Spooner et al., 2002; Farrell and Polycarpou, 2006). In this approach, engineers are trying to emulate the low-level biological functions of the brain and use these to solve challenging control problems. For instance, for some control problems we may train an artificial neural network to remember how to regulate a system by repeatedly providing it with examples of how to perform such a task. After the neural network has learned the task, it can be used to recall the control input for each value of the sensed output. Some other approaches to neural control, taken from Hunt et al. (1994 and the references therein), include a neural "internal model control" method and a "model reference structure" that is based on an approach to using neural networks for system identification.

Still other neural control approaches bear some similarities to the FMRLC in Figure 54.3 in the sense that they automatically learn how to control a system by observing the behavior from that system. For instance, in Figure 54.5 we show a "neural predictive control" approach from Hunt et al. (1994) where one neural network is used as an identifier (structure) for the plant and another is used as a feedback controller for the plant that is tuned online. This tuning proceeds at each step by having the "optimizer" specify an input u' for the neural model of the plant over some time interval. The predicted behavior of the plant y' is obtained and used by the optimizer, along with y_m to pick the best parameters of the neural controller so that the difference between the plant and reference model outputs is as small as possible (if y' predicts y well we would expect that the optimizer would be quite successful at tuning the controller). For more details on the multitude of techniques for using neural networks for control see Hunt et al. (1994).

54.2.4 Genetic Algorithms for Control

A genetic algorithm uses the principles of evolution, natural selection, and genetics from natural biological systems in a computer algorithm to simulate evolution (Goldberg, 1989). Essentially, the genetic algorithm performs a parallel, stochastic, but directed search to evolve the population that is most fit. It has been shown that a genetic algorithm can be used effectively in the (off-line) computer-aided-design of control systems because it can artificially "evolve" an appropriate controller that meets the performance specifications to the greatest extent possible. To do this, the genetic algorithm maintains a population of strings where each represents a different controller and it uses the genetic operators of "reproduction" (which represents the "survival of the fittest" characteristic of evolution), "crossover" (which represents "mating"),

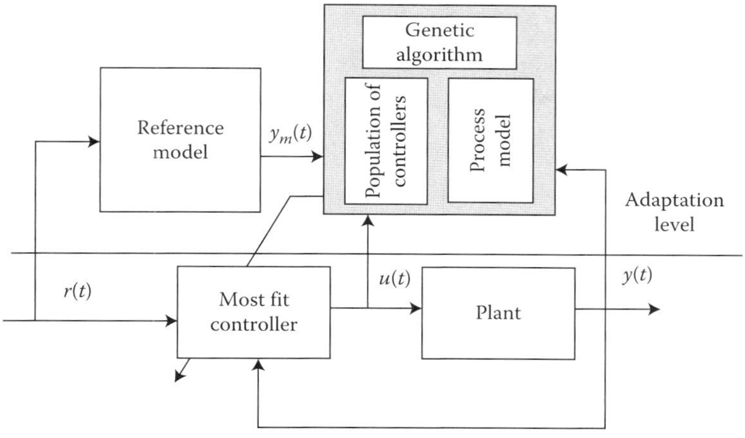

FIGURE 54.6 Genetic model reference adaptive controller.

and "mutation" (which represents the random introduction of new "genetic material"), coupled with a "fitness measure" (which often quantifies the performance objectives) to generate successive generations of the population. After many generations, the genetic algorithm often produces an adequate solution to a control design problem since the stochastic, but directed, search helps avoid locally optimal designs and seeks to obtain the best design possible.

Another more challenging problem is that of how to evolve controllers while the system is operating, rather than in off-line design. Progress in this direction has been made by the introduction of the genetic model reference adaptive controller (GMRAC) shown in Figure 54.6 (Porter and Passino, 1994). As in the FMRLC, the GMRAC uses a reference model to characterize the desired performance. For the GMRAC a genetic algorithm maintains a population of strings that represent candidate controllers. This genetic algorithm uses a process model (e.g., a linear model of the process) and data from the process to evaluate the fitness of each controller in the population at each time step. Using this fitness evaluation the genetic algorithm propagates controllers into the next generation via the standard genetic operators. The controller that is the most fit one in the population is used to control the system. This allows the GMRAC to automatically evolve a controller from generation to generation (i.e., from one time step to the next) and hence to tune a controller in response to changes in the process (e.g., due to temperature variations, parameter drift, etc.) or due to an online change of the specifications in the reference model. Early indications are that the GMRAC seems quite promising as a new technique for stochastic adaptive control since it provides a unique feature where alternative controllers can be quickly applied to the problem if they appear useful, and because it has some inherent capabilities to learn via evolution of its population of controllers. There is, however, a significant amount of comparative and nonlinear analysis that needs to be done to fully evaluate this approach to control.

54.3 Autonomous Control

The goal of the field of autonomous control is to design control systems that automate enough functions so that they can independently perform well under significant uncertainties for extended periods of time even if there are significant system failures or disturbances. Below, we overview some of the basic ideas from Antsaklis and Passino (1993) and Passino (2005) on how to specify controllers that can in fact achieve high levels of autonomy.

54.3.1 The Intelligent Autonomous Controller

Figure 54.7 shows a functional architecture for an intelligent autonomous controller with an interface to the process (plant) involving sensing (e.g., via conventional sensing technology, vision, touch, smell, etc.), actuation (e.g., via hydraulics, robotics, motors, etc.), and an interface to humans (e.g., a driver, pilot, crew, etc.) and other systems. The "execution level" has low-level numeric signal processing and control algorithms (e.g., PID, optimal, or adaptive control; parameter estimators, failure detection and identification (FDI) algorithms). The "coordination level" provides for tuning, scheduling, supervision, and redesign of the execution level algorithms, crisis management, planning and learning capabilities for the coordination of execution level tasks, and higher-level symbolic decision making for FDI and control algorithm management. The "management level" provides for the supervision of lower level functions and for managing the interface to the human(s). In particular, the management level will interact with the users in generating goals for the controller and in assessing capabilities of the system. The management level also monitors performance of the lower level systems, plans activities at the highest level (and in cooperation with the human), and performs high level learning about the user and the lower level algorithms. Applications that have used this type of architecture can be found in Antsaklis and Passino (1993), Valavanis and Saridis (1992), and Gazi et al. (2001).

Intelligent systems/controllers (fuzzy, neural, genetic, expert, etc.) can be employed as appropriate in the implementation of various functions at the three levels of the intelligent autonomous controller (adaptive fuzzy control may be used at the execution level, planning systems may be used at the management level for sequencing operations, and genetic algorithms may be used in the coordination level to pick an optimal coordination strategy). Hierarchical controllers composed of a hybrid mix of intelligent and conventional systems are commonly used in the intelligent control of complex dynamical systems. This is due to the fact that to achieve high levels of autonomy, we often need high levels of intelligence, which calls for incorporation of a diversity of decision-making approaches for complex dynamic reasoning.

There are several fundamental characteristics that have been identified for intelligent autonomous control systems (see Valavanis and Saridis, 1992; Antsaklis and Passino, 1993, and the references therein). For example, there is generally a successive delegation of duties from the higher to lower levels and the number of distinct tasks typically increases as we go down the hierarchy. Higher levels are often

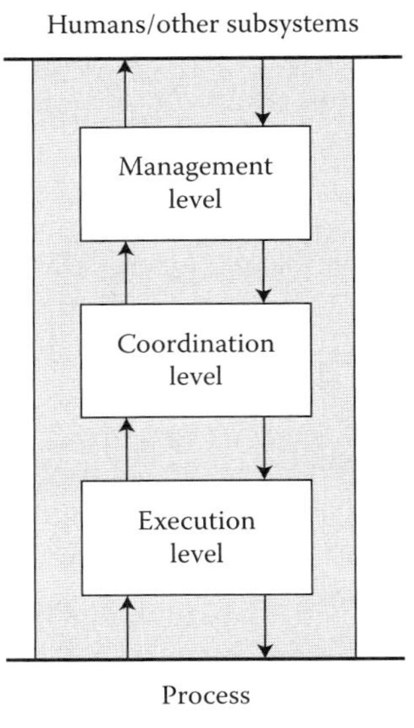

FIGURE 54.7 Intelligent autonomous controller.

concerned with slower aspects of the system's behavior and with its larger portions, or broader aspects. There is then a smaller contextual horizon at lower levels, that is, the control decisions are made by considering less information. Higher levels are typically concerned with longer time horizons than lower levels. It is said that there is "increasing intelligence with decreasing precision as one moves from the lower to the higher levels" (see Valavanis and Saridis, 1992, and the references therein). At the higher levels there is typically a decrease in time scale density, a decrease in bandwidth or system rate, and a decrease in the decision (control action) rate. In addition, there is typically a decrease in granularity of models used, or equivalently, an increase in model abstractness at the higher levels. Finally, we note that there is an ongoing *evolution* of the intelligent functions of an autonomous controller so that by the time one implements its functions they no longer appear intelligent—just algorithmic. It is this evolution principle and the fact that implemented intelligent controllers are nonlinear controllers that many researchers feel more comfortable focusing on *achieving autonomy* rather than whether the resulting controller is *intelligent*.

54.3.2 The Control-Theoretic View of Autonomy

Next, it is explained how to incorporate the notion of autonomy into the conventional manner of thinking about control problems. Consider the general control system shown in Figure 54.8 where P is a model of the plant, C represents the controller, and T represents specifications on how we would like the closed-loop system to behave (i.e., closed-loop specifications). For some classical control problems the scope is limited so that C and P are linear and T simply represents, for example, stability, rise time, overshoot, and steady-state tracking error specifications. In this case, intelligent control techniques may not be needed. As engineers, the simplest solution that works is the best one. We tend to need more complex controllers for more complex plants (where, for example, there is a significant amount of uncertainty) and more demanding closed-loop specifications T (see Valavanis and Saridis, 1992; Antsaklis and Passino, 1993, and the references therein).

Consider the case where

1. P is so complex that it is most convenient to represent it with ordinary differential equations and discrete event system (DES) models (or some other "hybrid" mix of models) and for some parts of the plant the model is not known (i.e., it may be too expensive to find).
2. T is used to characterize the desire to make the system perform well and act with *high degrees of autonomy* (i.e., so that the system performs well under significant uncertainties in the system and its environment for extended periods of time, and compensates for significant system failures without external intervention (Antsaklis and Passino, 1993).

The *general control problem* is how to construct C, given P, so that T holds. The intelligent autonomous controller described briefly in the previous section provides a general architecture for C to achieve highly autonomous behavior specified by T for very complex plants P.

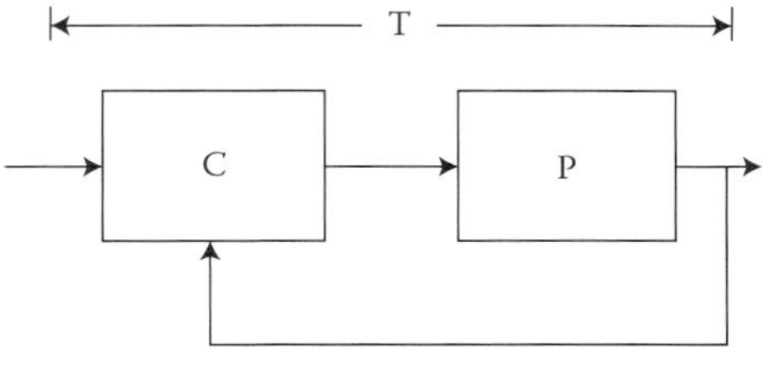

FIGURE 54.8 Control system.

54.3.3 Modeling and Analysis*

Conventional approaches to modeling P include the use of ordinary differential or difference equations, partial differential equations, stochastic models, models for hierarchical and distributed systems, and so on. However, often some portion of P is more easily represented with an automata model, Petri net, or some other DES model. Moreover, for analysis of the closed-loop system where there are a variety of intelligent and conventional controllers that are working together to form C (e.g., a planning system and a conventional adaptive controller) there is the need for "hybrid" modeling formalisms that can represent dynamics with differential equations and DES models. Control engineers use a model of the plant P to aid in the construction of the model of the controller C and this model is then implemented to control the real plant. In addition, models of C and P are used as formalisms to represent the dynamics of the closed-loop system so that analysis of the properties of the feedback system is possible before implementation (for redesign, verification, certification, and safety analysis). If the model of P is chosen to be too complex and C is very complex it will be difficult to develop and utilize mathematical approaches for the analysis of the resulting closed-loop system. Often we want the simplest possible model P that will allow for the development of the (simplest) controller C, and allow for it to be proven/demonstrated so that the closed-loop specifications T are met (of course, a separate more complex model P may be needed for simulation). Unfortunately, there is no clear answer to the question of how much or what type of modeling is needed for the plant P and there is no standardization of models for intelligent control in the way that there is for many areas of conventional control. Hence, although it is not exactly clear how to proceed with the modeling task, it is clear that knowledge of many different types of models may be needed, depending on the task at hand.

Given the model of the plant P and the model of the controller C, the next task often considered by a control engineer is the use of analysis to more fully understand the behavior of P or the closed-loop system, and to show that when C and P are connected, the closed-loop specifications T are satisfied. Formal mathematical analysis can be used to verify stability, controllability, and observability properties. There is, in fact, a growing amount of literature on nonlinear analysis (e.g., stability and describing function analysis) of fuzzy control systems (both direct and adaptive [Jenkins and Passino, 1999; Spooner et al., 2002; Farrell and Polycarpou, 2006]). There is a significant amount of activity in the area of nonlinear analysis of neural control systems and results in the past on nonlinear analysis of (numerical) learning control systems (see the later chapter in this section of this book and Hunt et al. (1994). There have already been some applications of DES theory to artificial intelligent (AI) planning systems and there have been recent results on stability analysis of expert control systems (Passino and Lunardhi, 1995). There has been some progress in defining models and developing approaches to analysis for some hybrid systems, but there is the need for much more work in this area. Many fundamental modeling and representation issues need to be reconsidered, different design objectives and control structures need to be examined, our repertoire of approaches to analysis and design needs to be expanded, and there is the need for more work in the area of simulation and experimental evaluation for hybrid systems. The importance of the solution to the hybrid control system analysis problem is based on the importance of solving the general control problem described above; that is, hybrid system analysis techniques could provide an approach to verifying the operation of intelligent controllers that seek to obtain truly autonomous operation. Finally, it must be emphasized that while formal verification of the properties of a control system is important, simulation and experimental evaluation always plays an especially important role also.

54.4 Concluding Remarks

We have provided a brief overview of the main techniques in the field of intelligent control and have provided references for the reader who is interested in investigating the details of any one of these techniques.

* The reader can find references for the work on modeling and analysis of intelligent control systems discussed in this section in [Passino, 2005].

The intent of this chapter was to provide the reader with an overview of a relatively new area of control, while the intent of the next two chapters is to provide an introduction to two of the more well-developed areas in intelligent control—fuzzy and neural control. In a chapter so brief it seems important to indicate what has been omitted. We have not discussed: (1) FDI methods that are essential for a truly autonomous controller, (2) reconfigurable (fault tolerant) control strategies that use conventional nonlinear robust control techniques and intelligent control techniques, (3) sensor fusion and integration techniques that will be needed for autonomous control, (4) architectures for intelligent and autonomous control systems (e.g., alternative ways to structure interconnections of intelligent subsystems), (5) distributed intelligent systems (e.g., multiagent systems), (6) attentional systems, and (7) applications.

54.5 Defining Terms

Expert system: A computer program designed to emulate the actions of a human who is proficient at some task. Often the expert system is broken into two components: a "knowledge-base" that holds information about the problem domain, and an inference mechanism (engine) that evaluates the current knowledge and decides what actions to take. An "expert controller" is an expert system that is designed to automate the actions of a human operator who controls a system.

Fuzzy systems: A type of knowledge-based system that uses fuzzy logic for knowledge representation and inference. It is composed of four primary components: the fuzzification interface, the rule-base (a knowledge-base that is composed of rules), an inference mechanism, and a defuzzification interface. A fuzzy system that is used to control a system is called a "fuzzy controller."

Planning system: Computer program designed to emulate human planning activities. These may be a type of expert system that has a special knowledge-base that has plan fragments and strategies for planning and an inference process that generates and evaluates alternative plans.

Neural network: Artificial hardware (e.g., electrical circuits) designed to emulate biological neural networks. These may be simulated on conventional computers or on specially designed "neural processors."

Genetic algorithm: A genetic algorithm uses the principles of evolution, natural selection, and genetics from natural biological systems in a computer algorithm to simulate evolution. Essentially, the genetic algorithm performs a parallel, stochastic, but directed search to evolve the most fit population.

Intelligent autonomous control system: A control system that uses conventional and intelligent control techniques to provide enough automation so that the system can independently perform well under significant uncertainties for extended periods of time even if there are significant system failures or disturbances.

Acknowledgment

Partial support for this work came from the National Science Foundation under grants IRI-9210332 and EEC-9315257.

References

Antsaklis, P.J., Passino, K.M., eds., *An Introduction to Intelligent and Autonomous Control*, Kluwer Academic Publishers, Norwell, MA, 1993.

Astrom, K.J., Anton, J.J., Arzen, K.E., Expert control, *Automatica*, Vol. 22, pp. 277–286, 1986.

Dean, T., Wellman, M.P., *Planning and Control*, Morgan Kaufman, CA, 1991.

Farrell, J.A., Polycarpou, M.M., *Adaptive Approximation Based Control: Unifying Neural, Fuzzy, and Traditional Approximation Based Approaches*, John Wiley, NJ, 2006.

Gazi, V., Moore, M.L., Passino, K.M., Shackleford, W., Proctor, F., and Albus, J.S., *The RCS Handbook: Tools for Real Time Control Systems Software Development*, John Wiley and Sons, New York, NY, 2001.

Goldberg, D.E., *Genetic Algorithms in Search, Optimization, and Machine Learning*, Addison-Wesley, Reading, MA, 1989.

Hunt, K.J., Sbarbaro, D., Zbikowski, R., Gawthrop, P.J., Neural networks for control systems—A survey, in M.M. Gupta and D.H. Rao, eds., *Neuro-Control Systems: Theory and Applications*, IEEE Press, New York, 1994.

Jenkins, D., Passino, K.M., An introduction to nonlinear analysis of fuzzy control systems, *Journal of Intelligent and Fuzzy Systems*, Vol. 7, No. 1, pp. 75–103, 1999.

Layne, J.R., Passino, K.M., Fuzzy model reference learning control for cargo ship steering, *IEEE Control Systems Magazine*, Vol. 13, No. 6, pp. 23–34, 1993.

Passino, K. M. and Yurkovich, S., *Fuzzy Control*, Addison-Wesley Longman, Menlo Park, CA, 1998

Passino, K. M., *Biomimicry for Optimization, Control, and Automation*, Springer-Verlag, London, UK, 2005.

Spooner, J. T., Maggiore, M., Ordonez, R., and Passino, K. M., *Stable Adaptive Control and Estimation for Nonlinear Systems: Neural and Fuzzy Approximator Techniques*, John Wiley and Sons, New York, NY, 2002.

Valavanis, K. P., Saridis, G. N., *Intelligent Robotic Systems: Theory, Design, and Applications*, Kluwer Academic Publishers, Norwell, MA, 1992.

For Further Information

While the books and articles referenced above (particularly, Passino, 2005) should provide the reader with an introduction to the area of intelligent control, there are other sources that may also be useful. For instance, there are many relevant conferences including: (1) *IEEE International Symposium on Intelligent Control*, (2) *American Control Conference*, (3) *IEEE Conference on Decision and Control*, (4) *IEEE Conference on Control Applications*, and (5) *IEEE International Conference on Systems, Man, and Cybernetics*. In addition, there are many conferences on fuzzy systems, expert systems, genetic algorithms, and neural networks where applications to control are often studied. There are many journals that cover the topic of intelligent control including: (1) *IEEE Control Systems Magazine*, (2) *IEEE Trans. on Control Systems Technology*, (3) *IEEE Trans. on Systems, Man, and Cybernetics*, (4) *IEEE Trans. on Fuzzy Systems*, (5) *IEEE Trans. on Neural Networks*, (6) *Engineering Applications of Artificial Intelligence*, (7) *Journal of Intelligent and Robotic Systems*, (8) *Applied Artificial Intelligence*, and (9) *Journal of Intelligent and Fuzzy Systems*. There are many other journals on expert systems, neural networks, genetic algorithms, and fuzzy systems where applications to control can often be found. The professional societies most active in intelligent control are the IEEE Control Systems Society, International Federation on Automatic Control, and the IEEE Systems, Man, and Cybernetics Society.

55 Fuzzy Control

Kevin M. Passino
The Ohio State University

Stephen Yurkovich
The Ohio State University

55.1 Introduction

When confronted with a control problem for a complicated physical process, the control engineer usually follows a predetermined design procedure, which begins with the need for understanding the process and the primary control objectives. A good example of such a process is that of an automobile "cruise control," designed with the objective of providing the automobile with the capability of regulating its own speed at a driver-specified set-point (e.g., 55 mph). One solution to the automotive cruise control problem involves adding an electronic controller that can sense the speed of the vehicle via the speedometer and actuate the throttle position so as to regulate the vehicle speed at the driver-specified value even if there are road grade changes, head-winds, or variations in the number of passengers in the automobile. Control engineers typically solve the cruise control problem by (1) developing a model of the automobile dynamics (which may model vehicle and power train dynamics, road grade variations, etc.), (2) using the mathematical model to design a controller (e.g., via a linear model develop a linear controller with techniques from classical control), (3) using the mathematical model of the closed-loop system and mathematical or simulation-based analysis to study its performance (possibly leading to redesign), and (4) implementing the controller via, for example, a microprocessor, and evaluating the performance of the closed-loop system (again possibly leading to redesign).

The difficult task of modeling and simulating complex real-world systems for control systems development, especially when implementation issues are considered, is well documented. Even if a relatively accurate model of a dynamical system can be developed, it is often too complex to use in controller development, especially for many conventional control design procedures that require restrictive assumptions for the plant (e.g., linearity). It is for this reason that in practice conventional controllers are often developed via simple crude models of the plant behavior that satisfy the necessary assumptions, and via the *ad hoc* tuning of relatively simple linear or nonlinear controllers. Regardless, it is well understood (although sometimes forgotten) that heuristics enter the design process when the conventional control design process is used as long as one is concerned with the actual implementation of the control system. It must be acknowledged, however, that conventional control engineering approaches that use appropriate heuristics to tune the design have been relatively successful (the vast majority of all controllers currently in operation are conventional PID controllers). One may ask the following questions: How much of the success can be attributed to the use of the mathematical model and conventional control design approach, and how much should be attributed to the clever heuristic tuning that the control engineer uses upon implementation? If we exploit the use of heuristic information throughout the entire design process can we obtain higher performance control systems?

Fuzzy control provides a formal methodology for representing, manipulating, and implementing a human's heuristic knowledge about how to control a system. Fuzzy controller design involves incorporating human expertise on how to control a system into a set of rules (a rule-base). The inference mechanism in the fuzzy controller reasons over the information in the knowledge-base, the process outputs, and the user-specified goals to decide what inputs to generate for the process so that the closed-loop fuzzy control system will behave properly (e.g., so that the user-specified goals are met). From the cruise control example discussed above, it is clear that anyone who has experience in driving a car can practice regulating the speed about a desired set-point and load this information into a rule-base. For instance, one rule that a human driver may use is "IF speed is lower than the set-point THEN press down further on the accelerator pedal." A rule that would represent even more detailed information about how to regulate the speed would be "IF speed is lower than the set-point AND speed is approaching the set-point very fast THEN release the accelerator pedal by a small amount." This second rule characterizes our knowledge about how to make sure that we do not overshoot our desired (goal) speed. Generally speaking, if we load very detailed expertise into the rule-base we enhance our chances of obtaining better performance. Overall, the focus in fuzzy control is on the use of heuristic knowledge to achieve good control, whereas in conventional control the focus is on the use of a mathematical model for control systems development and subsequent use of heuristics in implementation.

55.1.1 Philosophy of Fuzzy Control

Due to the substantial amount of hype and excitement about fuzzy control, it is important to begin by providing a sound control engineering philosophy for this approach. First, there is a need for the control engineer to assess what (if any) advantages fuzzy control methods have over conventional methods. Time permitting, this must be done by careful comparative analyses involving modeling, mathematical analysis, simulation, implementation, and a full engineering cost–benefit analysis (which involves issues of cost, reliability, maintainability, flexibility, lead-time to production, etc.). When making the assessment of what control technique to use, the engineer should be cautioned that most work in fuzzy control to date has only focused on its *advantages* and has not taken a critical look at what possible *disadvantages* there could be to using it. For example, the following questions are cause for concern:

- Will the behaviors observed by a human expert include all situations that can occur due to disturbances, noise, or plant parameter variations?
- Can the human expert realistically and reliably foresee problems that could arise from closed-loop system instabilities or limit cycles?

- Will the expert be able to effectively incorporate stability criteria and performance objectives (e.g., rise-time, overshoot, and tracking specifications) into a rule-base to ensure that reliable operation can be obtained?
- Can an effective and widely used synthesis procedure be devoid of mathematical modeling and subsequent use of proven mathematical analysis tools?

These questions may seem even more troublesome if: (1) the control problem involves a "critical environment" where the failure of the control system to meet performance objectives could lead to loss of human life or an environmental disaster (e.g., in aircraft or nuclear power plant control), or (2) if the human expert's knowledge implemented in the fuzzy controller is somewhat inferior to that of a very experienced specialist that we expect to have design the control system (different designers have different levels of expertise). Clearly, then, for some applications there is a need for a methodology to develop, implement, and evaluate fuzzy controllers to ensure that they are reliable in meeting their performance specifications.

As it is discussed above, the standard control engineering methodology involves repeatedly coordinating the use of modeling, controller (re)design, simulation, mathematical analysis, and experimental evaluations to develop control systems. What is the relevance of this established methodology to the development of fuzzy control systems? Engineering a fuzzy control system uses many ideas from the standard control engineering methodology, except that in fuzzy control it is often said that a formal mathematical model is assumed unavailable so that mathematical analysis is impossible. While it is often the case that it is difficult, impossible, or cost-prohibitive to develop an accurate mathematical model for many processes, it is almost always possible for the control engineer to specify some type of approximate model of the process (after all, we do know what physical object we are trying to control). Indeed, it has been our experience that most often the control engineer developing a fuzzy control system does have a mathematical model available. While it may not be used directly in controller design, it is often used in simulation to evaluate the performance of the fuzzy controller before it is implemented (and it is often used for rule-base redesign). Certainly there are some applications where one can design a fuzzy controller and evaluate its performance directly via an implementation. In such applications one may not be overly concerned with a high performance level of the control system (e.g., for some commercial products such as washing machines or a shaver). In such cases, there may thus be no need for conducting simulation-based evaluations (requiring a mathematical model) before implementation. In other applications there is the need for a high level of confidence in the reliability of the fuzzy control system before it is implemented (e.g., in systems where there is a concern for safety).

In addition to simulation-based studies, one approach to enhancing our confidence in the reliability of fuzzy control systems is to use the mathematical model of the plant and nonlinear analysis for (1) verification of stability and performance specifications and (2) possible redesign of the fuzzy controller (for an overview of the results in this area see [1]). Some may be confident that a true expert would never need anything more than intuitive knowledge for rule-base design, and therefore, never design a faulty fuzzy controller. However, a true expert will certainly use all available information to ensure the reliable operation of a control system including approximate mathematical models, simulation, nonlinear analysis, and experimentation. We emphasize that mathematical analysis cannot alone provide the definitive answers about the reliability of the fuzzy control system because such analysis proves properties about the model of the process, not the actual physical process. It can be argued that a mathematical model is never a perfect representation of a physical process; hence, while nonlinear analysis (e.g., of stability) may appear to provide definitive statements about control system reliability, it is understood that such statements are only accurate to the extent that the mathematical model is accurate. Nonlinear analysis does not replace the use of common sense and evaluation via simulations and experimentation; it simply assists in providing a rigorous engineering evaluation of a fuzzy control system before it is implemented.

It is important to note that the advantages of fuzzy control often become most apparent for very complex problems where we have an intuitive idea about how to achieve high performance control. In such control applications an accurate mathematical model is so complex (i.e., high order, nonlinear, stochastic, with many inputs and outputs) that it is sometimes not very useful for the analysis and design of conventional control systems (since assumptions needed to utilize conventional control design approaches are often violated). The conventional control engineering approach to this problem is to use an approximate mathematical model that is accurate enough to characterize the essential plant behavior, yet simple enough so that the necessary assumptions to apply the analysis and design techniques are satisfied. However, due to the inaccuracy of the model, upon implementation the developed controllers often need to be tuned via the "expertise" of the control engineer. The fuzzy control approach, where explicit characterization and utilization of control expertise is used earlier in the design process, largely avoids the problems with model complexity that are related to design. That is, for the most part fuzzy control system design does not depend on a mathematical model unless it is needed to perform simulations to gain insight into how to choose the rule-base and membership functions. However, the problems with model complexity that are related to analysis have not been solved (i.e., analysis of fuzzy control systems critically depends on the form of the mathematical model); hence, it is often difficult to apply nonlinear analysis techniques to the applications where the advantages of fuzzy control are most apparent. For instance, as shown in [1], existing results for stability analysis of fuzzy control systems typically require that the plant model be deterministic, satisfy some continuity constraints, and sometimes require the plant to be linear or "linear-analytic." The only results for analysis of steady-state tracking error of fuzzy control systems, and the existing results on the use of describing functions for analysis of limit cycles, essentially require a linear time-invariant plant (or one that has a special form so that the nonlinearities can be bundled into a separate nonlinear component in the loop).

The current status of the field, as characterized by these limitations, coupled with the importance of nonlinear analysis of fuzzy control systems, make it an open area for investigation that will help establish the necessary foundations for a bridge between the communities of fuzzy control and nonlinear analysis. Clearly fuzzy control technology is leading the theory; the practitioner will proceed with the design and implementation of many fuzzy control systems without the aid of nonlinear analysis. In the mean time, theorists will attempt to develop a mathematical theory for the verification and certification of fuzzy control systems. This theory will have a synergistic effect by driving the development of fuzzy control systems for applications where there is a need for highly reliable implementations.

55.1.2 Summary

The focus of this chapter is on providing a practical introduction to fuzzy control (a "users guide") in the style of the book [2] (which is available at Kevin M Passino's web site for a free download); hence, we omit discussions of mathematical analysis of fuzzy control systems and invite the interested reader to investigate this topic further by consulting the bibliographic references. The remainder of this chapter is arranged as follows. We begin by providing a general mathematical introduction to fuzzy systems in a tutorial fashion. Next, we introduce a rotational inverted pendulum "theme problem." Many details on control design using principles of fuzzy logic are presented via this theme problem. We perform comparative analyses for fixed (nonadaptive) fuzzy and linear controllers. Following this we introduce the area of adaptive fuzzy control and show how one adaptive fuzzy technique has proven to be particularly effective for balancing control of the inverted pendulum. In the concluding remarks we explain how the area of fuzzy control is related to other areas in the field of intelligent control and what research needs to be performed as the field of fuzzy control matures.

55.2 Introduction to Fuzzy Control

The functional architecture of the fuzzy system (controller)* is composed of a *rule-base* (containing a fuzzy logic quantification of the expert's linguistic description of how to achieve good control), an *inference mechanism* (which emulates the expert's decision-making in interpreting and applying knowledge about how to do good control), a *fuzzification* interface (which converts controller inputs into information that the inference mechanism can easily use to activate and apply rules), and a *defuzzification* interface (which converts the conclusions of the inference mechanism into actual inputs for the process). Here we describe each of these four components in more detail (see Section 55.4 for a block diagram) [3–5].

55.2.1 Linguistic Rules

For our purposes, a fuzzy system is a static nonlinear mapping between its inputs and outputs (i.e., it is not a dynamical system). It is assumed that the fuzzy system has inputs $u_i \in \mathcal{U}_i$ where $i = 1, 2, \ldots, n$ and outputs $y_i \in \mathcal{Y}_i$ where $i = 1, 2, \ldots, m$. The ordinary ("crisp") sets $\mathcal{U}_i$ and $\mathcal{Y}_i$ are called the "universes of discourse" for u_i and y_i, respectively (in other words they are their domains).

To specify rules for the rule-base the expert will use a "linguistic description"; hence, linguistic expressions are needed for the inputs and outputs and the characteristics of the inputs and outputs. We will use "linguistic variables" (constant symbolic descriptions of what are in general time-varying quantities) to describe fuzzy system inputs and outputs. For our fuzzy system, linguistic variables denoted by $\tilde{u}_i$ are used to describe the inputs u_i. Similarly, linguistic variables denoted by $\tilde{y}_i$ are used to describe outputs y_i. For instance, an input to the fuzzy system may be described as $\tilde{u}_i =$"velocity error" and an output from the fuzzy system may be $\tilde{y}_i =$"voltage in."

Just as u_i and y_i take on values over each universe of discourse $\mathcal{U}_i$ and $\mathcal{Y}_i$, respectively, linguistic variables $\tilde{u}_i$ and $\tilde{y}_i$ take on "linguistic values" that are used to describe characteristics of the variables. Let $\tilde{A}_i^j$ denote the jth linguistic value of the linguistic variable $\tilde{u}_i$ defined over the universe of discourse $\mathcal{U}_i$. If we assume that there exist many linguistic values defined over $\mathcal{U}_i$, then the linguistic variable $\tilde{u}_i$ takes on the elements from the set of linguistic values denoted by $\tilde{A}_i = \{\tilde{A}_i^j : j = 1, 2, \ldots, N_i\}$ (sometimes for convenience we will let the j indices take on negative integer values). Similarly, let $\tilde{B}_i^j$ denote the jth linguistic value of the linguistic variable $\tilde{y}_i$ defined over the universe of discourse $\mathcal{Y}_i$. The linguistic variable $\tilde{y}_i$ takes on elements from the set of linguistic values denoted by $\tilde{B}_i = \{\tilde{B}_i^p : p = 1, 2, \ldots, M_i\}$ (sometimes for convenience we will let the p indices take on negative integer values). Linguistic values are generally expressed by descriptive terms such as "positive large," "zero," and "negative big" (i.e., adjectives).

The mapping of the inputs to the outputs for a fuzzy system is in part characterized by a set of *condition* $\rightarrow$ *action* rules, or in modus ponens (*If*... *Then*) form,

$$\textbf{If } \text{(antecedent) } \textbf{Then } \text{(consequent).} \tag{55.1}$$

As usual, the inputs of the fuzzy system are associated with the antecedent, and the outputs are associated with the consequent. These *If*... *Then* rules can be represented in many forms. Two standard forms, multi-input multi-output (MIMO) and multi-input single output (MISO) are considered here. The MISO form of a linguistic rule is:

$$\textbf{If } \tilde{u}_1 \text{ is } \tilde{A}_1^j \textbf{ and } \tilde{u}_2 \text{ is } \tilde{A}_2^k \textbf{ and}, \ldots, \textbf{and } \tilde{u}_n \text{ is } \tilde{A}_n^l \textbf{ Then } \tilde{y}_q \text{ is } \tilde{B}_q^p. \tag{55.2}$$

It is a whole set of linguistic rules of this form that the expert specifies on how to control the system. Note that if $\tilde{u}_1$="velocity error" and $\tilde{A}_1^j$="positive large," then "$\tilde{u}_1$ is $\tilde{A}_1^j$," a single term in the antecedent of the rule, means "velocity error is positive large." It can be easily shown that the MIMO form for a rule

* Sometimes a fuzzy controller is called a "fuzzy logic controller" or even a "fuzzy linguistic controller" since, as we will see, it uses fuzzy logic in the quantification of linguistic descriptions.

(i.e., one with consequents that have terms associated with each of the fuzzy controller outputs) can be decomposed into a number of MISO rules (using simple rules from logic). We assume that there are a total of R rules in the rule-base numbered $1, 2, \ldots, R$. For simplicity we will use tuples $(j, k, \ldots, l; p, q)_i$ to denote the ith MISO rule of the form given in Equation 55.2. Any of the terms associated with any of the inputs for any MISO rule can be included or omitted. Finally, we naturally assume that the rules in the rule-base are distinct (i.e., there are no two rules with exactly the same antecedents and consequents).

55.2.2 Fuzzy Sets, Fuzzy Logic, and the Rule-Base

Fuzzy sets and fuzzy logic are used to heuristically quantify the meaning of linguistic variables, linguistic values, and linguistic rules that are specified by the expert. The concept of a fuzzy set is introduced by first defining a "membership function." Let $\mathcal{U}_i$ denote a universe of discourse and $\tilde{A}_i^j \in \tilde{A}_i$ denote a specific linguistic value for the linguistic variable $\tilde{u}_i$. The function $\mu(u_i)$ associated with $\tilde{A}_i^j$ that maps $\mathcal{U}_i$ to $[0, 1]$ is called a "membership function." This membership function describes the "certainty" that an element of $\mathcal{U}_i$, denoted u_i, with a linguistic description $\tilde{u}_i$, may be classified linguistically as $\tilde{A}_i^j$. Membership functions are generally subjectively specified in an *ad hoc* (heuristic) manner from experience or intuition. For instance, if $\mathcal{U}_i = [-150, 150]$, $\tilde{u}_i$="velocity error," and $\tilde{A}_i^j$="positive large," then $\mu(u_i)$ may be a bell-shaped curve that peaks at one at $u_i = 75$ and is near zero when $u_i < 50$ or $u_i > 100$. Then if $u_i = 75$, $\mu(75) = 1$ so that we are absolutely certain that u_i is "positive large." If $u_i = -25$ then $\mu(-25)$ is very near zero, which represents that we are very certain that u_i is not "positive large." Clearly, many other choices for the shape of the membership function are possible (e.g., triangular and trapezoidal shapes) and these will each provide a different meaning for the linquistics that they quantify. Below, we will show how to specify membership functions for a fuzzy controller for the rotational inverted pendulum.

Given a linguistic variable $\tilde{u}_i$ with a linguistic value $\tilde{A}_i^j$ defined on the universe of discourse $\mathcal{U}_i$, and membership function $\mu_{A_i^j}(u_i)$ (membership function associated with the fuzzy set A_i^j) that maps $\mathcal{U}_i$ to $[0, 1]$, a "fuzzy set" denoted with A_i^j is defined as

$$A_i^j = \{(u_i, \mu_{A_i^j}(u_i)) : u_i \in \mathcal{U}_i\}. \tag{55.3}$$

Next, we specify some set-theoretic and logical operations on fuzzy sets. Given fuzzy sets A_i^1 and A_i^2 associated with the universe of discourse $\mathcal{U}_i$ ($N_i = 2$), with membership functions denoted $\mu_{A_i^1}(u_i)$ and $\mu_{A_i^2}(u_i)$, respectively, A_i^1 is defined to be a "fuzzy subset" of A_i^2, denoted by $A_i^1 \subset A_i^2$, if $\mu_{A_i^1}(u_i) \leq \mu_{A_i^2}(u_i)$ for all $u_i \in \mathcal{U}_i$.

The intersection of fuzzy sets A_i^1 and A_i^2 which are defined on the universe of discourse $\mathcal{U}_i$ is a fuzzy set, denoted by $A_i^1 \cap A_i^2$, with a membership function defined by either:

$$\begin{aligned} \text{Minimum:} \quad & \mu_{A_i^1 \cap A_i^2} = \min\{\mu_{A_i^1}(u_i), \mu_{A_i^2}(u_i) : u_i \in \mathcal{U}_i\}, \\ \text{Algebraic Product:} \quad & \mu_{A_i^1 \cap A_i^2} = \{\mu_{A_i^1}(u_i)\mu_{A_i^2}(u_i) : u_i \in \mathcal{U}_i\}. \end{aligned} \tag{55.4}$$

Suppose that we use the notation $x * y = \min\{x, y\}$ or at other times we will use it to denote the product $x * y = xy$ ($*$ is sometimes called the "triangular norm"). Then $\mu_{A_i^1}(u_i) * \mu_{A_i^2}(u_i)$ is a general representation for the intersection of two fuzzy sets. In fuzzy logic, intersection is used to represent the "and" operation.

The union of fuzzy sets A_i^1 and A_i^2, which are defined on the universe of discourse $\mathcal{U}_i$, is a fuzzy set denoted $A_i^1 \cup A_i^2$, with a membership function defined by either:

$$\begin{aligned} \text{Maximum:} \quad & \mu_{A_i^1 \cup A_i^2}(u_i) = \max\{\mu_{A_i^1}(u_i), \mu_{A_i^2}(u_i) : u_i \in \mathcal{U}_i\}, \\ \text{Algebraic Sum:} \quad & \mu_{A_i^1 \cup A_i^2}(u_i) = \{\mu_{A_i^1}(u_i) + \mu_{A_i^2}(u_i) - \mu_{A_i^1}(u_i)\mu_{A_i^2}(u_i) : u_i \in \mathcal{U}_i\}. \end{aligned} \tag{55.5}$$

Suppose that we use the notation $x \oplus y = \max\{x, y\}$ or at other times we will use it to denote $x \oplus y = x + y - xy$ ($\oplus$ is sometimes called the "triangular conorm"). Then $\mu_{A_i^1}(u_i) \oplus \mu_{A_i^2}(u_i)$ is a general representation for the union of two fuzzy sets. In fuzzy logic, union is used to represent the "or" operation.

The intersection and union above are both defined for fuzzy sets that lie on the same universe of discourse. The fuzzy Cartesian product is used to quantify operations on many universes of discourse. If $A_1^j, A_2^k, \ldots, A_n^l$ are fuzzy sets defined on the universes of discourse $\mathcal{U}_1, \mathcal{U}_2, \ldots, \mathcal{U}_n$, respectively, their Cartesian product is a fuzzy set (sometimes called a "fuzzy relation"), denoted by $A_1^j \times A_2^k \times \cdots \times A_n^l$, with a membership function defined by

$$\mu_{A_1^j \times A_2^k \times \cdots \times A_n^l}(u_1, u_2, \ldots, u_n) = \mu_{A_1^j}(u_1) * \mu_{A_2^k}(u_2) * \cdots * \mu_{A_n}(u_n). \tag{55.6}$$

Next, we show how to quantify the linguistic elements in the antecedent and consequent of the linguistic *If*... *Then* rule with fuzzy sets. For example, suppose we are given the *If*... *Then* rule in MISO form in Equation 55.2. Define the fuzzy sets:

$$\begin{aligned} A_1^j &= \{(u_1, \mu_{A_1^j}(u_1)) : u_1 \in \mathcal{U}_1\} \\ A_2^k &= \{(u_2, \mu_{A_2^k}(u_2)) : u_2 \in \mathcal{U}_2\} \\ &\vdots \\ A_n^l &= \{(u_n, \mu_{A_n^l}(u_n)) : u_n \in \mathcal{U}_n\} \\ B_q^p &= \{(y_q, \mu_{B_q^p}(y_q)) : y_q \in \mathcal{Y}_q\}. \end{aligned} \tag{55.7}$$

These fuzzy sets quantify the terms in the antecedent and the consequent of the given *If*... *Then* rule, to make a "fuzzy implication"

$$\textbf{If } A_1^j \textbf{ and } A_2^k \textbf{ and}, \ldots, \textbf{and } A_n^l \textbf{ Then } B_q^p, \tag{55.8}$$

where the fuzzy sets $A_1^j, A_2^k, \ldots, A_n^l$, and B_q^p are defined in Equation 55.8. Therefore, the fuzzy set A_1^j is associated with, and quantifies, the meaning of the linguistic statement "$\tilde{u}_1$ is $\tilde{A}_1^j$" and B_q^p quantifies the meaning of "$\tilde{y}_q$ is $\tilde{B}_q^p$." Each rule in the rule-base $(j, k, \ldots, l; p, q)_i$, $i = 1, 2, \ldots, R$ is represented with such a fuzzy implication (a fuzzy quantification of the linguistic rule). The reader who is interested in more mathematical details on fuzzy sets and fuzzy logic should consult [6].

55.2.3 Fuzzification

Fuzzy sets are used to quantify the information in the rule-base, and the inference mechanism operates on fuzzy sets to produce fuzzy sets; hence, we must specify how the fuzzy system will convert its numeric inputs $u_i \in \mathcal{U}_i$ into fuzzy sets (a process called "fuzzification") so that they can be used by the fuzzy system. Let $\mathcal{U}_i^*$ denote the set of all possible fuzzy sets that can be defined on $\mathcal{U}_i$. Given $u_i \in \mathcal{U}_i$, fuzzification transforms u_i to a fuzzy set denoted by $\hat{A}_i^{\text{fuz}}$ defined* over the universe discourse $\mathcal{U}_i$. This transformation is produced by the fuzzification operator $\mathcal{F}$ defined by $\mathcal{F} : \mathcal{U}_i \to \mathcal{U}_i^*$, where $\mathcal{F}(u_i) := \hat{A}_i^{\text{fuz}}$. Quite often "singleton fuzzification" is used, which produces a fuzzy set $\hat{A}_i^{\text{fuz}} \in \mathcal{U}_i^*$ with a membership function defined by

$$\mu_{\hat{A}_i^{\text{fuz}}}(x) = \begin{cases} 1 & x = u_i, \\ 0 & \text{otherwise} \end{cases} \tag{55.9}$$

(any fuzzy set with this form for its membership function is called a "singleton"). Singleton fuzzification is generally used in implementations since, without the presence of noise, we are absolutely certain

* In this section, as we introduce various fuzzy sets. We will always use a hat over any fuzzy set whose membership function changes dynamically over time as u_i changes.

that u_i takes on its measured value (and no other value) and since it provides certain savings in the computations needed to implement a fuzzy system (relative to, e.g., "Gaussian fuzzification" which would involve forming bell-shaped membership functions about input points). Throughout the remainder of this chapter we use singleton fuzzification.

55.2.4 The Inference Mechanism

The inference mechanism has two basic tasks: (1) determining the extent to which each rule is relevant to the current situation as characterized by the inputs u_i, $i = 1, 2, \ldots, n$ (we call this task "matching"), and (2) drawing conclusions using the current inputs u_i and the information in the rule-base (we call this task an "inference step"). For matching note that $A_1^j \times A_2^k \times \cdots \times A_n^l$ is the fuzzy set representing the antecedent of the ith rule $(j, k, \ldots, l; p, q)_i$ (there may be more than one such rule with this antecedent). Suppose that at some time we get inputs u_i, $i = 1, 2, \ldots, n$, and fuzzification produces $\hat{A}_1^{\text{fuz}}, \hat{A}_2^{\text{fuz}}, \ldots, \hat{A}_n^{\text{fuz}}$, the fuzzy sets representing the inputs. The first step in matching involves finding fuzzy sets $\hat{A}_1^j, \hat{A}_2^k, \ldots, \hat{A}_n^l$ with membership functions

$$\begin{aligned}
\mu_{\hat{A}_1^j}(u_1) &= \mu_{A_1^j}(u_1) * \mu_{\hat{A}_1^{\text{fuz}}}(u_1)\\
\mu_{\hat{A}_2^k}(u_2) &= \mu_{A_2^k}(u_2) * \mu_{\hat{A}_2^{\text{fuz}}}(u_2)\\
&\vdots\\
\mu_{\hat{A}_n^l}(u_n) &= \mu_{A_n^l}(u_n) * \mu_{\hat{A}_n^{\text{fuz}}}(u_n)
\end{aligned}$$

(for all $j, k, \ldots, l$) that combine the fuzzy sets from fuzzification with the fuzzy sets used in each of the terms in the antecedents of the rules. If singleton fuzzification is used then each of these fuzzy sets is a singleton that is scaled by the antecedent membership function (e.g., $\mu_{\hat{A}_1^j}(\bar{u}_1) = \mu_{A_1^j}(\bar{u}_1)$ for $\bar{u}_1 = u_1$ and $\mu_{\hat{A}_1^j}(\bar{u}_1) = 0$ for $\bar{u}_1 \neq u_1$). Second, we form membership values $\mu_i(u_1, u_2, \ldots, u_n)$ for each rule that represent the overall certainty that rule i matches the current inputs. In particular, we first let

$$\bar{\mu}_i(u_1, u_2, \ldots, u_n) = \mu_{\hat{A}_1^j}(u_1) * \mu_{\hat{A}_2^k}(u_2) * \cdots * \mu_{\hat{A}_n^l}(u_n) \tag{55.10}$$

be the membership function for $\hat{A}_1^j \times \hat{A}_2^k \times \cdots \times \hat{A}_n^l$. Notice that since we are using singleton fuzzification, we have

$$\bar{\mu}_i(\bar{u}_1, \bar{u}_2, \ldots, \bar{u}_n) = \mu_{A_1^j}(\bar{u}_1) * \mu_{A_2^k}(\bar{u}_2) * \cdots * \mu_{A_n^l}(\bar{u}_n) \tag{55.11}$$

for $\bar{u}_i = u_i$, and $\bar{\mu}_i(\bar{u}_1, \bar{u}_2, \ldots, \bar{u}_n) := 0$ for $\bar{u}_i \neq u_i$, $i = 1, 2, \ldots, n$. Since the u_i are given, $\mu_{A_1^j}(u_1), \mu_{A_2^k}(u_2), \ldots, \mu_{A_n^l}(u_n)$ are constants. Define

$$\mu_i(u_1, u_2, \ldots, u_n) = \mu_{A_1^j}(u_1) * \mu_{A_2^k}(u_2) * \cdots * \mu_{A_n^l}(u_n), \tag{55.12}$$

which is simply a function of the inputs u_i. We use $\mu_i(u_1, u_2, \ldots, u_n)$ to represent the certainty that the antecedent of rule i matches the input information. This concludes the process of matching input information with the antecedents of the rules.

Next, the inference step is taken by computing, for the ith rule $(j, k, \ldots, l; p, q)_i$, the "implied fuzzy set" $\hat{B}_q^i$ with membership function

$$\mu_{\hat{B}_q^i}(y_q) = \mu_i(u_1, u_2, \ldots, u_n) * \mu_{B_q^p}(y_q). \tag{55.13}$$

The implied fuzzy set $\hat{B}_q^i$ specifies the certainty level that the output should be a specific crisp output y_q within the unverse of discourse $\mathcal{Y}_q$, taking into consideration only rule i. Note that since $\mu_i(u_1, u_2, \ldots, u_n)$

will vary with time so will the shape of the membership functions $\mu_{\hat{B}_q^i}(y_q)$ for each rule. Alternatively, the inference mechanism could, in addition, compute the "overall implied fuzzy set" $\hat{B}_q$ with membership function

$$\mu_{\hat{B}_q}(y_q) = \mu_{\hat{B}_q^1}(y_q) \oplus \mu_{\hat{B}_q^2}(y_q) \oplus \cdots \oplus \mu_{\hat{B}_q^R}(y_q) \tag{55.14}$$

that represents the conclusion reached considering all the rules in the rule-base at the same time (notice that determining $\hat{B}_q$ can, in general, require significant computational resources).

Using the mathematical terminology of fuzzy sets, the computation of $\mu_{\hat{B}_q}(y_q)$ is said to be produced by a "sup-star compositional rule of inference." The "sup" in this terminology corresponds to the $\oplus$ operation and the "star" corresponds to $*$. "Zadeh's compositional rule of inference" [7] is the special case of the sup-star compositional rule of inference when max is used for $\oplus$ and min is used for $*$. The overall justification for using the above operations to represent the inference step lies in the fact that *we can be no more certain about our conclusions than we are about our premises (antecedents).* The operations performed in taking an inference step adhere to this principle. To see this, study Equation 55.13 and note that the scaling from $\mu_i(u_1, u_2, \ldots, u_n)$ that is produced by the antecedent matching process ensures that $\sup_{y_q}\{\mu_{\hat{B}_q^i}(y_q)\} \leq \mu_i(u_1, u_2, \ldots, u_n)$ (a similar statement holds for the overall implied fuzzy set).

Up to this point we have used fuzzy logic to quantify the rules in the rule-base, fuzzification to produce fuzzy sets characterizing the inputs, and the inference mechanism to produce fuzzy sets representing the conclusions that it reaches considering the current inputs and the information in the rule-base. Next, we look at how to convert this fuzzy set quantification of the conclusions to a numeric value that can be input to the plant.

55.2.5 Defuzzification

A number of defuzzification strategies exist. Each provides a means to choose a single output (which we denote with y_q^{crisp}) based on either the implied fuzzy sets or the overall implied fuzzy set (depending on the type of inference strategy chosen). First, we present typical defuzzification techniques for the overall implied fuzzy set $\hat{B}_q$:

- *Max Criteria:* A crisp output y_q^{crisp} is chosen as the point on the output universe of discourse $\mathcal{Y}_q$ for which the overall implied fuzzy set $\hat{B}_q$ achieves a maximum that is,

$$y_q^{\text{crisp}} \in \left\{ \arg \sup_{\mathcal{Y}_q} \left\{ \mu_{\hat{B}_q}(y_q) \right\} \right\}. \tag{55.15}$$

  Since the supremum can occur at more than one point in $\mathcal{Y}_q$ one also needs to specify a strategy on how to pick only one point for y_q^{crisp} (e.g., choosing the smallest value). Often this defuzzification strategy is avoided due to this ambiguity.
- *Mean of Maximum:* A crisp output y_q^{crisp} is chosen to represent the mean value of all elements whose membership in $\hat{B}_q$ is a maximum. We define $\hat{b}_q^{\max}$ as the supremum of the membership function of $\hat{B}_q$ over the universe of discourse $\mathcal{Y}_q$. Moreover, we define a fuzzy set $\hat{B}_q^* \in \mathcal{Y}_q$ with a membership function defined as

$$\mu_{\hat{B}_q^*}(y_q) = \begin{cases} 1 & \mu_{\hat{B}_q}(y_q) = \hat{b}_q^{\max}, \\ 0 & \text{otherwise}, \end{cases} \tag{55.16}$$

  then a crisp output, using the mean of maximum method, is defined as

$$y_q^{\text{crisp}} = \frac{\int_{\mathcal{Y}_q} y_q \cdot \mu_{\hat{B}_q^*}(y_q) \cdot dy_q}{\int_{\mathcal{Y}_q} \mu_{\hat{B}_q^*}(y_q) \cdot dy_q}. \tag{55.17}$$

Note that the integrals in Equation 55.17 must be computed at each time instant since they depend on $\hat{B}_q$ which changes with time. This can require excessive computational resources; hence, this defuzzification technique is often avoided in practice.

- *Center of Area (COA):* A crisp output y_q^{crisp} is chosen as the COA for the membership function of the overall implied fuzzy set $\hat{B}_q$. For a continuous output universe of discourse $\mathcal{Y}_q$ the COA output is denoted by

$$y_q^{\text{crisp}} = \frac{\int_{\mathcal{Y}_q} y_q \cdot \mu_{\hat{B}_q}(y_q)\, dy_q}{\int_{\mathcal{Y}_q} \mu_{\hat{B}_q}(y_q)\, dy_q}. \tag{55.18}$$

Note that similar to the mean of the maximum method this defuzzification approach can be computationally expensive. Also, the fuzzy system must be defined so that $\int_{\mathcal{Y}_q} \mu_{\hat{B}_q}(y_q) dy_q \neq 0$ for all u_i.

Next, we specify typical defuzzification techniques for the implied fuzzy sets $\hat{B}_q^i$:

- *Centroid:* A crisp output y_q^{crisp} is chosen using the centers of each of the output membership functions and the maximum certainty of each of the conclusions represented with the implied fuzzy sets and is given by

$$y_q^{\text{crisp}} = \frac{\sum_{i=1}^{R} c_q^i \sup_{y_q}\{\mu_{\hat{B}_q^i}(y_q)\}}{\sum_{i=1}^{R} \sup_{y_q}\{\mu_{\hat{B}_q^i}(y_q)\}}, \tag{55.19}$$

where c_q^i is the COA of the membership function of B_q^p associated with the implied fuzzy set $\hat{B}_q^i$ for the ith rule $(j, k, \ldots, l; p, q)_i$. Notice that $\sup_{y_q}\{\mu_{\hat{B}_q^i}(y_q)\}$ is often very easy to compute since if $\mu_{B_q^p}(y_q) = 1$ for at least one y_q (which is the normal way to define membership functions), then for many inference strategies $\sup_{y_q}\{\mu_{\hat{B}_q^i}(y_q)\} = \mu_i(u_1, u_2, \ldots, u_n)$ which as already been computed in the matching process. Notice that the fuzzy system must be defined so that $\sum_{i=1}^{R} \sup_{y_q}\{\mu_{\hat{B}_q^i}(y_q)\} \neq 0$ for all u_i.

- *Center of Gravity (COG):* A crisp output y_q^{crisp} is chosen using the COA and area of each implied fuzzy set and is given by

$$y_q^{\text{crisp}} = \frac{\sum_{i=1}^{R} c_q^i \int_{\mathcal{Y}_q} \mu_{\hat{B}_q^i}(y_q)\, dy_q}{\sum_{i=1}^{R} \int_{\mathcal{Y}_q} \mu_{\hat{B}_q^i}(y_q)\, dy_q}, \tag{55.20}$$

where c_q^i is the COA of the membership function of B_q^p associated with the implied fuzzy set $\hat{B}_q^i$ for the ith rule $(j, k, \ldots, l; p, q)_i$. Notice that COG can be easy to compute since it is often easy to find closed-form expressions for $\int_{\mathcal{Y}_q} \mu_{\hat{B}_q^i}(y_q)\, dy_q$, which is the area under a membership function. Notice that the fuzzy system must be defined so that $\sum_{i=1}^{R} \int_{\mathcal{Y}_q} \mu_{\hat{B}_q^i}(y_q)\, dy_q \neq 0$ for all u_i.

Overall, we see that using the overall implied fuzzy set in defuzzification is often undesirable for two reasons: (1) the overall implied fuzzy set $\hat{B}_q$ is itself difficult to compute in general, and (2) the defuzzification techniques based on an inference mechanism that provides $\hat{B}_q$ are also difficult to compute. It is for this reason that most existing fuzzy controllers (including the ones in this chapter) use defuzzification techniques based on the implied fuzzy sets such as Centroid or COG.

55.3 Theme Problem: Rotational Inverted Pendulum

One of the classic problems in the study of nonlinear systems is that of the inverted pendulum. The primary control problem one considers with such a system is regulating the position of the pendulum (typically a rod with mass at the endpoint) to the vertical (up) position; that is, "balanced." A secondary problem is that of "swinging up" the pendulum from its rest position (vertical down). Often, actuation is accomplished either via a motor at the base of the pendulum (at the hinge), or via a cart through translational motion. In this example actuation of the pendulum is accomplished through *rotation* of a separate, attached link, referred to, henceforth, as the "base."

55.3.1 Experimental Apparatus

The test bed consists of three primary components: the plant, digital and analog interfaces, and the digital controller. The overall system is shown in Figure 55.1 where the three components can be clearly identified [8]. The plant is composed of a pendulum and a rotating base made of aluminum rods, two optical encoders as the angular position sensors with effective resolutions of 0.2 degrees for the pendulum and 0.1 degrees for the base, and a large, high-torque permanent-magnet DC motor (with rated stall torque of 5.15 N m). As the base rotates through the angle θ_0 the pendulum is free to rotate (high precision bearings are utilized) through its angle θ_1 made with the vertical.

Interfaces between the digital controller and the plant consist of two data acquisition cards and some signal conditioning circuitry, structured for the two basic functions of sensor integration and control signal generation. The signal conditioning is accomplished via a combination of several logic gates to filter quadrature signals from the optical encoders, which are then processed through a separate data acquisition card to utilize the four 16-bit counters (accessed externally to count pulses from the circuitry itself). Another card supplies the control signal interface through its 12-bit D/A converter (to generate the actual control signal), while the board's 16-bit timer is used as a sampling clock. The computer used for control is a personal computer with its Intel 80486DX processor operating at 50 MHz. The real-time codes for control are written in C.

55.3.2 Mathematical Model

For brevity, and because this system is a popular example for nonlinear control, we omit details of the necessary physics and geometry for modeling. The differential equations that approximately describe the dynamics of the plant are given by

$$\ddot{\theta}_0 = -a_p\dot{\theta}_0 + K_p v_a, \tag{55.21}$$

$$\ddot{\theta}_1 = -\frac{C_1}{J_1}\dot{\theta}_1 + \frac{m_1 g \ell_1}{J_1}\sin(\theta_1) + K_1\ddot{\theta}_0, \tag{55.22}$$

where, again, θ_0 is the angular displacement of the rotating base, $\dot{\theta}_0$ is the angular speed of the rotating base, θ_1 is the angular displacement of the pendulum, $\dot{\theta}_1$ is the angular speed of the pendulum, v_a is the motor armature voltage, K_p and a_p are parameters of the DC motor with torque constant K_1, g is the acceleration due to gravity, m_1 is the pendulum mass, ℓ_1 is the pendulum length, J_1 is the pendulum inertia, and C_1 is a constant associated with friction (actual parameter values appear in [8]).

For controller synthesis (and model linearization) we will require a state variable description of the system. This is easily done by defining state variables $x_1 = \theta_0$, $x_2 = \dot{\theta}_0$, $x_3 = \theta_1$, $x_4 = \dot{\theta}_1$, and control signal $u = v_a$. Linearization of these equations *about the vertical position* (i.e., $\theta_1 = 0$), and using the system physical parameters [8] results in the following linear, time invariant state variable description:

$$\begin{bmatrix} \dot{x}_1 \\ \dot{x}_2 \\ \dot{x}_3 \\ \dot{x}_4 \end{bmatrix} = \begin{bmatrix} 0 & 1 & 0 & 0 \\ 0 & -33.04 & 0 & 0 \\ 0 & 0 & 0 & 1 \\ 0 & 49.30 & 73.41 & -2.29 \end{bmatrix} \begin{bmatrix} x_1 \\ x_2 \\ x_3 \\ x_4 \end{bmatrix} + \begin{bmatrix} 0 \\ 74.89 \\ 0 \\ -111.74 \end{bmatrix} u. \tag{55.23}$$

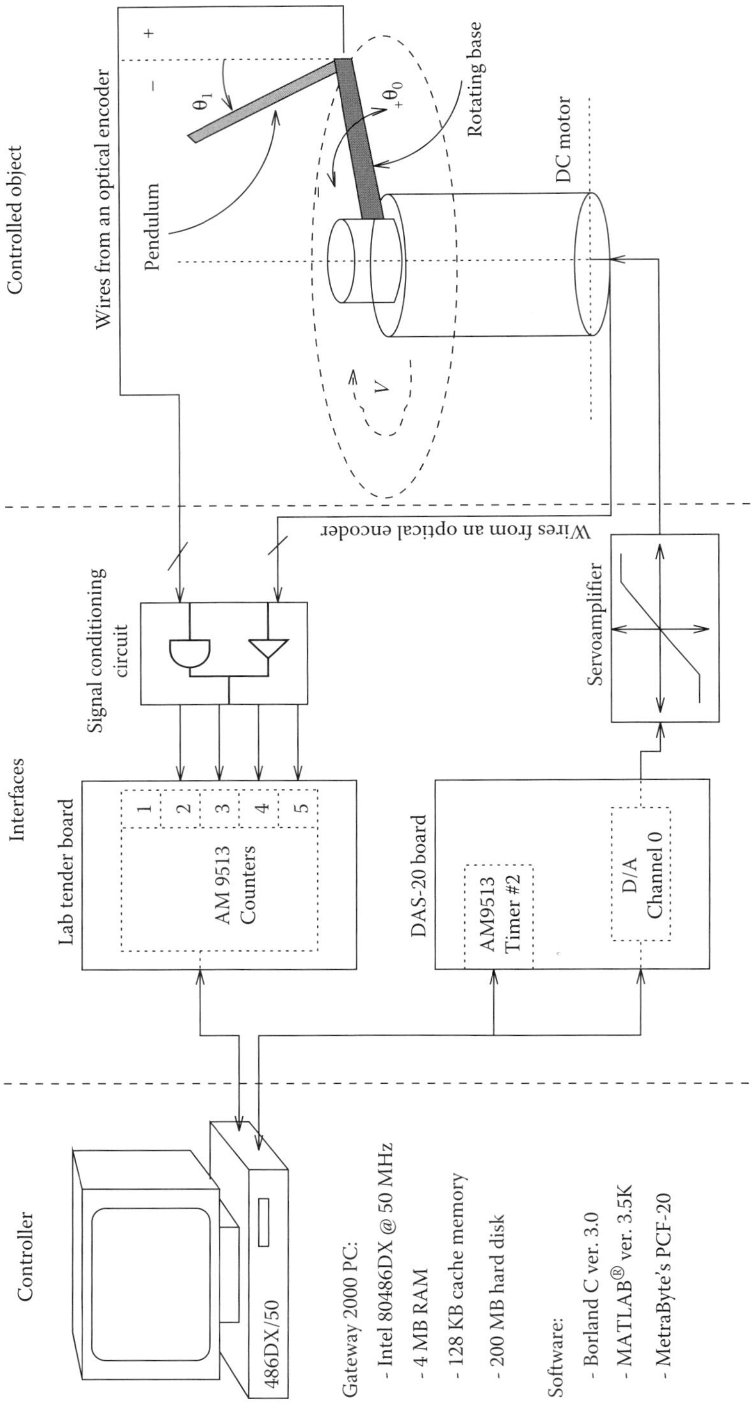

FIGURE 55.1 Hardware setup.

55.3.3 Swing-Up Control

Because we intend to develop control laws that will be valid in regions about the vertical position ($\theta_1 = 0$), it is crucial to swing the pendulum up so that it is near vertical at near zero (angular) velocity. Elaborate schemes can be used for this task (such as those employing concepts from differential geometry), but for the purposes of this example we choose to use a simple heuristic procedure based on an "energy pumping strategy" proposed in [9] for a similar *under-actuated* system. The goal of this simple swing-up control strategy is to "pump" energy into the pendulum link in such a way that the energy or magnitude of each swing increases until the pendulum approaches its inverted position. To apply such an approach, we simply consider how one would (intuitively) swing the pendulum from its hanging position ($\theta_1 = \pi$) to its upright position. If the rotating base is swung to the left and right continually at an appropriate frequency, the magnitude of the pendulum at each swing will increase.

The control scheme we will ultimately employ consists of two main components: the "scheduler" (which we will also call a "supervisor") observes the position of the pendulum relative to its stable equilibrium point ($\theta_1 = \pi$), then schedules the transitions between two reference positions of the rotating base (${\theta_0}^{ref} = \pm\Gamma$); and, the "positioning control" regulates the base to the desired reference point. These two components compose a closed-loop planning algorithm to command the rotating base to move in a certain direction based on the position of the pendulum. In effect, the human operator acts as the supervisor in tuning the positioning control (through trial and error on the system).

For simplicity, a proportional controller will be used as the positioning control. The gain K_p is chosen just large enough so that the actuator drives the base fast enough without saturating the control output; after several trials, K_p was set to 0.5. The parameter Γ determines how far the base is allowed to swing; larger swings transfer more energy to swinging up the pendulum. The swing-up motion of the pendulum can be approximated as an exponentially growing cosine function. The parameter Γ significantly affects the "negative damping" (i.e., exponential growth) of the swing-up motion. By tuning Γ, one can adjust the motion of the pendulum in such a way that the velocity of the pendulum and the control output are minimum when the pendulum reaches its inverted position (i.e., the pendulum has the largest potential energy and the lowest kinetic energy). Notice that if the dynamics of the pendulum are changed (e.g., adding extra weight to the endpoint of the pendulum), then the parameter Γ must be tuned. In [8] it is shown how a rule-based system can be used to effectively automate the swing-up control by implementing fuzzy strategies in the supervisor portion of the overall scheme.

55.3.4 Balancing Control

Synthesis of the fuzzy controllers to follow is aided by (1) a good understanding of the pendulum dynamics (the analytical model and intuition related to the physical process), and (2) experience with performance of linear control strategies. Although numerous linear control design techniques have been applied to this particular system, here we consider the performance of only one linear strategy (the one tested) as applied to the experimental system: the linear quadratic regulator (LQR). Our purpose is twofold. First, we form a baseline for comparison to fuzzy control designs to follow, and second, we provide a starting point for synthesis of the fuzzy controller. It is important to note that extensive simulation results (on the nonlinear model) were carried out prior to application to the laboratory apparatus; designs were carried out on the linearized model of the system. Specifics of the design process for the LQR and other applicable linear design techniques may be found in other chapters of this volume.

Because the linearized system is completely controllable and observable, state feedback strategies, including the optimal strategies of the LQR, are applicable. Generally speaking, the system performance is prescribed via the optimal performance index

$$J = \int_0^{\infty} (x(t)^T Q x(t) + u(t)^T R u(t))\, dt, \qquad (55.24)$$

where Q and R are the weighting matrices corresponding to the state x and input u, respectively. Given fixed Q and R, the feedback gains that optimize the function J can be uniquely determined by solving an algebraic Riccati equation. Because we are more concerned with balancing the pendulum than regulating the base, we put the highest priority in controlling θ_1 by choosing the weighting matrices $Q = \text{diag}(1, 0, 5, 0)$ and $R = [1]$. For a 10 ms sampling time, the discrete optimal feedback gains corresponding to the weighting matrices Q and R are $k_1 = -0.9$, $k_2 = -1.1$, $k_3 = -9.2$, and $k_4 = -0.9$. Although observers may be designed to estimate the states $\dot{\theta}_1$ and $\dot{\theta}_0$, we choose to use an equally effective and simple first-order approximation for each derivative.

Note that this controller is designed in simulation for the system as modeled (and subsequently linearized). When the resulting controller gains (k_1 through k_4) are implemented on the actual system, some "trial-and-error" tuning is required (due primarily to modeling uncertainties), which amounted to adjusting the designed gain by about 10% to obtain performance matching the predicted results from simulation. Moreover, it is critical to note that the design process (as well as the empirical tuning) has been done for the "nominal" system (i.e., the pendulum system with no additional mass on the endpoint).

Using a swing-up control strategy tuned for the nominal system, the results of the LQR control design are given in Figure 55.2 for the base angle (top plot), pendulum angle (center plot), and control output (bottom plot).

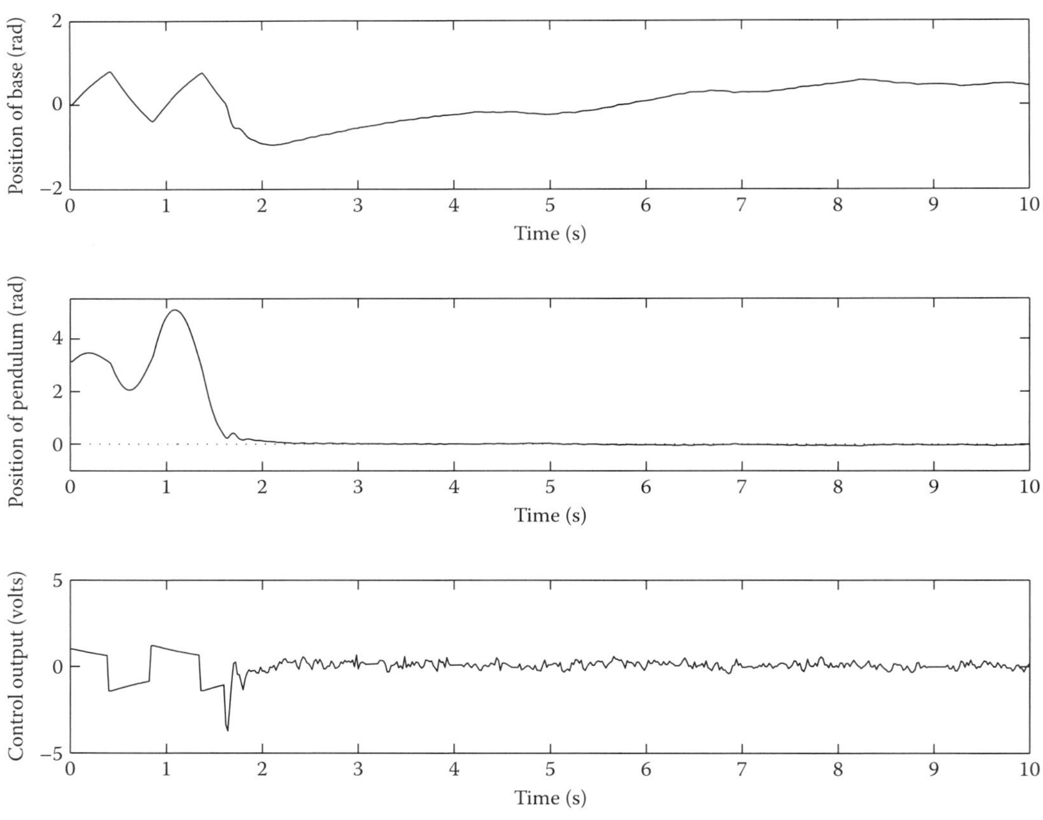

FIGURE 55.2 LQR on the *nominal* system.

55.4 Fuzzy Control for the Rotational Inverted Pendulum

Aside from serving to illustrate procedures for synthesizing a fuzzy controller, several reasons arise for considering the use of a nonlinear control scheme for the pendulum system. Because all linear controllers are designed based on a linearized model of the system, they are inherently valid only for a region about a specific point (in this case, the vertical, $\theta_1 = 0$ position). For this reason, such linear controllers tend to be very sensitive to parametric variations, uncertainties, and disturbances. This is indeed the case for the experimental system under study; when an extra weight or *sloshing liquid* (using a water-tight bottle) is attached at the endpoint of the pendulum, the performance of all linear controllers degrades considerably, often resulting in unstable behavior. Thus, to enhance the performance of the balancing control, one naturally turns to some nonlinear control scheme that is expected to exhibit improved performance in the presence of disturbances and uncertainties in modeling. Two such nonlinear controllers will be investigated here: in the next section, a direct fuzzy controller is constructed and later an adaptive version of this same controller is discussed.

55.4.1 Controller Synthesis

For simplicity, the controller synthesis example explained next will utilize singleton fuzzification and symmetric, "triangular" membership functions on the controller inputs and output (they are, in fact, very simple to implement in real-time code). We choose to use seven membership functions for each input, uniformly distributed across their universes of discourse (over crisp values of each input e_i) as shown in Figure 55.3. The linguistic values for the ith input are denoted by $\tilde{E}_i^r$ where $r \in \{-3, -2, -1, 0, 1, 2, 3\}$. Linguistically, we would therefore define $\tilde{E}_i^{-3}$ as "negative large," $\tilde{E}_i^{-2}$ as "negative medium," $\tilde{E}_i^0$ as "zero," and so on. Note also that a "saturation nonlinearity" is built in for each input in the membership functions corresponding to the outermost regions of the universes of discourse. We use min to represent the antecendent (i.e., "$*$" from Section 55.2 is min) and COG defuzzification.

To synthesize a fuzzy controller for our example system, we pursue the idea of seeking to "expand" the region of operation of the fixed (nonadaptive) controller. In doing so, we will utilize the results of the LQR design presented in Section 55.3 to lead us in the design. A block diagram of the fuzzy controller is shown in Figure 55.4. Similar to the LQR, the fuzzy controller for the inverted pendulum system will have four inputs and one output. The four (crisp) inputs to the fuzzy controller are the position error of the base e_1, its derivative e_2, the position error of the pendulum e_3, and its derivative e_4.

The *normalizing* gains g_i essentially serve to expand and compress the universes of discourse to some predetermined, uniform region, primarily to standardize the choice of the various parameters in synthesizing the fuzzy controller. A crude approach to choosing these gains is strictly based on intuition and does not require a mathematical model of the plant. In that case the input normalizing gains are chosen in such a way that all the desired operating regions are mapped into $[-1, +1]$. Such a simple approach in design works often for a number of systems, as witnessed by the large number of applications documented in the open literature. For complicated systems, however, such a procedure can be very difficult to implement because there are many ways to define the linguistic values and linguistic rules; indeed, it can be extremely difficult to find a viable set of linguistic values and rules just to maintain stability. Such was the case for this system.

What we propose here is an approach based on experience in designing the LQR controller for the linearized model of the plant, leading to a mechanized procedure for determining the normalizing gains, output membership functions, and rule-base. Recall from our discussion in Section 55.2 that a fuzzy system is a static nonlinear map between its inputs and output. Certainly, therefore, a linear map such as the LQR can be easily approximated by a fuzzy system (for small values of the inputs to the fuzzy system). Two components of the LQR are the optimal gains and the summer; the optimal gains can be replaced with the normalizing gains of a fuzzy system, and the summer can essentially be incorporated into the rule-base of a fuzzy system. By doing this, we can effectively utilize a fuzzy system implementation to expand

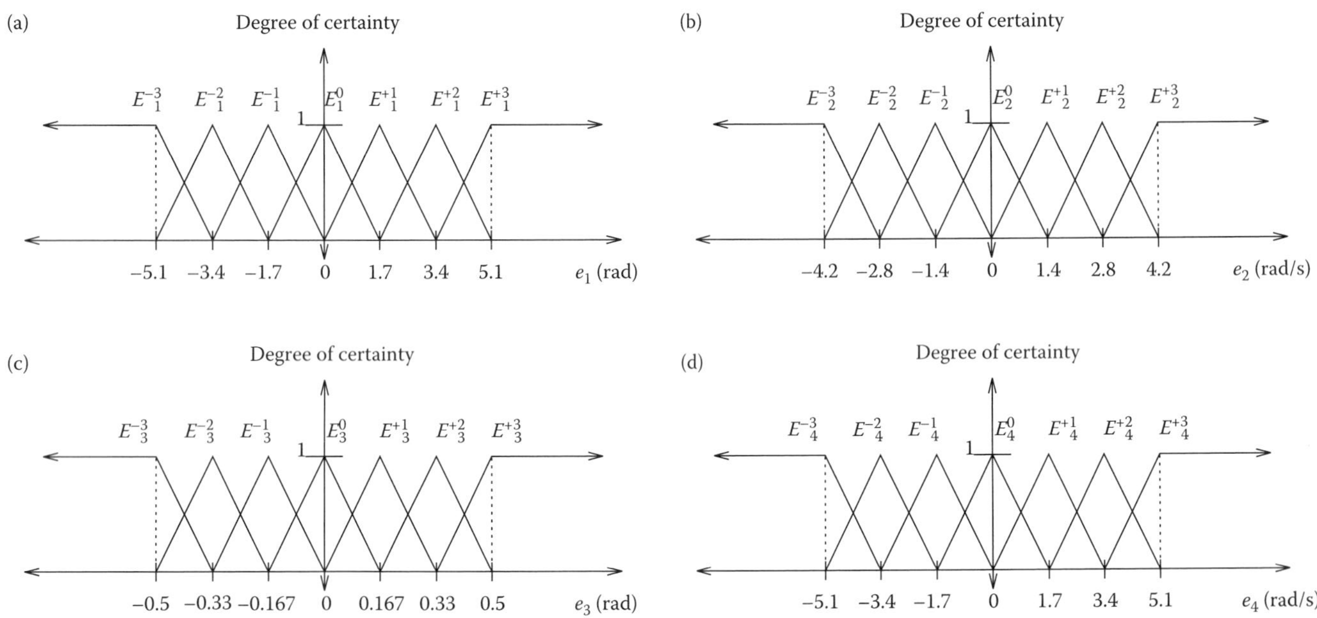

FIGURE 55.3 Four sets of input membership functions: (a) "base position error" ($\tilde{\tilde{e}}_1$), (b) "base derivative error" ($\tilde{\tilde{e}}_2$), (c) "pendulum position error" ($\tilde{\tilde{e}}_3$), and (d) "pendulum derivative error" ($\tilde{\tilde{e}}_4$).

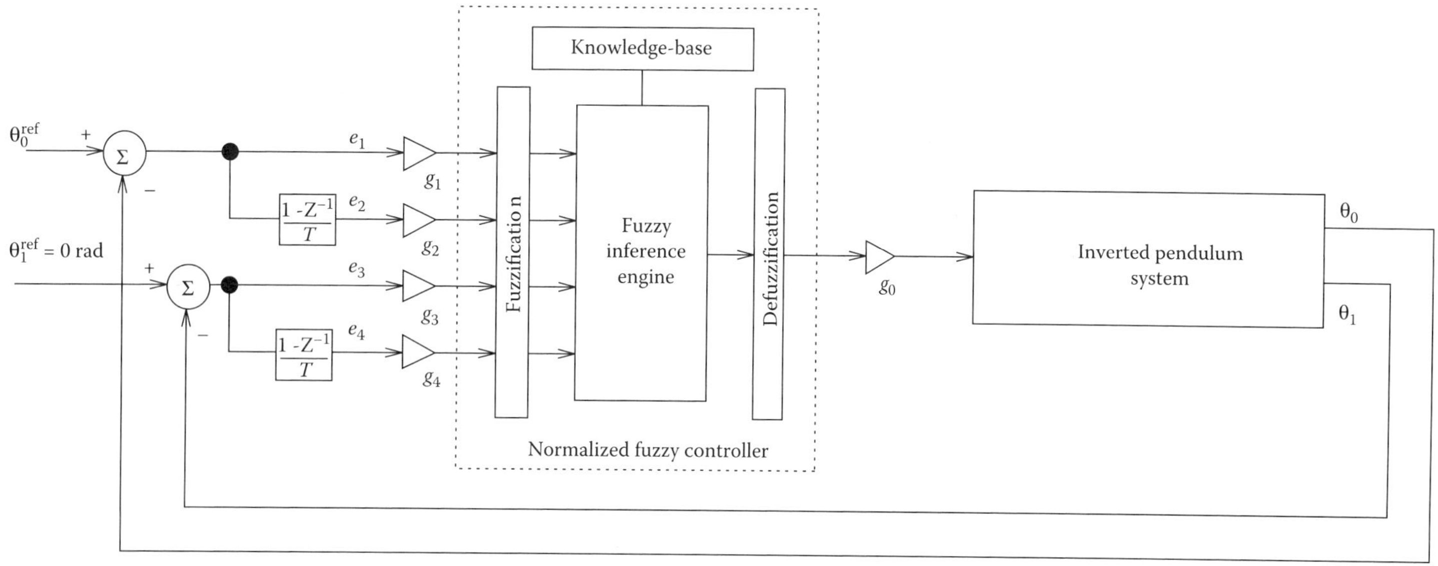

FIGURE 55.4 Block diagram of a direct fuzzy controller.

the region of operation of the controller beyond the "linear region" afforded by the linearization/design process. Intuitively, this is done by making the "gain" of the fuzzy controller match that of the LQR when the fuzzy controller inputs are small, while shaping the nonlinear mapping representing the fuzzy controller for larger inputs (in regions further from zero).

As pointed out in Section 55.2, the rule-base contains information about the relationships between the inputs and output of a fuzzy system. Within the controller structure chosen, recall that we wish to construct the rule-base to perform a weighted sum of the inputs. The summation operation is straightforward; prior to normalization, it is simply a matter or arranging each *If...Then* rule such that the antecedent indices sum to the consequent indices. Specification of the normalizing gains is explained next.

The basic idea [10] in specifying the $g_0 - g_4$ is so that for "small" controller inputs (e_i) the local slope (about zero) of the input–output mapping representing the controller will be the same as the LQR gains (i.e., the k_i). As alluded to above, the normalizing gains $g_1 - g_4$ transform the (symmetric) universes of discourse for each input (see Figure 55.3) to $[-1, 1]$. For example, if $[-\beta_i, \beta_i]$ is the interval of interest for input i, the choice $g_i = 1/\beta_i$ would achieve this normalization, whereas the choice $g_0 = \beta_0$ would map the output of the normalized fuzzy system to the real output to achieve a corresponding interval of $[-\beta_0, \beta_0]$. Then, assuming the fuzzy system provides the summation operation, the "net gain" for the ith input–ouput pair is $g_i g_0$. Finally, therefore, this implies that $g_i g_0 = k_i$ is required to match local slopes of the LQR controller and the fuzzy controller (in the sense of input–output mappings).

We are now in position to summarize the gain selection procedure. Recalling (Section 55.3) that the optimal feedback gains based on the LQR approach are $k_1 = -0.9$, $k_2 = -1.1$, $k_3 = -9.2$, and $k_4 = -0.9$, transformation of the optimal LQR gains into the normalizing gains of the fuzzy system is achieved according to the following simple scheme:

- Choose the controller input which most greatly influences plant behavior and overall control objectives; in our case, we choose the pendulum position θ_1. Subsequently, we specify the operating range of the this input (e.g., the interval $[-0.5, +0.5]$ radians, for which the corresponding normalizing input gain $g_3 = 2$).
- Given g_3, the output gain of the fuzzy controller is calculated according to $g_0 = k_3/g_3 = -4.6$.
- Given the output gain g_0, the remaining input gains can be calculated according to $g_j = k_j/g_0$, where $j \in \{1, 2, 3, 4\}$, $j \neq i$ (note that $i = 3$). For $g_0 = -4.6$, the input gains g_1, g_2, g_3, and g_4 are 0.1957, 0.2391, 2, and 0.1957, respectively.

Determination of the controller output universe of discourse and corresponding normalizing gain is dependent on the structure of the rule-base. A nonlinear mapping can be used to rearrange the output membership functions (in terms of their centers) for several purposes, such as to add higher gain near the center, to create a dead zone near the center, to eliminate discontinuities at the saturation points, and so on. This represents yet another area where intuition (i.e., knowledge about how to best control the process) may be incorporated into the design process. In order to preserve behavior in the "linear" region (i.e., the region near the origin) of the LQR-extended controller, but at the same time provide a smooth transition from the linear region to its extensions (e.g., regions of saturation), we choose an arctangent-type mapping to achieve this rearrangement. Because of the "flatness" of such a mapping near the origin, we expect the fuzzy controller to behave like the LQR when the states are near the process equilibrium.

It is important to note that, unlike the input membership function of Figure 55.3, the output membership functions at the outermost regions of the universe of discourse do *not* include the saturating effect; rather, they return to zero value which is required so that the fuzzy controller mapping is well-defined. In general, for a fuzzy controller with n inputs and one output, the center of the controller output fuzzy set Y^s would be located at where $s = j + k + \cdots + l$ is the index of the output fuzzy set Y^s (and the output linguistic value), $\{j, k, \ldots, l\}$ are the indices of the input fuzzy sets (and linguistic values), N is the number of membership functions on each input, and n is the number of inputs. Note that we must nullify the effect of divisor n by multiplying the output gain g_0 by the same factor.

55.4.2 Performance Evaluation

55.4.2.1 Simulation

Some performance evaluation via simulation is prudent to investigate the effectiveness of the strategies employed in the controller synthesis. Using the complete nonlinear model, the simulated responses of the direct fuzzy controller with seven membership functions on each input indicate that the fuzzy controller successfully balances the pendulum, but with a slightly degraded performance as compared to that of the LQR (e.g., the fuzzy controller produces undesirable high-frequency "chattering" effects over a bandwidth that may not be realistic in implementation).

One way to increase the "resolution" of the fuzzy controller is to increase the number of membership functions. As we increase the number of membership functions on each input to 25, responses using the fuzzy controller become smoother and closer to that of the LQR. Additionally, the control surface of the fuzzy controller also becomes smoother and has a much smaller gain near the center. As a result, the control output of the fuzzy controller is significantly smoother. On the other hand, the direct fuzzy controller, with 25 membership functions on each input comes with increased complexity in design and implementation (e.g., a four-input, one-output fuzzy system with 25 membership functions on each input has $25^4 = 390{,}625$ linguistic rules).

55.4.2.2 Application to Nominal System

Given the experience of the simulation studies, the final step is to implement the fuzzy controller (with seven membership functions on each input) on the actual apparatus. For comparative purposes, we again consider application to the *nominal* system, that is, the pendulum alone with no added weight or disturbances. With the pendulum initialized at its hanging position ($\theta_1 = \pi$), the swing-up control was tuned to give the best swing-up response, as in the case of the LQR results of Section 55.3. The sampling time was set to 10 ms (smaller sampling times produced no significant difference in responses for any of the controllers tested on this apparatus). The only tuning required for the fuzzy control scheme (from simulation to implementation in experimentation) was in adjusting the value for g_3 upward to improve the performance; recall that the gain g_3 is critical in that it essentially determines the other normalizing gains.

Figure 55.5 shows the results for the fuzzy controller on the laboratory apparatus; the top plot shows the base position (angle), the center plot shows the pendulum position (angle), and the bottom plot shows the controller output (motor voltage input). The response is comparable to that of the LQR controller (compare to Figure 55.2), in terms of the pendulum angle (ability to balance in the vertical position). However, some oscillation is noticed (particularly in the controller output, as predicted in simulation studies), but any difference in the ability to balance the pendulum is only slightly discernible in viewing the operation of the system.

55.4.2.3 Application to Perturbed System

When the system experiences disturbances and changes in dynamics (by attaching additional weight to the pendulum endpoint, or by attaching a bottle half filled with liquid), degraded responses are observed for these controllers. Such experiments are also informative for considerations of *robustness* analysis, although here we regard such perturbations on the nominal system as probing the limits of linear controllers (i.e., operating outside the linear region).

As a final evaluation of the performance of the fuzzy controller as developed above, we show results when a container half-filled with water was attached to the pendulum endpoint. This essentially gives a "sloshing liquid" effect, because the additional dynamics associated with the sloshing liquid are easily excited. In addition, the added weight shifted the pendulum's center of mass away from the pivot point; as a result, the natural frequency of the pendulum decreased. Furthermore, the effect

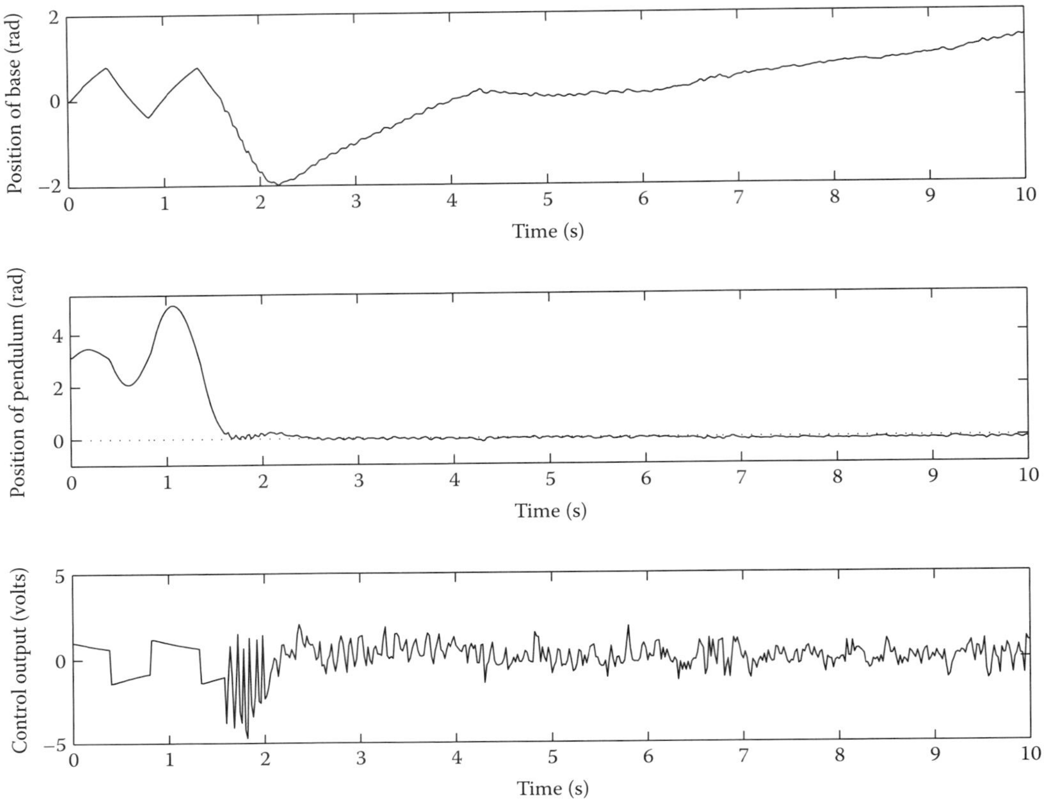

FIGURE 55.5 Direct fuzzy control on the *nominal* system.

of friction becomes less dominant because the inertia of the pendulum increases. These effects obviously come to bear on the balancing controller performance, but also significantly affect the swing-up controller as well. From the present chapter, we note that the swing-up control scheme requires tuning, once additional weight is added to the endpoint, and we refer the interested reader to [8] for details of a *supervisory fuzzy controller* scheme where tuning of the swing-up controller is carried out autonomously.

With the sloshing liquid added to the pendulum endpoint, the LQR controller (and, in fact, other linear control schemes we implemented on this system) produced an unstable response (was unable to balance the pendulum). Of course, the linear control schemes can be tuned to improve the performance *for the perturbed system*, at the expense of degraded performance for the nominal system. Moreover, it is important to note that tuning of the LQR type controller is difficult and *ad hoc* without additional modeling to account for the added dynamics. Such an attempt on this system produced a controller with stable but poor performance.

The fuzzy controller, on the other hand, because of its expanded region of operation (in the sense that it acts like the LQR for small inputs and induces a nonlinearity for larger signals), was able to maintain stability in the presence of the additional dynamics and disturbances caused by the sloshing liquid, *without tuning*. These results are shown in Figure 55.6 where some degradation of controller performance is apparent. Such experiments may also motivate the need for a controller, which can adapt to changing dynamics during operation; this issue is discussed later when we address adaptation in fuzzy controllers.

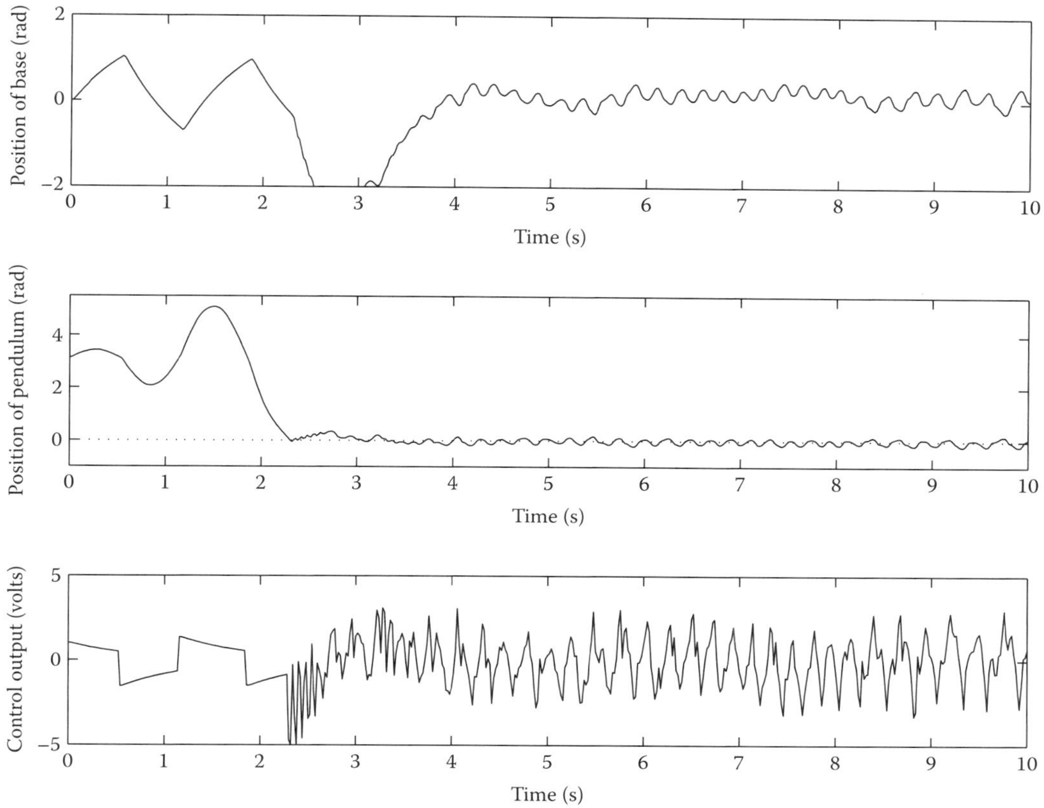

FIGURE 55.6 Direct fuzzy control on the pendulum with sloshing liquid at its endpoint.

55.5 Adaptive Fuzzy Control

55.5.1 Overview

While fuzzy control has, for some applications, emerged as a practical alternative to classical control schemes, there exist rather obvious drawbacks. We will not address all of these drawbacks here (such as stability which is a current research direction); rather, we will focus on an important emerging topical area within the realm of fuzzy control, that of adaptive fuzzy control, as it relates to some of these drawbacks.

The point that is probably most often raised in discussion of controller synthesis using fuzzy logic is that such procedures are usually performed in an *ad hoc* manner, where mechanized synthesis procedures, for the most part, are nonexistent (e.g., it is often not clear exactly how to justify the choices for many controller parameters, such as membership functions, defuzzification strategy, and inference strategy). On the other hand, some mechanized synthesis procedures do exist for particular applications, such as the one discussed above for the balancing control part of the inverted pendulum problem where a conventional LQR scheme was utilized in the fuzzy control design. Typically such procedures arise primarily out of necessity because of system complexity (such as when many inputs and multiple objectives must be achieved). Controller adaptation, in which a form of *automatic controller synthesis* is achieved, is one way of attacking this problem, when no other "direct" synthesis procedure is known.

Another, perhaps equally obvious requirement in the design and operation of any controller (fuzzy or otherwise) involves questions of system robustness. For instance, we illustrated in our theme example that the performance of the direct fuzzy controller constructed for the nominal plant may degrade if significant and unpredictable plant parameter variations, structural changes, or environmental disturbances occur. Clearly, controller adaptation is one way of overcoming these difficulties to achieve reliable controller performance in the presence of unmodeled parameter variations and disturbances.

Some would argue that the solution to such problems is always to incorporate more expertise into the rule-base to enhance the performance; however, there are several limitations to such a philosophy, including: (1) the difficulties in developing (and characterizing in a rule-base) an accurate intuition about how to best compensate for the unpredictable and significant process variations that can occur for all possible process operating conditions; and (2) the complexities of constructing a fuzzy controller that potentially has a large number of membership functions and rules. Experience has shown that it is often possible to tune fuzzy controllers to perform very well if the disturbances are known. Hence, the problem does not result from a lack of basic expertise in the rule-base, but from the fact that there is no facility for automatically redesigning (i.e., retuning) the fuzzy controller so that it can appropriately react to unforeseen situations as they occur.

There have been many techniques introduced for adaptive fuzzy control. For instance, one adaptive fuzzy control strategy that borrows certain ideas from conventional "model reference adaptive control" (MRAC) is called "fuzzy model reference learning control" (FMRLC) [11]. The FMRLC can automatically synthesize a fuzzy controller for the plant and later tune it if there are significant disturbances or process variations. The FMRLC has been successfully applied to an inverted pendulum, a ship-steering problem [11], antiskid brakes [12], reconfigurable control for aircraft [10], and in implementation for a flexible-link robot [13]. Modifications to the basic FMRLC approach have been studied in [14]. The work on the FMRLC and subsequent modifications to it tend to follow the main focus in fuzzy control where one seeks to employ heuristics in control. There are other techniques that take an approach that is more like conventional adaptive control in the sense that a mathematical model of the plant and a Lyapunov-type approach is used to construct the adaptation mechanism. Such work is described in [3]. There are many other "direct" and "indirect" adaptive fuzzy control approaches that have been used in a wide variety of applications (e.g., for scheduling manufacturing systems [15]). The reader should consult the references in the papers cited above for more details.

Another type of system adaptation, where a significant amount and variety of knowledge can be loaded into the rule-base of a fuzzy system to achieve high-performance operation, is the *supervisory fuzzy controller*, a two-level hierarchical controller which uses a higher-level fuzzy system to supervise (coordinate or tune) a lower-level conventional or fuzzy controller. For instance, an expert may know how to control the system very well in one set of operating conditions, but if the system switches to another set of operating conditions, the controller may be required to behave differently to again achieve high-performance operation. A good example is the PID controller which is often designed (tuned) for one set of plant operating conditions, but if the operating conditions change the controller will not be properly tuned. This is such an important problem that there is a significant amount of expertise on how to manually and automatically tune PID controllers. Such expertise may be utilized in the development of a supervisory fuzzy controller, which can observe the performance of a low-level control system and automatically tune the parameters of the PID controller. Many other examples exist of applications where the control engineer may have a significant amount of knowledge about how to tune a controller. One such example is in aircraft control when controller gains are scheduled based on the operating conditions. Fuzzy supervisory controllers have been used as schedulers in such applications. In other applications we may know that conventional or fuzzy controllers need to be switched on based on the operating conditions (see the work in [16] for work on fuzzy supervision of conventional controllers for a flexible-link robot) or a supervisory fuzzy controller may be used to tune an adpative controller (see the work in [10] where a fuzzy supervisor is used to tune an adaptive fuzzy controller that is used as a reconfigurable controller for an aircraft). It is this concept of *monitoring* and *supervising* lower-level controllers (possibly fuzzy, possibly conventional) that defines the supervisory control scheme. Indeed, in this sense supervisory and adaptive systems can be described as special cases of one another.

55.5.2 Auto-Tuning for Pendulum Balancing Control

Many techniques exist for automatically tuning a fuzzy controller in order to meet the objectives mentioned above. One simple technique we present next, studied in [8,14,15], expands on the idea of increasing the "resolution" of the fuzzy controller in terms of the characteristics of the input membership functions. Recall from Section 55.4 for our theme problem that when we increased the number of membership functions on each input to 25, improved performance (and smoother control action) resulted. Likewise, we suspect that increasing the resolution would result in improved performance for the perturbed pendulum case.

To increase the resolution of the direct fuzzy controller with a limited number of membership functions (as before, we will impose a limit of seven), we propose an "auto-tuned fuzzy control." To gain insight on how the auto-tuned fuzzy control works, consider the idea of a "fine controller," with smooth interpolation, achieved using a fuzzy system where the input and output universes of discourse are narrow (i.e., the input normalizing gains are large, and the output gain is small). In this case there are many membership functions on a small portion of the universe of discourse (i.e., "high resolution"). Intuitively, we reason that as the input gains are increased and the output gain is decreased, the fuzzy controller will have better resolution. However, we also conjecture that to obtain the most effective control action the input universes of discourse must also be large enough to avoid saturation. This obviously raises a question of trying to satisfy two opposing objectives. The answer is to adjust the gains based on the current operating states of the system. For example, if the states move closer to the center, then the input universe of discourse should be *compressed* to obtain better resolution, yet still cover all the active states. If the states move away from the center, then the input universe of discourse must expand to cover these states, at the expense of lowering the resolution.

The input–output gains of the fuzzy controller can be tuned periodically (e.g., every 50 samples) using the auto-tuning mechanism shown in Figure 55.7. Ideally, the auto-tuning algorithm should not alter the nominal control algorithm near the center; we therefore do not adjust each input gain independently. We can, however, tune the most significant input gain, and then adjust the rest of the gains based on this gain. For the inverted pendulum system, the most significant controller input is the position error of the pendulum, $e_3 = \theta_1$.

The input–output gains are updated every n_s samples in the following manner:

- Find the maximum e_3 over the most recent n_s samples and denote it by $e_3^{\max}$.
- Set the input gain $g_3 = 1/|e_3^{\max}|$.
- Recalculate the remaining gains using the technique discussed in Section 55.4 so as to preserve the nominal control action near the center.

We note that the larger n_s is, the slower the updating rate is, and that too fast an updating rate may cause instability. Of course, if a large enough buffer were available to store the most recent n_s samples of the input, the gains could be updated at every sample (utilizing an average); here we minimized the usage of memory and opted for the procedure mentioned above (finding the maximum value of e_3).

Simulation tests (with a 50-sample observation window and normalizing gain $g_3 = 2$) reveal that when the fuzzy controller is activated (after swing up), the input gains gradually increase while the output gain decreases, as the pendulum moves closer to its inverted position. As a result, the input and output universes of discourses contract, and the resolution of the fuzzy system increases. As g_3 reaches its maximum value*, the control action near $\theta_1 = 0$ is smoother than that of direct fuzzy control with 25 membership functions (as investigated previously via simulation), and very good balancing performance is achieved.

When turning to actual implementation on the laboratory apparatus, some adjustments were done in order to optimize the performance of the auto-tuning controller. As with the direct fuzzy controller, the value of g_3 was adjusted upward, and the tuning (window) length was increased to 75

* In practice, it is important to constrain the maximum value for g_3 (for our system, to a value of 10) because disturbances and inaccuracies in measurements could have have adverse effects.

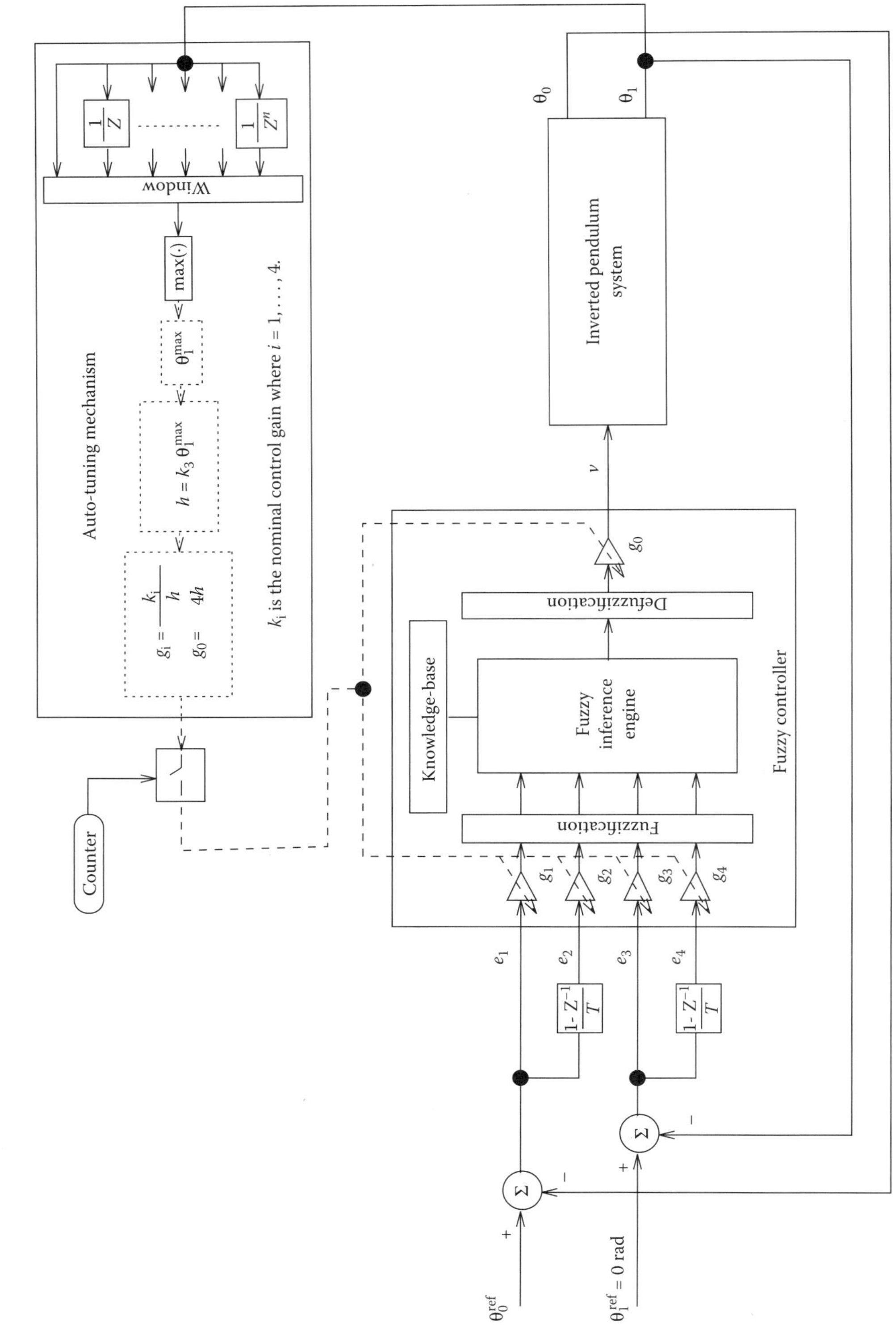

FIGURE 55.7 Auto-tuned fuzzy control.

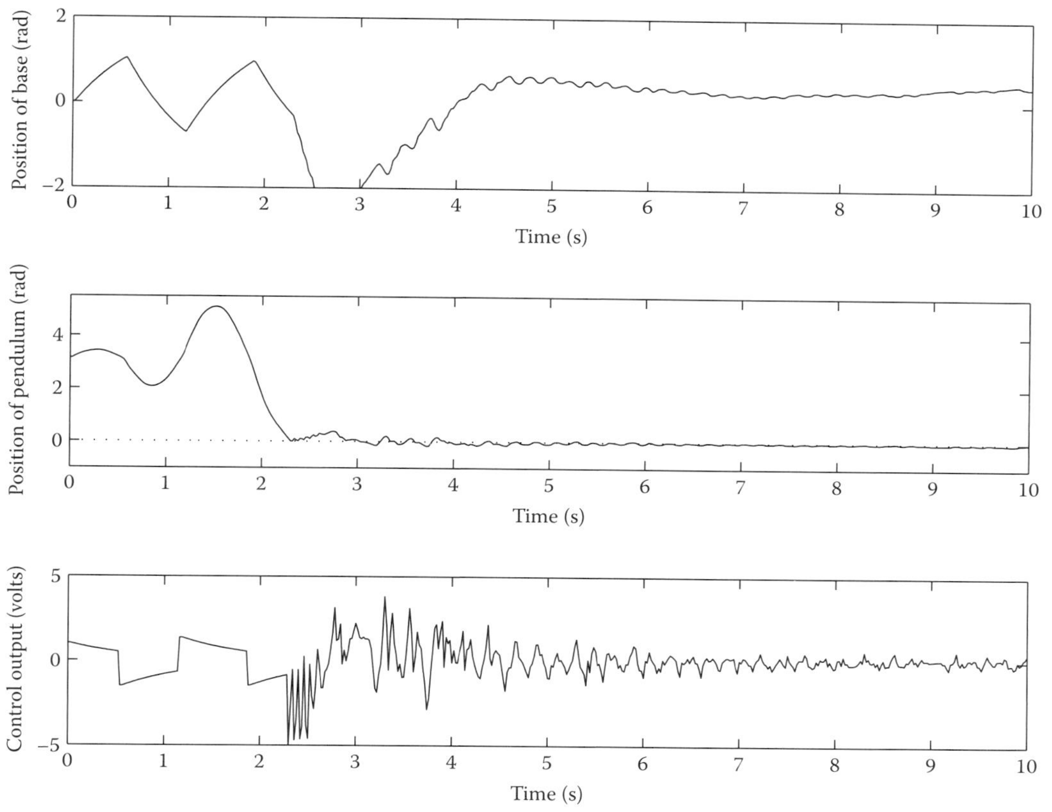

FIGURE 55.8 Auto-tuned fuzzy control on the pendulum with sloshing liquid at its endpoint.

samples. The first test was to apply the scheme to the nominal system; the auto-tuning mechanism improved the response of the direct fuzzy controller (Figure 55.5) by varying the controller resolution online. That is, as the resolution of the fuzzy controller increased over time, the high-frequency effects diminished.

The true test of the adaptive (auto-tuning) mechanism is to evaluate its ability to adapt its controller parameters as the process dynamics change. Once again we investigate the performance when the "sloshing liquid" dynamics (and additional weight) are appended to the endpoint of the pendulum. As expected from simulation exercises, the tuning mechanism, which "stretches" and "compresses" the universes of discourse on the input and output, not only varied the resolution of the controller but also effectively contained and suppressed the disturbances caused by the sloshing liquid, as clearly shown in Figure 55.8.

55.6 Concluding Remarks

We have introduced the general fuzzy controller, shown how to design a fuzzy controller for our rotational inverted pendulum theme problem, and compared its performance to a nominal LQR controller. We overviewed supervisory and adaptive fuzzy control techniques and presented a particular adaptive scheme that worked very effectively for the balancing control of our rotational inverted pendulum theme problem. Throughout the chapter we have emphasized the importance of comparative analysis of fuzzy controllers with conventional controllers with the hope that more attention will be given to detailed engineering cost–benefit analyses rather than the hype that has surrounded fuzzy control in the past.

There are close relationships between fuzzy control and several other intelligent control methods. For instance, expert systems are generalized fuzzy systems since they have a knowledge-base (a generalized version of a rule-base where information in the form of general rules or other representations may be used), an inference mechanism that utilizes more general inference strategies, and are constructed using the same general approach as fuzzy systems. There are close relationships between some types of neural networks (particularly radial basis function neural networks) and fuzzy systems and, while we did not have the space to cover it here, fuzzy systems can be trained with numerical data in the same way that neural networks can (see [3] for more details). Genetic algorithms provide for a stochastic optimization technique and can be useful for computer-aided-design of fuzzy controllers, training fuzzy systems for system identification, or tuning fuzzy controllers in an adaptive control setting.

While there exist such relationships between fuzzy control and other techniques in intelligent control, the exact relationships between all the techniques have not been established. Research along these lines is progressing but will take many years to complete. Another research area that is gaining increasing attention is the nonlinear analysis of fuzzy control systems and the use of comparative analysis of fuzzy control systems and conventional control systems to determine the advantages and disadvantages of fuzzy control. Such work, coupled with the application of fuzzy control to increasingly challenging problems, will help establish the technique as a viable control engineering approach.

55.7 Defining Terms

Rule-base: A part of the fuzzy system that contains a set of If...Then rules that quantify a human expert's knowledge about how to best control a plant. It is a special form of a knowledge-base that only contains If...Then rules that are quantified with fuzzy logic.

Inference Mechanism: A part of the fuzzy system that reasons over the information in the rule-base and decides what actions to take. For the fuzzy system the inference mechanism is implemented with fuzzy logic.

Fuzzification: Converts standard numerical fuzzy system inputs into a fuzzy set that the inference mechanism can operate on.

Defuzzification: Converts the conclusions reached by the inference mechanism (i.e., fuzzy sets) into numeric inputs suitable for the plant.

Supervisory Fuzzy Controller: A two-level hierarchical controller which uses a higher-level fuzzy system to supervise (coordinate or tune) a lower level conventional or fuzzy controller.

Adaptive Fuzzy Controller: An adaptive controller that uses fuzzy systems either in the adaptation mechanism or as the controller that is tuned.

Acknowledgments

This work was supported in part by National Science Foundation Grants IRI 9210332 and EEC 9315257. A large portion of the ideas, concepts, and results reported in this chapter have evolved over several years from the work of many students under the direction of the authors. We, therefore, gratefully acknowledge the contributions of A. Angsana, E. Garcia-Benitez, S. Gyftakis, D. Jenkins, W. Kwong, E. Laukonen, J. Layne, W. Lennon, A. Magan, S. Morris, V. Moudgal, J. Spooner, and M. Widjaja. We would like to, in particular, acknowledge the work of J. Layne and E. Laukonen for their assistance in writing earlier versions of Section 55.2 of this chapter and M. Widjaja who produced the experimental results for the rotational inverted pendulum.

References

1. D. Jenkins and K. Passino, An introduction to nonlinear analysis of fuzzy control systems, *Journal of Intelligent and Fuzzy Systems*, vol. 7, no. 1, pp. 75–103, 1999.
2. K. Passino and S. Yurkovich, *Fuzzy Control*. Menlo Park, CA: Addison Wesley Longman, 1998. This book available for free download at author's Web site.
3. L.-X. Wang, *Adaptive Fuzzy Systems and Control: Design and Stability Analysis*. Englewood Cliffs, NJ: Prentice-Hall, 1994.
4. D. Driankov, H. Hellendoorn, and M. Reinfrank, *An Introduction to Fuzzy Control*. New York, NY: Springer-Verlag, 1993.
5. W. Pedrycz, *Fuzzy Control and Fuzzy Systems*. New York, NY: Wiley, 2nd ed., 1993.
6. G. Klir and T. Folger, *Fuzzy Sets, Uncertainty and Information*. Englewood Cliffs, NJ: Prentice-Hall, 1988.
7. L. Zadeh, Outline of a new approach to the analysis of complex systems and decision processes, *IEEE Transactions on Systems, Man, and Cybernetics*, vol. 3, no. 1, pp. 28–44, 1973.
8. M. Widjaja and S. Yurkovich, Intelligent control for swing up and balancing of an inverted pendulum system, in *Proceedings of the IEEE Conference on Control Applications* (Albany, NY), pp. 534–542, September 1995.
9. M. Spong, Swing up control of the acrobot, in *Proceedings of the IEEE Conference on Robotics and Automation* (San Diego, CA), pp. 2356–2361, May 1994.
10. W. Kwong, K. Passino, E. Laukonen, and S. Yurkovich, Expert supervision of fuzzy learning systems for fault tolerant aircraft control, in *Proceedings of the IEEE, Special Issue on Fuzzy Logic in Engineering Applications*, vol. 83, pp. 466–483, March 1995.
11. J. Layne and K. Passino, Fuzzy model reference learning control for cargo ship steering, *IEEE Control Systems Magazine*, vol. 13, pp. 23–34, 1993.
12. J. Layne, K. Passino, and S. Yurkovich, Fuzzy learning control for anti-skid braking systems, *IEEE Transactions on Control Systems Technology*, vol. 1, pp. 122–129, 1993.
13. V. Moudgal, W. Kwong, K. Passino, and S. Yurkovich, Fuzzy learning control for a flexible-link robot, *IEEE Transactions on Fuzzy Systems*, vol. 3, pp. 199–210, 1995.
14. W. Kwong and K. Passino, Dynamically focused fuzzy learning control, *IEEE Transactions on Systems, Man, and Cybernetics*, vol. 26, pp. 53–74, 1996.
15. A. Angsana and K. Passino, Distributed fuzzy control of flexible manufacturing systems, *IEEE Transactions on Control Systems Technology*, vol. 2, pp. 423–435, 1994.
16. E. Garcia-Benitez, S. Yurkovich, and K. Passino, Rule-based supervisory control of a two-link flexible manipulator, *Journal of Intelligent and Robotic Systems*, vol. 7, no. 2, pp. 195–213, 1993.

Further Reading

Other introductions to fuzzy control are given in [2–5]. The book [2] is available for a free download (as a .pdf file) from http://www.ece.osu.edu/˜passino/. To study other applications of fuzzy control the reader should consult the references in the papers and books cited above. There are several conferences that have sessions or papers that study various aspects of fuzzy control. For instance, see the *Proceedings of FuzzIEEE* (in the World Congress on Computational Intelligence), *IEEE International Symposium on Intelligent Control, Proceedings of the IEEE Conference on Decision and Control, the American Control Conference*, and many others. The best journals to consult are the *IEEE Transactions on Fuzzy Systems, Fuzzy Sets and Systems*, and the *IEEE Transactions on Systems, Man, and Cybernetics*. Also, there are many short courses that are given on the topic of fuzzy control.

56

Neural Control

Marios M. Polycarpou
University of Cyprus

Jay A. Farrell
University of California, Riverside

56.1 Introduction

In control, filtering, and prediction applications the available *a priori* model information is sometimes not sufficiently accurate for a fixed design to satisfy all of the design specifications. In these circumstances, the designer may

- Reduce the specified level of required performance.
- Expend additional efforts to reduce the level of model uncertainty.
- Design a system to adjust itself online to increase its performance level as the accumulated input and output data are used to reduce the amount of model uncertainty during operation.

Controllers (filters and predictors can be implemented by similar techniques) that implement the last approach are the topic of this chapter. Such techniques are suitable when additional *a priori* effort in modeling and validation is unacceptable because of cost and feasibility, and the desired level of performance must be maintained.

Adaptive linear control mechanisms (see [17] and Chapters 34 and 35 of this book) have been developed for systems that, by suitable limitation of their operating domain, can be adequately modeled as linear.

The online system identification and control techniques developed for nonlinear systems can be classified into two categories, parametric and nonparametric. Parametric techniques assume that the functional form of the unknown nonlinearity is known, based on physical modeling principles, but that the model parameters are unknown. Nonparametric techniques are necessary when the functional form of the unknown nonlinearity is itself unknown. In this case a general family of function approximators is selected based on the known properties of the approximation family and the characteristics of the application. Similar to parametric techniques, the objective of the nonparametric approach is still to select a set of parameters to achieve an optimal approximation of the unknown function.

Learning and neural control (and system identification) [13,14,36] are nonparametric techniques developed to enhance the performance of those poorly modeled nonlinear systems for which a suitable model structure is not available. This performance improvement is achieved by exploiting experiences obtained through online interactions with the actual plant. *Adaptive approximation-based control* is an alternative name used in the literature for this class of techniques, and it includes approximators such as neural networks, fuzzy systems and traditional approximation approaches. Neural control is the name applied in the literature when a *neural network* is selected as the function approximator. To be both concise and general, adaptive approximation-based control will be the name used throughout this chapter.

Adaptive approximation-based control has an important role to play in the development of advanced control systems. This class of techniques has become feasible in recent decades due to the rapid advances that have occurred in computing technologies. Inexpensive desktop computing has inspired many *ad hoc* approximation-based control approaches. In addition, similar approaches in different communities (e.g., neural, fuzzy) have been derived and presented using different nomenclature yet nearly identical theoretical results. Our objective in this chapter is to treat such approaches within a unifying framework so that the resulting presentation encompasses a variety of approximation structures with different properties.

The three terms, adaptation, learning, and self-organization, are used with different meanings by different authors in the literature. In this chapter, we will use *adaptation* to refer to temporal changes. For example, adaptive control is applicable when the estimated parameters are slowly varying functions of time. We will use *learning* to refer to methods that retain information as a function of measured variables. Herein, learning is implemented via function approximation. Therefore, learning has a spatial connotation whereas adaptation refers to temporal effects. The process of learning requires adaptation, but the retention of information as a function of other variables in learning implies that learning is a higher level process than is adaptation. Self-organization refers to methods that adapt the function approximation structure during online operation [11,38]. Such methods are not covered in this chapter.

This chapter will focus on the motivation for, and implementation of, adaptive approximation-based control systems. Many of the concepts and results from function approximation and data interpolation are important relative to adaptive approximation-based control. Section 56.2 discusses various properties of function approximators as they relate to adaptive function approximation for control purposes. Section 56.3 presents a few function approximation structures that have been considered for implementation of adaptive approximation-based controllers. Section 56.4 focuses on issues related to parameter estimation. Section 56.5 reviews various nonlinear control system design methodologies. In Section 56.6 we bring together the ideas of function approximation, parameter estimation, and nonlinear control in the design of adaptive approximation-based control systems. Section 56.7 presents briefly the concept of approximate dynamic programming (ADP), which constitutes an important class of neural control schemes that has attracted a lot of attention in the literature. Finally, Section 56.8 contains some concluding remarks.

56.2 Approximator Properties

This section discusses issues and trade-offs related to the selection of a function approximation structure. We consider the situation where a control problem and solution have been defined that include a function $h(z)$. If the function h were known, then the control approach could be directly implemented. When

h is unknown, then the control law is rewritten using $\hat{h}(z;\theta)$, which is intended to approximate the unknown function $h(z)$ well for some optimal value $\theta = \theta^*$, which will be estimated online. In this discussion, the symbol z represents a dummy vector variable that may contain portions of the state vector or exogenous input variables. To proceed, it is necessary that the vector z be known or measured. We are interested in those applications where $h(z)$ is unknown, but we have information available to us that allows approximation of $h(z)$ for $z \in \mathcal{D}$, where $\mathcal{D}$ is a domain of interest.

A few examples are useful to clarify the meaning of the function h and the argument z. For example, $h(z)$ could represent the system dynamics $f(x, u)$ where z would then represent the vector $[x, u]$. In an adaptive gain scheduling approach, $h(z)$ might represent a mapping from the current operating condition represented by z to linear control system parameters.

It is almost always desirable to have theoretical results showing that the control law performance using $\hat{h}(z;\theta)$ is asymptotically equivalent to that of the control law using $h(z)$. From the control theory point of view, it is necessary to show boundedness of all signals in the control loop and asymptotic convergence of certain signals (e.g., tracking errors) toward zero. Preferably, the previously mentioned bounds can be made as small as desired, without increasing the control system gains (i.e., bandwidth). In addition, it is sometimes of interest to analyze whether $\hat{h}(z;\theta)$ converges to $h(z)$ as $t \to \infty$ for all $z \in \mathcal{D}$. It should be noted that the control performance goal can be achieved without convergence of the approximator.

Note that we have converted the function approximation problem to a parameter estimation problem, by assuming a known structure for the approximator. Selecting this structure is a critical design decision.

56.2.1 The Domain of Approximation

Whether convergence of the function approximator is required or not, proper definition of the function approximation problem is important. The optimal function approximator will minimize some measure of the error between $\hat{h}(z;\theta)$ and $h(z)$. Although the function $h(z)$ may be defined for $z \in \Re^m$, the function approximation problem can only be well defined over some compact set denoted by $\mathcal{D}$. For example, the problem of finding

$$\theta^* = argmin_{\theta} J(\theta), \tag{56.1}$$

where

$$J(\theta) = \int_D \|h(z) - \hat{h}(z;\theta)\|^2 \, dz \tag{56.2}$$

is well defined* for compact $\mathcal{D}$ and $h, \hat{h} \in C(\mathcal{D})$. To see the necessity of the constraint that the optimization be over a compact set $\mathcal{D}$, consider the simple problem of finding the optimal constant approximation $\hat{h}(z:\theta) = \theta$ to the quadratic function $h(z) = z^2$. While this problem has a well-defined solution over every compact subset of $\Re$, there is no solution over the set of real numbers $\Re$.

While a few papers fail to state assumptions on the domain of approximation $\mathcal{D}$, the vast majority of the papers in the literature work with a fixed domain of approximation that is defined *a priori*. At this point in time, the authors are unaware of articles that define the region $\mathcal{D}$ during operation in a rigorous fashion, which is an issue closely related to the topic of self-organization of the function approximation structure.

Finally, with the approximator defined over a domain $\mathcal{D}$, the designer must be concerned with two situations related to the time evolution of the argument $z(t)$ as a function of time. First, the initial value $z(0)$ may not be in $\mathcal{D}$. In this case, the designer must ensure that $z(t)$ enters $\mathcal{D}$ in finite time so that the benefits of the adaptive approximation can be realized. Second, once $z(t)$ enters $\mathcal{D}$, the designer must ensure either that it does not leave $\mathcal{D}$ or that it returns to $\mathcal{D}$ in finite time [32].

* The notation $C(\mathcal{D})$ is read as "the set of functions continuous on domain $\mathcal{D}$."

56.2.2 Approximation Error

When $h(z) \in C(\mathcal{D})$ and $\hat{h}(z;\theta) \in C(\mathcal{D})$ for all θ, then for any θ, the approximation error

$$\epsilon(z;\theta) = h(z) - \hat{h}(z;\theta) \tag{56.3}$$

is continuous and bound for $z \in \mathcal{D}$. Therefore, the minimization defined in Equation 56.2 is well defined and the optimal error function

$$\epsilon^*(z) = \epsilon(z;\theta^*) = h(z) - \hat{h}(z;\theta^*) \tag{56.4}$$

is also a continuous bounded function for $z \in \mathcal{D}$. The function $\epsilon^*(z)$ will be referred to as the minimum function approximation error (MFAE).

Many articles in neural control include an assumption similar to the following.

Assumption 56.1:

For the functions $h(z) \in C(\mathcal{D})$ and $\hat{h}(z;\theta) \in C(\mathcal{D})$, there exists $\theta^* \in \Re^p$ and $\bar{\epsilon} > 0$ such that $\|\epsilon(z;\theta^*)\| \leq \bar{\epsilon}$, $\forall z \in \mathcal{D}$.

The boundedness of the approximation error—for a continuous function $h(z)$, a continuous approximator $\hat{h}(z;\theta)$, and a compact domain $\mathcal{D}$—is a property of the problem statement. The assumption is only giving a name $\bar{\epsilon}$ to the bound. Ideally, the designer expects the bound $\bar{\epsilon}$ to be small, but the size of the bound will be directly affected by the designer's choice of the function approximation structure $\hat{h}(z;\theta)$. Because the approximation error may significantly affect the control performance, the choice of the approximation structure is very important.

56.2.3 The Context

Function approximation problems are well studied, especially in the case where $h(z)$ and $\hat{h}(z;\theta)$ are known analytically and $\mathcal{D}$ is given. Examples include the decomposition of a known function into its Fourier or Taylor series representation. Such problems are often studied in purely mathematical contexts [22,29].

There is also vast literature related to the development of functions to either interpolate or approximate a given batch of data. Online approximation applications are different in that the data are not a fixed batch, but are arriving continuously as the system operates. In addition, there is no guarantee that the data that arrive will represent the region $\mathcal{D}$ in a statistically uniform sense. Such problems are sometimes referred to as incremental, scattered data, approximation problems.

In the applications that are of interest to this chapter, $h(z)$ is not known as an analytic formula; therefore, Equation 56.2 cannot be optimized analytically. Instead, a typical approach is to optimize a cost function based on online measurements that depend on the function (i.e., $h(z(t))$), where $z(t)$ is defined by the system trajectory. When the values of $h(z(t))$ are available as a function of time, the cost function is effectively

$$J(\theta) = \frac{1}{t}\int_0^t \left\| h(z(\tau)) - \hat{h}(z(\tau);\theta) \right\|^2 d\tau. \tag{56.5}$$

Note, that while the cost function of Equation 56.2 is uniformily weighted over the region $\mathcal{D}$, the cost function of Equation 56.5 is biased toward those portions of $\mathcal{D}$ wherein the system operates most often. Therefore, the two cost functions can have significantly different minimizing values of θ.

Finally, it is often the case that the values of $h(z(t))$ are not directly available; therefore, even Equation 56.5 cannot be directly evaluated. Instead, adjustments to the parameter estimate $\hat{\theta}$ will be inferred, possibly based on stability-based formulations using the tracking error or identification error.

56.2.4 Active and Passive Learning

The fact that the sample locations $z(t) \in \mathcal{D}$ are determined by the system trajectory leads to an interesting distinction between active and passive learning scenarios. *Active learning* describes those situations in which the designer has the freedom to control the training sample distribution. *Passive learning* describes those situations in which the training sample density is defined by some external mechanism. Online applications usually involve passive learning because the plant is performing some useful function.

When active learning is possible, the designer will have the additional task to define the trajectory $z(t)$ to appropriately trade-off control performance relative to achieving accurate function approximation over $\mathcal{D}$.

56.2.5 Local versus Global Models

In the selection of a suitable model structure, it is useful to consider the domain over which the model is expected to apply. In particular, the following two definitions are of interest.

Definition 56.1: Local Approximation Structure

A parametric model $\hat{h}(z;\theta)$ is a local approximation to $h(z)$ at $z_0 \in \mathcal{D}$ if, for any $\varepsilon > 0$, there exist θ and δ so that $\|h(z) - \hat{h}(z;\theta)\| \leq \varepsilon$ for all $z \in \mathcal{B}(z_0, \delta)$ where $\mathcal{B}(z_0, \delta) = \{z \mid \|z - z_0\| < \delta\}$.

Definition 56.2: Global Approximation Structure

A parametric model $\hat{h}(z;\theta)$ is a global approximation to $h(z)$ over domain $\mathcal{D}$ if for any $\varepsilon > 0$ there exists θ so that $\|h(z) - \hat{h}(z;\theta)\| \leq \varepsilon$ for all $z \in \mathcal{D}$.

The following items are of interest relative to the definition of local and global models:

- Physical models derived from first principles are expected to be global models (i.e., valid over the domain of operation $\mathcal{D}$).
- Whether a given approximation structure is local or global depends on the system that is being modeled and the domain $\mathcal{D}$.
- If a parameter vector θ exists satisfying Definition 56.2 for a particular ε, then this θ also satisfies Definition 56.1 for the same ε at each $x_0 \in \mathcal{D}$. Therefore, the set of global models is a strict subset of the set of local models.
- As the desired level of accuracy increases (i.e., ε decreases), the dimension of the parameter vector or the complexity of the model structure will usually have to increase.

56.2.6 Adaptation versus Learning

As the operating point $z(t)$ moves through $\mathcal{D}$, a local approximation structure could maintain accuracy in the vicinity of the operating point by adjusting the parameter vector through time (i.e., adaptation). Alternatively, the parameter vector could be stored as a function of the operating point (i.e., learning). The former approach is typical of linear adaptive control methodologies in nonlinear control applications. The latter approach is one implementation of learning control [12] that constructs a global approximation structure by connecting several local approximating structures together in a smooth fashion [3].

The conceptual differences between adaptive linear control and learning control are of interest. In a nonlinear application, adaptive linear control methodologies can be developed to maintain a locally accurate model or control law at the current operating point z_0. With the assumption that the linearized model will change over time, as the operating point changes, the adaptive mechanism will adjust the parameter estimates to maintain the accuracy of the local fit. Such an approach can provide satisfactory performance if the locally linearized model changes slowly, but will have to estimate the same model parameters repetitively if the operating point moves repetitively over a set of standard operating points. A more ambitious proposal is to develop a global model of the dynamics that stores model information locally as a function of operating condition. One mechanism for this approach is to store the local linear models or controllers as a function of the operating point.

For example,

$$\hat{h}(z;\theta) = \sum_{i=1}^{m} \left(A_i(z - z_i) + B_i\right) w\left(\frac{\|z - z_i\|}{\mu}\right), \tag{56.6}$$

where $\mu > 0$; $w(\lambda) : \Re^+ \mapsto \Re^+$ is continuous and positive function that is non-zero only for $\lambda \in [0, 1]$ with $\sum_{i=1}^{m} w\left(\frac{\|z-z_i\|}{\mu}\right) = 1$ for all $z \in \mathcal{D}$; and $A_i \in \Re^{n\times n}$ and $B_i \in \Re^n$ are unknown parameters. Given these definitions, with appropriately chosen parameters, the function $A_i(z - z_i) + B_i$ is the linearized model at operating point z_i. The support of each linearized model is $S_i = \mathcal{B}(z_i, \mu)$. By the assumptions stated, $\mathcal{D} \subset \cup_{i=1}^{m} S_i$. If μ is selected small enough so that each linear model achieves approximation accuracy ε over its support region, then the approximator $\hat{h}$ defined in Equation 56.6 will achieve accuracy ε over $\mathcal{D}$. If a parameter adjustment mechanism is designed such that A_i and B_i are only adjusted for $z \in S_i$, then the model will be adjusted locally and retained locally; therefore, learning (i.e., retention of information within a recallable context) will have been achieved.

The name *Learning Control* stems from the idea that past performance information is used to estimate and store information (relevant to an operating point) for use when the system operates in the vicinity of the same operating point in the future. Learning Control requires more extensive use of computer memory and computational effort than adaptive or fixed parameter control. The required amount of memory and computation will be determined predominantly by the selection of function approximation and parameter estimation (training) techniques.

56.2.7 Universal Approximation

Given the problem of approximating the unknown function $h(z) \in C(\mathcal{D})$ over the compact domain $\mathcal{D}$, the structure of the approximator $\hat{h}(z;\theta)$ should be carefully considered. Ideally, the designer should know that if the dimension of θ is increased, in some appropriate way, then it is possible to make

$$\max_{z\in\mathcal{D}} |\epsilon^*(z)| = \max_{z\in\mathcal{D}} |h(z) - \hat{h}(z;\theta^*)|$$

sufficiently small. Such results are referred to as *universal approximation* (UA) theorems and the approximators that satisfy the theorems are referred to as *universal approximators.*

Several UA results have been derived in the literature, for example, [16,26]. A typical result is presented below for discussion. Prior to the statement of the theorem, we require a definition for $\Sigma\Pi$ networks.

Definition 56.3:

The family of r input, N node, single hidden layer ($\Sigma\Pi$) network approximators associated with nodal processor $g(\cdot)$ is defined by

$$S_{rN} = \left\{ h(z;\theta,w,b) = \sum_{i=1}^{N} \theta_i \prod_{j=1}^{q} g\left(w_{ij}^{\top} z + b_{ij}\right) \right\}.$$

Theorem 56.1: [16]

Let $\mathcal{D} \subset \Re^r$ be compact and $g : \Re \mapsto \Re$ be any continuous nonconstant function. Then, the set S_{rN} with nodal processor g has the property that for any $h \in C(\mathcal{D})$ and any $\varepsilon > 0$, there exists N (sufficiently large) such that $\max_{z\in\mathcal{D}} |\epsilon^*(z)| < \varepsilon$.

This theorem indicates that if g is a nodal function satisfying certain technical assumptions and ε is a specified approximation accuracy, then any given continuous function can be approximated over the compact set $\mathcal{D}$ to the desired degree of accuracy if the number of nodes N is sufficiently large. Such UA results are powerful, but should be interpreted with caution.

First, such theorems are very general. They do not state that any specific type of neural network has superior abilities for function approximation. In fact, the constraints on the nodal function g are quite lax.

Second, UA requires that the number of nodes per layer be expandable. In most applications, the number of nodes N is fixed *a priori.* Once each N is fixed, the network can no longer approximate an arbitrary continuous function to a specified accuracy ε; hence, approximation structures that allow criteria for the network structure specification are beneficial. Also, UA is achieved by making N "sufficiently" large. Starting *a priori* with a small N defeats the objective of using a universal approximator.

Third, UA results state existence. They are not constructive. In particular, neither the required value of N for a particular nodal processor, nor the approximator structure are defined by the theorems.

56.2.8 Generalization

For the applications discussed, the parameters of the approximation structure will be estimated using the training set $\mathcal{T}_t = \{z(\tau) \text{ for } \tau \in [0, t]\}$ defined by the system trajectory. However, the approximating function may be evaluated at any point $z \in D$. Therefore, it is desirable that the resulting approximation converge to a form that is accurate throughout $\mathcal{D}$, not only at the points $z \in \mathcal{T}_t$. The ability of parametric approximators to provide answers—good or bad—at points outside the training set is referred to in the neural network literature as *generalization.*

The fact that an approximation structure is capable of *generalizing from the training data* is not necessarily a beneficial feature. Note that the approximator will output results $\hat{h}(z;\theta)$ at any evaluation point z. The user must take appropriate steps to verify or ensure the accuracy of the resulting approximation at that point. Families of approximators that allow this assessment to occur online are desirable.

Generalization of training data may also be described as interpolation or extrapolation of the training data. Interpolation refers to the process of filling in the holes between nearby training samples. Interpolation is a desirable form of generalization. Most approximators interpolate well. Some approximators will allow online analysis of the interpolation accuracy. Extrapolation refers to the process of providing answers in regions of the learning domain $\mathcal{D}$ that the training set $\mathcal{T}_t$ does not adequately represent. Extrapolation from the training data is risky when the functional form is unknown by assumption. Clearly, it is desirable to know whether the approximator is interpolating between the training data or extrapolating from the training data.

56.2.9 Parameter (Non)Linearity

A general representation of function approximators is

$$\hat{h}(z;\theta,\sigma) = \phi(z;\sigma)^\top\theta, \tag{56.7}$$

where the adjustable parameters are represented by θ, which appears linearly, and σ, which appears nonlinearly. For example, in the approximator

$$\hat{h}(z;\theta,\sigma) = \sum_{i=1}^{N} \theta_i \exp\left(-\left(z-\sigma_i\right)^2\right). \tag{56.8}$$

the coefficients θ_i appear linearly while the center location parameters σ_i appear nonlinearly. The design must determine whether to adjust both sets of parameters or just those that appear linearly. If all the nonlinear parameters are fixed during the design stage, then a linear-in-the-parameters (LIP) approximator is obtained. Nonlinear in the parameter approximators are more flexible, but also are more difficult to analyze theoretically and to tune in an online fashion.

The relative drawbacks of LIP approximators are discussed, for example, in [4] where it is shown that under certain technical assumptions, nonlinear in the parameter approximators have squared errors of order $\mathcal{O}\left(1/N\right)$, whereas LIP approximators cannot have approximation errors less than $\mathcal{O}\left(1/N^{2/n}\right)$ (N is the number of parameters and n is the dimension of $\mathcal{D}$).

Function approximators that are LIP can be represented as

$$\hat{h}(z;\theta) = \phi(z)^\top\theta, \tag{56.9}$$

where θ is the unknown parameter vector, and the regressor vector $\phi(z)$ is a known, usually nonlinear, vector function of z. The powerful parameter estimation and performance analysis techniques that exist for LIP approximators (see [15]) make this a beneficial property.

By substituting Equation 56.3 into Equation 56.2, we obtain

$$J(\theta) = \int_{\mathcal{D}} \|\epsilon(z;\theta)\|^2\, dz, \tag{56.10}$$

which is strictly convex in ϵ. For LIP approximators, defining $\tilde{\theta} = \theta - \theta^*$, by Equations 56.3 and 56.4 it is clear that

$$\epsilon(z;\theta) = -\phi(z)^\top\tilde{\theta} + \epsilon^*(z),$$

which is linear in the parameter error vector. Therefore, $J(\theta)$ is convex in the parameter error vector. Note that for a fixed value of z, there is a linear subspace of values Θ defined as $\Theta = \{\theta : \phi(z)^\top\theta = 0\}$. If z varies sufficiently, then this subspace will shrink down to the single point θ^*. The fact that J is convex in ϵ, which is linear in θ, ensures that Θ is the unique equivalence set of global minima. When the parameters appear in a nonlinear fashion, then numerous local minima may also exist.

As discussed in Section 56.2.3, when a sample error cost function such as Equation 56.5 is used in place of the cost function of Equation 56.2, the optimal parameter estimate that results will depend on the distribution of the training samples in the training set $\mathcal{T}_t$. The fact that different parameter estimates result from different sets of training samples is *not* the result of multiple local minima; it is, instead, the result of different weightings by the sample distribution in the cost function.

56.2.10 Coverage

Coverage ensures that the value of the approximator can be adjusted at any point $z \in \mathcal{D}$. This property can be stated formally as follows: for any $z \in D$, there exists at least one θ_j such that $|\partial h(z;\theta_j)/\partial\theta| \neq 0$. If this property does not apply, then there exists some point $z \in D$ for which the approximation cannot be changed. This is obviously an undesirable situation.

56.2.11 Localization

The localization property is stated formally as follows: for θ_j, if $\left|\partial h(z;\theta)/\partial\theta_j\right|$ is nonzero in the vicinity of $z_o \in \mathcal{D}$, then it must be zero outside the ball $\mathcal{B}(z_o,\delta)$ for some $\delta > 0$. The smallest such δ is the radius of support.

With the localization property, the effects of changing any single parameter are limited to radius of support of that parameter. Thus, experience and consequent learning in one part of the input domain cannot positively nor negatively affect the knowledge previously accrued in other parts of the mapping domain. A polynomial or Fourier series approximation does not have the localization property. It is easy to see that adjusting any parameter of such an approximator will affect the approximation throughout $\mathcal{D}$.

The fact that a limited subset of the model parameters is relevant for a given point in the learning domain also makes it possible to reduce significantly the required amount of computation per sample interval. The trade-off is that an approximator with the localization property may require a higher number of parameters (more memory) than an approximator without the property. For example, if the domain $\mathcal{D}$ has dimension d, and m parameters are necessary for a given local approximator per input dimensions, then on the order of m^d parameters will generally be required to approximate the $\mathcal{D}$ dimensional function. This exponential increase in the memory requirements with input dimension is referred to in the literature as *the curse of dimensionality*.*

56.3 Function Approximators

Examples of several classes of function approximators are given in this section. This list of examples is not meant to be comprehensive. Many additional classes of approximators exist. See for example [13].

56.3.1 Sigmoidal Neural Networks

A strict mathematical definition of what is or is not a neural network does not exist; however, the class of sigmoidal neural networks has become widely accepted as function approximators.

Single hidden-layer neural networks have the form

$$\hat{h}(z) = \sum_{j=1}^{N} \theta_j \phi(\sigma_{j0} - [\sigma_{j1}, \ldots, \sigma_{jM}]z), \tag{56.11}$$

where usually both the linear parameters $\theta \in \Re^N$ and the nonlinear parameters $\sigma \in \Re^{N\times(M+1)}$ are adjusted, and the nodal processor ϕ is a scalar function of a single variable. Most often, the nodal processor has a sigmoidal shape similar to $\phi(v) = atan(v)$ or $\phi(v) = 1/(1+e^{-v})$.

Multiple layer networks are constructed by cascading single layer networks together, with the output from one network serving as the input to the next.

56.3.2 Orthogonal Polynomial Series

A set of polynomial functions $\phi_i(z), i = 1, 2, \ldots$ is orthogonal with respect to a nonnegative weight function $w(t)$ over the interval $[a, b]$, if

$$\int_a^b w(z)\phi_i(z)\phi_j(z)\,dz = \begin{cases} r_i, & i = j, \\ 0, & i \neq j, \end{cases}$$

for some nonzero constants r_i. When a function $h(z)$ satisfies certain integrability conditions, it can be expanded as

$$h(z) = \sum_{n=0}^{\infty} \theta_n \phi_n(z),$$

* The term *curse of dimensionality* was originally used by Bellman to refer to the increasing complexity of the solution to dynamic programming problems with increasing input dimension.

where

$$\theta_n = \frac{1}{r_n}\int_a^b w(z)h(z)\phi_j(z)\,dz.$$

The mth order finite approximation of $h(z)$ with respect to the polynomial series $\phi_i(z)$ is given by

$$\hat{h}(z) = \sum_{n=0}^{m-1} \theta_n\phi_n(z). \tag{56.12}$$

The integral of the error between $h(z)$ and $\hat{h}(z)$ over $[a, b]$ converges to zero as m approaches ∞. When $h(z)$ is unknown, it may be reasonable to approximate the θ_n in Equation 56.12 using online data.

56.3.3 Localized Basis Influence Functions

Due to the usefulness of the interconnection of local models to generate global models, the class of *Basis Influence Functions* [25] is presented. The definition is followed by examples of several approximation architectures that satisfy the definition. The purpose of the examples is to demonstrate the concept and to illustrate that several approximators often discussed independently can be studied as a class within the setting of the definition.

Definition 56.4: Localized-Basis Influence Functions

A function approximator is of the BI Class if, and only if, it can be written as

$$\hat{h}(z;\hat{\theta}) = \sum_i h_i(z;\hat{\theta}, z_i)\Gamma_i(z; z_i), \tag{56.13}$$

where each $h_i(z;\hat{\theta}, z_i)$ is a local approximation to $h(z)$ for all $z \in \mathcal{B}(z_i, \delta)$; $\Gamma(z; z_i)$ has local support $S_i = \{z : \Gamma(z; z_i) \neq 0\}$, which is a subset of $\mathcal{B}(z_i, \delta)$; and $\mathcal{D} \subseteq \bigcup_i S_i$.

In this framework, the $h_i(z;\hat{\theta}, z_i)$ are the basis functions and the $\Gamma_i(z; z_i)$ are the influence functions. It is interesting to note the similarity between BI class approximations, as described above, and some fuzzy approximators. In the language of fuzzy systems, a basis function is called a *rule* and an influence function is called a *membership function*.

BOXES. One of the simplest approximation structures that satisfies Definition 56.4 is the piecewise constant function

$$h(z;\theta) = \sum_i \theta_i\Gamma_i,$$

where

$$\Gamma_i = \begin{cases} 1, & \text{if } z \in D_i, \\ 0, & \text{otherwise,} \end{cases}$$

where $\cup_i D_i = \mathcal{D}$ and $D_i \cap D_j = \emptyset$. Often, the D_i are selected in a rectangular grid covering $\mathcal{D}$. In this case, very efficient code can be written to calculate the approximation and to update its parameters. Generalizations of this concept include using more complex basis elements (e.g., parameterized linear functions) in place of the constants θ_i, and interpolating between several neighboring D_i to produce a more continuous approximation. A disadvantage is that the influence functions are not continuous.

56.3.4 Spline-Like Functions

Interpolating between regions in the BOXES approach can result in spline functions, such as (for a one dimensional domain)

$$h(z;\theta) = \lambda_i\theta_i + (1-\lambda_i)\theta_{i+1},$$

where

$$\lambda_i = \frac{z_{i+1} - z}{z_{i+1} - z_i} \tag{56.14}$$

for $z_{i+1} > z \geq z_i$ and θ_i is the value of $h(z;\theta)$ at z_i. When the knots (i.e., the z_i for $i = 1, \ldots, N$) are equally spaced, Equation 56.14 can be rewritten to satisfy Definition 56.4 with constant basis functions and influence functions of the form

$$\Gamma_i(z) = \begin{cases} 1 - \dfrac{|z - z_i|}{z_{i+1} - z_i}, & \text{if } z_{i-1} \leq z \leq z_{i+1}, \\ 0, & \text{otherwise.} \end{cases} \tag{56.15}$$

Note that such schemes can be extended to both multidimensional inputs and outputs, at the expense of increased memory and computational requirements. For example, a multidimensional output could be the state feedback gain as a function of operating condition.

56.3.5 Radial Basis Functions

An approximation structure defined as

$$\hat{h}(z,\hat{\theta}) = \sum_i \theta_i \Gamma(z;z_i),$$

where $\Gamma(z;z_i)$ is a function such as $exp(-\|z - z_i\|^2)$, which changes with the distance of z from z_i, is called a Radial Basis Function (RBF) Approximator. In the case where $\Gamma(z;z_i)$ has only local support, the RBF also satisfies Definition 56.4, with the basis elements being constant functions. Of course, the basis functions can again be generalized to be more inclusive functions, at the expense of more involved computations.

56.4 Parameter Estimation

Section 56.2 has presented various key properties of function approximators and Section 56.3 described several neural and other approximation structures that can be used to adaptively approximate unknown nonlinearities in uncertain dynamical systems. The present section addresses the problem of designing and analyzing parameter estimation algorithms with certain stability and robustness properties.

Unlike standard parameter estimation problems, in the case of nonlinear dynamical systems with unknown nonlinearities, there is a need to first derive a suitable parametric model before designing and analyzing the learning scheme. The overall procedure is referred to as adaptive approximation problem, which is summarized as follows.

56.4.1 Adaptive Approximation Problem

Given an input/output system containing unknown nonlinear functions, the adaptive approximation problem deals with the design of online learning schemes and parameter adaptive laws for approximating the unknown nonlinearities.

The overall design procedure for solving the adaptive approximation problem consists of the following three steps:

1. Derive a *parametric model* by rewriting the dynamical system in the form
$$\chi(t) = W(s)[\hat{f}(z(t); \theta^*, \sigma^*)] + \delta(t), \tag{56.16}$$
where $\chi(t) \in \Re^n$ is a vector that can be computed from available signals, $W(s)$ is a known transfer function (in the Laplace s-domain) of dimension $n \times p$, the vector function $\hat{f} : \Re^m \times \Re^{q_\theta} \times \Re^{q_\sigma} \mapsto \Re^p$ represents an adaptive approximator, $z(t) \in \Re^m$ is the input to the adaptive approximator, $\theta^* \in \Re^{q_\theta}$ and $\sigma^* \in \Re^{q_\sigma}$ are unknown "optimal" weights for the adaptive approximator, and $\delta(t) \in \Re^n$ is a possibly filtered version of the MFAE $\epsilon_f^*(x(t))$.
2. Design a *learning scheme* of the form
$$\hat{\chi}(t) = \mathcal{L}(z(t), \hat{\theta}(t), \hat{\sigma}(t)),$$
$$e(t) = \hat{\chi}(t) - \chi(t),$$
where $\hat{\theta}(t), \hat{\sigma}(t)$ are adjustable weights of the adaptive approximator, $\mathcal{L}$ is the structure of the learning scheme, and $\hat{\chi}(t)$ is an estimate of $\chi(t)$, which is used to generate the output estimation error $e(t)$. The output estimation error $e(t)$ provides a measure of how well the estimator approximates the unknown nonlinearities, and therefore is utilized in updating the parameter adaptive laws.
3. Design a *parameter adaptive law* for updating $\hat{\theta}(t)$ and $\hat{\sigma}(t)$, of the form
$$\dot{\hat{\theta}}(t) = \mathcal{A}_\theta(z(t), \chi(t), \hat{\chi}(t), \hat{\theta}(t)),$$
$$\dot{\hat{\sigma}}(t) = \mathcal{A}_\sigma(z(t), \chi(t), \hat{\chi}(t), \hat{\sigma}(t)),$$
where $\mathcal{A}_\theta$ and $\mathcal{A}_\sigma$ represent the right-hand side of the adaptive law for $\hat{\theta}(t)$ and $\hat{\sigma}(t)$), respectively.

The role of the filter $W(s)$ will become clear in the subsequent presentation. For some applications, the form of the filter $W(s)$ is imposed by the structure of the problem. In other applications, the structure of the problem may purposefully be manipulated to insert the filter in order to take advantage of its beneficial noise reduction properties.

The analysis of the learning scheme consists of proving (under reasonable assumptions) the following properties:

Stable Adaptation Property: In the case of zero mismatch error (i.e., $\delta(t) = 0$), the estimation error $e(t) = \hat{\chi}(t) - \chi(t)$ remains bounded and asymptotically approaches zero (or a small neighborhood of zero).

Stable Learning Property: In the case of zero mismatch error (i.e., $\delta(t) = 0$), the function approximation error $\hat{f}(z(t), \hat{\theta}(t), \hat{\sigma}(t)) - \hat{f}((z(t), \theta^*, \sigma^*)$ remains bounded for all z in some domain of interest $\mathcal{D}$ and asymptotically approaches zero (or is asymptotically less than some threshold ϵ over $\mathcal{D}$).

Robust Adaptive and Learning Properties: In the case of nonzero mismatch error (i.e., $\delta(t) \neq 0$), the function approximation error $\hat{f}(z(t), \hat{\theta}(t), \hat{\sigma}(t)) - \hat{f}((z(t), \theta^*, \sigma^*)$, and the estimation error $e(t) = \hat{\chi}(t) - \chi(t)$ remain bounded for all z in some domain of interest $\mathcal{D}$ and satisfy a small-in-the-mean-square property [17] with respect to the magnitude of the mismatch error.

It is important for the reader to note that the parameter estimation methodologies for neural control and adaptive approximation-based control did not develop in a research vacuum, but they are the extension of a large number of parameter estimation results. Parametric estimation is a well-established field in science and engineering since it is one of the key components in developing models from observations. Several books are available for parameter estimation in the context of system identification [21], adaptive control [17] and time series analysis [10]. The parameter estimation methods presented herein are based on these standard estimation techniques but with special emphasis on the adaptive approximation problem.

A significant number of results have been developed for offline parameter estimation, where all the data are first collected and then processed to fit an assumed model. Both frequency and time-domain approaches can be used, depending on the nature of the input–output data. Moreover, stochastic techniques have been extensively used to deal with measurement noise and other types of uncertainty. A key component in offline parameter estimation is the selection of the norm, which determines the objective function to be minimized.

Adaptive approximation-based control requires the use of online parameter estimation methods; that is, techniques that are based on the idea of first choosing an initial estimate for the unknown parameter vector, then recursively updating the estimate based on the current set of measurements. One of the key characteristics of online parameter estimation methods is that as streaming data become available in real-time, it is processed, via updating the of parameter estimates, and then discarded. Therefore, the presented techniques require no data storage during real-time processing applications, except possibly for some buffering window that can be used to filter measurement noise. In general, the information presented by the past history of measurements (in time and/or space) is encapsulated by the current value of the parameter estimate. Adaptive parameter estimation methods are used extensively in various applications, especially those dealing with time-varying systems or unstable open-loop systems. It is also used as a way of avoiding long delays and high costs that result from offline system identification methods.

As discussed in Section 56.2, adaptive approximators can be classified into two categories of interest: linearly parameterized and nonlinearly parameterized. In the case of linearly parameterized approximators, the parameters denoted by σ are selected *a priori* and remain fixed. Therefore, the remaining adaptable weights $\hat{\theta}$ appear linearly. For nonlinearly parameterized approximators, both θ and σ weights are updated online. The case of linearly parameterized approximators allows the derivation of stronger analytical results for stability and convergence.

The adaptive parameter estimation problem can be formulated both in a continuous-time as well as a discrete-time framework. In practical applications, the actual plant typically evolves in continuous-time, while data processing (parameter estimation, monitoring, etc.) and feedback control is implemented in discrete-time using computing devices. Therefore, real-time applications yield so-called *hybrid systems*, where both continuous-time and discrete-time signals are intertwined [1].

It is important to keep in mind that different applications may have different objectives relevant to parameter convergence. In most control applications that focus on accurate tracking of reference input signals, the main objective is not necessarily to make the parameter estimates $\hat{\theta}(t)$ and $\hat{\sigma}(t)$ converge to the optimal values θ^* and σ^*, respectively, since accurate tracking performance can be achieved without convergence of the parameters. In general, parameter convergence is a strong requirement. In applications where parameter convergence is desired, the input to the approximator, denoted by $z(t)$, must also satisfy a so-called *persistency of excitation* condition. The structure of the persistence of excitation condition can be strongly affected by the choice of function approximator.

56.5 Nonlinear Control

Adaptive approximation-based control is based on the concept of designing control algorithms for systems with unknown nonlinearities. To understand adaptive approximation-based control methods, it is important to start by considering the simpler problem of designing of nonlinear control algorithms for systems with *known* nonlinearities. In other words, if the nonlinearities were known, how would we go about designing a control system? This section provides a general overview of various approaches for nonlinear control design. An excellent treatment of nonlinear systems and control methods is given in [18].

56.5.1 Small-Signal Linearization and Gain Scheduling

A traditional approach for dealing with nonlinear systems is to linearize the system around an equilibrium point or around a trajectory. This is referred to as *small-signal linearization*. The main idea behind small-signal linearization is to approximate the nonlinear system by a linear model of the form $\dot{x} = Ax + Bu$, where A and B are matrices of appropriate dimension. The term *small-signal linearization* is used to characterize the fact that the linear model is an accurate approximation of the nonlinear system only in a neighborhood of the point around which the linearization took place.

A key limitation of the small-signal linearization approach is the fact that the linear model is accurate only in a neighborhood around the operating (linearizing) point. To alleviate this limitation, the *gain scheduling* control approach is based on the idea of linearizing around multiple operating points. For each linear model there corresponds a feedback controller, thus creating a family of feedback control laws, each applicable in the neighborhood of a specific operating point. The family of feedback controllers can be combined into a single control whose parameters are changed by a scheduling scheme based on the trajectory or some other scheduling variables [34].

The performance and robustness of the gain scheduling control approaches is impacted by the fact that the controller parameters are precomputed offline for each operating condition. Hence, during operation the controller is fixed, even though the linear control gains are changing as the operating conditions change. In the presence of modeling errors or changes in the system dynamics, the gain scheduling controller may result in deterioration of the performance since the method does not provide any learning capability to correct—during operation—any inaccurate schedules. Another possible drawback of the gain scheduling approach is that it requires considerable offline effort to derive a reliable gain schedule for each possible situation that the plant will encounter.

The gain scheduling approach can be conveniently viewed as a special case of the adaptive approximation approach since a local linear model is an example case of a local approximation function. One of the key differences between the standard gain scheduling technique and the adaptive approximation-based control approach is the ability of the latter to adjust certain parameters (weights) during operation. Unlike gain scheduling, adaptive approximation is designed around the principle of "learning" and thus reduces the amount of modeling effort that needs to be expended offline. Moreover, it allows the control scheme to deal with unexpected changes in plant dynamics due to faults or severe disturbances.

56.5.2 Feedback Linearization

Feedback linearization is one of the most powerful and commonly found techniques in nonlinear control. This approach is based on cancelling the nonlinearities by the combined use of feedback and change of coordinates. A nonlinear system

$$\dot{x} = f(x) + G(x)u \tag{56.17}$$

is said to be *input-state feedback linearizable* if there exists a diffeomorphism $z = T(x)$, with $T(0) = 0$, such that

$$\dot{z} = Az + B\beta^{-1}(z)\left[u - \alpha(z)\right], \tag{56.18}$$

where (A, B) is a controllable pair and $\beta(z)$ is an invertible matrix for all z in a domain of interest $D_z \subset \Re^n$. Similarly, input–output feedback linearization describes input–output systems that can be linearized by the use of feedback.

Therefore, we see that the class of feedback linearizable systems includes not only systems that can be directly transformed to a linear system by feedback, but also systems that can be transformed into that form by a nonlinear state transformation. Determining whether a given nonlinear system is feedback linearizable and what is an appropriate diffeomorphism are not obvious issues, and in fact they can be extremely difficult since in general they involve solving a set of partial differential equations.

One of the key drawbacks of feedback linearization is that it depends on exact cancellation of nonlinear functions. If one of the functions is uncertain then cancellation is not possible. This is one of the motivations for adaptive approximation-based control. Another possible difficulty with feedback linearization is that not all systems can be transformed to a linearizable form. Another technique, referred to as *Backstepping* [19], can be applied to a class of systems that may not be feedback linearizable.

56.5.3 Robust Nonlinear Control

The methodologies introduced so far in this section were based on the key assumption that the control designer exactly knows the system nonlinearities. In practice, this is not a realistic assumption. Another class of nonlinear control design tools are based on the principle of assuming that the unknown components of the nonlinearities are bounded in some way by a known function. If this assumption is satisfied then it is possible to derive nonlinear control schemes that utilize these known bounding functions instead of the unknown nonlinearities.

This class of tools is referred to as robust nonlinear control design methods [13,18], and it includes: (1) bounding control, (2) sliding mode control, (3) Lyapunov redesign method, (4) nonlinear damping, and (5) adaptive bounding. Although these techniques have been extensively studied in the nonlinear control literature, they tend to yield conservative control laws, especially in cases where the uncertainty is significant. The term "conservative" is used among control engineers to indicate the fact that due to the uncertainty the control effort applied is more than needed. As a result, the control signal $u(t)$ may be large with rapid temporal variations (due to high-gain feedback). These characteristics may cause several problems, such as saturation or excessive wear of the actuators, large error in the presence of measurement noise, excitation of unmodeled dynamics, and large transient errors.

The robust nonlinear control design methods provide an interesting contrast to adaptive approximation-based control. Specifically, adaptive approximation-based control can be viewed as a way of reducing uncertainty during operation such that the need for conservative robust control can be eliminated or reduced. Another reason for studying these techniques in the context of adaptive approximation is their utilization to guarantee closed-loop stability outside of the approximation region $\mathcal{D}$.

56.6 Adaptive Approximation-Based Control

Sections 56.2 and 56.3 have presented approximator properties and structures, respectively. Section 56.4 discussed methods for parameter estimation and issues related to adaptive approximation, while Section 56.5 reviewed various nonlinear control design methods. This section brings these different topics together in the synthesis of adaptive approximation-based control systems.

The role of adaptive approximation-based control is to estimate unknown nonlinear functions and cancel their effect using the feedback control signal. Cancelling the estimated nonlinear function allows accurate tracking to be achieved with a smoother control signal, as compared to the robust nonlinear control design methods discussed in the previous section. The trade-off is that the adaptive approximation-based controller will typically have much higher state dimension (with the approximator adaptive parameters considered as states). This trade-off has become significantly more feasible over the past two decades, since controllers are frequently implemented via digital computers that have increased remarkably in memory and computational capabilities over this recent time span.

To illustrate some of the key concepts in the design and analysis of adaptive approximation-based control schemes, let us consider a simple scalar system

$$\dot{x} = (f_o(x) + f^*(x)) + (g_o(x) + g^*(x))u, \tag{56.19}$$

where $u(t)$ is the control input, and $y(t) = x(t)$ is the output. The functions $f_o(x)$ and $g_o(x)$ are the known components of the dynamics and $f^*(x)$ and $g^*(x)$ are the unknown parts of the dynamics. The objective is to achieve tracking of a certain signal $y_d(t)$ by the output $y(t)$.

The unknown portions of the model will be approximated over the compact region $\mathcal{D}$. This region is sometimes referred to as the safe operating envelope. For any system, the region $\mathcal{D}$ is physically determined at the design stage. For example, an electrical motor is designed to operate within certain voltage, current, torque, and speed constraints. If these constraints are violated, then the electrical or mechanical components of the motor may fail; therefore, the controller must ensure that the safe physical limits of the system represented by $\mathcal{D}$ are not violated.

The state Equation 56.19 can be expressed as

$$\dot{x} = \left(f_o(x) + \hat{f}(x;\theta_f^*)\right) + \epsilon_f^*(x) + \left(g_o(x) + \hat{g}(x;\theta_g^*)\right) u + \epsilon_g^*(x)u, \tag{56.20}$$

where θ_f^* and θ_g^* are the unknown "optimal" weight vectors, and ϵ_f^* and ϵ_g^* represent the MFAE as defined in Equation 56.4. The MFAE is a critical quantity, representing the minimum possible deviation between the unknown function f^* and the input/output function of the adaptive approximator $\hat{f}(x;\hat{\theta}_f)$. In general, increasing the number of adjustable weights reduces the MFAE. The UA results indicate that any specified approximation accuracy ϵ can be attained uniformly on the compact region $\mathcal{D}$ if the number of weights is sufficiently large.

The optimal weight vectors θ_f^* and θ_g^* are unknown quantities required only for analytical purposes. Typically, θ_f^* (similarly for θ_g^*) is defined as the value of θ_f that minimizes some cost function as defined in Equation 56.2. A special case, for the ∞-norm yields the uniform network approximation error over all $x \in \mathcal{D}$:

$$\theta_f^* := \arg \min_{\hat{\theta}_f \in \mathcal{R}^{q_f}} \left\{ \sup_{x\in\mathcal{D}} \left| f^*(x) - \hat{f}(x,\hat{\theta}_f) \right| \right\}. \tag{56.21}$$

The approximation-based feedback linearizing control law is summarized as

$$u_a = -a_m\tilde{x} + \dot{y}_d - f_o(x) - \phi_f(x)^\top \hat{\theta}_f - v_f, \tag{56.22}$$

$$u = \frac{u_a}{g_o(x) + \phi_g(x)^\top \hat{\theta}_g + v_g}, \tag{56.23}$$

$$\dot{\hat{\theta}}_f = \Gamma_f \phi_f \tilde{x}, \qquad \text{for } x \in \mathcal{D}, \tag{56.24}$$

$$\dot{\hat{\theta}}_g = \mathcal{P}_S\left(\Gamma_g \phi_g \tilde{x} u\right), \quad \text{for } x \in \mathcal{D}, \tag{56.25}$$

where $\tilde{x} = x - y_d$ is the tracking error, $a_m > 0$ is a positive design constant, Γ_f and Γ_g are positive-definite matrices, and $\mathcal{P}_S$ is the projection operator that will be used to ensure the stabilizability condition on $\hat{\theta}_g$. The auxiliary terms v_f and v_g are included to ensure that the state remains within the approximation region $\mathcal{D}$.

In the ideal case of zero MFAE, with $v_f = v_g = 0$, the adaptive approximation-based control scheme with linearly parameterized approximators guarantees that all the signals remain bounded and the tracking error $\tilde{x}(t)$ converges to zero. The stability proof is based on the use of Lyapunov stability theory [27].

In the case where the MFAE is nonzero (i.e., the adaptive approximation model is not able to approximate exactly the unknown nonlinearities within the region $\mathcal{D}$), it is possible for the parameter estimates to become unbounded in their attempt to estimate a function that cannot be physically estimated with the current number of weights. To prevent this from happening, the parameter update laws are modified by the use of the so-called dead-zone or σ-modification algorithms [17].

Another issue that needs to be addressed is what happens if the trajectory goes outside the operating envelope $\mathcal{D}$, which is a physically defined region over which it is safe and desirable for the system to operate. The trajectory generation system ensures that the desired state remains in $\mathcal{D}$. The control designer must ensure that the actual state converges to $\mathcal{D}$. Within $\mathcal{D}$ the objective is high accuracy trajectory tracking; therefore, the designer will select the approximator structure to provide confidence about the capability of the approximators $\hat{f}$ and $\hat{g}$ to approximate the unknown functions f^* and g^* accurately for $x \in \mathcal{D}$.

The control algorithms developed earlier had assumed that if $x(t)$ leaves the region $\mathcal{D}$, then the auxiliary control terms v_f and v_g are able to bring the state back within $\mathcal{D}$. Here, we show how to ensure that the design of the auxiliary terms v_f and v_g achieves the objective of bringing the trajectory within $\mathcal{D}$. Let

$\overline{\mathcal{D}} = \mathcal{R}^n - \mathcal{D}$; that is, $\overline{\mathcal{D}}$ is the region outside of $\mathcal{D}$. We assume that outside of $\mathcal{D}$, the unknown functions $f^*(x)$ and $g^*(x)$ are bounded by known nonlinearities as follows:

$$f_L(x) \le f^*(x) \le f_U(x), \quad x \in \overline{\mathcal{D}},$$
$$0 < g_L(x) \le g^*(x) \le g_U(x), \quad x \in \overline{\mathcal{D}}.$$

The control design for $x \in \mathcal{D}$ has already been considered. For $x \in \overline{\mathcal{D}}$, the adaptation of the parameter estimates $\hat{\theta}_f$ and $\hat{\theta}_g$ is stopped and $\phi_f(x) = 0$, $\phi_g(x) = 0$; that is, no basis functions are placed in $\overline{\mathcal{D}}$. Therefore, for $x \in \overline{\mathcal{D}}$, the auxiliary terms v_f and v_g are selected as follows:

$$v_f = \begin{cases} f_U(x), & \text{if } e \ge 0, \\ f_L(x), & \text{if } e \le 0, \end{cases} \tag{56.26}$$

$$v_g = \begin{cases} g_U(x), & \text{if } eu_a \ge 0, \\ g_L(x), & \text{if } eu_a \le 0, \end{cases} \tag{56.27}$$

where $e(t) = \tilde{x}(t)$ is the tracking error. The stability of the closed-loop system for $x \in \overline{\mathcal{D}}$ is obtained again by using Lyapunov stability theory [13].

The functions v_f and v_g are not Lipschitz functions. Their simplicity facilitates a clear discussion of methods to enforce convergence to $\mathcal{D}$. Usually these functions are smoothed across the boundary of $\mathcal{D}$ for practical implementations.

56.7 Reinforcement Learning

Optimal control theory was developed formally in the 1950s, following the contributions of L. S. Pontryagin and R. Bellman. Pontryagin introduced the minimum principle [28], which gave necessary conditions for the existence of optimal trajectories. Bellman introduced the concept of dynamic programming [7,8] that led to the notion of the celebrated Hamilton–Jacobi–Bellman (HJB) partial differential equation, which has the *value function* V^* as its solution.

Consider the nth order nonlinear system

$$\dot{x} = f(x) + g(x)u, \tag{56.28}$$

where $x = [x_1, \ldots, x_n]^\top \in R^n$ is the state and $u \in R^n$ is the control input. The optimal control problem is to find the control signal u to minimize a cost function $J(x, u)$ subject to the constraint on the trajectory evolution defined in Equation 56.28.

When the system model of Equation 56.28 is linear and and the cost function is quadratic in x and u, for example:

$$J_\infty = \int_{t_0}^{\infty} (x^\top(\tau)Qx(\tau) + \|u(\tau)\|^2)\, d\tau, \tag{56.29}$$

where Q is a positive-definite matrix, this becomes the Linear Quadratic Gaussian (LQG) problem [2], (i.e., the $\mathcal{H}_2$ optimal control problem) and the HJB partial differential equation becomes two separate Riccati equations, which can be solved very efficiently. For general nonlinear systems, it is extremely difficult to solve the HJB equation to obtain an optimal controller.

If $f(x)$ is nonlinear and known, a standard dynamic programming argument reduces the above optimal control problem to finding the value function V^* solving the HJB partial differential equation

$$V_x^* f - \frac{1}{4}\left(V_x^* g g^\top V_x^{*\top}\right) + x^\top Q x = 0, \tag{56.30}$$

where V_x^* denotes $\partial V^*/\partial x$. If there exists a continuously differentiable positive-definite solution V^* to Equation 56.30, then the optimal controller can be expressed as

$$u = -\frac{1}{2} g^\top V_x^{*\top}. \tag{56.31}$$

There are two obstacles that complicate the solution to the optimal control problem. The first challenge is that it can be extremely difficult to solve the HJB partial differential Equation 56.30 when the nonlinear function $f(x)$ is known. The second challenge is that $f(x)$ may be only partially unknown.

Given the above challenges, many approximate optimal control techniques have been developed for nonlinear systems. For example, control approaches such as receding horizon control [23], approximation of value functions [6,20,33,37], ADP [9,30,31] have been developed. Many ADP approaches address situations where the model is at least partially unknown and therefore address applications similar to those discussed in the main body of this chapter. There are numerous varieties of ADP, too many to address in this chapter. The interested reader can consult [24,30,35]. As one example of ADP, adaptive critic designs (ACD) are based on an algorithm that cycles between a policy-improvement routine (i.e., control) and a value-determination (i.e., estimation of V^*) operation [5,31]. At each optimizing cycle, the algorithm approximates the optimal control law and the value function based on the status of the system; therefore, there are at least two function approximation problems being solved during the system operation. Theoretically, when system dynamics are time-invariant, as the final time approaches infinity, the method yields the optimal control law.

56.8 Concluding Remarks

This chapter has discussed techniques for using online function approximation to improve control or state estimation performance. Successful implementations require proper selection of an objective function, an approximation structure, a training algorithm, and analysis of these three choices under the appropriate experimental conditions.

Several characteristics of function approximators have been discussed. In the literature, much attention is devoted to selecting certain function approximators because as a class they have the UA property. By itself, this condition is not sufficient. First, the set of function approximators, that are dense on the set of continuous functions, is larger than the set of universal function approximators. Second, universal function approximation requires that the approximator has an arbitrarily large size (i.e., width). Once the size of the approximator is limited, a limitation is also placed on the set of functions that can be approximated to within a given ϵ. In a particular application, the designer should instead consider which approximation structure can most efficiently represent the class of expected functions. Efficiency should be measured both in terms of the number of parameters (i.e., computer memory) and the amount of computation (i.e., required computer speed). Due to the inexpensive cost of computer memory and the hard constraints on the amount of possible computation due to sampling frequency considerations, computational efficiency is often more important in applications.

Although the goal of many applications presented in the literature is to approximate a given function, the approximation accuracy is rarely evaluated directly—even in simulation examples. Instead, performance is often evaluated by analyzing graphs of the training error. This approach can be misleading because such graphs indicate only the accuracy of the approximation at the pertinent operating points. It is quite possible to have a small sample error over a given time span without ever accurately approximating the desired function over the desired region. If the goal is to approximate a function over a given region, then some analysis other than monitoring the sample error during training must be performed to determine the success level. In simulation, the approximated and actual functions can be compared directly. Successful simulation performance is necessary to provide the confidence for implementation. In implementations, where the actual function is not known, a simple demonstration that the function has been approximated correctly is to show that the performance does not degrade when the parameter update law is turned off.

Acknowledgments

The authors gratefully acknowledge the support of the National Science Foundation (NSF) under Grant No. ECCS-0701791 and the Research Promotion Foundation (RPF) of Cyprus. Any opinions, findings, and conclusions or recommendations expressed in this chapter are those of the authors and do not necessarily reflect the views of the NSF and RPF.

References

1. P. J. Antsaklis, W. Kohn, A. Nerode, and S. Sastry, *Hybrid Systems II*, Lecture Notes in Computer Science, vol. 999. New York: Springer-Verlag, 1995.
2. M. Athans, The role and use of the stochastic linear-quadratic-Gaussian problem in control system design, *IEEE Transactions on Automatic Control*, vol. 16, no. 6, pp. 529–552, 1971.
3. C. G. Atkeson, A. W. Moore, and S. Schaal, Locally weighted learning, *Artificial Intelligence Review*, vol. 11, pp. 11–73, 1997.
4. A. Barron, Universal approximation bounds for superpositions of a sigmoidal function, *IEEE Transactions on Information Theory*, vol. 39, no. 3, pp. 930–945, 1993.
5. A. G. Barto, R. S. Sutton, and C. W. Anderson, Neuronlike adaptive elements that can solve difficult learning control problems, *IEEE Transactions on Systems, Man, and Cybernetics*, vol. 13, no. 5, pp. 834–846, 1983.
6. R. Beard, G. Saridis, and J. Wen, Galerkin approximations of the generalized Hamilton-Jacobi–Bellman equation, *Automatica*, vol. 33, no. 12, pp. 2159–2177, 1997.
7. R. Bellman, The theory of dynamic programming, *Proceedings of the National Academy of Science, USA*, vol. 38, 1952.
8. R. Bellman, *Dynamic Programming*. Princeton University Press, Princeton, NJ, 1957.
9. D. P. Bertsekas, *Dynamic Programming and Optimal Control*, 3rd ed. Athena Scientific, Nashua, NH, 2007.
10. G. Box, G. M. Jenkins, and G. Reinsel, *Time Series Analysis: Forecasting and Control*, 3rd ed. Englewood Cliffs, NJ: Prentice-Hall, 1994.
11. M. Cannon and J.-J. Slotine, Space–frequency localized basis function networks for nonlinear system estimation and control, *Neurocomputing*, vol. 9, pp. 293–342, 1995.
12. J. A. Farrell and W. Baker, Learning control systems, in *Intelligent-Autonomous Control Systems*, P. Antsaklis and K. Passino, Eds. Dordrecht: Kluwer Academic, 1993.
13. J. A. Farrell and M. M. Polycarpou, *Adaptive Approximation Based Control: Unifying Neural, Fuzzy, and Traditional Adaptive Approximation Approaches*. Hoboken, NJ: Wiley-Interscience, 2006.
14. K. Fu, Learning control systems—review and outlook, *IEEE Transactions on Pattern Analysis and Machine Intelligence*, vol. 8, no. 3, pp. 327–342, 1986.
15. G. C. Goodwin and K. S. Sin, *Adaptive Filtering Prediction and Control*. Englewood Cliffs, NJ: Prentice-Hall, 1984.
16. K. Hornik, M. Stinchcombe, and H. White, Multilayer feedforward networks are universal approximators, *Neural Networks*, vol. 2, pp. 359–366, 1989.
17. P. A. Ioannou and J. Sun, *Robust Adaptive Control*. Upper Saddle River, NJ: Prentice-Hall, 1996.
18. H. Khalil, *Nonlinear Systems*. Englewood Cliffs, NJ: Prentice-Hall, 1996.
19. M. Krstic, I. Kanellakopoulos, and P. Kokotovic, *Nonlinear and Adaptive Control Design*. New York: Wiley, 1995.
20. R. J. Leake and R.-W. Liu, Construction of suboptimal control sequences, *Journal of SIAM Control*, vol. 5, no. 1, pp. 54–63, 1967.

21. L. Ljung, *System Identification: Theory for the User*, 2nd ed. Englewood Cliffs, NJ: Prentice-Hall, 1999.
22. G. Lorentz, *Approximation of Functions*. New York: Holt, Rinehart, and Winston, 1966.
23. D. Q. Mayne, J. B. Rawlings, C. V. Rao, and P. O. M. Scokaert, Constrained model predictive control: Stability and optimality, *Automatica*, vol. 36, pp. 789–814, 2000.
24. W. T. Miller, R. S. Sutton, and P. J. Werbos, *Neural Networks for Control*. Cambridge, MA: MIT Press, 1990.
25. P. Millington, Associative reinforcement learning for optimal control, Master's thesis, Department of Aeronautics and Astronautics, MIT, Cambridge, MA, 1991.
26. J. Park and I. Sandberg, Universal approximation using radial basis function networks, *Neural Computation*, vol. 3, no. 2, pp. 246–257, 1991.
27. M. M. Polycarpou, Stable adaptive neural control scheme for nonlinear systems, *IEEE Transactions on Automatic Control*, vol. 41, no. 3, pp. 447–451, 1996.
28. L. S. Pontryagin, Optimal control processes, *Uspehi Mat. Nauk (in Russian)*, vol. 14, pp. 3–20, 1959.
29. M. Powell, *Approximation Theory and Methods*. Cambridge, UK: Cambridge University Press, 1981.
30. W. B. Powell, *Approximate Dynamic Programming: Solving the Curses of Dimensionality*. Hoboken, NJ: Wiley & Sons Inc., 2007.
31. D. V. Prokhorov and D. C. Wunsch, Adaptive critic designs, *IEEE Transactions on Neural Networks*, vol. 8, no. 5, pp. 997–1007, 1997.
32. R. Sanner and J. Slotine, Gaussian networks for direct adaptive ccontrol, *IEEE Transactions on Neural Networks*, vol. 3, no. 6, pp. 837–863, 1992.
33. G. N. Saridis and C.-S. Lee, An approximation theory of optimal control for trainable manipulators, *IEEE Transactions on Systems, Man, and Cybernetics*, vol. 9, no. 3, pp. 152–158, 1979.
34. J. S. Shamma and M. Athans, Analysis of gain scheduled control for nonlinear plants, *IEEE Transactions on Automatic Control*, vol. 35, no. 8, pp. 898–907, 1990.
35. J. Si, A. Barto, W. Powell, and D. Wunsch, Eds., *Handbook of Learning and Approximate Dynamic Programming*. Hoboken, NJ: Wiley-Interscience, 2004.
36. Y. Tsypkin, *Foundations of the Theory of Learning Systems*. New York, NY: Academic Press, 1971.
37. D. Vrabie and F. Lewis, Adaptive optimal control algorithm for continuous-time nonlinear systems based on policy iteration, *Proc. of the 47th IEEE Conf. on Decision and Control*, pp. 73–79, 2008.
38. Y. Zhao and J. A. Farrell, Self-organizing approximation based control for higher order systems, *IEEE Transactions on Neural Networks*, vol. 18, no. 4, pp. 1220–1231, 2007.

IX

System Identification

57

System Identification

Lennart Ljung
Linköping University

57.1 The Basic Ideas

57.1.1 Ten Basic Questions about System Identification

1. **What is system identification?**
   System Identification allows you to build mathematical models of a dynamic system based on measured data.

2. **How is that done?**
   Essentially by adjusting parameters within a given model until its output coincides as well as possible with the measured output.
3. **How do you know if the model is any good?**
   A good test is to take a close look at the model's output compared to the measurements on a data set that was not used for the fit ("Validation Data").
4. **Can the quality of the model be tested in other ways?**
   It is also valuable to look at what the model could not reproduce in the data ("the residuals"). There should be no correlation with other available information, such as the system's input.
5. **What models are most common?**
   The techniques apply to very general models. Most common models are difference equations descriptions, such as ARX and ARMAX models, as well as all types of linear state-space models. Lately, black box nonlinear structures, such as Artificial Neural Networks, Fuzzy models, and so on, have been much used.
6. **Do you have to assume a model of a particular type?**
   For parametric models, you have to specify the structure. However, if you just assume that the system is linear, you can directly estimate its impulse or step response using Correlation Analysis or its frequency response using Spectral Analysis. This allows useful comparisons with other estimated models.
7. **How do you know what model structure to try?**
   Well, you do not. For real life systems there is never any "true model," anyway. You have to be generous at this point, and try out several different structures.
8. **Can nonlinear models be used in an easy fashion?**
   Yes. Most common model nonlinearities are such that the measured data should be nonlinearly transformed (like squaring a voltage input if you think that it is the power i.e., the stimulus). Use physical insight about the system you are modeling and try out such transformations on models that are linear in the new variables, and you will cover a lot.
9. **What does this article cover?**
   After reviewing an archetypical situation in this section, we describe the basic techniques for parameter estimation in arbitrary model structures. Section 57.3 deals with linear models of black box structure, and Section 57.5 deals with particular estimation methods that can be used (in addition to the general ones) for such models. Physically parameterized model structures are described in Section 57.6, and nonlinear black box models (including neural networks) are discussed in Section 57.7. The final Section 57.8 deals with the choices and decisions the user is faced with.
10. **Is this really all there is to system identification?**
    Actually, there is a huge amount written on the subject. Experience with real data is the driving force to further understanding. It is important to remember that any estimated model, no matter how good it looks on your screen, has only picked up a simple reflection of reality. Surprisingly often, however, this is sufficient for rational decision making.

57.1.2 Background and Literature

System identification has its roots in standard statistical techniques and many of the basic routines have direct interpretations as well known statistical methods such as least squares and maximum likelihood. The control community took an active part in the development and application of these basic techniques to dynamic systems right after the birth of "modern control theory" in the early 1960s. Maximum likelihood estimation was applied to difference equations (ARMAX models) by Åström and Bohlin (1965) and thereafter a wide range of estimation techniques and model parameterizations flourished. By now, the area is well matured with established and well-understood techniques. Industrial use and application of the techniques has become standard. See Ljung (2007) for a common software package.

The literature on system identification is extensive. For a practical user oriented introduction we may mention Ljung and Glad (1994). Texts that go deeper into the theory and algorithms include Ljung (1999), and Söderström and Stoica (1989). A classical treatment is Box and Jenkins (1970).

These books all deal with the "mainstream" approach to system identification, as described in this article. In addition, there is a substantial literature on other approaches, such as "set membership" (compute all those models that reproduce the observed data within a certain given error bound), estimation of models from given frequency response measurement (Pintelon and Schoukens, 2001), online model estimation (Ljung and Söderström, 1983), nonparametric frequency domain methods (Brillinger, 1981), gray-box identification (Bohlin, 2006) and so on. To follow the development in the field, the IFAC series of Symposia on System Identification (Saint Malo, 2009; Newcastle, 2006) is a good source. A recent perspective paper is Ljung (2010).

57.1.3 An Archetypical Problem: ARX Models and the Linear Least-Squares Method

57.1.3.1 The Model

We shall generally denote the system's input and output at time t by $u(t)$ and $y(t)$, respectively. Perhaps, the most basic relationship between the input and output is the *linear difference equation*

$$y(t) + a_1 y(t-1) + \cdots + a_n y(t-n) = b_1 u(t-1) + \cdots + b_m u(t-m) \tag{57.1}$$

We have chosen to represent the system in *discrete time*, primarily since observed data are always collected by sampling. It is thus more straightforward to relate observed data to discrete time models. Nothing prevents us however from working with continuous time models: we shall return to that in Section 57.6.

In Equation 57.1, we assume the *sampling interval* to be one time unit. This is not essential, but makes notation easier.

A pragmatic and useful way to see Equation 57.1 is to view it as a way of *determining the next output value* given previous observations:

$$y(t) = -a_1 y(t-1) - \cdots - a_n y(t-n) + b_1 u(t-1) + \cdots + b_m u(t-m) \tag{57.2}$$

For more compact notation we introduce the vectors

$$\theta = [a_1, \ldots, a_n\ b_1, \ldots, b_m]^T \tag{57.3}$$

$$\varphi(t) = [-y(t-1) \cdots - y(t-n)\ u(t-1) \cdots u(t-m)]^T \tag{57.4}$$

With these Equation 57.2 can be rewritten as

$$y(t) = \varphi^T(t)\theta$$

To emphasize that the calculation of $y(t)$ from past data Equation 57.2 indeed depends on the parameters in θ, we shall rather call this calculated value $\hat{y}(t|\theta)$ and write

$$\hat{y}(t|\theta) = \varphi^T(t)\theta \tag{57.5}$$

57.1.3.2 The Least-Squares Method

Now suppose for a given system that we do not know the values of the parameters in θ, but that we have recorded inputs and outputs over a time interval $1 \leq t \leq N$:

$$Z^N = \{u(1), y(1), \ldots, u(N), y(N)\} \tag{57.6}$$

An obvious approach is then to select θ in Equations 57.1 through 57.5 so as to fit the calculated values $\hat{y}(t|\theta)$ as well as possible to the measured outputs by the least-squares method:

$$\min_{\theta} V_N(\theta, Z^N) \tag{57.7}$$

where

$$\begin{aligned} V_N(\theta, Z^N) &= \frac{1}{N}\sum_{t=1}^{N}(y(t) - \hat{y}(t|\theta))^2 \\ &= \frac{1}{N}\sum_{t=1}^{N}(y(t) - \varphi^T(t)\theta)^2 \end{aligned} \tag{57.8}$$

We shall denote the value of θ that minimizes Equation 57.7 by $\hat{\theta}_N$:

$$\hat{\theta}_N = \arg\min_{\theta} V_N(\theta, Z^N) \tag{57.9}$$

("arg min" means the minimizing argument, i.e., that value of θ which minimizes V_N.)

Since V_N is quadratic in θ, we can find the minimum value easily by setting the derivative to zero:

$$0 = \frac{d}{d\theta} V_N(\theta, Z^N) = \frac{2}{N}\sum_{t=1}^{N} \varphi(t)(y(t) - \varphi^T(t)\theta)$$

which gives

$$\sum_{t=1}^{N} \varphi(t)y(t) = \sum_{t=1}^{N} \varphi(t)\varphi^T(t)\theta \tag{57.10}$$

or

$$\hat{\theta}_N = \left[\sum_{t=1}^{N} \varphi(t)\varphi^T(t)\right]^{-1} \sum_{t=1}^{N} \varphi(t)y(t) \tag{57.11}$$

Once the vectors $\varphi(t)$ are defined, the solution can easily be found by modern numerical software, such as MATLAB®.

Example 57.1: First-Order Difference Equation

Consider the simple model

$$y(t) + ay(t-1) = bu(t-1).$$

This gives us the estimate according to Equations 57.3, 57.4, and 57.11

$$\begin{bmatrix} \hat{a}_N \\ \hat{b}_N \end{bmatrix} = \begin{bmatrix} \sum y^2(t-1) & -\sum y(t-1)u(t-1) \\ -\sum y(t-1)u(t-1) & \sum u^2(t-1) \end{bmatrix}^{-1} \begin{bmatrix} -\sum y(t)y(t-1) \\ \sum y(t)u(t-1) \end{bmatrix}$$

All sums are from $t = 1$ to $t = N$. A typical convention is to take values outside the measured range to be zero. In this case we would thus take $y(0) = 0$.

The simple model (Equation 57.1) and the well-known least-squares method (Equation 57.11) form the archetype of System Identification. Not only that—they also give the most commonly used parametric identification method and are much more versatile than perhaps perceived at first sight. In particular, one should realize that Equation 57.1 can directly be extended to several different inputs (this just calls for a redefinition of $\varphi(t)$ in Equation 57.4) and that the inputs and outputs do not have to be the raw measurements. On the contrary—it is often most important to think over the physics of the application and come up with suitable inputs and outputs for Equation 57.1, formed from the actual measurements.

Example 57.2: An Immersion Heater

Consider a process consisting of an immersion heater immersed in a cooling liquid. We measure:

- $v(t)$: The voltage applied to the heater
- $r(t)$: The temperature of the liquid
- $y(t)$: The temperature of the heater coil surface

Suppose we need a model for how $y(t)$ depends on $r(t)$ and $v(t)$. Some simple considerations based on common sense and high school physics ("Semi-physical modeling") reveal the following:

- The change in temperature of the heater coil over one sample is proportional to the electrical power in it (the inflow power) minus the heat loss to the liquid
- The electrical power is proportional to $v^2(t)$
- The heat loss is proportional to $y(t) - r(t)$

This suggests the model

$$y(t) = y(t-1) + \alpha v^2(t-1) - \beta(y(t-1) - r(t-1))$$

which fits into the form

$$y(t) + \theta_1 y(t-1) = \theta_2 v^2(t-1) + \theta_3 r(t-1))$$

This is a two input (v^2 and r) and one output model, and corresponds to choosing

$$\varphi(t) = [-y(t-1) \quad v^2(t-1) \quad r(t-1)]^T$$

in Equation 57.5.

57.1.3.3 Some Statistical Remarks

Model structures, such as Equation 57.5 that are linear in θ are known in statistics as *linear regression* and the vector $\varphi(t)$ is called the *regression vector* (its components are the *regressors*). "Regress" here alludes to the fact that we try to calculate (or describe) $y(t)$ by "going back" to $\varphi(t)$. Models such as Equation 57.1 where the regression vector—$\varphi(t)$—contains old values of the variable to be explained—$y(t)$—are then partly *auto-regressions*. For that reason the model structure Equation 57.1 has the standard name ARX-model (Auto-regression with extra inputs).

There is a rich statistical literature on the properties of the estimate $\hat{\theta}_N$ under varying assumptions. See, for example, Draper and Smith (1981). So far we have just viewed Equations 57.7 and 57.8 as "curve-fitting." In Section 57.2.2, we shall deal with a more comprehensive statistical discussion, which includes the ARX model as a special case. Some direct calculations will be done in the following subsection.

57.1.3.4 Model Quality and Experiment Design

Let us consider the simplest special case, that of a Finite Impulse Response (FIR) model. That is obtained from Equation 57.1 by taking $n = 0$:

$$y(t) = b_1 u(t-1) + \cdots + b_m u(t-m) \tag{57.12}$$

Suppose that the observed data really have been generated by a similar mechanism

$$y(t) = b_1^0 u(t-1) + \cdots + b_m^0 u(t-m) + e(t) \tag{57.13}$$

where $e(t)$ is a white noise sequence with variance λ, but otherwise unknown. (i.e., $e(t)$ can be described as a sequence of independent random variables with zero mean values and variances λ.) Analogous to

Equation 57.5, we can write this as

$$y(t) = \varphi^T(t)\theta_0 + e(t) \tag{57.14}$$

We can now replace $y(t)$ in Equation 57.11 by the above expression, and obtain

$$\begin{aligned}\hat{\theta}_N &= \left[\sum_{t=1}^{N} \varphi(t)\varphi^T(t)\right]^{-1} \sum_{t=1}^{N} \varphi(t)y(t) \\ &= \left[\sum_{t=1}^{N} \varphi(t)\varphi^T(t)\right]^{-1} \left[\sum_{t=1}^{N} \varphi(t)\varphi^T(t)\theta_0 + \sum_{t=1}^{N} \varphi(t)e(t)\right]\end{aligned}$$

or

$$\tilde{\theta}_N = \hat{\theta}_N - \theta_0 = \left[\sum_{t=1}^{N} \varphi(t)\varphi^T(t)\right]^{-1} \sum_{t=1}^{N} \varphi(t)e(t) \tag{57.15}$$

Suppose that the input u is independent of the noise e. Then φ and e are independent in this expression, so it is easy to see that $E\tilde{\theta}_N = 0$, since e has zero mean. The estimate is consequently *unbiased*. Here E denotes *mathematical expectation*.

We can also form the expectation of $\tilde{\theta}_N\tilde{\theta}_N^T$, that is, the covariance matrix of the parameter error. Denote the matrix within brackets by R_N. Take expectation with respect to the white noise e. Then R_N is a deterministic matrix and we have

$$P_N = E\tilde{\theta}_N\tilde{\theta}_N^T = R_N^{-1} \sum_{t,s=1}^{N} \varphi(t)\varphi^T(s)Ee(t)e(s)R_N^{-1} = \lambda R_N^{-1} \tag{57.16}$$

since the double sum collapses to λR_N.

We have thus computed the covariance matrix of the estimate $\hat{\theta}_N$. It is determined entirely by the input properties and the noise level. Moreover, define

$$\bar{R} = \lim_{N\to\infty} \frac{1}{N} R_N \tag{57.17}$$

This will be the *covariance matrix* of the input, that is, the $i-j$-element of $\bar{R}$ is $R_{uu}(i-j)$, as defined by Equation 57.104 later on.

If the matrix $\bar{R}$ is nonsingular, we find that the covariance matrix of the parameter estimate is approximately (and the approximation improves as $N \to \infty$)

$$P_N = \frac{\lambda}{N}\bar{R}^{-1} \tag{57.18}$$

A number of things follow from this. All of them are typical of the general properties to be described in Section 57.2.2:

- The covariance decays like $1/N$, so the parameters approach the limiting value at the rate $1/\sqrt{N}$.
- The covariance is proportional to the noise-to-signal ratio. That is, it is proportional to the noise variance and inversely proportional to the input power.
- The covariance does not depend on the input's or noise's signal shapes, only on their variance/covariance properties.
- Experiment design, that is, the selection of the input u, aims at making the matrix $\bar{R}^{-1}$ "as small as possible." Note that the same $\bar{R}$ can be obtained for many different signals u.

57.1.4 The Main Ingredients

The main ingredients for the System Identification problem are as follows:

- The data set Z^N
- A class of candidate model descriptions; *a Model Structure*
- A criterion of fit between data and models
- Routines to validate and accept resulting models

We have seen in Section 57.1.3, a particular model structure, the ARX-model. In fact the major problem in system identification is to select a good model structure, and a substantial part of this article deals with various model structures. See Sections 57.3, 57.6, and 57.7, which all concern this problem. Generally speaking, a model structure is a parameterized mapping from past inputs and outputs Z^{t-1} (*cf* Equation 57.6) to the space of the model outputs:

$$\hat{y}(t|\theta) = g(\theta, Z^{t-1}) \tag{57.19}$$

Here θ is the finite dimensional vector used to parameterize the mapping.

Actually, the problem of fitting a given model structure to measured data is much simpler, and can be dealt with independently of the model structure used. We shall do so in the following section.

The problem of assuring a data set with adequate information contents is the problem of *experiment design*, and it will be described in Section 57.8.1.

Model validation is both a process to discriminate between various model structures and the final quality control station, before a model is delivered to the user. This problem is discussed in Section 57.8.2.

57.2 General Parameter Estimation Techniques

In this section, we shall deal with issues that are independent of model structure. Principles and algorithms for fitting models to data, as well as the general properties of the estimated models are all model-structure independent and equally well applicable to, say, ARMAX models and Neural Network models.

The section is organized as follows. In Section 57.2.1, the general principles for parameter estimation are outlined. Sections 57.2.2 and 57.2.3 deal with the asymptotic (in the number of observed data) properties of the models, while algorithms, both for online and off-line use are described in Section 57.2.5.

57.2.1 Fitting Models to Data

In Section 57.1.3, we showed one way to parameterize descriptions of dynamical systems. There are many other possibilities and we shall spend a fair amount of this contribution to discuss the different choices and approaches. *This is actually the key problem in system identification.* No matter how the problem is approached, the bottom line is that such a model parameterization leads to a predictor

$$\hat{y}(t|\theta) = g(\theta, Z^{t-1}) \tag{57.20}$$

that depends on the unknown parameter vector and past data Z^{t-1} (see Equation 57.6). This predictor can be linear in y and u. This in turn contains several special cases both in terms of black box models and physically parameterized ones, as will be discussed in Sections 57.3 and 57.6, respectively. The predictor could also be of general, nonlinear nature, as discussed in Section 57.7.

In any case *we now need a method to determine a good value of* θ, based on the information in an observed, sampled data set (Equation 57.6). It suggests itself that the basic least-squares like approach Equations 57.7 through 57.9 still is a natural approach, even when the predictor $\hat{y}(t|\theta)$ is a more general function of θ.

A procedure with some more degrees of freedom is the following one

1. From observed data and the predictor $\hat{y}(t|\theta)$ form the sequence of prediction errors,

$$\varepsilon(t,\theta) = y(t) - \hat{y}(t|\theta), \quad t = 1, 2, \ldots N \tag{57.21}$$

2. Possibly filter the prediction errors through a linear filter $L(q)$,

$$\varepsilon_F(t,\theta) = L(q)\varepsilon(t,\theta) \tag{57.22}$$

(here q denotes the shift operator, $qu(t) = u(t+1)$) so as to enhance or depress interesting or unimportant frequency bands in the signals.

3. Choose a scalar valued, positive function $\ell(\cdot)$ so as to measure the "size" or "norm" of the prediction error:

$$\ell(\varepsilon_F(t,\theta)) \tag{57.23}$$

4. Minimize the sum of these norms:

$$\hat{\theta}_N = \arg\min_{\theta} V_N(\theta, Z^N) \tag{57.24}$$

where

$$V_N(\theta, Z^N) = \frac{1}{N}\sum_{t=1}^{N} \ell(\varepsilon_F(t,\theta)) \tag{57.25}$$

This procedure is natural and pragmatic—we can still think of it as "curve-fitting" between $y(t)$ and $\hat{y}(t|\theta)$. It also has several statistical and information theoretic interpretations. Most importantly, if the noise source in the system (like in Equation 57.61 below) is supposed to be a sequence of independent random variables $\{e(t)\}$ each having a probability density function $f_e(x)$, then Equation 57.24 becomes the maximum likelihood estimate (MLE) if we choose

$$L(q) = 1 \quad \text{and} \quad \ell(\varepsilon) = -\log f_e(\varepsilon) \tag{57.26}$$

The MLE has several nice statistical features and thus gives a strong "moral support" for using the outlined method. Another pleasing aspect is that the method is independent of the particular model parameterization used (although this will affect the actual minimization procedure). For example, the method of "back propagation" often used in connection with neural network parameterizations amounts to computing $\hat{\theta}_N$ in Equation 57.24 by a recursive gradient method. We shall deal with these aspects in Section 57.2.5.

57.2.2 Model Quality

An essential question is, of course, what properties will the estimate have resulting from Equation 57.24. These will naturally depend on the properties of the data record Z^N defined by Equation 57.6. It is in general a difficult problem to characterize the quality of $\hat{\theta}_N$ exactly. One normally has to be content with the asymptotic properties of $\hat{\theta}_N$ as the number of data, N, tends to infinity.

It is an important aspect of the general identification method Equation 57.24 that the asymptotic properties of the resulting estimate can be expressed in general terms for arbitrary model parameterizations.

The first basic result is the following one:

$$\hat{\theta}_N \to \theta^* \quad \text{as } N \to \infty \quad \text{where} \tag{57.27}$$

$$\theta^* = \arg\min_{\theta} E\ell(\varepsilon_F(t,\theta)) \tag{57.28}$$

That is, as more and more data become available, the estimate converges to that value θ^*, that would minimize the expected value of the "norm" of the filtered prediction errors. This is, in a sense *the best* *possible approximation* of the true system that is available within the model structure. The expectation E in Equation 57.28 is taken with respect to all random disturbances that affect the data and it also includes averaging over the input properties. This means in particular that θ^* will make $\hat{y}(t|\theta^*)$ a good approximation of $y(t)$ with respect to those aspects of the system that are enhanced by the input signal used.

The second basic result is the following one: If $\{\varepsilon(t,\theta^*)\}$ is approximately white noise, then the covariance matrix of $\hat{\theta}_N$ is approximately given by

$$E(\hat{\theta}_N - \theta^*)(\hat{\theta}_N - \theta^*)^T \sim \frac{\lambda}{N}[E\psi(t)\psi^T(t)]^{-1} \tag{57.29}$$

where

$$\lambda = E\varepsilon^2(t,\theta^*) \tag{57.30}$$

$$\psi(t) = \frac{d}{d\theta}\hat{y}(t|\theta)|_{\theta=\theta^*} \tag{57.31}$$

Think of ψ as the sensitivity derivative of the predictor with respect to the parameters. Then Equation 57.29 says that the covariance matrix for $\hat{\theta}_N$ is proportional to the inverse of the covariance matrix of this sensitivity derivative. This is a quite natural result.

Note: For all these results, the expectation operator E can, under general conditions, be replaced by the limit of the sample mean, that is

$$E\psi(t)\psi^T(t) \leftrightarrow \lim_{N\to\infty}\frac{1}{N}\sum_{t=1}^{N}\psi(t)\psi^T(t) \tag{57.32}$$

The results, Equations 57.27 through 57.31 are general and hold for all model structures, both linear and nonlinear ones, subject only to some regularity and smoothness conditions. They are also fairly natural, and will give the guidelines for all user choices involved in the process of identification. See Ljung (1999) for more details around this.

57.2.3 Measures of Model Fit

Some quite general expressions for the expected model fit, that are independent of the model structure, can also be developed.

Let us measure the (average) fit between any model (Equation 57.20) and the true system as

$$\bar{V}(\theta) = E|y(t) - \hat{y}(t|\theta)|^2 \tag{57.33}$$

Here expectation E is over the data properties (i.e., expectation over "Z^∞" with the notation Equation 57.6). Recall that expectation also can be interpreted as sample means as in Equation 57.32.

Before we continue, let us note the very important aspect that the fit $\bar{V}$ will depend, not only on the model and the true system, *but also on data properties*, like input spectra, possible feedback, and so on. We shall say that the fit depends on the *experimental conditions*.

The estimated model parameter $\hat{\theta}_N$ is a random variable, because it is constructed from observed data, that can be described as random variables. To evaluate the model fit, we then take the expectation of $\bar{V}(\hat{\theta}_N)$ with respect to the estimation data. That gives our measure

$$F_N = E\bar{V}(\hat{\theta}_N) \tag{57.34}$$

In general, the measure F_N depends on a number of things:

- The model structure used
- The number of data points N
- The data properties for which the fit $\bar{V}$ is defined
- The properties of the data used to estimate $\hat{\theta}_N$

The rather remarkable fact is that if the two last data properties coincide, then, asymptotically in N [see, e.g., Ljung (1999), Chapter 16]

$$F_N \approx \bar{V}_N(\theta^*)\left(1 + \frac{dim\,\theta}{N}\right) \tag{57.35}$$

Here θ^* is the value that minimizes the expected criterion (Equation 57.28). The notation $dim\,\theta$ means the number of estimated parameters. The result also assumes that the criterion function $\ell(\varepsilon) = \|\varepsilon\|^2$, and that the model structure is successful in the sense that $\varepsilon_F(t)$ is approximately white noise.

Despite the reservations about the formal validity of Equation 57.35, it carries a most important conceptual message: If a model is evaluated on a data set with the same properties as the estimation data, then *the fit will not depend on the data properties*, and it will depend on the model structure *only in terms of the number of parameters used and of the best fit offered within the structure.*

The expression can be rewritten as follows. Let $\hat{y}_0(t|t-1)$ denote the "true" one step ahead prediction of $y(t)$, and let

$$W(\theta) = E|\hat{y}_0(t|t-1) - \hat{y}(t|\theta)|^2 \tag{57.36}$$

and let

$$\lambda = E|y(t) - \hat{y}_0(t|t-1)|^2 \tag{57.37}$$

Then λ is the *innovations* variance, that is, that part of $y(t)$ that cannot be predicted from the past. Moreover, $W(\theta^*)$ is the *bias error*, that is, the discrepancy between the true predictor and the best one available in the model structure. Under the same assumptions as above, Equation 57.35 can be rewritten as

$$F_N \approx \lambda + W(\theta^*) + \lambda\frac{dim\,\theta}{N} \tag{57.38}$$

The three terms constituting the model error then have the following interpretations:

- λ is the unavoidable error, stemming from the fact that the output cannot be exactly predicted, even with perfect system knowledge.
- $W(\theta^*)$ is the bias error. It depends on the model structure, and on the experimental conditions. It will typically decrease as $dim\,\theta$ increases.
- The last term is the *variance error*. It is proportional to the number of estimated parameters and inversely proportional to the number of data points. It does not depend on the particular model structure or the experimental conditions (as long as these are the same for the estimation and evaluation).

57.2.4 Model Structure Selection

The most difficult choice for the user is no doubt to find a suitable model structure to fit the data to. This is of course a very application-dependent problem, and it is difficult to give general guidelines. (Still, some general practical advice will be given in Section 57.8.)

At the heart of the model structure selection process is to handle the trade-off between bias and variance, as formalized by Equation 57.38. The "best" model structure is the one that minimizes F_N, the fit between the model and the data for a *fresh* data set—one that was not used for estimating the model. Most procedures for choosing the model structures are also aiming at finding this best choice.

57.2.4.1 Cross Validation

A very natural and pragmatic approach is *cross validation*. This means that the available data set is split into two parts, *estimation data*, $Z_{\text{est}}^{N_1}$ that is used to estimate the models:

$$\hat{\theta}_{N_1} = \arg\min V_{N_1}\left(\theta, Z_{\text{est}}^{N_1}\right) \tag{57.39}$$

and *validation data*, $Z_{\text{val}}^{N_2}$ for which the criterion is evaluated:

$$\hat{F}_{N_1} = V_{N_2}(\hat{\theta}_{N_1}, Z_{\text{val}}^{N_2}) \tag{57.40}$$

Here V_N is the criterion Equation 57.25. Then $\hat{F}_N$ will be an unbiased estimate of the measure F_N, defined by (Equation 57.34), which was discussed at length in the previous section. The procedure would the be to try out a number of model structures, and choose the one that minimizes $\hat{F}_{N_1}$.

Such cross validation techniques to find a good model structure has an immediate intuitive appeal. We simply check if the candidate model is capable of "reproducing" data have it has not yet seen. If that works well, we have some confidence in the model, regardless of any probabilistic framework that might be imposed. Such techniques are also the most commonly used ones.

A few comments could be added. In the first place, one could use different splits of the original data into estimation and validation data. For example, in statistics, there is a common cross validation technique called "leave one out." This means that the validation data set consists of one data point "at a time," but successively applied to the whole original set. In the second place, the test of the model on the validation data does not have to be in terms of the particular criterion (Equation 57.40). In system identification it is common practice to simulate (or predict several steps ahead) the model using the validation data, and then visually inspect the agreement between measured and simulated (predicted) output.

57.2.4.2 Estimating the Variance Contribution: Penalizing the Model Complexity

It is clear that the criterion (Equation 57.40) has to be evaluated on the validation data to be of any use—it would be strictly decreasing as a function of model flexibility if evaluated on the estimation data. In other words, the adverse effect of the dimension of θ shown in Equation 57.38 would be missed. There are a number of criteria—often derived from entirely different viewpoints—that try to capture the influence of this variance error term. The two best-known ones are *Akaike's information theoretic criterion* (AIC), which has the form (for Gaussian disturbances)

$$\tilde{V}_N(\theta, Z^N) = \left(1 + \frac{2dim\,\theta}{N}\right)\frac{1}{N}\sum_{t=1}^{N}\varepsilon^2(t,\theta) \tag{57.41}$$

and *Rissanen's minimum description length criterion* (MDL) in which $dim\,\theta$ in the expression above is replaced by $\log N\,dim\,\theta$. See Akaike (1974a) and Rissanen (1978).

The criterion $\tilde{V}_N$ is then to be minimized both with respect to θ and to a family of model structures. The relation to the expression (Equation 57.35) for F_N is obvious.

57.2.5 Algorithmic Aspects

In this section, we discuss how to achieve the best fit between observed data and the model, that is, how to carry out the minimization of Equation 57.24. For simplicity, we here assume a quadratic criterion and set the prefilter L to unity:

$$V_N(\theta) = \frac{1}{2N}\sum_{t=1}^{N}|y(t) - \hat{y}(t|\theta)|^2 \tag{57.42}$$

No analytic solution to this problem is possible unless the model $\hat{y}(t|\theta)$ is linear in θ, so the minimization has to be done by some numerical search procedure. A classical treatment of the problem of how to minimize the sum of squares is given in Dennis and Schnabel (1983).

Most efficient search routines are based on iterative local search in a "downhill" direction from the current point. We then have an iterative scheme of the following kind

$$\hat{\theta}^{(i+1)} = \hat{\theta}^{(i)} - \mu_i R_i^{-1} \hat{g}_i \tag{57.43}$$

Here $\hat{\theta}^{(i)}$ is the parameter estimate after iteration number i. The search scheme is thus made up of the three entities

- μ_i step size
- $\hat{g}_i$ an estimate of the gradient $V_N'(\hat{\theta}^{(i)})$
- R_i a matrix that modifies the search direction

It is useful to distinguish between two different minimization situations

i. *Off-line* or *batch:* The update $\mu_i R_i^{-1} g_i^1$ is based on the whole available data record Z^N.
ii. *Online* or *recursive:* The update is based only on data up to sample i (Z^i), (typically done so that the gradient estimate $\hat{g}_i$ is based only on data just before sample i).

We discuss these two modes separately below. First, some general aspects will be treated.

57.2.5.1 Search Directions

The basis for the local search is the gradient

$$V_N'(\theta) = \frac{dV_N(\theta)}{d\theta} = -\frac{1}{N}\sum_{t=1}^{N}(y(t) - \hat{y}(t|\theta))\psi(t,\theta) \tag{57.44}$$

where

$$\psi(t,\theta) = \frac{\partial}{\partial\theta}\hat{y}(t|\theta) \tag{57.45}$$

The gradient ψ is in the general case a matrix with *dim* θ rows and *dim* y columns. It is well known that gradient search for the minimum is inefficient, especially close to the minimum. Then it is optimal to use the *Newton search direction*

$$R^{-1}(\theta)V_N'(\theta) \tag{57.46}$$

where

$$\begin{aligned} R(\theta) = V_N''(\theta) &= \frac{d^2 V_N(\theta)}{d\theta^2} \\ &= \frac{1}{N}\sum_{t=1}^{N}\psi(t,\theta)\psi^T(t,\theta) + \frac{1}{N}\sum_{t=1}^{N}(y(t) - \hat{y}(t|\theta))\frac{\partial^2}{\partial\theta^2}\hat{y}(t|\theta) \end{aligned} \tag{57.47}$$

The true Newton direction will thus require that the second derivative

$$\frac{\partial^2}{\partial\theta^2}\hat{y}(t|\theta)$$

be computed. Also, far from the minimum, $R(\theta)$ need not be positive semidefinite. Therefore, alternative search directions are more common in practice:

- *Gradient direction:* Simply take

$$R_i = I \tag{57.48}$$

- *Gauss–Newton direction:* Use

$$R_i = H_i = \frac{1}{N}\sum_{t=1}^{N}\psi(t,\hat{\theta}^{(i)})\psi^T(t,\hat{\theta}^{(i)}) \tag{57.49}$$

- *Levenberg–Marquard direction:* Use

$$R_i = H_i + \delta I$$

where H_i is defined by Equation 57.49.
- *Conjugate gradient direction:* Construct the Newton direction from a sequence of gradient estimates. Loosely, think of V_N'' as constructed by difference approximation of d gradients. The direction in Equation 57.46 is however, constructed directly, without explicitly forming and inverting V''.

It is generally considered (Dennis and Schnabel, 1983), that the Gauss–Newton search direction is to be preferred. For ill-conditioned problems, the Levenberg–Marquard modification is recommended.

57.2.5.2 Online Algorithms

The expressions (Equations 57.44 and 57.47) for the Gauss–Newton search clearly assume that the whole data set Z^N is available during the iterations. If the application is of an off-line character, that is, the model $\hat{g}_N$ is not required during the data acquisition, this is also the most natural approach.

However, many adaptive situations require online (or recursive) algorithms, where the data are processed as they are measured. (Such algorithms are in Neural Network contexts often also used in off-line situations. Then the measured data record is concatenated with itself several times to create a (very) long record that is fed into the online algorithm.) We may refer to Ljung and Söderström (1983) as a basic source of information for recursive identification techniques.

It is natural to consider the following algorithm as the basic one:

$$\hat{\theta}(t) = \hat{\theta}(t-1) + \mu_t R_t^{-1} \psi(t, \hat{\theta}(t-1)) \varepsilon(t, \hat{\theta}(t-1)) \tag{57.50}$$

$$\varepsilon(t, \theta) = y(t) - \hat{y}(t|\theta) \tag{57.51}$$

$$R_t = R_{t-1} + \mu_t [\psi(t, \hat{\theta}(t-1)) \psi^T(t, \hat{\theta}(t-1)) - R_{t-1}] \tag{57.52}$$

The reason is that if $\hat{y}(t|\theta)$ is linear in θ, then Equations 57.50 through 57.52, with $\mu_t = 1/t$, provides the analytical solution to the minimization problem Equation 57.42. This also means that this is a natural algorithm close to the minimum, where a second order expansion of the criterion is a good approximation. In fact, it is shown in Ljung and Söderström (1983), that Equations 57.50 through 57.52 in general gives an estimate $\hat{\theta}(t)$ with the same ("optimal") statistical, asymptotic properties as the true minimum to Equation 57.42.

It should be mentioned that the quantities $\hat{y}(t|\hat{\theta}(t-1))$ and $\psi(t, \hat{\theta}(t-1))$ would normally (except in the linear regression case) require the whole data record to be computed. This would violate the recursiveness of the algorithm. In practical implementations these quantities are therefore replaced by recursively computed approximations. The idea behind these approximations is to use the defining equation for $\hat{y}(t|\theta)$ and $\psi(t, \theta)$ (which typically are recursive equations), and replace any appearance of θ with its latest available estimate. See Ljung and Söderström (1983) for more details.

Some averaged variants of Equations 57.50 through 57.52 have also been discussed:

$$\hat{\hat{\theta}}(t) = \hat{\hat{\theta}}(t-1) + \mu_t R_t^{-1} \psi(t, \hat{\theta}(t-1)) \varepsilon(t, \hat{\theta}(t-1)) \tag{57.53}$$

$$\hat{\theta}(t) = \hat{\theta}(t-1) + \rho_t [\hat{\hat{\theta}}(t) - \hat{\theta}(t-1)] \tag{57.54}$$

The basic algorithm of Equations 57.50 through 57.52 then corresponds to $\rho_t = 1$. Using $\rho_t < 1$ gives a so called "accelerated convergence" algorithm. It was introduced by Polyak and Juditsky (1992) and has then been extensively discussed by Kushner and Yang (1993), and others. The remarkable thing with this averaging is that we achieve the same asymptotic statistical properties of $\hat{\theta}(t)$ by Equations 57.53 and 57.54 with $R_t = I$ (gradient search) as by Equations 57.50 through 57.52 if

$$\rho_t = 1/t \quad \mu_t >> \rho_t \quad \mu_t \to 0$$

It is thus an interesting alternative to Equations 57.50 through 57.52, in particular if dim θ is large so R_t is a big matrix.

57.2.5.3 Local Minima

A fundamental problem with minimization tasks like Equation 57.42 is that $V_N(\theta)$ may have several or many local (nonglobal) minima, where local search algorithms may get caught. There is no easy solution to this problem. It is usually well worth the effort to find a good initial value $\theta^{(0)}$ where to start the iterations. Other than that, only various global search strategies are left, such as random search, random restarts, simulated annealing, and the genetic algorithm.

57.3 Linear Black Box Systems

57.3.1 Linear System Descriptions in General

57.3.1.1 A Linear System with Additive Disturbances

A linear system with additive disturbances $v(t)$ can be described by

$$y(t) = G(q)u(t) + v(t) \tag{57.55}$$

Here $u(t)$ is the input signal, and $G(q)$ is the transfer function from input to output $y(t)$. The symbol q is the shift operator, so Equation 57.55 should be interpreted as

$$y(t) = \sum_{k=0}^{\infty} g_k u(t-k) + v(t) = \left(\sum_{k=0}^{\infty} g_k q^{-k}\right) u(t) + v(t) \tag{57.56}$$

The disturbance $v(t)$ can in general terms be characterized by its *spectrum*, which is a description of its frequency content. It is often more convenient to describe $v(t)$ as being (thought of as) obtained by filtering a white noise source $e(t)$ through a linear filter $H(q)$:

$$v(t) = H(q)e(t) \tag{57.57}$$

This is, from a linear identification perspective, equivalent to describing $v(t)$ as a signal with spectrum

$$\Phi_v(\omega) = \lambda |H(e^{i\omega})|^2 \tag{57.58}$$

where λ is the variance of the noise source $e(t)$. We shall assume that $H(q)$ is normalized to be monic, that is,

$$H(q) = 1 + \sum_{k=1}^{\infty} h_k q^{-k} \tag{57.59}$$

Putting all of this together, we arrive at the standard linear system description

$$y(t) = G(q)u(t) + H(q)e(t) \tag{57.60}$$

57.3.1.2 Parameterized Linear Models

Now, if the transfer functions G and H in Equation 57.60 are not known, we would introduce parameters θ in their description that reflect our lack of knowledge. The exact way of doing this is the topic of this section as well as of Section 57.6.

In any case the resulting, parameterized model will be described as

$$y(t) = G(q,\theta)u(t) + H(q,\theta)e(t) \tag{57.61}$$

The parameters θ can then be estimated from data using the general procedures described in Section 57.2.

57.3.1.3 Predictors for Linear Models

Given a system description of Equation 57.61 and input–output data up to time $t-1$,

$$y(s), \quad u(s) \;\; s \le t-1 \tag{57.62}$$

how shall we predict the next output value $y(t)$?

In the general case of Equation 57.61 the prediction can be deduced in the following way: Divide Equation 57.61 by $H(q,\theta)$:

$$H^{-1}(q,\theta)y(t) = H^{-1}(q,\theta)G(q,\theta)u(t) + e(t)$$

or

$$y(t) = [1 - H^{-1}(q,\theta)]y(t) + H^{-1}(q,\theta)G(q,\theta)u(t) + e(t) \tag{57.63}$$

In view of the normalization of Equation 57.59 we find that

$$1 - H^{-1}(q,\theta) = \frac{H(q,\theta)-1}{H(q,\theta)} = \frac{1}{H(q,\theta)}\sum_{k=1}^{\infty} h_k q^{-k}$$

The expression $[1 - H^{-1}(q,\theta)]y(t)$ thus only contains old values of $y(s)$, $s \le t-1$. The right side of Equation 57.63 is thus known at time $t-1$, with the exception of $e(t)$. The prediction of $y(t)$ is simply obtained from Equation 57.63 by deleting $e(t)$:

$$\hat{y}(t|\theta) = [1 - H^{-1}(q,\theta)]y(t) + H^{-1}(q,\theta)G(q,\theta)u(t) \tag{57.64}$$

This is a general expression for how ready-made models predict the next value of the output, given old values of y and u. Inserting the predictor in Equation 57.25 (with $\ell(\varepsilon) = \varepsilon^2$) give the identification criterion

$$\frac{1}{N}\sum_{t=1}^{N} |H^{-1}(q,\theta)(y(t) - G(q,\theta)u(t))|^2 \tag{57.65}$$

57.3.1.4 A Characterization of the Limiting Model in a General Class of Linear Models

Let us apply the general limit result in Equations 57.27 and 57.28 to the linear model structure (Equation 57.61 or 57.64). If we choose a quadratic criterion $\ell(\varepsilon) = \varepsilon^2$ (in the scalar output case) then this result tells us, in the time domain, that the limiting parameter estimate is the one that minimizes the filtered prediction error variance (for the input used during the experiment). Suppose that the data actually have been generated by

$$y(t) = G_0(q)u(t) + v(t) \tag{57.66}$$

Let $\Phi_u(\omega)$ be the input spectrum and $\Phi_v(\omega)$ be the spectrum for the additive disturbance v. Then the filtered prediction error can be written

$$\begin{aligned} \varepsilon_F(t,\theta) &= \frac{L(q)}{H(q,\theta)}[y(t) - G(q,\theta)u(t)] \\ &= \frac{L(q)}{H(q,\theta)}[(G_0(q) - G(q,\theta))u(t) + v(t)] \end{aligned} \tag{57.67}$$

By Parseval's relation, the prediction error variance can also be written as an integral over the spectrum of the prediction error. This spectrum, in turn, is directly obtained from Equation 57.67, so the limit estimate θ^* in Equation 57.28 can also be defined as

$$\begin{aligned} \theta^* = \arg\min_{\theta} \Bigg[&\int_{-\pi}^{\pi} |G_0(e^{i\omega}) - G(e^{i\omega},\theta)|^2 \frac{\Phi_u(\omega)|L(e^{i\omega})|^2}{|H(e^{i\omega},\theta)|^2} d\omega \\ &+ \int_{-\pi}^{\pi} \Phi_v(\omega)|L(e^{i\omega})|^2/|H(e^{i\omega},\theta)|^2 d\omega \Bigg] \end{aligned} \tag{57.68}$$

If the noise model $H(q,\theta) = H_*(q)$ does not depend on θ [as in the output error model (Equation 57.75)] the expression (Equation 57.68) thus shows that the resulting model $G(e^{i\omega},\theta^*)$ will give that frequency function in the model set that is closest to the true one, in a quadratic frequency norm with weighting function

$$Q(\omega) = \Phi_u(\omega)|L(e^{i\omega})|^2/|H_*(e^{i\omega})|^2 \tag{57.69}$$

This shows clearly that the fit can be affected by the choice of prefilter L, the input spectrum Φ_u and the noise model H_*.

57.3.2 Linear and Ready-Made Models

Sometimes we are faced with systems or subsystems that cannot be modeled based on physical insights. The reason may be that the function of the system or its construction is unknown or that it would be too complicated to sort out the physical relationships. It is then possible to use standard models, which by experience are known to be able to handle a wide range of different system dynamics. Linear systems constitute the most common class of such standard models. From a modeling point of view these models thus serve as *ready-made models*: tell us the size (model order), and it should be possible to find something that fits (to data).

57.3.2.1 A Family of Transfer Function Models

A very natural approach is to describe G and H in Equation 57.61 as rational transfer functions in the shift (delay) operator with unknown numerator and denominator polynomials.

We would then have

$$G(q,\theta) = \frac{B(q)}{F(q)} = \frac{b_1 q^{-nk} + b_2 q^{-nk-1} + \cdots + b_{nb} q^{-nk-nb+1}}{1 + f_1 q^{-1} + \cdots + f_{nf} q^{-nf}}, \tag{57.70}$$

Then

$$\eta(t) = G(q,\theta)u(t) \tag{57.71}$$

is a shorthand notation for the relationship

$$\eta(t) + f_1\eta(t-1) + \cdots + f_{nf}\eta(t-nf) = b_1 u(t-nk) + \cdots + b_{nb}(t-(nb+nk-1)) \tag{57.72}$$

There is also a time delay of nk samples. We assume assume, for simplicity, that the sampling interval T is one time unit.

In the same way the disturbance transfer function can be written

$$H(q,\theta) = \frac{C(q)}{D(q)} = \frac{1 + c_1 q^{-1} + \cdots + c_{nc} q^{-nc}}{1 + d_1 q^{-1} + \cdots + d_{nd} q^{-nd}} \tag{57.73}$$

The parameter vector θ thus contains the coefficients b_i, c_i, d_i, and f_i of the transfer functions. This ready-made model is thus described by five structural parameters: nb, nc, nd, nf, and nk. When these have been chosen, it remains to adjust the parameters b_i, c_i, d_i, and f_i to data. This is done with the methods of Section 57.2. The ready-made model Equations 57.70 through 57.73 gives

$$y(t) = \frac{B(q)}{F(q)} u(t) + \frac{C(q)}{D(q)} e(t) \tag{57.74}$$

and is known as the *Box–Jenkins (BJ) model*, after the statisticians G. E. P. Box and G. M. Jenkins.

An important special case is when the properties of the disturbance signals are not modeled, and the noise model $H(q)$ is chosen to be $H(q) \equiv 1$; that is, $nc = nd = 0$. This special case is known as an *output error (OE) model* since the noise source $e(t)$ will then be the difference (error) between the actual output and the noise-free output:

$$y(t) = \frac{B(q)}{F(q)} u(t) + e(t) \tag{57.75}$$

A common variant is to use the same denominator for G and H:

$$F(q) = D(q) = A(q) = 1 + a_1 q^{-1} + \cdots + a_{na} a^{-na} \tag{57.76}$$

Multiplying both sides of Equation 57.74 by $A(q)$ then gives

$$A(q)y(t) = B(q)u(t) + C(q)e(t) \tag{57.77}$$

This ready-made model is known as the *ARMAX model*. The name is derived from the fact that $A(q)y(t)$ represents an AutoRegression and $C(q)e(t)$ a Moving Average of white noise, while $B(q)u(t)$ represents an eXtra input (or with econometric terminology, an eXogenous variable).

Physically, the difference between ARMAX and BJ models is that the noise and input are subjected to the same dynamics (same poles) in the ARMAX case. This is reasonable if the dominating disturbances enter early in the process (together with the input). Consider for example an airplane where the disturbances from wind gusts give rise to the same type of forces on the airplane as the deflections of the control surfaces.

Finally, we have the special case of Equation 57.77 that $C(q) \equiv 1$, that is, $nc = 0$

$$A(q)y(t) = B(q)u(t) + e(t) \tag{57.78}$$

which with the same terminology is called an *ARX model*, and which we discussed at length in Section 57.1.3. Figure 57.1 shows the most common model structures.

To use these ready-made models, decide on the orders na, nb, nc, nd, nf, and nk and let the computer pick the best model in the class thus defined. The obtained model is then scrutinized, and it might be found that other order must also be tested.

A relevant question is how to use the freedom that the different model structures give. Each of the BJ, OE, ARMAX, and ARX structures offer their own advantages, and we will discuss them in Section 57.8.2.

57.3.2.2 Prediction

Starting with model (Equation 57.74), it is possible to predict what the output $y(t)$ will be, based on measurements of $u(s)$, $y(s)$ $s \leq t-1$, using the general formula (Equation 57.64). It is easiest to calculate

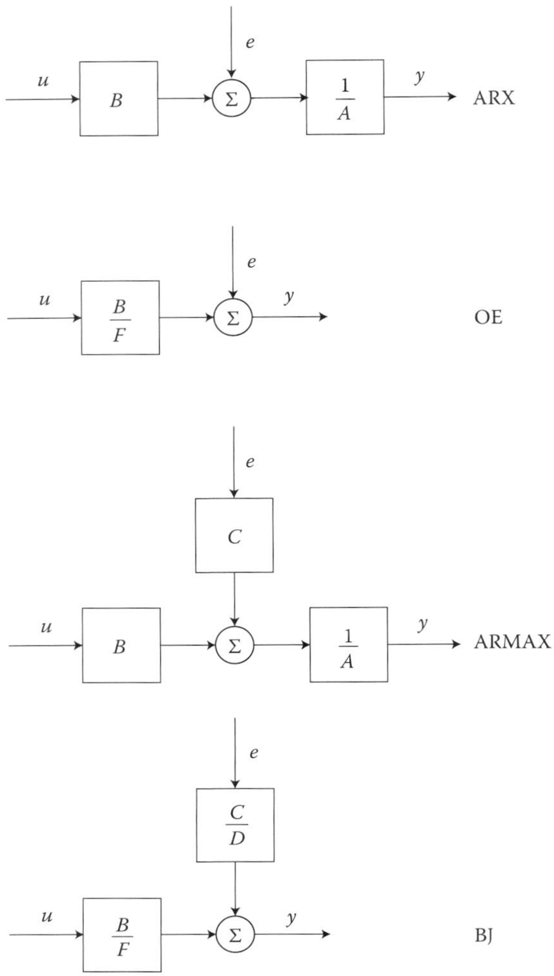

FIGURE 57.1 Model structures.

the prediction for the OE-case, $H(q, \theta) \equiv 1$, when we obtain the model

$$y(t) = G(q, \theta)u(t) + e(t)$$

with the natural prediction $(1 - H^{-1} = 0)$

$$\hat{y}(t|\theta) = G(q, \theta)u(t) \tag{57.79}$$

From the ARX case (Equation 57.78) we obtain

$$y(t) = -a_1 y(t-1) - \cdots - a_{na} y(t-na) + b_1 u(t-nk) + \cdots + b_{nb} u(t-nk-nb+1) + e(t) \tag{57.80}$$

and the prediction (delete $e(t)$!)

$$\hat{y}(t|\theta) = -a_1 y(t-1) - \cdots - a_{na} y(t-na) + b_1 u(t-nk) + \cdots + b_{nb} u(t-nk-nb+1) \tag{57.81}$$

Note the difference between Equations 57.79 and 57.81. In the OE model the prediction is based entirely on the input $\{u(t)\}$, whereas the ARX model also uses old values of the output.

57.3.2.3 Linear Regression

Both tailor-made and ready-made models describe how the predicted value of $y(t)$ depends on old values of y and u and on the parameters θ. We denote this prediction by

$$\hat{y}(t|\theta)$$

See Equation 57.64. In general this can be a rather complicated function of θ. The estimation work is considerably easier if the prediction is a linear function of θ:

$$\hat{y}(t|\theta) = \theta^T \varphi(t) \tag{57.82}$$

Here θ is a column vector that contains the unknown parameters, while $\varphi(t)$ is a column vector formed by old inputs and outputs. Such a model structure is called a *linear regression*. We discussed such models in Section 57.1.3, and noted that the ARX model (Equation 57.78) is one common model of the linear regression type. Linear regression models can also be obtained in several other ways. See Example 57.2.

57.4 Frequency Domain Techniques for Linear Models

It gives a useful complementary view as well as algorithms to consider estimating a linear model (Equation 57.61) using frequency domain data. Such techniques are carefully and comprehensively described in Schoukens and Pintelon (1991) and Pintelon and Schoukens (2001).

57.4.1 Frequency Domain Data

The data could either be Fourier transforms of inputs and outputs or estimates of the frequency function:

Frequency domain data from sampled measurements:

$$Z^N = \{U_N(e^{i\omega_1 T}), Y_N(e^{i\omega_1 T}), \ldots, U_N(e^{i\omega_N T}), Y_N(e^{i\omega_N T})\} \tag{57.83}$$

where U_N and Y_N are the discrete time Fourier transforms of sampled inputs and outputs, with sampling interval T.

$$U_N(e^{i\omega T}) = \frac{1}{\sqrt{N}} \sum_{k=1}^{N} u(kT)e^{-i\omega kT} \tag{57.84}$$

If these transforms are computed on the "DFT-grid"

$$\omega_k = 2\pi k/(NT), \quad k = 0, 1, \ldots, N-1 \tag{57.85}$$

the data (Equation 57.83) will become the DFT (discrete Fourier transform) of the time domain data (with a special normalization).

It is interesting to note (see, e.g., Theorem 14.25 in Pintelon and Schoukens (2001)) that under weak assumptions $U_N(\omega)$ and $Y_N(\omega)$ will have an asymptotically (as $N \to \infty$) normal distribution. The values will also be (asymptotically) independent at different values of ω_k on the DFT grid.

Measurements of the sampled frequency response function:

We assume that we have measurements of a sampled-data frequency function $G(e^{i\omega T})$:

$$Z^N = \{G_m(e^{i\omega_1 T}), \ldots, G_m(e^{i\omega_N T})\} \tag{57.86a}$$

$$Z_U^N = \{W(i\omega_1), \ldots, W(i\omega_N)\} \tag{57.86b}$$

where the values W are some kind of reliability measure of measurements. The frequency function estimates G_m in Equation 57.86 can be directly measured by certain hardware equipment, *frequency analyzers*. Such an equipment could implement Fourier analysis as in Equation 57.87 below, or could rely upon the definition of frequency responses by directly measuring phase and amplitude shifts for a number of different sinusoidal inputs.

The frequency responses can also be estimated/constructed from measured data either in the time or the frequency domain. This is the topic of *Spectral Analysis*, which is further dealt with in Section 57.5.2. Let us comment on the simplest case of spectral analysis, the *Empirical Transfer Function Estimate, (ETFE)*. It is formed as the ratio of the output and input Fourier transforms

$$\hat{\hat{G}}_N(e^{i\omega T}) = \frac{Y_N(e^{i\omega T})}{U_N(e^{i\omega T})} \tag{57.87}$$

in the discrete time case. If the observations y and u have been obtained from a noise-corrupted linear system with frequency function $G_0(i\omega)$ it can be shown that the ETFE has the following statistical properties: Lemma 6.1 in Ljung (1999).

$$E\hat{\hat{G}}_N(i\omega) = G_0(i\omega) + \frac{\rho_1}{\sqrt{N}U_N(i\omega)} \tag{57.88a}$$

$$E\left|\hat{\hat{G}}_N(i\omega) - G_0(i\omega)\right|^2 = \frac{\Phi_v(\omega)}{|U_N(i\omega)|^2} + \frac{\rho_2}{N|U_N(i\omega)|^2} \tag{57.88b}$$

Here $\Phi_v(\omega)$ is the spectrum of the additive noise (at the output of the system) and ρ_i are constant bounds that depend on the impulse response of the system, the bound on the input, and the covariance function of the noise.

All this means that we can think of the ETFE as a "noisy measurement" of the frequency function:

$$\hat{\hat{G}}_N(i\omega_k) = G_0(i\omega_k) + v_k \tag{57.89}$$

with v_k being a zero mean random variable with variance $\Phi_v(\omega_k)/|U_N(\omega_k)|^2$. We have then ignored the terms with ρ in the expressions above. Note that the variance of v_k would correspond to the uncertainty estimate $W(i\omega_k)$ in Equation 57.86b.

Something must also be said about the frequency grid in Equation 57.89: If the Fourier transforms are obtained by DFT of equidistantly sampled data, the natural frequencies to use in Equation 57.89 are the DFT grid:

$$\omega_k = 2k\pi/(NT); \quad k = 0, \ldots, N-1 \tag{57.90}$$

This gives two advantages:

- Frequencies in between these grid points carry no extra information: they are merely (trigonometric) interpolations of the values on the DFT grid. This also determines the maximum frequency resolution of the frequency function.
- v_k are (asymptotically) uncorrelated on this grid.

In the case of p outputs, v_k is a column vector and Φ_v is a $p \times p$ matrix.

The expression (Equation 57.87) assumes that u is a scalar (single input system). Formulas for multi-input systems are given in Ljung (2003b).

57.4.2 Fitting Models to Fourier Transformed Data

Consider now the case that we are given Fourier transform values of the inputs and the outputs as in Equation 57.83. The relationship between these values is obtained from Equation 57.55 (with G_0 denoting the "true value"):

$$Y_N(e^{i\omega T}) = G_0(e^{i\omega T})U_N(e^{i\omega T}) + V_N(e^{i\omega T})$$

$$E\left|V_N(e^{i\omega T})\right|^2 = \Phi_v(\omega) \tag{57.91}$$

where V is the transform of the noise, corresponding to Equation 57.84. The relationship is not exact, only approximate, since there are transients and deviations due to the fact that the data may not be periodic.

The relation (Equation 57.91) is like a measurement equation with uncertainties

$$y(t) = \theta\phi(t) + e(t), \quad \text{Var}(e(t)) = \alpha_t \tag{57.92}$$

and applying least squares, with (optimal) weights being inverse error variances gives the estimation method

$$\min_{\theta} \sum |y(t) - \theta\phi(t)|^2/\alpha_t \tag{57.93}$$

In our case we obtain a criterion

$$V_N(\theta, Z^N) = \sum_{k=1}^{N} \left| Y_N(e^{i\omega_k T}) - G(e^{i\omega_k T}, \theta) U_N(e^{i\omega_k T}) \right|^2 \Big/ \Phi_v(\omega_k) \tag{57.94}$$

In case the spectrum for v is not given, but parameterized as in Equation 57.58, we get a parameterized weighting, which should be balanced that as in:

$$\begin{aligned} V_N(\theta, Z^N) &= \sum_{k=1}^{N} |Y_N^k - G(e^{i\omega_k T}, \theta) U_N^k|^2 / (\lambda |H(e^{i\omega_k T}, \theta)|^2) \\ &\quad + \sum_{k=1}^{N} \log \lambda |H(e^{i\omega_k T}, \theta)|^2 \end{aligned} \tag{57.95}$$

where, for short, $Y_N^k = Y(e^{i\omega_k T})$. Indeed, this will be the true log-likelihood function in case Y_N^k are Gaussian distributed and independent for different k.

Remark

It may be noted that

$$\int_{-\pi}^{\pi} \log |H(e^{i\omega})| = 0 \tag{57.96}$$

for any monic, stable, and inversely stable transfer function H. This means that the last sum in Equation 57.95 is almost θ-independent for large N and equidistant frequency points.

57.4.3 Fitting to Frequency Response Data

Suppose now that the data is given in terms of measured frequency-response function values, Equation 57.86. A clear-cut curve-fitting approach to estimating the model would be to form the analog of Equation 57.24:

$$V_N(\theta, Z^N) = \sum_{k=1}^{N} |G_m e^{(i\omega_k)} - G(e^{i\omega_k}, \theta)|^2 / W(i\omega_k) \tag{57.97}$$

where we used the uncertainty measure in Equation 57.86b for the weights.

In the case that the frequency function estimate is an ETFE as in Equation 57.87 we would use the uncertainty measure Equation 57.88b. This gives

$$\begin{aligned} |G_m(i\omega) - G(i\omega, \theta)|^2 / W(i\omega) &= \left| \frac{Y_N(i\omega)}{U_N(i\omega)} - G(i\omega, \theta) \right|^2 \frac{|U_N(i\omega)|^2}{\Phi_v(\omega)} \\ &= |Y_N(i\omega) - G(i\omega) U_N(i\omega)|^2 \big/ \Phi_v(\omega) \end{aligned}$$

This means that for these frequency function estimates, the criterion Equation 57.97 exactly coincides with Equation 57.65.

57.4.4 Connections between Time and Frequency Domains

Let us show the relationship between frequency domain fit (Equation 57.94) and the time domain fit (Equation 57.65) using Parseval's relationship. The Fourier transform (Equation 57.84) of the prediction error is (neglecting transients or assuming periodic data):

$$E_N(e^{i\omega T},\theta) = H^{-1}(e^{i\omega T},\theta)(Y_N(e^{i\omega T}) - G(e^{i\omega T},\theta)U_N(e^{i\omega T}))$$

Applying Parseval's relationship to Equation 57.65 and ignoring transient effects (or assuming periodic data) now gives for this criterion

$$V(\theta, Z^N) = \sum_{k=1}^{N} |Y_N(e^{i\omega_k T}) - G(e^{i\omega_k T},\theta)U_N(e^{i\omega_k T})|^2 / |H(e^{i\omega_k T},\theta)|^2 \tag{57.98}$$

Dividing this expression by λ and using Equation 57.58 we see that this expression is exactly equal to Equation 57.94. Consequently, also the time domain expression (Equation 57.65) can be interpreted as curve fitting the parameterized model to the ETFE. We have also displayed the nature of the noise model in Equation 57.55: it just provides the weighting in this fit. See McKelvey (2000) and the special session Schoukens et al. (2004), Ljung (2004) for a closer discussion on time and frequency domain connections.

57.5 Special Estimation Techniques for Linear Black-Box Models

An important feature of a linear, time invariant system is that it is entirely characterized by its *impulse response*. So if we know the system's response to an impulse, we will also know its response to any input. Equivalently, we could study the *frequency response*, which is the Fourier transform of the impulse response.

In this section, we shall consider estimation methods for linear systems, that do not use particular model parameterizations. First, in Section 57.5.1, we shall consider direct methods to determine the impulse response and the frequency response, by simply applying the definitions of these concepts.

In Section 57.5.2, spectral analysis for frequency function estimation will be discussed. Finally, in Section 57.5.3, a recent method to estimate general linear systems (of given order, by unspecified structure) will be described.

57.5.1 Transient and Frequency Analysis

57.5.1.1 Transient Analysis

The first step in modeling is to decide which quantities and variables are important to describe what happens in the system. A simple and common kind of experiment that shows how and in what time span various variables affect each other is called *step-response analysis* or *transient analysis*. In such experiments the inputs are varied (typically one at a time) as a step: $u(t) = u_0$, $t < t_0$; $u(t) = u_1$, $t \geq t_0$. The other measurable variables in the system are recorded during this time. We thus study the *step response* of the system. An alternative would be to study the impulse response of the system by letting the input be a pulse of short duration. From such measurements, information of the following nature can be found:

1. The variables affected by the input in question. This makes it easier to draw block diagrams for the system and to decide which influences can be neglected.
2. The time constants of the system. This also allows us to decide which relationships in the model can be described as static (i.e., they have significantly faster time constants than the time scale we are working with).

3. The characteristic (oscillatory, poorly damped, monotone, and the like) of the step responses, as well as the levels of static gains. Such information is useful when studying the behavior of the final model in simulation. Good agreement with the measured step responses should give a certain confidence in the model.

57.5.1.2 Frequency Analysis

If a linear system has the transfer function $G(q)$ and the input is

$$u(t) = u_0 \cos \omega kT, \quad (k-1)T \leq t \leq kT \tag{57.99}$$

then the output after possible transients have faded away will be

$$y(t) = y_0 \cos(\omega t + \varphi), \quad \text{for } t = T, 2T, 3T, \ldots \tag{57.100}$$

where

$$y_0 = |G(e^{i\omega T})| \cdot u_0 \tag{57.101}$$

$$\varphi = \arg G(e^{i\omega T}) \tag{57.102}$$

If the system is driven by the input (Equations 57.99) for a certain u_0 and ω_1 and we measure y_0 and φ from the output signal, it is possible to determine the complex number $G(e^{i\omega_1 T})$ using Equation 57.101 and Equation 57.102. By repeating this procedure for a number of different ω, we can get a good estimate of the frequency function $G(e^{i\omega T})$. This method is called *frequency analysis*. Sometimes it is possible to see or measure u_0, y_0, and φ directly from graphs of the input and output signals. Most of the time, however, there will be noise and irregularities that make it difficult to determine φ directly. A suitable procedure is then to correlate the output with $\cos \omega t$ and $\sin \omega t$.

57.5.2 Estimating the Frequency Response by Spectral Analysis

57.5.2.1 Definitions

(In this section, the sampling interval T is assumed to be one time unit; $T = 1$.) The *cross spectrum* between two (stationary) signals $u(t)$ and $y(t)$ is defined as the Fourier transform of their cross covariance function, provided this exists:

$$\Phi_{yu}(\omega) = \sum_{\tau=-\infty}^{\infty} R_{yu}(\tau) e^{-i\omega\tau} \tag{57.103}$$

where $R_{yu}(\tau)$ is defined by

$$R_{yu}(\tau) = E(y(t) - Ey(t))(u(t-\tau) - Eu(t-\tau)) \tag{57.104}$$

This cross covariance function is typically estimated as

$$\hat{R}_{yu}^{N}(\tau) = \frac{1}{N} \sum_{t=1}^{N} y(t) u(t-\tau) \tag{57.105}$$

The (auto) *spectrum* $\Phi_u(\omega)$ of a signal u is defined as $\Phi_{uu}(\omega)$, that is, as its cross spectrum with itself.

The spectrum describes the frequency contents of the signal. The connection to more explicit Fourier techniques is evident by the following relationship:

$$\Phi_u(\omega) = \lim_{N \to \infty} |U_N(\omega)|^2 \tag{57.106}$$

where U_N is the discrete time Fourier transform (Equation 57.84). The relationship (Equation 57.106) is shown in Ljung and Glad (1994).

Consider now the general linear model (Equation 57.55):

$$y(t) = G(q)u(t) + v(t)$$

It is straightforward to show that the relationships between the spectra and cross spectra of y and u (provided u and v are uncorrelated) is given by

$$\Phi_{yu}(\omega) = G(e^{i\omega})\Phi_u(\omega) \tag{57.107}$$

$$\Phi_y(\omega) = |G(e^{i\omega})|^2\Phi_u(\omega) + \Phi_v(\omega) \tag{57.108}$$

It is easy to see how the transfer function $G(e^{i\omega})$ and the noise spectrum $\phi_v(\omega)$ can be estimated using these expressions, if only we have a method to estimate cross spectra.

57.5.2.2 Estimation of Spectra

The spectrum is defined as the Fourier transform of the correlation function. A natural idea would then be to take the transform of the estimate $\hat{R}^N_{yu}(\tau)$ in Equation 57.105. That will not work in most cases, though. The reason could be described as follows: the estimate $\hat{R}^N_{yu}(\tau)$ is not reliable for large τ, since it is based on only a few observations. These "bad" estimates are mixed with good ones in the Fourier transform, thus creating an overall bad estimate. It is better to introduce a weighting, so that correlation estimates for large lags τ carry a smaller weight:

$$\hat{\Phi}^N_{yu}(\omega) = \sum_{\ell=-\gamma}^{\gamma} \hat{R}^N_{yu}(\ell) \cdot w_\gamma(\ell)e^{-i\ell\omega} \tag{57.109}$$

This spectral estimation method is known as the The *Blackman–Tukey approach*. Here $w_\gamma(\ell)$ is a window function that decreases with $|\tau|$. This function controls the trade-off between *frequency resolution* and *variance of the estimate*. A function that gives significant weights to the correlation at large lags will be able to provide finer frequency details (a longer time span is covered). At the same time it will have to use "bad" estimates, so the statistical quality (the variance) is poorer. We shall return to this trade-off in a moment. How should we choose the shape of the window function $w_\gamma(\ell)$? There is no optimal solution to this problem, but the most common window used in spectral analysis is the *Hamming window*:

$$\begin{aligned} w_\gamma(k) &= \frac{1}{2}\left(1 + \cos\frac{\pi k}{\gamma}\right) \quad |k| < \gamma \\ w_\gamma(k) &= 0|k| \geq \gamma \end{aligned} \tag{57.110}$$

From the spectral estimates Φ_u, Φ_y and Φ_{yu} obtained in this way, we can now use Equation 57.107, to obtain a natural estimate of the frequency function $G(e^{i\omega})$:

$$\hat{G}_N(e^{i\omega}) = \frac{\hat{\Phi}^N_{yu}(\omega)}{\hat{\Phi}^N_u(\omega)} \tag{57.111}$$

Furthermore, the disturbance spectrum can be estimated from Equation 57.108 as

$$\hat{\Phi}^N_v(\omega) = \hat{\Phi}^N_y(\omega) - \frac{|\hat{\Phi}^N_{yu}(\omega)|^2}{\hat{\Phi}^N_u(\omega)} \tag{57.112}$$

To compute these estimates, the following steps are performed:

Algorithm SPA (57.113)

1. Collect data $y(k)$, $u(k)$ $k = 1, \ldots, N$.
2. Subtract the corresponding sample means form the data. This will avoid bad estimates at very low frequencies.

3. Choose the width of the lag window $w_\gamma(k)$.
4. Compute $\hat{R}_y^N(k)$, $\hat{R}_u^N(k)$, and $\hat{R}_{yu}^N(k)$ for $|k| \leq \gamma$ according to Equation 57.105.
5. Form the spectral estimates $\hat{\Phi}_y^N(\omega)$, $\hat{\Phi}_u^N(\omega)$, and $\hat{\Phi}_{yu}^N(\omega)$ according to Equation 57.109 and analogous expressions.
6. Form Equation 57.111 and possibly also Equation 57.112.

The user only has to choose γ. A good value for systems without sharp resonances is $\gamma = 20$ to 30. Larger values of γ may be required for systems with narrow resonances.

57.5.2.3 Quality of the Estimates

The estimates $\hat{G}_N$ and $\hat{\Phi}_w^N$ are formed entirely from estimates of spectra and cross spectra. Their properties will therefore be inherited from the properties of the spectral estimates. For the Hamming window with width γ, it can be shown that the frequency resolution will be about

$$\frac{\pi}{\gamma\sqrt{2}} \quad \text{radians/time unit} \tag{57.114}$$

This means that details in the true frequency function that are finer than this expression will be smeared out in the estimate. It is also possible to show that the estimate's variances satisfy

$$\text{Var } \hat{G}_N(i\omega) \approx 0.7 \cdot \frac{\gamma}{N} \cdot \frac{\Phi_v(\omega)}{\Phi_u(\omega)} \tag{57.115}$$

and

$$\text{Var } \hat{\Phi}_v^N(\omega) \approx 0.7 \cdot \frac{\gamma}{N} \cdot \Phi_v^2(\omega) \tag{57.116}$$

["Variance" here refers to taking expectation over the noise sequence $v(t)$.] Note that the relative variance in Equation 57.115 typically increases dramatically as ω tends to the Nyquist frequency. The reason is that $|G(i\omega)|$ typically decays rapidly, while the noise-to-signal ratio $\Phi_v(\omega)/\Phi_u(\omega)$ has a tendency to increase as ω increases. In a Bode diagram the estimates will thus show considerable fluctuations at high frequencies. Moreover, the constant frequency resolution (Equation 57.114) will look thinner and thinner at higher frequencies in a Bode diagram due to the logarithmic frequency scale. See Ljung and Glad (1994) for a more detailed discussion.

57.5.2.4 Choice of Window Size

The choice of γ is a pure trade-off between frequency resolution and variance (variability). For a spectrum with narrow resonance peaks it is thus necessary to choose a large value of γ and accept a higher variance. For a more flat spectrum, smaller values of γ will do well. In practice a number of different values of γ are tried out. Often we start with a small value of γ and increase it successively until an estimate is found that balances the trade-off between frequency resolution (true details) and variance (random fluctuations). A typical value for spectra without narrow resonances is $\gamma = 20$–30.

57.5.3 Subspace Estimation Techniques for State-Space Models

A linear system can always be represented in state-space form as

$$\begin{aligned} x(t+1) &= Ax(t) + Bu(t) + w(t) \\ y(t) &= Cx(t) + Du(t) + v(t) \end{aligned} \tag{57.117}$$

with white noises w and v. Alternatively, we could just represent the input–output dynamics as in

$$\begin{aligned} x(t+1) &= Ax(t) + Bu(t) \\ y(t) &= Cx(t) + Du(t) + v(t) \end{aligned} \tag{57.118}$$

where the noise at the output, v, very well could be colored. It should be noted that the input–output dynamics could be represented with a lower order model in Equation 57.118 than in Equation 57.117 since describing the noise character might require some extra states.

To estimate such a model, the matrices can be parameterized in ways that are described in Section 57.6—either from physical grounds or as black boxes in canonical forms.

However, there are also other possibilities: We assume that we have no insight into the particular structure, and we would just estimate any matrices A, B, C, and D that give a good description of the input–output behavior of the system. Since there are an infinite number of such matrices that describe the same system (the similarity transforms), we will have to fix the coordinate basis of the state-space realization.

Let us for a moment assume that not only are u and y measured, but also the sequence of state vectors x. This would, by the way, fix the state-space realization coordinate basis. Now, with known u, y, and x, the model (Equation 57.117) becomes a linear regression: the unknown parameters, all of the matrix entries in all the matrices, mix with measured signals in linear combinations. To see this clearly, let

$$Y(t) = \begin{bmatrix} x(t+1) \\ y(t) \end{bmatrix} \quad \Theta = \begin{bmatrix} A & B \\ C & D \end{bmatrix}$$

$$\Phi(t) = \begin{bmatrix} x(t) \\ u(t) \end{bmatrix} \quad E(t) = \begin{bmatrix} w(t) \\ v(t) \end{bmatrix}$$

Then, Equation 57.117 can be rewritten as

$$Y(t) = \Theta\Phi(t) + E(t) \tag{57.119}$$

From this, all the matrix elements in Θ can be estimated by the simple least squares method (which in the case of Gaussian noise and known covariance matrix coincides with the maximum likelihood method), as described in Section 57.1.3 The covariance matrix for $E(t)$ can also be estimated easily as the sample sum of the squared model residuals. That will give the covariance matrices as well as the cross covariance matrix for w and v. These matrices will, among other things, allow us to compute the Kalman filter for Equation 57.117. Note that all of the above holds without changes for multivariable systems, that is, when the output and input signals are vectors.

The only remaining problem is where to get the state vector sequence x from? It has long been known (Rissanen, 1974; Akaike, 1974b), that all state vectors $x(t)$ that can be reconstructed from input–output data, in fact, are linear combinations of the components of the n k-step ahead output predictors

$$\hat{y}(t+k|t), \quad k = \{1, 2, \ldots, n\} \tag{57.120}$$

where n is the model order (the dimension of x). See also Appendix 4. A in Ljung (1999). We could then form these predictors, and select a basis among their components:

$$x(t) = L\hat{Y}_r(t) \tag{57.121}$$

$$\hat{Y}_t(t) = \begin{bmatrix} \hat{y}(t+1|t) \\ \vdots \\ \hat{y}(t+r|t) \end{bmatrix} \tag{57.122}$$

The choice of L will determine the basis for the state-space realization, and is done in such a way that it is well conditioned. The predictor $\hat{y}(t+k|t)$ is a linear function of $u(s), y(s), 1 \leq s \leq t$ and can efficiently be determined by linear projections directly on the input–output data. (There is one complication in that $u(t+1), \ldots, u(t+k)$ should not be predicted, even if they affect $y(t+k)$.)

For practical reasons the predictor is approximated so that it only depends on a fixed and finite amount of past data, like the s_1 past outputs and the s_2 past inputs. This means that it takes the form

$$\hat{y}(t+k-1|t-1) = \alpha_1 y(t-1) + \cdots + \alpha_{s_1} y(t-s_1) + \beta_1 u(t-1) + \cdots + \beta_{s_2} u(t-s_2) \tag{57.123}$$

This predictor can then efficiently be determined by another linear least-squares projection directly on the input–output data. That is, set up the model

$$y(t+k-1) = \theta_k^T \varphi_s(t) + \gamma_k^T U(t) + \varepsilon(t+k-1) \tag{57.124}$$

where θ and φ are the variables in Equation 57.123 and U accounts for the influence of $u(t+k-j)$, $j \le k$) on $y(t+k-1)$ which should not be modeled in the predictor.

For large enough s, this will give a good approximation of the true predictors.

The method thus consists of the following steps:

Basic Subspace Algorithm (57.125)

1. Choose s_1, s_2, and r and form $\hat{Y}_r(t)$ in Equations 57.123 and 57.122 and $\mathbf{Y}$ as in:
$$\mathbf{Y} = \left[\hat{Y}_r(1) \ \ldots \ \hat{Y}_r(N)\right]. \tag{57.126}$$
2. Estimate the rank n of $\mathbf{Y}$ and determine L in Equation 57.121 so that $x(t)$ corresponds to a well-conditioned basis for it.
3. Estimate A, B, C, D and the noise covariance matrices by applying the LS method to the linear regression (Equation 57.119).

What we have described now is the *subspace projection* approach to estimating the matrices of the state-space model (Equation 57.117), including the basis for the representation and the noise covariance matrices. There are a number of variants of this approach. See among several references, for example, Van Overschee and DeMoor (1996), Larimore (1983), Verhaegen (1994) or Sections 7.3 or 10.6 in Ljung (1999).

The approach gives very useful algorithms for model estimation, and is particularly well suited for multivariable systems. The algorithms also allow numerically very reliable implementations, and typically produce estimated models with good quality. If desired, the quality may be improved by using the model as an initial estimate for the prediction error method (Equation 57.24).

The algorithms contain a number of choices and options, like how to choose s_i and r, and also how to carry out step number 3. There are also several "tricks" to do step 3 so as to achieve consistent estimates even for finite values of s_i. Accordingly, several variants of this method exist.

57.6 Physically Parameterized Models

So far, we have treated the parameters θ only as vehicles to give reasonable flexibility to the transfer functions in the general linear model (Equation 57.61). This model can also be arrived at from other considerations.

Consider a continuous time state-space model

$$\dot{x}(t) = A(\theta)x(t) + B(\theta)u(t) \tag{57.127a}$$

$$y(t) = C(\theta)x(t) + v(t) \tag{57.127b}$$

Here $x(t)$ is the state vector and typically consists of physical variables (such as positions and velocities etc.). The state-space matrices A, B, and C are parameterized by the parameter vector θ, reflecting the physical insight we have into the process. The parameters could be physical constants (resistance, heat transfer coefficients, aerodynamical derivatives, etc.) whose values are not known. They could also reflect other types of insights into the system's properties.

Example 57.3 An Electric Motor

Consider an electric motor with the input u being the applied voltage and the output y being the angular position of the motor shaft.

A first, but reasonable approximation of the motor's dynamics is as a first-order system from voltage to angular velocity, followed by an integrator:

$$G(s) = \frac{b}{s(s+a)}$$

If we select the state variables

$$x(t) = \begin{bmatrix} y(t) \\ \dot{y}(t) \end{bmatrix}$$

we obtain the state-space form

$$\begin{aligned} \dot{x} &= \begin{bmatrix} 0 & 1 \\ 0 & -a \end{bmatrix} x + \begin{bmatrix} 0 \\ b \end{bmatrix} u \\ y &= \begin{bmatrix} 1 & 0 \end{bmatrix} x + v \end{aligned} \tag{57.128}$$

where v denotes disturbances and noise. In this case, we thus have

$$\begin{aligned} \theta &= \begin{bmatrix} a \\ b \end{bmatrix} \\ A(\theta) &= \begin{bmatrix} 0 & 1 \\ 0 & -a \end{bmatrix} \quad B(\theta) = \begin{bmatrix} 0 \\ b \end{bmatrix} \\ C &= \begin{bmatrix} 1 & 0 \end{bmatrix} \end{aligned} \tag{57.129}$$

The parameterization reflects our insight that the system contains an integration, but is in this case not directly derived from detailed physical modeling. Basic physical laws would in this case have given us how θ depends on physical constants, such as resistance of the wiring, amount of inertia, friction coefficients and magnetic field constants.

Now, how do we fit a continuous-time model (Equation 57.127a) to sampled observed data? If the input $u(t)$ has been piecewise constant over the sampling interval

$$u(t) = u(kT) \quad kT \leq t < (k+1)T$$

then the states, inputs and outputs at the sampling instants will be represented by the discrete time model

$$\begin{aligned} x((k+1)T) &= \bar{A}(\theta)x(kT) + \bar{B}(\theta)u(kT) \\ y(kT) &= C(\theta)x(kT) + v(kT) \end{aligned} \tag{57.130}$$

where

$$\bar{A}(\theta) = e^{A(\theta)T}, \quad \bar{B}(\theta) = \int_0^T e^{A(\theta)\tau}B(\theta)\,d\tau \tag{57.131}$$

This follows from solving Equation 57.127 over one sampling period. We could also further model the added noise term $v(kT)$ and represent the system in the innovations form

$$\begin{aligned}\bar{x}((k+1)T) &= \bar{A}(\theta)\bar{x}(kT) + \bar{B}(\theta)u(kT) + \bar{K}(\theta)e(kT) \\ y(kT) &= C(\theta)\bar{x}(kT) + e(kT)\end{aligned} \tag{57.132}$$

where $\{e(kT)\}$ is white noise. The step from Equations 57.130 through 57.132 is really a standard Kalman filter step: $\bar{x}$ will be the one-step ahead predicted Kalman states. A pragmatic way to think about it is as follows: In Equation 57.130 the term $v(kT)$ may not be white noise. If it is colored we may separate out that part of $v(kT)$ that cannot be predicted from past values. Denote this part by $e(kT)$: it will be the *innovation*. The other part of $v(kT)$—the one that can be predicted—can then be described as a combination of earlier innovations, $e(\ell T)$ $\ell < k$. Its effect on $y(kT)$ can then be described via the states, by changing them from x to $\bar{x}$, where $\bar{x}$ contains additional states associated with getting $v(kT)$ from $e(\ell T)$, $k \leq \ell$.

Now Equation 57.132 can be written in input–output form (let $T = 1$)

$$y(t) = G(q,\theta)u(t) + H(q,\theta)e(t) \tag{57.133}$$

with

$$\begin{aligned}G(q,\theta) &= C(\theta)(qI - \bar{A}(\theta))^{-1}\bar{B}(\theta) \\ H(q,\theta) &= I + C(\theta)(qI - \bar{A}(\theta))^{-1}\bar{K}(\theta)\end{aligned} \tag{57.134}$$

We are thus back at the basic linear model (Equation 57.61). The parameterization of G and H in terms of θ is, however, more complicated than the ones we discussed in Section 57.3.2.

The general estimation techniques, model properties [including the characterization (Equation 57.68)], algorithms, and so on, apply exactly as described in Section 57.2.

From these examples it is also quite clear that nonlinear models with unknown parameters can be approached in the same way. We would then typically arrive at a structure

$$\begin{aligned}\dot{x}(t) &= f(x(t), u(t), \theta) \\ y(t) &= h(x(t), u(t), \theta) + v(t).\end{aligned} \tag{57.135}$$

In this model, all noise effects are collected as additive output disturbances $v(t)$ which is a restriction, but also a very helpful simplification. If we define $\hat{y}(t|\theta)$ as the simulated output response to Equation 57.135, for a given input, ignoring the noise $v(t)$, everything that was said in Section 57.2 about parameter estimation, model properties, and so on, is still applicable.

57.7 Nonlinear Black-Box Models

In this section, we describe the basic ideas behind model structures that have the capability to cover any nonlinear mapping from past data to the predicted value of $y(t)$. Recall that we defined a general model structure as a parameterized mapping in Equation 57.19:

$$\hat{y}(t|\theta) = g(\theta, Z^{t-1}) \tag{57.136}$$

We consequently allow quite general nonlinear mappings g. This section will deal with some general principles for how to construct such mappings, and will cover Artificial Neural Networks (ANN) as a special case. See Sjöberg et al. (1995) and Juditsky et al. (1995) for comprehensive surveys.

57.7.1 Nonlinear Black-Box Structures

Now, the model structure family (Equation 57.136) is really too general, and it turns out to be useful to write g as a concatenation of two mappings: one that takes the increasing number of past observations Z^{t-1} and maps them into a finite dimensional vector $\varphi(t)$ of fixed dimension and one that takes this vector to the space of the outputs:

$$\hat{y}(t|\theta) = g(\theta, Z^{t-1}) = g(\varphi(t), \theta) \tag{57.137}$$

where

$$\varphi(t) = \varphi(Z^{t-1}) \tag{57.138}$$

Let the dimension of φ be d. As before, we shall call this vector the *regression vector* and its components will be referred to as the *regressors*. We also allow the more general case that the formation of the regressors is itself parameterized:

$$\varphi(t) = \varphi(Z^{t-1}, \eta) \tag{57.139}$$

which we, for short, write $\varphi(t, \eta)$. For simplicity, the extra argument η will, however, be used explicitly only when essential for the discussion.

The choice of the nonlinear mapping in Equation 57.136 has thus been reduced to two partial problems for dynamical systems:

1. How to choose the nonlinear mapping $g(\varphi)$ from the regressor space to the output space (i.e., from R^d to R^p).
2. How to choose the regressors $\varphi(t)$ from past inputs and outputs.

The second problem is the same for all dynamical systems, and it turns out that the most useful choices of regression vectors are to let them contain past inputs and outputs, and possibly also past predicted/simulated outputs. The regression vector will thus be of the character in (Equation 57.4). We now turn to the first problem.

57.7.2 Nonlinear Mappings: Possibilities

57.7.2.1 Function Expansions and Basis Functions

The nonlinear mapping

$$g(\varphi, \theta) \tag{57.140}$$

goes from R^d to R^p for any given θ. At this point it does not matter how the regression vector φ is constructed. It is just a vector that lives in R^d.

It is natural to think of the parameterized function family as function expansions:

$$g(\varphi, \theta) = \sum \theta(k) g_k(\varphi) \tag{57.141}$$

where g_k are the *basis functions* and the coefficients $\theta(k)$ are the "coordinates" of g in the chosen basis.

Now, the only remaining question is: how to choose the basis functions g_k? Depending on the support of g_k [i.e., the area in R^d for which $g_k(\varphi)$ is (practically) nonzero] we shall distinguish between three types of basis functions

- Global basis functions
- Semiglobal or ridge-type basis functions
- Local basis functions

57.7.2.2 Global Basis Function

A typical and classical global basis function expansion would then be the Taylor series, or polynomial expansion, where g_k would contain multinomials in the components of φ of total degree k. Fourier series are also relevant examples. We shall, however, not discuss global basis functions here any further. Experience has indicated that they are inferior to the semilocal and local ones in typical practical applications.

57.7.2.3 Local Basis Functions

Local basis functions have their support only in some neighborhood of a given point. Think (in the case of $p = 1$) of the indicator function for the unit cube:

$$\kappa(\varphi) = 1 \;\text{ if } |\varphi_k| \leq 1 \;\forall k, \quad \text{and} \quad 0 \text{ otherwise} \tag{57.142}$$

By scaling the cube and placing it at different locations we obtain the functions

$$g_k(\varphi) = \kappa(\alpha_k * (\varphi - \beta_k)) \tag{57.143}$$

By allowing α to be a vector of the same dimension as φ and interpreting the multiplication $*$ as component-wise multiplication (like ".*" in MATLAB) we may also reshape the cube to be any parallelepiped. The parameters α are thus *scaling* or *dilation* parameters while β determine *location* or *translation*. For notational convenience we write

$$g_k(\varphi) = \kappa(\alpha_k * (\varphi - \beta_k)) = \kappa(\rho_k \cdot \varphi) \tag{57.144}$$

where

$$\rho_k = [\alpha_k, \alpha_k * \beta_k]$$

In the last equality, with some abuse of notation, we expanded the regression vector φ to contain some "1":s. This is to stress the point that the argument of the basic function κ is bilinear in the scale and location parameters ρ_k and in the regression vector φ. The notation $\rho_k \cdot \varphi$ indicates this.

This choice of g_k in Equation 57.141 gives functions that are piecewise constant over areas in R^d that can be chosen arbitrarily small by proper choice of the scaling parameters. It should be fairly obvious that such functions g_k can approximate any reasonable function arbitrarily well.

Now it is also reasonable that the same will be true for any other localized function, such as the Gaussian bell function:

$$\kappa(\varphi) = e^{-|\varphi|^2} \tag{57.145}$$

57.7.2.4 Ridge-Type Basis Functions

A useful alternative is to let the basis functions be local in one direction of the φ-space and global in the others. This is achieved quite analogously to Equation 57.143 as follows. Let $\sigma(x)$ be a local function from R to R. Then form

$$g_k(\varphi) = \sigma(\alpha_k^T(\varphi - \beta_k)) = \sigma(\alpha_k^T\varphi + \gamma_k) = \sigma(\rho_k \cdot \varphi) \tag{57.146}$$

where the scalar $\gamma_k = -\alpha_k^T\beta_k$, and

$$\rho_k = [\alpha_k, \gamma_k]$$

Note the difference with Equation 57.143! The scalar product $\alpha_k^T\varphi$ is constant in the subspace of R^d that is perpendicular to the scaling vector α_k. Hence, the function $g_k(\varphi)$ varies like σ in a direction parallel to α_k and is constant across this direction. This motivates the term *semiglobal* or *ridge-type* for this choice of functions.

As in Equation 57.143 we expanded in the last equality in Equation 57.146 the vector φ with the value "1," again just to emphasize that the argument of the fundamental basis function σ is bilinear in ρ and φ.

57.7.2.5 Connection to "Named Structures"

Here we briefly review some popular structures, other structures related to interpolation techniques are discussed in Sjöberg et al. (1995), Juditsky et al. (1995).

57.7.2.5.1 *Wavelets*

The local approach corresponding to Equations 57.141 and 57.143 has direct connections to wavelet networks and wavelet transforms. The exact relationships are discussed in Sjöberg et al. (1995). Loosely, we note that via the dilation parameters in ρ_k we can work with different scales simultaneously to pick up both local and not-so-local variations. With appropriate translations and dilations of a single suitably chosen function κ (the "mother wavelet"), we can make the expansion (Equation 57.141) orthonormal. This is discussed extensively in Juditsky et al. (1995).

57.7.2.5.2 *Wavelet and Radial Basis Networks*

The choice (Equation 57.145) without any orthogonalization is found in both wavelet networks (Zhang and Benveniste, 1992) and radial basis neural networks (Poggio and Girosi, 1990).

57.7.2.5.3 *Neural Networks*

The ridge choice (Equation 57.146) with

$$\sigma(x) = \frac{1}{1+e^{-x}}$$

gives a much used neural network structure, namely, the *one hidden layer feedforward sigmoidal net.*

57.7.2.5.4 *Hinging Hyperplanes*

If instead of using the sigmoid σ function we choose "V-shaped" functions (in the form of a higher-dimensional "open book") Breiman's *hinging hyperplane* structure is obtained, (Breiman 1993). Hinging hyperplanes model structures (Breiman 1993) have the form

$$g(x) = \max\left\{\beta^+ x + \gamma^+, \beta^- x + \gamma^-\right\} \quad \text{or} \quad g(x) = \min\left\{\beta^+ x + \gamma^+, \beta^- x + \gamma^-\right\}.$$

It can be written in a different way:

$$g(x) = \frac{1}{2}[(\beta^+ + \beta^-)x + \gamma^+ + \gamma^-] \pm \frac{1}{2}|(\beta^+ - \beta^-)x + \gamma^+ - \gamma^-|.$$

Thus a hinge is the superposition of a linear map and a semiglobal function. Therefore, we consider *hinge* functions as semiglobal or ridge-type, though it is not in strict accordance with our definition.

57.7.2.5.5 *Nearest Neighbors or Interpolation*

By selecting κ as in Equation 57.142 and the location and scale vector ρ_k in the structure (Equation 57.143), such that exactly one observation falls into each "cube," the nearest-neighbor model is obtained: just load the input–output record into a table, and, for a given φ, pick the pair $(\widehat{y}, \widehat{\varphi})$ for $\widehat{\varphi}$ closest to the given φ, $\widehat{y}$ is the desired output estimate. If one replaces Equation 57.142 by a smoother function and allows some overlapping of the basis functions, we get interpolation type techniques such as kernel estimators.

57.7.2.5.6 *Fuzzy Models*

Also so called *fuzzy models* based on fuzzy set membership belong to the model structures of the class (Equation 57.141). The basis functions g_k then are constructed from the fuzzy set membership functions and the inference rules. The exact relationship is described in Sjöberg et al. (1995).

57.7.3 Estimating Nonlinear Black-Box Models

The model structure is determined by the following choices:

- The regression vector (typically built up from past inputs and outputs)
- The basic function κ (local) or σ (ridge)
- The number of elements (nodes) in the expansion Equation 57.141

Once these choices have been made $\hat{y}(t|\theta) = g(\varphi(t), \theta)$ is a well-defined function of past data and the parameters θ. The parameters are made up of coordinates in the expansion Equation 57.141, and from location and scale parameters in the different basis functions.

All the algorithms and analytical results of Section 57.2 can thus be applied. For Neural Network applications these are also the typical estimation algorithms used, often complemented with *regularization*, which means that a term is added to the criterion (Equation 57.24), that penalizes the norm of θ. This will reduce the variance of the model, in that "spurious" parameters are not allowed to take on large, and mostly random values. See, for example, (Sjöberg et al. 1995).

For wavelet applications it is common to distinguish between those parameters that enter linearly in $\hat{y}(t|\theta)$ (i.e., the coordinates in the function expansion) and those that enter nonlinearly (i.e., the location and scale parameters). Often the latter are seeded to fixed values and the coordinates are estimated by the linear least squares method. Basis functions that give a small contribution to the fit (corresponding to nonuseful values of the scale and location parameters) can then be trimmed away ("pruning" or "shrinking").

57.8 User's Issues

57.8.1 Experiment Design

It is desirable to affect the conditions under which the data are collected. The objective with such *experiment design* is to make the collected data set Z^N as informative as possible with respect to the models to be built using the data. A considerable amount of theory around this topic can be developed and we shall here just review some basic points. An inspiring discussion on experiment design is given in Hjalmarsson (2005).

The first and most important point is the following one:

1. *The input signal u must be such that it exposes all the relevant properties of the system.* It must thus not be too "simple." For example, a pure sinusoid

$$u(t) = A \cos \omega t$$

will only give information about the system's frequency response at frequency ω. This can also be seen from Equation 57.68. The rule is that

- The input must contain at least as many different frequencies as the order of the linear model to be built.
- To be on the safe side, a good choice is to let the input be random (such as filtered white noise). It then contains all frequencies.

Another case where the input is too simple is when it is generated by feedback such as

$$u(t) = -Ky(t) \tag{57.147}$$

If we would like to build a first-order ARX model

$$y(t) + ay(t-1) = bu(t-1) + e(t)$$

we find that for any given α all models such that

$$a + bK = \alpha$$

will give identical input–output data. We can thus not distinguish between these models using an experiment with Equation 57.147. That is, we cannot distinguish between any combinations of "a" and "b" if they satisfy the above condition for a given "α." The rule is

- If closed-loop experiments have to be performed, the feedback law must not be too simple. It is to be preferred that a set-point in the regulator is being changed in a random fashion.

The second main point in experimental design is

2. *Allocate the input power to those frequency bands where a good model is particularly important.* This is also seen from the expression (Equation 57.68).

If we let the input be filtered white noise, this gives information how to choose the filter. In the time domain it is often useful to think like this:

- Use binary (two-level) inputs if linear models are to be built: this gives maximal variance for amplitude-constrained inputs.
- Check that the changes between the levels are such that the input occasionally stays on one level so long that a step response from the system has time, more or less, to settle. There is no need to let the input signal switch so quickly back and forth that no response in the output is clearly visible.

Note that the second point is really just a reformulation in the time domain of the basic frequency domain advice: let the input energy be concentrated in the important frequency-bands.

A third basic piece of advice about experiment design concerns the choice of sampling interval.

3. *A typical good sampling frequency is 10 times the bandwidth of the system.* That corresponds roughly to 5–7 samples along the rise time of a step response.

57.8.2 Model Validation and Model Selection

The system identification process has, as we have seen, these basic ingredients

- The set of models
- The data
- The selection criterion

Once these have been decided upon, we have, at least implicitly, defined a model: the one in the set that best describes the data according to the criterion. It is thus, in a sense, the best available model in the chosen set. But is it good enough? It is the objective of *model validation* to answer that question. Often the answer turns out to be "no," and we then have to go back and review the choice of model set, or perhaps modify the data set. See Figure 57.2.

How do we check the quality of a model? The prime method is to investigate how well it is capable of reproducing the behavior of a new set of data (*the validation data*) that was not used to fit the model. That is, we simulate the obtained model with a new input and compare this simulated output. One may then use one's eyes or numerical measurements of fit to decide if the fit in question is good enough. Suppose we have obtained several different models in different model structures (say a fourth order ARX model, a second order BJ model, a physically parameterized one, etc.) and would like to know which one is best. The simplest and most pragmatic approach to this problem is then to simulate each one of them on validation data, evaluate their performance, and pick the one that shows the most favorable fit to measured data. (This could indeed be a subjective criterion!)

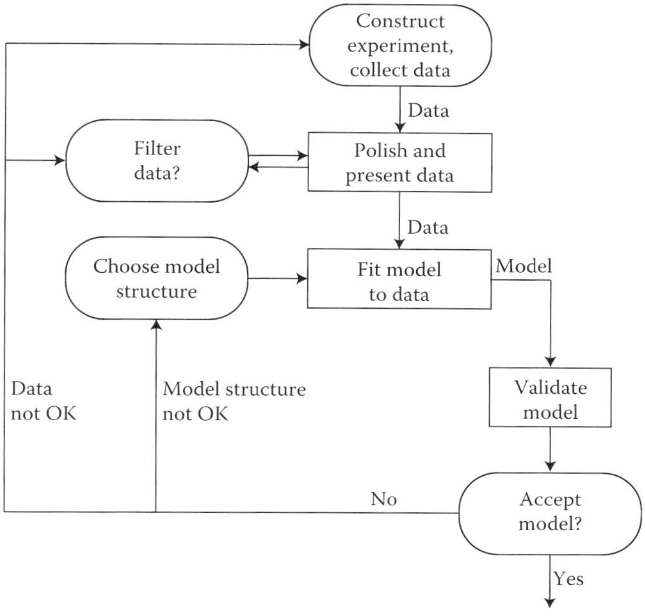

FIGURE 57.2 The identification loop.

The second basic method for model validation is to examine the residuals ("the leftovers") from the identification process. These are the prediction errors

$$\varepsilon(t) = \varepsilon(t, \hat{\theta}_N) = y(t) - \hat{y}(t|\hat{\theta}_N)$$

that is, what the model could not "explain." Ideally these should be independent of information that was at hand at time $t - 1$. For example, if $\varepsilon(t)$ and $u(t - \tau)$ turn out to be correlated, then there are things in $y(t)$ that originate from $u(t - \tau)$ but have not been properly accounted for by $\hat{y}(t|\hat{\theta}_N)$. The model has then not squeezed out all relevant information about the system from the data.

It is good practice to always check the residuals for such (and other) dependencies. This is known as *residual analysis.*

57.8.3 Software for System Identification

In practice System Identification is characterized by some quite heavy numerical calculations to determine the best model in each given class of models. This is mixed with several user choices, trying different model structures, filtering data, and so on. In practical applications we will thus need good software support. There are now many different packages for identification available, such as Mathwork's System Identification Toolbox (Ljung, 2007), Matrix$_x'$s System Identification Module (*MATRIX$_x$*, 1991), PIM (Landau, 1990), UNIT (Ninness and Wills, 2006) and CONTSID (Garnier and Mensler, 2000). They all have in common that they offer the following routines:

A *Handling of data, plotting, and so on.*
Filtering of data, removal of drift, choice of data segments, and so on.

B *Nonparametric identification methods*
Estimation of covariances, Fourier transforms, correlation-, and spectral-analysis, and so on.

C *Parametric estimation methods*
Calculation of parametric estimates in different model structures.

D *Presentation of models*
Simulation of models, estimation and plotting of poles and zeros, computation of frequency functions, and plotting Bode diagrams, and so on.

E *Model validation*
Computation and analysis of residuals ($\varepsilon(t, \hat{\theta}_N)$). Comparison between different models' properties, and so on.

The existing program packages differ mainly in various user interfaces and by different options regarding the choice of model structure according to C above. For example, MATLAB's Identification Toolbox (Ljung, 2007) covers all linear model structures discussed here, including arbitrarily parameterized linear models in continuous time.

Regarding the user interface, there is now a clear trend to make it graphically oriented. This avoids syntax problems and relies more on "click and move," at the same time as tedious menu-labyrinths are avoided. More aspects of CAD tools for system identification are treated in Ljung (2003a).

57.8.4 The Practical Side of System Identification

It follows from our discussion that the most essential element in the process of identification—once the data have been recorded—is to try out various model structures, compute the best model in the structures, using Equation 57.24, and then validate this model. Typically this has to be repeated with quite a few different structures before a satisfactory model can be found.

The difficulties of this process should not be underestimated, and it will require substantial experience to master it. Here follows, however, a procedure that could prove useful to try out.

Step 1: Looking at the data. Plot the data. Look at them carefully. Try to see the dynamics with your own eyes. Can you see the effects in the outputs of the changes in the input? Can you see nonlinear effects, like different responses at different levels, or different responses to a step up and a step down? Are there portions of the data that appear to be "messy" or carry no information. Use this insight to select portions of the data for estimation and validation purposes.
Do physical levels play a role in your model? If not, detrend the data by removing their mean values. The models will then describe how changes in the input give changes in output, but not explain the actual levels of the signals. This is the normal situation. The default situation, with good data, is that you detrend by removing means, and then select the first two-thirds or so of the data record for estimation purposes, and use the remaining data for validation. (All of this corresponds to the "Data Quickstart" in the MATLAB Identification Toolbox.)

Step 2: Getting a feel for the difficulties. Compute and display the spectral analysis frequency response estimate, the correlation analysis impulse response estimate as well as a fourth-order ARX model with a delay estimated from the correlation analysis and a default order state-space model computed by a subspace method. (All of this corresponds to the "Estimate Quickstart" in the MATLAB Identification Toolbox.) This gives three plots. Look at the agreement between the

- Spectral analysis estimate and the ARX and state-space models' frequency functions.
- Correlation analysis estimate and the ARX and state-space models' transient responses.
- Measured Validation Data output and the ARX and state-space models' simulated outputs. We call this the *Model Output Plot*.

If these agreements are reasonable, the problem is not so difficult, and a relatively simple linear model will do a good job. Some fine tuning of model orders, and noise models have to be made and you can proceed to Step 4. Otherwise go to Step 3.

Step 3: Examining the difficulties. There may be several reasons why the comparisons in Step 2 did not look good. This section discusses the most common ones, and how they can be handled:

- **Model unstable:** The ARX or state-space model may turn out to be unstable, but could still be useful for control purposes. Then change to a 5- or 10-step ahead prediction instead of simulation in the Model Output Plot.

- **Feedback in data:** If there is feedback from the output to the input, due to some regulator, then the spectral and correlations analysis estimates are not reliable. Discrepancies between these estimates and the ARX and state-space models can therefore be disregarded in this case. In residual analysis of the parametric models, feedback in data can also be visible as correlation between residuals and input for negative lags.
- **Noise model:** If the state-space model is clearly better than the ARX model at reproducing the measured output this is an indication that the disturbances have a substantial influence, and it will be necessary to carefully model them.
- **Model order:** If a fourth-order model does not give a good Model Output plot, try eighth order. If the fit clearly improves, it follows that higher-order models will be required, but that linear models could be sufficient.
- **Additional inputs:** If the Model Output fit has not significantly improved by the tests so far, think over the physics of the application. Are there more signals that have been, or could be, measured that might influence the output? If so, include these among the inputs and try again a fourth-order ARX model from all the inputs. (Note that the inputs need not at all be control signals, anything measurable, including disturbances, should be treated as inputs).
- **Nonlinear effects:** If the fit between measured and model output is still bad, consider the physics of the application. Are there nonlinear effects in the system? In that case, form the nonlinearities from the measured data. This could be as simple as forming the product of voltage and current measurements, if you realize that it is the electrical power that is the driving stimulus in, say, a heating process, and temperature is the output. This is of course application dependent. It does not cost very much work, however, to form a number of additional inputs by reasonable nonlinear transformations of the measured ones, and just test if inclusion of them improves the fit. See Example 57.2.
- **Still problems?** If none of these tests leads to a model that is able to reproduce the Validation Data reasonably well, the conclusion might be that a sufficiently good model cannot be produced from the data. There may be many reasons for this. The most important one is that the data simply do not contain sufficient information, for example, due to bad signal to noise ratios, large and nonstationary disturbances, varying system properties, and so on. The reason may also be that the system has some quite complicated nonlinearities, which cannot be realized on physical grounds. In such cases, nonlinear, black-box models could be a solution. Among the most used models of this character are the ANN. See Section 57.7.

Otherwise, use the insights on which inputs to use and which model orders to expect and proceed to Step 4.

Step 4: Fine tuning orders and noise structures. For real data there is no such thing as a "correct model structure." However, different structures can give quite different model quality. The only way to find this out is to try out a number of different structures and compare the properties of the obtained models. There are a few things to look for in these comparisons:

- **Fit between simulated and measured output:** Look at the fit between the model's simulated output and the measured one for the Validation Data. Formally, you could pick that model, for which this number is the lowest. In practice, it is better to be more pragmatic, and also take into account the model complexity, and whether the important features of the output response are captured.
- **Residual analysis test:** You should require of a good model so that the cross correlation function between residuals and input does not go significantly outside the confidence region. A clear peak at lag k shows that the effect from input $u(t-k)$ on $y(t)$ is not properly described. A rule of thumb is that a slowly varying cross correlation function outside the confidence region is an indication of too few poles, while sharper peaks indicate too few zeros or wrong delays.

- **Pole zero cancellations:** If the pole–zero plot (including confidence intervals) indicates pole–zero cancellations in the dynamics, this suggests that lower-order models can be used. In particular, if it turns out that the order of ARX models has to be increased to get a good fit, but that pole–zero cancellations are indicated, then the extra poles are just introduced to describe the noise. Then try ARMAX, OE, or BJ model structures with an A or F polynomial of an order equal to that of the number of noncanceled poles.

What Model Structures Should be Tested? Well, you can spend any amount of time to check out a very large number of structures. It often takes just a few seconds to compute and evaluate a model in a certain structure, so that you should have a generous attitude to the testing. However, experience shows that when the basic properties of the system's behavior have been picked up, it is not of much use to fine-tune orders in absurdum just to improve the fit by fractions of percents. For ARX models and state-space models estimated by subspace methods there are also efficient algorithms for handling many model structures in parallel.

Multivariable Systems: Systems with many input signals and/or many output signals are called multivariable. Such systems are often more challenging to model. In particular systems with several outputs could be difficult. A basic reason for the difficulties is that the couplings between several inputs and outputs lead to more complex models: The structures involved are richer and more parameters will be required to obtain a good fit.

Generally speaking, it is preferable to work with state-space models in the multivariable case, since the model structure complexity is easier to deal with. It is essentially just a matter of choosing the model order.

Working with Subsets of the Input–Output Channels: In the process of identifying good models of a system it is often useful to select subsets of the input and output channels. Partial models of the system's behavior will then be constructed. It might not, for example, be clear if all measured inputs have a significant influence on the outputs. That is most easily tested by removing an input channel from the data, building a model for how the output(s) depend on the remaining input channels, and checking if there is a significant deterioration in the model output's fit to the measured one. See also the discussion under Step 3. Generally speaking, the fit gets better when more inputs are included and worse when more outputs are included. To understand the latter fact, you should realize that a model that has to explain the behavior of several outputs has a tougher job than one that just must account for a single output. If you have difficulties to obtain good models for a multioutput system, it might thus be wise to model one output at a time, to find out which are the difficult ones to handle. Models that just are to be used for simulations could very well be built up from single-output models, for one output at a time. However, models for prediction and control will be able to produce better results if constructed for all outputs simultaneously. This follows from the fact that knowing the set of all previous output channels gives a better basis for prediction, than just knowing the past outputs in one channel.

Step 5: Accepting the model. The final step is to accept, at least for the time being, the model to be used for its intended application. Recall the answer to question 10 in the introduction: *No matter how good an estimated model looks on your screen, has only picked up a simple reflection of reality. Surprisingly often, however, this is sufficient for rational decision making.*

References

Akaike, H. 1974a. A new look at the statistical model identification. *IEEE Transactions on Automatic Control*, AC-19:716–723.

Akaike, H. 1974b. Stochastic theory of minimal realization. *IEEE Transactions on Automatic Control*, AC-19:667–674.

Åström, K. J. and Bohlin, T. 1965. Numerical identification of linear dynamic systems from normal operating records. In *IFAC Symposium on Self-Adaptive Systems*, Teddington, England.

Bohlin, T. 2006. *Practical Grey-Box Process Identification.* Springer-Verlag, London.

Box, G. E. P. and Jenkins, D. R. 1970. *Time Series Analysis, Forcasting and Control.* Holden-Day, San Francisco.

Breiman, L. 1993. Hinging hyperplanes for regression, classification and function approximation. *IEEE Trans. Inf. Theory*, 39:999–1013.

Brillinger, D. 1981. *Time Series: Data Analysis and Theory.* Holden-Day, San Francisco.

Dennis, J. E. and Schnabel, R. B. 1983. *Numerical Methods for Unconstrained Optimization and Nonlinear Equations.* Prentice-Hall, Englewood Cliffs, NJ.

Draper, N. and Smith, H. 1981. *Applied Regression Analysis,* Wiley, New York, NY.

Garnier, H. and Mensler, M. 2000. The CONTSID toolbox: A MATLAB toolbox for CONtinuous-Time System IDentification. In *Proceedings of 12th IFAC Symposium on Identification*, Santa Barbara, CA, USA.

Hjalmarsson, H. 2005. From experiment design to closed loop control. *Automatica*, 41(3):393–438.

Juditsky, A., Hjalmarsson, H., Benveniste, A., Delyon, B., Ljung, L., Sjöberg, J., and Zhang, Q. 1995. Nonlinear black-box modeling in system identification: Mathematical foundations. *Automatica*, 31(12):1724–1750.

Kushner, H. J. and Yang, J. 1993. Stochastic approximation with averaging of the iterates: Optimal asymptotic rate of convergence for general processes. *SIAM Journal of Control and Optimization*, 31(4):1045–1062.

Landau, I. D. 1990. *System Identification and Control Design Using P.I.M. + Software.* Prentice-Hall, Englewood Cliffs, NJ.

Larimore, W. E. 1983. System identification, reduced order filtering and modelling via canonical variate analysis. In *Proceedings of 1983 American Control Conference*, San Francisco, CA.

Ljung, L. 1999. *System Identification—Theory for the User*, 2nd edn. Prentice-Hall, Upper Saddle River, NJ.

Ljung, L. 2003a. Educational aspects of identification software user interfaces. In P. van der Hof, B. Wahlberg, S. W., Ed. *Proceedings of 13th IFAC Symposium on System Identification*, pp. 1590–1594, Rotterdam, The Netherlands.

Ljung, L. 2003b. Linear system identification as curve fitting. In *New Directions in Mathematical Systems Theory and Optimization, Springer Lecture Notes In Control and Information*, volume 286, pp. 203–215. Springer-Verlag, Berlin.

Ljung, L. 2004. State of the art in linear system identification: Time and frequency domain methods. In *Proceedings of American Control Conference*, Boston, MA.

Ljung, L. 2007. *System Identification Toolbox for Use with MATLAB. Version* 7th edn., The MathWorks, Inc, Natick, MA.

Ljung, L. 2010. Pespectives on system identification. *IFAC Annual Reviews*, Spring Issue, 34(1): 1–12.

Ljung, L. and Glad, T. 1994. *Modeling of Dynamic Systems.* Prentice Hall, Englewood Cliffs, NJ.

Ljung, L. and Söderström, T. 1983. *Theory and Practice of Recursive Identification.* MIT Press, Cambridge, MA.

MATRIXx 1991. *MATRIXx Users Guide.* Integrated Systems Inc., Santa Clara, CA.

McKelvey, T. 2000. Frequency domain identification. In Smith, R. and Seborg, D., Ed., *Proceedings of 12th IFAC Symposium on System Identification*, Plenary Paper, Santa Barbara, CA, USA.

Ninness, B. and Wills, A. 2006. An identification toolbox for profiling novel techniques. In *14th IFAC Symposium on System Identification*, Newcastle, Australia. http://sigpromu.org/idtoolbox/.

Pintelon, R. and Schoukens, J. 2001. *System Identification—A Frequency Domain Approach.* IEEE Press, New York, NY.

Poggio, T. and Girosi, F. 1990. Networks for approximation and learning. *Proceedings of IEEE*, 78:1481–1497.

Polyak, B. T. and Juditsky, A. B. 1992. Acceleration of stochastic approximation by averaging. *SIAM Journal of Control and Optimization*, 30:838–855.

Rissanen, J. 1974. Basis of invariants and canonical forms for linear dynamic systems. *Automatica*, 10:175–182.

Rissanen, J. 1978. Modelling by shortest data description. *Automatica*, 14:465–471.

Schoukens, J. and Pintelon, R. 1991. *Identification of Linear Systems: A Practical Guideline to Accurate Modeling.* Pergamon Press, London, UK.

Schoukens, J., Pintelon, R., and Rolain, Y. 2004. Time domain identification, frequency domain identification, equivalences! differences? In *Proceedings of American Control Conference*, Boston, MA.

Sjöberg, J., Zhang, Q., Ljung, L., Benveniste, A., Delyon, B., Glorennec, P., Hjalmarsson, H., and Juditsky, A. 1995. Nonlinear black-box modeling in system identification: A unified overview. *Automatica*, 31(12):1691–1724.

Söderström, T. and Stoica, P. 1989. *System Identification.* Prentice-Hall Int., London.

Van Overschee, P. and DeMoor, B. 1996. *Subspace Identification of Linear Systems: Theory, Implementation, Applications.* Kluwer Academic Publishers, Dordrecht.

Verhaegen, M. 1994. Identification of the deterministic part of MIMO state space models, given in innovations form from input–output data. *Automatica*, 30(1):61–74.

Zhang, Q. and Benveniste, A. 1992. Wavelet networks. *IEEE Trans. Neural Networks*, 3:889–898.

X

Stochastic Control

58

Discrete Time Markov Processes

Adam Shwartz
Technion–Israel Institute of Technology

58.1 Caveat

What follows is a quick survey of the main ingredients in the theory of discrete-time Markov processes. It is a bird's view, rather than the definitive "state of the art." To maximize accessibility, the nomenclature of mathematical probability is avoided, although rigor is not sacrificed. To compensate, examples (and counterexamples) abound and the references are annotated. Relevance to control is discussed in Section 58.9.

58.2 Introduction

Discrete time Markov processes, or Markov chains, are a powerful tool for modeling and analysis of discrete time systems, whose behavior is influenced by randomness. A Markov chain is probably the simplest object, which incorporates both dynamics (i.e., notions of "state" and time) and randomness. Let us illustrate the idea through a gambling example.

Example 58.1:

A gambler bets one dollar on "red" at every turn of a (fair) game of roulette. Then, at every turn he gains either one dollar (win, with probability $1/2$) or (-1) (lose). In an ideal game, the gains form a sequence of independent, identically distributed (i.i.d.) random variables. We cannot predict the outcome of each bet, although we do know the odds. Denote by X_t the total fortune the gambler has at time t. If we know X_s, then we can calculate the distribution of X_{s+1} (i.e., the probability that $X_{s+1} = y$, for all possible values of y), and even of X_{s+k} for any $k > 0$. The variable X_t serves as a "state" in the following sense: given X_s, knowledge of X_t for $t < s$ is irrelevant to the calculation of the distribution of future values of the state.

This notion of a state is similar to classical "state space" descriptions. Consider the standard linear model of a dynamical system

$$x_{t+1} = Ax_t + v_t, \tag{58.1}$$

or the more general nonlinear model

$$x_{t+1} = f(x_t, v_t). \tag{58.2}$$

It is intuitively clear that, given the present state, the past does not provide additional information about the future, as long as v_t is not predictable (this is why deterministic state-space models require that v_t be allowed to change arbitrarily at each t). This may remain true even when v_t is random: for example, when they are i.i.d., for then the past does not provide additional information about future v_s's. As we shall see in Theorem 58.1, in this case Equations 58.1 and 58.2 define Markov chains.

In the next section we give the basic definitions and describe the dynamics of Markov chains. We assume that the state space **S** is countable. We restrict our attention to time-homogeneous dynamics (see comment following Theorem 58.1), and discuss the limiting properties of the Markov chain. Finally, we shall discuss extensions to continuous time and to more general state spaces. We conclude this section with a brief review of standard notation. All of our random variables and events are defined on a probability space Ω with a collection $\mathcal{F}$ of events (subsets of Ω) and probability P. The "probability triple" $(\Omega, \mathcal{F}, P)$ is fixed, and we denote expectation by E. For events A and B with $P(B) > 0$, the basic definition of a conditional probability (the multiplication rule) is

$$P(A|B) \triangleq \frac{P(A \bigcap B)}{P(B)}. \tag{58.3}$$

The abbreviation i.i.d. stands for *independent, identically distributed*, and random variables are denoted by capital letters. The identity matrix is denoted by I and $\mathbf{1}_A$ is the indicator function of A, that is, $\mathbf{1}_A(\omega) = 1$ if $\omega \in A$ and $= 0$ otherwise.

58.3 Definitions and Construction

Let $X_0, X_1, \ldots$ be a sequence of random variables, with values in a state space **S**. We assume that **S** is finite or countable, and for convenience we usually set $\mathbf{S} = \{1, 2, \ldots\}$.

Definition 58.1:

*A sequence $X_0, X_1, \ldots$ on **S** is a Markov chain if it possesses the Markov property, that is, if for all $t > 0$ and all $i_{t-1}, \ldots, i_0$,*

$$P(X_t = j | X_{t-1} = i_{t-1}, X_{t-2} = i_{t-2}, \ldots, X_0 = i_0) = P(X_t = j | X_{t-1} = i_{t-1}).$$

A Markov chain is called homogeneous if $P(X_t = j|X_{t-1} = i)$ does not depend on t. In this case we denote the transition probability from i to j by

$$\mathrm{p}_{ij} \triangleq P(X_t = j | X_{t-1} = i).$$

The (possibly infinite) matrix $P \triangleq \{\mathrm{p}_{ij},\ i, j = 1, \ldots\}$ is called the transition matrix.

The restriction to homogeneous chains is not significant: if we define a new state $\tilde{x} \triangleq \{t, x\}$, then it is not hard to see that we can incorporate explicit time dependence within a homogeneous chain, and the new state space is still countable. Henceforth, we restrict our attention to homogeneous chains.

Using the Markov property, a little algebra with the definition of conditional probability gives the more general Markov property: if $t_1 \leq t_2 \leq \cdots \leq t_k \leq \cdots \leq t_\ell$ then

$$\begin{aligned} P\left(X_{t_\ell} = j_\ell, X_{t_{\ell-1}} = j_{\ell-1}, \ldots, X_{t_k} = j_k | X_{t_{k-1}} = j_{k-1}, \ldots, X_{t_1} = j_1\right) \\ = P\left(X_{t_\ell} = j_\ell, X_{t_{\ell-1}} = j_{\ell-1}, \ldots, X_{t_k} = j_k | X_{t_{k-1}} = j_{k-1}\right). \end{aligned} \tag{58.4}$$

This is a precise statement of the intuitive idea given in the introduction: given the present state $X_{t_{k-1}} = j_{k-1}$, the past does not provide additional information.

A chain is called finite if **S** is a finite set. An alternative name for "homogeneous Markov chain" is "Markov chain with stationary transition probabilities." There are those who call a homogeneous Markov chain a "stationary Markov chain." However, since "stationary process" means something entirely different, we shall avoid such usage (see Example 58.6).

Suppose that a process is a Markov chain according to Definition 58.1, and that its initial distribution is given by the row vector $\mu(0)$, that is,

$$P\left(X_0 = i\right) = \mu_i(0), \quad i = 1, 2, \ldots.$$

Then, we can calculate the joint probability distribution at times 0 and 1 from the definition of conditional probability,

$$P\left(X_0 = i, X_1 = j\right) = P\left(X_1 = j | X_0 = i\right) \cdot P\left(X_0 = i\right) = \mu_i(0) \cdot p_{ij}.$$

More generally, using the Markov property:

$$P\left(X_0 = j_0, X_1 = j_1, \ldots, X_t = j_t\right) = \mu_{j_0}(0) \cdot \prod_{s=1}^{t} p_{j_{s-1}j_s}.$$

So, the probability distribution of the whole process can be calculated from the two quantities: the initial probability distribution, and the transition probabilities. In particular, we can calculate the probability distribution at any time t

$$\begin{aligned} \mu_j(t) &= P\left(X_t = j\right) \\ &= \sum_{j_0, j_1, \ldots, j_{t-1}} \mu_{j_0}(0) \prod_{s=1}^{t} p_{j_{s-1}j_s} \end{aligned} \tag{58.5}$$

where the sum is over all states in **S**, that is, each index is summed over the values $1, 2, \ldots$. In vector notation,

$$\mu(t) = \mu(t-1)P = \cdots = \mu(0)P^t. \tag{58.6}$$

(If **S** is countable then, of course, the vector μ and matrix P are infinite, but their product is defined in exactly the same way as for finite ones.) Thus, the probability *distribution* of a Markov chain evolves as a linear dynamical system, even when its evolution equation 58.2 is nonlinear. The one-dimensional probability distribution and the transition probabilities clearly satisfy

$$\begin{cases} \mu_j(t) \geq 0 \\ \sum_{j \in \mathbf{S}} \mu_j(t) = 1 \end{cases} \qquad \begin{cases} p_{ij} \geq 0 \\ \sum_{j \in \mathbf{S}} p_{ij} = 1. \end{cases} \tag{58.7}$$

That is, the rows of the transition matrix P sum to one. Thus, P is a *stochastic matrix*: its elements are nonnegative and its rows sum to one.

If we denote by $p_{ij}^{(n)} \triangleq P\left(X_{m+n}=j|X_m=i\right)$ the *n step transition probability* from *i* to *j*, then we obtain from the definition of conditional probability and the Markov property (Equation 58.4) the Chapman–Kolmogorov equations

$$p_{ij}^{(n+m)} = \sum_{k\in \mathbf{S}} p_{ik}^{(n)} p_{kj}^{(m)}$$

or equivalently

$$P^{n+m} = P^n P^m. \tag{58.8}$$

Therefore, the $p_{ij}^{(n)}$ are the elements of the matrix P^n. This matrix notation yields a compact expression for expectations of functions of the state. To compute $Eg(X_t)$ for some function g on $\mathbf{S}=\{1,2,\ldots\}$, we represent g by a column vector $\underline{g} \triangleq \{g(1), g(2), \ldots\}^T$ (where T denotes transpose). Then

$$Eg(X_t) = \mu(0)\cdot P^t \cdot \underline{g}\,.$$

Note that this expression does not depend on the particular **S**: since the state space is countable we can, by definition, relabel the states so that the state space becomes $\{1,2,\ldots\}$. Let us now summarize the connection between the representations (Equations 58.1 and 58.2) and Markov chains.

Theorem 58.1:

*Let $V_0, V_1, \ldots$ be a sequence of i.i.d. random variables, independent of X_0. Then for any (measurable) function f, the sequence $X_0, X_1, \ldots$ defined through Equation 58.2 is a Markov chain. Conversely, let $\tilde{X}_0, \tilde{X}_1, \ldots$ be a Markov chain with values in **S**. Then there is a (probability triple and a measurable) function f and a sequence $V_0, V_1, \ldots$ of i.i.d. random variables so that the process $X_0, X_1, \ldots$ defined by Equation 58.2 with $X_0=\tilde{X}_0$ has the same probability distribution as $\tilde{X}_0, \tilde{X}_1, \ldots$ that is, for all t and $j_0, j_1, \ldots, j_t$,*

$$P\left(X_0=j_0, X_1=j_1, \ldots, X_t=j_t\right) = P\left(\tilde{X}_0=j_0, \tilde{X}_1=j_1, \ldots, \tilde{X}_t=j_t\right).$$

Note that, whether the system (Equation 58.2) is linear or not, the evolution of the probability distribution (Equations 58.5 and 58.6) is always linear.

We have seen that a Markov chain defines a set of transition probabilities. The converse is also true: given a set of transition probabilities and an initial probability distribution, it is possible to construct a stochastic process which is a Markov chain with the specified transitions and probability distribution.

Theorem 58.2:

If $X_0, X_1, \ldots$ is a homogeneous Markov chain then its probability distribution and transition probabilities satisfy Equations 58.7 and 58.8. Conversely, given $\mu(0)$ and a matrix P that satisfy Equation 58.7, there exists a (probability triple and a) Markov chain with initial distribution $\mu(0)$ and transition matrix P.

Example 58.2:

Let $V_0, V_1, \ldots$ be i.i.d. and independent of X_0. Assume both V_t and X_0 have integer values. Then

$$X_{t+1} \triangleq X_t + V_t = X_0 + \sum_{s=0}^{t} V_s \tag{58.9}$$

defines a Markov chain called a chain with *stationary independent increments*, with state space $\ldots, -1, 0, 1, \ldots$. The transition probability p_{ij} depends only on the difference $j-i$. It turns out [1] that the converse is also true: if the transition probabilities of a Markov chain depend only on the difference $j-i$ then the process can be obtained via Equation 58.9 with i.i.d. V_t.

A *random walk* is a process defined through Equation 58.9, but where the V_t are not necessarily integers—they are real valued.

58.4 Properties and Classification

Given two states i and j, it may or may not be possible to reach j from i. This leads to the notion of classes.

Definition 58.2:

We say a state i leads to j if

$$P\left(x_t = j \text{ for some } t | x_0 = i\right) > 0.$$

This holds if and only if $p_{ij}^{(t)} > 0$ *for some t. We say states i and j communicate, denoted by* $i \leftrightarrow j$*, if i leads to j and j leads to i.*

Communication is a property of pairs: it is obviously symmetric ($i \leftrightarrow j$ if and only if $j \leftrightarrow i$) and is transitive ($i \leftrightarrow j$ and $j \leftrightarrow k$ implies $i \leftrightarrow k$) by the Chapman–Kolmogorov equations 58.8. By convention, $i \leftrightarrow i$. By these three properties, $\leftrightarrow$ defines an equivalence relation. We can therefore partition $\mathbf{S}$ into nonempty *communicating classes* $\mathbf{S}_0, \mathbf{S}_1, \ldots$ with the properties

1. Every state i belongs to exactly one class
2. If i and j belong to the same class then $i \leftrightarrow j$
3. If i and j belong to different classes then i and j do not communicate

We denote the class containing state i by $\mathbf{S}(i)$. Note that if i leads to j but j does not lead to i then i and j *do not* communicate.

Definition 58.3:

A set of states C is closed, or absorbing, if $p_{ij} = 0$ *whenever* $i \in C$ *and* $j \notin C$*. Equivalently,*

$$\sum_{j \in C} p_{ij} = 1 \quad \text{for all } i \in C.$$

If a set C is not closed, then it is called open. A Markov chain is irreducible if all states communicate. In this case its partition contains exactly one class. A Markov chain is indecomposable if its partition contains at most one closed class.

Define the *incidence matrix* $\mathcal{I}$ as follows:

$$\mathcal{I}_{ij} = \begin{cases} 1 & \text{if } p_{ij} > 0 \\ 0 & \text{otherwise.} \end{cases}$$

We can also define a directed graph whose nodes are the states, with a directed arc between any two states for which $\mathcal{I}_{ij} = 1$. The communication properties can obviously be extracted from $\mathcal{I}$ or from the directed graph. The chain is irreducible if and only if the directed graph is connected in the sense that, going in the direction of the arcs, we can reach any node from any other node. Closed classes can also be defined in terms of the incidence matrix or the graph. The classification leads to the following maximal decomposition.

Theorem 58.3:

By reordering the states, if necessary, we can put the matrix P into the block form

$$P = \begin{bmatrix} P_1 & 0 & \dots & 0 & 0 \\ 0 & P_2 & \dots & 0 & 0 \\ \vdots & \vdots & & \vdots & \vdots \\ 0 & 0 & \dots & P_m & 0 \\ R_1 & R_2 & \dots & R_m & Q \end{bmatrix} \tag{58.10}$$

where the blocks P_i correspond to closed irreducible classes. It is maximal in the sense that smaller classes will not be closed, and no subset of states corresponding to Q is a closed irreducible class.

If **S** is countable, then the number of classes and the size of some blocks may be infinite. Note that all the definitions in this section apply when we replace **S** with any closed class.

Much like other dynamical systems, Markov chains can have cyclic behavior, and can be unstable. The relevant definitions are

---

Definition 58.4:

The period d of a state i is the greatest common divisor of the set $\{t : \mathrm{p}_{ii}^{(t)} > 0\}$. If d is finite and $d > 1$ then the state is called periodic; otherwise it is aperiodic.

Definition 58.5:

A state i is called recurrent if the probability of starting at i and returning to i in finite time is 1. Formally, if

$$P\left(X_t = i \text{ for some } t > 1 | x_1 = i\right) = 1.$$

Otherwise it is called transient.

Example 58.3:

In the chain on $\mathbf{S} = \{1, 2\}$ with $\mathrm{p}_{ij} = 1$ if and only if $i \neq j$, both states are periodic with period $d = 2$ and both states are recurrent. The states communicate, and so **S** contains exactly one class, which is therefore closed. Consequently, the chain is irreducible and indecomposable. However, if $\mathrm{p}_{12} = \mathrm{p}_{22} = 1$ then the states are not periodic, state 2 is recurrent and state 1 is transient. In this case, the partition contains two sets: the closed set $\{2\}$, and the open set $\{1\}$. Consequently, the chain is not irreducible, but it is indecomposable.

When $\mathbf{S}$ is finite, then either it is irreducible or it contains a closed proper subset.

Example 58.4:

Let $\mathbf{S} = \{1, 2, \ldots\}$. Suppose $p_{ij} = 1$ if and only if $j = i + 1$. Then all states are transient, and $\mathbf{S}$ is indecomposable but not irreducible. Every set of the form $\{i : i \geq k\}$ is closed, but in the partition of the state space each state is the only member in its class. Suppose now $p_{11} = 1$ and for $i > 1$, $p_{ij} = 1$ if and only if $j = i - 1$. Then state 1 is the only recurrent state, and again each state is alone in its class.

Theorem 58.4:

Let $\mathbf{S}_k$ be a class. Then either all states in $\mathbf{S}_k$ are recurrent, or all are transient. Moreover, all states in $\mathbf{S}_k$ have the same period d.

58.5 Algebraic View and Stationarity

The matrix P is positive, in the sense that its entries are positive. When $\mathbf{S}$ is finite, the Perron–Frobenius theorem implies [3]

Theorem 58.5:

Let $\mathbf{S}$ be finite. Then P has a nonzero left eigenvector π whose entries are nonnegative, and $\pi \cdot P = \pi$, that is, the corresponding eigenvalue is 1. Moreover, $|\lambda| \leq 1$ for all other eigenvalues λ. The multiplicity of the eigenvalue 1 is equal to the number of irreducible closed subsets of the chain. In particular, if the entries of P^n are all positive for some n, then the eigenvalue 1 has multiplicity 1. In this case, the entries of π are all positive and $|\lambda| < 1$ for all other eigenvalues λ.

If the entries of P^n are all positive for some n then the chain is irreducible and aperiodic, hence the second part of the theorem. If the chain is irreducible and periodic with period d, then the d roots of unity are left eigenvalues of P, each is of multiplicity 1 and all other eigenvalues have strictly smaller modulus. The results for a general finite chain can be obtained by writing the chain in the block form (Equation 58.10).

Definition 58.6:

Let $\mathbf{S}$ be finite or countable. A probability distribution μ satisfying $\mu \cdot P = \mu$ is called invariant (under P) or stationary.

Theorem 58.5 thus implies that every finite Markov chain possesses at least one invariant probability distribution. For countable chains, Example 58.4 shows that this is not true.

Example 58.5:

Returning to Example 58.3, in the first case $(1/2, 1/2)$ is the only invariant probability distribution, while in the second case $(0, 1)$ is the only invariant probability distribution. In Example 58.4, in the first case there is no invariant probability distribution, while in the second case $(1, 0, 0, \ldots)$ is the only invariant probability distribution. Finally, if $P = I$, the 2×2 identity matrix, then $\pi \triangleq (p, 1-p)$ is invariant for any $0 \le p \le 1$.

Example 58.6:

Recall that a process $X_0, X_1, \ldots$ is called stationary if, for all positive t and s, the distribution of $\{X_0, X_1, \ldots, X_t\}$ is the same as the distribution of $\{X_s, X_{1+s}, \ldots, X_{t+s}\}$. From the definitions it follows that a (homogeneous) Markov chain (finite or not) is stationary if and only if $\mu(0)$ is invariant.

A very useful tool in the calculation of invariant probability distributions is the "balance equations":

$$\pi_i = \sum_{j:j\to i} \pi_j p_{ji} = \pi_i \sum_{j:i\to j} p_{ij} \tag{58.11}$$

where the first equality is just a restatement of the definition of invariant probability, and the second follows since by Equation 58.7, the last sum equals 1. The intuition behind these equations is very useful: in steady state, the rate at which "probability mass enters" must be equal to the rate it "leaves." This is particularly useful for continuous-time chains. More generally, given any set S, the rate at which "probability mass enters" the set (under the stationary distribution) equals the rate it "leaves":

Theorem 58.6:

Let S be a set of states and π invariant under P. Then

$$\sum_{i\in S}\sum_{j:j\to i} \pi_j p_{ji} = \sum_{i\in S} \pi_i \sum_{j:i\to j} p_{ij}.$$

Example 58.7:

Random walk with a reflecting barrier. This example models a discrete-time queue where, at each instance, either arrival or departure occurs. The state space $\mathbf{S}$ is the set of nonnegative integers (including 0), and

$$p_{00} = 1-p, \quad p_{i(i+1)} = p, \quad p_{i(i-1)} = 1-p \quad \text{for } i \ge 1.$$

Then all states communicate so that the chain is irreducible, the chain is aperiodic and recurrent. From Equation 58.11 we obtain

$$\pi_0 = \pi_0 p_{00} + \pi_1 p_{10}$$
$$\pi_i = \pi_{i-1} p_{(i-1)i} + \pi_{i+1} p_{(i+1)i}, \quad i \ge 1.$$

When $p < 1/2$, this and Equation 58.7 imply that $\pi_i = [(1-2p)/(1-p)]p^i/(1-p)^i$ for $i \ge 0$. When $p > 1/2$ the equations imply that any solution must be increasing in i and so cannot be a distribution. Indeed, in this case the chain is transient.

Example 58.8:

Birth-death process. A Markov chain on $\mathbf{S} = \{0, 1, \ldots\}$ is a birth-death process if $p_{ij} = 0$ whenever $|i-j| \ge 2$. If X_t is the number of individuals alive at time t then, at any point in time, this number can increase by one (birth), decrease by one (death) or remain constant (simultaneous birth and death). Unlike Example 58.7, here the probability of a change in size may depend on the state.

58.6 Random Variables

In this section we shift our emphasis back from algebra to the stochastic process. We define some useful random variables associated with the Markov chain. It will be convenient to use P_j for the probability conditioned on the process starting at state j. That is, for an event A,

$$P_j(A) \triangleq P\left(A|X_0 = j\right),$$

with a similar convention for expectation E_j. The Markov property implies that the past of a Markov chain is immaterial given the present. But suppose we observe a process until a random time, say the time a certain event occurs. Is this property preserved? The answer is positive, but only for nonanticipative times:

Definition 58.7:

Let S be a collection of states, that is, a subset of **S**. *The hitting time τ_S of S is the first time the Markov chain visits a state in S. Formally,*

$$\tau_S = \inf\{t > 0 : X_t \in S\}.$$

Note that by convention, if X_t never visits S then $\tau_S = \infty$. The initial time, here $t = 0$, does not qualify in testing whether the process did or did not visit S. By definition, hitting times have the following property. In order to decide whether or not $\tau_S = t$, it suffices to know the values of $X_0, \ldots, X_t$. This gives rise to the notion of Markov time or stopping time.

Definition 58.8:

A random variable τ with positive integer values is called a stopping time, or a Markov time (with respect to the process $X_0, X_1, \ldots$) if one of the following equivalent conditions hold. For each $t \geq 0$

1. *It suffices to know the values of $X_0, X_1, \ldots, X_t$ in order to determine whether the event $\{\tau = t\}$ occurred or not*
2. *There exists a function f_t so that*

$$\mathbf{1}_{\tau=t}(\omega) = f_t\left(X_0(\omega), \ldots, X_t(\omega)\right).$$

An equivalent, and more standard definition is obtained by replacing $\tau = t$ by $\tau \leq t$. With respect to such times, the Markov property holds in a stronger sense.

Theorem 58.7: Strong Markov Property

If τ is a stopping time for a homogeneous Markov chain $X_0, X_1, \ldots$ then

$$\begin{aligned} &P\left(X_{\tau+1} = j_1, X_{\tau+2} = j_2, \ldots, X_{\tau+m} = j_m | X_t = i_t,\ t < \tau,\ X_\tau = i^*\right) \\ &\quad = P\left(X_1 = j_1, X_2 = j_2, \ldots, X_m = j_m | X_0 = i^*\right). \end{aligned}$$

We can now rephrase and complement the definition of recurrence. We write τ_j when we really mean $\tau_{\{j\}}$.

Definition 58.9:

The state j is recurrent if $P_j(\tau_j < \infty) = 1$. It is called positive recurrent if $E_j\tau_j < \infty$, and null-recurrent if $E_j\tau_j = \infty$.

If state j is recurrent, then the hitting time of j is finite. By the strong Markov property, when the processes hits j for the first time, it "restarts": therefore, it will hit j again! and again! So, let N_j be the number of times the process hits state j:

$$N_j \triangleq \sum_{t=1}^{\infty} \mathbf{1}_{X_t=j}.$$

Theorem 58.8:

1. *If a state is positive recurrent, then all states in its class are positive recurrent. The same holds for null recurrence.*
2. *Suppose j is recurrent. Then $P_j(N_j = \infty) = 1$, and consequently $E_jN_j = \infty$. Moreover, for every state i,*

$$\begin{aligned} P_i\left(N_j = \infty\right) &= P_i\left(\tau_j < \infty\right) \cdot P_j(N_j = \infty) \\ &= P_i\left(\tau_j < \infty\right), \end{aligned}$$

   and if $P_i\left(\tau_j < \infty\right) > 0$ then $E_iN_j = \infty$.
3. *Suppose j is transient. Then $P_j(N_j < \infty) = 1$, and for all i,*

$$E_iN_j = \frac{P_i\left(\tau_j < \infty\right)}{1 - P_j\left(\tau_j < \infty\right)}.$$

To see why the last relation should hold, note that by the strong Markov property,

$$\begin{aligned} &P_i(\tau_j < \infty \text{ and a second visit occurs}) \\ &\quad = P_i(\tau_j < \infty) \cdot P_j(\tau_j < \infty), \end{aligned}$$

and similarly for later visits. This means that the distribution of the number of visits is geometric: with every visit we get another chance, with equal probability, to revisit. Therefore,

$$\begin{aligned} E_iN_j &= P_i\left(\tau_j < \infty\right) + P_i\left(\tau_j < \infty \text{ and a second visit occurs}\right) + \cdots \\ &= P_i\left(\tau_j < \infty\right)\left(1 + P\left(\text{a second visit occurs} | \tau_j < \infty\right)\right) + \cdots. \\ &= P_i\left(\tau_j < \infty\right)\left(1 + P_j\left(\tau_j < \infty\right) + \cdots\right) \end{aligned}$$

which is what we obtain if we expand the denominator. A similar interpretation gives rise to

$$E_i\tau_j = 1 + \sum_{k:k \neq j \in \mathbf{S}} p_{ik}E_k\tau_j.$$

We have a simple criterion for recurrence in terms of transition probabilities, since

$$E_i N_j = E_i \sum_{t=1}^{\infty} \mathbf{1}_{X_t = j} = \sum_{t=1}^{\infty} P_i\left(X_t = j\right) = \sum_{t=1}^{\infty} \mathrm{p}_{ij}^{(t)}.$$

58.7 Limit Theorems: Transitions

Classical limit theorems concern the behavior of t-step transition probabilities, for large t. Limits for the random variables are discussed in Section 58.8.

Theorem 58.9:

For every Markov chain, the limit

$$P^* \triangleq \lim_{t\to\infty} \frac{1}{t} \sum_{s=0}^{t-1} P^s \tag{58.12}$$

exists and satisfies

$$P^* \cdot P = P \cdot P^* = P^* \cdot P^* = P^*.$$

If **S** *is finite then* P^* *is a stochastic matrix.*

1. *Suppose the Markov chain is indecomposable, recurrent, and nonperiodic. Then, for all states* i, k,

$$\lim_{t\to\infty} \sum_{j\in\mathbf{S}} |\mathrm{p}_{ij}^{(t)} - \mathrm{p}_{kj}^{(t)}| = 0.$$

2. *An irreducible chain is positive recurrent if and only if it has an invariant probability distribution* π. *If it is positive recurrent and nonperiodic then* $\lim_{t\to\infty} \mathrm{p}_{ij}^{(t)} = \pi(j)$ *for all* i, j. *If it is positive recurrent with period* $d > 1$ *then* $\lim_{t\to\infty} \mathrm{p}_{ij}^{(t)} + \cdots + \mathrm{p}_{ij}^{(t+d-1)} = d \cdot \pi(j)$ *for all* i, j. *If it is null recurrent then for all* i, j, $\lim_{t\to\infty} \mathrm{p}_{ij}^{(t)} = 0$. *If state* j *is transient then* $\sum_t \mathrm{p}_{ij}^{(t)} < \infty$.

Since a finite Markov chain always contains a finite closed set of states, there always exists an invariant distribution. Moreover, if a set is recurrent, then it is positive recurrent.

Example 58.9:

Example 58.3 Continued. For the periodic chain, $\mathrm{p}_{ij}^{(t)}$ clearly does not converge. However, $\mathrm{P}_{ij}^* = 1/2$ for all i, j, and the rows define an invariant measure.

Example 58.10:

Example 58.8 Continued. Assume that for the birth–death process $\mathrm{p}_{i(i+1)} > 0$ and $\mathrm{p}_{(i+1)i} > 0$ for all $i \geq 0$ and $\mathrm{p}_{ii} > 0$ for some i. Then the chain is obviously irreducible, and aperiodic (if $\mathrm{p}_{ii} = 0$ for all i then $d = 2$). Using Equation 58.11 we obtain that an invariant probability distribution, if it exists, must satisfy

$$\pi_i = \frac{p_{01} \cdots p_{(i-1)i}}{p_{10} \cdots p_{i(i-1)}} \cdot \pi_0. \tag{58.13}$$

Therefore, any invariant probability must satisfy $\pi_i > 0$ for all i, and in particular $\pi_0 > 0$. So, we can invoke Equation 58.7 to obtain the following dichotomy. Either

$$Z \triangleq \sum_{i \in \mathbf{S}} \frac{p_{01} \cdots p_{i-1i}}{p_{10} \cdots p_{ii-1}} < \infty, \tag{58.14}$$

in which case Equations 58.13 and 58.14 determine the unique invariant probability, and we conclude that the Markov chain is positive recurrent. Or $Z = \infty$, in which case there is no invariant probability and the chain is not positive recurrent.

In terms of the transition matrix P, if a chain is nonperiodic, indecomposable, and recurrent then the matrix converges (uniformly over rows) to a matrix having identical rows, which are either all zeroes (null-recurrent case), or equal to the invariant probability distribution. Here are the missing cases from Theorem 58.9. Denote the mean hitting time of state j starting at i by $m_{ij} = E_i\tau_j$. Clearly m_{jj} is infinite if j is not positive recurrent, and we shall use the convention that $a/\infty = 0$ whenever a is finite.

Theorem 58.10:

If a state j is transient then $\lim_{t\to\infty} p_{ij}^{(t)} = 0$. If j is recurrent with period d then $\lim_{t\to\infty} p_{jj}^{(nd)} = \frac{d}{m_{jj}}$. If j is nonperiodic, this remains true with $d = 1$, so that (by Theorem 58.9) $\pi(j) \cdot m_{jj} = 1$.

The last statement should be intuitive: the steady state probability of visiting state j is a measure of how often this state is "visited," and this is inversely proportional to the mean time between visits. The rate at which convergence takes place depends on the second largest eigenvalue of P. Therefore, if the Markov chain is finite, indecomposable, and aperiodic with invariant probability distribution π, then

$$\left| p_{ij}^{(t)} - \pi_j \right| \le R\rho^t \quad \text{for all } i, j$$

with $\rho < 1$. This of course implies that the one dimensional distributions converge geometrically fast. On the other hand, the Markov structure implies that if indeed the one-dimensional distributions converge, then the distribution of the whole process converges:

Theorem 58.11:

Suppose that for all i and j we have $\lim_{t\to\infty} p_{ij}^{(t)} = \pi_j$, for some probability distribution π. Then π is an invariant probability distribution, and for any i,

$$\lim_{t\to\infty} P_i\left(X_{t+1} = j_1, X_{t+2} = j_2 \ldots\right) = P_\pi\left(X_1 = j_1, X_2 = j_2 \ldots\right)$$

where P_π is obtained by starting the process with the distribution π. In fact, the distribution of the process converges in the sense that all finite dimensional distributions converge to the corresponding distributions under P_π.

58.8 Ergodic Theorems

We do not expect the Markov chain to converge: since transition probabilities are homogeneous, the probability of leaving a given state does not change in time. However, in analogy with i.i.d. random variables, there are limit theorems under the right scaling. The connection to the i.i.d. case comes from the following construction. Fix an arbitrary state j and define

$$\begin{aligned} R_1 &= T_1 = \tau_j \\ R_k &= \inf\{t > R_{k-1} : X_t = j, \quad k > 1\} \\ T_k &= R_k - R_{k-1}, \quad k > 1. \end{aligned}$$

Theorem 58.12:

If j is recurrent and the Markov chain starts at j (with probability one), then $T_1, T_2 \ldots$ is a sequence of i.i.d. random variables. Moreover, the random vectors

$$Z_k \triangleq \{X_{R_k}, X_{R_k+1}, \ldots, X_{R_{k+1}-1}\}$$

are independent and identically distributed (in the space of sequences of variable length!).

This is another manifestation of the fact that, once we know the Markov chain hits some state, future behavior is (probabilistically) determined. Fix a recurrent state i and a time t, and define

$$T^t \triangleq \max_k \{T_k : T_k \leq t\}.$$

Denote by $N_s(j)$ the number of times in $1, 2, \ldots, s$ that $X_u = j$. By Theorem 58.12, the random variables $\{N_{R_k+1}(j) - N_{R_k}(j), \ k = 1, 2, \ldots\}$ are independent, and (except possibly for $k = 1$) are identically distributed for each j. By the law of large numbers this implies the following.

Theorem 58.13:

Let i be recurrent. Then starting at i (P_i a.s.)

$$\lim_{t \to \infty} \frac{N_t(j)}{N_t(\ell)} = \frac{E_i \sum_{s=1}^{\tau_i} \mathbf{1}_{X_s = j}}{E_i \sum_{s=1}^{\tau_i} \mathbf{1}_{X_s = \ell}} = \frac{\pi_j}{\pi_\ell},$$

and π is an invariant probability distribution, concentrated on the closed class containing i, that is,

$$\pi_k = \sum_{j \in \mathbf{S}} \pi_j p_{jk}, \qquad \sum_{k : i \to k \in \mathbf{S}} \pi_k = 1$$

so that $\pi_k = 0$ if $i \not\to k$. Moreover, if we start in some state j then

$$\lim_{t \to \infty} \frac{N_t(i)}{t} = \frac{\mathbf{1}_{\tau_i < \infty}}{m_{ii}} \quad \text{with } P_j\text{-probability 1.}$$

The last relation implies Equation 58.12: taking E_j expectations, we obtain that if i is recurrent then $P^* = P_j(\tau_i < \infty)/m_{ii}$. From here follows a limit theorem for functions of a Markov chain.

Theorem 58.14:

Ergodic Theorem. *Let* **S** *be a single recurrent class, and assume* π *is an invariant probability distribution. Let* f *and* g *be functions such that* $E_\pi \left|f(X_0)\right| < \infty$ *and* $E_\pi \left|g(X_0)\right| < \infty$. *Then for an arbitrary starting state* i, *with probability one,*

$$\lim_{t\to\infty} \frac{\sum_{s=0}^{t} f(X_s)}{\sum_{s=0}^{t} g(X_s)} = \frac{E_i \sum_{s=1}^{\tau_i} f(X_s)}{E_i \sum_{s=1}^{\tau_i} g(X_s)} = \frac{E_\pi f(X_0)}{E_\pi g(X_0)}$$

provided not both numerator and denominator of the last terms are zero.

Setting $g \equiv 1$ we obtain a law of large numbers for a function of the Markov chain. Here is a statement of a Central Limit theorem and a Law of Iterated Logarithm.

Theorem 58.15:

Let **S** *be a single positive recurrent class with an invariant probability distribution* π. *Fix a function* f *and suppose that for some* i, $E_i\left[\tau_i^2\right] < \infty$ *and* $E_\pi f(X_0) = 0$ *and*

$$E_i\left[\left(\sum_{t=1}^{\tau_i} \left|f(x_t)\right|\right)^2\right] < \infty.$$

Define

$$\gamma^2 \triangleq \pi_i \cdot E_i\left[\left(\sum_{t=1}^{\tau_i} f(x_t)\right)^2\right].$$

Then $\gamma^2 < \infty$, *and if* $\gamma^2 > 0$ *then*

Central Limit Theorem

$$\frac{\sum_{s=0}^{t} f(X_s)}{\sqrt{t\cdot\gamma^2}}$$

converges in distribution (as $t \to \infty$*) to a standard Gaussian random variable. Moreover, the sum satisfies the Law of Iterated Logarithm, that is, the lim sup of*

Law of Iterated Logarithm

$$\frac{\sum_{s=0}^{t} f(X_s)}{\sqrt{2\gamma^2 t \log\log t}},$$

as $t \to \infty$, *is* 1, *and the lim inf is* (-1).

58.9 Extensions and Comments

Markov processes are very general objects, of which our treatment covered just a fraction. Many of the basic ideas extend, but definitely not all, and usually some effort is required. In addition, the mathematical difficulties rise exponentially fast. There are two obvious directions to extend: more general state spaces, and continuous time.

When the state space is not countable, the probability that an arbitrary point in the state space is "visited" by the process is usually zero. Therefore, the notions of communication, hitting, recurrence, and periodicity have to be modified. The most extensive reference here is [2].

In the discrete-space continuous time setting, the Markov property implies that if $x_t = i$ then the values of x_s, $s < t$ are not relevant. In particular, the length of time from the last change in state until t should be irrelevant. This implies that the distribution of the time between jumps (= change in state) should be exponential. If the only possible transition is then from i to $i+1$ and if all transition times have the same distribution (i.e., they are all exponential with the same parameter), then we obtain the Poisson process. More generally, we can describe most discrete state, continuous time Markov chains as follows. The process stays at state i an exponential amount of time with parameter $\lambda(i)$. It then jumps to the next state according to a transition probability p_{ij}, and the procedure is repeated (this is correct if, for example, $\lambda(i) > \lambda > 0$ for all i). This subject is covered, for example in [1] and in a new section in [2]. If we observe such a process at jump times, then we recover a Markov chain. This is one of the major tools in the analysis of continuous time chains.

Semi-Markov processes are a further generalization, where the time between events is drawn from a general distribution, which depends on the state, and possibly on the next state. This is no longer a Markov process; however, if we observe the process only at jump times, then we recover a Markov chain.

Finally, in applications, the information structure, and consequently the set of events, is richer: we can measure more than the values of the Markov chain. This is often manifested in a recursion of the type (Equation 58.2), but where the V_t are not independent. Do we still get a Markov chain? And in what sense? The rough answer is that, if the Markov property (Equation 58.4) holds, but where we condition on all the available information, then we are back on track: all of our results continue to hold. For this to happen we need the "noise sequence" $V_0, V_1, \ldots$ to be nonanticipative in a probabilistic sense.

Criteria for Stability: As in the case of dynamical systems, there are criteria for stability and for recurrence, based on Lyapunov functions. This is one of the main tools in [2]. These techniques are often the easiest and the most powerful.

Relevance to Control: Many models of control systems subject to noise can be modeled as Markov processes, and the discrete-time, discrete-space models are controlled Markov chains. Here the transition probabilities are parameterized by the control: see the section on Dynamic Programming. In addition, many filtering and identification algorithms give rise to Markov chains (usually with values in $\mathbb{R}^d$). Limit theorems for Markov chains can then be used to analyze the limiting properties of these algorithms. See, for example [2,4].

Example 58.11:

Extending Example 59.2, consider the recursion

$$X_{t+1} \overset{\triangle}{=} X_t + V_t + U_t$$

where U_t is a control variable. This is a simple instance of a controlled recursion of the ARMA type. Suppose that U_t can only take the values ± 1. Of course, we require that the control depends only on past information. If the control values U_t depend of the past states, then $X_0, X_1, \ldots$ may not be a Markov chain. For example, if we choose $U_t = \text{sign}(X_0)$, then the sequence $X_0, X_1, \ldots$ violates the Markov property (Definition 59.1). However, we do have a controlled Markov chain. This means that Definition 59.1 is replaced with the relation

$$\begin{aligned} &P(X_t = j | X_{t-1} = i_{t-1}, \ldots, X_0 = i_0, U_{t-1}, \ldots, U_0 = u_0) \\ &= P(X_t = j | X_{t-1} = i_{t-1}, U_{t-1} = u_{t-1}). \end{aligned}$$

This in fact is the general definition of a controlled Markov chain. If we choose a feedback control, that is, $U_t = f(X_t)$ for some function f, then $X_0, X_1, \ldots$ is again a Markov chain; but the transitions and the limit behavior now depend on the choice of f.

If we are interested in optimizing a functional of a controlled chain, we obtain a Markov Decision process. The evaluation of the long-time average cost is often that of functionals as in Theorem 58.14.

Such functionals also appear in learning theory, and for the same reason. For more information on controlled Markov chains, see the section on Dynamic Programming.

Finally, Markov chains have found important applications recently in the context of simulation—specifically in Markov Chain Monte Carlo simulation.

Bibliography

Çinlar, E., *Introduction to Stochastic Processes*, Prentice-Hall, Englewood Cliffs, NJ, 1975.
A precise, elegant and accessible book, covering the basics.

Kemeny, J.G., Snell, J., and Knapp, A.W., *Denumerable Markov Chains*, Van Nostrand, Princeton, NJ, 1966.
The most elementary in this list, but fairly thorough, not only in coverage of Markov chains, but also as introduction to Markov processes.

Nummelin, E., *General Irreducible Markov Chains and Non-Negative Operators*, Cambridge University Press, Cambridge, 1984.
More general than [3], treats general state spaces. Introduced many new techniques; perhaps less encyclopedic than [2], but fairly mathematical.

Orey, S., *Limit Theorems for Markov Chain Transition Probabilities*, Van Nostrand Reinhold, London, 1971.
A thin gem on limit theorems, fairly accessible.

Revuz, D., *Markov Chains*, North-Holland, Amsterdam, 1984.
A mathematical, thorough treatment.

References

1. Chung, K.L., *Markov Chains with Stationary Transition Probabilities*, Springer-Verlag, New York, 1967.
A classic on discrete space Markov chains, both discrete and continuous time. Very thorough and detailed, but mathematically not elementary.
2. Meyn, S.P. and Tweedie, R.L., *Markov Chains and Stochastic Stability*, 2nd edition. Cambridge University Press, Cambridge, 2009.
Deals with general discrete-time Markov chains, and covers the state of the art. It is therefore demanding mathematically, although not much measure theory is required. The most comprehensive book if you can handle it.
3. Seneta, E., *Non-Negative Matrices and Markov Chains*, Springer-Verlag, New York, 1981.
Gives the algebraic point of view on Markov chains.
4. Tijms, H.C., *Stochastic Modelling and Analysis: A Computational Approach*, John Wiley, New York, 1986.
Contains a basic introduction to the subject of Markov chains, both discrete and continuous time, with a wealth of examples for applications and computations.

59

Stochastic Differential Equations

John A. Gubner
University of Wisconsin–Madison

59.1 Introduction

This chapter deals with nonlinear differential equations of the form

$$\frac{dX_t}{dt} = a(t, X_t) + b(t, X_t)Z_t, \quad X_{t_0} = \Xi,$$

where Z_t is a Gaussian white noise driving term that is independent of the random initial state Ξ. Since the solutions of these equations are random processes, we are also concerned with the probability distribution of the solution process $\{X_t\}$. The classical example is the Langevin equation,

$$\frac{dX_t}{dt} = -\mu X_t + \beta Z_t,$$

where μ and β are positive constants. In this linear differential equation, X_t models the velocity of a free particle subject to frictional forces and to impulsive forces due to collisions. Here μ is the coefficient of friction, and $\beta = \sqrt{2\mu kT/m}$, where m is the mass of the particle, k is Boltzmann's constant, and T is the absolute temperature [7]. As shown at the end of Section 59.5, with a suitable Gaussian initial condition, the solution of the Langevin equation is a Gaussian random process known as the Ornstein–Uhlenbeck process.

The subject of stochastic differential equations is highly technical. However, to make this chapter as accessible as possible, the presentation is mostly on a heuristic level. On occasion, when deeper theoretical results are needed, the reader is referred to an appropriate text for details. Suggestions for further reading are given at the end of the chapter.

59.1.1 Ordinary Differential Equations

Consider a deterministic nonlinear system whose state at time t is $x(t)$. In many engineering problems, it is reasonable to assume that x satisfies an ordinary differential equation (ODE) of the form

$$\frac{dx(t)}{dt} = a\big(t, x(t)\big) + b\big(t, x(t)\big)z(t), \quad x(t_0) = \xi, \tag{59.1}$$

where $z(t)$ is a separately specified input signal. Note that if we integrate both sides from t_0 to t, we obtain

$$x(t) - x(t_0) = \int_{t_0}^{t} \big[a\big(\theta, x(\theta)\big) + b\big(\theta, x(\theta)\big)z(\theta)\big]\, d\theta.$$

Since $x(t_0) = \xi$, $x(t)$ satisfies the integral equation

$$x(t) = \xi + \int_{t_0}^{t} \big[a\big(\theta, x(\theta)\big) + b\big(\theta, x(\theta)\big)z(\theta)\big]\, d\theta.$$

Under certain technical conditions, e.g., [3], it can be shown that there exists a unique solution to Equation 59.1; this is usually accomplished by solving the corresponding integral equation.

Now suppose $x(t)$ satisfies Equation 59.1. If $x(t)$ is passed through a nonlinearity, say $y(t) := g\big(x(t)\big)$, then $y(t_0) = g(\xi)$, and by the chain rule, $y(t)$ satisfies the differential equation,

$$\begin{aligned}\frac{dy(t)}{dt} &= g'\big(x(t)\big)\frac{dx(t)}{dt} \\ &= g'\big(x(t)\big)a\big(t, x(t)\big) + g'\big(x(t)\big)b\big(t, x(t)\big)z(t),\end{aligned} \tag{59.2}$$

assuming g is differentiable.

59.1.2 Stochastic Differential Equations

If x models a mechanical system subject to significant vibration, or if x models an electronic system subject to significant thermal noise, it makes sense to regard $z(t)$ as a stochastic, or random process, which we denote by Z_t. (Our convention is to denote deterministic functions by lowercase letters with arguments in parentheses and to denote random functions by uppercase letters with subscript arguments.) Now, with a random input signal Z_t, the ODE of Equation 59.1 becomes the stochastic differential equation (SDE),

$$\frac{dX_t}{dt} = a(t, X_t) + b(t, X_t)Z_t, \quad X_{t_0} = \Xi, \tag{59.3}$$

where the initial condition Ξ is also random. As our notation indicates, the solution of an SDE is a random process. Typically, we take Z_t to be a white noise process; i.e,

$$\mathsf{E}[Z_t] = 0 \quad \text{and} \quad \mathsf{E}[Z_t Z_s] = \delta(t - s),$$

where E denotes expectation and δ is the Dirac delta function. In this discussion we further restrict attention to Gaussian white noise. The surprising thing about white noise is that it cannot exist as an ordinary random process (though it does exist as a generalized process [1]). Fortunately, there is a well-defined ordinary random process, known as the *Wiener process* (also known as *Brownian motion*), denoted by W_t, that makes a good model for integrated white noise, i.e., W_t behaves as if

$$W_t = \int_0^t Z_\theta \, d\theta,$$

or symbolically, $dW_t = Z_t dt$. Thus, if we multiply Equation 59.3 by dt, and write dW_t for $Z_t dt$, we obtain

$$dX_t = a(t, X_t) \, dt + b(t, X_t) \, dW_t, \quad X_{t_0} = \Xi. \tag{59.4}$$

To give meaning to Equation 59.4 and to solve Equation 59.4, we will always understand it as shorthand for the corresponding integral equation,

$$X_t = \Xi + \int_{t_0}^t a(\theta, X_\theta) \, d\theta + \int_{t_0}^t b(\theta, X_\theta) \, dW_\theta. \tag{59.5}$$

In order to make sense of Equation 59.5, we have to assign a meaning to integrals with respect to a Wiener process. There are two different ways to do this. One is due to Itô, and the other is due to Stratonovich. Since the Itô integral is more popular, and since the Stratonovich integral can be expressed in terms of the Itô integral [1], we restrict attention in our discussion to the Itô integral.

Now suppose X_t is a solution to the SDE of Equation 59.4, and suppose we pass X_t through a nonlinearity, say $Y_t := g(X_t)$. Of course, $Y_{t_0} = g(\Xi)$, but astonishingly, by the stochastic chain rule, the analog of Equation 59.2 is [1]

$$\begin{aligned} dY_t &= g'(X_t) \, dX_t + \frac{1}{2} g''(X_t) b(t, X_t)^2 \, dt \\ &= g'(X_t) a(t, X_t) \, dt + g'(X_t) b(t, X_t) \, dW_t + \frac{1}{2} g''(X_t) b(t, X_t)^2 \, dt, \end{aligned} \tag{59.6}$$

assuming g is twice continuously differentiable. Equation 59.6 is known as *Itô's rule*, and the last term in Equation 59.6 is called the *Itô correction term*. In addition to explaining its presence, the remainder of our discussion is as follows. Section 59.2 introduces the Wiener process as a model for integrated white noise. In Section 59.3, integration with respect to the Wiener process is defined, and a simple form of Itô's rule is derived. Section 59.4 focuses on SDEs. Itô's rule is derived for time-invariant nonlinearities, and its extension to time-varying nonlinearities is also given. In Section 59.5, Itô's rule is used to solve special forms of linear SDEs. ODEs are derived for the mean and variance of the solution in this case. When the initial condition is also Gaussian, the solution to the linear SDE is a Gaussian process, and its distribution is completely determined by its mean and variance. Nonlinear SDEs are considered in Section 59.6. The solutions of nonlinear SDEs are non-Gaussian Markov processes. In this case, we characterize their transition distribution in terms of the *Kolmogorov forward* (*Fokker–Planck*) and backward partial differential equations.

59.2 White Noise and the Wiener Process

A random process $\{Z_t\}$ is said to be a *white noise process* if

$$\mathsf{E}[Z_t] = 0 \quad \text{and} \quad \mathsf{E}[Z_t Z_s] = \delta(t - s), \tag{59.7}$$

where $\delta(t)$ is the Dirac delta, which is characterized by the two properties $\delta(t) = 0$ for $t \neq 0$ and $\int_{-\infty}^{\infty} \delta(t) \, dt = 1$.

Consider the *integrated white noise*

$$W_t = \int_0^t Z_u \, du. \tag{59.8}$$

We show that integrated white noise satisfies the following five properties. For $0 \le \theta \le \tau \le s \le t$,

$$W_0 = 0, \tag{59.9}$$

$$\mathsf{E}[W_t] = 0, \tag{59.10}$$

$$\mathsf{E}\big[(W_t - W_s)^2\big] = t - s, \tag{59.11}$$

$$\mathsf{E}\big[(W_t - W_s)(W_\tau - W_\theta)\big] = 0, \tag{59.12}$$

$$\mathsf{E}[W_t W_s] = \min\{t, s\}. \tag{59.13}$$

In other words:

- W_0 is a constant random variable with value zero.
- W_t has zero mean.
- $W_t - W_s$ has variance $t - s$.
- If $(\theta, \tau]$ and $(s, t]$ are nonoverlapping time intervals, then the increments $W_\tau - W_\theta$ and $W_t - W_s$ are uncorrelated.
- The correlation between W_t and W_s is $\mathsf{E}[W_t W_s] = \min\{t, s\}$.

(A process that satisfies Equation 59.12 is said to have orthogonal increments. A very accessible introduction to orthogonal increments processes can be found in [4].)

The property defined in Equation 59.9 is immediate from Equation 59.8. To establish Equation 59.10, write $\mathsf{E}[W_t] = \int_0^t \mathsf{E}[Z_u] \, du = 0$. To derive Equation 59.11, write

$$\begin{aligned}
\mathsf{E}\big[(W_t - W_s)^2\big] &= \mathsf{E}\left[\left(\int_s^t Z_u \, du\right)\left(\int_s^t Z_v \, dv\right)\right] \\
&= \int_s^t \left(\int_s^t \mathsf{E}[Z_u Z_v] \, du\right) dv \\
&= \int_s^t \left(\int_s^t \delta(u - v) \, du\right) dv \\
&= \int_s^t 1 \, dv \\
&= t - s.
\end{aligned}$$

To obtain Equation 59.12, write

$$\mathsf{E}[(W_t - W_s)(W_\tau - W_\theta)] = \mathsf{E}\left[\left(\int_s^t Z_u \, du\right)\left(\int_\theta^\tau Z_v \, dv\right)\right] = \int_\theta^\tau \left(\int_s^t \delta(u - v) \, du\right) dv.$$

Because the ranges of integration, which are understood as $(\theta, \tau]$ and $(s, t]$, do not intersect, $\delta(u - v) = 0$, and the inner integral is zero. Finally, the properties defined in Equations 59.9, 59.11, and 59.12 yield Equation 59.13 by writing, when $t > s$,

$$\begin{aligned}
\mathsf{E}[W_t W_s] &= \mathsf{E}[(W_t - W_s) W_s] + \mathsf{E}[W_s^2] \\
&= \mathsf{E}[(W_t - W_s) W_s] + s \\
&= \mathsf{E}[(W_t - W_s)(W_s - W_0)] + s \\
&= s.
\end{aligned} \tag{59.14}$$

If $t < s$, a symmetric argument yields $\mathsf{E}[W_t W_s] = t$. Hence, we can in general write $\mathsf{E}[W_t W_s] = \min\{t, s\}$.

It is well known that no process $\{Z_t\}$ satisfying Equation 59.7 can exist in the usual sense [1]. Hence, defining W_t by Equation 59.8 does not make sense. Fortunately, it is possible to define a random process $\{W_t, t \geq 0\}$ satisfying Equations 59.9 through 59.13 (as well as additional properties).

Definition 59.1:

The standard **Wiener process**, *or* **Brownian motion**, *denoted by* $\{W_t, t \geq 0\}$, *is characterized by the following four properties:*

W-1 $W_0 = 0$.

W-2 For $0 \leq s < t$, *the increment* $W_t - W_s$ *is a Gaussian random variable with zero mean and variance* $t - s$, *i.e.,*

$$\Pr(W_t - W_s \leq w) = \int_{-\infty}^{w} \frac{\exp\left(-\dfrac{x^2}{2(t-s)}\right)}{\sqrt{2\pi(t-s)}}\, dx.$$

W-3 $\{W_t, t \geq 0\}$ *has independent increments; i.e., if* $0 \leq t_1 \leq \cdots \leq t_n$, *then the increments*

$$(W_{t_2} - W_{t_1}), (W_{t_3} - W_{t_2}), \ldots, (W_{t_n} - W_{t_{n-1}})$$

are statistically independent random variables.

W-4 $\{W_t, t \geq 0\}$ *has continuous sample paths with probability 1.*

A proof of the existence of the Wiener process is given, for example, in [2] and in [4]. A sample path of a standard Wiener process is shown in Figure 59.1.

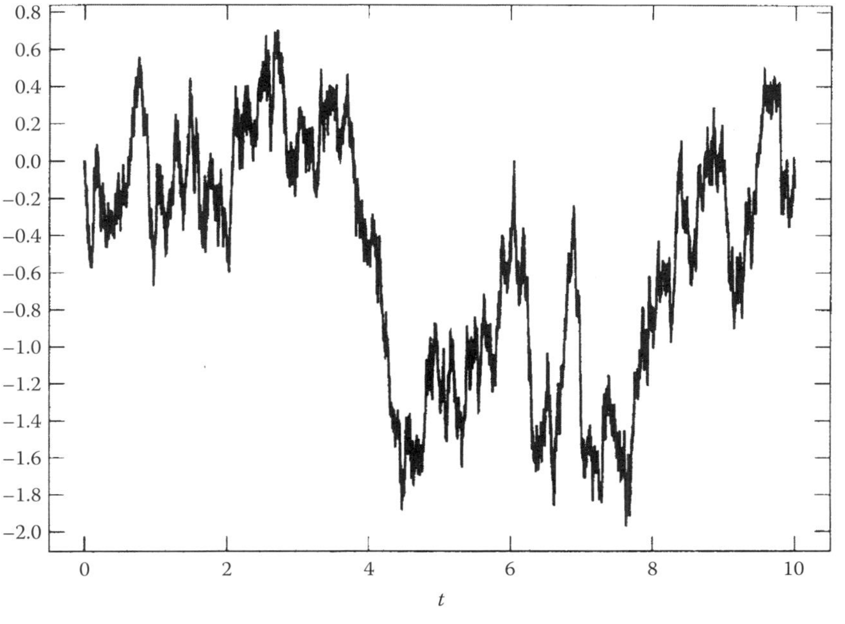

FIGURE 59.1 Sample path of a standard Wiener process.

We now show that properties W-1, W-2, and W-3 are sufficient to prove that the Wiener process satisfies Equations 59.9 through 59.13. Clearly, property W-1 and Equation 59.9 are the same. To establish Equation 59.10, put $s = 0$ in property W-2 and use property W-1. It is clear that property W-2 implies Equation 59.11. Also, Equation 59.12 is an immediate consequence of properties W-3 and W-2. Finally, since the Wiener process satisfies Equations 59.9, 59.11, and 59.12, the derivation in Equation 59.14 holds for the Wiener process, and thus Equation 59.13 also holds for the Wiener process.

Remark 59.1

From Figure 59.1, we see that the Wiener process has very jagged sample paths. In fact, if we zoom in on any subinterval, say [2,4] as shown in Figure 59.2, the sample path looks just as jagged. In other words, the Wiener process is continuous, but seems to have corners everywhere. In fact, it can be shown mathematically [2] that the sample paths of the Wiener process are nowhere differentiable. In other words, W_t cannot be the integral of any reasonable function, which is consistent with our earlier claim that continuous-time white noise cannot exist as an ordinary random process.

59.3 The Itô Integral

Let $\{W_t, t \geq 0\}$ be a standard Wiener process. The *history* of the process up to (and including) time t is denoted by $\mathcal{F}_t := \sigma(W_\theta, 0 \leq \theta \leq t)$. For our purposes, we say that a random variable X is $\mathcal{F}_t$*-measurable* if it is a function of the history up to time t; i.e., if there is a deterministic function h such that $X = h(W_\theta, 0 \leq \theta \leq t)$. For example, if $X = W_t - W_{t/2}$, then X is $\mathcal{F}_t$-measurable. Another example would be $X = \int_0^t W_\theta \, d\theta$; in this case, X depends on all of the variables $W_\theta, 0 \leq \theta \leq t$ and is $\mathcal{F}_t$-measurable. We also borrow the following notation from probability theory. If Z is any random variable, we write

$$\mathsf{E}[Z|\mathcal{F}_t] \quad \text{instead of} \quad \mathsf{E}[Z|W_\theta, 0 \leq \theta \leq t].$$

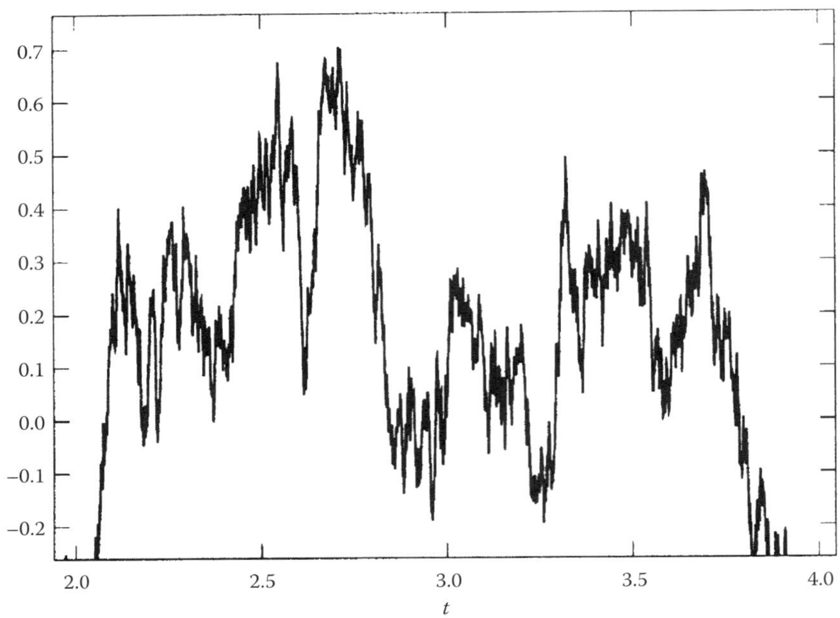

FIGURE 59.2 Closeup of sample path in Figure 59.1.

If Z is arbitrary but X is $\mathcal{F}_t$-measurable, then [2]

$$\mathsf{E}[XZ|\mathcal{F}_t] = X\,\mathsf{E}[Z|\mathcal{F}_t]. \tag{59.15}$$

A special case of Equation 59.15 is obtained if $Z = 1$; then $\mathsf{E}[Z|\mathcal{F}_t] = \mathsf{E}[1|\mathcal{F}_t] = 1$, and hence, if X is $\mathcal{F}_t$-measurable,

$$\mathsf{E}[X|\mathcal{F}_t] = X. \tag{59.16}$$

We also require the following results from probability theory [2], which we refer to as the *smoothing properties of conditional expectation*:

$$\mathsf{E}[Z] = \mathsf{E}\big[\mathsf{E}[Z|\mathcal{F}_t]\big], \quad t \geq 0, \tag{59.17}$$

and

$$\mathsf{E}[Z|\mathcal{F}_s] = \mathsf{E}\big[\mathsf{E}[Z|\mathcal{F}_t]\big|\mathcal{F}_s], \quad t \geq s \geq 0. \tag{59.18}$$

Definition 59.2:

A random process $\{H_t, t \geq 0\}$ is $\{\mathcal{F}_t\}$-adapted (or simply adapted if $\{\mathcal{F}_t\}$ is understood), if for each t, H_t is an $\mathcal{F}_t$-measurable random variable.

Obviously, $\{W_t, t \geq 0\}$ is $\{\mathcal{F}_t\}$-adapted. We now show that W_t is a *martingale*, i.e.,

$$\mathsf{E}[W_t|\mathcal{F}_s] = W_s, \quad t \geq s. \tag{59.19}$$

To see this, first note that since $W_0 = 0$,

$$\mathcal{F}_s := \sigma(W_\theta, 0 \leq \theta \leq s) = \sigma(W_\theta - W_0, 0 \leq \theta \leq s).$$

It follows, on account of the independent increments of the Wiener process, that $W_t - W_s$ is independent of the history $\mathcal{F}_s$ for $t \geq s$; i.e., for any function f,

$$\mathsf{E}[f(W_t - W_s)|\mathcal{F}_s] = \mathsf{E}[f(W_t - W_s)].$$

Then, since $W_t - W_s$ has zero mean,

$$\mathsf{E}[W_t - W_s|\mathcal{F}_s] = \mathsf{E}[W_t - W_s] = 0, \quad t \geq s, \tag{59.20}$$

which is equivalent to Equation 59.19.

Having shown that W_t is a martingale, we now show that $W_t^2 - t$ is also a martingale. To do this, we need the following three facts:

1. $W_t - W_s$ is independent of $\mathcal{F}_s$ with $\mathsf{E}[(W_t - W_s)^2] = t - s$
2. W_t is a martingale (*cf.* Equation 59.19)
3. Properties defined in Equations 59.15 and 59.16

For $t > s$, write

$$\begin{aligned}
\mathsf{E}[W_t^2|\mathcal{F}_s] &= \mathsf{E}[(W_t - W_s)^2 + 2W_tW_s - W_s^2|\mathcal{F}_s] \\
&= \mathsf{E}[(W_t - W_s)^2|\mathcal{F}_s] + 2W_s\mathsf{E}[W_t|\mathcal{F}_s] - W_s^2 \\
&= t - s + 2W_s^2 - W_s^2 \\
&= t + W_s^2 - s.
\end{aligned}$$

Rearranging, we have

$$\mathsf{E}[W_t^2 - t|\mathcal{F}_s] = W_s^2 - s, \quad t > s,$$

i.e., $W_t^2 - t$ is a martingale.

59.3.1 Definition of Itô's Stochastic Integral

We now define the Itô integral. As in the development of the Riemann integral, we begin by defining the integral for functions that are piecewise constant in time; i.e., we consider integrands $\{H_t, t \geq 0\}$ that are $\{\mathcal{F}_t\}$-adapted processes satisfying

$$H_t = H_{t_i} \quad \text{for } t_i \leq t < t_{i+1}, \tag{59.21}$$

for some breakpoints $t_i < t_{i+1}$. Thus, while $H_t = H_{t_i}$ on $[t_i, t_{i+1})$, the value H_{t_i} is an $\mathcal{F}_{t_i}$-measurable random variable. Without loss of generality, given any $0 \leq s < t$, we may assume $s = t_0 < \cdots < t_n = t$. Then the Itô integral is defined to be

$$\int_s^t H_\theta \, dW_\theta := \sum_{i=0}^{n-1} H_{t_i}(W_{t_{i+1}} - W_{t_i}). \tag{59.22}$$

To handle the general case, suppose H_t is a process for which there exists a sequence of processes H_t^k of the form of Equation 59.21 (where now n and the breakpoints $\{t_i\}$ depend on k) such that

$$\lim_{k\to\infty} \int_s^t \mathsf{E}[|H_\theta^k - H_\theta|^2] \, d\theta = 0.$$

Then we take $\int_s^t H_\theta \, dW_\theta$ to be the mean-square limit (which exists) of $\int_s^t H_\theta^k \, dW_\theta$; i.e., there exists a random variable, denoted by $\int_s^t H_\theta \, dW_\theta$, such that

$$\lim_{k\to\infty} \mathsf{E}\left[\left|\int_s^t H_\theta^k \, dW_\theta - \int_s^t H_\theta \, dW_\theta\right|^2\right] = 0. \tag{59.23}$$

See [10] for details. It should also be noted that when $H_t = h(t)$ is a deterministic function, the right-hand side of Equation 59.22 is a Gaussian random variable, and since mean-square limits of such quantities are also Gaussian [4], $\left\{\int_s^t h(\theta) \, dW_\theta, t \geq s\right\}$ is a Gaussian random process. When the integrand of an Itô integral is deterministic, the integral is sometimes called a *Wiener integral.*

59.3.2 Properties of the Itô Integral

The Itô integral satisfies the following three properties. First, the Itô integral is a zero-mean random variable, i.e.,

$$\mathsf{E}\left[\int_s^t H_\theta \, dW_\theta\right] = 0. \tag{59.24}$$

Second, the variance of $\int_s^t H_\theta \, dW_\theta$ is

$$\mathsf{E}\left[\left(\int_s^t H_\theta \, dW_\theta\right)^2\right] = \int_s^t \mathsf{E}[H_\theta^2] \, d\theta. \tag{59.25}$$

Third, if $X_t := \int_0^t H_\theta \, dW_\theta$, then X_t is a martingale, i.e., $\mathsf{E}[X_t|\mathcal{F}_s] = X_s$, or in terms of Itô integrals,

$$\mathsf{E}\left[\int_0^t H_\theta \, dW_\theta \middle| \mathcal{F}_s\right] = \int_0^s H_\theta \, dW_\theta, \quad t \geq s.$$

We verify these properties when H_t is of the form of Equation 59.21. To prove Equation 59.24, take the expectations of Equation 59.22 to obtain

$$\mathsf{E}\left[\int_s^t H_\theta \, dW_\theta\right] = \sum_{i=0}^{n-1} \mathsf{E}[H_{t_i}(W_{t_{i+1}} - W_{t_i})].$$

By the first smoothing property of conditional expectation, the fact that H_{t_i} is $\mathcal{F}_{t_i}$-measurable, and the properties defined in Equations 59.15 and 59.20,

$$\begin{aligned}\mathsf{E}[H_{t_i}(W_{t_{i+1}} - W_{t_i})] &= \mathsf{E}\big[\mathsf{E}[H_{t_i}(W_{t_{i+1}} - W_{t_i})|\mathcal{F}_{t_i}]\big] \\ &= \mathsf{E}\big[H_{t_i}\mathsf{E}[W_{t_{i+1}} - W_{t_i}|\mathcal{F}_{t_i}]\big] \\ &= 0.\end{aligned}$$

To establish Equation 59.25, use Equation 59.22 to write

$$\mathsf{E}\left[\left(\int_s^t H_\theta\, dW_\theta\right)^2\right] = \sum_{i=0}^{n-1}\sum_{j=0}^{n-1} \mathsf{E}[H_{t_i}H_{t_j}(W_{t_{i+1}} - W_{t_i})(W_{t_{j+1}} - W_{t_j})].$$

First consider a typical term in the double sum for which $j = i$. Then by the first smoothing property and Equation 59.15,

$$\begin{aligned}\mathsf{E}[H_{t_i}^2(W_{t_{i+1}} - W_{t_i})^2] &= \mathsf{E}\big[\mathsf{E}[H_{t_i}^2(W_{t_{i+1}} - W_{t_i})^2|\mathcal{F}_{t_i}]\big] \\ &= \mathsf{E}\big[H_{t_i}^2\mathsf{E}[(W_{t_{i+1}} - W_{t_i})^2|\mathcal{F}_{t_i}]\big].\end{aligned}$$

Since $W_{t_{i+1}} - W_{t_i}$ is independent of $\mathcal{F}_{t_i}$,

$$\mathsf{E}[(W_{t_{i+1}} - W_{t_i})^2|\mathcal{F}_{t_i}] = \mathsf{E}[(W_{t_{i+1}} - W_{t_i})^2] = t_{i+1} - t_i.$$

Hence,

$$\mathsf{E}[H_{t_i}^2(W_{t_{i+1}} - W_{t_i})^2] = \mathsf{E}[H_{t_i}^2](t_{i+1} - t_i).$$

For the terms with $j \neq i$, we can, without loss of generality, take $j < i$. In this case,

$$\begin{aligned}\mathsf{E}[H_{t_i}H_{t_j}(W_{t_{i+1}} - W_{t_i})(W_{t_{j+1}} - W_{t_j})] &= \mathsf{E}\big[\mathsf{E}[H_{t_i}H_{t_j}(W_{t_{i+1}} - W_{t_i})(W_{t_{j+1}} - W_{t_j})|\mathcal{F}_{t_i}]\big] \\ &= \mathsf{E}\big[H_{t_i}H_{t_j}(W_{t_{j+1}} - W_{t_j})\mathsf{E}[W_{t_{i+1}} - W_{t_i}|\mathcal{F}_{t_i}]\big] \\ &= 0, \qquad \text{by Equation 59.20.}\end{aligned}$$

Thus,

$$\mathsf{E}\left[\left(\int_s^t H_\theta\, dW_\theta\right)^2\right] = \sum_{i=0}^{n-1} \mathsf{E}[H_{t_i}^2](t_{i+1} - t_i) = \int_s^t \mathsf{E}[H_\theta^2]\, d\theta.$$

To show that $X_t := \int_0^t H_\theta\, dW_\theta$ is a martingale, it suffices to prove that $\mathsf{E}[X_t - X_s|\mathcal{F}_s] = 0$. Write

$$\begin{aligned}\mathsf{E}[X_t - X_s|\mathcal{F}_s] = \mathsf{E}\left[\int_s^t H_\theta\, dW_\theta\middle|\mathcal{F}_s\right] &= \mathsf{E}\left[\sum_{i=0}^{n-1} H_{t_i}(W_{t_{i+1}} - W_{t_i})\middle|\mathcal{F}_s\right] \\ = \sum_{i=0}^{n-1} \mathsf{E}[H_{t_i}(W_{t_{i+1}} - W_{t_i})|\mathcal{F}_s].\end{aligned}$$

Then by the second smoothing property of conditional expectation and the properties defined in Equations 59.15 and 59.20,

$$\mathsf{E}[X_t - X_s|\mathcal{F}_s] = \sum_{i=0}^{n-1} \mathsf{E}\big[H_{t_i}\mathsf{E}[W_{t_{i+1}} - W_{t_i}|\mathcal{F}_{t_i}]\big|\mathcal{F}_s\big] = 0.$$

59.3.3 A Simple Form of Itô's Rule

The following result is essential to derive Itô's rule [1].

Lemma 59.1:

For any partition of $[s, t]$, *say* $s = t_0 < \cdots < t_n = t$, *put* $\Delta := \max_{0 \le i \le n-1} |t_{i+1} - t_i|$. *If*

$$V := \sum_{i=0}^{n-1} (W_{t_{i+1}} - W_{t_i})^2, \tag{59.26}$$

then

$$\mathsf{E}[|V - (t-s)|^2] \le 2\Delta(t-s).$$

The importance of the lemma is that it implies that as the Δ of the partition becomes small, the sum of squared increments V converges in mean square to the length of the interval, $t - s$.

Proof 59.1. The first step is to note that $t - s = \sum_{i=0}^{n-1}(t_{i+1} - t_i)$. Next, let D denote the difference

$$D := V - (t-s) = \sum_{i=0}^{n-1} \Big\{ (W_{t_{i+1}} - W_{t_i})^2 - (t_{i+1} - t_i) \Big\}.$$

Thus, D is a sum of independent, zero-mean random variables. It follows that the expectation of the cross terms in D^2 vanishes, thus leaving

$$\mathsf{E}[D^2] = \sum_{i=0}^{n-1} \mathsf{E}\Big[\Big\{ (W_{t_{i+1}} - W_{t_i})^2 - (t_{i+1} - t_i) \Big\}^2 \Big].$$

Put $Z_i := (W_{t_{i+1}} - W_{t_i}) / \sqrt{t_{i+1} - t_i}$. Then Z_i is a zero-mean Gaussian random variable with variance 1 (which implies $\mathsf{E}[Z_i^4] = 3$). We can now write

$$\begin{aligned}
\mathsf{E}[D^2] &= \sum_{i=0}^{n-1} \mathsf{E}\Big[\Big\{ (Z_i^2 - 1)(t_{i+1} - t_i) \Big\}^2 \Big] \\
&= \sum_{i=0}^{n-1} \mathsf{E}[(Z_i^2 - 1)^2](t_{i+1} - t_i)^2 \\
&= \sum_{i=0}^{n-1} 2(t_{i+1} - t_i)^2 \\
&\le \sum_{i=0}^{n-1} 2\Delta(t_{i+1} - t_i) = 2\Delta(t-s).
\end{aligned}$$

Let $\{H_t, t \ge 0\}$ be a continuous $\{\mathcal{F}_t\}$-adapted process. It can be shown [1] that as the partition becomes finer,

$$\sum_{i=0}^{n-1} H_{t_i}(W_{t_{i+1}} - W_{t_i})^2 \to \sum_{i=0}^{n-1} H_{t_i}(t_{i+1} - t_i) \to \int_s^t H_\theta \, d\theta.$$

We now derive a special case of Itô's rule. Let $g(w)$ be a twice continuously differentiable function of w. If $Y_t = g(W_t)$, our intuition about the chain rule might lead us to write $dY_t = g'(W_t)\, dW_t$. As we now show, the correct answer is $dY_t = g'(W_t)\, dW_t + \frac{1}{2} g''(W_t)\, dt$.

Consider the Taylor expansion,

$$g(w_2) - g(w_1) \approx g'(w_1)(w_2 - w_1) + \frac{1}{2}g''(w_1)(w_2 - w_1)^2.$$

Suppose $Y_t = g(W_t)$. For $s = t_0 < \cdots < t_n = t$, write

$$\begin{aligned} Y_t - Y_s &= \sum_{i=0}^{n-1} g(W_{t_{i+1}}) - g(W_{t_i}) \\ &\approx \sum_{i=0}^{n-1} \left\{ g'(W_{t_i})(W_{t_{i+1}} - W_{t_i}) + \frac{1}{2}g''(W_{t_i})(W_{t_{i+1}} - W_{t_i})^2 \right\}. \end{aligned}$$

Note that $g'(W_{t_i})$ and $g''(W_{t_i})$ are $\mathcal{F}_{t_i}$-measurable. Hence, as the partition becomes finer, we obtain

$$Y_t - Y_s = \int_s^t g'(W_\theta)\, dW_\theta + \frac{1}{2}\int_s^t g''(W_\theta)\, d\theta.$$

Writing this in differential form, we have a special case of Itô's rule: If $Y_t = g(W_t)$, then

$$dY_t = g'(W_t)\, dW_t + \frac{1}{2}g''(W_t)\, dt.$$

Example 59.1:

As a simple application of this result, we show that

$$\int_0^t W_\theta\, dW_\theta = \frac{(W_t^2 - t)}{2}.$$

Take $g(w) = w^2$. Then $g'(w) = 2w$, and $g''(w) = 2$. The special case of Itô's rule gives

$$dY_t = 2W_t\, dW_t + 1\, dt.$$

Converting this to integral form and noting that $Y_0 = W_0^2 = 0$,

$$\begin{aligned} Y_t &= Y_0 + \int_0^t 2W_\theta\, dW_\theta + \int_0^t 1\, d\theta \\ &= 2\int_0^t W_\theta\, dW_\theta + t. \end{aligned}$$

Since $Y_t = g(W_t) = W_t^2$, the result follows. As noted earlier, integrals with respect to the Wiener process are martingales. Since $\int_0^t W_\theta\, dW_\theta = (W_t^2 - t)/2$, we now have an alternative proof to the one following Equation 59.20 that $W_t^2 - t$ is a martingale.

59.4 Stochastic Differential Equations and Itô's Rule

Suppose X_t satisfies the SDE

$$dX_t = a(t, X_t)\,dt + b(t, X_t)\,dW_t, \tag{59.27}$$

or equivalently, the integral equation,

$$X_t = X_s + \int_s^t a(\theta, X_\theta)\,d\theta + \int_s^t b(\theta, X_\theta)\,dW_\theta. \tag{59.28}$$

If $Y_t = g(X_t)$, where g is twice continuously differentiable, we show that Y_t satisfies the SDE

$$dY_t = g'(X_t)\,dX_t + \frac{1}{2}g''(X_t)b(t, X_t)^2\,dt. \tag{59.29}$$

Using the Taylor expansion of g as we did in the preceding section, write

$$Y_t - Y_s \approx \sum_{i=0}^{n-1}\Big\{g'(X_{t_i})(X_{t_{i+1}} - X_{t_i}) + \frac{1}{2}g''(X_{t_i})(X_{t_{i+1}} - X_{t_i})^2\Big\}.$$

From Equation 59.28 we have the approximation

$$X_{t_{i+1}} - X_{t_i} \approx a(t_i, X_{t_i})(t_{i+1} - t_i) + b(t_i, X_{t_i})(W_{t_{i+1}} - W_{t_i}).$$

Hence, as the partition becomes finer,

$$\sum_{i=0}^{n-1} g'(X_{t_i})(X_{t_{i+1}} - X_{t_i})$$

converges to

$$\int_s^t g'(X_\theta)a(\theta, X_\theta)\,d\theta + \int_s^t g'(X_\theta)b(\theta, X_\theta)\,dW_\theta.$$

It remains to consider sums of the form (*cf.* Equation 59.26)

$$\sum_{i=0}^{n-1} g''(X_{t_i})(X_{t_{i+1}} - X_{t_i})^2.$$

The ith term in the sum is approximately

$$g''(X_{t_i})\Big\{a(t_i, X_{t_i})^2(t_{i+1} - t_i)^2 + 2a(t_i, X_{t_i})b(t_i, X_{t_i})(t_{i+1} - t_i)(W_{t_{i+1}} - W_{t_i}) + b(t_i, X_{t_i})^2(W_{t_{i+1}} - W_{t_i})^2\Big\}.$$

Now with $\Delta = \max_{0 \le i \le n-1} |t_{i+1} - t_i|$,

$$\left|\sum_{i=0}^{n-1} g''(X_{t_i})a(t_i, X_{t_i})^2(t_{i+1} - t_i)^2\right| \le \Delta \sum_{i=0}^{n-1} |g''(X_{t_i})|a(t_i, X_{t_i})^2(t_{i+1} - t_i),$$

which converges to zero as $\Delta \to 0$. Also,

$$\sum_{i=0}^{n-1} g''(X_{t_i})a(t_i, X_{t_i})(t_{i+1} - t_i)(W_{t_{i+1}} - W_{t_i})$$

converges to 0. Finally, note that

$$\sum_{i=0}^{n-1} g''(X_{t_i})b(t_i, X_{t_i})^2(W_{t_{i+1}} - W_{t_i})^2$$

converges to $\int_s^t g''(X_\theta)b(\theta, X_\theta)^2\, d\theta$. Putting this all together, as the partition becomes finer, we have

$$Y_t - Y_s = \int_s^t g'(X_\theta)\big[a(\theta, X_\theta)\, dt + b(\theta, X_\theta)\, dW_\theta\big] + \frac{1}{2}\int_s^t g''(X_\theta)b(\theta, X_\theta)^2\, d\theta,$$

which is indeed the integral form of Itô's rule in Equation 59.29.

Itô's rule can be extended to handle a time-varying nonlinearity $g(t, x)$ whose partial derivatives

$$g_t := \frac{\partial g}{\partial t}, \quad g_x := \frac{\partial g}{\partial x}, \quad \text{and} \quad g_{xx} := \frac{\partial^2 g}{\partial x^2}$$

are continuous [1]. If $Y_t = g(t, X_t)$, where X_t satisfies Equation 59.27, then

$$\begin{aligned} dY_t &= g_t(t, X_t)\, dt + g_x(t, X_t)\, dX_t + \tfrac{1}{2} g_{xx}(t, X_t)b(t, X_t)^2 dt \\ &= g_t(t, X_t)\, dt + g_x(t, X_t)a(t, X_t)\, dt + g_x(t, X_t)b(t, X_t)\, dW_t + \frac{1}{2} g_{xx}(t, X_t)b(t, X_t)^2\, dt. \end{aligned} \tag{59.30}$$

Example 59.2:

Consider the Langevin equation

$$dX_t = -3X_t\, dt + 5\, dW_t.$$

Suppose $Y_t = \sin(tX_t)$. Then with $g(t, x) = \sin(tx)$, $g_t(t, x) = x\cos(tx)$, $g_x(t, x) = t\cos(tx)$, and $g_{xx}(t, x) = -t^2\sin(tx)$. By the extended Itô's rule,

$$dY_t = [(1 - 3t)X_t\cos(tX_t) - \frac{25}{2}t^2\sin(tX_t)]\, dt + 5t\cos(tX_t)\, dW_t.$$

59.5 Applications of Itô's Rule to Linear SDEs

Using Itô's rule, we can verify explicit solutions to linear SDEs. By a linear SDE we mean an equation of the form

$$dX_t = [\alpha(t) + c(t)X_t]\, dt + [\beta(t) + \gamma(t)X_t]\, dW_t, \quad X_{t_0} = \Xi. \tag{59.31}$$

If Ξ is independent of $\{W_t - W_{t_0}, t \geq t_0\}$, and if c, β, and γ are bounded on a finite interval $[t_0, t_f]$, then a unique continuous solution exists on $[t_0, t_f]$; if α, c, β, and γ are bounded on $[t_0, t_f]$ for every finite $t_f > t_0$, then a unique solution exists on $[t_0, \infty)$ [1].

59.5.1 Homogeneous Equations

A linear SDE of the form

$$dX_t = c(t)X_t\, dt + \gamma(t)X_t\, dW_t, \quad X_{t_0} = \Xi, \tag{59.32}$$

is said to be *homogeneous*. In this case, we claim that the solution is $X_t = \Xi\exp(Y_t)$, where

$$Y_t := \int_{t_0}^t \left[\frac{c(\theta) - \gamma(\theta)^2}{2}\right] d\theta + \int_{t_0}^t \gamma(\theta)\, dW_\theta,$$

or in differential form,

$$dY_t = \left[\frac{c(t) - \gamma(t)^2}{2}\right] dt + \gamma(t)\, dW_t.$$

To verify our claim, we follow [1] and simply apply Itô's rule of Equation 59.29:

$$\begin{aligned} dX_t &= \Xi \exp(Y_t)\, dY_t + \frac{1}{2}\Xi \exp(Y_t)\gamma(t)^2\, dt \\ &= X_t\left[\frac{c(t) - \gamma(t)^2}{2}\right] dt + X_t\gamma(t)\, dW_t + \frac{1}{2}X_t\gamma(t)^2\, dt \\ &= c(t)X_t\, dt + \gamma(t)X_t\, dW_t. \end{aligned}$$

Since $X_{t_0} = \Xi \exp(Y_{t_0}) = \Xi \exp(0) = \Xi$, we have indeed solved Equation 59.32.

Example 59.3:

Consider the homogeneous SDE

$$dX_t = \cos(t)X_t\, dt + 2X_t\, dW_t, \quad X_0 = 1.$$

For this problem, $Y_t = \sin(t) - 2t + 2W_t$, and the solution is $X_t = e^{Y_t} = e^{\sin(t) - 2(t - W_t)}$.

59.5.2 Linear SDEs in the Narrow Sense

A linear SDE of the form

$$dX_t = [\alpha(t) + c(t)X_t]\, dt + \beta(t)\, dW_t, \quad X_{t_0} = \Xi, \tag{59.33}$$

is said to be linear in the narrow sense because it is obtained by setting $\gamma(t) = 0$ in the general linear SDE in Equation 59.31. The solution of this equation is obtained as follows. First put

$$\Phi(t, t_0) := \exp\left(\int_{t_0}^{t} c(\theta)\, d\theta\right).$$

Observe that

$$\frac{\partial \Phi(t, t_0)}{\partial t} = c(t)\Phi(t, t_0), \tag{59.34}$$

$\Phi(t_0, t_0) = 1$, and $\Phi(t, t_0)\Phi(t_0, \theta) = \Phi(t, \theta)$. Next, let

$$Y_t := \Xi + \int_{t_0}^{t} \Phi(t_0, \theta)\alpha(\theta)\, d\theta + \int_{t_0}^{t} \Phi(t_0, \theta)\beta(\theta)\, dW_\theta,$$

or in differential form,

$$dY_t = \Phi(t_0, t)\alpha(t)\, dt + \Phi(t_0, t)\beta(t)\, dW_t.$$

Now put

$$\begin{aligned} X_t &:= \Phi(t, t_0)Y_t \\ &= \Phi(t, t_0)\Xi + \int_{t_0}^{t} \Phi(t, \theta)\alpha(\theta)\, d\theta + \int_{t_0}^{t} \Phi(t, \theta)\beta(\theta)\, dW_\theta. \end{aligned} \tag{59.35}$$

In other words, $X_t = g(t, Y_t)$, where $g(t, y) = \Phi(t, t_0)y$. Using Equation 59.34, $g_t(t, y) = c(t)\Phi(t, t_0)y$. We also have $g_y(t, y) = \Phi(t, t_0)$, and $g_{yy}(t, y) = 0$. By the extended Itô's rule of Equation 59.30,

$$dX_t = g_t(t, Y_t)\, dt + g_y(t, Y_t)\, dY_t + \frac{1}{2}g_{yy}(t, Y_t)\, dt$$

$$
\begin{aligned}
&= c(t)\Phi(t,t_0)Y_t\,dt + \Phi(t,t_0)[\Phi(t_0,t)\alpha(t)\,dt + \Phi(t_0,t)\beta(t)\,dW_t] \\
&= c(t)X_t\,dt + \alpha(t)\,dt + \beta(t)\,dW_t,
\end{aligned}
$$

which is exactly Equation 59.33.

Recalling the text following Equation 59.23, and noting the form of the solution in Equation 59.35, we see that $\{X_t\}$ is a Gaussian process if and only if Ξ is a Gaussian random variable. In any case, we can always use Equation 59.35 to derive differential equations for the mean and variance of X_t. For example, put $m(t) := \mathsf{E}[X_t]$. Since Itô integrals have zero mean (recall Equation 59.24), we obtain from Equation 59.35,

$$
m(t) = \Phi(t,t_0)\mathsf{E}[\Xi] + \int_{t_0}^{t} \Phi(t,\theta)\alpha(\theta)\,d\theta,
$$

and thus

$$
\begin{aligned}
\frac{dm(t)}{dt} &= c(t)\Phi(t,t_0)\mathsf{E}[\Xi] + \Phi(t,t)\alpha(t) + \int_{t_0}^{t} c(t)\Phi(t,\theta)\alpha(\theta)\,d\theta \\
&= c(t)\left\{\Phi(t,t_0)\mathsf{E}[\Xi] + \int_{t_0}^{t} \Phi(t,\theta)\alpha(\theta)\,d\theta\right\} + \alpha(t) \\
&= c(t)m(t) + \alpha(t),
\end{aligned}
$$

and $m(t_0) = \mathsf{E}[\Xi]$. We now turn to the covariance function of X_t, $r(t,s) := \mathsf{E}[(X_t - m(t))(X_s - m(s))]$. We assume that the initial condition Ξ is independent of $\{W_t - W_{t_0}, t \geq t_0\}$. Write

$$
X_t - m(t) = \Phi(t,t_0)(\Xi - \mathsf{E}[\Xi]) + \int_{t_0}^{t} \Phi(t,\theta)\beta(\theta)\,dW_\theta.
$$

For $s < t$, write

$$
\int_{t_0}^{t} \Phi(t,\theta)\beta(\theta)\,dW_\theta = \int_{t_0}^{s} \Phi(t,\theta)\beta(\theta)\,dW_\theta + \int_{s}^{t} \Phi(t,\theta)\beta(\theta)\,dW_\theta.
$$

Then

$$
r(t,s) = \Phi(t,t_0)\mathsf{E}[(\Xi - \mathsf{E}[\Xi])^2]\Phi(s,t_0) + \int_{t_0}^{s} \Phi(t,\theta)\beta(\theta)^2\Phi(s,\theta)\,d\theta.
$$

Letting $\operatorname{var}(\Xi) := \mathsf{E}[(\Xi - \mathsf{E}[\Xi])^2]$, for arbitrary s and t, we can write

$$
r(t,s) = \Phi(t,t_0)\operatorname{var}(\Xi)\Phi(s,t_0) + \int_{t_0}^{\min\{s,t\}} \Phi(t,\theta)\beta(\theta)^2\Phi(s,\theta)\,d\theta.
$$

In particular, if we put $v(t) := r(t,t) = \mathsf{E}[(X_t - m(t))^2]$, then a simple calculation shows

$$
\frac{dv(t)}{dt} = 2c(t)v(t) + \beta(t)^2, \quad v(t_0) = \operatorname{var}(\Xi).
$$

59.5.3 The Langevin Equation

If $\alpha(t) = 0$ and $c(t)$ and $\beta(t)$ do not depend on t, then the narrow-sense linear SDE in Equation 59.33 becomes

$$
dX_t = cX_t\,dt + \beta dW_t, \quad X_{t_0} = \Xi,
$$

which is the Langevin equation when $c < 0$ and $\beta > 0$. Now, since

$$
\Phi(t,t_0) = \exp\left(\int_{t_0}^{t} c\,d\theta\right) = e^{c[t-t_0]},
$$

the solution in Equation 59.35 simplifies to

$$X_t = e^{c[t-t_0]}\Xi + \int_{t_0}^{t} e^{c[t-\theta]}\beta\, dW_\theta.$$

Then the mean is

$$m(t) := \mathsf{E}[X_t] = e^{c[t-t_0]}\mathsf{E}[\Xi],$$

and the covariance is

$$\begin{aligned} r(t,s) &= e^{c[t-t_0+s-t_0]}\operatorname{var}(\Xi) + \int_{t_0}^{\min\{t,s\}} e^{c[t-\theta+s-\theta]}\beta^2\, d\theta \\ &= e^{c[t+s-2t_0]}\left[\operatorname{var}(\Xi) + \frac{\beta^2}{2c}\right] + e^{c|t-s|}\left(\frac{-\beta^2}{2c}\right). \end{aligned}$$

Now assume $c < 0$, and suppose that $\mathsf{E}[\Xi] = 0$ and $\operatorname{var}(\Xi) = -\beta^2/(2c)$. Then

$$\mathsf{E}[X_t] = 0 \quad \text{and} \quad r(t,s) = e^{c|t-s|}\left(\frac{-\beta^2}{2c}\right). \tag{59.36}$$

Since $\mathsf{E}[X_t]$ does not depend on t, and since $r(t,s)$ depends only on $|t-s|$, $\{X_t, t \geq t_0\}$ is said to be wide-sense stationary. If Ξ is also Gaussian, then $\{X_t, t \geq t_0\}$ is a Gaussian process known as the *Ornstein–Uhlenbeck process.*

59.6 Transition Probabilities for General SDEs

The transition function for a process $\{X_t, t \geq t_0\}$ is defined to be the conditional cumulative distribution

$$P(t,y|s,x) \;:=\; \Pr(X_t \leq y | X_s = x).$$

When X_t is the solution of an SDE, we can write down a partial differential equation that uniquely determines the transition function if certain hypotheses are satisfied. Under these hypotheses, the conditional cumulative distribution $P(t,y|s,x)$ has a density $p(t,y|s,x)$ that is uniquely determined by another partial differential equation.

To motivate the general results below, we first consider the narrow-sense linear SDE in Equation 59.33, whose solution is given in Equation 59.35. Since $\Phi(t,\theta) = \Phi(t,s)\Phi(s,\theta)$, Equation 59.35 can be rewritten as

$$\begin{aligned} X_t &= \Phi(t,s)\left[\Phi(s,t_0)\Xi + \int_{t_0}^{t} \Phi(s,\theta)\alpha(\theta)\, d\theta + \int_{t_0}^{t} \Phi(s,\theta)\beta(\theta)\, dW_\theta\right] \\ &= \Phi(t,s)\left[X_s + \int_{s}^{t} \Phi(s,\theta)\alpha(\theta)\, d\theta + \int_{s}^{t} \Phi(s,\theta)\beta(\theta)\, dW_\theta\right] \\ &= \Phi(t,s)X_s + \int_{s}^{t} \Phi(t,\theta)\alpha(\theta)\, d\theta + \int_{s}^{t} \Phi(t,\theta)\beta(\theta)\, dW_\theta. \end{aligned}$$

Now consider $\Pr(X_t \leq y | X_s = x)$. Since we are conditioning on $X_s = x$, we can replace X_s in the preceding equation by x. For notational convenience, let

$$Z \;:=\; \Phi(t,s)x + \int_{s}^{t} \Phi(t,\theta)\alpha(\theta)\, d\theta + \int_{s}^{t} \Phi(t,\theta)\beta(\theta)\, dW_\theta.$$

Then $\Pr(X_t \leq y | X_s = x) = \Pr(Z \leq y | X_s = x)$. Now, in the definition of Z, the only randomness comes from the Itô integral with deterministic integrand over $[s,t]$. The randomness in this integral comes only from the increments of the Wiener process on $[s,t]$. From Equation 59.35 with t replaced by s, we see that the only randomness in X_s comes from Ξ and from the increments of the Wiener process on $[t_0,s]$. Hence, Z and X_s are independent, and we can write $\Pr(Z \le y|X_s = x) = \Pr(Z \le y)$. Next, from the development of the Itô integral in Section 59.3, we see that Z is a Gaussian random variable with mean

$$m(t|s,x) := \Phi(t,s)x + \int_s^t \Phi(t,\theta)\alpha(\theta)\,d\theta$$

and variance

$$v(t|s) := \int_s^t [\Phi(t,\theta)\beta(\theta)]^2\,d\theta.$$

Hence, the transition function is

$$P(t,y|s,x) = \int_{-\infty}^{y} \frac{\exp\left(\frac{1}{2}[z - m(t|s,x)]^2/v(t|s)\right)}{\sqrt{2\pi v(t|s)}}\,dz,$$

and the transition density is

$$p(t,y|s,x) = \frac{\exp\left(\frac{1}{2}[y - m(t|s,x)]^2/v(t|s)\right)}{\sqrt{2\pi v(t|s)}}.$$

Example 59.4:

Consider the SDE $dX_t = -3X_t\,dt + 5\,dW_t$. Then $m(t|s,x) = e^{-3(t-s)}x$, and $v(t|s) = \frac{25}{6}[1 - e^{-6(t-s)}]$.

We now return to the general SDE,

$$dX_t = a(t,X_t)\,dt + b(t,X_t)\,dW_t, \quad X_{t_0} = \Xi, \tag{59.37}$$

where Ξ is independent of $\{W_t - W_{t_0}, t \ge t_0\}$. To guarantee a unique continuous solution on a finite interval, say $[t_0,t_f]$, we assume [1] that there exists a finite constant $K > 0$ such that for all $t \in [t_0,t_f]$ and all x and y, a and b satisfy the Lipschitz conditions

$$|a(t,x) - a(t,y)| \le K|x-y|,$$
$$|b(t,x) - b(t,y)| \le K|x-y|,$$

and the growth restrictions

$$|a(t,x)|^2 \le K^2(1+|x|^2),$$
$$|b(t,x)|^2 \le K^2(1+|x|^2).$$

If such a K exists for every finite $t_f > t_0$, then a unique solution exists on $[t_0,\infty)$ [1]. Under the above conditions for the general SDE in Equation 59.37, a unique solution exists, although one cannot usually give an explicit formula for it, and so one cannot find the transition function and density as we did in the narrow-sense linear case. However, if for some $b_0 > 0$, $b(t,x) \ge b_0$ for all t and all x, then [10] the transition function P is the unique solution of *Kolmogorov's backward equation*

$$\frac{1}{2}b(s,x)^2\frac{\partial^2 P(t,y|s,x)}{\partial x^2}+a(s,x)\frac{\partial P(t,y|s,x)}{\partial x}=-\frac{\partial P(t,y|s,x)}{\partial s},\quad t_0<s<t<t_f, \tag{59.38}$$

satisfying

$$\lim_{s\uparrow t} P(t,y|s,x)=\begin{cases}1, & y>x,\\ 0, & y<x.\end{cases}$$

Furthermore, P has a density,

$$p(t,y|s,x)=\frac{\partial P(t,y|s,x)}{\partial y},\quad t_0<s<t<t_f.$$

If $\partial a/\partial x$, $\partial b/\partial x$, and $\partial^2 b/\partial x^2$ also satisfy the Lipschitz and growth conditions above, and if $\partial b/\partial x \geq b_0 > 0$ and $\partial^2 b/\partial x^2 \geq b_0 > 0$, then the transition density p is the unique fundamental solution of *Kolmogorov's forward equation*

$$\frac{1}{2}\frac{\partial^2[b(t,y)^2p(t,y|s,x)]}{\partial y^2}-\frac{\partial[a(t,y)p(t,y|s,x)]}{\partial y}=\frac{\partial p(t,y|s,x)}{\partial t},\quad t_0<s<t<t_f, \tag{59.39}$$

satisfying

$$p(s,y|s,x)=\delta(y-x).$$

The forward partial differential equation is also known as the Fokker–Planck equation. Equation 59.39 is called the forward equation because it evolves forward in time starting at s; note also that x is fixed and y varies. In the backward equation, t and y are fixed, and x varies as s evolves backward in time starting at t.

Remark 59.2

If the necessary partial derivatives are continuous, then we can differentiate the backward equation with respect to y and obtain the following equation for the transition density p:

$$\frac{1}{2}b(s,x)^2\frac{\partial^2 p(t,y|s,x)}{\partial x^2}+a(s,x)\frac{\partial p(t,y|s,x)}{\partial x}=-\frac{\partial p(t,y|s,x)}{\partial s},\quad t_0<s<t<t_f.$$

Example 59.5:

In general, it is very hard to obtain explicit solutions to the forward or backward equations. However, when $a(t,x)=a(x)$ and $b(t,x)=b(x)$ do not depend on t, it is sometimes possible to obtain a limiting density $p(y)=\lim_{t\to\infty}p(t,y|s,x)$ that is independent of s and x. The existence of this limit suggests that for large t, $p(t,y|s,x)$ settles down to a constant as a function of t. Hence, the partial derivative with respect to t on the right-hand side of the forward equation in Equation 59.39 should be zero. This results in the ODE

$$\frac{1}{2}\frac{d^2[b(y)^2p(y)]}{dy^2}-\frac{d[a(y)p(y)]}{dy}=0. \tag{59.40}$$

For example, let μ and λ be positive constants, and suppose that

$$dX_t=-\mu X_t\,dt+\sqrt{\lambda(1+X_t^2)}\,dW_t.$$

(The case $\mu=1$ and $\lambda=2$ was considered by [10].) Then Equation 59.40 becomes

$$\frac{1}{2}\frac{d^2[\lambda(1+y^2)p(y)]}{dy^2}-\frac{d[-\mu y p(y)]}{dy}=0.$$

Integrating both sides, we obtain

$$\frac{\lambda}{2}\frac{d[(1+y^2)p(y)]}{dy}+\mu y p(y)=\kappa$$

for some constant κ. Now, the left-hand side of this equation is $(\lambda+\mu)yp(y)+\lambda(1+y^2)p'(y)/2$. If we assume that this goes to zero as $|y|\to\infty$, then $\kappa=0$. In this case,

$$\frac{p'(y)}{p(y)}=-\left(\frac{1+\mu}{\lambda}\right)\frac{2y}{1+y^2}.$$

Integrating from 0 to y yields

$$\ln\frac{p(y)}{p(0)}=-\left(\frac{1+\mu}{\lambda}\right)\ln(1+y^2),$$

or

$$p(y)=\frac{p(0)}{(1+y^2)^{1+\mu/\lambda}}.$$

Of course, $p(0)$ is determined by the requirement that $\int_{-\infty}^{\infty}p(y)\,dy=1$. For example, if $\mu/\lambda=1/2$, then $p(0)=1/2$, which can be found directly after noting that the antiderivative of $1/(1+y^2)^{3/2}$ is $y/\sqrt{1+y^2}$. As a second example, suppose $\mu/\lambda=1$. Then $p(y)$ has the form $p(0)f(y)^2$, where $f(y)=1/(1+y^2)$. If we let $F(\omega)$ denote the Fourier transform of $f(y)$, then Parseval's equation yields $\int_{-\infty}^{\infty}|f(y)|^2\,dy=\int_{-\infty}^{\infty}|F(\omega)|^2\,d\omega/2\pi$. Since $F(\omega)=\pi e^{-|\omega|}$, this last integral can be computed in closed form, and we find that $p(0)=2/\pi$.

59.7 Defining Terms

Adapted: A random process $\{H_t\}$ is $\{\mathcal{F}_t\}$-adapted, where $\mathcal{F}_t=\sigma(W_\theta,0\le\theta\le t)$, if for each t, H_t is $\mathcal{F}_t$-measurable. See measurable.

Brownian motion: Synonym for *Wiener process.*

Fokker–Planck equation: Another name for *Kolmogorov's forward equation.*

History: The history of a process $\{W_\theta,\theta\ge 0\}$ up to and including time t is denoted by $\mathcal{F}_t=\sigma(W_\theta,0\le\theta\le t)$. See also *measurable.*

Homogeneous: Linear stochastic differential equations (SDEs) of the form in Equation 59.32 are homogeneous.

Integrated white noise: A random process W_t that behaves as if it had the representation $W_t=\int_0^t Z_\theta\,d\theta$, where Z_θ is a white noise process. The Wiener process is an example of integrated white noise.

Itô correction term: The last term in Itô's rule in Equations 59.6 and 59.30. This term accounts for the fact that the Wiener process is not differentiable.

Itô's rule: A stochastic version of the chain rule. The general form is given in Equation 59.30.

Kolmogorov's backward equation: The partial differential Equation 59.38 satisfied by the transition function of the solution of an SDE.

Kolmogorov's forward equation: The partial differential Equation 59.39 satisfied by the transition density of the solution of an SDE.

Martingale: $\{W_t\}$ is an $\{\mathcal{F}_t\}$-martingale if $\{W_t\}$ is $\{\mathcal{F}_t\}$-adapted and if for all $t\ge s\ge 0$, $\mathsf{E}[W_t|\mathcal{F}_s]=W_s$, or equivalently, $\mathsf{E}[W_t-W_s|\mathcal{F}_s]=0$.

Measurable: See also history. Let $\mathcal{F}_t=\sigma(W_\theta,0\le\theta\le t)$. A random variable X is $\mathcal{F}_t$-measurable if it is a deterministic function of the random variables $\{W_\theta,0\le\theta\le t\}$.

Narrow sense: An SDE is linear in the narrow sense if it has the form of Equation 59.33.

Ornstein–Uhlenbeck process: A Gaussian process with zero mean and covariance function in Equation 59.36.

Smoothing properties of conditional expectation: See Equations 59.17 and 59.18.

Transition function: For a process $\{X_t\}$, the transition function is $P(t, y|s, x) := \Pr(X_t \leq y|X_s = x)$.

White noise process: A random process with zero mean and covariance $\mathsf{E}[Z_t Z_s] = \delta(t - s)$.

Wiener integral: An Itô integral with deterministic integrand. Always yields a Gaussian process.

Wiener process: A random process satisfying properties W-1 through W-4 in Section 59.2. It serves as a model for integrated white noise.

Acknowledgments

The author is grateful to Bob Barmish, Wei-Bin Chang, Majeed Hayat, Bill Levine, Raúl Sequeira, and Rajesh Sharma for reading the first draft of this chapter and for their suggestions for improving it.

References

1. Arnold, L., *Stochastic Differential Equations: Theory and Applications,* Wiley, New York, 1974, 48–56, 91, 98, 105, 113, 128, 137, 168.
2. Billingsley, P., *Probability and Measure*, 2nd ed., Wiley, New York, 1986, sect. 37, pp 469–470 (Equations 34.2, 34.5, 34.6).
3. Coddington, E. A. and Levinson, N., *Theory of Ordinary Differential Equations,* McGraw-Hill, New York, 1955, chap. 1.
4. Davis, M. H. A., *Linear Estimation and Stochastic Control,* Chapman and Hall, London, 1977, chap. 3, 52–53.
5. Elliott, R. J., *Stochastic Calculus,* Springer-Verlag, New York, 1982.
6. Ethier, S. N. and Kurtz, T. G., *Markov Processes: Characterization and Convergence,* Wiley, New York, 1986.
7. Karatzas, I. and Shreve, S. E., *Brownian Motion and Stochastic Calculus*, 2nd ed., Springer-Verlag, New York, 1991, 397.
8. Protter, P., *Stochastic Integration and Differential Equations,* Springer-Verlag, Berlin, 1990.
9. Segall, A., Stochastic processes in estimation theory, *IEEE Trans. Inf. Theory,* 22(3), 275–286, 1976.
10. Wong, E. and Hajek, B., *Stochastic Processes in Engineering Systems,* Springer-Verlag, New York, 1985, chap. 4, pp 173–174.

Further Reading

For background on probability theory, especially conditional expectation, we recommend [2].

For linear SDEs driven by orthogonal-increments processes, we recommend the very readable text by Davis [4].

For SDEs driven by continuous martingales, there is the more advanced book by Karatzas and Shreve [7].

For SDEs driven by right-continuous martingales, the theory becomes considerably more complicated. However, the tutorial paper by Segall [9], which compares discrete-time and continuous-time results, is very readable. Also, Chapter 6 of [10] is accessible.

Highly technical books on SDEs driven by right-continuous martingales include [5] and [8].

For the reader interested in Markov processes there is the advanced text of Ethier and Kurtz [6].

60

Linear Stochastic Input–Output Models

Torsten Söderström
Uppsala University

60.1 Introduction

Stationary stochastic processes are good ways of modeling random disturbances. The treatment here is basically for linear discrete-time input–output models. Most modern systems for control and signal processing work with sampled data; hence discrete-time models are of primary interest.

Properties of models, ways to calculate variances, and other second-order moments are treated. This chapter is organized as follows. Autoregressive moving average (ARMA) and ARMAX (ARMA with an eXogenous input) models are introduced in Section 60.2, while Section 60.3 deals with the effect of linear filtering. Spectral factorization, which has a key role when finding appropriate model representations for optimal estimation and control, is described in Section 60.4. Some ways to analyze stochastic systems by covariance calculations are presented in Section 60.5, while Section 60.6 gives a summary of results for continuous-time procecces.

In stochastic control, it is fundamental to predict future values of the process. A more general situation is the problem to estimate unmeasurable variables. Mean square optimal prediction is dealt with in Section 60.7, with minimal output variance control as a special application. In Section 60.8, a more general estimation problem is treated (covering optimal prediction, filtering, and smoothing), using Wiener filters, which are described in some detail.

This chapter is based on [6,7], where proofs and derivations can be found, as well as several extensions to the multivariable case, and to complex-valued signal processing problems. Other aspects, primarily related to control, can be found, for example, in [1–5].

60.2 ARMA and ARMAX Models

Wide sense stationary random processes are often characterized by their first- and second-order moments, that is by the mean value

$$m = Ex(t) \tag{60.1}$$

and the covariance function

$$r(\tau) \triangleq E[x(t+\tau) - m][x(t) - m], \tag{60.2}$$

where t, τ take integer values $0, \pm 1 \pm 2, \ldots$. For a wide sense stationary processes the expected values in Equations 60.1 and 60.2 are independent of t. As an alternative to the covariance function one can use its discrete Fourier transform, that is, the spectrum,

$$\phi(z) = \sum_{n=-\infty}^{\infty} r(n) z^{-n}. \tag{60.3}$$

Evaluated on the unit circle it is called the spectral density,

$$\phi(e^{i\omega}) = \sum_{n=-\infty}^{\infty} r(n) e^{-in\omega}. \tag{60.4}$$

As

$$r(\tau) = \frac{1}{2\pi} \int_{-\pi}^{\pi} \phi(e^{i\omega}) e^{i\tau\omega} d\omega = \frac{1}{2\pi i} \oint \phi(z) z^{\tau} \frac{dz}{z} \tag{60.5}$$

(where the last integration is counterclockwise around the unit circle) the spectral density describes how the energy of the signal is distributed over different frequency bands (set $\tau = 0$ in Equation 60.5).

Similarly, the cross-covariance function between two wide sense stationary processes $y(t)$ and $x(t)$ is defined as

$$r_{yx}(\tau) \triangleq E[y(t+\tau) - m_y][x(t) - m_x], \tag{60.6}$$

and its associated spectrum is

$$\phi_{yx}(z) = \sum_{n=-\infty}^{\infty} r_{yx}(n)\, z^{-n}. \tag{60.7}$$

A sequence of independent identically distributed (i.i.d.) random variables is called *white noise*. A white noise will have

$$r(\tau) = 0 \quad \text{for } \tau \neq 0. \tag{60.8}$$

Equivalently, its spectrum is constant for all z. Hence its energy is distributed evenly over all frequencies.

In order to simplify the development here, it is generally assumed that signals have zero mean. This is equivalent to considering only deviations of the signals from an operating point (given by the mean values).

Next, an important class of random processes, obtained by linear filtering of white noise, is introduced. Consider $y(t)$ given as the solution to the difference equation

$$y(t) + a_1 y(t-1) + \cdots + a_n y(t-n) = e(t) + c_1 e(t-1) + \cdots + c_m e(t-m), \tag{60.9}$$

where $e(t)$ is the white noise. Such a process is called an ARMA process. If $m = 0$, it is called an autoregressive (AR) process, and if $n = 0$ a moving average (MA) process.

Introduce the polynomials

$$\begin{aligned} A(z) &= z^n + a_1 z^{n-1} + \cdots + a_n, \\ C(z) &= z^m + c_1 z^{m-1} + \cdots + c_m, \end{aligned} \tag{60.10}$$

and the shift operator q, $qx(t) = x(t+1)$. The ARMA model (Equation 60.9) can then be written compactly as

$$A(q)y(t-n) = C(q)e(t-m). \tag{60.11}$$

As the white noise can be "relabeled" without changing the statistical properties of $y(t)$, (Equation 60.9) is much more frequently written in the form

$$A(q)y(t) = C(q)e(t). \tag{60.12}$$

Some illustrations of ARMA processes are given next.

Example 60.1:

Consider an ARMA process

$$A(q)y(t) = C(q)e(t).$$

The coefficients of the $A(q)$ and $C(q)$ polynomials determine the properties of the process. In particular, the roots of $A(z)$, which are called the poles of the process, determine the frequency contents of the process. The closer the poles are located toward the unit circle, the slower or more oscillating the process. Figure 60.1 illustrates the connections between the A and C coefficients, realizations of processes and their second-order moments as expressed by covariance function and spectral density.

As an alternative to q, one can use the *backward* shift operator, q^{-1}. An ARMA model would then be written as

$$\overline{A}(q^{-1})y(t) = \overline{C}(q^{-1})e(t),$$

where the polynomials are

$$\begin{aligned} \overline{A}(q^{-1}) &= 1 + a_1 q^{-1} + \cdots + a_n q^{-n}, \\ \overline{C}(q^{-1}) &= 1 + c_1 q^{-1} + \cdots + c_m q^{-m}. \end{aligned}$$

The advantage of using the q-formalism is that stability corresponds to the "natural" condition $|z| < 1$. An advantage of the alternative q^{-1}-formalism is that causality considerations (see below) become easier to handle. The q-formalism is used here.

In some situations, such as modeling a drifting disturbance, it is appropriate to allow $A(z)$ to have zeros on or even outside the unit circle. Then the process will not be wide sense stationary, but drift away. It should hence only be considered for a finite period of time. The special simple case $A(z) = z - 1$, $C(z) = z$ is known as a random walk.

If an input signal term is added to Equation 60.10, we obtain an ARMAX model

$$A(q)y(t) = B(q)u(t) + C(q)e(t). \tag{60.13}$$

The system must be *causal*, which means that $y(t)$ is not allowed to depend on *future* values of the input $u(t+\tau)$, $\tau > 0$. Hence it is required that $\deg A \geq \deg B$. Otherwise $y(t)$ would depend on *future* input values.

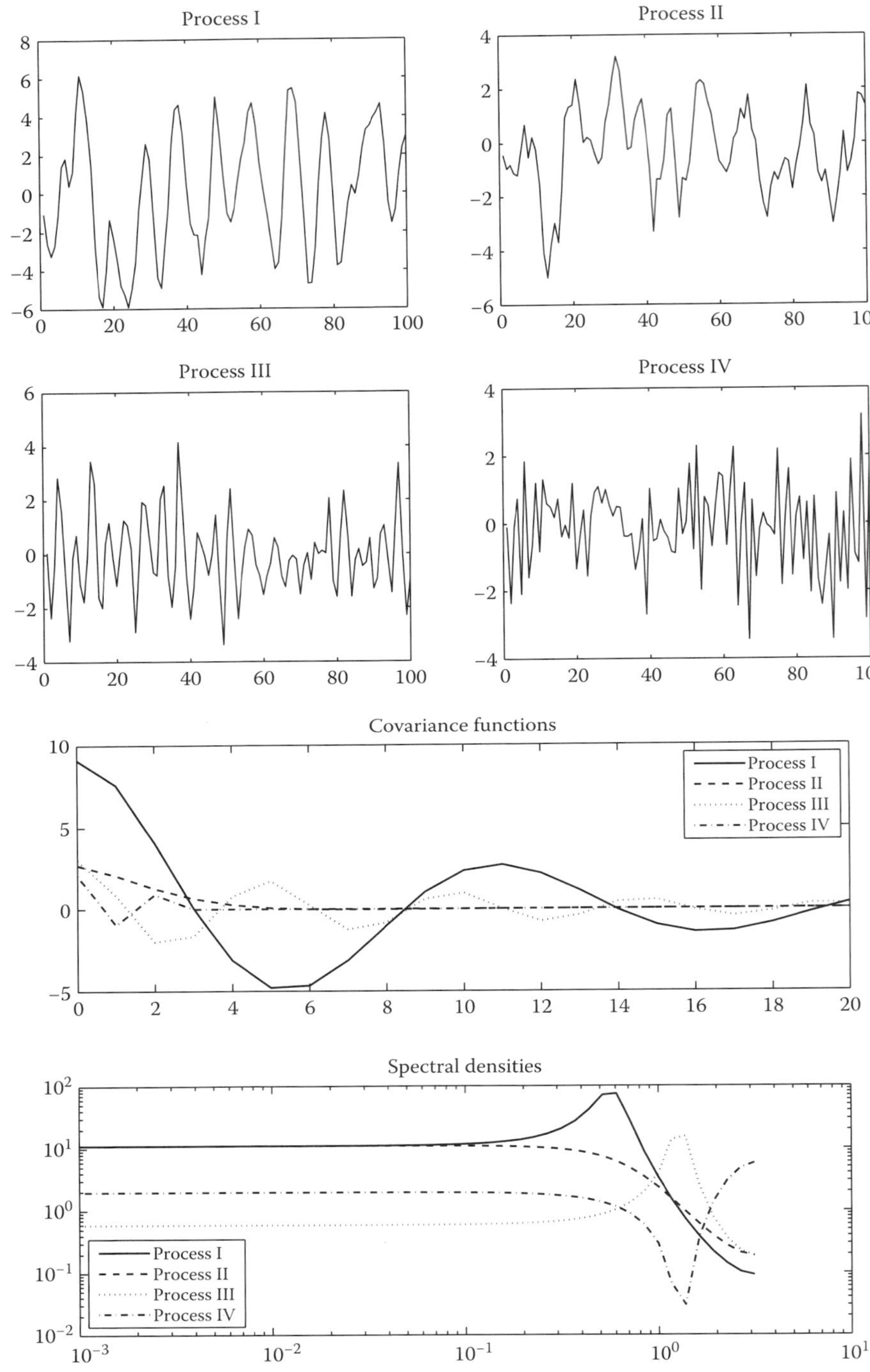

FIGURE 60.1 Illustration of some ARMA processes. (Process I: $A(q) = q^2 - 1.5q + 0.8$, $C(q) = 1$. Pole locations: $0.75 \pm i0.49$. Process II: $A(q) = q^2 - 1.0q + 0.3$, $C(q) = 1$. Pole locations: $0.50 \pm i0.22$. Process III: $A(q) = q^2 - 0.5q + 0.8$, $C(q) = 1$. Pole locations: $0.25 \pm i0.86$. Process IV: $A(q) = q^2$, $C(q) = q^2 - 0.5q + 0.9$. Pole locations: $0, 0$.)

Sometimes higher-order moments (i.e., moments of order higher than two) are useful. Such spectra are useful tools for the following:

- Extracting information due to deviations from a Gaussian distribution.
- Estimating the phase of a non-Gaussian process.
- Detecting and characterizing nonlinear mechanisms in time series.

To exemplify, we consider the bispectrum, which is the simplest form of higher-order spectrum. Bispectra are useful only for signals that do *not* have a probability density function that is symmetric around its mean value. For signals *with* such a symmetry, spectra of order at least four are needed.

Let $x(t)$ be a scalar stationary process of zero mean. Its *third moment sequence*, $R(m, n)$ is defined as

$$R(m,n) = Ex(t)x(t+m)x(t+n), \tag{60.14}$$

and satisfies a number of symmetry relations. The *bispectrum* is

$$B(z_1, z_2) \triangleq \sum_{m=-\infty}^{\infty} \sum_{n=-\infty}^{\infty} R(m,n) z_1^{-m} z_2^{-n}. \tag{60.15}$$

Let us consider two special cases:

- Let $x(t)$ be zero mean Gaussian. Then

$$R(m,n) = 0, \quad B(z_1, z_2) = 0. \tag{60.16}$$

- Let $x(t)$ be non-Gaussian *white* noise, so that $x(t)$ and $x(s)$ are independent for $t \neq s$, $Ex(t) = 0$, $Ex^2(t) = \sigma^2$, $Ex^3(t) = \beta$. Then

$$R(m,n) = \begin{cases} \beta & \text{if } m = n = 0, \\ 0 & \text{elsewhere.} \end{cases} \tag{60.17}$$

The bispectrum becomes a constant,

$$B(z_1, z_2) = \beta. \tag{60.18}$$

60.3 Linear Filtering

Let $u(t)$ be a stationary process with mean m_u and spectrum $\phi_u(z)$, and consider

$$y(t) = G(q)u(t) = \sum_{k=0}^{\infty} g_k u(t-k), \tag{60.19}$$

where $G(q) = \sum_{k=0}^{\infty} g_k q^{-k}$ is an asymptotically stable filter (e.g., it has all poles strictly inside the unit circle). Then $y(t)$ is a stationary process with mean

$$m_y = G(1) m_u, \tag{60.20}$$

and spectrum $\phi_y(z)$ and cross-spectrum $\phi_{yu}(z)$ given by

$$\phi_y(z) = G(z)G(z^{-1})\phi_u(z), \tag{60.21}$$

$$\phi_{yu}(z) = G(z)\phi_u(z). \tag{60.22}$$

The interpretation of Equation 60.20 is that the mean value m_u is multiplied by the static gain of the filter to obtain the output mean value m_y.

The following corollary is a form of Parseval's relation. Assume that $u(t)$ is a white-noise sequence with variance λ^2. Then

$$Ey^2(t) = \lambda^2 \sum_{k=0}^{\infty} g_k^2 = \frac{\lambda^2}{2\pi i} \oint G(z)G(z^{-1})\frac{dz}{z}. \tag{60.23}$$

Let $u(t)$ further have bispectrum $B_u(z_1, z_2)$. Then $B_y(z_1, z_2)$ can be found after straightforward calculation,

$$B_y(z_1, z_2) = G(z_1^{-1}z_2^{-1})G(z_1)G(z_2)B_u(z_1, z_2). \tag{60.24}$$

This is a generalization of Equation 60.21. Note that the spectral density (the power spectrum) does not carry information about the phase properties of a filter. In contrast to this, phase properties can be recovered from the bispectrum when it exists. This point is indirectly illustrated in Example 60.1.

60.4 Spectral Factorization

Let $\phi(z)$ be a scalar spectrum that is rational in z, that is, it can be written as

$$\phi(z) = \frac{\sum_{|k|\le m} \beta_k z^k}{\sum_{|k|\le n} \alpha_k z^k}. \tag{60.25}$$

Then there are two polynomials:

$$\begin{aligned} A(z) &= z^n + a_1 z^{n-1} + \cdots + a_n, \\ C(z) &= z^m + c_1 z^{m-1} + \cdots + c_m, \end{aligned} \tag{60.26}$$

and a positive real number λ^2 so that, (1) $A(z)$ has all zeros inside the unit circle, (2) $C(z)$ has all zeros inside or on the unit circle, and (3)

$$\phi(z) = \lambda^2 \frac{C(z)}{A(z)} \frac{C(z^{-1})}{A(z^{-1})}. \tag{60.27}$$

In the case where $\phi(e^{iw}) > 0$ for all ω, $C(z)$ will have no zeros on the circle.

Note that any continuous spectral density can be approximated arbitrarily well by a rational function in $z = e^{i\omega}$ as in Equations 60.25 through 60.27, provided that m and n are appropriately chosen. Hence, the assumptions imposed are not restrictive. Instead, the results are applicable, at least with a small approximation error, to a very wide class of stochastic processes.

It is an important implication that (as far as second-order moments are concerned) the underlying stochastic process can be regarded as generated by filtering white noise, that is, as an ARMA process

$$\begin{aligned} A(q)y(t) &= C(q)e(t), \\ Ee^2(t) &= \lambda^2. \end{aligned} \tag{60.28}$$

Hence, for describing stochastic processes (as long as they have rational spectral densities), there is no restriction to assume that the input signals are white noise. In the representation (Equation 60.28), the sequence $\{e(t)\}$ is called the *output innovations.*

Spectral factorization can also be viewed as a form of *aggregation of noise sources.* Assume, for example, that an ARMA process

$$A(q)x(t) = C(q)v(t) \tag{60.29}$$

is observed but the observations include measurement noise

$$y(t) = x(t) + e(t), \tag{60.30}$$

and that $v(t)$ and $e(t)$ are uncorrelated white-noise sequences with variances λ_v^2 and λ_e^2, respectively. As far as the second-order properties (such as the spectrum or the covariance function) are concerned, $y(t)$

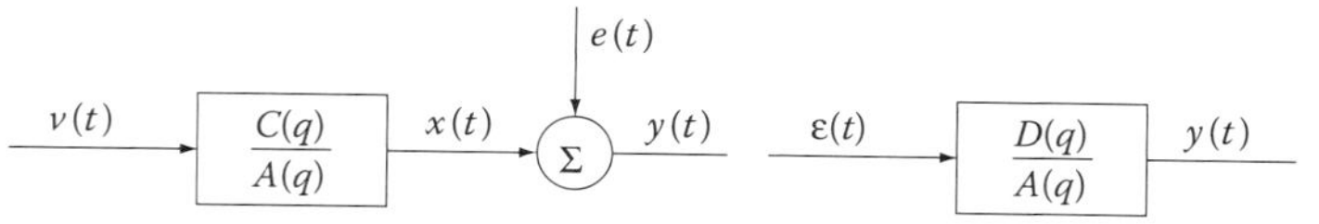

FIGURE 60.2 Two representations of an ARMA process with noisy observations.

can be viewed as generated from one single noise source:

$$A(q)y(t) = D(q)\varepsilon(t). \tag{60.31}$$

The polynomial $D(q)$ and the noise variance λ_ε^2 are derived as follows. The spectrum is, according to Equations 60.29 through 60.31,

$$\phi_y(z) = \lambda_v^2 \frac{C(z)C(z^{-1})}{A(z)A(z^{-1})} + \lambda_e^2 = \lambda_\varepsilon^2 \frac{D(z)D(z^{-1})}{A(z)A(z^{-1})}.$$

Equating these two expressions gives

$$\lambda_\varepsilon^2 D(z)D(z^{-1}) \equiv \lambda_v^2 C(z)C(z^{-1}) + \lambda_e^2 A(z)A(z^{-1}). \tag{60.32}$$

The two representations (Equations 60.29 and 60.30), and Equation 60.31 of the process $y(t)$, are displayed schematically in Figure 60.2.

Spectral factorization can also be performed using a state-space formalism. Then an algebraic Riccati equation (ARE) has to be solved. Its different solutions corresponds to different polynomials $C(z)$ satisfying Equation 60.27. The positive-definite solution of the ARE corresponds to the $C(z)$ polynomial with all zeros inside the unit circle.

One would expect that for a given process, the same type of filter representation will appear for the power spectrum and for the bispectrum. This is *not* so in general, as illustrated by the following example.

Example 60.2:

Consider a process consisting of the sum of two independent AR processes

$$y(t) = \frac{1}{A(q)} e(t) + \frac{1}{C(q)} v(t), \tag{60.33}$$

$e(t)$ being Gaussian white noise and $v(t)$ non-Gaussian white noise. Both sequences are assumed to have unit variance, and $Ev^3(t) = 1$.

The Gaussian process will not contribute to the bispectrum. Further, $R_v(z_1, z_2) \equiv 1$, and according to Equation 60.24 the bispectrum will be

$$B_y(z_1, z_2) = \frac{1}{C(z_1^{-1} z_2^{-1})} \frac{1}{C(z_1)} \frac{1}{C(z_2)}, \tag{60.34}$$

so

$$H(z) = \frac{1}{C(z)} \tag{60.35}$$

is the relevant filter representation as far as the bispectrum is concerned. However, the power spectrum becomes

$$\phi(z) = \frac{1}{A(z)A(z^{-1})} + \frac{1}{C(z)C(z^{-1})}, \tag{60.36}$$

and in this case it will have a spectral factor of the form

$$H(z) = \frac{B(z)}{A(z)C(z)}, \tag{60.37}$$

where

$$B(z)B(z^{-1}) \equiv A(z)A(z^{-1}) + C(z)C(z^{-1}) \tag{60.38}$$

due to the spectral factorization. Clearly, the two filter representations of Equations 60.35 and 60.37 differ.

60.5 Yule–Walker Equations and Other Algorithms for Covariance Calculations

When analyzing stochastic systems it is often important to compute variances and covariances between inputs, outputs, and other variables. This can mostly be reduced to the problem of computing the covariance function of an ARMA process. Some ways to do this are presented in this section.

Consider first the case of an AR process

$$y(t) + a_1 y(t-1) + \cdots + a_n y(t-n) = e(t), \quad Ee^2(t) = \lambda^2. \tag{60.39}$$

Note that $y(t)$ can be viewed as a linear combination of all the old values of the noise, that is, $\{e(s)\}_{s=-\infty}^{t}$. By multiplying $y(t)$ by a delayed value of the process, say $y(t-\tau)$, $\tau \geq 0$, and applying the expectation operator, one obtains

$$Ey(t-\tau)[y(t) + a_1 y(t-1) + \cdots + a_n y(t-n)] = Ey(t-\tau)e(t),$$

or

$$r(\tau) + a_1 r(\tau-1) + \cdots + a_n r(\tau-n) = \begin{cases} 0, & \tau > 0, \\ \lambda^2, & \tau = 0, \end{cases} \tag{60.40}$$

which is called a Yule–Walker equation. By using Equation 60.40 for $\tau = 0, \ldots, n$, one can construct the following system of equations for determining the covariance elements $r(0), r(1), \ldots, r(n)$:

$$\begin{pmatrix} 1 & a_1 & \cdots & & a_n \\ a_1 & 1+a_2 & & a_n & 0 \\ \vdots & & \ddots & & \vdots \\ a_n & a_{n-1} & \cdots & & 1 \end{pmatrix} \begin{pmatrix} r(0) \\ \vdots \\ \\ r(n) \end{pmatrix} = \begin{pmatrix} \lambda^2 \\ 0 \\ \vdots \\ 0 \end{pmatrix}. \tag{60.41}$$

Once $r(0), \ldots, r(n)$ are known, Equation 60.40 can be iterated (for $\tau = n+1, n+2, \ldots$) to find further covariance elements.

Consider next a full ARMA process

$$\begin{aligned} &y(t) + a_1 y(t-1) + \cdots + a_n y(t-n) = e(t) + c_1 e(t-1) + \cdots + c_m e(t-m), \\ &Ee^2(t) = \lambda^2. \end{aligned} \tag{60.42}$$

Now computing the cross-covariance function between $y(t)$ and $e(t)$ must involve an intermediate step. Multiplying Equation 60.42 by $y(t-\tau)$, $\tau \geq 0$, and applying the expectation operator, gives

$$r_y(\tau) + a_1 r_y(\tau-1) + \cdots + a_n r_y(\tau-n) = r_{ey}(\tau) + c_1 r_{ey}(\tau-1) + \cdots + c_m r_{ey}(\tau-m). \tag{60.43}$$

In order to obtain the output covariance function $r_y(\tau)$, the cross-covariance function $r_{ey}(\tau)$ must first be found. This is done by multiplying Equation 60.42 by $e(t-\tau)$, and applying the expectation operator, which leads to

$$r_{ey}(-\tau) + a_1 r_{ey}(-\tau+1) + \cdots + a_n r_{ey}(-\tau+n) = \lambda^2[\delta_{\tau,0} + c_1\delta_{\tau-1,0} + \cdots + c_m\delta_{\tau-m,0}], \qquad (60.44)$$

where $\delta_{t,s}$ is the Kronecker delta ($\delta_{t,s} = 1$ if $t = s$, and 0 elsewhere). As $y(t)$ is a linear combination of $\{e(s)\}_{s=-\infty}^{t}$, it is found that $r_{ey}(\tau) = 0$ for $\tau > 0$. Hence Equation 60.43 gives

$$r_y(\tau) + a_1 r_y(\tau-1) + \cdots + a_n r_y(\tau-n) = 0, \quad \tau > m. \qquad (60.45)$$

The use of Equations 60.43 through 60.45 to derive the autocovariance function is illustrated next by applying them to a first-order ARMA process.

Example 60.3:

Consider the ARMA process

$$y(t) + ay(t-1) = e(t) + ce(t-1), \quad Ee^2(t) = \lambda^2.$$

In this case, $n = 1, m = 1$. Equation 60.45 gives

$$r_y(\tau) + ar_y(\tau-1) = 0, \quad \tau > 1.$$

Using Equation 60.43 for $\tau = 0$ and 1 gives

$$\begin{pmatrix} 1 & a \\ a & 1 \end{pmatrix}\begin{pmatrix} r_y(0) \\ r_y(1) \end{pmatrix} = \begin{pmatrix} 1 & c \\ c & 0 \end{pmatrix}\begin{pmatrix} r_{ey}(0) \\ r_{ey}(-1) \end{pmatrix}.$$

Consider Equation 60.44 for $\tau = 0$ and 1, which gives

$$\begin{pmatrix} 1 & 0 \\ a & 1 \end{pmatrix}\begin{pmatrix} r_{ey}(0) \\ r_{ey}(-1) \end{pmatrix} = \lambda^2\begin{pmatrix} 1 \\ c \end{pmatrix}.$$

By straightforward calculations, it is found that

$$r_{ey}(0) = \lambda^2, \quad r_{ey}(-1) = \lambda^2(c-a),$$
$$r_y(0) = \frac{\lambda^2}{1-a^2}(1+c^2-2ac), \quad r_y(1) = \frac{\lambda^2}{1-a^2}(c-a)(1-ac),$$

and finally,

$$r_y(\tau) = \frac{\lambda^2}{1-a^2}(c-a)(1-ac)(-a)^{\tau-1}, \quad \tau \geq 1.$$

As an example of alternative approaches for covariance calculations, consider the following situation. Assume that two ARMA processes are given:

$$\begin{aligned} A(q)y(t) &= B(q)e(t), \\ A(q) &= q^n + a_1q^{n-1} + \cdots + a_n, \\ B(q) &= b_oq^n + b_1q^{n-1} + \cdots + b_n, \end{aligned} \qquad (60.46)$$

and

$$\begin{aligned} C(q)w(t) &= D(q)e(t), \\ C(q) &= q^m + c_1q^{m-1} + \cdots + c_m, \\ D(q) &= d_oq^m + d_1q^{m-1} + \cdots + d_m. \end{aligned} \qquad (60.47)$$

Assume that $e(t)$ is the same in Equations 60.46 and 60.47 and that it is white noise of zero mean and unit variance. The problem is to find the cross-covariance elements

$$r(k) = Ey(t+k)w(t) \tag{60.48}$$

for a number of arguments k. The cross-covariance function $r(k)$ is related to the cross-spectrum $\phi_{yw}(z)$ as (see Equations 60.7 and 60.22)

$$\phi_{yw}(z) = \sum_{k=-\infty}^{\infty} r(k)z^{-k} = \frac{B(z)}{A(z)}\frac{D(z^{-1})}{C(z^{-1})}. \tag{60.49}$$

Introduce the two polynomials

$$\begin{aligned} F(z) &= f_o z^n + f_1 z^{n-1} + \cdots + f_n, \\ G(z^{-1}) &= g_o z^{-m} + g_1 z^{-(m-1)} + \cdots + g_{m-1} z^{-1}, \end{aligned} \tag{60.50}$$

through

$$\frac{B(z)}{A(z)}\frac{D(z^{-1})}{C(z^{-1})} \equiv \frac{F(z)}{A(z)} + z\frac{G(z^{-1})}{C(z^{-1})}, \tag{60.51}$$

or, equivalently,

$$B(z)D(z^{-1}) \equiv F(z)C(z^{-1}) + zA(z)G(z^{-1}). \tag{60.52}$$

Since $zA(z)$ and $C(z^{-1})$ are coprime (i.e., these two polynomials have no common factor), Equation 60.52 has a unique solution. Note that as a linear system of equations, Equation 60.52 has $n+m+1$ equations and the same number of unknowns. The coprimeness condition ensures that a unique solution exists. Equations 60.49 and 60.51 now give

$$\sum_{k=-\infty}^{\infty} r(k)z^{-k} = \frac{F(z)}{A(z)} + z\frac{G(z^{-1})}{C(z^{-1})}. \tag{60.53}$$

The two terms in the right-hand side of Equation 60.53 can be identified with two parts of the sum. In fact,

$$\frac{F(z)}{A(z)} = \sum_{k=0}^{\infty} r(k)z^{-k}, \quad \frac{zG(z^{-1})}{C(z^{-1})} = \sum_{k=-\infty}^{-1} r(k)z^{-k}. \tag{60.54}$$

Equating the powers of z gives

$$\begin{cases} r_{yw}(0) = f_o, \\ r_{yw}(1) = f_1 - a_1 r(0), \\ \vdots \\ r_{yw}(k) = f_k - \sum_{j=1}^{k} a_j r(k-j), \quad (2 \le k \le n), \\ r_{yw}(k) = -\sum_{j=1}^{n} a_j r(k-j), \quad (k > n). \end{cases} \tag{60.55}$$

Note that the last part of Equation 60.55 is nothing but a Yule–Walker type of equation.

Similarly,

$$\begin{cases} r_{yw}(-1) = g_o, \\ r_{yw}(-k) = g_{k-1} - \sum_{j=1}^{k-1} c_j r(-k+j), \quad (2 \le k \le m), \\ r_{yw}(-k) = -\sum_{j=1}^{m} c_j r(-k+j), \quad (k > m). \end{cases} \tag{60.56}$$

Example 60.4:

Consider again a first-order ARMA process

$$y(t) + ay(t-1) = e(t) + ce(t-1), \quad Ee^2(t) = 1.$$

In this case, the autocovariance function is sought. Hence choose $z(t) \equiv y(t)$ and thus $A(q) = q + a$, $B(q) = q + c$, $C(q) = q + a$, and $D(q) = q + c$. The identity (Equation 60.52) becomes

$$(z + c)(z^{-1} + c) \equiv (f_o z + f_1)(z^{-1} + a) + z(z + a)(g_o z^{-1}).$$

Equating the powers of z leads to

$$f_o = \frac{1 + c^2 - 2ac}{1 - a^2}, \quad f_1 = c, \quad g_o = \frac{(c-a)(1-ac)}{1-a^2}.$$

Hence Equation 60.55 implies that

$$\begin{aligned} r(0) &= f_o = \frac{1 + c^2 - 2ac}{1 - a^2}, \\ r(1) &= f_1 - ar(0) = \frac{(c-a)(1-ac)}{1-a^2}, \\ r(k) &= (-a)^{k-1} r(1), \quad k \geq 1. \end{aligned}$$

while Equation 60.56 gives

$$\begin{aligned} r(-1) &= g_o = \frac{(c-a)(1-ac)}{1-a^2}, \\ r(-k) &= (-a)^{k-1} r(-1), \quad k \geq 1. \end{aligned}$$

Needless to say, these expressions for the covariance function are the same as those derived in the previous example.

60.6 Continuous-Time Processes

This section illustrates how some of the properties of discrete-time stochastic systems appear in an analog form for continuous-time models. However, white noise in continuous time leads to considerable mathematical difficulties that must be solved in a rigorous way. See [1] or [7] for more details on this aspect.

The covariance function of a process $y(t)$ is still defined as (*cf.* Equation 60.2)

$$r(\tau) = Ey(t + \tau)y(t), \tag{60.57}$$

assuming for simplicity that $y(t)$ has zero mean. The spectrum will now be

$$\phi(s) = \int_{-\infty}^{\infty} r(\tau) e^{-s\tau} d\tau, \tag{60.58}$$

and the spectral density is

$$\phi(i\omega) = \int_{-\infty}^{\infty} r(\tau) e^{i\omega\tau} d\tau. \tag{60.59}$$

The inverse relation to Equation 60.58 is

$$r(\tau) = \frac{1}{2\pi i} \int \phi(s) e^{s\tau} ds, \tag{60.60}$$

where integration is along the whole imaginary axis.

Consider a stationary stochastic process described by a spectral density $\phi(i\omega)$ that is a rational function of $i\omega$. By pure analogy with the discrete-time case it is found that

$$\phi(i\omega) = \frac{B(i\omega)B(-i\omega)}{A(i\omega)A(-i\omega)}, \tag{60.61}$$

where the polynomials

$$\begin{aligned} A(p) &= p^n + a_1 p^{n-1} + \cdots + a_n, \\ B(p) &= b_1 p^{n-1} + \cdots + b_n, \end{aligned} \tag{60.62}$$

have all their roots in the left half-plane (i.e., in the stability area). Here p is an arbitrary polynomial argument, but can be interpreted as the differentiation operator ($py(t) = \dot{y}(t)$).

The effect of filtering a stationary process, say $u(t)$, with an asymptotically stable filter, say $H(p)$, can be easily phrased using the spectra. Let the filtering be described by

$$y(t) = H(p)u(t). \tag{60.63}$$

Then

$$\phi_y(s) = H(s)H(-s)\phi_u(s), \tag{60.64}$$

again paralleling the discrete-time case. As a consequence, one can interpret any process with a rational spectral density (Equation 60.61) as having been generated by filtering as in Equation 60.63 by using

$$H(p) = \frac{B(p)}{A(p)}. \tag{60.65}$$

The signal $u(t)$ would then have a *constant* spectral density, $\phi_u(i\omega) \equiv 1$. As for the discrete-time case, such a process is called *white noise*. It will have a covariance function $r(\tau) = \delta(\tau)$ and hence in particular an infinite variance. This indicates difficulties to treat it with mathematical rigor.

60.7 Optimal Prediction

Consider an ARMA process,

$$A(q)y(t) = C(q)e(t), \quad Ee^2(t) = \lambda^2, \tag{60.66}$$

where $A(q)$ and $C(q)$ are of degree n, and have all their roots inside the unit circle. We seek a k-step predictor, that is, a function of available data $y(t), y(t-1), \ldots$ that will be close to the future value $y(t+k)$. In particular, we seek the predictor that is optimal in a mean square sense. The clue to finding this predictor is to rewrite $y(t+k)$ in two terms. The first term is a weighted sum of future noise values, $\{e(t+j)\}_{j=1}^{k}$. As this term is uncorrelated to all available data, it cannot be reconstructed in any way. The second term is a weighted sum of past noise values $\{e(t-s)\}_{s=0}^{\infty}$. By inverting the process model, the second term can be written as a weighted sum of output values, $\{y(t-s)\}_{s=0}^{\infty}$. Hence, it can be computed exactly from data.

In order to proceed, introduce the *predictor identity*

$$z^{k-1}C(z) \equiv A(z)F(z) + L(z), \tag{60.67}$$

where

$$F(z) = z^{k-1} + f_1 z^{k-2} + \cdots + f_{k-1}, \quad (60.68)$$

$$L(z) = \ell_o z^{n-1} + \ell_1 z^{n-2} + \cdots + \ell_{n-1}. \quad (60.69)$$

Equation 60.67 is a special case of a Diophantine equation for polynomials. A solution is always possible. This is analogous to the Diophantine equation

$$n = qm + r$$

for integers (for given integers n and m, there exist unique integers q and r). Now,

$$\begin{aligned} y(t+k) &= \frac{C(q)}{A(q)} e(t+k) = \frac{q^{k-1}C(q)}{A(q)} e(t+1) \\ &= \frac{A(q)F(q) + L(q)}{A(q)} e(t+1) = F(q)e(t+1) + \frac{qL(q)}{A(q)} e(t) \\ &= F(q)e(t+1) + \frac{qL(q)}{A(q)} \frac{A(q)}{C(q)} y(t) = F(q)e(t+1) + \frac{qL(q)}{C(q)} y(t). \end{aligned} \quad (60.70)$$

This is the decomposition mentioned above. The term $F(q)e(t+1)$ is a weighted sum of *future* noise values, while $qL(q)/C(q)y(t)$ is a weighted sum of *available* measurements Y^t. Note that it is crucial for stability that $C(q)$ has all zeros strictly inside the unit circle (but that this is not restrictive due to spectral factorization). As the future values of the noise are unpredictable, the *mean square optimal predictor* is given by

$$\hat{y}(t+k|t) = \frac{qL(q)}{C(q)} y(t), \quad (60.71)$$

while the associated prediction error is

$$\begin{aligned} \tilde{y}(t+k) &= F(q)e(t+1) \\ &= e(t+k) + f_1 e(t+k-1) + \cdots + f_{k-1} e(t+1), \end{aligned} \quad (60.72)$$

and has variance

$$E\tilde{y}^2(t+k) = \lambda^2(1 + f_1^2 + \cdots + f_{k-1}^2). \quad (60.73)$$

As a more general case, consider prediction of $y(t)$ in the ARMAX model

$$A(q)y(t) = B(q)u(t) + C(q)e(t). \quad (60.74)$$

In this case, proceeding as in Equation 60.70,

$$\begin{aligned} y(t+k) &= \frac{B(q)}{A(q)} u(t+k) + F(q)e(t+1) + \frac{qL(q)}{A(q)} \left[\frac{A(q)}{C(q)} y(t) - \frac{B(q)}{C(q)} u(t) \right] \\ &= F(q)e(t+1) + \frac{qL(q)}{C(q)} y(t) + \frac{qB(q)F(q)}{C(q)} u(t). \end{aligned} \quad (60.75)$$

We find that the prediction error is still given by Equation 60.72, while the optimal predictor is

$$\hat{y}(t+k|t) = \frac{qL(q)}{C(q)} y(t) + \frac{qB(q)F(q)}{C(q)} u(t). \quad (60.76)$$

This result can also be used to derive a minimum output variance regulator. That is, let us seek a feedback control for the process (Equation 60.74) that minimizes $Ey^2(t)$. Let $k = \deg A - \deg B$ denote the delay in the system. As $\tilde{y}(t+k)$ and $\hat{y}(t+k|t)$ are independent,

$$Ey^2(t+k|t) = E\hat{y}^2(t+k|t) + E\tilde{y}^2(t+k) \geq E\tilde{y}^2(t+k), \tag{60.77}$$

with equality if and only if $\hat{y}(t+k|t) = 0$, the regulator is

$$u(t) = -\frac{L(q)}{B(q)F(q)}y(t). \tag{60.78}$$

Optimal prediction can also be carried out using a state-space formalism. It will then involve computing the Kalman filter, and a Riccati equation has to be solved (which corresponds to the spectral factorization). See Chapter 17 for a treatment of linear quadratic stochastic control using state-space techniques.

60.8 Wiener Filters

The steady-state linear least-mean square estimate is considered in this section. It can be computed using a state-space formalism (like Kalman filters and smoothers), but here a polynomial formalism for an input–output approach is utilized. In case time-varying or transient situations have to be handled, a state-space approach must be used. See Chapter 12 for a parallel treatment of Kalman filters.

Let $y(t)$ and $s(t)$ be two correlated and stationary stochastic processes. Assume that $y(t)$ is measured and find a causal, asymptotically stable filter $G(q)$ such that $G(q)y(t)$ is the optimal linear mean square estimator, that is, it minimizes the criterion

$$V = E[s(t) - G(q)y(t)]^2. \tag{60.79}$$

This problem is best treated in the frequency domain. This implies in particular that data are assumed to be available since the infinite past $t = -\infty$. Introduce the estimation error

$$\tilde{s}(t) = s(t) - G(q)y(t). \tag{60.80}$$

The criterion V, Equation 60.79, can be rewritten as

$$V = \frac{1}{2\pi i}\oint \phi_{\tilde{s}}(z)\frac{dz}{z}. \tag{60.81}$$

Next note that

$$\phi_{\tilde{s}}(z) = \phi_s(z) - G(z)\phi_{ys}(z) - \phi_{sy}(z)G(z^{-1}) + G(z)G(z^{-1})\phi_y(z). \tag{60.82}$$

Now let $G(q)$ be the optimal filter and $G_1(q)$ any causal filter. Replace $G(q)$ in Equation 60.79 by $G(q) + \varepsilon G_1(q)$. As a function of ε, V can then be written as $V = V_0 + \varepsilon V_1 + \varepsilon^2 V_2$. For $G(q)$ to be the *optimal* filter it is required that $V \geq V_0$ for all ε, which leads to $V_1 = 0$, giving

$$0 = \text{tr}\frac{1}{2\pi i}\oint [G(z)\phi_y(z) - \phi_{sy}(z)]G_1(z^{-1})\frac{dz}{z}. \tag{60.83}$$

It is possible to give an interpretation and alternative view of Equation 60.83. For the optimal filter, the estimation error, $\tilde{s}(t)$, should be uncorrelated with all past measurements, $\{y(t-j)\}_{j=0}^{\infty}$. Otherwise there would be another linear combination of the past measurements giving smaller estimation error variance.

Hence,

$$E\tilde{s}(t)y(t-j) = 0, \quad \text{all } j \geq 0, \tag{60.84}$$

or

$$E\tilde{s}(t)[G_1(q)y(t)] = 0 \quad \text{for any stable and causal} \quad G_1(q). \tag{60.85}$$

This can be rewritten as

$$\begin{aligned} 0 &= E[s(t) - G(q)y(t)][G_1(q)y(t)] \\ &= \frac{1}{2\pi i}\oint [\phi_{sy}(z) - G(z)\phi_y(z)]G_1(z^{-1})\frac{dz}{z}, \end{aligned} \tag{60.86}$$

which is precisely (60.83).

From Equation 60.83, one easily finds the *unrealizable Wiener filter*. Setting the integrand to zero gives $G(z)\phi_y(z) = \phi_{sy}(z)$, and

$$G(z) = \phi_{sy}(z)\phi_y^{-1}(z). \tag{60.87}$$

The filter is not realizable since it relies (except in very degenerate cases) on all *future* data points of $y(t)$. Note, however, that when "deriving" Equation 60.87 from Equation 60.83, it was effectively required that Equation 60.83 holds for *any* $G_1(z)$. However, it is only required that Equation 60.83 holds for any *causal and stable* $G_1(z)$. Such an observation will eventually lead to the optimal *realizable* filter.

To proceed, let the process $y(t)$ have the *innovations representation* (remember that this is always possible as in Section 60.4)

$$\begin{aligned} y(t) &= H(q)e(t), \\ Ee(t)e(s) &= \lambda^2\delta_{t,s}, \quad H(0) = 1, \end{aligned} \tag{60.88}$$

$$H(q), H^{-1}(q) \text{ asymptotically stable.}$$

Then $\phi_y(z) = H(z)H(z^{-1})\lambda^2$. Further, introduce the *causal part* of an analytical function. Let

$$G(z) = \sum_{j=-\infty}^{\infty} g_j z^{-j}, \tag{60.89}$$

where it is required that the series converges in a strip that includes the unit circle. The *causal part* of $G(z)$ is defined as

$$[G(z)]_+ = \sum_{j=0}^{\infty} g_j z^{-j}, \tag{60.90}$$

and the *anticausal part* is the complementary part of the sum:

$$[G(z)]_- = \sum_{j=-\infty}^{-1} g_j z^{-j} = G(z) - [G(z)]_+. \tag{60.91}$$

It is important to note that the term $g_0 z^{-0}$ in Equation 60.89 appears in the causal part, $[G(z)]_+$. Note that the anticausal part $[G(z)]_-$ of a transfer function $G(z)$ has no poles inside or on the unit circle, and that a filter $G(z)$ is causal if and only if $G(z) = [G(z)]_+$. Using the conventions (Equation 60.90) and

(Equation 60.91), the optimality condition (Equation 60.83) can be formulated as

$$0 = \frac{1}{2\pi i}\oint \{G(z)H(z) - \phi_{sy}(z)\{H(z^{-1})\}^{-1}\lambda^{-2}\}\lambda^2 H(z^{-1})G_1(z^{-1})\frac{dz}{z}$$
$$= \frac{1}{2\pi i}\oint \{G(z)H(z) - [\phi_{sy}(z)\{H(z^{-1})\}^{-1}\lambda^{-2}]_+$$
$$- [\phi_{sy}(z)\{H(z^{-1})\}^{-1}\lambda^{-2}]_-\}\lambda^2 H(z^{-1})G_1(z^{-1})\frac{dz}{z}. \qquad (60.92)$$

The stability requirements imply that the function $H(z^{-1})G_1(z^{-1})$ does not have any poles inside the unit circle. The same is true for $[\phi_{sy}(z)\{H(z^{-1})\}^{-1}\lambda^{-2}]_-$, by construction. The latter function has a zero at $z = 0$. Hence, by the residue theorem,

$$\frac{1}{2\pi i}\oint [\phi_{sy}(z)\{H(z^{-1})\}^{-1}]_- H(z^{-1})G_1(z^{-1})\frac{dz}{z} = 0. \qquad (60.93)$$

The optimal condition of Equation 60.92 is therefore satisfied if

$$G(z) = \frac{1}{\lambda^2}[\phi_{sy}(z)\{H(z^{-1})\}^{-1}]_+ H^{-1}(z). \qquad (60.94)$$

This is the *realizable Wiener filter*. It is clear from its construction that it is a causal and asymptotically stable filter.

The Wiener filter will be illustrated by two examples.

Example 60.5:

Consider the ARMA process

$$A(q)y(t) = C(q)e(t),$$
$$Ee^2(t) = \lambda^2,$$

Treat the prediction problem

$$s(t) = y(t+k), \qquad k > 0,$$

In this case,

$$H(z) = \frac{C(z)}{A(z)},$$
$$\phi_{sy}(z) = z^k \phi_y(z).$$

The *unrealizable* filter (Equation 60.87) becomes, as before,

$$G(z) = z^k \phi_y(z)\phi_y^{-1}(z) = z^k,$$

meaning that

$$\hat{s}(t) = y(t+k).$$

Note that it is noncausal, but it is otherwise a perfect estimate since it is without error! Next, the *realizable* filter is calculated:

$$G(z) = \frac{1}{\lambda^2}\left[z^k \lambda^2 \frac{C(z)}{A(z)}\frac{C(z^{-1})}{A(z^{-1})}\frac{A(z^{-1})}{C(z^{-1})}\right]_+ \frac{A(z)}{C(z)} = \left[z^k \frac{C(z)}{A(z)}\right]_+ \frac{A(z)}{C(z)}.$$

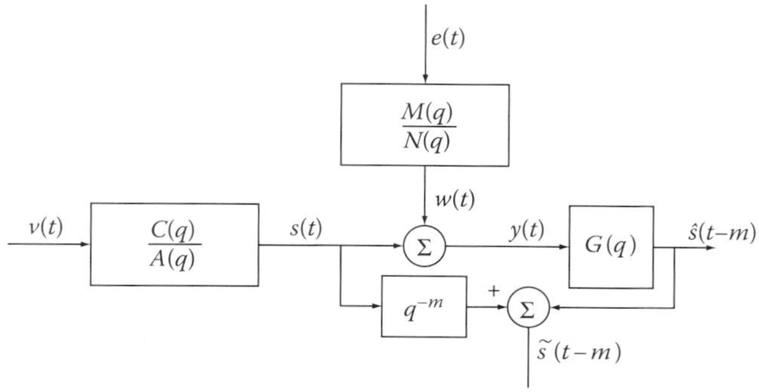

FIGURE 60.3 Setup for a polynomial estimation problem.

To proceed, let $A(z)$ and $C(z)$ have degree n, and introduce the polynomial $F(z)$ of degree $k-1$ and the polynomial $L(z)$ of degree $n-1$ by the *predictor identity* (Equation 60.67). This gives

$$G(z) = \left[\frac{zA(z)F(z) + zL(z)}{A(z)}\right]_+ \frac{A(z)}{C(z)} = \frac{zL(z)}{A(z)}\frac{A(z)}{C(z)} = \frac{zL(z)}{C(z)}.$$

The optimal predictor therefore has the form

$$\hat{s}(t) = \hat{y}(t+k|t) = \frac{qL(q)}{C(q)}y(t)$$

in agreement with Equation 60.71.

Example 60.6:

Consider the measurement of a random signal $s(t)$ in additive noise $w(t)$. This noise source need not be white, but it is assumed to be uncorrelated with the signal $s(t)$. Model $s(t)$ and $w(t)$ as ARMA processes; see Figure 60.3. Thus,

$$y(t) = s(t) + w(t), \quad s(t) = \frac{C(q)}{A(q)}v(t), \quad w(t) = \frac{M(q)}{N(q)}e(t),$$
$$Ev(t)v(s) = \lambda_v^2\delta_{t,s}, \quad Ee(t)e(s) = \lambda_e^2\delta_{t,s}, \quad Ee(t)v(s) = 0.$$

The polynomials in the model are

$$\begin{aligned} C(q) &= q^n + c_1q^{n-1} + \cdots + c_n, \\ A(q) &= q^n + a_1q^{n-1} + \cdots + a_n, \\ M(q) &= q^r + m_1q^{r-1} + \cdots + m_r, \\ N(q) &= q^r + n_1q^{r-1} + \cdots + n_r. \end{aligned}$$

Assume that all four polynomials have all their roots inside the unit circle. The problem to be treated is to estimate $s(t-m)$ from $\{y(t-j)\}_{j=0}^{\infty}$, where m is an integer. (By definition, $m=0$ gives filtering, $m>0$ smoothing, and $m<0$ prediction.)

To solve the estimation problem, first perform a spectral factorization of the output spectrum:

$$\phi_y(z) = \lambda_v^2\frac{C(z)C(z^{-1})}{A(z)A(z^{-1})} + \lambda_e^2\frac{M(z)M(z^{-1})}{N(z)N(z^{-1})} \equiv \lambda_\varepsilon^2\frac{B(z)B(z^{-1})}{A(z)N(z)A(z^{-1})N(z^{-1})},$$

requiring that $B(z)$ is a monic polynomial (i.e., it has a leading coefficient equal to one) of degree $n+r$, and that it has all zeros inside the unit circle. The polynomial $B(z)$ is therefore uniquely given by the identity

$$\lambda_\varepsilon^2 B(z)B(z^{-1}) \equiv \lambda_v^2 C(z)C(z^{-1})N(z)N(z^{-1}) + \lambda_e^2 A(z)A(z^{-1})M(z)M(z^{-1}).$$

Hence,

$$H(z) = \frac{B(z)}{A(z)N(z)}, \quad \lambda^2 = \lambda_\varepsilon^2, \quad \phi_{sy}(z) = z^{-m}\lambda_v^2 \frac{C(z)C(z^{-1})}{A(z)A(z^{-1})}.$$

According to Equation 60.94, the optimal filter becomes

$$\begin{aligned} G(z) &= \frac{1}{\lambda_\varepsilon^2}\left[z^{-m}\lambda_v^2 \frac{C(z)C(z^{-1})}{A(z)A(z^{-1})} \frac{A(z^{-1})N(z^{-1})}{B(z^{-1})}\right]_+ \frac{A(z)N(z)}{B(z)} \\ &= \frac{\lambda_v^2}{\lambda_\varepsilon^2}\left[z^{-m}\frac{C(z)C(z^{-1})N(z^{-1})}{A(z)B(z^{-1})}\right]_+ \frac{A(z)N(z)}{B(z)}. \end{aligned}$$

The causal part $[\ ]_+$ can be found by solving the Diophantine equation

$$z^{-m}C(z)C(z^{-1})N(z^{-1}) \equiv z^{\min(0,-m)}B(z^{-1})R(z) + z^{\max(0,-m)}A(z)L(z^{-1}),$$

where the unknown polynomials have degrees

$$\begin{aligned} \deg R &= n - \min(0,-m), \\ \deg L &= n + r - 1 + \max(0,-m). \end{aligned}$$

Note that the "-1" that appears in $\deg L$ has no direct correspondence in $\deg R$. The reason is that the direct term $g_0 z^{-0}$ in Equation 60.89 is associated with the causal part of $G(z)$.

The optimal filter is readily found:

$$\begin{aligned} G(z) &= \frac{\lambda_v^2}{\lambda_\varepsilon^2}\left(\left[\frac{z^{\min(0,-m)}B(z^{-1})R(z)}{A(z)B(z^{-1})}\right]_+ + \left[\frac{z^{\max(0,-m)}A(z)L(z^{-1})}{A(z)B(z^{-1})}\right]_+\right)\frac{A(z)N(z)}{B(z)} \\ &= \frac{\lambda_v^2}{\lambda_\varepsilon^2}\left[\frac{z^{\min(0,-m)}R(z)}{A(z)}\right]_+ \frac{A(z)N(z)}{B(z)} = \frac{\lambda_v^2}{\lambda_\varepsilon^2}\,\frac{z^{\min(0,-m)}R(z)N(z)}{B(z)}. \end{aligned}$$

References

1. K. J. Åström, *Introduction to Stochastic Control*. Academic Press, New York, 1970.
2. K. J. Åström and B. Wittenmark, *Computer Controlled Systems*, Prentice-Hall, Englewood Cliffs, NJ, (1984), 1990.
3. M. J. Grimble and M. A. Johnson, *Optimal Control and Stochastic Estimation*. John Wiley & Sons, Chichester, UK, 1988.
4. K. J. Hunt, Ed. *Polynomial Methods in Optimal Control and Filtering*. Peter Peregrinus Ltd, Stevenage, UK, 1993 (in particular, Chapter 6: A. Ahlén and M. Sternad: Optimal filtering problems.).
5. V. Kučera, *Discrete Linear Control*. John Wiley & Sons, Chichester, UK, 1979.
6. T. Söderström, *Discrete-Time Stochastic Systems: Estimation and Control*, Prentice-Hall International, Hemel Hempstead, UK, 1994.
7. T. Söderström, *Discrete-Time Stochastic Systems: Estimation and Control*, 2nd edition, Springer-Verlag, London, UK, 2002.

61

Dynamic Programming

P.R. Kumar
University of Illinois at Urbana–Champaign

61.1 Introduction

Dynamic programming is a recursive method for obtaining the optimal control as a function of the state in multistage systems. The procedure first determines the optimal control when there is only one stage left in the life of the system. Then it determines the optimal control where there are two stages left, etc. The recursion proceeds backward in time.

This procedure can be generalized to continuous-time systems, stochastic systems, and infinite horizon control. The cost criterion can be a total cost over several stages, a discounted sum of costs, or the average cost over an infinite horizon.

A simple example illustrates the main idea.

61.1.1 Example: The Shortest Path Problem

A bicyclist wishes to determine the shortest path from Bombay to the Indian East Coast. The journey can end at any one of the cities N_3, C_3, or S_3. The journey is to be done in 3 stages.

Stage zero is the starting stage. The bicyclist is in Bombay, labelled C_0 in Figure 61.1. Three stages of travel remain. The bicyclist has to decide whether to go north, center, or south. If the bicyclist goes north, she reaches the city N_1, after travelling 600 kms. If the bicyclist goes to the center, she reaches C_1, after travelling 450 kms. If she goes south, she reaches S_1 after travelling 500 kms.

At stage one, she will therefore be in one of the cities N_1, C_1, or S_1. From wherever she is, she has to decide which city from among N_2, C_2, or S_2 she will travel to next. The distances between cities N_1, C_1, and S_1 and N_2, C_2, S_2 are shown in Figure 61.1. At stage two, she will be in one of the cities N_2, C_2, or

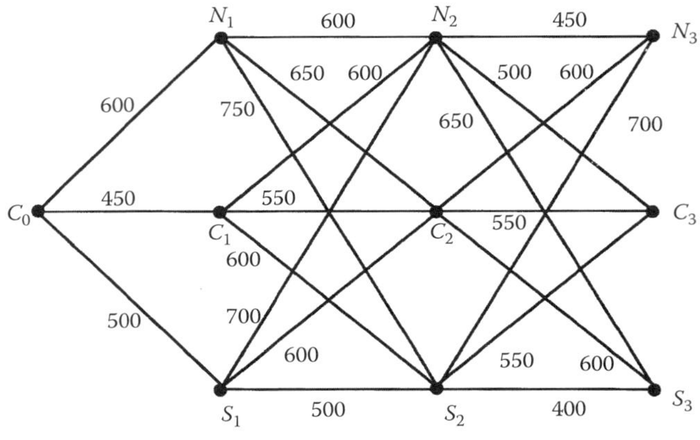

FIGURE 61.1 A shortest path problem.

S_2, and she will then have to decide which of the cities N_3, C_3, or S_3 to travel to. The journey ends in stage three with the bicyclist in one of the cities N_3, C_3 or S_3.

61.1.2 The Dynamic Programming Method

We first determine the optimal decision when there is only *one* stage remaining.

If the bicyclist is in N_2, she has three choices—north, center, or south, leading respectively to N_3, C_3 or S_3. The corresponding distances are 450, 500 and 650 kms. The best choice is to go north, and the shortest distance from N_2 to the East Coast is 450 kms.

Similarly, if the bicyclist is in C_2, the best choice is to go center, and the shortest distance from C_2 to the East Coast is 550 kms. From S_2, the best choice is to go south, and the shortest distance to the East Coast is 400 kms. We summarize the optimal decisions and the optimal costs, when only one stage remains, in Figure 61.2.

Now we determine the optimal decision when *two* stages remain.

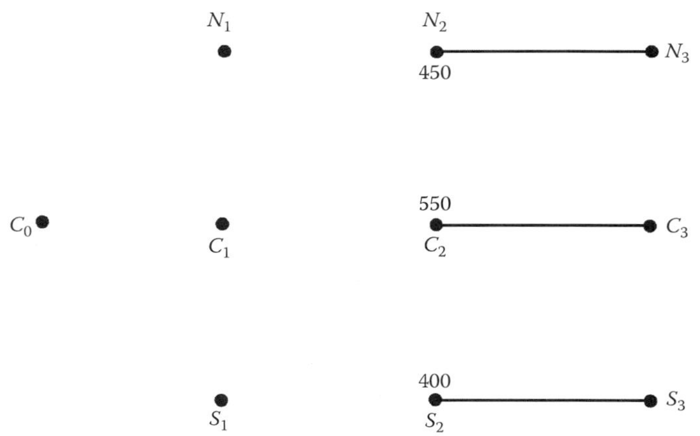

FIGURE 61.2 Optimal solution when one stage remains.

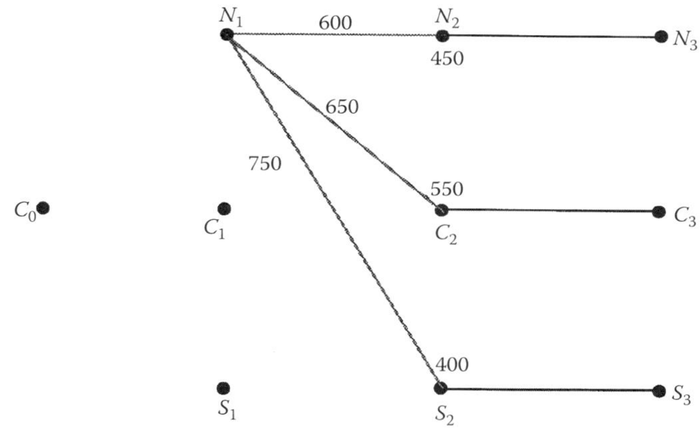

FIGURE 61.3 Making a decision in city N_1.

Suppose the bicyclist is in N_1. If she goes north, she will reach N_2 after cycling 600 kms. Moreover from N_2, she will have to travel a *further* 450 kms to reach the East Coast, as seen from Figure 61.2. From N_1, if she goes north, she will therefore have to travel a total of 1050 kms to reach the East Coast.

If instead she goes to the center from N_1, then she will have to travel 650 kms to reach C_2, and from C_2 she will have to travel a further 550 kms to reach the East Coast. Thus, she will have to travel a total of 1200 kms to reach the East Coast.

The only remaining choice from N_1 is to go south. Then she will travel 750 kms to reach S_2, and from there she has a further 400 kms to reach the East Coast. Thus, she will have to travel 1150 kms to reach the East Coast.

The consequences of each of these three choices are shown in Figure 61.3. Thus the optimal decision from N_1 is to go north and travel to N_2. Moreover, the shortest distance from N_1 to the East Coast is 1050 kms.

Similarly, we determine the optimal paths from C_1 and S_1, also, as well as the shortest distances from them to the East Coast. The optimal solution, when two stages remain, is shown in Figure 61.4.

Now we determine the optimal decision when there are *three* stages remaining.

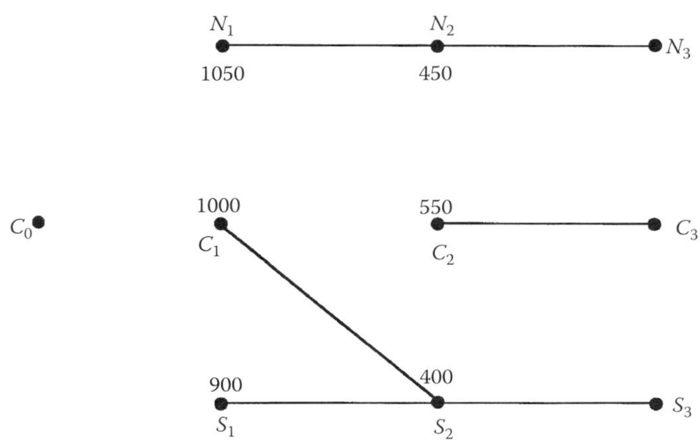

FIGURE 61.4 Optimal solution when two stages remain.

The bicyclist is in C_0. If she goes north, she travels 600 kms to reach N_1, and then a further 1050 kms is the minimum needed to reach the East Coast. Thus the total distance she will have to travel is 1650 kms. Similarly, if she goes to the center, she will travel a total distance of 1450 kms. On the other hand, if she goes south, she will travel a total distance of 1400 kms to reach the East Coast. Thus the best choice from C_0 is to travel south to S_1, and the shortest distance from C_0 to the East Coast is 1400 kms. This final result is shown in Figure 61.5.

61.1.3 Observations on the Dynamic Programming Method

We observe a number of important properties.

1. In order to solve the shortest path from just one city C_0 to the East Coast, we actually solved the shortest path from *all* of the cities $C_0, N_1, C_1, S_1, N_2, C_2, S_2$ to the East Coast.
2. Hence, dynamic programming can be *computationally intractable* if there are many stages and many possible decisions at each stage.
3. Dynamic programming determines the optimal decision as a function of the state and the number of stages remaining. Thus, it gives an optimal *closed-loop* control policy.
4. Dynamic programming proceeds *backward* in time. After determining the optimal solution when there are s stages remaining, it determines the optimal solution when there are $(s+1)$ stages remaining.
5. Fundamental use is made of the following relationship in obtaining the backward recursion:

Optimal distance from state x to the end

$$= \min_i \left\{ \begin{array}{l} \text{Distance travelled on current stage when decision } d_i \text{ is made + } \textit{optimal} \\ \text{distance from the state } x' \text{ to the end, where } x' \text{ is the state to which decision} \\ d_i \text{ takes you from state } x \end{array} \right\}$$

Hence, if one passes through state x' from x, an optimal continuation is to take the shortest path from state x' to the end. This can be summarized by saying "segments of optimal paths are optimal in themselves." This is called the *Principle of Optimality*.

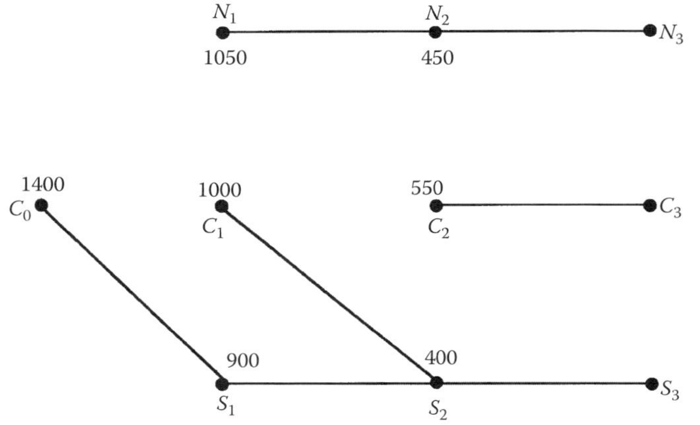

FIGURE 61.5 Optimal solution when three stages remain.

61.2 Deterministic Systems with a Finite Horizon

We can generalize the solution method to deterministic, time-varying control systems. Consider a system whose state at time t is denoted by $x(t) \in \mathcal{X}$. If a control input $u(t) \in \mathcal{U}$ is chosen at time t, then the next state $x(t+1)$ is determined as,

$$x(t+1) = f(x(t), u(t), t).$$

The system starts in some state $x(t_0) = x_0$ at time t_0. For simplicity, suppose that there are only a *finite* number of possible control actions, i.e., $\mathcal{U}$ is finite.

The cost incurred by the choice of control input $u(t)$ at time t is $c(x(t), u(t))$. We have a *finite time horizon* T. The goal is to determine the controls to minimize the cost function,

$$\sum_{t=t_0}^{T-1} c(x(t), u(t), t) + d(x(T)). \tag{61.1}$$

Theorem 61.1: Deterministic System, Finite horizon, Finite Control Set

Define

$$V(x, T) := d(x), \quad \textit{and} \tag{61.2}$$
$$V(x, t-1) := \min_{u \in \mathcal{U}} \left\{ c(x, u, t-1) + V(f(x, u, t-1), t) \right\}. \tag{61.3}$$

Let $u^*(x, t-1)$ be a value of u which minimizes the right-hand side (RHS) above. Then,

1. $V(x,t) = \min_{u(\cdot)} \sum_{n=t}^{T-1} c[x(n), u(n), n] + d[x(T)]$, *when* $x(t) = x$, *that is, it is the optimal cost-to-go from state x at time t to the end.*
2. $u^*(x, t)$ *is an optimal control when one is in state x at time t.*

Proof 61.1. Define $V(x,t)$ as the optimal cost-to-go from state x at time t to the end. $V(x, T) = d(x)$, because there is no decision to make when one is at the end. Now we obtain the *backward recursion* of dynamic programming. Suppose we are in state x at time $t-1$. If we apply control input u, we will incur an immediate cost $c(x, u, t-1)$, and move to the state $f(x, u, t-1)$, from which the optimal cost-to-go is $V(f(x, u, t-1), t)$. Because some action u has to be taken at time $t-1$,

$$V(x, t-1) \le \min_{u \in \mathcal{U}} \left\{ c(x, u, t-1) + V(f(x, u, t-1), t) \right\}.$$

Moreover, if $V(x, t-1)$ was strictly smaller than the RHS above, then one would obtain a contradiction to the minimality of $V(\cdot, t)$. Hence equality holds, and Equation 61.3 is satisfied.

By evaluating the cost of the control policy given by $u^*(x, t)$, one can verify that it gives the minimal cost $V(x_0, t_0)$.

We have proven the Principle of Optimality. We have also proven that, for the cost criterion Equation 61.1, there is an optimal control policy that takes actions based only on the value of current state and current time and does not depend on past states or actions. Such a control policy is called a *Markovian policy.*

61.2.1 Infinite Control Set $\mathcal{U}$

When the control set $\mathcal{U}$ is infinite, the RHS in Equation 61.3 may not have a minimizing u. Thus the recursion requires an "inf" in place of the "min." There may be no optimal control law, only nearly optimal ones.

61.2.2 Continuous-Time Systems

Suppose one has a continuous-time system

$$\dot{x}(t) = f(x(t), u(t), t), \tag{61.4}$$

and the cost criterion is

$$d(x(T)) + \int_{t_0}^{T} c(x(t), u(t), t)\, dt.$$

Several technical issues arise. The differential Equation 61.4 may not have a solution for a given $u(t)$, or it may have a solution but only until some finite time. Ignoring such problems, we examine what type of equations replace the discrete-time recursions. Clearly, $V(x, T) = d(x)$. Considering a control input u which is held constant over an interval $[t, t + \Delta)$, and approximating the resulting state (all possible only under more assumptions),

$$\begin{aligned} V(x,t) &= \inf_{u\in\mathcal{U}} \left\{c(x,u,t)\Delta + V(x + f(x,u,t)\Delta, t+\Delta) + o(\Delta)\right\}, \\ &= \inf_{u\in\mathcal{U}} \left\{c(x,u,t)\Delta + V(x,t) + \frac{\partial V}{\partial x}(x,t)f(x,u,t)\Delta + \frac{\partial V}{\partial t}(x,t)\Delta + o(\Delta)\right\}. \end{aligned}$$

Thus, one looks for the partial differential equation

$$\frac{\partial V}{\partial t}(x,t) + \inf_{u\in\mathcal{U}} \left\{\frac{\partial V}{\partial x}(x,t)f(x,u,t) + c(x,u,t)\right\} = 0, \tag{61.5}$$

to be satisfied by the optimal cost-to-go. This is called the *Hamilton–Jacobi–Bellman equation*. To prove that the optimal cost-to-go satisfies such an equation requires many technical assumptions.

However, if a smooth V exists that satisfies such an equation, then it is a lower bound on the cost, at least for Markovian control laws. To see this, consider any control law $u(x, t)$, and let $x(t)$ be the trajectory resulting from an initial condition $x(t_0) = x_0$ at time t_0. Then,

$$\frac{\partial V}{\partial t}(x(t),t) + \frac{\partial V}{\partial x}(x(t),t)f(x(t),u(x(t),t),t) + c(x(t),u(x(t),t)) \geq 0. \tag{61.6}$$

Hence $\frac{d}{dt}[V(x(t), t)] + c(x(t), u(x(t), t), t) \geq 0$, and so,

$$\begin{aligned} V(x(t_0), t_0) &\leq V(x(T), T) + \int_{t_0}^{T} c(x(t), u(t), t)\, dt \\ &= d(x(T)) + \int_{t_0}^{T} c(x(t), u(t), t)\, dt. \end{aligned}$$

Moreover, if $u^*(x, t)$ is a control policy which attains the "inf" in Equation 61.5, then it attains equality above and is, hence, optimal.

In many problems, the optimal cost-to-go $V(x, t)$ may not be sufficiently smooth, and one resorts to the Pontryagin Minimum Principle rather than the HJB Equation 61.5.

61.3 Stochastic Systems

Consider the simplest case where time is discrete, and both the state-space $\mathcal{X}$ and the control set $\mathcal{U}$ are finite. Suppose that the system is time invariant and evolves probabilistically as a *controlled Markov chain*, i.e.,

$$\text{Prob}[x(t+1) = j | x(t) = i, u(t) = u] = p_{ij}(u). \tag{61.7}$$

The elements of the matrix $P(u) = [p_{ij}(u)]$ are the *transition probabilities.*

Given a starting state $x(t_0) = x_0$ at time t_0, one wishes to minimize the *expected* cost

$$E\left[\sum_{t=t_0}^{T-1} c(x(t), u(t), t) + d(x(T))\right].$$

A *nonanticipative control policy* γ is a mapping from past information to control inputs, i.e., $\gamma = (\gamma_{t_0}, \gamma_{t_0+1}, \ldots, \gamma_{T-1})$ where $\gamma_t : (x(t_0), u(t_0), x(t_0+1), u(t_0+1), \ldots, x(t)) \mapsto u(t)$. Define the conditionally expected cost-to-go when policy γ is applied, as

$$V_t^\gamma := E\left[\sum_{n=t}^{T-1} c(x(n), u(n), n) + d(x(T)) \| x(t_0), u(t_0), \ldots, x(t)\right].$$

Analogously to the deterministic case, set Equation 61.2, and recursively define

$$V(x, t-1) = \min_{u \in \mathcal{U}} \left\{ c(x, u, t-1) + \sum_j p_{xj}(u) V(j, t) \right\}. \tag{61.8}$$

Now let us compare V_t^γ with $V(x(t), t)$ when policy γ is used. Clearly $V_T^\gamma = V(x(T), T) = d(x(T))$ a.s. Now suppose, by induction, that $V_t^\gamma \geq V(x(t), t)$ a.s. Then

$$\begin{aligned}
V_{t-1}^\gamma &= c(x(t-1), u(t-1), t-1) + E\left\{\sum_{n=t}^{T} c[x(n), u(n), n) + d(x(T)] | x(t_0), u(t_0), \ldots, x(t-1)\right\} \\
&= c(x(t-1), u(t-1), t-1) + E\left\{E\left[\sum_{n=t}^{T} c[x(n), u(n), n) + d(x(T)] | x(t_0), u(t_0), \ldots, x(t)\right]\right. \\
&\qquad \left. x(t_0), u(t_0), \ldots, x(t-1)\right\} \\
&= c(x(t-1), u(t-1), t-1) + E\left[V_t^\gamma | x(t_0), u(t_0), \ldots, x(t-1)\right] \\
&\geq c(x(t-1), u(t-1), t-1) + E\left[V(x(t), t) | x(t_0), u(t_0), \ldots, x(t-1)\right] \\
&= c(x(t-1), u(t-1), t-1) + \sum_j p_{x(t-1)j}[u(t-1)] V(j, t) \\
&\geq \min_{u \in \mathcal{U}} \left\{ c(x(t-1), u, t-1) + \sum_j p_{x(t-1)j}(u) V(j, t) \right\} \\
&= V(x(t), t) \text{ a.s.}
\end{aligned} \tag{61.9}$$

Thus $V(x, t)$ is a lower bound on the expected cost of any non anticipative control policy. Suppose, moreover, that $u^*(x, t)$ attains the minimum on the RHS of Equation 61.8 above. Consider the Markovian control policy $\gamma^* = (\gamma_{t_0}^*, \gamma_{t_1}^*, \ldots, \gamma_{T-1}^*)$ with

$$\gamma_t^*[x(t_0), u(t_0), \ldots, x(t)] = u^*[x(t), t]. \tag{61.10}$$

For γ^* it is easy to verify that the inequalities in Equation 61.9 are equalities, and so it is optimal.

Theorem 61.2: Stochastic Finite State, Finite Control System, Finite Horizon

Recursively define $V(x, t)$ from Equations 61.2 and 61.8. Let $u^*(x, t-1)$ attain the minimum on the RHS of Equation 61.8, and consider the Markovian control policy γ^* defined in Equation 61.10. Then,

i. *$V(x_0, t_0)$ is the optimal cost.*
ii. *The Markovian control policy γ^* is optimal over the class of all nonanticipative control policies.*

61.3.1 Countably Infinite State and Control Spaces

The result (i) can be extended to countably infinite state and control policies by replacing the "min" above with an "inf". However, the "inf" need not be attained, and an optimal control policy may not exist.

If one considers uncountably infinite state and control spaces, then further highly technical issues arise. A policy needs to be a *measurable* map. Moreover, $V(x, t)$ will also need to be a measurable function because one must take its expected value. One must impose appropriate conditions to insure that the "inf" over an uncountable set still gives an appropriately measurable function, and further that one can synthesize a minimizing $u(x, t)$ that is measurable.

61.3.2 Stochastic Differential Systems

Consider a system described by a *stochastic differential equation,*

$$dx(t) = f(x(t), u(t), t)\, dt + dw(t),$$

where w is a standard Brownian motion, with a cost criterion

$$E\{d[x(T)] + \int_{t_0}^{T} c[x(t), u(t), t]\}.$$

If $V(x, t)$ denotes the optimal cost–to–go from a starting state x when there are t time units remaining, one expects from Ito's differentiation rule that it satisfies the stochastic version of the Hamilton-Jacobi-Bellman equation:

$$\frac{\partial V(x,t)}{\partial t} + \frac{1}{2}\frac{\partial^2 V}{\partial x^2}(x,t) + \inf_{u \in \mathcal{U}} \left\{ \frac{\partial V}{\partial x}(x,t) f(x,u,t) + c(x,u,t) \right\} = 0.$$

The existence of a solution to such partial differential equations is studied using the notion of *viscosity* solutions.

61.4 Infinite Horizon Stochastic Systems

We will consider finite state, finite control, controlled Markov chains, as in Equation 61.7. We study the infinite horizon case.

61.4.1 The Discounted Cost Problem

Consider an *infinite horizon total discounted cost* of the form

$$E \sum_{t=0}^{+\infty} \beta^t c(x(t), u(t)), \tag{61.11}$$

where $0 < \beta < 1$ is a *discount factor*. The discounting guarantees that the summation is finite. We are assuming that the one-stage cost function c does not change with time. Let $W(x, \infty)$ denote the optimal value of this cost, starting in state x. The "∞" denotes that there is an infinity of stages remaining.

Let $W(x,N)$ denote the optimal value of the finite horizon cost $E\sum_{t=0}^{N-1}\beta^t c[x(t),u(t)]$, when starting in state x at time 0. Note that the index N refers to the number of *remaining* stages. From the finite horizon case, we see that

$$W(x,N) = \min_{u\in\mathcal{U}}\left\{c(x,u) + \sum_j p_{xj}(u)\beta W(j,N-1)\right\}. \tag{61.12}$$

Note the presence of the factor β multiplying $W(j,N-1)$ above.

Denote by $R^{\mathcal{X}}$ the class of all real valued functions of the state, i.e., $V\in R^{\mathcal{X}}$ is a function of the form $V:\mathcal{X}\to R$. Define an operator $T: R^{\mathcal{X}}\to R^{\mathcal{X}}$ by its action on V as

$$TV(x) := \min_{u\in\mathcal{U}}\left\{c(x,u) + \beta\sum_j p_{xj}(u)V(j)\right\} \quad \text{for all } x\in\mathcal{X}. \tag{61.13}$$

Using the operator T, one can rewrite Equation 61.12 as $W(\cdot,N) = TW(\cdot,N-1)$. Also, $W(\cdot,0)=O$, where O is the identically zero function, i.e., $O(x)\equiv 0$ for all x. Note that T is a monotone operator, i.e., if $V(x)\le\bar V(x)$ for all x, then $TV(x)\le T\bar V(x)$ for all x. Let us suppose now that $\bar c\ge c(x,u)\ge 0$ for all x, u. (This can be achieved by simply adding a large enough constant to each original one-step cost.) Due to this assumption, $TO\ge 0$. Hence by monotonicity, $T^{(N)}O\ge T^{(N-1)}O\ge\cdots\ge TO\ge 0$. Thus $T^{(N)}O(x)$ converges, for every x, to a finite number $\left(\text{since } T^{(N)}O\le\frac{\bar c}{1-\beta}\right)$. Let

$$W(x) := \lim_{N\to\infty} T^{(N)}O(x).$$

Now $W(x,\infty)\ge W(x,N)$ for all N, because $c(x,u)\ge 0$. Hence $W(x,\infty)\ge\lim_{N\to\infty}W(x,N) = W(x)$. Moreover $W(x,\infty)\le W(x,N)+\frac{\bar c\beta^N}{1-\beta}$, because one can employ an optimal policy for an N-stage problem. Thus $W(x,\infty)\le\lim_{N\to\infty}\left[W(x,N)+\frac{\bar c\beta^N}{1-\beta}\right] = W(x)$. Hence $W(x)=W(x,\infty)$, and is, therefore, the *optimal infinite horizon cost.*

Hence, we would like to characterize $W(x)$. Since T is continuous, $TW(x) = \lim_{N\to\infty}T^{N+1}O(x) = W(x)$. Hence W is a *fixed point* of T, i.e.,

$$W(x) = \min_{u\in\mathcal{U}}\left\{c(x,u)+\beta\sum_j p_{xj}(u)W(j)\right\}. \tag{61.14}$$

Simple calculations show that T is a contraction with respect to the $\|\cdot\|_\infty$ norm, i.e., $\max_x|TV(x)-T\bar V(x)|\le\beta\max_x|V(x)-\bar V(x)|$. Thus T has a unique fixed point. Since T is a contraction, we also know that $\lim_{N\to\infty}T^{(N)}V = W$ for *any* V.

Suppose now that $u^*(x)$ attains the minimum in Equation 61.14 above. Consider the policy γ^* which always chooses control $u^*(x)$ whenever the state is x. Such a control policy is called *stationary.* If one applies the stationary control policy γ^*, then the expected cost over N days, starting in state x, is $T^{(N)}O(x)$. Thus the infinite horizon cost of γ^* is $\lim_{N\to\infty}T^NO = W$.

Theorem 61.3: Stochastic Finite State, Finite Control System, Discounted Cost Criterion

Let $T: R^{\mathcal{X}}\to R^{\mathcal{X}}$ *denote the operator (Equation 61.13).*

1. T *is a contraction.*
2. *Let* $W=\lim_{N\to\infty}T^{(N)}V$ *for any* V. *Then* W *is the unique solution of Equation 61.14.*
3. $W(x)$ *is the optimal cost when starting in state* x.

4. *Let $\gamma^*(x)$ be the value of u which attains the minimum on the RHS in Equation 61.14. Then γ^* is a stationary control policy which is optimal in the class of all nonanticipative policies.*

The procedure for determining $W(x)$ as $\lim_{N\to\infty} T^{(N)}O$ is called *value iteration*. Another procedure which determines the optimal policy in a finite number of steps is *policy iteration*.

61.4.1.1 Policy Iteration Procedure

For a stationary policy γ, define $T_\gamma : R^{\mathcal{X}} \to R^{\mathcal{X}}$ by, $T_\gamma V(x) = c(x, \gamma(x)) + \beta \sum_j p_{xj}(\gamma(x))V(j)$.

1. Let γ_0 be any stationary policy.
2. Solve the linear system of equations $T_{\gamma_0} W_{\gamma_0} = W_{\gamma_0}$ to determine its cost W_{γ_0}.
3. If $TW_{\gamma_0} \neq W_{\gamma_0}$, then let $\gamma_1(x)$ be the value of u which attains the minimum in $\min_{u\in\mathcal{U}}\left\{c(x,u) + \beta \sum_j p_{xj}(u) W_{\gamma_0}(j)\right\}$.
4. Then γ_1 is a strict improvement of γ_0 (since $W_{\gamma_1} = \lim_{N\to\infty} T_{\gamma_1}^{(N)} W_{\gamma_0} \lneqq W_{\gamma_0}$).
5. By repeating this procedure, one obtains a sequence of strictly improving policies, that must terminate in a finite number of steps, because the total number of stationary policies is finite.

61.4.2 The Average Cost Problem

We consider the *average cost per unit time* over an infinite horizon,

$$\limsup_{N\to\infty} \frac{1}{N} E \sum_{t=0}^{N-1} c[x(t), u(t)].$$

Then the dynamic programming equation needs to be modified slightly.

Theorem 61.4:

Suppose a constant J^* exists, and a function $V : \mathcal{X} \to R$ exists so that

$$J^* + V(x) = \min_{u\in\mathcal{U}}\left\{c(x,u) + \sum_j p_{xj}(u)V(j)\right\}. \tag{61.15}$$

1. *Then J^* is the optimal value of the average cost criterion, starting from any state x.*
2. *Let $\gamma^*(x)$ be a value of u which minimizes the RHS in Equation 61.15 above. Then γ^* is a stationary policy which is optimal in the class of all nonanticipative policies.*

Proof 61.2. Consider any nonanticipative policy. Then

$$J^* + V(x(t)) \leq c(x(t), u(t)) + \sum_j p_{x(t)j}(u(t))V(j).$$

Noting that $E \sum_j p_{x(t)j}(u(t))V(j) = E\{V(x(t+1))\}$, we see that

$$E(c(x(t), u(t)) \geq J^* + E(V(x(t)) - E\{V(x(t+1))\}.$$

Hence

$$\limsup_{N\to\infty} \frac{1}{N} E \sum_{t=0}^{N-1} c(x(t), u(t)) \geq J^* + \limsup_N \frac{V(x(0)) - E\{V(x(N))\}}{N} = J^*.$$

Thus J^* is a lower bound on the average cost, starting from *any* state x. Moreover, if γ^* is the policy under consideration, then equality holds above, and so it is indeed optimal.

The question, still a topic of active research, is when does a solution $[J^*, V(\cdot)]$ exist for Equation 61.15? Let us consider the following simplifying assumption. For *every* stationary policy γ, the Markov chain $P_\gamma = [p_{ij}(\gamma(i))]$ is *irreducible*. By this is meant that there exists a *unique* steady state distribution π_γ, satisfying $\pi_\gamma P_\gamma = \pi_\gamma, \quad \pi_\gamma(i) \geq 0, \quad \sum_i \pi_\gamma(i) = 1$, which is *strictly positive*, i.e., $\pi_\gamma(i) > 0$ for all i. Then the average cost J_γ starting from any state is a constant and satisfies, $J_\gamma = \sum_i \pi_\gamma(i)c[i, \gamma(i)] = \pi_\gamma c_\gamma$ (where $c_\gamma := [c_\gamma(1), \ldots, c_\gamma(1x)]$). Hence if $e = (1, \ldots, 1)^T$, $\pi_\gamma(J_\gamma e - c_\gamma) = 0$. Hence $(J_\gamma e - c_\gamma)$ is orthogonal to the null space of $(P_\gamma - I)^T$, and is, therefore, in the range space of $(P_\gamma - I)$. Thus a V_γ exists so that $J_\gamma e - c_\gamma = (P_\gamma - I)V_\gamma$, which simply means that

$$J_\gamma + V_\gamma(x) = c[x, \gamma(x)] + \sum_j p_{xj}[\gamma(x)]V_\gamma(j). \tag{61.16}$$

Note that $V_\gamma(\cdot)$ is *not* unique because $V_\gamma(\cdot) + a$ is also a solution for any a. Let us therefore fix $V_\gamma(\bar{x}) = 0$ for some $\bar{x}$. One can obtain a policy iteration algorithm, as well as prove the existence of J^* and $V(\cdot)$, as shown below.

61.4.2.1 Policy Iteration Procedure

1. Let γ_0 be any stationary policy.
2. Solve Equation 61.16, with γ replaced by γ_0, to obtain $(J_{\gamma_0}, V_{\gamma_0})$. If $(J_{\gamma_0}, V_{\gamma_0})$ does *not* satisfy Equation 61.15, then let $\gamma_1(x)$ attain the minimum in $\min_{u \in \mathcal{U}} \left\{ c(x, u) + \sum_j p_{xj}(u) V_{\gamma_0}(x) \right\}$.

Then γ_1 is a strict improvement of γ_0. (This follows because $\pi_{\gamma_1}(i) > 0$ for all i, and so

$$\begin{aligned} J_{\gamma_0} + \pi_{\gamma_1} V_{\gamma_0} &= \pi_{\gamma_1}(J_{\gamma_0} e + V_{\gamma_0}) > \pi_{\gamma_1} c_{\gamma_1} + \pi_{\gamma_1} P_{\gamma_1} V_{\gamma_0} \\ &= J_{\gamma_1} + \pi_{\gamma_1} V_{\gamma_0}, \text{and so } J_{\gamma_0} > J_{\gamma_1}). \end{aligned}$$

Because the policy space is finite, this procedure terminates in a finite number of steps. At termination, Equation 61.15 is satisfied.

61.4.3 Connections of Average Cost Problem with Discounted Cost Problems and Recurrence Conditions

The average cost problem can be regarded as a limit of discounted cost problems when $\beta \nearrow 1$. We illustrate this for systems with *countable state space* and finite control set.

Theorem 61.5: Connection between Discounted and Average Cost

Let $W_\beta(x)$ denote the optimal discounted cost $E \sum_{t=0}^{+\infty} \beta^t \, c(x(t), u(t))$ when starting in the state x. Suppose that $|W_\beta(x) - W_\beta(x')| \leq M$ all x, x', and all $\beta \in (1 - \epsilon, 1)$, for some $\epsilon > 0$. For an arbitrary state $\bar{x} \in \mathcal{X}$, let $\beta_n \nearrow 1$ be a sub-sequence so that the following limits exist:

$$\lim_{n \to \infty} (1 - \beta_n) W_{\beta_n}(x) =: J^*, \quad \textit{and} \quad \lim_{n \to \infty} (W_{\beta_n}(x) - W_{\beta_n}(\bar{x})) =: V(x).$$

Then,

1. $J^* + V(x) = \min_{u \in \mathcal{U}} \left\{ c(x, u) + \sum_j p_{xj}(j) V(j) \right\}$.

2. *If a stationary policy γ^* is optimal for a sequence of discount factors β_n with $\beta_n \nearrow 1$, then γ^* is optimal for the average cost problem.*

Proof 61.3. The dynamic programming Equation 61.14 for the discounted cost problem can be rewritten as,

$$(1-\beta)W_\beta(x) = \min_{u\in\mathcal{U}}\left\{c(x,u) + \beta\sum_j p_{xj}[W_\beta(j) - W_\beta(x)]\right\}.$$

Taking limits along $\beta_n \nearrow 1$ yields the results.

The existence of $(J^*, V(\cdot))$ satisfying the average cost dynamic programming Equation 61.15 is guaranteed under certain uniform recurrence conditions on the controlled Markov chain.

Theorem 61.6: Uniformly Bounded Mean First Passage Times

Let τ denote the first time after time 1 that the system enters some fixed state $\bar{x}$, i.e.,

$$\tau = \min\left\{t \geq 1 : \quad x(t) = \bar{x}\right\}.$$

Suppose that the mean first passage times are uniformly bounded, i.e.,

$$E(\tau \mid x(0) = x \text{ and } \gamma \text{is used}) \leq M < +\infty,$$

for all states x and all stationary policies γ. Then a solution $(J^*, V(\cdot))$ exists to Equation 61.15.

Proof 61.4. Under a stationary policy γ_β which is optimal for the discount factor β,

$$\begin{aligned} W_\beta(x) &= E\left[\sum_{t=0}^{\tau-1}\beta^t c(x(t),u(t)) + \beta^\tau W_\beta(\bar{x}) \mid x(0) = x\right] \\ &\leq \bar{c}E[\tau \mid x(0) = x] + W_\beta(\bar{x}) \leq \bar{c}M + W_\beta(\bar{x}). \end{aligned}$$

Moreover, by Jensen's inequality,

$$\begin{aligned} W_\beta(x) &\geq E[\beta^\tau W_\beta(\bar{x}) \mid x(0) = x] \\ &= W_\beta(\bar{x})E[\beta^\tau \mid x(0) = x] \geq W_\beta(\bar{x})\beta^M. \end{aligned}$$

Hence, $-\bar{c}M \leq \frac{\bar{c}(\beta^M - 1)}{1-\beta} \leq W_\beta(\bar{x})(\beta^M - 1) \leq W_\beta(x) - W_\beta(\bar{x}) \leq \bar{c}M$, and the result follows from the preceding Theorem.

61.4.4 Total Undiscounted Cost Criterion

Consider a *total* infinite horizon cost criterion of the form

$$E\sum_{t=0}^{+\infty} c[x(t), u(t)].$$

In order for the infinite summation to exist, one often assumes that either

$$\begin{aligned} &c(x,u) \geq 0 \quad \text{for all } x, u \quad \text{(the positive cost case), or} \\ &c(x,u) \leq 0 \quad \text{for all } x, u \quad \text{(the negative cost case).} \end{aligned}$$

These two cases are rather different. In both cases, one exploits the monotonicity of the operator T.

References

1. Blackwell, D., Discounted dynamic programming. *Ann. of Math. Statist.* 36, 226–335, 1965.
2. Strauch, R., Negative dynamic programming. *Ann. Math. Statist.*, 37, 871–890, 1966.
3. Blackwell, D., Positive dynamic programming. *Proceedings of the 5th Berkeley Symposium on Mathematical Statistics and Probability*, 415–418, 1965.
4. Blackwell, D., On stationary policies. *J. Roy. Statist. Soc. Ser. A*, 133, 33–38, 1970.
5. Ornstein, D., On the existence of stationary optimal strategies. *Proc. Amer. Math. Soc.*, 20, 563–569, 1969.
6. Blackwell, D., Discrete dynamic programming. *Ann. of Math. Statist.*, 33, 719–726, 1962.
7. Hernandez-Lerma, O. and Lasserre, J. B., Weak conditions for average optimality in Markov control processes. *Syst. and Contr. Lett.*, 22, 287–291, 1994.
8. Bertsekas, D. P. and Shreve, S. E., *Stochastic Optimal Control: The Discrete Time Case*, Academic, New York, 1978.
9. Lions, P. L., Optimal control of diffusion processes and Hamilton-Jacobi equations, Part I–the dynamic programming principle and applications. *Comm. Partial Differential Equations*, 10, 1101–1174, 1983.
10. Crandall, M., Ishii, H. and Lions, P. L., User's guide to viscosity solutions of second order partial differential equations. *Bull. Amer. Math. Soc.* 27, 1–67, 1990.
11. Bellman, R., *Dynamic Programming*. Princeton University Press, Princeton, NJ, 1957.
12. Bertsekas, D. P., *Dynamic Programming: Deterministic and Stochastic Models*. Prentice Hall, Englewood Cliffs, NJ, 1987.
13. Kumar, P. R. and Varaiya, P. P., *Stochastic Systems: Estimation, Identification and Adaptive Control*. Prentice Hall, Englewood Cliffs, NJ, 1986.
14. Ross, S. M., *Applied Probability Models with Optimization Applications*. Holden-Day, San Francisco, 1970.

Further Reading

For discounted dynamic programming, the classic reference is Blackwell [1]. For the positive cost case, we refer the reader to Strauch [2]. For the negative cost case, we refer the reader to Blackwell [3,4] and Ornstein [5]. For the average cost case, we refer the reader to Blackwell [6] for early fundamental work, and to Hernandez-Lerma and Lasserre [7] and the references contained there for recent developments. For a study of the measurability issues which arise when considering uncountable sets, we refer the reader to Bertsekas and Shreve [8]. For continuous time stochastic control of diffusions we refer the reader to Lions [9], and to Crandall, Ishii and Lions [10] for a guide to viscosity solutions.

For the several ways in which dynamic programming can be employed, see Bellman [11].

Some recent textbooks which cover dynamic programming are Bertsekas [12], Kumar and Varaiya [13], and Ross [14].

62

Approximate Dynamic Programming

Draguna Vrabie
The University of Texas at Arlington

Frank L. Lewis
The University of Texas at Arlington

62.1 Introduction to Approximate Dynamic Programming

62.1.1 Approximate Dynamic Programming in Relation to Modern Control Approaches

The regular approach to *control system design* is a two-step procedure as follows:

1. Use modeling and/or identification techniques, of the sort presented in this handbook, to determine a most accurate mathematical description of the system.

2. Employ a controller design method to determine, based on the identified model, a control strategy such that the prescribed performances are obtained.

This approach at deriving control strategies is successful under the assumption that the system dynamics can be modeled and are not changing over time. *Optimal controllers* fall in this category as they are generally determined considering a fixed given model of the system. The goal of an optimal control strategy is the minimization of a cost index, which reflects the amount of energy used for control purposes and the distance between the present and desired performance of the controlled system. Though they have good robustness properties relative to possible changes in the system dynamics, optimal controllers are neither adaptive nor are they determined considering possible unmodeled dynamics. From this perspective one can generally say that an optimal controller is just as close to optimality as the model of the system, used during the design phase, is close to the real plant to be controlled.

The class of *adaptive controllers* has been developed to confront the variations that occur in the system dynamics. In this case adaptation algorithms modify the structure and/or parametric description of the controller such that the performances prescribed for the control system are closely satisfied in the presence of variations in the system dynamics. Regular adaptive control methods are not optimal in the sense of minimizing a cost functional of the sort considered by optimal control.

To confront optimality requirements in the presence of unknown or uncertain system dynamics a new class of control strategies has been introduced. *approximate dynamic programming* (ADP), [24], also addressed as *adaptive dynamic programming*, [13], or *neuro dynamic programming*, [7], is a hybrid approach to *online adaptive optimal control*. It combines elements of adaptive control with modeling and identification techniques for the purpose of obtaining, in an online fashion, optimal control strategies, with respect to a measurable cost index, for general systems.

The objective of this chapter is twofold:

- To motivate the ADP approach to controller design in the context of other control approaches, and introduce the main ingredients of ADP algorithms.
- To overview the main results and provide practical ideas for the implementation of ADP algorithms in a comparative approach relative to known optimal control results.

62.1.2 Reinforcement Learning

ADP methods have their roots in the *Reinforcement Learning* (RL) approach to optimization. RL refers to the capability of an active agent, able to interact with its environment, to modify its behavior (i.e., learn and adapt), based on measurable reward or punishment stimuli form the environment. Such stimuli reflect the distance between the performances associated with the present behavior strategy, relative to a desired performance index. This approach to learning is constructed on the idea that successful actions should be remembered, based on reinforcement information, such that they are more likely to be used a second time.

These methods were introduced in the computational intelligence community with the purpose of providing means of improvement for the behavior of an agent, which acts in an unknown and hostile environment. The idea originates from experimental animal learning where it has been observed that the dopamine neurotransmitter in the basal ganglia acts as a reinforcement signal, which favors learning at the level of the neuron, while encoding information on the difference between the actual and expected result of a performed action.

ADP algorithms were mostly developed and successfully implemented in the framework of Markov decision processes (MDP), which are of interest for the computational intelligence community. In the MDP formulation the spaces of states of the environment and actions of the agent are discrete and finite. Also, the transitions between different states, conditioned by a choice of an action, are performed according to a probability distribution at discrete moments in time. In control engineering the interest shifts toward developing controllers for man-engineered systems, which evolve in continuous state and action spaces in a deterministic fashion. Such systems will be the object of the ADP methods presented in this chapter.

In the remaining part of this section we will present the main components of ADP methods in a comparative (side-by-side) treatise for discrete-time and continuous-time systems. Discussions that refer to discrete-time or continuous-time formulations are indicated by "DT" and "CT" at the beginning of the paragraphs. Sections 62.2 and 62.3 will outline several ADP methods while giving practical implementation ideas. In Section 62.2 of this chapter we will focus our attention on ADP for discrete-time systems, as the implementation tools for discrete-time state-feedback controllers are more at hand in control engineering practice. We will then shift the focus, in Section 62.3, on the ADP methods developed in a continuous-time framework. Further reading suggestions are included, after the bibliographical references, for those interested in this new and fast-growing research area.

62.1.3 System/Environment and Control/Behavior Policy

Most ADP algorithms were developed while considering affine in the inputs systems, which unfold their dynamics in discrete-time. Such systems are described by the set of state-space difference equations

$$x_{k+1} = f(x_k) + g(x_k)u_k \tag{62.1}$$

where k denotes the discrete-time index, $x_k \in \mathbb{R}^n$ and $u_k \in \mathbb{R}^m$, and $f(.)$, $g(.)$ are nonlinear functions.

For systems that have a continuous-time evolution of the state vector, the description is given in terms of a set of ordinary differential equations

$$\dot{x} = f(x) + g(x)u \tag{62.2}$$

where $f(.)$, $g(.)$ are nonlinear functions, $x \in \mathbb{R}^n$ and $u \in \mathbb{R}^m$.

The goal of the ADP methods will be one of calculating optimal state-feedback controllers that map the state space in the action space, $h : \mathbb{R}^n \to \mathbb{R}^m$, and have the form

$$u_k = h(x_k) \tag{62.3}$$

and respectively

$$u = h(x). \tag{62.4}$$

A state-feedback control function will be addressed as a control policy. Given the control policy in Equation 62.3, the solution of the differential equation 62.1, considering initial conditions x_0 at time $k = 0$, will be denoted by $\varphi_k^h = \varphi(k; x_0, h(.))$. Similarly, the solution of Equation 62.2 with control policy (Equation 62.4) will be $\varphi^h(t) = \varphi(t; x_0, h(.))$.

In order to have a well-posed problem, that is, a optimal control problem which admits a stabilizing control solution, we assume the system is stabilizable on some set $\Omega \in R^n$ that includes the equilibrium origin. Thus, we assume that there exists a control policy of the form (Equation 62.3 or 62.4), such that the closed-loop system $x_{k+1} = f(x_k) + g(x_k)h(x_k)$ or $\dot{x} = f(x) + g(x)h(x)$ for the continuous-time case, is asymptotically stable on Ω.

Generally, for a given system, a state-feedback control policy can be determined using various methods of design presented in this handbook. In line with the RL approach to control, where the control policy is learned in real time based on stimuli from the environment, the goal of the ADP methods of design presented in this chapter is to determine optimal state-feedback policies based on online measured performance information.

Model Free Feature of RL Algorithms Desired for ADP Methods

In RL the agent learns the control policy, based on reinforcement stimuli, and makes use of this mapping to determine and perform the control actions, relative to each location in the state space of the environment, such that the highest amount of reward is achieved over time. Everything outside the agent is considered to be part of the unknown environment. For this reason in standard RL algorithms an explicit model of the system is never required. This is an attractive feature of RL algorithms and is desired to be maintained by the ADP algorithms developed in the control systems community. By considering the system that has to be controlled as being part of the unknown environment, such ADP algorithms would be model-independent approaches to optimal control solutions.

In general the ADP methods will be developed considering a model of the system. This is desired for theoretical analysis in order to provide proofs of stability and performance. Nonetheless, as we shall see in the next sections, for the implementation of the ADP algorithms this requirement can be removed in some cases, the result being a model-free or partially model-free adaptive controller.

62.1.4 Control System Performance and Bellman's Optimality Principle

62.1.4.1 Cost/Value Function

The notion of optimal performance of a certain control system is captured by defining a performance measure of the sort used for optimal control purposes.

DT

In a discrete-time framework the *cost function* is defined as

$$V_h(x_k) = \sum_{i=k}^{\infty} \gamma^{i-k} r(x_i, u_i) \tag{62.5}$$

where $u_k = h(x_k)$ is a feedback control policy, $r(x_k, u_k)$ is a function which quantizes the one-step reinforcement received from the environment and γ, which satisfies $0 < \gamma \leq 1$, is a discount factor. This cost function is also known as the *cost-to-go* and is a sum of discounted future *reinforcement* values $r(x_k, u_k)$ from the current time k into the infinite horizon future. The discount factor reflects the fact that reinforcement values, further into the future, are less important than immediate reinforcements. For simplicity of the derivation we will consider in the following that the discount factor is $\gamma = 1$.

The reinforcement function $r(x_k, u_k)$ is also known as the *utility*, and is a measure of the one-step cost of the control policy. A standard form for the reinforcement function is the quadratic energy function $r(x_k, u_k) = x_k^T Q x_k + u_k^T R u_k$, or the more general form

$$r(x_k, u_k) = Q(x_k) + u_k^T R u_k. \tag{62.6}$$

In Equation 62.6 the matrix valued function $Q(x)$ is such that $\forall x \neq 0, Q(x) > 0$ and $x = 0 \Rightarrow Q(x) = 0$, and the matrix R is positive-definite such that the cost function (Equation 62.5) is well defined.

CT

In a continuous-time framework, the infinite horizon integral cost associated with the control policy $u = h(x)$ is

$$V^h(x(t)) = \int_t^{\infty} r(x(\tau), u(\tau))d\tau \tag{62.7}$$

where $x(\tau)$ denotes the solution of Equation 62.2 for initial condition $x(t)$ and input $\{u(\tau) = h(x(\tau)); \tau \geq t\}$. The instantaneous reinforcement function takes the form $r(x, u) = Q(x) + u^T Ru$, with conditions $Q(x) > 0, R > 0$ and $x = 0 \Rightarrow Q(x) = 0$.

Admissible control policies are those maps $h : \mathbb{R}^n \rightarrow \mathbb{R}^m$ that guarantee stability properties for the closed-loop system while resulting in finite value of the cost indices, $V_h(x_k)$, or respectively $V^h(x(t))$, which reflect the performance of the control system. We denote with $\Psi(\Omega)$ the set of all admissible control policies defined on a set $\Omega \subset \mathbb{R}^n$.

We have to note here that in the cases in which the reinforcement function is interpreted as a *reward* then the cost function is called *value function* and the objective is to maximize it. Though in this chapter we will only discuss minimization problems, the extension to maximization problems is straightforward. However, since the ADP methods originated in the computational intelligence community where maximization problems are generally solved, we choose to use the term of *value function* also in reference to the *cost function*, which is to be minimized. In the following the two terms will be used interchangeably when in fact referring to the cost function associated with the minimization problems, which are interesting features of this chapter.

62.1.4.2 Optimal Control Objective and Hamiltonian Functions

DT

The objective of optimal controller design is to determine the policy that minimizes the cost (Equation 62.5), to obtain

$$V^*(x_k) = \min_{h(.)}\left(\sum_{i=k}^{\infty} r(x_i, h(x_i))\right), \tag{62.8}$$

which is the optimal cost. The optimal control policy is then

$$h^*(x_k) = \arg\min_{h(.)}\left(\sum_{i=k}^{\infty} r(x_i, h(x_i))\right). \tag{62.9}$$

A difference equation equivalent to Equation 62.5 is given by

$$V_h(x_k) = r(x_k, h(x_k)) + V_h(x_{k+1}), \quad V_h(0) = 0. \tag{62.10}$$

This equation is referred to as *Bellman's equation* and is a consistency relation that provides the interconnection between the one-step reward function and the cost function associated with a given control policy at successive time steps. Bellman's equation can be solved for the cost function associated with any given admissible control policy, instead of evaluating the infinite sum given in Equation 62.5. Writing this equation for linear systems with quadratic cost functions one obtains a Lyapunov equation. *Evaluating the value of a current policy using this equation is a key concept in developing RL techniques.* We shall show how to solve the Bellman equation on-line in real time using observed data from the system trajectories.

The discrete-time Hamiltonian function can be defined as

$$H(x_k, h(x_k), \Delta V_k) = r(x_k, h(x_k)) + V_h(x_{k+1}) - V_h(x_k) \tag{62.11}$$

where $\Delta V_k = V_h(x_{k+1}) - V_h(x_k)$ is the forward difference operator. The Hamiltonian function captures the energy content along the trajectories of a system as reflected in the desired optimal performance. The Hamiltonian function must be equal to zero for the value associated with a given control policy. We shall see later how the discrete-time Hamiltonian, being related to the so called *temporal difference (TD) error*, provides means for online solution for the value function associated with a certain control policy.

CT

In the continuous-time case the expressions of the optimal cost and optimal control policy are similar to Equations 62.8 and 62.9 and follow directly when the sum operator is replaced by the integral. Similar to Equation 62.10, in the continuous-time case one obtains the infinitesimal version of Equation 62.7

$$0 = r(x, h(x)) + (\nabla V^h)^T(f(x) + g(x)h(x)), \quad V^h(0) = 0 \tag{62.12}$$

where ∇V^h (a column vector) denotes the gradient of the value function $V^h(x)$, as the value function does not depend explicitly on time. Equation 62.12 is a Lyapunov equation for nonlinear systems which, given the controller $h(x) \in \Psi(\Omega)$, can be solved for the value function $V^h(x)$ associated with it. Given that $h(x)$ is an admissible control policy, if $V^{\mu}(x)$ satisfies (Equation 62.12), with $r(x, h(x)) \geq 0$, then $V^{\mu}(x)$ is a Lyapunov function for the system (Equation 62.2) with control policy $h(x)$.

The Hamiltonian function is defined as

$$H(x, h, \nabla V^h) = r(x, h(x)) + (\nabla V^h)^T (f(x) + g(x)h(x)) \tag{62.13}$$

and must be equal to zero for the value associated with a given control policy. One can also show that, for a given admissible control policy $u = h(x)$, the equation $H(x, h, \nabla V^h) = 0$ has the same cost function solution as

$$\int_t^{t+T} r(x, h(x))\mathrm{d}\tau + V^h(x(t+T)) - V^h(x(t)) = 0, \quad V^h(0) = 0, \tag{62.14}$$

$\forall x(t) \in \Omega$ initial condition, and $\forall T > 0$, where $x(t+T) = \varphi^h(t+T) = \varphi(t+T; x(t), h(.))$.

62.1.4.3 Bellman's Optimality Principle

Bellman's optimality principle [6], a cornerstone of optimal control, is the central idea of the dynamic programming algorithm providing means for the calculation of sequences of optimal control actions (i.e., an optimal-action policy). The principle states that an optimal policy has the property that no matter what the previous control decisions have been, the remaining decisions determined based on that policy will be optimal relative to the state resulting from the previous decisions.

DT

Bellman's optimality principle means that

$$V^*(x_k) = \min_{h(.)} \left(r(x_k, h(x_k)) + V^*(x_{k+1}) \right). \tag{62.15}$$

This equation is known as the *Bellman optimality equation*, or the discrete-time Hamilton–Jacobi–Bellman (HJB) equation. The optimal policy is

$$h^*(x_k) = \arg \min_{h(.)} \left(r(x_k, h(x_k)) + V^*(x_{k+1}) \right). \tag{62.16}$$

Since one must know the optimal policy at time $k+1$ to use Equation 62.15 to determine the optimal policy at time k, Bellman's optimality principle yields a *backwards-in-time* procedure for solving optimal control problems (i.e., one has to start from the end-goal and work his way backwards through time to determine the optimal control sequence that would provide the smallest cost over the entire time interval). For this reason dynamic programming is by nature an off-line planning method that requires the full knowledge of the system dynamics.

CT

For the continuous-time case the optimal cost function $V^*(x)$ satisfies the continuous-time HJB equation

$$0 = \min_{u \in \Psi(\Omega)} [H(x, u, \nabla V^*)]. \tag{62.17}$$

Assuming that the minimum on the right-hand side of Equation 62.17 exists and is unique, and the reward function is given by $r(x, u) = Q(x) + u^T Ru$, then the infinite horizon optimal control solution for the given problem is

$$u^*(x) = -\frac{1}{2}R^{-1}g^T(x)\nabla V^*. \tag{62.18}$$

Inserting this optimal control policy in the Hamiltonian the continuous-time HJB equation takes the form

$$0 = Q(x) + (\nabla V^*)^T f(x) - \frac{1}{4}(\nabla V^*)^T g(x)R^{-1}g^T(x)\nabla V^*; \quad V^*(0) = 0. \tag{62.19}$$

In order to find the optimal control solution for the problem one only needs to solve the HJB equation 62.19 for the value function and then substitute the solution in Equation 62.18 to obtain the optimal control. Solving this nonlinear HJB equation is generally difficult and requires complete knowledge of the system dynamics.

We shall see in the following how Bellman's equation and optimality principle can be used in ADP methods to solve the optimal control problem, both in continuous-time and discrete-time frameworks, in an online fashion, based on data measured in real time from the system, in a *forward-in-time* procedure, and, at times, without requiring knowledge on the dynamics of the system to be controlled.

62.1.5 ADP and Adaptive Critics

62.1.5.1 ADP

ADP names the class of RL algorithms that provide in an iterative manner an approximate solution of optimal control problems. These methods consist in a series of iterations between the steps of *policy evaluation/value function update* and *policy update/improvement*. Successful completion of the policy evaluation step leads to starting the policy update step. We can say that a policy is an improved one in comparison with another one, if it has a smaller associated cost. The iterations end when a policy update step no longer leads to an improvement of the control policy. This means that the policy with the smallest associated cost, that is, optimal control policy, has been found.

62.1.5.2 Adaptive Critics

The structural representation of ADP methods is given by the actor–critic interconnection [5,9,15], presented in a general form in Figure 62.1. The actor structure plays the role of the controller, which computes the control signal, according to the present control policy, based on the measurements that define the present state of the environment/system. The critic structure has the purpose of evaluating the performance of the present action policy. Based on the critic's assessment on the value of the present policy, various adaptation schemes can be used to modify the control strategy such that the actor's new policy will yield a value, which is an improvement over the previous one for the entire state space of interest (i.e., the new value has to be smaller in the case when minimization of the cost is desired, or higher for the case of maximization purposes).

It is in order here to discuss the concept behind the ADP methods in contrast with the classical model-based adaptive optimal control mechanisms.

In the case of model based adaptive optimal controllers (or indirect adaptive optimal control)

- First an approximation structure, serving as a model of the system, is trained to approximate (i.e., learns using an identification procedure) the system dynamics.
- Then this model is used in combination with a controller design method to determine a control policy such that the desired optimal performances of the closed-loop system are satisfied.

These two steps of system identification and optimal controller design can be iterated such that in the case in which the system dynamics change, a new control policy can be determined such that the desired closed-loop optimal performances can be maintained. One can immediately see that the controller

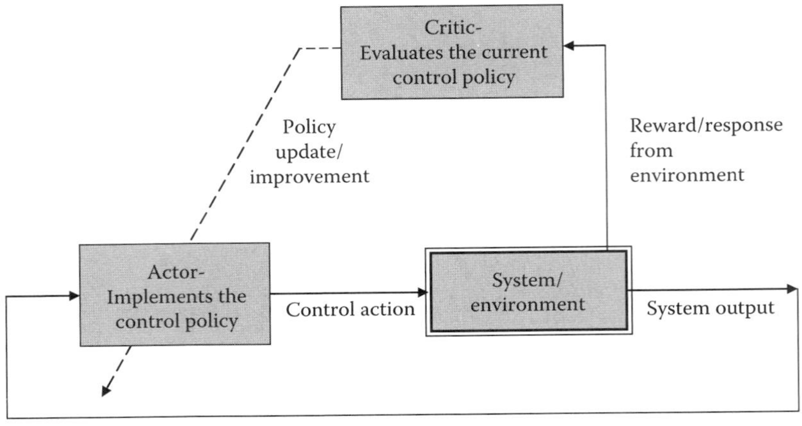

FIGURE 62.1 Actor–critic structure for the implementation of approximate dynamic programming algorithms.

calculated in this manner brings the closed-loop system close to optimality in direct relation with the differences between the approximate model of the system dynamics and the true dynamics of the plant. Simply put, the controller is just as "optimal" as the model is "accurate."

In the case of ADP algorithms the learning process is moved to a higher level, having no longer as object of interest the details of a system's dynamics but a performance index that quantifies how close to optimality the closed-loop control system operates. From this perspective ADP methods or RL in general, provide means of learning optimal behaviors (i.e., adaptation of the controller structure to provide an optimal control policy) by observing the response from the environment to nonoptimal control policies. ADP methods search for the admissible control policy which optimizes the measurable performance. In this case the resulting controller is just as "optimal" as the representation of the cost is "accurate."

62.1.6 Value Function and Q-Function

The value function associated with a given control policy has been defined and discussed in Section 62.1.4. This is a functional which maps the state space of the system into the space of real numbers.

Value functions

- Provide the means of comparison between control policies (e.g., in the case of a minimization problem, a policy is better than other if it has a smaller value function for all the states in the region of interest in the state space).
- Are required for the calculation of the optimal control policy.

When solving the minimization in Equation 62.16 one sees that in order to calculate the optimal control policy knowledge of the system dynamics is required for representation of x_{k+1}. For this reason such equation can not be used to solve the optimal control problem under a model-free condition. It was thus of interest to determine a different function that encapsulates the information of the value while at the same time can be minimized with respect to the control policy without maintaining the requirement of known system dynamics in a similar fashion to the regular RL methods.

Since the value function is part of the Hamiltonian function, which is to be minimized when solving for the optimal control policy, based on Equation 62.16 a so called *Q-function* was introduced [23], also known as an *action dependent value function* [24]. As we shall see, the defined Q-function maps the space of states and actions into the space of real numbers, and can be minimized with respect to the control policy such that it directly provides the optimal controller without necessity of the system dynamics. In effect the definition of the Q-function was the key for a completely model-free ADP algorithm, as promised by the RL methods, an algorithm formulated in the regular terms of a standard optimal control problem.

DT

Starting from the difference Equation 62.10 the Q-function is defined as

$$Q_h(x_k, u_k) = r(x_k, u_k) + V_h(x_{k+1}) \tag{62.20}$$

where $x_{k+1} = f(x_k) + g(x_k)u_k$ and $u_k \in \mathbb{R}^m$. The Q-function represents the value associated with using

- The control input u_k as one first-step in the control sequence followed by
- The control policy $u_k = h(x_k)$ for all the future steps $\forall i \geq k+1$.

The relation between the Q-function and the optimal value is given by

$$V^*(x_k) = \min_{u_k} Q^*(x_k, u_k) \tag{62.21}$$

where

$$Q^*(x_k, u_k) = r(x_k, u_k) + V^*(x_{k+1}). \tag{62.22}$$

One can also write Equation 62.20 only in terms of the Q-function as

$$Q_h(x_k, u_k) = r(x_k, u_k) + Q_h(x_{k+1}, h(x_{k+1})). \tag{62.23}$$

Equation 62.23, written in terms of the Q-function, is equivalent to Equation 62.10 written in terms of the value function. Thus, given a control policy $u_k = h(x_k)$, Equation 62.23 can be solved for the Q-function associated with it.

In terms of Q^* the optimal control is

$$\left(h^*(x_k) = \arg\min_{u_k} \left(Q^*(x_k, u_k) \right) \right). \tag{62.24}$$

In the absence of control constraints, one obtains the optimal value by solving

$$\frac{\partial}{\partial u} Q^*(x_k, u) = 0. \tag{62.25}$$

CT

The definition of a Q-function in a continuous-time framework is not straight-forward and will not be discussed now. Different definitions were introduced from the perspective of the two equivalent Equations 62.12 and 62.14 in [12,14].

62.1.7 Key Ingredients for Online Implementation of ADP Methods: Value Function Approximation and TD Error

The key aspects of the online implementation of the ADP algorithms are related to the way in which the value function is defined and the manner in which the critic structure learns to approximate the value function. This requires answers to the following questions:

Q1: Which value function does the critic approximate?
A1: The options are

- Value function
- Action dependent value function (i.e., Q-function)
- Gradient of the value function or gradient of the Q-function.

Q2: What is an appropriate parametric representation for the chosen value function?

A2: In [7,9,24,15] the critic function is represented as a neural network, which is known to have the universal approximation property. In this chapter we will discuss the case in which the critic can be represented as a linear combination of a set of basis functions, which spans the space of value functions to be approximated.

To motivate this approach, let us determine the value function approximator for the familiar linear quadratic regulation (LQR) problem in the discrete-time formulation. In LQR the value associated with any admissible control policy $u_k = -Kx_k$ is quadratic in the state (i.e., $V_K(x_k) = x_k^T P x_k$ with P a positive-definite matrix). Using the Kronecker product notation, the value function can be represented as

$$V_K(x_k) = x_k^T P x_k = (vec(P))^T (x_k \otimes x_k) \equiv \bar{p}^T \bar{x}_k \tag{62.26}$$

where $\otimes$ denotes the Kronecker product, $vec(P) \equiv \bar{p}$ is a column vector formed by stacking the columns of the matrix P, and $\bar{x}_k = x_k \otimes x_k$ is the quadratic polynomial vector containing all possible products of the n components of x_k. Noting that P is symmetric and has only $n\left(n+1\right)/2$ independent elements, one can remove the redundant terms in $x_k \otimes x_k$ to define a quadratic basis set $\bar{x}_k$ with $n\left(n+1\right)/2$ independent elements. The unknown parameter vector, which will adapt during learning, is $\bar{p}$ that contains the elements of matrix P.

Value Function Approximation

In the case in which the critic function takes a general nonlinear form, assuming that the value function is sufficiently smooth, then, according to the Weierstrass higher-order approximation theorem, there exists a dense basis set of functions $\{\varphi_i(x)\}$ such that the value of any admissible control policy can be represented as

$$V_h(x) = \sum_{i=1}^{\infty} w_i \varphi_i(x) = \sum_{i=1}^{L} w_i \varphi_i(x) + \sum_{i=L+1}^{\infty} w_i \varphi_i(x) \equiv W^T \phi(x) + \varepsilon_L(x) \tag{62.27}$$

with the basis vector $\phi(x) = \begin{bmatrix}\varphi_1(x) & \varphi_2(x) & \cdots & \varphi_L(x)\end{bmatrix} : R^n \rightarrow R^L$ such that $W^T\phi(x)$ is positive-definite over the set Ω, and where the error $\varepsilon_L(x)$ converges uniformly to zero as $L \rightarrow \infty$. It is standard usage to choose a polynomial basis set. Results have been shown for other basis sets including sigmoid, hyperbolic tangent, Gaussian radial basis functions, and so on. such that the approximation error $\varepsilon_L(x)$ is bounded by a constant on the compact set Ω. One can obtain intuition on choosing the basis set for the approximation of the value function knowing that they are positive-definite and they will play the role of Lyapunov functions for the closed-loop system.

62.1.7.1 Q-Function or Action-Dependent Value Function Approximation

Similar to the value function approximation case, general Q-functions for nonlinear systems can have a parametric representation in the form

$$Q_h(x, u) = W^T \phi(z) + \varepsilon_L(z) \tag{62.28}$$

where $z^T = [x^T \;\; u^T]$, $\phi(z)$ is a basis set of activation functions and W is the vector of parameters of the Q-function, such that $W^T\phi(x, u)$ is positive definite for all $x \in \Omega$ considering all admissible control policies $u \in \Psi(\Omega)$.

Q3: What is the error function which has to be minimized?

A3: The error function which is generally minimized during the Critic learning phase of an ADP algorithm is the *TD error* defined as

$$e_k = r(x_k, h(x_k)) + V_h(x_{k+1}) - V_h(x_k). \tag{62.29}$$

One sees that the right-hand side of this equation is the discrete-time Hamiltonian. Also, if Bellman's equation holds then the TD error is zero. Therefore, to find the value $V_h(.)$ of a fixed control policy $u_k = h(x_k)$ one can solve the equation $e_k = 0$, for all $x_k \in \Omega$.

Q4: Which identification method is employed for determining online the parameters of the approximate representation of the value function?

A4: To answer this question we consider the discrete-time and continuous-time cases separately.

DT

Considering the chosen parametric representation for the value function one can write the TD error equation in the form

$$e_k = r(x_k, h(x_k)) + W_h^T \phi(x_{k+1}) - W_h^T \phi(x_k). \tag{62.30}$$

a. Based on measurements of the states at successive moments in time x_k, x_{k+1} and of the rewards $r(x_k, h(x_k))$, taken from the system/environment while using the control policy $u_k = h(x_k)$, one can set up an online parameter identification procedure, of the sort described in this handbook, to solve for the parameters W_h of the cost function associated with the given control policy such that the TD error is minimized in the least-squares sense over the entire set Ω.
Solving on-line for the parameters of the value function is equivalent to obtaining on-line an approximate solution of a nonlinear Lyapunov equation. The solution is obtained in this case without using knowledge on the dynamics of the system, while only using data measured along the state trajectories produced by the specified admissible control policy.

b. Another approach to the solution starts with the observation that Equation 62.10 is a fixed point equation. The solution of this equation can then be reached by means of the contraction

$$V_h^j(x_k) = r(x_k, h(x_k)) + V_h^{j-1}(x_{k+1}), \tag{62.31}$$

starting with an initial guess $V_h^0(.)$. As $j \to \infty$ then $V_h^j(.) \to V_h(.)$ uniformly over Ω. Based on this, in terms of the parametric approximation of the value function, the approximate solution of $e_k = 0$ can be obtained by solving iteratively for the parameters W_j, for $j \geq 1$, starting with an initial guess W_0. At each iteration step j, the error

$$e_k(W_j) = r(x_k, h(x_k)) + W_{j-1}^T \phi(x_{k+1}) - W_j^T \phi(x_k) \tag{62.32}$$

has to be minimized in the least-squares sense while making use of online measured data.
The error $e_k(W_j)$ is called *TD prediction error* as it can be interpreted as an error between

- The new overall value represented as $W_j^T \varphi(x_k)$.
- The predicted performance corrected by observation $r(x_k, h(x_k)) + W_{j-1}^T \varphi(x_{k+1})$ obtained in response to an action applied to the system.

The TD error expressed in terms of the Q-function approximation is

$$e_k(W) = r(x_k, h(x_k)) + W^T \phi(z_{k+1}) - W^T \phi(z_k). \tag{62.33}$$

This equation can be solved online, in a similar fashion as described above, for the values of the parameters of the Q-function.

We now discuss a potential problem that appears when solving on-line for the parameters of the Q-function, and then we provide the solution. To make the issue clear, let us take again, as example, the familiar LQR case.

The reinforcement function has the form $r(x_k, u_k) = x_k^T Q_d x_k + u_k^T R u_k$. The Q-function associated with a control policy $u_k = h(x_k)$ is

$$Q_h(x_k, u_k) = x_k^T Q_d x_k + u_k^T R u_k + x_{k+1}^T P_h x_{k+1} \tag{62.34}$$

and inserting the system dynamics equation $x_{k+1} = Ax_k + Bu_k$ one obtains

$$Q_h(x_k, u_k) = x_k^T Q_d x_k + u_k^T R u_k + (Ax_k + Bu_k)^T P_h (Ax_k + Bu_k) \tag{62.35}$$

which can be arranged in the form

$$Q_h(x_k, u_k) = \begin{bmatrix} x_k \\ u_k \end{bmatrix}^T \begin{bmatrix} Q_d + A^T P_h A & A^T P_h B \\ B^T P_h A & R + B^T P_h B \end{bmatrix} \begin{bmatrix} x_k \\ u_k \end{bmatrix} \tag{62.36}$$

Then the Q-function will be written as

$$Q_h(x_k, u_k) = z_k^T H z_k = \begin{bmatrix} x_k \\ u_k \end{bmatrix}^T \begin{bmatrix} H_{xx} & H_{xu} \\ H_{ux} & H_{uu} \end{bmatrix} \begin{bmatrix} x_k \\ u_k \end{bmatrix} \tag{62.37}$$

where $H_{xx} = Q_d + A^T P_h A$, $H_{xu} = A^T P_h B$, $H_{ux} = B^T P_h A$ and $H_{uu} = R + B^T P_h B$.

In this case, when it comes to learning online the parameters of the Q-function based on measured data from the system, one encounters a problem: Due to the fact that the control policy signal $u_k = Kx_k$ is a linear combination of the values of the state, the persistence of excitation condition required for the regression vector $\varphi(z_k) - \varphi(z_{k+1})$ does no longer hold.

This problem is solved by adding persistently an exciting probing noise to the control inputs that were calculated based on the state-feedback control policy [8]. In this case the control inputs that are sent to the system during the online identification procedure of the Q-function are $u_k = Kx_k + n_k$. In [1] it is shown that this does not result in any bias in the Q-function estimates.

CT

The value function associated with the performance of a continuous-time state-feedback control policy can be similarly determined using a parameter identification procedure. The only difference consists in the fact that the one-step reinforcement signal $r(x_k, h(x_k))$, measured from the system in the discrete-time case, is now replaced by the measurement of the integral term $\int_t^{t+T} r(x(\tau), h(x(\tau)))d\tau$.

The TD error is defined in this case as

$$e(x(t), T, h) = \int_t^{t+T} r(x, h(x))d\tau + V^h(x(t+T)) - V^h(x(t)). \tag{62.38}$$

Q5: When is the value function identification procedure (i.e., policy evaluation step) finished, such that the policy update step can be started?

A5: The value function identification procedure ends when the convergence of the critic parameters has been obtained, in the sense that the predictions of the critic match closely, within some bounds that depend on the selected model, number of parameters to be trained and measurement errors, the measurements of the reinforcement received from the environment.

Q6: Is the online identification of the value function always possible?

A6: All ADP algorithms are developed on the assumption that correct estimation of the value function is possible. This means that, in order to successfully apply the online learning algorithm, enough excitation must be present in the system to guarantee correct estimation of the value function at each value update step. In relation to this, one must also note that if the policies obtained at the policy update steps are admissible policies then they may not provide the necessary excitation for the success of the value function learning phase. This is the well-known exploration/exploitation dilemma [17], which characterizes adaptive controllers that have simultaneous conflicting goals such as optimal control and fast and effective adaptation/learning. Solutions to this problem can be imagined (e.g., adding persistently exciting components to the signals computer by the controller).

62.1.8 Policy Update

DT

Given the value function associated with an admissible control policy, say $V_h(x_k)$ is the value associated with the policy $u_k = h(x_k)$, then the update of the control policy is given by the equation

$$\mu(x_k) = \arg \min_{\nu(.)\in\Psi(\Omega)} \left(r(x_k, \nu(x_k)) + V_h(x_{k+1})\right). \tag{62.39}$$

The new resulting control policy will be improved compared with the old one in the sense that it will have a smaller associated value.

For the case in which the Critic learns the Q-function, the policy update step is based on

$$\frac{\partial}{\partial u} Q(x, u) = 0. \tag{62.40}$$

Using the Q-function approximation, Equation 62.40 becomes

$$\frac{\partial}{\partial u} Q(x, u) = \frac{\partial}{\partial u} W^T \phi(x, u) = 0. \tag{62.41}$$

Since the Q-function depends explicitly on the control action u, the derivatives can be computed without reference to further details such as the system dynamics. Thus, in order to obtain an explicit updated policy $u_k = h(x_k)$ one requires application of the implicit function theorem to the Q-function approximation structure.

For the LQR case the Q-function is given by Equation 62.37 and the policy update step becomes

$$u_k = -\left(H_{uu}\right)^{-1} H_{ux} x_k. \tag{62.42}$$

From this equation one sees that in this case, since the kernel matrix H has been found using online learning, knowledge on the system dynamics is not needed for this policy improvement step.

CT

In the continuous-time case the policy update is performed according to the equation

$$\mu = \arg \min_{\nu\in\Psi(\Omega)} [H(x, \nu, \nabla V^h)]. \tag{62.43}$$

Considering the approximate representation of the value function $V^h(x) = W^T\phi(x)$ associated with an admissible control policy $u = h(x)$, and the particular form of the reinforcement $r(x, u) = Q(x) + u^T Ru$, one obtains explicitly the updated control policy as

$$\mu(x) = -R^{-1} g^T(x) \left(\frac{\partial \phi(x)}{\partial x}\right)^T W. \tag{62.44}$$

62.1.9 Features Which Provide Differentiation between ADP Algorithms

The differences between the ADP algorithms are related to

- The performance function, which has to be approximated through learning by the critic structure
  - *Value function:* The algorithms which use the value function as means of evaluation of the control performance are also referred to as V-learning. These are the default case of ADP.
  - *Action dependent value function (i.e., Q-function)*: In this case the algorithms are referred to as *Q-learning* or *action dependent-learning*. Due to the virtues of the Q-function such algorithms have the desired model-free property.
  - *Gradient of the value function:* In this case the algorithms are referred to as *Dual-learning*.
  - *Gradient of the Q-function:* These ADP are known as action dependent dual-learning.
- The conditions in which the control policy is updated
  - Thus, if the policy update is performed before the critic's parameters have converged to the values associated with the value of the present control policy then we obtain *value iteration* or *heuristic dynamic programming* (HDP) algorithms.
  - If the policy update is performed after convergence of the critic's parameters has been obtained such that the critic function represents the value associated with the current control policy we obtain *policy iteration* algorithms.

62.1.10 Terminology—Revisited

Reinforcement learning: Class of methods that provide solution, in an online fashion, to optimal control problems by means of a reinforcement scalar signal measured from the environment, which indicates the level of control performance.

Behavior policy: The sum of all rules that are followed by an active agent while interacting with its environment.

Control policy: A function which maps the state space of the system to be controlled onto the space of control actions that can be applied to the system.

Value function: A function which maps the state space of a system onto the set of real numbers such that it represents the amount of rewards obtained in response to performed control actions, while starting from a given initial state in the state space. Value functions provide an index of performance of a certain control policy. Control policies that have higher value functions are better than others. The term is used in relation to performance functions associated with maximization problems.

Cost function: A function which maps the state space of a system onto the set of real numbers, such that it reflects the amount of resources used for control purposes when starting from a specific state of the system. Control policies with smaller costs are better than others. As minimization problems are converse to maximization problems, with slight abuse of language cost functions are also referred to as value functions.

Q-function: A function which maps the space of states of a system and actions that can be performed, onto the space of real numbers such that it reflects the value associated with using an initial action (located in the action space) at a specific initial state of the system followed by optimal actions for all the subsequent control steps.

ADP: Class of algorithms which provide online solution to optimal control problems by using approximate representations of the value function to be minimized and of the control algorithm to be performed, and employing Bellman's optimality principle, central in dynamic programming, to provide means for training on-line the two approximation structures based on measured data from the system. Being mathematically formulated, such algorithms allow development of rigorous proofs of convergence for the approximation based approaches.

Actor–critic structure: The structural representation of ADP algorithms. It reflects the informational interconnection between

- The actor, which reacts in real-time to measurements from the system, and learns to adapt based on performance information from the critic.
- The critic, which learns to approximate a value function based on performance data and state data measured from the system, and provides performance information relative to the presently used control policy to the actor.

Adaptive critics: All algorithms which provide means for learning optimal control policies in an online fashion while using an actor–critic structure.

Bellman's optimality principle: The central idea of the dynamic programming algorithm, which states that an optimal policy has the property that no matter what the previous control decisions have been, the remaining decisions determined based on that policy will be optimal, relative to the state resulting from the previous decisions.

Bellman's equation: The mathematical expression of the relation between the one-step reward function and the infinite horizon cost function associated with a given control policy.

Value function approximation: The class of techniques used for online calculation of the parameters of an approximate expression of the value function associated with a given control policy.

TD error: The error difference between

- The expected infinite horizon cost predicted at the previous time step.
- The summation between the predicted future cost at the present time and the obtained reward over the time interval between the previous time step and the present time moment (i.e., prediction corrected by observation).

62.2 ADP Algorithms for Discrete-Time Systems

In this section are outlined the ADP algorithms developed considering the discrete-time formulation of the system to be controlled, of the control policy to be learned, and of the reinforcement signal, which provides means for critic learning. The connection between the ADP algorithms and optimal control is given at the end of this section, where we provide the equivalent algorithms for the well-known LQR problem in terms of iterations on discrete-time algebraic Riccati equations.

62.2.1 Policy Iteration

The policy iteration technique is applicable to a broad class of systems and it is guaranteed to converge to the optimal control with stepwise stability. The convergence guarantee is maintained also for the ADP online version of the policy iteration technique, which uses a value function approximator in the form of the critic structure.

62.2.1.1 V-Learning

62.2.1.1.1 V-Learning Policy Iteration Algorithm

Initialize: Select any admissible control policy $h_0(x_k)$ and let $j = 0$.

Policy Evaluation Step: Determine the value of the current policy using Bellman's equation

$$V_{h_j}(x_k) = r(x_k, h_j(x_k)) + V_{h_j}(x_{k+1}) \tag{62.45}$$

Policy Improvement Step: Determine an improved policy using

$$h_{j+1}(x_k) = \arg\min_{h(.)} \left(r(x_k, h(x_k)) + V_{h_j}(x_{k+1}) \right) \tag{62.46}$$

Stop Algorithm Step: If

$$\left\| h_{j+1}(x_k) - h_j(x_k) \right\| < \varepsilon_s \tag{62.47}$$

then STOP, else let $j = j + 1$ and go to the policy evaluation step.

As $j \to \infty$, this algorithm converges to the optimal value (Equation 62.8) and optimal control policy (Equation 62.9). The algorithm can be stopped when the norm of the difference between the two successive controllers is smaller than a specified error ε_s. This means that an approximate optimal control policy has been determined.

If the reinforcement is given in the special form $r(x_k, u_k) = Q(x_k) + u_k^T R u_k$ then the policy improvement step becomes explicitly

$$h_{j+1}(x_k) = -R^{-1} g^T(x_k) \nabla V_{h_j}(x_{k+1}) \tag{62.48}$$

where $\nabla V(x) = (\partial V(x)/\partial x)$, a column vector, is the gradient of the value function.

The online version of this policy iteration algorithm implemented using value function approximation on an actor–critic structure is now given.

62.2.1.1.2 Online V-Learning Policy Iteration Algorithm

Initialize: Select any admissible control policy $h_0(x_k)$ and let $j = 0$.
Policy Evaluation Step: Determine online the parameters of the critic W_j such that

$$W_j^T \left(\phi(x_k) - \phi(x_{k+1}) \right) = r(x_k, h_j(x_k)) \tag{62.49}$$

is solved in the least-squares sense.
Policy Improvement Step: Determine an improved policy using

$$h_{j+1}(x_k) = \arg\min_{h(.)} \left(r(x_k, h(x_k)) + W_j^T \phi(x_{k+1}) \right). \tag{62.50}$$

Stop Algorithm Step: If

$$\left\| W_j - W_{j-1} \right\| < \varepsilon_{Ws} \tag{62.51}$$

then STOP, else let $j = j + 1$ and go to the policy evaluation step.

The algorithm can be stopped when the norm of the difference between the two successive value of the critic parameters is smaller than a specified error ε_{Ws}.

Any online parameter identification algorithm can be used for the policy evaluation step such that the critic parameters can be determined, in the least-squares sense, based on online measured data of the system states x_k, x_{k+1} and the reward $r(x_k, h_j(x_k))$. The convergence of online parameter identification algorithm requires that the regression vector $\left(\phi(x_k) - \phi(x_{k+1}) \right)$ is persistently exciting.

If the reward function has the form $r(x_k, u_k) = Q(x_k) + u_k^T R u_k$ then the improved policy is given by

$$h_{j+1}(x_k) = -\frac{1}{2} R^{-1} g^T(x_k) \nabla \phi^T(x_{k+1}) W_j \tag{62.52}$$

where $\nabla \phi(x) = (\partial \phi(x)/\partial x) \in R^{L \times n}$ is the Jacobian of the vector of basis functions.

One sees that even if the policy evaluation step can be executed without using knowledge on the system dynamics, based only on measurements form the system, the policy improvement step of the algorithm *requires complete knowledge on the system dynamics* since they appear in the expression of x_{k+1}, that is, $x_{k+1} = f(x_k) + g(x_k)h(x_k)$.

62.2.1.2 Q-Learning

The policy iteration algorithm on the Q-function has been introduced in [8] to solve the LQR problem.

62.2.1.2.1 Q-Learning Policy Iteration Algorithm

Initialize: Select any admissible control policy $h_0(x_k)$ and let $j = 0$.

Policy Evaluation Step: Determine the value of the current policy using Bellman's equation

$$Q_{h_j}(z_k) = r(x_k, h_j(x_k)) + Q_{h_j}(z_{k+1}) \tag{62.53}$$

Policy Improvement Step: Determine an improved policy using

$$h_{j+1}(x_k) = \arg\min_{u(.)} \left(Q_{h_j}(x_k, u(x_k))\right) \tag{62.54}$$

Stop Algorithm Step: If

$$\left\| h_{j+1}(x_k) - h_j(x_k) \right\| < \varepsilon_s \tag{62.55}$$

then STOP, else let $j = j + 1$ and go to the policy evaluation step.

62.2.1.2.2 Online Q-Learning Policy Iteration Algorithm

Initialize: Select any admissible control policy $h_0(x_k)$ and let $j = 0$.

Policy Evaluation Step: Determine the parameters of the critic W_j, by solving online, in the least-squares sense, the equation

$$W_j^T \left(\phi(z_k) - \phi(z_{k+1})\right) = r(x_k, h_j(x_k)). \tag{62.56}$$

Policy Improvement Step: Determine an improved policy using

$$h_{j+1}(x_k) = \arg\min_{u(.)} \left(W_j^T \phi(x_k, u)\right) \tag{62.57}$$

Stop Algorithm Step: If

$$\left\| W_j - W_{j-1} \right\| < \varepsilon_{Ws} \tag{62.58}$$

then STOP, else let $j = j + 1$ and go to the policy evaluation step.

Any online parameter identification algorithm can be used for the policy evaluation step such that the critic parameters can be determined, in the least-squares sense, based on online measured data of the system states x_k, x_{k+1} and the reward $r(x_k, h_j(x_k))$.

In contrast to the online V-learning algorithm, for the online Q-learning algorithm the policy update step can be executed without using any knowledge on the system dynamics. This is by virtue of the formulation of the Q-function. One sees now that the policy iteration algorithm where the Critic structure learns the Q-function can be executed without using knowledge on the system dynamics. This means that it is a completely model-free method, which provides online, in a stepwise adaptive manner, with guaranteed stepwise stability, an approximate optimal controller.

62.2.2 Value Iteration

Here we present value iteration algorithms. The convergence of value iteration algorithms, while using linear systems with quadratic cost indices, has been proven in [11]. In [1] the convergence of these algorithms was shown while solving H-infinity problems. The convergence proof for nonlinear value iteration was given in [2].

62.2.2.1 V-Learning

62.2.2.1.1 Value Iteration Algorithm (HDP)

Initialize: Select any control policy $h_0(x_k)$, not necessarily admissible or stabilizing, and $V_0(x_k)$ such that $V_0(0) = 0$, and let $j = 0$.

Value Update Step: Update the value using

$$V_{j+1}(x_k) = r(x_k, h_j(x_k)) + V_j(x_{k+1}). \tag{62.59}$$

Policy Improvement Step: Determine an improved policy using

$$h_{j+1}(x_k) = \arg\min_{h(.)} \left(r(x_k, h(x_k)) + V_{j+1}(x_{k+1})\right) \tag{62.60}$$

Stop Algorithm Step: If

$$\left\| h_{j+1}(x_k) - h_j(x_k) \right\| < \varepsilon_s \tag{62.61}$$

then STOP, else let $j = j + 1$ and go to the value update step.

It is important to note that the value iteration algorithm does not require an initial stabilizing policy.

Writing together the two steps of the value iteration algorithm one obtains a recursion on Bellman's optimality equation. This shows that the convergence of the value iteration algorithm is based on the fact that Bellman's optimality equation is a fixed point equation. The interleaved steps of value update and policy improvement are the means of using the contraction map such that the fixed point solution of this map, that is, the optimal cost function, is obtained iteratively. For the case of the LQR problem the value iteration algorithm is equivalent with iterations on the discrete-time algebraic Riccati equation.

62.2.2.1.2 Online Value Iteration Algorithm

Initialize: Select W_0 and any control policy $h_0(x_k)$, not necessarily admissible or stabilizing, and let $j = 0$.

Value Update Step: Determine online the parameters of the critic W_{j+1} such that

$$W_{j+1}^T \phi(x_k) = r(x_k, h_j(x_k)) + W_j^T \phi(x_{k+1}). \tag{62.62}$$

is solved in the least-squares sense.

Policy Improvement Step: Determine an improved policy using

$$h_{j+1}(x_k) = \arg\min_{h(.)} \left(r(x_k, h(x_k)) + W_{j+1}^T \phi(x_{k+1})\right). \tag{62.63}$$

Stop Algorithm Step: If

$$\left\| W_{j+1} - W_j \right\| < \varepsilon_{Ws} \tag{62.64}$$

then STOP, else let $j = j + 1$ and go to the value update step.

If the reward function has the form $r(x_k, u_k) = Q(x_k) + u_k^T R u_k$ then the improved policy is given by

$$h_{j+1}(x_k) = -\frac{1}{2} R^{-1} g^T(x_k) \nabla\phi^T(x_{k+1}) W_{j+1} \tag{62.65}$$

where $\nabla\phi(x) = (\partial\phi(x)/\partial x) \in R^{L\times n}$ is the Jacobian of the vector of basis functions.

Note that, when solving the value update step, the regression vector is $\phi(x_k)$, which must be persistently exciting for convergence to the least-squares to be achieved when using a recursive method of identification.

62.2.2.2 Q-Learning

Value Iteration Algorithm (Action Dependent Heuristic Dynamic Programming):

In a similar manner as for the HDP method, it can be derived an action-dependent heuristic dynamic programming (ADHDP) method, in which the critic learns the Q-function, with the fixed point equation

$$Q_h(x_k, h(x_k)) = r(x_k, h(x_k)) + Q_h(x_{k+1}, h(x_{k+1})). \tag{62.66}$$

62.2.2.3 Dual Learning and Dual Action-Dependent Learning

In [24] it is shown how RL techniques can be used while the critic evaluates the costate function

$$\lambda_k = \frac{\partial V_h(x_k)}{\partial x_k}, \tag{62.67}$$

which is the gradient of the value function. This function carries more information about the cost than the value function alone. The ADP algorithms where the critic learns the costate function are called *dual heuristic programming* (DHP).

The fixed point equation which allows implementation of value iteration on the dual value function is

$$\frac{\partial}{\partial x_k} V_h(x_k) = \frac{\partial}{\partial x_k} r(x_k, h(x_k)) + \frac{\partial}{\partial x_k} V_h(x_{k+1}) \tag{62.68}$$

or, explicitly,

$$\lambda_k = \frac{\partial r(x_k, u_k)}{\partial x_k} + \left[\frac{\partial u_k}{\partial x_k}\right]^T \frac{\partial r(x_k, u_k)}{\partial u_k} + \left[\frac{\partial x_{k+1}}{\partial x_k} + \frac{\partial x_{k+1}}{\partial u_k}\frac{\partial u_k}{\partial x_k}\right]^T \lambda_{k+1}, \tag{62.69}$$

for a prescribed policy $u_k = h(x_k)$. A critic structure can be used to approximate λ_k and online learning can be implemented is a similar fashion as discussed before.

Since in Equation 62.69 $(\partial x_{k+1}/\partial x_k) = f(x_k)$ and $(\partial x_{k+1}/\partial u_k) = g(x_k)$, one sees that any ADP scheme based on this fixed point equation requires knowledge of the full plant dynamics. Moreover, it is clear that in this case the online implementation of the value function identification techniques is computationally intensive as the costate function is a vector.

Similarly, there have been formulated algorithms based on the gradient of the Q-function. The resulting ADDHP algorithm has the same features noted for DHP.

62.2.3 Policy Iterations and Value Iterations on the Algebraic Riccati Equation of the LQR Problem

We have seen how to implement V-learning and Q-learning on-line for nonlinear systems using value function approximators. In the case of Q-learning, no knowledge of the system dynamics is needed for online RL and convergence to the optimal control solution. In this subsection we derive the equivalent algorithms, which provide solution for the discrete-time LQR problem, in terms of standard ideas from systems theory. The algorithms presented next are underlying the ADP methods, which have just been discussed. The reader is, however, advised that these algorithms are not for implementation purposes, since they require knowledge of the full system dynamics.

62.2.3.1 Discrete-Time LQR Problem

The system dynamics are given by

$$x_{k+1} = Ax_k + Bu_k \tag{62.70}$$

where $x_k \in \mathbb{R}^n$, $u_k \in \mathbb{R}^m$, $A \in \mathbb{R}^{n\times n}$, and $B \in \mathbb{R}^{n\times m}$.

The objective is to calculate the optimal state-feedback control policy $u_k = K^* x_k$, which minimizes the infinite horizon quadratic cost functional

$$V(x_k) = \sum_{i=k}^{\infty} x_i^T Q_d x_i + u_i^T R u_i \tag{62.71}$$

where $Q_d \in \mathbb{R}^{n\times n}$, $Q_d \geq 0$, and $R \in \mathbb{R}^{m\times m}$, $R > 0$.

It is known that the optimal cost is quadratic in the state and is given by

$$V^*(x_k) = x_k^T P x_k \tag{62.72}$$

with $P \geq 0$.

62.2.3.2 V-Learning Policy Iteration

Algorithm

Initialize: Select any admissible control policy $u_k = h_0(x_k) = K_0 x_k$ and let $j = 0$.

Policy Evaluation Step: Determine the value of the current policy, that is, the value of the matrix P_{h_j}, using Bellman's equation

$$x_k^T P_{h_j} x_k = x_k^T Q_d x_k + u_k^T R u_k + x_{k+1}^T P_{h_j} x_{k+1}. \tag{62.73}$$

Equation 62.73 is explicitly

$$x_k^T P_{h_j} x_k = x_k^T Q_d x_k + u_k^T R u_k + x_k^T (A + BK_j)^T P_{h_j} (A + BK_j) x_k \tag{62.74}$$

or, simply the discrete-time Lyapunov equation

$$P_{h_j} = Q_d + K_j^T R K_j + (A + BK_j)^T P_{h_j} (A + BK_j). \tag{62.75}$$

Policy Improvement Step: Determine an improved policy using

$$K_{j+1} = \arg\min_K \left(Q_d + K^T R K + (A + BK)^T P_{h_j} (A + BK) \right) \tag{62.76}$$

which explicitly is

$$K_{j+1} = -(R + B^T P_{h_j} B)^{-1} B^T P_{h_j} A. \tag{62.77}$$

Stop Algorithm Step: If

$$\left\| K_{j+1} - K_j \right\| < \varepsilon_{Ks} \tag{62.78}$$

then STOP, else let $j = j + 1$ and go to the policy evaluation step.

62.2.3.3 V-Learning Value Iteration

Algorithm

Initialize: Select any control policy $u_k = h_0(x_k) = K_0 x_k$ and $P_0 \geq 0$, and let $j = 0$.

Value Update Step: Update the value, that is, determine the value of the matrix P_{j+1}, using

$$x_k^T P_{j+1} x_k = x_k^T Q_d x_k + u_k^T R u_k + x_{k+1}^T P_j x_{k+1}. \tag{62.79}$$

which is explicitly

$$P_{j+1} = Q_d + K_j^T R K_j + (A + BK_j)^T P_j (A + BK_j). \tag{62.80}$$

Policy Improvement Step: Determine an improved policy using

$$K_{j+1} = \arg\min_{K}\left(Q_d + K^T RK + (A+BK)^T P_{j+1}(A+BK)\right) \tag{62.81}$$

which explicitly is

$$K_{j+1} = -(R + B^T P_{j+1}B)^{-1}B^T P_{j+1}A. \tag{62.82}$$

Stop Algorithm Step: If

$$\left\|K_{j+1} - K_j\right\| < \varepsilon_{Ks} \tag{62.83}$$

then STOP, else let $j = j+1$ and go to the policy evaluation step.

The test (Equation 62.83) can be replaced by the test $\left\|P_{j+1} - P_j\right\| < \varepsilon_{Ps}$ which is in terms of the value function.

62.2.3.4 Q-Learning Policy Iteration

Algorithm

Initialize: Select any admissible control policy $u_k = h_0(x_k) = K_0 x_k$ and let $j = 0$.

Policy Evaluation Step: Determine the value of the current policy, that is, the value of the matrix H_{h_j}, using Bellman's equation

$$\begin{bmatrix} x_k \\ u_k \end{bmatrix}^T H_{h_j} \begin{bmatrix} x_k \\ u_k \end{bmatrix} = \begin{bmatrix} x_k \\ u_k \end{bmatrix}^T \begin{bmatrix} Q_d & 0 \\ 0 & R \end{bmatrix} \begin{bmatrix} x_k \\ u_k \end{bmatrix} + \begin{bmatrix} x_{k+1} \\ K_j x_{k+1} \end{bmatrix}^T H_{h_j} \begin{bmatrix} x_{k+1} \\ K_j x_{k+1} \end{bmatrix}. \tag{62.84}$$

or, the discrete-time Lyapunov equation

$$H_{h_j} = \begin{bmatrix} Q_d & 0 \\ 0 & R \end{bmatrix} + \begin{bmatrix} A & B \\ K_j A & K_j B \end{bmatrix}^T H_{h_j} \begin{bmatrix} A & B \\ K_j A & K_j B \end{bmatrix}. \tag{62.85}$$

Policy Improvement Step: Determine an improved policy using

$$K_{j+1} = \arg\min_{K}\left(\begin{bmatrix} x_k \\ Kx_k \end{bmatrix}^T H_{h_j} \begin{bmatrix} x_k \\ Kx_k \end{bmatrix}\right) \tag{62.86}$$

which explicitly is

$$K_{j+1} = -(H_{uuh_j})^{-1} H_{uxh_j}. \tag{62.87}$$

Stop Algorithm Step: If

$$\left\|K_{j+1} - K_j\right\| < \varepsilon_{Ks} \tag{62.88}$$

then STOP, else let $j = j+1$ and go to the policy evaluation step.

62.2.3.5 Q-Learning Value Iteration

Algorithm

Initialize: Select any control policy $u_k = h_0(x_k) = K_0 x_k$ and $H_0 \geq 0$, and let $j = 0$.

Value Update Step: Update the value, that is, determine the value of the matrix H_{j+1}, using

$$H_{j+1} = \begin{bmatrix} Q_d & 0 \\ 0 & R \end{bmatrix} + \begin{bmatrix} A & B \\ K_j A & K_j B \end{bmatrix}^T H_j \begin{bmatrix} A & B \\ K_j A & K_j B \end{bmatrix}. \tag{62.89}$$

Policy Improvement Step: Determine an improved policy using

$$K_{j+1} = -(H_{uu\,j+1})^{-1} H_{ux\,j+1}. \tag{62.90}$$

Stop Algorithm Step: If

$$\left\| K_{j+1} - K_j \right\| < \varepsilon_{Ks} \tag{62.91}$$

then STOP, else let $j = j + 1$ and go to the policy evaluation step.

The test (Equation 62.83) can be replaced by $\left\| H_{j+1} - H_j \right\| < \varepsilon_{Hs}$, which is in terms of the parameters of the Q-function.

62.3 ADP Algorithms for Continuous-Time Systems

Similar to the ADP methods presented above, which were developed considering discrete-time system dynamics, there have also been developed algorithms, which aim at solving the optimal control problem for systems with continuous-time dynamics. This is strongly motivated by the fact that the dynamics of a large class of human-engineered systems unfold in continuous-time and an approximate discrete-time formulation, obtained by discretization techniques, would only result in suboptimal control strategies for the considered systems. The formulation of continuous-time ADP methods starts from the fact that the infinite horizon cost associated with the use of a certain control policy $u = h(x)$ can be written as

$$V^h(x(t)) = \int_t^{t+T} r(x(\tau), h(x(\tau)))d\tau + V^h(x(t+T)), \tag{62.92}$$

for any time interval $T > 0$. This equation has the same value function solution as

$$0 = H(x, h, \nabla V^h). \tag{62.93}$$

It has been shown in [18] that Equation 62.92 is a fixed point equation for continuous-time systems, similar with Bellman's equation defined for the discrete-time case.

According to Bellman's optimality principle, the optimal cost satisfies

$$V^*(x(t)) = \min_{u(\tau);\, t \le \tau < t+T} \left(\int_t^{t+T} r(x(\tau), u(\tau))d\tau + V^*(x(t+T)) \right), \tag{62.94}$$

while the HJB equation is

$$0 = \min_{u \in \Psi(\Omega)} [H(x, u, \nabla V^*)]z. \tag{62.95}$$

The optimal control policy is then given by

$$h^*(x(t)) = \arg\min_{h(x)} \left(\int_t^{t+T} r(x(\tau), h(x(\tau)))d\tau + V^*(x(t+T)) \right). \tag{62.96}$$

In the continuous-time case the TD error over any interval $T > 0$ is defined as

$$e(x(t), T, h) = \int_t^{t+T} r(x, h(x))d\tau + V(x(t+T)) - V(x(t)). \tag{62.97}$$

It is now direct to formulate the policy iteration and value iteration algorithms for continuous-time systems given in [19,20]. In Section 62.3.3 we provide the connection between these algorithms and classical results from optimal control theory by presenting their underlying formulation while solving the well-known continuous-time LQR problem.

62.3.1 Policy Iteration

Algorithm

Initialize: Select any admissible control policy $h^{(0)}(x)$ and let $j = 0$.
Policy Evaluation Step: Solve for $V^{h^{(j)}}(x(t))$ using

$$V^{h^{(j)}}(x(t)) = \int_t^{t+T} r(x(\tau), h^{(j)}(x(\tau)))d\tau + V^{h^{(j)}}(x(t+T)) \quad \text{with } V^{h^{(j)}}(0) = 0 \tag{62.98}$$

Policy Improvement Step: Determine an improved policy using

$$h^{(j+1)} = \arg\min_u [H(x, u, \nabla V^{h^{(j)}})] \tag{62.99}$$

Stop Algorithm Step: If

$$\left\| h^{(j+1)}(x) - h^{(j)}(x) \right\| < \varepsilon_s \tag{62.100}$$

then STOP, else let $j = j + 1$ and go to the policy evaluation step.

The algorithm converges to the optimal controller as $j \to \infty$.

For the special case in which the instantaneous reward function is given by $r(x, u) = Q(x) + u^T Ru$ and the system is Equation 62.2 then the improved state-feedback control policy is explicitly written as

$$h^{(j+1)}(x) = -\frac{1}{2} R^{-1} g^T(x) \nabla V^{h^{(j)}}. \tag{62.101}$$

Online Policy Iteration Algorithm

Initialize: Select any admissible control policy $h^{(0)}(x)$ and let $j = 0$.
Policy Evaluation Step: Solve online for the parameters of the critic W_j such that

$$W_j^T [\phi(x(t)) - \phi(x(t+T))] = \int_t^{t+T} r(x(\tau), h^{(j)}(x(\tau)))d\tau \quad \text{where } \phi(0) = 0 \tag{62.102}$$

is solved in the least-squares sense.
Policy Improvement Step: Determine an improved policy using

$$h^{(j+1)}(x) = \arg\min_u \left[H\left(x, u, \left(\frac{\partial \phi(x)}{\partial x} \right)^T W_j \right) \right] \tag{62.103}$$

Stop Algorithm Step: If

$$\| W_j - W_{j-1} \| < \varepsilon_{Ws} \tag{62.104}$$

then STOP, else let $j = j + 1$ and go to the policy evaluation step.

If $r(x, u) = Q(x) + u^T Ru$ then the policy improvement step becomes

$$h^{(i+1)}(x) = -\frac{1}{2} R^{-1} g^T(x) \left(\frac{\partial \phi(x)}{\partial x} \right)^T W_j. \tag{62.105}$$

Policy iteration becomes Newton's method [10] for solving the continuous-time algebraic Riccati equation associated with finding the solution of the LQR problem. One can now see that, using the RL approach, Newton's algorithm can be implemented using only information about the input to state dynamics of the system. That is, continuous-time policy iteration on the value function solves the continuous time algebraic Riccati equation without knowing the system internal dynamics by using data measured online alongthe system's state trajectories.

62.3.2 Value Iteration

Algorithm

Initialize: Select any control policy $h^{(0)}(x)$, not necessarily stabilizing, $V^0(x)$ such that $V^0(0) = 0$, and let $j = 0$.

Value Update Step: Solve for V^{j+1} using

$$V^{j+1}(x(t)) = \int_t^{t+T} r(x(s), h^{(i)}(x(s)))ds + V^j(x(t+T)) \text{ with } V^{j+1}(0) = 0 \tag{62.106}$$

Policy Improvement Step: Determine an improved policy using

$$h^{(j+1)} = \arg\min_u [H(x, u, \nabla V^{j+1})] \tag{62.107}$$

Stop Algorithm Step: If

$$\left\| h^{(j+1)}(x) - h^{(j)}(x) \right\| < \varepsilon_s \tag{62.108}$$

then STOP, else let $j = j + 1$ and go to the value update step.

In the case of $r(x, u) = Q(x) + u^T Ru$ the policy improvement step is explicitly

$$h^{(i+1)}(x) = -\frac{1}{2} R^{-1} g^T(x) \nabla V^{j+1}. \tag{62.109}$$

Online Value Iteration Algorithm

Initialize: Select any control policy $h^{(0)}(x)$, not necessarily stabilizing, any vector of parameters W_0, and let $j = 0$.

Value Update Step: Solve on-line for the parameters of the critic, W_{j+1}, such that

$$W_{j+1}^T \phi(x(t)) = \int_t^{t+T} r(x(s), h^{(j)}(x(s)))ds + W_j^T \phi(x(t+T)) \text{ with } \phi(0) = 0 \tag{62.110}$$

is satisfied in the least-squares sense.

Policy Improvement Step: Determine an improved policy using

$$h^{(j+1)} = \arg\min_u [H(x, u, \left(\frac{\partial \phi(x)}{\partial x}\right)^T W_{j+1})] \tag{62.111}$$

Stop Algorithm Step: If

$$\| W_{j+1} - W_j \| < \varepsilon_{Ws} \tag{62.112}$$

then STOP, else let $j = j + 1$ and go to the value update step.

It is observed that both continuous-time versions of the policy iteration and the value iteration do not require knowledge about the internal dynamics of the system, that is, the function $f(x)$. Both these algorithms can be implemented online using online identification techniques for the parameters of the control policy (i.e., actor) and value function (i.e., critic) structures. The actor–critic structure for the implementation of these algorithms is discussed in the next section.

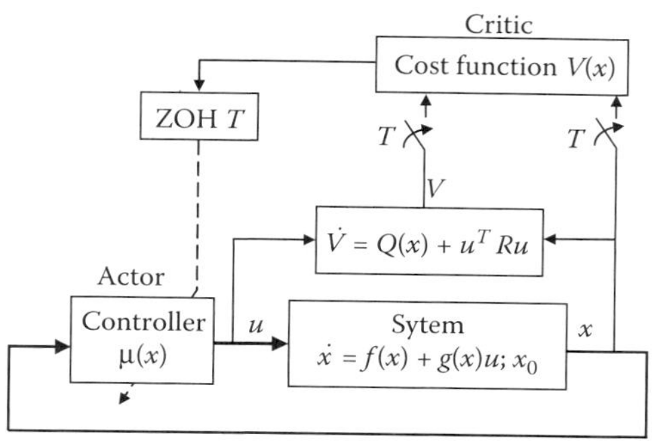

FIGURE 62.2 Structure of the system with adaptive controller for the implementation of continuous-time ADP methods.

62.3.3 Actor–Critic Structure for the Continuous-Time ADP Algorithms

We conclude the presentation of ADP algorithms for continuous-time systems with a discussion of the actor–critic structure used in this case. The structure of the system with optimal adaptive controller is presented in Figure 62.2.

The ADP algorithms require only measurements of the states at discrete moments in time and knowledge of the observed value function over the interval $[t, t+T]$. We denote with

$$d(t, T, u(.)) = \int_t^{t+T} r(x(\tau), u(x(\tau)))d\tau \qquad (62.113)$$

the reinforcement signal measured over the time interval T. Thus, the data measured at each time increment, which is required for the critic structure to learn the value function, is formed by the sets of triplets $\left(x(t), x(t+T), d(t, T, u(.))\right)$. Since the fixed point equations involved in the formulation of the continuous-time ADP algorithms are valid for any positive value of the time interval T, the reinforcement time interval T need not have the same value at every step of the iteration. The value of T can be adapted online in relation to the amount of time required to measure meaningful information from the system.

In order to extract reward information regarding the cost associated with the given policy the system was augmented with an additional state $V(t)$, with dynamics $\dot{V} = Q(x) + u^T Ru$. The update of both the actor and the critic is performed at discrete moments in time based on sampled information, while the actor performs continuous-time state-feedback control. Since the critic learns the value function based on observations of the continuous-time value over a finite interval, the algorithm converges to the solution of the continuous-time optimal control problem.

62.3.4 Policy Iterations and Value Iterations and the LQR Problem

We present next, in terms of standard ideas from optimal control theory, the iterative algorithms equivalent with the online ADP methods, which provide solution for the continuous-time LQR problem. We must again specify that these algorithms, which underlie the iterative ADP methods, are not for implementation purposes, since they require knowledge of the full system dynamics.

62.3.4.1 Continuous-Time LQR Problem

The system dynamics are given by

$$\dot{x} = Ax + Bu \qquad (62.114)$$

where $x \in \mathbb{R}^n$, $u \in \mathbb{R}^m$, $A \in \mathbb{R}^{n\times n}$, and $B \in \mathbb{R}^{n\times m}$.

It is desired to determine the optimal state-feedback control policy $u = K^*x$, which minimizes the infinite horizon quadratic cost functional

$$V(x(t), u(.)) = \int_t^\infty (x^T Qx + u^T Ru)d\tau \quad (62.115)$$

where $Q \in \mathbb{R}^{n\times n}$, $Q \geq 0$, and $R \in \mathbb{R}^{m\times m}$, $R > 0$.

It is known that the optimal cost is quadratic in the state and is given by

$$V^*(x(t)) = x(t)^T P^* x(t) \quad (62.116)$$

with $P^* \geq 0$.

62.3.4.2 Policy Iteration

Algorithm

Initialize: Select any admissible control policy $h^{(0)}(x) = K^0 x$ and let $j = 0$.

Policy Evaluation Step: Solve for $V^{h^{(j)}}(x(t))$ using

$$V^{h^{(j)}}(x(t)) = \int_t^{t+T} (x^T Qx + h^{(j)T}(x)Rh^{(j)}(x))d\tau + V^{h^{(j)}}(x(t+T)) \text{ with } V^{h^{(j)}}(0) = 0 \quad (62.117)$$

which can be written as

$$\int_t^{t+T} \frac{d(x^T P^{h^{(j)}} x)}{d\tau} d\tau = -\int_t^{t+T} [x^T(Q + K^{jT} RK^j)x]d\tau \quad (62.118)$$

Taking the derivative and using the system dynamics in (62.118) it becomes the Lyapunov equation

$$(A + BK^j)^T P^{h^{(j)}} + P^{h^{(j)}}(A + BK^j) = -(Q + K^{jT} RK^j)$$

which can be solved for $P^{h^{(j)}}$.

Policy Improvement Step: Determine an improved policy using

$$h^{(j+1)}(x) = -R^{-1} B^T P^{h^{(j)}} x \quad (62.119)$$

Stop Algorithm Step: If

$$\left\| h^{(j+1)}(x) - h^{(j)}(x) \right\| < \varepsilon_s \quad (62.120)$$

then STOP, else let $j = j + 1$ and go to the policy evaluation step.

Initialize: Select any control policy $h^{(0)}(x) = K^0 x$ and $P_0 \geq 0$, and let $j = 0$.

Value Update Step: Update the value, that is, determine the value of the matrix P_{j+1}, using

$$x^T(t)P_{j+1}x(t) = \int_t^{t+T} [x^T(Q + K^{jT} RK^j)x]d\tau + x^T(t+T)P_j x(t+T). \quad (62.121)$$

Differentiating Equation 62.121 with respect to time along the trajectories given by the controller $h^{(j)}(x) = K^j x$ one obtains the Lyapunov equation

$$(A + BK^j)^T P_{j+1} + P_{j+1}(A + BK^j) + K^{jT} RK^j + Q = (e^{(A+BK_j)T})^T((A + BK^j)^T P_j + P_j(A + BK^j) + K^{jT} RK^j + Q)e^{(A+BK_j)T}. \quad (62.122)$$

Adding and subtracting $A_i^T P_i + P_i A_i$ and using the notation

$$Ric(P_i) = A^T P_i + P_i A + Q - P_i BR^{-1} B^T P_i \quad (62.123)$$

Equation 62.122 becomes

$$A_i^T(P_{i+1} - P_i) + (P_{i+1} - P_i)A_i = -Ric(P_i) + e^{A_i TT} Ric(P_i)e^{A_i T}. \quad (62.124)$$

Policy Improvement Step: Determine an improved policy using

$$h^{(j+1)}(x) = K^{j+1}x = -R^{-1}B^T P_{j+1}x. \tag{62.125}$$

Stop Algorithm Step: If

$$\left\| K^{j+1} - K^j \right\| < \varepsilon_{Ks} \tag{62.126}$$

then STOP, else let $j = j + 1$ and go to the value update step.

The test (Equation 62.126) can be replaced by the test $\left\| P_{j+1} - P_j \right\| < \varepsilon_{Ps}$, which is in terms of the value function.

62.4 ADP-Based Optimal Controller Design: Examples

We now present optimal controller design results that were obtained considering a power system with a linear model [3]. For the same system we will use both a discrete-time and a continuous-time approach to ADP to obtain the state-feedback controller, which optimizes a quadratic cost index. This is the familiar LQR problem.

We will first present the results obtained using the continuous-time HDP approach. We then show that the same results will be obtained using the discrete-time ADHDP algorithm, when considering a very small sampling period.

62.4.1 Power System and Cost Function

Even though power systems are characterized by nonlinearities, linear state-feedback control is regularly employed for load-frequency control at a certain nominal operating points that are characterized by small variations of the system load around a constant value. Although this assumption seems to have simplified the design problem of a load-frequency controller, a new problem appears from the fact that the parameters of the actual plant are not precisely known. For this reason it is particularly advantageous to apply model free ADP methods to obtain the optimal LQR controller for a given operating point of the power system.

The continuous-time linear model of the system that is considered here, [21], is characterized by the realistic values

$$A = \begin{bmatrix} -0.665 & 8 & 0 & 0 \\ 0 & -3.663 & 3.663 & 0 \\ -6.86 & 0 & -13.736 & -13.736 \\ 6 & 0 & 0 & 0 \end{bmatrix} \tag{62.127}$$
$$B^T = \begin{bmatrix} 0 & 0 & 13.736 & 0 \end{bmatrix}$$

One can obtain the discrete version of this model by discretization using a zero-order hold method with the sample time T=0.01s.

The infinite horizon cost function is defined for the continuous-time case as

$$V = \int_t^{\infty} [x^T(\tau)Qx(\tau) + u^T(\tau)Ru(\tau)]\mathrm{d}\tau \tag{62.128}$$

where the matrices Q and R were chosen to be identity matrices of appropriate dimensions. For the case of the discrete-time ADP algorithm the cost function is

$$V = \sum_{i=k}^{\infty} (x_i^T Q_d x_i + u_i^T R_d u_i) \tag{62.129}$$

where the parameters, given by the matrices Q_d and R_d, were chosen as $Q * T$ and $R * T$, where T denotes the sampling period. One sees that for the closed-loop system with state-feedback controller in the form $u = -Kx$ the two cost functions can be expressed as $V(x(t)) = x^T(t)Px(t)$ and $V(x_k) = x_k^T Px_k$, respectively.

62.4.2 Continuous-Time Value Iteration

Online Value Iteration Algorithm

Initialize: A restriction on the initial matrix P_0 such that the corresponding $K_0 = -R^{-1}B^T P_0$ be a stabilizing controller is not required. Thus, we choose to start the ADP algorithm considering the case in which the system operates without controller (i.e., $P_0 = 0_{n\times n}$ and $h^{(0)}(x) = K_0 x = -R^{-1}B^T P_0 x$). Let $j = 0$.

Value Update Step: Solve for the parameters given by the matrix P_{j+1} using

$$x^T(t)P_{j+1}x(t) = \int_t^{t+T} (x^T Qx + (h^{(j)}(x))^T Rh^{(j)}(x))\mathrm{d}\tau + x^T(t+T)P_j x(t+T). \quad (62.130)$$

The value function update amounts to the update of the kernel matrix P_j.

Policy Update Step: Determine an improved policy using

$$h^{(j+1)}(x) = -R^{-1}B^T P_{j+1}x = K_{j+1}x \quad (62.131)$$

Stop Algorithm Step: If

$$\left\| P_{j+1} - P_j \right\| < \varepsilon_{Ps} \quad (62.132)$$

then STOP, else let $j = j+1$ and go to the value update step.

To implement this iteration scheme, one only needs to know the value of the matrix, which is required explicitly in the policy update step. The information on is not required at any step of the algorithm. The states, $x(t)$ and $x(t+T)$, and the reward over the time interval $[t, t+T]$, $\mathrm{d}(t, T, h(.)) = \int_t^{t+T} r(x(\tau), h(x(\tau)))\mathrm{d}\tau$, are observed on-line. Representing the value function as $V(x) = x^T Px = (vec(P))^T(x \otimes x) \equiv \bar{p}^T\bar{x}$ then the value update step can be written as

$$\bar{p}_{j+1}^T\bar{x}(t) = \mathrm{d}(t, T, h^{(j)}(.)) + \bar{p}_j^T\bar{x}(t+T) \quad (62.133)$$

Using this equation one can now solve for the parameters $\bar{p}_{j+1}$, in the least-squares sense, after collecting online enough state-trajectory points defined by the triplets $\left(x(t), x(t+T), d(t, T, h^{(j)}(.))\right)$ to set up a solvable problem. The target will be the right-hand side of Equation 62.133 and the least-squares solution will provide the parameters of the matrix P_{j+1}.

Since the symmetric matrix P_j has a number of 10 different parameters, each iteration step requires collecting at least 10 state-trajectory points in order to set up a solvable least-squares problem. We will choose to solve for the parameters of P_j using a batch least-squares method considering 15 data points collected along the state trajectory of the system. This is because we intend to compare the results obtained with the continuous-time HDP algorithm with the ones resulting from the application of the discrete-time Q-learning algorithm (i.e., ADHDP strategy) and the Q-function in this case has a number of 15 parameters (which require the measurement of 15 state points in order to solve a least-squares problem).

For the implementation of the continuous-time value iteration algorithm the state measurements were taken at each 0.1 s time period, such that a cost function update was performed at each 1.5 s. Over 60 s a number of 40 iterations of value function and control policy updates were performed. The convergence of some of the Critic parameters, namely $P(1,1)$, $P(1,3)$, $P(2,4)$, and $P(4,4)$, is presented in Figure 62.3.

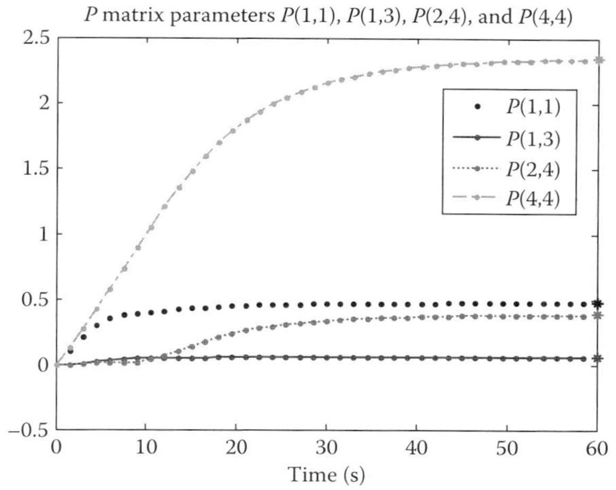

FIGURE 62.3 Convergence of the P matrix parameters obtained online using the continuous-time HDP algorithm.

The optimal control solution obtained after running the algorithm (without using knowledge regarding the system matrix A) is

$$P_{CT-HDP} = \begin{bmatrix} 0.4753 & 0.4771 & 0.0602 & 0.4770 \\ 0.4771 & 0.7838 & 0.1238 & 0.3852 \\ 0.0602 & 0.1238 & 0.0513 & 0.0302 \\ 0.4770 & 0.3852 & 0.0302 & 2.3462 \end{bmatrix} \quad (62.134)$$

while the exact solution of the continuous-time algebraic Riccati equation is

$$P_{CARE} = \begin{bmatrix} 0.4750 & 0.4766 & 0.0601 & 0.4751 \\ 0.4766 & 0.7831 & 0.1237 & 0.3829 \\ 0.0601 & 0.1237 & 0.0513 & 0.0298 \\ 0.4751 & 0.3829 & 0.0298 & 2.3370 \end{bmatrix}. \quad (62.135)$$

Comparing the values of the two matrices given by Equations 62.134 and 62.135 one sees that the solution obtained online using the ADP algorithm and without using any knowledge on the internal dynamics of the system specified by matrix A is very close to the exact solution obtained by solving the algebraic Riccati equation associated with the infinite horizon optimal control problem.

62.4.3 Discrete-Time Action-Dependent Value Iteration

Discrete-Time Action-Dependent Value Function–Q-Function

The Q-function for the LQR case is defined as

$$Q^*(x_k, u_k) = x_k^T Q_d x_k + u_k^T R_d u_k + x_{k+1}^T P^* x_{k+1} \quad (62.136)$$

where $x_{k+1} = Ax_k + Bu_k$, and P^* denotes the parameters of the optimal value function.

With the notation $z_k^T = [x_k^T \; u_k^T]$ the Q-function will have the form

$$Q^*(x_k, u_k) = z_k^T H z_k = \begin{bmatrix} x_k \\ u_k \end{bmatrix}^T \begin{bmatrix} H_{xx} & H_{xu} \\ H_{ux} & H_{uu} \end{bmatrix} \begin{bmatrix} x_k \\ u_k \end{bmatrix} \quad (62.137)$$

where $H_{xx} = Q_d + A^T P^* A$, $H_{xu} = A^T P^* B$, $H_{ux} = B^T P^* A$, and $H_{uu} = R_d + B^T P^* B$.

The optimal Q-function is equal with the optimal value function when u_k is the optimal policy

$$V^*(x_k) = \min_{u_k} Q^*(x_k, u_k) = \min_{u_k} \begin{bmatrix} x_k^T & u_k^T \end{bmatrix} H \begin{bmatrix} x_k^T & u_k^T \end{bmatrix}^T. \tag{62.138}$$

Given the optimal Q-function then the optimal control policy is obtained from

$$\frac{\partial Q^*(x_k, u_k)}{\partial u_k} = 0, \tag{62.139}$$

which implies that $2H_{ux}x_k + 2H_{uu}u_k = 0$, thus

$$u_k^* = H_{uu}^{-1} H_{ux} x_k. \tag{62.140}$$

The second-order sufficiency condition for the minimization is

$$\frac{\partial^2 Q^*(x_k, u_k)}{\partial u_k^2} > 0, \quad \text{that is } H_{uu} > 0, \tag{62.141}$$

and it is satisfied since $H_{uu} = R_d + B^T P^* B$, with $R_d > 0$ and $P^* > 0$.

In terms of the optimal Q-function one has the following recurrence relation

$$Q^*(x_k, u_k) = x_k^T Q_d x_k + u_k^T R_d u_k + \min_{u_{k+1}} Q^*(x_{k+1}, u_{k+1}), \tag{62.142}$$

which is a fixed-point equation. Based on this equation, and using the representation of the optimal Q-function (Equation 62.137), the online Q-learning algorithm is now formulated.

Online Q-Learning Algorithm

Initialize: A restriction on the initial matrix H_0 such that the corresponding $h_0(x_k) = H_{uu0}^{-1} H_{ux0} x_k = K_0 x_k$ is a stabilizing controller is not required. Thus, we choose to start the ADP algorithm considering the case in which the system operates without controller (i.e., $H_0 = 0_{(n+m)\times(n+m)}$ and $h_0(x_k) = H_{uu0}^{-1} H_{ux0} x_k$). Let $j = 0$.

Value Update Step: Determine online, the parameters of the Q-function, given by the matrix H_{j+1}, such that the equation

$$z_k^T H_{j+1} z_k = r(x_k, u_k) + z_{k+1}^T H_j z_{k+1}, \tag{62.143}$$

where $u_k = h_j(x_k) = K_j x_k$ and $u_{k+1} = h_j(x_{k+1}) = K_j x_{k+1}$, is solved in the least-squares sense. The value function update amounts to the update of the kernel matrix H_j.

Policy Update Step: Determine an improved policy using

$$h_{j+1}(x_k) = H_{uuj+1}^{-1} H_{uxj+1} x_k = K_{j+1} x_k. \tag{62.144}$$

Stop Algorithm Step: If

$$\left(\left\| H_{j+1} - H_j \right\| < \varepsilon_{Hs} \right) \tag{62.145}$$

then STOP, else let $j = j + 1$ and go to the value update step.

We now present details related to the online implementation of the value update step.

In a similar representation with the one given for the value function in Equation 62.26, the Q-function can be represented as

$$Q(z_k) = z_k^T H z_k = \bar{H}^T \bar{z}_k \tag{62.146}$$

where $\bar{H} = vec(H)$ and $\bar{z}_k = z_k \otimes z_k$.

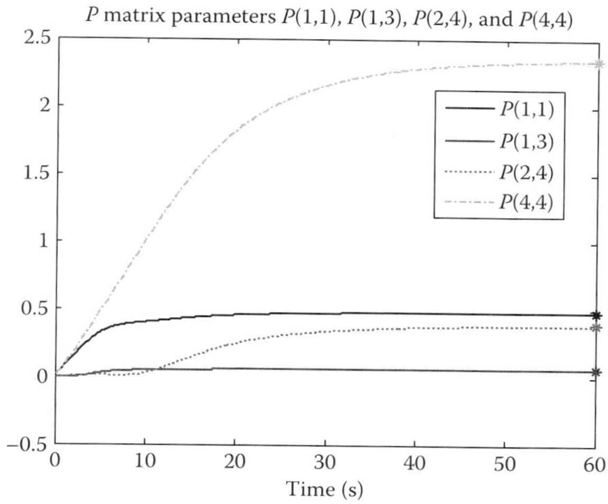

FIGURE 62.4 Convergence of the P matrix parameters for online discrete-time ADHDP algorithm.

Then the value update step becomes

$$\bar{H}_{j+1}^T \bar{z}_k = r(x_k, u_k) + \bar{H}_j^T \bar{z}_{k+1}. \tag{62.147}$$

The vector of parameters of the Q-function $\bar{H}_{j+1}$ is found by minimizing the error

$$e_k = r(x_k, u_k) + \bar{H}_j^T \bar{z}_{k+1} - \bar{H}_{j+1}^T \bar{z}_k \tag{62.148}$$

in least-squares sense over the compact set Ω.

The relation (Equation 62.144) shows that the control signal u_k is a linear combination of the state vector. Thus, not all the elements of the vector z_k are linearly independent. This fact causes the problem of insolvability of Equation 62.144.

This problem is removed by adding exploration noise to the control signal. Thus, let $u_{ke} = u_k + n_k$ where $n_k(0, \sigma)$is exploration noise with zero mean and variance σ. With this modification the vector z_k becomes

$$z_{ke} = [x_k^T \; u_{ke}^T]^T = [x_k^T \; (u_k + n_k)^T]^T, \tag{62.149}$$

and now obtaining the online solution of Equation 62.147 is guaranteed by the excitation condition.

In Figure 62.4 is presented the convergence of the critic parameters when the discrete-time ADHDP algorithm was used. The duration of the simulation was 60 s, during which a number of 400 cost function (Q-function) updates took place. Each update required 15 measurements of the state of the system, which were taken with sample time of $T = 0.01$ s, since there are 15 independent parameters of the Q-function, given by the elements of the symmetric matrix H.

The optimal value function can be calculated based on the obtained optimal Q-function and the obtained optimal control policy using the equation

$$P^* = [I_n \; K^{*T}]H^* \begin{bmatrix} I_n \\ K^* \end{bmatrix} \tag{62.150}$$

In our example the solution for the parameters of the optimal value function obtained after running the algorithm is given in the matrix

$$P_{DT-ADHDP} = \begin{bmatrix} 0.4802 & 0.4768 & 0.0603 & 0.4754 \\ 0.4768 & 0.7887 & 0.1239 & 0.3834 \\ 0.0603 & 0.1239 & 0.0567 & 0.0300 \\ 0.4754 & 0.3843 & 0.0300 & 2.3433 \end{bmatrix}. \quad (62.151)$$

Comparing the results presented in Figure 62.4 with the ones given in Figure 62.3 one sees that the discrete-time online algorithm converges to the optimal solution of the continuous-time algebraic Riccati equation for this case in which a very small sampling period was considered. It is clear that fairly equal amounts of time are required for both the continuous-time and the discrete-time ADP algorithms to converge, however, the discrete-time ADHDP algorithm is computationally more intensive (400 required iterations) than the continuous-time HDP (40 iterations). The advantage of using the discrete-time algorithm consists in its model-free characteristic while the presented continuous-time algorithm still requires knowledge of the system's B matrix. This observation motivates further research for finding a continuous-time model-free ADP approach to solve the optimal control problem.

Further Reading

For those interested in applying ADP methods and/or developing new ADP algorithms, we suggest for further reading the recently published papers [4,22], and the papers referred therein. The consistent lists of references are accurately encompassing the numerous results in this fast growing research field.

References

1. Al-Tamimi A., Lewis F. L., and Abu-Khalaf M., Model-free Q-learning designs for discrete-time zero-sum games with application to H-infinity control, *Automatica*, 43, 473–481, 2007.
2. Al-Tamimi A. and Lewis, F. Discrete-time nonlinear HJB solution using approximate dynamic programming: Convergence proof, *Proc. of IEEE Symposium on Approximate Dynamic Programming and Reinforcement Learning ADPRL'07*, 38–43, Hawaii, 2007.
3. Al-Tamimi A., Vrabie D., Lewis F. L., and Abu-Khalaf M., Model-free approximate dynamic programming schemes for linear systems, *Proc. of IJCNN'07*, 371–378, Orlando, 2007.
4. Balakrishnan S.N., Ding J., and Lewis F.L., Issues on stability of ADP feedback controllers for dynamical systems, *Trans. Systems, Man, Cybernetics, Part B: Special issue on ADP/RL*, 38(4), 913–917, 2008.
5. Barto A. G., Sutton R. S., and Anderson C., Neuron-like adaptive elements that can solve difficult learning control problems, *IEEE Trans. on Systems, Man, and Cybernetics*, SMC-13, 834–846, 1983.
6. Bellman R., *Dynamic Programming*, Princeton University Press, Princeton, NJ, 1957.
7. Bertsekas D. P. and J. N. Tsitsiklis D. P., *Neuro-Dynamic Programming*, Athena Scientific, MA, 1996.
8. S. Bradtke S., B. Ydstie B., and A. Barto A., Adaptive linear quadratic control using policy iteration, *Proc. of American Control Conference*, 3, 3475–3479, 1994.
9. T. Hanselmann T., L. Noakes L., and Zaknich A., Continuous-time adaptive critics, *IEEE Trans. on Neural Networks*, 18(3), 631–647, 2007.
10. D. Kleinman D., On an iterative technique for Riccati equation computations, *IEEE Trans. on Automatic Control*, 13, 114–115, 1968.
11. T. Landelius T., *Reinforcement Learning and Distributed Local Model Synthesis*, PhD Dissertation, Linkoping University, Sweden, 1997.
12. Mehta P. and Meyn S., Q-learning and Pontryagin's minimum principle, *Proc. of IEEE Conference on CDC/CCC*, 3598–3605, 2009.
13. Murray J. J., Cox C. J., Lendaris G. G., and Saeks R., Adaptive dynamic programming, *IEEE Trans. on Systems, Man and Cybernetics*, 32(2), 140–153, 2002.
14. Pareigis S., Numerical schemes for the continuous Q-function of reinforcement learning, Technical report, Bericht 97–15, *Lehrstuhl Praktische Mathematik*, Kiel University, 1997.

15. Prokhorov D. and Wunsch D., Adaptive critic designs, *IEEE Trans. on Neural Networks*, 8(5), 997–1007, 1997.
16. Si J., Barto A., Powell W., and Wunsch D., *Handbook of Learning and Approximate Dynamic Programming*, John Wiley, New Jersey, NJ, 2004.
17. Sutton R. S. and Barto A. G., *Reinforcement Learning—An Introduction*, MIT Press, Cambridge, MT, 1998.
18. Vrabie D. and Lewis F., Generalized policy iteration for continuous-time systems, *Proc. of IJCNN'09*, Atlanta, 2009.
19. Vrabie D. and Lewis F.L., Neural network approach to continuous-time direct adaptive optimal control for partially-unknown nonlinear systems, *Neural Networks—special issue: Goal-Directed Neural Systems*, 22(3), 237–246, 2009.
20. Vrabie D. Abu-Khalaf M., Lewis F.L., and Wang Y., Continuous-time ADP for linear systems with partially unknown dynamics, *Proc. of Symposium on Approximate Dynamic Programming and Reinforcement Learning (ADPRL)*, 247–253, Hawaii, 2007.
21. Wang Y., Zhou R., and Wen C., Robust load-frequency controller design for power systems, *IEE Proc.-C*, 140(I), 1993.
22. Wang F. Y., Zhang H., and Liu D., Adaptive dynamic programming: An introduction, *IEEE Comp. Intelligence Magazine*, 39–47, May 2009.
23. Watkins C. J. C. H. and Dayan P., Q-learning, *Machine Learning*, 8, 279–292, 1992.
24. Werbos P. J., Approximate dynamic programming for real-time control and neural modeling, *Handbook of Intelligent Control*, eds. D.A. White and D.A. Sofge, Van Nostrand Reinhold, New York, NY, 1992.

63

Stability of Stochastic Systems

Kenneth A. Loparo
Case Western Reserve University

63.1 Introduction

In many applications where dynamical system models are used to describe the behavior of a real world system, stochastic components and random noises are included in the model to capture uncertainties in the operating environment and the system structure of the physical process being studied. The analysis and control of such systems then involves evaluating the stability properties of a random dynamical system. Stability is an important property of a system and we know from classical control theory that input–output stability is a necessary condition for control system design, but are aware that analytic techniques for evaluating stability are often restricted to linear systems, or special classes of nonlinear systems. The general study of the stability properties of stochastic dynamical systems is important and considerable effort has been devoted to the study of stochastic stability. Significant results have been reported in the literature with applications to physical and engineering systems.

A comprehensive survey on the topic of stochastic stability was given by [27], and since that time there have been many significant developments of the theory and its applications in science and engineering. In this chapter, we present some basic results on the study of stability of stochastic systems. Because of limited space, only selected topics are presented and discussed. We begin with a discussion of the basic definitions of stochastic stability and the relationships among them. Kozin's survey provides an excellent introduction to the subject and a good explanation of how the various notions of stochastic stability are related.

It is not necessary that readers have an extensive background in stochastic processes or other related mathematical topics. In this chapter, the results and methods will be stated as simply as possible and no proofs of the theorems are given. When necessary, important mathematical concepts that are the foundation to some of the results, will be discussed briefly. This will provide a better appreciation of the results, the application of the results, and the key steps required to develop the theory further. Those readers interested in a particular topic or result discussed in this chapter are encouraged to go to the original papers and the references therein for more detailed information.

There are at least three times as many definitions for the stability of stochastic systems as there are for deterministic systems. This is because in a stochastic setting there are three basic types of convergence: convergence in probability, convergence in mean (or moment), and convergence in an almost sure (sample path, probability one) sense. Kozin [27] presented several definitions for stochastic stability, and these definitions have been extended in subsequent research works. Readers are cautioned to examine carefully the definition of stochastic stability that is being used when interpreting any stochastic stability results.

We begin with some preliminary definitions of stability concepts for a deterministic system. Let $x(t; x_0, t_0)$ denote the trajectory of a dynamic system initial from x_0 at time t_0.

Definition 63.1: Lyapunov Stability

The equilibrium solution, assumed to be 0 unless stated otherwise, is said to be stable if, given $\varepsilon > 0$, $\delta(\varepsilon, t_0) > 0$ exists so that, for all $\|x_0\| < \delta$,

$$\sup_{t \geq t_0} \|x(t; x_0, t_0)\| < \varepsilon.$$

Definition 63.2: Asymptotic Lyapunov Stability

The equilibrium solution is said to be asymptotically stable if it is stable and if $\delta' > 0$ exists so that $\|x_0\| < \delta'$, guarantees that

$$\lim_{t \to \infty} \|x(t; x_0, t_0)\| = 0.$$

If the convergence holds for all initial times, t_0, it is referred to as uniform asymptotic stability.

Definition 63.3: Exponential Lyapunov Stability

The equilibrium solution is said to be exponentially stable if it is asymptotically stable and if there exists a $\delta > 0$, an $\alpha > 0$, and a $\beta > 0$ so that $\|x_0\| < \delta$ guarantees that

$$\|x(t; x_0 t_0)\| \leq \beta \|x_0\| \exp^{-\alpha(t-t_0)}.$$

If the convergence holds for all initial times, t_0, it is referred to as uniform exponential stability.

These deterministic stability definitions can be translated into a stochastic setting by properly interpreting the notion of convergence, that is, in probability, in moment, or almost surely. For example, in Definition 63.1 for Lyapunov stability, the variable of interest is $sup_{t \geq t_0} \|x(t; x_0, t_0)\|$, and we have to study the various ways in which this (now random) variable can converge. We denote the fact that the variable is random by including the variable ω, that is, $x(t; x_0, t_0, \omega)$, and we will make this more precise later. Then,

---

Definition 63.4: I_p: Lyapunov Stability in Probability

The equilibrium solution is said to be stable in probability if, given ε, $\varepsilon' > 0$, $\delta(\varepsilon, \varepsilon', t_0) > 0$ exists so that for

all $\|x_0\| < \delta$,

$$P\{\sup_{t\geq t_0} \|x(t; x_0, t_0, \omega)\| > \varepsilon'\} < \epsilon.$$

Here, P denotes probability.

Definition 63.5: I_m: Lyapunov Stability in the *p*th Moment

The equilibrium solution is said to be stable in pth moment, $p > 0$ *if, given* $\varepsilon > 0$, $\delta(\varepsilon, t_0) > 0$ *exists so that* $\|x_0\| < \delta$ *guarantees that*

$$E\{\sup_{t\geq t_0} \|x(t; x_0, t_0, \omega)\|^p\} < \epsilon.$$

Here, E denotes expectation.

Definition 63.6: $\text{I}_{\text{a.s.}}$: Almost Sure Lyapunov Stability

The equilibrium solution is said to be almost surely stable if

$$P\{\lim_{\|x_0\|\to 0} \sup_{t\geq t_0} \|x(t; x_0, t_0, \omega)\| = 0\} = 1.$$

Note, almost sure stability is equivalent to saying that, with probability one, all sample solutions are Lyapunov stable.

Similar statements can be made for asymptotic stability and for exponential stability. These definitions are introduced next for completeness and because they are often used in applications.

Definition 63.7: II_p: Asymptotic Lyapunov Stability in Probability

The equilibrium solution is said to be asymptotically stable in probability if it is stable in probability and if $\delta' > 0$ *exists so that* $\|x_0\| < \delta'$ *guarantees that*

$$\lim_{\delta\to\infty} P\{\sup_{t\geq\delta} \|x(t; x_0, t_0, \omega)\| > \varepsilon\} = 0.$$

If the convergence holds for all initial times, t_0, it is referred to as uniform asymptotic stability in probability.

Definition 63.8: II_m: Asymptotic Lyapunov Stability in the *p*th Moment

The equilibrium solution is said to be asymptotically pth moment stable if it is stable in the pth moment and if $\delta' > 0$ *exists so that* $\|x_0\| < \delta'$ *guarantees that*

$$\lim_{\delta\to\infty} E\{\sup_{t\geq\delta} \|x(t; x_0, t_0, \omega)\|\} = 0.$$

Definition 63.9: $\text{II}_{\text{a.s.}}$: Almost Sure Asymptotic Lyapunov Stability

The equilibrium solution is said to be almost surely asymptotically stable if it is almost surely stable and if $\delta' > 0$ *exists so that* $\|x_0\| < \delta'$ *guarantees that, for any* $\varepsilon > 0$,

$$\lim_{\delta \to \infty} \{\sup_{t \geq \delta} \|x(t; x_0, t_0, \omega)\| > \varepsilon\} = 0.$$

Weaker versions of the stability definitions are common in the stochastic stability literature. In these versions the stochastic stability properties of the system are given in terms of particular instants, t, rather than the seminfinite time interval $[t_0, \infty)$ as given in the majority of the definitions given above. Most noteworthy, are the concepts of the pth moment and almost sure exponential stability:

Definition 63.10: III_{m}: pth Moment Exponential Lyapunov Stability

The equilibrium solution is said to be pth moment exponentially stable if there exists a $\delta > 0$, *an* $\alpha > 0$, *and a* $\beta > 0$ *so that* $\|x_0\| < \delta$ *guarantees that*

$$E\{\|x(t; x_0, t_0, \omega)\|\} \leq \beta \|x_0\| \exp^{-\alpha(t-t_0)}.$$

Definition 63.11: $\text{III}_{\text{a.s.}}$: Almost Sure Exponential Lyapunov Stability

The equilibrium solution is said to be almost surely exponentially stable if there exist a $\delta > 0$, *an* $\alpha > 0$, *and a* $\beta > 0$ *so that* $\|x_0\| < \delta$ *guarantees that*

$$P\{\{\|x(t; x_0, t_0, \omega)\|\} \leq \beta \|x_0\| \exp^{-\alpha(t-t_0)}\} = 1$$

Example 63.1:

Consider the scalar Ito equation

$$dx(t) = ax(t)dt + \sigma x(t)\, dw(t) \tag{63.1}$$

where $\{w(t)\}_{t \geq 0}$ is a standard Wiener process. The infinitesimal generator for the system is given by

$$\mathcal{L} = \frac{1}{2}\sigma^2 x^2 \frac{d^2}{dx^2} + ax\frac{d}{dx}. \tag{63.2}$$

The solution process x_t for $t \geq 0$ is given by

$$x(t) = e^{(a - \frac{1}{2}\sigma^2)t} e^{\sigma \int_0^t dw(s)} x_0.$$

Hence,

$$\log \frac{x(t)}{x_0} = (a - \frac{1}{2}\sigma^2)t + \sigma \int_0^t dw_s$$

and the asymptotic exponential growth rate of the process is

$$\lambda = \lim_{t\to\infty} \frac{1}{t} \log \frac{x(t)}{x_0} = (a - \frac{1}{2}\sigma^2) + \lim_{t\to\infty} \frac{\sigma}{t} \int_0^t dw(s)s = a - \frac{1}{2}\sigma^2$$

The last equality follows from the fact that the Wiener process, with increment $dw(t)$, is a zero-mean ergodic process. We then conclude that the system is almost surely exponentially stable in the sense that

$$P_{x_0}\left\{\lim_{t\to\infty} x(t) = 0, \text{at an exponential rate a.s.}\right\} = 1,$$

if, and only if, $a < \frac{1}{2}\sigma^2$.

Next we compare this with the second-moment stability result; see also Example 63.2 in the next section where the same conclusion follows from a Lyapunov stability analysis of the system. From our previous calculation, x_t^2 for $t \geq 0$ is given by

$$x_t^2 = e^{2(a-\frac{1}{2}\sigma^2)t} e^{2\sigma\int_0^t dw_s} x_0^2.$$

Then,

$$E\{x_t^2\} = e^{(2a+\sigma^2)t} E\{x_0^2\},$$

and we conclude that the system is exponentially second-moment stable if, and only if, $(a + \frac{1}{2}\sigma^2) < 0$, or $a < -\frac{1}{2}\sigma^2$. Therefore, unlike deterministic systems, even though moment stability implies almost sure (sample path) stability, almost sure stability need not imply moment stability of the system. For the system Equation 63.1, the pth moment is exponentially stable if and only if $a < \frac{1}{2}\sigma^2(1-p)$ where $p = 1, 2, 3, \ldots$.

In the early stages of the development of stability criteria for stochastic systems, investigators were primarily concerned with moment stability and stability in probability; the mathematical theory for the study of almost sure (sample path) stability was not yet fully developed. During this initial development period there was some confusion about the basic stability concepts, their usefulness in applications, and the relationship among the different stability concepts. Kozin's survey clarified some of the confusions and provided a good foundation for further work. During the past 25 years, almost sure (sample path) stability studies have attracted increasing attention of researchers. This is not surprising because it is the sample paths rather than moments that are observed in real systems, and the stability properties of the sample paths can be most closely related to their deterministic counterpart, as argued by Kozin [27]. Practically speaking, moment stability criteria, when used to infer sample path stability, are often too conservative to be useful in applications.

One of the most fruitful and important advances in the study of stochastic stability is the development of the theory of Lyapunov exponents for stochastic systems. This is the stochastic counterpart of the notion of characteristic exponents introduced in Lyapunov's original work on asymptotic (exponential) stability. This approach provides necessary and sufficient conditions for almost sure asymptotic (exponential) stability, but significant computational problems must be solved. The Lyapunov exponent method uses sophisticated tools from stochastic process theory and other related branches of mathematics, and has the potential of providing testable conditions for almost sure asymptotic (exponential) stability for stochastic systems. We will provide an introduction to this approach for analyzing the stability of stochastic systems.

In this chapter, we divide the results into two categories, the Lyapunov function method and the Lyapunov exponent method.

63.2 Lyapunov Function Method

The Lyapunov function method, known as Lyapunov's second (direct) method, provides a powerful tool for the study of stability properties of dynamic systems because the technique does not require solving the system equations explicitly. For deterministic systems, this method can be interpreted briefly as described in the following paragraph.

Consider a nonnegative continuous function $V(x)$ on $\mathbb{R}^n$ with $V(0) = 0$ and $V(x) > 0$ for $x \neq 0$. Suppose for some $m \in \mathbb{R}$, the set $Q_m = \{x \in \mathbb{R}^n : V(x) < m\}$ is bounded and $V(x)$ has continuous first partial derivatives in Q_m. Let the initial time $t_0 = 0$ and let $x(t) = x(t, x_0)$ be the unique solution of the initial value problem:

$$\begin{cases} \dot{x}(t) = f(x(t)), & t \geq 0, \\ x(0) = x_0 \in \mathbb{R}^n, & f(0) = 0, \end{cases} \tag{63.3}$$

for $x_0 \in Q_m$. Because $V(x)$ is continuous, the open set Q_r for $r \in (0, m]$ defined by $Q_r = \{x \in \mathbb{R}^n : V(x) < r\}$, contains the origin and monotonically decreases to the singleton set $\{0\}$ as $r \to 0^+$. The total derivative $\dot{V}(x)$ of $V(x)$ (along the solution trajectory $x(t, x_0)$) is given by

$$\dot{V}(x) = \frac{dV(x)}{dt} = f^T(x) \cdot \frac{\partial V}{\partial x} \stackrel{\text{def}}{=} -k(x). \tag{63.4}$$

If $-k(x) \leq 0$ for all $x \in Q_m$, with $k(x)$ continuous, then $V(x_t)$ is a nonincreasing function of t, and $V(x_0) < m$ implies $V(x(t)) < m$ for all $t \geq 0$. Equivalently, $x_0 \in Q_m$ implies that $x_t \in Q_m$, for all $t \geq 0$. This establishes the stability of the zero solution of Equation 63.3 in the sense of Lyapunov, and $V(x)$ is called a Lyapunov function for Equation 63.3. Let us further assume that $k(x) > 0$ for $x \in Q_m \backslash \{0\}$. Then $V(x(t))$, as a function of t, is strictly monotone decreasing. In this case, $V(x(t)) \to 0$ as $t \to +\infty$ from Equation 63.4. This implies that $x(t) \to 0$ as $t \to +\infty$. This fact can also be seen through an integration of Equation 63.4, that is,

$$0 < V(x_0) - V(x(t)) = \int_0^t k(x_s)\, ds < +\infty \quad \text{for } t \in [0, +\infty). \tag{63.5}$$

It is evident from Equation 63.5 that $x(t) \to \{0\} = \{x \in Q_m : k(x) = 0\}$ as $t \to +\infty$. This establishes the asymptotic stability of the trajectories of system Equation 63.3.

The Lyapunov function $V(x)$ can be interpreted as a generalized energy function of the system Equation 63.3, and the above argument illustrates the physical intuition that if the energy of a physical system is always decreasing near an equilibrium state, then the equilibrium state is stable.

Since Lyapunov's original work, this direct method for stability study has been extensively investigated. The main advantage of the method is that one can obtain considerable information about the stability of a given system without explicitly solving the system equation. One major drawback of this method is that for general classes of nonlinear systems a systematic method for constructing a suitable Lyapunov function does not exist, and stability criteria (usually sufficient conditions) determined using the method depend critically on the Lyapunov function that is chosen.

The first attempts to generalize the Lyapunov function method to stochastic stability studies were made by [8] and [24]. A systematic treatment of this topic was later given by [22,30,31] (primarily for white-noise stochastic systems). The key idea of the Lyapunov function approach for a stochastic system is the following:

Consider the stochastic system defined on a probability space $(\Omega, \mathcal{F}, P)$, where Ω is the set of elementary events (sample space), $\mathcal{F}$ is the σ field that consists of all measurable subsets (events) of Ω, and P is a probability measure:

$$\begin{cases} \dot{x}(t) = f(x(t), \omega), & t \geq 0, \\ x(0) = x_0. \end{cases} \tag{63.6}$$

It is not reasonable to require that $\dot{V}(x(t)) \leq 0$ for all ω, where $x(t) = x(t, x_0, \omega)$ is a sample solution of Equation 63.6 initial from x_0. Note, when t is fixed, $x(t, x_0, \omega)$ is a random variable on $(\Omega, \mathcal{F}, P)$, and when ω is fixed ($\omega = \omega^*$), $x(t, x_0, \omega^*)$ is a deterministic time function. The basic idea to insure "stability" of the system is that the time derivative of the expectation of $V(x(t))$, denoted by $\mathcal{L}V(x(t))$, is nonpositive. Here, $\mathcal{L}$ is the infinitesimal generator of the process $x(t)$. Suppose that the system is Markovian so that the solution process is a strong, time homogeneous Markov process. Then $\mathcal{L}$ is defined by

$$\mathcal{L}V(x_0) = \lim_{\Delta t \to 0} \frac{E_{x_0}(V(x_{\Delta t})) - V(x_0)}{\Delta t} \tag{63.7}$$

where the domain of $\mathcal{L}$ is defined as the space of functions $V(x)$ for which Equation 63.7 is well-defined. This is a natural analog of the total derivative of $V(x)$ along the solution trajectory $x(t)$ in the deterministic case. Now suppose that, for a Lyapunov function $V(x)$ that satisfies the conditions stated above, $\mathcal{L}V(x) \le -k(x) \le 0$. It follows that

$$0 \le V(x_0) - E_{x_0} V(x(t)) = E_{x_0} \int_0^t k(x(s))\, ds = -E_{x_0} \int_0^t \mathcal{L}V(x(s))\, ds < +\infty \tag{63.8}$$

and for $t, s > 0$

$$E_{x(s)} V(x(t+s)) - V(x(s)) \le 0 \quad \text{a.s.} \tag{63.9}$$

Equation 63.9 means that $V(x(t))$ is a supermartingale, and, by the martingale convergence theorem, we expect that $V(x(t)) \to 0$ a.s. (almost surely) as $t \to +\infty$. This means that $x(t) \to 0$ a.s. as $t \to +\infty$. A similar argument can be obtained by using Equation 63.8 an analog of Equation 63.5. It is then reasonable to expect from Equation 63.8 that $x(t) \to \{x \in \mathbb{R}^n : k(x) = 0\}$ almost surely. These are the key ideas behind the Lyapunov function approach to the stability analysis of stochastic systems.

Kushner [30–32] used the properties of strong Markov processes and the martingale convergence theorem to study the Lyapunov function method for stochastic systems with solution processes which are strong Markov processes with right continuous sample paths. A number of stability theorems were developed in these works and the references therein. Kushner also presented various definitions of stochastic stability. The reader should note that his definition of stability "with probability one" is equivalent to the concept of stability in probability introduced earlier here. Also, asymptotic stability "with probability one" means stability "with probability one" and sample path stability, that is, $x(t) \to 0$ a.s. as $t \to +\infty$. The key results of Kushner are based on the following supermartingale inequality that follows directly from Equation 63.8 where x_0 is given:

$$P_{x_0}\{ \sup_{0 \le t < +\infty} V(x(t)) \ge \epsilon\} \le \frac{V(x_0)}{\epsilon}. \tag{63.10}$$

From this, the following typical results were obtained by Kushner. For simplicity, we assume that $x(t) \in Q_m$ almost surely for some $m > 0$ and state these results in a simpler way.

Theorem 63.1: Kushner

1. *Stability "with probability one":*
   If $\mathcal{L}V(x) \le 0$, $V(x) > 0$ for $x \in Q_m \backslash \{0\}$, then the origin is stable "with probability one".
2. *Asymptotic stability "with probability one":*
   If $\mathcal{L}V(x) = -k(x) \le 0$ with $k(x) > 0$ for $x \in Q_m \backslash \{0\}$ and $k(0) = 0$, and if for any $d > 0$ small, $\epsilon_d > 0$ exists so that $k(x) \ge d$ for $x \in \{Q_m : \|x\| \ge \epsilon_d\}$, then the origin is stable "with probability one" and

   $$P_{x_0}\{x(t) \to 0 \quad \text{as } t \to +\infty\} \ge 1 - \frac{V(x_0)}{m}.$$

   In particular, if the conditions are satisfied for arbitrarily large m, then the origin is asymptotically stable "with probability one."

3. *Exponential asymptotic stability "with probability one":*
   If $V(x) \geq 0$, $V(0) = 0$ and $\mathcal{L}V(x) \leq -\alpha V(x)$ on Q_m for some $\alpha > 0$, then the origin is stable "with probability one," and

$$P_{x_0}\left\{\sup_{T \leq t < +\infty} V(x(t)) \geq \lambda\right\} \leq \frac{V(x_0)}{m} + \frac{V(x_0)e^{-\alpha T}}{\lambda}, \quad \forall T \geq 0.$$

   In particular, if the conditions are satisfied for arbitrarily large m, then the origin is asymptotically stable "with probability one," and

$$P_{x_0}\left\{\sup_{T \leq t < +\infty} V(x(t)) \geq \lambda\right\} \leq \frac{V(x_0)e^{-\alpha T}}{\lambda}.$$

Many interesting examples were also developed in Kushner's work to demonstrate the application of the stability theorems. These examples also illustrated some construction procedures for Lyapunov functions for typical systems.

Example 63.2: Kushner

Consider the scalar Ito equation

$$dx(t) = ax(t)\,dt + \sigma x(t)\,dw(t) \tag{63.11}$$

where $w(t)$ is a standard Wiener process. The infinitesimal generator for the system is given by

$$\mathcal{L} = \frac{1}{2}\sigma^2 x^2 \frac{d^2}{dx^2} + ax\frac{d}{dx}. \tag{63.12}$$

If the Lyapunov function candidate is $V(x) = x^2$, then

$$\mathcal{L}V(x) = (\sigma^2 + 2a)x^2. \tag{63.13}$$

If $\sigma^2 + 2a < 0$, then with $Q_m = \{x : x^2 < m^2\}$, from (1) of the previous theorem, the zero solution is stable "with probability one." Let $m \to +\infty$. By (2) of the theorem,

$$\lim_{t\to\infty} x(t) = 0 \qquad \text{a.s.} \tag{63.14}$$

where $x(t)$ is the solution process of Equation 63.11. By (3) of the theorem,

$$P_{x_0}\{\sup_{T \leq t < +\infty} x_t^2 \geq \lambda\} \leq \frac{V(x_0)e^{-\alpha T}}{\lambda} \tag{63.15}$$

for some $\alpha > 0$.

Remark 63.1

As calculated in the previous section, the asymptotic exponential growth rate of the process is

$$\lambda = \lim_{t\to\infty} \frac{1}{t}\log\frac{x(t)}{x_0} = (a - \frac{1}{2}\sigma^2) + \lim_{t\to\infty}\frac{\sigma}{t}\int_0^t dw(s) = a - \frac{1}{2}\sigma^2.$$

We conclude that the system is almost surely exponentially stable in the sense that

$$P_{x_0}\left\{\lim_{t\to\infty} x(t) = 0, \text{at an exponential rate a.s.}\right\} = 1,$$

if, and only if, $a < \frac{1}{2}\sigma^2$. Compare this with the stability result $a < -\frac{1}{2}\sigma^2$, given above, using the Lyapunov function method. Note that $a < -\frac{1}{2}\sigma^2$ is actually the stability criterion for second-moment stability, which is a conservative estimate of the almost sure stability condition $a < \frac{1}{2}\sigma^2$.

Has'minskii [22] and the references cited therein, provide a comprehensive study of the stability of diffusion processes interpreted as the solution process of a stochastic system governed by an Ito stochastic differential equation of the form,

$$\begin{cases} dx(t) = b(t,x)\,dt + \sum_{r=1}^{k} \sigma_r(t,x)\,d\zeta_r(t), & t \geq s \\ x(s) = x_s, \end{cases} \tag{63.16}$$

where $\zeta_r(t)$ are independent standard Wiener processes and the coefficients $b(t,x)$ and $\sigma_r(t,x)$ satisfy Lipschitz and growth conditions. In this case, the infinitesimal generator $\mathcal{L}$ of the system (associated with the solution process) is a second-order partial differential operator on functions $V(t,x)$ that are twice continuously differentiable with respect to x and continuously differentiable with respect to t. $\mathcal{L}$ is given by

$$\mathcal{L}V(t,x) = \frac{\partial V}{\partial t} + \sum_{i=1}^{n} b_i(t,x)\frac{\partial V}{\partial x_i} + \frac{1}{2}\sum_{i,j=1}^{n} a_{ij}(t,x)\frac{\partial^2 V}{\partial x_i \partial x_j}. \tag{63.17}$$

The key idea of Has'minskii's approach is to establish an inequality like Equation 63.10 developed in Kushner's work. Below are some typical results obtained by Has'minskii; the reader is referred to [22] for a more detailed development.

Let U be a neighborhood of 0 and $U_1 = \{t > 0\} \times U$. The collection of functions $V(t,x)$ defined in U_1, that are twice continuously differentiable in x except at the point $x = 0$ and continuously differentiable in t, are denoted by $C_2^0(U_1)$. A function $V(t,x)$ is said to be positive definite in the Lyapunov sense if $V(t,0) = 0$ for all $t \geq 0$ and $V(t,x) \geq \omega(x) > 0$ for $x \neq 0$ and some continuous function $\omega(x)$.

Theorem 63.2: Has'minskii

1. *The trivial solution of Equation 63.16 is stable in probability (same as our definition) if there exists $V(t,x) \in C_2^0(U_1)$, positive definite in the Lyapunov sense, so that $\mathcal{L}V(t,x) \leq 0$, for $x \neq 0$.*
2. *If the system Equation 63.16 is time homogeneous, that is, $b(t,x) = b(x)$ and $\sigma_r(t,x) = \sigma_r(x)$ and if the nondegeneracy condition,*

$$\sum_{i,j=1}^{n} a_{ij}(x)\lambda_i\lambda_j > m(x)\sum_{i=1}^{n} \lambda_i^2, \quad \text{for } \lambda = (\lambda_1 \ldots \lambda_n)^T \in \mathbb{R}^n, \tag{63.18}$$

   is satisfied with continuous $m(x) > 0$ for $x \neq 0$, then a necessary and sufficient condition for the trivial solution to be stable in probability is that a twice continuously differentiable function $V(x)$ exists, except perhaps at $x = 0$, so that

$$\mathcal{L}_0 V(x) = \sum_{i=1}^{n} b_i(x)\frac{\partial V}{\partial x_i} + \sum_{i,j=1}^{n} a_{ij}(x)\frac{\partial^2 V(x)}{\partial x_i \partial x_j} \leq 0$$

   where $\mathcal{L}_0$ is the infinitesimal generator of the time homogeneous system.
3. *If the system Equation 63.16 is linear, that is, $b(t,x) = b(t)x$ and $\sigma_r(t,x) = \sigma_r(t)x$, then the system is exponentially p-stable (the* pth *moment is exponentially stable), that is,*

$$E_{x_0}\{\|x(t,x_0,s)\|^p\} \leq A \cdot \|x\|^p \exp\{-\alpha(t-s)\}, \quad p > 0,$$

   for some constant $\alpha > 0$ if, and only if, a function $V(t,x)$ exists, homogeneous of degree p in x, so that for some constants $k_i > 0$, $i = 1,2,3,4$,

$$k_1\|x\|^p \leq V(t,x) \leq k_2\|x\|^p, \quad \mathcal{L}V(t,x) \leq -k_3\|x\|^p,$$

and

$$\left\|\frac{\partial V}{\partial x}\right\| \leq k_4 \|x\|^{p-1}, \quad \left\|\frac{\partial^2 V}{\partial x^2}\right\| \leq k_4 \|x\|^{p-2}.$$

In addition to the stability theorems, Has'minskii also studied other asymptotic properties of stochastic systems and presented many interesting examples that illustrate the stabilizing and destabilizing effects of random noise in stochastic systems.

Just as in the case of deterministic systems, the Lyapunov function approach has the advantage that one may obtain considerable information about the qualitative (asymptotic) behavior of trajectories of the system, in particular stability properties of the system that are of interest, without solving the system equation. However, no general systematic procedure exists to construct a candidate Lyapunov function. So even though the theorems, like Has'minskii's, provide necessary and sufficient conditions for stability, one may never find the "appropriate" Lyapunov function in a given application to determine the stability (or instability) of the system. Further, the nature of the stability condition obtained using the Lyapunov method depends critically on the choice of Lyapunov function that is used. Various techniques have been proposed by investigators to construct a suitable family of Lyapunov functions to obtain the "best" stability results. In the following, we summarize some of these efforts.

There are several works related to the stability of linear systems of the form,

$$\begin{cases} \dot{x}(t) = [A + F(t)]x(t); \quad t \geq 0, \\ x(0) = x_0 \in \mathbb{R}^n, \end{cases} \tag{63.19}$$

with A a stable (Hurwitz) matrix and $F(t)$ a stationary and ergodic matrix-valued random process. This problem was first considered by Kozin [26] by using the Gronwall–Bellman inequality rather than a Lyapunov function technique. Kozin's results were found to be too conservative. Caughey and Gray [12] were able to obtain better results by using a very special type of quadratic form Lyapunov function. Later, Infante [23] extended these stability theorems by using the extremal property of the so-called regular pencil of a quadratic form. The basic idea behind Infante's work is the following:

Consider the quadratic form Lyapunov function $V(x) = x'Px$. Then the time derivative along the sample paths of the system Equation 63.19 is

$$\dot{V}(x(t)) = x(t)'(F'(t)P + PF(t))x(t) - x(t)'Qx(t) \tag{63.20}$$

where Q is any positive-definite matrix and P is the unique solution of the Lyapunov equation

$$A'P + PA = -Q\,. \tag{63.21}$$

If $\lambda(t) = \dot{V}(x(t))/V(x(t))$, then

$$V(x(t)) = V(x_0)\exp\{\int_0^t \lambda(s)\,ds\}. \tag{63.22}$$

From the extremal properties of matrix pencils,

$$\lambda(t) \leq \lambda_{\max}[(A + F(t))' + Q(A + F(t))Q^{-1}]. \tag{63.23}$$

Here, $\lambda_{\max}(K)$ denotes the largest magnitude of the eigenvalues of the matrix K. By the ergodicity property of the random matrix process $F(t)$, we have the almost sure stability condition

$$\lim_{t\to\infty}\frac{1}{t}\int_0^t \lambda(\tau)\,d\tau = E\{\lambda(t)\} \leq E\{\lambda_{\max}[(A + F(t))' + Q(A + F(t))Q^{-1}]\} < 0.$$

Man [39] tried to generalize the results of Infante. However, several obvious mistakes can be observed in the derivation of the theorem and the two examples given in this work.

Following Infante et al. [29] used distributional properties of the random coefficient matrix $F(t)$ to obtain improved results for two specific second-order systems in the form,

$$\begin{cases} \ddot{x}(t) + 2\beta\dot{x}(t) + [c + f(t)]x(t) = 0, \\ \ddot{x}(t) + [2\beta + g(t)]\dot{x}(t) + cx(t) = 0, \end{cases} \tag{63.24}$$

where $f(t)$ and $g(t)$ are stationary and ergodic random processes.

Parthasarthy and Evan-Zwanoskii [48] presented an effective computational procedure, using an optimization technique, to apply the Lyapunov type procedure to higher order systems in the form of Equation 63.19 with $F(t) = k(t)G$. Here G is a constant matrix and $k(t)$ is a scalar (real-valued) stationary and ergodic random process. After proper parameterization of the quadratic form Lyapunov function, the Fletcher–Powell–Davidson optimization algorithm was used to optimize the stability region that depends on the system data and $k(t)$. A fourth-order system was studied using the suggested procedure, and various simulation results were used to show that this procedure yielded a stability region that was not unduly conservative. Because an optimization procedure was used and the solution of a Lyapunov equation was required, this procedure required an extensive computational effort.

Wiens and Sinha [56] proposed a more direct method for higher order systems defined as an interconnection of a set of second-order subsystems. Consider the system,

$$M\ddot{x}(t) + [C_0 + C(t)]\dot{x}(t) + [K_0 + K(t)]x(t) = 0, \tag{63.25}$$

where M, C_0, and K_0 are nonsingular $n \times n$ matrices and $C(t), K(t)$ are $n \times n$ stationary and ergodic matrix-valued, random processes. The technique for constructing a Lyapunov function suggested by Walker [55] for deterministic systems was used to construct a quadratic form Lyapunov function $V(\tilde{x})$ for the deterministic counterpart of Equation 63.25,

$$\begin{pmatrix} \dot{x}_1 \\ \dot{x}_2 \end{pmatrix} = \begin{pmatrix} 0 & I \\ -M^{-1}K_0 & -M^{-1}C_0 \end{pmatrix} \begin{pmatrix} x_1 \\ x_2 \end{pmatrix}; \quad \tilde{x} \triangleq \begin{pmatrix} x_1 \\ x_2 \end{pmatrix} \in \mathbb{R}^{2n\times 2n}, \tag{63.26}$$

that is

$$V(\tilde{x}) = \tilde{x}' \begin{bmatrix} P_1 & \frac{1}{2}P_3' \\ \frac{1}{2}P_3 & P_2 \end{bmatrix} \tilde{x}; \quad V(\tilde{x}) \triangleq \tilde{x}'P\tilde{x}, \quad P_i' = P_i > 0, \quad \text{for } i = 1, 2,$$

with a time derivative along sample paths,

$$\dot{V}(\tilde{x}) = \tilde{x}'A_0\tilde{x},$$

where A_0 is properly defined and the following choices are made:

$$\begin{aligned} P_3 &= P_2M^{-1}C_0, \\ P_1 &= P_2M^{-1}K_0 + \frac{1}{2}(M^{-1}C_0)'P_2(M^{-1}C_0). \end{aligned}$$

Then, Infante's approach was used to obtain the following result:

Theorem 63.3: Wiens and Sinha

The system Equation 63.25 is almost surely asymptotically stable (Definition $II_{a.s.}$ in Section 63.1) in the large if a positive definite matrix P_2 exists so that

1. *$P_2M^{-1}K_0$ is positive definite,*
2. *The symmetric parts of $P_2M^{-1}C_0$ and $(M^{-1}C_0)'P_2(M^{-1}K_0)$ are positive definite, and*
3. $E\{\lambda \max[(A_0 + C(t) + K(t))P^{-1}]\} < 0$,

where $C(t)$ and $K(t)$ are given by

$$C(t) = \begin{bmatrix} 0 & \frac{1}{2}(M^{-1}C_0)'P_2M^{-1}C(t) \\ \frac{1}{2}(M^{-1}C_0)'P_2M^{-1}C(t) & \{P_2M^{-1}(H)\} + \{P_2M^{-1}C(t)\}' \end{bmatrix}$$

and

$$K(t) = \begin{bmatrix} \frac{1}{2}\{(M^{-1}C_0)'P_2M^{-1}K(t)\} + \frac{1}{2}\{M^{-1}C_0)'P_2M^{-1}K(t)\}' & \{P_2M^{-1}K(t)\}' \\ P_2M^{-1}K(t) & 0 \end{bmatrix}.$$

When applying their results to the second-order system Equation 63.24, Wiens and Sinha [56] obtained the stability criteria,

$$E\{f^2(t)\} < 4\beta^2 c, \quad \text{and} \tag{63.27}$$

$$E\{g^2(t)\} < 4\beta^2 c/(c + 2\beta^2). \tag{63.28}$$

These results are similar to those obtained by Caughey and Gray [12]. Equation 63.27 is the "optimal" result of Infante, but Equation 63.28 is not. This result is not surprising, because no optimization procedure was used in deriving the stability criteria. The usefulness of the Wiens and Sinha theorem was demonstrated by applying their method to higher-order systems ($n = 4$ and 6) that yielded stability regions of practical significance.

Another research direction for application of the Lyapunov function method is the study of stochastic feedback systems where the forward path is a time-invariant linear system and the random noise appears in the feedback path as a multiplicative feedback gain. In this work, Lyapunov functions are constructed by analogous methods for deterministic systems, for example, the Lyapunov equation, the path-integral technique and the Kalman–Yacubovich–Popov method. However, most of the results obtained can be derived by directly using a quadratic form Lyapunov function together with the associated Lyapunov equation.

Kleinman [25] considered a stochastic system in the form,

$$dx(t) = Ax(t)\,dt + Bx(t)\,d\zeta(t), \tag{63.29}$$

where $\zeta(t)$ is a scalar Wiener process with $E\{[\zeta(t) - \zeta(\tau)]^2\} = \sigma^2 \mid t - \tau \mid$. By using the moment equation and a quadratic Lyapunov function, the following necessary and sufficient condition for 0 to be "stable with probability one" (this is the same as Kushner's definition) was derived:

$$I \otimes A + A \otimes I + \sigma^2 B \otimes B$$

is a stable matrix where "$\otimes$" denotes the Kronecker product of matrices. However, this result should be carefully interpreted as discussed by Willems [58].

Willems [58,59] studied the feedback system in the form,

$$\begin{cases} dx(t) = (A_0x(t) - kbcx(t))\,dt - bcx\,d\zeta(t), \\ x(0) = x_0 \in \mathbb{R}^n, \end{cases} \tag{63.30}$$

with $\zeta(t)$ a scalar-valued Wiener process with $E\{[\zeta(t) - \zeta(\tau)]^2\} = \sigma^2 \mid t - \tau \mid$, that is, a system consisting of a linear time-invariant plant in the forward path with rational transfer function $H(s)$ and minimal realization (A_0, b, c), and a multiplicative feedback gain that is the sum of a deterministic constant k and a stochastic noise component $\zeta(t)$. Suppose the system is written in companion form with state $x(t) = (y(t), Dy(t), \ldots D^{n-1}y(t))$ where $D = d/dt$ and $y(t) = cx(t)$ is the scalar output. The closed-loop transfer function is

$$G(s) = c(sI - A_0 - kbc)^{-1}b = \frac{q(s)}{p(s)} \tag{63.31}$$

with $p(s)$ and $q(s)$ relatively prime because the realization is minimal. Then, the input–output relation can be written in the form,

$$p(D)y(t)\,dt + q(D)y(t)\,d\zeta(t) = 0. \tag{63.32}$$

Willems observed that if a positive-definite quadratic form $V(x)$ was used as a Lyapunov function, then, following Kushner, $\mathcal{L}V(x)$ is the sum of a "deterministic part" obtained by setting the noise equal to zero and the term $\frac{1}{2}\sigma^2(q(D)y)^2(\partial^2 V/\partial x_n^2)$ $(x_n = D^{n-1}y)$. Here, $\mathcal{L}$ is the infinitesimal generator of the system. To guarantee the negativity of $\mathcal{L}V(x)$ to assure stability, using Equation 63.32, Willems used a construction technique for a Lyapunov function that was originally developed by Brockett [10], that is

$$V(x) = \int_{t(0)}^{t(x)} [p(D)yh(D)y - (q(D)y)^2]\,dt$$

where the polynomial $h(s) = s^n + h_{n-1}s^{n-1} + \cdots + h_0$ is the unique solution (assuming $p(s)$ is strictly Hurwitz) of

$$\frac{1}{2}[h(s)p(-s) + h(-s)p(s)] = q(s)q(-s)$$

so that

$$\mathcal{L}V(x) = -(q(D)y)^2 + \frac{1}{2}h_{n-1}\sigma^2(q(D)y)^2.$$

By applying Kushner's results, Willems obtained the following theorem:

Theorem 63.4: Willems

1. *The origin is mean-square stable in the large (in the sense that $R(t) = E\{x(t)x'(t)\} < M < +\infty$ for $t \geq 0, x_0 \in \mathbb{R}^n$ and $\sup_{t\geq 0} \|R(t)\| \to 0$ as $\|R(0)\| \to 0$) and stable "with probability one" (in Kushner's sense) in the large, if $p(s)$ is strictly Hurwitz and*

   $$\frac{\sigma^2 h_{n-1}}{2} \leq 1. \tag{63.33}$$

   Moreover, the following identity holds:

   $$\frac{\sigma^2 h_{n-1}}{2} = \sigma^2 \int_0^{\infty} [g(t)]^2\,dt = \frac{\sigma^2}{2\pi} \int_{-\infty}^{+\infty} |\,G(i\omega)\,|^2\,dw$$

   where $g(t)$ is the impulse response of the closed-loop (stable) deterministic system and $g(t)$ and $G(s)$ are Laplace transform pairs.
2. *If the inequality Equation 63.33 holds, the stability of the origin as indicated in condition 1 above is asymptotic.*

Brockett and Willems [11] studied the linear feedback system given by

$$\dot{x}(t) = Ax(t) - BK(t)Cx(t) \tag{63.34}$$

where (A, B, C) is a completely symmetric realization, that is, $A^T = A \in \mathbb{R}^{n\times n}$, $B = C' \in \mathbb{R}^{n\times m}$, and $K(t) \in \mathbb{R}^{m\times n}$ is a stationary ergodic matrix process. By using a quadratic Lyapunov function, the specific properties of a completely symmetric system, and the well-known Kalman–Yacubovich–Popov Lemma, they obtained the following stability theorem:

Theorem 63.5: Brockett and Willems

For the system Equation 63.34,

1. *If $K(t) = K'(t)$ almost surely and*

$$\bar{\lambda}_{\max} = E\{\lambda_{\max}(A - BK(t)C)\} < 0,$$

then the origin is almost surely asymptotically stable (in the sense that $\lim_{t\to+\infty} x(t) = 0$ a.s.). *In particular, if $m = 1$, which is analogous to a single input–single output system, then*

$$\bar{\lambda}_{\max} = \int_{Z_1}^{\infty} \sigma p\left(-\frac{1}{g(\sigma)}\right)\left|\frac{\partial g(\sigma)/\partial\sigma}{g^2(\sigma)}\right| d\sigma$$

where Z_1 is the largest zero of the open-loop transfer function $g(s) = C(sI - A)^{-1}B$ and $p(\cdot)$ is the density function of $K(0)$.

2. *If $m = 1$ and*

$$g(s) = C(sI - A)^{-1}B = \frac{q_{n-1}s^{n-1} + \cdots + q_0}{s^n + p_{n-1}s^{n-1} + \cdots + p_0},$$

the origin is almost surely asymptotically stable if a constant β exists so that
   a. $E\{\min(\beta, K(t))\} > 0$,
   b. *The poles of $g(s)$ lie in $Re\{s\} < -q_{n-1}\beta$,*
   c. *The locus of $G(i\omega - q_{n-1}\beta)$, $-\infty < \omega < +\infty$, does not encircle or intersect the closed disc centered at $(-1/2\beta, 0)$ with radius $1/2\beta$ in the complex plane.*

Mahalanabis and Purkayastha [38] as well as Socha [53], applied techniques similar to Willems and Willems and Brockett to study nonlinear stochastic systems. Readers should consult their papers and the references cited therein for more details on this approach.

An extension of Kushner's and Has'minskii's work on the Lyapunov function approach uses the Lyapunov function technique to study large-scale stochastic systems. The development of large-scale system theory in the past several decades was an impetus for these studies. The Lyapunov function technique of Kushner and Has'minskii is based on the use of scalar, positive-definite functions and is not effective for large-scale systems. As in the deterministic case, the difficulties are often overcome, if a vector positive-definite function is used. Michel and Rasmussen [40,41,51] used vector valued positive-definite functions to study various stability properties of large-scale stochastic systems. Their approach was to construct a Lyapunov function for the complete system from those of the subsystems. Stability properties were studied by investigating the stability properties of the lower order subsystems and the interconnection structure. Ladde and Siljak in [33] and Siljak [52] established quadratic mean stability criteria by using a positive-definite vector Lyapunov function. This is an extension of the comparison principle developed by Ladde for deterministic systems to the stochastic case. In these works, a linear comparison system was used. Bitsris [9] extended their work by considering a nonlinear comparison system. White noise systems were studied in all of the above works. Socha [54] investigated a real noise system where the noise satisfied the law of large numbers. Has'minskii's result [22] was extended to a large-scale system in this work. Interested readers are referred to the original papers and references cited therein.

63.3 The Lyapunov Exponent Method and the Stability of Linear Stochastic Systems

One of the major advances in the study of stochastic stability during the past several decades is the application of the Lyapunov exponent concept to stochastic systems. This method uses sophisticated mathematical techniques to study the sample behavior of stochastic systems and often yields necessary and sufficient conditions for almost sure (sample) stability in the sense that

$$\lim_{t\to\infty} \|x(t, x_0, \omega)\| = 0 \quad \text{a.s.} \tag{63.35}$$

Because this method focuses on the sample path behavior of a stochastic system, rather than the moments, it has the greatest potential for applications in science and engineering. In this section we present a summary of this method along with selected results.

After the introduction of the concept of Lyapunov exponents by Lyapunov [37], this concept has formed the foundation for many investigations into the stability properties of deterministic dynamical systems. However, it is only recently that the Lyapunov exponent method has been used to study almost sure stability of stochastic systems. The setup is as follows:

Consider the continuous time linear stochastic system defined by

$$\begin{cases} \dot{x}(t) = A(t,\omega)x(t), & t \geq 0, \\ x(0) = x_0 \in \mathbb{R}^d. \end{cases} \qquad (\Sigma_c)$$

Let $x(t, x_0, \omega)$ denote the unique sample solution of (Σ_c) initial from x_0 for almost all $\omega \in \Omega$. (We always denote the underlying probability space by $(\Omega, \mathcal{F}, P)$). The Lyapunov exponent $\overline{\lambda}_\omega(x_0)$ determined by x_0 is defined by the random variable,

$$\overline{\lambda}_\omega(x_0) = \overline{\lim}_{t\to+\infty} \frac{1}{t} \log \|x(t, x_0, \omega)\|, \tag{63.36}$$

for (Σ_c). Although x_0 can be a random variable, for simplicity we will concern ourselves primarily with the case where x_0 is nonrandom and fixed.

In the case $A(t, \omega) = A(t)$ is a deterministic $\mathbb{R}^{d\times d}$-valued continuous bounded function, Lyapunov [37] proved the following fundamental result for the exponent $\overline{\lambda}(x_0)$ for the system,

$$\dot{x}(t) = A(t)x(t): \qquad (\Sigma)$$

1. $\overline{\lambda}(x_0)$ is finite for all $x_0 \in \mathbb{R}^d\backslash\{0\}$.
2. The set of real numbers which are Lyapunov exponents for some $x_0 \in \mathbb{R}^d\backslash\{0\}$ is finite with cardinality p, $1 \leq p \leq d$;
$$-\infty < \lambda_1 < \lambda_2 < \cdots < \lambda_p < +\infty, \quad \lambda_i \in \mathbb{R}\forall i.$$
3. $\overline{\lambda}(cx_0) = \overline{\lambda}(x_0)$ for $x_0 \in \mathbb{R}^d\backslash\{0\}$ and $c \in \mathbb{R}\backslash\{0\}$. $\overline{\lambda}(\alpha x_0 + \beta y_0) \leq \max\{\overline{\lambda}(x_0), \overline{\lambda}(y_0)\}$ for $x_0, y_0 \in \mathbb{R}^d\backslash\{0\}$ and $\alpha, \beta \in \mathbb{R}$ with equality if $\overline{\lambda}(x_0) < \overline{\lambda}(y_0)$ and $\beta \neq 0$. The sets $\mathcal{L}_i = \{x \in \mathbb{R}^d\backslash\{0\}\colon \overline{\lambda}(x) = \lambda_i\}$, $i = 1, 2, \ldots, p$, are linear subspaces of $\mathbb{R}^d$, and $\{\mathcal{L}_i\}_{i=0}^p$ is a filtration of $\mathbb{R}^d$, that is,
$$\{0\} \triangleq \mathcal{L}_0 \subset \mathcal{L}_1 \subset \cdots \subset \mathcal{L}_p = \mathbb{R}^d$$
where $d_i \triangleq \dim(\mathcal{L}_i) - \dim(\mathcal{L}_{i-1})$ is called the *multiplicity* of the exponent λ_i for $i = 1, 2, \ldots, p$ and the collection $\{(\lambda_i, d_i)\}_{i=1}^p$ is referred to as the *Lyapunov spectrum* of the system (Σ). We have the

relation

$$\sum_{i=1}^{p} d_i\lambda_i \leq \underline{\lim}_{t\to+\infty}\frac{1}{t}\log\|\Phi(t)\| \leq \overline{\lim}_{t\to+\infty}\frac{1}{t}\log\|\Phi(t)\| \tag{63.37}$$

where $\Phi(t)$ is the transition matrix of (Σ). The system is said to be (forward) regular if the two inequalities in Equation 63.37 are equalities. For a forward regular system the $\underline{\lim}$ and $\overline{\lim}$ can be replaced by lim .

For the stochastic system (Σ_c) with $\omega \in \Omega$ fixed, the relationship given in Equation 63.36 implies that if $\overline{\lambda}_\omega(x_0) < 0$, then the sample solution $x(t, x_0, \omega)$ will converge to zero at the exponential rate $|\overline{\lambda}_\omega(x_0)|$ and, if $\overline{\lambda}_\omega(x_0) > 0$, then the sample solution $x(t, x_0, \omega)$ cannot remain in any bounded region of $\mathbb{R}^d$ indefinitely. From this we see that $\overline{\lambda}_\omega(x_0)$ contains information about the sample path stability of the system and as we will see later, in many cases a necessary and sufficient condition for sample path (almost sure) stability is obtained.

Arnold and Wihstutz [5] have given a detailed survey of research work on Lyapunov exponents. The survey is mathematically oriented and presents a summary of general properties and results on the topic. Readers are encouraged to refer to Arnold and Wihstutz [5] for more details. Our focus, however, is more on the application of Lyapunov exponents to the study of stability of stochastic systems.

The fundamental aspects of the Lyapunov exponent method were applied to stochastic systems by Furstenberg, Oseledec, and Has'minskii. We will briefly review the work of Oseledec and Furstenberg and then focus attention on the work of Has'minskii, that best illustrates the application of the Lyapunov exponent to the study of stochastic stability.

The random variables $\overline{\lambda}_\omega(x_0)$ given in Equation 63.36 are simple, nonrandom constants under certain conditions, for example, stationarity and ergodicity. This is a major consequence of the multiplicative ergodic theorem of Oseledec [46] that establishes conditions for the regularity of stochastic systems. In his original work, Lyapunov used the regularity of (Σ) to determine the stability of a perturbed version of the system (Σ) from the stability of (Σ). Although regularity is usually very difficult to verify for a particular system, Oseledec proved an almost sure statement about regularity. Because of our interest in the stability of stochastic systems, the theorem can be stated in the following special form for (Σ_c), see Arnold et al. [2].

Theorem 63.6: Multiplicative Ergodic Theorem: Oseledec

Suppose $A(t, \omega)$ is stationary with finite mean, that is, $E\{A(0, \omega)\} < \infty$. Then for (Σ_c), we have

1. *State-space decomposition:*
   For almost all $\omega \in \Omega$, an integer $r = r(\omega)$ exists with $1 \leq r(\omega) \leq d$, real numbers $\lambda_1(\omega) < \lambda_2(\omega) < \cdots < \lambda_r(\omega)$, and linear subspaces (Oseledec spaces) $E_1(\omega), \ldots, E_r(\omega)$ with dimension $d_i(\omega) = \dim[E_i(\omega)]$ so that

$$\mathbb{R}^d = \bigoplus_{i=1}^{r} E_i(\omega)$$

and

$$\lim_{t\to+\infty}\frac{1}{t}\log\|\Phi(t,\omega)x_0\| = \lambda_i(\omega), \quad \text{if } x_0 \in E_i(\omega)$$

where $\Phi(t, \omega)$ is the transition matrix of (Σ_c) and "$\bigoplus$" denotes the direct sum of subspaces.

2. *Domain of attraction of* $E_i(\omega)$*:*

$$\lim_{t\to+\infty}\frac{1}{t}\log\|\Phi(t,\omega)x_0\| = \lambda_i(\omega), \quad \textit{iff}\ x_0 \in \mathcal{L}_i(\omega)\backslash\mathcal{L}_{i-1}(\omega),$$

where $\mathcal{L}_i(\omega) = \bigoplus_{j=1}^{i} E_j(\omega)$.

3. *Center of gravity of exponents:*

$$\sum_{i=1}^{r(\omega)} d_i(\omega)\lambda_i(\omega) = \lim_{t\to+\infty}\frac{1}{t}\log\big|\det\Phi(t,\omega)\big| = trE\{A_0(\omega)\big|\widetilde{\mathcal{F}}\}$$

where $\widetilde{\mathcal{F}}$ *is the* σ *algebra generated by the invariant sets of* $A(t,\omega)$.

4. *Invariance property: If* $A(t,\omega)$ *is ergodic as well, then the random variables* $r(\omega)$, $\lambda_i(\omega)$, *and* $d_i(\omega)$, $i = 1, 2, \ldots, r$, *are independent of* ω *and are nonrandom constants.*

Note that under the current assumptions, all $\overline{\lim}$ are actually lim and (3) is equivalent to the almost sure regularity of the stochastic system (Σ_c). Oseledec's theorem is a very general result and the above is a special version for the system (Σ_c). A detailed statement of the theorem is beyond the scope of this chapter. As far as sample path (almost sure) stability is concerned, the sign of the top exponent λ_r is of interest. We present a simple example to illustrate the application of the theorem.

Example 63.3:

Consider the randomly switched linear system

$$\begin{cases}\dot{x}(t) = \frac{1}{2}(1-y(t))A_{-1}x(t) + \frac{1}{2}(1+y(t))A_1x(t), \quad t \geq 0,\\ x(0) = x_0,\end{cases} \tag{63.38}$$

where

$$A_{-1} = \begin{bmatrix}-a & 0\\ 0 & -b\end{bmatrix}, \quad A_1 = \begin{bmatrix}c & 1\\ 0 & c\end{bmatrix}, \quad 0 < b < a < +\infty \ \ 0 < c < +\infty, \tag{63.39}$$

and $y(t) \in \{-1, +1\}$ is the random telegraph process with mean time between jumps $\alpha^{-1} > 0$. Then,

$$e^{A_{-1}t} = \begin{bmatrix}e^{-at} & 0\\ 0 & e^{-bt}\end{bmatrix}, \tag{63.40}$$

and

$$e^{A_1t} = \left[\begin{pmatrix}e^{ct} & te^{ct}\\ 0 & e^{ct}\end{pmatrix}\right]. \tag{63.41}$$

The phase curves of these two linear systems are as shown in Figure 63.1.

It is easy to verify that, for $x_0 \in \mathbb{R}^2\backslash\{0\}$,

$$\overline{\lambda}_\omega(x_0) = \lim_{t\to+\infty}\frac{1}{t}\log\|x(t,x_0,\omega)\| = \alpha\cdot\lim_{k\to\infty}\frac{1}{k}\log\|e^{A_{y(k)}\tau^{(k)}}\cdots e^{A_{y(1)}\tau^{(1)}}x_0\| \text{ a.s.} \tag{63.42}$$

and

$$\overline{\lambda}_\omega(\mathbb{R}^2) \stackrel{\text{def}}{=} \lim_{t\to+\infty}\frac{1}{t}\log\|\det\Phi(t,\omega)\| = \alpha\lim_{k\to\infty}\frac{1}{k}\log\|\det\left(e^{A_{y(k)}\tau^{(k)}}\cdots e^{A_{y(1)}\tau^{(1)}}\right)\| \text{ a.s.} \tag{63.43}$$

where $\{y(k) : k \geq 1\}$ is the embedded Markov chain of $y(t)$ and $\{\tau^{(k)} : k \geq 1\}$ is a sequence of i.i.d. random variables exponentially distributed with parameter α, and independent of $\{y(k) : k \geq 1\}$. Because $y(t)$ is

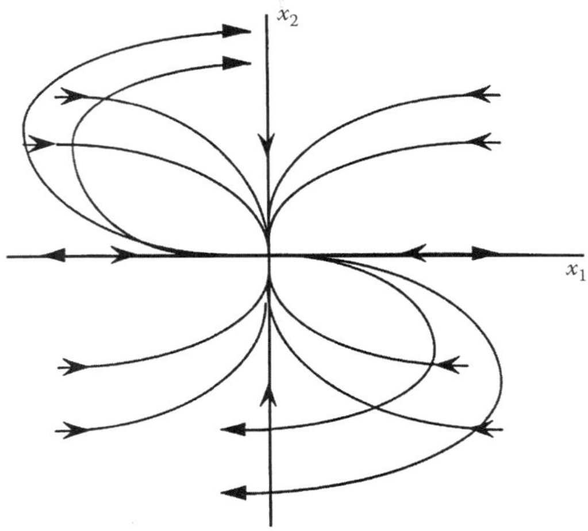

FIGURE 63.1 Phase curves.

stationary and ergodic with unique stationary (invariant) distribution $P\{y(t)=1\}=P\{y(t)=-1\}=\frac{1}{2}$, Oseledec's theorem applies and a $\mathcal{F}$-set N_0 exists with $P(N_0)=0$ and two real numbers λ_1 and λ_2 (with possibly $\lambda_1=\lambda_2$) so that

$$\overline{\lambda}_\omega(x_0)\in\{\lambda_1,\lambda_2\}\quad \text{for } x_0\in\mathbb{R}^2\backslash\{0\}\quad \text{and}\quad \omega\in\Omega\backslash N_0 \tag{63.44}$$

From Equations 63.40, 63.41, and 63.43 and the law of large numbers,

$$\begin{aligned}\overline{\lambda}_\omega(\mathbb{R}^2)&=\alpha\cdot\lim_{k\to\infty}\frac{1}{k}\log e^{-(a+b)(\tau^{(1)}+\cdots+\tau^{(i_k)})}\cdot e^{2c(\tau_{(1)}+\cdots+\tau_{(k-i_k)})},\\&=-(a+b)\alpha\lim_{k\to\infty}\frac{1}{k}(\tau^{(1)}+\cdots+\tau^{(i_k)})+2c\alpha\lim_{k\to\infty}\frac{1}{k}(\tau_{(1)}+\cdots+\tau_{(k-i_k)}),\\&=-\frac{(a+b)}{2+c}\quad\text{a.s.}\end{aligned}\tag{63.45}$$

Here i_k is the number of times that $y(j)$ takes the value -1 in k steps and $\{\tau^{(i)},\tau_{(j)};\ i,j\geq 1\}$ is a sequence of i.i.d. random variables exponentially distributed with parameter α.

Because $E=\text{span}\{e_1=(1,0)^T\}$ is a common eigenspace of $e^{A_{-1}t}$ and e^{A_1t}, by the law of large numbers, it is clear that for any $x_0\in E\backslash\{0\}$,

$$\overline{\lambda}_\omega(x_0)=-\frac{1}{2}a+\frac{1}{2}c\quad\text{a.s.}\quad\text{for } x_0\in E\backslash\{0\}. \tag{63.46}$$

From (3) of the theorem, Equations 63.45 and 63.46, we obtain

$$\lambda_1=\frac{1}{2}c-\frac{1}{2}a\quad\text{and}\quad\lambda_2=\frac{1}{2}c-\frac{1}{2}b \tag{63.47}$$

(with $\lambda_1<\lambda_2$). We can identify the following objects in Oseledec's theorem:

$$\begin{cases}r=2, & E_1(\omega)=E,\quad \text{for } \omega\in\Omega\backslash N_1,\\ d_1=1, & E_2(\omega)=E^{\perp}=\text{span}\{e_2=(0,1)^T\},\quad \text{for } \omega\in\Omega\backslash N_1,\\ d_2=1, & \mathcal{L}_1(\omega)=E,\ \mathcal{L}_2(\omega)=\mathbb{R}^2,\quad \text{for } \omega\in\Omega\backslash N_1,\end{cases}\tag{63.48}$$

where N_1 is a $\mathcal{F}$-set with $P(N_1) = 0$ and

$$\begin{cases} \overline{\lambda}_\omega(x_0) = \lambda_1, & \text{for } x_0 \in E\backslash\{0\} \quad \text{and} \quad \omega \in \Omega\backslash N_1, \\ \overline{\lambda}_\omega(x_0) = \lambda_2, & \text{for } x_0 \in \mathbb{R}^2\backslash E \quad \text{and} \quad \omega \in \Omega\backslash N_1. \end{cases} \tag{63.49}$$

From Equation 63.49 it follows that the system Equation 63.38 is almost surely exponentially stable in the sense that

$$P\{\|x(t, \omega, x_0)\| \to 0 \text{ as } t \to +\infty, \text{ at an exponential rate }\} = 1$$

if and only if $\lambda_2 < 0$ (iff $c < b$). Note that the system is almost surely exponentially unstable if and only if $\lambda_1 > 0$ (iff $c > 0$).

We remark here that we can compute λ_1 and λ_2 directly using Equation 63.46 and the center of gravity of the exponent relation for this simple system. An interesting question is what happens if $b > a$. In this case, the top exponent is $\frac{1}{2}c - \frac{1}{2}a$ and only the top exponent in Oseledec's theorem is physically "realizable" in the sense that there exists a $x_0 \in \mathbb{R}^2$ so that

$$P\{\overline{\lambda}_\omega(x_0) = \frac{1}{2}c - \frac{1}{2}a\} > 0.$$

Actually, it follows from Feng and Loparo [14,15] that $\overline{\lambda}_\omega(x_0) = \frac{1}{2}c - \frac{1}{2}a$ a.s. for any $x_0 \in \mathbb{R}^2\backslash\{0\}$ in this case.

Motivated by Bellman [7], Furstenberg and Kesten [16], as a starting point, studied the product of random matrices and generalized the classical law of large numbers of an i.i.d. sequence of random variables. They showed that under some positivity conditions of the entries of the matrices, the random variable $n^{-1} \log \|X_n \ldots X_1\|$ tended to a constant almost surely for a sequence of i.i.d. random matrices $\{X_k\}_{k=1}^{\infty}$. Furstenberg [17,18] went one step further and showed that, if $X_i \in G$, for all i, where G is a noncompact semisimple connected Lie group with finite center, there exists a finite-dimensional vector space of functions ($\psi(\cdot)$, rather than just $\log \|\cdot\|$) so that $n^{-1}\psi(X_n \ldots X_1)$ tended to a constant α_ψ (known as Furstenberg's constant) almost surely as $n \to +\infty$. These studies also considered the action of the group G on certain manifolds, for example, $\mathbb{P}^{d-1} \subset \mathbb{R}^d$ (the projective sphere in $\mathbb{R}^d$, that is, $\mathbb{P}^{d-1}$ is obtained by identifying s and $-s$ on the unit sphere S^{d-1} in $\mathbb{R}^d$). Furstenberg showed that under a certain transitivity or irreducibility condition, there exists a unique invariant probability measure ν for the Markov chain $Z_n = X_n * \ldots * X_1 * Z_0$, which evolves on $\mathbb{P}^{d-1}$, with common probability measure μ for $X_i \in G$ (here, " $*$ " denotes the action of G on $\mathbb{P}^{d-1}$) and for any $x_0 \in \mathbb{R}^d\backslash\{0\}$, the Lyapunov exponent

$$\overline{\lambda}_\omega(x_0) = \overline{\lim_{n\to\infty}} \frac{1}{n} \log \|X_n * \cdots * X_1 * x_0\| = \int\int_{G\times\mathbb{P}^{d-1}} \log \|g * z\| \mu(dg)\nu(dz). \tag{63.50}$$

Furstenberg's works developed deep and significant results and Equation 63.50 is a prototype for computing the Lyapunov exponent in most of the later work that followed. The following simple discrete-time example due to Has'minskii best illustrates Furstenberg's idea.

Example 63.4: Has'minskii

Consider a discrete time system, $x_{n+1} = A_n(\omega)x_n$, with $\{A_n(\omega)\}_{n=1}^{\infty}$, a sequence of i.i.d. random matrices. Let

$$\begin{cases} \rho_n = \log \|x_n\|, \quad \text{and} \\ \varphi_n = \|x_n\|^{-1} x_n \in S^{d-1}. \end{cases}$$

Then,

$$\rho_n = \log \|x_n\| = \rho_{n-1} + \log \|A_n \rho_{n-1}\| = \rho_0 + \sum_{k=1}^{n} \log \|A_k \varphi_{k-1}\|.$$

Because $\{A_n\}_{n=1}^{\infty}$ are independent, $\{A_n, \varphi_{n-1}\}_{n=1}^{\infty}$ forms a Markov chain in $\mathbb{R}^{d\times d}\times S^{d-1}$. If $\{A_n\varphi_{n-1}\}_{n=1}^{\infty}$ is ergodic, the law of large number gives

$$\begin{aligned}\bar{\lambda}_\omega(x_0) &= \overline{\lim_{n\to\infty}}\frac{1}{n}\rho_n = \lim_{n\to\infty}\frac{1}{n}\sum_{i=1}^{n}\log\|A_i\varphi_{i-1}\| = E\{\log\|A\varphi\|\}\\ &= \int\int_{\mathbb{R}^{d\times d}\times S^{d-1}}\log\|A\varphi\|\mu(dA)\nu(d\varphi)\quad \text{a.s.}\end{aligned}$$

where μ is the common probability measure of $\{A_i\}_{i=1}^{\infty}$ and ν is the unique invariant probability measure of φ_n with respect to μ, (i.e., ν is such that if φ_0 is distributed according to ν on S^{d-1}, then $\varphi_1 = \|A_1\varphi_0\|^{-1}A_1\varphi_0$ has the same distribution ν on S^{d-1}).

Has'minskii [20] generalized Furstenberg's idea to a linear system of Ito stochastic differential equations and obtained a necessary and sufficient condition for the almost sure stability of the system:

$$\begin{cases} dx(t) = Ax(t)\,dt + \sum_{i=1}^{m} B_i x(t)\,d\xi_i(t), & t\ge 0\\ x(0) = x_0 \end{cases} \tag{63.51}$$

where $\xi_i(t)$ are independent standard Wiener processes. The process $x(t)$ is then a Markov diffusion process with the infinitesimal generator

$$\mathcal{L}u = \langle Ax, u_x\rangle + \frac{1}{2}\sum_{i,j=1}^{d}\sigma_{ij}(x)\frac{\partial^2 u}{\partial x^i\partial x^j} \tag{63.52}$$

where u is a real-valued twice continuously differentiable function, $u_x = \partial u/\partial x$, $\langle\cdot,\cdot\rangle$ is the standard inner product on $\mathbb{R}^d$, and

$$\Sigma(x) = (\sigma_{ij}(x))_{d\times d} = \sum_{i=1}^{m} B_i x x' B_i' \tag{63.53}$$

By introducing the polar coordinates, $\rho = \log\|x\|$ and $\varphi = \|x\|^{-1}x$ for $x \neq 0$, and applying Ito's differentiation rule,

$$d\rho(t) = Q(\varphi(t))\,dt + \sum_{i=1}^{m}\varphi(t)'B_i\varphi(t)\,d\xi_i(t) \tag{63.54}$$

$$d\varphi(t) = H_0(\varphi(t))\,dt + \sum_{j=1}^{m} H_j(\varphi)\,d\xi_j(t) \tag{63.55}$$

where

$$Q(\varphi(t)) = \varphi(t)'A\varphi(t) + \frac{1}{2}tr\Sigma(\varphi(t)) - \varphi(t)'\Sigma(\varphi(t))\varphi(t)$$

and $H_0(\varphi(t))$, $H_i(\varphi(t))$ are projections of the linear vector fields Ax and B_ix on $\mathbb{R}^d\backslash\{0\}$ onto S^{d-1}. Then from Equation 63.55 we see that $\varphi(t)$ is independent of $\rho(t)$ and $\varphi(t)$ is a Markov diffusion process on S^{d-1}. Has'minskii assumed a nondegeneracy condition of the form,

$$(H)\quad \alpha'\Sigma(x)\alpha \ge m\|\alpha\|\cdot\|x\|,\quad \forall\alpha\in\mathbb{R}^m \quad \text{and} \quad x\in\mathbb{R}^d$$

that guaranteed that the angular process $\varphi(t)$ is an ergodic process on S^{d-1} with unique invariant probability measure ν. Then the time average of the right-hand side of Equation 63.54 is equal to the ensemble average, that is,

$$\overline{\lambda}_\omega(x_0) = \overline{\lim_{t\to+\infty}} \frac{1}{t}\int_0^t Q(\varphi(t))\,dt = E\{\varphi(t)\} = \int_{S^{d-1}} Q(\varphi)\nu(d\varphi) = J \quad \text{a.s.} \tag{63.56}$$

independent of $x_0 \in \mathbb{R}^d \backslash \{0\}$.

Theorem 63.7: Has'minskii

Under condition (H), $J < 0$ implies that the system Equation 63.51 is almost surely stable. If $J > 0$, then the system Equation 63.51 is almost surely unstable and

$$P\{\lim_{t\to+\infty} \|x(t,x_0,\omega)\| = +\infty\} = 1.$$

If $J = 0$, then

$$P\left\{\overline{\lim_{t\to+\infty}} \|x(t,x_0,\omega)\| = +\infty\right\} > 0,$$

$$P\left\{\underline{\lim_{t\to+\infty}} \|x_t(x_0,\omega)\| < +\infty\right\} > 0.$$

To determine the sign of J, the invariant probability measure ν must be found. This can be accomplished by solving the so-called Fokker–Planck (forward) equation for the density μ of the measure ν. For simplicity, we consider the two dimensional case ($d = 2$). It can be shown from Equation 63.55 that the angular process $\varphi(t)$ satisfies the Ito equation

$$d\varphi(t) = \Phi(\varphi(t))\,dt + \Psi(\varphi(t))\,dx_i(t) \tag{63.57}$$

with $x_i(t)$ a standard Wiener process and Φ and Ψ appropriate functions of $\varphi(t)$. $\varphi(t)$ has generator

$$\mathcal{L}_\varphi = \Phi(\varphi)\frac{d}{d\varphi} + \frac{1}{2}\Psi^2(\varphi)\frac{d^2}{d\varphi^2}. \tag{63.58}$$

Then, the Fokker–Planck equation for μ is

$$\mathcal{L}^*_\varphi\mu = \frac{1}{2}\frac{d^2}{d\varphi^2}(\Psi^2(\varphi)\mu) - \frac{d}{d\varphi}(\Phi(\varphi)\mu) = 0 \tag{63.59}$$

with normalization and consistency constraints

$$\int_0^{2\pi} \mu(\varphi)d\varphi = 1, \quad \mu(0) = \mu(2\pi). \tag{63.60}$$

Condition (H) guarantees that Equation 63.59 is nonsingular in the sense that $\Psi(\varphi) \neq 0$ for any $\varphi \in S^{d-1}$, and thus admits a unique solution satisfying Equation 63.60. However, even for simple but nontrivial systems, an analytic solution for the invariant density μ is very difficult to obtain. The problem is even more complicated if (H) is not satisfied, and this is more often than not the case in many applications.

In this situation, singularity on S^1 is a result of $\Psi(\varphi) = 0$ for some $\varphi \in S^1$ and the Markov diffusion process $\varphi(t)$ may not be ergodic on the entire manifold S^1. One must then examine each ergodic component of S^1 to determine a complete description of the invariant probability measure. The law of large numbers is then used to compute the exponent for each x_0 belonging to the ergodic components. Has'minskii [19–22] presented two second-order examples to illustrate the treatment of ergodic components and the usefulness of the theorem. In one of the examples it was shown that an unstable linear deterministic system could be stabilized by introducing white noise into the system. This settled a long standing conjecture about the possibility of stabilizing an unstable linear system using noise. Has'minskii's work was fundamental to the later developments of the Lyapunov exponent method for studying stochastic stability.

Following Has'minskii, Kozin and his coworkers studied the case when the nondegeneracy condition (H) fails. This work made extensive use of one-dimensional diffusion theory to obtain analytical and numerical results for second-order linear white-noise systems. Kozin and Prodromou [28] applied Has'minskii's idea to the random harmonic oscillator system in the two forms,

$$\frac{d^2u(t)}{dt^2} + (1 + \sigma\dot{B}(t))u(t) = 0, \tag{63.61}$$

and

$$\frac{d^2u(t)}{dt^2} + 2\xi \cdot \frac{du(t)}{dt} + (1 + \sigma\dot{B}(t))u(t) = 0, \tag{63.62}$$

where $\dot{B}(t)$ represents a Gaussian white-noise process. After transforming the systems into an Ito representation, they observed that condition (H) is not satisfied with singularities $\Psi(\pm\frac{\pi}{2}) = 0$. Kozin and Prodromou then used one-dimensional diffusion theory to study the singularities and sample path behavior of the angular process $\varphi(t)$ near the singularities. The angular process traversed the circle in clockwise direction and was neither trapped at any point, nor did it remain at a singularity $+\frac{\pi}{2}$ or $-\frac{\pi}{2}$ for any positive time interval. Thus, the angular process $\varphi(t)$ was ergodic on the entire circle S^1. From the law of large numbers, they obtained

$$\bar{\lambda}_\omega(x_0) = J = E\{Q(\varphi)\} = \frac{E\left\{\int_0^{\mathcal{T}_r} Q(\varphi(s))\,ds\right\} + E\{\int_0^{\mathcal{T}_\ell} Q(\varphi(s))\,ds\}}{E\{\mathcal{T}_r\} + E\{\mathcal{T}_\ell\}} \quad \text{a.s.}$$

independent of x_0. $\mathcal{T}_r$ and $\mathcal{T}_\ell$ are Markov times for $\varphi(t)$ to travel the right and left half circles, respectively. Using the so-called speed and scale measures, they were able to show the positivity of J for Equation 63.61 for $\sigma^2 \in (0, +\infty)$. This proved the almost sure instability of the random harmonic oscillator with white noise parametric excitation. For the damped oscillator Equation 63.62, they failed to obtain analytical results, and the almost sure stability region was determined numerically in terms of the parameters (ξ, σ).

Mitchell [42] studied two second-order Ito differential equations where the condition (H) is satisfied. By solving the Fokker–Planck equation, he obtained necessary and sufficient conditions for almost sure stability in terms of Bessel functions. Mitchell and Kozin [43] analyzed the linear second-order stochastic system in canonical form,

$$\begin{cases} \dot{x}_1(t) = x_2(t), \quad \text{and} \\ \dot{x}_2(t) = -\omega^2 x_1(t) - 2\xi\omega x_2(t) - f_1(t)x_1(t) - f_2(t)x_2(t). \end{cases} \tag{63.63}$$

Here the $f_i(t)$ are stationary Gaussian random processes with wide bandwidth power spectral density. After introducing a correction term due to Wong and Zakai, Equation 63.63 was written as an Ito stochastic differential equation and Has'minskii's idea was again applied. Because the angular process $\varphi(t)$ is singular, one-dimensional diffusion theory was used to give a careful classification of the singularities on S^1 and the behavior of the sample path of $\varphi(t)$ near the singularities. The ergodic components were also examined in detail. The difficulty in the analytical determination of the invariant measures corresponding to the ergodic components of the angular process was again experienced. An extensive numerical simulation was conducted and the almost sure stability regions were represented graphically for different parameter sets of the system. In this work, the authors provided a good explanation of the ergodic properties of the angular process $\varphi(t)$ and also illustrated various concepts, such as stabilization and destabilization of a linear system by noise, by using examples corresponding to different parameter sets. Interestingly, they provided an example where the sample solutions of the system behaved just like a deterministic system. Examples were also used to show that the regions for second moment stability may be small when compared with the regions for a.s. stability, that is, moment stability may be too conservative to be practically useful. This is one of the reasons why a.s. stability criteria are important.

A more detailed study of the case when the nondegeneracy condition is not satisfied was presented by Nishioka [45] for the second-order white-noise system given by

$$dx(t) = Ax(t)\,dt + B_1x(t)\,d\xi_1(t) + B_2x(t)\,d\xi_2(t) \tag{63.64}$$

with $\xi_i(t), i = 1, 2$, independent standard Wiener processes. Results similar to Kozin and his coworkers were obtained.

Has'minskii's results were later refined by Pinsky [49] using the so-called Fredholm alternative and the stochastic Lyapunov function $f(\rho,\varphi) = \rho + h(\varphi)$ in the following manner. From the system Equation 63.51 with the generator Equation 63.52, Pinsky observed that

$$\mathcal{L}f(\rho,\varphi) = \mathcal{L}_\rho(\rho) + \mathcal{L}_\varphi[h(\varphi)] \tag{63.65}$$

that is, the action of $\mathcal{L}$ on $f(\rho,\varphi)$ separates $\mathcal{L}$ into the radial part $\mathcal{L}_\rho$ and the angular part $\mathcal{L}_\varphi$. Furthermore, $\mathcal{L}_\rho(\rho) = v(\varphi)$ is a function of φ only, and $\mathcal{L}_\varphi$ is a second-order partial differential operator that generates a Markov diffusion process on S^{d-1}. If Has'minskii's nondegeneracy condition (H) is satisfied, then $\mathcal{L}_\varphi$ is ergodic and satisfies a Fredholm alternative, that is, the equation

$$\mathcal{L}_\varphi h(\varphi) = g(\varphi) \tag{63.66}$$

has a unique solution, up to an additive constant, provided that

$$\int_{S^{d-1}} g(\varphi)m(d\varphi) = 0 \tag{63.67}$$

with m being the unique invariant probability measure of the process $\varphi(t)$ on S^{d-1}. Define

$$q = \int_{S^{d-1}} v(\varphi)m(d\varphi). \tag{63.68}$$

Choose $h(\varphi)$ as a solution of $\mathcal{L}_\varphi h(\varphi) = q - v(\varphi)$. From Ito's formula, Pinsky obtained

$$\rho(t) + h(\varphi(t)) = \rho_0 + h(\varphi_0) + \int_0^t \mathcal{L}f(\rho(s),\varphi(s))\,ds + M_t = \rho_0 + h(\varphi_0) + qt + M_t \tag{63.69}$$

where

$$M_t = \int_0^t H(\varphi(s))\,d\xi(s))$$

for an appropriate function H on S^{d-1}. M_t is a zero-mean martingale with respect to the σ-algebra generated by the process $\{\varphi_t,\ t \geq 0\}$ and $\lim_{t\to+\infty} \frac{1}{t}M_t = 0$ a.s. Thus, upon dividing both sides of Equation 63.69

by t and taking the limit as $t \to +\infty$,

$$\overline{\lambda}_{\omega}(x_0) = \lim_{t \to +\infty} \frac{1}{t}\rho(t) = q \quad \text{a.s.} \tag{63.70}$$

Loparo and Blankenship [35], motivated by Pinsky, used an averaging method to study a class of linear systems with general jump process coefficients

$$\dot{x}(t) = Ax(t) + \sum_{j=1}^{m} y_j(t)B_j x(t) \tag{63.71}$$

with $A' = -A$ and $y(t) = (y_1(t) \ldots y_m(t))'$ is a vector-valued jump process with bounded generator Q. Using the transformation $z_t = e^{-At}x(t)$, $\rho(t) = \log \|z(t)\| = \log \|x(t)\|$, and $\varphi(t) = \|z(t)\|^{-1}z(t)$, the system was transformed into a system of equations involving $(\rho(t), \varphi(t))$ that are homogeneous in $y = (y_1(t) \ldots y_m(t))'$ but inhomogeneous in time. Thus, an artificial process $\tau(t)$ was introduced so that $[\rho(t), \varphi(t), \tau(t), y(t)]$ is a time-homogeneous Markov process. Using the averaging method and the Fredholm alternative, a second-order partial differential operator $\overline{\mathcal{L}}$ was constructed. $\overline{\mathcal{L}}$ is the generator of a diffusion process on $\mathbb{R} \times S^{d-1}$. Applying Pinsky's idea they obtained

$$q = \int_{S^{d-1}} \overline{\mathcal{L}}_{\rho}(\rho)m(d\theta) = \int_{S^{d-1}} \nu(\theta)m(d\theta) \tag{63.72}$$

where m is the unique invariant probability measure on S^{d-1} for the diffusion φ_t generated by $\overline{\mathcal{L}}_{\varphi}$. Unfortunately, q is not the top Lyapunov exponent for the system and cannot be used to determine the a.s. stability properties of the system. This can be easily verified by considering the random harmonic oscillator

$$\ddot{u}(t) + k^2(1 + by(t))u(t) = 0 \tag{63.73}$$

with $y(t) \in \{-1, +1\}$ a telegraph process with mean time between jumps $\lambda^{-1} > 0$. The formula for q is

$$q = \frac{k^2 \lambda b^2}{8(k^2 + \lambda^2)} \quad \text{a.s.} \tag{63.74}$$

independent of $x_0 \in \mathbb{R}^2 \backslash \{0\}$. This is only the first term in an expansion for the Lyapunov exponent $\overline{\lambda}_{\omega}(x_0)$; see the discussion below. Thus Equations 63.72 and 63.74 do not provide sufficient conditions for almost sure stability or instability.

Following Loparo and Blankenship's idea, Feng and Loparo [13,36] concentrated on the random harmonic oscillator system Equation 63.73. By integration of the Fokker–Planck equation, the positivity of $\overline{\lambda}_{\omega}(x_0) = \overline{\lambda}$ was established for any $k, \lambda \in (0, +\infty)$ and $b \in (-1, 1)\backslash\{0\}$. Thus, the system Equation 63.73 is almost surely unstable for any k, λ, and b as constrained above. To compute the Lyapunov exponent $\overline{\lambda}$ explicitly (it is known that $\overline{\lambda}_{\omega}(x_0) = \overline{\lambda} =$ constant a.s. for any $x_0 \in \mathbb{R}^2 \backslash \{0\}$), the following procedure was used. As usual, introducing polar coordinates $\rho = \log \|x\|$, $\varphi = \|x\|^{-1}x$, where $x = (ku, \dot{u})'$, the system becomes

$$\begin{cases} \dot{\rho}(t) = y(t)g_0(\varphi(t)) \quad \text{and} \\ \dot{\varphi}(t) = h(\varphi(t), y(t)) \end{cases} \tag{63.75}$$

for some smooth functions g_0 and h on S^1. The Markov process $(\rho(t), \varphi(t), y(t))$ has generator

$$\mathcal{L} = Q + A\varphi \cdot \frac{\partial}{\partial \varphi} + y\left[g_0(\varphi)\frac{\partial}{\partial \rho} + h(\varphi)\frac{\partial}{\partial \varphi}\right] \triangleq \mathcal{L}_1 + \mathcal{L}_2$$

where $\mathcal{L}_1 = Q + A\varphi \cdot \partial/\partial\varphi$ satisfies a Fredholm alternative. When $\mathcal{L}$ acts on the function, $F = \rho + h(\varphi) + f_1(y, \varphi)$, where $f_1(y, \varphi)$ is a correction term introduced in the work of Blankenship and Papanicolaou, see [13] for details,

$$\begin{cases} \mathcal{L}F = q_0 + \mathcal{L}_2 h(\varphi) \quad \text{and} \\ q_0 = \overline{\pi}\mathcal{L}_2 f_1(y,\varphi) = \overline{\pi}[\nu(\varphi)]. \end{cases} \tag{63.76}$$

In Equation 63.76 the correction term is chosen as the solution of $\mathcal{L}_1 f_1 = -yg_0(\varphi)$, and $\mathcal{L}_2 f_1 = \nu(\varphi)$ is a function of φ only. Here, $\overline{\pi}$ is the uniform measure on S^1, and $h(\varphi)$ is the solution of $A\varphi \cdot \partial h/\partial\varphi = [\nu(\varphi) - \overline{\pi}\nu(\varphi)]$. It follows that $\mathcal{L}_2 h(\varphi) = yg_1(\varphi)$ and a martingale convergence argument gives

$$\overline{\lambda} = \lim_{t\to+\infty} \frac{1}{t}\rho(t), = q_0 + \lim_{t\to+\infty} \frac{1}{t}\int_0^t (\mathcal{L}_2 h)(s)(\varphi(s))\,ds, = q_0 + \overline{\lambda}_1,$$

where $\overline{\lambda}_1$ is the Lyapunov exponent of the system,

$$\begin{cases} \dot{\rho}(t) = \mathcal{L}_2 h(\varphi(t)) = y(t)g_1(\varphi(t)) \quad \text{and} \\ \dot{\varphi}(t) = h(\varphi(t), y(t)). \end{cases} \tag{63.77}$$

Noting the similarity between the systems given by Equations 63.75 and 63.77, the above procedure can be repeated for Equation 63.77. Hence,

$$\overline{\lambda} = q_0 + \overline{\lambda}_1 = q_0 + (q_1 + \overline{\lambda}_2) = \sum_{k=0}^{\infty} q_k$$

The absolute convergence of the above series for any $k, \lambda \in (0, +\infty)$, and $b \in (-1, 1)$ was proved by using Fourier series methods, and a formula for the general term q_k was obtained. To third-order (b^6), Feng and Loparo obtained

$$\begin{aligned} \overline{\lambda} = {} & \frac{k^2\lambda b^2}{8(k^2+\lambda^2)} + \frac{5k^4\lambda b^4}{64(k^2+\lambda^2)^2} + \frac{k^4\lambda b^6}{8\times 16^2(\lambda^2+k^2)^2}\left[k^2\left(\frac{75}{k^2+\lambda^2} + \frac{160}{\lambda^2+rk^2} + \frac{21}{\lambda^2+9k^2}\right)\right. \\ & \left. - \lambda^2\left(\frac{25}{k^2+\lambda^2} + \frac{32}{\lambda^2+4k^2} + \frac{3}{\lambda^2+9k^2}\right)\right] + \sum_{n=3}^{\infty} q_n. \end{aligned} \tag{63.78}$$

In the formulas for the Lyapunov exponent obtained by most researchers, for example, Has'minskii for Equation 63.56, computation of an invariant probability measure was always required to determine the exponent. This is a very difficult problem that was overcome in the work of Feng and Loparo by sequentially applying the Fredholm alternative for the simple, but nontrivial, random harmonic oscillator problem. Due to the difficulty involved with the analytical determination of the invariant probability measure, several researchers have attacked the problem by using an analytic expansion and perturbation technique; some of these efforts have been surveyed by Wihstutz [57].

Auslender and Mil'shtein [6] studied a second-order system perturbed by a small white-noise disturbance, as given by

$$dx^{\epsilon}(t) = Bx^{\epsilon}(t) + \epsilon\sum_{j=1}^{k}\sigma_j x^{\epsilon}(t)\,d\xi_j(t)) \tag{63.79}$$

where $\xi_j(t)$ are independent, standard Wiener processes and $0 < \epsilon < 1$ models the small noise. Assuming that Has'minskii's nondegeneracy condition (H) is satisfied, the angular process $\varphi^{\epsilon}(t) = \|x^{\epsilon}(t)\|^{-1}x^{\epsilon}(t)$ is ergodic on S^1, and the invariant density μ^{ϵ} satisfies the Fokker–Planck Equation 63.58, and the exponent $\lambda(\epsilon)$ is given by Equation 63.56. In this case a small parameter ϵ is involved. Auslender and Mil'shtein computed $\lambda(\epsilon)$ to second order (ϵ^2) and estimated the remainder term to obtain an expansion for $\lambda(\epsilon)$ to order ϵ^2 as $\epsilon \downarrow 0^+$. For different eigenstructures of B, they obtained the following results:

1. $B = \begin{bmatrix} a & 0 \\ 0 & b \end{bmatrix}$, $a > b$, two distinct real eigenvalues.

$$\lambda(\epsilon) = a - \frac{\epsilon^2}{2} \sum_{r=1}^{k} (\sigma_r^{11})^2 + \epsilon^4 \rho(\epsilon) + \rho_0(\epsilon)$$

where $\sigma_r = (\sigma_r^{ij})2 \times 2$, $|\rho(\epsilon)| \leq m < \infty$ and $|\rho_0(\epsilon)| \leq ce^{-c_1/\epsilon^2}$ for some constants c and $c_1 > 0$.

2. $B = \begin{bmatrix} a & b \\ -b & a \end{bmatrix}$, $a, b > 0$, a complex conjugate pair of eigenvalues.

$$\lambda(\epsilon) = a + \frac{\epsilon^2}{8} \sum_{r=1}^{k} \left[\left(\sigma_r^{12} - \sigma_r^{21}\right)^2 + \left(\sigma_r^{11} + \sigma_r^{22}\right)^2 \right] + \epsilon^4 \mathcal{R}(\epsilon)$$

where $|\mathcal{R}(\epsilon)| \leq m < +\infty$.

3. $B = \begin{bmatrix} a & 0 \\ 0 & a \end{bmatrix}$, one real eigenvalue of geometric multiplicity 2.

$$\lambda(\epsilon) = a + \epsilon^2 \int_0^{2\pi} \mathcal{L}(\varphi)\mu(\varphi)\, d\varphi$$

where

$$\mathcal{L}(\varphi) = \frac{1}{2} \sum_{r=1}^{k} \langle \sigma_r \lambda(\varphi), \sigma_r \lambda(\varphi) \rangle - \sum_{r=1}^{k} \langle \sigma_r \lambda(\varphi), \lambda(\varphi) \rangle^2,$$

$\lambda(\varphi) = (\cos\varphi, \sin\varphi)'$, and $\mu(\varphi)$ is the density determined by a Fokker–Planck equation and is independent of ϵ.

4. $B = \begin{bmatrix} a & 1 \\ 0 & a \end{bmatrix}$, one real eigenvalue of geometric multiplicity 1.

$$\lambda(\epsilon) = a + \epsilon^{2/3} \frac{\pi^{1/2}}{\Gamma(1/6)} \left[\frac{3}{4} \sum_{r=1}^{k} \left(\sigma_r^{21}\right)^2 \right]^{1/3} + O(\epsilon)$$

where $\Gamma(x)$ is the gamma function and $O(\epsilon)$ denotes a quantity of the same order of ϵ (i.e., $\lim_{\epsilon \to 0} O(\epsilon)/\epsilon =$ constant).

In the above work, an intricate computation is required to obtain $\lambda(\epsilon)$ as an expansion of powers of ϵ. A much easier way to obtain an expansion is the direct use of perturbation analysis of the linear operator associated with the forward equation $\mathcal{L}_\varphi^* \mu = 0$. Pinsky [50] applied this technique to the random oscillator problem,

$$\ddot{u}(t) + \{\gamma + \sigma F(\xi(t))\} u(t) = 0 \tag{63.80}$$

where γ is a positive constant that determines the natural frequency of the noise-free system, σ is a small parameter that scales the magnitude of the noise process, and $\xi(t)$ is a finite-state continuous time Markov process with state space $M = \{1, 2, \ldots, N\}$. $F(\cdot)$ is a function satisfying $E\{F(\xi(t))\} = 0$. After introducing polar coordinates in the form $u\sqrt{\gamma} = \rho\cos\varphi$ and $\dot{u} = \rho\sin\varphi$, Pinsky obtained

$$\begin{cases} \dot{\varphi}(t) = h(\varphi(t), \xi(t)) = -\sqrt{\gamma} + \dfrac{\sigma F(\xi(t))}{\sqrt{\gamma}} \cos^2\varphi(t) \\ \dot{\rho}/\rho = q(\varphi(t), \xi(t)) = \dfrac{\sigma F(\xi(t))}{2\sqrt{\gamma}} \sin 2\varphi(t) \end{cases} \tag{63.81}$$

where $(\varphi(t), \xi(t))$ is a time-homogeneous Markov process with generator,

$$\mathcal{L} = Q + h\frac{\partial}{\partial\varphi},$$

and Q is the generator of the process $\xi(t)$. The Fokker–Planck equation for the density $p_\sigma(\varphi, \xi)$, the invariant probability density of $(\varphi(t), \xi(t))$, is given by

$$0 = \mathcal{L}^* p_\sigma$$

where $\mathcal{L}^*$ is the formal adjoint of $\mathcal{L}$. The exponent is computed by

$$\lambda(\sigma) = \int\limits_{S^1 \times M} q(\varphi, \xi) p_\sigma(\varphi, \xi)\, d\varphi\, d\xi. \tag{63.82}$$

Note that

$$\mathcal{L} = Q + h\frac{\partial}{\partial\varphi} = Q - \sqrt{\gamma}\frac{\partial}{\partial\varphi} + \sigma\frac{F(\xi)}{\sqrt{\gamma}} \cos^2\varphi \frac{\partial}{\partial\varphi} \triangleq \mathcal{L}_0 + \sigma\mathcal{L}_1.$$

Assume an approximation of p_σ in the form

$$p_\sigma = p_0 + \sigma^1 p_1 + \cdots + \sigma^n p_n.$$

Then, it follows from

$$0 = \mathcal{L}^* p_\sigma = (\mathcal{L}_0^* + \sigma\mathcal{L}_1^*)(p_0 + \sigma^1 p_1 + \cdots + \sigma_n p_n) \tag{63.83}$$

that

$$\mathcal{L}_0^* p_i + \mathcal{L}_1^* p_{i-1} = 0, \quad \text{for } i = 1, 2, \ldots, n, \tag{63.84}$$

by setting the coefficients of the term σ^i in Equation 63.83 equal to zero. Pinsky showed that the expansion satisfying Equation 63.84 can be obtained by taking

$$p_0 = \frac{1}{2\pi} \quad \text{and} \quad \int\limits_{S^1 \times M} p_n = 0 \tag{63.85}$$

and proved the convergence of the series expansion by using properties of the finite-state Markov process to show that

$$|p_\sigma - (p_0 + \sigma^1 p_1 + \cdots + \sigma^n p_n)| \le c\sigma^{n+1} \max|\mathcal{L}_1^* p_n|.$$

By evaluating p_1 from Equations 63.84 and 63.85, Pinsky obtained

$$\lambda(\sigma) = \frac{\sigma^2}{4\gamma} \sum_{k=2}^{N} \frac{\lambda_k < F_1, \psi_k >^2}{\lambda_k^2 + 4\gamma} + O(\sigma^3), \quad \sigma \downarrow 0^+ \tag{63.86}$$

where λ_k^{-1} is the mean sojourn time of $\xi(t)$ in state k and $\{\psi_i; i = 1, 2, \ldots, N\}$ are normalized eigenfunctions of the linear operator Q. For the case when $F(\xi) = \xi$ is a telegraph process, Pinsky obtained a refined expansion of Equation 63.86 consisting of the first two terms in Equation 63.78 when $\sigma \downarrow 0^+$.

Contemporaneous with Pinsky, Arnold, Papanicolaou, and Wihstutz [4] used a similar technique to study the random harmonic oscillator in the form,

$$-\ddot{y}(t) + \sigma F\left(\xi\left(\frac{t}{\rho}\right)\right) y(t) = \gamma y(t), \tag{63.87}$$

where γ, σ and ρ are parameters modeling the natural frequency, the magnitude of the noise, and the time scaling of the noise. $\xi(t)$ is only assumed to be an ergodic Markov process on a smooth connected Riemannian manifold M, and $F(\cdot)$ is a function satisfying $E\{F(\xi(t))\} = 0$. A summary of their results is given next.

1. Small noise $(\sigma \downarrow 0^+,\ \gamma > 0,\ \rho = 1)$:

$$\lambda(\sigma) = \sigma^2 \frac{\pi}{4\gamma} \hat{f}(2\sqrt{\gamma}) + O(\sigma^3), \quad \sigma \downarrow 0^+$$

   where $\hat{f}(\omega)$ is the power spectral density of $F(\xi_t)$.

2. Large noise $(\sigma \uparrow +\infty,\ \gamma = \gamma_0 + \sigma\gamma_1,\ \rho = 1)$:

$$\lambda(\sigma) = \frac{\sqrt{\gamma_1}}{4\pi} \int_0^{2\pi} d\varphi \int_M \nu(d\xi) \frac{\sqrt{\gamma_1 - F(\xi)}}{\gamma_1 - F(\xi)\cos^2\varphi} Q(\log(\gamma_1 - F(\xi)\cos^2\varphi)) + O\left(\frac{1}{\sqrt{\sigma}}\right), \quad \sigma \uparrow +\infty$$

   where Q is the generator of ξ and ν is the unique invariant probability measure of ξ on M.

3. Fast noise $(\rho \downarrow 0^+,\ \sigma$ and ν fixed): if $\gamma > 0$,

$$\lambda(\rho) = \rho \cdot \frac{\sigma^2 \pi}{4\gamma} \hat{f}(0) + O(\rho^2), \quad \rho \downarrow 0^+, \quad \text{and}$$

   if $\gamma < 0$,

$$\lambda(\rho) = \sqrt{-\gamma} + \rho\, \frac{\sigma^2 \pi}{4\gamma} \hat{f}(0) + O(\rho^2), \quad \rho \downarrow 0^+.$$

4. Slow noise $(\rho \uparrow +\infty,\ \sigma$ and γ fixed): if $\gamma > \sigma \max(F)$,

$$\lambda(\rho) = \frac{1}{\rho} \frac{\sqrt{\gamma}}{4\pi} \int_0^{2\pi} d\varphi \int_M \nu(d\xi) \frac{\sqrt{\gamma - \sigma F(\xi)}}{\gamma - \sigma F(\xi)\cos^2\varphi} Q[\log(\gamma - \sigma F(\xi)\cos^2\varphi)] + O\left(\frac{1}{\rho^2}\right), \quad \rho \uparrow +\infty$$

   if $\gamma < \sigma \min(F)$,

$$\lambda(\rho) = \int_M \nu(d\xi) \sqrt{\sigma F(\xi) - \gamma} + \frac{1}{\rho} \int_M \nu(d\xi) Q\big[(\log \sin(\varphi + \psi(\xi))\big]\Big|_{\varphi = \psi(\xi)} + O\left(\frac{1}{\rho^2}\right), \quad \rho \uparrow +\infty,$$

   where $\psi(\xi) = \tan^{-1}\left((\sigma F(\xi) - \gamma)/\sqrt{-\gamma}\right)$.

Pardoux and Wihstutz [47] studied the two-dimensional linear white-noise system Equation 63.79 by using similar perturbation techniques. Instead of using the Fokker–Planck equation, perturbation analysis was applied to the backward (adjoint) equation and a Fredholm alternative was used. Pardoux and Wihstutz were able to obtain a general scheme for computing the coefficients of the series expansion of the exponent for all powers of ϵ. Their results are the same as (1), (2), and (3) of Auslender and Mil'shtein, if only the first two terms in the expansion are considered. There is another approach for studying the problem based on a differential geometric concept and nonlinear control theory, that has been suggested by Arnold and his coworkers. For a linear system, after introducing polar coordinates, the angular process $\varphi(t)$ is separated from the radial process and is governed by the nonlinear differential equation

$$\dot{\varphi}(t) = h(\xi(t), \varphi(t)), \quad \varphi(t) \in S^{d-1} \tag{63.88}$$

where $\xi(t)$ is the noise process and h is a smooth function of ξ and φ. To determine the Lyapunov exponent $\overline{\lambda}_\omega(x_0)$, the ergodicity of the joint process $(\xi(t), \varphi(t))$ needs to be determined. This question is naturally related to the concept of reachable sets of the random process $\varphi(t)$ and invariant sets of Equation 63.88 on S^{d-1} for certain "controls" $\xi(t)$. It is from this perspective that geometric nonlinear control theory is applied.

Arnold et al. [3] considered the system given by

$$\begin{cases} \dot{x}(t) = A(\xi(t))x(t) \quad \text{and} \\ x(0) = x_0 \in \mathbb{R}^d \end{cases} \tag{63.89}$$

where $A : M \to \mathbb{R}^{d\times d}$ is an analytic function with domain M, an analytic connected Riemannian manifold that is the state space of a stationary ergodic diffusion process $\xi(t)$ satisfying the Stratonovich equation,

$$d\xi(t) = X_0(\xi(t))\, dt + \sum_{j=1}^{r} X_j(\xi(t))\, d\xi_j(t) \tag{63.90}$$

where $\xi_j(t)$ are independent, standard Wiener processes. The following Lie algebraic conditions on the vector fields were imposed by Arnold et al.:

a. $\dim LA(X_1, \dots, X_r)(\xi) = \dim M, \quad \forall \xi \in M.$
b. $\dim LA(h(\xi, \cdot), \xi \in M)(\varphi) = d - 1, \quad \forall \varphi \in \mathbb{P}^{d-1}.$

Here $h(\xi, \psi)$ is the smooth function given in Equation 63.89 for the angular process, and $\mathbb{P}^{d-1}$ is the projective sphere in $\mathbb{R}^d$. Condition (a) guarantees that a unique invariant probability density ρ of ξ on M exists solving the Fokker–Planck equation $Q^*\rho = 0$, with Q the generator of $\xi(t)$. Condition (b) is equivalent to an accessibility condition for the angular process φ governed by Equation 63.88 or the fact that the system group,

$$G = \left\{ \Pi_{i=1}^{n} e^{t_i A(\xi_i)}; t_i \in \mathbb{R}, \xi_i \in M, \text{ n finite} \right\},$$

acts transitively on S^{d-1}, that is, for any $x, y \in S^{d-1}$, $g \in G$ exists so that $g * x = y$. Under these conditions, they showed that a unique invariant control set C in $\mathbb{P}^{d-1}$ exists and that this is an ergodic set; all trajectories $\varphi(t)$ entering C remain there forever with probability one. Henceforth, a unique invariant probability measure μ of (ξ, φ) on $M \times \mathbb{P}^{d-1}$ exists with support $M \times C$. The following result was given.

Theorem 63.8: Arnold, Kilemann, and Oeljeklaus

For the system defined by Equations 63.89 and 63.90, suppose (a) and (b) above hold, then

1.

$$\lambda = \int_{M\times C} q(\xi, \varphi)\mu(d\xi, d\varphi), \quad q(\xi, \varphi) = \varphi' A(\xi)\varphi \tag{63.91}$$

is the top exponent in Oseledec's theorem.

2. *For each*

$$x_0 \neq 0, \quad \overline{\lambda}_\omega(x_0) = \lambda \text{ a.s.} \tag{63.92}$$

3. *For the fundamental matrix* $\Phi(t)$ *of Equation 63.89,*

$$\lim_{t\to+\infty} \frac{1}{t} \log \|\Phi(t)\| = \lambda \quad a.s. \tag{63.93}$$

Equation 63.91 is in the same form as given by Has'minskii. Note that, even though Oseledec's theorem states that only $\overline{\lambda}_{\omega}(x_0) = \lambda = \lambda_{\max}$, the top exponent, is realizable, that is, with probability one can be observed from sample solutions, the difficulty is still in determining the invariant density μ.

Arnold [1,3], motivated by the results for the undamped random oscillator obtained by Molchanov [44] studied the relationship between sample and moment stability properties of systems defined by Equations 63.89 and 63.90 and obtained some very interesting results. Besides conditions (a) and (b) above, another Lie algebraic condition was needed to guarantee that the generator $\mathcal{L}$ of (ξ, φ) is elliptic. Define the Lyapunov exponent for the pth moment, for $p \in \mathbb{R}$, as

$$g(p, x_0) = \overline{\lim}_{t\to+\infty} \frac{1}{t} \log E(\|x(t, x_0, \omega)\|^p), \quad x_0 \in \mathbb{R}^d\backslash\{0\}. \tag{63.94}$$

By Jenson's inequality $g(p, x_0)$ is a finite convex function of p with the following properties:

1. $| g(p, x_0) | \leq k | p |, \ k = \max_{\xi \in M} \|A(\xi)\|$.
2. $g(p, x_0) \geq \lambda p\,, \ \overline{\lambda}_{\omega}(x_0) = \lambda$ a.s.
3. $g(p, x_0)/p$ is increasing as a function of p with $x_0 \in \mathbb{R}^d\backslash\{0\}$ fixed.
4. $g'(0^-, x_0) \leq \lambda \leq g'(0^+, x_0)$, here, g' denotes the derivative of g with respect to p.

Linear operator theory was then used to study the strongly continuous semigroup $T_t(p)$ of the generator $\mathcal{L}(p) \triangleq \mathcal{L} + pq(\xi, \varphi)$. Here, $\mathcal{L}$ is the generator of (ξ, φ) and $q(\xi, \varphi) = \varphi' A(\xi)\varphi$. Arnold showed that $g(p, x_0)$ is actually independent of x_0, that is, $g(p, x_0) = g(p)$, and $g(p)$ is an analytic function of p with $g'(0) = \lambda$ and $g(0) = 0$. $g(p)$ can be characterized by the three figures shown in Figure 63.2 (excluding the trivial case, $g(p) \equiv 0$).

If $g(p) \equiv 0$, then $\lambda = 0$. If $g(p) \not\equiv 0$, besides 0, $g(p)$ has at most one zero $p_0 \neq 0$, and it follows that

1. $\lambda = 0$ iff $g(p) > 0, \quad \forall p \neq 0$.
2. $\lambda > 0$ iff $g(p) < 0, \quad$ for some $p < 0$.
3. $\lambda < 0$ iff $g(p) < 0, \quad$ for some $p > 0$.

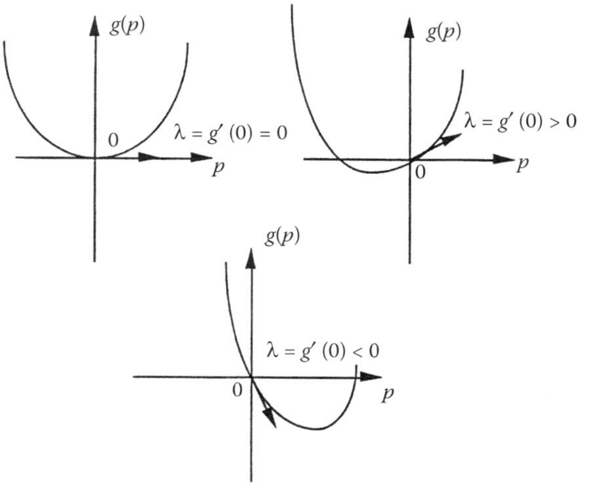

FIGURE 63.2 pth moment Lyapunov exponent.

In the case $trA(\xi) \equiv 0$, more information can be obtained for the second zero $p_0 \neq 0$ of $g(p)$. Under conditions that $\xi(t)$ is a reversible process along with reachability of $\varphi(t)$ on $\mathbb{P}^{d-1}$, Arnold showed that $p_0 = -d$, where d is the system dimension. However, if $trA(\xi) \not\equiv 0$, then the system Equation 63.89 is equivalent to

$$\dot{x}(t) = d^{-1} trA(\xi(t))x(t) + A_0(\xi(t))x(t) \tag{63.95}$$

where $trA_0(\xi(t)) \equiv 0$. Observing that $d^{-1}trA(\xi(t))I$ commutes with $A_0(\xi(t))$, it follows that

$$g(p) = \alpha_0 p + g_0(p) \tag{63.96}$$

where $\alpha_0 = d^{-1}tr\{EA[\xi(0)]\}$ and $g_0(p)$ is the exponent for the pth moment of $\dot{y}(t) = A_0(\xi(t))y(t)$. Therefore, $g_0(-d) = 0$. Applying the results to the damped harmonic oscillator,

$$\ddot{y}(t) + 2\beta\dot{y}(t) + (1 + \xi(t))y(t) = 0, \tag{63.97}$$

with $x = (y, \dot{y})' \in \mathbb{R}^2$, Arnold obtained

$$g(p) = -\beta p + g_0(p) \tag{63.98}$$

where $g_0(p)$ is the moment exponent for the undamped oscillator with $g_0(-2) = 0$. Thus, it follows that

1. For the undamped oscillator Equation 63.97, if $\beta = 0$, then $\lambda > 0$ and $g(p) > 0$ for $p > 0$. This implies almost sure sample instability and pth moment instability for $p > 0$.
2. The undamped harmonic oscillator can be stabilized by introducing positive damping so that with $\lambda > 0$ and $g(p) < 0$ for $p \in (0, p_1)$, where p_1 is a positive real number.

We remark here that the random harmonic oscillator problem has attracted considerable attention from researchers and Equation 63.97 occurs in many science and engineering applications, such as mechanical or electrical circuits, solid state theory, wave propagation in random media, and electric power systems. It is also of particular interest from a theoretical viewpoint because of the simple structure of the model. From the result (1) above, we see that pth moment stability for some $p > 0$ will imply almost sure sample stability. One natural question to ask is the converse question, that is, does almost sure sample stability have any implication for the pth moment stability? The result (2) above for the random oscillator gives a partial answer to this question. One may ask whether p_1 in (2) can be arbitrarily large or, equivalently, when does the sample stability ($\lambda < 0$) imply pth moment stability for all $p > 0$? If

$$0 < \gamma \triangleq \lim_{p \to +\infty} \frac{1}{p} g(p) < +\infty,$$

then we may choose $\beta > \gamma$ sufficiently large so that $g(p) < 0$ for all $p > 0$. Hence, γ is the quantity that characterizes the the property that is of interest.

Another interesting question in the stability study of stochastic systems is when can an unstable linear system be stabilized by introducing noise into the system? This question has attracted the attention of researchers for a long time. Has'minskii [20] presented an example giving a positive answer to the question. Arnold et al. [2] presented a necessary and sufficient condition for stabilization of a linear system $\dot{x}(t) = A(t)x(t)$ by stationary ergodic noise $F(t) \in \mathbb{R}^{d \times d}$ in the sense that the new system $\dot{x}(t) = [A(t) + F(t)]x(t)$ will have a negative top exponent and hence is almost surely stable. This result is presented next.

Theorem 63.9: Arnold, Crauel, and Wihstutz

Given the system $\dot{x}(t) = Ax(t)$ with $A \in \mathbb{R}^{d \times d}$ a constant matrix, the perturbed system is given by $\dot{x}(t) = [A + F(t)]x(t)$ where $F(t)$ is a stationary, ergodic, and measurable stochastic process in $\mathbb{R}^{d \times d}$ with finite mean. Then

1. *For any choice of $F(t)$ with $E\{F(t)\} = 0$, the top Lyapunov exponent $\lambda_{max}[A + F(t)]$ of the perturbed system satisfies*
$$d^{-1}tr(A) \leq \lambda_{max}(A + F(t)).$$
2. *For any $\epsilon > 0$ fixed, an $F(t)$ exists with $E\{F(t)\} = 0$ so that*
$$d^{-1}tr(A) \leq \lambda_{max}[A + F(t)] \leq d^{-1}tr(A) + \epsilon.$$

In particular, the linear system $\dot{x}(t) = Ax(t)$ can be stabilized by a zero-mean stochastic process if, and only if, $tr(A) < 0$.

Note that (2) above implies that the undamped, random harmonic oscillator cannot be stabilized by a random linear feedback control of the ergodic type.

Li and Blankenship [34] generalized Furstenberg's results on the product of random matrices and studied linear stochastic system with Poisson process coefficients of the form,

$$dx(t) = Ax(t)\,dt + \sum_{i=1}^{m} B_i x(t)\,dN_i(t), \tag{63.99}$$

where $N_i(t)$ are independent Poisson (counting) processes. The solution of Equation 63.99 can be written as a product of i.i.d. matrices on $\mathbb{R}^{d\times d}$ acting on the initial point $x_0 \in \mathbb{R}^d\backslash\{0\}$:

$$x(t) = \exp\left[A(t - t_{N(t)})\right] D_{\tau_{N(t)}}, \ldots, D_{\tau_1} \exp(A\tau_1)x_0,$$

for $t \in [t_{N(t)}, t_{N(t+1)})$, where $D_i = I + B_i$, for $i = 1, 2, \ldots, m$, and $N(t) = N_1(t) + \cdots + N_m(t)$ is also a Poisson process with mean interarrival time λ^{-1}, $\{\tau_j;\ \ j \geq 1\}$ are the interarrival times of $N(t)$, $t_j \triangleq \tau_1 + \cdots + \tau_j$ are the occurrence times of $N(t)$.

Using the fact that $\{X_j(\omega) = D_{\tau_j} \exp(A_{\tau_j})\}_{j=1}^{\infty}$ is an i.i.d. sequence, they generalized Furstenberg's results from semisimple Lie groups to general semigroups (note that D_i may be singular) in the following sense. Let $M = S^{d-1} \cup \{0\}$, and let μ be the probability distribution of X_1 induced by

$$\mu(\Gamma) = P\{D_{\mu_1} e^{A_{\tau_1}} \in \Gamma :\ \ \Gamma \in B(\mathbb{R}^{d\times d})\}$$

where $B(\mathbb{R}^{d\times d})$ is the Borel σ algebra on $\mathbb{R}^{d\times d}$. Let ν be an invariant probability measure on M with respect to μ, that is, if $x_0 \in M$, and is distributed according to ν, then $x_1 = \|X_1x_0\|^{-1}X_1x_0$ has the same distribution ν. If Q_0 denotes the collection of so-called extremal invariant probabilities on M with respect to μ, then Li and Blankenship obtained the following result:

Theorem 63.10: Li and Blankenship

For all $\nu \in Q_0$,

$$r_\nu \triangleq \sum_{i=1}^{m} \lambda_i \int_M \int_0^{\infty} \log \|D_i \exp(At)s\| e^{-\lambda t}\,dt\nu(ds) < +\infty$$

and

$$\overline{\lambda}_\omega(x_0) = \lim_{t\to+\infty} \frac{1}{t} \log \frac{\|x(t, x_0)\|}{\|x_0\|} = \lambda \cdot r_\nu \quad \text{a.s.}$$

for all $x_0 \in E_\nu^0$, where λ_i^{-1} is the mean interarrival time of $N_i(t)$, λ^{-1} is the mean interarrival time of N_t, and E_ν^0 is an ergodic component corresponding to $\nu \in Q_0$. There are only a finite number of different values of r_ν, say, $r_1 < r_2 < \cdots < r_\ell$, $\ell \leq d$. Furthermore, if $\bigcup_{\nu\in Q_0} E_\nu^0$ contains a basis for $\mathbb{R}^d$, then the system

Equation 63.99 is asymptotically almost surely stable if $r_\ell < 0$, *and Equation 63.99 is almost surely unstable if* $r_1 > 0$. *In the case where* $r_1 < 0$ *and* $r_\ell > 0$, *then the stability of the system depends on the initial state* $x_0 \in \mathbb{R}^d \backslash \{0\}$.

Here, the difficulty is still in determining the extremal invariant probabilities. Li and Blankenship also discussed large deviations and the stabilization of the system by linear state feedback.

Motivated by Oseledec and Furstenberg's work, Feng and Loparo [14,15] studied the linear system given by

$$\begin{cases} \dot{x}(t) = A(y(t))x(t), t \geq 0, \quad \text{and} \\ x_0 \in \mathbb{R}^d, \end{cases} \tag{63.100}$$

where $y(t) \in N = \{1, 2, \ldots, n\}$ is a finite-state continuous time Markov process with infinitesimal generator

$$Q = (q_{ij})_{n \times n}$$

and initial probabilities $(p_1, \ldots, p_n)$, and $A(i) = A_i \in \mathbb{R}^{d \times d}$. By using properties of finite-state Markov processes and a sojourn time description of the process $y(t)$, they were able to relate the Oseledec spaces of the system to invariant subspaces of the constituent linear differential equations $\dot{x}(t) = A_i x(t)$ and obtained a spectrum theorem for the Lyapunov exponents of the stochastic system Equation 63.100. The exponents in the spectrum theorem given below are exactly the exponents given in Oseledec's theorem that are physically realizable. It was also shown that the spectrum obtained is actually independent of the choice of the initial probabilities $(p_1, \ldots, p_n)$ of $y(t)$. Thus, stationarity is not required. The theorem is stated next.

Theorem 63.11: Feng and Loparo

For the system Equation 63.100, suppose $p_i > 0$ *and* $q_{ij} > 0$ *for all* $i, j \in N = \{1, 2, \ldots, n\}$. *Then there exists* k *real numbers,* $k \leq d$, *constants* $\lambda_i, 1 = 1, \ldots, k$ *with*

$$-\infty < \lambda_1 < \lambda_2 < \cdots < \lambda_k < +\infty,$$

an orthonormal basis for $\mathbb{R}^d$

$$\{e_1^{(1)} \ldots e_{i_1}^{(1)} | e_1^{(2)} \ldots e_{i_2}^{(2)} | \ldots | e_1^{(k)} \ldots e_{i_k}^{(k)}\}$$

with $i_1 + \cdots + i_k = d$ *and* $i_j \geq 1$ *for* $j = 1, 2, \ldots, k$ *so that, if*

$$E^{i_j} = span\ \{e_1^{(j)} \ldots e_{i_j}^{(j)}\}$$

is an i_j*-dimensional subspace of* $\mathbb{R}^d$ *and*

$$\begin{cases} \mathcal{L}_0 = \{0\} \\ \mathcal{L}_j = \oplus_{\ell=1}^{j} E^{i_\ell} \quad j = 1, 2, \ldots, k \end{cases}$$

which is a filtration of $\mathbb{R}^d$, *that is,*

$$\{0\} \triangleq \mathcal{L}_0 \subset \mathcal{L}_1 \subset \ldots \subset, \mathcal{L}_{,k} = \mathbb{R}^d$$

where "$\oplus$*" denotes the direct sum of subspaces, then*

1. $\mathcal{L}_j = \{x \in \mathbb{R}^d \backslash \{0\} : \overline{\lambda}_\omega(x) \leq \lambda_j$ *a.s.* $\}$ *and* $\overline{\lambda}_\omega(x) = \lambda_j$ *a.s. iff* $x \in \mathcal{L}_j \backslash \mathcal{L}_{j-1}$ *for* $j = 1, 2, \ldots, k$.

2. *$\mathcal{L}_j$ is an A_i-invariant subspace of $\mathbb{R}^d$ for all $i = 1, 2, \ldots, n$ and $j = 1, 2, \ldots, k$.*
3. *All of the results above are independent of the initial probabilities $(p_1, \ldots, p_n)$ chosen.*
4. *If $y(t)$ is stationary, λ_k is the top exponent in Oseledec's theorem.*

Stability results can be obtained directly from the theorem, for example, if $\lambda_i < 0$ and $\lambda_{i+1} > 0$, then

$$P\{\overline{\lim_{t\to+\infty}} \|x(t, x_0, \omega)\| = 0\} = 1, \quad \text{if } x_0 \in \mathcal{L}_i,$$

and

$$P\{\overline{\lim_{t\to+\infty}} \|x(t, x_0, \omega)\| = +\infty\} = 1, \quad \text{if } x_0 \in \mathbb{R}^d \backslash \mathcal{L}_i.$$

63.4 Conclusions

In this chapter we have introduced the different concepts of stability for stochastic systems and presented two techniques for stability analysis. The first extends Lyapunov's direct method of stability analysis for deterministic systems to stochastic systems. The main ingredient is a Lyapunov function and, under certain technical conditions, the stability properties of the stochastic system are determined by the "derivative" of the Lyapunov function along sample solutions of the stochastic system. As in the deterministic theory of Lyapunov stability, a major difficulty in applications is constructing a proper Lyapunov function for the system under study.

Because it is well known that even though moment stability criteria for a stochastic system may be easier to determine, moment stability does not necessarily imply sample path stability and that moment stability criteria can be too conservative to be practically useful, it is important to determine the sample path (almost sure) stability properties of stochastic systems. It is the sample paths, not the moments, that are observed in applications. Focusing primarily on linear stochastic systems, we presented the concepts and theory of the Lyapunov exponent method for sample path stability of stochastic systems. Computational difficulties in computing the top Lyapunov exponent, or its algebraic sign, must still be resolved before this method will see more widespread applications in science and engineering.

Dedication

The chapter is dedicated to the memory of Dr. Xiangbo Feng, my former student, colleague, and friend, who passed away since the first version of this chapter was published.

References

1. Arnold, L., A formula connecting sample and moment stability of linear stochastic systems, *SIAM J. Appl. Math.*, 44, 793–802, 1984.
2. Arnold, L., Crauel, H., and Wihstutz, V., Stabilization of linear system by noise. *SIAM J. Control Optim.*, 21(3), 451–461, 1983.
3. Arnold, L., Kliemann, W., and Oeljeklaus, E., Lyapunov exponents of linear stochastic systems. In *Lecture Notes in Math.*, No. 1186, Springer, Berlin, 1985.
4. Arnold, L., Papanicolaou, G., and Wihstutz, V., Asymptotic analysis of the Lyapunov exponent and rotation number of the random oscillator and application. *SIAM J. App. Math.*, 46(3), 427–449, 1986.
5. Arnold, L. and Wihstutz, V., Eds., *Lyapunov Exponents*, Lecture Notes in Math., No. 1186, Springer, Berlin, 1985.
6. Auslender, E.I. and Mil'shtein, G.N., Asymptotic expansion of Lyapunov index for linear stochastic system with small noise. *Prob. Math. Mech. USSR*, 46, 277–283, 1983.
7. Bellman, R., Limit theorem for noncommutative operations-I. *Duke Math. J.*, 21, 491–500, 1954.

8. Bertram, J.E. and Sarachik, P.E., Stability of circuits with randomly time-varying parameters. *Trans. IRE,* PGIT-5, Special Supplement, p. 260, 1959.
9. Bitsris, G., On the stability in the quadratic mean of stochastic dynamical systems. *Int. J. Control,* 41(4), 1061–1075, 1985.
10. Brockett, R.W., *Finite Dimensional Linear Systems,* John Wiley & Sons, New York, 1970.
11. Brockett, R.W. and Willems, J.C., Average value criteria for stochastic stability. In *Stability of Stochastic Dynamical Systems,* Lecture Notes in Math., No. 294, Curtain, R.F., Ed., Springer, New York, 1972.
12. Caughey, T.K. and Gray, A.H., Jr., On the almost sure stability of linear dynamic systems with stochastic coefficients. *J. Appl. Mech.,* 32, 365, 1965.
13. Feng, X. and Loparo, K.A., Almost sure instability of the random harmonic oscillator. *SIAM J. Appl. Math.,* 50(3), 744–759, 1990.
14. Feng, X. and Loparo, K.A., A nonrandom spectrum for Lyapunov exponents of linear stochastic systems. *Stochastic Anal. Appl.,* 9(1), 25–40, 1991.
15. Feng, X. and Loparo, K.A., A nonrandom spectrum theorem for products of random matrices and linear stochastic systems. *J. Math. Syst., Estimation Control,* 2(3), 323–338, 1992.
16. Furstenberg, H. and Kesten, H., Products of random matrices. *Ann. Math. Statist.,* 31, 457–469, 1960.
17. Furstenberg, H., Noncommuting random products. *Trans. Am. Math. Soc.,* 108, 377–428, 1963.
18. Furstenberg, H., A Poisson formula for semi-simple Lie group. *Ann. Math.,* 77, 335–386, 1963.
19. Has'minskii, R.Z., A limit theorem for solutions of differential equations with random right-hand sides. *Theory Probl. Appl.,* 11, 390–406, 1966.
20. Has'minskii, R.Z., Necessary and sufficient condition for the asymptotic stability of linear stochastic systems. *Theory Prob. Appl.,* 12, 144–147, 1967.
21. Has'minskii, R.Z., *Stability of Systems of Differential Equations Under Random Perturbation of Their Parameters,* (in Russian), Nauka Moscow, 1969.
22. Has'minskii, R.Z., *Stochastic Stability of Differential Equations,* Sijthoff and Noordhoff, Ma., 330–332, 1980.
23. Infante, E.F., On the stability of some linear non-autonomous random system. *J. Appl. Mech.,* 35, 7–12, 1968.
24. Kats, I.I. and Krasovskii, N.N., On the stability of systems with random parameters. *Prkil. Met. Mek.,* 24, 809, 1960.
25. Kleinman, D.L., On the stability of linear stochastic systems. *IEEE Trans. Automat. Control,* AC-14, 429–430, 1969.
26. Kozin, F., On almost sure stability of linear systems with random coefficients. *M.I.T. Math. Phys.,* 43, 59, 1963.
27. Kozin, F., A survey of stability of stochastic systems. *Automatica,* 5, 95–112, 1969.
28. Kozin, F. and Prodromou, S., Necessary and sufficient condition for almost sure sample stability of linear Ito equations. *SIAM J. Appl. Math.,* 21(3), 413–424, 1971.
29. Kozin, F. and Wu, C.M., On the stability of linear stochastic differential equations. *J. Appl. Mech.,* 40, 87–92, 1973.
30. Kushner, H.J., *Stochastic Stability and Control,* Academic Press, New York, 1967.
31. Kushner, H.J., *Introduction to Stochastic Control Theory,* Holt, Rinehart and Winston, New York, 1971.
32. Kushner, H.J., Stochastic stability. In *Stability of Stochastic Dynamical Systems,* Lecture Notes in Math., No. 249, Curtain, R.F., Ed., Springer, New York, pp. 97–124, 1972.
33. Ladde, G.S. and Siljak, D.D., Connective stability of large scale systems. *Int. J. Syst. Sci.,* 6(8), 713–721, 1975.
34. Li, C.W. and Blankenship, G.L., Almost sure stability of linear stochastic system with Poisson process coefficients. *SIAM J. Appl. Math.,* 46(5), 875–911, 1986.
35. Loparo, K.A. and Blankenship, G.L., Almost sure instability of a class of linear stochastic system with jump parameter coefficients. In *Lyapunov Exponents,* Lecture Notes in Math., No. 1186, Springer, Berlin, 1985.
36. Loparo, K.A. and Feng, X., Lyapunov exponent and rotation number of two-dimensional linear stochastic systems with telegraphic noise. *SIAM J. Appl. Math.,* 53(1), 283–300, 1992.
37. Lyapunov, A.M., Probléme générale de la stabilité du muvement. *Comm. Soc. Math. Kharkov, 2,* 1892,*3,* 1893. Reprint *Ann. Math. Studies,* 17, Princeton University Press, Princeton, 1949.
38. Mahalanabis, A.K. and Parkayastha, S., Frequency domain criteria for stability of a class of nonlinear stochastic systems. *IEEE Trans. Automat. Control,* AC-18(3), 266–270, 1973.
39. Man, F.T., On the almost sure stability of linear stochastic systems. *J. Appl. Mech.,* 37(2), 541, 1970.
40. Michel, A.N., Stability analysis of stochastic large scale systems. *Z. Angew. Math. Mech.,* 55, 93–105, 1975.

41. Michel, A.N. and Rasmussen, R.D., Stability of stochastic composite systems. *IEEE Trans. Automat. Control,* AC-21, 89–94, 1976.
42. Mitchell, R.R., Sample stability of second order stochastic differential equation with nonsingular phase diffusion. *IEEE Trans. Automat. Control,* AC-17, 706–707, 1972.
43. Mitchell, R.R. and Kozin, F., Sample stability of second order linear differential equation with wide band noise coefficients. *SIAM J. Appl. Math.,* 27, 571–605, 1974.
44. Molchanov, S.A., The structure of eigenfunctions of one-dimensional unordered structures. *Math. USSR Izvestija,* 12, 69–101, 1978.
45. Nishioka, K., On the stability of two-dimensional linear stochastic systems. *Kodai Math. Sem. Rep.,* 27, 221–230, 1976.
46. Oseledec, V.Z., A multiplicative ergodic theorem Lyapunov characteristic number for dynamical systems. *Trans. Moscow Math. Soc.,* 19, 197–231, 1969.
47. Pardoux, E. and Wihstutz, V., Two-dimensional linear stochastic systems with small diffusion. *SIAM J. Appl. Math.,* 48, 442–457, 1988.
48. Parthasarathy, A. and Evan-Zwanowskii, R.M., On the almost sure stability of linear stochastic systems. *SIAM J. Appl. Math.,* 34(4), 643–656, 1978.
49. Pinsky, M.A., Stochastic stability and the Dirichlet problem. *Common. Pure. Appl. Math.,* 27, 311–350, 1974.
50. Pinsky, M.A., Instability of the harmonic oscillator with small noise. *SIAM J. Appl. Math.,* 46(3), 451–463, 1986.
51. Rasmussen, R.D. and Michel, A.N., On vector Lyapunov functions for stochastic dynamic systems. *IEEE Trans. Automat. Control,* AC-21, 250–254, 1976.
52. Siljak, D.D., *Large-Scale Dynamic Systems.* North-Holland, Amsterdam, 1978.
53. Socha, L., Application of Yakubovich criteria for stability of nonlinear stochastic systems. *IEEE Trans. Automat. Control,* AC-25(2), 330–332, 1980.
54. Socha, L., The asymptotic stochastic stability in large of the composite systems. *Automatica,* 22(5), 605–610, 1986.
55. Walker, J.A., On the application of Lyapunov's direct method to linear lumped-parameter elastic systems. *J. Appl. Mech.,* 41, 278–284, 1974.
56. Wiens, G.J. and Sinha, S.C., On the application of Lyapunov direct method to discrete dynamic systems with stochastic parameters. *J. Sound. Vib.,* 94(1), 19–31, 1984.
57. V. Wihstutz, V. Parameter dependence of the Lyapunov exponent for linear stochastic system. A survey. In *Lyapunov Exponents*, Lecture Notes in Math., No. 1186, Springer, Berlin, 1985.
58. Willems, J.L., Lyapunov functions and global frequency domain stability criteria for a class of stochastic feedback systems. In *Stability of Stochastic Dynamical Systems*, Lecture Notes in Math., No. 294, Curtain, R.F., Ed., Springer, New York, 1972.
59. Willems, J.L., Mean square stability criteria for stochastic feedback systems. *Int. J. Syst. Sci.,* 4(4), 545–564, 1973.

64

Stochastic Adaptive Control for Continuous-Time Linear Systems

T.E. Duncan
University of Kansas

B. Pasik-Duncan
University of Kansas

Another important and commonly used class of systems is the class of continuous-time linear systems. The models are assumed to evolve in continuous time rather than discrete time because this assumption is natural for many models and it is important for the study of discrete-time models when the sampling rates are large and for the analysis of numerical round-off errors. The stochastic systems are described by linear stochastic differential equations. It is assumed that there are complete observations of the state.

The general approach to adaptive control that is described here exhibits a splitting or separation of the problems of identification of the unknown parameters and adaptive control. Maximum likelihood (or equivalently least-squares) estimates are used for the identification of the unknown constant parameters. These estimates are given recursively and are shown to be strongly consistent. The adaptive control is usually constructed by the so-called certainty equivalence principle, that is, the optimal stationary controls are computed by replacing the unknown true parameter values by the current estimates of these values. Since the optimal stationary controls can be shown to be continuous functions of the unknown parameters, the self-tuning property is verified. It is shown that the family of average costs using the control from the certainty equivalence principle converges to the optimal average cost. This verifies the self-optimizing property.

A model for the adaptive control of continuous-time linear stochastic systems with complete observations of the state can be described by the following stochastic differential equation

$$\mathrm{d}X(t) = (A(\alpha)X(t) + BU(t))\mathrm{d}t + \mathrm{d}W(t) \tag{64.1}$$

where $X(t) \in \mathbb{R}^n$, $U(t) \in \mathbb{R}^m$.

$$A(\alpha) = A_0 + \sum_{i=1}^{p} \alpha^i A_i \tag{64.2}$$

$A_i \in \mathcal{L}(\mathbb{R}^n)$ $i = 0, \ldots, p$, $B \in \mathcal{L}(\mathbb{R}^m, \mathbb{R}^n)$, $(W(t), t \in \mathbb{R}_+)$ is a standard $\mathbb{R}^n$-valued Wiener process and $X_0 \equiv a \in \mathbb{R}^n$. It is assumed that

(3.A1) $\mathcal{A} \subset \mathbb{R}^p$ is compact and $\alpha \in \mathcal{A}$.

(3.A2) $(A(\alpha), B)$ is reachable for each $\alpha \in \mathcal{A}$.
(3.A3) The family $(A_i,\ i = 1, \ldots, p)$ is linearly independent.

Let $(\mathcal{F}_t, t \in \mathbb{R}_+)$ be a filtration such that X_t is measurable with respect to $\mathcal{F}_t$ for all $t \in \mathbb{R}_+$ and $(W(t), \mathcal{F}_t, t \in \mathbb{R}_+)$ is a Brownian martingale. The ergodic, quadratic control problem for Equation 64.1 is to minimize the ergodic cost functional

$$\limsup_{t \to \infty} \frac{1}{t} J(X_0, U, \alpha, t) \tag{64.3}$$

where

$$J(X_0, U, \alpha, t) = \int_0^t [\langle QX(s), X(s)\rangle + \langle PU(s), U(s)\rangle]\, \mathrm{d}s \tag{64.4}$$

and $t \in (0, \infty]$, $X(0) = X_0$, $Q \in \mathcal{L}(\mathbb{R}^n)$, and $P \in \mathcal{L}(\mathbb{R}^m)$ are self-adjoint and P^{-1} exists, $(X(t), t \in \mathbb{R}_+)$ satisfies Equation 64.1 and $(U(t), t \in \mathbb{R}_+)$ is adapted to $(\mathcal{F}_t, t \in \mathbb{R}_+)$. It is well known [9] that if α is known then there is an optimal linear feedback control such that

$$U^*(t) = KX(t) \tag{64.5}$$

where $K = -P^{-1}B^*V$ and V is the unique, symmetric, nonnegative definite solution of the algebraic Riccati equation

$$VA + A^*V - VB^*P^{-1}BV + Q = 0. \tag{64.6}$$

For an unknown α the admissible adaptive control policies $(U(t), t \in \mathbb{R}_+)$ are linear feedback controls

$$U(t) = K(t)X(t) = \tilde{K}(t, X(u),\ u \le t - \Delta)X(t) \tag{64.7}$$

where $(K(t), t \ge 0)$ is an $\mathcal{L}(\mathbb{R}^n, \mathbb{R}^m)$-valued process that is uniformly bounded and there is a fixed $\Delta > 0$ such that $(K(t), t \ge 0)$ is measurable with respect to $\sigma(X_u, u \le t - \Delta)$ for each $t \ge \Delta$ and $(K(t), t \in [0, \Delta))$ is a deterministic function. For such an adaptive control, it is elementary to verify that there is a unique strong solution of Equation 64.1. The delay $\Delta > 0$ accounts for some time that is required to compute the adaptive control law from the observation of the solution of Equation 64.1.

Let $(U(t), t \ge 0)$ be an admissible adaptive control and let $(X(t), t \ge 0)$ be the associated solution of Equation 64.1. Let $\mathcal{A}(t) = (a_{ij}(t))$ and $\tilde{\mathcal{A}}(t) = (\tilde{a}_{ij}(t))$ be $\mathcal{L}(\mathbb{R}^p)$-valued processes such that

$$a_{ij}(t) = \int_0^t \langle A_iX(s), A_jX(s)\rangle \mathrm{d}s,$$
$$\tilde{a}_{ij}(t) = \frac{a_{ij}(t)}{a_{ii}(t)}.$$

To verify the strong consistency of a family of least-squares estimates it is assumed that

(3.A4) $\liminf_{t \to \infty} |\det \tilde{\mathcal{A}}(t)| > 0$ a.s.

The estimate of the unknown parameter vector at time t, $\hat{\alpha}(t)$, for $t > 0$ is the minimizer for the quadratic functional of α, $L(t, \alpha)$, given by

$$L(t, \alpha) = -\int_0^t \langle (A(\alpha) + BK(s))X(s), \mathrm{d}X(s)\rangle + \frac{1}{2}\int_0^t |(A(\alpha) + BK(s))X(s)|^2 \mathrm{d}s \tag{64.8}$$

where $U(s) = K(s)X(s)$ is an admissible adaptive control. The following result [6] gives the strong consistency of these least-squares estimators.

Theorem 64.1:

Let $(K(t), t \geq 0)$ be an admissible adaptive feedback control law. If (3.A1–3.A4) *are satisfied and $\alpha_0 \in \mathcal{A}^\circ$, the interior of $\mathcal{A}$, then the family of least-squares estimates $(\hat{\alpha}(t), t > 0)$, where $\hat{\alpha}(t)$ is the minimizer of Equation 64.8, is strongly consistent, that is,*

$$P_{\alpha_0}\left(\lim_{t\to\infty} \hat{\alpha}(t) = \alpha_0\right) = 1 \tag{64.9}$$

where α_0 is the true parameter vector.

The family of estimates $(\hat{\alpha}(t), t > 0)$ can be computed recursively because this process satisfies the following stochastic equation:

$$\mathrm{d}\hat{\alpha}(t) = \mathcal{A}^{-1}(t)\langle \mathbb{A}(t)X(t), \mathrm{d}X(t) - A(\hat{\alpha}(t))X(t)\,\mathrm{d}t - BU(t)\,\mathrm{d}t\rangle, \tag{64.10}$$

where $\langle \mathbb{A}(t)x, y\rangle = (\langle A_i x, y\rangle)\ i = 1, \ldots, p$.

Now the performance of some admissible adaptive controls is described.

Proposition 64.1:

Assume that (3.A1–3.A4) *are satisfied and that*

$$\lim_{t\to\infty} \frac{1}{t}\langle VX(t), X(t)\rangle = 0 \quad \text{a.s.}, \tag{64.11}$$

$$\limsup_{t\to\infty} \frac{1}{t}\int_0^t |X(t)|^2\,\mathrm{d}s < \infty \quad \text{a.s.}, \tag{64.12}$$

where $(X(t), t \geq 0)$ is the solution of Equation 64.1 with the admissible adaptive control $(U(t), t \geq 0)$ and $\alpha = \alpha_0 \in \mathcal{K}$ and V is the solution of the algebraic Riccati equation 64.6 with $\alpha = \alpha_0$. Then

$$\liminf_{T\to\infty} \frac{1}{T} J(X_0, U, \alpha_0, T) \geq \ tr\ V \quad \text{a.s.}, \tag{64.13}$$

If U is an admissible adaptive control $U(t) = K(t)X(t)$ such that

$$\lim_{t\to\infty} K(t) = k_0 \qquad \text{a.s.}, \tag{64.14}$$

where $k_0 = -P^{-1}B^*V$, then

$$\lim_{T\to\infty} \frac{1}{T} J(X_0, U, \alpha_0, T) = \ tr\ V \quad \text{a.s.}, \tag{64.15}$$

Corollary 64.1:

Under the assumptions of Proposition 64.1, if Equation 64.14 is satisfied, then Equations 64.11 and 64.12 are satisfied.

The previous results can be combined for a complete solution to the stochastic adaptive control problem (Equations 64.1 and 64.3) [6].

Theorem 64.2:

Assume that (3.A1–3.A4) are satisfied. Let $(\hat{\alpha}(t), t > 0)$ be the family of least-squares estimates where $\hat{\alpha}(t)$ is the minimizer of Equation 64.8. Let $(K(t), t \geq 0)$ be an admissible adaptive control law such that

$$K(t) = -P^{-1}B^* V(\hat{\alpha}(t - \Delta))$$

where $V(\alpha)$ is the solution of Equation 64.6 for $\alpha \in \mathcal{A}$. Then the family of estimates $(\hat{\alpha}(t), t > 0)$ is strongly consistent,

$$\lim_{t\to\infty} K(t) = k_0 \quad \text{a.s.,} \tag{64.16}$$

where $k_0 = -P^{-1}B^* V(\alpha_0)$ and

$$\lim_{T\to\infty} \frac{1}{T} J(X_0, U, \alpha_0, T) = tr\, V \quad \text{a.s.,} \tag{64.17}$$

Now a second formulation for the stochastic adaptive control of an unknown continuous time linear system with a quadratic cost is given. In this description a fixed delay $\Delta > 0$ is introduced for measurability and computability of an optimal adaptive control. Let $(X(t), t \geq 0)$ be a controlled linear diffusion, that is, a solution of the stochastic differential equation

$$\mathrm{d}X(t) = AX(t)\,\mathrm{d}t + BU(t)\,\mathrm{d}t + CW(t), \quad X(0) = X_0, \tag{64.18}$$

where $X(t) \in \mathbb{R}^n$, $U(t) \in \mathbb{R}^m$, and $(W(t), t \geq 0)$ is a standard p-dimensional Wiener process. The probability space is $(\Omega, \mathcal{F}, P)$ and $(\mathcal{F}_t, t \geq 0)$ is an increasing family of sub-σ-algebras of $\mathcal{F}$ such that $\mathcal{F}_0$ contains all P-null sets, $(W(t), \mathcal{F}_t, t \geq 0)$ is a continuous martingale and $X(t) \in \mathcal{F}_t$ for all $t \geq 0$. The linear transformations A, B, and C are assumed to be unknown. Since the adaptive control does not depend on C it suffices to estimate the pair (A, B). For notational simplicity let $\theta^T = [A, B]$.

For the adaptive control problem, it is required to minimize the ergodic cost functional

$$\limsup_{t\to\infty} J(t, U) = \limsup_{t\to\infty} \frac{1}{t} \int_0^t (X^T(s)Q_1X(s) + U^T(s)Q_2U(s))\,\mathrm{d}s \tag{64.19}$$

where $Q_1 \geq 0$ and $Q_2 > 0$ and U is an admissible control.

Since both A and B are unknown, it is necessary to ensure sufficient excitation in the control to obtain consistency of a family of estimates. This is accomplished by a diminishing excitation control (dither) that is asymptotically negligible for the ergodic cost functional Equation 64.19. Let $(v_n, n \in \mathbb{N})$ be a sequence of $\mathbb{R}^m$-valued independent, identically distributed random variables that is independent of the Wiener process $(W(t), t \geq 0)$. It is assumed that $E[v_n] = 0$, $E\left[v_n v_n^T\right] = I$ for all $n \in \mathbb{N}$, and there is a $\sigma > 0$ such that $\|v_n\|^2 \leq \sigma$ a.s. for all $n \in \mathbb{N}$. Let $\varepsilon \in \left(0, \frac{1}{2}\right)$ and fix it. Define the $\mathbb{R}^m$-valued process $(V(t), t \geq 0)$ as

$$V(t) = \sum_{n=0}^{[\frac{t}{\Delta}]} \frac{v_n}{n^{\varepsilon/2}} 1_{[n\Delta,(n+1)\Delta)}(t) \tag{64.20}$$

where $\Delta > 0$ is fixed. A family of least-squares estimates $(\theta(t), t \geq 0)$ is used to estimate the unknown $\theta = [A, B]^T$. The estimate $\theta(t)$ is given by

$$\theta(t) = \Gamma(t) \int_0^t \varphi(s)\,\mathrm{d}X^T(s) + \Gamma(t)\Gamma^{-1}(0)\theta(0), \tag{64.21}$$

$$\Gamma(t) = \left(\int_0^t \varphi(s)\varphi^T(s)\,\mathrm{d}s + aI\right)^{-1}, \tag{64.22}$$

$$\varphi(s) = [X^T(s) \quad U^T(s)]^T, \tag{64.23}$$

where $\theta(0)$ and $a > 0$ are arbitrary. The diminishingly excited control is

$$U(t) = U^{\mathrm{d}}(t) + V(t) \tag{64.24}$$

where U^{d} is a "desired" control.

For A stable, (A, C) controllable and some measurability and asymptotic boundedness of the control, the family of estimates $(\theta(t), t \geq 0)$ is strongly consistent [2].

Theorem 64.3:

Let $\varepsilon \in \left(0, \frac{1}{2}\right)$ be given in Equation 64.20. For Equation 64.18, if A is stable, (A, C) is controllable and the control $(U(t), t \geq 0)$ is given by Equation 64.24, where $U^{\mathrm{d}}(t) \in \mathcal{F}_{(t-\Delta)\vee 0}$ for $t \geq 0$ and $\Delta > 0$ is fixed and

$$\int_0^t \|U^{\mathrm{d}}(s)\|^2 \mathrm{d}s = O(t^{1+\delta}) \quad \text{a.s.,} \tag{64.25}$$

as $t \to \infty$ for some $\delta \in [0,\ 1 - 2\varepsilon)$ then,

$$\|\theta - \theta(t)\|^2 = O\left(\frac{\log t}{t^{\alpha}}\right) \quad \text{a.s.,} \tag{64.26}$$

as $t \to \infty$ for each $\alpha \in \left(\frac{1+\delta}{2}, 1 - \varepsilon\right)$, where $\theta = [A \quad B]^T$ and $\theta(t)$ satisfies Equation 64.21.

Now a self-optimizing adaptive control is constructed for the unknown linear stochastic system (Equation 64.18) with the quadratic ergodic cost functional (Equation 64.19). The adaptive control switches between a certainty equivalence control and the zero control. The family of admissible controls $\mathcal{U}(\Delta)$ is defined as follows:

$$\begin{aligned}\mathcal{U}(\Delta) = \Big\{U : U(t) &= U^{\mathrm{d}}(t) + U^1(t), U^{\mathrm{d}}(t) \in \mathcal{F}_{(t-\Delta)\vee 0} \text{ and} \\ & U^1(t) \in \sigma(V(s), (t-\Delta) \vee 0 \leq s \leq t) \text{ for all } t \geq 0 \\ |X(t)\|^2 &= \mathrm{o}(t) \text{ a.s. and } \int_0^t (\|U(s)\|^2 + \|X(s)\|^2)\, \mathrm{d}s \\ &= \mathrm{O}(t) \text{ a.s. as } t \to \infty\Big\}.\end{aligned} \tag{64.27}$$

Define the $\mathbb{R}^m$-valued process $(U^0(t), t \geq \Delta)$ using the equation

$$\begin{aligned}U^0(t) &= -Q_2^{-1}B^T(t-\Delta)P(t-\Delta) \\ &\quad \times \left(e^{\Delta A(t)}X(t-\Delta) + \int_{t-\Delta}^t e^{(t-s)A(t-\Delta)}B(t-\Delta)U^{\mathrm{d}}(s)\,\mathrm{d}s\right),\end{aligned} \tag{64.28}$$

where $A(t)$ and $B(t)$ are the least-squares estimates of A and B given by Equation 64.21 and $P(t)$ is the minimal solution of the algebraic Riccati equation

$$A^T(t)P(t) + P(t)A(t) - P(t)B(t)Q_2^{-1}B^T(t)P(t) + Q_1 = 0, \tag{64.29}$$

if $A(t)$ is stable and otherwise $P(t) = 0$.

To define the switching in the adaptive control, the following two sequences of stopping times $(\sigma_n, n = 1, 2, \ldots)$ and $(\tau_n, n = 1, 2, \ldots)$ are given as follows:

$$\sigma_0 = 0,$$

$$\sigma_n = \sup\left\{t \geq \tau_n : \int_0^t \|U^0(r)\|^2 \mathrm{d}r \leq s\tau_n^{\delta},\ A(s - \Delta) \text{ is stable for all } s \in [\tau_n, t)\right\} \tag{64.30}$$

$$\tau_n = \inf\left\{t \geq \tau_{n-1} + 1 : \int_0^t \|U^0(r)\|^2 \mathrm{d}r \leq \frac{1}{2} t^{1+\delta}\ A(t - \Delta) \text{ is stable and } \|x(t - \Delta)\|^2 \leq t^{1+\delta/2}\right\}. \tag{64.31}$$

The adaptive control $(U^*(t), t \geq 0)$ is given by

$$U^*(t) = U^{\mathrm{d}}(t) + V(t), \tag{64.32}$$

where

$$U^{\mathrm{d}}(t) = \begin{cases} 0 & \text{if } t \in [\sigma_n, \sigma_{n+1}) \quad \text{for some } n \geq 0, \\ U^0(t) & \text{if } t \in [\tau_n, \sigma_n) \quad \text{for some } n \geq 1. \end{cases} \tag{64.33}$$

The adaptive control U^* is self-optimizing [2].

Theorem 64.4:

If A is stable and (A, C) is controllable, then the adaptive control $(U^*(t), t \geq 0)$ given by Equation 64.22 belongs to $\mathcal{U}(\Delta)$ and is self-optimizing for Equations 64.18 and 64.19, that is,

$$\inf_{U \in \mathcal{U}(\Delta)} \limsup_{t \to \infty} J(t, U) = \lim_{t \to \infty} J(t, U^*) = tr\,(C^T PC) + tr\,(B^T PR(\Delta)PBQ_2^{-1}) \text{ a.s.}, \tag{64.34}$$

where P is the minimal solution of the algebraic Riccati equation 64.29 using A and B and

$$R(\Delta) = \int_0^{\Delta} e^{tA} CC^T e^{tA^T} \mathrm{d}t.$$

A third adaptive linear-quadratic Gaussian (LQG) control problem is considered now. In this case only a delay $\Delta > 0$ is introduced for the measurability of the adaptive control. This problem is solved in [4]. In this case, the unknowns are A and B and there are only the usual assumptions of controllability and observability that are made for the LQG control problem when the system is known. Let $(X(t), t \geq 0)$ be the process that satisfies the stochastic differential equation

$$\mathrm{d}X(t) = AX(t)\,\mathrm{d}t + BU(t)\,\mathrm{d}t + D\mathrm{d}W(t), \quad X(0) = X_0, \tag{64.35}$$

where $X(t) \in \mathbb{R}^n$, $U(t) \in \mathbb{R}^m$, and $(W(t), \mathcal{F}(t), t \geq 0)$ is an $\mathbb{R}^p$-valued standard Wiener process, and $(U(t), \mathcal{F}(t), t \geq 0)$ is a control from a family that is specified subsequently. The random variables are defined on a fixed complete probability space $(\Omega, \mathcal{F}, \mathbb{P})$ and the filtration $(\mathcal{F}(t), t \geq 0)$ is defined on this space and specified subsequently. It is assumed that the matrices A and B are unknown.

The objective is to design an admissible control process $(U(t), \mathcal{F}(t), t \geq 0)$ so that the following ergodic cost functional for the system (Equation 64.35) is minimized

$$J(U) = limsup_{T\to\infty} \frac{1}{T} \int_0^T [X^T(t)Q_1X(t) + U^T(t)Q_2U(t)]\, dt, \tag{64.36}$$

where $Q_2 > 0$ and $Q_1 \geq 0$ are symmetric and a control $(U(t), t \geq 0)$ for the system Equation 64.35 is said to be *admissible* if it is adapted to $(\mathcal{F}(t), t \geq 0)$ and

$$limsup_{t\to\infty} \frac{\int_0^t |U(s)|^2 de}{\int_0^t |X(s)|^2 ds} < \infty \tag{64.37}$$

The following standard assumptions are made.

A1: (A, B) is controllable.
A2: $(A, Q_1^{1/2})$ is observable.

It is well known that under the assumptions A1 and A2, the optimal control for the known system in the family of admissible controls is a linear feedback expressed as

$$U^0(t) = -Q_2^{-1}B^TRX(t), \tag{64.38}$$

where R is the unique positive, symmetric solution of the following algebraic Riccati equation

$$A^TR + RA - ABQ_2^{-1}B^TR + Q_1 = 0. \tag{64.39}$$

The corresponding minimal cost is

$$J(U^0) = inf_U J(U) = tr(D^TRD) \qquad \text{a.s.}, \tag{64.40}$$

To describe the estimation problem in a standard form, let

$$\theta^T = [A \qquad B], \tag{64.41}$$

$$\phi(t)^T = [X^T(t) \qquad U^T(t)], \tag{64.42}$$

so that Equation 64.35 can be rewritten as a linear regression

$$dX(t) = \theta^T\phi(t)dt + DdW(t). \tag{64.43}$$

Now the faimily of continuous-time, weighted least-squares (WLS) estimates [1], $(\theta(t), t \geq 0)$ is given by

$$d\theta(t) = a(t)P(t)\phi(t)[dX^T(t) - \phi(t)\theta(t)\, dt], \tag{64.44}$$

$$dP(t) = -a(t)P(t)\phi(t)\phi^T(t)P(t)\, dt, \tag{64.45}$$

where $\theta(0) = [A_0, B_0]^T$ and $P(0) > 0$ are arbitrary deterministic values such that (A_9, B_0) is controllable and $(A_0, Q_1^{1/2})$ is observable

$$a(t) = \frac{1}{f(r(t))}, \quad r(t) = \|P^{-1}(0)\| + \int_0^t |\phi(s)|^2\, ds,$$

and $f \in \mathbb{F}$ with

$$\mathbb{F} = \{f | f : \mathbb{R}_+ \to \mathbb{R}_+ \text{ slowly increasing}, \int_c^\infty \frac{dx}{xf(x)} < \infty \text{ for some } c > 0\}$$

where a function f is called slowly increasing if it is increasing and satisfies $f \geq 1$ and $f(x^2) = O(f(x))$ as $x \to \infty$.

Remark

A necessary condition for a function $f \in \mathbb{F}$ is that $f(x) = o(logx))$ Some typical functions that are used in WLS algorithms are $log^{1+\delta}x$ and $(logx)(loglogx)^{1+\delta}$. In fact the family of weights $(a(t), t \geq 0)$ satisfies

$$a^{-1}(t) = f(r(t)) + O(log^k r(t))$$

for some $k > 0$ as $t \to \infty$.

It can be shown that the convergence rate of the WLS algorithm can be characterized by $P(t)$, that is,

$$\|\theta(t) - \theta\|^2 = O(\|P(t)\|)$$

The explicit solution of $P(t)$ is

$$P(t) = [P^{-1}(0) + \int_0^t a(s)\phi(s)\phi(s)^T \, ds]^{-1},$$

which clearly shows that $P(t)$ is positive and nonincreasing so $(P(t), t \geq 0)$ converges a.s. as $t \to \infty$. It is also worth noting that the standard least-squares algorithm corresponds to the choice $f(x) \equiv 1$ which is excluded for the WLS algorithm.

Definition 64.1:.

A family of linear system models $(A(t), B(t), t \geq 0)$ is said to be uniformly controllable if there is a $c > 0$ such that

$$\sum_{i=0}^{[t]-1} A^i(t)B(t)B^T(t)A^{iT}(t) \geq cI \tag{64.46}$$

for all $t \in [0, \infty)$.

A family of models $(A(t), C(t), t \geq 0)$ is said to be uniformly observable if $(A^T(t), C^T(t), t \geq 0)$ is uniformly controllable.

Let $(\eta_k, k \in \mathbb{N})$ be a sequence of independent, identically distributed $\mathcal{M}(n+m, n)$-valued random variables that is independent, identically distributed $\mathcal{M}(n+m, n)$-valued random varibles that is independent of $(W(t), t \geq 0)$ so that for each $k \in \mathbb{N}$ the random variable η_k is uniformly distributed on the unit ball for a norm of the matrices. The maximization procedure is recursively defined as

$$\beta_0 = 0 \tag{64.47}$$

$$\beta_k = \eta_k \quad \text{if} \, f(k, \eta_k) \geq (1 + \gamma f(k, \beta_{k-1})) \tag{64.48}$$

$$= \beta_{k-1} \quad \text{otherwise}$$

Finally, the family of continuous time estimates $(\hat{\theta}(t), t \geq 0)$ to be used for the adaptive control problem is simply a piecewise constant function induced by Equation 64.44

$$\hat{\theta}(t) = \bar{\theta}_k, \quad t \in (k, k+1], \tag{64.49}$$

$$\bar{\theta}_k = \theta(k) - P^{1/2}(k)\beta_k, \tag{64.50}$$

where $k \in \mathbb{N}$.

The above estimates are expressed as

$$\hat{\theta}^T(t) = [A(t) \quad B(t)]. \tag{64.51}$$

It can be verified that the family $(A(t), B(t), Q_1^{1/2}, t \geq 0)$ is uniformly controllable and observable. Thus, the following stochastic algebraic Riccati equation:

$$A^T(t)B(t) + R(t)A(t) - R(t)B(t)Q_2^{-1}B^T R(t) + Q_1 = 0 \tag{64.52}$$

has a unique, adapted, symmetric, positive solution $R(t)$ for each $t \in [0, \infty)$ a.s.

Using $R(t)$, define a lagged certainty equivalence LQG control by

$$U(t) = -Q_2^{-1}B^T(t)R(t)X(t) \tag{64.53}$$

The following theorem states that for the above lagged certainty equivalence control, the solution of Equation 64.35 is stable in the averaging sense.

Theorem 64.5:

The process $(X(t), t \geq 0)$ that is the solution of Equation 64.35 is stable in the sense that

$$limsup_{T\to\infty}\frac{1}{T}\int_0^T |X(s)|^2 ds < \infty \qquad \text{a.s.,} \tag{64.54}$$

To obtain the optimality of the quadratic cost functional, it is necessary to obtain the strong consistency for the family of estimates $(\hat{\theta}(t), t \geq 0)$. For this, a diminishing excitation is added to the adaptive control, Equation 64.53, that is

$$U^*(t) = L_k X(t) + \gamma_k[V(t) - V_k] \tag{64.55}$$

or

$$dU^*(t) = L_k dX(t) + \gamma_k dV(t) \tag{64.56}$$

for $t \in (k, k+1]$ and $k \in \mathbb{N}$, where $U^*(0) \in \mathbb{R}^m$ is an arbitrary deterministic vector,

$$L_k = -Q_2^{-1}B^T(k)R(k) \quad \text{and} \quad \gamma_k^2 = \frac{log(k)}{\sqrt{k}} \tag{64.57}$$

for $k \geq 1$. The process $(V(t), t \geq 0)$ is an $\mathbb{R}^m$-valued standard Wiener process that is independent of $(W(t), t \geq 0)$ and $(\eta_k, k \in \mathbb{N})$. Without loss of generality, the sub-σ-algebra $\mathcal{F}(t)$ is defined as the $\mathbb{P}$-completion of $\sigma(X_0, W(s), \eta_j, V(s), s \leq t, j \leq t)$.

The following theorem states that the family of regularized WLS estimates is strongly consistent using the lagged certainty equivalence control with diminishing excitation [4].

Theorem 64.6:

Let $(\hat{\theta}(t), t \geq 0)$ be the family of estimates given by Equation 64.49 using the control (Equation 64.55) in Equation 64.35. If A1 and A2 are satisfied, then

$$lim_{t\to\infty}\hat{\theta}(t) = \theta \qquad \text{a.s.,} \tag{64.58}$$

where θ is the true system parameters.

The following theorem states the self-optimality of the diminishingly excited lagged certainty equivalence control [4].

Theorem 64.7:

Let A1 and A2 be satisfied for the stochastic system (Equation 64.35) with the cost functional Equation 64.36 where A and B are unknown. Then the adaptive control defined by Equation 64.55 is admissible and optimal, that is,

$$limsup_{T\to\infty}\frac{1}{T}\int_0^T [x^T(t)Q_1X(t)+U^{*T}(t)Q_2U^*(t)]\mathrm{d}t = tr(D^TRD) \quad \text{a.s.,} \tag{64.59}$$

64.1 Adaptive Control for Continuous-Time Scalar Linear Systems with Fractional Brownian Motions

Brownian motion is often used to model some continuous perturbations of many physical systems because this process is Gauss–Markov and has independent increments. However, empirical measurements of many physical phenomena suggest that Brownian motion is inappropriate to use in mathematical models of these phenomena. A family of processes that has empirical evidence of wide physical applicability is fractional Brownian motion. Fractional Brownian motion is a family of Gaussian processes that was defined by Kolmogorov [13] in his study of turbulence [14,15]. While this family of processes includes Brownian motion, it also includes other processes that describe behavior that is bursty or has a long-range dependence. The first empirical evidence of the usefulness of these latter processes was made by Hurst [11] in his statistical analysis of rainfall along the Nile River. Mandelbrot used fractional Brownian motions to describe economic data and noted that Hurst's statistical analysis was identifying the appropriate fractional Brownian motion (FBM). Mandelbrot and van Ness [16] provided some of the initial theory for FBMs. Empirical justifications for modeling with FBMs have occurred in a wide variety of phenomena, such as, economic data , flicker noise in electronic devices, turbulence, internet traffic, biology, and medicine.

A real-valued process $(B(t), t \geq 0)$ on a complete probability space $(\Omega, \mathcal{F}, \mathbb{P})$ is called a (real-valued) standard fractional Brownian motion with the Hurst parameter $H \in (0, 1)$ if it is a Gaussian process with continuous sample paths that satisfies

$$\begin{aligned}\mathbb{E}\left[B(t)\right] &= 0\\ \mathbb{E}\left[B(s)B(t)\right] &= \frac{1}{2}\left(t^{2H}+s^{2H}-|t-s|^{2H}\right)\end{aligned} \tag{64.60}$$

for all $s, t \in \mathbb{R}_+$.

Let $H \in (1/2, 1)$ be fixed and B be a fractional Brownian motion with Hurst parameter H. For the applications given here, only a few results from a stochastic calculus for fractional Brownian motion are necessary. Let $f\colon [0, T] \to \mathbb{R}$ be a Borel measurable function. If f satisfies

$$|f|^2_{L^2_H} = \rho(H)\int_0^T \left(u_{1/2-H}(s)\left|I^{H-1/2}_{T-}\left(u_{H-1/2}f\right)(s)\right|\right)^2 \mathrm{d}s < \infty, \tag{64.61}$$

then $f \in L^2_H$ and $\int_0^T f\, \mathrm{d}B$ is a zero-mean Gaussian random variable with second moment [10]

$$\mathbb{E}\left[\left(\int_0^T f\, \mathrm{d}B\right)^2\right] = |f|^2_{L^2_H}, \tag{64.62}$$

where $u_a(s) = s^a$ for $a > 0$ and $s \geq 0$, $I_{T-}^{H-1/2}$ is a fractional integral [17] defined almost everywhere and given by

$$\left(I_{T-}^{H-1/2} f\right)(x) = \frac{1}{\Gamma(\alpha)} \int_x^T \frac{f(t)}{(t-x)^{3/2-H}}\, dt \tag{64.63}$$

for $x \in [0, T)$, $f \in L^1([0, T])$, and $\Gamma(\cdot)$ is the Gamma function and

$$\rho(H) = \frac{H\Gamma(H+1/2)\Gamma(3/2-H)}{\Gamma(2-2H)}.$$

Consider the optimal control problem where the scalar state X satisfies

$$dX(t) = \alpha_0 X(t)\, dt + bU(t)\, dt + dB(t), \quad X(t) = X_0, \tag{64.64}$$

and the ergodic (or average cost per unit time) cost function J is

$$J(U) = \limsup_{T\to\infty} \frac{1}{T} \int_0^T \left(qX^2(t) + rU^2(t)\right)\, dt, \tag{64.65}$$

where $q > 0$ and $r > 0$ are constants. The family $\mathcal{U}$ of admissible controls is all $(\mathcal{F}_t)$ adapted processes such that Equation 64.64 has one and only one solution.

To introduce some notation, recall the well-known solution with $H = 1/2$, that is, $(B(t), t \geq 0)$ is a standard Brownian motion. An optimal control is U^* given by

$$U^*(t) = -\frac{b}{r}\rho_0 X^*(t) \tag{64.66}$$

where $(X^*(t), t \geq 0)$ is the solution of Equation 64.64 with the control U^*, ρ_0 is the unique positive solution of the scalar algebraic Riccati equation

$$\frac{b^2}{r}\rho^2 - 2a\rho - q = 0, \tag{64.67}$$

hence

$$\rho_0 = \frac{r}{b^2}[\alpha_0 + \delta_0], \tag{64.68}$$

$$\delta_0 = \sqrt{\alpha_0^2 + \frac{b^2}{r}q}\,. \tag{64.69}$$

Furthermore,

$$J(U^*) = \rho_0 \qquad \text{a.s.}, \tag{64.70}$$

The following result is given in [12] and solves the analogous control problem for $H \in (1/2, 1)$.

Theorem 64.8:

Let $(U^*(t), t \geq 0)$ be the control given by

$$U^*(t) = -\frac{b}{r}\rho_0[X^*(t) + V^*(t)], \tag{64.71}$$

$$\begin{aligned} V^*(t) &= \int_0^t \delta_0 V^*(s)\,\mathrm{d}s + \int_0^t \left[\bar{k}(t,s) - 1\right]\left(\mathrm{d}X^*(s) - \alpha_0 X^*(s) - bU^*(s)\right)\mathrm{d}s \\ &= \int_t^\infty e^{-\delta_0(s-t)}\,\mathrm{d}B(s \mid t), \end{aligned} \tag{64.72}$$

where $(X^*(t), t \geq 0)$ is the solution of Equation 64.64 with the admissible control $(U^*(t), t \geq 0)$, ρ_0 and δ_0 are given in Equations 64.68 and 64.69 respectively, and

$$\begin{aligned} \bar{k}(t,s) &= -c_H^{-1} s^{1/2-H} \frac{d}{\mathrm{d}s} \int_s^t (r-s)^{1/2-H} \gamma(r,r)\,dr, \\ \gamma(t,s) &= \delta_0 e^{\delta_0 t} \int_0^\infty e^{\delta_0 \tau} K_H(\tau, s)\,d\tau, \\ K_H(t,s) &= H(2H-1) \int_s^\tau r^{H-1/2}(r-s)^{H-3/2}\,dr. \end{aligned}$$

$$\begin{aligned} B(s \mid t) &= \mathbb{E}[B(s) \mid \mathcal{F}_t] \\ &= B(t) + \int_0^t u_{1/2-H}(I_{t-}^{1/2-H}(I_{s-}^{H-1/2} 1_{[t,s)}))\,\mathrm{d}B \\ &= B(t) + \int_0^t u_{1/2-H}(I_{s-}^{H-1/2} u_{H-1/2} 1_{[t,s)})\,dW, \end{aligned} \tag{64.73}$$

where c_H is a constant that only depends on H, $u_a(s) = s^a$ for $s \geq 0$, $I^{H-1/2}$ is a fractional integral (Equation 64.63), $I^{1/2-H}$ is the fractional derivative and $(W(t), t \geq 0)$ is a standard Brownian motion (Wiener process) associated with $(B(t), t \geq 0)$ (e.g., [3]).

Then the control U^* is optimal in $\mathcal{U}$ and the optimal cost is

$$J(U^*) = \lambda \qquad \text{a.s.}, \tag{64.74}$$

where

$$\lambda = \frac{q\Gamma(2H+1)}{2\delta_0^{2H}}\left[1 + \frac{\delta_0 + \alpha_0}{\delta_0 - \alpha_0}\right]. \tag{64.75}$$

If α_0 is unknown, then it is important to find a family of strongly consistent estimators of the unknown parameter α_0 in Equation 64.64. A method is used in [7] that is called pseudo-least squares because it uses the least-squares estimate for α_0 assuming $H = 1/2$, that is, B is a standard Brownian motion in Equation 64.64. It is shown in [7] that the family of estimators $(\hat{\alpha}(t), t \geq 0)$ is strongly consistent for $H \in (1/2, 1)$ where

$$\hat{\alpha}(t) = \alpha_0 + \frac{\int_0^t X^0(s)\,\mathrm{d}B(s)}{\int_0^t (X^0(s))^2\,\mathrm{d}s}, \tag{64.76}$$

where

$$\mathrm{d}X^0(t) = \alpha_0 X^0(t)\,\mathrm{d}t + \mathrm{d}B(t), \quad X^0(0) = X_0. \tag{64.77}$$

This family of estimators can be obtained from Equation 64.64 by removing the control term by subtraction. The family of estimators $\hat{\alpha}$ is modified here using the fact that $\alpha_0 \in [a_1, a_2]$ as

$$\alpha(t) = \hat{\alpha}(t) 1_{[a_1,a_2]}(\hat{\alpha}(t)) + a_1 1_{(-\infty,a_1)}(\hat{\alpha}(t)) + a_2 1_{(a_2,\infty)}(\hat{\alpha}(t)) \tag{64.78}$$

for $t \geq 0$. $\hat{\alpha}(0)$ is chosen arbitrarily in $[a_1, a_2]$.

An adaptive control $(U^{\wedge}(t), t \geq 0)$, is obtained from the certainty equivalence principle, that is, at time t, the estimate $\alpha(t)$ is assumed to be the correct value for the parameter. Thus the stochastic equation for the system (Equation 64.64) with the control $U^{\wedge}$ is

$$\begin{aligned} dX^{\wedge}(t) &= (\alpha_0 - \alpha(t) - \delta(t))X^{\wedge}(t)\,dt - \frac{b\rho(t)}{r}V^{\wedge}(t)\,dt + dB(t) \\ &= (-\alpha_0 - \alpha(t) - \delta(t))X^{\wedge}\,dt - (\alpha(t) + \delta(t))V^{\wedge}(t)\,dt + dB(t), \\ X^{\wedge}(0) &= X_0, \end{aligned} \tag{64.79}$$

and

$$\delta(t) = \sqrt{\alpha^2(t) + \frac{b^2}{r}q} \tag{64.80}$$

$$U^{\wedge}(t) = -\frac{b\rho(t)}{r}\left[X^{\wedge}(t) + V^{\wedge}(t)\right], \tag{64.81}$$

$$\rho(t) = \frac{r}{b^2}\left[\alpha(t) + \delta(t)\right], \tag{64.82}$$

$$\begin{aligned} V^{\wedge}(t) &= \int_0^t \tilde{\delta}(s)V^{\wedge}(s)\,ds + \int_0^t [\tilde{k}(t,s) - 1][dX^{\wedge}(s) - \alpha(s)X^{\wedge}(s)\,ds - bU^{\wedge}(s)\,ds] \\ &= \int_0^t \tilde{\delta}(s)V^{\wedge}(s)\,ds + \int_0^t [\tilde{k}(t,s) - 1][dB(s) + (\alpha_0 - \alpha(t))X^{\wedge}(s)\,ds], \end{aligned} \tag{64.83}$$

$$\tilde{\delta}(t) = \delta(t) + \alpha(t) - \alpha_0, \tag{64.84}$$

and $\tilde{k}$ denotes the use of $\tilde{\delta}$ instead of δ_0 in $\bar{k}$. Note that $\delta(t) \geq -\alpha(t) + c$ for some $c > 0$ and all $t \geq 0$ so that

$$\alpha_0 - \alpha(t) - \delta(t) < -c.$$

The solution of the stochastic equation 64.79 is

$$X^{\wedge}(t) = e^{-\int_0^t \hat{\delta}}X_0 + \int_0^t e^{-\int_s^t \hat{\delta}}[-(\alpha(s) + \delta(s)]V^{\wedge}(s)\,ds + dB(s)]. \tag{64.85}$$

The following result [8] provides the solution of an adaptive control problem in the average sense instead of the pointwise sense.

Theorem 64.9:

Let the scalar-valued control system satisfy the Equation 64.64. Let $(\alpha(t), t \geq 0)$ be the family of estimators of α_0 given by Equation 64.78, let $(U^{\wedge}(t), t \geq 0)$ be the associated adaptive control in Equation 64.81, and let $(X^{\wedge}(t), t \geq 0)$ be the solution of Equation 64.64 with the control $U^{\wedge}$. Then

$$\lim_{t\to\infty} \frac{1}{t}\mathbb{E}\int_0^t \left|U^*(s) - U^{\wedge}(s)\right|^2\,ds = 0 \tag{64.86}$$

and

$$\lim_{t\to\infty} \frac{1}{t}\mathbb{E}\int_0^t \left|X^*(s) - X^{\wedge}(s)\right|^2\,ds = 0, \tag{64.87}$$

hence

$$\lim_{t\to\infty} \frac{1}{t}\mathbb{E}\int_0^t \left(q(X^{\wedge}(s))^2 + r(U^{\wedge}(s))^2\right)\,ds = \lambda, \tag{64.88}$$

where λ is given in Equation 64.75.

References

1. B. Bercu, Weighted estimation and tracking for ARMAX models, *SIAM J. Control Optim.* 33, 1995, 89–106.
2. H. F. Chen, T. E. Duncan, and B. Pasik-Duncan, Stochastic adaptive control for continuous-time linear systems with quadratic cost, *Appl. Math. Optim.*, 34, 1996, 113–138.
3. T. E. Duncan, Prediction for some processes related to a fractional Brownian motion, *Stat.Prob.Lett.* 76, 2006, 128–134.
4. T. E. Duncan, L. Guo, and B. Pasik-Duncan, Adaptive continuous-time linear quadratic Gaussian control, *IEEE Trans. Automat. Control* 44, 1999, 1653–1662.
5. T. E. Duncan, J. Jakubowski and B. Pasik-Duncan, Stochastic integration for fractional Brownian motions in Hilbert spaces, *Stochastics Dynamics*, 6, 2006, 53–75.
6. T. E. Duncan and B. Pasik-Duncan, Adaptive control of continuous-time linear stochastic systems, *Math. Control Signals Systems*, 3, 1990, 45–60.
7. T. E. Duncan and B. Pasik-Duncan, Parameter identification for some linear systems with fractional Brownian motion, *Proceedings of the Fifteenth Triennial World Congress IFAC 2002*, Barcelona, 2002.
8. T. E. Duncan and B. Pasik-Duncan, Adaptive control of a scalar linear stochastic system with a fractional Brownian motion, *Proc. IFAC World Congress*, Seoul 2002, 4096-4101.
9. W. H. Fleming and R. W. Rishel, *Deterministic and Stochastic Optimal Control*, Springer, New York, 1975.
10. L. Guo, Self-convergence of weighted least-squares with applications to stochastic adaptive control, *IEEE Trans. Automat. Control* 41, 1996, 79–89.
11. H. E. Hurst, Long-term storage capacity in reservoirs, *Trans. Am. Soc. Civil Eng.*, 116, 1951, 400–410.
12. M. L. Kleptsyna, A. Le Breton, and M. Viot, About the linear quadratic regulator problem under a fractional Brownian perturbation, *ESAIM Probab. Stat.* 9, 2003, 161–170.
13. A. N. Kolmogorov, Wienersche spiralen und einige andere interessante kurven in Hilbertschen Raum, *C.R. (Doklady) Acad. USSS (N.S.)*, 26, 1940, 115–118.
14. A. N. Kolmogorov, The local structure of turbulence in incompressible viscouse fluid for very large Reynolds numbers, *Dokl. Akdad. Nauk SSSR*, 30, 1941, 301–305.
15. A. N. Kolmogorov, Dissipation of energy in the locally isotropic turbulence, *Dokl. Adad. Nauk SSSSR*, 31, 1941, 538–540.
16. B. B. Mandelbrot and J. W. Van Ness, Fractional Brownian motions, fractional noises and applications, *SIAM Rev.*, 10, 1968, 422–437.
17. S. G. Samko, A. A. Kilbas, and O. I. Marichev, *Fractional Integrals and Derviatives*, Gordon and Breach, Yverdon, 1993.

65

Probabilistic and Randomized Tools for Control Design

Fabrizio Dabbene
Polytechnic University of Turin

Roberto Tempo
Polytechnic University of Turin

65.1 Motivations and Preliminaries

In this chapter, we study probabilistic and randomized methods for analysis and design of uncertain systems. This area is fairly recent, see [10,11], even though its roots lie in the robustness techniques for handling complex control systems developed in the 1980s. In contrast to these previous deterministic techniques, the main feature of these methods is the use of probabilistic concepts. One of the goals of this methodology is to provide a rapprochement between the classical stochastic and robust paradigms, combining worst-case bounds with probabilistic information, thus potentially reducing the conservatism inherent in the worst-case design. In this way, the control engineer gains additional insight that may help bridging the gap between theory and applications.

We consider an uncertain system affected by parametric uncertainty

$$q = [q_1 \ \cdots \ q_\ell]^\top$$

which is bounded in an hyperrectangle of radius $\rho \in [0, 1]$ centered at the nominal value $\bar{q}$

$$\mathbb{Q}_\rho \doteq \left\{q \in \mathbb{R}^\ell : q_i \in [(1-\rho)\,\bar{q}_i, (1+\rho)\,\bar{q}_i], i = 1, \ldots, \ell\right\}.$$

Henceforth, the objective is to design controller parameters θ in the presence of uncertain constraints of the form $f(\theta, q) \leq 0$; in Section 65.2 we provide a specific example of function f.

The algorithms derived in this probabilistic context are based on uncertainty randomization and are usually called *randomized algorithms*. That is, assuming that q is a random vector with given probability measure Pr, a controller is constructed utilizing a number of random samples $q^{(i)}$ of q. These algorithms provide controller parameters θ which probabilistically satisfy the system constraint $f(\theta, q) \leq 0$. In other words, a certain *probability of violation* $V(\theta, q)$ is associated to the controller θ, but this probability may be suitably bounded by given (probabilistic) accuracy $\epsilon \in (0, 1)$ and confidence $\delta \in (0, 1)$. The results presented in the literature for solving probabilistic design of uncertain systems can be divided into two main categories consisting of sequential and nonsequential methods, which are now described more precisely.

Sequential methods, see Section 65.3, generally build upon a convexity assumption regarding how the parameters θ enter into the function $f(\theta, q)$. The resulting algorithms are based on a probabilistic oracle and subsequent update rule. Various update rules have been derived, including gradient [3,9,27], ellipsoid [7], and cutting plane [4]. In this context, finite-time probabilistic properties have been obtained providing bounds on the maximum number of required iterations to guarantee a termination criterion with given probabilistic accuracy and confidence. Design of robust Linear Quadratic regulators, synthesis of linear parameter-varying (LPV) systems, and solution of uncertain linear matrix inequalities (LMIs) are successful examples of the efficacy of these methods.

Nonsequential methods, see Sections 65.4 and 65.5, are mainly based on statistical learning theory; see [13] for further details. From the control design point of view, the objective is to derive sample size bounds which guarantee uniform convergence properties for feasibility or optimization problems. The methods developed are very general and are not based on any convexity assumption. We present an algorithm which requires a one-shot (local) solution of an optimization problem subject to randomized constraints, whose number is dictated by the sample complexity previously obtained.

In the particular case of convex optimization with nonsequential methods, see Section 65.4, a successful paradigm has been introduced in [1]. In this approach, the original robust control problem is reformulated in terms of a single convex optimization problem with sampled constraints which are randomly generated. The main result of this line of research is to determine the sample complexity without resorting to statistical learning methods.

Once a probabilistic controller has been obtained either with sequential or nonsequential methods, the control engineer should perform an *a posteriori* analysis, both deterministic and probabilistic; see Section 65.6. In particular, regarding the former analysis, standard robustness methods can be used. Regarding the latter, in a classical Monte Carlo setting, we can establish the sample complexity for analysis so that the probability of violation is estimated with given accuracy and confidence. The probabilistic margin obtained with this approach is larger than that computed with deterministic methods at the expense of a "small" probability of violation.

To better illustrate these concepts, we resort to a motivating example, which will be revisited in various forms in this chapter.

Example 65.1: Design of a Lateral Motion Controller for an Aircraft

We consider a multivariable example given in [11] (see also [18] for a slightly different model and set of data) which studies the design of a controller for the lateral motion of an aircraft. The state-space equation consists of four states and two inputs and is given by

$$\dot{x}(t) = \begin{bmatrix} 0 & 1 & 0 & 0 \\ 0 & L_p & L_\beta & L_r \\ g/V & 0 & Y_\beta & -1 \\ N_{\dot\beta}(g/V) & N_p & N_\beta + N_{\dot\beta} Y_\beta & N_r - N_{\dot\beta} \end{bmatrix} x(t) + \begin{bmatrix} 0 & 0 \\ 0 & L_{\delta_a} \\ Y_{\delta_r} & 0 \\ N_{\delta_r} + N_{\dot\beta} Y_{\delta_r} & N_{\delta_a} \end{bmatrix} u(t), \tag{65.1}$$

where x_1 is the bank angle, x_2 its derivative, x_3 is the sideslip angle, x_4 the yaw rate, u_1 the rudder deflection, and u_2 the aileron deflection.

We consider the case when the 13 aircraft parameters entering into the state and input matrices of Equation 65.1 are uncertain. Hence, we consider the system

$$\dot{x}(t) = A(q)x(t) + B(q)u(t),$$

where we introduced the uncertainty vector $q = [q_1 \cdots q_\ell]^\top$ with $\ell = 13$. In particular, we consider the case when each parameter q_i is perturbed by a relative uncertainty equal to 10% around its nominal value $\bar{q}_i$, as reported in Table 65.1.

Formally, the vector q is assumed to range in the hyperrectangle centered at the nominal value $\bar{q}$, that is,

$$\mathbb{Q}_\rho = \left\{ q \in \mathbb{R}^\ell : q_i \in [0.9\,\bar{q}_i,\ 1.1\,\bar{q}_i], i = 1, \ldots, \ell \right\}. \tag{65.2}$$

We are interested in designing a state feedback controller $u = Kx$ that robustly stabilizes the system guaranteeing a desired decay rate $\alpha > 0$, which is equivalent to having all closed-loop eigenvalues with real part smaller than $-\alpha$. A sufficient condition [22] for the existence of such a controller requires finding a symmetric positive definite matrix $P \in \mathbb{R}^{4,4}$, and a matrix $W \in \mathbb{R}^{4,2}$ such that the following *quadratic performance* criterion is satisfied for all values of $q \in \mathbb{Q}_\rho$,

$$\Phi_{QP}(W, P, q) \doteq A(q)P + PA^\top(q) + B(q)W^\top + WB^\top(q) + 2\alpha P \preceq 0, \tag{65.3}$$

where the symbols $\succ$ and $\prec$ ($\succeq$ and $\preceq$) denote positive and negative definite (semidefinite) matrices. Further, if we find common $P \succ 0$ and W that simultaneously satisfy Equation 65.3 for all $q \in \mathbb{Q}_\rho$, then a control gain K robustly guaranteeing the desired decay rate can be recovered as $K = W^\top P^{-1}$. This approach has strong connections with classical optimal control, and in particular guaranteed-cost regulator design; see [11,18] for further details.

It should be noted that in many practical situations the requirement that Equation 65.3 is satisfied for all possible values of q may be too strict and may result in an overly conservative design. Suppose further that additional information on q is available, namely q has probabilistic nature, that is, it is a random vector distributed according to a probability measure Pr. In this case, it is reasonable to look for a design that guarantees that the closed-loop system behaves as desired with a given "high" probability. More precisely, fixing *a priori* a (small) probability level $\epsilon \in (0, 1)$, we look for $P \succ 0$, W that satisfies the probability constraint

$$\Pr\left\{ q \in \mathbb{Q}_\rho : \Phi_{QP}(P, W, q) \preceq 0 \right\} \geq 1 - \epsilon.$$

Note also that, if we introduce the function

$$f(P, W, q) = \lambda_{\max}\left(\Phi_{QP}(P, W, q)\right),$$

TABLE 65.1 Uncertain Parameters and Nominal Values

q	L_p	L_β	L_r	g/V	Y_β	$N_{\dot\beta}$	N_p	N_β	N_r	L_{δ_a}	Y_{δ_r}	N_{δ_r}	N_{δ_a}
$\bar{q}$	−2.93	−4.75	0.78	0.086	−0.11	0.1	−0.042	2.601	−0.29	−3.91	0.035	−2.5335	0.31

where $\lambda_{max}(\cdot)$ denotes the largest eigenvalue of its argument, then we may rewrite a probabilistic problem as that of finding $P \succ 0$, W such that

$$\Pr\left\{q \in \mathbb{Q}_\rho : f(P, W, q) \leq 0\right\} \geq 1 - \epsilon. \tag{65.4}$$

Computing a solution that satisfies Equation 65.4 amounts to solving a so-called chance-constrained semidefinite feasibility problem: a non-convex and very hard problem in general.

65.2 Probabilistic Design

The design techniques developed in the probabilistic framework are based on the interplay of random sampling in the uncertainty space and deterministic optimization in the design parameter space. Formally, let $\theta \in \mathbb{R}^{n_\theta}$ denote the vector of design variables. For instance, returning to the previous motivating example, we choose θ to represent the free parameters of the symmetric matrix P and the full matrix W, so that $n_\theta = 18$. Then, define a *performance function* that takes into account all the design and performance constraints related to the system with design parameters θ. These are rewritten in the form of the following (uncertain) inequality

$$f(\theta, q) = f(P, W, q) \leq 0, \tag{65.5}$$

where $f(\theta, q) : \mathbb{Q}_\rho \times \mathbb{R}^{n_\theta} \to \mathbb{R}$ is a scalar-valued function. The design vector θ such that the inequality (Equation 65.5) is satisfied "for most" (in a probabilistic sense) of the outcomes of q is called a probabilistic robust design. More specifically, define the *probability of violation* of the design θ as

$$V(\theta) \doteq \Pr\left\{q \in \mathbb{Q}_\rho : f(\theta, q) > 0\right\},$$

then, an ϵ-*probabilistic robust design* is a design guaranteeing

$$V(\theta) < \epsilon.$$

Equivalently, we define the *probability of performance*, or shortly *reliability*, of the design θ as

$$R(\theta) = 1 - V(\theta) = \Pr\left\{q \in \mathbb{Q}_\rho : f(\theta, q) \leq 0\right\}.$$

In other words, we say that a design is reliable if it guarantees a probability of performance of at least $(1 - \epsilon)$. Most of the results presented in the literature for solving the probabilistic design problem have been derived under the assumption that the function $f(\theta, q)$ is *convex* in θ for all $q \in \mathbb{Q}_\rho$. This assumption, which will be used in Sections 65.3 and 65.4, is now formally stated.

Convexity Assumption 65.1:

The function $f(\theta, q)$ is convex in θ for any fixed value of $q \in \mathbb{Q}_\rho$.

A standard example of a convex function f arises when considering performance requirements expressed by LMI conditions, as in the case of the motivating example. In particular, consider robust feasibility problem of the form

$$\text{Find } \theta \quad \text{such that } F(\theta, q) \preceq 0 \quad \forall q \in \mathbb{Q}_\rho,$$

where the constraint $F(\theta, q) \preceq 0$ is an LMI in θ for fixed q, that is,

$$F(\theta, q) = F_0(q) + \sum_{i=1}^{n_\theta} \theta_i F_i(q), \tag{65.6}$$

and $F_i(q)$, $i = 0, \ldots, n_\theta$, are symmetric real matrices of appropriate dimensions which depend in a generic and possibly nonlinear way on the uncertainty $q \in \mathbb{Q}_\rho$. Then, to rewrite this problem in the scalar-function framework, we simply set

$$f(\theta, q) \doteq \lambda_{\max}(F(\theta, q)).$$

Finally, we remark that considering scalar-valued constraint functions is without loss of generality, since multiple constraints

$$f_1(\theta, q) \leq 0, \ldots, f_{n_f}(\theta, q) \leq 0$$

can be immediately reduced to a single scalar-valued constraint by setting

$$f(\theta, q) = \max_{i=1,\ldots,n_f} f_i(\theta, q).$$

The randomized algorithms discussed in this chapter provide a numerically viable way to compute approximate probabilistic solutions for the problems previously described.

65.3 Sequential Methods for Design

In this section, we present randomized sequential methods for finding a probabilistic feasible solution θ to the uncertain inequality

$$f(\theta, q) \leq 0 \tag{65.7}$$

introduced in the previous section, under the Convexity Assumption. To this end, we first state the following formal definition.

Definition 65.1: r-Feasibility

For given $r > 0$, we say that the inequality (Equation 65.7) is r-feasible if the solution set

$$\mathcal{S} = \left\{\theta \in \mathbb{R}^{n_\theta} : f(\theta, q) \leq 0, \forall q \in \mathbb{Q}_\rho\right\}$$

contains a full-dimensional ball of radius r.

The algorithms presented in the literature for finding a probabilistic feasible solution (i.e., a solution that lies in $\mathcal{S}$ with high probability) are based on two fundamental ingredients: (1) an Oracle, which should check probabilistic feasibility of a candidate solution, and (2) an Update Rule, which exploits the convexity of the problem for constructing a new candidate solution based on the Oracle outcome. All the available algorithms can hence be recast in the form of the following meta-algorithm.

1. *Initialization:* Set $k = 0$ and choose an initial candidate solution θ_0.
2. *Oracle:* Invoke the Oracle with θ_k. The Oracle returns `true` if θ_k is a probabilistic robust design. In this case, Exit returning $\theta_{\text{seq}} = \theta_k$. Otherwise, the Oracle returns `false`, together with a *violation certificate* q_k, that is a realization of the uncertainty q such that $f(\theta_k, q_k) > 0$.
3. *Update:* Construct the new candidate solution θ_{k+1} based on θ_k and on the violation certificate q_k.
4. *Outer iteration:* Set $k = k + 1$ and Goto 2.

In the next sections, we describe in detail the two basic components of the previous algorithm: the Oracle and the Update Rule.

65.3.1 Probabilistic Oracle

The Oracle constitutes the randomized part of the algorithm, and its role is to decide on the probabilistic feasibility of the current solution. This decision is made based on random samples of the uncertainty. More precisely, a number N_k of independently identically distributed (i.i.d) uncertainty samples

$$q^{(1)}, \ldots, q^{(N_k)} \in \mathbb{Q}_\rho$$

are drawn according to the underlying distribution Pr, and the candidate design θ_k is deemed probabilistically robust if

$$f(\theta_k, q^{(i)}) \leq 0, \quad i = 1, 2, \ldots, N_k.$$

This leads to the following simple randomized scheme.

Algorithm 65.1: `Oracle`
Input: θ_k, N_k
Output: `feas (true/false)` and violation certificate q_k

for $i = 0$ to N_k **do**
  draw a random sample $q^{(i)}$

  Randomized Test

  if $f(q^{(i)}, \theta_k) > 0$ **then**
    set $q_k = q^{(i)}$, `feas=false`
    exit and return q_k
  end if
end for
set `feas=true`

Note that at step k the feasibility of the candidate solution θ_k is verified with respect to a number of samples N_k. If the test is passed, the solution is deemed feasible; otherwise, the uncertainty value q_k for which the randomized test failed is returned as a violation certificate. The *sample size* N_k depends on k, and has to be chosen to guarantee the desired probabilistic properties of the solution. Before discussing this issue in Section 65.3.3, we concentrate our attention on the outer loop of the algorithm, and show how one can update the current solution if the Oracle establishes its unfeasibility.

65.3.2 Update Rules

Various update rules have been proposed in the literature on randomized algorithms. For clarity of presentation, we discuss in detail only the simplest one, which is based on a (sub)gradient descent technique. In this approach, it is assumed that a *subgradient* $\partial_k(\theta)$ of the function $f(\theta, q)$ at the violation certificate q_k is computable. In the case when $f(\theta, q_k)$ is differentiable at θ, then $\partial_k(\theta)$ is simply the gradient of f, that is,

$$\partial_k(\theta) = \nabla_\theta f(\theta, q_k).$$

Then, the Update Rule consists of a classical gradient descent step. This is summarized in Algorithm 65.2. The main distinguishing feature of the method lies in the particular choice of the stepsize η_k, which is given by

$$\eta_k = \begin{cases} \dfrac{f(\theta_k, \delta_k)}{\|\partial_k(\theta_k)\|} + r & \text{if } \partial_k(\theta_k) \neq 0, \\ 0 & \text{otherwise,} \end{cases} \tag{65.8}$$

where r is a positive parameter.

Algorithm 65.2: `UpdateRule` (gradient method)

Input: θ_k, q_k

Output: θ_{k+1}

compute the subgradient $\partial_k(\theta)$ of $f(\theta, q_k)$.

compute the stepsize η_k according to (Equation 65.8).

set

$$\theta_{k+1} = \begin{cases} \theta_k - \eta_k \dfrac{\partial_k(\theta_k)}{\|\partial_k(\theta_k)\|} & \text{if } \partial_k(\theta_k) \neq 0 \\ \theta_k & \text{otherwise.} \end{cases}$$

Remark 65.1: Subgradient for LMIs

Note that, in the case of the LMI defined in Equation 65.6, a subgradient of the function $f(\theta, q_k) = \lambda_{\max}(F(\theta, q_k))$ at $\theta = \theta_k$ is readily computable as

$$\partial_k(\theta_k) = \left[\xi_{\max}{}^\top F_1(q_k)\xi_{\max} \cdots \xi_{\max}{}^\top F_{n_\theta}(q_k)\xi_{\max}\right]^\top,$$

where $\xi_{\max}$ is a unit norm eigenvector associated with the largest eigenvalue of $F(\theta_k, q_k)$.

The specific choice of stepsize in Equation 65.8, and in particular the selection of the parameter r, has the purpose of guaranteeing finite-time convergence, by imposing a termination condition to the sequential algorithm when the oracle cannot find any probabilistic solution. Note in fact that, in this situation, the meta-algorithm introduced in the previous section would run indefinitely. In the next section, we show how to avoid this behavior, by imposing a finite termination condition to the algorithm. Then, we analyze the probabilistic properties of the ensuing randomized algorithm.

65.3.3 Probabilistic Properties

Let us first define the quantity

$$N_{\text{outer}} = \left\lceil \frac{R^2}{r^2} \right\rceil, \tag{65.9}$$

where R is the distance of the initial solution θ_0 from the center of a ball of radius r contained in the solution set $\mathcal{S}$.

To study the probabilistic properties of the iterative scheme discussed in this section, we fix desired probabilistic levels $\epsilon, \delta \in (0, 1)$ and assume that at step k the Oracle is invoked with the sample size which has been derived in [8]

$$N_k \geq N_{\text{oracle}}(\epsilon, \delta, k) = \left\lceil \frac{\log \dfrac{\pi^2(k+1)^2}{6\delta}}{\log \dfrac{1}{1-\epsilon}} \right\rceil. \tag{65.10}$$

With these positions, the meta-algorithm can be formally rewritten as follows.

Algorithm 65.3: `SequentialDesign`

Input: $\epsilon, \delta \in (0, 1)$, N_{outer}

Output: θ_{seq}

Initialization

choose $\theta_0 \in \mathbb{R}^{n_\theta}$, set $k = 0$ and `feas=false`

Outer Iteration

while `feas=false` and $k < N_{\text{outer}}$ **do**

determine the sample size N_k according to Equation 65.10

Oracle

(`feas`, q_k) = `Oracle`(θ_k, N_k)

Update

if `feas=false` **then**

update $\theta_{k+1} =$ `UpdateRule`(θ_k)

else

set $\theta_{\text{seq}} = \theta_k$

end if

set $k = k + 1$

end while

The probabilistic properties of Algorithm 65.3 are formally stated in the next theorem.

Theorem 65.1: Probabilistic Properties

Let the Convexity Assumption hold and let $\epsilon, \delta \in (0, 1)$ be given probability levels. Then, the following statements hold:

1. The probability that Algorithm 65.3 terminates at some outer iteration $k < N_{\text{outer}}$ and returns θ_{seq} such that $V(\theta_{\text{seq}}) > \epsilon$ is less than δ.
2. If Algorithm 65.3 reaches the outer iteration N_{outer}, then the problem is not r-feasible.

Case 1 corresponds to a *successful exit* of the algorithm: the algorithm returned, with high probability, a probabilistic robust design θ_{seq}. Case 2 represents instead an *unsuccessful exit*: no solution has been found. In this situation, however, we have a certificate q_k which shows that the original problem is not r-feasible. We remark that, to determine N_{outer}, the quantities r and R have to be available. In particular,

r can be chosen to represent the desired level of r-feasibility, according to Definition 65.1. The radius R in Equation 65.9 can be instead replaced by an appropriate upper bound which can be easily estimated *a priori*.

We also remark that the sample size N_k provided in Equation 65.10 is independent of the number of uncertain and design parameters. This represents one of the key features of randomized algorithms: in general, the number of required computations is independent of the problem dimension, and therefore these algorithms are polynomial-time. For this reason, they are said to break the curse of dimensionality at the expense of a probability of violation; see [11].

65.3.4 Advanced Techniques

The update rule discussed in the previous section is one of the simplest methods that are currently available. More sophisticated techniques that still guarantee probabilistic properties of the ensuing solution can be developed. With these techniques, improved convergence rates are provided.

In particular, different techniques have been proposed falling in the class of the so-called localization methods. In these methods, the update rule is based on the computation, at each step k, of a center of a suitably defined localization set $\mathcal{L}_k$. The set $\mathcal{L}_k$ is guaranteed to contain the feasible set $\mathcal{S}$, that is, $\mathcal{S} \subseteq \mathcal{L}_k$. The set $\mathcal{L}_k$ is constructed based on the violation certificate q_k returned by the Oracle. The point q_k is used to construct a separating hyperplane $h_k \doteq \{\xi \in \mathbb{R}^{n_\theta} : a_k^\top \xi = b_k\}$ having the property that

$$a_k^\top \theta_k \geq b_k \quad \text{and} \quad a_k^\top \theta \leq b_k, \quad \forall \theta \in \mathcal{S}.$$

The separating hyperplane h_k indicates that the half-space $\{\theta : a_k^\top \theta > b_k\}$ cannot contain a feasible point and can therefore be eliminated (cut) in subsequent steps of the algorithm. In this case, we know that $\mathcal{S} \subseteq \mathcal{L}_k \cap \mathcal{H}_k$, where

$$\mathcal{H}_k \doteq \{\theta : a_k^\top \theta \leq b_k\},$$

and the algorithm constructs an updated localization set $\mathcal{L}_{k+1}$ such that

$$\mathcal{L}_{k+1} \supseteq \mathcal{L}_k \cap \mathcal{H}_k.$$

A new query point $\theta_{k+1} \in \mathcal{L}_{k+1}$ is then computed, and the process is repeated. This is summarized in the following scheme.

Algorithm 65.4: `UpdateRule` (localization methods)

Input: θ_k, q_k
Output: θ_{k+1}

compute the subgradient $\partial_k(\theta)$ of $f(\theta, q_k)$
construct the half-space $\mathcal{H}_k$ based on the subgradient
update the localization set $\mathcal{L}_{k+1} \supseteq \mathcal{L}_k \cap \mathcal{H}_k$
return $\theta_{k+1} = \texttt{Center}(\mathcal{L}_k)$

Different methods descend from different choices of the shape and description of the localization set. In particular, in probabilistic cutting plane methods, the localization set is a polytope, and the candidate solution θ_{k+1} is a center of this polytope (usually, the analytic center). In the probabilistic ellipsoid algorithm, the localization set is instead an ellipsoid and the candidate solution is the ellipsoid center.

It should be noted that Theorem 65.1 still holds also for the update rules introduced in this section, provided that N_{outer} is properly chosen. In particular, for the ellipsoid methods N_{outer} grows as

$O(n_\theta^2 \log(\sqrt{n_\theta} R/r))$, while for the best-known cutting-plane method it is of the order of $O(n_\theta \log^2(R/r))$, where in the first case R represents the radius of a ball and in the second case the radius of a cube, both inscribing the set $\mathcal{S}$, and r is the desired r-feasibility level.

Example 65.2: Probabilistic Sequential Design

To show how the sequential algorithm presented in this section can be applied to a specific design problem, we revisit the aircraft lateral motion design example. In particular, we set $\alpha = 0.5$ and seek a probabilistic feasible solution $\theta = \{P, W\}$ to the uncertain LMIs

$$P \succeq \beta I, \tag{65.11}$$

$$A(q)P + PA^\top(q) + B(q)W^\top + WB^\top(q) + 2\alpha P \preceq 0, \tag{65.12}$$

where the uncertainty q is assumed to vary in the set $\mathbb{Q}_\rho$ defined in Equation 65.2 and $\beta = 0.01$. We apply the algorithm `SequentialDesign` with ellipsoid update rule, and probability levels $\epsilon = 0.01$, $\delta = 10^{-6}$. With this setting, the algorithm is guaranteed to return (with 99.9999% probability) a solution P, W such that Equations 65.11 and 65.12 hold with 99% probability.

The algorithm was run with random initial candidate solution θ_0, and terminated after $k = 28$ outer iterations returning the solution

$$P_{\text{seq}} = \begin{bmatrix} 0.3075 & -0.3164 & -0.0973 & -0.0188 \\ -0.3164 & 0.5822 & -0.0703 & -0.0993 \\ -0.0973 & -0.0703 & 0.2277 & 0.2661 \\ -0.0188 & -0.0993 & 0.2661 & 0.7100 \end{bmatrix}, \tag{65.13}$$

$$W_{\text{seq}} = \begin{bmatrix} -0.0191 & 0.2733 \\ -0.0920 & 0.4325 \\ 0.0803 & -0.3821 \\ 0.4496 & -0.2032 \end{bmatrix}. \tag{65.14}$$

This solution was deemed probabilistically feasible by the `Oracle` after checking Equations 65.11 and 65.12 for $N_k = 2,089$ uncertainty samples. Then, the probabilistic controller is constructed as

$$K_{\text{seq}} = \begin{bmatrix} -2.9781 & -1.9139 & -3.2831 & 1.5169 \\ 7.3922 & 5.1010 & 4.1401 & -0.9284 \end{bmatrix}.$$

Further properties of this controller (worst-case and probabilistic) are studied in Section 65.6.

65.4 Scenario Approach to Optimization Problems

The sequential methods presented in the previous sections have been developed for feasibility problems. In this section, we present a nonsequential (batch) method for addressing in a probabilistic setting uncertain convex optimization problems, that is, optimization programs of the form

$$\min_{\theta \in \mathbb{R}^{n_\theta}} c^\top \theta \quad \text{subject to } f(\theta, q) \leq 0, \quad \text{for all } q \in \mathbb{Q}_\rho. \tag{65.15}$$

A probabilistic solution to this problem is found by replacing the semiinfinite set of constraints $f(\theta, q) \leq 0$ (one for every possible value of $q \in \mathbb{Q}_\rho$) with a finite number N of constraints $f(\theta, q^{(i)}) \leq 0, i = 1, \ldots, N$, one for each random sample $q^{(i)}$. This approach is summarized in the next algorithm.

Algorithm 65.5: `ScenarioDesign`

Input: $\epsilon, \delta, n_\theta$
Output: θ_{scen}

compute the sample size $N \geq N_{\text{scen}}(\epsilon, \delta, n_\theta)$ as in Equation 65.18
draw N i.i.d. samples $q^{(1)}, \ldots, q^{(N)}$

Scenario Problem

solve the convex optimization problem

$$\theta_{\text{scen}} = \arg \min_{\theta \in \mathbb{R}^{n_\theta}} c^\top \theta \quad \text{subject to } f(\theta, q^{(i)}) \leq 0, \quad i = 1, \ldots, N \tag{65.16}$$

To analyze the properties of the proposed algorithm, we consider for simplicity the case when the scenario problem (Equation 65.16) admits a feasible solution θ_{scen} and that this solution is unique.* We state here a result that was originally presented in [1], see also subsequent improvements in [26].

Theorem 65.2: Convex Scenario Design

Let the Convexity Assumption hold. Suppose that $N > n_\theta$, and $\epsilon, \delta \in (0, 1)$ satisfy the inequality

$$\binom{N}{n_\theta}(1 - \epsilon)^{N - n_\theta} \leq \delta, \tag{65.17}$$

then the probability that $V(\theta_{\text{scen}}) > \epsilon$ is at most δ.

The bound in Equation 65.17 provides an implicit relation between N, ϵ, δ, and n_θ. This relation can be made explicit to derive the *sample complexity* of the scenario approach. We now state a bound derived in [14], proving that the sample complexity is proportional to $1/\epsilon$ for fixed δ. In particular, it was shown that, for given $\epsilon, \delta \in (0, 1)$, Equation 65.17 holds if

$$N \geq N_{\text{scen}}(\epsilon, \delta, n_\theta) \doteq \left\lceil \frac{2}{\epsilon} \ln \frac{1}{2\delta} + 2n_\theta + \frac{2n_\theta}{\epsilon} \ln 4 \right\rceil. \tag{65.18}$$

Example 65.3: Scenario Design

To exemplify the sample complexity involved in a scenario design, we consider the problem of determining a probabilistic solution $\theta = \{P, W\}$ of the optimization problem

$$\min_{P,W} \operatorname{Tr} P \quad \text{subject to Equations 65.11 and 65.12}$$

where Tr denotes the trace of a matrix.

---

* Note that, if the scenario problem (Equation 65.16) is unfeasible, then clearly also the original robust convex program (Equation 65.15) is unfeasible. The assumption on uniqueness of the solution can be relaxed in most cases, as shown in Appendix A of [1].

In this case, if we set as previously $\epsilon = 0.01$, $\delta = 10^{-6}$, applying the bound (Equation 65.18) with $n_\theta = 18$, we obtain $N_{\text{scen}} = 7,652$. This means that we have to solve an optimization problem with 7,652 LMI constraints and 18 design variables. Then, Algorithm 65.5 returned the solution

$$P_{\text{scen}} = \begin{bmatrix} 0.1445 & -0.0728 & 0.0035 & 0.0085 \\ -0.0728 & 0.2192 & -0.0078 & -0.0174 \\ 0.0035 & -0.0078 & 0.1375 & 0.0604 \\ 0.0085 & -0.0174 & 0.0604 & 0.1975 \end{bmatrix},$$

$$W_{\text{scen}} = \begin{bmatrix} 0.0109 & 0.0908 \\ 7.2929 & 3.4846 \\ 0.0439 & -0.0565 \\ 0.6087 & -3.9182 \end{bmatrix}.$$

Then, a probabilistic controller is constructed as

$$K_{\text{scen}} = \begin{bmatrix} 20.0816 & 40.3852 & -0.4946 & 5.9234 \\ 10.7941 & 18.1058 & 9.8937 & -21.7363 \end{bmatrix}.$$

In the next section, we consider the more general case when the function $f(\theta, q)$ may be nonconvex in θ.

65.5 Learning Approach to Nonconvex Optimization

A nonsequential probabilistic approach for nonconvex control design is based on statistical learning theory; see [13] for further details. One of the objectives of this theory is to derive *uniform convergence laws*. Hence, from the control point of view, the main utility of statistical learning is to derive convergence results and compute the sample complexity which hold uniformly for all controller parameters θ. In turn, this leads to a powerful methodology for control synthesis which is not based on a convexity assumption on the controller parameters. The obtained sample complexity bounds are significantly larger than those derived in the convex case, even though the structure of the bounds show various similarities. We also remark that the setting of statistical learning may be seen as a major extension of the classical Monte Carlo approach which is useful only in the context of *a posteriori* analysis (see Section 65.6) where the controller parameters are fixed.

To treat the problem in full generality, we consider the following optimization problem

$$\min_{\theta \in \mathbb{R}^{n_\theta}} J(\theta) \quad \text{subject to } g(\theta, q) = 0, \quad \text{for all } q \in \mathbb{Q}_\rho, \tag{65.19}$$

where the cost function $J : \mathbb{R}^{n_\theta} \to \mathbb{R}$ and the binary function $g : \mathbb{R}^{n_\theta} \times \mathbb{Q}_\rho \to \{0, 1\}$ may be nonconvex. Note that constraints of the form (Equation 65.5) may be recast in this framework by setting

$$g(\theta, q) = \mathbb{I}(f(\theta, q) > 0),$$

where the indicator function $\mathbb{I}(\cdot)$ is one if the clause is true and zero otherwise. We consider here only the case when the binary function $g(\theta, q)$ can be written as a Boolean polynomial expression on the decision variable θ, as now formally stated.

Definition 65.2: (γ, m)-Boolean Function

The function $g : \mathbb{R}^{n_\theta} \times \mathbb{Q}_\rho \to \{0, 1\}$ is a (γ, m)-Boolean if, for fixed q, it can be written as a Boolean expression consisting of Boolean operators involving m polynomials

$$\beta_1(\theta), \ldots, \beta_m(\theta)$$

in the variables θ_i, $i = 1, \ldots, n_\theta$, and the degree with respect to θ_i of all these polynomials is no larger than $\gamma > 0$.

A probabilistic approximate solution to problem (Equation 65.19) may be found by means of the following algorithm, which has a similar structure of Algorithm 65.5.

Algorithm 65.6: `NonConvex`

Input: $\epsilon, \delta, n_\theta$
Output: θ_{ncon}

compute the sample size $N \geq N_{\text{ncon}}(\epsilon, \delta, n_\theta)$ as in Equation 65.21
draw N i.i.d. samples $q^{(1)}, \ldots, q^{(N)}$

Nonconvex Scenario Problem

solve the nonconvex optimization problem

$$\theta_{\text{ncon}} = \arg \min_{\theta \in \mathbb{R}^{n_\theta}} J(\theta) \quad \text{subject to } g(\theta, q^{(i)}) = 0, \quad i = 1, \ldots, N \tag{65.20}$$

Note that Equation 65.20 requires in general the solution of a nonconvex optimization problem. This usually leads only to a local minimum. However, the following theorem, stated in [14], guarantees that this minimum is ϵ-feasible.

Theorem 65.3: Nonconvex Learning-Based Design

Let $g(\theta, q)$ be (γ, m)-Boolean. Given $\epsilon \in (0, 0.14)$ and $\delta \in (0, 1)$, if

$$N \geq N_{\text{ncon}}(\epsilon, \delta, n_\theta) \doteq \left\lceil \frac{4.1}{\epsilon} \left(\ln \frac{21.64}{\delta} + 4.39 n_\theta \log_2 \left(\frac{8 e \gamma m}{\epsilon} \right) \right) \right\rceil, \tag{65.21}$$

where e is the Euler number, then the probability that $V(\theta_{\text{ncon}}) > \epsilon$ is less than δ.

Example 65.4: Nonconvex Learning-Based Design

In contrast with the previous examples which considered quadratic stability, we aim here at designing a controller K that minimizes (in probability) the decay rate α of Hurwitz stability. Moreover, as in [18], we want to impose a saturation constraint on the entries of the gain matrix K. That is, we aim at solving the following nonconvex optimization problem

$$\min_{\alpha, K} (-\alpha)$$

subject to

$$A(q) + B(q)K + \alpha I \text{ is Hurwitz for all } q \in \mathbb{Q}_\rho \tag{65.22}$$

$$-\bar{K}_{i,j} \leq K_{i,j} \leq \bar{K}_{i,j}, \quad i = 1, 2, \; j = 1, 2, 3, 4 \tag{65.23}$$

where

$$\bar{K} = \begin{bmatrix} 5 & 0.5 & 5 & 5 \\ 5 & 2 & 20 & 1 \end{bmatrix}.$$

To use the framework developed in this section, we set the design parameters to $\theta = \{\alpha, K\}$. Then, we note that checking the constraint in Equation 65.22 can be performed using the classical Hurwitz test, which requires strict positivity of all the Hurwitz determinants $H_i(\theta, q)$, $i = 1, \ldots, n$. That is, Equation 65.22 is equivalent to the Boolean condition

$$\mathcal{B}_H(\theta, q) \doteq \{H_1(\theta, q) > 0\} \wedge \cdots \wedge \{H_{n_s}(\theta, q) > 0\},$$

where the symbol $\wedge$ stands for "and." Note now that $H_i(\theta, q)$ are n_s polynomials in θ, where $n_s = 4$ is the dimension of the state matrix, whose degree is at most $\gamma_i = i(i+1)/2$; see [12] for additional details. Hence, $\mathcal{B}_H(\theta, q)$ is (γ, n_s)-Boolean, with

$$\gamma = \max_{i=1,\ldots,n_s} \gamma_i = n_s(n_s+1)/2 = 10.$$

Additionally, constraint (65.23) is immediately rewritten as the Boolean condition

$$\mathcal{B}_K(\theta) \doteq \{K_{1,1} \geq -\bar{K}_{1,1}\} \wedge \{K_{1,1} \leq \bar{K}_{1,1}\} \wedge \cdots \wedge \{K_{2,4} \geq -\bar{K}_{2,4}\} \wedge \{K_{2,4} \leq \bar{K}_{2,4}\}$$

which is (1, 16)-Boolean. Hence, the binary function

$$g(\theta, q) \doteq \begin{cases} 0 & \text{if } \mathcal{B}_H(\theta, q) \wedge \mathcal{B}_K(\theta), \\ 1 & \text{otherwise}, \end{cases}$$

is (10, 20)-Boolean. Then, we let $\epsilon = 0.01$ and $\delta = 10^{-6}$, and, noting that $n_\theta = 9$, $\gamma = 10$, $m = 20$, we apply Theorem 65.3 obtaining $N_{\text{ncon}}(\epsilon, \delta, n_\theta) = 310{,}341$. A (local) solution of the ensuing nonconvex scenario problem (Equation 65.20) was computed by means of a numerical optimization method based on local linearization obtaining

$$\alpha_{\text{ncon}} = 3.7285,$$
$$K_{\text{ncon}} = \begin{bmatrix} 0.8622 & 0.2714 & -5.0000 & 2.7269 \\ 5.0000 & 1.4299 & 3.9328 & -1.0000 \end{bmatrix}.$$

We note that three gains in the matrix K_{ncon} saturate to their extreme values.

65.6 *A Posteriori* Performance Analysis

In this section, we study the performance of the probabilistic controller selected with the sequential or nonsequential methods previously presented. Since the controller is fixed, this amounts to solving *a-posteriori* deterministic and probabilistic analysis problems, as discussed next.

65.6.1 Deterministic Analysis

For a fixed value of the design parameter θ, the deterministic, or worst-case, analysis consists in computing the *radius of deterministic performance* ρ_{wc}, which is the largest value of $\rho \in [0, 1]$ for which the constraint

$$f(\theta, q) \leq 0$$

is robustly satisfied for all q in the set $\mathbb{Q}_\rho$. Then, once this worst-case analysis is completed, the objective is to perform a probabilistic analysis beyond the radius of deterministic performance, that is, for values of $\rho \in [\rho_{\text{wc}}, 1]$.

Deterministic and probabilistic analysis are now illustrated more precisely for the example considered in this chapter. First, we perform a worst-case analysis of the controller θ_{seq} derived in Section 65.3.

Example 65.5: Deterministic Analysis

Note that in the example under consideration the entries of

$$A(q) = \begin{bmatrix} 0 & 1 & 0 & 0 \\ 0 & q_1 & q_2 & q_3 \\ q_4 & 0 & q_5 & -1 \\ q_4q_6 & q_7 & q_8 + q_5q_6 & q_9 - q_6 \end{bmatrix}, \quad B(q) = \begin{bmatrix} 0 & 0 \\ 0 & q_{10} \\ q_{11} & 0 \\ q_{12} + q_6q_{11} & q_{13} \end{bmatrix}$$

depend multiaffinely* on the uncertainty q. In this case, for fixed ρ, to detect quadratic performance of an uncertain system affected by multiaffine uncertainty, it is well-known (see, e.g., [20]), that it suffices to check the simultaneous satisfaction of the uncertain constraint (Equation 65.3) at the vertices of the hyperrectangle $\mathbb{Q}_\rho$. That is, for given $P \succ 0$ and W, if the inequality

$$\Phi_{\text{QP}}(P, W, v^i) = A(v^i)P + PA^\top(v^i) + B(v^i)W^\top + WB^\top(v^i) + 2\alpha P \preceq 0 \tag{65.24}$$

is satisfied for all vertices $v^i, i = 1, 2, \ldots, 2^\ell$ of the hyperrectangle $\mathbb{Q}_\rho$, then the uncertain constraint (Equation 65.3) is satisfied for all $q \in \mathbb{Q}_\rho$.

Then, computing the radius ρ_{wc} amounts to solving a one-dimensional problem in the variable ρ and, for each value of ρ, to verify if the inequality (Equation 65.24) is satisfied for all vertices of $\mathbb{Q}_\rho$. This problem can be solved, for example, using a bisection method, but for each value of ρ an exponential number of vertices 2^ℓ should be considered.

Performing this worst-case analysis on the design P_{seq} and W_{seq} derived in Equations 65.13 and 65.14, we compute

$$\rho_{\text{wc}} \approx 0.12.$$

We conclude that the controller derived for the aircraft model is robustly stable and attains robust quadratic performance for all values of $q \in \mathbb{Q}_\rho$, with $\rho \in [0, \rho_{\text{wc}}]$.

65.6.2 Probabilistic Analysis

In this case, the *a-posteriori* analysis consists of designing a Monte Carlo experiment which is based on random extractions of uncertainty samples. The result of the experiment is to return an estimated probability of satisfaction of the uncertain constraint under attention. The probability provided by this randomized algorithm is within an *a-priori* accuracy $\epsilon \in (0, 1)$ from the true probability, with confidence $\delta \in (0, 1)$. That is, the algorithm may indeed fail to return an approximately correct estimate, but the probability of failure is at most δ.

More precisely, let $q \in \mathbb{Q}_\rho$ represents the random uncertainty acting on the system. Then, for a given value of the design parameter θ, we aim at providing an estimate $\hat{R}_N$ of the probability of performance

$$R(\theta) = \Pr\left\{q \in \mathbb{Q}_\rho \,:\, f(\theta, q) \le 0\right\}$$

using N i.i.d. samples $q^{(i)}, i = 1, 2, \ldots, N$ of $q \in \mathbb{Q}_\rho$. The estimate $\hat{R}_N$, called the empirical probability, is given by

$$\hat{R}_N = \frac{1}{N} \sum_{i=1}^{N} \mathbb{I}[f(\theta, q^{(i)}) \le 0].$$

The sample complexity N is related to the accuracy and confidence and it can be determined using fairly classical large deviations inequalities. In particular, the so-called Chernoff Bound [5] can be readily used

* A function $f : \mathbb{R}^\ell \to \mathbb{R}$ is said to be multiaffine if the following condition holds: If all components $q_1, \ldots, q_\ell$ except one are fixed, then f is affine. For example, $f(q) = 3q_1q_2q_3 - 6q_1q_3 + 4q_2q_3 + 2q_1 - 2q_2 + q_3 - 1$ is multiaffine.

to determine the number of required samples. More precisely, given $\epsilon, \delta \in (0, 1)$, if

$$N \geq N_{\text{cher}}(\epsilon, \delta) = \left\lceil \frac{1}{2\epsilon^2} \ln \frac{2}{\delta} \right\rceil,$$

then, the probability inequality

$$|R(\theta) - \hat{R}_N| \leq \epsilon$$

holds with probability at least $1 - \delta$. We remark that the sample size N_{cher} provided by the Chernoff Bound is independent of the number ℓ of uncertain parameters q.

The following algorithm describes the Monte Carlo experiment discussed so far.

Algorithm 65.7: `Probabilistic Analysis`

Input: ϵ, δ, θ
Output: $\hat{R}_N$

compute the sample size $N \geq N_{\text{cher}}(\epsilon, \delta)$
draw N i.i.d. samples $q^{(1)}, \ldots, q^{(N)}$
return $\hat{R}_N = \frac{1}{N} \sum_{i=1}^{N} \mathbb{I}[f(\theta, q^{(i)}) \leq 0]$

Then, the next step is to construct the *probability degradation function*, which is the plot of the probability of stability as a function of the radius ρ. This plot may be compared with the radius of performance ρ_{wc}, to provide additional important information to the control engineer on the behavior of his/her design beyond the worst-case performance margin. This is illustrated in the next example.

Example 65.6: Probabilistic Analysis

For the controller θ_{seq}, we proceed with a probabilistic analysis for perturbations q whose radius goes beyond the deterministic radius ρ_{wc} previously computed. That is, we study how the probability of quadratic performance degrades with increasing radius ρ. More precisely, for fixed $\rho > \rho_{\text{wc}}$, we compute an estimate $\hat{R}_N$. Note that this analysis can be carried out using smaller values of ϵ, δ than those employed in the design phase. For instance, taking $\epsilon = 0.005$, $\delta = 10^{-6}$, by means of the Chernoff bound we obtain $N = 290,174$. Then, we estimated the probability degradation function for 100 equispaced values of ρ in the range [0.12, 0.5]. For each grid point the estimated probability of performance is computed by means of Algorithm 65.7. For each value of ρ, the accuracy of this estimate satisfies the inequality

$$|R(\theta_{\text{seq}}) - \hat{R}_N| \leq \epsilon$$

with probability at least $1 - \delta$.

The obtained results showing the estimated probability together with the deterministic radius ρ_{wc} are given in Figure 65.1. From this plot we observe, for instance, that if a 2% loss of probabilistic performance may be tolerated, then the performance margin may be increased by approximately 270% with respect to its deterministic counterpart. In fact, for $\rho = 0.34$, the estimated probability of performance is 0.98. In addition, we note that the estimated probability is equal to one for values of the radius up to $\rho \approx 0.26$.

In Figure 65.2, we plot the closed-loop eigenvalues for $\rho = 0.34$.

65.7 Randomized Algorithms Control Toolbox

The algorithms discussed in this chapter may be readily implemented in MATLAB®. However, while this may be straightforward in the case of analysis problems, it may require a nontrivial effort for the

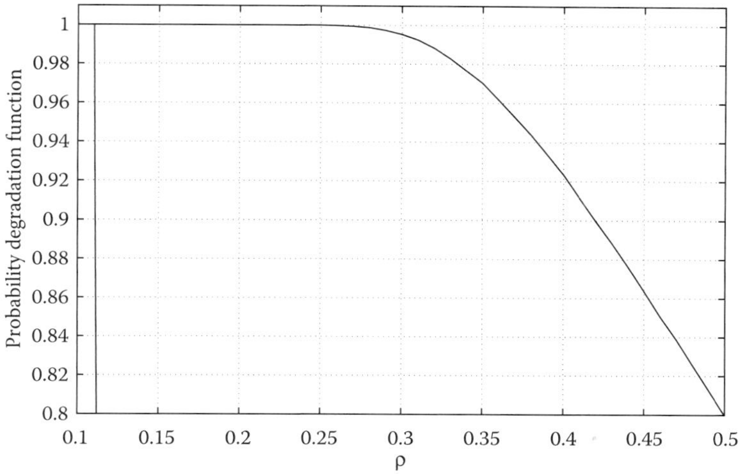

FIGURE 65.1 Probability degradation function.

design problems discussed in Sections 65.3 through 65.5. For this reason, to render these techniques easily accessible to the interested researchers, an effort was made to unify them into a coherent set of MATLAB routines. This effort led to the release of a Randomized Algorithms Control Toolbox (RACT); see [36].

This toolbox provides convenient uncertain object manipulation and implementation of randomized methods using state-of-the-art theoretical and algorithmic results. The two main features of the package

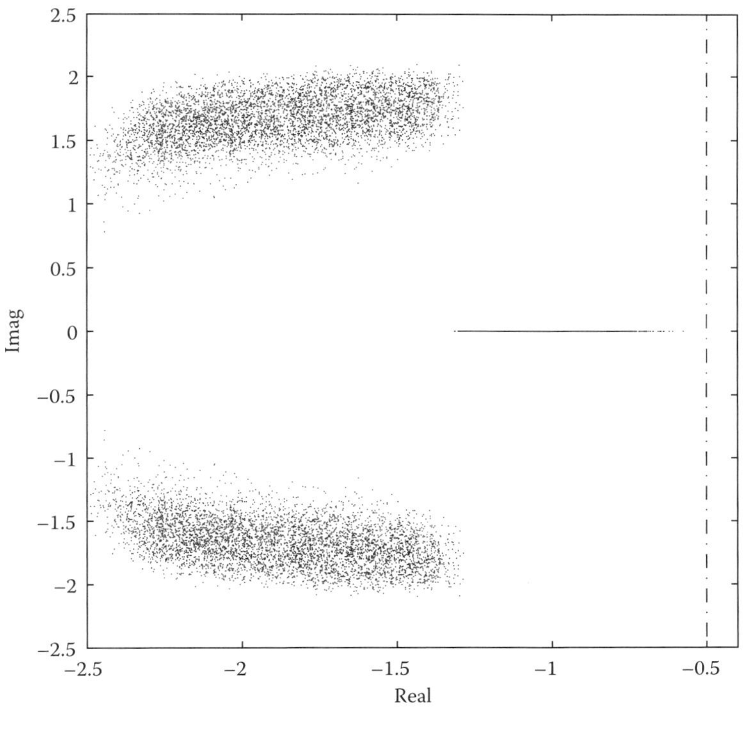

FIGURE 65.2 Closed-loop eigenvalues for $\rho = 0.34$.

are a functional approach with m-file templates, and a definition of design problems in generic LMI format using the widely used YALMIP syntax. This first release of the toolbox provides an easy-to-use interface of current randomized algorithms for control and is intended to be used by researchers, engineers, and students interested in robust control, uncertain systems, and optimization. The package can be freely downloaded from http://ract.sourceforge.net.

RACT features include

- Definition of a variety of uncertain objects: scalar, vector, and matrix uncertainties, with different density functions.
- Easy and fast sampling of uncertain objects of almost any type.
- Randomized algorithms for probabilistic performance verification and probabilistic worst-case performance.
- Randomized algorithms for feasibility of uncertain LMIs using stochastic gradient, ellipsoid, or cutting plane methods.
- Optimal design methods using scenario approach.

65.8 Miscellaneous Topics

65.8.1 General Uncertainty Description

In this chapter, we have studied the case when the uncertain parameters q are bounded in a hyperrectangle $\mathbb{Q}_\rho$. However, in some applications, different bounding sets, such as ellipsoids or diamonds; see, for example, [20], are more appropriate and should be studied. In other situations, the uncertainty affecting the system is of general type, that is, it consists of parametric and nonparametric uncertainty. In this case, various performance problems may be reformulated using the so-called $M - \Delta$ configuration and the related μ-theory, see [11], which is a quite flexible and general representation of linear uncertain systems. From the probabilistic point of view, structured uncertainty descriptions require to develop suitable randomized algorithms for generation of (vector and matrix) uncertainty samples within various bounding sets, see Section 65.8.2 for additional details.

Moreover, in the probabilistic setting described in this chapter, the probability measure Pr of the uncertainty q is assumed to be known. If this is not the case, then clearly the probability of violation depends on the specific choice of the measure Pr. The result is that, in extreme cases, the probability of violation may vary between zero and one when taking different measures. Therefore, without any guideline on the choice of the measure Pr, the obtained probability of violation, or its estimate, may be meaningless. In various applications, the probability measure may be estimated directly from available data, but in general the selection of the "right" measure should be performed with great care. This problem has been studied in [19,21]. In particular, in [21], the theory of distribution-free robustness, which has the objective to determine the worst-case measure in a given class of bell-shaped distributions, is discussed.

65.8.2 Sample Generation Algorithms

All the previously described randomized methods critically rely on efficient techniques for random sample generation. The interested reader may refer to [11] for a general discussion on the topic. In this section, we briefly discuss the problem of generating uniform i.i.d. samples in the ℓ_p norm-ball of radius ρ

$$\mathcal{B}_\rho^p \doteq \left\{ q \in \mathbb{R}^\ell : \|q\|_p \leq \rho \right\}.$$

In particular, we report an algorithm, presented in [23], that returns a real random vector q uniformly distributed in the norm-ball $\mathcal{B}_\rho^p$. This algorithm is based on the Generalized Gamma density $\bar{G}_{a,c}(x)$, defined as

$$\bar{G}_{a,c}(x) = \frac{c}{\Gamma(a)} x^{ca-1} e^{-x^c}, \quad x \geq 0,$$

where a and c are given parameters and $\Gamma(a)$ is the Gamma function.

Algorithm 65.8: `Uniform Generation`

Input: ρ, p

Output: q uniformly distributed in $\mathcal{B}_\rho^p$

generate ℓ independent random real scalars $\xi_i \sim \overline{G}_{1/p,p}$
construct the vector $x \in \mathbb{R}^\ell$ of components $x_i = s_i\xi_i$, where s_i are i.i.d. random signs
generate $z = w^{1/\ell}$, where w is uniform in $[0, 1]$
return $q = \rho z \frac{x}{\|x\|_p}$

Figure 65.3 visualizes the main steps of this algorithm in the simple case of sample generation of two-dimensional real vectors in a circle of radius one ($\ell = 2$, $p = 2$, $\rho = 1$). First, we note that for $p = 2$ the Generalized Gamma density $\bar{G}_{\frac{1}{p},p}(x)$ is related to the Gaussian density function. The random samples drawn from a Gaussian distribution (step 1 in the figure) are radially symmetric with respect to the ℓ_2 norm. Roughly speaking, this means that their level curves are ℓ_2-spheres. Secondly, the samples are normalized obtaining random vectors uniformly distributed on the boundary of the circle (step 2), and then injected according to the volumetric factor z (step 3).

We remark that in [11] a similar algorithm for complex vectors is presented. The sample generation problem becomes much harder when we are interested in the uniform generation of real and complex matrix samples Δ bounded in the induced ℓ_p-norm. Specific algorithms are presented in [11].

65.8.3 Mixed Deterministic and Probabilistic Setting

We remark that the setup considered here can be easily extended to mixed deterministic and random uncertainties. To this end, consider our motivating example. In this example, we treated all the parameters as random ones, but we could easily deal with the case when we want to robustly guarantee performance against some of them.

Just for exemplification, assume for instance that the vector q is partitioned as $q = \{q_{\text{prob}}, q_{\text{det}}\}$, so that the first m uncertain parameters $q_{\text{prob}} = \left[q_1 \ \cdots \ q_m\right]^\top$ have probabilistic nature with uniform distribution in the hyperrectangle $\mathbb{Q}_{\text{prob}}$, and the parameters $q_{\text{det}} = \left[q_{m+1} \ \cdots \ q_\ell\right]^\top$ are deterministic uncertainties, unknown but bounded in the hyperrectangle $\mathbb{Q}_{\text{det}}$, where the sets $\mathbb{Q}_{\text{prob}}, \mathbb{Q}_{\text{det}}$ are defined accordingly.

In this case, the present approach can be extended considering the following set of LMIs which depend on q_{prob}

$$\begin{gathered} A(q_{\text{prob}}, v^1)P + PA^\top(q_{\text{prob}}, v^1) + B(q_{\text{prob}}, v^1)W^\top + WB^\top(q_{\text{prob}}, v^1) + 2\alpha P \preceq 0 \\ \vdots \\ A(q_{\text{prob}}, v^M)P + PA^\top(q_{\text{prob}}, v^M) + B(q_{\text{prob}}, v^M)W^\top + WB^\top(q_{\text{prob}}, v^M) + 2\alpha P \preceq 0 \end{gathered} \tag{65.25}$$

where v^i, $i = 1, \dots, M \doteq 2^{\ell-m}$ are the vertices of the hyperrectangle $\mathbb{Q}_{\text{det}}$.

Then, an ϵ-feasible solution to the uncertain constraint (Equation 65.25) can be found using the tools described in Sections 65.2 through 65.4. Note that this solution would guarantee, as desired, *robust* performance with respect to q_{det} and *probabilistic* performance with respect to q_{prob}.

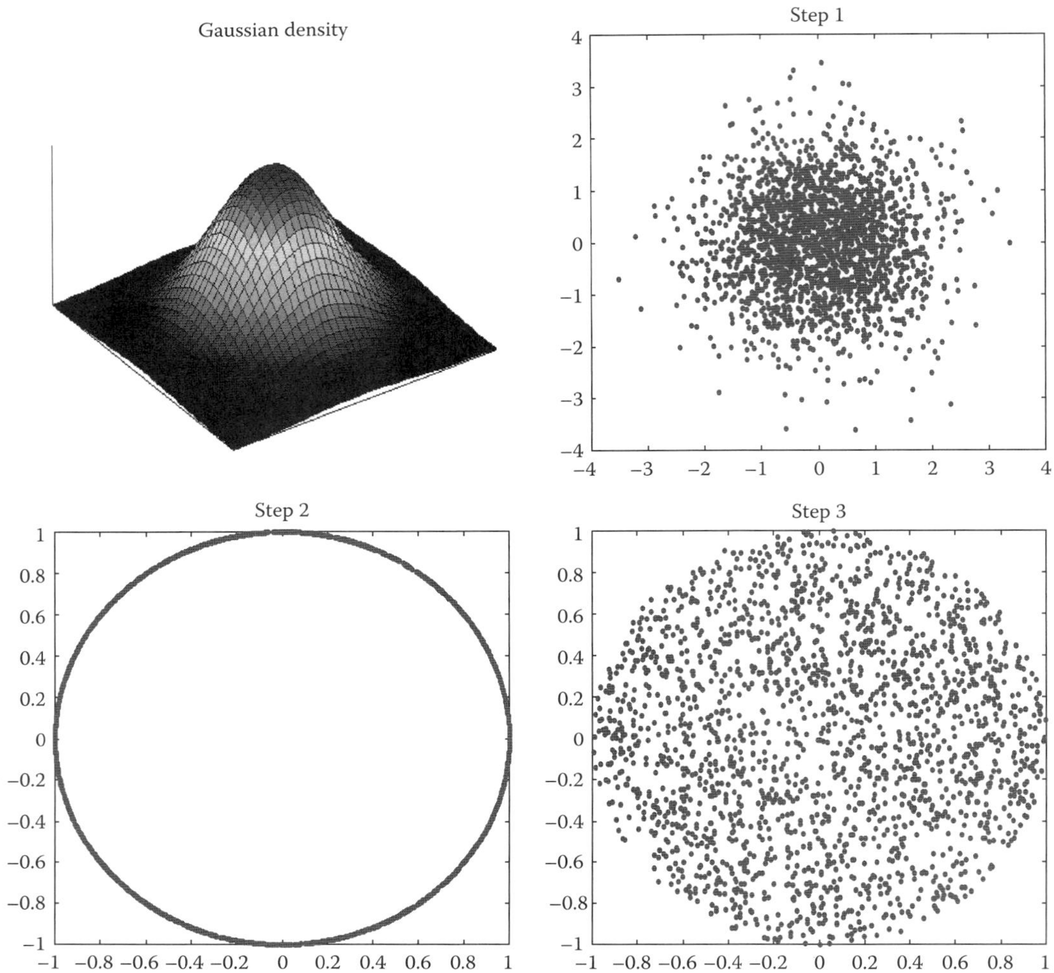

FIGURE 65.3 Generation of random vectors uniformly distributed in a circle.

65.8.4 Linear Parameter-Varying Systems

LPVs provide a good starting point for control design of fairly general classes of gain scheduling problems. This methodology is suitable and frequently used, for example, in aerospace control and in other applications as well, due to the time variations of the operating conditions. In this context, the system matrices depend on an unknown scheduling parameter vector $\xi(t)$ which can be measured online by the controller at each time instant t. More precisely, the LPV system is described by the equations

$$\dot{x}(t) = A(\xi(t))x(t) + B(\xi(t))u(t),$$
$$y(t) = C(\xi(t))u(t)$$

and the feedback controller is of the form

$$u(t) = K(\xi(t))y(t).$$

Since the original motivation for introducing gain scheduling is to handle plant nonlinearities, a critical issue of the resulting LPV model is to properly deal with the time-varying parameters $\xi(t)$. An LPV performance problem can be easily reformulated using parameter-dependent LMIs, but these inequalities are nonlinear with respect to ξ and it is practically impossible to solve them exactly. To this end, three different approaches have been developed in recent years and are now described.

The first approach (approximation LPV) studies a specific class of functions of the scheduling parameters. For example, the entries of the matrices of the LPV model can be taken as multiaffine functions, or linear fractional transformations, of ξ. This technique reduces the design problem to tractable formulae which involve a finite number of LMIs, but some conservatism is introduced in the approximation.

The second method (gridding LPV) requires to grid the parameters ξ which are assumed to be bounded within a given set Ξ. In this case, control design requires the solution of a finite set of LMIs. However, the number of LMIs depends on the grid points and it exponentially increases with the number of scheduling parameters. In addition, the satisfaction of the LMI conditions at the grid points does not provide any guarantee of their satisfaction for the entire set Ξ.

In the third technique (probabilistic LPV), the scheduling parameters ξ entering into the LMI equations are treated as random variables. Therefore, to compute a probabilistic controller, a randomized sequential algorithm of the form of Algorithm 65.3 can be derived. Clearly, in contrast to the previous deterministic methods, the obtained results enjoy probabilistic properties similar to those stated in Theorem 65.1. In [6], a specific technique for LPV systems, based on the simultaneous update of two gradient iterations similar to those in Algorithm 65.2, is developed. We refer to this chapter for further details and specific numerical results for feedback design of the lateral motion of an aircraft with nine scheduling parameters.

65.8.5 Systems and Control Applications

Several applications of probabilistic and randomized methods have been studied. In particular, we recall the following:

- *Aerospace control:* Applications of randomized strategies for the design of control algorithms in the field of aeronautics and aerospace was initiated by Stengel (see, e.g., [34]). In [32], a modern approach based on ellipsoid techniques is proposed for the design of an LPV control of an F-16 aircraft.
- *Flexible and truss structures:* Probabilistic robustness of flexible structures consisting of a mass-spring-damper model affected by random bounded uncertainty with force actuators and position sensors. Comparisons with standard robustness techniques are made [2]. In the field of truss topology optimization, a scenario-based approach was proposed in [24].
- *Model (in)validation:* A computationally efficient algorithm for model (in)validation in the presence of structured uncertainty and robust performance over finite horizons was proposed in [35].
- *Adaptive control:* A methodology for the design of cautious adaptive controllers based on a two-step procedure is introduced in [25]. First, a probability measure is updated online based on observations; then a controller with certain robust control specifications is tuned to this updated probability by means of randomized algorithms.
- *Switched systems:* Randomized algorithms for synthesis of multimodal systems with state-dependent switching rules. Convergence properties (with probability one) of nonconvex sequential methods are analyzed. Simulations show the efficacy of the method for various practical problems [29].
- *Network control:* Congestion control of high-speed communication networks by means of randomized algorithms. Various methods are developed and compared using different network topologies, including Monte Carlo and Quasi-Monte Carlo techniques [17].
- *Automotive:* A randomization-based methodological approach for validation of advanced driver assistance systems is studied in [28]. The case study also points out some characteristic properties of randomized algorithms regarding the necessary sample complexity, and the sensitivity to model uncertainty.
- *Model predictive control (MPC), fault detection, and isolation (FDI):* Sequential methods (ellipsoid-based) are derived in [30] to design robustly stable finite-horizon MPC schemes for linear uncertain

systems, when the uncertainty is not restricted to some specific class. In [33], a risk-adjusted approach based on randomization is proposed for robust simultaneous fault detection and isolation of MIMO systems.

- *Circuits and embedded systems:* Performance of electric circuits subject to uncertain components introduced by the manufacturing process. The objective is to evaluate the probability that a given "system property" holds providing "hard" (deterministic) bounds [31]. In [15,16], randomized techniques are applied to estimate the performance degradation of digital architectures and embedded systems subject to various sources of perturbation.

Acknowledgments

The authors are grateful to professor Teodoro Alamo for providing the code for the example on nonconvex learning-based design.

References

1. G. Calafiore and M.C. Campi. The scenario approach to robust control design. *IEEE Transactions on Automatic Control*, 51(5):742–753, 2006.
2. G. Calafiore, F. Dabbene, and R. Tempo. Randomized algorithms for probabilistic robustness with real and complex structured uncertainty. *IEEE Transactions on Automatic Control*, 45:2218–2235, 2000.
3. G. Calafiore and B.T. Polyak. Stochastic algorithms for exact and approximate feasibility of robust LMIs. *IEEE Transactions on Automatic Control*, 46:1755–1759, 2001.
4. G.C. Calafiore and F. Dabbene. A probabilistic analytic center cutting plane method for feasibility of uncertain LMIs. *Automatica*, 43:2022–2033, 2007.
5. H. Chernoff. A measure of asymptotic efficiency for tests of a hypothesis based on the sum of observations. *Annals of Mathematical Statistics*, 23:493–507, 1952.
6. Y. Fujisaki, F. Dabbene, and R. Tempo. Probabilistic robust design of LPV control systems. *Automatica*, 39:1323–1337, 2003.
7. S. Kanev, B. De Schutter, and M. Verhaegen. An ellipsoid algorithm for probabilistic robust controller design. *Systems and Control Letters*, 49:365–375, 2003.
8. Y. Oishi. Polynomial-time algorithms for probabilistic solutions of parameter-dependent linear matrix inequalities. *Automatica*, 43(3):538–545, 2007.
9. B.T. Polyak and R. Tempo. Probabilistic robust design with linear quadratic regulators. *Systems and Control Letters*, 43:343–353, 2001.
10. R. Tempo, E.-W. Bai, and F. Dabbene. Probabilistic robustness analysis: Explicit bounds for the minimum number of samples. *Systems and Control Letters*, 30:237–242, 1997.
11. R. Tempo, G. Calafiore, and F. Dabbene. *Randomized Algorithms for Analysis and Control of Uncertain Systems*. Communications and Control Engineering Series. Springer-Verlag, London, 2005.
12. M. Vidyasagar. Statistical learning theory and randomized algorithms for control. *IEEE Control Systems Magazine*, 18:69–85, 1998.
13. M. Vidyasagar. *Learning and Generalization: With Applications to Neural Networks* (2nd edn.). Springer-Verlag, New York, NY, 2002.

Additional References

14. T. Alamo, R. Tempo, and E.F. Camacho. Statistical learning theory: A pack-based strategy for uncertain feasibility and optimization problems. In V.D. Blondel, S.P. Boyd, and H. Kimura, Eds, *Recent Advances in Learning and Control*. Springer-Verlag, London, 2008.

15. C. Alippi. A probably approximately correct framework to estimate performance degradation in embedded systems. *IEEE Transactions on Computer-Aided Design of Integrated Circuits and Systems*, 21(7):749–762, 2002.
16. C. Alippi. Randomized algorithms: A system-level, poly-time analysis. *IEEE Transactions on Computers*, 51(7):740–749, 2002.
17. T. Alpcan, T. Başar, and R. Tempo. Randomized algorithms for stability and robustness analysis of high speed communication networks. *IEEE Transactions on Neural Networks*, 16:1229–1241, 2005.
18. B.D.O. Anderson and J.B. Moore. *Optimal Control: Linear Quadratic Methods*. Prentice-Hall, Englewood Cliffs, NJ, 1990.
19. E.-W. Bai, R. Tempo, and M. Fu. Worst-case properties of the uniform distribution and randomized algorithms for robustness analysis. *Mathematics of Control, Signals, and Systems*, 11:183–196, 1998.
20. B.R. Barmish. *New Tools for Robustness of Linear Systems*. MacMillan, New York, NY, 1994.
21. B.R. Barmish and C.M. Lagoa. The uniform distribution: A rigorous justification for its use in robustness analysis. *Mathematics of Control, Signals, and Systems*, 10:203–222, 1997.
22. S. Boyd, L. El Ghaoui, E. Feron, and V. Balakrishnan. *Linear Matrix Inequalities in System and Control Theory*. SIAM, Philadelphia, 1994.
23. G. Calafiore, F. Dabbene, and R. Tempo. Radial and uniform distributions in vector and matrix spaces for probabilistic robustness. In D.E. Miller and L. Qiu, Eds, *Topics in Control and its Applications*, p. 17–31. Springer-Verlag, New York, NY, 1999.
24. G.C. Calafiore and F. Dabbene. Optimization under uncertainty with applications to design of truss structures. *Structural and Multidisciplinary Optimization*, 35(3):189–200, 2008.
25. M.C. Campi and M. Prandini. Randomized algorithms for the synthesis of cautious adaptive controllers. *Systems and Control Letters*, 49:21–36, 2003.
26. M.C. Campi and S. Garatti. The exact feasibility of randomized solutions of robust convex programs. *SIAM Journal on Optimization*, 19(3):1211–1230, 2008.
27. Y. Fujisaki and Y. Oishi. Guaranteed cost regulator design: A probabilistic solution and a randomized algorithm. *Automatica*, 43:317–324, 2007.
28. O.J. Gietelink, B. De Schutter, and M. Verhaegen. Probabilistic approach for validation of advanced driver assistance systems. *Transportation Research Record*, (1910):20–28, 2005.
29. H. Ishii, T. Basar, and R. Tempo. Randomized algorithms for synthesis of switching rules for multimodal systems. *IEEE Transactions on Automatic Control*, 50:754–767, 2005.
30. S. Kanev and M. Verhaegen. Robustly asymptotically stable finite-horizon MPC. *Automatica*, 42(12):2189–2194, 2006.
31. C. Lagoa, F. Dabbene, and R. Tempo. Hard bounds on the probability of performance with application to circuit analysis. *IEEE Transactions on Circuits and Systems I*, 55:3178–3187, 2008.
32. B. Lu and F. Wu. Probabilistic robust linear parameter-varying control of an F-16 aircraft. *Journal of Guidance, Control, and Dynamics*, 29(6):1454–1460, 2006.
33. W. Ma, M. Sznaier, and C.M. Lagoa. A risk adjusted approach to robust simultaneous fault detection and isolation. *Automatica*, 43(3):499–504, 2007.
34. C.I. Marrison and R.F. Stengel. Design of robust control systems for a hypersonic aircraft. *Journal of Guidance, Control, and Dynamics*, 21(1):58–63, 1998.
35. M. Sznaier, C.M. Lagoa, and M.C. Mazzaro. An algorithm for sampling subsets of H_∞ with applications to risk-adjusted performance analysis and model (in)validation. *IEEE Transactions on Automatic Control*, 50(3):410–416, 2005.
36. A. Tremba, G. Calafiore, F. Dabbene, E. Gryazina, B.T. Polyak, P.S. Shcherbakov, and R. Tempo. RACT: Randomized algorithms control toolbox for MATLAB. In *Proceedings of the IFAC World Congress*, Seoul, 2008.

66

Stabilization of Stochastic Nonlinear Continuous-Time Systems

Miroslav Krstić
University of California, San Diego

Shu-Jun Liu
Southeast University

66.1 Introduction

Stochastic differential equations of Itô's kind are a suitable model for the study of nonlinear dynamic systems subject to random disturbances, for example, marine vehicles in the presence of wave forcing. For Itô stochastic systems, Khas'minskiĭ [8] and others have developed Lyapunov techniques for stability analysis of these systems. Since the mid-1990s this class of systems has been a fertile ground for feedback design, particularly for stabilization [1,2], stochastic disturbance attenuation [3], and adaptive control [3].

In this chapter we review various fundamental results in the area of control of nonlinear continuous-time stochastic systems. Our designs are presented through several worked examples. The emphasis is on the construction of feedback laws, without theorem statements.

66.2 Stability of Stochastic Systems

We consider stochastic systems of the form

$$dx = f(t,x)\,dt + h(t,x)\,dw, \tag{66.1}$$

where $x \in \mathbb{R}^n$ is the state, $x_0 = x(0)$ is the initial state, w is an r-dimensional standard Brownian motion defined on the complete probability space $(\Omega, \mathcal{F}, \{\mathcal{F}_t\}_{t\geq 0}, P)$, with Ω being a sample space, $\mathcal{F}$ being a σ-field, $\{\mathcal{F}_t\}_{t\geq 0}$ being a filtration, and P being a probability measure; $f : \mathbb{R}_+ \times \mathbb{R}^n \to \mathbb{R}^n$ and $h : \mathbb{R}_+ \times \mathbb{R}^n \to \mathbb{R}^{n\times r}$.

For any given function $V(t,x) \in \mathcal{C}^{1,2}(\mathbb{R}_+ \times \mathbb{R}^n; \mathbb{R}_+)$, associated with the stochastic differential equation 66.1, we define the differential operator $\mathcal{L}$ as follows ([8]):

$$\mathcal{L}V = \frac{\partial V}{\partial t} + \frac{\partial V}{\partial x} f + \frac{1}{2}\mathrm{Tr}\left\{h^T \frac{\partial^2 V}{\partial x^2} h\right\}. \tag{66.2}$$

Let τ denote the maximal interval of existence of the solution process $x(t)$ of the system (Equation 66.1). We say that the solution $x(t)$ is *regular* if $P\{\tau = \infty\} = 1$.

For system (Equation 66.1), suppose that $f(t,x)$ and $h(t,x)$ are locally Lipschitz and $f(t,0), h(t,0)$ are bounded uniformly in t. If there exists a nonnegative function $V(t,x) \in \mathcal{C}^{1,2}(\mathbb{R}_+ \times \mathbb{R}^n; \mathbb{R}_+)$ such that for some $c > 0$,

$$\begin{aligned} &\mathcal{L}V \leq cV, \\ &V_R = \inf_{\|x\|>R} V(t,x) \to \infty, \quad \text{as } R \to \infty, \end{aligned}$$

then system (Equation 66.1) has an almost surely unique (strong) solution on $[0,\infty)$.

66.2.1 Boundedness in Probability

A stochastic process $\{x(t), t \geq 0\}$ is said to be bounded in probability if

$$\lim_{c\to\infty} \sup_{0\leq t<\infty} P\{\|x(t)\| > c\} = 0.$$

For system (Equation 66.1), if there exist functions $V(t,x) \in \mathcal{C}^{1,2}(\mathbb{R}_+ \times \mathbb{R}^n; \mathbb{R}_+)$, $\mu_1(\cdot), \mu_2(\cdot) \in \mathcal{K}_\infty$ such that $\mu_1(\|x\|) \leq V(t,x) \leq \mu_2(\|x\|)$, and constants $c \geq 0, c_1 > 0$ such that $\mathcal{L}V \leq -c_1 V + c$, then the solution process is bounded in probability.

Compared with deterministic systems, there is more than one way to define stability for stochastic systems because in probability theory there are several concepts of convergence. Generally, these alternative stability definitions are not equivalent. Some common stability notions and (related Lyapunov-type tests) are given next. These basic notions are

- Boundedness in probability (defined above)
- Stochastic stability with "almost sure" convergence
- Stability in probability
- Noise-to-state stability (with respect to unknown but deterministic covariance)
- Stochastic input-to-state stability (ISS) (with respect to a stochastic input)
- Practical stochastic stability

66.2.2 Stability Notion 1

For system (Equation 66.1) with $f(t,0) \equiv 0, h(t,0) \equiv 0$, the solution $x(t) \equiv 0$ is said to be *stochastically asymptotically stable in the large* if for any $\epsilon > 0$,

$$\lim_{\|x_0\| \to 0} P\left\{\sup_{t \geq 0} \|x(t)\| \geq \epsilon\right\} = 0$$

and for any initial condition x_0,

$$P\{\lim_{t \to \infty} x(t) = 0\} = 1.$$

This form of stability, introduced in [8], is not an immediate analog of the standard deterministic stability concepts. In order to establish a clearer connection between deterministic stability results in the style of Hahn [6] or Khalil [7] and stochastic stability results, we introduce the following stability concepts from Krstic and Deng [9], which are based on class $\mathcal{K}$ functions [6] rather than on the $\epsilon - \delta$ formalism.

66.2.3 Stability Notion 2

For system (Equation 66.1) with $f(t,0) \equiv 0, h(t,0) \equiv 0$, the solution $x = 0$ is

- *Globally stable in probability* if for all $\epsilon > 0$ there exists a function $\gamma(\cdot) \in \mathcal{K}$ such that

$$P\{\|x(t)\| < \gamma(\|x_0\|)\} \geq 1 - \epsilon, \quad \forall t \geq 0, \ \forall x_0 \in \mathbb{R}^n \backslash \{0\};$$

- *Globally asymptotically stable in probability* if for all $\epsilon > 0$ there exists a function $\beta(\cdot,\cdot) \in \mathcal{KL}$ ([6]) such that

$$P\{\|x(t)\| < \beta(\|x_0\|, t)\} \geq 1 - \epsilon, \quad \forall t \geq 0, \ \forall x_0 \in \mathbb{R}^n \backslash \{0\}.$$

For these stability notions, we have the following test.

66.2.3.1 Test for Stability A

Consider system (Equation 66.1) with $f(t,0) \equiv 0, h(t,0) \equiv 0$. If there exist functions $V(t,x) \in \mathcal{C}^{1,2}(\mathbb{R}_+ \times \mathbb{R}^n; \mathbb{R}_+)$, $\mu_1(\cdot), \mu_2(\cdot) \in \mathcal{K}_\infty$, and a continuous and nonnegative function $W(x)$, such that

$$\mu_1(\|x\|) \leq V(t,x) \leq \mu_2(\|x\|),$$
$$\mathcal{L}V \leq -W(x),$$

then the equilibrium is globally stable in probability and $P\{\lim_{t\to\infty} W(x) = 0\} = 1$. Moreover, if $W(x)$ is positive definite, the solution $x \equiv 0$ is stochastically asymptotically stable in the large and also globally asymptotically stable in probability.

The notion of deterministic ISS introduced by Sontag [15] plays an important role in nonlinear system analysis and synthesis. For stochastic nonlinear systems, the analogous ISS (see [14,16,17]) is a richer property as it can be considered relative to more than one class of inputs.

66.2.4 Stability Notion 3

Consider the system

$$dx = f(t,x)\,dt + h(t,x)\Sigma(t)\,dw, \tag{66.3}$$

where $\Sigma : \mathbb{R}_+ \to \mathbb{R}^{r \times r}$ is Borel measurable and bounded, and $\Sigma(t)\Sigma^T(t)$ is the infinitesimal covariance function of driving noise $\Sigma(t)\,dw$. System (Equation 66.3) is said to be *noise-to-state stable (NSS)* if for any $\epsilon > 0$, there exist functions $\beta(\cdot,\cdot) \in \mathcal{KL}$ and $\gamma(\cdot) \in \mathcal{K}$, such that*

$$P\{\|x(t)\| < \beta(\|x_0\|, t) + \gamma\left(\sup_{0\le s\le t} \|\Sigma(s)\Sigma(s)^T\|_F\right)\} \ge 1-\epsilon, \quad \forall t \ge 0, \ \forall x_0 \in \mathbb{R}^n\backslash\{0\}.$$

66.2.4.1 Test for Stability B

Consider system (Equation 66.3) and suppose there exist functions $V(t,x) \in \mathcal{C}^{1,2}(\mathbb{R}_+ \times \mathbb{R}^n; \mathbb{R}_+)$, $\alpha_1(\cdot), \alpha_2(\cdot), \rho(\cdot) \in \mathcal{K}_\infty$, and $\alpha_3(\cdot) \in \mathcal{K}$, such that

$$\alpha_1(\|x\|) \le V(t,x) \le \alpha_2(\|x\|),$$
$$\|x\| \ge \rho(\|\Sigma\Sigma^T\|_F) \Rightarrow \mathcal{L}V \le -\alpha_3(\|x\|).$$

Then the system (Equation 66.1) is NSS.

To introduce a different ISS notion, referred to as *stochastic input-to-state stability*, consider the following system:

$$dx = f(x,v)\,dt + g(x,v)\,dw, \tag{66.4}$$

where $v = v(x,t) : \mathbb{R}^n \times \mathbb{R}_+ \to \mathbb{R}^m$ is the input; w is an r-dimensional standard Brownian motion defined on the complete probability space $(\Omega, \mathcal{F}, \{\mathcal{F}_t\}_{t\ge 0}, P)$, with Ω being a sample space, $\mathcal{F}$ being a σ-field, $\{\mathcal{F}_t\}_{t\ge 0}$ being a filtration, and P being a probability measure; and $f : \mathbb{R}^n \times \mathbb{R}^m \to \mathbb{R}^n$ and $g : \mathbb{R}^n \times \mathbb{R}^m \to \mathbb{R}^{n\times r}$ are assumed to be locally Lipschitz in their arguments. Assume that for every initial condition x_0 and each essentially bounded measurable input v, system (Equation 66.4) has a unique solution $x(t)$ on $[0,\infty)$ which is $\mathcal{F}_t$-adapted, t-continuous, and measurable with respect to $\mathcal{B} \times \mathcal{F}$, where $\mathcal{B}$ denotes the Borel σ-algebra of $\mathbb{R}$.

66.2.5 Stability Notion 4

System (Equation 66.4) is *practically stochastically input-to-state stable* if for any given $\epsilon > 0$; there exist functions $\beta(\cdot,\cdot) \in \mathcal{KL}$, $\gamma(\cdot) \in \mathcal{K}$ and a constant $d \ge 0$ such that

$$P\left\{\|x(t)\| < \beta(\|x_0\|, t) + \gamma\left(\sup_{0\le s\le t} \|v_s\|\right) + d\right\} \ge 1-\epsilon, \quad \forall t \ge 0, \ \forall x_0 \in \mathbb{R}^n\backslash\{0\}, \tag{66.5}$$

where $\|v_s\| = \inf_{\mathcal{A}\subset\Omega, P(\mathcal{A})=0} \sup\{\|v(x(\omega,s),s)\| : \omega \in \Omega\backslash\mathcal{A}\}$. When $d = 0$ in Equation 66.5, system (Equation 66.4) is said to be *stochastically input-to-state stable (SISS).*

The property (Equation 66.5) ensures that when the input $v \equiv 0$ and $d = 0$, system (Equation 66.4) is globally asymptotically stable in probability. It also ensures that the solution process of system (Equation 66.4) is bounded in probability when the input v is bounded almost surely (in other words, there exists a constant $M > 0$ such that $P\{\sup_{t\ge 0} \|v(x(\omega,t),t)\| \le M\} = 1$). When $v(x,t) = v(t)$ is deterministic and $d = 0$, the above definition is a generalization of NSS.

A $\mathcal{C}^2$ function $V(x)$ is said to be a *practical SISS–Lyapunov function* for the system (Equation 66.4) if there exist $\mathcal{K}_\infty$ functions $\alpha_1, \alpha_2, \alpha, \chi$, and a constant $d \ge 0$ such that

$$\alpha_1(\|x\|) \le V(x) \le \alpha_2(\|x\|),$$
$$\mathcal{L}V \le \chi(\|v\|) - \alpha(\|x\|) + d.$$

When $d = 0$, the function V is said to be an *SISS–Lyapunov function* for system (Equation 66.4).

---

* For $X \in \mathbb{R}^{n\times m}$, $\|X\|_F$ presents the Frobenius form defined by $\|X\|_F = \sqrt{\mathrm{Tr}(X^TX)}$.

66.2.5.1 Test for Stability C

System (Equation 66.4) is practically SISS (resp. SISS) if there exists a practical SISS- (resp. SISS-) Lyapunov function.

For other versions of "stochastic" ISS, refer to [16,17].

66.3 Stabilization of Stochastic Systems: The Basics

Due to the Itô differentiation rule (Equation 66.2), additional quadratic terms, not encountered when working with deterministic systems, arise in the stochastic Lyapunov analysis. For this reason, a control law designed for a deterministic system does not necessarily result in a stabilizing design for the corresponding stochastic problem. Moreover, different notions of stability, different forms of Lyapunov functions, and different performance indices (in optimal control formulations) may be used in the stochastic case. Efforts toward Lyapunov stabilization of stochastic nonlinear systems started in the work of Florchinger [4,5] who suggested the use of control Lyapunov functions and Sontag's stabilization formula in the stochastic setting. This was followed, by various other authors, by extensions of these concepts to problems with unknown noise covariance, adaptive control formulations, differential game and inverse optimal control formulations, output-feedback designs, and numerous constructive recursive procedures employing "backstepping." These efforts are reviewed here.

Consider a stochastic system which, besides the noise input w, has a control input u:

$$dx = f(x)\,dt + g_1(x)\,dw + g_2(x)u\,dt, \tag{66.6}$$

where $f(0) = 0, g_1(0) = 0$, and $u \in \mathbb{R}^m$. We say that system (Equation 66.6) is *globally asymptotically stabilizable in probability* if there exists a control law $u = \alpha(x)$ continuous everywhere, with $\alpha(0) = 0$, such that the equilibrium $x = 0$ of the closed-loop system is globally asymptotically stable in probability.

A smooth positive definite radially unbounded function $V : \mathbb{R}^n \to \mathbb{R}_+$ is called a *stochastic control Lyapunov function (sclf)* for system (Equation 66.6) if it satisfies

$$\inf_{u\in\mathbb{R}^m}\left\{L_f V + \frac{1}{2}\mathrm{Tr}\left\{g_1^T\frac{\partial^2 V}{\partial x^2}g_1\right\} + L_{g_2}Vu\right\} < 0, \quad \forall x \neq 0,$$

where $L_f V = \frac{\partial V}{\partial x}f(x), L_{g_2}V = \frac{\partial V}{\partial x}g_2(x)$.

A positive definite radially unbounded function $V(x)$ is an sclf if and only if for all $x \neq 0$,

$$L_{g_2}V = 0 \;\Rightarrow\; L_f V + \frac{1}{2}\mathrm{Tr}\left\{g_1^T\frac{\partial^2 V}{\partial x^2}g_1\right\} < 0.$$

It is shown in [9] that *system (Equation 66.6) is globally asymptotically stabilizable in probability if there exists an sclf with the small control property.**

The efforts on constructive methods for stabilization of broad classes of stochastic nonlinear systems commenced with the result of Pan and Başar [11], who derived the first backstepping design for strict-feedback systems, employing a risk-sensitive cost criterion (this approach was revisited a decade later and extended in [13]). This result was immediately followed by extensive efforts by various authors employing Lyapunov techniques for stabilization of stochastic continuous-time systems. The key results

* An sclf $V(x)$ is said to satisfy the small control property if there exists a control law $\alpha_c(x)$ continuous on $\mathbb{R}^n$ such that

$$L_f V + \frac{1}{2}\mathrm{Tr}\left\{g_1^T\frac{\partial^2 V}{\partial x^2}g_1\right\} + L_{g_2}V\alpha_c < 0, \quad \forall x \neq 0.$$

have focused on three classes of systems, which Pan [10] characterized as canonical forms and presented coordinate-free differential geometric conditions for transforming a given stochastic system into a particular canonical form.

1. *Strict-feedback canonical form*

$$
\begin{aligned}
dx_1 &= (x_2 + a_1(x_1))\,dt + \sum_{i=1}^{s} g_{i1}(x_1)\,dw_i,\\
&\vdots\\
dx_{n-1} &= (x_n + a_{n-1}(x_1,\ldots,x_{n-1}))\,dt + \sum_{i=1}^{s} g_{i,n-1}(x_1,\ldots,x_{n-1})\,dw_i,\\
dx_n &= (a_n(x_1,\ldots,x_n) + b_n(x_1,\ldots,x_n)u)\,dt + \sum_{i=1}^{s} g_{in}(x_1,\ldots,x_n)\,dw_i.
\end{aligned}
\tag{66.7}
$$

2. *Observer canonical form*

$$
\begin{aligned}
dx_1 &= \left(x_2 + a_1(x_1) + \sum_{i=1}^{p} b_{i1}(x_1)u_i\right) dt + \sum_{i=1}^{s} g_{i1}(x_1)\,dw_i,\\
dx_2 &= \left(x_3 + a_2(x_1) + \sum_{i=1}^{p} b_{i2}(x_1)u_i\right) dt + \sum_{i=1}^{s} g_{i2}(x_1)\,dw_i,\\
&\vdots\\
dx_n &= \left(a_n(x_1) + \sum_{i=1}^{p} b_{in}(x_1)u_i\right) dt + \sum_{i=1}^{s} g_{in}(x_1)\,dw_i,\\
y &= x_1.
\end{aligned}
\tag{66.8}
$$

3. *Systems with zero dynamics*

$$
\begin{aligned}
dx_z &= a_z(x_z,x_1)\,dt + \sum_{i=1}^{s} g_{iz}(x_z)\,dw_i,\\
dx_1 &= (x_2 + a_1(x_z,x_1))\,dt + \sum_{i=1}^{s} g_{i1}(x_z,x_1)\,dw_i,\\
&\vdots\\
dx_{r-1} &= (x_r + a_{r-1}(x_z,x_1,\ldots,x_{r-1}))\,dt + \sum_{i=1}^{s} g_{i,r-1}(x_z,x_1,\ldots,x_{r-1})\,dw_i,\\
dx_r &= (a_r(x_z,x_1,\ldots,x_r) + b_r(x_z,x_1,\ldots,x_r)u)\,dt + \sum_{i=1}^{s} g_{ir}(x_z,x_1,\ldots,x_r)\,dw_i,\\
y &= x_1.
\end{aligned}
\tag{66.9}
$$

In the next three sections we present control designs for the three classes of stochastic systems. The following design tools are illustrated in the subsequent sections:

- Stochastic backstepping (Section 66.4)
- Observer-based output feedback for systems with stochastic forcing through nonlinear input vector fields, but with non-noisy output measurements (Section 66.5)
- Domination of uncertain and unmeasured zero dynamics (Section 66.6)
- Stochastic adaptive control subject to unknown noise covariance and other parametric uncertainties (Section 66.6)
- Inverse optimal redesign leading to cost functionals given as expectations of time integrals of positive definite functions of the state and control (Section 66.7)

66.4 Stabilization in Probability via Backstepping

66.4.1 Introductory Example

We introduce the design idea of stabilization in probability via backstepping for an example of a system in the strict-feedback canonical form. Consider the system

$$\begin{aligned} dx_1 &= x_2\,dt + \varphi_1^T(x_1)\,dw, \\ dx_2 &= u\,dt + \varphi_2^T(x_1, x_2)\,dw, \end{aligned} \tag{66.10}$$

where φ_1, φ_2 are r-dimensional vector-valued smooth nonlinear functions with $\varphi_1(0) = \varphi_2(0, 0) = 0$. Our goal is to stabilize the equilibrium $x_1 = 0, x_2 = 0$ in probability.

Define the backstepping change of variables

$$\begin{aligned} z_1 &= x_1, \\ z_2 &= x_2 - \alpha_1(x_1), \end{aligned} \tag{66.11}$$

where $\alpha_1(x_1)$ is a yet-to-design stabilizing function and z_2 is an error variable expressing the fact that x_2 is not the true control. By Itô's formula, differentiating z_1 and z_2 with respect to time, the complete system (Equation 66.10) is expressed in the error coordinates (Equation 66.11):

$$\begin{aligned} dz_1 &= dx_1 = x_2\,dt + \varphi_1^T dw = (z_2 + \alpha_1)\,dt + \varphi_1^T dw, \\ dz_2 &= dx_2 - d\alpha_1 = u\,dt + \varphi_2^T\,dw - \frac{\partial \alpha_1}{\partial x_1} dx_1 - \frac{1}{2}\frac{\partial^2 \alpha_1}{\partial x_1^2}\varphi_1^T \varphi_1\,dt \end{aligned} \tag{66.12}$$

$$= \left(u - \frac{\partial \alpha_1}{\partial x_1} x_2 - \frac{1}{2}\frac{\partial^2 \alpha_1}{\partial x_1^2}\varphi_1^T \varphi_1\right) dt + \left(\varphi_2^T - \frac{\partial \alpha_1}{\partial x_1}\varphi_1^T\right) dw. \tag{66.13}$$

The Lyapunov design for stochastic systems is more difficult than for deterministic systems because of the term $\frac{1}{2}\mathrm{Tr}\left\{h^T \frac{\partial^2 V}{\partial x^2} h\right\}$ in Equation 66.2, and it cannot be carried out by using the quadratic Lyapunov function $V = z_1^2 + z_2^2$ as in the deterministic case. We use the following *quartic* Lyapunov function:

$$V = \frac{1}{4} z_1^4 + \frac{1}{4} z_2^4.$$

With it, we obtain that

$$\mathcal{L}V = z_1^3(z_2 + \alpha_1) + \frac{3}{2} z_1^2 \varphi_1^T \varphi_1 + z_2^3\left(u - \frac{\partial \alpha_1}{\partial x_1} x_2 - \frac{1}{2}\frac{\partial^2 \alpha_1}{\partial x_1^2}\varphi_1^T \varphi_1\right) + \frac{3}{2} z_2^2 \left(\varphi_2^T - \frac{\partial \alpha_1}{\partial x_1}\varphi_1^T\right)\left(\varphi_2 - \frac{\partial \alpha_1}{\partial x_1}\varphi_1\right). \tag{66.14}$$

Since $\varphi_1(0) = 0, \alpha_1$ will vanish at $x_1 = z_1 = 0$. Thus, by the mean value theorem, $\alpha_1(x_1)$ can be expressed as

$$\alpha_1(x_1) = \alpha_1(z_1) = z_1 \alpha_{11}(z_1), \tag{66.15}$$

where $\alpha_{11}(z_1)$ is a smooth function. Thus, the functions $\varphi_1(x_1), \varphi_2(x_1, x_2)$ can be written as follows:

$$\varphi_1(x_1) = z_1\varphi_{11}(z_1) \triangleq z_1\psi_{11}(x_1), \tag{66.16}$$

$$\begin{aligned}\varphi_2(x_1, x_2) &= x_1\varphi_{21}(x_1, x_2) + x_2\varphi_{22}(x_1, x_2)\\ &= z_1\varphi_{21}(x_1, x_2) + (z_2 + \alpha_1)\varphi_{22}(x_1, x_2)\\ &= z_1\varphi_{21}(x_1, x_2) + z_2\varphi_{22}(x_1, x_2) + z_1\alpha_{11}(z_1)\varphi_{22}(x_1, x_2)\\ &\triangleq z_1\psi_{21}(x_1, x_2) + z_2\psi_{22}(x_1, x_2)\end{aligned} \tag{66.17}$$

where $\varphi_{ik}, \psi_{ik}, i, k = 1, 2$, are smooth functions. So, Equation 66.14 can be rewritten as

$$\begin{aligned}\mathcal{L}V = z_1^3 z_2 + z_1^3\left(\alpha_1 + \frac{3}{2}z_1\psi_{11}^T\psi_{11}\right) + z_2^3\left[u - \frac{\partial\alpha_1}{\partial x_1}x_2 - \frac{1}{2}\frac{\partial^2\alpha_1}{\partial x_1^2}\varphi_1^T\varphi_1\right]\\ + \frac{3}{2}z_2^2\left(\varphi_2^T - \frac{\partial\alpha_1}{\partial x_1}\varphi_1^T\right)\left(\varphi_2 - \frac{\partial\alpha_1}{\partial x_1}\varphi_1\right).\end{aligned} \tag{66.18}$$

To handle the first and the fourth terms on the right-hand side of Equation 66.18, we need Young's inequality

$$x^T y \leq \frac{\epsilon^p}{p}\|x\|^p + \frac{1}{q\epsilon^q}\|y\|^q,$$

where the constant $\epsilon > 0$, the constants $p > 1$ and $q > 1$ satisfy $(p-1)(q-1) = 1$, and $x, y \in \mathbb{R}^n$. Thus, for the first term of Equation 66.18, by Young's inequality, we have

$$z_1^3 z_2 \leq \frac{3}{4}\epsilon_1^{\frac{4}{3}} z_1^4 + \frac{1}{4\epsilon_1^4}z_2^4, \tag{66.19}$$

where the constant $\epsilon_i > 0$, and for the fourth term, we try to arrange it in the form

$$z_2^{\gamma}\eta(x_1, x_2), \quad \gamma \geq 3,$$

so that it can also be combined into the second and the third terms in Equation 66.18, and be canceled by α_1 and u. Substituting Equations 66.16 and 66.17 into the last term in Equation 66.18 yields

$$\frac{3}{2}z_2^2\left(\varphi_2^T - \frac{\partial\alpha_1}{\partial x_1}\varphi_1^T\right)\left(\varphi_2 - \frac{\partial\alpha_1}{\partial x_1}\varphi_1\right) = \frac{3}{2}z_2^2\left(\sum_{k=1}^{2} z_k\beta_{2k}\right)^T\left(\sum_{k=1}^{2} z_k\beta_{2k}\right), \tag{66.20}$$

where

$$\begin{aligned}\beta_{21}(x_1, x_2) &= \psi_{21} - \frac{\partial\alpha_1}{\partial x_1}\psi_{11},\\ \beta_{22}(x_1, x_2) &= \psi_{22}.\end{aligned}$$

The next major step is to separate z_1 and z_2; every term can be handled by either α_1 or hence u. Hence, we rearrange the right term of Equation 66.20 as

$$\frac{3}{2}z_2^2\left(\sum_{k=1}^{2} z_k\beta_{2k}\right)^T\left(\sum_{k=1}^{2} z_k\beta_{2k}\right) = \frac{3}{2}z_2^4\beta_{22}^T\beta_{22} + 3z_2^3\beta_{22}^T z_1\beta_{21} + \frac{3}{2}z_2^2(z_1\beta_{21})^T(z_1\beta_{21}). \tag{66.21}$$

The first two terms on the right-hand side of Equation 66.21 are already in the desired form. Now we concentrate on the last term in Equation 66.21:

$$\begin{aligned}\frac{3}{2}z_2^2(z_1\beta_{21})^T(z_1\beta_{21}) = z_2^2 z_1^2\frac{3}{2}\sum_{j=1}^{r}\beta_{21j}^2 \leq \frac{3}{4}\sum_{j=1}^{r}\left(\frac{1}{\epsilon_{211}^2}z_2^4\beta_{21j}^4 + \epsilon_{211}^2 z_1^4\right) = \frac{3}{4}\sum_{j=1}^{r}\frac{1}{\epsilon_{211}^2}z_2^4\beta_{21j}^4 + \frac{3}{4}\sum_{j=1}^{r}z_1^4\epsilon_{211}^2\\ = \frac{3}{4}z_2^4\sum_{j=1}^{r}\frac{1}{\epsilon_{211}^2}\beta_{21j}^4 + \frac{3r}{4}z_1^4\epsilon_{211}^2.\end{aligned} \tag{66.22}$$

where β_{2kj} is the jth component of the vector β_{2k}.

Substituting Equations 66.19 through 66.22 into Equation 66.18, we get

$$
\begin{aligned}
\mathcal{L}V &\leq \frac{3}{4}\epsilon_1^{\frac{4}{3}}z_1^4 + \frac{1}{4\epsilon_1^4}z_2^4 + z_1^3\left(\alpha_1 + \frac{3}{2}z_1\psi_{11}^T\psi_{11}\right) + z_2^3\left[u - \frac{\partial\alpha_1}{\partial x_1}x_2 - \frac{1}{2}\frac{\partial^2\alpha_1}{\partial x_1^2}\varphi_1^T\varphi_1\right] \\
&\quad + \frac{3}{2}z_2^4\beta_{22}^T\beta_{22} + 3z_2^3\beta_{22}^T z_1\beta_{21} + \frac{3}{4}z_2^4\left(\sum_{j=1}^{r}\frac{1}{\epsilon_{211}^2}\beta_{21j}^4\right) + \frac{3r}{4}z_1^4\epsilon_{211}^2 \\
&= z_2^3\left[u - \frac{\partial\alpha_1}{\partial x_1}(z_2+\alpha_1) - \frac{1}{2}\frac{\partial^2\alpha_1}{\partial x_1^2}\varphi_1^T\varphi_1 + \frac{1}{4\epsilon_1^4}z_2 + \frac{3}{2}z_2\beta_{22}^T\beta_{22} + 3\beta_{22}^T z_1\beta_{21}\right. \\
&\quad \left. + \frac{3}{4}z_2\sum_{j=1}^{r}\frac{1}{\epsilon_{211}^2}\beta_{21j}^4\right] + z_1^3\left(\alpha_1 + \frac{3}{2}z_1\psi_{11}^T\psi_{11} + \frac{3}{4}\epsilon_1^{\frac{4}{3}}z_1 + \frac{3r}{4}z_1\epsilon_{211}^2\right).
\end{aligned}
$$

At this point, all the terms can be canceled by u and α_1. If we choose them as

$$
\begin{aligned}
\alpha_1 &= -c_1z_1 - \frac{3}{4}\epsilon_1^{\frac{4}{3}}z_1 - \frac{3}{2}z_1\psi_{11}^T\psi_{11} - \frac{3r}{4}z_1\epsilon_{211}^2, \\
u &= -c_2z_2 + \frac{\partial\alpha_1}{\partial x_1}(z_2+\alpha_1) + \frac{1}{2}\frac{\partial^2\alpha_1}{\partial x_1^2}\varphi_1^T\varphi_1 - \frac{1}{4\epsilon_1^4}z_2 - \frac{3}{2}z_2\beta_{22}^T\beta_{22} - 3\beta_{22}^T z_1\beta_{21} - \frac{3}{4}z_2\sum_{j=1}^{r}\frac{1}{\epsilon_{211}^2}\beta_{21j}^2,
\end{aligned}
$$

where constants $c_i > 0$, then the infinitesimal generator of the closed-loop system is negative definite:

$$\mathcal{L}V \leq -\sum_{i=1}^{2}c_iz_i^4,$$

which means that the equilibrium $z = 0$ is globally asymptotically stable in probability. In view of Equation 66.11, the same is true about $x = 0$.

66.4.2 General Recursive Design Procedure

A systematic backstepping design has been developed for the class of nonlinear systems transformable into the strict-feedback form:

$$dx_i = x_{i+1}\,dt + \varphi_i^T(\bar{x}_i)\,dw, \quad i = 1, 2, \ldots, n-1 \tag{66.23}$$

$$dx_n = u\,dt + \varphi_n^T(\bar{x}_n)\,dw, \tag{66.24}$$

where $\bar{x}_i = [x_1, \ldots, x_i]^T$, and where $\varphi_i(\bar{x}_i)$ are vector-valued smooth functions with $\varphi_i(0) = 0$.

The general design is summarized in Table 66.1. We can show that the Lyapunov function $V_n = \frac{1}{4}\sum_{i=1}^{n}z_i^4$ satisfies

$$\mathcal{L}V_n \leq -\sum_{i=1}^{n}c_iz_i^4,$$

which guarantees that the equilibrium $z = 0$ is globally asymptotically stable in probability. Thus the equilibrium $x = 0$ is globally asymptotically stable in probability.

66.5 Output-Feedback Stabilization

For linear systems, a common solution to the output-feedback problem is a stabilizing state-feedback controller employing the state estimates from an exponentially converging observer or filter. However,

TABLE 66.1 Backstepping Design for General System (Equations 66.23 and 66.24)

$$z_i = x_i - \alpha_{i-1}, \quad i = 1, \ldots, n,$$

$$\psi_{ik} = \sum_{l=k}^{i} \alpha_{l-1,k}\varphi_{il}, \quad k = 1, \ldots, i,$$

$$\beta_{ik} = \psi_{ik} - \sum_{l=k}^{i-1} \frac{\partial\alpha_{i-1}}{\partial x_l}\psi_{lk}, \quad k = 1, \ldots, i,$$

$$\alpha_{ik} = \frac{1}{2}\sum_{p=k}^{i-1} \psi_{pk}^T \sum_{q=1}^{i-1} \frac{\partial^2\alpha_{i-1}}{\partial x_q \partial x_q} \sum_{j=1}^{q} z_j \psi_{qj} + \sum_{l=k-1}^{i-1} \frac{\partial\alpha_{i-1}}{\partial x_l}\alpha_{lk} - 3\beta_{ii}^T\beta_{ik}, \quad k = 1, \ldots, i-1,$$

$$\alpha_{ii} = -\left(c_i + \frac{3}{4}\epsilon_i^{\frac{4}{3}} + \frac{1}{4\epsilon_{i-1}^4} + \frac{3}{2}\beta_{ii}^T\beta_{ii} + \frac{3r}{4}\sum_{k=i+1}^{n}\sum_{l=1}^{k-1}\epsilon_{kil}^2 + \frac{3}{4}\sum_{j=1}^{r}\sum_{k=1}^{i-1}\sum_{l=1}^{i-1}\frac{1}{\epsilon_{ikl}^2}\beta_{ikj}^2\beta_{ilj}^2\right) + \frac{\partial\alpha_{i-1}}{\partial x_{i-1}},$$

$$\alpha_{i,i+1} = 1,$$

$$\alpha_i = \sum_{k=1}^{i} z_k \alpha_{ik}.$$

Control Law:

$$u = \alpha_n.$$

this approach ("certainty equivalence") is not applicable to nonlinear systems. The situation is even more difficult in the stochastic case where nonlinear exponentially convergent observers are very difficult to design.

Fortunately, classes of problems do exist where output-feedback stabilization can be solved. One such class is the output-feedback canonical form (Equation 66.8), for which we present a feedback design in this section.

Consider, for example, the nonlinear stochastic system given by

$$\begin{aligned} dx_1 &= x_2\,dt + f_1(y)\,dt + \varphi_1^T(y)\,dw, \\ dx_2 &= x_3\,dt + f_2(y)\,dt + b_1\beta(y)u\,dt + \varphi_2^T(y)\,dw, \\ dx_3 &= f_3(y)\,dt + b_0\beta(y)u\,dt + \varphi_3^T(y)\,dw, \\ y &= x_1, \end{aligned} \tag{66.25}$$

where only y is measured, the functions $f_i(y), i = 1, 2, 3$ are dependent on the output only and assumed to vanish at zero; the polynominal $b_1 s + b_0$ is assumed to be Hurwitz (i.e., $b_0 b_1 > 0$), and $\beta(y) \neq 0, \forall y \in \mathbb{R}$, which guarantees that the system has a well-defined relative degree $\rho = 3 - 1 = 2$.

The goal is to design the output feedback control law to guarantee that the solution process of the closed-loop system is bounded in probability and, when the diffusion terms $\varphi_i(0) = 0, i = 1, 2, 3$, the zero solution of the closed-loop system is asymptotically stable in the large.

Since the states x_2, x_3 are not measured, we first design an observer which would provide exponentially convergent estimates of the unmeasured states in the absence of noise. The observer is designed as

$$\begin{aligned} \dot{\hat{x}}_1 &= \hat{x}_2 + k_1(y - \hat{x}_1) + f_1(y), \\ \dot{\hat{x}}_2 &= \hat{x}_3 + k_2(y - \hat{x}_1) + f_2(y) + b_1\beta(y)u, \\ \dot{\hat{x}}_3 &= k_3(y - \hat{x}_1) + f_3(y) + b_0\beta(y)u, \end{aligned}$$

where k_1, k_2, and k_3 are parameters.

Let $\hat{x} = [\hat{x}_1, \hat{x}_2, \hat{x}_3]^T$, and denote the observer error by $\tilde{x} = x - \hat{x} = [\tilde{x}_1, \tilde{x}_2, \tilde{x}_3]^T$. Then we have

$$d\tilde{x} = \begin{bmatrix} -k_1 & 1 & 0 \\ -k_2 & 0 & 1 \\ -k_3 & 0 & 0 \end{bmatrix} \tilde{x}\, dt + \varphi^T(y)\, dw \triangleq A\tilde{x}\, dt + \varphi^T(y)\, dw,$$

where $\varphi(y) = [\varphi_1, \varphi_2, \varphi_3]$. The parameters k_1, k_2, and k_3 are chosen such that the matrix A is Hurwitz.

Define $\bar{\hat{x}} = [y, \hat{x}_2, \hat{x}_3]^T$, and let $g_{01}(y) = f_1(y), g_{0i}(y, \tilde{x}_1) = f_i(y) + k_i\tilde{x}_1, i = 2, 3$. Then we have the following dynamics for $\bar{\hat{x}}$:

$$d\bar{\hat{x}} = D\bar{\hat{x}}\, dt + G(y, \tilde{x}_1)\, dt + [1, 0, 0]^T \tilde{x}_2\, dt + \beta(y)Bu\, dt + H(y)\, dw,$$

where

$$D = \begin{bmatrix} 0 & 1 & 0 \\ 0 & 0 & 1 \\ 0 & 0 & 0 \end{bmatrix}, \quad G(y, \tilde{x}_1) \triangleq \begin{bmatrix} g_{01}(y) \\ g_{02}(y, \tilde{x}_1) \\ g_{03}(y, \tilde{x}_1) \end{bmatrix},$$

$$B = [0, b_1, b_0]^T, \quad H(y) = [\varphi_1(y), 0, 0]^T.$$

In the design, we employ two similarity transformations. With the first transformation we transform vector B into an input vector that has all zero elements except the element b_1. Taking the first transformation as

$$\xi = \begin{bmatrix} 1 & 0 & 0 \\ 0 & 1 & 0 \\ 0 & -\dfrac{b_0}{b_1} & 1 \end{bmatrix} \bar{\hat{x}} \triangleq T_1 \bar{\hat{x}},$$

we obtain

$$d\xi = D_1\xi\, dt + G_1(y, \tilde{x}_1)\, dt + [1, 0, 0]^T \tilde{x}_2\, dt + \beta(y)B_1 u\, dt + H_1(y)\, dw,$$

where

$$D_1 = T_1 D T_1^{-1} = \begin{bmatrix} 0 & 1 & 0 \\ 0 & d_{11} & 1 \\ 0 & d_{12} & -d_{11} \end{bmatrix},$$

$$d_{11} = \frac{b_0}{b_1} > 0, \quad d_{12} = -\left(\frac{b_0}{b_1}\right)^2 = -d_{11}^2 < 0,$$

$$G_1(y, \tilde{x}_1) = T_1 G(y, \tilde{x}_1) = \begin{bmatrix} g_{01} \\ g_{02} \\ -\dfrac{b_0}{b_1} g_{02} + g_{03} \end{bmatrix} \triangleq \begin{bmatrix} g_{11}(y) \\ g_{12}(y, \tilde{x}_1) \\ g_{13}(y, \tilde{x}_1) \end{bmatrix},$$

$$B_1 = T_1 B = [0, b_1, 0]^T, \quad H_1(y) = T_1 H(y) = H(y).$$

With the second transformation, we would like to transform the second column of the matrix D_1 into the first unit vector. Let $\eta = T_2\xi$, where

$$T_2 = \begin{bmatrix} 1 & 0 & 0 \\ -d_{11} & 1 & 0 \\ -d_{12} & 0 & 1 \end{bmatrix}.$$

Then we obtain that

$$d\eta = D_2\eta\, dt + G_2(y, \tilde{x}_1)\, dt + [1, -d_{11}, -d_{12}]^T \tilde{x}_2\, dt + \beta(y)B_2 u\, dt + H_2(y)\, dw,$$

where

$$D_2 = T_2 D_1 T_2^{-1} = \begin{bmatrix} d_{21} & 1 & 0 \\ d_{22} & 0 & 1 \\ d_{23} & 0 & -d_{11} \end{bmatrix}, \quad d_{21} = d_{11}, \; d_{22} = d_{12}, \; d_{23} = -d_{11}d_{12},$$

$$G_2(y,\tilde{x}_1) = T_2 G_1 = [g_{11}, -g_{11}d_{11} + g_{12}, -d_{12}g_{11} + g_{13}]^T \triangleq [g_{21}(y), g_{22}(y,\tilde{x}_1), g_{23}(y,\tilde{x}_1)]^T$$

$$B_2 = B_1, \quad H_2 = T_2 H = [\varphi_1(y), -d_{11}\varphi_1(y), -d_{12}\varphi_1(y)]^T \triangleq [h_1^T(y), h_2^T(y), h_3^T(y)]^T.$$

Noting that

$$T_2 T_1 = \begin{bmatrix} 1 & 0 & 0 \\ -d_{11} & 1 & 0 \\ d_{11}^2 & -d_{11} & 1 \end{bmatrix},$$

we obtain that

$$\begin{aligned} \eta_1 &= y \\ \eta_2 &= \hat{x}_2 - d_{11}y \\ \eta_3 &= \hat{x}_3 - d_{11}\hat{x}_2 + d_{11}^2 y. \end{aligned}$$

Denoting

$$\zeta = \eta_3,$$

the following dynamics of lower triangular form are obtained for the η-system:

$$\begin{aligned} dy &= d_{21}y\,dt + (\eta_2 + \tilde{x}_2)\,dt + g_{21}(y)\,dt + h_1(y)\,dw, \\ d\eta_2 &= (\zeta + d_{22}y + g_{22}(y,\tilde{x}_1) + b_1\beta(y)u - d_{11}\tilde{x}_2)\,dt + h_2(y)\,dw, \\ d\zeta &= -d_{11}\zeta\,dt + L_1 y\,dt + L_2\tilde{x}_2\,dt + \Omega(y,\tilde{x}_1)\,dt + \Phi(y)\,dw, \end{aligned}$$

where

$$\begin{aligned} L_1 &= d_{23} = -d_{11}^3, \\ L_2 &= -d_{12} = -d_{11}^2, \\ \Omega(y,\tilde{x}_1) &= g_{23}(y,\tilde{x}_1) = d_{11}^2 f_1(y) - d_{11}f_2(y) + f_3(y) + (k_3 - d_{11}k_2)\,\tilde{x}_1, \\ \Phi(y) &= h_3(y) = d_{11}^2\varphi_1^T(y). \end{aligned}$$

We are now ready to design the output feedback stabilizing controller via backstepping. Define the error variable

$$z = \eta_2 - \alpha(y) = \hat{x}_2 - (d_{11}y + \alpha(y)), \tag{66.26}$$

where $\alpha(y)$ is a smooth virtual control law to be defined. Then we have

$$\begin{aligned} d\tilde{x} &= A\tilde{x}\,dt + \varphi^T(y)\,dw, \\ d\zeta &= -d_{11}\zeta\,dt + L_1 y\,dt + L_2\tilde{x}_2\,dt + \Omega(y,\tilde{x}_1)\,dt + \Phi(y)\,dw, \\ dy &= M_1(y)\,dt + (z + \alpha(y) + \tilde{x}_2)\,dt + \Psi_1(y)\,dw, \\ dz &= [\zeta + M_2(\tilde{x}_1, y, \eta_2) + b_1\beta(y)u + N\tilde{x}_2]\,dt + \Psi_2(y)\,dw, \end{aligned}$$

where

$$\begin{aligned} M_1(y) &= d_{21}y + f_1(y), \\ M_2(\tilde{x}_1, y, \eta_2) &= g_{22}(y,\tilde{x}_1) + d_{22}y - \alpha'(y)[\eta_2 + g_{21}(y) + d_{21}y] - \frac{1}{2}\alpha''(y)\|h_1(y)\|^2, \\ \Psi_1(y) &= h_1(y) = \varphi_1^T(y), \quad \Psi_2(y) = N(y)\varphi_1^T(y), \\ N(y) &= -d_{11} - \alpha'(y). \end{aligned}$$

Since A is Hurwitz, there exists a positive-definite symmetric matrix P satisfying

$$A^T P + PA = -I.$$

Instead of employing a quartic Lyapunov function as in Section 66.4, here we illustrate the possibility of employing a Lyapunov function which is "nominally" quadratic, but it incorporates state-dependent weights on the quadratic terms. We postulate the following Lyapunov function candidate:

$$V_2 = \delta \tilde{x}^T P \tilde{x} + \frac{1}{2d_{11}} \zeta^2 + y^2 + \Xi(y) z^2,$$

where $\delta > 0$ is a constant to be determined, and smooth function $\Xi(y) > 0$ is a weighting function to be determined.

Then, we have

$$\begin{aligned}\mathcal{L}V_2 = & -\delta\|\tilde{x}\|^2 + \delta \mathrm{Tr}\{\varphi(y) P \varphi^T(y)\} - \zeta^2 + \frac{1}{2d_{11}} \mathrm{Tr}\{\Phi^T(y)\Phi(y)\} \\ & + \frac{1}{d_{11}} \zeta[L_1 y + L_2 \tilde{x}_2 + \Omega(y, \tilde{x}_1)] + \Xi'(y)[\eta_2 + M_1(y) + \tilde{x}_2] z^2 \\ & + 2y[z + \alpha + M_1(y) + \tilde{x}_2] + 2z\Xi(y)[\zeta + M_2(\tilde{x}_1, y, \eta_2) + b_1\beta(y) u + N\tilde{x}_2] \\ & + \Psi_1\Psi_1^T + \frac{1}{2}\mathrm{Tr}\left\{[\Psi_1^T, \Psi_2^T] \frac{\partial^2[\Xi(y)z^2]}{\partial(y,z)^2} [\Psi_1^T, \Psi_2^T]^T\right\}. \end{aligned} \tag{66.27}$$

The functions Ψ_i can be decomposed by using the mean value theorem as follows:

$$\begin{aligned}\Psi_1(y) &= h_1(y) = h_1(0) + \bar{h}_1(y) y, \\ \Psi_2(y) &= \Psi_2(0) + \overline{\Psi}_{21}(y) y.\end{aligned}$$

Then we have

$$\Psi_1\Psi_1^T = (h_1(0) + \bar{h}_1(y)y)(h_1(0) + \bar{h}_1(y)y)^T = h_1(0)h_1^T(0) + \left[2h_1(0)\bar{h}_1^T(y) + \bar{h}_1(y)\bar{h}_1^T(y)y\right] y, \tag{66.28}$$

$$\begin{aligned}\frac{1}{2}\mathrm{Tr}\left\{[\Psi_1^T, \Psi_2^T]\frac{\partial^2[\Xi(y)z^2]}{\partial(y,z)^2}[\Psi_1^T, \Psi_2^T]^T\right\} &= \frac{1}{2}\mathrm{Tr}\left\{\begin{bmatrix}\Psi_1 \\ \Psi_2\end{bmatrix}^T \begin{bmatrix}\Xi''(y)z^2 & 2\Xi'(y)z \\ 2\Xi'(y)z & 2\Xi(y)\end{bmatrix}\begin{bmatrix}\Psi_1 \\ \Psi_2\end{bmatrix}\right\} \\ &= \frac{1}{2}\mathrm{Tr}\left\{\begin{bmatrix}\Psi_1 \\ \Psi_2\end{bmatrix}^T \begin{bmatrix}\Xi''(y)z & 2\Xi'(y) \\ 2\Xi'(y) & 0\end{bmatrix}\begin{bmatrix}\Psi_1 \\ \Psi_2\end{bmatrix}\right\} z + \Xi(y)\Psi_2\Psi_2^T, \\ &= \frac{1}{2}\left(z\Xi''(y) + 4\Xi'(y)N(y)\right)\|\varphi_1(y)\|^2 z + \Xi(y)\Psi_2\Psi_2^T. \end{aligned} \tag{66.29}$$

In Equation 66.29, we can cancel out the first term by the choice of α, u. We cannot cancel out the second term, but we can bound it in terms of y^2, as follows:

$$\begin{aligned}\Xi\Psi_2\Psi_2^T &= \Xi[\Psi_2(0) + \overline{\Psi}_{21}(y)y] \times [\Psi_2(0) + \overline{\Psi}_{21}(y)y]^T \\ &\le 2\Xi\|\Psi_2(0)\|^2 + 2\Xi\|\overline{\Psi}_{21}(y)\|^2 y^2.\end{aligned}$$

We can design the weighting function $\Xi(y)$, such that $\Xi\|\overline{\Psi}_{21}\|^2$ is sufficiently small by selecting

$$\Xi(y) = \frac{\kappa}{1 + \|\overline{\Psi}_{21}(y)\|^2},$$

where κ is a design parameter.

Then we obtain that

$$\Xi\Psi_2\Psi_2^T \leq 2\kappa\|\Psi_2(0)\|^2 + 2\kappa y^2. \tag{66.30}$$

In order to deal with "second variation terms" in Equation 66.27, we employ the mean value theorem to obtain the representations

$$\varphi(y) = \varphi(0) + y\overline{\varphi}(y), \quad \Phi(y) = \Phi(0) + y\overline{\Phi}(y),$$

and

$$\begin{aligned} L_2\tilde{x}_2 + \Omega(y, \tilde{x}_1) &= L_2\tilde{x}_2 + \Omega(0,0) + y\overline{\Omega}(y) + L_3\tilde{x}_1 \\ &\triangleq [L_3, L_2]\overline{\tilde{x}}_2 + \Omega(0,0) + y\overline{\Omega}(y). \end{aligned}$$

Then, we have

$$\delta\mathrm{Tr}\{\varphi(y)P\varphi^T(y)\} = \delta\mathrm{Tr}\{\varphi(0)P\varphi^T(0)\} + \delta\mathrm{Tr}\{2\varphi(0)P\overline{\varphi}^T(y) + \overline{\varphi}(y)P\overline{\varphi}^T(y)y\}y, \tag{66.31}$$

$$\frac{1}{2d_{11}}\mathrm{Tr}\{\Phi^T(y)\Phi(y)\} = \frac{1}{2d_{11}}\mathrm{Tr}\{\Phi^T(0)\Phi(0)\} + \frac{1}{2d_{11}}\mathrm{Tr}\{2\Phi^T(0)\overline{\Phi}(y) + \overline{\Phi}^T(y)\overline{\Phi}(y)y\}y, \tag{66.32}$$

$$\begin{aligned} \frac{1}{d_{11}}\zeta\left[L_1 y + L_2\tilde{x}_2 + \Omega(y,\tilde{x}_1)\right] &= \frac{1}{d_{11}}\zeta\left[\left(L_1 + \overline{\Omega}(y)\right)y + [L_3, L_2]\overline{\tilde{x}}_2 + \Omega(0,0)\right] \\ &\leq (4d_{11}^2\varepsilon_0)^{-1}[L_1 + \overline{\Omega}(y)]^2 y^2 + \varepsilon_0\zeta^2 + \varepsilon_1\frac{L_2^2 + L_3^2}{4d_{11}^2}\zeta^2 \\ &\quad + \varepsilon_1^{-1}\|\overline{\tilde{x}}_2\|^2 + \varepsilon_2\zeta^2 + (4d_{11}^2\varepsilon_2)^{-1}\|\Omega(0,0)\|^2, \end{aligned} \tag{66.33}$$

$$2y\tilde{x}_2 = 2y\tilde{x}_2 \leq \varepsilon_3\tilde{x}_2^2 + \varepsilon_3^{-1}y^2, \tag{66.34}$$

$$\left(\Xi'(y)z^2 + 2\Xi(y)z\right)\tilde{x}_2 \leq \varepsilon_3\tilde{x}_2^2 + \varepsilon_3^{-1}\left(\Xi'(y)z + 2\Xi(y)\right)^2 z^2, \tag{66.35}$$

where $\varepsilon_i > 0, i = 1, 2, 3$, are analysis parameters to be chosen appropriately.

Combining Equation 66.27 with Equation 66.28 through 66.35 we arrive at

$$\begin{aligned} \mathcal{L}V_2 \leq & -\delta\|\tilde{x}\|^2 + \delta\mathrm{Tr}\{\varphi(0)P\varphi^T(0)\} + \frac{1}{2d_{11}}\mathrm{Tr}\{\Phi^T(0)\Phi(0)\} - \zeta^2 + 2\varepsilon_3\tilde{x}_2^2 \\ & + h_1(0)h_1^T(0) + \varepsilon_1\frac{L_2^2 + L_3^2}{4d_{11}^2}\zeta^2 + \varepsilon_1^{-1}\|\overline{\tilde{x}}_2\|^2 + \varepsilon_2\zeta^2 + (4d_{11}^2\varepsilon_1)^{-1}\|\Omega(0,0)\|^2 \\ & + \frac{1}{2d_{11}}\mathrm{Tr}\{2\Phi^T(0)\overline{\Phi}(y) + \overline{\Phi}^T(y)\overline{\Phi}(y)y\}y + \delta\mathrm{Tr}\{2\varphi(0)P\overline{\varphi}^T(y) + \overline{\varphi}(y)P\overline{\varphi}^T(y)y\}y \\ & + (2h_1(0)\overline{h}_1^T(y) + \overline{h}_1(y)\overline{h}_1^T(y)y)y + (4d_{11}^2\varepsilon_0)^{-1}[L_1 + \overline{\Omega}(y)]^2 y^2 + \varepsilon_0\zeta^2 \\ & + \varepsilon_3^{-1}y^2 + \varepsilon_3^{-1}\left(\Xi'(y)z + 2\Xi(y)\right)^2 z^2 + \Xi'(y)[\eta_2 + M_1(y)]z^2 + 2y[z + \alpha + M_1(y)] \\ & + 2z\Xi(y)[\zeta + M_2 + b_1\beta(y)u] + 2\kappa\|\Psi_2(0)\|^2 + \frac{1}{2}\left(z\Xi''(y) + 4\Xi'(y)N(y)\right)\|\varphi_1(y)\|^2 z + 2\kappa y^2. \end{aligned} \tag{66.36}$$

In the above differential inequality, we can select the control law (including virtual control law) and the design parameters to cancel out all the unfavorable terms except the constant bias terms. The virtual control α and the actual control u are chosen as

$$\begin{aligned}\alpha(y) = & -\left(\frac{\beta_1}{2}+\frac{1}{2\varepsilon_3}\right)y - M_1(y) - \frac{1}{4d_{11}}\mathrm{Tr}\{2\Phi^T(0)\overline{\Phi}(y)+\overline{\Phi}^T(y)\overline{\Phi}(y)y\} \\ & -\frac{1}{2}(2h_1(0)\overline{h}_1^T(y)+\overline{h}_1(y)\overline{h}_1^T(y)y) - \frac{\delta}{2}\mathrm{Tr}\{2\varphi(0)P\overline{\varphi}^T(y)+\overline{\varphi}(y)P\overline{\varphi}^T(y)y\} \\ & -(4d_{11}^2\varepsilon_0)^{-1}[L_1+\overline{\Omega}(y)]^2 y, \\ u = & -\frac{1}{b_1\beta(y)}\left\{\frac{1}{2}\beta_2 z+\zeta+M_2(\tilde{x}_1,y,\eta_2)+\frac{1}{\Xi(y)}y+\frac{1}{2\varepsilon_3\Xi(y)}\left(\Xi'(y)z+2\Xi(y)\right)^2 z\right. \\ & \left.+\frac{\Xi'(y)}{2\Xi(y)}[\eta_2+M_1(y)]z+\frac{1}{4\Xi(y)}\left(z\Xi''(y)+4\Xi'(y)N(y)\right)\|\varphi_1(y)\|^2\right\}.\end{aligned}$$

Choosing the analysis parameters $\varepsilon_0, \varepsilon_1, \varepsilon_2, \varepsilon_3$, and κ, sufficiently small, the analysis parameter δ sufficiently large, and the controller parameter β_1 sufficiently large, such that

$$\begin{aligned}\bar{c}_1 &\triangleq \delta - 2\varepsilon_3 - \varepsilon_1^{-1} > 0, \\ \bar{c}_2 &\triangleq 1 - \varepsilon_1\frac{L_2^2+L_3^2}{4d_{11}^2} - \varepsilon_0 - \varepsilon_2 > 0, \\ \bar{c}_3 &\triangleq \beta_1 - 2\kappa > 0,\end{aligned}$$

and by choosing the control parameter β_2 as positive, we obtain that

$$\mathcal{L}V_2 \le -c_1 V_2 + c_2, \tag{66.37}$$

where

$$\begin{aligned}c_1 &= \min\left\{\frac{\bar{c}_1}{\delta\lambda_{\max}(P)}, 2\bar{c}_2 d_{11}, \bar{c}_3, \beta_2\right\} > 0, \\ c_2 &= \delta\mathrm{Tr}\{\varphi(0)P\varphi^T(0)\} + \left(\frac{d_{11}}{2}+2\kappa N^2(0)\right)\|\varphi_1(0)\|^2 + (4d_{11}^2\varepsilon_1)^{-1}\|d_{11}^2 f_1(0) - d_{11}f_2(0) + f_3(0)\|^2 \ge 0.\end{aligned}$$

Thus, the solution process of the closed-loop system is bounded in probability and, moreover, when $\varphi_i(0) = 0, f_i(0) = 0, i = 1, 2, 3$, we obtain $c_2 = 0$ in Equation 66.37, which guarantees that the origin $x = \hat{x} = 0$ is stochastically asymptotically stable in the large (and, in fact, also globally asymptotically stable in probability).

For a recursive design procedure for the general output-feedback system in the observer canonical form, the reader is referred to [12], as well as to an earlier result in [9] where the output-feedback problem without zero dynamics was solved using quartic Lyapunov functions.

66.6 Adaptive Stabilization in Probability

In this section, we introduce the idea of stochastic adaptive control in the presence of functional and dynamic (nonlinear) uncertainties. Our design is presented in the output-feedback case, with nonlinear zero dynamics.

Consider for example the stochastic nonlinear system of relative degree two, given in the following form:

$$dx_z = f_0(x_z, y)\,dt + g_0^T(x_z)\,dw, \tag{66.38}$$

$$dx_1 = (x_2 + f_1(x_z, y))\,dt + g_1^T(x_z, y)\,dw, \tag{66.39}$$

$$dx_2 = (u + f_2(x_z, y))\,dt + g_2^T(x_z, y)\,dw, \tag{66.40}$$

$$y = x_1,$$

where $y \in \mathbb{R}$ is the measurable output; $x_z \in \mathbb{R}^m$ is the state of unmodeled dynamics, and $f_i, g_i, i = 0, 1, 2$, are uncertain locally Lipschitz functions.

We assume that the uncertain nonlinear functions satisfy the following linear parameterized upper bounds (where the parameters are unknown) and the stochastic unmodeled dynamics are stochastic SISS.

A1. $|f_i(x_z, y)| \le l_i^* \varphi_{i1}(\|x_z\|) + l_i^* \varphi_{i2}(|y|)$, $\|g_i(x_z, y)\| \le h_i^* \psi_{i1}(\|x_z\|) + h_i^* \psi_{i2}(|y|)$, $i = 1, 2$, $\forall (x_z, y) \in \mathbb{R}^m \times \mathbb{R}$, where φ_{i1}, φ_{i2}, ψ_{i1}, and ψ_{i2} are known smooth nonnegative functions, and $\varphi_{i1}(0) = 0$, $\psi_{i1}(0) = 0$, and l_i^*, h_i^*, $i = 1, 2$, are unknown positive constants.
For the noise of unknown covariance, if it can be formulated as $\Sigma(t)\,dw$, where $w(t)$ is a standard Brownian motion and $\Sigma(t)$ is a uniformly bounded deterministic (possibly unknown) function, then we can modify A1 as follows: $\|g_i(x_z, y)\Sigma(t)\| \le h_i^* \psi_{i1}(\|x_z\|) + h_i^* \psi_{i2}(|y|), i = 1, 2$.

A2. For Equation 66.38, there exists a C^2 function $V_z(x_z)$ and $\mathcal{K}_\infty$ functions $\alpha_1(\cdot)$, $\alpha_2(\cdot)$, $\alpha(\cdot)$, and $\gamma(\cdot)$ such as $\forall (x_z, y) \in \mathbb{R}^m \times \mathbb{R}$,

$$\alpha_1(\|x_z\|) \le V_z(x_z) \le \alpha_2(\|x_z\|), \tag{66.41}$$

$$\mathcal{L}V_z(x_z) \le \gamma(|y|) - \alpha(\|x_z\|). \tag{66.42}$$

The control objective is to design a smooth adaptive output-feedback controller

$$\dot{\chi} = \varpi(\chi, y), \quad u = \mu(\chi, y), \tag{66.43}$$

so that the solution process of the closed-loop systems (Equations 66.38 through 66.43) is bounded in probability and moreover, when the drift and diffusion vector fields vanish at the origin, the output can be regulated to the origin almost surely.

66.6.1 Backstepping Design

We view the overall control system as being composed of the unmodeled dynamics (Equation 66.38) and the rest of the controlled plants (Equations 66.39 and 66.40). For the subsystem (Equations 66.39 and 66.40), we will use the tuning function method to design an adaptive output-feedback controller with unknown gain function, where x_z is considered as a disturbance input.

First, we introduce a state-estimator for the subsystems (Equations 66.39 and 66.40):

$$\begin{aligned}\dot{\hat{x}}_1 &= \hat{x}_2 + a_1(y - \hat{x}_1),\\ \dot{\hat{x}}_2 &= u + a_2(y - \hat{x}_1),\end{aligned}$$

where a_1, a_2 are constants such that $s^2 + a_1 s + a_2$ is a Hurwitz polynomial. We denote $\hat{x} = [\hat{x}_1, \hat{x}_2]^T$, $\tilde{x} = \frac{x - \hat{x}}{l^*}$, $l^* = \max\{1, l_i^*, h_i^*, i = 1, 2\}$, and

$$\begin{aligned}A &= \begin{bmatrix} -a_1 & 1 \\ -a_2 & 0 \end{bmatrix},\\ F(x_z, y) &= [f_1(x_z, y), f_2(x_z, y)]^T,\\ G(x_z, y) &= [g_1(x_z, y), g_2(x_z, y)].\end{aligned}$$

The complete system is governed by the stochastic differential equations

$$dx_z = f_0(x_z, y)\,dt + g_0^T(x_z, y)\,dw, \tag{66.44}$$

$$d\tilde{x} = \left(A\tilde{x} + \frac{1}{l^*}F(x_z, y)\right)dt + \frac{1}{l^*}G^T(x_z, y)\,dw, \tag{66.45}$$

$$dy = (\hat{x}_2 + l^*\tilde{x}_2 + f_1(x_z, y))\,dt + g_1^T(x_z, y)\,dw, \tag{66.46}$$

$$d\hat{x}_2 = (u + a_2(y - \hat{x}_1))\,dt, \tag{66.47}$$

where Equation 66.44 is the zero dynamics, Equation 66.45 is the observer error system, and Equation 66.46 and 66.47 are the equations on which the backstepping design will be performed.

We start our design by introducing the backstepping transformation

$$z_1 = y, \quad z_2 = \hat{x}_2 - \phi_1(y, \hat{l}),$$

where $\phi_1(y, \hat{l})$ is a smooth virtual control law to be designed, and $\hat{l}$ is a parameter estimate. In the z-variables, the subsystems (Equations 66.46 and 66.47) is governed by

$$dz_1 = (z_2 + \phi_1 + l^*\tilde{x}_2 + f_1(x_z, y))\,dt + g_1^T(x_z, y)\,dw, \tag{66.48}$$

$$dz_2 = d\hat{x}_2 - \frac{\partial\phi_1}{\partial y}dy - \frac{\partial\phi_1}{\partial\hat{l}}\dot{\hat{l}}\,dt - \frac{1}{2}\frac{\partial^2\phi_1}{\partial y^2}g_1^T g_1\,dt = (u + \Omega_1 + \Omega_2 + \Omega_3 + \Omega_4)\,dt + \Phi^T\,dw, \tag{66.49}$$

where

$$\Omega_1 = -\frac{\partial\phi_1}{\partial y}\hat{x}_2 + a_2(y - \hat{x}_1), \quad \Omega_2 = -\frac{1}{2}\frac{\partial^2\phi_1}{\partial y^2}g_1^T g_1 - \frac{\partial\phi_1}{\partial y}f_1, \quad \Omega_3 = -\frac{\partial\phi_1}{\partial y}l^*\tilde{x}_2,$$

$$\Omega_4 = -\frac{\partial\phi_1}{\partial\hat{l}}\dot{\hat{l}}, \quad \Phi = -\frac{\partial\phi_1}{\partial y}g_1.$$

Next, we present our design in two steps.

Step 1. Since the polynomial $s^2 + a_1 s + a_2$ is Hurwitz, so is the matrix A. Thus, there exists a positive-definite matrix P such that $A^T P + PA = -I$. We start building our Lyapunov function by introducing

$$V_1 = \frac{\delta_1}{2}(\tilde{x}^T P\tilde{x})^2 + \frac{1}{4}y^4 + \frac{1}{2\lambda_0}(\hat{l} - l)^2,$$

where $\delta_1 > 0$ is an analysis parameter, $\lambda_0 > 0$ is the adaptation gain to be chosen freely by the designer, $l = \max\left\{l_1^{*\frac{4}{3}}, l^{*\frac{4}{3}}, h_1^{*4}\right\}$ is an unknown constant, and $\hat{l}(t)$ the estimate of l for which the update law is yet to be chosen.

By Itô formula and Equation 66.48, we obtain that

$$\begin{aligned}\mathcal{L}V_1 = &-\delta_1\tilde{x}^T P\tilde{x}|\tilde{x}|^2 + \frac{2\delta_1}{l^*}\tilde{x}^T P\tilde{x}(F^T P\tilde{x}) + \frac{\delta_1}{l^{*2}}\mathrm{Tr}\{G(2P\tilde{x}\tilde{x}^T P + \tilde{x}^T P\tilde{x}P)G^T\}\\ &+ y^3(z_2 + \phi_1 + l^*\tilde{x}_2 + f_1) + \frac{3}{2}y^2 g_1^T g_1 + \frac{1}{\lambda_0}(\hat{l} - l)\dot{\hat{l}}.\end{aligned} \tag{66.50}$$

We need to deal with the uncertain functions f_1, f_2, g_1, and g_2. Toward this end, with the help of the mean value theorem we observe that

$$\varphi_{i2}^4(|y|) = (\varphi_{i2}(0) + |y|\breve{\varphi}_{i2}(|y|))^4 \le 8\varphi_{i2}^4(0) + 8y^4\breve{\varphi}_{i2}^4(|y|) \le 8\varphi_{i2}^4(0) + 8y^4\breve{\varphi}_{i2}(y),$$

$$\psi_{i2}^4(|y|) = (\psi_{i2}(0) + |y|\breve{\psi}_{i2}(|y|))^4 \le 8\psi_{i2}^4(0) + 8y^4\breve{\psi}_{i2}^4(|y|) \le 8\psi_{i2}^4(0) + 8y^4\breve{\psi}_{i2}(y)$$

for some smooth nonnegative functions $\breve{\varphi}_{i2}, \breve{\psi}_{i2}, \breve{\varphi}_{i2}$, and $\breve{\psi}_{i2}$. In addition, with Young's inequality, we obtain

$$y^3 z_2 \le \frac{3}{4}\varepsilon_1^{\frac{4}{3}}y^4 + \frac{1}{4\varepsilon_1^4}z_2^4,$$

$$y^3 l^*\tilde{x}_2 \le \frac{3}{4}\varepsilon_1^{\frac{4}{3}}y^4 l + \frac{1}{4\varepsilon_1^4}\tilde{x}_2^4,$$

$$y^3 f_1(x_z, y) \le \left(\frac{3}{2}\eta_0^{\frac{4}{3}}l + \frac{2}{\eta_0^4}\breve{\varphi}_{12}\right)y^4 + \frac{1}{4\eta_0^4}\varphi_{11}^4 + \frac{2}{\eta_0^4}\varphi_{12}^4(0),$$

$$\frac{3}{2}y^2 g_1^T g_1 \le \frac{3}{\eta_1}y^4 l + \frac{3}{2}\eta_1\psi_{11}^4 + 12\eta_1\breve{\psi}_{12}y^4 + 12\eta_1\psi_{12}^4(0),$$

$$\frac{2\delta_1}{l^*}\tilde{x}^T P\tilde{x}(F^T P\tilde{x}) \leq \frac{3\delta_1}{2}\varepsilon^{\frac{4}{3}}\|P\|^2|\tilde{x}|^4 + \frac{8\delta_1}{\varepsilon^4}\|P\|^2\sum_{i=1}^{2}\varphi_{i1}^4 + \frac{64\delta_1}{\varepsilon^4}\|P\|^2\sum_{i=1}^{2}\breve{\varphi}_{i2}y^4$$
$$+\frac{64\delta_1}{\varepsilon^4}\|P\|^2\sum_{i=1}^{2}\varphi_{i2}^4(0),$$
$$\frac{\delta_1}{l^{*2}}\mathrm{Tr}\{G(2P\tilde{x}\tilde{x}^T P + \tilde{x}^T P\tilde{x}P)G^T\} \leq \frac{2\delta_1\|P\|^2 + \delta_1\|P\|\mathrm{Tr}\{P\}}{\varepsilon}\sum_{i=1}^{2}\psi_{i1}^4 + 4\varepsilon(2\delta_1\|P\|^2 + \delta_1\|P\|\mathrm{Tr}\{P\})|\tilde{x}|^4$$
$$+8\frac{2\delta_1\|P\|^2 + \delta_1\|P\|\mathrm{Tr}\{P\}}{\varepsilon}\sum_{i=1}^{2}\breve{\psi}_{i2}y^4$$
$$+8\frac{2\delta_1\|P\|^2 + \delta_1\|P\|\mathrm{Tr}\{P\}}{\varepsilon}\sum_{i=1}^{2}\psi_{i2}^4(0).$$

So, substituting these bounds into Equation 66.50, we obtain

$$\mathcal{L}V_1 \leq y^3\left(\phi_1 + \frac{3}{4}\varepsilon_1^{\frac{4}{3}}y + \frac{2}{\eta_0^4}\breve{\varphi}_{12}y + 12\eta_1\breve{\psi}_{12}y + \tilde{\varphi}(y) + \tilde{\psi}(y)\right) + l\left(\frac{3}{4}\varepsilon_1^{\frac{4}{3}}y^4 + \frac{3}{2}\eta_0^{\frac{4}{3}}y^4 + \frac{3}{\eta_1}y^4\right)$$
$$+\frac{1}{4\varepsilon_1^4}z_2^4 + \frac{1}{\lambda_0}(\hat{l}-l)\dot{\hat{l}} + \frac{1}{4\eta_0^4}\varphi_{11}^4 + \frac{3}{2}\eta_1\psi_{11}^4 + \tilde{\varphi}_1(|x_z|) + \frac{1}{4\varepsilon_1^4}\tilde{x}_2^4 - \delta_1\lambda_{\min}(P)|\tilde{x}|^4 + \tilde{\psi}_1(|x_z|)$$
$$+\tilde{c}|\tilde{x}|^4 + \tilde{\psi}_0 + \frac{2}{\eta_0^4}\varphi_{12}^4(0) + 12\eta_1\psi_{12}^4(0) + \tilde{\varphi}_0, \tag{66.51}$$

where

$$\tilde{\varphi}(y) = \frac{64\delta_1}{\varepsilon^4}\|P\|^2\sum_{i=1}^{2}\breve{\varphi}_{i2}(y)y, \tag{66.52}$$
$$\tilde{\psi}(y) = 8\frac{2\delta_1\|P\|^2 + \delta_1\|P\|\mathrm{Tr}\{P\}}{\varepsilon}\sum_{i=1}^{2}\breve{\psi}_{i2}(y)y, \tag{66.53}$$
$$\tilde{\varphi}_1(|x_z|) = \frac{8\delta_1}{\varepsilon^4}\|P\|^2\sum_{i=1}^{2}\varphi_{i1}^4(|x_z|), \tag{66.54}$$
$$\tilde{\psi}_1(|x_z|) = \frac{2\delta_1\|P\|^2 + \delta_1\|P\|\mathrm{Tr}\{P\}}{\varepsilon}\sum_{i=1}^{2}\psi_{i1}^4(|x_z|), \tag{66.55}$$
$$\tilde{c} = \frac{3\delta_1}{2}\varepsilon^{\frac{4}{3}}\|P\|^2 + 4\varepsilon\left(2\delta_1\|P\|^2 + \delta_1\|P\|\mathrm{Tr}\{P\}\right), \tag{66.56}$$
$$\tilde{\psi}_0 = 8\frac{2\delta_1\|P\|^2 + \delta_1\|P\|\mathrm{Tr}\{P\}}{\varepsilon}\sum_{i=1}^{2}\psi_{i2}^4(0), \tag{66.57}$$
$$\tilde{\varphi}_0 = \frac{64\delta_1}{\varepsilon^4}\|P\|^2\sum_{i=1}^{2}\varphi_{i2}^4(0), \tag{66.58}$$

and $\varepsilon, \varepsilon_1, \eta_0$, and η_1 are design parameters to be chosen later.

We are now ready to select the virtual update law and the virtual control as

$$\varpi_1(y,\hat{l}) = -\lambda_0\sigma\hat{l} + \lambda_0\left(\frac{3}{4}\varepsilon_1^{\frac{4}{3}}y^4 + \frac{3}{2}\eta_0^{\frac{4}{3}}y^4 + \frac{3}{\eta_1}y^4\right), \tag{66.59}$$

$$\begin{aligned}\phi_1(y,\hat{l}) = &-\beta_1 y - \nu_1(y^2)y - \frac{3}{4}\varepsilon_1^{\frac{4}{3}}y - \frac{2}{\eta_0^4}\breve{\varphi}_{12}(y)y - 12\eta_1\breve{\psi}_{12}(y)y \\ &-\hat{l}\left(\frac{3}{4}\varepsilon_1^{\frac{4}{3}}y + \frac{3}{2}\eta_0^{\frac{4}{3}}y + \frac{3}{\eta_1}y\right) - \tilde{\varphi}(y) - \tilde{\psi}(y),\end{aligned} \tag{66.60}$$

where $\sigma > 0$ is the parameter of the standard "sigma-modification/leakage" and $\nu_1(\cdot)$ is a smooth nonnegative function to be designed later to dominate various functional and dynamic uncertainties, including the effect of the uncertain and unmeasured zero dynamics. It now follows from Equations 66.51, 66.59 and 66.60 that

$$\begin{aligned}\mathcal{L}V_1 \le &-\beta_1 y^4 - \nu_1(y^2)y^4 + \frac{1}{\lambda_0}(\hat{l}-l)(\dot{\hat{l}} - \varpi_1) - \sigma(\hat{l}-l)\hat{l} + \frac{1}{4\varepsilon_1^4}z_2^4 + \frac{1}{4\eta_0^4}\varphi_{11}^4(|x_z|) \\ &+ \frac{3}{2}\eta_1\psi_{11}^4(|x_z|) + \tilde{\varphi}_1(|x_z|) + \frac{1}{4\varepsilon_1^4}\tilde{x}_2^4 + \tilde{\psi}_1(|x_z|) - \delta_1\lambda_{\min}(P)|\tilde{x}|^4 + \tilde{c}|\tilde{x}|^4 \\ &+ \frac{2}{\eta_0^4}\varphi_{12}^4(0) + 12\eta_1\psi_{12}^4(0) + \tilde{\varphi}_0 + \tilde{\psi}_0,\end{aligned} \tag{66.61}$$

which completes the first step of the design process.

Step 2. Consider the Lyapunov function candidate

$$V = V_1(\tilde{x}^T, y, \hat{l}) + \frac{1}{4}z_2^4.$$

By Equation 66.49, and Young's inequality, we obtain

$$\begin{aligned}\mathcal{L}V = &\mathcal{L}V_1 + z_2^3(u + \Omega_1 + \Omega_2 + \Omega_3 + \Omega_4) + \frac{3}{2}z_2^2\Phi_2^T\Phi_2 \\ \le &\mathcal{L}V_1 + z_2^3\left(u + \Omega_1 + \frac{1}{4\varepsilon_2^4}z_2\right) - z_2^3\frac{\partial\phi_1}{\partial\hat{l}}\dot{\hat{l}} \\ &+ l\left(\frac{1}{\epsilon_1}(\frac{\partial^2\phi_1}{\partial y^2})^2 z_2^6 + \frac{3}{2}\eta_0^{\frac{4}{3}}\left(\left(\frac{\partial\phi_1}{\partial y}\right)^2+1\right)^2 z_2^4 + \frac{3}{4}\varepsilon_2^{\frac{4}{3}}\left(\left(\frac{\partial\phi_1}{\partial y}\right)^2+1\right)^2 z_2^4 + \frac{3}{\eta_1}(\frac{\partial\phi_1}{\partial y})^4 z_2^4\right) \\ &+ \frac{\epsilon_1}{2}\psi_{11}^4 + \frac{1}{4\eta_0^4}\varphi_{11}^4 + \frac{2}{\eta_0^4}\breve{\varphi}_{12}y^4 + 4\epsilon_1\breve{\psi}_{12}y^4 + 12\eta_1\breve{\psi}_{12}y^4 + \frac{1}{4\varepsilon_2^4}\tilde{x}_2^4 \\ &+ \frac{3}{2}\eta_1\psi_{11}^4 + \frac{2}{\eta_0^4}\varphi_{12}^4(0) + 4\epsilon_1\psi_{12}^4(0) + 12\eta_1\psi_{12}^4(0) \\ \le &-\beta_1 y^4 - \nu_1(y^2)y^4 + \frac{1}{\lambda_0}(\hat{l}-l)(\dot{\hat{l}} - \varpi_1) - \sigma(\hat{l}-l)\hat{l} - \delta_1\lambda_{\min}(P)|\tilde{x}|^4 + \tilde{c}|\tilde{x}|^4 \\ &+ \frac{1}{4\varepsilon_1^4}z_2^4 + z_2^3\left(u + \Omega_1 + \frac{1}{4\varepsilon_2^4}z_2\right) - z_2^3\frac{\partial\phi_1}{\partial\hat{l}}\dot{\hat{l}} \\ &+ l\left(\frac{1}{\epsilon_1}(\frac{\partial^2\phi_1}{\partial y^2})^2 z_2^6 + \frac{3}{2}\eta_0^{\frac{4}{3}}\left(\left(\frac{\partial\phi_1}{\partial y}\right)^2+1\right)^2 z_2^4 + \frac{3}{4}\varepsilon_2^{\frac{4}{3}}\left(\left(\frac{\partial\phi_1}{\partial y}\right)^2+1\right)^2 z_2^4 + \frac{3}{\eta_1}(\frac{\partial\phi_1}{\partial y})^4 z_2^4\right) \\ &+ \frac{2}{\eta_0^4}\breve{\varphi}_{12}y^4 + 4\epsilon_1\breve{\psi}_{12}y^4 + \frac{1}{4\varepsilon_1^4}\tilde{x}_2^4 + \frac{1}{4\varepsilon_2^4}\tilde{x}_2^4 + 12\eta_1\breve{\psi}_{12}y^4 + \delta(|x_z|) + \Gamma_0,\end{aligned} \tag{66.62}$$

where

$$\delta(|x_z|) = \frac{1}{2\eta_0^4}\varphi_{11}^4(|x_z|) + 3\eta_1\psi_{11}^4(|x_z|) + \frac{\epsilon_1}{2}\psi_{11}^4(|x_z|) + \tilde{\varphi}_1(|x_z|) + \tilde{\psi}_1(|x_z|),$$

$$\Gamma_0 = \frac{4}{\eta_0^4}\varphi_{12}^4(0) + (4\epsilon_1 + 24\eta_1)\psi_{12}^4(0) + \tilde{\varphi}_0 + \tilde{\psi}_0, \tag{66.63}$$

and $\epsilon_1, \varepsilon_2$, and ϵ_2 are design parameters to be chosen later.

We now choose our final parameter update law and control law as

$$\dot{\hat{l}} = \varpi_2(y, z_2, \hat{l}) = \varpi_1(y, \hat{l}) + \lambda_0 z_2^4 \vartheta_2, \tag{66.64}$$

$$u = -\beta_2 z_2 - \Omega_1 - \frac{1}{4\varepsilon_1^4} z_2 - \hat{l} z_2 \vartheta_2 + \frac{\partial \phi_1}{\partial \hat{l}}\varpi_2, \tag{66.65}$$

where

$$\vartheta_2 = \frac{1}{\epsilon_1}\left(\frac{\partial^2\phi_1}{\partial y^2}\right)^2 z_2^2 + \frac{3}{2}\eta_0^{\frac{4}{3}}\left(\left(\frac{\partial\phi_1}{\partial y}\right)^2 + 1\right)^2 + \frac{3}{4}\varepsilon_2^{\frac{4}{3}}\left(\left(\frac{\partial\phi_1}{\partial y}\right)^2 + 1\right)^2 + \frac{3}{\eta_1}\left(\frac{\partial\phi_1}{\partial y}\right)^4. \tag{66.66}$$

The function $\nu_1(\cdot)$ is chosen as

$$\nu_1(y^2) = (12\eta_1 + 4\epsilon_1)\breve{\psi}_{12}(y) + \frac{2}{\eta_0^4}\breve{\varphi}_{12}(y) + \nu(y^2), \tag{66.67}$$

where $\nu(\cdot)$ is a smooth $\mathcal{K}_\infty$ function which is yet to be chosen in order to deal with the dynamic uncertainty coming from the zero dynamics.

With the choices (Equations 66.64 through 66.67), we obtain

$$\mathcal{L}V \le -\beta_1 y^4 - \beta_2 z_2^4 - \nu(y^2)y^4 - \sigma(\hat{l} - l)\hat{l} - \delta_1\lambda_{\min}(P)|\tilde{x}|^4 + \frac{1}{4\varepsilon_2^4}\tilde{x}_2^4 + \frac{1}{4\varepsilon_1^4}\tilde{x}_2^4 + \tilde{c}|\tilde{x}|^4 + \delta(|x_z|) + \Gamma_0. \tag{66.68}$$

Using the property of "sigma-modification" that

$$-\sigma(\hat{l} - l)\hat{l} \le -\frac{\sigma}{2}(\hat{l} - l)^2 + \frac{\sigma}{2}l^2, \tag{66.69}$$

we obtain

$$\mathcal{L}V \le -\beta_1 y^4 - \beta_2 z_2^4 - \nu(y^2)y^4 - \frac{\sigma}{2}(\hat{l} - l)^2 - \delta_1\lambda_{\min}(P)|\tilde{x}|^4 + \frac{1}{4\varepsilon_2^4}\tilde{x}_2^4 + \frac{1}{4\varepsilon_1^4}\tilde{x}_2^4 + \tilde{c}|\tilde{x}|^4 + \frac{\sigma}{2}l^2 + \delta(|x_z|) + \Gamma_0. \tag{66.70}$$

Recalling Equation 66.55, we obtain

$$\mathcal{L}V \le -\beta_1 y^4 - \beta_2 z_2^4 - \nu(y^2)y^4 - \frac{\sigma}{2}(\hat{l} - l)^2 - c_0|\tilde{x}|^4 + \frac{\sigma}{2}l^2 + \delta(|x_z|) + \Gamma_0, \tag{66.71}$$

where

$$c_0 \triangleq \delta_1\left(\lambda_{\min}(P) - \frac{3}{2}\varepsilon^{\frac{4}{3}}\|P\|^2 - 4\varepsilon(2\|P\|^2 + \|P\|\mathrm{Tr}\{P\})\right) - \sum_{i=1}^{2}\frac{1}{4\varepsilon_i^4}.$$

Choosing the control parameter ε sufficiently small, and then choosing the analysis parameter δ_1 sufficiently large (for given ϵ_2, ε_1, and ε_2 of any positive values), we obtain $c_0 > 0$. The design parameters η_0, η_1, ϵ_1, β_1, β_2, and λ_0 are chosen as positive, but otherwise of any value. Hence, we obtain

$$\mathcal{L}V \le -W(\tilde{x}, y, z_2, \hat{l}) - \nu(y^2)y^4 + \delta(|x_z|) + \Gamma_0 + \frac{\sigma}{2}l^2, \tag{66.72}$$

where $W(\tilde{x}, y, z_2, \hat{l}) = \sum_{i=1}^{2}\beta_i z_i^4 + c_0|\tilde{x}|^4 + \frac{\sigma}{2}(\hat{l} - l)^2$.

66.6.2 Dominating the Uncertain and Unmeasured Zero Dynamics

From Equations 66.42 and 66.72, we know that the subsystems (Equations 66.48, 66.49, 66.64, and 66.65) is practically SISS with respect to the state x_z of unmodeled dynamics, while the unmodeled dynamics (Equation 66.38) are assumed to be SISS with respect to the output y. We need to deal with this stochastic feedback connection and its effect on closed-loop stability, by making one final choice for our control law, the function $\nu(\cdot)$ in Equation 66.67.

Toward this end, we note that there exist known smooth nonnegative functions ψ_z and ψ_0 satisfying $|\frac{\partial V_z(x_z)}{\partial x_z}| \le \psi_z(|x_z|)$ and $\|g_0(x_z)\| \le \psi_0(|x_z|)$. If

$$\limsup_{s\to 0+} \frac{\varphi_{i1}^4(s)}{\alpha(s)} < \infty, \ \limsup_{s\to 0+} \frac{\psi_{i1}^4(s)}{\alpha(s)} < \infty, \ \limsup_{s\to 0+} \frac{\psi_z^2(s)\psi_0^2(s)}{\alpha(s)} < \infty, \quad i = 1, 2, \tag{66.73}$$

where the first two sets of terms appear in Equation 66.63 and α appears in Assumption A2, then we define continuous increasing functions $\xi(s) \ge 0$ and $\zeta(s) > 0$ on $[0, \infty)$ as $\xi(s) = 4\sup_{\tau\in(0,s]} \delta(\tau)/\alpha(\tau)$ for $s > 0$ and $\xi(0) = 4\limsup_{s\to 0+} \delta(\tau)/\alpha(\tau)$, and $\zeta(s) = 2\sup_{\tau\in(0,s]} \psi_z^2(\tau)\psi_0^2(\tau)/\alpha(\tau)$ for $s > 0$ and $\zeta(0) = 2\limsup_{s\to 0+} \psi_z^2(\tau)\psi_0^2(\tau)/\alpha(\tau)$. In this way we obtain $\delta(s) \le \xi(s)\alpha(s)/4$ and $\psi_z^2(s)\psi_0^2(s) \le \zeta(s)\alpha(s)/2$, which we shall use later in selecting the control function $\nu(\cdot)$.

If Equation 66.73 holds and if

$$\int_0^\infty [\xi(\alpha_1^{-1}(s))]' e^{-\int_0^s [\zeta(\alpha_1^{-1}(\tau))]^{-1} d\tau}\, ds < \infty, \tag{66.74}$$

then there exists a nondecreasing positive function $\rho \in \mathcal{C}^1[0, \infty)$ such that for all $x \in \mathbb{R}^m$,

$$\rho(V_z(x))\alpha(|x|) \ge 2\rho'(V_z(x))\psi_z^2(|x|)\psi_0^2(|x|) + 4\delta(|x|), \tag{66.75}$$

where $\delta(\cdot)$ is given as in Equation 66.63. The function ρ, which will be used to modify the Lyapunov function, can be constructed as, for example,

$$\rho(0) = \xi(0) + \int_0^\infty [\xi(\alpha_1^{-1}(s))]' e^{-\int_0^s q_1(\tau)d\tau}\, ds$$

and

$$\rho(s) = e^{\int_0^s q_1(\tau)d\tau}\Big[\rho(0) - \int_0^s q_2(u) e^{-\int_0^u q_1(\tau)d\tau}\, du\Big], \quad s > 0,$$

where $q_1(s) = \frac{1}{\zeta(\alpha_1^{-1}(s))}$, $q_2(s) = \frac{\xi(\alpha_1^{-1}(s))}{\zeta(\alpha_1^{-1}(s))}$.

We now consider the Lyapunov function

$$\overline{V}(\tilde{x}, x_z, y, z_2, \hat{l}) = \int_0^{V_z(x_z)} \rho(s)\, ds + V(\tilde{x}, y, \hat{l}, z_2).$$

By Itô's formula we have

$$\begin{aligned}
\mathcal{L}\overline{V} &= \rho(V_z)\mathcal{L}V_z + \frac{1}{2}\rho'(V_z)\left\|\left(\frac{\partial V_z}{\partial x_z}\right)^T g_0\right\|_F^2 - W(\tilde{x}, y, z_2, \hat{l}) - \nu(y^2)y^4 + \delta(|x_z|) + \Gamma_0 + \frac{\sigma}{2}l^2 \\
&\le \rho(V_z)[\gamma(|y|) - \alpha(|x_z|)] + \frac{1}{2}\rho'(V_z)\psi_z^2(|x_z|)\psi_0^2(|x_z|) - W(\tilde{x}, y, z_2, \hat{l}) - \nu(y^2)y^4 + \delta(|x_z|) + \Gamma_0 + \frac{\sigma}{2}l^2 \\
&\le \rho(\eta(|y|))\gamma(|y|) - \frac{1}{2}\rho(V_z)\alpha(|x_z|) + \frac{1}{2}\rho'(V_z)\psi_z^2(|x_z|)\psi_0^2(|x_z|) \\
&\quad - W(\tilde{x}, y, z_2, \hat{l}) - \nu(y^2)y^4 + \delta(|x_z|) + \Gamma_0 + \frac{\sigma}{2}l^2,
\end{aligned} \tag{66.76}$$

where $\eta = \alpha_2(\alpha^{-1}(2\gamma(\cdot))) \in \mathcal{K}_\infty$. From Equation 66.75, we have

$$\frac{1}{4}\rho(V_z)\alpha(|x_z|) - \delta(|x_z|) \geq \frac{1}{2}\rho'(V_z)\psi_z^2(|x_z|)\psi_0^2(|x_z|), \tag{66.77}$$

thus we get

$$\mathcal{L}\overline{V} \leq \rho(\eta(|y|))\gamma(|y|) - \frac{1}{4}\rho(V_z)\alpha(|x_z|) - W(\tilde{x}, y, z_2, \hat{l}) - \nu(y^2)y^4 + \Gamma_0 + \frac{\sigma}{2}l^2. \tag{66.78}$$

Assuming that the gain function $\gamma(\cdot)$ in Assumption A2 is locally quartic, that is

$$\limsup_{s\to 0+} \frac{\gamma(s)}{s^4} < \infty, \tag{66.79}$$

we choose $\nu(\cdot)$ as

$$\nu(y^2) = \rho(\alpha_2(\alpha^{-1}(2\gamma(y^2+1)))) \sup_{s\in(0,y^2+1]} \frac{\gamma(s)}{s^4}.$$

This, together with Equations 66.76 and 66.77, yields

$$\mathcal{L}\overline{V} \leq -\frac{1}{4}\rho(0)\alpha(|x_z|) - W(\tilde{x}, y, z_2, \hat{l}) + \Gamma_0 + \frac{\sigma}{2}l^2. \tag{66.80}$$

Define

$$W_1(\tilde{x}, x_z, y, z_2, \hat{l}) = \frac{1}{4}\rho(0)\alpha(|x_z|) + W(\tilde{x}, y, z_2, \hat{l}),$$

which is positive-definite and radially unbounded in $(\tilde{x}, x_z, y, z_2, \hat{l})$. Thus

$$\mathcal{L}\overline{V} \leq -W_1(\tilde{x}, x_z, y, z_2, \hat{l}) + \Gamma_0 + \frac{\sigma}{2}l^2 \leq -\beta(|(\tilde{x}, x_z, y, z_2, \hat{l})|) + \Gamma_0 + \frac{\sigma}{2}l^2, \tag{66.81}$$

where β is a class $\mathcal{K}_\infty$ function.

66.6.3 Boundedness, Stability, and Regulation

From Equation 66.81, by Test for Stability C, the closed-loop system is practically SISS (and hence the solution process is bounded in probability). Moreover, if $\varphi_{i2}(0) = 0$, $\psi_{i2}(0) = 0$, $i = 1, 2$, we obtain $\Gamma_0 = 0$. In this case, we take $\sigma = 0$ in Equation 66.59; hence from Equation 66.80 we obtain

$$\mathcal{L}\overline{V} \leq -W_2(\tilde{x}, x_z, y, z_2), \tag{66.82}$$

where $W_2(\tilde{x}, x_z, y, z_2) = \frac{1}{4}\rho(0)\alpha(|x_z|) + \sum_{i=1}^{2}\beta_i z_i^4 + c_0|\tilde{x}|^4$. Thus, by Test for Stability A, the equilibrium $(\tilde{x}, z, x_0, \hat{l}) = (0, 0, 0, l)$ is globally stable in probability, and in addition, the output is regulated to the origin almost surely, more precisely, $P\{\lim_{t\to\infty}(|y| + |x_z| + |x_1| + |x_2|) = 0\} = 1$.

In this section, we have used a different Lyapunov function from the previous two sections. The basic idea of construction comes from the method of changing the supply function of unmodeled dynamics (or zero dynamics). For further details and an extension to a general class of systems the reader is referred to [14].

66.7 Inverse Optimal Stabilization in Probability

The problem of *inverse optimal stabilization in probability* for system (Equation 66.6) is said to be solvable if there exists a $\mathcal{K}_\infty$ function γ_2 whose derivative γ_2' is also a $\mathcal{K}_\infty$ function, a matrix-valued function $R_2(x)$

such that $R_2(x) = R_2(x)^T > 0$ for all x, a positive definite function $l(x)$, and a feedback control law $u = \alpha(x)$ continuous everywhere with $\alpha(0) = 0$, which guarantees global asymptotic stability in probability of the equilibrium $x = 0$ and minimizes the cost functional

$$J(u) = E\left\{\int_0^\infty [l(x) + \gamma_2(|R_2(x)^{\frac{1}{2}}u|)]\, d\tau\right\}.$$

Next, we present a constructive solution to this nonlinear stochastic optimal control problem.

Consider the control law

$$u = \alpha(x) = -R_2^{-1}(L_{g_2}V)^T \frac{\ell\gamma_2(|L_{g_2}VR_2^{-1/2}|)}{|L_{g_2}VR_2^{-1/2}|^2}, \tag{66.83}$$

where $V(x)$ is a Lyapunov function candidate, γ_2 is a $\mathcal{K}_\infty$ function whose derivative is also a $\mathcal{K}_\infty$ function, $R_2(x)$ is a matrix-valued function such that $R_2(x) = R_2^T(x) > 0$, and

$$\ell\gamma(r) = \int_0^r (\gamma')^{-1}(s)\, ds$$

represents the Legendre–Fenchel transform. If the control law (Equation 66.83) achieves global asymptotic stability in probability for system (Equation 66.6) with respect to $V(x)$, then the control law

$$u^* = \alpha^*(x) = -\frac{\beta}{2}R_2^{-1}(L_{g_2}V)^T \frac{(\gamma_2')^{-1}(|L_{g_2}VR_2^{-1/2}|)}{L_{g_2}VR_2^{-1/2}}, \quad \beta \geq 2 \tag{66.84}$$

solves the problem of inverse optimal stabilization in probability for system (Equation 66.6) by minimizing the cost functional

$$J(u) = E\left\{\int_0^\infty \left[l(x) + \beta^2\gamma_2\left(\frac{2}{\beta}|R_2^{1/2}u|\right)\right] d\tau\right\},$$

where

$$l(x) = 2\beta\left[\ell\gamma_2(|L_{g_2}VR_2^{-1/2}|) - L_fV - \frac{1}{2}\mathrm{Tr}\left\{g_1^T\frac{\partial^2 V}{\partial x^2}g_1\right\}\right] - \beta(\beta-2)\ell\gamma_2(|L_{g_2}VR_2^{-1/2}|).$$

For the design procedures for control laws in the form (Equation 66.83), and hence of inverse-optimal-in-probability control laws in the form (Equation 66.84), the reader is referred to the book by Krstić and Deng [9].

66.8 Extensions

In the preceding sections, we introduced stabilization tools for three classes of canonical systems. These design ideas can be applied to the following more general systems.

Systems not in the output-feedback form:

$$\begin{aligned}
dx_z &= f_0(x_z, y)\, dt + g_0^T(x_z, y)\, dw,\\
dx_i &= (x_{i+1} + f_i(x_z, \overline{x}_i))\, dt + g_i^T(x_z, \overline{x}_i)\, dw, \quad i = 1, \ldots, n-1,\\
dx_n &= (u + f_n(x_z, \overline{x}_n))\, dt + g_n^T(x_z, \overline{x}_n)\, dw,\\
y &= x_1,
\end{aligned}$$

where nonlinear terms $f_i, g_i, i = 1, 2, \ldots, n$ depends on the unmeasurable states $x_2, \ldots, x_i$ besides the measured output y. To design the output-feedback stabilization controller, some extra growth conditions on these nonlinear terms are usually needed.

Unknown virtual control coefficients:

$$dx_i = (b_i x_{i+1} + f_i(\bar{x}_i))\,dt + g_i^T(\bar{x}_i)\,dw, \quad i = 1, \ldots, n-1,$$
$$dx_n = (b_n u + f_n(\bar{x}_n))\,dt + g_n^T(\bar{x}_n)\,dw,$$

where the constant coefficients b_i, $i = 1, \ldots, n$ are unknown, but the signs of b_i are assumed to be known. Similar to the deterministic case, using the tuning functions design which, in addition to estimating the b_i's, also estimates the inverse coefficients $\frac{1}{b_i}$, $i = 1, \ldots, n$, we construct adaptive control laws that achieve global stability in probability and "almost sure" regulation to the origin.

References

1. H. Deng and M. Krstić, Stochastic nonlinear stabilization-I: A backstepping design, *Systems and Control Letters*, vol. 32, pp. 143–150, 1997.
2. H. Deng and M. Krstić, Stochastic nonlinear stabilization-II: Inverse optimality, *Systems and Control Letters*, vol. 32, pp. 151–159, 1997.
3. H. Deng, M. Krstić, and R. J. Williams, Stabilization of stochastic nonlinear systems driven by noise of unknown covariance, *IEEE Transactions on Automatic Control*, vol. 8, pp. 1237–1253, 2001.
4. P. Florchinger, A universal formula for the stabilization of control stochastic differential equations, *Stochastic Analysis and Its Applications*, vol. 11, pp. 155–162, 1993.
5. P. Florchinger, Lyapunov-like techniques for stochastic stability, *SIAM Journal of Control and Optimization*, vol. 33, pp. 1151–1169, 1995.
6. W. Hahn, *Stability of Motion*, New York, NY: Springer-Verlag, 1967.
7. H. Z. Khalil, *Nonlinear Systems* (2nd edition) Upper Saddle River, NJ: Prentice-Hall, 1996.
8. R. Z. Khas'minskiĭ, *Stochastic Stability of Differential Equations*, Rockville, MD; S & N International Publisher, 1980.
9. M. Krstić and H. Deng, *Stabilization of Nonlinear Uncertain Systems*, London: Springer-Verlag, 1998.
10. Z. G. Pan, Canonical forms for stochastic nonlinear systems, *Automatica*, vol. 38, pp. 1163–1170, 2002.
11. Z. G. Pan and T. Başar, Backstepping controller design for nonlinear stochastic systems under a risk-sensitive cost criterion, *SIAM Journal on Control and Optimization*, vol. 37, pp. 957–995, 1999.
12. Z. G. Pan, Y. G. Liu, and S. J. Shi, Output feedback stabilization for stochastic nonlinear systems in observer canonical form with stable zero-dynamics, *Science in China (Series F)*, vol. 44, pp. 292–308, 2001.
13. Y. G. Liu and J. F. Zhang, Practical output-feedback risk-sensitive control for stochatic nonlinear systems with stable zero-dynamics, *SIAM Journal on Control and Optimization*, vol. 45, pp. 885–926, 2006.
14. S. J. Liu, J. F. Zhang, and Z. P. Jiang, Decentralized adaptive output-feedback stabilization for large-scale stochastic nonlinear systems, *Automatica*, vol. 34, pp. 238–251, 2007.
15. E. D. Sontag, Smooth stabilization implies coprime factorization, *IEEE Transactions on Automatic Control*, vol. 34, pp. 435–443, 1989.
16. J. Tsinias, The concept of Exponential input to state stability for stochastic systems and applications to feedback stabilization, *Systems and Control Letters*, vol. 36, pp. 221–229, 1999.
17. Z. J. Wu, X. J. Xie, and S. Y. Zhang, Adaptive backstepping controller design using stochastic small-gain theorem, *Automatica*, vol. 43(4), pp. 608–620, 2007.

XI

Control of Distributed Parameter Systems

67

Control of Systems Governed by Partial Differential Equations

Kirsten Morris
University of Waterloo

67.1 Introduction

In many applications, such as diffusion and structural vibrations, the physical quantity of interest depends on both position and time. Some examples are shown in Figures 67.1 through 67.3. These systems are modeled by partial differential equations (PDEs) and the solution evolves on an infinite-dimensional Hilbert space. For this reason, these systems are often called infinite-dimensional systems. In contrast, the state of a system modeled by an ordinary differential equation evolves on a finite-dimensional system, such as $\mathbb{R}^n$, and these systems are called finite-dimensional. Since the solution of the PDE reflects the distribution in space of a physical quantity such as the temperature of a rod or the deflection of a beam, these systems are often also called distributed-parameter systems (DPS). Systems modeled by delay differential equations also have a solution that evolves on an infinite-dimensional space. Thus, although the physical situations are quite different, the theory and controller design approach is quite similar to that of systems modeled by PDEs. However, delay differential equations will not be discussed directly in this chapter.

The purpose of controller design for infinite-dimensional systems is similar to that for finite-dimensional systems. Every controlled system must of course be stable. Beyond that, the goals are to improve the response in some well-defined manner, such as by solving a linear-quadratic (LQ) optimal control problem. Another common goal is design the controller to minimize the system's response to disturbances.

Classical controller design is based on an input/output description of the system, usually through the transfer function. Infinite-dimensional systems have transfer functions. However, unlike the transfer functions of finite-dimensional systems, the transfer function is not a rational function. If a closed-form expression of the transfer function of an infinite-dimensional system can be obtained, it may be possible to design a controller directly. This is known as the *direct controller design* approach. Generalizations of many well-known finite-dimensional stability results such as the small gain theorem and the Nyquist

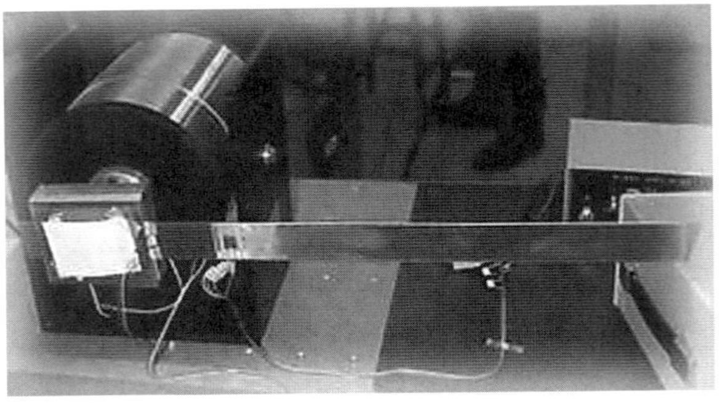

FIGURE 67.1 A flexible beam is the simplest example of transverse vibrations in a structure. It has relevance to control of flexible robots and space structures. This photograph shows a beam controlled by means of a motor at one end. (Photo courtesy of Prof. M.F. Golnaraghi, Simon Fraser University.)

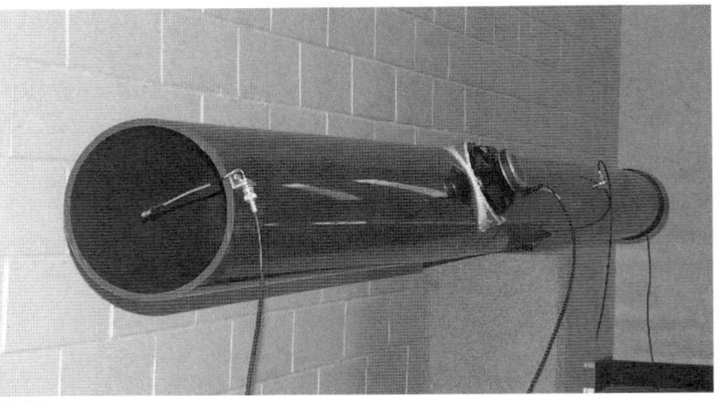

FIGURE 67.2 Acoustic noise in a duct. A noise signal is produced by a loudspeaker placed at one end of the duct. In this photo, a loudspeaker is mounted midway down the duct where it is used to control the noise signal. The pressure at the open end is measured by means of a microphone as shown in the photo. (Photo courtesy of Prof. S. Lipshitz, University of Waterloo.)

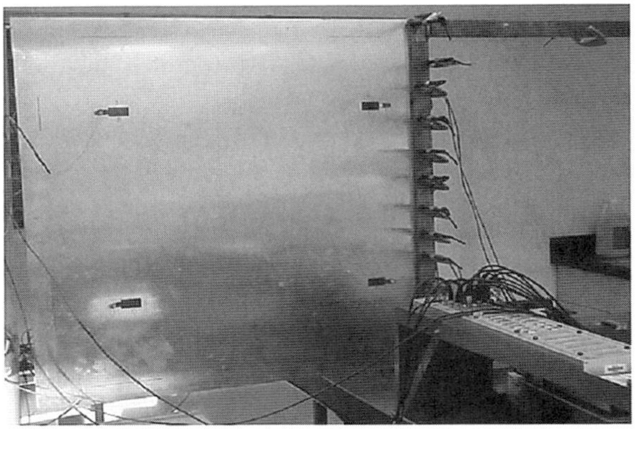

FIGURE 67.3 Vibrations in a plate occur due to various disturbances. In this apparatus the vibrations are controlled via the piezo-electric patches shown. (Photo courtesy of Prof. M. Demetriou, Worcester Polytechnic University.)

stability criterion exist; see [24,25]. Passivity generalizes in a straightforward way to irrational transfer functions [9, Chapters V, VI] and just as for finite-dimensional systems, any positive real system can be stabilized by the static output feedback $u = -\kappa y$ for any $\kappa > 0$. For some of these results it is not required to know the transfer function in order to ensure stability of the controlled system. It is only required to know whether the transfer function lies in the appropriate class. PI-control solutions to tracking problems for irrational transfer functions and the internal model principle is covered in [25,26,35]. For a high-performance controlled system, a model of the system needs to be used. H_∞-controller design has been successfully developed as a method for robust control and disturbance rejection in finite-dimensional systems. A theory of robust H_∞-control designs for infinite-dimensional systems using transfer functions is described in [11]. More recent results on this approach can be found in [19] and references therein.

The chief drawback of direct controller design is that an explicit representation of the transfer function is required. Another drawback of direct controller design is that, in general, the resulting controller is infinite-dimensional and must be approximated by a finite-dimensional system. For this reason, direct controller design is sometimes referred to as *late lumping* since a last step in the controller design is to approximate the controller by a finite-dimensional, or lumped parameter, system.

For many practical examples, controller design based on the transfer function is not feasible, since a closed-form expression for the transfer function may not be available. Instead, a finite-dimensional approximation of the system is first obtained and controller design is based on this finite-dimensional approximation. This approach is known as *indirect controller design*, or *early lumping*. This is the most common method of controller design for systems modeled by PDEs. The hope is that the controller has the desired effect on the original system. That this method is not always successful was first documented in Balas [1], where the term *spillover effect* was coined. Spillover refers to the phenomenon that a controller which stabilizes a reduced-order model need not necessarily stabilize the original model. Systems with infinitely many poles either on or asymptoting to the imaginary axis are notorious candidates for spillover effects. However, conditions under which this practical approach to controller design works have been obtained and are presented in this chapter.

In the next section a brief overview of the state-space theory for infinite-dimensional systems is given. Some issues associated with approximation of systems for the purpose of controller design are discussed in the following section. Results for the most popular methods for multi-input–multi-output controller design, LQ controller, and H_∞-controller design are then presented in Sections 67.4 and 67.5.

67.2 State-Space Formulation

Systems modeled by linear ordinary differential equations are generally written as a set of n first-order differential equations putting the system into the state-space form

$$\dot{z}(t) = Az(t) + Bu(t), \tag{67.1}$$

where $z(t) \in \mathbb{R}^n, A \in \mathbb{R}^{n\times n}, B \in \mathbb{R}^{n\times m}$, and m is the number of controls.

We write systems modeled by PDEs in a similar way. The main difference is that the matrices A becomes an operator acting, not on $\mathbb{R}^n$, but on an infinite-dimensional Hilbert space, $\mathcal{Z}$. Similarly, B maps the input space into the Hilbert space $\mathcal{Z}$. More detail on the systems theory described briefly in this section can be found in [8].

We first need to generalize the idea of a matrix exponential $\exp(At)$. Let $\mathcal{L}(\mathcal{X}_1, \mathcal{X}_2)$ indicates bounded linear operators from a Hilbert space $\mathcal{X}_1$ to a Hilbert space $\mathcal{X}_2$.

Definition 67.1:

A strongly continuous (C_0-) semigroup $S(t)$ on Hilbert space $\mathcal{Z}$ is a family $S(t) \in \mathcal{L}(\mathcal{Z}, \mathcal{Z})$ such that

1. $S(0) = I$,
2. $S(t)S(s) = S(t+s)$,
3. $\lim_{t\downarrow 0} S(t)z = z$, for all $z \in \mathcal{Z}$.

Definition 67.2:

The infinitesimal generator A of a C_0-semigroup on $\mathcal{Z}$ is defined by

$$Az = \lim_{t\downarrow 0} \frac{1}{t}(S(t)z - z)$$

with domain $\mathcal{D}(A)$ the set of elements $z \in \mathcal{Z}$ for which the limit exists.

The matrix exponential $\exp(At)$ is a special case of a semigroup, defined on a finite-dimensional space. Its generator is the matrix A. Note that we only have strong convergence of $S(t)$ to the identity I in Definition 67.1(3). Uniform convergence implies that the generator is a bounded operator defined on the whole space and that the semigroup can be defined as

$$\sum_{i=0}^{\infty} \frac{(At)^i}{i!}$$

just as for a matrix exponential. However, for PDEs the generator A is an unbounded operator and only strong convergence of $S(t)$ to I is obtained.

If A is the generator of a C_0-semigroup $S(t)$ on a Hilbert space $\mathcal{Z}$, then for all $z_0 \in D(A)$,

$$\frac{d}{dt}S(t)z_0 = AS(t)z_0 = S(t)Az_0.$$

It follows that the differential equation on $\mathcal{Z}$

$$\frac{dz(t)}{dt} = Az(t), \quad z(0) = z_0$$

has the solution

$$z(t) = S(t)z_0.$$

Furthermore, owing to the properties of a semigroup, this solution is unique, and depends continuously on the initial data z_0.

Thus, for infinite-dimensional systems, instead of Equation 67.1, we consider systems described by

$$\frac{dz}{dt} = Az(t) + Bu(t), \quad z(0) = z_0 \tag{67.2}$$

where A with domain $\mathcal{D}(A)$ generates a strongly continuous semigroup $S(t)$ on a Hilbert space $\mathcal{Z}$ and $B \in \mathcal{L}(\mathcal{U}, \mathcal{Z})$. We assume also that $\mathcal{U}$ is finite-dimensional (for instance, $\mathbb{R}^m$) as is generally the case in practice.

For some situations, such as control on the boundary of the region, typical models lead to a state-space representation where the control operator B is unbounded on the state space. More precisely, it

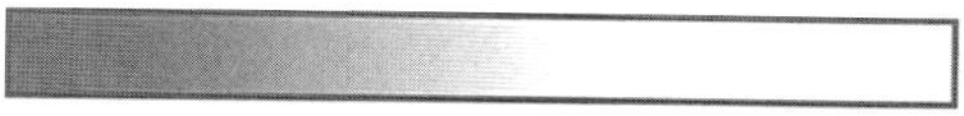

FIGURE 67.4 Heat flow in a rod. The regulation of the temperature profile of a rod is the simplest example of a control system modeled by a PDE.

is a bounded operator into a larger space than the state space. However, this complicates the analysis considerably. To simplify the exposition, this chapter considers only bounded B. Appropriate references are given for extension to unbounded operators where available. Note, however, that including a model for the actuator often changes a simple model with an unbounded actuator to a more complex model with bounded control; see, for instance, [17].

Example 67.1: Diffusion

Consider the temperature in a rod of length L with constant thermal conductivity K_0, mass density ρ and specific heat C_p (see Figure 67.4). Applying the principle of conservation of energy to arbitrarily small volumes in the bar leads to the following PDE for the temperature $z(x,t)$ at time t at position x from the left-hand end (see, e.g., [14, e.g., Section 1.3])

$$C_p\rho\frac{\partial z(x,t)}{\partial t} = K_0\frac{\partial^2 z(x,t)}{\partial x^2}, \quad x \in (0,L), \quad t \geq 0. \tag{67.3}$$

In addition to modeling heat flow, this equation also models other types of diffusion, such as chemical diffusion and neutron flux. To fully determine the temperature, one needs to specify the initial temperature profile $z(x,0)$ as well as the boundary conditions at each end. Assume Dirichlet boundary conditions:

$$z(0,t) = 0, \quad z(L,t) = 0.$$

and some initial temperature distribution $z(0) = z_0, z_0 \in \mathcal{L}_2(0,L)$. Define $Az = \frac{K_o}{C_p e}\partial^2 z/\partial x^2$. Since we cannot take derivatives of all elements of $\mathcal{L}_2(0,L)$ and we need to consider boundary conditions, define

$$\mathcal{D}(A) = \{z \in \mathcal{H}^2(0,L);\ z(0) = 0, z(L) = 0\},$$

where $\mathcal{H}^2(0,L)$ indicates the Sobolev space of functions with weak second derivatives [37]. We can rewrite the problem as

$$\dot{z}(t) = Az(t), \quad z(0,t) = z_0.$$

The operator A with domain $\mathcal{D}(A)$ generates a strongly continuous semigroup $S(t)$ on $\mathcal{Z} = \mathcal{L}_2(0,L)$. The state z, the temperature of the rod, evolves on the infinite-dimensional space $\mathcal{L}_2(0,L)$.

Suppose that the temperature on an unit interval [0,1] is controlled using an input flux $u(t)$

$$\frac{\partial z}{\partial t} = \frac{\partial^2 z}{\partial x^2} + Bu(t), \quad 0 < x < 1,$$

where $B \in \mathcal{L}(\mathbb{R}, \mathcal{Z})$ describes the distribution of applied energy, with the same Dirichlet boundary conditions. Since the input space is one-dimensional, B can be defined by $Bu = b(x)u$ for some $b(x) \in \mathcal{L}_2(0,1)$. This leads to

$$\dot{z}(t) = Az(t) + b(x)u(t).$$

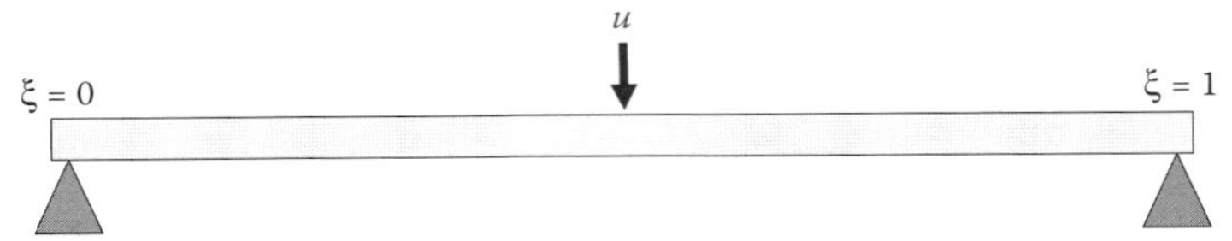

FIGURE 67.5 Simply supported beam with applied force.

Example 67.2: Simply Supported Beam

Consider a simply supported Euler-Bernoulli beam as shown in Figure 67.5 and let $w(x,t)$ denote the deflection of the beam from its rigid body motion at time t and position x. The control $u(t)$ is a force applied at the center with width δ. The analysis of beam vibrations is useful for applications such as flexible links in robots; but also in understanding the dynamics of more complex structures. If we normalize the variables, we obtain the PDE (see, e.g., [14, Chapter 6]):

$$\frac{\partial^2 w}{\partial t^2} + \frac{\partial^4 w}{\partial x^4} = b(x)u(t), \quad t \geq 0, \ 0 < x < 1,$$

$$b(x) = \begin{cases} \frac{1}{\delta}, & |x-0.5| < \frac{\delta}{2} \\ 0, & |x-0.5| \geq \frac{\delta}{2} \end{cases}$$

with boundary conditions

$$w(0,t)=0, \quad w_{xx}(0,t)=0, \quad w(1,t)=0, \ w_{xx}(1,t)=0. \tag{67.4}$$

This system is second order in time, and analogously to a simple mass–spring system, we define the state as $z(t) = [w(\cdot,t)\ \dot{w}(\cdot,t)]$. Let

$$H_s(0,1) = \{w \in \mathcal{H}^2(0,1), w(0)=0, w(1)=0\}$$

and define the state space $\mathcal{Z} = H_s(0,1) \times \mathcal{L}_2(0,1)$. A state-space formulation of the above PDE problem is

$$\frac{d}{dt}z(t) = Az(t) + Bu(t),$$

where

$$B = \begin{bmatrix} 0 \\ b(\cdot) \end{bmatrix}, \qquad A = \begin{bmatrix} 0 & I \\ -\frac{d^4}{dx^4} & 0 \end{bmatrix},$$

with domain

$$\mathcal{D}(A) = \{(\phi,\psi) \in H_s(0,1) \times H_s(0,1); \quad \phi_{xx} \in H_s(0,1)\}.$$

The operator A with domain $\mathcal{D}(A)$ generates a C_0-semigroup on $\mathcal{Z}$.

Even for finite-dimensional systems, the entire state cannot generally be measured. Measurement of the entire state is never possible for systems described by PDEs, and we define

$$y(t) = Cz(t) + Eu(t), \tag{67.5}$$

where $C \in \mathcal{L}(\mathcal{Z},\mathcal{Y})$, $E \in \mathcal{L}(\mathcal{U},\mathcal{Y})$, and $\mathcal{Y}$ is a Hilbert space. The expression (Equation 67.5) can also represent the cost in controller design. Note that as for the control operator, it is assumed that C is a bounded operator from the state space $\mathcal{Z}$. The operator E is a feedthrough term that is nonzero in some control configurations.

The state at any time t and control $u \in L_2(0,t;U)$ is given by

$$z(t) = S(t)z_0 + \int_0^t S(t-s)Bu(s)\,ds$$

and the output is

$$y = CS(t)z_0 + C\int_0^t S(t-\tau)Bu(\tau)\,d\tau,$$

or defining

$$g(t) = CS(t)B,$$
$$y(t) = CS(t)Bz_0 + (g * u)(t)$$

where $*$ indicates convolution.

The Laplace transform G of g yields the transfer function of the system: If $z(0) = 0$,

$$\hat{y}(s) = G(s)\hat{u}(s).$$

The transfer function of a system modeled by a system of ordinary differential equations is always rational with real coefficients; for example, $(2s+1)/(s^2+s+25)$. Transfer functions of systems modeled by PDEs are nonrational. There are some differences in the systems theory for infinite-dimensional systems [7]. For instance, it is possible for the transfer function to have different limits along the real and imaginary axes.

The definitions of stability for finite-dimensional systems generalize to infinite-dimensions.

Definition 67.3:

A system is externally stable or $\mathcal{L}_2$-stable if for every input $u \in \mathcal{L}_2(0,\infty;\mathcal{U})$, the output $y \in \mathcal{L}_2(0,\infty;\mathcal{Y})$. If a system is externally stable, the maximum ratio between the norm of the input and the norm of the output is called the L_2-gain.

Define

$$\mathbb{H}_\infty = \{G : \mathbb{C}_0^+ \to \mathbb{C} | G \quad \text{analytic and} \quad \sup_{\text{Re}s>0} |G(s)| < \infty\}$$

with norm

$$\|G\|_\infty = \sup_{\text{Re}s>0} |G(s)|.$$

Matrices with entries in $\mathbb{H}_\infty$ will be indicated by $M(\mathbb{H}_\infty)$. The H_∞-norm of matrix functions is

$$\|G\|_\infty = \sup_{\text{Re}s>0} \sigma_{\max}\big(G(s)\big).$$

The theorem below is stated for systems with finite-dimensional input and output spaces $\mathcal{U}$ and $\mathcal{Y}$ but it generalizes to infinite-dimensional $\mathcal{U}$ and $\mathcal{Y}$.

Theorem 67.1:

A linear system is externally stable if and only if its transfer function matrix $G \in M(\mathbb{H}_\infty)$. In this case, $\|G\|_\infty$ is the L_2-gain of the system and we say that G is a stable transfer function.

As for finite-dimensional systems, we need additional conditions to ensure that internal stability and external stability are equivalent.

Definition 67.4:

The semigroup $S(t)$ is exponentially stable if there is $M \geq 1, \alpha > 0$ such that $\|S(t)\| \leq Me^{-\alpha t}$ for all $t \geq 0$.

Definition 67.5:

The system (A, B, C) is internally stable if A generates an exponentially stable semigroup $S(t)$.

Definition 67.6:

The pair (A, B) is stabilizable if there exists $K \in \mathcal{L}(\mathcal{U}, \mathcal{Z})$ such that $A - BK$ generates an exponentially stable semigroup.

Definition 67.7:

The pair (C, A) is detectable if there exists $F \in \mathcal{L}(\mathcal{Y}, \mathcal{Z})$ such that $A - FC$ generates an exponentially stable semigroup.

Theorem 67.2: [18, Theorem 26, Corollary 27]

A stabilizable and detectable system is internally stable if and only if it is externally stable.

Now, let G be the transfer function of a given plant and let H be the transfer function of a controller, of compatible dimensions, arranged in the standard feedback configuration shown in Figure 67.6. This framework is general enough to include most common control problems. For instance, in tracking, r is the reference signal to be tracked by the plant output y_1. Since r can also be regarded as modeling sensor noise and d as modeling actuator noise, it is reasonable to regard the control system in Figure 67.6 as externally stable if the four maps from r, d to e_1, e_2 are in $M(\mathbb{H}_\infty)$. (Stability could also be defined in terms of the transfer matrix from (r, d) to (y_1, y_2): both notions of stability are equivalent.) Let (A, B, C, E) be a state-space realization for G and similarly let (A_c, B_c, C_c, E_c) be a state-space realization for H. If $I + EE_c$ is invertible, then the 2×2 transfer matrix $\Delta(G, H)$ which maps the pair (r, d) into the pair (e_1, e_2) is

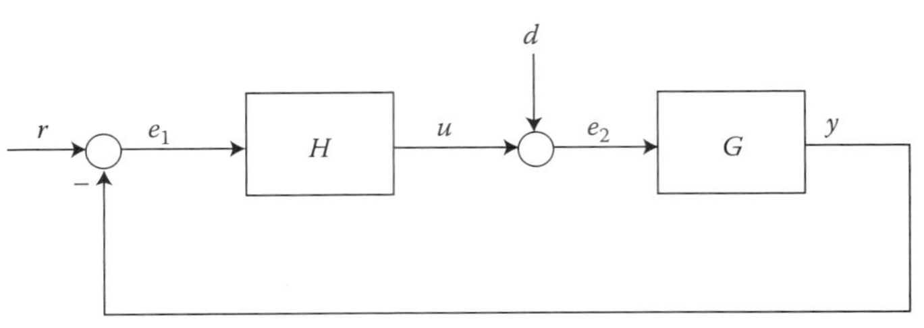

FIGURE 67.6 Standard feedback diagram.

given by

$$\Delta(G,H) = \begin{bmatrix} (I+GH)^{-1} & -G(I+HG)^{-1} \\ H(I+GH)^{-1} & (I+HG)^{-1} \end{bmatrix}.$$

Definition 67.8:

The feedback system (Figure 67.6), or alternatively the pair (G,H), is said to be externally stable if $I+EE_c$ is invertible, and each of the four elements in $\Delta(G,H)$ belongs to $M(\mathbb{H}_\infty)$.

Typically the plant feedthrough E is zero and so the invertibility of $I+EE_c$ is trivially satisfied. The above definition of external stability is sufficient to ensure that all maps from uncontrolled inputs to outputs are bounded. Furthermore, under the additional assumptions of stabilizability and detectability, external stability and internal stability are equivalent.

Theorem 67.3: [18, Theorem 35]

Assume that (A,B,C,E) is a jointly stabilizable/detectable control system and that a controller (A_c,B_c,C_c,E_c) is also jointly stabilizable/detectable.The closed-loop system is externally stable if and only if it is internally stable.

This equivalence between internal stability and external stability justifies the use of controller design techniques based on system input/output behavior for infinite-dimensional systems.

67.3 Issues in Controller Design

For most practical examples, a closed-form solution of the PDE or of the transfer function is not available and an approximation needs to be used. This approximation is generally calculated using one of the many standard methods, such as finite elements, developed for simulation of PDE models. The resulting system of ordinary differential equations is used in controller design. The advantage to this approach is that the wide body of synthesis methods available for finite-dimensional systems can be used.

The usual assumption made on an approximation scheme used for simulation are as follows. Suppose the approximation lies in some finite-dimensional subspace $\mathcal{Z}_n$ of the state space $\mathcal{Z}$, with an orthogonal projection $P_n : \mathcal{Z} \to \mathcal{Z}_n$ where for each $z \in \mathcal{Z}, \lim_{n\to\infty} \|P_n z - z\| = 0$. The space $\mathcal{Z}_n$ is equipped with the norm inherited from $\mathcal{Z}$. Define $B_n = P_n B, C_n = C|_{\mathcal{Z}_n}$ (the restriction of C_n to $\mathcal{Z}_n$) and define $A_n \in \mathcal{L}(\mathcal{Z}_n, \mathcal{Z}_n)$ using some method. This leads to a sequence of finite-dimensional approximations

$$\frac{dz}{dt} = A_n z(t) + B_n u(t), \quad z(0) = P_n z_0,$$
$$y(t) = C_n z(t).$$

Let S_n indicate the semigroup (a matrix exponential) generated by A_n. The following assumption is standard.

(A1) For each $z \in \mathcal{Z}$, and all intervals of time $[t_1, t_2]$,

$$\lim_{n\to\infty} \sup_{t\in[t_1,t_2]} \|S_n(t)P_n z - S(t)z\| = 0.$$

Assumption (A1) is required for convergence of initial conditions. Assumption (A1) is often satisfied by ensuring that the conditions of the Trotter–Kato Theorem hold; see, for instance, [34, Section 3.4].

Assumption (A1) implies that $P_n z \to z$ for all $z \in \mathcal{Z}$. The strong convergence $P_n z \to z$ for all $z \in \mathcal{Z}$ and the definitions of $B_n = P_n B$ and $C_n = C|_{\mathcal{Z}_n}$ imply that for all $u \in \mathcal{U}, z \in \mathcal{Z}, \|B_n u - Bu\| \to 0, \|C_n P_n z - Cz\| \to 0$.

The following result on open-loop convergence is straightforward.

Theorem 67.4:

Suppose that the approximating systems (A_n, B_n, C_n) satisfy assumption (A1). Then for each initial condition $z \in \mathcal{Z}$, and time $T > 0$ the uncontrolled approximating state $z(t)$ converges uniformly on bounded intervals of time $[0, T]$ to the exact state. Also, for each $u \in \mathcal{L}_2(0, T; \mathcal{U}), y_n \to y$ in the norm on $\mathcal{L}_2(0, T; \mathcal{Y})$.

Example 67.3: Indirect Controller Design for Simply Supported Beam. (Eg. 67.2 cont.)

Consider a simply supported Euler–Bernoulli beam (Figure 67.5). As shown in Example 67.2, a state-space formulation with state space $\mathcal{Z} = H_s(0,1) \times \mathcal{L}_2(0,1)$ is

$$\frac{d}{dt}z(t) = Az(t) + Bu(t),$$

where

$$A = \begin{bmatrix} 0 & I \\ -\frac{d^4}{dx^4} & 0 \end{bmatrix}, \quad B = \begin{bmatrix} 0 \\ b(\cdot) \end{bmatrix},$$

with domain

$$\mathcal{D}(A) = \{(\phi, \psi) \in H_s(0,1) \times H_s(0,1);\ \phi'' \in H_s(0,1)\}$$

and

$$b(x) = \begin{cases} \frac{1}{\delta}, & |x - 0.5| < \frac{\delta}{2} \\ 0, & |x - 0.5| \geq \frac{\delta}{2} \end{cases}.$$

Let $\phi_i(x)$ indicate the eigenfunctions of $\partial^4 w/\partial x^4$ with simply supported boundary conditions (Equation 67.4). Defining $\mathcal{X}_n$ to be the span of $\phi_i, i = 1 \ldots n$, we choose $\mathcal{Z}_n = \mathcal{X}_n \times \mathcal{X}_n$ and define P_n to be the projection onto $\mathcal{Z}_n$. Let $\langle \cdot, \cdot \rangle$ indicate the inner product on $\mathcal{Z}$ (and on $\mathcal{Z}_n$). Define the approximating system (A_n, B_n) by the Galerkin approximation

$$\langle \dot{z}(t), \phi_i \rangle = \langle Az(t), \phi_i \rangle + \langle b, \phi_i \rangle u(t), \quad i = 1 \ldots n.$$

This leads to the standard modal approximation

$$\dot{z}(t) = A_n z(t) + B_n u(t), \quad z_n(0) = P_n z(0). \tag{67.6}$$

This approximation scheme satisfies assumption (A1).

Consider LQ controller design [30]: find a control $u(t)$ so that the cost functional

$$J(u, z_0) = \int_0^\infty \langle z(t), z(t) \rangle + |u(t)|^2 dt$$

is minimized where $z(t)$ is determined by Equation 67.6. The resulting optimal controller is $u(t) = -K_n z(t)$, where $K_n = B_n^* \Pi_n z(t)$, and Π_n solves the algebraic Riccati equation

$$A_n^* \Pi_n + \Pi_n A_n - \Pi_n B_n B_n^* \Pi_n + I = 0$$

where M^* indicates the adjoint operator of M. For a matrix M, M^* is the complex conjugate transpose of M. Suppose we use the first three modes (or eigenfunctions) to design the controller. As expected, the controller stabilizes the model used in design. However, Figure 67.7b shows that if even one additional mode is added to the system model, the controller no longer stabilizes the system. This phenomenon is often called *spillover* [1]. Figure 67.8 shows the sequence of controllers obtained for increasing model order. The controller sequence is not converging to some controller appropriate for

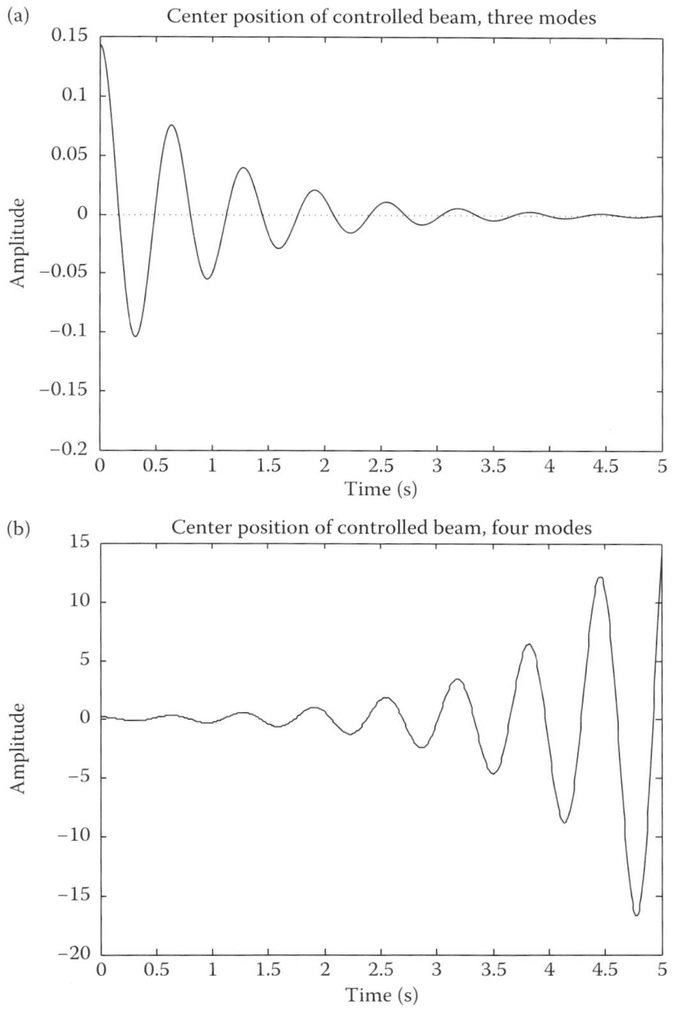

FIGURE 67.7 An LQ feedback controller K was designed for the beam in Example 67.3 using the first three modes (eigenfunctions). Simulation of the controlled system with (a) three modes and (b) four modes is shown. The initial condition in both cases is the first eigenfunction. Figure (b) illustrates that the addition of only one additional mode to the model leads to instability in the controlled system.

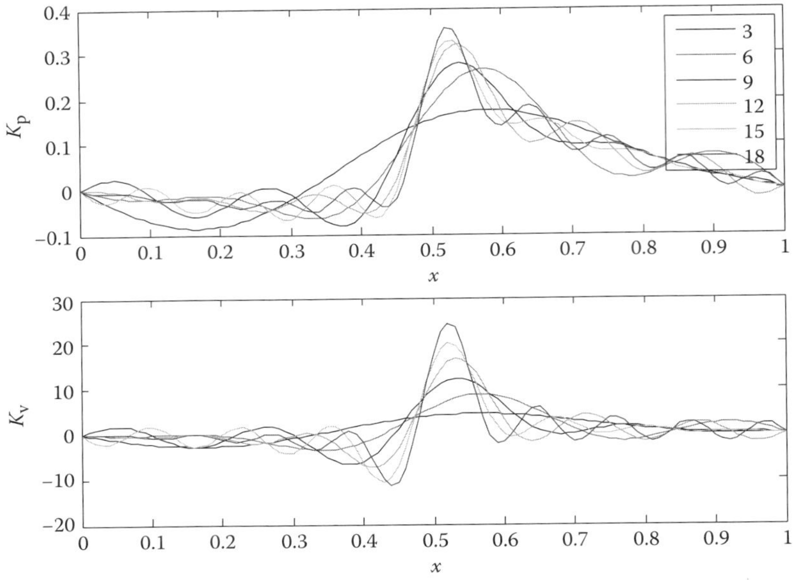

FIGURE 67.8 LQ optimal feedback for modal approximations of the simply supported beam in Example 67.3. Since the input space $U = \mathbb{R}$, the feedback operator K_n is a bounded linear functional and hence can be uniquely identified with a function, called the gain. The upper figure shows the feedback gain for the position of beam; the lower figure shows the velocity gains. Neither sequence is converging as the approximation order increases.

the original infinite-dimensional system. The increase in $\|K_n\|$ as approximation order increases suggests that the original system is not stabilizable. This is the case here. Although the approximations are stabilizable, the original model is not stabilizable [13].

As shown by the above example, and also by an example in [29], requirements additional to those sufficient for simulation are required when an approximation is used for controller design. The issues are illustrated in Figure 67.9. Possible problems that may occur are the following:

- The controlled system may not perform as predicted.
- The sequence of controllers for the approximating systems may not converge.
- The original control system may not be stabilizable, even if the approximations are stabilizable.
- The performance of the controlled infinite-dimensional system may not be close to that predicted by simulations with approximations (and may be unstable).

Although (A1) guarantees open-loop convergence, additional conditions are required in order to obtain closed loop convergence.

The *gap topology*, first introduced in [40], is useful in establishing conditions under which an approximation can be used in controller design. Consider a stable system G; that is, $G \in M(\mathbb{H}_\infty)$. A sequence G_n converges to G in the gap topology if and only if $\lim_{n\to\infty} \|G_n - G\|_\infty = 0$. The extension to unstable systems uses *coprime factorizations.* Let G be the transfer function of a system, with right coprime factorization $G = \tilde{N}\tilde{D}^{-1}$:

$$\tilde{X}\tilde{N} + \tilde{Y}\tilde{D} = I, \quad \tilde{X}, \tilde{N}, \tilde{Y}, \tilde{D} \in M(\mathbb{H}_\infty).$$

(If $G \in \mathbb{H}_\infty$, then we can choose $\tilde{N} = G, \tilde{D} = I, \tilde{X} = I, \tilde{Y} = 0$.) A sequence G_n converges to G in the gap topology if and only if for some right coprime factorization $(\tilde{N}_n, \tilde{D}_n)$ of G_n, $\tilde{N}_n$ converges to $\tilde{N}$ and $\tilde{D}_n$

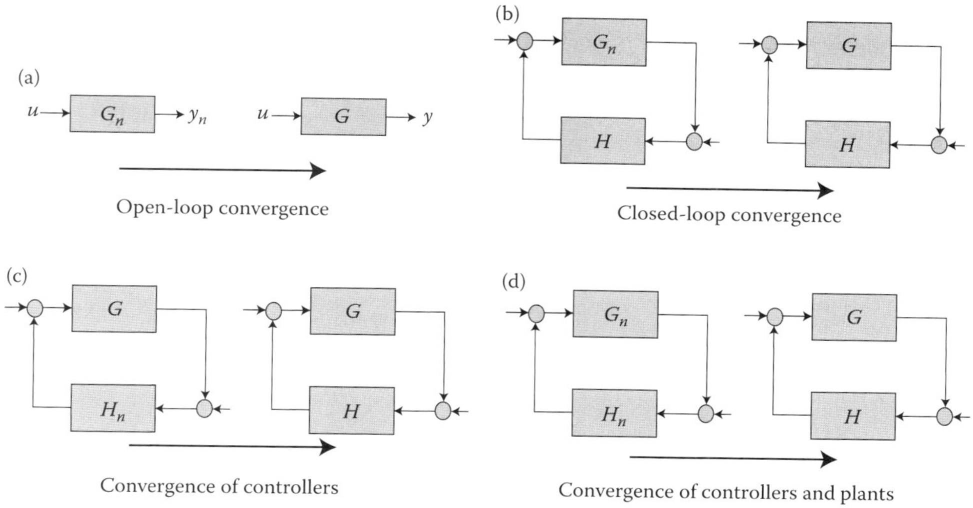

FIGURE 67.9 Typical approximation criteria ensure that the open loops converge. However, in controller design, the controller is generally implemented as a feedback controller and control of the resulting closed-loop system is needed. Furthermore, a sequence of controllers is produced by applying a controller synthesis technique to the approximations. This sequence of controllers should converge so that closed-loop performance with the original plant is similar to that predicted by simulations with the approximations.

converges to $\tilde{D}$ in the $\mathbb{H}_\infty$-norm. For details on the gap (or graph) topology, see [39,42]. The importance of the gap topology in controller design is stated in the following result.

Theorem 67.5: [39]

Let $G_n \in M(\mathbb{H}_\infty)$ be a sequence of system transfer functions.

1. *Suppose G_n converges to G in the gap topology. Then if H stabilizes G, there is an N such that H stabilizes G_n for all $n \geq N$ and the closed-loop transfer matrix $\Delta(G_n, H)$ converges to $\Delta(G, H)$ in the $\mathbb{H}_\infty$-norm.*
2. *Conversely, suppose that there exists an H that stabilizes G_n for all $n \geq N$ and so that $\Delta(G_n, H)$ converges to $\Delta(G, H)$ in the $\mathbb{H}_\infty$-norm. Then G_n converges to G in the gap topology.*

Thus, failure of a sequence of approximations to converge in the gap topology implies that for each possible controller H at least one of the following conditions holds:

- H does not stabilize G_n for all n sufficiently large.
- The closed-loop response $\Delta(G_n, H)$ does not converge to $\Delta(G_n, H)$.

On the other hand, if a sequence of approximations does converge in the gap topology, then the closed-loop performance $\Delta(G_n, H)$ converges and moreover, every H that stabilizes G also stabilizes G_n for sufficiently large approximation order.

The following condition, with assumption (A1) provides a sufficient condition for convergence of approximations in the gap topology. It was first formulated in [3] in the context of approximation of LQ regulators for parabolic PDEs.

Definition 67.9:

The control systems (A_n, B_n) *are uniformly stabilizable if there exists a sequence of feedback operators* $\{K_n\}$ *with* $\|K_n\| \leq M_1$ *for some constant* M_1 *such that* $A_n - B_nK_n$ *generate* $S_{Kn}(t)$ *where* $\|S_{Kn}(t)\| \leq M_2e^{-\alpha_2 t}, M_2 \geq 1, \alpha_2 > 0$.

Since $\mathcal{U}$ is assumed finite-dimensional, B is a compact operator. Thus, (A1) and uniform stabilizability imply stabilizability of (A, B) [15, Theorem 2.3] and in fact the existence of a uniformly stabilizing sequence K_n satisfying Definition 67.9 with $\lim_{n\to\infty} K_nP_nz = Kz$ for all $z \in \mathcal{Z}$. The beam in Example 67.3 is not stabilizable and therefore there is no sequence of uniformly stabilizable approximations.

Theorem 67.6: [29, Theorem. 4.2]

Consider a stabilizable and detectable control system (A, B, C, E), *and a sequence of approximations* (A_n, B_n, C_n, E) *that satisfy (A1). If the approximating systems are uniformly stabilizable then they converge to the exact system in the gap topology.*

Example 67.4: Diffusion (Eg. 67.1 cont.)

$$\frac{\partial z}{\partial t} = \frac{\partial^2 z}{\partial x^2} + b(x)u(t), \quad 0 < x < 1,$$
$$z(0,t) = 0, \qquad z(1,t) = 0,$$
$$y(t) = \int_0^1 c(x)z(x,t)\,dx$$

for some $b, c \in \mathcal{L}_2(0,1)$. This can be written as

$$\dot{z}(t) = Az(t) + Bu(t),$$
$$y(t) = Cz(t),$$

where A and B are as defined in Example 67.1 and

$$Cz = \int_0^1 c(x)z(x)\,dx.$$

The operator A generates an exponentially stable semigroup $S(t)$ with $\|S(t)\| \leq e^{-\pi^2 t}$ on the state-space $\mathcal{L}_2(0,L)$ [8] and so the system is trivially stabilizable and detectable. The eigenfunctions $\phi_i(x) = \sqrt{2}\sin(\pi i x), i = 1, 2, \ldots,$ of A form an orthonormal basis for $\mathcal{L}_2(0,L)$. Defining $\mathcal{Z}_n = \text{span}_{i=1\ldots n}\phi_i(x)$, and letting P_n be the projection onto $\mathcal{Z}_n$, define A_n by the Galerkin approximation

$$\langle A_n\phi_j, \phi_i\rangle = \langle A\phi_j, \phi_i\rangle \quad i = 1\ldots n, \ j = 1\ldots n.$$

It is straightforward to show that this set of approximations satisfies assumption (A1) and that the semigroup generated by A_n has bound $S_n(t) \leq e^{-\pi^2 t}$. Hence the approximations are uniformly stabilizable (using $K_n = 0$). Thus, the sequence of approximating systems converges in the gap topology and will yield reliable results when used in controller design.

It is easy to show that if the original problem is exponentially stable, and the eigenfunctions of A form an orthonormal basis for $\mathcal{Z}$, then any approximation formed using the eigenfunctions as a basis, as in the previous example, will both satisfy assumption (A1) and be trivially uniformly stabilizable and detectable. However, in practice, other approximation methods, such as finite elements, are often used.

Many generators A for PDE models can be described as follows. Let V be a Hilbert space that is dense in $\mathcal{Z}$. The notation $\langle \cdot, \cdot \rangle$ indicates the inner product on $\mathcal{Z}$, and $\langle \cdot, \cdot \rangle_V$ indicates the inner product on V. The norm on $\mathcal{Z}$ is indicated by $\| \cdot \|$ while the norm on V will be indicated by $\| \cdot \|_V$. Let the bilinear form $a : V \times V \mapsto C$ be such that for some $c_1 > 0$

$$|a(\phi, \psi)| \leq c_1 \|\phi\|_V \|\psi\|_V \tag{67.7}$$

for all $\phi, \psi \in V$. An operator A can be defined through this form by

$$\langle A\phi, \psi \rangle = -a(\phi, \psi), \quad \forall \psi \in V$$

with $\mathcal{D}(A) = \{\phi \in V | \, a(\phi, \cdot) \in \mathcal{Z}\}$. Assume that in addition to Equation 67.7, $a(\cdot, \cdot)$ satisfies Garding's inequality: there exists $k \geq 0$, such that for all $\phi \in V$

$$\text{Re } \mathrm{a}(\phi, \phi) + k\langle \phi, \phi \rangle \geq c\|\phi\|_V^2. \tag{67.8}$$

For example, in the diffusion example above, $V = \mathcal{H}_0^1(0, 1)$ and

$$a(\phi, \psi) = \int_0^1 \phi'(x)\psi'(x)\, dx.$$

The inequalities (Equations 67.7 and 67.8) guarantee that A generates a C_0- semigroup with bound $\|T(t)\| \leq e^{kt}$ [37, Section 4.6]. This framework includes many problems of practical interest such as diffusion and beam vibrations with Kelvin–Voigt damping [29].

Defining a sequence of finite-dimensional subspaces $\mathcal{Z}_n \subset V$, the approximating generator A_n is defined by

$$\langle A_n z_n, v_n \rangle = -a(z_n, v_n), \quad \forall z_n, v_n \in \mathcal{Z}_n. \tag{67.9}$$

This type of approximation is generally referred to as a *Galerkin* approximation and includes finite-elements as well as many other popular approximation methods. For such problems, the following result, which generalizes [3, Lemma 3.3] is useful. It applies to a number of common applications, such as the usual linear spline finite-elements for approximating the heat equation and other diffusion problems. Finite-element cubic spline approximations to damped beam vibrations are also included.

Theorem 67.7: [29, Theorem 5.2,5.3]

Let $\mathcal{H}_n \subset V$ be a sequence of finite-dimensional subspaces such that for all $z \in V$ there exists a sequence $z_n \in \mathcal{Z}_n$ with

$$\lim_{n \to \infty} \|z_n - z\|_V = 0. \tag{67.10}$$

If the operator A satisfies the inequalities (Equations 67.7 and 67.8) then

1. *Assumption (A1) is satisfied with $\|S_n(t)\| \leq e^{kt}$;*
2. *If $K \in \mathcal{L}(\mathcal{Z}, U)$ is such that $A - BK$ generates an exponentially stable semigroup then the semigroups $S_{nK}(t)$ generated by $A_n - B_n K P_n$ are uniformly exponentially stable. In other words, there exists $M \geq 1, \alpha > 0$ such that*

$$\|S_{nK}(t)\| \leq M e^{-\alpha t} \quad \forall n > N. \tag{67.11}$$

and the approximations (A_n, B_n) are thus uniformly stabilizable.

Example 67.5: Damped String

The wave equation

$$\frac{\partial^2 w(x,t)}{\partial t^2} = c^2 \frac{\partial^2 w(x,t)}{\partial x^2}, \quad 0 < x < 1, \ t \geq 0,$$

describes the deflection $z(x,t)$ of a vibrating string of unit length as well as many other situations such as acoustic plane waves, lateral vibrations in beams, and electrical transmission lines (see, e.g., [38, Chapter 1]). Suppose the ends are fixed:

$$w(0,t) = 0, \quad w(1,t) = 0.$$

Including control and observation, as well as the effect of some light damping [36] leads to the model

$$\frac{\partial^2 w(x,t)}{\partial t^2} + \epsilon \langle \frac{\partial w(\cdot,t)}{\partial t}, b(\cdot) \rangle b(x) = \frac{\partial^2 w}{\partial x^2} + b(x)u(t), \quad 0 < x < 1,$$

$$y(t) = \int_0^1 b(x) \frac{\partial w(x,t)}{\partial t}\, dx$$

where $\epsilon > 0$, and $b \in \mathcal{L}_2(0,1)$ describes both the control and observation action, which is a type of distributed colocation. The state space is $\mathcal{Z} = \mathcal{H}_0^1(0,1) \times \mathcal{L}_2(0,1)$ and the state-space equations are

$$\frac{d}{dt}\begin{bmatrix} z \\ \frac{dz}{dt} \end{bmatrix} = A \begin{bmatrix} z \\ \frac{dz}{dt} \end{bmatrix} + \begin{bmatrix} 0 \\ b(x) \end{bmatrix} u(t),$$

$$y(t) = C \begin{bmatrix} z \\ \frac{dz}{dt} \end{bmatrix}$$

where

$$A\begin{bmatrix} w \\ v \end{bmatrix} = \begin{bmatrix} 0 & I \\ \frac{\partial^2 w}{\partial x^2} & -\epsilon \langle v, b \rangle b(x) \end{bmatrix},$$

$$\mathcal{D}(A) = \left\{ (w,v) \in H_0^1(0,1) \times H_0^1(0,1) \right\},$$

$$C\begin{bmatrix} w \\ v \end{bmatrix} = \begin{bmatrix} 0 & \langle b(x), v \rangle \end{bmatrix}.$$

Suppose that

$$b(x) = \begin{cases} 1, & 0 < x < \frac{1}{2}, \\ 0, & \frac{1}{2} < x < 1 \end{cases}$$

The eigenvalues λ_n of A have all negative real parts, but asymptote to the imaginary axis so that $\sup_n \mathrm{Re}(\lambda_n) = 0$. The results in [18] (see also [8, Section 5.2]) imply that the system is not exponentially stabilizable. Thus, no sequence of approximations is uniformly stabilizable.

However, it is possible to construct a sequence of finite-dimensional approximations that converge in the gap topology. The transfer function is

$$G(s) = \frac{\frac{s}{2}\sinh(s) + 2\cosh(\frac{s}{2}) - 3\cosh^2(\frac{s}{2}) + 1}{s(s+\frac{\varepsilon}{2})\sinh(s) + \varepsilon(2\cosh(\frac{s}{2}) - 3\cosh^2(\frac{s}{2}) + 1)}.$$

The function $G \in \mathbb{H}_\infty$; so the system is L_2-stable, and furthermore,

$$\lim_{|s| \to \infty, \mathrm{Re}s > 0} G(s) = 0.$$

Thus, we can find a sequence of rational functions G_n so that

$$\lim_{n\to\infty} \|G_n(s) - G(s)\|_\infty = 0.$$

The state-space realizations corresponding to $\{G_n\}$ are finite-dimensional and thus there are finite-dimensional approximations that converge in the gap topology.

The above example illustrates that uniform stabilizability is a sufficient, not necessary, condition for convergence of approximations in the gap topology.

Once an approximation scheme that converges in the gap topology is found (typically, by finding one that satisfies (A1) and is uniformly stabilizable), the next step is controller design. The sequence of controllers designed using the approximations should converge to a controller for the original infinite-dimensional system that yields the required performance, as well as stability (see Figure 67.9).

A common procedure for controller design is to first design a state feedback controller: $u(t) = -Kz(t)$. Then, since the full state is not available, an estimator is designed to obtain an estimate of the state using knowledge of the output y and the input u. The controller is formed by using the state estimate as input to a state feedback controller. Controller design of this type for infinite-dimensional systems is described in [8, Section 5.3].

However, typically both the state feedback and the estimator are designed using a finite-dimensional approximation (A_n, B_n, C_n, E). Suppose $F_n \in \mathcal{L}(\mathcal{Y}, \mathcal{Z})$ is found so that all the eigenvalues of $A_n - F_nC_n$ have negative real parts, and similarly $K_n \in \mathcal{L}(\mathcal{Z}, \mathcal{U})$ is such that all the eigenvalues of $A_n - B_nK_n$ have negative real parts. The resulting finite-dimensional controller is

$$\begin{aligned}\dot{z}_c(t) &= A_n z_c(t) + B_n u(t) + F_n\big(y(t) - C_n z_c(t)\big)\\ y_c(t) &= -K_n z_c(t)\end{aligned}$$

This framework does not include the effect of disturbances to the controlled system, shown as r and v in Figure 67.6. To include these effects, and put the system into the standard framework, define an augmented system output

$$\tilde{y}(t) = \begin{bmatrix} y(t)\\ u(t)\end{bmatrix} = \begin{bmatrix} C\\ 0\end{bmatrix} z(t) + \begin{bmatrix} E\\ I\end{bmatrix} u(t)$$

Letting $e(t) = r - \tilde{y}$ indicate the controller input, the controller equations are then

$$\begin{aligned}\dot{z}_c(t) &= (A_n - F_nC_n)z_c(t) - [F_n \quad B_n]\, e(t),\\ y_c(t) &= -K_n z_c(t)\end{aligned} \tag{67.12}$$

and the plant input is $y_c + v$.

It is well known that such a controller stabilizes the approximation (A_n, B_n, C_n, E) (see, e.g., [30]). However, it must also stabilize the original system (A, B, C, E). For this to happen, the controller sequence must converge in some sense. For controller convergence, an assumption in addition to (A1) and uniform stabilizability is required.

Definition 67.10:

The observation systems (A_n, C_n) are uniformly detectable if there exists a sequence of operators $\{F_n\}$ with $\|F_n\| \le M_3$ for some constant M_3 such that $A_n - F_nC_n$ generates $S_{Fn}(t)$, where $\|S_{Fn}(t)\| \le M_4 e^{-\alpha_4 t}, M_4 \ge 1$, and $\alpha_4 > 0$.

The approximating systems (A, C) are uniformly detectable if and only if (A_n^*, C_n^*) is uniformly stabilizable and thus uniform detectability can be established using conditions for uniform stabilizability. In particular, if A is defined through a bilinear form satisfying Equations 67.7 and 67.8 and condition (Equation 67.10) in Theorem 67.7 is satisfied, then detectability of (A, C) implies uniform detectability of (A_n, C_n).

Theorem 67.8:

Assume that (A1) holds, that the operators K_n, F_n used to define the sequence of controllers (Equation 67.12) satisfy Definitions 67.9 and 67.10 respectively and there exists $K \in \mathcal{L}(\mathcal{U}, \mathcal{Z}), F \in \mathcal{L}(\mathcal{Y}, \mathcal{Z})$ such that $\lim_{n\to\infty} K_n P_n z = Kz$ for all $z \in \mathcal{Z}, \lim_{n\to\infty} F_n y = Fy$ for all $y \in \mathcal{Y}$. Indicating the controller transfer function by H_n, the controllers converge in the gap topology and so for sufficiently large n, the output feedback controllers (Equation 67.12) stabilize the infinite-dimensional system (Equations 67.2, 67.5). Furthermore, the closed loop systems $\Delta(G_n, H_n)$ converge uniformly to the closed system $\Delta(G, H)$ where H indicates the transfer function of the infinite-dimensional controller

$$\begin{aligned} \dot{z}_c(t) &= (A - FC)z_c(t) - [F \quad B]\, e(t), \\ y_c(t) &= -Kz_c(t). \end{aligned} \tag{67.13}$$

Proof. The assumptions on the controller imply that, as for the plant (Theorem 67.6) the controllers converge in the gap topology. See [28] for details. This implies that closed loops $\Delta(G, H_n)$ converge to $\Delta(G, H)$ and that the controllers stabilize the original system for large enough n. Since the assumptions also imply that the approximating systems G_n converge in the gap topology to G, the closed-loop systems $\Delta(G_n, H_n)$ converge uniformly to the closed system $\Delta(G, H)$. ■

Controller design is explored further in the next two sections for the synthesis methods most commonly used for multi-input–multi-output systems: LQ control and $\mathbb{H}_\infty$-control.

67.4 LQ Regulators

Consider the LQ controller design objective of finding a control $u(t)$ so that the cost functional

$$J(u, z_0) = \int_0^\infty \langle C_1 z(t), C_1 z(t)\rangle + \langle u(t), Ru(t)\rangle \, dt \tag{67.14}$$

is minimized where $R \in \mathcal{L}(\mathcal{U}, \mathcal{U})$ is a symmetric positive-definite operator weighting the control, $C_1 \in \mathcal{L}(\mathcal{Z}, \mathcal{Y})$ (with Hilbert space $\mathcal{Y}$) weights the state, and $z(t)$ is determined by Equation 67.2. The theoretical solution to this problem is similar in structure to that for finite-dimensional systems [8,12,22,23].

Definition 67.11:

The system (Equation 67.2) with cost (Equation 67.14) is optimizable if for every $z_0 \in \mathcal{Z}$ there exists $u \in L_2(0, \infty; \mathcal{U})$ such that the cost is finite.

Theorem 67.9: [8, Theorem 6.2.4, 6.2.7]

If Equation 67.2 with cost (Equation 67.14) is optimizable and (A, C_1) is detectable, then the cost function (Equation 67.14) has a minimum for every $z_0 \in \mathcal{Z}$. Furthermore, there exists a self-adjoint nonnegative *operator $\Pi \in \mathcal{L}(H, H)$ such that*

$$\min_{u \in L_2(0,\infty;\mathcal{U})} J(u, z_0) = \langle z_0, \Pi z_0 \rangle.$$

The operator Π is the unique nonnegative solution to the operator equation

$$\langle Az_1, \Pi z_2 \rangle + \langle \Pi z_1, Az_2 \rangle + \langle C_1 z_1, C_1 z_2 \rangle - \langle B^* \Pi z_1, R^{-1} B^* \Pi z_1 \rangle = 0 \quad z_1, z_2 \in \mathcal{D}(A). \tag{67.15}$$

Defining $K = R^{-1} B^* \Pi$, the corresponding optimal control is $u = -Kz(t)$ and $A - BK$ generates an exponentially stable semigroup.

It is straightforward to show that the assumption of optimizability in Theorem 67.9 is equivalent to stabilizability.

The Riccati operator equation 67.15 is equivalent to

$$\left(A^* \Pi + \Pi A - \Pi B R^{-1} B^* \Pi + C_1^* C_1\right) z = 0, \quad \forall z \in \mathcal{D}(A).$$

In practice, the operator Equation 67.15 cannot be solved and the control is calculated using an approximation. The cost functional becomes

$$J(u, z_0) = \int_0^\infty \langle C_{1n} z(t), C_{1n} z(t) \rangle + \langle u(t), Ru(t) \rangle \, dt \tag{67.16}$$

where $z(t)$ is the state of the approximating system

$$\dot{z}(t) = A_n z(t) + B_n u(t), \quad z(0) = P_n z_0,$$

on $\mathcal{Z}_n$ and $C_{1n} = C_1|_{\mathcal{Z}_n}$. If (A_n, B_n) is stabilizable and (A_n, C_{1n}) is detectable, then the cost functional has the minimum cost $\langle P_n z_0, \Pi_n P_n z_0 \rangle$ where Π_n is the unique nonnegative solution to the algebraic Riccati equation

$$A_n^* \Pi_n + \Pi_n A_n - \Pi_n B_n R^{-1} B_n^* \Pi_n + C_{1n}^* C_{1n} = 0 \tag{67.17}$$

on the finite-dimensional space $\mathcal{Z}_n$. The feedback control $K_n = R^{-1} B_n^* \Pi_n$ is used to control the original system (Equation 67.2).

The sequence of controllers K_n, along with the associated performance must converge in some sense in order for this approach to be valid. Assumption (A1), along with uniform stabilizability, guarantees convergence, of the approximating systems. However, in order to obtain controller convergence, a set of assumptions involving the dual system (A^*, B^*, C_1^*) is required.

(A1*) 1. For each $z \in \mathcal{Z}$, and all intervals of time $[t_1, t_2]$

$$\sup_{t \in [t_1, t_2]} \|S_n^*(t) P_n z - S^*(t) z\| \to 0;$$

2. For all $u \in U, y \in Y$, $\|C_{1n}^* y - C_1^* y\| \to 0$ and $\|B_n^* P_n z - B^* z\| \to 0$.

Theorem 67.10: [3, Theorem 6.9], [15, Theorem 2.1, Corollary 2.2]

If assumptions (A1), (A1) are satisfied, (A_n, B_n) is uniformly stabilizable and (A_n, C_{1n}) is uniformly detectable, then for each n, the finite-dimensional ARE (Equation 67.17) has a unique nonnegative solution Π_n with $\sup \|\Pi_n\| < \infty$. There exists constants $M_1 \geq 1, \alpha_1 > 0$, independent of n, such that the semigroup*

$S_{nK}(t)$ generated by $A_n - B_nK_n$ satisfy

$$\|S_{nK}(t)\| \leq M_1 e^{-\alpha_1 t}.$$

For sufficiently large n, the semigroups $S_{Kn}(t)$ generated by $A - BK_n$ are uniformly exponentially stable; that is there exists $M_2 \geq 1, \alpha_2 > 0$, independent of n, such that

$$\|S_{Kn}(t)\| \leq M_2 e^{-\alpha_2 t}.$$

Furthermore, letting Π indicate the solution to the infinite-dimensional Riccati equation 67.15, for all $z \in \mathcal{Z}$,

$$\lim_{n\to\infty} \|\Pi_n P_n z - \Pi z\| = 0$$

and

$$\lim_{n\to\infty} \|K_n P_n z - Kz\| = 0,$$

and the cost with feedback $K_n z(t)$ converges to the optimal cost:

$$J(-K_n z(t), z_0) \to \langle \Pi z_0, z_0 \rangle.$$

The assumption (A1*) implies open-loop convergence of the dual systems (A_n^*, C_{1n}^*, B_n^*). It is required since the optimal control Kz relates to an optimization problem involving the dual system. Note that (A, C_1) is uniformly detectable if and only if (A^*, C_1^*) is uniformly stabilizable, and so (A1*) along with uniform detectability can be regarded as dual assumptions to (A1) and uniform stabilizability. Since the operators B and C_1 are bounded, (A1*2) holds if both the input and output spaces are finite-dimensional. However, the satisfaction of (A1*1), strong convergence of the adjoint semigroups, is not automatic. A counterexample may be found in [5] where the assumptions except (A1*1) are satisfied and the conclusions of the above theorem do not hold. The conclusions of the above theorem, that is, uniform boundedness of Π_n and the uniform exponential stability of $S_{nK}(t)$, imply uniform stabilizability of (A_n, B_n). Although Example 67.5 illustrated that uniform stabilizability is not necessary for convergence of the approximating systems, it is necessary to obtain an LQ controller sequence that provides uniform exponential stability. The above result has been extended to unbounded control operators B for parabolic PDEs, such as diffusion problems [2,22].

Example 67.6: Damped Beam

As in Examples 67.2 and 67.3, consider a simply supported Euler–Bernoulli beam but now include viscous damping with parameter $c_d = 0.1$. We obtain the PDE

$$\frac{\partial^2 w}{\partial t^2} + c_d \frac{\partial w}{\partial t} + \frac{\partial^4 w}{\partial x^4} = b_r(x)u(t), \quad t \geq 0, 0 < x < 1,$$

with the same boundary conditions as before. However, we now consider an arbitrary location r for the control operator b so that

$$b_r(x) = \begin{cases} \frac{1}{\delta}, & |x - r| < \frac{\delta}{2} \\ 0, & |x - r| \geq \frac{\delta}{2} \end{cases}.$$

Recall that the state space is $\mathcal{Z} = H_s(0,1)(0,1) \times \mathcal{L}_2(0,1)$ with state $z(t) = (w(\cdot,t), \frac{\partial}{\partial t}w(\cdot,t))$. An obvious choice of weight for the state is $C_1 = I$. Since there is only one control, choose control

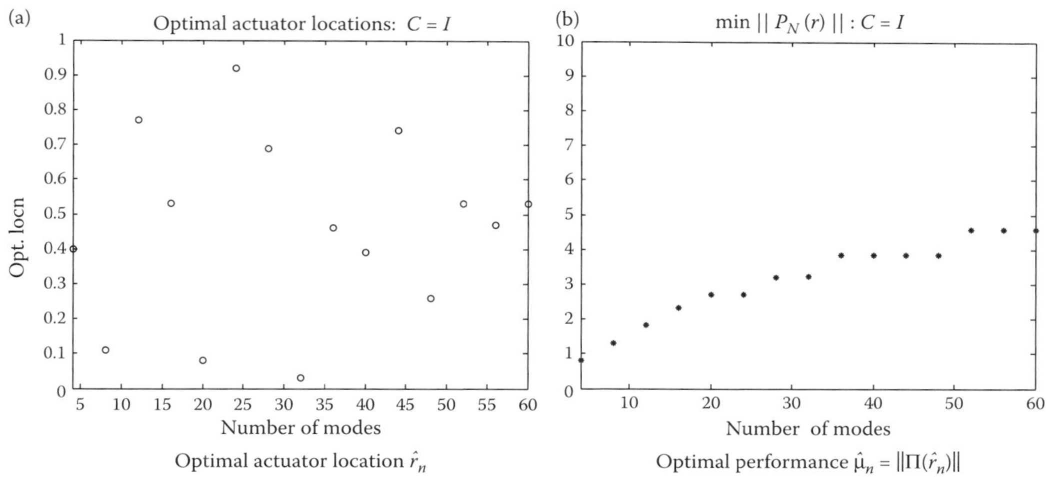

FIGURE 67.10 Optimal actuator location and performance for approximations of the viscously damped beam with weights $C_1 = I, R = 1$. No convergence of the optimal location or performance is seen as the approximation order is increased.

weight $R = 1$. We wish to choose the actuator location in order to minimize the response to the worst choice of initial condition. In other words, choose r in order to minimize

$$\max_{z_0 \in \mathcal{Z} \|z_0\|=1} \min_{u \in L_2(0,\infty;U)} J^r(u, z_0) = \|\Pi(r)\|.$$

The performance for a particular r is $\mu(r) = \|\Pi(r)\|$ and the optimal performance

$$\hat{\mu} = \inf_{r \in [0,1]} \|\Pi(r)\|.$$

This optimal actuator location problem is well-posed and a optimal location $\hat{r}$ exists [31, Theorem 2.6].

Let $\phi_i(x)$ indicate the eigenfunctions of $\partial^4 w/\partial x^4$ with simply supported boundary conditions. Defining X_n to be the span of $\phi_i, i = 1 \ldots n$, we choose $\mathcal{Z}_n = X_n \times X_n$. This approximation scheme satisfies all the assumptions of Theorem 67.10 and so the sequence of solutions Π_n to the corresponding finite-dimensional AREs converge strongly to the exact solution Π.

However, as shown in Figure 67.10, this does not imply convergence of the optimal actuator locations, or of the corresponding actuator locations.

The problem is that strong convergence of the Riccati operators is not sufficient to ensure that as the approximation order increases, the optimal cost $\hat{\mu}_n$ and a corresponding sequence of optimal actuator locations $\hat{r}_n$ converge. Since the cost is the norm of the Riccati operator, uniform convergence of the operators is required. That is,

$$\lim_{n \to \infty} \|\Pi_n P_n - \Pi\| = 0,$$

is needed in order to use approximations in determining optimal actuator location. The first point to consider is that since Π_n has finite rank, Π must be a compact operator in order for uniform convergence to occur, regardless of the choice of approximation method. Since the solution to an ARE is not always compact, it is not possible to compute the optimal actuator location for every problem.

Example 67.7: [8]

Consider any A, B, C_1 such that $A^* = -A$ and $C_1 = B^*$. Then $\Pi = I$ is a solution to the ARE

$$A^*\Pi + \Pi A - \Pi BB^*\Pi + C_1^*C_1 = 0.$$

The identity I is not compact on any infinite-dimensional Hilbert space.

Example 67.8:

This example is a generalization of [6, Example 1]. On the Hilbert space $\mathcal{Z} = \mathbb{R} \times H$ where H is any infinite-dimensional Hilbert space, define

$$A = \begin{bmatrix} -1 & 0 \\ 0 & -I \end{bmatrix}, \quad B = \begin{bmatrix} 1 \\ 0 \end{bmatrix}, \quad C_1 = \begin{bmatrix} \sqrt{3} & 0 \\ 0 & \sqrt{2}M \end{bmatrix}$$

where M is a bounded operator on H. The solution to the ARE

$$A^*\Pi + \Pi A - \Pi BB^*\Pi + C_1^*C_1 = 0$$

is

$$\Pi = \begin{bmatrix} 1 & 0 \\ 0 & M^2 \end{bmatrix}.$$

This operator is not compact if M is not a compact operator; for instance if $M = I$. This example is particularly interesting because A is a bounded operator and also generates an exponentially stable semigroup.

Theorem 67.11: [31, Theorem 2.9,3.3]

If B and C_1 are both compact operators, then the Riccati operator Π is compact. Furthermore, if a sequence of approximations satisfy (A1), (A1) and are uniformly stabilizable and detectable, then the minimal nonnegative solution Π_n to (Equation 67.17) converges uniformly to the nonnegative solution Π to (Equation 67.15): $\lim_{n\to\infty} \|\Pi_n - \Pi\| = 0$.*

Thus, if B and C_1 are compact operators, guaranteeing compactness of the Riccati operator, then any approximation method satisfying the assumptions of Theorem 67.10 will lead to a convergent sequence of optimal actuator locations. A finite-dimensional input space guarantees that B is a compact operator; and similarly a finite-dimensional output space will guarantee that C_1 is compact, although these assumptions are not necessary.

For an important class of problems the Riccati operator is compact, even if the observation operator C_1 is not compact.

Definition 67.12:

A semigroup $S(t)$ is analytic if $t \to S(t)$ is analytic in some sector $|\arg t| < \theta$.

Analytic semigroups have a number of interesting properties [34,37]. Recall that the solution $S(t)z \in \mathcal{D}(A), t \geq 0$, if $z \in \mathcal{D}(A)$. If S is an analytic semigroup, $S(t)z \in \mathcal{D}(A)$ for all $z \in \mathcal{Z}$. Also, the eigenvalues of the generator A of an analytic semigroup lie in a sector $|\arg\lambda| < \pi - \epsilon$, where $\epsilon > 0$. The heat equation and other parabolic PDEs lead to an analytic semigroup. Weakly damped wave and beam equations are not associated with analytic semigroups.

If A generates an analytic semigroup, uniform convergence can be obtained without compactness of the state weight C_1. The result [22, Theorem. 4.1], applies to operators B and C_1 that may be unbounded. It is stated below for bounded B and C_1.

Theorem 67.12:

Let A generate an analytic semigroup $S(t)$ with $\|S(t)\| \le Me^{\omega_0 t}$ and define $\hat{A} = (\omega I - A)$ for $\omega > \omega_0$. Assume that the system (A, B, C_1) has the following properties:

1. *(A, B) is stabilizable and (A, C_1) is detectable.*
2. *Either $C_1^* C_1 \ge rI, r > 0$, or for some $F \in \mathcal{L}(\mathcal{Y}, \mathcal{Z})$ such that $A - FC_1$ generates an exponentially stable semigroup, $\hat{A}^{-1}FC_1$ is compact.*
3. *Either $B^*\hat{A}^{-1}$ is compact or there exists a compact operator $K \in \mathcal{L}(\mathcal{Z},\mathcal{U})$ such that $A - BK$ generates an exponentially stable semigroup.*

Assume the following list of properties for the approximation scheme, where γ is any number $0 \le \gamma < 1$:

1. *For all $z \in \mathcal{Z}$, $\|P_n z - z\| \to 0$.*
2. *The approximations are uniformly analytic. That is, for some $\epsilon > 0$*

$$\|A_n^\theta e^{A_n t}\| \le \frac{M_\theta e^{(\omega_0+\epsilon)t}}{t^\theta}, \quad t > 0, \ 0 \le \theta \le 1.$$

3. *For some s and γ independent of n, $0 \le \gamma < 1$,*
   a. $\|\hat{A}^{-1} - \hat{A}_n^{-1} P_n\| \le \frac{M}{n^s}$.
   b. $\|B^* z - B_n^* P_n z\| \le M n^{s(\gamma-1)} \|z\|_{[\mathcal{D}(A^*)]}$, $z \in \mathcal{D}(A^*)$.

Then $\lim_{n\to\infty} \|\Pi_n P_n - \Pi\| = 0$.

The above conditions on the approximation scheme imply assumptions (A1) and (A1*) as well as uniform stabilizability and detectability.

Provided that Π_n converges to Π in operator norm at each actuator location, the sequence of optimal actuator locations for the approximations converges to the correct optimal location.

Theorem 67.13: [31, Theorem 3.5]

Let Ω be a closed and bounded set in $\mathbb{R}^N$. Assume that $B(r), r \in \Omega$, is compact and such that for any $r_0 \in \Omega$,

$$\lim_{r\to r_0} \|B(r) - B(r_0)\| = 0.$$

Assume also that $(A_n, B_n(r), C_{1n})$ is a family of uniformly stabilizable and detectable approximations satisfying (A1) and (A1) such that for each r,*

$$\lim_{n\to\infty} \|\Pi_n(r) - \Pi(r)\| = 0.$$

Let $\hat{r}$ be the optimal actuator location for $(A, B(r), C_1)$ with optimal cost $\hat{\mu}$; that is

$$\hat{\mu} = \inf_{r\in\Omega} \|\Pi(r)\| = \|\Pi(\hat{r})\|$$

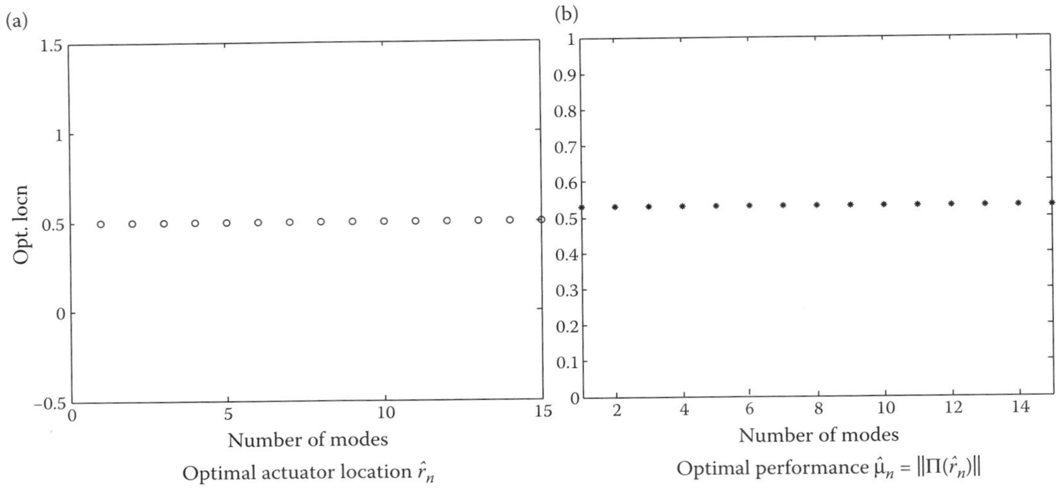

FIGURE 67.11 Optimal actuator location and performance for different approximations of the viscously damped beam with weights $C_1 = [I\ 0], R = 1$. Although the output space is infinite-dimensional, C_1 is a compact operator. This implies uniform convergence of the Riccati operators and thus convergence of both the optimal actuator locations $\hat{r}_n$ and optimal costs $\hat{\mu}_n$.

Defining similarly $\hat{r}_n, \hat{\mu}_n$, *it follows that*

$$\hat{\mu}_n \to \hat{\mu},$$
$$\hat{r}_n \to \hat{r}.$$

Example 67.9: Viscously Damped Beam, cont.

Consider the same viscously damped beam and control problem as in Example 67.6, except that now instead of trying to minimize the norm of the entire state, $C_1 = I$, we consider only the position. Choose the weight $C_1 = [I\ \ 0]$, where I here indicates the mapping from $\mathcal{H}_0^2(0,1)$ into $\mathcal{L}_2(0,1)$. Although the semigroup is not analytic, both B and C_1 are compact operators on $\mathcal{Z}$. Using the same modal approximations as before, we obtain convergence of the approximating optimal performance and the actuator locations. This is illustrated in Figure 67.11.

Example 67.10: Diffusion (Eg. 67.1 cont.)

$$\frac{\partial z}{\partial t} = \frac{\partial^2 z}{\partial x^2} + b_r(x)u(t) \quad 0 < x < 1,$$

$$b_r(x) = \begin{cases} \dfrac{1}{\delta}, & |x - r| < \dfrac{\delta}{2} \\ 0, & |x - r| \geq \dfrac{\delta}{2} \end{cases}$$

$$\frac{\partial z}{\partial x}(0,t) = 0, \quad \frac{\partial z}{\partial x}(1,t) = 0.$$

We wish to determine the best location r of the actuator to minimize

$$J^r(u, z_0) = \int_0^\infty 1000 \int_0^1 |z(x,t)|^2\, dx + |u(t)|^2\, dt$$

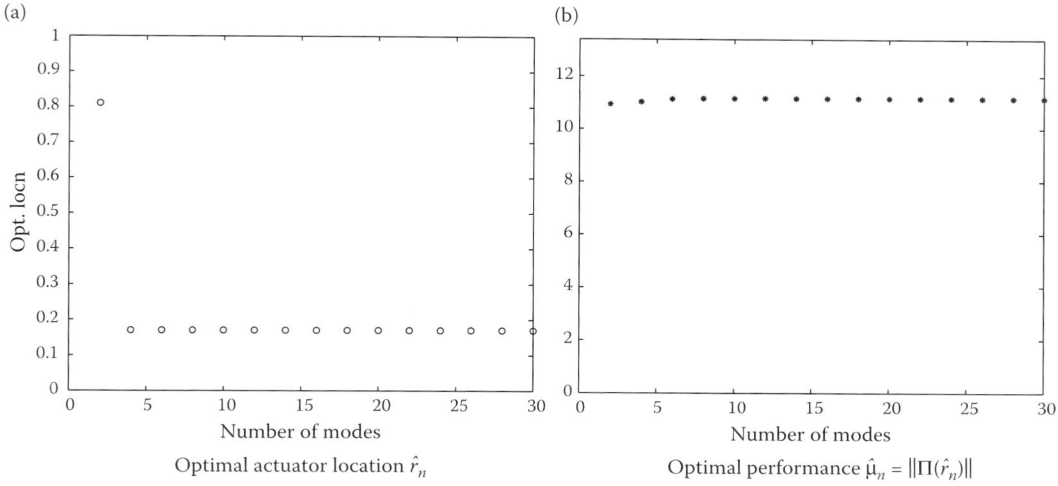

FIGURE 67.12 Optimal actuator location and performance for approximations of the of the diffusion equation, $C_1 = \sqrt{1000}I, R = 1$. Since the semigroup is analytic, uniform convergence of Π_n to Π is obtained, even for a noncompact C_1 such as used here. This leads to convergence of the optimal performance and of a corresponding sequence of actuator locations.

with respect to the worst possible initial condition. This means that we want to minimize $\|\Pi(r)\|$, where Π solves the ARE with $C_1 = \sqrt{1000}I$ and $R = 1$. Note that C_1 is not a compact operator. However, $A = \partial^2/dx^2$ with domain $\mathcal{D}(A) = \{z \in H_2(0,1) | z'(0) = z'(1) = 0\}$ generates an analytic semigroup on $\mathcal{L}_2(0,1)$. Defining $\mathcal{Z}_n$ to be the span of the first n eigenfunctions and defining the corresponding Galerkin approximation as in Example 67.4 leads to an approximation that satisfies the assumptions of Theorem 67.12 [22] and so $\lim_{n\to\infty} \|\Pi_n(r) - \Pi(r)\| = 0$ for each location r. Convergence of the optimal performance and of a corresponding sequence of actuator locations is shown in Figure 67.12.

Figure 67.13 illustrates that this optimal actuator location problem is nonconvex. We are only guaranteed to have convergence of a sequence of optimal actuator locations, not every sequence.

The discussion in this section has so far been concerned only with state feedback. However, in general, the full state z is not available and a measurement

$$y(t) = C_2 z(t), \tag{67.18}$$

where $C_2 \in \mathcal{L}(\mathcal{Z}, \mathcal{W})$ and $\mathcal{W}$ is a finite-dimensional Hilbert space, is used to estimate the state.

As for finite-dimensional systems, we construct an estimate of the state. Choose some $F \in \mathcal{L}(\mathcal{W}, \mathcal{Z})$ so that $A - FC_2$ generates an exponentially stable semigroup. This can be done, for instance, by solving a Riccati equation dual to that for control: $F = \Sigma C_2^* R_e^{-1}$ where for some $B_1 \in \mathcal{L}(\mathcal{Z}, \mathcal{V})$, $\mathcal{V}$ a Hilbert space and $R_e \in \mathcal{L}(\mathcal{W}, \mathcal{W})$ with $R_e > 0$,

$$\left(\Sigma A^* + A\Sigma + B_1 B_1^* - \Sigma C_2^* R_e^{-1} C_2 \Sigma\right) z = 0, \quad \forall z \in \mathcal{D}(A^*). \tag{67.19}$$

The estimator is

$$\dot{z}_e(t) = Az_e(t) + Bu(t) + F\big(y(t) - C_2 z_e(t)\big).$$

Stability of $A - FC_2$ guarantees that $\lim_{t\to\infty} \|z_e(t) - z(t)\| = 0$.

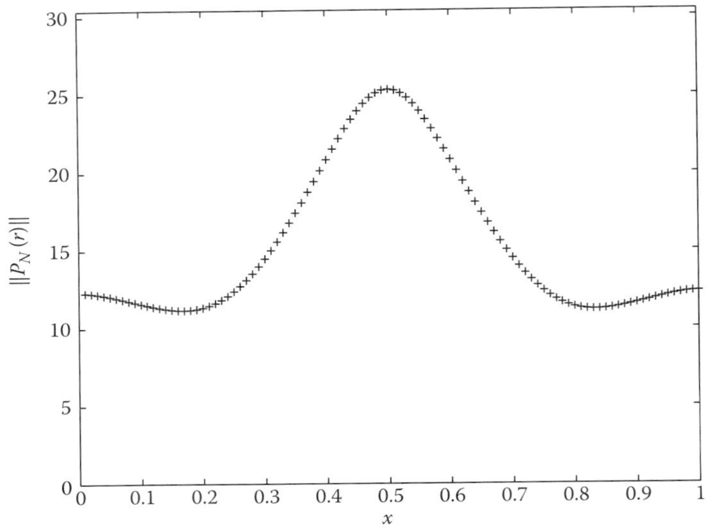

FIGURE 67.13 The cost $\|\Pi(r)\|$ as a function of r for the heat equation, $C = \sqrt{1000}I$, is not a convex function.

The controller is formed by using the state estimate as input to a state feedback controller. As explained in Section 67.3, this leads to the controller

$$\begin{aligned}\dot{z}_c(t) &= (A - FC_2)z_c(t) - [F \quad B]e(t),\\ y_c(t) &= -Kz_c(t),\end{aligned} \tag{67.20}$$

where $e(t) = r(t) - \tilde{y}(t)$ and

$$\tilde{y}(t) = \begin{bmatrix} y(t) \\ u(t) \end{bmatrix} = \begin{bmatrix} C_2 \\ 0 \end{bmatrix} z(t) + \begin{bmatrix} 0 \\ I \end{bmatrix} u(t).$$

As for finite-dimensional systems, if $A - FC_2$ and $A - BK$ each generate an exponentially stable semigroup, then the above controller stabilizes the infinite-dimensional system (Equations 67.2 and 67.18) [8, Section 5.3].

The controller (Equation 67.20) is infinite-dimensional. A finite-dimensional approximation to this controller can be calculated using a finite-dimensional approximation (A_n, B_n, C_{2n}) to the original system (A, B, C_2). Consider $F_n = \Sigma_n C_{2n}^* R_e^{-1}$, where Σ_n solves the ARE

$$\Sigma_n A_n^* + A_n \Sigma_n + B_{1n} B_{1n}^* - \Sigma_n C_{2n}^* R_e^{-1} C_{2n} \Sigma = 0. \tag{67.21}$$

Results for convergence of solutions Σ_n and the operators F_n follow from arguments dual to those for convergence of solutions to the control Riccati equation 67.17.

Theorem 67.14:

Assume that assumptions (A1) and (A1) hold for approximations of* $\left(A, [B \quad B_1], \begin{bmatrix} C_1 \\ C_2 \end{bmatrix}\right)$ *and that the approximations are also uniformly stabilizable and uniformly detectable. Then the operators* K_n, F_n *obtained by solving the Riccati equations 67.17 and 67.21 respectively converge to the operators* K *and* F *obtained* *by solving Equations 67.15 and 67.19. Furthermore, the sequence K_n uniformly stabilizes (A_n, B_n) and the sequence F_n is uniformly detectable for (A_n, C_{2n}).*

Defining

$$\tilde{y}(t) = \begin{bmatrix} y(t) \\ u(t) \end{bmatrix} = \begin{bmatrix} C_2 \\ 0 \end{bmatrix} z(t) + \begin{bmatrix} 0 \\ I \end{bmatrix} u(t)$$

and letting $e(t) = r - \tilde{y}$, the finite-dimensional controller is

$$\begin{aligned} \dot{z}_c(t) &= (A_n - F_n C_{2n}) z_c(t) - \begin{bmatrix} F_n & B_n \end{bmatrix} e(t), \\ y_c(t) &= -K_n z_c(t). \end{aligned} \tag{67.22}$$

It follows from Theorem 67.8 that the sequence of controllers (Equation 67.22) converge in the gap topology to Equation 67.20, and that they stabilize the original system for large enough n. Furthermore, the corresponding closed-loop systems converge.

67.5 $\mathbb{H}_\infty$ Control

Many applications involve an unknown and uncontrolled disturbance $d(t)$. An important objective of controller design in these situations is to reduce the system's response to the disturbance. The system (Equations 67.2 and 67.5) become

$$\frac{dz}{dt} = Az(t) + Bu(t) + Dd(t), \quad z(0) = 0 \tag{67.23}$$

with cost

$$y_1(t) = C_1 z(t) + E_1 u(t). \tag{67.24}$$

Since we are interested in reducing the response to the disturbance, the initial condition $z(0)$ is set to zero. We assume that $d(t) \in L_2(0, \infty; \mathcal{V})$, where $\mathcal{V}$ is a Hilbert space and that $D \in \mathcal{L}(\mathcal{V}, \mathcal{Z})$ is a compact operator. (This last assumption follows automatically if $\mathcal{V}$ is finite-dimensional.) The measured signal, or input to the controller is

$$y_2(t) = C_2 z(t) + E_2 d(t) \tag{67.25}$$

The operator $C_2 \in \mathcal{L}(\mathcal{Z}, \mathcal{W})$ for some finite-dimensional Hilbert space $\mathcal{W}$ and $E_2 \in \mathcal{L}(\mathcal{V}, \mathcal{W})$.

Let G denote the transfer function from d to y_1 and let H denote the controller transfer function:

$$\hat{u}(s) = H(s)\hat{y}_2(s).$$

The map from the disturbance d to the cost y_1 is

$$\begin{aligned} \hat{y}_1 &= C_1(sI - A)^{-1}(D\hat{d} + B\hat{u}) \\ &= C_1(sI - A)^{-1}(D\hat{d} + BH\hat{y}_2). \end{aligned}$$

Using Equations 67.23 and 67.25 to eliminate y_2, and defining

$$\begin{aligned} G_{11}(s) &= C_1(sI - A)^{-1}D, & G_{12}(s) &= C_1(sI - A)^{-1}B + E_1, \\ G_{21}(s) &= C_2(sI - A)^{-1}D + E_2, & G_{22}(s) &= C_2(sI - A)^{-1}B, \end{aligned}$$

we obtain the transfer function

$$\mathcal{F}(G, H) = G_{11}(s) + G_{12}(s)H(s)(I - G_{22}(s)H(s))^{-1}G_{21}(s)$$

from the disturbance d to the cost y_1.

The controller design problem is to find, for given $\gamma > 0$, a stabilizing controller H so that

$$\int_0^\infty \|y_1(t)\|^2 < \int_0^\infty \gamma^2 \|d(t)\|^2 \, dt.$$

If such a controller is found, the controlled system will then have L_2-gain less than γ.

Definition 67.13:

The system (Equations 67.23 through 67.25) is stabilizable with attenuation γ if there is a stabilizing controller H so that

$$\|\mathcal{F}(G,H)\|_\infty < \gamma.$$

To simplify the formulae, we make the assumptions that

$$E_1^* \begin{bmatrix} C_1 & E_1 \end{bmatrix} = \begin{bmatrix} 0 & I \end{bmatrix}, \quad \begin{bmatrix} D \\ E_2 \end{bmatrix} E_2^* = \begin{bmatrix} 0 \\ I \end{bmatrix}. \tag{67.26}$$

With these simplifying assumptions, the cost y_1 has norm

$$\int_0^\infty \|y_1(t)\|^2 \, dt = \int_0^\infty \|C_1 z(t)\|^2 + \|u(t)\|^2 \, dt$$

which is the LQ cost (Equation 67.14) with normalized control weight $R = I$. The difference is that here we are considering the effect of the disturbance d on the cost, instead of the initial condition $z(0)$. Also, the problem formulation (Equations 67.23 through 67.25) can include robustness and other performance constraints. For details, see, for example [30,41].

We assume throughout that (A, D) and (A, B) are stabilizable and that (A, C_1) and (A, C_2) are detectable. These assumptions ensure that an internally stabilizing controller exists; and that internal and external stability are equivalent for the closed loop if the controller realization is stabilizable and detectable.

Consider first the full information case:

$$y_2(t) = \begin{bmatrix} x(t) \\ d(t) \end{bmatrix} = \begin{bmatrix} I \\ 0 \end{bmatrix} x(t) + \begin{bmatrix} 0 \\ I \end{bmatrix} d(t). \tag{67.27}$$

An important characteristic of $\mathbb{H}_\infty$-disturbance attenuation is that, in general, a system is not stabilizable with attenuation γ for every γ. However, if it is stabilizable with attenuation γ, the attenuation (in the full-information case) can be achieved with constant state-feedback.

Definition 67.14:

The state feedback $K \in \mathcal{L}(\mathcal{Z},\mathcal{U})$ is γ-admissible if $A-BK$ generates an exponentially stable semigroup and the feedback $u(t) = -Kz(t)$ is such that γ-attenuation is achieved.

Theorem 67.15: [4,20]

Assume that (A, B) is stabilizable and (A, C_1) is detectable. For $\gamma > 0$, the following are equivalent:

- *The full-information system (Equations 67.23, 67.24, and 67.27) is stabilizable with disturbance attenuation γ.*

- *There exists a nonnegative, self-adjoint operator* Σ *on* $\mathcal{Z}$ *such that for all* $z \in \mathcal{D}(A)$,

$$\left(A^*\Pi + \Pi A + \Pi\left(\frac{1}{\gamma^2}DD^* - BB^*\right)\Pi + C_1^*C_1\right)z = 0 \tag{67.28}$$

and $A - BB^*\Pi + \frac{1}{\gamma^2}DD^*\Pi\frac{1}{\gamma^2}DD^*\Pi$ *generates an exponentially stable semigroup on* $\mathcal{Z}$.

In this case, $K = B^*\Pi$ *is a* γ*-admissible state feedback.*

Note that as $\gamma \to \infty$, the Riccati equation for the LQ problem is obtained. Since the operator Riccati equation 67.28 cannot be solved, a finite-dimensional approximation to the original infinite-dimensional system is used to approximate the exact control $K = B^*\Pi$. As in previous sections, let $\mathcal{Z}_n$ be a finite-dimensional subspace of $\mathcal{Z}$ and P_n the orthogonal projection of $\mathcal{Z}$ onto $\mathcal{Z}_n$. Consider a sequence of operators $A_n \in \mathcal{L}(\mathcal{Z}_n, \mathcal{Z}_n)$, $B_n = P_nB$, $D_n = P_nD$, and $C_{1n} = C_1|_{\mathcal{Z}_n}$. Assumptions similar to those used for LQ control are required.

Theorem 67.16: [16, Theorem 2.5, Corollary 2.6]

Assume a sequence of approximations satisfy (A1),(A1),* (A_n, B_n) *are uniformly stabilizable, and* (A_n, C_{1n}) *are uniformly detectable. Assume that the original problem is stabilizable with attenuation* γ*. For sufficiently large n the Riccati equation*

$$A_n^*\Pi_n + \Pi_nA_n + \Pi_n\left(\frac{1}{\gamma^2}D_nD_n^* - B_nB_n^*\right)\Pi_n + C_{1n}^*C_{1n} = 0, \tag{67.29}$$

has a nonnegative, self-adjoint solution Π_n*. For such n*

- *There exist positive constants* M_1 *and* ω_1 *such that the semigroup* $S_{n2}(t)$ *generated by* $A_n + \frac{1}{\gamma^2}D_nD_n^*\Pi_n - B_nB_n^*\Pi_n$ *satisfies* $\|S_{n2}(t)\| \leq M_1e^{-\omega_1 t}$.
- $K_n = (B_n)^*\Pi_n$ *is a* γ*-admissible state feedback for the approximating system and there exists* $M_2, \omega_2 > 0$ *such that the semigroup* $S_{nK}(t)$ *generated by* $A_n + B_nK_n$ *satisfies* $\|S_{nK}(t)\| \leq M_2e^{-\omega_2 t}$.

Moreover, for all $z \in \mathcal{Z}$, $\Pi_nP_nz \to \Pi z$ *as* $n \to \infty$ *and* $K_n = (B_n)^*\Pi_n$ *converges to* $K = B_2^*\Pi$ *in norm. For n sufficiently large,* K_nP_n *is a* γ*-admissible state feedback for the infinite-dimensional system.*

The optimal disturbance attenuation problem is to find

$$\hat{\gamma} = \inf \gamma$$

over all γ such that Equations 67.23, 67.24, and 67.27 are stabilizable with attenuation γ. Let $\hat{\gamma}_n$ indicate the corresponding optimal disturbance attenuation for the approximating problems. Theorem 67.16 implies that $\limsup_{n\to\infty} \hat{\gamma}_n \leq \hat{\gamma}$ but in fact convergence of the optimal disturbance attenuation can be shown.

Corollary 67.1: [16, Theorem 2.8]

With the same assumptions as Theorem 67.16, it follows that

$$\lim_{n\to\infty} \hat{\gamma}_n = \hat{\gamma}.$$

TABLE 67.1 Physical Constants

E	2.1×10^{11} N/m^2
I	1.2×10^{-10} m^4
ρ	3.0 kg/m
c_v	0.0010 N s /m^2
c_d	0.010 N s /m^2
L	7.0 m
I_h	39.0 kg m^2
ρ_d	0.12 1/m

Example 67.11: Flexible Slewing Beam

Consider a Euler–Bernoulli beam clamped at one end and free at other end. Let $w(x,t)$ denote the deflection of the beam from its rigid body motion at time t and position x. The deflection can be controlled by applying a torque at the clamped end ($x = 0$). We assume that the hub inertia I_h is much larger than the beam inertia, so that, letting $\theta(t)$ indicate the rotation angle, $u(t) = I_h\ddot{\theta}(t)$ is a reasonable approximation to the applied torque. The disturbance $d(t)$ induces a uniformly distributed load $\rho_d d(t)$. The values of the physical parameters used in the simulations are listed in Table 67.1. Use of the Kelvin–Voigt model for damping leads to the following description of the beam vibrations:

$$\rho\frac{\partial^2 w}{\partial t^2} + c_v\frac{\partial w}{\partial t} + \frac{\partial^2}{\partial x^2}\left[EI\frac{\partial^2 w}{\partial x^2} + c_d I\frac{\partial^3 w}{\partial x^2 \partial t}\right] = \frac{\rho x}{I_h}u(t) + \rho_d\, d(t), \quad 0 < x < L.$$

The boundary conditions are

$$w(0,t) = 0, \quad \frac{\partial w}{\partial x}(0,t) = 0,$$

$$\left[EI\frac{\partial^2 w}{\partial x^2} + c_d I\frac{\partial^3 w}{\partial x^2 \partial t}\right]_{x=L} = 0, \quad \left[EI\frac{\partial^3 w}{\partial x^3} + c_d I\frac{\partial^4 w}{\partial x^3 \partial t}\right]_{x=L} = 0.$$

Let $z(t) = (w(\cdot,t), \frac{\partial}{\partial t}w(\cdot,t))$, $H_f(0,L)$ be the closed linear subspace of $\mathcal{H}^2(0,L)$ defined by

$$H_f(0,L) = \left\{w \in \mathcal{H}^2(0,L) : w(0) = \frac{dw}{dx}(0) = 0\right\}.$$

With the state space $\mathcal{Z} = H_f(0,L) \times \mathcal{L}^2(0,L)$, a state-space formulation of the above PDE problem is

$$\frac{d}{dt}z(t) = Az(t) + Bu(t) + Dd(t),$$

where

$$B = \begin{bmatrix} 0 \\ \frac{x}{I_h} \end{bmatrix}, \quad D = \begin{bmatrix} 0 \\ \frac{\rho_d}{\rho} \end{bmatrix}, \quad A = \begin{bmatrix} 0 & I \\ -\frac{EI}{\rho}\frac{d^4}{dx^4} & -\frac{c_d I}{\rho}\frac{d^4}{dx^4} - \frac{c_v}{\rho} \end{bmatrix},$$

with, defining $M = EI\frac{d^2}{dx^2}\phi + c_d I\frac{d^2}{dx^2}\psi$, A has domain

$$\mathcal{D}(A) = \{(\phi,\psi) \in X : \psi \in H_f(0,L); \quad M \in \mathcal{H}^2(0,L) \text{ with } M(L) = \frac{d}{dx}M(L) = 0\}.$$

The operators B and D are clearly bounded operators from $\mathbb{R}$ to $\mathcal{Z}$.

Suppose the objective of the controller design is to reduce the effect of disturbances on the tip position:

$$y(t) = C_1 z(t) = w(L, t).$$

Sobolev's Inequality implies that evaluation at a point is bounded on $H_f(0, L)$ and so C_1 is bounded from $\mathcal{Z}$ to $\mathbb{R}$.

Define the bilinear form on $H_f(0, L) \times H_f(0, L)$

$$\sigma(\phi, \psi) = \int_0^L \frac{EI}{\rho} \frac{d^2}{dx^2}\phi(x) \frac{d^2}{dx^2}\psi(x)\, dx.$$

Define $\mathcal{V} = H_f(0, L) \times H_f(0, L)$ and define $a(\cdot, \cdot)$ on $\mathcal{V} \times \mathcal{V}$ by

$$a((\phi_1, \psi_1), (\phi_2, \psi_2)) = -\sigma(\psi_1, \phi_2) + \sigma(\phi_1 + \frac{c_d}{E}\psi_1, \psi_2) + \langle \frac{c_v}{\rho}\psi_1, \psi_2\rangle$$

for $(\phi_i, \psi_i) \in V$, $i = 1, 2$. Then A can be defined by

$$\langle Ay, z\rangle_{\mathcal{V}^* \times \mathcal{V}} = -a(y, z), \quad \text{for } y,\ z \in \mathcal{V}.$$

The form $a(\cdot, \cdot)$ satisfies inequality (Equation 67.7) and also (Equation 67.8) with $k < 0$ and so the operator A generates an exponentially stable semigroup on $\mathcal{Z}$.

Let $H_n \subset H_f(0, L)$ be a sequence of finite-dimensional subspaces formed by the span of n standard finite-element cubic $B-$spline approximations [33]. The approximating generator A_n on $\mathcal{Z}_n = H_n \times H_n$ is defined by the Galerkin approximation

$$\langle A_n y_n, z_n\rangle = -a(y_n, z_n), \quad \forall z_n,\ y_n \in \mathcal{Z}_n.$$

For all $\phi \in H_f(0, L)$ there exists a sequence $\phi_n \in H_n$ with $\lim_{n\to\infty} \|\phi_n - \phi\|_{H_f(0,L)} = 0$ [33]. It follows then from Theorem 67.7 and exponential stability of the original system that (A1) is satisfied and that the approximations are uniformly exponentially stabilizable (trivially, by the zero operator). The adjoint of A can be defined through $a(z, y)$ and (A1*) and uniform detectability also follow. Thus, Theorem 67.16 applies and convergence of the approximating feedback operators is obtained.

The corresponding series of finite-dimensional Riccati equation 67.29 were solved with $\gamma = 2.3$. Figure 67.14 compares the open- and closed-loop responses $w(L, t)$ to a disturbance consisting of a 100 second pulse, for the approximation with 10 elements. The feedback controller leads to a closed-loop system which is able to almost entirely reject this disturbance. Figure 67.15 compares the open- and closed-loop responses to the periodic disturbance $\sin(\omega t)$, where ω is the first resonant frequency: $\omega = \min_j |\text{Im}(\lambda_j(A_{10})|$. The resonance in the open loop is not present in the closed loop.

Figure 67.16 displays the convergence of the feedback gains predicted by Theorem 67.16. Since $\mathcal{Z}_n$ is a product space, the first and second components of the gains are displayed separately as displacement and velocity gains, respectively.

In general, of course, the full state is not measured and the measured output is described by Equation 67.25.

Theorem 67.17: [4,20]

The system (Equations 67.23 through 67.25) is stabilizable with attenuation $\gamma > 0$ if and only if the following two conditions are satisfied:

1. *There exists a nonnegative self-adjoint operator Π on $\mathcal{Z}$ satisfying the Riccati equation*

$$\left(A^*\Pi + \Pi A + \Pi\left(\frac{1}{\gamma^2}DD^* - BB^*\right)\Pi + C_1^*C_1\right)z = 0, \quad \forall z \in \mathcal{D}(A), \tag{67.30}$$

such that $A + (\frac{1}{\gamma^2}DD^* - BB^*)\Pi$ generates an exponentially stable semigroup on $\mathcal{Z}$;

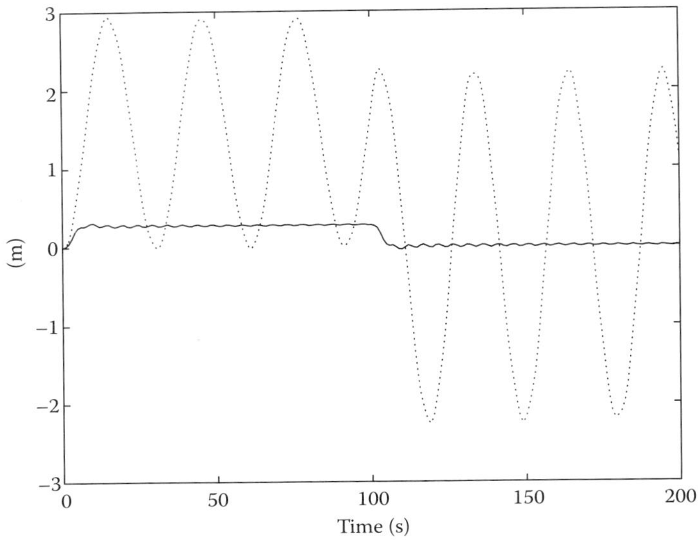

FIGURE 67.14 H_∞-state feedback for flexible slewing beam. Open-(..) and closed-loop (–) responses to a disturbance $d(t) = 1, t \leq 100$ *s*: 10 elements.

2. *Define* $\tilde{A} = A + \frac{1}{\gamma^2}DD^*\Pi$ *and* $K = B^*\Pi$. *There exists a nonnegative self-adjoint operator* $\tilde{\Sigma}$ *on* X *satisfying the Riccati equation*

$$(\tilde{A}\tilde{\Sigma} + \tilde{\Sigma}\tilde{A}^* + \tilde{\Sigma}\left(\frac{1}{\gamma^2}K^*K - C_2^*C_2\right)\tilde{\Sigma} + DD^*)z = 0, \quad \forall z \in \mathcal{D}(A^*), \tag{67.31}$$

such that $\tilde{A} + \tilde{\Sigma}\left(\frac{1}{\gamma^2}K^*K - C_2^*C_2\right)$ *generates an exponentially stable semigroup on* X.

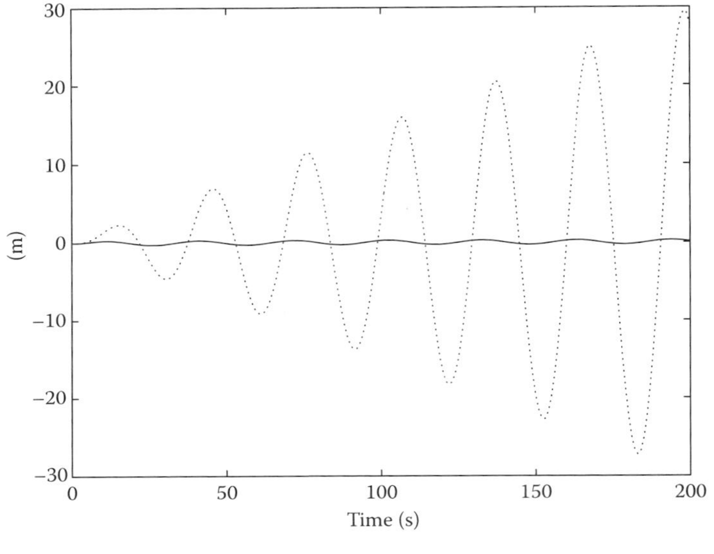

FIGURE 67.15 H_∞-state feedback for flexible slewing beam. Open-(..) and closed-loop (–) responses to a disturbance $d(t) = \sin(\omega t)$: 10 elements.

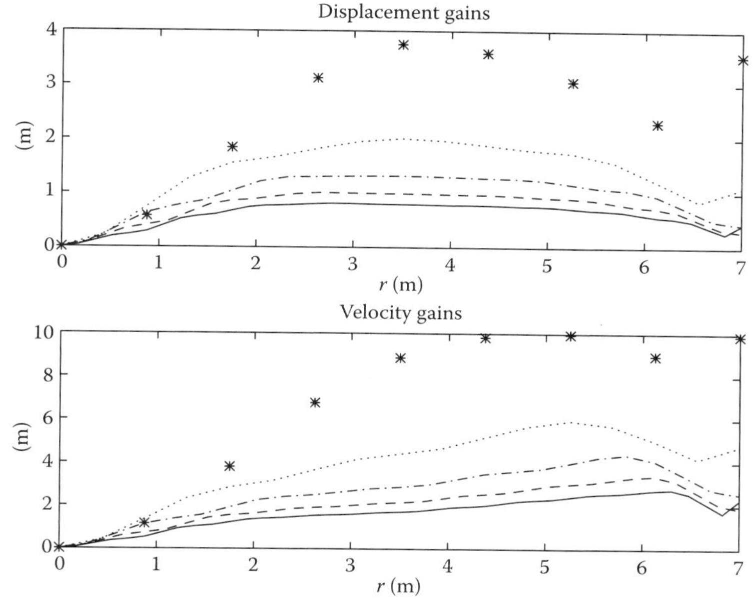

FIGURE 67.16 H_∞-state feedback for flexible slewing beam. Feedback gains for 2 elements *, 4 elements ....., 6 elements _._., 8 elements _ _, and 10 elements, __ are plotted. As predicted by the theory, the feedback operators are converging.

Moreover, if both conditions are satisfied, define $F = \tilde{\Sigma} C_2^*$ *and* $A_c = A + \frac{1}{\gamma^2} DD^* \Sigma - BK - FC_2$. *The controller with state-space description*

$$\dot{z}_c(t) = A_c z_c(t) + F y_2(t), \qquad (67.32)$$
$$u(t) = -K z_c(t)$$

stabilizes the system (Equations 67.23 through 67.25) with attenuation γ.

Condition (1) is simply the Riccati equation to be solved for the full-information state feedback controller. Condition (2) leads to an estimate of the state of the system controlled by $Kz(t)$. Unlike LQ control, the design of the state feedback and the estimator is coupled. Condition (2) above is more often written as the following two equivalent conditions:

a. There exists a nonnegative, self-adjoint operator Σ on X satisfying the Riccati equation

$$\left(A\Sigma + \Sigma A^* + \Sigma \left(\frac{1}{\gamma^2} C_1^* C_1 - C_2^* C_2 \right) \Pi + DD^* \right) z = 0, \quad \forall z \in \mathcal{D}(A^*) \qquad (67.33)$$

such that $A + \Pi(\frac{1}{\gamma^2} C_1^* C_1 - C_2^* C_2)$ generates an exponentially stable semigroup on $\mathcal{Z}$

b. $r(\Sigma\Pi) < \gamma^2$ where r indicates the spectral radius.

In the presence of condition (1) in Theorem 67.17, condition (2) is equivalent to conditions (a) and (b). Also $\tilde{\Sigma} = (I - \frac{1}{\gamma^2}\Sigma\Pi)^{-1}\Sigma = \Sigma(I - \frac{1}{\gamma^2}\Pi\Sigma)^{-1}$. The advantage of replacing condition (2) by conditions (a) and (b) is numerical. The Riccati equation in (2) is coupled to the solution of (1) whereas the Riccati equation in (a) is independent of the solution of (1). This theoretical result has been extended to a class of control systems with unbounded control and observation operators [20].

For bounded control and observation operators, a complete approximation theory exists. Define a sequence of approximations on finite-dimensional spaces $\mathcal{Z}_N$, as for the full information case, with the addition of $C_{2n} = C_2|_{\mathcal{Z}_n}$.

Strong convergence of solutions Σ_n to Riccati equations approximating Equation 67.33 will follow from Theorem 67.16 and a straightforward duality argument if (A1) and (A1*) hold, along with assumptions on uniform stabilizability of (A, D) and uniform detectability of (A, C_2). However, strong convergence of both $\Pi_n \to \Pi$ and $\Sigma_n \to \Sigma$ does not imply convergence (or even existence) of the inverse operator $(I - \frac{1}{\gamma^2}\Sigma_n\Pi_n)^{-1}$ so we cannot show controller convergence. Convergence of the solution $\tilde{\Sigma}_n$ to the estimation Riccati equation 67.31 can be proven.

Theorem 67.18: [27, Theorem 3.5]

Assume that (A1) and (A1) hold, that (A_n, D_n) are uniformly stabilizable, and that (A_n, C_{2n}) are uniformly detectable. Let $\gamma > 0$ be such that the infinite-dimensional problem is stabilizable. Let N be large enough that approximations to the full-information problem are stabilizable with attenuation γ, and let Π_n indicate the solution to the ARE (Equation 67.29) for $n > N$. Define $K_n = B_n^*\Pi_n$ and $\tilde{A}_n = A_n + \frac{1}{\gamma^2}D_nD_n^*\Pi_n$.*

For sufficiently large n the Riccati equation

$$\tilde{A}_n\tilde{\Sigma}_n + \tilde{\Sigma}_n\tilde{A}_n^* + \tilde{\Sigma}_n\left(\frac{1}{\gamma^2}K_n^*K_n - C_{2n}^*C_{2n}\right)\tilde{\Sigma}_n + D_nD_n^* = 0 \tag{67.34}$$

has a nonnegative, self-adjoint solution $\tilde{\Sigma}_n$. For such n there exist positive constants M_3 and ω_3 such that the semigroup $\tilde{S}_{n2}(t)$ generated by $\tilde{A}_n + \frac{1}{\gamma^2}\tilde{\Sigma}_nK_n^*K_n - \tilde{\Sigma}_nC_{2n}^*C_{2n}$ satisfies $\|\tilde{S}_{n2}(t)\| \leq M_3e^{-\omega_3 t}$. Moreover, for each $z \in \mathcal{Z}$, $\tilde{\Sigma}_nP_n z \to \tilde{\Sigma} z$ as $n \to \infty$ and $F_n = \tilde{\Sigma}_nC_{2n}^*$converges to $F = \tilde{\Sigma}C_2^*$ in norm.

Defining $A_{cn} = A_n + \frac{1}{\gamma^2}D_nD_n^*\Sigma - B_nK_n - F_nC_{2n}$, Theorems 67.16 and 67.18 imply convergence of the controllers

$$\begin{aligned} \dot{z}_c(t) &= A_{cn}z_c(t) + F_ny_2(t), \\ u(t) &= -K_nz_c(t) \end{aligned} \tag{67.35}$$

to the infinite-dimensional controller (Equation 67.32) in the gap topology. The same assumptions imply convergence of the plants which leads to the following result.

Theorem 67.19: [27, Theorem 3.6]

Let γ be such that the infinite-dimensional system is stabilizable with attenuation γ. Assume that (A1) and (A1) hold, that (A, B) and (A, D) are uniformly stabilizable, and that (A, C_1) and (A, C_2) are uniformly detectable. Then the finite-dimensional controllers (Equation 67.35) converge in the gap topology to the infinite-dimensional controller (Equation 67.32). For sufficiently large n, the finite-dimensional controllers (Equation 67.35) stabilize the infinite-dimensional system and provide γ-attenuation.*

Convergence of the optimal attenuation also holds for the output feedback problem.

Corollary 67.2: [27, Theorem 3.7]

Let $\hat{\gamma}$ indicate the optimal disturbance attenuation for the output-feedback problem (Equations 67.23 through 67.25), and similarly indicate the optimal attenuation for the approximating problems by $\hat{\gamma}_n$. With *the same assumptions as for Theorem 67.19,*

$$\lim_{n\to\infty} \hat{\gamma}_n = \hat{\gamma}.$$

67.6 Summary

For most practical systems, an approximation to the PDE must be used in controller design. Similarly, control for delay-differential equations often proceeds by using an approximation, although delay-differential equations were not discussed directly in this chapter. This has the advantage of making available the vast array of tools available for design of controllers for finite-dimensional systems. Since the underlying model is infinite-dimensional, this process is not entirely straightforward. However, there are a number of tools and techniques available for satisfactory controller design. This chapter has presented an overview of the main issues surrounding controller design for these systems. The key point is that the controller must control the original system. Sufficient conditions for satisfactory LQ controller design and H_∞-controller design were presented. Uniform stabilizability and detectability, along with convergence of the adjoint systems, are assumptions not required in simulation but key to obtaining satisfactory performance of a controller designed using a finite-dimensional approximation. There are results guaranteeing these assumptions for many problems. However, for problems where these assumptions are not known and proof is not feasible, the approximating systems should be checked numerically for uniform stabilizability and detectability. One test is to verify that the sequence of controlled systems is uniformly exponentially stable.

This chapter discussed only systems with bounded control and observation operators B and C. Introducing a better model for an actuator sometimes converts a control system with an unbounded control operator to a more complex model with a bounded operator, and similarly for sensor models, e.g. [17,43]. For some systems, however, the most natural model leads to unbounded operators. There are considerably fewer results for controller design for these systems. However, results do exist for LQ control of parabolic problems, such as diffusion [2,22].

Once a suitable approximation scheme, and controller synthesis technique, has been chosen, the problem of computation remains. The primary difficulty is solving the large Riccati equations that arise in LQ and H_∞-controller design. For problems where an approximation of suitable accuracy is of relatively low order (less than about a hundred), direct methods can be used to solve the Riccati equations. However, for larger problems, an iterative method is required. Probably the most popular method for the Riccati equations that arise in LQ control is the Newton–Kleinman method [21] and its variants—see, for example [10,32]. This method is guaranteed to converge, and has quadratic convergence. However, calculation of an H_∞-controller corresponds to calculation of a saddle point, not a minimum as in the case of LQ control. Suitable methods for design of H_∞-controllers for large-scale systems is an open problem at the present time.

References

1. M. J. Balas. Active control of flexible systems. *J. Optim. Theory Appl.*, 23(3):415–436, 1976.
2. H. T. Banks and K. Ito. Approximation in LQR problems for infinite-dimensional systems with unbounded input operators. *J. Math. Systems Estim. Control*, 7(1):1–34, 1997.
3. H. T. Banks and K. Kunisch. The linear regulator problem for parabolic systems. *SIAM J. Control Optim.*, 22(5):684–698, 1984.
4. A. Bensoussan and P. Bernhard. On the standard problem of H_∞-optimal control for infinite dimensional systems. In *Identification and Control in Systems Governed by Partial Differential Equations*, pp. 117–140. SIAM, Philadelphia, PA, 1993.
5. J. A. Burns, K. Ito, and G. Propst. On non-convergence of adjoint semigroups for control systems with delays. *SIAM J. Control Optim.*, 26(6):1442–1454, 1988.

6. J. A. Burns, E. W. Sachs, and L. Zietsman. Mesh independence of Kleinman-Newton iterations for Riccati equations in Hilbert space. *SIAM J. Control Optim.*, 47(5):2663–2692, 2008.
7. R. F. Curtain and K. A Morris. Transfer functions of distributed parameter systems: A tutorial. *Automatica*, 45(5):1101–1116, 2009.
8. R. F. Curtain and H. Zwart. *An Introduction to Infinite-Dimensional Linear Systems Theory*. Springer-Verlag, Berlin, 1995.
9. C. A. Desoer and M. Vidyasagar. *Feedback Systems: Input–Output properties*. Academic Press, New York, 1975.
10. F. Feitzinger, T. Hylla, and E. W. Sachs. Inexact Kleinman–Newton method for Riccati equations. *SIAM J. Matrix Anal. Appl.*, 21(2):272–288, 2009.
11. C. Foias, H. H. Ozbay, and A. Tannenbaum. *Robust Control of Infinite Dimensional Systems: Frequency Domain Methods LNCIS vol. 209*. Springer-Verlag, Berlin, 1995.
12. J. S. Gibson. The Riccati integral equations for optimal control problems on Hilbert spaces. *SIAM J. Control Optim.*, 17(4):637–665, 1979.
13. J. S. Gibson. A note on stabilization of infinite-dimensional linear oscillators by compact linear feedback. *SIAM J. Control Optim.*, 18(3):311–316, 1980.
14. R. B. Guenther and J. W. Lee. *Partial Differential Equations of Mathematical Physics and Integral Equations*. Prentice-Hall, Englewood Cliffs, NJ, 1988.
15. K. Ito. Strong convergence and convergence rates of approximating solutions for algebraic Riccati equations in Hilbert spaces. In W. Schappacher F. Kappel, and K. Kunisch (Eds), *Distributed Parameter Systems*. Springer-Verlag, Berlin, 1987.
16. K. Ito and K. A. Morris. An approximation theory for solutions to operator Riccati equations for H_∞ control. *SIAM J. Control Optim.*, 36(1):82–99, 1998.
17. B. Jacob and K. A. Morris. Second-order systems with acceleration measurements. Preprint, 2009.
18. C. A. Jacobson and C. N. Nett. Linear state-space systems in infinite-dimensional space: The role and characterization of joint stabilizability/detectability. *IEEE Trans. Automat. Control*, 33(6):541–549, 1988.
19. K. Kashima and Y. Yamamoto. On standard H_∞ control problems for systems with infinitely many unstable poles. *Systems Control Lett.*, 57(4):309–314, 2008.
20. B.V. Keulen. *H_∞-Control for Distributed Parameter Systems: A State-Space Approach*. Birkhauser, Boston, 1993.
21. D. Kleinman. On an iterative technique for Riccati equation computations. *IEEE Trans. Automat. Control*, 13:114–115, 1968.
22. I. Lasiecka and R. Triggiani. *Control Theory for Partial Differential Equations: Continuous and Approximation Theories*, Vol. I. Cambridge University Press, Cambridge, UK, 2000.
23. I. Lasiecka and R. Triggiani. *Control Theory for Partial Differential Equations: Continuous and Approximation Theories*, Vol. II. Cambridge University Press, Cambridge, UK, 2000.
24. H. Logemann. Stabilization and regulation of infinite-dimensional systems using coprime factorizations. In R. F. Curtain (Ed.), *Analysis and Optimization of Systems: State and Frequency Domain Approaches for Infinite-Dimensional Systems*, pp. 102–139. Springer-Verlag, Berlin, 1992.
25. H. L. Logemann. Circle criteria, small-gain conditions and internal stability for infinite-dimensional systems. *Automatica*, 27:677–690, 1991.
26. H. L. Logemann, E. P. Ryan, and S. Townley. Integral control of linear systems with actuator nonlinearities: Lower bounds for the maximal regulating gain. *IEEE Trans. Automat. Control*, 44:1315–1319, 1999.
27. K. A. Morris. H_∞ output feedback control of infinite-dimensional systems via approximation. *Systems Control Lett.*, 44(3):211–217, 2001.
28. K. A. Morris. Convergence of controllers designed using state-space methods. *IEEE Trans. Automat. Control*, 39:2100–2104, 1994.
29. K. A. Morris. Design of finite-dimensional controllers for infinite-dimensional systems by approximation. *J. Math. Systems, Estim. Control*, 4:1–30, 1994.
30. K. A. Morris. *An Introduction to Feedback Controller Design*. Harcourt-Brace Ltd., New York, 2001.
31. K. A. Morris. Convergence of LQ-optimal actuator locations for systems governed by partial differential equations. *IEEE Trans. Automat. Control*, 46:93–111, 2010.
32. K. A. Morris and C. Navasca. Approximation of low rank solutions for linear quadratic feedback control of partial differential equations. *Comp. Opt. App.*, 46:93–111, 2010.
33. N. Ottosen and H. Petersson. *Introduction to the Finite Element Method*. Prentice-Hall, Englewood Cliffs, NJ, 1992.

34. A. Pazy. *Semigroups of Linear Operators and Applications to Partial Differential Equations.* Springer-Verlag, New York, 1983.
35. R. R. Rebarber and G. Weiss. Internal model-based tracking and disturbance rejection for stable well-posed systems. *Automatica*, 39(9):1555–1569, 2003.
36. D. L. Russell. Linear stabilization of the linear oscillator in Hilbert space. *J. Math. Anal. App.*, 25(3): 663–675, 1969.
37. R. E. Showalter. *Hilbert Space Methods for Partial Differential Equations.* Pitman Publishing Ltd., London, 1977.
38. D. H. Towne. *Wave Phenomena.* Dover Publications, New York, 1988.
39. M. Vidyasagar. *Control System Synthesis: A Factorization Approach.* MIT Press, Cambridge, USA, 1985.
40. G. Zames and A. El-Sakkary. Unstable systems and feedback: The gap metric. In *Proc. Allerton Conf.*, pp. 380–385, 1980.
41. K. Zhou, J. C. Doyle, and K. Glover. *Robust and Optimal Control.* Prentice-Hall, Englewood Cliffs, NJ, 1996.
42. S. Q. Zhu. Graph topology and gap topology for unstable systems. *IEEE Trans. Automat. Control*, 34:848–855, 1989.
43. B. J. Zimmer, S. P. Lipshitz, K.A. Morris, J. Vanderkooy, and E. E. Obasi. An improved acoustic model for active noise control in a duct. *ASME J. Dynamic Systems, Measurement Control*, 125(3):382–395, 2003.

68

Controllability of Thin Elastic Beams and Plates

J. E. Lagnese
Georgetown University

G. Leugering
University of Bayreuth

68.1 Dynamic Elastic Beam Models

Consider the deformation of a thin, initially curved beam of length ℓ and constant cross-section of area A, which, in its undeformed reference configuration, occupies the region

$$\Omega = \{\mathbf{r} =: \mathbf{r}_0(x_1) + x_2\mathbf{e}_2(x_1) + x_3\mathbf{e}_3(x_1)|\; x_1 \in [0, \ell],$$
$$(x_2, x_3) := x_2\mathbf{e}_2(x_1) + x_3\mathbf{e}_3(x_1) \in A\}$$

where $\mathbf{r}_0 : [0, \ell] \to \mathbb{R}^3$ is a smooth function representing the *centerline*, or the *reference line*, of the beam at rest. The orthonormal triads $\mathbf{e}_1(\cdot), \mathbf{e}_2(\cdot), \mathbf{e}_3(\cdot)$ are chosen as smooth functions of x_1 so that $\mathbf{e}_1$ is the direction of the tangent vector to the centerline, i.e., $\mathbf{e}_1(x_1) = (d\mathbf{r}_0/dx_1)(x_1)$, and $\mathbf{e}_2(x_1), \mathbf{e}_3(x_1)$ span the orthogonal cross section at x_1. The meanings of the variables x_i are as follows: x_1 denotes arc length along the undeformed centerline, and x_2 and x_3 denote lengths along lines orthogonal to the reference line. The set Ω can then be viewed as obtained by translating the reference curve $\mathbf{r}_0(x_1)$ to the position $x_2\mathbf{e}_2 + x_3\mathbf{e}_3$ within the cross-section perpendicular to the tangent of $\mathbf{r}_0$.

At a given time t, let $\mathbf{R}(x_1, x_2, x_3, t)$ denote the position vector after deformation to the particle which is at $\mathbf{r}(x_1, x_2, x_3)$ in the reference configuration. We introduce the displacement vector by $\mathbf{V} := \mathbf{R} - \mathbf{r}$. The position vector $\mathbf{R}(x_1, 0, 0, t)$ to the deformed reference line $x_2 = x_3 = 0$ is denoted by $\mathbf{R}_0$. Accordingly, the displacement vector of a particle on the reference line is $\mathbf{W} := \mathbf{V}(x_1, 0, 0) = \mathbf{R}_0 - \mathbf{r}_0$. The position vector $\mathbf{R}$ may be approximated to first order by

$$\mathbf{R} = \mathbf{R}_0 + x_2\mathbf{E}_2 + x_3\mathbf{E}_3,$$

where $\mathbf{E}_i$ are the tangents at $\mathbf{R}$ with respect to x_i, respectively. Note, however, that the triad $\mathbf{E}_i$ *is not necessarily orthogonal*, due to shearing.

The deformation of $\mathbf{r}(\cdot)$ into $\mathbf{R}(\cdot, t)$ will be considered as a succession of two motions: (1) a rotation carrying the triad $\mathbf{e}_i(x_1)$ to an intermediate orthonormal triad $\hat{\mathbf{e}}_i(x_1, t)$, followed by (2) a deformation carrying $\hat{\mathbf{e}}_i(x_1, t)$ into the nonorthogonal triad $\mathbf{E}_i(x_1, t)$. The two triads $\hat{\mathbf{e}}_i$ and $\mathbf{E}_i$ then differ on account of a strain $\bar{\varepsilon}$ to be specified below. We choose to orient the intermediate (right-handed) triad $\hat{\mathbf{e}}_i$, which serves as a moving orthonormal reference frame, so that

$$\mathbf{E}_1 = (\mathbf{E}_1 \cdot \hat{\mathbf{e}}_1)\hat{\mathbf{e}}_1 = |\mathbf{E}_1|\hat{\mathbf{e}}_1, \quad \hat{\mathbf{e}}_2 \cdot \mathbf{E}_3 = \hat{\mathbf{e}}_3 \cdot \mathbf{E}_2.$$

A strain $\bar{\varepsilon}$, related to the deformation carrying the triad $\hat{\mathbf{e}}_i$ to the triad $\mathbf{E}_i$, is defined by

$$\begin{aligned}
\hat{\mathbf{e}}_1 \cdot \mathbf{E}_1 &=: 1 + \bar{\varepsilon}_{11}, \quad & \hat{\mathbf{e}}_1 \cdot \mathbf{E}_2 &=: 2\bar{\varepsilon}_{12}, \\
\hat{\mathbf{e}}_2 \cdot \mathbf{E}_2 &=: 1 + \bar{\varepsilon}_{22}, \quad & \hat{\mathbf{e}}_1 \cdot \mathbf{E}_3 &=: 2\bar{\varepsilon}_{13}, \\
\hat{\mathbf{e}}_3 \cdot \mathbf{E}_3 &=: 1 + \bar{\varepsilon}_{33}, \quad & \hat{\mathbf{e}}_2 \cdot \mathbf{E}_3 &=: \bar{\varepsilon}_{23}.
\end{aligned}$$

The remaining strains are defined by requiring the symmetry $\bar{\varepsilon}_{ij} = \bar{\varepsilon}_{ji}$. If distortion of the planar cross sections is neglected, then $\bar{\varepsilon}_{22} \approx \bar{\varepsilon}_{33} \approx \bar{\varepsilon}_{23} \approx 0$. The normal $\mathbf{N}$ to the cross section is then $\mathbf{N} = \mathbf{E}_2 \times \mathbf{E}_3 = \hat{\mathbf{e}}_1 - 2\bar{\varepsilon}_{21}\hat{\mathbf{e}}_2 - 2\bar{\varepsilon}_{31}\hat{\mathbf{e}}_3$.

Let Θ_i denote the angles associated with the orthogonal transformation carrying the orthonormal basis $\mathbf{e}_i$ into $\hat{\mathbf{e}}_i$, whereas the rotation of $\mathbf{e}_i$ into $\mathbf{E}_i$ is represented by the angles ϑ_i (dextral mutual rotations). Up to quadratic approximations we obtain

$$\begin{aligned}
\vartheta_1 &\doteq \Theta_1, \\
\vartheta_2 &\doteq \Theta_2 + 2\bar{\varepsilon}_{31} + 2\Theta_1\bar{\varepsilon}_{21}, \\
\vartheta_3 &\doteq \Theta_3 - 2\bar{\varepsilon}_{21} + 2\Theta_1\bar{\varepsilon}_{31}.
\end{aligned}$$

These angles are interpreted as the *global rotations*. It is obvious from these relations that the shear strains vanish if, and only if, the angles Θ_i, ϑ_i coincide, $i = 1, 2, 3$, or, what is the same, if the normal $\mathbf{N}$ to the cross section coincides with $\mathbf{E}_1$. This is what is known as the *Euler–Bernoulli hypothesis*.

To complete the representation of the reference strains in terms of the angles above and the displacements $W_i := \mathbf{W} \cdot \mathbf{e}_i$, we compute up to quadratic approximations in all rotations and linear approximations in all strains $\bar{\varepsilon}_{ij}$. To this end we introduce curvatures and twist for the undeformed reference configuration by Frénet-type formulae

$$\kappa_2 = \mathbf{e}_2 \cdot \mathbf{e}_{1,1}\,, \quad \kappa_3 = \mathbf{e}_3 \cdot \mathbf{e}_{1,1}\,, \quad \tau = \mathbf{e}_3 \cdot \mathbf{e}_{2,1}\,.$$

(The index separated by a comma indicates a partial derivative with respect to the corresponding variable x_i.) The reference strains are then approximated by

$$\begin{aligned}
\bar{\varepsilon}_{11} &= W_{1,1} - \kappa_2 W_2 - \kappa_3 W_3 + \frac{1}{2}((W_{3,1} + \kappa_3 W_1 + \tau W_2)^2 + (W_{2,1} + \kappa_2 W_1 - \tau W_3)^2),\\
\bar{\varepsilon}_{21} &= \frac{1}{2}(W_{2,1} - \vartheta_3 + \kappa_2 W_1 - \tau W_3 + \vartheta_1(\vartheta_2 + W_{3,1} + \kappa_3 W_1 + \tau W_2)),\\
\bar{\varepsilon}_{31} &= \frac{1}{2}(W_{3,1} + \vartheta_2 + \kappa_3 W_1 + \tau W_2 + \vartheta_1(\vartheta_3 - W_{2,1} - \kappa_2 W_1 + \tau W_3)),
\end{aligned}$$

whereas the approximate bending strains are

$$\begin{aligned}
\tilde{\kappa}_2 &= \vartheta_{3,1} + \kappa_3\vartheta_1 + \tau\vartheta_2,\\
\tilde{\kappa}_3 &= -\vartheta_{2,1} - \kappa_2\vartheta_1 + \tau\vartheta_3,\\
\tilde{\tau} &= \vartheta_{1,1} - \kappa_2\vartheta_2 - \kappa_3\vartheta_3.
\end{aligned}$$

The approximations given above comprise the theory of rods with *infinitesimal strains and moderate rotations*; see Wempner [27].

68.2 The Equations of Motion

Under the assumptions of the previous section, the total strain (or potential) energy is given by

$$\mathcal{U} = \int_0^\ell \left[\frac{EA}{2}\bar{\varepsilon}_{11}^2 + 2GA(\bar{\varepsilon}_{12}^2 + \bar{\varepsilon}_{13}^2) + \frac{EI_{22}}{2}\tilde{\kappa}_2^2 + \frac{EI_{33}}{2}\tilde{\kappa}_3^2 + \frac{GI}{2}\tilde{\tau}^2\right] dx,$$

where $E, G, I_{22}, I_{33}, I = I_{22} + I_{33}$ are Young's modulus, the shear modulus, and moments of the cross section, respectively. The kinetic energy is given by

$$\mathcal{K} = \int_0^\ell \|\dot{\mathbf{R}}\|^2 dx,$$

where $\dot{} = d/dt$. *Controls* may be introduced as distributed, pointwise, or boundary forces and couples $(\mathbf{F}, \mathbf{M}, \mathbf{f}, \mathbf{m})$ through the total work

$$\mathcal{W} = \int_0^\ell (\mathbf{F} \cdot \mathbf{W} + \mathbf{M} \cdot \vartheta)\, dx + \sum_{x=0,\xi,\ell} (\mathbf{f} \cdot \mathbf{W} + \mathbf{m} \cdot \vartheta).$$

Controls may also be introduced in geometric boundary conditions, but we shall not do so here. Typically, a beam will be rigidly clamped at one end (say at $x_1 = 0$) and simply supported or free at the other end, $x_1 = \ell$. If the space of test functions $V_0 := \{f \in H^1(0,\ell) | f(0) = 0\}$ is introduced, the requirement that the end $x_1 = 0$ be rigidly clamped is mathematically realized by the "geometric" constraints $W_i, \vartheta_i \in V_0$. If $x_1 = 0$ is simply supported, rather than clamped, the appropriate geometric boundary conditions are $W_i \in V_0$ for $i = 2$ and $i = 3$ only.

Let

$$\mathcal{L} = \int_0^T [\mathcal{K}(t) + \mathcal{W}(t) - \mathcal{U}(t)]\, dt$$

be the *Lagrangian*. Then, by *Hamilton's principle* (see, for example [26]), the dynamics of the deformation satisfy the stationarity conditions $\delta\mathcal{L}_W = 0$, $\delta\mathcal{L}_\vartheta = 0$, the variations being in V_0. The equations of motion then follow by integration by parts, collecting terms, etc., in the usual manner. Due to space limitations, neither the most general beam model nor the entirety of all meaningful boundary conditions can be described here. Rather, a partial list of beam models which, in part, have been studied in the literature and which can easily be extracted from the formulas above, will be provided. We focus on the typical situation of a beam which is clamped at $x_1 = 0$ and free (resp., controlled) at $x_1 = \ell$.

68.2.1 Initially Straight and Untwisted Linear Shearable 3-Dimensional Beams

68.2.1.1 Equations of Motion

$$\left.\begin{aligned}
m_0 \ddot{W}_1 &= [EAW_1]' && \text{(longitudinal motion)}\\
m_0 \ddot{W}_2 &= [GA(W_2' - \vartheta_3)]' && \text{(lateral motion)}\\
m_0 \ddot{W}_3 &= [GA(W_3 + \vartheta_2)]' && \text{(vertical motion)}\\
m_0 \ddot{\vartheta}_1 &= [GI\vartheta_1']' && \text{(torsional motion)}\\
m_0 \ddot{\vartheta}_2 &= [EI_{33}\vartheta_2']' - GA(W_3' + \vartheta_2), && \text{(shear around } \hat{\mathbf{e}}_2)\\
m_0 \ddot{\vartheta}_3 &= [EI_{22}\vartheta_3']' + GA(W_2' - \vartheta_3) && \text{(shear around } \hat{\mathbf{e}}_3)
\end{aligned}\right\} \tag{68.1}$$

where $' = d/dx_1$.

68.2.1.2 Boundary Conditions

The *geometric boundary conditions* are

$$W_i(0) = 0, \quad \vartheta_i(0) = 0, \quad i = 1, 2, 3. \tag{68.2}$$

The *dynamical boundary conditions* are

$$\left.\begin{aligned}
EAW_1'(\ell) &= f_1,\\
GA(W_2' - \vartheta_3)(\ell) &= f_2,\\
GA(W_3' + \vartheta_2)(\ell) &= f_3,\\
GI\vartheta_1'(\ell) &= m_1,\\
EI_{33}\vartheta_2'(\ell) &= m_2,\\
EI_{22}\vartheta_3'(\ell) &= m_3.
\end{aligned}\right\} \tag{68.3}$$

68.2.1.3 Initial Conditions

$$\begin{aligned}
&W_i(\cdot, t = 0) = W_{i0}(\cdot), \quad \dot{W}_i(\cdot, t = 0) = W_{i1}(\cdot),\\
&\vartheta_i(\cdot, t = 0) = 0, \quad \dot{\vartheta}_i(\cdot, t = 0) = 0, \quad i = 1, 2, 3.
\end{aligned}$$

68.2.2 Initially Straight and Untwisted, Nonshearable Nonlinear 3-Dimensional Beams

68.2.2.1 Equations of Motion

These are comprised of four equations which describe the longitudinal, lateral, vertical, and torsional motions, respectively.

$$\left.\begin{aligned} &m_0\ddot{W}_1 = \left[EA\left(W_1' + \frac{1}{2}(W_2')^2 + \frac{1}{2}(W_3')^2\right)\right]', \\ &m_0\ddot{W}_2 - [\rho_0 I_{22}\ddot{W}_2']' + [EI_{22}W_2'']'' = \left[\left(EA(W_1' + \frac{1}{2}(W_2')^2 + \frac{1}{2}(W_3')^2)W_2'\right)\right]', \\ &m_0\ddot{W}_3 - [\rho_0 I_{33}\ddot{W}_3']' + [EI_{33}W_3'']'' = \left[\left(EA(W_1' + \frac{1}{2}(W_2')^2 + \frac{1}{2}(W_3')^2)W_3'\right)\right]', \\ &\rho_0 I\ddot{\vartheta}_1 = [GI\vartheta_1']'. \end{aligned}\right\} \quad (68.4)$$

68.2.2.2 Boundary Conditions

The geometric boundary conditions are

$$\left.\begin{aligned} &W_i(0) = 0, \quad i = 1, 2, 3, \\ &W_2'(0) = W_3'(0) = \vartheta_1(0) = 0, \end{aligned}\right\} \quad (68.5)$$

while the dynamical boundary conditions are

$$\left.\begin{aligned} &EA\left(W_1' + \frac{1}{2}((W_2')^2 + (W_3')^2)\right)(\ell) = f_1, \\ &EI_{33}W_3''(\ell) = -m_2, \quad EI_{22}W_2''(\ell) = m_3, \\ &\left[EA(W_1' + \frac{1}{2}((W_2')^2 + (W_3')^2))\right]W_2'(\ell) - [EI_{22}W_2'']'(\ell) + \rho_0 I_{22}\ddot{W}_2'(\ell) = f_2, \\ &\left[EA(W_1' + \frac{1}{2}((W_2')^2 + (W_3')^2))\right]W_3'(\ell) - [EI_{33}W_3'']'(\ell) + \rho_0 I_{33}\ddot{W}_3'(\ell) = -f_3, \\ &GI\vartheta_1'(\ell) = m_1 \end{aligned}\right\} \quad (68.6)$$

68.2.2.3 Initial Conditions

$$\begin{aligned} &W_i(\cdot, t=0) = W_{i0}(\cdot), \quad \dot{W}_i(\cdot, t=0) = W_{i1}(\cdot), \\ &\vartheta_1(\cdot, t=0) = 0, \quad \dot{\vartheta}_1(\cdot, t=0) = 0, \quad i = 1, 2, 3. \end{aligned}$$

68.2.3 Nonlinear Planar, Shearable Straight Beams

68.2.3.1 Equations of Motion

The three equations which describe the longitudinal, vertical, and shear motions, respectively, are

$$\left.\begin{aligned} &m_0 h\ddot{W}_1 = \left[Eh(W_1' + \frac{1}{2}W_3^2)\right]', \\ &m_0\ddot{W}_3 = [Gh(\vartheta_2 + W_3')]' + \left[Eh(W_1' + \frac{1}{2}W_3^2)W_3'\right]', \\ &\rho_0 I_{33}\ddot{\vartheta}_2 = EI_{33}\vartheta_2'' - Gh(\vartheta_2 + W_3'). \end{aligned}\right\} \quad (68.7)$$

68.2.3.2 Boundary Conditions

The geometric boundary conditions are

$$W_i(0) = 0, \quad i = 1, 3, \quad \vartheta_2(0) = 0, \tag{68.8}$$

while the dynamical boundary conditions are

$$\left.\begin{aligned} &Eh\left[W_1' + \frac{1}{2}W_3^2\right](\ell) = f_1, \\ &Gh[\vartheta_2 + W_3'](\ell) + \left[Eh(W_1' + \frac{1}{2}W_3^2)W_3'\right](\ell) = f_2, \\ &EI_{33}\vartheta_2'(\ell) = m_1. \end{aligned}\right\} \tag{68.9}$$

68.2.3.3 Initial Conditions

$$\begin{aligned} W_i(\cdot, t = 0) = W_{i0}(\cdot), \quad \dot{W}_i(\cdot, t = 0) = W_{i1}(\cdot), \quad i = 1, 3, \\ \vartheta_2(\cdot, t = 0) = \vartheta_{20}, \quad \dot{\vartheta}_2(\cdot, t = 0) = \vartheta_{21}. \end{aligned}$$

Remark 68.1

The model in this section is attributed to Hirschhorn and Reiss [7]. If the longitudinal motion is neglected and the quadratic term in W_3 is averaged over the interval $[0, \ell]$, the model then reduces to a Woinowski–Krieger type. The corresponding partial differential equations are quasi-linear and hard to handle. If, however, one replaces Θ_2 by ϑ_2 in the expression for the strain $\bar{\varepsilon}_{11}$ (which is justified for small strains), then a semilinear partial differential equation with a cubic nonlinearity in ϑ_2 is obtained.

68.2.4 Planar, Nonshearable Nonlinear Beam

The equations of motion for the longitudinal and vertical motions, respectively, are

$$\left.\begin{aligned} m_0\ddot{W}_1 &= \left[EA(W_1' + \frac{1}{2}(W_3')^2)\right]', \\ m_0\ddot{W}_3 - [\rho_0 I_{33}\ddot{W}_3']' + [EI_{33}W_3'']'' &= \left[\left(EA(W_1' + \frac{1}{2}(W_3')^2)\right)W_3'\right]'. \end{aligned}\right\} \tag{68.10}$$

We dispense with displaying the boundary and initial conditions for this and subsequent models, as those can be immediately deduced from the previous ones. This model has been derived in Lagnese and Leugering [16].

The models above easily reduce to the classical beam equations as follows. We first concentrate on the nonshearable beams.

68.2.5 The Rayleigh Beam Model

Here the longitudinal motion is not coupled to the remaining motions. The equation for vertical motion is

$$m_0\ddot{W}_3 - [\rho_0 I_{33}\ddot{W}_3']' + [EI_{33}W_3'']'' = 0. \tag{68.11}$$

68.2.6 The Euler–Bernoulli Beam Model

This is obtained by ignoring the rotational inertia of cross sections in the Rayleigh model:

$$m_0\ddot{W}_3 + [EI_{33}W_3'']'' = 0. \tag{68.12}$$

With regard to shearable beams, two systems are singled out.

68.2.7 The Bresse System

This is a model for a planar, linear shearable beam with initial curvature involving couplings of longitudinal, vertical, and shear motions.

$$\left.\begin{aligned} m_0\ddot{W}_1 &= [Eh(W_1' - \kappa_3 W_3)]' - \kappa_3 Gh(\vartheta_2 + W_3' + \kappa_3 W_1), \\ m_0\ddot{W}_3 &= [Gh(\vartheta_2 + W_3' + \kappa_3 W_1)]' + \kappa_3 Eh[W_1' - \kappa_3 W_3], \\ \rho_0 I_{33}\ddot{\vartheta}_2 &= EI_{33}\vartheta_2'' - Gh(\vartheta_2 + W_3' + \kappa_3 W_1). \end{aligned}\right\} \tag{68.13}$$

This system was first introduced by Bresse [6].

68.2.8 The Timoshenko System

This is the Bresse model for a straight beam ($\kappa_3 = 0$), so that the longitudinal motion uncouples from the other two equations, which are

$$\left.\begin{aligned} m_0\ddot{W}_3 &= [Gh(\vartheta_2 + W_3')]' \\ \rho_0 I_{33}\ddot{\vartheta}_2 &= EI_{33}\vartheta_2'' - Gh(\vartheta_2 + W_3'). \end{aligned}\right\} \tag{68.14}$$

Remark 68.2

The models above can be taken to be the basic beam models. In applications it is also necessary to account for damping and various other (local or non-local) effects due to internal variables, such as viscoelastic damping of Boltzmann (non-local in time) or Kelvin–Voigt (local in time) types, structural damping, so-called shear-diffusion or spatial hysteresis type damping; see Russell [24] for the latter. It is also possible to impose large (and usually fast) rigid motions on the beam. We refer to [9]. A comprehensive treatment of elastic frames composed of beams of the types discussed above is given in [17]. With respect to control applications, one should also mention the modeling of beams with piezoceramic actuators; see Banks et al. [3].

68.3 Exact Controllability

68.3.1 Hyperbolic Systems

Consider the models (Equations 68.1, 68.7, 68.13, and 68.14). We concentrate on the linear equations first. All of these models can be put into the form

$$\mathbf{M}\ddot{\mathbf{z}} = [\mathbf{K}(\mathbf{z}' + \mathbf{C}\mathbf{z})]' - \mathbf{C}^T\mathbf{K}(\mathbf{z}' + \mathbf{C}\mathbf{z}) + \mathbf{f}, \tag{68.15}$$

$$\mathbf{z}(0) = 0, \quad \mathbf{K}(\mathbf{z}' + \mathbf{C}\mathbf{z})(\ell) = \mathbf{u}, \tag{68.16}$$

$$\mathbf{z}(\cdot, 0) = \mathbf{z}_0, \quad \dot{\mathbf{z}}(\cdot, 0) = \mathbf{z}_1, \tag{68.17}$$

with positive definite matrices $\mathbf{M}$,$\mathbf{K}$ depending continuously on x. In particular, for the model (Equation 68.1)

$$\begin{aligned} \mathbf{z} = (\mathbf{W}, \vartheta)^T, \quad \mathbf{M} &= \mathrm{diag}(m_0, m_0, m_0, \rho_0 I, \rho_0 I_{33}, \rho_0 I_{22}) \\ \mathbf{K} &= \mathrm{diag}(EA, GA, GA, GI, EI_{33}, EI_{22}), \\ C_{20} &= -1, \quad C_{35} = 1, \quad C_{ij} = 0 \quad \text{otherwise.} \end{aligned}$$

In the case of the Bresse beam,

$$\begin{aligned}\mathbf{z} = (W_1, W_3, \vartheta_2)^T, \quad \mathbf{M} &= \text{diag}(m_0, m_0, \rho_0 I_{33}),\\ \mathbf{K} &= \text{diag}(Eh, Gh, GA, EI_{33}),\\ C_{12} = -\kappa_3, \quad C_{21} &= \kappa_3, \quad C_{23} = 1, \quad C_{ij} = 0 \quad \text{otherwise.}\end{aligned}$$

The Timoshenko system is obtained by setting $\kappa_3 = 0$. If $\mathbf{C} = 0$, Equation 68.15 reduces to the one-dimensional wave equation.

We introduce the spaces

$$\mathbf{H} = L^2(0, \ell, \mathbb{R}^q), \quad \mathbf{V} = \{\mathbf{z} \in H^1(0, \ell, \mathbb{R}^q) : \quad \mathbf{z}(0) = 0\}, \tag{68.18}$$

where q is the number of state variables and $H^k(0, \ell, \mathbb{R}^q)$ denotes the Sobolev space consisting of $\mathbb{R}^q$ valued functions defined on the interval $(0, \ell)$ whose distributional derivatives up to order k are in $L^2(0, \ell; \mathbb{R}^q)$. (The reader is referred to [1] for general information about Sobolev spaces.) We further introduce the energy forms

$$(\mathbf{z}, \hat{\mathbf{z}})_{\mathbf{V}} := \frac{1}{2}\int_0^\ell \mathbf{K}(\mathbf{z}' + \mathbf{C}\mathbf{z}) \cdot (\hat{\mathbf{z}}' + \mathbf{C}\hat{\mathbf{z}})\, dx,$$

$$(\mathbf{z}, \hat{\mathbf{z}})_{\mathbf{H}} := \frac{1}{2}\int_0^\ell \mathbf{M}\mathbf{z} \cdot \hat{\mathbf{z}}\, dx.$$

Indeed, the norm induced by $\|\mathbf{z}\| = (\mathbf{z}, \mathbf{z})_{\mathbf{V}}^{1/2}$ is equivalent to the usual Sobolev norm of H^1 given by

$$\|\mathbf{z}\|_1 = \left(\int_0^\ell (|\mathbf{z}'|^2 + |\mathbf{z}|^2)\, dx\right)^{1/2}.$$

Let $\mathbf{z}_0 \in \mathbf{V}$, $\mathbf{z}_1 \in \mathbf{H}$, $\mathbf{f} \in L^2(0, T, \mathbf{H})$, $\mathbf{u} \in L^2(0, T, \mathbf{R}^q)$, $T > 0$. It can be proven that a unique function $\mathbf{z} \in C(0, T, \mathbf{V}) \cap C^1(0, T, \mathbf{H})$ exists which satisfies Equations 68.15, 68.16, and 68.17 in the following weak sense:

$$\frac{d^2}{dt^2}(\mathbf{z}(t), \phi)_{\mathbf{H}} + (\mathbf{z}(t), \phi)_{\mathbf{V}} = (\mathbf{M}^{-1}\mathbf{f}(t), \phi)_{\mathbf{H}} + \mathbf{u}(t) \cdot \phi(\ell), \quad \forall \phi \in \mathbf{V}, \tag{68.19}$$

and

$$(\mathbf{z}(0), \phi)_{\mathbf{H}} = (\mathbf{z}_0, \phi)_{\mathbf{H}}, \quad \frac{d}{dt}(\mathbf{z}(t), \phi)_{\mathbf{H}}|_{t=0} = (\mathbf{z}_1, \phi)_{\mathbf{H}}.$$

Because of space limitations, only boundary controls and constant coefficients will be considered. Distributed controls are easy to handle, while pointwise controls have much in common with boundary controls except for the liberty of their location. Thus we set $\mathbf{f} \equiv 0$ in Equation 68.15. The problem of *exact controllability* in its strongest sense can be formulated as follows: *Given initial data Equation* 68.17 *and final data* $(\mathbf{z}_{T0}, \mathbf{z}_{T1})$ *in* $\mathbf{V} \times \mathbf{H}$ *and given* $T > 0$, *find a control* $\mathbf{u} \in L^2(0, T, \mathbf{R}^q)$ *so that the solution* $\mathbf{z}$ *of Equation 68.15 satisfies Equations 68.16, 68.17 and* $\mathbf{z}(T) = \mathbf{z}_{T0}, \dot{\mathbf{z}}(T) = \mathbf{z}_{T1}$.

It is, in principle, possible to solve the generalized eigenvalue problem

$$\gamma^2 \frac{d^4}{dx^4}\phi = \lambda^2 (I - \gamma^2 \frac{d^2}{dx^2})\phi,$$

$$\phi(0) = \phi'(0) = \phi''(\ell) = (\phi''' + \lambda^2 \phi')(\ell) = 0,$$

and write the solution of Equations 68.15, 68.16, 68.17 using Fourier's method of separation of variables. The solution together with its time derivative can then be evaluated at T and the control problem reduces to a trigonometric (or to a complex exponential) moment problem. The controllability requirement is then equivalent to the base properties of the underlying set of complex exponentials $[exp(i\lambda_k t) | k \in \mathbf{Z}, t \in$

$(0, T)]$. If that set constitutes a Riesz basis in its $L^2(0, T)$-closure, then exact controllability is achieved. For conciseness, we do not pursue this approach here and, instead, refer to Krabs [12] for further reading. Rather, the approach we want to consider here, while equivalent to the former one, does not resort to the knowledge of eigenvalues and eigenelements. The controllability problem, as in finite dimensions, is a question of characterizing the image of a linear map, the control-to-state map. Unlike in finite dimensions, however, it is not sufficient here to establish a uniqueness result for the homogeneous adjoint problem, that is, to establish injectivity of the adjoint of the control-to-state map. In infinite-dimensional systems this implies only that the control-to-state map has dense range, which in turn is referred to as *approximate controllability*. Rather, we need some additional information on the adjoint system, namely, uniformity in the sense that the adjoint map is uniformly injective with respect to all finite energy initial (final) conditions. In particular, given a bounded linear map $\mathbf{L}$ between Hilbert spaces X, Y, the range of $\mathbf{L}$ is all of Y if, and only if, $\mathbf{L}^*$, the adjoint of $\mathbf{L}$, satisfies $\|\phi\| < \nu\|\mathbf{L}^*\phi\|$ for some positive ν and all $\phi \in Y$. This inequality also implies that the right inverse $\mathbf{L}^*(\mathbf{LL}^*)^{-1}$ exists as a *bounded* operator (see the finite-dimensional analog, where 'bounded' is generic). It is clear that this implies *norm-minimality* of the controls constructed this way. This result extends to more general space setups. It turns out that an inequality like this is needed to assure that the set of complex exponentials is a Riesz base in its L^2-closure. As will be seen shortly, such an inequality is achieved by nonstandard energy estimates, which constitute the basis for the so-called HUM method introduced by Lions. It is thus clear that this inequality is the crucial point in the study of exact controllability.

In order to obtain such estimates, we consider smooth enough solutions ϕ of the homogeneous adjoint final value problem,

$$\mathbf{M}\ddot{\phi} = [\mathbf{K}(\phi' + \mathbf{C}\phi)]' - \mathbf{C}^T\mathbf{K}(\phi' + \mathbf{C}\phi), \tag{68.20}$$

$$\phi(0) = 0, \quad \mathbf{K}(\phi' + \mathbf{C}\phi)(\ell) = 0, \tag{68.21}$$

$$\phi(\cdot, T) = \phi_0, \quad \dot{\phi}(\cdot, T) = \phi_1. \tag{68.22}$$

Let m be a smooth, positive, increasing function of x. Multiply Equation 68.20 by $m\phi'$, where $m(\cdot)$ is a smooth function in x, and integrate by parts over (x, t). After some calculus, we obtain the following crucial identity, valid for any sufficiently smooth solution of Equation 68.20:

$$\begin{aligned} 0 = &\int_0^\ell m\mathbf{M}\dot{\phi} \cdot (\phi' + \mathbf{C}\phi)|_0^T dx - \int_0^T m(x)e(x,t)|_{x=0}^\ell dt + \int_0^T \int_0^\ell m' e dx dt \\ &- \int_0^T \int_0^\ell mC^T\mathbf{M}\dot{\phi} \cdot \dot{\phi} dx dt + \int_0^T \int_0^\ell mC^T\mathbf{K}(\phi' + \mathbf{C}\phi) \cdot (\phi' + \mathbf{C}\phi)\, dx dt, \end{aligned} \tag{68.23}$$

where $\rho = \int_0^\ell m\mathbf{M}\dot{\phi} \cdot (\phi' + \mathbf{C}\phi)\, dx$ and $e(x, t)$ denotes the energy density given by

$$e = \frac{1}{2}[\mathbf{M}\dot{\phi} \cdot \dot{\phi} + \mathbf{K}(\phi' + \mathbf{C}\phi) \cdot (\phi' + \mathbf{C}\phi)].$$

Set the total energy at time t,

$$\mathcal{E}(t) = \int_0^\ell e(x, t)\, dx, \tag{68.24}$$

and denote the norm of the matrix $\mathbf{C}$ by ν. If we choose m in Equation 68.23 so that $m'(x) - \nu m(x) \geq c_0 > 0, \forall x \in (0, \ell)$, we obtain the estimates

$$\gamma \int_0^T |\dot{\phi}(\ell, t)|^2 dt \leq \mathcal{E}(0) \leq \Gamma \int_0^T |\dot{\phi}(\ell, t)|^2 dt, \tag{68.25}$$

for some positive constants $\gamma(T), \Gamma(T)$, where T is sufficiently large (indeed, $T > 2\times$ "optical length of the beam" is sufficient). The second inequality in Equation 68.25 requires the multiplier above and some estimation, whereas the first requires the multiplier $m(x) = -1 + 2x/\ell$, and is straightforwardly proved. These inequalities can also be obtained by the method of characteristics which, in addition, yields the smallest possible control time [17]. It is then shown that the norm of the adjoint to the control-to-state map (which takes the control $\mathbf{u}$ into the final values $\mathbf{z}(T), \dot{\mathbf{z}}(T)$ (for zero initial conditions)), applied to ϕ_0, ϕ_1, is exactly equal to $\int_0^T |\dot{\phi}(\ell,t)|^2 dt$. By the above argument, the original map is onto between the control space $L^2(0, T, \mathbb{R}^q)$ and the *finite energy space* $\mathbf{V} \times \mathbf{H}$.

Theorem 68.1:

Let $(\mathbf{z}_0, \mathbf{z}_1), (\mathbf{z}_{T0}, \mathbf{z}_{T1})$ be in $\mathbf{V} \times \mathbf{H}$ and $T > 0$ sufficiently large. Then a unique control $\mathbf{u} \in L^2(0, T, \mathbb{R}^q)$ exists, with minimal norm, so that $\mathbf{z}$ satisfies Equations 68.15, 68.16, 68.17 and $\mathbf{z}(T) = z_{T0}, \dot{\mathbf{z}}(T) = z_{T1}$.

Remark 68.3

Controllability results for the fully nonlinear planar shearable beam Equation 68.7, and also for the Woinowski–Krieger-type approximation, are not known at present. The semilinear model is locally exactly controllable, using the implicit function theorem. The argument is quite similar to the one commonly used in finite-dimensional control theory.

68.3.2 Quasi-Hyperbolic Systems

In this section we discuss the (linearized) nonshearable models with rotational inertia, namely, Equations 68.4, 68.10, 68.11. We first discuss the linear subsystems. Observe that in that situation all equations decouple into wave equations governing the longitudinal and torsional motion and equations of the type of Equation 68.11. Hence it is sufficient to consider the latter. For simplicity, we restrict ourselves to constant coefficients. The system is then

$$\begin{aligned} \rho h \ddot{W} - \rho I \ddot{W}'' + EIW'''' &= f, \\ W(0) = 0, \quad W'(0) &= 0, \end{aligned}$$

and

$$\begin{aligned} EIW''(\ell) = u_1, \quad (EIW''' - \rho I \ddot{W}')(\ell) = u_2, \\ W(\cdot, 0) = W_0, \quad \dot{W}(\cdot, 0) = W_1. \end{aligned} \tag{68.26}$$

Setting $\gamma^2 := I/A$ and rescaling by $t \to t\sqrt{\rho/E}$, this system can be brought into a nondimensional form. We define spaces,

$$\begin{aligned} \mathbf{H} &= [v \in H^1(0, \ell): \quad v(0) = 0], \\ \mathbf{V} &= [v \in H^2(0, \ell): \quad v(0) = v'(0) = 0], \end{aligned} \tag{68.27}$$

and forms,

$$(u, v) = \int_0^\ell uv dx, \quad (u, v)_{\mathbf{H}} = (u, v) + \gamma^2(u', v'),$$
$$(u, v)_{\mathbf{V}} = \gamma^2(u'', v'').$$

Let $W_0 \in \mathbf{V}$, $W_1 \in \mathbf{H}$, and $\mathbf{u} \in L^2(0, T, \mathbb{R}^2)$, $T > 0$. It may be proved that there is a unique $\mathbf{W} \in C(0, T, \mathbf{V}) \cap C^1(0, T, \mathbf{H})$ satisfying Equation 68.26 in an appropriate variational sense.

Remark 68.4

The nonlinear models can be treated using the theory of nonlinear maximal monotone operators; see Lagnese and Leugering [16].

To produce an energy identity analogous to Equation 68.23, we multiply the first equation of Equation 68.26 by $xW' - \alpha W$, where $\alpha > 0$ is a free parameter, and then we integrate over $(0, \ell) \times (0, T)$. If we introduce the auxiliary functions $\rho_1 = \int_0^\ell \dot{W}(xW' - \alpha W)\, dx$ and $\rho_2 = \gamma^2 \int_0^\ell \dot{W}'(xW' - \alpha W)'\, dx$, and $\rho = \rho_1 + \rho_2$, we find after some calculus

$$
\begin{aligned}
0 = \rho(T) - \rho(0) - \frac{\ell}{2}\int_0^T \{[\dot{W}(\ell,t)]^2 + \gamma^2[\dot{W}'(\ell,t)]^2\}\, dt \\
+ \gamma^2 \int_0^T [W''(\ell,t)]^2 dt + \int_0^T (\gamma^2 W''' - \gamma^2 \ddot{W}')(\ell,t)(\ell W' - \alpha W)(\ell,t))\, dt \\
+ \int_0^T \int_0^\ell \left[\left(\frac{1}{2} + \alpha\right)\dot{W}^2 + \gamma^2\left(\alpha - \frac{1}{2}\right)(\dot{W}')^2 + \gamma^2\left(\frac{3}{2} - \alpha\right)(W'')^2\right] dxdt.
\end{aligned}
$$

With the total energy now defined by

$$
\mathcal{E}(t) = \frac{1}{2}\{\|\dot{W}(t)\|_{\mathbf{H}}^2 + \|W\|_{\mathbf{V}}^2\}, \tag{68.28}
$$

this identity can now be used to derive the energy estimate,

$$
\begin{aligned}
\pi \int_0^T [\dot{\phi}^2(\ell,t) + \gamma^2(\dot{\phi}')^2(\ell,t)]\, dt \le \mathcal{E}(0) \\
\le \Pi \int_0^T [\dot{\phi}^2(\ell,t) + \gamma^2(\dot{\phi}')^2(\ell,t)]\, dt,
\end{aligned}
$$

for some positive constants $\pi(T)$ *and* $\Pi(T)$, which is valid for sufficiently smooth solutions ϕ to the homogeneous system, Equation 68.26, and for sufficiently large $T > 0$ (again T is related to the "optical length," i.e., to wave velocities). The first estimate is more standard and determines the regularity of the solutions. It is again a matter of calculating the control-to-state map and its adjoint. After some routine calculation, one verifies that the norm of the adjoint, applied to the final data for the backwards running homogeneous equation, coincides with the time integral in the energy estimate. This leads to the exact controllability of the system, Equation 68.26, in the space $\mathbf{V} \times H$, with $\mathbf{V}$ and $\mathbf{H}$ as defined in Equation 68.27, using controls $\mathbf{u} = (u_1, u_2) \in L^2(0, T, \mathbb{R}^2)$.

Remark 68.5

As in [18], for $\gamma \to 0$, the controllability results for Rayleigh beams carry over to the corresponding results for Euler-Bernoulli beams. It is however instructive and, in fact, much easier to establish controllability of the Euler-Bernoulli beam directly with control only in the shear force.

68.3.3 The Euler–Bernoulli Beam

We consider the nondimensional form of Equation 68.12, namely,

$$
\begin{aligned}
\ddot{W} + W'''' = 0 \\
W(0) = W'(0) = 0, \quad W''(\ell) = 0, \quad W'''(\ell) = u \\
W(\cdot, 0) = W_0, \quad \dot{W}(\cdot, \ell) = W_1.
\end{aligned} \tag{68.29}
$$

We introduce the spaces

$$
\mathbf{H} = L^2(0, \ell), \quad \mathbf{V} = \{v \in H^2 | v(0) = v'(0) = 0\} \tag{68.30}
$$

and the corresponding energy functional

$$\mathcal{E}(t) = \frac{1}{2}(\|\dot{W}(t)\|_{\mathbf{H}}^2 + \|W(t)\|_{\mathbf{V}}^2. \tag{68.31}$$

Again, we are going to use multipliers to establish energy identities. The usual choice is $m(x) = x$. Upon introducing $\rho = \int_0^\ell x\dot{W}W'\,dx$ and multiplying the first equation by mW', followed by integration by parts, we obtain

$$\rho(T) - \rho(0) + \frac{1}{2}\int_0^T\int_0^\ell \dot{W}^2 dxdt + \frac{3}{2}\int_0^T\int_0^\ell (W'')^2 dxdt = \frac{\ell}{2}\int_0^T \dot{W}^2 dt + \ell\int_0^T W'(\ell,t)u(t)\,dt. \tag{68.32}$$

Using this identity for the homogeneous system solved by ϕ, we obtain the energy estimates

$$\pi\int_0^T \dot{\phi}^2(\ell,t)\,dt \le \mathcal{E}(0) \le \Pi\int_0^T \dot{\phi}^2(\ell,t)\,dt$$

where again $\pi(T)$ and $\Pi(T)$ depend on $T > 0$, with T sufficiently large.

One way to obtain the adjoint control-to-state-map is to consider

$$\frac{d}{dt}\int_0^\ell \{\dot{W}\dot{\phi} + W''\phi''\}dx = -u(t)\dot{\phi}(\ell,t),$$

for W as above and ϕ solving the backwards running adjoint equation (i.e., Equation 68.29 with final conditions ϕ_{T0} and ϕ_{T1}). Integrating with respect to time over $(0, T)$ yields

$$(L_T(u), (\phi_{T0}, \phi_{T1}))_{\mathbf{V}\times\mathbf{H}} = -\int_0^T u(t)\dot{\phi}(\ell,t)\,dt.$$

The same argument as above yields the conclusion of exact controllability of the system (Equation 68.29), in the space $\mathbf{V} \times \mathbf{H}$, where $\mathbf{V}$ and $\mathbf{H}$ are defined in Equation 68.30, using controls $u \in L^2(0, T, \mathbb{R})$.

Remark 68.6

It may be shown that the control time T for the Euler-Bernoulli system can actually be taken arbitrarily small. That is typical for this kind of model (Petrovskii type systems) and is closely related to the absence of a uniform wave speed. The reader is referred to the survey article [15] for general background information on controllability and stabilizability of beams and plates.

68.4 Stabilizability

We proceed to establish uniform exponential decay for the solutions of the various beam models by *linear* feedback controls at the boundary $x = \ell$. There is much current work on nonlinear and constrained feedback laws. However, the results are usually very technical, and, therefore, do not seem suitable for reproduction in these notes. In the linear case it is known that for time reversible systems, exact controllability is equivalent to uniform exponential stabilizability. In contrast to the finite-dimensional case, however, we have to distinguish between various concepts of controllability, such as exact, spectral, or approximate controllability. Accordingly, we have to distinguish between different concepts of stabilizability, as uniform exponential decay is substantially different from nonuniform decay. Because of space limitations, we do not dwell on the relation between controllability, stabilizability, and even observability. The procedure we follow is based on Lyapunov functions, and is the same in all of the models. Once again the energy identities, Equations 68.23, 68.28, and 68.32 are crucial. We take the hyperbolic case as an exemplar and outline the procedure in that case.

68.4.1 Hyperbolic Systems

Apply Equation 68.23 to solve Equation 68.15 (with $\mathbf{f} = 0$) and Equation 68.16 with the control $\mathbf{u}$ in Equation 68.16 replaced by a linear feedback law $\mathbf{u}(t) = -k\mathbf{z}(\dot{\ell}, t)$, $k > 0$. Recall that $\rho = \int_0^\ell m\mathbf{M}\dot{\mathbf{z}} \cdot (\mathbf{z}' + \mathbf{C}\mathbf{z})\, dx(t)$. Then using Equation 68.23,

$$\dot{\rho} \leq \gamma|\dot{\mathbf{z}}(\ell, t)|^2 - c_0\mathcal{E}(t),$$

where $\mathcal{E}$ is given by Equation 68.24. Therefore, introducing the function $\mathcal{F}_\epsilon(t) := \mathcal{E}(t) + \epsilon\rho(t)$ one finds

$$\dot{\mathcal{F}}_\epsilon \leq \dot{\mathcal{E}}(t) + \gamma\epsilon|\dot{\mathbf{z}}(\ell, t)|^2 - \epsilon c_0\mathcal{E}(t).$$

However, $\dot{\mathcal{E}}(t) = -k|\dot{\mathbf{z}}(\ell, t)|^2$, and therefore the boundary term can be compensated for by choosing ϵ sufficiently small. This results in the estimate $\dot{\mathcal{F}}_\epsilon(t) \leq -c_1\epsilon\mathcal{E}(t)$ for some $c_1 > 0$, which in turn implies

$$\mathcal{F}_\epsilon(t) + c_1\epsilon\int_0^t \mathcal{E}(s)ds \leq \mathcal{F}(0).$$

It also straightforward to see that $\mathcal{F}(t)$ satisfies

$$\pi_\epsilon\mathcal{E}(t) \leq \mathcal{F}(t) \leq \Pi_\epsilon\mathcal{E}(t).$$

The latter implies $\int_t^\infty \mathcal{E}(s)ds \leq (1/\lambda)\mathcal{E}(0)$, with $\lambda = \Pi_\epsilon/(c_1\epsilon)$ and $t \geq 0$. One defines $\eta(t) := \int_t^\infty \mathcal{E}(s)ds$ and obtains a differential inequality $\dot{\eta} + \lambda\eta \leq 0$. A standard Gronwall argument implies $\eta(t) \leq \exp(-\lambda t)\eta(0)$, that is,

$$\int_t^\infty \mathcal{E}(s)ds \leq \frac{\exp(-\lambda t)}{\lambda}\mathcal{E}(0).$$

Now, because $\mathcal{E}(t)$ is nonincreasing,

$$\tau\mathcal{E}(\tau + t) \leq \left\{\int_t^{\tau+t} + \int_{\tau+t}^\infty\right\}\mathcal{E}(s)ds \leq \frac{\exp(-\lambda t)}{\lambda}\mathcal{E}(0)$$

and this, together with the choice $\tau = 1/\lambda$, gives

$$\mathcal{E}(t) \leq e\exp(-\lambda t)\mathcal{E}(0), \quad \forall t \geq \tau. \tag{68.33}$$

Hence, we have the following result.

Theorem 68.2:

Let $\mathbf{V}$ *and* $\mathbf{H}$ *be given by Equation* 68.18. *Given initial data* $\mathbf{z}_0, \mathbf{z}_1$ *in* $\mathbf{V} \times \mathbf{H}$, *the solution to the closed-loop system, Equations 68.15 through 68.17, with* $\mathbf{u}(t) = -k\dot{\mathbf{z}}(\ell, t)$, *satisfies*

$$\mathcal{E}(t) \leq M\exp(-\omega t)\mathcal{E}(0), \quad t \geq 0, \tag{68.34}$$

for some positive constants M *and* ω.

Remark 68.7

The linear feedback law can be replaced by a monotone nonlinear feedback law with certain growth conditions. The corresponding energy estimates, however, are beyond the scope of these notes. Ultimately, the differential inequality above is to be replaced by a nonlinear one. The exponential decay has then (in general) to be replaced by an algebraic decay, see [17]. The decay rate can be optimized using "hyperbolic estimates" as in [17]. Also the dependence on the feedback parameter can be made explicit.

68.4.2 Quasi-Hyperbolic Systems

We consider the problem of Equation 68.26 with feedback controls,

$$u_1(t) = -k_1\dot{W}'(\ell,t), \quad u_2(t) = k_2\dot{W}(\ell,t), \tag{68.35}$$

with positive feedback gains k_1 and k_2. The identity Equation 68.23 is used to calculate the derivative of the function $\rho(t)$. By following the same procedure as above, we obtain the decay estimate Equation 68.34 for the closed-loop system, Equations 68.26 and 68.35, where the energy functional $\mathcal{E}$ is given by Equation 68.28.

Remark 68.8

One can show algebraic decay for certain monotone nonlinear feedbacks. In addition, the nonlinear system, Equation 68.10, exhibits those decay rates as well; see Lagnese and Leugering [16].

68.4.3 The Euler–Bernoulli Beam

Here we consider the system Equation 68.29 and close the loop by setting $u(t) = -k\dot{W}(\ell,t), k > 0$. By utilizing the estimate Equation 68.32 and proceeding in much the same way as above, the decay estimate Equation 68.34 can be established for the closed-loop system, where $\mathcal{E}$ is given by Equation 68.31.

68.5 Dynamic Elastic Plate Models

Let Ω be a bounded, open, connected set in $\mathbb{R}^2$ with a Lipschitz continuous boundary consisting of a finite number of smooth curves. Consider a deformable three-dimensional body which, in equilibrium, occupies the region

$$[(x_1,x_2,x_3): \quad (x_1,x_2)\in\overline{\Omega}, \quad |x_3|\le h/2]. \tag{68.36}$$

When the quantity h is very small compared to the diameter of Ω, the body is referred to as a *thin plate of uniform thickness h* and the planar region,

$$[(x_1,x_2,0): (x_1,x_2)\in\overline{\Omega}]$$

is its *reference surface*.

Two-dimensional mathematical models describing the deformation of the three-dimensional body Equation 68.36 are obtained by relating the displacement vector associated with the deformation of each point within the body to certain *state variables* defined on the reference surface. Many such models are available; three are briefly described below.

68.5.1 Linear Models

Let $\mathbf{W}(x_1,x_2,x_3,t)$ denote the displacement vector at time t of the material point located at (x_1,x_2,x_3), and let $\mathbf{w}(x_1,x_2,t)$ denote the displacement vector of the material point located at $(x_1,x_2,0)$ in the reference surface. Further, let $\mathbf{n}(x_1,x_2,t)$ be the unit-normal vector to the deformed reference surface at the point $(x_1,x_2,0)+\mathbf{w}(x_1,x_2,t)$. The direction of $\mathbf{n}$ is chosen so that $\mathbf{n}\cdot\mathbf{k}>0$, where $\mathbf{i},\mathbf{j},\mathbf{k}$ is the natural basis for $\mathbb{R}^3$.

68.5.1.1 Kirchhoff Model

The basic kinematic assumption of this model is

$$\mathbf{W}(x_1,x_2,x_3,t) = \mathbf{w}(x_1,x_2,t) + x_3(\mathbf{n}(x_1,x_2,t)-\mathbf{k}), \tag{68.37}$$

which means that a filament in its equilibrium position, orthogonal to the reference surface, remains straight, unstretched, and orthogonal to the deformed reference surface. It is further assumed that the material is linearly elastic (Hookean), homogeneous and isotropic, that the transverse normal stress is small compared to the remaining stresses, and that the strains and the normal vector $\mathbf{n}$ are well-approximated by their linear approximations. Write $\mathbf{w} = w_1\mathbf{i} + w_2\mathbf{j} + w_3\mathbf{k}$. Under the assumptions above, there is no coupling between the in-plane displacements w_1, w_2 and the transverse displacement $w_3 := w$. The former components satisfy the partial differential equations of linear plane elasticity and the latter satisfies the equation

$$\rho h\frac{\partial^2 w}{\partial t^2} - I_\rho \Delta\frac{\partial^2 w}{\partial t^2} + D\Delta^2 w = F, \tag{68.38}$$

where $\Delta = \partial^2/\partial x_1^2 + \partial^2/\partial x_2^2$ is the harmonic operator in $\mathbb{R}^2$, ρ is the mass density per unit of reference volume, $I_\rho = \rho h^3/12$ is the polar moment of inertia, D is the modulus of flexural rigidity, and F is the transverse component of an applied force distributed over Ω. The "standard" Kirchhoff plate equation is obtained by omitting the term $I_\rho\Delta(\partial^2 w/\partial t^2)$, which accounts for the rotational inertia of cross sections, from Equation 68.38.

68.5.1.2 Reissner–Mindlin System

The basic kinematic assumption of this model is

$$\mathbf{W}(x_1, x_2, x_3, t) = \mathbf{w}(x_1, x_2, t) + x_3\mathbf{U}(x_1, x_2, t), \tag{68.39}$$

where $|\mathbf{U} + \mathbf{k}| = 1$. Equation 68.39 means that a filament in its equilibrium position, orthogonal to the reference surface, remains straight and unstretched but not necessarily orthogonal to the deformed reference surface. Write

$$\mathbf{U} = U_1\mathbf{i} + U_2\mathbf{j} + U_3\mathbf{k}, \quad \mathbf{u} = U_1\mathbf{i} + U_2\mathbf{j}.$$

In the linear approximation, $U_3 = 0$ so that the state variables of the problem are $\mathbf{w}$ and $\mathbf{u}$. The latter variable accounts for transverse shearing of cross sections. It is further assumed that the material is homogeneous and Hookean. The stress-strain relations assume that the material is isotropic in directions parallel to the reference surface but may have different material properties in the transverse direction. As in the previous case, there is no coupling between w_1, w_2, and the remaining state variables in the linear approximations. The equations of motion satisfied by $w_3 := w$ and $\mathbf{u}$, referred to as the Reissner or Reissner–Mindlin system, may be written

$$\left.\begin{aligned} &\rho h\frac{\partial^2 w}{\partial t^2} - Gh\,\mathrm{div}\,(\mathbf{u} + \nabla w) = F, \quad \text{and} \\ &I_\rho\frac{\partial^2 \mathbf{u}}{\partial t^2} - \frac{h^3}{12}\mathrm{div}\,\sigma(\mathbf{u}) + Gh(\mathbf{u} + \nabla w) = \mathbf{C}, \end{aligned}\right\} \tag{68.40}$$

where Gh is the shear modulus and $\sigma(\mathbf{u}) = (\sigma_{ij}(\mathbf{u}))$ is the stress tensor associated with $\mathbf{u}$, i.e.,

$$\sigma_{ij}(\mathbf{u}) = 2\mu\varepsilon_{ij}(\mathbf{u}) + \frac{2\mu\lambda}{2\mu + \lambda}\delta_{ij}\sum_{k=1}^{2}\varepsilon_{kk}(\mathbf{u}), \quad i, j = 1, 2.$$

$\varepsilon_{ij}(\mathbf{u})$ denotes the linearized strain

$$\varepsilon_{ij}(\mathbf{u}) = \frac{1}{2}\left(\frac{\partial U_i}{\partial x_j} + \frac{\partial U_j}{\partial x_i}\right),$$

λ and μ are the Lamé parameters of the material,

$$\mathrm{div}\,(\mathbf{u} + \nabla w) = \nabla\cdot(\mathbf{u} + \nabla w), \quad \mathrm{div}\,\sigma(\mathbf{u}) = \sum_{j=1}^{3}\frac{\partial}{\partial x_j}\sigma_{ij}(\mathbf{u}),$$

and $\mathbf{C} = C_1\mathbf{i} + C_2\mathbf{j}$ is a distributed force couple.

68.5.2 A Nonlinear Model: The von Kármán System

Unlike the two previous models, this is a "large deflection" model. It is obtained under the same assumptions as the Kirchhoff model except for the linearization of the strain tensor. Rather, in the general strain tensor,

$$\varepsilon_{ij}(\mathbf{W}) = \frac{1}{2}\left(\frac{\partial W_i}{\partial x_j} + \frac{\partial W_j}{\partial x_i}\right) + \frac{1}{2}\sum_{k=1}^{3}\frac{\partial W_k}{\partial x_i}\frac{\partial W_k}{\partial x_j},$$

the quadratic terms involving W_3 are retained, an assumption formally justified if the planar strains are small relative to the transverse strains. The result is a nonlinear plate model in which the in-plane components of displacement w_1 and w_2 are coupled to the transverse displacement $w_3 := w$. Under some further simplifying assumptions, w_1 and w_2 may be replaced by a single function G, called an Airy stress function, related to the in-plane stresses. The resulting nonlinear equations for w and G are

$$\left.\begin{aligned} &\rho h\frac{\partial^2 w}{\partial t^2} - I_\rho \Delta\frac{\partial^2 w}{\partial t^2} + D\Delta^2 w - [w, G] = F, \\ &\Delta^2 G + \frac{Eh}{2}[w, w] = 0, \end{aligned}\right\} \tag{68.41}$$

where E is Young's modulus and where

$$[\phi, \psi] = \frac{\partial^2\phi}{\partial x_1^2}\frac{\partial^2\psi}{\partial x_2^2} + \frac{\partial^2\psi}{\partial x_1^2}\frac{\partial^2\phi}{\partial x_2^2} - 2\frac{\partial^2\phi}{\partial x_1\partial x_2}\frac{\partial^2\psi}{\partial x_1\partial x_2}.$$

One may observe that $[w, w]/2$ is the Gaussian curvature of the deformed reference surface $x_3 = w(x_1, x_2)$. The "standard" dynamic von Kármán plate system is obtained by setting $I_\rho = 0$ in Equation 68.41.

Remark 68.9

For a derivation of various plate models, including thermoelastic plates and viscoelastic plates, see [18]. For studies of junction conditions between two or more interconnected (not necessarily co-planar) elastic plates, the reader is referred to the monographs [17,21] and references therein.

68.5.3 Boundary Conditions

Let Γ denote the boundary of Ω. The boundary conditions are of two types: geometric conditions, that constrain the geometry of the deformation at the boundary, and mechanical (or dynamic) conditions, that represent the balance of linear and angular momenta at the boundary.

68.5.3.1 Boundary Conditions for the Reissner–Mindlin System

Geometric conditions are given by

$$w = \bar{w}, \quad \mathbf{u} = \bar{\mathbf{u}} \quad \text{on } \Gamma, \ t > 0. \tag{68.42}$$

The case $\bar{w} = \bar{\mathbf{u}} = 0$ corresponds to a rigidly clamped boundary.

The mechanical boundary conditions are given by

$$\left.\begin{aligned} &Gh\nu \cdot (\mathbf{u} + \nabla w) = f, \\ &\frac{h^3}{12}\sigma(\mathbf{u})\nu = \mathbf{c} \quad \text{on } \Gamma, t > 0, \end{aligned}\right\} \tag{68.43}$$

where ν is the unit exterior pointing normal vector to the boundary of Ω, $\mathbf{c} = c_1\mathbf{i} + c_2\mathbf{j}$ is a boundary force couple and f is the transverse component of an applied force distributed over the boundary. The problem consisting of the system of Equations 68.40 and boundary conditions Equations 68.42 *or* 68.43, together with the *initial conditions*

$$\left.\begin{aligned} &w = w^0, \quad \frac{\partial w}{\partial t} = w^1, \\ &\mathbf{u} = \mathbf{u}^0, \quad \frac{\partial \mathbf{u}}{\partial t} = \mathbf{u}^1 \quad \text{at } t = 0, \end{aligned}\right\} \tag{68.44}$$

has a unique solution if the data of the problem is sufficiently regular. The same is true if the boundary conditions are Equation 68.42 on one part of Γ and Equation 68.43 on the remaining part, or if they consist of the first (resp., second) of the two expressions in Equation 68.42 and the second (resp., first) of the two expressions in Equation 68.43.

68.5.3.2 Boundary Conditions for the Kirchhoff Model

The geometric boundary conditions are

$$w = \bar{w}, \quad \frac{\partial w}{\partial \nu} = -\nu \cdot \bar{\mathbf{u}} \quad \text{on } \Gamma, t > 0, \tag{68.45}$$

and the mechanical boundary conditions may be written

$$\left.\begin{aligned} &\frac{h^3}{12}\nu \cdot \sigma(\nabla w)\nu = -\nu \cdot \mathbf{c}, \\ &\frac{\partial}{\partial \nu}\left(I_\rho \frac{\partial^2 w}{\partial t^2} - D\Delta w\right) - \frac{h^3}{12}\frac{\partial}{\partial \tau}[\tau \cdot \sigma(\nabla w)\nu] = \frac{\partial}{\partial \tau}(\tau \cdot \mathbf{c}) + f. \end{aligned}\right\} \tag{68.46}$$

68.5.3.3 Boundary Conditions for the von Kármán System

The state variables w and G are not coupled in the boundary conditions. The geometric and mechanical boundary conditions for w are those of the Kirchhoff model. The boundary conditions satisfied by G are

$$G = 0, \quad \frac{\partial G}{\partial \nu} = 0. \tag{68.47}$$

These arise if there are no in-plane applied forces along the boundary.

68.6 Controllability of Dynamic Plates

In the models discussed in the last section, some, or all, of the applied forces and moments $F, \mathbf{C}, f, \mathbf{c}$, and the geometric data $\bar{w}, \bar{\mathbf{u}}$, may be considered as *controls* which must be chosen in order to affect the transient behavior of the solution in some specified manner. These controls may either be *open loop*, or *closed loop*. Open-loop controls are usually associated with problems of controllability, which is that of steering the solution to, or nearly to, a specified state at a specified time. Closed-loop controls are usually associated with problems of stabilizability, that is, of asymptotically driving the solution towards an equilibrium state of the system.

In fact, for infinite-dimensional systems of which the above plate models are representative, there are various related but distinct concepts of controllability (spectral, approximate, exact) and of stabilizability (weak, strong, uniform), distinctions which disappear in finite-dimensional approximations of these models (see [2, Chapter 4]). Stabilizability problems will be discussed in the next section. With regard to controllability, *exact controllability* is the most stringent requirement because it requires a complete description of the configuration space (reachable set) of the solution. This is equivalent to steering any initial state of the system to any other permissible state within a specified interval of time. The notion of *spectral controllability* involves exactly controlling the span of any set of *finitely many* of the eigenmodes of the system. *Approximate controllability* involves steering an arbitrary initial state to a given, but arbitrary, neighborhood of a desired configuration within a specified time.

Among the possible controls, distinctions are made between *distributed controls* such as F and $\mathbf{C}$, which are distributed over all or a portion of the face of the plate, and *boundary controls,* such as $f, \mathbf{c}, \bar{w}, \bar{\mathbf{u}}$, which are distributed over all or a portion of the edge of the plate. Within the class of boundary controls, a further distinction is made between mechanical controls, f and $\mathbf{c}$, and geometric controls $\bar{w}$ and $\bar{\mathbf{u}}$. Because mechanical controls correspond to forces and moments, they are, in principle, physically implementable; these are the only types of controls which will be considered here. In addition, only boundary control problems will be considered in detail; however, some remarks regarding distributed control problems will be provided.

68.6.1 Controllability of Kirchhoff Plates

Assume that $\Gamma = \overline{\Gamma}_0 \cup \overline{\Gamma}_1$, where Γ_0 and Γ_1 are disjoint, relatively open subsets of Γ with $\Gamma_1 \neq \emptyset$. The problem under consideration consists of the partial differential equation 68.38, boundary conditions Equation 68.45 on Γ_0, boundary conditions Equation 68.46 on Γ_1, and initial conditions

$$w(x,0) = w^0(x), \quad \frac{\partial w}{\partial t}(x,0) = w^1(x), \quad x \in \Omega. \tag{68.48}$$

In this system, the distributed force F, the geometric quantities $\bar{w}, \bar{\mathbf{u}}$, and the initial data (w^0, w^1) are assumed as given data, while $f, \mathbf{c}$ are the controls, chosen from a certain class $\mathcal{C}$ of *admissible controls.* The *configuration space*, or the *reachable set*, at time T is

$$\mathcal{R}_T = \{(w(T), \dot{w}(T)) : (f, \mathbf{c}) \in \mathcal{C}\},$$

where, for example, $w(T)$ stands for the function $[w(x,T) : x \in \Omega]$ and where $\dot{w} = \partial w/\partial t$. If z denotes the solution of the uncontrolled problem, i.e., the solution with $f = 0$ and $\mathbf{c} = 0$, then

$$\mathcal{R}_T = \mathcal{R}_T^0 \oplus \{[z(T), \dot{z}(T)]\},$$

where $\mathcal{R}_T^0$ denotes the configuration space when all of the given data are zero. Therefore, to study the reachable set it may be assumed without loss of generality that the data $F, \bar{w}, \bar{\mathbf{u}}, w^0, w^1$ vanish. The problem under consideration is, therefore,

$$\rho h \frac{\partial^2 w}{\partial t^2} - I_\rho \Delta \frac{\partial^2 w}{\partial t^2} + D\Delta^2 w = 0, \tag{68.49}$$

$$w = 0, \quad \frac{\partial w}{\partial \nu} = 0 \quad \text{on } \Gamma_0, t > 0, \tag{68.50}$$

$$\left.\begin{aligned} &\frac{\partial}{\partial \nu}\left(I_\rho \frac{\partial^2 w}{\partial t^2} - D\Delta w\right) - \frac{h^3}{12}\frac{\partial}{\partial \tau}[\tau \cdot \sigma(\nabla w)\nu] = \frac{\partial}{\partial \tau}(\tau \cdot \mathbf{c}) + f, \\ &\frac{h^3}{12}\nu \cdot \sigma(\nabla w)\nu = -\nu \cdot \mathbf{c} \quad \text{on } \Gamma_1, t > 0, \end{aligned}\right\} \tag{68.51}$$

$$w(x,0) = \frac{\partial w}{\partial t}(x,0) = 0, \quad x \in \Omega. \tag{68.52}$$

If w is a solution of Equation 68.49, its *kinetic energy* at time t is

$$\mathcal{K}(t) = \frac{1}{2}\int_\Omega (\rho h \dot{w}^2 + I_\rho |\nabla \dot{w}|^2)\, d\Omega,$$

where the quantities in the integrand are evaluated at time t. The *strain energy* of this solution at time t is given by

$$\mathcal{U}(t) = \frac{1}{2}\frac{h^3}{12}\sum_{i,j=1}^{2}\int_{\Omega}\sigma_{ij}(\nabla w)\varepsilon_{ij}(\nabla w)\,d\Omega.$$

A pair of functions (w_0, w_1) defined on Ω is called a *finite energy pair* if

$$\sum_{i,j=1}^{2}\int_{\Omega}\sigma_{ij}(\nabla w_0)\varepsilon_{ij}(\nabla w_0)\,d\Omega < \infty,$$

$$\int_{\Omega}(\rho h w_1^2 + I_\rho|\nabla w_1|^2)\,d\Omega < \infty.$$

A solution w of Equation 68.49 is called a *finite energy solution* if $[w(t), \dot{w}(t)]$ is a finite energy pair for each $t \geq 0$ and is continuous with respect to t into the space of finite energy pairs. This means that the solution has finite kinetic and strain energies at each instant which vary continuously in time.

Many choices of the control space $\mathcal{C}$ are possible, each of which will lead to different configuration space $\mathcal{R}_T^0$. One requirement on the choice of $\mathcal{C}$ is that the solution w corresponding to given input, f, $\mathbf{c}$ be reasonably well behaved. Another is that the choice of $\mathcal{C}$ lead to a sufficiently rich configuration space. For the problem under consideration, it is very difficult to determine the precise relation between the control and configuration spaces. For example, there is no simple characterization of those inputs for which the corresponding solution has finite energy at each instant. On the other hand, when standard control spaces with simple structure, such as L^2 spaces, are utilized, the regularity properties of the solution are, in general, difficult to determine. (This is in contrast to the situation which occurs in the analogous boundary control problem for Rayleigh beams, where it is known that finite energy solutions correspond exactly to inputs which are L^2 in time.)

In order to make the ideas precise, it is necessary to introduce certain function spaces based on the energy functionals $\mathcal{K}$ and $\mathcal{U}$. Let $L^2(\Omega)$ denote the space of square integrable functions defined on Ω, and let $H^k(\Omega)$ be the Sobolev space consisting of functions in $L^2(\Omega)$ whose derivatives up to order k (in the sense of distributions) belong to $L^2(\Omega)$. Let

$$H = \{v \in H^1(\Omega) : v|_{\Gamma_0} = 0\}.$$

The quantity

$$\|v\|_H = \left(\int_{\Omega}(\rho h v^2 + I_\rho|\nabla v|^2)\,d\Omega\right)^{1/2}$$

defines a Hilbert norm on H which is equivalent to the standard induced $H^1(\Omega)$ norm. Similarly, define

$$V = \{v \in H : v \in H^2(\Omega),\ \left.\frac{\partial v}{\partial \nu}\right|_{\Gamma_0} = 0\}.$$

The quantity

$$\|v\|_V = \left(\int_{\Omega}\left(\frac{h^3}{12}\sigma_{ij}(\nabla v)\varepsilon_{ij}(\nabla v)\right)d\Omega\right)^{1/2}$$

defines a seminorm on V. In fact, as a consequence of Korn's lemma, $\|\cdot\|_V$ is actually a *norm* equivalent to the standard induced $H^2(\Omega)$ norm whenever $\Gamma_0 \neq \emptyset$. Such will be assumed in what follows to simplify the discussion. The Hilbert space V is dense in H and the injection $V \mapsto H$ is compact. Let H be identified with its dual space and let V^* denote the dual space of V. Then $H \subset V^*$ with compact injection. A finite energy solution of Equations 68.49 through 68.51 is characterized by the statements $w(t) \in V$, $\dot{w}(t) \in H$ for each t, and the mapping $t \mapsto (w(t), \dot{w}(t))$ is continuous into the space $V \times H$. The space $V \times H$ is sometimes referred to as *finite energy space*.

Write $\mathbf{c} = c_1\mathbf{i} + c_2\mathbf{j}$. In order to assure that the configuration space is sufficiently rich, the control space is chosen as

$$\mathcal{C} = \{[f, \mathbf{c}) : f \in L^2(\Gamma_1 \times (0,T)], \quad c_i \in L^2[\Gamma_1 \times (0,T)]\}. \tag{68.53}$$

The penalty for this simple choice is that the corresponding solution, which may be defined in a certain weak sense and is unique, is not necessarily a finite energy solution. In fact, it can be shown by variational methods that, if w is the solution of Equations 68.49 through 68.52 corresponding to an input $(f, \mathbf{c}) \in \mathcal{C}$, then $w(t) \in H$, $\dot{w}(t) \in V^*$ and the mapping $t \mapsto [w(t), \dot{w}(t)]$ is continuous into $H \times V^*$. (A more refined analysis of the regularity of the solution may be found in [20].)

68.6.1.1 Approximate Controllability

The system (Equations 68.49 through 68.52) is called *approximately controllable* at time T if $\mathcal{R}_T^0$ is dense in $H \times V^*$. To study this problem, introduce the *control-to-state map* C_T defined by

$$C_T : \mathcal{C} \mapsto H \times V^*, \quad C_T(f, \mathbf{c}) = [w(T), \dot{w}(T)].$$

Then the system, Equations 68.49 through 68.52, is *approximately controllable* at time T exactly when range(C_T) is dense in $H \times V^*$. The linear operator C_T is bounded, so, therefore, is its dual operator $C_T^* : H \times V \mapsto \mathcal{C}$. Thus, proving the approximate controllability of Equations 68.49 through 68.52 is equivalent to showing that

$$(\phi^1, \phi^0) \in H \times V, \quad C_T^*(\phi^1, \phi^0) = 0 \Rightarrow (\phi^1, \phi^0) = 0.$$

The quantity $C_T^*(\phi^1, \phi^0)$ may be explicitly calculated (see, [14]). It is given by the trace

$$C_T^*(\phi^1, \phi^0) = (\phi, \nabla\phi)|_{\Gamma_1 \times (0,T)},$$

where ϕ is the solution of the final value problem

$$\rho h \frac{\partial^2 \phi}{\partial t^2} - I_\rho \Delta \frac{\partial^2 \phi}{\partial t^2} + D\Delta^2 \phi = 0, \tag{68.54}$$

$$\phi = 0, \quad \frac{\partial \phi}{\partial \nu} = 0 \quad \text{on } \Gamma_0, 0 < t < T, \tag{68.55}$$

$$\left.\begin{aligned} &\frac{\partial}{\partial \nu}\left(I_\rho \frac{\partial^2 \phi}{\partial t^2} - D\Delta\phi\right) - \frac{h^3}{12}\frac{\partial}{\partial \tau}[\tau \cdot \sigma(\nabla\phi)\nu] = 0, \\ &\frac{h^3}{12}\nu \cdot \sigma(\nabla\phi)\nu = 0 \quad \text{on } \Gamma_1, 0 < t < T, \end{aligned}\right\} \tag{68.56}$$

$$\phi(x, T) = \phi^0, \quad \frac{\partial \phi}{\partial t}(x, T) = \phi^1, \quad x \in \Omega. \tag{68.57}$$

Therefore, the system, Equations 68.49 through 68.52, is approximately controllable if the only solution of Equations 68.54 through 68.57, which also satisfies

$$\phi|_{\Gamma_1 \times (0,T)} = 0, \quad \nabla\phi|_{\Gamma_1 \times (0,T)} = 0, \tag{68.58}$$

is the trivial solution. However, the boundary conditions, Equations 68.56 and 68.58, together, imply that ϕ satisfies Cauchy data on $\Gamma_1 \times (0, T)$, that is, ϕ and its derivatives up to order three vanish on $\Gamma_1 \times (0, T)$. If T is large enough, a general uniqueness theorem (Holmgren's theorem) then implies that $\phi \equiv 0$ in $\Omega \times (0, T)$. This implies approximate controllability in time T.

Theorem 68.3:

There is a $T_0 > 0$ so that the system Equations 68.49 through 68.52 is approximately controllable in time $T > T_0$.

Remark 68.10

The optimal time T_0 depends on the material parameters and the geometry of Ω and Γ_1. If Ω is convex, then $T_0 = 2\sqrt{I_\rho/D}\, d(\Omega, \Gamma_1)$, where

$$d(\Omega, \Gamma_1) = \sup_{x\in\Omega} \inf_{y\in\Gamma_1} |x - y|.$$

Remark 68.11

(Distributed control.) Consider the problem of approximate controllability using a distributed control rather than boundary controls. Let ω be a nonempty, open subset of Ω and let the control space be

$$\mathcal{C} = [F : F \in L^2(\Omega \times (0, T)), \quad F = 0 \text{ in } \Omega\backslash\omega].$$

Consider the system consisting of Equation 68.38 and (for example) the homogeneous boundary conditions,

$$w = \frac{\partial w}{\partial \nu} = 0 \quad \text{on } \Gamma, t > 0.$$

Assume that the initial data is zero and let

$$\mathcal{R}_T^0 = \{(w(T), \dot{w}(T)) : F \in \mathcal{C}\}.$$

For any input F taken from $\mathcal{C}$ the corresponding solution may be shown to be a finite energy solution. Let H and V be defined as above with $\Gamma_0 = \Gamma$. The control-to-state map $C_T : F \mapsto (w(T), \dot{w}(T))$ maps $\mathcal{C}$ boundedly into $V \times H$ and its dual is given by

$$C_T^*(\phi^1, \phi^0) = \phi|_{\omega\times(0,T)},$$

where ϕ is the solution of Equation 68.54 with final data, Equation 68.57, and boundary conditions

$$\phi = 0, \quad \frac{\partial \phi}{\partial \nu} = 0 \quad \text{on } \Gamma, 0 < t < T. \tag{68.59}$$

If $\phi|_{\omega\times(0,T)} = 0$ and T is sufficiently large, an application of Holmgren's theorem gives $\phi \equiv 0$ in $\Omega \times (0, T)$ and, therefore, the system is approximately controllable in time T. When Ω is convex, the optimal control time is $T_0 = 2\sqrt{I_\rho/D}\, d(\Omega, \omega)$.

68.6.1.2 Exact Controllability

Again consider the system Equations 68.49 through 68.52 with the control space given by Equation 68.53. If $\mathcal{D}$ is a subspace in $H \times V^*$, the system is *exactly controllable to* $\mathcal{D}$ at time T if $\mathcal{D} \subset \mathcal{R}_T^0$. The *exact controllability problem*, in the strictest sense, consists of explicitly identifying $\mathcal{R}_T^0$ or, in a less restricted sense, of explicitly identifying dense subspaces $\mathcal{D}$ of $H \times V^*$ contained in $\mathcal{R}_T^0$.

To obtain useful explicit information about $\mathcal{R}_T^0$ it is necessary to restrict the geometry of Γ_0 and Γ_1. It is assumed that there is a point $x_0 \in \mathbb{R}^2$ so that

$$(x - x_0) \cdot \nu \le 0, \quad x \in \Gamma_0. \tag{68.60}$$

Condition (Equation 68.60) is a "nontrapping" assumption of the sort found in early work on scattering of waves from a reflecting obstacle. Without some such restriction on Γ_0, the results described below would not be valid. It is further assumed that

$$\overline{\Gamma}_0 \cap \overline{\Gamma}_1 = \emptyset. \tag{68.61}$$

This is a technical assumption needed to assure that solutions of the uncontrolled problem have adequate regularity up to the boundary (*cf.* Remark 68.13 below).

Theorem 68.4:

Under the assumptions (Equations 68.60 and 68.61), there is a $T_0 > 0$ so that

$$\mathcal{R}_T^0 \supset V \times H \tag{68.62}$$

if $T > T_0$.

The inclusion stated in Equation 68.62 may be stated in equivalent form in terms of the control-to-state mapping C_T. In fact, let

$$\mathcal{C}_0 = \{(f, \mathbf{c}) \in \mathcal{C} : C_T(f, \mathbf{c}) \in V \times H\},$$

and consider the restriction of C_T to $\mathcal{C}_0$. This is a closed, densely defined linear operator from $\mathcal{C}_0$ into $V \times H$. The inclusion (Equation 68.62) is the same as the assertion $C_T(\mathcal{C}_0) = V \times H$, the same as proving that the dual of the operator has a bounded inverse from $V^* \times H$ to $\mathcal{C}_0$, that is,

$$\|(\phi^0, \phi^1)\|^2_{H \times V^*} \leq c \int_0^T \int_{\Gamma_1} (\phi^2 + |\nabla\phi|^2)\, d\Omega, \tag{68.63}$$

where ϕ is the solution of Equations 68.54 through 68.57. For large T, the "observability estimate" (Equation 68.63) was proved in [18] under the additional geometric assumption

$$(x - x_0) \cdot \nu > 0, \quad x \in \Gamma_1. \tag{68.64}$$

However, this hypothesis may be removed by application of the results of [20].

Remark 68.12

It is likely that the optimal control time T_0 for exact controllability is the same as that for approximate controllability, but that has not been proved.

Remark 68.13

The hypothesis (Equation 68.61) may be replaced by the assumption that the sets $\overline{\Gamma}_0$ and $\overline{\Gamma}_1$ meet in a strictly convex angle (measured in the interior of Ω).

Remark 68.14

The conclusion of Theorem 68.4 is false if the space of controls is restricted to finite-dimensional controllers of the form,

$$\mathbf{c}(x, t) = \sum_{i=1}^{N} \alpha_i(x)\mathbf{c}_i(t),$$

$$f(x, t) = \sum_{i=1}^{N} \beta_i(x) f_i(t), \quad x \in \Gamma_1, \quad t > 0,$$

where α_i, β_i are given $L^2(\Gamma_1)$ functions and $\mathbf{c}_i, f_i$ are $L^2(0, T)$ controls, $i = 1, \ldots, N$(see [25]).

Remark 68.15

Given a desired final state $(w_0, w_1) \in V \times H$, there are many ways of constructing a control pair $(f, \mathbf{c})$ so that $w(T) = w_0$ and $\dot{w}(T) = w_1$. The *unique* control of minimum $L^2(\Gamma_1 \times (0, T))$ norm may be constructed as follows. Set $\Sigma_1 = \Gamma_1 \times (0, T)$. Let ϕ be the solution of Equations 68.54 through 68.57 and let w be the solution of Equations 68.49 through 68.52 with

$$f = \phi|_{\Sigma_1}, \quad \text{and} \quad \mathbf{c} = -\nabla\phi|_{\Sigma_1}. \tag{68.65}$$

Then $[w(T), \dot{w}(T)]$ depends on (ϕ^0, ϕ^1). A linear mapping Λ is defined by setting

$$\Lambda(\phi^0, \phi^1) = [\dot{w}(T), -w(T)].$$

The inequality (Equation 68.63) may be used to show that, for any $(w_0, w_1) \in V \times H$, the pair (ϕ^0, ϕ^1) may be chosen so that $\Lambda(\phi^0, \phi^1) = (w_1, -w_0)$. The argument is based on the calculation (using integrations by parts)

$$\begin{aligned} 0 &= \int_0^T \int_\Omega \phi(\rho h \ddot{w} - I_\rho \Delta \ddot{w} + D\Delta^2 w)\, d\Omega dt, \\ &= \int_\Omega [\rho h \dot{w}(T)\phi^0 + I_\rho \nabla \dot{w}(T) \cdot \nabla\phi^0 - \rho h w(T)\phi^1 - I_\rho \nabla w(T) \cdot \nabla\phi^1]\, d\Omega - \int_{\Sigma_1} (\phi^2 + |\nabla\phi|^2)\, d\Sigma, \end{aligned}$$

that is,

$$(\Lambda(\phi^0, \phi^1), (\phi^0, \phi^1))_{H\times H} = \int_{\Sigma_1} (\phi^2 + |\nabla\phi|^2)\, d\Sigma. \tag{68.66}$$

According to Equation 68.63, for T large enough, the right-hand side of Equation 68.66 defines a Hilbert *norm* $\|(\phi^0, \phi^1)\|_F$ and a corresponding Hilbert space F which is the completion of sufficiently smooth pairs (ϕ^0, ϕ^1) with respect to $\|\cdot\|_F$. The identity (Equation 68.66) shows that Λ is exactly the Riesz isomorphism of F onto its dual space F^*. Because $(w_0, w_1) \in \mathcal{R}_T^0$ precisely when $(w_1, -w_0) \in \text{range}(\Lambda)$, it follows that

$$\mathcal{R}_T^0 = [(w_0, w_1) : (w_1, -w_0) \in F^*].$$

The inequality (Equation 68.63) implies that $H \times V^* \supset F$. Therefore $H \times V \subset F^*$, which is the conclusion of Theorem 68.4. If $(w_0, w_1) \in V \times H$, then the minimum norm control is given by Equation 68.65, where $(\phi^0, \phi^1) = \Lambda^{-1}(w_1, -w_0)$. This procedure for constructing the minimum norm control is the basis of the *Hilbert Uniqueness Method* introduced in [22,23].

Remark 68.16

In the situation where $I_\rho = 0$ in Equations 68.49 and 68.51, the only change is that $H = L^2(\Omega)$ rather than the space defined above and the optimal control time is known to be $T_0 = 0$, i.e., exact controllability holds in *arbitrarily* short time (*cf.* [28]).

Remark 68.17

If distributed controls are used rather than boundary controls as in Remark 68.11, Equation 68.62 is not true, in general, but is valid if ω is a neighborhood of Γ.

68.6.2 Controllability of the Reissner–Mindlin System

The controllability properties of the Reissner–Mindlin system are similar to those of the Kirchhoff system. As in the last subsection, only the boundary control problem is considered in detail. Again, we work within the context of L^2 controls and choose Equation 68.53 as the control space. As above, it may be assumed without losing generality that the data of the problem, $F, C, \bar{w}, \bar{\mathbf{u}}, w^0, w^1, \mathbf{u}^0, \mathbf{u}^1$ vanish. The problem under consideration is, therefore,

$$\left.\begin{aligned} &\rho h \frac{\partial^2 w}{\partial t^2} - Gh\text{div}\,(\mathbf{u} + \nabla w) = 0, \\ &I_\rho \frac{\partial^2 \mathbf{u}}{\partial t^2} - \frac{h^3}{12}\text{div}\,\sigma(\mathbf{u}) + Gh(\mathbf{u} + \nabla w) = 0, \end{aligned}\right\} \tag{68.67}$$

$$w = 0, \quad \mathbf{u} = 0 \quad \text{on } \Gamma_0, t > 0, \tag{68.68}$$

$$\left.\begin{aligned} &Gh\nu \cdot (\mathbf{u} + \nabla w) = f, \\ &\frac{h^3}{12}\sigma(\mathbf{u})\nu = \mathbf{c} \quad \text{on } \Gamma_1, t > 0, \end{aligned}\right\} \tag{68.69}$$

$$\begin{aligned} w(x,0) &= \frac{\partial w}{\partial t}(x,0) = 0, \\ \mathbf{u}(x,0) &= \frac{\partial \mathbf{u}}{\partial t}(x,0) = 0 \quad \text{in} \quad \Omega. \end{aligned} \tag{68.70}$$

For convenience, it is assumed that $\Gamma_0 \neq \emptyset$.

Set $\mathbf{w} = \mathbf{u} + w\mathbf{k}$ and introduce the configuration space (reachable set) at time T for this problem by

$$\mathcal{R}_T^0 = \{[\mathbf{w}(T), \dot{\mathbf{w}}(T)] : \quad (f, \mathbf{c}) \in \mathcal{C}\}.$$

To describe $\mathcal{R}_T^0$, certain function spaces based on the kinetic and strain energy functionals of the above problem must be introduced. Let

$$\begin{aligned} H &= [\mathbf{v} = u_1\mathbf{i} + u_2\mathbf{j} + w\mathbf{k} \overset{\text{def}}{=} \mathbf{u} + w\mathbf{k} : \ u_i, w \in L^2(\Omega)], \\ \|\mathbf{v}\|_H &= \left(\int_\Omega (\rho h w^2 + I_\rho |\mathbf{u}|^2)\, d\Omega\right)^{1/2}, \\ V &= \{\mathbf{v} \in H : \quad u_i, w \in H^1(\Omega), \quad \mathbf{v}|_{\Sigma_1} = 0\}, \\ \|\mathbf{v}\|_V &= \left(\int_\Omega \left(\frac{h^3}{12}\sum_{i,j=1}^{2} \sigma_{ij}(\mathbf{u})\varepsilon_{ij}(\mathbf{u}) + Gh|\mathbf{u} + \nabla w|^2\right) d\Omega\right)^{1/2}. \end{aligned}$$

It is a consequence of Korn's lemma and $\Gamma_0 \neq \emptyset$ that $\|\cdot\|_V$ is a norm equivalent to the induced $H^1(\Omega)$ norm. The space V is dense in H and the embedding $V \mapsto H$ is compact. A solution of Equations 68.67 through 68.70 is a *finite energy solution* if $\mathbf{w}(t) \in V, \dot{\mathbf{w}}(t) \in H$ for each t, and the mapping $t \mapsto (w(t), \dot{w}(t))$ is continuous into $V \times H$. This means that the solution has finite kinetic and strain energies at each instant. As with the Kirchhoff model, solutions corresponding to inputs taken from $\mathcal{C}$ are not necessarily finite energy solutions. However, it is true that $\mathbf{w}(t) \in H, \dot{\mathbf{w}}(t) \in V^*$ and the mapping $t \mapsto [w(t), \dot{w}(t)]$ is continuous into $H \times V^*$, where the concept of a solution is defined in an appropriate weak sense.

The system, (Equations 68.67 through 68.70) is called *approximately controllable* if $\mathcal{R}_T^0$ is dense in $H \times V^*$. The *exact controllability problem* consists of explicitly identifying dense subspaces of $H \times V^*$ contained in $\mathcal{R}_T^0$.

With this setup, Theorem 68.3 and a slightly weaker version of Theorem 68.4 may be proved for the Reissner-Mindlin system. The proofs again consist of an examination of the control-to-state map

$C_T : \mathcal{C} \mapsto H \times V^*$ defined by $C_T(f, \mathbf{c}) = [\mathbf{w}(T), \dot{\mathbf{w}}(T)]$. The dual mapping $C_T^* : H \times V \mapsto \mathcal{C}$ is given by

$$C_T^*(\Phi^1, \Phi^0) = \Phi|_{\Gamma_1 \times (0,T)},$$

where

$$\Phi = \phi + \psi \mathbf{k}, \quad \phi = \phi_1 \mathbf{i} + \phi_2 \mathbf{j},$$
$$\Phi^0 = \phi^0 + \psi^0 \mathbf{k}, \quad \Phi^1 = \phi^1 + \psi^1 \mathbf{k},$$

and ϕ, ψ satisfy

$$\left.\begin{aligned} &\rho h \frac{\partial^2 \psi}{\partial t^2} - Gh \mathrm{div}\,(\phi + \nabla \psi) = 0, \\ &I_\rho \frac{\partial^2 \phi}{\partial t^2} - \frac{h^3}{12} \mathrm{div}\,\sigma(\phi) + Gh(\phi + \nabla \psi) = 0, \end{aligned}\right\} \tag{68.71}$$

$$\psi = 0, \quad \phi = 0 \quad \text{on } \Gamma_0, 0 < t < T, \tag{68.72}$$

$$\left.\begin{aligned} &Gh\nu \cdot (\phi + \nabla \psi) = 0, \\ &\frac{h^3}{12} \sigma(\phi)\nu = 0 \quad \text{on } \Gamma_1, 0 < t < T, \end{aligned}\right\} \tag{68.73}$$

$$\left.\begin{aligned} &\psi(x, T) = \psi^0, \quad \frac{\partial \psi}{\partial t}(x, T) = \psi^1, \\ &\phi(x, T) = \phi^0, \quad \frac{\partial \phi}{\partial t}(x, T) = \phi^1 \quad \text{in } \Omega. \end{aligned}\right\} \tag{68.74}$$

Approximate controllability amounts to showing that $C_T^*(\Phi^1, \Phi^0) = 0$ only when $(\Phi^1, \Phi^0) = 0$. However, if Φ is the solution of Equations 68.71 through 68.74 and satisfies $\Phi|_{\Gamma_1 \times (0,T)} = 0$, then Φ and its first derivatives vanish on $\Gamma_1 \times (0, T)$. If T is large enough, Holmgren's theorem then implies that $\Phi \equiv 0$.

With regard to the exact controllability problem, to prove the inclusion, (Equation 68.62) for the Reissner system amounts to establishing the observability estimate (*cf.* Equation 68.63),

$$\|(\Phi^0, \Phi^1)\|^2_{H \times V^*} \le c \int_0^T \int_{\Gamma_1} |\Phi|^2 d\Gamma dt. \tag{68.75}$$

For sufficiently large T, this estimate has been proved in [18] under assumptions, Equations 68.60, 68.61, and 68.64. The following exact controllability result is a consequence of Equation 68.75.

Theorem 68.5:

Under assumptions Equations 68.60, 68.61, and 68.64, there is a $T_0 > 0$ so that

$$\mathcal{R}_T^0 \supset V \times H$$

if $T > T_0$.

Remark 68.18

If Ω is convex, then the optimal control time for approximate controllability is

$$T_0 = 2 \max(\sqrt{I_\rho / D}, \sqrt{\rho / G})\, d(\Omega, \Gamma_1).$$

The optimal control time for exact controllability is probably the same, but this has not been proved. See, however, [10,11, Chapter 5]. Assumption, Equation 68.64, is probably unnecessary, but this has not been established. Remark 68.13 is valid also for the Reissner–Mindlin system. The remarks concerning approximate and exact controllability of the Kirchhoff model utilizing distributed controls remain true for the Reissner–Mindlin system.

68.6.3 Controllability of the von Kármán System

The *global* controllability results which hold for the Kirchhoff and Reissner–Mindlin models cannot be expected to hold for nonlinear partial differential equations, in general, or for the von Kármán system in particular. Rather, only *local* controllability is to be expected (although global controllability results may obtain for certain *semilinear* systems; *cf.* [19]), that is, in general the most that can be expected is that the reachable set (assuming zero initial data) contains *some* ball S_r centered at the origin in the appropriate energy space, with the control time depending on r.

The problem to be considered is

$$\left.\begin{aligned} &\rho h\frac{\partial^2 w}{\partial t^2}-I_\rho\Delta\frac{\partial^2 w}{\partial t^2}+D\Delta^2 w-[w,G]=0,\\ &\Delta^2 G+\frac{Eh}{2}[w,w]=0, \end{aligned}\right\} \tag{68.76}$$

$$w=\frac{\partial w}{\partial \nu}=0, \quad \text{on } \Gamma_0, t>0, \tag{68.77}$$

$$\left.\begin{aligned} &\frac{\partial}{\partial \nu}\left(I_\rho\frac{\partial^2 w}{\partial t^2}-D\Delta w\right)-\frac{h^3}{12}\frac{\partial}{\partial \tau}[\tau\cdot\sigma(\nabla w)\nu]=\frac{\partial}{\partial \tau}(\tau\cdot\mathbf{c})+f,\\ &\frac{h^3}{12}\nu\cdot\sigma(\nabla w)\nu=-\nu\cdot\mathbf{c}, \quad \text{on } \Gamma_1, t>0, \end{aligned}\right\} \tag{68.78}$$

$$G=0, \quad \frac{\partial G}{\partial \nu}=0 \quad \text{on } \Gamma, t>0, \tag{68.79}$$

$$w(x,0)=\frac{\partial w}{\partial t}(x,0)=0, \quad x\in\Omega. \tag{68.80}$$

It is assumed that $\Gamma_0 \neq \emptyset$. Note that there is no initial data for G.

The above system is usually analyzed by uncoupling w from G. This is done by solving the second equation in Equation 68.79, subject to the boundary conditions, Equation 68.79, for G in terms of w. One obtains $G=-(Eh/2)\mathcal{G}[w,w]$, where $\mathcal{G}$ is an appropriate Green's operator for the biharmonic equation. One then obtains for w the following equation with a cubic nonlinearity:

$$\rho h\frac{\partial^2 w}{\partial t^2}-I_\rho\Delta\frac{\partial^2 w}{\partial t^2}+D\Delta^2 w-\frac{Eh}{2}[w,\mathcal{G}[w,w]]=0. \tag{68.81}$$

The problem for w consists of Equation 68.81 together with the boundary conditions, Equations 68.77, 68.78, and initial conditions Equation 68.80.

The function spaces H and V based on the kinetic and strain energy functionals, respectively, related to the transverse displacement w, are the same as those introduced in discussing controllability of the Kirchhoff model, as is the reachable set $\mathcal{R}_T^0$ corresponding to vanishing data. Let

$$S_r=\{(v,h)\in V\times H:\ (\|v\|_V^2+\|h\|_H^2)^{1/2}<r\}.$$

With this notation, a local controllability result analogous to Theorem 68.4 can be established.

Theorem 68.6:

Under assumptions, Equations 68.60, 68.61, and 68.64, there is an $r > 0$ and a time $T_0(r) > 0$ so that

$$\mathcal{R}_T^0 \supset S_r \tag{68.82}$$

if $T > T_0$.

Curiously, a result for the von Kármán system analogous to Theorem 68.3 is not known.

Theorem 68.6 is proved by utilizing the global controllability of the linearized (i.e., Kirchhoff) problem, together with the implicit function theorem in a manner familiar in the control theory of finite-dimensional nonlinear systems.

Remark 68.19

If the underlying dynamics are modified by introducing a dissipative term $b(x)\dot{w}$, $b(x) > 0$, into Equation 68.81, it may be proved, under assumptions, Equations 68.60 and 68.61, that the conclusion, Equation 68.82, is valid for *every* $r > 0$. However, the optimal control time T_0 will continue to depend on r, so that such a result is still local.

68.7 Stabilizability of Dynamic Plates

The problem of stabilization is concerned with the description of *feedback controls* which assure that the trajectories of the system converge asymptotically to an equilibrium state of the system. For infinite-dimensional systems in general and distributed parameter systems in particular, there are various distinct notions of asymptotic stability: weak, strong, and uniform (distinctions which, incidentally, disappear in finite-dimensional approximations of the system). The differences in the various types of stability are related to the topology in which convergence to an equilibrium takes place. The most robust notion of stability is that of uniform stability, which guarantees that all possible trajectories starting near an equilibrium of the system converge to that equilibrium *at a uniform rate*. In this concept, convergence is usually measured in the *energy norm* associated with the system. This is the classical viewpoint of stability. Strong stability, on the other hand, guarantees asymptotic convergence of each trajectory (in the energy norm) but at a rate which may become arbitrarily small, depending on the initial state of the system. The concept of weak stability is similar; however, in this case, convergence to an equilibrium takes place in a topology weaker than associated with the energy norm. In the discussion which ensues, only uniform and strong asymptotic stability will be considered.

68.7.1 Stabilizability of Kirchhoff Plates

Consider the Kirchoff system consisting of Equation 68.49, boundary conditions, Equations 68.50, and 68.51, and initial conditions

$$w(x,0) = w^0(x), \quad \frac{\partial w}{\partial t}(x,0) = w^1(x), \quad x \in \Omega. \tag{68.83}$$

It is assumed that $\Gamma_i \neq \emptyset$, $i = 0, 1$. The boundary inputs $\mathbf{c}, f$ are the controls. The *boundary outputs* are

$$y = \left.\frac{\partial w}{\partial t}\right|_{\Gamma_1 \times (0,\infty)}, \quad \mathbf{z} = \nabla\left.\left(\frac{\partial w}{\partial t}\right)\right|_{\Gamma_1 \times (0,\infty)}. \tag{68.84}$$

The problem is to determine the boundary inputs in terms of the boundary outputs to guarantee that the resulting closed-loop system is asymptotically stable in some sense.

The *total energy* of the system at time t is

$$\mathcal{E}(t) = \mathcal{K}(t) + \mathcal{U}(t),$$
$$= \frac{1}{2}\int_\Omega (\rho h \dot{w}^2 + I_\rho |\nabla \dot{w}|^2)\, d\Omega + \frac{1}{2}\frac{h^3}{12}\sum_{i,j=1}^{2}\int_\Omega \sigma_{ij}(\nabla w)\varepsilon_{ij}(\nabla w)\, d\Omega,$$

where $\dot{w} = \partial w/\partial t$. A direct calculation shows that

$$\frac{d\mathcal{E}}{dt} = \int_{\Gamma_1}\left[(-\nu\cdot\mathbf{c})\frac{\partial \dot{w}}{\partial \nu} + \left(\frac{\partial}{\partial \tau}(\tau\cdot\mathbf{c}) + f\right)\dot{w}\right] d\Gamma.$$

When the loop is closed by introducing the proportional feedback law

$$f = -k_0 y, \quad \mathbf{c} = k_1\mathbf{z}, \quad k_i \geq 0, \quad k_0 + k_1 > 0, \tag{68.85}$$

it follows that

$$\frac{d\mathcal{E}}{dt} = -\int_{\Gamma_1}\left[k_1\left(\frac{\partial \dot{w}}{\partial \nu}\right)^2 + k_0(\dot{w})^2 + k_1\left(\frac{\partial \dot{w}}{\partial \tau}\right)^2\right] d\Gamma,$$
$$= -\int_{\Gamma_1}[k_0(\dot{w})^2 + k_1|\nabla \dot{w}|^2]\, d\Gamma \leq 0.$$

Thus the feedback laws Equation 68.85 are dissipative with respect to the total energy functional $\mathcal{E}$.

Let H and V be the Hilbert spaces based on the energy functionals $\mathcal{K}$ and $\mathcal{U}$, respectively, as introduced above. If $(w^0, w^1) \in V \times H$, the Kirchoff system, Equations 68.49 through 68.51, with initial conditions Equation 68.83, boundary outputs Equation 68.84, and feedback law Equation 68.85, is well-posed: it has a unique finite energy solution w. The system is called *uniformly asymptotically stable* if there is a positive, real-valued function $\alpha(t)$ with $\alpha(t) \to 0$ as $t \to \infty$, so that

$$\|(w(t), \dot{w}(t))\|_{V\times H} \leq \alpha(t)\|(w^0, w^1)\|_{V\times H}.$$

Therefore, $\mathcal{E}(t) \leq \alpha^2(t)\mathcal{E}(0)$. If such a function α exists, it is necessarily exponential: $\alpha(t) = Ce^{-\omega t}$ for some $\omega > 0$. The system is *strongly asymptotically stable* if, for every initial state $(w^0, w^1) \in V \times H$, the corresponding solution satisfies

$$\lim_{t\to\infty} \|(w(t), \dot{w}(t))\|_{V\times H} = 0$$

or, equivalently, that $\mathcal{E}(t) \to 0$ as $t \to \infty$. Uniform and strong asymptotic stability are not equivalent concepts. Strong asymptotic stability *does not* imply uniform asymptotic stability. Strong stability has the following result.

Theorem 68.7:

Assume that $k_i > 0$, $i = 0, 1$. Then the closed-loop Kirchhoff system is strongly asymptotically stable.

The proof of this theorem amounts to verifying that, under the stated hypotheses, the problem has no spectrum on the imaginary axis. The latter is a consequence of the Holmgren uniqueness theorem (see [13, Chapter 4] for details).

For the closed-loop Kirchhoff system to be uniformly asymptotically stable, the geometry of Γ_i must be suitably restricted.

Theorem 68.8:

Assume that $k_i > 0$, $i = 0, 1$, and that Γ_i satisfy Equations 68.60 and 68.61. Then the closed-loop Kirchhoff system is uniformly asymptotically stable.

The proof of Theorem 68.8 follows from the estimate,

$$\mathcal{E}(T) \le C_T \int_0^T \int_{\Gamma_1} [k_0(\dot{w})^2 + k_1|\nabla \dot{w}|^2]\, d\Gamma dt, \quad T \text{ large.} \tag{68.86}$$

The proof of Equation 68.86 is highly nontrivial. From Equation 68.86 and the above calculation of $d\mathcal{E}/dt$, it follows that

$$\mathcal{E}(T) \le \frac{1}{1 + C_T}\mathcal{E}(0),$$

which implies the conclusion of the theorem.

Remark 68.20

Theorem 68.8 was first proved in [13] under the additional assumption Equation 68.64, but the latter condition may be removed by applying the results in [20]. Assumption, Equation 68.61, may be weakened; see Remark 68.13. If $I_\rho = 0$, the conclusion holds even when $k_1 = 0$.

Remark 68.21

In place of the linear relationship Equation 68.85, one may consider a nonlinear feedback law

$$y = -y(\dot{w}), \quad \mathbf{z} = \mathbf{z}(\nabla \dot{w}),$$

where $y(\cdot)$ is a real-valued function, $\mathbf{z}(\cdot) : \mathbb{R}^2 \mapsto \mathbb{R}^2$ and satisfies

$$xy(x) > 0, \quad \forall x \in \mathbb{R}\backslash\{0\},$$
$$\mathbf{x} \cdot \mathbf{z}(\mathbf{x}) > 0 \quad \forall \mathbf{x} \in \mathbb{R}^2\backslash\{0\}.$$

The closed-loop system is then dissipative. In addition, suppose that both $y(\cdot)$ and $\mathbf{z}(\cdot)$ are continuous, monotone increasing functions. The closed-loop system is then well-posed in finite energy space. Under some additional assumptions on the growth of $y(\cdot), \mathbf{z}(\cdot)$ at 0 and at ∞, the closed-loop system have a decay rate which, however, will be algebraic, rather than exponential, and will depend on a bound on the initial data; *cf.* [13, Chapter 5] and [16].

68.7.2 Stabilizability of the Reissner–Mindlin System

The system, (Equations 68.67 through 68.69) is considered, along with the initial conditions

$$\left.\begin{aligned} w(x,0) &= w^0(x), & \frac{\partial w}{\partial t}(x,0) &= w^1(x), \\ \mathbf{u}(x,0) &= \mathbf{u}^0(x), & \frac{\partial \mathbf{u}}{\partial t}(x,0) &= \mathbf{u}^1(x), \quad x \in \Omega. \end{aligned}\right\} \tag{68.87}$$

The boundary inputs $\mathbf{c}, f$ are the controls. The *boundary outputs* are

$$y = \left.\frac{\partial w}{\partial t}\right|_{\Gamma_1 \times (0,\infty)}, \quad \mathbf{z} = \left.\frac{\partial \mathbf{u}}{\partial t}\right|_{\Gamma_1 \times (0,\infty)}.$$

The *total energy* of the system at time t is

$$\mathcal{E}(t)=\mathcal{K}(t)+\mathcal{U}(t)=\frac{1}{2}\int_{\Omega}(\rho h\dot{w}^2+I_\rho|\mathbf{u}|^2)\,d\Omega+\frac{1}{2}\int_{\Omega}\left(\frac{h^3}{12}\sum_{i,j=1}^{2}\sigma_{ij}(\mathbf{u})\varepsilon_{ij}(\mathbf{u})+Gh|\mathbf{u}+\nabla w|^2\right)d\Omega.$$

Then

$$\frac{d\mathcal{E}}{dt}=\int_{\Gamma_1}(f\dot{w}+\mathbf{c}\cdot\dot{\mathbf{u}})\,d\Gamma,$$

which suggests that the loop be closed by introducing the proportional feedback law

$$f=-k_0y,\quad \mathbf{c}=-k_1\mathbf{z},\quad k_i\geq 0,\quad k_0+k_1>0,$$

so that the closed-loop system is dissipative. In fact, it may be proved that the conclusions of Theorems 68.7 and 68.8 above hold for this system; see [13], where this is proved under the additional geometric assumption, Equation 68.64.

68.7.3 Stabilizability of the von Kármán System

Consider the system consisting of Equation 68.81, boundary conditions Equations 68.77 and 68.78, and initial conditions Equation 68.83. The inputs, outputs, and total energy of this system are defined as for the Kirchhoff system, and the loop is closed using the proportional feedback law Equation 68.85. The following result has been proved in [4].

Theorem 68.9:

Assume that $k_i>0$, $i=0,1$, and that Γ_i satisfy Equations 68.60, 68.61, and 68.64. Then there is an $r>0$ so that

$$\mathcal{E}(t)\leq Ce^{-\omega t}\mathcal{E}(0)\qquad(68.88)$$

provided $\mathcal{E}(0)<r$, where $\omega>0$ does not depend on r.

Remark 68.22

If $I_\rho=0$ and $\Gamma_0=\emptyset$, the estimate Equation 68.88 was established in [13, Chapter 5] for every $r>0$, with constants C,ω independent of r, but under a modified feedback law for $\mathbf{c}$.

Remark 68.23

If the underlying dynamics are modified by introducing a dissipative term $b(x)\dot{w}$, $b(x)>0$, into Equation 68.81, it is proven in [5] that, under assumptions, Equations 68.60 and 68.61, the conclusion Equation 68.88 is valid for *every* $r>0$, where both constants C,ω depend on r. This result was later extended in [8] to the case of nonlinear feedback laws (*cf.* Remark 68.21).

References

1. Adams, R.A., *Sobolev Spaces*, Academic, New York, 1975.
2. Balakrishnan, A.V., *Applied Functional Analysis*, 2nd ed., Springer, New York, 1981.
3. Banks, H.T. and Smith, R.C., Models for control in smart material structures, in *Identification and Control in Systems Governed by Partial Differential Equations*, Banks, H.T., Fabiano, R.H., and Ito K., Eds., SIAM, 1992, pp 27–44.

4. Bradley, M. and Lasiecka, I., Local exponential stabilization for a nonlinearly perturbed von Kármán plate, *Nonlinear Analysis: Theory, Methods and Appl.*, 18, 333–343, 1992.
5. Bradley, M. and Lasiecka, I., Global decay rates for the solutions to a von Kármán plate without geometric constraints, *J. Math. Anal. Appl.*, 181, 254–276, 1994.
6. Bresse, J. A. C., *Cours de mechanique applique*, Mallet Bachellier, 1859.
7. Hirschhorn, M. and Reiss, E., Dynamic buckling of a nonlinear Timoshenko beam, *SIAM J. Appl.Math.*, 37, 290–305, 1979.
8. Horn, M. A. and Lasiecka, I., Nonlinear boundary stabilization of a von Kármán plate equation, *Differential Equations, Dynamical Systems and Control Science: A Festschrift in Honor of Lawrence Markus*, Elworthy, K. D., Everit, W. N., and Lee, E. B., Eds., Marcel Dekker, New York, 1993, pp 581–604.
9. Kane, T.R., Ryan, R.R., and Barnerjee, A.K., Dynamics of a beam attached to a moving base, *AIAA J. Guidance, Control Dyn.*, 10, 139–151, 1987.
10. Komornik, V., A new method of exact controllability in short time and applications, *Ann. Fac. Sci. Toulouse*, 10, 415–464, 1989.
11. Komornik, V., Exact controllability and stabilization, in *The Multiplier Method*, Masson - John Wiley & Sons, Paris, 1994.
12. Krabs, W., *On Moment Theory and Controllability of One-Dimensional Vibrating Systems and Heating Processes*, in Lecture Notes in Control and Information Sciences, Vol. 173, Springer, New York, 1992.
13. Lagnese, J. E., *Boundary Stabilization of Thin Plates*, Studies in Applied Mathematics, SIAM, Philadelphia, 1989, Vol. 10.
14. Lagnese, J. E., The Hilbert uniqueness method: A retrospective, in *Optimal Control of Partial Differential Equations*, Hoffmann, K. H. and Krabs, W., Eds., Springer, Berlin, 1991, pp 158–181.
15. Lagnese, J. E., Recent progress in exact boundary controllability and uniform stabilizability of thin beams and plates, in *Distributed Parameter Systems: New Trends and Applications* Chen, G., Lee, E. B., Littmann, W., and Markus, L., Eds., Marcel Dekker, New York, 1991, pp 61–112.
16. Lagnese, J. E. and Leugering, G., Uniform stabilization of a nonlinear beam by nonlinear boundary feedback, *J. Diff. Eqns.*, 91, 355–388, 1991.
17. Lagnese, J. E., Leugering, G., and Schmidt, E.J.P.G., *Modeling, Analysis and Control of Dynamic Elastic Multi-Link Structures*, Birkhäuser, Boston-Basel-Berlin, 1994.
18. Lagnese, J. E. and Lions, J. L., *Modelling, Analysis and Control of Thin Plates*, Collection RMA, Masson, Paris, 1988, Vol. 6.
19. Lasiecka, I. and Triggiani, R., Exact controllability of semilinear abstract systems with application to waves and plates boundary control problems, *Appl. Math. Opt.*, 23, 109–154, 1991.
20. Lasiecka, I. and Triggiani, R., Sharp trace estimates of solutions to Kirchhoff and Euler-Bernoulli equations, *Appl. Math. Opt.*, 28, 277–306, 1993.
21. LeDret, H., *Problèmes variationnels dans les multi-domains: Modélisation des jonctions et applications*, Collection RMA, Masson, Paris, 1991, vol. 19.
22. Lions, J. L., *Contrôlabilité exacte, perturbations et stabilisation de systèmes distribués, Tome I: Contrôlabilté exacte*, Collection RMA, Masson, Paris, 1988, Vol. 8.
23. Lions, J. L., Exact controllability, stabilization and perturbations for distributed systems, *SIAM Review*, 30, 1–68, 1988.
24. Russell, D. L., On mathematical models for the elastic beam with frequency-proportional damping, in *Control and Estimation in Distributed Parameter Systems*, Banks, H.T., Ed., SIAM, 1992, pp 125–169.
25. Triggiani, R., Lack of exact controllability for wave and plate equations with finitely many boundary controls, *Diff. Int. Equations*, 4, 683–705, 1991.
26. Washizu, K., *Variational Methods in Elasticity and Plasticity*, 3rd ed., Pergamon, Oxford, 1982.
27. Wempner, G., *Mechanics of Solids with Applications to Thin Bodies*, Sijthoff and Noordhoff, Rockville, MD, 1981.
28. Zuazua, E., Contrôlabilité exacte d'un modèle de plaques vibrantes en un temps arbitrairement petit, *C. R. Acad. Sci.*, 304, 173–176, 1987.

69

Control of the Heat Equation

Thomas I. Seidman
University of Maryland, Baltimore County

69.1 Introduction

We must begin by making an important distinction between the considerations appropriate to a great variety of practical problems, for example, controlled heat transfer or observed diffusion and those more theoretical considerations which arise when we wish to apply the essential ideas developed for the control theory of ordinary differential equations in the context of systems governed by partial differential equations—here, the linear *heat equation*

$$\frac{\partial v}{\partial t} = \frac{\partial^2 v}{\partial x^2} + \frac{\partial^2 v}{\partial y^2} + \frac{\partial^2 v}{\partial z^2} \quad \text{for } t > 0, \quad \mathbf{x} = (x, y, z) \in \Omega \subset \mathbb{R}^3. \tag{69.1}$$

Many of the former set of problems are ones of optimal design, rather than of dynamic control and many of the essential concerns are related to fluid flow in a heat exchanger or to phase changes (e.g., condensation) or to other issues which go well beyond the physical situations described by Equation 69.1. Some properties of Equation 69.1 are relevant for these problems and we shall mention these, but the essential concerns which dominate them are outside the scope of this chapter.

The primary focus of this chapter is, from the point of view of control theory, on the inherent distinctions one must make between "lumped parameter systems" (with finite-dimensional state space, governed by ordinary differential equations) and "distributed parameter systems" governed by partial differential equations such as Equation 69.1 so the state, for each t, is a function of position in the spatial region Ω.

While Equation 69.1 may be viewed abstractly as an ordinary differential equation*

$$\frac{\mathrm{d}v}{\mathrm{d}t} = \mathbf{\Delta} v + \psi \quad \text{for } t > 0, \tag{69.3}$$

it is important to realize that abstract ordinary differential equations such as Equation 69.3 are quite different in nature from the more familiar ordinary differential equations with finite-dimensional state; hence, one's intuition must be attuned to this situation. Further, the intuition appropriate to consideration of the parabolic partial differential equation 69.3 is quite different from what would be appropriate, say, for the *wave equation*

$$\frac{\mathrm{d}^2 w}{\mathrm{d}t^2} = \mathbf{\Delta} w + \psi_2 \quad \text{for } t > 0, \tag{69.4}$$

which describes a very different set of physical phenomena with very different properties (although in Section 69.5.3 we do mention an interesting relation for the corresponding theories of observation and control).

We restrict our attention largely to autonomous linear problems for which frequency-domain approaches involving matrix theory (transfer functions, etc.) and Riccati equations are well-known approaches in lumped parameter (finite-dimensional) settings. While we do note that, with appropriate technical conditions, these approaches generalize to infinite-dimensional settings (note the emphasis on operator transfer functions in [3] or on the Riccati equation in [1,6]), we will not consider those approaches here, but will concentrate on approaches which are distinctly relevant to considerations of partial differential equations, following the approaches of [4,5,7–10], for example.

One new consideration is that the geometry of the region Ω is relevant here. We here concentrate primarily on linear problems in which input/output interaction (for control and for observation) is restricted to the boundary—partly because this is physically reasonable and partly because it is only for a system governed by a partial differential equation that one could even consider "control via the boundary conditions."

The first two sections of this chapter provide, as background, some relevant properties of Equation 69.1, including the presentation of some examples and implications of these general properties for practical heat conduction problems. We then turn to the discussion of system-theoretic properties of Equation 69.1 or 69.3; note that this restricts our discussion to linear problems. We emphasize, in particular, the considerations which arise when the input/output occurs in a way which has no direct analog in the theory of lumped parameter systems—not through the equation itself, but through the boundary conditions which are appropriate to the partial differential equation 69.1. This mode of interaction is, of course, quite plausible for physical implementation since it is typically difficult to influence or to observe directly the behavior of the system in the interior of a solid spatial region.

69.2 Background: Physical Derivation

Unlike situations involving ordinary differential equations with finite-dimensional state space, it is almost impossible to work with partial differential equations without developing a deep appreciation for the characteristic properties of the particular kind of equation. For the classical equations, such as Equation 69.1, this is closely related to physical interpretations. Thus, we begin with a discussion of some interpretations of Equation 69.1 and only then note the salient properties which will be needed to understand its control.

* Now $v(t)$ denotes the state, viewed as an element of an infinite-dimensional space of functions on Ω, and $\mathbf{\Delta} = \vec{\nabla}^2$ is the *Laplace operator*, given in the 3-dimensional case by

$$\mathbf{\Delta} : v \mapsto \frac{\partial^2 v}{\partial x^2} + \frac{\partial^2 v}{\partial y^2} + \frac{\partial^2 v}{\partial z^2} \tag{69.2}$$

together with specification of the relevant boundary conditions.

While we speak of Equation 69.1 as the *heat equation*, governing conductive heat transfer, our intuition will be aided by noting also that this same equation also governs molecular diffusion for dilute solutions and certain dispersion phenomena as well as the evolution of the probability distribution in the stochastic theory of Brownian motion.

For heat conduction, we begin with the fundamental notions of *heat content* $\mathcal{Q}$ and *temperature*, related* by

$$[\text{heat content}] = [\text{heat capacity}] \cdot [\text{temperature}]. \tag{69.6}$$

or, in symbols,

$$Q = \rho c\, T \tag{69.7}$$

where Q is here the heat density (per unit volume), ρ is the mass density, T is the temperature, and c is the "incremental heat capacity" [amount of heat needed to raise the temperature of a unit mass by, say, 1°]. The well-known physics of the situation is that *heat will flow by conduction from one body to another at a rate proportional to the difference of their temperatures.* Within a continuum one has a *heat flux* vector $\vec{q}$ describing the heat flow: $\vec{q} \cdot \vec{n}\, dA$ is the rate (per unit time) at which heat flows through any (imaginary) surface element dA, oriented by its unit normal $\vec{n}$. This is now given by *Fourier's Law*:

$$\vec{q} = -k\, grad\, T = -k\vec{\nabla} T \tag{69.8}$$

with a (constant†) *coefficient of heat conduction* $k > 0$.

For any (imaginary) region $\mathcal{B}$ in the material, the total rate of heat flow out of $\mathcal{B}$ is then $\int_{\partial\mathcal{B}} \vec{q} \cdot \vec{n}\, dA$ (where $\vec{n}$ is the outward normal to the bounding surface $\partial\mathcal{B}$) and, by the Divergence Theorem, this equals the volume integral of div $\vec{q}$. Combining this with Equations 69.7 and 69.8—and using the arbitrariness of $\mathcal{B}$—this gives the governing‡ heat equation

$$\rho c \frac{\partial T}{\partial t} = \vec{\nabla} \cdot k\vec{\nabla} T + \psi \tag{69.9}$$

where ψ is a possible source term for heat.

Let us now derive the equation governing molecular diffusion, we consider the spread of some substance in another (e.g., a "solute" in a "solvent") caused, as discussed in one of Einstein's famous papers of 1905, by the random collisions of molecules. Assuming a dilute enough solution that one can neglect the volume fraction occupied by the solute in comparison with the solvent, we present our analysis simply in terms of the concentration (relative density) C of the relevant chemical component. One has, entirely analogous

---

* More precisely, since the *mass density* ρ and the *incremental heat capacity* c (i.e., the amount of heat needed to raise the temperature of a unit mass of material by, e.g., 1°C when it is already at temperature ϑ) are each temperature dependent, the heat content in a region $\mathcal{R}$ with temperature distribution $T(\cdot)$ is given by

$$Q = Q(\mathcal{B}) = \int_{\mathcal{B}} \int_0^T [\rho c](\vartheta)\, d\vartheta\, dV. \tag{69.5}$$

For our present purposes we are assuming that (except, perhaps, for the juxtaposition of regions with dissimilar materials) we may take ρc to be effectively constant. Essentially, this means that we assume, the temperature variation is not so large as to force us to work with the more complicated nonlinear model implied by Equation 69.5. In particular, it means that we will not treat situations involving phase changes such as vaporization, condensation, or melting.

† This coefficient k is, in general, also temperature dependent as well as a material property. Our earlier assumption in connection with ρc is relevant here also to permit us to take k to be a constant.

‡ It is essential to realize that $\vec{q}$, as given in Equation 69.8, refers to heat flow *relative to the material.* If there is spatial motion of the material itself, then this argument remains valid provided the regions $\mathcal{B}$ are taken as moving correspondingly—that is, Equation 69.3 holds in material coordinates. When this is referred to stationary coordinates, we view the heat as transported in space by the material motion—that is, we have advection as well as diffusion; the *Peclet number* indicates the relative importance of these transport mechanisms.

to the previous derivation, a *material flux* vector $\vec{J}$ which is now given by *Fick's Law*:

$$\vec{J} = -D\vec{\nabla}C \tag{69.10}$$

where $D > 0$ is the *diffusion coefficient**. As in deriving Equation 69.9, this law for the flux immediately leads to the conservation equation

$$\frac{\partial C}{\partial t} = \vec{\nabla} \cdot D\vec{\nabla}C + \psi, \tag{69.11}$$

where ψ is now a source term for this component—say, by some chemical reaction.

A rather different mechanism for the spread of some substance in another depends on the effect of comparatively small relative velocity fluctuations of the medium—for example, gusting in the atmospheric spread of the plume from a smokestack or the effect of path variation through the interstices of packed soil in considering the spread of a pollutant in groundwater flow. Here one again has a material flux for the concentration—given now by *Darcy's Law*, which appears identical to Equation 69.10. The situation can be more complicated here, however, since one may well have anisotropy (D is then a matrix) and/or various forms of degeneracy (e.g., D becoming 0 when $C = 0$); nevertheless, we still obtain Equation 69.11 with this *dispersion coefficient* D. Here, as earlier, we focus on settings where we may take a constant scalar $D > 0$ and neglect advection.

As we are assuming constant coefficients in each case, we may simplify Equation 69.9 or 69.11 by writing these as

$$v_t = D\Delta v + \psi \tag{69.12}$$

where v stands either for the temperature T or the concentration C, subscript t denotes a partial derivative, and, in considering Equation 69.9, D stands for the *thermal diffusivity* $\alpha = k/\rho c$. We may, of course, always choose units to make $D = 1$ in Equation 69.12; hence, it becomes precisely Equation 69.3. It is interesting and important for applications to have some idea of the wide range of magnitudes of the coefficient D in Equation 69.12 in fixed units—say, cm^2/s.—for various situations. For heat conduction, typical values of the coefficient $D = \alpha$ are, quite approximately:

> 8.4 for heat conduction in diamond; 1.1 for copper; 0.2–0.6 for steam (rising with temperature); 0.17 for cast iron and 0.086 for bronze; 0.011 for ice; 7.8×10^{-3} for glass; 4×10^{-3} for soil; $1.4\text{–}1.7 \times 10^{-3}$ for water; 6.2×10^{-4} for hard rubber; and so on.

For molecular diffusion, typical figures for D might be

> around 0.1 for many cases of gaseous diffusion; 0.28 for the diffusion of water vapor in air and 2×10^{-5} for air dissolved in water; 2×10^{-6} for a dilute solution of water in ethanol and 8.4×10^{-6} for ethanol in water; 1.5×10^{-8} for solid diffusion of carbon in iron and 1.6×10^{-10} for hydrogen in glass, and so on.

Finally, for example, the dispersion of a smoke plume in mildly stable atmosphere (say, a 15 *mph breeze*) might, on the other hand, have D approximately 10^6 cm^2/s.—as might be expected, dispersion is a far more effective spreading mechanism than molecular diffusion, but the same mathematical description covers both.

Assuming that one knows the initial state of the physical system

$$v(x, t = 0) = v_0(x) \quad \text{on } \Omega, \tag{69.13}$$

where Ω is the region of $\mathbb{R}^3$ we wish to consider, we still cannot expect to determine the system evolution unless we also know (or can determine) the source term ψ and, unless Ω would be all of $\mathbb{R}^3$, can furnish

* More detailed treatments might consider the possibility that D depends on the temperature, and so on, of the solvent and is quite possibly also dependent on the existing concentration, even for dilute concentrations. As earlier, we neglect these effects as insignificant for the situations under consideration and take D to be constant.

adequate information about the interaction at the boundary $\partial\Omega$. The simplest setting is that there is to be no such interaction at all: the physical system is to be *insulated* from the rest of the universe and thus there is no flux across the boundary. Formally, this means $\vec{q}\cdot\vec{n}=0$, where $\vec{n}$ is now the unit normal to $\partial\Omega$ or, from Equation 69.11 or 69.8 with the scaling of Equation 69.12,

$$-D\frac{\partial v}{\partial n} = -D\vec{\nabla}v\cdot\vec{n} = 0. \tag{69.14}$$

More generally, the flux might be more arbitrary but known; hence we have the *inhomogeneous Neumann condition*:

$$-D\frac{\partial v}{\partial n} = g_1 \quad \text{on } \Sigma = (0,T)\times\partial\Omega. \tag{69.15}$$

An alternative* set of data would involve knowing the temperature (concentration) at the boundary, that is, having the *Dirichlet condition*:

$$v = g_0 \quad \text{on } \Sigma = (0,T)\times\partial\Omega. \tag{69.17}$$

The mathematical theory supports our physical interpretation:

> *If we have Equation 69.1 on $\mathcal{Q}=(0,T)\times\Omega$ with ψ specified on $\mathcal{Q}$ and the initial condition Equation 69.13 specified on Ω, then either† of the boundary conditions Equation 69.15 or 69.17 suffices to determine the evolution of the system on $\mathcal{Q}$, that is, for $0<t\le T$.*

We refer to either of these as the *direct problem*. An important property of this problem is that it is *well-posed*, that is, a unique solution exists for each choice of the data and small changes in the data produce ‡ correspondingly small changes in the solution.

69.3 Background: Significant Properties

In this section we note some of the characteristic properties of the "direct problem" for the partial differential equation 69.1 and, related to these, introduce the representation formulas underlying the mathematical treatment.

69.3.1 The Maximum Principle and Conservation

One characteristic property, going back to the physical derivation, is that Equation 69.1 is a *conservation equation*. In the simplest form, when heat or material is neither created nor destroyed in the interior

* Slightly more plausible, physically, would be to assume that the ambient temperature or concentration would be known or determinable to be g "just outside" $\partial\Omega$ and then to use the flux law (proportionality to the difference) directly:

$$-D\frac{\partial v}{\partial n} = \vec{q}\cdot\vec{n} = \lambda(v-g) \quad \text{on } \Sigma \tag{69.16}$$

with a *flux transfer coefficient* $\lambda>0$. Note that, if $\lambda\approx 0$ (negligible heat or material transport), then we effectively get Equation 69.14. On the other hand, if λ is very large ($v-g=-(D/\lambda)\partial v/\partial n$ with $D/\lambda\approx 0$), then v will immediately tend to match g at $\partial\Omega$, giving Equation 69.17; see Section 69.4.1

† We may also have, more generally, a partition of $\partial\Omega$ into $\Gamma_0\cup\Gamma_1$ with data given in the form Equation 69.17 on $\Sigma_0=(0,T)\times\Gamma_0$ and in the form Equation 69.15 on $\Sigma_1=(0,T)\times\Gamma_1$.

‡ We note that making this precise—that is, specifying the appropriate meanings of "small"—becomes rather technical and, unlike the situation for ordinary differential equations, can be done in several ways which each may be useful for different situations. We will see, on the other hand, that some other problems which arise in system-theoretic analysis turn out to be "ill-posed", that is, not to have this well-posedness property; see, for example, Section 69.4.3.

($\psi \equiv 0$) and if the region is insulated Equation 69.14, then [total heat or material] $= \int_\Omega v \, dV$ is constant in time. More generally, we have

$$\frac{d}{dt}\left[\int_\Omega v \, dV\right] = \int_{\partial\Omega} g_1 \, dA + \int_\Omega \psi \, dV \tag{69.18}$$

for v satisfying Equations 69.3 through 69.15.

Another important property is the Maximum Principle:

> *Let v satisfy Equation 69.3 with $\psi \geq 0$ on $\mathcal{Q}_\tau := (0, \tau) \times \Omega$. Then the minimum value of $v(t, \mathbf{x})$ on $\overline{\mathcal{Q}_\tau}$ is attained either initially ($t = 0$) or at the boundary ($\mathbf{x} \in \partial\Omega$). Unless v is a constant, this value cannot also occur in the interior of $\mathcal{Q}_\tau$; if it is a boundary minimum with $t > 0$, then one must have $\partial v/\partial \vec{n} > 0$ at that point. Similarly, if v satisfies Equation 69.3 with $\psi \leq 0$, then its maximum is attained for $t = 0$ or at $\mathbf{x} \in \partial\Omega$, and so on.*

One simple argument for this rests on the observation that at an interior minimum one would necessarily have $v_t = \partial v/\partial t \leq 0$ and also $\Delta v \geq 0$.

The Maximum Principle shows, for example, that the mathematics of Equation 69.1 is consistent with the requirement for physical interpretation that a concentration cannot become negative and the fact that, since heat flows "from hotter to cooler," it is impossible to develop a "hot spot" except by providing a heat source.

69.3.2 Smoothing and Localization

Perhaps the dominant feature of Equation 69.1 is that solutions rapidly smooth out, with peaks and valleys of the initial data flattening out. We will see this in more mathematical detail later, but comment now on three points:

- Approach to steady state
- Infinite propagation speed
- Localization and geometric reduction

The first simply means that if neither ψ nor the data g_0 would vary in time, then the solution v of Equations 69.3 through 69.17 on $(0, \infty) \times \Omega$ would tend, as $t \to \infty$, to the unique solution $\bar{v}$ of the (elliptic) *steady-state equation*

$$-\left[\frac{\partial^2 \bar{v}}{\partial x^2} + \frac{\partial^2 \bar{v}}{\partial y^2} + \frac{\partial^2 \bar{v}}{\partial z^2}\right] = \psi, \quad \bar{v}\Big|_{\partial\Omega} = g_0. \tag{69.19}$$

The timescale of this transient is given by the lowest eigenvalue of the Laplace operator here. Essentially the same would hold if we were to use Equation 69.15 rather than Equation 69.17 except that, as is obvious from Equation 69.18, we must then impose a consistency condition that

$$\int_{\partial\Omega} g_1 \, dA + \int_\Omega \psi \, dV = 0$$

for there to be a steady state at all—and then must note that the solution of the steady-state equation

$$-\left[\frac{\partial^2 \bar{v}}{\partial x^2} + \frac{\partial^2 \bar{v}}{\partial y^2} + \frac{\partial^2 \bar{v}}{\partial z^2}\right] = \psi, \quad \partial\bar{v}/\partial\vec{n} = g_1 \tag{69.20}$$

only becomes unique when one supplements Equation 69.20 by specifying, from the initial conditions (Equation 69.13), the value of $\int_\Omega \bar{v} \, dV$.

Unlike the situation with the wave equation 69.4, the mathematical formulation (Equation 69.1), and so on, implies an infinite propagation speed for disturbances—for example, the effect of a change in the boundary data $g_0(t, \mathbf{x})$ at some point $\mathbf{x}_* \in \partial\Omega$ occurring at a time $t = t_*$ is immediately felt throughout the region, affecting the solution for every $\mathbf{x} \in \Omega$ at every $t > t_*$. One can see that this is necessary to have the Maximum Principle, for example, but it is certainly nonphysical. This phenomenon is a consequence of idealizations in our derivation and becomes consistent with our physical intuition when we note that this "immediate influence" is extremely small: there is, indeed, a noticeable delay before a perturbation will have a noticeable effect at a distance.

Consistent with the last observation, we note that the behavior in any subregion will, to a great extent, be affected only very slightly (in any fixed time) by what happens at parts of the boundary which may be very far away; this is a sort of "localization" principle. For example, if we are only interested in what is happening close to one part of the boundary, then we may effectively treat the far boundary as "at infinity." To the extent that there is little spatial variation in the data at the nearby part of the boundary, we may then approximate the solution quite well by looking at the solution of the problem considered on a half-space with spatially constant boundary data, dependent only on time. Taking coordinates so that the boundary becomes the plane "$x = 0$," one easily sees that this solution will be independent of the variables y, z if the initial data and source term are. Equation 69.1 then reduces to a one-dimensional form

$$\frac{\partial v}{\partial t} = \frac{\partial^2 v}{\partial x^2} + \psi(t, x) \tag{69.21}$$

for $t > 0$ and, now, $x > 0$ with, for example, specification of $v(t, 0) = g_0(t)$ and of $v(0, x) = v_0(x)$. Similar dimensional reductions occur in other contexts—one might obtain Equation 69.21 for $0 < x < L$, where L gives the thickness of a slab in appropriate units or one might get a two-dimensional form corresponding to a body which is long compared to its constant constant cross-section and with data which is relatively constant longitudinally. In any case, our equation will be Equation 69.3, with the dimensionally suitable interpretation of the Laplace operator. Even if the initial data does depend on the variables to be omitted, our first property asserts that this variation will tend to disappear; hence we may still get a good approximation after waiting through an initial transient. On the other hand, one usually cannot accept this approximation near, for example, the ends of the body where "end effects" due to those boundary conditions may become significant.

69.3.3 Linearity

We follow Fourier in using the *linearity* of the heat equation, expressed as a "superposition principle" for solutions, to obtain a general representation for solutions as an infinite series. Let $\{[e_k, \lambda_k] : k = 0, 1, \ldots\}$ be the pairs of *eigenfunctions* and *eigenvalues* for $-\boldsymbol{\Delta}$ on Ω, that is,

$$-\boldsymbol{\Delta} e_k = \lambda_k e_k \quad \text{on } \Omega \text{ (with BC)} \quad \text{for } k = 0, 1, \ldots \tag{69.22}$$

where "BC" denotes one of the homogeneous conditions

$$e_k = 0 \quad \text{or} \quad \frac{\partial e_k}{\partial \vec{n}} = 0 \quad \text{on } \partial\Omega \tag{69.23}$$

according as we are considering Equation 69.17 or 69.15. It is always possible to take these so that

$$\int_\Omega |e_k|^2 \, dV = 1, \qquad \int_\Omega e_i e_k \, dV = 0 \quad \text{for } i \neq k, \tag{69.24}$$

with $0 \leq \lambda_0 < \lambda_1 \leq \cdots \to \infty$; we have $\lambda_0 > 0$ for Equation 69.17 and $\lambda_0 = 0$ for Equation 69.15.

One sees immediately from Equation 69.22 that each function $e^{-\lambda_k t} e_k(\mathbf{x})$ satisfies Equation 69.1; hence superposing, we see that

$$v(t, \mathbf{x}) = \sum_k c_k e^{-\lambda_k t} e_k(\mathbf{x}) \tag{69.25}$$

gives the "general solution" with the coefficients (c_k) obtained from Equation 69.13 by

$$c_k = \langle e_k, v_0 \rangle \quad \text{so } v_0(\cdot) = \sum_k c_k e_k(\cdot), \tag{69.26}$$

assuming Equation 69.24. Note that $\langle \cdot, \cdot \rangle$ denotes the $L^2(\Omega)$ inner product: $\langle f, g \rangle = \int_\Omega f(\mathbf{x}) g(\mathbf{x})\, d^m\mathbf{x}$ (for m-dimensional Ω—with, physically, $m = 1, 2, 3$). The expansion (Equation 69.26, and so Equation 69.25), is valid if the function v_0 is in the Hilbert space $L^2(\Omega)$, that is, if $\int_\Omega |v_0|^2 < \infty$. Note that the series (Equation 69.26) need not converge pointwise unless one assumes more smoothness for v_0 but, since it is known that, asymptotically as $k \to \infty$, one has

$$\lambda_k \sim Ck^{2/m} \quad \text{with } C = C(\Omega), \tag{69.27}$$

the factors $e^{-\lambda_k t}$ decrease quite rapidly for any fixed $t > 0$ and Equation 69.25 then converges nicely to a smooth function. Indeed, this is just the "smoothing" noted above: this argument can be used to show that solutions of Equation 69.1 are analytic (representable locally by convergent power series) in the interior of Ω for any $t > 0$ and we note that this does not depend on having homogeneous boundary conditions.

Essentially the same approach can be used when there is a source term ψ as in Equation 69.9 but we still have homogeneous boundary conditions as, for example, $g_0 = 0$ in Equation 69.17. We can then obtain the more general representation

$$\begin{aligned} v(t, \mathbf{x}) &= \sum_k \gamma_k(t) e_k(\mathbf{x}), \quad \text{where} \\ \gamma_k(t) &= c_k e^{-\lambda_k t} + \int_0^t e^{-\lambda_k (t-s)} \psi_k(s)\, ds, \\ c_k &= \langle e_k, v_0 \rangle, \qquad \psi_k(t) = \langle e_k, \psi(t, \cdot) \rangle \end{aligned} \tag{69.28}$$

for the solution of Equation 69.9. When ψ is constant in t, this reduces to

$$\gamma_k(t) = \psi_k/\lambda_k + [c_k - \psi_k/\lambda_k]\, e^{-\lambda_k t} \longrightarrow \psi_k/\lambda_k$$

which not only shows that $v(t, \cdot) \to \bar{v}$, as in Equation 69.19 with $g_0 = 0$, but also demonstrates the exponential rate of convergence with the transient dominated by the principal terms, corresponding to the smaller eigenvalues. This last must be modified slightly when using Equation 69.15, since one then has $\lambda_0 = 0$.

Another consequence of linearity is that the effect of a perturbation is simply additive: if $\hat{v}$ is the solution of Equation 69.9 with data $\hat{\psi}$ and $\hat{v}_0$ and one perturbs this to obtain a new perturbed solution $\tilde{v}$ for the data $\hat{\psi} + \psi$ and $\hat{v}_0 + v_0$ (and unperturbed boundary data), then the solution perturbation $v = \tilde{v} - \hat{v}$ itself satisfies Equation 69.9 with data ψ and v_0 and homogeneous boundary conditions. If we now multiply the partial differential equation by v and integrate, we obtain

$$\frac{d}{dt}\left(\frac{1}{2}\int_\Omega |v|^2\right) + \int_\Omega |\vec{\nabla} v|^2 = \int_\Omega v\psi,$$

using the divergence theorem to see that $\int v\Delta v = -\int |\vec{\nabla} v|^2$ with no boundary term since the boundary conditions are homogeneous. The Cauchy–Schwarz inequality gives $\left|\int v\psi\right| \leq \|v\|\, \|\psi\|$, where $\|\cdot\|$ is the $L^2(\Omega)$-norm: $\|v\| = \left[\int_\Omega |v|^2\right]^{1/2}$ and we can then apply the Gronwall Inequality* to obtain, for example, the *energy inequality*

$$\|v(t)\|^2,\ 2\int_0^t \|\vec{\nabla} v\|^2\, ds \leq \left(\|v_0\|^2 + \int_0^t \|\psi\|^2\, ds\right) e^t. \tag{69.29}$$

This is one form of the well-posedness property asserted at the end of the last section.

* *If a function $\varphi \geq 0$ satisfies $\varphi(t) \leq C + M\int_0^t \varphi(s)\, ds$ for $0 \leq t \leq T$, then it satisfies: $\varphi(t) \leq Ce^{Mt}$ there.*

69.3.4 Autonomy, Similarity, and Scalings

Two additional useful properties of the heat equation are *autonomy* and *causality*. The first just means that the equation itself is time independent; hence a time-shifted setting just gives the time-shifted solution. For the pure initial-value problem—that is, Equation 69.1 with $g = 0$ in Equation 69.17 or 69.15—"causality" means that $v(t, \cdot)$ is determined by its "initial data" at *any* previous time t_0; hence we may write

$$v(t, \cdot) = \mathbf{S}(t - t_0)\, v(t_0, \cdot) \tag{69.30}$$

where $\mathbf{S}(\tau)$ is the *solution operator* for Equation 69.1 for elapsed time $\tau \geq 0$. This operator $\mathbf{S}(\tau)$ is a nice linear operator in a variety of settings, for example, $L^2(\Omega)$ or the space $C(\bar{\Omega})$ of continuous functions with convergence of functions meaning uniform convergence. A comparison with Equation 69.25 shows that

$$\mathbf{S}(t) : e_k \mapsto e^{-\lambda_k t} e_k \quad \text{so } \mathbf{S}(t)\left[\sum_k c_k e_k\right] = \left[\sum_k c_k e^{-\lambda_k t} e_k\right]. \tag{69.31}$$

From Equation 69.30 one obtains the fundamental "semigroup property"

$$\mathbf{S}(s + t) = \mathbf{S}(t) \circ \mathbf{S}(s) \quad \text{for } t, s \geq 0. \tag{69.32}$$

This only means that, if one initiates Equation 69.1 with any initial data v_0 at time 0 and so obtains $v(s, \cdot) = \mathbf{S}(s)v_0$ after a time s and $v(s + t, \cdot) = \mathbf{S}(s + t)v_0$ after a longer time interval of length $s + t$, as in Equation 69.30, "causality" gives $v(s + t, \cdot) = \mathbf{S}(t)v(s, \cdot)$. It is possible to verify that this operator function is strongly continuous at $t = 0$:

$$\mathbf{S}(t)v_0 \to v_0 \quad \text{as } t \to 0 \quad \text{for each } v_0$$

and is differentiable for $t > 0$: Equation 69.1 just tells us that

$$\frac{\mathrm{d}}{\mathrm{d}t}\mathbf{S}(t) = \mathbf{\Delta}\mathbf{S}(t), \tag{69.33}$$

where the Laplace operator $\mathbf{\Delta}$ here includes specification of the appropriate boundary conditions; we refer to $\mathbf{\Delta}$ in Equation 69.33 as "the infinitesimal generator of the semigroup $\mathbf{S}(\cdot)$."

In terms of $\mathbf{S}(\cdot)$ we obtain a new solution representation for Equation 69.9:

$$v(t, \cdot) = \mathbf{S}(t)v_0 + \int_0^t \mathbf{S}(t - s)\psi(s, \cdot)\, \mathrm{d}s. \tag{69.34}$$

Note that $\mathbf{S}$ precisely corresponds to the "Fundamental Solution of the homogeneous equation" for ordinary differential equations and Equation 69.35 is just the usual "variation of parameters" solution for the inhomogeneous equation 69.9; compare also with Equation 69.33. We may also treat the system with inhomogeneous boundary conditions by introducing the *Green's operator* $\mathbf{G} : g \mapsto w$, defined by solving

$$-\mathbf{\Delta}w = 0 \text{ on } \Omega, \quad \mathbf{B}w = g \text{ at } \partial\Omega. \tag{69.35}$$

with $\mathbf{B}w$ either w or $\partial w/\partial \vec{n}$, according as we consider Equation 69.17 or 69.15. Since $u = v - w$ then satisfies $u_t = \mathbf{\Delta}u + (\psi - w_t)$ with homogeneous boundary conditions, we may use Equations 69.33 and 69.34 to obtain, after an integration by parts,

$$v(t, \cdot) = \mathbf{S}(t)v_0 + \mathbf{G}[g_0(t) - g_0(0)]\, \mathrm{d}s + \int_0^t \mathbf{S}(t - s)\psi(s, \cdot)\, \mathrm{d}s - \int_0^t \mathbf{\Delta}\mathbf{S}(t - s)\mathbf{G}g_0(s)\, \mathrm{d}s. \tag{69.36}$$

The autonomy/causality above corresponds to the invariance of Equation 69.1 under time-shifting and we now note the invariance under some other transformations. For this, we temporarily ignore considerations related to the domain boundary and take Ω to be the whole 3-dimensional space $\mathbb{R}^3$.

It is immediate that in considering Equation 69.12 (with $\psi = 0$) with constant coefficients we have ensured that we may shift solutions arbitrarily in space. Not quite as obvious mathematically is the physically obvious fact that we may rotate in space. In particular, we may consider solutions which, spatially, depend only on the distance from the origin; so $v = v(t,r)$ with $r = |\mathbf{x}| = \sqrt{x^2+y^2+z^2}$. Equation 69.12 with $\psi = \psi(t,r)$ is then equivalent to

$$\frac{\partial v}{\partial t} = D\left[\frac{\partial^2 v}{\partial r^2} + \frac{2}{r}\frac{\partial v}{\partial r}\right] + \psi, \tag{69.37}$$

which involves only a single spatial variable. For the two-dimensional setting $\mathbf{x} = (x,y)$ as in Section 69.3.2, this becomes

$$\frac{\partial v}{\partial t} = D\left[\frac{\partial^2 v}{\partial r^2} + \frac{1}{r}\frac{\partial v}{\partial r}\right] + \psi. \tag{69.38}$$

More generally, for a d-dimensional case, Equations 69.37 and 69.38 can be written as

$$v_t = r^{-(d-1)}\left(r^{d-1}v_r\right)_r + \psi. \tag{69.39}$$

The apparent singularity of these equations as $r \to 0$ is, of course, only an effect of the use of polar coordinates. As in Section 69.3.3, we may seek a series representation like Equation 69.25 for solutions of Equation 69.39 with the role of the eigenfunction equation 69.22 now played by Bessel's equation; we then obtain an expansion in Bessel functions with the exponentially decaying time dependence $e^{-\lambda_k t}$ as earlier.

Finally, we may also make a combined scaling of both time and space. If, for some constant c, we set

$$\hat{t} = c^2 Dt, \quad \hat{\mathbf{x}} = c\mathbf{x}, \tag{69.40}$$

then, for any solution v of Equation 69.12 with $\psi = 0$, the function $\hat{v}(\hat{t},\hat{\mathbf{x}}) = v(t,\mathbf{x})$ will satisfy Equation 69.1 in the new variables. This corresponds to the earlier comment that we may make $D = 1$ by appropriate choice of units.

Closely related to the above is the observation that the function

$$k(t,\mathbf{x}) = (4\pi Dt)^{-d/2}e^{-|\mathbf{x}|^2/4Dt} \tag{69.41}$$

satisfies Equation 69.39 for $t > 0$ while a simple computation shows* that

$$\int_{\mathbb{R}^d} k(t,\mathbf{x})\, d_d\mathbf{x} = 1 \quad \text{for each } t > 0; \tag{69.42}$$

thus $k(t,\cdot)$ becomes a δ-function as $t \to 0$. So, $k(t-s, \mathbf{x}-\mathbf{y})$ is the *impulse response function* for an impulse at $(s,\mathbf{y})$. Taking $d = 3$, we note that

$$v(t,\mathbf{x}) = \int_{\mathbb{R}^3} k(t,\mathbf{x}-\mathbf{y})v_0(\mathbf{y})\, d_3\mathbf{y} \tag{69.43}$$

is a superposition of solutions (now by integration, rather than by summation); hence, linearity ensures that v is itself a solution; we also have

$$v(t,\cdot) \longrightarrow v_0 \quad \text{as } t \to 0, \tag{69.44}$$

where the specific interpretation of this convergence depends on how smooth v_0 is assumed to be. Thus, Equation 69.43 provides another solution representation–although, as noted, it ignores the effect of the boundary for a physical region which is not all of $\mathbb{R}^3$. For practical purposes, following the ideas of Section 69.3.2, the formula (Equation 69.43) will be a good approximation to the solution so long as $\sqrt{2Dt}$ is quite small† as compared to the distance from the point $\mathbf{x}$ to the boundary of the region.

---

* We may observe that $k(t,\cdot)$ is a multivariate normal distribution (Gaussian) with standard deviation $\sqrt{2Dt} \to 0$ as $t \to 0$.

† When $\mathbf{x}$ is too close to the boundary for this to work well, it is often plausible to think of $\partial\Omega$ as "almost flat" on the relevant spatial scale and then to extend v_0 by reflection across it—as an odd function if one were using Equation 69.17

69.4 Some Control-Theoretic Examples

In this section we provide three comparatively elementary examples to see how the considerations above apply to some control-theoretic questions. The first relates to a simplified version of a quite practical heat transfer problem and is treated with the use of rough approximations, essentially to see how such heuristic treatment can be used to obtain practical results. The second describes the problem of control to a specified terminal state—which would be a standard problem in the case of ordinary differential equations but which involves some new considerations in this distributed parameter setting. The final example is a "coefficient identification" problem: using interaction (input/output) at the boundary to determine the function $q = q(x)$ in an equation of the form $u_t = u_{xx} - qu$, generalizing Equation 69.21.

69.4.1 A Simple Heat Transfer Problem

We consider a slab of thickness a and diffusion coefficient D within which heat is generated at constant rate ψ. On the one side, this is insulated ($v_x = 0$) and on the other, it is in contact with a stream of coolant (diffusion coefficient D') moving in an adjacent duct with constant flow rate F in the y-direction. Thus, the slab occupies $\{(x,y) : 0 < x < a,\ 0 < y < L\}$ and the duct occupies $\{(x,y) : a < x < \bar{a},\ 0 < y < L\}$ with $a, \bar{a} \ll L$ and no dependence on z.

If the coolant enters the duct at $y = 0$ with input temperature u_0, our problem is to determine how hot the slab will become. For this purpose, we assume that we are operating in steady state, that the coolant flow is turbulent enough to ensure perfect mixing (and so constant temperature) across the duct, and that—to a first approximation—the longitudinal transfer of heat is entirely by the coolant flow so we may consider conduction in the slab only in the transverse direction ($0 < x < a$).

The source term ψ in the slab gives heat production $a\psi$ per unit distance in y and this must be carried off by the coolant stream to have a steady state. We might, as noted earlier, shift to material coordinates in the stream to obtain an equation there but, more simply, we just observe that when the coolant has reached the point y it must have absorbed the amount $a\psi y$ of heat per second and, for a flow rate F (choosing units so that ρc in Equation 69.7 is 1) this will have raised the coolant temperature from u_0 to $[u_0 + a\psi y/F] =: u(y)$.

Now consider the transverse conduction in the slab. We have there $v_t = Dv_{xx} + \psi$ with $v_t = 0$ for steady state. As $v_x = 0$ at the outer boundary $x = 0$, the solution has the form $v = v^* - (\psi/2D)x^2$, where v^* is exactly what we wish to determine. If we assume a simple temperature match of slab to coolant ($v = u(y)$ at $x = a$), then this gives $v^*(y) - (\psi/2D)a^2 = v(a,y) = u(y) = u_0 + a\psi y/F$; so

$$\begin{aligned} v^* = v^*(y) &= u_0 + \left[\frac{y}{F} + \frac{a}{2D}\right] a\psi, \\ v &= u_0 + \left[\frac{y}{F} + \frac{a}{2D}\left(1 - \left[\frac{x}{a}\right]^2\right)\right] a\psi. \end{aligned} \tag{69.46}$$

A slight correction of this derivation is worth noting: for the coolant flow we expect a boundary layer (say, of thickness δ) of "stagnant" coolant at the duct wall and within this layer we have $u_x \approx \text{constant} = -\left[v(a) - u\,|_{\text{flow}}\right]/\delta$, while also $-D'u_x = \text{flux} = a\psi$ by Fourier's Law; so, instead of matching $v(a) = u(y)$, we obtain $v(a) = u(y) + (\delta/D')a\psi$ which also increases v^*, v by $(\delta/D')a\psi$ as a correction to Equation 69.46; effectively, this correction notes the reduction of heat transfer through

---

with $g_0 = 0$ or as an even function if one were using Equation 69.15 with $g_1 = 0$. For Equation 69.17 with, say, $g_0 = g_0(t)$ locally, there would then be a further correction by adding

$$\int_0^t \hat{k}(t-s,x) g_0(s)\,ds, \quad \hat{k}(\tau,x) = \frac{x}{\tau} k_1(\tau,x) = 2D\frac{\partial k_1}{\partial x} \tag{69.45}$$

where $k_1 = k_1(\tau,x)$ is as in Equation 69.41 for $d = 1$ and x is here the distance from $\mathbf{x}$ to the boundary; compare Equation 69.36. There are also comparable correction formulas for more complicated settings.

replacing the boundary conditions (Equation 69.17) by (Equation 69.16) with $\lambda = D'/\delta$. Much more complicated corrections would be needed if one would have to consider conduction within the duct, especially with a velocity profile other than the plug flow assumed here.

We also note that in this derivation we neglected longitudinal conduction in the slab, essentially omitting the v_{yy} term in Equation 69.12. Since Equation 69.46 gives $v_{yy} = 0$, this is consistent with the equation. It is, however, inconsistent with reasonable boundary conditions at the ends of the slab ($y = 0, L$) and one would expect "end effects" as well as some evening out of v^*.

We note that, although this was derived in steady state, we could think of using Equation 69.46 for an optimal control problem (especially if ψ would be time dependent, but slowly varying) with the flow rate F as control.

69.4.2 Exact Control

We consider the problem of using ψ as control to reach a specified "target state," $\omega = \omega(x)$ at time T. We base the discussion on the representation* (Equation 69.28), which permits us to treat each component independently: the condition that $v(T,\cdot) = \omega$ becomes the sequence of "moment equations"

$$\begin{aligned} \gamma_k(T) &= c_k e^{-\lambda_k T} + \int_0^T e^{-\lambda_k (T-s)} \psi_k(s)\, ds \\ &= \omega_k := \langle e_k, \omega \rangle \end{aligned} \tag{69.48}$$

for each k. This does not determine the control uniquely, when one exists, so we select by optimality, minimizing the norm of ψ in $L^2(\mathcal{Q})$ with $\mathcal{Q} = (0, T) \times \Omega$. This turns out to be equivalent to requiring that $\psi_k(t)$ should be a constant times $e^{\lambda_k (T-t)}$; so, noting that $\int_0^T \left| e^{-\lambda_k (T-s)} \right|^2 ds = \left[1 - e^{-2\lambda_k T}\right]/2\lambda_k$, the conditions (Equation 69.48) give us the formula

$$\psi(t, \mathbf{x}) = 2 \sum_k \lambda_k \left(\frac{\omega_k - c_k e^{-\lambda_k T}}{1 - e^{-2\lambda_k T}} \right) e^{-\lambda_k (T-t)} e_k(\mathbf{x}). \tag{69.49}$$

This formula converges if (and only if) the specified target state ω is, in fact, attainable by some control in $L^2(\mathcal{Q})$. So far, so good!

Let us now see what happens when we actually attempt to implement the use of Equation 69.49. Adding a touch of realism, one must truncate the expansion (say, at $k = K$) and one must then find each coefficient $\alpha_k = \omega_k - c_k e^{-\lambda_k T}$ with an error bound ε_k by using some algorithm of numerical integration on Ω to compute the inner products $\langle e_k, \omega \rangle$. [For simplicity we assume that we would already know exactly the relevant eigenvalues and eigenfunctions, as is the case for Equation 69.21 and for a variety of higher-dimensional geometries.] Denoting the optimal control by Ψ and the approximation obtained by Ψ_K, we can bound the total error by

$$\|\Psi - \Psi_K\|_{\mathcal{Q}}^2 \le 2 \sum_{k=1}^{K} \left[\frac{\lambda_k \varepsilon_k^2}{1 - e^{-2\lambda_k T}} \right] + 2 \sum_{k>K} \left[\frac{\lambda_k \alpha_k^2}{1 - e^{-2\lambda_k T}} \right]. \tag{69.50}$$

For an attainable target ω the second sum is small for large K, corresponding to convergence of Equation 69.49. The use of a fixed error bound $|\varepsilon_k| \le \varepsilon$ for the coefficient computation would make the first sum of the order of $K^{1+(2/d)}\varepsilon$ by Equation 69.27 so this sum would become large as K increased.

---

* For definiteness, one may think of the 1-dimensional heat equation 69.21 with homogeneous Dirichlet boundary conditions at $x = 0, 1$. The eigenvalues and normalized eigenfunctions are then

$$\lambda_k = k^2 \pi^2, \quad e_k(x) = \sqrt{2} \sin k\pi x; \tag{69.47}$$

so the expansions, starting at $k = 1$, for convenience, are standard Fourier sine series.

To make the total error (Equation 69.50) small requires picking K and then choosing ε dependent on this choice—or using a relative error condition: $|\varepsilon_k| \leq \varepsilon|\alpha_k|$. This last seems quite plausible for numerical integration with floating point arithmetic—but one trap remains! Neglecting v_0, a plausible form of the error estimate for a method of numerical integration might be

$$|\varepsilon_k| \leq C_\nu h^\nu \|\omega e_k\|_{[\nu]} \sim C'_\nu h^\nu \lambda_k^{\nu/2} \|\omega\|_{[\nu]},$$

where h characterizes a mesh size and the subscript on $\|\cdot\|_{[\nu]}$ indicates consideration of derivatives of order up to ν, with ν depending on the choice of integration method; we have noted that $\|e_k\|_{[\nu]} \sim \lambda_k^{\nu/2}$ since the differential operator $\mathbf{\Delta}$ is already of order 2. This means that one might have to refine the mesh progressively to obtain such a uniform relative error for large k.

69.4.3 System Identification

Finally, we consider a 1-dimensional example governed by an equation known to have the form*

$$\frac{\partial v}{\partial t} = D\frac{\partial^2 v}{\partial x^2} - q(x)v, \tag{69.51}$$

but with D and the specific coefficient function $q(\cdot)$ unknown or known with inadequate accuracy. We will assume here that $u = 0$ at $x = 1$, but that interaction is possible at the end $x = 0$ where one can both manipulate the temperature and observe the resulting heat flux; for simplicity, we assume that $v_0 = 0$. Thus, we consider the input/output pairing: $g \mapsto f$, defined through Equation 69.51 with

$$\begin{aligned} u(t,1) = 0 \qquad u(t,0) &= g(t) \\ f(t) := -Du_x(t,0). \end{aligned} \tag{69.52}$$

By linearity, causality, and autonomy of Equation 69.51, we see that this pairing takes the convolution form

$$f(t) = \int_0^t \sigma(t-s)g(s)\,\mathrm{d}s = \int_0^t \sigma(s)g(t-s)\,\mathrm{d}s \tag{69.53}$$

where $\sigma(\cdot)$ is a kind of impulse response function. Much as we obtained Equations 69.25 and 69.36, we obtain

$$\begin{aligned} \sigma(t) &= \sum_k \sigma_k e^{-\lambda_k t} \\ \text{with} \quad \sigma_k &:= -D\lambda_k e'_k(0)\langle z, e_k\rangle \\ Dz'' - qz = 0 \quad & z(0) = 1,\ z(1) = 0, \\ -De''_k + qe_k = \lambda_k e_k \quad & e_k(0) = 0 = e_k(1), \end{aligned} \tag{69.54}$$

noting that z and $\{(\lambda_k, e_k)\}$ are unknown since q is unknown.

Viewing Equation 69.53 as an integral equation for σ, it can be shown that Equation 69.53 determines σ for appropriate choices of the input $g(\cdot)$—simplest would be if we could take g to be a δ-function (impulse) so that the observed f would just be σ: otherwise we must first solve a Volterra equation of first kind, which is already an ill-posed problem. The function $\sigma(\cdot)$ contains all the information about the unknown q which we can get and it is possible to show that quite large differences for q may produce only very small perturbations of σ; thus, this identification problem cannot be "well-posed," regardless of $g(\cdot)$.

None of the coefficients σ_k will be 0; so, given $\sigma(\cdot)$, Equation 69.54 uniquely determines the eigenvalues $\{\lambda_k\}$ which appear there as exponents. We note that $\lambda_k \sim D\pi^2 k^2$; so $D = \lim_k \lambda_k/\pi^2 k^2$ is then determined.

* For example, such an equation might arise for a rod reduced to a simplified 1-dimensional form with heat loss to the environment appearing through a boundary condition at the surface of the rod as in Equation 69.16, with g = constant and λ, here q, varying along the rod.

It is then possible to show (by an argument involving analytic continuation, Fourier transforms, and properties of the corresponding wave equation) that $\sigma(\cdot)$ uniquely determines $q(\cdot)$, as desired.

The discussion above gives no suggestion as to how to compute D, $q(\cdot)$ from the observations. Typically, one seeks nodal values for a discretization of q. This can be done, for example, by *history matching*, an approach often used for such identification problems, in which one solves the direct problem with a guessed q to obtain a resulting "$f = f(q)$" and proceeds to find the q which makes this best match the observed f. It is a useful viewpoint to consider the guessed q as a control so that the idea of 'best match' makes this an optimal control problem

With some further *a priori* information about the function q—say, a known bound on the derivative q'—the uniqueness result, although itself nonconstructive, serves to ensure convergence for these computations to the correct result as the discretization is refined. Note that it is the auxiliary *a priori* information which converts this to a well-posed problem, although one which will be quite badly conditioned; hence the practical difficulties do not entirely disappear. A frequently used approach to this is the use of *regularization*, for example, including in the optimization criterion for "best match" some penalty for undesirable oscillations of q, trading resolution for stability of the computations.

69.5 More Advanced System Theory

In this section we consider the system–theoretic results available for the heat equation, especially regarding observability and controllability. Our emphasis is on how, although the relevant questions are quite parallel to those standard in "lumped parameter" control theory, one has new technical difficulties which can occur only because of the infinite-dimensional state space; this will also mean that this section describes in more detail the results of Functional Analysis* and special mathematical results for the partial differential equations involved.

69.5.1 The Duality of Observability/Controllability

For the finite-dimensional case, controllability for a problem and observability for the adjoint problem are dual—essentially, one can control $\dot{x} = Ax + Bg$ ($g(\cdot) =$ control) from one arbitrary state to another if and only if only the trivial solution of the adjoint equation $-\dot{y} = A^*y$ can give [observation] $= B^*y(\cdot) \equiv 0$. Something similar holds for the heat equation (e.g., with boundary I/O), but we must be rather careful in our statement.

We begin by computing the relevant adjoint problem, taking the boundary control problem as

$$u_t = \mathbf{\Delta} u \quad \text{on } \mathcal{Q} \text{ with } \mathbf{B}u = g \text{ on } \Sigma \text{ and } u\Big|_{t=0} = u_0 \tag{69.55}$$

in which the control function g is the data for the boundary conditions, defined on $\Sigma = (0, T) \times \partial\Omega$. As for Equation 69.35, the operator $\mathbf{B}$ will correspond to either Equation 69.17 or 69.15; we may now further include in the specification of $\mathbf{B}$ a requirement that $g(\cdot)$ is restricted to be 0 outside some fixed "patch"—that is, a relatively open subset $\mathcal{U} \subset \partial\Omega$, viewed as an "accessible" portion of $\partial\Omega$—and then refer to this as a problem of "boundary patch control." Note that

$$u_T := u(T, \cdot) = \mathbf{S}(T)u_0 + \mathbf{L}g(\cdot) \tag{69.56}$$

where Equation 69.36 gives

$$\mathbf{L} : g(\cdot) \mapsto \mathbf{G}[g(T) - g(0)] - \int_0^T \mathbf{\Delta S}(T - s)\mathbf{G}g(s)\, ds.$$

* We note [1,2] as possible general references for Functional Analysis, specifically directed toward distributed parameter system theory.

For the adjoint problem, we consider

$$-v_t = \mathbf{\Delta} v \quad \mathbf{B}v = 0; \; \varphi = [\hat{\mathbf{B}}v]\Big|_{\mathcal{U}}, \tag{69.57}$$

where $\hat{B}$ gives the "complementary" boundary data: $\hat{B}v := \partial v/\partial \vec{n}$ if $\mathbf{B}$ corresponds to Equation 69.17 and $\hat{\mathbf{B}}v := v\Big|_{\partial\Omega}$ if $\mathbf{B}$ corresponds to Equation 69.15. We then have, using the Divergence Theorem,

$$\begin{aligned}\left[\int_\Omega u_T v_T\right] - \left[\int_\Omega u_0 v_0\right] &= \int_{\mathcal{Q}} (uv)_t \;= \int_{\mathcal{Q}}[(\vec{\nabla}^2 u)v - u(\vec{\nabla}^2 v)] \\ &= \int_\Sigma [u_{\vec{n}} v - u v_{\vec{n}}] \;= -\int_{\mathcal{U}} g\varphi,\end{aligned}$$

where we write v_T, v_0 for $v(T,\cdot), v(0,\cdot)$, respectively, and set $\mathcal{Q} = (0,T)\times\Omega$. Thus, with subscripts indicating the domain for the inner product of $L^2(\cdot)$, we have the identity

$$\langle u_T, v_T\rangle_\Omega + \langle g, \varphi\rangle_{\mathcal{U}} = \langle u_0, v_0\rangle_\Omega \tag{69.58}$$

from which we wish to draw conclusions.

First, consider the *reachable set* $\mathcal{R} = \{u_T : g = \text{any} \in L^2(\mathcal{U}); u_0 = 0\}$, which is just the range of the operator $\mathbf{L} : L^2(\mathcal{U}) \to L^2(\Omega)$. If this were not dense, that is, if we did not have $\overline{\mathcal{R}} = L^2(\Omega)$, then (by the Hahn–Banach Theorem) there would be some nonzero v_T^* orthogonal to all $u_T \in \overline{\mathcal{R}}$; so Equation 69.58 would give $\langle g, \varphi^*\rangle_{\mathcal{U}} = 0$ for all g, whence $\varphi^* = 0$, violating *detectability* (i.e., that $\varphi^* = 0$ only if $v^* = 0$). Conversely, a violation of detectability would give a nonzero v^* with $v_T^* \neq 0$ orthogonal to $\overline{\mathcal{R}}$. Thus, detectability is equivalent to *approximate controllability*. This last means that one could control arbitrarily closely to any target state, even if it cannot be reached exactly—a meaningless distinction for finite-dimensional linear systems although significant for the heat equation since, as we have already noted, solutions of Equation 69.1 are very smooth (analytic) in the interior of Ω; so only very special targets could be exactly reachable.

Detectability means that the map $v_T \mapsto v \mapsto \varphi$ is 1–1; hence, inverting, $\varphi \mapsto v_T \mapsto v_0$ is well-defined: one can predict (note the time-reversal in Equation 69.57) v_0 from observation of φ on $\mathcal{U}$. In the finite-dimensional case, any linear map such as $\mathbf{A} : \varphi \mapsto v_0$ would necessarily be continuous (bounded), but here this is not automatically the case; note that the natural domain of $\mathbf{A}$ is the range of $v_T \mapsto \varphi$ and, if one had continuity, this would extend to the closure $\mathcal{M} = \mathcal{M}_{\mathcal{U}} \subset L^2(\mathcal{U})$. For bounded $\mathbf{A} : \mathcal{M} \to L^2(\Omega)$ there is a bounded adjoint operator $\mathbf{A}^* : L^2(\Omega) \to \mathcal{M}$ and, if we were to set $g = \mathbf{A}^* u_0$ in Equation 69.55, we would obtain

$$\langle u_T, v_T\rangle_\Omega = \langle u_0, \mathbf{A}\varphi\rangle_\Omega - \langle \mathbf{A}^* u_0, \varphi\rangle_{\mathcal{U}} = 0 \quad \text{for every } \; v_T \in L^2(\Omega).$$

This would imply $u_T = 0$; hence $g = \mathbf{A}^* u_0$ is a nullcontrol from u_0—indeed, it turns out that this g is the *optimal* nullcontrol in the sense of minimizing the $L^2(\mathcal{U})$-norm. Conversely, if there is some nullcontrol $\tilde{g}$ for each u_0, there will be a minimum-norm nullcontrol g and the map $\mathbf{C} : u_0 \mapsto g$ is then linear and is continuous by the Closed Graph Theorem. Further, its adjoint $\mathbf{A} = \mathbf{C}^*$ is just the observation operator: $\varphi \mapsto v_T$ whence bounded observability for the adjoint problem is equivalent to nullcontrollability* for Equation 69.55 which, from Equation 69.56, is equivalent to knowing that the range of $\mathbf{L}$ contains the range of $\mathbf{S}(T)$.

Suppose we have nullcontrollability for arbitrarily small $T > 0$, always taking $\mathcal{U} = [0,T] \times U$ for some fixed patch $U \subset \partial\Omega$. A simple argument shows that the reachable set $\mathcal{R}$ must then be entirely independent of T and of the initial state u_0. No satisfactory characterization of $\mathcal{R}$ is available, although there are various known sufficient considerations to have some $\omega \in \mathcal{R}$.

* This observation is the heart of the Hilbert Uniqueness Method (HUM) introduced by J.-L. Lions. If one has detectability one might use the great freedom Hilbert space theory gives for selecting norms to find one making the space of observations complete so $\mathbf{C}^*$ would be bounded; then, if the resultant dual space can be suitably characterized, one will have found an appropriate context for nullcontrollability.

69.5.2 The One-Dimensional Case

From the discussion above, it will clearly be sufficient to prove bounded observability for the one-dimensional heat equation to have nullcontrollability also. (As a historical note, this equivalence was not realized at the time these results were first proved; so originally they were proved independently.) We will consider specifically the observability problem with $\Omega = (0, 1), \mathcal{U} = (0, T) \times \{0\}$ and

$$v_t = v_{xx} \quad v(t,0) = v(t,1) = 0; \quad \varphi(t) := v_x(t,0), \tag{69.59}$$

for which we explicitly know Equation 69.47. From Equation 69.25, we obtain

$$\varphi(t) = \sum_k \tilde{c}_k e^{-\lambda_k t}, \tag{69.60}$$

$$v(T,\cdot) = \sum_k \frac{\tilde{c}_k}{k\pi} e^{-\lambda_k T} e_k(\cdot). \tag{69.61}$$

[Note that, for convenience, we have re-reversed time in comparison with Equation 69.57 and that, from Equation 69.25, we have $\tilde{c}_k = \sqrt{2}k\pi \int_0^1 v_0(x) \sin k\pi x \, dx$—although we will have no need of any explicit information about v_0.]

The form of Equation 69.60 is a *Dirichlet series*; note that this becomes a power series in $\xi = e^{-\pi^2 t}$ with only the k^2 powers appearing: $e^{-\lambda_k t} = e^{-k^2 \pi t} = \xi^{k^2}$. The theory of such series centers on the Müntz–Szász Theorem (extending the Weierstrass Approximation Theorem) which, for our purposes, shows that only quite special functions can have L^2-convergent expansions (Equation 69.60) when $\Sigma 1/\lambda_k < \infty$. One has estimates for Equation 69.60 of the form

$$|\tilde{c}_k| \le \beta_k \|\varphi\|_{L^2(0,\infty)} \tag{69.62}$$

with the values of β_k explicitly computable as an infinite product

$$\beta_k = \sqrt{1 + 2\lambda_k} \prod_{i \ne k} \left| 1 + \frac{1 + 2\lambda_k}{\lambda_i - \lambda_k} \right| \tag{69.63}$$

(convergent when $\sum_k 1/\lambda_k$ is convergent); note that $1/\beta_k$ is the distance in $L^2(0,\infty)$ from $\exp[-\lambda_k t]$ to span $\{\exp[-\lambda_i t] : i \ne k\}$. L. Schwarz has further shown that for functions given as in Equation 69.60 one has

$$\|\varphi\|_{L^2(0,\infty)} \le \Gamma_T \|\varphi\|_{L^2(0,T)}. \tag{69.64}$$

Combining these estimates shows that

$$\|v(T,\cdot)\|_{L^2(0,1)} \le C_T \Gamma_T \|\varphi\|_{L^2(0,T)} \quad \left(C_T^2 := \sum_k \left[\frac{\beta_k}{k\pi} \right]^2 e^{-2k^2\pi^2 T} \right). \tag{69.65}$$

The sequence β_k increases moderately rapidly as $k \to \infty$ but the exponentials $\exp[-k^2\pi^2 T]$ decay even more rapidly; so the sum giving C_T^2 is always convergent and Equation 69.65 provides a bound ($\|\mathbf{A}\| \le \Gamma_T C_T < \infty$) for the observation operator* $\mathbf{A} : \mathcal{M} = \mathcal{M}_{[0,T]} \to L^2(\Omega) : \varphi \mapsto v(T,\cdot)$, when $T > 0$ is arbitrarily small.

* We note at this point that an estimate by Borwein and Erdélyi makes it possible to obtain comparable results when $\mathcal{U}$ has the form $U = \mathcal{E} \times \{0\}$ with $\mathcal{E}$ any subset of $[0, T]$ having positive measure; one consequence of this is a proof of the bang-bang principle for time-optimal constrained boundary control.

A somewhat different way of looking at this is to note that the linear functional: $\mathcal{M} \to \mathbb{R} : \varphi \mapsto \tilde{c}_k$ must be given by a function $g_k \in L^2(0, T)$ such that

$$\int_0^T g_k(t)e^{-\lambda_i t}\,dt = \delta_{i,k} := \begin{cases} 0 & \text{if } i \neq k, \\ 1 & \text{if } i = k. \end{cases} \tag{69.66}$$

If we think of $g_k(\cdot)$ as defined on $\mathbb{R}$ (0 off $[0, T]$), we may take the Fourier transform and note that Equation 69.66 just asserts the "interpolation conditions"

$$\hat{g}_k(-j\lambda_i) = \sqrt{2\pi}\delta_{i,k} \quad (j = \sqrt{-1}); \tag{69.67}$$

so it is sufficient to construct functions $\hat{g}_k$ satisfying Equation 69.67, together with the properties required by the Paley–Wiener Theorem to get the inverse Fourier transform in $L^2(0, T)$ with $\|g_k\| = \beta_k$. This approach leads to the sharp asymptotic estimate

$$\ln \|\mathbf{A}\| = \mathcal{O}(1/T) \quad \text{as } T \to 0, \tag{69.68}$$

showing how much more difficult* observability or controllability becomes for small T, even though one does have these for every $T > 0$.

A variant of this considers an interior point observation $\varphi(t) := v(t, a)$. The observability properties now depend on number-theoretic properties of $0 < a < 1$ — for rational $a = m/n$, one obtains no information at all about c_k when k is a multiple of n, since then $\sin k\pi a = 0$, and there is difficulty if a is approximable by rationals too rapidly. It can be shown that one does have bounded observability (with arbitrarily small $T > 0$) for a in a set of full measure whereas the complementary nullset for which this fails is uncountable in each subinterval.†

Finally, we note that an essentially identical treatment for all of the material of this subsection would work more generally for Equation 69.21 and with other boundary conditions.

69.5.3 Higher-Dimensional Geometries

We have already noted that the geometry may be significant here in ways which have no finite-dimensional parallel. For higher-dimensional cases, we note first that we can obtain observability for any "cylindrical" region $\Omega := (0, 1) \times \hat{\Omega} \subset \mathbb{R}^d$ with $\mathcal{U} = (0, T) \times [0 \times \hat{\Omega}]$ by using the method of "separation of variables" to reduce this to a sequence of independent one-dimensional problems: noting that we have here

$$\lambda_{k,\ell} = k^2\pi^2 + \hat{\lambda}_\ell \quad e_{k,\ell}(x, \hat{\mathbf{x}}) = [\sqrt{2}\sin k\pi x]\hat{e}_\ell(\hat{\mathbf{x}}),$$

we obtain

$$\mathbf{A}v_x(\cdot, 0, \cdot) = \sum_\ell [\mathbf{A}_1\varphi_\ell](x)e^{-\hat{\lambda}_\ell T}\hat{e}_\ell(\hat{\mathbf{x}}),$$
$$\text{with} \quad \varphi_\ell(t) := e^{\hat{\lambda}_\ell t}\langle v_x(t, 0, \cdot), \hat{e}_\ell\rangle_{\hat{\Omega}}$$

where $\mathbf{A}_1$ is the observability operator for Equation 69.59. It is easy to check that this gives $\|\mathbf{A}\| \leq \|\mathbf{A}_1\| < \infty$ and we have nullcontrollability by duality.

For more general regions, when $\mathcal{U}$ is all of $\Sigma := (0, T) \times \partial\Omega$ we may shift to the context of nullcontrollability for Equation 69.55 and rely on a simple geometric observation. Suppose we have $\Omega \subset \tilde{\Omega} \subset \mathbb{R}^d$,

---

* This may be compared to the corresponding estimate: $\|\mathbf{A}\| = \mathcal{O}\left(T^{-(K+1/2)}\right)$ for the finite-dimensional setting, with K, the minimal index, giving the rank condition there.

† Since the "bad set" has measure 0, one might guess that observation using local integral averages (as a "generalized thermometer") should always work but, somewhat surprisingly, this turns out to be false.

where $\tilde{\Omega}$ is some conveniently chosen region (e.g., a cylinder, as above) for which we already know that we have nullcontrollability for Equation 69.55, that is, with $\mathcal{Q}$, $\mathcal{U}$ replaced by $\tilde{\mathcal{Q}} = (0, T) \times \tilde{\Omega}$ and $\tilde{\mathcal{U}} = \tilde{\Sigma} = (0, T) \times \partial\tilde{\Omega}$, respectively. Given any initial data $u_0 \in L^2(\Omega)$ for Equation 69.55, we extend it as 0 to all of $\tilde{\Omega}$ and, as has been assumed possible, let $\tilde{u}$ be a (controlled) solution of Equation 69.55, vanishing on all of $\tilde{\Omega}$ at time T. The operator $\mathbf{B}$ acting (at Σ) on $\tilde{u}$ will have *some* value, which we now call "g" and using this in Equation 69.55 necessarily (by uniqueness) gives the restriction to $\mathcal{Q}$ of $\tilde{u}$ which vanishes at T. Thus, this g is a nullcontrol for Equation 69.55. As already noted, once we have a nullcontrol g for each u_0, it follows, as noted earlier, that the nullcontrol operator $\mathbf{C}$ for this setting is well defined and continuous; by duality, one also has bounded observability.

At this point we note an irreversible deep connection [8] between the control theories for the wave and heat equations:

> *observability for the wave equation* $w_{tt} = \mathbf{\Delta} w$ *for some* $[\Omega, U]$ *and some* $T^* > 0$ *implies observability for the heat equation* $u_t = \mathbf{\Delta} u$ *for arbitrary* $T > 0$ *in the same geometric setting* $[\Omega, U]$

with observability results for the wave equation obtainable [8,10] from known Scattering Theory results. From this, one obtains observability/controllability for the heat equation for a variety of geometric settings, but there is a price in terms of significant geometric restrictions on $[\Omega, U]$ related to wave propagation.

On the other hand, it has now been shown that use of an arbitrary patch $U \subset \Omega$ or $U \subset \partial\Omega$ suffices for observation or control of Equation 69.1. This uses the technique of Carleman estimates* which is quite technical; hence we have given here only a very brief sketch of how it applies to Equation 69.1. Given, for example, an interior patch $U \subset \Omega$; so we would be considering the restriction to U as the observation operator in Equation 69.57, one begins with a positive function ϕ on Ω vanishing on $\partial\Omega$ and with $|\vec{\nabla}\phi| > 0$ outside the patch U and then, with parameters r, s, one sets

$$\sigma = \frac{e^{2s\,\phi_{max}} - e^{s\phi(x)}}{(t/T)(1-t/T)T}, \quad \rho = \frac{rs\, e^{s\phi(x)}}{(t/T)(1-t/T)T}, \quad v = \frac{e^{-r\sigma}}{\sqrt{\rho}} u(t,x). \tag{69.69}$$

After integration by parts of products coming from relevant derivatives, one obtains an estimate of the form

$$s\int_0^T \int_\Omega \left[|v_t|^2 + |\mathbf{\Delta} v|^2 + |\rho\vec{\nabla} v|^2 + |\rho^2 v|^2\right] \le C_\phi \int_0^T \int_U \rho |u|^2$$

for large enough s and then $r = r(s)$. From this one can obtain the desired observability estimate $\|u_T\|_{\mathcal{Q}} \le C\|u\|_{\mathcal{U}}$ and so also has nullcontrollability using controls supported in this interior patch. (Further, the structure of Equation 69.69 gives the asymptotics Equation 69.68 in this context as well.)

For boundary control supported in a patch $U \subset \partial\Omega$, one can use a trick: extend the region Ω to a larger $\hat{\Omega}$ by a small bulge sitting on U and select a (very small) patch $\hat{U}$ in the interior of the bulge. Applying the patch control result above to the setting $(\hat{\Omega}, \hat{U})$, one can get a nullcontrol supported in $\hat{U}$ for $\hat{\Omega}$; the trace of this solution on the part of U contained in $\hat{\Omega}$ (extended as 0 on the rest of $\partial\Omega$) will be a nullcontrol, obviously supported on U, for the problem on Ω as desired.

As final comments we note concern for similar problems with nonlinearities in the equation or in the boundary conditions. Perhaps still more interesting is concern for systems in which the heat equation is coupled with other components. For example, one such (linear) example which has been given consideration by several authors is a thermoelastic plate model

$$w_{tt} + \mathbf{\Delta}^2 w = \alpha\mathbf{\Delta}\vartheta, \quad \vartheta_t = \mathbf{\Delta}\vartheta - \alpha\mathbf{\Delta} w_t, \tag{69.70}$$

coupling an Euler–Bernoulli plate with heat conduction. The interesting problem here is observability/nullcontrollability with observation or control limited not only to a patch U but also to a single

* The Carleman estimate technique applies not only to the heat equation, but also to similar problems for wave and elliptic partial differential equations, and so on; see, for example, [4].

component (temperature ϑ or displacement w or momentum w_t). (As of this writing, this is known for an interior patch U, but not for a boundary patch since the trick above does not apply when one wishes to consider only one component of a system.)

References

1. A. Bensoussan, G. Da Prato, M. C. Delfour, S. K. Mitter, *Representation and Control of Infinite Dimensional Systems,* 2nd edition, Birkhäuser, Boston, 2007.
2. R.F. Curtain and A.J. Pritchard, *Functional Analysis in Modern Applied Mathematics*, Academic Press, New York, 1977.
3. R.F. Curtain and H. Zwart, *An Introduction to Infinite-Dimensional Linear Systems Theory,* Springer-Verlag, New York, 1995.
4. A.V. Fursikov and O.Yu. Imanuvilov, *Controllability of Evolution Equations*, Lecture Notes Series 34, Seoul National University, Research Institute of Mathematics, Seoul, 1996.
5. W. Krabs, *On Moment Theory and Controllability of One-Dimensional Vibrating Systems and Heating Processes*, Lecture Notes in Control and Inf. Sci. # 173, Springer-Verlag, Berlin, 1992.
6. I. Lasiecka and R. Triggiani, *Differential and Algebraic Riccati Equations with Application to Boundary/Point Control Problems: Continuous Theory and Approximation Theory*, Lecture Notes in Control and Inf. Sci. # 164, Springer-Verlag, Berlin, 1991.
7. J.-L. Lions, Exact controllability, stabilization, and perturbations, *SIAM Review* 30, pp. 1–68, 1988.
8. D.L. Russell, A unified boundary controllability theory for hyperbolic and parabolic partial differential equations, *Stud. Appl. Math.* LII, pp. 189–211, 1973.
9. T.I. Seidman, Boundary observation and control for the heat equation, in *Calculus of Variations and Control Theory* D.L. Russell, (Ed.), Academic Press, New York, NY, pp. 321–351, 1976.
10. T.I. Seidman, Exact boundary controllability for some evolution equations, *SIAM J. Control Opt.* 16, pp. 979–999, 1978.

70

Observability of Linear Distributed-Parameter Systems

David L. Russell
Virginia Polytechnic Institute and State University

70.1 Comparison with the Finite Dimensional Case

The general question of *observability* in the finite dimensional setting concerns a system

$$\dot{x} = f(x, \ldots), \quad x \in \mathbf{R}^n,$$

where the ellipsis indicates that the system might involve additional parameters, controls, disturbances, etc., and an *output (measurement, observation)* function

$$y = g(x, \ldots), \quad y \in \mathbf{R}^m.$$

Observability theory is concerned, first of all, with the question of *distinguishability*, i.e., whether distinct system trajectories $x(t), \hat{x}(t)$ necessarily give rise to distinct outputs $y(t), \hat{y}(t)$ over a specified time interval. The *observability* question, properly speaking, concerns actual *identification* of the trajectory $x(t)$, equivalently, the initial state x_0, from the available observations on the system with an ultimate view to the possibility of *reconstruction* of the trajectories, or perhaps the initial or current states, which, in application, are generally not directly available from the outputs, typically consisting of a fairly small set of recorded instrument readings on the system. In the linear case, observability and distinguishability are equivalent, and both questions can be treated in the context of a vector/matrix system

$$\dot{x} = A(t)x, \quad x \in \mathbf{R}^n, \; A(t) \in \mathbf{R}^{n \times n}, \tag{70.1}$$

together with an output vector y related to x via

$$y = C(t)x, \quad y \in \mathbf{R}^m, \; C(t) \in \mathbf{R}^{m \times n}, \tag{70.2}$$

and it is a standard result [1] that observability obtains on an interval $[0, T]$, $T > 0$, just in case $y \equiv 0$ implies that $x \equiv 0$ on that interval. This observability property, in turn, is equivalent to the positive definiteness of the matrix integral

$$Z(T) = \int_0^T \Phi(t,0)^* C(t)^* C(t) \Phi(t,0)\, dt, \tag{70.3}$$

where $\Phi(t,s)$ is the fundamental solution matrix of the system with $\Phi(s,s) = I$. The initial state x_0 can then be reconstructed from the observation $y(t)$ by means of the *reconstruction operator* (cf. [18])

$$x_0 = Z(T)^{-1} \int_0^T \Phi(t,0)^* C(t)^* y(t)\, dt. \tag{70.4}$$

In the constant coefficient case $A(t) \equiv A$, observability (on any interval of positive length) is equivalent to the algebraic condition that no eigenvector of A should lie in the null space of C; there are many other equivalent formulations.

In the (infinite dimensional) *distributed-parameter* setting, it is not possible to provide any comparably concise description of the general system to be studied or universal definition of the terms involved, but we will make an attempt in this direction later in the chapter. To set the stage for that, let us begin with a very simple example that illustrates many of the complicating factors involved.

The term *distributed-parameter system* indicates a system whose state parameters are distributed over a spatial region rather than constituting a discrete set of dependent variables. For our example, we take the spatial region to be the interval $[0, 1]$ and represent the state by a function $w(x,t), x \in [0,1], t \in [0,\infty)$. Let us think of $w(x,t)$ as representing some physical property (temperature, concentration of a dissolved chemical, etc.) in a fluid moving from left to right through a conduit whose physical extent corresponds to $0 \le x \le 1$ with a uniform unit velocity. If we assume no diffusion process is involved, it is straightforward to see that $w(x,t)$ will be, in some sense that we do not elaborate on at the moment, a solution of the first-order *partial differential equation* (PDE)

$$\frac{\partial w}{\partial t} + \frac{\partial w}{\partial x} = 0, \tag{70.5}$$

to which we need to adjoin a *boundary condition*

$$w(0,t) = v(t), t \ge 0, \tag{70.6}$$

and an *initial condition* or *initial state* given, without loss of generality, at $t = 0$,

$$w(x,0) = w_0(x), x \in [0,1]. \tag{70.7}$$

It can be shown (cf. [4], e.g.) that with appropriate regularity assumptions on $v(t)$ and $w_0(x)$ there is a unique solution $w(x,t)$ of Equations 70.5 through 70.7 for $x \in [0,1], t \in [0,\infty)$.

We will consider two different types of measurement. The first is a *point* measurement at a given location x_0,

$$y(t) \equiv w(x_0, t), \tag{70.8}$$

while the second is a *distributed* measurement, which we will suppose to have the form

$$y(t) \equiv \int_0^1 c(x) w(x,t)\, dx, \tag{70.9}$$

for some piecewise continuous function $c(x)$, defined and not identically equal to zero on $[0, 1]$.

Let us suppose that we have an initial state $w_0(x)$ defined on $[0, 1]$, while the boundary input (Equation 70.6) is identically equal to 0. Let us examine the simplest form of observability, *distinguishability*, for the resulting system. In this case, the solution takes the form, for $t \geq 0$,

$$w(x,t) \equiv \begin{cases} w_0(x-t), & t \leq x \leq 1, \\ 0, x < t. \end{cases}$$

For a point observation at x_0, the output obtained is clearly

$$y(t) = w(x_0,t) = \begin{cases} w_0(x_0-t), & 0 \leq t \leq x_0, \\ 0, t > x_0. \end{cases}$$

We see that if $x_0 < 1$, the data segment consisting of the values

$$w_0(x), \quad x_0 \leq x \leq 1$$

is lost from the data. Consequently, we do not have distinguishability in this case because initial states $w_0(x)$, $\tilde{w}_0(x)$ differing only on the indicated interval cannot be distinguished on the basis of the observation $y(t)$ on any interval $[0, T]$. On the other hand, for $x_0 = 1$ the initial state is simply rewritten, in reversed order, in the values of $y(t), 0 \leq t \leq 1$ and, thus, we have complete knowledge of the initial state $w_0(x)$ and, hence, of the solution $w(x, t)$ determined by that initial state, provided the length of the observation interval is at least unity. If the length of the interval is less than one, we are again lacking some information on the initial state. It should be noted that this is already a departure from the finite dimensional case; we have here a time-independent system for which distinguishability is dependent on the length of the interval of observation.

Now let us consider the case of a distributed observation; for definiteness, we consider the case wherein $c(x)$ is the characteristic function of the interval $[1-\delta, 1]$ for $0 \leq \delta \leq 1$. Thus,

$$y(t) = \int_{1-\delta}^{1} w(x,t)\,dx = \int_{min\{0,1-(t+\delta)\}}^{min\{0,1-t\}} w_0(x)\,dx$$

If $y(t) \equiv 0$ on the interval $[0, 1]$, then by starting with $t = 1$ and decreasing to $t = 0$ it is easy to see that $w_0(x) \equiv 0$ and thus $w(x, t) \equiv 0$; thus, we again have the property of distinguishability if the interval of observation has length ≥ 1. But now an additional feature comes into play, again not present in the finite dimensional case. If we consider trigonometric initial states

$$w_0(x) = \sin \omega x, \quad \omega > 0$$

we easily verify that $|y(t)| \leq \frac{\delta}{\omega}, t \geq 0$, a bound tending to zero as $\omega \to \infty$. Thus, it is not possible to bound the supremum norm of the initial state in terms of the corresponding norm of the observation, whatever the length of the observation interval. Thus, even though we have the property of distinguishability, we lack observability in a stronger sense to the extent that we cannot reconstruct the initial state from the indicated observation in a continuous (i.e., bounded) manner. This is again a departure from the finite dimensional case wherein we have noted, for linear systems, that distinguishability/observability is equivalent to the existence of a bounded reconstruction operator.

70.2 General Formulation in the Distributed-Parameter Case

To make progress toward some rigorous definitions we have to introduce a certain degree of precision into the conceptual framework. In doing this we assume that the reader has some background in the basics of functional analysis [17]. Accordingly, then, we assume that the process under study has an associated *state space*, W, which we take to be a *Banach space* with *norm* $\|w\|_W$ [in specific instances, this is often strengthened to a *Hilbert space* with *inner product* $(w, \hat{w})_W]$. The process itself is described by an operator differential equation in W,

$$\dot{w} = Aw, \tag{70.10}$$

where A is a (typically unbounded, differential) closed linear operator with domain $\mathcal{D}(A)$ constituting a dense subspace of W satisfying additional conditions (cf. [6], e.g.) so that the *semigroup of bounded operators* e^{At} is defined for $t \geq 0$, *strongly continuous* in the sense that the state trajectory associated with an initial state $w_0 \in W$,

$$w(t) = e^{At}w_0, \tag{70.11}$$

is continuous in W for $t \geq 0$. Many time-independent PDEs can be represented in this way, with A corresponding to the "spatial" differential operator appearing in the equation (e.g., in the case of Equation 70.1 we could take W to be $C[0, 1]$, A to be the operator defined by $Aw = -\partial w/\partial x$ and $\mathcal{D}(A) = C_0^1[0, 1]$, the subspace of $C[0, 1]$ consisting of continuously differentiable functions on $[0, 1]$ with $w(0) = 0$). It should be noted that the range of e^{At}, and hence $w(t)$, is not, in general, in $\mathcal{D}(A)$. It can be shown that this is the case if $w_0 \in \mathcal{D}(A)$.

Now let Y be a second Banach space, the *output*, or *measurement* space and let $C : W \to Y$ be the observation operator; in general, unbounded but with domain including $\mathcal{D}(A)$. Further, given an observation interval $[0, T]$, $T > 0$, let $\mathcal{Y}_T$, be the *space of observations*; e.g., when Y is a Banach space we might let $\mathcal{Y}_T = C([0, T]; Y)$, the space of Y continuous functions on $[0, T]$ with the supremum (Y) norm, or, in the case where Y is a Hilbert space, we might wish to take $\mathcal{Y}_T = L^2([0, T]; Y)$, the space of norm square integrable functions with range in Y. This space is generally defined so that, for an initial state $w_0 \in \mathcal{D}(A)$, which results, via Equation 70.11, in a trajectory $w(t) \in \mathcal{D}(A)$, the observation function is

$$y(t) = Cw(t) \in \mathcal{Y}_T. \tag{70.12}$$

The observation operator C is said to be *admissible* if the linear map from $\mathcal{D}(A)$ to $\mathcal{Y}_T, w_0 \to y(\cdot)$, has a continuous extension to a corresponding map from W to $\mathcal{Y}_T$; we will continue to describe this map via Equation 70.12 even though $Cw(t)$ will not, in general, be defined for general $w_0 \in W$. This whole process may seem very complicated but it cannot be avoided in many, indeed, the most important, examples. In fact, it becomes necessary in the case of point observations on Equation 70.1 if that system is posed in the space $W = L^2[0, 1]$ because, although the state $w(\cdot, t)$ is defined as an element of $L^2[0, 1]$ for each t, the value $w(1, t)$ may not be defined for certain values of t.

With this framework in place we can introduce some definitions.

Definition 70.1:

The linear observed system of Equations 70.10 and 70.12 is distinguishable on an interval $[0, T]$, $T \geq 0$, *if and only if* $y(\cdot) = 0$ *in* $\mathcal{Y}_T$ *implies that* $w_0 = 0$ *in* W.

Definition 70.2:

Given $T \geq 0$ *and* $\tau \in [0, T]$, *the linear observed system of Equations 70.10 and 70.12 is* τ*-observable on* $[0, T]$ *if and only if there exists a positive number* $\gamma \geq 0$ *such that, for every initial state* $w_0 \in W$ *and resulting state (via Equation 70.11)* $w(\tau)$ *in* W, *we have*

$$\|y(\cdot)\|_{\mathcal{Y}_T} \geq \|w(\tau)\|_W, \tag{70.13}$$

$y(\cdot)$ *being the observation obtained via Equations 70.11 and 70.12.*

Remark 70.1

In the finite dimensional context, for $\tau_1, \tau_2 \in [0, T]$, τ_1-observability is equivalent to τ_2-observability, though the (largest) corresponding values of γ may be different. We will see, in a *heat conduction* example to be discussed later, that this need no longer be the case for distributed-parameter systems. It does remain true for *time-reversible* distributed-parameter systems of the type discussed here, corresponding, e.g., to the *wave equation* [19], or in the case of the *elastic beam equation*, which we discuss at some length later.

Let us note that, just as in the finite dimensional case, the theory of observability for distributed-parameter systems forms the basis for *observer theory* and *state estimation theory* (*cf.* [15,18]) in the distributed-parameter context. Observability also enters into the question of asymptotic stability for certain linear distributed-parameter systems via its connection with the *La Salle Invariance Principle* [12,20]. The question of observability also arises in *parameter identification* studies [3]. Distributed parameter observability plays a dual role to *controllability* for the *dual control system*, but the relation between the two is not quite as simple as it is in the corresponding finite dimensional context. The reader is referred to [5] for details of this relationship.

In the case of finite dimensional linear systems, a weaker concept than observability, *detectability*, is often introduced. The constant coefficient version of Equations 70.1 and 70.2 is detectable (on the interval $[0, \infty)$) just in case $y(t) \equiv 0 \rightarrow x(t) \rightarrow 0$ as $t \rightarrow \infty$. This concept is not as useful in the distributed-parameter context because, unlike the constant coefficient linear case, those components of the solution tending to zero do not necessarily tend to zero at a uniform exponential rate (see, e.g., [20]). A more useful concept is that of γ-*detectability* for a given $\gamma \geq 0$; the system of Equations 70.10 and 70.12 enjoys this property on the interval $[0, \infty)$ just in case $y(t) \equiv 0$ implies that, for some $M \geq 0$, $\|w(\cdot, t)\|_W \leq Me^{\gamma t}$, $t \geq 0$.

70.3 Observation of a Heat Conduction Process

Let us consider an application involving PDEs of *parabolic type* in several space dimensions. In the steel industry, it is important that the temperature distribution in a steel ingot in preparation for rolling operations should be as uniform as possible. It is clearly difficult, if not impossible, to determine the temperature distribution in the interior of the ingot directly, but the measurement of the surface temperature is routine. We therefore encounter the problem of the observability of the temperature distribution throughout the ingot from the available surface measurements.

In order to analyze this question mathematically, we first require a model for the process. If we represent the spatial region occupied by the ingot as a region $\Omega \in \mathbf{R}^3$ with smooth boundary Γ and suppose the measurement process takes place over a time interval $0 \leq t \leq T$, we are led by the standard theory of heat conduction to consider the parabolic PDE

$$\rho \frac{\partial w}{\partial t} = k\left(\frac{\partial^2 w}{\partial x^2} + \frac{\partial^2 w}{\partial y^2} + \frac{\partial^2 w}{\partial z^2}\right), \tag{70.14}$$

where ρ is the specific heat of the material and k is its thermal conductivity, both assumed constant here. The rate of heat loss to the exterior environment is $k\frac{\partial w}{\partial \nu}$, where ν denotes the unit normal vector to Γ, external with respect to Ω, and this rate is proportional to the difference between the surface temperature w at the same point on the surface and the ambient temperature, which we will assume, for simplicity, to be zero. The system under study therefore consists of Equation 70.14 together with a Dirichlet-Neumann boundary condition

$$k\frac{\partial w}{\partial \nu} = \sigma w, \tag{70.15}$$

where σ is a positive constant of proportionality. Using X to stand for the triple x, y, z, the (unknown) initial condition is

$$w(X, 0) = w_0(X). \tag{70.16}$$

The standard theory of parabolic PDEs [22] guarantees the existence of a unique solution $w(X,t), X \in \Omega, t \in [0,T]$ of Equation 70.14 with the boundary/initial data of Equations 70.15 and 70.16. The available measurement data are

$$y(X,t) = w(X,t), \quad X \in \Gamma, \quad t \in [0,T]. \tag{70.17}$$

The question then is whether it is possible to reconstruct the temperature distribution $w(X,\tau), X \in \Omega$ at a particular instant $\tau \in [0,T]$ on the basis of the measurement (Equation 70.18). This question has been extensively studied, sometimes indirectly via the dual question of boundary controllability. Brevity requirements constrain us to cite only [13], [14] here, but we will indicate below some of the mathematical issues involved.

The Laplacian operator appearing on the right-hand side of Equation 70.14, defined on a dense domain in $L^2(\Omega)$ incorporating the boundary condition of Equation 70.15, is known to be a positive self-adjoint differential operator with positive eigenvalues $\lambda_k, k = 1,2,3,\ldots$ and has corresponding normalized eigenfunctions $\phi_k, k = 1,2,3,\ldots$ forming an orthonormal basis for the Hilbert space $L^2(\Omega)$. An initial state w_0 in that space has an expansion, convergent in that space,

$$w_0 = \sum_{k=1}^{\infty} c_k \phi_k. \tag{70.18}$$

Corresponding to this expansion, the system of Equations 70.14 through 70.16 has the solution

$$w(X,t) = \sum_{k=1}^{\infty} c_k e^{-\lambda_k t} \phi_k(X), \tag{70.19}$$

with the corresponding measurement, or observation

$$y(X,t) = \sum_{k=1}^{\infty} c_k e^{-\lambda_k t} \phi_k(X), \quad X \in \Gamma, \quad t \in [0,T]. \tag{70.20}$$

involving known λ_k, ϕ_k, but unknown c_k.

Let us first consider the question of distinguishability: Can two solutions $w(X,t), \tilde{w}(X,t)$, corresponding to initial states $w_0, \tilde{w}_0$ produce the same observation $y(X,t)$ via Equation 70.17? From the linear homogeneous character of the system it is clear that this is equivalent to asking whether a nonzero initial state (Equation 70.18), i.e., such that not all $c_k = 0$, can give rise to an observation (Equation 70.20) that is identically zero. This immediately requires us to give attention to the boundary values $\eta_k(X,t) \equiv e^{-\lambda_k t}\phi_k(X)$ corresponding to $w_0(X) = \phi_k(X)$. Clearly, the boundary observation corresponding to a general initial state (Equation 70.18) is then

$$y(X,t) = \sum_{k=1}^{\infty} c_k \eta_k(X,t), \quad X \in \Gamma, \quad t \in [0,T], \tag{70.21}$$

Can $y(X,t)$, taking this form, be identically zero if the c_k are not all zero? The statement that this is not possible, hence that the system is distinguishable, is precisely the statement that the $\eta_k(X,t), X \in \Gamma, t \in [0,T]$ are *weakly independent* in the appropriate boundary space, for simplicity, say $L^2(\Gamma \times [0,T])$. As a result of a variety of investigations [9], we can assert that this is, indeed, the case for any $T \geq 0$. In fact, these investigations show that a stronger result, *spectral observability*, is true. The functions $\eta_k(X,t)$ are actually *strongly independent* in $L^2(\Gamma \times [0,T])$, by which we mean that there exist *biorthogonal* functions, not necessarily unique, in $L^2(\Gamma \times [0,T])$ for the $\eta_k(X,t)$, i.e., functions $\zeta_k(X,t) \in L^2(\Gamma \times [0,T])$,

$k = 1, 2, 3, \ldots$, such that

$$\int_{\Gamma\times[0,T]} \zeta_k(X,t)\,\eta_j(X,t)\,dX\,dt = \begin{cases} 0, & k \neq j, \\ 1, & k = j. \end{cases} \tag{70.22}$$

The existence of these biorthogonal functions implies that any finite number of the coefficients c_k can be constructed from the observation $y(X, t)$ via

$$c_k = \int_{\Gamma\times[0,T]} \zeta_k(X,t)\,y(X,t)\,dX\,dt. \tag{70.23}$$

Indeed, we can construct a map from the output space, here assumed to be $L^2(\Gamma \times [0, T])$, namely,

$$S_{\tau,K} Y = \sum_{k=1}^{K} e^{-\lambda_k \tau} \phi_k(X) \int_{\Gamma\times[0,T]} \zeta_k(X,t)\,y(X,t)\,dX\,dt, \tag{70.24}$$

carrying the observation Y into the "K-approximation" to the state $w(\cdot, \tau), \tau \in [0, T]$. Since the sum (Equation 70.18) is convergent in $L^2(\Omega)$, this property of spectral observability is a form of *approximate observability* in the sense that it permits reconstruction of the initial state of Equation 70.18, or a corresponding subsequent state $w(\cdot, \tau)$, within any desired degree of accuracy. Unfortunately, there is, in general, no way to know how large K should be in Equation 70.24 in order to achieve a specified accuracy, nor is there any way to obtain a uniform estimate on the effect of errors in measurement of Y.

The question of τ-observability, for $\tau \in [0, T]$, is a more demanding one; it is the question as to whether the map $S_{\tau,K}$ defined in Equation 70.24 extends by continuity to a bounded linear map S_τ taking the observation Y into the corresponding state w_0. It turns out [21] that this is possible for any $\tau > 0$, but it is not possible for $\tau = 0$; i.e., the initial state can never be continuously reconstructed from measurements of this kind. Fortunately, it is ordinarily the terminal state $w(\cdot, T)$ that is the more relevant, and reconstruction is possible in this case. The proof of these assertions (*cf.* [7,21], e.g.) relies on delicate estimates of the norms of the biorthogonal functions $\zeta_k(X, t)$ in the space $L^2(\Gamma \times [0, T])$. The boundedness property of the operator S_τ is important not only in regard to convergence of approximate reconstructions of the state $w(\cdot, \tau)$ but also in regard to understanding the effect of an error δY. If such an error is present, the estimate obtained for $w(\cdot, \tau)$ will clearly be

$$\tilde{w}(\cdot, \tau) = S_\tau(Y + \delta Y) = w(\cdot, \tau) + S_\tau \delta Y, \tag{70.25}$$

and the norm of the reconstruction error thus does not exceed $\|S_\tau\|\|\delta Y\|$. A further consequence of this clearly is the importance, since reconstruction operators are not, in general, unique, of obtaining a reconstruction operator of least possible norm. If we denote the subspace of $L^2(\Gamma \times [0, T])$ spanned by the functions $\eta_k(X, t)$, as described following Equation 70.21, by $E(\Gamma \times [0, T])$, it is easy to show that the biorthogonal functions ζ_k described via Equation 70.22 are unique and have least possible norm if we require that they should lie in $E(\Gamma \times [0, T])$. The particular reconstruction operator $\hat{S}_\tau$, constructed as the limit of operators (Equation 70.24) with the least norm biorthogonal functions $\hat{\zeta}_k$, may then be seen to have least possible norm. In applications, reconstruction operators of the type we have described here are rarely used; one normally uses a state estimator (*cf.* [18], e.g.) that provides only an asymptotic reconstruction of the system state, but the performance of such a state estimator is still ultimately limited by considerations of the same sort as we have discussed here.

It is possible to provide similar discussions for the "wave" counterpart of Equation 70.14, i.e.,

$$\rho\frac{\partial^2 w}{\partial t^2} = k\left(\frac{\partial^2 w}{\partial x^2} + \frac{\partial^2 w}{\partial y^2} + \frac{\partial^2 w}{\partial z^2}\right), \tag{70.26}$$

with a variety of boundary conditions, including Equation 70.15. A very large number of such studies have been made, but they have normally been carried out in terms of the dual control system [19] rather than in terms of the linear observed system. In some cases, methods of harmonic analysis similar to those just described for the heat equation have been used [8], but the most definitive results have been obtained using methods derived from the *scattering theory* of the wave equation and from related methods such as *geometrical optics* [2] or *multiplier methods* [11]. These studies include treatment of cases wherein the observation/(dual) control process is restricted to a subset $\Gamma_1 \subset \Gamma$ having certain geometrical properties. Other contributions [21] have shown the study of the wave equation to be pivotal in the sense that results for related heat and elastic processes can be inferred, via harmonic analysis, once the wave equation results are in place.

70.4 Observability Theory for Elastic Beams

There are several different models for elastic beams, even when we restrict attention to small deformation linear models. These include the Euler–Bernoulli, Rayleigh and Timoshenko models. The oldest and most familiar of these is the Euler–Bernoulli model, consisting of the PDE

$$\rho(x)\frac{\partial^2 w}{\partial t^2} = \frac{\partial^2}{\partial x^2}\left(EI(x)\frac{\partial^2 w}{\partial x^2}\right), \quad x \in [0, L], \; t \in [0, \infty), \tag{70.27}$$

wherein $\rho(x)$ denotes the mass per unit length and $EI(x)$ is the so-called *bending modulus*. We are concerned with solutions in a certain *weak* sense, which we will not elaborate upon here, corresponding to a given initial state

$$w(\cdot, 0) = w_0 \in H^2[0, L], \quad \frac{\partial w}{\partial t}(x, 0) = v_0 \in L^2[0, L], \tag{70.28}$$

where $H^2[0, L]$ is the standard *Sobolev space* of functions with square integrable second derivatives on the indicated interval. Additionally, one needs to give boundary conditions at $x = 0$ and at $x = L$; these vary with the physical circumstances. For the sake of brevity, we will confine our discussion here to the *cantilever* case, where the left-hand end is assumed "clamped" while the right-hand end is "free"; the appropriate boundary conditions are then

$$w(0, t) \equiv 0, \quad \frac{\partial w}{\partial x}(0, t) \equiv 0, \tag{70.29}$$

$$\frac{\partial^2 w}{\partial x^2}(L, t) \equiv 0, \quad \frac{\partial}{\partial x}\left(EI(x)\frac{\partial^2 w}{\partial x^2}\right)(L, t) \equiv 0. \tag{70.30}$$

In many applications a mechanical structure, such as a manipulator arm, is clamped to a rotating base that points the arm/beam in various directions in order to carry out particular tasks. Each "slewing" motion results in a degree of vibration of the structure, which, for most practical purposes, can be thought of as taking place within the context of the model of Equations 70.27 through 70.30. In order to attenuate the undesired vibration, it is first of all necessary to carry out an observation procedure in order to determine the oscillatory state of the system preparatory to, or in conjunction with, control operations. A number of different measurement options exist whose feasibility depends on the operational situation in hand. We will cite three of these. In the first instance, one might attach a *strain gauge* to the beam near the clamped end. The extension or compression of such a (normally piezoelectric) device provides a scaled physical realization of the mathematical measurement

$$y(t) = \frac{\partial^2 w}{\partial x^2}(0, t). \tag{70.31}$$

Alternatively, one can place an *accelerometer* near the free end of the beam to provide a measurement equivalent to

$$y(t) = \frac{\partial^2 w}{\partial t^2}(L, t), \tag{70.32}$$

or one can use a laser device, relying on the Doppler effect, to measure

$$y(t) = \frac{\partial w}{\partial t}(L, t). \tag{70.33}$$

Each of these measurement modes provides a scalar valued function $y(t)$ carrying a certain amount of information on the system state $w(x, t)$; the problem, as before, is to reconstruct the initial state, or the current state at time T, from the record $y(t), 0 \leq t \leq T$. The mathematical theory of this reconstruction is in many ways similar to the one we have just described for the heat equation, but with some significant differences. The most notable of these is immediately apparent from the equation itself; it is invariant under reversal of the time direction. This means that there is no inherent difference between initial and terminal states or, indeed, any intermediate state. We should expect this to show up in the mathematics, and it does.

The differential operator defined by

$$(Aw) \equiv -\frac{1}{\rho}\frac{\partial^2}{\partial x^2}\left(EI(x)\frac{\partial^2 w}{\partial x^2}\right), \tag{70.34}$$

on the subspace of $H^4[0, L]$ resulting from imposition of the cantilever boundary conditions, is an unbounded positive self-adjoint operator with positive eigenvalues, listed in increasing order, $\lambda_k, k = 1, 2, 3, \ldots$. The corresponding normalized eigenfunctions $\phi_k(x)$ form an orthonormal basis for $L^2[0, L]$. Defining $\omega_k = \sqrt{\lambda_k}$, the solution of Equations 70.27 through 70.30 takes the form

$$w(x, t) = \sum_{k=1}^{\infty}\left(c_k e^{i\omega_k t} + d_k e^{-i\omega_k t}\right)\phi_k(x), \tag{70.35}$$

where, with w_0 and v_0 as in Equation 70.28

$$w_0(x) = \sum_{k=1}^{\infty} w_{0,k}\phi_k(x), \quad v_0(x) = \sum_{k=1}^{\infty} v_{0,k}\phi_k(x), \tag{70.36}$$

with

$$w_{0,k} = c_k + d_k, \quad v_{0,k} = i\omega_k(c_k - d_k). \tag{70.37}$$

The norm of the state w_0, v_0 in the state space $H^2[0, L] \otimes L^2[0, L]$ is equivalent to the norm of the double sequence $\{w_{0,k}, v_{0,k}\}$ in the Hilbert space, which we call ℓ^2_ω. That norm is

$$\|w_{0,k}, v_{0,k}\| = \left[\sum_{k=1}^{\infty}\left(\omega_k w_{0,k}^2 + v_{0,k}^2\right)\right]^{\frac{1}{2}}.$$

Any one of the measurement modes discussed earlier now takes the form

$$y(t) = \sum_{k=1}^{\infty}\left(\gamma_k c_k e^{i\omega_k t} + \delta_k d_k e^{-i\omega_k t}\right), \tag{70.38}$$

with γ_k and δ_k depending on the particular measurement mode being employed. Thus, in the cases of Equations 70.31, 70.32, and 70.33, respectively, we have

$$\gamma_k = \delta_k = \frac{d^2\phi_k}{dx^2}(0), \tag{70.39}$$

$$\gamma_k = \delta_k = -\lambda_k\phi_k(L), \tag{70.40}$$

$$\gamma_k = i\omega_k\phi_k(L), \quad \delta_k = -i\omega_k\phi_k(L). \tag{70.41}$$

Just as in the earlier example of the heat equation, but now we are concerned only with the *scalar* exponential functions

$$e^{i\omega_k t}, e^{-i\omega_k t}, k = 1, 2, 3, \ldots, \tag{70.42}$$

everything depends on the *independence* properties of these functions in $L^2[0, T]$, where T is the length of the observation interval, in relation to the asymptotic growth of the ω_k, $k \to \infty$, the basic character of which is that the ω_k are distinct, increasing with k, if ordered in the natural way, and

$$\omega_k = \mathcal{O}(k^2), k \to \infty. \tag{70.43}$$

The relationship between properties of the functions of Equation 70.42 and the asymptotic and/or separation properties of the exponents ω_k is one of the questions considered in the general topic of *nonharmonic Fourier series*, whose systematic study began in the 1930s with the work of Paley and Wiener [16]. The specific result that we make use of is due to A. E. Ingham [10]. Combined with other, more or less elementary, considerations, it implies that if the ω_k are real and satisfy a separation condition, for some positive integer K and positive number Γ,

$$\omega_{k+1} \geq \omega_k + \Gamma, k \geq K, \tag{70.44}$$

then the functions of Equation 70.42 are *uniformly independent* and *uniformly convergent* in $L^2[0, T]$, provided that $T \geq \frac{2\pi}{\Gamma}$, which means there are numbers $b, B, 0 < b < B$, such that, for any square summable sequences of coefficients $c_k, d_k, k = 1, 2, 3, \ldots$, we have

$$b\sum_{k=1}^{\infty}\left(|c_k|^2 + |d_k|^2\right) \leq \| \sum_{k=1}^{\infty}\left(c_k e^{i\omega_k t} + d_k e^{-i\omega_k t}\right) \|^2_{L^2[0,T]} \leq B\sum_{k=1}^{\infty}\left(|c_k|^2 + |d_k|^2\right) \tag{70.45}$$

Since the asymptotic property of Equation 70.43 clearly implies, for any $\Gamma > 0$, that Equation 70.44 is satisfied if $K = K(\Gamma)$ is sufficiently large, inequalities of Equation 70.45, with $b = b(\Gamma), B = B(\Gamma)$, must hold for any $T > 0$. It follows that the linear map, or operator,

$$C : \sum_{k=1}^{\infty}\left(c_k e^{i\omega_k t} + d_k e^{-i\omega_k t}\right) \to \{c_k, d_k | k = 1, 2, 3, \ldots\} \in \ell^2, \tag{70.46}$$

defined on the (necessarily closed, in view of Equation 70.45) subspace of $L^2[0, T]$ spanned by the exponential functions in question, is both bounded and boundedly invertible. Applied to the observation $y(t), t \in [0, T]$, corresponding to an initial state of Equation 70.36, the boundedness of C, together with the relationship of Equation 70.37 between the $w_{0,k}, v_{0,k}$ and the c_k, d_k and the form of the ℓ^2_ω norm, it is not hard to see that the boundedness of the linear operator C implies the existence of a bounded reconstruction operator from the observation (Equation 70.38) to the initial state (Equation 70.36) provided the coefficients δ_k, γ_k in Equation 70.38 satisfy an inequality of the form

$$|c_k| \geq C\omega_k, \quad |d_k| \geq D\omega_k, \tag{70.47}$$

for some positive constants C and D. Using the trigonometric/exponential form of the eigenfunctions $\phi_k(x)$, one easily verifies this to be the case for each of the observation modes of Equations 70.39 through 70.41; indeed, the condition is overfulfilled in the case of the accelerometer measurement (Equation 70.40). Thus, bounded observability of elastic beam states via scalar measurements can be considered to be typical.

At any later time $t = \tau$, the system state resulting from the initial state of Equation 70.36 has the form

$$w_\tau(x) = \sum_{k=1}^{\infty}\left(\cos \omega_k t w_{0,k} + \frac{1}{\omega_k} \sin \omega_k t v_{0,k}\right)\phi_k(x),$$
$$v_\tau(x) = \sum_{k=1}^{\infty}\left(-\omega_k \sin \omega_k t w_{0,k} + \cos \omega_k t v_{0,k}\right)\phi_k(x). \tag{70.48}$$

Using Equation 70.48 with the form of the norm in ℓ^2_ω, one can see that the map from the initial state $w_0, v_0 \in H^2[0,L] \otimes L^2[0,L]$ to w_τ, v_τ in the same space is bounded and boundedly invertible; this is another way of expressing the time reversibility of the system. It follows that the state w_τ, v_τ can be continuously reconstructed from the observation $y(t), t \in [0,T]$ in precisely the same circumstances as we can reconstruct the state w_0, v_0.

References

1. Anderson, B.D.O. and Moore, J.B., *Linear Optimal Control*, Elect. Eng. Series, Prentice Hall, Englewood Cliffs, NJ, 1971, chap. 8.
2. Bardos, C., LeBeau, G., and Rauch, J., Contrôle et stabilisation dans des problèmes hyperboliques, in *Controlabilité Exacte, Perturbations et Stabilisation de Systèmes Distribués*, Lions, J.-L., Masson, Paris, 1988, appendix 2.
3. Beck, J.V. and Arnold, K.J., *Parameter Estimation in Engineering and Science*, John Wiley & Sons, New York, 1977.
4. Courant, R. and Hilbert, D., *Methods of Mathematical Physics; Vol. 2: Partial Differential Equations*, Interscience Publishers, New York, 1962.
5. Dolecki, S. and Russell, D.L., A general theory of observation and control, *SIAM J. Control Opt.*, 15, 185–220, 1977.
6. Dunford, N. and Schwartz, J.T., *Linear Operators; Vol. 1: General Theory*, Interscience, New York, 1958.
7. Fattorini, H.O. and Russell, D.L., Exact controllability theorems for linear parabolic equations in one space dimension, *Arch. Ration. Mech. Anal.*, 4, 272–292, 1971.
8. Graham, K.D. and Russell, D.L., Boundary value control of the wave equation in a spherical region, *SIAM J. Control*, 13, 174–196, 1975.
9. Ho, L.F., Observabilité frontière de l'équation des ondes, *Cah. R. Acad. Sci.*, Paris, 302, 1986.
10. Ingham, A.E., Some trigonometrical inequalities with applications to the theory of series, *Math. Z.*, 41, 367–379, 1936.
11. Lagnese, J., Controllability and stabilizability of thin beams and plates, in *Distributed Parameter Control Systems*, Chen, G., Lee, E.B., Littman, W., Markus, L., Eds., Lecture Notes in Pure and Applied Math., Marcel Dekker, New York, 1991, 128.
12. LaSalle, J.P. and Lefschetz, S., *Stability by Lyapunov's Direct Method, with Applications*, Academic Press, New York, 1961.
13. Lions, J.L., *Optimal Control of Systems Governed by Partial Differential Equations*, Grund. Math. Wiss. Einz., Springer-Verlag, New York, 1971, 170.
14. Lions, J.L., *Controlabilité Exacte, Perturbations et Stabilisation de Systèmes Distribués*, Tomes 1,2, *Recherches en Mathematiques Appliquées*, Masson, Paris, 1988, 8, 9.
15. Luenberger, D.G., An introduction to observers, *IEEE Trans. Autom. Control*, 22, 596–602, 1971.
16. Paley, R.E.A.C. and Wiener, N., *Fourier Transforms in the Complex Domain*, Colloq. Pub. 19, American Mathematical Society, Providence, RI, 1934.
17. Pazy, A., *Semigroups of Linear Operators and Applications to Partial Differential Equations*, Applied Mathematical Sciences 44, Springer-Verlag, New York, 1983.
18. Russell, D.L., *Mathematics of Finite Dimensional Control Systems: Theory and Design*, Marcel Dekker, Inc., New York, 1979.
19. Russell, D.L., Controllability and stabilizability theory for linear partial differential equations: Recent progress and open questions, *SIAM Rev.*, 20, 639–739, 1978.

20. Russell, D.L., Decay rates for weakly damped systems in Hilbert space obtained with control-theoretic methods, *J. Diff. Eq.*, 19, 344–370, 1975.
21. Russell, D.L., A unified boundary controllability theory for hyperbolic and parabolic partial differential equations, *Stud. Appl. Math.*, LII, 189–211, 1973.
22. Showalter, R.E., *Hilbert Space Methods for Partial Differential Equations*, Pitman Publishing Ltd., San Francisco, 1977.

71

Boundary Control of PDEs: The Backstepping Approach

Miroslav Krstić
University of California, San Diego

Andrey Smyshlyaev
University of California, San Diego

71.1 Introduction

Aerodynamic fluid flows, flexible structures, plasmas in electromagnetic fields, acoustic and optical waves, thermal dynamics, and chemical processes—distributed parameter systems—are so ubiquitous in control applications that they are the norm, rather than an exception. Although finite-dimensional techniques often suffice in solving such control problems, there are significant advantages in considering such systems in their full partial differential equation (PDE) format, which has driven the development of the field of control of distributed parameter systems since the 1960s.

Numerous papers, surveys, and even books exist on control of distributed parameter systems [2,7,8, 17,18], dedicated to solving problems of controllability and optimal control. The latter are particularly of interest as they produce feedback laws, which afford a degree of robustness to various uncertainties. Optimal control laws for PDEs require solutions to operator Riccati equations—infinite dimensional nonlinear equations which are in general very difficult to solve. In this chapter we present the recently emerged approach, inspired by the finite-dimensional "backstepping" design for nonlinear systems [12], which avoids the requirement to solve Riccati equations. Instead of Riccati equations, our design problem consists in solving only a linear hyperbolic PDE in the spatial variables, which is performed once, offline. This simplification of the design problem is achieved by foregoing optimality. The approach consists in constructing a "coordinate transformation" and a control law, to cast the closed-loop system in a desirable form, which we refer to as the "target system." This is similar to the approach employed in *feedback linearization* and backstepping. The choice of a target system is different for each of the different classes of PDEs. This factor requires a deeper level of involvement on the part of the control designer, who needs to have a certain level of insight into stability properties of different classes of PDEs. However, for this additional analytical effort, the designer is rewarded with a control law which is much simpler than optimal controllers employing Riccati equations, and with a dynamic behavior which is physically intuitive because it follows the behavior of well-understood classes of physical PDEs.

This line of research at present deals only with the problems of *boundary control*. Many solutions exist, at least in principle, for control of PDEs both with interior or distributed controls, as well as with boundary controls. It is, however, generally recognized that boundary control problems, where the actuation is applied only through the boundary conditions of a PDE, are not only the most realistic in many applications (e.g., this is the only way to control fluid flows using micro-electro-mechanical systems [MEMS] actuators), but are analytically the most challenging because the input and output operators are unbounded. On the intuitive level, the boundary control problems are always characterized by the state space having a spatial dimension at least one higher than that of the actuator(s). For instance, the system evolving over a rectangle and controlled only through the sides of the rectangle. In boundary control problems one necessarily has an order of magnitude fewer control inputs than the states. Despite this "extreme" character of the boundary control problem, the backstepping approach happens to be perfectly suited for it as it is based on propagating a control action through a sequence of ordered subsystems [12].

71.2 Unstable Heat Equation

To introduce the basic idea of the backstepping control design, let us start with the simplest unstable PDE, reaction–diffusion equation

$$u_t(x,t) = u_{xx}(x,t) + \lambda u(x,t), \quad x \in (0,1) \tag{71.1}$$

$$u(0,t) = 0, \tag{71.2}$$

$$u(1,t) = U(t), \tag{71.3}$$

where λ is an arbitrary constant and $U(t)$ is the input. The open-loop system (Equations 71.1 and 71.2) (with $U(t) = 0$) is unstable with arbitrarily many unstable eigenvalues for sufficiently large λ.

The main idea of the backstepping method is to use the coordinate transformation

$$w(x,t) = u(x,t) - \int_0^x k(x,y)u(y,t)\,dy \tag{71.4}$$

and the control law

$$U(t) = \int_0^1 k(1,y)u(y,t)\,dy \tag{71.5}$$

to transform the systems (Equations 71.1 through 71.3) into the heat equation

$$w_t(x,t) = w_{xx}(x,t) \tag{71.6}$$

$$w(0,t) = 0, \tag{71.7}$$

$$w(1,t) = 0, \tag{71.8}$$

which is exponentially stable.

To derive the equation for $k(x,y)$ we first substitute Equation 71.4 into Equation 71.6:

$$\begin{aligned} w_t(x,t) &= u_t(x,t) - \int_0^x k(x,y)(u_{yy}(y,t) + \lambda u(y,t))\,dy \\ &= u_{xx}(x,t) + \lambda u(x,t) - k(x,x)u_x(x,t) + k(x,0)u_x(0,t) + k_y(x,x)u(x,t) \\ &\quad - \int_0^x (k_{yy}(x,y) + \lambda k(x,y))u(y,t)\,dy, \end{aligned} \tag{71.9}$$

$$\begin{aligned} w_{xx}(x,t) &= u_{xx}(x,t) - k(x,x)u_x(x,t) - k_x(x,x)u(x,t) \\ &\quad - \frac{d}{dx}k(x,x)u(x,t) - \int_0^x k_{xx}(x,y)u(y,t)\,dy. \end{aligned} \tag{71.10}$$

Combining Equations 71.9 and 71.10 and using Equation 71.1 gives

$$\begin{aligned} 0 &= w_t(x,t) - w_{xx}(x,t) + cw(x,t) \\ &= \int_0^x \left\{k_{xx}(x,y) - k_{yy}(x,y) - \lambda k(x,y)\right\} u(y,t)\,dy \\ &\quad + \left(2\frac{d}{dx}k(x,x) + \lambda\right) u(x,t) + k(x,0)u_x(0,t). \end{aligned} \tag{71.11}$$

We can see now that the kernel $k(x,y)$ must satisfy the following hyperbolic PDE:

$$k_{xx}(x,y) - k_{yy}(x,y) = \lambda k(x,y), \quad (x,y) \in \mathcal{T}, \tag{71.12}$$

$$k(x,0) = 0, \tag{71.13}$$

$$k(x,x) = -\frac{\lambda x}{2}. \tag{71.14}$$

where $\mathcal{T} = \{x, y : 0 < y < x < 1\}$. The solution to this PDE is [21]

$$k(x,y) = -\lambda y \frac{I_1\left(\sqrt{\lambda(x^2-y^2)}\right)}{\sqrt{\lambda(x^2-y^2)}}, \tag{71.15}$$

where I_1 is a first-order modified Bessel function. The controller (Equation 71.5) is therefore,

$$U(t) = -\int_0^1 \lambda y \frac{I_1\left(\sqrt{\lambda(1-y^2)}\right)}{\sqrt{\lambda(1-y^2)}} u(y,t)\,dy. \tag{71.16}$$

One can show that the transformation (Equation 71.4) is invertible and its inverse is

$$u(x,t) = w(x,t) - \int_0^x \lambda y \frac{J_1\left(\sqrt{\lambda(x^2-y^2)}\right)}{\sqrt{\lambda(x^2-y^2)}} w(y,t)\,dy. \tag{71.17}$$

where J_1 is the usual (nonmodified) Bessel function of the first order. The smoothness of the kernels of the direct and inverse transformations in x and y establishes the equivalence of norms of u and w in both

L^2 and H^1. Therefore, from the properties of the heat equations 71.6 through 71.8 we conlude that the closed-loop system is exponentially stable in L^2 and H^1.

Note that the systems (Equations 71.1 through 71.3 and 71.16) are not only well posed but their solutions are explicitly available. To find this solution, we first solve the heat equation (Equations 71.6 through 71.8):

$$w(x,t) = 2\sum_{n=1}^{\infty} e^{-\pi^2 n^2 t} \sin(\pi n x) \int_0^1 w_0(\xi) \sin(\pi n \xi)\, d\xi. \tag{71.18}$$

The initial condition w_0 can be calculated explicitly from u_0 via Equations 71.4 and 71.15. Substituting the result into Equation 71.17, changing order of integration, and calculating one of the integrals we obtain the explicit solution to closed-loop systems (Equations 71.1 through 71.3 and 71.16):

$$u(x,t) = \sum_{n=1}^{\infty} e^{-\pi^2 n^2 t} \phi_n(x) \int_0^1 \psi_n(\xi) u_0(\xi)\, d\xi, \tag{71.19}$$

where

$$\psi_n(x) = \sin(\pi n x) + \int_x^1 \lambda x \frac{I_1\left(\sqrt{\lambda(\xi^2 - x^2)}\right)}{\sqrt{\lambda(\xi^2 - x^2)}} \sin(\pi n \xi)\, d\xi,$$

$$\phi_n(x) = \frac{2\pi n}{\sqrt{\lambda + \pi^2 n^2}} \sin\left(\sqrt{\lambda + \pi^2 n^2}\, x\right). \tag{71.20}$$

The above control design is easily extended for the Neumann type of actuation, one only needs to replace the boundary condition (Equation 71.8) with the Neumann one. The control law is

$$u_x(1,t) = U(t) = k(1,1)u(1,t) + \int_0^1 k_x(1,y) u(y,t)\, dy, \tag{71.21}$$

where $k(x,y)$ is given by Equation 71.15.

For the rest of the chapter we will not repeat the procedure of derivation of a PDE for the kernel and the arguments of the previous paragraph proving stability and well posedness.

71.3 Observers

The stabilizing controller developed in the previous section requires complete measurements from the interior of the domain which are usually unavailable. So we look for the observers that estimate $u(x,t)$ inside the domain. Consider the plant

$$u_t(x,t) = u_{xx}(x,t) + \lambda u(x,t), \tag{71.22}$$

$$u_x(0,t) = 0, \tag{71.23}$$

$$u(1,t) = U(t). \tag{71.24}$$

Suppose the only available measurement of our system is at $x = 0$, the opposite end to actuation. To estimate the state inside the domain, we introduce the following observer:

$$\hat{u}_t(x,t) = \hat{u}_{xx}(x,t) + \lambda \hat{u}(x,t) + p_1(x)[u(0,t) - \hat{u}(0,t)], \tag{71.25}$$

$$\hat{u}_x(0,t) = p_{10}[u(0,t) - \hat{u}(0,t)], \tag{71.26}$$

$$\hat{u}(1,t) = U(t). \tag{71.27}$$

Here $p_1(x)$ and p_{10} are output injection functions (p_{10} is a constant) *to be designed.* Note that we introduce output injection not only in Equation 71.25 but also at the boundary where the measurement is available.

The observer (Equations 71.25 through 71.27) is in the standard form of "copy of the system plus injection of the output estimation error." This standard form allows us to pursue duality between the observer and the controller design, similar to the way duality is used to find the gains of a Luenberger observer based on the pole-placement control algorithm, or to the way duality is used to construct a Kalman filter based on the LQR design.

The observer error $\tilde{u} = u - \hat{u}$ satisfies the PDE

$$\tilde{u}_t(x,t) = \tilde{u}_{xx}(x,t) + \lambda\tilde{u}(x,t) - p_1(x)\tilde{u}(0,t), \tag{71.28}$$

$$\tilde{u}_x(0,t) = -p_{10}\tilde{u}(0,t), \tag{71.29}$$

$$\tilde{u}(1,t) = 0. \tag{71.30}$$

Observer gains $p_1(x)$ and p_{10} should be now chosen to stabilize the error system. We look for a coordinate transformation

$$\tilde{u}(x,t) = \tilde{w}(x,t) - \int_0^x p(x,y)\tilde{w}(y,t)\,dy \tag{71.31}$$

that transforms the system (Equations 71.28 through 71.30) into the exponentially stable system

$$\tilde{w}_t(x,t) = \tilde{w}_{xx}(x,t), \tag{71.32}$$

$$\tilde{w}_x(0,t) = 0, \tag{71.33}$$

$$\tilde{w}(1,t) = 0. \tag{71.34}$$

By substituting Equation 71.31 into Equations 71.28 through 71.30 we obtain a set of conditions on the kernel $p(x,y)$ in the form of the hyperbolic PDE

$$p_{yy}(x,y) - p_{xx}(x,y) = \lambda p(x,y) \tag{71.35}$$

for $0 < y < x < 1$, with the boundary conditions

$$\frac{d}{dx}p(x,x) = \frac{\lambda}{2}, \tag{71.36}$$

$$p(1,y) = 0. \tag{71.37}$$

and observer gains given by

$$p_1(x) = \varepsilon p_y(x,0), \quad p_{10} = p(0,0). \tag{71.38}$$

Let us make a change of variables

$$\breve{x} = 1 - y, \quad \breve{y} = 1 - x, \quad \breve{p}(\breve{x},\breve{y}) = p(x,y). \tag{71.39}$$

It can be verified that in these new variables the problems (Equations 71.35 through 71.37) become exactly the same as Equations 71.12 through 71.14 for $k(x,y)$, and therefore,

$$\breve{p}(\breve{x},\breve{y}) = -\lambda\breve{y}\frac{I_1\left(\sqrt{\lambda(\breve{x}^2 - \breve{y}^2)}\right)}{\sqrt{\lambda(\breve{x}^2 - \breve{y}^2)}}. \tag{71.40}$$

Using Equation 71.38 we obtain the observer gains

$$p_1(x) = \frac{\lambda(1-x)}{x(2-x)}I_2\left(\sqrt{\lambda x(2-x)}\right), \quad p_{10} = -\frac{\lambda}{2}. \tag{71.41}$$

The design procedure presented above is easily extended to other input/output setups, such as Neumann sensing ($u_x(0,t)$ is measured) and sensor/actuator located at the same boundary (e.g., $u(1,t)$ is measured, $u_x(1,t)$ is controlled); see [22] for details.

Since the observer design is dual to control design, only control designs will be presented in the rest of the chapter.

71.4 Other Parabolic Plants

71.4.1 Reaction–Advection–Diffusion Systems

Consider the reaction–advection–diffusion PDE

$$u_t(x,t) = \varepsilon(x)u_{xx}(x,t) + b(x)u_x(x,t) + \lambda(x)u(x,t), \tag{71.42}$$

$$u_x(0,t) = -qu(0,t), \tag{71.43}$$

$$u(1,t) = U(t). \tag{71.44}$$

This equation describes a variety of systems with thermal, fluid, and chemically reacting dynamics. The spatially varying coefficients come from applications with nonhomogenous materials, unusually shaped domains, and can also arise from the linearization.

Using the so-called gauge transformation, it is possible to convert this system into the one with constant diffusion and zero advection terms. Consider a coordinate change

$$v(z,t) = \varepsilon^{-1/4}(x)u(x,t)e^{\int_0^x \frac{b(s)}{2\varepsilon(s)}\,ds}, \tag{71.45}$$

where

$$z = \sqrt{\varepsilon_0}\int_0^x \frac{ds}{\sqrt{\varepsilon(s)}}, \quad \varepsilon_0 = \left(\int_0^1 \frac{ds}{\sqrt{\varepsilon(s)}}\right)^{-2}. \tag{71.46}$$

Then the new state v satisfies the following PDE:

$$v_t(z,t) = \varepsilon_0 v_{zz}(z,t) + \lambda_0(z)v(z,t), \tag{71.47}$$

$$v_z(0,t) = -q_0 v(0,t), \tag{71.48}$$

$$v(1,t) = U(t)\varepsilon^{-1/4}(1)e^{\int_0^1 \frac{b(s)}{2\varepsilon(s)}\,ds}, \tag{71.49}$$

where

$$\lambda_0(z) = \lambda(x) + \frac{\varepsilon''(x)}{4} - \frac{b'(x)}{2} - \frac{3}{16}\frac{(\varepsilon'(x))^2}{\varepsilon(x)} + \frac{1}{2}\frac{b(x)\varepsilon'(x)}{\varepsilon(x)} - \frac{1}{4}\frac{b^2(x)}{\varepsilon(x)}, \tag{71.50}$$

$$q_0 = q\sqrt{\frac{\varepsilon(0)}{\varepsilon_0}} - \frac{b(0)}{2\sqrt{\varepsilon_0\varepsilon(0)}} - \frac{\varepsilon'(0)}{4\sqrt{\varepsilon_0\varepsilon(0)}}. \tag{71.51}$$

We use the transformation (Equation 71.4) (with x replaced by z) to map the modified plant into the target system

$$w_t(z,t) = \varepsilon_0 w_{zz}(z,t) - cw(z,t), \tag{71.52}$$

$$w_z(0,t) = w(1,t) = 0. \tag{71.53}$$

Here c is a design parameter that determines the decay rate of the closed-loop system. The transformation kernel is found as a solution of the following PDE:

$$k_{zz}(z,y) - k_{yy}(z,y) = \frac{\lambda_0(y)+c}{\varepsilon_0}k(z,y), \tag{71.54}$$

$$k_y(z,0) = -q_0 k(z,0), \tag{71.55}$$

$$k(z,z) = -q_0 - \frac{1}{2\varepsilon_0}\int_0^z (\lambda_0(y)+c)\,dy. \tag{71.56}$$

This PDE is well posed but cannot be solved in closed form. However, it is easy to solve it symbolically (by successive approximations series) or numerically, substantially easier than solving a Ricatti equation [21]. Note that for constant ε, b, λ, and q, the kernel PDE can be solved explicitly. The controller is

$$U(t) = \varepsilon^{1/4}(1)\sqrt{\varepsilon_0}\int_0^1 k\left(1, \int_0^y \frac{\sqrt{\varepsilon_0}\,ds}{\sqrt{\varepsilon(s)}}\right)\varepsilon^{-3/4}(y)\,e^{-\int_y^1 \frac{b(s)}{2\varepsilon(s)}ds}\,u(y,t)\,dy. \quad (71.57)$$

71.4.2 Other Spatially Causal Plants

Consider the spatially causal system

$$u_t(x,t) = u_{xx}(x,t) + g(x)u(0,t) + \int_0^x f(x,y)u(y,t)dy, \quad (71.58)$$

where the boundary conditions are $u_x(0,t) = 0$ and $u(1,t) = U(t)$. This equation is partly motivated by the model of unstable burning in solid propellant rockets [6]. Another reason to consider this plant is that it often appears as a part of the control design for more complicated systems (see Section 71.8). Without the derivation we present the PDE for the kernel of the transformation (Equation 71.4) that maps it into the heat equations 71.6 through 71.8 [21]:

$$k_{xx}(x,y) - k_{yy}(x,y) = -f(x,y) + \int_y^x k(x,\xi)f(\xi,y)d\xi, \quad (71.59)$$

$$k_y(x,0) = g(x) - \int_0^x k(x,y)g(y)dy, \quad (71.60)$$

$$k(x,x) = 0. \quad (71.61)$$

This PDE can be solved using the method of successive approximations. The controller is given by Equation 71.5.

71.5 Complex-Valued and Coupled PDEs

The design scheme presented in previous sections extends to complex-valued PDEs in a straightforward way. The only difference is that the gain of the transformation (Equation 71.4) becomes complex-valued.

As an example, consider the linear Schrödinger equation

$$u_t(x,t) = -ju_{xx}(x,t), \quad (71.62)$$

$$u_x(0,t) = 0, \quad (71.63)$$

$$u(1,t) = U(t). \quad (71.64)$$

Note that $u(x,t)$ is a complex-valued function so that Equations 71.62 through 71.64 can be viewed as two coupled real-valued PDEs. Without control, this system displays oscillatory behavior and is not asymptotically stable. Let us think of $-j$ as the diffusion coefficient in the heat equation and apply the transformation (Equation 71.4) to map the plant into the target system

$$w_t(x,t) = -jw_{xx}(x,t) - cw(x,t), \quad (71.65)$$

$$w_x(0,t) = w(1,t) = 0. \quad (71.66)$$

It is easy to check that this system is well damped for $c > 0$. The gain kernel for the controller (Equation 71.5) is found as

$$\begin{aligned} k(1,y) &= -cj\frac{I_1\left(\sqrt{cj(1-y^2)}\right)}{\sqrt{cj(1-y^2)}} \\ &= \sqrt{\frac{c}{2\left(1-y^2\right)}}\left[(1+j)\mathrm{bei}_1\left(\sqrt{c\left(1-y^2\right)}\right)+(j-1)\mathrm{ber}_1\left(\sqrt{c\left(1-y^2\right)}\right)\right], \end{aligned} \tag{71.67}$$

where $\mathrm{ber}_1(\cdot)$ and $\mathrm{bei}_1(\cdot)$ are Kelvin functions. The controller is given by (Equation 71.5).

A more general version of the Schrödinger equation is a Ginzburg–Landau equation, which is a popular simplified model for studying vortex shedding in the flows past submerged obstacles. In [1], a flow around a two-dimensional circular cylinder was considered and the backstepping method was successfully applied to design the output-feedback boundary controllers with both sensing and actuation on the cylinder surface.

Another application which involves dealing with coupled equations is a thermal convection loop [5], which describes the dynamics of a heated fluid between two concentric cylinders in a gravity field. In [27], the backstepping design was combined with the singular perturbation approach to solve this problem.

71.6 First-Order Hyperbolic Systems

We now turn our attention to hyperbolic PDEs. Before considering oscillatory systems such as strings and beams we deal with first-order hyperbolic equations. These equations describe quite different physical problems such as traffic flows, chemical reactors, and heat exchangers. They can also serve as a model for delays.

The general first-order hyperbolic equation that can be handled by the backstepping method is

$$u_t(x,t) = u_x(x,t) + \lambda u(x,t) + g(x)u(0,t) + \int_0^x f(x,y)u(y,t)\,dy, \tag{71.68}$$

$$u(1,t) = U(t). \tag{71.69}$$

Note that unlike in second-order (in space) PDEs, here we specify only one boundary condition. For g or f positive and large, the uncontrolled plant is unstable.

Following our general procedure, we use the transformation (Equation 71.4) to convert the plant (Equation 71.68) into the target system

$$w_t(x,t) = w_x(x,t), \tag{71.70}$$

$$w(1,t) = 0. \tag{71.71}$$

This system has the solution

$$w(x,t) = \begin{cases} w_0(t+x) & 0 \le t < 1, \\ 0 & t \ge 1, \end{cases} \tag{71.72}$$

where $w_0(x)$ is the initial condition. We see that this system acts as a pure delay and converges to zero in finite time.

One can derive the following well-posed kernel PDE from Equations 71.68 through 71.71:

$$k_x(x,y) + k_y(x,y) = \int_y^x k(x,\xi)f(\xi,y)d\xi - f(x,y), \tag{71.73}$$

$$k(x,0) = \int_0^x k(x,y)g(y)dy - g(x). \tag{71.74}$$

This PDE has a closed-form solution for constant g and f but needs to be solved numerically in general. The controller is given by Equation 71.5.

71.7 Wave Equation

71.7.1 Undamped Wave Equation

Consider the PDE that describes a vibrating string on a finite interval

$$u_{tt}(x,t) = u_{xx}(x,t), \tag{71.75}$$

$$u_x(0,t) = 0, \tag{71.76}$$

$$u(1,t) = 0. \tag{71.77}$$

The boundary conditions correspond to the situation where the end of the string at $x = 1$ is "pinned" and the other end is "free."

One classical method of stabilizing this system is to add a damping term to the boundary. Specifically, the boundary condition (Equation 71.76) becomes

$$u_x(0,t) = c_0 u_t(0,t). \tag{71.78}$$

The boundary control provided by Equation 71.78 has the capacity to add considerable damping to the system, however it requires actuation on the free end $x = 0$, which is not always feasible.

Let us now consider the wave equation with damping boundary control at the constrained end

$$u_{tt}(x,t) = u_{xx}(x,t) \tag{71.79}$$

$$u_x(0,t) = 0 \tag{71.80}$$

$$u_x(1,t) = U(t). \tag{71.81}$$

The attempt to use the controller $U(t) = -c_1 u_t(1,t), c_1 > 0$, fails because in that case the system has an arbitrary constant as an equilibrium profile. To deal with this multitude of equilibria a more sophisticated controller at $x = 1$ (e.g., backstepping) is needed if the boundary condition at $x = 0$ is to remain free.

The backstepping transformation

$$w(x,t) = u(x,t) + c_0 \int_0^x u(y,t)\,dy \tag{71.82}$$

maps Equations 71.79 through 71.81 the plant into the target system

$$w_{tt}(x,t) = w_{xx}(x,t), \tag{71.83}$$

$$w_x(0,t) = c_0 w(0,t), \tag{71.84}$$

$$w_x(1,t) = -c_1 w_t(1,t). \tag{71.85}$$

The idea is to use a large c_0 in the boundary condition at $x = 0$ to make $w_x(0,t) = c_0 w(0,t)$ behave like $w(0,t) = 0$.

To establish stability of the target system, consider the Lyapunov function

$$V = \frac{1}{2}\left(\|w_x\|^2 + \|w_t\|^2 + c_0 w^2(0)\right) + \delta \int_0^1 (1+x) w_x(x) w_t(x)\, dx. \tag{71.86}$$

One can show that for sufficiently small δ, there exist $m_1, m_2 > 0$ such that

$$m_1 U \leqslant V \leqslant m_2 U, \quad U = \|w_x\|^2 + \|w_t\|^2 + w^2(0), \tag{71.87}$$

which shows that V is positive definite. Straightforward calculation gives

$$\dot{V} = -\left(c_1 - \delta(1+c_1^2)\right) w_t^2(1) - \frac{\delta}{2}\left[\|w_x\|^2 + \|w_t\|^2\right] - \frac{\delta}{2}\left(w_t^2(0) + c_0^2 w^2(0)\right), \tag{71.88}$$

which is negative definite for sufficiently small $\delta < \frac{c_1}{1+c_1^2}$. From Equation 71.87 it follows that

$$U(t) \leq M e^{-t/M} U(0)$$

for some (possibly large) M, and therefore the target system is exponentially stable.

The resulting Neumann backstepping controller is given by

$$U(t) = -c_1 u_t(1,t) - c_0 u(1,t) - c_1 c_0 \int_0^1 u_t(y,t) dy, \tag{71.89}$$

where c_0 is large and c_1 is around 1. Some insight into the properties of the above controller may be obtained by observing the individual terms in Equation 71.89. The term $-c_1 u_t(1,t)$ provides basic damping and the action of $-c_1 u_t(1,t) - c_0 u(1,t)$ is similar to PD control. The last term is a spatially averaged velocity and is the backstepping term that allows actuation at the constrained end.

71.7.2 Wave Equation with Kelvin–Voigt Damping

A totally different backstepping design is possible when the wave equation has a small amount of "Kelvin–Voigt" damping (internal material damping, present in all realistic materials):

$$u_{tt}(x,t) = (1 + d\partial_t) u_{xx}(x,t), \tag{71.90}$$

$$u_x(0,t) = 0, \tag{71.91}$$

$$u(1,t) = U(t), \tag{71.92}$$

where ∂_t is the time derivative and d is a small positive constant. Using the usual backstepping transformation (Equation 71.4) with the kernel

$$k(x,y) = -cx \frac{I_1\left(\sqrt{c(x^2-y^2)}\right)}{\sqrt{c(x^2-y^2)}} \tag{71.93}$$

and the controller

$$U(t) = -\int_0^1 c \frac{I_1\left(\sqrt{c(1-y^2)}\right)}{\sqrt{c(1-y^2)}} u(y,t)\, dy, \tag{71.94}$$

we map the plant into the target system

$$w_{tt}(x,t) = (1 + d\partial_t)(w_{xx}(x,t) - cw(x,t)), \tag{71.95}$$

$$w_x(0,t) = 0, \tag{71.96}$$

$$w(1,t) = 0. \tag{71.97}$$

The nth pair of eigenvalues σ_n of this system satisfies the quadratic equation

$$\sigma_n^2 + d\left[c + \left(\frac{\pi}{2} + \pi n\right)^2\right]\sigma_n + \left[c + \left(\frac{\pi}{2} + \pi n\right)^2\right] = 0, \tag{71.98}$$

where $n = 0, 1, 2, \ldots$. There are two sets of eigenvalues: for lower n's the eigenvalues reside on the circle

$$\left(\mathrm{Re}(\sigma_n) + \frac{1}{d}\right)^2 + (\mathrm{Im}(\sigma_n))^2 = \frac{1}{d^2}, \tag{71.99}$$

and for higher n's the eigenvalues are real, with one branch accumulating toward $-1/d$ as $n \to \infty$ and the other branch converging to $-\infty$ on real axis. Increasing c moves the eigenvalues along the circle in the negative real direction and decreases the number of them on the circle (ultimately they become real). This increases the damping ratio of the nth conjugate complex eigenvalue pair of the closed-loop system by a factor of $\sqrt{1 + \frac{4}{\pi^2(1+2n)^2}c}$ and moves it leftward in the complex plane by a factor of $\left(1 + \frac{4}{\pi^2(1+2n)^2}c\right)$. With a very high value of c all of the eigenvalues can be made real. While possibly, this would not necessarily be a good idea, neither for servo response, nor for disturbance attenuation, and certainly not from the point of view of control effort. Thus, the flexibility to improve the damping using the backstepping transformation and controller should be used judiciously, with lower values of c if d is already relatively high.

71.8 Beams

The most general 1D beam model—the Timoshenko model—can be represented by the two coupled wave equations

$$\epsilon u_{tt} = u_{xx} - \alpha_x, \tag{71.100}$$

$$\mu\epsilon\alpha_{tt} = \epsilon\alpha_{xx} + a(u_x - \alpha), \tag{71.101}$$

where $u(x, t)$ denotes the displacement, $\alpha(x, t)$ denotes the angle of rotation, and the positive constants ϵ, μ, d, and a represent physical parameters of the beam. We consider a beam which is free at the end $x = 0$, that is,

$$u_x(0, t) = \alpha(0, t), \tag{71.102}$$

$$\alpha_x(0, t) = 0 \tag{71.103}$$

and which is controlled at the end $x = 1$ through the boundary conditions $u_x(1, t)$ and $\alpha(1, t)$.

For the case of small μ, we obtain the so-called "slender beam." When μ is set to zero, the Timoshenko equations 71.100 and 71.101 reduce to the "shear beam" model which we will consider in this section:

$$\epsilon u_{tt} = u_{xx} - \alpha_x, \tag{71.104}$$

$$0 = \alpha_{xx} - b^2\alpha + b^2 u_x, \tag{71.105}$$

where $b = \sqrt{a/\epsilon}$. The solution to the α ODE (Equation 71.105) is

$$\alpha(x, t) = \cosh(bx)\alpha(0, t) + b\sinh(bx)u(0, t) - b^2\int_0^x \cosh(b(x - y))u(y, t)\,dy. \tag{71.106}$$

The boundary value $\alpha(0, t)$ can be expressed in terms of $\alpha(1, t)$ by setting $x = 1$ in Equation 71.106. In order to eliminate the nonstrict feedback integral term (and thus set $\alpha(0, t)$ to zero), let us choose our first control to be

$$\alpha(1,t) = U_1(t) = b\sinh(b)u(0,t) - b^2\int_0^1 \cosh(b(1-y))u(y,t)\,dy. \tag{71.107}$$

Substituting Equation 71.106 with $\alpha(0,t) = 0$ into Equation 71.104 gives

$$\epsilon u_{tt}(x,t) = u_{xx}(x,t) + b^2u(x,t) - b^2\cosh(bx)u(0,t) + b^3\int_0^x \sinh(b(x-y))u(y,t)\,dy, \tag{71.108}$$

where this PDE is now of the form similar to Equation 71.58. Now consider the transformation (Equation 71.4) to the exponentially stable target system

$$\epsilon w_{tt}(x,t) = w_{xx}(x,t), \tag{71.109}$$

$$w_x(0,t) = c_0w(0,t), \tag{71.110}$$

$$w_x(1,t) = -c_1w_t(1,t), \tag{71.111}$$

where c_0 and c_1 are design parameters. The transformation kernel can be shown to satisfy the following PDE

$$k_{xx}(x,y) = k_{yy}(x,y) + b^2k(x,y) - b^3\sinh(b(x-y)) + b^3\int_y^x k(x,\xi)\sinh(b(\xi-y))d\xi, \tag{71.112}$$

$$k_y(x,0) = b^2\int_0^x k(x,y)\cosh(by)dy - b^2\cosh(bx), \tag{71.113}$$

$$k(x,x) = -\frac{b^2}{2}x - c_0. \tag{71.114}$$

The second boundary controller (the first one is given by Equation 71.107) is obtained by differentiating the transformation (Equation 71.4) with respect to x and setting $x = 1$:

$$\begin{aligned} u_x(1,t) = U_2(t) = &-\left(\frac{b^2}{2} + c_0\right)u(1,t) + \int_0^1 k_x(1,y)u(y,t)\,dy \\ &- c_1u_t(1,t) + c_1\int_0^1 k(1,y)u_t(y,t)\,dy. \end{aligned} \tag{71.115}$$

71.9 Adaptive Control of Parabolic PDEs

In systems with thermal, fluid, or chemically reacting dynamics, which are usually modeled by parabolic PDEs, physical parameters are often unknown. Thus a need exists for adaptive controllers that are able to stabilize a potentially unstable, parametrically uncertain plant. While adaptive control of finite-dimensional systems is a mature area that has produced adaptive control methods for most LTI systems of interest [10], adaptive control techniques have been developed for only a few of the classes of PDEs for which nonadaptive controllers exist. The prebackstepping results [3,4,9] focus on model reference adaptive control (MRAC) type schemes and the control action distributed in the PDE domain; see [11] for a more detailed literature review. The backstepping controllers presented in previous sections are explicitly parametrized in physical parameters of the corresponding plants, making them an ideal choice for nominal controllers in adaptive schemes. In the rest of the chapter we overview three different design methods based on those explicit controllers—the Lyapunov method, and certainty equivalence approaches with passive and swapping identifiers. For tutorial reasons, the presentation proceeds through a series of one-unknown-parameter benchmark examples with sketches of the stability proofs. The designs are then extended to to reaction–advection–diffusion plants in 2D and plants with spatially varying (functional) parametric uncertainties. The chapter ends with the *output*-feedback adaptive design for reaction–advection–diffusion systems with only boundary sensing and actuation. These systems have an infinite relative degree, infinitely many unknown parameters and are open-loop unstable, representing the ultimate challenge in adaptive control for PDEs. The detailed proofs for the designs presented here are given in [13,23–26].

71.10 Lyapunov Design

Consider the plant

$$u_t(x,t) = u_{xx}(x,t) + \lambda u(x,t), \quad 0 < x < 1, \tag{71.116}$$

$$u(0,t) = 0, \tag{71.117}$$

$$u_x(1,t) = U(t), \tag{71.118}$$

where λ is an unknown constant parameter. We use the Neumann boundary controller (Equation 71.21) with $k(x,y)$ given by Equation 71.15 and λ replaced by its estimate $\hat{\lambda}$:

$$U(t) = -\frac{\hat{\lambda}}{2}u(1,t) - \hat{\lambda}\int_0^1 \xi \frac{I_2\left(\sqrt{\hat{\lambda}(1-\xi^2)}\right)}{1-\xi^2}u(\xi,t)d\xi. \tag{71.119}$$

Straightforward calculation shows that invertible change of variables

$$w(x,t) = u(x,t) - \int_0^x \hat{k}(x,\xi)u(\xi,t)\,d\xi, \quad \hat{k}(x,\xi) = -\hat{\lambda}\xi\frac{I_1\left(\sqrt{\hat{\lambda}(x^2-\xi^2)}\right)}{\sqrt{\hat{\lambda}(x^2-\xi^2)}} \tag{71.120}$$

maps Equations 71.116 through 71.119 into

$$w_t(x,t) = w_{xx}(x,t) + \dot{\hat{\lambda}}\int_0^x \frac{\xi}{2}w(\xi,t)\,d\xi + \tilde{\lambda}w(x,t), \tag{71.121}$$

$$w(0,t) = w_x(1,t) = 0, \tag{71.122}$$

where $\tilde{\lambda} = \lambda - \hat{\lambda}$ is the parameter estimation error.

Consider the Lyapunov function*

$$V = \frac{1}{2}\log\left(1+\|w\|^2\right) + \frac{1}{2\gamma}\tilde{\lambda}^2. \tag{71.123}$$

Its time derivative along the solutions of Equations 71.121 and 71.122 is

$$\dot{V} = -\frac{\|w_x\|^2}{1+\|w\|^2} + \frac{\dot{\hat{\lambda}}}{2}\frac{\int_0^1 w(x)\left(\int_0^x \xi w(\xi)d\xi\right)dx}{1+\|w\|^2} + \tilde{\lambda}\left(-\frac{1}{\gamma}\dot{\hat{\lambda}} + \frac{\|w\|^2}{1+\|w\|^2}\right) \tag{71.124}$$

(the calculation involves integration by parts). Based on Equation 71.124, we choose the update law

$$\dot{\hat{\lambda}} = \gamma\frac{\|w\|^2}{1+\|w\|^2}, \quad 0 < \gamma < 1. \tag{71.125}$$

We are going to show that with this update law, $u(x,t)$ is regulated to zero for all $x \in [0,1]$, for arbitrarily large initial data $u(x,0)$ and for an arbitrarily poor initial estimate $\hat{\lambda}(0)$.

* In Sections 71.10 through 71.15, we suppress time dependence to simplify the notation, that is, $w(0) \equiv w(0,t)$, $\|w\| \equiv \|w(\cdot,t)\|$, etc.

Note that V depends only on time and contains the logarithm of the spatial L_2 norm [19]. This results in the normalized update law (Equation 71.125) and slows down the adaptation, which is necessary because the control law (Equation 71.119) is of certainty equivalence type. An additional measure of preventing overly fast adaptation in Equation 71.125 is the restriction on the adaptation gain ($\gamma < 1$).

Using Cauchy and Poincare inequalities, one obtains

$$\left|\int_0^1 w(x)\left(\int_0^x \xi w(\xi)d\xi\right)dx\right| \le \frac{2}{\sqrt{3}}\|w_x\|^2. \tag{71.126}$$

Substituting Equations 71.126 and 71.125 into Equation 71.124 and using the fact that $|\dot{\hat{\lambda}}| < \gamma$ (see Equation 71.125), we obtain

$$\dot{V} \le -\left(1-\frac{\gamma}{\sqrt{3}}\right)\frac{\|w_x\|^2}{1+\|w\|^2}. \tag{71.127}$$

This implies that $V(t)$ remains bounded for all time whenever $0 < \gamma \le \sqrt{3}$. From the definition of V it follows that $\|w\|$ and $\hat{\lambda}$ remain bounded for all time. To show that $w(x,t)$ is bounded for all time and for all x, we estimate (using Agmon, Young, and Poincare inequalities):

$$\begin{aligned}\frac{1}{2}\frac{d}{dt}\|w_x\|^2 &= -\|w_{xx}\|^2 + \tilde{\lambda}\|w_x\|^2 + \frac{\dot{\hat{\lambda}}}{4}\left(w(1)^2 - \|w\|^2\right)\\ &\le -(1-\gamma)\|w_{xx}\|^2 + \tilde{\lambda}\|w_x\|^2 \le \tilde{\lambda}\|w_x\|^2.\end{aligned} \tag{71.128}$$

Integrating the last inequality, we obtain

$$\|w_x(t)\|^2 \le \|w_x(0)\|^2 + 2\sup_{0\le\tau\le t}|\tilde{\lambda}(\tau)|\int_0^t \|w_x(\tau)\|^2 d\tau. \tag{71.129}$$

Using Equation 71.127 and the fact that $\|w\|$ is bounded, we obtain

$$\int_0^t \|w_x(\tau)\|^2 d\tau \le (1+C)\int_0^t \frac{\|w_x(\tau)\|^2}{1+\|w(\tau)\|^2}d\tau < \infty, \tag{71.130}$$

where C is the bound on $\|w\|^2$. From Equations 71.129 and 71.130, we get that $\|w_x\|^2$ is bounded. Since $\max_{x\in[0,1]}|w(x)|^2 \le \|w_x\|^2$, we observe that $w(x,t)$ is bounded for all x and t.

Next, we prove regulation of $w(x,t)$ to zero. Using Equations 71.121 and 71.122, it is easy to show that

$$\frac{1}{2}\left|\frac{d}{dt}\|w\|^2\right| \le \|w_x\|^2 + \left(|\tilde{\lambda}| + \frac{\gamma}{4\sqrt{3}}\right)\|w\|^2. \tag{71.131}$$

Since $\|w\|$ and $\|w_x\|$ have been proven bounded, it follows that $\frac{d}{dt}\|w\|^2$ is bounded, and thus $\|w(t)\|$ is uniformly continuous. From Equation 71.130 and Poincare inequality, we get that $\|w\|^2$ is integrable in time over the infinite time interval. By Barbalat's lemma, it follows that $\|w\| \to 0$ as $t \to \infty$. The regulation in the maximum norm follows from Agmon's inequality.

Having proved the boundedness and regulation of w, we now set out to establish the same for u. We start by noting that the inverse transformation to Equation 71.120 is [21]

$$u(x) = w(x) + \int_0^x \hat{l}(x,\xi)w(\xi)d\xi, \quad \hat{l}(x,\xi) = -\hat{\lambda}\xi\frac{J_1\left(\sqrt{\hat{\lambda}(x^2-\xi^2)}\right)}{\sqrt{\hat{\lambda}(x^2-\xi^2)}}. \tag{71.132}$$

Since $\hat{\lambda}$ is bounded, the function $\hat{l}(x,\xi)$ has bounds $L_1 = \max_{0\le\xi\le x\le 1}\hat{l}(x,\xi)^2$, $L_2 = \max_{0\le\xi\le x\le 1}\hat{l}_x(x,\xi)^2$. It is straightforward to show that

$$\|u_x\|^2 \le 2\left(1+\hat{\lambda}^2+4L_2\right)\|w_x\|^2, \tag{71.133}$$

Noting that $u(x,t)^2 \leq \|u_x\|^2$ for all $(x,t) \in [0,1] \times [0,\infty)$ and using the fact that $\|w_x\|$ is bounded, we get uniform boundedness of u. To prove regulation of u, we estimate from Equation 71.132

$$\|u\|^2 \leq 2(1+L_1)\|w\|^2. \tag{71.134}$$

Since $\|w\|$ is regulated to zero, so is $\|u\|$. By Agmon's inequality $u(x,t)^2 \leq 2\|u\|\|u_x\|$, where $\|u_x\|$ is bounded. Therefore $u(x,t)$ is regulated to zero for all $x \in [0,1]$.

The Lyapunov design incorporates all the states of the closed-loop system into a single Lyapunov function and therefore Lyapunov adaptive controllers possess the best transient performance properties. However, this method is not applicable as broadly as the *certainty equivalence* approaches, which we consider next.

71.11 Certainty Equivalence Design with Passive Identifier

Consider the plant

$$u_t(x,t) = u_{xx}(x,t) + \lambda u(x,t), \tag{71.135}$$

$$u(0,t) = 0, \tag{71.136}$$

$$u(1,t) = U(t), \tag{71.137}$$

where a constant parameter λ is *unknown*. We use the Dirichlet controller (Equation 71.16) with λ replaced by its estimate $\hat{\lambda}$:

$$U(t) = -\hat{\lambda}\int_0^1 \xi \frac{I_1\left(\sqrt{\hat{\lambda}(1-\xi^2)}\right)}{\sqrt{\hat{\lambda}(1-\xi^2)}} u(\xi,t)\,d\xi. \tag{71.138}$$

Following the certainty equivalence principle, first we need to design an identifier which will provide the estimate $\hat{\lambda}$.

Consider the following system:

$$\hat{u}_t = \hat{u}_{xx} + \hat{\lambda}u + \gamma^2(u-\hat{u})\int_0^1 u^2(x)\,dx, \tag{71.139}$$

$$\hat{u}(0) = 0, \tag{71.140}$$

$$\hat{u}(1) = u(1). \tag{71.141}$$

Such a system is often called an "observer," even though it is not used here for state estimation (the entire state u is available for measurement in our problem). The purpose of this "observer" is to help identify the unknown parameter. This identifier employs a copy of the PDE plant and an additional nonlinear term. We will refer to the systems (Equations 71.139 through 71.141) as a "passive identifier." The term "passive identifier" comes from the fact that an operator from the parameter estimation error $\tilde{\lambda} = \lambda - \hat{\lambda}$ to the inner product of u with $u - \hat{u}$ is strictly passive. The additional term in Equation 71.139 acts as nonlinear damping whose task is to slow down the adaptation.

Let us introduce the error signal $e = u - \hat{u}$. Using Equations 71.135 and 71.136 and Equations 71.139 through 71.141, we obtain

$$e_t = e_{xx} + \tilde{\lambda}u - \gamma^2 e\|u\|^2, \tag{71.142}$$

$$e(0) = e(1) = 0. \tag{71.143}$$

With the Lyapunov function

$$V = \frac{1}{2}\int_0^1 e^2(x)\,dx + \frac{\tilde{\lambda}^2}{2\gamma}, \tag{71.144}$$

we obtain

$$\dot{V} = -\|e_x\|^2 - \gamma^2\|e\|^2\|u\|^2 + \tilde{\lambda}\int_0^1 e(x)u(x)\,dx - \frac{\tilde{\lambda}\dot{\hat{\lambda}}}{\gamma}. \tag{71.145}$$

Choosing the update law

$$\dot{\hat{\lambda}} = \gamma\int_0^1 (u(x) - \hat{u}(x))u(x)\,dx, \tag{71.146}$$

we obtain

$$\dot{V} = -\|e_x\|^2 - \gamma^2\|e\|^2\|u\|^2, \tag{71.147}$$

which implies $V(t) \le V(0)$ and from the definition of V we get that $\tilde{\lambda}$ and $\|e\|$ are bounded functions of time. Integrating Equation 71.147 with respect to time from zero to infinity we get that the spatial norms $\|e_x\|$ and $\|e\|\|u\|$ are square integrable over infinite time. From the update law (Equation 71.146) we have $|\dot{\hat{\lambda}}| \le \gamma\|e\|\|u\|$, which shows that $\dot{\hat{\lambda}}$ is also square integrable in time.

One can show that the transformation

$$\hat{w}(x) = \hat{u}(x) - \int_0^x \hat{k}(x,y)\hat{u}(y)\,dy \tag{71.148}$$

with $\hat{k}$ given by Equation 71.120, maps Equations 71.139 through 71.141 into the following "target system"

$$\hat{w}_t = \hat{w}_{xx} + \dot{\hat{\lambda}}\int_0^x \frac{\xi}{2}\hat{w}(\xi)\,d\xi + (\hat{\lambda} + \gamma^2\|u\|^2)e_1, \tag{71.149}$$

$$\hat{w}(0) = \hat{w}(1) = 0, \tag{71.150}$$

where e_1 is the transformed estimation error

$$e_1(x) = e(x) - \int_0^x \hat{k}(x,y)e(y)\,dy. \tag{71.151}$$

We observe that in comparison to nonadaptive target system (plain heat equation) two additional terms appeared in Equation 71.149, one is proportional to $\dot{\hat{\lambda}}$ and the other to the estimation error e. The identifier guarantees that both of these terms are square integrable in time.

Since $\hat{\lambda}$ is bounded, and the functions $\hat{k}(x,y)$ and $\hat{l}(x,y)$ are twice continuously differentiable with respect to x and y, there exist constants M_1, M_2, and M_3 such that

$$\|e_1\| \le M_1\|e\|, \quad \|u\| \le \|\hat{u}\| + \|e\| \le M_2\|\hat{w}\| + \|e\| \tag{71.152}$$

$$\|u_x\| \le \|\hat{u}_x\| + \|e_x\| \le M_3\|\hat{w}_x\| + \|e_x\|. \tag{71.153}$$

Using Equation 71.152, Young's, Cauchy–Schwarz, and Poincare inequalities along with the identifier properties, one can obtain the following estimate:

$$\frac{1}{2}\frac{d}{dt}\|\hat{w}\|^2 \le -\frac{1}{16}\|\hat{w}\|^2 + l_1\|\hat{w}\|^2 + l_2, \tag{71.154}$$

where l_1, l_2 are some integrable functions of time on $[0,\infty)$. Using Lemma B.6 from [12] we get boundedness and square integrability of $\|\hat{w}\|$. From Equation 71.152 we get boundedness and square integrability of $\|\hat{u}\|, \|u\|$, which also proves boundedness of $\dot{\hat{\lambda}}$.

In order to get pointwise in x boundedness, one estimates

$$\frac{1}{2}\frac{d}{dt}\int_0^1 \hat{w}_x^2\,dx \le -\frac{1}{8}\|\hat{w}_x\|^2 + \frac{|\dot{\hat{\lambda}}|^2\|\hat{w}\|^2}{4} + (\lambda_0 + \gamma^2\|u\|^2)^2 M_1\|e\|^2, \tag{71.155}$$

$$\frac{1}{2}\frac{d}{dt}\int_0^1 e_x^2\,dx \le -\frac{1}{8}\|e_x\|^2 + \frac{1}{2}|\tilde{\lambda}|^2\|u\|^2. \tag{71.156}$$

Since the right-hand sides of Equations 71.155 and 71.156 are integrable, using Lemma B.6 from [12] we get boundedness and square integrability of $\|\hat{w}_x\|$, $\|e_x\|$, and $\|\hat{u}_x\|$, $\|u_x\|$ (from Equation 71.153). By Agmon's inequality, we get that $\hat{u}$ and u are bounded for all $x \in [0, 1]$.

To show the regulation of u to zero, note that

$$\left|\frac{1}{2}\frac{d}{dt}\|e\|^2\right| \le |\tilde{\lambda}|\|e\|\|u\| < \infty. \tag{71.157}$$

The boundedness of $|(d/dt)\|w\|^2|$ follows from Equation 71.154. By Barbalat's lemma, we obtain $\|\hat{w}\| \to 0$, $\|e\| \to 0$ as $t \to \infty$. Since the transformation (Equation 71.148) is invertible, we have $\|\hat{u}\| \to 0$ and from Equation 71.152 $\|u\| \to 0$. Using Agmon's inequality and the fact that $\|u_x\|$ is bounded, we obtain

$$\lim_{t\to\infty}\max_{x\in[0,1]}|u(x,t)| = 0. \tag{71.158}$$

71.12 Certainty Equivalence Design with Swapping Identifier

The certainty equivalence design with *swapping* identifer is perhaps the most common method of parameter estimation in adaptive control. Filters of the "regressor" and of the measured part of the plant are implemented to convert a dynamic parameterization of the problem (given by the plant's dynamic model) into a static parametrization where standard gradient and least-squares estimation techniques can be used.

Consider the plant

$$u_t(x,t) = u_{xx}(x,t) + \lambda u(x,t), \tag{71.159}$$

$$u(0,t) = 0, \tag{71.160}$$

$$u(1,t) = U(t) \tag{71.161}$$

with unknown constant parameter λ. We start by employing two filters: the state filter

$$v_t(x,t) = v_{xx}(x,t) + u(x,t), \tag{71.162}$$

$$v(0,t) = v(1,t) = 0, \tag{71.163}$$

and the input filter

$$\eta_t(x,t) = \eta_{xx}(x,t), \tag{71.164}$$

$$\eta(0,t) = 0, \tag{71.165}$$

$$\eta(1,t) = U(t). \tag{71.166}$$

The "estimation" error

$$e = u - \lambda v - \eta \tag{71.167}$$

is then exponentially stable:

$$e_t(x,t) = e_{xx}(x,t), \tag{71.168}$$

$$e(0,t) = e(1,t) = 0. \tag{71.169}$$

Using the static relationship (Equation 71.167) as a parametric model, we implement a "prediction error" as

$$\hat{e} = u - \hat{\lambda} v - \eta, \quad \hat{e} = e + \tilde{\lambda} v. \tag{71.170}$$

We choose the gradient update law with normalization

$$\dot{\hat{\lambda}} = \gamma \frac{\int_0^1 \hat{e}(x) v(x)\, dx}{1 + \|v\|^2}. \tag{71.171}$$

With the Lyapunov function

$$V = \frac{1}{2}\int_0^1 e^2\, dx + \frac{1}{8\gamma}\tilde{\lambda}^2 \tag{71.172}$$

one obtains

$$\dot{V} \leq -\frac{1}{2}\|e_x\|^2 - \frac{1}{8}\frac{\|\hat{e}\|^2}{1 + \|v\|^2}. \tag{71.173}$$

This gives the following properties:

$$\frac{\|\hat{e}\|}{\sqrt{1 + \|v\|^2}}, \ \tilde{\lambda}, \ \dot{\hat{\lambda}} \ \text{ are bounded,} \tag{71.174}$$

$$\frac{\|\hat{e}\|}{\sqrt{1 + \|v\|^2}}, \ \dot{\hat{\lambda}} \ \text{ are square integrable in time over } [0, \infty). \tag{71.175}$$

In contrast with the passive identifier, the normalization in the swapping identifier is employed in the update law. This makes $\dot{\hat{\lambda}}$ not only square integrable but also bounded.

We use the controller (Equation 71.138) with the state u replaced by its estimate $\hat{\lambda} v + \eta$:

$$U(t) = -\hat{\lambda}\int_0^1 \xi \frac{I_1\left(\sqrt{\hat{\lambda}(1 - \xi^2)}\right)}{\sqrt{\hat{\lambda}(1 - \xi^2)}} (\hat{\lambda} v(\xi, t) + \eta(\xi, t))\, d\xi. \tag{71.176}$$

The transformation (Equation 71.148) is also modified to use the estimate of the state:

$$\hat{w}(x) = \hat{\lambda} v(x) + \eta(x) - \int_0^x \hat{k}(x, \xi)(\hat{\lambda} v(\xi) + \eta(\xi))\, d\xi, \tag{71.177}$$

where $\hat{k}(x, \xi)$ is given by Equation 71.120. Using Equations 71.162 through 71.166 and the inverse transformation

$$\hat{\lambda} v(x) + \eta(x) = \hat{w}(x) + \int_0^x \hat{l}(x, \xi)\hat{w}(\xi)\, d\xi \tag{71.178}$$

with $\hat{l}(x, \xi)$ given by Equation 71.132, one can obtain the following PDE for $\hat{w}$:

$$\hat{w}_t = \hat{w}_{xx} + \dot{\hat{\lambda}} v + \dot{\hat{\lambda}}\int_0^x \left(\frac{\xi}{2}\hat{w}(\xi) - \hat{k}(x, \xi) v(\xi)\right) d\xi + \hat{\lambda}\left(\hat{e}(x) - \int_0^x \hat{k}(x, \xi)\hat{e}(\xi)\, d\xi\right) \tag{71.179}$$

$$\hat{w}(0) = \hat{w}(1) = 0. \tag{71.180}$$

Rewriting the filters (Equations 71.162 through 71.163) as

$$v_t = v_{xx} + \hat{e} + \hat{w} + \int_0^x \hat{l}(x, \xi)\hat{w}(\xi)\, d\xi, \tag{71.181}$$

$$v(0) = v(1) = 0, \tag{71.182}$$

we obtain two interconnected systems for v and $\hat{w}$, Equations 71.179 through 71.182, which are driven by the signals $\dot{\hat{\lambda}}$ and $\hat{e}$ with properties (Equations 71.174 and 71.175). Note that the situation here is more complicated than in the passive design where we had to analyze only the $\hat{w}$-systems (Equations 71.149 through 71.150). While the signal v in Equations 71.179 through 71.180 is multiplied by a square-integrable signal $\dot{\hat{\lambda}}$, the signal $\hat{w}$ in the v-systems (Equations 71.181 through 71.182) is multiplied by a bounded but possibly large gain $\hat{l}$. Therefore, to show boundedness of $\|\hat{w}\|$ and $\|v\|$, we use a weighted Lyapunov function

$$W = A\|\hat{w}\|^2 + \|v\|^2, \tag{71.183}$$

where A is a large enough constant. One can then show that

$$\dot{W} \leq -\frac{1}{4A}W + l_3 W, \tag{71.184}$$

where $l_3(t)$ is an integrable function of time over $[0, \infty)$, and with the help of Lemma B.6 from [12] we get the boundedness of $\|\hat{w}\|$ and $\|v\|$. Using this result it can be shown that

$$\frac{d}{dt}\left(\|\hat{w}_x\|^2 + \|v_x\|^2\right) \leq -\|\hat{w}_{xx}\|^2 - \|v_{xx}\|^2 + l_4 \tag{71.185}$$

with integrable $l_4(t)$, which proves that $\|\hat{w}_x\|$ and $\|v_x\|$ are bounded. From Agmon's inequality we get that $\hat{w}$ and v are bounded pointwise in x. By Barbalat's lemma, $\|\hat{w}\| \to 0$, $\|v\| \to 0$ as $t \to \infty$. From Equations 71.178 and 71.167 we get the pointwise boundedness of η and u and $\|u\| \to 0$. Finally, the pointwise regulation of u to zero follows from Agmon's inequality.

The swapping method uses the highest order of dynamics of all identifier approaches. Lyapunov design has the lowest dynamic order as it only incorporates the dynamics of the parameter update, and the passivity-based method is better than the swapping method because it uses only one filter, as opposed to "one-filter-per-unknown-parameter" in the case of the swapping approach. Despite its high dynamic order, the swapping approach is popular because it is the most transparent (its stability proof is the simplest due to the static parametrization) and it is the only method that incorporates least-squares estimation.

71.13 Extension to Reaction–Advection–Diffusion Systems in Higher Dimensions

All the approaches presented in Sections 71.10 through 71.12 can be readily extended to reaction–advection–diffusion plants and higher dimensions (2D and 3D). As an illustration, consider a 2D plant with four unknown parameters ε, b_1, b_2, and λ:

$$u_t = \varepsilon(u_{xx} + u_{yy}) + b_1 u_x + b_2 u_y + \lambda u \tag{71.186}$$

on the rectangle $0 \leq x \leq 1, 0 \leq y \leq L$ with actuation applied on the side with $x = 1$ and Dirichlet boundary conditions on the other three sides.

We choose to design the scheme with passive identifier. We introduce the following "observer"

$$\hat{u}_t = \hat{\varepsilon}(\hat{u}_{xx} + \hat{u}_{yy}) + \hat{b}_1\hat{u}_1 + \hat{b}_2\hat{u}_2 + \hat{\lambda}u + \gamma^2(u - \hat{u})\|\nabla u\|^2, \tag{71.187}$$

$$\hat{u} = 0, \quad (x, y) \in \{[0, 1] \times [0, 1]\}\backslash\{x = 1\}, \tag{71.188}$$

$$\hat{u} = u, \quad x = 1, \ \ 0 \leq y \leq 1. \tag{71.189}$$

There are two main differences compared to 1D case with one parameter considered in Section 71.11. First, since the unknown diffusion coefficient ε is positive, we must use projection to ensure $\hat{\varepsilon} > \underline{\varepsilon} > 0$:

$$\text{Proj}_{\underline{\varepsilon}}\{\tau\} = \begin{cases} 0 & \hat{\varepsilon} = \underline{\varepsilon} \quad \text{and} \quad \tau < 0, \\ \tau & \text{else.} \end{cases} \tag{71.190}$$

Although this operator is discontinuous, it can be easily modified to avoid dealing with Filippov solutions and noise due to frequent switching of the update law; see [13] for more details. However, we use Equation

71.190 here for notational clarity. Note that $\hat{\varepsilon}$ does not require the projection from above and all other parameters do not require projection at all.

Second, we can see in Equation 71.187 that while the diffusion and advection coefficients multiply the operators of $\hat{u}$, the reaction coefficient multiplies u in the observer. This is necessary in order to eliminate any λ-dependence in the error system so that it is stable.

The update laws are

$$\dot{\hat{\varepsilon}} = -\gamma_0 \mathrm{Proj}_{\underline{\varepsilon}}\left\{\int_0^1\int_0^1 u_x(u_x - \hat{u}_x) + u_y(u_y - \hat{u}_y)\,dx\,dy\right\}, \tag{71.191}$$

$$\dot{\hat{b}}_1 = \gamma_1 \int_0^1\int_0^1 (u - \hat{u})u_x\,dx\,dy, \quad \dot{\hat{b}}_2 = \gamma_1 \int_0^1\int_0^1 (u - \hat{u})u_y\,dx\,dy, \tag{71.192}$$

$$\dot{\hat{\lambda}} = \gamma_2 \int_0^1\int_0^1 (u - \hat{u})u\,dx\,dy, \tag{71.193}$$

and the controller is

$$u(1, y, t) = -\int_0^1 \frac{\hat{\lambda}}{\hat{\varepsilon}}\xi e^{-\frac{\hat{b}_1(1-\xi)}{2\hat{\varepsilon}}} \frac{I_1\left(\sqrt{\frac{\hat{\lambda}}{\hat{\varepsilon}}(1-\xi^2)}\right)}{\sqrt{\frac{\hat{\lambda}}{\hat{\varepsilon}}(1-\xi^2)}}\hat{u}(\xi, y, t)\,d\xi. \tag{71.194}$$

The results of the simulation of the above scheme are presented in Figures 71.1 and 71.2. The true parameters are set to $\varepsilon = 1, b_1 = 1, b_2 = 2, \lambda = 22$, and $L = 2$. With this choice the open-loop plant has

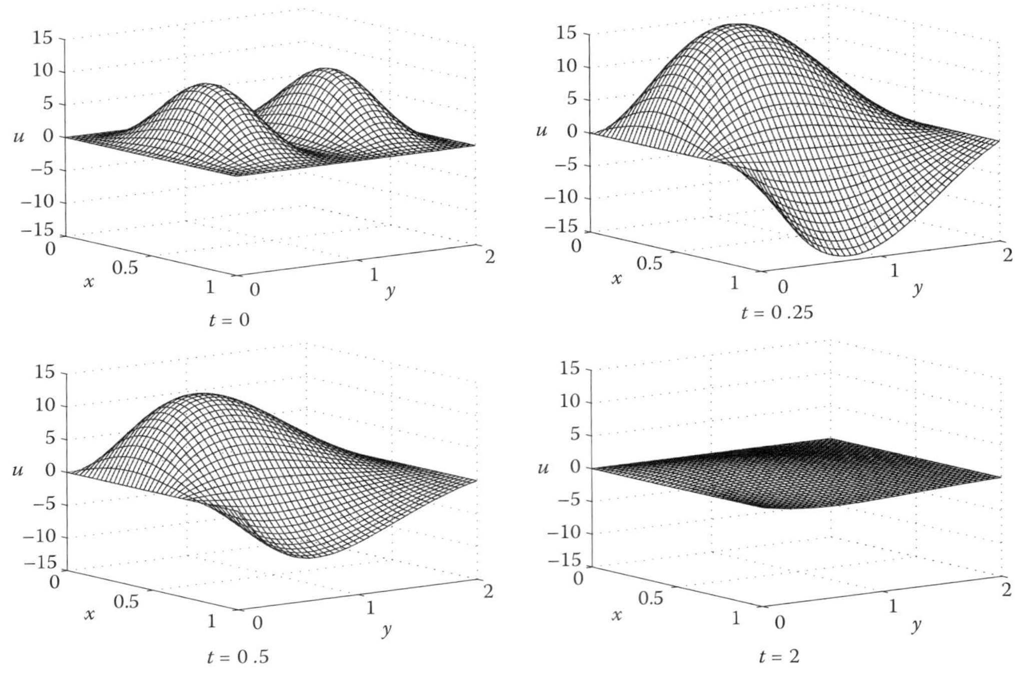

FIGURE 71.1 The closed-loop state for the plant (Equation 71.186) with adaptive controller (Equation 71.194) and passive identifiers (Equations 71.187 through 71.193) at different times.

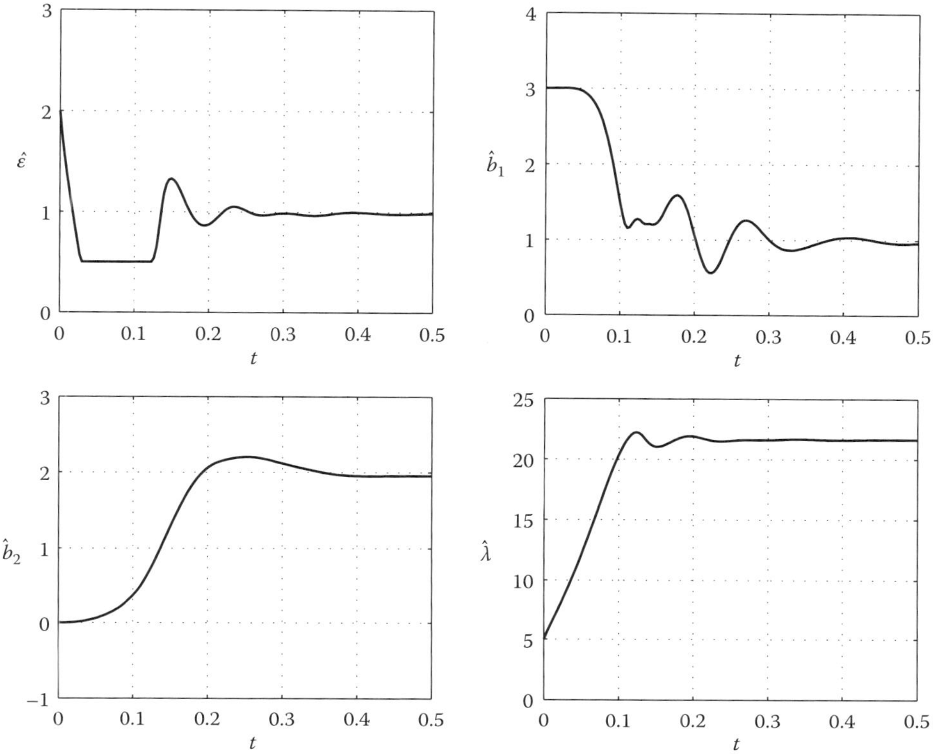

FIGURE 71.2 The parameter estimates for the plant (Equation 71.186).

two unstable eigenvalues at 8.4 and 1. All estimates come close to the true values at approximately $t = 0.5$ and after that the controller stabilizes the system.

71.14 Plants with Spatially Varying Uncertainties

The designs presented in Sections 71.10 through 71.12 can be extended to plants with spatially varying unknown parameters. For example, for the plant

$$u_t(x,t) = u_{xx}(x,t) + \lambda(x)u(x,t), \tag{71.195}$$

$$u_x(0,t) = 0, \tag{71.196}$$

the Lyapunov adaptive controller would be

$$u(1,t) = \hat{k}(1,1)u(1,t) + \int_0^1 \hat{k}_x(1,\xi)u(\xi,t)\,d\xi \tag{71.197}$$

with

$$\hat{\lambda}_t(x,t) = \gamma \frac{u(x,t)\left(w(x,t) - \int_x^1 \hat{k}(\xi,x)w(\xi,t)d\xi\right)}{1 + \|w(t)\|^2},$$

where $\hat{\lambda}(t,x)$ is the online functional estimate of $\lambda(x)$, $w(x) = u(x) - \int_0^x \hat{k}(x,\xi)u(\xi)d\xi$, and the kernel $\hat{k}(x,\xi) = \hat{k}_n(x,\xi)$ is obtained recursively from

$$\hat{k}_0(x,y) = -\frac{1}{2}\int_{\frac{x-\xi}{2}}^{\frac{x+\xi}{2}} \hat{\lambda}\left(\zeta\right) d\zeta, \tag{71.198}$$

$$\hat{k}_{i+1}(x,y) = \hat{k}_i(x,y) + \int_{\frac{x-\xi}{2}}^{\frac{x+\xi}{2}} \int_0^{\frac{x-\xi}{2}} \hat{\lambda}\left(\zeta-\sigma\right) \hat{k}_i\left(\zeta+\sigma,\zeta-\sigma\right) d\sigma\, d\zeta, \quad i = 0, 1, \ldots, n, \tag{71.199}$$

for each new update of $\hat{\lambda}(x,t)$. Stability is guaranteed for sufficiently small γ and sufficiently high n. The recursion (Equation 71.199) was proved convergent in [21]. The certainty equivalence designs with passive and swapping identifiers can also be extended to the case of functional unknown parameters using the same recursive procedure. For further details, the reader is referred to [23].

71.15 Output-Feedback Design

Consider the plant

$$u_t(x,t) = u_{xx}(x,t) + \lambda(x)u(x,t), \tag{71.200}$$

$$u_x(0,t) = 0, \tag{71.201}$$

$$u(1,t) = U(t), \tag{71.202}$$

where $\lambda(x)$ is an unknown continuous function and only the boundary value $u(0,t)$ is measured.

The key step in the output-feedback design is the transformation of the original plant (Equations 71.200 through 71.202) into a system in which unknown parameters multiply the measured output.

71.15.1 Transformation to Observer Canonical Form

One can show that the transformation

$$v(x,t) = u(x,t) - \int_0^x p(x,y)u(y,t)\, dy, \tag{71.203}$$

where $p(x,y)$ is a solution of the PDE

$$p_{xx}(x,y) - p_{yy}(x,y) = \lambda(y)p(x,y), \tag{71.204}$$

$$p(1,y) = 0, \tag{71.205}$$

$$p(x,x) = \frac{1}{2}\int_x^1 \lambda(s)\, ds, \tag{71.206}$$

maps the systems (Equations 71.200 through 71.202) into

$$v_t(x,t) = v_{xx}(x,t) + \theta(x)v(0,t), \tag{71.207}$$

$$v_x(0,t) = \theta_1 v(0,t), \tag{71.208}$$

$$v(1,t) = U(t), \tag{71.209}$$

where

$$\theta(x) = -p_y(x,0), \quad \theta_1 = -p(0,0) \tag{71.210}$$

are the new unknown functional parameters.

The systems (Equations 71.207 through 71.209) is the PDE analog of observer canonical form. Note from Equation 71.203 that $v(0) = u(0)$ and therefore $v(0)$ is measured. The transformation

(Equation 71.203) is invertible so that stability of v implies stability of u. Therefore, it is enough to design the stabilizing controller for v–system and then use the condition $u(1) = v(1)$ (which follows from Equation 71.205) to obtain the controller for the original system. The new unknown parameters $\theta(x)$ and θ_1 are going to be estimated directly, therefore there is no need to solve the PDE (Equations 71.204 through 71.206) for the control scheme implementation.

71.15.2 Filters

The unknown parameters θ and $\theta(x)$ enter the boundary condition and the domain of the v–system. Therefore, the following output filters are needed:

$$\phi_t(x,t) = \phi_{xx}(x,t), \tag{71.211}$$

$$\phi_x(0,t) = u(0,t), \tag{71.212}$$

$$\phi(1,t) = 0, \tag{71.213}$$

and

$$\Phi_t(x,t,\xi) = \Phi_{xx}(x,t,\xi) + \delta(x-\xi)u(0,t), \tag{71.214}$$

$$\Phi_x(0,t,\xi) = \Phi(1,t,\xi) = 0. \tag{71.215}$$

Here the filter $\Phi = \Phi(x,t,\xi)$ is parametrized by $\xi \in [0,1]$ and $\delta(x-\xi)$ is the delta function. The reason for this parametrization is the presence of the functional parameter $\theta(x)$ in the domain. Therefore, loosely speaking, we need an infinite "array" of filters, one for each $x \in [0,1]$ (since the swapping design normally requires one filter per unknown parameter). We also introduce the input filter

$$\psi_t(x,t) = \psi_{xx}(x,t), \tag{71.216}$$

$$\psi_x(0,t) = 0, \tag{71.217}$$

$$\psi(1,t) = U(t). \tag{71.218}$$

It is straightforward to show now that the error

$$\bar{e}(x,t) = v(x,t) - \psi(x,t) - \theta_1\phi(x,t) - \int_0^1 \theta(\xi)\Phi(x,t,\xi)\,d\xi \tag{71.219}$$

satisfies the exponentially stable PDE

$$\bar{e}_t(x,t) = \bar{e}_{xx}(x,t), \tag{71.220}$$

$$\bar{e}_x(0,t) = \bar{e}(1,t) = 0. \tag{71.221}$$

Typically, the swapping method requires one filter per unknown parameter and since we have functional parameters, infinitely many filters are needed. However, it is possible to reduce their number down to only two by representing the state $\Phi(x,\xi,t)$ algebraically through $\phi(x,t)$ at each instant t. Comparing the explicit solutions of the systems (Equations 71.211 through 71.213) and (Equations 71.214 through 71.215), one can show that Φ can be represented as $\Phi = F + \Delta F$, where ΔF satisfies

$$\Delta F_t = \Delta F_{xx}, \tag{71.222}$$

$$\Delta F_x(0,\xi,t) = \Delta F(1,\xi,t) = 0, \tag{71.223}$$

and $F(x,\xi,t)$ is obtained from the filter ϕ through the solution of the PDE

$$F_{xx}(x,\xi) = F_{\xi\xi}(x,\xi), \tag{71.224}$$

$$F(0,\xi) = -\phi(\xi), \tag{71.225}$$

$$F_x(0,\xi) = F_\xi(x,0) = F(x,1) = 0. \tag{71.226}$$

It is then easy to show that the signal

$$e(x) = v(x) - \psi(x) - \theta_1\phi(x) - \int_0^1 \theta(\xi)F(x,\xi)\,d\xi \tag{71.227}$$

is governed by the exponentially stable heat equation:

$$e_t(x,t) = e_{xx}(x,t), \tag{71.228}$$

$$e_x(0,t) = e(1,t) = 0. \tag{71.229}$$

Therefore, instead of solving the infinite "array" of parabolic equations 71.214 through 71.215, one can solve only two dynamic equations for filters ϕ and ψ and compute the solution of the standard wave equation 71.224 through 71.226 at each time step.

71.15.3 Update Laws

The following equation serves as a parametric model:

$$e(0) = v(0) - \psi(0) - \theta_1\phi(0) + \int_0^1 \theta(\xi)\phi(\xi)\,d\xi. \tag{71.230}$$

The estimation error is

$$\hat{e}(0) = v(0) - \psi(0) - \hat{\theta}_1\phi(0) + \int_0^1 \hat{\theta}(\xi)\phi(\xi)\,d\xi. \tag{71.231}$$

We employ the gradient update laws with normalization

$$\hat{\theta}_t(x,t) = -\gamma(x)\frac{\hat{e}(0)\phi(x)}{1+\|\phi\|^2+\phi^2(0)}, \quad \dot{\hat{\theta}}_1 = \gamma_1\frac{\hat{e}(0)\phi(0)}{1+\|\phi\|^2+\phi^2(0)}, \tag{71.232}$$

where $\gamma(x)$ and γ_1 are positive adaptation gains.

With the Lyapunov function

$$V = \frac{1}{2}\|e\|^2 + \frac{1}{2\gamma_1}\tilde{\theta}_1^2 + \int_0^1 \frac{\tilde{\theta}^2(x)}{2\gamma(x)}\,dx \tag{71.233}$$

one obtains

$$\dot{V} \le -\frac{1}{2}\|e_x\|^2 - \frac{1}{2}\frac{\hat{e}^2(0)}{1+\|\phi\|^2+\phi^2(0)}, \tag{71.234}$$

which gives the following properties of the adaptive laws:

$$\frac{\hat{e}(0)}{\sqrt{1+\|\phi\|^2+\phi^2(0)}},\ \|\hat{\theta}_t\|,\ \dot{\hat{\theta}}_1,\ \|\tilde{\theta}\|,\ \tilde{\theta}_1 \ \text{are bounded,} \tag{71.235}$$

$$\frac{\hat{e}(0)}{\sqrt{1+\|\phi\|^2+\phi^2(0)}},\ \|\hat{\theta}_t\|,\ \dot{\hat{\theta}}_1 \ \text{are square integrable in time.} \tag{71.236}$$

71.15.4 Controller

The PDE for the estimated gain kernel for the plant (Equations 71.207 through 71.209) is

$$\hat{k}_{xx}(x,y) = \hat{k}_{yy}(x,y), \tag{71.237}$$

$$\hat{k}_y(x,0) = \hat{\theta}_1\hat{k}(x,0) + \hat{\theta}(x) - \int_0^x \hat{k}(x,y)\hat{\theta}(y)\,dy, \tag{71.238}$$

$$\hat{k}(x,x) = \hat{\theta}_1. \tag{71.239}$$

The solution to this PDE is $\hat{k}(x,y) = \kappa(x-y)$, where κ satisfies the simple integral equation

$$\kappa(x) = \hat{\theta}_1 - \int_0^x \hat{\theta}(y)\,dy - \int_0^x \left[\hat{\theta}_1 - \int_0^{x-y} \hat{\theta}(s)\,ds\right]\kappa(y)\,dy, \tag{71.240}$$

which has to be solved online.

The controller is given by

$$U = \int_0^1 \kappa(1-y)\left(\psi(y) + \hat{\theta}_1\phi(y) + \int_0^1 F(y,\xi)\hat{\theta}(\xi)\,d\xi\right)dy. \tag{71.241}$$

Using the transformation

$$\begin{aligned} w(x) &= \psi(x) + \hat{\theta}_1\phi(x) + \int_0^1 F(x,\xi)\hat{\theta}(\xi)\,d\xi \\ &\quad - \int_0^x \kappa(x-y)\left[\psi(y) + \hat{\theta}_1\phi(y) + \int_0^1 F(y,\xi)\hat{\theta}(\xi)\,d\xi\right]dy \end{aligned} \tag{71.242}$$

and analyzing stability properties of the interconnected PDEs for w and ϕ similarly to the way it is done in Section 71.12, one can show that all the signals in the closed-loop system are bounded and $u(x,t)$ is regulated to zero uniformly in $x \in [0,1]$.

The design methodology presented above can also be applied to general reaction–advection–diffusion systems

$$u_t = \varepsilon(x)u_{xx} + b(x)u_x + \lambda(x)u + g(x)u(0) + \int_0^x f(x,y)u(y)\,dy \tag{71.243}$$

$$u_x(0) = -qu(0), \tag{71.244}$$

$$u(1) = U, \tag{71.245}$$

where $\varepsilon(x)$ is known and $b(x), \lambda(x), g(x), f(x,y)$, and q are unknown parameters.

Further Reading

Due to space constraints, the material presented in this chapter does not include all the backstepping-based designs developed up to the time of writing this chapter. Most of those can be accessed in two recently published books, [15,28].

In particular, the stabilizing controllers in Sections 71.2 through 71.4 can be modified so that a certain meaningful cost functional is minimized, providing stability margins [21] (the so-called inverse-optimal design).

In addition to the design for the shear beam in Section 71.8, which is presented in more details in [16], backstepping controllers have also been developed for the Euler–Bernoulli beam [20].

Designs for nonlinear parabolic PDEs are presented in [29,30].

Finally, the backstepping method is particularly suited for dealing with PDE–ODE or PDE–PDE cascades, for example, systems with actuator/sensor dynamics in the form of the pure delay [14].

References

1. O. M. Aamo, A. Smyshlyaev, and M. Krstic. Boundary control of the linearized Ginzburg–Landau model of vortex shedding. *SIAM Journal of Control and Optimization*, 43:1953–1971, 2005.

2. A. Bensoussan, G. Da Prato, M. C. Delfour, and S. K. Mitter. *Representation and Control of Infinite-Dimensional Systems*. Birkhäuser, Boston, 2006.
3. J. Bentsman and Y. Orlov. Reduced spatial order model reference adaptive control of spatially varying distributed parameter systems of parabolic and hyperbolic types. *International Journal of Adaptive Control Signal Process.*, 15:679–696, 2001.
4. M. Bohm, M. A. Demetriou, S. Reich, and I. G. Rosen. Model reference adaptive control of distributed parameter systems. *SIAM Journal of Control and Optimization*, 36:33–81, 1998.
5. D. M. Boskovic and M. Krstic. Nonlinear stabilization of a thermal convection loop by state feedback. *Automatica*, 37:2033–2040, 2001.
6. D. M. Boskovic and M. Krstic. Stabilization of a solid propellant rocket instability by state feedback. *International Journal of Robust and Nonlinear Control*, 13:483–495, 2003.
7. P. Christofides. *Nonlinear and Robust Control of Partial Differential Equation Systems: Methods and Applications to Transport-Reaction Processes*. Birkhäuser, Boston, 2001.
8. R. F. Curtain and H. J. Zwart. *An Introduction to Infinite Dimensional Linear Systems Theory*. Springer-Verlag, 1995.
9. K. S. Hong and J. Bentsman. Direct adaptive control of parabolic systems: algorithm synthesis and convergence and stability analysis. *IEEE Transactions on Automatic Control*, 39:2018–2033, 1994.
10. P. Ioannou and J. Sun. *Robust Adaptive Control*. Prentice-Hall, Englewood cliffs, NJ, 1996.
11. M. Krstic. Systematization of approaches to adaptive boundary stabilization of PDEs. *International Journal of Robust and Nonlinear Control*, 16:812–818, 2006.
12. M. Krstic, I. Kanellakopoulos, and P. Kokotovic. *Nonlinear and Adaptive Control Design*. Wiley, New York, 1995.
13. M. Krstic and A. Smyshlyaev. Adaptive boundary control for unstable parabolic PDEs—Part I: Lyapunov design. *IEEE Transactions on Automatic Control*, 53:1575–1591, 2008.
14. M. Krstic and A. Smyshlyaev. Backstepping boundary control for first order hyperbolic PDEs and application to systems with actuator and sensor delays. *Systems and Control Letters*, 57:750–758, 2008.
15. M. Krstic and A. Smyshlyaev. *Boundary Control of PDEs: A Course on Backstepping Designs*. SIAM, 2008.
16. M. Krstic, A. Smyshlyaev, and A.A. Siranosian. Backstepping boundary controllers and observers for the slender Timoshenko beam: Part I—Design. In *Proceedings of the 2006 American Control Conferencel*, 2006.
17. I. Lasiecka and R. Triggiani. *Control Theory for Partial Differential Equations: Continuous and Approximation Theories*. Cambridge Univ. Press, Cambridge, UK, 2000.
18. Z. H. Luo, B. Z. Guo, and O. Morgul. *Stability and Stabilization of Infinite Dimensional Systems with Applications*. Springer-Verlag, 1999.
19. L. Praly. Adaptive regulation: Lyapunov design with a growth condition. *International Journal of Adaptive Control Signal Processing*, 6:329–351, 1992.
20. A. Smyshlyaev, B. Z. Guo, and M. Krstic. Boundary controllers for Euler–Bernoulli beam with arbitrary decay rate. In *Proceedings of the 2008 IEEE Conference on Decision and Control*, 2008.
21. A. Smyshlyaev and M. Krstic. Closed form boundary state feedbacks for a class of 1D partial integro-differential equations. *IEEE Transactions on Automatic Control*, 49:2185–2202, 2004.
22. A. Smyshlyaev and M. Krstic. Backstepping observers for a class of parabolic PDEs. *Systems and Control Letters*, 54:613–625, 2005.
23. A. Smyshlyaev and M. Krstic. Lyapunov adaptive boundary control for parabolic PDEs with spatially varying coefficients. In *Proceedings of 2006 American Control Conference*, 2006.
24. A. Smyshlyaev and M. Krstic. Output-feedback adaptive control for parabolic PDEs with spatially varying coefficients. In *Proceedings of 2006 IEEE Conference on Decision and Control*, 2006.
25. A. Smyshlyaev and M. Krstic. Adaptive boundary control for unstable parabolic PDEs—Part II: Estimation-based designs. *Automatica*, 43:1543–1556, 2007.
26. A. Smyshlyaev and M. Krstic. Adaptive boundary control for unstable parabolic PDEs—Part III: Output feedback examples with swapping identifiers. *Automatica*, 43:1557–1564, 2007.
27. R. Vazquez and M. Krstic. Explicit integral operator feedback for local stabilization of nonlinear thermal convection loop PDEs. *Systems and Control Letters*, 55:624–632, 2006.
28. R. Vazquez and M. Krstic. *Control of Turbulent and Magnetohydrodynamic Channel Flows*. Birkhäuser, Boston, 2007.
29. R. Vazquez and M. Krstic. Control of 1-D parabolic PDEs with Volterra nonlinearities—Part I: Design. *Automatica*, 44:2778–2790, 2008.
30. R. Vazquez and M. Krstic. Control of 1-D parabolic PDEs with Volterra nonlinearities—Part II: Analysis. *Automatica*, 44:2791–2803, 2008.

72

Stabilization of Fluid Flows

Miroslav Krstić
University of California, San Diego

Rafael Vazquez
University of Seville

72.1 Introduction

Recent years have been marked by dramatic advances in the field of *active flow control.* This development can be credited not only to advances in the various fields that intersect through this discipline (such as control theory, fluid mechanics, PDE theory, and numerical methods), but also to technological developments such as Micro-Electro-Mechanical Systems (MEMS) sensors and actuators and the ever-increasing prowess of last-generation computers, that have augmented the possibilities of effective implementation in both real-life and numerical experiments. However, the area is far from being mature with still many opportunities and challenging open problems.

Efforts over the last few years have led to a wide range of developments in many different directions, reflecting the interdisciplinary character of the research community. However, most implementable developments so far have been obtained using discretized versions of the plant models and finite-dimensional control techniques. In contrast, in this chapter we present a design method that is based on the "continuum" version of the backstepping approach, applied to the partial differential equation (PDE) model of the flow. The postponement of the spatial discretization until the implementation stage offers advantages that range from numerical to analytical. In fact, this design method offers a unique physical intuition by forcing the closed-loop system to dynamically behave as a set of well-damped heat equation PDEs.

In the next sections we present an example of feedback control and observer design for an incompressible 3D magnetohydrodynamic (MHD) channel flow, also known as the Hartmann flow, a benchmark model for applications such as cooling systems (computer systems, fusion reactors), hypersonic flight, and propulsion. In this flow, an electrically conducting fluid moves between parallel plates and is affected by an imposed transverse magnetic field. The velocity and electromagnetic fields are mathematically described by the MHD equations, which are the Navier–Stokes equation coupled with the Maxwell equations. For zero magnetic field or nonconducting fluids, the plant reduces to the 3D Navier–Stokes channel flow, and hence, the control and observer designs apply to the 3D Navier–Stokes system when the magnetic field parameter is set to zero.

Owing to the high complexity of the Navier–Stokes problem (further aggravated by the presence of Maxwell's equations), most derivations and proofs are skipped or at best only sketched. Further details and methods for flow control can be found in the recent book by Vazquez and Krstic [26]. The main method used in the designs is the backstepping method for PDEs. Its basics are outlined in Chapter 71 on "Boundary control for PDEs" by Krstic and Smyshlyaev in this Handbook. Some very elementary background on Lyapunov stability (see Chapter 43 by Khalil in this Handbook) and basic PDE theory is assumed in this chapter, but no previous familiarity with flow control is necessary.

In Section 72.2, we present the mathematical model of the Hartmann flow, and then study its equilibrium profile. We linearize the system around the equilibrium and derive the plant equations in Fourier space. We follow Section 72.3, where we design (in Fourier space) a feedback control law that stabilizes the linearized system using the backstepping method. In Section 72.4 we study the problem of designing a nonlinear estimator, with linear output injection, using the backstepping observer design method. The observer and controller, combined together, provide a stabilizing output feedback control law. We finish with a brief literature review in Section 72.5.

72.2 Channel with MHD Flow

We consider a benchmark 3D MHD channel flow, known as the Hartmann flow. This flow consists of an incompressible conducting fluid enclosed between two parallel plates, separated by a distance L_p, under the influence of a pressure gradient ∇P parallel to the walls and a magnetic field B_0 normal to the walls, as shown in Figure 72.1. Under the assumption of a very small magnetic Reynolds number

$$Re_M = \mu_m \sigma U_c L_p \ll 1, \tag{72.1}$$

where μ_m is the magnetic permeability of the fluid, σ the conductivity of the fluid, and U_c the reference velocity (maximum velocity of the Hartmann equilibrium profile), the dynamics of the magnetic field can be neglected and the dimensionless velocity and electric potential field are governed by the inductionless MHD equations [17].

We set nondimensional coordinates (x, y, z), where x is the streamwise direction (parallel to the pressure gradient), y the wall normal direction (parallel to the magnetic field), and z the spanwise

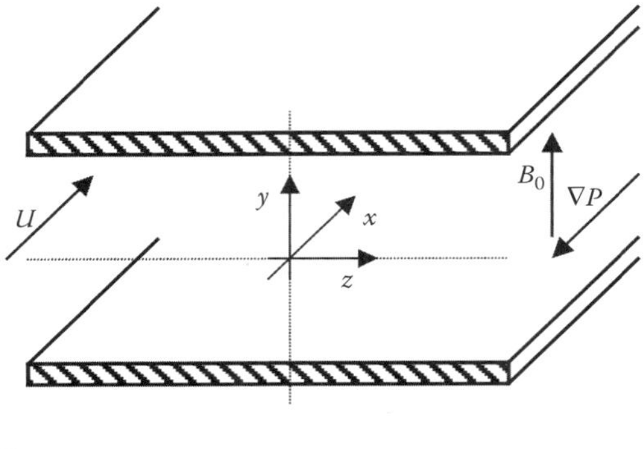

FIGURE 72.1 Hartmann flow.

direction, so that $(x, y, z) \in (-\infty, \infty) \times [0, 1] \times (-\infty, \infty)$. The governing equations are

$$U_t = \frac{\Delta U}{Re} - UU_x - VU_y - WU_z - P_x + N\phi_z - NU\,, \tag{72.2}$$

$$V_t = \frac{\Delta V}{Re} - UV_x - VV_y - WV_z - P_y\,, \tag{72.3}$$

$$W_t = \frac{\Delta W}{Re} - UW_x - VW_y - WW_z - P_z - N\phi_x - NW\,, \tag{72.4}$$

$$\Delta\phi = U_z - W_x\,, \tag{72.5}$$

where U, V, and W denote, respectively, the streamwise, wall-normal and spanwise velocities, P the pressure, ϕ the electric potential, $Re = (U_c L_p)/\nu$ is the Reynolds number, $N = (\sigma L_p B_0^2)/(\rho U_c)$ the Stuart number, and $\Delta = \partial_{xx} + \partial_{yy} + \partial_{zz}$ denotes the standard Laplacian operator. The notations U_t and U_x (and other subscripts such as y, z, xx, etc.) denote the partial derivatives of U with respect to t and x, respectively. Since the fluid is incompressible, the continuity equation is verified

$$U_x + V_y + W_z = 0\,. \tag{72.6}$$

The boundary conditions for the velocity field are

$$U(t, x, 0, z) = U(t, x, 1, z) = U_c(t, x, z), \tag{72.7}$$

$$V(t, x, 0, z) = V(t, x, 1, z) = V_c(t, x, z), \tag{72.8}$$

$$W(t, x, 0, z) = W(t, x, 1, z) = W_c(t, x, z), \tag{72.9}$$

where $U_c(t, x, z)$, $V_c(t, x, z)$, and $W_c(t, x, z)$ denote, respectively, the actuators for streamwise, wall-normal, and spanwise velocity in the upper wall. Assuming perfectly conducting walls, the electric potential must verify

$$\phi(t, x, 0, z) = 0, \; \phi(t, x, 1, z) = \Phi_c(t, x, z), \tag{72.10}$$

where $\Phi_c(t, x, z)$ is the imposed potential (electromagnetic actuation) in the upper wall.

We assume that all actuators can be independently actuated for every $(x, z) \in \mathbb{R}^2$. Note that no actuation is done inside the channel or at the bottom wall.

Remark 72.1

If we set $N = 0$ (zero magnetic field, or nonconducting fluid) in Equations 72.2 through 72.5, they reduce to the classical Navier–Stokes Equations without body forces. Then Equations 72.2 through 72.4, 72.6, and Equations 72.7 through 72.9 describe a pressure-driven channel flow, known as the Poiseuille flow.

72.2.1 Hartmann Equilibrium Profile

The equilibrium profile for systems (Equations 72.2 through 72.5) with no control can be found explicitly (as is also the case with the Poiseuille solution for Navier–Stokes channel flow). We assume a steady solution with only one nonzero nondimensional velocity component, U^e, that depends only on the y coordinate. Substituting U^e in Equation 72.2, one finds that it verifies the following equation:

$$0 = \frac{U^e_{yy}(y)}{Re} - P^e_x - NU^e(y)\,, \tag{72.11}$$

whose nondimensional solution is, setting P^e such that the maximum velocity (centerline velocity) is unity,

$$U^e(y) = \frac{\sinh(H(1-y)) - \sinh H + \sinh(Hy)}{2\sinh H/2 - \sinh H}, \quad (72.12)$$

$$V^e = W^e = \phi^e = 0, \quad (72.13)$$

$$P^e = \frac{N \sinh H}{2\sinh H/2 - \sinh H} x, \quad (72.14)$$

$$j^{xe} = j^{ye} = 0, j^{ze} = U^e(y). \quad (72.15)$$

where $H = \sqrt{ReN} = B_0 L_P \sqrt{\frac{\sigma}{\rho\nu}}$ is the Hartmann number. In Figure 72.2 (left) we show $U^e(y)$ for different values of H. Since the equilibrium profile is nondimensional the centerline velocity is always 1. For $H = 0$, the Poiseuille equilibrium profile is obtained, which is parabolic. In Figure 72.2 (right) we show $U^e_y(y)$, proportional to shear stress, whose maximum is reached at the boundaries and grows with H.

72.2.2 The Plant in Wave Number Space

Define the fluctuation variables

$$u(t,x,y) = U(t,x,y) - U^e(y), \quad (72.16)$$

$$p(t,x,y) = P(t,x,y) - P^e(y), \quad (72.17)$$

where $U^e(y)$ and $P^e(y)$ are, respectively, the equilibrium velocity and pressure given in Equations 72.12 and 72.14. The linearization of Equations 72.2 through 72.4 around the Hartmann equilibrium profile, written in the fluctuation variables (u, V, W, p, ϕ), is

$$u_t = \frac{\Delta u}{Re} - U^e(y)u_x - U^e_y(y)V - p_x + N\phi_z - Nu, \quad (72.18)$$

$$V_t = \frac{\Delta V}{Re} - U^e(y)V_x - p_y, \quad (72.19)$$

$$W_t = \frac{\Delta W}{Re} - U^e(y)W_x - p_z - N\phi_x - NW. \quad (72.20)$$

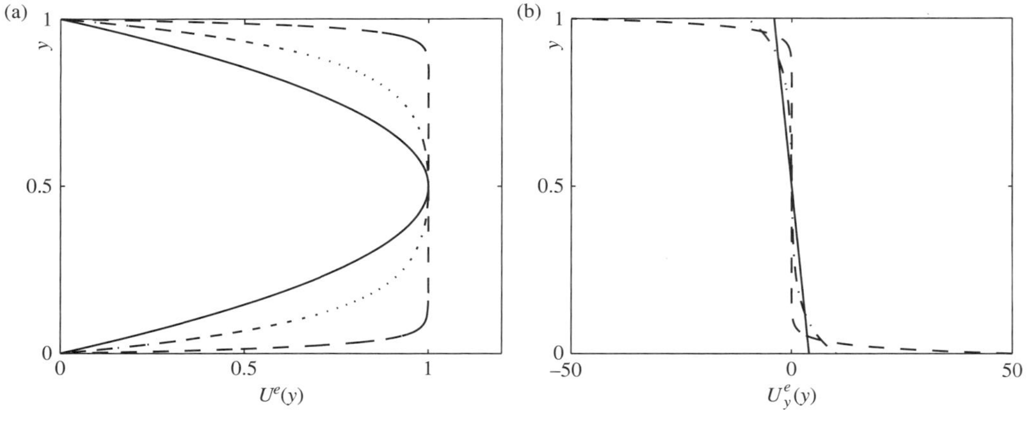

FIGURE 72.2 Streamwise equilibrium velocity $U^e(y)$ (left) and $U^e_y(y)$ (right), for different values of H. Solid, $H = 0$; dash-dotted, $H = 10$; dashed, $H = 50$.

The equation for the potential is

$$\Delta\phi = u_z - W_x\,, \tag{72.21}$$

and the fluctuation velocity field verifies the continuity equation,

$$u_x + V_y + W_z = 0\,, \tag{72.22}$$

and the following boundary conditions:

$$u(t,x,0,z) = W(t,x,0,z) = V(t,x,0,z) = 0, \tag{72.23}$$
$$u(t,x,1,z) = U_c(t,x,z), \tag{72.24}$$
$$V(t,x,1,z) = V_c(t,x,z), \tag{72.25}$$
$$W(t,x,1,z) = W_c(t,x,z), \tag{72.26}$$
$$\phi(t,x,0,z) = 0,\ \phi(t,x,1,z) = \Phi_c(t,x,z). \tag{72.27}$$

Since the plant is linear and invariant to shifts in the streamwise (x) and spanwise (z) directions, we use a Fourier transform in x and z coordinates. The transform pair (direct and inverse transform) is defined as

$$f(k_x,y,k_z) = \int_{-\infty}^{\infty}\int_{-\infty}^{\infty} f(x,y,z)\mathrm{e}^{-2\pi i(k_x x+k_z z)}\,dz\,dx, \tag{72.28}$$
$$f(x,y,z) = \int_{-\infty}^{\infty}\int_{-\infty}^{\infty} f(k_x,y,k_z)\mathrm{e}^{2\pi i(k_x x+k_z z)}\,dk_z\,dk_x. \tag{72.29}$$

Note that we use the same symbol f for both the original $f(x,y,z)$ and the transform $f(k_x,y,k_z)$. The quantities k_x and k_z are referred to as the "wave numbers."

Denoting $\alpha^2 = 4\pi^2(k_x^2+k_z^2)$ and $\beta = 2\pi i k_x U^e$, the plant equations in Fourier space are

$$u_t = \frac{-\alpha^2 u + u_{yy}}{Re} - \beta u - U_y^e V - 2\pi k_x i p + 2\pi k_z i N\phi - Nu, \tag{72.30}$$
$$V_t = \frac{-\alpha^2 V + V_{yy}}{Re} - \beta V - p_y, \tag{72.31}$$
$$W_t = \frac{-\alpha^2 W + W_{yy}}{Re} - \beta W - 2\pi k_z i p - 2\pi k_x i N\phi - NW. \tag{72.32}$$

The continuity equation in wave number space is expressed as

$$2\pi i k_x u + V_y + 2\pi k_z W = 0, \tag{72.33}$$

and the equation for the potential is

$$-\alpha^2\phi + \phi_{yy} = 2\pi i\left(k_z u - k_x W\right). \tag{72.34}$$

The boundary conditions are

$$u(t,k_x,0,k_z) = W(t,k_x,0,k_z) = V(t,k_x,0,k_z) = \phi(t,k_x,0,k_z) = 0, \tag{72.35}$$
$$u(t,k_x,1,k_z) = U_c(t,k_x,k_z),\ V(t,k_x,1,k_z) = V_c(t,k_x,k_z), \tag{72.36}$$
$$W(t,k_x,1,k_z) = W_c(t,k_x,k_z),\ \phi(t,k_x,1,k_z) = \Phi_c(t,k_x,k_z). \tag{72.37}$$

72.3 Boundary Control Design

The Hartmann and Poiseuille flows are unstable for large Reynolds numbers. To guarantee stability, our design task is to design feedback laws U_c, V_c, W_c, and Φ_c, so that the origin of the velocity fluctuation system is exponentially stable. Full state knowledge is assumed in this section.

We design the controller in wave number space. Note that Equations 72.30 through 72.37 are uncoupled for each wave number. Therefore, the set of wave numbers $k_x^2+k_z^2 \leq M^2$ (for some large $M>0$), which we refer to as the *controlled* wave number range, and the range $k_x^2+k_z^2 > M^2$, the *uncontrolled* wave number range, can be studied separately. If stability for all wave numbers is established, stability in physical space follows. The number M, which will be computed in Section 72.3.2, is a parameter that ensures stability for uncontrolled wave numbers.

We define χ, a *truncating* function, as

$$\chi(k_x,k_z)=\begin{cases}1, & k_x^2+k_z^2\leq M^2,\\ 0, & \text{otherwise.}\end{cases} \tag{72.38}$$

Then, we reflect that we only use control for small (controlled) wave numbers by setting

$$\begin{pmatrix}U_c(t,x,z)\\ V_c(t,x,z)\\ W_c(t,x,z)\\ \Phi_c(t,x,z)\end{pmatrix}=\int_{-\infty}^{\infty}\int_{-\infty}^{\infty}\chi(k_x,k_z)\begin{pmatrix}U_c(t,k_x,k_z)\\ V_c(t,k_x,k_z)\\ W_c(t,k_x,k_z)\\ \Phi_c(t,k_x,k_z)\end{pmatrix}e^{2\pi i(k_x x+k_z z)}\,dk_z\,dk_x. \tag{72.39}$$

Next, we find control laws for small wave numbers and study uncontrolled wave numbers.

72.3.1 Controlled Velocity Wave Number Analysis

Consider $k_x^2+k_z^2\leq M^2$. Then $\chi=1$, so control is applied. Using the continuity Equation 72.33 and taking divergence of Equations 72.30 through 72.32, a Poisson Equation for the pressure is derived,

$$-\alpha^2 p+p_{yy}=-4\pi k_x i U_y^e(y)V+NV_y. \tag{72.40}$$

Evaluating Equation 72.31 at $y=0$ one finds that

$$\begin{aligned}p_y(k_x,0,k_z)&=\frac{V_{yy}(k_x,0,k_z)}{Re}\\ &=-2\pi i\frac{k_x u_{y0}+k_z W_{y0}}{Re},\end{aligned} \tag{72.41}$$

where we use Equation 72.33 for expressing V_{yy} at the bottom in terms of the boundary variables $u_{y0}=u_y(k_x,0,k_z)$ and $W_{y0}=W_y(k_x,0,k_z)$. Similarly, evaluating Equation 72.31 at $y=1$ we obtain

$$\begin{aligned}p_y(k_x,1,k_z)&=\frac{V_{yy}(k_x,1,k_z)}{Re}-(V_c)_t-\alpha^2\frac{V_c}{Re}\\ &=-2\pi i\frac{k_x u_{y1}+k_z W_{y1}}{Re}-(V_c)_t-\alpha^2\frac{V_c}{Re},\end{aligned} \tag{72.42}$$

where we use Equation 72.33 for expressing V_{yy} at the top wall in terms of $u_{y1}=u_y(k_x,1,k_z)$ and $W_{y1}=W_y(k_x,1,k_z)$ and the controller V_c.

Equation 72.40 can be solved in terms of integrals of the state and the boundary terms appearing in Equations 72.41 and 72.42.

$$
\begin{aligned}
p = & -\frac{4\pi k_x i}{\alpha} \int_0^y U_y^e(\eta) \sinh(\alpha(y-\eta)) V(k_x,\eta,k_z)\, d\eta + N \int_0^y \frac{\sinh(\alpha(y-\eta))}{\alpha} V_y(k_x,\eta,k_z)\, d\eta \\
& + 2\pi i \frac{\cosh(\alpha(1-y))}{\alpha \sinh\alpha} \frac{k_x u_{y0} + k_z W_{y0}}{Re} - N \frac{\cosh(\alpha y)}{\alpha \sinh\alpha} \int_0^1 \cosh(\alpha(1-\eta)) V_y(k_x,\eta,k_z)\, d\eta \\
& + \frac{4\pi k_x i \cosh(\alpha y)}{\alpha \sinh\alpha} \int_0^1 U_y^e(\eta) \cosh(\alpha(1-\eta)) V(k_x,\eta,k_z)\, d\eta \\
& - 2\pi i \frac{\cosh(\alpha y)}{\alpha \sinh\alpha} \frac{k_x u_{y1} + k_z W_{y1}}{Re} - \frac{\cosh(\alpha y)}{\alpha \sinh\alpha} \left((V_c)_t + \alpha^2 \frac{V_c}{Re} \right).
\end{aligned} \tag{72.43}
$$

We use the controller V_c, which appears *inside* the pressure solution (Equation 72.43), to make the pressure strict-feedback [16] (spatially causal in y), for applying a backstepping boundary controller as in [27]. Since the first two lines in Equation 72.43 are already spatially causal, we need to cancel the third, fourth, and fifth lines of Equation 72.43. Set

$$
\begin{aligned}
(V_c)_t = & \alpha^2 \frac{V_c}{Re} + 2\pi i \frac{k_x(u_{y0} - u_{y1}) + k_z(W_{y0} - W_{y1})}{Re} - N \int_0^1 \cosh(\alpha(1-\eta)) V_y(k_x,\eta,k_z)\, d\eta \\
& + 4\pi k_x i \int_0^1 U_y^e(\eta) \cosh(\alpha(1-\eta)) V(k_x,\eta,k_z)\, d\eta,
\end{aligned} \tag{72.44}
$$

which can be written as

$$
\begin{aligned}
(V_c)_t = & \alpha^2 \frac{V_c}{Re} + 2\pi i \frac{k_x(u_{y0} - u_{y1}) + k_z(W_{y0} - W_{y1})}{Re} - N V_c \\
& + \int_0^1 \cosh(\alpha(1-\eta)) V(k_x,\eta,k_z) \left(N + 4\pi k_x i U_y^e(\eta) \right) d\eta.
\end{aligned} \tag{72.45}
$$

Then, the pressure is written in terms of a strict-feedback integral of the state V and the boundary terms u_{y0}, W_{y0} (proportional to the skin friction at the bottom) as follows:

$$
\begin{aligned}
p = & -\frac{4\pi k_x i}{\alpha} \int_0^y U_y^e(\eta) \sinh(\alpha(y-\eta)) V(k_x,\eta,k_z)\, d\eta + N \int_0^y \frac{\sinh(\alpha(y-\eta))}{\alpha} V_y(k_x,\eta,k_z)\, d\eta \\
& - 2\pi i \frac{\cosh(\alpha y) - \cosh(\alpha(1-y))}{Re\, \alpha \sinh\alpha} \left(k_x u_{y0} + k_z W_{y0} \right).
\end{aligned} \tag{72.46}
$$

Similarly, solving for ϕ in terms of the control Φ_c and the right-hand side of its Poisson equation 72.34,

$$
\begin{aligned}
\phi = & \frac{2\pi i}{\alpha} \int_0^y \sinh(\alpha(y-\eta)) \left(k_z u(k_x,\eta,k_z) - k_x W(k_x,\eta,k_z) \right) d\eta + \frac{\sinh(\alpha y)}{\sinh\alpha} \Phi_c(k_x,k_y) \\
& - \frac{2\pi i \sinh(\alpha y)}{\alpha \sinh\alpha} \int_0^1 \sinh(\alpha(1-\eta)) \left(k_z u(k_x,\eta,k_z) - k_x W(k_x,\eta,k_z) \right) d\eta.
\end{aligned} \tag{72.47}
$$

As in the pressure Equation 72.43, an actuator (Φ_c in this case) appears inside the solution for the potential. The second line of Equation 72.47 is a nonstrict-feedback integral and needs to be cancelled to apply the backstepping method. For this we use Φ_c by setting

$$\Phi_c(k_x,k_y)=\frac{2\pi i}{\alpha}\int_0^1 \sinh\left(\alpha(1-\eta)\right)\left(k_z u(k_x,\eta,k_z)-k_x W(k_x,\eta,k_z)\right)d\eta. \tag{72.48}$$

Then the potential can be expressed as a strict-feedback integral of the states u and W as follows:

$$\phi=\frac{2\pi i}{\alpha}\int_0^y \sinh\left(\alpha(y-\eta)\right)\left(k_z u(k_x,\eta,k_z)-k_x W(k_x,\eta,k_z)\right)d\eta. \tag{72.49}$$

Introducing the expressions (Equations 72.46 and 72.49) in Equations 72.30 and 72.32, we obtain

$$\begin{aligned}
u_t=&\frac{-\alpha^2u+u_{yy}}{Re}-\beta u-U_y^e(y)V-4\pi^2k_x\frac{\cosh(\alpha y)-\cosh(\alpha(1-y))}{Re\,\alpha\sinh\alpha}\left(k_xu_{y0}+k_zW_{y0}\right)\\
&-Nu-\frac{8\pi k_x^2}{\alpha}\int_0^y U_y^e(\eta)\sinh\left(\alpha(y-\eta)\right)V(k_x,\eta,k_z)\,d\eta\\
&-2\pi ik_xN\int_0^y\frac{\sinh\left(\alpha(y-\eta)\right)}{\alpha}V_y(k_x,\eta,k_z)\,d\eta\\
&-\frac{4\pi^2k_zN}{\alpha}\int_0^y\sinh\left(\alpha(y-\eta)\right)\left(k_zU(k_x,\eta,k_z)-k_xW(k_x,\eta,k_z)\right)\,d\eta,
\end{aligned} \tag{72.50}$$

$$\begin{aligned}
W_t=&\frac{-\alpha^2W+W_{yy}}{Re}-\beta W-NW-4\pi^2k_z\frac{\cosh(\alpha y)-\cosh(\alpha(1-y))}{Re\,\alpha\sinh\alpha}\left(k_xu_{y0}+k_zW_{y0}\right)\\
&-\frac{8\pi k_xk_z}{\alpha}\int_0^y U_y^e(\eta)\sinh\left(\alpha(y-\eta)\right)V(k_x,\eta,k_z)\,d\eta\\
&-2\pi ik_zN\int_0^y\frac{\sinh\left(\alpha(y-\eta)\right)}{\alpha}V_y(k_x,\eta,k_z)\,d\eta\\
&+\frac{4\pi^2k_xN}{\alpha}\int_0^y\sinh\left(\alpha(y-\eta)\right)\left(k_zU(k_x,\eta,k_z)-k_xW(k_x,\eta,k_z)\right)\,d\eta.
\end{aligned} \tag{72.51}$$

We have omitted the equation for V since, from Equation 72.33 and using the fact that $V(k_x,0,k_z)=0$, V is computed as

$$V=-2\pi i\int_0^y\left(k_xU(k_x,\eta,k_z)+k_zW(k_x,\eta,k_z)\right)d\eta. \tag{72.52}$$

Now we use the following change of variables and its inverse:

$$Y=2\pi i\left(k_xu+k_zW\right),\quad \omega=2\pi i\left(k_zu-k_xW\right), \tag{72.53}$$

$$u=\frac{2\pi i}{\alpha^2}\left(k_xY+k_z\omega\right),\quad W=\frac{2\pi i}{\alpha^2}\left(k_zY-k_x\omega\right). \tag{72.54}$$

Defining $\epsilon=\frac{1}{Re}$ and the following functions:

$$f=4\pi ik_x\left\{\frac{U_y^e}{2}+\int_\eta^y U_y^e(\sigma)\frac{\sinh\left(\alpha(y-\sigma)\right)}{\alpha}d\sigma\right\}+N\alpha\sinh\left(\alpha(y-\sigma)\right), \tag{72.55}$$

$$g=-\alpha\frac{\cosh(\alpha y)-\cosh(\alpha(1-y))}{Re\sinh\alpha}, \tag{72.56}$$

$$h_1=2\pi ik_zU_y^e, \tag{72.57}$$

$$h_2=-N\alpha\sinh\left(\alpha(y-\eta)\right), \tag{72.58}$$

Equations 72.50 and 72.51 expressed in terms of Y and ω are

$$Y_t = \epsilon\left(-\alpha^2 Y + Y_{yy}\right) - \beta Y - NY + gY_{y0} + \int_0^y f(k_x, y, \eta, k_z) Y(k_x, \eta, k_z)\, d\eta\,, \tag{72.59}$$

$$\omega_t = \epsilon\left(-\alpha^2 \omega + \omega_{yy}\right) - \beta\omega - N\omega + h_1 \int_0^y Y(k_x, \eta, k_z)\, d\eta + \int_0^y h_2(y, \eta)\omega(k_x, \eta, k_z)\, d\eta, \tag{72.60}$$

where we have used the inverse change of variables (Equation 72.54) to express u_{y0} and W_{y0} in terms of $Y_{y0} = Y_y(k_x, 0, k_z)$ as follows:

$$Y_{y0} = 2\pi i\left(k_x u_{y0} + k_z W_{y0}\right), \tag{72.61}$$

with boundary conditions

$$Y(t, k_x, 0, k_z) = \omega(t, k_x, 0, k_z) = 0, \tag{72.62}$$

$$Y(t, k_x, 1, k_z) = Y_c(t, k_x, k_z), \quad \omega(t, k_x, 1, k_z) = \omega_c(t, k_x, k_z), \tag{72.63}$$

where

$$Y_c = 2\pi i\left(k_x U_c + k_z W_c\right), \quad \omega_c = 2\pi i\left(k_z U_c - k_x W_c\right). \tag{72.64}$$

Equations 72.59 and 72.60 are a coupled, strict-feedback plant, with integral and reaction terms. A variant of the backstepping control design presented in [21] can be used to stabilize the system considered here, which consists of a pair of coupled PDEs, by using a *pair* of backstepping transformations. The transformations map, for each k_x and k_z, the variables (Y, ω) into the variables (Ψ, Ω), that verify the following pair of heat equations (parameterized in k_x, k_z):

$$\Psi_t = \epsilon\left(-\alpha^2\Psi + \Psi_{yy}\right) - \beta\Psi - N\Psi, \tag{72.65}$$

$$\Omega_t = \epsilon\left(-\alpha^2\Omega + \Omega_{yy}\right) - \beta\Omega - N\Omega, \tag{72.66}$$

with boundary conditions

$$\Psi(k_x, 0, k_z) = \Psi(k_x, 1, k_z) = \Omega(k_x, 0, k_z) = \Omega(k_x, 1, k_z) = 0. \tag{72.67}$$

The transformation is defined as follows:

$$\Psi = Y - \int_0^y K(k_x, y, \eta, k_z) Y(k_x, \eta, k_z)\, d\eta, \tag{72.68}$$

$$\Omega = \omega - \int_0^y \Gamma_1(k_x, y, \eta, k_z) Y(k_x, \eta, k_z)\, d\eta - \int_0^y \Gamma_2(k_x, y, \eta, k_z)\omega(k_x, \eta, k_z)\, d\eta. \tag{72.69}$$

The kernel functions $K(k_x, y, \eta, k_z)$, $\Gamma_1(k_x, y, \eta, k_z)$, and $\Gamma_2(k_x, y, \eta, k_z)$ are found as the solution of the following partial integro-differential equations:

$$\epsilon K_{yy} = \epsilon K_{\eta\eta} + \left(\beta(y) - \beta(\eta)\right) K - f + \int_\eta^y f(\eta, \xi) K(y, \xi)\, d\xi, \tag{72.70}$$

$$\epsilon\Gamma_{1yy} = \epsilon\Gamma_{1\eta\eta} + \left(\beta(y) - \beta(\eta)\right)\Gamma_1 - h_1 + \int_\eta^y \Gamma_2(y, \xi) h_1(\xi)\, d\xi + \int_\eta^y f(\eta, \xi)\Gamma_1(y, \xi)\, d\xi, \tag{72.71}$$

$$\epsilon\Gamma_{2yy} = \epsilon\Gamma_{2\eta\eta} + \left(\beta(y) - \beta(\eta)\right)\Gamma_2 - h_2 + \int_\eta^y h_2(\xi, \eta)\Gamma_2(y, \xi)\, d\xi. \tag{72.72}$$

Equations 72.70 through 72.72 are hyperbolic partial integro-differential equation in the region $\mathcal{T} = \{(y, \eta) : 0 \le y \le 1, 0 \le \eta \le y\}$. Their boundary conditions are

$$K(y, y) = -\frac{g(0)}{\epsilon}, \; K(y, 0) = \frac{\int_0^y K(y, \eta) g(\eta)\, d\eta - g(y)}{\epsilon}, \tag{72.73}$$

$$\Gamma_1(y, y) = 0, \; \Gamma_1(y, 0) = \frac{\int_0^y \Gamma_1(y, \eta) g(\eta)\, d\eta}{\epsilon}, \tag{72.74}$$

$$\Gamma_2(y, y) = 0, \; \Gamma_2(y, 0) = 0. \tag{72.75}$$

Remark 72.2

Equations 72.70 through 72.75 are well-posed and can be solved symbolically, by means of a successive approximation series, or numerically [21]. Note that Equations 72.70 and 72.72 are autonomous. Hence, one must solve first for $K(k_x, y, \eta, k_z)$ and $\Gamma_2(k_x, y, \eta, k_z)$. Then the solution for Γ_2 is plugged in Equation 72.71 which then can be solved for $\Gamma_1(k_x, y, \eta, k_z)$.

Control laws Y_c and W_c are found evaluating Equations 72.68 and 72.69 at $y = 1$ and using Equations 72.63 and 72.67 which yield

$$Y_c(t, k_x, k_z) = \int_0^1 K(k_x, 1, \eta, k_z) Y(k_x, \eta, k_z)\, d\eta, \tag{72.76}$$

$$\omega_c(t, k_x, k_z) = \int_0^1 \Gamma_1(k_x, 1, \eta, k_z) Y(k_x, \eta, k_z)\, d\eta + \int_0^1 \Gamma_2(k_x, 1, \eta, k_z) \omega(k_x, \eta, k_z)\, d\eta. \tag{72.77}$$

Using Equations 72.53 and 72.54 to write Equations 72.76 and 72.77 in(u, W), we obtain

$$U_c = \int_0^1 K^{Uu}(k_x, 1, \eta, k_z) u(k_x, \eta, k_z)\, d\eta + \int_0^1 K^{UW}(k_x, 1, \eta, k_z) W(k_x, \eta, k_z)\, d\eta, \tag{72.78}$$

$$W_c = \int_0^1 K^{Wu}(k_x, 1, \eta, k_z) u(k_x, \eta, k_z)\, d\eta + \int_0^1 K^{WW}(k_x, 1, \eta, k_z) W(k_x, \eta, k_z)\, d\eta, \tag{72.79}$$

where

$$\begin{pmatrix} K^{Uu} \\ K^{UW} \\ K^{Wu} \\ K^{WW} \end{pmatrix} = \mathbf{A} \begin{pmatrix} K(k_x, y, \eta, k_z) \\ \Gamma_1(k_x, y, \eta, k_z) \\ 0 \\ \Gamma_2(k_x, y, \eta, k_z) \end{pmatrix}, \tag{72.80}$$

and where the matrix $\mathbf{A}$ is defined as

$$\mathbf{A} = -\frac{4\pi^2}{\alpha^2} \begin{pmatrix} k_x^2 & k_x k_z & k_x k_z & k_z^2 \\ k_x k_z & k_z^2 & -k_x^2 & -k_x k_z \\ k_x k_z & -k_x^2 & k_z^2 & -k_x k_z \\ k_z^2 & -k_x k_z & -k_x k_z & k_x^2 \end{pmatrix}. \tag{72.81}$$

Stability in the controlled wave number range follows from stability of Equations 72.65 and 72.66, and the invertibility of the transformations (Equations 72.68 and 72.69). We obtain the following result.

Proposition 72.1:

For $k_x^2 + k_z^2 \leq M^2$, the equilibrium $u \equiv V \equiv W \equiv 0$ of systems (Equations 72.30 and 72.37) with control laws (Equations 72.45, 72.48, 72.78 and 72.79) is exponentially stable in the L^2 norm, that is,

$$\int_0^1 \left(|u|^2 + |V|^2 + |W|^2\right)(t, k_x, y, k_z)\, dy \leq C_1 e^{-2\epsilon t} \int_0^1 \left(|u|^2 + |V|^2 + |W|^2\right)(0, k_x, y, k_z)\, dy, \tag{72.82}$$

where $C_1 \geq 0$.

Proof. From Equations 72.65 and 72.66 we obtain, using a standard Lyapunov argument,

$$\int_0^1 \left(|\Psi|^2 + |\Omega|^2\right)(t, k_x, y, k_z)\, dy \le e^{-2\epsilon t} \int_0^1 \left(|\Psi|^2 + |\Omega|^2\right)(0, k_x, y, k_z)\, dy, \tag{72.83}$$

and then from the transformation (Equations 72.68 and 72.69) and its inverse (which is guaranteed to exist [21]), we obtain

$$\int_0^1 \left(|Y|^2 + |\omega|^2\right)(t, k_x, y, k_z)\, dy \le C_0 e^{-2\epsilon t} \int_0^1 \left(|Y|^2 + |\omega|^2\right)(0, k_x, y, k_z)\, dy, \tag{72.84}$$

where $C_0 > 0$ is a constant depending on the kernels K, Γ_1, and Γ_2 and their inverses. Then writing (u, W) in terms of (Y, ω) and bounding the norm of V by the norm of Y (using $Y = -V_y$ and Poincare's inequality), the result follows.

72.3.2 Uncontrolled Velocity Wave Number Analysis

When $k_x^2 + k_z^2 > M$, the plant verifies the following equations:

$$u_t = \frac{-\alpha^2 u + u_{yy}}{Re} - \beta u - U_y^e(y)V - 2\pi k_x ip + 2\pi k_z iN\phi - Nu\,, \tag{72.85}$$

$$V_t = \frac{-\alpha^2 V + V_{yy}}{Re} - \beta V - p_y\,, \tag{72.86}$$

$$W_t = \frac{-\alpha^2 W + W_{yy}}{Re} - \beta W - 2\pi k_z ip - 2\pi k_x iN\phi - NW\,, \tag{72.87}$$

the Poisson equation for the potential

$$-\alpha^2\phi + \phi_{yy} = 2\pi i\left(k_z u - k_x W\right) \tag{72.88}$$

the continuity equation

$$2\pi ik_x u + V_y + 2\pi k_z W = 0, \tag{72.89}$$

and Dirichlet boundary conditions

$$u(t, k_x, 0, k_y) = V(t, k_x, 0, k_y) = W(t, k_x, 0, k_y) = 0, \tag{72.90}$$

$$u(t, k_x, 1, k_y) = V(t, k_x, 1, k_y) = W(t, k_x, 1, k_y) = 0, \tag{72.91}$$

$$\phi(t, k_x, 0, k_y) = \phi(t, k_x, 1, k_y) = 0. \tag{72.92}$$

Using the transformation (Equation 72.53) to write the system in (Y, ω) coordinates, one gets the following equations for Y and ω:

$$Y_t = \epsilon\left(-\alpha^2 Y + Y_{yy}\right) - \beta Y - 2\pi k_x iU_y^e V + \alpha^2 p - NY, \tag{72.93}$$

$$\omega_t = \epsilon\left(-\alpha^2\omega + \omega_{yy}\right) - \beta\omega - 2\pi k_z iU_y^e V - \alpha^2 N\phi - N\omega. \tag{72.94}$$

The Poisson equation for the potential is, in terms of ω,

$$-\alpha^2\phi + \phi_{yy} = \omega. \tag{72.95}$$

Consider the Lyapunov function

$$\Lambda = \int_0^1 \frac{|u|^2 + |V|^2 + |W|^2}{2}\, dy, \tag{72.96}$$

where we write $\int_0^1 f = \int_0^1 f(k_x, y, k_z)\, dy$. The function Λ is the L^2 norm (kinematic energy) of the velocity field.

Substituting Y and ω from Equation 72.54 into Equation 72.96, we obtain

$$\begin{aligned}\Lambda &= \int_0^1 4\pi^2 \left[\frac{k_x^2|Y|^2 + k_z^2|\omega|^2 + k_xk_z(\bar{Y}\omega + Y\bar{\omega})}{2\alpha^4} + \frac{k_z^2|Y|^2 + k_x^2|\omega|^2 - k_xk_z(\bar{Y}\omega + Y\bar{\omega})}{2\alpha^4}\right] dy \\ &+ \int_0^1 \frac{|V|^2}{2}\, dy \\ &= \int_0^1 \frac{|Y|^2 + |\omega|^2 + \alpha^2|V|^2}{2\alpha^2}\, dy. \end{aligned} \tag{72.97}$$

Define then a new Lyapunov function,

$$\Lambda_1 = \alpha^2\Lambda = \int_0^1 \frac{|Y|^2 + |\omega|^2 + \alpha^2|V|^2}{2}\, dy. \tag{72.98}$$

The time derivative of Λ_1 can be estimated as follows:

$$\begin{aligned}\dot{\Lambda}_1 &= -2\epsilon\alpha^2\Lambda_1 - \epsilon\int_0^1 \left(|Y_y|^2 + |\omega_y|^2 + \alpha^2|V_y|^2\right) - N\int_0^1 \left(|Y|^2 + |\omega|^2\right) \\ &- \int_0^1 \pi i U_y^e(y) V(2k_x\bar{Y} + k_z\bar{\omega}) + \int_0^1 \pi i U_y^e(y)\bar{V}(2k_xY + k_z\omega) \\ &- \alpha^2 N\int_0^1 \frac{\bar{\phi}\omega + \phi\bar{\omega}}{2} + \alpha^2\int_0^1 \frac{\bar{P}Y + P\bar{Y} - \bar{P}_yV - P_y\bar{V}}{2}. \end{aligned} \tag{72.99}$$

For bounding Equation 72.99, we use the following two lemmas.

Lemma 72.1:

$$-\alpha^2\int_0^1 \frac{\bar{\phi}\omega + \phi\bar{\omega}}{2} \le \int_0^1 |\omega|^2. \tag{72.100}$$

Lemma 72.2:

$$|U_y^e(y)| \le 4 + H. \tag{72.101}$$

Integrating by parts and applying Lemma 72.1,

$$\begin{aligned}\dot{\Lambda}_1 &\le -2\epsilon\alpha^2\Lambda_1 - \epsilon\int_0^1 \left(|Y_y|^2 + |\omega_y|^2 + \alpha^2|V_y|^2\right) + \int_0^1 \pi i U_y^e(y)\bar{V}(k_xY + k_z\omega) \\ &- \int_0^1 \pi i U_y^e(y) V(k_x\bar{Y} + k_z\bar{\omega}) - N\int_0^1 |Y|^2. \end{aligned} \tag{72.102}$$

Using Lemma 72.2 to bound U_y^e in Equation 72.102,

$$\begin{aligned}\dot{\Lambda}_1 &\le -2\epsilon\left(1+\alpha^2\right)\Lambda_1 - N\int_0^1 |Y|^2 dy + 2\pi\left(4+H\right)\int_0^1 \left(|V|(|k_x||Y| + |k_z||\omega|\right) dy \\ &\le \left(4 + H - 2\epsilon\left(1+\alpha^2\right)\right)\Lambda_1 \end{aligned} \tag{72.103}$$

where we have applied Young's and Poincare's inequalities. Hence, if $\alpha^2 \geq \frac{4+H}{2\epsilon}$,

$$\dot{\Lambda}_1 \leq -2\epsilon\Lambda_1. \tag{72.104}$$

Dividing Equation 72.104 by α^2 and using Equation 72.98, we obtain

$$\dot{\Lambda} \leq -2\epsilon\Lambda, \tag{72.105}$$

and stability in the uncontrolled wave number range follows when $k_x^2 + k_z^2 \geq M^2$ for M (conservatively) chosen as

$$M \geq \frac{1}{2\pi}\sqrt{\frac{(H+4)Re}{2}}. \tag{72.106}$$

We summarize the result in the following proposition.

Proposition 72.2:

For $k_x^2 + k_z^2 \geq M^2$, where $M \geq \frac{1}{2\pi}\sqrt{\frac{(H+4)Re}{2}}$, the equilibrium $u \equiv V \equiv W \equiv 0$ of the uncontrolled systems (Equations 72.85 through 72.92) is exponentially stable in the L^2 sense, that is,

$$\int_0^1 \left(|u|^2 + |V|^2 + |W|^2\right)(t, k_x, y, k_z)\, dy \leq e^{-2\epsilon t} \int_0^1 \left(|u|^2 + |V|^2 + |W|^2\right)(0, k_x, y, k_z)\, dy. \tag{72.107}$$

72.3.3 Closed-Loop Stability Properties

Substituting Equations 72.45, 72.48, and Equations 72.78, 72.79 into Equation 72.39, and using the Fourier convolution theorem, we obtain the control laws in physical space, which can be expressed compactly as

$$\begin{pmatrix} U_c \\ W_c \\ \Phi_c \end{pmatrix} = \int_{-\infty}^{\infty}\int_0^1\int_{-\infty}^{\infty} \mathbf{\Sigma}(x-\xi, \eta, z-\zeta)\begin{pmatrix} u(\xi,\eta,\zeta) \\ W(\xi,\eta,\zeta)\end{pmatrix} d\xi\, d\eta\, d\zeta, \tag{72.108}$$

where

$$\mathbf{\Sigma}(\xi,\eta,\zeta) = \int_{-\infty}^{\infty}\int_{-\infty}^{\infty} \mathbf{\Sigma}(k_x, \eta, k_z)\chi(k_x,k_z)e^{2\pi i(k_x\xi + k_z\zeta)}\, dk_z\, dk_x, \tag{72.109}$$

and

$$\mathbf{\Sigma} = \begin{pmatrix} K^{Uu}(k_x, 1, \eta, k_z) & K^{UW}(k_x, 1, \eta, k_z) \\ K^{Wu}(k_x, 1, \eta, k_z) & K^{WW}(k_x, 1, \eta, k_z) \\ \frac{2\pi i k_z \sinh(\alpha(1-\eta))}{\alpha} & -\frac{2\pi i k_k \sinh(\alpha(1-\eta))}{\alpha} \end{pmatrix}, \tag{72.110}$$

where the kernels appearing in Equation 72.110 were defined in Equation 72.80. Control law V_c is a dynamic feedback law computed as the solution of the following forced parabolic equation:

$$(V_c)_t = \frac{(V_c)_{xx} + (V_c)_{zz}}{Re} - NV_c + g(t,x,z), \tag{72.111}$$

where $g(t,x,z)$ is defined as

$$g = \int_{-\infty}^{\infty}\int_{-\infty}^{\infty}\Bigg[\int_0^1 g_V(x-\xi,\eta,z-\zeta)V(\xi,\eta,\zeta)\, d\eta + g_W(x-\xi,z-\zeta)\left(W_y(\xi,0,\zeta)\right.$$
$$\left. - W_y(\xi,1,\zeta)\right) + g_u(x-\xi,z-\zeta)\left(u_y(\xi,0,\zeta) - u_y(\xi,1,\zeta)\right)\Bigg] d\xi\, d\zeta, \tag{72.112}$$

and

$$g_u = \int_{-\infty}^{\infty}\int_{-\infty}^{\infty} 2\pi i \frac{k_x}{Re}\chi(k_x,k_z)e^{2\pi i(k_x\xi+k_z\zeta)}\,dk_z\,dk_x, \tag{72.113}$$

$$g_V = \int_{-\infty}^{\infty}\int_{-\infty}^{\infty} \cosh\left(\alpha(1-\eta)\right)\left(N + 4\pi k_x i U_y^e(\eta)\right)\chi(k_x,k_z)e^{2\pi i(k_x\xi+k_z\zeta)}dk_z\,dk_x, \tag{72.114}$$

$$g_W = \int_{-\infty}^{\infty}\int_{-\infty}^{\infty} 2\pi i \frac{k_z}{Re}\chi(k_x,k_z)e^{2\pi i(k_x\xi+k_z\zeta)}dk_z\,dk_x. \tag{72.115}$$

Considering all wave numbers and using Propositions 72.1 and 72.2, the following result holds regarding the convergence of the closed-loop system.

Theorem 72.1:

Consider the systems (Equations 72.18 through 72.27) with control laws (Equations 72.108 through 72.115). Then the equilibrium profile $u \equiv V \equiv W \equiv 0$ is asymptotically stable in the L^2 norm, that is,

$$\int_{-\infty}^{\infty}\int_0^1\int_{-\infty}^{\infty}\left(u^2+V^2+W^2\right)(t,x,y,z)\,dx\,dy\,dz$$
$$\leq C_2 e^{-2\epsilon t}\int_{-\infty}^{\infty}\int_0^1\int_{-\infty}^{\infty}\left(u^2+V^2+W^2\right)(0,x,y,z)\,dx\,dy\,dz. \tag{72.116}$$

where $C_2 = \max\{C_1, 1\} \geq 0$.

Remark 72.3

In case that $N = 0$, meaning that either there is no imposed magnetic field or the fluid is nonconducting, Equations 72.2 through 72.4 are the Navier–Stokes equations and our controller solves the stabilization problem for a 3D channel flow. Some physical insight can be gained analyzing this case. In the context of hydrodynamic stability theory, the linearized system written in (Y, ω) variables verify equations analogous to the classical Orr-Sommerfeld–Squire equations. These are Equations 72.59 and 72.60 for controlled wave numbers and Equations 72.93 and 72.94 for uncontrolled wave numbers. Note that we use the backstepping transformations (Equations 72.68 and 72.69) not only to stabilize (using gain K) but also to decouple the system (using gains Γ_1, Γ_2) in the small wave number range, where nonnormality effects are more severe. Even if the linearized system is stable, nonnormality produces large transient growths [19], which enhanced by nonlinear effects may allow the velocity field to wander far away from the origin. This warrants the use of extra gains to map the system into two uncoupled heat equations 72.65 and 72.66.

72.4 Observer Design

In this section we design an observer for the MHD channel flow introduced in Section 72.2. Our observer generates estimates of the velocity, pressure, electric potential, and current fields in the whole domain, derived only from wall measurements. Obtaining such an estimate can be of interest in itself, depending on the application. For example, the absence of effective state estimators modeling turbulent fluid flows is considered one of the key obstacles to reliable, model-based weather forecasting. In other engineering applications in which active control is needed, such as drag reduction or mixing enhancement for cooling systems, designs usually assume unrealistic full state knowledge, therefore, a state estimator is necessary for effective implementation.

72.4.1 Observer Structure

For simplicity, we first design an estimator for the linearized system. In Section 72.4.6, using the linear gains, we present a nonlinear observer.

We employ the fluctuation variables around the equilibrium of the Hartmann flow, u and p, which were defined in Equations 72.16 and 72.17; the linearized equations written in fluctuation variables are given in Equations 72.18 through 72.27.

The observer consists of a copy of the linearized equations, to which we add output injection of the pressure p, the potential flux ϕ_y (proportional to current), and both the streamwise and spanwise velocity gradients, u_y and W_y, (proportional to friction) at the bottom wall.

Denoting the observer (estimated) variables by a hat, the equations for the estimated velocity field are

$$\hat{u}_t = \frac{\Delta \hat{u}}{Re} - U^e(y)\hat{u}_x - U^e_y(y)\hat{V} - \hat{p}_x + N\hat{\phi}_z - N\hat{u} - Q^U, \tag{72.117}$$

$$\hat{V}_t = \frac{\Delta \hat{V}}{Re} - U^e(y)\hat{V}_x - \hat{p}_y - Q^V, \tag{72.118}$$

$$\hat{W}_t = \frac{\Delta \hat{W}}{Re} - U^e(y)\hat{W}_x - \hat{p}_z - N\hat{\phi}_x - N\hat{W} - Q^W. \tag{72.119}$$

The additional Q terms in the observer equation are related to output injection and are defined as follows:

$$\begin{pmatrix} Q^U \\ Q^V \\ Q^W \end{pmatrix} = \int_{-\infty}^{\infty}\int_{-\infty}^{\infty} \mathbf{L}(x-\xi, y, z-\zeta) \begin{pmatrix} p(\xi,0,\zeta) - \hat{p}(\xi,0,\zeta) \\ u_y(\xi,0,\zeta) - \hat{u}_y(\xi,0,\zeta) \\ W_y(\xi,0,\zeta) - \hat{W}_y(\xi,0,\zeta) \\ \phi_y(\xi,0,\zeta) - \hat{\phi}_y(\xi,0,\zeta) \end{pmatrix} d\xi\, d\zeta, \tag{72.120}$$

where $\mathbf{L}$ is an output injection kernel matrix, defined as

$$\mathbf{L} = \begin{pmatrix} L^{UP} & L^{UU} & L^{UW} & L^{U\phi} \\ L^{VP} & L^{VU} & L^{VW} & L^{V\phi} \\ L^{WP} & L^{WU} & L^{WW} & L^{W\phi} \end{pmatrix}, \tag{72.121}$$

whose entries will be designed to ensure observer convergence. The estimated potential is computed from

$$\Delta\hat{\phi} = \hat{u}_z - \hat{W}_x, \tag{72.122}$$

and the observer verifies the continuity equation,

$$\hat{u}_x + \hat{V}_y + \hat{W}_z = 0, \tag{72.123}$$

and the same boundary conditions as the plant,

$$\hat{u}(t,x,0,z) = \hat{W}(t,x,0,z) = \hat{V}(t,x,0,z) = \hat{\phi}(t,x,0,z) = 0, \tag{72.124}$$

$$\hat{u}(t,x,1,z) = U_c,\ \hat{W}(t,x,1,z) = W_c, \tag{72.125}$$

$$\hat{V}(t,x,1,z) = V_c,\ \hat{\phi}(t,x,1,z) = \Phi_c. \tag{72.126}$$

Remark 72.4

Note that the observer Equations 72.117 through 72.126 can be regarded as forced MHD equations, with the output injection acting as a body force. This means that any standard DNS solver for the forced MHD equations can be used to implement the observer without the need of major modifications.

As inputs to the observer, appearing in Equation 72.120, one needs measurements of pressure, skin friction, and current on the lower wall. For obtaining these measurements, pressure, skin friction, and current sensors have to be embedded into one of the walls. Pressure and skin friction sensors are common in flow control, while for current measurement one could use an array of discrete current sensors, as depicted in Figure 72.3.

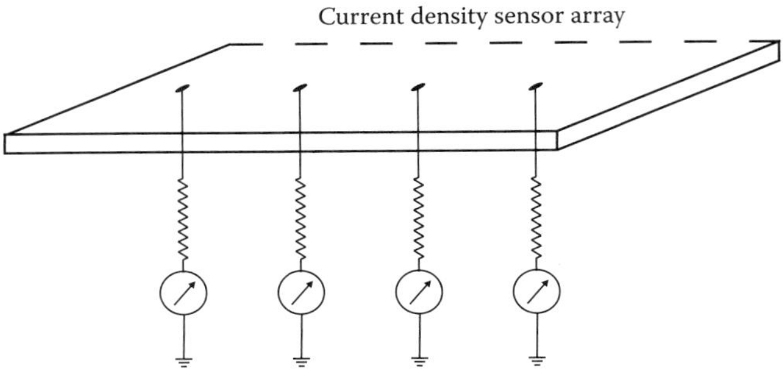

FIGURE 72.3 An array of current sensors in the lower wall.

72.4.2 Observer Gain Design and Convergence Analysis

Substracting the observer equations from the linearized plant equations we obtain the error equations, with states $\tilde{U} = u - \hat{u} = U - \hat{U}$, $\tilde{V} = V - \hat{V}$, $\tilde{W} = W - \hat{W}$, $\tilde{P} = p - \hat{p}$, and $\tilde{\phi} = \phi - \hat{\phi}$,

$$\tilde{U}_t = \frac{\triangle \tilde{U}}{Re} - U^e(y)\tilde{U}_x - U^e_y(y)\tilde{V} - \tilde{P}_x + N\tilde{\phi}_z - N\tilde{U} + Q^U, \tag{72.127}$$

$$\tilde{V}_t = \frac{\triangle \tilde{V}}{Re} - U^e(y)\tilde{V}_x - \tilde{P}_y + Q^V, \tag{72.128}$$

$$\tilde{W}_t = \frac{\triangle \tilde{W}}{Re} - U^e(y)\tilde{W}_x - \tilde{P}_z - N\tilde{\phi}_x - N\tilde{W} + Q^W. \tag{72.129}$$

The observer error verifies the continuity equation,

$$\tilde{U}_x + \tilde{V}_y + \tilde{W}_z = 0, \tag{72.130}$$

while the potential error is governed by

$$\triangle\tilde{\phi} = \tilde{U}_z - \tilde{W}_x. \tag{72.131}$$

The boundary conditions for the error states are

$$\tilde{U}(t,x,0,z) = \tilde{V}(t,x,0,z) = \tilde{W}(t,x,0,z) = 0, \tag{72.132}$$

$$\tilde{U}(t,x,1,z) = \tilde{V}(t,x,1,z) = \tilde{W}(t,x,1,z) = 0, \tag{72.133}$$

$$\tilde{\phi}(t,x,0,z) = \tilde{\phi}(t,x,1,z) = 0. \tag{72.134}$$

To guarantee observer convergence, our design task is to design the output injection gains **L** defined in Equation 72.120 that appear in Q^U, Q^V, and Q^W, so that the origin of the error system is exponentially stable.

Using the Fourier transform in x and z as defined in Equations 72.28 and 72.29, we obtain the observer error equations in Fourier space, which are

$$\begin{aligned}\tilde{U}_t = {} & \frac{-\alpha^2\tilde{U} + \tilde{U}_{yy}}{Re} - \beta\tilde{U} - U^e_y\tilde{V} - 2\pi k_x i\tilde{P} + 2\pi k_z iN\tilde{\phi} - N\tilde{U} \\ & + L^{UP}P_0 + L^{UU}U_{y0} + L^{UW}W_{y0} + L^{U\phi}\phi_{y0},\end{aligned} \tag{72.135}$$

$$\tilde{V}_t = \frac{-\alpha^2\tilde{V} + \tilde{V}_{yy}}{Re} - \beta\tilde{V} - \tilde{P}_y + L^{VP}P_0 + L^{VU}U_{y0} + L^{VW}W_{y0} + L^{V\phi}\phi_{y0}, \tag{72.136}$$

$$\begin{aligned}\tilde{W}_t = {} & \frac{-\alpha^2\tilde{W} + W_{yy}}{Re} - \beta\tilde{W} - 2\pi k_z i\tilde{P} - 2\pi k_x iN\tilde{\phi} - N\tilde{W} \\ & + L^{WP}P_0 + L^{WU}U_{y0} + L^{WW}W_{y0} + L^{W\phi}\phi_{y0},\end{aligned} \tag{72.137}$$

where we have used the definition (Equation 72.120) of the output injection terms as convolutions, which become products in Fourier space. We have written for short $P_0 = \tilde{P}(k_x, 0, k_z)$, $U_{y0} = \tilde{U}_y(k_x, 0, k_z)$, $W_{y0} = \tilde{W}_y(k_x, 0, k_z)$, and $\phi_{y0} = \tilde{\phi}_y(k_x, 0, k_z)$.

The continuity equation in Fourier space is expressed as

$$2\pi i k_x \tilde{U} + \tilde{V}_y + 2\pi k_z \tilde{W} = 0, \tag{72.138}$$

and the equation for the potential is

$$-\alpha^2 \tilde{\phi} + \hat{\phi}_{yy} = 2\pi i \left(k_z \tilde{U} - k_x \tilde{W}\right). \tag{72.139}$$

Note that Equations 72.138 through 72.143 is uncoupled for each wave number. Therefore, as in Section 72.3, we define the range $k_x^2 + k_z^2 \leq M^2$ as the *observed* wave number range, and the range $k_x^2 + k_z^2 > M^2$ as the *unobserved* wave number range, and study them separately. If stability for all wave numbers is established, stability in physical space follows as in Section 72.3.3. A bound for the number M was given by Equation 72.106 to ensure stability for the unobserved wave number range.

We emphasize that we do not use output injection for unobserved wave numbers by writing

$$\mathbf{L}(k_x, y, k_z) = \chi(k_x, y, k_z)\mathbf{R}(k_x, y, k_x), \tag{72.140}$$

where χ was defined in Equation 72.38. Then $\mathbf{L}$ can be written in physical space, using the definition of the Fourier transform and the convolution theorem as

$$\mathbf{L}(x, y, z) = \int_{-\infty}^{\infty}\int_{-\infty}^{\infty} \chi(k_x, y, k_z)\mathbf{R}(k_x, y, k_z)e^{2\pi i(k_x x + k_z z)}\, dk_z\, dk_x. \tag{72.141}$$

The matrix $\mathbf{R}$ is defined as

$$\mathbf{R} = \begin{pmatrix} R^{UP} & R^{UU} & R^{UW} & R^{U\phi} \\ R^{VP} & R^{VU} & R^{VW} & R^{V\phi} \\ R^{WP} & R^{WU} & R^{WW} & R^{W\phi} \end{pmatrix}, \tag{72.142}$$

and using $\mathbf{R}$ we can write the observer error equations as

$$\begin{aligned} \tilde{U}_t = {} & \frac{-\alpha^2 \tilde{U} + \tilde{U}_{yy}}{Re} - \beta \tilde{U} - U_y^e \tilde{V} - 2\pi k_x i \tilde{P} + 2\pi k_z i N \tilde{\phi} - N\tilde{U} \\ & + \chi(k_x, k_z)\left\{R^{UP} P_0 + R^{UU} U_{y0} + R^{UW} W_{y0} + R^{U\phi}\phi_{y0}\right\}, \end{aligned} \tag{72.143}$$

$$\tilde{V}_t = \frac{-\alpha^2 \tilde{V} + \tilde{V}_{yy}}{Re} - \beta \tilde{V} - \tilde{P}_y + \chi(k_x, k_z)\left\{R^{VP} P_0 + R^{VU} U_{y0} + R^{VW} W_{y0} R^{V\phi}\phi_{y0}\right\}, \tag{72.144}$$

$$\begin{aligned} \tilde{W}_t = {} & \frac{-\alpha^2 \tilde{W} + W_{yy}}{Re} - \beta \tilde{W} - 2\pi k_z i \tilde{P} - 2\pi k_x i N \tilde{\phi} - N\tilde{W} \\ & + \chi(k_x, k_z)\left\{R^{WP} P_0 + R^{WU} U_{y0} + R^{WW} W_{y0} + R^{W\phi}\phi_{y0}\right\}. \end{aligned} \tag{72.145}$$

72.4.3 Observed Wave Number Analysis

Consider $k_x^2 + k_z^2 \leq M^2$. Then $\chi = 1$, so output injection is present. Using the continuity Equations 72.138 and 72.143 through 72.145, the following Poisson equation for the pressure is derived:

$$-\alpha^2 \tilde{P} + \tilde{P}_{yy} = \Upsilon - 4\pi k_x i U_y^e(y)\tilde{V} + NV_y, \tag{72.146}$$

where Υ contains all the terms due to output injection,

$$\begin{aligned} \Upsilon = {} & P_0\left(2\pi i k_x R^{UP} + R_y^{VP} + 2\pi k_z R^{WP}\right) + U_{y0}\left(2\pi i k_x R^{UU} + R_y^{VU} + 2\pi k_z R^{WU}\right) \\ & + W_{y0}\left(2\pi i k_x R^{UW} + R_y^{VW} + 2\pi k_z R^{WW}\right) + \phi_{y0}\left(2\pi i k_x R^{U\phi} + R_y^{V\phi} + 2\pi k_z R^{W\phi}\right). \end{aligned} \tag{72.147}$$

We want to make Equation 72.146 independent of the output injection gains, for which we need $\Upsilon = 0$. Hence, we set

$$R^{VP}(k_x, y, k_z) = R^{VP}(k_x, 0, k_z) - 2\pi i \int_0^y \left(k_x R^{UP} + k_z R^{WP}\right)(k_x, \eta, k_z)\, d\eta, \tag{72.148}$$

$$R^{VU}(k_x, y, k_z) = R^{VU}(k_x, 0, k_z) - 2\pi i \int_0^y \left(k_x R^{UU} + k_z R^{WU}\right)(k_x, \eta, k_z)\, d\eta, \tag{72.149}$$

$$R^{VW}(k_x, y, k_z) = R^{VW}(k_x, 0, k_z) - 2\pi i \int_0^y \left(k_x R^{UW} + k_z R^{WW}\right)(k_x, \eta, k_z)\, d\eta, \tag{72.150}$$

$$R^{V\phi}(k_x, y, k_z) = R^{V\phi}(k_x, 0, k_z) - 2\pi i \int_0^y \left(k_x R^{U\phi} + k_z R^{W\phi}\right)(k_x, \eta, k_z)\, d\eta, \tag{72.151}$$

which means that, in physical space, $\nabla \cdot \mathbf{L} = 0$. Hence, as Equations 72.146 is derived by taking divergence of Equations 72.143 through 72.145, the output injection terms cancel away.

Expression 72.146 can be solved in terms of the values of the pressure at the bottom wall.

$$\begin{aligned}\tilde{P} = & -\frac{4\pi k_x i}{\alpha} \int_0^y U_y^e(\eta) \sinh\left(\alpha(y-\eta)\right) \tilde{V}(k_x, \eta, k_z)\, d\eta + \cosh\left(\alpha y\right) P_0 \\ & + \frac{\sinh\left(\alpha y\right)}{\alpha} \tilde{P}_y(k_x, 0, k_z) + N \int_0^y \frac{\sinh\left(\alpha(y-\eta)\right)}{\alpha} \tilde{V}_y(k_x, \eta, k_z)\, d\eta.\end{aligned} \tag{72.152}$$

Evaluating Equation 72.144 at $y = 0$, one finds that

$$\tilde{P}_y(k_x, 0, k_z) = \Upsilon_0 - 2\pi i \frac{k_x U_{y0}(k_x, 0, k_z) + k_z \tilde{W}_{y0}}{Re}, \tag{72.153}$$

where we have used Equation 72.138 for writing $\tilde{V}_{yy}$ at the bottom in terms of measurements. In Equation 72.153,

$$\Upsilon_0 = P_0 R^{VP}(k_x, 0, k_z) + U_{y0} R^{VU}(k_x, 0, k_z) + W_{y0} R^{VW}(k_x, 0, k_z) + \phi_{y0} R^{V\phi}(k_x, 0, k_z), \tag{72.154}$$

and as before we force the pressure to be independent of any gains. Hence, we set

$$R^{VP}(k_x, 0, k_z) = R^{VU}(k_x, 0, k_z) = R^{VW}(k_x, 0, k_z) = R^{V\phi}(k_x, 0, k_z) = 0. \tag{72.155}$$

Then, the pressure can be expressed independently of the output injection gains in terms of a strict-feedback integral of the state $\tilde{V}$ and measurements,

$$\begin{aligned}\tilde{P} = & -\frac{4\pi k_x i}{\alpha} \int_0^y U_y^e(\eta) \sinh\left(\alpha(y-\eta)\right) \tilde{V}(k_x, \eta, k_z)\, d\eta + \cosh\left(\alpha y\right) P_0 \\ & - 2\pi i \frac{\sinh\left(\alpha y\right)}{Re\alpha} \left(k_x U_{y0} + k_z W_{y0}\right) + N \int_0^y \frac{\sinh\left(\alpha(y-\eta)\right)}{\alpha} \tilde{V}_y(k_x, \eta, k_z)\, d\eta.\end{aligned} \tag{72.156}$$

Similarly, solving for ϕ in terms of the measurement ϕ_{y0} and the right-hand side of Equation 72.139,

$$\tilde{\phi} = \frac{2\pi i}{\alpha} \int_0^y \sinh\left(\alpha(y-\eta)\right) \left(k_z \tilde{U}(k_x, \eta, k_z) - k_x \tilde{W}(k_x, \eta, k_z)\right) d\eta + \frac{\sinh\left(\alpha y\right)}{\alpha} \phi_{y0}. \tag{72.157}$$

Introducing the expressions (Equations 72.156 and 72.157) in Equations 72.143 and 72.145, we obtain

$$\begin{aligned}\tilde{U}_t =& \frac{-\alpha^2\tilde{U}+\tilde{U}_{yy}}{Re} - \beta\tilde{U} - U_y^e(y)\tilde{V} - N\tilde{U} + U_{y0}\left(R^{UU} - \frac{4\pi^2k_x^2}{\alpha Re}\sinh\left(\alpha y\right)\right)\\ &+ P_0\left(R^{UP} - 2\pi k_x i\cosh\left(\alpha y\right)\right) + W_{y0}\left(R^{UW} - \frac{4\pi^2k_xk_z}{\alpha Re}\sinh\left(\alpha y\right)\right)\\ &+ \phi_{y0}\left(R^{U\phi} + N\frac{2\pi k_z i}{\alpha}\sinh\left(\alpha y\right)\right) - \frac{8\pi k_x^2}{\alpha}\int_0^y U_y^e(\eta)\sinh\left(\alpha(y-\eta)\right)\tilde{V}(k_x,\eta,k_z)\,d\eta\\ &- \frac{4\pi^2k_zN}{\alpha}\int_0^y \sinh\left(\alpha(y-\eta)\right)\left(k_z\tilde{U}(k_x,\eta,k_z) - k_x\tilde{W}(k_x,\eta,k_z)\right)d\eta\\ &- 2\pi i k_x N\int_0^y \frac{\sinh\left(\alpha(y-\eta)\right)}{\alpha}\tilde{V}_y(k_x,\eta,k_z)\,d\eta, \end{aligned} \tag{72.158}$$

$$\begin{aligned}\tilde{W}_t =& \frac{-\alpha^2\tilde{W}+W_{yy}}{Re} - \beta\tilde{W} - N\tilde{W} + U_{y0}\left(R^{WU} - \frac{4\pi^2k_xk_z}{\alpha Re}\sinh\left(\alpha y\right)\right)\\ &+ P_0\left(R^{WP} - 2\pi k_z i\cosh\left(\alpha y\right)\right) + W_{y0}\left(R^{WW} - \frac{4\pi^2k_z^2}{\alpha Re}\sinh\left(\alpha y\right)\right)\\ &+ \phi_{y0}\left(R^{W\phi} - N\frac{2\pi k_x i}{\alpha}\sinh\left(\alpha y\right)\right) - \frac{8\pi k_xk_z}{\alpha}\int_0^y U_y^e(\eta)\sinh\left(\alpha(y-\eta)\right)\tilde{V}(k_x,\eta,k_z)\,d\eta\\ &+ \frac{4\pi^2k_xN}{\alpha}\int_0^y \sinh\left(\alpha(y-\eta)\right)\left(k_z\tilde{U}(k_x,\eta,k_z) - k_x\tilde{W}(k_x,\eta,k_z)\right)d\eta\\ &- 2\pi i k_z N\int_0^y \frac{\sinh\left(\alpha(y-\eta)\right)}{\alpha}\tilde{V}_y(k_x,\eta,k_z)\,d\eta. \end{aligned} \tag{72.159}$$

Note that we have omitted the equation for $\tilde{V}$ since, from Equation 72.138 and using the fact that $\tilde{V}(k_x,0,k_z)=0$, $\tilde{V}$ is computed from $\tilde{U}$ and $\tilde{W}$:

$$\tilde{V} = -2\pi i\int_0^y \left(k_x\tilde{U}(k_x,\eta,k_z) + k_z\tilde{W}(k_x,\eta,k_z)\right)d\eta. \tag{72.160}$$

We now set the output injection terms to directly cancel the boundary terms coming from Equations 72.156 and 72.157, while still leaving some additional gains for stabilization. Thus, we define

$$R^{UP} = 2\pi k_x i\cosh\left(\alpha y\right),\ R^{WP} = 2\pi k_z i\cosh\left(\alpha y\right), \tag{72.161}$$

$$R^{UU} = \frac{4\pi^2k_x^2}{\alpha Re}\sinh\left(\alpha y\right) + \Pi_1(k_x,y,k_z),\ R^{WU} = \frac{4\pi^2k_xk_z}{\alpha Re}\sinh\left(\alpha y\right) + \Pi_2(k_x,y,k_z), \tag{72.162}$$

$$R^{UW} = \frac{4\pi^2k_xk_z}{\alpha Re}\sinh\left(\alpha y\right) + \Pi_3(k_x,y,k_z),\ R^{WW} = \frac{4\pi^2k_z^2}{\alpha Re}\sinh\left(\alpha y\right) + \Pi_4(k_x,y,k_z), \tag{72.163}$$

$$R^{U\phi} = -N\frac{2\pi k_z i}{\alpha}\sinh\left(\alpha y\right),\ R^{W\phi} = N\frac{2\pi k_x i}{\alpha}\sinh\left(\alpha y\right), \tag{72.164}$$

where the gains Π_1, Π_2, Π_3, and Π_4 are to be defined later. From Equations 72.148 through 72.151, 72.155 and 72.161 through 72.164, we obtain an explicit expression for the remaining entries of **R**,

$$R^{VP} = \alpha\sinh\left(\alpha y\right),\ R^{V\phi} = 0, \tag{72.165}$$

$$R^{VU} = 2\pi i(k_x+k_z)\frac{1-\cosh\left(\alpha y\right)}{Re} - 2\pi i\int_0^y\left(k_x\Pi_1(k_x,\eta,k_z) + k_z\Pi_2(k_x,\eta,k_z)\right)d\eta, \tag{72.166}$$

$$R^{VW} = 2\pi i(k_x+k_z)\frac{1-\cosh\left(\alpha y\right)}{Re} - 2\pi i\int_0^y\left(k_x\Pi_3(k_x,\eta,k_z) + k_z\Pi_4(k_x,\eta,k_z)\right)d\eta. \tag{72.167}$$

Introducing Equations 72.161 through 72.167 in Equations 72.158 and 72.159, we obtain

$$\begin{aligned}\tilde{U}_t =& \frac{-\alpha^2\tilde{U}+\tilde{U}_{yy}}{Re} - \beta\tilde{U} - U^e_y(y)\tilde{V} - N\tilde{U} + \Pi_1 U_{y0} + \Pi_3 W_{y0}\\ &- \frac{4\pi^2 k_z N}{\alpha}\int_0^y \sinh\left(\alpha(y-\eta)\right)\left(k_z\tilde{U}(k_x,\eta,k_z) - k_x\tilde{W}(k_x,\eta,k_z)\right)d\eta\\ &- \frac{8\pi k_x^2}{\alpha}\int_0^y U^e_y(\eta)\sinh\left(\alpha(y-\eta)\right)\tilde{V}(k_x,\eta,k_z)\,d\eta\\ &- 2\pi i k_x N\int_0^y \frac{\sinh\left(\alpha(y-\eta)\right)}{\alpha}\tilde{V}_y(k_x,\eta,k_z)\,d\eta,\end{aligned} \tag{72.168}$$

$$\begin{aligned}\tilde{W}_t =& \frac{-\alpha^2\tilde{W}+W_{yy}}{Re} - \beta\tilde{W} - N\tilde{W} + \Pi_2 U_{y0} + \Pi_4 W_{y0}\\ &- \frac{8\pi k_x k_z}{\alpha}\int_0^y U^e_y(\eta)\sinh\left(\alpha(y-\eta)\right)\tilde{V}(k_x,\eta,k_z)\,d\eta\\ &+ \frac{4\pi^2 k_x N}{\alpha}\int_0^y \sinh\left(\alpha(y-\eta)\right)\left(k_z\tilde{U}(k_x,\eta,k_z) - k_x\tilde{W}(k_x,\eta,k_z)\right)d\eta\\ &- 2\pi i k_z N\int_0^y \frac{\sinh\left(\alpha(y-\eta)\right)}{\alpha}\tilde{V}_y(k_x,\eta,k_z)\,d\eta.\end{aligned} \tag{72.169}$$

Now, we introduce the following change of variables and its inverses:

$$Y = 2\pi i\left(k_x\tilde{U} + k_z\tilde{W}\right), \quad \omega = 2\pi i\left(k_z\tilde{U} - k_x\tilde{W}\right), \tag{72.170}$$

$$U = \frac{2\pi i}{\alpha^2}\left(k_x Y + k_z\omega\right), \quad W = \frac{2\pi i}{\alpha^2}\left(k_z Y - k_x\omega\right). \tag{72.171}$$

Then using the definitions in Equations 72.55 through 72.58, Equations 72.168 and 72.169 expressed in terms of Y and ω are

$$\begin{aligned}Y_t =& \epsilon\left(-\alpha^2 Y + Y_{yy}\right) - \beta Y - NY - \frac{4\pi^2}{\alpha^2}\left(k_x^2\Pi_1 + k_x k_z\Pi_2 + k_x k_z\Pi_3 + k_z^2\Pi_4\right)Y_{y0}\\ &- \frac{4\pi^2}{\alpha^2}\left(k_x k_z\Pi_1 + k_z^2\Pi_2 - k_x^2\Pi_3 - k_x k_z\Pi_4\right)\omega_{y0} + \int_0^y f(k_x,y,\eta,k_z)Y(k_x,\eta,k_z)\,d\eta,\end{aligned} \tag{72.172}$$

$$\begin{aligned}\omega_t =& \epsilon\left(-\alpha^2\omega + \omega_{yy}\right) - \beta\omega - N\omega - \frac{4\pi^2}{\alpha^2}\left(k_x k_z\Pi_1 - k_x^2\Pi_2 + k_z^2\Pi_3 - k_x k_z\Pi_4\right)Y_{y0}\\ &- \frac{4\pi^2}{\alpha^2}\left(k_z^2\Pi_1 - k_x k_z\Pi_2 - k_x k_z\Pi_3 + k_x^2\Pi_4\right)\omega_{y0} + h_1(y)\int_0^y Y(k_x,\eta,k_z)\,d\eta\\ &+ \int_0^y h_2(y,\eta)\omega(k_x,\eta,k_z)\,d\eta,\end{aligned} \tag{72.173}$$

where we have used the inverse change of variables (Equation 72.171) to express U_{y0} and W_{y0} in terms of $Y_{y0} = Y(k_x, 0, k_z)$ and $\omega_{y0} = \omega(k_x, 0, k_z)$. We define now the output injection gains Π_1, Π_2, Π_3, and Π_4 in the following way:

$$\begin{pmatrix}\Pi_1\\ \Pi_2\\ \Pi_3\\ \Pi_4\end{pmatrix} = \mathbf{A}^{-1}\begin{pmatrix} l(k_x,y,0,k_z)\\ 0\\ \theta_1(k_x,y,0,k_z)\\ \theta_2(k_x,y,0,k_z)\end{pmatrix}, \tag{72.174}$$

where the matrix $\mathbf{A}$ was defined in Equation 72.81. Note that since $\det(A) = -1$ its inverse appearing in Equation 72.174 is well defined. The functions $l(k_x,y,\eta,k_z)$, $\theta_1(k_x,y,\eta,k_z)$, and $\theta_2(k_x,y,\eta,k_z)$

in Equation 72.174 are to be found. Using Equation 72.174, Equations 72.172 and 72.173 become

$$Y_t = \epsilon\left(-\alpha^2 Y + Y_{yy}\right) - \beta Y - NY + l(k_x, y, 0, k_z)Y_{y0} + \int_0^y f(k_x, y, \eta, k_z)Y(k_x, \eta, k_z)\,d\eta, \tag{72.175}$$

$$\begin{aligned}\omega_t &= \epsilon\left(-\alpha^2\omega + \omega_{yy}\right) - \beta\omega - N\omega + \theta_1(k_x, y, 0, k_z)Y_{y0} + \theta_2(k_x, y, 0, k_z)\omega_{y0}\\ &\quad + h_1\int_0^y Y(k_x, \eta, k_z)\,d\eta + \int_0^y h_2(y, \eta)\omega(k_x, \eta, k_z)\,d\eta.\end{aligned} \tag{72.176}$$

Equations 72.175 and 72.176 are a coupled, strict-feedback plant, with integral and reaction terms. A variant of the design presented in [22] for anticollocated systems can be used to design the gains $l(k_x, y, 0, k_z)$, $\theta_1(k_x, y, 0, k_z)$, and $\theta_2(k_x, y, 0, k_z)$ using a pair of backstepping transformations. The transformation maps, for each k_x and k_z, the variables (Y, ω) into the variables (Ψ, Ω), that verify the following pair of heat equations (parameterized in k_x, k_z):

$$\Psi_t = \epsilon\left(-\alpha^2\Psi + \Psi_{yy}\right) - \beta\Psi - N\Psi, \tag{72.177}$$

$$\Omega_t = \epsilon\left(-\alpha^2\Omega + \Omega_{yy}\right) - \beta\Omega - N\Omega, \tag{72.178}$$

with boundary conditions

$$\Psi(k_x, 0, k_z) = \Psi(k_x, 1, k_z) = \Omega(k_x, 0, k_z) = \Omega(k_x, 1, k_z) = 0. \tag{72.179}$$

The transformation is defined as follows:

$$Y = \Psi - \int_0^y l(k_x, y, \eta, k_z)\Psi(k_x, \eta, k_z)\,d\eta, \tag{72.180}$$

$$\omega = \Omega - \int_0^y \theta_1(k_x, y, \eta, k_z)\Psi(k_x, \eta, k_z)\,d\eta - \int_0^y \theta_2(k_x, y, \eta, k_z)\Omega(k_x, \eta, k_z)\,d\eta. \tag{72.181}$$

We find the kernel functions $l(k_x, y, \eta, k_z)$, $\theta_1(k_x, y, \eta, k_z)$, and $\theta_2(k_x, y, \eta, k_z)$ by solving the following partial integrodifferential equations:

$$\epsilon l_{\eta\eta} = \epsilon l_{yy} - \left(\beta(y) - \beta(\eta)\right) l - f + \int_\eta^y f(y, \xi)l(\xi, \eta)\,d\xi, \tag{72.182}$$

$$\epsilon\theta_{1\eta\eta} = \epsilon\theta_{1yy} - \left(\beta(y) - \beta(\eta)\right)\theta_1(y, \eta) - h_1 + h_1\int_\eta^y l(\xi, \eta)\,d\xi + \int_\eta^y h_2(y, \xi)\theta_1(\xi, \eta)\,d\xi, \tag{72.183}$$

$$\epsilon\theta_{2\eta\eta} = \epsilon\theta_{2yy} - \left(\beta(y) - \beta(\eta)\right)\theta_2 - h_2 + \int_\eta^y h_2(y, \xi)\theta_2(\xi, \eta)\,d\xi. \tag{72.184}$$

Equations 72.182 through 72.184 are hyperbolic partial integro-differential equation in the region $\mathcal{T} = \{(y, \eta) : 0 \le y \le 1, 0 \le \eta \le y\}$. Their boundary conditions are

$$l(k_x, y, y, k_z) = \theta_1(k_x, y, y, k_z) = \theta_2(k_x, y, y, k_z) = 0, \tag{72.185}$$

$$l(k_x, 1, \eta, k_z) = \theta_1(k_x, 1, \eta, k_z) = \theta_2(k_x, 1, \eta, k_z) = 0. \tag{72.186}$$

Remark 72.5

Equations 72.182 through 72.186 are well-posed and can be solved symbolically, by means of a successive approximation series, or numerically [21]. Note that both Equation 72.182 and Equation 72.184 are autonomous. Hence, one must solve first for $l(k_x, y, \eta, k_z)$ and $\theta_2(k_x, y, \eta, k_z)$. Then the solution for l is plugged in Equation 72.183 which then can be solved for $\theta_1(k_x, y, \eta, k_z)$. The observer gains are then found just by setting $\eta = 0$ in the kernels $l(k_x, y, \eta, k_z)$, $\theta_2(k_x, y, \eta, k_z)$, and $\theta_1(k_x, y, \eta, k_z)$.

Stability in the observed wave number range follows from stability of Equations 72.177 and 72.178 and the invertibility of the transformations (Equations 72.180 and 72.181), as in Proposition 72.1.

72.4.4 Unobserved Wave Number Analysis

When $k_x^2 + k_z^2 > M$, there is no output injection, as $\chi = 0$, and the linearized observer error verifies the following equations:

$$\tilde{U}_t = \frac{-\alpha^2 \tilde{U} + \tilde{U}_{yy}}{Re} - \beta \tilde{U} - U_y^e(y)\tilde{V} - 2\pi k_x i \tilde{P} + 2\pi k_z i N \tilde{\phi} - N\tilde{U}, \tag{72.187}$$

$$\tilde{V}_t = \frac{-\alpha^2 \tilde{V} + \tilde{V}_{yy}}{Re} - \beta \tilde{V} - \tilde{P}_y, \tag{72.188}$$

$$\tilde{W}_t = \frac{-\alpha^2 \tilde{W} + W_{yy}}{Re} - \beta \tilde{W} - 2\pi k_z i \tilde{P} - 2\pi k_x i N \tilde{\phi} - N\tilde{W}, \tag{72.189}$$

the Poisson equation for the potential Equation 72.139 and the continuity Equation 72.138.

Note that Equations 72.187 through 72.189 are the same as Equations 72.85 through 72.87. Hence, the analysis of Section 72.3.2 can be applied, obtaining a result similar to Proposition 72.2. Hence, stability in the unobserved wave number range follows when $k_x^2 + k_z^2 \geq M^2$ for M as in Equation 72.106.

72.4.5 Observer Convergence Properties

Considering all wave numbers, the following holds regarding the convergence of the observer.

Theorem 72.2:

Consider the systems (Equations 72.18 through 72.27), and the systems (Equations 72.117 through 72.126), and suppose that both have classical solutions. Then, the L^2 norms of $\tilde{U}$, $\tilde{V}$, and $\tilde{W}$ converge to zero, that is,

$$\lim_{t\to\infty} \int_{-\infty}^{\infty} \int_0^1 \int_{-\infty}^{\infty} \left(\tilde{U}^2 + \tilde{V}^2 + \tilde{W}^2\right)(t, x, y, z)\, dx\, dy\, dz = 0. \tag{72.190}$$

Remark 72.6

The convergence result stated in Theorem 72.2 guarantees asymptotic convergence of the estimated states to the actual values of the *linearized* plant. For this to be true for the *nonlinear* plant we need additional conditions. Namely, the estimates have to be initialized close enough to the real initial values and the MHD system has to stay in a neighborhood of the equilibrium at all times.

Remark 72.7

In case that $N = 0$, meaning that either there is no imposed magnetic field or the fluid is nonconducting, Equations 72.18 through 72.20 are the linearized Navier–Stokes equations and the observer reduces to a velocity/pressure estimator for a 3D channel flow. This is a result that can be seen as dual to the 3D channel flow control problem. See Remark 72.2 for some physical insight for this case.

Remark 72.8

We obtain an output feedback law that stabilizes the plants (Equations 72.18 through 72.27) using only wall measurements. Such a control law uses the estimates ($\hat{u}$, $\hat{V}$, $\hat{W}$) from the observers (Equations 72.117 through 72.126) to replace the real states (u, V, W) in the control laws (Equations 72.108 through 72.115). Then, using Theorems 72.1 and 72.2 and standard arguments for linear output feedback controllers, a similar result holds guaranteeing the L^2 stability of the closed-loop output feedback system.

72.4.6 A Nonlinear Estimator with Boundary Sensing

We present a nonlinear observer to obtain convergence in open-loop estimation problems (the linear estimator achieves convergence in feedback problems, where both the state and the estimate are driven to zero, however, the linear observer, which neglects the nonlinear terms in the model, cannot capture open-loop dynamics, and hence cannot achieve convergence to the solutions of an uncontrolled system). The nonlinear observer has the same structure and gains as the linear observer, but the nonlinear terms are added. In our design, we follow an approach similar to an Extended Kalman Filter, in which gains are deduced for the linearized plant and then used for a nonlinear observer.

The nonlinear observer equations are the following:

$$\hat{U}_t = \frac{\triangle\hat{U}}{Re} - \hat{U}\hat{U}_x - \hat{V}\hat{U}_y - \hat{W}\hat{U}_z - \hat{P}_x + N\hat{\phi}_z - N\hat{U} - Q^U, \quad (72.191)$$

$$\hat{V}_t = \frac{\triangle\hat{V}}{Re} - \hat{U}\hat{V}_x - \hat{V}\hat{V}_y - \hat{W}\hat{V}_z - \hat{P}_y - Q^V, \quad (72.192)$$

$$\hat{W}_t = \frac{\triangle\hat{W}}{Re} - \hat{U}\hat{W}_x - \hat{V}\hat{W}_y - \hat{W}\hat{W}_z - \hat{P}_z - N\hat{\phi}_x - N\hat{W} - Q^W. \quad (72.193)$$

The estimated potential is computed from

$$\triangle\hat{\phi} = \hat{U}_z - \hat{W}_x, \quad (72.194)$$

and the observer verifies the continuity equation,

$$\hat{U}_x + \hat{V}_y + \hat{W}_z = 0, \quad (72.195)$$

and the same boundary conditions as in Equations 72.124 through 72.126.

In Equations 72.191 through 72.193, the Q terms are the same as for the linear observer. Hence, the observer is designed for the linearized plant and then the linear gains are used for the nonlinear observer. Such a nonlinear observer will produce closer estimates of the states in a larger range of initial conditions.

Using the fluctuation variable and the observer error variables, we can write the nonlinear observer velocity field error equations as follows:

$$\tilde{U}_t = \frac{\triangle\tilde{U}}{Re} - U^e(y)\tilde{U}_x + \mathcal{N}^U(\tilde{U},\tilde{V},\tilde{W},u,V,W) - U^e_y(y)\tilde{V} - \tilde{P}_x + N\tilde{\phi}_z - N\tilde{U} + Q^U, \quad (72.196)$$

$$\tilde{V}_t = \frac{\triangle\tilde{V}}{Re} - U^e(y)\tilde{V}_x + \mathcal{N}^V(\tilde{U},\tilde{V},\tilde{W},u,V,W) - \tilde{P}_y + Q^V, \quad (72.197)$$

$$\tilde{W}_t = \frac{\triangle\tilde{W}}{Re} - U^e(y)\tilde{W}_x + \mathcal{N}^W(\tilde{U},\tilde{V},\tilde{W},u,V,W) - \tilde{P}_z - N\tilde{\phi}_x - N\tilde{W} + Q^W, \quad (72.198)$$

where we have introduced

$$\mathcal{N}^U = \tilde{U}\tilde{U}_x - u\tilde{U}_x - \tilde{U}u_x + \tilde{V}\tilde{U}_y - V\tilde{U}_y - \tilde{V}u_y + \tilde{W}\tilde{U}_z - W\tilde{U}_z - \tilde{W}u_z, \quad (72.199)$$

$$\mathcal{N}^V = \tilde{U}\tilde{V}_x - u\tilde{V}_x - \tilde{U}V_x + \tilde{V}\tilde{V}_y - V\tilde{V}_y - \tilde{V}V_y + \tilde{W}\tilde{V}_z - W\tilde{V}_z - \tilde{W}V_z, \quad (72.200)$$

$$\mathcal{N}^W = \tilde{U}\tilde{W}_x - u\tilde{W}_x - \tilde{U}W_x + \tilde{V}\tilde{W}_y - V\tilde{W}_y - \tilde{V}W_y + \tilde{W}\tilde{W}_z - W\tilde{W}_z - \tilde{W}W_z, \quad (72.201)$$

Assuming, for the purposes of observer design and analysis, that the observer state $(\hat{U},\hat{V},\hat{W})$ is close to the actual state (U,V,W) (i.e., the error state is close to zero), and that the fluctuation (u,V,W) around the equilibrium state is small, then $\mathcal{N}_U(\tilde{U},\tilde{V},\tilde{W},u,V,W)$, $\mathcal{N}_V(\tilde{U},\tilde{V},\tilde{W},u,V,W)$, and $\mathcal{N}_W(\tilde{U},\tilde{V},\tilde{W},u,V,W)$ are small and dominated by the linear terms in the equations, so they can be neglected. The linearized error equations are then

$$\tilde{U}_t = \frac{\triangle\tilde{U}}{Re} - U^e(y)\tilde{U}_x - U^e_y(y)\tilde{V} - \tilde{P}_x + N\tilde{\phi}_z - N\tilde{U} + Q^U, \quad (72.202)$$

$$\tilde{V}_t = \frac{\triangle \tilde{V}}{Re} - U^e(y)\tilde{V}_x - \tilde{P}_y + Q^V, \tag{72.203}$$

$$\tilde{W}_t = \frac{\triangle \tilde{W}}{Re} - U^e(y)\tilde{W}_x - \tilde{P}_z - N\tilde{\phi}_x - N\tilde{W} + Q^W, \tag{72.204}$$

which are the same as Equations 72.127 through 72.129. Thus, as expected, the error equations for the observer designed for the linearized plant, and the linearized error equations for the nonlinear observer are the same; this is the main reason why the same gains derived in Section 72.4.2 are used.

Remark 72.9

Following [10], we may consider the *mean turbulent profile* instead of considering the exact *laminar* equilibrium profile. This amounts to changing U^e in definition (Equation 72.12). Since U^e appears in the coefficients of the output injection gain PDEs (Equations 72.182 through 72.184), the observer gains will change (quantitatively). However, Theorem 72.1 still holds and guarantees convergence of estimates, but for these estimates to be good enough it is required that the state has to stay close enough to the mean turbulent profile at all times.

72.5 For Further Information

We survey some representative results in flow control, a recent but rapidly growing field.

By far the most studied problem in flow control is channel flow stabilization for large Reynolds numbers. There are many complex issues underlying this problem [13,14,19], making it extremely challenging. Optimal control has so far been the most successful technique for addressing channel flow stabilization [13], in a (streamwise- and spanwise-) periodic setting, by using a discretized version of the equations and employing high-dimensional algebraic Riccati equations for computation of gains. Using a Lyapunov/passivity approach, another control design [1,3] was developed for stabilization of the (periodic) channel flow. Other works make use of nonlinear model reduction techniques to solve the problem, though they employ in-domain actuation [5]. Boundary controllers using spectral decomposition and pole-placement methods have been developed [25]. Other techniques include separation control [2] and turbulence suppression by using transverse wall oscillations [15].

Observer design has been so far a largely neglected area in flow control. For channel flows it is known that pressure and skin friction at one of the walls completely determine the flow inside the domain. For these reason, they have been called the "footprints" of turbulence [8]. Based on these measurements, designs have been done in the form of Extended Kalman Filter for the spatially discretized Navier–Stokes equations [10,11].

The area of conducting fluids moving in magnetic fields, even though rich in applications, has only been recently considered. Recent results in stabilization of MHD flows make use of nonlinear model reduction [4], open-loop control [7], and optimal control [12]. Applications include, for instance, drag reduction [18], or mixing enhancement for cooling systems [20]. Some experimental results are available as well, showing that control of such flows is technologically feasible [9,18,24]. Mathematical studies of controllability have been done, although they do not provide explicit controllers [6,23].

For a more detailed presentation of flow control, the reader is referred to the monograph *Control of Turbulent and Magnetohydrodynamic Channel Flows* by Vazquez and Krstic [26]. The book introduces new constructive design methods (based on the backstepping technique) for boundary stabilization and boundary estimation for several benchmark problems in flow control, with potential applications to turbulence control, weather forecasting, and plasma control. It contains the details of methods presented here, and includes an introduction on spatial invariance, Fourier series and transforms, the stability of infinite-dimensional systems, and the backstepping method for PDEs. It also covers other topics not mentioned here, such as well-posedness, H^1 and H^2 stability, or Poiseuille flow transfer.

References

1. O. M. Aamo and M. Krstic, *Flow Control by Feedback: Stabilization and Mixing*. Berlin: Springer, 2002.
2. M.-R. Alam, W.-J. Liu and G. Haller, Closed-loop separation control: An analytic approach, *Phys. Fluids*, 18:043601, 2006.
3. A. Balogh, W.-J. Liu and M. Krstic, Stability enhancement by boundary control in 2D channel flow, *IEEE Trans. Automat. Control*, 46:1696–1711, 2001.
4. J. Baker, A. Armaou, and P. D. Christofides, Drag reduction in incompressible channel flow using electromagnetic forcing, *Proc. 2000 ACC*, 4269–4273, 2000.
5. J. Baker, A. Armaou, and P. D. Christofides, Nonlinear control of incompressible fluid flow: Application to Burgers' equation and 2D channel flow, *J. Math. Anal. Appl.*, 252:230–255, 2000.
6. V. Barbu, C. Popa, T. Havarneanu, and S. S. Sritharan, Exact controllability of magneto-hydrodynamic equations, *Commun Pure App. Math.*, 56(6):732–783, 2003.
7. T. W. Berger, J. Kim, C. Lee, and J. Lim, Turbulent boundary layer control utilizing the Lorentz force, *Phys. Fluids*, 12:631, 2000.
8. T. R. Bewley and B. Protas, Skin friction and pressure: The 'footprints' of turbulence, *Physica D*, 196:28–44, 2004.
9. K. S. Breuer and J. Park, Actuation and control of a turbulent channel flow using Lorentz forces, *Phys. Fluids*, 16(4):897, 2004.
10. M. Chevalier, J. Hoepffner, T. R. Bewley, and D. S. Henningson, State estimation in wall-bounded flow systems. Part 2. Turbulent flows. *J. Fluid Mech.*, 552:167–187, 2006.
11. J. Hoepffner, M. Chevalier, T. R. Bewley, and D. S. Henningson, State estimation in wall-bounded flow systems. Part 1. Perturbed laminar flows. *J. Fluid Mech.*, 534:263–294, 2005.
12. K. Debbagh, P. Cathalifaud, and C. Airiau, Optimal and robust control of small disturbances in a channel flow with a normal magnetic field, *Phys. Fluids*, 19(1):014103–014103-14, 2007.
13. M. Hogberg, T. R. Bewley, and D. S. Henningson, Linear feedback control and estimation of transition in plane channel flow, *J. Fluid Mech.*, 481:149–175, 2003.
14. M. R. Jovanovic and B. Bamieh, Componentwise energy amplification in channel flows, *J. Fluid Mech.*, 543:145–183, 2005.
15. M. R. Jovanovic, Turbulence suppression in channel flows by small amplitude transverse wall oscillations, *Phys. Fluids*, 20(1):014101, 2008.
16. M. Krstic, I. Kanellakopoulos, and P. V. Kokotovic, *Nonlinear and Adaptive Control Design*, New York, NY: Wiley, 1995.
17. D. Lee and H. Choi, Magnetohydrodynamic turbulent flow in a channel at low magnetic Reynolds number, *J. Fluid Mech.*, 439:367–394, 2001.
18. J. Pang and K.-S. Choi, Turbulent drag reduction by Lorentz force oscillation, *Phys. Fluids*, 16(5):L35, 2004.
19. P. J. Schmid and D. S. Henningson, *Stability and Transition in Shear Flows*, Springer, Berlin, 2001.
20. E. Schuster, L. Luo, and M. Krstic, MHD channel flow control in 2D: Mixing enhancement by boundary feedback, *Automatica*. 44:2498–2507, 2008.
21. A. Smyshlyaev and M. Krstic, Closed form boundary state feedbacks for a class of partial integro-differential equations, *IEEE Trans. Automat. Control*, 49:2185–2202, 2004.
22. A. Smyshlyaev and M. Krstic, Backstepping observers for parabolic PDEs, *Systems Control Lett.*, 54:1953–1971, 2005.
23. S. S. Sritharan, V. Barbu, T. Havarneanu and C. Popa, Local controllability for the magnetohydrodynamic equations revisited, *Adv. Diff. Eq*, 10(5):481–504, 2005.
24. J.-P. Thibault and L. Rossi, Electromagnetic flow control: Characteristic numbers and flow regimes of a wall-normal actuator, *J. Phys. D: Appl. Phys.*, 36:2559–2568, 2003.
25. R. Triggiani, Stability enhancement of a 2-D linear Navier–Stokes channel flow by a 2-D, wall-normal boundary controller, *Disc. Cont. Dyn. Sys.—B*, 8(2), 2007.
26. R. Vazquez and M. Krstic, *Control of Turbulent and Magnetohydrodynamic Channel Flows*, Boston: Birkhauser, 2007.
27. R. Vazquez and M. Krstic, A closed-form feedback controller for stabilization of the linearized 2D Navier–Stokes Poiseuille flow, *IEEE Trans. Automat. Control*, 52:2298–2312, 2007.

XII

Networks and Networked Controls

73

Control over Digital Networks

Nuno C. Martins
University of Maryland

73.1 Introduction

The advent of digital communication networks has created a new subfield of control engineering broadly identified as networked control. It addresses the analysis and design of control systems whose components are not colocated, and for which the dissemination of information requires a communication network. As evidenced in [2], the vast current and future applications of networked control systems, and the lack of systematic analysis and design methods, warrants a significant research effort.

Two major challenges in networked control research are the development of analysis and design tools that account for the deterioration of performance that digital communication networks cause when they are used to connect two or more modules of the feedback loop. This performance degradation is due to the typical detrimental features of digital networks such as the finite bit-rate limits, which result from the use of data packets carrying a finite number of bits, and erasures or packet losses due to fading, interference or congestion. Currently, most research in networked control can be categorized in two types. The first focuses on the effects of erasures by assuming that the number of bits carried in each data packet is large enough to support the hypothesis that any resulting quantization noise is negligible in light of other sources of noise. The second type addresses not only erasures, but it also acknowledges that the effects of bit-rate limits and the associated quantization distortion need to be analyzed. In this chapter, we provide a brief discussion on the latter, which we complement by citing only a few representative references, on a need basis. An informative survey comprising a much richer collection of references, which are up to date until 2007, can be found in [3]. An extensive and comprehensive discussion can also be found in [31].

73.1.1 Chapter Organization

This chapter is centered on the stabilizability of unstable finite dimensional linear time invariant (LTI) systems via data-rate-limited feedback. The general formulation is given in Section 73.2, while Sections 73.3 and 73.4 focus on internal and external stabilization, respectively.

73.2 Basic Framework

A prototypical formulation for a networked control system is one where the sensors and the controller are not collocated. Here, we assume that these blocks are interconnected by a digital communication channel, as shown in Figure 73.1. In our framework the plant, the channel and the remaining blocks operate synchronously in discrete time.

Throughout this chapter, we assume that the digital channels function as random digital links, which transmit $\mathbf{R}_k$ bits at each time instant k. Here, $\mathbf{R}_k$ is a random process taking values in the set $\{0, 1, \ldots, \bar{R}\}$, and whose statistics reflect the relevant detrimental effects of the underlying network, such as fading and congestion [5]. The signal $\mathbf{S}_k$ is an ordered $\bar{R}$-tuple taking values in $\{0, 1\}^{\bar{R}}$ and it represents the data that is placed for transmission through the channel. The channel operates by transmitting only the first $\mathbf{R}_k$ bits of the word placed for transmission. Hence, the output of the channel, denoted as $\mathbf{V}_k$, is a truncated version of $\mathbf{S}_k$ where the last $\bar{R} - \mathbf{R}_k$ bits get dropped. This model was adopted in [7] to study the effects of channel randomness in the robust stability of networked control systems. The authors of [9] have also proposed this type of abstraction in an information theoretic setting. We should, however, note that this class of channel models represents a particular case, in light of other more general analog and discrete models that represent certain channels more accurately. There is also a significant body of work for Gaussian channels, from some of the pioneering work of [12,21] to more recent work addressing the difficult problem of networked control over Gaussian channels with memory [15]. A comprehensive study for general channels can be found in [8].

The two blocks $\mathcal{S}$ and $\mathcal{K}$ in Figure 73.1 represent the sensing and controller blocks, respectively. The controller produces a control signal $\mathbf{U}_k$ based on causal processing of the output of the channel, while the sensing block accepts causal measurements of the output of the plant and uses them to decide when and what should be placed for transmission through the channel. The measurement and process exogenous

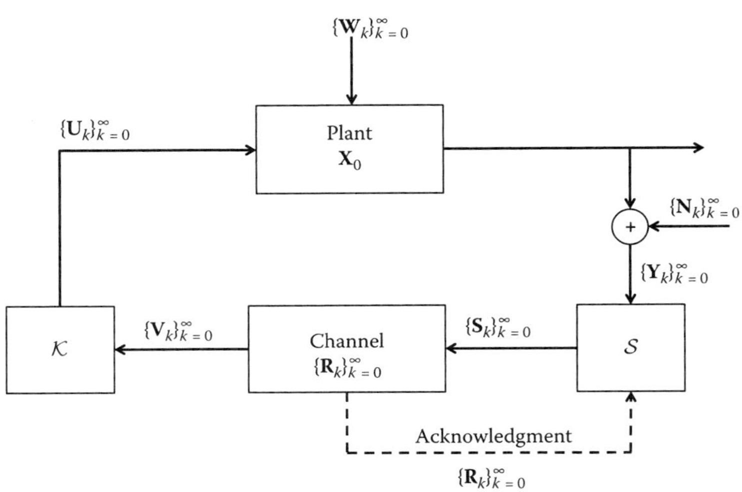

FIGURE 73.1 Depiction of a networked control system.

signals are represented by $\mathbf{N}_k$ and $\mathbf{W}_k$, respectively, while the initial condition of the plant is denoted by $\mathbf{X}_0$. Throughout this paper, we assume that $\{\mathbf{R}_k\}_{k=0}^{\infty}$ and $\mathbf{X}_0$ are independent.

73.3 Internal Stabilization

We consider that the plant in Figure 73.1 is finite-dimensional LTI, and that it evolves in discrete time. The state-space representation of the plant is given as follows:

$$\mathbf{X}_{k+1} = A\mathbf{X}_k + B\mathbf{U}_k, \quad k \geq 0 \tag{73.1}$$

$$\mathbf{Y}_k = C\mathbf{X}_k + D\mathbf{U}_k, \quad k \geq 0 \tag{73.2}$$

where the initial condition $\mathbf{X}_0$ is a random variable. In general, $\mathbf{X}_k$, $\mathbf{Y}_k$, and $\mathbf{U}_k$ are vector valued and A, B, and C are the matrices of appropriate dimension.

We refer to the feedback loop in Figure 73.1 as internally stabilizing if the state of the plant (Equation 73.1) is stable in the presence of the initial condition $\mathbf{X}_0$. Stability in the presence of external inputs is investigated in Section 73.4.

73.3.1 Deterministic Channel Model

The simplest case to analyze is when the channel is deterministic, implying that $\mathbf{R}_k = \bar{R}$ holds for all k, where $\bar{R}$ is a predefined positive constant. From the pioneering work in [10,11] it follows that, for a scalar LTI plant, internal stabilization via the scheme of Figure 73.1 implies that the following condition must hold:

$$\bar{R} \geq \max\{0, \log_2 |a|\} \tag{73.3}$$

where a defines the state-space representation of the scalar plant via:

$$\mathbf{X}_{k+1} = a\mathbf{X}_k + B\mathbf{U}_k, \quad k \geq 0 \tag{73.4}$$

We proceed by giving a simple explanation of Equation 73.3, where, for now, the feedback loop is qualified as internally stabilizing if the following holds with probability one:

$$|\mathbf{X}_k| \leq \nu < \infty, \quad k \geq 0 \tag{73.5}$$

for some positive real β.

Now consider the following estimate of $\mathbf{X}_0$:

$$\hat{\mathbf{Z}}_k = -\sum_{j=0}^{k-1} a^{-j-1} B\mathbf{U}_j, \quad k \geq 1 \tag{73.6}$$

which can be used, via the convolution formula, to express $\mathbf{X}_k$ as:

$$\mathbf{X}_k = a^k \left(\mathbf{X}_0 - \hat{\mathbf{Z}}_k\right), \quad k \geq 1 \tag{73.7}$$

If follows from Equations 73.7 and 73.5 that internal stabilization implies:

$$|\mathbf{X}_0 - \hat{\mathbf{Z}}_k| \leq \nu |a|^{-k}, \quad k \geq 1 \tag{73.8}$$

Hence, we conclude from Equation 73.8 that $\hat{\mathbf{Z}}_k$ constitutes an estimate of $\mathbf{X}_0$ whose error decreases exponentially with time. If $|a| > 1$ holds then the fact that Equation 73.8 holds, regardless of the probability distribution of $\mathbf{X}_0$, implies that there exists information about $\mathbf{X}_0$ flowing to $\hat{\mathbf{Z}}_k$ at every time step. In particular, with $a = 2$ the error in Equation 73.8 is reduced by half at every time step, and as a result, there is at least one bit of information about $\mathbf{X}_0$ being transmitted to $\hat{\mathbf{Z}}_k$ at every time step. Since $\hat{\mathbf{Z}}_k$ is constructed solely from $\mathbf{U}_k$ and all information about $\mathbf{X}_0$ that it contains must be conveyed through the channel, we conclude that for $a = 2$ it must be true that $\bar{R} \geq 1$. Since for $a = 2$ it holds that $\bar{R} \geq 1$, it is natural that for arbitrary a the necessary condition becomes Equation 73.3. This argument is explained rigorously in [10,11], and from it one can conclude that Equation 73.3 is a necessary condition for any arbitrary memoryless channel provided that $\bar{R}$ is replaced with Shannon's channel capacity, which is a measure of the highest information rate attainable by a given channel. As it is shown in [10,11], the inequality in Equation 73.3 is also a sufficient condition for the existence of stabilizing $\mathcal{S}$ and $\mathcal{K}$. Hence, from Equation 73.3, we can clearly conclude that larger $|a|$, or the more unstable is the system, the larger is the required rate $\bar{R}$ for stabilizability. However, as it was pointed out in [8], if the channel is not deterministic then finer notions of stochastic stability are needed, for which the inequality in Equation 73.3, with $\bar{R}$ replaced with capacity, might no longer be a sufficient condition for the existence of stabilizing $\mathcal{S}$ and $\mathcal{K}$. We will explore an instance of this remark in Section 73.3.2.

Using an appropriate modal decomposition technique and counting arguments applied to plants of arbitrary dimension, the authors of [17] proved that the following is a necessary condition for the existence of stabilizing $\mathcal{S}$ and $\mathcal{K}$:

$$\bar{R} \geq \sum_{i=1}^{n} \max\{0, \log|\lambda_i(A)|\} \tag{73.9}$$

where n is the dimension of the plant and $\{\lambda_i(A)\}_{i=1}^{n}$ are the eigenvalues of A. In [17] it is shown that the condition (Equation 73.9) is also sufficient if the plant is stabilizable and detectable. Although proving this result for the general case is nontrivial, one can gain some insight by noticing that the necessity and sufficiency of Equation 73.3, for the stabilization of scalar systems, can be used to derive the corresponding condition (Equation 73.9) for multistate plants, for the particular case where A is diagonal, or diagonalizable with eigenvalues that are powers of 2.

73.3.2 Random Channel Model

In order to access stability in the presence of a random channel, that is, when $\mathbf{R}_k$ is no longer deterministic, we adopt two notions: almost sure stability and moment stability.

A feedback system is almost surely stable, or a.s.s. for short, if the following holds with probability one:

$$|\mathbf{X}_k| \leq \nu_{ass} < \infty \tag{73.10}$$

Given a positive integer m, an alternative notion is one where a system is mth moment stable if the following holds:

$$E[|\mathbf{X}_k|^m] \leq \nu_m < \infty, \quad k \geq 0 \tag{73.11}$$

where $E[\cdot]$ denotes statistical expectation with respect to the randomness introduced by the channel and the initial condition $\mathbf{X}_0$.

The authors of [18] have shown that the following is a necessary condition for stabilization in both the a.s.s. and the mth moment sense:

$$C \geq \sum_{i=1}^{n} \max\{0, \log|\lambda_i(A)|\} \tag{73.12}$$

Here C is defined as follows:

$$C \stackrel{def}{=} \underbrace{\liminf}_{\text{in probability}} \frac{\sum_{j=0}^{k} \mathbf{R}_j}{k} \tag{73.13}$$

where lim inf above is assumed to hold in probability. It follows from Equation 73.13 that, for the channel model adopted here, C coincides with Shannon's capacity as defined in the information theory literature.

The authors of [19] show that for stabilizable and detectable plants, the condition in Equation 73.12 is sufficient for stabilizability, provided that $\mathbf{R}_k$ is a time-varying deterministic sequence. Here, the assumption that $\mathbf{R}_k$ is deterministic guarantees that $\mathcal{S}$ knows what is effectively transmitted to $\mathcal{K}$ through the channel. This predictability underpins the development of algorithms executed at $\mathcal{S}$ that can track and reproduce any action taken at $\mathcal{K}$, a feature that dramatically simplifies the development of stabilizing schemes. This desirable attribute, which we qualify as classical information pattern [1], may not hold for random channels. If the channel is random then the information pattern is classical if the channel sends an acknowledgement signal to $\mathcal{S}$ containing $\mathbf{R}_k$. Here, the acknowledgement signal effectively indicates how many bits are successfully transmitted through the channel from $\mathcal{S}$ to $\mathcal{K}$. This type of acknowledgement signal may be conveyed via a dedicated link as in Figure 73.1, or it can be signaled by $\mathcal{K}$ through the plant in certain cases. For channels that are supported on a computer network, one can associate classical information patterns with TCP protocols, while UDP protocols are nonclassical.

73.3.2.1 Classical Information Pattern

If the channel is memoryless, that is, $\mathbf{R}_k$ is white, and the plant is stabilizable and detectable and in the presence of acknowledgement signals then the necessary condition in Equation 73.12 may also be sufficient for the existence of stabilizing $\mathcal{S}$ and $\mathcal{K}$, in the a.s.s. sense. Sufficiency in the absence of process noise is proved in [18]. Other results for the multistate case can be found in [4] and references therein, where the authors provide explicit stabilizing schemes that rely on the predictability guaranteed by the classical information pattern hypothesis. It is interesting to notice that the condition (Equation 73.12) may not be sufficient in the presence of process and measurement noises.

As it is remarked in [8], even if the information pattern of the feedback loop is classical, the necessary condition in Equation 73.12 is no longer sufficient to guarantee internal stabilization in the mth moment sense. In fact, for scalar plants and memoryless stationary channels, and under a classical information pattern, the necessary and sufficient condition for mth moment stabilizability can be cast as follows [7]:

$$\mathcal{C} \geq \max\{0, \log_2 |a|\} + \beta(m) \tag{73.14}$$

where $\beta(m)$ is a positive increasing function of m. The necessary and sufficient condition for stabilization for the aforementioned formulation can also be expressed as follows:

$$a^m E\left[2^{-m\mathbf{R}_k}\right] \leq 1, \quad k \geq 0. \tag{73.15}$$

We show next that the condition in Equation 73.15 becomes particularly simple when $\mathbf{R}_k$ is governed by a Bernoulli process that selects either 0 or $\bar{R}$ signifying an erasure or a successful transmission, respectively. More specifically, if p_e denotes the probability of erasure then Equation 73.15 becomes $a^m\left(p_e + 2^{-m\bar{R}}(1-p_e)\right) \leq 1$. As pointed out in [14], it is interesting to note that as $\bar{R}$ tends to infinity, the aforementioned condition becomes $a^m p_e \leq 1$, which coincides with the necessary and sufficient condition for stabilizability derived in [20] for the packet-drop model. Indeed, one should expect this consistency in the limit when $\bar{R}$ tends to infinity in light of the formulation in [20], which adopts a channel described by a Bernoulli-driven device that either causes an erasure (packet drop) or enables the transmission of a real number. Here, the transmission of a real number can be viewed as the transmission of a data packet with an infinite number of bits. It follows from the aforementioned argument for scalar plants, and by using modal decomposition, that a necessary and sufficient condition for the stabilization of stabilizable and detectable plants in the mth moment sense with $\bar{R}$ tending to infinity is given by the following inequality:

$$\varrho(A)^m p_e \leq 1 \tag{73.16}$$

where $\varrho(A)$ denotes the spectral radius of A. Here, once again, Equation 73.16 coincides with the analogous condition derived under the packet drop model [20].

There are other interesting limiting cases of Equation 73.15 studied in [7], such as when m tends to zero for which we obtain Equation 73.3 or when m tends to infinity for which we arrive at $R_{min} \geq \max\{0, \log_2 |a|\}$, where R_{min} is the minimum value attained by $\mathbf{R}_k$ with probability one. Similar limiting cases were investigated for the multistate case in [14].

Existing work on necessary and sufficient conditions for the existence of a stabilizing scheme for a plant with multiple states, for the case when $\mathbf{R}_k$ is random [14], are not as elegant and simple as their counterparts (Equation 73.12) for deterministic channels. A reason for this difficulty is that, as pointed out in [7], for multistate plants the blocks $\mathcal{S}$ and $\mathcal{K}$ have to implement allocation algorithms that dictate what information is sent at any particular time instant through the channel. Recent results in [14] provide necessary and sufficient conditions for which there is a gap within which the existence of a stabilizing scheme cannot be determined. This gap is well characterized in [14], including examples where it is zero. The work in [14] also provides explicit algorithms and a thorough statistical analysis.

Remark

Some of the necessary and sufficient conditions presented above can be used to determine the stabilizability of the plant in the presence of measurement and process noises, provided that they are modified to require that the corresponding inequalities hold strictly.

73.3.2.2 Nonclassical Information Pattern

In general, if the channel is random and acknowledgement signals from $\mathcal{K}$ to $\mathcal{S}$ are not available then the analysis and design of stabilizing schemes becomes rather involved. This is the case because $\mathcal{S}$ does not have perfect information about what is successfully transmitted to $\mathcal{K}$. The analysis for this case, accompanied by the description of appropriate codes for the stabilization of scalar plants, can be found in Section V-A of [8]. Tight necessary and sufficient conditions for multistate plants are, in general, not yet known. Clear progress has been made on obtaining necessary and sufficient conditions [16,33] for the multistate case provided that the channel is Gaussian additive.

73.3.3 Decentralized Networked Control

Decentralized networked control refers to the case in which the plant comprises several dynamically coupled subsystems with a subcontroller at each. Here one needs to clearly specify the information pattern that dictates how information is disseminated among subcontrollers. An information pattern is decentralized if each subcontroller can act directly only on its own subsystem, but has access to only partial and possibly degraded information about the state of the other subsystems. The framework described in Section 73.2 involves one controller, one plant and one point-to-point channel. Hence, a natural extension to the decentralized case is obtained when certain pairs of subcontrollers are "connected" via point-to-point channels to exchange measurement as well as other data that is relevant for distributed control. The stabilizability via decentralized control schemes must be determined with respect to the properties of the plant, the channel, and the information pattern. This problem is unsolved for the general case, but significant progress has been made for deterministic channels [26,27,29,30].

73.3.4 Notes on Optimality

Consider the typical optimal control formulation specified by the quintessential cost comprising of the expectation of quadratic forms of the state and the control. Under a classical framework, where acknowledgments are present, the separation principle holds and it is possible to generate an optimal control signal at $\mathcal{K}$ based on the optimal state estimate that it constructs from the information transmitted through the channel. This is possible because $\mathcal{S}$ can use the acknowledgment signals to determine the information received by $\mathcal{K}$ and use it to indirectly compute the control signal. Hence, there is no incentive to use control for signaling and in theory it is possible to determine the optimal $\mathcal{K}$ and $\mathcal{S}$. However, unfortunately, the optimal estimator at $\mathcal{K}$ is, in general, nonlinear and it may be infinite-dimensional and time varying. Here, suboptimal solutions might be obtained by propagating an approximate and truncated conditional distribution and constructing the corresponding estimator. For the nonclassical case the problem can be formulated but, in general, even approximate solutions are elusive. Interesting structural results and optimal policies for particular cases have been proposed in [1]. A new characterization of nonclassical information structures and its implications for the design of decentralized decision systems is discussed in [28].

73.4 Input to State Stability

Consider the following modification of Equation 73.1 so as to the exogenous signals $\mathbf{N}_k$ and $\mathbf{W}_k$:

$$\mathbf{X}_{k+1} = A\mathbf{X}_k + B\mathbf{U}_k + \mathbf{W}_k, \quad k \geq 0, \quad X_0 = 0 \tag{73.17}$$

$$\mathbf{Y}_k = C\mathbf{X}_k + D\mathbf{U}_k + \mathbf{N}_k, \quad k \geq 0 \tag{73.18}$$

The so-called finite gain input to state stability is a mainstay notion in many paradigms of modern control, such as $\mathcal{H}_\infty$ controller design. A feedback loop stabilizes (Equation 73.17) in the finite gain sense if and only if there exists a positive real gain η such that the following inequality holds for every possible realization of $\{\mathbf{N}_k\}_{k=0}^{\infty}$, $\{\mathbf{W}_k\}_{k=0}^{\infty}$ and $\{\mathbf{X}_k\}_{k=1}^{\infty}$:

$$\|X\|_s \leq \eta \|(N, W)\|_e \tag{73.19}$$

where $V \overset{def}{=} \{V_k\}_{k=0}^{\infty}$, $W \overset{def}{=} \{W_k\}_{k=0}^{\infty}$ and $X \overset{def}{=} \{X_k\}_{k=1}^{\infty}$ represent realizations of $\{\mathbf{N}_k\}_{k=0}^{\infty}$, $\{\mathbf{W}_k\}_{k=0}^{\infty}$ and $\{\mathbf{X}_k\}_{k=1}^{\infty}$, respectively. Here $\|\cdot\|_s$ and $\|\cdot\|_e$ are preselected norms, which are appropriately chosen to quantify the desired notion of stability.

Although most results discussed in Section 73.3.2.1 hold in the presence of process and measurement noise with a variety of statistical descriptions, they do not address the input–output stability as defined in Equation 73.19. In fact, it has been shown in [32] that finite gain stability is not feasible for unstable plants that are controlled over a feedback loop comprising one or more bit-rate constrained blocks. Hence, in particular, finite gain stability is impossible under the scheme of Figure 73.1. In fact, as it is shown in [32], the ratio between any norm of the state and any norm of the external noises becomes unbounded in the limits when the exogenous signals are arbitrarily small or arbitrarily large. As pointed out in [32], these facts substantially limit the robustness of the resulting networked control systems and they also indicate that other notions of stability are required for the performance analysis of these systems.

The following inequality is a relaxation of Equation 73.19:

$$\|X\|_s \leq \mathcal{B}\left(\|(N, W)\|_e\right) \tag{73.20}$$

where $\mathcal{B} : \mathbb{R}_{\geq 0} \rightarrow \mathbb{R}_{\geq 0}$. If the feedback loop is such that $\mathcal{B}$ can be selected to be continuous, strictly increasing, unbounded and $\mathcal{B}(0) = 0$ holds then we qualify it as input to state stable (ISS), a concept discovered and coined in the foundational work in [23]. Using this concept, the authors of [22] show that if the plant is stabilizable and detectable and $\mathbf{R}_k$ is a constant satisfying (Equation 73.12) then there exists $\mathcal{S}$ and $\mathcal{K}$ for which the feedback loop is ISS. An important characteristic of the solution presented in [22] is that the algorithms executed at $\mathcal{S}$ and $\mathcal{K}$ do not have prior knowledge of an upper bound on $\|(N, W)\|_e$. This feature is, in fact, essential to guarantee that the feedback loop is ISS because ISS requires that (Equation 73.20) holds for every possible realization of X, N, and W. A key idea in [22] is that of designing $\mathcal{S}$ as a time-varying quantizer with adaptable resolution, whereby large signal levels are met with coarser quantization cells, which can then be made finer when the amplitude of the state lessens. This formulation was initially proposed in [24] and its extensions for the ISS case, including robustness analysis, are now fully developed in [25]. These results are remarkable in light of the effects that state quantization may bring about on the behavior of feedback systems [6].

References

1. A. Mahajan and D. Teneketzis, Optimal performance of networked control systems with non-classical information structures, *SIAM Journal of Control and Optimization*, 48(3), 1377–1404, 2009.
2. R. M. Murray (Ed.), Control in an information rich world: Report of the panel on future directions in control, dynamics, and systems, SIAM Report, 2003.
3. G. N. Nair, F. Fagnani, S. Zampieri, and R. J. Evans, Feedback control under data rate constraints: An overview, *Proceedings of the IEEE*, 95(1), 108–137, 2007.
4. A. S. Matveev and A. V. Savkin, Shannon zero error capacity in the problems of state estimation and stabilization via noisy communication channels, *International Journal of Control*, 80(2), 241–255, 2007.
5. L. L. Peterson and B. S. Davie, *Computer Networks: A Systems Approach*, 4th ed., San Mateo, CA: Morgan Kaufmann, 2007.
6. D. F. Delchamps, Stabilizing a linear system with quantized state feedback, *IEEE Transactions on Automatic Control*, 35(9), 916–924, 1990.
7. N. C. Martins, M. A. Dahleh, and N. Elia, Feedback stabilization of uncertain systems in the presence of a direct link, *IEEE Transactions on Automatic Control*, 51(3), 438–447, 2006.
8. A. Sahai and S. Mitter, The necessity and sufficiency of anytime capacity for stabilization of a linear system over a noisy communication link; Part I: Scalar systems, *IEEE Transactions on Information Theory*, 52(8), 3369–3395, 2006.
9. D. Tse and R. Yates, Fading broadcast channels with state information at the receivers, arXiv:0904.3165v1 [cs.IT], *IEEE Transactions on Information Theory*, 2009, submitted for publication.
10. W. S. Wong and R. W. Brockett, Systems with finite communication bandwidth constraints II: Stabilization with limited information feedback, *IEEE Transactions on Automatic Control*, 44(5), 1049–1053, 1999.
11. J. Baillieul, Feedback designs in information-based control, in *Stochastic Theory and Control Proceedings of a Workshop*, B. Pasik-Duncan, Ed. Lawrence, KS, New York: Springer-Verlag, pp. 35–57, 2001.
12. V. S. Borkar and S. K. Mitter, LQG control with communication constraints in *Communications, Computation, Control and Signal Processing*, A. Paulraj, V. Roychowdhury, C.D. Schaper, Eds. Dordrecht: Kluwer, 365–373, 1997.
13. G. N. Nair, R. J. Evans, I. M. Y. Mareels, and W. Moran, Topological feedback entropy and nonlinear stabilization, *IEEE Transactions on Automatic Control*, 49(9), 1585–1597, 2004.
14. P. Minero, M. Franceschetti, S. Dey, and G. N. Nair, Data rate theorem for stabilization over time-varying feedback channels, *IEEE Transactions on Automatic Control*, 54(2), 243–255, 2009.
15. R. H. Middleton, A. J. Rojas, J. S. Freudenberg, and J. H. Braslavsky, Feedback stabilization over a first order moving average Gaussian noise channel, *IEEE Transactions on Automatic Control*, 54(1), 163–167, 2009.
16. J. H. Braslavsky, R. H. Middleton, and J. S. Freudenberg, Feedback stabilization over signal-to-noise ratio constrained channels, *IEEE Transactions on Automatic Control*, 52(8), 1391–1403, 2007.
17. S. Tatikonda and S. Mitter, Control under information constraints, *IEEE Transactions on Automatic Control*, 49(7), 1056–1068, 2004.
18. S. Tatikonda and S. Mitter, Control over noisy channels, *IEEE Transactions on Automatic Control*, 49(7), 1196–1201, 2004.
19. G. N. Nair and R. J. Evans, Stabilizability of stochastic linear systems with finite feedback data rates, SIAM Journal on Control and Optimization, *Society for Industrial and Applied Mathematics, USA*, 43(2), 413–436, 2004.
20. V. Gupta, B. Hassibi, and R. M. Murray, Optimal LQG control across packet-dropping links, *System and Control Letters*, 56(6), 439–446, 2007.
21. J. H. Braslavsky, R. H. Middleton, and J. S. Freudenberg, Feedback stabilization over signal-to-noise ratio constrained channels automatic control, *IEEE Transactions on Automatic Control*, 52(8), 1391–1403, 2007.
22. D. Liberzon and D. Nešić, Input to state stabilization of linear systems with quantized state measurements, *IEEE Transactions on Automatic Control*, 52(5), 2007.
23. E. D. Sontag, *Mathematical Control Theory*, 2nd ed., Berlin: Springer, 1998.
24. R. Brockett and D. Liberzon, Quantized feedback stabilization of linear systems, *IEEE Transactions on Automatic Control*, 45(7), 1279–1289, 2000.
25. D. Nešić and D. Liberzon, A unified framework for design and analysis of networked and quantized control systems (with D. Nesic), *IEEE Transactions on Automatic Control*, 54(4), 732–747, 2009.
26. G. N. Nair and R. J. Evans, Cooperative networked stabilizability of linear systems with measurement noise, *In Proc. 15th IEEE Mediterranean Conf. Control and Automation*, Athens, Greece, 2007.

27. M. Rotkowitz and G. Nair, An LP for stabilization over networks with rate constraints, *Proceedings of the International Symposium on Mathematical Theory of Networks and Systems*, Blacksburg, Virginia, 2008.
28. S. Yuksel, Stochastic nestedness and the belief sharing information pattern, *IEEE Trans. Automat. Control*, 55, 2773–2786, 2009.
29. S. Yuksel and T. Basar, Optimal signaling policies for decentralized multi-controller stabilizability over communication channels, *IEEE Trans. Automat. Control*, 52, 1969–1974, 2007.
30. S. Yuksel and T. Basar, Communication constraints for decentralized stabilizability with time-invariant policies, *IEEE Trans. Automat. Control*, 52, 1060–1066, 2007.
31. A. S. Matveev and A. V. Savkin, *Estimation and Control over Communication Networks (Control Engineering)*, Boston: Birkhauser, 2008.
32. N. C. Martins, Finite gain LP stability requires analog control, systems and control letters, 55, 949–954, 2006.
33. S. Tatikonda, A. Sahai, and S. Mitter, Stochastic linear control over a communication channel, *IEEE Trans. Automat. Control*, 49(9), 1549–1561, 2004.

74

Decentralized Control and Algebraic Approaches

Michael C. Rotkowitz
The University of Melbourne

74.1 Introduction

This chapter addresses the problem of decentralized control, where multiple controllers have an access to different information but need to achieve or optimize a global objective. Most of conventional control analysis breaks down when information is decentralized, even in the simplest possible scenarios (Witsenhausen, 1968).

This chapter addresses decentralized control problems in a simple unified framework. This framework is introduced in Section 74.2, where we see that a standard controls framework may be utilized, but with the addition of a particular constraint on the controller that needs to be designed. Section 74.3 then briefly reviews the parametrization of stabilizing controllers for centralized control, when one does not have this decentralization constraint.

Section 74.4 introduces quadratic invariance—an algebraic condition under which decentralized control problems may be cast as convex optimization problems. Section 74.5 looks at particular classes of problems to see when this condition holds, and to get some intuition behind when decentralized control problems may be tractable. Section 74.6 then discusses the perfectly decentralized control problem, a problem which is often of interest yet which typically does not satisfy this condition. Finally, while the rest of this chapter focuses on the case where both the system to be controlled and the possible controllers are all linear, Section 74.7 discusses some related results for nonlinear control.

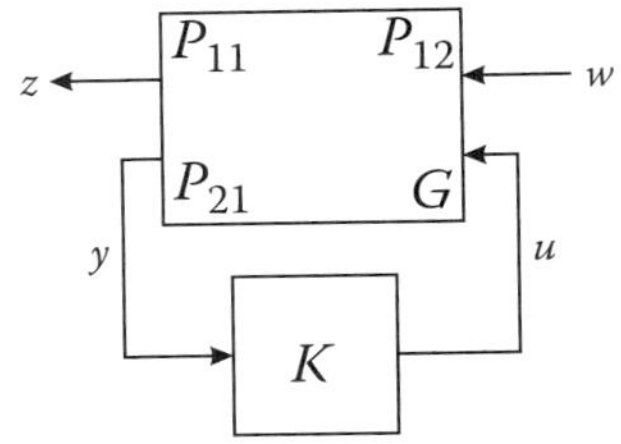

FIGURE 74.1 Standard feedback control framework.

74.2 Framework and Setup

We introduce a unified framework for studying optimal feedback control problems subject to decentralized information constraints.

74.2.1 Standard Framework

We first review a standard framework for centralized control synthesis.

Figure 74.1 represents the standard design framework of modern control theory, and is used in many other chapters. The signal w represents the vector of exogenous inputs, those the designer has no control over, such as wind gusts if one is considering an example in aerospace, and z represents everything the designer would like to keep small, which would typically include deviations from a desired state or trajectory, or a measure of control effort, for example. The signal y represents the vector of measurements that the controller K has access to, and u is the vector of inputs from the controller that feeds back into the plant. The plant is subdivided into four blocks that maps w and u into z and y. The block which maps the controller input u to the measurements y is simply referred to as G, since it corresponds to the plant of classical control analysis, and so that we can later refer to its subdivisions without any ambiguity.

The design objective is to construct a controller K to keep a measure of the size of the mapping from w to z, known as the *closed-loop map*, as small as possible. There are many ways in which one can measure the size of a mapping, and thus this basic setup underpins much of modern controls including $\mathcal{H}_2$-control and $\mathcal{H}_\infty$-control. In this framework, a decentralized information structure may be viewed as a constraint on the structure of the controller K, as now illustrated by examples.

74.2.2 Information Constraint

We now illustrate why, in this framework, decentralization may be simply encapsulated as a constraint that the controller lies in a particular subspace.

The diagram in Figure 74.2 represents three different subsystems, each of which may effect its neighbors, and each of which has its own controller, which only has access to measurements coming from its own subsystem. In this case, if we look at the system as a whole, we need to design a controller K that can be

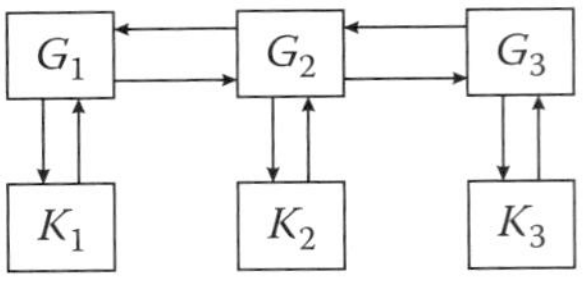

FIGURE 74.2 Perfectly decentralized control.

written as

$$\begin{bmatrix} u_1 \\ u_2 \\ u_3 \end{bmatrix} = \underbrace{\begin{bmatrix} K_1 & 0 & 0 \\ 0 & K_2 & 0 \\ 0 & 0 & K_3 \end{bmatrix}}_{K} \begin{bmatrix} y_1 \\ y_2 \\ y_3 \end{bmatrix}$$

since each controller input may only depend upon the measurement from its corresponding subsystem. In other words, we need to design the best possible K which is block diagonal. The overall problem can be viewed as minimizing the size of the closed-loop map subject to the additional constraint that $K \in S$, where S is the set of all block diagonal controllers. This concept readily extends to any type of structural constraint we may need to impose in formulating an optimal control problem for controller synthesis. For instance, if in the above example, each controller was able to share information with its neighbors, then we would end up with a constraint set S which is tri-diagonal. In general, the ijth component of the controller is held to 0 if the ith controller is unable to see the jth measurement y_j.

If controllers were instead allowed to communicate with each other, but with some delays, this too could be reflected in another constraint set S. This situation is represented in Figure 74.3, where the controller for a given subsystem i can see the information from another subsystem j after a transmission delay of t_{ij}. In this case, if we look at the system as a whole, we need to design a controller K that can be written as

$$\begin{bmatrix} u_1 \\ u_2 \\ u_3 \end{bmatrix} = \underbrace{\begin{bmatrix} D_{t_{11}}\tilde{K}_{11} & D_{t_{12}}\tilde{K}_{12} & D_{t_{13}}\tilde{K}_{13} \\ D_{t_{21}}\tilde{K}_{21} & D_{t_{22}}\tilde{K}_{22} & D_{t_{23}}\tilde{K}_{23} \\ D_{t_{31}}\tilde{K}_{31} & D_{t_{32}}\tilde{K}_{32} & D_{t_{33}}\tilde{K}_{33} \end{bmatrix}}_{K} \begin{bmatrix} y_1 \\ y_2 \\ y_3 \end{bmatrix}$$

where $D_{t_{ij}}$ represents a delay of t_{ij}, and $\tilde{K}_{ij}$ represents the part of the controller which we are free to design, since each subsystem controller must wait the proscribed amount of time before it can use information from each of the other controllers.

The set S above is called the *information constraint,* as it captures the information available to various parts of the controller. The overarching point is that the objective of decentralized control may be considered to be the minimization of a closed-loop map subject to an information constraint $K \in S$. The approach is extremely broad, as it seamlessly incorporates any type of decentralization, any control objective, and heterogeneous subsystems. It has thus come to be accepted as the canonical problem one would like to solve in decentralized control.

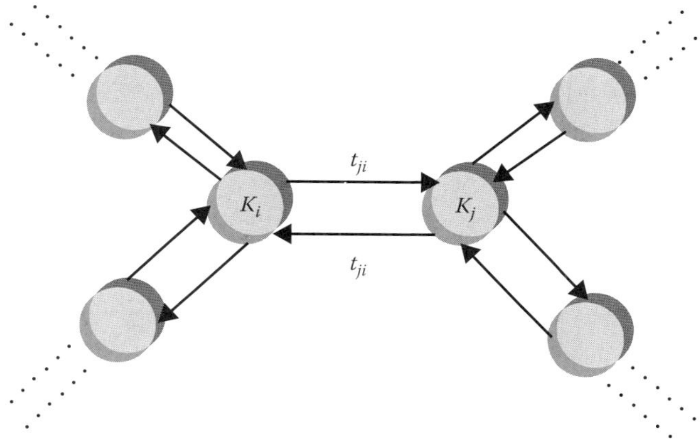

FIGURE 74.3 Network with delays.

74.2.3 Problem Formulation

The mapping from w to z that we wish to keep small in Figure 74.1, the closed-loop map, can be written as $f(P,K) = P_{11} + P_{12}K(I - GK)^{-1}P_{21}$. The problem that we would like to address may then be formulated as

$$\begin{aligned} \text{minimize} \quad & ||f(P,K)|| \\ \text{subject to} \quad & K \text{ stabilizes } P \\ & K \in S. \end{aligned} \tag{74.1}$$

The norm $(|| \cdot ||)$ is any appropriate system norm chosen, based on the particular performance objectives, which could be the $\mathcal{H}_2$-norm or $\mathcal{H}_\infty$-norm as described in detail in other chapters. The information constraint S is the subspace of admissible controllers which encapsulates the decentralized nature of the system, as exemplified above. The stabilization constraint is needed in the most typical case where the signals lie in extended spaces and the plant and controller are rational proper systems whose interconnections may thus be unstable. It may not be necessary, or another technical condition may be necessary such as the invertibility of $(I - GK)$, for other spaces of interest, such as Banach spaces with bounded linear operators (Rotkowitz and Lall, 2002, 2006a).

74.3 Stabilizing Controller Parametrization

If the plant to be controlled is stable, we could use the following change of variables:

$$Q = -K(I - GK)^{-1} \iff K = -Q(I - GQ)^{-1} \tag{74.2}$$

and then allowing the new parameter Q to be stable is equivalent to the controller K stabilizing the plant P, and the set of all achievable closed-loop maps (ignoring the information constraint) is then given as

$$\{P_{11} - P_{12}QP_{21} \mid Q \text{ stable}\}. \tag{74.3}$$

This is generalized by the Youla–Kucera or YJBK parametrization (Youla, Jabr, and Bonjiorno, 1976), which gives a similar change of variables for unstable plants such as allowing the new (Youla) parameter Q to vary over all stable systems is still equivalent to considering all stabilizing controllers K, and the set of all achievable closed-loop maps is then given by

$$\{T_1 - T_2QT_3 \mid Q \text{ stable}\} \tag{74.4}$$

where T_1, T_2, T_3 are other stable systems.

We see that these parametrizations allow the set of achievable closed-loop maps to be expressed as an affine function of a stable parameter, and thus allow our objective function in our main problem (Equation 74.1) to be cast as a convex function of that parameter. However, the information constraint $K \in S$ will typically not be simple to express in the new parameter, and this will ruin the convexity of the optimization problem.

74.4 Quadratic Invariance

We have seen that we can employ a change of variables that will make our objective convex, but that will generally cause the information constraint to no longer be affine. We thus seek to characterize problems for which the information constraint may be written as an affine constraint in the Youla parameter, such that a convex reformulation of our main problem will result.

The following property, first introduced in (Rotkowitz and Lall, 2002), provides that characterization.

Definition 74.4.1

The set S is ***quadratically invariant*** *with respect to G if*

$$KGK \in S \quad \text{for all } K \in S$$

In other words, given any admissible controller K, the composition KGK has to be admissible as well. When this condition holds, it follows that a controller being admissible is further equivalent to the linear-fractional transformation we encountered earlier lying in the constraint set (Rotkowitz and Lall, 2006a, 2006b):

$$K \in S \iff K(I - GK)^{-1} \in S \tag{74.5}$$

We can see immediately from Equation 74.2 that for the stable case this results in the equivalence of enforcing the information constraint on the controller or on the new parameter:

$$K \in S \iff Q \in S \tag{74.6}$$

and it can be shown that when the plant is unstable, another change of variables can be made such that this equivalence still holds (Rotkowitz and Lall, 2006b).

Thus when the information constraint S is quadratically invariant with respect to the plant G, the optimal decentralized control problem (Equation 74.1) may be recast as the following:

$$\begin{aligned} \text{minimize} \quad & \|T_1 - T_2QT_3\| \\ \text{subject to} \quad & Q \text{ stable} \\ & Q \in S \end{aligned} \tag{74.7}$$

which is a convex optimization problem.

74.5 Examples

This section looks at particular classes of information constraints to see when this quadratic invariance condition holds, to identify those decentralized problems which are amenable to convex synthesis. We see that this algebraic condition often has intuitive interpretations for specific classes of problems.

74.5.1 Structural Constraints

We first look at structural constraints, or sparsity constraints, where each subcontroller can see the measurements from some subsystems but not from others. This structure can be represented with a binary matrix K^{bin}. For instance, $K_{kl}^{\text{bin}} = 1$ if the kth control input u_k is allowed to be a function of the lth measurement y_l, and $K_{kl}^{\text{bin}} = 0$ if it cannot see that measurement. The information constraint S is then the set of all controllers which have the structure proscribed by K^{bin}; that is, all of the controllers such that none of the subcontrollers use information which they cannot see.

A binary matrix G^{bin} can similarly be used to give the structure of the plant. For instance, $G_{ij}^{\text{bin}} = 1$ if G_{ij} is nonzero and the ith measurement y_i is affected by the jth control input u_j, and $G_{ij}^{\text{bin}} = 0$ if it is unaffected by that input. Given this representation of the structure of the plant and the controller constraints, we have the following result:

S is quadratically invariant with respect to G if and only if

$$K_{ki}^{\text{bin}} G_{ij}^{\text{bin}} K_{jl}^{\text{bin}} (1 - K_{kl}^{\text{bin}}) = 0 \quad \text{for all } i, j, k, l. \tag{74.8}$$

Figure 74.4 illustrates this condition. The condition in Equation 74.8 requires that, for arbitrary i, j, k, l, if the three blocks on the bottom are all nonzero (or allowed to be chosen nonzero), then the top

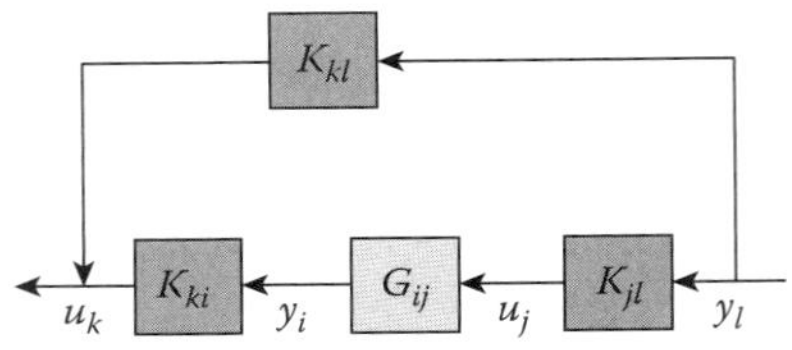

FIGURE 74.4 Structural quadratic invariance.

block must be allowed to be nonzero as well. In other words, if there is an indirect connection from a measurement to a control input, then there has to be a direct connection as well.

When this condition is met, the problem is quadratically invariant, and we can recast our optimal decentralized control problem as the convex optimization problem in Equation 74.7.

74.5.2 Symmetry

We briefly consider the problem of symmetric synthesis. Suppose that we need to design the best symmetric controller; that is, the best controller such that $K_{kl} = K_{lk}$ for all k, l, and that the information constraint S is the set of all such symmetric controllers. If the plant is also symmetric; that is, if $G_{ij} = G_{ji}$ for all i, j, then KGK is symmetric for any symmetric K. Thus, $KGK \in S$ for all $K \in S$, the problem is quadratically invariant, and the optimal symmetric control problem may be recast as Equation 74.7.

74.5.3 Delays

We now return to the problem of Figure 74.3, where we have multiple nodes/subsystems, each with its own controller, and each subsystem i, can see the information from another subsystem j after a transmission delay of t_{ij}.

We similarly consider that the inputs to a given subsystem j may affect other subsystems after some delay, and denote the amount of time after which it may affect another subsystem i by the propagation delay p_{ij}.

The overall problem of controlling such a network with propagation delays, with controllers that may communicate with transmission delays, is depicted in Figure 74.5.

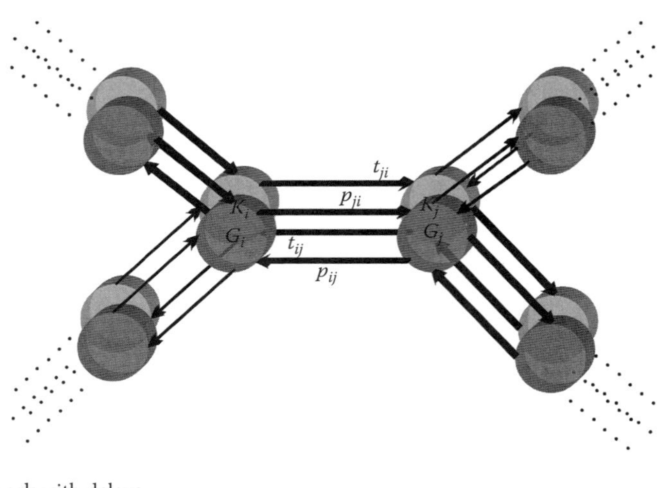

FIGURE 74.5 Network with delays.

When this problem is tested for quadratic invariance, one first finds that the following condition is necessary and sufficient:

$$t_{ki} + p_{ij} + t_{jl} \geq t_{kl} \quad \text{for all } i, j, k, l. \tag{74.9}$$

This is reminiscent of condition (Equation 74.8) for structural constraints, as it similarly requires that any direct path from y_l to u_k must be at least as fast as any indirect path through the plant. This condition can be further reduced to a very simple intuitive condition (Rotkowitz, Cogill, and Lall, 2005), as long as we may assume that the transmission delays themselves satisfy the triangle inequality; that is,

$$t_{ki} + t_{ij} \geq t_{kj} \quad \text{for all } k, i, j. \tag{74.10}$$

This is typically a very reasonable assumption, as it states that information is transmitted between nodes in the quickest manner available through the network. If the inequality failed for some k, j, one would want to reroute the transmissions from j to k along the faster route such that the inequality would then hold.

If the triangle inequality among transmissions does hold, then condition (Equation 74.9), and thus quadratic invariance, is reduced to simply

$$p_{ij} \geq t_{ij} \quad \text{for all } i, j. \tag{74.11}$$

In other words, for any pair of nodes, information needs to be transmitted faster than the dynamics propagate. When this simple condition holds, the problem is quadratically invariant, and the optimal decentralized control problem may be recast as the convex problem (Equation 74.7).

This very intuitive result has a counterintuitive complement when one considers computational delays as well. Suppose now that the ith controller cannot use a measurement from the jth subsystem until a pure transmission delay of $\tilde{t}_{ij}$, representing the time it takes to send the information from one subsystem to the other, as well as a computational delay of c_i, representing the time it takes to process the information once it is received.

While intuition might suggest that these two quantities would end up being added and then replacing the right-hand side of Equation 74.11, if we now assume that the pure transmission delays satisfy the triangle inequality, the condition for quadratic invariance becomes

$$p_{ij} + c_j \geq \tilde{t}_{ij} \quad \text{for all } i, j \tag{74.12}$$

with the computational delay on the other side of the inequality.

This shows that, regardless of computational delay, if information can be transmitted faster than dynamics propagate, then the optimal decentralized control problem can be reformulated as a convex optimization problem. If we consider a problem with multiple aerial vehicles, for example, where dynamics between any pair of subsystems will propagate at the speed of sound, this tells us that transmissions just have to be faster than that threshold for the optimal control problem to be recast as (Equation 74.7).

These results have also been extended to spatio-temporal systems (Rotkowitz, Cogill, and Lall, 2010), including the special case of spatially invariant systems.

74.6 Perfectly Decentralized Control

We now revisit the problem of Figure 74.2, where each controller can use only the measurements from its own subsystem, and thus the information constraint is block diagonal. This problem is never quadratically invariant, and will never satisfy condition (Equation 74.8), except for the case where the subsystems do not affect one another; that is, except for the case where G is block diagonal as well.

In all other cases where subsystems may have some affect on others, we thus cannot parametrize all of the admissible stabilizing controllers in a convex fashion, and cannot cast the optimal decentralized control problem as a convex problem such as in Equation 74.7. However, a Youla parametrization can similarly be used, and while Equation 74.6 does not hold, as the information constraint on the controller is not equivalent to enforcing it on the Youla parameter, it is instead equivalent to a quadratic equality constraint on the parameter (Manousiouthakis, 1993)

$$K \in S \iff W_1 + W_2Q + QW_3 + QW_4Q = 0 \tag{74.13}$$

for stable operators W_1, W_2, W_3, W_4. When returning to the optimal decentralized control problem, this equality constraint replaces the final $Q \in S$ constraint of Equation 74.7. The problem is no longer convex due to the quadratic term, but the overall difficulty is transformed to one well-understood type of constraint, for which many methods exist to approximate optimal solutions.

Other structural constraints, which are neither block diagonal nor quadratically invariant, can be similarly parametrized by first converting them to a perfectly decentralized problem (Rotkowitz, 2010).

74.7 Nonlinear Decentralized Control

The framework discussed thus far assumes that the operators, both the plant to be controlled and the possible controllers that we may design for it, are all linear, and when applicable, time-invariant as well. A similar convex parametrization of stabilizing decentralized controllers exists even when the systems are possibly nonlinear and possibly time-varying (NLTV) (Rotkowitz, 2006). The condition allowing for the parametrization then becomes

$$K_1(I \pm GK_2) \in S \quad \text{for all } K_1, K_2 \in S.$$

When the plant is stable, the stabilizing controllers may be parametrized similarly to Equation 74.3 (Desoer and Liu, 1982), and when the plant is unstable, the stabilizing controllers may typically be parametrized similarly to Equation 74.4 (Anantharam and Desoer, 1984). Similar to quadratic invariance, the above condition then yields the equivalence of the controller and the feedback map satisfying the information constraint (Equation 74.5), which then gives the equivalence of the controller and the parameter satisfying the constraint as in Equation 74.6. The convex parametrization of all causal stabilizing decentralized controllers then results, analogous to the linear case with quadratic invariance.

While this condition may appear quite different from quadratic invariance, they actually both reduce to the same conditions when one considers the classes of sparsity constraints or delay constraints, and so these results extend to all of the cases covered in Sections 74.5.1 and 74.5.3.

References

V. Anantharam and C. A. Desoer. On the stabilization of nonlinear systems. *IEEE Transactions on Automatic Control*, 29(6):569–572, 1984.

C. A. Desoer and R. W. Liu. Global parametrization of feedback systems with nonlinear plants. *Systems and Control Letters*, 1(4):249–251, 1982.

V. Manousiouthakis. On the parametrization of all stabilizing decentralized controllers. *Systems and Control Letters*, 21(5):397–403, 1993.

M. Rotkowitz. Information structures preserved under nonlinear time-varying feedback. In *Proc. American Control Conference*, Minneapolis, Minnesota, pp. 4207–4212, June 2006.

M. Rotkowitz. Parametrization of all stabilizing controllers subject to any structural constraint. In *Proc. IEEE Conference on Decision and Control*, Atlanta, Georgia, December 2010.

M. Rotkowitz, R. Cogill, and S. Lall. A simple condition for the convexity of optimal control over networks with delays. In *Proc. IEEE Conference on Decision and Control*, Seville, Spain, pp. 6686–6691, December 2005.

M. Rotkowitz, R. Cogill, and S. Lall. Convexity of optimal control over networks with delays and arbitrary topology. *International Journal of Systems, Control, and Communications*, 2:30–54, 2010.

M. Rotkowitz and S. Lall. Decentralized control information structures preserved under feedback. In *Proc. IEEE Conference on Decision and Control*, Las Vegas, Nevada, pp. 569–575, December 2002.

M. Rotkowitz and S. Lall. Affine controller parameterization for decentralized control over Banach spaces. *IEEE Transactions on Automatic Control*, 51(9):1497–1500, 2006a.

M. Rotkowitz and S. Lall. A characterization of convex problems in decentralized control. *IEEE Transactions on Automatic Control*, 51(2):274–286, 2006b.

H. S. Witsenhausen. A counterexample in stochastic optimum control. *SIAM Journal of Control*, 6(1):131–147, 1968.

D.C. Youla, H.A. Jabr, and J.J. Bonjiorno Jr. Modern Wiener–Hopf design of optimal controllers: Part II. *IEEE Transactions on Automatic Control*, 21(3):319–338, 1976.

75

Estimation and Control across Analog Erasure Channels

Vijay Gupta
University of Notre Dame

75.1 Introduction

In this chapter, we will adopt the analog erasure model to describe the communication channel present inside a control loop. This model, also referred to as the packet erasure or packet loss model, can be described as follows [8]. The channel operates in discrete time steps. At every time step, the channel accepts as input a finite-dimensional real vector $r(k)$. The value of the output of the channel $y(k)$ is chosen according to an *erasure process*. At every time step, the erasure process assumes either the value T or the value R. If the value at time k is T, $y(k+1) = r(k)$ and a successful transmission is said to have occurred. Otherwise, $y(k+1) = \phi$ and an erasure event, or a packet loss, is said to have occurred at time k. The symbol ϕ denotes that the receiver does not receive any data; however, the receiver is aware that an erasure event has occurred at that time. Note that we have assumed that the channel introduces a constant delay of one time step.

The analog erasure model aims to capture data loss due to a communication channel. Due to effects such as interference and fading in wireless channels, congestion in shared networks, or even overload and interrupts at a microcontroller, various parts of a control loop between the sensor and controller, or the controller and actuator, may exhibit information loss. By considering the idealization in which every successful transmission results in the communication of a real vector of a bounded dimension, such situations can be modeled using an analog erasure representation. While an analog erasure model has an infinite capacity in an information theoretic sense, it is often a useful representation for the cases when the communication protocols allow for large data packets to be transmitted at every time step. For instance, the minimum size of an ethernet data packet is 72 bytes. This is much more space for carrying information than usually required inside a control loop. If the data packets allow for transmission of control and sensing data to a high fidelity, the quantization effects are often ignored and an analog erasure model is adopted.

Various descriptions of the erasure process are possible. The following two models are the most popular:

1. *Maximum Allowable Transmit Interval (MATI)-based models*: This model (see, e.g., [3]) is described using two integer values n_1 and n_2. For an (n_1, n_2) model, in any interval of length n_1, at most n_2 erasure events can occur. Apart from this constraint, the erasure process is arbitrary.
2. *Stochastic erasure event-based models*: In this model (see, e.g., [12]), the erasure process is a random process. The simplest case is when the erasure events are independent and identically distributed at different time steps. In such a case, the erasure process is described by the erasure probability $p \overset{def}{=} \text{Prob}(y(k) = \phi)$, at any time step k. More complicated models when the erasure process can be described by, for example, a Markov chain (possibly on the lines of the Gilbert–Eliot channel model [4]) can also be considered.

In this chapter, we will concentrate on the stochastic erasure event-based models. As with any networked control system, two questions can be asked.

1. What is the effect of introducing channels on the structure of the estimators and controllers?
2. What are the optimal encoders and decoders that transmit the maximum amount of control relevant information to achieve the fundamental limits of performance in such systems?

These questions are answered in the next two sections, respectively. Section 75.4 lists some extensions to the simple model considered here for pedagogical ease, and points out some open research directions. Some results on Markovian jump linear systems (MJLS) that are used in the chapter are presented in Section 75.5.

75.2 Effect on Estimation and Control Performance

75.2.1 Estimation

We begin with the problem of estimating a linear time-invariant process across an analog erasure channel. Consider the setup as shown in Figure 75.1. The process with state $x(k) \in \mathbf{R}^n$ evolves as

$$x(k+1) = Ax(k) + w(k),$$

where $w(k)$ is the process noise modeled as white, zero mean, Gaussian with covariance $R_w > 0$. The initial condition $x(0)$ is assumed to be drawn from a Gaussian distribution with zero mean and covariance $\Pi(0)$. The state is observed by a sensor of the form

$$y(k) = Cx(k) + v(k),$$

where $v(k) \in \mathbf{R}^p$ is measurement noise modeled by white, zero mean, Gaussian with covariance $R_v > 0$. We assume that the sources of randomness $x(0)$, $\{w(j)\}$ and $\{v(j)\}$, are mutually independent. The sensor transmits its measurements to an estimator across an analog erasure channel with erasure probability p at every time step. The pair (A, C) is assumed to be observable.

The estimator receives those measurements that are transmitted successfully across the channel. It has access to the constant matrices A, C, R_w, R_v and $\Pi(0)$. The estimator aims at generating the minimum

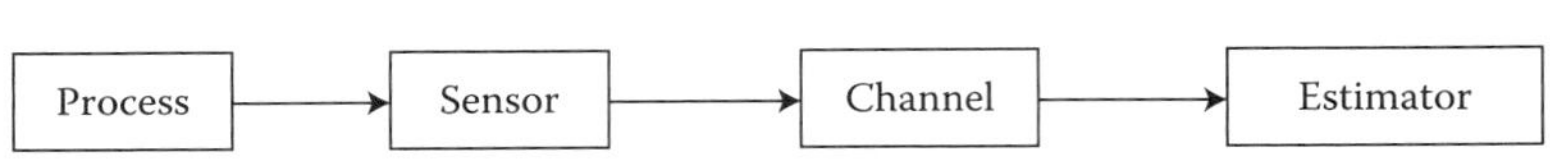

FIGURE 75.1 Basic setup of the estimation problem.

mean-squared error (MMSE) estimate $\hat{x}(k)$ of the process state $x(k)$ based on the information it has access to at time k. If no erasure events were to occur, the optimal estimate is given by the Kalman filter, and the estimate error covariance evolves according to a discrete Riccati recursion. In particular, for the assumptions as made above, the error covariance is stable in the sense of being bounded as time k increases. We wish to extend this analysis to the case when erasure events occur with a probability p at every time step.

Since the event of a packet drop is known to the estimator, the problem is equivalent to estimation of the following Markov jump linear system with the jumps occurring according to a Bernoulli process:

$$x(k+1) = Ax(k) + w(k)$$
$$y(k) = C_{r(k)}x(k) + v(k),$$

where $r(k)$ is the Markov state such that $r(k) = 1$ with probability $1 - p$ and $r(k) = 2$ otherwise. Moreover, $C_1 = C$ and $C_2 = 0$. We can thus utilize the standard results from MJLS theory*. In particular, from Corollary 75.2 we obtain that the optimal estimator for such a system is provided by a time-varying Kalman filter. Due to probabilistic erasure events, the estimate error covariance $\Pi(k)$ evolves as a random variable. An upper bound of the expected estimate error covariance $E[\Pi(k)]$ is provided by the quantity $V(k)$ that evolves as

$$V(k+1) = AV(k)A^T + R_w - (1-p)AV(k)C^T\left(CV(k)C^T + R_v\right)^{-1}CV(k)A^T,$$

with the initial condition $V(0) = \Pi(0)$. A sufficient condition for stabilizability can also be obtained through Corollary 75.2. We can also express the condition as a Linear Matrix Inequality (LMI) [14]. The system is stabilizable if there exists a matrix $X > 0$ and a matrix K such that

$$\begin{bmatrix} X & \sqrt{(1-p)}(XA + KC) & \sqrt{p}XA \\ \sqrt{(1-p)}(A^TX + C^TK^T) & X & 0 \\ \sqrt{p}A^TX & 0 & X \end{bmatrix} > 0.$$

For our problem, we can also derive a lower bound on $E[\Pi(k)]$ as follows. At time step k, the error covariance $\Pi(k)$ is lower bounded by R_w if a packet is received by the estimator, and is equal to $A\Pi(k-1)A^T + R_w$ if a packet is not received. Thus, $E[\Pi(k)]$ is lower bounded by $S(k)$ which evolves as

$$S(k+1) = p\left(AS(k)A^T + R_w\right) + (1-p)R_w = pAS(k)A^T + R_w.$$

This is a discrete algebraic Lyapunov recursion. By considering the convergence properties of the recursion, we obtain that a necessary and sufficient condition for stability of $S(k)$ (and, thus, a necessary condition for stability of $E[\Pi(k)]$) is given by $p\rho(A)^2 < 1$, where $\rho(A)$ is the spectral radius of A.

75.2.2 Control

We now move on to the control problem considered, for example, in [9,14]. To begin with, consider the setup shown in Figure 75.2 that has only one channel in the control loop, present between the sensor and the controller. Such a situation can arise, for example, when the controller is colocated with the process and the sensor is remote, or if the controller has access to large transmission power. The linear

* A brief review of such results is provided in Section 75.5.

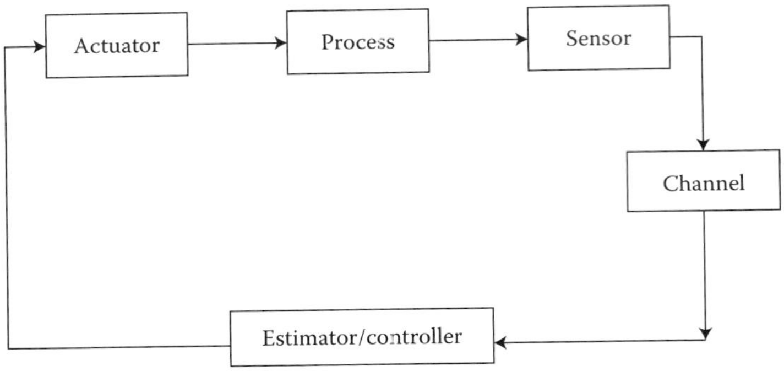

FIGURE 75.2 Basic setup of the control problem with a single channel present between the sensor and the controller.

time-invariant process now evolves as

$$x(k+1) = Ax(k) + Bu(k) + w(k),$$

where the additional variable $u(k) \in \mathbf{R}^m$ is the control input chosen to minimize the cost

$$J_{LQG} = E\left[\sum_{k=1}^{K}\left(x^T(k)Qx(k) + u^T(k)Ru(k)\right) + x^T(K+1)P(K+1)x(K+1)\right],$$

where the expectation at time k is taken with respect to the future values of the packet erasure events, the initial condition, and the measurement and process noises. Further, the matrices $P(K+1)$, Q and R are all assumed to be positive definite. The pair (A, B) is assumed to be controllable.

We can utilize the Markov state representation to solve the Linear Quadratic Gaussian (LQG) problem as well. The system can again be written as a Markov jump linear system of the form

$$x(k+1) = Ax(k) + Bu(k) + w(k),$$
$$y(k) = C_{r(k)}x(k) + v(k),$$

where $r(k)$ is the Markov state such that $r(k) = 1$ with probability $1 - p$ and $r(k) = 2$ otherwise. Moreover, $C_1 = C$ and $C_2 = 0$. Thus, we can utilize the separation principle from Section 75.5.3 to obtain the optimal controller as the combination of the LQ optimal control, with the MMSE estimate of the state used in place of the state value. Note that the MMSE estimate can be calculated as in Section 75.2.1. Moreover, since neither the matrix A nor the matrix B depend on the Markov state, the LQ optimal control corresponds to the control input for the system

$$x(k+1) = Ax(k) + Bu(k),$$

that minimizes the cost function

$$J = \left[\sum_{k=1}^{K}\left(x^T(k)Qx(k) + u^T(k)Ru(k)\right) + x^T(K+1)P(K+1)x(K+1)\right],$$

when the controller has full state information, that is, to calculate the input $u(k)$ at time k, the state $x(0), x(1), \ldots, x(k)$ is available.

We can also consider the case of two channels being present. Consider the system setup shown in Figure 75.3, with the process and the sensor as described above. In addition to the sensor–controller channel, there is an additional channel between the controller and the actuator. Assume that the erasures

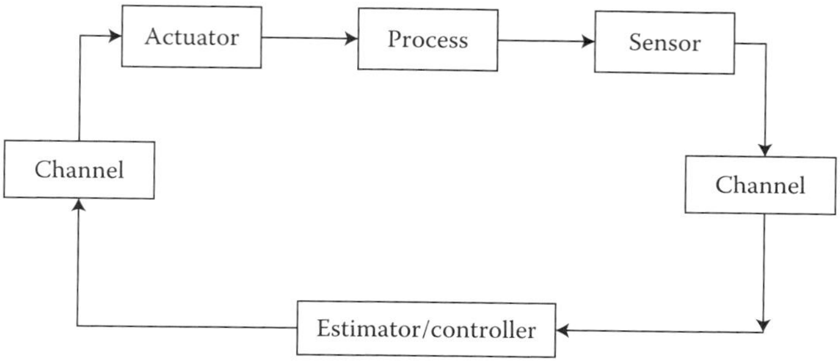

FIGURE 75.3 Setup of the control problem with two channels.

on the controller–actuator channel occur in an i.i.d. fashion, with probability of erasure q at any time step. Moreover, the erasure events on this channel are independent of all other random variables in the system.

In this case, it is also important to specify the action that the actuator takes when it does not receive a packet. The action depends on the amount of processing, memory, and information about the process that is assumed to be available at the actuator. We consider the simplest choice, which is to apply zero control input if no packet was received. Other actions by the actuator can be treated in a similar fashion. For this case, the Markov jump linear system representation of the system is now given by

$$\begin{aligned} x(k+1) &= Ax(k) + B_{r(k)}u(k) + w(k) \\ y(k) &= C_{r(k)}x(k) + v(k), \end{aligned}$$

where $r(k)$ is the Markov state that can take values 1, 2, 3, and 4 with probabilities $(1-p)(1-q)$, $p(1-q)$, $(1-p)q$, and pq respectively. The system matrices are given by $B_1 = B_2 = B$, $B_3 = B_4 = 0$, $C_1 = C_3 = C$ and $C_2 = C_4 = 0$. The solution of the LQG problem requires one additional assumption. The separation result in Theorem 75.6 assumes that the controller at time k has knowledge of the control inputs that were previously applied till time $k-1$. Indeed, if the past control inputs are not available at the controller, then the control will have a dual effect. For our problem, this implies that the controller must know whether or not the transmission over the controller actuator channel has been successful. To provide this information to the controller, we will assume a perfect acknowledgment from the actuator to the controller for any data packet received by the actuator. This is often called the TCP-like case. The case when acknowledgements are not available is termed as the UDP-like case. For the UDP-like case, the separation principle does not hold and the form of the optimal controller is unknown, in general.

Note that since the controller can detect any packet drops on the sensor–controller channel, and receives acknowledgments about transmissions over the controller–actuator channel, the controller has access to the Markov states $r(0), \ldots, r(k-1)$ at time k, but not $r(k)$. In particular, at time step k, the controller does not know the value of $B_{r(k)}$. While in general Theorem 75.6 requires knowledge of the current Markov state at the controller, in this particular case, the problem is still solvable. To see this, assume that the controller uses the value $B_{r(k)} = B$ to calculate the optimal control input. If indeed $r_k = 1$ or 2, the actuator successfully receives this packet and the control input is optimal. If $r_k = 3$ or 4, the optimal control input should be zero since for those states $B_{r(k)} = 0$. However, at these time steps, the transmission by the controller is not successful and the actuator applies zero as the control input. Thus, once again, the optimal control input is applied. Thus, we see that the LQG problem can be solved for this case using Theorem 75.6. We can also identify stability conditions and performance bounds by utilizing the MJLS theory.

75.3 Optimal Coding

We now turn to the more general question of identifying fundamental limits on the performance of a system being controlled across an analog erasure channel, and the design of encoders and decoders to achieve such limits as discussed, for example, in [7]. To proceed, we must define the class of encoders that we will consider. The information theoretic capacity of an analog erasure channel is infinite. Thus, the only constraints we impose on the encoder are that the transmitted vector is some causal (possibly time-varying) function of the measurements available to the encoder until time k and that the dimension of the vector is finite. The encoder is collocated with the sensor, while the decoder is located at the estimator/controller. We will sometimes refer to the encoder as an encoding algorithm.

We begin by considering the system setup in Figure 75.2 and the associated assumptions about the process, the sensor and the cost function in Section 75.2.2. However, instead of transmitting measurements $y(k)$ at every time step k, the sensor can now calculate a vector $s(k) = f(k, \{y(j)\}_{j=0}^{k})$ and transmit it. Note that we have not assumed that the encoder has access to any acknowledgments from the decoder about which transmissions have been successful. However, we will show that the presence of such acknowledgements does not improve the optimal performance achievable by a suitable encoder.

Denote by $\mathcal{I}(k)$ the information set that the decoder can utilize to calculate the control at time k. As an example, if no erasures were happening, $\mathcal{I}(k) = \{y(0), y(1), \dots, y(k-1)\}$. More generally, given any packet erasure pattern, we can define a time stamp $t_s(k)$ at every time step k such that the erasures did not allow any information transmitted by the encoder after time $t_s(k)$ to reach the decoder. Without loss of generality, we can restrict attention to information-set feedback controllers. For a given information set, $\mathcal{I}(.)$ denote the minimal value of the cost J_{LQG} that can be achieved with the optimal controller design by $J^{\star}_{LQG}(\mathcal{I})$, and the smallest sigma algebra generated by the information set as $\mathbf{I}(.)$. If two information sets $I^1(.)$ and $I^2(.)$ are such that $\mathbf{I}^1(k) \subseteq \mathbf{I}^2(k)$, we have $J^{\star}_{LQG}(\mathcal{I}^2) \leq J^{\star}_{LQG}(\mathcal{I}^1)$.

Consider an algorithm $\mathcal{A}_1$ in which at every time step k, the sensor transmits all measurements $y(0), y(1), \dots, y(k)$ to the decoder. Note that this algorithm is not a valid encoding algorithm since the dimension of the transmitted vector is not bounded, as k increases. However, with this algorithm, for any drop sequence, the decoder has access to an information set of the form $\mathcal{I}^{\max}(k) = \{y(0), y(1), \dots, y(t_s(k))\}$, where $t_s(k) \leq k-1$ is the time stamp defined above. This is the maximal information set that the decoder can have access to with any algorithm in the sense that $\mathbf{I}(k) \subseteq \mathbf{I}^{\max}(k)$, for any other algorithm that yields the information set $\mathcal{I}(k)$. Thus, one way to achieve the optimal value of the cost function is to utilize an algorithm that makes $\mathcal{I}^{\max}(k)$ available to the sensor at every time k along with a controller that optimally utilizes this set. Further, one such encoder algorithm is $\mathcal{A}_1$. However, as discussed above, $\mathcal{A}_1$ is not a valid encoding algorithm. Surprisingly, as shown below, we can achieve the same performance with an algorithm that transmits a vector with finite dimension.

We begin with the following separation principle when the decoder has access to the maximal information set. Denote by $\hat{\alpha}(k|\beta(k))$ the MMSE estimate of the random variable $\alpha(k)$ based on the information $\beta(k)$.

Theorem 75.1: Separation Principle with Maximal Information Set

Consider the control problem as defined above, when the decoder has access to the maximal information set $\mathcal{I}^{\max}(k)$ at every time step. Then, the optimal control input is given by

$$u(k) = \hat{u}_{LQ}\left(k|\mathcal{I}^{\max}(k), \{u(j)\}_{j=0}^{k-1}\right),$$

where $u_{LQ}(k)$ is the optimal LQ control law.

The proof of this result is similar to the standard separation principle (see, e.g., [10, Chapter 9]) and is omitted here. For our setting, the importance of this result lies in the fact that it recognizes that $\hat{u}_{LQ}\left(k|\mathcal{I}^{\max}(k),\{u(j)\}_{j=0}^{k-1}\right)$ (or, in turn, $\hat{x}_{LQ}\left(k|\mathcal{I}^{\max}(k),\{u(j)\}_{j=0}^{k-1}\right)$) is a sufficient statistic to calculate the control input that achieves the minimum possible cost for *any* encoding algorithm. Utilizing the fact that the optimal MMSE estimate of $x(k)$ is linear in the effects of the maximal information set and the previous control inputs, we can identify the quantity that the encoder should transmit that depends only on the measurements. We have the following result.

Theorem 75.2: Separation of the Effect of the Control Inputs

The quantity $\hat{x}_{LQ}\left(k|\mathcal{I}^{\max}(k),\{u(j)\}_{j=0}^{k-1}\right)$ *can be calculated as*

$$\hat{x}_{LQ}\left(k|\mathcal{I}^{\max}(k),\{u(j)\}_{j=0}^{k-1}\right)=\bar{x}_{LQ}\left(k|\mathcal{I}^{\max}(k)\right)+\psi(k),$$

where $\bar{x}_{LQ}\left(k|\mathcal{I}^{\max}(k)\right)$ *depends only on* $\mathcal{I}^{\max}(k)$ *but not on the control inputs and* $\psi(k)$ *depends only on the control inputs* $\{u(j)\}_{j=0}^{k-1}$. *Further both* $\bar{x}_{LQ}\left(k|\mathcal{I}^{\max}(k)\right)$ *and* $\psi(k)$ *can be calculated recursively.*

Proof. The proof follows readily from noting that $\hat{x}_{LQ}\left(k|\mathcal{I}^{\max}(k),\{u(j)\}_{j=0}^{k-1}\right)$ can be obtained from the Kalman filter which is affine in both measurements and control inputs. We can identify

$$\begin{aligned}\bar{x}_{LQ}\left(k|\mathcal{I}^{\max}(k)\right)&=A^{k-t_s(k)-1}\breve{x}(t_s(k)+1|t_s(k)),\\ \psi(k)&=A^{k-t_s(k)-1}\breve{\psi}(t_s(k)+1)+\sum_{i=0}^{k-t_s(k)-2}A^iBu(k-i-1),\end{aligned}$$

where $\breve{x}(j+1|j)$ evolves as

$$\begin{aligned}M^{-1}(j|j)&=M^{-1}(j|j-1)+C^TR_v^{-1}C,\\ M^{-1}(j|j)\breve{x}(j|j)&=M^{-1}(j|j-1)\breve{x}(j|j-1)+C^TR_v^{-1}y(j),\\ M(j|j-1)&=AM(j-1|j-1)A^T+R_w,\\ \breve{x}(j|j-1)&=A\breve{x}(j-1|j-1),\end{aligned}$$

with the initial conditions $\breve{x}(0|-1)=0$ and $M(0|-1)=\Pi(0)$, and $\breve{\psi}(j)$ evolves as

$$\begin{aligned}\breve{\psi}(j)&=Bu(j-1)+\Gamma(j-1)\breve{\psi}(j-1),\\ \Gamma(j)&=AM^{-1}(j-1|j-1)M(j-1|j-2),\end{aligned}$$

with the initial condition $\breve{\psi}(0)=0$. ■

Now consider the following algorithm $\mathcal{A}_2$. At every time step k, the encoder calculates and transmits the quantity $\breve{x}(k|k)$ using the algorithm in the above proof. The decoder calculates the quantity $\psi(k)$.

If the transmission is successful, the decoder calculates

$$\hat{x}_{LQ}\left(k+1|\mathcal{I}^{\max}(k+1),\{u(j)\}_{j=0}^{k}\right)=\bar{x}_{LQ}\left(k+1|\mathcal{I}^{\max}(k+1)\right)+\psi(k)$$
$$=A\check{x}(k|k)+\psi(k).$$

If the transmission is unsuccessful, the decoder calculates

$$\hat{x}_{LQ}\left(k+1|\mathcal{I}^{\max}(k+1),\{u(j)\}_{j=0}^{k}\right)=A^{k-t_s(k)}\bar{x}_{LQ}\left(k+1|\mathcal{I}^{\max}(t_s(k)+1)\right)+\psi(k),$$

where the quantity $\bar{x}_{LQ}(k+1|\mathcal{I}^{\max}(t_s(k)+1))$ is stored in the memory from the last successful transmission (note that only the last successful transmission needs to be stored). Using the Theorems 75.1 and 75.2 clearly allows us to state the following result.

Theorem 75.3: Optimality of the Algorithm $\mathcal{A}_2$

Algorithm $\mathcal{A}_2$ is optimal in the sense that it allows the controller to calculate the control input $u(k)$ that minimizes J_{LQG}.

Proof. At every time step, the algorithm $\mathcal{A}_2$ makes $\hat{x}_{LQ}(k+1|\mathcal{I}^{\max}(k+1),\{u(j)\}_{j=0}^{k})$ available to the controller. Thus, the controller can calculate the same control input as with the algorithm $\mathcal{A}_1$ which together with an LQ controller yields the minimum value of J_{LQG}. ■

Note that the optimal algorithm is nonlinear (in particular, it is a switched linear system). This is not unexpected, in view of the nonclassical information pattern in the problem.

Remarks

- *Boundedness of the Transmitted Quantity:* It should be emphasized that the quantity $\check{x}(k|k)$ that the encoder transmits is *not* the estimate of $x(k)$ (or the state of some hypothetical open-loop process) based only on the measurements $y(0),\ldots,y(k)$. In particular, under the constraint on the erasure probability that we derive later, the state $x(k)$ is stable and hence the measurements $y(k)$ are bounded. *Thus, the quantity $\check{x}(k|k)$ is bounded.* This can also be seen from the recursive filter used in the proof of Theorem 75.2. If the closed-loop system $x(k)$ is unstable due to high erasure probabilities, $\check{x}(k|k)$ would, of course, not be bounded. However, the optimality result implies that the system cannot be stabilized by transmitting any other bounded quantity (such as measurements).
- *Optimality for any Erasure Pattern and the "Washing Away" Effect:* The optimality of the algorithm required no assumption about the erasure statistics. The optimality result holds for an arbitrary erasure sequence, and at every time step (not merely in an average sense). Moreover, any successful transmission "washes away" the effect of the previous erasures in the sense that it ensures that the control input is identical to the case as if all previous transmissions were successful.
- *Presence of Delays:* We assumed that the communication channel introduces a constant delay of one time step. However, the same algorithm continues to remain optimal even if the channel introduces larger (or even time-varying) delays, as long as there is the provision of a time stamp from the encoder regarding the time it transmits any vector. The decoder uses the packet it receives at any time step only if it was transmitted later than the quantity it has stored from the previous time steps. If this is not true due to packet reordering, the decoder continues to use the quantity stored from previous time steps. Further, if the delays are finite, the stability conditions derived below remain unchanged. Infinite delays are equivalent to packet erasures, and can be handled by using the same framework.

Stability and Performance

Both the stability and performance of the system with this optimal coding algorithm in place can be analyzed by assuming specific models for the erasure process. For pedagogical ease, we adopt the i.i.d. erasure model, with an erasure occurring at any time step with probability p. Due to the separation principle, to obtain the stability conditions, we need to consider the conditions under which the LQ control cost for the system, and the covariance of the estimation error between the state of the process $x(k)$ and the estimate at the controller $\hat{x}(k)$ remain bounded, as time k increases. Under the controllability and observability assumptions, the LQ cost remains bounded, if the control value does. Define the estimation error and its covariance as

$$e(k) = x(k) - \hat{x}(k),$$
$$P(k) = E\left[e(k)e^T(k)\right],$$

where the expectation is taken with respect to the process and measurement noises, and the initial condition (but not the erasure process). Due to the "washing away" effect of the algorithm, the error of the estimate at the decoder evolves as

$$e(k+1) = \begin{cases} \bar{e}(k+1) & \text{no erasure} \\ Ae(k) & \text{erasure event,} \end{cases}$$

where $\bar{e}(k)$ is the error between $x(k)$ and the estimate of $x(k)$ given all control inputs $\{u(j)\}_{j=0}^{k-1}$ and measurements $\{y(j)\}_{j=0}^{k-1}$. Thus, the error covariance evolves as

$$P(k+1) = \begin{cases} M(k+1) & \text{with probability } 1-p \\ AP(k)A^T + R_w & \text{with probability } p, \end{cases}$$

where $M(k)$ is the covariance of the error $\bar{e}(k)$. Thus, we obtain

$$E[P(k+1)] = (1-p)M(k+1) + pR_w + pAE[P(k)]A^T,$$

where the extra expectation for the error covariance is taken over the erasure process in the channel. Since the system is observable, $M(k)$ converges geometrically to a steady-state value $M^{\star}$. Thus, the necessary and sufficient condition for the convergence of the above discrete algebraic Lyapunov recursion is

$$p\rho(A)^2 < 1,$$

where $\rho(A)$ is the spectral radius of A. Due to the optimality of the algorithm considered above, this condition is necessary for stability of the system with any causal encoding algorithm. In particular, for the strategy of simply transmitting the latest measurement from the sensor as considered in Section 75.2, this condition turns out to be necessary for stability (though not sufficient for a general process model). For achieving stability with this condition, we require an encoding strategy, such as the recursive algorithm provided above.

This analysis can be generalized to more general erasure models. For example, for a Gilbert–Eliot type channel model, the necessary and sufficient condition for stability is given by

$$q_{00}\rho(A)^2 < 1,$$

where q_{00} is the conditional probability of an erasure event at time $k+1$, provided an erasure occurs at time k. In addition, by calculating the terms $E[P(k)]$ and the LQ control cost of the system with full state information, the performance J_{LQG} can also be calculated through the separation principle proved above. The value of the cost function thus achieved provides a lower bound to the value of the cost function achievable using any other encoding or control algorithm, for the same probability of erasure. An alternative viewpoint is to consider the encoding algorithm above as a means for transmitting data with lesser frequency to achieve the same level of performance than, for example, transmitting measurements to the controller.

Higher-Order Moments

It can be seen that the treatment above can be extended to consider the stability of higher-order moments of the estimation error, or the state value. In fact, the entire steady-state probability distribution function of the estimation error can be calculated.

75.4 Extensions and Open Questions

The above framework was explained for a very simple setup of an LQG problem. It is natural to consider its generalization to other models by removing various assumptions. We consider some of these assumptions below. We also point out some of the open questions.

- *Channel between the Controller and the Actuator:* The encoding algorithm presented above continues to remain optimal when a channel is present between the controller and the actuator (as considered in Figure 75.3) as long as there is a provision for acknowledgment from the actuator to the controller for any successful transmission, and the protocol that the actuator follows in case of an erasure is known at the controller. This is because these two assumptions are enough for the separation principle to hold. If no such acknowledgment is available, the control input begins to have a dual effect and the optimal algorithm is still unknown. Moreover, the problem of designing the optimal encoder for the controller–actuator channel can also be considered. This design will intimately depend on the information that is assumed to be known at the actuator (e.g., the cost function, the system matrices, etc.). Algorithms that optimize the cost function for such information sets are largely unknown. A simpler version of the problem would involve either
    - Analyzing the stability and performance gains for given encoding and decoding algorithms employed by the controller and the actuator, respectively.
    - Considering algorithms that are stability optimal, in the sense of designing recursive algorithms that achieve the largest stability region for any possible causal encoding algorithm.

  Both these directions have seen research activity. For the first direction, algorithms typically involve transmitting some future control inputs at every time step, or the actuator using some linear combination of past control inputs if an erasure occurs. The second direction has identified the stability conditions that are necessary for any causal algorithm. Moreover, recursive designs that can achieve stability when these conditions are satisfied have also been identified. Surprisingly, the design is in the form of a *universal actuator* that does not require access to the model of the plant. Even if such knowledge were available, the stability conditions do not change. Thus, the design is stability optimal.
- *Presence of a Communication Network:* So far we have concentrated on the case when the sensor and the controller are connected using a single communication channel. A typical scenario, particularly in a wireless context, would instead involve a communication network with multiple such channels. If no encoding algorithm is implemented, and every node in the network (including the sensor) transmits simply the measurements, the network can be replaced by a giant erasure channel with the equivalent erasure probability being some measure of the reliability of the network. The analysis in Section 75.2 carries over to this case; however, the performance degrades rapidly as the network size increases. If encoding is permitted, such an equivalence breaks down. The optimal algorithm is an extension of the single channel case, and is provided in [6]. The stability and performance

calculations are considerably more involved. However, the stability condition has an interesting interpretation in terms of the capacity for fluid networks. The necessary and sufficient condition for stability can be expressed as the inequality

$$p_{\text{max-cut}}\rho(A)^2 < 1,$$

where $p_{\text{max-cut}}$ is the max-cut probability calculated in a manner similar to the min-cut capacity of fluid networks. We construct cut-sets by dividing the nodes in the network into two sets with one set containing the sensor, and the other containing the controller. For each cut-set, we calculate the cut-set erasure probability by multiplying the erasure probabilities of all the channels from the set containing the sensor to the set containing the controller. The maximum such cut-set erasure probability (over all possible cut-sets) denotes the max-cut probability of the network. The improvement in the performance and stability region of the system by using the encoding algorithm increases drastically with the size and the complexity of the network.

- *Multiple Sensors:* Another direction in which the above framework can be extended is to consider multiple sensors observing the same process. As with the case with one sensor, one can identify the necessary stability conditions and a lower bound for the achievable cost function with any causal coding algorithm. These stability conditions are also sufficient and recursive algorithms for achieving stability when these conditions are satisfied have been identified. These conditions are a natural extension of the stability conditions for the single sensor case. As an example, for the case of two sensors described by sensing matrices C_1 and C_2 that transmit data to the controller across erasure channels for which erasure events are i.i.d. with probabilities p_1 and p_2 respectively, the stability conditions are given by the set

$$p_2\rho(A_1)^2 < 1, \quad p_1\rho(A_2)^2 < 1, \quad p_1p_2\rho(A)^2 < 1,$$

where $\rho(A_i)$ denotes the spectral radius of the unobservable part of the system matrix A, when the pair (A, C_i) is represented in the observability cano + nical form. However, the problem of identifying distributed encoding algorithms to be followed at each sensor for achieving the lower bounds on the achieved cost function remains largely open. This problem is related to the track fusion problem that considers identifying algorithms for optimal fusion of information from multiple sensors that interact intermittently (e.g., see [1]). That transmitting estimates based on local data from each sensor is not optimal is long known. While algorithms that achieve a performance close to the lower bound of the cost function have been identified, a complete solution is not available.
- *Inclusion of More Communication Effects:* Our discussion has focussed on modeling the loss of data transmitted over the channel. In our discussion of the optimal encoding algorithms, we also briefly considered the possibility of data being delayed or received out of order. An important direction for future work is to consider other effects due to communication channels. Both from a theoretical perspective, and for many applications such as underwater systems, an important effect is to impose a limit on the number of bits that can be communicated for every successful transmission. Some recent work [11,13] has considered the analog digital channel in which the channel supports n bits per time step and transmits them with a certain probability p at every time step. Stability conditions for such a channel have been identified and are a natural combination of the stability conditions for the analog erasure channel above and the ones for a noiseless digital channel, as considered elsewhere in the book. The performance of optimal encoding algorithms and the optimal performance that is achievable remain unknown. Another channel effect that has largely been ignored is the addition of channel noise to the data received successfully.
- *More General Performance Criteria:* Our treatment focussed on a particular performance measure—a quadratic cost, and the stability notions emanating from that measure. Other cost functions may be relevant in applications. Thus the cost function may be related to target tracking, measures such as H_2 or H_∞ [15], or some combination of communication and control costs. The analysis and

optimal encoding algorithms for such measures are expected to differ significantly. An an example, for target tracking, the properties of the reference signal that needs to be tracked can be expected to play a significant role. Similarly, for H_∞ related costs, the sufficient statistic, and hence the encoding algorithms to transmit it, may be vastly different than the LQG case. Finally, a distributed control problem with multiple processes, sensors, and actuators is a natural direction to consider.

- *More General Plant Dynamics:* The final direction is to consider plant dynamics that are more general than the linear model that we have considered. Moving to models such as jump linear systems, hybrid systems, and general nonlinear systems will provide new challenges and results. As an example, for nonlinear plants, concepts such as spectral radius no longer hold. Thus, the analysis techniques are likely to be different and measures such as Lyapunov exponents and the Lipschitz constant for the dynamics will likely become important.

75.5 Some Results on Markovian Jump Linear Systems

We present a short overview of Markov jump linear systems. A more thorough and complete treatment is given in [2]. Consider a discrete-time discrete-state Markov process with state $r(k) \in \{1, 2, \cdots, m\}$ at time k. Denote the transition probability $\text{Prob}(r(k+1) = j | r(k) = i)$ by q_{ij}, and the resultant transition probability matrix by Q. Also denote

$$\text{Prob}(r(k) = j) = \pi_j(k),$$

with $\pi_j(0)$ as given. The evolution of MJLS, denoted by $\mathcal{S}_1$ for future reference, can be described by the following equations:

$$\begin{aligned} x(k+1) &= A_{r(k)}x(k) + B_{r(k)}u(k) + F_{r(k)}w(k), \\ y(k) &= C_{r(k)}x(k) + G_{r(k)}v(k), \end{aligned} \tag{75.1}$$

where $w(k)$ is zero mean white Gaussian noise with covariance R_w, $v(k)$ is zero mean white Gaussian noise with covariance R_v and the notation $X_{r(k)}$ implies that the matrix $X \in \{X_1, X_2, \ldots, X_m\}$ with the matrix X_i being chosen when $r(k) = i$. The initial state $x(0)$ is assumed to be a zero mean Gaussian random variable with variance $\Pi(0)$. For simplicity, we will consider $F_{r(k)} = G_{r(k)} \equiv I$ for all values of $r(k)$ in the sequel. We also assume that $x(0)$, $\{w(k)\}$, $\{v(k)\}$ and $\{r(k)\}$ are mutually independent.

75.5.1 LQ Control

The Linear Quadratic Regulator (LQR) problem for the system $\mathcal{S}_1$ is posed by assuming that the noises $w(k)$ and $v(k)$ are not present. Moreover, the matrix $C_{r(k)} \equiv I$ for all choices of the state $r(k)$. The problem aims at designing the control input $u(k)$ to minimize the finite horizon cost function

$$J_{LQR} = \sum_{k=1}^{K} \left(E_{\{r(j)\}_{j=k+1}^{K}} \left[x^T(k)Qx(k) + u^T(k)Ru(k) \right] \right) + x^T(K+1)P(K+1)x(K+1),$$

where the expectation at time k is taken with respect to the future values of the Markov state realization, and $P(K+1)$, Q and R are all assumed to be positive definite. The controller at time k has access to control inputs $\{u(j)\}_{j=0}^{k-1}$, state values $\{x(j)\}_{j=0}^{k}$, and the Markov state values $\{r(j)\}_{j=0}^{k}$. Moreover, the system is said to be stabilizable if the infinite horizon cost function $J_\infty \stackrel{def}{=} \lim_{K\to\infty} \frac{J_{LQR}}{K}$ is finite.

The solution to this problem can readily be obtained through dynamic programming arguments. The optimal control is given by the following result.

Theorem 75.4:

Consider the LQR problem posed above for the system $\mathcal{S}_1$.

- *At time k, if $r(k) = i$, then the optimal control input is given by*

$$u(k) = -\left(R + B_i^T P_i(k+1) B_i\right)^{-1} B_i^T P_i(k+1) A_i x(k),$$

where for $j = 1, 2, \ldots, m$,

$$P_j(k) = \sum_{t=1}^{m} q_{tj} \left(Q + A_t^T P_t(k+1) A_t - A_t^T P_t(k+1) B_t \left(R + B_t^T P_t(k+1) B_t\right)^{-1} B_t^T P_t(k+1) A_t\right),$$

and $P_j(K+1) = P(K+1), \forall j = 1, 2, \ldots, m$.

- *Assume that the Markov states reach a stationary probability distribution. A sufficient condition for stabilizability of the system is that there exist m positive-definite matrices $X_1, X_2, \ldots, X_m$ and m^2 matrices $K_{1,1}, K_{1,2}, \ldots, K_{1,m}, K_{2,1}, \ldots, K_{m,m}$ such that for all $j = 1, 2, \ldots, m$,*

$$X_j > \sum_{i=1}^{m} q_{ij} \left((A_i^T + K_{i,j} B_i^T) X_i (A_i^T + K_{i,j} B_i^T)^T + Q + K_{ij} R K_{ij}^T\right).$$

Note that the sufficient condition can be cast in alternate forms as linear matrix inequalities, which can be efficiently solved. We omit such representations. A special case of Markov jump linear systems is when the discrete states are chosen independently from one time step to the next. Since this is the case we have concentrated on in this chapter, we summarize the results pertaining to this case below.

Corollary 75.1:

Consider system $\mathcal{S}_1$ with the additional assumption that the Markov transition probability matrix is such that for all states i and j, $q_{ij} = q_i$ (in other words, the states are chosen independently and identically distributed from one time step to the next). Consider the LQR problem posed above for the system $\mathcal{S}_1$.

- *At time k, if $r(k) = i$, then the optimal control input is given by*

$$u(k) = -\left(R + B_i^T P(k+1) B_i\right)^{-1} B_i^T P(k+1) A_i x(k),$$

where

$$P(k) = \sum_{t=1}^{m} q_t \left(Q + A_t^T P(k+1) A_t - A_t^T P(k+1) B_t \left(R + B_t^T P(k+1) B_t\right)^{-1} B_t^T P(k+1) A_t\right).$$

- *Assume that the Markov states reach a stationary probability distribution. A sufficient condition for stabilizability of the system is that there exists a positive-definite matrix X, and m matrices K_1, $K_2, \ldots, K_m$ such that*

$$X > \sum_{i=1}^{m} q_i \left((A_i^T + K_i B_i^T) X (A_i^T + K_i B_i^T)^T + Q + K_i R K_i^T\right).$$

75.5.2 MMSE Estimation

The MMSE estimate problem for the system $\mathcal{S}_1$ is posed by assuming that the control $u_{r(k)}$ is identically zero. The objective is to identify at every time step k an estimate $\hat{x}(k+1)$ of the state $x(k+1)$ that minimizes the mean-squared error covariance

$$\Pi(k+1) = E_{\{w(j)\},\{v(j)\}}\left[(x(k+1)-\hat{x}(k+1))(x(k+1)-\hat{x}(k+1))^T\right],$$

where the expectation is taken with respect to the process and measurement noises (but not the Markov state realization). The estimator at time k has access to observations $\{y(j)\}_{j=0}^{k}$ and the Markov state values $\{r(j)\}_{j=0}^{k}$. Moreover, the error covariance is said to be stable if the expected steady-state error covariance $\lim_{k\to\infty} E_{\{r(j)\}_{j=0}^{k-1}}[\Pi(k)]$ is bounded, where the expectation is taken with respect to the Markov process. The estimator at time k has access to the measurements $y(0), y(1), \ldots, y(k)$, and Markov state values $r(0), r(1), \ldots, r(k)$.

Since the estimator has access to the Markov state values till time k, the optimal estimate can be calculated through a time-varying Kalman filter. Thus, if at time k, $r_k = i$, the estimate evolves as

$$\hat{x}(k+1) = A_i\hat{x}(k) + K(k)\left(y(k) - C_i\hat{x}(k)\right),$$

where

$$K(k) = A_i\Pi(k)C_i^T\left(C_i\Pi(k)C_i^T + R_v\right)^{-1}$$

$$\Pi(k+1) = A_i\Pi(k)A_i^T + R_w - A_i\Pi(k)C_i^T\left(C_i\Pi(k)C_i^T + R_v\right)^{-1}C_i\Pi(k)A_i^T.$$

The error covariance $\Pi(k)$ is available through the above calculations. However, calculating $E_{\{r(j)\}_{j=0}^{k-1}}[\Pi(k)]$ seems to be intractable. Instead, we present an upper bound to this quantity* that will also help in obtaining sufficient conditions for the error covariance to be stable.

The intuition behind obtaining the upper bound is simple. The optimal estimator presented above optimally utilizes the information about the Markov states till time k. Consider an alternate estimator that at every time step k averages over the values of the Markov states $r_0, \ldots, r_{k-1}$. Such an estimator is suboptimal and the error covariance for this estimator forms an upper bound for $E_{\{r(j)\}_{j=0}^{k-1}}[\Pi(k)]$. A formal proof using Jensen's inequality is given in [5, Theorem 5]. We present the statement below while omitting the proof.

Theorem 75.5:

The term $E_{\{r(j)\}_{j=0}^{k-1}}[\Pi(k)]$ *obtained from the optimal estimator is upper bounded by* $M(k) = \sum_{j=1}^{m} M_j(k)$ *where*

$$M_j(k) = \sum_{t=1}^{m} q_{tj}\left(R_w + A_tM_t(k-1)A_t^T - A_tM_t(k-1)C_t^T\left(R_v + C_tM_t(k-1)C_t^T\right)^{-1}C_tM_t(k-1)A_t^T\right),$$

with $M_j(0) = \Pi(0)\ \forall j$. *Moreover, assume that the Markov states reach a stationary probability distribution. A sufficient condition for stabilizability of the system is that there exist* m *positive-definite matrices* $X_1, X_2, \ldots, X_m$ *and* m^2 *matrices* $K_{1,1}, K_{1,2}, \ldots, K_{1,m}, K_{2,1}, \ldots, K_{m,m}$ *such that for all* $j = 1, 2, \ldots, m$,

$$X_j > \sum_{i=1}^{m} q_{ij}\left((A_i + K_{i,j}C_i)X_i(A_i + K_{i,j}C_i)^T + R_w + K_{ij}R_vK_{ij}^T\right).$$

---

* We say that A is upperbounded by B if $B - A$ is positive semi definite.

We can once again consider the special case of states being chosen in an independent and identically distributed fashion.

Corollary 75.2:

Consider the estimation problem posed above for the system $\mathcal{S}_1$ with the additional assumption that the Markov transition probability matrix is such that for all states i and j, $q_{ij} = q_i$ (in other words, the states are chosen independently and identically distributed from one time step to the next). The term $\mathrm{E}_{\{r(j)\}_{j=0}^{k-1}}[\Pi(k)]$ obtained from the optimal estimator is upper bounded by $M(k)$, where

$$M(k) = \sum_{t=1}^{m} q_t \left(R_w + A_t M(k-1) A_t^T - A_t M(k-1) C_t^T \left(R_v + C_t M(k-1) C_t^T \right)^{-1} C_t M(k-1) A_t^T \right),$$

with $M(0) = \Pi(0)$. Further, a sufficient condition for stabilizability of the system is that there exists a positive-definite matrix X, and m matrices $K_1, K_2, \ldots, K_m$ such that

$$X > \sum_{i=1}^{m} q_i \left((A_i + K_i C_i) X (A_i + K_i C_i)^T + R_w + K_i R_v K_i^T \right).$$

75.5.3 LQG Control

The LQG problem for the system $\mathcal{S}_1$ aims at designing the control input $u(k)$ to minimize the finite horizon cost function

$$J_{LQG} = \mathrm{E}\left[\sum_{k=1}^{K} \left(x^T(k) Q x(k) + u^T(k) R u(k) \right) + x^T(K+1) P(K+1) x(K+1) \right],$$

where the expectation at time k is taken with respect to the future values of the Markov state realization, and the measurement and process noises. Further, the matrices $P(K+1)$, Q and R are all assumed to be positive definite. The controller at time k has access to control inputs $\{u(j)\}_{j=0}^{k-1}$, measurements $\{y(j)\}_{j=0}^{k}$ and the Markov state values $\{r(j)\}_{j=0}^{k}$. The system is said to be stabilizable if the infinite horizon cost function $J_\infty \stackrel{def}{=} \lim_{K\to\infty} \frac{J_{LQG}}{K}$ is finite.

The solution to this problem is provided by a separation principle and using Theorems 75.4 and 75.5. We have the following result.

Theorem 75.6:

Consider the LQG problem for the system $\mathcal{S}_1$. At time k, if $r(k) = i$, then the optimal control input is given by

$$u(k) = -\left(R + B_i^T P_i(k+1) B_i \right)^{-1} B_i^T P_i(k+1) A_i \hat{x}(k),$$

where for $P_i(k)$ is calculated as in Theorem 75.4 and $\hat{x}(k)$ is calculated using a time-varying Kalman filter.

We can also obtain the conditions for stabilizability of the system by utilizing Theorems 75.4 and 75.5.

References

1. K. C. Chang, R. K. Saha, and Y. Bar-Shalom, On optimal track-to-track fusion, *IEEE Trans. Aerospace Electronic Systems*, AES-33(4):1271–1276, 1997.
2. O. L. V. Costa, M. D. Fragoso, and R. P. Marques, Discrete-time Markov jump linear systems, Springer, Berlin, 2006.
3. D. Dacic and D. Nesic, Quadratic stabilization of linear networked control systems via simultaneous controller and protocol synthesis, *Automatica*, 43(7):1145–1155, 2007.
4. E. O. Elliott, Estimates of error rates for codes on burst-noise channels, *Bell Systems Tech. Jo.* 42: 1977–1997, 1963.
5. V. Gupta, T. Chung, B. Hassibi, and R. M. Murray, On a stochastic sensor selection algorithm with applications in Sensor scheduling and dynamic sensor coverage, *Automatica*, 42(2):251–260, 2006.
6. V. Gupta, A. F. Dana, J. Hespanha, R. M. Murray, and B. Hassibi, Data transmission over networks for estimation and control, *IEEE Trans. Automat. Control*, 54(8):1807–1819, August 2009.
7. V. Gupta, D. Spanos, B. Hassibi, and R. M. Murray, Optimal LQG control across a packet-dropping link, *Systems and Controls Letters*, 56(6):439–446, 2007.
8. J. Hespanha, P. Naghshtabrizi, and Y. Xu, A Survey of recent results in networked control systems. *Proc. of IEEE, Special Issue on Techno. Networked Control Systems*, 95(1): 138–162, 2007.
9. O.C. Imer, S. Yuksel and T. Basar, Optimal control of LTI systems over communication networks, *Automatica*, 42(9):1429–1440, 2006.
10. T. Kailath, A.H. Sayed and B. Hassibi, *Linear Estimation*, Prentice-Hall, Englewood Cliffs, NJ, 2000.
11. N. C. Martins and M. A. Dahleh, Feedback control in the presence of noisy channels: 'Bode-like' fundamental limitations of performance, *IEEE Transac. Automat. Control*, 52(7):1604–1615, 2008.
12. A. Matveev and A. Savkin, The problem of state estimation via asynchronous communication channels with irregular transmission times, *IEEE Trans. Automat. Control*, 48(4):670–676, 2003.
13. P. Minero, M. Franceschetti, S. Dey, and G. Nair, Data rate theorem for stabilization over time-varying feedback channels, *IEEE Trans. Automat. Control*, 54(2): 243–255, February 2009.
14. L. Schenato, B. Sinopoli, M. Franceschetti, K. Poolla, and S. S. Sastry, foundations of control and estimation over Lossy networks, *Proc. IEEE*, 95(1):163–187, 2007.
15. P. Seiler and R. Sengupta, An H_∞ approach to networked control, *IEEE Trans. Automat. Control*, 50(3):356–364, 2005.

76

Passivity Approach to Network Stability Analysis and Distributed Control Synthesis

Murat Arcak
University of California, Berkeley

76.1 Introduction

This chapter reviews a passivity-based approach for the design and analysis of networked nonlinear systems. This approach exploits the structure of the network and breaks up the design and analysis procedures into two levels: At the network level, one represents the components with input/output properties, such as passivity and other forms of dissipativity, as abstractions of their complex dynamic models and determines which input/output properties guarantee stability and other desirable properties for the given network structure. This means that the study of the network does not rely on detailed knowledge of the components and does not require the components to be homogeneous. At the component level, one studies the individual dynamic models and verifies or assigns appropriate input/output properties without relying on further knowledge of the network.

The question of when a network of dissipative dynamic systems is stable was studied in the early Refs. [1,2] and several tests were developed. From today's standpoint, these studies are encompassed by the elegant and unifying framework of Integral Quadratic Constraints [3] discussed elsewhere in this handbook. The objectives of this review are

1. To present recurrent interconnection structures in several modern networks and to show that a key input/output property compatible with these structures is *passivity*.
2. To illustrate the verification and assignment of passivity properties in these network models.

The second task is particularly challenging in applications where the equilibrium of the network model and other global parameters are not available to the components.

We illustrate the passivity approach with case studies in communication networks, motion coordination of autonomous agents, and biochemical reaction networks. In Section 76.2 we study an interconnection structure that arises in decentralized resource allocation algorithms in communication networks, including congestion control for the Internet discussed in Section 76.2.1 and power control for wireless networks discussed in Section 76.2.2. This structure exhibits a symmetry in the coupling of the components, which implies that a passivity property of the update algorithms (for packet rates, transmission power levels, etc.) guarantees stability of the network. With this observation as a starting point, we present a systematic framework which, rather than giving specific algorithms, prescribes a passivity constraint that new classes of algorithms must satisfy.

In Section 76.3, we show that a similar symmetric-coupling structure arises in motion coordination when the information flow between the agents is bidirectional. Passivity-based design for motion coordination is related to potential function-based schemes, such as those in [4–6]. By making explicit the inherent passivity property in these schemes, in Section 76.3.1 we present a systematic framework that allows complex, nonlinear agent models and that offers additional design flexibility. The ability of the passivity framework to address complex agent dynamics is illustrated with an attitude coordination design for rigid body models in Section 3.2. The design flexibility is illustrated with an adaptive redesign in Section 76.3.3.

In Section 76.4, we present a passivity-based analysis of biochemical reaction networks. In Section 76.4.1 we discuss a cyclic interconnection structure that is common in biological systems and show how the passivity-based approach recovers and significantly strengthens the *secant criterion* [7,8] used by mathematical biologists for studying the stability of such networks. After identifying passivity as the core of the secant criterion, in Section 76.4.2 we generalize this criterion to other topologies using the concept of *diagonal stability* [9,10]. Finally, in Section 76.4.3 we employ passivity properties to ensure robustness against the destabilizing effect of diffusion in spatially distributed models.

Following [11], we say that a dynamic system

$$\dot{x} = f(x, u), \quad y = h(x, u), \tag{76.1}$$

where input u and output y are of the same dimension, is *passive* if there exists a continuously differentiable *storage function* $S(x) \geq 0$ such that

$$\dot{S} = \nabla S(x) f(x, u) \leq u^T y. \tag{76.2}$$

The Passivity Theorem [12], when adapted to systems in state-space form, states that the negative feedback interconnection of two passive blocks as in Figure 76.1a with positive definite storage functions is stable. Indeed, the sum of the storage functions of individual blocks serves as a composite Lyapunov function for the interconnected system. A further property of passive systems which is particularly useful in this article is that postmultiplication of the output y by a matrix and premultiplication of the input u by the transpose of the same matrix as in Figure 76.1b maintain passivity with respect to the new input–output pair $(\bar{u}, \bar{y})$.

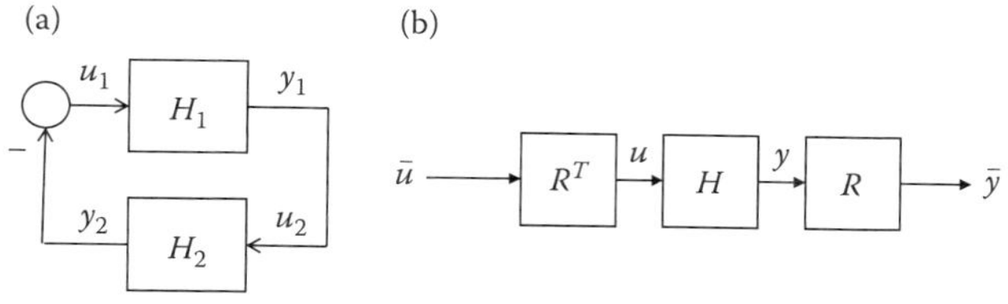

FIGURE 76.1 (a) Negative feedback interconnection of two passive systems. (b) Postmultiplication of the output y by a matrix and premultiplication of the input u by the transpose of the same matrix maintain passivity with respect to the new input–output pair $(\bar{u}, \bar{y})$.

76.2 Resource Allocation in Communication Networks

76.2.1 A Unifying Passivity Framework for Internet Congestion Control

Congestion control algorithms aim to maximize network throughput while ensuring an equitable allocation of bandwidth to the users. In a decentralized congestion control scheme, each link increases its packet drop or marking probability (interpreted as the "price" of the link) as the transmission rate approaches the capacity of the link. Sources then adjust their sending rates based on the aggregate price feedback they receive in the form of dropped or marked packets. Reference [13] showed that a passivity-based design is particularly suitable for this decentralized feedback structure and developed a unifying design methodology.

To see the interconnection structure of sources and links, consider a network where packets from sources $i = 1, \ldots, N$ are routed through links $\ell = 1, \ldots, L$ according to an $L \times N$ routing matrix R in which the (ℓ, i) entry is 1 if source i uses link ℓ and 0 otherwise. Because the transmission rate y_ℓ of link ℓ is the sum of the sending rates x_i of sources using that link, the vectors of link rates y and source rates x are related by

$$y = Rx. \tag{76.3}$$

Likewise, since the total price feedback q_i received by source i is the sum of the prices p_ℓ of the links on its path, the vectors q and p are related by

$$q = R^T p. \tag{76.4}$$

The resulting feedback loop is depicted in Figure 76.2 where the blocks Σ_i represent decentralized update algorithms for source rates x_i, and Λ_ℓ represent decentralized update algorithms for link prices p_ℓ.

The congestion control problem as formulated by Kelly et al. [14] and studied by numerous other authors (see the excellent reviews [15,16]) is to design these algorithms in such a way that the closed-loop system is stable and the network equilibrium (x^*, y^*) solves the optimization problem:

$$\max_{x_i \geq 0} \sum_{i=1}^{N} U_i(x_i) \quad \text{s.t.} \quad y_\ell \leq c_\ell, \tag{76.5}$$

where $U_i(\cdot)$ is a concave utility function for source i and c_ℓ is the capacity of link ℓ. The key observation in [13] is that postmultiplication by R and premultiplication by R^T in Figure 76.2 preserve passivity properties and thus, a design that renders Σ_i passive from input $q_i - q_i^*$ to output $-(x_i - x_i^*)$ and Λ_ℓ passive from input $y_\ell - y_\ell^*$ to output $p_\ell - p_\ell^*$ guarantees stability of the network according to the Passivity Theorem.

The passivity framework streamlines the design process by eliminating the need for a separate network stability analysis for every new algorithm. Indeed, numerous existing update rules already satisfy the

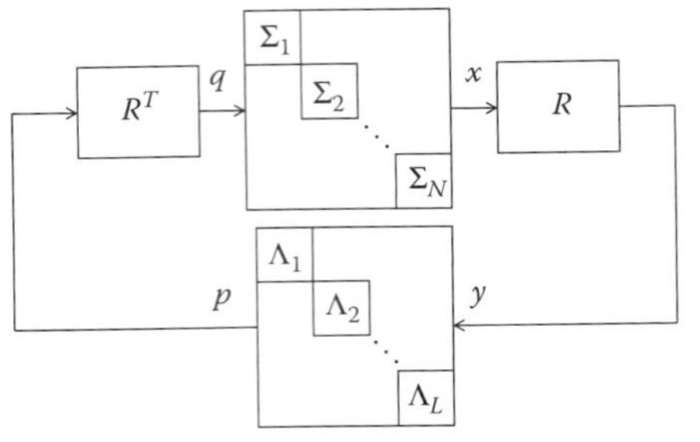

FIGURE 76.2 The feedback structure arising in decentralized congestion control. R is the routing matrix, Σ_i blocks represent the source algorithms for updating sending rates x_i, and Λ_l blocks represent the link algorithms for updating link prices p_l.

desired passivity property, such as Kelly's *primal* algorithm:

$$\dot{x} = k_i(U_i'(x_i) - q_i), \quad p_\ell = h_\ell(y_\ell), \tag{76.6}$$

where $k_i > 0$, $U_i'(\cdot)$ is the derivative of the utility function $U_i(\cdot)$, and $h_\ell(\cdot)$ is a penalty function that grows rapidly as y_ℓ approaches the link capacity c_ℓ. The network equilibrium resulting from this algorithm approximates the solution of the Kuhn–Tucker optimality conditions for Equation 76.5 and the stability of this equilibrium follows from passivity properties established in [13] by exploiting the monotone decreasing property of $U_i'(\cdot)$ and monotone increasing property of $h_\ell(\cdot)$.

As an illustration of the design flexibility offered by the passivity framework [13] presented new classes of algorithms, including the following extension of the source control algorithm in Equation 76.6

$$\dot{\xi}_i = A_i\xi_i + B_i(U_i'(x_i) - q_i), \tag{76.7}$$

$$\dot{x}_i = C_i\xi_i + D_i(U_i'(x_i) - q_i), \tag{76.8}$$

where the matrices A_i, B_i, C_i, D_i must be selected such that the transfer function $H_i(s) = C_i(sI - A_i)^{-1}B_i + D_i$ is *strictly positive-real* (SPR). The benefit of increasing the dynamic order of Equation 76.6 is demonstrated in [13] with an example where the SPR filter design adds phase lead to counter time delays.

A passivity analysis and a dynamic extension was also pursued in [13] for Kelly's *dual* algorithm where, in contrast to the *primal* algorithm (Equation 76.6), the source controller is static and the link controller is dynamic. In addition, the passivity approach has made stability proofs and systematic Lyapunov function constructions possible for *primal-dual* algorithms where both source and link controllers are dynamic. As an illustration, for the following algorithm in [16]

$$\dot{x}_i = f_i(x_i)(U_i'(x_i) - q_i), \tag{76.9}$$

$$\dot{p}_\ell = g_\ell(p_\ell)(y_\ell - c_\ell), \tag{76.10}$$

where $f_i(\cdot)$ and $g_\ell(\cdot)$ are positive-valued functions, a Lyapunov function constructed from a sum of storage functions is

$$V = \sum_{i=1}^{N} \int_{x_i^*}^{x_i} \frac{x - x_i^*}{f_i(x)}\,dx + \sum_{\ell=1}^{M} \int_{p_\ell^*}^{p_\ell} \frac{p - p_\ell^*}{g_\ell(p)}\,dp. \tag{76.11}$$

Lyapunov functions obtained within the passivity framework have also been instrumental in robust redesigns against disturbances, time delays, and uncooperative users [17,18].

76.2.2 CDMA Power Control Game

Another important resource allocation problem for communication networks is uplink transmission power control in code division multiple access (CDMA) systems. Increased power levels ensure longer transmission distance and higher data transfer rate, but also increase battery consumption and interference to neighboring users. A particularly elegant approach to this problem is a game-theoretic formulation [19] where the cost function for the ith mobile is

$$J_i = P_i(p_i) - U_i(\gamma_i(p)), \quad i = 1, \ldots, N. \tag{76.12}$$

$P_i(\cdot)$ is a penalty function for the power level p_i and $U_i(\cdot)$ is a utility function for the signal-to-interference ratio (SIR), given by

$$\gamma_i(p) = \frac{Lh_i p_i}{\sigma^2 + \sum_{k \neq i} h_k p_k}, \tag{76.13}$$

where h_i is the channel gain between the ith mobile and the base station, L is the spreading gain and σ^2 is the noise variance.

To develop a distributed power update law, [20] employed the logarithmic utility function

$$U_i(\gamma_i) = \log(L + \gamma_i) \tag{76.14}$$

and studied a first-order gradient algorithm which, upon algebraic manipulations, is given by the expression

$$\dot{p}_i = -\lambda_i \frac{\partial J_i}{\partial p_i} = -\lambda_i P_i'(p_i) + \lambda_i \frac{h_i}{\sigma^2 + \sum_k h_k p_k}. \tag{76.15}$$

This algorithm is to be implemented by the mobiles with the help of the feedback signal:

$$u_i = \frac{h_i}{\sigma^2 + \sum_k h_k p_k} \tag{76.16}$$

received from the base station. The resulting feedback structure of the network is depicted in Figure 76.3 where Σ_i represents the update law (Equation 76.15) for p_i, h denotes the column vector of channel gains, and Equation 76.16 is represented as a function of $y := \sum_k h_k p_k$ in the feedback block.

The equilibrium p^* of Equation 76.15 is unique when $P_i(\cdot)$ is strictly convex and coincides with the Nash equilibrium for the game defined by the cost function (Equation 76.12). Global stability of this equilibrium follows from the Passivity Theorem because premultiplication by h and postmultiplication by h^T in Figure 76.3 preserve passivity of the Σ_i blocks in the feedforward path from input $u_i - u_i^*$ to output $y_i - y_i^*$. Likewise, the function $\phi(y)$ is strictly increasing and thus, the feedback block is passive from input $y - y^*$ to output $\phi(y) - \phi(y^*)$.

Reference [20] uses this observation as a starting point to develop classes of passive power update laws and base-station algorithms that include Equations 76.15 and 76.16 as special cases. It further employs the

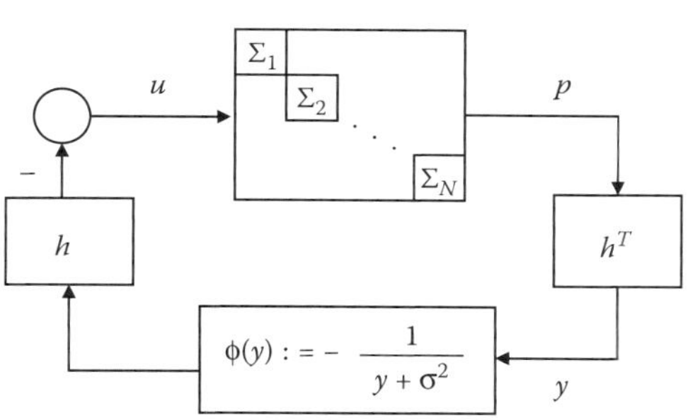

FIGURE 76.3 The feedback structure arising from the distributed power update law (Equation 76.15). The Σ_i blocks represent the update laws $\dot{p}_i = -\lambda_i P_i'(p_i) + \lambda_i u_i$, h denotes the column vector of channel gains, and Equation 76.16 is represented as a function of $y := \sum_k h_k p_k$ in the feedback block.

Lyapunov function obtained from the Passivity Theorem to study robustness of power control algorithms against a time-varying channel gain $h(t)$ which is assumed to be constant in the nominal stability analysis. A related paper [21] studies a team-optimization approach to power control rather than the game-theoretic formulation discussed above and pursues a passivity-based design.

76.3 Distributed Feedback Design for Motion Coordination

76.3.1 Passivity-Based Design Procedure for Position Coordination

We represent a group of agents and their communication structure with a graph that consists of N nodes connected by M links. The presence of a link between nodes i and j means that agents i and j have access to the relative distance information $x_i - x_j$. We assign an orientation to each link and recall that the $N \times M$ *incidence matrix* D of the graph is defined as

$$d_{ik} = \begin{cases} +1 & \text{if node } i \text{ is the positive end of link } k \\ -1 & \text{if node } i \text{ is the negative end of link } k \\ 0 & \text{otherwise.} \end{cases} \tag{76.17}$$

The assignment of orientation is for analysis only, and the particular choice does not change the results. Our objective is to develop distributed feedback laws that obey the information structure defined by this graph and that guarantee the following group behaviors:

P1: The velocity of each node approaches a common velocity vector $v^d(t)$ prescribed for the group; that is, $\lim_{t\to\infty}(\dot{x}_i - v^d(t)) = 0,\ i = 1, \ldots, N.$

P2: If nodes i and j are neighbors connected by link k, then the difference variable

$$z_k := \sum_{\ell=1}^{N} d_{\ell k} x_\ell = \begin{cases} x_i - x_j & \textit{if } i \textit{ is the positive end of link } k \\ x_j - x_i & \textit{if } i \textit{ is the negative end of link } k \end{cases} \tag{76.18}$$

converges to a prescribed set $\mathcal{A}_k$, $k = 1, \ldots, M$.

Examples of sets $\mathcal{A}_k$ include the origin in a rendezvous problem, or a sphere if the positions of agents must maintain a prescribed distance in a formation. From Equation 76.18, the concatenated vectors

$$x := [x_1^T \cdots x_N^T]^T, \quad z := [z_1^T \cdots z_M^T]^T \tag{76.19}$$

satisfy

$$z = (D^T \otimes I)x, \tag{76.20}$$

where I is an identity matrix with dimension consistent with that of x_i and "$\otimes$" represents the Kronecker product.

To achieve P1 and P2 [22] presented a two-step design procedure: Step 1 is to design an internal feedback loop for each node that achieves passivity from an external feedback signal u_i^{ext}, left to be designed in Step 2, to the velocity error $y_i := \dot{x}_i - v^d(t)$. This design step is depicted with a block diagram in Figure 76.4, where H_i represents the open-loop dynamic model of agent i and $\tilde{H}_i$ represents the passive block obtained from the internal feedback design. Step 2 is to design an external feedback law of the form

$$u_i^{\text{ext}} := -\sum_{k=1}^{M} d_{ik}\, \psi_k(z_k), \tag{76.21}$$

where z_k's are the relative distance variables as in Equation 76.18 and the multivariable nonlinearities $\psi_k(\cdot)$ are to be designed. The feedback law (Equation 76.21) is implementable with locally available signals because $d_{ik} \neq 0$ only for links k that are connected to node i.

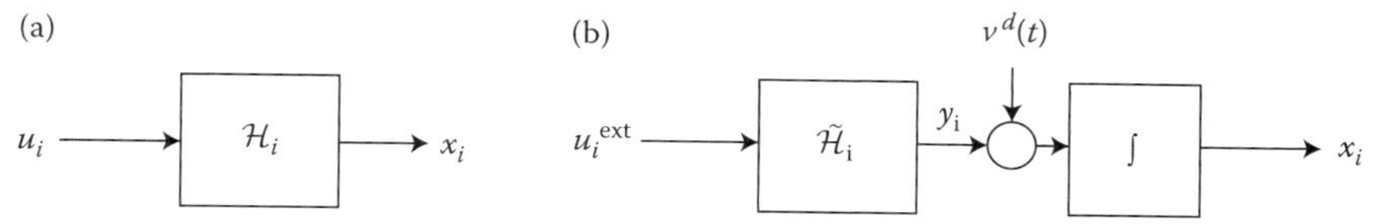

FIGURE 76.4 The task of the internal feedback design is to render the plant (a) passive from the external signal u_i^{ext} to the velocity error $y_i := \dot{x}_i - v^d(t)$. With this internal feedback, the dynamics take the form of (b) where the $\tilde{\mathcal{H}}_i$ block is passive.

The combination of the internal and external feedback loops result in the interconnected system in Figure 76.5, where asymptotic stabilization of the set

$$\mathcal{A} = \left\{(z, y) | \; y = 0, \; z \in \{\mathcal{A}_1 \times \cdots \times \mathcal{A}_M\} \cap \mathcal{R}(D^T \otimes I)\right\} \tag{76.22}$$

is synonymous to achieving objectives P1–P2 above. To accomplish this stabilization task, we exploit the interconnection structure in Figure 76.5 where premultiplication by $D^T \otimes I$ and postmultiplication by its transpose $D \otimes I$ preserve passivity properties of the feedforward path. The input $1_N \otimes v^d(t)$ does not affect the feedback loop in Figure 76.5 because it lies in the null space of $D^T \otimes I$. We design the nonlinearities ψ_k to be passive when cascaded with an integrator as in Figure 76.5 so that the feedforward path is passive from input y to output $-u_{ext}$ and, thus, the feedback loop is stable from the Passivity Theorem.

To guarantee passivity of the nonlinearity $\psi_k(z_k)$ preceded by an integrator, we let

$$\psi_k(z_k) = \nabla P_k(z_k), \tag{76.23}$$

where $P_k(z_k)$ is a nonnegative and sufficiently smooth function defined on an open set $\mathcal{G}_k$ in which z_k is allowed to evolve. To steer z_k to $\mathcal{A}_k$ while keeping it within $\mathcal{G}_k$, we construct the function $P_k(z_k)$ to grow unbounded as z_k approaches the boundary of $\mathcal{G}_k$, and let $P_k(z_k)$ and its gradient $\nabla P_k(z_k)$ vanish on the set $\mathcal{A}_k$. Using this construction and the passivity properties of the feedback and feedforward paths in Figure 76.5, [22] proves asymptotic stability of the set $\mathcal{A}$.

As an illustration, consider the point mass model

$$\ddot{x}_i = u_i, \tag{76.24}$$

where $x_i \in \mathbb{R}^2$ is the position of each mass and $u_i \in \mathbb{R}^2$ is the force input. The internal feedback

$$u_i = -K_i(\dot{x}_i - v^d(t)) + \dot{v}^d(t) + u_i^{ext}, \quad K_i = K_i^T > 0 \tag{76.25}$$

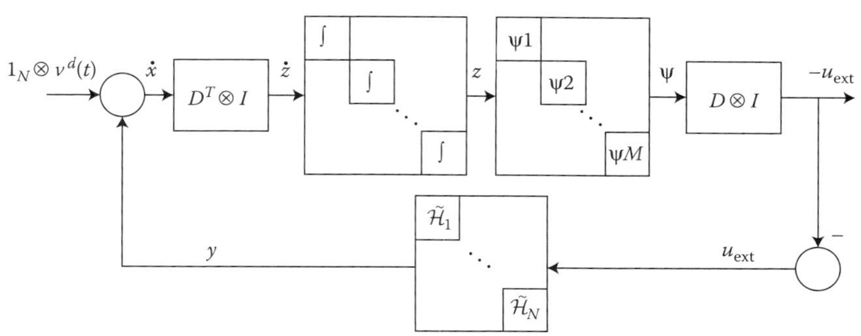

FIGURE 76.5 A block diagram representation for the interconnection of the subsystems in Figure 76.4 via the external feedback (Equation 76.21).

and the change of variables $y_i = \dot{x}_i - v^d(t)$ bring Equation 76.24 to the form

$$\dot{x}_i = y_i + v^d(t), \tag{76.26}$$

$$\dot{y}_i = -K_i y_i + u_i^{\text{ext}}, \tag{76.27}$$

where the y_i-subsystem with input u_i^{ext} plays the role of the passive block $\tilde{\mathcal{H}}_i$ in Figure 76.4.

To create and stabilize an equilateral triangle formation with unit side lengths while avoiding collisions, we let $\mathcal{A}_k$ be the unit circle, $\mathcal{G}_k = \mathbb{R}^2 - \{0\}$, and let the potential functions be of the form

$$P_k(z_k) = \int_1^{|z_k|} \sigma_k(s)ds, \quad k = 1, 2, 3, \tag{76.28}$$

where $\sigma_k : \mathbb{R}_{>0} \to \mathbb{R}$ is a continuously differentiable, strictly increasing function such that

$$\sigma_k(1) = 0 \quad \lim_{s\to\infty} \sigma_k(s) = -\infty, \quad \lim_{s\to 0} \sigma_k(s) = -\infty \tag{76.29}$$

and such that, as $|z_k| \to 0$, $P_k(z_k) \to \infty$. Then, the interaction forces

$$\psi_k(z_k) = \nabla P_k(z_k) = \sigma_k(|z_k|)\frac{1}{|z_k|}z_k \quad z_k \neq 0 \tag{76.30}$$

guarantee asymptotic stability of the desired formation. In particular, $\sigma_k(|z_k|)$ creates an attraction force when $|z_k| > 1$ and a repulsion force when $|z_k| < 1$.

76.3.2 From Point Mass to Rigid-Body Models

A key advantage of the passivity framework is its ability to address high-order and complex agent dynamics by exploiting their inherent passivity properties. As an illustration, we now study a rigid-body model and design a controller that achieves identical orientation and synchronous rotation of the agents. This means that the objectives P1 and P2 in Section 76.3.1 above must now be modified as

A1: The angular velocity of each agent converges to the group reference $\omega^d(t)$; that is, $\lim_{t\to\infty} ({}^i\omega_i - \omega^d(t)) = 0$, $i = 1, \ldots N$, where ${}^i\omega_i$ denotes the angular velocity of agent i in the ith body frame.

A2: Each agent achieves the same attitude as its neighbors in the limit; that is, the orientation matrices R_i satisfy $\lim_{t\to\infty} R_i^T R_j = I$, $i, j = 1, \ldots, N$.

To achieve objectives A1 and A2 we first design an internal feedback loop τ_i for each agent $i = 1, \ldots, N$ that renders the attitude dynamics

$$\mathcal{G}_i : {}^i\mathcal{I}_i\, {}^i\dot{\omega}_i + {}^i\omega_i \times {}^i\mathcal{I}_i\, {}^i\omega_i = \tau_i \tag{76.31}$$

passive from an external input signal τ_i^{ext} left to be designed, to the angular velocity error

$$\Delta\omega_i := {}^i\omega_i - \omega^d(t). \tag{76.32}$$

One such controller is

$$\tau_i = {}^i\mathcal{I}_i\, \dot{\omega}^d + \omega^d \times {}^i\mathcal{I}_i\, {}^i\omega_i - f_i\Delta\omega_i + \tau_i^{\text{ext}}, \quad f_i > 0, \tag{76.33}$$

which indeed achieves strict passivity from τ_i^{ext} to $\Delta\omega_i$ in the error dynamics system:

$$\tilde{\mathcal{G}}_i : {}^i\mathcal{I}_i \Delta\dot{\omega}_i + \Delta\omega_i \times {}^i\mathcal{I}_i\, {}^i\omega_i = -f_i\Delta\omega_i + \tau_i^{\text{ext}}. \tag{76.34}$$

Other designs (possibly using dynamic controllers) that achieve passivity from τ_i^{ext} to $\Delta\omega_i$ may also be employed in this framework. This design flexibility is illustrated in the next section with an adaptive redesign.

To achieve objective A2 using only relative attitude information, we parameterize the relative orientation matrix

$$\tilde{R}^k := \begin{cases} R_i^T R_j & \text{if node } i \text{ is the positive end of link } k \\ R_j^T R_i & \text{if node } i \text{ is the negative end of link } k \end{cases} \tag{76.35}$$

using one of the standard parameterizations of $SO(3)$, such as the *unit quaternion* representation $q^k = \begin{bmatrix} q_0^k & q_v^{k\,T} \end{bmatrix}^T$, where q_v^k denotes the vector part of the quaternion. We then design an external torque feedback of the form

$$\tau_i^{\text{ext}} = \sum_{l \in \mathcal{N}_i^+} q_v^l - \sum_{p \in \mathcal{N}_i^-} q_v^p, \tag{76.36}$$

where $\mathcal{N}_i^+$ denotes the set of links for which node i is the positive end and $\mathcal{N}_i^-$ is the set of links for which node i is the negative end. To synthesize this external feedback signal, agent i obtains its neighbors' relative attitudes with respect to its own frame, parameterizes them by unit quaternions, and adds up their vector parts.

The closed-loop system resulting from the internal and external feedback laws (Equations 76.33 and 76.36) is depicted in Figure 76.6, where the J_k blocks represent the quaternion kinematics, which possess passivity properties from the relative angular velocity $\tilde{\omega}^k$ to the vector component q_v^k of the unit quaternion representation [23]. To incorporate the rotation matrices between the body frames, we replace the matrix $D \otimes I$ in Figure 76.5 with a new $3N \times 3M$ *rotational incidence matrix* $\bar{D}$, which consists of the 3×3 sub-blocks:

$$\bar{d}_{ik} =: \begin{cases} -I & \text{if node } i \text{ is the positive end of link } k \\ (\tilde{R}^k)^T & \text{if node } i \text{ is the negative end of link } k \\ \mathbf{0} & \text{otherwise.} \end{cases} \tag{76.37}$$

As shown in [24], stability of the closed-loop system in Figure 76.6 follows from the Passivity Theorem because premultiplication by $\bar{D}^T$ and postmultiplication by $\bar{D}$ preserve passivity of the feedforward path, and because the feedback path is passive by the internal feedback design. Although the foregoing arguments are based on the unit quaternion representation of $SO(3)$, they have been generalized in [24] to other parameterizations.

76.3.3 Adaptive Redesign

Thus far we assumed that the reference velocities $v^d(t)$ and $\omega^d(t)$ in objectives P1 and A1 above are available to each agent in the group. A more realistic situation is when a leader in the group possesses or autonomously determines this information, while others have access only to the relative distance and relative orientation with respect to their neighbors. Can the agents estimate $v^d(t)$ and $\omega^d(t)$ online from

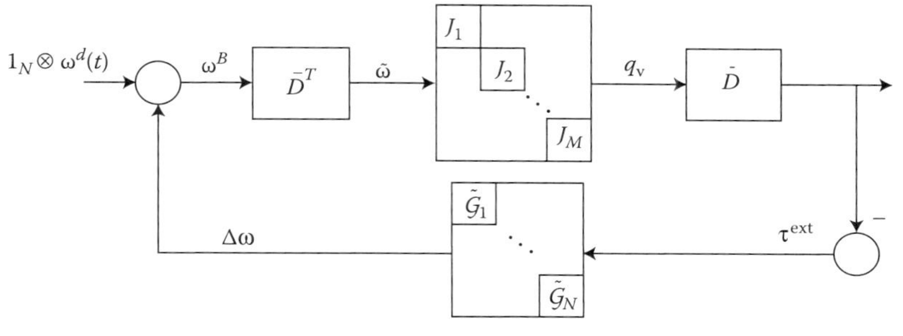

FIGURE 76.6 A block diagram representation for the network resulting from the attitude coordination scheme (Equations 76.33 and 76.36). The concatenated vector ω^B consists of the angular velocities of the agents in their own body frames and the *rotational incidence matrix* $\bar{D}$ is as defined in Equation 76.37.

this relative distance and orientation information? We now present an adaptive redesign from [25] that accomplishes this task by relying only on the connectivity of the graph and the passivity properties of the interconnected system.

This adaptive redesign modifies the internal feedback loop to assign the estimate $\hat{v}_i$ and ${}^i\hat{\omega}_i$ to agent i, instead of the unknown references $v^d(t)$ and $\omega^d(t)$. To develop update laws for $\hat{v}_i$ and ${}^i\hat{\omega}_i$, we parameterize $v^d(t)$ and $\omega^d(t)$ as

$$v^d(t) = \sum_j \phi^j(t)\theta^j, \quad \omega^d(t) = \sum_j \gamma^j(t)\beta^j, \tag{76.38}$$

where $\phi^j(t)$ and $\gamma^j(t)$ are scalar base functions known to each agent, and θ^j and β^j are vectors available only to the leader. Agent i estimates the unknown θ^j and β^j by $\hat{\theta}_i^j$ and $\hat{\beta}_i^j$, and reconstructs $\hat{v}_i(t)$ and ${}^i\hat{\omega}_i(t)$ from

$$\hat{v}_i(t) = \sum_j \phi^j(t)\hat{\theta}_i^j, \quad {}^i\hat{\omega}_i(t) = \sum_{j=1} \gamma^j(t)\hat{\beta}_i^j. \tag{76.39}$$

The update laws proposed in [25] are of the form

$$\dot{\hat{\theta}}_i = \Lambda_i(\Phi(t) \otimes I)u_i^{\text{ext}}, \quad \dot{\hat{\beta}}_i = \Delta_i(\Gamma(t) \otimes I)\tau_i^{\text{ext}}, \tag{76.40}$$

where $\Lambda_i = \Lambda_i^T > 0$ and $\Delta_i = \Delta_i^T > 0$ are adaptation gain matrices, $\Phi(t)$, $\Gamma(t)$, $\hat{\theta}_i$, $\hat{\beta}_i$ denote concatenations of $\phi^j(t)$, $\gamma^j(t)$, $\hat{\theta}_i^j$, $\hat{\beta}_i^j$ respectively, and u_i^{ext} and τ_i^{ext} are the external force and torque feedback laws. The adaptation of $\hat{\theta}_i$ and $\hat{\beta}_i$ continues until the group reaches the desired position and orientation configuration, in which case the external force and torque feedback signals employed in Equation 76.40 vanish.

Reference [25] proves that this adaptive scheme indeed stabilizes the desired configuration by exploiting the passivity of the adaptation algorithm, which is depicted in Figure 76.7 for position coordination as an additional module in the existing passive feedback loop. Parameter convergence is established under additional conditions. One situation in which parameter convergence is guaranteed is when the reference velocity v^d is constant and the graph is connected, in which case the proof in [25] makes use of the Krasovskii–LaSalle Invariance Principle. Another situation is when the target sets in P2 are $\mathcal{A}_k = \{0\}$ and

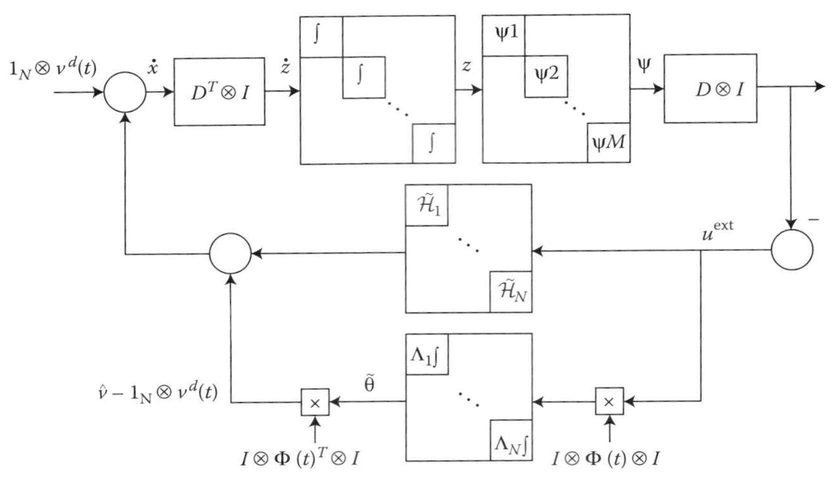

FIGURE 76.7 Block diagram for the adaptive scheme (Equations 76.21, 76.39, and 76.40) for position coordination. Λ_i must be interpreted as zero for the leader. $\tilde{\theta}$ denotes the concatenation of the vectors $\hat{\theta}_i - \theta$. The adaptive module in the feedback path preserves the passivity properties of the closed-loop system.

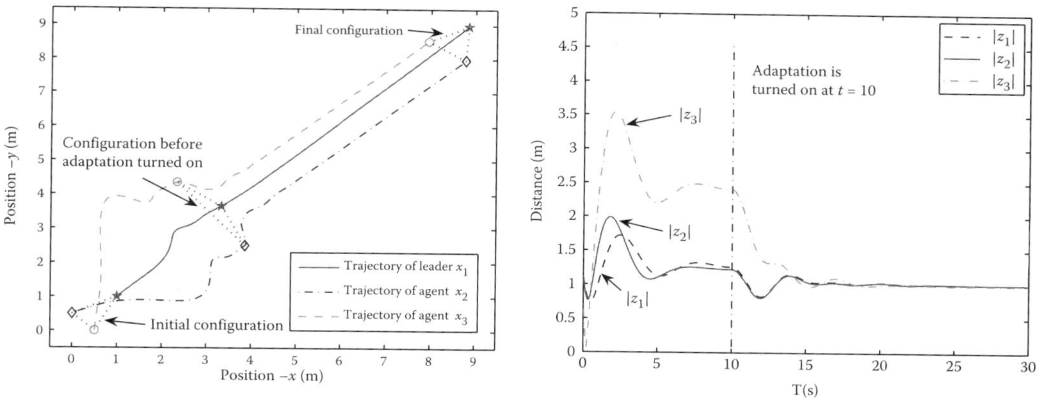

FIGURE 76.8 Left: Snapshots of the formation in the adaptive design with constant reference velocity v^d. The adaptation is turned on at $t = 10$, after which point the trajectories converge to the desired formation. Right: The relative distance variables z_k plotted as a function of time.

the *regressor vector* $\Phi(t)$ is *persistently exciting*, in which case parameter convergence is established with an application of the Teel–Matrosov Theorem [26].

As an illustration of the adaptive redesign, we now revisit the example (Equation 76.24) and suppose that $v^d(t)$ is available only to agent 1. We modify the feedback law (Equation 76.25) for the agents $i = 2, 3$ as

$$u_i = -K_i(\dot{x}_i - \hat{v}_i) + \dot{\hat{v}}_i + u_i^{\text{ext}}, \quad K_i = K_i^T > 0, \tag{76.41}$$

where the signal $\hat{v}_i$ and its derivative $\dot{\hat{v}}_i$ are available for implementation from the parametrization (Equation 76.39) and the update law (Equation 76.40). In the simulation presented in Figure 76.8, we take the constant reference velocity $v^d = [0.2 \ \ 0.2]^T$ and start with the adaptation turned off. Since agents 2, 3 possess incorrect information about v^d and since there is no adaptation, the relative distances z_k do not converge to their prescribed sets $\mathcal{A}_k = \{z_k : |z_k| = 1\}$. At $t = 10$, we turn on the adaptation for agents 2 and 3, which results in convergence to the desired distances $|z_k| = 1$ asymptotically.

As a case study for the adaptive redesign, [27] investigated a *gradient climbing* problem in which the leader performs *extremum seeking* to reach the minima or maxima of a field distribution and the other agents maintain a formation with respect to the leader. To incorporate an extremum seeking algorithm in the motion, [27] let the reference velocity $v^d(t)$ be determined autonomously by the leader, in the form of segments $v_k^d(t)$, $t \in [t_k, t_{k+1}]$, that are updated in every iteration according to the next Newton step. To calculate this Newton step, the leader performs a dither motion from which it collects samples of the field and generates finite-difference approximations for the gradient and the Hessian. The adaptive redesign discussed above treats the Newton direction as the unknown parameter vector in Equation 76.38 and allows the followers to estimate this direction. This case study raised several new problems which led to a refinement of the basic adaptive procedure. Among these problems is a judicious tuning of the design parameters to ensure that the followers respond to the Newton motion while filtering out the dither component.

76.4 Passivity Approach to Biochemical Reaction Networks

76.4.1 The Secant Criterion for Cyclic Networks

Reference [28] proposed a passivity-based analysis technique for *cyclic* biochemical reaction networks, where the end product of a sequence of reactions inhibits the first reaction upstream. This technique recovered the *secant criterion* [7,8] developed earlier by mathematical biologists for the local stability of

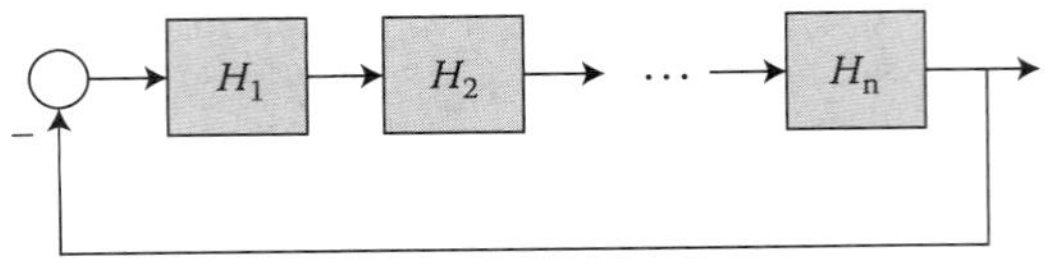

FIGURE 76.9 Cyclic feedback interconnection of dynamic blocks $H_1, \ldots, H_n$.

such reactions and strengthened it to become a global stability test. Cyclic reaction networks are of great interest because, as surveyed in [29], they are widespread in gene regulation, cell signaling, and metabolic pathways. Unlike positive feedback systems which constitute a subclass of *monotone systems* [30], the negative feedback due to inhibition gives rise to the possibility of attractive periodic orbits. Indeed, a Poincaré–Bendixson Theorem proven in [31] for negative feedback cyclic systems of arbitrary order shows that bounded trajectories converge either to fixed points or to periodic orbits. Stability criteria for cyclic networks are thus important for determining which parameter regimes guarantee convergence to fixed points and which regimes yield oscillations.

To evaluate local stability properties of negative feedback cyclic systems, [7,8] analyzed the Jacobian linearization, represented in Figure 76.9 as the feedback interconnection of first-order linear blocks $H_i(s) = \gamma_i/(\tau_i s + 1)$. They then proved that the interconnected system is Hurwitz if the dc gains γ_i satisfy the secant criterion:

$$\gamma_1 \cdots \gamma_n < \sec(\pi/n)^n. \tag{76.42}$$

In contrast to a small-gain condition which would restrict the right-hand side of Equation 76.42 to be 1, the secant criterion exploits the phase properties of the feedback loop and allows the gain to be arbitrarily large when $n = 2$, and to be as high as 8 when $n = 3$.

Local stability of the equilibrium proven in [7,8], however, does not rule out the possibility of periodic orbits as shown in [29] with the example:

$$\dot{x}_1 = -x_1 + \varphi(x_3), \quad \dot{x}_2 = -x_2 + x_1, \quad \dot{x}_3 = -x_3 + x_2, \tag{76.43}$$

where the nonlinearity $\varphi(x_3) = e^{-10(x_3-1)} + 0.1\text{sat}(25(x_3 - 1))$ has a negative slope of magnitude $\gamma_3 = 7.5$ at the unique equilibrium $x_1 = x_2 = x_3 = 1$. With $\gamma_1 = \gamma_2 = 1$, the local secant criterion (Equation 76.42) guarantees asymptotic stability of the equilibrium as in Figure 76.10a. However, the numerical simulation in Figure 76.10b indicates that an attractive periodic orbit exists in addition.

To develop a global stability test for cyclic networks, [28] exploited a passivity property that is implicit in the local secant criterion [32]. To make this property explicit, the first step in [28] is to break down the network into n nonlinear subsystems representing the dynamics of each species, interconnected according to the cyclic structure in Figure 76.5. The second step is to verify, for each block H_i, the *output strict passivity* (OSP) property [33,34]:

$$\dot{S}_i \leq -y_i^2 + \gamma_i u_i y_i, \tag{76.44}$$

where u_i and y_i denote the input and the output of H_i, and $S_i(x_i - x_i^*)$ is a positive definite *storage function* of the deviation of the concentration x_i from its equilibrium value x_i^*. The third step is to construct a Lyapunov function for the network from a weighted sum of these storage functions S_i and to prove that a set of weights that render its derivative negative definite exists if and only if the secant criterion (Equation 76.42) holds. In this new procedure, the first-order blocks $H_i(s) = \gamma_i/(\tau_i s + 1)$ employed in the local secant criterion are replaced by nonlinear passive systems with *OSP gain* γ_i as in Equation 76.44 and the secant condition (Equation 76.42) guarantees global asymptotic stability for the network.

How does one verify the OSP property (Equation 76.44) and estimate the gain γ_i? Although this task may appear intractable for highly uncertain biological models, [28,35] gave procedures to verify OSP without explicit knowledge of the nonlinearities and the equilibrium value x_i^*. Instead, one verifies OSP by qualitative arguments that exploit the monotone increasing or decreasing properties of the nonlinearities,

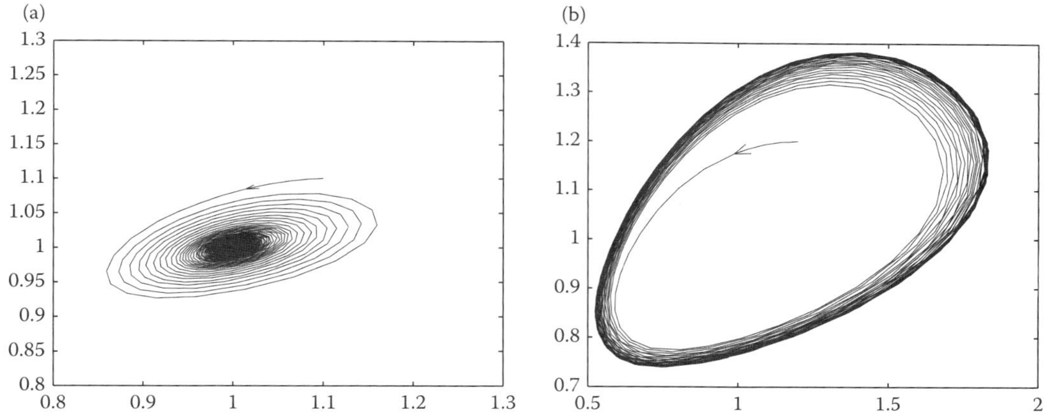

FIGURE 76.10 The trajectories of Equation 76.43 starting from initial conditions (a) $x = [1.1\ 1.1\ 1]$, and (b) $x = [1.2\ 1.2\ 1.2]$, projected onto the $x_1 - x_2$ plane. The trajectory in (a) converges to the equilibrium $x_1 = x_2 = x_3 = 1$, while the trajectory in (b) converges to a periodic orbit.

such as Michaelis–Menten and Hill equations that arise in activation and inhibition models in enzyme kinetics [36]. Once OSP is verified, an upper bound on the gain γ_i is obtained by inspecting the maximum slope of the *steady-state characteristic curve* $\bar{y}_i = k_i(\bar{u}_i)$, where $\bar{y}_i$ denotes the steady-state response of the output y_i to a constant input $u_i = \bar{u}_i$.

As an illustration of the global secant test, consider the following simplified model of a mitogen activated protein kinase (MAPK) cascade with inhibitory feedback, proposed in [37,38]:

$$\dot{x}_1 = -\frac{b_1 x_1}{c_1 + x_1} + \frac{d_1(1-x_1)}{e_1 + (1-x_1)}\frac{\mu}{1+kx_3} \tag{76.45}$$

$$\dot{x}_2 = -\frac{b_2 x_2}{c_2 + x_2} + \frac{d_2(1-x_2)}{e_2 + (1-x_2)}x_1 \tag{76.46}$$

$$\dot{x}_3 = -\frac{b_3 x_3}{c_3 + x_3} + \frac{d_3(1-x_3)}{e_3 + (1-x_3)}x_2, \tag{76.47}$$

where the variables $x_i \in [0, 1]$ denote the concentrations of the active forms of the proteins, and the terms $1 - x_i$ correspond to the inactive forms (after an appropriate nondimensionalization that scales the sum of the active and inactive concentrations to 1). The inhibition of x_1 by x_3 in this model is due to the decreasing function $\mu/(1 + kx_3)$ in Equation 76.45, the steepness of which is determined by the parameter k. With the coefficients $b_1 = e_1 = c_1 = b_2 = 0.1$, $c_2 = e_2 = c_3 = e_3 = 0.01$, $b_3 = 0.5$, $d_1 = d_2 = d_3 = 1$, $\mu = 0.3$, we obtained the OSP gains γ_i numerically for various values of k. This numerical experiment showed that the secant condition $\gamma_1\gamma_2\gamma_3 < 8$ is satisfied in the range $k \leq 4.35$, which reduces the gap between the small-gain estimate $k \leq 3.9$ given in [39] and the Hopf bifurcation value $k = 5.1$. The secant test further guarantees global asymptotic stability which cannot be ascertained from a bifurcation analysis.

76.4.2 Generalization to Other Network Topologies

The passivity-based analysis outlined above makes the key property behind the local secant criterion explicit, extends it to be a global stability test, and further opens the door to a generalization of this test to network topologies other than the cyclic structure. Reference [35] achieved this generality by representing the reaction network with a directed graph and by making use of the concept of *diagonal stability* [9,10] to expand the passivity-based analysis to such a graph. The nodes $x_1, \ldots, x_n$ in this directed graph denote the concentrations of the species and the links $k = 1, \ldots, m$ represent the reactions, as in Figure 76.11. In particular, solid links represent activation terms which play the role of positive feedback and dashed links represent inhibitory terms which act as negative feedback.

To derive a stability test that mimics the secant criterion, [35] breaks down the network into m subsystems each associated with a link and forms an $m \times m$ *dissipativity matrix* E of the form

$$e_{lk} = \begin{cases} -1/\gamma_l & \text{if } k = l \\ \text{sign(link } k) & \text{if source}(l) = \text{sink}(k) \\ 0 & \text{otherwise,} \end{cases} \tag{76.48}$$

where γ_l is the OSP gain for lth subsystem as in Equation 76.44 and the sign of a link is $+1$ or -1 depending on whether it represents positive or negative feedback. This matrix incorporates information about the passivity properties of the subsystems, the interconnection structure of the network, and the signs of the interconnection terms.

To determine the stability of the network, [35] checks the *diagonal stability* of E; that is, the existence of a diagonal solution $P > 0$ to the Lyapunov equation $E^T P + PE < 0$. When such a P exists, its diagonal entries serve as the weights of the storage functions in a composite Lyapunov function. It is important to note that here diagonal stability is not used as a local stability test, but is employed to construct a composite Lyapunov function as in the large-scale systems literature [2,40,41]. By taking into account the signs of the off-diagonal terms in Equation 76.48, however, the diagonal stability test exploits the phase properties of the feedback loops in the network and avoids the conservatism of small-gain-type dominance approaches prevalent in large-scale systems studies.

This diagonal stability test encompasses the secant criterion because, as shown in [28], when E is constructed according to the cyclic interconnection topology, its diagonal stability is equivalent to the secant condition (Equation 76.42). For other practically important network structures, [35] obtained variants of the secant criterion by investigating when the dissipativity matrix E is diagonally stable. As an illustration, for the feedback configurations (a) and (b) in Figure 76.11, the dissipativity matrices obtained according to Equation 76.48 are

$$E_a = \begin{bmatrix} -\frac{1}{\gamma_1} & 0 & 0 & -1 \\ 1 & -\frac{1}{\gamma_2} & -1 & 0 \\ 0 & 1 & -\frac{1}{\gamma_3} & 0 \\ 0 & 1 & 0 & -\frac{1}{\gamma_4} \end{bmatrix}, \quad E_b = \begin{bmatrix} -\frac{1}{\gamma_1} & 0 & 0 & 1 \\ 1 & -\frac{1}{\gamma_2} & -1 & 0 \\ 0 & 1 & -\frac{1}{\gamma_3} & 0 \\ 0 & 1 & 0 & -\frac{1}{\gamma_4} \end{bmatrix}. \tag{76.49}$$

As shown in [35], matrix E_a is diagonally stable if $\gamma_1\gamma_2\gamma_4 < 8$, and E_b is diagonally stable if $\gamma_1\gamma_2\gamma_4 < 1$. For the feedback configuration in Figure 76.11c, the dissipativity matrix is

$$E_c = \begin{bmatrix} -\frac{1}{\gamma_1} & 0 & 0 & 1 & 0 \\ 1 & -\frac{1}{\gamma_2} & -1 & 0 & 0 \\ 0 & 1 & -\frac{1}{\gamma_3} & 0 & 1 \\ 0 & 1 & 0 & -\frac{1}{\gamma_4} & 1 \\ 0 & 0 & 0 & 1 & -\frac{1}{\gamma_5} \end{bmatrix} \tag{76.50}$$

and a necessary condition for its diagonal stability is

$$\gamma_1\gamma_2\gamma_4 + \gamma_4\gamma_5 < 1. \tag{76.51}$$

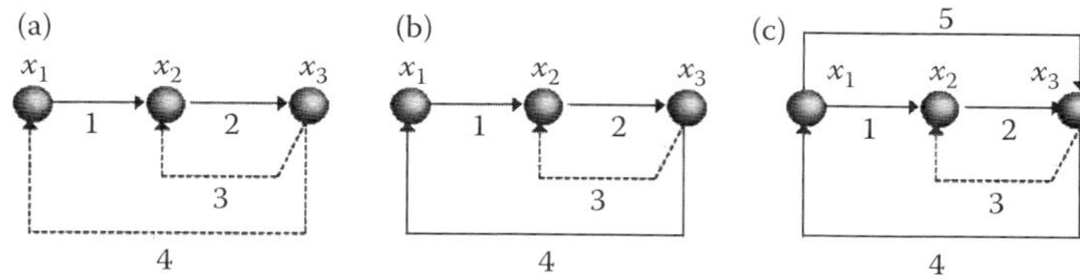

FIGURE 76.11 Three feedback configurations proposed in [42] for MAPK networks in PC-12 cells. The nodes x_1, x_2, and x_3 represent Raf-1, Mek1/2, and Erk1/2, respectively. The dashed links indicate negative feedback signals. Depending on whether the cells are activated with (a) epidermal or (b) neuronal growth factors, the feedback from Erk1/2 to Raf-1 changes sign. (c) An increased connectivity from Raf-1 to Erk1/2 is noted in [42] when neuronal growth factor activation is observed over a longer period.

Although the necessary condition (Equation 76.51) does not depend on γ_3, a numerical investigation shows that, unlike the feedback configurations (a) and (b), γ_3 is implicated in the diagonal stability for the configuration in Figure 76.11c.

76.4.3 Passivity as a Certificate of Robustness to Diffusion

Thus far we assumed a well-mixed reaction environment and represented the concentration of each species $i = 1, \ldots, n$ with a lumped variable x_i. The study of spatially distributed reaction models is of interest because the presence of diffusion in the spatial domain can lead to subtle instability mechanisms in an otherwise stable reaction system [43]. In contrast, the passivity-based stability tests presented in Sections 76.4.1 and 76.4.2 rule out such mechanisms and guarantee robustness against diffusion. References [29,35] studied both reaction–diffusion PDE models and compartmental ODE models where the compartments represent a discrete set of spatial domains as further described below, and proved that the secant condition and its generalization in Section 76.4.2 above guarantee global asymptotic stability of the spatially homogeneous equilibrium. This homogeneous behavior is illustrated in Figure 76.12 on the MAPK example (Equations 76.45 through 76.47) where the concentrations $x_i(t, \xi)$ are now functions of the spatial variable ξ, and the dynamic equation for each $i = 1, 2, 3$ is augmented with diffusion terms.

The structural property that ensures robustness against diffusion is particularly transparent in a compartmental ODE model in which the state vector X^j incorporates the concentrations x_i^j of species $i = 1, \ldots, n$ in compartment $j = 1, \ldots, N$. To make the structure of this $N n$th order ODE explicit we introduce a new, undirected, graph $\mathcal{G}$ in which the nodes represent the compartments and the links describe the interconnection of the compartments. Denoting by $L_{\mathcal{G}}$ the Laplacian matrix for this graph, we obtain the block diagram in Figure 76.13, where the feedforward blocks Σ_j are copies of the lumped

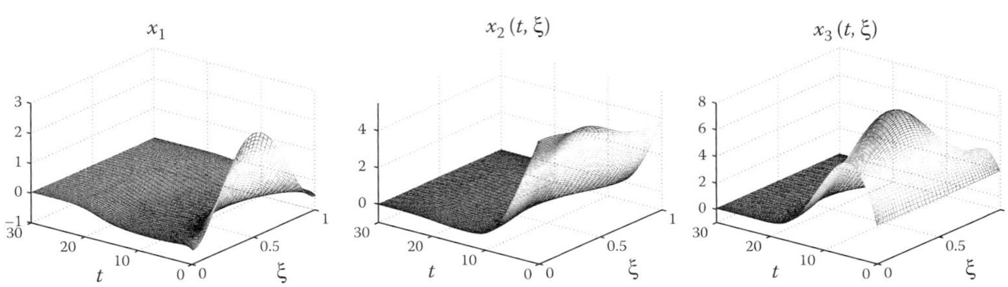

FIGURE 76.12 Solutions of the MAPK model (Equations 76.45 through 76.47) augmented with diffusion terms: ξ represents the spatial coordinate and the solutions converge to the spatially homogeneous equilibrium.

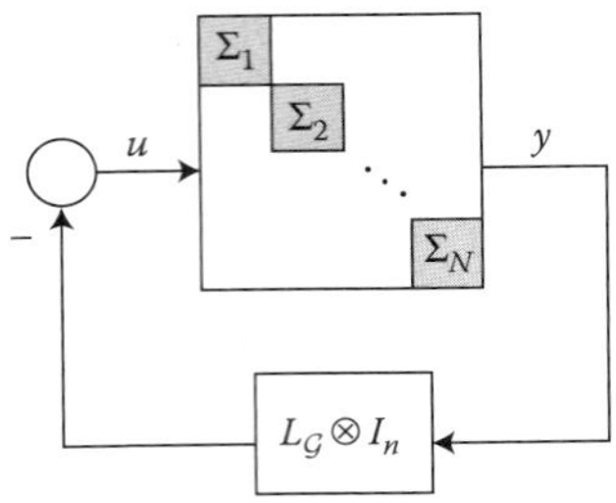

FIGURE 76.13 Diffusive coupling between the compartments Σ_j in Equation 76.52. $L_{\mathcal{G}}$ is the Laplacian matrix for the graph $\mathcal{G}$ representing the interconnection of the compartments, and the concatenated vectors u and y denote $u = [u_1^T \cdots u_N^T]^T$ and $y = [y_1^T \cdots y_N^T]^T$.

reaction model perturbed by the diffusion input u_j:

$$\Sigma_j : \dot{X}^j = F(X^j) + u_j, \quad y_j = CX^j, \quad u_j, y_j, X^j \in \mathbb{R}^n, \tag{76.52}$$

and C is a diagonal matrix whose entries are the diffusion coefficients of the species.

Stability of the interconnection in Figure 76.13 then follows from the results of [44] on positive operators with repeated monotone nonlinearities, extended to multivariable nonlinearities in [45]. If the decoupled system (Equation 76.52) with $u_j = 0$ admits a Lyapunov function $V(X^j)$, then, for the coupled system, the sum of these Lyapunov functions for each compartment satisfies

$$\frac{d}{dt}\sum_{j=1}^{N} V(X^j) = \sum_{j=1}^{N} \nabla V(X^j)F(X^j) - [\nabla V(X^1) \cdots \nabla V(X^N)](L_{\mathcal{G}} \otimes C)\begin{bmatrix} X^1 \\ \vdots \\ X^N \end{bmatrix}, \tag{76.53}$$

where the first term on the right-hand side is negative definite. The second term is due to the coupling of the compartments and includes the repeated nonlinearity $\nabla V(\cdot)$. Because the graph Laplacian matrix $L_{\mathcal{G}}$ is *doubly hyperdominant with zero excess*, it follows from [44,45] that the second term on the right-hand side of Equation 76.53 is nonpositive if $\nabla V(C^{-1}\cdot)$ is a *monotone mapping* as defined in [45]. Indeed, under mild additional assumptions, the Lyapunov functions $V(X^j)$ constructed in [28,29,35] consist of a sum of convex storage functions of x_i^j which guarantee the desired monotonicity property, thus proving negative definiteness of Equation 76.53.

76.5 Conclusions and Future Topics

The notion of passivity emerged from energy conservation and dissipation concepts in electrical and mechanical systems [11] and became a fundamental tool for nonlinear system design and analysis [33,34]. In this chapter we showed that passivity is a powerful design and analysis approach for several types of networks. We have further identified recurrent interconnection structures in these networks, such as the *symmetric-coupling* structure in Figures 76.2, 76.3, 76.5 through 76.7 and 76.13 and the cyclic structure in Figure 76.9, which are well suited to this passivity approach.

Verification and assignment of passivity properties were hampered by the unavailability of the network equilibrium to the components in the communication and biological network examples in Sections 76.2 and 76.4. This difficulty was overcome by exploiting monotone increasing or decreasing properties of nonlinearities which led to *incremental* forms of passivity that do not depend on the equilibrium location. Likewise, the unavailability of the network reference velocity in the motion coordination study of Section 76.3 was overcome with an adaptive redesign. Despite these encouraging results, further studies are needed for achieving passivity of the components when global network parameters are unavailable.

References

1. P.J. Moylan and D.J. Hill. Stability criteria for large-scale systems. *IEEE Transactions on Automatic Control*, 23(2):143–149, 1978.
2. M. Vidyasagar. *Input–Output Analysis of Large Scale Interconnected Systems*. Springer-Verlag, Berlin, 1981.
3. A. Megretski and A. Rantzer. System analysis via integral quadratic constraints. *IEEE Transactions on Automatic Control*, 42:819–830, 1997.
4. P. Ögren, E. Fiorelli, and N.E. Leonard. Cooperative control of mobile sensor networks: Adaptive gradient climbing in a distributed network. *IEEE Transactions on Automatic Control*, 49(8):1292–1302, 2004.
5. V. Gazi and K.M. Passino. Stability analysis of social foraging swarms. *IEEE Transactions on Systems, Man, and Cybernetics—Part B: Cybernetics*, 34(1):539–557, 2004.
6. H.G. Tanner, A. Jadbabaie, and G.J. Pappas. Flocking in fixed and switching networks. *IEEE Transactions on Automatic Control*, 52(5):863–868, 2007.
7. J.J. Tyson and H.G. Othmer. The dynamics of feedback control circuits in biochemical pathways. In *Progress in Theoretical Biology*, R. Rosen and F.M. Snell, Ed., Vol. 5, pp. 1–62. Academic Press, New York, NY, 1978.
8. C.D. Thron. The secant condition for instability in biochemical feedback control—Parts I and II. *Bulletin of Mathematical Biology*, 53:383–424, 1991.
9. R. Redheffer. Volterra multipliers—Parts I and II. *SIAM Journal on Algebraic and Discrete Methods*, 6(4):592–623, 1985.
10. E. Kaszkurewicz and A. Bhaya. *Matrix Diagonal Stability in Systems and Computation*. Birkhauser, Boston, 2000.
11. J.C. Willems. Dissipative dynamical systems. Part I: General theory; Part II: Linear systems with quadratic supply rates. *Archive for Rational Mechanics and Analysis*, 45:321–393, 1972.
12. G. Zames. On the input–output stability of time-varying nonlinear feedback systems—Parts I and II. *IEEE Transactions on Automatic Control*, 11:228–238 and 465–476, 1966.
13. J.T. Wen and M. Arcak. A unifying passivity framework for network flow control. *IEEE Transactions on Automatic Control*, 49(2):162–174, 2004.
14. F.P. Kelly, A. Maulloo, and D. Tan. Rate control in communication networks: Shadow prices, proportional fairness and stability. *Journal of the Operational Research Society*, 49:237–252, 1998.
15. S.H. Low, F. Paganini, and J.C. Doyle. Internet congestion control. *IEEE Control Systems Magazine*, 22(1):28–43, 2002.
16. R. Srikant. *The Mathematics of Internet Congestion Control*. Birkhauser, Boston, 2004.
17. X. Fan, M. Arcak, and J.T. Wen. Robustness of network flow control against disturbances and time delays. *Systems and Control Letters*, 53(1):13–29, 2004.
18. X. Fan, K. Chandrayana, M. Arcak, S. Kalyanaraman, and J.T. Wen. A two-time-scale design for edge-based detection and rectification of uncooperative flows. *IEEE/ACM Transactions on Networking*, 14(6):1313–1322, 2006.
19. T. Alpcan, T. Başar, R. Srikant, and E. Altman. CDMA uplink power control as a noncooperative game. *Wireless Networks*, 8:659–669, 2002.
20. X. Fan, T. Alpcan, M. Arcak, J.T. Wen, and T. Basar. A passivity approach to game-theoretic CDMA power control. *Automatica*, 42(11):1837–1847, 2006.
21. T. Alpcan, X. Fan, T. Basar, M. Arcak, and J.T. Wen. Power control for multicell CDMA wireless networks: A team optimization approach. *Wireless Networks*, 14(5):647–657, 2008.
22. M. Arcak. Passivity as a design tool for group coordination. *IEEE Transactions on Automatic Control*, 52(8):1380–1390, 2007.
23. F. Lizarralde and J.T. Wen. Attitude control without angular velocity measurement: A passivity approach. *IEEE Transactions on Automatic Control*, 41(3):468–472, 1996.
24. H. Bai, M. Arcak, and J. Wen. Rigid body attitude coordination without inertial frame information. *Automatica*, 44(12):3170–3175, 2008.
25. H. Bai, M. Arcak, and J.T. Wen. Adaptive design for reference velocity recovery in motion coordination. *Systems and Control Letters*, 57(8):602–610, 2008.
26. A. Loría, E. Panteley, D. Popović, and A.R. Teel. A nested Matrosov theorem and persistency of excitation for uniform convergence in stable nonautonomous systems. *IEEE Transactions on Automatic Control*, 50(2):183–198, 2005.
27. E. Biyik and M. Arcak. Gradient climbing in formation via extremum-seeking and passivity-based coordination rules. *Asian Journal of Control*, 10(2):201–211, 2008.

28. M. Arcak and E.D. Sontag. Diagonal stability of a class of cyclic systems and its connection with the secant criterion. *Automatica*, 42(9):1531–1537, 2006.
29. M. Jovanović, M. Arcak, and E. Sontag. A passivity-based approach to stability of spatially distributed systems with a cyclic interconnection structure. *IEEE Transactions on Automatic Control*, 53(1):75–86, 2008.
30. H. Smith. *Monotone Dynamical Systems: An Introduction to the Theory of Competitive and Cooperative Systems*. American Mathematical Society, Providence, RI, 1995.
31. J. Mallet-Paret and H.L. Smith. The Poincaré–Bendixson theorem for monotone cyclic feedback systems. *Journal of Dynamics and Differential Equations*, 2:367–421, 1990.
32. E.D. Sontag. Passivity gains and the "secant condition" for stability. *Systems Control Letters*, 55(3):177–183, 2006.
33. A. J. van der Schaft. *$\mathcal{L}_2$-gain and Passivity Techniques in Nonlinear Control*, 2nd edn. Springer-Verlag, New York, 2000.
34. R. Sepulchre, M. Janković, and P. Kokotović. *Constructive Nonlinear Control*. Springer-Verlag, New York, NY, 1997.
35. M. Arcak and E. Sontag. A passivity-based stability criterion for a class of biochemical reaction networks. *Mathematical Biosciences and Engineering*, 5(1):1–19, 2008.
36. J. Keener and J. Sneyd. *Mathematical Physiology*. Springer, New York, NY, 2004.
37. B.N. Kholodenko. Negative feedback and ultrasensitivity can bring about oscillations in the mitogen-activated protein kinase cascades. *European Journal of Biochemistry*, 267:1583–1588, 2000.
38. S.Y. Shvartsman, M.P. Hagan, A. Yacoub, P. Dent, H.S. Wiley, and D.A. Lauffenburger. Context-dependent signaling in autocrine loops with positive feedback: Modeling and experiments in the EGFR system. *American Journal of Physiology—Cell Physiology*, 282:C545–C559, 2001.
39. E.D. Sontag. Asymptotic amplitudes and Cauchy gains: A small-gain principle and an application to inhibitory biological feedback. *Systems and Control Letters*, 47:167–179, 2002.
40. D.D. Šiljak. *Large-Scale Systems: Stability and Structure*. North-Holland, New York, NY, 1978.
41. A.N. Michel and R.K. Miller. *Qualitative Analysis of Large Scale Dynamical Systems*. Academic Press, New York, 1977.
42. S.D.M. Santos, P.J. Verveer, and P.I.H. Bastiaens. Growth factor induced MAPK network topology shapes Erk response determining PC-12 cell fate. *Nature Cell Biology*, 9:324–330, 2007.
43. A. Turing. The chemical basis of morphogenesis. *Philosophical Transactions of Royal Society of London*, B273:37–72, 1952.
44. J.C. Willems. *The Analysis of Feedback Systems*. MIT Press, Cambridge, MA, 1971.
45. R. Mancera and M. Safonov. Stability multipliers for MIMO monotone nonlinearities. In *Proceedings of the 2003 American Control Conference*, pp. 1861–1866, Denver, CO, 2003.

Index

Note: n = Footnote

A

B

C

D

E

F

I

J

K

M

N

O

P

Q

R

U

V

W

Y

Z